THE INSECTS
OF AUSTRALIA

A Male of *Lissopimpla excelsa* (Casta) (Hymenoptera–Ichneumonidae; length 11 mm) attempting to copulate with the orchid *Cryptostylis erecta*.

B Workers of *Myrmecia forficata* (F.) (Hymenoptera–Formicidae; length *ca* 20 mm) drinking honey-dew from Psyllidae (Hemiptera–Homoptera) on a eucalypt leaf.

C Disruptively patterned grasshopper, *Tinzeda eburneata* (Walk.) (Orthoptera–Tettigoniidae; length 45 mm).

D *Amenia imperialis dubitalis* Mall. (Diptera–Calliphoridae; length 15 mm) on *Leptospermum* blossom.

E *Tisiphone abeona* (Don.) (Lepidoptera–Nymphalidae; length of fore wing *ca* 30 mm).

F Larva of *Papilio aegeus* Don. (Lepidoptera–Papilionidae; length 50 mm) with osmeterium extruded.

THE INSECTS OF AUSTRALIA

A Textbook for Students and Research Workers

SPONSORED BY

THE DIVISION OF ENTOMOLOGY

COMMONWEALTH SCIENTIFIC AND

INDUSTRIAL RESEARCH ORGANIZATION

CANBERRA

MELBOURNE UNIVERSITY PRESS

First published 1970

Printed in Hong Kong
by Dai Nippon Co. (International) Ltd,
colour plates printed
by Wilke and Co. Ltd, Clayton, Victoria 3168
for Melbourne University Press, Carlton, Victoria 3053

SBN 522 83837 5
Dewey Decimal Classification Number 595.709945
Aus 67–1962

Text set in Times Roman Type

AUTHORS

E. B. BRITTON, D.Sc., F.R.E.S.,
Division of Entomology, CSIRO, Canberra; *formerly Department of Entomology British Museum (Natural History), London* (Chapter 30)

WILLIAM L. BROWN Jr, B.S., Ph.D.,
Professor of Entomology, Cornell University, U.S.A. (Chapter 37)

J. H. CALABY,
Division of Wildlife Research, CSIRO, Canberra (Chapter 25)

D. H. COLLESS, B.Sc.Agr., Ph.D., F.R.E.S.,
Division of Entomology, CSIRO, Canberra (Chapter 34)

I. F. B. COMMON, M.A., M.Agr.Sc.,
Division of Entomology, CSIRO, Canberra (Chapter 36)

M. F. DAY, B.Sc., Ph.D., F.A.A.,
formerly Division of Entomology, CSIRO, Canberra (Chapter 2)

G. M. DUNNET, B.Sc., Ph.D.,
Culterty Field Station, University of Aberdeen, Scotland (Chapter 33)

V. F. EASTOP, M.Sc., Ph.D., F.R.E.S.,
Department of Entomology, British Museum (Natural History), London (Chapter 26)

J. W. EVANS, M.A., Sc.D., D.Sc., F.R.E.S.,
formerly Director, The Australian Museum, Sydney (Chapter 26)

F. J. GAY, B.Sc., D.I.C.,
Division of Entomology, CSIRO, Canberra (Chapter 15)

E. T. GILES, M.Sc., Ph.D., D.I.C., F.R.E.S.,
Associate Professor of Zoology, University of New England, Armidale, N.S.W. (Chapter 19)

D. GILMOUR, D.Sc.,
Division of Entomology, CSIRO, Sydney, N.S.W. (Chapter 2)

H. E. HINTON, Ph.D., Sc.D., F.R.S., F.R.E.S.,
Professor of Entomology, University of Bristol, England (Chapter 4)

K. H. L. KEY, D.Sc., Ph.D., D.I.C., F.A.A.,
Division of Entomology, CSIRO, Canberra (Chapters 6, 16, 18, 21, 22)

D. K. MCALPINE, M.Sc., F.R.E.S.,
The Australian Museum, Sydney (Chapter 34)

I. M. MACKERRAS, F.R.A.C.P., M.C.P.A., F.A.A., F.R.E.S.,
Research Fellow, Division of Entomology, CSIRO, Canberra
(Chapters 1, 4, 7, 9, 10)

M. JOSEPHINE MACKERRAS, D.Sc., M.B., M.C.P.A.,
Division of Entomology, CSIRO, Canberra (Chapter 14)

CHARLES D. MICHENER, Ph.D.,
Watkins Distinguished Professor of Entomology, University of Kansas, U.S.A.
(Chapter 37)

K. R. NORRIS, D.Sc.,
Division of Entomology, CSIRO, Canberra (Chapter 5)

A. F. O'FARRELL, B.Sc., A.R.C.S., F.R.E.S.,
Professor of Zoology, University of New England, Armidale, N.S.W. (Chapter 13)

E. M. REED, B.Sc.,
Division of Entomology, CSIRO, Canberra (Chapter 27)

E. F. RIEK, M.Sc.,
Division of Entomology, CSIRO, Canberra
(Chapters 8, 12, 20, 28, 29, 31, 32, 35, 37)

EDWARD S. ROSS, Ph.D.,
California Academy of Sciences, San Francisco, U.S.A. (Chapter 23)

C. N. SMITHERS, M.Sc., F.R.E.S.,
The Australian Museum, Sydney (Chapters 17, 24)

R. W. TAYLOR, M.Sc., Ph.D.,
Division of Entomology, CSIRO, Canberra; *formerly Department of Biology, Harvard University, U.S.A.* (Chapter 37)

M. M. H. WALLACE, B.Sc., F.R.E.S.,
Division of Entomology, CSIRO, Perth, W.A. (Chapter 10)

D. F. WATERHOUSE, D.Sc., F.R.S., F.A.A., F.R.A.C.I.,
Chief, Division of Entomology, CSIRO, Canberra (Chapter 2)

J. A. L. WATSON, B.Sc., Ph.D.,
Division of Entomology, CSIRO, Canberra; *formerly Queen Elizabeth II Fellow*
(Chapter 11)

M. J. D. WHITE, D.Sc., F.R.S., F.A.A.,
Professor of Genetics, University of Melbourne, Victoria (Chapter 3)

T. E. WOODWARD, M.Sc., Ph.D., D.I.C., F.R.E.S.,
Department of Entomology, University of Queensland, Brisbane (Chapter 26)

FOREWORD

The first important general work on the insects of this continent, *Australian Insects* by W. W. Froggatt, was published in 1907. It was written primarily from the point of view of the field naturalist, and there is no doubt that it proved most useful to the student of those days. Nineteen years later, the classical *Insects of Australia and New Zealand* by R. J. Tillyard appeared, marking a major advance both in the content of accumulated knowledge and in the wealth of biology and comparative morphology that it presented. More than forty years have passed since then, a vast number of new facts have been gathered, points of view have changed, and 'Tillyard' has long been out of print. A few elementary or specialized books have since been published, but there has been an increasing requirement for something more comprehensive. 'Tillyard' was such an individual kind of book, with its own indefinable capacity for imparting enthusiasm, that no one but the original author could have revised it effectively. Furthermore, it is too much nowadays to expect an individual to possess the breadth of knowledge that is necessary to produce a modern work at the same level single-handed. Consequently, to meet the obvious need, the Division of Entomology eventually decided to sponsor a co-operative venture. However, the present work is not to be regarded as a second edition of 'Tillyard', although it has followed a generally similar plan of presentation. We sincerely hope that it will nevertheless generate the same sort of enthusiasm in its readers.

The expansion in our knowledge of Australian insects over the past forty years is reflected in a 45 per cent increase in the number of known species (37,300 in Tillyard, 54,071 in this book). In the same time the number of families has increased from 401 to 574. This is principally due to subdivision of large families, but it nevertheless reflects progress in the study of the class. The chief additions of known forms have been in the Diptera and Hymenoptera (in each more than 4,000 new species), followed by Lepidoptera and Coleoptera. These numbers reflect rather more the number of taxonomists working in the various major groups than the amount still to be achieved, which may well be more than has already been accomplished.

Tillyard's book suffered from the fact that, like any other entomologist even in his time, he could not write with equal authority on all the orders, and the same would be even truer of anyone who undertook such a great task today. It was therefore agreed that the new book would have many authors, and it was particularly fortunate that Dr I. M. Mackerras became available at that time to act as editor. His task proved to be much more arduous than we had expected, and he has discharged it in a scholarly fashion and with great care and thoroughness.

The task of the authors has been to give an account of the insects of Australia primarily from the systematic point of view, but with some account of their morphology, where they live, and what they do. To achieve even this limited objective in a single volume of reasonable size has involved rigorous selection of subject matter and imposed severe limitations on its presentation. We were indeed faced at one stage with the choice between eliminating the introductory chapters or reducing the space allocated to the orders. We chose the latter, although for anyone but the student we may have been wrong. I expect that, in consequence, few authors of the systematic chapters are really satisfied with the adequacy with which the allocated space has permitted them to cover their orders. There is, however, a good deal of information in the nine introductory chapters that helps in the general understanding of the insects as a class and which should therefore be of interest and value to specialist and student alike. Moreover, much of it is presented in an entirely new fashion, and some of it would be difficult to find in other texts. Apart from considerations of space it has been necessary for all authors to conform rather closely to an established pattern; this they have accepted for the most part cheerfully, although they would sometimes have preferred to deviate to suit their own particular needs.

In the teaching of Zoology the trend for some years has been away from a taxonomic approach as a major foundation on which to build other studies. Although this may mean that books like the present will progressively occupy a less and less central position in the formal training of students, they will continue to be invaluable sources of information and to provide an entry into the relevant but widely scattered literature. Indeed it is not too much to hope that a work of this sort will stimulate interest in a more broadly based taxonomy than has been general in Australia in the past. The keys—many of which are new—the abundance of new illustrations, and the wealth of previously unpublished biological information should certainly provide the student with a much better foundation to the subject than has been available for many years.

D. F. WATERHOUSE
Chief,
Division of Entomology,
CSIRO

Canberra
December, 1967

ACKNOWLEDGMENTS

The indebtedness of individual authors to those who helped them in their own work is acknowledged separately in the appropriate chapters. Here we wish to record our appreciation and thanks to those who contributed in more general ways.

As the book was intended primarily to be a textbook, it was important to present it in a way that would be most useful to students. For this reason, we are particularly indebted to Mr F. A. Perkins, formerly Reader in Entomology, Dr T. E. Woodward, the staff, and the students of the Department of Entomology, University of Queensland, who read and commented on several of the chapters and, in particular, tested many of the keys in the course of regular class work. Professor T. O. Browning, University of Adelaide, Professor A. F. O'Farrell and Associate Professor E. T. Giles, University of New England, Dr R. D. Hughes, then Australian National University, Canberra, and Associate Professor D. J. Lee, School of Public Health and Tropical Medicine, University of Sydney, also gave helpful advice from the same point of view. The drafts of Chapters 1 to 9 were sent to all authors, and several responded with useful comments.

The illustrations are as important as the text in a work of this kind. Most of those in this volume are original, and nearly all of the drawings were made, under the guidance of the authors, by the artists whose names appear in square brackets in the legends. Photographs and borrowed illustrations are acknowledged similarly. We owe a special debt to Mr Frank Nanninga, who painted the coloured plates and drew a great many of the figures in the text. Among those not so specifically acknowledged, Mr L. A. Marshall and Mrs G. C. Palmer prepared maps and diagrams, and they and Messrs C. Lourandos and John Green were responsible for much of the final technical preparation of the illustrations for the press. The Duplicating Service of the Division and the Editorial and Publications Section of CSIRO also gave valuable assistance.

Even with all this effort, there remained some figures that had to be derived directly from already published sources. These are acknowledged individually in the legends, and we wish to thank the following publishers for permission to make use of them:

The McGraw-Hill Book Company, New York, for Figs. 4.5 to 4.8 from Johannsen and Butt (1941);

The Ronald Press Company, New York, for Figs. 4.1 and 4.2 from Hagan (1951);

The Melbourne University Press, Melbourne, for Figs. 3.3 and 9.15 from Leeper (1962);

The Royal Entomological Society of London, for Fig. 4.15 from Hinton (1963b);

The Company of Biologists, Cambridge, for Fig. 4.16 from Savage (1956);

The Editors of *Science Progress*, London, and the *Entomologist's Monthly Magazine*, Oxford, for Fig. 4.12 from Hinton (1958b,c);

The Editor of the *Journal of Insect Physiology*, London, for Fig. 4.4A from Hinton (1967);

The Editor-in-Chief, Editorial and Publications Section, CSIRO, Melbourne, for Figs. 3.1 from White and Key (1957), 3.2 from Martin (1963), 3.6 from White, Cheney and Key (1963), and 26.38 from Eastop (1966).

All the references were checked from the original publications, and we are most grateful for the unstinted help in this sometimes difficult task that we received from the staff of the Canberra Library of CSIRO, Dr Elizabeth Exley, University of Queensland, and Mrs C. A. Gosney, British Museum (Natural History), London.

Preparation of such a large manuscript in good order for the Press is an arduous business, and we are indebted to Mrs Geraldine Davy, Miss Dilys Ward, and especially Miss Robin Holland, on whom the greatest burden fell, for the care and accuracy with which they completed the work. Integration of the manuscripts, so far as that could be done, was the responsibility of Dr I. M. Mackerras, and he shared with Miss Holland and Dr K. R. Norris the task of checking text and illustrations at all stages of production. Miss Holland and Miss J. Cardale gave valuable help in the preparation of the index.

Finally, the Chief and the editorial group in the Division would like to record with particular appreciation the uniformly cordial and understanding relationship that has existed between them and the Director and staff of the Press over what has now proved to have been quite a long period.

CONTENTS

PLATES

(Plates 1–8 painted by F. Nanninga)

INTRODUCTION

Insects are amongst the most abundant and successful of terrestrial animals. They include about three-fourths of all the described species of animals, and they have become adapted to a great range of environments, from high latitudes to the equator, from rain forest to desert, from mountains to the shore, and to varied ways of living—phytophagous, carnivorous, saprophagous, parasitic. Many have become aquatic during part or the whole of their life history. The breadth of their adaptive radiations has brought some of them, though only a very small proportion of the total, into close contact with human populations. On the one hand, they damage or destroy crops and domestic animals, transmit diseases of plants and animals, including man himself, and damage or destroy his habitations. On the other hand, man has added bees and silkworms to his flocks, and has learned to use other species to control noxious insects and weeds. Moreover, they have provided him with some of his most valuable tools of biological research in many fields from the broadest aspects of evolution to the detailed mechanisms of inheritance.

It is natural, then, that entomology should have shared in the steadily increasing volume of research that has made it so difficult to be both brief and comprehensive when reviewing any scientific discipline. Selection is unavoidable, and the purpose of these notes is to set out the scope of this book and the ways in which problems of presentation and compression have been met.

1. It has been necessary to assume that the reader would have a background of general knowledge equivalent to what might reasonably be expected at the end of the first year of a university course in Pure Science. This may impose some hardship on the amateur entomologist who may wish to use the book, but it is hoped that he will find sufficient explanation in the text and illustrations to meet his needs. There is no glossary. Terms with which the reader might not be familiar are defined in the text and entered in the index.

2. Chapters 1 to 9 have been reduced to the minimum that is considered essential for the general student of entomology as a background for the chapters that follow. Chapter 1 has presented particular difficulty in this respect, because the anatomical terminology used in it has been based, as far as possible, on usage among morphologists, whereas the writers on many of the orders carry the historical burden of a terminology that has grown up with little or no reference to what has been done outside those orders. There seems to be no way to avoid this unfortunate situation. There is the further minor difficulty that the earlier chapters must anticipate Chapter 7, but a general picture of the classification

used may be obtained by referring to Table 8.1 on p. 169.

3. The presentation of the remaining chapters is essentially systematic, an arrangement that serves to emphasize the evolutionary perspective into which any study of insects should be fitted. The systematic statements have been used as pegs on which to hang short accounts of where the insects live and what they do. Wider problems are touched on when describing insects that illustrate them particularly well.

4. The book is designed to cover Australian insects only. The insects of New Zealand were included in Tillyard (1926b), and the literature on them up to 1952 was made accessible by D. Miller (1956). The insects of New Guinea and the Pacific show predominantly Oriental relationships (chapter 9), and they are referred to here only when they are relevant to particular problems that are being discussed. However, an attempt has been made to put the Australian fauna into perspective with that of the rest of the world by including at least a mention of important groups that do not occur here.

5. In the systematic chapters (except 11), the censuses include undescribed species and unrecorded synonymy known to the authors, but the completeness of this information differs from order to order, so the figures should be taken as approximations. Their reliability as indices of the total number of species in the country varies greatly. Thus, it is reasonable to suppose that most of the species of Australian butterflies and mosquitoes are already known, but probably less than two-thirds of the other Lepidoptera and Diptera, and still smaller proportions of some of the less-studied groups.

6. The keys to families and higher taxa have been based primarily, though not exclusively, on Australian material. Some of them may have a wider application, but they should be used with reserve for placing specimens from other regions. There has not been room for more than a few keys below the family level.

7. Common names have been used sparingly, and only for widely known species. A full list, with standard abbreviations of the names of authors of species, is given by Gay (1966). The techniques of collecting, preserving, and studying insects have been described by Norris (1966).

8. The following abbreviations have been used, where appropriate, for political divisions of Australia that have compound names:

N.S.W. New South Wales;
S.A. South Australia;
W.A. Western Australia;
N.T. Northern Territory;
A.C.T. Australian Capital Territory.

9. References have presented a particular problem. It is impossible to give extensive bibliographies in the space available, and reliance has been placed as much as possible on references to reviews and monographic works which provide a recent, readily available entry into the major literature. Papers in the *Annual Review of Entomology* have proved particularly useful for this purpose, as have the bibliographies in the latest edition of Imms's well-known textbook, as revised by Richards and Davies (1960). It must be stressed that this in no way implies any lack of appreciation of earlier classical papers. Direct study of these works is indispensable to any serious student, and many of them have been referred to in general terms when the full references are included in a publication that is cited. References in round brackets relate to the statements in which they occur, those in square brackets usually to the group as a whole. Papers on Australian insects up to 1930 have been listed in the bibliography by Musgrave (1932).

1

SKELETAL ANATOMY

by I. M. MACKERRAS

Arthropods are metamerically segmented invertebrates with articulated exoskeletons, which not only provide support for soft tissues, attachments for muscles, and protection from physical stresses, but determine the external appearance of the animals as well. Consequently, this chapter is concerned primarily with the external anatomy of the adult insect. It includes the infoldings and appendages, but does not take the detailed structure of the cuticle (p. 29) into account, and it is intended to provide no more than a background for later chapters. Fuller descriptions, with references to literature, will be found in Snodgrass (1935, 1952) and Richards and Davies (1960).

A major difficulty facing the student is the plethora of anatomical terminologies that are used in the different orders. To some extent this is inevitable, because the comparative anatomy and embryology of some groups have not been studied in sufficient detail for firm conclusions to be drawn. This kind of difficulty is well illustrated by the external genitalia (Tuxen, 1956). When these legitimate reasons for disparities have been taken into account, there remain a still greater number that are due to the fact that specialists on different orders have developed their own terminologies without reference to homologies in which they were not interested. The problem seems to be insoluble without international meetings to agree on a standard anatomical nomenclature.

ORIENTATION AND RELATIONSHIP OF PARTS

It is necessary to define some general terms at the outset. An insect is, basically, a bilaterally symmetrical, horizontally oriented,

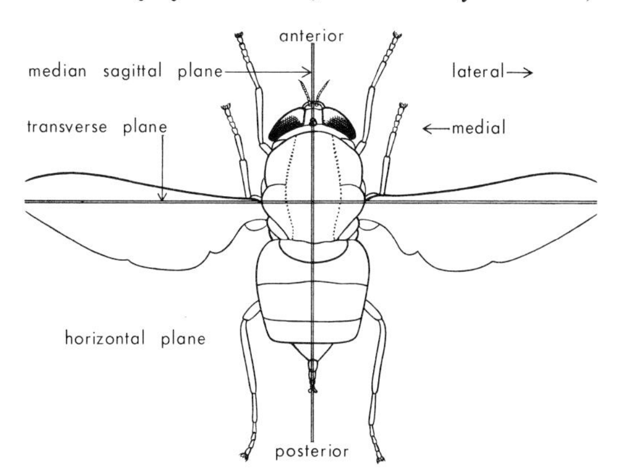

Fig. 1.1. A generalized fly (*Pelecorhynchus fusconiger*, ♀) to show orientation and anatomical planes. [B. Rankin]

forwardly progressing animal, and its body can consequently be divided by three sets of primary anatomical planes at right angles to each other (Fig. 1.1): vertical *sagittal* planes in the direction of its long axis, the one passing through the central axis of the body being the *median sagittal plane*; *horizontal* planes, also parallel with the long axis; and *transverse* planes at right angles to its long axis and to the other two planes. The head end of the body is *anterior*, *cranial*, or *cephalic*, the hind end is *posterior* or *caudal*, and the antero-posterior relationships of parts are described by these adjectives. The upper surface is *dorsal* and the lower surface *ventral*. A line traversing the surface of the body in the median sagittal plane is the *median line*, and an area symmetrically disposed about it the *median area*; an intermediate zone may be termed *sublateral*; and the outer zone, including the side of the insect, is *lateral*. Structures lying further from, or nearer to, the median sagittal plane than other structures are referred to as being *lateral*, or *medial*, to them. Similarly, parts of appendages (or of other attached structures) that lie nearer to the body are referred to as *proximal* and those further from the body as *distal*. There seems to be no good reason to perpetuate the use of such terms as 'mesal' for 'medial' or the usually adverbial suffix '–ad'. On the other hand, combined terms, such as dorsolateral, anteroventral, etc., are often convenient and clearly descriptive.

All these terms are used in relation to the morphologically horizontal position of the insect, no matter what attitude it may take up. Some care in interpretation is consequently necessary, particularly in defining the surfaces of the leg segments (p. 13). It cannot be stressed too much that precise terms should always be used to describe anatomical relationships.

GENERAL ORGANIZATION

Insects are composed of twenty original

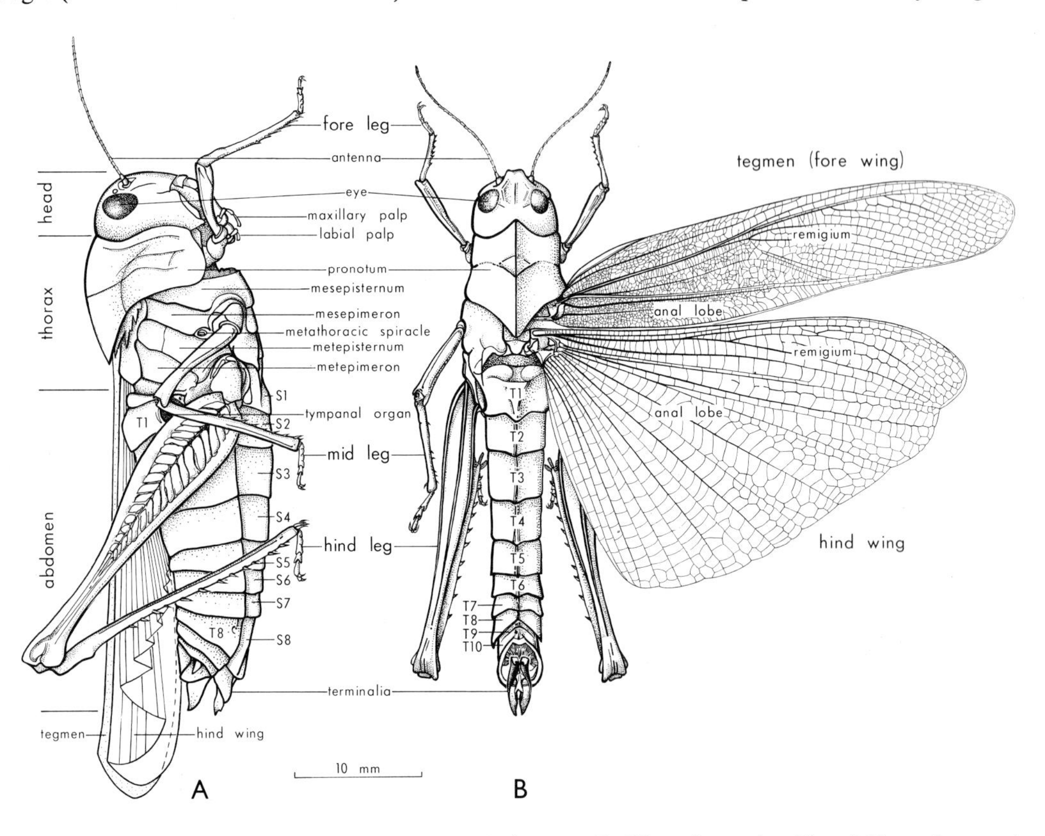

Fig. 1.2. A common locust (*Gastrimargus musicus*, Orthoptera-Caelifera, ♀–see also Plate 2,F) to show main anatomical divisions and landmarks. [T. Nolan]

somites. They show considerable advance on their presumed myriapodan ancestors in having developed a much greater degree of *tagmosis*, which is the co-ordination or amalgamation of segments to perform some particular group of functions. As Tiegs and Manton (1958) have stressed, tagmosis has been vitally important in the evolution of insects, because, originally developed in response to some immediate need, each episode of tagmosis has opened the way for further adaptive radiation. The result has been the division of the insect body into the three familiar regions of *head*, *thorax*, and *abdomen*, each with its specially modified appendages (Fig. 1.2).

The head consists of three primitive preoral somites (with the antennae and eyes), to which three postoral appendage-bearing somites have become fused to provide the mouth-parts. The thorax consists of three somites, which are still more or less clearly distinguishable as separate segments. Originally developed in response to the adoption of a hexapod gait, it provided the point of balance at which wings could function effectively, and became further enlarged and modified when they evolved on its second and third segments. The abdomen remained relatively unmodified, except that the appendages of most of its segments atrophied more or less completely, although they can still be seen in Apterygota and the embryos of higher insects. It consisted originally of 11 segments, and there is a minor degree of tagmosis in the terminal segments.

The primitive metameric segmentation is indicated in unsclerotized larvae by the insertions of the longitudinal segmental muscles and by corresponding transverse grooves on the surface of the body (Fig. 1.3A). When sclerotization develops, it does so in dorsal and ventral plates extending from just anterior to each intersegmental groove *(antecostal suture)* for a varying distance towards the posterior end of the segment. The inflected part is the *antecosta*; the strip anterior to it is the *acrotergite* dorsally and the *acrosternite* ventrally; and the posterior part of the plate is the body of the *tergum* dorsally and of the *sternum* ventrally (Fig. 1.3B).* The unsclerotized part of the

* It is usual to refer to the whole segmental plate as tergum or sternum and to its parts as tergites or sternites. When, however, a single plate is known, or suspected, to represent only a portion of the original whole, many authors prefer to call it a tergite or sternite, in order to avoid the implication that it is complete.

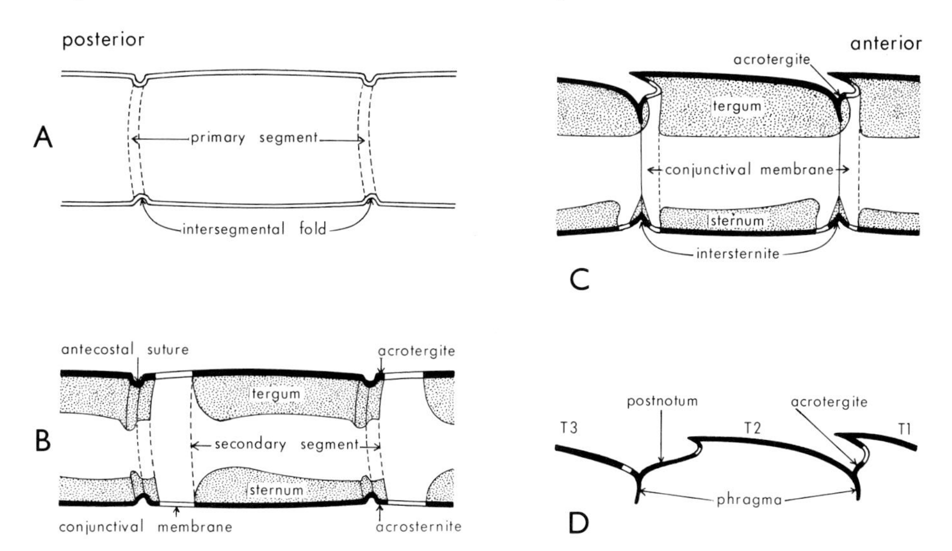

Fig. 1.3. Diagrams of primary and secondary segmentation (based on Snodgrass, 1935): A, primary; B, simple secondary; C, more advanced secondary; D, dorsal sclerites of thorax in section.

segmental cuticle posterior to the tergal and sternal plates is the so-called *conjunctival* or *intersegmental* membrane, which generally becomes tucked in as the plates extend (Fig. 1.3C).

In this way, a secondary functional segmentation is imposed on the primary metameric segmentation. In the abdomen the sides *(pleura)* of the body remain membranous; but in the thorax, where greater rigidity is required, strengthening sclerites are laid down in the pleural membrane, and become associated with the tergal and sternal plates to form a box with a precisely limited capacity for distortion. In the head, where no flexibility at all is needed, all the sclerites become fused into a single strong capsule. In addition, there are sclerotized infoldings *(apodemes)* projecting into the body to add strength and provide attachments for muscles.

Finally, the appendages mostly consist of segmented tubes or plates with flexible articulations at the joints;* the wings are specialized outgrowths of the thoracic terga; and the tracheae of the respiratory apparatus open at segmental *spiracles* on each side. Various parts of the body may also be more or less covered with hairs of different kinds—*macrotrichia*, which may be modified into spines or scales, and *microtrichia*, which may become specialized into the fine pile of a plastron (p. 39). The arrangement of the hairs *(trichiation, chaetotaxy)* is often of taxonomic value.

* Anatomically, a *joint* is an articulation, the parts joined together being *segments* or *articles*.

HEAD

Cranium

The skull of an insect (Fig. 1.4) is a hard, usually more or less globular capsule, which is incomplete below (where the preoral cavity is partly closed by the mouth-parts), and opens posteriorly by the *occipital foramen*, through which the nerve cords, oesophagus, aorta, salivary ducts, and a pair of tracheae enter or leave the neck. The part dorsal and anterior to the occipital foramen is the *epicranium*, which is divided into the *occiput* posteriorly, the *vertex* dorsally, the *frontoclypeus* anteriorly, and the *genae* (cheeks) laterally, although the limits of these areas are often not clearly defined.

The inflected skeleton is formed by a series of apodemes, of which the most important are the anterior, posterior, and sometimes dorsal *tentorial* arms. These usually unite to form a *frontal plate* anteriorly and a *corpus tentorii* posteriorly, with a more or less extensive gap or foramen between them (Fig. 1.5). Other internal ridges occur less consistently, for example between frons

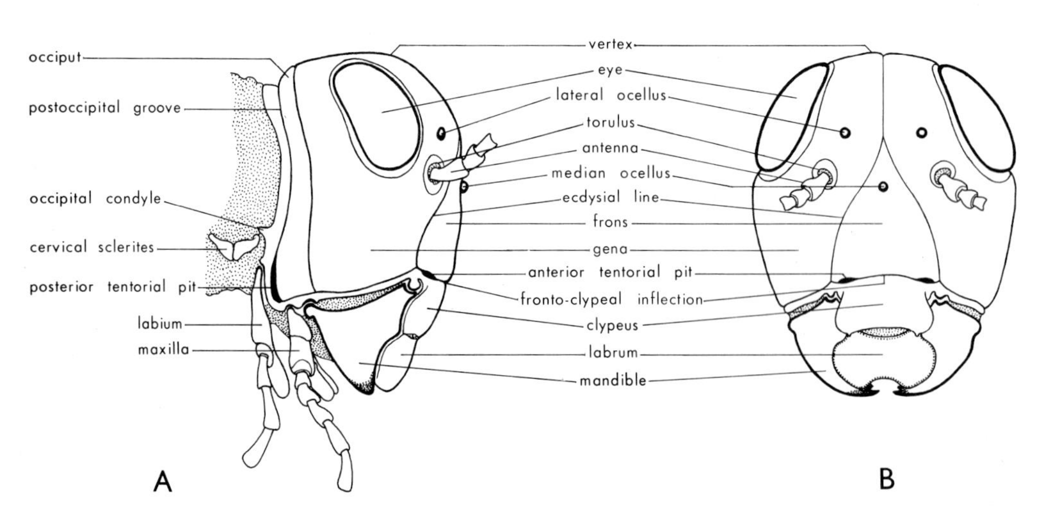

Fig. 1.4. A generalized cranium (after Snodgrass, 1935): A, lateral; B, anterior.

and clypeus, between both of these and the genae, around the eyes, or around the antennae. All may show indications of their presence in the form of external grooves or 'sutures', and thus provide useful anatomical landmarks.

Compound eyes are present in most adult insects. They usually occupy a considerable area on each side of the head. When so large that they meet in the median line, they are termed *holoptic*; when separated, they

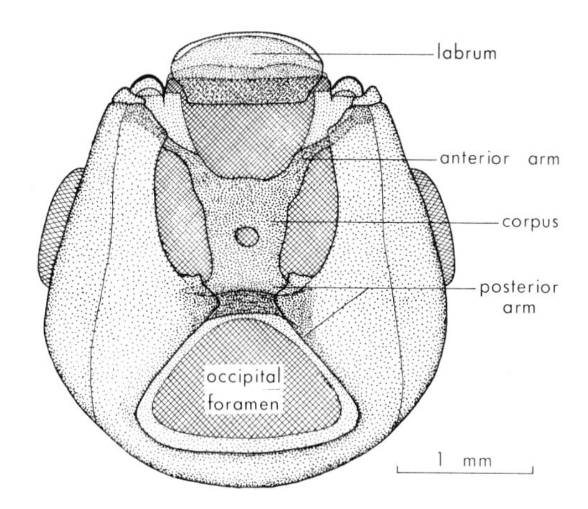

Fig. 1.5. Tentorial skeleton of *Mastotermes darwiniensis*, Isoptera, ♂, from a cleared preparation.

are *dichoptic* (Fig. 1.6).* Externally, the eye consists of a large number of hexagonal *facets* (or corneal lenses) formed of transparent cuticle, each lens covering a single eye element (p. 45). Sometimes the upper facets are much larger than the lower, and occasionally the eyes are divided into separated dorsal and ventral parts (e.g. Fig. 34.18D), or they may fuse across the mid-dorsal line. They may show a pattern of bands or patches of contrasting colour in life, and the interfacettal junctions are often provided with fine hairs, which may be dense enough to give the eyes a distinctive appearance.

Three *ocelli*, typically arranged in an isosceles triangle on the vertex, are present in so many insects that they must be presumed to have occurred in the ancestors of all the

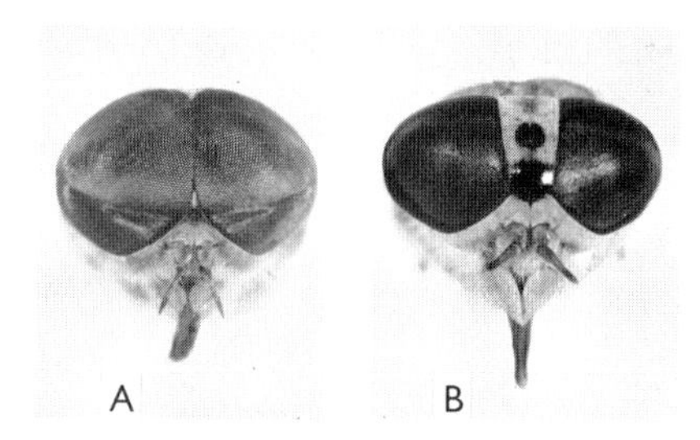

Fig. 1.6. Frontal view of head of *Tabanus particaecus*, Diptera, to show dimorphism in eyes: A, ♂; B, ♀. [Photos by J. Green]

orders. Each *lateral* (or posterior) ocellus has a single lens, but differs from an ommatidium of a compound eye in that the lens covers a number of internal eye elements. The *median* (or anterior) ocellus was apparently formed from two separate ocelli which became fused together, and it is innervated from both sides of the deutocerebrum.

With the main areas and landmarks of the cranium thus broadly defined, we may turn to the problems presented by the so-called sutures of the head (see DuPorte, 1957; Hinton, 1958a).

In the first place, the only lines that may represent original segmental divisions of the cranium are the *postoccipital groove* just anterior to the occipital foramen in many insects (Fig. 1.4A), which may delimit the labial segment, and a groove anterior to this in Archaeognatha, which may mark off the maxillary segment (Snodgrass, 1935). The gnathal segments are more clearly defined in the larvae of some Thysanura and in the Palaeozoic Monura. Other morphologically significant lines are the fronto-clypeal inflection (see below), which marks the division between the primitively preoral and postoral parts of the cranium, and the division between the clypeus and labrum anteriorly.

Secondly, there are two kinds of 'sutures'. The first is represented by the dorsal and ventral ecdysial lines, along which the capsule splits or hinges in immature insects, and which persist as unpigmented lines in some adults. The dorsal lines, known as *epicranial*

* Strictly, these terms should refer to the insect, but they are often conveniently applied to the eyes, as here.

sutures, typically form an inverted Y, with the median *coronal* arm extending forward from the occiput and the two lateral, or *frontal*, arms diverging between the ocelli (Fig. 1.4B). The ventral lines may be single or double, and they may enclose a median area, the *ventral apotome*, the position and extent of which are of functional rather than morphological significance (Hinton, 1963a). All these lines vary in position, and they may become incorporated in, or confused with, other depressions, so that they are less useful for morphological analysis than had been supposed.

The second kind of suture is formed by inflections of the surface which are normally associated with apodemes. They are generally better defined than the ecdysial lines, but their positions are also determined primarily by functional requirements, and they do not necessarily mark off morphologically identical areas in different insects. The most constant and conspicuous inflections are the *anterior tentorial pits*, which mark the origins of the anterior tentorial arms. There is a less conspicuous pair of *posterior tentorial pits* marking the origin of the posterior tentorial arms on either side of the occiput, and these may lie on an inflection dividing the occiput from the posterior genae behind the eyes. Anteriorly, the *fronto-clypeal* inflection joins the tentorial pits, and separates the *frons* above from the *clypeus* below. It may be displaced upwards or disappear, so the original position of the mouth can sometimes be determined best by the level of the tentorial pits (DuPorte, 1957). Laterally, the frons is usually separated from the genae by a deep *fronto-genal* inflection on each side, and this is continued below the tentorial pits as an equally deep *clypeo-genal* inflection.

While the frons may be regarded, basically, as extending from the vertex to the fronto-clypeal inflection or, in its absence, to a horizontal line joining the anterior tentorial pits, the landmarks are not always clear, and there may be additional inflections on the frontal or clypeal areas. Consequently, some systematists adopt purely topographical definitions, limiting the frons to the area dorsal to the insertions of the antennae, and calling the whole or part of the central area bounded by the antennal insertions dorsally, the labrum ventrally, and the fronto-genal and clypeo-genal inflections laterally the *face*.

When a median ventral sclerite is differentiated behind the labium, especially in prognathous insects, it is usually termed the *gula*, or gular plate.

Antennae

These are a pair of mobile, segmented appendages, which are inserted in the head between the eyes. The articulation to the skull often consists of a sclerotized ring

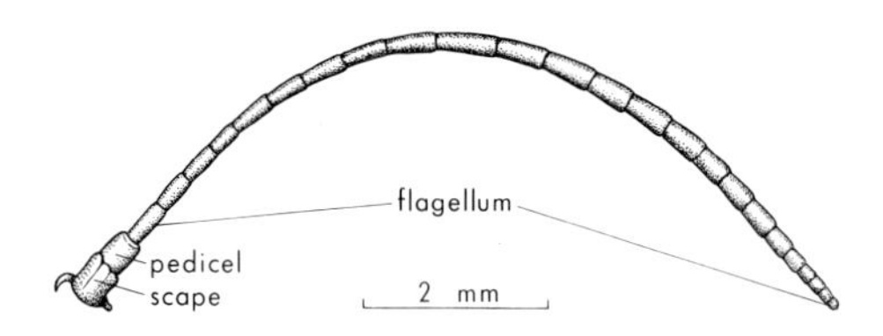

Fig. 1.7. Antenna of *Gastrimargus musicus*, ♂. [B. Rankin]

forming a socket *(torulus)*, and there may be a projecting peg which provides for free movement of the antenna in all directions. Three principal divisions of the antenna may usually be recognized (Fig. 1.7): the *scape*, or basal segment, often longer or larger than any of the succeeding segments, and containing the intrinsic muscles; the *pedicel*, or second segment, which is filled with a mass of sensory cells called Johnston's organ (p. 44); and the *flagellum*, which forms the remainder of the antenna and is usually multisegmented. Antennae are of many diverse forms, which are described under the various orders. Functionally, they are organs of special sense, and they often show important secondary sexual characters, as, for example, in some Diptera–Nematocera (Fig. 34.2A,B) and Lepidoptera; those of the males are occasionally modified for holding the females.

Mouth-parts

The *preoral cavity*, which opens posteriorly

into the definitive mouth, is bounded anteriorly by the labrum, posteriorly by the labium, and laterally by the lower margins of the genae; the mandibles and first maxillae are articulated at its sides (Fig. 1.8). In contrast with the other hexapods described in Chapter 10, the mouth-parts of true insects are primitively exposed *(ectognathous)* and dependent *(hypognathous)*, although they have become secondarily *prognathous* in some specialized groups (e.g. Fig. 30.3). There are many variations in detail, depending on the method of feeding, and the following general description (see Fig. 1.9) is based on an insect with mouth-parts adapted for biting and chewing.

The *labrum* is normally a movable plate attached to the lower margin of the clypeus. Its outer surface is generally strongly sclerotized, and its distal margin sharply defined; its internal surface, the *epipharynx*, is membranous and furnished with small tactile hairs and taste organs. The complete structure is often termed the labrum-epipharynx.

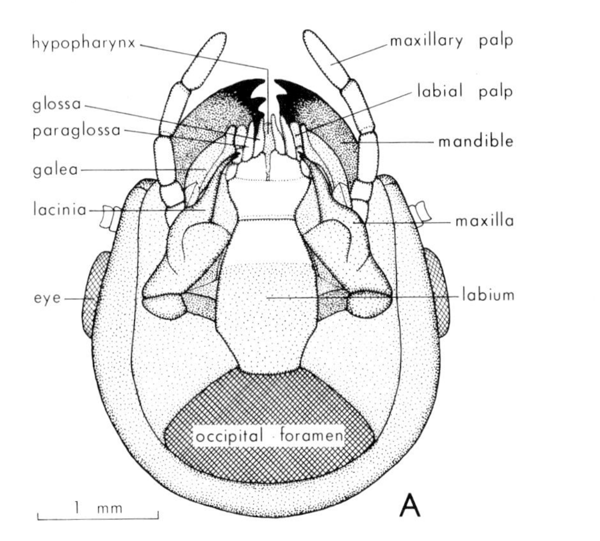

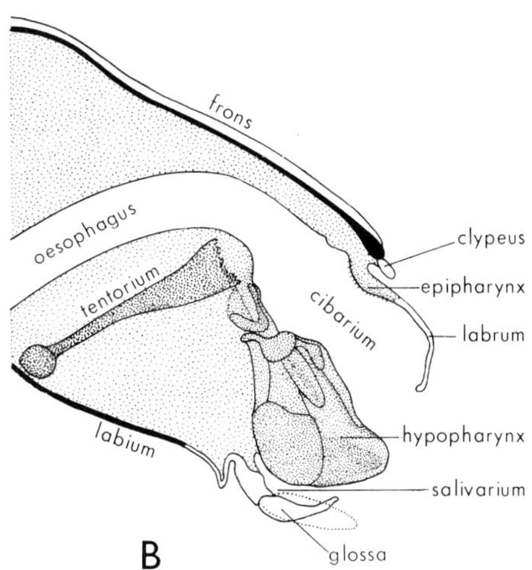

Fig. 1.8. Head of *Mastotermes darwiniensis*, ♂: A, posteroventral aspect to show relationships of mouth-parts B, paramedian sagittal section (semidiagrammatic).

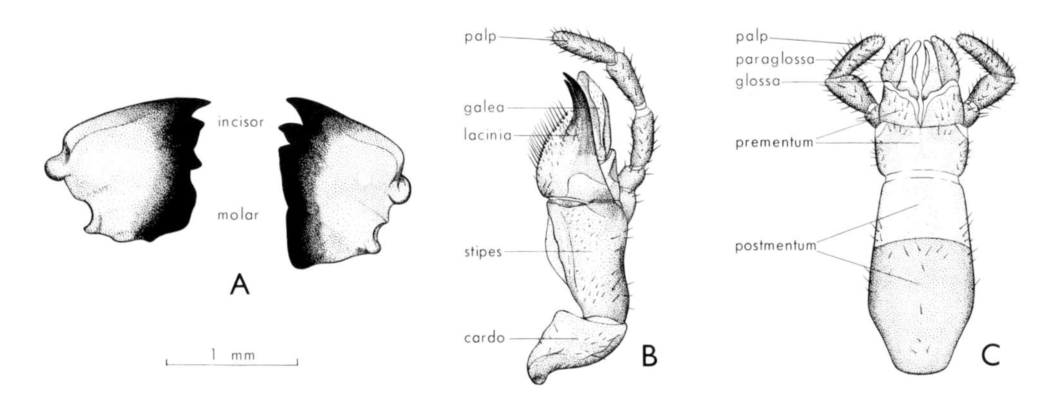

Fig. 1.9. Mouth-parts of *Mastotermes darwiniensis*, ♂: A, mandibles; B, right maxilla; C, labium. [B. Rankin]

The *mandibles* are a pair of strongly sclerotized, usually toothed jaws situated immediately posterior to the labrum. Except in Archaeognatha (p. 217), they articulate with a process of the clypeus anteriorly by a ginglymus (hinge) joint and with the gena posteriorly by a condyle (ball), and they are operated by the most powerful muscles in the head. They are the principal feeding organs, being used primitively to bite off and chew the food. They are not segmented, although there is an accessory *prostheca* ('lacinia mobilis') in Ephemeroptera (as in some Diplura), and they do not have a palp.

The *maxillae* (first maxillae) lie postero-ventral to the mandibles and anterodorsal to the labium. They differ from the mandibles in showing evidence of having originated by modification of walking legs, and they are usually divided into several parts. The first segment is the *cardo*, which is attached to the head proximally and to a longer second segment, the *stipes*, distally. The stipes bears two lobes, the lateral being the *galea* and the medial the *lacinia*. The maxillary *palpus*, or *palp*, usually of one to seven segments, is attached laterally to the distal part of the stipes, sometimes by a separated section, the *palpifer*. There is considerable variation, the galea sometimes being two-segmented and the lacinia spined or toothed on its medial border. The maxillae serve as accessory jaws, the laciniae helping to hold the food when the mandibles are extended, as well as assisting in mastication, and the galea and palp assisting to select the food by touch and taste.

The *labium* consists of the fused second maxillae. It is attached to the ventral surface of the cranium (to the gular plate, if that sclerite is developed), is bilaterally symmetrical, and is divided into the following parts, the corresponding names for the first maxillae being included in brackets to show the homologies: *postmentum* (cardo) proximally; *prementum* (stipes) more distally; two distal processes articulated to the prementum on each side, the *glossa* (lacinia) medially and *paraglossa* (galea) laterally; and a pair of labial palps, normally of one to four segments, arising from a lateral part of the prementum which is sometimes differentiated as the *palpiger*. When the prementum is divided transversely into two parts, the distal portion bearing the glossae and paraglossae is known as the *ligula*.

In addition to the paired mouth-parts, there is usually a median, unpaired, tongue-like organ, the *hypopharynx*, projecting forward from the back of the preoral cavity, and dividing it into a dorsal *cibarium*, which serves as a food-pouch, and a ventral *salivarium*, into which the salivary ducts open (Fig. 1.8B). Sometimes the hypopharnyx is trilobed, the median *lingua* bearing a pair of lateral *superlinguae*; sometimes it is a simple lobe supported by sclerotized plates; and sometimes it is produced as a stylet.

There are a great many variations from this basic pattern. In lucanid beetles (p. 549) there is marked sexual dimorphism in the size and shape of the mandibles. In sucking insects (e.g. Figs. 34.3, 4) both mandibles and maxillae may be transformed into spear-like organs, or the mandibles may disappear; the maxillary stylet may be derived from the lacinia, or the galea, or (in Hemiptera; Fig. 26.2A) the stipes, the other parts being reduced or absent; in Lepidoptera (Fig. 36.1) the laciniae disappear and the two galeae together form the haustellum or coiled sucking tube. The labium may be drawn out into an elongate, sometimes segmented trough to hold the other mouth-parts, and its palps may be absent (Hemiptera) or reduced to a pair of labella at its distal end (Diptera). The hypopharynx, too, may be prolonged into a strong, tapering spear carrying the orifices of the salivary ducts to its apex, and it may sometimes fuse with a strongly sclerotized, pointed labium.

THORAX

The head is joined to the thorax by a flexible, membranous neck, which is usually quite short, and generally strengthened by small sclerites to which the muscles that control the movements of the head are attached.

The segments of the thorax are the *prothorax*, *mesothorax*, and *metathorax* (Fig.

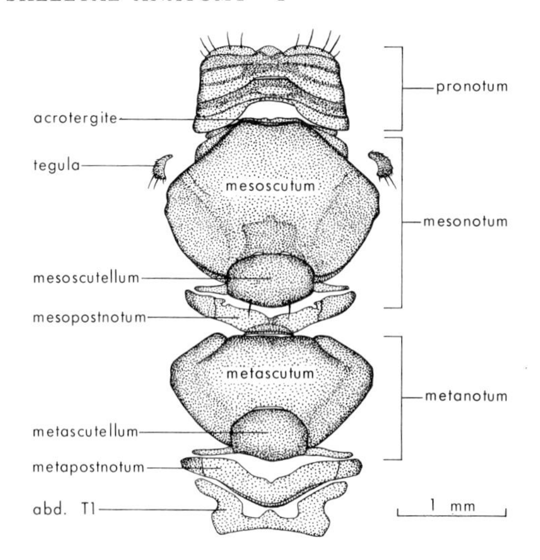

Fig. 1.10. Dorsal thoracic sclerites of *Chorista australis*, Mecoptera, ♀, from a cleared preparation. The prescuta are not clearly defined.

1.10), and their sclerites carry the same prefixes, for example, prosternum, mesepimeron; these are not to be confused with the prefixes pre- and post- which are used to define particular parts of each segment. Primitively, the three segments are more or less equal and similar, but in the winged insects the mesothorax and metathorax are enlarged and more closely united to form a relatively rigid *pterothorax*. The prothorax is sometimes reduced to a small annulus, but in some insects, such as cockroaches (e.g. Fig. 14.11), its dorsal part is developed into a shield. *Paranotal lobes* are present on the prothorax of a few insects (e.g. Fig. 26.13). The mesothorax is largest in those insects in which the fore wings are the stronger, notably in the Diptera, whereas those that fly with their hind wings or have powerful hind legs generally have a large metathorax.

The conjunctival membrane is reduced, and the sclerites cover most, sometimes all, of the thorax. Dorsally, each tergal plate is known as a *notum*, which is usually more or less clearly subdivided into *prescutum*, *scutum*, and *scutellum* (Fig. 1.10). Additional rigidity is provided by the antecostae and acrotergites. When the acrotergites remain narrow and retain their association with the notum posterior to them, they also retain their name; when they develop into wider plates associated with the notum anterior to them (to which they really belong morphologically), they are called *postnota* (Fig. 1.3D). Thus the metathorax has both acrotergite and postnotum in the Orthoptera and Coleoptera. On wing-bearing segments of Neoptera, there are also two lateral projections, the *anterior* and *posterior notal processes*, which provide hinge-joints for the bases of the wings.

The pleural sclerites laterally (Fig. 1.11) are thought to have been derived from subcoxal elements of the ancestral legs, which became incorporated into the lateral body wall. They

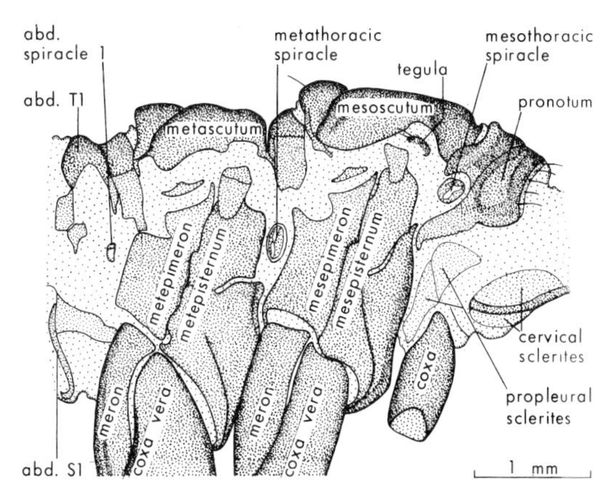

Fig. 1.11. Thorax of *Chorista australis*, ♀, lateral, from a cleared preparation.

are relatively small and variable in the Apterygota, but more extensive in the winged insects, on each segment of which they consist of an anterior *episternum* separated by the *pleural suture* from a posterior *epimeron*. There is considerable variation, and the episternum may be divided into an upper *anepisternum* and lower *katepisternum*, while similar divisions of the epimeron have been named *anepimeron* (or pteropleuron) and *katepimeron*. Sometimes the episternum (or its katepisternum) is fused below with the sternum, and the combined plate is known as the *sternopleuron*. Another variation, seen most often in primitive orders and in the mesothorax of Hymenoptera, is a separation of the anterior part of the episternum as a distinct plate, the *prepectus* (Fig. 37.5C). The wing-bearing segments also have a *pleural wing process* for articulation with the base of the wing, together with one or more small *basalar* sclerites anterior to the process and *subalar* sclerites posterior to it, which serve for attachment of some of the direct wing muscles. There are only two pairs of *thoracic spiracles*, mesothoracic and metathoracic, in true insects. They normally lie in or near the upper part of the divisions between the segments.

The sternal plates *(eusterna)* may be simple, or divided into three sclerites, *presternum*, *basisternum*, and *sternellum* (Fig. 1.12), and in generalized insects the basisternum is

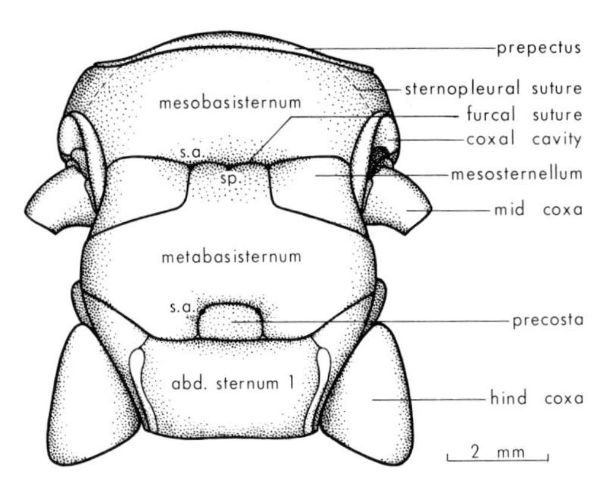

Fig. 1.12. Pterothoracic sterna of *Gastrimargus musicus*, ♂. [B. Rankin]
s.a., *sp.*, pits marking inflection of furcal arms and spina, respectively.

separated from the sternellum by a transverse suture connecting the apodemal pits at the origin of the furcal arms. The *spinasternum (post-sternellum)* follows the sternellum, and corresponds to the postnotum; it is often derived from an intersternite between the segments rather than from an acrosternite. Separate *laterosternites* are sometimes found at the sides of the eusternum, and fusion of the sternal and pleural regions may result in the formation of precoxal and postcoxal bridges. The sterna may be narrow, or largely unsclerotized; they may be marked by a median longitudinal inflection; or they may become incorporated in a sternopleuron, as in the mesothorax of many Diptera (p. 661).

The principal inflected skeleton of the thorax (Fig. 1.13) consists of *phragmata (endotergites)*, extending inwards from the antecostae dorsally; lateral apodemes *(endo-*

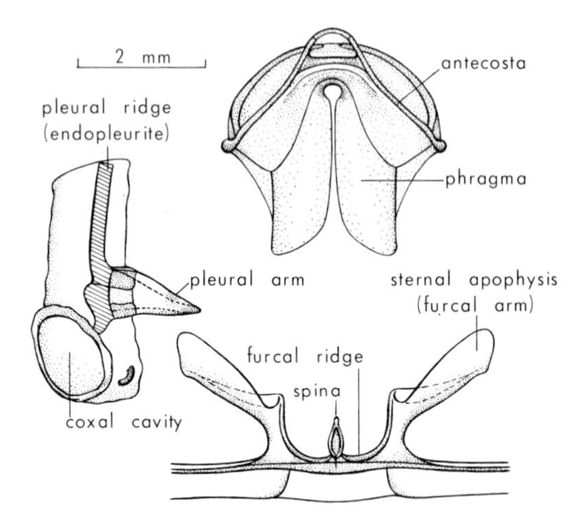

Fig. 1.13. Inflected mesothoracic skeleton of *Gastrimargus musicus*, ♂, posterior aspect, 'exploded'. [B. Rankin]

pleurites), formed in the wing-bearing segments by infolding along the pleural suture, and sometimes produced into medial arms which may fuse with the furcae ventrally; and *furcae (endosternites)*, arising from the apodemal pits between the basisterna and sternella, and generally forming Y-shaped bodies internal to the ventral body wall. In generalized insects there is also a median *spina* projecting into the body from the spinasternum, but it is usually lost or consolidated with the furca in higher forms. These apodemes provide surfaces for attachment of the large thoracic muscles.

LEGS

The insect leg (Fig. 1.14) consists, basically, of six segments: coxa, trochanter, femur, tibia, tarsus, and pretarsus bearing the claws; but the tarsus is usually subdivided into several segments, although it is still moved as a whole by a single pair of muscles

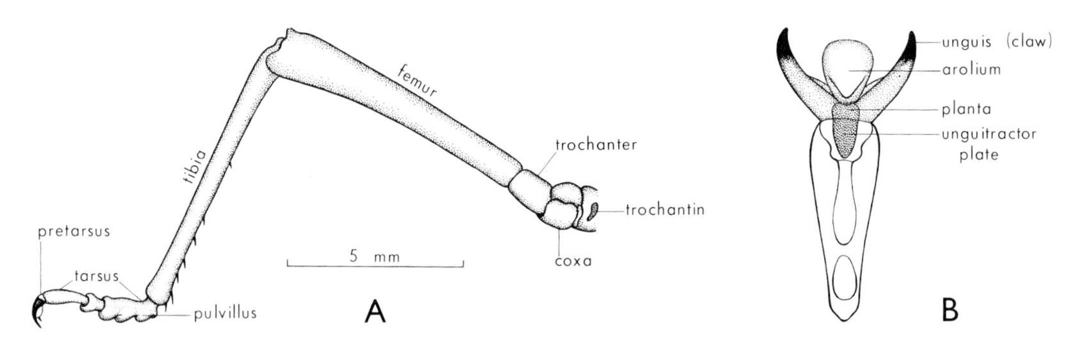

Fig. 1.14. Mid leg of *Gastrimargus musicus*, ♂: A, anterior aspect; B, last tarsal segment and pretarsus, ventral. [B. Rankin]

arising from the distal end of the tibia, and the pretarsus is not generally counted as a separate segment. For purposes of orientation, the legs are treated as if they were extended horizontally at right angles to the body, so that the segments present dorsal, ventral, anterior, and posterior aspects for examination.

The *coxa* is usually short and stout; it is often divided into an anterior *coxa vera* and a posterior lobe, the *meron*. It articulates with the thorax by a *coxal process* at the ventral end of the pleural suture, and sometimes also with a separate plate *(trochantin)* or with the sternum, either serving to restrict its range of movement. Other articular plates occur in some insects.

The *trochanter* is usually a short segment interposed between the coxa and the femur, sometimes freely articulated with both, occasionally divided into two parts, but often firmly attached to the femur, for which it provides a coxal articulation.

The *femur* is generally the stoutest, strongest, and sometimes the longest segment of the leg. It is sometimes armed with strong spines, but not with movable spurs.

The *tibia* is often longer than the femur, but nearly always more slender; it may be spiny, and is often armed with one or more articulated subapical spurs, which are occasionally very large.

The *tarsus* is usually divided into five segments *(tarsomeres)* of which the first is often the longest and is sometimes called the basitarsus or metatarsus. The number of tarsal segments varies from five down to one in different groups. The *pretarsus* is closely associated with the distal end of the last tarsal segment. It normally consists of a ventral *unguitractor plate*, a pair of claws, which vary in size and are sometimes toothed, and a projecting central lobe, the *arolium*. In the Diptera, there may be a pair of pads, the *pulvilli*, between the claws, and the arolium is usually replaced by a more ventrally situated *empodium*. In some insects, pulvilli *(euplantulae)* occur on the ventral surface of other tarsal segments.

Normally, all three pairs of legs are used for running (and are referred to as *cursorial*) or for walking *(gressorial)*; but they are employed only for perching and seizing prey in the Odonata, mainly for clinging to flowers in nectar-feeding Diptera, and are specialized for swimming *(natatorial)* in many aquatic Coleoptera and Hemiptera. The mid legs usually remain relatively simple, but the fore or hind may become extremely specialized, the fore legs for seizing prey *(raptorial)* or for digging *(fossorial)*, and the hind enlarged for jumping *(saltatorial)*. The mid or hind legs are raptorial in some Diptera and the hind in some Mecoptera. The fore legs of male Ephemeroptera are characteristically lengthened to hold the female in copulation, whereas the mid and hind may be more or less atrophied.

WINGS

It is generally believed that wings evolved from paranotal lobes which had developed on the thoracic terga of the ancestral insects (p. 159). Problems of balance would have arisen as these lobes became more aerodynamically effective, and those on the pronotum could have taken no further part in the development unless there were an associated shift in the centre of gravity, which apparently did not occur in any insect.

Almost endless possibilities of further modification arose once the wings became functional. Thus, the mesothoracic pair became modified into protective *tegmina* in the blattoid–orthopteroid orders, *hemelytra* in heteropteran Hemiptera, and *elytra* in the Coleoptera. This division of function led to concentration of the propulsive drive and at least most of the lift in the hind wings, and the process was carried further in groups with reduced tegmina or elytra. On the other hand, reduction of the hind wings in many insects led to improved aerodynamic efficiency in the fore wings, and the same advantage underlies the evolution of various methods of coupling the fore and hind wings together. Nevertheless, the ancient Odonata became highly efficient fliers without coupling, by developing excellent aerofoils and a remarkable degree of neuro-muscular control. Still another series of specializations is seen in small insects that take advantage of rising air currents and are dispersed in an 'aerial plankton' (Wigglesworth, 1963a). They tend to have small wings, reduced venation, and marginal fringes of long hairs. Probably the most remarkable modification of all is the transformation of the wings into paddles in the males of a marine midge (p. 114). Finally, wings may lack selective value in insects that have adopted a sedentary, cryptic, or parasitic way of life, or that live in montane, insular, or high-latitude habitats, and we find *brachypterous* (short-winged) and *apterous* (wingless) forms in almost every order of the Neoptera.

Apart from, or perhaps associated with, their great adaptive significance, the wings of insects show so many characters of value in classification that they have been used more extensively than any other structure in comparative studies. Their importance is the greater, because they are usually the only remains of insects that are recognizably preserved in fossils, which often show essential details, even down to the trichiation, with remarkable clarity.

Structure and Topography of the Wings

The wings develop as expansions of the terga to which they belong. During development, they are flat, lightly sclerotized bags, lined by epidermis, filled with blood, and containing nerve fibres and tracheae. The veins are laid down as bands of pigment in the cuticle. Those on the morphologically dorsal surface give rise to 'convex' veins which come to lie on ridges in the completed wing, and those on the ventral surface give rise to 'concave' veins lying in hollows or furrows. When the wing has expanded and dried after emergence of the adult, it becomes a membranous sheet, with the two layers of cuticle closely approximated, except where they are separated by the sclerotized tubes that form the strengthening and supporting veins. The more or less regular alternation between convexity and concavity in the arrangement of the veins is a further strengthening mechanism.

The membrane may be clear or marked by a pattern of pigment, and there is often a sharply defined, opaque or pigmented area, the *pterostigma*, near the distal end of the anterior margin. Hairs on the wings are of two kinds (Fig. 1.15): *microtrichia (aculeae)*, which are very small, and irregularly scattered

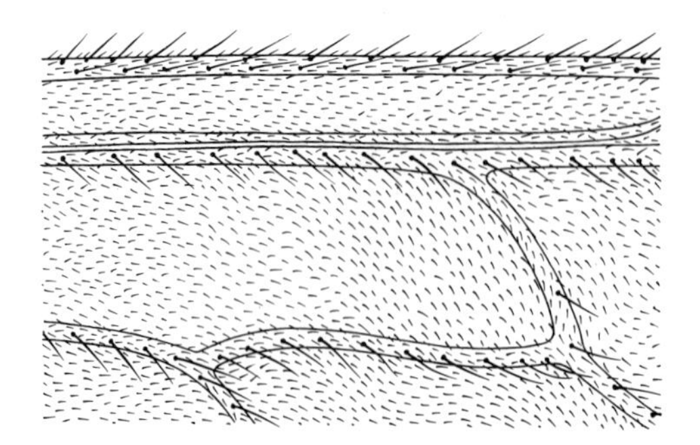

Fig. 1.15. Portion of fore wing of *Caecilius* sp., Psocoptera, showing macrotrichia and microtrichia. [B. Rankin]

over membrane and veins or gathered into diffuse patches; and *macrotrichia*, which are larger, provided with sockets, and often restricted to the veins. The *scales* on the wings of Lepidoptera, Trichoptera, and some Diptera are flattened and striated macrotrichia.

The wing is treated morphologically as being held in the horizontal plane at right angles to the long axis of the body. It is approximately triangular in shape, and its borders can be defined as *anterior* (or costal), *lateral* (or termen) from the apex to the anal angle (tornus), and *anal* (or ano-jugal) from there to the base (Fig. 1.16). A third *(humeral)* angle may be recognized when the costa curves sharply backwards at the base of the wing anteriorly. The wing may also be divided into four topographical areas (Snodgrass, 1935). At the base, there is a triangular *axillary* area containing the articular sclerites. The main part of the wing chiefly responsible for maintaining and propelling the insect in flight is called the *remigium*. It is separated by an *anal fold* from the more or less triangular *anal* (or vannal) *area*. The fourth area, less constantly present, is a much smaller *jugal area (neala)*, separated by a *jugal fold*, and lying between the anal area and the base of the wing posteriorly. These areas and the divisions between them are important in relation to the folding of the wing at rest, as well as to the functions of its different parts in flight, and they are also useful in defining the limits of the venational fields.

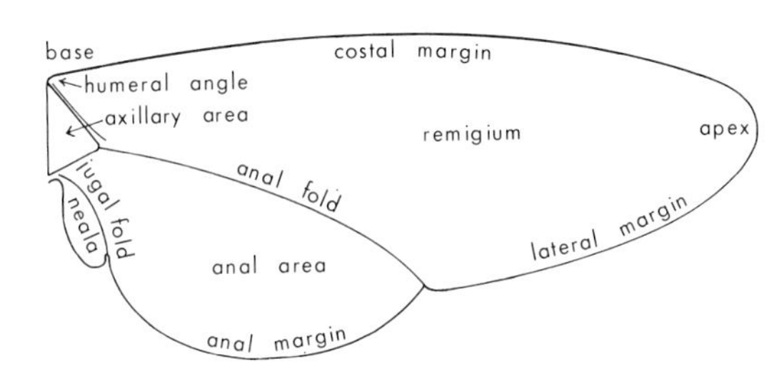

Fig. 1.16. Topography of an insect wing, diagrammatic. The 'lateral' margin is often better referred to as posterior or posterolateral.

Basal Attachment

The attachment of the wings to the thorax is somewhat complicated, in conformity with the fact that their movements are rather precisely limited and are produced in two different ways. Vertical flapping movements are generally produced by distortion of the notum through the action of the indirect muscles (p. 51); other movements, including antero-posterior and rotational movements, as well as those that bring the wing into the resting position in the Neoptera, are performed by the direct muscles that are inserted on some of the basal sclerites of the wing and

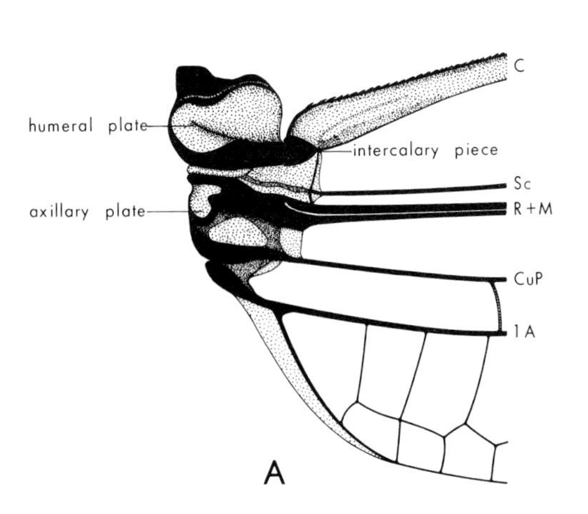

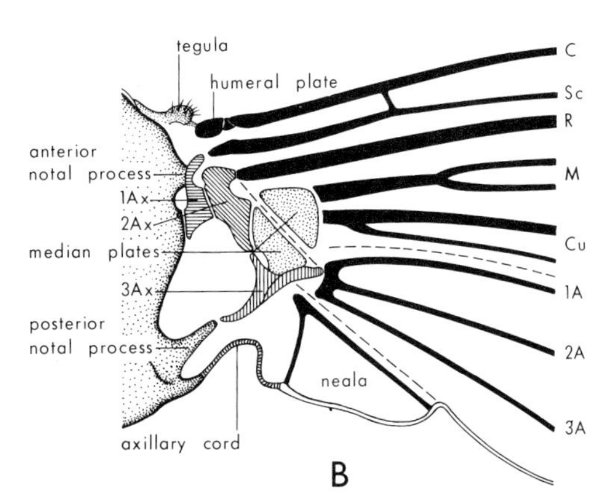

Fig. 1.17. Basal articulation of wings: A, Palaeoptera (*Hemianax papuensis*, Odonata, ♀); B, generalized Neoptera, diagrammatic (based on Snodgrass, 1935).

subalar pleura. Moreover, the mechanisms of attachment differ appreciably in the Palaeoptera and Neoptera.

In the Odonata (Fig. 1.17A) the wing has two large basal plates, the *humeral plate* anteriorly, to which the costa is articulated by a small intercalated piece, and the *axillary plate*, to which the remaining main veins are fused. These two plates articulate, somewhat loosely, with each other, with separate zones on the lateral edge of the notum dorsally, and with separate fulcral arms of the pleural wing process ventrally. As a result, flapping movement is unimpeded, deflection of the costal margin (important in flight control) is moderately free, and the small movements in other directions that are needed for stability and manoeuvrability can be made; but the wings can be closed (in nearly all Zygoptera and a very few Anisoptera) only by rotating them upwards over the back, where they usually come to lie in a more or less anteroposterior direction (Fig. 13.13). The Ephemeroptera appear to have rudimentary axillary sclerites, much as in Neoptera, but incorporated in a somewhat sclerotized area, and the wings can be closed only by apposition above the body (Fig. 12.1A).

The Neoptera can perform the same flight movements as the Palaeoptera, but (except in a few forms with secondarily limited movement) they can also fold the wings back along the abdomen, sometimes with extensive pleating of an enlarged anal area. The wing is hinged dorsally on the anterior and posterior notal processes of the thorax, and articulated ventrally with the pleural wing process at the upper margin of the episternum, which serves as a fulcrum. The linkage with these processes is mediated by a series of separate movable sclerites *(pteralia)* in the base of the wing (Fig. 1.17B), consisting of: the *tegula*, a scale-like sclerite basal to the costa of each fore wing; a *humeral plate* between the tegula and the base of the costa; the *first axillary* between the anterior notal process and the base of the subcosta; a *second axillary* articulating partly with the first axillary, partly with the base of the radius, and ventrally with the pleural wing process; and a *third axillary* articulating with the posterior notal process and the base of the anal veins. There may be a *fourth axillary* between the posterior notal process and the third axillary, and a pair of *median plates* distal to the second and third axillaries. The membrane at the base of the wing posteriorly is usually thickened to form a strengthening *axillary cord*. The costa can rotate backwards at the joints provided by the humeral plate, and the axillaries can angulate on one another, producing a sharp fold between the median plates and a crumpling of the wing basal to them, so that this part of the wing folds up as the rest of it is drawn back.

Coupling Mechanisms

A slow, irregular flight of unpredictable direction has certain advantages in protecting against fast-moving predators, as anyone who has chased butterflies knows; but there are also advantages in improved aerodynamic efficiency leading to higher speeds and a greater range of flight in relation to available energy reserves. As already mentioned, this has been achieved in various ways, and we are concerned here with the methods of coupling the fore and hind wings together that have been evolved independently in several orders (e.g. Figs. 24.2; 35.4; 36.4, 11, 13; 37.8G).

In the Hymenoptera a sclerotized fold along the posterior margin of the fore wing is engaged by a row of small hooks (hamuli) on the anterior margin of the hind wing, giving the *hamulate* type of coupling, which is also more or less developed in some Trichoptera and a few other groups. In some Hemiptera the wings are held together by a variety of small hooks or folds along the margins, and in the Psocoptera the costa of the hind wing is held by a spiny or hooked process at the node in the fore wing where vein CuP reaches the margin. The most complex and varied couplings are developed in the panorpoid orders (Tillyard, 1918–19). The jugal area of the fore wing is often developed into a *jugal lobe* ('fibula' in Lepidoptera, 'alula' in Diptera), and the anterior margin of the hind wing, when present, is produced into a small *humeral lobe* near its base. In the simplest condition these lobes merely overlap, and an extension of this process over a wider area produces the *amplexiform* method of coupling. In Mecoptera both lobes may bear long bristles, the jugal bristles lying on top of the hind wing in flight, while the humeral bristles form the *frenulum* which presses against the underside of the fore wing. In some Trichoptera and lower Lepidoptera the jugal lobe is produced into a projecting *jugum*. (For the varied methods of coupling in Lepidoptera see pp. 769–70.)

Wing Venation

The patterns of wing venation and the homologies of the veins were elucidated progressively in the classical studies by Comstock and Needham, Tillyard, Lameere, and Martynov; and it is a highly significant result of their work that the varied venations of all the orders of winged insects can be brought into a common basic plan. It is not difficult to imagine that an ability to fly might have evolved several times, as it has in the vertebrates, but it is unlikely that the same pattern of venation would have evolved more than once, and this is perhaps the strongest evidence in favour of a monophyletic origin of the Pterygota.*

Four criteria have been used in assessing the homologies of the veins.

1. The classical method was to follow the tracheation in the developing wing-sheath. In many insects the main veins develop on the courses of tracheae, which enter the wing in two main trunks, an anterior *costo-radial* and a posterior *cubito-anal*, and then divide to form the venational fields. The limitation of this method is that tracheation is incomplete or distorted in some orders.

2. Tillyard (1918) made use of the fact that macrotrichia are often limited to main longitudinal veins and their branches. There are many exceptions, and the method seems to have most use in indicating where veins may have disappeared.

3. There is a strong tendency for main veins to be alternately 'convex' and 'concave' (Fig. 1.18C), especially towards their bases, and they are often denoted, respectively, by plus and minus signs. This condition is undoubtedly primitive, and it is of considerable practical value in identifying veins and determining their homologies. Its limitation is that it was originally a response to mechanical need, and the level of a vein, particularly of its distal part, may have changed in response to changing needs in the course of evolution or as a result of capture or replacement by a vein of opposite sign.

4. The final court of appeal is a careful study of comparative morphology within groups and between groups, including the evidence that can be obtained from fossils. Generalized members of an order rarely present serious difficulty; it is when the venation is reduced or distorted that it may become difficult or impossible to apply any consistent notation to it.

It seems likely that the expanding paranotal lobes of the ancestors of the winged insects first became strengthened by a meshwork of thickenings, some of which has persisted as an *archedictyon* in some primitive groups (e.g. Fig. 14.10, fore wing), and that the main supporting veins were formed by condensation of the archedictyon on dorsal and ventral lines of stress as the lobes grew longer and thinner. It is widely held, too, that the most ancient winged insects had a fully developed venation, not unlike that of Palaeodictyoptera (Fig. 8.3), but Riek (unpublished) considers it more probable that the venation of the first winged insects was relatively simple (Fig. 1.18B), with the concave veins originally intercalated between the convex veins, like the triads of recent Palaeoptera (p. 19). An important consequence of this theory is that it would derive the Odonata from a more ancient stock than the Palaeodictyoptera.

The differences in interpretation that have developed over the years are of great theoretical importance, but they have involved only minor changes in notation. It is proposed to follow the system of Martynov (1930), except

* But see I. M. Mackerras, *J. Aust. ent. Soc.* **6**: 3–11, 1967.

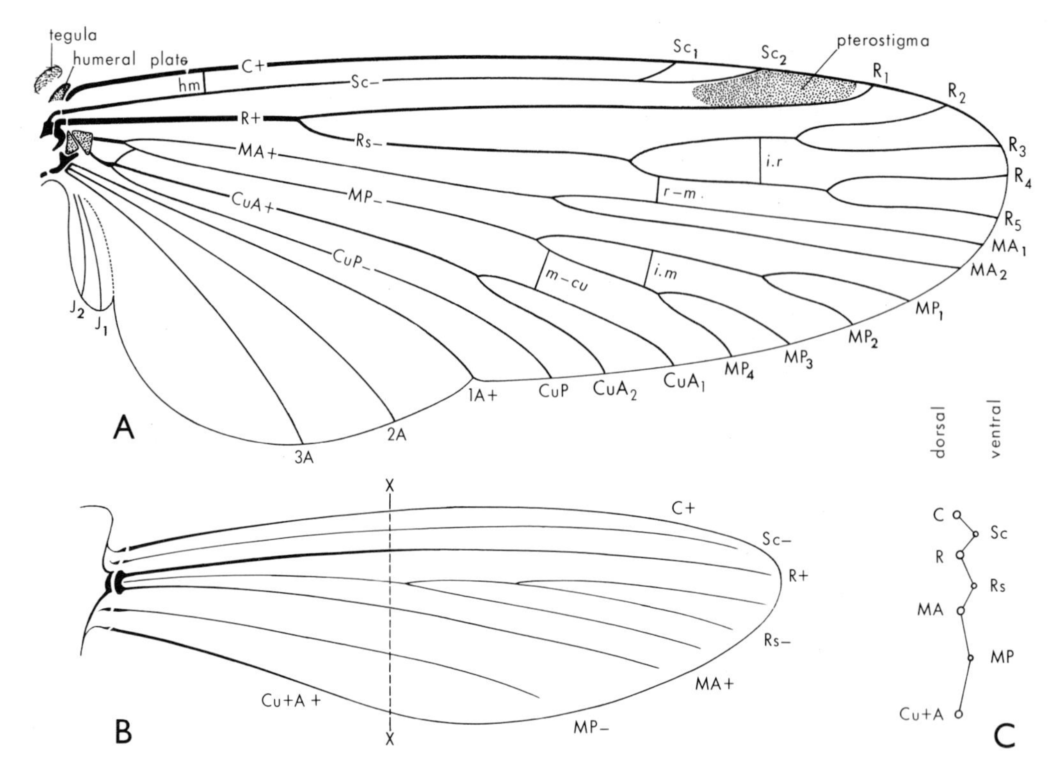

Fig. 1.18. Wing venation, diagrammatic: A, general plan to illustrate the modified Martynov notation; B, possible venational plan of early pterygotes with undeveloped cubito-anal field (from E. F. Riek, unpublished); C, section at X–X showing primary 'convexity' and 'concavity' of veins.

for the conventional division of R and Rs into R_1 to R_5. The primarily concave veins nearly always appear as branches of the convex vein anterior to them in recent insects, so six groups of veins, or *venational fields*, are usually recognized (Fig. 1.18A): costal, subcostal, radial, median, cubital, and anal.

The *costa* (C) is convex, usually strong, and usually marginal; it generally extends to the apex of the wing, and is often continuous with a delicate *ambient vein* around the lateral and anal margins.

The *subcosta* (Sc) is a concave, usually rather delicate vein lying in what is often a deep furrow between the costa and the radius; it is sometimes branched.

The *radius* (R) is often the largest vein of the wing, and, with Rs, covers the largest field. It is strongly convex, continued as R_1, and often unbranched. Rs generally arises as a posterior, concave branch from R (but sometimes from M); it normally divides twice to give four terminal veins, R_2 to R_5. At least R_1, and usually the anterior branches of Rs, reach the margin anterior to the apex and the posterior branches usually behind it.

The *media* (M) is usually not as strong as the radius, but it also is primitively two-branched, the *media anterior* (MA) being convex and normally single or two-branched, and the *media posterior* (MP), which corresponds to Rs, being concave and branching twice. MA is present, sometimes alone, in Palaeoptera and most exopterygote Neoptera; it may have disappeared in Hemiptera; and it probably became fused with Rs, MP alone occupying the median field, in endopterygote Neoptera. Where this has happened, or the identity of the stem is uncertain (as in Hemiptera and some other orders with reduced venation), it is customary to refer to the branches simply as M_1 to M_4 (e.g. Figs. 26.4, 34.7).

The *cubitus* (Cu) is another strong vein, balancing R (though on a smaller scale) behind the middle of the wing. Again there are two primary branches, a strongly convex CuA, which usually divides distally into two branches called CuA_1 and CuA_2, and a concave, usually unbranched CuP, which often lies in the anal fold of the wing, and is sometimes reduced to a mere line.

The *anal* veins (A) generally arise serially from a common base. Usually there are three, which are alternately convex and concave (though all sometimes appear to be convex), and they are designated 1A, 2A, 3A. In hind wings with a very wide, sharply defined anal area, as in most orthopteroid orders, they may be very numerous, giving a fan-like appearance (Fig. 1.2), and on this account they have been called vannal veins by Snodgrass (1935). He also regards the vein here designated 1A as independent from the others, and calls it the postcubitus (Pcu).

Lastly, there are often one or two small veins in the jugal area at the base of the wing, and these have sometimes been called first and second *jugals*, or arcuate and cardinal veins respectively.

The generalized arrangement shown in Figure 1.18A involves a simple *dichotomous* branching of the veins, but two other types of branching also occur (Fig. 1.19). One is *pectinate*, in which a number of branches come by migration to lie serially along a single stem (see also Fig. 29.7). The other, called *triadic* by Tillyard, represents a primitive condition that is extensively developed only in the Palaeoptera amongst recent insects. It consists in the interpolation of a longitudinal vein between the pairs of branches of the main veins. The intercalated vein is of opposite sign to those between which it lies, and the group is called a positive triad when a concave vein is intercalated between two convex branches, and a negative triad when a convex vein is intercalated between two concave ones.

In addition, there may be considerable terminal branching beyond the basic amount already described, and short branches *(veinlets)* may diverge from the main veins. The *humeral vein* (hm) near the base of the costal area of the wing and the costal series of veinlets that may follow it are the most important examples of this class. *Cross-veins* are transverse struts, which are rarely preceded by a trachea, and which strengthen the wing by joining longitudinal veins together. They vary in number and position, and are indicated by lower-case italic letters hyphenated together (for example, *r–m* between the radial and median fields); only those that are most constant in occurrence and position are shown in Figure 1.18A.

Finally, the areas into which the wing membrane is divided by the longitudinal and cross-veins, or by confluence of two longitudinal veins, are called *cells*, and they are designated by letters and figures corresponding to the veins or portions of veins behind which they lie, as, for example, cell R_4 behind vein R_4 at the apex of the wing. Special cells, such as the 'median' (or 'discal') and 'basal' cells, are usually indicated by name, and it is not to be forgotten that these terms are used for different cells in different orders.

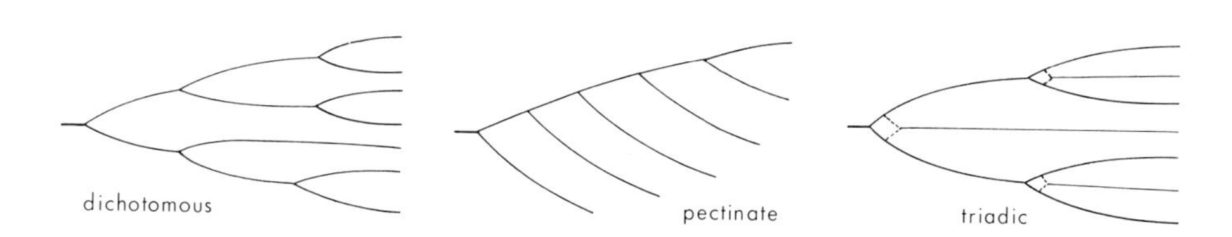

Fig. 1.19. Types of branching of wing veins.

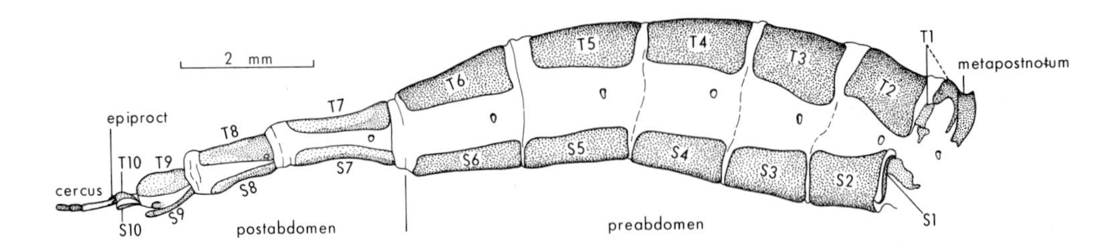

Fig. 1.20. Abdomen of *Chorista australis*, ♀, from a cleared preparation with segments fully extended.

ABDOMEN

The abdomen (Fig. 1.20) is the least specialized part of the insect. It consists primitively of eleven segments, although the first is often reduced, sometimes absent or incorporated in the thorax, and there may also be reduction at the caudal end. The first eight segments are generally of broadly similar architecture. Each is strengthened by a more or less arched tergal plate and a smaller, flatter sternal plate, separated from each other by a relatively wide expanse of pleural membrane, and the conjunctival membrane between consecutive plates is also relatively extensive. Consequently the abdomen is normally more mobile, and certainly more distensible, than the other parts of the body. There are no appendages on these segments, except in the Apterygota, although remnants of larval gill filaments are retained in Plecoptera. Spiracles are normally present on the pleura of segments 1 to 8, but may be reduced, or incorporated in lateral extensions of the terga. The endoskeleton consists of tergal phragmata and sternal furcae, which are sometimes highly developed. In some groups the first four or five segments are developed into a more or less globular *preabdomen* with rather closely united terga, the remainder forming a tubular, retractile *postabdomen*. The Hymenoptera show some remarkable modifications in the size and shape of the segments, and the first is fused with the thorax in the Apocrita.

Terminalia

The segments from 8 or 9 (occasionally from 7) to the end of the abdomen are modified and more or less combined to form an ano-genital tagma, which may be referred to as the *terminalia*, the genital parts being the *genitalia*. Descriptions of these structures in the different orders have been published by a number of authors in Tuxen (1956). The male terminalia have been widely used in taxonomy, particularly as aids in distinguishing between species, and in some genera the only clearly recognizable specific characters are to be found in them. They are simpler and less varied in the females, and their value in this sex seems usually to lie more in providing a guide to the phylogeny of a group than in the differentiation between species.

Excretory and sensory adaptations of the terminalia do not, in general, differ greatly between the sexes or in different groups of insects, but the reproductive adaptations do. Insects almost certainly evolved from ancestors in which internal fertilization was achieved by indirect transfer, the females picking up spermatophores deposited by the males. This method has survived in the Apterygota. True copulation (i.e. direct transfer of spermatophores or free sperms by apposition of specialized genital structures) evolved at least twice in the Pterygota: by the development of an intromittent organ on abdominal sternum 2 of the male in Odonata; and by specialization of the terminalia in all other orders. The primitive mating position in these orders appears to have been with the male beneath (or at the side), but the evolution of male dominance in the act led to the adoption of other positions, and to the development of efficient holding and intromittent organs (Alexander, in Highnam, 1964).

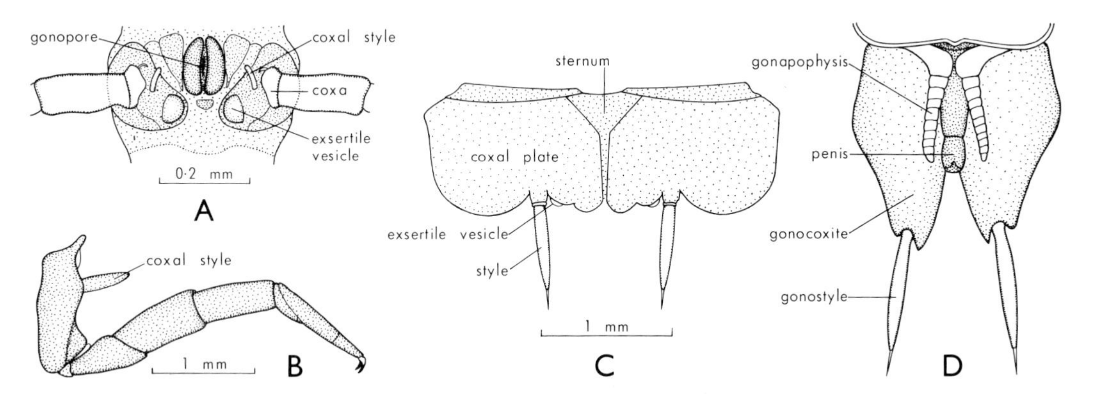

Fig. 1.21. A, 4th trunk segment of *Hanseniella* sp., Symphyla, ♂, ventral, with functional limbs, coxal styles, and exsertile vesicles; B, mid leg of *Allomachilis froggatti*, Archaeognatha, ♀, posterior aspect, with coxal style (no exsertile vesicles on thorax); C, mid-abdominal segment of *A. froggatti*, ventral, with coxites becoming incorporated in sternum, styles (rudimentary distal segments of limb), and exsertile vesicles; D, 9th (genital) segment of *Machilis variabilis*, ♂, with coxites, gonostyles, and gonapophyses, but no exsertile vesicles (after Snodgrass, 1957).

The accessory structures that were available to become adapted for reproductive functions can be appreciated best on the theory, for which there is a good deal of evidence, that the ancestors of the insects had reduced abdominal limbs, with coxal styles at their bases and eversible vesicles medial to them, the basic plan being similar to that of the Symphyla (Fig. 1.21A). In Archaeognatha segments 8 and 9 of the abdomen have well-defined coxites *(gonocoxites)* and often two pairs of styliform appendages, but apparently no vesicles. There has been considerable confusion, because the term 'style' has been used indiscriminately for both kinds of appendage, but the most plausible explanation is that the distal, more lateral pair on each segment are rudimentary limbs *(telopodites)* and the medial pair are true coxal styles. The latter may be distinguished as *gonapophyses*, but the term 'style' (or *gonostyle*) has become so widely used for the former that its continued employment cannot now be avoided. No definite appendages have been detected on segment 10 of any primitive insect.

Female

The functions of the terminalia are to receive sperms, to deposit eggs in appropriate situations, sometimes (Blattodea, Mantodea) to form an ootheca, occasionally (viviparous species) to accommodate a developing embryo or larva, to dispose of waste products, and to serve as a base for posterior sense organs. Modifications to aid the male in clasping are not uncommon on other parts of the body, but are generally inconspicuous in the genital segments. The *gonopore*, which is the opening of the distal (usually secondary) part of the common oviduct, is in the median line behind sternum 7 in Ephemeroptera and Dermaptera, 8 or 9 in other insects, sometimes on the surface, but commonly at the bottom of a more or less deeply invaginated *genital atrium* or *vagina*, the external opening

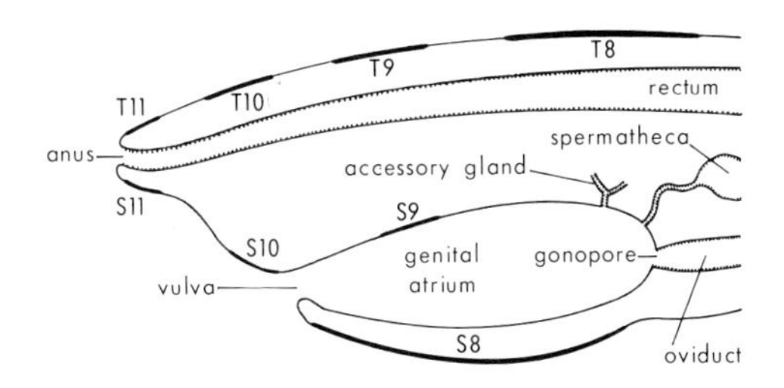

Fig. 1.22. Terminal abdominal segments of a generalized ♀ neopteron, diagrammatic median sagittal section with appendages omitted (cf. Figs. 4.1, 14.3A).

then being known as the *vulva* (Fig. 1.22). The spermathecal duct and accessory glands (p. 69) usually open into the atrium. The *anus* is on segment 11, which may become incorporated in 10.

In Apterygota (Fig. 1.23), segments 8 and 9 have well developed coxites, styles, and gonapophyses, the two pairs of gonapophyses being closely applied together to form an *ovipositor*; segment 10 is complete; segment 11 is small, the tergum forming a mid-dorsal *epiproct*, and the sternum being

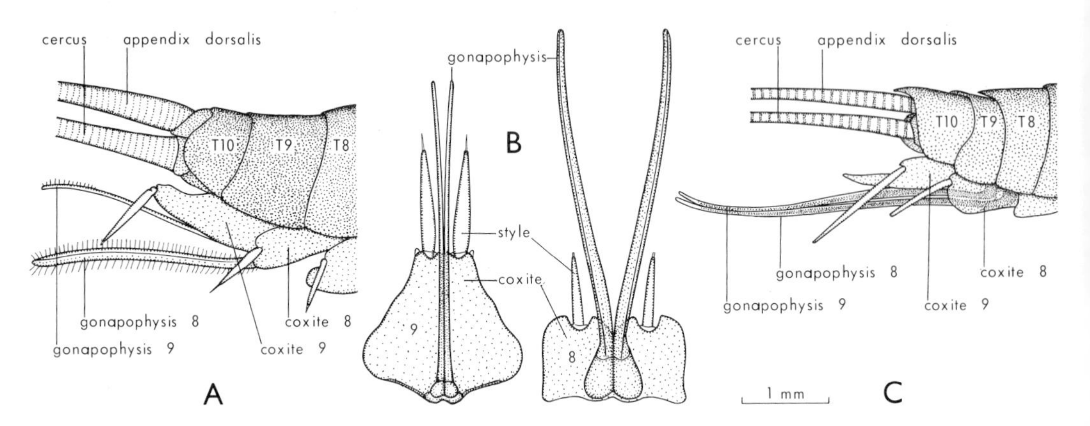

Fig. 1.23. Female terminalia of Apterygota: A, *Allomachilis froggatti*, Archaeognatha, lateral (for normal position of gonapophyses see Fig. 7.5); B, appendages of 9th and 8th segments of same, dorsal aspect; C, *Ctenolepisma longicaudata*, Thysanura, lateral. All at same magnification.

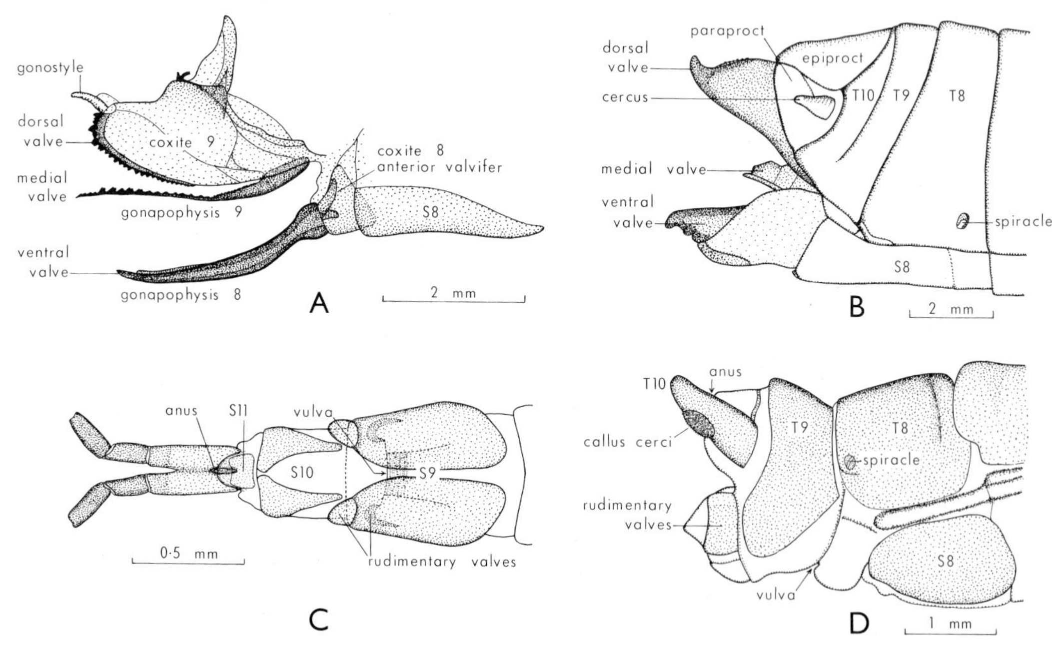

Fig. 1.24. Female terminalia of some Pterygota: A, ovipositor of *Synlestes tillyardi*, Odonata, dissected, lateral–the gonapophyses form the terebra (for normal position of the parts see Fig. 13.6); B, *Gastrimargus musicus*, Orthoptera-Caelifera, lateral; C, *Chorista australis*, Mecoptera, ventral–the rudimentary appendages of S9 presumably represent the medial and dorsal valves; D, *Archichauliodes guttiferus*, Megaloptera, showing reduction and compaction. All except B from cleared preparations.

partly divided into a pair of ventral *paraprocts*; the anus is surrounded by three small papillae. The *cerci* are articulated to segment 11 laterally, and a median *appendix dorsalis* ('caudal style') arises from the distal end of the epiproct just above the dorsal anal papilla. This is a remarkable appendage, which is found only in the Apterygota, some Palaeodictyoptera, and most Ephemeroptera. There is no trace of it in Neoptera, such median outgrowths as occur being secondary structures, nor in more primitive hexapods (Collembola, Protura, Diplura). Its significance in the evolution of the insects is obscure.

In the Pterygota, the appendages of segments 8 and 9 are reduced, and the ovipositor consists of three pairs of *valves*. In some Odonata (Fig. 1.24A), for example, the gonocoxites of segment 8 are reduced to a small sclerite (*anterior* or first *valvifer*) on each side, with a long process (*ventral*, anterior, or first valve) representing the gonapophysis, and no gonostyle. The gonocoxites of segment 9 are elongate; their bases are the *posterior* or second valvifers, their distal parts the *dorsal*, lateral, or third valves, with a subapical style, and the gonapophyses the *medial*, posterior, or second valves.* Scudder (1961) has noted that a detached piece (gonangulum), probably derived from the base of the 9th-segment coxite, is constantly present, and replaces the 8th-segment coxite as the anterior valvifer in some orders; he also proposed the term gonoplac for the dorsal valves. Except for loss of the style, the structure is basically similar, however widely the functional requirements may vary, in Orthoptera (Fig. 1.24B) and other groups (homopteran Hemiptera, terebrantian Thysanoptera, Hymenoptera) that have retained a well-developed ovipositor. In Blattodea, a large genital atrium is formed by infolding of sternum 8, the surface of which is modified as a mould for the ootheca, and the valves are considerably reduced (Figs. 14.3A, B).

The segmental structures are retained only to the extent that they perform a specific function. Thus, the appendages of segments 8 and 9 are greatly reduced or lost in Ephemeroptera, many Odonata, Plecoptera, Isoptera, most hemipteroids, and most endopterygotes; segments 10 and 11 are compacted in representatives of many orders, though the epiproct and paraprocts usually remain recognizable; and cerci are absent in hemipteroids, and reduced or rudimentary in endopterygotes. The type illustrated in Figure 1.24C is fairly typical of many panorpoid terminalia.

In the Ephemeroptera, sternum 7 is sometimes produced posteriorly to form a *subgenital plate*. The subgenital plate is formed by sternum 7 in Blattodea also (8 being inflected), but by 8 in most other groups in which it is differentiated. Tergum 10 or 9 (or both) may be divided into *hemitergites* by a median cleft, either may form a *supra-anal plate* over the anal segment dorsally, and the corresponding sternites may be reduced or absent. An endoskeletal *furca* may also be developed as a separate internal sclerite. An extreme modification is seen in the Lepidoptera, in which there may be two genital openings (one for copulation, one for oviposition), or they may be combined with the rectum in a common cloaca opening at the end of the body (p. 772).

Male

The functions of the terminalia are to maintain apposition with the genital opening of the female during copulation, often to construct spermatophores, to inject spermatophores or semen into the genital tract (occasionally into the haemocoele) of the female, and to perform the same excretory and sensory functions as in the female. So far as the structures are concerned, there is endless confusion in terminology (see Tuxen, 1956) and sharp disagreement on homologies. It is consequently difficult to present a clear picture of evolutionary sequences. The Odonata, as already mentioned have a different method of copulation from other insects, and their genitalia are described separately on p. 246.

* The terms that appear to be most commonly used are italicized.

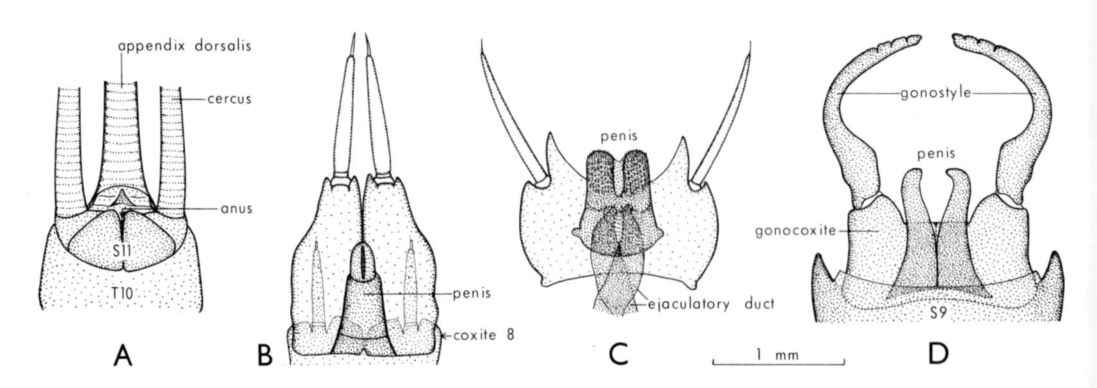

Fig. 1.25. Male terminalia: A, terminal segments of *Allomachilis froggatti*, Archaeognatha, ventral, with S10 and underlying structures removed; B, 9th-segment appendages and penis of *A. froggatti*, dorsal aspect; C, same of *Ctenolepisma longicaudata*, Thysanura; D, same of *Mirawara aapta*, Ephemeroptera, with fused gonocoxites.

The males of a few Archaeognatha (Fig. 1.21D) have gonocoxites, gonostyles, and gonapophyses on both segments 8 and 9, and are consequently like females, except that a membranous median *penis* (or phallus) projects backwards between the posterior gonapophyses. In most Apterygota (Figs. 1.25B,C) the 8th-segment coxites are relatively small and without gonapophyses, and the penis of Thysanura tends to be bilobed and sometimes contains separate ejaculatory ducts. The terminal segments and appendages are similar to those of the female. That the differences between the sexes are so slight is presumably associated with the absence of complete copulation, the male producing a thread beaded with sperm droplets, and simply guiding the vulva of the female into contact with it by means of an antenna or cercus.

The Ephemeroptera (Fig. 1.26) probably show the most primitive method of aerial mating in recent Pterygota (Brinck, 1957). The male approaches the female from below, hooks his long, 'double-jointed' fore tarsi over the bases of her wings, which he steadies with his cerci, grasps her abdomen with his 'forceps', and inserts his penes into her vulva from below and behind. There are no appendages on segment 8 (nor in any pterygote males); the gonocoxites of 9 often have segmented styles (Fig. 1.25D); there are usually two penes containing separate ejaculatory ducts; and there are sometimes associated processes which may be gonapophyses. Segment 10 is well developed, 11 is represented by a small epiproct and separate paraprocts, and the cerci and appendix dorsalis are usually long. The similarity of the apparatus to that of the Apterygota is evident.

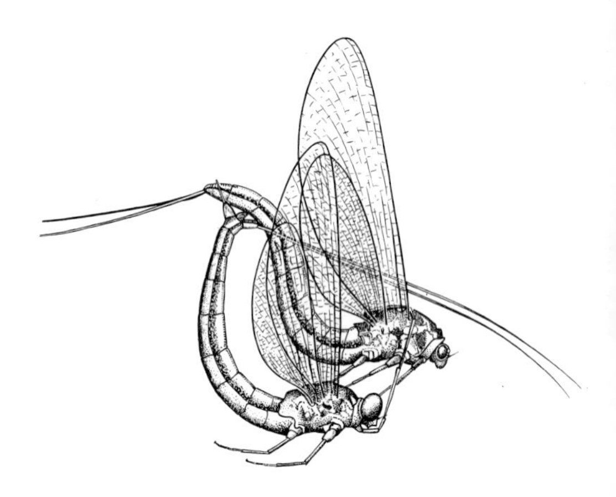

Fig. 1.26. Aerial mating of Ephemeroptera (after Brinck, 1957). [T. Nolan]

A marked contrast is provided by the blattoid–orthopteroid orders. There is little doubt that terrestrial mating was established very early in their history, and Alexander has illustrated the divergences from the primitive mating position that developed as the males became dominant. In the simplest (e.g. in

Plecoptera and acridoid Orthoptera), the male mounts on the female, grasps her thorax with his fore and mid legs, curls the end of his abdomen under hers, and applies his genitalia to hers in the same relationship as in the original male-beneath position (Fig. 1.27A). Contact can be maintained by general muscular action, and position by the peg-like action of his phallic organs in her

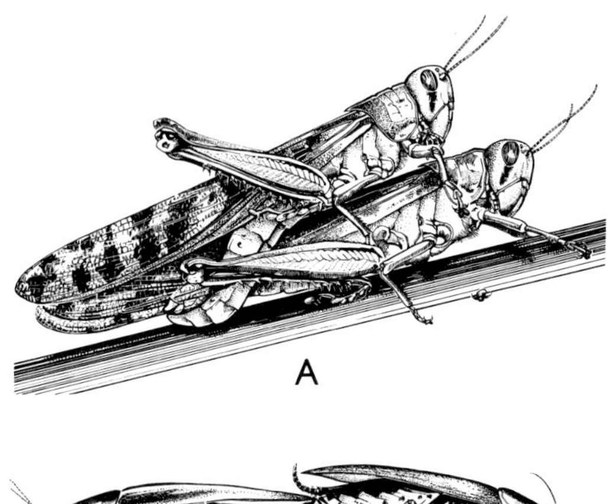

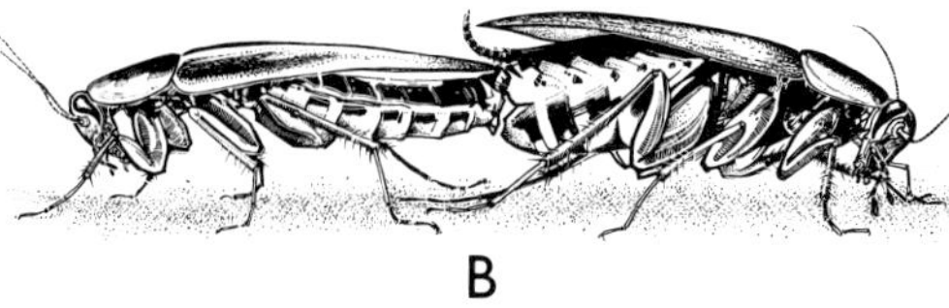

Fig. 1.27. Terrestrial mating of: A, a locust, Orthoptera (after Lecomte, 1964); B, *Blattella germanica*, Blattodea (after Roth and Willis, 1954). [T. Nolan]

genital chamber. The other principal modification may be illustrated by the Blattodea (Fig. 1.27B). The male attracts the female on to his back by means of a dorsal gland (or other tergal structures), joins his genitalia to hers, and the insects then rotate until they point in opposite directions without twisting the terminalia.

The mechanical requirements are entirely different from those of the Ephemeroptera. Strong abdominal claspers are not needed, and would be an impediment to the kinds of rotation that occur in these orders. Small gonocoxites are preserved in Grylloblattodea (Fig. 1.28A), but in the other orders they are absorbed into sternum 9, which becomes a *subgenital plate* underlying a large genital cavity. Their styles remain recognizable only in Grylloblattodea, Blattodea, Mantodea, some Isoptera, and some ensiferan Orthoptera. On the other hand, the phallic organs rising from the base of the genital chamber are often strongly and variously developed. Primitively, there appear to be three, a median penis or penile lobe and a *phallomere* (Figs. 1.28A,B) on each side of it; but there is great variation, the penis sometimes acquiring additional sclerotizations, as in Acridoidea, and the phallomeres becoming modified or distorted in a variety of ways (Snodgrass, 1937). Thus, in Blattodea (Fig. 14.3C) there is gross asymmetry, the penis is replaced by a

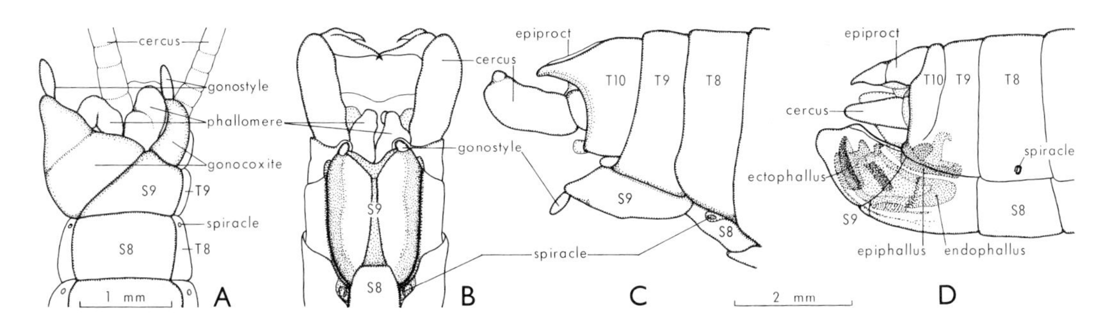

Fig. 1.28. Male terminalia of orthopteroids: A, *Grylloblatta campodeiformis*, Grylloblattodea, ventral–asymmetrical coxites and styles present; B, *Euconocephalus* sp., Orthoptera-Ensifera, posteroventral–coxites absorbed in S9, styles present; C, same, lateral; D, *Gastrimargus musicus*, Orthoptera-Caelifera, cleared preparation with phallic complex seen through cuticle–S9 divided transversely, but no coxites or styles. B–D at same magnification.

'ventral phallomere', and the apex of the left or right phallomere is developed into a strong process, which hooks under the edge of the subgenital plate of the female and retains a hold while the bodies move round. Conversely, in some Plecoptera (which do not rotate) the phallic parts are reduced, and the whole genital chamber may be eversible, or a secondary intromittent organ may be developed from the epiproct (Brinck, 1956). The Dermaptera (p. 308) have an elongate, sometimes double penis, but no phallomeres.

So far, the picture is reasonably clear, and it is now necessary to examine two important questions before proceeding further. These concern the composition of the intromittent organs and the homologies of the clasping organs in hemipteroid and endopterygote insects.

The intromittent organ of most higher insects is a more complex structure than the simple penis of Apterygota or Ephemeroptera. It develops postembryonically by fusion of initially paired primary genital papillae to form a median, tubular, often eversible *endophallus* (Snodgrass, 1957), with

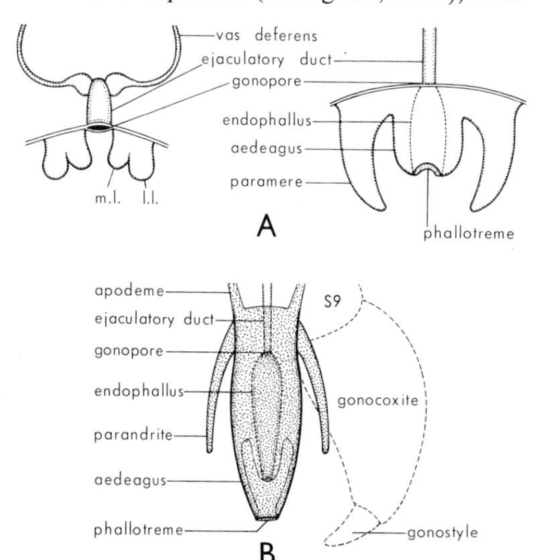

Fig. 1.29. A, development of the external genitalia of the male (after Snodgrass, 1957)–left, division of the primary genital papillae into medial and lateral lobes (*m.l.*, *l.l.*); right, fusion of the medial lobes to form the intromittent organ. B, structure of a generalized aedeagus with a pair of parandrites ('parameres').

the conjoined ejaculatory ducts opening by the gonopore at its base, and itself opening distally by a *phallotreme* (Fig. 1.29A). Its external walls may become sclerotized or modified in a wide variety of ways. The whole organ is commonly known as the *aedeagus*, and its functions are not always exclusively intromittent. It develops from the tissues behind sternum 9, and some authors regard it as belonging to segment 9, whereas others (e.g. Snodgrass, 1957) consider that it is derived from 10. There is even wider difference of opinion on the origins of its sclerites, and especially on the nature of a pair of processes (sometimes called 'parameres' or 'penis valves') that may be present at its sides or incorporated in its lateral walls (Fig. 1.29B). They may not be homologous with the 'parameres' discussed below, and *parandrite* (Crampton, 1938) is probably the best term to use for them when they are present.

The clasping organs of the higher insects have generally been regarded as homologous with the gonocoxites and gonostyles of Ephemeroptera, their reduction in some groups being parallel to what occurs in the blattoid–orthopteroid orders. This theory is based primarily on the comparative anatomy of the adults. It carries the implication that these appendages are 9th-segment structures, and the suggestion that the ancestors of these insects may have mated in the air, as Ephemeroptera do today. Moreover, strong external clasping would preclude the kind of rotation that is possible in the blattoid–orthopteroid orders, so that deviations from the primitive mating position would be likely to produce twisting of the terminal segments, a phenomenon that is not uncommon in endopterygotes.

On the other hand, Crampton (1938) and Snodgrass (1957) have maintained that these organs are derived from the phallic complex and not the clasping complex of primitive insects, because they arise by subdivision of the primary genital papilla on each side. As the term 'paramere' was first used for the structures in beetles that they believe to be homologous with the clasping organs of other endopterygotes, they applied it to all. It follows from this theory that the 'parameres' may be 10th-segment structures; that the ancestors of both hemipteroid and endopterygote insects mated on the ground or vegetation, and had already lost all trace of the 9th-segment coxites and styles; and that unsegmented 'parameres' close to the aedeagus (as in beetles; Fig. 1.31c) represent the primitive condition, their development into efficient, two-segmented claspers in the panorpoid orders being a response to increasing mobility of the insects and exposure to hazards during mating.

It is unlikely that hemipteroids and endopterygotes evolved from the same ancestral stock, so neither theory need be true for both groups. There is some collateral evidence to support the paramere theory in the hemipteroids, but the weight of probability still lies with the gonocoxite theory for the endopterygotes, and it would seem undesirable to depart from it for either group without further evidence.

End-to-end mating without torsion of the parts seems to be common in the hemipteroid orders, and the genitalia are usually narrow, with approximation of the gonocoxites to the aedeagus and loss of the gonostyle (Fig. 1.30). There is often asymmetry, and in cimicoid bugs this is associated with development of one gonocoxite into a piercing organ for use in haemocoelic insemination (see Hinton, in Highnam, 1964).

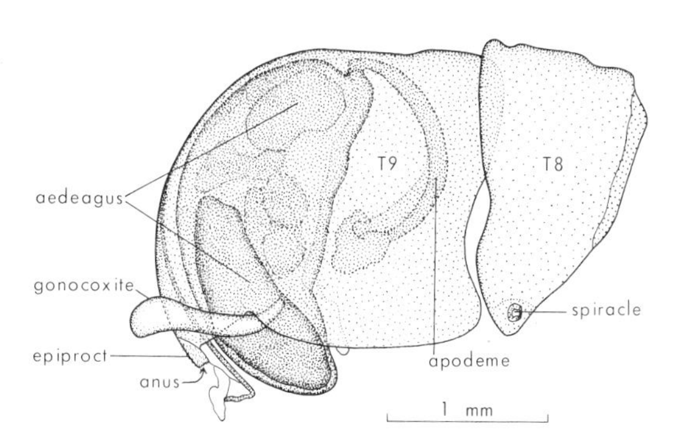

Fig. 1.30. Male terminalia of *Pristhesancus papuensis*, Hemiptera-Heteroptera, cleared preparation with aedeagus seen through cuticle of the pygofer (T9).

Methods of mating are more varied in the

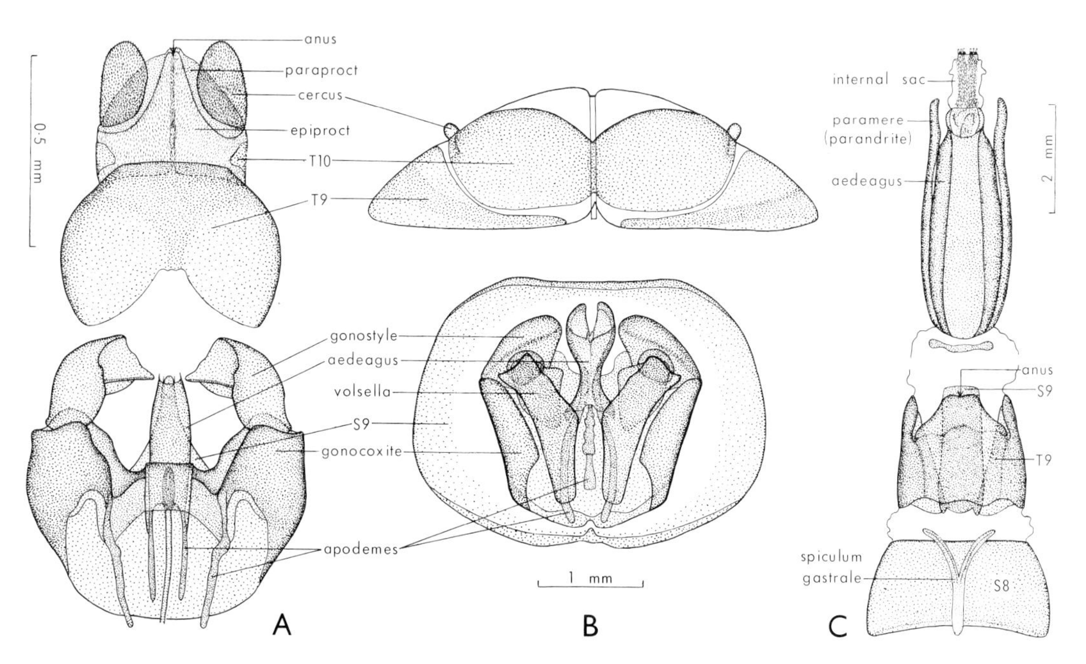

Fig. 1.31. Male terminalia of some endopterygotes, dissected: A, *Chiromyza* sp., Diptera; B, *Perga affinis*, Hymenoptera-Symphyta; C, *Ptomaphila lacrymosa*, Coleoptera, T8 removed and the aedeagus withdrawn from its resting position within segments 8 and 9.

endopterygotes, including primitive side-by-side orientation, as well as male-above, end-to-end, and even venter-to-venter in a few Diptera and Hymenoptera. Permanent torsion of the terminal segments is common in Diptera and occasional in other orders. There is a trend towards displacement and reduction of parts in Megaloptera, Neuroptera, and Coleoptera, and to loss of styles but development of parandrites on the aedeagus and accessory appendages ('volsellae' or 'claspettes') on the coxites in Hymenoptera (Fig. 1.31B). The panorpoid orders show the greatest variation, from diagrammatic simplicity in some generalized Diptera to complicated arrangements involving, not only the presence of parandrites and claspettes, but various displacements, and the development of a wide variety of accessory lobes or processes, culminating in the complexity seen in the Lepidoptera (p. 771). There may also be strong apodemes* extending into the body from the base of the endophallus, aedeagus, or coxites. Segments 10 and 11 and the cerci are often reduced or partly absorbed, but the appearance of the cerci and paraprocts usually remains characteristic, and is often a valuable aid in orientation when rotation is suspected.

* These are not limited to endopterygotes; they occur also, for example, in Dermaptera.

2

GENERAL ANATOMY AND PHYSIOLOGY

by D. GILMOUR

in association with D. F. WATERHOUSE *and* M. F. DAY

This chapter extends the survey of insect anatomy to the internal organs, and relates the anatomy of the organ systems to their physiological function, to the chemistry of the tissues, and to the flow of metabolism. References are given to some selected publications. For more extensive general treatments the books by Snodgrass (1935), Wigglesworth (1965), Roeder (1953), Rockstein (1964), and Gilmour (1965) should be consulted, as well as those by Bullock and Horridge (1965) on the nervous system, Dethier (1963) on the special senses, and Davey (1965) on the reproductive system.

THE INTEGUMENT

Structure and Function

The integument (Fig. 2.1) comprises a single layer of epidermal (or 'hypodermal') cells and the cuticle of varying thickness which they secrete. The *epidermis* is bordered on its inner side by a thin sheet of connective tissue, the *basement membrane*. The *cuticle* covers the whole external surface, is continued into the fore and hind guts, and lines the ducts of the dermal glands and the entire tracheal system. As the cuticle serves not only as a barrier against the external world but also as the insect's skeleton, it may be extremely thick and hard in some regions, although in other areas, such as the intersegmental membranes or most of the body surface of many larvae, it is relatively soft and flexible. Two major subdivisions of the cuticle are recognized on the basis of chemical constitution: a thin, outer, non-chitinous *epicuticle*, and a thicker, inner, chitinous *procuticle*. The latter is traversed in a direction normal to the surface by fine *pore canals*, which may contain protoplasmic extensions of the epidermal cells. The outer region of the procuticle normally undergoes hardening after a moult, thus differentiating this layer into an outer, dense, dark-coloured *exocuticle*, and an inner, more transparent *endocuticle*. In section, the endocuticle is seen to be made up of regularly alternating laminae of differing refractive index. The surface of the cuticle may be sculptured in various ways. It may bear hairs or scales (p. 6), or minute mushroom-shaped projections, or enclose a sponge-like system of air-filled galleries, as in the plastron of aquatic species (p. 39).

The Epicuticle. The epicuticle normally has at least two major components: an inner lipoprotein *(cuticulin)* layer and an outer *wax layer*. The wax layer is responsible

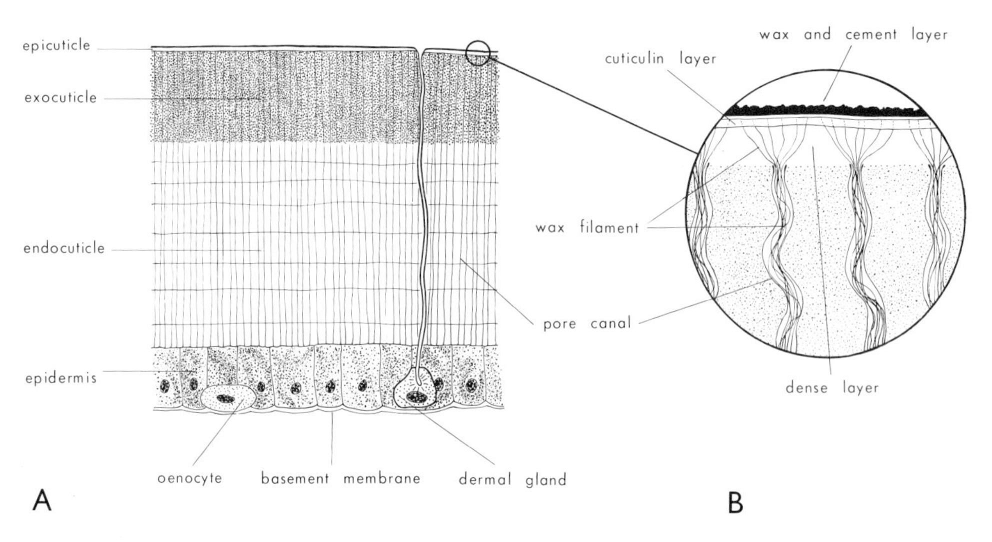

Fig. 2.1. Structure of the integument: A, generalized diagram; B, enlargement of outer region (after Locke 1961).

for the great resistance of the cuticle to movement of water from the insect's body to the environment, a property of critical importance to insects living in dry terrestrial habitats. Insect cuticular waxes are mixtures of straight-chain paraffins and esters of straight-chain fatty acids and primary alcohols, along with smaller amounts of free acids and alcohols. Chain lengths of the paraffins and fatty acids vary from 12 to 34 carbon atoms, while the alcohols have from 20 to 36 carbon atoms. The waxes contain varying amounts of unsaturated constituents and chemically reactive compounds, such as aliphatic aldehydes. These have lower melting points than the saturated components, and, when present in appreciable quantities, may produce a soft mobile wax, as in the cockroach, in contrast to the more usual solid wax layer (Gilby, 1965). The more reactive molecules tend to polymerize and to combine with other cuticular components, thus stabilizing the wax layer after its secretion. Relatively large amounts of polymerized lipids in the form of resins, which are chemically more resistant than the other wax components, occur in some cuticles. Such resins are probably the main constituents of the *cement layers* seen in some epicuticles, usually as an external covering for the wax.

It has been shown that the inner layers of the wax covering are most important in establishing the cuticle's resistance to water loss, and it has been postulated that the significant element in waterproofing is an oriented monolayer of hydrocarbon molecules with their polar groups directed towards the hydrated protein component of the cuticle. Not only do terrestrial insects resist water loss; the cuticles of some are able to absorb liquid water or water vapour from humid air. That is, the integument acts as a valve permitting water movement in one direction only. This phenomenon is an expression of the metabolic activity of the epidermal cells; it disappears when the insect is anaesthetized. The formation of oriented monolayers in the wax layer, as well as metabolic control of the state of hydration of the protein layers of the cuticle, have been considered to play a part in the operation of this water 'pump' (Beament, 1964). Another explanation suggests that the minute wax-filled canals that traverse the cuticulin layer (Fig. 2.1B) are the route of water entry. These canals are thought to contain lipid-water liquid crystals, in which a change of phase with an increase in water content allows the inward flow of water (Locke, 1965).

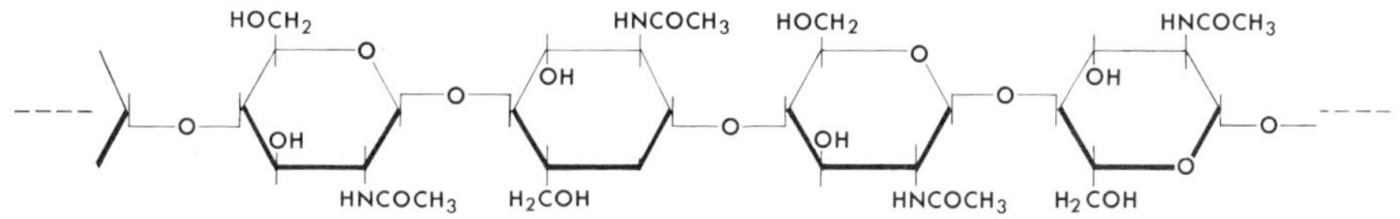

HOCH2 HNCOCH3 HOCH2 HNCOCH3 OH OH OH OH HNCOCH3 H2COH HNCOCH3 H2COH

Fig. 2.2. Structure of chitin.

The Procuticle. The procuticle contains protein and chitin, usually in about equal amounts. *Chitin* is a polysaccharide of large molecular weight in which the unit sugar is *N*-acetylglucosamine (Fig. 2.2). It is usually secreted in close association with protein, and some evidence suggests that it is linked with protein in the form of a stable copolymer, or glycoprotein. The laminations seen in most endocuticles do not express the separation of chitin and protein layers, but are more probably the result of different orientation of microfibrils containing both components. In a few specialized regions which possess a highly elastic cuticle, however, the epidermis secretes chitin and protein sequentially, so that the two constituents are separated in different layers.

The proteins extracted from the cuticle over the general body area have a fairly high content of tyrosine. It is thought that this amino acid participates, along with the free amino groups of the protein, in the hardening process which is described below. The regions of elastic cuticle have been found to contain a unique rubber-like protein, *resilin*. These areas are particularly important in the articulation of the wings, where they form hinges and elastic ligaments which aid energy conversion in flight. Resilin is secreted in the form of a completely insoluble protein with apparently no polypeptide chains with free end-groups. It is believed that this structure is achieved by the reaction of end-groups with quinones secreted at the time of deposition of the protein. Resilin has relatively few tyrosine residues, which, like amino end-groups, would be likely to react with quinones, so only a small number of cross-linkages are formed between chains. The result is a molecular structure of long, freely rotating polypeptide chains held together by infrequent stable linkages (Fig. 2.3), a structure that is responsible for resilin's rubber-like ability to recover completely from gross plastic deformation (Andersen and Weis-Fogh, 1964). Untanned inner layers of the general cuticle may also exhibit some elasticity or plasticity, but whether this is due to the presence of resilin is not known.

Changes in the Integument at Moulting

The cuticulin layer of the epicuticle is inextensible, and any increase in size of an insect between moults is achieved by smoothing out folds in the epicuticle, not by stretching it. The process of moulting, that is, the periodic replacement of the old cuticle by a new one, is described on p. 92. Here we are concerned with the changes in the integument that accompany this renewal. The beginning of the moult (apolysis) is marked by separation of the old cuticle from the epidermal cells, and by changes in the epidermis which establish the external form of the new instar. The refashioned epidermis secretes a thin lipoprotein cuticle, which becomes the cuticulin layer of the new epicuticle. This epicuticular layer is fully permeable to molecules in solution, since the products of the digestion of the old cuticle pass through it into the epidermis. The *moulting fluid* secreted between old and new cuticles contains enzymes which digest the protein of the old endocuticle to amino acids and the chitin to individual sugar molecules or small oligosaccharides. These are resorbed by the epidermis, and may be used immediately in the formation of new cuticle, although much

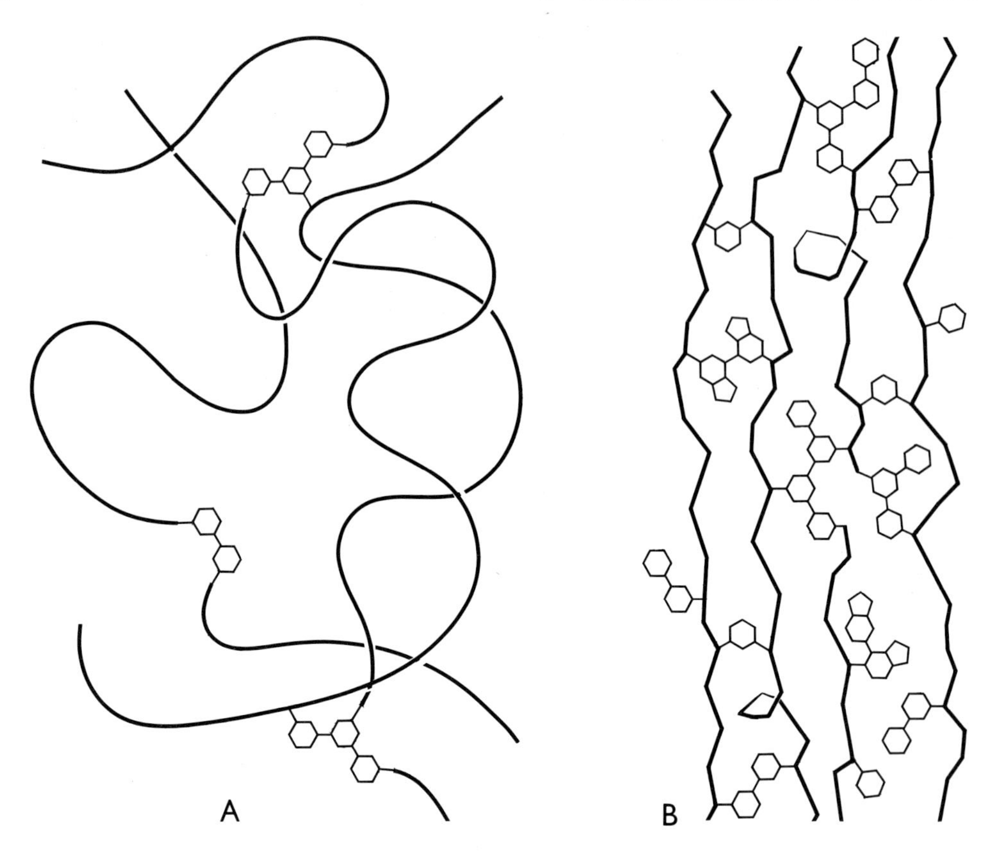

Fig. 2.3. Structure of cuticular proteins (diagrammatic): A, resilin; B, sclerotin.

cuticular material passes into the haemolymph and is stored in the fat body. Chitin, for instance, may be degraded all the way to glucose, which is then built up again into trehalose in the blood and glycogen in the fat body.

When the digestion and absorption of the old endocuticle is complete, the epidermis secretes the wax layers of the new cuticle, thus rendering it impermeable, and the remainder of the old cuticle is usually cast off at ecdysis. The cuticle of the newly emerged insect is generally soft and colourless, but it soon hardens and darkens, and it continues to grow in thickness, new layers of endocuticle being added throughout the course of the instar.

Hardening and Darkening

The fibrils of protein and chitin of the new cuticle are laid down in a fully hydrated state. They are converted into the horny insect exoskeleton by reaction of the protein chains with quinones generated in the cuticle by the action of specific enzymes. This process, which is known as *sclerotization*, is analogous to the tanning of mammalian skin proteins in the formation of leather. The quinones are derived ultimately from tyrosine, which accumulates in the blood before ecdysis, and then moves into the cuticle at the time of hardening. The immediate precursor of the tanning quinone in the formation of the blow-fly puparium, and probably in other cuticular hardening processes, is the diphenol *N*-acetyldopamine, which is derived from tyrosine by the series of reactions shown in Figure 2.4 (Karlson, 1963). The quinone is formed in the presence of oxygen by the action of phenolase, which is secreted into or on to the cuticle along with the diphenol.

Quinone molecules react with the free amino groups of the protein of the epicuticle and the outer layers of the procuticle, causing extensive cross-bonding and dehydration, thus converting the soft covering of the newly emerged insect into a tough exoskeleton.

Other diphenols derived from tyrosine have also been detected in the cuticle. They include a number of deaminated derivatives which may have a role in sclerotization in some species and also dihydroxyphenylalanine ('dopa'). Quinones derived from dopa by the action of phenolase condense together to form the polymerized

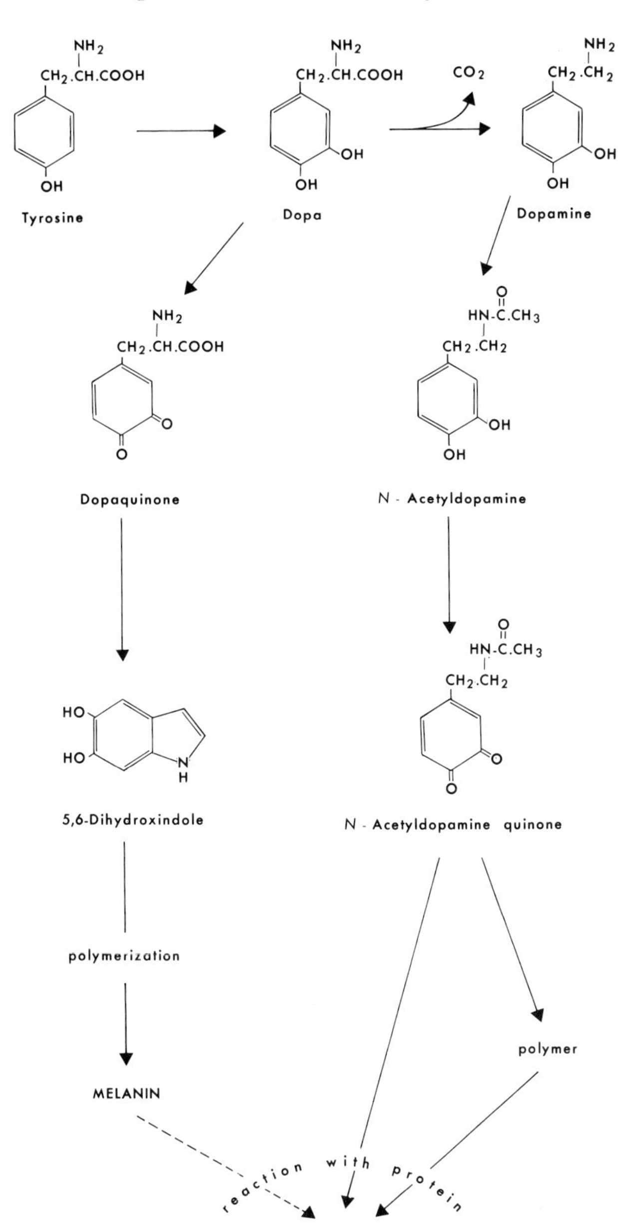

Fig. 2.4. Reaction pathways in the formation of sclerotin.

indole pigment *melanin* (Fig. 2.4). In most animals melanin occurs in discrete granules in the epidermal cells, but in insects the dark brown or black pigments are built into the cuticular structure by cross-reactions with the proteins. Although indole pigments have not been positively identified in insect cuticle, it is generally agreed that melanin participates in the darkening process.

Tyrosine residues of the cuticular proteins may also be oxidized *in situ* by the phenolase, and thus become further points of potential cross-linkage of chains. Moreover, the quinones, of whatever origin, tend to condense with one another, forming polymers of varying size, which fill the spaces between protein chains formerly occupied by water molecules. As all these quinone derivatives are coloured, the fully hardened cuticle is usually dark, although in some cuticles sclerotization proceeds only to the production of a stable cross-linked structure which is still flexible and light in colour. It is also probable, as indicated above, that variations in the kind as well as the quantity of the quinone may be responsible for differences in the pigmentation of different areas. The cross-linked, usually pigmented protein of the exocuticle has been named *sclerotin*. It is probable that the differences in its properties from those of the transparent resilin are due to its higher content of phenolic residues and the presence of condensed quinonoid systems (Fig. 2.3B).

Integumentary Colours

The pigmented protein of the cuticle produced by the action of quinones is responsible for the common brown and black coloration of the insect integument. But this is only one factor of the many that go to make the varied hues of insects. In many insects the colours are iridescent, changing with the angle of view. These are interference effects produced by the interaction of light with fine periodic structures in the cuticle or cuticular scales, including multiple thin films separated by material of different refractive index and fine striations on reflecting surfaces. The external colours of other insects are the result of the accumulation of pigment beneath a fairly transparent cuticle, either in the blood or in the epidermal cells. The non-cuticular pigments are varied in chemistry and metabolic origin. Some are simply concentrated from the diet. These include the carotenoids, flavonoids, and anthraquinones of plant origin, and some porphyrins formed by the degradation of either the plant pigment chlorophyll or the animal pigment haemoglobin. Carotenoids are particularly important in the coloration of plant feeders, such as Orthoptera and larvae of Lepidoptera. Coccids and other plant bugs tend to accumulate anthraquinones, a notable example being the deep red pigment of the cochineal insect of Central America. *Insectoverdin*, a common green pigment of the blood and epidermis of insects, is a protein conjugated with two coloured compounds, one blue, one yellow. The blue compound is usually a tetrapyrrole, but may be an anthocyanin, whereas the yellow pigment is usually a carotenoid.

Two groups of pigments synthesized by insects are particularly important in their metabolism. These are the *ommochromes* and the *pterins* (Fig. 2.5). The ommochromes, which are derived from the amino acid tryptophan, are responsible for red and yellow colours on the bodies and wings of insects

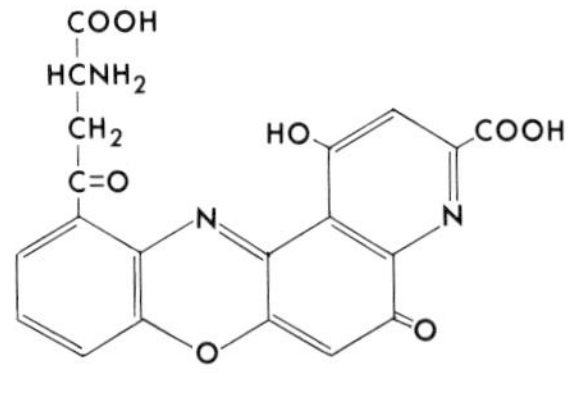

Xanthommatin

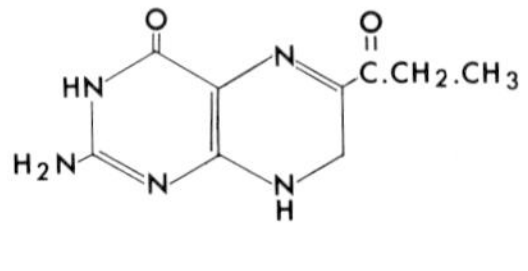

Sepiapterin

Fig. 2.5. Insect pigments.

of several orders, particularly Hymenoptera and Lepidoptera. Perhaps their most important function, however, is that of masking pigments in the compound eyes. The pterins are responsible for a wide range of integumentary colours, but their biological importance is much greater than this might suggest, for they have a general role as coenzymes in cellular metabolism. Ommochromes always occur in association with pterins in pigment granules, and it is possible that the pterins may play some part in ommochrome synthesis.

The compounds that contribute to the external colour of insects may also include a number of by-products of metabolism. These metabolites, which accumulate in the scales of some butterfly wings, include degradation products of tryptophan and some purines, including uric acid, which is the usual vehicle of nitrogen excretion in insects.

Some insects are able to change colour in response to a variety of environmental stimuli. Such colour changes, which may be the result of migration of pigment granules within epidermal cells, or of the migration of pigmented mesodermal chromatocytes, or of alterations in pigment metabolism, are described on p. 138.

Dermal Glands

Ectodermal glands associated with the integument (Fig. 2.1) produce a wide variety of secretions with important functions in the biology of insects. One is wax. Whereas in many insects the cuticular wax is secreted generally by the epidermal cells, in others wax production is the function of special glands. Even when the whole epidermis participates in wax secretion, certain layers of the epicuticle, such as the cement layer, may be derived from dermal glands. Wax glands are particularly important in insects which secrete quantities of wax far in excess of that needed to form the normal waterproofing coat. The wax of the honey bee, for instance, secreted in the form of plates or scales by ventral abdominal glands, is used in the construction of the comb. Many coccoids secrete a tent-like covering of wax underneath which they live.

Dermal glands associated with the mouthparts are usually digestive (p. 61), but they may have other functions. In social species the cephalic glands of some members of the colony may be specialized for the production of brood food or of pheromones (see below). In the silkworm the labial glands generate silk. *Fibroin*, the major constituent of the silk of *Bombyx*, is a unique protein with an unusual preponderance of the simpler amino acids glycine and alanine. It is secreted as a viscous fluid which solidifies as it passes through the narrow orifice of the *spinneret*, emerging as two semi-crystalline continuous threads, one derived from each of the two lobes of the gland. A second protein, *sericin*, with a high content of the amino acid serine is secreted at the same time. Sericin hardens in air to form a tough glue surrounding the two fibroin threads and binding them together. Dermal glands on other parts of the body are responsible for silk production in other insects (p. 133). These silks differ in amino acid composition from that of *Bombyx*, but all agree in having a high content of the simpler amino acids.

An important function of a number of dermal glands is to produce *pheromones* (Karlson and Butenandt, 1959) by which individuals communicate with, or influence the behaviour or metabolism of, other individuals of the same species. The pheromones are particularly important in promoting co-operative behaviour in social insects, but they may play a part in the lives of solitary species also. Several pheromones have been isolated and identified. Some of them are related chemically to the aliphatic acids and alcohols of the cuticular wax; others are more complex compounds such as terpenoids, many of which are responsible for the fragrance of plant oils.

Some female moths secrete volatile compounds which are capable of attracting males from astonishing distances. Two such sex attractants have been isolated, one from the silk moth and one from the gypsy moth; both are long-chain unsaturated aliphatic

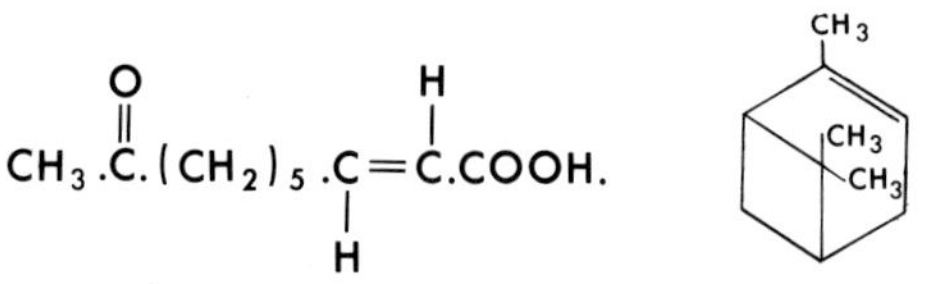

$$CH_3.\overset{O}{\overset{\|}{C}}.(CH_2)_5.\underset{H}{\underset{|}{C}}=\overset{H}{\overset{|}{C}}.COOH.$$

9-Ketodecenoic acid

CH_3, CH_3, CH_3

α-Pinene

Fig. 2.6 Insect pheromones.

alcohols. A pheromone with profound metabolic effects is 9–ketodecenoic acid (Fig. 2.6), the 'queen substance' of the honey bee. This compound, which is the active principle of the secretion of the mandibular gland of the queen, suppresses ovarian development in the workers that imbibe it (p. 137). Pheromones with comparable metabolic effects must be passed between individuals of other social insects, since it is known, for instance, that the presence of an active king and queen in a termite community inhibits the development of other reproductives except in remote parts of the colony. Several other pheromones involved in communication between members of a colony have been identified. They include geraniol, which honey bees deposit at an abundant source of nectar as a marker for other bees, and pinene (Fig. 2.6), the alarm substance which attracts and arouses aggressive behaviour in the soldiers of Australian termites of the genus *Nasutitermes*. Both these compounds are volatile terpenoids.

Finally, completing the list of known dermal-gland secretions, we have the defensive substances (Roth and Eisner, 1962). Evil-smelling or toxic secretions are a means of defence for many species, and provide an effective offensive armament for others. Such secretions may simply render the species repugnant to birds or other predators, or they may be expelled with force and aimed to repel attackers. Still others provide predacious insects with a means of subduing their prey. Defensive glands occur on almost any part of the body, but they are often found at the posterior end of the abdomen, where they are known as *anal* or *pygidial* glands. In many Hymenoptera the accessory glands of the female reproductive system are modified to form venom glands. These may be associated with an exoskeletal sting, which is a modified ovipositor. In the formicine ants, which have no sting, the contents of the venom glands are simply squirted on to an attacker, usually over the wounds inflicted by the ant's mandibles.

Many different volatile and reactive compounds have been identified in defensive secretions. Quinones, which are secreted by some cockroaches, as well as by carabid and tenebrionid beetles, are the active components of the strangest defensive system of all, that of the bombardier beetle (p. 529). The defensive glands of a bombardier beetle produce hydrogen peroxide, together with diphenols, the most abundant of which is hydroquinone. When the beetle is alarmed, the mixture is impelled into a central chamber, where a violent enzymic reaction takes place, the end products being quinones and oxygen gas. The rising pressure of oxygen blows out the solution of quinones from the orifice of the gland in a series of audible explosions.

Aliphatic acids and aldehydes are also commonly used in defence. They range from the simplest member, formic acid, to the unsaturated aldehyde decenal. Formic acid may be produced in relatively enormous quantities in the venom glands of formicine ants. Decenal and aldehydes of shorter chain length are found in the stink glands of many bugs. Aromatic aldehydes, such as benzaldehyde and salicylaldehyde, are secreted by some beetles. It is not known by what means insects are able to secrete and store high concentrations of these compounds, many of which are potent cellular poisons. Terpenoids, some of which are of a complex cyclic structure, occur in the defensive secretions of dolichoderine ants, and are also found in other orders. The tendency of reactive terpenoids to polymerize has been exploited by some species. Thus soldiers of *Nasutitermes* eject a sticky polymer of pinene from the horn-like frontal organ. The polymer is accompanied by the volatile monomer, which, as already noted, acts as an alarm substance for these insects. Moreover, evaporation of the monomer quickly hardens the secretion, effectively gumming up the limbs of an attacker on to which it has been squirted.

The venoms associated with insect stings contain several compounds which induce pharmacological reactions in mammals. These include histamine, acetylcholine, and serotonin. Venoms may also include enzymes, which aid the penetration of the toxic principle through the tissues of the victims, or generate further toxic substances within the tissues. Most interesting of all the insect toxins are the components, believed to be proteins and often highly specific in action, which induce prolonged narcosis of the prey of those wasps and bees that store up victims as food for the larvae (p. 129).

THE OENOCYTES

The oenocytes are groups of cells of ectodermal origin which are differentiated early in embryonic life. They have been identified in all but a few primitive groups. A segmental arrangement of the cells is often retained in postembryonic life, and in many forms they remain in close association with the epidermis (Fig. 2.1). In others, however, they may be scattered through the body, and often occur among the cells of the fat body. The oenocytes of many larvae are conspicuous lobed cells, and are typically few in number; in the adult they are usually smaller and more numerous.

Oenocytes undergo cyclic changes at the time of moulting, which suggests that they have a secretory role in the formation of new cuticle. They increase in size and their cytoplasm often becomes vacuolated just before the new cuticle appears, then they decrease again after it is formed. The cytoplasm in the enlarged phase reacts to lipid stains in the same way as does the lipoprotein of the epicuticle, suggesting that secretion of cuticulin is their main function. In some insects they may also play a part in secreting the wax layer of the epicuticle. In the adult, secretory activity of the oenocytes is associated with egg production, indicating a possible role in the secretion of egg-shell material.

THE RESPIRATORY SYSTEM

The Tracheal System

In insects, as in many other terrestrial arthropods, oxygen is conveyed to the tissues, and carbon dioxide is removed, through a tracheal system (Fig. 2.7). This is a set of very fine tubes, ramifying to almost every cell of the body, and opening to the external atmosphere through the *spiracles*. The *tracheae* running from them are interconnected by a series of longitudinal *tracheal trunks*, which communicate with one another by transverse connectives. The tracheae become progressively narrower at successive branchings, until they end in the very fine *tracheoles*, which are small enough to penetrate individual cells. A trachea may branch abruptly into a group of fine

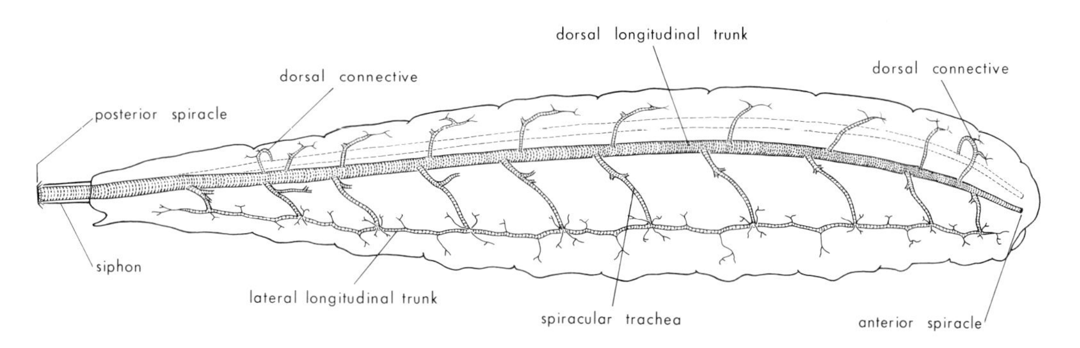

Fig. 2.7. Tracheal system of a syrphid larva with amphipneustic arrangement of spiracles. [R. Dencio]

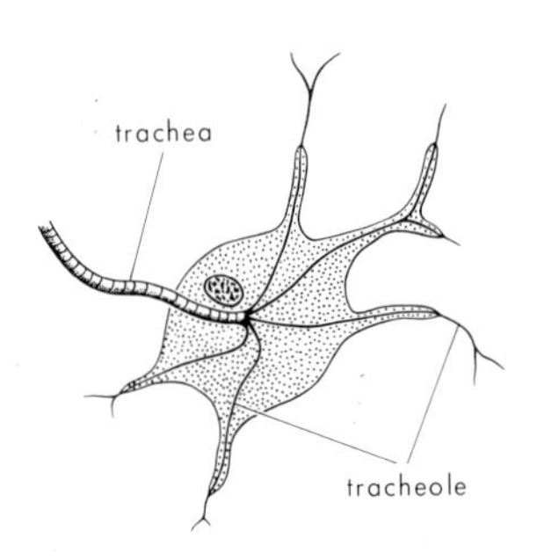

Fig. 2.8 Tracheal end cell (after Bongardt, 1903).

tracheoles enclosed in a large stellate cell, called a *tracheal end cell* (Fig. 2.8), but in other places the transition may be more gradual. The cuticular lining of the tracheal system is continuous with the external cuticle, and is secreted by an investing sheath of cells which is continuous with the epidermis. It consists of a mixture of chitin and protein, except in the finer tracheal branches, from which chitin is absent. The cuticle of the larger tracheae is shed at each moult, specific points of rupture in the walls of trunks allowing the withdrawal of the old lining from the tracheae of the emerging instar. In some species the opening from which the old tracheal lining is withdrawn becomes the new spiracle, but in others a new spiracle is formed either around or to one side of the old opening, which then closes to form the *ecdysial scar*.

The cuticular walls of the tracheae are strengthened by spiral thickenings which prevent them from collapsing, but the system also includes a number of relatively large, thin-walled dilatations, the *tracheal air sacs*, which may occupy a large part of the body cavity. These aid ventilation (see below), and also allow growth of the internal organs within a rigid exoskeleton. The net-like masses of tracheae have another non-respiratory function, in that they provide support for the internal organs, thereby assuming much of the role of the fibrous connective tissue of higher animals.

Spiracles vary greatly in form. They may be simple external openings, but more frequently are situated at the base of a spiracular pit or *atrium*, of often complex structure (Fig. 2.9). In many insects the spiracles are provided with a muscular closing apparatus which controls movement of gas through the opening. The distribution of functional spiracles varies considerably in different groups of insects, particularly among endopterygote larvae, and it may differ in different instars of the same species. The positions of the non-functional spiracles are usually marked by detectable scars, and they still have an important function, because they continue to provide a mechanism for withdrawal of the old tracheal linings at ecdysis. The patterns of spiracular arrangement have been classified in the following way.

HOLOPNEUSTIC: All ten pairs of spiracles (2 thoracic, 8 abdominal) functional—primitive.

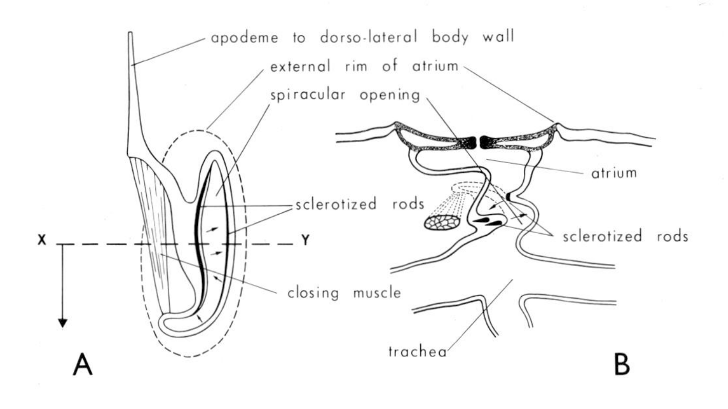

Fig. 2.9. Spiracle of moth larva, showing closing apparatus: A, in plan, from within; B, section at X–Y, viewed in the direction of the arrow. Contraction of the closing muscle causes the edges of the spiracle to move in the directions shown by the small arrows. The sclerotized rods are in cuticle of the spiracle rim.

HEMIPNEUSTIC: One or more pairs non-functional.

Peripneustic: Most spiracles functional—usually only metathoracic or metathoracic and last abdominal non-functional.

Amphipneustic: Only 'prothoracic' (= mesothoracic) and last abdominal spiracles functional.

Propneustic: Only 'prothoracic' (= mesothoracic) spiracles functional.

Metapneustic: Only last abdominal spiracles functional.

APNEUSTIC: Without functional spiracles—respiration cutaneous.

Mechanisms of Respiration

Aerial Respiration. The entry of oxygen and the elimination of carbon dioxide in insects depends essentially on gaseous diffusion. Although the cuticle of some insects living in damp situations may be relatively permeable to gases, that of the great majority of terrestrial insects is almost completely impermeable. Gaseous exchange in these types must take place entirely through the spiracles and the tracheal system. Because of the small distances involved, diffusion through the spiracles provides an adequate supply of oxygen to the tissues of many species, but in the more active flying insects diffusion is supplemented by forced ventilation of parts of the system. Numerous small tracheal sacs attached to the sides of the flight muscles (Fig. 2.23) may be deformed by the contraction of the muscles, thus producing a flow of air into and out of the sacs. In addition, the large tracheal sacs of the abdomen are ventilated by rhythmic contractions of the abdominal walls. This process may be accompanied by co-ordinated opening and closing of anterior and posterior spiracles to produce a directed flow of air through the tracheal trunks.

Since conservation of water is of major importance to most terrestrial insects, and the main pathway for loss of water is through the tracheal system, the spiracles remain open only for the minimum time required to meet the respiratory needs. Oxygen lack, the accumulation of CO_2 in the blood and tissues, and the state of desiccation of the insect, all play a part in controlling spiracular movement. Although the most important stimulus for the opening of the spiracles is the accumulation to a critical level of CO_2 in the form of bicarbonate, the critical level itself may vary with the water content of the insect. In some pupae the spiracles open widely for only very short intervals, separated by periods of several hours during which they are almost completely closed. During the open periods, CO_2 which has been stored in the blood and tissues is ventilated off rapidly, so that CO_2 production by these insects seems to occur in discontinuous 'bursts'.

Aquatic Respiration. Many aquatic insects establish periodic communication with the air, and their respiratory mechanisms do not differ from those of aerial species, except that functional spiracles are usually reduced to a single pair. Various adaptations to this type of respiration are described on p. 112. In some common water beetles and bugs the time interval between visits to the surface has been lengthened by the insect carrying a bubble or surface film of air down with it. The bubble or film acts as a hydrostatic organ, as an expendable store, and as a temporary physical gill, at least some of the oxygen used by the insect being replaced by diffusion from water, while the nitrogen diffuses outward as the result of its increased partial pressure, until the bubble is too small to be effective, when it must be replenished (Thorpe, 1950).

Typical spiracular respiration is also the rule even in some species which can remain submerged indefinitely. The spiracles in such species are in contact with a layer of air held permanently in a mat of hydrofuge hairs or in a fine cuticular meshwork. Surface tension stabilizes the air-water interface across which gaseous exchange takes place. The structure, which is known as a *plastron*, thus acts as an external physical gill. Perhaps its most extreme development is found in the 'spiracular gills' of the pupae of some flies and beetles. These are large, feathery, plastron-covered outgrowths of the pupal epidermis. The gas space of the plastron of *Simulium* is in contact with a pupal spiracle and with a spiracle of the pharate adult (p. 93) below the pupal cuticle, and closure of the pupal

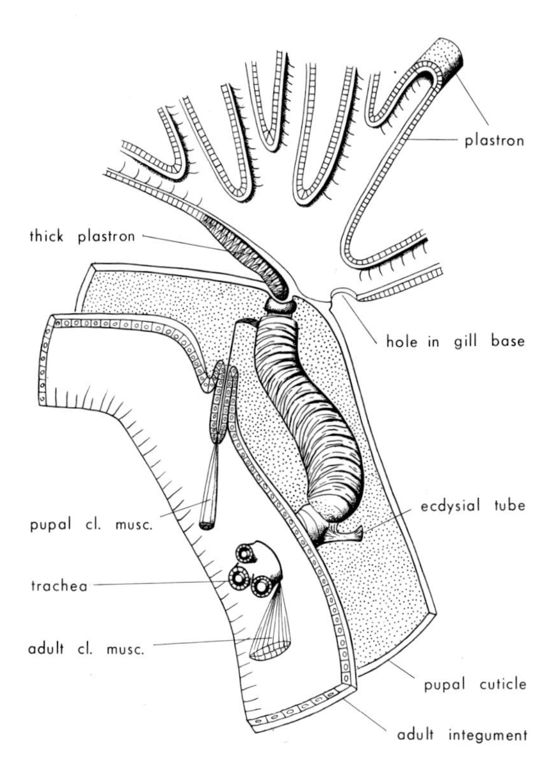

Fig. 2.10. Base of the spiracular gill of a simuliid in the pharate-adult phase of development (after Hinton, 1957).
adult cl. musc., closing muscle of adult spiracle; *pupal cl. musc.*, closing muscle in adult that regulates pupal spiracle. The area of thick plastron communicates with the pupal spiracle.

spiracle is controlled by a muscle of the adult (Fig. 2.10; see Hinton, 1957). Plastron respiration is a unique adaptation to life in variable streams or in situations that are intermittently flooded, since it can be combined with a completely waterproofed cuticle and functions equally well in air as in water. It is found also in many insect eggs that may need to withstand occasional flooding (Hinton, 1962b).

Other aquatic insects have a closed tracheal system with no functional spiracles. The cuticle is permeable to gases in solution, either over the whole body surface, or, more usually, in specialized areas which act as true gills. The gills are filamentous or leaf-like outgrowths of the body wall in most aquatic larvae, but in anisopteran Odonata they are developed in an enlargement of the rectum, and are aerated by the rhythmic intake and expulsion of water. Most gills are well supplied with tracheae into which oxygen, which enters the body in solution, is released in gaseous form. The tracheal system thus retains much of its role in gas transport within the body. Other gills are filled with blood, and have few or no tracheae. Here the blood must have a more important role in oxygen transport.

Oxygen Transport in the Blood. Although many cells are in contact with tracheoles, others receive their oxygen supply from the blood. The diffusion pathway is extremely short, however, and as a rule no oxygen-carrying pigment is found, gases being transported in simple solution. The few exceptions to this rule include a number of chironomid larvae (Diptera), in which haemoglobin is present in the haemolymph, as well as some aquatic Hemiptera and the larvae of bot flies in which the haemoglobin occurs in special groups of cells provided with many tracheae. The haemoglobin of these insects is capable of combining with oxygen at very low partial pressures, thus enabling them to trap oxygen in the poorly aerated environments in which they all live.

Anaerobic Mechanisms. Most insects can survive for several hours in the absence of oxygen, although it is usually observed that they become motionless under anaerobic conditions. A few species, such as chironomid larvae that live at the bottom of lakes, may be able to carry on active life in the complete absence of oxygen, but this has not been established with certainty. Most insect cells, in common with animal cells in general, are able to derive energy anaerobically for a short period by the conversion of glucose to lactic acid. Alternative or additional metabolic pathways, combined with the excretion of potentially harmful end products, are found in true anaerobes, such as some parasitic worms. It seems possible that such pathways may also occur in some insects, but so far they have not been identified.

THE NERVOUS SYSTEM

The central nervous system of insects comprises a *supra-oesophageal ganglion*, or brain, connected by *para-oesophagael commissures* with a *sub-oesophagael ganglion*, from which a pair of *ventral nerve cords* runs throughout the length of the body, connecting the segmental ganglia. Three divisions are recognized in the brain: the *protocerebrum*, which bears the optic lobes; the *deutocerebrum*, which innervates the antennae; and the *tritocerebrum*, which is formed by the

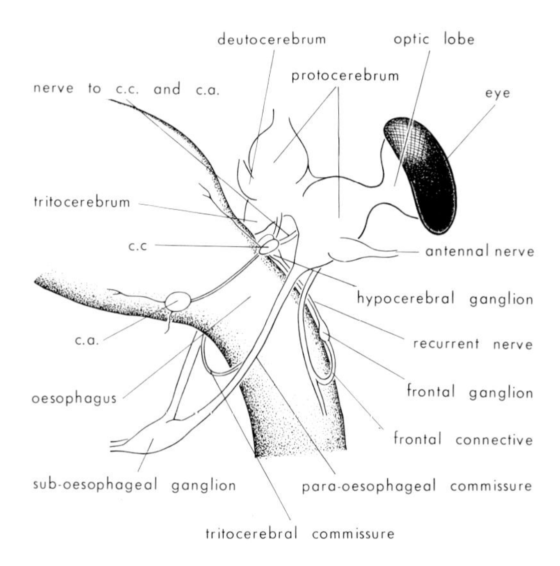

Fig. 2.11. Brain and associated organs of a grasshopper. The frontal ganglion, hypocerebral ganglion, and their associated nerves comprise the anterior division of the autonomic nervous system.
c.a., corpus allatum; *c.c.*, corpora cardiaca.

forward migration of the ganglia of the first postoral segment, evidence of which is seen in the existence of a commissure connecting its two halves and passing below the anterior end of the alimentary canal (Fig. 2.11). The sub-oesophageal ganglion is formed by the fusion of the ganglia of the three gnathal segments.

The primitively paired ganglia of the succeeding segments are fused in the mid-line, although the nerve cord remains paired. In the primitive condition, a ganglion is present in each of the three thoracic and the anterior eight of the 11 abdominal segments. There is evidence that the eighth abdominal ganglion is the product of fusion of segmental ganglia 8-10. In more highly evolved insects various additional degrees of fusion are found. In some groups all of the ganglia posterior to the sub-oesophageal are fused into a single mass, and in some higher Diptera the entire central nervous system is concentrated into a single ganglionic mass through which the oesophagus passes (Fig. 2.12).

Insects also possess an autonomic nervous system innervating involuntary muscles. It has an anterior division consisting of several ganglia and nerves connected directly with the brain and innervating a large part of the alimentary canal, the dorsal aorta, and several endocrine organs (Fig. 2.11), and a posterior division of nerves and ganglia connected with the posterior segmental ganglia and innervating the gonads, the posterior intestine, and the spiracles.

Sensory neurones are peripheral. They usually have a short distal process associated with a sense organ, and a longer proximal process to a ganglion (Fig. 2.13). Others have several branched processes, which end on the inner surface of the epidermis, the somatic muscles, or internal organs. The *motor neurones* and *associative neurones* are situated in the ganglia, where their branched processes, or *dendrites*, form synapses with the branched ends of the sensory fibres or with one another.

Electrical activity associated with conduction is similiar in insect nerve to that in vertebrates, and consists of a propagated wave of depolarization of the fibre membrane. Transmission of excitation across synapses in the central nervous system is thought to involve the release and breakdown of acetylcholine, a system which is the point of attack of many insecticides. Other transmitter substances may have a role in other parts of the nervous system, but their identity is still uncertain.

One of the outstanding features of the insect nervous system is its ability to conduct in an ionic environment which varies widely from that known to be necessary for conduction in vertebrates. This environment is

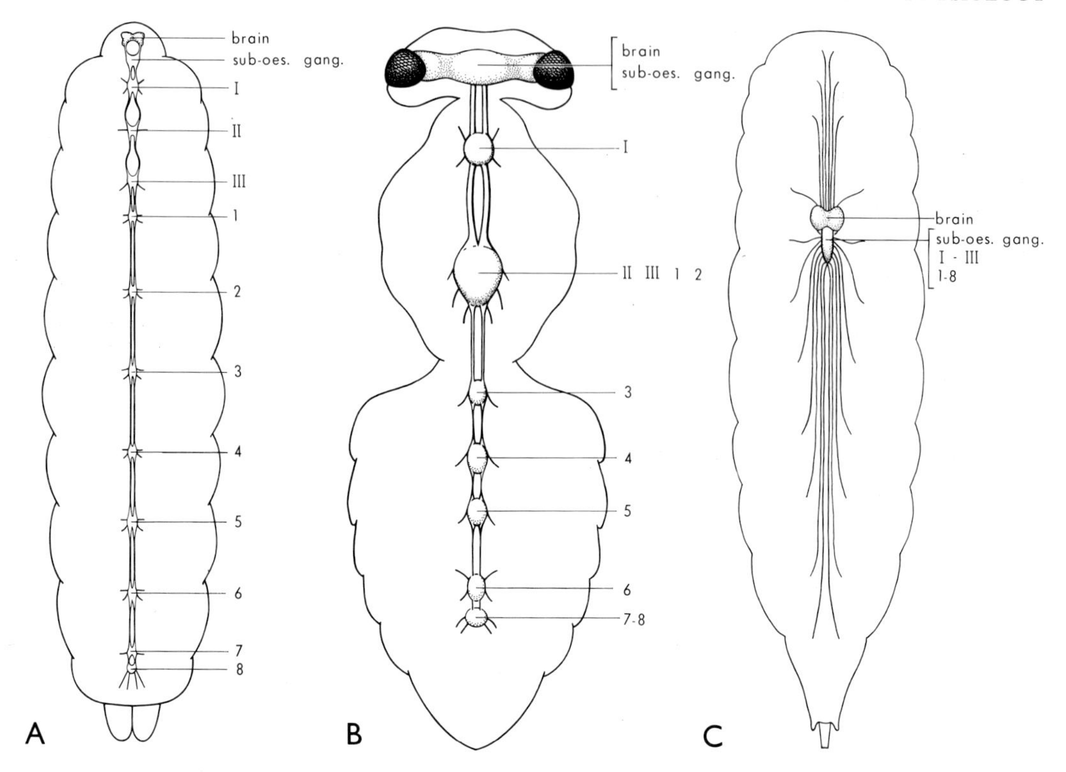

Fig. 2.12. Degrees of fusion of ganglia of the central nervous system: A, moth larva; B, honey bee adult; C, syrphid larva.
I–III, thoracic ganglia; *1-8*, abdominal ganglia; *sub-oes. gang.*, sub-oesophageal ganglion.

the haemolymph, the ionic constitution of which is poorly controlled in many insects, and in phytophagous species often reflects that of the food plant. In these insects the concentration of potassium ions, and in particular the K^+:Na^+ ratio, is much greater than that which causes complete inhibition of vertebrate nerve. The ability to conduct in such an environment is ex-

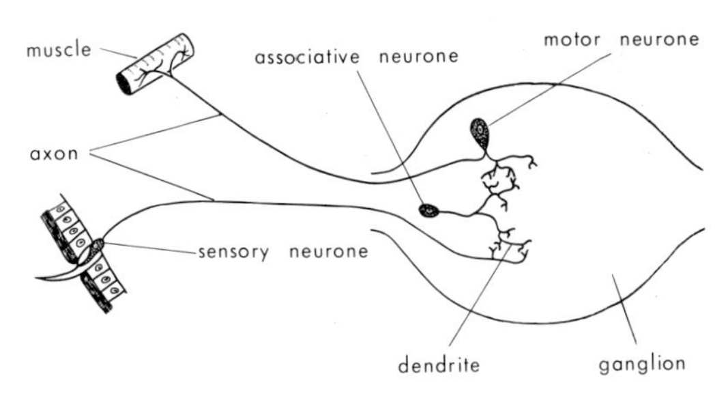

Fig. 2.13. Diagram of reflex arc (after Wigglesworth, 1965).

plained partly by the presence of a fibrous sheath around the nervous system of insects. By sequestering certain ions, this sheath maintains an environment immediately around the nerve with a higher concentration of Na^+ than the surrounding haemolymph, and one less subject to the violent fluctuations of the outer medium. The protection afforded by the neural sheath does not completely explain the properties of insect nerves, however, and it is possible that future work may reveal mechanisms of ion transport unknown in other animals.

Insect muscles are small, but their component fibres are not reduced in size. Consequently many insect muscles have very few fibres. This presents a problem in fine nervous control that has been overcome by some unique features of the neuromuscular mechanism. In the first place, innervation is *multiterminal*—that is, each axon branches to

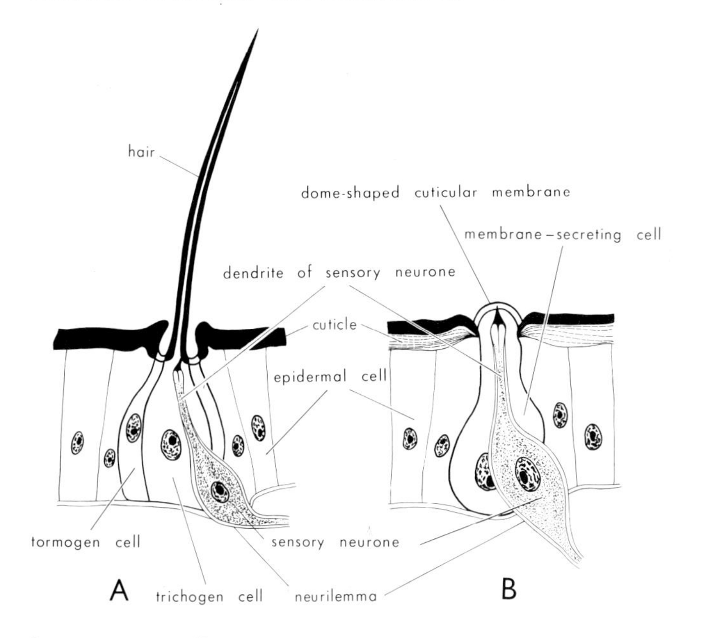

Fig. 2.14. Mechanoreceptors (diagrammatic): A, tactile sensillum; B, campaniform sensillum. In A, the tormogen cell secretes the hair socket and the trichogen cell the hair itself.

form many endings on the one fibre, instead of the single end-plate of vertebrates. Secondly, the innervation is *polyneural*, many fibres receiving terminals from two or three axons. One of these (the fast axon) elicits a rapid twitch; another (the slow axon) produces a slow contraction with repetitive stimulation; the function of the third is uncertain, although it may be inhibitory. Finally, the muscle membrane, unlike mammalian muscle, shows a graded depolarization to nervous stimuli of varying strength.

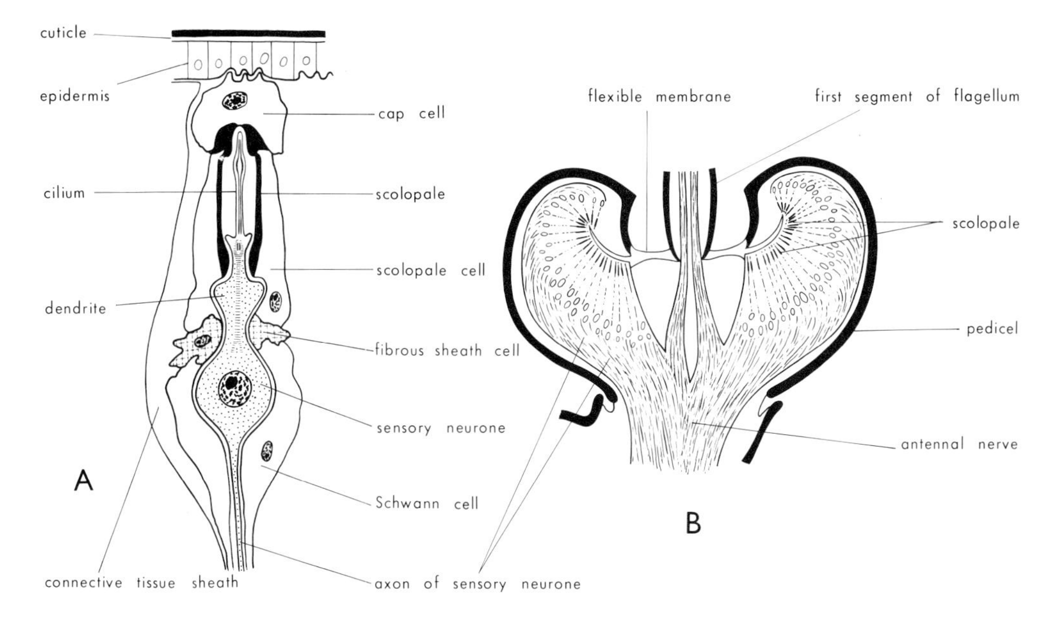

Fig. 2.15. Scolopophorous organs; A, scolopophorous sensillum (after Gray, 1960); B, Johnston's organ (after Child, 1894).

THE SENSE ORGANS

Touch and Hearing

The simplest type of peripheral sense organ is the *tactile sensillum*. This consists of an articulated cuticular hair, with the distal process of the sense cell attached to its base (Fig. 2.14A). In the *campaniform sensilla* (Fig. 2.14B), the process of the nerve cell ends below a dome-shaped structure on the cuticle. These respond to pressure and strains in the cuticle. The *scolopophorous sensilla* (Fig. 2.15A) usually lie more deeply, the sensory cell being separated from the body wall by one or more accessory cells. Its dendrite, which terminates in a cilium, is enveloped by the *scolopale cell*, within which is a dense body, the *scolopale*. A cap cell distal to the scolopale cell is attached to the epidermis, usually in some elastic membranous region of the cuticle. The scolopophorous sensilla are presumably stimulated by vibrations of the elastic regions to which they are attached. Sensilla of this type are commonly found in the antennae, particularly in the pedicel, where a number of them may be grouped together to form *Johnston's organ* (Fig. 2.15B). This organ, quite small in some species, reaches its highest development in Culicidae and Chironomidae (Diptera), which have swollen pedicels packed with scolopophorous sensilla and their accessory cells. The cells are attached to the flexible membrane separating the pedicel from the first flagellar segment, so that any movement in the flagellum causes activation of some sensilla.

Another type of mechanoreceptor is the *stretch receptor*, a sensory neurone with several dendrites enclosed in a strand of connective tissue, which may be attached to two points on the body wall or internal organs, or run between the integument and the viscera. The neurones respond to changes in tension in the connective-tissue strands.

In the *auditory organs* the distal processes of a group of scolopophorous sensilla are attached to a thin, drum-like area of cuticle, the *tympanum*, which vibrates with the impinging sound waves (Fig. 2.16). Such tympanal organs occur on the abdomen of cicadas and locusts, on the fore tibiae of most Ensifera, and on the thorax and abdomen of some moths. Communication by sound is an important factor in the behaviour of many insects, and it is possible that groups of scolopophorous sensilla, even when not associated with a tympanum, may act as primitive auditory organs in some species.

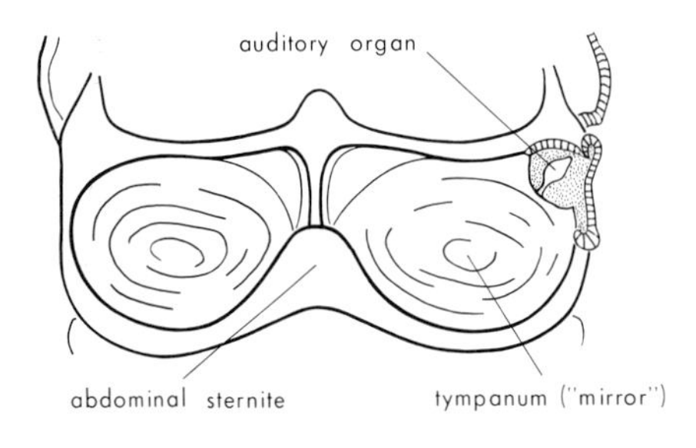

Fig. 2.16. Auditory apparatus of a cicada, ventral view with cuticle cut away at one side to show auditory organ.

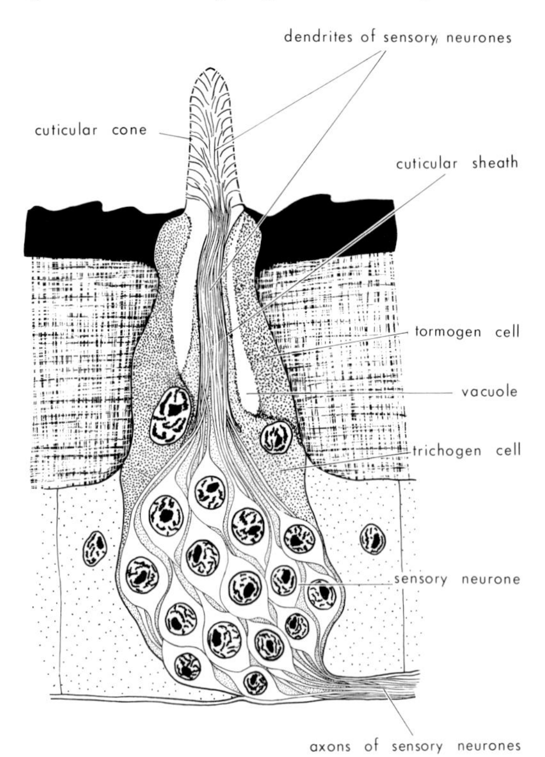

Fig. 2.17. Basiconic sensillum (after Slifer, Prestage and Beams, 1959). The cuticular cone has many pores.

Smell and Taste

Sensilla that respond to chemical stimuli usually have a group of nerve cells with their distal processes ending at openings in the cuticle. In some the cuticular part of the organ is a seta, which differs from the hairs of the tactile sensilla in having a minute opening at the tip. In others the cuticular parts are rounded lobes or flat plates. These are known as *basiconic* (Fig. 2.17) and *plate* sensilla. Dendrites from the sense cells send branches to numerous minute pores in the cuticle of the lobes. Chemosensory sensilla may be extremely sensitive and highly specific in their response to stimulation by a particular compound or ion. Some are stimulated by water, and thus endow the insect with a specific water taste. Chemoreceptors are present on the antennae and mouth-parts of many insects, the tarsi of bees, flies, and Lepidoptera, and the ovipositors of some wasps.

Sight

Light-sensitive organs consist essentially of a transparent area of cuticle, the *cornea*, which may be lens-like in shape, and a *retina* of sense cells. Below the cornea may be a secondary lens structure, the *crystalline cone*, which is derived from special epidermal cells. One border of the sense cells has a laminated structure in which the light-sensitive pigment is situated. This may be either the distal border, or the lateral borders of adjacent sense cells, where the laminated regions of several contiguous sense cells together form the *rhabdome* (Fig. 2.19). In addition, various pigment cells may be present, either with a masking function, or in the form of a *tapetum* or reflective layer.

The simple eyes are classified into *stemmata*, which are the visual organs of most larvae, and *ocelli*, which occur along with compound eyes in adult insects. Stemmata vary in form from very simple light-sensitive structures to eyes with multicomponent retinas, the structure characteristic of ocelli (Fig. 2.18). Simple eyes, in general, are of quite primitive structure, but some, at least, appear to be capable of rudimentary form perception and colour discrimination.

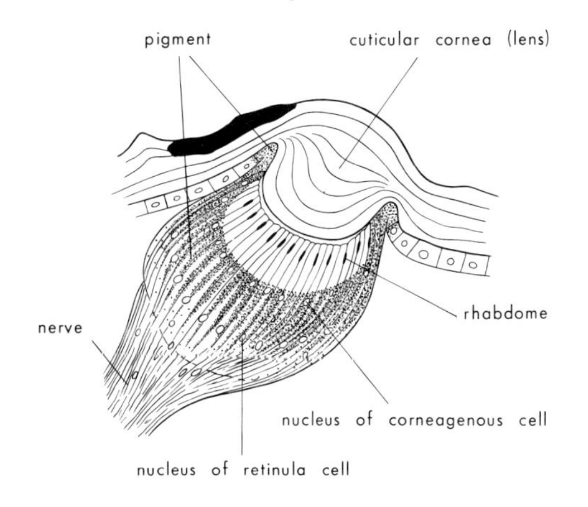

Fig. 2.18. Ocellus of a pentatomid bug (after Link, 1909).

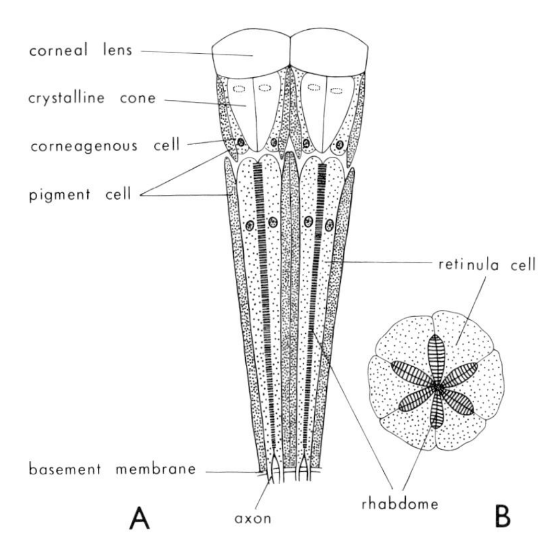

Fig. 2.19. Structure of the compound eye (diagrammatic): A, two ommatidia sectioned in a plane normal to the surface; B, transverse section of a group of retinula cells showing the orientation of the micro-tubules of the rhabdome.

The main visual organs are the *compound eyes*, in which a number of individual photoreceptive units, or *ommatidia*, are grouped together. Externally, the eye is seen to be composed of a number of facets, usually tightly packed and hexagonal in shape, which are the corneas of the individual ommatidia. The number of ommatidia varies from less than ten in some insects to

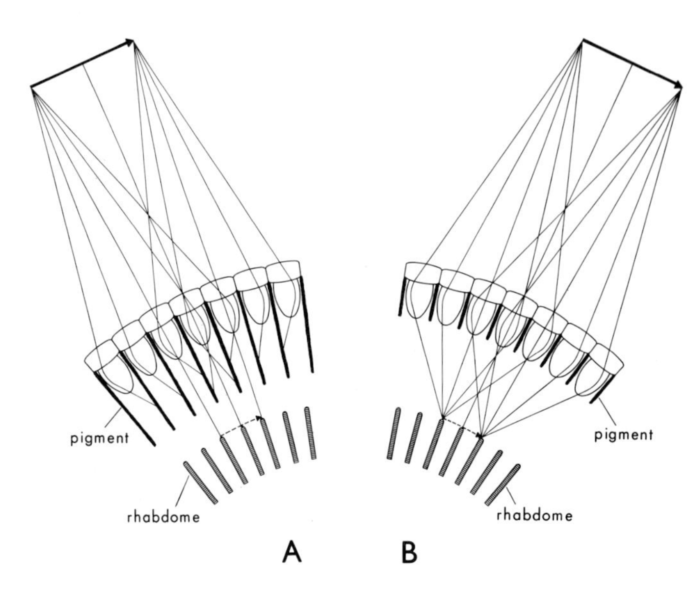

Fig. 2.20. Theory of image formation in the compound eye: A, apposition image, in strong light–pigment extended so that only rays parallel to the ommatidial axis penetrate to the retina; B, superposition image, in weak light–pigment withdrawn so that groups of ommatidia can co-operate in image formation.

many thousands in others. Each ommatidium (Fig. 2.19) comprises a cornea, a crystalline cone, and a group of usually 6 or 8 *retinula* cells. Each retinula contains a *rhabdomere*, which under the electron microscope has the appearance of many minute tubules stacked in layers normal to the axis of the ommatidium. Often the rhabdomeres occupy the inner borders of the retinulae, grouped along the central axis to form the rhabdome, but they may lie more deeply in the cells. Flanking each ommatidium and separating it from its neighbours are pigment cells.

Controversy still exists about how the compound eye works. One theory, which has dominated thought on the subject for many years, holds that the ommatidia function independently of one another, each responding to the narrow cone of light received by the lens. The whole eye would thus reproduce the surrounding world in the form of lighter and darker patches, which together would form a *mosaic* image. Ideally, mosaic vision would require the complete optical separation of the ommatidia, and this seems to be the rule in the eyes of some diurnal insects, in which the pigment cells extend distally to the region of the crystalline cones. In other insects, usually night-flying or crepuscular species, the ommatidia are not so isolated, and the retina lies at some distance from the lens system, so that each rhabdome is able to receive light from several facets (Fig. 2.20). Groups of facets thus co-operate to form a geometric image (probably of poor definition) on the retina. Reduced pigmentation results in relatively greater illumination of each rhabdome, thus enhancing sensitivity. In some eyes of this type, pigment granules migrate within the pigment cells in response to the intensity of illumination, being concentrated distally to produce the true mosaic or *apposition* image in strong light and allowing an overlapping or *superposition* image in weak light. Mosaic vision is well adapted for the perception of movement, and, indeed, behavioural studies have shown that, whereas static form perception is relatively poor in insects as compared with mammals, insect eyes are particularly sensitive to movement.

There have been some challenges to the mosaic theory. Thus, Burtt and Catton

(1962) have suggested that the physical mechanism of the compound eye is the formation of *diffraction* images. According to this idea, the confluence of light reflected from a small object and passing through a group of facets produces a series of diffraction images at different depths in the retina. It is suggested that the retinulae respond at the level of these sharp images, transmitting the appropriate information to the brain.

There is still insufficient information to discriminate between these theories. Furthermore, even if we could be sure of the physical laws governing image formation in the compound eye, there would still remain the problem of how electrical information transmitted from the retinulae is integrated within the optic lobes of the brain. Does it, in fact, produce anything like the sensation we recognize as vision? Probably it does, but we are a long way from appreciating an insect's view of the outside world.

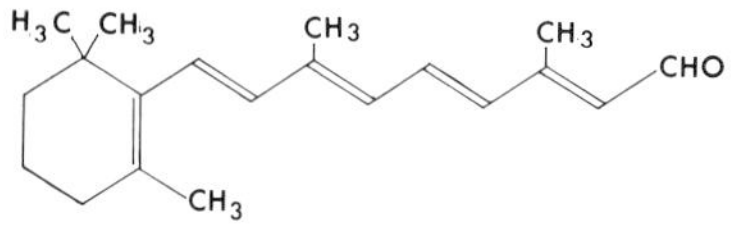

Fig. 2.21. $Retinene_1$.

The molecular mechanism of photoreception may be the same in insects as it is in mammals. The photosensitive compound $retinene_1$ (Fig. 2.21), which is the aldehyde derivative of vitamin A_1, has been found in insect eyes. $Retinene_1$ is also the chromophore of the mammalian visual pigment rhodopsin. Absorption of light by rhodopsin results in bleaching of the pigment, dissociation of retinene from its protein partner, and a change in the configuration of the chromophore molecule, events which in some way trigger the excitation process. Similar events presumably occur in insect eyes, but few details have yet been established. Retinene is believed to be conjugated with several different proteins to produce pigments sensitive to different wave lengths, since insects are known to be able to discriminate colours. Indeed, the spectral sensitivity of many insects extends into the ultraviolet. Electrical recordings from single retinula cells have revealed as many as four different classes of cells with four different wave-length sensitivities.

Some night-flying moths respond to infrared radiations, and it has been suggested that they 'see' with these wave lengths. No molecular mechanism for such a visual process has been proposed, however. Other insects are able to discriminate the plane of polarization of incoming light. The cornea apparently does not act as a polarization analyser, and it is thought that the constant angular inclination of the tubules in the rhabdomeres over a series of ommatidia may constitute the receptive apparatus.

Behaviour

The complex and highly sensitive sensory equipment of insects directs a wealth of information about the external environment into the central nervous system. Indeed, the perceptive range of insect senses collectively far exceeds that of man. Thus, some moths 'hear' the echo-locating noises emitted by bats, which are outside our auditory range, and the band of wave lengths visible to insects may extend beyond both ends of the spectrum we see. Appreciation of the plane of polarization of the light coming from unclouded sky by honey bees and other insects equips them with a directional sense that operates whether the sun is visible or not.

The co-ordination of this rich and varied sensory input with motor output results in precise and rapid muscular responses, which in some areas of activity, such as flight, may be unmatched anywhere in the animal kingdom. Yet the sum total of such activity, which we call behaviour, tends, in general, to be rigidly conditioned or stereotyped. Thus, all the individuals of a species may react in a fixed pattern to a given stimulus, regardless of whether the reaction aids or jeopardizes their survival as individuals. Moreover, much of this behaviour is genetically fixed or 'instinctive', as, for instance, in many caterpillars, which are conditioned from hatching to eat a certain plant and will starve to death in the presence of food that is nutritionally identical but lacks a particular compound that triggers feeding behaviour. The biochemical basis of this preconditioning is unknown; but one may speculate that reflex pathways that can be facilitated by frequent use in other animals (as in Pavlov's dogs) are in insects facilitated in response to genetic information when the nervous system is being formed in the embryo or reorganized at metamorphosis (p. 98). Consequently, alternative reflex arcs are never used, and the animal never has the opportunity to exercise a 'choice' over the motor response in the light of modifying sensory information.

However, the range of variation is wide, and there are some insects that evince an ability to learn and a considerable degree of flexibility of behaviour. The network of synapses between sensory and motor neurones and of integrative cross-links within the central nervous system of an insect is quite complex, and, for the most part, differs quantitatively rather than qualitatively from that of man. It seems possible that the complexity may already be such that the higher integrative functions that we call consciousness and reasoning are present to some degree. To what degree will probably always be the subject of argument; but it must be recognized that, if ability to communicate with one's fellows is a measure of intelligence, then the honey bee, at least, rates highly in the animal kingdom. The language of the honey bee, which includes sounds, odours, and various dances (von Frisch, 1954), is a very effective medium of communication. Moreover, worker bees can not only establish direction by their ability to recognize the plane of polarization of light, they also make allowance for elapsed time in communicating this information to their fellows by means of an oriented dance. Although we should not be carried away by these evidences of 'intelligence', we should also be wary of ascribing all insect behaviour to mere automatism.

ENDOCRINE ORGANS

Chemical regulation of growth, maturation, metabolism, and behaviour is well developed in insects (Wigglesworth, 1954, 1964; Schneiderman and Gilbert, 1964). The hormones that establish this control are released into the blood from a variety of endocrine organs. As in other animals, the nervous system is a major source of hormones. Neurones that are modified for hormone production *(neurosecretory cells)* are usually enlarged and have their cytoplasm packed with material recognized as 'secretory granules'. Hormones may pass into the haemolymph directly, or move in the form of secretory granules down the axons of the modified neurones to a storage organ, from which they are released on an appropriate stimulus. Other sources of hormones are the endocrine glands, the secretions of which may also be released into the blood and distributed generally through the body, or travel in granular form through channels in connective-tissue membranes directly to the target organ.

The Brain-Cardiacum System

Neurosecretory cells in the protocerebrum are the source of a secretion which passes down a pair of nerves to the *corpora cardiaca*, two bodies lying in or on the wall of the aorta (Fig. 2.11). Axons from neurosecretory cells also pass to the *corpora allata* (see below), and branches carrying neurosecretory material have been traced back along the gut. Nevertheless, the corpora cardiaca are probably the main recipients of secretions originating in the brain. They also contain glandular tissue which is believed to elaborate intrinsic hormones.

The neurosecretory cells of the protocerebrum are the source of the *brain hormone*, which stimulates the prothoracic glands to produce the moulting hormone, which, in its turn, acts on the epidermis to initiate moulting. The brain thus exerts an over-riding control over periodic growth by moulting. Nervous stimuli and possibly reciprocal influences from other endocrines or general body tissues (feed-back) combine to determine the initiation of cycles of activity in the neurosecretory cells. The brain hormone may also have a direct effect on the body tissues, since in its absence there is a general cessation of protein synthesis throughout the body—the condition known as diapause (p. 108). In diapause not only the epidermis but most of the other tissues are dormant, and some, such as the wing muscles, may actually degenerate. Some tissues, such as heart muscle, are immune to these effects, however, and protein synthesis may be reawakened locally for the healing of wounds.

Hormones of the corpora cardiaca also control a number of metabolic functions. One regulates the concentration of trehalose in the blood by promoting its formation from fat-body glycogen (p. 58). Another controls the rate of heart beat by stimulating the formation in the pericardial cells of a compound (probably a tryptophan derivative) with a pharmacodynamic action on the heart.

Other Neural Endocrine Organs

The sub-oesophageal ganglion of the female silk moth produces a diapause hormone. Presence or absence of this hormone determines whether embryonic development in the eggs laid by the female will be interrupted by diapause or not. Urinary excretion in the bug *Rhodnius* is controlled by a diuresis hormone produced from the ventral ganglionic mass. Other effects of the same sort have been described and more doubtless await discovery.

The Prothoracic Glands

The endocrine organs usually called prothoracic glands arise as a mass of ectodermal cells in the labial segment of the embryo. In some insects they remain in the head and are called 'ventral glands', but in many others they form diffuse masses of tissue in the larval thorax. In Diptera–Brachycera they are incorporated with the corpora cardiaca and corpora allata in a ring of tissue (the *ring gland*) which surrounds the anterior end of the aorta. Prothoracic glands degenerate at the end of pre-adult life, except in the Apterygota.

The product of the prothoracic glands (moulting hormone, or *ecdysone*) acts primarily on the epidermis, initiating the formation of the new cuticle and the production of the enzymes that bring about hardening of the cuticle after ecdysis or of the puparium in some flies. Ecdysone is probably not a general growth hormone, however: some tissues, such as the fat body and the imaginal discs, grow continuously throughout larval life, and are apparently unaffected by the cycles of activity in the prothoracic glands. Moreover, removal of the glands does not prevent moulting in all insects. Presumably the degree of co-ordination of cyclic growth exercised by them varies within the class. The injection of ecdysone does terminate pupal diapause in moths, but one cannot be sure that the renewal of epidermal growth has not reawakened neurosecretory activity in the brain. Growth and differentiation in insects is controlled by such a finely balanced and mutually interacting system of endocrine organs and body tissues that establishment of the actions of individual hormones is rarely precise.

The Corpora Allata

The corpora allata are small spherical glands, arising from the maxillary segment, and situated just behind the corpora cardiaca on the wall of the aorta or the side of the oesophagus (Fig. 2.11). During larval life the corpora allata produce a hormone, known as the *juvenile hormone*, that controls the way in which the larval cells differentiate at each moult. As long as they produce this hormone, larval form is retained; when production of juvenile hormone ceases, adult form is expressed, either by cells which are derived directly by mitosis from larval cells, or by rapid growth of the imaginal discs after breakdown of the larval tissues (p. 97). Thus, although the general body form arising from each moult is controlled by a quantitative relationship between the amounts of ecdysone and juvenile hormone in circulation, the response of individual larval cells to this balance may be different—some change the direction of their metabolism, others cease all activity and disintegrate.

The corpora allata resume secretion in the adult, producing a hormone which seems to be identical with the juvenile hormone, but which now acts as a gonadotropin, stimulating the ovaries, for instance, to absorb protein from the blood and store it in the eggs as yolk. The corpora allata are also said to have a general growth-promoting effect, but this may be a consequence of the storage of brain hormone, transported from the brain via nerve axons, rather than an effect of juvenile hormone.

Nature and Mode of Action of Hormones

The brain hormone has not been identified. The moulting hormone, ecdysone, is a steroid with a multiplicity of hydroxyl

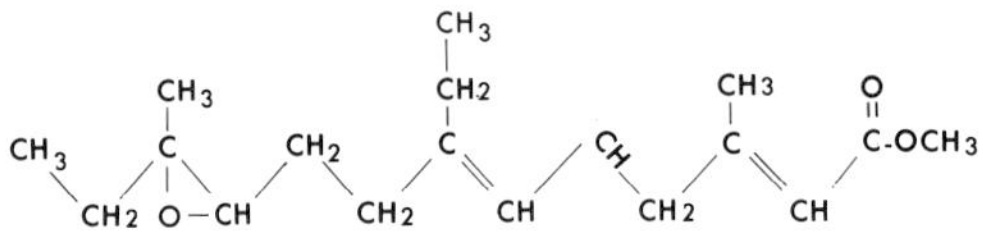

The Juvenile Hormone

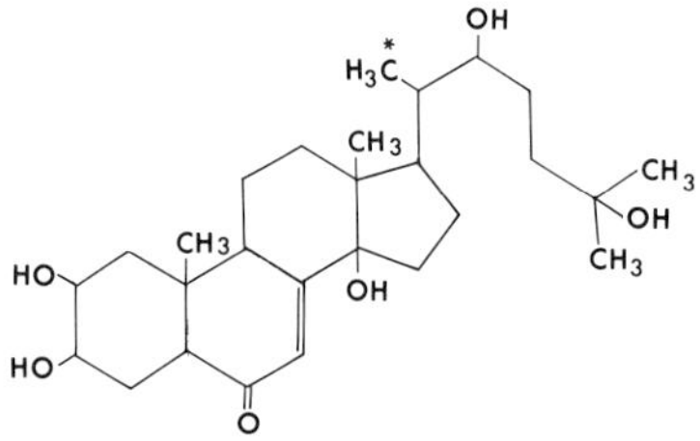

Ecdysone

Fig. 2.22. Insect hormones.

groups (Fig. 2.22). Since insects cannot synthesize sterols from simpler precursors, they must modify preformed sterols obtained in the diet to produce this vital hormone. The earliest observed effect of the injection of ecdysone into fly larvae deprived of their ring glands is the formation of 'puffs' at one or more specific points on the giant chromosomes of the salivary glands (Beermann and Clever, 1964). These puffs are regions where the desoxyribonucleic-acid (DNA) strands of the chromosomes are separated and partly unwound, and where the synthesis of ribonucleic acid (RNA) is proceeding. They are recognized as centres of gene activity—points at which the DNA template is reproduced in RNA molecules, which later move from the nucleus to the cytoplasm, where they initiate and code for the synthesis of specific proteins. The effect of ecdysone on the salivary-gland chromosomes suggests that the hormone, which has been shown to induce the formation of several of the enzymes involved in the hardening of the new cuticle (Karlson, 1963), may exert its effect by the direct activation of the genes responsible for these enzymes.

The juvenile hormone (Fig. 2.22), identified as the methyl ester of a branched-chain acid of 17 carbon atoms (Röller *et al.*, 1967), may also act directly on the nucleus. It could be thought to promote the activity of one whole set of genes, those producing larval form, while suppressing the genes that carry the code for adult form. It is possible that those cells that do not survive larval life have permanent suppressors of the adult genes, so that they run down and die when the activator of the larval genes is no longer present.

THE MUSCULAR SYSTEM

Anatomy and Histology

The somatic muscles of insects are derived embryonically from the paired mesodermal somites. Although their origin is intrasegmental, the development of body-wall sclerites which do not conform with the segmentation means that longitudinal muscles are functionally intersegmental. Several main groups of muscles may be recognized. They are the *dorsal*, *tergo-pleural*, *tergo-sternal*, *pleuro-sternal*, and *ventral* muscles. The cranial muscles that operate the mouth-parts and pharyngeal structures are more complex; but the main musculature of the thorax is often simpler, the dorsal and tergo-sternal muscles occupying most of the pterothorax of flying insects, although there is also a series of pleural muscles inserted on the wing articulations and on the coxae of the legs. Movement of the legs is achieved by sets of opposing muscles located in the segments proximal to their points of insertion. Thus, in the jumping leg of a grasshopper the tibial muscles are attached to the walls of the enlarged femur, which they fill.

In addition to the voluntary segmental muscles, there is an involuntary musculature associated with the viscera and innervated by the autonomic nervous system.

Insect muscles contain little connective tissue, and do not have a continuous connective-tissue sheath which is extended into a tendon, as in vertebrate muscle. Instead, the myofibrils are attached to the cuticle by fibres of unknown composition *(tonofibrillae)*, which pass through the epidermal cells into the cuticular substance. The insertions may be on the body wall itself, or on inflected apodemes, which may be rigid and heavily sclerotized or relatively long, flexible, and tendon-like.

Both voluntary and involuntary insect muscles contain striated fibres. They thus resemble the skelctal muscle, not the involuntary muscle, of vertebrates. Three histological types of insect skeletal muscle have been recognized (Pringle, 1957a). The commonest type has fibres of relatively small diameter, which appear tubular in cross-section, because they have a central core of cytoplasm containing nuclei but no fibrils, and an outer region in which the fibrils are arranged radially. It is characteristic of most body muscles, and is also found in the flight muscles of some primitive insects, such as cockroaches. The two other histological types occur in the flight muscles of other groups. In those of higher orthopteroids and the Lepidoptera, the fibres are still of relatively small diameter, but are tightly packed with fibrils and mitochondria, with the nuclei external to the fibrils. In the so-called 'fibrillar' flight muscle, which is found in the Coleoptera, Diptera, and Hymenoptera, the number of fibres is reduced and their diameter is greatly increased. Regular invaginations of the cell membrane (sarcolemma), which carry nerves and tracheoles into the body of the fibre, break it up into segments that resemble the fibres of other muscles. Within the fibres the individual fibrils are separated from one another by columns of very large mitochondria.

The Physiology of Flight

The power for flight was probably provided primitively by muscles attached directly to sclerites at the wing bases, and such direct muscles are still responsible for much of the motive power in Odonata, blattoids, many orthopteroids, and beetles (Pringle, 1957a). In all other orders, however, the predominant flight muscles are dorsal longitudinal and tergo-sternal (Fig. 2.23). Development of this indirect musculature is associated with the evolution of mechanisms for coupling fore and hind wings or the reduction of

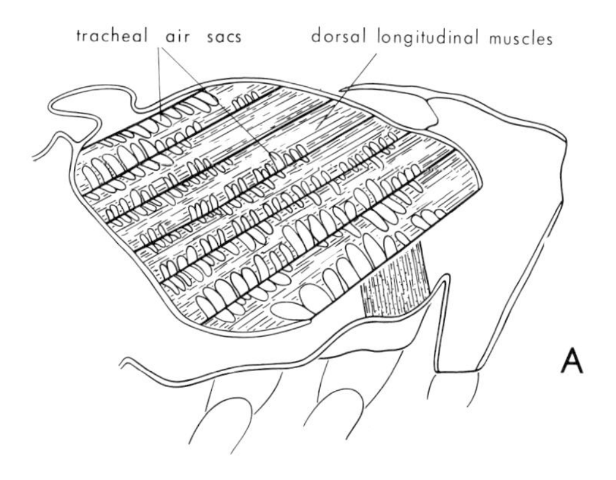

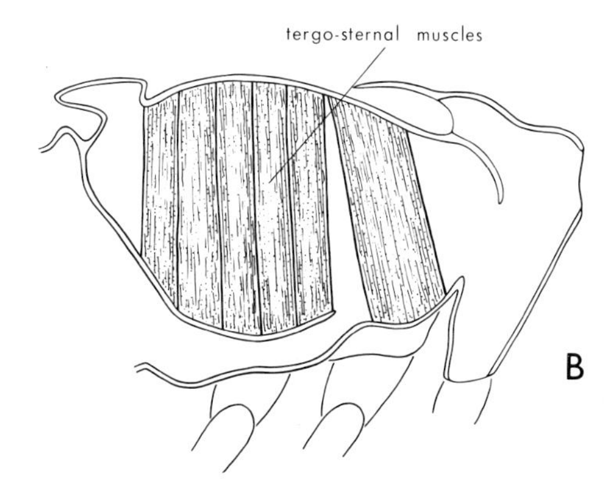

Fig. 2.23. Flight muscles of a cicada: A, median sagittal section of thorax; B, same after removal of longitudinal muscles.

functional wings to a single pair. The indirect flight muscles move the wings by deforming the elastic box formed by the exoskeleton of the thorax, the smaller muscles attached to the wing bases being used to alter the angle of attack of the wings, or to fold them in the Neoptera.

Contraction of the indirect flight muscles is translated into wing movement through a complicated cuticular hinge which constitutes the so-called 'click' mechanism. Contraction of the dorsal longitudinal muscles produces a rising tension in the thorax, but little or no wing movement until the tension reaches a critical level, at which point the hinge clicks over to its down position, at the same time releasing the tension. A similar mechanism in the reverse direction, with the dorsoventral muscles contracting, causes the up-stroke of the wings. The down-stroke is also aided by an elastic component in the hinge which is stretched in the up position. The value of this component is that it stores energy during the up-stroke, when movement of the wings is aided by the lift they generate, allowing this energy to be released during the down-stroke when the wings must do positive work against the lift. By this conservation of energy, the insect achieves a high rate of power output from its energy expenditure during flight (Weis-Fogh, 1961).

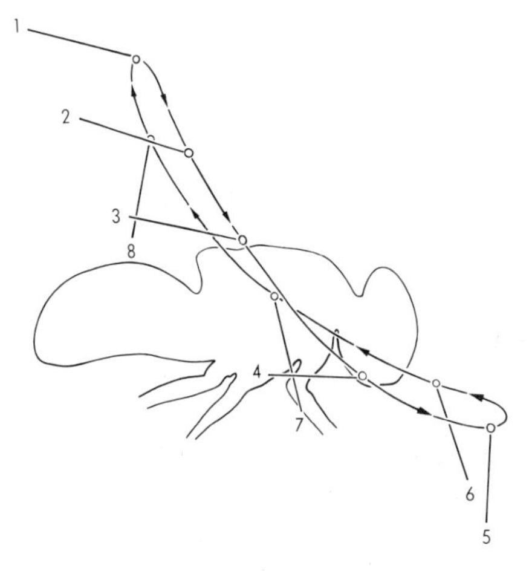

Fig. 2.24. Figure-eight course of the wing tip during flight (after Chadwick, in Roeder, 1953). The straight lines numbered 1–8 show the inclination of the wing at successive positions in the stroke.

Although some large-winged insects, such as butterflies, flap their wings relatively slowly (4 to 20 strokes per second), others attain extraordinarily high frequencies of wing movement—200 to 300 strokes per second for the house fly, and as high as 1,000 per second for some other flies. The wing tip describes a figure eight in the up and down sweep of the wings, as is shown in Figure 2.24. The natural curvature of the wings and the resistance of the air produce the aerofoil surface that generates lift. Horizontal movement and steering are achieved by twisting the wings and changing the plane of the figure-eight movement on one or both sides. This flexible arrangement endows the insect with great manoeuvrability in the air.

The click mechanism, together with some unique properties of the flight muscles, explains the high frequencies of wing beat attained in flight, frequencies much higher than would be possible for a muscle repetitively stimulated to perform the normal contraction-relaxation cycle. Myoneural studies have shown, in fact, that during flight the muscles are repetitively stimulated by their motor nerves, but at a frequency much lower than that of the wing beat. At the beginning of the down-stroke of the wing the longitudinal muscles are fully activated and exerting tension. When this tension reaches the critical level, the wing clicks over to the down position, the tension is suddenly released, and as a consequence the muscle is momentarily deactivated. In this condition it can be stretched by contraction of the opposing dorsoventral muscles, which, in turn, go through a cycle of rising tension followed by deactivation produced by the quick release of tension. An oscillation is thus set up, with a frequency depending largely on the mechanical properties of the thorax, in spite of the fact that the muscles can be considered to be in a state of continuous stimulation.

Sound Production

Many insects produce sound, usually by rubbing one part of the body over an adjacent

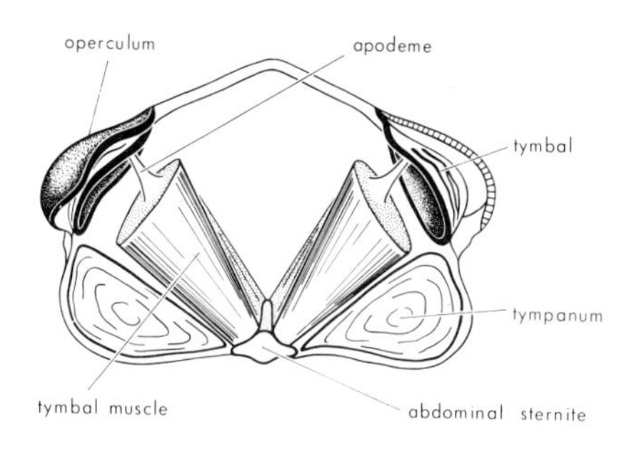

Fig. 2.25. Sound-producing apparatus of a cicada; transverse section of the anterior end of the abdomen, looking caudally. [R. Dencio]

roughened area at high frequency. Sound may be used as a mating call, or to communicate other information between individuals, or as a defence mechanism. In Australia probably the best known of the sound-producing organs are the *tymbals*, or drums, of the cicadas. These consist of a pair of ridged areas of cuticle, each often covered by a flap or *operculum*, on the dorsolateral part of the abdomen (Fig. 2.25). A strong muscle, which has its origin on the ventral abdominal wall, is attached by an apodeme to the inner surface of each tymbal. The muscles are surrounded by tracheal dilatations, and thus lie in an air space. This organ also acts by a 'click' mechanism. Contraction of the muscle builds up tension in the tymbal to a critical level at which it suddenly deforms, emitting a pulse of sound, releasing tension in the muscle, and temporarily deactivating it. The elasticity of the tymbal then causes it to return to its original shape, whereupon the muscle once more starts to exert tension. The oscillation thus set up results in the characteristic loud song of the cicada (Pringle, 1957b). The thin drum-like areas of cuticle adjacent to the tymbals on the ventral surface of the abdomen of the cicada are the tympana of the auditory organs (p. 44).

Metabolism of Muscle

The mechanical strength of insect muscle is about equivalent to that of mammalian muscle. Its resistance to passive stretch is higher, however, in spite of its smaller content of connective tissue, and its elasticity is probably a factor in the special mechanics of flight muscle. An outstanding feature of the metabolism of insect muscle is the tremendous rate of energy conversion during flight. Insect flight muscles, with their efficient respiratory system and their high concentrations of the cytochromes and associated oxidative enzymes, are particularly well adapted for aerobic energy production. The oxidative enzymes are located in the mitochondria, which are always a prominent feature of flight muscle. The presence of such a high concentration of the haem cytochrome pigments imparts a pink colour to many insect muscles, even though they do not contain the red oxygen-carrying pigments haemoglobin and myoglobin that are present in mammalian muscle.

Insect flight muscle is well equipped for the total oxidation of glucose and other substrates to CO_2 and water. That of the house fly is specialized for the oxidation of carbohydrates, using either glycogen stored in the muscle or trehalose from the blood, both of which yield glucose on hydrolysis. The migratory locust can burn either carbohydrate or fat, although it uses mostly fat on long migratory flights. The flight muscles of some butterflies oxidize fat exclusively, whereas the amino acid proline, which is present in high concentration in the blood, is the energy source for flight in the tsetse fly.

Not all insect muscle is so well adapted for aerobic energy production. The jumping muscle of the locust, for instance, like most mammalian skeletal muscle, obtains its energy by the conversion of glycogen to lactic acid, bypassing the mitochondrial oxidative enzymes. Lactic acid production (glycolysis) is an adaptation found in muscles in which the oxygen supply cannot keep pace with energy demands.

THE BLOOD AND CIRCULATION

Organs of Circulation

The blood, or *haemolymph*, occupies all the space *(haemocoele)* between the internal organs. Its circulation is controlled by a tubular dorsal *heart* (Fig. 2.26), which consists

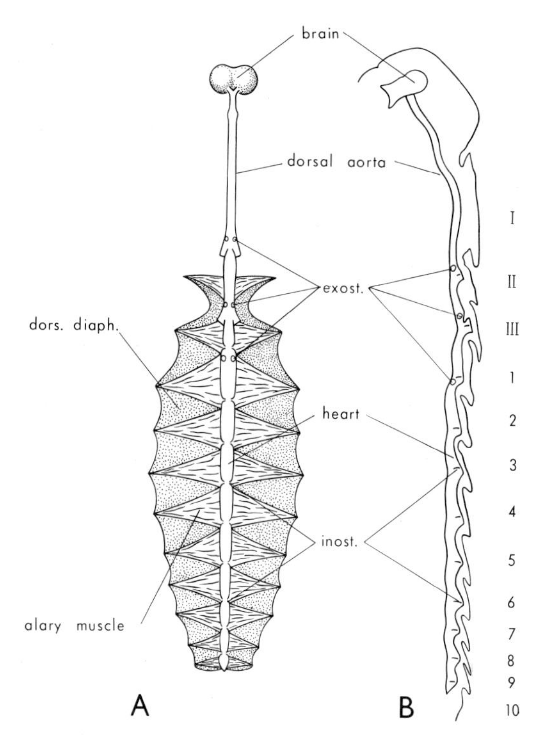

Fig. 2.26. Heart and aorta of a cricket (after Nutting, 1951): A, in plan from the ventral side; B, in sagittal section.
dors. diaph., dorsal diaphragm; *exost.*, excurrent ostium; *inost.*, incurrent ostium.

primitively of a series of segmentally arranged pulsatile chambers located in the abdomen and continued in some lower insects (Blattodea, Dermaptera) into the thorax. The heart is supported by the wing-like *alary muscles* which arise from the dorsal body wall and are attached to the sides of the chambers. Blood enters the heart through paired slit-like openings, the *ostia*, which are located towards the rear of each chamber. The segmental arrangement of chambers has been lost in most insects. In only a few does the chambered portion of the dorsal vessel extend further forward than the second abdominal segment, and in some species the heart is reduced to a single chamber, although the presence of more than one pair of ostia or alary muscles reveals its multiple origin.

The heart is continued into the *dorsal aorta*, which extends forward to the head. A membrane of connective tissue, the *dorsal diaphragm*, lies just ventral to the heart and aorta, forming an incomplete septum which divides off a *dorsal sinus* from the general body cavity. A ventral diaphragm enclosing a ventral sinus is also found in some insects. Movement of the blood is achieved by peristaltic contractions of the wall of the heart. Its direction is usually toward the head, although in some species it regularly flows posteriorly, and in others the direction of flow is reversed at intervals. Circulation in some insects is aided by small, accessory, pulsatile organs, which may be located in the thorax, legs, or head and promote blood flow through the appendages.

The pulse rate is under endocrine control (p. 49), and varies with temperature and activity. Rates as high as 140 per minute have been recorded in a sphinx moth during activity, and as low as one per hour in chilled diapausing pupae of the cecropia silk moth.

Composition of the Blood

The blood volume may be as high as 40 per cent of the total body volume in some larvae, but is usually much less than this in adults (6 per cent in the cockroach). The blood is an aqueous solution of inorganic ions and organic molecules, in which several kinds of cells are usually, although not invariably, suspended. The concentrations of the biologically important ions, Na^+, K^+, Mg^{++}, and Ca^{++}, vary greatly from species to species, and are even far from constant during the life of an individual. Although insect tissues are clearly much more tolerant of changes in their ionic environment than are those of mammals, nevertheless mechanisms for ionic regulation do exist, and are particularly well developed in aquatic insects (Shaw and Stobbart, 1963). The excretory system may selectively eliminate certain ions (p. 67), and selective absorption from the environment is also possible, as, for example, by the anal papillae of mosquito larvae. The osmotic pressure of the blood, which is commonly higher than that of mammalian blood, may also vary over a considerable range, but here again there are regulatory mechanisms that control the flow of water and salts into or out of the organism.

Insect blood contains numerous organic constituents, many of which make a substantial contribution to the osmotic pressure and ionic balance. The great variety and high concentration of amino acids is a characteristic feature. The blood provides a convenient store of amino acids needed for special metabolic events, such as tyrosine for cuticular hardening, several amino acids for silk formation, and even proline as an energy source, but whether the blood amino acids have any general significance beyond this is not known.

The major blood sugar of insects is believed to be the disaccharide trehalose (Fig. 2.27), although recent work has indicated that other disaccharides, such as maltose and cellobiose,

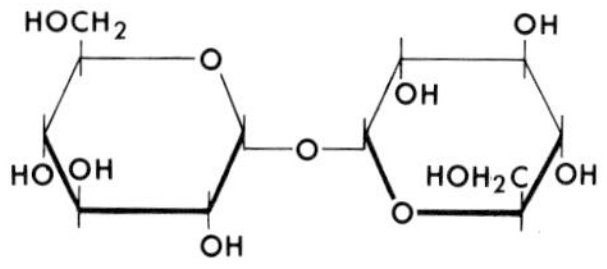

Fig. 2.27. Trehalose.

may be prominent in some species. The monosaccharides glucose and fructose are usually only minor constituents, although they are present in high concentration in a few insects, such as the honey bee. Trehalose is the labile energy source for muscular contraction in many species. Its concentration is under hormonal control (p. 49), as is blood glucose in man. Insect blood contains an enzyme, trehalase, which splits trehalose to two molecules of glucose. The activity of this enzyme is presumably also subject to some kind of metabolic control. Several organic acids (citric, succinic, malic, etc.), which are intermediates in glucose oxidation, are commonly found in high concentrations in insect blood, but whether this accumulation has any metabolic significance is not known.

Fats are conjugated with specific proteins as a preliminary to transport in the blood. The lipid parts of such lipoproteins are believed to be diglycerides, that is, esters of glycerol with two fatty acids. Pigments that colour the blood yellow or green occur in many species (p. 34).

Blood Cells

The blood cells, or *haemocytes*, are derived from scattered groups of mesodermal cells which remain in the haemocoele after the other tissues are formed. Numerous cells circulate in the blood of many insects, but they have a tendency to adhere to surfaces, and in some species are entirely sessile. The haemocytes produced during postembryonic life probably arise by division from pre-existing blood cells. Sessile groups of blood cells engaged in cell division may act as a haemopoietic (haemocyte-producing) organ.

Various histological types of blood cells have been distinguished (Fig. 2.28). Differences between species and different nomenclatures employed by investigators have hindered attempts to arrive at a general classification, but in recent years several cell types have been recognized as occurring widely (Jones, 1962). *Prohaemocytes*, small cells with very little cytoplasm, appear to be precursors of the numerous amoeboid cells with cytoplasm of uniform density known as *plasmatocytes*. Cells in which the cytoplasm contains inclusions of different shapes and sizes may be plasmatocytes engaged in special functions, such as the transport of metabolites. For instance, *adipohaemocytes*, which contain lipid inclusions, are known to be derived from the plasmatocytes in the wax moth *Galleria mellonella*. They differentiate

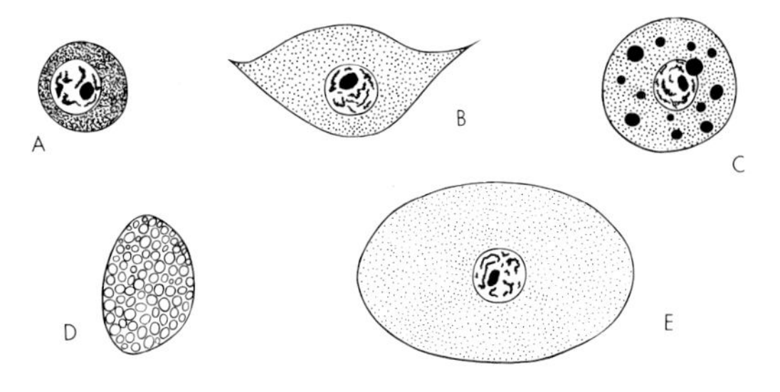

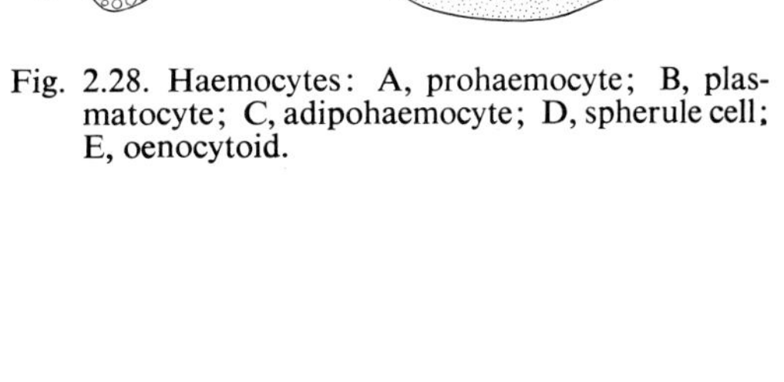

Fig. 2.28. Haemocytes: A, prohaemocyte; B, plasmatocyte; C, adipohaemocyte; D, spherule cell; E, oenocytoid.

towards the end of larval life, and may possibly be engaged in the storage and transport of lipids released by the autolysis of larval cells.

Another granular cell, called a *spherule* cell, is apparently not derived from a plasmatocyte. Spherule cells, which are of varied shape, are packed with small round inclusions of uniform size. Tyrosinase activity has been demonstrated in the spherule cells of some species, and they are also thought to be involved in the transport of material for silk production in Lepidoptera. A cell type of puzzling origin is the *oenocytoid*, in which the relatively large volume of cytoplasm commonly becomes hyaline after the cells have been removed from the body. This 'hyaline transformation', although characteristic of oenocytoids, occurs also in other cell types. As their name suggests, the oenocytoids resemble oenocytes, and have been thought to be derived from them, although there is no positive evidence for this view. It is possible, however, that oenocytoids, like the oenocytes, may have some role in moulting, as they have been reported to accumulate before a moult and later to be destroyed phagocytically by plasmatocytes. In some species tyrosinase activity is found in the oenocytoids rather than in spherule cells.

The haemocytes have a major role in immunity in insects, which do not possess the antigen-antibody mechanism of vertebrates. They engulf pathogens, such as bacteria and viruses, and may adhere to and encapsulate larger foreign bodies, such as the eggs of parasites. They may also play a part in the phagocytic destruction of larval tissues at pupation, although in some species this destruction is entirely autolytic. Blood cells have also been implicated in tissue formation. It is known that haemocytes may deliver mucopolysaccharide for inclusion in developing connective-tissue membrane, and some organs may be formed at metamorphosis from undifferentiated cells in the haemocoele (p. 99). In many insects a type of blood cell which has not been positively identified morphologically is involved in blood coagulation. These cells adhere to surfaces where the blood is in contact with air, and presumably release an agent that induces gel formation in the lymph in their vicinity. Some of the cells that engage in clot formation throw out numerous fine pseudopodia, and eventually disintegrate into a series of coagulated protoplasmic strands which help to stabilize the clot.

THE PERICARDIAL CELLS

The pericardial cells have an origin similar to that of haemocytes, but, instead of circulating in the blood stream, they form a diffuse tissue closely associated with the heart. They may lie on its outer or inner wall, on the alary muscles, or be clustered around connective-tissue strands of the dorsal diaphragm. Because of their ability to absorb dyes and other colloidal material injected into the blood, the pericardial cells have been thought to have an excretory function, and were called nephrocytes. More recently, they have been equated with the reticulo-endothelial system of vertebrates, and are thought to supplement the functions of the haemocytes in immunity, absorbing foreign proteins and other potentially harmful material of a non-particulate nature. An endocrine function for pericardial cells has also been established (p. 49).

THE FAT BODY

The fat body is a diffuse tissue derived from the mesoderm of the embryo. It usually consists of sheets or clumps of cells, supported by tracheae, lying mostly in the abdominal cavity. It grows in size during larval life, may provide much of the material for the formation of new tissues during pupation, and is commonly reduced to quite small size in the adult, when its stored reserves are used in egg production. Most fat-body cells contain numerous fat globules, as well as protein and glycogen granules. Others enclose crystals of urates, and still others may be specialized in some insects for the accommodation of intracellular symbiotic micro-organisms.

The fat body is important as a storage organ, and also as a site of intermediate metabolism in insects. It thus assumes many

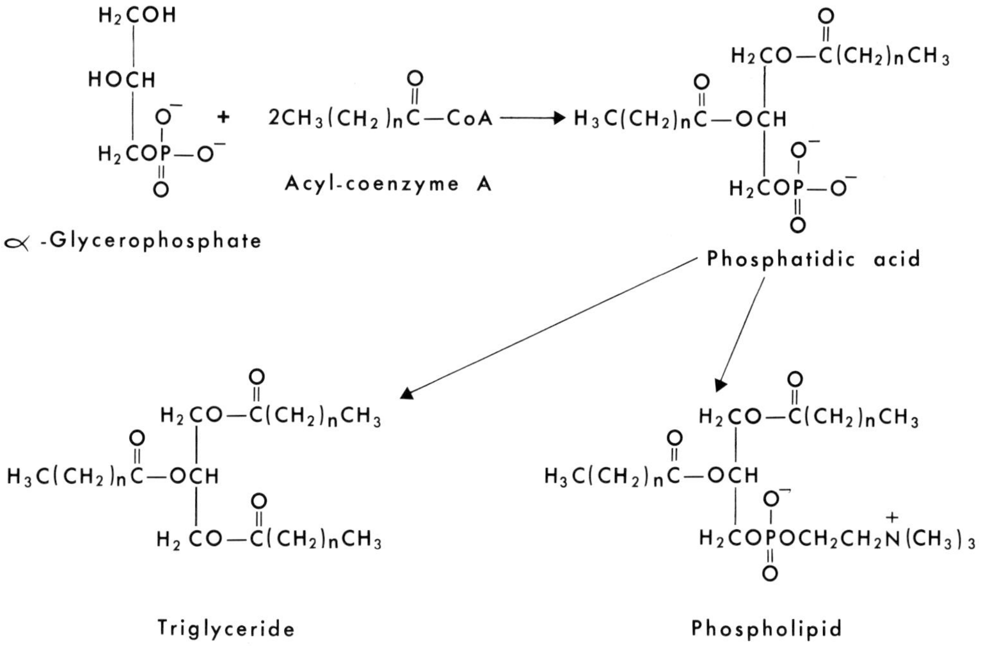

Fig. 2.29. Synthesis of lipids.

of the functions of the mammalian liver. The presence of uric acid may be the result of the accumulation of waste products (storage excretion), or it may represent an additional means of nitrogen storage, as there is some evidence that uric acid is returned to the metabolic pool during pupation. The symbiotes found in the fat body of many insects supplement their hosts' metabolism, synthesizing vitamins or essential amino acids which other insects must acquire from their diet. These symbiotes are transferred from generation to generation in, or on the surface of, the eggs. The metabolic functions of the fat body have been reviewed by Kilby (1963).

Fat Metabolism

As might be expected, the cells of the fat body contain the enzymes that catalyse the synthesis of fats. Long-chain fatty acids are formed from two-carbon fragments derived ultimately from acetate, although the ester of malonic acid with coenzyme A (malonyl-coA) is the immediate donor of 2-carbon units. Fatty acyl-coenzyme A derivatives condense with glycerol-1-phosphate to form phosphatidic acids, which yield triglycerides when the phosphate is replaced by a third fatty acid, or phospholipids when the phosphate group enters into a second ester linkage with another molecule, such as choline, inositol, etc. (Fig. 2.29). Storage of reserves in the form of fat is favoured in many larval insects, in which the fat body is able to convert both carbohydrates and amino acids to fat, presumably through the intermediate formation of acetic acid.

Most storage fats of insects are in the form of triglycerides; the more complex lipids, such as the phospholipids, enter into the protoplasmic architecture of cells. The constituent fatty acids of storage fats vary widely from group to group. In some, as in most animal fats, the predominant acids are the 16- and 18-carbon saturated palmitic and stearic acids and the 18-carbon mono-unsaturated oleic acid. But many insect fats resemble plant oils in having a much higher proportion of unsaturated fatty acids, including members with two, three, or more double bonds. In a number of aphids, shorter-chain saturated fatty acids predominate, with chain lengths ranging down to the 4-carbon butyric acid.

The fat body is capable of releasing fat into the haemolymph, probably in the form of diglycerides conjugated with specific proteins. Breakdown of fats may also occur within the fat body. Lipases capable of splitting the ester linkages of glycerides to form free fatty acids and glycerol are found in these cells, as well as enzymes of the fatty-acid oxidation sequence which convert long-chain fatty acids to acetic acid. A necessary preliminary to this oxidation is the conjugation of the fatty acid with carnitine, which is synthesized by almost all animals, but is a vitamin for a few species of beetles (p. 65).

Carbohydrate Metabolism

Glucose molecules are built up into glycogen polymers in the fat body, after prior condensation with the nucleotide uridine triphosphate (UTP) to form uridine diphosphate glucose (UDP-glucose). A similar reaction leads to the synthesis of trehalose, but here the UDP-glucose conjugate is condensed, not with the end of a growing glycogen chain, but with a molecule of glucose-6-phosphate (Fig. 2.30). The resulting trehalose-6-phosphate is dephosphorylated by a specific enzyme. Glucoside formation in the fat body by conjugation of various molecules with UDP-glucose is an important detoxication mechanism in insects. Poisonous compounds, such as phenols, whether absorbed from the environment or produced by the insect's own metabolism, may be detoxified in this way. Detoxication by glucoside formation is another biochemical peculiarity which insects share with plants. Comparable detoxifying mechanisms in most other animals involve conjugation with glucuronic acid.

Glycogen is broken down in the fat body by phosphorylase, which condenses a phosphate radicle with each glucose unit removed from the glycogen chain. Trehalose, on the

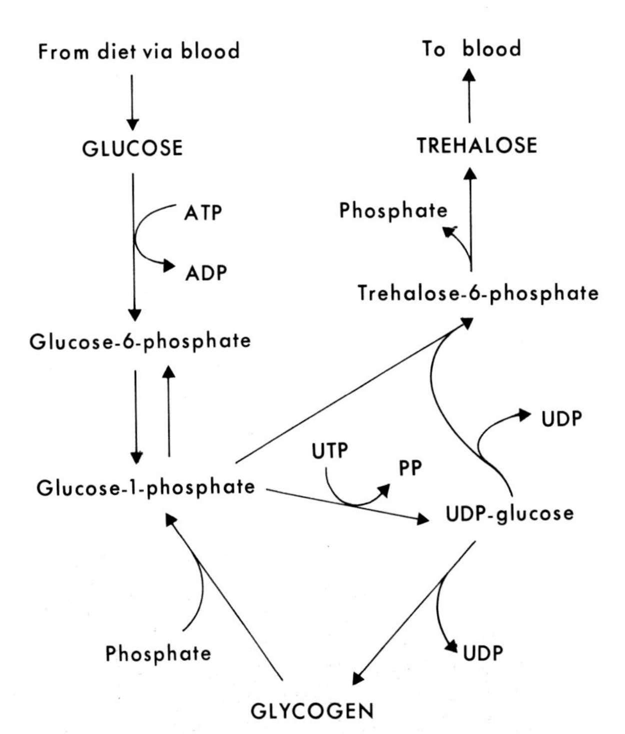

Fig. 2.30. Carbohydrate metabolism of the fat body.

other hand, is hydrolysed directly by trehalase. It has already been pointed out (p. 49) that the trehalose concentration of the blood is in dynamic equilibrium with the glycogen of the fat body.

Amino-acid and Protein Metabolism

Transaminases, which are important in amino-acid synthesis, are active in the fat body. Glutamic acid is the most effective donor of amino groups, and the most rapid transaminase reaction is:

glutamate + oxaloacetate ⇋ aspartate + α-ketoglutarate.

Other keto acids can accept amino groups, although rates are lower than in the above reaction. Fat body also contains an active glutamic acid dehydrogenase which oxidatively deaminates glutamic acid, transferring hydrogen to the coenzyme nicotinamide adenine dinucleotide (NAD):

glutamate + H_2O + NAD → α-ketoglutarate + NH_3 + $NADH_2$

This enzyme, acting in conjunction with the transaminases, provides a pathway for the deamination of most amino acids. Other enzymes involved in the intermediate metabolism of amino acids have been demonstrated in the fat body. They include the enzyme responsible for the synthesis of serine from glycine and formate, and that which forms glutamine from glutamate and ammonia.

Proteins are synthesized and stored in the fat body, and may be released from there into the blood. Fat body is probably the main source of the protein that is incorporated into the yolk of developing eggs.

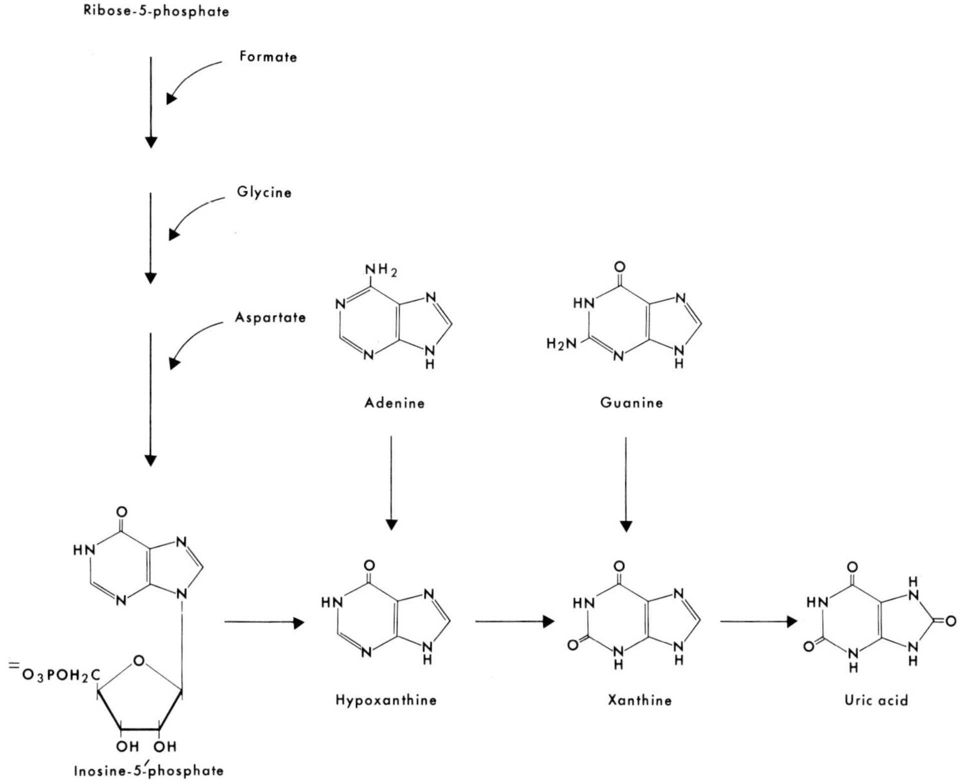

Fig. 2.31. Purine metabolism of the fat body.

Purine Metabolism

Uric acid, which is the main vehicle of nitrogen excretion in insects, is synthesized in the fat body by a pathway that incorporates carbon and nitrogen into the molecule from a variety of sources (formate, glycine, aspartate, etc.). The starting point of this pathway is the phosphate ester of the sugar ribose, and the final product is probably the purine ribotide inosine-5′-phosphate (Fig. 2.31). Hypoxanthine derived from inosine-5′-phosphate is converted to xanthine and uric acid by xanthine dehydrogenase. The fat body contains enzymes that deaminate the amino purines, which can thus also be degraded to uric acid A number of insects excrete degradation products of uric acid as well as the purine itself (p. 67), and in some of these the enzymes of uric acid breakdown are present in the fat body.

Pigment Metabolism

Pigments of plant origin, such as carotene, are frequently stored in the fat body, and may undergo some metabolism there. The fat body is also a major site of synthesis of the ommochromes from tryptophan. In some insects (e.g. *Drosophila*) only the conversion of tryptophan to kynurenine occurs in the fat body. Kynurenine accumulates during larval life, and is used by the pharate adult for the formation of the ommochrome eye pigments. However, in some Lepidoptera in which ommochromes occur as general body pigments, the whole process of pigment synthesis takes place within the fat body. The pterins, which always seem to be associated with the ommochromes, are prominent in the fat body, which may be an important site of the synthesis of these pigments.

LIGHT-PRODUCING ORGANS

Of the many unrelated organisms that produce light, several insects are among the most conspicuous. The widely distributed fireflies, which are mostly adult lampyrid and elaterid beetles (p. 512), are well known for their bright flashing lights, and the cave-dwelling glow-worms (mycetophilid larvae) also produce spectacular effects. Other glow-worms are the larvae or larviform females of fireflies. Different tissues have been specialized for light production in insects. In the majority, the fat body is responsible. The entire tissue may glow in some species and transmit light through a uniformly transparent body wall. In most of the fireflies the luminous organs are specialized regions of fat body situated under localized areas of transparent epidermis. In the Central American beetle *Phengodes* the luminous organs resemble, and may be derived from, the oenocytes; whereas in the Australian and New Zealand mycetophilid glow-worms of the genus *Arachnocampa* light is produced from the enlarged tips of the Malpighian tubes, which are applied to the ventral surface of the rectum and are visible through the cuticle of the last abdominal segment. Light serves as a lure for prey in the mycetophilid glow-worms, but in the adult fireflies it acts as a sex attractant and mating signal.

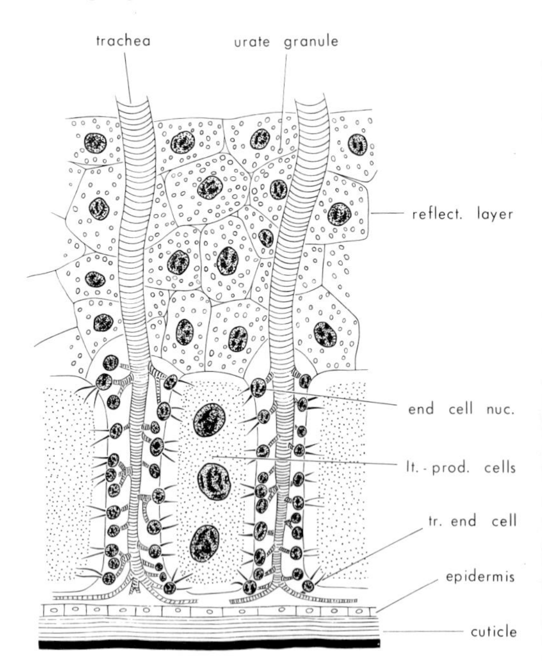

Fig. 2.32. Vertical section of firefly lantern (after Hess, 1922).
end cell nuc., end cell nucleus; *lt.-prod. cells*, light-producing cells; *reflect. layer*, reflective layer; *tr. end cell*, tracheal end cell.

The photogenic organ of the firefly consists of a layer of light-emitting cells, below which is a reflective layer of cells packed with urate crystals (Fig. 2.32). The light-producing cells have a massive tracheal supply, and tracheal end cells are prominent in those species that produce brief intense flashes. The organs are innervated, and these beetles are able to exert a considerable degree of physiological control over light production.

As in other organisms, the chemistry of light emission in the firefly *Photinus pyralis* involves the oxidation of an organic substrate by an enzyme called luciferase. In *Photinus* the substrate, adenyl-luciferin, is formed by reaction of the nucleotide adenosine triphosphate with an aromatic carboxylic acid (McElroy and Seliger, 1962). Oxidation produces an excited intermediate, which decays with a loss of energy in the form of light. The quantum efficiency is 1; that is, one quantum of light is produced for every mole of substrate oxidized. *Photinus* produces greenish yellow light with a wave length of maximal emission of 5500 Å. Other insects produce light of different colours, and presumably oxidize different luciferins from that of *Photinus*. It is clear that the mechanism of luminescence provides several bases for the establishment of physiological control. For instance, control could be exercised by regulating the supply of oxygen, or by relieving inhibition of luciferase, which is easily inhibited by the products of its action, but no overriding evidence in support of any of the possible control mechanisms has yet been found.

THE ALIMENTARY CANAL

Structure

The alimentary canal consists of ectodermal *fore* and *hind guts* joined by an endodermal *mid gut* (Fig. 2.33). The gut epithelium, one cell in thickness, rests on a basement membrane, external to which are circular and longitudinal muscle fibres innervated by the autonomic nervous system. The muscles control the movement of the gut contents. Circular muscles are strongly developed in some localized areas to form sphincters. The epithelium has a cuticular covering on its inner surface in the fore and hind gut regions, but not in the mid gut.

At the anterior end of the fore gut is the preoral or *buccal cavity*, into which the ducts of one or more pairs of dermal glands associated with the mouth-parts may open. These *salivary glands* are labial in most groups, although they may be mandibular in lepidopterous larvae in which the labial glands are used for silk production. The salivary glands may lie partly or wholly in the cephalic,

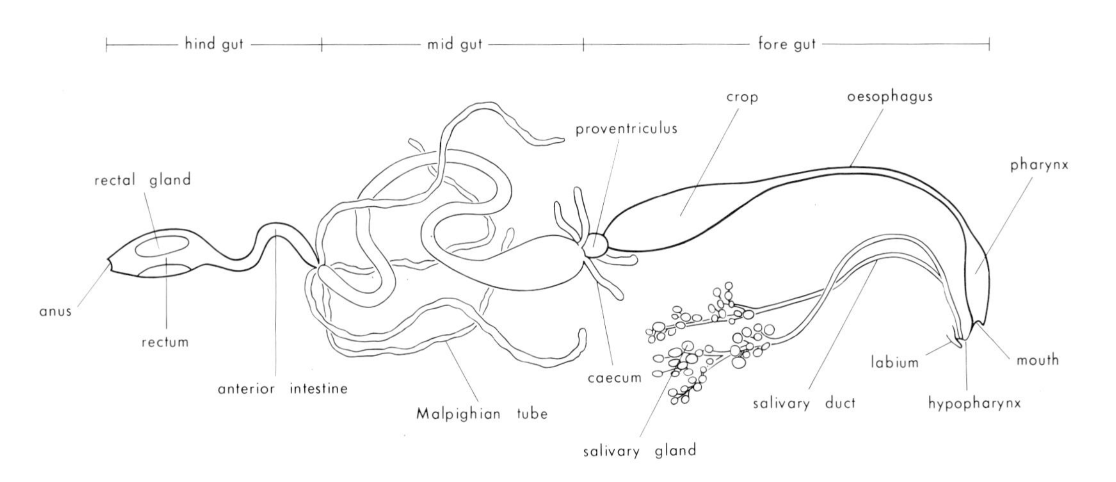

Fig. 2.33. Generalized diagram of the alimentary canal.

thoracic, or abdominal regions of the body cavity (Fig. 2.33); their ducts unite to form a median salivary duct which opens between the hypopharynx and the labium. Mechanisms for the forcible ejection of saliva are present in several groups. In the Hemiptera the apparatus consists of a cuticular and muscular *salivary pump* (Fig. 2.34) which lies in a diverticulum rom the median salivary duct. The pump is filled by contraction of the muscles, and emptied by the elastic recoil of the resilin walls. Cuticular flaps in the ducts act as valves controlling the direction of flow.

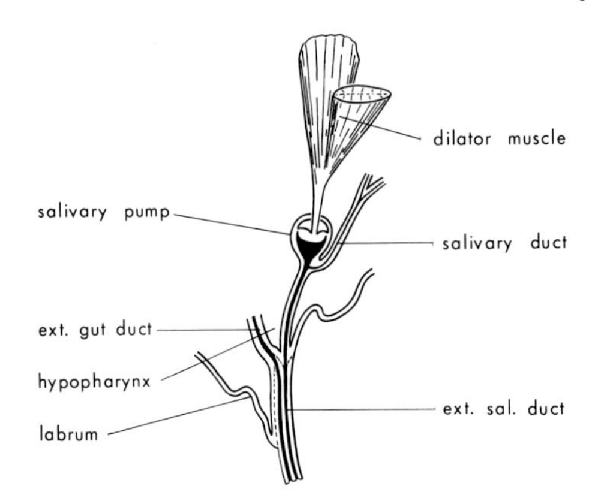

Fig. 2.34. Salivary pump of a cicada (after Snodgrass, 1935).
ext. gut duct, external duct leading to gut; *ext. sal. duct*, external duct of the salivary system.

The saliva contains not only a variety of digestive enzymes, the functions of which are discussed on p. 64, but also mucopolysaccharides. It thus has a general function in moistening the food and lubricating its passage, in addition to initiating the digestive process. The saliva also performs a number of special functions in certain groups. In blood-sucking forms it contains an anticoagulant. In plant bugs it is responsible for the formation of the stylet sheath, which surrounds the mouth-parts, and forms a more or less permanent duct between the plant surface and the phloem vessels from which the bugs draw the sap. The salivary glands of these bugs contain a lipoprotein which gels rapidly on release from the glands. Tyrosine and a phenolase are also present, and probably aid sheath formation by tanning the extruded lipoprotein. Penetration of the stylets into the plant is aided by the secretion of a salivary pectinase which hydrolyses pectin, the polysaccharide cementing substance between plant-cell walls. A similar function is served by hyaluronidase in insects that inject saliva into the bodies of captured prey. Hyaluronidase breaks down the polysaccharide ground substance of connective tissue, aids the penetration of the saliva, and assists in liquefying the tissues of the victim.

The first section of the fore gut is the *pharynx*. It is usually strong-walled and muscular, forming a powerful sucking organ in insects that ingest liquid food. Movements of the pharynx are produced by its own muscular coat and by muscles which have their origin on the exoskeleton of the head and are inserted on the pharynx wall. The epithelium of the pharynx is flattened and pavement-like. Its cuticular lining may be quite thick, and often bears hardened ridges and teeth which assist in grinding the food.

Posterior to the pharynx is the *oesophagus*, which may take the form of a narrow tube, although in many species it is dilated into a *crop* for the storage of food, or may possess sac-like diverticula with the same function. The epithelium of the oesophagus is also flattened, and the cuticle often thin, although crops sometimes have thick cuticular linings and walls thrown into folds which can be smoothed out when the crop is distended with food. The posterior end of the oesophagus projects into the anterior end of the mid gut to form the *oesophageal invagination* (Fig. 2.35). In insects that eat solid food, the region just anterior to the oesophageal invagination is differentiated into a muscular *proventriculus*. The cuticular lining is very thick in this region, and often has hardened, tooth-like or hair-like projections (Fig. 14.3D). Circular muscles are strongly developed. The proventriculus has some function in grinding the food; it also acts as a valve and a sieve, allowing only the finest food particles to pass into the mid gut.

The *mid gut* is the main organ of digestion and absorption of food products. The cells

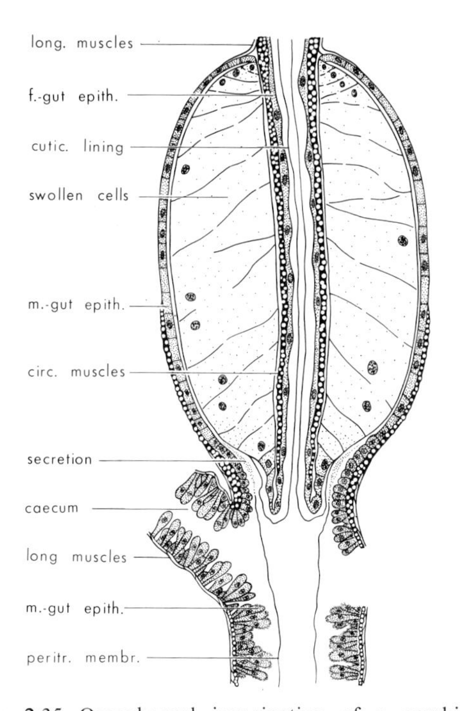

Fig. 2.35. Oesophageal invagination of a syrphid larva.
circ. muscles, circular muscles; *cutic. lining*, cuticular lining of fore gut; *f.-gut epith.*, fore-gut epithelium; *long. muscles*, longitudinal muscles; *m.-gut epith.*, mid-gut epithelium. The secretion is transformed into the peritrophic membrane *(peritr. membr.)*. The swollen cells of the fore gut form the peritrophic membrane press.

of its epithelium are large and columnar, and possess hair-like brush borders on their inner surfaces. They show signs of intense secretory activity. At the end of their lifetime, which is usually short, they disintegrate, and are replaced by mitosis from groups of undifferentiated cells, which may be scattered through the epithelium or localized in special pockets or crypts. The mid gut usually possesses a number of diverticula or *caeca*, which are typically located at the anterior end near the oesophageal invagination, but may occur in other regions (Fig. 30.10). It may be a simple tube, or differentiated into a series of regions of differing morphology, epithelial histology, and functions.

In all insects except those that live on a purely liquid diet, the contents of the mid gut are separated from the epithelium by a thin, transparent *peritrophic membrane*. This structure appears as a series of concentric lamellae in some groups, since it is formed by the whole mid-gut epithelium and periodically separated from the cells. In other species it is formed as a single continuous tube by specialized cells in the crypt formed by the oesophageal invagination (Fig. 2.35). The peritrophic membrane encloses the food during its passage through the mid gut, usually remains intact in the hind gut, and can often be seen investing the faecal pellets. Although it prevents the passage of particulate material, it is freely permeable to molecules in solution. Its function seems to be protective, and thus parallels that of the mucous secretions of vertebrate stomachs. Its composition is partly proteinaceous, partly chitinous, and also includes mucopolysaccharides such as hyaluronic acid.

The junction between mid and hind gut is often marked by a sphincter, or *pyloric valve*. The Malpighian tubes (p. 66) also open at this point. The *hind gut* is usually differentiated into a narrow, tubular *anterior intestine* and a more bulbous and muscular *posterior intestine* or *rectum*. In plant-sucking bugs a loop of the anterior intestine is held in permanent association with the anterior end of the mid gut to form the *filter*

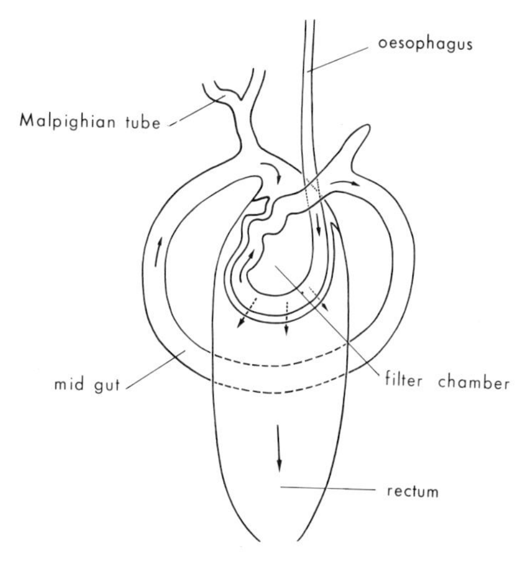

Fig. 2.36. Alimentary canal of a coccid (after Weber, 1933).

chamber, which aids in concentrating the highly liquid diet (Fig. 2.36). In most insects the rectum has a number of regions with a high columnar epithelium, contrasting with the generally flattened cell layer of the hind gut, and known as *rectal glands*. They also are concerned with water economy (p. 66).

Digestion

The diet of insects is so varied that it is no surprise to find a wide range of enzymes and digestive mechanisms within the class (Waterhouse, 1957). In general, the digestive enzymes are adapted to the normal food. Omnivorous insects possess a wide range of enzymes; those that live on a restricted diet have a narrower range.

The salivary glands secrete digestive enzymes in most species. Salivary carbohydrases are common, but proteases and lipases may also occur. In some of the insects that practise external digestion by salivating on the surface of food, or injecting saliva into a victim and sucking up the soluble products, the salivary glands may be the sole source of enzymes. In most insects, however, the salivary secretion merely supplements the activity of the mid gut, which is the main producer of digestive enzymes. In the larvae of dytiscid beetles, for instance, midgut digestive fluid passes forward through channels in the mandibles into the body of the victim to initiate digestion. Digestion by the salivary enzymes in insects that eat solid food proceeds while the food is in the fore gut and during storage in the crop, and is also aided in some species by the action of enzymes passed forward from the mid gut. In most species digestion is complete by the time the food leaves the mid gut, but digestion in the hind gut is the rule in some wood-eating termites.

Most insects secrete enzymes capable of hydrolysing the simpler carbohydrates, such as sucrose, maltose, etc., as well as the starch-digesting amylase. Other digestive carbohydrases of a less usual kind are cellulase and chitinase. Many insects feed on wood, but not all are capable of digesting the cellulose of cell walls, and some live on the stored starch and other metabolites in the wood. Of those that do digest cellulose, only a few species (beetle larvae, silverfish) are known to synthesize a cellulase themselves. Others have an intestinal flora of bacteria which carry out the digestion, and in many termites the hind gut is crammed with a varied fauna of protozoa which have the same function. Digestive chitinases have been identified in several insects (predacious dragonflies, the omnivorous cockroach), and may be of quite usual occurrence in insectivorous species.

Digestive lipases, which are common in insects, split fatty acids from triglycerides. Beeswax, which is not digested by most animals, figures prominently in the diet of larvae of the wax moth, *Galleria mellonella*. Paraffin constituents of the wax are not digested by the moth's own enzymes, however, but by symbiotic bacteria that live in the gut.

Most insect proteases work at pH above 7·0, and are thus similar to mammalian trypsin. Pepsin-like enzymes with pH optima as low as 2·4 have, however, been reported in one or two species. Although the gut contents are acid in these species, nothing similar to the acid-secreting stomach of mammals is known in insects. The pH of the gut contents is near neutrality in most insects, but is regularly alkaline (pH 9–10) in lepidopterous larvae. Some insects digest strongly cross-linked proteins which are resistant to attack by normal animal proteases, including the collagen of connective tissue and the keratin of wool, hair, and feathers. Blow-fly larvae produce a collagenase which can degrade pure collagen, while larvae of the clothes moth have a keratinase which can initiate the splitting of keratin. Digestion of wool by the clothes moth is aided by the maintenance of extremely reducing conditions in the gut. Two enzymes probably have a role in maintaining these conditions. They are cystine reductase, which reduces cystine released from the wool to cysteine, and cysteine desulphydrase, which splits hydrogen sulphide from cysteine. Cysteine and hydrogen sulphide are powerful reducing agents capable of reducing the disulphide bridges of wool in the alkaline gut contents, thereby rendering it more susceptible to attack by the gut protease. In addition to proteases, which act on whole proteins, insects secrete a number of peptidases, which split specific linkages between amino acids in peptides.

Absorption

The cuticular lining of the fore gut is relatively impermeable, and little absorption occurs there. The absorption of fat from the crop has been reported in the cockroach, however, and it may be that lipids pass more easily than water-soluble compounds through the fore-gut cuticle, as they do through the external cuticle. The mid gut is the main site of absorption, although the hind-gut cuticle is freely permeable and an appreciable amount of absorption occurs there. Since the peritrophic membrane prevents particles from reaching the mid-gut cells, all absorption is of substances in solution, as, for example, single sugars derived from carbohydrates, amino acids from proteins, and fatty acids from fats. Free fatty acids may induce the absorption of unchanged fats by emulsifying them into droplets small enough to pass the peritrophic membrane and enter the cells. Histochemical methods have shown that in some insects parts of the mid gut are specialized for the absorption of particular substances—inorganic ions by some cells, fats by others, and so on—but it seems likely that the mid-gut cells of other insects combine all these absorptive functions, as well as secreting enzymes.

Intermediate Metabolism in the Gut

The alimentary canal has always been considered to play a role in intermediate metabolism complementary to that of the fat body. It is clear that mid-gut cells of many insects are capable of synthesizing the protein and polysaccharide of the peritrophic membrane. Some facets of amino-acid metabolism have also been identified in the mid gut, but evidence in support of a general role in intermediate metabolism is scanty.

Nutrition

Although the three main classes of foodstuffs—carbohydrates, fats, and proteins—are interconvertible within the animal body, most animals require a reasonable balance of all three in their diet for optimum growth. The same is true of some insects, but many species live on remarkably restricted diets, and show metabolic specializations appropriate to this restriction. Thus, carbohydrates are essential elements in the diets of several insects, but others, such as the clothes moth, can live on a wholly protein diet. Termites can survive for some time on pure cellulose, but cannot grow on this diet, as they need organic nitrogen, which is normally supplied by micro-organisms, fungi, and organic debris within the wood they eat.

In spite of the extraordinary specialization of the diet in some species, insects, in general, have the same requirements for vitamins and essential amino acids as mammals. When these are not present in the normal food, they are supplied by micro-organisms which the insects harbour within their own bodies. These organisms may live in the gut, where they may play a part in digestion as well as synthesizing essential organic molecules, or they may be internal symbiotes, often accommodated in special cells of the fat body. Some flour beetles, for instance, can live on extracted flour which lacks several of the water-soluble vitamins, but soon die on this diet when separated from their symbiotic micro-organisms. The cockroach *Blattella* has a plentiful supply of symbiotes which render it independent of many vitamins and several amino acids that are essential to the diet of other animals.

Insects deprived of their symbiotes need the water-soluble vitamins thiamin, riboflavin, pyridoxine, nicotinic acid, pantothenic acid, biotin, folic acid, and choline. There seems to be no requirement for cobalamin (vitamin B_{12}), and ascorbic acid is known to be needed by only a few species. The essential metabolic role of carnitine, which is involved in fatty-acid oxidation (p. 58), was first demonstrated in insects, as a few tenebrionid beetles are the only animals known to require this compound in their diets. Of the fat-soluble vitamins, either carotene or vitamin A is presumed to be essential, since the visual pigment of insects is a derivative of vitamin A, and they are not known to synthesize carotenoids; but attempts to demonstrate a need for carotenoids in the diet have so far been successful with only a few phytophagous species.

An unusual dietary requirement of insects is for sterols. These important compounds, which form part of the lipoprotein structure of cells, and which act as hormones in insects, are synthesized by most other animals.

The amino acids which are essential elements in the diet of insects are the same ten that are needed by most animals. They are arginine, histidine, isoleucine, leucine, lysine, methionine, phenylalanine, threonine, tryptophan, and valine. In addition, the simplest amino acid, glycine, seems to be a dietary requirement for the Diptera, which are also unusual in needing purines or nucleic acid in the diet. Poly-unsaturated fatty acids are a dietary requirement for some insects, as they are for the rat. This need may be more general in insects than has so far been established. Insects deficient in unsaturated fatty acids are unable to separate themselves completely from the old cuticle at ecdysis.

THE EXCRETORY SYSTEM

The alimentary canal has a major function in excretion, removing metabolites from the blood and releasing them in the gut lumen. Moreover, the periodic renewal of the gut epithelium at moulting, or the continuous regeneration that occurs in many insects, provides a means by which unwanted metabolites accumulated in gut cells may be passed to the exterior. However, in all insects except some aphids, diverticula of the alimentary canal are responsible for most of the excretory function. These specialized excretory organs are the *Malpighian tubes* (Fig. 2.33), which open at or near the junction of mid and hind guts. Many insects retain the 4 or 6 primary Malpighian tubes throughout their lives; in others secondary tubes arise either in late embryonic or larval development, and the final number may be large (p. 101). The tubes usually lie freely in the abdominal cavity, but in the so-called 'cryptonephridial' arrangement of some Coleoptera and Lepidoptera their blind ends are applied closely to the rectal epithelium, and held in permanent association by connective tissue. Some are uniform throughout their length, but others show anatomical differentiation, typically into a swollen terminal portion and a more slender proximal portion. The epithelium is a single layer of cells with a brush border on their inner surfaces. It rests on a basement membrane. Occasional muscle fibres are responsible for writhing movements.

Water Balance

The excretory system has a major role in the water economy of terrestrial insects and osmoregulation in aquatic species. Water is taken in by aquatic insects with the food and also absorbed through permeable regions of the cuticle. It is eliminated largely through the Malpighian tubes. Urine in the form of a colourless fluid collects in the lumen of the tubes, and passes down them and through the hind gut to the exterior. Most terrestrial insects avail themselves of any opportunity to drink, but conservation of water is of great importance. There is extensive resorption of water during the passage of urine down the Malpighian tubes and hind gut, so that the product that finally passes out with the excreta is extremely concentrated and may even be dry and solid. The cells of the proximal regions of the Malpighian tubes and of the hind gut may all play a part in this resorption, but the rectal glands are responsible for most of it. In those species in which the tips of the Malpighian tubes are applied to the rectal epithelium the circulation may be more direct, as some of the resorbed water could pass immediately into the ends of the tubes rather than into the haemolymph. Absorption of water by the rectal glands is an active process taking place against a considerable osmotic gradient.

In some insects there is a need for water excretion rather than conservation, a need that is satisfied in plant bugs by the development of the filter chamber (p. 63), a means whereby water can be extracted from the liquid food in the anterior part of the mid gut and transferred directly to the hind gut. Whether the filter chamber allows for the passive transfer of water along a gradient or involves active absorption and secretion is unknown. Since the phloem sap which the bugs tap with their stylets is under positive hydrostatic pressure, the filter chamber could act to some extent as a purely physical sieve to hold back larger organic molecules. The amount of fluid that can be excreted by this system in Hemiptera is astonishingly large, as any small boy who has hunted cicadas knows.

Ionic Regulation

The excretory system is largely responsible for ionic regulation, a function which it shares with the external cuticle in aquatic insects (Shaw and Stobbart, 1963). Ions move through the cuticle of aquatic insects either passively or by active absorption, as by the anal papillae of mosquito larvae. Most inorganic ions apparently pass into the Malpighian tubes along electrochemical gradients, but the ionic composition of the urine is altered from that of the haemolymph by selective absorption of ions in either the proximal parts of the Malpighian tubes or the hind gut. The potassium ion is an exception to this rule, being actively secreted into the Malpighian-tube lumen against a concentration gradient, probably carrying water with it and thus inducing the flow of liquid urine. Some potassium is resorbed in the hind gut, although usually not to the same extent as is sodium.

Nitrogen Excretion

Some aquatic insects excrete excess nitrogen as ammonia. Ammonia is toxic to cells in low concentrations, and can only be excreted in a copious dilute urine. Whether ammonia is formed in these insects by the direct deamination of amino acids (which would represent nitrogen excretion in its most primitive form), or whether it is produced by the secondary breakdown of other excretory products, is not known. There is evidence that the ammonia excreted by blow-fly larvae is derived from adenosine by a deaminase, suggesting that ammonia excretion in these insects is a secondary modification of the purine synthetic pathway of nitrogen excretion characteristic of most insects. The other main excretory product of blow-fly larvae is allantoin, which is an oxidation product of uric acid.

Many insects excrete small quantities of urea, and possess the enzyme arginase which splits urea from arginine. No evidence of the existence of the complete ornithine cycle of nitrogen incorporation into urea has been found, however, so it seems that the small amount of urea excretion may be a vestige of an ancestral excretory mechanism. Amino acids are also regular components of the urine in some species.

Most insects excrete the bulk of their excess nitrogen as uric acid, which is formed in the fat body, and extracted from the haemolymph by the Malpighian tubes and alimentary canal. It is not known whether any mechanism of active secretion is involved in this process. A number of insects excrete breakdown products of uric acid as well as the purine itself, and have enzymes in the Malpighian tubes which carry out the degradation. These are uricase, which oxidizes uric acid to allantoin, allantoinase, which converts allantoin to allantoic acid, and, in one insect at least, allantoicase, which splits allantoic acid to urea and glyoxyllic acid

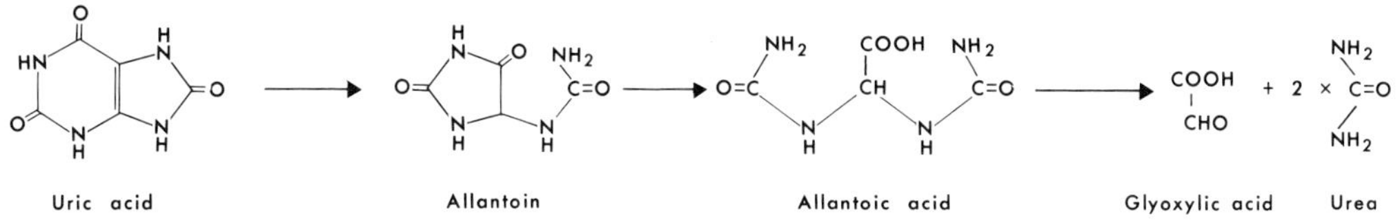

Uric acid → Allantoin → Allantoic acid → Glyoxylic acid + 2 × Urea

Fig. 2.37. Metabolism of uric acid.

(Fig. 2.37). Other insects may lack xanthine dehydrogenase, and thus excrete hypoxanthine and xanthine in place of uric acid. All these uric-acid derivatives are more soluble than uric acid itself, and their presence in the urine presumably indicates some relaxation of selection pressures in favour of water retention.

End products of nitrogen metabolism may also be stored in a variety of tissues, and may remain there throughout life, or be voided all at once, as, for example, in the *meconium* excreted by newly emerged adults. Uric acid may be stored in the urate cells of the fat body, in pericardial cells, in epidermal cells, and in the wing scales of butterflies (p. 35). Parts of the Malpighian tubes may also be used for storage of excretory products, particularly of calculi formed from calcium salts.

Silk Production by the Malpighian Tubes

In a few insects of widely separate orders (Neuroptera, Coleoptera, Hymenoptera) the cocoon silk is produced by the Malpighian tubes. The tubes, which presumably function as excretory organs during larval life, change over to silk production just before pupation, secreting proteinaceous material which emerges as a thread from the anus.

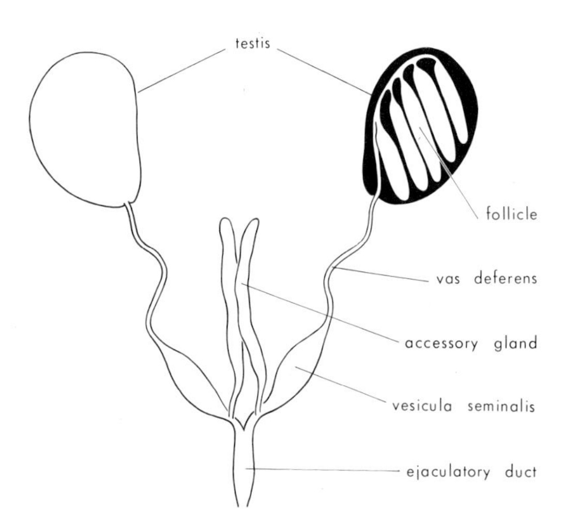

Fig. 2.38. Generalized diagram of the male organs of reproduction.

THE REPRODUCTIVE SYSTEM

The reproductive organs comprise paired gonads containing the germ cells enclosed in mesodermal tissue, paired mesodermal gonoducts, and an ectodermal median gonoduct lined with cuticle and opening at the gonopore.

The Male Organs

Each *testis* (Fig. 2.38) contains a number of tubular *follicles* held together in a common connective-tissue sheath. They communicate by paired *vasa deferentia* with the median *ejaculatory duct*, which has muscular walls and a cuticular lining. The vasa deferentia may be provided with enlargements or diverticula for sperm storage, the *vesiculae seminales*. *Accessory glands* discharge their secretion into the ejaculatory duct, which usually opens at the end of a penis or aedeagus (p. 26).

Spermatogenesis. The apical germ cells, or *spermatogonia*, become enclosed in individual cysts of epithelial cells. Each spermatogonium divides repeatedly to form a large number of *spermatocytes* within the walls of the cyst. By a further two divisions, one of which is meiotic, each spermatocyte gives rise to four *spermatids*. Finally, the spermatids transform into the elongate *spermatozoa*. These events proceed as the cysts move down the follicle, although elongation of the spermatozoa may force them back up towards the apex, from which they then migrate down again. The spermatozoa break free from the cysts at the base of the follicle, and move into the vasa deferentia. Those derived from a single cyst are often held together by a hyaline cap, which usually disappears when the sperms enter the vas, although it may persist until they are transferred to the female. Spermatozoa are typically very long and narrow, with no obvious differentiation into head and tail.

The Female Organs

Each *ovary* (Fig. 2.39) consists usually of a cluster of egg tubes or *ovarioles*. These vary in number from two to some hundreds, except in a few viviparous flies and aphids

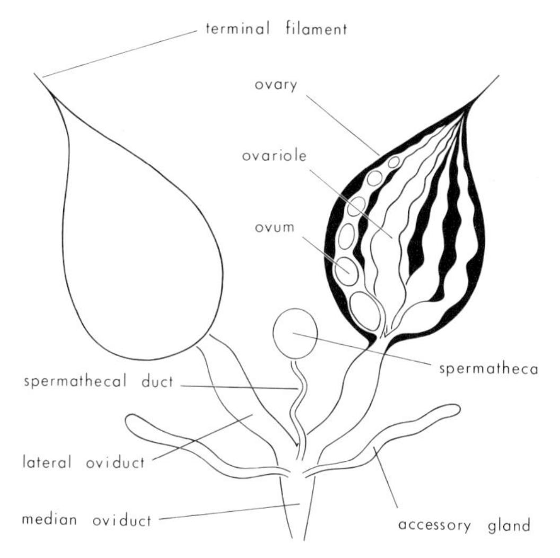

Fig. 2.39. Generalized diagram of the female organs of reproduction.

which have only one. Each ovariole is enclosed in a connective-tissue sheath, which tapers apically to form a *terminal filament*. The ends of the filaments are usually attached to the body wall or dorsal diaphragm. The ovarioles are connected by stalks to the paired *oviducts*, which open into the *median oviduct*. The ducts of paired *accessory (colleterial) glands* open either into the median oviduct or the genital atrium. An organ for the storage of sperm, the *spermatheca*, is also connected by a duct, generally long and narrow, with the median oviduct or genital atrium. Two or more spermathecae may be present in some species. The inner cuticular lining of the spermatheca is often sclerotized and its walls may be glandular, or there may be a separate gland attached to it. Both the spermatheca and its duct are invested by a muscular coat. The median oviduct, which also has a well-developed muscular coat, may be enlarged to form a *uterus* in viviparous insects. In most species it accommodates the aedeagus during copulation, and may therefore be referred to as a *vagina* or *bursa copulatrix*. In a number of unrelated groups, however, there is a separate bursa opening to the exterior by a *copulatory aperture* and communicating with the median oviduct by a *seminal duct* (Fig. 2.40).

Oogenesis. The germarium is situated at the apex of each ovariole. Its cells, the *oogonia*, differentiate to form *oocytes* in some species, or oocytes plus *nurse cells* in other species. The oocytes move down the ovariole as they are differentiated, so that each ovariole contains a series of oocytes which are progressively larger and older the further removed they are from the apex. Each oocyte is enclosed in an envelope of *follicle cells* derived from the epithelium of the ovariole. When nurse cells are present, they may travel down with the oocytes, each cluster of nurse cells being connected with the oocyte and with each other by protoplasmic channels through which nutrients enter the oocyte. In other species the nurse cells may remain together near the apex of

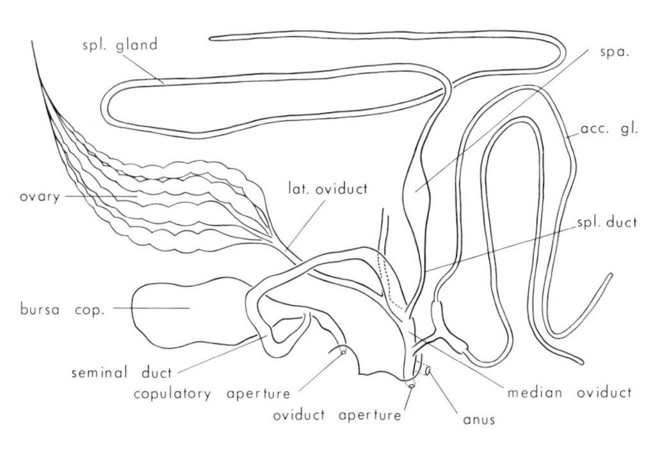

Fig. 2.40. Female reproductive organs of a butterfly (after Ehrlich and Davidson, 1961).
acc. gl., accessory gland; *bursa cop.*, bursa copulatrix; *spa.*, spermatheca; *spl. duct*, spermathecal duct; *spl. gland*, spermathecal gland.

the ovariole, but still retain their protoplasmic connections with the oocytes as the latter move in series down the ovariole. Ovarioles are termed *panoistic* when they lack nurse cells, *polytrophic* when the nurse cells travel with the oocytes, *acrotrophic* when they remain near the apex of the ovariole.

The nurse cells are centres of intense synthetic activity. Their nuclei synthesize ribonucleic acid, which passes to the cytoplasm and takes part in the synthesis of enzymes and granules of storage protein. Metabolites of many kinds flow from the nurse cells through the protoplasmic channels to form the generally massive yolk of the developing egg. The follicle cells also contribute to the nutrition of the oocytes, and are wholly responsible for it in insects that lack nurse cells. They absorb protein, fat, and carbohydrate from the blood and pass them on to the oocyte. Finally, the follicle cells secrete the egg shell, or *chorion*, just before the egg passes out into the oviduct. In its passage to the oviduct the egg breaks through the plug of follicle cells at the base of the ovariole. The empty follicle then disintegrates, making way for the descent of the next oldest oocyte.

The Transfer of Sperm

The male accessory glands produce a seminiferous secretion and also, in many species, a protein which hardens to form a capsule, or *spermatophore*, to enclose the semen. Fluid semen may be injected by the male directly into the spermatheca, but more often is deposited in the vagina or bursa. Spermatophores are also usually inserted by the male into the vagina or bursa, but they may be formed during copulation around and largely outside the genital opening of the female, or even (in Apterygota and other primitive hexapods) deposited on the ground to be picked up by the female. The sharp aedeagus of male cimicids penetrates the abdominal body wall of the female, depositing the sperm either directly into the haemocoele or into a special group of receptive cells, the *spermalège*, from which they migrate to the oviducts (Fig. 26.12).

Spermatozoa enclosed in a spermatophore are released, either by rupture of the case by spines in the vagina, or, possibly, by enzymic digestion by a secretion of the female, or they may migrate or be ejected from the neck of a flask-shaped spermatophore. The spermatozoa, which were quiescent in the male, become active in the female, and remain so for the often lengthy period of storage in the spermatheca. The reason for this difference in activity is uncertain, although in *Rhodnius*, at least, it seems that a change in pH between the semen and the spermathecal secretion is responsible. Movement into the spermatheca is apparently effected largely by muscular contractions in the ducts. The secretion of the accessory gland of the male *Periplaneta americana* includes a neurohormone (probably a tryptophan derivative) which induces peristaltic contractions in the female oviduct. Empty spermatophores may be dissolved and absorbed in the vagina, or in some species may be eaten by the female.

Fertilization and Oviposition

The egg is usually fertilized as it passes down the median oviduct with its micropyle oriented towards the opening of the spermathecal duct. In some species there is an extraordinary conservation of sperm, only one being extruded for each egg. The accessory glands of the female usually have a function in oviposition. In many species they secrete a glue which cements the eggs to a surface (hence the term 'colleterial'). The secretion may be a silk protein in some Neuroptera, which attach each egg to a surface by means of a delicate stalk. In Blattodea and Mantodea the glands secrete the material of an *ootheca* that encloses a cluster of eggs. In several species of cockroach, for instance, the left accessory gland produces a structural protein, a phenolase, and a phenyl glucoside. The right gland produces a glucosidase. When the two secretions are mixed together and extruded during the formation of the ootheca, the glucosidase frees the phenol, which is oxidized to a quinone by the phenolase. The quinone reacts with the protein, tanning it to form the tough ootheca wall. The glands may also produce salts of organic acids, such as calcium oxalate in one species and calcium citrate in another, which appear as crystals in the finished ootheca. In viviparous species the accessory glands may manufacture nutritive secretions to feed the larva within the uterus.

ACKNOWLEDGMENTS. We are greatly indebted to Drs L. B. Barton-Browne, R. J. Bartell, R. H. Hackman, and J. A. L. Watson, of the Division of Entomology, CSIRO, for supervising the preparation of the illustrations.

3

CYTOGENETICS

by M. J. D. WHITE

Chromosome cytology is the study of the chromosome sets of animal and plant species, whereas cytogenetics is more concerned with analytical studies of chromosomal mechanisms regarded as the physical basis of genetic systems. Cytotaxonomy, which includes both the use of these studies in elucidating the relationships of higher taxa and comparison of the chromosome sets of related species, is sometimes considered to be a distinct field of its own. Obviously these various fields intergrade with one another and with genetics and cell physiology. They involve, in fact, the whole of the genetic apparatus on which speciation, evolution, and adaptive radiation depend.

Cytogenetics is of exceptional significance in entomology, because of the great variety of genetic systems found in the Insecta. Some of these, like the male haploidy of the Hymenoptera, characterize whole orders. Others are found in smaller subdivisions such as tribes or families. Thus, there are several distinct types of chromosome cycle in the scale insects (Coccoidea). Other special chromosomal mechanisms are peculiar to single species.

THE KARYOTYPE

When considering the chromosome set, or *karyotype*, of a species we may ask: (1) how many chromosomes are there in the male and female cells; (2) what are the sizes and shapes of these chromosomes, and what special 'markers' in the form of nucleoli, constrictions, specially condensed regions, etc., do they bear; (3) how much desoxyribonucleic acid (DNA) do they contain (in picograms or gm^{-12})? The chromosome numbers of several thousands of insect species are now known (Makino, 1951; Smith, 1953, 1960), and illustrations of the karyotypes of most of these have been published. Information on the amounts of DNA in these karyotypes exists for only about 25 species, however, and most of it is in the form of relative rather than absolute amounts.

The haploid chromosome numbers of insect species range from 2 in certain simuliid and chironomid midges, the iceryine coccids, and a few Phthiraptera–Mallophaga to 190–191 in the butterfly *Lysandra nivescens* from Spain, which is the highest number known in the animal kingdom. In certain groups the chromosome number is extremely constant, whereas in others it is highly variable. Diptera show quite low numbers, Lepidoptera generally have high ones. Members of certain groups, such as the orthopteroid orders, have large chromosomes, whereas others, such as the Thysanoptera and Phthiraptera, have very minute ones.

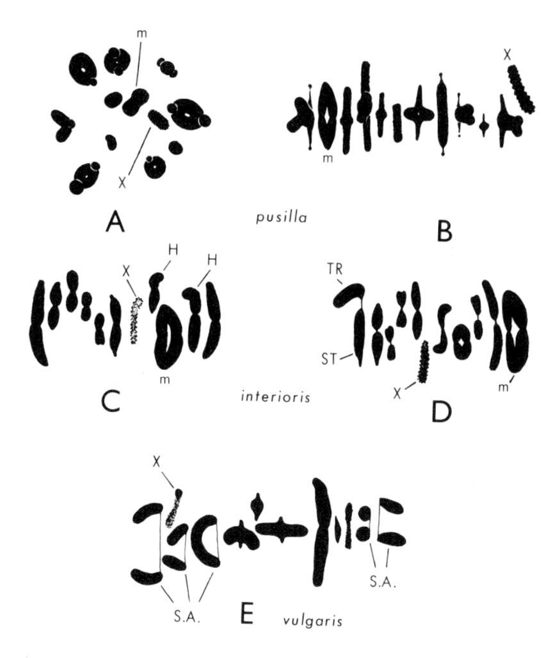

Fig. 3.1. First meiotic metaphase chromosomes in males of 3 species of the grasshopper genus *Austroicetes*: A, B, *A. pusilla*, polar and side views; C, D, *A. interioris*, side views of metaphases from two individuals; E, *A. vulgaris*, side view. In *pusilla* the chiasmata are non-localized; in *interioris* they show strong distal localization and are almost all completely terminalized by first metaphase; in *vulgaris* there is a strong tendency for the chiasmata to be located in the short arms of acrocentric bivalents (*S.A.* in E). In C two bivalents are heterozygous for pericentric inversions; in D the bivalent on the left is heterozygous for the Trangie and Standard sequences. *m* indicates metacentric bivalents; the one in *pusilla* has arisen by a pericentric inversion for which all individuals of the species are homozygous, whereas the one in *interioris* has arisen by a fusion. [Mainly from White and Key, 1957, by permission of the publishers]

Many insects have anomalous mechanisms of meiosis. In some of them the chiasmata (points of genetic crossing-over) may be localized in various ways (Fig. 3.1), thereby canalizing and restricting genetic recombination. Throughout the higher Diptera (with the possible exception of the Phoridae), the male bivalents lack chiasmata, and there is no genetic crossing-over in that sex. 'Achiasmatic' meiosis also occurs in the males of some families of Nematocera (e.g. Mycetophilidae, Bibionidae), in the mecopteran genus *Panorpa*, in a few African eumastacid grasshoppers, and in several genera of mantids, including, in all probability, all the Australian Perlamantinae and Iridopteryginae. According to some authorities, there are no chiasmata in oogenesis of Lepidoptera, although they are undoubtedly present in the male.

Two fundamentally different types of chromosomes exist in insects: the *monocentric* type with a single *centromere* or attachment to the mitotic spindle, and the *holocentric* type which is attached to the spindle throughout its whole length. It is still uncertain whether holocentric chromosomes have numerous centromeres at intervals along their length or a 'diffuse centromere activity'. Monocentric chromosomes seem to occur in the orthopteroid orders, Neuroptera, Coleoptera, Mecoptera, Diptera, and Hymenoptera, and almost certainly in the Odonata although suggestions to the contrary have been made. The Hemiptera have holocentric chromosomes, and so, in all probability, do the Phthiraptera. The other group which probably has holocentric chromosomes is the Lepidoptera, but, in spite of the large amount of work that has been done on this order, complete certainty has not been reached.

The position of the centromere is revealed in monocentric chromosomes by a non-staining gap or constriction, and by the fact that the chromosome is frequently bent at this point when attached to the spindle, especially when chromosomes are pulled apart at anaphase. Holocentric chromosomes are never bent or angled in this manner at anaphase. The location of the centromere is shown particularly clearly in cells that have been exposed to certain drugs, such as colchicine. In most cases it is constant for a particular chromosome, but it may be changed by inversions or other structural rearrangements. If it is situated approximately midway along the length of the chromosome, the latter is said to be *metacentric*. Chromosomes with the centromere extremely close to one end, so that the two arms are of very unequal length, one of them being quite minute, are called *acrocentric*. Chromosomes with strictly terminal centromeres *(telocentrics)* do not seem to occur naturally in insects. Some species have all their chromosomes metacentric or acrocentric but one often finds both types coexisting in the same karyotype. And intermediate kinds of chromosomes (sub-acrocentric, sub-metacentric, or 'J-shaped' elements) occur in many groups.

Certain chromosomes or chromosomal regions are out of phase with the rest of the karyotype in their cycle of condensation at mitosis. At a stage when most of the karyotype is in a diffuse state, these regions may be condensed and compact, and may stain darkly with basic dyes. This kind of behaviour is called *positive heteropycnosis*. The opposite condition, in which a chromosome or chromosome segment is diffuse or undercondensed by comparison with the rest of the karyotype, is *negative heteropycnosis*. Chromosomes or regions that exhibit either type of behaviour are said to be *heterochromatic* or composed of *heterochromatin*, the rest of the karyotype being made of *euchromatin*. The X-chromosomes of the Acridoidea show positive heteropycnosis at certain stages of spermatogenesis and negative heteropycnosis at other stages. Heterochromatin has been spoken of as genetically inert, but its 'inertness' is relative rather than absolute.

SOMATIC POLYPLOIDY

The somatic cells in many insect tissues lose the ability to divide by mitosis at a certain stage of development, and become polyploid as a result of chromosomal replication without nuclear division. This process is known as *endomitosis*, and the cells that have undergone one or more endomitotic cycles are said to be *endopolyploid*. Classical examples of endopolyploidy are the salivary-gland cells of the pond-skater *Gerris* studied

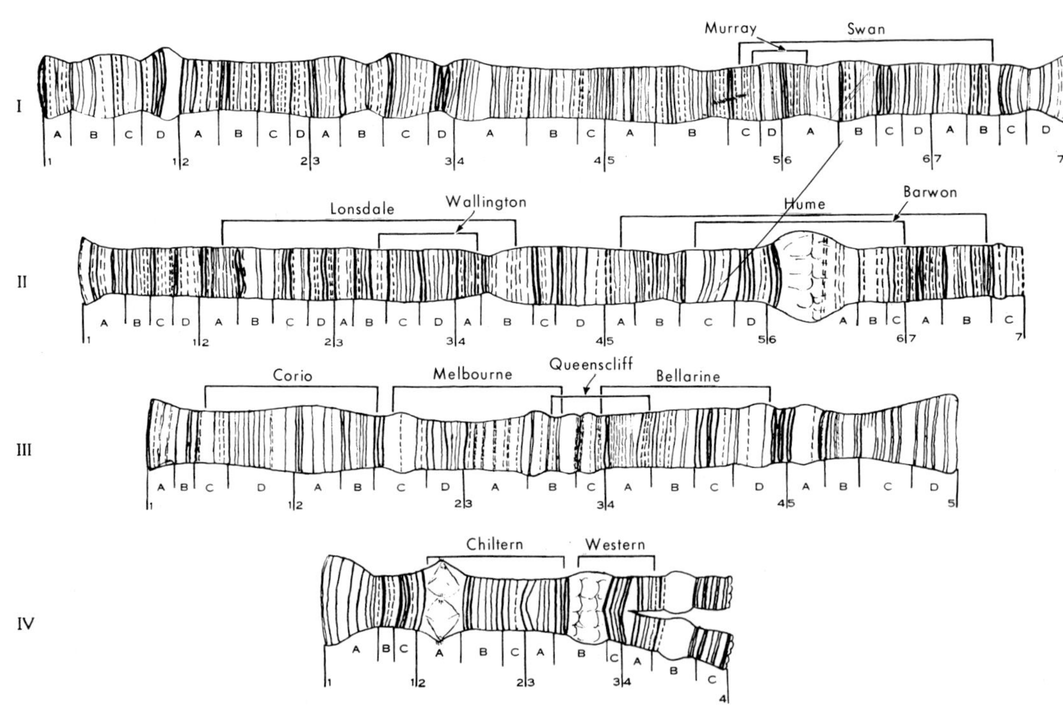

Fig. 3.2. Maps of the 4 polytene salivary-gland chromosomes of *Chironomus intertinctus* (Diptera), showing the standard banding pattern and the various inversions that have been detected in this species. Hume and Barwon are 'included' inversions, Chiltern and Western are 'independent', while Queenscliff and Bellarine are 'overlapping'. [From J. Martin, 1963, by permission of the publishers]

by Geitler (1953), the branched nuclei of the spinning-gland cells of Lepidoptera and Trichoptera, and the ileal epithelium of mosquito larvae. Endopolyploidy is, however, found in most tissues of adult insects. Some insect tissues consist of a mosaic of large and small cells of different 'ploidies', and are said to be *mixoploid.*

A special case of endopolyploidy is provided by the *polytene* nuclei of many species of Diptera, which usually attain their maximum size in the salivary-gland cells. In these giant nuclei the chromosomes have undergone repeated cycles of replication without separation of the resulting strands, so that they come to resemble many-stranded telephone cables (Fig. 3.2). While the great thickness of these chromosomal elements is due to repeated replication, their great length is due to much of the coiling that is normally present being unwound. The polytene chromosomes show numerous dark-staining bands, each due to partial fusion of chromomeres, one in each of the hundreds or thousands of strands of which the polytene 'cable' is formed. Certain of these bands, which represent genetic loci, manifest their biosynthetic activities by swelling up to form 'puffs' at specific stages of development and in particular tissues (Beermann, 1962).

The maternal and paternal polytene chromosomes are usually closely paired or synapsed in the nucleus. Thus rearrangements, duplications, or deletions of even very minute regions down to single bands can be directly observed under the microscope. The polytene chromosomes of the Diptera are hence a unique cytological tool of enormous value. They are not known outside this order of insects, and are not found in a fully developed form in all Diptera.

CHROMOSOMAL POLYMORPHISM

Simple mutations at a single locus cannot be detected cytologically, but other changes often can. Chromosomes can be broken by radiation, by abnormal stretching on the spindle, or by various chemical treatments. They also undergo breakage 'spontaneously' (i.e. from unknown causes) from time to time. A chromosome break leads to the production of two 'sticky' ends. It is a property of such artificial ends that they will tend to fuse with other sticky ends. This is the basis of all structural rearrangements in which the sequence of the genetic loci is altered. Natural chromosome ends *(telomeres)* are not 'sticky', i.e. they will not fuse with one another or with sticky ends. The basic rules governing the occurrence and survival of chromosomal rearrangements were formulated by Muller (1938).

Chromosomal rearrangements occur spontaneously with a relatively high frequency; in the Australian grasshopper *Keyacris scurra* approximately 1 in 750 individuals is heterozygous for a newly arisen rearrangement. Stone (1962) estimated that something like 350 million inversions and about four times as many translocations have occurred in all the individuals of *Drosophila* that have ever lived. Many of these rearrangements, particularly those that involve losses or gains of chromosomal regions (deletions and duplications) are lethal—either to the cell or the individual. Others, however, especially inversions (segments of chromosome turned through 180°), have become established in many species. If the original type of chromosome and the new type both persist in the population, we have a situation of *chromosomal polymorphism.* On the other hand, the new type may reach a frequency of 100 per cent in the species as a whole or in one geographic race. The karyotype thus undergoes periodic changes in the course of evolution, and in most groups there are visible differences between the karyotypes of related species.

Some of the structural rearrangements that occur in evolution involve changes in chromosome number. There are two main types of these: *centric fusions* of two acrocentrics to give a metacentric, and *dissociations* of a metacentric to give two acrocentrics. These changes are slightly more complex than simple fusion and fragmentation. In a centric fusion one acrocentric breaks close to the centromere in the long arm, and the other breaks in the minute short arm. After re-joining has occurred, we get a monocentric metacentric and a minute monocentric, which will contain very little genetic material and must be expected to get lost in the course of a few generations. A dissociation is, in a sense, the opposite of a centric fusion. Before it can occur, a minute fragment-chromosome (with a centromere and two telomeres) has to be present. A break in this and a break very close to the centromere in a metacentric member of the normal karyotype may give rise, after rejoining, to two viable acrocentric chromosomes.

Centric fusions and dissociations (sometimes called Robertsonian changes) have been detected in almost all animal groups that have been studied, and no less than 34 centric fusions and 20 dissociations are known to have occurred in the evolution of the Australian morabine grasshoppers. However,

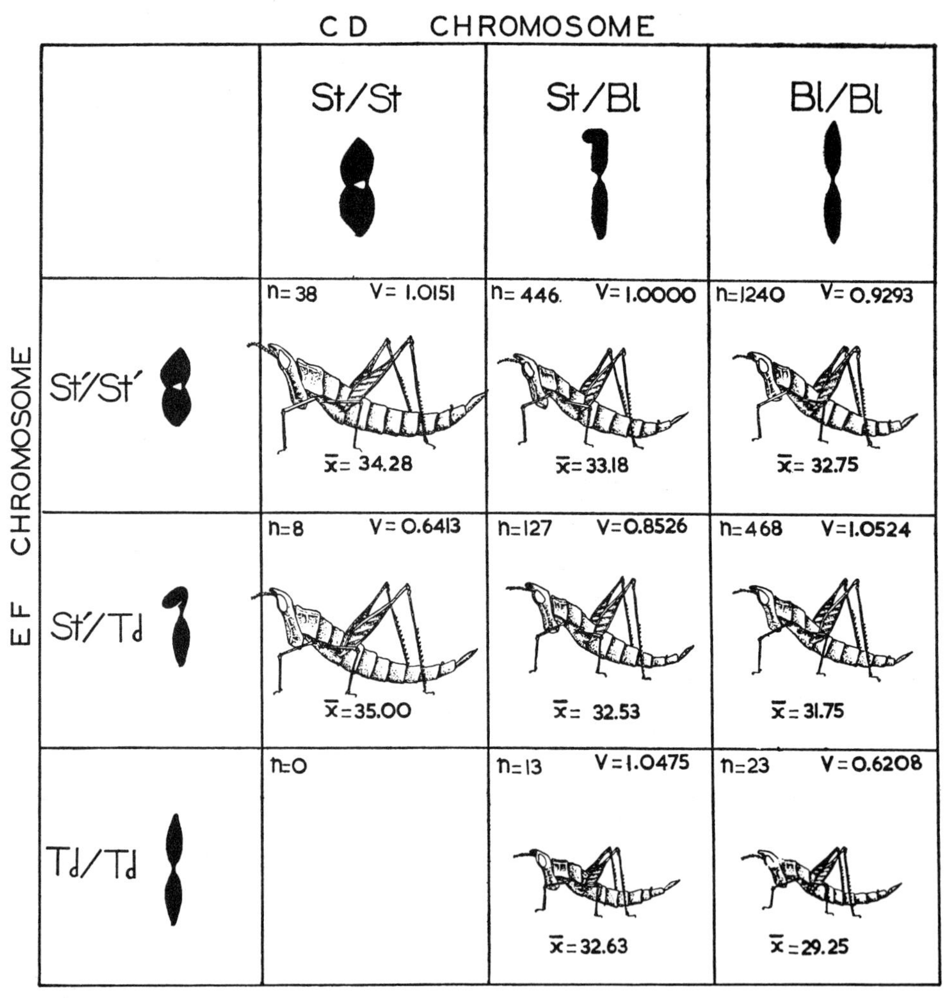

Fig. 3.3. Diagram showing the calculated effects of the pericentric inversions of the morabine grasshopper *Keyacris scurra* on relative viability and size in the male sex. The diagram is based on a sample from a population at Wombat, N.S.W. The CD bivalent may be Standard/Standard (St/St), Standard/Blundell (St/Bl), or Blundell/Blundell (Bl/Bl); similarly, the EF bivalent can be Standard/Standard (St′/St′), Standard/Tidbinbilla (St′/Td), or Tidbinbilla/Tidbinbilla (Td/Td). [From White and Andrew, in Leeper, 1962, by permission of the publishers] *n*, number of individuals of each karyotype in the sample; *V*, calculated viability; $\bar{x}$, mean live weight. The figures of grasshoppers show in a somewhat exaggerated manner the 'size effect' of the inversion sequences.

dissociations have not been unequivocally demonstrated in the genus *Drosophila* nor in grasshoppers of the family Acrididae.

Different geographic races of a species frequently differ cytologically, either in chromosome number or in chromosomal rearrangements, such as inversions, for which the populations may be polymorphic or monomorphic. Thus *Keyacris scurra* on the southern tableland of N.S.W. has an eastern race with 2n♂ = 15, 2n♀ = 16, and a western race with 2n♂ = 17, 2n♀ = 18; in the latter a large metacentric chromosome has undergone dissociation into two acrocentrics. The evidence suggests that the zone of overlap between these two races was only a few hundred metres wide before the habitat of the species was destroyed by grazing (White and Chinnick, 1957). Within each of these races there is a large number of local micro-races characterized by different frequencies of various inversions. For example, populations of the eastern race in the Canberra–Lake George area are highly polymorphic for inversions, whereas Victorian populations show little or no chromosomal polymorphism. *Moraba viatica* also shows a very narrow zone of overlap between an eastern race (with 2n♂ = 19) and a western one (2n♂ = 17) in the vicinity of Keith, South Australia (White, Blackith, Blackith and Cheney, 1967). In both species interbreeding and chromosome-number heterozygotes have occurred in the overlap zone.

Inversion polymorphism occurs in most species of *Drosophila*. The biological significance of this phenomenon has been investigated especially by Dobzhansky and his school (Da Cunha, 1960). Without going into the complexities, we may say that, in general, heterozygotes for naturally occurring inversions exhibit *heterosis*, i.e. they show a higher 'selective value' than either of the homozygous genotypes. But genetic interactions between inversions in different regions of the same chromosome also occur; they have been studied in the tropical Australian species *Drosophila rubida* by Mather (1963). Chromosomal inversions have all sorts of subtle effects on the physiological properties of the individuals; for example, those of *K. scurra* have effects on the overall size and relative viability of the insects (Fig. 3.3).

Inversion polymorphism has also been found in most other families of Diptera in which the existence of polytene chromosomes renders its detection easy. It is common in the Chironomidae, and has been studied in Australian species by J. Martin (1962, 1963), who has also found non-random associations of inversions in the same chromosome, indicative of genetic interactions having a powerful effect on viability. Inversion polymorphism is also well known in some species of *Simulium* and *Anopheles*.

Inversions in the Diptera are almost all *paracentric*—i.e. the inverted region does not contain the centromere. *Pericentric* inversions, in which the centromere lies within the inverted region, occur in a number of species of grasshoppers, and are easy to detect even in the absence of polytene chromosomes. Among Australian grasshoppers, *Cryptobothrus chrysophorus*, *Austroicetes interioris*, and about 15 species of Morabinae are known to show pericentric-inversion polymorphism.

Apart from inversions, the natural populations of many insect species are polymorphic for so-called *supernumerary chromosomes*, which are present in certain individuals, but not in all, and are hence not necessary for life. Their behaviour at both mitosis and meiosis is apt to deviate from the usual, and, as a result of directed or random mitotic non-disjunction (both daughter chromosomes passing to the same pole at anaphase), mosaic individuals with different numbers of supernumeraries in their various tissues may be produced. Supernumerary chromosomes are known in several species of Australian grasshoppers, but their role in the population genetics of the insects is not understood.

It is evident from the foregoing that chromosomal polymorphisms of several kinds are very frequent in insect populations. Not all of them will lead to speciation, but all cytotaxonomic differences between species must have arisen from such polymorphisms. It is of considerable evolutionary significance, too, that the amount of recombination that can occur varies greatly from species to species and from group to group. To take an extreme example, Mallophaga with only two pairs of chromosomes and few points of crossing-over must exhibit very tight linkage, whereas certain Lepidoptera with over a hundred chromosome pairs probably have fifty to a hundred times as much genetic recombination.

SEX DETERMINATION

With the exception of those groups in which the males arise from unfertilized eggs (p. 80) and a few other special cases, sex determination in insects depends on sex chromosomes. Typically these form a pair of similar chromosomes (XX) in one sex and a pair of dissimilar elements (XY) in the other (Fig. 3.4). The XX sex is said to be *homogametic*, the XY one *heterogametic*. In female heterogamety some authors use the letters W and Z to designate the sex chromosomes, so that we have ZW (♀): ZZ (♂). All species of Lepidoptera and Trichoptera apparently show female heterogamety, whereas male heterogamety occurs in all other insects that have been studied, including Apterygota and Palaeoptera.

Sex determination in *Drosophila* depends on a balance between female-determining genes in the X-chromosomes and male-determining genes in certain of the autosomes. In a normal female the ratio of X's to haploid sets of autosomes is 1·0, whereas in males it is 0·5. Individuals with 2 X's and 3 haploid sets of autosomes (X:A ratio 0·67) are intersexes. The Y in *Drosophila melanogaster* has no direct role in sex determination; individuals with one X and no Y (termed XO flies) are sterile males with abnormal spermatogenesis, successful sperm formation taking place only when several factors located in the Y are present. In the silkworm *Bombyx mori*, on the other hand, any individual with a Y (=W) chromosome is necessarily female, and all moths without Y's are male.

Although XO individuals of *D. melanogaster* are sterile, there are many groups of insects (including five species of *Drosophila*) in which the males are normally XO, so that the 2n number in the heterogametic sex is uneven. The XO condition seems to be the primitive one in the Odonata and in all the orthopteroid orders. There are also numerous species of Hemiptera and Coleoptera with XO males.

The evolution of the sex chromosomes may be regarded as a progressive genetic divergence of the X and Y. In very primitive sex-chromosome mechanisms, such as occur in the chironomid midges, the X and Y are genetically alike throughout their length, except for a sex locus, usually or always terminal in location. In slightly more advanced systems the X and Y may carry different inversion sequences, or the same inversions may be present in a polymorphic condition on both X and Y, but with quite different frequencies (Martin, 1962). The X of *Drosophila* species carries many sex-linked genes affecting development in a great variety of ways. The Y, on the other hand, is genetically rather inactive, except for the factors involved in the determination of male fertility.

Although the Y was 'lost' at an early stage in the phylogeny of several insect orders, it has been regained in a considerable number of species of grasshoppers and in members of other groups that usually have XO males. In such cases we may speak of neo-XY sex-chromosome mechanisms (Fig. 3. 4).

The usual way in which neo-XY systems arise from the XO condition is through a fusion between an acrocentric X and an acrocentric autosome. This creates a metacentric neo-X, one limb of which is genetically homologous with the unfused autosome, which now becomes confined to the heterogametic sex and may be spoken of as a neo-Y. This process has occurred in the Australian grasshoppers *Stenocatantops angustifrons* and *Tolgadia* spp. It has also happened nine times in the phylogeny of the morabine grasshoppers and in the tettigoniid *Yorkiella picta*. Neo-XY mechanisms are also known in non-Australian species of Coleoptera, Odonata, and Hemiptera–

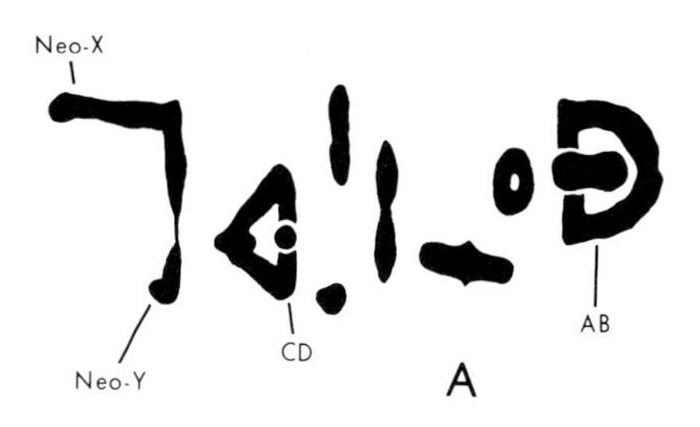

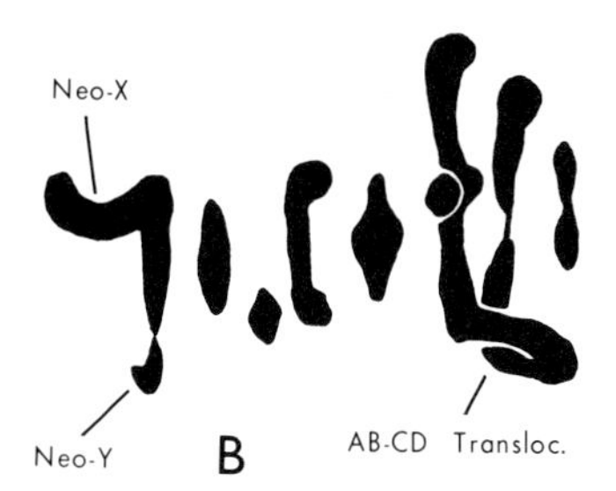

Fig. 3.4. First metaphases of meiosis in males of the morabine grasshopper species 'P52a', a species from southern Queensland with a neo-XY sex-chromosome mechanism: A, first metaphase from a normal individual; B, from an individual heterozygous for a translocation between the 'AB' and 'CD' chromosomes, which consequently form a chain of four at meiosis.

Heteroptera. The origin of a neo-XY system involves a decrease in the chromosome number (−1 in the heterogametic sex, −2 in the homogametic one).

In some members of many insect orders, XY : XX and XO : XX mechanisms have given rise, in the course of evolution, to complex mechanisms involving several kinds of X's (X_1, X_2, X_3, . . .) or Y's (Y_1, Y_2, . . .). These systems depend on the principle that in the heterogametic sex all the chromosomes designated by the same letter behave as a unit at meiosis and are inherited together. Thus in most X_1X_2Y species, the three sex-chromosomes will form a trivalent, X_1 and X_2 will pass to the same pole, while the Y passes to the other. In most species of Heteroptera with complex sex-chromosome mechanisms, however, the X's and Y's merely approach one another briefly before separating again to the poles at the second meiotic anaphase.

Multiple sex-chromosome systems seem to have arisen by three kinds of mechanisms: (1) fusions of acrocentric autosomes with acrocentric X or Y chromosomes (XY → either X_1X_2Y or XY_1Y_2); (2) reciprocal translocations between a metacentric X and a metacentric autosome (in this case XO → X_1X_2Y without passing through an XY stage); (3) 'dissociation' of an X in either an XO or an XY system or of a Y in the latter—this mechanism requires a minute supernumerary fragment-chromosome to supply the necessary telomeres and (in monocentric chromosomes) a centromere. The first kind of transformation has occurred five times in the Australian morabine grasshoppers; the second has occurred in a large group of mantid genera, including the well-known Australian *Orthodera*, *Tenodera*, *Sphodropoda*, *Hierodula*, and *Rhodomantis* (Fig. 3.5); and the third type is known to have occurred in many Heteroptera (review in Manna, 1958).

In the Sciaridae and Cecidomyiidae (Diptera) there are remarkable differences between chromosome numbers of the somatic cells and those of the germ-line. Certain chromosomes (called L-chromosomes in the Sciaridae and E-chromosomes in the Cecidomyiidae) are eliminated from the future somatic cells during the cleavage divisions in the embryo, and sex chromosomes may be eliminated as well. Thus, in *Sciara coprophila* there are 3 L-chromosomes, 3 X's, and 3 pairs of autosomes in the fertilized egg (i.e. 12 chromosomes altogether); but the male somatic cells have 3 pairs of autosomes and a single X, the female ones 3 pairs of autosomes and 2 X's, while the spermatogonia and oogonia contain 3 pairs of autosomes, 2 X's, and 2 L-chromosomes. The cecidomyiid *Trishormomyia helianthi* has 24 chromosomes in spermatogonia and oogonia, of which 16 are E-chromosomes; the male somatic nuclei contain 6 chromosomes and the female ones 8, there being an X_1X_2O sex-chromosome mechanism. In most (but not quite all) cecidomyiids the E-chromosomes are inherited maternally through the egg, and the sperm only carries a haploid set of the chromosomes present in the soma

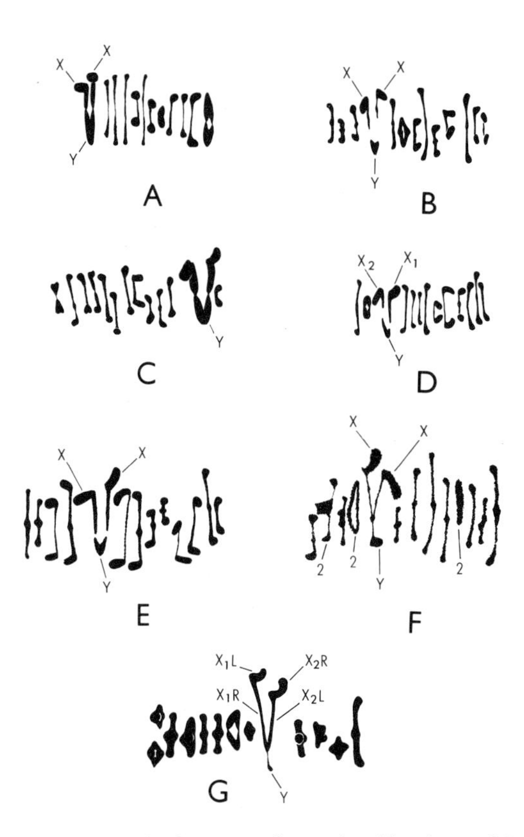

Fig. 3.5. Male first metaphases, in side view, of 7 species of Australian mantids with X_1X_2Y sex-chromosome mechanisms: A, *Orthodera gunnii*; B, *Sphodropoda tristis*; C, *Mantis octospilota*; D, *Archimantis sobrina*; E, *Sphodropoda* sp.; F, *Rhodomantis pulchella*; G, *Rhodomantis* sp. 3. All have 2n♂ = 27, except *Orthodera* which has 2n♂ = 25. In all genera except *Rhodomantis* the 'right' limb of X_1 (X_1R) and the 'left' limb of X_2 (X_2L) are associated with *different* limbs of the Y, but in *Rhodomantis* spp., as a result of a translocation, they pair with the *same* limb of the Y, leaving the other limb free as a small knob. (After White, 1962a, 1965.)

(S-chromosomes). Highly anomalous meiotic mechanisms exist in these groups. Peculiar chromosome cycles also occur in many scale insects, and important studies of this group have been carried out by S. W. Brown (e.g. 1965).

PARTHENOGENESIS

In the entire order Hymenoptera males arise from unfertilized eggs, and are hence fatherless ('impaternate'). In sexual reproduction females are produced from fertilized eggs; they may also be produced by diploid parthenogenesis in many species. The workers of the social Hymenoptera are simply sterile females which do not differ chromosomally from the fertile queens. Because of the way they arise, the males are genetic haploids and have a single chromosome set in their germ-line (most of the somatic tissues are polyploid, as in other insects). The genetic system of the Hymenoptera is consequently referred to as male haploidy or haplo-diploidy. The associated allelic controls of sex determination have been discussed by Whiting (1945) and Kerr (1962). Outside the Hymenoptera, haplo-diploidy occurs in the coccids of the tribe Iceryini, in some, at least, of the Aleyrodidae and Thysanoptera, and in the beetle *Micromalthus debilis*. It is also known in some mites.

A very peculiar form of haplo-diploidy occurs in the armoured scale insects (Diaspididae). Both males and females arise from fertilized eggs, but in the males the chromosome set contributed by the sperm is expelled from the nucleus at certain of the cleavage mitoses, so that it is not represented in the adult (Brown and Bennett, 1957).

The only normally hermaphroditic insects are a few Phoridae (Diptera) and certain species of coccids belonging to the tribe Iceryini, including the well-known pest *Icerya purchasi*. This is a self-fertilizing hermaphrodite, which resembles a female externally, but has an ovotestis. The ovarian part of this gonad is diploid (2n = 4), while the testicular part is haploid (n = 2), having undergone a reduction division during the embryonic stage of development (Hughes-Schrader, 1948). Some other species of Iceryini have normal diploid females and haploid males which are produced from unfertilized eggs. Haploid males are occasionally found also in *I. purchasi*, so that cross-fertilization does occur in this species, although rarely.

Thelytoky is the system of obligatory parthenogenesis in which a population consists exclusively of females, or of females and a few functionless males. The concept of thelytoky thus excludes occasional parthenogenesis in normally bisexual species, haplodiploidy, and cyclical parthenogenesis.

Thelytoky is the only genetic system in which fertilization has been completely abolished and genetic recombination is consequently absent. Thus, whatever the precise mechanism, it differs radically from all other genetic systems. Thelytokous forms can evolve only by accumulation of mutations *seriatim*; two favourable mutations present in different individuals cannot be brought together in a single individual as a result of a reproductive process—only mutations in individuals in a single line of descent will be accumulated.

There are two main types of thelytoky—*automictic* and *apomictic*. In the automictic type meiosis still occurs; the reduction in chromosome number is compensated for either by a pre-meiotic doubling of the number, a fusion of two of the four nuclei resulting from meiosis, or a fusion of embryonic nuclei in pairs. In apomictic thelytoky no meiosis occurs, so that no compensatory doubling process is required. Whereas bisexual species of animals are almost always diploids or haplo-diploids, many thelytokous forms are triploids, tetraploids, or in a few instances higher polyploids. Polyploid thelytokous biotypes have been studied in weevils especially by Suomalainen (1962), in Simuliidae by Basrur and Rothfels (1959), and in psychid Lepidoptera by Seiler (1963).

Relatively few Australian insects are known to reproduce by thelytoky. Two that have been investigated are the eumastacid grasshopper *Moraba virgo* by White, Cheney and Key (1963) and the chironomid midge *Lundstroemia parthenogenetica* by Edward (1963). The former is a diploid with a peculiar karyotype (2n♀ = 15), and there is a doubling of the chromosome number in the egg just before synapsis, leading to the formation of 15 bivalents (Fig. 3.6). The mechanism is hence an automictic one; two meiotic divisions reduce the chromosome number to 15 once more. *M. virgo* is heterozygous for a number of chromosomal rearrangements. It and many other thely-

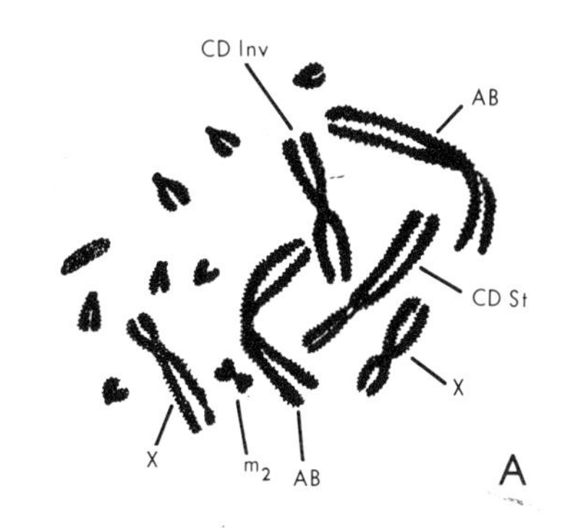

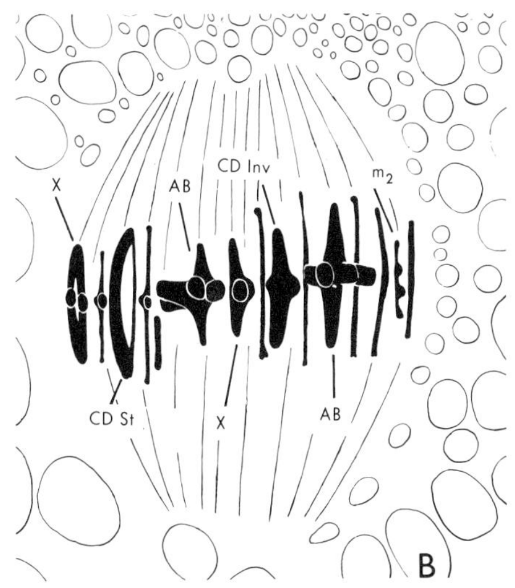

Fig. 3.6. A, somatic metaphase, following colcemid injection in the parthenogenetic grasshopper *Moraba virgo*. All individuals have 2n♀ = 15, the CD chromosome pair being heterozygous for a pericentric rearrangement; there are 9 small chromosomes, one of which (m_2) is a metacentric. B, first metaphase in an egg of *Moraba virgo*. There are 15 bivalents, which are invariably structurally homozygous, e.g. an inverted CD is never synapsed with a standard CD. [From White, Cheney and Key, 1963, by permission of the publishers]

tokous forms probably reap the advantages of heterosis without having to carry the 'genetic load' which a bisexual species would have to bear in the form of biologically inferior homozygotes produced in each generation by segregation.

Cyclical parthenogenesis is the system in which there is an alternation of sexual and parthenogenetic generations. It may thus combine the reproductive advantages of thelytoky during periods of abundant food with the evolutionary advantages of genetic recombination at other times, and it may result in highly complicated life cycles, as in some aphids (Fig. 5. 45; pp. 421-3).

TAXONOMIC APPLICATIONS

If we knew just how the major differences in genetic mechanisms had arisen in the course of evolution, we might use this knowledge in establishing the relationships of the insect orders and other higher categories on a firmer foundation. But in most cases the origins of the main types of genetic systems are as unknown as the origins of the groups themselves. Nevertheless, useful indications of the status and relationships of certain orders have been obtained (White, 1957). For example, the universality of male haploidy (which occurs only sporadically in other orders) sets the Hymenoptera apart among the endopterygotes; the Anoplura and Mallophaga share a unique type of chromosome cycle in spermatogenesis that supports the anatomical evidence of their close relationship to one another; and the female heterogamety of Lepidoptera and Trichoptera strongly suggests a monophyletic origin of these two orders. On the other hand, genetic mechanisms are extremely varied in Hemiptera and lower Diptera, and their taxonomic usefulness would fall mostly at lower levels in these orders. Nevertheless, at any level the evidence from comparative morphology can often be checked and supplemented by cytogenetic studies.

At the specific level, detailed comparison of karyotypes has not only thrown a great deal of light on genetic polymorphism, it has also proved to be a valuable tool in distinguishing between closely similar *(sibling)* or variable species. Thus, in the *pusilla* group of the grasshopper genus *Austroicetes* studied by White and Key (1957), the specific status of two supposed races with overlapping phenotypes could not have been revealed without the cytological evidence. This kind of study can be undertaken only in groups with chromosomes that are not too small, not too numerous, and not too uniform within the group. It has been particularly fruitful in Diptera that have polytene chromosomes, especially in *Drosophila* (see Mather, 1955–6, for a review of Australian species) and Simuliidae. It may fairly be said that no difficult taxonomic problem at the specific level should be studied today without an attempt to obtain essential cytological information. Some of the simpler techniques have been outlined in Norris (1966).

4

REPRODUCTION AND METAMORPHOSIS

by H. E. HINTON *and* I. M. MACKERRAS

This chapter is concerned with the series of events that occur between fertilization (or cleavage of the ovum in parthenogenetic species) and the emergence of the adult insect. As in the preceding chapters, it will be limited to what happens in the true insects (for notes on the entognathous classes, see pp. 155–7).

METHODS OF REPRODUCTION

Most insects reproduce bisexually, but parthenogenesis is fairly common, and hermaphroditism is known in a few species (p. 80). Fertilization of the ovum usually occurs as it passes the orifice of the spermathecal duct on its way down the genital tract, but it may take place in the ovariole, and occasionally in the haemocoele. The resulting progeny are produced in a variety of ways, which are summarized below (see Hagan, 1951, for a classification of viviparity).

OVIPARITY. The eggs hatch after being laid. Oviposition generally occurs soon after fertilization, but it may be delayed until embryonic development is more or less advanced. This is the common method in most orders.

OVOVIVIPARITY. The eggs are retained, usually in the vagina, until they hatch, and the young larvae then deposited; but the term is also used when hatching occurs immediately after deposition. The vaginal wall may be richly tracheated to provide for embryonic respiration, irrespective of whether eclosion takes place before or after deposition, and there is no physiological difference between these two conditions. Ovoviviparity occurs in some species of most orders.

ADENOTROPHIC VIVIPARITY. After hatching the larvae are retained in the enlarged vagina ('uterus'), where they feed on maternal secretions and grow to full size; they pupate soon after being deposited. This occurs only in some Diptera (p. 674).

PSEUDOPLACENTAL VIVIPARITY. The egg is deficient in yolk, and usually lacks a chorion. It develops in an enlarged part of the vagina or genital pouch, the embryo being nourished by a combination of specialized maternal and embryonic tissues which may be fused or in close contact. Some Blattodea (Fig. 4.1), Dermaptera, Psocoptera, and Hemiptera reproduce this way.

HAEMOCOELIC VIVIPARITY. There are no oviducts. The eggs become dispersed in the haemocoele, lack a chorion, and are nourished by a *trophic membrane* derived from the mother. It is known only in the Strepsiptera, in which the larvae escape through the

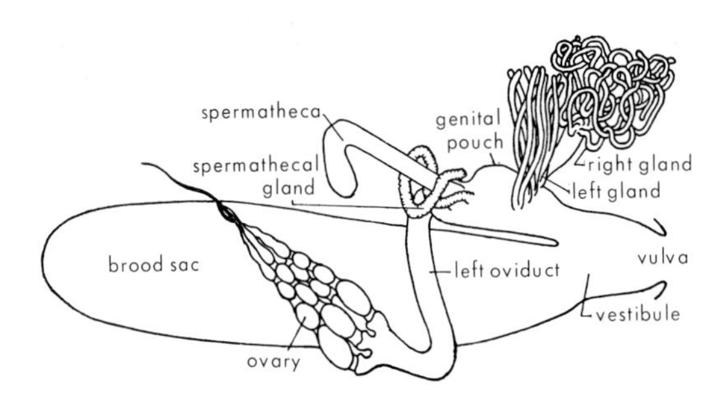

Fig. 4.1. Female reproductive system of *Diploptera dytiscoides*, a viviparous cockroach with a brood pouch developed from the floor of the genital atrium–cf. Fig. 14.3A.

[From Hagan, 1951, by permission of the publishers]

maternal brood canal, and in paedogenetic Cecidomyiidae (Diptera), in which the larvae break through the integument after devouring the maternal tissues.

POLYEMBRYONY. This might be described as uniovular twinning carried to excess, and it has been observed only in parasitic species. The eggs are microlecithal, and a polar nucleus, formed by fusion of the first and second polar bodies, divides to form an investing membrane *(trophamnion)* around the egg. The ovum proper divides to form two to many morulae, each of which develops into a complete embryo in which the true amnion may be suppressed. An adventitious sheath of cells, derived from the host, may encapsulate the whole. Up to 3,000 parasites may be produced from the ramifying 'embryo chains' in a single host. The phenomenon occurs commonly in the Hymenoptera, but rarely in other orders.

PAEDOGENESIS. In a beetle *(Micromalthus)* and some Cecidomyiidae (Fig. 4.2), the ovaries become functional in the larva, and the eggs develop parthenogenetically. Some

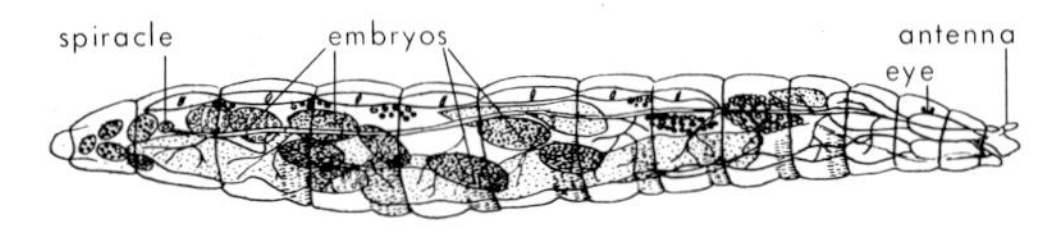

Fig. 4.2. Haemocoelic embryos in larva of *Miastor metroloas*, Diptera-Cecidomyiidae.

[From Hagan, 1951, by permission of the publishers]

of the resulting larvae also reproduce parthenogenetically, whereas others may develop into adults. In some parthenogenetic aphids embryonic development begins precociously, but the larvae are not deposited until the female matures. Similarly, some polyctenid bugs are fertilized and embryonic development begins before the final ecdysis. On the other hand, oviposition by the 'pupa' of *Tanytarsus* (Diptera) is an example of parthenogenetic reproduction by a pharate adult (Hinton, 1946c).

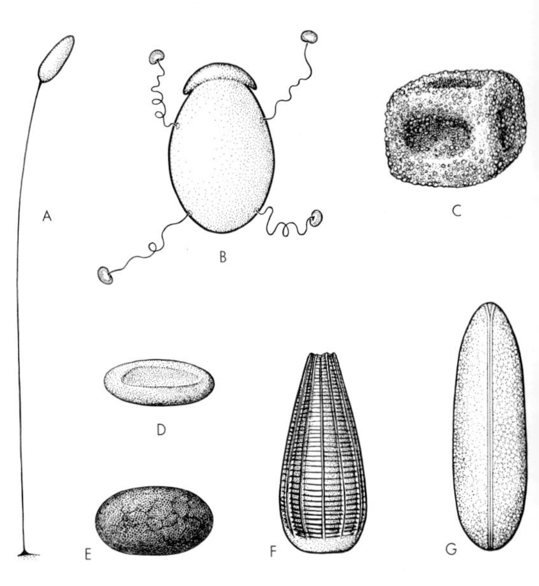

Fig. 4.3. Diverse eggs of insects: A, Neuroptera-Chrysopidae (stalk *ca* 5 mm); B, Ephemeroptera-Baetidae (0·18 × 0·09 mm); C, *Harpobittacus* sp., Mecoptera-Bittacidae (0·8 × 0·6 mm); D, *Stenoperla* sp., Plecoptera-Eustheniidae (0·3 × 0·1 mm); E, *Coscinocera hercules*, Lepidoptera-Saturniidae (3·5 × 2 mm); F, *Pieris rapae*, Lepidoptera-Pieridae (height *ca* 2 mm); G, *Lucilia cuprina*, Diptera-Calliphoridae (1·0 × 0·3 mm). [S. Curtis]

THE EGG

Insects eggs are usually more or less round or ovate, but some are stalked, or have conspicuous lateral flanges, or long respiratory horns, or a complex cap at the anterior end (Fig. 4.3). They may be laid singly or in groups. They are frequently cemented to the substrate by secretions from the accessory glands. In many aquatic forms, such as

Trichoptera and chironomid Diptera, the accessory glands secrete a hygroscopic protein in which the eggs are enveloped. In Blattodea, Mantodea, and *Mastotermes* (Isoptera) the accessory glands secrete a protein that is subsequently tanned and forms an ootheca in which the eggs are contained.

The general structure of the egg is shown in Figure 4.5A. The outer protective layers consist of (1) those secreted by the follicle cells, i.e. the layers that constitute the *chorion* proper, and (2) those secreted by the oocyte and later by the zygote and embryo. The latter consist of the *subchoral membranes* and the so-called 'true' embryonic or *serosal* cuticles.

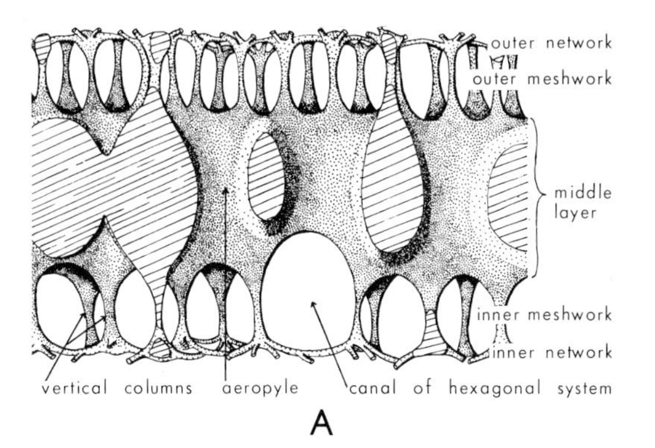

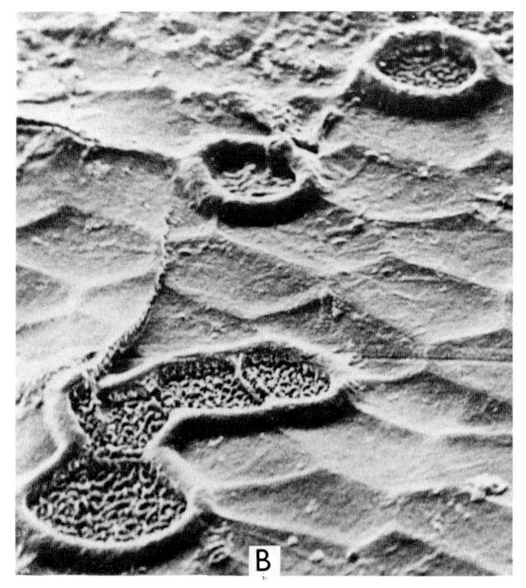

Fig. 4.4. A, diagrammatic section through the egg-shell of a muscid fly, outside the hatching lines–depth of section *ca* 0·01 mm [from Hinton, 1967, by permission of the publishers]; B, middle sides of the egg-shell of *Musca vetustissima*, Diptera-Muscidae, showing limitation of plastron to discrete 'craters'–width of photo *ca* 0·06 mm (stereoscan electron micrograph by H. E. Hinton).

The chorion consists of a protein that is often called chorionin, although we do not know whether the composition of the different chorionins is similar. It usually has two distinct layers, but there are more in many insects. It may be thick or thin, and its outer surface is often strikingly sculptured. The micropylar area is usually at or near the anterior end of the egg, and one or more micropyles extend, usually diagonally, right through the chorion. In a few aquatic and many terrestrial eggs the chorion has a respiratory system that usually consists of a thin layer of air held between the interstices of vertical columns or meshwork in the inner chorion (Fig. 4.4A). Exchanges between the air layer in the chorion and the atmosphere are effected through small canals or holes called *aeropyles*. In some aquatic and many terrestrial eggs part or all of the outer chorion may consist of an air-containing meshwork that functions as a plastron (Fig. 4.4B) when the egg is submerged in water. Sometimes the plastron is restricted to a limited part of the outer surface and sometimes to the surface of respiratory horns, e.g. in some Hemiptera, Hymenoptera, and Diptera. The respiratory systems of insect eggs have been reviewed by Hinton (1962b).

There has been no adequate comparative work on the subchoral membranes of insect eggs. In some a *primary wax layer* is secreted by the oocyte through the vitelline membrane. Before fertilization, it is present as a thin film across the inner opening of the micropyles. After fertilization, a fertilization membrane, which prevents the entry of additional spermatozoa, is secreted in some, and perhaps all, eggs. After oviposition, more lipid, the *secondary wax*, may be added to the primary wax layer and the fertilization membrane. In many eggs no fatty-acid layers exist between the vitelline membrane and the chorion. In some, e.g. *Calliphora*, two subchoral membranes are present, but neither is particularly impermeable, and it therefore appears that neither has a continuous layer of fatty acids. In many eggs a *serosal hydropyle* is developed, which functions to absorb water during development. In some, e.g. the Nepidae, the serosal hydropyle is formed against the inner face of a *chorionic hydropyle*.

DEVELOPMENT OF THE EMBRYO

Cleavage is meroblastic, an amnion is formed, there is no true ventral flexure, and the embryo goes through a complicated migration in the egg (blastokinesis) in most

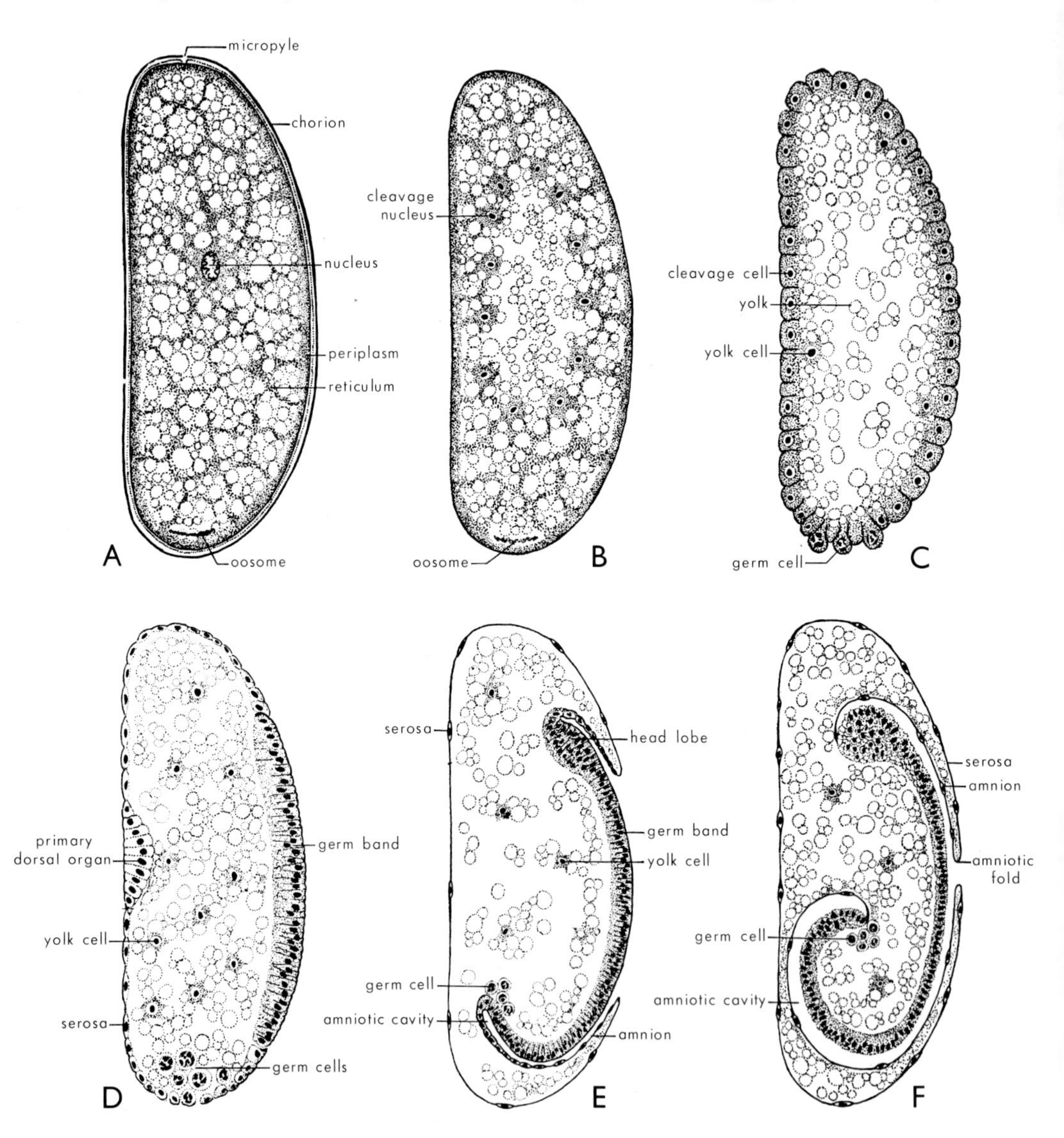

Fig. 4.5. Median sagittal sections of egg and embryos: A, egg; B, cleavage; C, formation of blastoderm; D, formation of germ band; E, F, development of amnion.

[From Johannsen and Butt, 1941, by permission of the publishers] true insects. There are, however, many variations in detail, and Johannsen and Butt (1941) have given a comprehensive account of embryonic development in all the orders. The sequence of events up to blastokinesis is broadly that given below.

Formation of the Blastoderm. The zygote (or parthenogenetic) nucleus divides repeatedly, many of the resulting nuclei migrate into the peripheral cytoplasm, cell walls begin to appear between them, and the surface of the ovum becomes covered by a single layer of cells, the *blastoderm* (Fig. 4.5C). Some of the nuclei remain in, or migrate back into, the yolk, where they enlarge, acquire cytoplasm, and become *vitellophages*. A few migrating towards the posterior end pass through a specialized area of cytoplasm and become the primordial germ cells, which remain recognizable during later stages of development. At about this time, the ventral cells of the blastoderm become thickened and columnar to form the *germ band* (Fig. 4.5D), whereas the more lateral and dorsal cells become flattened to form the *serosa*. A transitory thickening that sometimes forms on the dorsal surface of the serosa is the *primary dorsal organ*.

Organization of the embryo is determined by this stage in many insects, two controlling centres being demonstrable by experimental manipulation but not histologically. The first is the *activation centre*, near the posterior pole, which determines the formation of the germ band. The second is the *differentiation centre*, in the position of the future thorax, which comes into operation when the germ band reaches that area and determines its subsequent differentiation. The embryo becomes a mosaic when this centre has done its work, and subsequent interference with any part merely results in a local defect.

Formation of the Amnion. The anterior end of the germ band becomes thickened to form the *head lobe*, and the posterior end curls into the yolk, drawing a pocket of marginal germ band and serosa with it. The thicker cells become the *amnion*, which subsequently flattens, and the thin ones remain serosal. Similar folds are formed anteriorly and laterally, their margins grow together, and they finally fuse to form the *amniotic cavity* (Fig. 4.5F), within which the growing embryo is cushioned in fluid during the earlier stages of differentiation.

Possession of an amniotic cavity is an important characteristic of insects. There is none in myriapods, Collembola, or Diplura, in all of which the ventral flexure may serve for some of its functions. It is incomplete in Apterygota, forming a deep furrow in Archaeognatha *(Machilis)*, and being connected to the surface by a pore in Thysanura *(Lepisma)*. It is complete in nearly all other insects, although varying in size and orientation.

'Gastrulation' and Differentation of Mesoderm. While the amnion is forming, a longitudinal mid-ventral cleft appears in the germ band, which has now become the definitive embryo. Its walls invaginate to form a flattened tube, and the lumen disappears to leave an 'inner layer' two or three cells deep lying dorsal to the once more continuous ectoderm (Fig. 4.6A). The inner layer divides into a median strand of 'endodermal' cells and a pair of lateral bands which form the mesoderm (Fig. 4.6B). A little later, when segmentation is beginning, the *stomodaeum* forms as an invagination of the ventral surface near the anterior end of the embryo, and the *proctodaeum* subsequently appears as a similar invagination at the posterior end. They are both capped internally by a mass of 'secondary' endoderm cells (Fig. 4.7A) continuous with the median strand.

Differentiation of the mesoderm also begins after segmentation has commenced. Clefts appear in the lateral parts of the inner layer, and enlarge to form the primitive coelomic sacs, which separate the mesoderm into a ventral *somatic* layer and a dorsal *splanchnic* or visceral layer on each side (Fig. 4.6C). The yolk also retracts from the ingrowing splanchnic layer to form the *epineural sinus*, which subsequently coalesces with the coelomic sacs laterally to form the haemocoele. The somatic mesoderm divides into a denser ventrolateral layer, from which

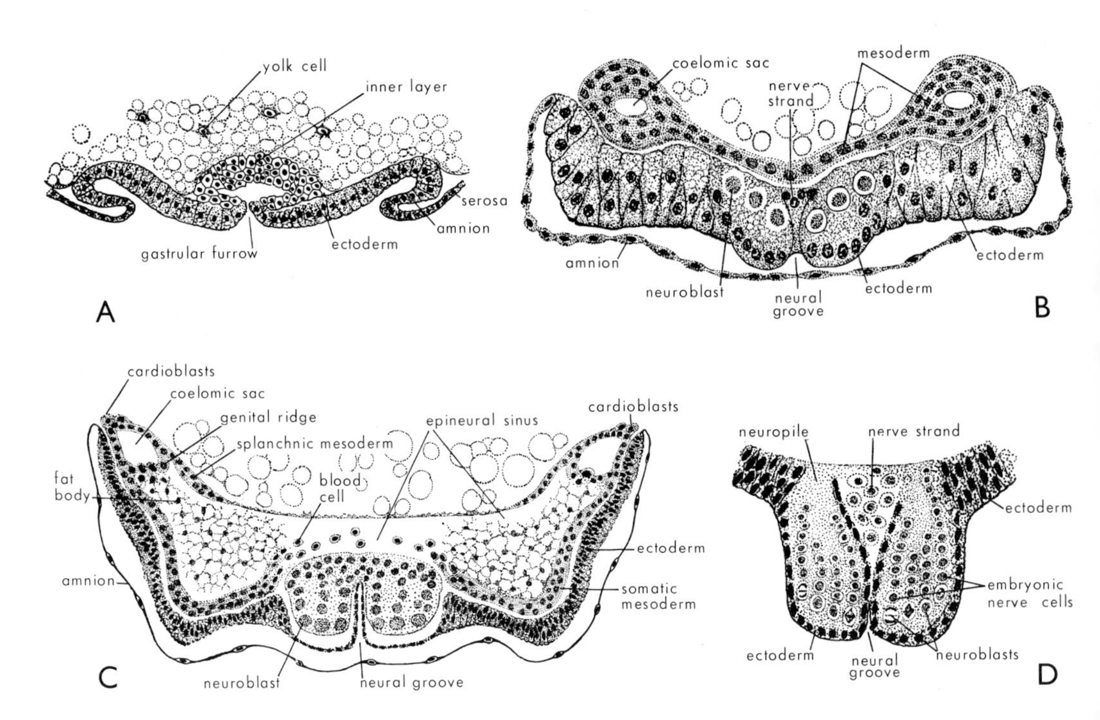

Fig. 4.6. Transverse sections of embryos: A, 'gastrulation' and formation of inner layer; B, formation of neural groove and early differentiation of mesoderm; C, later differentiation of mesoderm; D, further development of ventral nerve cord. [From Johannsen and Butt, 1941, by permission of the publishers]

the skeletal muscles develop, and a dorsomedial layer, at first limited to the coelomic sacs, which is the rudiment of the fat body. The splanchnic layer also divides into two sheets, a *splanchnic lamella* dorsal to the epineural sinus, which gives origin to the visceral connective tissues, and a *ventrolateral lamella*, which joins the fat body, and thickens to form the *genital ridge* into which the primordial germ cells migrate. The *cardioblasts* are groups of cells with large nuclei, situated at the dorsolateral corners of the coelomic sacs where the primary layers of mesoderm meet.

The median strand begins to break up when the epineural sinus reaches the mid-line, and its cells disperse, some possibly to form endoderm, some possibly to form blood cells, and some to degenerate in the yolk.

The Neural Groove. After the inner layer has separated, the neural groove appears as a second median longitudinal cleft in the ectoderm (Fig. 4.6B). It is bounded laterally by thickened *neural crests*, which are the forerunners of the central nervous system. There are thus two successive mid-ventral invaginations of the embryonic area, one to form mesoderm and possibly endoderm and the other to form nerve tissue.

Segmentation of the Embryo. Division of the mesoderm into somites begins as soon as the bands of mesoderm have become defined, and it may be preceded or followed by corresponding divisions in the ectoderm. It usually progresses in stages. The first is into a *protocephalic* or primary head region, which will include the three divisions of the brain, and a *protocormic* or primary trunk region for the rest of the body. The three gnathal segments are then added to the head from the protocorm, the three thoracic

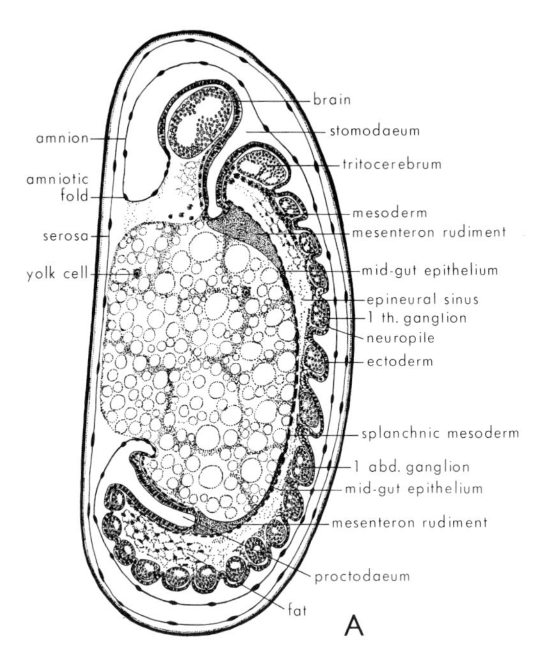

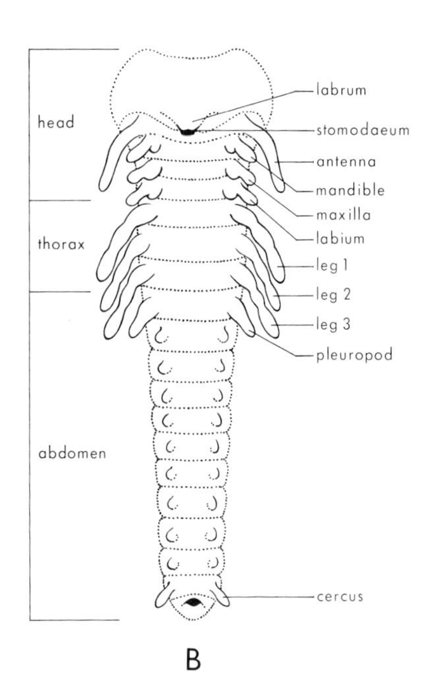

Fig. 4.7. Segmentation of the embyro: A, paramedian sagittal section [from Johannsen and Butt, 1941, by permission of the publishers]; B, ventral aspect, showing appendages (after Snodgrass, in Johannsen and Butt, 1941).

segments appear next, and the abdominal segments finally separate, usually from before backwards. In more primitive insects there are 11 abdominal segments, each with its pair of coelomic sacs and ganglia (Fig. 4.7A); but there are often fewer, and the coelomic sacs may be incomplete or absent. An unpaired lobe anterior to the stomodaeum is the rudiment of the labrum, and there is an unpaired periproct ('telson') at the posterior end of the embryo.

The appendages (Fig. 4.7B) then appear, growing out as buds from the ventrolateral aspects of the somites, and they may contain extensions of the coelomic sacs. The first protocephalic ('ocular') segment has no appendages, the antennae develop on the second but later migrate forward, and the third (intercalary) has a transitory pair of rudimentary appendages. The next three are the mandibles, maxillae, and labium, and the following three are the thoracic legs, which often become larger than the others at a fairly early stage. The abdominal appendages are more variable, but they are all usually distinct in more primitive forms, and the first is often enlarged to form a glandular organ, the *pleuropod.* The last pair may persist as cerci, but the others disappear before hatching in nearly all insects.

Blastokinesis. Most embryos migrate through the yolk in a remarkable fashion (Fig. 4.8). The movement begins when the caudal end sinks into the yolk, continues as the embryo sinks further during development of the amnion, and is completed by it rotating *(anatrepsis)* until its ventral surface lies against the dorsal surface of the egg and then returns *(katatrepsis)* to its original position. There are many variations in the path followed, and the significance of the manoeuvre is not known.

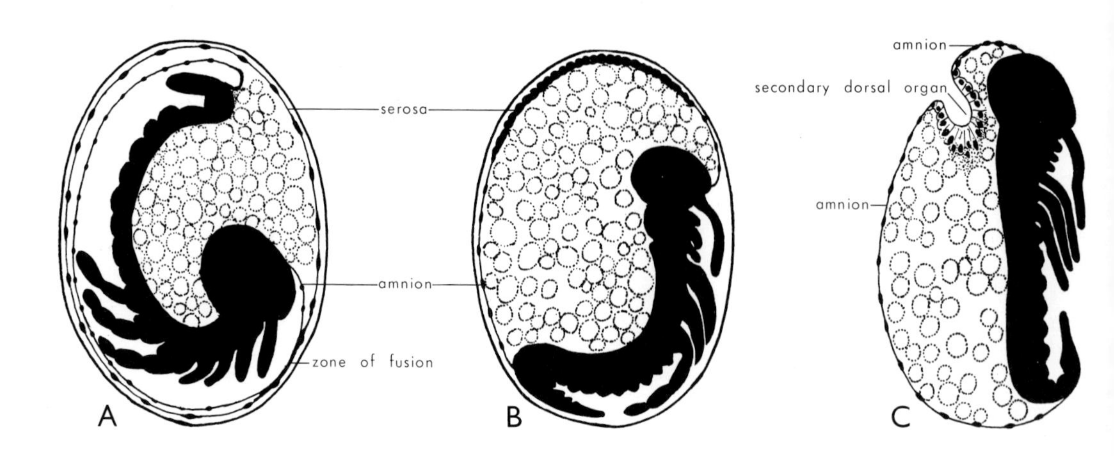

Fig. 4.8. Blastokinesis: A, anatrepsis; B, katatrepsis; C, completion of katatrepsis and provisional dorsal closure. [From Johannsen and Butt, 1941, by permission of the publishers]

Dorsal Closure. As development advances, the amnion and serosa fuse at the head end, a fissure appears, and the membranes retract to the dorsal surface, where the serosa sinks into the yolk to form the *secondary dorsal organ* (Fig. 4.8C). The remaining amnion forms a provisional dorsal closure at this stage, but the ectoderm has been growing upward in the meantime, and it ultimately completes the definitive closure, the secondary dorsal organ being absorbed in the yolk. The walls of the mid gut have also been extending, and embryonic development is completed when they finally enclose the residue of the yolk. Apolysis (p. 93) occurs, and the organism is then a pharate first-instar larva until it hatches.

There are considerable variations in the method of closure, and several types have been described, depending mainly on whether provisional closure is by amnion, or serosa, or both (see Richards and Davies, 1960). In any case, the steadily decreasing mass of yolk comes to be enclosed on its dorsal aspect successively by blastoderm, serosa, residual membranes, ectoderm, and mid gut in the course of development, although some of it may remain external and be eaten by the larva before hatching.

It remains to complete the story by a preliminary account of organogeny (see also pp. 97–103).

Nervous System. The neural ridges become segmented, forming the ganglia and the connectives. The first three ganglia fuse to form the brain, those in the three gnathal segments form the suboesophageal ganglion, and the remainder may also tend to fuse in groups which vary with the orders. The anterior (stomatogastric) part of the autonomic nervous system arises from invaginations of the dorsal wall of the stomodaeum, the remainder from segmental ganglia. The peripheral sense cells are also ectodermal in origin.

Endocrine Organs. The corpora cardiaca develop as invaginations of the stomodaeum, often with the unpaired hypocerebral ganglion, from which they may later separate more or less completely. The corpora allata originate as ectodermal invaginations at the base of the maxillary segment, and later migrate to a position which is generally above the oesophagus. It has been claimed that in *Pieris* (Lepidoptera) they arise from the mandibular segment, and that in *Sitophilus* (Coleoptera) they originate from the mesoderm of the antennal segment. The prothoracic glands arise as ectodermal invaginations of the labial segment.

Respiratory System. The tracheae are formed by paired tubular ectodermal invaginations that appear on the mesothorax, metathorax, and first eight abdominal segments soon after the neural ridges have become segmented. Their openings are the future spiracles and the tubes divide into anterior and posterior branches which fuse to form the longitudinal tracheal trunks. Further branching produces the transverse trunks and the ramifying system of tracheae and tracheoles throughout the body.

Circulatory System. The strands of cardioblasts, at first lateral, are pushed upward round the yolk by the growth of body wall and mesoderm on each side, until they meet in the mid-dorsal line and fuse to form the tubular dorsal vessel. The most anterior part of the dorsal vessel arises independently, from the medial walls of the antennal coelomic sacs, in some insects. As indicated earlier, the blood cells may arise from the median strand but it has also been suggested that they may be derived from the ventro-lateral corners of the coelomic sacs.

Alimentary Canal. The stomodaeal and proctodaeal invaginations deepen, and the masses of 'secondary' endoderm cells, the *mesenteron rudiments* (Fig. 4.7A), begin to proliferate, forming paired ribbons of cells, which spread towards each other and fuse in a continuous band between the yolk and the splanchnic mesoderm. Some of the cells of the original median strand (and possibly also some vitellophages) may contribute to the process. The band grows out around the yolk, and fuses dorsally to complete the mid gut, continuity through the tract being established when the blind ends of the stomodaeum and proctodaeum break down at the end of embryonic life. The salivary glands arise as a pair of ectodermal ingrowths of the labial segment.

Malpighian Tubes. The primary Malpighian tubes arise as two or three pairs of diverticula from the cap of endoderm cells at the anterior end of the proctodaeum. The origin of secondary tubes in late embryonic and postembryonic stages is described on p. 101.

Reproductive System. This contains the cells of the germ line supported and nourished by mesoderm. The genital ridges, into which the primordial germ cells have migrated, extend at first from the posterior part of the thorax nearly to the end of the abdomen, but later become restricted to those segments in which the gonads and the mesodermal portions of their ducts will develop. Formation of the ectodermal parts of the ducts is usually postembryonic.

HATCHING

Hatching may be delayed for a considerable period after embryonic development has been completed. The shell is normally burst open by the muscular activity of the larva, which may swallow air or amniotic fluid, and thus increase its volume and the pressure it exerts. Hatching spines or egg-bursters are present on various parts of the body of many larvae. They may be either on the cuticle of the first-instar larva or on the embryonic cuticle between the larval cuticle and the chorion (summary in van Emden, 1946). Sometimes the shell has a special preformed line of weakness along which it splits when enough pressure is exerted by the larva, e.g. in many Hemiptera–Heteroptera, Phthiraptera, and Diptera–Cyclorrhapha. Some larvae, such as caterpillars, eat their way out of the shell. The larvae of exopterygotes usually hatch while still enclosed in the embryonic cuticle, and its removal is often assisted by gluing it at some point to the inside of the shell. For instance, in most Blattodea and Mantodea 'silk glands' that open on the cerci glue the cuticle at the tips of the embryonic cerci to the inside of the shell.

GROWTH

As the epicuticle is inextensible, growth within the covering it provides is limited to the amount that can be accommodated by pressing out the folds and wrinkles that were formed when it was laid down. A new cuticle is then secreted, the old one is sloughed off, and a new phase of growth begins, often accompanied by more or less evident alteration in body form. Growth thus proceeds in a series of *stages* or *instars*, each separated by a *moult*.

The number of larval instars varies considerably, generally being greater in more primitive groups (e.g. 20 or more in Ephemeroptera) than in more specialized groups (e.g. 3 or 4 in many endopterygotes). It is sometimes constant in an order or suborder, but may vary even within species. The Apterygota continue to moult throughout their lives (as do other primitive hexapods), but the Pterygota do not, the last instar being the fully winged adult *(imago)* or its apterous equivalent. The Ephemeroptera, however, have two fully winged instars, subimago and imago.

Moulting Cycle. The moulting cycle is initiated by ecdysone (p. 50). The flattened epidermis thickens, DNA synthesis commences, generally followed by mitosis, and the epidermis retracts from the old cuticle. It then secretes the new epicuticle and the *moulting fluid*, which dissolves the old endocuticle, the space so formed being traversed by the tonofibrillae (p. 51). The innermost layers of the endocuticle may remain as a *moulting membrane*. As the new procuticle is deposited, the digested products of the old endocuticle, and eventually the moulting fluid itself, are resorbed and replaced by air.

Fig. 4.9. A tettigoniid grasshopper emerging at night from the last larval cuticle.
[Photo by A. J. Nicholson]

It only remains for the old cuticle to be sloughed off. When this process, known as *ecdysis*, is about to begin, the insect usually becomes quiescent, and its abdominal muscles contract, increasing the blood pressure. It may also distend itself by swallowing air or water. In due course, the old cuticle splits along the back. The most constant ecdysial lines are the mid-dorsal line on the thorax, often extending on to the abdomen, and the Y-shaped line on the head. The emerging instar then works its way out of the slit by muscular action, sometimes aided by gravity as it hangs from a convenient support (Fig. 4.9). As it emerges, the old linings of the tracheae, but not of the tracheoles, and of the fore and hind guts are drawn out and remain attached to the old cuticle, the whole commonly being called the *exuviae*. The new cuticle is nearly always soft and pale at this stage. It is usually expanded by swallowing additional air or water, and generally hardens and darkens fairly rapidly. The epidermis continues to secrete endocuticle after ecdysis, and it may go on doing so throughout the life of the instar. In exopterygotes and many endopterygotes the layers of the endocuticle deposited during the night differ from those deposited during the day.

In the conventional interpretation the word 'moult' has been used variously, sometimes being applied to ecdysis, but sometimes covering the entire moulting cycle described above. In either case, the instar has been regarded as beginning at ecdysis, and ending at ecdysis.

However, it has been pointed out (Snodgrass, 1935; Hinton, 1946c, 1958c) that the

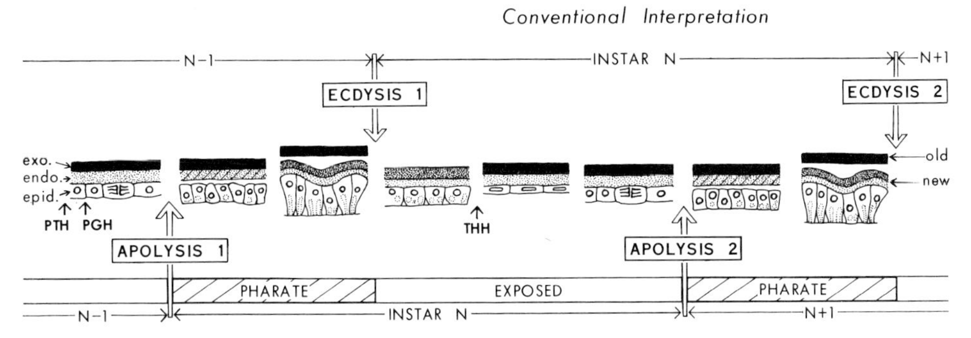

Fig. 4.10. The moulting cycle of an insect, showing the relationship between apolysis and ecdysis, when ecdysis is not delayed or suppressed (after a diagram by P. M. Jenkin). The stages of growth and morphogenesis (instars) are considered to extend from ecdysis to ecdysis in the 'conventional' interpretation, from apolysis to apolysis in the Snodgrass–Hinton interpretation.

form of the new stage is determined about the time of epidermal retraction, when the epidermis starts to deposit new cuticle. The term *apolysis* has been proposed for this retraction (Jenkin and Hinton, 1966), and the instar is recognized as lasting from one retraction to the next (Fig. 4.10). Apolysis is followed by a 'cloaked' or *pharate* phase (Fig. 4.11), in which the new stage is enclosed by the cuticle of the old. The pharate phase usually ends with ecdysis, although an instar may remain pharate throughout its life.

Many writers refer to a 'prepupal' stage of development. This sometimes corresponds to the pharate pupa, sometimes to the last part of the last larval instar *plus* the pharate pupa. In sawflies, the 'eonymph' is simply the last larval instar and the 'pronymph' the pharate pupal stage.

The additional precision which this terminology permits in describing developmental events, or the behaviour of insects in particular stages of growth, makes its use desirable in these contexts. However, as Hinton (1958c) pointed out, there are sometimes practical difficulties in identifying, without dissection, the stage of development of the organism inside a cuticle.

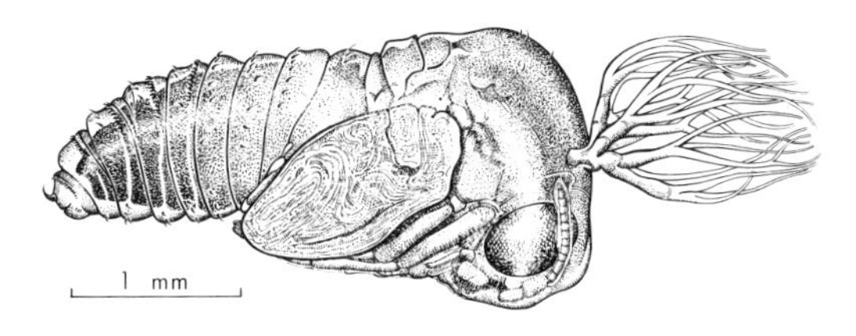

Fig. 4.11. Pharate adult of *Cnephia tonnoiri fuscoflava*, Diptera-Simuliidae, showing through the pupal cuticle. [T. Nolan]

The pharate phase is often very brief, as in most larval stages of both exopterygotes and endopterygotes; but it may be more or less prolonged, as in the first-instar larvae of many exopterygotes, adults of aquatic exopterygotes, and the pupal and adult stages of endopterygotes. This flexibility in its duration has been of great evolutionary significance, because it has provided a ready-made set of conditions in which the profound changes that follow the larval-pupal and pupal-adult apolyses in the endopterygotes could take place with the least hazard to the insects. Time is needed for these changes if the insect is not to be launched into the world part made, and time is what the pharate phase can provide. The dissociation between larval and adult structure and modes of life on which the success of the endopterygotes has depended would not have been possible without it.

A more or less prolonged pharate phase may also have direct adaptive value, and the old cuticle may still subserve a number of

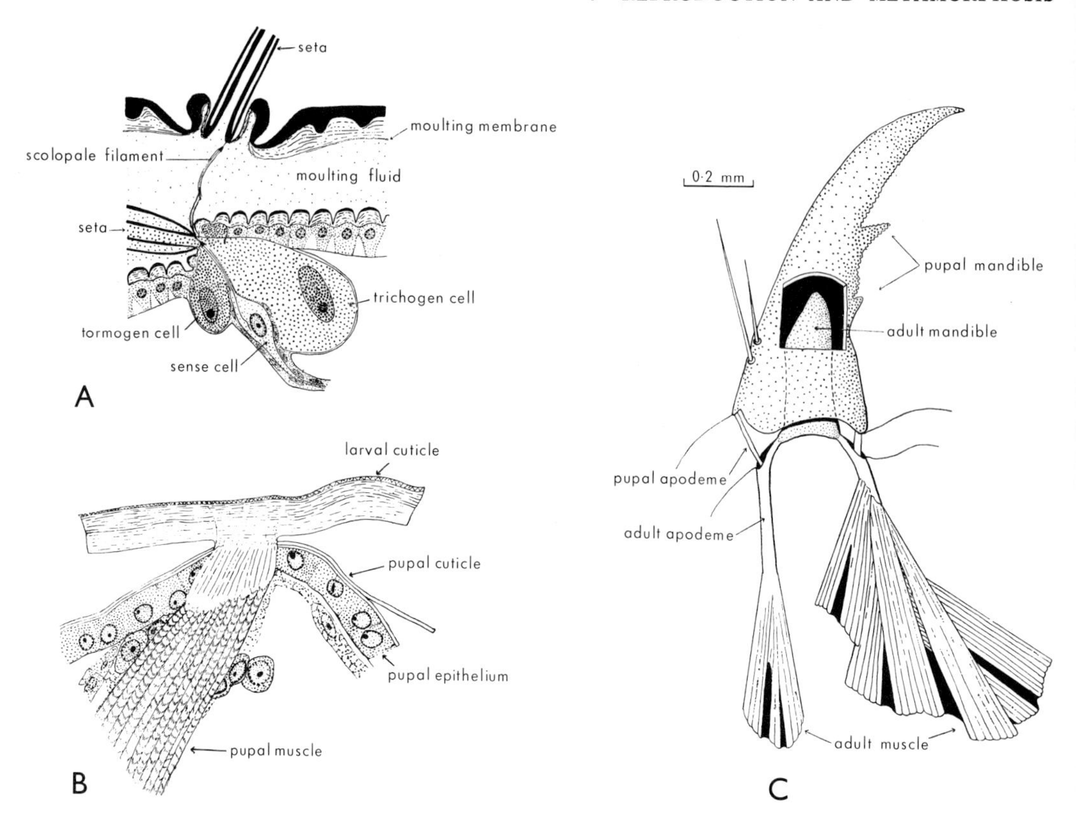

Fig. 4.12. Some connections of pharate phases with the old cuticle: A, scolopale receptor of a pharate larva linked with filament and seta of previous instar; B, pharate pupa of *Simulium ornatum*, Diptera, showing connection of an abdominal muscle with the larval cuticle; C, pupal mandible of *Rhyacophila dorsalis*, Trichoptera, enclosing, and actuated by, the mandible of the pharate adult.

[From Hinton, 1958b,c, by permission of the publishers]

important functions. It limits the shape of the body, and continues to have protective functions while the new one is developing. Its mechanoreceptors function for the new instar through connecting scolopale filaments, which persist until shortly before ecdysis. The tonofibrillar connections of many muscles to the old cuticle may remain intact until just before ecdysis. Further, as new apodemes ensheath the old ones (Fig. 4.12c), the muscles of the pharate insect can move such appendages as the mandibles of the earlier stage. It follows that many activities that have been attributed to the last larval and pupal stages of endopterygotes are, in fact, carried out by the pharate pupal and adult stages respectively. Thus, the transition from water to air by species with aquatic pupae is usually made by the pharate adult (as it is also in exopterygotes with aquatic larvae); the active 'gill-spot larvae' of Simuliidae are pharate pupae (Hinton, 1958b); and hibernating 'pupae' are usually pharate adults.

Changes during Growth. The most obvious change is a great increase in size, some adult endopterygotes weighing more than 10,000 times as much as their eggs. This is brought about progressively in each instar, primarily by division or increase in the size *(hypertrophy)* of the cells. The epidermal cells become free to reorganize as soon as they separate from the old cuticle at apolysis, and they may show bursts of mitotic activity and cell degeneration. When this 'fever of growth and decay' is over, the remaining cells and nuclei assume the orderly arrangement needed to define the form of the next instar (Wigglesworth, 1954).

Growth of the softer parts of the body may appear to be continuous, the discontinuities being evident only in sclerotized parts, such as the head capsule. When these are measured they generally show constant proportional increments (usually of about 1·4:1) in linear dimensions in successive instars (Dyar's 'law'). Some insects show *allometric* growth of particular parts, such as the heads of some termite soldiers and worker ants, which grow at a progressively changing ratio to the increments of other parts or of the body as a whole. Essentially, normal growth gives a straight line when plotted on a semi-log. grid and allometric growth a straight line on a log.-log. grid.

There are also changes in form and structure that will be discussed in the next section, but it is necessary first to mention important differences that occur in the development of the wings. In the Palaeoptera and more primitive Neoptera the wings develop, usually progressively, in external sheaths (Fig. 4.13A) that lie on the dorsal surface of the body of the larva, which ultimately transforms directly into an adult. This type of wing growth is known as *exopterygote*, and the larva is often called a *nymph*. In the higher orders of Neoptera, in which transformation to the adult takes place through a pupal stage, the wings develop in pockets in the integument (Fig. 4.13B), and are everted after the larval-pupal apolysis. This type of development is known as *endopterygote*. These differences, and the quantitative differences in metamorphosis generally associated with them, were formerly used as a basis for primary taxonomic subdivision of the winged insects (p. 161).

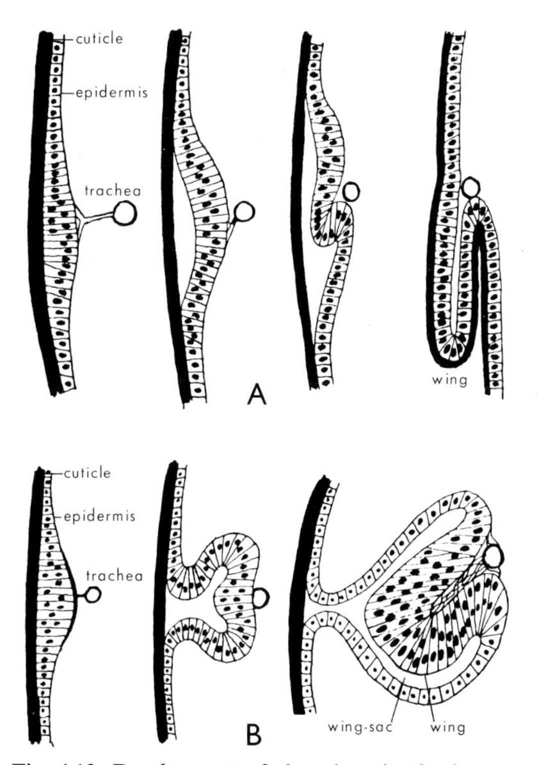

Fig. 4.13. Development of the wings in the larva: A, of an exopterygote insect; B, of an endopterygote insect (after McBride, 1914).

METAMORPHOSIS

The word *metamorphosis* means change of form, and it is used only for postembryonic changes of form. The expression 'at metamorphosis' is generally used for the changes that occur after the larval-adult apolysis of the exopterygotes and after both the larval-pupal and pupal-adult apolyses in endopterygotes. Several different terms have been used to distinguish between what appear to be different degrees of metamorphosis. Thus the Apterygota change little during growth, and are said to be 'ametabolous'. In the great majority of the exopterygotes (e.g. Fig. 14.4) the external form of the larva gradually approaches that of the adult in successive larval instars, and the last larval instar resembles the adult. These are said to be 'hemimetabolous'. Some writers (e.g. Snodgrass, 1954) use the term 'paurometabolous' instead of hemimetabolous, restricting the latter term to Ephemeroptera, Odonata (Fig. 4.14A), Plecoptera, and some Homoptera, in which the last larval instar is distinctly different from the adult. In nearly all endopterygotes and a few exopterygotes, e.g. male Coccidae, the larval instars do not resemble the adult, and there is therefore a

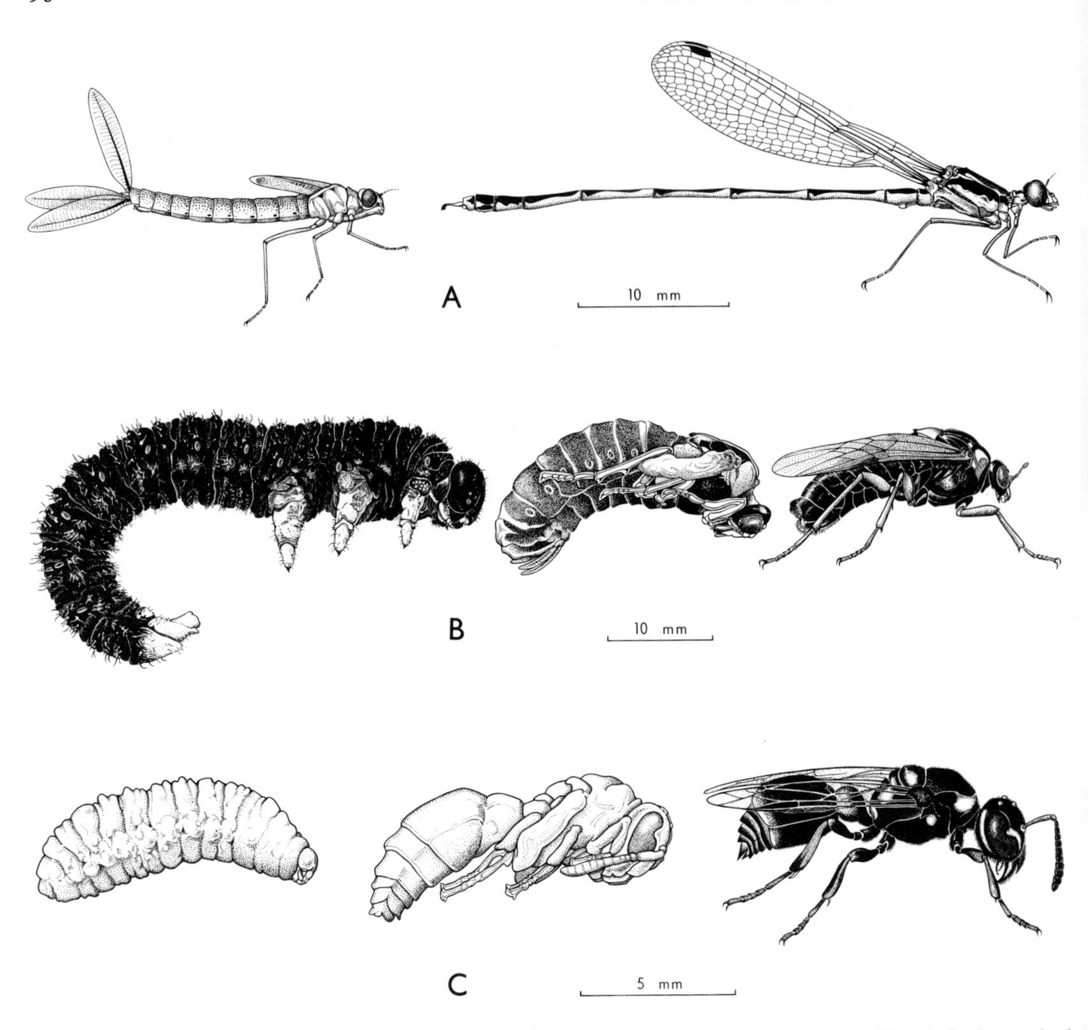

Fig. 4.14. Metamorphosis: A, *Austrolestes analis*, Odonata-Zygoptera, an exopterygote with relatively marked divergence between larva and adult; B, *Perga affinis*, Hymenoptera-Symphyta, a relatively primitive endopterygote; C, *Paralastor* sp., Hymenoptera-Apocrita, a relatively specialized endopterygote.
[T. Nolan]

striking change in external form at metamorphosis (Figs. 4.14B, C). Such insects are called 'holometabolous'.

The different types of metamorphosis described refer to the amount of change between the gross external appearance of the larva and that of the adult. A characteristic feature of hemimetabolous insects is also the large number of larval organs that are carried over more or less intact into the adult stage. However, after the larval-adult apolysis, some may be reconstituted in varying degree; some may be destroyed and replaced by organs subserving the same function; some may be destroyed and not replaced; and some are formed that have no counterpart in the larval stage. The essential distinctions between the 'Hemimetabola' and 'Holometabola' are quantitative and not qualitative. In holometabolous insects a far higher proportion of functional larval tissues and organs is generally destroyed at metamorphosis, and in some highly specialized endopterygotes the pupa may be almost entirely formed from embryonic replacement cells or groups of such cells. In these insects there is an almost total destruction of previously functional larval cells. Most insects are mixtures of tissues and organs that develop in a 'hemimetabolous' way and others that develop in a 'holometabolous' way. This is clear from the survey given below, which is largely taken from Hinton (in preparation) where references to the literature will be given. (For the endocrine organs see p. 48).

Epidermis. During larval life, the epidermis of the body wall may grow either (1) by cell multiplication with relatively little increase in cell size, or (2) by an increase in cell size without cell division. The first method is the primitive one; the second is the specialized one characteristic of the Diptera–Cyclorrhapha and many Hymenoptera.

When growth during the larval stages is by cell multiplication, the same process occurs at metamorphosis; the larval cells divide, and the cells that secrete the cuticle of the new stage are the descendants of the cells that secreted the cuticle of the previous stage. If some of the epidermal cells of the last larval instar prove to be carried over into the next stage without division in some insects, it would have to be recognized that a cell specialized for the secretion of larval cuticle could also secrete the different cuticle of the next stage without going through a process of dedifferentiation and division.

When the second type of growth occurs, renovation of the epidermis at metamorphosis is by proliferation from isolated groups of embryonic replacement cells called *imaginal* or *pupal discs*. In this case, cells that function in two different stages—larval and adult (exopterygotes) or larval and pupal (endopterygotes)—exist side by side in at least the later larval instars.

There is great variation in detail. Thus, although in some highly specialized endopterygotes epidermal renovation after the larval-pupal apolysis is exclusively of the specialized type, renovation after the pupal-adult apolysis is of the primitive type. In some Diptera–Cyclorrhapha the epidermis of the head and thorax is renovated from imaginal discs after the larval-pupal apolysis, whereas that of the abdomen is renovated in a similar manner after the pupal-adult apolysis. Also, in exopterygotes and those endopterygotes in which the epidermis of the general body wall is renovated by cell multiplication, the epidermis of organs that do not function in the larval stage, e.g. the wings, is nevertheless proliferated from imaginal or pupal discs, but destruction of previously existing larval cells is not involved. Moreover, particular parts may grow only by an increase in cell size and be renovated at metamorphosis by proliferation from discs, e.g. the fore gut, colon, anal gills, and main tracheal trunks of the mosquito *Aedes aegypti*. Again, in some endopterygotes with functional larval legs renovation of the epidermis after the larval-pupal apolysis is of the primitive type, whereas in others certain more or less restricted areas of the leg epidermis may thicken and the cells of these areas proliferate.

Oenocytes. These arise in the embryo from epidermal cells close to and usually behind the abdominal spiracles at about the time of the first tracheal invaginations. The definitive number of oenocytes may be present on hatching or may be attained during the first larval instar; but in many insects, e.g. some Hemiptera, increases in their number occur after each larval apolysis or even throughout life. The larval oenocytes usually degenerate in endopterygotes, and a new generation of adult oenocytes is differentiated from the epidermis. Sometimes, as in the chalcid wasp *Nasonia* (=*Mormoniella*), cells destined to become the imaginal oenocytes are present as clusters of small cells in the first-instar larva. These grow slowly, and in the mature larva are no more than about 15μ in diameter, whereas the larval oenocytes are three times as large. At metamorphosis all of the larval oenocytes disappear, and the imaginal oenocytes are carried over into the adult.

Nervous System. During postembryonic growth the ganglia of the ventral cord tend to become increasingly fused in successive larval instars, but exceptions are known in which they become slightly more separated during growth. At metamorphosis of many of the higher endopterygotes all of the thoracic and abdominal ganglia may be fused into a single mass. Such anterior displacements of the ganglia do not affect the points of origin of the peripheral nerves nor the ganglia which they enter. It is probable that the primitive method of growth of the nervous system involves division of the cells of the cortex and neuropile (Fig. 4.6D), and this method occurs even in some highly specialized endopterygotes. In others, however, the central nervous system may undergo a metamorphosis after the larval-pupal apolysis that is no less profound than that of many other organs or tissues of these insects.

The brain and ventral cord are enveloped by a mesodermal sheath, the *perilemma*, which consists of a layer of cells, the *perineurium*, and an outer acellular layer, the *neural lamella*. During larval growth the perineurial cells increase in size, but do not divide. They degenerate at the larval-pupal apolysis, and the perineurium is replaced by proliferation of embryonic replacement cells that were present amongst the larval cells. The replacement by the new perineurium may take place in such a way that the continuity of the perilemma is preserved and blood cells do not enter the nerve cord. At the same time the glia and nerve cells may be more or less completely replaced by proliferation of embryonic replacement cells. In specialized endopterygotes in which this happens, e.g. *Nasonia*, metamorphosis of the peripheral nerves is similar. Their mesodermal covering is renovated without discontinuity, and in due course the axons of the new nerve cells extend within the reconstituted sheaths towards the central nervous system and finally establish connections with the newly differentiated association cells in the brain and ventral cord.

Sense Organs. In general, larvae of exopterygotes and Mecoptera have compound eyes like those of the adults; other endopterygote larvae are either blind or have lateral ocelli, the compound eyes developing only at metamorphosis. New sense organs may be differentiated in the epidermis after apolysis in both exo- and endopterygotes. When this occurs, the growing axon of the new nerve cell wanders about amongst the epidermal cells on the outer side of the basement membrane until it meets with a previously existing mesodermal sheath already containing one or more axons. It then grows inwards within the sheath until it enters the central system. Sense cells that differentiate in postembryonic stages do not initiate new sheaths through the haemocoele, but apparently depend upon existing ones.

Muscles. The degree of metamorphosis undergone by muscles increases as the difference in form between the larva and adult increases. In most exopterygotes the great majority of the larval muscles are carried over into the adult stage with no more than relatively slight changes in fibre length and thickness, sometimes no greater than those that occur between one larval instar and another. When the adult has structures that are either not present or not functional in the larva, the muscles required for them may be larval muscles that have altered their position, or they may be entirely new muscles, as are, for instance, some of the ovipositor muscles of thrips. If skeletal structures are lost or very much altered after the larval-adult apolysis, the muscles concerned in their functioning may be totally destroyed, e.g. two of the muscles of the larval labium in the dragonfly *Aeshna*. The new muscles that are produced after the larval-adult apolysis are formed either from myoblasts that are part of the larval muscle-fibre syncytium or from scattered myoblasts. For instance, in the male of the scale insect *Pseudococcus*, seven of the 20 new skeletal muscles of the adult arise in the former manner and 13 in the latter.

In the endopterygotes the only functional larval skeletal muscles that persist in the adult are some of those of the abdomen, all others being formed from imaginal discs or myoblasts, and in some highly specialized endopterygotes even the abdominal muscles of the larva are all destroyed. In both exo- and endopterygotes some larval muscles, called *caducous* muscles, may persist for a short time in the adult, and may play a significant role until they are destroyed, e.g. some of the abdominal muscles of moths, which persist throughout the pharate adult stage, and do not degenerate until about two days after emergence of the moth.

It is important to note that, as with other tissues, there is no difference between exo- and endopterygotes in the kind or in the degree of metamorphosis of the muscles, but only in the extent of the more drastic changes. This is clearly shown by the growth and metamorphosis of the indirect flight muscles (review by Hinton, 1959). For instance, in the Odonata, Blattodea, Mantodea, Phasmatodea, and some Hemiptera, the rudimentary fibres (imaginal discs) of the indirect flight muscles are present when the larva hatches, the fibres grow by division of their nuclei, and there is no incorporation of free myoblasts. Precisely the same type of growth occurs in the Diptera–Nematocera (Fig. 4.15). In the Diptera–Cyclorrhapha, on the other hand, all myoblasts of the indirect flight muscles are contained within the larval muscle, as in most endopterygotes, e.g. Neuroptera, Coleoptera, Mecoptera, and Lepidoptera.

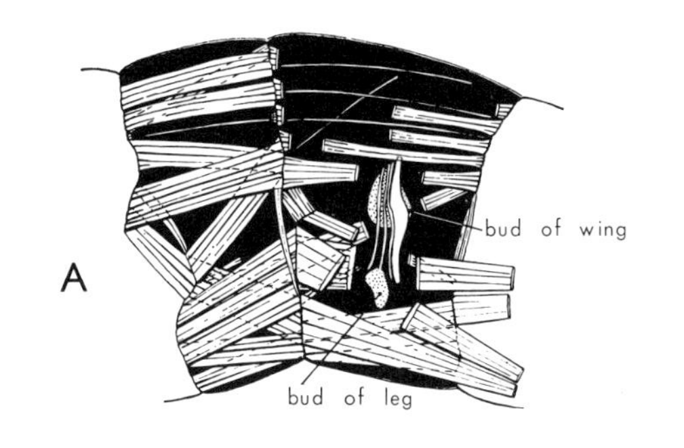

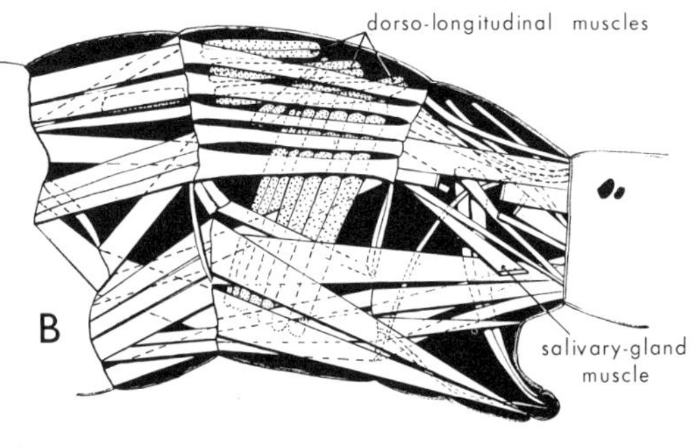

Fig. 4.15. Meso- and metathorax of *Simulium ornatum*, Diptera: A, penultimate larval instar, the fibre rudiments of the indirect flight muscles and the tergal depressor of the trochanter unshaded; B, pharate pupa, the fibre rudiments of the same muscles stippled.
[From Hinton, 1963b, by permission of the publishers]

Circulatory System. There are usually various fibro-muscular septa and ampullae or accessory hearts in the circulatory system of the adult that are not present in the larva. These new organs are formed at metamorphosis from undifferentiated blood cells. The dorsal vessel generally undergoes some modification during metamorphosis. For instance, in some dragonflies the aortic diverticula increase in size in successive larval instars and again at metamorphosis. In many exo- and endopterygotes pulsatile organs develop in the mesothorax at metamorphosis, but it is not known whether they are formed by division of larval cells or proliferation of embryonic replacement cells. Metamorphosis of the dorsal vessel may be considerably greater amongst some exopterygotes than amongst some endopterygotes, such as some Diptera–Nematocera and the weevil *Sitophilus*, in which the larval dorsal vessel is apparently carried over directly into the adult without change (in *Sitophilus* growth in all postembryonic stages is only by increase in cell size). On the other hand, the adult heart of *Nasonia*, for instance, is formed in the pupal stage by proliferation of embryonic replacement cells, and all of the larval cells degenerate and are absorbed.

Blood Cells. The haemocyte may well be the only class of cell that divides during larval growth and at metamorphosis in all insects. Mitoses, especially of the freely circulating blood cells, are generally most numerous at the times of the moults. Quite often a high proportion, sometimes all, of the blood cells are attached to other organs and tissues, and sometimes they appear to form more or less discrete haemopoietic organs in the thorax or abdomen. However mitosis of the circulating blood cells often occurs even in those insects in which conspicuous haemopoietic organs have been reported.

The haemocytes tend to be very labile in shape, but a number of types have been recognized (p. 55). From a general point of view, it is useful to think of the young cell that usually becomes a haemocyte as a kind of undifferentiated mesenchyme cell more or less totipotent for the formation of a wide variety of other tissues—connective tissue, basement membrane, fat body, contractile fibres, skeletal muscle, and so on.

Pericardial Cells. These are usually bi- or trinucleate cells. In many Diptera, some beetles, and presumably many exopterygotes they are carried over into the adult stage without evident change at metamorphosis. In other insects, e.g. *Sitophilus*, most may degenerate at metamorphosis. Some writers believe that the pericardial cells give rise to ordinary blood cells at metamorphosis.

Fat Body. The primitive metameric arrangement of the fat body is often lost by secondary aggregations early in embryonic development, especially in the more highly specialized insects. Sometimes cells that are functionally and histologically indistinguishable from those of normal fat body arise from the epidermis (Cynipidae) or blood cells. The fat-body cells are generally in well-defined lobes or ribbons, each aggregation being bounded by a membrane composed of fibrillae in a ground substance, the 'tunica propria' of writers. In a number of insects, however, they are free in the body cavity, and it is then difficult to distinguish them from blood cells.

During larval growth the number of fat-body cells usually increases by division, and at metamorphosis the cells of the final larval instar are carried over into the adult either with or without an increase in their number. When this happens, e.g. in many Coleoptera, the form of the fat body of the adult may be very similar to that of the larva. It has recently been shown that in some Diptera (Simuliidae, Thaumaleidae) the pigmented cells of the fat body (chromatocytes) do not divide at metamorphosis, but persist in the pupa and adult, in which they may migrate extensively and form new colour patterns. In *Nasonia* the fat-body cells of the first instar persist into the adult, and do not divide. In at least some Diptera–Cyclorrhapha there appears to be a total destruction of the larval fat body, and a new fat body is formed at metamorphosis from undifferentiated mesenchyme (blood) cells.

Alimentary Canal. At metamorphosis the alimentary canal of many insects, especially endopterygotes, undergoes great changes in outward form and in the relative lengths of its three principal parts. These changes are brought about either by a transformation of the functional larval epithelium, or by destruction of the epithelium and formation of the new parts by proliferation of groups of replacement cells (imaginal discs, imaginal rings).

In the Diptera–Cyclorrhapha and other specialized endopterygotes in which the body-wall epidermis is renovated entirely from discs at metamorphosis, the epithelium of the fore and hind guts is renovated in the same way. When renovation of the epidermis is by cell multiplication, renovation of the fore- and hind-gut epithelium is often, probably usually, by cell multiplication too, e.g. in many Coleoptera and Lepidoptera. However, exceptions are known, as in the mosquito *Aedes aegypti*, in which the epithelium of the fore gut of the larva is destroyed and replaced by a new fore gut proliferated from an imaginal ring.

More is known about the renovation of the mid gut. In all Antennata except Pauropoda the epithelium is renovated in one of two ways: (1) the old epithelium is replaced by proliferation of scattered groups of embryonic replacement cells, which either rest on the basement membrane between the functional epithelial cells or are contained in *regenerative crypts* which may project outwards between the gut muscles; or (2) the old epithelium may be partly or entirely replaced by one that arises by proliferation from one or other of the imaginal rings.

These two types of mid-gut renovation can occur after different apolyses in the same insect; and in a few insects, e.g. *Nasonia*, the mid gut may be renovated in both ways after the same apolysis. Renovation may be total or partial immediately following an apolysis, and sometimes, as in the caterpillars of *Galleria* and *Achroia*, there is a continuous partial renovation between apolyses. In some insects renovation may be entirely suppressed in the larval instars, and there is then a total renovation after the larval-adult (exopterygotes) or larval-pupal apolysis.

In most Insecta (as in Chilopoda, Collembola, and Diplopoda), whenever renovation of the mid gut occurs, it is from embryonic replacement cells that are scattered in groups between the functional epithelial cells. This seems to be the only type of renovation known in Apterygota and the exopterygotes. Sometimes, as in some Apterygota, it is partial during the period between apolyses, but total immediately after the apolyses. In some Odonata *(Aeshna)* it is reported that renovation is entirely suppressed until the larval-adult apolysis.

In most endopterygotes the mid gut is renovated after the larval apolyses precisely as in the exopterygotes; and the changes that occur after the larval-pupal and pupal-adult apolyses differ in no way from those that occur after the larval-adult apolysis of the exopterygotes. Sometimes the pupal mid gut persists into the adult stage, but as a general rule it is destroyed and a new adult mid gut is formed. When renovation is only partial after the larval-larval apolyses, the epithelial cells that persist from instar to instar become progressively more hypertrophied. When renovation is totally suppressed during the larval instars, the mid gut grows by an increase in cell size, as in Diptera–Cyclorrhapha, some beetles, and some Hymenoptera. In many beetles and some Hymenoptera the mid gut is renovated after the larval-pupal apolysis by posterior extension from the anterior imaginal ring, the new epithelium stripping off the old, which is then broken down in the lumen of the gut. It is of interest to note that a similar type of mid-gut renovation, but from the posterior imaginal ring, occurs in the Symphyla. In *Nasonia* the anterior half of the mid gut is renovated after the larval-pupal apolysis from the anterior imaginal ring, from which it grows posteriorly until it joins the posterior half, which has been renovated in the meantime from scattered groups of embryonic replacement cells.

To understand the functional significance of the renovation of the mid gut, it is necessary to see the process not only as a way of growth but also as a method of excretion: degeneration of the epithelial cells results in the translocation of excretory products into the lumen of the gut. It is characteristic of the subphylum Antennata that during adult life there is a continual renovation of the mid gut without an increase in size: the mid gut may be considered to be the most universal excretory organ of these animals.

Malpighian Tubes. The Malpighian tubes are appendages of the mid gut. In the Lepidoptera–Ditrysia, however, the posterior imaginal ring develops in an unusual way during embryogenesis, and the tubes of the larva come to arise from the distal ends of two common ducts of which the proximal parts are ectodermal and the distal parts endodermal. Following the larval-pupal apolysis there is a complete re-formation of this region: the new Malpighian tubes now arise from the mid gut anterior to its junction with the hind gut, as in larvae of the Lepidoptera–Monotrysia as well as most, if not all, larvae and adults of other insects.

The primitive number of Malpighian tubes in the class Insecta is six. Most insects have six or four primary tubes, but some Hymenoptera and many exopterygotes have large numbers of secondary tubes: the greatest number is probably found in mayflies, some of which have up to 600 tubes. Cell division in the primary tubes normally appears to be restricted to the first half of embryonic life, and thereafter growth is by an increase in cell size. In the earwig *Forficula auricularia*, for instance, there are 20 Malpighian tubes in the adult, arranged in four groups of five each. One member of each group is a primary tube proliferated in the embryo. Its cell number does not increase in postembryonic life. In each of the four larval instars a

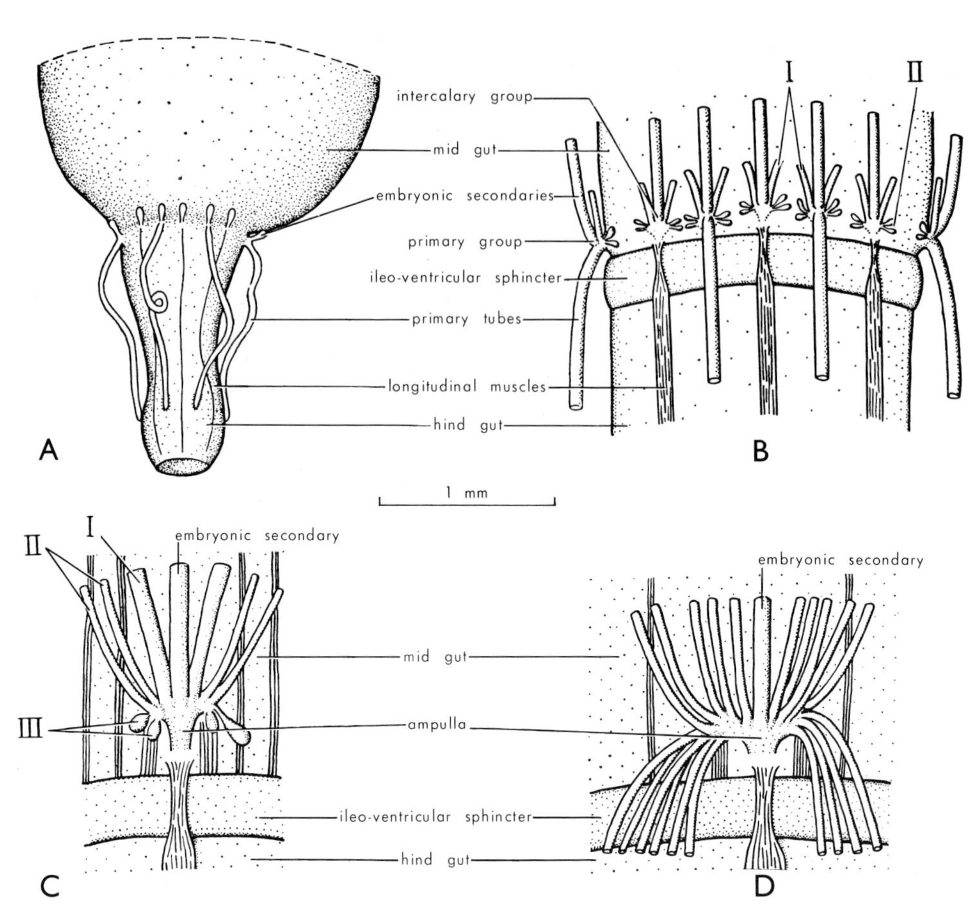

Fig. 4.16. Development of the Malpighian tubes in *Schistocerca gregaria*, Orthoptera: A, in a 9-day embryo; B, in an early second-instar larva; C, dorsal intercalary group in an early third instar; D, same in an adult. [From Savage, 1956, by permission of the publishers]

secondary tube is proliferated from the imaginal ring within a few days of ecdysis. Growth of these secondary tubes is at first by cell division and later only by an increase in cell size. The tubes of the last larval instar are carried over into the adult. In the locust *Schistocerca gregaria* (Fig. 4.16), on the other hand, there are two generations of tubes produced in the embryo, the first composed of six primary tubes and the second of 12 embryonal secondaries. In each of the five larval instars a variable number of secondaries is added, until in the last larval instar there are about 250 tubes arranged in 12 groups. The development of each tube conforms to the normal pattern: a period of initiation, a period of mitosis and elongation, and a period of differentiation. Mitoses cease in the primaries before the development of the embryonal secondaries, and mitoses at each succeeding stage also end before the appearance of a new generation of secondaries.

In those Lepidoptera–Ditrysia in which the primary tubes persist into the adult stage, it is only the middle sections that do so without drastic alteration. In a few others, e.g. many Tineidae and some Pyralidae, there is a total destruction of the larval Malpighian tubes at the larval-pupal apolysis, and a new generation of tubes is proliferated from the posterior imaginal ring at metamorphosis. In these species, therefore, the larval tubes are primary and the adult tubes are secondary, as they are in many Hymenoptera–Apocrita. On the other hand, in some Hymenoptera, e.g. some Chalcidoidea, the larva lacks Malpighian tubes, and the primary adult tubes arise from the posterior imaginal ring at metamorphosis. In a number of Coleoptera, e.g. some Chrysomelidae and Curculionidae, the tubes are renovated at metamorphosis from scattered embryonic replacement cells precisely as in the mid gut of many insects.

ORIGIN OF THE ENDOPTERYGOTE PUPA

The origin and functional significance of the pupal stage have been the subject of controversy almost since man first began to be interested in insects, and the views that have been expressed have been summarized by Hinton (1963b). The pupa of the endopterygotes is like the adult in most respects, and its external features generally bear no resemblance to those of the preceding larval instar. The origin of the endopterygotes from some

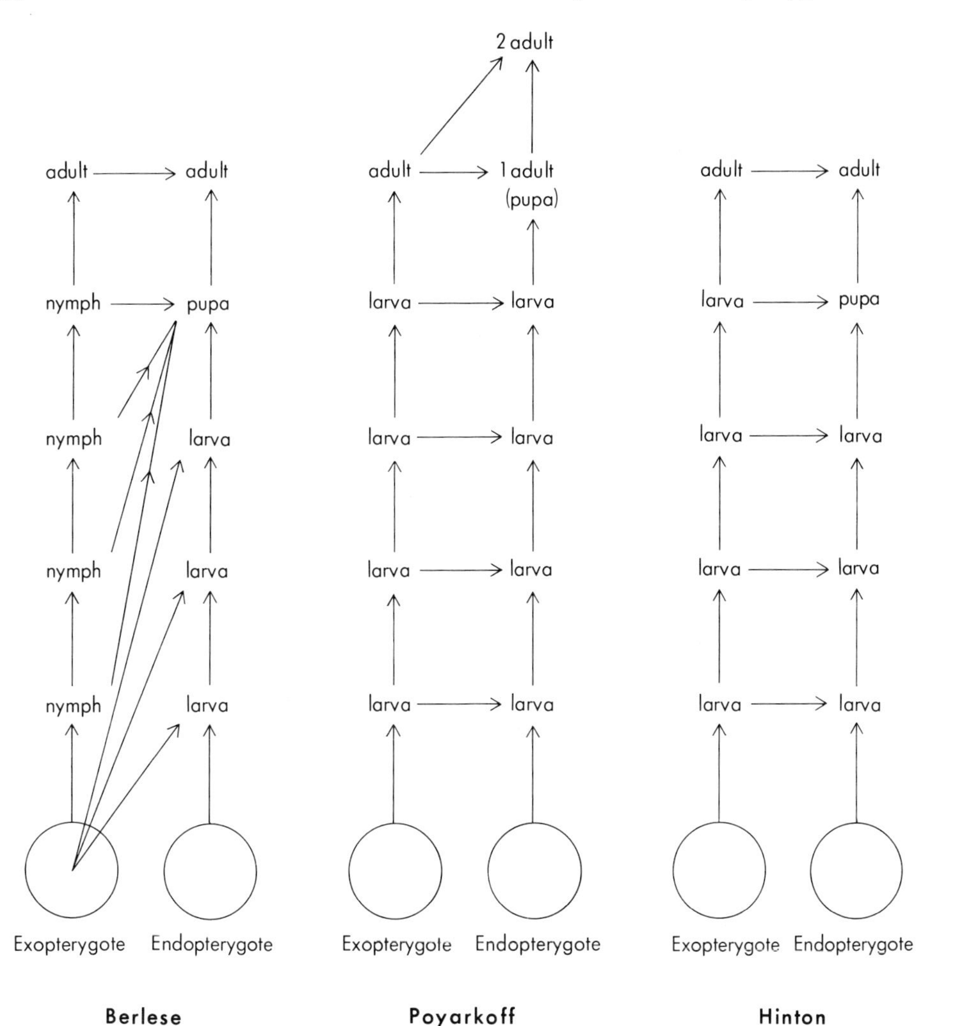

Fig. 4.17. Diagram showing the relationships between the theories of Berlese, Poyarkoff, and Hinton concerning the origin of the pupal stage (after Hinton, 1963b).

group of exopterygotes is not seriously disputed, and, as new developmental stages arise only by modification of previously existing stages, it is not surprising that there has been much speculation about the exopterygote stage that gave rise to the pupa. Three quite different theories have been propounded, and the differences between them are shown diagrammatically in Figure 4.17.

There are now known to be serious difficulties in accepting either Berlese's or Poyarkoff's theory (Hinton, 1963b). On the other hand, the fact that the pupa resembles the adult so much more closely than does the final larval instar has a simple explanation if it originated as a specialization of the last larval instar of the exopterygote ancestors of the endopterygotes. *All that has to be assumed is that the last larval stage with external wings was retained in the endopterygotes to bridge the difference between a larva with internal wings and the adult.* If, in time, such a stage became less and less mobile as the feeding larval stages became more and more specialized, and therefore more and more different from the adult, we might expect that the feeding larval stages might ultimately cease to resemble the last larval instar, now the pupa, the relative immobility of which would have disrupted many of its former relations with the external environment. In many exopterygotes the last larval instar resembles the adult quite as much as the pupa of the endopterygotes resembles the adult endopterygote.

It is supposed that there has been a functional subdivision of the larval stages into a series chiefly concerned with feeding and a final larval instar, now a pupa, chiefly concerned with bridging the gap between the specialized feeding stages and an adult specialized for reproduction and distribution. In this connection, it may be noted that subdivision of the larval stages into two series specialized for very different modes of life has occurred independently on a number of occasions in both primitive and advanced endopterygotes (p. 106).

The pupal stage is one of the most vulnerable in the life of the endopterygotes, and it would therefore appear that its compensating advantages must be enormous. These would seem to lie primarily in the fact that possession of a pupal stage has made it possible for the form and structure of endopterygote larvae to become completely dissociated from those of the adults, and has thereby enabled them to invade a large number of habitats from which the exopterygotes are excluded. This radiation was facilitated both by the existence of pharate phases that could be prolonged (p. 93) and by the absence of the impediments provided by the external wing sheaths.

Some writers believe that the pupal stage is not especially vulnerable, but a kind of encystment stage, the initial selective advantage of which was to resist the rigours of winter. But only a few of the most highly specialized endopterygotes (e.g. some of the more advanced Lepidoptera–Ditrysia) pass the winter in the true pupal stage. The impression that many species do so has been largely due to a failure to distinguish between the pupa and the pharate adult.

Because of the space occupied in the larval thorax by muscles and other organ systems, there is insufficient room for the wing *anlage* to develop to the size necessary for the wings to be effective, even allowing for some growth after their evagination. One apolysis is required to evaginate the wings. Following it there is considerable growth of their epidermis, and large pupal wing-cases are formed. A second apolysis is then required, at which the epidermis is freed from the pupal cuticle and the number of its cells can increase greatly, with the result that the definitive adult epidermis is often much folded within the pupal cuticle before secretion of the adult wing cuticle is completed. The growth of the wing *anlage* after the larval-pupal apolysis is enormous. For instance, at the beginning of the last larval instar of *Ephestia kuehniella*, the *anlage* of the hind wing has about 400 cells, not counting those of the epithelial sac in which it is contained. Following the larval-pupal apolysis there is an outburst of mitosis, and by the time the wing is evaginated the number of cells is about 38,000. By the end of the pharate pupal period there are nearly 70,000 cells, and there is another prolonged outburst of mitosis following the pupal-adult apolysis, which in this insect occurs after the larval-pupal ecdysis.

THE KINDS OF ENDOPTERYGOTE PUPAE

The pupal and pharate adult stages of all primitive endopterygotes are passed in a cell or cocoon of some kind, and it is only some specialized Coleoptera, Hymenoptera, Lepidoptera, and Diptera that are known not to have one. In the Diptera–Cyclorrhapha and some others the last larval cuticle *(puparium)* serves as a cocoon. The habit of making a cell or cocoon could scarcely have been evolved without the simultaneous development of a method of escape for the adult; and it has been shown (Hinton, 1946b) that the chief structural variations of pupae can be related to the manner in which the adult effects its escape from its cell or cocoon.

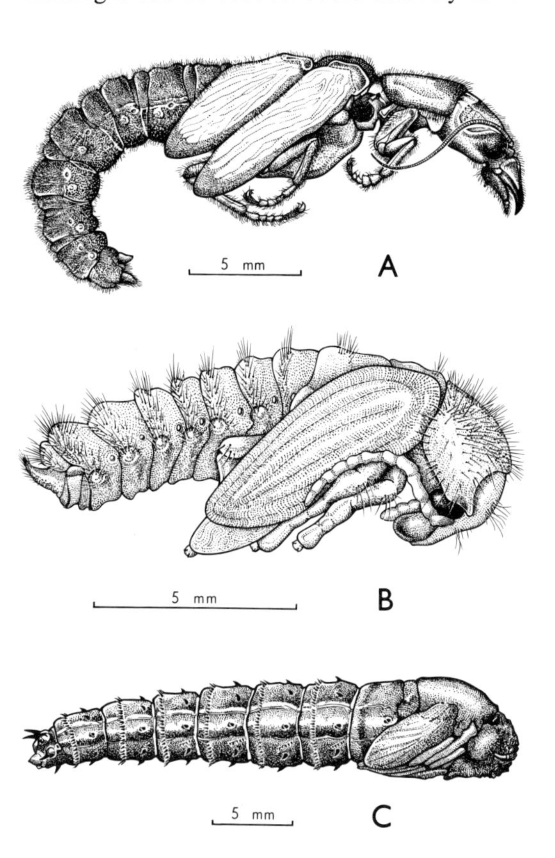

Fig. 4.18. The three kinds of pupae: A, *Archichauliodes guttiferus*, Megaloptera, decticous exarate; B, *Paropsis atomaria*, Coleoptera, adecticous exarate; C, *Pelecorhynchus fulvus*, Diptera, adecticous obtect. [T. Nolan]

In primitive endopterygotes, such as Megaloptera, Neuroptera, and Mecoptera, as well as in many others, the pupal cuticle is not generally shed by the adult until it has escaped from the cell or cocoon with the aid of the pupal mandibles. Even amongst the few (some Neuroptera) that shed the pupal cuticle within the cocoon, the pupal mandibles are still used to cut it open. The pupae of all these have strongly sclerotized mandibles, each of which has a dorsal and ventral articulation to the anterior margin of the cranium. A pupa with articulated mandibles is known as a *pupa dectica* (Fig. 4.18A). A pupa without articulated mandibles, and therefore one in which the mandibles cannot be used by the pharate adult, is known as a *pupa adectica.*

The pharate adult of those groups that have decticous pupae not only uses the mandibles of the pupa in escaping but also its feet. It may wander far from the site of the cell or cocoon before it sheds the pupal cuticle. Its journey may involve burrowing through considerable depths of soil or, as in most Trichoptera, swimming through the water.

There are two kinds of adecticous pupae, the exarate and the obtect. An *exarate* pupa (Fig. 4.18B) is one in which the appendages are free and are not cemented to the body wall. An *obtect* pupa (Fig. 4.18C) is one in which the appendages are more or less strongly cemented to the body, presumably by tanning of a protein contained in the moulting fluid. The cuticle of obtect pupae is usually much more strongly sclerotized than that of exarate pupae, and there is rarely any practical difficulty in distinguishing between the two. Decticous pupae are never obtect: if the appendages were cemented to the body the pupal legs could not be used in locomotion.

The distinction between adecticous exarate

and adecticous obtect pupae is comparatively trivial, and both categories are polyphyletic. The decticous pupa may evolve directly into an adecticous exarate pupa, as have those of some genera of the trichopterous family Phryganeidae and most Coleoptera, Hymenoptera, and possibly Siphonaptera. The available evidence suggests that the obtect pupae of both the Lepidoptera and the Diptera arose directly from decticous pupae without an intervening adecticous exarate stage. The few obtect pupae in the Coleoptera and Hymenoptera are derived from adecticous exarate pupae, whereas the exarate pupae of the cyclorrhaphous Diptera–Brachycera are clearly derived from adecticous obtect pupae.

The methods of emergence from the pupal cell or cocoon of pharate adults with adecticous pupae are very varied. When the pupa is exarate, the pupal cuticle is usually shed in the cell and the adult bites its way out with its own mandibles, as in most Coleoptera and Hymenoptera. When the pupa is obtect, the pharate adult may utilize backwardly directed spines on the pupal cuticle to force its way out of the cell or cocoon. Such pupae often have cocoon-cutters on the head, as in many Lepidoptera and Diptera. The many departures from these relatively primitive methods of emergence have been summarized by Hinton (1946b).

HETEROMORPHOSIS

Endopterygote insects in which two or more of the successive larval instars are specialized for different modes of life and differ greatly in form are said to undergo hypermetamorphosis, but it is best to call this phenomenon *larval heteromorphosis* (Snodgrass, 1954). It is best known in parasitic insects that lay their eggs far from the hosts. The first-instar larva, generally called a *planidium* (triungulin in Coleoptera, triungulinid in Strepsiptera) is a free-living, active creature (e.g. Fig. 31.5), which is often able to stand erect on its hind end, and is sometimes capable of jumping to reach a host or an insect that will carry it to a host. There is remarkable convergence in form between the planidia of different groups. Active life ceases after the host is found, and the subsequent instars normally become stout and grub-like. Heteromorphosis of this type occurs in the Neuroptera (Mantispidae), Coleoptera (some Carabidae and Staphylinidae, all Meloidae and Rhipiphoridae, Drilidae, a few Colydiidae), all Strepsiptera, Diptera (Acroceridae, Nemestrinidae, Bombyliidae, some Tachinidae, a few Sarcophagidae), Lepidoptera (e.g. Cyclotornidae and Epipyropidae which are ectoparasites of Homoptera), and Hymenoptera (Perilampinae, Eucharitinae).* Examples of most of these are given in later chapters.

Larval heteromorphosis may also occur when the eggs are laid on or in the host, as in some agromyzid Diptera and many Hymenoptera. In these the first instar is not a planidium. In some Hymenoptera it is virtually an embryo with an unsegmented abdomen and undeveloped nervous and respiratory systems. It is seen, too, in some non-parasitic insects, such as the paedogenetic Micromalthidae. A few other beetles, e.g. some Cerambycidae and Bruchidae, have a first instar so different from subsequent instars that they might well be considered to be heteromorphic. In a few leaf-mining Lepidoptera, e.g. some Phyllocnistidae, Gracillariidae, and Lithocolletidae, the first instars are strongly flattened and adapted for mining, whereas the final instar is like a normal caterpillar and its chief function seems to be to spin the cocoon. Thus, when all groups are considered, there is a complete series of intermediates between the normal and the heteromorphic.

* Also in a specialized flea (Figs. 33.3B, C).

5

GENERAL BIOLOGY

by K. R. NORRIS

It has been emphasized in the Introduction that insects have been highly successful animals, and the purpose of this chapter is to give some account of their more striking adaptations to many different kinds of existence, using Australian examples wherever possible. It should be read in conjunction with chapter 2, to which it is, in many respects, a sequel.

ADAPTATIONS TO TERRESTRIAL EXTREMES

There are few niches on the surface of the land that have not been occupied by some group of insects, and few climates to which none has become adapted. In the Subarctic and Arctic many species live in areas that are too cold for activity during most of the year. Their growth and reproduction are crowded into the very few warm months. Opportunities for insects are fewer in the south, and only two littoral chironomid flies are known from the Antarctic continent, although Collembola and mites have been found less than 350 miles from the pole (Wise, 1964). The sucking lice of seals (Fig. 5.1) also have to cope with severe conditions (Murray, Smith and Soucek, 1965), but the biting lice of birds are better protected, and a fair number of species have been recorded. The Subantarctic islands have a more varied fauna, chiefly of Coleoptera, Diptera, and Lepidoptera. Many of the species on these islands are brachypterous or apterous (Gressitt, 1961; Fig. 5.2), a tendency which is explained by some as due to selection against flying insects in a stormy environment, and by others as due to the absence of ants and other predators of apterous insects.

Conditions in Australia do not call for exceptional cold tolerance, the extremes being represented in Tasmania and the higher parts of the Australian Alps. A wingless mecopteron, *Apteropanorpa tasmanica*, has been found active on snow; but the summer fauna of these areas exhibits a rich variety

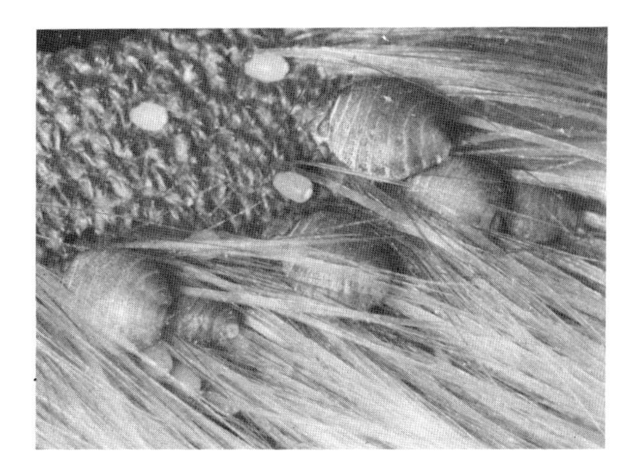

Fig. 5.1. *Antarctophthirus ogmorhini* (Phthiraptera-Echinophthiriidae) and eggs on Weddell seal, exposed by plucking hair; largest adults 3 mm long. [Photo by I. Roper]

of nectar-seeking insects, stream-haunting species with aquatic immature stages, mountain cockroaches and grasshoppers (Plate 2, B, H, J, L), blood-sucking Diptera, and blow flies. Some, like pelecorhynchid flies, have the developmental cycle extended, with larval growth interrupted each winter, whereas the grasshoppers overwinter in the egg stage and the blow flies as full-grown larvae buried in the snow-covered soil. Mountain insects, in general, concentrate all their feeding and reproductive activities into the relatively brief, warm part of the year.

By contrast, the mountains may afford a haven in summer for some species which find the lowland conditions too extreme at this time. The outstanding example is the bogong moth, *Agrotis infusa*. In early summer many of these moths migrate from the larval feeding areas scattered throughout N.S.W. and southern Queensland to mountain peaks above 1,500 m in the Australian Alps, and undergo diapause there, clustered together in rock crevices (Fig. 5.3). During this period, which lasts from November to early April, the moths subsist largely on fat reserves, and do not mate. During autumn they disperse over wide areas to mate and lay their eggs in the pastures, where dicotyledonous annuals suitable as larval food are now available in place of the unsuitable grasses that dominated the pastures in the hotter months. By a combination of migration and diapause this species is able to survive each year over a period when environmental conditions in the breeding grounds are unfavourable (Common, 1954).

Diapause, indeed, is a potent factor in the adaptation of insects to fluctuating environments, as it enables a stage with little or no food requirements to tide the species over periods when the habitat is inhospitable or, at most, favourable for periods too brief for the completion of development. It is to be distinguished from simple *quiescence*, in which development or activity merely ceases during adverse (usually cold) periods and is resumed immediately conditions again become favourable. In *diapause* a particular set of environmental stimuli operates through the endocrine system to inhibit development in a particular stage in the life cycle, and development is not resumed, even in the presence of otherwise favourable conditions,

Fig. 5.3. *Agrotis infusa* (Lepidoptera-Noctuidae) aestivating in a rock crevice on Mt Gingera, A.C.T.
[Photo by I. F. B. Common and M. S. Upton]

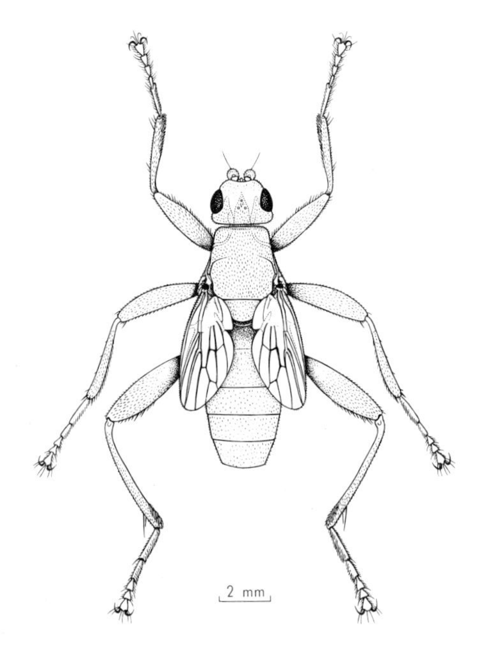

Fig. 5.2. Brachypterous fly, *Baeopterus robustus* (Coelopidae), ♂, from Campbell I. (52° 30′ S).
[F. Nanninga]

until a second set of environmental stimuli has operated, again through the endocrine system, to 'break' the diapause. Diapause and quiescence may, however, operate effectively together, as in the example quoted below (Andrewartha and Birch, 1954).

The grasshopper *Austroicetes cruciata* lives in a part of South Australia with a cool, wet winter and a hot, dry summer (Fig. 5.4). The eggs are laid in the soil in the early summer. They undergo partial development to a stage at which they resist water loss, and then enter diapause, in which they remain even if unseasonable rains should fall during the summer. In the winter the low temperatures progressively break the diapause, but at the same time retard development of the embryo until the temperature rises again in the following spring. Diapause and cold quiescence therefore ensure that only one generation of the grasshopper occurs each year, and that the eggs hatch when conditions are most suitable for growth and reproduction of the active stages.

Australia is rich in insects with special adaptations to escape the hot, dry conditions that prevail for long periods over most of the continent.

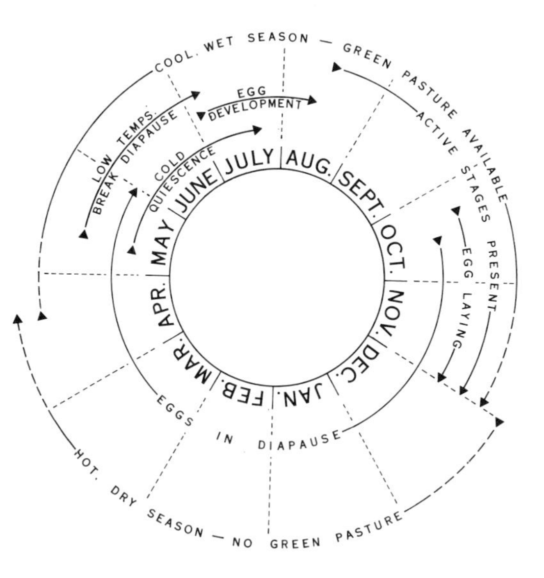

Fig. 5.4. The annual life cycle of *Austroicetes cruciata* (Orthoptera-Acrididae) in South Australia.

Fig. 5.5. Distended workers of a honeypot ant displayed by their Aboriginal gatherer in central Australia. [Photo by C. P. Mountford]

The drought-enduring species [of grasshoppers] manage to survive prolonged periods of low moisture availability (usually combined with high temperatures and low shelter availability) by various adaptations of physiology, behaviour, and structure. Dormancy, ability to restrict water loss, ability to tolerate water loss, ability to survive on dew-soaked plant debris, special sheltering behaviour patterns, or a habit of feeding on the foliage of evergreen shrubs or trees, may be involved. Typically the capacity for drought resistance is well developed in only one life-cycle stage, either the egg or the adult, so that the problem of survival resolves itself very largely into the problem of ensuring that the insect will be in, or will rapidly enter, the appropriate resistant stage when the dry period commences. (Key, in Keast *et al.*, 1959).

The last statement is the complement of the last one in the preceding paragraph.

Storage of requirements is a further mechanism for drought endurance. In central Australia workers of the honeypot ants (*Melophorus bagoti* and *Camponotus* spp.) gather honey-dew from scale insects and psyllids, and feed it to other workers, which become mere nectar-storage vessels, with terga and sterna appearing as dark patches in the greatly stretched membrane of the abdomen (Fig. 5.5). These helpless replete ants, which regurgitate some of their nectar when solicited by other workers, are kept safe from the effects of drought in special galleries, often deep underground (Fig. 5.35).

Similarly, seed harvesting enables many inland Australian ants to take advantage of food supplies afforded by the temporary abundance of ephemeral plants or irregular fruiting of shrubs after rain. Many species of termites living in dry areas have developed the habit of harvesting and storing grass, there being limited opportunities for wood-eating species in the area.

The other major environmental extreme, the truly wet tropics, is limited in Australia to patches of rain forest on the north-east coast. It has an abundant insect fauna, but only a few groups (especially Coleoptera and Lepidoptera) show something of the rich coloration and diversity of form that is usual in wet–tropical insects. A more general response, well seen, for example, in mosquitoes, is acceleration of life cycles and often smaller size at maturity than their relatives in more temperate climates.

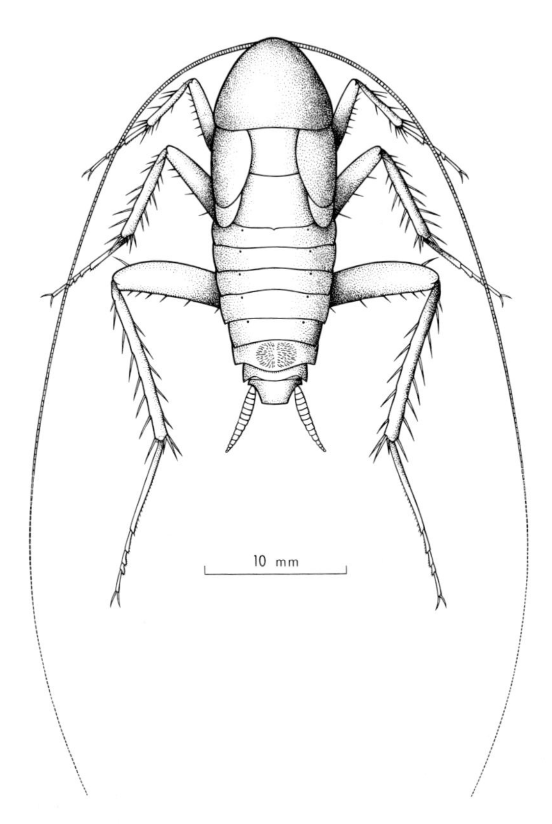

Fig. 5.6. A blind cavernicolous cockroach, *Troglo-blattella nullarborensis* (Blattellidae), ♂.
[F. Nanninga]

The last extreme to be noted is the uniformly dark, cold, moist, still interior of limestone caves, to life in which many of the permanent inhabitants *(troglobites)* show remarkable structural and physiological adaptations. So far, only two Australian insects, a carabid beetle and a cockroach (Fig. 5.6), are known as obligate cave inhabitants, although a number of other beetles and some cave crickets are facultatively cavernicolous troglophiles (Moore, 1964).

ADAPTATIONS TO AQUATIC LIFE

Many groups of insects have become adapted to life in water, as may be seen from the following summary.

Early stages aquatic: all Ephemeroptera, Odonata, Plecoptera, Australian Megaloptera, and Australian Trichoptera; many Hemiptera, Coleoptera, and Diptera; a few Neuroptera, Mecoptera, Lepidoptera, and Hymenoptera.

All stages aquatic, though adults can usually leave the water: some Hemiptera and Coleoptera, a few members of other orders.

To these should, perhaps, be added the occasional semiaquatic Blattodea, Orthoptera, etc., as well as Phthiraptera and adult Siphonaptera on aquatic hosts.

Australian aquatic species show the same kinds of adaptations that have been observed in other parts of the world. The Hydrometridae and Gerridae (Hemiptera; Fig. 5.9) are specialized to live exclusively on the water surface, supported by coverings of hydrofuge hairs on the appendages and body. A further step is shown by the Gyrinidae (Coleoptera), which have very short antennae and highly specialized legs, and are able to dive below the surface to elude enemies or to lay eggs. Many groups of insects have become more definitely adapted to life beneath the surface, and this has involved them in problems of locomotion, respiration, and emergence.

Thus, the capacity for dispersive flight is usually retained by adults, but their legs are often modified into paddles by flattening or development of fringes of stiff hairs (Fig. 5.7),

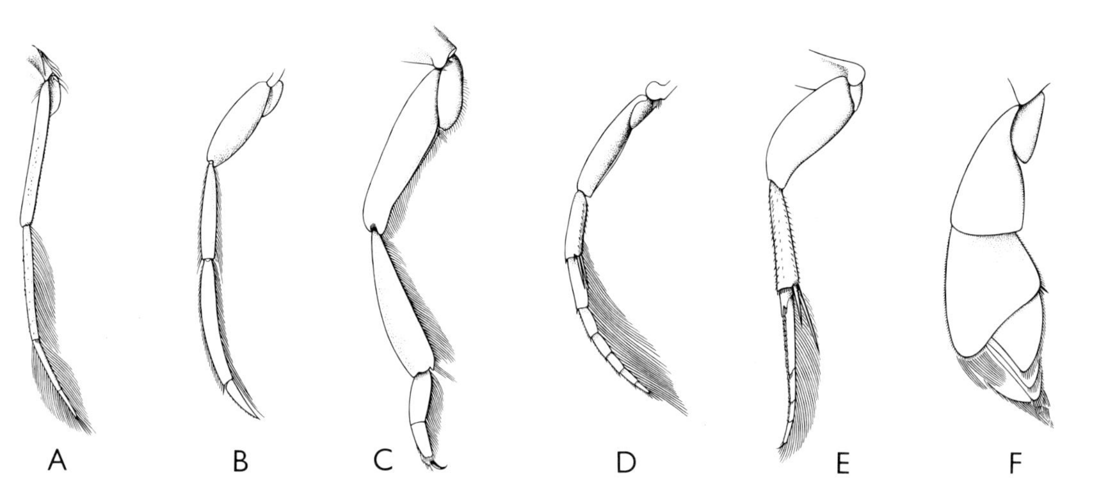

Fig. 5.7. Hind legs modified for movement on or in water. Hemiptera: A, Notonectidae; B, Corixidae; C, Belostomatidae. Coleoptera: D, Dytiscidae; E, Hydrophilidae; F, Gyrinidae. [F. Nanninga]

as in the Gyrinidae already mentioned, Dytiscidae, and Hydrophilidae (Coleoptera) and the Belostomatidae and Corixidae (Hemiptera). Alternatively, the brushes of hairs may permit locomotion on the surface of the water, as in the Notonectidae and some Gerridae (Hemiptera). Elongate larvae, such as those of zygopteran dragonflies and mosquitoes, swim by a fish-like wriggling, whereas the larvae of anisopteran dragonflies can progress by forcing jets of water from the rectum. Reduction or recessing of antennae is usual, and streamlining of the body a prominent feature of fast-moving forms.

The great majority of aquatic insects live in relatively still waters, and few are found in places where wave action is strong. Running streams, in which the disadvantages of turbulence appear to be offset by better aeration and food supplies, are an exception, and representatives of several orders have colonized fast-moving, often torrential streams and the splash zones of waterfalls. Most (e.g. larvae of Odonata, Ephemeroptera, Plecoptera, Megaloptera, Trichoptera, and adult bugs and beetles) live under stones or in other sheltered situations on the bottom; but some face the full force of the stream, and may show special adaptations in flattened bodies, strong legs for clinging, or the development of special adhesive organs, as in the dipterous families Blephariceridae and Simuliidae. They are all able to progress slowly while still maintaining a safe hold on the substrate.

There is reason to believe that all aquatic insects are descendants of ancestors that had impermeable cuticles and breathed air by means of spiracles which were too small to separate a significant amount of oxygen by diffusion from the water. The attendant difficulties in adaptation to aquatic respiration have been solved in a variety of ways, which have been reviewed by Hinton (1953). In many species the spiracles are simply brought into direct contact with the atmosphere. Thus, Nepidae (Hemiptera) replenish their air supply through a long respiratory siphon formed by the apposition of a pair of processes at the end of the body, the tip being kept in the air while the bug forages on the bottom of shallow water. Some Syrphidae (Diptera) show a parallel adaptation in a flexible extension of the abdomen ('rat-tail'), with the functional spiracles at its tip in direct contact with the atmosphere. In mosquito larvae the functional eighth abdominal spiracles are commonly at the apex of a more rigid siphon, by which the larva

Fig. 5.8. Life history of the mosquito, *Culex fatigans*. From left to right: pupa, larva, adult emerging from pupal skin at surface, and egg raft; larva 6 mm long. [Photo by A. J. Nicholson]

hangs from the surface as it breathes (Fig. 5.8). Their pupae, and those of some other Nematocera, respire similarly by means of thoracic trumpets. The larvae of a few Diptera (e.g. *Mansonia* in the mosquitoes, *Chrysogaster* in the Syrphidae) and exotic beetles *(Donacia)* have strong, sharply pointed siphons, which they thrust into the submerged tissues of plants to obtain air directly from intercellular spaces. The pupae of *Mansonia* use their trumpets in the same way, and the cocoons of *Donacia* are kept in communication with the intercellular air through holes bitten in the epidermis of the plant by the full-grown larvae. Indeed, all aquatic pupae respire directly from a gaseous atmosphere, except those of Trichoptera, some Diptera, and some psephenid beetles. More complex adaptations, including the use of an expendable store of air, plastron respiration, and respiration by means of gills, are described on pp. 39–40.

Most species with aquatic pre-imaginal stages leave the water before emergence of the adult. In the exopterygotes it is generally the pharate adult that crawls out on to rocks or vegetation, where the exuviae may often be found in abundance; but in most endopterygotes the last larval instar leaves the water to construct the pupal cell on land. However, Trichoptera, many Diptera, and a few beetles pupate in the water. In Trichoptera the pharate adult subsequently cuts its way out of the cocoon by means of the pupal mandibles, swims to the surface with the specialized pupal legs, and emerges there from the floating skin (Hinton, 1958c). Some Diptera (e.g. mosquitoes, chironomids) also use the pupal exuviae as a platform on which to emerge and expand (Fig. 5.8). A remarkable adaptation is shown by the black flies (Simuliidae). The pharate pupa spins an incomplete silken cocoon under the water, in which it transforms. When ready to emerge, the adult secretes gas beneath the pupal skin, and rises to the surface fully expanded in the bubble that is set free when the skin splits.

Exposure to the air presents as great a hazard to aquatic insects as ephemeral floods do to terrestrial ones. Temporary drying can be withstood by species that have retained an impervious cuticle and spiracular respiration, and others may survive by retracting permeable parts of the cuticle under impermeable parts, or by tolerating considerable reduction in the water content of their tissues. When, however, really arid conditions have to be faced for long periods, there appear to be only two effective kinds of adaptation. One is the possession of an impervious life-history stage which can enter prolonged quiescence or diapause. Some inland mosquitoes of the genus *Aedes* lay their eggs in drying depressions in the ground, where they remain dormant for many months. When rains come again, the eggs hatch and give rise to multitudes of adults before the new pools have had time to dry out (Fenner and Ratcliffe, 1965). Similarly, larvae of the black fly *Austrosimulium pestilens* can develop only in the turbulent waters of flooding inland streams, swarms of blood-hungry adults emerging during the second week after the floods come down. The intervals between plagues may extend to years, and again it seems likely that the egg is the resistant stage (Mackerras and Mackerras, 1948b).

The other special adaptation is *cryptobiosis*, which has so far been studied in detail

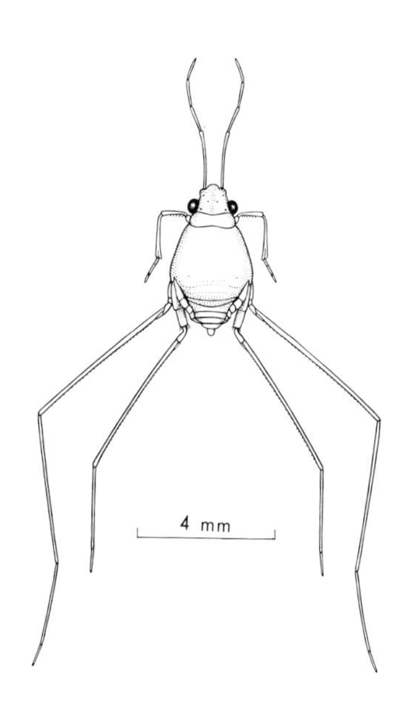

Fig. 5.9. *Halobates mjobergi* (Hemiptera-Gerridae), one of the few kinds of insects able to exist on the open ocean. [F. Nanninga]

only in an African chironomid, though it is known to occur also in unrelated species of the same family in W.A. Larvae of the African *Polypedilum vanderplanki* live in rock-pools that are subject to drying and intense insolation, and they have been found experimentally to tolerate complete desiccation for periods up to several years, quickly rehydrating and becoming active again when immersed in water (Hinton, 1960b). In the dried state all cellular activities appear to be suppressed, and the larvae can withstand extremes (e.g. +102 to −270°C) that would be immediately fatal to normal insects.

Colonization of Marine Habitats

Although they have wider osmotic tolerances than vertebrates (p. 54), aquatic insects appear to be severely limited by the osmotic pressure of the water in which they live. Some halophilous species have met this problem by evolving a hydrofuge covering or a cuticle impermeable to ions. Otherwise it has been met by the elimination of osmoregulatory structures, reduction of chloride intake through the gut, and excretion of hypertonic solutions by the Malpighian tubes and rectal glands. Nevertheless, even inland saline waters have a very impoverished insect fauna consisting mainly of Hemiptera, Coleoptera, Diptera, and Trichoptera. In the sea there are additional hazards (Mackerras, 1950), including turbulence and changing direction of currents, varying hydrostatic pressure in the intertidal zone, abundance of predators, and competition with a long-established fauna that includes many other arthropods.

Estuaries are a transition zone between fresh and salt water, and gerrid bugs of the genus *Halobates* (Fig. 5.9) are widely distributed. The fauna of the mangrove zone is more conspicuous, because it includes viciously biting mosquitoes and Ceratopogonidae, and there is also a small beach fauna of seaweed breeders, ceratopogonids, and chironomids.

On the ocean beaches pale, effacingly coloured, quick-moving Diptera (mainly Asiloidea and Tabanidae) and cicindeline beetles occur on the drier sand, and the larvae of at least some of the Diptera live in the sand between the tide-marks. Near the water the zone of decaying seaweed is occupied by a more numerous 'wrack fauna' consisting mostly of cyclorrhaphous Diptera, carabid and staphylinid beetles, and heteropteran bugs. All these insects have become more or less adapted to contact with salt water.

A different fauna is found on the rocky coastal platforms. In the supralittoral zone clefts in the cliffs are inhabited by *Allomachilis* (Archaeognatha) and the rock-pools of extremely varying salinity by the larvae of a few Diptera, especially those of the mosquito *Aedes australis*, which is widespread geographically, but strictly limited to this zone. The more truly marine insects are in the intertidal zones where the pools are in regular communication with the sea. There are a few bugs (including *Halobates*) and beetles, a caddis-fly, *Philanisus plebeius*, which breeds among coralline algae where its early stages are well hidden, and Tipulidae (Diptera) in associated habitats. The adult

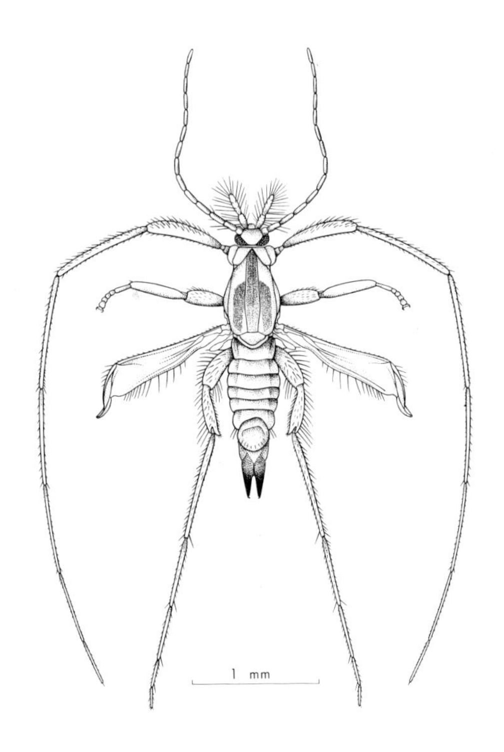

Fig. 5.10. Male of *Pontomyia* sp. (Diptera-Chironomidae), which skates on the sea, propelled by its modified wings. [F. Nanninga]

caddises and tipulids flit about among the rocks on the platform. The most characteristic species, however, are clunionine Chironomidae, the larvae and pupae of which live among the algae in the pools. The adults are usually found close to the pool margins, and they show a marked trend towards aptery in the females and brachyptery in the males. Other groups of terrestrial arthropods have also invaded the intertidal zones—mites, a spider *(Desis)*, and at least three genera of Collembola in Australia, and also a pseudoscorpion and chilopods in other parts of the world.

The coral reefs north of the tropic have an essentially similar insect fauna to the rocky platforms, differing mainly in the presence of minute bugs and beetles in the soft coral rock and the prominence of the extraordinary chironomine flies of the genus *Pontomyia* (Fig. 5.10). The larvae and pupae of these midges usually occur in coral or sandy pools among *Halophila* or *Zostera* (themselves among the few flowering plants that have colonized the sea). The adults emerge from the pupal cuticle at the water surface, where the males skate around, propelled by their modified wings and carrying the wingless, almost legless females during mating (Tokunaga, 1932).

In the open sea we find only *Halobates* and related genera skating on the surface, where they feed on animal remains and lay their eggs in floating debris, and *Pontomyia* males occasionally off the Queensland coast.

ASSOCIATIONS WITH PLANTS

Plants provide food for an enormous fauna of insects, which are described as *polyphagous* if they attack a wide range of species, *oligophagous* if they feed on only a few species, and *monophagous* if they have only one species of food plant. An example of polyphagy is afforded by the scale insect *Ceroplastes rubens* feeding on hundreds of different host plants. *Paropsis atomaria* (Coleoptera) would qualify as an oligophagous species, as it feeds on a limited range of species of *Eucalyptus*, whereas the larva of the butterfly *Thysonotis*

Fig. 5.11. *Eucalyptus radiata* (foreground) defoliated by *Didymuria violescens* (Phasmatodea-Phasmatidae); Victoria. [Photo by Z. Mazanek]

hymetus is monophagous, in that it eats only 'red ash', *Alphitonia excelsa.*

Plants of all types are hosts to leaf-feeding insects, the commonest form of attack probably being the piecemeal consumption of leaves. Some leaf-feeders are spectacular defoliators: *Didymuria violescens* (Phasmatodea) frequently strips thousands of acres of eucalypt forest in N.S.W. (Fig. 5.11), the migrating swarms of *Chortoicetes terminifera* (Orthoptera) make great inroads into pastures and crops in eastern Australia, and *Hednota pedionoma* (Lepidoptera) may consume up to 50 per cent of pastures in parts of W.A. Disturbance to the environment by the activities of man are thought to be responsible for some defoliators assuming plague proportions, but other cases are not understood. It is generally even more difficult to say why the activities of the vast majority of leaf-eaters remain relatively inconspicuous, but we have clear indications for some species. Thus, the weevil *Gonipterus scutellatus* feeds on eucalypts, but attracts little attention in Australia. When it was accidentally introduced to South Africa, however, it caused havoc in the eucalypt plantations, until it was brought under control by an egg-parasite, *Patasson nitens* (Hymenoptera), which was introduced in turn from Australia. The weevil is obviously maintained in balance with its host plant by parasitization, but in other cases other factors, such as host resistance, may play a part.

Some phytophagous insects, like *Caliroa cerasi* (Hymenoptera) and the earlier instars of limacodid moths, do not consume the entire thickness of the leaf, but browse on one surface, leaving the other intact. As a further specialization, insects of many types feed only between the intact upper and lower epidermis of leaves, in a 'mine' which has a characteristic appearance for the species concerned. Leaf-miners are found chiefly in the Lepidoptera, Coleoptera, Hymenoptera, and Diptera, and the larvae (except in the Diptera) tend to become flattened in accommodation to the dimensions of the leaf. A notable Australian example is an undescribed moth, the larvae of which mine in the leaves of *Eucalyptus*

Fig. 5.12. Blotch mines of the jarrah leaf miner (Lepidoptera-Incurvariidae) in leaves of *Eucalyptus marginata*. The holes (mean length 4 mm) are sites from which larvae cut discs for pupal cases. [Photo by M. M. H. Wallace]

marginata (Fig. 5.12) in such numbers that vast areas of the Western Australian jarrah forest frequently appear as if scorched by fire.

The feeding of Hemiptera involves negligible mechanical injury, and may be so precisely directed that the stylets penetrate only by a microscope channel to the phloem. However, the effects are sometimes greatly aggravated by the toxic action of salivary secretions, and considerable leaf injury may result. Sucking mouth-parts are also well adapted to the transmission of virus diseases, and some species are consequently of more importance as plant-disease vectors than for the direct damage they inflict by feeding. For instance, the leafhopper *Austroagallia torrida* transmits the disease known as rugose leaf curl to plants of eight families. The females convey the infection transovarially to their offspring, which are infective on hatching.

Thrips, which exhibit a type of feeding intermediate between chewing and sucking, may cause extensive direct damage to plants, and *Chaetanaphothrips signipennis* is a major pest of bananas in Queensland.

There is a credit side to the picture of insect attack on foliage, in the beneficial results derived from introducing oligophagous or monophagous insects from the homeland of plants that have become weeds in Australia. The classic example is the moth *Cactoblastis cactorum* (Fig. 36.39I), which freed some 50,000,000 acres of Queensland (Fig. 5.13) from virtual dominance by prickly pear *(Opuntia inermis)*. Beetles of the genus *Chrysomela* have also materially reduced stands of St John's wort *(Hypericum perforatum)* in Victoria (F. Wilson, 1960).

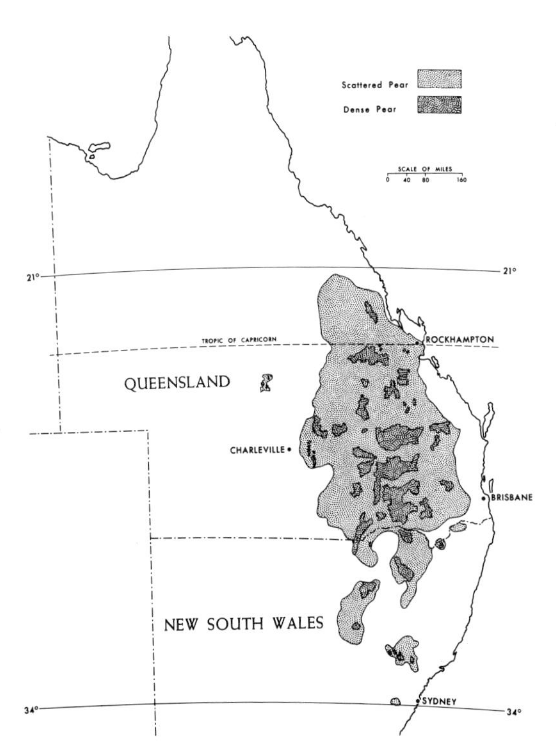

Fig. 5.13. Parts of Queensland and N.S.W. formerly occupied by *Opuntia inermis*. The activity of *Cactoblastis* restored land use over almost all of this area.

Fig. 5.14. *Apiomorpha pedunculata* (Hemiptera-Eriococcidae), ♀ from a eucalypt, and its gall; insect 25 mm long. [Photo by A. J. Nicholson]

Many phytophagous insects induce the production of abnormal growth reactions ('galls'; Fig. 5.14) in the tissues of the host plants, inside which they live and feed. Galls may occur in leaves, stems, flowers, or roots, and their shape and location are often characteristic of the plant and insect species concerned. The gall is entirely a product of the plant, developing in response to a chemical stimulus from the secretions of the insect. Gall insects are found principally in the Cecidomyiidae (Diptera) and Cynipidae (Hymenoptera), but in Australia some striking galls are formed by the endemic Apiomorphinae (Hemiptera), and a few gall-formers occur in the Aphididae (Hemiptera), Thysanoptera, Coleoptera, and Lepidoptera. The Agaonidae (Hymenoptera) form hidden galls in the flowers of figs, and there are many other gall-forming Chalcidoidea.

Many galls on leaves, stem tips, leaf buds, and flower buds of *Eucalyptus* are caused by the combined action of nematodes *(Fergusobia)* and the larvae of small flies of the genus *Fergusonina* (Fig. 5.15). Fertilized

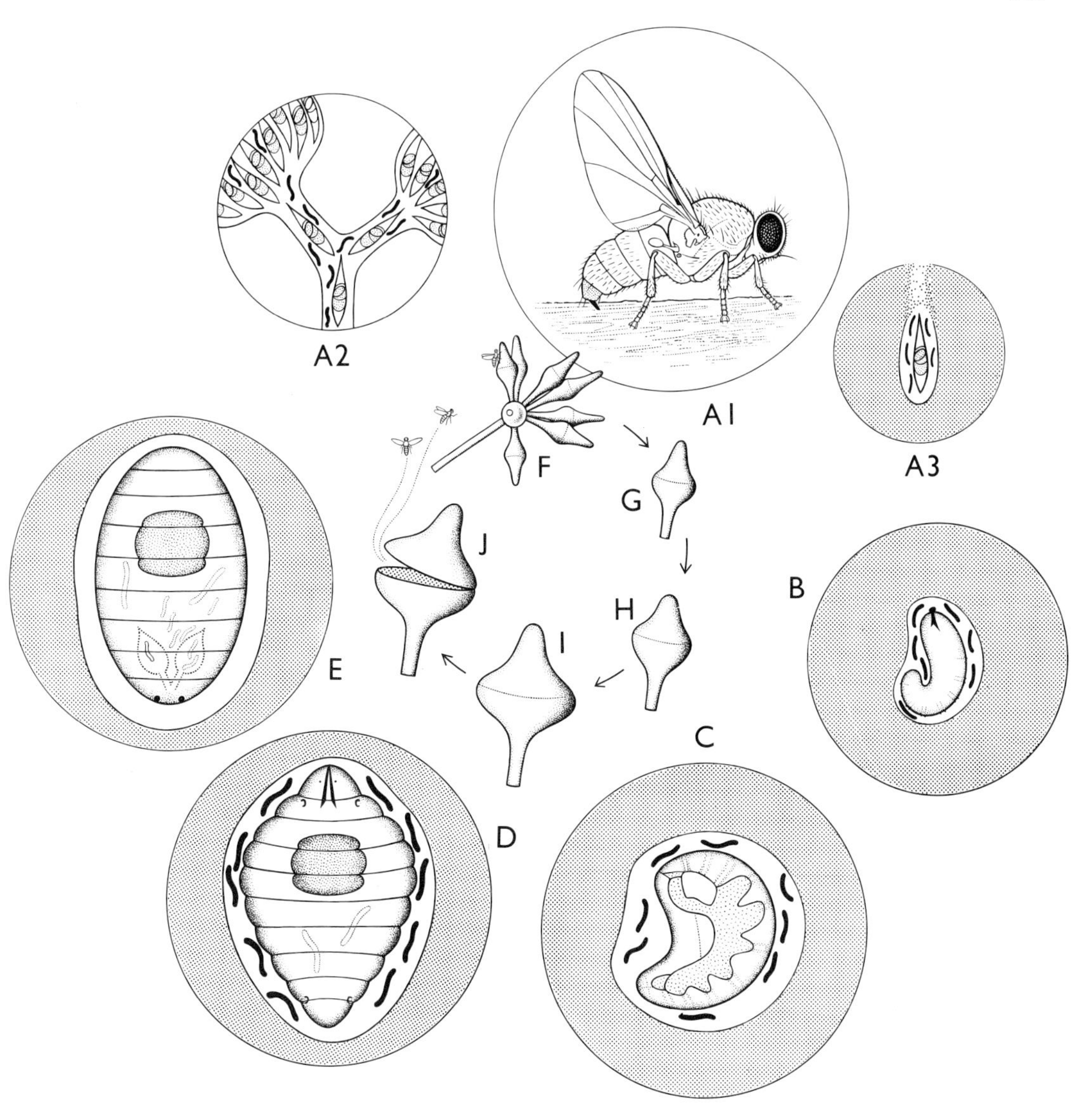

Fig. 5.15. Life cycle of *Fergusonina nicholsoni* (Diptera-Fergusoninidae) on *Eucalyptus macrorhyncha*: A1, female laying in flower bud; A2, eggs and nematode larvae in oviducts and vagina; A3, egg and nematode larvae in plant tissue; B and C, first and second instar larvae, respectively, with parthenogenetic commensal nematodes; D, third instar larva with sexually reproducing nematodes–two fertilized females (dotted) have entered body of the larva; E, puparium, in which larval nematodes are invading the ovary rudiments (dotted) of the pharate adult; F–J, corresponding stages in the galling of the flower bud–in J the operculum of the bud has lifted, permitting the adult flies to escape; flower buds in F 12 mm long. [F. Nanninga]

female nematodes enter the body cavity of advanced female fly larvae, and give rise to larval progeny, which invade the oviducts, and so pass out with the eggs when they are laid into eucalypt tissues. The nematodes then invade the tissues of the plant with the fly larvae, and multiply parthenogenetically inside the developing gall. Males are produced in autumn and winter, and fertilize females, which then complete the cycle by

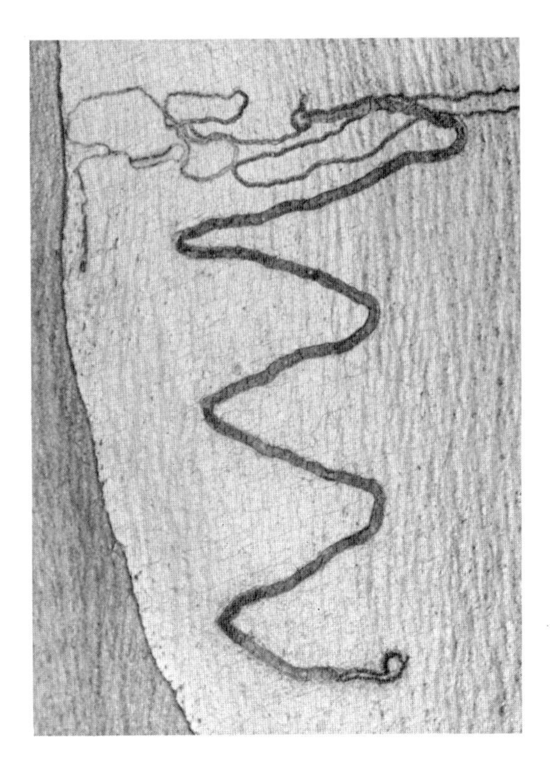

Fig. 5.16. The 'scribble' (mine) of *Ogmograptis scribula* (Lepidoptera-Yponomeutidae) on eucalypt bark; overall height of scribble 11 cm. [Photo by M. S. Upton]

invading the body cavity of a fly larva. This association may be regarded as a true symbiosis (Currie, 1937).

The bark and timber of living trees are attacked by a wide range of insects. The characteristic zig-zag 'scribbles' (Fig. 5.16) which are seen on the trunks of *Eucalyptus haemastoma* and a few other smooth-barked eucalypts are the empty tunnels of the larvae of a small moth, *Ogmograptis scribula*, which had burrowed at the junction between the old and new season's bark before the old bark was shed. The ambrosia beetles (p. 512) are also responsible for some very characteristic tunnelling systems between the bark and sapwood of eucalypts. The parent female excavates an egg-tunnel, lays eggs in niches chewed at regular intervals along its walls, and each larva forms a burrow extending at right angles to the oviposition tunnel. The parents remain with the growing brood, and cultivate an ambrosia fungus, which spreads into the galleries and serves as food for the larvae.

The eggs of insects with wood-boring larvae may be laid into the wood by means of a piercing ovipositor (Hymenoptera), or on the bark, sometimes in excavations specially chewed by the female (Coleoptera), or else they may be laid into exit holes of an earlier generation (Lepidoptera). Wood-boring larvae necessarily have very strong mandibles and a well-sclerotized head capsule. In keeping with their burrow-dwelling habit, they tend to be pale and relatively hairless. Adult beetles can usually chew their way out of the burrows after emergence from the pupal cuticle; but the larvae of Lepidoptera make an exit hole, lightly plugged with web and frass, out of which the pharate adult wriggles far enough to permit emergence. It is not uncommon to find empty hepialid and cossid pupal sheaths protruding from the openings of the burrows in tree trunks (Fig. 5.17). The abundance and long-term availability of food provided by tree trunks permits growth to a large size, and the larvae of *Xyleutes boisduvali* (Cossidae) which burrow in the trunks of eucalypts, attain a length of 18 cm. In addition, the low

Fig. 5.17. Pupal exuviae of *Xyleutes* sp. (Lepidoptera-Cossidae) protruding from larval burrow in a wattle; newly emerged adult moth is about 50 mm long. [Photo by I. F. B. Common]

nutritive value of wood tends to result in long life cycles.

Living trees are also attacked by some species of termites, of which the most devastating is *Coptotermes acinaciformis*. Newly mated royal pairs gain access to a living tree through damaged areas, and the colony develops in an excavation in the heartwood, into which soil may be carried as a supplementary building material. Ventilation is achieved by shafts emerging on protuberances on the bark, and underground tunnels radiate to other living trees, which are then subject to attack by way of their root systems, even if completely free from mechanical or fire damage.

One of the most significant functions of insects in relation to plants is the part they play in pollination. Insects (like nectar-feeding birds) have evolved along with the flowering plants, each influencing the evolution of the other. The role of bees and other Hymenoptera in cross-fertilizing flowers is well known, but an important part is also played by a wide variety of nectar-seeking insects, such as blow flies, beetles, and moths, and the gall-forming Agaonidae mentioned earlier play a vital part in fertilizing the figs they inhabit. Plants show a remarkable range of adaptations to ensure fertilization. Orchids of the genus *Cryptostylis*, for example, apparently emit an odour so closely resembling the sex attractant of *Lissopimpla* species that the male wasps eagerly attempt to copulate with the flower (Frontispiece). In so doing they become loaded with adhesive pollinia (Fig. 5.18), from which pollen is transferred to other flowers during subsequent attempts at copulation (Coleman, 1928).

Fig. 5.19. *Cephalotus follicularis*, an insectivorous plant from south-western Australia.
[Photo by Vincent Serventy]

Fig. 5.18. Male of *Lissopimpla excelsa* (Hymenoptera-Ichneumonidae) bearing pollinia of *Cryptostylis erecta*, on which it is resting after having attempted copulation; male 11 mm long, excluding antennae.
[Photo by A. J. Nicholson]

Some plants, or parts of them, serve as shelters for insects, one of the most striking being the ant plant, *Myrmecodia beccarii*, an epiphyte on *Melaleuca*. It has large pseudobulbs containing cavities communicating with the exterior by small holes. The cavities are nearly always occupied by colonies of ants of various species, but the significance of this to the plant is obscure. They may also provide shelter for the myrmecophilous larvae of the butterfly *Hypochrysops apollo*.

Perhaps in response to the abundance of

insects, diverse kinds of plants have become insectivorous. *Cephalotus follicularis*, a small plant of rosette habit which grows in swampy places in south-western Australia, is an example. Some of its leaves are in the form of lidded pitchers up to 5 cm in height (Fig. 5.19). These serve as efficient traps, as they have glandular areas attractive to insects, which then find difficulty in escaping past an inturned collar and claw-like projections. Many are eventually drowned and digested in liquid at the bottom of the pitcher (Lloyd, 1942). Leaves of the common sundews (Droseraceae; Fig. 5.20) bear numerous small tentacles secreting droplets of insect-snaring mucilage which enfold and digest the victims. *Byblis gigantea* (Byblidaceae) in south-western Australia also has pedunculate mucilage glands for entrapping insects. The plasticity of the insects is, however, reflected in the fact that some species have become adapted to exploit the food sources afforded by the prey of carnivorous plants. For instance, *B. gigantea* is inhabited by a flightless, predatory bug, *Setocoris bybliphilus*, which moves around with impunity among the mucilage droplets and sucks the juices of trapped flies (Lloyd, 1942). The larvae of some pterophorid moths feed on sundews, and the true pitcher plants (*Nepenthes* spp.) of Cape York not only capture prey but provide breeding grounds for unusual mosquitoes and other Diptera.

Fig. 5.20. A sundew, *Drosera peltata*; unopened leaf buds at top; extended leaf right; 'tentacles' of leaf on left have closed over a trapped midge; height 26 mm. [Photo by J. Green]

Finally, insects play a prominent part in the disintegration and decomposition of dead wood, plant debris, fungi, and animal faeces, and hence in the cycling of plant nutrients. Embioptera, larvae and adults of beetles, larvae of Lepidoptera, and many other insects live on leaf mould, and so speed up the destruction of vegetable matter. Their tunnelling aerates the soil, improves its moisture-holding capacity, and assists in the burial of the frass (see also p. 124).

ASSOCIATIONS WITH VERTEBRATES

Insects are an important item in the diet of a considerable number of vertebrates; but their consumption, deliberate or accidental, can also be harmful. Certain species of cockroaches and cockchafer beetles, for

Fig. 5.21. Sores on throat, behind shoulder, and on lower flank of the cow are due largely to rubbing and licking to alleviate irritation of buffalo-fly bites; Malanda, Queensland. [Photo by K. R. Norris]

instance, are intermediate hosts for nematode worms of domestic animals (see also p. 122). Moreover, many cattle in Queensland die as a result of eating the larvae of a sawfly, *Lophyrotoma interrupta*, which contain a toxic substance. Large numbers of these larvae sometimes perish from starvation at the foot of their food tree, *Eucalyptus melanophloia*, after stripping it of leaves, and cattle consume them avidly, either alive or at any stage of decomposition (Roberts, 1952).

Blood-feeding is a habit that has evolved independently in more than a dozen different groups of insects. It is a feature particularly of sucking lice, fleas, and various adult Diptera, but the larvae of one Australian fly, *Passeromyia longicornis*, suck the blood of nestling birds, and the Cimicidae and some exotic Reduviidae (Hemiptera) are noted blood-suckers. Haematophagous insects include some serious pests, chiefly on account of the irritation caused by the bites rather than the loss of blood involved. Great damage is caused in this way by the introduced, dung-breeding buffalo fly, *Haematobia exigua*. During the wet season in northern Australia, every beast in a herd may carry thousands of flies, which bite several times a day, and the resulting irritation leads to the formation of sores (Fig. 5.21) and causes great

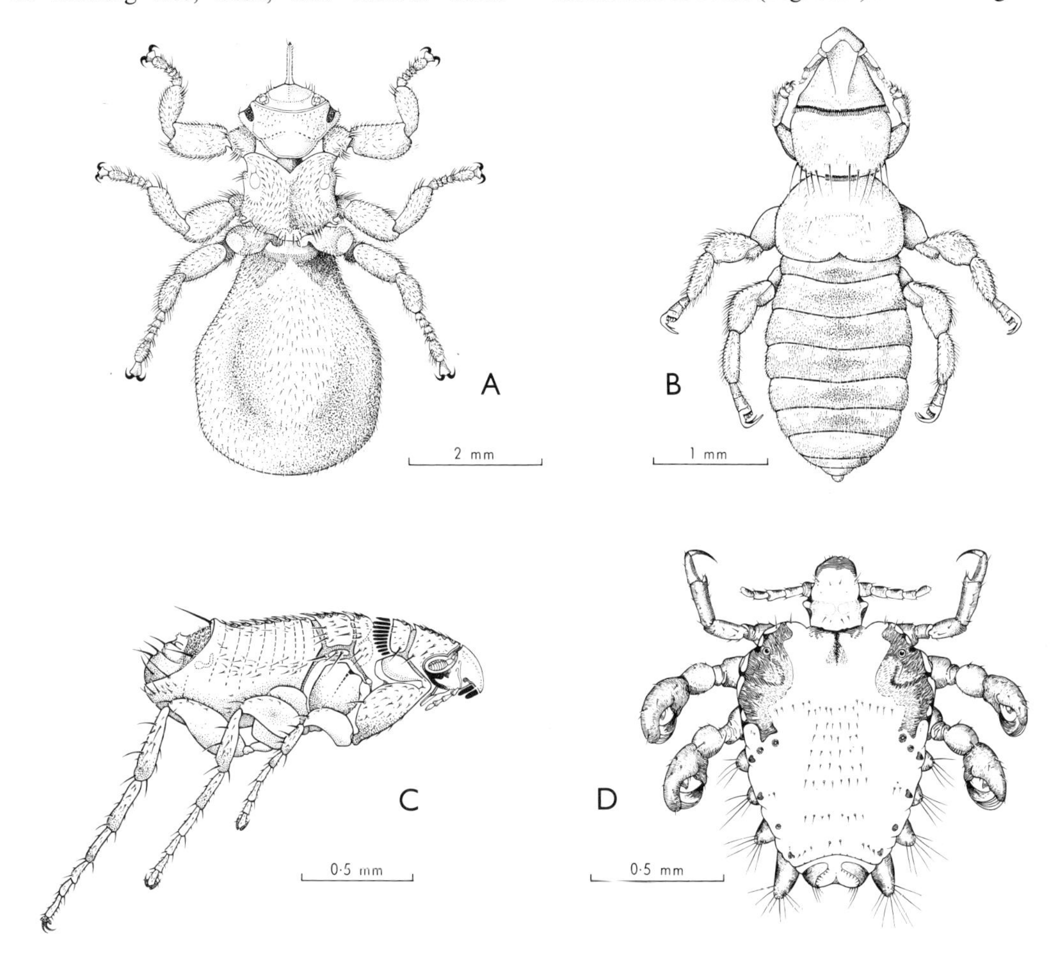

Fig. 5.22. Typical ectoparasites: A, *Melophagus ovinus* (Diptera-Hippoboscidae); B, *Adroctenes magnus* (Hemiptera-Polyctenidae); C, *Lagaropsylla mera* (Siphonaptera-Ischnopsyllidae); D, *Pthirus pubis* (Phthiraptera-Pediculidae). [T. Nolan]

loss of condition. Similarly, the bites of *Austrosimulium pestilens* have been known to cause deaths of stock and wild mammals in central and western Queensland.

The blood-sucking habit provides an opportunity for insects to become vectors of a fairly wide range of blood and tissue parasites. Mosquitoes provide the classical examples, although they play only a minor role as disease transmitters in Australia. There appear still to be a few centres of endemic malaria in the Northern Territory and the Torres Strait Is.; filariasis has almost disappeared; the dengue virus is an occassional visitor; the MVE (Murray Valley encephalitis) virus spreads south in favourable seasons; and several other arthropod-borne viruses of undetermined significance have been discovered in Queensland (Doherty, 1964). In these infections there is some form of cyclical development of the parasite in the mosquito. An important example of mechanical transmission (not involving development in the vector) is myxomatosis of rabbits, which is spread by several species of mosquito and sometimes by other blood-suckers (Fenner and Ratcliffe, 1965). Rat fleas spread murine typhus in country centres, and are potential plague hazards in seaports; but neither lice nor *Phlebotomus* (Diptera) are vectors in this country. There is usually a considerable degree of specificity in these host-parasite associations, an interesting example being that one filarial worm of the dog *(Dirofilaria immitis)* is transmitted by mosquitoes, whereas the other *(Dipetalonema repens)* is transmitted by fleas, a fact that caused confusion in the literature for many years.

Many ectoparasitic insects have lost the power of flight, and wings are absent in parasitic earwigs, the bat-parasitic polyctenid bugs, lice, fleas, and some parasitic Diptera. There is also a tendency to the development of spines and very large claws or pincer-like legs for maintaining a grip on the host (Fig. 5.22). The antennae are often reduced in size and may be recessed into grooves, so that they do not impede the movement of the parasite through feathers or hair. The eyes also are reduced in parasites that do not leave their hosts.

Ectoparasites (and free-living blood-suckers too) show a considerable degree of host specificity, and in lice this may even extend to a predilection for one part of the body. For example, man is subject to infestation by two species of lice, *Pediculus humanus* and *Pthirus pubis*. The former has two races, one of which infests the hair of the head and the other the clothing, whereas *P. pubis* is adapted to life in the coarse hair of the pubic region. Sheep may be infested by three species of lice, which prefer the trunk, face, and legs, respectively. Not a great deal is known of the limitations of the ectoparasites of Australian birds and mammals, but two native species of 'stickfast' fleas, *Echidnophaga myrmecobii* and *E. perilis*, have

Fig. 5.23. Eggs of pigeon body louse, *Hohorstiella lata* (Menoponidae); eggs 0·9 mm long.
[Photo by I. Roper]

become pests of cats and dogs in inland Australia. The specificity of fleas, in general, is not as marked as that of lice.

Life cycles show a variety of adaptations to parasitism. Lice cement their eggs firmly to the vestiture of the host (Fig. 5.23), and the larvae are parasitic from the moment of hatching. This is true of an Australian species of flea also (p. 651), but most fleas deposit their eggs in the lair or nest of the host, where the apodous larvae live on detritus containing much undigested blood passed in the faeces of the adult. The reproductive cycle of the European rabbit flea is geared to that of its host by hormones taken up with the blood meal (Rothschild, 1965). The 'pupiparous' Diptera have reconciled mobile hosts with legless larvae by becoming viviparous. One larva at a time is nourished inside the body of the female and deposited when ready to pupate. This habit enables the group to utilize large, free-ranging ungulates as hosts, a niche virtually denied to the fleas.

In fleas and hippoboscids the parental food is also utilized for the nourishment of the young. The reverse may occur in insects with larvae that are endoparasites of large mammals. In at least four families of higher Diptera the larvae accumulate sufficient food reserves from the generous amounts of host tissue available to them to enable the free-living adults to reproduce without the need for seeking food. In fact, adult mouth-parts

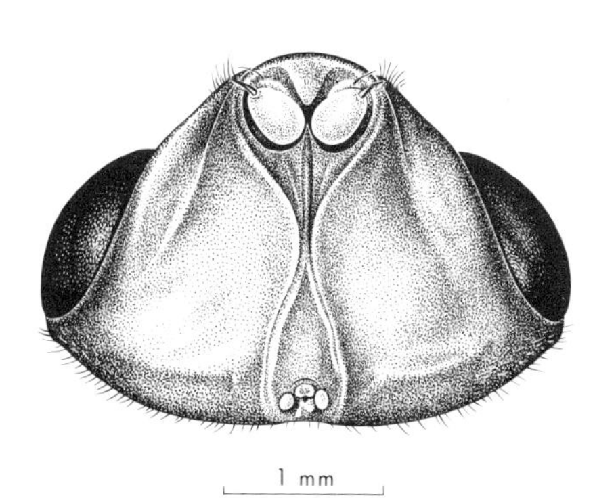

Fig. 5.24. Head of *Tracheomyia macropi* (Diptera-Oestridae) showing vestigial mouth-parts. [F. Nanninga]

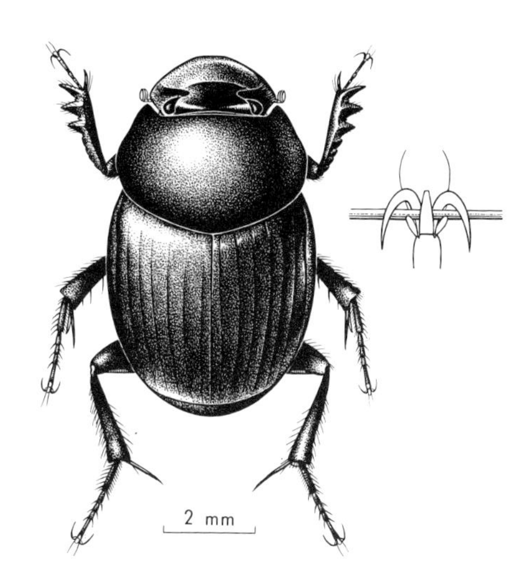

Fig. 5.25. *Macropocopris* sp. (Coleoptera-Scarabaeidae), with detail of claws showing specialization for grasping hairs. [F. Nanninga]

in these groups have become atrophied, and feeding is no longer possible. The only native fly of this type is *Tracheomyia macropi* (Fig. 5.24) the larvae of which live in the trachea of kangaroos, but the introduced *Oestrus ovis* of sheep and *Gasterophilus* spp. of horses afford other examples.

In all these groups, therefore, the host provides nutriment for both the immature stages and the adults. The same is true of the free-living buffalo flies which alight on cow dung to lay their eggs within seconds of its being dropped. Scarabaeid beetles of the genus *Macropocopris* (Fig. 5.25) are another example. They have enlarged claws with which they cling to the fur near the anus of kangaroos and wallabies, where they feed on the secretions. They are thus in a position to mount the droppings as they are extruded, and to commence burying them immediately as food for their larvae (Arndt, unpublished).

Larvae of chloropid flies (*Batrachomyia* spp.) are subcutaneous parasites of a number of species of frogs in Australia, but when they leave the host to pupate the frog may be little the worse for the infestation. The blow fly *Chrysomya bezziana*, which attacks stock in New Guinea, is exclusively parasitic on

Fig. 5.26. Larvae of *Lucilia cuprina* (Diptera-Calliphoridae) in fleece over a lesion they have caused on a living sheep; larvae 12 mm long.
[Photo by C. Lourandos]

mammals, making lesions so extensive that the death of the host may ensue. Something in the nature of a half-way stage to tissue parasitism is shown by the sheep blow flies. They deposit their eggs or larvae on the fleece where it has become soiled by faeces, urine, or merely rain or dew, and the larvae attack the skin and underlying tissues of the sheep. The Australian species of *Calliphora* do this comparatively seldom, as they breed principally in carrion, but in pastoral country the introduced *Lucilia cuprina* breeds more on living sheep (Fig. 5.26) than on carrion, and may be regarded as evolving rapidly towards obligate parasitism (Norris, in Keast *et al.*, 1959).

Consideration of myiasis leads to the question of carcass decomposition, in which insects play an important role. In Australia, where carrion-eating birds and mammals are unimportant, the blow flies, chiefly species of *Calliphora* in southern Australia and of *Chrysomya* in tropical areas, are the predominant primary destroyers of cadavers, although there is a long succession of other flies, beetles, etc., before the skeleton alone is left (Fuller, 1934b).

INSECTS AS HOSTS AND PARASITOIDS

Insects provide a rich field for parasitization, and most of their parasites are themselves insects. Much of this parasitism of insects by insects is of a special type, which results in the death of the host and the total consumption of its internal tissues, in contrast with most other kinds of parasitism, in which it is advantageous to the parasite for the host to survive. It is, in effect, midway between predation and true parasitism, and organisms that exhibit it are known as *parasitoids*. Usually only the larval stages are parasitic, but there are exceptions, like the Strepsiptera, most of which have degenerate females that remain parasitic even when adult.

Parasites of insects may be other insects, arachnids, helminths, protozoa, fungi, and a wide range of bacteria, rickettsias, and viruses (Steinhaus, 1963). Conversely, hosts of parasitic insects include other insects, spiders, ticks, chilopods, terrestrial isopods, earthworms (Fig. 5.27), and snails.

The two great groups of insects that contain entomophagous parasitoids (i.e. those that attack other insects) are the Hymenoptera and the Diptera; but the Strepsiptera are almost exclusively parasitic on insects, and there are entomophagous parasitoids in the Coleoptera and Lepidoptera. The great majority of the hosts of parasitic Hymenoptera and Diptera are Coleoptera and

Fig. 5.27. Earthworm being consumed alive by the third-instar larva of *Calliphora tibialis* (Diptera-Calliphoridae); arrow indicates approximate position of head of larva inside tissues of worm; remains of worm 55 mm long.
[Photo by C. Lourandos]

Lepidoptera, but there is also a wide parasitization of many other orders, including the Hymenoptera and Diptera themselves (Clausen, 1940; Sweetman, 1958). Larvae are the chief hosts, with pupae, eggs, and adults in descending order of frequency.

Some parasitoids lay their eggs in the eggs of the host. In some minute chalcidoid and proctotrupoid wasps a single host egg supplies sufficient nourishment for one or more parasites to reach maturity. In larger forms oviposition in eggs permits the parasite to hatch immediately after eclosion of the host larva, and so to gain competitive advantage over other parasites. *Opius oophilus*, for example, which was introduced into Australia in an attempt to control the fruit fly *Strumeta tryoni*, lays its eggs in the eggs of fruit flies, but the parasite does not hatch until the fruit-fly larva begins to grow rapidly.

Eggs laid on the bodies of hosts may be stalked or sessile, but they are always firmly cemented to the integument. More usually the ovipositor is used to insert the egg into the body cavity of the host through unsclerotized areas of the integument. The ovipositor is sometimes of relatively great length to enable the female to penetrate a considerable distance into the material sheltering the host (Fig. 5.28).

Some parasitoids merely deposit eggs in a place where they are likely to be eaten by the host. Thus, some Trigonalidae (Hymenoptera) lay their eggs in the tissues of leaves on which sawfly larvae feed. Alternatively, the first-instar larvae may be active planidia or triungulins (p. 106), which eventually find their way to a host, sometimes employing a winged insect as a vehicle, as when triungulinids of Strepsiptera gain access to larval Hymenoptera by riding on the adult wasps tending the nests.

Insect parasitoids may be ectoparasitic, when the larva is external to the body of its host but attached to it by the mouth-parts.

Fig. 5.28. *Rhyssa persuasoria* (Hymenoptera-Ichneumonidae) probing a pine tree in search of *Sirex* larvae (Hymenoptera-Siricidae); the disengaged ovipositor sheath points downward; the entire length of the ovipositor (35 mm) can be sunk into the timber. [Photo by K. L. Taylor]

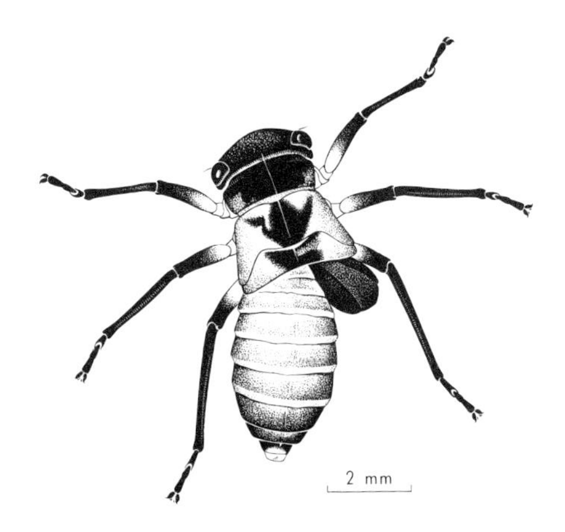

Fig. 5.29. Larva of a dryinid wasp (Hymenoptera) ectoparasitic upon a nymph of a eurymelid (Hemiptera-Homoptera). [F. Nanninga]

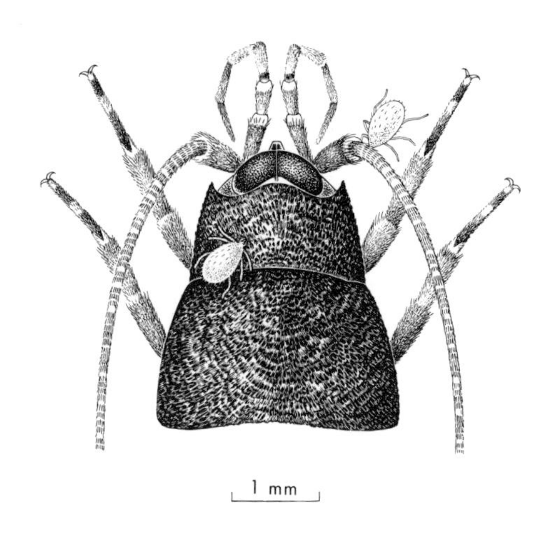

Fig. 5.30. Larval erythraeid mites (Acarina) ectoparasitic upon *Allomachilis froggatti* (Archaeognatha); mites about 0·75 mm long.
[F. Nanninga]

They are usually protected by being inside the cocoon or pupal cell of the host, but in some forms, such as dryinid wasps which parasitize Homoptera, most of the body of the larva is exposed on the surface of the free-living host (Fig. 5.29). The larvae of most parasitoids, however, live inside the body cavity of the host, sometimes establishing a respiratory connection with the atmosphere by a special opening in its integument, or else perforating one of its larger tracheal trunks with a respiratory siphon. Parasitoids usually avoid the vital organs in the earlier stages of feeding, so that the host remains alive, and even grows, until they reach the final larval instar, when all remaining host tissues are consumed.

Parasitoids are themselves frequently subject to parasitization, or, as it is called, *hyperparasitization*. Thus, the actual (primary) host of the larva of some species of *Taeniogonalos* (Hymenoptera) is the larva of a hymenopterous parasite (*Enicospilus* sp.) living in the larva of *Perga dorsalis* (Hymenoptera), which is regarded as the secondary host of the *Taeniogonalos*.

Arachnid parasites of insects are chiefly mites (Acarina). The first-instar larvae of Erythraeidae are ectoparasitic on various insects. These small red mites attach by their mouth-parts to the host (Fig. 5.30), but drop off to moult when fully fed; later instars are free-living. Other parasitic mites found on terrestrial insects commonly belong to the Gamasidae, whereas aquatic insects are parasitized by larvae of Hydrachnellidae. A common mite on insect larvae, such as those of the grain moth, *Sitotroga cerealella*, is the cosmopolitan *Pyemotes ventricosus*. The females of this species normally engorge on the larvae, and eventually give birth to sexually mature mites; but people handling infested hay may also receive irritating bites.

Some Acarina use insects as dispersive vehicles only, a phenomenon known as *phoresis*. Thus, beetles not uncommonly carry gamasid mites, and certain species of Tyroglyphidae and other families have resistant, non-feeding, sometimes even non-motile, dispersive stages which attach themselves to insects by suckers or secretions. Occasionally pseudoscorpions also attach to insects for transport, and, on the other hand, the larvae of some African black flies are phoretic on freshwater crabs.

Insects serve as intermediate hosts of a few trematodes and cestodes, Acanthocephala, and many nematodes. Some nematodes are important parasites of insects, usually of the larvae, and several species have been used in the United States for the control of scarab beetles. Larval mermithoids are best known in aquatic larvae of Diptera, but they also parasitize Orthoptera and ants, and two species have been recorded from aestivating adults of the bogong moth (p. 781). The even more conspicuous larvae of gordioids (Nematomorpha) are found in terrestrial Blattodea, Orthoptera, and Coleoptera, although their adults, like those of most mermithoids, are aquatic. There has been a considerable amount of parallel evolution in the two groups of worms.

Insects have an intricate web of relationships with micro-organisms, ranging from symbiotes that provide essential nutrients (p. 65) to virulent pathogens. A few protozoa, (e.g. *Nosema*) kill their hosts, and some bacteria and viruses may cause spectacular

Fig. 5.31. *Musca vetustissima* (Diptera-Muscidae) killed by the fungus *Empusa muscae*; conidiophores have protruded through the intersegmental membranes and thrown off the surrounding deposit of conidia; length of fly 5·5 mm. [Photo by J. Green]

mortalities. *Bacillus thuringiensis* is a highly pathogenic bacterium, specific to insects, which is cultured commercially in the United States for the control of pests. In the Australian bush, natural infections with polyhedral viruses are not uncommon in larvae of the moths *Doratifera casta* and *Antheraea eucalypti*. Infected larvae at first become inactive, then flaccid, and die with the body contents liquefied. Fungal parasites are also conspicuous. One of the commonest is *Empusa muscae* (Fig. 5.31), which parasitizes house flies and other Diptera. The mycelium ultimately bloats the fly, so that the stretched intersegmental membranes appear as yellowish stripes, and the conidia form a halo around its body. Species of *Cordyceps* attack chiefly terrestrial larvae of Lepidoptera and Coleoptera. *C. aphodii* causes extensive mortality of *Aphodius howitti*, the slender fruiting bodies growing up several centimetres above the dead beetle larvae in the ground.

Social parasitism, which occurs in the nests of ants and termites, is a different kind of relationship. A wide range of insects of various orders live as *inquilines* in the nests, sometimes as spongers specially adapted to solicit food from the workers, sometimes as thieves taking food regurgitated by one worker for another, sometimes as scavengers, and sometimes as predators of the brood. All types, even the predators, may be tolerated by the ants or termites in return for secretions supplied. Some, however, lack chemical persuasion, and have evolved protective devices to counter the hostility they are likely to experience. For instance, the larvae of *Liphyra brassolis* (Lepidoptera), which are predacious on the brood of the green tree ant, *Oecophylla smaragdina*, in north Queensland, are subject to attack by the ants. To counter this, the larvae have a very tough integument, and the pupae and pharate adults remain within the last larval skin. The emerging butterflies have special deciduous scales that clog the mandibles of attacking ants long enough to permit them to escape from the nest. In addition to these more diverse types of association, some termites are specialized inquilines in the nests of other termites, bees in the nests of other bees, and ants in the nests of other ants.

INSECTS AS PREDATORS AND PREY

A precise definition of predation is almost impossible to frame; but generally a carnivorous organism is rated as *predacious* if it consumes two or more of its prey during its trophic phase, as opposed to the one, or part of one, consumed by a parasitoid, or the one, usually surviving, host attacked by a parasite. Predation is a widespread mode of life in the insects (Sweetman, 1958), taking various forms according to the group in which it occurs. All Odonata and Mantodea are predacious in all postembryonic stages, and predacious adults or larvae occur in about half of the remaining orders. There are even carnivorous larvae in diverse families of the Lepidoptera, which are generally regarded as the prime example of a phytophagous order.

The simplest form of predation is the capture and piecemeal consumption of the prey, well exemplified by Orthoptera of the genus *Paragryllacris*, which simply spring upon other insects and devour them while

Fig. 5.32. *Neoaratus hercules* (Diptera-Asilidae) sucking the juices from a honey bee which it has captured on the wing; fly about 35 mm long. [Photo by A. J. Nicholson]

holding them in their strongly spined legs. More subtle predation is exhibited by the Mantodea, which wait, motionless and inconspicuous, until their prey approaches close enough to be seized by their raptorial fore legs. An analogous mechanism of capture is found in the larvae of Odonata, in which the labium can be shot out and the prey seized in the strong terminal hooks.

Predation on the wing is less common. It reaches its zenith in the Odonata, in which the legs are virtually useless for locomotion, but serve as a very efficient basket for seizing insects on the wing and carrying them to the mouth-parts. Other insects may capture prey while in flight, but they generally alight to devour it, the outstanding examples being some Mecoptera and the Asilidae (Fig. 5.32) and Empididae among the Diptera.

Abundant, sluggish, soft-bodied forms, like aphids and coccids, virtually invite predation, and insects with a wide variety of mouth-parts and body forms take advantage of such freely available and rapidly reproducing food supplies. Some are normal mandibulate forms, like the adults of Hemerobiidae and Chrysopidae (Neuroptera); in others, such as the larvae of the same families, the mouth-parts may be modified into piercing stylets, through which digestive enzymes are pumped into the victim so that its contents may be sucked out. Aphids are also subject to predation by the larvae of Syrphidae (Diptera), which have no particular specialization of the mouth-parts for the purpose, and they may be fed upon by dolichopodid flies, such as *Heteropsilopus cingulipes*, which crush them between the labella.

On the other hand, the mouth-parts of many predators are modified into stout, piercing beaks, as in the Asilidae among the Diptera. The prey is seized by the prehensile legs, perforated by the beak, and the contents of the body are then drained after predigestion. A similar method of feeding is found in the predacious Hemiptera. The beak is particularly powerful in the Reduviidae, as, for example, in *Pristhesancus papuensis* (Fig. 7.7), a large bug which impales honey bees and other visitors to flowers. Predacious insects may inject a toxin which stops the struggling of their prey, and some reduviids have abdominal glands which are believed to exude a secretion avidly sought by ants. This narcotizes the ants, and the bug then feeds on them at leisure (N.Miller, 1956). The striking yellow and black *Ptilocnemus femoratus* of Australia is stated by McKeown (1942) to prey on ants in this fashion. Most aquatic bugs are also predacious, perhaps the

Fig. 5.33. Snare of *Arachnocampa* larva (Diptera-Mycetophilidae); the larva is in upper background in a horizontal position; full-grown larva may be 30 mm long. [Photo by A. J. Nicholson]

Fig. 5.34. Ant-lion pits (Neuroptera-Myrmeleontidae), Katherine, N.T.; in north Australia the long dry season permits the larvae to construct pits in the open; in the south they are generally under cover; largest pits about 35 mm across. [Photo by K. R. Norris]

most formidable being the 'fish-killers', *Lethocerus* spp. (Belostomatidae). These monsters, up to 7 cm long, capture insects, fish, and frogs with their raptorial fore legs, and impale them with their powerful beaks.

It is a curious fact that, although many insects spin web, unlike the spiders few have put it to use in snare construction. Some caddis-fly larvae (Hydropsychidae) construct conical snares of silk opening upstream and serving to entrap small organisms brought down by the current, and a few aquatic dipterous larvae spin rudimentary snares; but almost the only terrestrial snare-spinners in Australia are the larvae of some Mycetophilidae (Diptera). Perhaps the best known are the luminous larvae of *Arachnocampa tasmaniensis*, which live in caves and other damp, dark places, where they spin snares of hanging silken threads studded with mucoid droplets (Fig. 5.33) to entrap small insects attracted by the light produced by the larva.

Another type of snare is constructed by the 'ant-lion' larvae of certain Myrmeleontidae (Neuroptera). They excavate conical pits (Fig. 5.34) in dry, sandy places, by backing into the sand in a circular pattern and flicking off the material that falls on to their head and jaws, so that the sides come to lie at the angle of repose of the sand. Ants or other small insects that tumble into the pitfall are usually seized immediately by the ant-lion, which lies concealed in the sand at the bottom of the pit with its mouth-parts agape. However, if the victim attempts to escape up the sides of the pit, sand is flicked up and either dislodges it or causes the walls of the pit to slip beneath it. Vermilionine Rhagionidae (Diptera) have the same habit in the northern hemisphere, a remarkable example of convergent evolution.

Predation of a different type is exhibited by eumenid and sphecoid wasps, which are not themselves primarily eaters of protein food, but capture it for their larvae. The prey (flies, leafhoppers, caterpillars, grasshoppers, cicadas, or spiders) are stung by the ovipositor, and paralysed by injection of toxic secretions. Insects so treated do not die, but remain fresh long enough to enable the larvae of the wasps to complete their development on them. Vespid wasps, on the other hand, chew up captured insects, and feed them progressively to the larvae in the nest, imbibing some of the juices for their own nourishment as they do so.

Insects provide living food for a wide variety of other animals, including many freshwater and terrestrial vertebrates. Thus, among the Australian mammals there are exclusively or largely insectivorous species in the monotremes (*Tachyglossus*), marsupials (e.g. *Myrmecobius*, *Antechinus*, *Isoodon*), rodents (e.g. *Rattus assimilis*), and bats.

Fig. 5.35. Central Australian Aborigines digging for honeypot ants. [Photo by C. P. Mountford]

Even man exploits insects when they are large or abundant: the aestivating assemblages of bogong moths were sought after eagerly by the Australian Aborigines, who also relished the larvae of wood-boring and root-feeding beetles and moths ('bardee' and 'witchety' grubs), and dug up the replete workers of honeypot ants (Figs. 5.5, 35) to obtain the stored nectar. Invertebrates also exploit insects as food. Spiders, for example, subsist largely on insects, and the great evolutionary radiation of the orb-web spinners in particular was made possible by the development of flight in insects. Many other arachnids and myriapods are predacious largely upon insects.

All this predation, particularly that by vertebrates, has led to the evolution of a variety of mechanisms for avoiding or misleading the hunters. Warning and concealing coloration are dealt with in the next section, but there are other methods of escape. *Thanatosis* is a state of feigned death that is well exhibited by many weevils. If molested, the insect merely drops to the ground, and remains for a period immobile and consequently difficult to detect. *Autotomy*, or the reflex shedding of appendages which have been seized, is an escape mechanism exhibited, for example, by phasmatids. Lepidopterous larvae (e.g. Tortricidae) commonly wriggle violently when disturbed, and either drop to the ground or descend rapidly on threads, remaining suspended for a time before hauling themselves up again. Chemical means of defence are also common. Ants and termites afford well known examples. Coccinellid larvae (Coleoptera) exude distasteful fluids through pores near their tibio-femoral articulation. Bombardier beetles (Carabidae) discharge an explosive vapour (p. 36) from the anus. Larvae of sawflies of the genus *Perga* regurgitate a pungent, acrid concentrate of eucalypt oils when molested (Fig. 5.36). Some bugs secrete repellent substances from thoracic or abdominal glands, and in certain Coreidae these can actually be squirted a centimetre or so. Many lepidopterous larvae have urticating hairs, which are grouped on large protuberances in the Limacodidae.

Fig. 5.36. *Perga* larvae (Hymenoptera-Pergidae), 40 mm long, reacting to rapping on the branch; right-hand specimen has a large globule of regurgitated defensive fluid on the front of the head. [Photo by C. Lourandos]

WARNING AND CONCEALING COLORATION AND MIMICRY

Very important in offence and defence are the devices by which insects advertise their distastefulness or conceal their edibility. These are adaptive responses to predation by animals, especially birds, that hunt by sight rather than by smell or other senses.

The type of coloration, structure, and behaviour that aids concealment of an animal is termed *cryptic*. If the concealment aids the insect to escape the attention of a predator, it is known as *procryptic*, an instance being the stick-like appearance and posture of many Phasmatodea. Procryptic insects may be so coloured and shaped that they merge into the background which they commonly inhabit, and they generally contribute to the illusion by appropriate behaviour. Some moths, for instance, have intricate, disruptive wing markings of wavy lines, which blend with the bark of tree trunks when the moths rest, as they always do, with their wings pressed to the surface and their pattern aligned with the pattern of the bark (e.g. Nicholson, 1927, Plates 13, 14). Procrypsis may also be brought about by resemblance to some object in which the predator is not interested. Thus, young larvae of the butterfly *Papilio aegeus* are olive black with white markings and strongly resemble bird droppings, their sedentary day-time habits contributing to the deception. Larvae of geometrid moths commonly resemble twigs of bushes on which they feed, and their posture heightens the resemblance. The larvae of *Eucyclodes* (Lepidoptera) have several laterally expanded segments which give them a procryptic resemblance to the bipinnate leaves of the *Acacia* on which they feed.

When concealing coloration and behaviour enable a predator to approach its prey unobserved, or, more usually, allow the prey to approach the predator without seeing it, the phenomenon is known as *anticrypsis.* It is exhibited by mantids, which look much like twigs or leaves when quietly awaiting the approach of prey. However, it is likely that this adaptation also enables the mantids to escape bird predation, and is consequently as much procryptic as anticryptic.

An interesting sidelight to procrypsis is that many insects possessing it have, in effect, a second line of defence. If they are forced to break their cryptic posture by the close approach of a predator, they may startle it by a sudden switch to the display of a normally concealed pattern of bright colour. For example, if the mantids are provoked until they are forced to move, many of them then briefly exhibit blazes of colour on the wings or inner surfaces of the fore femora. The female of the mountain grasshopper, *Acripeza reticulata*, is quite inconspicuous in repose, but raises its tegmina when disturbed,

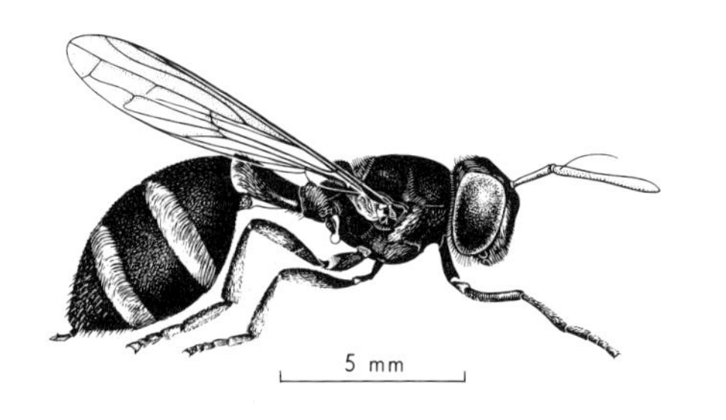

Fig. 5.37. Wasp-mimicking syrphid fly, *Microdon variegatus* (from a photo by A. J. Nicholson).
[F. Nanninga]

displaying a strikingly banded abdomen (Plate 2, L). The extrusible, often forked and brightly coloured organs that occur on the prothorax of the larvae of Papilionidae (Frontispiece), Notodontidae, and some other Lepidoptera have a disconcerting effect, but many of them probably dispense volatile repellent substances as well.

Thus, not all warning coloration is necessarily bluff, and many insects advertise genuine distastefulness or injurious characteristics by permanently displayed bold patterns and flamboyant colours. These *aposematic* patterns, as they are called, are well shown by many pompilid, vespid, and eumenid wasps, which have such potent stings that few vertebrates require many lessons before associating colour pattern with pain. Similarly, it has been demonstrated (see Sheppard, in Kennedy, 1961) that certain brightly coloured northern-hemisphere butterflies are genuinely distasteful to birds, and that the birds learn to recognize them. Consequently, relatively few individuals have to be sacrificed before the species achieves significant protection from attack by any particular predator. Sometimes groups of distasteful or injurious species exhibit a common pattern of warning coloration, which further reduces the sacrifices necessary to educate predators. This 'pooling of resources' is known as *Müllerian mimicry*.

A remarkable consequence of aposematism is the kind of mimicry called *Batesian*, in which the individuals of a species or morph (p. 140) that is devoid of distasteful and injurious properties derive protection by mimicking the colour, form, and behaviour of an aposematically coloured injurious insect (Plate 6). To be successful in such deceptions, a mimetic species must be much less common than the model, so that the protection does not break down through unconditioned predators finding a sufficiently high percentage of their 'warningly' coloured prey to be in fact palatable.

A number of Australian species of Syrphidae and Asilidae (Diptera) exhibit faithful mimicry of stinging wasps (Nicholson, 1927). The colour patterns are reproduced with great exactitude, and the 'waist' of the wasp is often simulated, either by actual constriction of the body, or by countershading to give the impression of constriction. The antennae are usually long and held as in the wasps, and even the longitudinally folded wings of the Eumenidae and Vespidae may be imitated. There is also usually a behavioural mimicry, extending to flicking of the wings and vibration of the antennae which are common in aposematically coloured Hymenoptera. It takes an entomologist with keen eyesight to be sure that the rare individuals of such species as *Microdon variegatus* (Fig. 5.37) encountered in nature are not stinging Hymenoptera.

Other wasp mimics occur in the Cerambycidae (Coleoptera). The long antennae of these forms constitute a good starting point (a basic general resemblance of model and mimic is presumed to be a prerequisite for evolution towards more detailed resemblance), which is heightened by the great abbreviation of the elytra. When the insect is not in flight, these cover the bases of the permanently extended hind wings, and convey the impression of the bulky thorax of a wasp.

The aposematically coloured, orange and black beetles of the genus *Metriorrhynchus* (Lycidae; Plate 6, A), which are distasteful to predators, are abundant in Australia, and serve as models for a wide variety of non-distasteful mimics, especially in the Coleoptera and Lepidoptera. Colour pattern, form, and the pectinate antennae are often faithfully imitated. Many ants are distasteful, or capable of biting or stinging severely, and so it is not surprising that ant mimics occur. For instance both immature and adult stages of the lygaeid *Daerlac tricolor* mimic the ant *Acanthoclinea doriae*, though the deception is produced by quite different anatomical modifications in the nymph and adult (Nicholson, 1927). Other ant mimics occur in Phasmatodea, Orthoptera, and Lepidoptera.

INSECTS AS BUILDERS

One of the most widely used construction materials employed by insects is silk—a secretion that hardens so rapidly on extrusion that it may be produced as a long continuous thread. A very tough structure results when layer upon layer is progressively built up by the spinning insect. Silk secretion has been evolved independently by a number of groups of insects. It may be produced by dermal

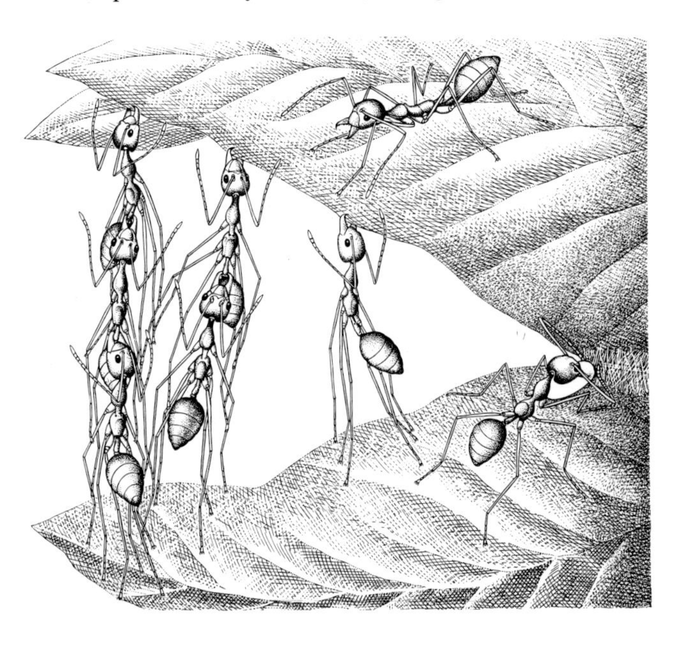

Fig. 5.38. Workers of *Oecophylla smaragdina* (Hymenoptera-Formicidae) drawing together leaves, which are bound with silk from larva held by worker on right; ants 8 mm long; (reconstructed from photos by A. J. Nicholson and R. W. Taylor). [F. Nanninga]

glands opening on the abdomen (some Coleoptera), or on the fore tarsi (Embioptera, some empidid Diptera), or by the Malpighian tubes discharging at the anus (Neuroptera, some Coleoptera and Hymenoptera). Most web-spinners, however, discharge the silk from the mouth cavity, the glands usually being modified salivary glands (Psocoptera, Siphonaptera, some Diptera, Trichoptera, Lepidoptera, Hymenoptera).

Silk is used extensively by larvae of Lepidoptera in construction of shelters. These may either be fixed abodes, like those of many Tortricoidea and Tineoidea which tie leaves together, or they may be portable houses, like those of the Psychidae and some Tineidae. The 'cases' of Psychidae (Fig. 36.21) often incorporate twigs and dried leaves. Shelters and silken cocoons of Lepidoptera often include shed larval hairs, which, if urticating, serve as additional protection from vertebrate predators. Thus, the communal homes of the larvae of the bag-shelter moth, *Ochrogaster contraria*, cause considerable irritation to people cutting the acacias in which the colonies live. Some other cocoons, such as those of the Limacodidae (cup moths), are like fibre-board in texture, the silk being consolidated after spinning by a brown, varnish-like liquid produced by the larva.

Embioptera weave silken tunnels (Fig. 23.5) in which they are protected from desiccation and predation. However, perhaps the most striking example of the use of silk in construction of terrestrial shelters is found in the green tree ant, *Oecophylla smaragdina*. The workers draw the leaves of plants together, often forming living chains of ants clinging to each other with jaws and legs. Other workers then bring advanced larvae from existing nests, and use them as living shuttles to bind the edges of the leaves together with the silk that they secrete (Fig. 5.38).

Among aquatic insects, larvae of most Trichoptera construct portable cases under water, gluing twigs, leaves, small pebbles, or sand together with a silky or gelatinous secretion. Others make fixed houses of pebbles, or gelatinous tunnels, sometimes reinforced with algae or debris. The larvae or pharate pupae of aquatic nematocerous Diptera may also spin underwater silk, as in the simuliid cocoons already mentioned.

Homopterous bugs produce structures of a different kind by secretion. Many cercopoid ('cuckoo-spit') larvae live in a foam of their own production. Extensions of the abdominal terga form a ventral pocket, into which the spiracles open, and from which air is blown through a soapy secretion exuded from the anus and posterior abdominal glands. Some cercopoids (Machaerotidae) live in tubes formed of a secretion from the Malpighian tubes. The formation of protective envelopes is also a prominent feature in the biology of Australian Coccidae, Aleyrodidae, and Psyllidae. Many psyllid larvae, in particular, construct delicate and complex tests ('lerps') from a carbohydrate secretion from the anus. As the larva grows, the edge of the test is lifted periodically and the structure extended.

Fig. 5.39. 'Decorated' nest entrance of *Camponotus nigriceps* (Hymenoptera-Formicidae), Alice Springs, N.T.; leaves of mulga *(Acacia aneura)* have been added to the mound around the entrance; coin is a 20-cent piece.
[Photo by K. R. Norris]

Earth and clay are in widespread use as building materials by many kinds of insects. In the nests of some ants the structures above ground appear to be spoil heaps incidental to subterranean excavation, but other species incorporate twigs, leaves, and litter into the mounds, which then become special structures functioning in the defence of the nest opening and protecting it against flooding (Fig. 5.39).

Termitaria are more imposing structures associated with ramifying underground tunnel systems. They are of intricate construction, usually having a solid outer wall and well-defined inner zones of galleries (Fig. 15.8). The outer wall is built largely of soil particles, each transported by a single termite, which supplies its own mortar in the form of saliva and faeces. The walls of the innermost galleries may be made of a substance termed 'carton', consisting largely of semi-digested wood or grass and organic matter. Termitaria include by far the largest structures built by insects, the grass-eating *Nasutitermes triodiae* of northern Australia constructing mounds 7 m or more in height.

Clay is employed by eumenid wasps in building nests which they stock with paralysed caterpillars. Wasps of several sphecoid families also adopt this habit. They select crevices, knot-holes, and keyholes, making good any gaps in the walls with clay, and sealing the cavity with clay after it has been stocked with paralysed prey.

The social wasps (Vespidae) construct their nests of a papery substance made by chewing up weathered wood and mixing it thoroughly with saliva. The nests consist of regularly arranged symmetrical cells, and hang with the openings downwards (Fig. 5.40). Bees of various families employ clay, resins, and a variety of vegetable material in nest construction; but the honey bee, *Apis mellifera*, although using 'propolis' compounded of plant resins in sealing and weather-proofing its dwellings, relies exclusively on wax secreted from dermal glands for building the comb that houses the brood and the stores of honey and pollen.

Fig. 5.40. Workers of *Polistes tasmaniensis* (Hymenoptera-Vespidae), about 12 mm long, on their nest; eggs and larvae in cells top left; sealed pupal cells below; wasp emerging centre.
[F. Nanninga]

SOCIAL INSECTS

The insects include a number of truly social species, some of which exhibit an amazing intricacy of community organization. Insect societies are essentially founded on the family, and are never based on aggregations of individuals of varied parentage. There are also various grades of subsocial family organization, and these are worth surveying for the light they may throw on the evolution of the social habit. Gregariousness, as shown by locusts, is a different phenomenon which is considered on p. 139.

Protection of the eggs by the females is a very ancient habit, occurring in myriapods and some entognathous hexapods. In the insects, subsocial groups in which there is a temporary cohesion of an actual family occur in Blattodea, Dermaptera, Embioptera, Hemiptera, Coleoptera, and Hymenoptera. Cockroaches of the genera *Panesthia* and *Macropanesthia* often live in family communities consisting of a pair and their immature progeny; the female of the sawfly *Perga lewisi* remains with her eggs until they hatch, and stands guard over the young larvae for some time (Fig. 5.41); females of Embioptera guard their eggs and first-instar offspring in their silken tunnels until the young spin tunnels of their own nearby; the female of the large bug *Tectocoris diophthalmus* stands over her egg clusters and guards them until they hatch; fungus-eating beetles of the subfamily Platypodinae show a degree of care for the young; some of the sphecid wasps referred to earlier feed their larvae each day until they are fully grown, but the young

Fig. 5.41. Female of *Perga lewisi* (Hymenoptera-Pergidae) brooding over her newly hatched larvae; female 16 mm long.

[Photo by A. J. Nicholson]

wasps show no further association with the mother after emergence.

The truly social insects are those in which the mother founds a colony by remaining with her eggs and feeding the young, which later co-operate with her in caring for her subsequent broods. This type of organization has arisen independently in termites, ants, some wasps, and some bees. A necessary prerequisite for colonial development is that the mother should remain active long enough to raise her brood and then to specialize completely in reproduction when they have relieved her of the burden of caring for the nest and young. In all the social groups there is a special reproductive caste, usually consisting of the founder of the colony, perhaps accompanied by her mate, whereas the insects that carry out the work of the colony are sterile. Reproductive specialization would thus appear to be an important factor in the success of social insects.

Michener and Michener (1951) cite a special type of nervous system as a prerequisite for social insects. They must be capable of the complex activities required of their particular social organization, which, despite the importance of inbuilt patterns of behaviour, invariably involve some degree of learning. Colonial insects also need to have mouth-parts that permit them to move objects around. Effective defence mechanisms are also necessary for the protection of the colony and its accumulated food supplies.

Termites are the most primitive of the social insects. There is, nevertheless, a clearly defined division of labour in their societies, and a correspondingly sharp division of the insects into specialized castes (Fig. 5.42) adapted to carry out particular functions. The community consists essentially of the pair of dealated reproductives that founded the colony and their progeny of sterile workers and soldiers, but there may also be wingless secondary reproductives ('neoteinics') and more than one kind of worker or soldier. Being exopterygotes, the immature stages have mandibulate mouth-parts and ambulatory legs, and so can carry out some or all of the duties of the workers, which may not be differentiated as a separate adult caste. In the more primitive termites caste determination is mediated by the endocrine system under pheromonal control by the royal pair, but in the higher Termitidae this plasticity no longer remains, and the worker, for instance, lacks the potential for

Fig. 5.42. Workers and soldiers of *Coptotermes lacteus* (Isoptera-Rhinotermitidae); parasitic mites on head capsules of two workers; overall length of soldiers 4·5 mm.

[Photo by C. Lourandos]

caste alteration (E. Wilson, 1965). An important unifying force in the social life of termites is the habit of oral and proctodaeal feeding of one individual by another. By this means, too, the symbiotic protozoa and bacteria of the termite gut are spread through the community.

The social Hymenoptera share a number of features which distinguish them sharply from the termites. The males ('drones') are produced parthenogenetically, and are invariably fully winged; their only function is to fertilize the females. All the workers are sterile females, which may be specially modified for the performance of domestic duties. The larvae are helpless grubs, and the pupae equally helpless.

In the Vespidae, as represented by *Polistes* (Michener and Michener, 1951), fertilized queens emerge from hibernation in spring, and, working alone, construct paper cells in which eggs are laid. The larvae are supplied regularly with masticated caterpillars, and, when fully fed, spin a silken cocoon which seals the opening of the cell. The wasps that emerge from these are sterile workers, stunted in comparison with the queen, probably because of food limitation. They take over the feeding of the brood, which the queen now abandons to them. The larvae yield a secretion, which is actively sought by the workers and probably acts as a unifying force in the colony. The wasps solicit the secretion so avidly that the growth of the larvae may be stunted until the colony has become large and prosperous late in the season. Males are then produced, and some of the workers secure enough food to enable them to develop into queens, which mate and are ready for hibernation.

The honey bee (Fig. 5.43) is the most highly specialized of the Apoidea, and has the most complex social organization. Mating takes place on a special flight, on which the queen is accompanied by a swarm of males. In mating, the successful male suffers a fatal loss of his genitalia. The fertilized queen then either returns to the hive or founds a new colony, in which she assumes her duty as the sole female reproductive. She differs from the workers in lacking the anatomical specializations necessary for honey and pollen gathering, and also in being equipped with a smooth sting, in contrast to the barbed sting that the workers employ in the defence of the colony. All larvae from fertilized eggs are capable of developing into queens, if fed throughout their lives on a special secretion ('royal jelly') by the worker nurses. However, only those that hatch from eggs laid in the large, specially shaped queen cells are treated thus; those from eggs laid in

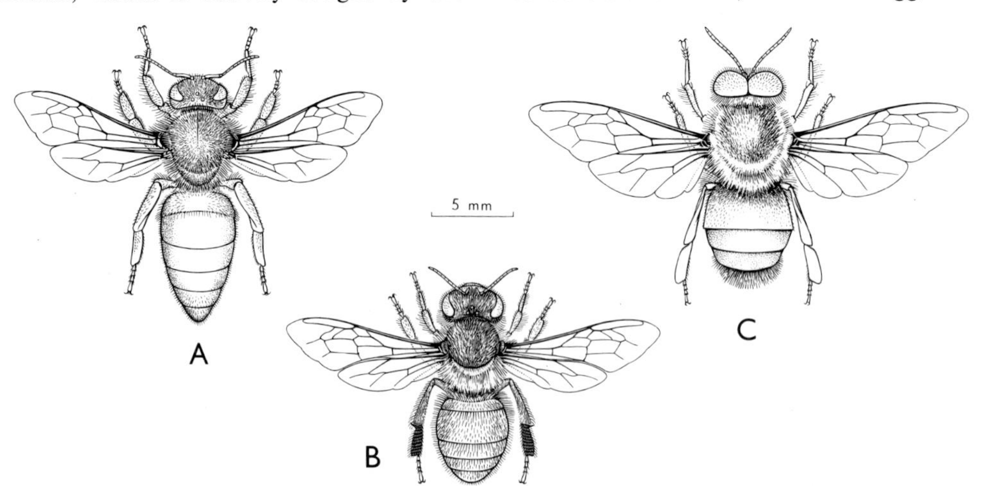

Fig. 5.43. *Apis mellifera* (Hymenoptera-Apidae): A, queen; B, worker (sterile ♀); C, drone (♂). [F. Nanninga]

ordinary cells are changed from the royal jelly to a diet of 'bee bread' (honey and pollen) after the third day, and these produce the future workers.

Unlike other colonial Hymenoptera, the larvae of honey bees do not return any secretion in exchange for the nourishment provided by the nurse bees. However, other influences serve to integrate the colony, the most potent being a pheromone secreted by the queen and known as 'queen substance' (p. 36), which is eagerly sought by the workers, and suppresses ovarian development in those that imbibe it. The action of this in colony cohesion is supplemented by food interchange between adult bees, by the need to conserve warmth resulting in clustering, and by visual factors and responses to vibration and odours. The complex behaviour patterns of bees have been described by von Frisch (1954).*

New colonies are generally established by the old queen departing with a swarm of workers to seek a fresh nest site. Other swarms may accompany emigrating daughter queens. A remaining daughter queen which has participated in a mating flight then carries on the hive after destroying any possible competitors that may remain.

Ant colonies are established after a mating flight, in which the queen is fertilized by one or more males. The males are short-lived, and the queen initiates the colony alone, immuring herself and shedding her wings, then waiting passively while the eggs develop. The wing muscles are resorbed to sustain her, and she nourishes the first worker brood on salivary secretions. As soon as the first workers emerge, they take over the task of foraging and caring for the brood, and the queen then specializes entirely on egg production. The presence of the queen and the habit of mutual feeding are important factors in the integration of ant colonies, and specific colony odours may also be significant. Workers may be of different kinds ('phases'), and there is considerable obscurity as to the mechanisms determining caste. In some species it is certainly determined in the egg, and in others it is known to be influenced by the feeding of the larvae during growth and also by seasonal temperature effects.

COLOUR CHANGES AND POLYMORPHISM

No species is completely uniform and some are highly variable; others change in various

* See also M. Lindauer, *A. Rev. Ent.* **12**: 439–70, 1967.

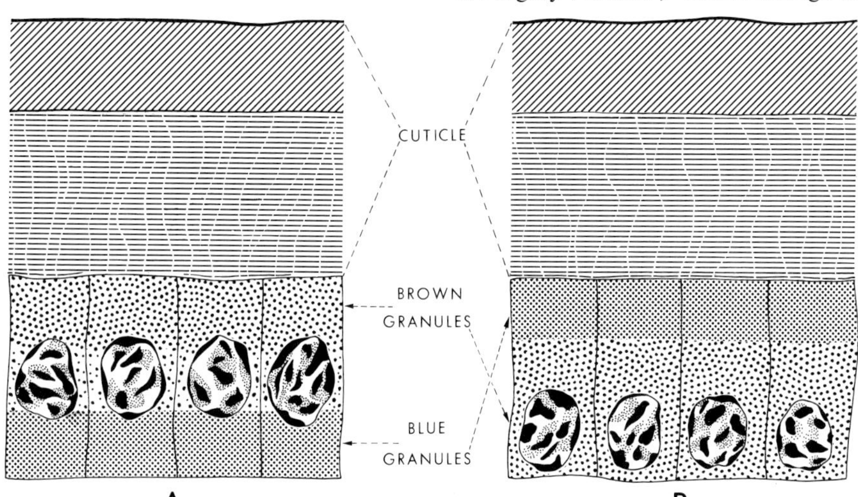

Fig. 5.44. Changes in epidermal cells of *Kosciuscola tristis* (Orthoptera-Acrididae) associated with temperature-induced body colour changes (see Plate 2,J): A, dark condition; B, blue condition (after Key and Day, 1954a).

ways from place to place and from time to time. Analysis of all this variability has revealed several rather distinctive phenomena, and it is with these that we are concerned here (see Kennedy, 1961, for a more detailed discussion).

Adaptive Colour Changes. These are of several kinds (review in Hinton, 1960a). In one series, called 'physiological', the changes develop rapidly in response to stimuli of relatively short duration, and are usually produced by the migration of pigment granules within the epidermal cells. By this means males of the Australian alpine grasshopper, *Kosciuscola tristis* (Fig. 5.44; Plate 2, J), change from almost black to blue as the temperature rises between 15 and 25°C. Several species of damselflies also change from dull purplish or greyish to brighter bluish between 10 and 15°C (O'Farrell, 1964). The stimulus acts directly in the grasshopper, and it has been suggested that the advantage

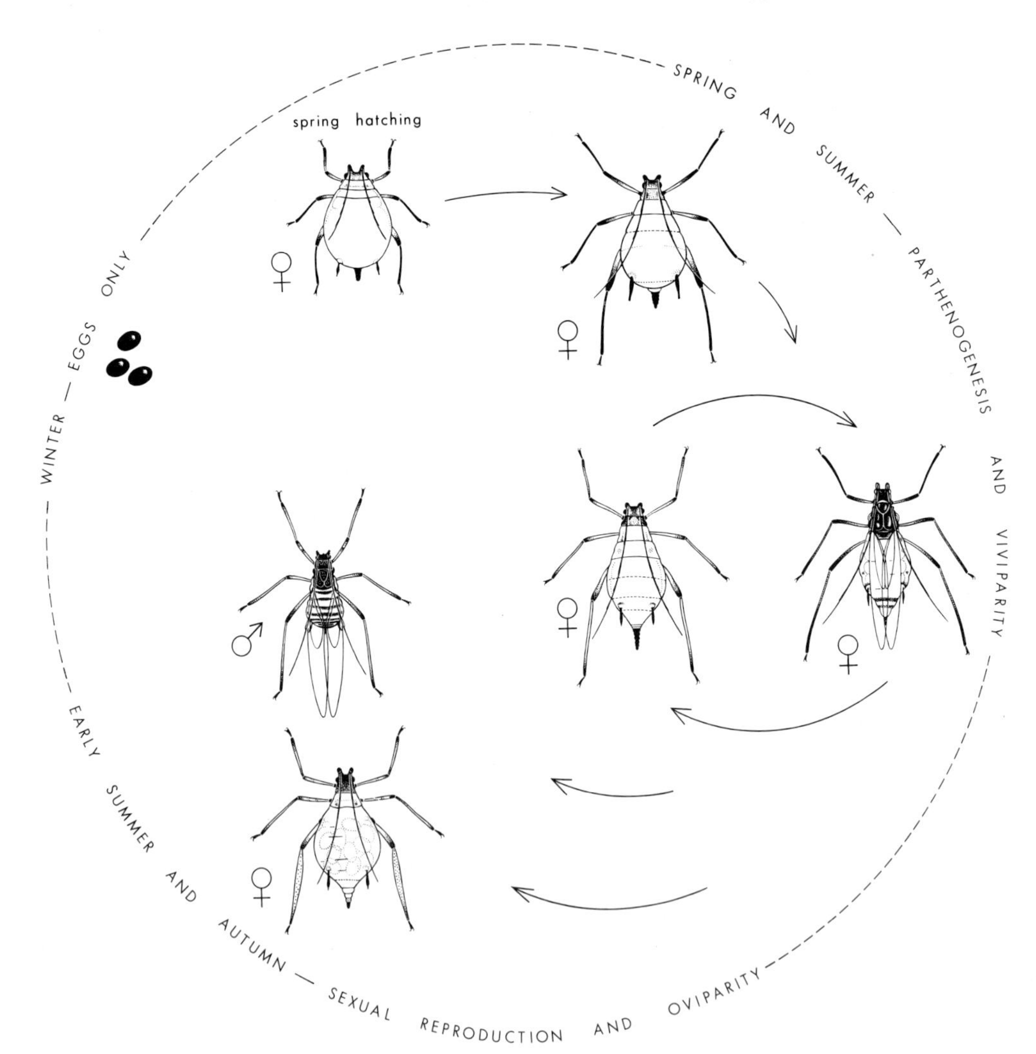

Fig. 5.45. Annual succession of morphs in *Megoura viciae* (Hemiptera-Aphididae) under British conditions; the winged parthenogenetic females at right of diagram colonize new host plants (based on Lees, in Kennedy, 1961).

of these responses may lie in quick heating of the body early in the day and perhaps reduction of overheating by reflection of radiant energy later. Similar but more complex changes in phasmatids (p. 354) have a diurnal rhythm, and are under endocrine control.

A remarkable 'physiological' change is produced by migration and changes of shape of mesodermal chromatocytes in the transparent larvae of *Chaoborus* (Diptera). When the larva is in the darker zone deep in the water, the chromatocytes spread over and hide the reflective tracheal air sacs, whereas they retract away from the dorsal surfaces of the sacs when the larva is in more brightly illuminated water nearer the surface. Thus the visibility of the larva is minimized according to the light intensity it is in.

Other kinds of change, called 'morphological', are the result of synthesis or destruction of pigment, and they are consequently slower to develop but more lasting than the 'physiological' changes. Some of the pigments are cuticular, and are deposited when new cuticle is being laid down; others are in or beneath the epidermal cells, and these are visible externally only where the cuticle is more or less transparent. The dark cuticular pigments are the products of tanning; the subcuticular pigments of various colours may be metabolic, or derived from the food, or a combination of both. Any or all of these mechanisms may contribute to the progressive adaptation of insects to different backgrounds, as in mantids (p. 298), Acridoidea (p. 329), and possibly the seasonal forms of some adult Lepidoptera (p. 778). In some insects the changes may be related to changes in environment at different stages in the life history, and in some locusts the *solitaria* phase (see below) has relatively little cuticular pigment but complete subcuticular insectoverdin (p. 34), whereas the *gregaria* phase has a strong cuticular pattern but only carotenoids beneath the cuticle (Kennedy). The mesodermal chromatocytes of some nematocerous Diptera are a special case, because they may provide a constant cryptic pattern through the larval instars followed by marked changes in the pupa and pharate adult as the chromatocytes are rearranged at metamorphosis (Hinton).

Kentromorphism. The changes here are more complex, but they too are triggered by environmental stimuli. They occur in phasmatids (p. 354), locusts (p. 329), the larvae of some Lepidoptera, and a few other insects. The essential feature that they share in common is that they are responses to population density, one form *(phase)* being characteristic of low-density populations, the other of high densities, and intermediate forms occurring at intermediate or changing densities. The two phases differ from one another in coloration and pattern, at least in the immature stages, and often also in anatomical proportions, physiology, and behaviour. The phasmatids simply become more crowded as their numbers increase, whereas the locusts and caterpillars become both positively gregarious and migratory. The low and high density phases of locusts and some caterpillars are known, respectively, as *solitaria* and *gregaria*. The migrating swarms of millions of locusts in the *gregaria* phase, devouring every scrap of edible plant tissue as they go, have been known from antiquity, and have been the subject of a great deal of research. The interactions between individuals that induce kentromorphic changes are partly sensory and partly pheromonal, and the endocrine system plays a major mediating role (Haskell, 1962).

Cyclical Polymorphism (cyclomorphosis). The colonies of social insects described in the previous section are examples of polymorphic families, in which the various castes are produced, more or less contemporaneously, from the eggs of one female. Many species of Aphididae show polymorphism of another sort, different forms appearing successively in different generations as the annual cycle proceeds. These forms, of which there are more than twenty in some species, differ from one another in a variety of anatomical features, and in being either parthenogenetic and viviparous or sexually reproducing and oviparous. Thus, in *Megoura viciae* (Fig. 5.45) a wingless female hatches in spring from a fertilized winter egg. She produces parthenogenetic daughters, which have relatively long legs, antennae, and dorsal cornicles, and also a colour pattern different from her own. Parthenogenetic reproduction proceeds until the later descendants in the summer are even more distinct from the original parent, and they may also be either wingless or winged. In late summer, sexually reproducing forms appear, winged males and apterous females which deposit the fertilized overwintering eggs (Lees, in Kennedy, 1961; see also pp. 421–3).

The polymorphism of aphids is thus a cyclic phenomenon, which appears to be regulated by endocrine switch-systems under the influence of such factors as photoperiod, temperature, degree of crowding, and innate time-interval mechanisms. The physiological changes at least give the species evident advantages in coping with marked seasonal changes, though the significance of some of the anatomical changes is obscure. If the environmental conditions permit, all forms may occur annually in the descendants of a single female ('clonal polymorphism').

Genetic Polymorphism. This is an entirely different phenomenon, which is not only adaptive in the short term, but also of considerable importance in speciation and longer-term evolution. It has been defined (see Ford, in Kennedy, 1961) as 'the occurrence together in the same habitat of two or more discontinuous forms of a species in such proportions that the rarest of them cannot be maintained merely by recurrent mutation.' It is usually 'balanced', the different phenotypes *(morphs)* that are the expressions of the genetic heterogeneity being maintained by natural selection in approximately constant proportions at any one time and place, though often in very different proportions in different parts of the geographical or seasonal range of the species. Survival of the less successful morphs is ensured by adaptive superiority of the heterozygote, and their relative proportions by their relative selective advantages under any particular set of environmental pressures. Genetic polymorphism thus favours the establishment of locally adapted subpopulations.

Sexual dimorphism is an almost universal expression of genetic polymorphism in insects, but other discontinuities are common, and have been reported in all the major orders. They may be independent of sex, or restricted to one sex, which is often the female. Probably the most spectacular examples are in tropical butterflies, especially mimetic species. In Australia colour morphs of Odonata are not uncommon, particularly in the Coenagrionidae (p. 247), and pattern morphs of acridid Orthoptera have been described by Key (1954). Chromosomal polymorphism is a special case in which obvious manifestations may be limited to the karyotype, but there is evidence that it too has adaptive value (p. 77).

ACKNOWLEDGMENTS. The author wishes to thank numerous colleagues, in particular Dr I. M. Mackerras, for considerable assistance in the preparation of this chapter; and also Dr A. J. Nicholson and Mr C. Lourandos for many of the photographs that illustrate it.

6

PRINCIPLES OF CLASSIFICATION AND NOMENCLATURE

by K. H. L. KEY

CLASSIFICATION OF ANIMALS

Since there are several million different species of living animals, with insects comprising some three-quarters of the total, it can be seen that their classification must entail major difficulties. This task is the province of *taxonomy*, or *systematics*. Although 'special' classifications may be employed for special purposes—e.g. classifications by geographical region or by economic significance—the classification with which we are concerned is the 'general' or 'natural' one, of which the foundations were laid by Carolus Linnaeus in the eighteenth century.

Taxonomic Categories and Taxa

The classification of the animal kingdom is hierarchical, first-order classes being divided into more numerous second-order classes, these in turn into third-order classes, and so on, so that each grouping may be said to have a *rank*. Groups of the same rank are said to belong to the same *taxonomic category*, and each category has received a name. Thus the animals are classified within a framework of standard categories, some of which, however, are 'optional'. The obligate (capitals) and the more usual optional categories, in descending order of rank, are:

ANIMAL KINGDOM
PHYLUM
Subphylum
CLASS
Subclass
Superorder
ORDER
Suborder
Superfamily
FAMILY
Subfamily
Tribe
Subtribe
GENUS
Subgenus
SPECIES
Subspecies

An individual animal belongs to a particular *species*, the species to a particular *genus*, and so on up the scale. These particular, concrete groupings of animals are given the general designation of *taxonomic units*, or *taxa* (singular, *taxon*), and each of them bears a distinctive name. The logical structure of the classification is thus that each taxon is included in a particular taxon of higher rank, but is also a member of the taxonomic category embracing all other taxa of the same rank (Buck and Hull, 1966). The insects as a whole constitute the Class Insecta within the Phylum Arthropoda.

Although it is often convenient to visualize taxonomic classification as a system of progressive subdivision of the animal kingdom, in principle it should be regarded as derived, not by this analytical method, but synthetically, by grouping individuals into species, 'related' species into genera, and so on.* Except at the species level, the 'scope', or 'limits', of the various taxa are to some extent arbitrary. For example, 12 related species might all be placed in one genus, or in two or three related genera, depending upon the personal outlook of the taxonomist. If we were able to plot the 12 species as points on a plane, separated from one another by distances inversely proportional to their levels of overall similarity, we might get the result shown in Figure 6.1. The diagram leaves no doubt as to which species should be grouped together for a three-genus or a two-genus solution, but it gives no guidance as to whether a one-, two-, or three-genus solution is to be preferred.

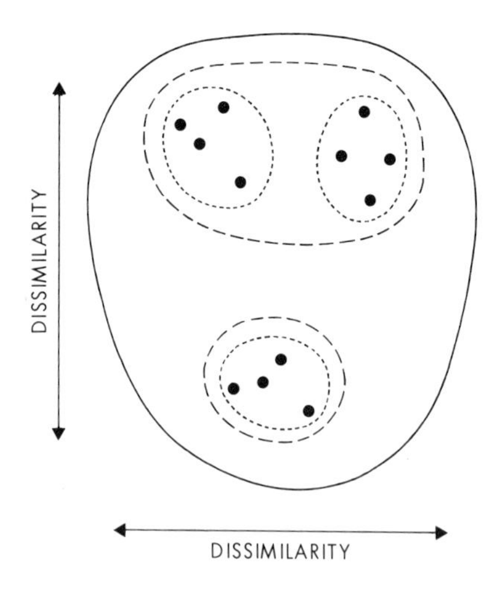

Fig. 6.1. A hypothetical two-dimensional phenetic-distance plot of 12 related species, showing possible groupings into one, two, or three genera.
——— One-genus solution. - - - - Two-genus solution. . . . . Three-genus solution.

* Throughout this chapter the words 'related', 'relationship', and 'affinity' are used in the sense of degree of similarity, rather than in the sense of familial relationship.

Similar considerations apply to the delimitation of taxa in all the higher categories. A guiding principle should be that the magnitude of the differences between taxa of any one rank should be approximately standard within a given group of animals. Unfortunately, historical reasons have led to the adoption of rather widely divergent standards as between major groups, e.g. between different orders of insects.

Principles of Classification

Linnaeus's classification of the animals known to him has been followed by two major revolutions in theoretical biology, the first arising from Darwin's theory of evolution and the second from the development of genetics. Although the bearing of these developments upon taxonomy is still controversial, it is surprising what comparatively small effect they have had on classification.

In principle, the viewpoint of the school of 'phenetic', or 'numerical' taxonomy seems to be irrefutable (see Sokal and Sneath, 1963). It argues that what the taxonomist actually does—and the only logical procedure for him to adopt—is to estimate the level of overall similarity ('phenetic affinity') between different organisms, based on as many characters as possible of all kinds, without weighting the characters in accordance with judgments of their relative 'importance'. It regards 'fundamental' or 'key' characters as indicators that are recognizable as such only when this overall comparison has been carried out, and whose uncritical use thereafter in unanalysed situations is fraught with danger. It disposes similarly of the supposed necessity for seeking other than empirical criteria for distinguishing between homologous and non-homologous structures in estimating relationship. Finally, it insists that our classification is not, and logically cannot be, based on phylogeny, but rather that our ideas of phylogeny rest, often very precariously, on inferences drawn from estimates of overall similarity, even in those instances where the fossil record is most adequate. In the insects, probably not one phyletic line in a hundred has any supporting fossil evidence.

The 'phylogenetic' taxonomists take a different point of view on many of these issues. They consider that classification should be based on phylogeny, with 'propinquity of descent' the principal criterion for delimiting and ranking taxa. They believe that phylogenies can be derived with a sufficient degree of probability (even in the absence of fossils) by assessing the 'primitive' versus 'specialized' status of the characters of the organisms concerned and by giving priority or special weight to characters deemed to be of major phylogenetic significance. In their view, circular reasoning need not be involved in this process. The viewpoint of the phylogenetic taxonomists is expounded by Hennig (1957, 1965) and in a broader setting by Simpson (1961).

The contribution of phenetic taxonomy to the logical methodology of taxonomy is certainly as significant as the various mathematical techniques it employs. The latter are principally applicable to groups in which the majority of the species are known, but in which the generic or higher grouping requires clarification. This situation is often not realized among the insects, and rarely among the Australian insects. On the other hand, the empirical approach, including the rejection of character weighting and the relegation of 'key' characters and phylogeny to their subordinate position, can be usefully applied even in the traditional process of non-numerical estimation of overall similarity.

Extension of the concepts of phenetic taxonomy to the recognition and delimitation of species tends to disregard the rather fundamental difference, in bisexual organisms, between species and groups of other rank within the classification (see below); its application at the species level will probably be somewhat limited.

The Species

Apart from optional infraspecific categories, the species is the lowest category in the hierarchy. It is also a particularly critical category, because the grouping of individuals into species is the first and most essential step in classification.

In asexual, self-fertilizing, and parthenogenetic organisms, the limits of species may be more or less arbitrary, as in taxa of other ranks, and the methods of numerical taxonomy are applicable at the species level. However, in bisexual organisms these limits can be established with greater objectivity, at least for the present fauna and contemporaneous past ones. The basic consideration in the recognition and delimitation of species has always been the possibility of demonstrating discontinuities between their ranges of variation. That is, however widely two related contemporaneous species may vary, there is nearly always at least one character in which their ranges of variation are clearly separated, i.e. in which they differ consistently. Population genetics emphasizes the reason for this. Among the individuals comprising a species there is no bar to interbreeding. There is a common 'gene pool', and in principle every gene is free to spread eventually throughout the species. By contrast, different species are reproductively isolated from one another: a gene becoming established in one of them is not free to spread to others, so that genetic differences accumulate and ultimately express themselves in detectable character differences. Thus the discontinuity in characters results from the genetic discontinuity, which, in turn, is maintained by the reproductive isolation (see Mayr, 1963).

These considerations have led to the adoption of the 'biological' definition of a species as 'a group of actually or potentially interbreeding natural populations reproductively isolated from other such groups' (Mayr, 1942). Apart from emphasizing the crucial factor of reproductive isolation, this definition gets away from the concept of the species as an idealized model and substitutes the concept of a concrete population in space and time.

In practice we seldom have direct evidence on the reproductive isolation of putative species, but must infer it from discontinuities in characters and from other leads. Failure to produce fertile hybrids in the laboratory (under conditions where both parental forms breed freely) is good evidence of reproductive isolation, but populations may be reproductively isolated in the field even though they can be induced to hybridize in the laboratory. Given enough material, it is usually possible to reach a definite conclusion on the status of putative species whose areas of distribution overlap, i.e. that are *sympatric*. In the case of *allopatric* populations (occupying disjunct areas) the decision must usually be arbitrary if they are characterized by slight but constant differences.

Characters of Species. Every piece of information about a species in any stage of development—about its external and internal anatomy and coloration; its cytology; its physiology, biochemistry, and serology; its behaviour and ecology—constitutes a character* of that species and ideally should be taken into account in assessing its relationships. Often, however, the available characters are restricted to those of the external anatomy and coloration, as when dried specimens are the only material available. Particularly useful characters are often afforded by the structure of the genitalia, satisfactory preparations of which can usually be made from dried material. Unlike most other characters, the genitalia are in the main not directly adaptive to the external environment; the selection pressures that have moulded them have had a different source, so that the evidence they offer is to a degree independent of that of other characters.

While, in the broad sense, the characters of a species include all our knowledge about it, the term is often used in the narrower sense of 'diagnostic character,' i.e. a character found to be useful in distinguishing the species from its relatives. A difficulty here is to know just which characters may prove to be needed for diagnostic purposes when all related species have been discovered.

Once a species has been recognized on the basis of character discontinuity and inferences as to reproductive isolation, as many as possible of its characters must be ascertained; the formal itemizing of these constitutes 'description' of the species. Here special, but by no means exclusive, attention needs to be focussed on diagnostic characters, which may also be employed in the construction of keys for the separation of related species. Occasionally there is conclusive biological evidence that two populations are reproductively isolated, although no obvious character difference can be found between them. Such species, in which character differences are exceedingly slight, or ascertainable only by special procedures, have been termed *sibling* species.

Variability of Species. Most characters of all species are in some degree variable. Where such variation is appreciable, its range needs to be indicated in characterizing the species.

Several different types of variation may be distinguished. One of the most widespread is the quantitative, continuous variation to which most characters are subject and which is polygenic in origin. Then there is *polymorphism*, or the occurrence of sharply contrasted types of individual which are nevertheless freely interbreeding (pp. 75, 140). The differences between such *morphs* are usually due to one or a few genes only, and their proportions in any population vary widely. Some degree of dimorphism is always to be noted between the sexes.

In addition to variability of genetic origin, all sorts of differences between individuals of the same species, especially in size, coloration, and behaviour, may be induced by direct effects of the environment on the phenotype (p. 138). Factors producing such effects are various climatic elements, the quantity and quality of the food supply, and the density of the population. Caste differentiation among social insects is a special case of this kind.

Variation from all these genetic and environmental sources may be quite bewildering. Often parallel, or 'homologous', variations occur in related species, and the differences between extremes in any one species may be more striking than the constant differences separating the species from one another. In such circumstances careful

* The valuable analysis of D. H. Colless (*Syst. Zool.* **16**: 6–27, 1967) of the concept of 'character' was received too late to be exploited here.

discrimination is necessary in delimiting and then characterizing the species.

Although all these variants may coexist and interbreed within a single local population of a species, it also frequently happens that populations occupying different regions, or different types of habitat, or feeding on different hosts, differ from one another in some character or characters, or in the relative frequencies of certain morphs. Such populations may be sufficiently different to be distinguished as geographical or ecological 'races'. Moreover, in species with wide distributions we often find that one or more characters will show a gradient of progressive change in a particular direction through the territory occupied; e.g. there may be a steady increase in average size from north to south. Such character gradients are termed *clines*. The absence of reproductive isolation in all these instances is indicated by the existence of a chain of intermediate individuals in the intervening regions between divergent local populations.

Adequate description of the variability of species calls for the use of statistical techniques (references in Simpson, 1961).

Polytypic Species. Species in which geographical variation is marked, and especially where several races of distinctive appearance occupy substantial areas, have been termed *polytypic* species. Their component races may be treated as subspecies by those who employ this category. Where what are considered to be subspecies or races of a polytypic species are isolated from one another by barriers to migration, e.g. on islands, a decision as to their specific versus subspecific rank is necessarily uncertain, unless they can be studied experimentally. However, where distributions are continuous, subspecies reveal their true nature by intergrading with neighbouring subspecies. Thus subspecies are to be distinguished from morphs by their occupation of distinct, though often contiguous, territories, and from species by interbreeding with their neighbours.

The differences between subspecies are due to selection from the common gene pool in response to the different environments encountered in different parts of the distribution area of the species, or to such phenomena as the 'founder principle' (see Mayr, 1963). Occasionally it happens that a species has extended its distribution along a more or less circular route around some barrier, differentiating into a chain of subspecies on the way, until, when the advance-guard again reaches the original home of the species, it is no longer capable of interbreeding with the populations there. It is then a matter for arbitrary decision whether we treat the terminal populations as species because they do not interbreed, or as subspecies because they are connected, around the circuitous route, by an unbroken chain of freely interbreeding populations.

Like the supraspecific taxa, subspecies are arbitrary units—but they are the result of subdivision, not synthesis. There is no limit to the possibilities of subdivision, for no two local populations of a species are identical. As a formal category, the subspecies has come under criticism for this reason, and also because there is often poor concordance between the geographical variation of different characters. The critics advocate abandonment of the category and substitution of a description of the general pattern of clines shown by the species.

Speciation

In spite of the strictures that have been made upon the claim that phylogeny can be the basis of classification, we know that the animals we classify today are the product of a long evolutionary history. In this process populations gradually change with time under the influence of selection, whose strength and direction are in turn determined by alterations in the general environment. However, such changes would not in themselves bring about diversity among living things, and it is diversity that raises the need for classification.

The critical step in the evolutionary induction of diversity is the development of reproductive isolation between populations, which thereafter are free to follow independent directions of change that open up to them new ecological niches and expose them to new selection pressures, until such time as they in turn bud off reproductively isolated daughter populations. This is the process of *speciation*, a process that must be supposed to have been at work from the earliest times and is still active (Dobzhansky, 1951; Mayr, 1942, 1963). Every bifurcation in the 'phylogenetic tree' is a point at which speciation occurred in the past.

An understanding of speciation is important to the taxonomist in dealing with problems at the specific and infraspecific levels. It is now widely accepted that an almost invariable precondition for speciation is the spatial isolation of some part of the ancestral population, usually at the periphery of its distribution. If this enforced isolation from reproductive contact with the main population is maintained long enough, the isolated part will have time to diverge genetically under locally directed selection pressures. If the two populations later come into contact again, the divergence may have proceeded far enough to cause significantly reduced hybrid viability or fertility. Once this happens, it has been argued (Dobzhansky, 1951) that selection will tend to favour the evolution of behavioural or other impediments to cross-mating, until in time full reproductive isolation is attained and hence full specific status.

The original spatial isolation may be brought about by various kinds of barrier, according to the properties of the organisms concerned (p. 202). Barriers must be adequate to prevent gene exchange almost completely, i.e. they must constitute unacceptable territory that can rarely be crossed. It is generally considered that a difference in the preferred habitat of two portions of a population does not in itself constitute effective isolation. Rare cases of speciation without spatial isolation, involving special mechanisms, are known. One such mechanism would be the mutational origin of a successful parthenogenetic form.

The Genus

The genus is the next obligate taxonomic category above the species. As we have seen, the limits of genera are to a degree arbitrary. The manner in which related species should be grouped to bring out their overall similarities can be made evident by the techniques of numerical taxonomy, or by similar less rigorous methods, but the point at which divergence should be considered to have reached the generic level remains a matter of opinion. We have seen that efforts should be made to maintain uniform standards within, say, orders. Apart from this, it should be noted that a classification that produces a large proportion of genera with only one species largely defeats the purpose of the genus, which is to provide a useful first grouping of the enormous number of animal species. On the other hand, a genus containing as many as 100 species is inconveniently large and leads to a shift of emphasis

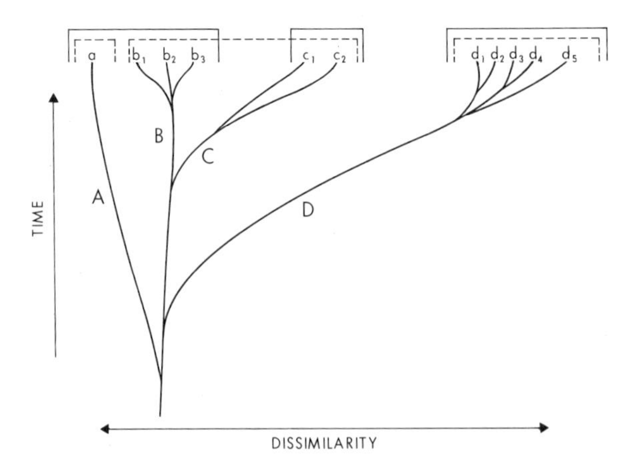

Fig. 6.2. A hypothetical phylogenetic tree, illustrating problems of generic classification.
A, B, C, D. Evolutionary lines, terminating in the living species a, b1–b3, c1 and c2, and d1–d5.
——— Generic limits as they might be drawn by a phenetic taxonomist.
– – – – Generic limits as they might be drawn by a phylogenetic taxonomist.

on to optional categories between genus and species. It is clearly awkward, too, to have genera with such narrow limits that they are determinable only on characters of one sex. A good generic classification will take account of all these considerations and work out a reasonable compromise between them.

It is at the generic level that the conflict in methodology between the phenetic taxonomists and the phylogeneticists becomes most evident. The problems involved may be illustrated by Figure 6.2, which depicts a phylogenetic tree showing four evolutionary lines; the tips of the ultimate branches represent living species and are visualized as lying in a horizontal plane at distances from one another inversely proportional to their overall similarity. For the purposes of the argument, we will assume that the tree is a true representation of the phylogenetic events. The phenetic taxonomist would be inclined to recognize the three genera enclosed by the solid lines, the one encompassing a and b_1-b_3 being 'polyphyletic'. The phylogenetic taxonomist might also have three genera, but his would be more likely to be those enclosed by the broken lines. He would regard the one encompassing b_1-b_3 and c_1-c_2 as somewhat heterogeneous, but justified by 'propinquity of descent'. Putting the matter more generally, the phenetic taxonomist would hold that the classification should be based on the degree of evolutionary divergence, rather than upon the paths along which this developed, or the time at which it commenced. We are not concerned here with the classification of extinct members of this hypothetical group, which raises additional problems.

An optional category below the genus is the subgenus. The use made of this varies widely from group to group. Where it is extensively used, as in the mosquitoes, the genera tend to be large and heterogeneous, and the subgenus fills the role played by the genus in other groups. The differences in practice seem to be largely historical in origin. Informal categories between species and genus are the 'superspecies' (see Mayr, 1942) and the 'species group'.

The Higher Categories

Little needs to be said specifically about the higher categories. The tendency has been to introduce more and more of them, by interpolation between existing ones. This process has obvious limits, and more use could well be made of informal, more elastic groupings, such as the 'hemipteroid orders' (cf. p. 162).

A problem is created for the higher classification by the practice of some taxonomists who elevate existing taxa so as to facilitate their subdivision, thus crowding the ranks above them or forcing a general elevation. In many such cases the solution is rather to make use of existing possibilities of subdivision without elevation, to use informal categories, or simply to recognize a heterogeneity which it is impracticable to express, for no system of classification can be expected to reflect all the nuances of the interrelations of a million species.

ZOOLOGICAL NOMENCLATURE

Zoological nomenclature has earned an unenviable reputation for complexity. This could hardly be otherwise, when one considers that some millions of units had to be named, and that these have come to our knowledge over a period of 200 years, during which great changes took place both in our views of the relationships of the taxa to one another and in nomenclatural practices.

Linnaeus, the father of animal classification, also laid the foundations of the system of nomenclature employed today. Indeed, the 'starting point' of zoological nomenclature has been fixed at the tenth edition of his *Systema Naturae*, published in 1758; i.e. names published before that date are disregarded. In this work Linnaeus first consistently employed names consisting of a combination of two Latin words—the 'generic' and 'specific' names—for his species. This practice has persisted and has led to the designation 'binominal nomenclature', although today, if we include the optional subgeneric and subspecific names, it is potentially quadrinominal. The use of two words rather than one for the names of species greatly reduces the number of different words required to name all animals, since the same specific name can be used in any number of different genera. However, it has the disadvantage that the name of a species (the binominal combination, or *binomen*) changes whenever, as a result of new knowledge of its characters and relationships, it comes to be classified in a different genus; some authors have recently advocated a uninominal system for this reason.

Requirements of a System of Nomenclature

To be most effective for its purpose, a nomenclature should be so designed that each taxon will be known by only one name at all times and places; i.e. the nomenclature should be stable and universal. Furthermore, no two taxa should bear the same name; i.e. the nomenclature should be unequivocal. Some device should be employed to establish an indissoluble link between each name and the particular taxon to which it refers. Finally, the form of a name should indicate in some way the rank of the taxon named, e.g. whether it is a species or a family. These requirements are not fully compatible. For example, instability is inevitably introduced by changing concepts of the limits and rank of particular taxa.

The requirements of a reasonably satisfactory nomenclature can be met only by the universal acceptance of a set of rules stipulating how taxa should be named and what should be done in the case of infractions. Realization of the need for such a set of rules came only gradually, and by the time the first international *Règles* were published in 1905, zoological nomenclature was already in great confusion. Early editions of the rules gave taxonomists little more than general guidance, but the completely revised 1961 edition, entitled *International Code of Zoological Nomenclature*, covers most of the situations likely to arise in the nomenclature of taxa between the ranks of subspecies and superfamily. Other ranks are still unregulated.

The International Code

The provisions of this important document (International Congresses of Zoology, 1961–64) are set out in 87 Articles grouped into 18 Chapters. The Code consists of mandatory rules, some of them operating only from specified dates, together with non-mandatory recommendations. The taxonomic categories covered by the Code are treated in three groups: the family group (superfamily, family, subfamily, tribe, etc.), the genus group (genus and subgenus), and the species group (species and subspecies). Names of taxa within any one of these groups are 'co-ordinate', i.e. subject to essentially the same rules.

Formation of Names. The Code provides that all names shall be 'Latin or latinized', or combinations of letters capable of being treated as such. Names in the genus and species groups are usually printed in italics, those in the family group in Roman type. The name of the original author (sometimes abbreviated) and the date of publication may be cited in conjunction with the name of a taxon, e.g. *Podacanthus wilkinsoni* Macleay, 1881, but they do not form part of the name.

Family-group names are based on the name of a genus included in the family-group taxon, by adding the suffix '-idae' to the stem of the generic name in the case of a family and '-inae' in the case of a subfamily; it is recommended that the suffix '-oidea' be used for superfamilies and '-ini' for tribes. Thus 'Muscinae' is the subfamily of flies containing the genus *Musca*. All these names are plural in form and are written with a capital initial letter.

Generic and subgeneric names consist of a noun in the nominative singular, or are treated as such. Both must be written with a capital initial letter. If a subgeneric name is to be cited in association with a binomen, the Code requires it to be placed in brackets between the generic and specific names, e.g. *Aedes (Finlaya) notoscriptus*. However, the subgenus does not have to be cited.

Specific and subspecific names must consist of a word that is, or is treated as, (a) an adjective in the nominative singular, agreeing in gender with the generic name with which it is at any time combined; (b) a noun in the nominative singular, standing in apposition to the generic name; (c) a noun in the genitive case; or (d) an adjective used as a noun in the genitive case and derived from the specific name of an organism with which the animal in question is associated. They must be written with a lower-case initial letter. Specific and subspecific names may only be used in conjunction with a generic name, although the latter may be abbreviated to its initial letter where confusion is impossible, or, in informal use, may even be left to be understood. Subspecies names are trinominal, e.g. *Anoplognathus macleayi aurora*.

A consequence of the co-ordinate status of names within each of the three groups is that any name is competent, immediately upon its establishment, to serve for taxa in each of the categories included within the group. For example, when the grasshopper genus *Monistria* was published by Ståal in 1873, a subgenus of the same name (the *nominate* subgenus) was deemed to have been established at the same time by the same author—although the subgeneric name was not required for use until, in 1953, Rehn established the new subgenera *Yeelanna* and *Cygniterra*. The genus *Monistria* then comprised the three subgenera *Monistria*, *Yeelanna*, and *Cygniterra*. Similarly, one of the subspecies of a species must always bear the same name as the species, and one of the tribes and subfamilies of a family must bear the same name as the family—with the requisite change of suffix.

Availability of Names. The word 'available' is used as a technical term to indicate that a name satisfies the various requirements of the Code. The chief of these are that the name must have been correctly formed, as outlined above (certain infractions do not prevent availability from the original author and date, but must be corrected), and it must have been 'published' (in accordance with a definition of publication given in the Code) together with information that purports to define the taxon to which the name is being applied. The information required in the case of names published today comprises essentially a diagnostic description and, for generic names, the designation of a *type species* (p. 150).

Although a name may be 'available' for nomenclatural use in appropriate circumstances, it may not be the 'valid' name for the taxon for which it was proposed, nor, for that matter, for any taxon. The reason for this will shortly become evident. An unavailable name is termed a *nomen nudum*.

The Law of Priority. Although a requirement of an effective system of nomenclature is that each taxon should have only one name, it often happens that more than one comes to be applied to it. This can result from the inadvertent naming of a taxon that has already been named, or from the naming of two or more supposedly distinct taxa which future study shows to be one and the same. Two or more available names applied to the same taxon are termed *synonyms*. The Law of Priority states that the 'valid' name for such a taxon is the synonym with the earliest date of publication as an available name; the others are 'junior' synonyms. A junior synonym may be resurrected and used again if the senior synonym is invalidated as a homonym (below), or if the junior name is found to refer after all to a taxon distinct from that bearing the senior name. As between co-ordinate taxa, the priority of a name is not affected by elevation or reduction in the rank of the taxon to which it refers.

The principle of priority is applied in certain other situations, as detailed in the Code, but it is also subject to limitations.

The Law of Homonymy. The requirement that a given name should be used for only one taxon may also be unfulfilled in practice. This happens if an author publishes a name for a new taxon, in ignorance of the fact that the same name is already in use for a different taxon of the same rank. The ostensibly different and independently published, but nevertheless identical names are termed *homonyms*. The Law of Homonymy states that a junior homonym of an available name must be rejected and replaced, by its oldest available synonym, if it has one, or otherwise by a new name.

Names in the species group represent a

special case, for these can be homonymous only if they are originally published in the same genus ('primary' homonyms), or subsequently brought together in the same genus ('secondary' homonyms). A junior primary homonym must be permanently rejected; a junior secondary homonym must be rejected by those who believe that the taxa in question are 'congeneric', i.e. belong in the same genus. Certain specified variant spellings are disregarded for the purpose of determining whether species-group names are identical.

Types. The necessary permanent connection between scientific names and the taxa to which they refer is established by the method of 'types'. The principles of this method may be illustrated by reference to the species group.

An author describing a new species or subspecies actually bases the name and description on one or more specimens. If he has only one, then that specimen, under the Code, is the *holotype*. If he has more than one, he should select one (and only one) as the holotype and label it as such. The holotype is the ultimate standard of reference that determines the application of the name. It is not necessarily 'typical' of the species in any ordinary sense; it merely serves to register that the name applies to the species of which the holotype is a member. Views may differ (on taxonomic grounds) as to what other specimens should be included in the species and therefore bear the name, but there can be no dispute as to the name of the holotype.

The earlier authors did not usually follow this practice, and the Code indicates how a type specimen should be selected in that case. If it can be established which specimens the original author had before him, then those specimens, subject to certain restrictions, constitute *syntypes*; a subsequent author may select one of these to be the *lectotype*, and the first such selection to be published must be accepted. The lectotype serves exactly the same purpose as a holotype. If, through loss or destruction, no holotype, lectotype, or syntype exists, some other specimen may be designated a *neotype*, provided that certain rather exacting preconditions are satisfied; the neotype also serves the same purpose as a holotype. The type specimen of a species, whether it be holotype, lectotype, or neotype, should be lodged in a museum or similar institution, where it can be safely preserved and made available for purposes of research.

Specimens other than the holotype which the original author considered to belong to the same species are termed *paratypes*; and syntypes other than the lectotype are termed *paralectotypes*. Neither of these constitutes type specimens in the strict sense, and it is always possible that they may prove to belong to a different species.

The type of a genus or subgenus is not a specimen, but a species, termed the *type species*. A generic name published after 1930 is not available unless the author designated a type species for it. If no type species was designated by the author of a name published before 1931, the Code states in detail how its type species should be determined. Failing certain positive indications—e.g. if only one species was referred to the genus by its author, that is the type species—a later author is empowered to select as type species any of the species listed by the original author, subject to certain limitations and conditions. As with the type specimen of a species, the type species of a genus is not necessarily typical of the genus. It is, however, the criterion by which the application of the generic name must be judged: the peg on which the name is hung. The limits of the genus bearing that name may be broadened or narrowed according to the taxonomic views of individual zoologists, but the type species must always remain in it.

The type of a taxon in the family group is a genus, the *type genus*. This is defined as the genus whose name is incorporated in the name of the family-group taxon.

If a taxon has subordinate taxa—e.g. subgenera in the case of a genus, subspecies in the case of a species—its type is automatically also the type of the nominate subordinate taxon and vice versa. For example, the type specimen of the butterfly subspecies *Papilio aegeus aegeus* is the same as that of the species *P. aegeus*. Demonstration that two nominally different taxa have the same type immediately establishes them as *objective* synonyms; nominal taxa that may be considered synonymous although having different types are *subjective* synonyms. A new name published to replace a junior homonym takes the type of the name it replaces.

The International Commission

The International Code aims to provide comprehensive guidance in the nomenclatural field. However, from time to time cases arise in which the correct procedure is open to doubt. To meet this situation, the International Commission on Zoological Nomenclature is empowered to issue authoritative interpretations of the Code, and the Code itself requires that in certain specified situations a decision should be sought from the Commission.

Many names that are invalid under the present Code have been in general use for many years, sometimes for taxa cited in dozens of major textbooks. Since the principal aim of the Code is to promote stability and universality of nomenclature, it is clearly undesirable that it should itself be a cause of name-changing in such cases. The International Commission is therefore empowered also to suspend the operation of the Code in individual instances brought before it for decision, and to make whatever ruling will best serve the interests of stability and universality of nomenclature.

Decisions of the Commission are published as Opinions or Declarations. Every name that the Commission validates is placed on one of the Official Lists of names in zoology, of which one is devoted to family-group, one to genus-group, and one to species-group names. There is also a series of corresponding Official Indexes, on which names ruled invalid are inscribed.

Nomenclature of Higher Taxa

Although names above the family group are not covered by the Code, there is a general tendency to follow similar nomenclatural practices in this field. All such names are latinized plurals, with capital initial letter, and are normally printed in Roman type. Priority is recognized in a loose sort of way and homonymy is rectified. However, the type concept is applied only sporadically and there are no distinctive suffixes for taxa of different rank, although in the insects many ordinal names end in '-ptera' and many subordinal ones in '-odea'.

GENERAL

The enormous number of animal taxa and the archival nature of publications dealing with their classification and nomenclature give the literature special importance in taxonomy. Apart from the International Code and the Official Lists and Indexes, great importance attaches to generic nomenclators such as that of Neave (1939–50), comprehensive world catalogues of major groups, monographic revisions, etc. At the other end of the scale, the working taxonomist has to keep in touch with the current output of new names in his group, which may be scattered over hundreds of journals in many languages. Here some of the biological abstracting and bibliographical journals are of great assistance, and above all the *Zoological Record*, published annually by the Zoological Society of London.

7

EVOLUTION AND CLASSIFICATION OF THE INSECTS

by I. M. MACKERRAS

The insects might be defined as labiate, hexapod, terrestrial arthropods, but that definition would ignore several important questions that have been the subject of study and speculation for more than a hundred years. Three are crucial. From what ancestry did they evolve? Are they monophyletic? How may they be subdivided so as to reflect the evolutionary cleavages that have occurred in them during their long history from the Palaeozoic to the present day? We shall be concerned with these questions, because all modern classifications of the insects and their relatives, at the levels to be considered here, have been based on attempts to answer them. This is the antithesis of the point of view developed in the preceding chapter. There is a fourth question that has received less attention, namely: how far may some of the major taxonomic divisions represent grades of organization (p. 161) similar to those that have been recognized in the vertebrates? It, too, must be considered, if we are to understand the evolutionary implications of the classification we are using.

Keys to orders have no more than ephemeral value for users of a book of this kind. With a little experience, the great majority of insects can be placed in an order at a glance, whereas the few that cannot require detailed comparative study that is beyond the scope of any manageable key. Consequently, the later sections of the chapter will consist mainly of short notes on the salient features and relationships of the surviving orders, brought together to provide an introductory conspectus and to facilitate comparison. The extinct orders are described in chapter 8, and the complete classification is set out in Table 8.1.

The Ancestry of Insects

This problem has been admirably reviewed by Tiegs and Manton (1958). Five major theories had been proposed up to 1930.

1. By Brauer, in 1869, that *Campodea* is close to the ancestral stock of the insects, and is itself derivable from chilopod ancestors. The second part of the theory has been rejected (most recently by Manton, 1964); the first part will require further consideration.

2. By Packard, in 1873, that the insects had a common origin with a group of small myriapods, which Ryder named the Symphyla in 1880 to mark 'the singular combination of myriapodous, insectan and thysanurous characters which it presents.' This also will require further consideration.

3. By Hansen, in 1893, that the insects evolved from Crustacea. This view was elaborated into a wider theory of arthropod phylogeny by Lankester

in 1904, and, in various forms, it received strong support from G. Carpenter, Snodgrass, and others.

4. By Handlirsch, in 1908, that insects evolved from trilobites, with a corollary (also in Lankester's theory) that the original insects had wings.

5. By Tillyard, in 1930, who demonstrated the untenability of both the crustacean and trilobite theories. He nearly accepted the symphylan theory, rejecting it only on what was then believed to be a fundamental division between the progoneate Symphyla and the opisthogoneate insects. He met this difficulty by proposing a hypothetical, few-segmented 'protapteron' (analogous with the *Nauplius* larva of Crustacea) as a common ancestor for myriapods and insects.

Tillyard's views were strongly contested. Calman and Imms, both in 1936, returned to the symphylan theory, and the recapitulation theory of Haeckel, on which Tillyard depended, was under heavy fire from de Beer and others. It remained, however, for Tiegs (1940, 1945) to attack the problem by seeking for essentially new evidence. In an elegant study of the embryology and postembryonic development of an Australian symphylan, *Hanseniella agilis*, he demonstrated that the gonoduct was a secondary epithelial invagination, showed that the distinction between progoneates and opisthogoneates was not a fundamental one, found additional features to link the Symphyla with the Diplura, and added to the evidence indicating affinity between Symphyla and Onychophora. He amplified these findings in a study of *Pauropus* in 1947.

His findings considerably reinforced Packard's theory, and we may therefore pause at this point to look at the Symphyla. They are small, slender, unpigmented, eyeless, agile myriapods (Fig. 7.1A), which live in soil, leaf litter, and rotten logs. Their skeletal anatomy has been described by Snodgrass (1952) and the systematics of Australian species reviewed by Scheller (1961), except for a Tasmanian genus described by Chamberlin (1920). The head is prognathous; antennae long, moniliform, and containing intrinsic muscles in all segments but the last; post-antennal organ of Tömösvary present; mouth-parts ectognathous and trignathous, the mandibles with a movable gnathal lobe, and the labium of peculiar form (Fig. 7.1C). There are 14 postcephalic segments: 1 and its appendages more or less reduced; 2 to 12 with short, 5-segmented legs, and with styles at the bases of the coxae and eversible vesicles medial to them on most segments (Fig. 1.21A); 13 with legs replaced by a pair of short cerci containing the ducts of silk

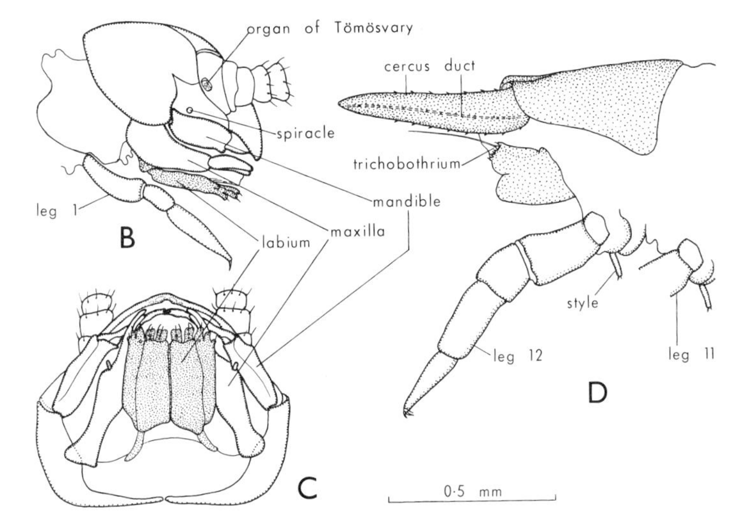

Fig. 7.1. *Hanseniella* sp., Symphyla-Scutigerellidae: A, live adult–length of body 6 mm [photo by C. Lourandos]; B, head, lateral; C, head, ventral; D, terminal segments, lateral. Labium stippled in B and C.

glands; 14 bearing the anus and a pair of trichobothria (Fig. 7.1D). The nervous, digestive, and circulatory systems are generalized; there are 2 long Malpighian tubes; tracheae are few and short, and open by a pair of spiracles on the head; the gonads have parietal germaria; and the gonopore is between the 4th pair of legs.

Cleavage of the ovum is total; there is early ventral flexure of the germ band; an amnion is not formed; and the embryo has a pair of large premandibular glands, a distinctive type of dorsal organ, and a series of segmental 'ventral organs'. The larva hatches with 6 or 7 pairs of legs; postembryonic growth is *anamorphic*, additional segments being differentiated between the preanal and anal segments, accompanied by successive replacement of the cerci. The ectodermal gonoducts begin to develop in larvae with 9 or 10 pairs of legs. Growth and ecdysis continue throughout life.

There is little room to doubt that the insects evolved from terrestrial ancestors that had antennae, ectognathous, labiate mouthparts, tracheal respiration, and, in all probability, 14 postcephalic segments, 12 pairs of legs, coxal (or subcoxal) styles and eversible vesicles on most segments, and a pair of terminal or subterminal cerci. Nevertheless, there are two evident difficulties in accepting the Symphyla as being really close to those ancestors. One, as Tiegs appreciated, is the considerable assemblage of derived *(apomorphic)* myriapod characters that they show. The other is that they lead more directly towards the Diplura than to the true insects (as represented by the Apterygota), and it seems impossible to derive the true insects from any recognizably protodipluran stock.

These difficulties (and others that were not so apparent) have been resolved by Manton (1964) in a detailed comparative study of mandibular mechanisms in the Arthropoda. Those of her conclusions that are immediately relevant to the present discussion are:

1. The terrestrial onychophoran–myriapod–hexapod evolutionary line of organisms with jaws (i.e. mandibles) derived from whole limbs was completely separate from the crustacean and trilobite–chelicerate lines with jaws derived from the bases of the limbs.

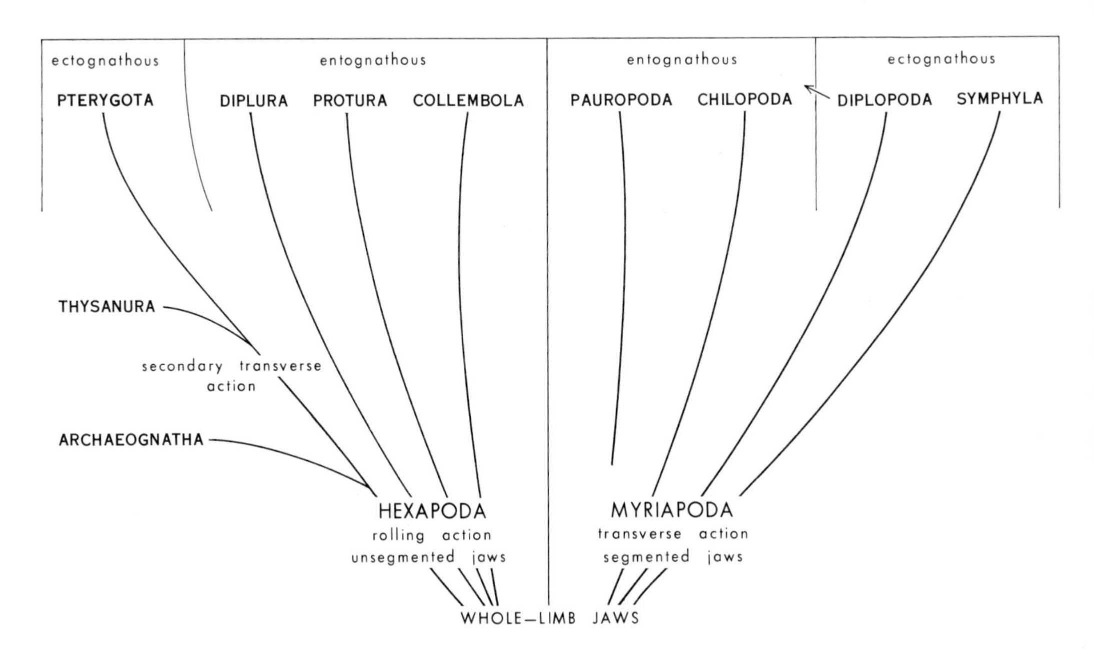

Fig. 7.2. The presumed evolutionary relationships of the hexapods and myriapods (based on Manton, 1964).

2. There was an early division of the animals with whole-limb jaws into two lines: one with unsegmented jaws and a rolling biting mechanism, from which the hexapods evolved; and the other with segmented jaws and a transversely biting mechanism, from which the myriapods evolved.

3. The jaw mechanisms of Symphyla are of the myriapod type, and they therefore could not be relicts of stock from which the hexapods had also evolved. Their resemblance to some primitive hexapods is due, partly to retention of primitive *(plesiomorphic)* characters that were possessed by their common ancestors, and partly to parallel evolution or convergence, as with the fusion of the 2nd maxillae into a labium.

4. The unsegmented rolling jaws of *Petrobius* (Archaeognatha) are not far removed from the primitive type, from which the entognathy of the Collembola, Protura, and Diplura could have evolved on the one hand, and the ectognathous, secondarily transversely biting jaws of the Thysanura and winged insects on the other.

Occasionally one finds that an analysis based on a particular set of structures provides a pattern into which the less conclusive, sometimes apparently conflicting evidence that had previously been available immediately fits in a convincingly coherent way. A comparison with alternative classifications that have been proposed (for example, that of Remington, 1955) suggests that this has happened here. Consequently, we may return to the classical Myriapoda and Hexapoda with increased confidence in the phylogenetic justification for treating them as separate superclasses within the subphylum Antennata. Their relationships are shown diagrammatically in Figure 7.2. The major divisions must have been established quite early in the development of a land fauna, because myriapods, Collembola, and a winged insect are known from the Devonian.

Sharov (1966), whose book was in the press when Manton's paper appeared, holds different views. He would follow Snodgrass in deriving the myriapod–hexapod line of evolution ('Atelocerata') from crustacean ancestors, and Tiegs in including Symphyla with the hexapods (as 'Dimalata' = Labiata of Tiegs) rather than with the other myriapods ('Monomalata'). He does not, however, bring any fresh evidence to bear on the problem.

Superclass HEXAPODA

Terrestrial or secondarily aquatic arthropods. Normally with antennae; mouthparts trignathous, comprising unsegmented whole-limb mandibles, 1st maxillae, and labium; head with anterior and posterior tentorial apodemes (Manton); postcephalic segments divided into a thorax of 3 segments, normally with a pair of legs on each, and an abdomen of 6 to 12 segments, with rudiments of limbs on the first 9 in primitive forms; *opisthogoneate* (i.e. gonopore near posterior end of body). Respiration normally tracheal; gonads normally with apical germaria. Cleavage of ovum normally meroblastic; postembryonic development normally *epimorphic* (i.e. there is no increase in the number of body segments during larval growth).

This assemblage includes the Collembola, Protura, Diplura, the Apterygota, and the winged insects, the first three being entognathous and the last two ectognathous.

The entognathous groups have the preoral cavity enclosed laterally by pleural folds which grow down from the sides of the head to fuse with the labium (Fig. 7.3). The mandibles are very mobile, and adapted for piercing, cutting, and triturating small particles, which can be well salivated and sucked up efficiently through the enclosed space; the hypopharynx is well developed, but both maxillary and labial palpi are reduced or absent. These groups have other features in common. Their heads are prognathous;

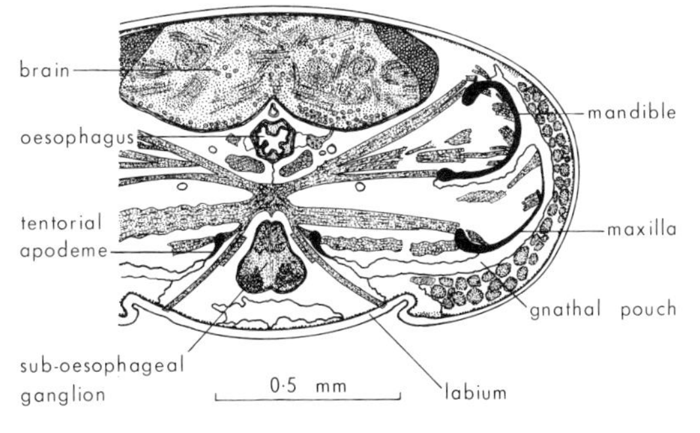

Fig. 7.3. Transverse section of head of *Heterojapyx evansi*, Diplura-Japygidae, to illustrate entognathy (from a preparation by Miss B. J. Bancroft, Australian National University).

Collembola and Diplura have intrinsic muscles in flagellar antennal segments (antennae absent in Protura); eyes are rudimentary (some Collembola) or absent; the thorax is feebly developed; the legs generally short and relatively inefficient; tarsi not more than 1-segmented; abdominal appendages markedly reduced or modified, and none on the scarcely modified genital segments; tracheae weakly developed; Malpighian tubes rudimentary or absent; insemination (where known) by indirect transfer of spermatophores; embryo with symphylan-like dorsal organ and no amnion in Collembola and Diplura (unknown in Protura).

This combination of characteristics might suggest a monophyletic origin of these three orders, which could thus be brought together in a class Entognatha, as Tuxen (1959) suggested. But some of them are simply ancestral features, and many, including the entognathy, are plainly adaptations to life in confined, humid habitats, from which only the Collembola have become partly emancipated. Moreover, the hexapod condition could be polyphyletic (Tiegs and Manton, 1958), and there are differences between the jaw mechanisms of Collembola and Diplura that suggest that they may have evolved independently from protomachiloid ancestors (Manton, 1964). Perhaps most important, there are significant differences between these three groups, and between any of them and the true insects, so it seems appropriate to treat them all as separate classes. It does not, however, appear necessary to change the familiar ordinal names when they are raised in rank.

Class and Order COLLEMBOLA

Small to minute, entognathous hexapods (Fig. 10.1). Antennae with few segments; organ of Tömösvary often present; legs 4-segmented, the distal segment a tibiotarsus; abdomen with never more than 6 segments, even in embryo, usually with specialized appendages on 1, 3, and 4; gonopore on 5; no cerci or appendix dorsalis; spiracular opening, when present, on neck. Gonads usually with parietal germaria; cleavage of ovum holoblastic; larval development epimorphic. There are many 'myriapodan' features in the embryology, and it seems that the Collembola must have diverged far back in the ancestry of the hexapods.

Class and Order PROTURA

Minute, elongate, entognathous hexapods (Fig. 10.9). Antennae absent; 'pseudoculi' may represent antennal bases or organs of Tömösvary; legs 5-segmented, fore pair tactile, mid and hind ambulatory; abdomen 12-segmented in adult, with rudimentary limbs on 1–3; gonopore between 11 and 12; no cerci, but segment 12 with median prolongation; spiracles, when present, on thoracic terga. Gonads with apical germaria; embryology unknown; larva hatches with 9 abdominal segments, and growth is anamorphic. The Protura show a baffling mixture of characters (Tuxen, 1958), and are best treated as distinct from both Diplura and insects until more is known about them.

Class and Order DIPLURA

Small to large, symphylan-like or dermapteron-like, entognathous hexapods (Figs. 7.4, 10.15). Antennae long, moniliform; no organ of Tömösvary; mandible often with a

Fig. 7.4. *Campodea* sp., Diplura-Campodeidae, live adult–length of body 2·5 mm. [Photo by C. Lourandos]

movable prostheca; legs 5-segmented; abdomen 10-segmented, usually with rudimentary limbs and eversible vesicles on most of first 7; gonopore between 8 and 9; cerci present, sometimes long and moniliform, sometimes short and forcipate, and sometimes containing the duct of a silk gland; no appendix dorsalis; tracheae relatively well developed, with 2 to 4 pairs of spiracles on thoracic pleura and none or 7 on abdominal pleura. Gonads with apical germaria; cleavage of ovum meroblastic; larval development epimorphic.

The Diplura have been included in the Apterygota by many writers, but their entognathy alone would remove them from the direct line of insect ancestry. Taking all their characters into account, they might be an offshoot from protomachiloid stock, but the converse could not be true. They seem more likely to have diverged from earlier protohexapods that failed to escape from the ancestral environment.

Class INSECTA

Minute to very large, very varied, ectognathous hexapods. Head normally hypognathous; compound eyes and ocelli normally present; antennae with intrinsic muscles in scape only; no organ of Tömösvary; maxillary and labial palpi normally well developed; thorax normally well developed; legs normally with more than 5 segments (i.e. tarsi more than 1-segmented); abdomen primitively 11-segmented; gonopore almost always on 8 in female, on 9 in male; usually 8 and 9 of female and 9 of male (as well as some of preceding segments in very primitive forms) with appendages; cerci usually present, sometimes forcipate, but never containing the duct of a silk gland; an appendix dorsalis in some primitive forms; tracheae normally well developed, with pleural spiracles (or their rudiments) on 2nd and 3rd thoracic and first 8 abdominal segments. Malpighian tubes almost always well developed; gonads with apical germaria; cleavage of ovum normally meroblastic; no strong ventral flexure of germ band; embryo normally enclosed ventrally in an amniotic cavity covered by amnion and serosa; dorsal organs unlike those of Symphyla; larval development epimorphic.

This definition is designed to emphasize the basic differences from the other hexapod classes, and takes little account of the immense diversity of structure in the insects, including the development of wings. The essential feature of insect evolution has been emancipation from ancestral habitats, and that depended on three evolutionary developments.

1. Reduction of functional walking legs to three pairs divided the body into thorax and abdomen, which could become specialized for locomotion and for trophic and reproductive functions, respectively. As with so many critical steps in evolution, the immediate advantages were probably small, and the potentialities of the new structural foundation could not be realized until the second step had been taken.

2. This was emergence into the open air on the exposed surface of the ground, and it depended primarily on control of water loss. Reduction of cuticular permeability, elaboration of the respiratory system for better ventilation and control of evaporation, development of an impervious chorion on the egg, and evolution of an amnion to protect the embryo were probably the most important basic modifications to this end.

3. Once free on the surface, but not until then, selection pressures for speed and agility would have come into operation, leading to the development of longer, stronger legs carried on a larger, more powerful thorax. This provided a point of balance in the anterior third of the body; while raising the anterior end off the substrate provided a basis for evolution of the varied types of hypognathous mouth-parts that occur in the insects.

The most primitive surviving insects, the Archaeognatha (Fig. 7.5), have reached this stage of evolution, and it may be taken as a critical point of demarcation between the insects and their protohexapod progenitors. Nevertheless, the Archaeognatha show some extremely primitive features, so their separa-

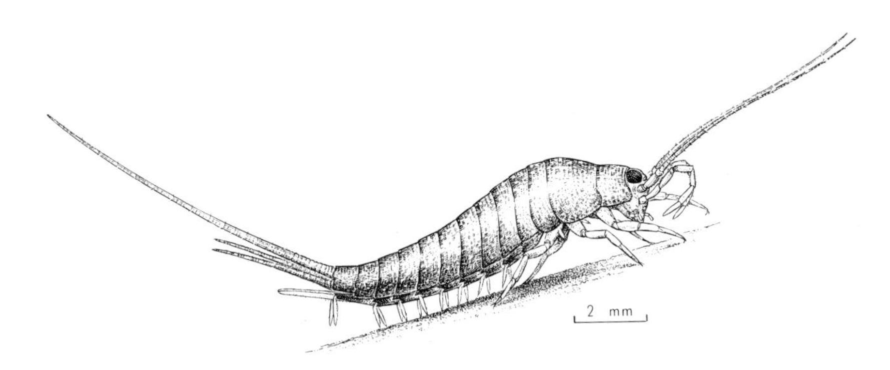

Fig. 7.5. *Allomachilis froggatti*, Archaeognatha-Meinertellidae, ♀, running on a rough, sloping surface (from a slow-motion cinematograph film by C. Lourandos). [M. Quick]

tion must have occurred very early in the history of the hexapod line. Thus, they have a rolling type of mandibular mechanism and efficient hydraulic suction in the preoral cavity (Manton, 1964), coxal styles on some thoracic legs, both rudimentary limbs and eversible vesicles on abdominal segments to the 9th, a primitive spiracular apparatus, and an incompletely closed amniotic cavity. They also have an appendix dorsalis as well as cerci. The Monura (Fig. 8.1) were even more primitive than the Archaeognatha in having distinguishable tergal rudiments of the mandibular to labial segments in the adult, less differentiated thoracic segments, 1-segmented tarsi, and less reduced abdominal limbs. They had a long appendix dorsalis, but apparently no cerci, a curious arrangement for which there is no obvious explanation.

The principal further developments in apterygote evolution, shown by the Thysanura, are: conversion of the rolling mandible into a hinged, transversely biting jaw, capable of a wide gape and able to deal with relatively large masses of hard food, though at the cost of decreased hydraulic efficiency (Manton); loss of styles on the thoracic limbs; and progressive reduction of the abdominal limb rudiments, except those on segments 8 and 9 of the female and 9 of the male. In these respects, the Thysanura forecast conditions found in the winged insects.

The Archaeognatha and Thysanura show few indications of regression associated with their mode of life, and it is now generally agreed that they are truly primitive insects. Consequently, the accepted primary division of the class is into Apterygota and Pterygota.

Subclass APTERYGOTA

Primitively wingless, terrestrial insects. Thorax with three subequal, separate segments; pleural sclerites small; abdomen 11-segmented, with limb rudiments on at least some of the pre-genital segments; long, multisegmented appendix dorsalis present, and also cerci in the recent orders; spiracles without regulatory apparatus operated by muscles. Mating by indirect transfer of spermatophores; amniotic cavity of embryo incompletely closed; development of larva direct, with little change other than attainment of sexual maturity; ecdysis repeated throughout life.

Order ARCHAEOGNATHA (Figs. 7.5, 11.2; 3 Australian spp.)

Mostly medium-sized, subcylindrical, rather laterally compressed apterygotes; living under bark, in litter, and in rock crevices (often on coastal cliffs), and able to jump considerable distances by sudden flexion of the abdomen. Mandibles monocondylar; eyes and ocelli large; thorax strongly arched, coxal styles often present; abdominal styles well developed, spiracles on segments 2–8, appendix dorsalis longer than cerci.

Order THYSANURA (Silverfish; Fig. 11.4; 23 Australian spp.)

Small to medium-sized, fusiform or cylindrical, rather dorsoventrally compressed apterygotes; living in forest litter, rotten logs, under stones, in houses, or as inquilines in ants' or termites' nests; can run quickly, but do not jump. Mandibles dicondylar; eyes small or absent; ocelli absent, except in the primitive Lepidothrichidae; thorax not strongly arched, no coxal styles; abdominal styles generally reduced in size and number, spiracles on segments 1–8, appendix dorsalis and cerci usually subequal in length.

Subclass PTERYGOTA

Winged or secondarily apterous, terrestrial or aquatic insects. Thoracic segments usually large, mesothorax and metathorax generally more or less united to form a pterothorax; pleural sclerites well developed; abdomen with 8 to 11 distinguishable segments; pregenital segments without appendages; terminalia often modified in relation to habits of mating or oviposition; spiracles with regulatory apparatus provided with muscles (secondarily lost in Diptera). Mating always by copulation; amniotic cavity closed (absent in a few Hymenoptera); metamorphosis from larva to adult slight to great; ecdysis ceases on attainment of sexual maturity.

The development of wings was the outstanding event in insect evolution, because it provided a means to escape from enemies, facilitated dispersal, and so opened up new environmental niches for the pterygotes to exploit. Something like the hexapod shape, with its muscular thoracic tagma and potentially aerodynamic balance, was a prerequisite for true flight; but there is no direct evidence to show how the wings developed on that foundation. However, it would appear from a recent discussion (Wigglesworth and others, 1963) that there is fairly wide agreement on three sets of propositions:

1. That wings developed in a single, terrestrial, apterygote stock (though some believe that they evolved more than once, and a few that the pre-pterygotes were aquatic).
2. That they developed from paranotal processes on the sides of the thorax.
3. That there were three principal stages in their evolution: (a) development of the processes for some function other than flight; (b) their use to aid in some aerial operation; (c) development of a basal articulation and flapping mechanism.

Paranotal processes (or equivalent structures) occur in a variety of flightless arthropods, including Thysanura, so it is not necessary to postulate that the pre-pterygotes had none. The difficulty is to visualize the aerial operations for which they were first used and the selection pressures that favoured their enlargement during that stage. Wigglesworth (1963a, and in the discussion) suggested that they aided light-bodied insects to disperse in an aerial plankton, Hinton (in the discussion) that they provided medium-sized insects with attitudinal control in landing from a fall and thereby favoured escape from predators. Flower (1964) has examined the aerodynamic implications of both suggestions, and concluded that rudimentary wings would have no selective advantages for small insects dispersing in air currents, whereas they would have unique advantages for controlling attitude and prolonging glide in insects 1–2 cm long. Once control had been achieved, the subsequent evolution of gliding planes would have been a relatively simple matter; and several authors have pointed out that the insects alone among arthropods are structurally 'pre-adapted', so that the transition from flexing to flapping movement of the planes would also have been a relatively simple one.

Gliding would have aided local dispersal and true flight wider dispersal, and there is little doubt that these were potent influences in the evolution of efficient wings. The early flights increased the ability of the insects to find habitats to which they were already adapted, but they would not have involved exploitation of previously unoccupied ecological niches. That came later, and it produced new selection pressures, because the permanently extended wings of the earlier groups would have been an impediment to movement in confined places, and there was probably also increasing need to protect the soft sides of the body. The primitive wing articulation became modified to permit the wings to be folded back along the body, and so, it is usually believed, arose the primary division of the Pterygota into Palaeoptera and Neoptera that was first recognized by both Martynov and Crampton in 1924.

On this view, Neoptera were derived from Palaeoptera. Some Russian workers, on the other hand (see Martynova, 1961), consider that the Neoptera were the more ancient. Sharov (1966) holds that neither group could have been derived directly from the other. He (and also Riek, unpublished) believes that the expanding paranotal lobes of the ancestral pterygotes did not grow out laterally, but obliquely backwards as in the larvae of many recent and extinct exopterygotes. He includes the Devonian Archaeoptera (p. 176) and the Carboniferous Paoliidae (Protoptera) in an infraclass Archoptera characterized by an oblique position of the wings at rest, from the paoliid section of which the Palaeoptera and Neoptera evolved independently in adaptation to different kinds of habitat and ways of life. This theory has attractive features, but it is founded on as yet slender fossil evidence, and its aerodynamic implications have not been examined.

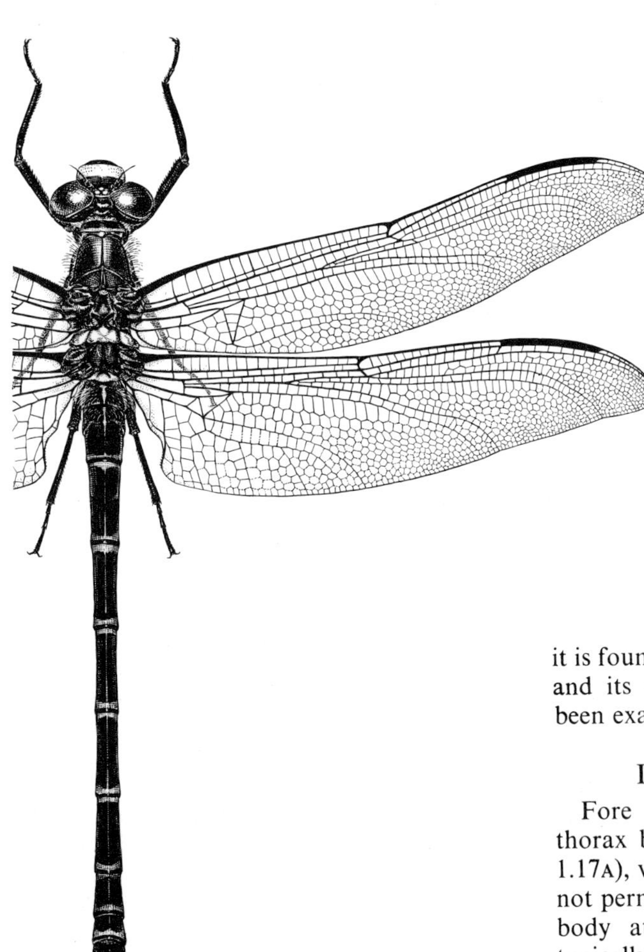

Fig. 7.6. *Petalura ingentissima*, Odonata-Petaluridae, ♂, the largest Australian dragonfly–natural size. [M. Quick]

Infraclass PALAEOPTERA

Fore and hind wings articulated to the thorax by humeral and axillary plates (Fig. 1.17A), which are fused with the veins and do not permit the wings to be folded against the body at rest; venation of recent orders typically with triadic veins and a network of cross-veins; abdomen with at least rudiments of cerci and sometimes an appendix dorsalis.

Larvae of recent orders aquatic; wings develop in external sheaths; change of form at ecdysis to imago (or subimago) considerable.

The two surviving orders belong to widely different lineages, which must have developed independently from a very early stage in pterygote evolution.

Order EPHEMEROPTERA (Mayflies; Fig. 12.1; 124 Australian spp.)

Delicate, short-lived insects, with vestigial mouth-parts, wide fore wings, reduced (or absent) hind wings, long cerci, and usually a long appendix dorsalis. Larvae machiliform, but with segmentally arranged tracheal gills. Mayflies are always associated with water, and are the only recent insects with two imaginal winged instars. They are, in many ways, the most primitive surviving pterygotes.

Order ODONATA (Dragonflies, damselflies; Plate 1; Fig. 7.6; 248 Australian spp.)

Active, usually strongly flying, predacious insects, with large eyes, biting mouth-parts, greatly reduced cerci, and no unequivocal appendix dorsalis; distinguished by their narrow, subequal, uncoupled, gauzy fore and hind wings, unique method of sperm transfer (p. 252), and association with water. Larvae with a protrusible, raptorial labium, the costal borders of the wing-sheaths apposed, and distinctive types of tracheal gills. The Odonata are ancient but specialized insects that have made efficient use of a basically primitive wing venation.

Infraclass NEOPTERA

Wings articulated by discrete axillaries (Fig. 1.17B) that permit them to be folded back along the body (except in secondarily specialized forms); venation with few or no triadic veins and usually a less dense reticulation of cross-veins than in the Palaeoptera; abdomen usually with cerci, but without appendix dorsalis. Larvae of varied form and habitat; metamorphosis slight to complete; no subimaginal winged instar.

The Pterygota were formerly divided into Exopterygota (or Hemimetabola), in which the wing-sheaths develop externally on the larva (usually known as a 'nymph'), and Endopterygota (or Holometabola), in which the wing-sheaths develop inside the larva and are everted after the penultimate (larval-pupal) apolysis (p. 95). The Palaeoptera were included in the Exopterygota, so the arrangement is inconsistent with the evolutionary sequence outlined above. Nevertheless, it has a broad general significance, and the designations *exopterygote* and *endopterygote* are useful descriptive terms.

They represent, in fact, *grades of organization* in the sense of de Beer (1954) and Huxley (1958).* A grade is not necessarily a monophyletic assemblage. Its essential feature is that the structural advances that characterize it, not only meet immediate adaptive needs, but collectively provide a foundation for a new series of adaptive radiations. Thus, attainment of the hexapod grade conferred immediate advantages in itself (Manton, 1958), and it incidentally provided a structural basis for the evolution of flight, but that much more important development was possible only for those hexapods that also developed efficient water-regulatory mechanisms. On criteria of this kind the recent insects could be divided into four progressive evolutionary grades: hexapod and apterygote; palaeopterous and exopterygote; neopterous and exopterygote; neopterous and endopterygote.

Grades are characterized by structural and functional developments; but they are recognized by the biological success of the animals that are included in them, and this may be measured, broadly, by the variety of environmental niches they occupy, by the aptness of their adaptations to those niches, by their population densities, and usually by their taxonomic diversity. The bursts of adaptive radiation that followed the emergence of new grades in the evolution of the insects have been shown diagrammatically by Smart (1963), while, in groups as old as these, the numbers of species also provide a useful index of their capacity to diversify. The following figures for the principal supraordinal divisions are therefore illuminating.

* See also I. M. Mackerras, *J. Aust. ent. Soc.* **6**: 3–11, 1967.

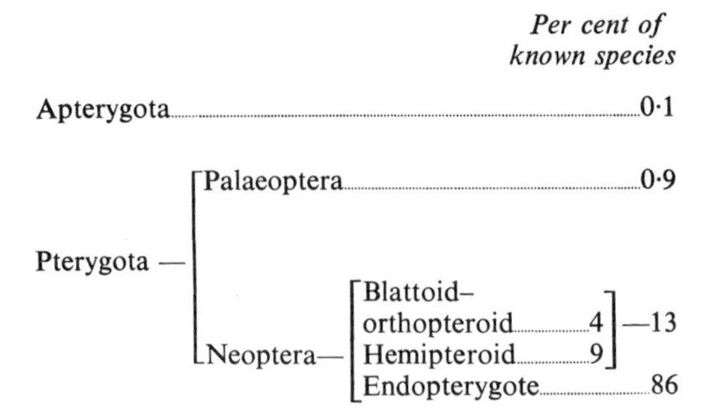

There have been various attempts to divide the Neoptera into monophyletic groups, and there seems to be a consensus of opinion among recent writers (Jeannel, 1949a; H. Ross, 1955; Hennig, 1962; also Richards and Davies, 1960, when their primary division is omitted) that there were three main lines of neopterous evolution, which probably arose independently, can be distinguished from each other by broad morphological features, and can be regarded as monophyletic in the sense of Simpson (1961), in that, at the least, each arose from a group of related ancestors. Martynov gave these divisions the names Polyneoptera, Paraneoptera, and Oligoneoptera, the first two being exopterygote and the last endopterygote. However, the position of a few orders is still obscure, and there is still insufficient agreement about superordinal groupings within the divisions, so we prefer to treat them informally as the 'blattoid–orthopteroid', 'hemipteroid', and 'endopterygote' groups of orders, respectively.

The Blattoid–Orthopteroid Orders

Mouth-parts of generalized mandibulate type; fore wing usually thickened to form a tegmen, hind wing usually with a large, fan-like ano-jugal lobe; venation usually complex; cerci present; terminalia of male often asymmetrical, and usually with coxites atrophied or absorbed into sternum 9; Malpighian tubes normally numerous; ventral nerve cord normally with several discrete ganglia. Larvae exopterygote; metamorphosis gradual and relatively slight.

The group is a very ancient one, possibly extending back into the Devonian. It can be divided into two long-separated series, the Blattodea, Isoptera, Mantodea, and probably Zoraptera forming one, and the Grylloblattodea, Dermaptera, Plecoptera, Orthoptera, Phasmatodea, and Embioptera the other. The position of the Dermaptera was disputed, but Giles (1963) has marshalled evidence to show that they are most nearly related to the Grylloblattodea.

Order BLATTODEA (Cockroaches; Plate 2, B–D; 439 Australian spp.)

Flattened, usually rather soft-bodied, terrestrial insects, with shield-like pronotum, cursorial legs, and the wings, when present, folded flat on the body, the fore wings forming overlapping tegmina; eggs laid in an ootheca. Mostly active and fast-moving, but cryptic, living under bark, logs, etc., sometimes on vegetation, and a few species serious domestic pests.

Order ISOPTERA (Termites; Figs. 5.42, 15.1; 182 Australian spp.)

Relatively small, soft-bodied, polymorphic, social insects, the winged caste with subequal, untegminized wings which are discarded after the dispersal flight. Distinguished especially by their habits of constructing galleries and nests (Figs. 15.8, 9) and of subsisting on wood or dried grass; many species are of economic importance. Termites are little more than greatly modified cockroaches.

Order MANTODEA (Praying mantids; Plate 2, A; 118 Australian spp.)

Distinctive predatory insects, with raptorial fore legs and a characteristic jugal lobe in the more or less tegminized fore wings; eggs laid in an ootheca. Generally found on vegetation. They are often combined with Blattodea in an order Dictyoptera, but are probably older and more widely separated than the Isoptera.

Order ZORAPTERA (Fig. 17.1)

Very small, winged or apterous insects, with no ovipositor; winged forms distinguished in the group by their reduced venation and small hind wings. Found under bark, in decaying wood, etc. Not known from Australia.

Order GRYLLOBLATTODEA (Fig. 18.1)

Eyeless, wingless, cryptic, primitive insects, with unmodified legs and a well-developed ovipositor. Not known from Australia.

Order DERMAPTERA (Earwigs; Plate 2, E; 60 Australian spp.)

Flattened, often apterous insects, with short legs and a long, flexible abdomen which almost always ends in a pair of strong forcipate cerci. Usually found under stones, logs, or bark. Winged forms can be distinguished from staphylinid beetles by their forceps and rounded hind wings with characteristic radiating venation; wingless forms from large Diplura *(Heterojapyx)* by their eyes, ectognathous mouth-parts, and 3-segmented tarsi.

Order PLECOPTERA (Stoneflies; Fig. 20.1; 84 Australian spp.)

Elongate, soft-bodied insects, with clearly separated thoracic segments, cursorial legs, and usually untegminized fore wings, which are characteristically folded straight back and closely applied to the abdomen (a ready distinction from most Trichoptera (Fig. 7.10) in the same situations); cerci usually long. Always associated with water; the only orthopteroid order with aquatic larvae, which may be distinguished from those of Ephemeroptera by their large prothorax and absence of an appendix dorsalis. The Plecoptera are, in some ways, the most primitive surviving orthopteroid order.

Order ORTHOPTERA (Locusts, grasshoppers, crickets; Plate 2, F–L; 1,513 Australian spp.)

An abundant, widespread order of usually medium-sized to large, winged, brachypterous, or apterous insects, with large pronotum bent downwards to form lateral lobes, hind legs usually saltatorial, an ovipositor, and short cerci; often with specialized stridulatory and auditory organs. Orientation of wing-sheaths reversed during growth of larva. Orthoptera occur in all environments, from mountain tops to desert, usually amongst vegetation or on open ground, but some under logs or debris; many species are of economic importance.

Order PHASMATODEA (Stick-insects; Figs. 22.4–9; 132 Australian spp.)

Large to very large, winged, brachypterous, or apterous insects, often narrow and cylindrical, sometimes flat and leaf-like; distinguished by their usually elongate mesonotum, gressorial legs, short tegmina, and small ovipositor. Closely associated with vegetation; some species damage forest trees.

Order EMBIOPTERA (Web-spinners; Fig. 23.1; 65 Australian spp.)

Small, slender species, with winged or apterous males and apterous females, living in communities in silken galleries (Fig. 23.5); distinguished by the greatly swollen basal segment of their fore tarsi, which contains a silk gland, and their narrow wings with reduced venation.

The Hemipteroid Orders

Mouth-parts specialized, usually suctorial; winged forms with whole or part of fore wings usually thickened and sclerotized to form tegmina or hemelytra, coupling mechanism often developed, venation generally reduced, and anal lobe small but sharply defined; cerci absent; terminalia of male usually symmetrical, less reduced than in the blattoid–orthopteroid orders; few Malpighian tubes; ventral nervous system more concentrated than in the previous orders except Zoraptera. Larvae exopterygote; metamorphosis slight to considerable, occasionally including resting 'pupal' instars.

These orders seem, for the most part to be fairly closely knit, in spite of the diversity in the structure of the mouth-parts, and the modifications that are associated with parasitic life in the Phthiraptera. They are essentially terrestrial, although a few families of Hemiptera have become aquatic. The Psocoptera and Phthiraptera form one pair of related orders and the Hemiptera and Thysanoptera another.

Order PSOCOPTERA (Psocids, booklice; Figs. 24.1, 6; 120 Australian spp.)

Usually very small, soft-bodied, sometimes brachypterous or apterous insects, with asymmetrical mandibles, specialized maxillae, and characteristically specialized venation in winged forms. Mostly living on bark or in debris, with the domestic booklice minor pests of museums and libraries.

Order PHTHIRAPTERA (Lice; Fig. 25.1; 208 Australian spp.)

Very small, flattened, wingless ectoparasites of birds and mammals; some are of veterinary and public-health importance. There are three suborders, of which the surface-feeding Mallophaga and blood-sucking Anoplura differ from one another in many important respects. They are sometimes thought to have evolved separately from psocopteroid ancestors, but there is cytogenetic evidence that they are closely related (White, 1957), and it does not seem necessary to treat them as separate orders in the present state of knowledge.

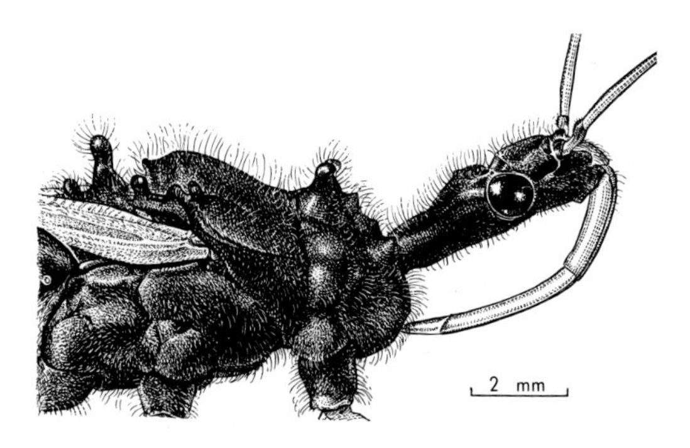

Fig. 7.7. *Pristhesancus papuensis*, Hemiptera-Reduviidae, ♀, to illustrate the hemipteran rostrum. [M. Quick]

Order HEMIPTERA (Bugs; Plate 3; 3,661 Australian spp.)

Large to small insects, characterized by the development of the mouth-parts into a strong, backwardly-directed beak (Fig. 7.7); prothorax large, mesothorax with a well-developed scutellum; fore wing usually longer and narrower than the hind, and usually more or less extensively tegminized. Metamorphosis usually gradual, but sometimes considerable. A successful order of very varied structure and habits, mostly associated with flowering plants (to which they may cause economically important damage), but some predatory (including a few blood-suckers), and some aquatic. The two suborders, Homoptera and Heteroptera, are sometimes treated as separate orders, but their close relationship to each other does not seem to be disputed.

Order THYSANOPTERA (Thrips; Fig. 27.4; 287 Australian spp.)

Small, slender, often apterous insects, with asymmetrical, piercing mouth-parts, and the wings, when present, narrow, with greatly reduced venation and a wide marginal fringe of hairs. Metamorphosis gradual, but with one or two pre-imaginal resting instars. Associated with flowering plants, and sometimes of economic importance.

The Endopterygote Orders

Mouth-parts basically mandibulate, but of very varied structure in conformity with the varied feeding habits of the insects; pterothorax generally strongly developed; wings primitively subequal, the fore wings sclerotized to form elytra in the Coleoptera, and the hind wings tending towards reduction in most other orders; a coupling mechanism often present; venation generalized or reduced; cerci short or reduced to a group of sensilla; terminalia of male normally symmetrical and with well-developed coxites and aedeagus. Larvae of different form and habits from the adults, most of the metamorphosis taking place in a resting pupal instar, in which the previously internal wing-buds are everted on to the surface of the body.

The evolution of a pupal stage within which a great deal of reconstruction was possible provided an effective mechanism for dissociating adaptive radiation, and therefore structure, of the larvae from those of the adult. This would appear to have been the primary basis for the success of the group as a whole, although not all of the included assemblages have been equally successful. Thus, the

Megaloptera seem scarcely to have advanced beyond the initial stages, and are now a relict group (as are the Mecoptera), the fleas have got into an evolutionary *cul-de-sac*, and the Trichoptera have become (or remained) tied to water; whereas the Coleoptera, Diptera, Lepidoptera, and Hymenoptera have become significantly more abundant and diverse than the Orthoptera and Hemiptera, the most successful representatives of the other groups of orders.

There is evidence (p. 170) that the endopterygote grade developed early in the history of the Neoptera, and it has consequently had a long time in which to reach its present eminence. There is no evidence (other, possibly, than the relative isolation of the Hymenoptera) that it was polyphyletic in origin, although it can be divided into three long-established subgroups, which are treated as superorders by many authors. These are:

NEUROPTEROIDEA	PANORPOIDEA	HYMENOPTEROIDEA
Megaloptera	Mecoptera	Hymenoptera
(Raphidioptera)	Siphonaptera	
Neuroptera	Diptera	
Coleoptera	Trichoptera	
Strepsiptera	(Zeugloptera)	
	Lepidoptera	

The Neuroptera are somewhat anomalous. They have retained many primitive characters, which is why Tillyard (1918–19) included them in his 'panorpoid complex', but they show other features that ally them more definitely with the megalopteran–coleopteran line, and Hinton (1958a) excluded them from the Panorpoidea; in some respects (for example, in wing venation), they are probably not far removed from the ancestral form of both lines. The Trichoptera also show some megalopteran features, but their larval anatomy, cytogenetics, and the annectent position of the Zeugloptera (Hinton, 1958a) link them closely to the Lepidoptera.

Order MEGALOPTERA (Alderflies; Fig. 28.1; 16 Australian spp.)

Mostly rather large, soft-bodied insects, with subequal, membranous wings, and cerci reduced to sensilla; distinguished from most Neuroptera by lacking end-twigging of the wing veins, and from Mecoptera by having short, unmodified mouth-parts and different terminalia in both sexes. Larvae of Australian species aquatic; distinguished from those of Gyrinidae (Coleoptera) by having a distinct labrum and 3-segmented palpi. The terrestrial Raphidioptera (which do not occur in Australia) are treated by some authors (e.g. on p. 182) as a separate order.

Order NEUROPTERA (Lacewings; Fig. 29.2; 396 Australian spp.)

Large to small, mostly slow-flying insects, with usually subequal membranous wings; distinguished by having unmodified mouth-parts, wings generally with numerous long veins and cross-veins and much end-twigging, and cerci reduced to patches of sensilla. Larvae mostly terrestrial, usually active and predacious; with mandibles and maxillae combined to form a pair of distinctive, needle-like or forcipate, piercing and sucking jaws.

Order COLEOPTERA (Beetles; Plates 4, 5; 19,219 Australian spp.)

Terrestrial or secondarily aquatic insects, almost always easily recognized by the modification of the fore wings into sclerotized elytra, which normally meet edge to edge and cover the abdomen at rest, but are sometimes greatly reduced. Larvae of varied form, but usually strongly mandibulate, and usually with 3 pairs of legs (a few are legless) but no abdominal prolegs (a caudal pair in Gyrinidae). A highly successful order, containing about 40 per cent of the known species of insects, and occupying a wide range of environments; many species of economic importance.

Order STREPSIPTERA (Figs. 31.1–3; 93 Australian spp.)

Very small, specialized endoparasites of Hymenoptera, Hemiptera, and occasionally other orders. Males and triungulinid larvae (and females of one family) free-living, older larvae and larviform females parasitic; males with greatly reduced, club-shaped elytra.

Closely related to beetles; Crowson (1960) and several other workers include them in the Coleoptera (see p. 518).

Order MECOPTERA (Scorpion-flies; Figs. 7.8, 32.1; 20 Australian spp.)

Slender, usually slow-flying insects, normally with subequal membranous wings, rarely apterous; distinguished by having the mouth-parts modified into a beak, primitive wing venation, short, 1- to 3-segmented cerci, and bulbous male terminalia. The larvae, alone among endopterygotes, have true compound eyes; those of Australian species are predacious, and terrestrial or aquatic. An ancient order that appears to have undergone most of its evolution in the Permian and Mesozoic.

Order SIPHONAPTERA (Fleas; Figs 7.9, 33.1; 68 Australian spp.)

Laterally compressed, apterous, blood-sucking ectoparasites of warm-blooded vertebrates; some species important as transmitters of disease. Mouth-parts suctorial; hind legs saltatorial. Larvae apodous, dipteron-like. The ancestry of the fleas is obscure, but Hinton has suggested that they may have evolved from the same stem as the Neomecoptera, which he treats as a separate order from the Mecoptera.

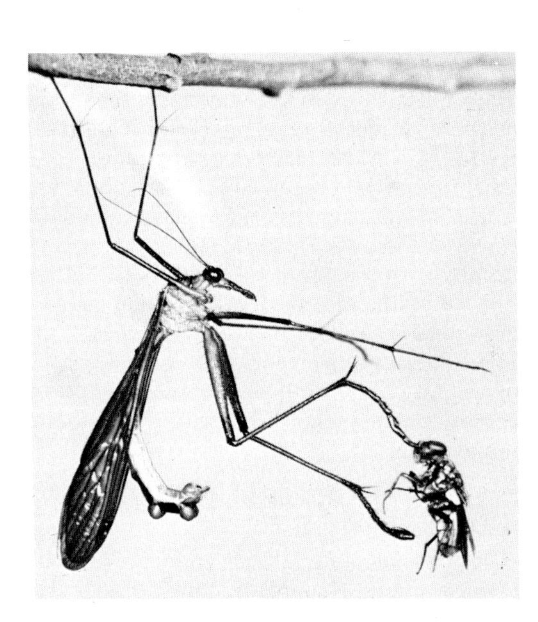

Fig. 7.8. *Harpobittacus australis*, Mecoptera-Bittacidae, ♂, with abdominal glands extruded–1·5 times natural size.
[Photo by G. F. Bornemissza and C. Lourandos]

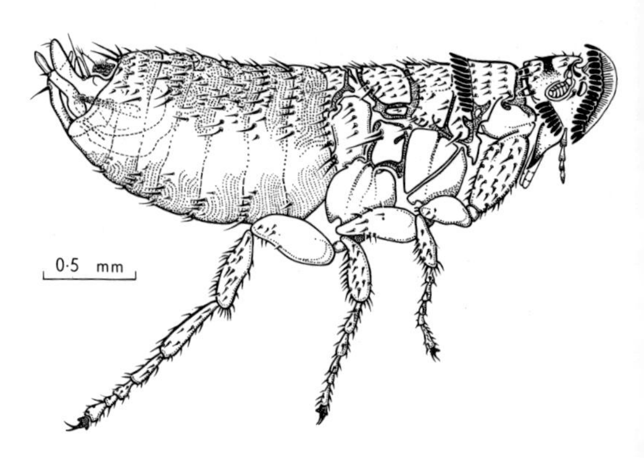

Fig. 7.9. *Stephanocircus pectinipes*, Siphonaptera-Stephanocircidae, ♂. [T. Nolan]

Order DIPTERA (Two-winged flies; Plates 5, 6; 6,256 Australian spp.)

All but apterous forms easily recognized by the reduction of the hind wings to form a pair of clubbed halteres. Mouth-parts suctorial; mesothorax large, prothorax and metathorax very small and fused with it; venation generalized or reduced; cerci short, 1- or 2-segmented. Larvae terrestrial, aquatic, or parasitic; apodous, but sometimes with prolegs. A successful order, with a wide range of types from primitively mecopteron-like to highly specialized; some species of economic, public-health, and veterinary importance.

Order TRICHOPTERA (Caddis-flies; Figs. 7.10, 35.1; 260 Australian spp.)

Small to medium-sized, moth-like insects, which are always associated with water. Distinguished by their setaceous, usually long antennae; reduced mouth-parts; membranous, coupled, more or less densely hairy (occasionally scaly) wings, which are usually held in a characteristic roof-like manner at rest; and short, 1-segmented cerci. Larvae nearly always aquatic, usually living in a case, the last abdominal segment with characteristically hooked caudal appendages.

Order LEPIDOPTERA (Moths, butterflies; Plates 7, 8; 11,221 Australian spp.)

Insects of very varied size and form, usually distinguished by a dense covering of scales on the body, legs, and at least parts of the wings. Mouth-parts mandibulate in a few primitive forms, generally modified into a specialized sucking proboscis, occasionally atrophied; venation not greatly reduced, but characteristically modified and with relatively few cross-veins. Larvae are caterpillars, usually with abdominal prolegs. A highly successful order; many species of economic importance. The Micropterigidae are treated here as a suborder Zeugloptera rather than as a separate order.

Order HYMENOPTERA (Wasps, bees, ants; Plates 5, 6; 8,834 Australian spp.)

An isolated assemblage of usually distinctive, strongly sclerotized insects, some with a well developed social organization. Wings membranous, hind reduced, except in some Symphyta; fore and hind interlocked by hamuli; venation usually reduced and highly specialized; abdomen with segment 1 incorporated in thorax and 2 constricted basally, except in Symphyta; females with an ovipositor adapted for sawing, piercing, or stinging; the only endopterygote order with numerous Malpighian tubes in the adult; males haploid. Larvae of Symphyta mostly caterpillar-like, with locomotor appendages, phytophagous; of Apocrita generally apodous, and parasitic or provided with food in cells. A very successful order, containing many beneficial and a few harmful species.

Fig. 7.10. *Kosrheithrus tillyardi*, Trichoptera-Philorheithridae, ♀, in resting position.
[M. Quick]

8

FOSSIL HISTORY

by E. F. RIEK

Most of our knowledge of fossil insects has come from the northern hemisphere, particularly from Russia, northern Europe, and North America. Not a great many are known from Australia, and too few from other southern countries to permit useful regional comparisons. The only discoveries so far made in Antarctica appear to have been two wings from Permian *Glossopteris* beds illustrated by Plumstead (1962) and Jurassic beetles recorded by Zeuner (1959).

The great majority of the remains consist of wings or portions of wings, some of which are beautifully preserved. This is fortunate, because the wings are used extensively in analysing the relationships of insects, so that the classification of the fossils can usually proceed along lines comparable to those that are applied in Recent species. Only rarely are the wings attached to bodies, and when this does occur the preservation is often unsatisfactory, except in Palaeoptera in which the wings are spread out. For this reason, descriptions and reconstructions of the bodies must be taken as generally much less reliable than those of the wings. Immature stages are rare as fossils, most of those found so far being nymphs of presumed aquatic species or their exuviae. The best documented are the nymphs of the Paraplecoptera from the Lower Permian of Kansas, the Plecoptera-like nymphs from the Permian of Russia, and those of Mesozoic Ephemeroptera and Plecoptera.

Jeannel (1949a) and Laurentiaux (1953) have given illustrated accounts of the major extinct groups of insects, Carpenter (in Brues, Melander and Carpenter, 1954) has published definitions, and the more recent literature has been reviewed by Martynova (1961).

THE GENERAL FOSSIL RECORD

Any discussion of fossil history must necessarily be made against a background of the geological time scale and the phylogenetic relationships of the organisms, as inferred from the comparative morphology of both fossil and Recent forms. These two components, so far as they are known for the insects, are brought together in Table 8.1. Concordance is not complete, but that merely reflects the fact that the history of the class can be seen but dimly. It is necessary to remember this when considering the varied opinions that have been expressed about the content and relationships of orders that have not survived to the present day.

Devonian. The earliest known hexapod is a probable collembolan, *Rhyniella praecursor*, from the Middle Devonian of Scotland; and the earliest known insect is *Eopterum devonicum* (Fig. 8.6C), from the Upper Devonian of Russia. Tillyard (1928) had described

TABLE 8.1 *The Geological History of the Insects*

General history indicated by continuous lines; spots are records of occurrences in Australia. Recent orders in bold face type.

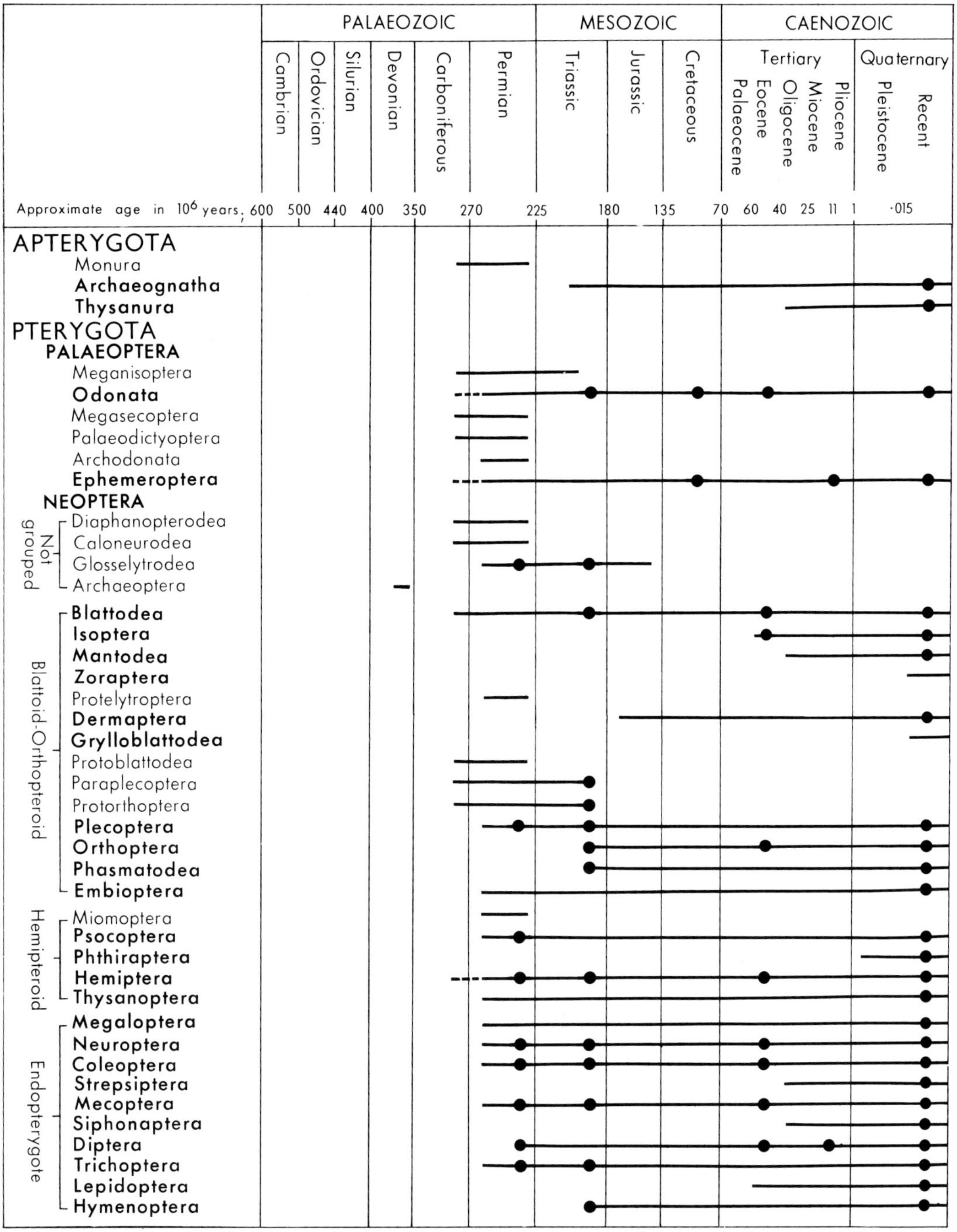

		PALAEOZOIC						MESOZOIC			CAENOZOIC						
		Cambrian	Ordovician	Silurian	Devonian	Carboniferous	Permian	Triassic	Jurassic	Cretaceous	Tertiary: Palaeocene	Eocene	Oligocene	Miocene	Pliocene	Quaternary: Pleistocene	Recent
	Approximate age in 10^6 years	600	500	440	400	350	270	225	180	135	70	60	40	25	11	1	·015
APTERYGOTA																	
	Monura																
	Archaeognatha																
	Thysanura																
PTERYGOTA																	
PALAEOPTERA																	
	Meganisoptera																
	Odonata																
	Megasecoptera																
	Palaeodictyoptera																
	Archodonata																
	Ephemeroptera																
NEOPTERA																	
Not grouped	Diaphanopterodea																
	Caloneurodea																
	Glosselytrodea																
	Archaeoptera																
Blattoid-Orthopteroid	**Blattodea**																
	Isoptera																
	Mantodea																
	Zoraptera																
	Protelytroptera																
	Dermaptera																
	Grylloblattodea																
	Protoblattodea																
	Paraplecoptera																
	Protorthoptera																
	Plecoptera																
	Orthoptera																
	Phasmatodea																
	Embioptera																
Hemipteroid	Miomoptera																
	Psocoptera																
	Phthiraptera																
	Hemiptera																
	Thysanoptera																
Endopterygote	**Megaloptera**																
	Neuroptera																
	Coleoptera																
	Strepsiptera																
	Mecoptera																
	Siphonaptera																
	Diptera																
	Trichoptera																
	Lepidoptera																
	Hymenoptera																

Age estimations by Bureau of Mineral Resources, Geology and Geophysics, Canberra, 1960.

Rhyniognatha hirsti from the same beds as *Rhyniella*, on a pair of mandible-like structures which he thought might have belonged to 'some minute early progenitor of the mandibulate Orthopteroid insects'.

Carboniferous. The next appearance of insect remains is at the base of the Upper Carboniferous. *Erasipteron larischi* (Meganisoptera) belonged to the Palaeoptera, *Stygne roemeri* (orthopteroid) to the Neoptera, and it has been suggested that the fragmentary *Metropator pusillus* might have been an endopterygote, though most authors doubt that it was. By the end of the period the insects had radiated widely and ten orders had become established, the Blattodea being the dominant group. Raphidioptera have also been recorded, but the remains may have been those of specialized Paraplecoptera. Although neither hemipteroids nor endopterygotes are known with certainty, both groups had radiated so widely by the Lower Permian that they must have been present in the Carboniferous.

Permian. All the Carboniferous orders were still present, and others had appeared, including many that have survived. Even allowing for imperfections in the record, the table shows clearly what a great burst of adaptive radiation occurred at about this time, when land plants were also developing rapidly. The Palaeoptera reached their climax, if they had not already done so during the Carboniferous, and the blattoid–orthopteroid orders also attained their greatest diversity. Nevertheless, these groups no longer dominated the fauna, because Psocoptera were well established and homopteran Hemiptera were abundant and diverse, as were Mecoptera and Coleoptera, although other endopterygote orders seem to have been rather sparsely represented.

Mesozoic. This was an era of change and development. Of the 12 orders now extinct, nine do not seem to have survived beyond the Permian, the Meganisoptera and Paraplecoptera disappeared in the Triassic, and the Glosselytrodea in the Jurassic. On the other hand, there was diverse evolution within the orders that did survive, the Neuroptera, Mecoptera, and Diptera providing notable examples. Suborders and numerous families were established, many of which survived. Some orders—Archaeognatha, Dermaptera, Orthoptera, Phasmatodea, Hymenoptera—appear in the record for the first time, though there is little doubt that most of them were really older.

Tertiary. The Tertiary insects are essentially modern. This is well illustrated by the many Recent families and genera that have been found among the remarkably preserved insects in Baltic amber, but is also seen in other beds. Of the six orders recorded with certainty for the first time from Tertiary beds, the Isoptera, Strepsiptera, and Siphonaptera may have been relatively late developments, but the others were probably more ancient.

THE AUSTRALIAN FOSSIL RECORD

About 350 fossil species, distributed in 19 orders, have been described from Australia. They cover a period of time extending from Upper Permian to probably Pliocene.

Permian. The earliest occurrence is in a rich horizon of the Upper Permian Newcastle Coal Measures at Belmont and Warner's Bay in N.S.W. The beds consist of a hard, fine-grained chert, in which the insects are associated with phyllopod Crustacea, fish scales in some areas, and a *Glossopteris* flora. The insect fauna is curiously unbalanced as compared with that of the corresponding period in the northern hemisphere. On the one hand, there are no Palaeoptera except an undescribed meganisopteron, no blattoids, and no orthopteroids other than a single species of Plecoptera. There is one species of Glosselytrodea. On the other hand, there are many Homoptera, several Psocoptera, three families of Neuroptera, a few Coleoptera and Trichoptera, an abundance of Mecoptera, and a few probable ancestors of the Diptera.

Triassic. Insect remains have been obtained from several localities in the Australian Triassic. The best known are Mt Crosby and Denmark Hill in south-eastern Queensland and Brookvale, near Sydney, in N.S.W. A few insects are also recorded from the

Wianamatta Shales around Sydney, and from Tasmania and W.A.

The Mt Crosby bed, at the base of the Upper Triassic Ipswich Series, has yielded more than a thousand recognizable insects. The cockroaches dominate the fauna, which is surprising in view of their absence from the Upper Permian at Belmont. The Homoptera and certain panorpoid orders constitute the other main elements, and there are smaller numbers of Orthoptera and Coleoptera, whilst Odonata, Paraplecoptera, Plecoptera, and Hymenoptera are each represented by one or two specimens.

The Denmark Hill fauna, from the top of the Ipswich Series, differs considerably from that of Mt Crosby. This, no doubt, reflects local differences in ecology and preservation as much as differences in fauna, for the two localities are not widely separated. About half the fossils are beetles, with the Blattodea and Homoptera as the other important components. There are a few Orthoptera, Neuroptera, and Mecoptera. The Odonata, Glosselytrodea, Paraplecoptera, Phasmatodea, and Trichoptera are each represented by one or two specimens.

The remains at Brookvale have been found in a lenticular shale in the Middle Triassic Hawkesbury Sandstone. They are mostly beautifully preserved wings, often of large insects, and some clearly showing a pigmented pattern. Several orders are represented, mainly orthopteroids, with one common mecopteron, a number of Homoptera, and some Blattodea. The orthopteroid fauna is dominated by the large, highly modified *Clatrotitan* (Fig. 8.11), which had an unusual stridulatory mechanism in the fore wing of the male. The occurrence of Protorthoptera and Paraplecoptera is also noteworthy, because it is unusual to find these two orders persisting beyond the Permian.

The Wianamatta Shales extend for about 100 m above the Hawkesbury Sandstone, and are considered to be of Upper Triassic or possibly Jurassic age. The fossil insects found so far are beetle elytra, some large, poorly preserved orthopteroid wings, a cockroach that is possibly related to the Mesoblattinidae of other Australian Triassic strata, a mecopteron, and the apex of an homopteran wing.

Fossil insects have been obtained in Tasmania from a plant-bearing horizon in the New Town Coal Measures at Hobart (Riek, 1962c) and from the Mt Nicholas Coal Measures at Fingal. At each locality there is the tegmen of a cockroach of the genus *Triassoblatta*, which is common in the Triassic of Queensland. Portion of the wing of a large scytinopterid homopteron was also obtained at the Hobart locality. Beetle elytra and part of a tegmen of a mesoblattinid cockroach have been found at Hill River in W.A.

Jurassic. A specimen from near Mudgee, N.S.W., was described (wrongly) as a cicada and referred to the Jurassic. A wing recorded by Etheridge from South Gippsland probably came from younger strata.

Cretaceous. Two hind wings of an anisopteran dragonfly belonging to the Mesozoic family Aeschnidiidae are known from the marine limestones of the Flinders River beds of north Queensland. There are abundant, very well preserved, undescribed nymphs of Ephemeroptera (Baetidae–Siphlonurinae) and a few Diptera, Mecoptera, Plecoptera, and Odonata, as well as a restricted representation of the terrestrial insect fauna, in the freshwater Lower Cretaceous of Victoria in association with abundant fish and phyllopod Crustacea.

Tertiary. There are only a few known Tertiary insect deposits in Australia. The best documented occur in the lower Tertiary near Redbank, southern Queensland, and in the possibly Pliocene deposits at Vegetable Creek, near Emmaville in northern N.S.W.

At Redbank Plains and Dinmore (a few miles away) the insects are associated with other freshwater fossils. The beds at Redbank Plains have yielded fossil fish, Dipnoi, reptilian skin, ostracods, and Cladocera. The insects are predominantly Homoptera and Coleoptera, with a few Blattodea, Hemiptera–Heteroptera, Neuroptera, Mecoptera, and Diptera. At Dinmore the insects are associated with fossil plants and unionid lamellibranchs. There is a wing belonging to the Orthoptera, one of Isoptera, the almost complete hind wing of a cicada differing little from Recent forms, and a fragment of the fore wing of an anisopteran dragonfly.

The insects at Vegetable Creek are in the youngest Tertiary stanniferous lode of the field. They are mostly well-preserved nymphs and larvae of aquatic insects. The nymphs are Ephemeroptera belonging to the Leptophlebiidae and Baetinae, the association suggesting sluggish or standing water. The larvae were described as those of a lampyrid beetle, but they all appear to have been chironomid Diptera. *Chironomus venerabilis*, from the same beds, may have been the adult of the larvae.

At Duaringa in central Queensland, two zygopteran dragonfly nymphs, said to differ little from nymphs of Recent Lestidae, were obtained from a bore core. The age of the sediments is not known, but they are assumed to be Tertiary. There are also localities in Victoria from which a few insect remains, mainly beetle elytra, have been obtained.

HISTORY OF THE ORDERS

APTERYGOTA

MONURA (Carb.–Perm.). This primitive order was based on the genus *Dasyleptus* (Fig. 8.1), described from the Upper Carboniferous of France, and subsequently found by Sharov (1957) to occur also in the Lower Permian of Russia. Its salient features have been noted on p. 158.

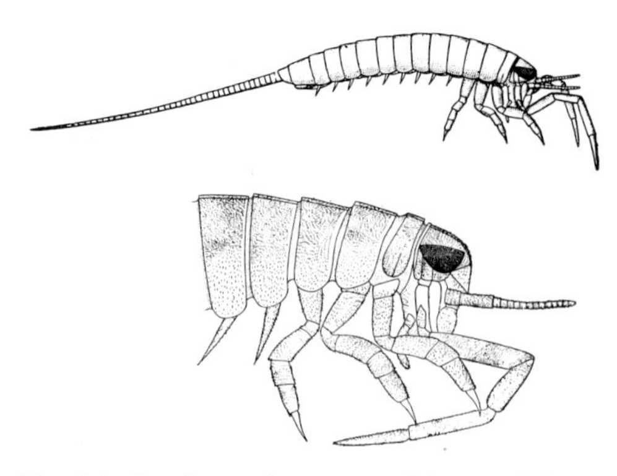

Fig. 8.1. *Dasyleptus brongniarti*, Monura (after Sharov, in Rohdendorf *et al.*, 1961).

ARCHAEOGNATHA (Trias.–Recent). The earliest record of the order is the family Triassomachilidae from Russia. More modern Machilidae are known from Oligocene Baltic amber.

THYSANURA (Oligocene–Recent). Lepidothrichidae have been found in Oligocene Baltic amber.

PALAEOPTERA

The Palaeodictyoptera are usually regarded as the most primitive winged insects so far known to have existed; but there were two early evolutionary lines—odonate and palaeodictyopteran—and the wing venation of the odonate line (e.g. Fig. 8.2) can be derived more directly than that of the palaeodictyopteran line from the postulated ancestral type shown in Figure 1.18B.

MEGANISOPTERA (Carb.–Trias.). These were large, often huge, probably predacious insects, with a wing expanse of 12 to 75 cm. They have been found in the Upper Carboniferous of France, the Lower Permian of North America, the Permian of Russia, and there is an undescribed species in the Upper Permian of Australia; one species has been recorded from the Triassic of France. They resembled dragonflies in having long narrow bodies, prominent eyes and mandibles, oblique thoracic segments, and spiny legs. The wing venation of the Meganeuridae (Fig. 8.2A) is considered to have been extremely primitive, in that all the main veins except Rs had separate origins at the base and the cubito-anal field was represented by a single vein. There was a dense reticulation of cross-veins, and triads were present.

ODONATA (Perm.–Recent). There is general agreement that the Odonata and Meganisoptera were derived from the same stock. Excluding the Upper Carboniferous Campylopteridae, the position of which is disputable, the earliest records of Odonata are from the Permian of North America and Russia, where the main lines of zygopterous and anisopterous descent were already differentiated. Six suborders are recognized, of which three have survived. The Protozygoptera, best known from *Kennedya* (Fig.

8.2B) and its allies, may have been ancestral to the Zygoptera, and the Protanisoptera (Fig. 8.2C), through the Anisozygoptera which replaced them in the Mesozoic, to the Anisoptera; the Mesozoic Archizygoptera appear to have been a divergent line that has left no descendants.

Few fossil Odonata are known from Australia. The fragmentary *Mesophlebia* from the Triassic at Denmark Hill was included by Tillyard in the Anisozygoptera, and *Triassagrion* from the same beds has been placed

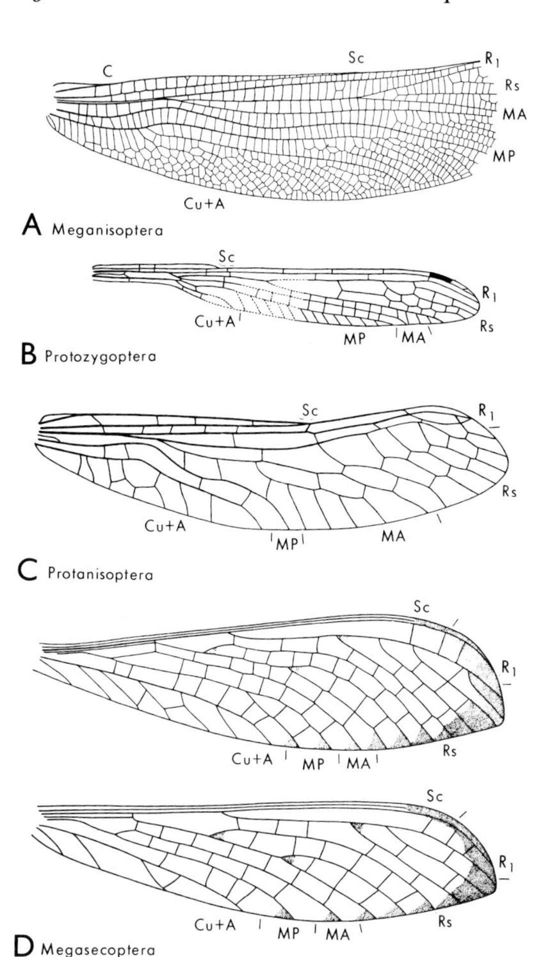

Fig. 8.2. A, *Oligotypus tillyardi*, fore wing (after Carpenter, 1947); B, *Kennedya mirabilis*, fore wing; C, *Ditaxineura anomalostigma*, fore wing; D, *Aspidothorax triangularis*, fore and hind wings. (B–D after Carpenter, 1931.)

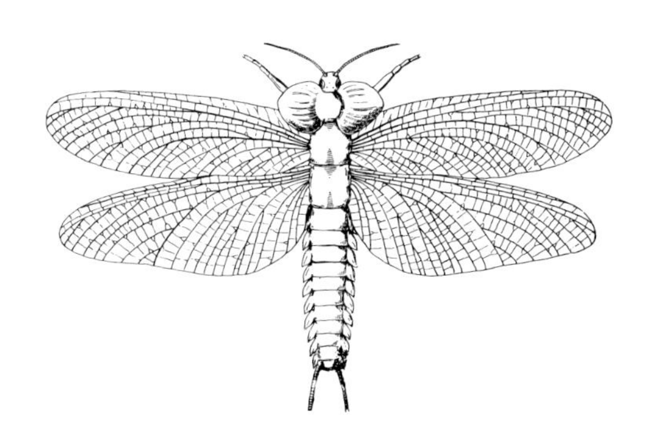

Fig. 8.3. *Lithomantis carbonaria*, Palaeodictyoptera-Lithomantidae, reconstruction (after Jeannel, 1949a).

in the Archizygoptera. *Aeschnidiopsis* (Anisoptera–Aeschnidiidae), from the Cretaceous of northern Queensland, is close to Recent Aeshnidae, and a few Tertiary fossils have been noted on p. 172.

MEGASECOPTERA (Carb.–Perm.). These were medium-sized to large (12 cm) insects, with rather slender bodies and very long cerci but no caudal style (Fig. 8.2D). Most workers relate them to the Palaeodictyoptera, but their cubito-anal venation resembles that of the meganisopteran-odonate line. They appear to have left no descendants except possibly the Diaphanopterodea (p. 175).

PALAEODICTYOPTERA (Carb.–Perm.). This was a large, diverse order of mostly large (expanding up to 20 or even 50 cm), mostly rather broadly built insects, which were widespread in the Upper Carboniferous. The antennae, where known, were setose, prothoracic paranota usually large (Fig. 8.3) and sometimes with rudiments of venation, thoracic segments subequal and discrete, legs slender, abdominal segments usually with prominent lateral lobes, cerci rather short, and no caudal style. Some species were more slender, with reduced lateral processes and long cerci (Fig. 8.4). The wings were subequal, sometimes widened, occasionally triangular. An archedictyon was present as an irregular polygonal meshwork, but was often more or less completely replaced by a regular system of cross-veins. The venation was

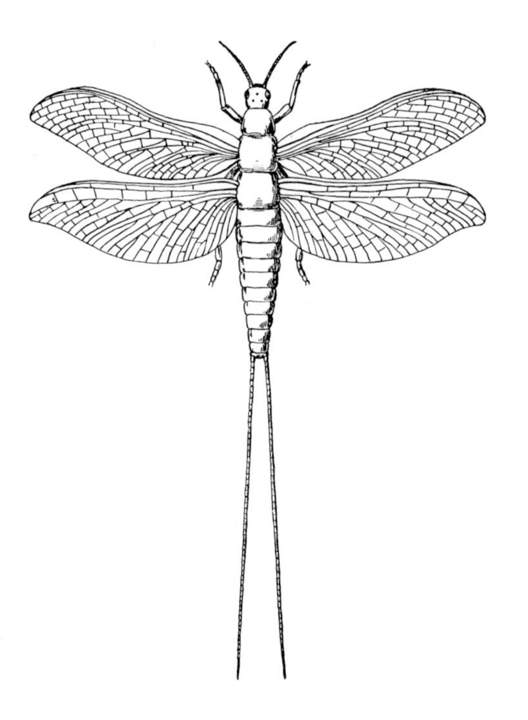

Fig. 8.4. *Homaloneura ornata*, Palaeodictyoptera-Spilopteridae, reconstruction (after Jeannel, 1949a).

complete, typically with the concave components (Rs, MP, CuP) arising as posterior branches from the convex components, and with a fully developed cubito-anal field which was sometimes more or less expanded. There were two suborders: Eupalaeodictyoptera with presumably mandibulate mouth-parts; and Protohemiptera with the mouth-parts modified into a long, suctorial rostrum, which was not the forerunner of the hemipteran rostrum.

ARCHODONATA (Perm.). This is a small order recorded from North America and Russia. The mesothorax was greatly developed, the fore wing had a well-developed pterostigma and few cross-veins (Fig. 8.5A), and the abdomen ended in a pair of long cerci. It has been compared with the Megasecoptera, but was more probably a two-winged development from the Palaeodictyoptera. It has left no recognizable descendants.

EPHEMEROPTERA (Perm.–Recent). The Ephemeroptera are closely related to the Palaeodictyoptera, and presumably developed from some ancestral form that had not lost the

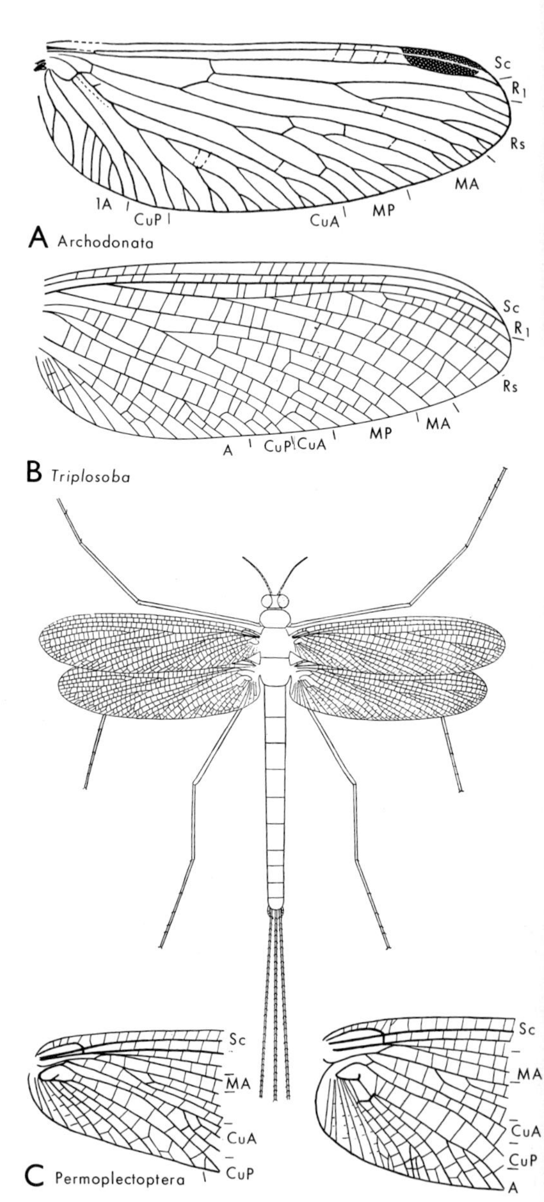

Fig. 8.5. A, *Permothemis libelluloides*, fore wing (modified from Jeannel, 1949a); B, *Triplosoba pulchella*, wing (after Edmunds and Traver, 1954); C, *Protereisma permianum*, restoration and base of fore and hind wings (after Carpenter, 1933).

caudal style. The Carboniferous *Triplosoba* (Fig. 8.5B) had the general facies of an ephemeropteron, and was thought by some authors to be a connecting link, but its unbranched MA and pectinately branched Rs would exclude it from the direct line of mayfly descent. The order was, however, well represented in the Permian of North America and Russia by the suborder Permoplectoptera (Fig. 8.5C), in which fore and hind wings were subequal in size. They differed from some more slender Palaeodictyoptera chiefly in venational characters and in possessing a long caudal style. Apart from a few doubtful nymphs, there is no further record of the order until the Jurassic, when two families of the surviving suborder Euephemeroptera appeared. Of these, the Mesephemeridae retained the nearly homonomous wings of the Permoplectoptera, whereas the Paedephemeridae had considerably reduced hind wings. Later fossils belong to Recent groups, as do the few recorded from Australia.

NOT GROUPED

Four small extinct orders of obscure relationships are placed here for convenience, because at least some of them appear to represent separate evolutionary lines from any of the Neoptera described below.

DIAPHANOPTERODEA (Carb.–Perm.). This order occurred in the Carboniferous and Permian of Russia and the Permian of North America. It was a neopterous development from (or similar to) the Megasecoptera, and was quite independent of other Neoptera; Carpenter treated it as a suborder of Megasecoptera. The fore and hind wings were similar, R was bent back at the base (less so in the hind wings), and the main veins were closely aligned in the basal part of the wing. The mechanism of folding is not clear.

CALONEURODEA (Carb.–Perm.). This order reached its greatest development in the Permian of North America and Russia. It included small to large insects, with long antennae, no pronotal lobes, slender legs, subequal membranous wings, slender abdomen, and short cerci. At least some of the species could fold their wings, and most authors associate them with the orthopteroids. However, they had a distinctive venation (Fig. 8.6A), with a double stem to Rs, incompletely developed cubito-anal field, and abundant convex cross-veins, so it seems likely that they were an independent neopterous development parallel to the Diaphanopterodea.

GLOSSELYTRODEA (Perm. – Juras.). These were small insects that resembled, and may have been related to, the Caloneurodea. Most species had a well-marked basal expansion enclosing rows of cellules in the costal area of the tegminous fore wing (Fig. 8.6B). Two genera are known from Australia, the generalized *Permoberothella* from Belmont and the more advanced Triassic *Polycytella* from Denmark Hill.

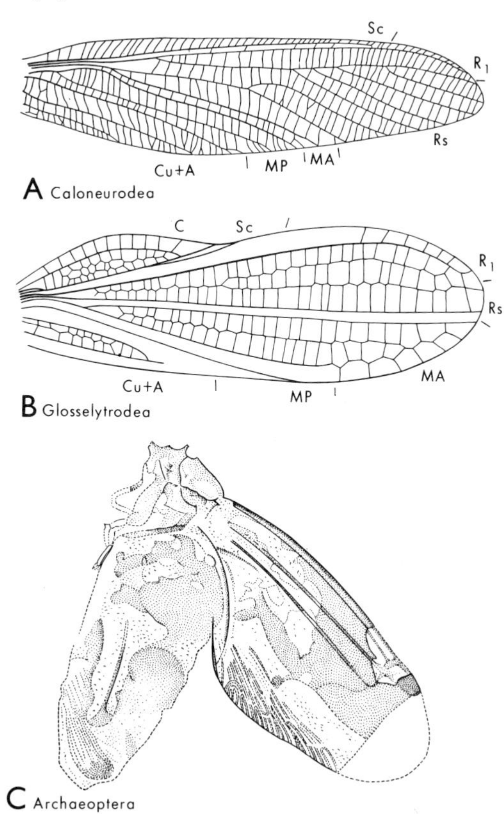

Fig. 8.6. A, *Caloneura dawsoni*, fore wing (after Carpenter, 1943); B, *Eoglosselytron perfectum*, fore wing (after Martynova, in Rohdendorf *et al.*, 1961); C, *Eopterum devonicum* (after Rohdendorf, 1961).

ARCHAEOPTERA (Devon.). *Eopterum devonicum*, on which the order is based, was described from what appear to be two wings and fragments of a body (Fig. 8.6C). It was included in the Neoptera by Rohdendorf (1961), but its preservation is poor, and its relationships cannot be regarded as established.

NEOPTERA

Most workers believe that the Neoptera were derived from the Palaeodictyoptera. Some would limit the derivation to a single blattoid line, from which, in turn, the orthopteroids, hemipteroids, and endopterygotes evolved (see Martynova, 1961). A close comparison of wing venation suggests, however, that the blattoid–orthopteroid and endopterygote groups, and possibly the hemipteroid line also, may have arisen separately from palaeodictyopteran ancestors.

There is no fossil record of Zoraptera or Grylloblattodea, and of Phthiraptera only from the Pleistocene.

Blattoid–Orthopteroid Orders

The extinct orders of this group form a confusing assemblage, about the content and arrangement of which there is little agreement. The writer considers that the lower Paraplecoptera could have been derived from the Palaeodictyoptera, and may have been ancestral to both the blattoid and orthopteroid series of orders.

BLATTODEA (Carb.–Recent). This is the most ancient of the surviving orders. The Palaeozoic species were remarkably like those of today, although a few had well-developed ovipositors, and the wing-sheaths of the nymphs (Fig. 8.7A) were larger and their cerci often longer than in Recent nymphs. The most primitive family was the Archimylacridae (Fig. 8.7B), which some authors consider to have been ancestral to the other major neopterous groups. Two genera of the extinct family Mesoblattinidae have been recorded from the Triassic in Australia and some unidentified remains from the Tertiary.

ISOPTERA (Eocene–Recent). The primitive Recent family Mastotermitidae, in which the hind wing has retained a small ano-jugal lobe (Fig. 15.2), occurred in the lower Tertiary of England, North America, and Australia, the Australian representative *(Blattotermes)* being from the Dinmore bed. Other Recent families have been recorded from Oligocene Baltic amber. The order may have evolved during the later part of the Mesozoic.

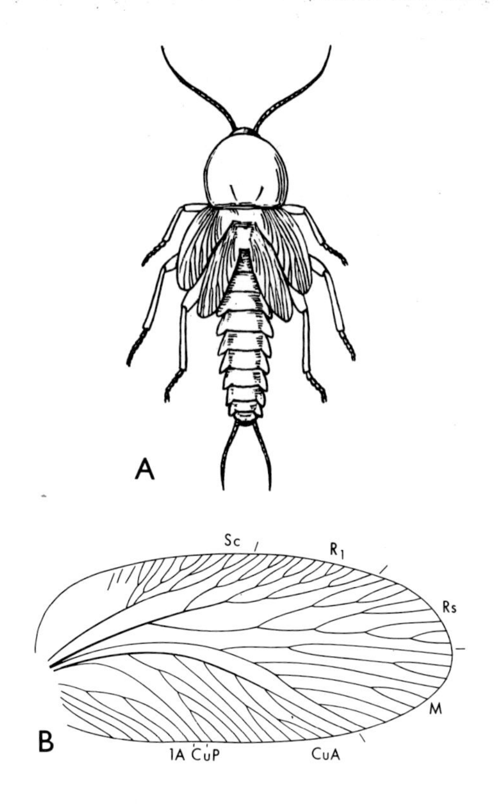

Fig. 8.7. Blattodea: A, nymph of primitive archimylacrid, restoration (after Jeannel, 1949a); B, *Archimylacris lerichei*, fore wing (after Tillyard, 1937).

MANTODEA (Oligocene–Recent). The earliest known mantids are modern types from Baltic amber, the Australian *Triassomantis* (p. 177) now being excluded. The venation of the mantid fore wing differs from that of Blattodea in important respects, and it seems likely that the order was not a relatively late derivative from the cockroaches, but may have arisen independently from Paraplecoptera-like ancestors. That no early fossils have been recognized may be due to the resemblance of the distal parts of the wings to those of Orthoptera, with which they could have been confused.

PROTELYTROPTERA (Perm.). These were small insects (Fig. 8.8A), which had short, thick antennae, robust legs with five tarsal segments, fore wings typically elytriform but showing traces of venation, hind wings larger and with longitudinal and transverse folds in the expanded anal area, broad abdomen, and short cerci. They are usually considered to have been ancestral to the Dermaptera.

DERMAPTERA (Juras.–Recent). *Protodiplatys* (Fig. 8.8B), from the Middle Jurassic of Turkestan, is the oldest known dermapteron. It resembled modern earwigs in many ways, but had 5-segmented tarsi (at least in the hind legs) and segmented cerci; it is placed in a separate suborder Archidermaptera. *Semenoviola*, also from the Jurassic, had the cerci transformed into forceps. Modern Labiduridae and Forficulidae have been recorded from Tertiary strata in North America and Europe.

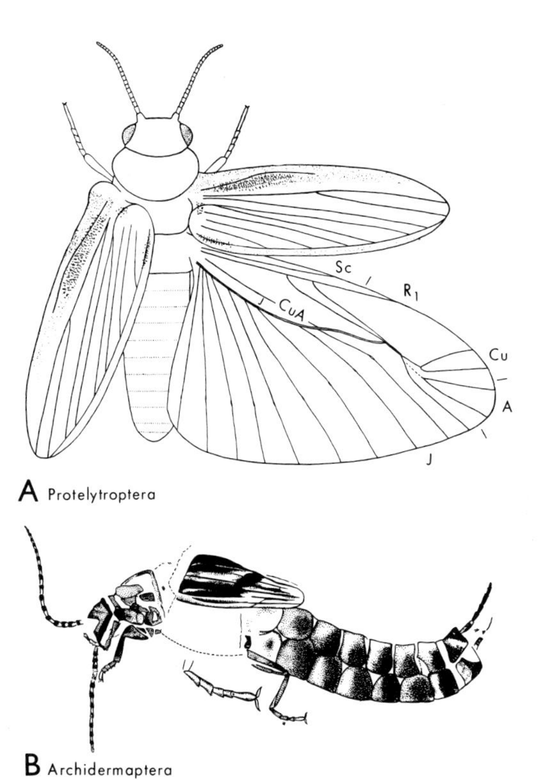

Fig. 8.8. A, *Protelytron permianum*, reconstruction (after Carpenter, 1933); B, *Protodiplatys fortis* (after Martynov, 1925).

PROTOBLATTODEA (Carb.–Perm.). These insects were not like cockroaches. They were actively flying, probably carnivorous, with small heads, projecting mandibles, often unenlarged pronota, strong legs, fore wings only lightly tegminized, and short cerci. Their venation also differed significantly from that of the Blattodea, and they were probably more nearly related to the orthopteroids, in which Carpenter included them.

PARAPLECOPTERA and PROTORTHOPTERA (Carb.–Juras.). This is a large, diverse, widely distributed assemblage. Martynova divided it into two orders: Paraplecoptera containing the bulk of the species, and Protorthoptera with two families of saltatorial species. Carpenter also recognized two orders: Protorthoptera (the older name) substantially for both orders of Martynova, and Protoperlaria for the Lemmatophoridae (Fig. 8.9) which had presumably aquatic nymphs. The writer would follow Martynova, but add a third family to the Protorthoptera.

Generalized Paraplecoptera (Fig. 8.9C) can be distinguished from Palaeodictyoptera by little more than the neopterous folding of the wings and the development of an ano-jugal fan in the hind wing. Some retained an extensive archedictyon, the stems of the main veins were free to the base, and both MA and MP were developed. Some had large pronotal paranota. Many were rather plecopteron-like, with long or short antennae and cerci, and the hind legs were usually long. The known Australian species were all Triassic. They comprise several diverse genera, including *Triassomantis* which is not closely related to the Mantodea.

The Protorthoptera are distinguished by having MP fused with CuA to form an oblique strut from MA to CuA (Fig. 8.10A), and by the development of a precostal

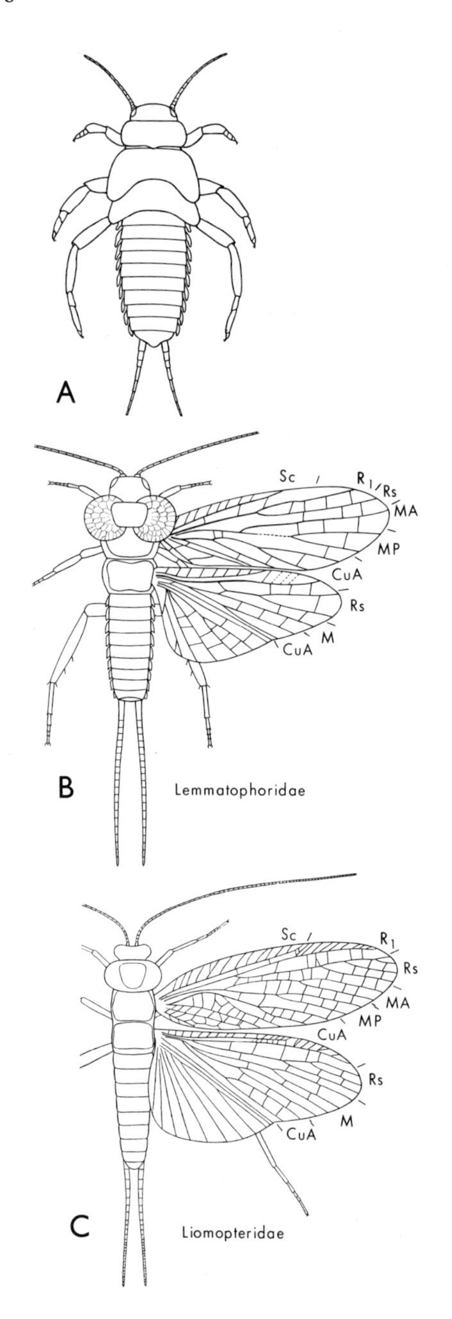

Fig. 8.9. Paraplecoptera, reconstructions: A, nymph; B, *Lemmatophora typa*; C, *Liomopterum ornatum*. (All after Carpenter, 1935.)

expansion at the base of the wing. They are represented in Australia by the Triassic *Mesacridites*.

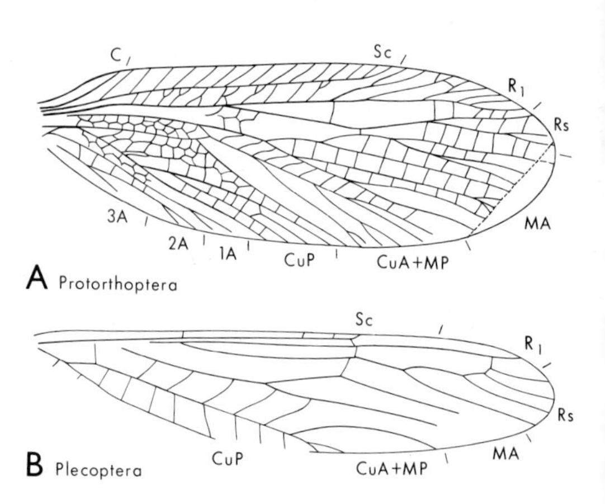

Fig. 8.10. A, *Sthenaropoda bruesi*, fore wing (after Zeuner, 1939); B, *Stenoperlidium triassicum*, fore wing (after Riek, 1956).

PLECOPTERA (Perm.–Recent). The oldest known Plecoptera, from the Permian of Russia and Australia, did not differ greatly from those of today. For example, the Australian *Stenoperlidium* (Fig. 8.10B), which has also been found in the Triassic, had a venation like that of the Recent *Stenoperla*, although the cross-veins appear to have been more reduced. The surviving subfamily Taeniopteryginae is known from the Jurassic of Turkestan, Gripopterygidae from the Lower Cretaceous of Victoria, and other Recent families from Oligocene Baltic amber. Nymphs have been recorded from the Permian, Jurassic, and Cretaceous.

ORTHOPTERA (Trias.–Recent). Ensifera and Caelifera are thought to have arisen separately from protorthopteran ancestors, from which they are distinguished mainly by reduction in the number of branches of MA and CuA and by the development of a stridulatory apparatus in many groups. Both suborders were well established in the Australian Triassic. The Haglidae (Fig. 8.12A) with MA 2-branched are the most primitive known Ensifera, and three gryllacridoid genera have been recorded from Mt Crosby and Denmark Hill. The Clatrotitanidae (Fig. 8.11), from the Triassic at Brookvale, may represent an unsuccessful offshoot from the Ensifera. Elsewhere a

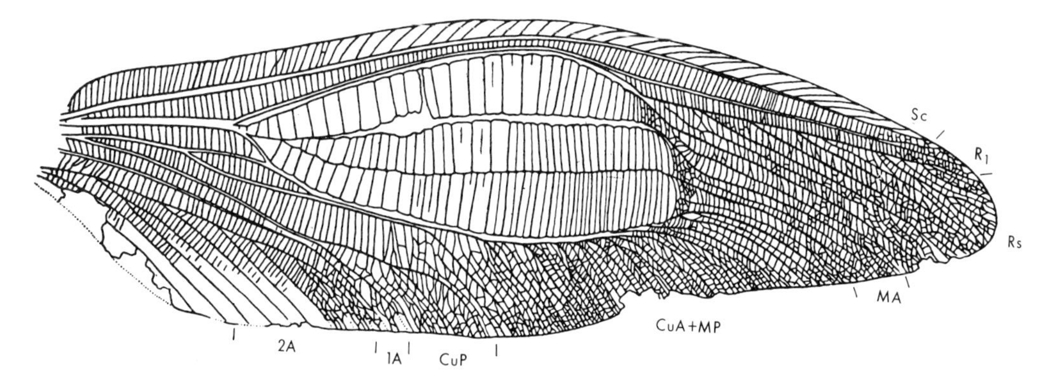

Fig. 8.11. *Clatrotitan scullyi*, Orthoptera, fore wing, ♂ (after McKeown, 1937).

grylloid has been found in the South African Triassic and Tettigonioidea in the northern Jurassic. The oldest known caeliferan is *Triassolocusta* (Fig. 8.12B) from the Queensland Triassic, but two other extinct families of the suborder have been recorded in the northern hemisphere, and Acrididae in the Tertiary.

PHASMATODEA (Trias.–Recent). The oldest known genus is the Australian *Aeroplana* (Fig. 8.12C), which Tillyard had thought to be a meganisopteron. It and three northern Mesozoic families differed from Recent families in having had long fore wings, with the longitudinal veins parallel and connected by many cross-veins; there was a large anal fan in the hind wing. These Mesozoic families are sometimes placed in a separate suborder Chresmododea.

EMBIOPTERA (Perm.–Recent). It has been suggested that the embiids may have evolved from some cursorial orthopteroid of the *Tillyardembia* type, although both *Tillyardembia* and *Protembia* are now excluded from the order. However, the genus *Sheimia* (Fig. 8.12D), described by Martynova from the Permian of Europe, does appear to be a true embiid. Apart from it, all the known fossils are Tertiary, mostly from amber, and all except *Burmitembia* from Eocene amber in Burma are similar to modern forms.

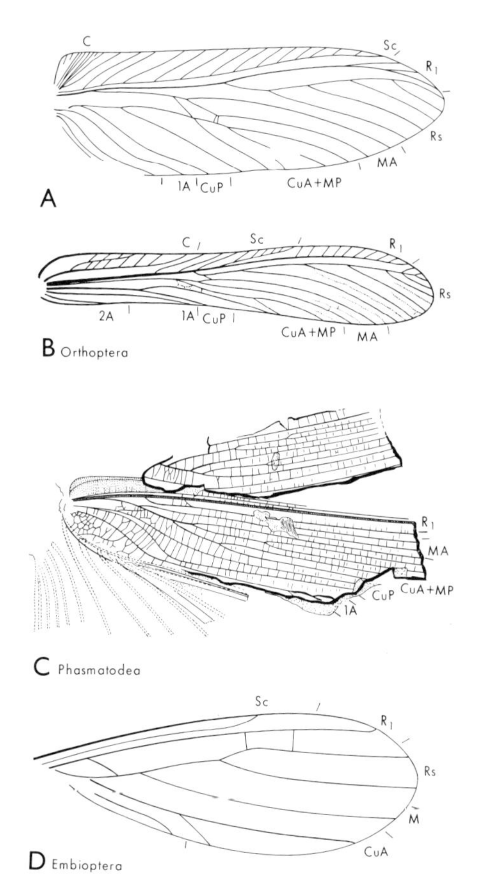

Fig. 8.12. A, *Prohagla superba*, Orthoptera, fore wing without cross-veins (after Riek, 1954d); B, *Triassolocusta leptoptera*, fore wing (after Tillyard, 1922b); C, *Aeroplana mirabilis*, fore and hind wings (after Martynov, 1928a); D, *Sheimia sojanensis*, fore wing (after Martynova, 1958).

Hemipteroid Orders

MIOMOPTERA (Perm.). This is a small order of small insects with short bodies, long

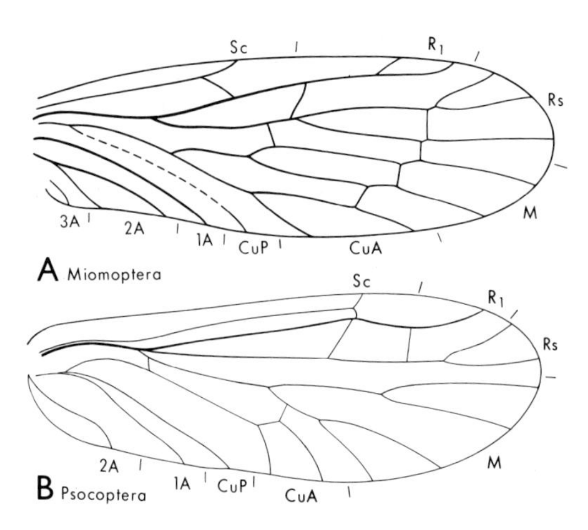

Fig. 8.13. A, *Delopterum kaltanicum*, fore wing (after Martynova, in Rohdendorf *et al.*, 1961); B, *Permopsocus latipennis*, Permopsocida, fore wing (after Carpenter, 1932).

wings, and rather simplified venation (Fig. 8.13A). Its status and content have been disputed, and it is now limited (see Martynova, 1961) to two families which seem to the writer to lie near the base of the hemipteroid stem.

PSOCOPTERA (Perm.–Recent). This order had radiated more widely before the end of the Palaeozoic than might have been expected from the residue surviving today. The Lower Permian suborder Permopsocida had subequal wings; quite generalized venation (Fig. 8.13B), with Rs 2-branched, M 4-branched, CuA 2-branched, and no fusion between the middle sections of the veins; and tarsi 4-segmented, at least in the fore legs. The Lophioneuridae and Zoropsocidae, from the Upper Permian of Australia and Russia, had reduced hind wings, but were divergent in having M 2-branched and CuA unbranched in the fore wing. *Zygopsocus*, also from the Australian Permian, was more like Recent psocids, but had Rs unbranched. Modern forms are known from Oligocene Baltic amber.

HEMIPTERA (Perm.–Recent). There has been less agreement about the classification of fossil Hemiptera than of any other order, and the phylogenetic problems they present have been reviewed by Evans (1963,

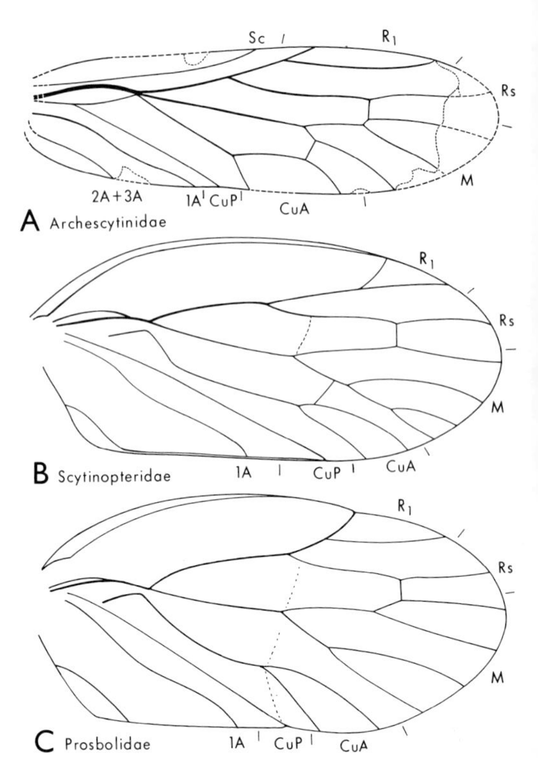

Fig. 8.14. Hemiptera: A, *Archescytina permiana*, fore wing (after Tillyard, 1926a); B, *Permocicada borealis*, fore wing (after Evans, 1956); C, *Prosbole reducta*, fore wing (after Evans, 1956).

1964). While the order is undoubtedly ancient, the Upper Carboniferous Blattoprosboloidea, which are considered to be Homoptera by the Russian workers, are best referred to the blattoid–orthopteroid complex. By the Lower Permian, however, there were at least three families of undoubted Homoptera, and twelve or more were known by the end of the period, many of them occurring at Belmont. They were a diverse assemblage (e.g. Figs. 8.14, 15), but most of the main lines of homopteran evolution can be recognized.

Several of the Permian families persisted into the Mesozoic, and ten others (e.g. Figs. 8.14, 15) appeared in the Australian Triassic, some of which extended into the Lower Jurassic of the northern hemisphere. Several modern groups (e.g. Cercopoidea, Cicadelloidea) appeared in Australia and others

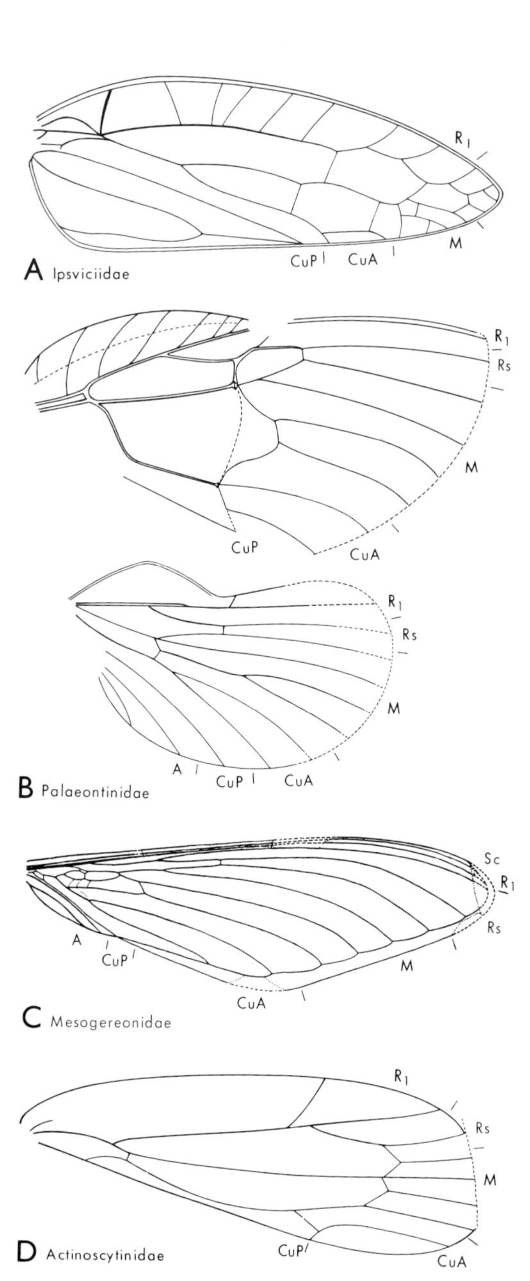

Fig. 8.15. Hemiptera: A, *Ipsvicia jonesi*, fore wing (after Evans, 1956); B, *Fletcheriana triassica*, fore and hind wings (fore wing after Evans, 1956, hind wing reconstruction); C, *Mesogereon superbum*, fore wing, reconstruction (after Tillyard, 1921b); D, *Actinoscytina belmontensis*, fore wing (after Evans, 1956).

(e.g. Tettigarctidae, Aphidoidea) a little later in Europe. Only a few homopteran fossils are known from the Cretaceous. Of the superfamilies not definitely recognized up to this time, the Peloridioidea are known with certainty only from the Recent fauna, the Aleyrodoidea may possibly be represented by some Permian nymphs, and the Coccoidea by a mealy bug-like fossil from the Permian (Evans, 1963). Modern forms, including coccids, are known from the Tertiary.

The Heteroptera present an even more difficult problem. Evans believed, on morphological grounds, that they probably separated from the Homoptera before the beginning of the Permian, whereas the writer thinks it more likely that they diverged much later. Several families of undoubted, mostly aquatic Heteroptera are known from the Lower Jurassic (Laurentiaux, 1953), but earlier records are less definite, the best documented being the Actinoscytinidae (Fig. 8.15D) from the Upper Permian and Triassic of Australia, Triassic of Asia, and Lower Jurassic of Germany (Wootton, 1963).

THYSANOPTERA (Perm.–Recent). The fossil record is scanty. *Permothrips*, from the Upper Permian of Russia, had the distinctive head and narrow, strap-like wings of the order. *Mesothrips*, from the Jurassic of Turkestan, had a body length of only 2 to 2·5 mm.

Endopterygote Orders

Martynova holds that the Neuroptera evolved from blattoid ancestors and gave rise to the panorpoid orders, from one branch of which the Hymenoptera also arose. There are, however, morphological difficulties in accepting a blattoid origin for the endopterygotes, whereas the venation of the Megaloptera, which are in some ways more primitive than the Neuroptera, can be compared directly with that of such Palaeodictyoptera as the Spilopteridae (Fig. 8.4). It is difficult to distinguish between primitive Megaloptera, Neuroptera, Mecoptera, and even Trichoptera, on wing venation, so allocation of some of the early genera to particular orders depends, to an extent, on the point of view of the worker. The Protocoleoptera, described originally from Belmont, are excluded from the account, because the fossil appears to have been an imprint of plant tissue.

MEGALOPTERA (Perm.–Recent). Three Permian families—Permosialidae (Fig. 8.16A), Tomiochoristidae, and Choristosialidae—are probably to be regarded as primitive Megaloptera rather than Protomecoptera; but there appears to be no further record of this order until modern Corydalidae appear in the Oligocene Baltic amber.

RAPHIDIOPTERA (Juras.–Recent). Excluding certain Carboniferous and Permian genera that seem to be more properly placed in the Paraplecoptera, the first record of Raphidioptera is from the Jurassic (Fig. 8.16B), and modern genera are known from Baltic amber and the Miocene of U.S.A.

NEUROPTERA (Perm.–Recent). Two groups of families were established in the Permian. The Sialidopsidae (Fig. 8.17A) and Palaemerobiidae had Sc and R_1 separate to the wing-margin, and are distinguishable from Megaloptera by little more than the greater amount of end-twigging of their veins. They can be compared with the Recent family Dilaridae. The Permithonidae (which were probably not directly ancestral to the Recent Ithonidae), Permopsychopidae, and Archaeosmylidae had Sc and R_1 fused distally (Fig. 8.17B). In the Triassic, the Archaeosmylidae were still present, with their descendants the surviving Osmylidae. The Osmylopsychopidae (Fig. 8.17C), their possible descendants the surviving Psychopsidae, and the Mesoberothidae (which were probably ancestral to the later Berothidae) also appeared. There was further development in the Jurassic: Osmylopsychopidae persisted; Prohemerobiidae and Kaligrammatidae prob-

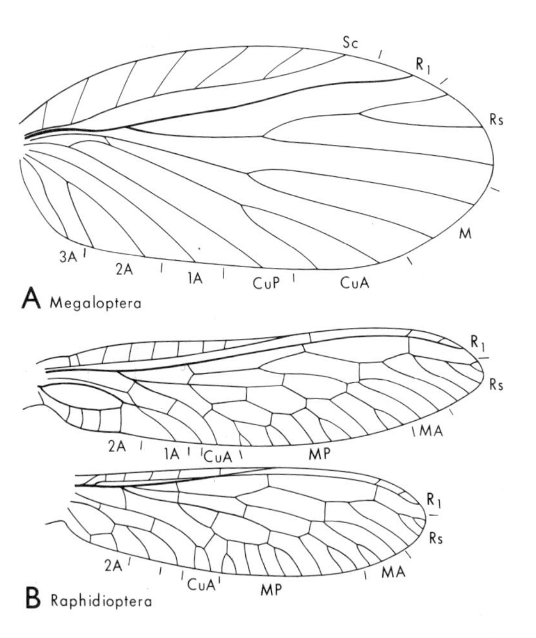

Fig. 8.16. A, *Permosialis immaculata*, fore wing (after Martynova, 1952); B, *Mesoraphidia inaequalis*, fore and hind wings (after Martynov, 1925).

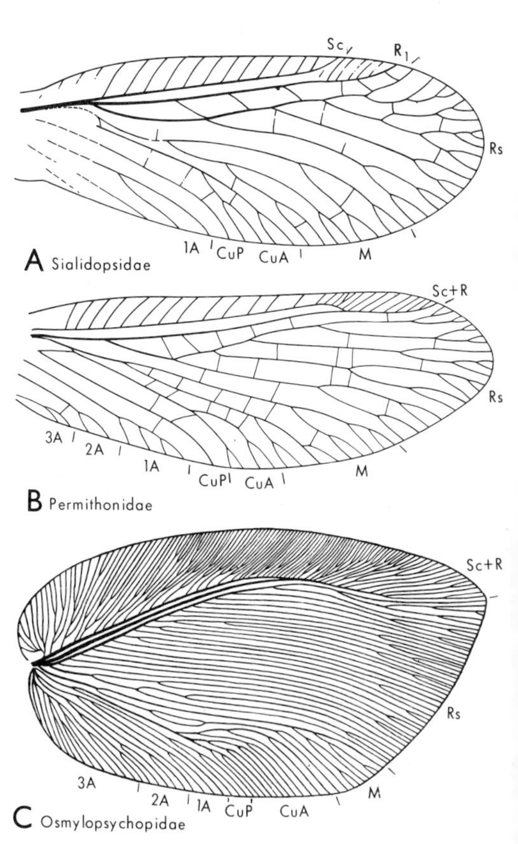

Fig. 8.17. Neuroptera: A, *Sialidopsis kargalensis*, fore wing (after Martynov, 1928b); B, *Permithone belmontensis*, fore wing (after Riek, 1953); C, *Osmylopsychops spillerae*, fore wing (after Tillyard, 1923b).

ably belonged to the hemerobiid line; the Nymphitidae may have represented a side-branch; and genera at least close to Recent Hemerobiidae, Polystoechotidae, Nymphidae, and Myrmeleontidae occurred. Thus, although there is no record of Coniopterygidae, almost the whole radiation of the order was completed by the middle of the Mesozoic.

The Australian fauna was relatively rich. The three Permian families that had Sc fused with R_1 occurred here, and there were five (Archaeosmylidae, Osmylopsychopidae, Mesoberothidae, Osmylidae, and Psychopsidae) in the Triassic. An osmylid, *Euporismites*, related to the Recent *Kempynus*, has been found in the lower Tertiary at Redbank.

COLEOPTERA (Perm.–Recent). Four families (or groups of families) of beetles are known from Permian elytra, which may be classified broadly as: cupedoid, primitive forms with remnants of a venation and a meshwork of cells (Fig. 8.18A); curculiopsid, with rows of tubercles; schizocoleid, with a single longitudinal groove; and permosynid, with striae (Fig. 8.18B). A mordellid of relatively modern facies has been recorded from the Jurassic of Turkestan. The Australian Permian beetles were mostly

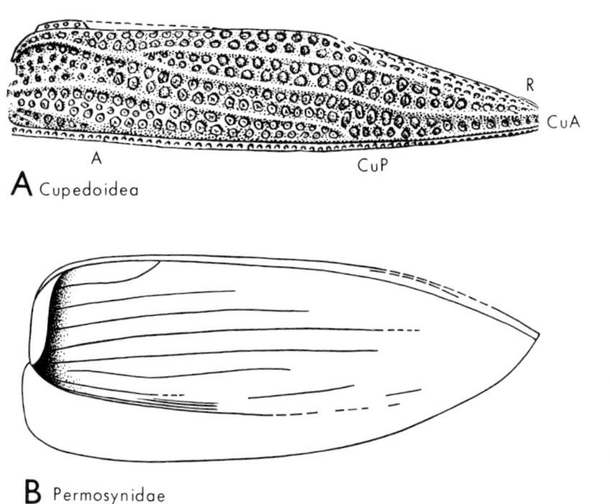

Fig. 8.18. Elytra of Coleoptera: A, *Archicupes jacobsoni* (after Rohdendorf, in Rohdendorf *et al.*, 1961); B, *Permocrossos elongatus* (after Jeannel, 1949a).

small permosynids *(Permosyne)*. In the Triassic, on the other hand, there were some cupedoids, such as *Simmondsia*, and many permosynids *(Ademosyne)*. A number of modern families are represented in the Lower Cretaceous of Victoria. Many unidentified elytra have also been obtained from the Redbank Tertiary.

STREPSIPTERA (Oligocene–Recent). The earliest record is of the surviving family Mengeidae from Baltic amber.

MECOPTERA (Perm.–Recent). Three suborders are known from the Permian. The Protomecoptera, with a regular series of costal cross-veins and CuA forked in both wings, were represented in the northern hemisphere by Kaltanidae (Fig. 8.19A), which are difficult to distinguish from Megaloptera; the suborder is not known again until it reappears as the Recent Meropeidae and Notiothaumidae. The Paramecoptera, with few costal cross-veins and CuA forked in the fore wing but not in the hind (Fig. 8.19B), are known only from Belmont. The Eumecoptera, normally with few costal cross-veins and with CuA simple in both wings, were represented by three families with Sc of the fore wing 3-branched and five families with it 2-branched. The first group died out in the Triassic, but the others had a more varied history. Two of them (Permopanorpidae and Mesopanorpodidae) disappeared by the middle of the Mesozoic; the Mesochoristidae (Fig. 8.20A) merged into the Recent Choristidae, and may have been ancestral to other Recent families; and the Nannochoristidae have survived to the present. The fifth family, Permotipulidae, was included with the Permotanyderidae in a suborder Protodiptera on account of the reduction in the cubito-anal venation and other dipterous features that they showed. However, the group is composite. Of the five Australian genera that comprised it, *Robinjohnia* should be transferred to the Nannochoristidae, *Permotipula* retained in a family of its own near the Nannochoristidae, the Triassic *Mesotanyderus* transferred to the Mesopanorpodidae, and the other two to the Diptera.

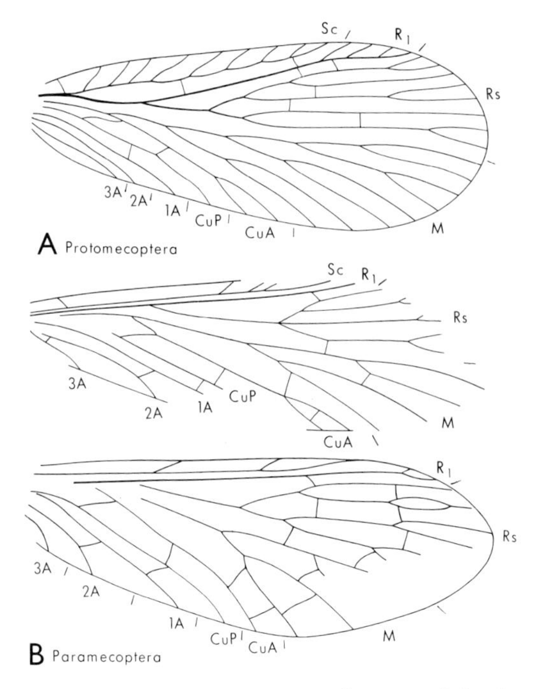

Fig. 8.19. Mecoptera: A, *Pinnachorista sarbalensis*, fore wing (after Martynova, in Rohdendorf *et al.*, 1961); B, *Belmontia mitchelli*, fore and hind wings (after Riek, 1953).

The changes in the Mesozoic included disappearance of Permian families; the appearance and disappearance of the suborder Paratrichoptera (Fig. 8.20B), which was distinguished from Eumecoptera by a rather distinctive alignment of the veins and a tendency to looping of the anals; and the appearance of the Neorthophlebiidae and Orthophlebiidae, which were probably derived from Mesochoristidae and evolved, respectively, into the Recent Bittacidae and Panorpidae. Though more diverse than today, the fauna had begun to take on a modern facies. Recent families only—Bittacidae, Choristidae, Panorpidae—are known from the Tertiary. There is no indication of the origin of the Boreidae.

The Australian fauna was rich and varied. In the Permian there were Paramecoptera (Belmontiidae), two families (Agetopanorpidae and Xenochoristidae) of the first division of the Eumecoptera, and four (Mesochoristidae, Mesopanorpodidae, Nannochoristidae, and Permotipulidae) of the second division. In the Triassic there was one family of Paratrichoptera (Mesopsychidae), Xenochoristidae were still present, and the second division of the Eumecoptera was represented by Permopanorpidae, Mesopanorpodidae, Mesochoristidae, Neorthophlebiidae, and Orthophlebiidae. *Chorista* occurred in the Lower Cretaceous of Victoria. In the Tertiary a species of *Chorista* and one of *Austropanorpa* (Panorpidae) occurred at Redbank.

SIPHONAPTERA (Oligocene – Recent). The only record is of a modern hystrichopsyllid *(Palaeopsylla)* from Baltic amber, but there is an undescribed species in the Lower Cretaceous of Victoria.

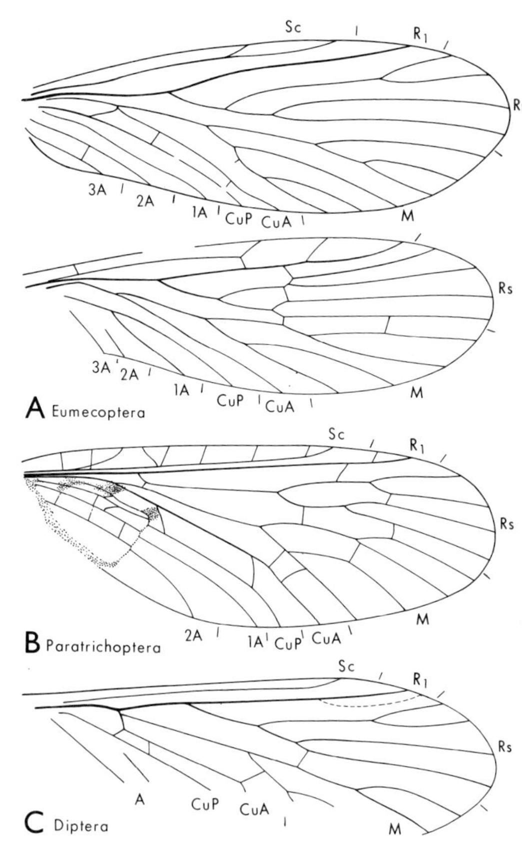

Fig. 8.20. Mecoptera: A, *Mesochorista australica*, fore and hind wings (after Riek, 1953); B, *Mesopsyche triareolata*, fore wing (after Tillyard, 1919a). Diptera: C, *Permotanyderus ableptus* fore wing (after Riek, 1953).

DIPTERA (Perm.–Recent). The Permotanyderidae, now restricted to the two Australian Permian genera *Permotanyderus* (Fig. 8.20C) and *Choristotanyderus*, may be included in the Diptera, because at least the former shows the characteristic kink in the stem of R, near the brace between R and M, that appears to provide the most reliable venational distinction between Diptera and Mecoptera. CuP is, however, relatively strong in both genera. It may be necessary to establish a separate group for them if they prove to have had four wings.

The only records for the Triassic appear to be the 'Archidiptera' and 'Eudiptera' recently described by Rohdendorf from wings of doubtfully dipterous affinities. However, well-developed Nematocera and some Brachycera were present in the Lower Jurassic (Hennig, 1954). Most of the Nematocera were Tipulomorpha of quite modern appearance; forms apparently related to Ptychopteridae; and Bibionomorpha of obscure relationship, but at least some belonging (or ancestral) to the Bibionidae, Mycetophilidae, and Anisopodidae, including what may be a representative of the modern genus *Olbiogaster*. There were also a few tanyderid- and psychodid-like forms, and a probable chironomid. The Jurassic Brachycera included several families of probable Tabanoidea, a few possible Asiloidea, and a possible cyclorrhaphan *(Archiphora)*. Ceratopogonidae and Chironomidae have been found in Cretaceous Canadian amber.

By the Eocene the fauna was quite modern, though with relatively few Schizophora. Many existing genera were present in northern-hemisphere strata, including, rather surprisingly, the tsetse flies (Glossininae). Acalyptrate Schizophora and Hippoboscidae appeared in the Oligocene and Miocene, and the only major families not recognized by the end of the Tertiary are the higher muscoids, such as Calliphoridae and Tachinidae, of which at least the Tachinidae seem to be still in the midst of a major radiation. The only fossil Diptera known from Australia after the Permian are a tipulid, some Mycetophilidae, and a possible muscoid from the lower Tertiary at Redbank, a chironomid from Vegetable Creek, and undescribed Chironomidae, Culicidae, and Simuliidae from the Lower Cretaceous of Victoria.

TRICHOPTERA (Perm.–Recent). The Platychoristidae (Fig. 8.21A) from the Lower Permian of North America and the Permomeropidae from Belmont are considered to be ancestral Trichoptera, rather than Protomecoptera, on account of their looped anal veins, although they had costal cross-veins and more terminal branches to Rs and M than is usual in the order. The Microptysmatidae (Fig. 8.21B) from the Permian of Russia and Cladochoristidae from Belmont provided transitions to more typical Trichoptera. The Cladochoristidae continued into the Triassic in Australia in company with more advanced Prorhyacophilidae, and the equally advanced Necrotauliidae appeared in the Lower Jurassic of England, Germany, and possibly South

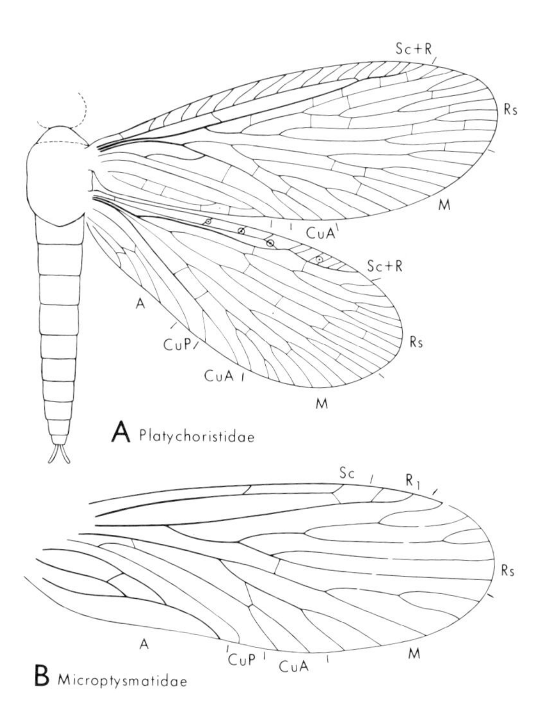

Fig. 8.21. Trichoptera: A, *Platychorista venosa*, reconstruction (after Carpenter, 1930); B, *Microptysma sibericum*, fore wing (after Martynova, in Rohdendorf *et al.*, 1961).

America. As with the Neuroptera and Mecoptera, the foundations of the modern families appear to have been laid before the end of the Mesozoic.

LEPIDOPTERA (Eocene–Recent). No extinct family is definitely known. Handlirsch had included the northern Jurassic Palaeontinidae and the Australian Triassic Mesogereonidae (Fig. 8.15c) in the Lepidoptera, but these are now regarded as cicadoid Homoptera. Tindale described the Australian Triassic *Eoses* in a new suborder Eoneura of the Lepidoptera, but it was later transferred to the Mecoptera. The earliest undoubted Lepidoptera are modern families from the Eocene of Europe. Zeugloptera, Dachnonypha, Monotrysia, and many families of Ditrysia, including butterflies, have been found in the Oligocene of Europe and North America. It is remarkable that so little is known of the history of such a successful order.

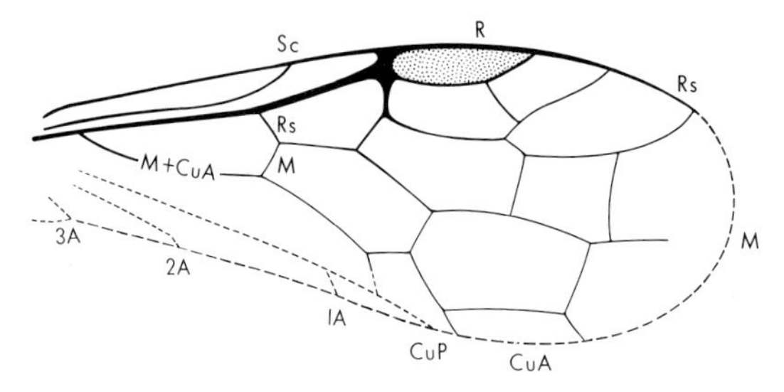

Fig. 8.22. *Archexyela crosbyi*, Hymenoptera-Symphyta, fore wing (after Riek, 1955j).

HYMENOPTERA (Trias.–Recent). The oldest Symphyta are the Xyelidae from the Lower Triassic of Central Asia (Rasnitsyn, 1964) and *Archexyela crosbyi* (Fig. 8.22) from the Upper Triassic at Mt Crosby. Other Symphyta have been described from the northern Jurassic. Two surviving families of Apocrita have been found in the Jurassic, and Ichneumonoidea, Cynipoidea, Proctotrupoidea, Chalcidoidea, and a few other groups in the Cretaceous. Undescribed Sphecoidea and parasitic species are present in the Lower Cretaceous of Victoria. It seems evident that the evolution of the order must have begun before the end of the Palaeozoic. Formicidae (including workers) and Vespidae are known from the Eocene, and bees from Oligocene Baltic amber.

9

COMPOSITION AND DISTRIBUTION OF THE FAUNA

by I. M. MACKERRAS

The materials of zoogeography are the present and past distributions of animal groups; its purpose is to throw light on the evolution of those groups in time and space, and to explain how they came to occupy the parts of the world in which they have been found. There are other useful functions for studies of distribution—to learn where vectors of disease may occur, where pests may have come from, where natural enemies of either may be found, and so forth—but enquiries of these kinds help only incidentally to solve the wider problems to be examined here.

The study of *evolutionary terrestrial zoogeography*, to which this chapter is limited, involves integration of knowledge from several different fields, and it may be practised at three overlapping but broadly distinguishable levels:

1. Relationships between the faunas of different parts of the world.
2. Faunal divisions within a region, or a country, and the reasons for their existence.
3. Geographic speciation within animal groups.

Each of these will be considered in turn, and it will be apparent from what follows that no apology is needed for restricting the present enquiry to insects or for choosing examples mostly from a few groups with which the author is familiar. Many more in other orders will be found in later chapters; the wider literature has been reviewed by Gressitt (1958) and Munroe (1965). Two digressions will be necessary in order to make the subject clear: one into the principles of zoogeographical research, and the other into the array of evidence that is now available to support the hypothesis of continental drift.

GENERAL PRINCIPLES

1. The basic premise of evolutionary zoogeography is that every group of animals, of whatever taxonomic rank, began as a single species in some limited part of the world, whence it might spread as it evolved and speciated in turn. If grades (p. 161) are of multiple ancestry, their monophyletic components form the zoogeographical units.

2. The groups of animals we can use are defined by their classification, so taxonomy is necessarily the basic tool to be employed in the study. The foundations were laid on empirical classifications, which provided the basis for defining the zoogeographical regions and formulating hypotheses to account for the faunal resemblances and differences between them. Faunal lists were used extensively, even up to the time of Harrison (1928). But questions were being

asked. Did the classification really reflect, not only monophyletic origins, but the relative propinquity of descent of the groups that were being compared? Was degree of resemblance a reliable guide, or did it conceal fallacies resulting from convergence or parallel evolution? These questions were vital, and it is clear today (e.g. Hennig, 1960; Brundin, 1965) that only strictly phylogenetic classifications, based on critical comparative morphology, are valid for studies of the kind discussed here. It follows that the zoogeographer must usually be his own taxonomist, and that the groups he can use are limited, for by no means all are equally amenable to phylogenetic analysis.

An important warning is necessary. *The whole of the analysis of descent and relation-*

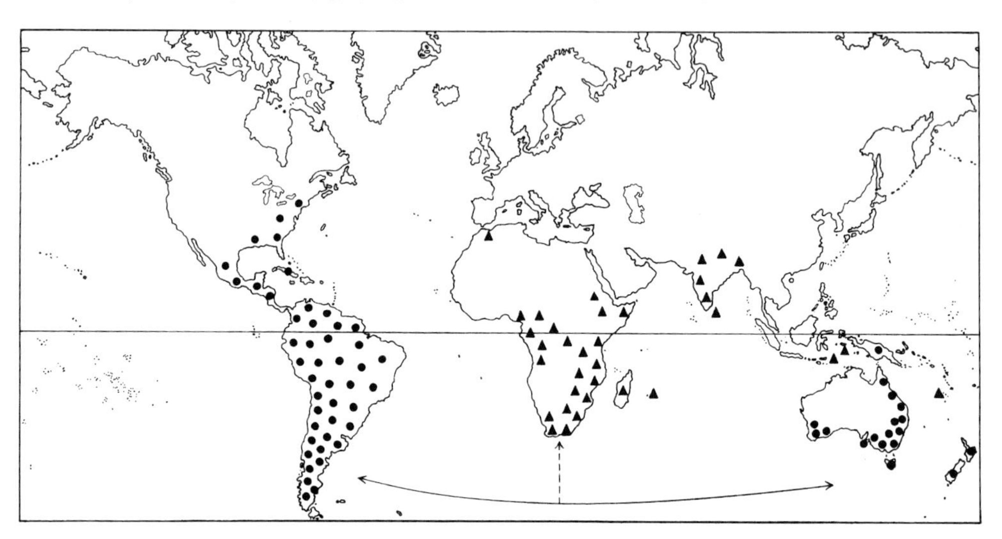

Fig. 9.1. Distribution of two older tribes of Tabanidae (Diptera). The Scionini (spots) and Philolichini (triangles, see also Fig. 9.5) are sister groups derived from a common 'protoscionine' ancestor.

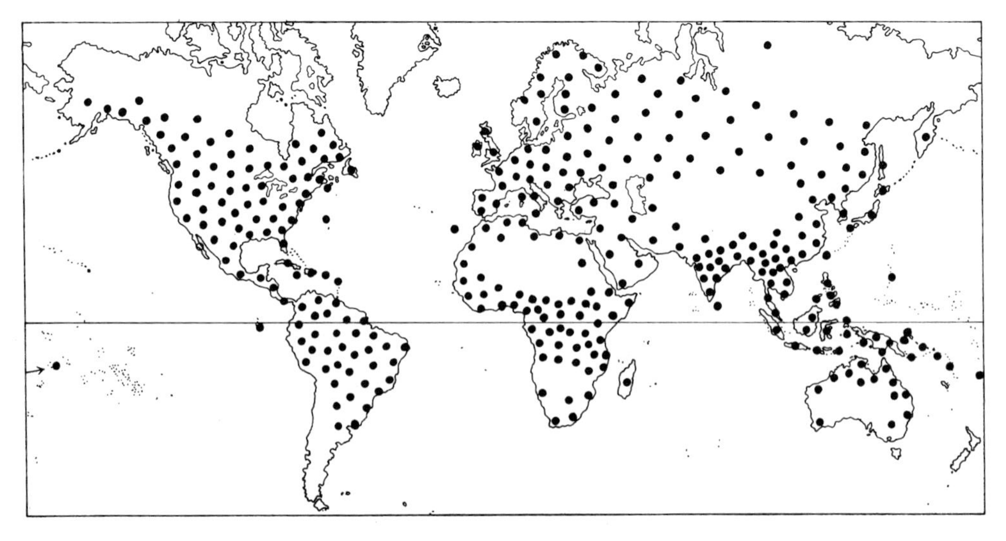

Fig. 9.2. Distribution of the Tabanini, a younger tribe of Tabanidae. The resemblance to the distribution of eutherian mammals is striking.

ship must be morphological and genetical: it is indefensible to use distributional data for this purpose, except at the subspecific level. An anomalous distribution should make us check our morphological or genetical evidence, but it should not be allowed to influence our ultimate decision.

3. The next step is to plot the distributions of the selected groups, in order to discover whether they show any coherent patterns. Three such distributions are shown in Figures 9.1 and 9.2. Some writers attempt to achieve greater precision by tabulating the number of species (or of some higher taxon) in different regions or parts of a region; but this approach, though useful, can be misleading, because the conditions that influence speciation and diversification are not necessarily the same as those that determine origins and directions of dispersal. There may be secondary centres of evolution, especially in the tropics, or, on the other hand, a widely distributed group may occasionally be reduced to relict status in its original home. An attempt should always be made, therefore, to determine the sequences of evolutionary development within the chosen group, and this cannot be done without careful phylogenetic analysis. Maps become much more significant when it is possible to plot arrows as well as spots on them, as in Figures 9.5 and 9.6.

4. As it cannot be assumed that centres of radiation and pathways of dispersal have remained the same from the Palaeozoic to the present day, the group used must be old enough (or young enough) to have relevance for the particular problem that is being investigated. The *relative* antiquity of related groups can often be estimated from morphological evidence, but it is also necessary to have some fixed points for reference. Consequently, we must know something of the geological history of the group we are studying. Sufficient fossil Diptera, for example, are known to justify the conclusion that the Nematocera were diversified and at least the more primitive families of Brachycera–Orthorrhapha established before the middle of the Mesozoic, whereas the Cyclorrhapha–Schizophora seem to have radiated rather suddenly in the early Tertiary. One could use the first two for studying earlier radiations, the last for later ones.

That is not the whole story, for many ancient groups continued to evolve and diversify, and they did so unevenly, so that one can distinguish between components of various age within them (for example, the tribes plotted in Figures 9.1 and 9.2). It follows that the taxonomic level to be used must be chosen as critically as the whole group, if one is to avoid confusing solutions to different problems.

5. The fossil evidence may also provide valuable information about the past distribution of a group of animals. Negative evidence is useless, and positive evidence must be used with caution, because specimens are usually scanty (and therefore doubtfully representative), the characters available for study are limited, and the age of fossilization may not be accurately known. Nevertheless, even a single fragment that can be *unequivocally* identified provides positive evidence that a given group occurred in a particular area at a particular time. No one, for example, would doubt that Mecoptera occurred in the Nearctic, Palaearctic, and Australian regions in the Permian, although there might be considerable room for argument about the detailed relationships between the three faunas.

6. We must know something of the 'vagility', or capacity to disperse, of the group we are using. Small, light insects (even if apterous) that form a frequent component of the aerial plankton are excluded from the analysis, as are parasites of wide-ranging hosts, and species that travel with man.* On the other hand, groups, such as mayflies, with short-lived dispersal stages, and especially those, like helminthid beetles, with narrowly limited ecological tolerances (Hinton, 1965), are particularly useful. In general, the reliability of a group as a zoogeographical

* Actually, if the fauna of an isolated area of land is composed *entirely* of these elements, that is evidence of previous extinction of other elements, as in Antarctica, or that the area was never connected to a major land-mass, as with many oceanic islands (see also Hinton, 1965).

tool varies directly with the difficulty one has in imagining that it could spread without something approximating to direct land connections. There are exceptions, and 'island hopping' appears to have been an effective method of dispersal in the Moluccan area and the Pacific, although it also had a powerful filtering action.

7. Lastly, we must also have other information about the ecological requirements and limitations of the group we are using. In the first place, although progressive diffusion and adaptation may occur at a margin, it matters little how good an animal's capacity to disperse may be, if it does not find a niche that it can colonize at the end of a journey over hostile terrain (see, especially, Gressitt, 1961). Secondly, ecological limitations determine what terrestrial conditions form barriers to the dispersal of different groups of animals, and are particularly important therefore in relation to filtration at regional margins and to speciation. Thirdly, there is evidence (also summarized by Gressitt) that the macroclimatic requirements of some groups of animals have not changed greatly over long geological periods, and it is sometimes possible to correlate past and present distributions with independent palaeoclimatological findings. Broadly, the distribution of groups of animals that can disperse readily is often determined primarily by their requirements, but in most groups it is the result of interaction between historical and ecological factors.

8. Most of the foregoing has been concerned with the restrictions that apply to the selection of zoogeographical indicators. Having chosen an indicator, its pattern of distribution is worked out and the implications noted. That is only one step. The same process must be repeated with other, equally carefully chosen, independent groups of different ancestry, geological age, and adaptive properties. Discrepancies are analysed, and the individual patterns are progressively integrated into a more general pattern of animal distribution in time and space, which can then be integrated with the patterns of distribution of land plants and marine life into a coherent theory of biogeography (see Evans, in Carey, 1958). Modern zoogeography is thus based not so much on whole faunas as on progressive sampling, which becomes limited in turn by the principle of diminishing returns.

9. Finally, it cannot be emphasized too strongly that zoologists, from their own data, can only state problems of evolutionary zoogeography; they cannot solve them without other data provided by the earth scientists. Moreover, the solutions arrived at can satisfy only in so far as the statements of the zoologists are not coloured by palaeogeographical preconceptions and the findings of the earth scientists are based primarily on non-biological evidence. A great deal of zoogeographical nonsense has resulted from ignoring this principle.

REGIONAL RELATIONSHIPS

The Zoogeographical Regions

The year 1858 was a vintage one for the Linnean Society of London, for its *Proceedings* contained not only the epoch-making papers on natural selection by Darwin and Wallace but also the first clear definition of the zoogeographical regions by Sclater. Darwin discussed the distribution of animals and plants from the evolutionary point of view in *The Origin of Species* in 1859, and Wallace (1876) considerably extended both Sclater's and Darwin's analyses. His great work remains the most valuable single treatise on the subject that has yet been written, and his arrangement of regions (Fig. 9.3) and subregions has been but little changed since.

The rather static treatment of most early workers has been criticized on the grounds that it took insufficient account of the facts that the faunas are made up of components *(faunal elements)* of varied origin and evolutionary history, that different groups of animals have different capacities to disperse, and that there are important transition zones between regions. Nevertheless, the regional faunas are, in fact, reasonably distinct from one another, and the classical regions do form useful geographical units for analysis and comparison. Schmidt (1954) proposed a more

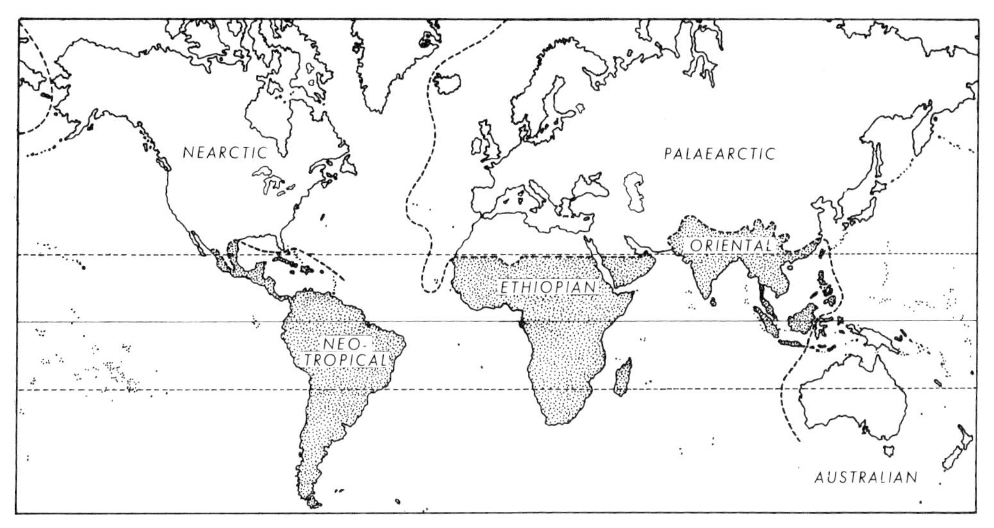

Fig. 9.3. The zoogeographical regions (after Wallace, 1876).

complex arrangement, based primarily on Tertiary radiations, but it has little value outside that context.

Modern descriptions of the regions and their faunas can be found in a number of general works (e.g. Darlington, 1957). Briefly the *Palaearctic* and *Nearctic* faunas are nearly related, and often combined as a single *Holarctic* fauna, which contains some older components and elements of both Ethiopian and Neotropical origin, but is characterized mainly by great Tertiary radiations of many groups. The *Neotropical* fauna is distinctive, with a strong southern element related to Australian and southern Ethiopian elements, tropical radiations of mixed relationships, and more recent Nearctic intrusions from the north. The *Ethiopian* fauna comprises a southern element, a northern (Palaearctic) element, tropical radiations derived from both, and an Oriental intrusion in the east. The *Oriental* fauna can be regarded as consisting mostly of a tropical Palaearctic expansion, with Gondwanaland relicts, a significant Ethiopian intrusion from the west, and a smaller Australian one in the east. It enters the Pacific through a wide zone of transition and filtration, lying approximately between Wallace's and Weber's lines (Fig. 9.4), and appropriately named *Wallacea*. The *Australian* region has been taken as extending south and east to a greater or lesser extent from Wallace's or Weber's line; its limits are discussed later.

The Australian Faunal Elements

The early naturalists were amazed at the novelty of the Australian animals (and plants) they studied. That impression of faunal distinctiveness and isolation still remains valid, but its impact has been progressively modified by the gradual perception of relationships with other faunas and the development of coherent theories of faunal origins. The trend took definite form between 1888 and 1896, when the studies of Tate, Hedley, and Baldwin Spencer led to the recognition of three faunal (and floral) elements—'Autochthonian', 'Euronotian', and 'Papuan'—to which Harrison (1928) added a fourth 'Pantropical' element. These have been considerably modified, and it is necessary to include also a small, but sometimes obtrusive, mixed, more recent element in order to complete the picture of the fauna as it exists today.

The 'Autochthonian' element may be disposed of before proceeding further. An *autochthonous* group is one that evolved within the country, the homopteran family Eurymelidae being a good example (Evans,

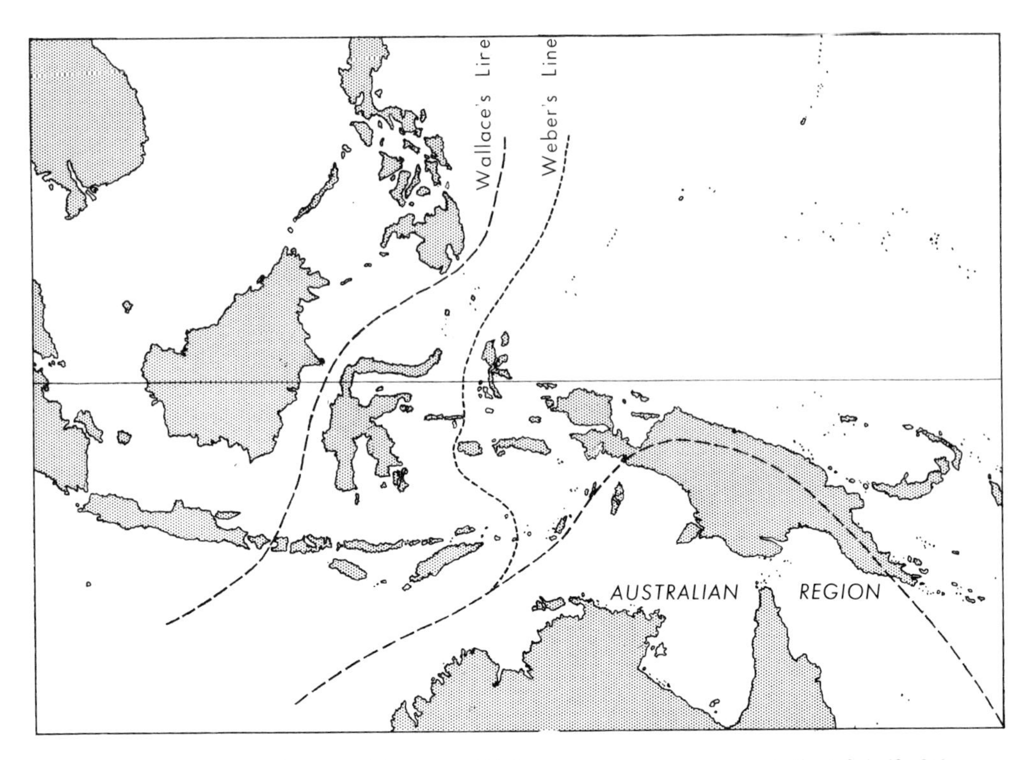

Fig. 9.4. The Oriental-Australian transition zone. Wallacea occupies approximately the left half of the map.

in Keast *et al.*, 1959). The designation is usually applied to higher taxa, and it is to be used with caution, because it often conceals ignorance of origins and relationships. An *endemic* (or *precinctive*) group is one that is known only from a particular country, irrespective of whether it evolved there or not. Thus, a group might have evolved in area A, extended to B, died out in A, and remained 'endemic' in B. *Degree of endemicity* is often defined as the percentage of the local members of a taxon (usually species of a genus) that is limited to a given country or faunal area. It is an unreliable statistic unless it is accompanied by an analysis of the evolutionary factors that have operated. In spite of their limitations, the phenomena underlying these terms often do provide useful indications of geographical isolation.

The Archaic Element. ('Euronotian' in part). This is a small element, consisting of primitive animals that have evidently survived with little change since Palaeozoic or early Mesozoic times. It is most clearly represented by the relict groups, each consisting of a few archaic species scattered over the world in a disjunct, indeterminate fashion. The Dipnoi are a classical example in the vertebrates and the Onychophora in the invertebrates. Examples in the insects include the mecopteran family Meropeidae (North America, Australia) and the dipteran genus *Nemopalpus* (Neotropical, Canary Is., South Africa, Malaysia, Australia, New Zealand, Baltic amber). Some apparent relicts are known from only a single region, the curious doubtfully nemestrinid genus *Exeretoneura* (p. 704) being an Australian example. This element merges into the southern element, and it is sometimes difficult to decide whether a particular group of insects should be allotted to the one or the other.

Some very ancient groups are not relicts. The cockroaches are an example, and it may be that their obscure zoogeographical patterns are due as much to the fact that they have changed remarkably little since the Carboniferous as to the still incompletely resolved phylogeny of the families and genera into which they have been divided.

The Southern Element. ('Antarctic', 'Antarctogaean', 'southern Gondwanaland', 'Euronotian' in part). This is one of the two, conspicuous, well-defined elements that together make up the bulk of the Australian fauna. Whatever the theories of its origin, it forms a significant faunal component of a number of groups of terrestrial and freshwater invertebrates (Harrison, 1928), of most orders of insects from mayflies to beetles (see later chapters) and of four classes of vertebrates—freshwater fish, amphibia, reptiles, and marsupials. Darlington (1965) and most other northern-hemisphere zoogeographers have not appreciated its magnitude (e.g. Fig. 9.1), diversity, and at the same time its unity.

It has two basic characteristics, which are clearly demonstrated in the insects. Firstly, it represents a rather, though not extremely, early level in the evolution of most of the orders in which it has been recognized. Thus, in the Diptera one finds its representatives in the more primitive sections of nearly all the families of Nematocera and Orthorrhapha, but very few in the Cyclorrhapha–Schizophora. The fossil evidence suggests that it is composed of groups that had probably evolved into recognizable entities by about the middle of the Mesozoic. Its long establishment in Australia is attested by the extensive adaptive radiations of many of its components.

Secondly, it has a characteristic pattern of distribution: in Australia, southern South America, southern Africa, and New Zealand (see Brundin, 1965, Fig. 2*). Most of its component groups are shared between Australia and South America, fewer with South Africa or New Zealand. A few extend into temperate Holarctica, but it does not appear to be represented in Oligocene Baltic amber (Colless, 1964; Illies, 1965). There are small extensions from Australia into Wallacea and the Pacific, but it is absent from the Oriental region, where its place is taken by the older northern element.

The Older Northern Element. ('Pantropical', 'Lemurian', 'northern Gondwanaland'). This also is a relatively primitive element, related to the southern, but usually distinguishable from it taxonomically as well as in distribution. Thus, Evans (in Keast *et al.*, 1959) regards the ledrine leafhoppers as indicating a very ancient connection between India and Australia. Somewhat better known is the Lemurian element, which has been studied by Harrison (1928) in various groups of invertebrates and by Mackerras (e.g. 1964) in the Tabanidae. This consists of a band of species-groups or higher taxa, which have centres of evolution in Africa or Madagascar, extend round or across the Indian Ocean, and

* Also L. Brundin, *A. Rev. Ent.* **12:** 149–68, 1967.

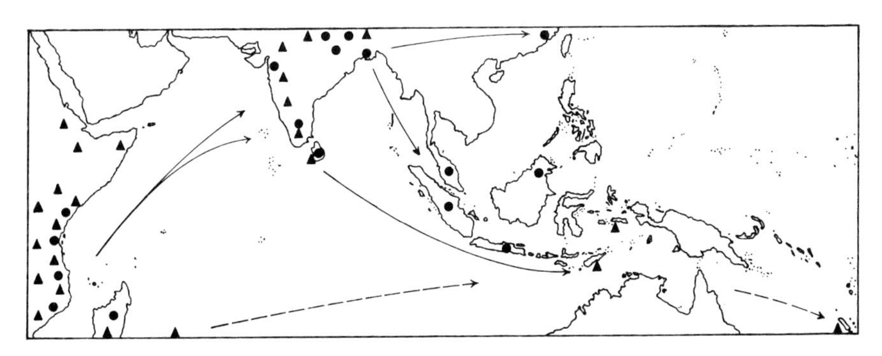

Fig. 9.5. 'Lemurian' distributions, illustrated by the tabanid tribes Philolichini (triangles) and Rhinomyzini (spots). The lines and arrows indicate relationships, not necessarily pathways of distribution.

sweep through New Guinea deeply into the Pacific (Fig. 9.5). There is also evidence of relationship with part of the older Neotropical fauna, which is why Harrison gave it the wider name Pantropical. Extensions of this element into Australia are mostly small and not always clearly definable, but some of them have diversified to a degree that suggests that they have been in the country for a considerable time.

The Younger Northern Element. ('Oriental', 'Indo-Malayan', 'Papuan' *sensu* Hedley). Like the southern element, this one is prominent, clearly defined, and of unequivocal relationships. It forms the most highly evolved and often the most conspicuous element in almost every order of insects. It stems back to the Holarctic Tertiary radiation, but is generally called Oriental or Indo-Malayan, because its immediate relationships are with the most prominent component in the Oriental fauna. Its pathway into Australia lay partly through New Guinea, but probably also from points in the Indonesian chain further west. It has undergone considerable speciation here, but comparatively little radiation at higher taxonomic levels.

Recent Elements. These may be divided into three groups. First, there are the small hexapods, such as Collembola, aphids and some chironomid Diptera, that are carried widely in the aerial plankton, a method of dispersal that has been studied by Gressitt and his colleagues in high southern latitudes. Second, there are larger insects (often single species of normally rather static groups) that have become recognized as 'good travellers'. *Tabanus ceylonicus*, distributed from Ceylon to the Philippine Is., Solomon Is., and north Queensland without indication of developing into distinguishable local populations, is an example.

The third group, of considerable practical importance, includes insects that have been spread by the agency of man. *Musca domestica*, *Culex fatigans*, and *Pulex irritans* may well have travelled as 'commensals' in the First Fleet, as *Pediculus humanus* and *Pthirus pubis* certainly did. Buffalo flies came in with introduced stock, although, remarkably, the pest species of *Hippobosca* from the same parts of Indonesia apparently did not. Many pests of introduced plants were brought in unwittingly with their hosts. Hive bees were introduced deliberately, and there have been many, much later, deliberate introductions of beneficial insects to control weeds or insect pests. In fact, most of the attention of economic entomologists in this country is devoted to introduced insects, useful or harmful, and it has proved a sound investment to spend considerable sums on quarantine measures to prevent further unwanted introductions.

Enigmatic Distributions. Some distributions do not fit readily into any of the patterns described above. The mosquito genus *Culiseta*, which is rather strictly adapted to cool temperate climates and found mainly in Holarctica, southern Australia, and Tasmania, is one example, and Illies' (1965) suggested dispersal of the plecopteran subfamily Notonemourinae from Australia to South America is another. Anomalies like these will ultimately have to be fitted into a pattern, but they should not be allotted arbitrarily to an element in the meantime.

Conclusion. Integration of inferences about phylogeny with the known facts of distribution strongly suggests that there were three major phases in the distribution of animals over the world: an early one, in which dispersal was more or less random; a middle one, in which dispersal was predominantly in wide latitudinal bands corresponding broadly with the present northern and southern land masses, the southern band being further divisible into temperate and tropical belts; and a later one, in which dispersal was predominantly meridional, dominated by the Holarctic radiation, but with subsidiary centres of radiation also, especially in the tropics and Australia. Correlation with the available palaeontological evidence suggests that the transition from the first to the second phase began in the later Palaeozoic and from the second to the third in late Mesozoic. There is also evidence that Australia was almost completely isolated from other regions throughout most of the Tertiary.

The Limits of the Australian Region

This problem brings up two important points. One, mentioned earlier, is that different groups of animals have different capacities to disperse and colonize, and are stopped by different barriers, so that transition zones like Wallacea provide filters or 'sweepstake routes' rather than complete barriers. The other is that there is good evidence that both pathways and barriers have varied considerably during geological history, so the time at which a group reached a regional or subregional margin may have determined whether it passed through or was stopped. There are other influences too. Thus, aerial plankton may be dispersed mainly by prevailing winds, but more powerful insects only in conditions of unusual turbulence, which may come from other directions. The end result is that regions lack marginal precision, so that arguments about their precise limits are largely futile.

Australia and Tasmania form the core of the Australian region, and few zoologists would question their overall faunal distinctiveness. To the north-west, recognizably Australian elements fall off rapidly beyond Weber's line, which lies close to the continental shelf (Fig. 9.15A) and could be taken as a convenient boundary. But, coming from the other direction, a flood of older and younger Oriental groups extends right across, and into, Australia, so that the boundary between the regions might be said to lie between Australia and New Guinea, or even within Australia itself (Gressitt, 1961). So far as New Guinea is concerned, this point of view is supported by the evidence that much of its Oriental component has been there long enough to radiate extensively and develop a recognizably 'Papuan' facies in many groups of insects, whereas the Australian component appears to consist almost entirely of relatively recent offshoots from mainland stocks. That, however, is not the whole story, because, although the insects of

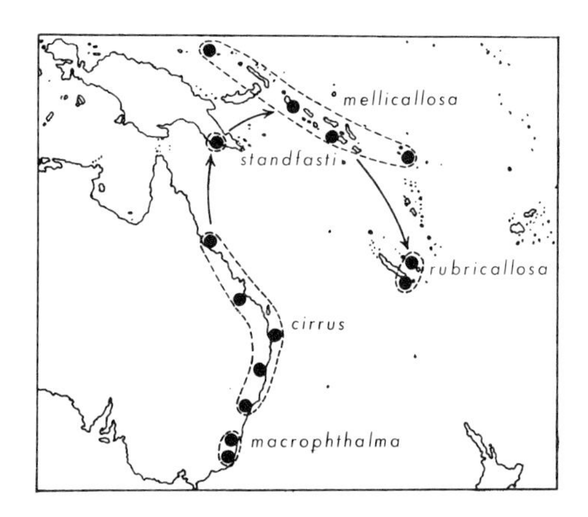

Fig. 9.6. An Australian contribution to the Pacific, illustrated by five nearly related coastal species of *Dasybasis* (Tabanidae).

New Guinea could be regarded as Oriental with Australian intrusions, the mammals, for example, could equally be regarded as Australian with Oriental intrusions. For present purposes, we may delimit the region conservatively as shown in Figure 9.4.

To the east, the Papuan fauna extends, progressively impoverished and with a good deal of speciation, along the Melanesian arc and into the western Pacific (Fig. 9.2). There are also some small Australian contributions that can be traced along the same arc to New Caledonia (Fig. 9.6), but they are relatively insignificant. All these areas, then, might reasonably be excluded from the Australian region.

Lord Howe and Norfolk Is. could be accepted, partly for convenience, as Australian, but New Zealand presents a different problem. Its endemic fauna appears to have been derived partly from the north (mainly through the New Caledonian part of the arc, along which there has probably been some traffic both ways), partly from the south (with relationships in some groups, e.g. Tabanidae, rather closer to South America than to Australia), and partly, but to a much smaller extent, by direct migration from Australia (as in the bird genus *Zosterops* as well as some insects). Its impoverishment in higher taxa may be the combined result of

Fig. 9.7. The faunal provinces in Australia.

isolation, limited environments, and destruction during Pleistocene glaciations. It is evident that the regional concept tends to lose force in the south-western Pacific area.

FAUNAL PROVINCES WITHIN AUSTRALIA

When Tate, Hedley, and Spencer recognized the faunal elements, they appreciated that they tend to occupy different parts of the country, and Spencer's division of the continent into Bassian, Torresian, and Eyrean faunal provinces (Fig. 9.7) has been widely accepted since. Tate's additional Autochthonian province in the south-west has been discarded, because its fauna, although rather distinctive, proved to be predominantly Bassian in relationships. Some authors have proposed further subdivisions or alternative names, but neither action has had more than limited usefulness.

The provinces can be defined broadly by both ecological and faunal characteristics (see papers in Keast *et al.*, 1959). The *Bassian* province has a moist-temperate climate, a vegetation ranging from southern rain and wet-sclerophyll forest to alpine herb fields, and an insect fauna that consists predominantly of the southern element. The archaic element, though always a minor one, is also mostly Bassian in distribution. The *Torresian* province has a wet to moist, tropical to subtropical climate, patches of tropical rain forest, a general vegetation ranging from wet-sclerophyll forest to open grassland, and an insect fauna dominated by the older and especially the younger northern elements, with recognizably Papuan intrusions (in the restricted sense) in the north-east. The *Eyrean* province comprises the whole of the semi-arid to arid interior within about the 500-mm (20-inch) isohyet. It has an impoverished insect fauna, not characterized by dominance of any element, but derived as progressively adapted offshoots from all. Acridid grasshoppers and some families of Neuroptera, for example, have radiated widely in dry country. Another series, consisting of individual species of a genus, is shown in Figure 9.9, but here the extensions are along major river systems, and the adaptations to aridity are not as great as they might appear from the map.

There are many divergences from the patterns outlined above. Thus, the Bassian province should really be shown as extending in a discontinuous northern tongue along the higher parts of the Dividing Range, and a few representatives of the southern element are even found in the highlands of New Guinea. Moreover, many otherwise typically 'Bassian' (i.e. southern-element) species can tolerate relatively hot conditions during their immature stages, so that adults are abundant in the winter or spring in coastal districts where adults of the northern elements dominate the scene in summer. Progressive adaptation may also occur, and Figure 9.8 shows the distribution of a more than usually successful 'Bassian' subgenus. Conversely, northern elements extend more or less deeply into Bassian territory (e.g. Fig. 9.9), a few of them even into Tasmania. Again, some of the older autochthonous groups of less evident origin, such as the short-horned grasshoppers (Key, in Keast *et al.*, 1959) appear to have evolved within Australia into series adapted to the different provincial environments. These phenomena complicate the picture; but they are illuminating rather than confusing when looked at from a dynamic point of view, and they do not detract from the general significance of the provincial divisions.

Fig. 9.8. The distribution of the subgenus *Scaptia* (Tabanidae) in Australia (cf. Fig. 9.1).

Fig. 9.9. The distribution of *Tabanus* in Australia (cf. Fig. 9.2).

PALAEOGEOGRAPHY

The essence of what has gone before is that there are four faunal elements in Australia that cannot be readily accounted for by casual means of transport: an old one with varied regional relationships; old southern and old northern of clearly southern-hemisphere latitudinal relationships; and a younger northern of equally clearly meridional northern-hemisphere relationships. The problem is to account for their occurrence in this country.

Older authors (e.g. Tillyard) simply raised and sank land-bridges at will, but that procedure was looked on coldly by the geologists. Then Wegener proposed his theory of drifting continents (the English edition in 1924), and it seemed to solve all the major problems of southern zoogeographers (see Harrison, 1928). His concept of continental blocks composed of a lighter, harder sial floating in isostatic (equivalent to hydrostatic) balance on a denser, more viscous sima was acceptable to most geologists, but they could not visualize the forces that would move the continents. Many zoogeographers, preoccupied with Tertiary radiations which nobody questions, accepted a static world geography, and only a few, especially in the southern hemisphere, found it impossible to account for the abundant southern elements as marginal relicts, or by convergent evolution, or by trans-oceanic transport. Even the demonstration that Antarctica supported a substantial flora from the Devonian until at least the middle of the Tertiary did not solve the problem.

In the last ten years, the wheel has turned almost full circle. Continental drift has come again into the picture, backed not only by new evidence from various sources, but by an independently supported theory of convection in the mantle that would make drift not only possible but inevitable. It is therefore worth closer examination.

The Evidence

This is of two kinds, relating directly to drift, and bearing primarily on the occurrence of convection cells in the mantle. The two are often inter-related, but it is possible to make a broad distinction between them. The geological evidence has been reviewed by Carey (1958) and King (1962), the geophysical evidence in Runcorn (1962) and more simply by J. Wilson (1963), the palaeoclimatological evidence by Opdyke (in Runcorn, 1962; see also Schwarzbach, 1963), and the palaeobotanical evidence by Plumstead (1962). The viewpoint of the sceptic has been put by Darlington (1965).

1. The classical basis for the theory of drift was the way in which the margins (not shore lines) of continental blocks can be fitted together, particularly on opposite sides of the Atlantic Ocean. This was reinforced by stratigraphic and tectonic continuities, sometimes of remarkable precision, on opposing aspects.

2. The mid-ocean ridges (again especially in the Atlantic), lateral ridges, and increasing age of islands with distance from the central ridges form patterns that conform with the supposed movements, and indicate a rate of movement (about 5 cm per year) of the same order as that deduced from geophysical studies (Wilson).

3. Some rift valleys (e.g. in Africa) show evidence of displacements that would conform with the hypothesis that existing continents are breaking up. Patterns of mountain building can also be accounted for by drift, but it appears that other mechanisms might be equally effective.

4. The evidence from palaeomagnetism is particularly cogent. Many rocks contain ferromagnetic minerals, of which the permanent magnetism was oriented in the direction of the earth's magnetic field when the rocks were formed, and some or all of the magnetic components commonly remain preserved indefinitely. Assuming the earth has always been a dipole, and subject to appropriate precautions, these magnetic components can therefore be used to determine the orientation of the field at the time of origin of the rocks. Palaeolatitudes are indicated by the dip, and *relative* palaeolongitudes by variation in the horizontal plane. Polar positions determined for any one continent are concordant, and indicate a particular pattern of 'polar wandering' for that continent. But, and this is the crucial point, the patterns of 'polar wandering' are not concordant for different continents, and they can be brought together only by moving the continents relative to one another, both in position and orientation, on essentially Wegenerian lines.

5. The palaeoclimatological evidence is also of several kinds, and much of it is independent of palaeontological evidence. Some of the indicators used are: $0^{16}:0^{18}$ ratios as measures of temperature; cross-stratification of aeolian sandstones produced in former trade-wind belts; evaporites (gypsum, etc.) deposited in hot, dry climates; bauxite, a product of deep weathering in a tropical or subtropical climate with contrasting wet and dry seasons; tillites and other stigmata of

glaciations; structural adaptations to wet or dry, hot or cold environments in fossil plants and animals, particularly in plants. The positions of the climatic equator deduced from these data cannot be reconciled with the present positions of the continents on any coherent theory of atmospheric circulation; but they do agree reasonably well with those of the terrestrial equator deduced from palaeomagnetism (Opdyke).

6. The crucial palaeobotanical evidence has been the identity of the distinctive *Glossopteris* flora, which flourished from late Palaeozoic to early Mesozoic times in South America, southern Africa, India, Australia, and Antarctica. It was also related geographically to the Permian glaciations in the same countries, and it occupied what must have been a fairly wide, cold to temperate, circumpolar belt. It thus supplements the palaeoclimatological evidence. It was not strictly contemporaneous in all the countries, and this has been interpreted as evidence that the South Pole wandered in a remarkable fashion; but it is more simply explained by assuming that the continents were close together and drifting collectively in relation to the geographical pole during that period, an explanation that is also in accord with the palaeomagnetic data. Plumstead (1962) has now shown that a close floral relationship between Antarctica and the other southern continents extended from the Devonian into the Tertiary.

7. The theory that there are convection currents in the earth's mantle, driven by the heat generated by radioactive decay, moving at the rate of a few centimetres per year, and forming cells that change in a predictable manner, is now supported both by mathematical analysis of theoretical models and by a considerable body of seismic, gravimetric, and thermal evidence. Many of the papers in Runcorn (1962) are devoted, wholly or in part, to these subjects, and only a few examples can be given here. Thus, temperature anomalies suggest that the mid-ocean ridges are lines of upwelling, and gravitational anomalies in trenches that currents are turning downwards in them (Fig. 9.10);

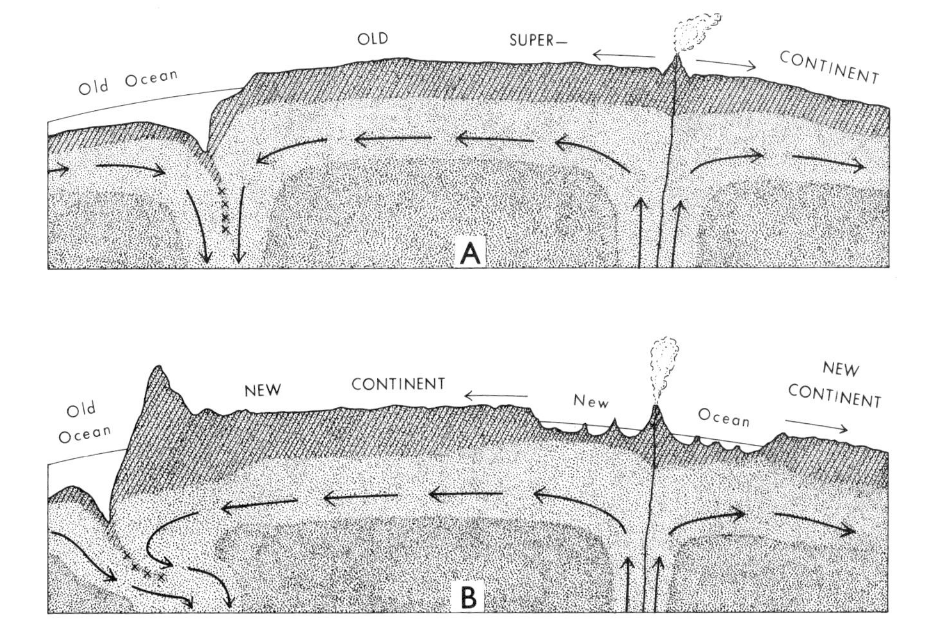

Fig. 9.10. Section of the earth showing convection in the mantle (based on Wilson, 1963).

rates of movement measured in rift valleys or calculated from other kinds of information are within the range of theoretical expectation; and calculation of the viscosity of the mantle, based on the rate of postglacial isostatic recoil in northern Europe, show that it would have to be vastly harder than it is in order to stop convection. If convection can be taken as proved, some form of drift would necessarily appear to follow from the known properties of the outer layers of the earth.

Patterns of World Geography

Not all the lines of evidence outlined above are equally strong. Some—particularly the palaeomagnetic, palaeoclimatic, and palaeobotanical data—are difficult to reconcile with any hypothesis other than drift; others are weaker, in that they could be susceptible to other explanations. Nevertheless, they all combine in an impressive way to produce geographical models that are broadly similar to those postulated by Wegener in 1912 and du Toit in 1937, though differing in some important respects. The theory cannot yet be accepted as final—it would be as indefensible for a zoologist to allow his taxonomic analysis to be influenced by it as by static continents—but it is the best supported today, and it fits the biogeographical data better than any other that has been put forward.

Wegener believed that the land surface of the world originally formed a single mass, Pangaea. By Permo-Carboniferous times, when many of the orders of insects were established (Table 8.1), it is generally visualized as being divided more or less completely into two supercontinents, a northern Laurasia, lying largely in tropical and subtropical latitudes, and a southern Gondwanaland, extending nearly to the pole, and including India as well as Antarctica and Australia in its most southern part. The fossil floras suggest that the two blocks were almost completely separated by the Tethys Sea (Plumstead), the insects that a fair amount of exchange occurred between them. Jeannel has suggested that the Palaeoptera and Blattodea evolved in Laurasia, the Hemiptera, Coleoptera, and panorpoid orders in Gondwanaland. Both blocks were moving with a northerly component, the southern impinging progressively on the northern, and they had broken into fragments corresponding to the existing continents by about the

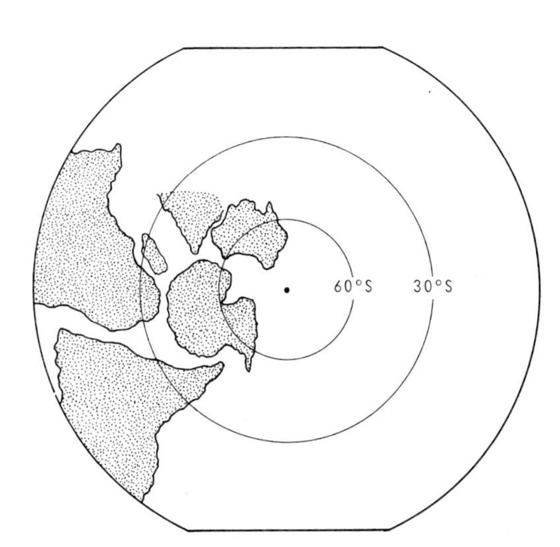

Fig. 9.11. The southern continents separating in the Mesozoic (after Runcorn, 1962, with Madagascar added).

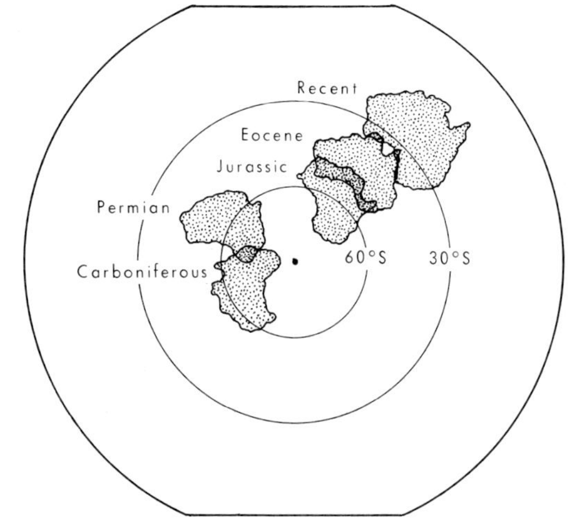

Fig. 9.12. Some positions of Australia since the Palaeozoic, as indicated by palaeomagnetism (based on Runcorn, 1962).

middle of the Mesozoic (Fig. 9.11). Thereafter they moved slowly into their present positions, India meeting Asia and pushing up the Himalayan chain, and Australia breaking into the Indonesian–Melanesian arc.

The associated climatic changes were important too. The pattern of atmospheric circulation is determined primarily by the rotation and inclination of the earth, so it would seem that the climate of the world must always have been zoned, although the intensity, width, and position of the zones have varied considerably from time to time. Laurasia was largely tropical in the Carboniferous; the Boreal Sea was too wide for glacial effects to be significant on land during the Permian; continental effects produced extensive aridity as the land-mass drifted through middle latitudes in the late Permian and Triassic; and its northern shores reached high latitudes in time to meet the impact of the Pleistocene glaciations. In contrast, Gondwanaland extended through all zones, which became intensified during the southern Permian glacial cycles, and then relaxed again until the Quaternary, when most of the southern lands were too far from the pole to be much affected by the glaciations. It is important to appreciate these differences, when comparing the northern and southern

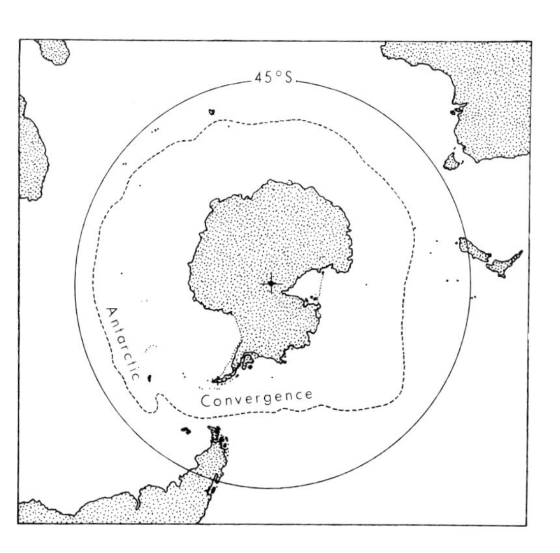

Fig. 9.13. South-polar projection (present).

fossil faunas described in the previous chapter.

The History of Australia

A series of palaeomagnetically determined positions of Australia are shown in Figure 9.12. Some writers hold that drift ceased early in the Tertiary, but Runcorn (1963) has reported evidence that there has been appreciable northward movement of both India and Australia during the past million years. At first Australia formed an integral part of southern Gondwanaland, with a cold to frigid but progressively ameliorating climate, and a fauna and flora shared with the other southern countries. When the mass broke up, South America, Antarctica, and Australia may have remained connected together for a time before separating into the three present blocks (Fig. 9.13). Australia was completely isolated thereafter until it impinged on the Indonesian–Melanesian arc, probably in the Pliocene.

There were significant changes within the continental area too. First the Permian glaciations may well have stimulated speciation in the same way that the later Pleistocene glaciations appear to have done (p. 203). Then the country had a warm climate in the Cretaceous, and was divided into two or three large islands by epicontinental seas, which were later replaced by extensive lakes. In the Tertiary it was reduced to a peneplain with an equable climate, and any speciation that occurred must have depended on distance rather than barriers. The first clearly defined arid belt to be recognized was in the Pliocene in Queensland; mountain building culminated in the Kosciusko uplift; and the climatic zoning of the Pleistocene developed. All these changes have influenced the evolution of the fauna.

It is important to appreciate that both the archaic and older southern components were inhabitants of southern Gondwanaland, and must have been adapted to cool environments. Some Gondwanaland groups adapted to warmer or drier climates (e.g. Nemestrinidae, Apioceridae) could have extended into the area before separation was complete, but the

Eyrean province could not have begun to exercise intensive selection pressures until very much later. The older northern element is more difficult to account for. That part of it that was not established early (as suggested by Evans) apparently could not have entered the country until a few million years ago. If so, some of it must have radiated with remarkable rapidity here, or done so undetected on the way. This is a problem that still requires elucidation.

The faunal relationship between Australia and New Zealand is made clearer by the theory of drift, because it consists largely of shared southern and northern Gondwanaland elements. New Guinea presents a more difficult problem. Traditionally it has been regarded as part of Australia, and has been thought to have moved with Australia. However, Good (in Gressitt, 1963) has suggested that most of New Guinea formed part of an Asiatic arc into which Australia impinged (see also Cheesman, 1951). This is an attractive hypothesis; but there were wide epicontinental seas that separated the island from the mainland for a long time, and these could explain why it seems to have served as a more effective earlier corridor into the Pacific than into Australia (Carey, *in litt.*).

GEOGRAPHIC SPECIATION

'That geographic speciation is the almost exclusive mode of speciation among animals', to quote Mayr (1963), has already been emphasized in Chapter 6. Its investigation therefore forms an important part of zoogeographical research. Essentially, we look

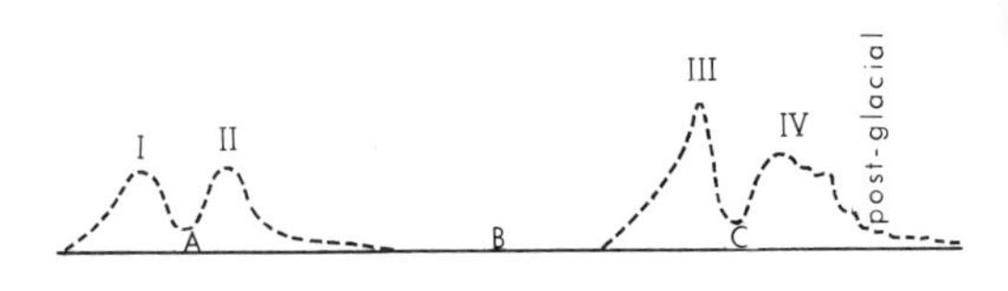

Fig. 9.14. Glacial cycles in the Pleistocene: I–IV, glaciations; A–C, interglacials (based on Mackerras, in Leeper, 1962).

for the occurrence of barriers—of various kinds, depending on the kinds of animals we are working with—that could have divided populations into discrete sections for long enough periods to permit the evolution of reproductive isolation. Existing barriers may be producing incipient speciation, for which evidence can often be found in the occurrence of subspecies; but we must find barriers that existed in the past and lasted, in general, for many thousands of years in order to account for the definitive species that we see around us today.

The Pleistocene epoch was a time of profound climatic and environmental fluctuations occurring just when they would have made the most significant impression on modern faunas, and it has received the most attention. It lasted about a million years, and included four major cycles of glaciation separated by three longer interglacial periods (Fig. 9.14). The last glaciation ended about 10,000 years ago, there were minor fluctuations since then, and we are now said to be about two-thirds of the way into an interglacial. During glaciations, ice-caps extended

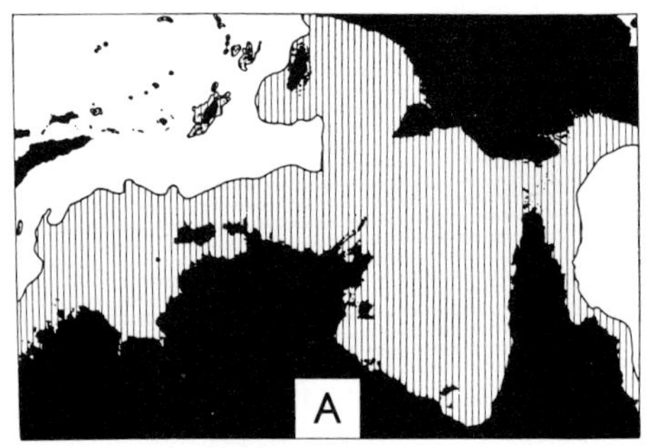

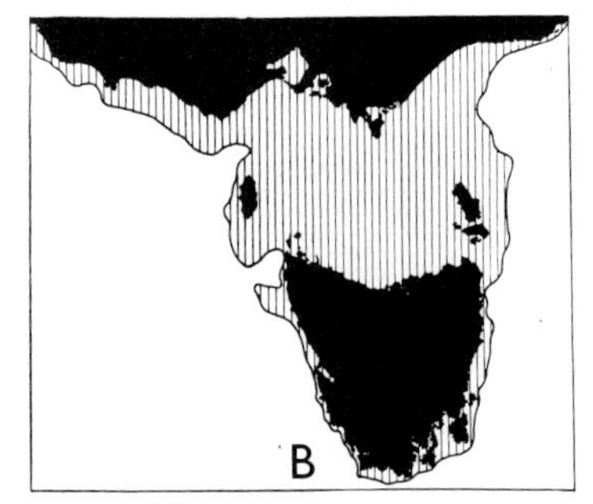

Fig. 9.15. A, northern, and B, southern, extensions of Australia during glacial phases of the Pleistocene, as indicated by the 100 m line.
[From Mackerras, in Leeper, 1962, by permission of the publishers]

far from the poles, sea-level was lowered by the water locked up in ice, and atmospheric circulation was intensified, resulting in pluvial conditions over extensive areas, with the arid belts apparently narrowed and displaced towards the equator. In interglacials, the ice disappeared, sea-level rose, low-pressure systems were displaced towards the poles, rainfall decreased over previously pluvial areas, and the arid belts became wider and more diffuse.

In the northern hemisphere glaciers bit deeply into the northern parts of the continents, and their effects in breaking up populations of animals have been studied extensively (see, e.g., Illies, 1965). In Africa other influences have been more important (see, especially, Moreau, 1963). In Australia, too, glaciation has been slight, though not negligible, and the effects of aridity have received particular attention (e.g. papers by Main, Mackerras, and others, in Leeper, 1962). Thus, it is clear that a fertile corridor north of the Great Australian Bight has been opened and closed at least three times, the resultant speciation contributing materially to the distinctiveness of the western fauna. Land connections (Fig. 9.15) between Australia and New Guinea in the north and Tasmania in the south have also been made and broken by the world-wide fluctuations in sea-level, and evidence of the resulting exchanges and speciation can be clearly seen in the Tabanidae. There is evidence, too, that speciation was promoted in at least one littoral group by the barrier presented by the extensive Tasmanian peninsula during glacial phases.

Other mechanisms must be sought for in arid-adapted groups and in those with a capacity for wide dispersal, but the kinds of study outlined can be very fruitful when they are undertaken with groups that are not too difficult taxonomically and have appropriately limited ecological ranges.

ACKNOWLEDGMENT. I am greatly indebted to Professor S. Warren Carey, University of Tasmania, for guidance on the geological and geophysical problems discussed.

COLOUR PLATES

Plate 1

ODONATA

(Males)

A *Synthemis eustalacta* (Burm.), Synthemidae
B *Nannophya dalei occidentalis* (Till.), Libellulidae
C *Austrogomphus guerini* (Ramb.), Gomphidae
D *Cordulephya pygmaea* Selys, Corduliidae
E *Diphlebia euphaeoides coerulescens* Till., Amphipterygidae
F *Synlestes tillyardi* Fraser, Chlorolestidae
G *Telephlebia godeffroyi* Selys, Aeshnidae
H *Austrolestes cingulatus* (Burm.), Lestidae
I *Nososticta solida* Selys, Protoneuridae
J *Austrocnemis splendida* (Martin), Coenagrionidae
K *Pseudagrion aureofrons* Till., Coenagrionidae

(All life size, except J, which is magnified 1·5 times)

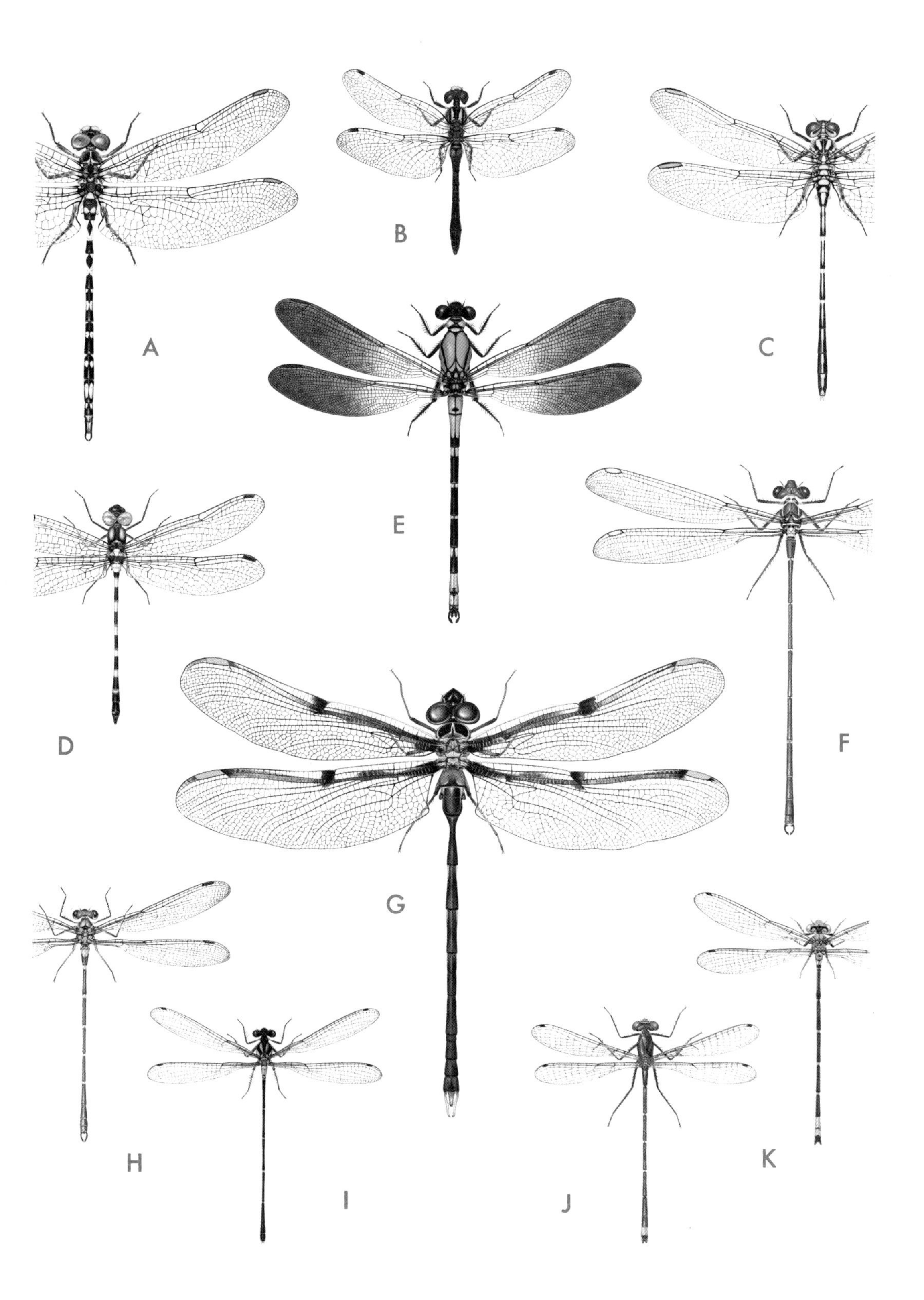
A
B
C
D
E
F
G
H
I
J
K

Plate 2

MANTODEA

A *Tenodera australasiae* (Leach), ♀, Mantidae

BLATTODEA

B *Polyzosteria viridissima* Shelf., ♂, Blattidae
C *Anamesia polyzona* (Walk.), ♂, Blattidae
D *Ellipsidion magnificum* Heb., ♀, Blattellidae

DERMAPTERA

E *Apachyus peterseni* Borelli, ♂, Labiduridae

ORTHOPTERA

F *Gastrimargus musicus* (F.), ♂, Acrididae
G *Froggattina australis* (Walk.), ♂, Acrididae
H *Monistria pustulifera* (Walk.), ♂, Pyrgomorphidae
I *Moraba virgo* Key, ♀, Eumastacidae
J *Kosciuscola tristis* Sjöst., ♂ (dark and pale phases), Acrididae
K *Alectoria superba* Brunn., ♀, Tettigoniidae
L *Acripeza reticulata* Guér., ♀, Tettigoniidae

(All life size)

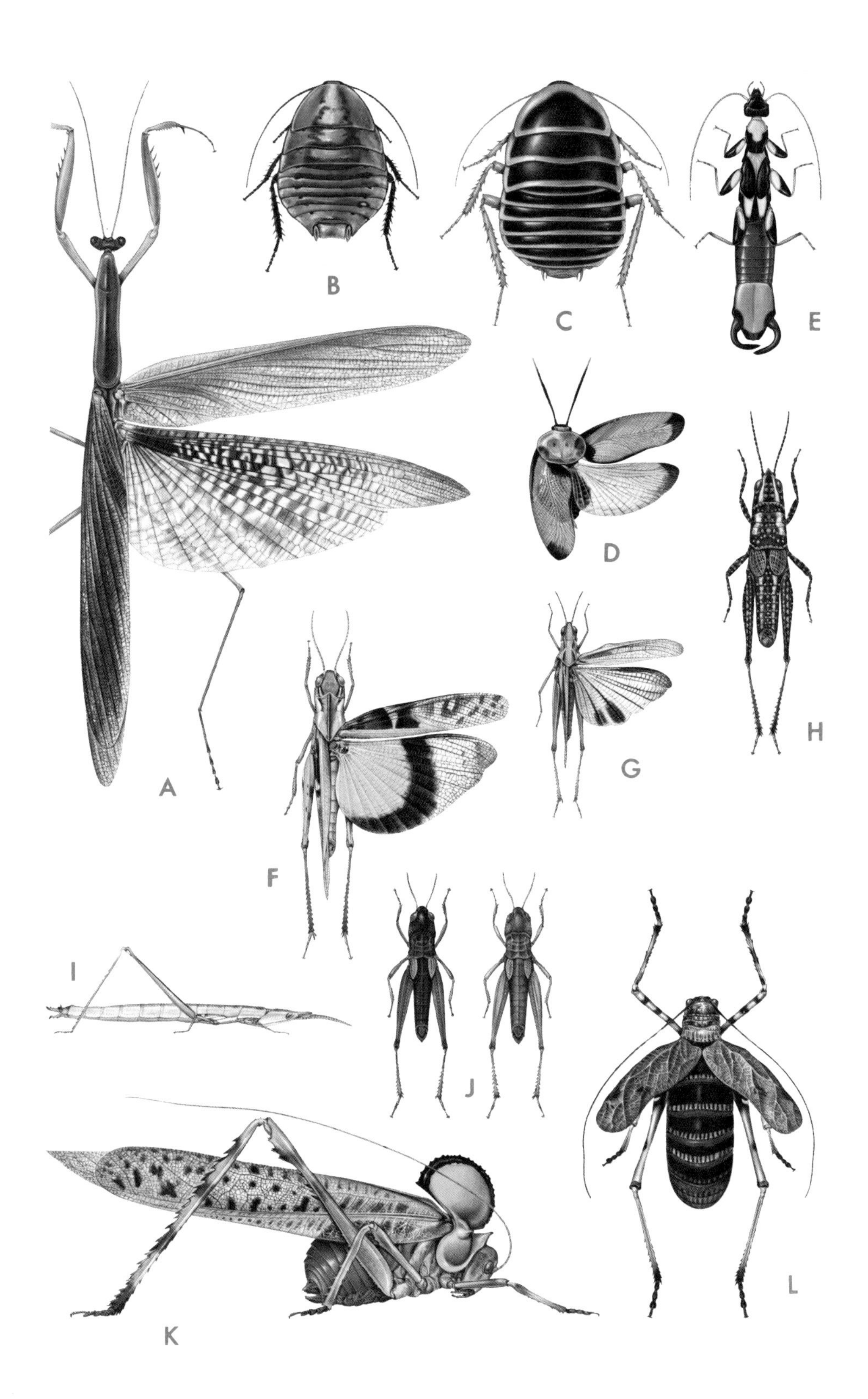
A
B
C
D
E
F
G
H
I
J
K
L

PLATE 3

HEMIPTERA

HOMOPTERA

A *Desudaba danae* (Gerst.), Fulgoridae
B *Achilus flammeus* Kirby, Achilidae (× 1·5)
C *Siphanta acuta* (Walk.), Flatidae (× 1·5)
D *Platybrachys leucostigma* (Walk.), Eurybrachyidae (× 1·5)
E *Gelastopsis insignis* Kirk., Eurybrachyidae (× 1·5)
F *Eoscarta carnifex* (F.), Cercopidae (× 1·5)
G *Cicadetta rubricincta* (Goding & Froggatt), Cicadidae
H *Eurymelops bicolor* (Burm.), Eurymelidae

HETEROPTERA

I *Ectomocoris decoratus* Stål, Reduviidae (× 1·5)
J *Ectomocoris ornatus* Stål, Reduviidae
K *Ptilocnemus femoratus* Horv., Reduviidae (× 2·0)
L *Morna florens* (Walk.), Pentatomidae (× 1·5)
M *Eumecopus armatus* (F.), Pentatomidae
N *Catacanthus nigripes* (Sulzer), Pentatomidae
O *Commius elegans* (Don.), Pentatomidae (× 1·5)
P *Tectocoris diophthalmus* (Thunb.), Scutelleridae
Q *Scutiphora pedicellata* (Kirby), Scutelleridae (× 1·5)
R *Stauralia chloracantha* Dallas, Acanthosomatidae (× 1·5)
S *Oncomeris* sp., Tessaratomidae (New Guinea)
T *Peltocopta crassiventris* (Bergr.), Tessaratomidae, ♂
U *Philia senator* (F.) Scutelleridae (× 1·5)
V *Melanerythrus mactans* (Stål), Lygaeidae (× 1·5)
W *Spilostethus hospes* (F.), Lygaeidae (× 1·5)
X *Dysdercus cingulatus* (F.), Pyrrhocoridae
Y *Physopelta famelica* Stål, Largidae
Z *Aulacosternum nigrorubrum* Dallas, Coreidae (× 1·5)

(Life size, unless otherwise indicated)

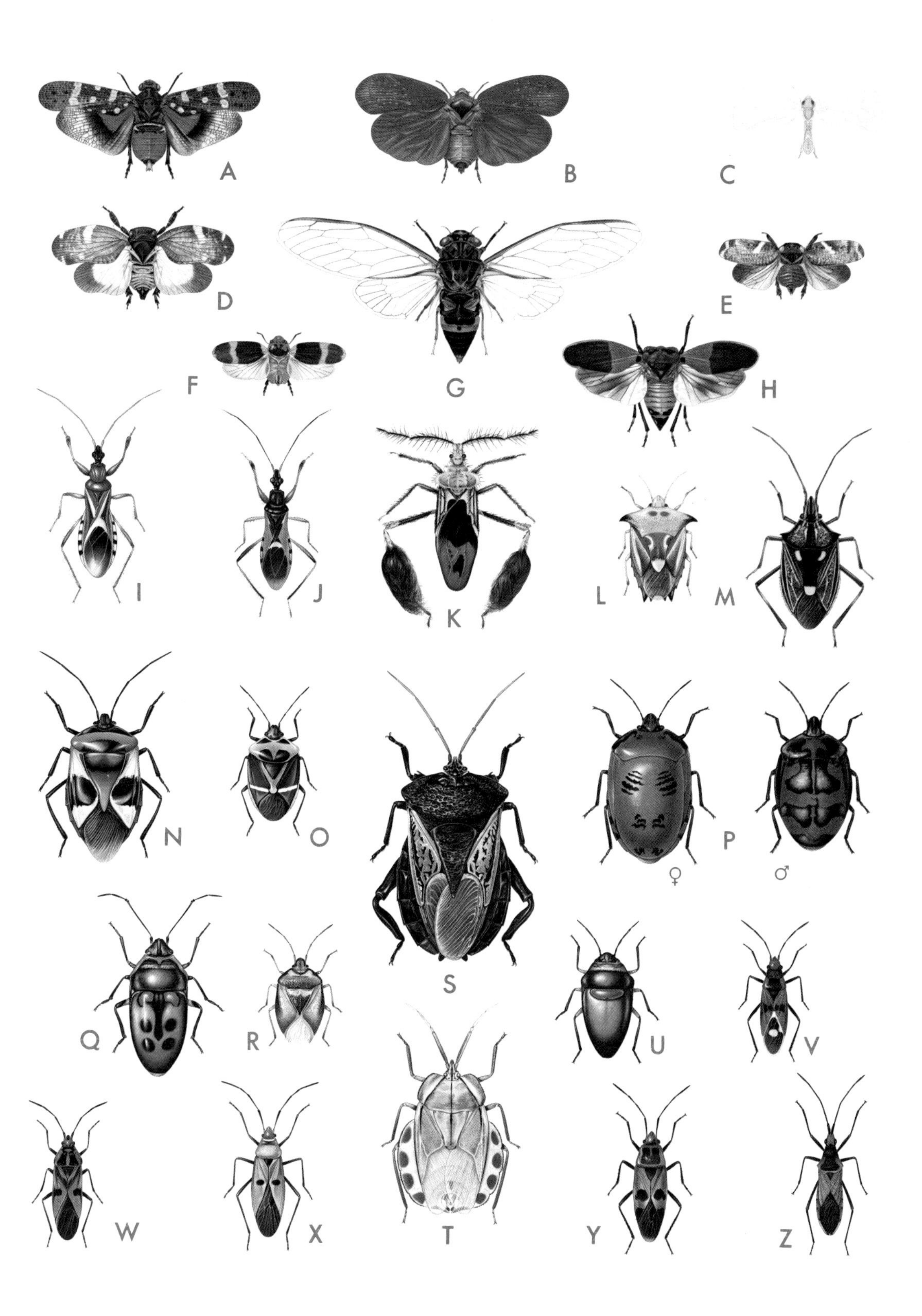
A
B
C
D
E
F
G
H
I
J
K
L
M
N
O
P
♀
♂
Q
R
S
U
V
W
X
T
Y
Z

Plate 4

COLEOPTERA

A *Megacephala australis* Chaud., Carabidae–Cicindelinae
B *Calosoma schayeri* Erichs., Carabidae–Carabinae
C *Prophanes mastersi* Pasc., Tenebrionidae–Cyphaleinae
D *Anoplognathus aureus* Waterh., Scarabaeidae–Rutelinae
E *Anoplognathus smaragdinus* Ohaus, Scarabaeidae–Rutelinae
F *Anoplognathus smaragdinus*, colour variety
G *Rhipidocerus australasiae* Westw., Cerambycidae–Prioninae
H *Paracalais gibboni* (Newm.), Elateridae
I *Eurhamphus fasciculatus* Shuck., Curculionidae–Curculioninae
J *Rhytiphora dallasi* Pasc., Cerambycidae–Lamiinae
K *Penthea pardalis* (Newm.), Cerambycidae–Lamiinae
L *Anoplognathus viriditarsis* Leach, Scarabaeidae–Rutelinae
M *Anoplostethus laetus* Rothsch. & Jord., colour variety, Scarabaeidae–Rutelinae
N *Phalacrognathus muelleri* Macl., Lucanidae
O *Ischiopsopha yorkiana* (Jans.), Scarabaeidae–Cetoniinae
P *Eupoecila australasiae* (Don.), Scarabaeidae–Cetoniinae
Q *Sagra papuana* Jac., Chrysomelidae–Sagrinae
R *Uracanthus triangularis* Hope, Cerambycidae–Cerambycinae
S *Carenum sumptuosum* Westw., Carabidae–Scaritinae
T *Stigmodera gratiosa* Chevr., Buprestidae–Stigmoderinae
U *Lamprima aurata* Latr., Lucanidae
V *Cyphogastra pistor* (Cast. & Gory), Buprestidae–Chalcophorinae
W *Stigmodera chevrolati* Géhin, Buprestidae–Stigmoderinae
X *Stigmodera amabilis* Cast. & Gory, Buprestidae–Stigmoderinae
Y *Pseudotaenia quadrisignata* (Saund.), Buprestidae–Chalcophorinae
Z *Stigmodera alternata* Lumholz, Buprestidae–Stigmoderinae
ZA *Calodema regalis* Cast. & Gory, Buprestidae–Stigmoderinae

(All life size)

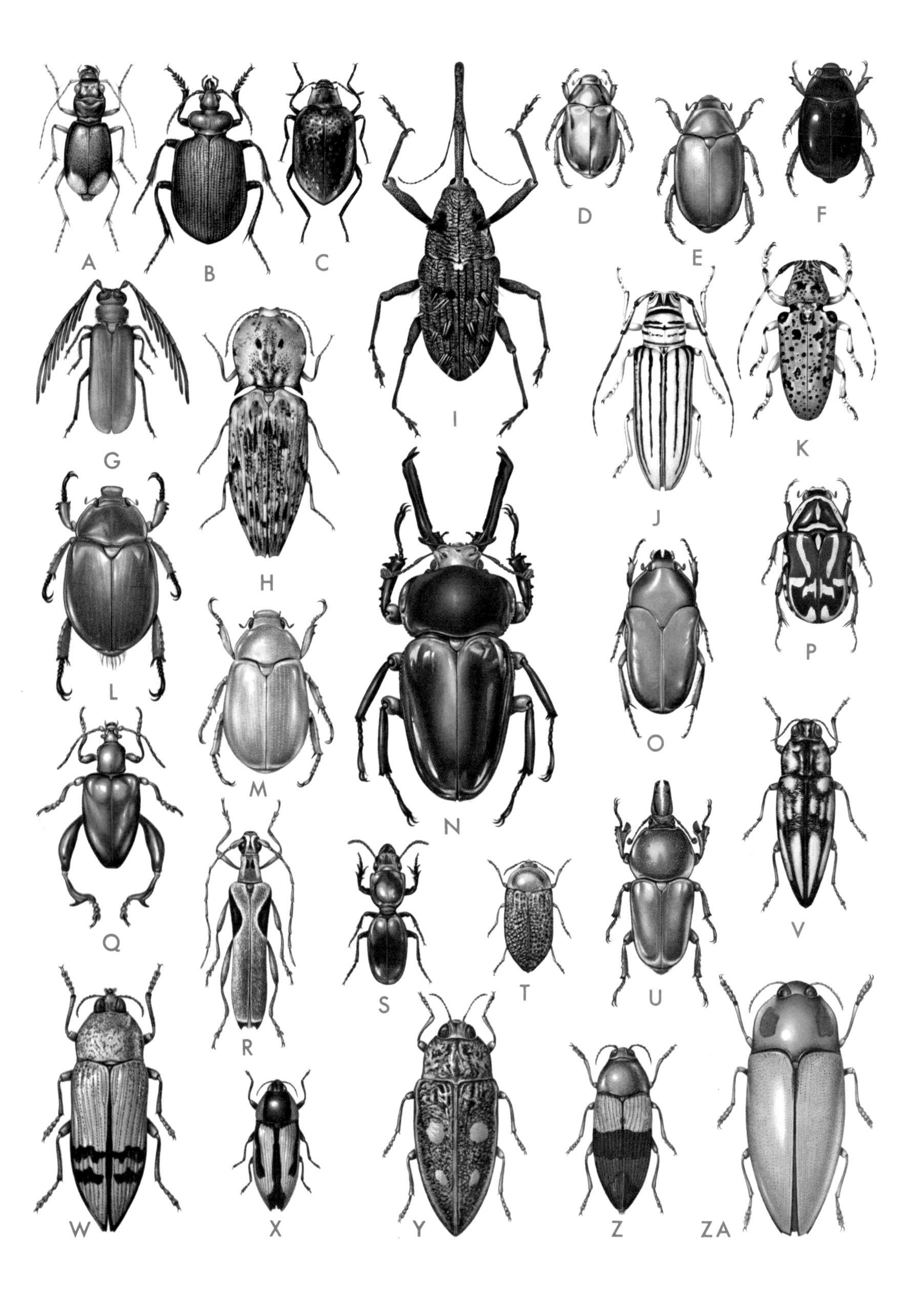
A
B
C
D
E
F
G
H
I
J
K
L
M
N
O
P
Q
R
S
T
U
V
W
X
Y
Z
ZA

PLATE 5

COLEOPTERA

A *Chrysolopus spectabilis* (F.), Curculionidae–Aterpinae, blue form
B *Episcaphula pictipennis* Crotch, Erotylidae
C *Spilopyra sumptuosa* Baly, Chrysomelidae–Eumolpinae
D *Lemodes mastersi* Macl., Anthicidae
E *Apterotheca besti* (Blackb.), Tenebrionidae–Cnodaloninae
F *Rupilia ruficollis* Clark, Chrysomelidae–Galerucinae
G *Encymon clavicornis* (Blackb.), Endomychidae
H *Pedilophorus gemmatus* Lea, Byrrhidae
I *Diphucephala colaspidoides* (Gyll.), Scarabaeidae–Melolonthinae
J *Paederus cruenticollis* Germ., Staphylinidae–Paederinae

(A life size, B–J magnified 3·0 times)

DIPTERA

K *Clytocosmus helmsi* Skuse, Tipulidae
L *Lamprogaster violacea* Mall., Platystomatidae
M *Laphria hirta* Ricardo, Asilidae
N *Scaptia auriflua* (Don.), ♂, Tabanidae
O *Formosia speciosa* (Erichs.), Tachinidae
P *Rutilia formosa* Rob.–Desv., Tachinidae

(All life size)

HYMENOPTERA

Q *Exeirus lateritius* Shuck., ♀, Sphecidae–Nyssoninae
R *Diamma bicolor* Westw., ♀, Tiphiidae–Thynninae
S *Myrmecia nigrocincta* Smith, worker, Formicidae–Myrmeciinae
T *Pompilus* sp., ♀, Pompilidae
U *Lestis aeratus* Smith, ♂, Anthophoridae–Xylocopinae
V *Stilbum splendidum* F., ♀, Chrysididae

(All life size)

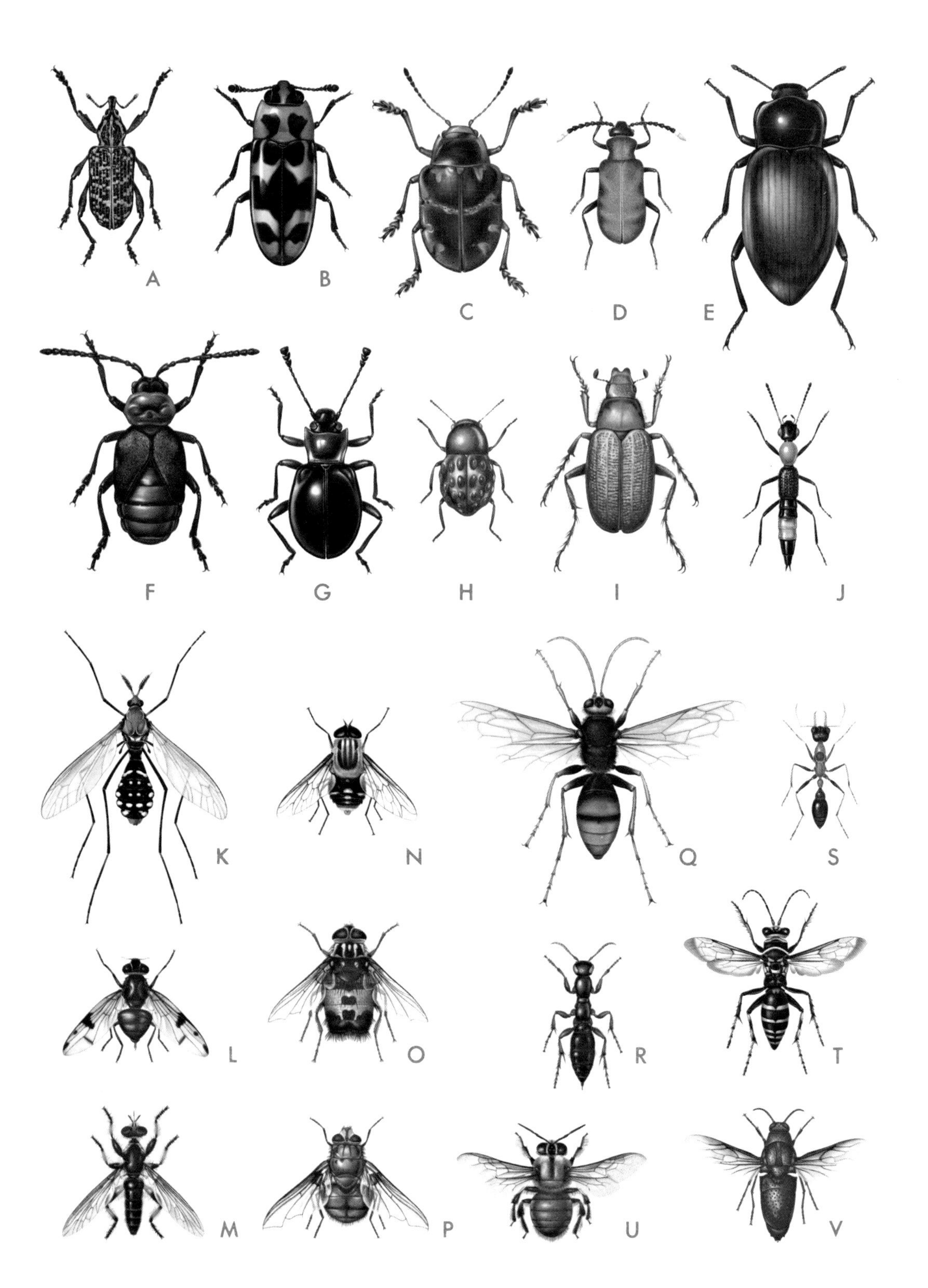
A
B
C
D
E
F
G
H
I
J
K
N
Q
S
L
O
R
T
M
P
U
V

PLATE 6

BATESIAN MIMICRY

The examples (all approximately life size) are mostly of group rather than specific resemblances. In Lycidae and Braconidae the models are presumably distasteful; in the other Hymenoptera illustrated they have a powerful sting.

A *Metriorrhynchus rhipidius* (Macl.), Coleoptera–Lycidae (model)
B *Eroschema poweri* Pasc., Coleoptera–Cerambycidae (mimic)
C *Tmesidera rufipennis* Westw., Coleoptera–Meloidae (mimic)
D *Rhinotia haemoptera* Kirby, Coleoptera–Belidae (mimic)
E *Stigmodera nasuta* Saund., Coleoptera–Buprestidae (mimic)
F *Snellenia lineata* (Walk.), Lepidoptera–Stathmopodidae (mimic)

G *Hyleoides concinna* (F.), Hymenoptera–Colletidae, a Müllerian mimic of many stinging wasps (model)
H *Codula limbipennis* Macq., Diptera–Asilidae (mimic)
I *Syndipnomyia auricincta* Kert., Diptera–Stratiomyidae (mimic)
J *Hesthesis variegata* (F.), Coleoptera–Cerambycidae (mimic)

K *Rhynchium abispoides* M.-W., Hymenoptera–Eumenidae (model)
L *Chrysopogon crabroniformis* Röder, Diptera–Asilidae (mimic)
M *Hesthesis ferruginea* (Boisd.), Coleoptera–Cerambycidae (group mimic)

N *Pseudozethus* sp., Hymenoptera–Eumenidae (model)
O *Hesthesis cingulata* (Kirby), Coleoptera–Cerambycidae (mimic)

P *Williamsita smithiensis* Leclercq, Hymenoptera–Sphecidae (model)
Q *Conops* sp., Diptera–Conopidae (mimic)

R *Sceliphron laetum* Sm., Hymenoptera–Sphecidae (model)
S *Systropus flavoornatus* Rob., Diptera–Bombyliidae (mimic)

T *Paralastor constrictus* Perk., Hymenoptera–Eumenidae (model)
U *Cerioides macleayi* Ferg., Diptera–Syrphidae (mimic)

V *Phanagenia* sp., Hymenoptera–Pompilidae (model)
W *Agapophytus flavicornis* Mann, Diptera–Therevidae (mimic)

X *Miscothyris* sp., Hymenoptera–Sphecidae (model)
Y *Leucopsina odyneroides* Westw., Diptera–Acroceridae (mimic)

Z A braconine wasp, Hymenoptera–Braconidae (model)
ZA *Plecia* sp., Diptera–Bibionidae (apparent mimic)

ZB A braconine wasp, Hymenoptera–Braconidae (model)
ZC *Hestiochora tricolor* (Walk.), Lepidoptera–Zygaenidae (apparent mimic)
ZD *Coracistis erythrocosma* Meyr., Lepidoptera–Stathmopodidae (apparent mimic)

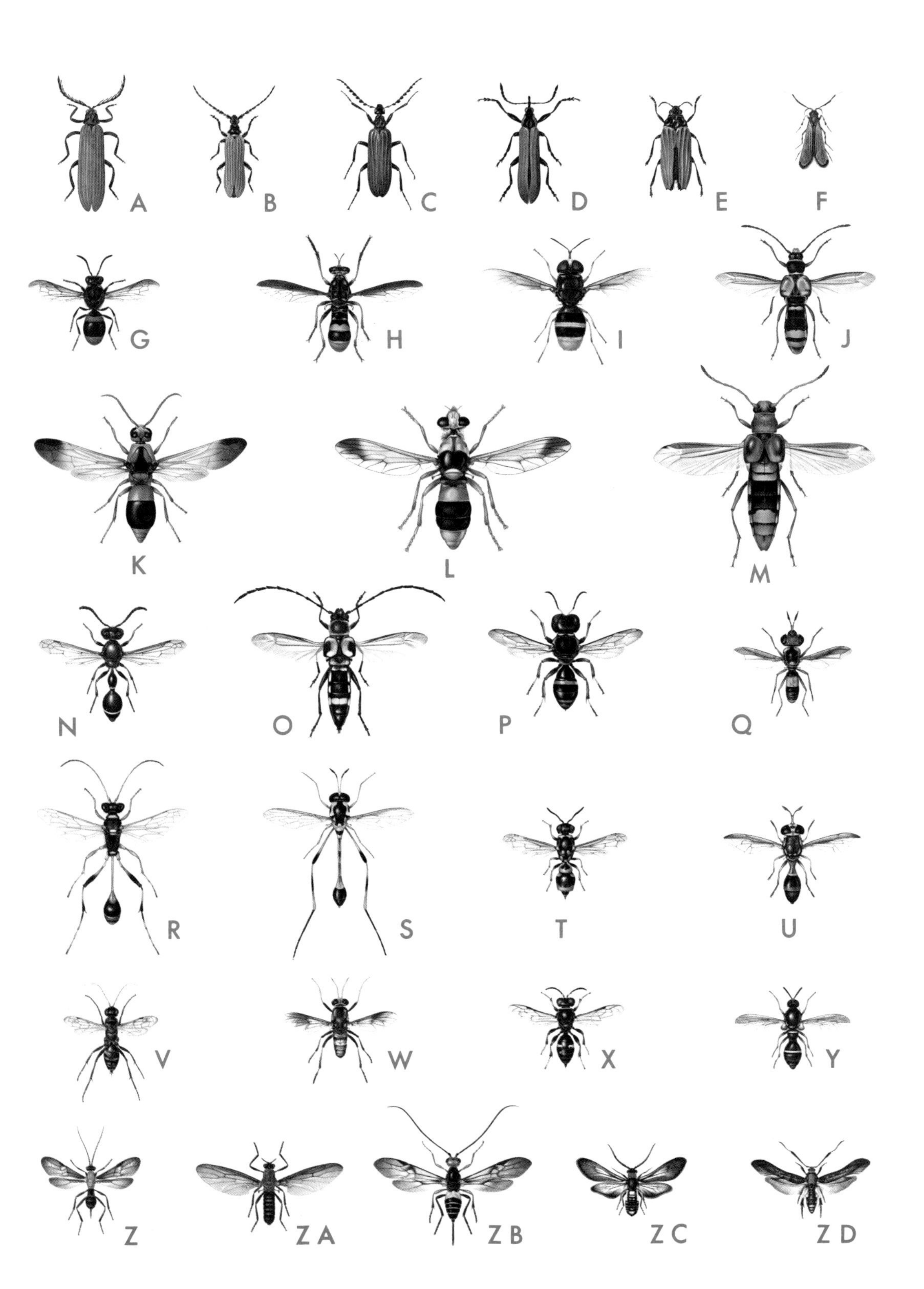
A
B
C
D
E
F
G
H
I
J
K
L
M
N
O
P
Q
R
S
T
U
V
W
X
Y
Z
ZA
ZB
ZC
ZD

PLATE 7

LEPIDOPTERA

A *Macarostola formosa* (Staint.), Gracillariidae
B *Wingia rectiorella* (Walk.), Oecophoridae
C *Heliodines princeps* Meyr., Heliodinidae
D *Synanthedon chrysophanes* (Meyr.), Aegeriidae
E *Snellenia lineata* (Walk.), Stathmopodidae
F *Macarangela leucochrysa* Meyr., Yponomeutidae
G *Oenistis entella* (Cram.), Arctiidae
H *Telecrates laetiorella* (Walk.), Xyloryctidae
I *Nemophora sparsella* (Walk.), Incurvariidae
J *Comana miltocosma* (Turn.), Limacodidae
K *Atasthalistis ochreoviridella* (Pag.), Gelechiidae
L *Atteva niphocosma* Turn., Yponomeutidae
M '*Margaronia*' *glauculalis* (Guen.), Pyralidae
N *Pollanisus trimaculus* (Walk.), Zygaenidae

(Each scale = 2 mm)

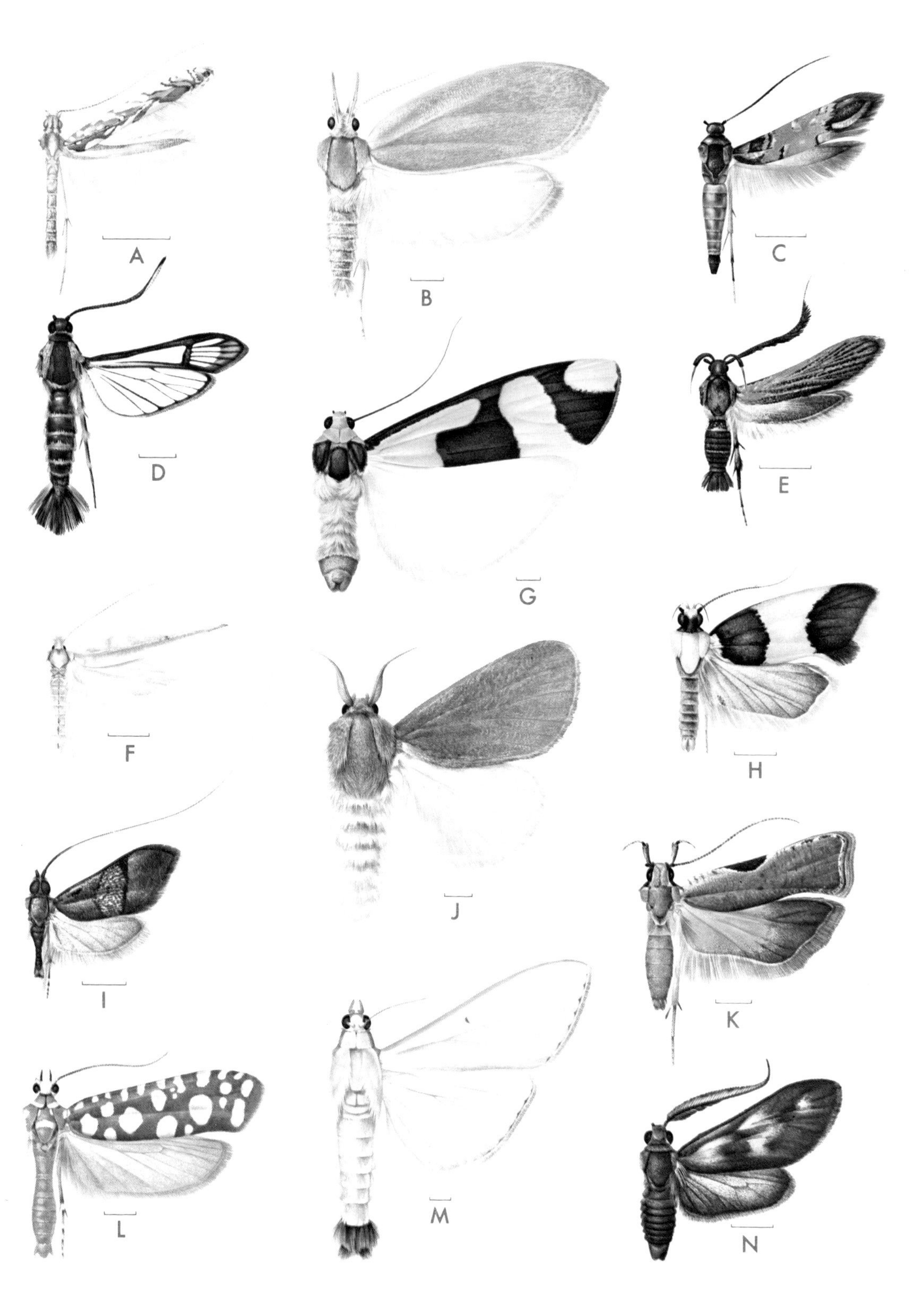
A
B
C
D
G
E
F
H
J
I
K
L
M
N

Plate 8

LEPIDOPTERA

A *Aenetus eximius* (Scott), ♂, Hepialidae
B *Synemon magnifica* Strand, Castniidae
C *Hypochrysops apelles* (F.), Lycaenidae
D *Dudgeonea actinias* Turn., Cossidae
E *Ogyris genoveva* Hewitson, ♂, Lycaenidae
F *Coscinocera hercules* (Miskin), ♂, Saturniidae
G *Cethosia chrysippe* (F.), Nymphalidae
H *Agape chloropyga* (Walk.), Hypsidae
I *Alcides zodiaca* (Butl.), Uraniidae
J *Euchromia creusa* (L.), Amatidae
K *Papilio ulysses* (L.), Papilionidae
L *Eucyclodes pieroides* (Walk.), Geometridae
M *Hecatesia fenestrata* Boisd., Agaristidae
N *Euschemon rafflesia* (Macl.), Hesperiidae

(Each scale = 10 mm)

A
B
C
D
E
F
G
H
I
J
K
L
M
N

10

THE ENTOGNATHOUS HEXAPODS

by M. M. H. WALLACE *and* I. M. MACKERRAS

This assemblage of primitive wingless arthropods includes the Collembola, Protura, and Diplura. Their evolutionary status and relationships to the class Insecta have been discussed on p. 156. Their zoogeographical patterns are obscure, and will not be considered here.

Class COLLEMBOLA

(Springtails)

Small to minute, pigmented or unpigmented hexapods; rudimentary eyes or ocelli often present; antennae 4- to 6-segmented, with intrinsic muscles in basal 3; thorax variously developed; legs 4-segmented; abdomen 6-segmented, with specialized appendages on 1, 3, and 4, gonopore on 5; cerci absent. Larval development epimorphic.

The springtails (Fig. 10.1) are soft-bodied, generally rather compact arthropods, usually about 2–3 mm long, but a few up to 10 mm. Their colour varies from white to almost black, and some species show characteristic colour patterns. Their bodies may be clothed in hairs or scales, and *trichobothria* (long sensory setae each supported on a tubercle or boss) are usually present. Between 1,500 and 2,000 species have been described. Their common name is derived from the fact that most species can spring considerable distances when disturbed. For accounts of their biology see Paclt (1956) and Christiansen (1964). The literature has been indexed by Salmon (1964), and the Australian species reviewed by Womersley (1939).

Anatomy of Adult

Head. Primitively prognathous, but more or less hypognathous in Symphypleona. Antennae with 4 (sometimes 6) segments, filiform, varying from shorter than the head *(Neelus)* to longer than the body (Entomobryidae); sometimes modified in the male as grasping organs (Fig. 10.3A). *Post-antennal organ of Tömösvary* ('pseudocellus') usually present, varying in form from a simple, ring-like structure to a multituberculate sensory area. Eyes, when present, composed of 8 or fewer ommatidia on each side. Mouth-parts (Fig. 10.2) elongate, adapted for triturating and sucking; mandibles slender; maxillae complex; hypopharynx trilobed; labium reduced by encroachment of the cranial folds; no labial or maxillary palpi, except in embryo.

Thorax. Segments clearly distinguishable in Arthropleona, fused with the abdomen in Symphypleona. Legs divided into coxa, trochanter, femur, and tibio-tarsus. The pretarsus bears a single claw with a sheath-like tunica and sometimes with basal tooth-like formations or 'pseudonychia'; empodium absent or claw-like.

Abdomen. Six-segmented; segments 1–4 form the pregenital region, in which 1 carries the *ventral tube*, 3 the *retinaculum*, and 4 the *furcula* or spring (Fig. 10.1). The ventral tube consists of a basal column containing a pair

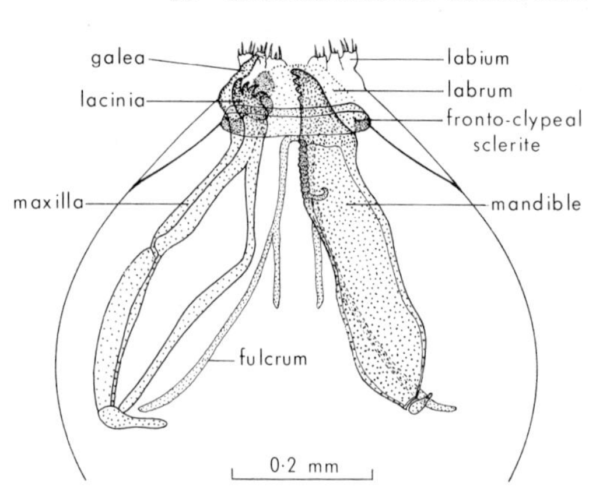

Fig. 10.2. Mouth-parts of *Sminthurus viridis*, dorsal, left mandible and right maxilla omitted, labrum darkened.

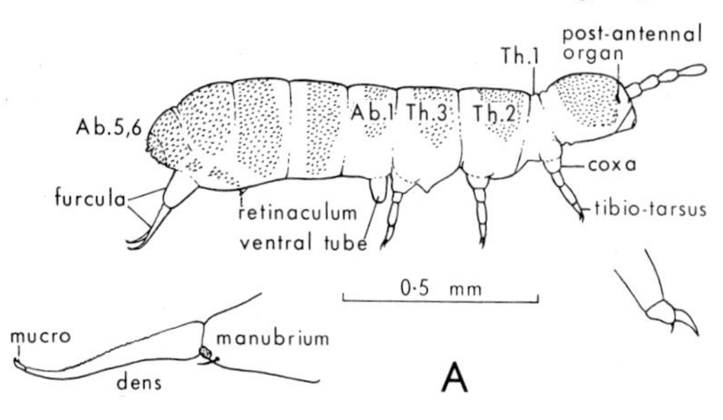

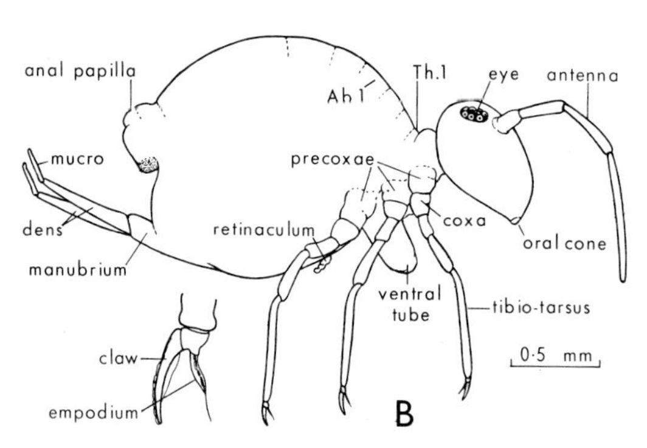

Fig. 10.1. General structure of Collembola (setae omitted): A, an isotomid, Arthropleona (furcula and fore pretarsus enlarged); B, *Sminthurus viridis*, Symphypleona (hind pretarsus enlarged).

of sacs which can be everted by blood pressure. It seems to serve as an adhesive organ enabling the collembolan to stick to smooth surfaces; but it also plays a part in moisture intake, and can often be seen in use when the animal is on a moist surface. The furcula consists of a single basal *manubrium* from which emerge two *dentes* terminating in lamellate *mucrones*. When not in use, it is tucked under the body and held in place by the retinaculum. Gonopore on segment 5, usually a horizontal slit in the female, a circular hole or vertical slit in the male, often at the apex of a spined hump (Fig. 10.3B). Anus on segment 6. No cerci, but females usually have a pair of stout subanal appendages.

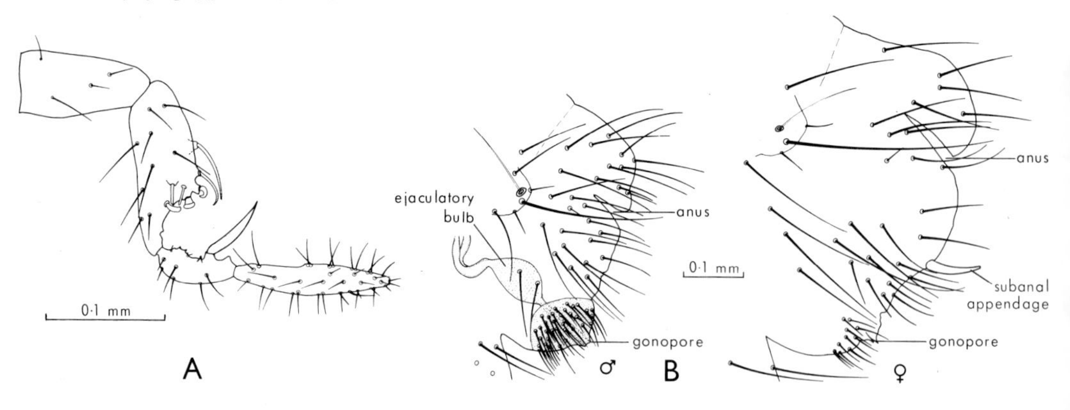

Fig. 10.3. A, antenna of ♂ *Sminthurides violacea*, showing modification for clasping ♀ (cf. Fig. 10.8B); B, terminal segments of *Sminthurus viridis*.

Internal Anatomy. Alimentary canal a simple tube with a large mid gut; no Malpighian tubes. Excretion is performed partly by the hind gut, partly by the mid gut, and partly by the fat bodies, which contain concretions of uric acid. Nervous system concentrated, with 3 thoracic but no separate abdominal ganglia. Some Symphypleona have tracheae opening by a single pair of spiracles between head and prothorax, but respiration is entirely cutaneous in most species. Reproductive organs usually simple or folded tubes; germaria generally, though not always, parietal as in myriapods; the female has a pair of small spermathecae; in some species the male has a complex sclerotized ejaculatory apparatus.

Immature Stages

The eggs are mostly spherical, pale, and with a smooth or finely sculptured chorion. Segmentation is complete; there is no amnion; a ventral flexure develops; the chorion ruptures precociously; and the embryo has a well-developed dorsal organ. These features suggest relationship with myriapods rather than with insects. The young are like their parents. There are several pre-imaginal instars, and adults may moult up to 50 times, usually without much additional growth. Males often have fewer ecdyses than females, and do not live as long.

Biology

Collembola are found most commonly on or near the soil surface, especially in decaying vegetable matter, and often in rotten logs. Some occur on the sea-shore, some on reefs between tide marks, and some on the surface of fresh water and even on snow. A few are found in ant and termite nests. They are rare in dry situations, but are present in all parts of the world from Subarctic regions to Antarctica. They are readily airborne.

Most species feed on decaying vegetable matter. Other important sources of food are algae, fungi, and lichens. Many Sminthuridae prefer fresh plant material. Spores and pollen are often found in the gut contents. Decomposing animal matter is sometimes eaten, particularly by littoral species. *Hypogastrura purpurascens* (Lubb.) feeds not only on decaying vegetable matter, but also on mycelia and spores of fungi, soft animal matter, such as dead flies, collembolans, earthworms, etc., and also on its own exuviae. Young larvae of *Sminthurus viridis* (L.) may feed actively upon recently dead larvae and adults, and in so doing ingest substances that hasten death. This process plays an important part in regulating the population density of this pest. In England *Podura aquatica* L. has been observed to engage in combat, especially at high densities. Cannibalism and predation have also been recorded in *Hypogastrura* and *Friesea* spp., whilst *Isotoma grandiceps* Reut. is said to attack and eat other species of Collembola.

Reproduction. Males attach globular spermatophores by a stalk to the substrate, usually on the soil surface or on vegetation near the ground (Fig. 10.4B). They are picked up by the female, which straddles the spermatophore and catches the sperm mass in her genital aperture. Parthenogenesis has been recorded in a few species. The eggs are deposited singly or in batches, and in *S. viridis* each egg is covered with freshly eaten soil voided through the anus at oviposition (Fig. 10.4C).

The life histories of few Collembola have been studied in detail. In Australia the introduced lucerne flea, *S. viridis*, has received greatest attention. It occurs in the winter-rainfall areas of southern Australia. The aestivating eggs hatch with autumn rains; the larvae pass through 4 instars, and usually become adult about 3 to 5 weeks after hatching, depending on the temperature. The females lay up to 3 batches of eggs between ecdyses. There are 3 to 4 generations from autumn to late spring, when high temperatures and low rainfall bring about the disappearance of the active stages. Eggs in diapause are laid in the spring, and diapause development is completed in the field by midsummer, after which the eggs hatch when rainfall and temperature again become favourable.

Abundance. Some Holarctic Collembola, such as *Hypogastrura longispina* Tullb., may

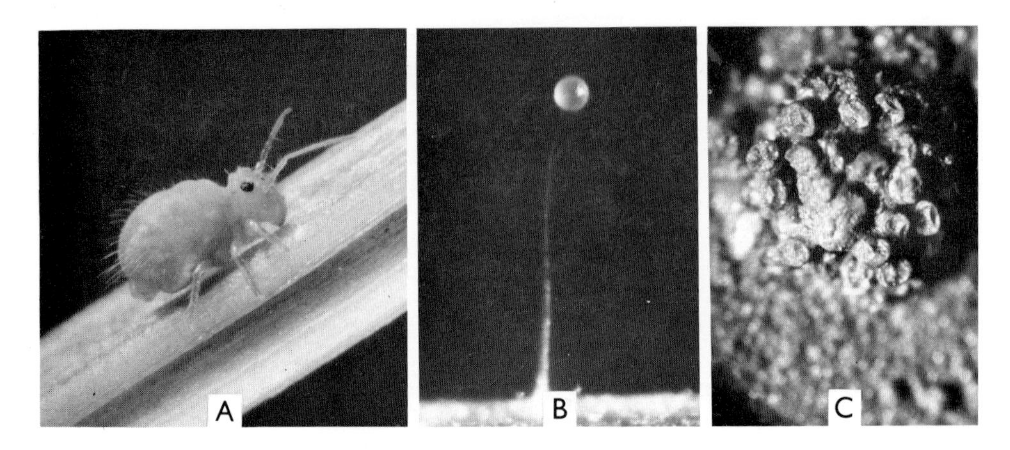

Fig. 10.4. *Sminthurus viridis*: A, female (length of body 3 mm); B, spermatophore (height 0·8 mm); C, egg mass (diameter 2·5 mm). [Photos by C. Lourandos and M. M. H. Wallace]

form layers up to 8 mm thick on the ground or on the surface of water. *Onychiurus alborufescens* (Volger) in Alpine countries produces a 'red snow' when it occurs in large numbers. Similar mass appearances have not been recorded in Australia, but large dark patches containing many thousands of individuals of an unidentified collembolan have been seen on Mt Wellington in Tasmania. More than 10,000 *Isotomurus chiltoni* (Carp.) and up to 60,000 *S. viridis* per square metre have been counted in pastures in W.A.

Natural Enemies. Collembola seem to be the main diet of the predatory Bdellidae and Cunaxidae (Acarina). In Australia the bdellid, *Bdellodes lapidaria* (Kram.), preys actively upon *S. viridis*, and has an important role in the biological control of this pest. Trombiculidae and Gamasidae also prey upon Collembola, as do various spiders and beetles. Application of insecticides to pastures often destroys the natural enemies, and so permits rapid increase of Collembola. Nematodes have been found in large numbers in the mid gut of *Entomobrya* sp.

Economic Significance. *S. viridis* is by far the most damaging collembolan in Australia through its destruction of clover pastures and leguminous crops, especially lucerne. Womersley (1939) listed 49 other species that might be classified as injurious, of which 38 occur also in other countries.

CLASSIFICATION

Class and Order COLLEMBOLA
(215 Australian spp.)

Suborder ARTHROPLEONA (157)

PODUROIDEA (51)	ENTOMOBRYOIDEA (106)
1. Poduridae (41)	3. Isotomidae (40)
2. Onychiuridae (10)	4. Entomobryidae (66)

Suborder SYMPHYPLEONA (58)

5. Sminthuridae (58)

The classification adopted is essentially that of Gisin (1960), who separates the order into five families all of which are represented in Australia.

Key to the Families of Collembola

1. Body globose; thorax and first 4 abdominal segments fused .... SYMPHYPLEONA. **Sminthuridae**
   Body elongate; most of the thoracic and abdominal segments distinctly separated, at most the last 2–3 fused .... ARTHROPLEONA. 2
2. Prothorax with several small setae .... PODUROIDEA. 3
   Prothorax completely hairless, often concealed under mesothorax .... ENTOMOBRYOIDEA. 4

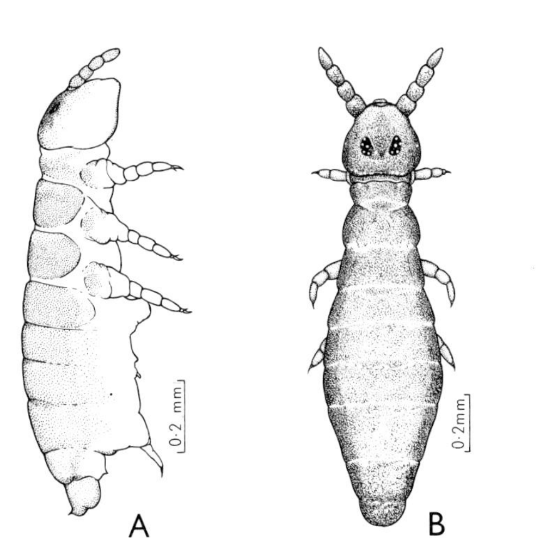

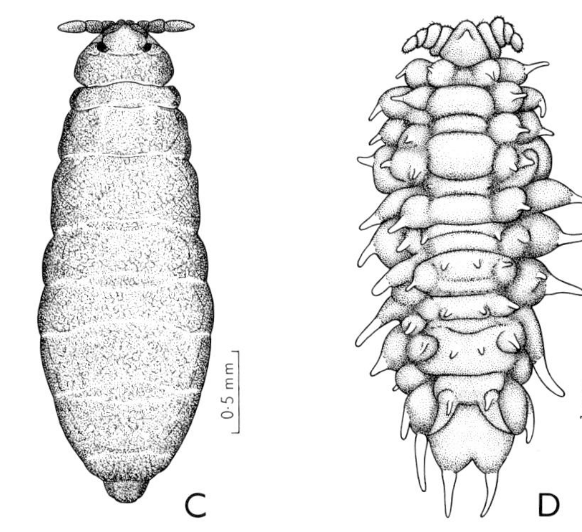

Fig. 10.5. Poduridae: A, *Hypogastrura manubrialis*; B, *Xenylla littoralis*; C, *Brachystomella fungicola*; D, *Ceratrimeria dendyi* (after Womersley, 1939). [B. Rankin]

3. Third antennal segment without unusually large sensilla, 4th with large retractile terminal vesicles; without pseudocelli .... **Poduridae**
   Third antennal segment with 2 (sometimes 3) large sensilla which are mostly protected by high papillae, 4th with or without minute terminal vesicles; pseudocelli present .... **Onychiuridae**
4. Body without scales, or with sparsely distributed, simple, weakly ciliated but not club-shaped setae; segments 3 and 4 of abdomen not very different in length; furcula often reduced .... **Isotomidae**
   Body scaled, or at least with dense, ciliated, club-shaped setae; segment 4 of abdomen usually much longer than 3; furcula always well developed ...... **Entomobryidae**

Suborder ARTHROPLEONA

1. Poduridae (Fig. 10.5). Fifteen genera are represented in Australia. One of the commonest species is the introduced *Hypogastrura armata* Nic. found in manure and decaying humus. *Xenylla littoralis* Wom. occurs under stones between tide marks in W.A. and S.A. During high tide, it buries itself in the sand, and probably utilizes the air occluded amongst its dorsal setae. *Pseudachorutes* sp. has been recorded from the intertidal zone on a coral reef in Queensland. *Brachystomella parvula* (Schaef.) is a cosmopolitan species occurring throughout most of the agricultural areas of Australia. It appears to feed largely on humus and manure, and often collects in pasture drains in enormous numbers. *Ceratrimeria* contains species up to 10 mm long with paratergal regions swollen and often spinose. Two of the 3 species are known only from Tasmania.

2. Onychiuridae (Fig. 10.6A). Delicate, elongate, usually whitish forms inhabiting soil and humus. *Onychiurus armatus* (Tullb.) is common in W.A., S.A., and Queensland;

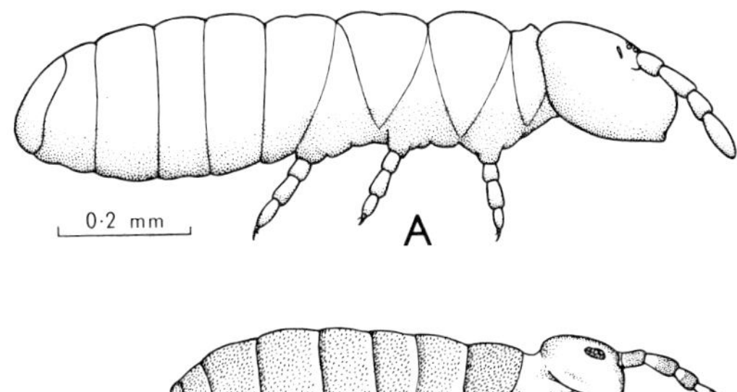

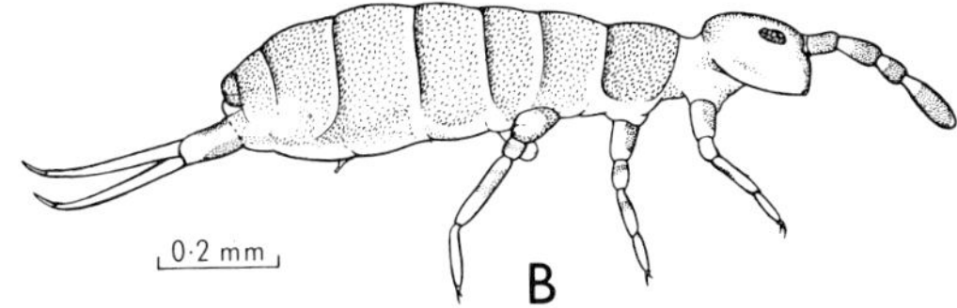

Fig. 10.6. A, *Onychiurus fimetarius*, Onychiuridae; B, *Isotoma tridentifera*, Isotomidae. [B. Rankin]

it is often a pest on the roots of plants. Other genera in Australia are *Tullbergia*, *Stenaphorura*, *Mesaphorura*, and *Dinaphorura*.

3. Isotomidae (Fig. 10.6B). There are 14 genera in Australia, 9 of them represented by a single species each. Most are as yet known only from isolated localities, but several are widespread. *Isotomurus chiltoni* occurs in pastures and garden plots from W.A. to Victoria. Species of *Proisotoma* are abundant in moss and in cultivated soil. *Axelsonia littoralis* Moniez is a reef inhabitant. *Isotoma* (12 spp.) is widespread, and some of the species are known also from Europe or the Subantarctic islands.

4. Entomobryidae (Fig. 10.7). A large family of elongate species that have the pronotum greatly reduced. *Entomobrya* is represented by 13 species, several of which are numerous in cultivated land in all States. *E. maritima* Wom. is common on sand and under stones between tide marks on the South Australian coast. *Mesira australica* (Schott) is a widely distributed species found under the loose bark of *Eucalyptus* trees. *Lepidocyrtinus domesticus* (Nic.) is well known in dry locations in houses, museums, etc. *Lepidocyrtus cyaneus* Tullb. is a common inhabitant of cultivated ground in Adelaide and Perth. *Cyphoderus* is usually associated with ants and termites.

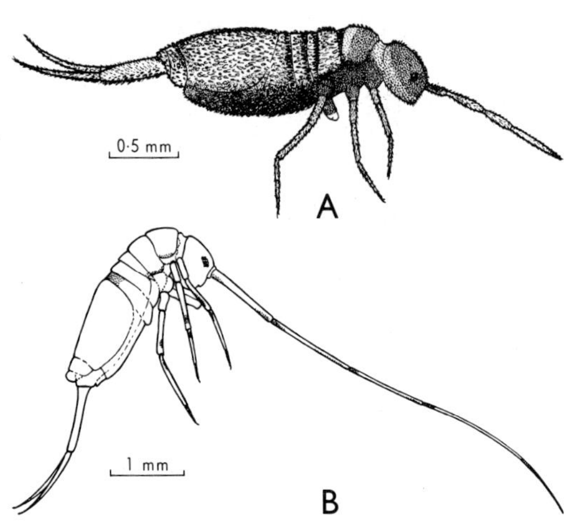

Fig. 10.7. Entomobryidae: A, *Lepidosira sagmarius*; B, *Mesira longicornis*. [B. Rankin]

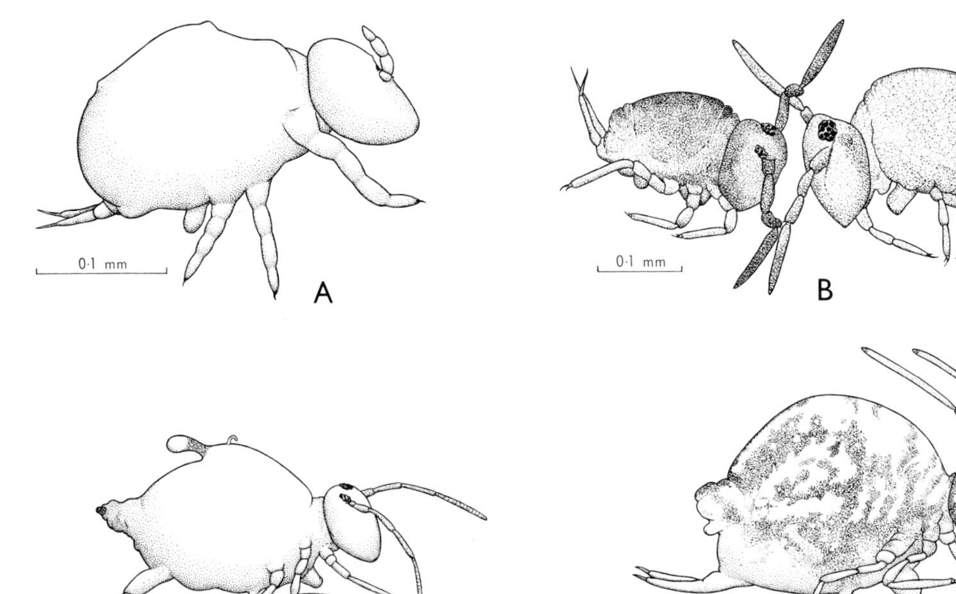

Fig. 10.8. Sminthuridae: A, *Megalothorax swani*; B, *Sminthurides pumilio*, ♂ and ♀; C, *Corynephoria* sp.; D, *Katianna oceanica*. [B. Rankin]

Suborder SYMPHYPLEONA

5. Sminthuridae (Figs.10.1B, 4, 8). A large family containing several injurious species. The best known is *Sminthurus viridis*, which is an important pest in southern Australia. *S. denisi* Wom. is common in the Western Australian bush, and easily recognized by its very long antennae. *Deuterosminthurus dromedarius* Wom. is also common in native bush. *Katianna australis* Wom. feeds extensively on clovers and other legumes, and is sometimes confused with *S. viridis*. The introduced *Bourletiella arvalis* Fitch and *hortensis* Fitch are both common in pastures throughout southern Australia. *Sminthurides stagnalis* Wom. and *aquaticus* (Bourlet) are found in small numbers on the surface of ponds and stagnant water. In this genus the modified male antennae (Fig. 10.3A) are used to grip the antennae of the female, which then carries the male around with her. *Neelus* and *Megalothorax* are represented by one species each, *M. swani*. Wom. being recorded from W.A. to Victoria.

Class PROTURA

Delicate, elongate, mostly unpigmented hexapods; eyes and antennae absent; thorax slightly developed; legs 5-segmented; abdomen 12-segmented in adult, with styles on segments 1–3, gonopore between 11 and 12; no cerci. Larval development anamorphic.

Protura are pale, slender, cryptic arthropods, less than 2 mm long (Fig. 10.9A). They are quite common, but are rarely seen except in concentrates from a 'Berlese' funnel. They occur in all zoogeographical regions, about 170 species being known in the world. Tuxen (1967) has reviewed the Australian species.

Anatomy. Head prognathous, pyriform or egg-shaped; smooth and almost completely enclosed ventrally, with the basal part of the labium eliminated, and only the tips of the mouth-parts showing anteriorly. No antennae or eyes, but a pair of 'pseudoculi', which may be rudiments of antennae and may function as humidity receptors. Mouth-parts (Fig. 10.10) slender and pointed; no hypopharynx; maxillary palpi 4-segmented, labial palpi 2-segmented, both better developed

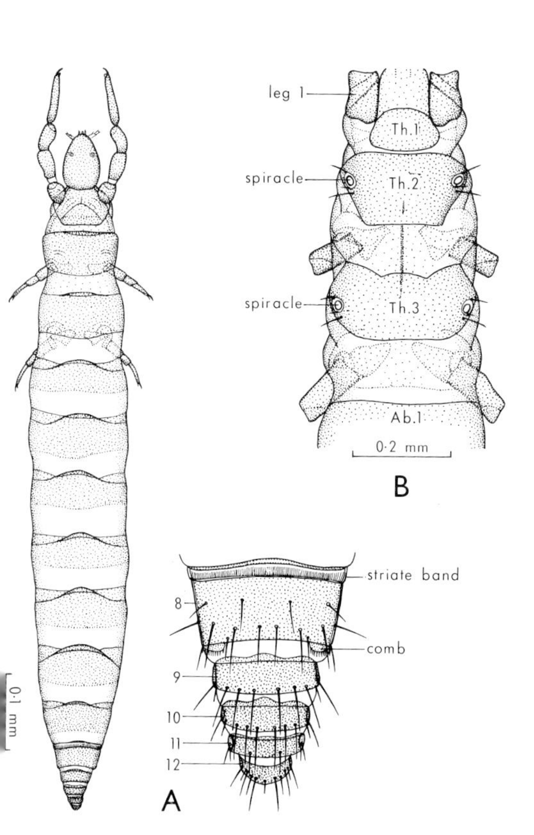

Fig. 10.9. Protura: A, *Acerentulus sexspinatus*, Acerentomidae, ♀, dorsal (setae omitted), and distal abdominal segments enlarged; B, thorax of *Eosentomon* sp., Eosentomidae, dorsal.

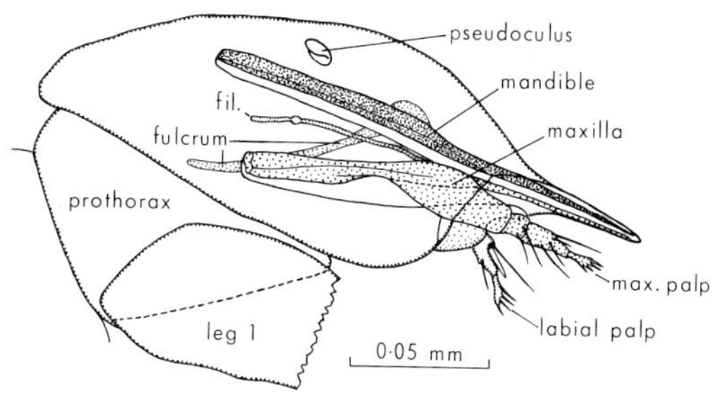

Fig. 10.10. Head of an *Acerentomon* in lateral optical section (after Tuxen, 1959).
fil., filamento di sostegno.

than in the other entognathous classes. There is a Y-shaped *fulcrum* similar to that of Collembola and Diplura, and a sclerotized thread, the 'filamento di sostegno', extends posteriorly from the lacinia on each side.

Thoracic segments with relatively strong terga, the propleural area flexible, and the prosternum extending anteriorly under the head. In *Eosentomon* (Fig.10.9B) there are spiracles on the meso- and metaterga. Legs 5-segmented, the pretarsus ending in a single claw and a bristle-like empodium. The mid and hind legs are used for walking; the fore legs, which are longer than the others, are held forward in front of the head and used as antennae.

Abdomen 12-segmented in adult, tapering, with well-developed terga and sterna. Small, 2- or 1-segmented styles, usually with apical exsertile vesicles, at the posterior corners of sterna 1–3 (Fig. 10.11). Thoracic and abdominal apodemes often conspicuous. Cerci absent; anus terminal. A deep genital pouch opens between sterna 11 and 12, and encloses the protrusible genital armature (Fig. 10.12). This differs only in detail in the two sexes, and the structures cannot be readily homologized with those of insects (Tuxen, 1956).

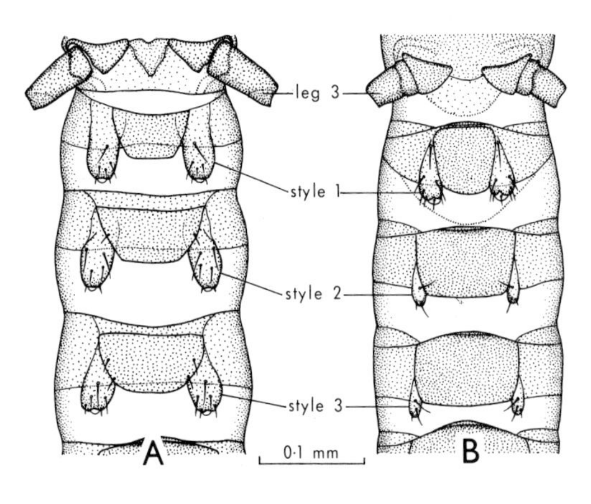

Fig. 10.11. Basal abdominal segments, ventral, showing rudimentary legs (styles) of: A, *Eosentomon* sp., Eosentomidae; B, *Australentulus* sp., Acerentomidae.

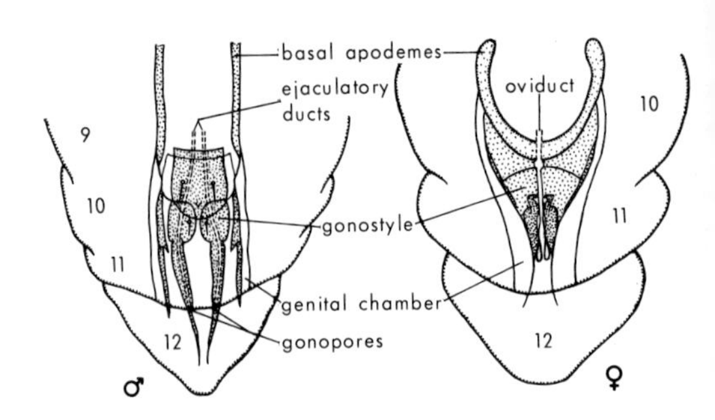

Fig. 10.12. Genital armature of an *Eosentomon* (after Tuxen, 1956).

There is a simple gut; paired maxillary and labial glands; Malpighian tubes represented by two groups of 3 papillae. Nervous system generalized, with 7 discrete abdominal ganglia. Lateral tracheal trunks extend almost the full length of the head and body in *Eosentomon*. The ovaries are large, panoistic, with apical germaria, and the oviducts unite to form a short common duct. In the male, the testes are joined anteriorly about the middle of the thorax, and the ducts are paired throughout.

Immature Stages. The embryology is unknown. There is little change of form during growth, except for anamorphic increase in the number of abdominal segments by interpolation between the last and penultimate segments. The larva ('prenymph') hatches with 9 abdominal segments and incomplete appendages, and the 4th stage, with 12 segments but incomplete chaetotaxy and no genital armature, is known as 'maturus junior'. It is not known whether the adult continues to moult throughout its life.

Biology. Protura occur in more or less damp situations in soil, leaf litter, moss, sometimes under stones, and occasionally under bark. They are said to feed mainly on vegetable material. Life cycles of Australian species have not been studied; but in Europe the eggs are apparently laid in early spring, and the adults hibernate through the winter; there may be more than one generation in a year.

CLASSIFICATION

Class and Order PROTURA

(30 Australian spp.)

1. Eosentomidae (9)
2. Protentomidae (1)
3. Acerentomidae (20)

Key to the Families of Protura

1. Tracheae and thoracic spiracles present *(Eosentomon)* .... **Eosentomidae**
   Tracheal system absent 2
2. Second pair of abdominal styles with terminal vesicles *(Protentomon)* ...... **Protentomidae**
   Second pair of abdominal styles rudimentary, without vesicles (4 genera in Australia) .... **Acerentomidae**

The species have been found in the southern and eastern parts of Australia from Perth to north Queensland, with one recorded from north-western Australia. Four have widely scattered distributions in other countries.

Class DIPLURA

Small to large, narrow-bodied hexapods, mostly unpigmented (except caudal segments of Japygidae); eyes absent; antennae moniliform, with intrinsic muscles in the flagellar segments; thorax usually weakly developed; legs 5-segmented; abdomen 10-segmented, with styles and exsertile vesicles on at least some of the segments, gonopore between 8 and 9; cerci variously developed. Larval development epimorphic.

This is a cosmopolitan order of 659 species (Paclt, 1957). They are mostly inhabitants of damp soil under logs or stones, and they are the most insect-like of the entognathous classes. They range from symphylan-like campodeids less than 5 mm long to dermapteron-like *Heterojapyx* up to 50 mm in length. The skeletal anatomy of *Heterojapyx* has been described by Snodgrass (1952), and the Australian species by Womersley (1939, 1945), Silvestri (1947), Pagés (1952), and Bornemissza (1957).

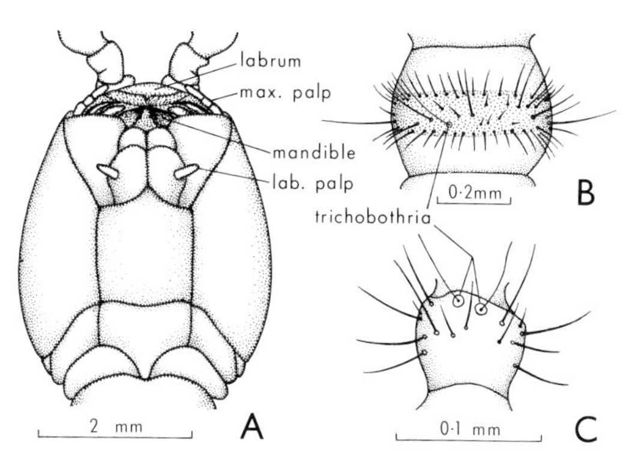

Fig. 10.13. A, head of *Heterojapyx evansi*, Japygidae, ventral; B, 7th segment of right antenna of *H. evansi*, dorsal; C, 6th segment of right antenna of *Tricampa* sp., Campodeidae, dorsal.

Anatomy of Adult

Head (Fig. 10.13). An egg-shaped, prognathous capsule, with indications of cleavage lines dorsally. No eyes. Antennae moniliform, often with trichobothria on some segments. Ventral surface almost totally enclosed by fusion of the cranial folds with the lateral margins of the wide labium on almost its whole length, so that only the tips of the mandibles and maxillae show anteriorly. The fulcrum, as in Collembola and presumably in Protura, corresponds with the posterior tentorial arms of the Insecta (Manton, 1964). Apart from the mobility of the mandibles associated with the entognathy, the most striking features of the mouth-parts are the presence of a prostheca (cf. Ephemeroptera) on the mandibles in all families except Japygidae, and the marked reduction of both maxillary and labial palpi. There is a large trilobed hypopharynx.

Thorax. Tergal and sternal plates well developed. Two to four laterally placed thoracic spiracles (2 in Parajapyginae, 3 in Campodeinae and Projapyginae, and 4 in Japyginae and Heterojapyginae). Legs usually short, the pretarsus with a pair of claws and sometimes a median claw also.

Abdomen. Ten-segmented, with well-developed terga and sterna. Sterna 2–7

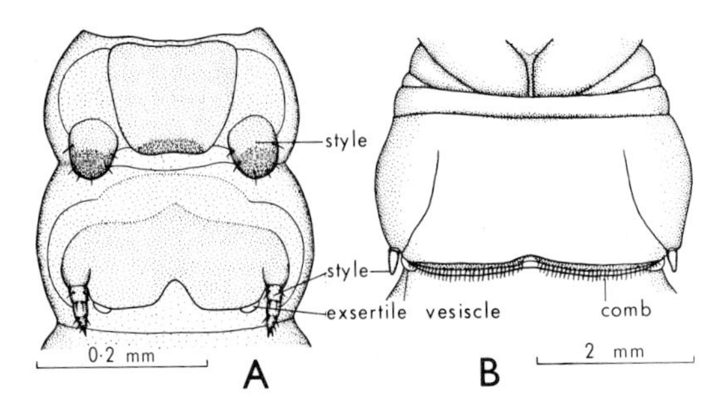

Fig. 10.14. Abdominal segments, ventral: A, 1st and 2nd of *Tricampa* sp.; B, 1st of *Heterojapyx evansi*.

usually with small lateral styles and exsertile vesicles (Fig. 10.14); the arrangement on sternum 1 varies in the families, and it also has a median glandular organ in many Japygidae. Gonopore in both sexes in a pouch behind sternum 8; usually no genital armature, although males of Japygidae have a pair of short appendages; anus terminal. The cerci differ markedly in the three major groups. In Campodeidae they are long, filiform, many-segmented, and rather resemble the antennae; in Projapygidae and Procampodeidae they are shorter, annulate, and contain the ducts of silk glands; in Japygidae they are reduced to a pair of strong, 1-segmented, pigmented forceps, which are held open by elastic tension at rest and closed by strong muscles. There are 7 pairs of abdominal spiracles in Projapygidae and Japygidae, but none in Campodeidae.

Internal Anatomy. Alimentary canal simple; Malpighian tubes reduced to 6 small papillae or absent. Usually 8 discrete abdominal ganglia, the 7th and 8th fused in Campodeidae. The tracheal systems of opposite sides apparently do not anastomose. Seven, apparently metamerically arranged pairs of ovarioles in *Japyx* and *Heterojapyx*, but only a single pair in campodeids; germaria apical; the genital tubes fuse to form a median duct in both sexes.

Immature Stages

The egg is heavily yolked, cleavage is meroblastic, there is no amnion, and the embryo has a large dorsal organ similar to that of Symphyla and Collembola. Post-embryonic development is epimorphic, with little change, except for increase in the number of antennal segments in some groups, progressive development of the chaetotaxy, and sometimes changes in the cerci. Development is slow, and moulting continues throughout life, about 30 ecdyses having been observed in *Campodea*. Sexual maturity appears to be reached at the stage when the chaetotaxy is complete.

Biology

The less aggressive Diplura, like the Symphyla, often occur in groups or small 'colonies'. A few species have been found in ants' nests and termitaria, but the association appears to be casual. *Campodea* has moderately effective legs, but *Heterojapyx* can do no more than scrabble in the soil, and progresses almost entirely by a worm-like action of its abdominal segments. The food of campodeids seems mostly to be vegetable, but japygids are carnivorous, and wait buried in the soil with only the forceps on the surface ready to seize any small arthropod that comes in contact with them.

Fertilization, where known, is by means of a spermatophore attached to the substrate by a short stalk. The smooth, spherical eggs are laid in clumps, often attached by a stalk, within rotting vegetation or in cracks in the soil. The females of some japygids guard them and the young larvae. The young members of a family group of *Heterojapyx* have been observed to devour the female parent, and cannibalism also occurs in the older stages in captivity. Isolated adults have been kept alive for more than a year.

CLASSIFICATION

Class and Order DIPLURA
(32 Australian spp.)

1. Campodeidae (11)
   Procampodeidae (0)
2. Projapygidae (2)
3. Japygidae (19)

We have followed Paclt (1957) in recognizing only four families, a convenient arrangement in the present meagre knowledge of the Australian fauna. The current tendency, however, is to divide the order into suborders and superfamilies, so we have included subfamilies in the key, in the expectation that at least some of them will eventually be given family rank.

Key to the Families and Australian Subfamilies of Diplura

1. Cerci 1-segmented, forcipate; mandible without prostheca .... **Japygidae.** 2
   Cerci multisegmented; mandible with a prostheca 4
2. No trichobothria on antenna; no labial palpi, but a pair of robust setae in their place; 2 pairs of thoracic spiracles .... PARAJAPYGINAE
   Trichobothria present on antenna; labial palpi present; 4 pairs of thoracic spiracles 3
3. Trichobothria on segments 4–6 of antenna; pretarsus with 2 lateral claws and a median unpaired claw; small to medium-sized species .... JAPYGINAE
   Trichobothria on segments 4–13 of antenna; pretarsus with 2 subequal claws, each of which has a conical empodium on the ventral surface of its base; very large species .... HETEROJAPYGINAE
4. Cerci long, without ducts of silk glands; trichobothria on segments 3–6 of antenna; maxillae without combs; abdominal sternum 1 with exsertile vesicles not developed, styles fleshy, rounded. (CAMPODEINAE) .... **Campodeidae**
   Cerci short, containing ducts of silk glands; 3rd antennal segment without trichobothria; maxillae with combs; abdominal sternum 1 with exsertile vesicles replaced by a pair of subcylindrical or elongate papillae, styles normal (PROJAPYGINAE) .... **Projapygidae**
   Cerci short, containing ducts of silk glands; trichobothria on segments 3–7 of antenna; maxillae without combs, neither vesicles nor styles on abdominal sternum 1 .... (Procampodeidae)

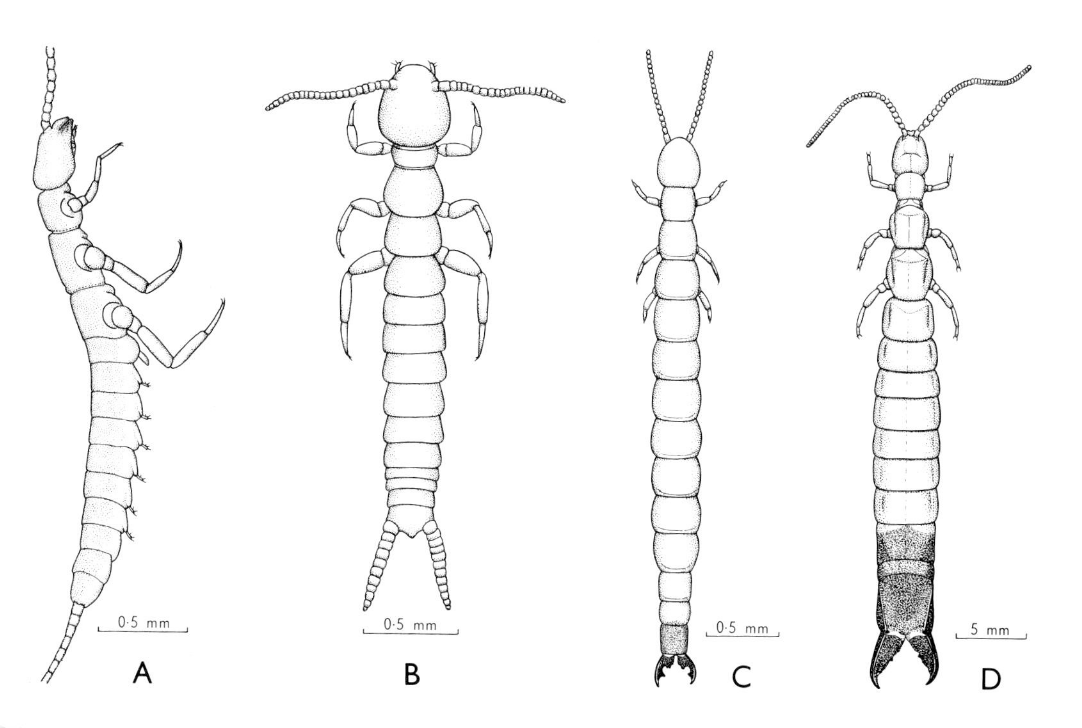

Fig. 10.15. Diplura (setae omitted): A, *Campodea* sp. (same as in Fig. 7.4), Campodeidae, lateral; B, *Symphylurinus* sp., Projapygidae, dorsal (based on Bornemissza, 1957); C, *Parajapyx swani*, Japygidae, dorsal (after Womersley, 1939); D, *Heterojapyx evansi*, Japygidae, dorsal. [B. Rankin]

1. Campodeidae (Fig. 10.15A). The Australian species may be recognized by their small size (2–5 mm long), whitish colour, and long, moniliform antennae and cerci. Three genera have been recorded: *Campodea*, with 5 species, of which *fragilis* Mein. is cosmopolitan and *tillyardi* Silv. occurs in all States and New Zealand; *Tricampa* (*Metriocampa* in Womersley, 1939), with 4 species; and *Campodella*, with only 3 known species in the world, *tiegsi* Wom. from Victoria, an undescribed species in W.A., and *clavigera* Silv. from South Africa.

2. Projapygidae (Fig. 10.15B). Small, primitive species, which can be distinguished from campodeids by their larger heads, more japygid-like build, and shorter, thicker cerci containing the ducts of silk glands. The 2 known Australian species of *Symphylurinus* appear to be rare, *swani* Wom. from north Queensland, and an undescribed species from north-western Australia.

3. Japygidae (Figs. 10.15C, D). The species of this family vary considerably in size and structure, but can be readily recognized by their forcipate cerci. *Heterojapyx* (HETEROJAPYGINAE, 4 spp.) is the most distinctive genus in the order, if only by reason of the remarkable robustness of the species, which are found mostly in the mountains in eastern Australia. The genus has also been recorded from New Zealand, Madagascar, Pamir, and Tibet. In contrast, the 3 Australian species of *Parajapyx* (PARAJAPYGINAE) are small (about 3–5 mm long) and widely scattered in north Queensland, north-western Australia, S.A., and south-western W.A. The Australian JAPYGINAE are now distributed in 5 genera, of which *Japygianus* (monotypic for *wheeleri* Silv., Queensland), with its short, squat cerci and long abdominal segment 10, is the most distinctive. The others are *Indjapyx*, *Teljapyx*, *Burmjapyx*, and *Notojapyx*. Most of the species are small (5–8 mm), but two are 13–15 mm long, and *B. glauerti* (Wom.) from W.A. is 25 mm, so it could be mistaken for a small *Heterojapyx*.

11

APTERYGOTA

(Primitively wingless insects)

by J. A. L. WATSON

The Apterygota comprise the primitively wingless insects as defined on p. 158. They are mostly free-living, about 3–15 mm long, and have elongate or oval bodies, ectognathous mouth-parts, long filiform antennae and cerci, an appendix dorsalis, vestiges of abdominal limbs, and well-developed ovipositors in the females. They were formerly treated as a single order, which included two evolutionary lines, machiloid and lepismatoid; but the similarities between the lines are due principally to the retention of primitive characters, whereas the differences in their jaw mechanisms are ancient and of great evolutionary significance (Manton, 1964; but cf. J. Watson, 1965). The fossil record is meagre. The upper Palaeozoic Monura (Fig. 8.1) are allied to the machiloids (Sharov, 1966), but no early lepismatoids are known, although dicondylar mandibles must have antedated the winged insects of the Upper Devonian. It is therefore appropriate to treat the two surviving apterygote lines as separate orders, Archaeognatha (Microcoryphia) and Thysanura.

Denis (1949), Paclt (1956), Delany (1957), and Sharov (1966) have reviewed the subclass, and Womersley (1939) the Australian fauna, which is very poorly known.

Order ARCHAEOGNATHA

Fusiform subcylindrical Apterygota, with the ability to jump. Appendix dorsalis longer than cerci; body bearing pigmented scales; compound eyes large, contiguous; ocelli present; mandibles with a single articulation; maxillary palp long, 7-segmented; thorax strongly arched, the terga extending over pleura; styles often present on mid and hind coxae; abdominal segments 2–9 with ventral styles, 1–7 generally with one or two pairs of exsertile vesicles.

This small, homogeneous, and cosmopolitan order includes about 250 known species in one extinct and two living families.

Anatomy of Adult

Head (Figs. 11.1A–E). Hypognathous. Compound eyes large, contiguous. Ocelli present, the lateral pair near mid-line below anterior margin of eyes and transversely elongate. Antennae multisegmented, elongate. Mandibles (Fig. 11.1B) extremely primitive; long, monocondylar, apical incisor process widely separated from molar process, operating with a rolling motion. Hypopharynx (Fig. 11.1E) trilobed; lingua lying between labium and maxilla and superlinguae between maxilla and mandible. Maxilla and labium normal; maxillary palp

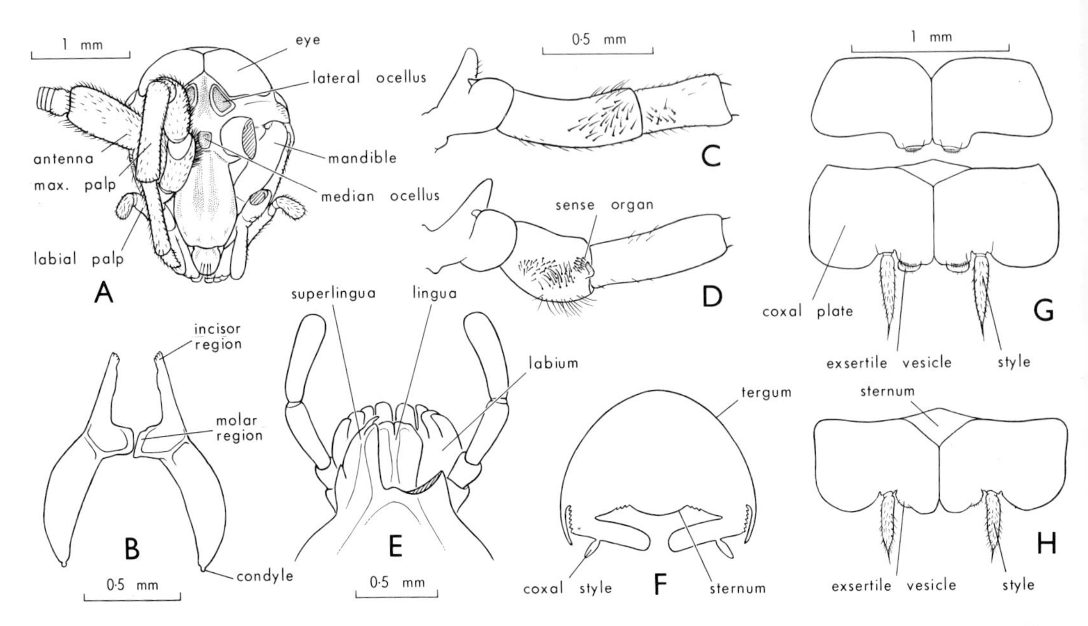

Fig. 11.1. Archaeognatha: A, head of *Allomachilis froggatti*, ♂, frontal view with left antenna and maxillary palp removed; B, mandibles of *Nesomachilis australicus*; C, base of male maxillary palp of *A. froggatti*; D, same of *N. australicus*; E, labium and hypopharynx of *A. froggatti* with right lateral lobe of hypopharynx removed; F, diagrammatic transverse section through thorax of *Allomachilis*; G, sternal regions of abdominal segments 1 and 2 of *N. australicus* with segments separated to show sternum 2; H, same of segment 2 of *A. froggatti*. Scales and many setae omitted throughout. [M. Quick]

long, 7-segmented; labial palp short, 3-segmented. Anterior arms of tentorium discrete.

Thorax. Terga entire, strongly arched, and produced into large lateral lobes (Fig. 11.1F). Pleura represented by one small sclerite in mesothorax, one in metathorax, and several in prothorax, covered by tergal lobes. Sterna small. Meso- and metathoracic spiracles present.

Legs. Coxa large, trochanter 2-segmented, tarsi 3- or rarely 2-segmented, the pretarsus with 2 claws. Mid and hind coxae often bearing styles (Fig. 11.2B), which lack muscles and are not homologous with abdominal styles. A plane of autotomy present between trochanter and femur.

Abdomen. Tapering, the contour continuous with that of thorax. Terga large, extending around side of body. Sternal region comprising a basal sternum, greatly reduced in the Meinertellidae, and postero-lateral coxal plates (Figs. 11.1G, H). Coxal plates 2–9 bearing large styles actuated by muscles, apparently representing reduced limbs. Coxal plates 1–7 generally with 1 or 2 pairs of exsertile vesicles medial to the styles, often inconspicuous. Spiracles present on segments 2–8. The terminalia are described on pp. 22, 24.

Internal Anatomy. Alimentary canal simple; crop small, proventriculus absent; enteric caeca present; 12–20 well-developed Malpighian tubes. Salivary glands and labial kidneys (Wigglesworth, 1965) present. Nervous system generalized, with 3 thoracic and 8 abdominal ganglia, the connectives appearing double throughout. Tracheal system well developed, but lacking inter- or intrasegmental anastomoses. Ovaries with 7, rarely 5, panoistic ovarioles; spermatheca and accessory glands apparently lacking. Testes with 3 or 4 follicles; vas deferens of each side double, the dual channels being interconnected by several transverse tubes; median reservoir present.

Immature Stages

Eggs globular, soft and orange when first laid, conforming to crevices in which they are placed, later hardening and blackening (Delany, 1957). Embryonic development typically insectan, but amniotic cavity incompletely closed. Larval development 'ametabolous', the larva closely resembling the adult. First-instar larva with abdominal styles and inner series of exsertile vesicles complete; scales appearing at second ecdysis; outer series of vesicles (if present), coxal styles, and gonapophyses developing progressively later. Sexual maturity attained in 8th or 9th instar in *Petrobius*.

Biology

The Archaeognatha are free-living and nocturnal, hiding by day under bark, in litter, or in rock crevices, and coming out to feed at night. The diet includes algae, lichens, and vegetable debris, the tips of the mandibles serving as augers. Several species eat their exuviae, and fragments of other arthropods have been found in the gut. *Allomachilis*, like *Petrobius* in the northern hemisphere, lives in crannies on coastal cliffs not far above high tide. Archaeognatha run nimbly, though not as quickly as Thysanura, and when disturbed escape by jumping, often more than 10 cm. The jumps are induced by abrupt downward flexure of the abdomen, the legs and styles playing only a minor role. *Allomachilis* can also jump short distances from the surface of water. The abdominal styles, at least in *Allomachilis*, appear to supplement the legs on slopes.

Reproduction is generally sexual and seasonal, and the transfer of sperm is primitive and indirect. In some machilids the male produces small droplets of sperm on a thread, which is attached to the ground and gathered by the female. Other machilids produce larger droplets ('spermatophores'), 0·5 mm or more long. In a few species the apparent absence or scarcity of males suggests that parthenogenesis may occur. The females lay batches of 2–30 eggs, commonly about 15, placing them deeply in crevices or in holes dug by the ovipositor. One batch is laid in each instar.

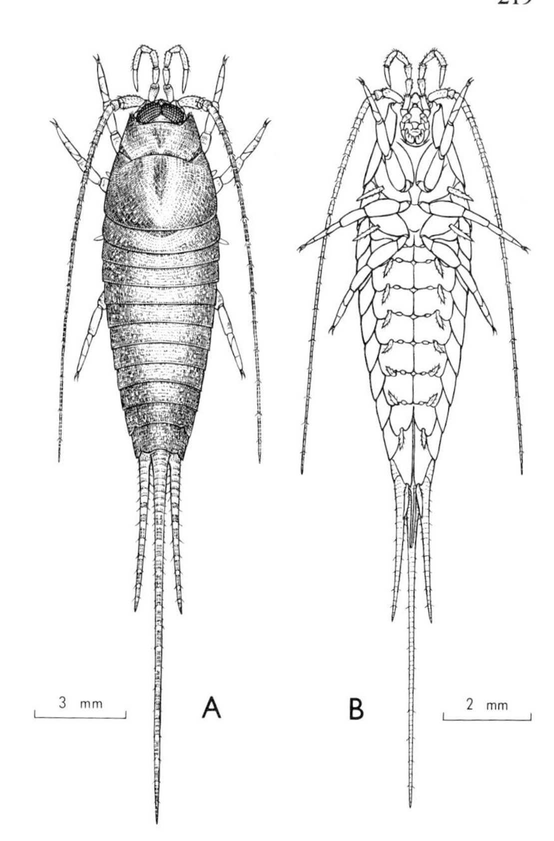

Fig. 11.2. Archaeognatha: A, *Allomachilis froggatti*, ♂, dorsal; B, *Nesomachilis australicus*, ♀, ventral with vesicles exserted (scales omitted). [M. Quick]

The eggs overwinter in diapause. Post-embryonic development is slow, the larvae taking from 6 months to 2 years to mature. Moulting continues throughout life.

Natural Enemies. Archaeognatha commonly harbour gregarine Sporozoa in the gut. They may also become infested with larval erythraeid mites, which attach themselves to the cuticle (Fig. 5.30). Several families of spiders are the main predators, but centipedes and carab beetles may also feed on them.

Economic Significance. None.

Special Features of the Australian Fauna

The Australian fauna is evidently meagre, as compared, for example, with the 19 species known in South Africa. All the species belong to the Meinertellidae, the more primitive living family, which is essentially southern; the Machilidae are found principally in the northern hemisphere (Denis, 1949). *Allomachilis* is endemic, and *Nesomachilis* is restricted to Australia, New Zealand, and Melanesia including New Guinea. *Machiloides*, to which *Nesomachilis* is closely allied, is principally South African, Madagascan, and South American (Wygodzinsky, 1948, 1955).

CLASSIFICATION

Order ARCHAEOGNATHA

(3 described Australian spp.)

1. Meinertellidae (3) Machilidae (0)

1. Meinertellidae (Figs. 7.5, 11.2). Distinguished by having very small abdominal sternites which protrude slightly, if at all, between the coxal plates, and by lacking scales on the legs, scape, and pedicel. The best known Australian species is *Allomachilis froggatti* Silv., up to 15–18 mm long, which occurs on the coastal cliffs of eastern, southern, and south-western Australia. It has small exsertile vesicles on abdominal segments 2–4 (Fig. 11.1H), and the 2nd segment of the male maxillary palp bears simple sensory setae (Fig. 11.1C). The other Australian species are smaller, up to 7–10 mm, have larger vesicles on abdominal segments 1–7 (Fig. 11.1G), and the male maxillary palp bears an elaborate sense organ (Fig. 11.1D). *Nesomachilis australicus* (Wom.) is abundant in rain forest in Queensland and N.S.W., occurring also in wet and dry sclerophyll forest. *Machiloides hickmani* Wom., with dumb-bell-shaped ocelli, is recorded from beach tussocks in Tasmania, and an allied *Machiloides* occurs near Canberra. Undetermined Archaeognatha are also known from Victorian forests.

Order THYSANURA

(Silverfish)

More or less flattened, cursorial Apterygota. Cerci and appendix dorsalis generally subequal in length, sometimes short; body scaled or bare, with or without pigment; compound eyes reduced or absent, never contiguous; ocelli absent, except in Lepidothrichidae; mandible dicondylar; maxillary palp 5-segmented; thorax not strongly arched, pleura exposed; coxae lacking styles, tarsi 2- to 5-segmented; styles present at most on abdominal segments 2–9, commonly less; 2–7 occasionally with exsertile vesicles.

The order includes some 330 species in 5 living families. It is more diverse than the Archaeognatha, both structurally and ecologically, and is of some economic importance.

Anatomy of Adult

Head (Fig. 11.3A). More prognathous than in Archaeognatha. Compound eyes reduced, the component ommatidia somewhat isolated, or absent. Median and lateral ocelli present in Lepidothrichidae, reduced to median frontal organ in Lepismatidae, otherwise absent. Antennae long. Mandibles dicondylar (Fig. 11.3B), biting transversely as in higher insects; molar and incisor regions contiguous. Hypopharynx simple. Maxilla and labium of generalized form, the maxillary palp 5-segmented, the last segment sometimes secondarily subdivided; labial palp shorter, 4-segmented. Anterior arms of tentorium fused into large central plate, approaching the condition found in orthopteroids.

Thorax (Figs. 11.3C–E). Terga entire, not strongly arched, sometimes produced laterally into lobes. Pleural region with 3 small sclerites. Sterna generally small, but produced into large posterior lobe in Lepismatidae (Fig. 11.4C). Mesothoracic spiracles sometimes appearing prothoracic, metathoracic spiracles normal.

Legs. Coxa large, flattened (Fig. 11.4C), trochanter simple, tarsi 2- to 5-segmented, the pretarsus bearing 2 lateral and a variable median claw. Plane of autotomy present between trochanter and femur.

Abdomen. Tapering, rarely much narrower than thorax. Terga less arched than in Archaeognatha. Sternal region in Lepidothrichidae and some Nicoletiidae divided into

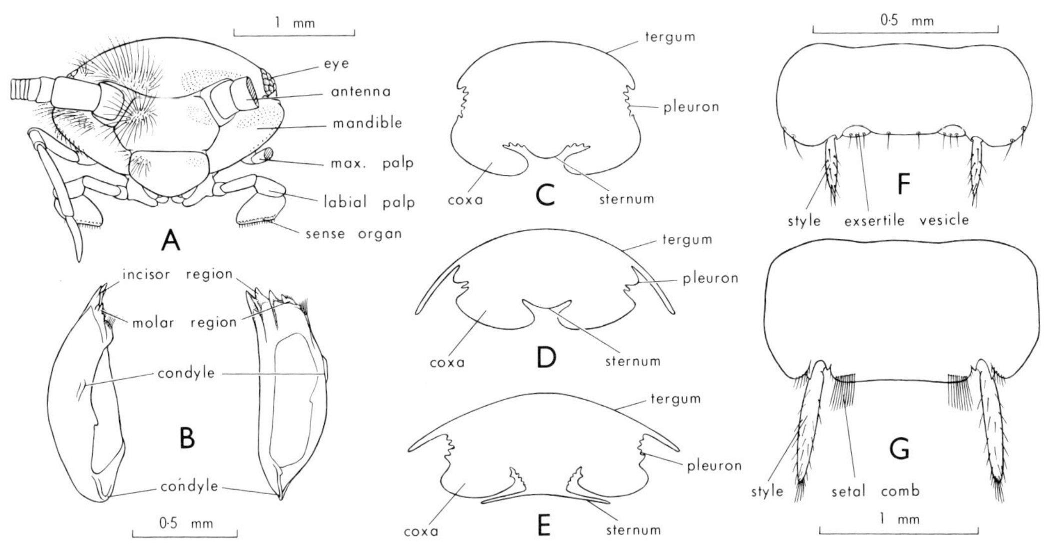

Fig. 11.3. Thysanura: A, head of *Ctenolepisma longicaudata*, frontal view with left antenna and maxillary palp removed; B, mandibles of *Ct. urbana*, upper and inner aspects; C, diagrammatic transverse section through thorax of *Nicoletia*; D, same of *Atopatelura*; E, same of *Lepismodes*; F, abdominal 'sternum' 6 of *Atopatelura hartmeyeri*; G, abdominal 'sternum' 7 of *Ct. longicaudata*. Scales and many setae omitted throughout. [M. Quick]

basal sternum and lateral coxal plates, otherwise fused into single 'sternum' (Figs. 11.3F, G). Abdominal styles variously distributed, at most on segments 2–9, commonly on 7–9, occasionally absent. A single pair of exsertile vesicles on segments 2–7 in lepidothrichids and some nicoletiids, otherwise reduced, and absent in Lepismatidae and Maindroniidae. Spiracles present on segments 1–8. The terminalia are described on pp. 22, 24.

Internal Anatomy. Alimentary canal simple; crop well developed; proventriculus absent in Lepidothrichidae and Nicoletiidae, reduced in Ateluridae, well developed in Lepismatidae; generally 4–8 Malpighian tubes. Salivary glands and labial kidneys present. Nervous system much as in Archaeognatha. Tracheal system lacking large longitudinal trunks, but intra- and intersegmental anastomoses present. Ovaries with 7 panoistic ovarioles in Lepidothrichidae, 2–5 in other families; spermatheca and accessory glands present. Testes with 2–7 follicles in lepismatids and numerous follicles in lepidothrichids; male tract often highly modified for production of spermatophores.

Immature Stages

Eggs pale brown, oval, but often distorted. Embryonic development as in Archaeognatha, the amnion closing except for a small pore. Frontal egg-tooth present in first instar, as are traces of separate cephalic tergites (Sharov, 1966). Larval development much as in Archaeognatha. Scales appear at third ecdysis; abdominal styles incomplete in first-stage lepismatids, appearing progressively during larval and adult life; both styles and vesicles complete in newly-hatched *Nicoletia*. Sexual maturity attained between 10th and 14th instar in lepismatids (Delany, 1957).

Biology

Most Thysanura are free-living, and all are extremely agile. The Lepidothrichidae and most Lepismatidae are cryptozoic, living under bark or litter, but some silverfish can withstand considerable desiccation (Beament *et al.*, 1964). Nicoletiids are principally subterranean or cavernicolous, and are vegetarian. Silverfish are otherwise omnivorous, and some can secrete an intrinsic cellulase (p. 64). The Ateluridae include the smallest Thysanura, and are inquilines in nests of ants and termites, as are a few lepismatids. Some dozen species of silverfish live in association with vertebrates, most commonly with man.

Reproduction is generally sexual, although *Nicoletia* may reproduce parthenogenetically. As in the Archaeognatha, transfer of sperm is indirect, the male placing a flask-shaped spermatophore on the ground, the female picking it up and transferring its contents to the spermatheca. Female lepismatids mate and lay one batch of eggs in each instar, whereas some nicoletiids lay individual eggs over much of the stadium. The eggs hatch in 10 to 60 days. Postembryonic development may be slow, but larvae of *Lepismodes* (= *Thermobia*) reach sexual maturity in 2–3 months. Moulting continues throughout the long adult life of up to 4 years, and the reproductive potential is great.

Natural Enemies. Many species of Thysanura harbour gregarine Sporozoa. The primitive Mengeidae (Strepsiptera) parasitize

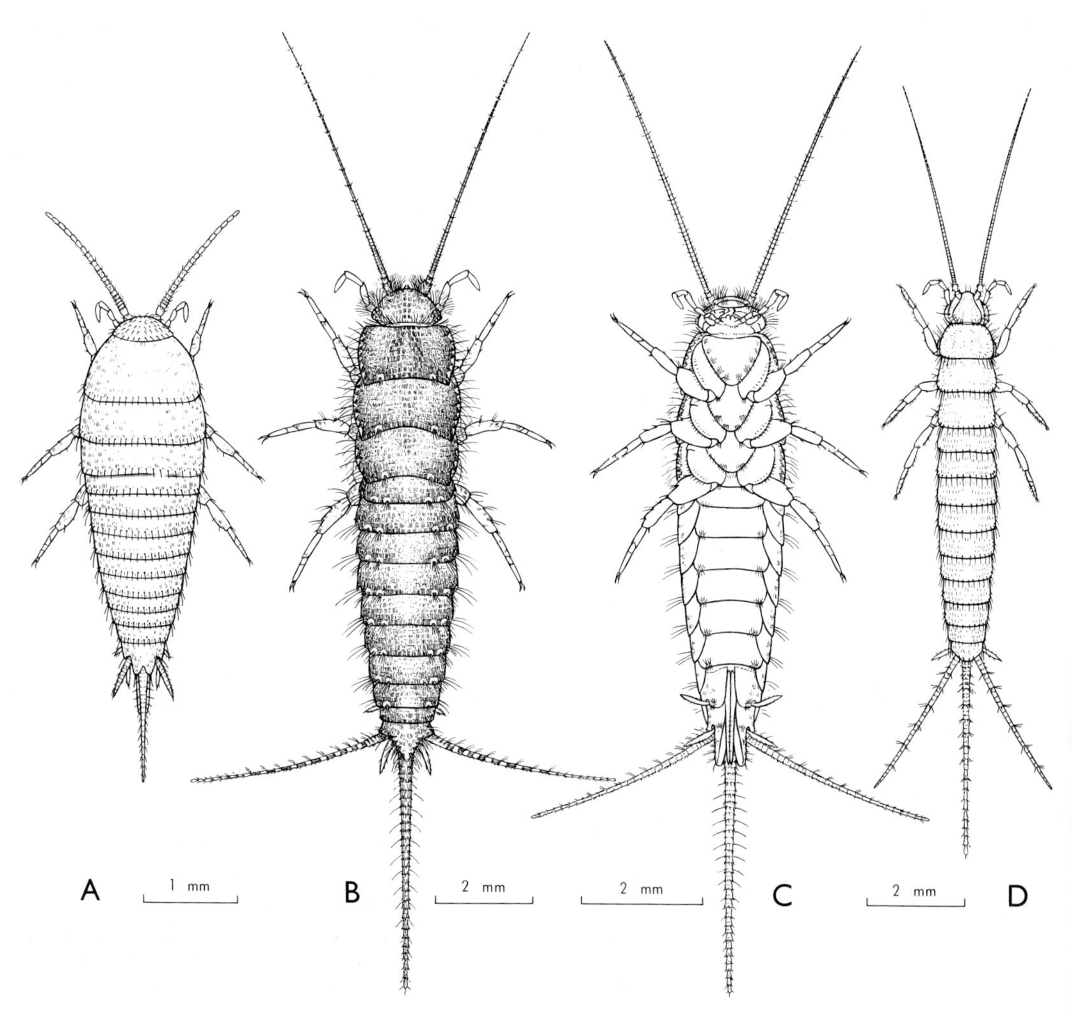

Fig. 11.4. Thysanura: A, *Atopatelura michaelseni*, Ateluridae, ♀ dorsal; B, C, *Acrotelsella devriesiana*, Lepismatidae, ♂ dorsal and ♀ ventral (scales omitted); D, *Trinemura excelsa*, Nicoletiidae, ♀ dorsal. [M. Quick]

lepismatids. As with the Archaeognatha, the principal predators appear to be spiders.

Economic Significance. A few nicoletiids are plant pests, and various lepismatids are pests in human habitations, where they feed on paper, glues, sizings, and scraps; none is known to transmit disease. Five pest species, all cosmopolitan, are known from Australia: *Lepisma saccharina* L., *Lepismodes inquilinus* Newm. (= *Thermobia domestica* (Pack.)), *Ctenolepisma lineata* (F.), *Ct. longicaudata* Esch., and *Ct. urbana* Slab. * Womersley (1939) keyed the first 4 species, *Ct. urbana* being recorded by Greaves (1947), who also discussed their control. The commonest are *Ct. longicaudata* and *Ct. urbana*, grey or silver species up to 15 mm long, the former having 9–12 and the latter 2–5 sensilla on the labial palp.

Special Features of the Australian Fauna

The three introduced gènera are cosmopolitan, and *Heterolepisma* is almost cosmopolitan, whereas *Acrotelsella* is confined to Australia and the Marquesas Is. The atelurid *Atopatelura* is restricted to Africa, Australia, and possibly South America; *Atelurodes* is Australian and Asian (Paclt, 1963).

CLASSIFICATION

Order THYSANURA

(23 described Australian spp.)

	Lepidothrichidae (0)	2.	Ateluridae (7)
1.	Nicoletiidae (2)	3.	Lepismatidae (14)
			Maindroniidae (0)

The classification adopted is basically that of Remington (1954) and Paclt (1963),† although the short-bodied, inquiline Ateluridae are separated from the elongate, free-living Nicoletiidae. The Lepidothrichidae are the most primitive dicondylous insects known, and hence closest to the origin of the winged insects (Wygodzinsky, 1961); the other families are more specialized. The lepidothrichids and maindroniids are small families of restricted distribution.

* A further pest species, *Acrotelsa collaris* (F.) is now known from Darwin (J. A. L. Watson and C. S. Li, *J. Aust. ent. Soc.* **6**: 89–90, 1967).

† And J. Paclt, *Genera Insect.* **218**: 1–86, 1967.

Key to the Families of Thysanura Known in Australia

1. Eyes present; body with scales, generally pigmented .... **Lepismatidae**
   Eyes absent; body with or without scales, white or golden 2
2. Body short, oval; scales always present; appendix dorsalis and cerci short, generally much less than half body length, appendix dorsalis sometimes longer than cerci; inquiline species .... **Ateluridae**
   Body subcylindrical, elongate, parallel-sided; scales generally absent; appendix dorsalis and cerci subequal, exceeding half the body length; not inquiline species .... **Nicoletiidae**

1. Nicoletiidae (Fig. 11.4D). The described Australian nicoletiids both belong to the genus *Trinemura* (treated as a synonym of *Nicoletia* by Paclt, 1963); a third species occurs in Melanesia. *T. novaehollandiae* Silv. is recorded from W.A. and *T. excelsa* Silv. from S.A. and Victoria. However, nicoletiids also occur in N.S.W. and Queensland.

2. Ateluridae (Fig. 11.4A). Four genera of this inquiline family are recorded from Australia. The species are all small, up to 3-6 mm long. *Allatelura hilli* Silv. is associated with *Mastotermes darwiniensis* Frogg. in northern Australia, and *Gastrotheus disjunctus* (Silv.) has been collected with termites and ants in S.A. and W.A. At least 3 of the 4 species of *Atopatelura* occur with both ants and termites, and the host of *Atelurodes similatus* (Silv.) is unknown. There are several undescribed species.

3. Lepismatidae (Figs. 11.4B, C). Five of the species are cosmopolitan household pests. The 4 recorded species of *Heterolepisma* and 5 of *Acrotelsella* are mostly found under bark, rocks, and litter in southern and central Australia, and *H. michaelseni* Silv. has also been collected in ants' nests.

12

EPHEMEROPTERA

(Mayflies)

by E. F. RIEK

Palaeoptera with mouth-parts mandibulate, but reduced to non-sclerotized vestiges in the adult; abdomen of both nymph and adult ending in two or three long caudal styles (cerci and appendix dorsalis). Almost invariably with subimaginal and adult winged stages.

This is a relatively small order of stream-frequenting insects. The adults (Fig. 12.1A) are short-lived (a few hours to a few days), and do not move far from water. They take no food, the alimentary canal is inflated with air, and they have a characteristic dancing flight. The immature stages of all species are aquatic. They occur in a wide variety of situations, but are most abundant under cool, clear-water conditions.

Anatomy of Adult

Head (Fig. 12.1B). Appears triangular when viewed from above, due mainly to the strong development of the posterolaterally situated eyes, especially in the males, in which they are often closely approximated, although they are more or less rounded and widely separated in the females. The eyes are usually divided transversely into an upper part with larger facets and a lower part with smaller, usually darker facets. Baetid males have the eyes divided into a lower rounded portion and a larger, separated, turbanate portion (p. 235) capping it. This exceptional development of the male eye is thought to be correlated with the fact that the male approaches the female from below in the mating flight. Three ocelli. Antennae shorter than the head, filiform, multisegmented, the scape short and thick. Mouth-parts vestigial but externally visible, asymmetrical in form, lacking sclerotization, and shrunken closely together in a single whitish mass beneath the clypeus.

Thorax. Highly specialized for flight. Prothorax reduced, meso- and metathorax large and fused together. Mesothorax most strongly developed, its notum deeply grooved and strongly convex dorsally. Metanotum quadrangular and transversely ridged, sometimes produced posteriorly.

Legs. Highly modified; adult mayflies walk very little. In some non-Australian groups the legs, except the fore legs of the male, have become vestigial, and the adult remains on the wing during the whole of its short imaginal life. Fore legs of males elongate, with a reversible joint at the base of the tarsus as an adaptation for seizing the female during the mating flight. Tarsi normally 5-segmented, but the number of tarsal segments may be reduced by fusion of the basal segments with the tibia in more specialized mayflies.

Wings (Figs. 12.1A, 4). Somewhat triangular, and both pairs held rigidly upright when

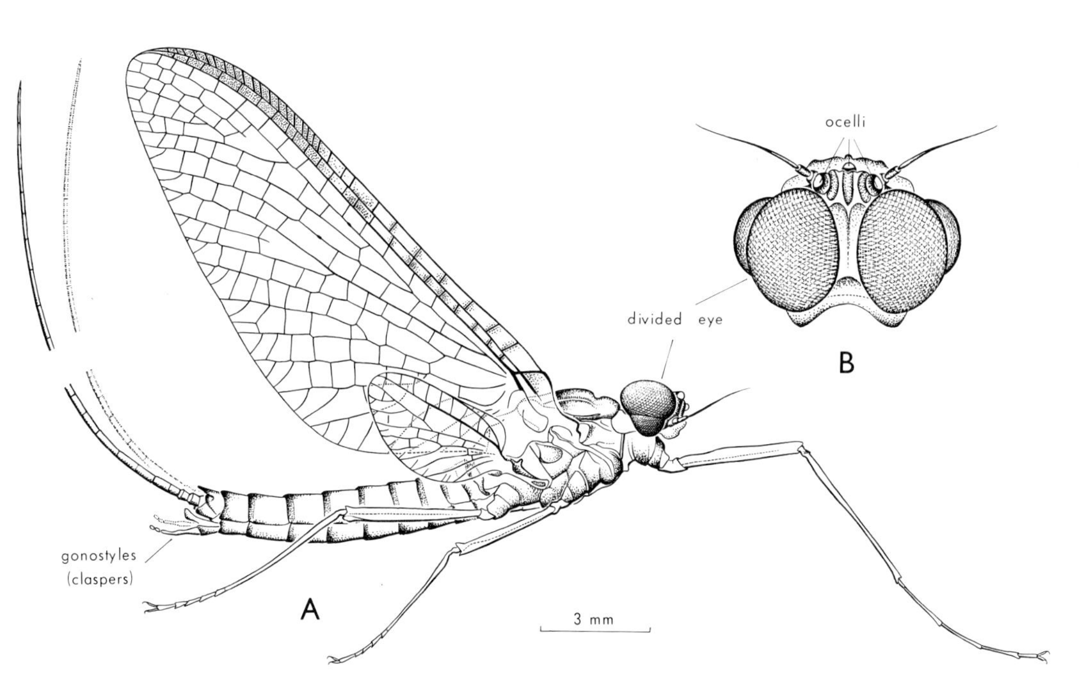

Fig. 12.1. *Atalophlebia* sp., ♂, Leptophlebiidae: A, lateral view; B, head, dorsal. [T. Binder]

at rest. With increasing specialization they become longer and narrower, except in genera with greatly reduced hind wings, in which there is a tendency for secondary development of a triangular or rounded wing. Scalloping of the wing margin occurs in Papuan Palingeniinae. The wings are clear or sometimes with a slight pattern, rarely deeply infuscated except along the costal margin. Mayflies are unique among pterygote insects in the completeness of the fluting of the wings, there being a regular alternation of convex and concave veins and numerous triadic veins (p. 19). When the hind wing is well developed, the wings are coupled in flight. In *Mirawara* the veins of the posterior edge of the fore wing and the corresponding costal veinlets of the hind wing weaken just before the margin, allowing the edge of the fore wing to roll downwards and forwards to effect coupling.

The venation is sometimes very reduced, and in Caenidae and some Tricorythinae the fluting of the wing is also lost behind R. The hind wing is absent in all Caenidae, most Tricorythinae, a large part of the Baetidae, and in diverse genera of the Leptophlebiidae.

Abdomen. Ten-segmented, with a short postabdomen fused with segment 10. Each segment is ring-like, composed of an enlarged tergite which extends down the sides and a small transverse sternite. In some species there are posterolateral spines on some segments, especially posteriorly, and occasionally they are very pronounced. The multi-segmented cerci and median caudal style, which is usually present, articulate with the composite segment 10. Spiracles present on segments 1 to 8.

The male terminalia (Figs. 1.25D, 12.2B) are remarkable in usually having segmented gonostyles, and in having a double penis. In more advanced genera, there appears to be a fusion of segments of the gonostyles, culminating in claspers consisting of a single segment in the Caenidae. The females are also remarkable, because the oviducts open separately by a pair of gonopores placed

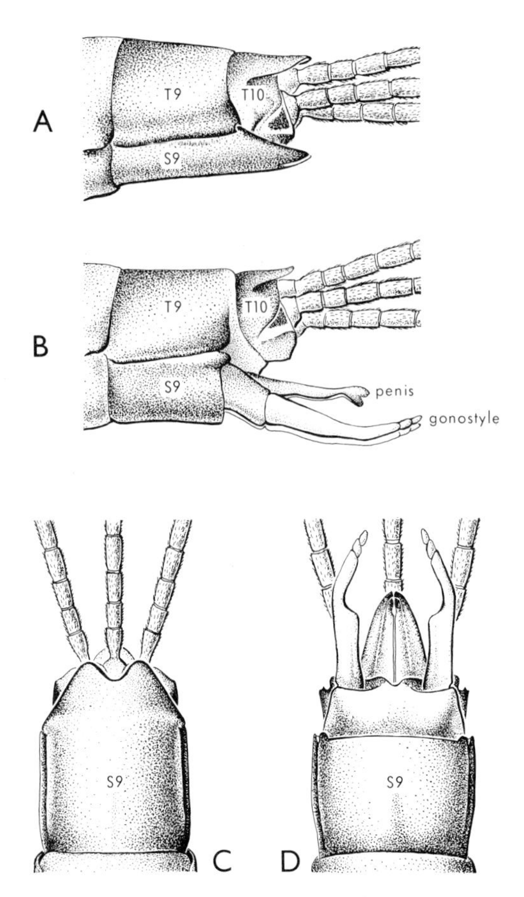

Fig. 12.2. Terminalia of *Atalophlebia* sp., Leptophlebiidae: A, ♀, lateral; B, ♂, lateral; C, ♀, ventral; D, ♂, ventral. [R. Ewins]

behind sternite 7 (Fig. 12.2A), which is often developed into a strong subgenital plate. The distal ends of sternites 7 and 8 are extended into 'egg-valves' in many mayflies, and further prolonged to form a simple ovipositor in some Leptophlebiidae. Its role in egg-laying by these species is not known.

Internal Anatomy. The alimentary tract is greatly modified, lined with a flattened epithelium which has lost its capacity to secrete enzymes or absorb nutrients, and provided with numerous dilator muscles. It functions as a reservoir for air, which is said to be drawn in and expelled through the mouth, the impulses contributing to the dancing quality of the flight. Malpighian tubes usually very numerous. Nervous, circulatory, and reproductive systems primitive. In the male the testes join separate vasa deferentia which end in separate ejaculatory ducts; in the female the numerous panoistic ovarioles similarly join a separate oviduct on each side; there are no accessory glands.

Immature Stages

Egg. The eggs are small (0·1 to about 0·5mm in diameter), of varied form, chorionic sculpturing, and colour; with or without an evident micropylar apparatus. They are generally provided with long coiled adhesive filaments, which serve to attach them to various substrates.

Nymph. The body form differs considerably from that of the adult, and there is wide variation within the order (Figs. 12.5–7, 9). The most noticeable differences are in the presence of fully developed mouth-parts and of abdominal gills. Some nymphs resemble the apterygotes quite closely.

The head (Fig. 12.3) has large eyes, and the sexual dimorphism in eye development is evident in immature nymphs; 3 ocelli; antennae multisegmented, filiform. The mouth-parts (Fig. 12.3) are of normal mandibulate structure. Mandibles with two articular condyles, and usually with distinct incisor and molar areas; the incisor area with one fixed and one movable tooth; the molar surface different in the two mandibles; a prostheca ('lacinia mobilis') of unknown function is usually present on inner side of the movable incisor. Maxilla with most of the components fused into a single unit; palp normally 2- or 3-segmented, with the segments sometimes greatly lengthened, but occasionally the palp vestigial or absent; there is generally a comb of feeding bristles at the apex of the maxilla and sometimes also at the apex of the palpi. Hypopharynx well developed. Labium usually with segmentation distinct; palpi well developed, normally 2- or 3- segmented. In certain carnivorous nymphs the 2nd segment of the maxillary and labial palp is secondarily subdivided (p. 234).

Prothorax larger than in the imago; meso- and metathorax closely fused. There is a median longitudinal ecdysial line on the thorax. Legs shorter and much stouter than in the adult, especially the femora. Tarsi simple, except in the mature nymphs of a few species (e.g. *Mirawara*), and ending in a single claw. There is a correlation between claw size and habitat. Species living in ponds or slow-flowing water or in sand have thin, attenuated claws without teeth or with a comb of very fine teeth below; those living in swifter currents have short, thick claws; and nymphs from rapids often have strong teeth on the underside of the claws. For the most part the size of the tarsal claws is fairly uniform within a genus.

The abdomen consists of 10 distinct segments, and ends in two or three long caudal filaments, which may be considerably longer than the body, though they are generally considerably shorter. In the free-swimming and darting nymphs (Figs. 12.5A, B, 6A), there is a great development of hairs medially on the cerci and laterally on the caudal style; they overlap to form a paddle, so that the nymphs are able to move quickly through the

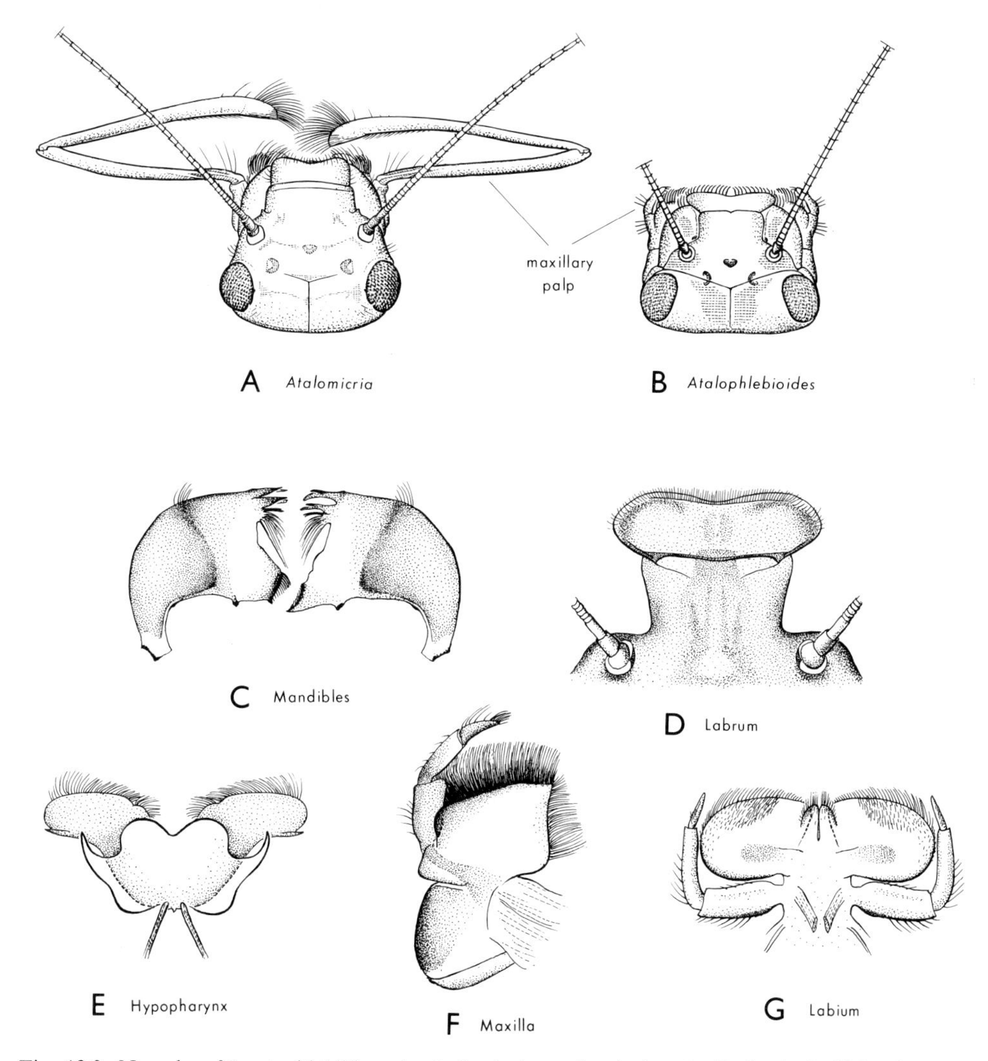

Fig. 12.3. Nymphs of Leptophlebiidae: A, *Atalomicria* sp., head, dorsal; B–G, *Atalophlebioides* sp., head and mouth-parts. [M. Quick, R. Ewins]

water by rapidly flicking their abdomens and straightening their tails. In species that do not swim (Fig. 12.7), the tails may be almost bare and often very long and thin.

Mayfly nymphs respire primarily by means of 4 to 7 pairs of articulated gills situated at the junction of tergum and sternum of some or all of the first 7 abdominal segments. The gills of actively swimming nymphs are often lamellate with a strengthened fore margin, whereas those of species living in standing water of low oxygen content are often large and plumose (Fig. 12.8). They may appear to be lateral, dorsal, or ventral, and in some one branch of the gill may be dorsal and the other ventral, especially the first gill, or when the gills are developed into a ventral suction disc. Gill filaments are developed on the anterior parts of the body in some genera. There are simple maxillary and labial gills in *Coloburiscoides*, and more varied arrangements in some non-Australian mayflies. Nymphs do not have open spiracles.

Biology

Adults. Some adults live only a few hours, but others may survive for more than a week, especially in cold dull weather. Their dispersal is generally very limited, but at times adults may be found at distances of several kilometres from the breeding grounds. Swarming depends to some extent on climate. In warmer regions adults tend to emerge throughout the year, and for this reason there are no great swarms. In colder regions many species collect in very large swarms to carry on their mating flight. Swarming often takes place over land adjoining the water, rather than over the water itself. Cold, rainy, or windy weather inhibits swarming flights. The burrowing mayflies are more or less seasonal, and emerge in great numbers over a short period. In Australia, the burrowing Caenidae form conspicuous swarms. In some swarms each male patrols a stretch of the stream. The rapid flight of the males and the great number of individuals may give the flight a somewhat frantic appearance. Similar patrolling of a stretch of the stream occurs in some species *(Atalophlebia)* that do not form dense swarms. In some species there is a change in the eye-colour of the adult dependent on light intensity.

Reproduction. Copulation (Fig. 1.26) takes place in flight. The male approaches the female from below, and his long, upwardly extended fore legs are securely wrapped over her mesothorax. His tarsi are crossed over one another across her mesonotum, passed around the anterior edge of the base of her fore wings, and the tarsal claws are hooked into the pleural wing recess near the base of each wing. Copulation is accomplished when the abdomen of the male is curved upward and the abdomen of the female is encircled, usually about segment 9, by his forceps. It may last for only a few seconds. The males of the non-Australian *Homoeoneuria* (Oligoneuriidae) have no forceps, and the abdomen of the female is held by the modified hind legs with their strongly bowed coxa and femur. In this genus, too, the fore legs of the male are shorter than the mid and hind legs. Parthenogenesis occurs in a few species.

The eggs are deposited in water in a number of ways. Usually the egg-mass is extruded in flight, hangs in a ribbon or ball from near the tip of the abdomen, and is washed off as the female dips the abdomen on to the surface of the water. When the egg-mass meets the water, it disintegrates, and the eggs sink freely to the bottom, where they become attached to stones and pebbles. In some species the eggs are extruded a few at a time as the abdomen touches the water. In others the adult female enters the water to lay, either placing the tip of the abdomen below the surface (some species of *Atalophlebia*), or crawling beneath the surface to oviposit on stones or other substrates on the bottom (some species of *Baetis*). In some exotic species *(Ephoron)*, the females fall to the surface of the water and drown during oviposition.

Normally, when mayfly eggs come in contact with water, they either form a protective jelly-like layer around themselves or release their attaching filaments. The tiny threads of the polar caps uncoil, and the eggs stick to any object that they strike. Some species of *Callibaetis* and *Cloeon* are 'ovoviviparous'. The fertilized eggs are retained within the abdomen of the female for several days, during which the embryos develop. Then, when the female alights on the surface of a suitable body of water, the eggs are expelled, and the nymphs hatch within a few minutes.

Immature Stages. Eggs of most mayflies hatch in from one to two weeks, depending on temperature or possibly light. In a few non-Australian genera the eggs deposited in late spring or summer do not hatch until early in the following spring. They undergo incomplete development in the autumn, and then go into a diapause which is broken only by low temperature. These eggs can withstand desiccation.

Most mayflies have a relatively long nymphal period. The majority have an annual life cycle, although there are two generations a year in some species, one in early spring and the other in late summer. Some of the larger mayflies take two years to reach maturity, but there are emergences each year due to overlapping broods. There are a large number of nymphal instars, up to 27 having been recorded.

Each species is normally restricted to a definite type of habitat, ranging from lakes and large rivers down to temporary ponds and creeks. A very few occur in stagnant, weed-choked backwaters and bogs. Swarms of mayfly larvae occur on submerged stones of the highest, cold, snow-melt streams. In Australia, the mayfly nymphs that occur in standing waters are usually the small, shrimp-like nymphs of *Cloeon* (Baetinae) or the large, flattened nymphs of the *australasica* group of *Atalophlebia* (Leptophlebiidae).

Burrowing nymphs in other countries exhibit some of the most remarkable adaptive characteristics. The legs have often become flattened digging structures, the fore legs being used for pushing the silt sideways and the hind legs for passing it back out of the burrow. Frontal processes of the head and long tusks of the mandibles may be developed to loosen the silt. The gills are often feathery and well supplied with tracheae. They are kept in constant motion during the digging, producing a current of water which passes posteriorly over the body and out through the burrow opening. These burrowing types of nymphs feed selectively on the detritus they excavate for the burrow. There are few burrowing types of nymphs in Australia. *Jappa* burrows under stones in silty sections of sluggish streams, and has long setae on the gill margins. The Caenidae and *Tasmanophlebia* (Siphlonurinae) have semi-burrowing nymphs that settle into sand or silt until they are almost completely hidden.

With few exceptions, the nymphs are herbivores or scavengers, living on vegetable detritus and microscopic aquatic organisms, principally diatoms. A few are partly predacious, while others have long sharp mandibles which would seem to indicate that they are entirely predacious in habit. Carnivorous nymphs with unusual mouth-parts have developed in one group of the Siphlonurinae in Australia, New Zealand, and South America, and a similar type of nymph is known to occur in the Cretaceous of Australia. Any mayfly nymphs held in a confined space will eat one another.

Subimago. Ephemeroptera emerge as winged subimagos, which usually moult to produce the imagos, or true adults. They are the only insects that undergo an ecdysis after having acquired functional wings. Some nymphs moult into the subimago on the surface film of the water in a few seconds; others crawl out on to stones and sticks, and the subimago emerges relatively slowly, taking up to five minutes. Emergence occurs throughout the day, with different times favoured by different species. Caenids normally emerge shortly after dawn, whereas those leptophlebiids that emerge on the surface film of lakes prefer the warmer middle portion of the day. Some siphlonurine nymphs crawl out of the water at dusk or during the night, and the subimagos, clinging to rocks, are ready to fly with the first light.

The subimaginal stage may last for as little as five minutes (Caenidae), but it is usually of one day, though the period may be extended in dull cold weather until conditions are suitable for swarming. It lasts 2–4 days in some of the larger Australian Siphlonurinae—they have been kept for up to a week in a refrigerator, and when removed completed ecdysis to the adult. Humidity may determine the success or failure of the moulting process, but temperature is a more important controlling factor initiating ecdysis, which does not take place at very low temperatures.

In the Papuan *Plethogenesia papuana* the subimagos do not moult, but breed and die in that stage, and in a few other exotic species only the males become imagos. Some show exuviation in part, retaining the subimaginal cuticle on the wings, while in others *(Caenis)* the subimaginal skin may remain on the caudal filaments in the females. Usually exuviation of the subimago takes place with the insect resting on some object, though some exotic forms exuviate in flight, and in these exuviation takes place from all parts of the body except the wings. In some caenids exuviation may occupy less than a minute, and the insect often takes flight again before the tails become fully free, the last of the cuticle being usually dislodged in flight. Exuviation in *Tasmanocoenis* usually takes from 10 to 20 seconds from the time the subimago alights to the time the imago flies off to rejoin the dancing swarm of mixed adults and subimagos.

Natural Enemies. Mayflies, both nymphs and adults, form an important item in the diet of freshwater fish. Many nymphs are eaten as they swim to the surface, or on the surface as they transform to the subimago. Many are also eaten by the carnivorous nymphs and larvae of other aquatic insects (Odonata, Plecoptera, Coleoptera). Adults are ensnared in spider webs spun over or along the stream, and others fall prey to insectivorous birds and the large hawking dragonflies (Aeshnidae).

The nymphs act as intermediate hosts for certain flukes that infest fish. Nematodes occur in the abdomens of both nymphs and adults. Certain tube-dwelling chironomid larvae are external parasites of the nymphs, but others are merely commensals. The commensal larvae have been observed attached to the pronotum and head of almost full-grown nymphs of *Coloburiscoides*. Sessile Protozoa may form large colonies on nymphs of *Atalophlebia*.

Economic Significance. Mayflies occupy an important place in the economy of aquatic communities. The nymphs almost invariably feed on vegetable matter, converting it to animal tissues. They are practically defenceless, but make up in numbers for their vulnerability. Since they are so readily available, mayfly nymphs are preyed upon by practically every aquatic predator. The adults of a few species may swarm to lights and become a nuisance in towns near lakes, but on the whole they are harmless.

Special Features of the Australian Fauna

There is a marked similarity between the mayfly faunas of Australia, New Zealand, and southern South America. It is evident from the tabular statement that the Australian fauna is quite restricted, most of the species belonging to the more generalized families. The large and diverse Ephemeroidea are represented by only a few species of the small, highly specialized, cosmopolitan family Caenidae, which is often placed in a separate superfamily because of its reduced venation. Of the other ephemeroid families, the Ephemeridae are known from New Zealand (Ephemerinae) and New Guinea (Palingeniinae), and the Neoephemeridae are present in Java. The Leptophlebioidea and Baetoidea are well represented, the Leptophlebiidae being the dominant family of Australian Ephemeroptera. Baetidae are represented by Baetinae and Siphlonurinae, but the Ametropodinae are absent.

Mayflies are most abundant in the highlands of south-eastern Australia and in Tasmania, but there are numerous species in the colder sections of the coastal streams of the whole east coast. There are few species in S.A., and in W.A. they are restricted to the wetter south-western corner. There are only a few widely distributed species in the slow-flowing inland streams of the eastern States. The large and more spectacular mayflies of the subfamily Siphlonurinae occur in Tasmania, Victoria, and the higher regions of southern N.S.W., but one genus has a discontinuous distribution to north Queensland.

CLASSIFICATION

Order EPHEMEROPTERA
(124 Australian spp.)

BAETOIDEA (37)
1. Baetidae (37)
Oligoneuriidae (0)
EPHEMEROIDEA (13)
2. Caenidae (13)
Ephemeridae (0)
Neoephemeridae (0)
Behningiidae (0)
PROSOPISTOMATOIDEA (0)
Prosopistomatidae (0)
Baetiscidae (0)
HEPTAGENIOIDEA (0)
Heptageniidae (0)
LEPTOPHLEBIOIDEA (74)
3. Leptophlebiidae (73)
4. Ephemerellidae (1)

There has been little agreement about the classification of this order (Needham, Traver and Hsu, 1935; Burks, 1953; Edmunds and Traver, 1954; Demoulin, 1958; Edmunds, 1962). The one adopted here recognizes 19 subfamilies and 11 families distributed in 5 superfamilies in the world fauna. The Heptageniidae were incorrectly recorded in the Australian fauna by Harker (1950, 1954).

Keys to the Families of Ephemeroptera Known in Australia

ADULTS

1. Cubital intercalary veins absent (though marginal intercalaries are present in this sector of the wing); cross-veins in disc of fore wing weak and net-like .... (PROSOPISTOMATOIDEA)
   Cubital intercalary veins present (except in Oligoneuriidae); cross-veins in disc of wing not net-like, sometimes absent 2
2. Posterior branch of MP of fore wing arising close to the base, sharply bent towards CuA near its base, running parallel with it in this area, or fusing with it for a short distance (when the hind wing is absent this branch of MP is straight) .... EPHEMEROIDEA. **Caenidae**
   Posterior branch of MP of fore wing arising some distance from base, straight throughout its length, or only slightly angled at base, or MP absent as a distinct vein 3
3. CuP strongly curved to the rear in fore wing. [At most with 2 long intercalaries between CuA and CuP in Australian species; tornus, when present, midway between CuA and CuP] .... LEPTOPHLEBIOIDEA. 4
   CuP straight or only slightly curved in fore wing 5
4. Fore wing with one or two long intercalary veins between MP and CuA, and usually with detached marginal intercalaries (absent in most Tricorythinae) .... **Ephemerellidae**
   Fore wing without long intercalary veins between MP and CuA, and without detached marginal intercalaries .... **Leptophlebiidae**
5. Four long intercalaries between CuA and CuP; tornus midway between CuA and CuP .... (HEPTAGENIOIDEA)
   Never more than 2 long intercalaries between CuA and CuP (and then only free and distinct when the cross-venation is partly reduced); tornus, when present, close to apex of CuA in Australian species .... BAETOIDEA. **Baetidae**

NYMPHS

1. Gills on abdominal segments 2–6 concealed under a carapace-like projection from the thoracic notum; segment 1 almost completely fused with the thorax .... (PROSOPISTOMATOIDEA)
   Abdominal gills, when present, not concealed by such a projection; segment 1 of abdomen free ... 2
2. Gill on abdominal segment 1 reduced to a simple filament (except in Behningiidae); gills with a dense margining of long fine filaments (Fig. 12.6B) .... EPHEMEROIDEA. **Caenidae**
   Gill on abdominal segment 1 similar in form to those on succeeding segments, though sometimes reduced, but occasionally enlarged or developed into an operculate gill (e.g. *Tasmanophlebia*), or gill absent (Ephemerellidae); gills without a dense margining of long fine filaments, except possibly on the fore margin and narrowed apical portion (Fig. 12.8G), or margining extremely fine and short .... 3...

3. Labrum almost or completely concealed under the projecting, rounded, anterior margin of the head. [Nymphs flattened] .... (HEPTAGENIOIDEA)
   Labrum exposed 4
4. Cerci bearing a dense row of setae only on the medial margin (Fig. 12.6A) (some setae on lateral margin in a few genera); nymphs shrimp-like, mostly active swimmers . BAETOIDEA. **Baetidae**
   Cerci uniformly clothed with short setae (Fig. 12.7); nymphs flattened for running on the substrate .... LEPTOPHLEBIOIDEA. 5
5. Gills present on abdominal segments 1–7 (Fig. 12.7) or 2–7 .... **Leptophlebiidae**
   Gills absent from segment 1 or very reduced; never more than 5 pairs of gills (Fig. 12.9) .... **Ephemerellidae**

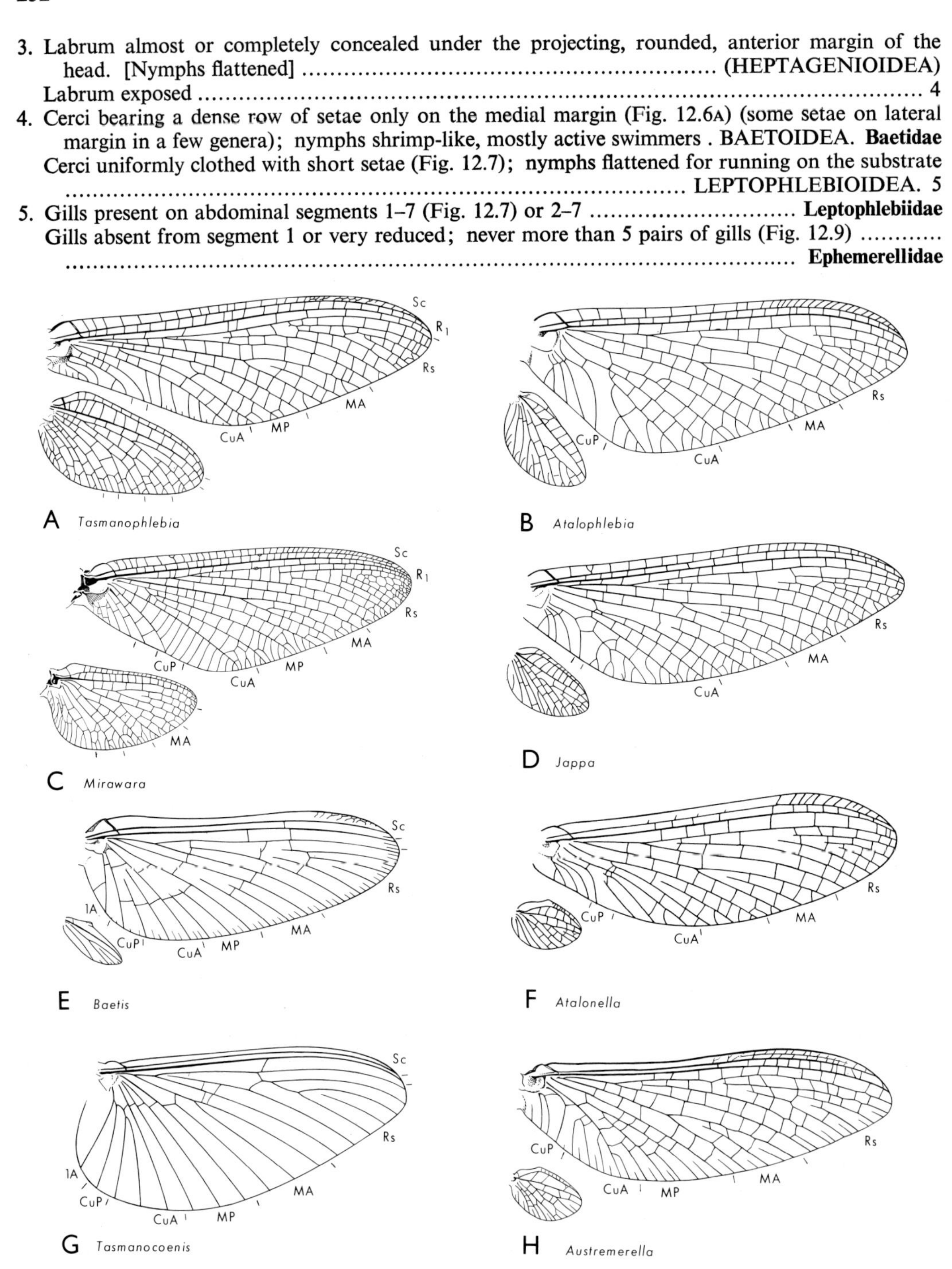

Fig. 12.4. Wing venation of: A, C, Baetidae-Siphlonurinae; B, D, F, Leptophlebiidae; E, Baetidae-Baetinae; G. Caenidae; H, Ephemerellidae. [M. Quick]

Superfamily BAETOIDEA

1. Baetidae. The two subfamilies that occur in Australia may be distinguished by the following keys. Although the nymphs are very similar in structure, they differ markedly in size.

ADULTS

Posterior branch of MP of fore wing not detached at base from stem of MP (Figs. 12.4A, C); hind wing well developed; hind tarsus with 4 clearly defined segments; ♂♂ with the two portions of the eye not distinctly separated .... SIPHLONURINAE

Posterior branch of MP of fore wing detached at base from stem of MP (Fig. 12.4E); hind wing greatly reduced or absent; ♂♂ with turbanate eyes .... BAETINAE

NYMPHS

Posterolateral angles of distal abdominal tergites prolonged as thin flat spines (Fig. 12.5); labrum with anterior margin entire, or with a broad median V-shaped notch .... SIPHLONURINAE

Posterolateral angles of distal abdominal tergites not prolonged as spines (Fig. 12.6A); labrum with a square median notch .... BAETINAE

The SIPHLONURINAE (13 spp.) are large mayflies, spanning up to 45 mm, with well-developed hind wings, and almost invariably with only 2 caudal filaments. Wings with numerous cross-veins; in the fore wing the cubital intercalaries form a series of parallel, often sinuate, but usually not branched, veins extending from CuA to the anal margin. The wing coupling is well developed, especially in *Mirawara*. The fore tarsus of the male is always much longer than the fore tibia. The subfamily has been revised by Riek (1955h).

The adults of Australian species appear in summer and early autumn. The subimagos of most species emerge at night from nymphs that have crawled out of the water. They normally leave the stream within an hour of sunrise, and settle in high trees. *Mirawara* emerges from the water surface. The sub-imago stage usually lasts for 2 days, but may be longer. Mating flights take place at different times. *Coloburiscoides* adults swarm at dusk where the broken water of the riffles gives place to unbroken water. *Mirawara* mates during the brightest and warmest part of the day, and groups of males may be seen flying upwind towards a female settled on the marginal vegetation.

The eggs of *Mirawara* are laid during the day in smooth-water sections of the stream. The females of *M. aapta* lay in the open sections of the stream, but the other species prefer shaded pools. The females always prefer the lower end of the pool above places where the water flows into the riffles. The eggs are shed a few at a time by touching the surface of the water with the apex of the abdomen. The female gradually works up the pool with each successive dip to the surface, turns, flies towards the lower end of the pool, and recommences egg-laying.

The nymphs of most species are strong and rapid swimmers. They are almost always found in rapidly flowing, clear, cold-water streams, but some species occur in small subalpine lakes. The rather large nymphs of Australian species of *Mirawara* and *Ameletoides* rest on rock surfaces in the pools of smoother water, those of *Tasmanophlebia* burrow in sand, and those of *Coloburiscoides* are limited to the broken water of the rapids. *Coloburiscoides* nymphs possess large spinose abdominal gills, and are not active swimmers. At times the nymphs of *Mirawara* can be observed out of the water on rocks that are wet by the spray of waterfalls.

Gills, almost invariably plate-like except in *Coloburiscoides*, are borne only on the abdomen. The tarsal claws of the nymphs are long and slender, but are always shorter than the tibiae. Each cercus has long setae on the inner side only, except to some extent in *Tasmanophlebia*. *Mirawara* is unusual in having 4-segmented tarsi in the mature nymph (Fig. 12.5A) and 3-segmented tarsi in younger nymphs. All other species have unsegmented tarsi (Fig. 12.5C), though the segmentation of the subimago may show through the skin of the mature nymph.

The 4 Australian genera belong to different tribes, which are sometimes regarded as subfamilies. *Mirawara* (Ameletopsini, 3 spp.) has peculiar carnivorous nymphs with elon-

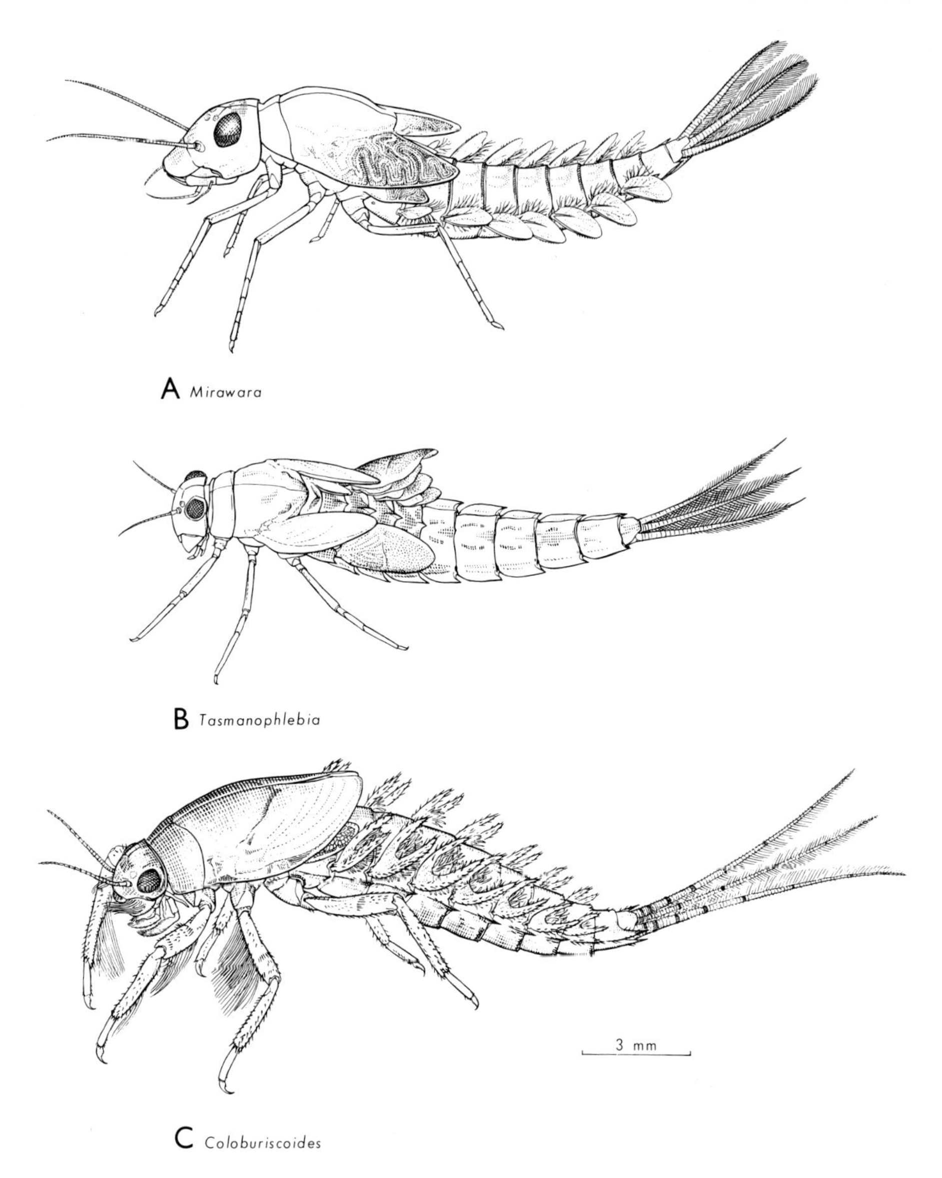

Fig. 12.5. Nymphs of Baetidae-Siphlonurinae.
[M. Quick]

gate mandibles and maxillae, thread-like, multisegmented labial and maxillary palpi, and segmented tarsi. They are distributed from the Victorian Alps to north Queensland. The nymphs usually occur at the margins of slowly flowing pools. Females are commonly observed laying, but males are less commonly seen. *M. aapta* from the upper Murrumbidgee area has clear wings, except for the costal margin. The other two species have females with purplish wings, although the males have clear wings.

Coloburiscoides (Coloburiscini, 5 spp.) nymphs (Fig. 12.5c) have long fringes of setae on the fore and mid legs, and maxillary and labial gills. They occur only in the riffles of fast-flowing streams. The species are restricted to south-eastern Australia, and have very short emergence periods in mid-summer.

Tasmanophlebia (Oniscigastrini, 4 spp.) nymphs (Fig. 12.5B) have the gills on abdominal segment 1 enlarged into gill covers. They are semi-burrowers in sand. This genus is well represented in the lakes and streams at high altitudes from Tasmania to the Blue Mts in N.S.W. The adults have a wing-span of 25–30 mm, and clear wings with some yellowish infuscation at the base.

Ameletoides (Siphlonurini, 1 sp.) nymphs are typical for the subfamily. They occur in the small, cold, feeder streams and in cold lakes. The species is known only from the Kosciusko area and Mt Buller in Victoria. The imago has clear wings (those of the subimago are attractively mottled) and 3 well-developed caudal filaments, a character unusual for the subfamily.

The BAETINAE (24 spp.) include some of the most simplified mayflies. The adults are small and delicate, with clear wings, sometimes slightly darkened along the anterior margin. The venation (Fig. 12.4E) is reduced through loss or fusion of some veins and atrophy in part of those that persist. The fore wings have few cross-veins, usually arranged in gradate series. Short marginal intercalaries are well developed. The number present in each interspace (1 or 2) is normally constant for each genus. The hind wings are very small and narrow, or sometimes absent. Adults have only 2 caudal filaments. The eyes of the males are greatly enlarged and divided into two distinct parts, a lower and outer oval or rounded part, usually darkly pigmented, and a much larger upper and inner portion, usually pale and with large facets, which is raised on a broad stalk and is known as a turbanate eye from its resemblance to a turban. The eyes in the female are relatively small and simple. The male terminalia are greatly reduced, the penis lobes being virtually internal membranes without form. There is often considerable sexual dimorphism in colouring and size; also, the males often have the central portion of the abdomen translucent, and the females the anterior margin of the wing darkened.

Mating flights take place over the water, or over grassy areas adjoining the stream or pond. Females lay 500 or more eggs in many instances. The length of the life-cycle of Australian species is not known. Some are seasonal in their occurrence, but others occur throughout most of the year.

The nymphs (Fig. 12.6A) are small (up to 9 mm long), slender, streamlined forms that swim actively with a jerky, irregular action. They are common in the rocky sections of flowing streams, but occur also amongst the water-weeds of ponds, dams, and slow-flowing streams and backwaters. They differ from siphlonurine nymphs mainly in size and in the structure of the labrum. The head is flexed downwards in front of the humped thorax. The legs have slender, denticulate, single tarsal claws. Plate-like, usually single and simple gills are present on abdominal segments 1–7. Abdomen with 3 caudal filaments in all known Australian nymphs, the cerci with a long hair-fringe only on the inner border. The immature nymphs often have contrasting transverse bands of black and white, but these are nearly always lost in the mature nymphs.

This is a difficult subfamily, in which it is often not possible to separate the females to species. The cosmopolitan genus *Baetis* is represented by 9 species, which occur only in the clear water of cold streams from Tasmania to south-eastern Queensland. They are amongst the earliest of mayflies to emerge, some appearing on warm days at the end of winter. In summer they form dense mating swarms dancing at the lower end of long, quiet pools where the pool joins broken water. The flights, usually over the water, often take place while the air is still cool and crisp. The 4 species of *Centroptilum* are generally larger than the *Baetis*, and they often occur in association with them or under similar conditions, but none has been recorded from Queensland.

The 2 species of *Pseudocloeon* are amongst the smallest of the family in Australia. The adults are almost invisible when on the wing.

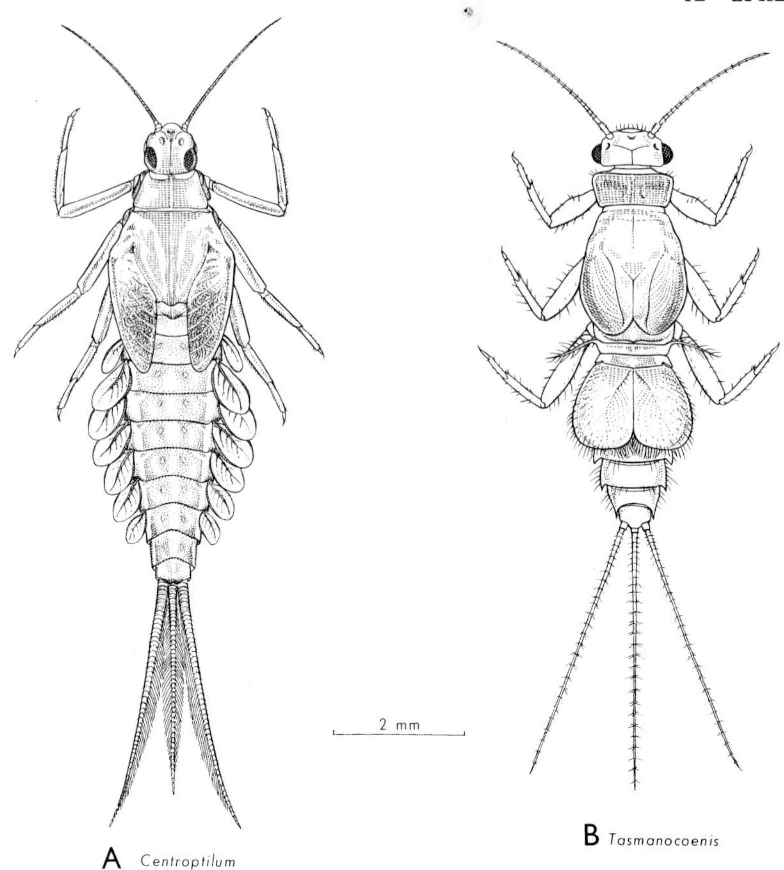

Fig. 12.6. Nymphs of: A, Baetidae-Baetinae; B, Caenidae. [M. Quick]

They occur throughout most of the year in south-eastern Queensland, and nymphs are known only from the mountain streams of coastal Queensland and N.S.W. One of the species occurs also in Java.

Cloeon (8 spp.) is widespread. The Australian species normally inhabit the standing waters of ponds and billabongs, where the nymphs live amongst the water-weeds. The males have clear wings, but the costal and subcostal spaces are usually infuscated in the females, which have relatively large eyes, and are usually the only sex attracted to light. The subimago emerges in the early afternoon, moults to the imago, and swarms the following morning. Most of the species occur in Queensland and inland N.S.W., but there is one in Tasmania and another in north-western Australia.

Superfamily EPHEMEROIDEA

2. Caenidae. These are amongst the smallest and most distinct of the mayflies. The sexes are almost identical in appearance, even the eyes being of similar size. The lateral ocellus is at least half as large as one of the compound eyes. The thorax is greatly developed, whereas the abdomen is relatively small. The fore wing (Fig. 12.4G) is broad, expanded in the anal region, with few cross-veins, often cloudy or milky, and the marginal cilia are numerous in the imago as well as in the subimago. The hind wing is absent. The tarsi are 5-segmented. There are 3 well-developed caudal filaments, with widely spaced joints; they bear prominent setae in the subimagos, but are bare in the adult males, although the females retain the subimaginal setae partly or completely. The

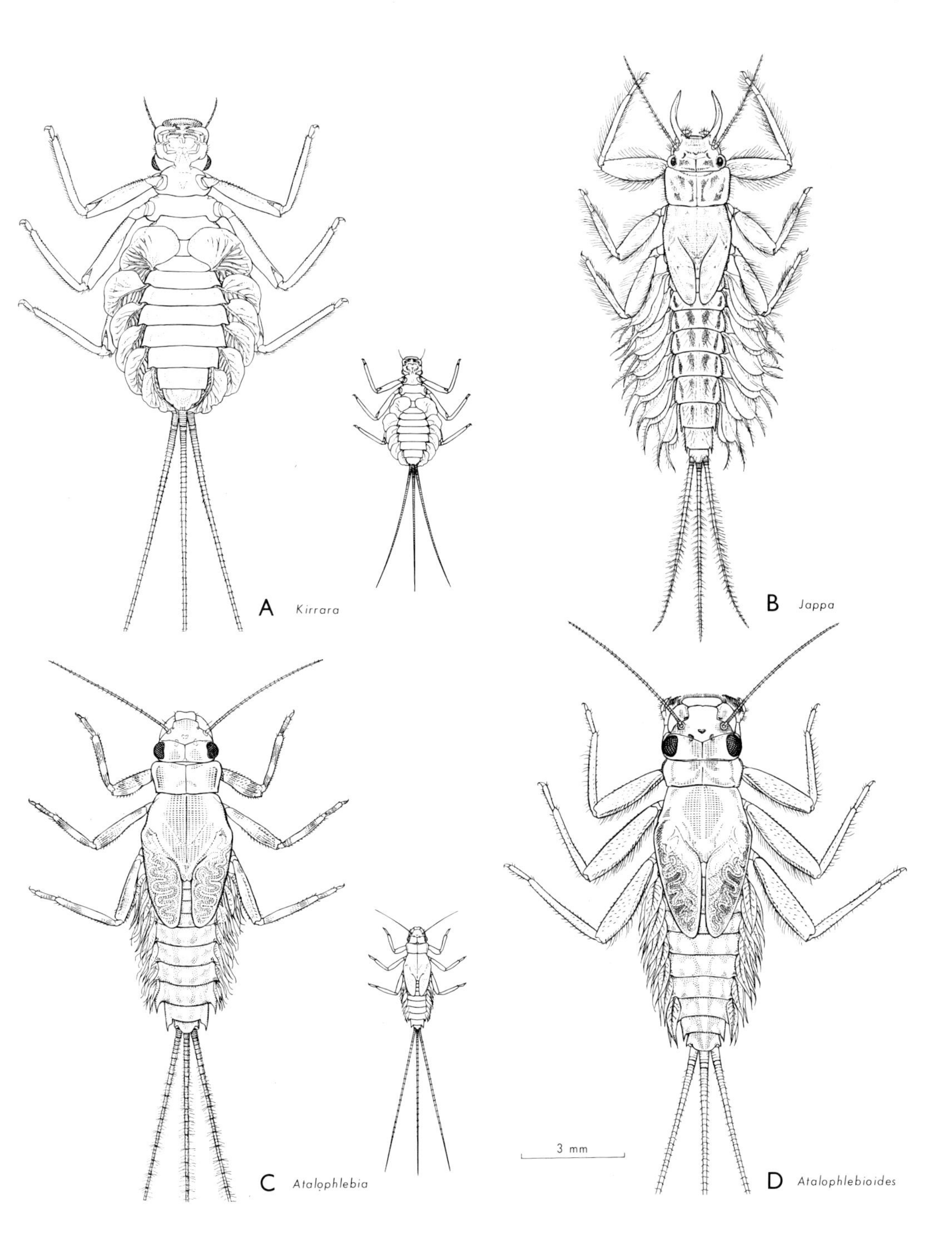

Fig. 12.7. Nymphs of Leptophlebiidae.
[M. Quick]

small, hairy, semi-burrowing nymphs (Fig. 12.6B) appear somewhat flattened, due to the strong development of the second pair of abdominal gills into operculate plates which cover the gills of the succeeding shortened segments 3–6; segment 1 has a pair of single filamentous gills. The lateral margins of the segments are produced, and there are 3 well-developed caudal filaments bearing setae on both sides.

The nymphs burrow into the mud and sediment on the bottom of ponds and standing rock pools, as well as in slow-moving streams. They are rarely found in swift-flowing waters. The subimagos emerge from the surface film of the water early in the morning, and quickly gather into large swarms. They soon settle, moult in less than a minute, and take to the wing again to mate and then die on the surface of the water. The only Australian genus, *Tasmanocoenis*, is found in Tasmania and on the mainland from the coastal plains to the subalpine zone of the Kosciusko plateau; one species exists in W.A.

Superfamily LEPTOPHLEBIOIDEA

3. Leptophlebiidae. This is the dominant family of Australian mayflies. The species are adapted to various habitats, from the warm standing waters of coastal waterholes to the melted snow of the subalpine areas. Some are rather large, but others are no bigger than the small baetids and, like them, have clear, glassy wings so that they are almost invisible during flight. The eyes in the males are composed of a large upper portion of comparatively large facets and a small lower portion of smaller, darker facets; these parts are distinctly separated, but the upper portion is not set on a well-developed stalk. The fore tarsus of the male has 5 segments, but the mid and hind and all tarsi in the female have

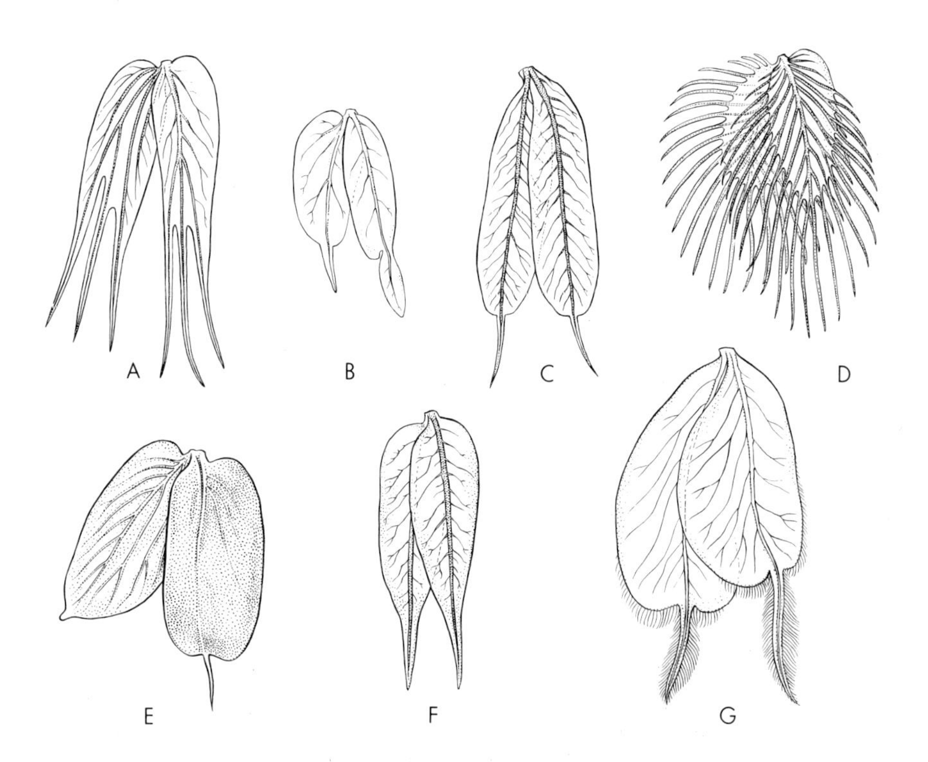

Fig. 12.8. Abdominal gills of leptophlebiid nymphs: A, *Atalophlebia* sp. of *australis* group; B, *Atalonella* sp.; C, *Atalomicria* sp.; D, *Atalophlebia* sp. of *costalis* group; E, genus nr *Massartella*; F, *Atalophlebioides* sp.; G, *Jappa* sp. [M. Quick]

only 4 distinctly differentiated segments (1st segment partly fused with tibia). The 2 claws on each tarsus may be similar or dissimilar. The hind wings (Figs. 12.4B, D, F) are rather reduced in most genera, absent in a few. The male forceps are 4-segmented, but the basal 2 segments are often almost completely fused; the 2nd segment is elongate.

The nymphs (Fig. 12.7) are typically slender and somewhat flattened. Those living in swift waters have a broad, very flattened body, with the eyes situated dorsally on the large head, and in *Kirrara* (Fig. 12.7A) the abdominal gills have ventral lobes which combine to form a large suction disc. There are always 3 long caudal filaments, all with relatively inconspicuous setae, and abdominal segments 1–7 bear lateral gills of various form (Fig. 12.8). The structure of the tarsal claws varies (p. 227). *Kirrara*, *Atalophlebioides*, and less so *Atalonella*, which occur in fast, cold riffles, have a few large ventral denticles on the claw; *Atalophlebia*, from standing to moderately flowing water, has a comb of many fine denticles; and *Atalomicria*, from moderately flowing small streams in heavily forested areas, has no denticles. The burrowing *Jappa* has a comb of many fine denticles.

The subimagos of many species, especially those that crawl out of the water, transform early in the evening, and the imagos emerge the next day. Other species that rise to the surface of the lake or stream to transform to the subimago do so during daylight, especially on warm, bright days.

Atalophlebia (25 spp.) contains all the common large mayflies (wing-span around 25 mm) of slower streams and lakes. The males of several species have only 2, very long (up to 35 mm), white-tipped caudal filaments. As they fly just above the water, one is more conscious of the moving dots of white than of the rest of the insect. The females of all species have 3 filaments. The subimagos of most species have an attractive, specifically diagnostic wing pattern. The genus is Australia-wide, and occurs from billabongs of the dry inland to subalpine streams.

Atalonella (13 spp.) includes the small representatives of the family, with clear wings in the adults and uniformly grey to almost black wings in the subimagos; there is a slight wing pattern in one large species from Kosciusko. There is marked sexual dimorphism in wing venation, the costal cross-veins being well developed in the female, but lacking in the male except at the pterostigma. The individuals are abundant, and the species occur in running waters in Tasmania and the whole of eastern Australia. The genus was described from South America, and one species of a closely allied genus occurs in W.A.

The 3 species of *Thraulophlebia* differ from small *Atalonella* mainly in having a distinctive angular projection on the anterior margin of the hind wing. They are widespread, but not generally collected because of their small size.

The 5 species of *Atalomicria* differ greatly in size. Their wing venation has a characteristic delicacy, the cross-veins being openly spaced and often arranged in several gradate series; the costal and subcostal spaces are pigmented. The subimago has a grey infuscation outlining the venation. The nymphs occur only in cool mountain streams bordered by dense vegetation. They have conspicuous, greatly elongate maxillary palpi (Fig. 12.3A). The species are found mostly in Queensland, but extend to southern N.S.W.

Atalophlebioides (10 spp.) includes some of the most common Australian mayflies. It occurs in the eastern States, Tasmania, and New Zealand. Adults form dense swarms during the summer, flying in the hottest parts of the day. All are of medium to small size, with clear wings and rather flesh-coloured bodies. The wings are broader at the base, less rounded at apex, and somewhat larger than in *Atalonella*. The subimagos are uniformly grey. The flattened nymphs (Fig. 12.7D) occur only in swiftly flowing water.

Of the 7 species of *Jappa*, the large species of the *strigata* group are widespread from central Queensland to southern N.S.W., where the nymphs (Fig. 12.7B) with large frontal horns are found burrowing below rocks in the slower-flowing streams. The nymphs of other species, all of which occur in the eastern States, are without large frontal horns.

Kirrara contains 2 large, handsome species with red-brown costal and subcostal spaces. The wing-span is similar to that of the Siphlonurinae, but the hind wing is quite reduced. The nymphs (Fig. 12.7A) are found only in the swiftest, cold waters of the Kosciusko plateau, New England area, and south-eastern Queensland.

There are a number of undescribed genera, at least two of which are similar to South American genera (Demoulin, 1955).

4. Ephemerellidae. The family has an almost world-wide distribution. The adults have small hind wings and 3 caudal filaments. The fore wing (Fig. 12.4H) has 1 or 2 long intercalary veins between MP and CuA, and there are detached marginal intercalaries between most of the veins. The males have large, rather indistinctly divided eyes that are almost or quite holoptic. The claspers are 3-segmented, with a very long 2nd segment. The nymphs (Fig. 12.9), which are often strikingly and cryptically coloured, require somewhat rapid, clear streams that are cool throughout the year, or else small clear lakes. The cuticle is more heavily sclerotized than in most other mayfly nymphs. One species of TELOGANODINAE has been found in south-eastern Queensland (Riek, 1963), the sub-family being known from South Africa, through India, Ceylon, and southern China to Australia.

ACKNOWLEDGMENTS. I am grateful to Professor George F. Edmunds Jr and Dr William L. Peters, University of Utah, U.S.A., for helpful discussions of the manuscript.

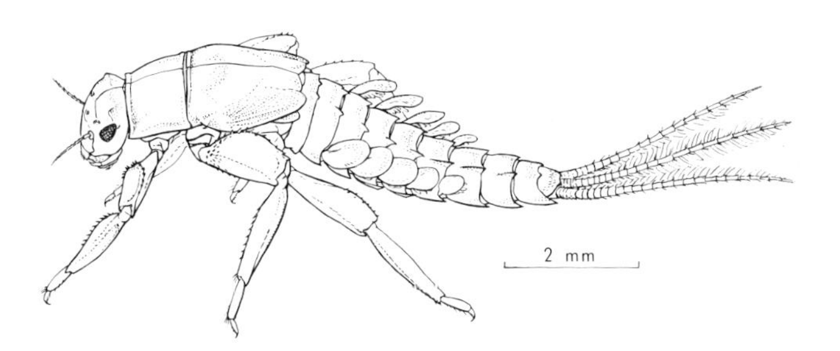

Fig. 12.9. Nymph of *Austremerella picta*, Ephemerellidae. [M. Quick]

13

ODONATA

(Dragonflies and damselflies)

by A. F. O'FARRELL

Predacious Palaeoptera with two equal or subequal pairs of wings; complex accessory genitalia developed from abdominal sterna 2 and 3 of male. Nymphs aquatic, having elongate prehensile labium modified for seizing prey, and respiring by tracheal gills developed either as external caudal appendages or from internal folds of the rectal wall.

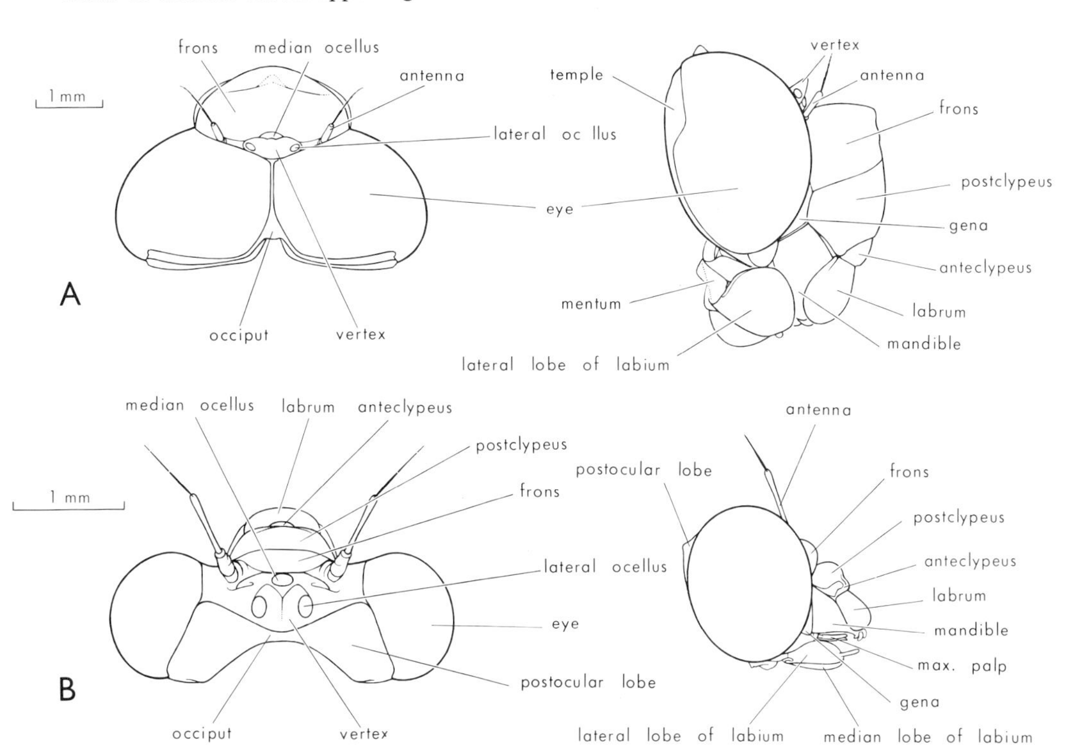

Fig. 13.1. Dorsal and lateral views of heads of ♂♂: A, *Acanthaeschna parvistigma*, Anisoptera-Aeshnidae; B, *Synlestes tillyardi*, Zygoptera-Chlorolestidae. [F. Nanninga]

This ancient and unique order includes some of the world's most spectacular insects. Only about 5 per cent of the known species are Australian, but these include some remarkable forms. Adult Odonata usually have more or less elongate, slender bodies and highly developed powers of flight, hunting by sight and seizing their prey (flying insects) on the wing. Mostly between 30 and 90 mm long, they include some huge forms exceeding 150 mm and small fragile species of less than 20 mm. Almost always aquatic, the nymphs lie in wait for, or sometimes actively stalk, small animals, which are seized by shooting out the long, prehensile labium in which the palpi are modified for grasping.

Recent general accounts of the order include those of Fraser (1957) on systematics and Corbet (1962) on biology. For the Australian species, see Fraser (1960), J. Watson (1962), and, for important data on distribution, Lieftinck (1949, 1951).

Anatomy of Adult

Head (Fig. 13.1). Large, concave behind, on a flexible, slender neck. Occipital region in female Anisoptera sometimes modified to articulate with male anal appendages in pairing. Region behind eyes in both sexes sometimes enlarged, forming postocular lobes or 'temples'. Compound eyes conspicuous as lateral swellings on transversely elongate head of Zygoptera, or largely covering more or less spherical head of other suborders. Three ocelli always present; vertex sometimes reduced to a small tubercle in front of contiguous compound eyes. Antennae minute; scape relatively large, pedicel variable, flagellum thin, at most 5-segmented. Frons prominent, projecting forwards, then downwards, in front of eyes. Clypeus large, divided transversely into a larger postclypeus *(nasus)* and a smaller anteclypeus *(rhinarium)*. Labrum conspicuous, sloping obliquely forwards in Zygoptera, but vertical in other suborders; genae small; large areas of mandibular bases exposed. Mouth-parts (Fig. 13.2) of modified biting type, adapted for predation; wide gape, strongly toothed mandibles, spined maxillae with unsegmented palpi.

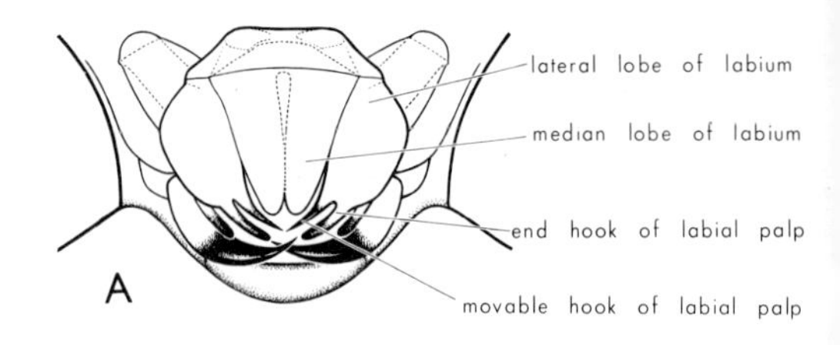

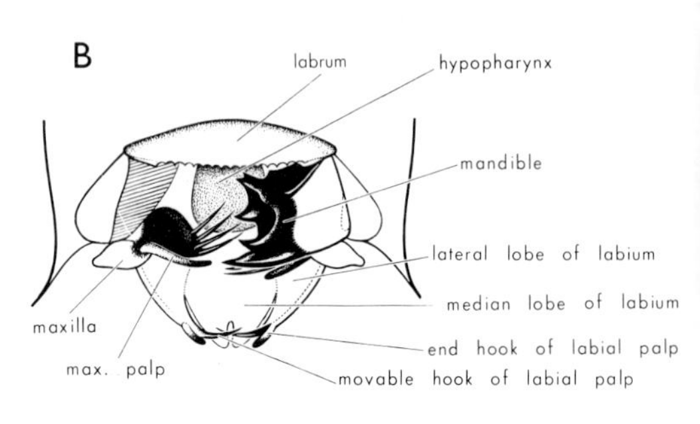

Fig. 13.2. Mouth-parts of ♂ *Synlestes tillyardi*, Zygoptera-Chlorolestidae: A, ventral view; B, anterior view, labrum and right mandible removed. [F. Nanninga]

Labial palpi modified into large lateral lobes, each bearing near its apex a movable hook and a spine or end-hook.

Thorax (Fig. 13.3). Prothorax small, transversely elongate, mobile on the large, rigid pterothorax. Pronotum (modified in female Zygoptera to articulate with male appendages in pairing) often complex in shape, sometimes with prominent posterior lobe. Propleura small, sutures obscure; mesothoracic spiracles just behind them, on anterolateral borders of mesothorax. Meso- and metathorax fused into a large rigid pterothorax ('synthorax'), strongly convex in profile, with legs placed far forward and wings set well back. Sterna and nota small; region between wing-bases sloping obliquely down and backwards. Pleura huge; mesepisterna enlarged anterodorsally, meeting in mid-dorsal line along dorsal carina to form extensive 'shoulders' in front of prealar ridges. Mesopleural (humeral) and metapleural (second lateral) sutures usually conspicuous. Interpleural (first lateral) suture, between mesepimeron and metepisternum,

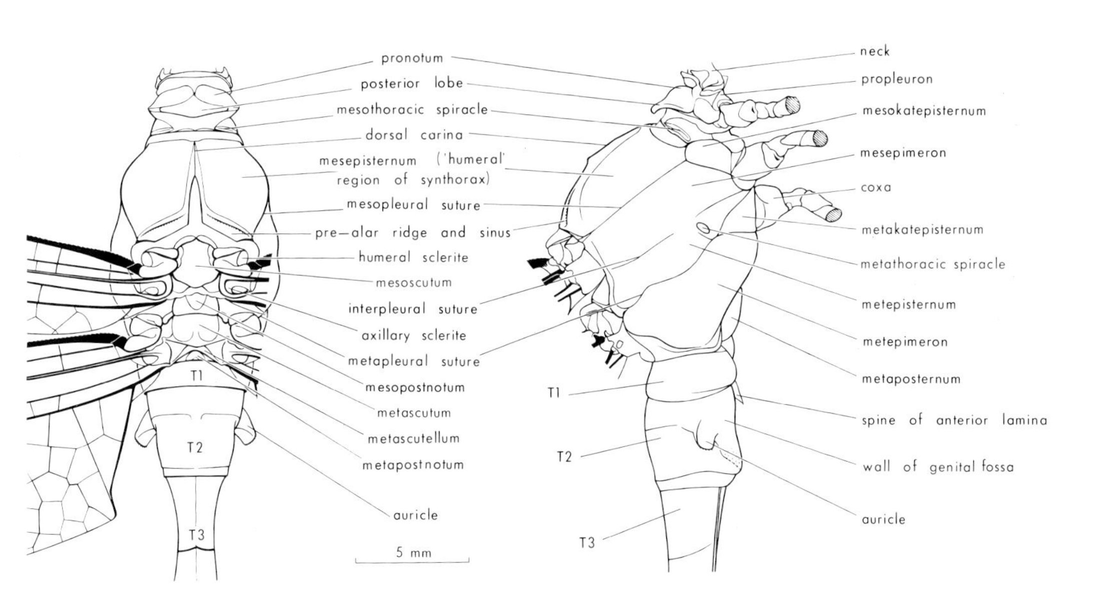

Fig. 13.3. Dorsal and lateral views of thorax and anterior abdominal segments of ♂ *Acanthaeschna parvistigma*, Anisoptera-Aeshnidae. [F. Nanninga]

sometimes obscure at dorsal end, but distinct below, with metathoracic spiracle just behind it. Katepisterna border coxae laterally below episterna.

Legs. Rather short, weak, adapted for seizing and holding prey or clinging and scrambling upon a suitable perch, but not for walking. Trochanters subdivided near base; femora strong, sometimes spiny; tibiae slender, no spurs, but usually one or more rows of comb-like spines (fused in male Corduliidae and Synthemidae into tibial keels); tarsi 3-segmented, longest segment distal, with a pair of claws, each usually armed with a tooth (claw-hook) about half-way along its length.

Wings. Humeral and axillary plates at the base of C and R+M articulate with notal sclerites of the pterothorax (Fig. 1.17A). The wings are membranous, usually hyaline, occasionally with large areas of pigment or structural colour, rarely opaque waxy patches; often with basal yellow pigment ('saffroning'). Veins and pterostigma usually black, brown, or yellowish, sometimes brightly coloured. Nearly all Anisoptera and some Zygoptera rest with wings spread horizontally or somewhat depressed; others hold them vertically side by side above the back, or in some intermediate position. The wings are not coupled, though their movements are clearly co-ordinated in flight.

Venation (Fig. 13.4) complex, with much secondary reticulation (there may be over 3,000 cells in a single wing), and subject to various interpretations. The modified Tillyard–Fraser system adopted here gives weight to fossil evidence, and is more readily comparable with other insects than the Needham notation, which was based mainly on nymphal tracheation (Table 13.1). The convex or concave position of the long veins is useful in identifying them, and is indicated by + or − signs in Figure 13.4 and the following description.

The main stems are strongly pleated at the base of the wing: C (+) on anterior margin; Sc (−); a strong R+M (+); CuP (−,

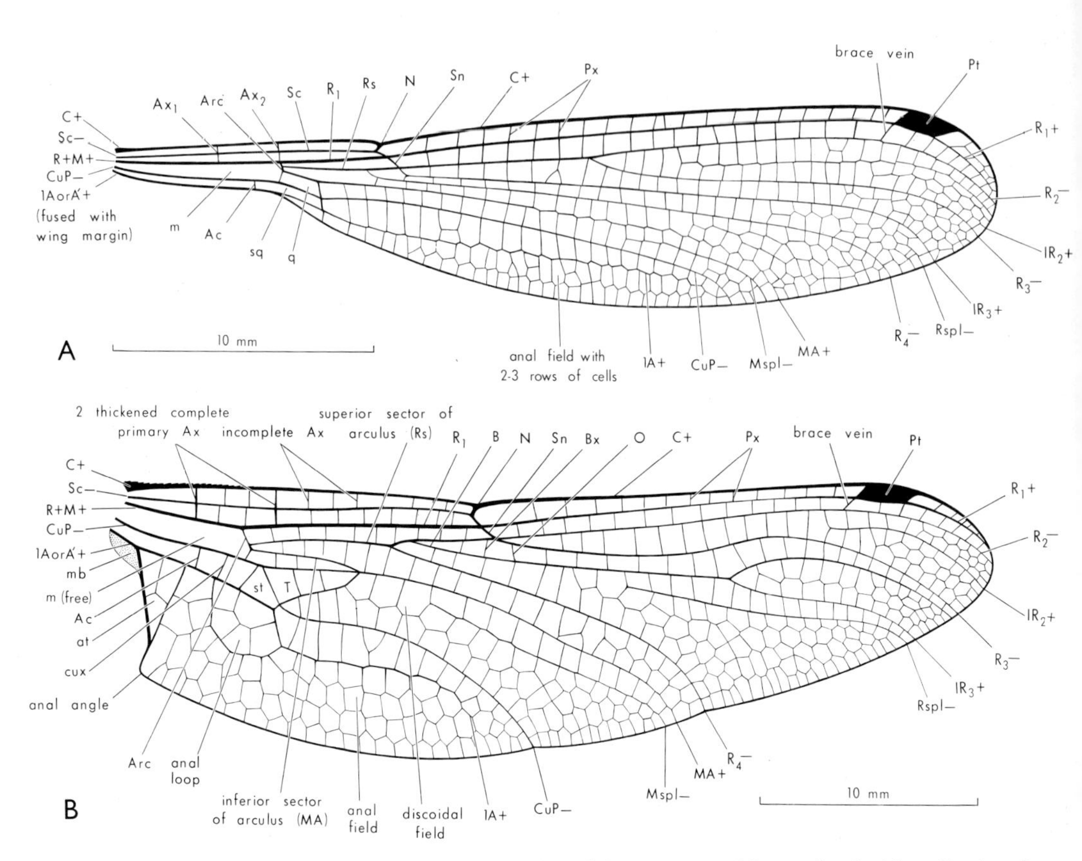

Fig. 13.4. Hind wings of ♂♂: A, *Argiolestes icteromelas nobilis*, Zygoptera-Megapodagrionidae; B, *Acanthaeschna parvistigma*, Anisoptera-Aeshnidae. [F. Nanninga]

usually termed Cu_2); and 1A (+), on posterior margin of petiolate portion in Zygoptera (except *Hemiphlebia* and a few others), separate from the margin in Anisoptera. MP and CuA are regarded as absent. Sc joins C at the *nodus* (N). R_1 (+), continuous with R+M, runs parallel to C, and reaches margin well beyond pterostigma (Pt). Rs+MA runs back obliquely to meet a cross-vein coming obliquely forwards from CuP, thus forming a distally convex bow, the *arculus* (Arc). Rs (−) and MA (+) ('sectors of Arc') may originate from Arc separately (Fig. 13.4) or by a common stem (Figs. 13.21, 23). R_2 (−) parallel to R_1, is joined across R_1 to N by thick oblique *subnodus* (Sn), and often joined to inner corner of Pt by 'brace' vein. R_3 (−) and R_4 (−) arise pectinately from R_2, and sweep to margin behind apex. IR_2 (+) and IR_3 (+) are intercalary veins behind R_2 and R_3 respectively. Rspl (−) is a secondary intercalary ('supplement', unrepresented by any nymphal trachea) between IR_3 and R_4. IR_3 is a strong vein; its basal part, up to oblique vein 0 (Fig. 13.4B), being a proximal extension of the original nymphal trachea in Anisoptera, is called the *bridge* (B).

MA (+) and CuP (−), unbranched, with the supplement MAspl (−) between them, enclose the discoidal field (d), which ends proximally in the *discoidal cell* (dc) immediately distal to the posterior segment of Arc. This cell is never divided longitudinally in Zygoptera, in which it is the *quadrilateral* (q), sometimes having a *subquadrilateral* (sq) lying behind it between CuP and 1A (Fig. 13.19). In Anisoptera, CuP bends sharply backward at the base of dc, which is divided by a longitudinal cross-vein into the characteristic *triangle* (T) and *supra-* or *hypertriangle* (ht, Figs. 13.20-23). Especially in primitive forms, T may appear four-sided, with anterior margin bent (Plate 1, B, D). If a cross-vein runs from CuP to 1A *at or distal to* level of Arc, it cuts off a *subtriangle* (st or ti, Figs. 13.20, 21) proximal to T.

In Zygoptera, except in *Lestoidea* and most Protoneuridae (Figs. 13.14, 15), in which it is greatly reduced, 1A leaves wing margin usually near distal end of petiole, and encloses a narrow anal field (a). The *anal crossing* (Ac) runs from the stem of CuP to 1A, often proximal to the departure of the latter from wing margin (Figs. 13.18, 19). In Anisoptera, portion of 1A proximal to Ac is sometimes designated 'A″', and the vein joins or closely approaches CuP at posterior corner of T. From here, in most hind wings and some fore wings, its conspicuous anterior branch, sometimes designated 'AA', runs roughly parallel to CuP. The anal field, usually much more extensive in the hind wing than in fore wing, contains, in all but some primitive forms, an *anal loop* (al, Figs. 13.4B, 21) formed by a convex branch of 1A swerving back just beyond Ac and then forward again to complete the loop. In more advanced Libelluloidea, this becomes an elaborate 'stocking-shaped' formation with a strong midrib (Figs. 13.22, 23). Behind 1A at extreme base of wing, a small opaque *membranule* (mb) may occur in both sexes; and, in males only, a well-defined basal or *anal triangle* (at) may be present (Figs. 13.4B, 20, 21).

Some cross-veins are of taxonomic importance. *Antenodals (Ax)* join C to R_1 ('complete') or to Sc only ('incomplete') proximal to N; *postnodals (Px)* join C to R_1 beyond N. Number and arrangement of cross-veins *Bx* in space between R_3, B, and 0; *mx* in median space (m) between R+M and stem of CuP; *cux* in cubital space (cu) behind stem of CuP; and, if present, in at, st, T, or ht, may be important. So, too, may be degree of alignment of cross-veins of different series, presence of extra intercalaries at wing margin, and number and arrangement of cells in (e.g.) discoidal or anal fields. A space not traversed by any cross-vein is called 'free' as distinct from 'crossed'.

TABLE 13.1

Alternative Terms for Some Features of the Odonate Wing

Tillyard–Fraser notation	Needham notation	de Selys and 19th century workers
R_1	R_1	Median nervure
R_2	M_1	Principal sector
IR_2	M_{1a}	Postnodal sector
R_3	M_2	Nodal sector
IR_3	Rs	Subnodal sector
R_4	M_3	Median sector
MA	M_4	Lower sector of arculus
CuP*	Cu and Cu_1	Submedian nervure and superior sector of triangle
1A	Cu_2	Inferior sector of triangle
A	—	Postcostal nervule

* Termed Cu_2 by Tillyard and Fraser.

Abdomen. Usually elongate, cylindrical, often constricted at one or more points, tapered or dorsoventrally flattened. Ten complete, flexibly articulated segments, 1–8 bearing spiracles. Terga strongly sclerotized, arched to embrace flat, narrow sterna; pleura narrow, membranous. Segments 1–2 and 8–10 or 9–10 usually shorter than rest. Males with sterna 2–3 modified to form elaborate secondary genitalia, not homologous with paired appendages; and, in some Anisoptera, with ventrolateral outgrowths of

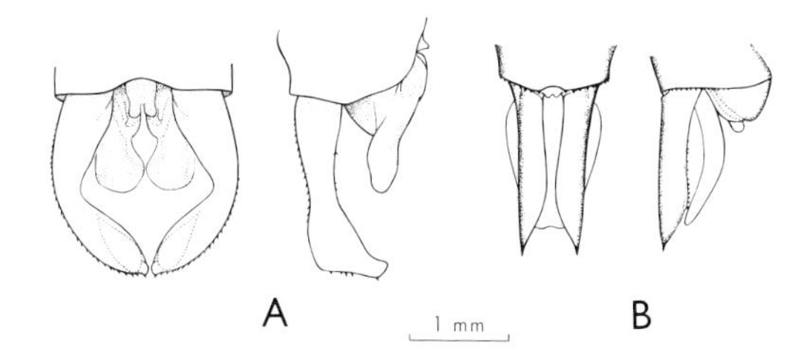

Fig. 13.5. Dorsal and lateral views of anal appendages of ♂♂: A, *Diphlebia nymphoides*, Zygoptera-Amphipterygidae; B, *Diplacodes bipunctata*, Anisoptera-Libellulidae. [F. Nanninga]

tergum 2 (*auricles* or oreillets, Figs. 13.3, 8). Segment 10 in both sexes with paired, unsegmented, superior anal appendages, probably cerci (homology disputable). Paired inferior anal appendages also present in male Zygoptera; in males of other suborders, an unpaired median inferior appendage (Fig. 13.5). The latter is sometimes so deeply cleft as to appear double (e.g. in some Gomphidae).

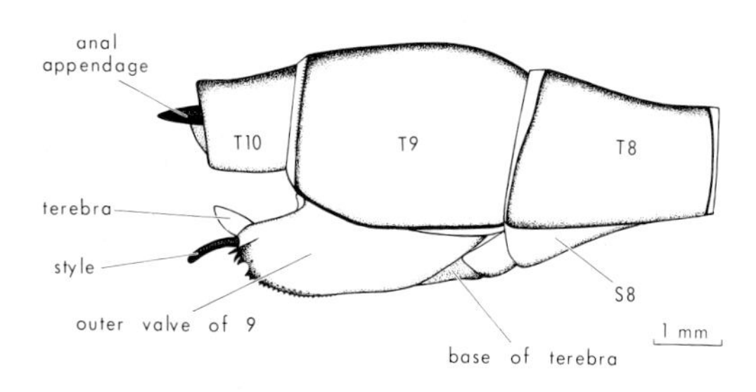

Fig. 13.6. Terminalia of ♀ *Synlestes tillyardi*, Zygoptera-Chlorolestidae, lateral. [F. Nanninga]

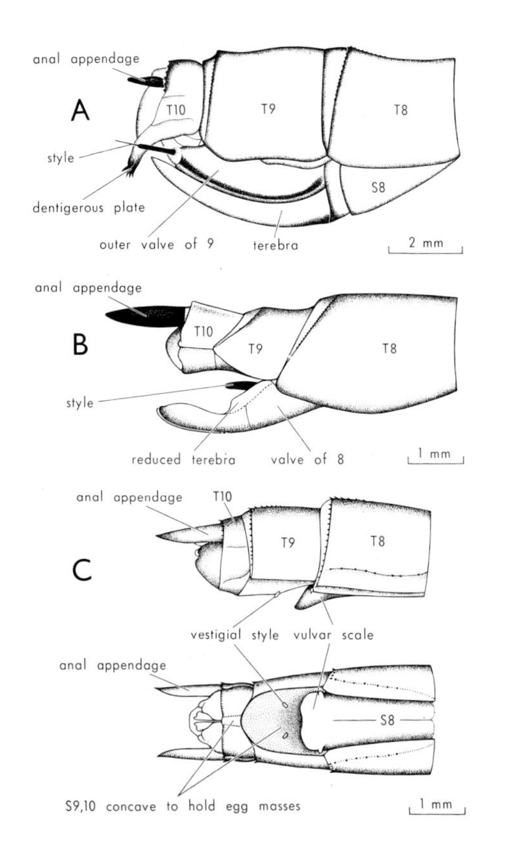

Fig. 13.7 Terminalia of ♀♀ Anisoptera: A, *Acanthaeschna unicornis*, Aeshnidae, lateral; B, *Synthemis eustalacta*, Synthemidae, lateral; C, *Diplacodes bipunctata*, Libellulidae, lateral and ventral. [F. Nanninga]

Female Genitalia (Figs. 13.6, 7). Gonopore behind sternum 8. Ovipositor complete in Zygoptera, Anisozygoptera, Aeshnidae, and Petaluridae, in which gonapophyses of segments 8–9 form a cutting, piercing, or sawing *terebra* ensheathed by the valves of 9, which bear a sometimes segmented sensory apical style and may have cutting edges or teeth. Ovipositor in other Anisoptera often reduced to a small, sometimes bivalved, vulvar scale; abdominal sterna 9–10 sometimes excavated to hold egg-masses extruded from the gonopore. More rarely, ovipositor only partially reduced, or functionally replaced by secondary structures.

Male Genitalia (Fig. 13.8). No external genitalia on segment 9, where gonopore opens. Forward ventral flexure of abdomen transfers semen to a reservoir *(penis vesicle)* on anterior end of sternum 3. Sternum 2 much modified to form *genital fossa*, within which lies the 'penis', and on either side of it clasping organs used in copulation to hold and guide female genitalia; these bear two pairs of processes *(hamules)* but, except in Aeshnidae, one pair is reduced. 'Penis' in Anisoptera 3-segmented, with orifice on its convex surface, folded with apex ventral, and partly covered anteriorly by backwardly directed *ligula* (penis sheath); points somewhat anteriorly when erected. Unsegmented, but otherwise similar, in Anisozygoptera. 'Penis' in Zygoptera probably homologous with anisopteran ligula; without an orifice and directed posteriorly, the tip usually concealed in the large penis vesicle.

Coloration. Cuticular pigments usually black, brown, or yellowish. Epidermal pigments (cream, yellow, green, blue, red) show through light or unpigmented cuticle, often in patterns of spots, stripes, or bands; metallic effects sometimes produced by structural coloration of cuticular surface; or surface covered by waxy, whitish to powder-

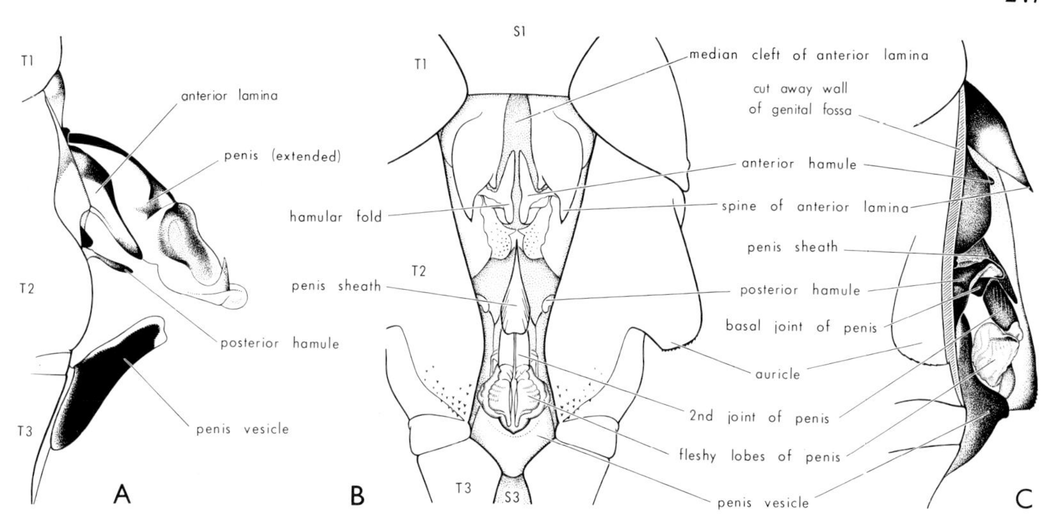

Fig. 13.8. Male accessory genitalia: A, *Synlestes tillyardi*, Zygoptera-Chlorolestidae, lateral: B, *Acanthaeschna parvistigma*, Anisoptera-Aeshnidae, ventral; C, *A. parvistigma*, from right side with right wall of genital fossa cut away. [F. Nanninga]

blue secretion (pruinescence). Short-term 'physiological' colour change from rapid responses of epidermal pigments to temperature has recently been demonstrated in some Australian species (O'Farrell, 1964). Females often resemble immature males in colour, and are usually less brilliant than mature males. Especially in Coenagrionidae, there may be several female colour morphs, usually including 'andromorph' like mature male and one or more 'heteromorphs' unlike male. Coloration fades in dried specimens, limiting its taxonomic value.

Internal Anatomy. Alimentary canal long and straight; salivary glands small; long narrow oesophagus extends into base of abdomen, where it dilates into a crop; gizzard small and weak in adults, but with complex internal folding and dentition (of taxonomic importance) in nymphs; mid gut long, simple, without caeca; hind gut short, rectum with 6 longitudinal papillae in adults, but only 3 in nymphs. Malpighian tubes 50–70, usually arranged in groups of 5 or 6. Central and sympathetic nervous systems and retrocerebral complex well developed; brain large, optic lobes very large and complex; 3 thoracic ganglia widely separated; 1st abdominal ganglion almost fused on to metathoracic ganglion, remaining 7 located anteriorly in each of abdominal segments 2 to 8; a ventral blood sinus is associated with the nerve-cord in adults, but apparently not in nymphs. Respiratory system with 3 main pairs of longitudinal tracheal trunks, opening in adults by 10 pairs of spiracles, but closed in nymphs, which respire by tracheal gills. The nymphal mesostigma, however, is well-developed and is often used in aerial respiration long before metamorphosis. Enormous air-sacs are developed in the pterothorax and dorsal abdomen of adults of many families. The dorsal vessel and pericardial membrane often lie beneath the dorsal abdominal air-sacs, which largely conceal them from above.

Gonads elongate in both sexes, often extending when mature through abdominal segments 2–7 (ovary) or 4–8 (testis). Ovarioles panoistic, numerous; oviducts very short, leading into large spermathecal pouch, which also receives duct from accessory glands. Testes with numerous follicles, each producing a radially arranged, rounded, sticky sperm-mass adapted for easy transfer from gonopore to secondary genitalia. Sperm masses awaiting transfer are stored in a sperm-sac, formed by large dorsal dilation of the common sperm-duct into which the short

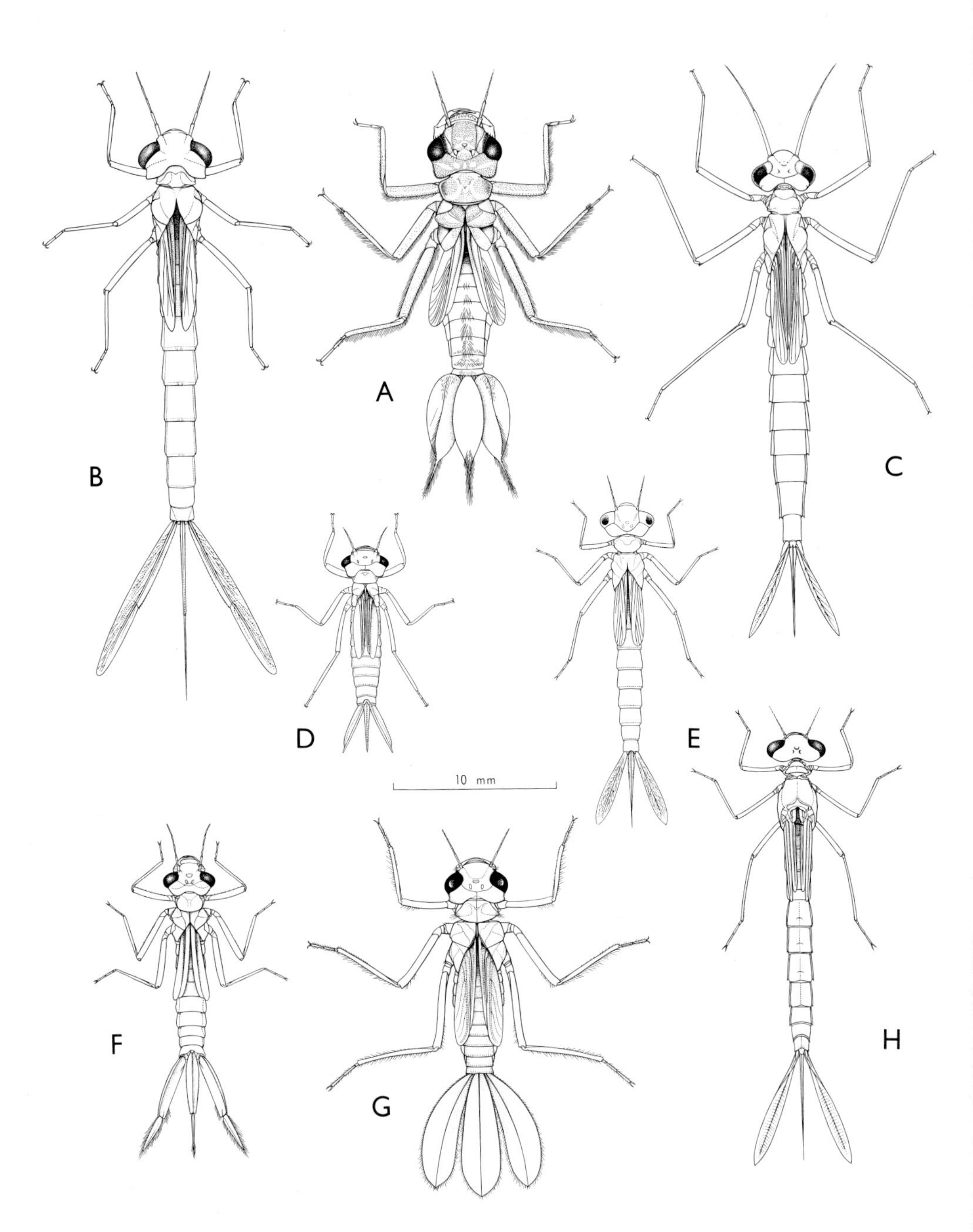

Fig. 13.9. Nymphs of Zygoptera: A, *Diphlebia euphaeoides euphaeoides*, Amphipterygidae; B, *Caliagrion billinghursti*, Coenagrionidae; C, *Synlestes tillyardi*, Chlorolestidae; D, *Nososticta solida*, Protoneuridae; E, *Ischnura heterosticta*, Coenagrionidae; F, *Austrosticta fieldi*, Protoneuridae; G, *Argiolestes icteromelas icteromelas*, Megapodagrionidae; H, *Austrolestes annulosus*, Lestidae. [F. Nanninga]

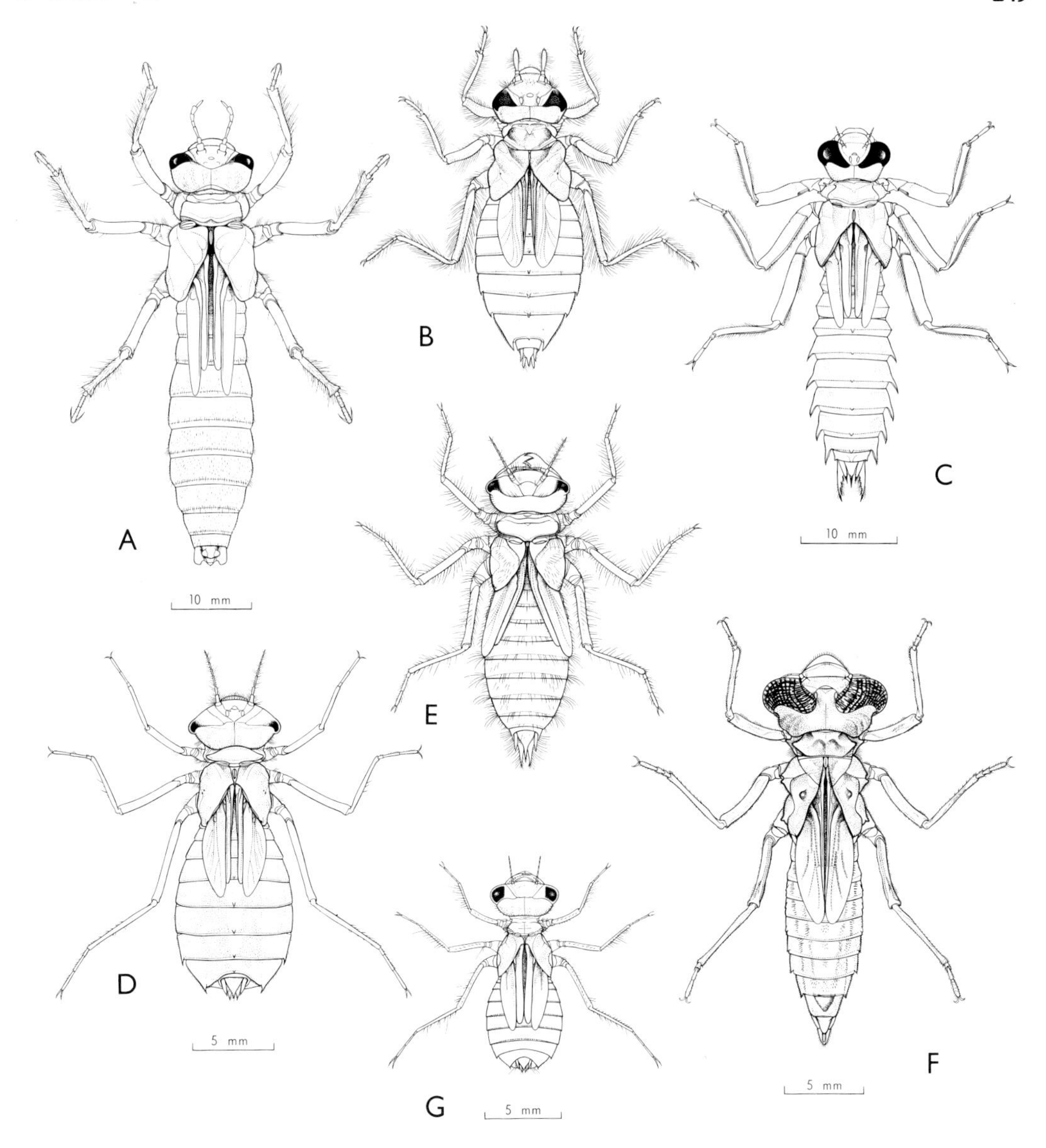

Fig. 13.10. Nymphs of Anisoptera: A, *Petalura hesperia*, Petaluridae; B, *Austrogomphus guerini*, Gomphidae; C, *Notoaeschna sagittata*, Aeshnidae; D, *Procordulia jacksoniensis*, Corduliidae; E, *Synthemis macrostigma occidentalis*, Synthemidae; F, *Telephlebia brevicauda*, Aeshnidae; G, *Diplacodes bipunctata*, Libellulidae. [F. Nanninga]

vasa deferentia discharge. Filling of the sperm-sac may coincide with completion of development of mature coloration.

Immature Stages

Egg. Initially creamy white, usually turning red-brown within 24 hours after fertilization. Elongate, smooth 'endophytic' eggs are inserted in plant tissues; ovoid or spherical 'exophytic' type, often with sculptured chorion and gelatinous investments or appendages for anchorage, are freely scattered into water. Embryonic development can be regulated until after germ-band forms; the embryonic envelopes arise by invagination of the germ-band into the yolk. The unsegmented embryonic cuticle ensheathing the folded appendages of the newly-hatched 'pronymph' is sometimes retained for several minutes or longer after eclosion from the egg.

Nymph. Shorter and stouter than adult, usually with relatively larger antennae and smaller eyes which are always well separated (Figs. 13.9, 10). Mouth-parts like those of adult, but labium (Fig. 13.11) elaborated as extensible grasping organ; labial palpi adapted for seizing prey. Hinged between pre- and postmentum, labium (or 'mask') largely covers face anteroventrally. Rapidly extended by local increase in blood pressure caused by sharp contraction of a diaphragm in abdominal segments 4–5, it is returned to the normal, ventrally folded position by retractor muscles. Prothorax larger and pterothorax less convex than in adult; external wing rudiments extending backwards to cover first few abdominal segments, and reversed so that hind wings overlie the fore wings; identification may be difficult before they appear. Legs normally placed, adapted for walking, clinging, or burrowing.

In Zygoptera, abdomen bears 3 (rarely only 2) large caudal tracheal gills, which may be used in swimming by lashing abdomen from side to side. They are usually plate-like (lamellate), but sometimes saccoid (Fig. 13.9A) in form; of taxonomic value provided damaged or regenerated gills are recognized as such. In other suborders, an elaborate

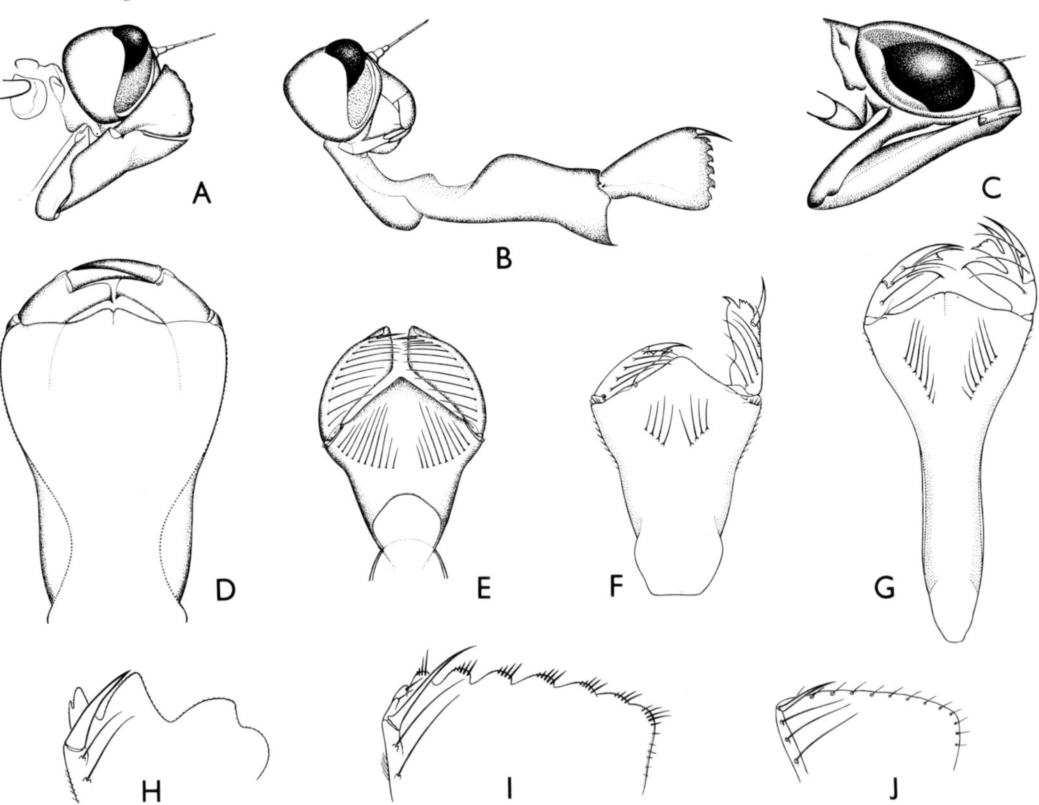

Fig. 13.11. Labia of nymphs: A, B, lateral views of head of *Synthemis eustalacta*, Synthemidae, showing labium retracted and extended; C, lateral view of head, with labium retracted, of *Aeshna brevistyla*, Aeshnidae; D, dorsal view of labium of *A. brevistyla*; E, same of *Diplacodes bipunctata*, Libellulidae; F, same of *Ischnura heterosticta*, Coenagrionidae; G, same of *Austrolestes analis*, Lestidae; H, distal border of labial palp of *Synthemis macrostigma orientalis*, Synthemidae; I, same of *Hemicordulia tau*, Corduliidae; J, same of *Diplacodes bipunctata*. [F. Nanninga]

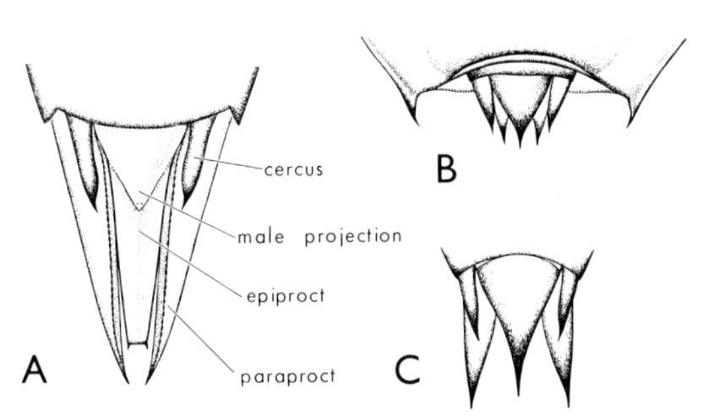

Fig. 13.12. Anal pyramids of nymphs of Anisoptera, dorsal: A, *Aeshna brevistyla*, Aeshnidae, ♂; B, *Procordulia jacksoniensis*, Corduliidae; C, *Orthetrum caledonicum*, Libellulidae.

[F. Nanninga]

tracheal meshwork (branchial basket) in the rectal wall replaces external gills. Intake and expulsion of water through valves guarding anus serves both for respiration and rapid 'jet-propulsion' swimming. Anal pyramid (Fig. 13.12) forms a spiny armature round anus; it is sometimes used as offensive weapon. There are 5 components: unpaired dorsal epiproct ('appendix dorsalis'), paired ventro-lateral paraprocts ('cerci' of some authors), and paired dorso-lateral cerci ('cercoids') which are absent in very young nymphs.

Female nymphs show external rudiments of the ovipositor unless adult has no ovipositor, and males rudiments of genitalia on abdominal segments 2–3. In Anisoptera, 'male projection' above epiproct is rudiment of male inferior anal appendage (Fig. 13.12A). Nymphal coloration is usually dull; mainly cuticular pigments, sometimes brightened by underlying brighter hues, especially greens. Pattern is often distinctive, sometimes changing progressively during development, or sometimes changing at an ecdysis in response to environmental conditions during period immediately before moulting. Shape and proportions of body and its appendages, and of dorsal or lateral abdominal hooks or spines, or specialized local areas of spines or setae, are among useful taxonomic features.

Biology

Adults. Of exceptional ethological interest, Odonata are sometimes referred to as 'bird-watchers' insects'. The adults often have complex patterns of territorial, sexual, and other behaviour, very largely based on visual stimuli, and the methods of the ornithologist rather than the entomologist are required to study these. Patience and a good pair of field-glasses are the essential items of equipment.

Typically diurnal, but sometimes crepuscular, even nocturnal, they fly in temperate Australia mainly from September to March, although in warmer areas some fly all the year round. Others are on the wing for only a few weeks. Onset of tropical rains, dry season, or winter cold may determine flight season. Little is known of causes or extent of occasional large-scale mass migrations over long distances. Metamorphosis usually occurs at night, or about dawn. The 'maiden' flight, from a few to 1,500 metres in length, is directed away from the breeding-place. Maturation of gonads, and of adult colouring may take days or weeks. Immature ('teneral') adults tend to avoid water; their cuticle is soft and the wings have a glassy sheen.

Both sexes in most Zygoptera congregate when mature, often in vast numbers, at suitable breeding-places, hovering or skimming over or near the water; males (e.g. *Diphlebia*) may occupy territories, but this is rarely obvious. In Anisoptera, most females and some males only visit breeding-places occasionally to mate and oviposit; most males spend much time over or near water, with territorial behaviour based on visual cues. Each male occupies a restricted area, where he attacks and drives off intruders of his own or other species; copulation usually follows attack on a 'willing' female, 'reluctant' females having characteristic responses. Smaller species seldom defeat larger ones in territorial combat. Most Aeshnidae and Corduliidae hawk and hover over elongate territories, but many Gomphidae and Libellulidae cover circular areas by short darting flights from a selected, centrally placed perch.

Reproduction. Both sexes may mate several times in one day. The accessory male genitalia are charged with semen either before

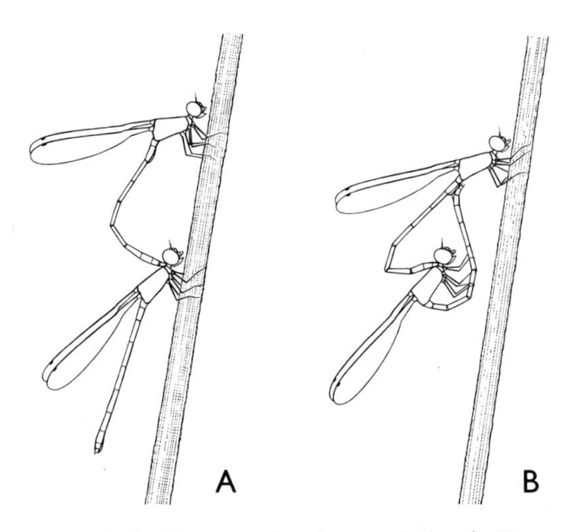

Fig. 13.13. Diagrams showing copulation in Zygoptera: A, tandem position; B, copulation. [F. Nanninga]

or after seizure of the female occiput (Anisoptera) or prothorax (Zygoptera) by the male anal appendages. The partners then bring their genitalia in contact by flexing their abdomens (Fig. 13.13). Visual courtship displays may precede mating, but tactile and chemical stimuli involved are little known. Prolonged flight in the 'tandem position' without genital contact may precede and follow copulation, which may occur in flight or at rest. The male often escorts the ovipositing female; he may hold her in tandem even during underwater oviposition. Aggressive species sometimes attempt interspecific or homosexual matings.

In endophytic species, eggs are inserted into tissues of submerged or emergent aquatic plants, terrestrial plants on the bank, or (notably in some Aeshnidae) into soil above water but below flood level; such eggs are best obtained by collecting material in which oviposition has been seen to occur. In exophytic species, without a functional ovipositor, eggs are extruded individually or in adhesive strings or masses from the gonopore, and often washed off by the female sharply flicking the water with the tip of her abdomen. They are easily collected in a vial of water. Smooth surfaces resembling water may stimulate exophytic oviposition—e.g. *Diplacodes* may attempt to oviposit over smooth bitumen roads.

Immature Stages. Although particular species have specialized habitats, nymphs of Odonata occur in most environments in fresh and even brackish water, including waterfalls, torrents, permanent and intermittent streams, lakes, ponds, temporary rain pools, swamps, bogs, and even estuaries. Some cling to submerged plants, often preferring a particular zone; others inhabit debris and detritus, or burrow into the bottom; torrent dwellers may shelter among debris or be adapted to clinging to bare rocks. Many nymphs tolerate prolonged exposure to humid air, and some species of *Megalagrion* in Hawaii are fully terrestrial. Adequate sampling of a major habitat requires varied collecting methods. Floods may concentrate nymphs into backwaters or overflow pools, where collecting is often good.

The numbers (usually 10–15) and intervals of nymphal ecdyses vary widely between and within species, according to temperature and food supply. 'Opportunist' species which colonize temporary pools, e.g. *Ischnura aurora* (Br.), *Orthetrum caledonicum* (Br.), *Diplacodes* spp., may develop from egg to adult in 8–10 weeks. Some species may take several years, but a range of 6–18 months is more usual. Diapause in eggs, nymphs, and adults is well known in some Odonata, although not yet in Australian species. Nymphal diapause may result in simultaneous emergence of adults in localities where the favourable season is brief.

The full-fed nymph moves towards the edge or surface of the water. Transition from aquatic to aerial respiration follows. The pharate adult then crawls out, anchors itself firmly by its tarsal claws to a suitable support, and the imago emerges. Cast nymphal exuviae can be collected to obtain information about life histories and even estimates of populations.

Natural Enemies. Fish, frogs, birds, and reptiles prey heavily on nymphs and on teneral and ovipositing adults. Often themselves voracious cannibals, nymphs and adults are attacked by predacious arthropods, including aquatic insects, spiders, wasps, mantids, etc. Hymenopterous egg-parasites, ectoparasitic aquatic mites, parasitic worms (including trematodes of aquatic birds), and sundry micro-organisms are also associated with Odonata.

Economic Significance. There is inadequate evidence to assert that Odonata play a major part in 'biological control' of mosquitoes, etc., although considerable nymphal and adult predation occurs. Edible fish, notably trout, some aquatic game birds, and (in parts of Asia) man himself feed extensively on nymphs and adults. The larger nymphs may attack young fish, including trout fry, and some adult Aeshnidae cause damage by preying on hive bees. The value of the order is mainly scientific and aesthetic.

Special Features of the Australian Fauna

Strictly confined to the Australian continent are the curious archaic groups Hemiphlebioidea, Lestoideidae, Chorismagrioninae (Chlorolestidae), and Cordulephyinae (Corduliidae). Extending into Papua and New Caledonia, the Synthemidae are almost certainly Australian in origin; but the Isostictinae (Protoneuridae) of similar distribution are more doubtful. Tropical Calopterygidae and Chlorocyphidae doubtfully extend into northern Australia, and Libellulidae, Coenagrionidae, and Megapodagrionidae include few genera peculiar to Australia. Other families, however, are represented on the continent wholly or largely by endemic genera.

Tasmania has two endemic cold-water genera, *Synthemiopsis* (Synthemidae) and *Archipetalia* (Aeshnidae–Neopetaliinae). Southwestern Australia also has two endemic cold-water genera, *Hesperocordulia* and *Lathrocordulia* (Corduliidae–Gomphomacromiinae). Other cold-water species in both these isolated areas tend to be rather distantly related to their congeners of the south-eastern mainland, whereas the more adaptable species are identical or closely allied to eastern mainland forms. With other distributional evidence, including the occurrence of forms with mainly South American affinities (e.g. the neopetaliine Aeshnidae and the gomphomacromiine Corduliidae), this suggests a dual origin for the Australian fauna. Probably many forms associated with cold permanent waters, found mainly in mountainous country, are relicts of an ancient southern invasion. Others, adaptable to warmer and drier conditions, seem to be of more recent tropical origin, entering Australia from the north; some of these may still be extending their range southwards and inland. The Anisozygoptera, now apparently confined to Japan and the Himalayas, may have been represented in the Triassic of Queensland (p. 173).

CLASSIFICATION

Order ODONATA (248 Australian spp.)

Suborder ZYGOPTERA (92)

COENAGRIONOIDEA (63)
- Platystictidae (0)
- 1. Protoneuridae (13)
- Platycnemididae (0)
- 2. Lestoideidae (1)
- 3. Coenagrionidae (29)
- Pseudostigmatidae (0)
- 4. Megapodagrionidae (20)

LESTINOIDEA (20)
- 5. Lestidae (13)
- 6. Chlorolestidae (7)

HEMIPHLEBIOIDEA (1)
- 7. Hemiphlebiidae (1)

CALOPTERYGOIDEA (8)
- 8. Amphipterygidae (6)
- 9. Chlorocyphidae (1)
- Heliocharitidae (0)
- Polythoridae (0)
- Epallagidae (0)
- 10. Calopterygidae (1)

Suborder ANISOPTERA (156)

AESHNOIDEA (64)
- 11. Gomphidae (22)
- 12. Petaluridae (4)
- 13. Aeshnidae (38)
- Cordulegasteridae (0)

LIBELLULOIDEA (92)
- 14. Synthemidae (20)
- 15. Corduliidae (25)
- 16. Libellulidae (47)

Suborder ANISOZYGOPTERA (0)

Epiophlebiidae (0)

KEY TO THE SUBORDERS OF ODONATA

Fore and hind wings similar in shape and venation; discoidal cell quadrangular, never longitudinally divided; eyes far apart; ♂ inferior appendages paired. Nymphs usually slender, with 3 (exceptionally only 2) large caudal gills .... ZYGOPTERA

Fore and hind wings dissimilar in venation and usually in shape; discoidal cell divided longitudinally into triangle and supratriangle; eyes touching or only moderately separated; one median inferior appendage in ♂. Nymphs stout, without caudal gills, but with an anal pyramid; antennae with either 4 or 6–7 segments .... ANISOPTERA

Facies of Anisoptera, but wings zygopteron-like (although discoidal cell differs between fore and hind wing). Nymphs like Anisoptera, but with 5-segmented antennae. [Japan and the Himalayas] .... (ANISOZYGOPTERA)

The subfamilies shown in the keys that follow are not all equivalent in taxonomic status, but represent a convenient division of the order, modified from that of Fraser (1957).

Suborder ZYGOPTERA

Keys to the Families and Subfamilies of Zygoptera Known in Australia

ADULTS

1. Wings with 5 or more *Ax* 2
   Wing with only 2 *Ax* (exceptionally 3) 3
2. *Ax* few, all incomplete except 2 conspicuous thickened primaries in each wing (Fig. 13.19) .... **Amphipterygidae**
   Ax many, but primaries still obvious .... **Chlorocyphidae**
   Ax many, primaries not distinguishable .... **Calopterygidae**
3. *Px* not at all in line with cross-veins behind them; tiny metallic insects .... **Hemiphlebiidae**
   Px mostly aligned with cross-veins behind them 4
4. CuP arching forwards as it leaves the distal end of the quadrilateral (Fig. 13.18) . **Chlorolestidae.** 5
   CuP not arching forwards in such a way 6
5. Quadrilateral in fore wing open basally .... CHORISMAGRIONINAE
   Quadrilateral in fore wing closed basally .... CHLOROLESTINAE
6. 1A reduced: IR_3 and R_4 arising about midway between levels of arculus and subnodus (Fig. 13.15) .... **Lestoideidae**
   1A normal; IR_3 and R_4 arising much nearer to level of arculus than that of subnodus (Fig.13.17) .... **Lestidae.** 7
   1A variable; IR_3 and R_4 arising near level of subnodus 8
7. Quadrilaterals similar in shape in all wings .... LESTINAE
   Quadrilaterals of fore and hind wings dissimilar in shape; wings closed when at rest .... SYMPECMATINAE
8. 1A wholly fused with wing margin (Fig. 13.14) .... **Protoneuridae.** 9
   1A well developed as a separate vein 10
9. CuP short, reaching $\frac{1}{2}$ to $2\frac{1}{2}$ cells beyond quadrilateral; ♂ anal appendages not forceps-like .... PROTONEURINAE
   CuP short as above, or long (3–8 cells beyond quadrilateral); ♂ anal appendages forceps-like .... ISOSTICTINAE
10. Distal wing-margin with supplementary veins reaching inward at least to level of pterostigma (Fig. 13.4A) .... **Megapodagrionidae**
    Supplements of distal wing-margin weak .... **Coenagrionidae.** 11
11. Arculus well distal to level of Ax_2 .... AGRIOCNEMINAE
    Arculus at most slightly distal to level of Ax_2 12
12. Pterostigmas dissimilar in ♂ fore and hind wings; a posterior mid-ventral vulvar spine on abdominal segment 8 of ♀; 1A (except in *Aciagrion*) leaving wing-margin proximal to level of Ac (Fig. 13.16B) .... ISCHNURINAE
    Not having the above combination of characters 13

13. 1A leaving wing-margin proximal to Ac .... COENAGRIONINAE
    1A leaving wing-margin at level of Ac; pterostigmas of normal size (Fig. 13.16A) .... PSEUDAGRIONINAE
    Either 1A leaves wing-margin well distal to Ac *(Teinobasis)*, or pterostigmas are minute *(Archibasis)* .... AMPHICNEMINAE

NYMPHS (families only)

1. Median gill reduced to a mere spine .... **Chlorocyphidae**
   Median gill lamellate, laterals triquetral .... **Calopterygidae**
   Median and lateral gills similar in form 2
2. Gills saccoid or lamellate, with constriction or node much nearer to apex than to base (Fig. 13.9F) .... some **Protoneuridae**
   Gills saccoid, sharply pointed, without a node 3
   Gills lamellate; node, if present, about halfway along 4
3. Gills about half as long as body (Fig. 13.9A) .... **Amphipterygidae**
   Gills much less than half as long as body .... **Lestoideidae**
4. Labium with paraglossae; very small nymphs .... **Hemiphlebiidae**
   Paraglossae absent; various sizes 5
5. Gills broad, rounded, sometimes with long terminal filaments; held horizontally (Fig. 13.9G) .... **Megapodagrionidae**
   Gills narrower, more tapered, without long terminal filaments; held vertically 6
6. Tracheal branches at 90° to long axis of gill (Fig. 13.9H) .... **Lestidae**
   Tracheal branches at acute angle to long axis of gill 7
7. Antennae large; gills small and compact; general form rather elongated (Fig. 13.9C) .... **Chlorolestidae**
   Not as above; gills relatively large 8
8. Labium bearing setae on mentum (Fig. 13.11F) .... **Coenagrionidae**
   Mental setae absent .... some **Protoneuridae**

Superfamily COENAGRIONOIDEA

1. Protoneuridae. Of tropical origin, our species of these slender, stream-frequenting insects occur in the north and east. *Nososticta solida* Selys (Plate 1, I), black and orange, widespread, and the northern species of *Notoneura*, with no orange colour, represent the PROTONEURINAE. Among ISOSTICTINAE, short superior anal appendages in the male distinguish the rare northern genus *Austrosticta* (CuP long) and the more widespread *Isosticta* (CuP short), of which *I. simplex* Martin occurs in the south-east. Long superior appendages occur in *Neosticta canescens* Till. (CuP long), pruinescent, locally common in the south-east, as well as in the very slender northern *Oristicta filicicola* Till. and the rarer, more robust *Phasmosticta* spp. (CuP short). Protoneurid nymphs (Figs. 13.9D, F) are peculiarly difficult to collect, and many are undescribed.

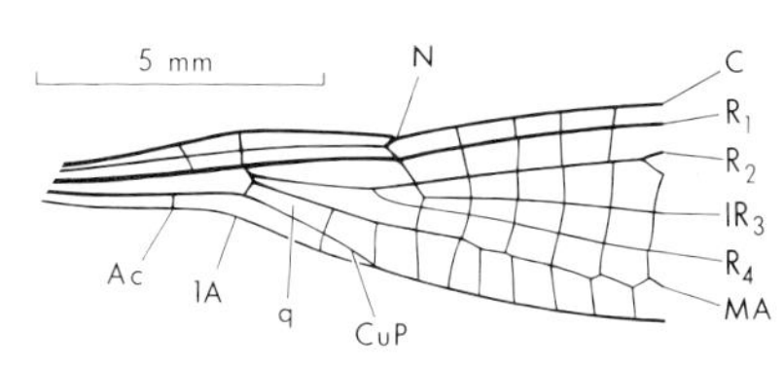

Fig. 13.14. Base of hind wing of *Nososticta solida*, Protoneuridae. [F. Nanninga]

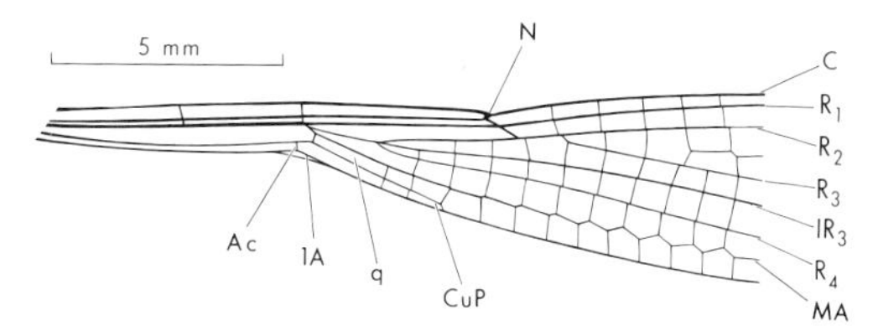

Fig. 13.15. Base of hind wing of *Lestoidea barbarae*, Lestoideidae. [F. Nanninga]

2. Lestoideidae. The genus *Lestoidea*, of north Queensland, combines protoneurid, lestid, megapodagrionid, and even amphipterygid features; Fraser treats it as an aberrant pseudolestid calopterygoid, but it certainly has strong coenagrionoid affinities.*

* A second species, *Lestoidea barbarae*, has been described by J. A. L. Watson (*J. Aust. ent. Soc.* **6**: 77–80, 1967).

3. Coenagrionidae. Adults typically skim low over static waters where nymphs live among aquatic plants. The mainly tropical AGRIOCNEMINAE include some of the world's smallest Odonata in the large genus *Agriocnemis*; the tiny, long-legged *Austrocnemis splendida* (Martin) (Plate 1, J) runs over lily-pads on tropical swamps, while the much larger, reddish *Argiocnemis rubescens* Selys is common on east coast streams. Heteromorphic females occur in this and the related cosmopolitan subfamily ISCHNURINAE, which includes some rarer northern forms and two of our commonest Zygoptera—*Ischnura heterosticta* (Burm.), males black and blue, and the much smaller *I. aurora* (Br.), males largely red. The latter is wind-borne all over S.E. Asia and the S.W. Pacific, occurring on oceanic islands and desert water-holes. The widespread eastern *Coenagrion lyelli* (Till.) and rare *C. brisbanense* (Till.) (COENAGRIONINAE) are our only species in this cosmopolitan group. Our AMPHICNEMINAE are exclusively tropical. The PSEUDAGRIONINAE include the widespread red, black, and blue *Xanthagrion erythroneurum* Selys, which forms vast colonies. *Caliagrion billinghursti* (Martin) is a large south-eastern form with blue and black male, distinguished by the proximal position of Ac. A common reddish northern species is *Ceriagrion aeruginosum* (Br.). *Pseudagrion*, eastern and northern, includes *P. ignifer* Till., dark, pruinescent, on fast streams; *P. aureofrons* Till., with spectacular blue and gold male (Plate 1, K); and others with blue and black males distinguished by the notched tips of the superior anal appendages. *Austroagrion*, probably annectent with Coenagrioninae, has western, northern, and eastern species, all small, with blue and black males.

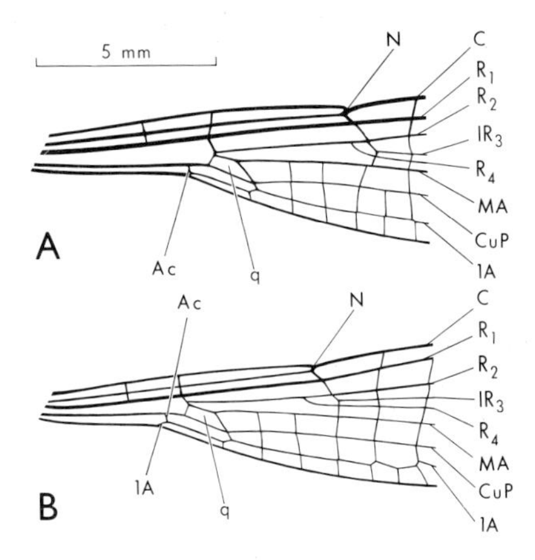

Fig. 13.16. Bases of hind wings of Coenagrionidae: A, *Pseudagrion ignifer*; B, *Ischnura heterosticta*. [F. Nanninga]

4. Megapodagrionidae. Of tropical origin, breeding in bog-holes as well as fast streams, our fauna includes 2 northern species of *Podopteryx*, black and pink, over 90 mm span, and numerous smaller forms (30–70 mm span) in *Argiolestes* (sometimes split into several subgenera). *A. icteromelas* Selys, blackish, and *A. griseus* Selys, pruinescent, both with several 'subspecies' or morphs of uncertain status, are our commonest eastern forms. Some pretty species with red or orange markings occur in rain forests.

Superfamily LESTINOIDEA

5. Lestidae. *Austrolestes* (SYMPECMATINAE, given family rank by Fraser), Australia-wide, includes several pretty bronze or black and blue southern species (Plate 1, H) and some rather duller northern ones, sometimes forming huge colonies. Our only species in the cosmopolitan LESTINAE is the northern *Lestes concinnus* Selys. Lestid nymphs are typically found among vegetation in still water.

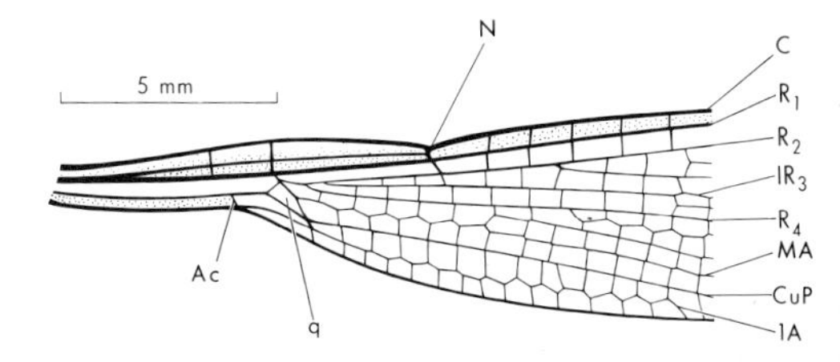

Fig. 13.17. Base of hind wing of *Austrolestes cingulatus*, Lestidae. [F. Nanninga]

6. Chlorolestidae. We have 3 distinctive genera of this mainly tropical family: the unique *Chorismagrion* (CHORISMAGRIONINAE) in north Queensland, and 6 elongate metallic

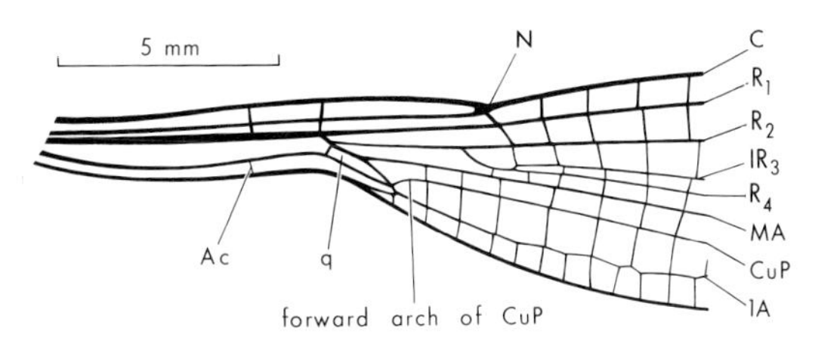

Fig. 13.18. Base of hind wing of *Synlestes tillyardi*, Chlorolestidae. [F. Nanninga]

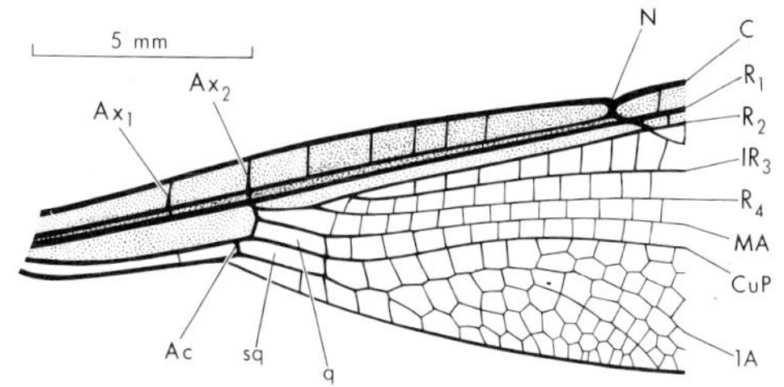

Fig. 13.19. Base of hind wing of *Diphlebia nymphoides*, Amphipterygidae. [F. Nanninga]

eastern species of CHLOROLESTINAE, *Episynlestes albicauda* (Till.) with white anal appendages, and 5 *Synlestes* species, of which *S. tillyardi* Fraser (Plate 1, F), common in N.S.W., seems to have been confused with *S. weyersi* Selys, more plentiful in Victoria. All species breed in running water.

Superfamily HEMIPHLEBIOIDEA

7. Hemiphlebiidae. The unique *Hemiphlebia mirabilis* Selys, with primitive nymphal and adult features, is very local on reedy backwaters in Victoria, possibly also further north.

Superfamily CALOPTERYGOIDEA

8. Amphipterygidae. Our only genus of this mainly tropical family, *Diphlebia*, has 6 large, robust, eastern species with vivid blue males (Plate 1, E) and dull females, resting with wide-open wings like Anisoptera. The stout, clumsy nymphs, with huge saccoid gills (Fig. 13.9A), cling to rocks in fast streams.

9, 10. Chlorocyphidae, Calopterygidae. *Rhinocypha tincta semitincta* Selys and *Neurobasis australis* Selys, respectively, represent these families in northern Australia, where they are said to occur rarely.

Suborder ANISOPTERA

Keys to the Families and Subfamilies of Anisoptera Known in Australia

ADULTS

1. Triangles of fore and hind wings dissimilar; eyes far apart; anal appendages broadly flattened, leaf-like; ♀ with ovipositor; huge insects (Fig. 7.6) .... **Petaluridae**
   Triangles in all wings similar; facies not as above; *Ax*, except 2 distinct primaries, mostly incomplete 2
   Triangles dissimilar in fore and hind wings; eyes touching; *Ax* mostly complete, primaries not always recognizable 7
2. Anal loop ill-defined (Fig. 13.20); eyes far apart .... **Gomphidae**
   Anal loop small but well-defined (Fig. 13.4B); eyes usually touching dorsally or close together .... **Aeshnidae.** 3
3. Several reddish spots on costa of each wing .... NEOPETALIINAE
   Costa, even if coloured or banded, without such spots 4
4. MA unbroken distally and running parallel to, or even diverging from, R_4 .... BRACHYTRONINAE
   MA degenerate distally and converging upon R_4 5
5. Anal appendages very long and thin; ♀ dentigerous plate a 2- or 3-pronged digging-fork .... GYNACANTHAGINAE
   Anal appendages and dentigerous plate not as above 6
6. R_3 smooth distally; anal triangle in hind wing of ♂ well developed .... AESHNINAE
   R_3 curving forward distally; anal triangle absent .... ANACTINAE
7. Primary *Ax* quite distinct; median space crossed (Fig. 13.21); ♂♂ with auricles and tibial keels .... **Synthemidae**

Primary *Ax* indistinct; median space free (Fig. 13.22); ♂♂ with tibial keels and (except *Procordulia* and *Hemicordulia*) conspicuous auricles and angulated anal margin to hind wing; posterior margin of eye with a slight sinuous projection near middle .... **Corduliidae.** 8
No auricles or tibial keels; hind wing (Fig. 13.23) rounded anally in both sexes; eyes globular, not sinuous behind .... **Libellulidae.** 11
8. Wings closed over back when resting, all similar in shape; no anal loop; triangle in fore wing 4-sided .... CORDULEPHYINAE
Wings not as above 9
9. 3 or more *cux* in hind wing; span 100 mm .... EPOPHTHALMIINAE
Fewer than 3 *cux* in hind wing; size moderate or small 10
10. Triangle in fore wing usually crossed; anal loop large, with a strong, well-defined midrib .... CORDULIINAE
Triangle in fore wing free; anal loop variable, midrib ill-defined and zigzagged or even absent .... GOMPHOMACROMIINAE
11. 5–8 *Ax*, all complete, in fore wing; Rs and MA with at most a very short common stem; posterior lobe of prothorax small, not conspicuously hairy .... UROTHEMISTINAE
Not having the above combination of characters 12
12. Hind wing very greatly broadened basally 13
Hind wing at most moderately broadened basally 14
13. Size moderate; fore wing with 7½–10½ *Ax* .... RHYOTHEMISTINAE
Size larger; fore wing with 12½ *Ax* or more, the last one sometimes complete .... PANTALIINAE
14. Eyes very large, broadly confluent dorsally; anal loop large, confluent with wing margin .... ZYXOMMATINAE
Eyes normal; anal loop closed before reaching wing margin 15
15. Triangle of fore wing 4-sided; anal loop of 7–8 cells; slender black forms with yellow markings .... TETRATHEMISTINAE
Not having the above combination of characters 16
16. Hind wing little broadened basally, quite similar in shape to fore wing; 12 or more *Ax* in fore wing, of which the last may or may not be complete .... LIBELLULINAE
Hind wing obviously broader basally than fore wing; *Ax* in fore wing variable, but less than 10 in smaller forms 17
17. Costal side of triangle in fore wing very short and usually quite straight .... most SYMPETRINAE
Costal side of triangle in fore wing about half as long as either of the other two sides, and sometimes bent 18
18. Abdomen excessively broad and flattened .... TRITHEMISTINAE
Abdomen, even if stout, not excessively broad and flat 19
19. Posterior lobe of prothorax small and not conspicuously hairy; anal loop many-celled *(Neurothemis)* .... SYMPETRINAE
Posterior lobe of prothorax large and hairy 20
20. Anal loop normal or *(Nannophya)* absent .... BRACHYDIPLACINAE
Anal loop, of 10 or more cells, sharply truncated, its distal end almost straight *(Nannodiplax)* .... SYMPETRINAE

NYMPHS (families only)

1. Labium flat, shallow; palpi usually without setae (Figs. 13.11C, D) 2
Labium deeply concave, ladle-shaped; palpi much broadened distally and armed with setae (Figs. 13.11A, B, E) 4
2. Antennae 4-segmented; fore tarsi 2-segmented .... **Gomphidae**
Antennae of 6 or 7 segments; all tarsi 3-segmented 3
3. Antennae and movable hooks of labium slender .... **Aeshnidae**
Antennae and movable hooks stout .... **Petaluridae**
4. Wing rudiments very strongly divergent (Fig. 13.10E); distal borders of labial palpi usually without setae (Fig. 13.11H) .... **Synthemidae**
Wing rudiments lying parallel to one another; distal borders of labial palpi usually with setae (Figs. 13.11I, J) 5

5. Labial palpi toothed distally (Fig. 13.11I); anal pyramid short, cerci usually at least half as long as paraprocts (Fig. 13.12B) .... **Corduliidae**
   Labial palpi not toothed distally (Figs. 13.11E, J), or, if toothed, then anal pyramid is long; cerci usually less than half as long as paraprocts (Fig. 13.12C) .... **Libellulidae**

Superfamily AESHNOIDEA

11. Gomphidae. Cosmopolitan, swift, slender forms, dark with yellow, cream, or greenish markings, frequenting running water. Fraser's revised classification is unworkable for Australian forms, including two widespread endemic genera, *Hemigomphus*, in

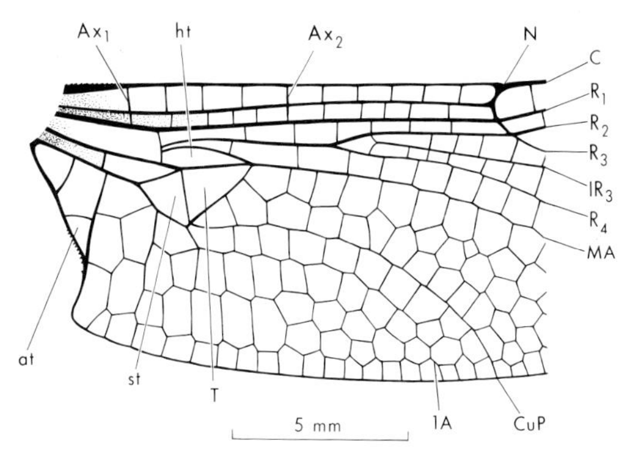

Fig. 13.20. Base of hind wing of *Austrogomphus guerini*, Gomphidae. [F. Nanninga]

which the ventrobasal branches of the superior anal appendages of the male lie outside the branches of the deeply cleft inferior appendage, and *Austrogomphus* (Plate 1, C), a heterogeneous group in which this situation is reversed. *Ictinogomphus australis* (Selys), our largest species, with abdominal segment 8 much expanded, represents a wide-ranging tropical group. Gomphid nymphs, with curious flattened antennae (Fig. 13.10B) and occasionally with the abdomen elongated posteriorly as a respiratory tube, are bottom-dwellers, often burrowers.

12. Petaluridae. An archaic relict of a once widespread fauna, our only genus, *Petalura* (Fig. 7.6), discontinuously distributed from S.W. to N.E. Australia, has 4 huge species, of which *P. ingentissima* Till. spans over 160 mm. The curious 'grub-like' nymphs (Fig. 13.10A) burrow in swamps and bogs.

13. Aeshnidae. Our commonest species are *Aeshna brevistyla* Ramb., the only Australian species in the cosmopolitan subfamily AESHNINAE, and *Hemianax papuensis* (Burm.) (ANACTINAE, represented also by *Anax* spp. in the north). *Austropetalia patricia* (Till.) (S.E. Australia) and *Archipetalia auriculata* Till. (Tasmania) are archaic NEOPETALIINAE, with their only living relatives in South America. Dull-coloured *Gynacantha* and *Agyrtacantha* species in the north, with the brighter *Austrogynacantha heterogena* Till. extending south to Brisbane, represent the tropical subfamily GYNACANTHAGINAE. The BRACHYTRONINAE have many eastern and one western species; the largest genus is *Acanthaeschna* (formerly called *Austroaeschna*), found mainly on running water in forest country, and distinguished by having 2 or more thin *Ax* between the primaries; *Notoaeschna sagittata* (Martin) has only one, while *Dendroaeschna conspersa* (Till.) has the median space crossed. All these have conspicuous pale dorsal spots on the abdomen; such spots are absent and cloudy bands along the costa usually present in species of *Telephlebia* (Plate 1, G), also in *Austrophlebia costalis* (Till.), spanning over 120 mm, found in eastern rain forests. Of unmistakable facies, aeshnid nymphs have diverse habits; e.g. *Notoaeschna* (Fig. 13.10C) clings to rocks in torrents, *Telephlebia* (Fig. 13.10F) may be semi-terrestrial in splash areas round waterfalls, and *Hemianax* clings to vegetation in almost stagnant water.

Superfamily LIBELLULOIDEA

14. Synthemidae. Confined to the S.W. Pacific, these graceful, superficially gomphid-like dragonflies, with yellow, creamy, or greenish markings on a dark ground, frequent swamps and running streams. *Synthemiopsis* (Tasmania) has black blotches at the nodus of all wings. *Choristhemis* (eastern) has 2 thin *Ax* proximal to the first primary.

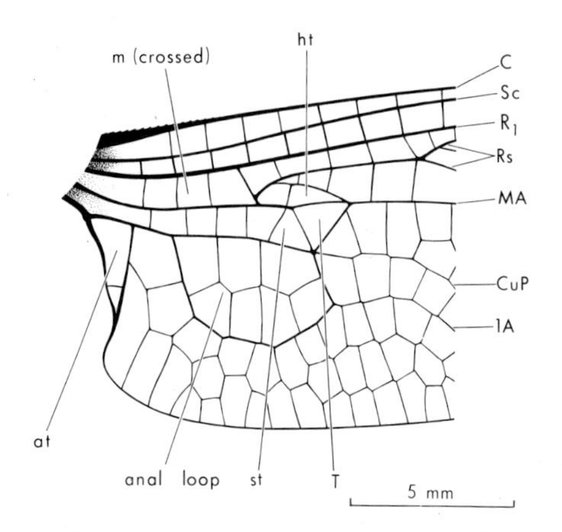

Fig. 13.21. Base of hind wing of *Synthemis eustalacta*, Synthemidae. [F. Nanninga]

Eusynthemis, also eastern, has very short anal appendages; the female has no ovipositor. Our remaining genus, *Synthemis* (Plate 1, A), with distinct forms in east and west, has long sinuous anal appendages and the female has a small ovipositor. Synthemid nymphs (Fig. 13.10E) are bottom-dwellers, and may bury themselves very deeply in time of drought.

15. Corduliidae. The primitive subfamily CORDULEPHYINAE, confined to eastern Australia, includes one genus, *Cordulephya* (Plate 1, D), of small black and yellow corduliids resting with closed wings after the manner of a zygopteron. In strong contrast are the robust tropical *Macromia* spp. (EPOPHTHALMIINAE). The cosmopolitan subfamily CORDULIINAE is represented by *Procordulia* (*P. jacksoniensis* (Ramb.) in eastern and *P. affinis* (Selys) in western Australia) and by several species of *Hemicordulia*, among which *H. tau* Selys and *H. australiae* (Ramb.) are ubiquitous and sometimes migrate in swarms. Other Corduliinae are rare, as are almost all the GOMPHOMACROMIINAE, of which we have several monospecific genera, including some obscure north-eastern forms, as well as the better-known western *Lathrocordulia* and *Hesperocordulia*. Corduliid nymphs occur, often among vegetation, in habitats ranging from swamps to clear fast streams. Often longer-legged and more 'spidery' in appearance than the nymphs of libellulids (Figs. 13.10,D, G), they are nevertheless not easy to distinguish from them without thorough examination.

16. Libellulidae. Cosmopolitan, of tropical origin, this family includes some of our commonest and most conspicuous dragonflies, often with red or pruinescent blue males,

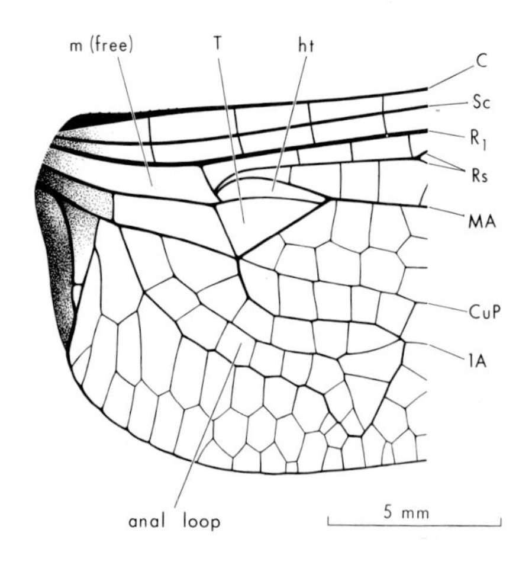

Fig. 13.22. Base of hind wing of *Procordulia jacksoniensis*, Corduliidae. [F. Nanninga]

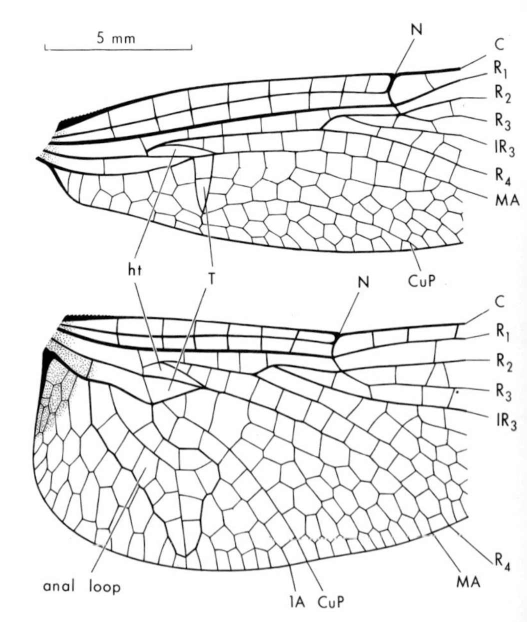

Fig. 13.23. Base of fore and hind wings of *Diplacodes bipunctata*, Libellulidae. [F. Nanninga]

yellowish females, and short, stout nymphs found mainly in still water. Possibly annectent with Corduliidae, the UROTHEMISTINAE (ranked as family Macrodiplactidae by Fraser) include, in the north-east, the attractive little *Aethriamanta circumsignata* Selys and the larger *Urothemis aliena* Selys, the females of which have a well-developed secondary ovipositor. *Macrodiplax cora* (Br.) has been recorded sporadically from most parts of Australia. Among the PANTALIINAE, *Hydrobasileus brevistylus* (Br.) is a handsome black and yellow northern form; several species of *Trapezostigma* (formerly *Tramea*) range widely over the continent, as does an almost cosmopolitan species, *Pantala flavescens* (F.), distinguished by the strong convergence of MA and CuP; *Camacinia othello* Till. is a gigantic northern form. The RHYOTHEMISTINAE, with short metallic body, often beautifully pigmented wings, and soaring flight, include several species of *Rhyothemis*, mainly northern. *Tholymis* and *Zyxomma* (ZYXOMMATINAE) are tropical genera, of crepuscular habit. *Tetrathemis* and *Nannophlebia*, small black and yellow forms found in the north and east, are our representatives of the primitive subfamily TETRATHEMISTINAE.

Among LIBELLULINAE, *Orthetrum caledonicum* (Br.), with powder-blue male, is one of our commonest dragonflies; other species of this cosmopolitan genus, together with *Agrionoptera*, *Lathrecista*, and *Potamarcha*, are found mainly in the north. The large subfamily SYMPETRINAE includes the widespread and conspicuous *Diplacodes bipunctata* (Br.) and *D. haematodes* (Burm.) with bright red males; other species occur in the north and east. Larger forms include *Rhodothemis lieftincki* Fraser (male red) and *Crocothemis nigrifrons* (Kirby) (male powder-blue); the latter seems to be extending its range southwards down both the east and west coasts. *Nannodiplax rubra* Br., mainly northern, is among the smallest known Anisoptera. *Neurothemis s. stigmatizans* (F.), with red-pigmented wings in mature males, is one of our commonest and most attractive tropical libellulids. Among BRACHYDIPLACINAE, *Brachydiplax* and the much rarer *Raphismia* are northern, but the tiny, wasp-like species of *Nannophya* (Plate 1, B) are very widespread although local. Also widespread, but extremely local, is *Austrothemis nigrescens* (Martin), our only representative of the TRITHEMISTINAE.

ACKNOWLEDGMENTS. I am indebted to Dr M. A. Lieftinck and Dr J. A. L. Watson for valuable comments and criticisms on the draft of this chapter, and to Messrs C. Sourry, C. W. Frazier, and J. Overell for help in obtaining living material for the illustrations. Dr Watson also kindly provided material and supervised the completion of the illustrations.

14

BLATTODEA

(Cockroaches)

by M. JOSEPHINE MACKERRAS

Exopterygote Neoptera with dorsoventrally compressed bodies; legs cursorial; fore wings, when present, modified into tegmina; male genitalia complex, asymmetrical, concealed by abdominal sternum 9; female with reduced ovipositor concealed by sternum 7; cerci with one to numerous segments; specialized stridulatory and auditory organs absent. Eggs contained in an ootheca.

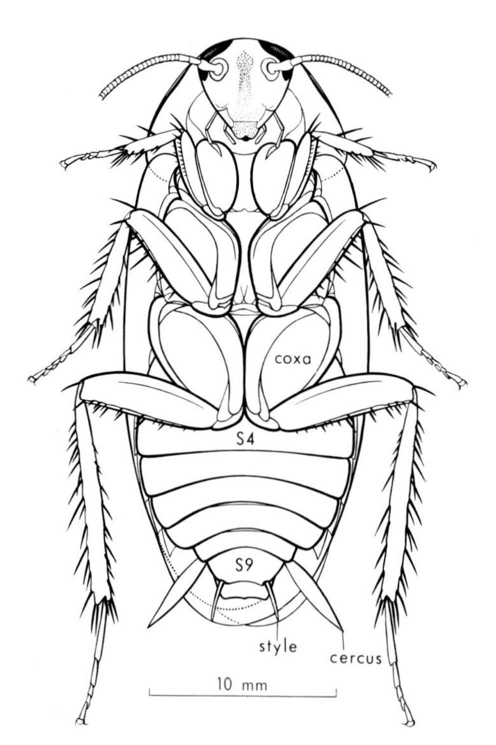

Fig. 14.1 *Methana marginalis*, ♂, ventral view. [F. Nanninga]

This is a widely distributed order of easily recognized terrestrial insects, which were formerly associated with mantids, grasshoppers, crickets, and stick-insects in the old order Orthoptera. It has been estimated to contain about 3,500 species, but Roth and Willis (1960) consider that at least 4,000 remain to be described. These authors give a full account of the bionomics, and their bibliography also includes most of the taxonomic literature on Australian groups. * The anatomy has been described by Snodgrass (1952).

Anatomy of Adult

Head (Fig. 14.2A). Hypognathous; relatively primitive, with most of the sclerites well defined. Compound eyes nearly always developed; lateral ocelli represented by two pale areas known as ocelliform spots. Antennae filiform, with numerous segments. Mandibles strong and toothed; maxillary palps 5-segmented; labial palps 3-segmented; hypopharynx large.

* See also K. Princis (1962–6). Blattaria. In M. Beier (ed.), *Orthopterorum Catalogus*, partes 3, 4, 6, 7, 8 (s'Gravenhage: Junk).

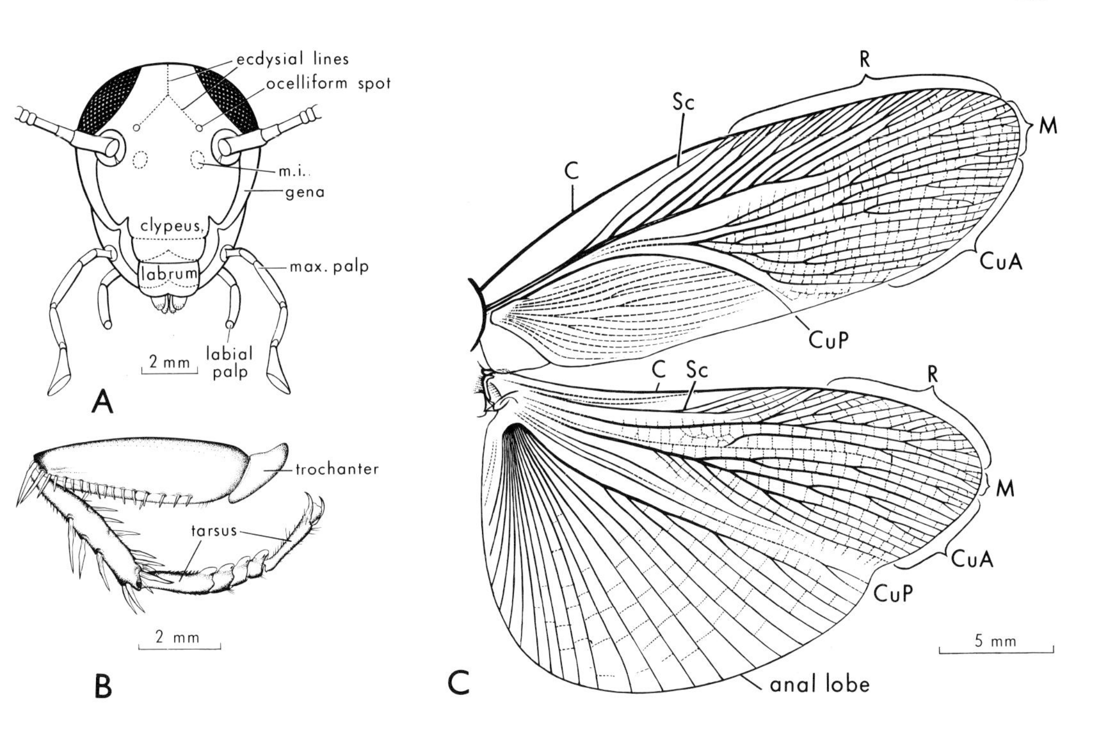

Fig. 14.2. A, head of *Polyzosteria limbata*, semidiagrammatic; B, fore leg of *Periplaneta brunnea*; C, right tegmen and wing of *P. brunnea*. [F. Nanninga]
m.i., impression of transverse frontal muscle.

Thorax. Prothorax with a large, shield-like pronotum, often overlapping the head; mesonotum and metanotum more or less rectangular and similar to each other. Pleura and sterna weakly sclerotized, somewhat compressed, mostly hidden by the coxae; meso- and metathoracic spiracles present.

Legs. (Figs. 14.1, 2B). Coxae closely approximated, very wide, often flanged; other segments unmodified, often spinose; tarsi 5-segmented, segments 1–4 usually with plantar pulvilli, terminal segment with claws and usually with arolium.

Wings. (Fig. 14.2C). Tegmina rather strongly sclerotized, protecting the membranous hind wings. C of tegmen marginal, Sc short, and Rs usually with numerous anterior pectinate branches; MA and CuA occupy a large part of the tegmen, and the short, curved CuP cuts off a distinctively shaped clavus. Hind wings with large anal lobe, folded at rest between CuP and 1A, and usually further folded in a fan-like manner.

Abdomen. Ten segments distinguishable by their terga, 11 being absorbed into 10 which forms the supra-anal plate. The tergal integument is soft in alate forms, tough and sclerotized in wingless species. Sternum 1 small or absent, 11 represented by the paraprocts, the subgenital plate concealing most of the terminal structures ventrally. Cerci inserted at the base of tergum 10, usually many-segmented, but may be reduced to a few segments, or even one (Panesthiinae). Spiracles of segments 2 to 7 open on the pleural membrane, those of 1 and 8 are attached to the lateral margins of the terga; a small plate surrounding each spiracle is visible from the ventral aspect in many species. A bilobed sternal scent gland opens between sterna 6 and 7 in some Blattidae (Figs. 14.5B, C, *gld.*), and paired tergal glands between terga 5 and 6. Males of several genera have well-developed dorsal glands

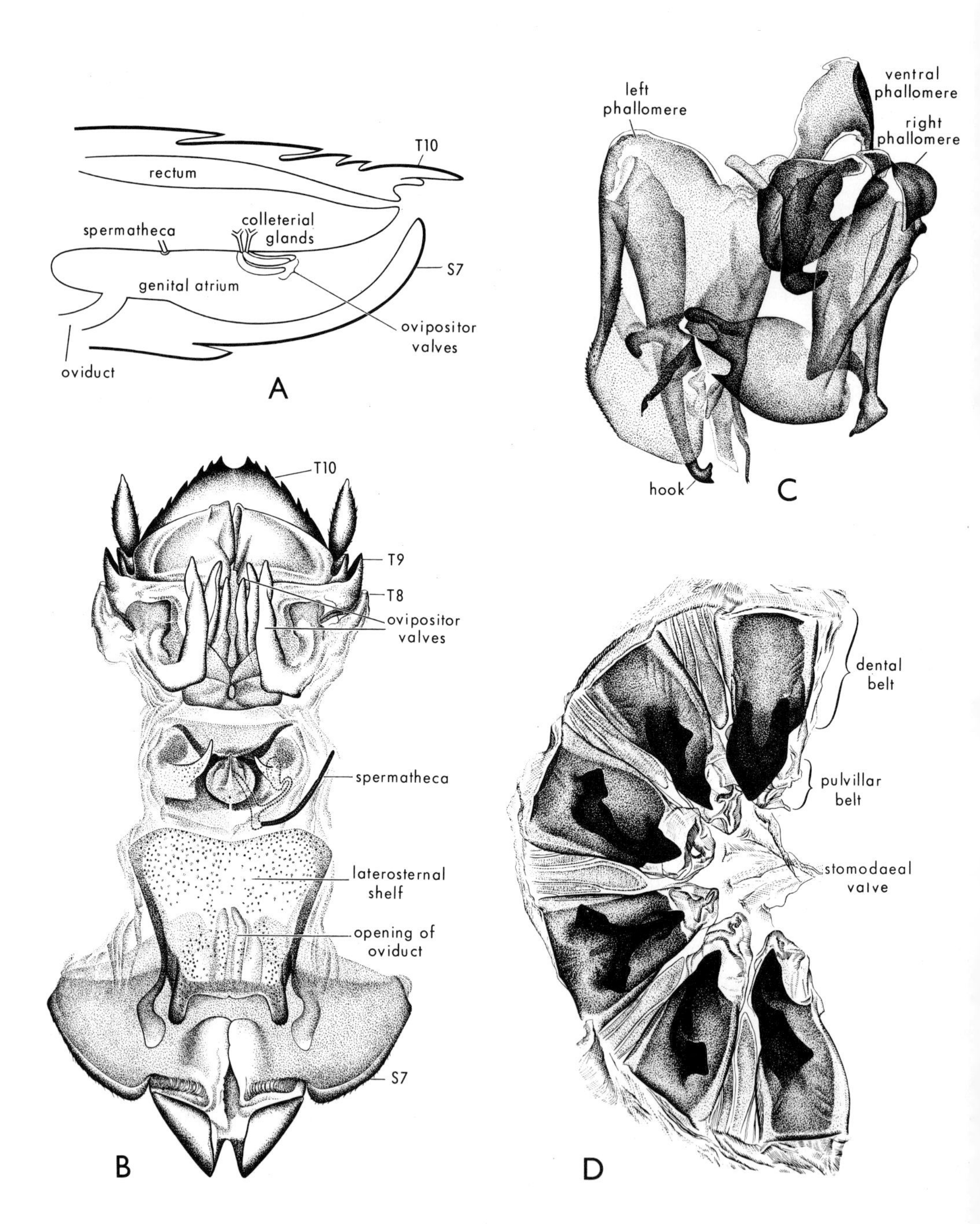

Fig. 14.3. A, median sagittal section of ♀ genital atrium, diagrammatic; B, genital atrium of *Platyzosteria melanaria*, ♀ (the lateral walls have been cut and the roof reflected upwards); C, genitalia of *Polyzosteria viridissima*, ♂ (the ventral phallomere reflected anteriorly); D, proventriculus of *Polyzosteria limbata* (slit longitudinally). [F. Nanninga]

opening on specially modified areas bearing tufts of fine hairs on tergum 1 (Fig. 14.8F), 7, or 8 (Fig. 14.6A, *p.*), which apparently secrete a substance licked by the female before copulation.

In the male, the subgenital plate is formed by sternum 9, usually bearing a pair of styles. The genitalia, concealed in the genital pouch, comprise a group of 2 or 3, asymmetrical, sclerotized or membranous phallomeres, with the gonopore opening between them. In the Blattidae both lateral phallomeres are complex, being divided into several sclerites, and the ejaculatory duct lies dorsal to a smooth lobiform ventral phallomere; a strong ventral sclerite of the left phallomere forms an eversible process known as the 'hook' (Fig. 14.3C). In the Blattellidae and Blaberidae the phallomeres are simpler, and either the right or left forms a 'hook' which is retractable into a pouch; there is no ventral phallomere, but a membranous penis, with the ejaculatory duct lying below it and accompanied in some subfamilies by a conspicuous spine-like sclerite known as the *virga*.

In the female the subgenital plate is formed by sternum 7, 8 is inflected and, with 9 and the styles, is absorbed into the wall of a large genital atrium (Fig. 14.3A). The ovipositor consists of three pairs of small, finger-like valves inside the chamber (Fig. 14.3B).

Internal Anatomy. Alimentary canal usually long and sinuous. Salivary glands large; crop well developed; proventriculus present, provided with large or medium-sized denticles (Fig. 14.3D), except in Blaberidae in which they are very small; mid gut with 8 caeca; about 80–100 Malpighian tubes. Nervous system with 3 thoracic and 4 to 6 separate abdominal ganglia. Fat body containing symbiotes in some species. Each testis consists of 4 or more follicles. The vasa deferentia join the ductus ejaculatorius, from which arise one or more pairs of seminal vesicles and numerous tubular accessory glands which secrete the material that forms the spermatophore. An unpaired conglobate gland opens between the phallomeres. Each ovary comprises a number of panoistic ovarioles. The common oviduct opens into the genital atrium on the reduced sternum 8. The spermatheca may be single or paired, and its orifice is on the dorsal wall of the atrium between sterna 8 and 9; paired accessory (colleterial) glands open separately (Fig. 14.3A), and secrete the materials from which the ootheca is formed.

Immature Stages

Egg and Ootheca. As each ovum emerges from the common oviduct, it is fertilized, then guided by the valves of the ovipositor, and held in position in the genital atrium by the folds of the intersternal membrane, as in a mould, while the secretions of the colleterial glands flow around it. The left gland secretes material containing protein and the right a diphenolic substance which forms a quinonoid tanning agent (p. 70). The eggs are placed alternately to the right and left of the mid-line, and gradually moved backwards, so that the developing ootheca protrudes from the genital opening. At first it is pale, but in oviparous species it hardens and darkens on exposure to the air. The completed ootheca (Figs. 14.4A, 8A, I) then consists of a tough outer case containing two rows of pockets divided by a median partition. Each egg occupies a pocket containing an air space with a duct in the dorsal ridge or keel leading to the external air. This ridge shows a number of serrations corresponding to the number of eggs present. Oothecae of different species are distinctive, differing in size, shape, surface ornamentation, and number of contained eggs, which vary from 12 to 40 in the Australian species studied (Pope, 1953). In ovoviviparous species the ootheca remains pale and soft, and is gradually withdrawn into the brood sac, an extensible pouch developed in the anteroventral wall of the genital chamber.

Nymph. Nymphs (Figs. 14.4B–D) resemble adults in general structure, but often differ considerably in colour and texture. All very young nymphs resemble males in having

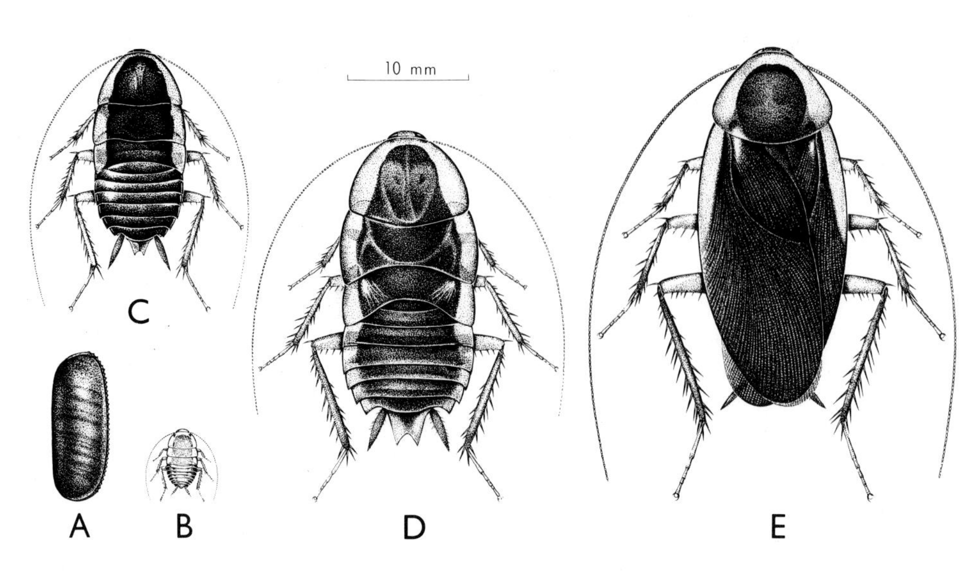

Fig. 14.4. Stages of development of *Methana marginalis*: A, ootheca; B, young nymph; C, medium-sized nymph; D, large nymph; E, adult. [F. Nanninga]

abdominal sterna 8 and 9 visible and 9 provided with styles; these sterna gradually disappear from view in female nymphs, and are incorporated into the genital atrium. The number of segments in the antennae and cerci may increase from stage to stage. In alate species the wing buds appear late in development, and do not undergo reversal during growth. The number of ecdyses is difficult to determine accurately. In *Blatta orientalis* L. 7 to 9 have been observed, in *Periplaneta americana* (L.) 11 in the female and 12 in the male, and in *Blattella germanica* (L.) 6 or 7 in the male and 7 in the female.

Biology

Adults. Most cockroaches are nocturnal. They are usually found near the ground, hiding under bark, logs, or stones during the day-time, but some are diurnal. *Ellipsidion* spp. are arboreal and run about amongst the foliage in bright daylight. Some brightly coloured diurnal species, e.g. *Polyzosteria viridissima* Shelf. (Plate 2, B), are surprisingly inconspicuous in the field. *P. limbata* Burm. also basks in the sun on bright days. Some cockroaches are adapted to arid conditions (e.g. several species of *Polyzosteria* and *Euzosteria*), others to tropical forests (e.g. numerous species of *Balta*). They occur from sea level to high country (about 1,800 m). A blind cavernicolous species (Fig. 5.6) occurs on the Nullarbor Plain. * Amphibious species of Epilamprinae occur in Central America, Burma, and Thailand, but no species adapted to this habitat has been reported in Australia. Many species, particularly the winged males of *Calolampra*, *Balta*, etc., are attracted to light. The food of native species is unknown, except that some Panesthiinae eat rotten wood. Domestic species appear to be omnivorous, and breed well on a diet of 'mouse cubes' and water. Some native species (e.g. *Methana* spp.) also thrive on a similar diet, but, although adults of most species will live on artificial food, their nymphs may prove difficult to rear, starving to death in the midst of a variety of foodstuffs. The first-instar nymphs apparently need some particular stimulus to induce feeding.

Reproduction. In the species so far studied, mating is preceded by a simple courtship behaviour, and the sexes become linked end to end. Copulation extends over a con-

* *Trogloblattella nullarborensis* Mackerras (*J. Aust. ent. Soc.* **6:** 39–44, 1967).

siderable time, sometimes many hours, a spermatophore being formed and transferred to the female genital atrium. The empty spermatophore is dropped after a day or two.

Three modes of reproduction may be distinguished: (1) oviparous species (Blattidae, Blattellidae) form a hardened ootheca which they carry for various lengths of time, sometimes (*Blattella germanica*) for the whole of the incubation period; (2) ovoviviparous species (Blaberidae) form a membranous ootheca, which they incubate internally in the brood sac, e.g. *Calolampra* spp., *Laxta* spp., *Panesthia* spp.; (3) the non-Australian blaberid, *Diploptera punctata* Esch., is viviparous, and the membranous ootheca is incomplete, allowing exchange of nutriment between the mother and the developing young. Oviparous species differ in the manner of depositing the ootheca. In captivity some simply drop it, others glue it to bark (*Ellipsidion* spp.), others conceal it by gluing it to the substrate and covering it with foreign material (e.g. *Methana marginalis* (Sauss.) and *Polyzosteria limbata*). Probably most species attempt to conceal their oothecae under natural conditions, and McKittrick (1964) has described many of the procedures followed.

Parthenogenesis occurs regularly in strains of *Pycnoscelus surinamensis* (L.) which have become established in Europe, North America, and Australia, whereas the strain in the Indo-Malaysian area and Hawaii is bisexual (Roth and Willis, 1961). These workers also proved that *Periplaneta americana* can reproduce parthenogenetically.

Life History. When incubation is complete, the young nymphs swallow air, and the expansion splits the ootheca along the dorsal edge; they emerge still enclosed in the embryonic cuticle. The nymphs live in the same situations as adults. Native cockroaches occur singly or in loose aggregations, and species of *Panesthia* and *Laxta* are usually found living in groups of various ages in rotting logs. Growth is relatively slow; small species usually mature more rapidly and have shorter lives than large species. Pope (1953) found that, in Brisbane, the incubation period of *Periplaneta* spp. and *Methana* spp. usually ranged from 5 to 14 weeks, and nymphal development from 4 months to over a year, whereas the corresponding periods for *Blattella germanica* and *Ellipsidion* spp. were 3 to 7 weeks and 2 to 8 months. The smaller species and *Methana* spp. lived for 1 to 2 years, and *Periplaneta* spp. from 2 to 4 years. *P. americana* proved a particularly long-lived insect, many surviving over 3 years, with a maximum recorded life span of 4·1 years. It was the most fecund species studied, females producing on an average over 50 oothecae, each containing 16 eggs. *Bl. germanica* produced fewer oothecae, the maximum number recorded being 5, but the large number of eggs (40) in each ootheca and the capacity of the female to carry it during incubation compensated for the smaller number produced.

Natural Enemies. Hymenopterous egg-parasites belonging to the Evaniidae, Cleonyminae, Eupelminae, other Encyrtidae, and Eulophidae occur in Australia. *Ellipsidion* spp., from their habit of depositing their oothecae in exposed places, have frequently been recorded as hosts. Parasitic Rhipiphoridae (Coleoptera) were studied by Riek (1955a), who found ground-dwelling cockroaches parasitized by several species, there being some correlation between family of host and genus of parasite. Cockroaches are preyed upon by many other terrestrial arthropods, as well as by frogs, reptiles, and insectivorous birds and mammals. Captive bandicoots and marsupial mice eat cockroaches readily.

Parasites. Many protozoan and helminth parasites of cockroaches are listed by Roth and Willis (1960), but there appear to be few records in Australia. Day (1950) recorded the presence of a large species of amoeba, ciliates belonging to the Clevelandellidae, and nematodes in the hind gut of *Macropanesthia rhinoceros* Sauss. Gordiid worms (Nematomorpha) may be found in the body cavity. Fielding, in 1926, described the transmission of the eye worm of poultry, *Oxyspirura mansoni*, by *Pycnoscelus surinamensis*. *Blattophila sphaerolaima* (an oxyurid nematode) was described from *Panesthia laevicollis* Sauss. in N.S.W.

Economic Significance. Nine cosmopolitan species have been introduced into Australia, where they have become pests in dwellings, markets, storehouses, etc. *Blatta orientalis* is present in the southern part of the continent. *Periplaneta americana* and *P. australasiae* (F.) are widespread, troublesome insects. *P. brunnea* Burm. (syn. *P. ignota* Shaw; Fig. 14.8B) occurs in Queensland and N.T., and *Neostylopyga rhombifolia* (Stoll) in N.T. These are all Blattidae–Blattinae. *Blattella germanica* and *Supella supellectilium* (Serv.) (Blattellidae) are widespread domestic pests. *Nauphoeta cinerea* (Oliv.) and *Pycnoscelus surinamensis* (Blaberidae) are more usually found in fowl-houses or grain stores. A few native species sometimes enter houses, and *Shawella couloniana* (Sauss.) (Fig. 14.8I) may be able to adapt itself to an indoor life.

Any man-made structure, where there is a suitable range of temperature and humidity and access to food, is liable to be occupied by cockroaches, and an estimate of the degree of infestation can be made only after dark, when they emerge from their hiding places. Most of the pest species prefer a tropical climate, but can withstand short exposures to extreme cold. They not only destroy foodstuffs by eating them, but foul whatever they have access to with their excreta. They will eat labels off containers and the binding of books. They are abundant in earth-closets, and congregate in countless multitudes in sewers in the warmer parts of the country. Some species have been found to harbour organisms pathogenic to man, e.g. poliomyelitis viruses in U.S.A. (Roth and Willis, 1960) and species of *Salmonella* in Australia (Mackerras and Mackerras, 1948a). Thus, their association with human populations, as well as their general ecology, is strikingly parallel to that of the rodent pests of cities.

Special Features of the Australian Fauna

The Australian fauna is rich in some groups of cockroaches. The Blattidae are well represented, especially in arid country, by numerous genera of medium-sized to large, flightless forms. The Blattellidae are also well represented by small, active, light coloured, winged or brachypterous species. Only 2 subfamilies of the Blaberidae have been recorded—Epilamprinae, with a few genera but numerous species, and Panesthiinae with 5 genera. The Polyphagidae may be more abundant than appears from published records and museum collections. In Africa and America, members of this family live in deserts, or are commensals in the nests of termites, ants, and wasps, situations which do not seem to have been sufficiently explored in Australia. The monotypic family Cryptocercidae is confined to North America and eastern Asia.

CLASSIFICATION

Order BLATTODEA (439 Australian spp.)

BLATTOIDEA (200)	BLABEROIDEA (239)	
1. Blattidae (200)	2. Polyphagidae (4)	4. Blaberidae (95)
Cryptocercidae (0)	3. Blattellidae (140)	

Cockroaches are a very ancient, well-defined group of insects, but there is no general agreement on their taxonomic treatment. Earlier classifications, culminating in that of Princis in 1960, depended on various external characters, and all tend to break down at some point, because the different groups have undergone parallel and convergent evolution, and the relationships suggested by the external characters are not always supported by the internal anatomy and structure of the genitalia. McKittrick (1964) considers that there are 5 clearly defined families which may be arranged in 2 superfamilies. Her classification, which is based on a comparative study of the male genitalia, the female genitalia and their musculature, the egg-laying behaviour, the structure of the proventriculus, and their correlation with external characters, has been followed here.

Key to the Families of Blattodea Known in Australia (Adults)

1. Anal area of hind wing not folded fanwise in repose; thickened clypeal shield sometimes present. .... **Polyphagidae**
   Anal area of hind wing folded fanwise in repose: without thickened clypeal shield 2
2. Styles of ♂ simple, slender, usually symmetrical; both lateral phallomeres complex, ventral phallomere present; sternum 7 of ♀ bivalvular .... **Blattidae**
   Styles of ♂ usually asymmetrical or absent; one of the lateral phallomeres forming a strong, eversible hook, the other often blunt, no ventral phallomere; sternum 7 of ♀ not bivalvular 3
3. Size usually small to medium; antennae longer than half body length; legs relatively long, slender, spiny; cerci fairly long, tapering, carried at right angles to body .... **Blattellidae**
   Size usually medium to large, ♀♀ always broad-bodied; antennae often less than half body-length; legs usually relatively short and stout, femora and tarsi sometimes smooth; cerci often short, not projecting much beyond the supra-anal plate .... **Blaberidae**

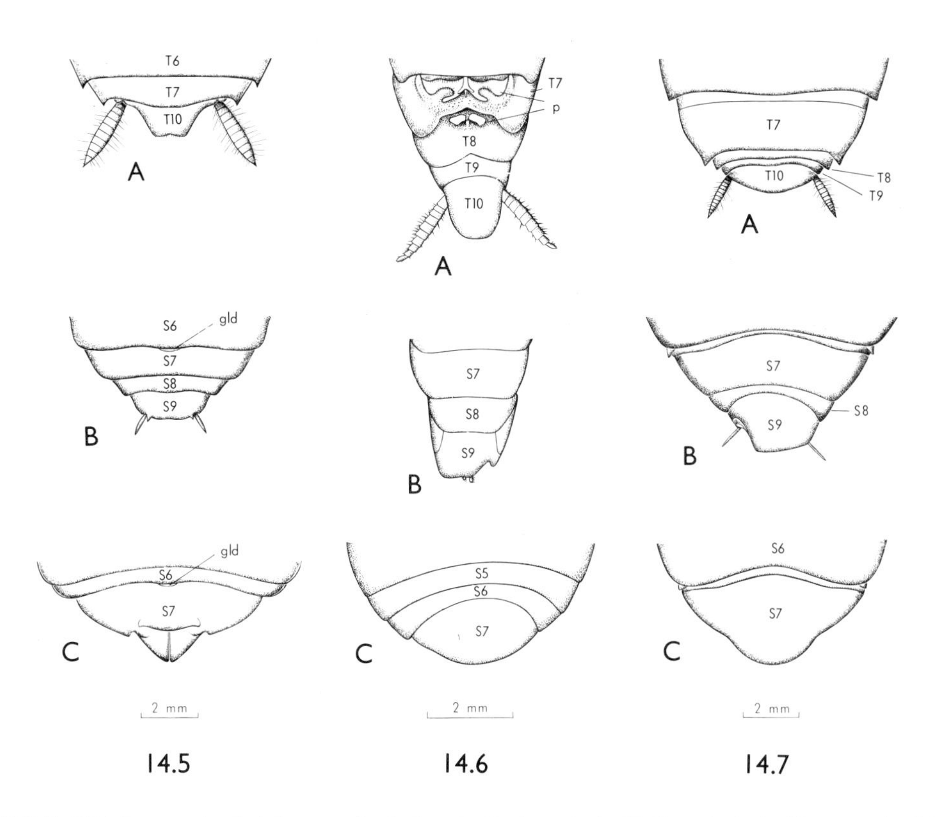

Fig. 14.5–7. Distal segments of abdomen: A, ♂, dorsal; B, ♂, ventral; C, ♀, ventral. 5, *Platyzosteria nitidella*, Blattidae; 6, *Blattella germanica*, Blattellidae; 7, *Laxta granicollis*, Blaberidae. [F. Nanninga] *gld*, opening of sternal scent gland; *p*, specialized pits on terga 7 and 8.

Superfamily BLATTOIDEA

1. Blattidae. Fore femur with numerous spines on anterior margin but few on posterior margin, mid and hind with some spines on both margins. Wings without apical triangle; R with numerous branches often subdivided; CuA with numerous branches running towards apical margin (Fig. 14.2c). Sternum 9 of male often rectangular, styles symmetrical, spineless, placed laterally

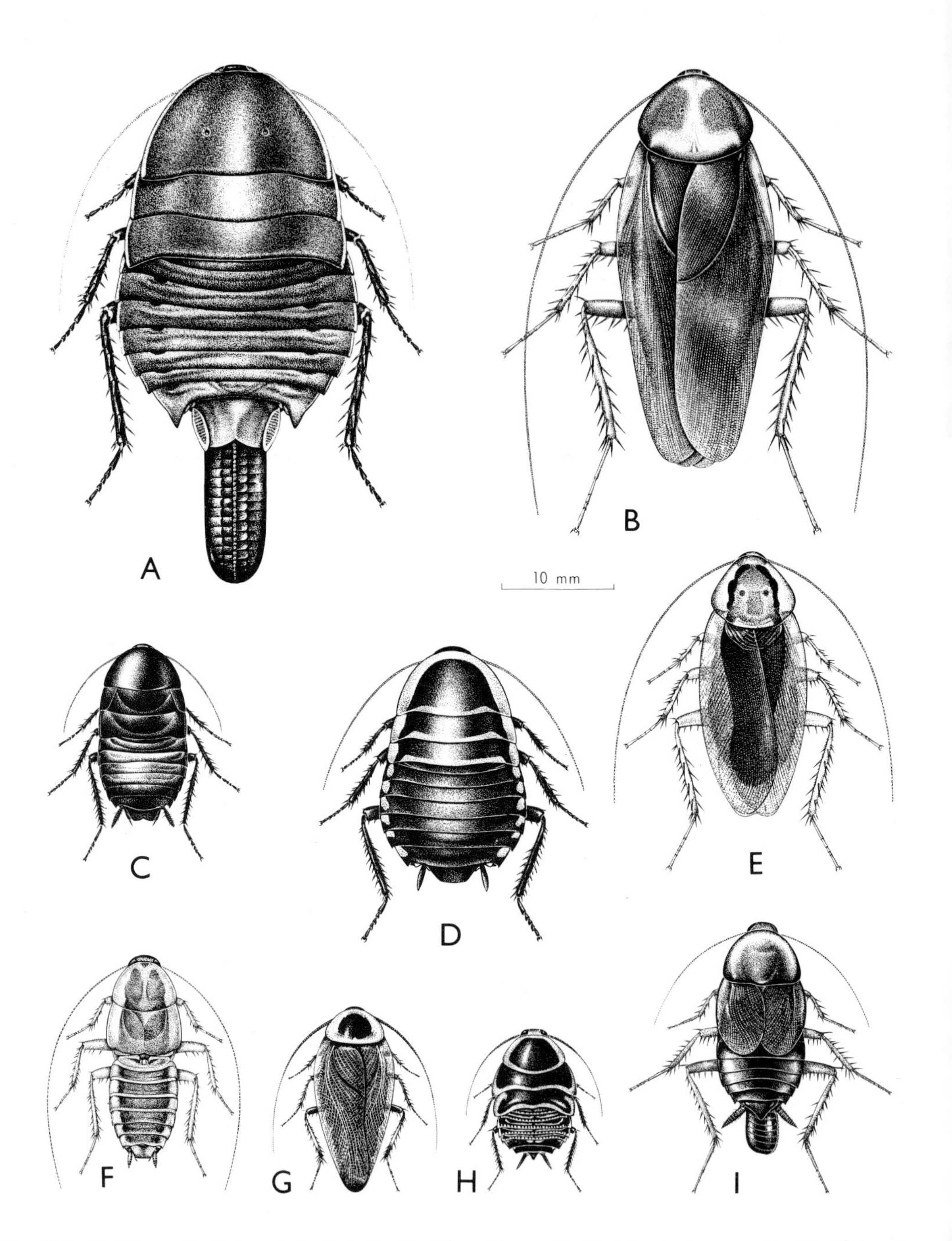

Fig. 14.8. Blattidae: A, *Polyzosteria limbata*, ♀ with ootheca; B, *Periplaneta brunnea*, ♂; C, *Platyzosteria nitidella*, ♂; D, *Cosmozosteria subzonata*, ♀; E, *Methana curvigera*, ♂; F, *Temnelytra truncata*, ♂. Blattellidae: G, *Ellipsidion australe*, ♂; H, *E. australe*, nymph; I, *Shawella couloniana*, ♀ with ootheca.

[F. Nanninga]

(Fig. 14.5B); paraprocts unmodified. Sternum 7 of female large, its posterior part modified to form two lobes or valves, which are united medially by an extensible membrane and partly separated from the remainder of the sternum by a transverse suture (Fig. 14.5C). Tergum 10 of both sexes rectangular, rounded, or triangular; cerci often short and flattened, but may be tapering and project well beyond the tergum. Oviparous species, which always carry the ootheca with the suture dorsal (Fig. 14.8A).

This family is world-wide in distribution. Our 5 cosmopolitan species of the subfamily BLATTINAE (p. 268) may be distinguished by having spines on the ventral surface of all 1st tarsal segments and on the 2nd segment of mid and hind tarsi. The Australian genera belong to the POLYZOSTERIINAE and TRYONICINAE. Of these, *Methana* is also found in the Oriental region, *Platyzosteria* in the Oriental region, New Zealand, and New Caledonia, and *Polyzosteria* has been reported in New Caledonia. All the others appear to be confined to Australia. They are particularly abundant in the southern, central, and western parts of the continent, where many species have become adapted in various ways to live under extremely harsh climatic conditions. All except *Methana* (Figs. 14.4, 8E) are flightless, but *Scabina* and *Temnelytra* (Fig. 14.8F) have short, quadrate tegmina, and some species of *Platyzosteria* (Fig. 14.8C), and *Tryonicus* have vestigial tegmina. *Polyzosteria* (Fig. 14.8A), *Cosmozosteria* (Fig. 14.8D), and other genera lack even vestigial organs of flight. The integument is tough; smooth or pitted and wrinkled, sometimes extremely rough and nodular. Some species have a covering of minute scales or hairs, particularly well developed in *Polyzosteria pubescens* Tepper and related species. Coloration may be black, green, brown, red, or yellow. *Polyzosteria mitchelli* (Angas) has pale diverging stripes on a green or brown dorsum, and some species of *Anamesia* (Plate 2, C) and *Cosmozosteria* have striking yellow bands or spots on a reddish brown ground. Most species of *Platyzosteria* are uniformly black or reddish brown, but some have pale borders or stripes. Australian species have been described by Tepper, Shelford, Shaw, Princis, and by Mackerras (1965–6).

Superfamily BLABEROIDEA

2. Polyphagidae. Spines may be absent on anterior and posterior margins of mid and hind femora. Wings with large preaxillary area and small, unfolded anal lobe. Tegmen sometimes with reduced venation, or with a network of veins. Sternum 7 of female sometimes bivalvular. This family is well represented in other parts of the world, but is little known in Australia. *Tivia australica* Princis, a small species with a speckled brown male, has been described from Western Australia; a nymph was also taken from an ants' nest. There are at least 3 undescribed species in eastern Australia; all are small, delicate insects with winged males that are attracted to light.

3. Blattellidae. Legs long, slender, spiny; ventral surfaces of tarsi spined. An apical triangle may be present in the wing; R usually well developed, with simple, regularly spaced, anterior rami; CuA usually reduced or with relatively few branches running towards apical margin (Fig. 14.9). Many species are fully alate, but there are also many with reduced wings in one or both sexes. Sternum 9 of male usually asymmetrical (Fig. 14.6B), with 1 or 2 spiny styles; paraprocts often spiny and asymmetrical. Sternum 7 of female broad, rounded (Fig. 14.6C), never bivalvular. Tergum 10 of both sexes

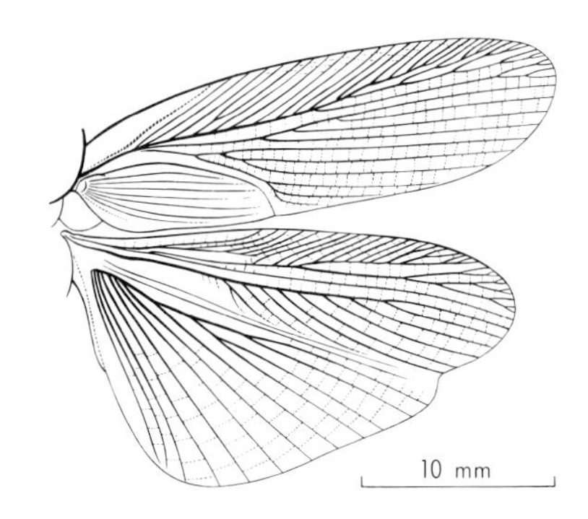

Fig. 14.9. Right tegmen and wing of *Gislenia australis*, Blattellidae. [F. Nanninga]

often triangular, with long, tapering cerci showing well-defined segmentation. Oviparous species, some of which rotate the ootheca and carry it with the suture lateral (Fig. 14.8I).

The family is abundant in all faunal regions; but, except for 2 cosmopolitan species, the Australian element seems to be rather isolated, only some of the genera occurring elsewhere. The family as a whole has never been reviewed in Australia, although Hebard (1943) published a revision of genera allotted at that time to the Ectobiinae and Chorisoneurinae. There are numerous species which are difficult to classify, so that it is more convenient to recognize groups of genera than subfamilies or tribes in the present state of our knowledge. These may be distinguished by the following key.

Key to the Generic Groups of Blattellidae

1. Ventral margins of hind femora armed with spines 2
   Ventral margins of hind femora unarmed ... 3
2. Antero-ventral margins of fore femora with small piliform spines, sometimes a few strong spines proximally .... *Balta* group
   Antero-ventral margins of fore femora strongly spined .... *Blattella* group
3. Sc of tegmen nearly parallel to costal margin .... *Mediastinia* group
   Sc of tegmen reaching the costal margin near its mid point .... *Ectoneura* group

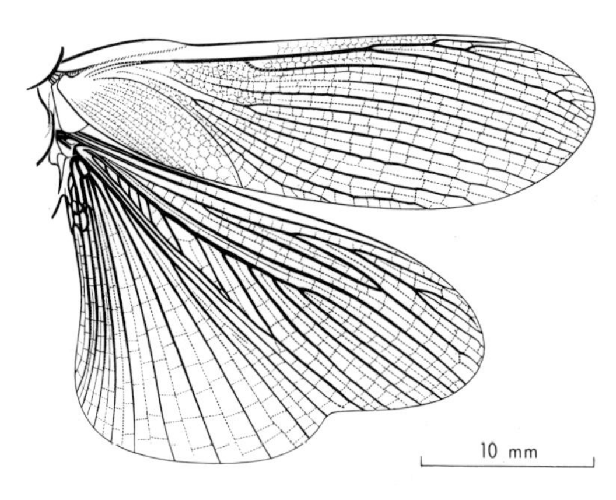

Fig. 14.10. Right tegmen and wing of *Panesthia australis*, Blaberidae. [F. Nanninga]

The *Mediastinia* group includes only 3 small northern species of *Mediastinia*. The *Ectoneura* group includes *Ectoneura*, *Stenectoneura*, *Choristomodes*, and *Choristima*; these are small, light-coloured, winged species. The *Balta* group includes *Balta*, *Megamareta*, and *Ellipsidion*; the first two are mostly pale insects, but many species of *Ellipsidion* are richly marked with orange and have characteristic black sterna with white edges (Plate 2, D; Figs. 14.8 G, H). The *Blattella* group includes numerous genera which will require careful study to determine their relationships.

4. Blaberidae. Legs often relatively short; femora and tarsi sometimes without spines. Wing with branches of R sometimes reduced to 5 or less; CuA with numerous branches, many going to the plical fold (Fig. 14.10). Sternum 9 of male rounded and more or less asymmetrical, bearing a pair of short slender styles (Fig. 14.7B), except in Panesthiinae which lack styles; right phallomere forming the 'hook'. Sternum 7 of female broad, rounded, notched, or truncate, never bivalvular (Fig. 14.7C). Cerci usually short. Ovoviviparous species which rotate the ootheca and incubate it internally. This family is world-wide in distribution, and much work is needed to elucidate the relationships of the Australian species.

The PANESTHIINAE are Oriental and Palaearctic, with an extension into Australia, where some specialized genera have developed. They are stoutly built insects with hard, pitted integument, and are adapted for burrowing. The legs are characteristic: coxae and femora hairy, and the femora almost devoid of large spines; tibiae strongly spined, sometimes fossorial; tarsi relatively short, without spines, segments 2–4 subequal, terminal segment long, with symmetrical claws; arolia absent. When wings are present, R has not more than 5 branches (Fig. 14.10). The number of abdominal segments appears to be reduced, as in many species only terga 1 to 7 and 10 can be counted. However, in very young nymphs and in adults of some species, terga 8 and 9 can be detected as narrow strips between 7 and 10. In the male the styles disappear, and

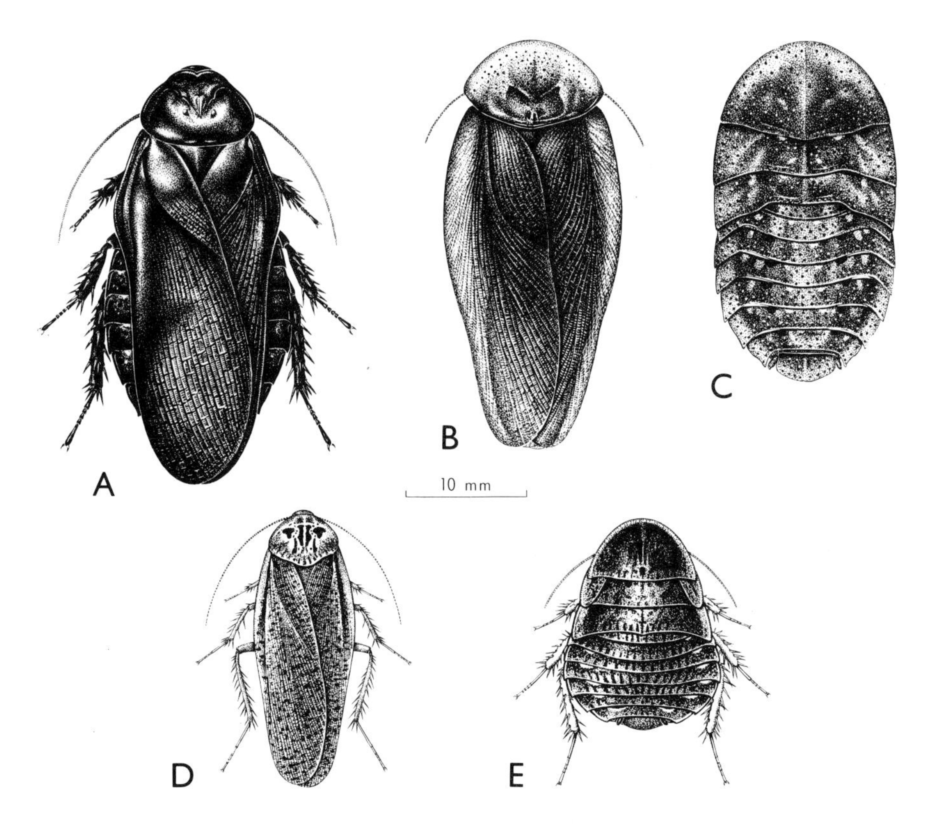

Fig. 14.11. Blaberidae: A, *Panesthia australis*. ♀; B, *Laxta granicollis*, ♂; C, *L. granicollis*, ♀; D, *Calolampra irrorata*, ♂; E, *C. irrorata*, ♀. [F. Nanninga]

sternum 9 is greatly reduced and overlapped by 7, which completely covers the vestigial 8. Cerci 1-segmented. *Panesthia* is the most widespread and abundant genus, two fairly large black species, *P. laevicollis* Sauss. and *P. australis* Brunn. (Fig. 14.11A), being common in rotting logs. Both sexes of several species have well-developed wings, but these are broken off close to the base soon after the final ecdysis. *Macropanesthia* and *Geoscapheus* contain large apterous species which burrow into the soil. *Hemipanesthia* and *Salganea* are represented by one species each.

The EPILAMPRINAE are widely distributed in Asia and America, but comparatively few genera have been recorded in Australia. *Laxta (= Oniscosoma)* contains very flattened insects; dorsum usually dark brown, and covered with irregular nodules; head completely concealed by pronotum; tergal angles usually sharply produced; mid and hind femora without spines on the ventral margins. The males may be fully winged, brachypterous, or apterous, but all known females are apterous. *L. granicollis* (Sauss.) (Figs. 14.11 B, C) is a medium-sized insect, common under bark or in leaf litter, the males possessing large wings. *Calolampra* includes plump, soft-bodied insects; the dorsal surface smooth and characteristically speckled; head not completely concealed by the pronotum; tergal angles slightly or not at all produced; mid and hind femora armed with spines. Males are usually alate and females have lobiform tegmina, but both sexes may be alate or apterous. These are abundant insects, with numerous species. *C. irrorata* (F.) (Figs. 14.11D, E) was one of the first Australian insects to be described, being collected in 1770 during Cook's first voyage. The females are usually taken in the soil, but males fly about, and are frequently collected in light traps. Other genera are *Molytria* and *Ataxigamia*. *M. inquinata* (Stål) is a large, speckled insect occurring in N.S.W. *A. tatei* Tepper has remarkably long, silky hairs on the thorax; males are attracted to light; its relationships are unknown. There are many undescribed species of *Laxta* and *Calolampra* and several undescribed genera.

15

ISOPTERA

(Termites)

by F. J. GAY

Polymorphic, mandibulate, exopterygote Neoptera, living in social units composed of a limited number of reproductive forms associated with numerous wingless sterile soldiers and workers. Antennae filiform or moniliform; wings, when present, elongate, membranous, held flat over the body at rest, and capable of being shed by means of basal sutures; cerci short; external genitalia rudimentary or wanting.

This is a relatively small order, closely related to the Blattodea, and occurring mainly in tropical and subtropical regions. It contains over 2,000 species, of which 145 described and at least 30 undescribed species occur in Australia. They are soft-bodied insects with cryptic habits. The alates have four wings, and vary in length from 6–7 mm (small *Amitermes* and *Microcerotermes*) to 17–18 mm *(Mastotermes)* and in wing-span from 12 mm (small *Amitermes*) to 50 mm *(Mastotermes)*. The soldiers and workers vary in body length from 2·5 mm (small *Tumulitermes*) to 15 mm (*Porotermes* and *Neotermes*). The world literature on the order has been catalogued by Snyder (1949–61), and the Australian species were monographed by Hill (1942).

Every termite colony contains a number of different castes which exhibit both anatomical and physiological specialization (Fig. 15.1). Four distinct castes are recognizable: (1) the primary reproductives (kings and queens), which are fully sclerotized individuals derived from alates, with the wing remnants in the form of small triangular scales; (2) supplementary reproductives (also known as *neotein-*

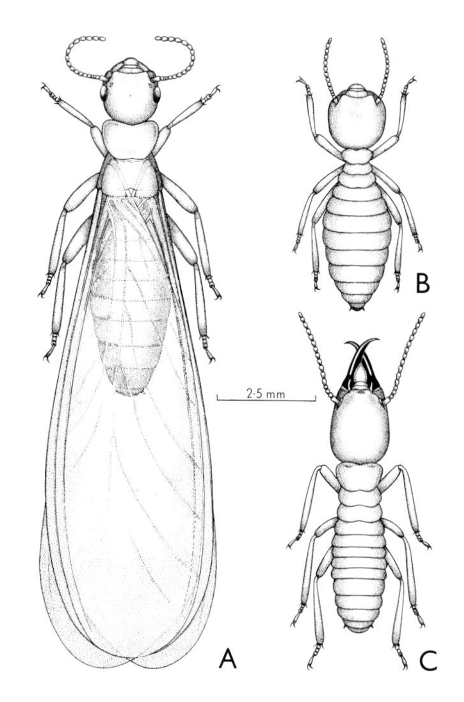

Fig. 15.1. Castes of *Coptotermes acinaciformis*, Rhinotermitidae: A, winged reproductive, or alate; B, worker; C, soldier. [B. Rankin]

ics, or second and third form reproductives), which are less heavily sclerotized, and are either without any trace of wing elements, or with rounded wing buds of variable size; (3) soldiers, which are sterile males and females, with heavily sclerotized and greatly modified heads; and (4) workers, which are sterile males and females, only lightly sclerotized, and without special modifications. Immature individuals of all these castes may also be present.

Anatomy of Adult

Reproductives

Head. Ovoid or rounded and somewhat flattened, often with distinct ecdysial lines. Compound eyes present (generally somewhat reduced in supplementaries). Ocelli present or absent; when present, only two (median ocellus absent). Antennae moniliform, of 10 to 32 segments; number of segments variable within a species and even from side to side of an individual. Labrum well developed. Clypeus divided into distinct ante- and postclypeus. Mouth-parts of typical mandibulate blattoid type. Mandibles of alates and workers of the same genus are virtually similar, and afford valuable taxonomic characters at the generic level (M. Ahmad, 1950). Large basal plate of the labium, known as the *gula* or gulamentum, also provides useful taxonomic characters because of variations in shape. Fontanelle (p. 278) present in Termitidae and Rhinotermitidae.

Thorax. Prothorax with a well-defined pronotum, the shape of which affords useful taxonomic characters. Meso- and metanotum subequal and without distinctive features. Nota well developed. Sterna membranous.

Legs. All three pairs of legs very similar, with large broad coxae, and long slender tibiae armed with terminal spines. Tarsi 4-segmented, except in *Mastotermes* in which they are 5-segmented. Arolium present in some of the primitive termites.

Wings. Two pairs of elongate, membranous wings; fore and hind similar in shape, except in *Mastotermes* in which the hind wing has a pronounced anal lobe of blattoid type (Fig. 15.2). Anterior veins strongly sclero-

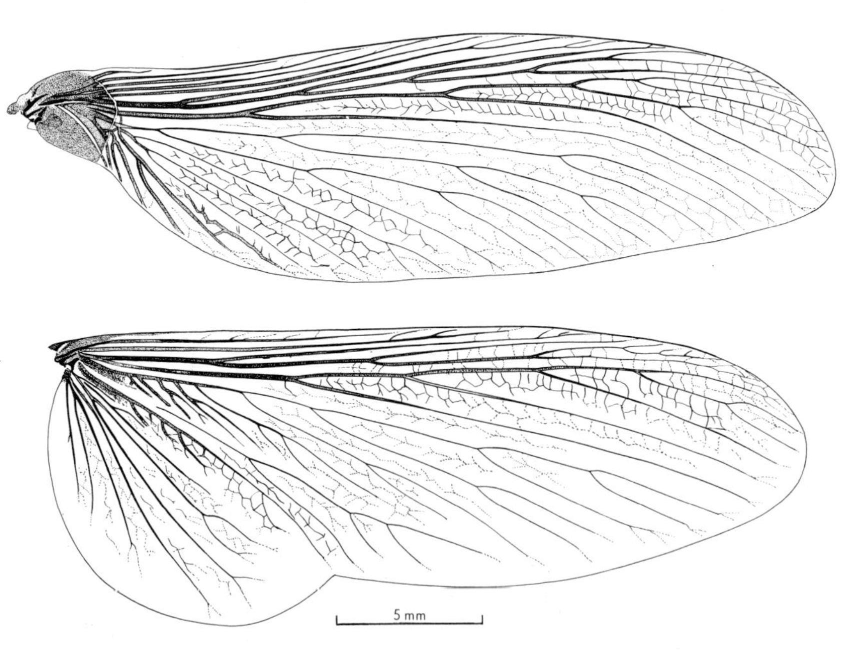

Fig. 15.2. Fore and hind wings of *Mastotermes darwiniensis*, Mastotermitidae. [N. Key]

tized, others poorly developed. No cross-veins, but wing membrane sometimes reticulate. Venation reduced and simple (Fig. 15.3), except in *Mastotermes* which has a primitive venational pattern with all veins present. All wings normally have a basal or humeral suture, which permits shedding of the wing, and leaves a triangular stump, the *scale*, attached to the thorax. This suture is present only on the fore wings of *Mastotermes*.

Abdomen. Of 10 evident segments, each with tergum and sternum, except that sternum 1 is absent. In males all sterna are visible and entire, except 9 which is divided longitudinally in some of the higher termites. In females sternum 7 is enlarged to form a subgenital plate which obscures the remaining sterna (Fig. 15.4). Terminal segment with a pair of short, 1- to 5-segmented cerci, which are present in all castes. A pair of small, unsegmented styles present on the posterior border of sternum 9 in the males of most species. External genitalia absent, except in *Mastotermes*, in which the female has a reduced ovipositor of blattoid type and the male a median, membranous, copulatory organ.

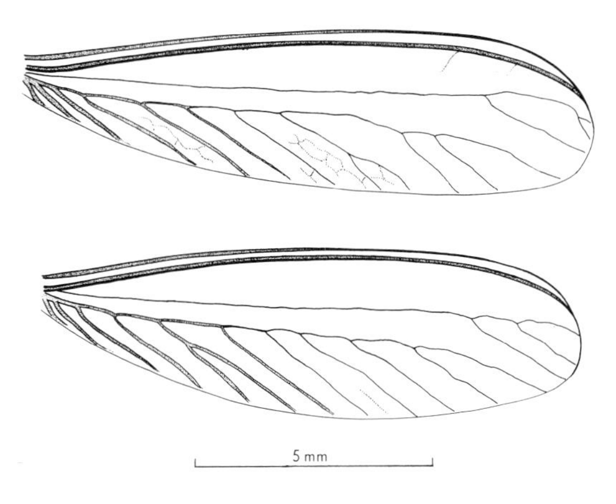

Fig. 15.3. Fore and hind wings of *Nasutitermes dixoni*, Termitidae. [N. Key]

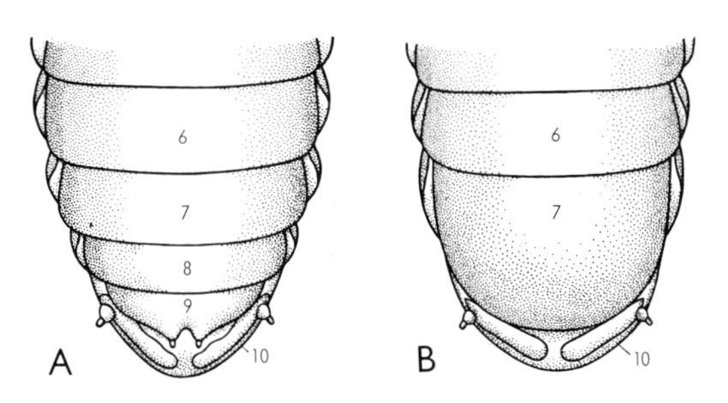

Fig. 15.4. Distal abdominal sterna of alates of *Coptotermes lacteus*, Rhinotermitidae: A, ♂; B, ♀. [B. Rankin]

Fig. 15.5. Physogastric queen of *Nasutitermes exitiosus*, Termitidae–length *ca* 30 mm. [Photo by C. Lourandos]

In most of the Rhinotermitidae and Termitidae, the primary, and to a lesser extent the supplementary, female reproductives exhibit the phenomenon of *physogastry*. This is a pronounced swelling of the abdomen resulting from the proliferation of the ovaries and fat body. The enlargement of these organs is accommodated by the expansion of intersegmental membranes, so that the originally contiguous terga and sterna of the abdomen appear as isolated islands of sclerotized tissue (Fig. 15.5).

The sole function of both primary and supplementary reproductives is to maintain the strength of the colony by providing a regular supply of spermatozoa and eggs.

Soldier

Structurally very specialized. Head greatly developed, often oblong or pyriform, and at times exceeding in size the rest of the body. Obsolescent or weakly developed compound eyes present in some primitive genera, absent elsewhere. Antennae moniliform or filiform, of 10 to 26 segments. Soldiers are genetically male and female, but secondary sexual

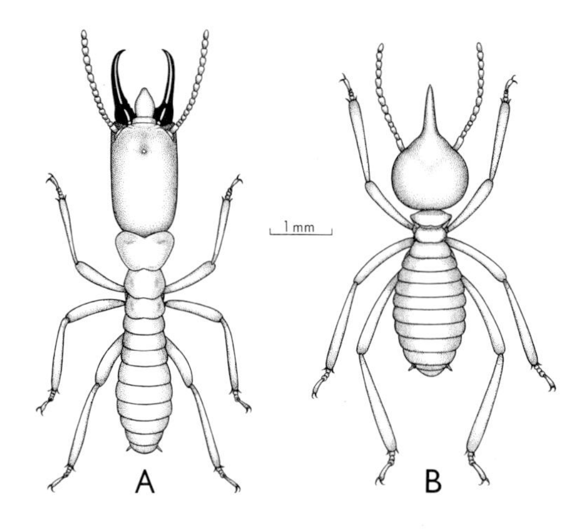

Fig. 15.6. A, mandibulate soldier of *Heterotermes ferox*, Rhinotermitidae; B, nasute soldier of *Nasutitermes exitiosus*, Termitidae. [B. Rankin]

characters are recognizable only in some primitive genera. Reproductive organs are wanting or vestigial.

Two distinct types of soldiers are recognizable: (1) mandibulate, with well-developed jaws often of very grotesque form; (2) nasute, with the head produced anteriorly into a rostrum, and with small or vestigial jaws (Fig. 15.6). In addition, two classes of soldier may occur within the same species, differing from one another not only in size but also in structure. This feature is found in Australian species of *Tumulitermes*, *Occultitermes*, and *Schedorhinotermes*.

A well-developed frontal gland is present in the Rhinotermitidae and Termitidae; it occupies a variable portion of the head, and in *Coptotermes* also extends well into the abdomen. The secretion from this gland is exuded through a pore on the frontal region of the head, known as the *fontanelle*. In the nasute soldiers the duct from the frontal gland opens at the tip of the rostrum, and the secretion is emitted either as a droplet or as a fine jet. In the minor soldiers of *Schedorhinotermes* the secretion is reported to volatilize rapidly from the fringe of hairs on the anterior border of the labrum.

The primary function of the soldier caste is to protect the colony from enemies. This may be achieved mechanically by biting, as in mandibulate soldiers; by blocking access galleries with their heavily sclerotized heads, which are specifically adapted for this purpose (*phragmotic* heads) in one or two genera; or by chemical means in those forms that have functioning frontal glands. The frontal-gland secretion may be toxic or repellent to invading insects, such as ants, or may merely be tacky and entangle their legs and antennae. The protective activities of soldiers are most noticeable during building or repair operations on the nest, or during the release of winged reproductives from the parent nest. At such times conspicuous numbers are present at all breaches of the nest or gallery system to prevent the entry of intruders.

Worker

Structurally unspecialized. Body usually pale and only lightly sclerotized. Mandibles strongly sclerotized and similar in pattern to those of the reproductives (Fig. 15.12). Compound eyes usually absent. Pronotum similar in form to that of soldier. Both sexes are represented, but reproductive organs are non-functional, and external sexual characters scarcely discernible.

Workers are numerically the largest caste in the colony. Their functions include foraging for food; feeding the young, the soldiers, the maturing reproductive nymphs, and the reproductives themselves; tending the eggs; and repairing and enlarging the nest and gallery system.

The worker caste is absent in the Kalotermitidae and Termopsidae. In these families the worker functions are performed by the early reproductive nymphal stages, or by 'pseudergates', which are late-stage nymphs that have lost the ability to develop into alates. In *Mastotermes* the colonies contain a high proportion of individuals that behave and look like workers, but their precise status is still not clear.

Internal Anatomy

The only organ system that exhibits obvious special features, certain of which are associated with the insects' way of life, is the alimentary canal. This is a convoluted tube of moderate length, similar to that of the Blattodea, and characterized by: (1) a muscular proventriculus bearing an internal armature of sclerotized plates or teeth for grinding food; (2) presence of a number of caeca at the junction of the proventriculus and mid gut in the more primitive termites; (3) reduction in the number of Malpighian tubes (2–8); and (4) in wood-eating species of all families but the Termitidae, considerable enlargement of the colonic region of the hind gut to form a large rectal pouch. Termites have well-developed salivary glands opening at the base of the hypopharynx.

Little is known about sex determination, other than the fact that both sexes are diploid. Parthenogenesis occurs very rarely, and results in wholly female progeny. Chromosome numbers appear to be high, the diploid numbers being 42 and 52 in two exotic species that have been studied.

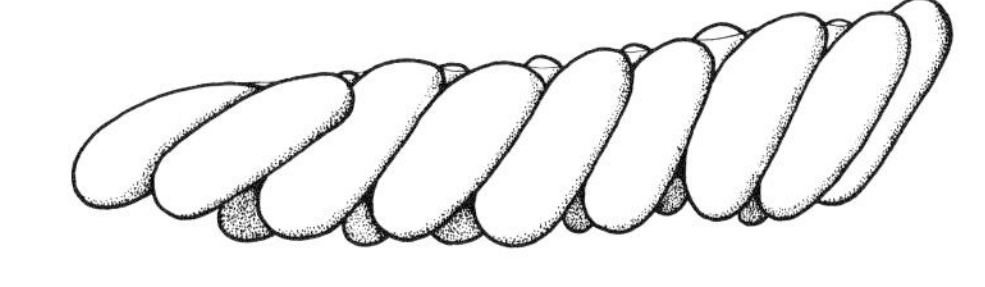

Fig. 15.7. Egg mass of *Mastotermes darwiniensis*. [B. Rankin]

Immature Stages

Egg. Yellowish or white in colour, reniform, and about 0·75 to 1·25 mm in length. Eggs are normally laid singly, and may be collected by the workers into irregular masses in the nest; but in *Mastotermes* they are laid in batches of up to 24 arranged in two regular rows in a structure resembling the ootheca of cockroaches (Fig. 15.7).

Nymph. On hatching the nymphs are of the same general form as their parents, and they undergo a number of ecdyses, varying from 4 to 10 or more, but commonly 7, before reaching maturity as workers, soldiers, or alates. There is similar variation in the number of ecdyses before a particular caste is differentiated, from as few as 2 for workers up to 6 or more for alates. The final ecdysis to a soldier is always preceded by an instar in which the nymph shows distinct evidence of transition from worker to soldier form, and is referred to as a 'white soldier' or 'pseudosoldier'. Nymphs destined to become alates may develop wing-pads as early as the 3rd instar; the pads increase in size at successive ecdyses, and the fully winged form may appear after the 7th or a subsequent ecdysis.

Biology

The Colony. Colony founding is normally initiated by the release of large numbers of alates from the parent colonies at a particular time of the year, and even at a particular time of the day. In Australia there are two main flight periods, late spring to early summer, and autumn. The release of winged reproductives of both sexes may be spread over a period of several weeks during which small batches of alates emerge at irregular intervals, or it may be concentrated into a brief period of a few days with most of the alates departing in a single colonizing flight. It is not uncommon for all nests of a particular species in one district to release their alates simultaneously. They are weak fliers, and the colonizing flight is generally of short duration and distance, unless assisted by wind. The wings are shed at the basal suture soon after alighting, and the de-alated females attract males to themselves by elevating the abdomen and presumably emitting a sex-attractant odour. Pairs of males and females then form tandems which seek out suitable sites for new colonies in soil or wood. The royal pair excavate a small nuptial chamber within which they seal themselves, after which copulation takes place. Colonies may also be founded by a group of nymphs splitting off from the main nest group, followed by the development of supplementary reproductives. This procedure is quite common in the more primitive termites which have a diffuse nest and gallery system, and appears to be usual in *Mastotermes*.

The primary queen may begin to lay eggs within a week of mating, but the initial batch is small and rarely numbers more than 20 to 30. These hatch after some weeks, and this primary brood is tended by the king and queen during its early development. Oviposition is resumed, generally after a lapse of some months, and the fecundity of the queen begins to rise as the progeny develop the ability to maintain themselves. Ultimately egg production becomes an almost continuous process, and mature physogastric queens of such species as *Nasutitermes exitiosus* (Hill) and *Coptotermes lacteus* (Frogg.) are capable of laying 2,000-3,000 eggs per day.

Only workers and soldiers are produced in young colonies, and the early broods are also distinctly undersized ('nanitic'); alates do not appear until the colony is several years old. Mature colonies, which take many years, or decades, to develop, may range in size from a few hundred individuals, as in many of the more primitive genera, to upwards of a million, as in *Mastotermes*, some species of *Coptotermes*, and some Termitidae.

Both of the primary reproductives may live for many years, possibly from 15 to 50 in the higher termites, and during this time the queen is periodically fertilized by the king. Normally there is only a single primary king and queen per colony, but instances of multiple primary queens are known in *Nasutitermes exitiosus*, *N. walkeri* (Hill), and *N. graveolus* (Hill). These are probably derived from fusion of adjacent colonies.

The colony is potentially immortal in species that have the ability to produce supplementary reproductives. These are always less fecund than the primary queens, so that multiple replacements are generally needed to maintain the strength of the colony. More than 100 supplementary queens have been recorded from a colony of *Amitermes laurensis* Mjöb. and over 50 from a colony of *Protocapritermes krisiformis* (Frogg.). Supplementary reproductives are not normally found in colonies in which the original primary reproductives are still present.

In mature colonies, the proportion of soldiers to workers, or to nymphal forms performing worker functions, varies considerably. In many Kalotermitidae and in some species of *Amitermes* it is less than 1 per cent, while in some species of *Nasutitermes* it may be as high as 15 per cent. These sterile castes are comparatively long-lived; soldiers and workers of *N. exitiosus* are known to live at least two years, and those of *Coptotermes acinaciformis* (Frogg.) up to four years.

Termites feed on sound and decayed wood, grass, fungi, and other materials of vegetable origin, such as dead leaves, bark, humus, and the dung of herbivores. The principal material utilized by the wood-eaters is cellulose, which, in all families but the Termitidae, is converted to an assimilable form by the action of a rich fauna of symbiotic flagellate Protozoa in the rectal dilatation of the hind gut. There is evidence to suggest that certain wood-eating Termitidae which digest cellulose without the aid of symbiotic Protozoa either rely on cellulases produced by the bacterial flora in their guts, or secrete their own cellulase. Species which feed on material other than wood either have a very reduced protozoan fauna or none at all.

Most termites consume their food *in situ*, but many species that feed on grass or plant litter harvest their food and carry it back to the nests. This occurs throughout the genus *Tumulitermes* so far as is known, in both species of *Drepanotermes*, some species of *Amitermes*, and some of *Nasutitermes*. The harvesting species of *Tumulitermes*, *Drepanotermes*, and *Nasutitermes* come out and forage on the surface on humid overcast days or at night, and the workers and soldiers of these species are generally heavily pigmented. In *Amitermes* harvesting is done under the protective cover of thin mud sheeting built over the grass or litter. Most of the harvesting species build mound nests which serve largely as food stores, but a few species (e.g. *Amitermes neogermanus* Hill) are subterranean nest builders.

Irrespective of the nature of the food, the location of new food sources depends very largely on random foraging. Termites carry out this exploratory activity and eventually return to the central nest with the aid of scent trails, which consist of minute amounts of volatile odoriferous materials produced by sternal glands and deposited on the substrate (Moore, 1966).

Both stomodaeal and proctodaeal feeding are common in termites. This exchange of food *(trophallaxis)* represents the normal manner in which soldiers, young nymphs, reproductive nymphs, and the reproductives themselves obtain their food and, to some extent, pheromones. In species with symbiotic Protozoa proctodaeal feeding is essential for the refaunation of individuals that have lost their intestinal fauna during ecdysis.

Caste Determination. It is now generally believed that every newly hatched nymph emerges from the egg with equal potentialities for developing into any caste, and the particular potentiality that is realized depends on extrinsic factors operating on the young nymphs. There is, however, no general agreement on the nature of these factors, nor is it certain that the same factor operates for all species of termites (see p. 135). Three types of controlling factors have been proposed. One is 'hormonal' control, which depends upon the circulation of a pheromone within the colony. This secretory material, which is ephemeral and must be produced continuously, is passed from individual to individual by licking or by food exchange, and is able to induce specific actions, such as stimulating the development of reproductive organs or inhibiting the development of particular structural characters. This mode of caste determination has been demonstrated experimentally in connexion with the production of supplementary sexuals in some primitive genera (*Kalotermes* and *Zootermopsis*). The second is nutritional control, in which differences in the amount and type of food given to young nymphs result in caste differences. There is a limited amount of experimental evidence from a species of *Reticulitermes* (Rhinotermitidae) to support this view. Thirdly, there is sensory control, in which tactile or olfactory stimuli between individuals within the colony are believed to be involved in caste determination.

Intercastes in termites are comparatively rare, and are of two types only: forms combining characters of the soldier and reproductive castes; and forms showing soldier and worker characters. Examples of the former type have been recorded in the Kalotermitidae, and in species of *Nasutitermes* and *Microcerotermes* (Termitidae). The second type has been observed in species of *Reticulitermes* (Rhinotermitidae) and *Nasutitermes*. No intercastes combining characters of the reproductive and worker castes are known, and this supports the widely held view that the worker caste has been derived phylogenetically from the soldier and not directly from the reproductive caste.

Nests. Termites build various types of nests. The simplest is that found in the Kalotermitidae and Termopsidae, in which the whole colony lives in a series of galleries and chambers excavated in moist or dry wood and is surrounded by its food. The subterranean forms construct a more or less complex central nest, either in the soil or in a log or stump, from which subterranean galleries or covered runways extend to food sources below or above ground. This type of nest habit is found in many genera, notably *Mastotermes*, *Heterotermes*, *Schedorhinotermes*, and some species of *Amitermes*. The most elaborate nest structures are the termitaria built by some species of *Coptotermes* and many of the Termitidae.

About one-fifth of all Australian species build termitaria, often of very distinctive form. Generally, the outer portion of the nest is hard to very hard; built mainly of clay or earthy material, it contains very few galleries or chambers. The inner region is much softer, and is generally composed of woody material fashioned into a complex system of galleries and chambers (Fig. 15.8). Here are found the eggs and young nymphal stages, as well as the royal pair often in a specialized cell. Mound-building species also reach their food sources by subterranean galleries and covered runways. The gallery system of a single colony may exploit all the food sources over as much as one hectare (2·47 acres), with individual galleries extending from 75 to 100 m in length. In at least some of the Australian species the internal environment of the termitarium is considerably more stable than that outside. Relative humidity is maintained close to saturation, and the daily oscillation of

Fig. 15.8. Mound of *Coptotermes lacteus*, Rhinotermitidae: A, external appearance; B, sectioned to show internal structure. [Photos by F. J. Gay]

temperature is damped down to only a few degrees. Such a combination provides ideal conditions for the continuous rearing of brood.

The size and shape of the termitarium ranges from the low, domed mounds built by a number of species to the tall, columnar structures of *Nasutitermes triodiae* (Frogg.) which may be more than 7 m high (Fig. 15.9C), and the curious wedge-shaped mounds of *Amitermes meridionalis* (Frogg.) in which the long axis is always oriented on a north-south line (Figs. 15.9A, B). In general, each species builds its own characteristic termitarium, but some, such as *N. triodiae* and *Drepanotermes rubriceps* (Frogg.), build distinctly different termitaria in different parts of their range, probably in response to variations in soil, vegetation, and climatic factors.

Although termitaria are normally built on the ground, two species of *Microcerotermes* commonly construct their clay-covered nests on poles, posts, or the trunks of trees. True tree nests of the 'nigger-head' type, consisting of dark, carton-like material, are built by two species of *Nasutitermes* (Fig. 15.9D).

The outer wall of the termitarium offers a suitable nesting place for other termites, and up to seven species have been collected from a single mound of *C. acinaciformis*. Other animals also make use of termitaria. Thus, three tropical species of the parrot genus *Psephotus* almost invariably excavate their nests in the walls of termite mounds, and kingfishers in the genera *Dacelo*, *Halcyon*, and *Tanysiptera* commonly exhibit the same habit. In addition, there are several records of the monitor lizard, *Varanus varius*, laying its eggs in mounds, particularly those of *N. exitiosus*.

Termitophiles. Termite nests harbour a rich fauna of insects and other arthropods. Some of these termitophiles are active predators feeding on eggs and young brood, some are merely scavengers feeding on nest debris, and some show a more intimate association with their hosts by providing attractive secretions in return for which they receive food. These last frequently show considerable structural adaptations, such as physogastry and highly developed exudatory organs, associated with this trophallactic exchange. Termitophiles recorded from Australia include representatives of the following orders and families: COLLEMBOLA: Entomobryidae. INSECTA: Thysanura—Nicoletiidae; Hemiptera—Aphididae; Coleoptera

Fig. 15.9. Mounds or nests of some Australian Termitidae: A, *Amitermes meridionalis*, north-south aspect; B, same, east-west aspect; C, *Nasutitermes triodiae*; D, *Nasutitermes walkeri*; E, *Amitermes vitiosus*. [Photos by G. F. Hill, F. J. Gay, and Australian News and Information Bureau]

—Carabidae, Curculionidae, Pselaphidae, Scarabaeidae, Staphylinidae, Tenebrionidae; Diptera—Phoridae, Sciaridae. CRUSTACEA: Isopoda—Armadillididae.

Natural Enemies. The most important natural enemies of termites are predators of various kinds. Winged reproductives emerging on their colonizing flight are destroyed in large numbers by lizards, snakes, frogs, insectivorous and omnivorous birds, ants, and other predatory insects, especially Odonata. Workers and soldiers of a wide range of species form an important part of the diet of the echidna *(Tachyglossus aculeatus)*, which has strong, long-clawed feet with which it attacks termitaria and subterranean galleries. The marsupial ant-eater *(Myrmecobius fasciatus)* is a specialized termite predator, and uses its strong, burrowing fore legs to open up subsurface soil galleries. Termites are significant items of food for other marsupials, particularly small insectivorous species of Dasyuridae, which prey on foraging columns on the surface, and for at least one species of bandicoot (Peramelidae) which has strong, digging fore feet. Workers and soldiers from underground galleries are eaten by typhlopid snakes, which are commonly found in the soil at the base of termite nests or under rotten stumps inhabited by termites, and geckoes have been observed preying on foraging columns moving in the open at night. Termites, particularly alates, appear to be a regular dietary item of many species of leptodactylid frogs, particularly in the drier regions of Australia. Rain showers which commonly stimulate colonizing flights of termites also bring the burrowing frogs to the surface for feeding or breeding. One leptodactylid, *Myobatrachus gouldi* of south-western Australia, is a specialized termite predator; it burrows in the soil, and locates the termites in their subterranean galleries.

Economic Significance. Termites are responsible for a considerable amount of damage to poles, posts, railway sleepers, bridgework, and other constructional timber throughout Australia. The annual loss from attacks of this kind is estimated to exceed $4,000,000. Damage to living trees in the indigenous hardwood forests, particularly of eastern Australia, is believed to be of the same order. Affected trees contain extensive gallery systems, which may extend out to the sapwood and frequently surround large central pipes. The resultant damage is a combination of actual loss of merchantable timber plus degradation in quality. In drier inland areas, fruit trees, ornamentals, vines, and vegetable crops are frequently damaged or destroyed. Other materials attacked by termites include fibre and particle boards, lead-sheathed telephone cables, plastic piping, and plastic-sheathed cables.

The economic significance of the activities of the many widely-distributed grass-eating species has not been fully assessed. It appears, however, that the complete removal of plant cover by these termites may accelerate sheet erosion by wind and water.

On the other hand, termites play a significant role in soil formation in many parts of Australia. This is particularly true of the species that do not build mounds, but live in complex subterranean systems of galleries and chambers. It is almost impossible to turn over a spadeful of earth without revealing evidence of their tunnelling activities, which facilitate the penetration of air and water to the deeper layers of the soil. Thus, in addition to their role in the breakdown of cellulose and its return to the soil, termites bring subsoil to the surface and, in our arid and semi-arid regions, fulfil some of the functions carried out by earthworms elsewhere.

Special Features of the Australian Fauna

The Australian termite fauna contains representatives of all known families but one. Half of the genera are in 4 primitive families and half in the Termitidae, in which several endemic genera have developed in 2 highly specialized groups, one with soldiers having asymmetrical, snapping mandibles and the other with nasute soldiers. Australia has more genera in common with the Ethiopian than any other region, but, if the relict Termopsidae are excluded, the Australian fauna is equally related to those of the Papuan and

Indo-Malayan subregions (Emerson, 1955).

The most distinctive feature of the fauna is the presence of relict primitive genera in the Mastotermitidae and Termopsidae. The Australian species of *Mastotermes*, the most primitive living termite, is the only surviving member of a family that is represented in other continents by fossil species from Tertiary deposits. In the Termopsidae, *Stolotermes*, with 3 mainland species and one in Tasmania, is represented elsewhere by a single species in New Zealand and one in South Africa, and *Porotermes* has one species in south-eastern Australia, one in Chile, and one in South Africa (Calaby and Gay, in Keast *et al.*, 1959). Notable absences from the fauna are the true harvesters belonging to the Hodotermitidae and fungus-growing termites belonging to the Macrotermitinae.

Two aspects of termite distribution within Australia are of special interest. The first is the paucity of species in rain-forest areas (in which only four species regularly occur). This is in marked contrast to the extremely rich termite faunas of other tropical rain forests, such as the Congo and Guyana, and may be due to the fact that Australian rain forest has never been a large and important community which could promote active speciation. The second feature is the absence of some species of termites from certain types of soil. The black earths of inland north-eastern Australia are virtually devoid of termites, although adjacent sandy-desert steppe soils have an abundant fauna. It is thought that the physical characteristics of the heavy soils, which crack deeply and widely in dry conditions and become waterlogged after rain, do not favour termite survival. *Mastotermes darwiniensis* Frogg., which is widely distributed across tropical Australia, is absent from rain forest and soils which once carried rain forest (Ratcliffe, Gay and Greaves, 1952), nor does it occur on the extensive bauxite soils of Cape York Peninsula.

As far as is known, only one exotic species occurs in Australia, the Philippines drywood termite *Cryptotermes dudleyi* Banks, which is established in Darwin. On the other hand several Australian species have become established overseas. *M. darwiniensis* occurs at Lae in New Guinea, and no less than 6 species have been introduced into New Zealand, namely, *Neotermes insularis* (Walk.), *Bifiditermes condonensis* (Hill), *Glyptotermes tuberculatus* Frogg., *Coptotermes acinaciformis*, *C. lacteus* (Frogg.), and *C. frenchi* Hill.

CLASSIFICATION

Order ISOPTERA (182 Australian spp.)

1. Mastotermitidae (1)
2. Kalotermitidae (20)
3. Termopsidae (5)
   Hodotermitidae (0)
4. Rhinotermitidae (19)
5. Termitidae (137)

The classification is essentially that proposed by Harris (1961), which recognizes the Termopsidae as of family rank and divides the order into 6 families.

Keys to the Families of Isoptera Known in Australia

ALATES

1. Tarsi distinctly 5-segmented, with arolium; antenna with 29–32 segments; hind wing with anal lobe .... **Mastotermitidae**
   Tarsi 4-segmented, viewed from above; antennae rarely with more than 22 segments; hind wing without anal lobe 2
2. Anterior wing scales short, not reaching to base of posterior scales; wings not reticulate . **Termitidae**
   Anterior wing scales covering at least the base of the posterior scales; wings reticulate 3
3. Ocelli absent .... **Termopsidae**
   Ocelli present 4
4. Fontanelle present .... **Rhinotermitidae**
   Fontanelle absent .... **Kalotermitidae**

SOLDIERS

1. Tarsi 5-segmented .... **Mastotermitidae**
   Tarsi 4-segmented (rarely with a rudimentary 5th segment) 2
2. Cerci long, 4- or 5-segmented .... **Termopsidae**
   Cerci short, 2-segmented 3
3. Fontanelle absent .... **Kalotermitidae**
   Fontanelle present 4
4. Pronotum flat, without anterior lobes .... **Rhinotermitidae**
   Pronotum saddle-shaped, with anterior lobes .... **Termitidae**

1. Mastotermitidae. This family is represented by only a single, relict species, *Mastotermes darwiniensis*, which was confined to tropical Australia, but has recently become established in New Guinea. It presents a remarkable assemblage of primitive characters. The tarsi are 5-segmented in all castes. The alate, which is the largest of all the Australian termites, has 29- to 32-segmented antennae, large eyes and ocelli, and reticulate

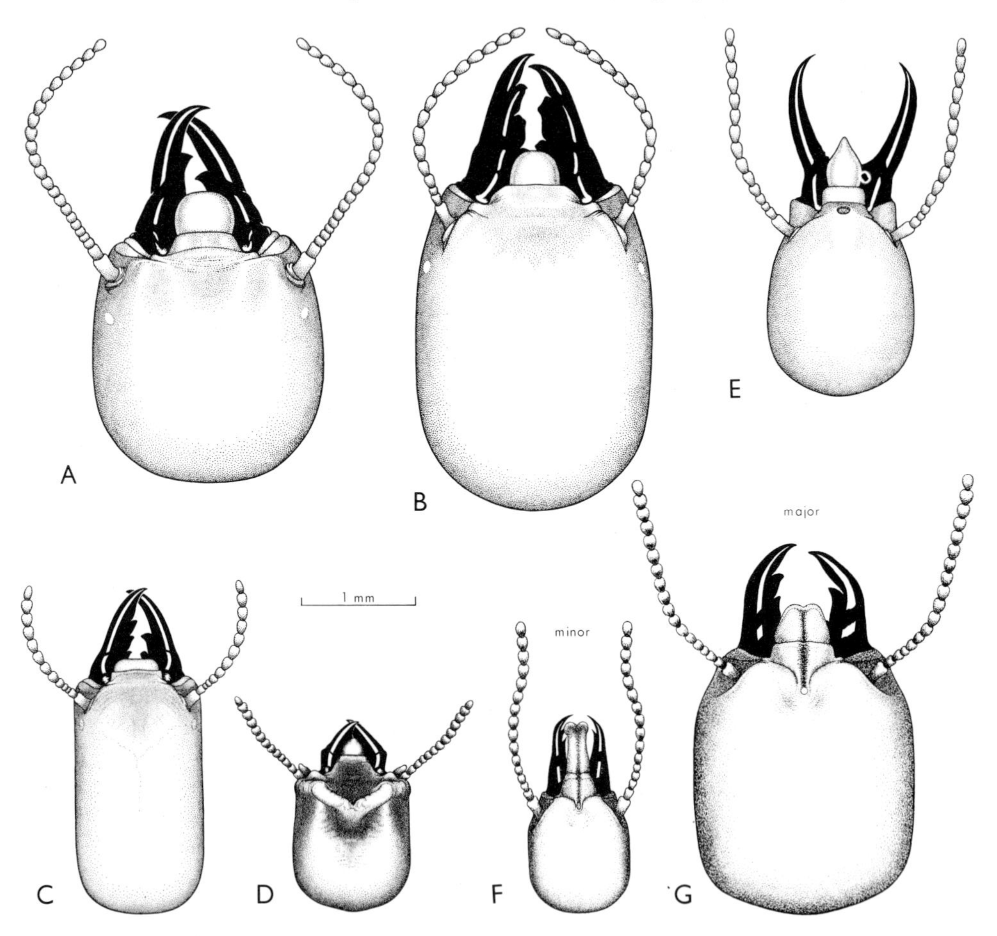

Fig. 15.10. Heads of soldiers of: A, *Mastotermes*, Mastotermitidae; B, *Porotermes*, Termopsidae; C, *Kalotermes*, Kalotermitidae; D, *Cryptotermes*, Kalotermitidae; E, *Coptotermes*, Rhinotermitidae; F, G, *Schedorhinotermes*, Rhinotermitidae. [B. Rankin]

wing venation; the hind wings (Fig. 15.2) lack the usual humeral suture, but have a distinct anal lobe, both blattoid characters. The soldiers have 20- to 26-segmented antennae and short, stout mandibles, each with a pointed tooth a little in advance of the middle (Fig. 15.10A). In both alates and soldiers, the fore tibiae have 3, and the mid and hind tibiae 4 apical spurs. The bulk of each colony is made up of individuals which fulfil the function of the worker caste, although their precise status is uncertain. Eggs are laid in pods (Fig. 15.7) homologous with the oothecae of Blattodea; this method of oviposition is found nowhere else in the Isoptera.

M. darwiniensis occurs widely in the tropical areas of Queensland, N.T., and W.A. It is not a mound-builder, and normally nests in the boles of trees or in logs or stumps. Under natural conditions these colonies are seldom very populous, but in centres of settlement they frequently attain great size (well in excess of a million individuals). Colonies are normally headed by numerous apterous reproductives, and primary reproductives appear to be very rare. Neither the supplementary nor primary reproductives are significantly physogastric. *M. darwiniensis* has a very high destructive potential, and, in addition to its severe and often rapid attacks on structural timber of all kinds, it causes serious damage to living trees and crop plants.

2. Kalotermitidae. Alates with small ocelli close to the eyes. Fontanelle absent. Antennae 11- to 21-segmented. Left mandible with an apical tooth, 2 distinct, almost equal marginal teeth, and a molar plate. Right mandible with 2 marginal teeth and molar plate (Figs. 15.12B, C). Pronotum flat, as broad as, or broader than, head. Tarsi 4-segmented. Fore-wing scale large and overlapping hind-wing scale; membrane of wings reticulate. Fore wing with Sc, R, and Rs sclerotized; M varying in position; 1A absent. Hind wing with Sc absent, short 1A present. Cerci short, 2-segmented. Soldiers with robust heads and well-developed mandibles. Head generally long, but phragmotic in some genera with sculptured frontal area (Figs. 15.10C, D). Eyes rudimentary, occasionally pigmented. Antennae 10- to 19-segmented. Mandibular dentition varied. A true worker caste is absent in this family.

The family, which has been studied critically by Krishna (1961), is represented in Australia by 6 genera. *Neotermes*, with the single, somewhat variable species *insularis*, occurs in coastal forests from Victoria to Torres Strait and across to Darwin. Its soldiers are larger than those of any other Australian termite, including *Mastotermes*. It forms moderately large colonies in systems of concentric galleries, generally in the upper portions of living eucalypts; it has also been recorded from hoop pine (*Araucaria Cunninghamii*) and a number of ornamental trees. Although this species produces no central pipe in the trees that it attacks, the extensive gallery systems are occasionally the cause of some economic loss.

Kalotermes (7 spp.) has small, darkly pigmented alates with the wing membrane densely covered with small, pimple-like, pigmented nodules; antennae 12- to 17-segmented; all species with an arolium on the tarsus. The soldiers, with the exception of *K. atratus* (Hill), have long, narrow heads, small, unpigmented eyes, strong mandibles with varied dentition, and 11- to 18-segmented antennae. All the species are wood-dwellers, and form small colonies in exposed dead wood, branch stubs, and fire scars on species of *Eucalyptus*, *Banksia*, and *Casuarina*. Except for *K. hilli* Emerson, which occurs in south-western Australia, they are restricted to the coastal and adjacent highland areas of south-eastern Australia and Tasmania.

Ceratokalotermes is represented by a single species, *C. spoliator* (Hill), which occurs in coastal and adjacent highland areas from southern N.S.W. to northern Queensland. It forms small colonies in and around branch stubs in a number of species of *Eucalyptus*.

Glyptotermes (5 spp.) has reddish brown to dark brown alates with smoky brown wings; Rs and M are both heavily sclerotized, Rs is unbranched, and M running closely parallel to it. The soldiers have a more or less elongated head which is distinctly to faintly bilobed in the frontal area; eyes unpigmented; 10- to 16-segmented antennae; mandibles short and broad. The species are restricted to the coastal and adjacent tableland areas of eastern and south-eastern Australia from north Queensland to Adelaide. They form small to moderate-sized colonies in dead wood and adjacent sound wood of trees belonging to a number of genera.

Bifiditermes (2 spp.) occurs in coastal areas from south Queensland to W.A. The alates have hyaline wings, 16- to 20-segmented antennae, and lack an arolium on the tarsus. The soldiers have long, thick, parallel-sided heads; antennae 11- to 18-segmented, with the 3rd segment enlarged and clavate. *B. condonensis* forms small to moderate-sized colonies which occupy irregular gallery systems in dead and sound wood of a number of species of *Eucalyptus*.

Cryptotermes (4 spp.) is characterized by soldiers with dark, phragmotic heads, and with a vertical or steeply sloping frontal area (Fig. 15.10D); antennae 11- to 15-segmented; mandibles short and humped basally, and bent sharply near the middle. The species are found in coastal and subcoastal areas from Darwin to Victoria. They form small colonies in gallery systems in the exposed dead wood of living trees or in stumps.

3. Termopsidae. Alates without ocelli or fontanelle. First marginal tooth of right mandible with a small subsidiary tooth on the distal edge (Fig. 15.12D). Pronotum flat and narrower than the head. Scales of fore wing larger than those of hind wing. Tarsi 4-segmented. Sternum 9 of males with styles. These are the true damp-wood termites; all species are found only in wood, principally in standing trees or fallen logs, less commonly in timber in service. There is no worker caste. The family is divided into 3 subfamilies, of which 2 occur in Australia, each being represented by a single relict genus.

In the POROTERMITINAE, the alates have 5-segmented cerci, and the soldiers have long, rectangular heads, with stout, toothed mandibles about one-third the length of the head (Fig. 15.10B). *Porotermes adamsoni* (Frogg.), the only representative of this subfamily, occurs in coastal and adjacent highland areas from south Queensland to Tasmania and S.A. It is mainly found in hardwood forests, where it forms moderately large colonies in both dead and living trees as well as in logs. Colonies in trees construct gallery systems, which may extend from the roots to the main branches, and frequently include a large central pipe which is packed with moist, clay-like, faecal material. They are responsible for economic damage to commercial-timber trees, and also to poles and other structural timber in moist situations near Melbourne.

In the STOLOTERMITINAE, the alates have 4-segmented cerci, and the soldiers have markedly flattened bodies and heads, with strongly toothed mandibles that are at least half as long as the head. *Stolotermes* is represented by 4 species: one in Tasmania; *S. victoriensis* Hill in mountainous country from Victoria to southern Queensland, and two on the Atherton Tableland in north Queensland. All form small to moderate-sized colonies in rotten logs on the ground or in pockets of rot in living trees.

4. Rhinotermitidae. Alates generally with ocelli (sometimes absent in *Heterotermes*). Fontanelle present. Antennae 14- to 22-segmented. Left mandible with 1 apical and 3 marginal teeth; right mandible with a small subsidiary tooth at base of the 1st marginal (Figs. 15.12E, F). Pronotum more or less flattened. Tarsi 4-segmented. Scale of fore wing large; wings often reticulate. Cerci 2-segmented. Soldiers various, generally without eyes. Mandibles well-developed, with or without marginal teeth. Fontanelle present. Antennae 12- to 18-segmented. Pronotum flattened. The Rhinotermitidae possess a true worker caste. All species are wood-eaters and subterranean in habit, although some construct conspicuous mounds. The family is divided into 6 subfamilies, 3 of which occur in Australia.

In the HETEROTERMITINAE, the alates have elongate-oval heads with flattened sides, and the wing membrane is slightly reticulate. The soldiers have long, rectangular heads, with sabre-like mandibles devoid of obvious marginal teeth. *Heterotermes* is represented by 8 species. They are all small termites, and are widely distributed throughout Australia in both coastal and inland areas. The colonies are usually small, and are found in the soil, in or under logs and stumps, and in the mounds of other species of termites. All species are wood-eaters, and some are of minor economic importance.

In the COPTOTERMITINAE, the alates have circular heads, small clypeus, relatively broad pronotum, and the wing membrane is not reticulate. The soldiers have pyriform heads with a conspicuous fontanelle, and sabre-shaped mandibles without obvious marginal teeth (Fig. 15.10E). *Coptotermes* (6 spp.) is widely distributed throughout the mainland, and readily recognized by the soldiers' habit of exuding a drop of milky fluid from the frontal gland when disturbed. All the species appear to form large colonies, some of which may exceed a million individuals. The nests are subterranean, or in stumps or hollowed trunks of dead or living trees, or in domed or conical mounds up to 3 m high. This mound-building habit is unusual, for, although the genus is widespread in the warmer regions of the world, it is only in Australia that mounds are built. The mounds are characterized by a tough, clay outer wall up to 30 cm thick, enclosing a tough, woody, honeycomb interior. All the species are wood-eaters, and some of them cause severe damage to timber in service and to living trees. They constitute the most important economic group of termites in Australia. *C. acinaciformis* occurs in all States, but is a mound-builder only in the northern and south-western portions of its range. *C. frenchi* is restricted to eastern and southern Australia, and builds mounds

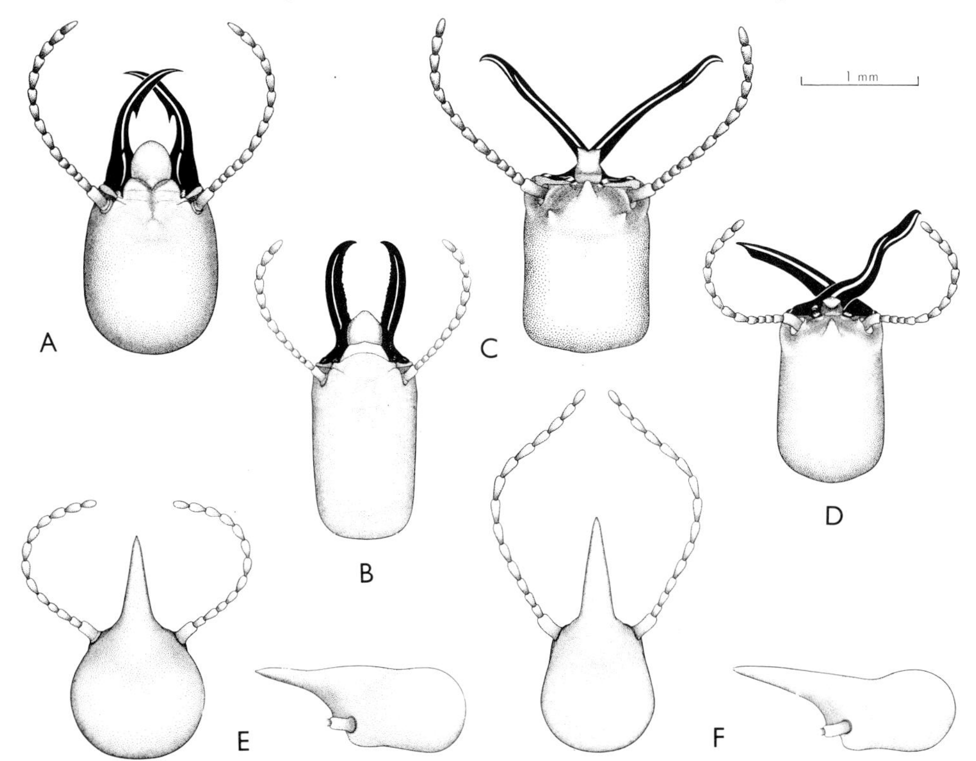

Fig. 15.11. Heads of soldiers of some genera of Australian Termitidae: A, *Amitermes*; B, *Microcerotermes*; C, *Termes*; D, *Paracapritermes*; E, *Nasutitermes*, plan and profile views; F, *Tumulitermes*, plan and profile views.
[B. Rankin]

only in the south. *C. michaelseni* Silv. in south-western Australia does not build mounds. *C. brunneus* Gay from W.A. and *C. lacteus* from south-eastern Australia appear to be almost obligate mound-builders.

The RHINOTERMITINAE are characterized by alates with circular heads; clypeus well developed and inflated, with a distinct median groove from the fontanelle to the labrum; and wings with distinct reticulation. The soldiers have an elongate, medially grooved labrum, conspicuous fontanelle, and well-developed mandibles with large marginal teeth. *Schedorhinotermes* (4 spp.) is unusual in possessing two distinct types of soldiers, differing in size and in the shape of heads and mandibles (Fig. 15.10F, G). They are all wood-eaters. Colonies are small to moderate in size, and are located in stumps, in and under logs, or in the soil. Frequently the attack on posts, logs, or dead branches is carried out under the protection of a fragile, roof-like plastering. *Parrhinotermes* is represented by a single species, *P. queenslandicus* Mjöb., found only in moist rotten logs in rain forests in north Queensland.

5. Termitidae. Alates with ocelli, and usually with a fontanelle (absent or vestigial in *Microcerotermes*). Antennae 13- to 18-segmented. Mandibular dentition various. Tarsi 4-segmented. Fore-wing scale about as large as, or only slightly larger than, hind-wing scale. M closer to Cu than to Rs. Soldiers either with rounded or rectangular heads bearing obvious mandibles with or without teeth, or with nasute-type head and vestigial mandibles. Antennae 10- to 20-segmented. Pronotum saddle-shaped. Workers with 12–19 antennal segments. Postclypeus generally inflated. Pronotum

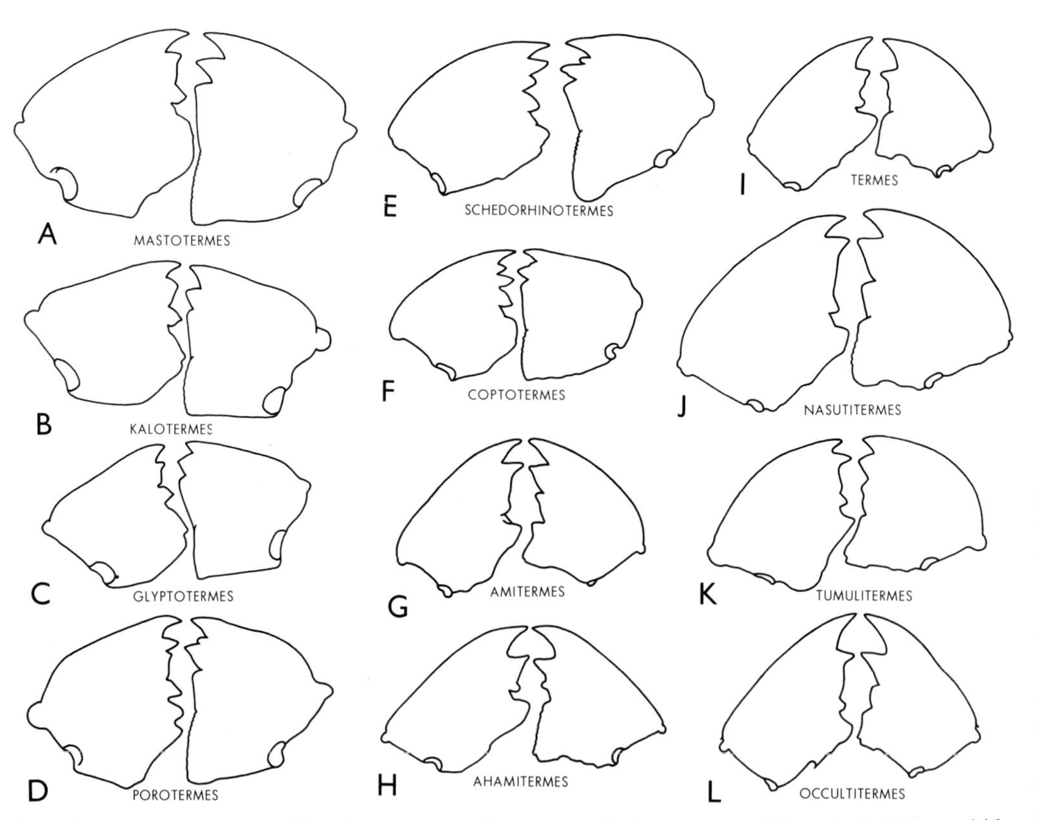

Fig. 15.12. Imago-worker mandibles in some Australian genera of: A, Mastotermitidae; B, C, Kalotermitidae; D, Termopsidae; E, F, Rhinotermitidae; G–L, Termitidae. [N. Key]

saddle-shaped. This is the largest family in the order. It is divided into 5 subfamilies, of which 3 occur in Australia. The species are mainly wood- or grass-eaters with subterranean habits; several are mound-builders, and a few construct arboreal nests. Fungus-growing termites (Macrotermitinae) are not represented in our fauna.

In the AMITERMITINAE, the alates have broadly oval heads, and the labrum is broader than long or as broad as long; antennae 14- to 18-segmented. The soldiers have oval or rectangular heads, and the mandibles are sabre-shaped with a more or less distinct tooth on each mandible, or rod-like with serrated inner margins.

Amitermes (50 spp.) is the largest Australian genus. It is characterized by soldiers with conspicuous and generally incurved mandibles each of which bears a single tooth, occasionally concealed by the labrum (Fig. 15.11A). Representatives of the genus are found all over Australia, but are particularly abundant in the drier inland areas of northern and western Australia. Most of the species are completely subterranean, and form small to moderate-sized colonies which feed on grass or vegetable debris and are also commonly found in and under heaps of animal dung; two species appear to be of slight economic importance because of attacks on timber in service. Five species are known to build mounds, some of which may contain very populous colonies, the best known being that of *A. meridionalis* in the N.T. (Figs. 15.9A, B).

In the endemic *Ahamitermes* (3 spp.), the mandibles of the soldiers are short and stout with vestigial teeth or none at all. All the species are obligate nest-eating inquilines in nests of *Coptotermes*. The derivative and highly specialized genus *Incolitermes* is represented by a single species, *I. pumilus* (Hill). The nests of this species are completely surrounded by those of its host, and it has lost the power of releasing its own alates on the colonizing flight, this being done by the workers of the host species when they release their own alates.

Drepanotermes (2 spp.) is also an endemic genus. *D. rubriceps* is a grass-eater, which is widely distributed, mainly in the drier inland areas, of all mainland States except Victoria. It forms populous colonies, which nest either in the soil or in conspicuous mounds up to 3 m high. The nests are commonly stocked with masses of chaffed grass, cut into short lengths about 1 cm long. Soldiers of *Drepanotermes* are moderately large, and vary in head colour from pale orange to almost black. They have long, sickle-shaped mandibles, each with a conspicuous, generally hatchet-shaped tooth.

Microcerotermes (11 spp.) is found all over the continent except in the south-eastern corner. The fontanelle is indistinct or wanting in the alates, and the soldiers have long, rectangular heads, with long slender mandibles which are generally obviously serrate on their inner margins (Fig. 15.11B). All species are wood-eaters and form moderate to very populous colonies. They are generally subterranean, but four species regularly build small, conical or domed mounds either on the ground, stumps, logs, or on the trunks of dead and living trees, and two of them occasionally damage timber in service.

In the TERMITINAE, the alate-worker mandibles have the apical tooth much larger than the 1st marginals and separated from them by an unusually long gap (Fig. 15.12I). The soldiers are equipped with biting mandibles, or with symmetrical or asymmetrical mandibles used for snapping (Figs. 15.11C, D).

Termes (19 spp.) is mostly confined to tropical Australia. The soldiers are characterized by rectangular or quadrate heads, with long, symmetrical, rod-like mandibles specialized for snapping, and a prominent frontal tubercle which conceals the fontanelle. Three species are known to build small, conical or domed mounds, but mostly they are subterranean or live in the nests of other termites. They form small to moderately populous colonies, and feed on weathered timber, bark, or litter. Two species occasionally damage timber in service.

In *Protocapritermes* (1 sp.), the mandibles of the soldier are symmetrical and resemble those of *Termes*, but there is no frontal tubercle concealing the fontanelle. *P. krisiformis* builds small blackish mounds only a few cm high in coastal districts from southern N.S.W. to south Queensland.

Paracapritermes (2 spp.) has soldiers with asymmetrical, rod-like mandibles adapted for snapping, and a well-developed frontal tubercle concealing the fontanelle. *P. kraepelinii* (Silv.) is frequently found in association with *Termes* in the nests of other termites in south-western Australia.

In the NASUTITERMITINAE, the alates have a long, narrow labrum, and on the left mandible the apical and 1st marginal teeth are about equal in size and the 2nd marginal is reduced to a long, blade-like area (Figs. 15.12J, K). The soldiers have pyriform heads and greatly reduced mandibles, and the fontanelle opening is at the end of a distinct frontal tube traversing the nasus or rostrum (Figs. 15.11E, F).

Nasutitermes (19 spp.) is widely distributed throughout Australia, particularly in the northern half of the continent. The soldiers are the typical nasute type, generally of only one size, and the head is without an evident constriction when viewed from above. The genus includes subterranean species which form small to moderately sized colonies, as well as mound- and tree-nest-builders which form colonies of 1–2 million individuals. Food materials include grass, vegetable debris, rotten and sound wood. *N. graveolus*, which occurs in coastal and near coastal districts from Townsville to Darwin, and *N. walkeri* in similar districts from Sydney to Cairns both construct 'nigger-head' nests on the trunks and branches of dead and living trees (Fig. 15.9D); they feed mainly on bark and weathered wood. *N. magnus* (Frogg.) is a grass-eating species which occurs commonly in coastal and near coastal districts from Grafton to Cairns. It builds low, domed mounds, up to 2 m in basal diameter and 1 m in height, which are generally abundantly stocked with chaffed grass. *N. triodiae*, a grass-eater found in northern Queensland, N.T., and north-western Australia, constructs large mounds of varying types, the most spectacular being massive columnar structures more than 7 m high. These huge nests are unique for the genus, which has a wide distribution in the warmer regions of the world. *N. exitiosus*, which extends from just north of the Queensland–N.S.W. border, through Victoria and S.A., across to W.A., is of considerable economic importance because of the damage it causes to timber in service. Throughout most of its range it builds low, dome-shaped mounds, about 1 m in basal diameter and 50 cm high.

In *Tumulitermes* (25 spp.), the soldier is nasutiform, with a relatively long, slender rostrum, and the head is generally obviously constricted behind the antennae (Fig. 15.11F). Several species have two distinct classes of soldier which differ in size and occasionally in structure. This is an endemic genus, which is widely distributed throughout Australia with the exception of the south-eastern region. For the most part they are completely subterranean, but five species are known to construct mounds. They all appear to be surface foragers, and their nests are stocked with stores of grass and other vegetable matter. *T. hastilis* (Frogg.) builds tall narrow mounds up to 2 m high in inland Queensland, N.T., and W.A. In some districts these mounds are so numerous as to constitute a distinctive feature of the landscape.

The remaining Australian genera in this subfamily are also endemic. *Occasitermes occasus* (Silv.) is recorded only from S.A. and south-western W.A.; the soldiers are of one size only, the head is not constricted, and the rostrum is long, thick, and conical. This species is subterranean, and feeds on rotten or weathered wood. *Occultitermes occultus* (Hill) in the N.T. has dimorphic soldiers which lack points on their mandibles and have either 11 or 12 antennal segments. *Australitermes dilucidus* (Hill) is found in south-eastern Queensland. The alates are characterized by a very high left mandible index* and a vestigial 2nd marginal tooth on the right

* The left mandible index is the linear distance between the pointed tips of the apical and 1st marginal teeth divided by the linear distance between the pointed tips of the 1st and 3rd marginal teeth. The 2nd marginal tooth of the left mandible is reduced and fused with the posterior cutting edge of the 1st (Emerson, 1960).

mandible (Fig. 15.12L). The soldiers are monomorphic, with a very slightly constricted head and antennae of 12 segments. *Macrosubulitermes* is represented by 2 uncommon species, one from N.T. and one from northern Queensland. The alates have a left mandible index close to 1·0, and the monomorphic soldiers have heads devoid of constriction and antennae of 12 segments. Little is known of the biology of the last three genera. The species are all subterranean, occurring either in the outer wall region of the nests of other termites or in gallery systems in the soil.

16

MANTODEA

(Praying mantids)

by K. H. L. KEY

Mandibulate, predacious, exopterygote Neoptera, having raptorial fore legs with large mobile coxae; pronotum without large, descending lateral lobes; wing rudiments of nymph not reversing their orientation in the later instars; specialized auditory and saltatorial organs lacking. Eggs enclosed in an ootheca.

Mantodea are rather large, terrestrial insects found in all the warmer parts of the world, especially the tropics. Their peculiar supplicatory posture has attracted wide attention and given rise to the popular name of 'praying mantis'. They were classified originally as a family of the Orthoptera. More recently, they have usually been treated as a suborder of 'Dictyoptera', along with the cockroaches, or as a separate order, which is the course adopted here (p. 162). General appearance and habits are rather uniform throughout the order. Some 1,800 species are known.

A general account of the group is given by Chopard (1949a) and by Beier (1964).

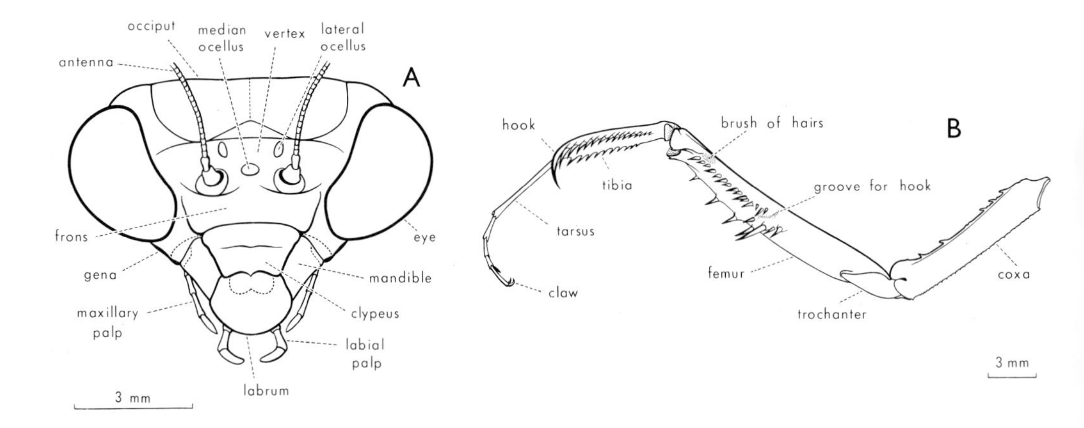

Fig. 16.1. ***Archimantis latistyla*, Mantidae, ♀: A, head, viewed normal to face; B, right fore leg, ventrocephalic (ventrointernal) view.** **[F. Nanninga]**

Anatomy of Adult

Head. Freely movable on a slender neck and rarely inserted to any extent into the pronotum; typically hypognathous, flattened anteroposteriorly, subtriangular, the large compound eyes occupying the dorsolateral corners and the narrow mouth the ventral apex (Fig. 16.1A). Antennae slender, many-segmented. Ocelli three, larger in the male than in the female. Mouth-parts comprising strong, toothed mandibles; maxillae with 5-segmented palp, soft galea, and toothed lacinia; and labium with 3-segmented palp and well-developed glossae and paraglossae. Hypopharynx large.

Thorax. Prothorax usually narrow and elongate, movable on the mesothorax; pronotum transversely arched, but without descending lateral lobes, not produced posteriorly over the wing bases; prosternum sclerotized, with short basisternum and long sternellum; propleura greatly reduced, the pronotum for most of its length connecting directly with the sternum. Meso- and metathorax similar, not elongated, rigidly connected; the sterna largely sclerotized, with large basisternum and small sternellum; pleura oblique, the epimera very narrow. Meso- and metathoracic spiracles present.

Legs. Mid and hind legs slender and unspecialized, with large coxae usually closely approximated ventrally. Fore legs (Fig. 16.1B) raptorial; coxae elongate and mobile; femora robust, spined ventrally, and with a brush of short hairs distally on the inner face; tibiae in all Australian species with ventral spines and a strong, sharp terminal hook on the inner margin; when the tibia is flexed against the femur, the spines enmesh and the hook fits into a groove on the inner face of the femur. Tarsi of all legs slender, almost always 5-segmented, with a pair of terminal claws but no arolium, those of the fore legs articulated to the tibia laterally before the apex.

Wings. In the male usually fully functional (Fig. 16.2); in the female of most Australian species reduced or absent. Fore wings narrow, usually tougher and more opaque than the hind wings, in the closed position overlying them along the abdomen, the one broadly overlapping its fellow except at the extreme base, and each with a small jugal lobe, which is folded under when the wings are closed. Hind wings much broader, membranous, except often for a narrow zone along the anterior margin, with a large anal area folding fanwise in repose. Venation (Ragge, 1955a) with C strongly developed along the whole anterior margin of both wings; Sc unbranched; R situated close behind Sc, divided near the base in the hind wing, and usually, but more distally, in the fore wing, into R_1 and Rs, which in the hind wing run parallel and unbranched to the wing apex; MA just behind R and branched or un-

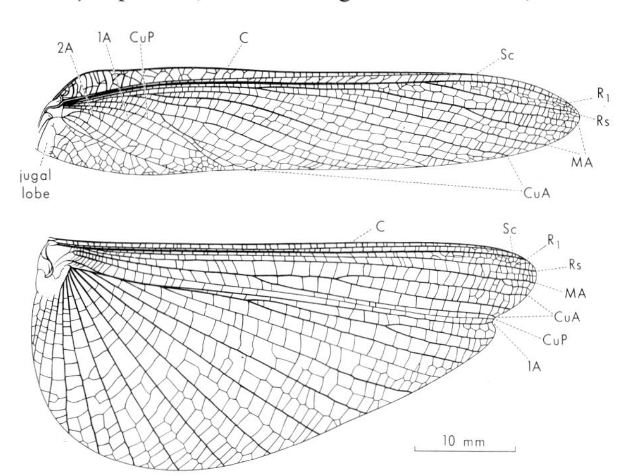

Fig. 16.2. *Archimantis latistyla*, Mantidae, wings of ♂. [F. Nanninga]

branched; CuA branched, CuP unbranched; several anals in the fore wing and many in the hind.

Abdomen. Dorsoventrally compressed, consisting of 11 segments. Tergum 10 constitutes the supra-anal plate. Sternum 1 reduced and usually not identifiable; sternum 11 represented by the paraprocts. Cerci segmented. Spiracles situated in the pleural membrane of tergum 1 and on the ventral borders of terga 2–8. In the male (Snodgrass, 1937), sternum 9 is the subgenital plate (Fig. 16.3A); it usually bears a pair of styles and partly conceals the genitalia, which comprise a complex, strongly asymmetrical, partly sclerotized group of structures of value in classification; sternum 10 is reduced and internal. In the female, sternum 7 is the subgenital plate (Fig. 16.3B); it narrows and divides posteriorly, enveloping the ovipositor, which consists of three pairs of small valves (Marks and Lawson, 1962); sterna 8–10 reduced and internal.

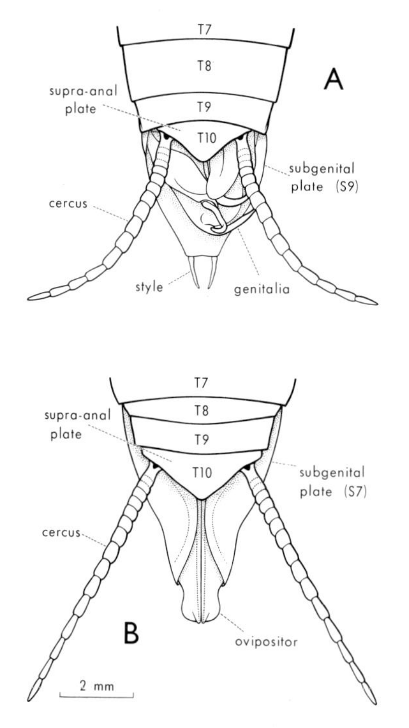

Fig. 16.3. *Tenodera australasiae*, Mantidae, dorsal view of terminalia: A, ♂: B, ♀.
[F. Nanninga]

Internal Anatomy. The alimentary canal is relatively straight, with a large crop, conical gizzard with 6 internal teeth (Judd, 1948), and short mid gut bearing 8 caeca; complex, paired salivary glands lie along the gut, and there are about 100 Malpighian tubes. The central nervous system includes 3 thoracic and 7 abdominal ganglia. The male reproductive system consists of large paired testes opening into the ejaculatory duct, to which are connected paired vesiculae seminales and a mass of tubular accessory glands, whose secretions are involved in the formation of the spermatophore. In the female, ovaries composed of numerous panoistic ovarioles discharge into a common oviduct; the spermatheca and the voluminous accessory glands, whose secretions form the ootheca, open independently into the genital atrium.

Karyotype. The Mantodea have a chromosome set with 2n♂ ranging from 15 to 39 in the species studied (White, 1965), all or most of the chromosomes being usually metacentric. The male is usually XO, but in many genera there is an X_1X_2Y sex mechanism.

Immature Stages

The egg is cylindrical and thin-shelled. Upon hatching, the young make their way out of the ootheca (p. 298) via the mid-dorsal apertures. Each is enclosed in the embryonic cuticle, which bears minute backwardly-directed spines on the dorsal surface and a dark, strongly sclerotized, spindle-shaped plate anteriorly; the appendages are held against the body, pointing backwards. Immediately after emergence, while the hind end is still lodged in one of the apertures—or, in some species, while suspended from the ootheca by silken threads—the insect casts the embryonic cuticle and the appendages can be moved freely. This brief first stage has been termed the 'pronymph', or 'vermiform larva', and the ecdysis by which it is terminated the 'intermediate moult'; it is not regarded as one of the nymphal instars.*

* The term 'instar' is used in this chapter in the conventional sense of a developmental stage delimited by ecdysis, which in turn is treated as the final and definitive episode of the moult.

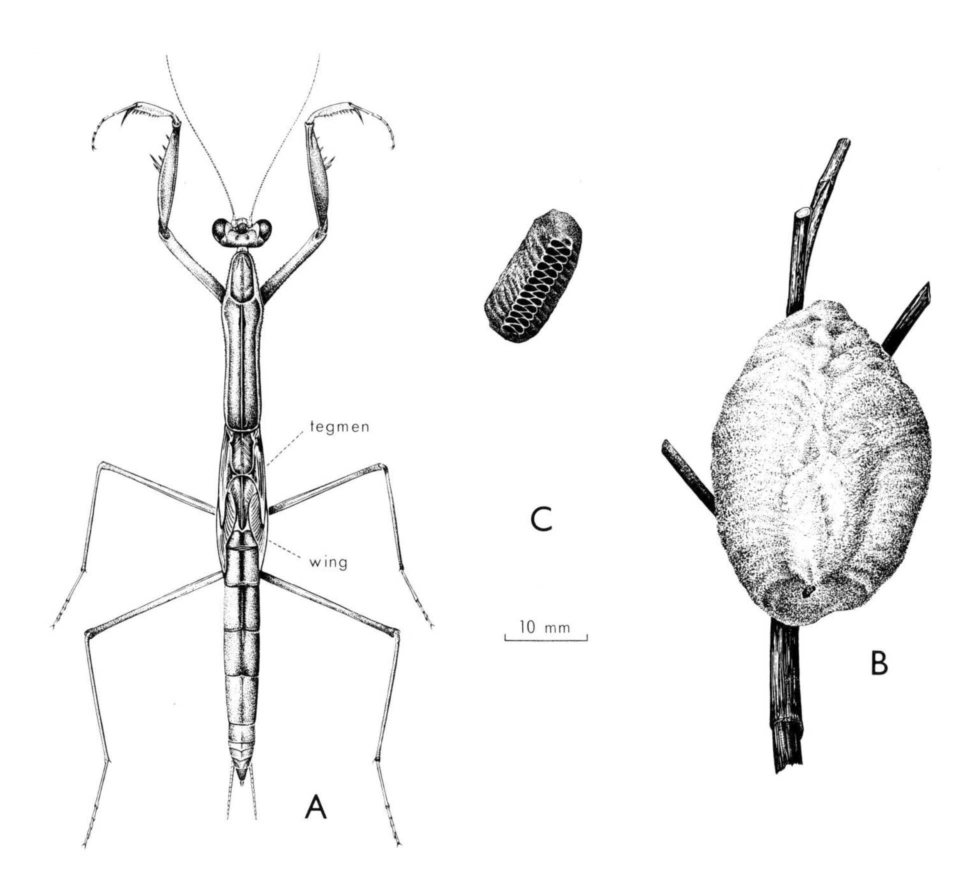

Fig. 16.4. Mantidae: A, *Tenodera australasiae*, late nymph; B, ootheca of *Archimantis latistyla*; C, ootheca of *Orthodera ministralis*. [F. Nanninga]

The nymphs (Fig. 16.4A) show only minor morphological differences from adults, involving principally the rudimentary wings, immature terminalia, and lower number of antennal segments. They pass through several ecdyses (usually fewer in the male than in the female), which take place in an inverted position from some support. The wing rudiments and terminalia increase in size from instar to instar, the former not reversing their orientation in the later instars. All pre-maturation divisions are usually completed in the testis by the time of the last ecdysis.

Biology

Much information on the biology of the order is given by Chopard (1938).

Adults. The Mantodea are solitary insects, frequenting principally shrubs, tall herbs, or the trunks of trees, but a few species live on the ground. Most of their time is spent at rest, with the fore legs raised in the supplicatory attitude. If disturbed, they may run rapidly or take to flight; many species can make short jumps, and some carry out gentle swaying movements of the body, reminiscent of the Phasmatodea. Some species, especially those with brightly coloured wings, will stand their ground when approached, raise their wings, and adopt an aggressive 'frightening' attitude, while striking out with the spined fore legs, which are capable of inflicting a painful prick (Crane, 1952). Autotomy of the mid and hind legs may occur, rarely of the fore legs. Males of many species fly to light on warm evenings. The females are often flightless, the wing area being insufficient to bear the weight of the body.

All Mantodea are strictly carnivorous, feeding on various insects and sometimes spiders, which they catch and hold with their raptorial fore legs. These are so constructed that they can be thrust forward, seizing the prey in the pincer-like grip of the spined tibia and femur; it can then be carried to the mouth by retraction of the limb towards the body and held there until consumed. The predator usually remains quite still until the prey comes within striking range. Many species possess procryptic adaptations of colour or structure (p. 131); it is not known whether this is merely a protection against their own enemies, as in many non-predacious orthopteroids, or whether it also has the function of concealing them from potential prey. Black or coloured markings are often present on the inner face of the fore coxa and femur, and are displayed in certain attitudes, but their function is not certainly known.

Reproduction. Copulation takes place with the male mounted on the female and his abdomen curved under hers to permit engagement of the copulatory organs from below. He transfers a small spermatophore containing the spermatozoa. In some species the female may attack and devour the male during the courtship approach.

The female deposits her eggs in an ootheca (Figs. 16.4B, C), formed from a frothy material produced by the accessory glands (p. 70) and moulded by movements of the ovipositor valves. This hardens upon drying to form a horny capsule, often encased in a voluminous spongy envelope. Ranged along the mid-dorsal line of the ootheca is a series of covered apertures leading into subvertical chambers containing the eggs. The size and shape of the ootheca varies with the species, as does the situation chosen for its reception. It may be attached to a relatively flat substrate, such as the trunk of a tree, a rock, or a wall; it may partly envelop a twig or group of grass stems; or it may be deposited in the soil. Commonly a female will construct several oothecae, each of which may contain from about 10 to 400 eggs, according to the species; in a few species she mounts guard over the ootheca until the young nymphs hatch.

Obligatory parthenogenesis occurs in an American mantid, in which males are unknown (White, 1951).

Nymph. After the intermediate moult, the nymph behaves in essentially the same way as the adult, feeding upon small, soft-bodied insects. Regeneration of lost appendages may occur. The nymphal period lasts some weeks or months.

Polymorphism, Colour Adaptation. Limited genetic colour polymorphism seems to occur in some species, involving especially the alternatives of a predominantly green, or predominantly straw or buff general coloration. Adaptation of the colour of the insect to that of the substrate during the life of the individual may also occur (Ergene, 1953).

Ecological Features. The density of mantodean populations is low, possibly because of the high incidence of egg parasitism and the vulnerability of the early nymphal instars. Cannibalism is frequent in cultures, and could be expected to occur in the field as densities increase, quite apart from its known occurrence in mating pairs. Indeed, some form of territorialism seems probable. Feeding is not concentrated upon particular prey species and in general prey is not actively sought out. In these circumstances the predator could not be expected to bear heavily upon any prey species, and its own population regulation is likely to be brought about in the main by agencies other than food shortage.

Natural Enemies. The eggs of Mantodea are often heavily parasitized by proctotrupoid and chalcidoid Hymenoptera. In Australia the latter are represented especially by species of *Podagrion.* Their small circular emergence holes may often be seen studding the surface of oothecae. Chloropid flies are often bred from oothecae, but whether they are parasites or only scavengers is not clear; in Australia, numerous *Botanobia tonnoiri* Mall. have been reared from oothecae of *Tenodera australasiae* (Leach). Crickets are also said to prey upon the oothecae. The active stages are subject to predation by birds, lizards, insectivorous mammals, and sphecid wasps, and to parasitism by mites and mermithid nematodes.

Economic Significance. It is doubtful whether Mantodea have any economic significance.

Special Features of the Australian Fauna

The mantodean fauna of Australia is characterized by the small number of families and subfamilies represented. It includes some 35 genera, two-thirds of which are endemic, although none of the subfamilies is. All the subfamilies and many of the genera have an almost Australia-wide distribution.

CLASSIFICATION

Order MANTODEA (118 Australian spp.)

1. Amorphoscelidae (38) 2. Mantidae (80) 6 exotic families (0)

The classification of Beier (1964) is followed here, except for minor details. He recognizes 8 families and 32 subfamilies, of which 2 families and 4 subfamilies are here accepted as Australian. At the generic and specific levels, the more significant works bearing on the taxonomy of the Australian Mantodea are those of Giglio-Tos (1927), Tindale (1923–4), Beier (1935), and White (1965).

Key to the Families of Mantodea Known in Australia

Ventroexternal margin of fore tibia without spines .... **Amorphoscelidae**
Ventroexternal margin of fore tibia with a row of spines .... **Mantidae**

1. Amorphoscelidae. This family attains its greatest development in Australia, where it is represented by several genera in the subfamily PARAOXYPILINAE. All the species are small and most of them frequent the trunks of rough-barked eucalypts, where their mottled procryptic pattern renders them inconspicuous. Several have various prominences or spines on the head and on the short pronotum. In *Paraoxypilus* (Fig. 16.5) and *Cliomantis* (Fig. 16.6A) the males are fully winged and the females apterous.

2. Mantidae. The Mantidae are by far the largest family of the order. We recognize only 3 Australian subfamilies, Beier's Compsothespinae being treated as not Australian, and the Australian genera in his Liturgusinae being transferred to Iridopteryginae.

The IRIDOPTERYGINAE are mainly small, delicate mantids with only 3 'discoidal' (i.e. mid-ventral) spines on the fore femur. The

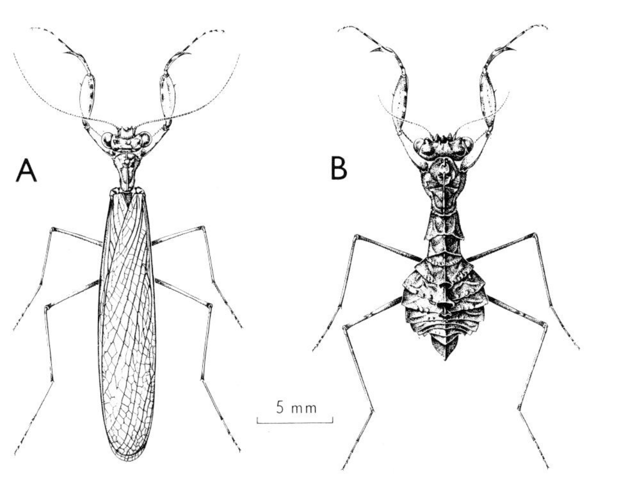

Fig. 16.5. *Paraoxypilus tasmaniensis*, Amorphoscelidae: **A**, ♂; **B**, ♀. [F. Nanninga]

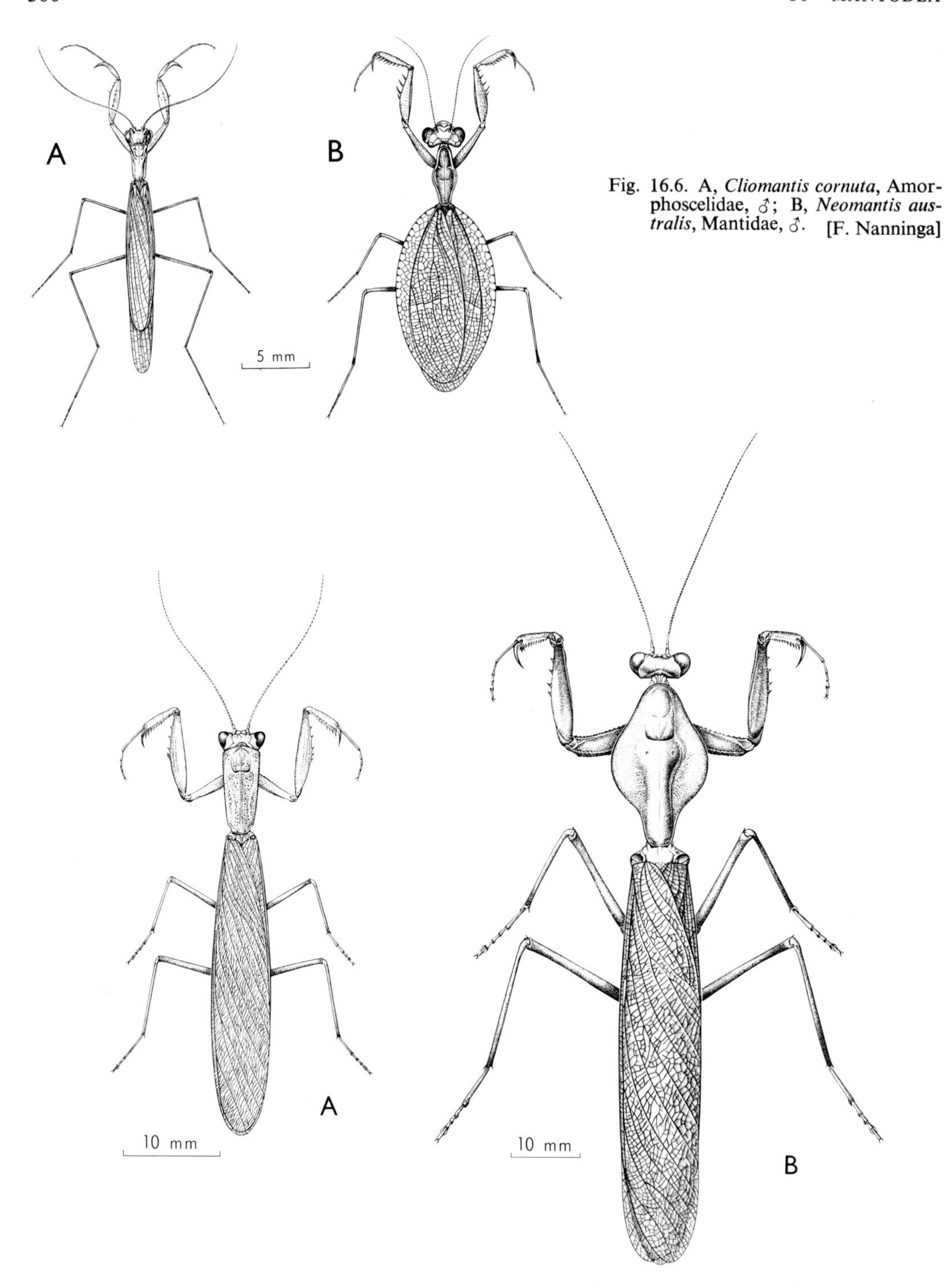

Fig. 16.6. A, *Cliomantis cornuta*, Amorphoscelidae, ♂; B, *Neomantis australis*, Mantidae, ♂. [F. Nanninga]

Fig. 16.7. Mantidae: A, *Orthodera ministralis*, ♀; B, *Hierodula tamolana*, ♂. [F. Nanninga]

genus *Bolbe* has fully winged males, which are often attracted to lights, and apterous females sometimes seen running rapidly on the ground. *B. pallida* Tind. has an XO sex-determining mechanism and achiasmatic meiosis (White, 1965); 2n♂ = 25. *Neomantis australis* (Sauss. & Zhnt.) is a small, pale green, northern species (Fig. 16.6B) with the general appearance of a hemerobiid (Neuroptera).

The ORTHODERINAE are medium-sized, green mantids, with the face strongly oblique to horizontal and the lateral margins of the pronotum nearly straight and somewhat expanded, especially anteriorly, where it is almost as wide as the head. The fore wing has a uniform, dense, and predominantly longitudinal reticulation between the main veins. *Orthodera ministralis* (F.) (Fig. 16.7A) may frequently be found on shrubs and eucalypt regrowth in all parts of Australia. Its horny oothecae (Fig. 16.4C) are commonly attached to fence posts and walls; their chemical composition has been studied by Hackman and Goldberg (1960). *Orthodera* possesses an X_1X_2Y sex mechanism, with 2n♂ = 25.

The largest subfamily is the MANTINAE, characterized by 4 discoidal spines on the fore femur, and a pronotum that narrows anteriorly from its widest point above the insertion of the fore coxae. All members of the group whose karyotype has been studied have the X_1X_2Y sex-determining mechanism and all Australian genera investigated have 2n♂ = 27. The genus *Mantis* is represented by *M. octospilota* Westw., and another cosmopolitan genus, *Tenodera*, by 2 species, of which *T. australasiae* (Plate 2, A) is one of the best known Australian mantids. The endemic genus *Archimantis* includes the large east coastal species *A. latistyla* (Serv.) (Fig. 16.8); females in this genus have abbreviated wings. *Hierodula tamolana* (Brancs.) is a large, heavily built species with the pronotum strongly expanded in the middle region (Fig. 16.7B). The endemic genus *Rhodomantis* comprises several very slender, dun-coloured species, found at or near ground level in open country in inland Australia. In the female the abbreviated hind wings are mainly black with a strong violet to blue metallic iridescence. The ootheca is known for a number of Australian Mantinae; it usually has a well developed spongy envelope (Fig. 16.4B); in *Rhodomantis* it is formed beneath the surface of sandy soil.

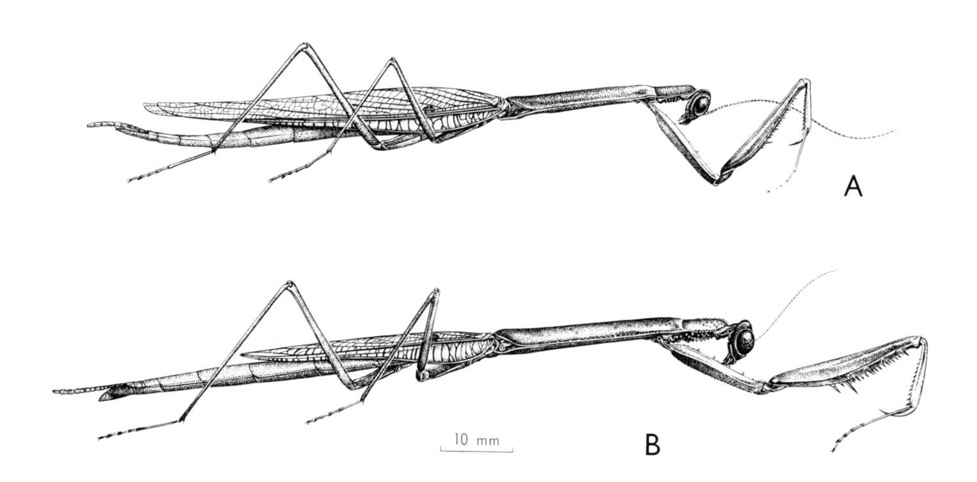

Fig. 16.8. *Archimantis latistyla*, Mantidae: **A**, ♂; **B**, ♀.
[F. Nanninga]

17

ZORAPTERA

by C. N. SMITHERS

Small exopterygote Neoptera; with moniliform antennae; Y-shaped epicranial suture; mandibulate mouth-parts; tarsi 2-segmented; wings, when present, membranous, with reduced venation, aptery common; cerci short; females with ovipositor greatly reduced or absent.

This is an order of 22 described species of one genus, *Zorotypus* (Fig. 17.1), recorded from the Ethiopian, Oriental, Nearctic, Neotropical, and Pacific regions; recently found in New Guinea, but not yet in Australia. They are less than 3 mm long, polymorphic, and have been found living gregariously under bark, in rotting wood, and in termites' nests. There are two schools of thought about their relationships, one associating them with the Psocoptera (i.e. with the hemipteroid orders), the other with the blattoid–orthopteroid orders. The head and mouth-parts are more blattoid than hemipteroid, and the structure of the thorax is reminiscent of that of the Isoptera. The presence of cerci is a blattoid feature; they are lacking even in fossil psocids. The male genitalia of Zoraptera also appear to be blattoid. On the other hand, apparent hemipteroid features in the internal anatomy and the superficial resemblances to psocids could be the result of convergence, so the weight of evidence would seem to favour a blattoid ancestry.

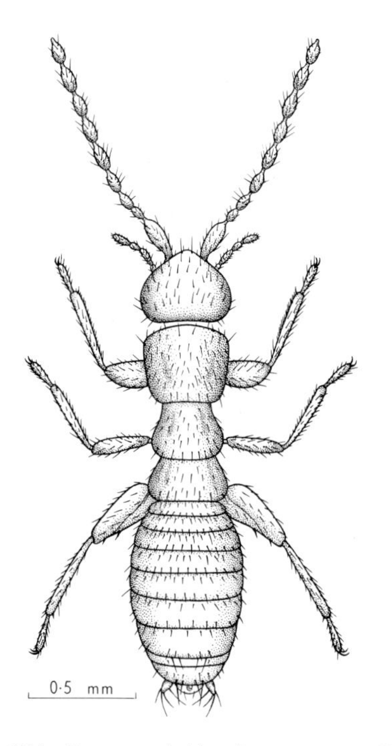

Fig. 17.1. *Zorotypus hubbardi*, apterous ♂, U.S.A. [B. Rankin]

Useful accounts of the order have been given by Tillyard (1926b), Gurney (1938), and Bolivar y Pieltain and Coronado-G (1963).

Anatomy of Adult

Head. Hypognathous; epicranial suture distinct; antennae 9-segmented; mandibles strong, the left bearing a group of hairs;

maxillae with lacinia fused to stipes, palpi 5-segmented; labium with divided prementum, palpi 3-segmented.

Thorax. Prothorax well developed, with undifferentiated notum; pterothoracic terga of winged forms with differentiated prescutum, scutum, scutellum, and postnotum; apterous forms with simple terga. Pleura of alates and apterous forms essentially similar, and with usual sclerites recognizable, but with a differentiated lateropleurite and laterosternite in the alates; apterous forms with pleural suture oblique, anteriorly displaced dorsally. Sterna broad, intersegmental membranes extensive.

Legs. Coxae large, femora fairly stout, tibiae cylindrical, 1st tarsal segment shorter than 2nd.

Wings. Hind wings smaller than fore wings; venation of both simplified (Fig. 17.2). The wings can be shed at basal fractures, the stumps remaining.

Abdomen. Eleven-segmented, with unsegmented cerci. The male genitalia are often asymmetrical, but the homologies of their parts are obscure. Delamare-Deboutteville (in Tuxen, 1956) has compared them with those of mantids.

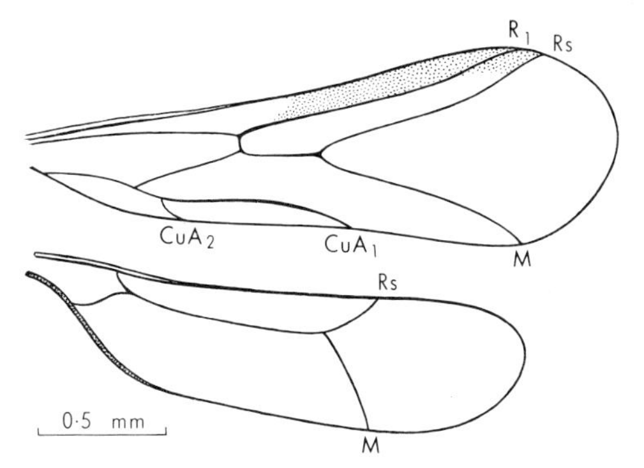

Fig. 17.2. Wing venation of *Zorotypus hubbardi*.

Internal Anatomy. Alimentary canal with a large crop extending well into abdomen; mid gut short, hind gut convoluted; 6 rectal papillae and 6 Malpighian tubes present. Nervous system concentrated, with 3 discrete thoracic and 2 abdominal ganglia. Testes round or ovoid, connected to a large seminal vesicle from which runs the ejaculatory duct; a pair of accessory glands present. Female reproductive system of 4–6 panoistic ovarioles and a spermatheca, leading by a narrow tube to gonopore behind sternum 8.

Immature Stages

The eggs are ovoid, with a finely granular chorion. Hatching is assisted by an egg-burster on the embryonic cuticle, which is shed immediately on eclosion. The nymphs are of two forms, those with wing-buds giving rise to pigmented winged adults, those without giving rise to pale apterous adults.

Biology

Little is known of the life histories of these gregarious insects, but there does not seem to be any social organization. They appear to be mainly fungivorous, but mite remains have also been found in the gut. Two forms of adults have been described for most species: a pigmented form, winged, with eyes and ocelli; and a pale form, apterous and blind.

ACKNOWLEDGMENT. I am grateful to Dr A. B. Gurney, U.S. National Museum, for helpful criticism of the manuscript.

18

GRYLLOBLATTODEA

by K. H. L. KEY

Apterous, mandibulate, exopterygote Neoptera, having all legs cursorial, with large coxae; pronotum without descending lateral lobes; auditory organs lacking; cerci long and flexible, segmented; sternum 9 of male bearing articulated coxites; ovipositor strongly projecting. Eggs free; young terrestrial.

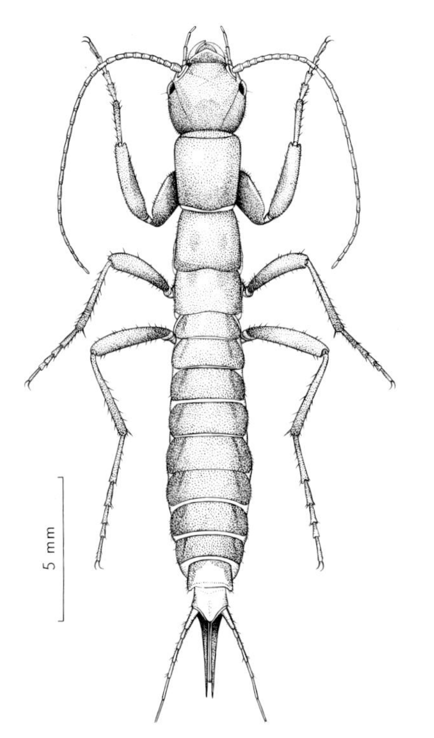

Fig. 18.1. *Grylloblatta campodeiformis*, ♀, U.S.A. [F. Nanninga]

The Grylloblattodea are a small group of soft-bodied, apterous, terrestrial insects found in cold, wet habitats in North America, Siberia, and Japan. The known forms are slender, somewhat depressed, finely pubescent, and 2–3 cm long (Fig. 18.1). They combine apparently primitive characters of both blattoid and orthopteroid orders with specializations associated with an unusual mode of life. Zeuner (1939) suggests they are living representatives of the fossil order Protorthoptera. Little more than a dozen species, in 3 genera and a single family, have been described. It is unlikely that the order will be found to occur in Australia. A bibliography is given by Gurney (1948) and later work noted by Gurney (1961).

Anatomy of Adult

Head. Flattened, prognathous. Antennae filiform, with numerous segments, inserted anterolaterally close to the base of the mandibles. Eyes small to absent; ocelli wanting. Mandibles toothed, without a molar region; maxillae with 5-segmented palp, 2-segmented galea, and toothed lacinia; labium with 3-segmented palps, paraglossae, and smaller glossae. Hypopharynx broad, flattened.

Thorax. The 3 segments similar, free, the prothorax the largest. Terga flattened; the pronotum without descending lateral lobes.

Sternal region comprising several separate sclerites joined by membrane. Pleura oblique; in the prothorax small, in meso- and metathorax weakly sclerotized. Meso- and metathoracic spiracles present.

Legs. All pairs similar, cursorial. Coxae large, but separated ventrally. Tarsi 5-segmented, with terminal claws, but no arolium.

Abdomen. Consisting of 11 segments, of which the first is quite free from the metathorax and bears ventrally an eversible sac. Tergum 10 is the supra-anal plate, the epiproct (tergum 11) being depressed and vestigial; paraprocts poorly developed. Cerci long and flexible, 5- to 9-segmented. Spiracles in the pleural membrane of segments 1–8. In the male, the large sternum 9 constitutes the subgenital plate; it bears a broad, triangular, scoop-shaped left coxite and a narrower, more digitiform right one, each with a terminal style (Fig. 1.28A). The asymmetrical genitalia, freely exposed between the coxites, comprise principally a right lobe bearing sclerotized processes and a left lobe with an eversible, membranous sac. In the female, sternum 8 is not extended backwards to form a subgenital plate; the strongly projecting ovipositor consists of 3 pairs of slender, tapering, partly free valves.

Internal Anatomy. Alimentary canal with a large crop and muscular gizzard; mid gut short and broad, with 2 ill-defined caeca anteriorly. Salivary glands closely apposed to the oesophagus, without reservoir. Malpighian tubes 12–24 in *Grylloblatta campodeiformis* Walk., opening independently into the intestine. Central nervous system with 3 thoracic and 7 discrete abdominal ganglia.

Immature Stages

The eggs of *G. campodeiformis* are large and black (Walker, 1937). The young closely resemble the adult, except for the undeveloped state of the reproductive organs and the smaller number of antennal and cercal segments and of ommatidia in the eyes. There are 8 preimaginal instars, of which the last is completely white except for the eyes.

Biology

The Grylloblattodea are cryptozoic insects, confined to cold, wet situations—beneath stones on the fringe of snow patches, in the rock crevices of moist talus slopes, in ice caves, or in rotting logs, usually at high altitude. It is believed they may descend to depths of several feet during the warmer months of the year. They have a thin, permeable integument, and a narrow temperature tolerance with an optimum only a few degrees above freezing. They appear to be comparatively rare, probably largely nocturnal insects; when disturbed during the day-time they run rapidly for shelter. Both moss and insects, especially if dead or immobile, may be eaten.

Rates of metabolism, food consumption, and development are low. In *G. campodeiformis*, eggs are first laid about a year after the final moult. They are inserted singly into soil or moss. The incubation period is said to be about a year, and the 8 juvenile instars require 5 years for their development, giving a total life cycle of 7 years. The unpigmented final instar is probably never exposed to the light.

19

DERMAPTERA

(Earwigs)

by E. T. GILES

Elongate, prognathous, winged or apterous, exopterygote Neoptera, with cerci modified into terminal forceps; thorax with many free sclerites; fore wings reduced to small tegmina; hind wings large, membranous, semicircular, almost entirely made up of anal fan, and almost completely folded beneath tegmina at rest; legs relatively short, cursorial, tarsi 3-segmented; abdomen long, freely movable, telescopic.

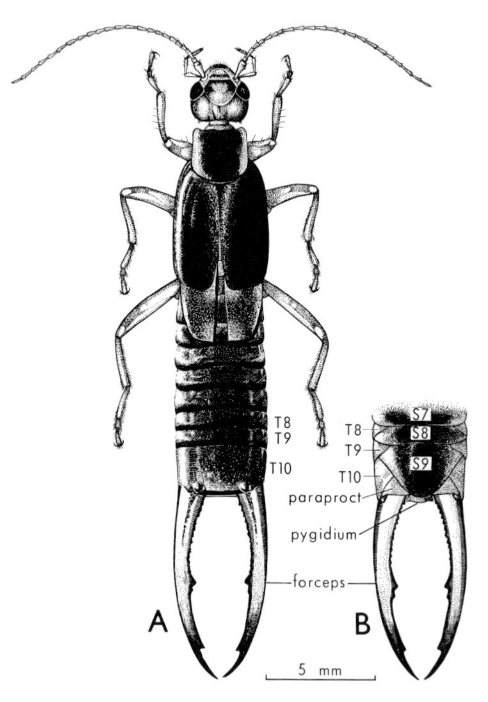

Fig. 19.1. *Labidura riparia truncata*, Labiduridae, ♂: A, dorsal; B, terminal segments, ventral. [T. Nolan]

Nearly 1,200 species of Dermaptera have been described from all parts of the world (except the polar regions), but only about 60 are known from Australia. All are elongate and flattened, the mobile, telescopic abdomen ending in a pair of forceps (Fig. 19.1). They range from approximately 7 to 50 mm long, and vary in colour from buff to black. They favour damp, confined spaces and, although nocturnal, are attracted to lights. Their relationships have been investigated by Giles (1963), who concluded that they are closer to the primitive Grylloblattodea than to any other orthopteroid order.

Earwigs are highly distinctive, but may be confused with Japygidae (Diplura) or Staphylinidae (Coleoptera). The Japygidae (p. 216) burrow into forest leaf litter. They can be distinguished by the creamy white colour of the body (excluding brown terminal segments and forceps), the absence of eyes, the entognathous mouth-parts, 1-segmented tarsi, and the abdominal segments not overlapping. The Staphylinidae (p. 544) are active ground scavengers in which the end of the abdomen bears bristles and straight processes instead of the forceps of earwigs.

Anatomy of Adult

Head. Broad, flattened, freely movable on neck; prognathous, with typical mandibulate mouth-parts. Compound eyes large in Forficulina, small or wanting in other recent suborders; ocelli wanting. Genae much reduced in front of eyes, but much inflated behind; Y-shaped ecdysial line usually conspicuous on vertex. Antennae of short annulate type. Clypeus divided into sclerotized postclypeus and membranous anteclypeus. Mandibles with 2 apical teeth; inner edge with distal incisor region and proximal molar area, except in Arixeniina and Hemimerina. Parts of maxillae much divided and primitive. Hypopharynx with 3 distal lobes. Labium large, primitive, without glossae, but with gula attached and almost covering ventral surface of head. Neck cylindrical, unprotected, with 3 series of sclerites.

Thorax. Segments free; pro- and mesothorax subequal, metathorax largest; generally primitive, with many free sclerites, Fully winged species have large pronotum, with ecdysial line; mesonotum small, comprised mainly of scutum; metanotum large, also comprised mainly of scutum, with 2 lines of bristles to lock tegmina at rest. Apterous species have large pronotum, smaller mesonotum, and still smaller metanotum, all with ecdysial line. Pleura always exposed, with many free sclerites and fairly extensive membrane; two pairs of spiracles in membrane. Sterna increase in size from in front backwards, metasternum very large; overlap from front to rear; intersegmental sclerites small.

Legs. Fairly short, subequal. Coxae wide apart and short. Tarsi 3-segmented, 1st and 3rd tarsomeres long and 2nd very short. Euplantulae, pulvilli, arolium, and empodium usually lacking. Pretarsal claws long.

Wings. Tegmina small, smooth, lacking definite veins. Hind wings large, almost entirely membranous, semicircular (Fig. 19.2). Pre-anal veins very much reduced and confined to partly sclerotized remigium. Flying membrane the very large anal fan supported by 10 radiating branches of 1A; 2A and 3A short, simple; intercalaries between branches of 1A and between 2A and 3A; cross-vein links all elements of anals. At rest, folded fanwise and then longitudinally beneath tegmen, except for sclerotized squama.

Abdomen. Long, freely movable, telescopic, depressed. Ten segments visible in male, but 8 in female, in which terga and sterna 8 and 9 are reduced and fused to tergum 10; forceps usually more curved in male (Fig. 19.1) than in female (Fig. 19.3). Terga overlap for about one-third their lengths; tergum 10 strongly sclerotized, rectangular. Sterna overlap similarly; sternum 1 wanting. Subgenital (or subanal) plate is sternum 9 in male and 7 in

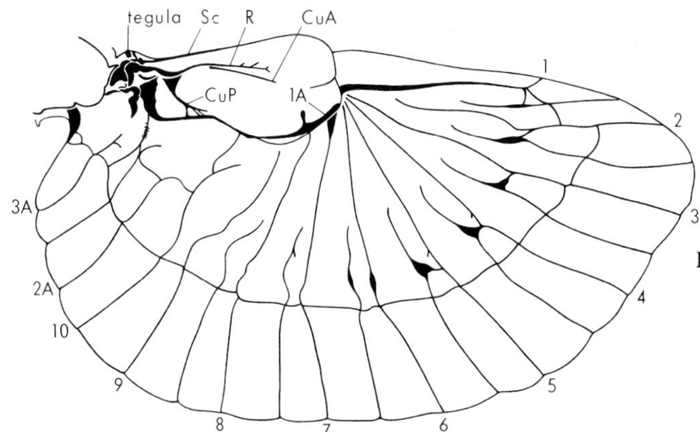

Fig. 19.2. Hind wing of *Echinosoma* sp., Pygidicranidae (after Giles, 1963).

female; large, triangular in both (Figs. 19.1, 3); in male sclerotized, wire-like loop, the *manubrium*, developed in middle of anterior edge for attachment of genitalic muscles. Pleural sclerites wanting; tergum overlaps sternum laterally in each segment; 8 pairs of spiracles in membrane. Triangular paraprocts at base of forceps, which are separated by epiproct of 2 plates: an anterior dorsal *pygidium* and posterior ventral *metapygidium*.

GENITALIA. In male (Fig. 19.4) lie in genital chamber above subgenital plate, and consist of large, subcylindrical, partly sclerotized and partly cuticular organ, with proximal single part within body wall and distal paired section outside. Each distal part consists of a penis, itself single basally and double apically. In primitive Labiduroidea each penis further divided into cuticular *medial lobe* traversed by *virga* (sclerotized terminal portion of the ejaculatory duct) and heavily sclerotized *lateral lobe*; in more specialized Forficuloidea (Fig. 19.4), Arixeniina, and Hemimerina one medial lobe lost and associated virga aborted to varying degree. Number and ornamentation of medial lobes, shape of lateral lobes, and structural details of virga all of systematic importance. In female

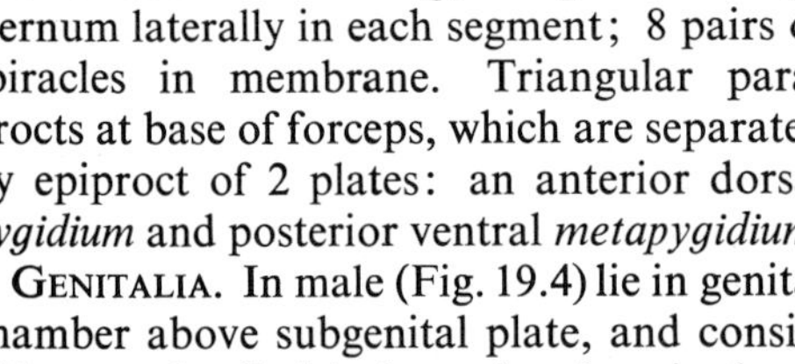

Fig. 19.3. Abdomen of *Labidura riparia truncata*, ♀: A, dorsal; B, ventral. [T. Nolan]

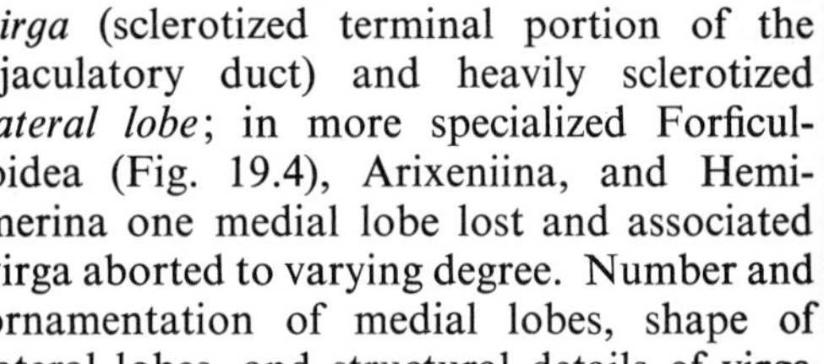

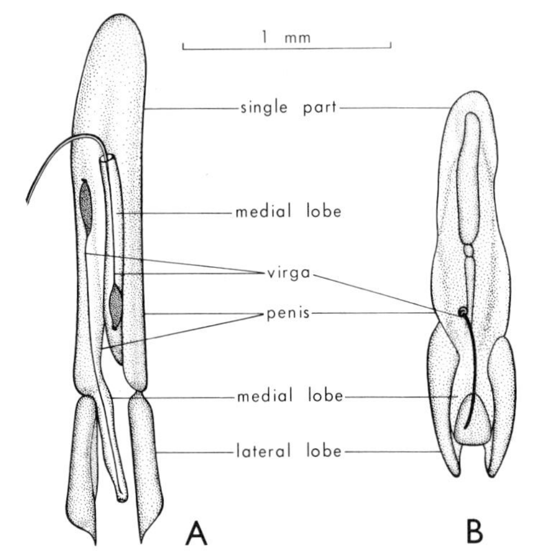

Fig. 19.4. Male genitalia, diagrammatic, of: A, *Labidura riparia truncata*; B, *Forficula auricularia*, Forficulidae. [B. Rankin]

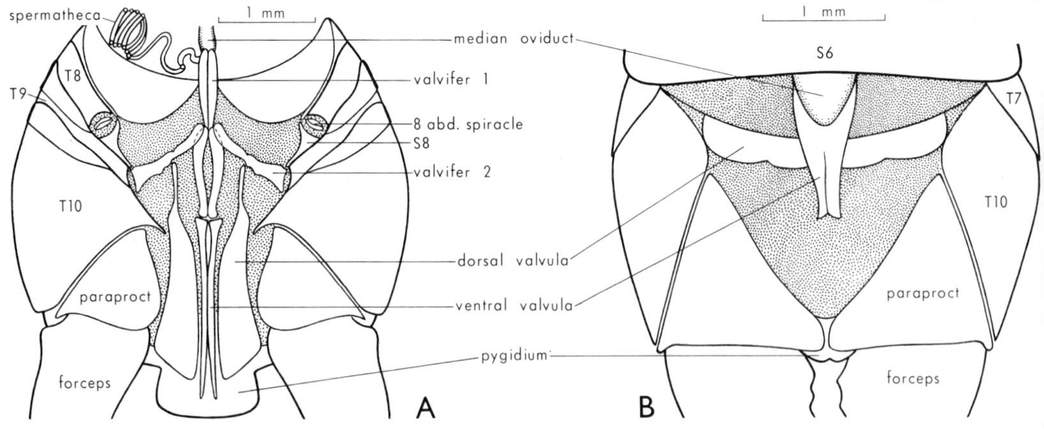

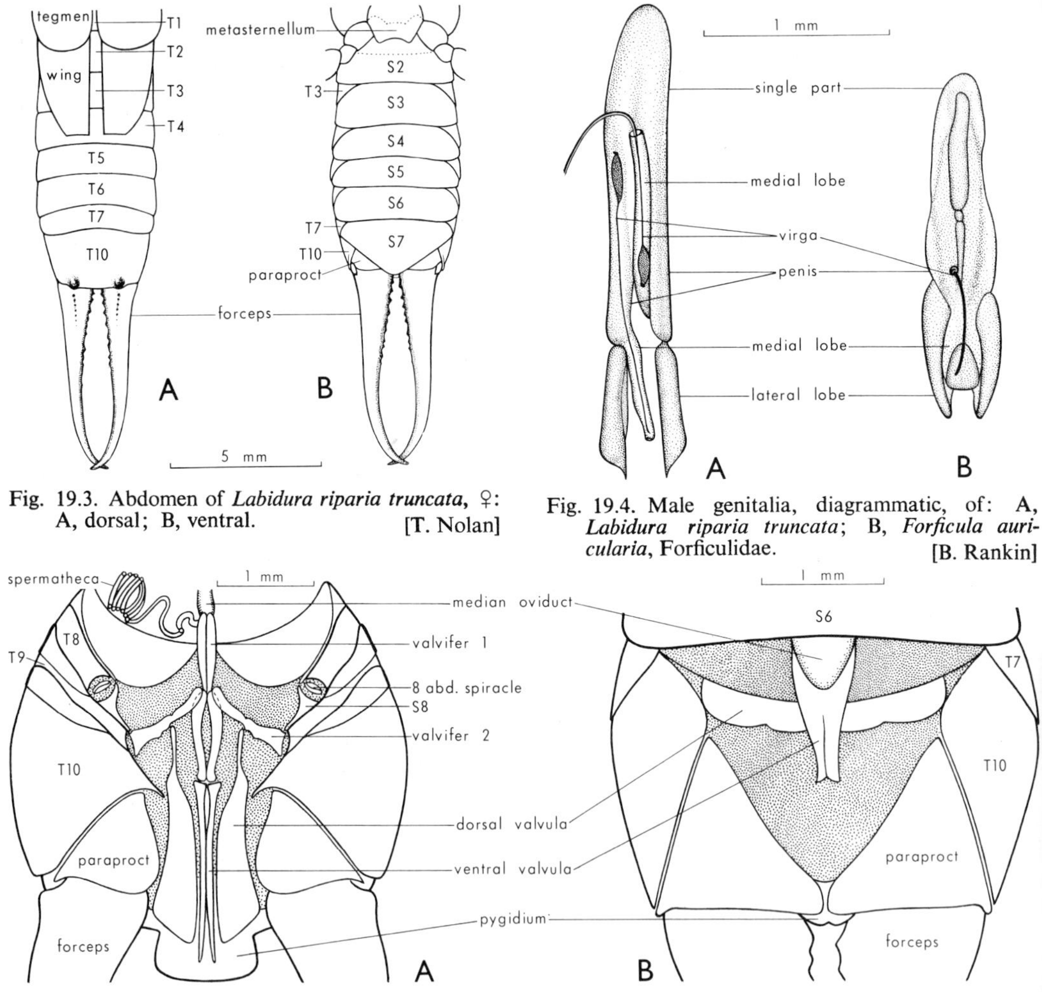

Fig. 19.5. Female genitalia, diagrammatic ventral view with subgenital plate (sternum 7) removed, of: A, *Echinosoma* sp. (after Giles, 1963); B, *Labidura riparia truncata*. [B. Rankin]

(Fig. 19.5) ovipositor always much reduced and concealed by subgenital plate, except in most Pygidicranidae in which valves project slightly; comprised of 2 pairs of valves, flattened laterals and rod-like medials (probably dorsal and ventral valvulae respectively). In both sexes the genitalia are unique among orthopteroids.

Internal Anatomy. Alimentary canal typical of orthopteroid insects; shows all usual regions, except salivary glands reduced and gastric caeca absent. Central nervous system with brain, sub-oesophageal ganglion, 3 thoracic, and 6 abdominal ganglia, all joined by thick connectives. Tracheal system without air sacs. Malpighian tubes long, slender, numerous in haemocoele, opening into commencement of hind gut in 4 groups; fat body diffuse and large, particularly just before breeding season. Female reproductive organs simple, comprising paired ovaries and lateral oviducts, median oviduct, spermatheca, and genital chamber; single genital aperture between abdominal segments 7 and 8. The polytrophic ovarioles have two types of arrangement in the ovaries: first (as in *Forficula auricularia* L.), numerous short ovarioles in 3 series around oviduct; second (as in *Labidura riparia truncata* Kirby), a few long ovarioles branching together in linear series from end of lateral oviduct. Forficulina oviparous, but Arixeniina and Hemimerina exhibit pseudoplacental viviparity. Male reproductive organs complex; portions are paired, but extent varies even within families; composed of paired testes, paired vasa deferentia, paired or single vesicula seminalis, and paired or single common ejaculatory duct ending in the sclerotized virga.

The anatomy of certain Dermaptera has been worked out by Giles (e.g. 1963, 1965). Earlier literature was reviewed in 1963, and most of the figures used there apply equally well here, for the order is notably uniform. Chopard (1949c) has given a good general account of the order.

Immature Stages

Egg. Ovoid, with thin, semitransparent, smooth, creamy white chorion; up to 2 mm long in larger species.

Nymph. Apart from size, nymphs generally resemble adults. In winged species, wing pads (always small) appear in 2nd or 3rd instar. Always distinguishable from adults by lighter colour, shorter antennae, conspicuous ecdysial line on head and thoracic terga, and male-type 10-segmented abdomen with straighter female-type forceps. Sexes cannot be distinguished externally. In *Diplatys* and *Bormansia* all nymphs have long, multi-articulate cerci in place of forceps.

Biology

Adults. Earwigs occur principally in the tropics and warmer temperate zones, although many are found in cooler regions. They frequent crevices under the bark of trees and in fallen logs, and crawl beneath all kinds of debris on the ground. They are thigmotactic, nocturnal, and are attracted to light. Their food consists of a wide range of living and dead plant and animal matter. The forceps are used for the capture of prey, for offence and defence, and occasionally for assisting in folding the hind wings beneath the tegmina.

Reproduction. Mating and egg-laying by tropical species seem to take place at any time of the year, but in temperate zones, although mating occurs throughout the year, egg-laying is restricted to the summer months. Spermatozoa are retained in a quiescent but viable condition in the female spermatheca from the time of copulation until the eggs are fertilized, which may be many months later. This is undoubtedly correlated with winter population trends (see below). A short courtship display 'dance' by the male precedes copulation, which takes place with the pair facing in opposite directions with the ends of their abdomens in contact, male uppermost. Depending upon the species, the pair may also grip each other with their forceps and either move about or remain still.

Earwigs are well known for the maternal care shown by the female for her eggs and young. The female excavates a short burrow in the soil beneath some debris, and at the bottom lays the batch of eggs (20 to 50 according to species). She remains in the burrow during incubation (from 2 to 3 weeks depending upon the climate), turning the eggs, protecting them from predators, and apparently cleaning them of fungi, etc. After about 12 days in temperate regions the pulsations of the dorsal vessel of the embryo can be easily seen, and a couple of days later the black compound eyes are clearly visible. There is no operculum, and the chorion splits lengthwise to allow the nymph to emerge.

Life History. For the first week or two of life the nymphs are also protected, but after that they must fend for themselves as the female becomes notably cannibalistic. Nymphs then have the same way of life as adults. According to the species, a nymph moults either 4 or 5 times before reaching adulthood. However, the length of the nymphal life varies considerably. In the tropics growth is continuous and development takes about 4 weeks, but in the temperate zones it ceases during winter, resuming months later in spring, in time for maturity to be reached before the breeding season.

Population Trends. For the common European earwig, *Forficula auricularia*, there is good evidence that there are fewer males than females in the population during the winter, and the same trend has been noted in *Anisolabis littorea* (White), the New Zealand littoral species. It is not known whether this is also a feature of tropical Dermaptera. For insects which copulate months before oviposition, it is of obvious selective advantage if males are fewer in winter, when living conditions become most severe.

Natural Enemies. Birds are well-known predators of earwigs, and European naturalists have recorded bats taking them. The level of natural parasitism is not high (probably below 10 per cent), and includes certain Tachinidae (Diptera), *Mermis* spp. (Nematoda), and *Hymenolepis* sp. (Cestoda). The harmless sporozoan *Gregarina ovata* infests up to 50 per cent of some populations of *Forficula auricularia*.

Economic Significance. It is doubtful if any of the endemic Australian species are of significant economic importance as pests. However, *Forficula auricularia*, which is nowadays virtually cosmopolitan in cooler regions, can be a serious nuisance if present in large numbers. It invades houses, and eats pieces out of flower buds and orchard fruits. On the other hand, *Labidura riparia truncata* has been observed in the A.C.T. readily attacking codling moth larvae searching for cocooning sites.

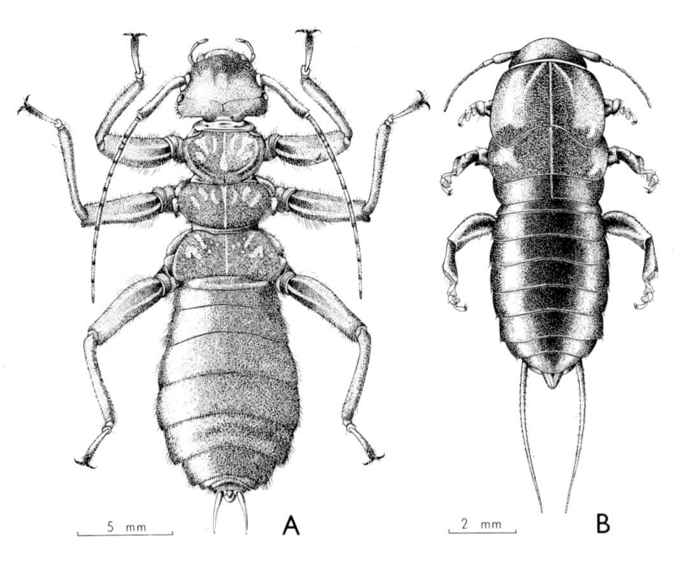

Fig. 19.6. A, *Arixenia esau*, Arixeniina, ♀; B, *Hemimerus vosseleri*, Hemimerina, ♀. [T. Nolan]

Special Features of the Australian Fauna

The primitive Labiduroidea are widely distributed in Australia, but the more advanced Forficuloidea are more restricted. The number of Australian species that have been described is not as large as might be expected, almost certainly as a result of inadequate collecting rather than of a poor fauna. Most have been found in the wetter regions, but *Labidura riparia truncata* and *Nala lividipes* (Dufour) have been collected throughout the continent. It is apparent that there is a considerable Indo-Malayan element, particularly in northern tropical regions. There is also a fair degree of correspondence between the faunas of north Queensland on the one hand and New Guinea, the Bismark Archipelago, and the Solomon Is. on the other. Most Australian species, nevertheless, are endemic. There are noticeable differences between the known faunas of the east and the west of the continent, undoubtedly another consequence of spasmodic collecting. Except for a few short papers by Burr, Mjöberg, Hebard, and Hincks, referred to in Hincks (1954), little has been written about Australian Dermaptera.

CLASSIFICATION

Order DERMAPTERA (60 Australian spp.)

Suborder FORFICULINA (60)

LABIDUROIDEA (33)
1. Pygidicranidae (6)
2. Labiduridae (27)

FORFICULOIDEA (27)
3. Labiidae (15)
4. Chelisochidae (8)
5. Forficulidae (4)

Suborder ARIXENIINA (0)

Suborder HEMIMERINA (0)

Of the 3 modern suborders, only the Forficulina have, up to the present, been found in Australia. The Arixeniina (Fig. 19.6A) include only 2 species, which are associated with 2 species of bats in Malaya, Indonesia, and the Philippines. They either live on the bats, or frequent their roosts, but are not necessarily parasitic. Both species are robust, hairy, apterous, viviparous, and almost blind. The forceps are rod-like, except in the males of one species. As close relatives of one of the bats occur in New Guinea and near Cape York, it seems possible that Arixeniina may be found in these localities in the future. The Hemimerina (Fig. 19.6B) comprise about 10 species, all of which are true ectoparasites under the fur of various species of *Cricetomys*, the giant rat of southern Africa. Apparently there is considerable host–parasite specificity. These insects are streamlined, depressed, and smooth for rapid movement through the fur, and the tarsal segments have specially large pads for gripping the hairs. They are blind and apterous, and the forceps are cylindrical and rod-like. The remaining suborder, the Jurassic Archidermaptera, is described on p. 177.

The last revision of the order was by Burr in 1911. The same author (1915, 1916) developed the work of Verhoeff and Zacher on the use of the male genitalia in classification, but failed to assign all known species. Hincks revised the family Pygidicranidae in 1955 and 1959, and the arrangement used here follows that proposed by him. Classification is based entirely on adults, and chiefly males; unassociated females are often impossible to identify. After the completion of this chapter, Popham (1965) proposed a new classification, which it is not possible to consider here.

Key to the Superfamilies and Families of Forficulina

1. ♂ genitalia with paired medial lobes; rarely one lobe reduced or vestigial, in which case genitalia asymmetrical .... LABIDUROIDEA. 2
   ♂ genitalia with single medial lobe .... FORFICULOIDEA. 3
2. ♂ genitalia with both medial lobes bent forward at rest, rarely one lobe reduced, lateral lobes usually with teeth or processes; femora often compressed and keeled; ♀ usually with discernible genitalia. Some nymphs with segmented cerci .... **Pygidicranidae**
   ♂ genitalia with one medial lobe bent forward and the other straight backward at rest, rarely one lobe vestigial, lateral lobes usually without teeth or processes; femora never compressed and keeled; ♀ without readily discernible genitalia. Nymphs with forcipate cerci .... **Labiduridae**

3. Second tarsal segment simple (Fig. 19.7A) .... **Labiidae**
   Second tarsal segment produced or expanded (Figs. 19.7B,C) 4
4. Second tarsal segment produced below 3rd, but not expanded laterally .... **Chelisochidae**
   Second tarsal segment produced below 3rd and expanded laterally .... **Forficulidae**

Superfamily LABIDUROIDEA

1. Pygidicranidae. This is considered to be the most primitive family of living Dermaptera. Members occur, nowhere commonly, in eastern Australia and W.A., are about 10 to 35 mm long, light to dark brown, winged, and have long antennae. The PYGIDICRANINAE (*Cranopygia*, Fig. 19.8A; *Dacnodes*) are fairly slender and smooth, whereas the ECHINOSOMATINAE *(Echinosoma)* are notably stout and clothed in short, stiff bristles.

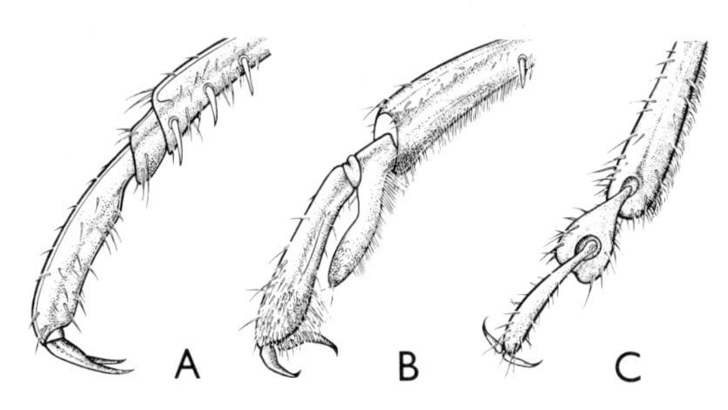

Fig. 19.7. Tarsi, to illustrate couplets 3 and 4 of key: A, Labiidae; B, Chelisochidae; C, Forficulidae. [T. Nolan]

2. Labiduridae. This cosmopolitan family of rather primitive earwigs is fairly well represented, largely by red-brown, apterous species from 10 to 45 mm long, in all parts of Australia. The classification is confused at the generic and specific levels, particularly in the Carcinophorinae, but it seems best at present to follow Burr's last revision. Five subfamilies have been recorded. All CARCINOPHORINAE are wingless, and have the manubrium at least as long as the subgenital plate. *Titanolabis* has the meso- and metasternum lobed posteriorly. The abdomen of *Anisolabis* is feebly dilated, but that

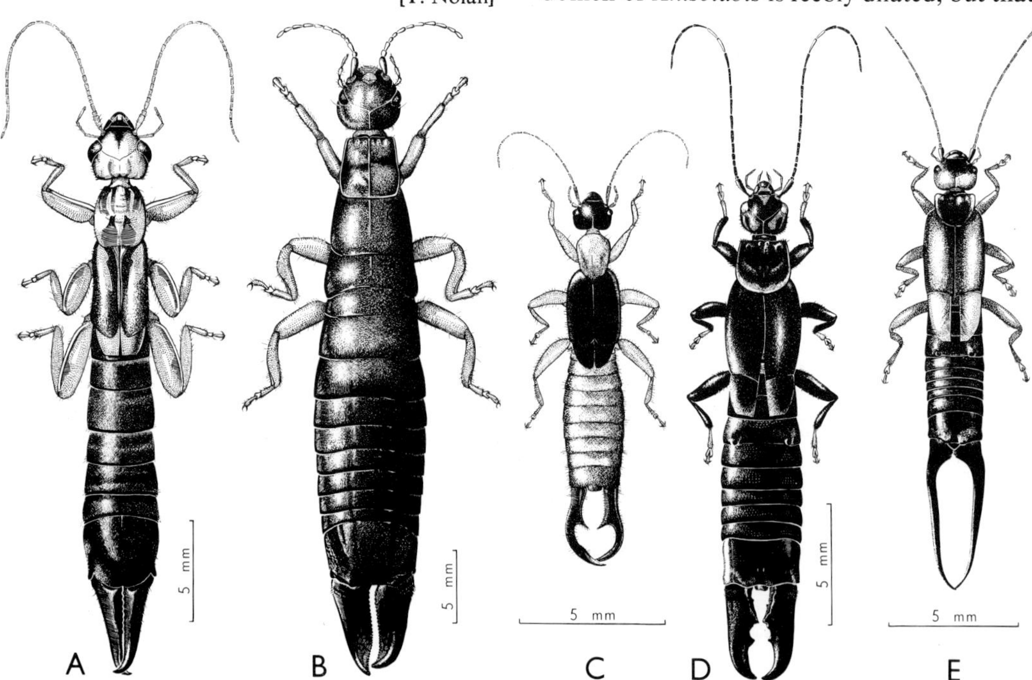

Fig. 19.8. A, *Cranopygia daemeli*, Pygidicranidae, ♀; B, *Titanolabis colossea*, Labiduridae, ♂; C, *Chaetospania brunneri*, Labiidae, ♂; D, *Chelisoches morio*, Chelisochidae, ♂; E, *Elaunon bipartitus*, Forficulidae, ♂. [T. Nolan]

of the three closely related genera *Gonolabis*, *Mongolabis*, and *Notolabis* is noticeably broadest towards the apex; however, these may be separated on characters of the male genitalia. *Titanolabis colossea* (Dohrn) is very large, robust, red-brown in colour, and not uncommon in the rain forests and wet sclerophyll bushland of eastern Australia; exceptional specimens reach a total length of 55 mm. It is the largest known earwig (Fig. 19.8B). The APACHYINAE (Plate 2, E) are large, winged, extremely flattened, and strikingly coloured. All 4 species of *Apachyus* are found on the eastern coastal plains and tablelands under the bark of trees in the rain forest and wet sclerophyll bushland; their incredible thinness fits them ideally for this habitat. All BRACHYLABINAE are apterous, dull, dark brown to black, often thickly covered with fine bristles, and have short antennae. *Antisolabis*, *Brachylabis*, and *Nannisolabis* may be separated by the lateral keel of the mesonotum and the form of the proximal antennal segments. LABIDURINAE are winged and glossy, and have longer antennae (more than 20 segments). *Labidura* is large, long-legged, and the ecdysial line is obsolete. *L. riparia truncata* is distributed all over Australia, particularly in sandy habitats, and is by far the commonest species. It is about 35 mm long, dull brown with straw-coloured markings, and the male has long slender forceps, widely spaced basally, with a prominent tooth near the middle of the inner edge. *Nala lividipes* is also common throughout, and is especially attracted to lights. The PARISOLABINAE generally resemble the Brachylabinae, but may be separated from the latter by the truncate posterior dorsal tergum. *Parisopsalis* is the only Australian genus.

Superfamily FORFICULOIDEA

3. Labiidae. In Australia this large and widespread family is rather poorly represented. All are winged, glossy brown to black, and small (usually under 15 mm). They are nowhere common, but have been collected principally on the east coast and tablelands. LABIINAE have a narrow head and small eyes. *Chaetospania* (Fig. 19.8C) is decidedly flattened and has the posterior margin of the head subsinuate, but in *Labia* the head is truncate posteriorly and femora are slighter. SPONGIPHORINAE have a broad head with prominent ecdysial line and large eyes. *Apovostox* and *Marava* are very close, but may be separated by the male genitalia. In the NESOGASTRINAE the elytra are keeled; *Nesogaster* is the sole genus.

4. Chelisochidae. This small, mainly tropical family is represented in northern Australia by dark brown to black earwigs from 15 to 25 mm long. *Hamaxas* is small, and has pubescent legs with tibiae smooth above. *Chelisoches* (Fig. 19.8D) is larger, robust, and has tibiae flattened and sulcate above; *Proreus* resembles *Chelisoches*, but is more slender, particularly in the antennae; *Lamprophorella* has the body strongly depressed.

5. Forficulidae. This large, cosmopolitan family of rather specialized earwigs is very poorly represented in Australia. All are brown, winged, and from 15 to 25 mm long. *Forficula* has the abdomen broadest in the middle and the forceps depressed and basally dilated; *Doru* and *Elaunon* (Fig. 19.8E) have the abdomen almost parallel-sided, but the pygidium of *Doru* bears a spine. *Forficula auricularia* is well established as a garden pest in cooler districts.

20

PLECOPTERA

(Stoneflies)

by E. F. RIEK

Mandibulate exopterygote Neoptera, with membranous wings, and aquatic nymphs.

This is a relatively small order of stream-frequenting insects. It contains rather more than 1,000 species, of which 50 described and at least 34 undescribed species occur in Australia. They are soft-bodied, four-winged insects, which vary in length from 4–5 mm (small *Nemoura* and *Leptoperla*) to 40–50 mm (large Eustheniidae). Most species are fully winged, but a few are brachypterous, and wingless adults, generally only the males, have been reported for several genera.

Stoneflies (Fig. 20.1) resemble most other orthopteroid orders in general form, and in having long, flexible, filiform antennae and an anal fan in the hind wing except in a few specialized genera; but differ in their distinctive venation, in the comparative softness of their integument, and in there being only a slight difference in texture between the fore and hind wings. They also differ markedly in having truly aquatic nymphal stages which usually possess tracheal gills, although a few exotic species have terrestrial nymphs that live under damp cool conditions. Adult stoneflies can be distinguished from the Embioptera, with which they might be confused, by their venation and the structure of their mouth-parts and by possessing 2–3 ocelli.

Anatomy of Adult

Head. Sessile on the broad prothorax. Compound eyes well developed; 2 or 3 ocelli always present. Antennae long and setaceous, filiform, many-segmented. Mouth-parts mandibulate, usually weak; mandibles well formed in all Australian species; maxillae well developed, with galea, lacinia, and 5-segmented palpi; labium complete, with 3-segmented palpi.

Thorax. The three segments free; the prothorax large and mobile, with wide, flat pronotum; meso- and metathorax subequal, both with normally placed spiracles.

Legs. Strongly built, with short coxae, strong, more or less flattened femora, slender tibiae without spurs, and 3-segmented tarsi with a pair of terminal claws and an empodium.

Wings. Membranous, very unequal, with weak coupling between fore and hind wings. Venation basically primitive, but often considerably specialized, as illustrated for the different families (Fig. 20.5); M always two-branched. At rest, the wings normally fold down closely over the dorsum, with the anal lobe of the hind wing folded fanwise against the body, and one fore wing closely enwrap-

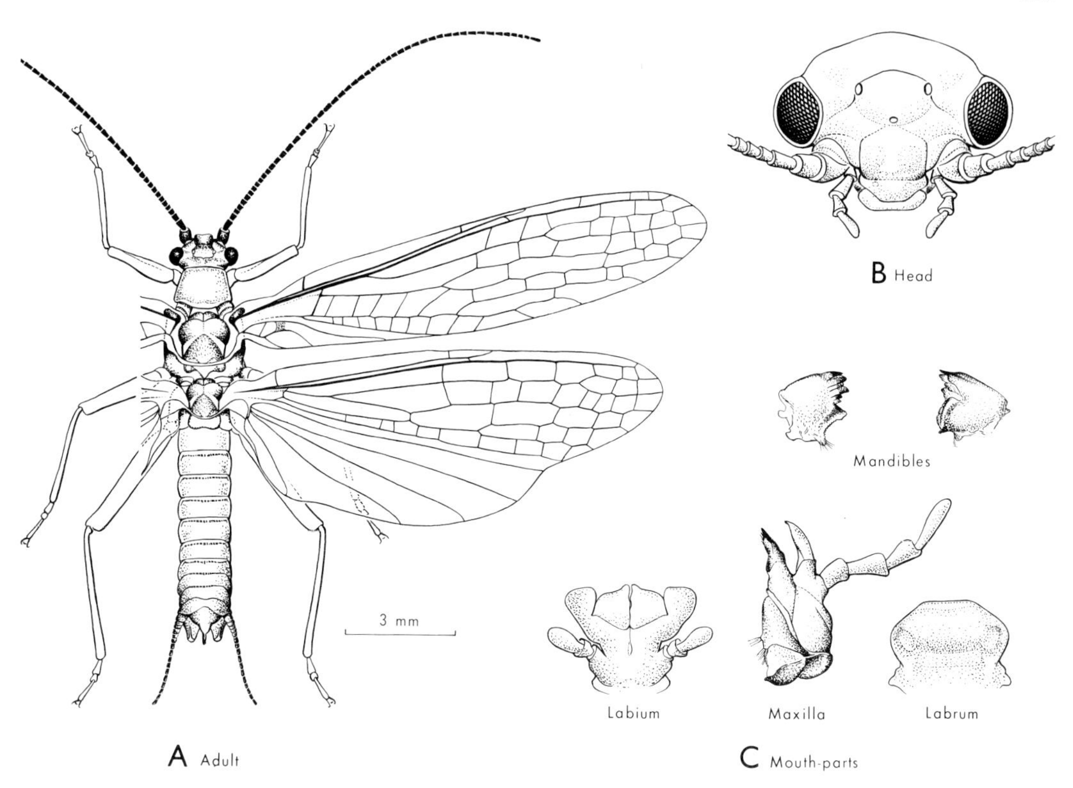

Fig. 20.1. Adult *Trinotoperla* sp., Gripopterygidae. [T. Binder]

ping all but the basal portion of the other. The folded wings generally project beyond the apex of the abdomen.

Abdomen. Soft, cylindrical or somewhat flattened, with 10 complete segments, and a pair of cerci with from one to usually many segments. Spiracles on segments 1–8. Male (Figs. 20.2A, B) with sternite 9 undivided and without styles; segment 10 complete, usually annular; tergite 10 and supra-anal plate often more or less fused, the latter often modified to form a hook-like copulatory organ, which bends forward over the terminal segments, and sometimes occupies a groove or pocket in tergite 10; paraprocts usually large, often fused with bases of cerci, frequently armed with copulatory hooks; phallus present or absent, sometimes with phallomeres. Female (Fig. 20.2C) generally without ovipositor.

Internal Anatomy. Oesophagus very long; gizzard rudimentary or absent; mid and hind guts both short; a pair of salivary glands present; about 20 to 60 Malpighian tubes. The first abdominal ganglion is fused with the metathoracic; the others show progressive fusion of the distal ganglia, culminating in fusion of 6 to 8. In the male the normally conjoined testes join separate vasa deferentia, which usually end in a median ejaculatory duct, though occasionally they remain paired up to the median gonopore. In the female the numerous panoistic ovarioles normally arise from a common duct joining the oviduct of each side; a spermatheca of variable form is usually present.

Immature Stages

Egg. Usually spherical in Australian species, and with an outer coating which swells markedly and become sticky when moistened, thus enabling it to adhere to the substrate. The flattened, elliptical eggs of *Stenoperla* have an adhesive disc on one of the flattened surfaces; they are about 0·4 mm long, and are laid in the one- or two-cell stage of development. Those of medium-sized *Leptoperla* are flat discs about 0·25 mm in diameter by 0·1 mm thick.

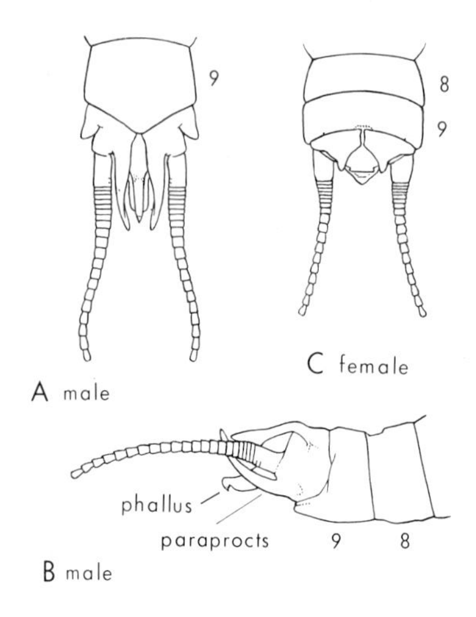

Fig. 20.2. Terminalia: A, ♂, ventral; B, ♂, lateral; C, ♀, ventral. [M. Quick]

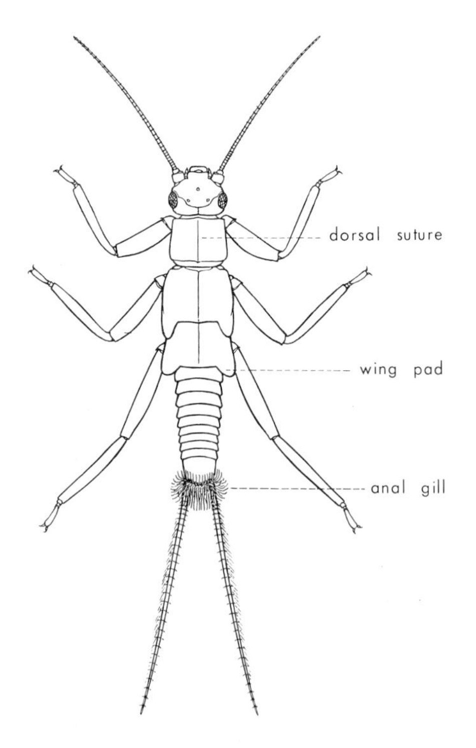

Fig. 20.3. A gripopterygid nymph. [M. Quick]

Nymph. The body form (Figs. 20.3, 6) closely resembles that of the adult. Nymphs may have external gills located on the mentum, submentum, neck, thoracic segments, the first few abdominal segments, subanal plates, or extruded from the anus, or they may lack external gills. Remnants of the nymphal gills are retained in the adults, and are of importance in classification. Abdominal or anal gills are present on all Australian species with the exception of the Nemouridae.

Biology

Adults. Stoneflies are found only near fresh water, usually along running streams or the margins of lakes, especially in high country. They are generally sluggish, of irregular, slow flight, usually blend well with the substrate on which they rest, and often elude capture by running instead of flying. They may be collected early in the morning or on dull days as they emerge on the rocks protruding from the stream, or at other times by beating the foliage along the stream margin. Many hide under the loose bark of trees and logs that border the stream. A few species, mostly of Setipalpia, are attracted to artificial lights at night. There is one apterous species of Nemouridae that apparently passes its whole life history at depths of more than 60 m in a lake in U.S.A.; it has no external gills on either nymph or adult.

There is a marked seasonal and altitudinal succession in the emergence of stoneflies, and in many parts of the country adults of different species may be collected throughout the year. Many nemourids, in particular, emerge from late autumn to early spring. Each eustheniid species has a very short emergence period in summer or, in *Thaumatoperla*, in the late autumn.

Reproduction. Most stoneflies mate during daylight, but the few nocturnal species mate only at night. They pair at rest, with the male superimposed on the female. The eggs are usually shed during flight when the abdomen touches the water. Thus, the females of *Dinotoperla* hover quite high over the water, and then suddenly glide down to the surface, where the eggs are washed off the tip of the abdomen. On the other hand, some species allow their eggs to fall freely, whereas others, such as *Thaumatoperla*, deposit theirs under water from a protruding rock or log. A few species have a long ovipositor, but the manner of depositing their eggs is unknown. The extruded eggs form a ball on the underside of the tip of the abdomen, which turns up, so that it appears that the egg-mass is placed dorsally. Egg-laying may occur once or several times. The total number of eggs deposited often exceeds 1,000, and *Stenoperla* lays that number in a single batch.

Immature Stages. Most stonefly nymphs live in waters with a gravel bottom. They require cool, well-aerated water for development. In a few species the nymphs develop in the shallow, wave-affected margins of cold lakes, as on the Kosciusko plateau, but most pass their immature stages in the shallow upper reaches of streams. A few occur in rather sluggish streams where the bottom is covered with detritus. The number of instars in species that have been reared from the egg has been found to vary from 22 to 33 (Claassen, 1931). The nymphal stage lasts about a year in most species, but some require two or three years for development. Some nymphs *(Stenoperla)* can live for periods in moist terrestrial habitats. Nymphs of certain species undergo diapause, usually at the fourth or fifth instar.

Nymphs are easily collected by overturning rocks, and by stirring gravel in stream beds upstream from the opening of a fine-meshed sieve. Nymphs of some species move away rapidly when disturbed, and hide under other rocks and debris. Those of a few species swim actively with an undulating action of the whole body. Others are quite sluggish, and cling to the overturned rocks. The mature nymph generally crawls on to rocks or debris at the water's edge, where the adult emerges.

Natural Enemies. There are no known parasites of any stage in the life history. Predators on the nymphs include the larger nymphs of Odonata and larvae of aquatic Coleoptera, as well as fish. The adults are eaten by small insectivorous birds as they rise from the stream. Many eustheniid adults collected from rock crevices around the margins of alpine lakes where crows had been seen foraging had the apices of the wings damaged, as though pecked by the birds in efforts to dislodge them.

Economic Significance. Stoneflies are one of the important orders of insects in the diet of freshwater fish, particularly of trout. The Filipalpia form a significant link in the food chain, for they utilize some of the plant tissues that are available in the stream.

Special Features of the Australian Fauna

The most striking features of the Australian stonefly fauna are the absence of one whole suborder and the relationship of 3 of the 4 families with New Zealand and southern South America; the fourth occurs in all regions, but the subfamily in which the Australian species is placed has a southern distribution. No family is entirely endemic. The local distribution is essentially cool temperate, with 17 species in Tasmania, 40 in south-eastern Australia, 8 in N.S.W. outside the highlands, 12 in Queensland, 5 in S.A., and 2 in south-western Australia. There are no records from the vast area of central, north, and north-western Australia, although there are probably some areas that would be suitable for stoneflies.

CLASSIFICATION

Order PLECOPTERA (84 Australian spp.)

Suborder SETIPALPIA (0)

Suborder FILIPALPIA (84)

1. Eustheniidae (14)
   Diamphipnoidae (0)
   Pteronarcidae (0)
2. Austroperlidae (5)
   Peltoperlidae (0)
   Scopuridae (0)
3. Gripopterygidae (50)
4. Nemouridae (15)

The classification adopted has been modified from that proposed by Ricker (1950, 1952), the order being divided into two suborders* which are distinguished as follows.

Paraglossae much longer than the glossae (Fig. 20.4A). Small to large species; nymphs usually carnivorous, and adults of most genera do not feed; generally commoner in warm waters. (Three families, none in Australia) .... SETIPALPIA

Paraglossae and glossae of about equal size, or at least not very unequal (Fig. 20.4B). Generally small to medium-sized species; nymphs and adults of many genera primarily vegetarian; commoner in colder waters .... FILIPALPIA

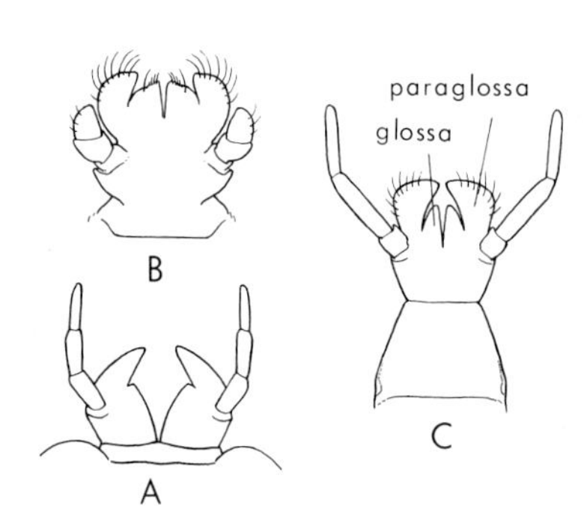

Fig. 20.4. Labia of nymphs: A, *Isoperla patricia*, Setipalpia (after Frison, 1942); B, *Trinotoperla* sp., Gripopterygidae; C, *Stenoperla* sp., Eustheniidae. [M. Quick]

Suborder FILIPALPIA

Keys to the Families of Filipalpia Known in Australia

ADULTS

1. Ano-jugal fan of hind wing with its marginal contour continuous with that of the rest of the wing (Figs. 20.5A,B); cross-veins present all over ano-jugal fan .... **Eustheniidae**
   Ano-jugal fan of hind wing, when present, with its marginal contour not continuous with that of the rest of the wing, forming a re-entrant angle at the end of 1A (Figs. 20.5C,D); cross-veins generally absent on ano-jugal fan (rarely a few present) 2
2. Cross-veins absent from the distal half of the wings (Fig. 20.5G). [CuA of fore wing simple; 2A of fore wing forked after the anal cell; 6 veins in ano-jugal fan of hind wing] .... **Nemouridae**
   Cross-veins present, normally numerous, in the distal half of the wings 3
3. 2A and 3A of fore wing separating at or before the apex of the anal cell; 7 distinct veins in ano-jugal fan of hind wing (Fig. 20.5C); pronotum rectangular, with the corners somewhat produced. [CuA of fore wing branched towards wing margin; 2A of hind wing normally branched] .... **Austroperlidae**
   2A and 3A of fore wing separating after the apex of the anal cell; 5 or 6 distinct veins in ano-jugal fan of hind wing (Figs. 20.5D,E); pronotum with convex margins. [CuA of fore wing simple or forked] .... **Gripopterygidae**

NYMPHS

1. With 5 or 6 pairs of lateral filamentous gills on segments 1–5 or 1–6 of abdomen (Fig. 20.6A). [Nymph very active] .... **Eustheniidae**
   Never with paired lateral gills on abdominal segments 2
2. With 3 or 5 simple anal gills (Fig. 20.6B); pronotum rectangular, with the corners somewhat produced. [Nymph sluggish] .... **Austroperlidae**
   With a tuft or rosette of filiform anal gills (Fig. 20.6C); pronotum normally with rounded margins .... **Gripopterygidae**
   Without external gills (Fig. 20.6D) .... **Nemouridae**

1. Eustheniidae. A small family confined to Australia, New Zealand, and South America. It includes large, mostly brightly coloured stoneflies, with abundant cross-veins on all parts of the wings. The adults emerge over a short period in summer and autumn. They do not move very far from the lake or stream, and hide during the day in rock crevices or rest amongst foliage. The family is well represented in the highlands of Tasmania and south-eastern Australia, with one species extending

* Illies (1960, 1965) considers the Eustheniidae and Diamphipnoidae to constitute a third suborder, Archiperlaria, but this separation is not adopted.

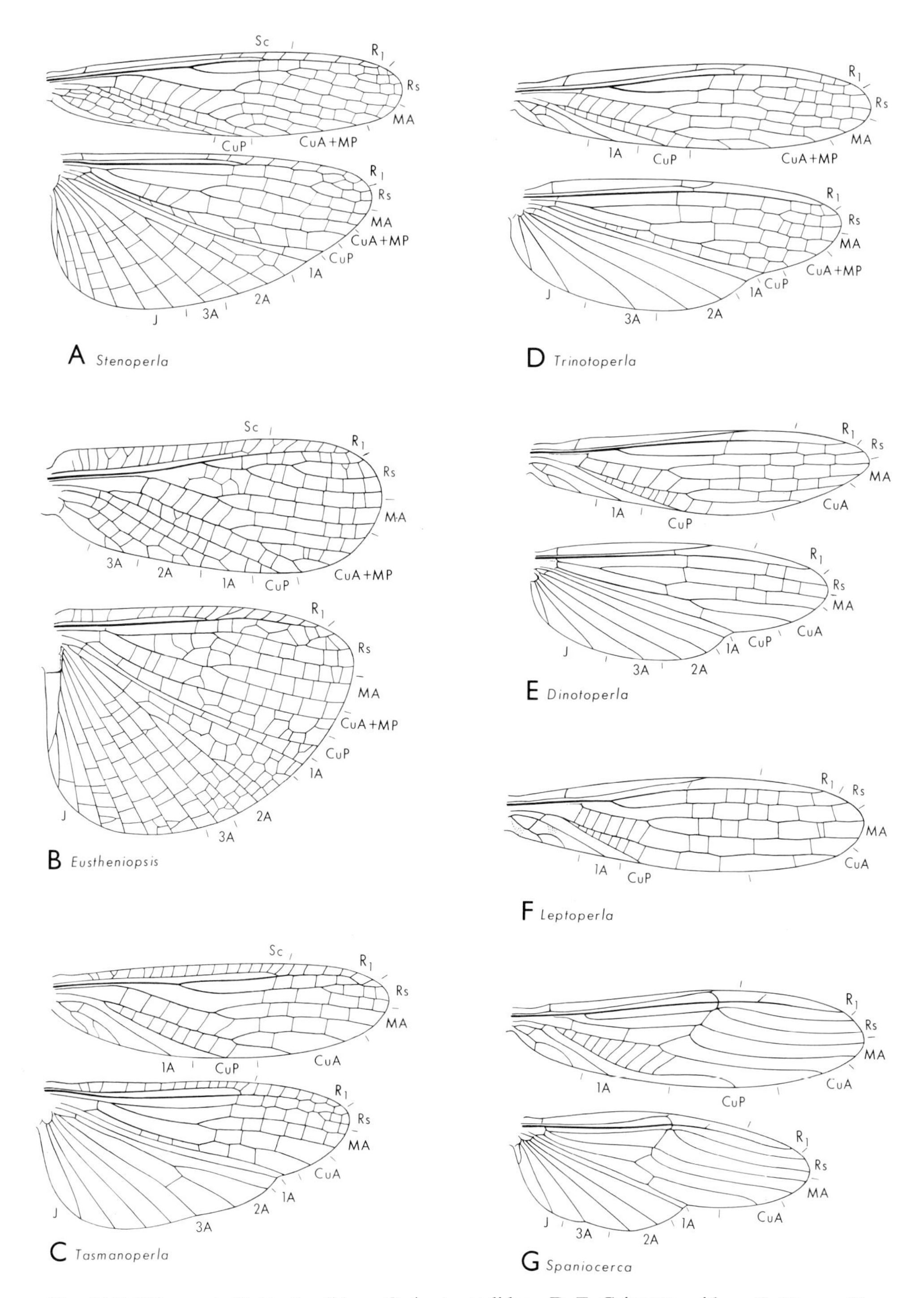

Fig. 20.5. Wings: A, B, Eustheniidae; C, Austroperlidae; D–F, Gripopterygidae; G, Nemouridae. [M. Quick]

into Queensland. The abdominal gills of the nymphs are folded obliquely but not tightly beneath the flat ventral surface. The nymphs run actively when disturbed, but are also capable of swimming with an undulating action.

There are 4 Australian genera (Tillyard, 1921a). *Eusthenia* (5 spp.) is restricted to Tasmania. The species all have a large brilliant orange mark over the basal half of the hind wing, and are otherwise variously marked with orange or purple. The nymphs have rather long fringes of hairs on the legs. *Eustheniopsis* (2 spp.) occurs in Victoria and the Kosciusko area. The wings (Fig. 20.5B) are shorter and broader than in *Eusthenia*, and the hind wings, in particular, are deep purple in colour. The Kosciusko species is unable to fly. The nymph is very similar to that of *Eusthenia*, but lacks the pronounced hair fringes on the legs. *Thaumatoperla* (3 spp.), which has the shortest and broadest wings of all, occurs in the high mountains of Victoria (Burns and Neboiss, 1957). The adults cannot fly. The species have dark to almost black wings, but the body may have yellowish markings. The reticulate venation of these three genera is considered to be a secondarily derived character.

Stenoperla (4 spp.) is the most generalized genus of the family. It is rather unlike the other genera in appearance. The single described species occurs in the upper reaches of the coastal streams of south-eastern Queensland. There is an undescribed species on the Atherton Tableland, another ranging from Barrington Tops to the Sydney area, and a widespread species in the highlands and cold streams of south-eastern Australia. The wings (Fig. 20.5A) are relatively narrow, and resemble those of the large Gripopterygidae, but are distinguished by the abundant cross-veins in the anal field and entire margin of the hind wing. The adults can fly, and are sometimes attracted to lights. The nymph differs from other Australian eustheniids in having only 5 pairs of abdominal gills.

2. Austroperlidae*. A small family confined to Australia, New Zealand, and South America. Easily distinguished from similar-sized *Trinotoperla* (Gripopterygidae) by the sharply rectangular pronotum. They are medium-sized insects (wing-span 20–35 mm), with dark bodies, and the fore margins of the wings reddish brown or the wings mottled. The adults emerge in spring or summer, and are often found some distance from water. The family occurs in Tasmania and in the highlands of south-eastern Australia at least as far north as the Barrington plateau.

Tasmanoperla is the only Australian genus. The closely related *Austroperla* occurs in New Zealand. The two Tasmanian species have mottled wings and pale legs, and emerge in summer. The described species from the mainland has the costal margin of the wings more or less brownish and the legs dark, and the adults emerge in early to late spring. The nymphs of *Tasmanoperla* (Fig. 20.6B) are sluggish, and often congregate in large numbers under moist rocks and detritus at the edges of cool mountain streams. They usually have 5 simple, white, anal gills, which are clearly visible, though somewhat shorter than the short, thin, mostly white cerci which probably function as additional gills. The rectangular pronotum bears relatively large, lateral, flange-like projections somewhat similar to wing pads. Immature nymphs occur in the streams at the same time as adults are on the wing, so the life history probably occupies two or more years. In undescribed Australian genera, there may be only 3 anal gills and also a dorsolateral tubercle on each segment of the abdomen.

3. Gripopterygidae. These are the dominant stoneflies of Australia. The family occurs also in New Zealand, Fiji, and South America. They are mostly uniformly dull-coloured, and of very small to medium size. Adults of some of the species can be collected throughout the summer, though those of many of the small species emerge early in spring. In north Queensland, there is an autumn emergence. Because adults of many species appear so early in spring, it is possible that less than half the Australian species are described, for very little collecting is done at that time of the

* The Penturoperlidae of Illies (1960), based on South American species, is synonymous with the Austroperlidae.

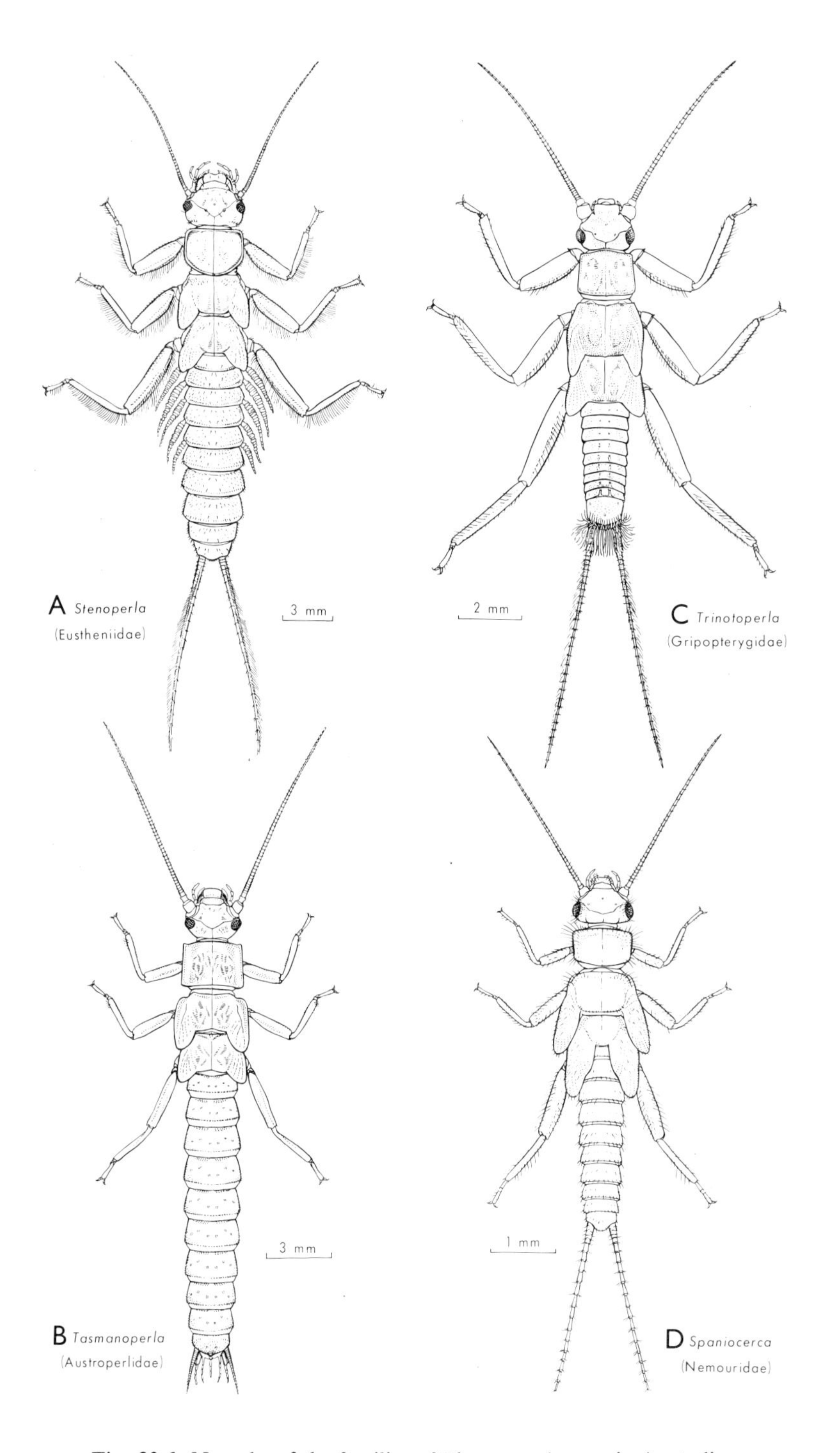

Fig. 20.6. Nymphs of the families of Plecoptera known in Australia. [M. Quick]

year. The nymphs (Fig. 20.6C) have a clearly visible tuft or rosette of filiform anal gills, whitish, pinkish, or pale mauve in colour, and the cerci are multisegmented. They are rather sluggish, and are found clinging to the undersurface of rocks and debris in swift water. A few species cover the body with detritus. The Australian species have been reviewed by Kimmins (1951).

There are 5 Australian genera. The single described species of *Eunotoperla* occurs in Victoria and southern N.S.W. It is large, dark, and similar in size to a *Stenoperla* (Eustheniidae), but is readily distinguished by the distinct angle between the border of the anal fan and the rest of the hind wing.

Trinotoperla (20 spp.) occurs from Victoria to north Queensland, with one species in South Australia. These are the common, large stoneflies of the family (wing-span generally more than 30 mm). Most species are uniformly dull, with a slight darkening at the bases of the wings. The genus has been reviewed by F. Perkins (1958). *Aldia* (3 spp.), from the Kosciusko area and Tasmania, is similar, but the fore wings show a distinct light mottling, and the nymphs are very different.

Dinotoperla (11 spp.) is widespread in Tasmania and eastern Australia. Specimens occur in warmer, more slowly flowing waters, as well as in the colder streams to which *Trinotoperla* is restricted. The species are all small, with a wing-span of about 20 mm. They are coloured much as in *Trinotoperla*, though one Queensland species has obvious transverse banding of the fore wing, and a Tasmanian species has mottled fore wings. The cerci of the adults are quite short and inconspicuous, as they are in *Trinotoperla*. This is the only genus of stoneflies recorded from South Australia, though undescribed species of *Trinotoperla* and *Leptoperla* are known to occur there. Several different nymphal types are known in this genus.

Leptoperla (15 spp.) occurs from Tasmania to north Queensland, and there are 2 species in W.A., where it is the only genus of stoneflies recorded. The species are all small, and the genus contains the smallest Australian stoneflies, with a wing-span of only about 10 mm. The species are mostly uniformly dull in colour. This is a large genus with numerous undescribed species. Many of them emerge early in spring, along with the nemourids, and so are not often collected. Members of the genus are easily recognized by the long cerci, which distinguish them from the smaller species of *Dinotoperla*.

The number of veins present at the wing margin is closely correlated with size in this family. Although the genera are based primarily on venation, there are corresponding genitalic differences.

4. Nemouridae. This large family, which is easily recognized by the absence of cross-veins in the distal half of the wings (Fig. 20.5G), is mostly Holarctic (Capniinae, Taeniopteryginae, and most Leuctrinae), but there are representatives on all continents. The only subfamily in Australia is the NOTONEMOURINAE, which occurs also in New Zealand, South Africa, and South America, and in which Sc arches up, often to meet or almost meet the costa, and the basal segment of the tarsus is as long as the apical. They are all small, insignificant, dark-coloured species. The adults generally emerge in winter and early spring (August–September). The family is well represented in Tasmania, several species occur in the highlands of south-eastern Australia, and there is a described species from south Queensland. The small, burrowing nymphs (Fig. 20.6D) are without external gills. The segmented cerci are about as long as the antennae, but they often break off at the basal stump to give the appearance of short, truncate, unsegmented cerci; each segment has an apical whorl of long hairs. The distal segments of the cerci are shed before the adult emerges. The thoracic segments of the nymph are clearly separated by constrictions. Both the described genera, *Spaniocerca* and *Kimminsoperla*, occur in Tasmania and on the mainland.

ACKNOWLEDGMENT. I am grateful to Mr A. Neboiss, National Museum of Victoria, Melbourne, for helpful discussions of the manuscript.

21

ORTHOPTERA

(Grasshoppers, locusts, crickets)

by K. H. L. KEY

Mandibulate, exopterygote Neoptera, having the hind legs usually saltatorial, with the hind coxae nearly always small and well separated; pronotum with large descending lateral lobes; wing rudiments of nymph reversing their orientation in the later instars.

The Orthoptera comprise the mainly medium-sized to large, terrestrial insects commonly known as long- and short-horned grasshoppers, locusts (in the strict sense: not the cicadas), and crickets, as well as related groups without common names. They occur over all but the coldest parts of the earth's surface, but are best developed in the tropics. They are often abundant as individuals, forming a characteristic and striking component of the fauna in many parts of the world. The order is perhaps best known for the power of jumping possessed by nearly all species, and for the 'singing' that many indulge in, especially at night. The locusts have become a byword for the devastation caused by their migrating swarms. They are mentioned in early writings of the Mediterranean and Chinese civilizations, while in the present century they have provoked one of the most massive concentrations of research ever achieved in the field of entomology. Biological types include phytophilous, geophilous, cavernicolous, myrmecophilous, and burrowing; diurnal and nocturnal; vegetarian and carnivorous. More than 20,000 species are known. Comprehensive accounts of the order are given by Chopard (1938, 1949b) and Beier (1955).

Anatomy of Adult

Head. Typically hypognathous, usually inserted to some extent into the pronotum, anterior part of vertex often projecting forward beyond the eyes to form a *fastigium*, face then usually reclinate; sometimes the whole of the anterior part of the head has a strongly projecting, elongate-conical form. Antennae short to several times as long as the body, with from 7 to very numerous segments, in a few burrowing forms greatly reduced. Eyes usually of medium to large size, but strongly reduced or absent in some cavernicolous and subterranean forms. Ocelli usually 3, sometimes fewer or absent. Mandibles large, somewhat asymmetrical, in predacious forms more slender and pointed, in phytophagous ones with broad grinding surfaces; exceptionally, but often only in the male, they may be grotesquely enlarged. Maxillae with 5- or rarely 6-segmented palp, large galea, and toothed lacinia. Labium with 3-segmented palp and large paraglossae; glossae reduced or absent. Hypopharynx present.

Thorax. Prothorax large, usually with only limited movement on the mesothorax.

Pronotum nearly always larger than the other nota, with large descending lateral lobes, which are usually subvertical and form the sides of the prothorax. Often the free margins of the pronotum are formed by a fold projecting beyond the lines of articulation with adjacent sclerites, especially posteriorly, where a broad extension commonly covers the wing bases. In a few burrowing forms the lateral lobes curve round ventrally, even meeting on the mid-ventral line. Spines, crests, or foliaceous outgrowths, some of which evidently have a procryptic significance, are sometimes developed. Prosternum sclerotized, very variably constructed; propleural sclerites often vestigial, at most their ventral extremities visible below the pronotum. Meso- and metathorax similar, rigidly connected, the latter usually the larger; the sterna sclerotized, usually broad, comprised principally of a large basisternum to which a smaller sternellum is partly fused; pleura well developed, similar, usually somewhat oblique, episternum and epimeron subequal. Meso- and metathoracic spiracles present.

Legs. Fore and mid legs usually similar and gressorial, but some predacious forms have raptorial fore legs provided with rows of spines on the ventral sides of femur and tibia; fore legs of burrowing forms sometimes comprehensively modified for digging. Hind legs nearly always modified for jumping, larger than the others, the femur thickened to accommodate powerful muscles, and the tibia provided distally with articulated spurs; occasionally they may be further adapted for life on sand by enlargement of the tibial spurs, for swimming by a broadening and flattening of the tibia and tarsus, or for movement on water surfaces by the development of long, hairy spines on the tibia. Coxae of hind legs nearly always small and well separated. Tarsi 1- to 4-segmented, nearly always with a pair of terminal claws and usually with an arolium. Fore tibia of most Ensifera bearing a pair of auditory tympanal organs near the base, the tympanum being exposed, or largely concealed by an integumental fold. Hind femur of some Caelifera equipped on its inner face with a ridge, or row of minute pegs, that plays a part in stridulation.

Wings. Most Orthoptera are fully winged in both sexes, but brachypterous or apterous forms are numerous in all the larger families, and some families are apterous throughout. Reduction usually affects both fore and hind wings; occasionally the hind may be lost, but the fore unaffected; and in two superfamilies the fore wings are always reduced, even when the hind are fully developed. Commonly the female is affected more than the male, rarely vice versa; the male may be fully winged and the female apterous. Reduction may be arrested when just enough of the fore wing remains to permit the performance of secondary functions, e.g. stridulation in Ensifera, protection of the abdominal tympanal organ in Acridoidea.

When fully developed, fore wings usually narrow, tegminized, in the closed position overlying the hind wings along the abdomen and overlapping across the mid-line; sometimes broader and modified to resemble leaves. Hind wings broad, largely or wholly membranous, with a large anal area; in repose folded fanwise, either throughout, or only in the anal or cubito-anal region. Venation (Ragge, 1955a) very varied (cf. Figs. 1.2, 21.1). C submarginal in the fore wing, at least towards the base, or absent, the more proximal part of the anterior wing margin unsupported, or strengthened by an ambient vein; in the hind wing marginal. MA usually present. Cu dividing near the wing base into CuA, often branched, and an almost always unbranched CuP. Anals numerous in the hind wing. Cross-veins and intercalary veins are generally present and an archedictyon may occur over part or all of the fore wing. The venation of the fore wing, and less often of the hind, may show a variety of stridulatory modifications often restricted to, or better developed in, the male.

Abdomen. Consists of 11 segments; usually smooth, but sometimes sculptured in brachypterous and apterous species. Terga 9 and 10 of male sometimes bear integumental outgrowths; 11 forms the supra-anal plate, which is sometimes fused with 10. Sterna

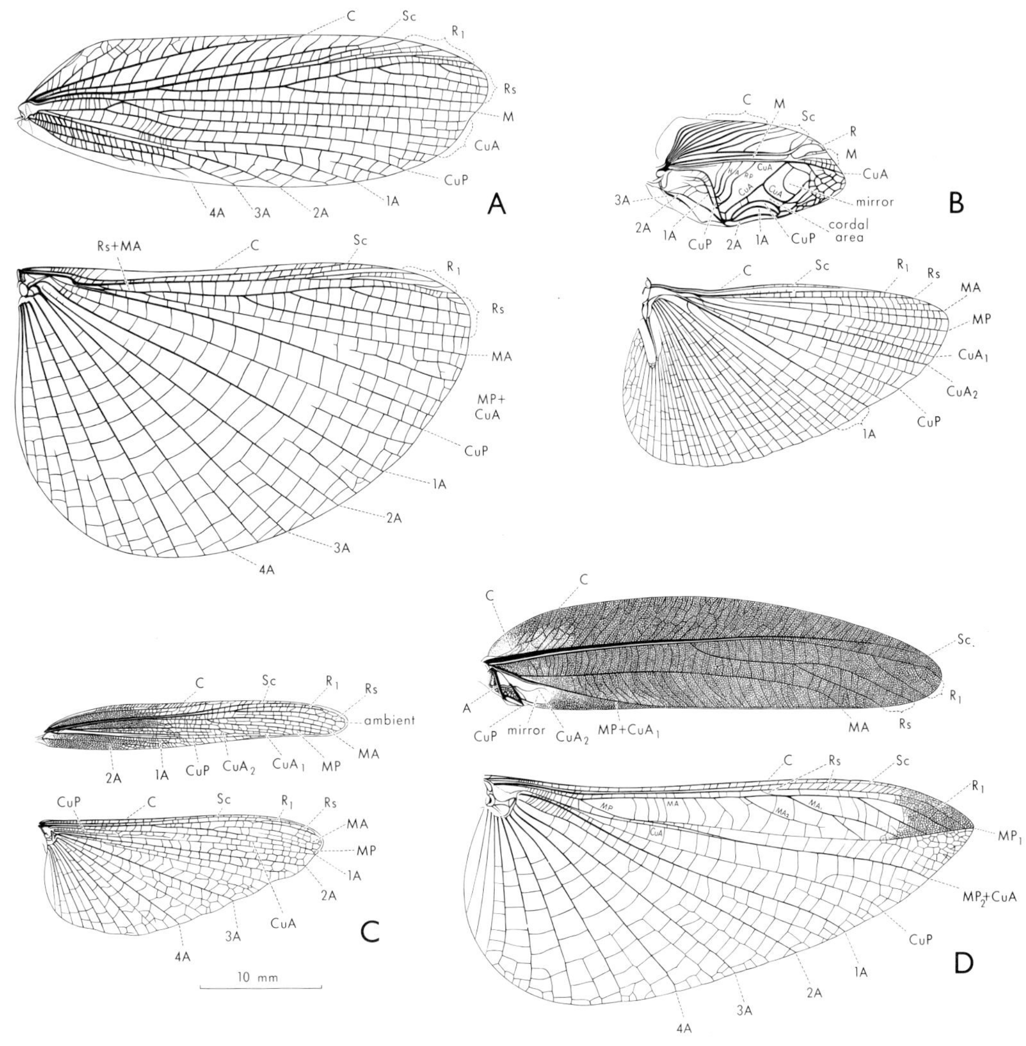

Fig. 21.1. Male wings: A, *Hadrogryllacris* sp., Gryllacrididae; B, *Teleogryllus commodus*, Gryllidae; C, *Bermius brachycerus*, Acrididae; D, *Torbia viridissima*, Tettigoniidae. [F. Nanninga]

fully sclerotized, or represented by sternites and laterosternites surrounded by membrane; 1 often closely associated with the metathoracic sternum; 11 represented by the paraprocts. Cerci usually unsegmented, of varied length, usually larger in male, where they may be modified as clasping organs. Spiracles present on segments 1 to 8. Sometimes one of the anterior terga bears a series of stridulatory ridges laterally. In most Acridoidea tergum 1 bears a conspicuous lateral tympanal organ, which is covered by the fore wing except in flight.

The male terminalia (cf. Fig. 1.28B–D) have been described by Snodgrass (1937). Sternum 9 (or part of it) forms the subgenital plate, which sometimes has a pair of styles, or is provided with a posterior lobe, which may be elongate or otherwise modified. Sternum 10 concealed. The genitalia, concealed by the subgenital plate, are diversely constructed, often largely sclerotized, phallic and periphallic structures of uncertain homology; they are often useful in classification.

In the female (cf. Fig. 1.24B), the subgenital plate is formed by sternum 8 or 9; 10 (or 9 and 10) reduced and concealed. The ovipositor (sometimes obsolete) comprises 3 pairs of valves, of which the inner pair may be greatly reduced. It may consist of an ensiform or cylindrical tube (most Ensifera), formed by the tongue-and-groove articulation of 2 or all of the valves and sometimes exceeding the rest of the body in length; or of a dorsal and ventral pair of free prongs, representing the anterior and posterior valves, hinged at their base and capable of being approximated and separated like jaws (most Caelifera).

Internal Anatomy. Alimentary canal straight or convoluted, with a large crop, variably developed gizzard, and mid gut with 2 to 6 caeca anteriorly. Salivary glands variably developed. Malpighian tubes numerous, entering the gut separately, or by one or more common trunks. The central nervous system includes 3 thoracic and 3 to 7 discrete abdominal ganglia. The male reproductive organs comprise a pair of testes, discrete or fused, connected to the ejaculatory duct by paired, simple or epididymis-like vasa deferentia; tubular accessory glands open into the duct, usually a pair of vesiculae seminales, and, in the Ensifera, paired or unpaired globular 'prostate' glands. In the female, the ovaries comprise a number of panoistic ovarioles, arranged serially along the lateral oviducts or arising in a cluster; the lateral oviducts join to form a short common oviduct, or occasionally open independently into the genital atrium, as does the paired, or usually unpaired spermatheca. Accessory glands open at the base of the ovipositor in some Ensifera, while in most Caelifera the anterior extremities of the oviducts are glandular and secrete the material that goes to form the egg pod. In some acridids a pair of 'Comstock–Kellogg' glands, thought to produce a sex-attractant substance, open into the genital atrium. Many Orthoptera possess integumental glands, of the most diverse types and location, from which repugnatorial fluids may be released, while similar glands in males of some species have a sex-attractant secretion.

Karyotype. The chromosome set has 2n♂ ranging from 8 to 57 in the species studied (White, 1951), while 2n = 68 in the parthenogenetic tetraploid *Saga pedo* Pall. (Tettigoniidae); supernumerary chromosomes sometimes occur. The chromosome number and morphology are almost constant throughout some large groups, but highly diversified in others. Sometimes the karyotype affords the best specific characters and may reveal the existence of sibling species; intraspecific chromosomal polymorphisms and chromosome races are also common (p. 77). The male is usually XO, but XY and X_1X_2Y sex mechanisms also occur.

Immature Stages

Egg. The eggs may be oval, elliptical, or cylindrical, sometimes curved, and sometimes strongly flattened; shell thin, pale, and smooth, or thick, pigmented, and minutely sculptured.

Nymph. The just-hatched young is enclosed in the embryonic cuticle, i.e. it is a 'pronymph', or 'vermiform larva'. At hatching the egg shell is fractured by pulsations of an extrusible *cervical ampulla* in the dorsal membrane of the neck, usually assisted by the cutting action of a ridge, or row of teeth situated on the front of the head. The ampulla also plays a part in the emergence of the pronymph from the egg repository in soil or plant tissue, and in the 'intermediate moult', by which the embryonic cuticle is cast immediately the insect is freed. The nymph differs from the adult mainly in the rudimentary reproductive organs and wings (in species possessing them), in the less elabor-

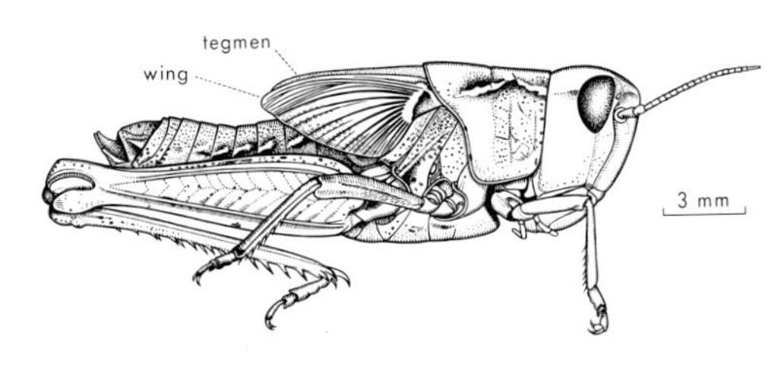

Fig. 21.2. *Chortoicetes terminifera*, Acrididae, last nymphal instar. [F. Nanninga]

ated pronotum, and usually in the smaller number of antennal segments. It undergoes several moults (usually fewer in the male than the female), which usually take place in an inverted position from some support, the cervical ampulla again coming into play in rupturing the old cuticle. The wing rudiments and terminalia increase in size from instar to instar (footnote on p. 296 applies to this chapter also). The former are initially lateral in position, with the costal margin ventral, as in the adult, but at one of the later ecdyses—usually the antepenultimate one—they rotate about their basal attachment and assume a reversed, more dorsal position, in which the costal margin is dorsal, the morphologically ventral surface external, and the hind wing overlaps the fore wing (Fig. 21.2). The number of antennal segments may increase by progressive division.

Biology

Adults. Orthoptera are solitary, or sometimes gregarious, mostly active insects, usually progressing by walking, but capable of powerful jumps. Jumping occurs usually as part of the escape reaction, or as a means of launching the insect for flight. In some small groups the capacity has been largely or wholly lost. Flight may be weak and of a fluttering or blundering character, or powerful and sustained, leading to extensive migrations. The Ensifera are predominantly nocturnal, the Caelifera diurnal, but many species of both groups are attracted to lights at night.

A characteristic element in the behaviour of many Orthoptera is stridulation. The sounds produced, as well as the mechanisms involved, are highly diversified (Busnel, 1955). The mechanisms fall into two principal categories: friction between specialized veins in the proximal half of the two fore wings, in the course of their rapid reciprocal motion (usual in Ensifera); and friction between a ridge, or row of minute teeth, on the hind femur and one or more specialized veins of the fore wing (common in Acridoidea). However, there are many others: the hind femur or tibia may engage with a variously placed series of proximal abdominal ridges; the tibia may strike across a series of evenly spaced cross-veins of the fore wing; the hind wings, alone or in conjunction with the fore wings, may produce in various ways a crackling or buzzing sound in flight; and the mouth-parts may be rubbed together to produce a sort of squeak. Stridulation is usually associated with courtship or sexual excitement, and is better developed in the male, or confined to that sex. The 'songs' produced are nearly always characteristic for the species; often they appear to constitute the principal isolating mechanisms between closely related species and may provide the best taxonomic characters for separating them (Alexander, 1962). However, stridulation in flight, which is often associated with coloured hind wings, or with a zigzag flight path, probably serves to disorient bird predators (see Cott, 1940). Mouth-part stridulation is commonly a reaction to being seized.

Many Orthoptera have excellent sight or hearing and are wary insects, difficult to catch. Others are sluggish, relying for protection upon procryptic or distasteful properties, or hiding in inaccessible places. If seized, almost all species will kick out with their usually spiny hind tibiae and regurgitate the acrid contents of the crop; most carnivorous species will bite without hesitation and can draw blood. If the insect is seized by a hind leg, the leg will generally be autotomized between femur and trochanter, where there is an occlusive diaphragm (cf. p. 353). Various species, often aposematically patterned, will squirt or exude repugnatorial secretions from integumental glands.

Most species are phytophagous, feeding especially on the foliage of higher plants, but some on roots, cryptogams, or unicellular organisms ingested with mud. Others are 'omnivorous', while a few substantial groups are predacious on other insects, which they seize with their spiny fore legs in the manner of Mantodea.

Four broad biological types may be distinguished: (1) those species whose active stages spend all their time in relatively exposed situations on plants or on the ground; (2) cryptozoic forms, spending much of their time in burrows in soil or rotting wood, or concealed beneath stones, debris, or loose bark, but also coming out and moving about freely in the open; (3) cavernicolous species, confined to the darkness and dampness of caves, but moving freely over the surface of walls, floor, or ceiling; and (4) wholly subterranean species. The number of species in these categories declines markedly from the first to the last, the great majority being in the first two.

Species in the first category show many features of structure and pattern that are of visual significance, primarily in relation to vertebrate predators, but sometimes to their fellows. Thus we find elaborate procryptic resemblances to leaves, twigs, bark, or stones, involving structural modifications of various parts of the body, combined with camouflaging colours and patterns and with behaviour appropriate to these adaptations (p. 130). Mimicry of other insects is also found. Aposematic colouring, associated with distasteful properties, sluggish behaviour, occupancy of conspicuous situations, and sometimes limited gregariousness, is not infrequent (Plate 2, H, L). Bright colours on parts of the body that are concealed at rest may be in the category of 'flash', or frightening colours (Plate 2, F); they are commonly different in related species and are much used by the taxonomist. A few species may bury themselves in sand for short periods, or overwinter in soil cracks or under litter, even though in general they live fully exposed lives. Some are semiaquatic, living on the hygrophilous vegetation or damp soil fringing bodies of water (or even on floating hydrophytes) and swimming freely, or skating over the surface film; these commonly show limited structural adaptation for aquatic life (p. 342).

The cryptozoic forms lack bright colours and procryptic adaptations. They are mostly pale, drab, or sometimes black insects, very smooth or with a short velvety pile of hairs, and mostly more or less cylindrical or dorsoventrally compressed; fore legs or mandibles may be adapted for digging. The cavernicolous forms have drab pigmentation, smooth, thin integument, long delicate appendages, and small eyes. The very few permanently subterranean species are pale, cylindrical, larviform creatures, with vestigial eyes, reduced antennae, non-saltatorial hind legs, and powerful burrowing adaptations; or small, depressed myrmecophiles.

Reproduction. Diverse attitudes are adopted in copulation. The sperm are transferred in spermatophores, which vary greatly in size and complexity, but consist essentially of a vesicle, with a narrow opening usually situated at the end of a tubular extension. In the Ensifera a single spermatophore is transferred at each copulation and the vesicle remains attached externally to the vulva; it may be eaten by the female after emptying. In the Acridoidea several small spermatophores may be inserted into the female tract, or the tubular part of a single spermatophore may penetrate the spermathecal duct, while the vesicle remains in the phallus.

The eggs are usually laid in the soil, but many Ensifera insert them into stems or leaves, and a few cement them in rows to twigs; a few Acridoidea lay in pithy stems, soft spots in dead timber, or grass tussocks, and the semiaquatic species may cement the eggs to water plants below the surface. Burrowing forms usually deposit theirs within the burrow, often in special chambers. In Ensifera only the slender ovipositor is inserted into the substrate; entrance is effected by alternate small penetrations of the valves, which slide longitudinally on each other. In Caelifera the pronged ovipositor penetrates by alternate opening and closing, drawing the abdomen after it; in the Acridoidea the latter may be stretched to twice its normal length in this way. Most Ensifera lay their eggs singly. The Caelifera lay theirs in batches of less than 10 to about 200, each batch being usually loosely held together by a proteinaceous foam, which at the same time often serves to cement the surrounding soil particles together to form a more or less discrete capsule, or 'pod'; several batches are commonly produced, at intervals of some days. Total egg production in the Orthoptera may reach many hundreds.

Facultative parthenogenesis is not uncommon, although fecundity is then usually much reduced and the viability of the progeny impaired. A very few species show obligatory thelytoky, with males unknown or extremely rare (p. 80).

Nymphs. After the intermediate moult, the nymph in general behaves in the same way as the adult. Sometimes, however, the first instar is more divergent, in both appearance and behaviour; thus in a few tettigoniids it may mimic an ant or a cicindeline beetle, but the resemblance diminishes in later instars and the adult is procryptic. Regeneration of damaged or autotomized legs is rare. The nymphal stage lasts some weeks or months.

Variation, Colour Adaptation. Intraspecific genetic variation in conspicuous characters is widespread, especially in Caelifera, and includes polymorphism and continuous variation within populations, as well as geographical variation. Polymorphism may involve such structural features as wing length, pronotal proportions, and surface sculpturing. Strikingly different colour patterns may occur as multiple alleles, or there may be polymorphism or continuous variation in colour of wing, hind tibia, etc. Geographical variation may involve similar features, along with percentage frequency of different morphs, and physiological characters such as diapause. Variation in the karyotype has been discussed on p. 75.

Environmental factors may have a direct effect on the phenotype. Thus dryness and high temperature tend to increase wing length; high temperature leads to a paler general coloration; and abundant succulent food may induce a green pigmentation. 'Morphological' colour adaptation to the background is widespread, especially among geophilous Acridoidea. A few species of Acridoidea show a physiological colour change under the control of temperature (pp. 138, 354). The male of the Australian alpine grasshopper *Kosciuscola tristis* Sjöst. (Plate 2, J) is bright blue above 25°C and nearly black below 15°C. The reaction is due to a direct influence of temperature on the epidermal cells (Key and Day, 1954 a, b).

Kentromorphism. Kentromorphism (p. 139) is exhibited by a few Tettigoniidae and a considerable number of Acridoidea, including at least 8 Australian species. Locusts are the classical subject of kentromorphic research and a large literature on their phases has developed (references in Uvarov, 1966). In locusts the high-density phase is characterized by strongly developed gregariousness and migration, so that it has been termed the 'phase *gregaria*', in distinction to 'phase *solitaria*' and 'phase *transiens*' (intermediates). Much has been written on the relation between phase transformation and locust outbreaks. Today it seems that the phases must be seen primarily as a consequence of the population fluctuations and only secondarily, and in a minor degree, as contributing to them.

Phase studies on Australian Orthoptera are limited to those of Key (1954) and Blackith (1957) on the locust *Chortoicetes terminifera* (Walk.) and two grasshoppers of the genus *Austroicetes*, and of Common (1948) on the locust *Gastrimargus musicus* (F.); but phases also occur in Australian populations of *Locusta migratoria* L. and in species of *Austracris* and *Valanga*.

Ecological Features. The order is ecologically heterogeneous. Such common features as it possesses are associated with the relatively large size of the insects, and their biting mouth-parts, terrestrial exopterygote nymphs, and saltatorial specializations. Taken together, these features imply a life in the open; i.e., the first of our biological types (p. 327) is the basic one and the others must be regarded as secondary developments. Leaving aside the small proportion of cavernicolous and wholly subterranean forms, the relative exposure of the Orthoptera means that much importance must attach both to the gross physical factors of the environment and to vertebrate predation.

The cryptozoic forms are generally hygrophilous and nocturnal. The wholly exposed forms can be grouped mainly into hygrophilous, phytophilous, and often nocturnal species (mostly Tettigoniidae) on the one hand, and more xerophilous, often geophilous, very largely diurnal species (mostly Caelifera) on the other. The first are much less under stress from physical factors or food shortage, and predation may be expected to be a more important mortality factor; their density seems to be relatively low and stable. The second are subject to marked density fluctuation, are often very abundant, and may be gregarious and/or migratory. Physical factors, especially rainfall and temperature, are very important in their lives, both directly and through their effect on the herbaceous vegetation, which usually supplies both food and shelter. Their distributions usually show a wet or a dry climatic limit, or both. The former is probably due to fungal and bacterial diseases that become prominent under humid conditions, the latter mainly to the effect of drought on food and shelter. The more xerophilous species show various drought-resisting adaptations, often concentrated in a particular resting stage.

Outbreaks of locusts result from the occurrence of favourable weather in areas (known as *outbreak areas*) possessing features of topography, soil, and vegetation favourable for survival and multiplication of the insects (Dempster, 1963). Details vary with the species of locust, but the outbreak areas usually display a patchwork distribution of habitat suitable for oviposition and habitat affording food and shelter. Contraction of the area occupied by either habitat occurs under appropriate weather conditions and may lead to enforced concentration of locusts and their subsequent gregarization. Emigration of the resulting swarms may be regarded as an ecological device for relaxing pressure on the food supply before a shortage develops (cf. Wynne-Edwards, 1962). At the same time, the increased mobility of the insects places more widely spaced favourable habitats at their disposal.

Natural Enemies. The list of natural enemies of Orthoptera includes many vertebrates and invertebrates, but the information available relates principally to the Acridoidea. In Australia, the straw-necked ibis, *Threskiornis spinicollis*, is an important predator of locusts and grasshoppers (Carrick, 1959). Sphecids of the genus *Priononyx* and asilids of the genus *Bathypogon* also prey on various Acridoidea, while the ant *Iridomyrmex purpureus* has been observed carrying large numbers of just-hatched locust nymphs to its nest. The sarcophagids *Blaesoxipha pachytyli* and *Taylorimyia iota* and the nemestrinid *Trichopsidea oestracea* are internal parasites. Ectoparasitic mites are common, especially species of the erythraeid genus *Leptus*. The eggs of the grasshopper *Austroicetes cruciata* (Sauss.) are preyed upon by the bombyliid *Cyrtomorpha flaviscutellaris* and those of various other Acridoidea are parasitized by mostly host-specific species of *Scelio*. Species of *Praxibulus* (Acridoidea) are particularly susceptible to parasitic worms, which are responsible for the development of intersexual adults, and to a fungus disease, which, from the characteristic attitude of the dead host, is probably due to *Entomophthora* sp.

Economic Significance. The Orthoptera are of great economic importance, but this is due overwhelmingly to the large number of pests in the Acridoidea. Virtually all the more arid regions of the world harbour several species of locust or grasshopper of major importance to agricultural or pastoral production. The inhabitants of many of these areas practise subsistence farming, and locust outbreaks have often caused famines. In more advanced countries, including Australia, locusts and grasshoppers may cause a recurrent financial drain, through crop losses and the cost of control measures, which has been estimated to average many millions of dollars per annum over the world as a whole.

A few tettigoniids are significant pests of crops and pastures in various parts of the world. Several mole crickets are also rated as pests and in Australia a field cricket, while in some countries tree crickets may cause damage to woody plants at times of heavy oviposition in their stems.

Special Features of the Australian Fauna

The Australian orthopteran fauna reflects the world spectrum in number of species in the 5 larger superfamilies. Estimates show that the percentage of the world total occurring in Australia ranges from 6 to 9 per cent, except in the Gryllacridoidea, where it reaches 13 per cent. The percentage endemism varies widely. At the generic level it is only 25 per cent in the Grylloidea, but as high as 90 per cent in the Acridoidea, with the other 3 superfamilies ranging from about 50 to over 60. The high figure for the Acridoidea may be due in part to their xerophily, which would hinder penetration of the equatorial rain forests lying athwart the main post-Pliocene route of entry into Australia. In this superfamily the predominant element is ancient and autochthonous (Key, in Keast *et al.*, 1959). In the others the relationships are principally with the Indo-Malayan fauna. Evidence of Antarctic relationships is restricted (Rhaphidophoridae, Cylindrachetidae).

CLASSIFICATION

Order ORTHOPTERA (1,513 Australian spp.)

Suborder ENSIFERA (615)

GRYLLACRIDOIDEA (136)
1. Stenopelmatidae (15)
2. Gryllacrididae (101)
3. Rhaphidophoridae (20)
   Schizodactylidae (0)

TETTIGONIOIDEA (300)
4. Tettigoniidae (300)
   Prophalangopsidae (0)

GRYLLOIDEA (179)
5. Gryllidae (169)
6. Myrmecophilidae (3)
7. Gryllotalpidae (7)

Suborder CAELIFERA (898)

ACRIDOIDEA (819)
8. Eumastacidae (200)
9. Pyrgomorphidae (35)
10. Acrididae (584)
    Other exotic families (0)

TETRIGOIDEA (70)
11. Tetrigidae (70)

TRIDACTYLOIDEA (9)
12. Tridactylidae (4)
13. Cylindrachetidae (5)

In spite of difficulties of formal diagnosis, resulting from a small percentage of aberrant forms, the suborders are well differentiated and may be distinguished as follows.

KEY TO THE SUBORDERS OF ORTHOPTERA

Antenna with well over 30 segments. [Auditory organs, when present, located on fore tibia; stridulatory specializations of the fore wings, when present, located in the overlapping, horizontal part of the wings in their folded position; ovipositor, when present, having the valves articulated along their length to form an ensiform or stilettiform structure] .... ENSIFERA

Antenna with less than 30 segments. [Auditory organs, when present, located on abdominal tergum 1; stridulatory specializations of the fore wings, when present, located in the lateral, subvertical part of the wings in their folded position; ovipositor, when present, consisting mainly of 4, separate, prong-like valves with a basal hinge articulation] .... CAELIFERA

The classification adopted here is nearest to that of Beier (1955), from which it differs in the separate superfamily ranking accorded to the Tetrigidae and the elevation of the Pyrgomorphinae (Acridoidea) and Myrmecophilinae (Grylloidea) to family rank, while the characterization of the Gryllacridoidea is facilitated by regarding the non-Australian Prophalangopsidae as Tettigonioidea. At the generic and specific levels, the catalogue of Kirby (1906, 1910), though out of date, is still the most comprehensive reference work.

Suborder ENISFERA

Key to the Superfamilies and Australian Families of Ensifera

1. Tarsi 3-segmented .... GRYLLOIDEA. 2
   Tarsi 4-segmented* 4

* In all Australian species.

2. Fore legs fossorial, with broad, flat femur and tibia, and with powerful teeth on both tibia and tarsus .... **Gryllotalpidae**
   Fore legs gressorial, of substantially normal form 3
3. Eyes greatly reduced; hind coxae closely approximated ventrally. [Small depressed ant inquilines] .... **Myrmecophilidae**
   Eyes not reduced; hind coxae well separated ventrally .... **Gryllidae**
4. Fore wings, when present, usually tough, tegminized, usually with stridulatory specializations in the ♂; fore and mid tibiae very rarely with ventral articulated spines .... TETTIGONIOIDEA. **Tettigoniidae**
   Fore wings, when present, soft, not tegminized, without stridulatory specializations; fore and mid tibiae usually with ventral articulated spines .... GRYLLACRIDOIDEA. 5
5. Tarsi depressed. [First tarsal segment with plantulae; auditory tympana absent] . **Gryllacrididae**
   Tarsi compressed 6
6. First tarsal segment with plantulae; auditory tympana present* .... **Stenopelmatidae**
   First tarsal segment without plantulae;* auditory tympana absent .... **Rhaphidophoridae**

* In all Australian species.

Superfamily GRYLLACRIDOIDEA

The Gryllacridoidea are a primitive superfamily with about 1,000 species. They are cricket-like in general appearance and include the 'king crickets' and 'cave crickets' among others. The following characters are additional to those given in the key. Antennae usually much longer than the body. Fore wings, when fully developed (cf. Fig. 21.1A), pliable, rather broad, wrapping closely around each other and the hind wings in repose, the longitudinal veins running largely parallel, archedictyon absent; hind wings usually as broad as long. C often with several anterior accessory branches arising near its base in the fore wing; Sc running parallel to C in both wings; R_1 usually with anterior branches, Rs with posterior; M often fused to R for part of its length, especially in the hind wing; CuA in the hind wing often fused at the base with R+M. Abdomen with only the central portion of the sterna sclerotized, sternum 8 in the female and 9 in the male constituting the subgenital plate; styles often present in the male. Ovipositor present in all Australian species, ensiform, usually with all 3 valves well developed.

Most of the species are pale to dark brown insects, hiding during the day in tree-holes, hollow logs, under bark or debris, or in burrows in the ground, and coming out at night, when they may be attracted to light. They occur in both humid and arid environments. Some are cavernicolous, and a few exotic species wholly subterranean. Some live a more exposed life on foliage, where they may construct shelters of rolled leaves with the aid of an oral secretion. Both phytophagous and predacious feeding habits are represented. Beier recognizes 4 families (apart from the Prophalangopsidae, here allotted to the Tettigonioidea), of which 3 are represented in Australia. There is no comprehensive work on the Australian fauna. Nearly two-thirds of the genera are endemic.

1. Stenopelmatidae. Mostly large wingless species, sometimes with the mandibles much enlarged in the male; one Indian genus blind and wholly subterranean. Fore coxa usually with a strong spine. Fore tibia with auditory tympana in all Australian species, the dorsal face usually with one or more spines in addition to the terminal ones. Tarsi compressed, 1st segment with plantulae. Of 4 subfamilies, 2, the HENICINAE and DEINACRIDINAE, occur in Australia. The latter is represented by species of the apterous genus *Australostoma* (Fig. 21.3A), found in rotting logs and similar situations along the eastern coast.

2. Gryllacrididae. This is the largest unit in the superfamily, with a world list of some 600 species. Most of them are fully winged, some brachypterous or apterous. Fore coxa usually with a strong spine. Fore tibia with only terminal spines on the dorsal face; auditory tympana wanting. Tarsi depressed, 1st segment with plantulae. Only one subfamily is recognized. The largest Australian

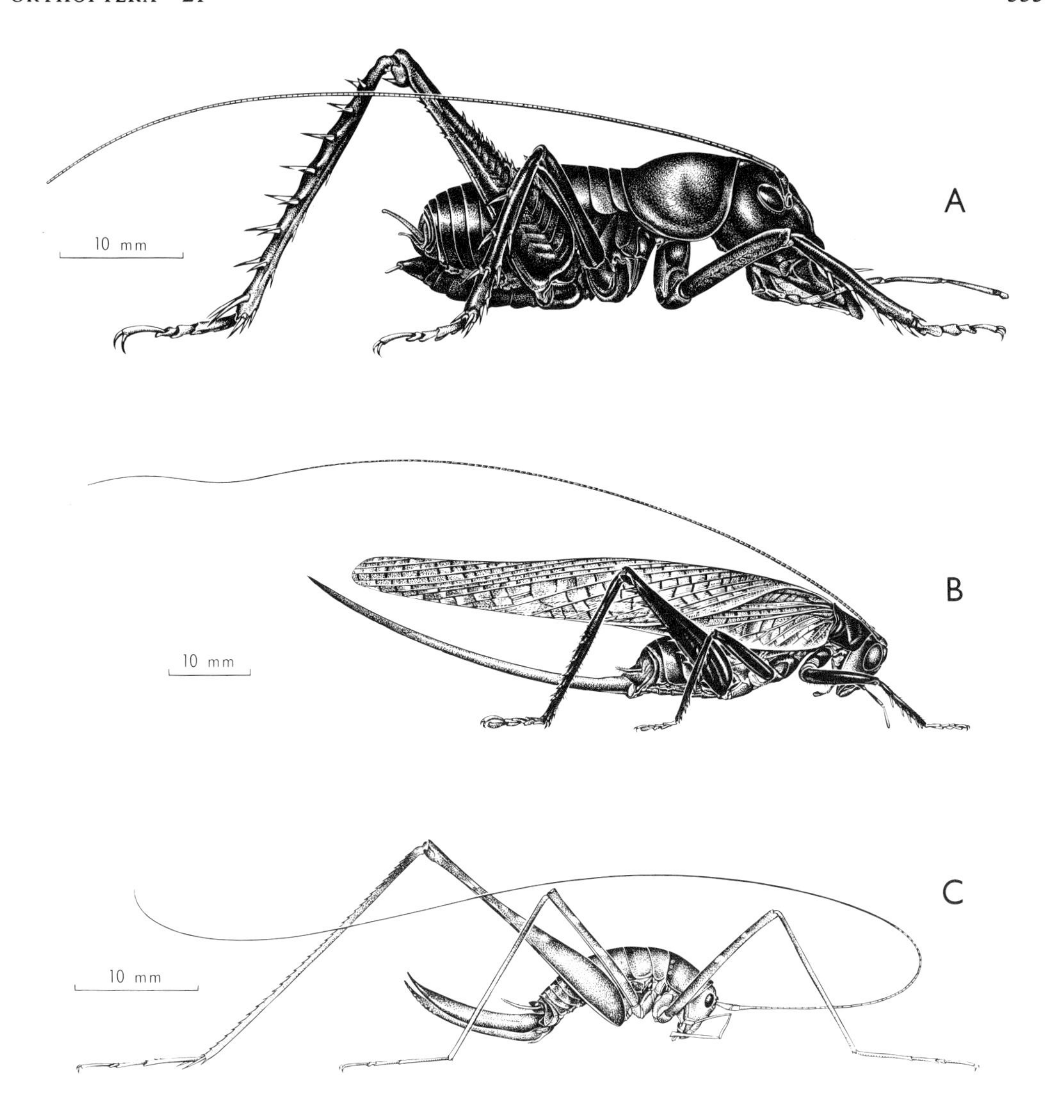

Fig. 21.3. A, *Australostoma opacum*, Stenopelmatidae, ♂; B, *Hadrogryllacris magnifica*, Gryllacrididae, ♀; C, *Micropathus tasmaniensis*, Rhaphidophoridae, ♀. [F. Nanninga]

genus is *Hadrogryllacris*, with about 25 described species. *H. magnifica* (Brunn.) (Fig. 21.3B), from the arid interior, has the head, pronotum, and legs shiny black with a metallic iridescence.

3. Rhaphidophoridae. Apterous insects, usually with slender, elongate appendages. Fastigium of vertex narrow and usually divided at the apex. Tarsi compressed, without plantulae. Auditory tympana lacking. All species are hygrophilous and many cavernicolous; they are probably in the main omnivorous scavengers. Chopard recognizes 3 subfamilies, of which one, the MACROPATHINAE, with a circum-Antarctic distribution, occurs in Australia. The Australian species are very similar in general appearance, with a soft integument and mottled, drab pattern (Fig. 21.3C); most of them have been taken in caves.

Superfamily TETTIGONIOIDEA

The Tettigonioidea, or long-horned grasshoppers, are the largest superfamily of the Ensifera, with some 5,000 species. They are characterized by the following features additional to those given in the key. Antennae usually longer than the body. Fore tibia with auditory tympana, except in the Australian subfamily Phasmodinae. Fore wing rarely wholly wanting in the male. CuA with its most anterior branch (or the vein as a whole) fusing with MP for part of its length, in the fore wing at least; CuP unbranched, running straight to the margin in the hind wing. The male fore wing usually has the cubital region specialized for stridulation; CuP, the vein principally affected, is bent towards the posterior margin of the wing before resuming its longitudinal course, the reflexed portion being thickened and toothed on the underside. Abdomen with only the central region of the sterna sclerotized, sternum 8 in the female and 9 in the male constituting the subgenital plate; styles usually present in the male. Ovipositor present, though sometimes small, laterally compressed, with all 3 valves well developed.

All members of the group belong to our first biological type. They are mostly phytophilous and hygrophilous, frequenting the foliage or stems of woody or herbaceous plants; a few are geophilous. They usually show procryptic adaptations, occasionally aposematic, and may be either diurnal or nocturnal, being often attracted to light. Most feed on foliage, but some are predacious on other insects. The Tettigonioidea have been divided into numerous, widely accepted, subordinate groupings, which, however, have been variously ranked by different workers. Here we recognize 2 families, of which the primitive Prophalangopsidae include only 3 living species (none of which is Australian), but a number of fossils.

4. Tettigoniidae. The Tettigoniidae are distinguished from the Prophalangopsidae by the more advanced stridulatory specialization. The fore wing folds along a line between CuA and CuP, the horizontal part being relatively small in fully winged species and the left wing always overlapping the right; hind wing with the first fold in the MP area. An archedictyon is usually present over the whole of the fore wing and sometimes on a small distal portion of the hind. MA almost always fused to Rs for a short distance in the hind wing and occasionally in the fore wing; CuA with its most anterior branch (or the vein as a whole) usually fused with MP over the whole of its distal portion. The stridulatory specialization of the male fore wing (cf. Fig. 21.1D) has been described by Ragge. In the right wing a posterior branch of CuA bends towards the posterior margin in the proximal region of the wing and divides to form the anterior and posterior borders of a usually membranous area, the 'mirror', whose proximal border is formed by CuP, likewise reflexed. In the left wing these veins take a corresponding course, but in addition CuP is raised and serrulate on its underside, forming the 'file', which in stridulation is scraped over the raised posterior margin of the right wing.

Of 19 subfamilies, to which Beier provides a key, 11 occur in Australia; 2 of these are endemic, as well as more than half the genera. There is no monograph on the Australian fauna. The subfamilies PSEUDOPHYLLINAE, MECOPODINAE, and PHYLLOPHORINAE are represented in Australia by only a few species each, the predacious DECTICINAE by some 25 species, distributed over most of the continent, and the small predacious subfamily SAGINAE by nearly 20 species, all brachypterous and mostly from W.A. The endemic ZAPROCHILINAE comprise 2 species in the remarkable genus *Zaprochilus*. *Z. australis* (Brullé) (Fig. 21.4A) has a strongly prognathous head and very narrow wings, which rise in a stick-like roll at a considerable angle to the lie of the abdomen. The closely related endemic PHASMODINAE are known at present only from females of a single Western Australian species, *Phasmodes ranatriformis* Westw. (Fig. 21.4D). These are apterous, exceedingly attenuate, and look like a phasmatid. The fore tibia lacks auditory tympana, and the hind femur is of uniform width throughout and clearly not saltatorial.

The large subfamily CONOCEPHALINAE is well represented with about 18 genera and

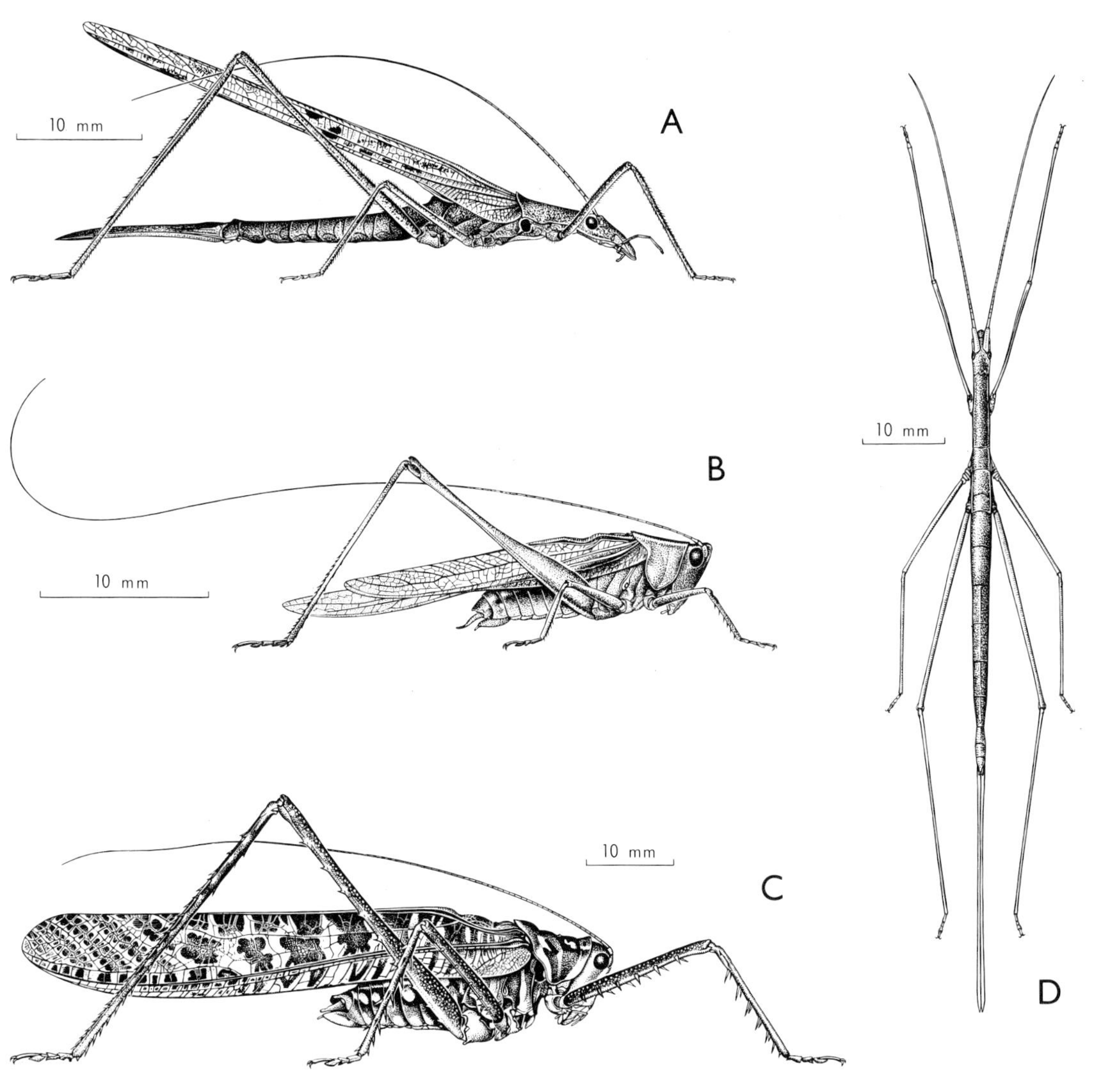

Fig. 21.4. Tettigoniidae: A, *Zaprochilus australis*, ♀; B, *Conocephalus* sp., ♂; C, *Yorkiella picta*, ♂; D, *Phasmodes ranatriformis*, ♀. [F. Nanninga]

many species. *Conocephalus* has more than 20 abundant, diurnal, grass-frequenting species (Fig. 21.4B). The predacious LISTROSCELINAE are represented by at least 10 species. The large, slender, very spiny *Yorkiella picta* Carl (Fig. 21.4C), beautifully mottled in olive-green and white, may be found on shrubs in the interior. In the small subfamily TYMPANOPHORINAE, represented by 2 species of *Tympanophora*, the thoracic structure of the male (Fig. 21.5A) has been extensively modified for stridulation, with inflated nota and fore wings, in which the stridulatory area is very large. The female (Fig. 21.5B) is apterous, with a small pronotum, emarginate behind. The phytophagous PHANEROPTERINAE, largest subfamily of Tettigoniidae, account for nearly half the Australian tettigoniid fauna. The largest genus is *Caedicia* (Fig. 21.5C). *Alectoria superba* Brunn. (Plate 2, K) has 3 strong pronotal spines and an enormous semicircular

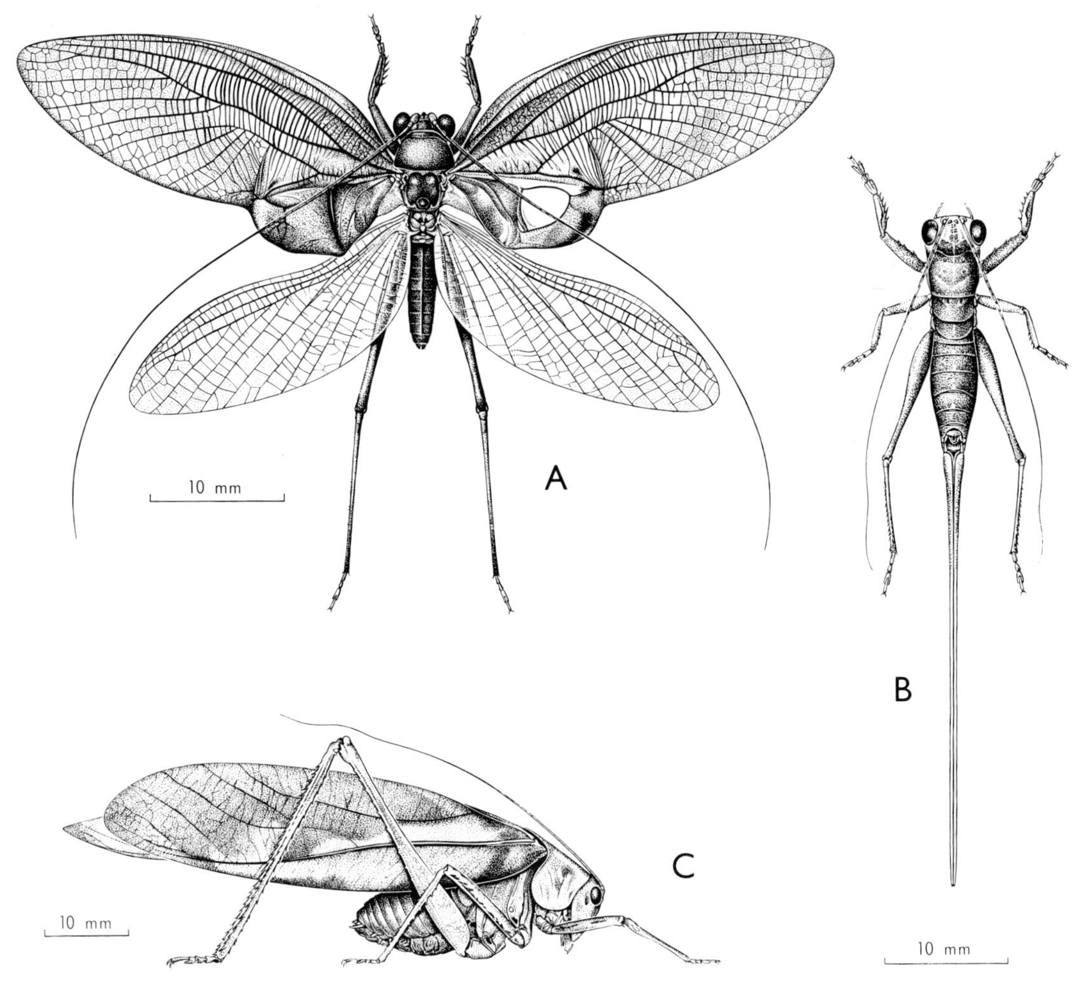

Fig. 21.5. Tettigoniidae: A, *Tympanophora uvarovi*, ♂; B, *T. pellucida*, ♀; C, *Caedicia major*, ♀. [F. Nanninga]

pronotal crest. *Acripeza reticulata* Guér. is almost black in the male, with fully developed wings; in the female (Plate 2, L), hind wings are absent and the shell-like fore wings can be raised to reveal brightly coloured bands on the abdomen, which presumably serve an aposematic function.

Superfamily GRYLLOIDEA

The Grylloidea, or true crickets (including the tree crickets and mole crickets), comprise some 2,000 species. Fore tibia bearing auditory tympana in species that stridulate. All tarsi 3-segmented. Fore wings, when fully developed, relatively broad, usually tegminized, folded near their mid-line, in the region of the median vein, to form a box-like cover for meso- and metathorax and the proximal part of hind wings and abdomen; horizontal part relatively much larger than in the Tettigonioidea, in the male most of it usually modified for stridulation in almost identical fashion in the two fore wings; right wing usually overlapping left. Hind wing folding throughout to form an attenuate spike usually projecting well beyond the fore wing. Many species are brachypterous or apterous, or polymorphic for wing length.

C usually absent or reduced in fore wing; Sc lying well behind anterior margin in fore wing, with numerous parallel, sinuous, anterior accessory branches, in the hind wing running just behind C. R, M, and CuA closely approximated to Sc along the mid line of the fore wing, with M usually branching in its distal part; CuA in the male fore wing giving rise to posterior accessory branches which contribute to the stridulatory apparatus, in the hind wing dividing at the base into 2 parallel branches which run straight to the wing margin; CuP in the male fore wing bending strongly towards the posterior margin before continuing towards the apex, with stridulatory teeth on ventral surface of bent section; 3 anals in the fore wing, which in the male fuse with each other and CuP and then separate again, and in the female run separate and parallel to the margin; anals numerous in the hind wing. Abdominal sterna fully sclerotized, sternum 8 in the female and 9 in the male constituting the subgenital plate, which lacks styles. Cerci similar in the sexes, long and flexible. Ovipositor, when present, usually stilettiform, composed of 2 pairs of valves, the 3rd being greatly reduced.

Most crickets are pale, drab, or sometimes black insects, usually hiding during the day under logs, stones, or debris, or in burrows in the ground, and coming out at night. They typically frequent humid environments, stridulating loudly on warm nights, especially after rain. Tree crickets, however, live exposed on foliage. Most Grylloidea are omnivorous. About 15 major groups are recognized, to

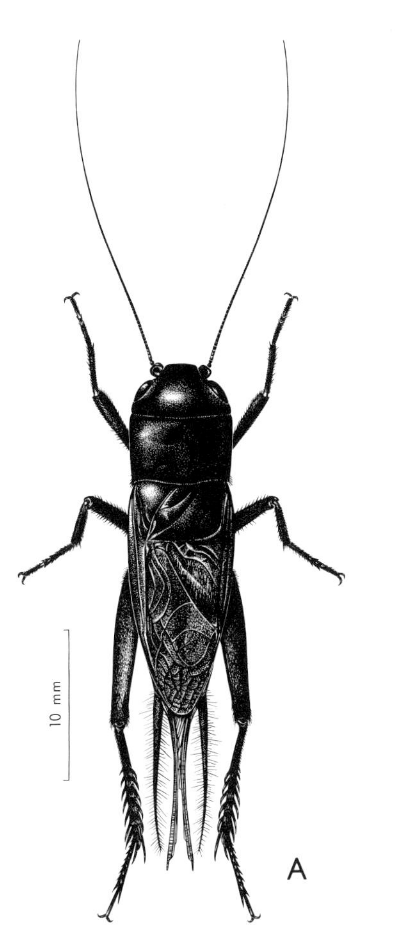

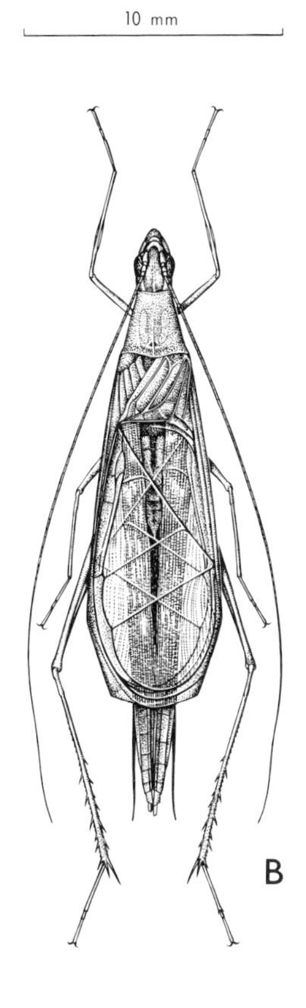

Fig. 21.6. Gryllidae: A, *Teleogryllus commodus*, ♂; B, *Oecanthus rufescens*, ♂. [F. Nanninga]

most of which Chopard gives family and Beier subfamily rank. We recognize 3 families, Gryllidae, Myrmecophilidae, and Gryllotalpidae.

5. Gryllidae. This is much the largest family. Some species have the fore tibia somewhat modified for digging, but the femur and tarsus are not affected and the leg remains primarily gressorial. The stridulatory apparatus of the fore wing (cf. Fig. 21.1B) typically includes four areas: the mirror, cordal area, harp, and basal area. Three posterior branches of CuA delimit the mirror and form the boundary between cordal area and harp, while the proximal boundary of the last is formed by CuP. The Australian gryllids have been monographed by Chopard (1951, see also 1961). Of his 13 major units (here treated as subfamilies), 11 are represented in Australia. About a quarter of the genera and many of the species are endemic, but none of the subfamilies.

The large subfamily GRYLLINAE includes numerous winged and some wingless Australian species. The black *Teleogryllus commodus* (Walk.) (Figs. 21.1B, 6A) is injurious to pastures and crops in limited areas of south-eastern Australia. Investigations of its biology and ecology have been carried out by Browning, Hogan, and Bigelow and Cochaux (1962). The NEMOBIINAE have 3 Australian genera, the MOGOPLISTINAE 2, and the PENTACENTRINAE 2 species. The PHALANGOPSINAE are represented principally by some dozen species of *Endacusta*, with brachypterous males and apterous females. There are 4 species of OECANTHINAE, or tree crickets (Fig. 21.6B). These are pale and slender, with prognathous head and very large membranous fore wings in the male; they live exposed on foliage and lay their eggs in the stems of plants. In the TRIGONIDIINAE are numerous, very small, mainly tropical species. The largest genus is *Metioche* (Fig. 21.7A), with the fore wings somewhat elytriform, similar in the sexes, and usually

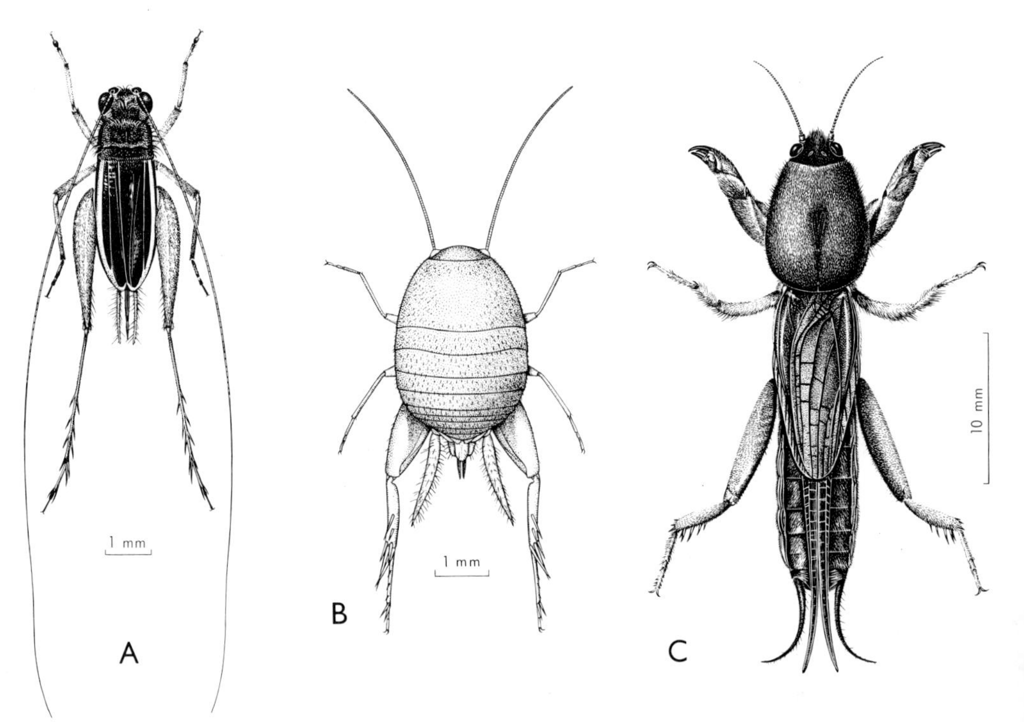

Fig. 21.7. A, *Metioche albovittata*, Gryllidae, ♀; B, *Myrmecophila testacea*, Myrmecophilidae, ♀; C, *Gryllotalpa africana*, Gryllotalpidae, ♀. [F. Nanninga]

with no trace of stridulatory apparatus. ENEOPTERINAE are represented by a number of species, and the Indo-Malayan ITARINAE by a single species. One of the largest groups in Australia is the PODOSCIRTINAE, with mainly rather large, fusiform species.

6. Myrmecophilidae. This is a small family of very small, apterous, depressed crickets, which live as inquilines in ants' nests. Eyes reduced. All coxae large and closely approximated ventrally. Cerci with a pseudo-segmentation. Ovipositor short and stout, strongly and rigidly descending, supported at the base by long descending processes of terga 8 and 9. These little crickets live on secretions produced by the ants, which they lick with the hypopharynx; they may also be directly fed by them. They participate in the movements of the ants within the nest and are apparently taken for other ants by their hosts. Only a very few, large eggs mature at one time. Some species are parthenogenetic. Three Australian species have been described, in the genus *Myrmecophila* (Fig. 21.7B).

7. Gryllotalpidae. The Gryllotalpidae, or mole crickets, comprise some 50 species of rather large, burrowing Grylloidea, most of which are winged and may fly to lights at night. Pronotum with the lateral lobes bending around ventrally, but not approximated on the mid-ventral line as in the superficially similar Cylindrachetidae (p. 347). Fore leg greatly modified for digging; both femur and tibia broad and compressed, the latter with greatly enlarged terminal teeth, or 'dactyls'; tarsus attached to outer face of tibia, its 2 proximal segments produced below to form supplementary dactyls, which participate in the digging process (cf. Cylindrachetidae). Hind legs relatively small, but the femur considerably larger than the mid femur. Fore wing of male with harp and basal area, delimited as in Gryllidae, but with no differentiation of mirror and cordal area, only the most proximal branch of CuA being reflexed. Ovipositor absent. Mole crickets make deep permanent burrows, as well as superficial foraging galleries. They may be heard stridulating at the entrance of the burrow, especially after rain. The eggs are laid in an underground chamber. Tindale (1928a) has monographed the Australian species and given some information on their biology; all belong to the cosmopolitan genus *Gryllotalpa*. *G. africana* Beauv. (Fig. 21.7C), a minor pest of crops, stores seeds underground.

Suborder CAELIFERA

Key to the Superfamilies and Australian Families of Caelifera

1. Tarsi of all legs 3-segmented .... ACRIDOIDEA. 2
   Fore and mid tarsi at most 2-segmented 4
2. Head with posterior margin more or less emarginate dorsally, exposing the (often sclerotized) cervical membrane;* hind tibia with the 2 ventral terminal spurs much reduced, or absent* .... **Eumastacidae**
   Head with posterior margin entire, not exposing cervical membrane; hind tibia with all 4 terminal spurs well developed 3
3. Fastigium of vertex with a mediolongitudinal sulcus extending backwards for a short distance from its anterior extremity .... **Pyrgomorphidae**
   Fastigium of vertex without a mediolongitudinal sulcus anteriorly .... **Acrididae**
4. Pronotum produced posteriorly so as to overlie dorsally the rest of the thorax and at least the first few abdominal segments; hind tarsus 3-segmented .... TETRIGOIDEA. **Tetrigidae**
   Pronotum covering at most the rest of the thorax; hind tarsus 1-segmented . TRIDACTYLOIDEA. 5
5. Pronotum with lateral lobes widely separated ventrally; hind legs saltatorial, the femur much larger than the mid femur .... **Tridactylidae**
   Pronotum with lateral lobes curving around ventrally and approximated on the mid-ventral line; hind legs not saltatorial, the femur scarcely larger than the mid femur .... **Cylindrachetidae**

* In all Australian species.

Superfamily ACRIDOIDEA

The Acridoidea are the largest superfamily of Orthoptera, with a world tally of more than 10,000 species. They comprise the short-horned grasshoppers and locusts. Locusts are species that form dense, strongly migrating swarms at times. Barely 20 are known; they are all Acrididae, but distributed in different subfamilies.

General characters of Acridoidea, additional to those given in the key, are: Antennae in all Australian species less than half the length of the body. Tarsi with an arolium. Fore wing (when fully developed) with the more proximal parts of Sc, R, and M (but not Cu) closely approximated (cf. Fig. 21.1c). Hind wing with only the anal area folding in repose; R and M fused, at least towards the base. Both wings with CuP tending to reduction. Abdomen with sterna fully sclerotized, lacking separate laterosternites; 8 visible in the female, of which the last is the subgenital plate; sternum 9 of male without styles, but bearing a posterior lobe, which envelops the genitalia and to which the term 'subgenital plate' is restricted. Cerci short, unsegmented. Paraprocts without cerciform processes. Male genitalia (Dirsh, 1956) usually strongly sclerotized, comprising an epiphallus and an intromittent phallus. Ovipositor always present.

All species belong to our first biological type (p. 327). They are mainly geophilous, or inhabitants of the herbaceous stratum, and more or less xerophilous. Although they are essentially diurnal, many species undertake long flights on hot evenings, and a few oviposit at night. They are exclusively phytophagous and normally feed upon green foliage; some include dew-moistened dead leaves in their diet. Copulation takes place with the male mounted on the female, although in some species he may become detached from that position and be dragged around by the female. Uvarov (1966) gives a comprehensive account of the anatomy and some aspects of the biology. The biology of Australian species has been reviewed by Key (1958, and in Keast *et al.*, 1959).

Dirsh (1961) recognizes 14 families, of which only 3 are Australian; none is endemic. The principal taxonomic works on the Australian Acridoidea are two monographs by Sjöstedt in 1921 and 1935 and the revisions of Rehn (1952–7), but these cannot be recommended to the student seeking identifications.

8. Eumastacidae. This is one of the more primitive families. The characters given in the key permit immediate recognition of the Australian representatives, but are not diagnostic for the family as a whole. The following additional characters may be noted: Antenna with one of the more distal segments usually bearing a small tubercle, the 'antennal organ', on its ventral surface. Hind legs in the resting position commonly extended outwards at a large angle to the body, the femur being rotated so that its outer surface faces the substrate, the tibia flexed against it. Hind tarsus with the basal segment usually bearing one or more spines or tubercles dorsally. Fore wing (when fully developed) with CuA unbranched; hind wing with M unbranched, CuA reduced, and an ambient vein around the anal area. Abdominal spiracles situated in the pleural membrane. Auditory tympanum absent. Male genitalia (Dirsh, 1961) very diverse, valuable in classification. Hind gut with 6 caeca in the species studied. Dirsh recognizes 20 subfamilies, of which only 2 are Australian: the Biroellinae, best developed in New Guinea, and the Morabinae, so far as known endemic to Australia.

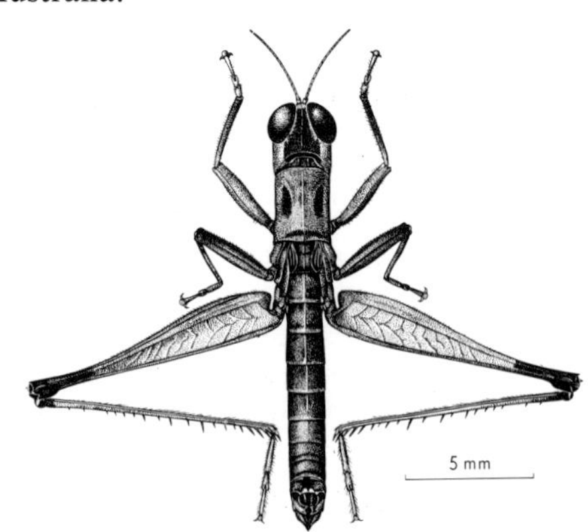

Fig. 21.8. *Biroella* sp., Eumastacidae, ♂. [F. Nanninga]

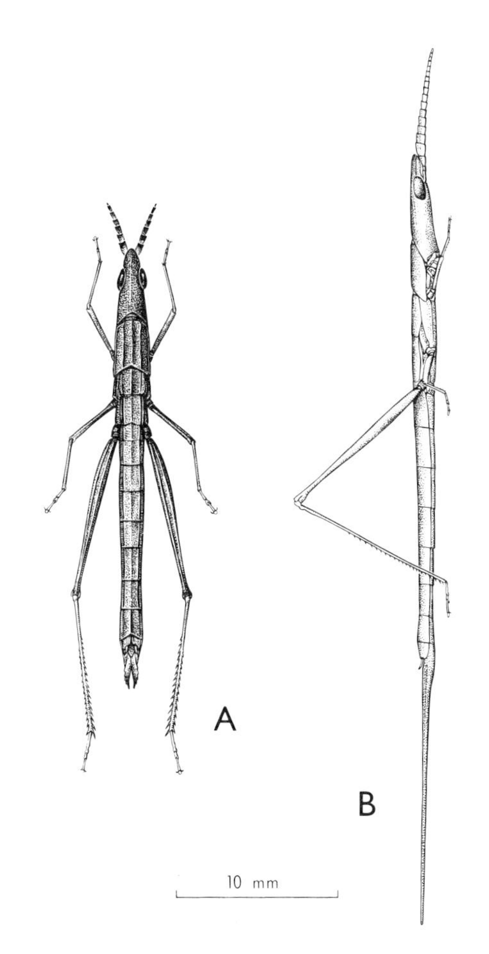

Fig. 21.9. Eumastacidae: A, *Keyacris scurra*, ♀; B, *Warramunga desertorum*, ♂. [F. Nanninga]

The BIROELLINAE comprise the single genus *Biroella* (Fig. 21.8), of which 5 species are known from north-eastern Queensland. They are rather small, brachypterous or apterous, with short filiform antennae, and the fastigium hardly projecting beyond the eyes; they frequent low regrowth on the margins of rain forest.

The MORABINAE comprise at least 195 species of apterous, usually elongate grasshoppers, which may be found in almost all parts of Australia. The antennae are ensiform or subensiform and the fastigium extends well beyond the eyes; the male subgenital plate nearly always bears a slender cultriform process at its extremity. The insects mostly frequent shrubs and feed on dicotyledons, but some live and feed on grasses and a few on ferns. Blackith and Blackith (1966) describe their anatomy. Extensive studies on the chromosomes of the morabine grasshoppers have been undertaken by White and his associates (see White, 1962b), working especially with *Keyacris scurra* (Rehn) (Fig. 21.9A). The basic chromosome number seems to be 2n♂ = 17, but numbers from 13 to 21 are known. *Moraba virgo* Key (Plate 2, I) is the only member of the Acridoidea known to reproduce by obligatory parthenogenesis (p. 81). The genus *Warramunga* (Fig. 21.9B) comprises a number of very attenuate species specialized for life on spiny grasses of the genus *Triodia*.

9. Pyrgomorphidae. Apart from the features given in the key, the Pyrgomorphidae are characterized by a conical head and an elevated median process on the prosternum. The fastigium often has the region near the margins slightly raised and uneven, and separated from the central part by furrows, which join the median furrow anteriorly, forming an inverted Y. The basic chromosome number is 2n♂ = 19. Most species feed on dicotyledons; many are aposematic and some produce repugnatorial secretions.

In Australia, 2 species of *Atractomorpha* (Fig. 21.10A), uniformly green or light brown grasshoppers with pale pink hind wings, represent the furthest penetration of a widespread palaeotropical genus. They feed on dicotyledonous herbs in moist situations. *Desmoptera* (Fig. 21.10B) and *Desmopterella*, with one species each, are similarly outliers of a larger distribution centred on New Guinea. They are found on litter in the rain forests of north-eastern Queensland. *Psednura* (Fig. 21.10C) and *Propsednura* comprise very attenuate, apterous or subapterous species found on plants with terete green leaves or stems, such as rushes, sedges, and *Xanthorrhoea*. The large endemic genus *Monistria* (Plate 2,H) has an Australia-wide distribution, usually on shrubs. The species are normally brachypterous, although several produce occasional macropterous individuals; all are aposematically spotted or striped with

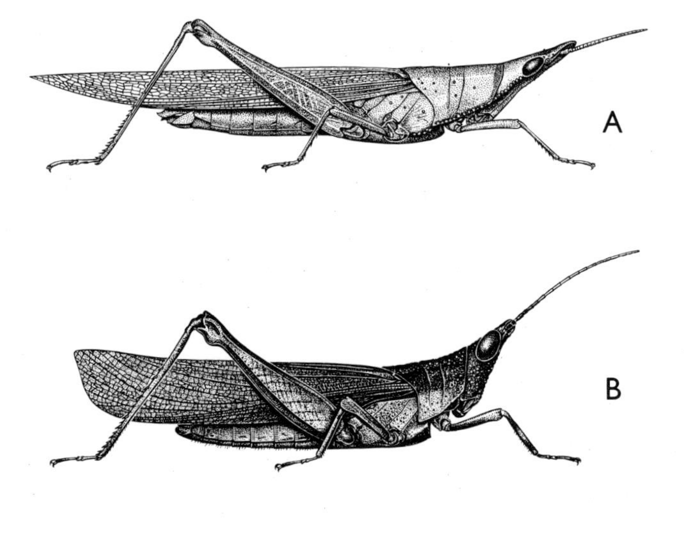

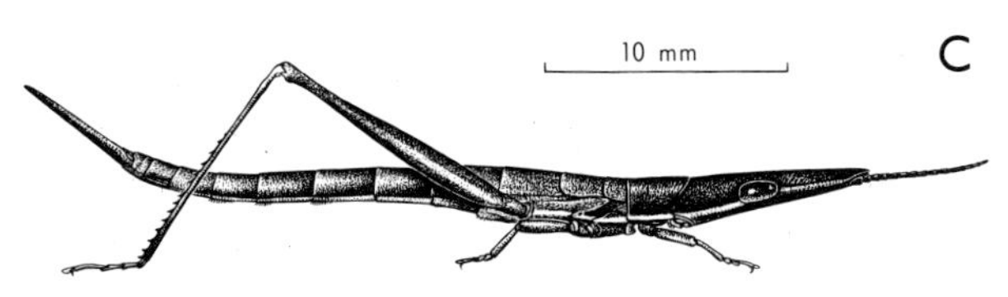

Fig. 21.10. Pyrgomorphidae: A, *Atractomorpha crenaticeps*, ♀; B, *Desmoptera truncatipennis*, ♂; C, *Psednura* sp., ♂. [F. Nanninga]

yellow to red and have scarlet hind wings, which the insect exposes by raising the fore wings when disturbed.

10. Acrididae. This is by far the largest family of Caelifera. It is characterized principally by lack of the various specializations shown by the other families of Acridoidea and is itself highly heterogeneous. The basic chromosome number is 2n♂ = 23; this represents also the upper limit, but numbers as low as 8 are known. Dirsh recognizes 17 subfamilies, with 6 occurring in Australia, but for present purposes we may reduce these to 4, constituted somewhat differently from his. There is about 90 per cent endemism at the generic level.

The OXYINAE include about a dozen Australian genera of Indo-Malayan origin. *Bermiella acuta* (Stål) (Fig. 21.11A), which occurs on rushes, sedges, and hygrophilous grasses near water, has a number of structural adaptations for a semiaquatic life, including expanded hind tibiae, dense patches of hairs on the more distal abdominal sterna and on the fore wing, and an air chamber formed by doming of the costal area of the fore wing over the first abdominal spiracle. The genus *Kosciuscola* (Plate 2,J) is of interest both for its colour response (p. 329) and also because, in spite of its Indo-Malayan origin, it has adapted better than any other Australian genus of Acrididae to the alpine environment.

Some 90 per cent of the Australian acridids belong to the CATANTOPINAE—a preponderance not attained in any other region. For the most part, the Australian representatives comprise an ancient autochthonous group, which has developed numerous strongly divergent lines. A very typical Australian genus is *Goniaea* (Fig. 21.11B), with several species occurring among fallen leaves under eucalypts; they feed upon the dead leaves and retreat beneath them at night. The more slender and procryptically mottled *Coryphistes* (Fig. 21.11C) is equally characteristic on tree trunks and fallen branches. *Ecphantus quadrilobus* Stål (Fig. 21.11D), a velvety, green or light brown species, is restricted to plants of the genus *Sida*. *Phaulacridium* is the only genus common to Australia and New Zealand. *P. vittatum* (Sjöst.) (Fig. 21.12A), with brachy-

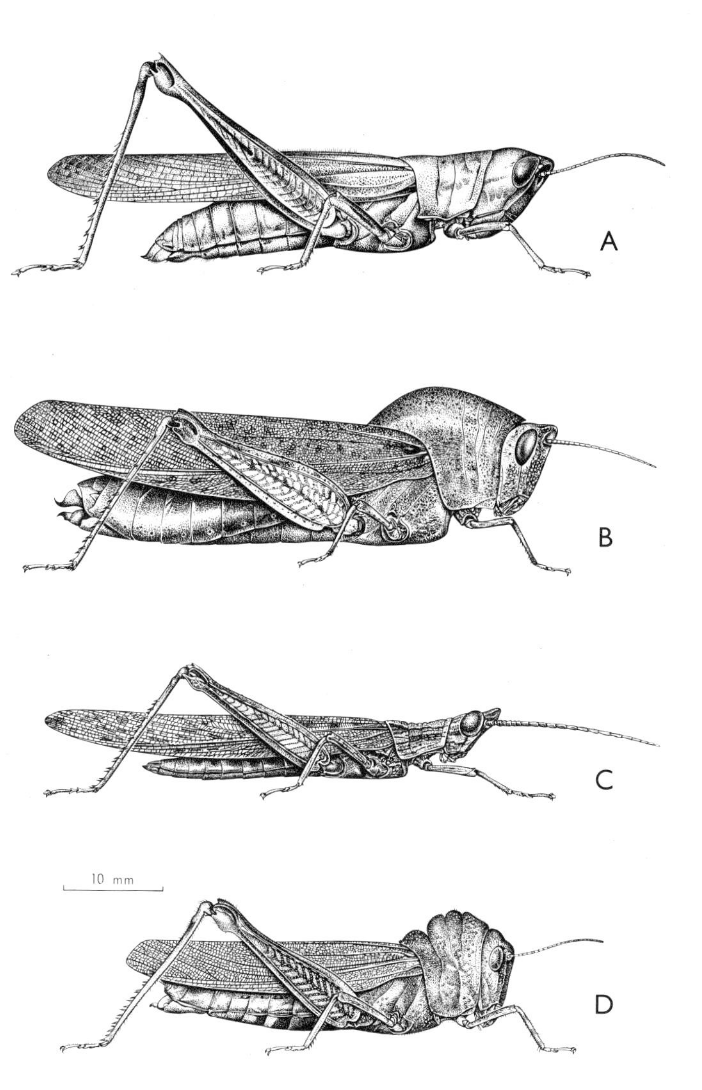

Fig. 21.11. Acrididae: A, *Bermiella acuta*, ♀; B, *Goniaea australasiae*, ♀; C, *Coryphistes ruricola*, ♂; D, *Ecphantus quadrilobus*, ♀. [F. Nanninga]

pterous and macropterous morphs, is of economic importance in southern Australia. *Raniliella* (Fig. 21.12B) has an enormously enlarged, rounded pronotum, giving the insect a close resemblance to the pebbles among which it lives in the semi-desert. Another desert genus, *Urnisiella* (Fig. 21.12C), has exceptionally long mid legs, which assist it to maintain a hold on shifting sand and to bury itself just beneath the surface.

The CYRTACANTHACRIDINAE comprise mainly very large tropical species, most of which are powerful fliers. They include several of the world's most serious locust pests. In Australia there are about a dozen, fully winged species in 2 genera, *Austracris* and *Valanga*, which have otherwise an Indo-Malayan and Pacific distribution. *Austracris guttulosa* (Walk.) (Fig. 21.12D), with an almost Australia-wide range, behaves as a locust in northern Australia and is of some economic significance. The nymphs show kentromorphic pattern differentiation. *Valanga irregularis* (Walk.) is the largest Australian acridoidean and one of the largest in the world. It feeds on the leaves of trees and shrubs in the moister tropics and subtropics, and may become of local horticultural importance.

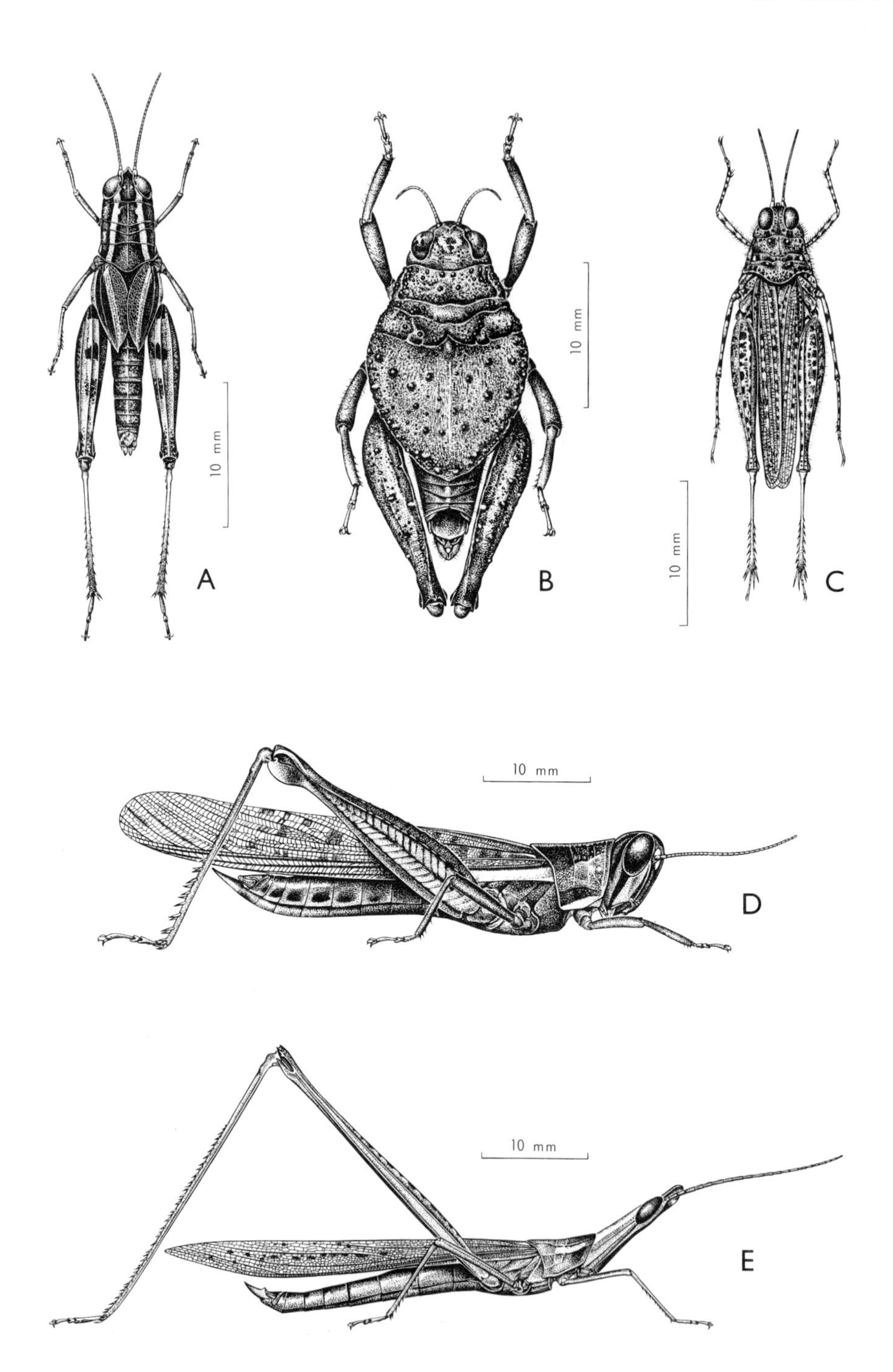

Fig. 21.12. Acrididae: A, *Phaulacridium vittatum*, ♀, brachypterous form; B, *Raniliella* sp., ♂; C, *Urnisiella rubropunctata*, ♀; D, *Austracris guttulosa*, ♂; E, *Acrida conica*, ♂. [F. Nanninga]

The ACRIDINAE are a large subfamily of almost world-wide distribution, including several locusts. Phylogenetically advanced, almost all are primarily grass feeders. In Australia there are only some 30 species in 21 genera. A third are large Old-World genera (e.g. *Acrida*, Fig. 21.12E; *Gastrimargus*) represented in Australia by a single species and apparently of geologically recent entry. However, many species of this restricted fauna are abundant as individuals. *Chortoicetes terminifera* (Figs. 21.13, 2) is the principal Australian locust pest, on which there is now a substantial literature (Key, 1958; Casimir, 1962). It is distinguished by the black spot at the tip of the hind wing. The swarming grasshopper *Austroicetes cruciata* and the locust *Gastrimargus musicus* (Plate 2, F; see Common, 1948) are of less importance. The last produces a characteristic clicking sound in flight. *Locusta migratoria*, one of the most serious of the Old-World locusts, occurs in Australia, but swarms only very occasionally. Males of *Froggattina australis* (Walk.) (Plate 2, G) have thickened anal veins in the hind wing and make a buzzing sound in flight.

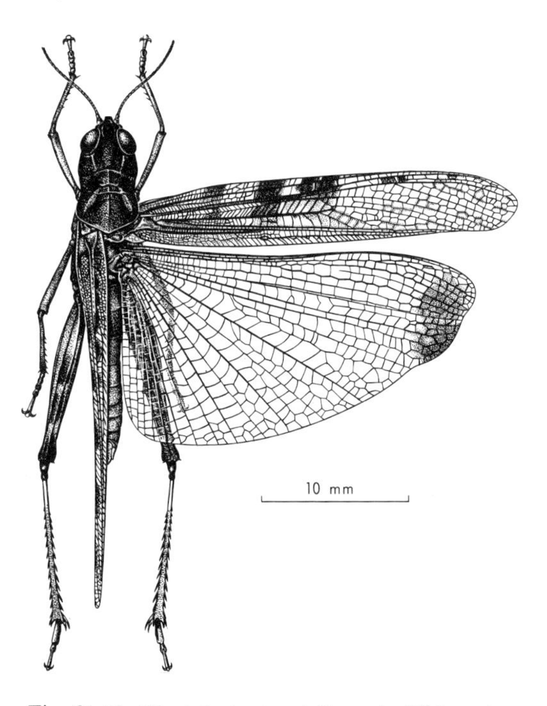

Fig. 21.13. *Chortoicetes terminifera*, Acrididae, ♂. [F. Nanninga]

Superfamily TETRIGOIDEA

The Tetrigoidea include some 1,000 species of small, dun-coloured grasshoppers, sometimes called 'grouse-locusts'. They are characterized by the following distinctive features. Frons below median ocellus bearing a single median carina, which divides ventrally, forming, with the fronto-clypeal suture, a 'supra-clypeal triangle'. Pronotum usually reaching or surpassing apex of abdomen, shielding the wings (if present), sometimes modified in bizarre fashion. Prosternum expanded along its anterior edge to form a collar around the mouth-parts and directly connected to the pronotum by a precoxal bridge. Fore and mid tarsi 2-segmented, hind tarsus 3-segmented; arolia absent. Hind legs always saltatorial. Fore wing reduced to a small, usually oval, lateral scale, which is sharply divided by the strong radial vein into an exposed, sclerotized, morphologically anterior region, and a membranous posterior region covered by the lateral edge of the pronotum; all other veins greatly reduced or absent. Hind wing with all veins unbranched, except for the basal division of Cu; remigium reduced to a narrow band, with marginal C, M fused with R for all or nearly all its length, and for much of its length closely pressed against Sc, CuA very short; anal fan large, 1A closely pressed against CuP for almost all its length. Both wings often absent, or concealed and functionless. Abdomen with discrete laterosternites; the female subgenital plate constituted by sternum 8, that of the male by a lobe of 9; without styles. Cerci short, unsegmented. Paraprocts without cerciform processes. Male genitalia simpler and quite differently constructed from those of the Acridoidea (Widdows and Wick, 1959). Ovipositor as in Acridoidea. Auditory and stridulatory organs absent. Abdominal spiracles located in the pleural membrane. So far as known, all Tetrigoidea have a karyotype with 2n♂ = 13, all the chromosomes being acrocentric.

Complicated pattern polymorphism is characteristic of many Tetrigoidea, as well as polymorphism in pronotal length, with which differences in rugosity and other structural features are correlated. Some flightless species are neoteinic in pronotal development. The insects are found exposed on bare soil or sand, less often on short turf, and usually in damp situations. Many species swim freely, and may remain for some time under water, resting on the bottom or clinging to water plants. Tetrigoidea feed principally on algae and other cryptogams, or on mud, from which they presumably obtain living or dead vegetable matter. In species studied, the eggs are laid in small batches in the soil, without any bonding secretion. They are cylindrical, with a slender appendage at the anterior pole.

11. Tetrigidae. Only one family, the Tetrigidae, is recognized. This has been divided into 9 or 10 often poorly delimited 'sections', which are sometimes treated as subfamilies; Rehn (1952) recognizes 4 in Australia. About half the genera are endemic. *Paratettix* (Fig. 21.14B), with several Australian species, has a wide extra-Australian distribution.

Superfamily TRIDACTYLOIDEA

This superfamily comprises only a few genera and some 90 species. Its components were formerly classified under the Grylloidea, species of *Tridactylus* being sometimes called 'pigmy mole crickets'. The true relationships of the group (Ander, 1934) have been shown to lie with the Tetrigoidea. There is, however, a remarkable parallelism between the Tridactyloidea and the Grylloidea, the Tridactylidae corresponding to the Gryllidae, and the Cylindrachetidae to the Gryllotalpidae. The following features are common to the two constituent families. Prosternum directly connected to the pronotum by a precoxal bridge. Fore tibia in all Australian species strongly fossorial: expanded and equipped with powerful teeth. Fore tarsus 1- or 2-segmented, inserted on inner face of tibia (cf. Gryllotalpidae); mid tarsus 2-segmented; hind 1-segmented; arolia absent. Abdomen with 9 fully sclerotized sterna in both sexes, sternum 9 constituting the subgenital plate, which lacks both styles and sternal lobe. Paraproct bearing a sclerotized hook in the male. Ovipositor absent in all Australian species. The principal work on the Australian representatives is that of Tindale (1928a), who treated them as gryllotalpids.

12. Tridactylidae. The tridactylids are very small, usually smooth and shiny, black or variegated Orthoptera. Eyes well developed. Pronotum overlapping the mesonotum, its lateral lobes well separated by a broad prosternum. Fore femur little expanded, tibia moderately so, tarsus 2-segmented. Hind legs saltatorial, with greatly enlarged femur; tibia long and slender, with the 4 terminal spurs long, especially the ventral pair, and several of the more distal spines along the dorsal edges often replaced by articulated lamellae; tarsus sometimes greatly reduced. Fore wing tegminized, falling well short of abdominal apex (even when the hind wing much surpasses the abdomen), bearing 2 veins (probably Sc and R) and sometimes a third (probably 1A). Hind wing, when fully developed, with the remigium reduced to a narrow sclerotized band, all veins unbranched, except Cu; C weak, supporting proximal part of anterior margin; R, M, and CuA fused or closely associated in basal third of wing; anal fan large, with very numerous anals, traversed by a single arc of cross-veins. A putative stridulatory file is sometimes located ventrally on the fore wing, along the distal part of Sc. Cercus 2-segmented in all Australian species. Paraproct bearing a cerciform appendage. Karyotypes with 2n♂ = 13 and 15 have been reported.

These little insects usually frequent the margins of water bodies, where they construct galleries in sandy ground. However, they spend much of their time crawling about on the surface and must be referred to our second biological type. The hind legs are kept flexed during walking; they are used only for making extremely powerful jumps when disturbed, and for swimming on and beneath the surface of the water, when the tibial lamellae presumably come into action. Food apparently consists of unicellular algae and particles of vegetable matter in or on the soil. The eggs are laid in batches in underground chambers. Of 2 subfamilies, only the TRIDACTYLINAE are found in Australia, with the cosmopolitan *Tridactylus* (Fig. 21.14A) the principal genus.

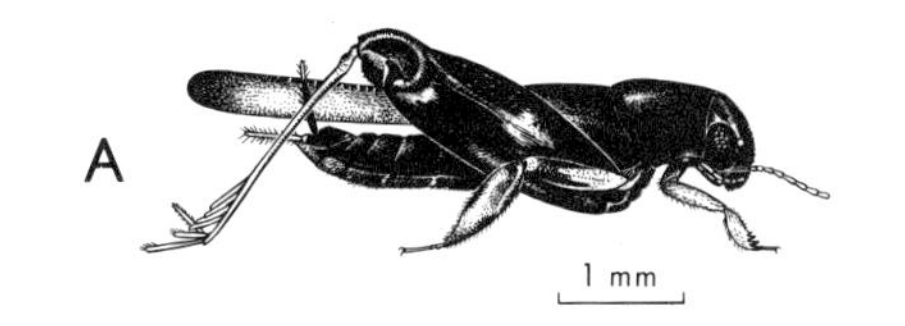

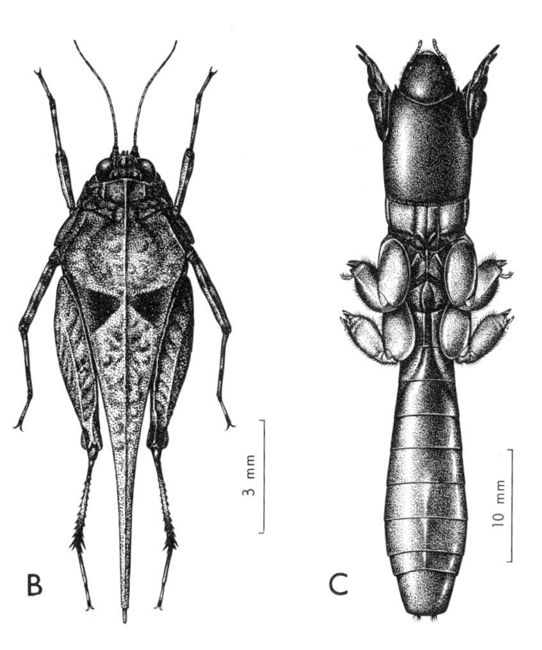

Fig. 21.14. A, *Tridactylus australicus*, Tridactylidae, ♂; **B**, *Paratettix argillaceus*, Tetrigidae, ♀; **C**, *Cylindracheta psammophila*, Cylindrachetidae, ♀. F. Nanninga]

13. Cylindrachetidae. This is a family of medium-sized, cylindrical, larviform, wholly subterranean insects comprising 7 described species, of which 5 are Australian, one New Guinean, and one Patagonian. Eyes reduced, ocelliform. Mandibles and maxillary palps modified for stridulation. Pronotum not overlapping mesonotum, its lateral lobes curving around ventrally and approximated on the mid-ventral line (cf. Gryllotalpidae, p. 339). Fore femur and tibia greatly flattened and expanded, the latter with very large dactyls; the tarsus 1- or 2-segmented, slender, without dactyls (cf. Gryllotalpidae). Hind legs not saltatorial; femur scarcely larger than the mid femur; tibia short and stout, with stout terminal spurs and no dorsal spines or lamellae. Wings absent. Abdomen with terga 9 and 10 partly fused; supra-anal plate flexed into a position ventral to tergum 10. Cerci stout, unsegmented, ventral in position, often directed forward. Paraproct without a cerciform appendage. There are 2 genera, *Cylindracheta* (Fig. 21.14C) in Australia and New Guinea, and *Cylindroryctes* in Patagonia. The first is unique in Orthoptera in possessing a thin, tubular, strongly projecting mesothoracic collar, which envelops the hind part of the pronotum, while allowing it free movement in all directions. The insects are found tunnelling in sand in both humid and arid regions.

22

PHASMATODEA

(Stick-insects)

by K. H. L. KEY

Mandibulate, phytophagous, exopterygote Neoptera, having all legs gressorial, with small, well separated coxae; pronotum without large descending lateral lobes; wing rudiments of nymph not reversing their orientation in the later instars; specialized auditory organs lacking; eggs free, thick-shelled, provided with a conspicuous operculum.

The Phasmatodea are large to very large terrestrial insects, inhabiting chiefly the warmer parts of the world, especially the tropics. They have attracted wide attention because of their size—a few species exceeding a foot in length—and the remarkable resemblance of most of them to sticks or leaves. The wingless, parthenogenetic *Carausius morosus* Brunn. is a widely studied laboratory animal. The stick-insects were long treated as a family within the Orthoptera. Basic structure and habits are rather uniform. More than 2,500 species are known.

Anatomy of Adult

A general account of the anatomy may be found in the comprehensive work of Beier (1957). Males of most species are markedly smaller and more slender than the females and may differ from them in many striking structural features.

Head. Typically prognathous, rectangular to oval, sometimes horned or spined. Antennae short to very long and slender, with 8 to more than 100 segments. Eyes rather small, situated anterolaterally. Ocelli present only in some winged species, then normally 3, larger in the male or confined to that sex. Mouth-parts comprising strong, cutting mandibles; maxillae with 5-segmented palp, 2-segmented galea, and a toothed lacinia bearing bristles on its inner face; and labium with 3-segmented palp, well developed paraglossae, and smaller glossae. Hypopharynx present.

Thorax. Prothorax shorter than the other segments, transverse to little more than twice as long as broad, movable on the mesothorax; pronotum transversely arched, without strongly descending lateral lobes, usually with a transverse and a median sulcus; prosternum sclerotized, comprising large, well separated basisternum and sternellum of about equal size; propleura well developed, the episternum and epimeron subequal. Mesothorax reaching well forward from the level of attachment of the fore wings, longer than the other segments, ranging from only slightly so to very elongate, often spiny or granulose; mesonotum transversely arched. Metathorax rigidly connected to the mesothorax, often elongate in apterous species. Meso- and metathoracic sterna comprising a large basisternum, with a much smaller,

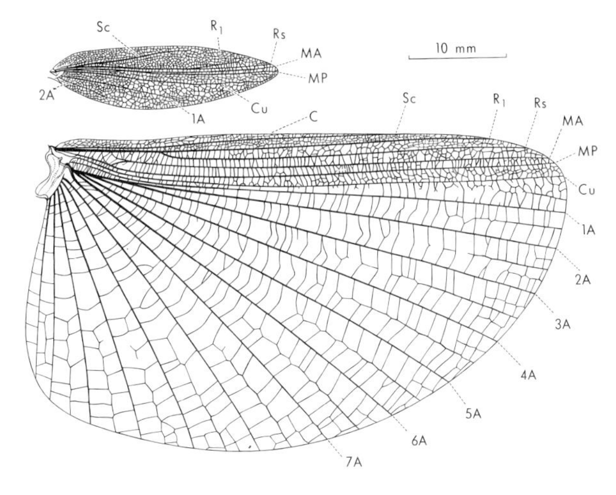

Fig. 22.1. *Podacanthus wilkinsoni*, wings of ♂. [F. Nanninga]

bilobed sternellum and a small spinasternum; pleura strongly oblique to longitudinal, with long and sometimes slender episternum and smaller, sometimes very small, epimeron, those of the metathorax extending posteriorly well beyond the extremity of the notum and often reaching that of the first abdominal tergum. Spiracles situated anterior to the meso- and metathoracic episterna.

Legs. All three pairs similar, gressorial, usually long and slender, often spined, occasionally bearing lobes or broad expansions. Coxae rather small, usually well separated. Fore femora usually curved and compressed proximally to accommodate the head when the legs are extended forwards in the mid-line. Tarsi of all Australian species 5-segmented (except in regenerated legs, where they are nearly always 4-segmented), with terminal claws and arolium.

Wings. Most Australian species are apterous in both sexes. When wings are present, they are usually fully developed and functional in the male (Fig. 22.1), but frequently reduced in the female. Fore wings tough opaque tegmina, nearly always short and covering only the base of the hind wings at rest, overlapping each other across the mid-line, usually with a knob-like dorsal eversion in the proximal half, which accommodates a prominence of the hind wing-base in the folded position; occasionally reduced to spines or absent, even when the hind wings are present. Hind wings broad, remigium tough and opaque like the fore wing, the large anal area membranous, in repose folded fanwise and overlain by the protective remigium. Venation (Ragge, 1955b) very uniform, with restricted branching; C absent in fore wing, in the hind wing marginal, unbranched, and usually weak; Sc unbranched; R usually unbranched in the fore wing, but in the hind usually breaking into R_1 and Rs, either of which may occasionally be itself bifurcate; M usually bifurcate in the fore wing, in the hind bifurcate in the basal part of the wing, or occasionally trifurcate; Cu unbranched; anals unbranched, usually one in the fore wing and many in the hind, of which the relatively weak first arises separately, with the second to seventh in one group and the rest in another.

Abdomen. Cylindrical or dorsoventrally compressed, sometimes bearing spines or lobular outgrowths, consisting of 11 segments. Tergum 1 (often termed the 'median segment') rigidly connected to the metanotum in all Australian species, the intervening suture sometimes lost; tergum 10 well developed, in the male often emarginate posteriorly and with the ventroposterior angles produced and sometimes caliper-like, their inner extremities usually provided with small teeth or spines, the lateral walls of the tergum sometimes curving round ventrally, or even meeting on the mid-ventral line anteriorly; tergum 11

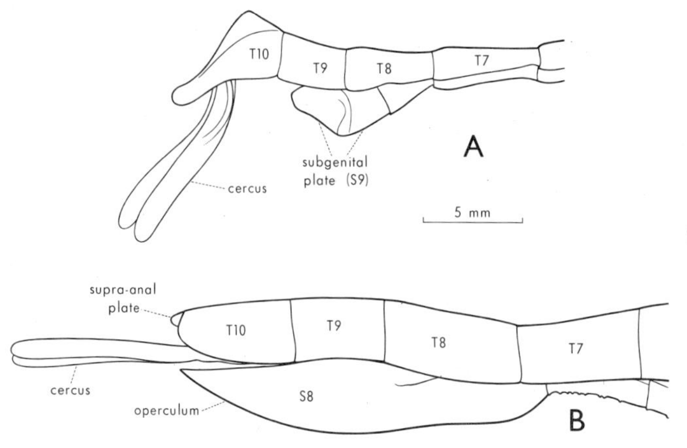

Fig. 22.2. *Podacanthus wilkinsoni*, apex of abdomen in lateral view: **A**, ♂; **B**, ♀. [F. Nanninga]

constituting the supra-anal plate, which is often concealed beneath tergum 10, especially in the male, but in the female may be elongate and either articulated on tergum 10 or completely fused to it. Sternum 1 reduced and usually not identifiable; sternum 10 usually well developed, fused posteriorly to the paraprocts, which represent the divided sternum 11. Cerci unsegmented, long or concealed beneath tergum 10, occasionally modified as claspers in the male. Spiracles situated in the pleural membrane near the ventroanterior angle of terga 2 to 8 and in a small sclerite below tergum 1.

In the male (Fig. 22.2A), sternum 9 constitutes the subgenital plate; it lacks styles, but bears a somewhat swollen, cup-shaped, distal lobe, sometimes called the 'poculum'. Sternum 10 is provided in some groups with a backwardly directed, sclerotized, usually prong-shaped copulatory process, the *vomer*. Genitalia (Snodgrass, 1937) concealed by the subgenital plate, comprising an asymmetrical group of largely membranous lobes.

In the female (Fig. 22.2B), sternum 8 is the subgenital plate; it consists of a scoop- or keel-shaped structure known as the 'operculum', whose free posterior end may project beyond the abdomen. The ovipositor, enveloped below by the operculum, consists of 3 pairs of slender valves.

Internal Anatomy. Alimentary canal straight, with a large crop, of which the posterior portion functions as a gizzard; mid gut long, in its anterior part provided with circular folds and in its narrower posterior part with numerous external glandular papillae, each bearing a terminal filament; caeca absent or rudimentary. Salivary glands large, bilobed. Numerous Malpighian tubes, arranged in two groups, open into the intestine by a common duct. The central nervous system includes 3 thoracic and 7 discrete abdominal ganglia. The male reproductive system comprises a pair of narrow, elongate testes, with a tubular seminal vesicle opening into each vas deferens and a number of tubular accessory glands into the short common ejaculatory duct. In the female, the ovaries are composed of free, well spaced, panoistic ovarioles strung along the lateral oviducts and prolonged into fine suspensory ligaments; there is usually a large bursa copulatrix with an independent opening into the genital atrium above that of the common oviduct; accessory glands and single or paired spermathecae may also be present. A pair of glands, which in some species at least have a repugnatorial function, is usually present in the prothorax, opening to the exterior in a notch at the anterolateral angle of the notum.

Karyotype. The karyotypes studied have 2n♂ ranging from 21 to about 100 (White, 1951). The higher numbers occur in parthenogenetic species and it has been suggested that some of these may be polyploid. The male

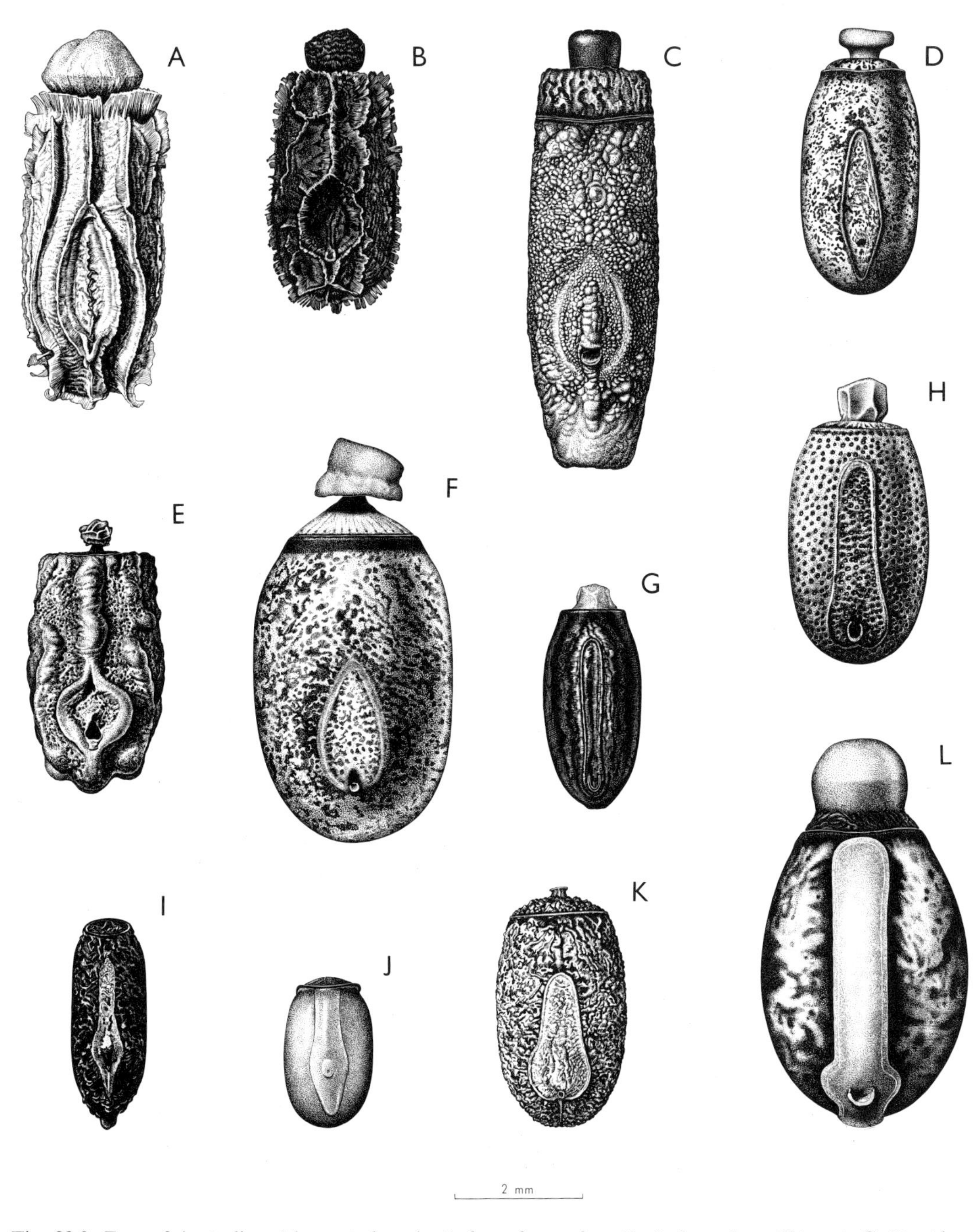

Fig. 22.3. Eggs of Australian: Phasmatodea: A, *Podacanthus typhon*; B, *Podacanthus wilkinsoni*; C, *Tropidoderus childrenii*; D, *Didymuria violescens*; E, *Ctenomorpha chronus*; F, *Eurycnema goliath*; G, *Ctenomorphodes tessulatus*; H, *Acrophylla titan*; I, *Hyrtacus tuberculatus*; J, *Sipyloidea filiformis*; K, *Pachymorpha squalida*; L, *Extatosoma tiaratum*. [F. Nanninga]

usually has an XO constitution, but a few XY species are known (Hughes-Schrader, 1959).

Immature Stages

Egg. The eggs (Fig. 22.3) are usually oval or barrel-shaped and show a striking resemblance to seeds. The shell is hard, and may be smooth and shiny, or heavily sculptured in various ways, and sometimes patterned. The anterior pole is truncated and fitted with a variably shaped operculum, while dorsally, associated with the micropyle, there is a longitudinal, scar-like area, the 'micropylar plate', which strongly resembles the hilum of many seeds. The eggs provide good taxonomic characters; they are known for some 25 Australian species.

The embryology is described by Beier (1957). In many species development has an appearance of capriciousness, owing to variability in diapause manifestations resulting from the interaction of genetic and environmental causes. Eggs laid by a single female may produce hatchings in both the first and second following springs, or even in the third, and each hatching period

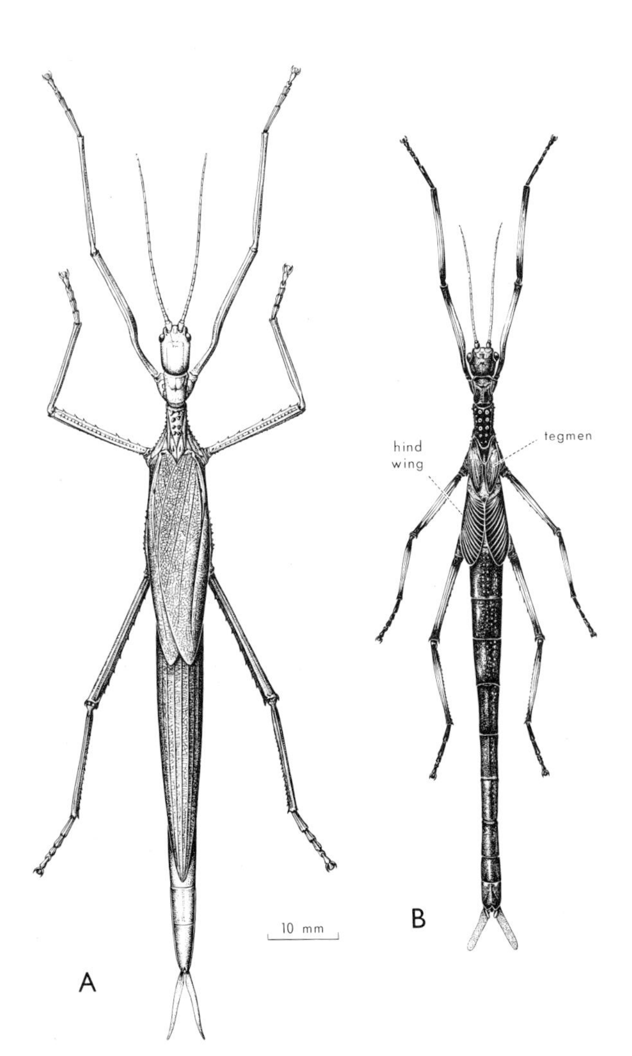

Fig. 22.4. *Podacanthus wilkinsoni*: A, adult ♀; B, last-instar ♂ nymph, high-density phase. [F. Nanninga]

may last for several weeks.

Nymph. Hatching occurs by detachment of the operculum. A 'pronymph', corresponding to that of Mantodea and Orthoptera, does not occur. The nymph (Fig. 22.4B) shows only minor anatomical differences from the adult, involving principally the fewer antennal segments and the rudimentary reproductive organs and wings (in those species possessing them). It undergoes several ecdyses (usually fewer in the male than in the female), hanging in an inverted position from an elevated support; the exuviae are commonly eaten. The wing rudiments and terminalia increase in size from instar to instar (the footnote on p. 296 applies to this chapter also), the former not reversing their original orientation as they do in Orthoptera. The number of antennal segments increases by progressive division.

Biology

General information on the biology of the Phasmatodea is given by Chopard (1938).

Adults. Most Phasmatodea are apparently rare insects, even allowing for the special features that make detection difficult. On the other hand, a few species may become extremely abundant, giving rise to outbreaks or plagues, in which, however, true gregariousness does not seem to be involved. The insects most characteristically frequent the foliage, or sometimes the trunks or stems, of trees and shrubs, but many species are found on plants of the herbaceous stratum, or even on litter at ground level. They spend much of their time immobile, often in extraordinary attitudes that must materially assist their procryptic adaptations of structure and colour in deceiving predators. From time to time, especially upon some slight disturbance, they may perform rhythmical swaying or rocking movements. In locomotion they are generally sluggish and clumsy. They will ascend any vertical support within their field of vision, constantly feeling for new footholds by wide, sweeping movements of the fore legs. In Australia, almost all the species frequenting trees and large shrubs are fully winged, at least in the male. Those found on low herbs are apterous, while those on the lower shrubs and on tall grasses include both types, as well as intermediate ones. Males of fully winged species sometimes fly to lights; most females are flightless. Many stick-insects are largely nocturnal, but this is not true of the commoner Australian species.

When suddenly approached, stick-insects often drop to the ground and lie motionless in a state of catalepsy, which may last for hours if contact with the tarsi is prevented. If the limbs are seized, reflex autotomy may occur between femur and trochanter, at which point the lumen of the leg is almost completely obstructed by a haemostatic diaphragm. When grasped, some species flex their spiny legs spasmodically over the fingers, or raise their wings with a rustling motion. Rarely will any attempt be made to escape by taking to the wing. Many species regurgitate the contents of the crop when handled and some are capable of squirting an irritant fluid from the repugnatorial glands of the prothorax. In two such instances (*Anisomorpha buprestoides* (Stoll) and *Megacrania wegneri* Will.) the insects have an aposematic colour pattern, while in *Anisomorpha* they are at times gregarious. In *Phyllium* spp. and *Extatosoma tiaratum* (Macl.), the first instar only is aposematic (p. 357). A few species have been reported as stridulating, by friction between fore and hind wing, or between the enlarged 3rd segments of the antennae. The hind wings are often coloured or mottled on the anal area, while the remigium commonly has a brightly coloured patch, distinctive for the species, in the region covered by the tegmen when the wings are closed.

All Phasmatodea feed upon foliage. They are rarely confined to a single host species, but the range of acceptable hosts is commonly limited. In Australia several species feed indifferently upon such divergent genera as *Eucalyptus* and *Acacia*, while most eucalypt feeders accept any species of that genus, at least in the later instars.

Reproduction. In copulation, the male is mounted on the female, with the posterior part of his abdomen curving downward and forward through 180°, so that its ventral face is apposed to the ventral face of the female. The ventroposterior angles of tergum 10 of the male clasp either the base of the female's operculum, or her sternum 7 towards its posterior end, and the intromittent organ is inserted from the side, over the edge of the depressed operculum. Copulation is often prolonged, the sperm being transferred in small spermatophores. The eggs are usually laid singly at the rate of one to several per day. Upon leaving the oviduct they may be retained for some time within the ovipositor or the operculum; finally they are either passively released, or projected with some force by a sharp flick of the abdomen, and fall to the ground amongst the litter. Some species deviate from the general pattern by inserting the eggs into the soil, or cementing them, singly or in groups, to a support. Normal egg production per female seems to range from about 100 to 1,300, according to species.

Parthenogenesis is widespread in the order. In some exotic species, including *Carausius morosus*, it is obligatory, with males unknown or exceedingly rare. More usually it is facultative, unfertilized eggs producing only females; but the parthenogenetic production of males as well as females has been reported (Hadlington and Shipp, 1961) for the Australian *Ctenomorphodes tessulatus* (Gray).

Nymphs. The nymph on hatching ascends the first upright support it encounters and continues upwards until it finds suitable foliage. At this stage it is dependent upon the softer parts and may have a more restricted host range than the adult. It is vulnerable to desiccation and predators, especially between hatching and establishment on suitable food; in some species hatching occurs in the evening, permitting establishment during the hours of darkness and high humidity. The nymphal stage lasts some weeks or months. Regeneration of damaged or autotomized appendages may occur, so that individuals with short or otherwise abnormal legs are often seen. The behaviour of nymphs is in general closely similar to that of adults; in a few species they are gregarious.

Variation, Colour Adaptation. Marked genetic variation in anatomical features occurs in many species. Horns, thoracic and abdominal spines and granulations, and lobes on abdomen or legs, may be variably developed or absent in the same population. Some species have green and non-green morphs. Many of these characters may also vary geographically, together with overall size and degree of attenuation, relative wing length, and colour of hind wing.

'Morphological' colour adaptation to the background has been recorded in a few species. More remarkable are the mechanisms of 'physiological' colour change (p. 138) found in *Carausius morosus* and some other species. Darkness and pallor are induced by the migration of dark granules within the epidermal cells. When the granules are clumped together near the basal end of the cells, the insect is pale; when they move distally and spread out to form a continuous curtain, the insect becomes dark. In *C. morosus* there is a diurnal rhythm, the insect being dark at night and pale in the day-time, basically in response to light intensity. Apart from this rhythm, and overriding it, darkening may be induced by about an hour's exposure to low temperature or high humidity, and pallor by the reverse conditions. In the temperature response, the epidermis acts as an independent effector, so that local pallor follows local application of heat to the integument; light and humidity operate through a nervous-endocrine mechanism.

Kentromorphism. Kentromorphic phase differences (p. 139) have been reported (Key, 1957) in the Australian phasmatids *Podacanthus wilkinsoni* Macl., *Didymuria violescens* (Leach), and *Ctenomorphodes tessulatus* (Gray), all of which sometimes reach high population densities. In the nymph, the procryptic low-density phase is rather uniform and usually green, whereas the conspicuous, supposedly aposematic, high-density phase is patterned with black, yellow, and sometimes white (Fig. 22.4B). In *P. wilkinsoni* a mean density of one insect per eucalypt branchlet is sufficient to induce the extreme high-density pattern, while the low-density extreme occurs at less than one per 20 branchlets. Intermediate patterns appear at intermediate densities, or in response to density change. All three species also show morphometric phase differences analogous to those of locusts. There is apparently no correlation between density and activity, and no overt gregariousness.

Ecological Features. The rarity of many Phasmatodea may be expected to impose difficulties in the way of sexual contact. These may be offset by the sedentary habits of the female and the behaviour of the young nymph, which tend to ensure that the progeny of a given female will find themselves on the same plant, or on adjacent plants in contact. They may be relieved, also, by the substantial powers of flight of the males of most tree-frequenting species, especially if the female produces a sex-attractant substance, as some observations suggest. However, the occurrence of facultative parthenogenesis is an insurance against failure of sexual contact. This view of the situation implies a patchy distribution of the rare species.

It is not known what determines the generally low density of stick-insect populations. Food is usually available in abundance, although the quantity or accessibility of young foliage may be inadequate for the needs of the first-instar nymph. On the other hand, in the species liable to outbreaks all edible foliage may be consumed over the greater part of a high-density area. Since the dispersal powers of the insects are low, only those around the fringes survive. Important mortality from predators and parasites is suffered in both the egg and active stages. Fire must have drastic effects, its severity and time of incidence determining whether it will disadvantage principally the phasmatids or their natural enemies (Campbell, 1961).

Although human modification of the habitat may have increased the frequency and severity of outbreaks, yet the capacity for kentromorphic change that has been evolved in all three of the Australian plague species indicates that they have always been subject to recurrent high density. In *Podacanthus wilkinsoni*, *Didymuria violescens*, and the American *Diapheromera femorata* (Say), outbreak densities are usually reached only every alternate year over an outbreak period, the incidence of high density having somehow come into relation with the predominantly two-year life cycle, perhaps through the proportionately greater pressure of predation on the low-density broods (Readshaw, 1965).

Natural Enemies. The procryptic adaptations and other defensive mechanisms of all stages of Phasmatodea are probably directed mainly against bird predators. In Australia, Readshaw (1965) states that large nymphs and adults of *Didymuria violescens* form at times the principal food of such birds as *Strepera graculina* (pied currawong), *Coracina novaehollandiae* (black-faced cuckoo-shrike), and *Anthochaera carunculata* (red wattle bird). Young nymphs are subject to attack by small birds, and also by ants (*Myrmecia* and *Iridomyrmex* spp.) and probably spiders. Eggs are eaten by a number of general predators on the forest floor, including ants, birds, and marsupial mice (*Antechinus* spp.).

Parasitism is also frequent. The active stages are attacked by tachinid flies, while erythraeid mites of the genus *Charletonia* are sometimes abundant as ectoparasites. The eggs are extensively parasitized by minute cleptid wasps. In Australia several species are involved, mainly in the genus *Myrmecomimesis.* The wingless female wasp chews a hole in the thick shell of the egg and deposits her own egg through this.

Economic Significance. Only a few species are rated of economic importance. In Australia, *Podacanthus wilkinsoni*, *Didymuria violescens*, and, to a less extent, *Ctenomorphodes tessulatus* are responsible for extensive defoliation of eucalypt forests. Some of the most valuable timber species are especially sensitive, a single defoliation causing a high percentage of deaths in both *Eucalyptus regnans* and *E. delegatensis.* Even moderate defoliation causes reduction in growth increment. Repeated defoliations have been associated with accelerated soil erosion on steeply sloping catchments. Papers on the biology of the three pest species are listed by Readshaw (1965).

Special Features of the Australian Fauna

The principal feature of the Australian phasmatodean fauna is the very poor representation of the family Phylliidae. The Phasmatidae have most of their subfamilies represented. More than half of about 50 genera, and nearly all the species, are endemic, but none of the subfamilies.

CLASSIFICATION

Order PHASMATODEA (132 Australian spp.)

1. Phylliidae (2) 2. Phasmatidae (130)

The classification of Günther (1953) is followed here, except for the subfamily assignment of a few genera. He recognizes two families: Phylliidae and Phasmatidae. Both are widely distributed in the warmer parts of the world, and both exhibit an extraordinary diversification in body form and integumental processes, many of the trends showing marked parallelism. In the Phylliidae the ventrodistal extremity of the tibiae, at least of the mid and hind legs, has a triangular area of the integument somewhat impressed relative to the surrounding surface, or at least demarcated from it by a shallow marginal furrow; the Phasmatidae lack this feature.

1. Phylliidae. The 'leaf insects' of south-east Asia and New Guinea, in which the body is strongly flattened dorsoventrally and the abdomen and legs bear broad lamellate expansions, are represented by a species of *Phyllium* in the rain forests of northern Queensland. There is also an early record of the spiny genus *Haaniella*.

2. Phasmatidae. Günther (1953) gives keys to the subfamilies, of which 8 out of 11 are represented in Australia. The taxonomy and nomenclature of the injurious Australian species and some of their allies have been discussed by Key (1957, 1960).

The subfamily PODACANTHINAE is the most characteristically Australian group, with a

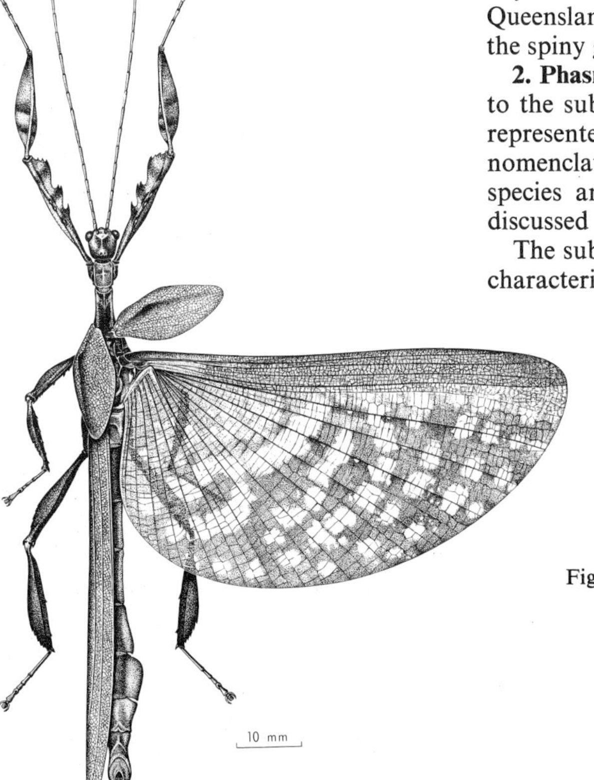

Fig. 22.5. *Extatosoma tiaratum*, ♂. [F. Nanninga]

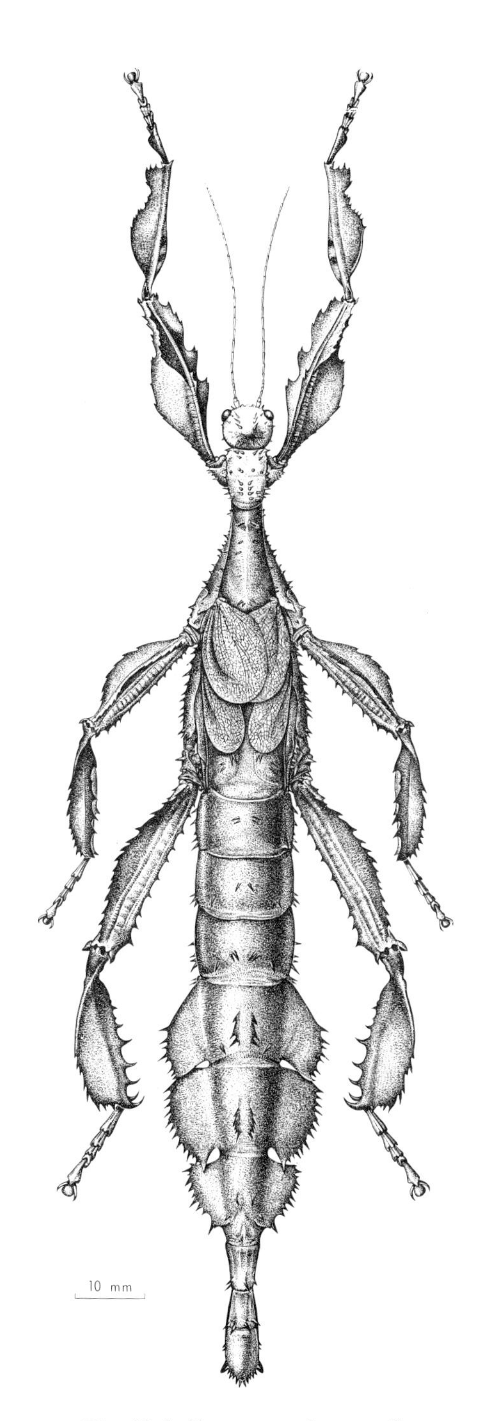

Fig. 22.6. *Extatosoma tiaratum*, ♀. [F. Nanninga]

continent-wide distribution. It comprises mainly large, winged species, which resemble leaves rather than twigs, although they are not markedly flattened. *Podacanthus* includes 4 species with rather regularly spined mesonotum, long cerci, and fully winged females, of which 2 are injurious. *P. wilkinsoni* (Fig. 22.4A) is distinguished by the orange-yellow proximal patch on the remigium of the hind wing. *Extatosoma tiaratum* (Figs. 22.5, 6; see Gurney, 1947) is remarkable for the numerous spines and integumental expansions on the body and legs, including a tuft of spines on the conical occiput of the hypognathous head. The male has large mottled wings, while the female is heavy-bodied and brachypterous. The insect often hangs inverted amongst foliage, with its highly procryptic abdomen curled over its back. The first instar has an aposematic pattern of black, with orange head and whitish collar; in its colouring, posture, and movements it appears to mimic ants of the genus *Leptomyrmex*. *Didymuria violescens* (Fig. 22.7) is fully winged in the male, which has inflated hind femora bearing two or three large black spines ventrally; the female is shorter-winged and flightless, with the hind femora unspecialized.

The PHASMATINAE are another prominent group with an Australia-wide distribution. The species are mainly large, winged, and stick-like. *Ctenomorphodes tessulatus* (Fig. 22.8) is a medium-sized, slender phasmatid, with spiny mesonotum and strongly tessellated hind-wing pattern. The males are fully winged, the females short-winged and flightless. The longest Australian phasmatid, which reaches a body length of 25 cm, is the spiny *Acrophylla titan* (Macl.) of the eastern coast, with large mottled wings in both sexes and wavy-margined three-ridged cerci. Somewhat shorter, but more heavily built, is the mainly tropical *Eurycnema goliath* (Gray), in which the female is conspicuously banded with green and yellow, with the fore wings and the remigium of the bluish-veined hind wings a bright red beneath.

The EURYCANTHINAE are mainly Papuan in distribution, but a few species have been

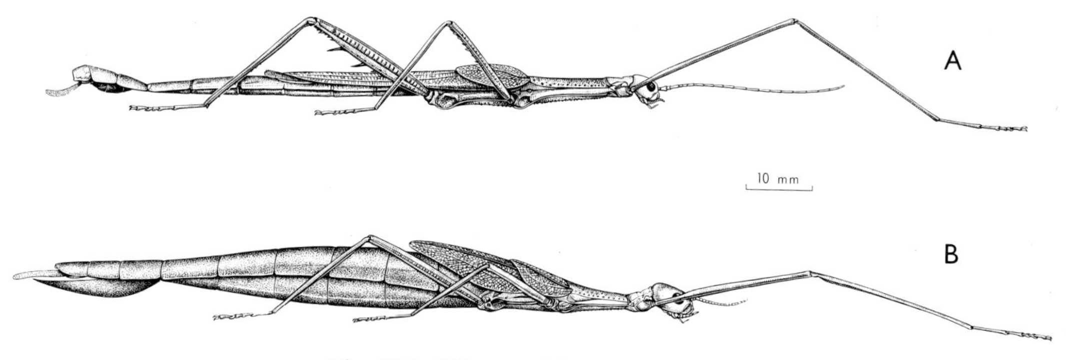

Fig. 22.7. ***Didymura violescens***: A, ♂; B, ♀.
[F. Nanninga]

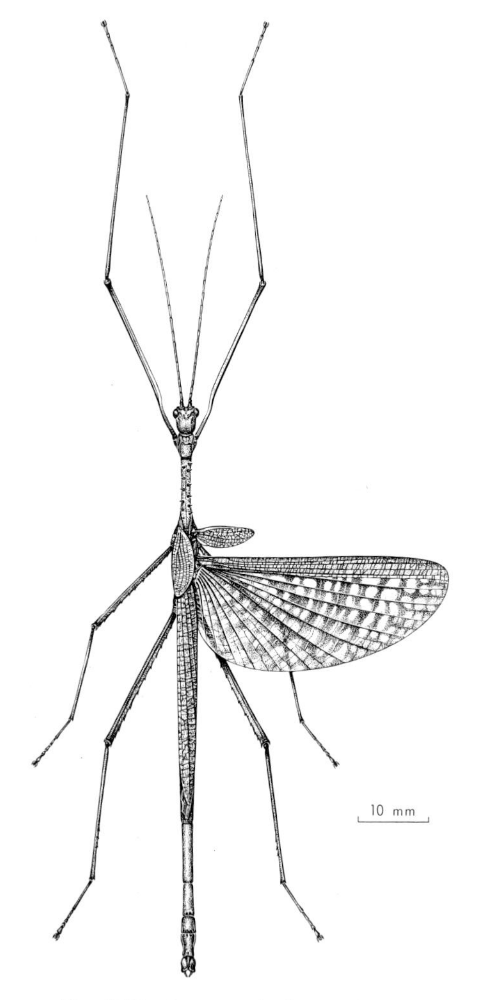

Fig. 22.8. *Ctenomorphodes tessulatus*, ♂.
[F. Nanninga]

recorded from the Cape York area. However, the most interesting Australian representative is the almost extinct *Dryococelus australis* (Montr.). This is a shiny, heavy-bodied, reddish brown, wingless species, the male having enormously thickened hind femora with a few heavy spines beneath (Gurney, 1947). It was formerly very abundant on Lord Howe I., where it frequented large cavities in the trunks of living trees, coming out at night to feed. It has been exterminated on the main island by rats, which were accidentally introduced in the early part of the century, but still occurs on an associated small island known as Ball's Pyramid.

The subfamily XERODERINAE, also mainly Papuan and Melanesian in distribution, is represented by two little-known species from northern Queensland. The African PALOPHINAE also have two or three Australian outliers.

The NECROSCIINAE are a large, principally south-east-Asian group of medium-sized phasmatids, with long slender antennae, very small fore wings (in the winged forms), and a prominent vomer in the male. In Australia the winged genus *Sipyloidea* (Fig. 22.9A) has a number of species frequenting tall tropical grasses. *Parasipyloidea*, with many apterous species, occurs on plants of the herbaceous stratum.

The PACHYMORPHINAE are rather small, dun-coloured, apterous phasmatids with short antennae, of which *Pachymorpha*, with several species (Fig. 22.9B), is widespread in Australia. The insects are usually found on the lower part of tree trunks, or on ground litter around their base.

A subfamily of particular interest in Australia is the apterous LONCHODINAE, which are widely distributed over the continent, as well as in New Guinea and south-east Asia. They frequent plants of the herbaceous stratum, or low shrubs. Several exceedingly slender ones occur on tall tropical grasses, or on the hummocks of arid spiny grasses of the genus *Triodia*. In *Hyrtacus* (Fig. 22.9C) the female supra-anal plate is more or less elongate, but clearly separated from tergum 10. In *Marcenia* it is elongate and fused to that tergum.

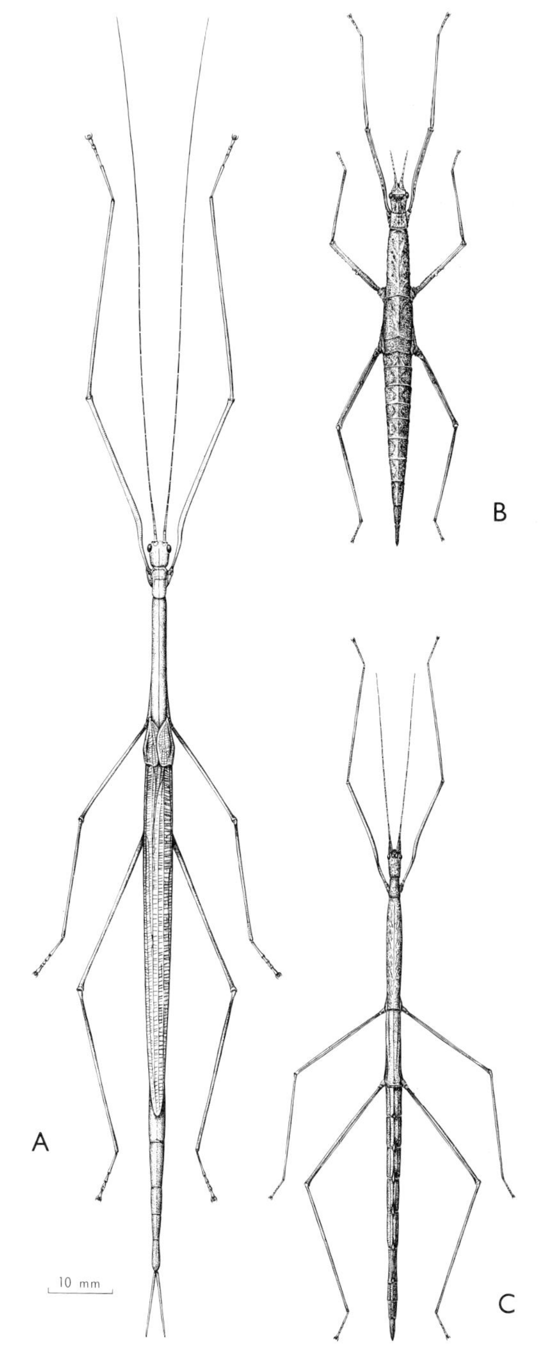

Fig. 22.9. A, *Sipyloidea filiformis*, ♀; B, *Pachymorpha squalida*, ♀; C, *Hyrtacus tuberculatus*, ♀. [F. Nanninga]

23

EMBIOPTERA

(Embiids, web-spinners, foot-spinners)

by EDWARD S. ROSS

Mandibulate exopterygote Neoptera, with apterous females and winged or apterous males; living in silken galleries; fore basitarsi globose.

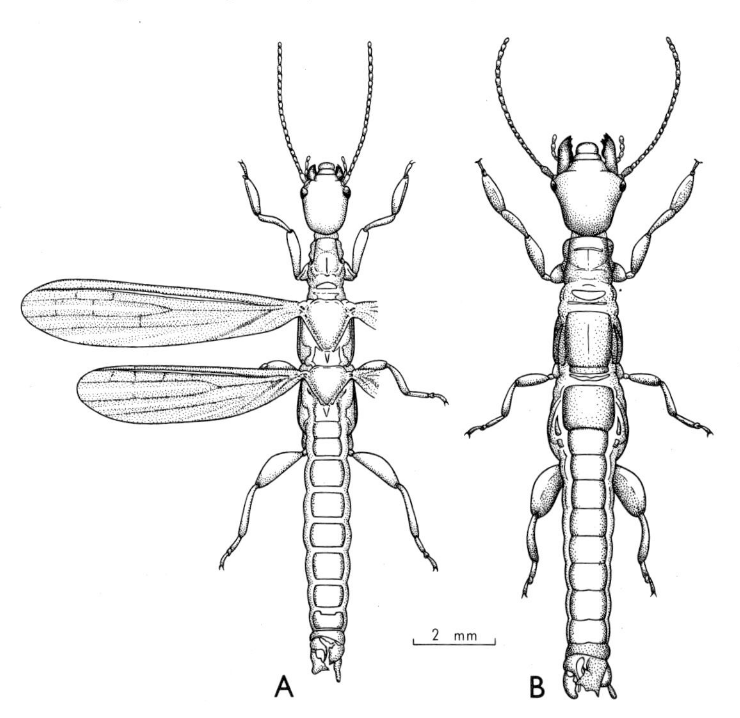

Fig. 23.1. Adult embiids, dorsal: A, *Notoligotoma nitens*, Notoligotomidae, ♂; B, *Metoligotoma septentrionalis*, Australembiidae, ♂.
[B. Rankin]

The Embioptera comprise one of the smaller, lesser-known orders of insects. Like their relatives, termites and earwigs, embiids are essentially tropical, but some species occur in warm-temperate climates. The order is well represented throughout Australia and Tasmania, two species even occurring above 2,000 m in the Australian Alps. Less than 200 species are recorded for the world, but recent field work indicates that as many as

2,000 may exist. Because of almost complete confinement of activity to their silk galleries (Fig. 23.5), embiids are seldom seen by the non-specialist, except for alate males attracted to light. They are small to medium-sized, narrow-bodied insects (Fig. 23.1), easily recognized by the greatly swollen fore tarsi which are packed with silk glands. They may be aberrant derivatives of the Protorthoptera or Protoperlaria (i.e. the Paraplecoptera of chapter 8), owing their persistence to the survival potential afforded by life in silk galleries.

The Australian fauna has been described by Tillyard (1923a), Davis (1936–44, 1944a),

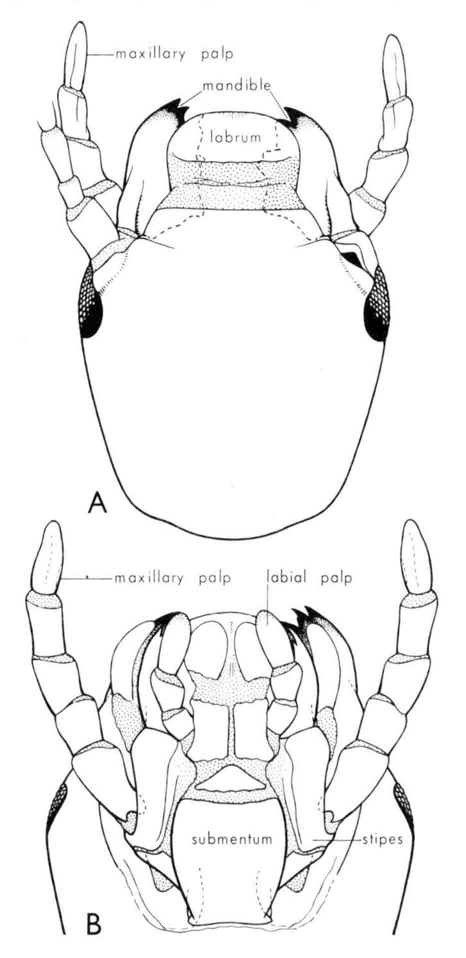

Fig. 23.2. Head of *Australembia incompta*, Australembiidae: A, dorsal; B, ventral.

and E. Ross (1963), while Barth (1954) has given an excellent account of silk production.

Anatomy of Adult

Head (Fig. 23.2). Strongly prognathous, always with a sclerotized gular bridge between submentum and occipital foramen. Antennae filiform 12- to 32-segmented. Eyes reniform, compound; often large in male, always small in female; ocelli absent. Mandibles of adult males usually flattened, elongate, often with only a few inner-apical dentations. Submentum of males often large, sclerotic, shield-like.

Thorax. Female without even rudimentary wings; ventral thoracic sclerites separated by membrane. Apterous males with thorax similar to that of females, or with short wing buds. Pterothorax of winged males flattened, rigid; nota triangular; axillary cords very long; scutellum small; ventral sclerites fused.

Legs. Very short; tarsi 3-segmented; basal segment of fore tarsus distended by silk glands (Fig. 23.4); mid legs relatively small; hind legs with enlarged femora due to large size of tibial depressor muscles.

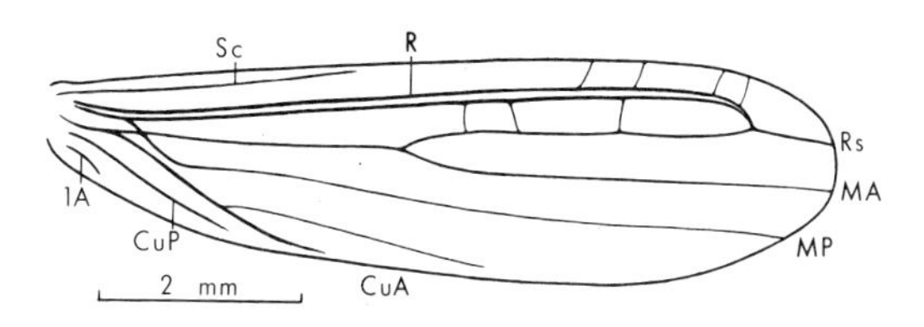

Fig. 23.3. Venation of fore wing of the oligotomoid type characteristic of all alate Australian Embioptera; *Aposthonia* sp. of *gurneyi* complex, Oligotomidae.

Wings. Elongate, subequal; with characteristic, pigmented, venal stripes alternating with hyaline stripes. R_1 a broad, inflatable blood sinus with granular, pink borders; other veins normal or represented only by a line of setae and a pigment stripe; basal half of Rs and MA closely parallel, appearing as one vein; few cross-veins (Fig. 23.3).

Abdomen. Elongate, with 10 well-defined, subequal segments; cylindrical in female, dorsoventrally flattened in male; cerci 2-segmented, tactile. External female genitalia represented only by slightly modified sterna 8 and 9 and rudimentary valvifer lobes. Male genitalia (Fig. 23.6) complex, asymmetrical; tergum 10 cleft, bearing complex processes and flaps; left cercus modified by lobes and/or by segment-fusion to function as a clasper in copulation.

Internal Anatomy. Alimentary canal simple, with large salivary glands and 6 rectal papillae; 20–30 Malpighian tubes in adult. Nervous system with 3 thoracic and 7 discrete abdominal ganglia. Reproductive system with indications of a segmental arrangement in the five pairs of serially arranged panoistic ovarioles in the female and five pairs of testes in the male. The tarsal silk glands are described below.

Immature Stages

The eggs are elongate and with a rimmed, circular operculum. Nymphs are similar to adult females, but those of males destined to have wings develop external wing pads.

Biology

The outstanding feature of embiids is an ability to spin galleries of silk with their fore tarsi. The silk glands contained in the enlarged basal tarsal segment are irregularly shaped globules tightly packed like seeds of a pomegranate fruit, as many as 200 occurring in a single tarsus (Fig. 23.4). A gland is a single layer of syncytial cells enclosing a large lumen which stores the liquid silk secretion of the gland wall. Single ducts conduct the secretion to hollow, seta-like silk ejectors on the ventral surface of the tarsus, where it issues involuntarily as the ejectors brush against a surface. The spinning action of the fore legs resembles the movements of a shadow boxer, and embiids, unless the temperature is too low, constantly extend their galleries to new food sources, or add silk layers to older galleries. Freshly-spun galleries are often blue-white or violaceous, but old, multi-layered silk is chalk-white.

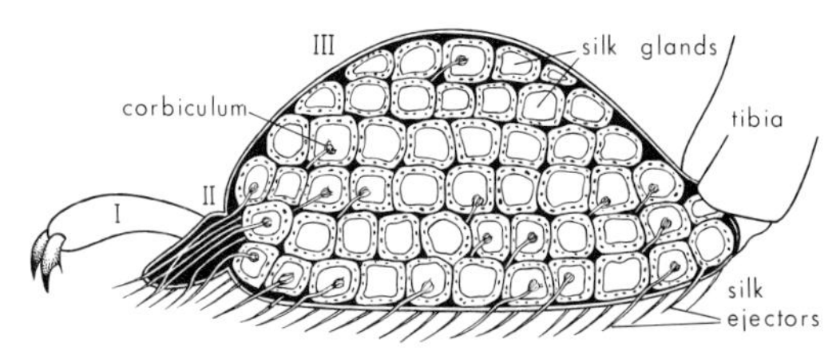

Fig. 23.4. Longitudinal section of fore tarsus, showing silk glands and ejectors, schematic. [B. Rankin]

The distinctive, enlarged fore tarsi serve as a universal recognition character of the order in all stages of the life history. It is remarkable that even first-instar nymphs spin silk, and that the ability is life-long. In most insects silk production seldom extends into the adult stage.

As an embiid spins it rotates its body, and forms a silken tube narrow enough to permit constant contact of its body hairs with the gallery wall. In humid regions, many species produce exposed galleries on bark surfaces, but others hide themselves in crevices and bark flakes with little exposure of silk. In some species faeces and masticated vegetable fragments are added to silk surfaces to provide additional cover, but in others the surface may be bare. In dry regions, heat, desiccation, and bush fires are evaded in galleries penetrating cracks in the soil (Fig. 23.5A), crevices in rock or termite mounds, or ramifying beneath stones. Some species live in old fence posts (Fig. 23.5B) or in leaf litter. In tropical cloud forests up to 4,500 m, hanging moss is often an embiid habitat.

All distinctive features of embiids—the elongate, supple body, short legs, prognathism, aptery, efficient reverse locomotion, peculiar wings, and sensitive cerci—are adaptations to life in silk galleries. The galleries are as much an embiid element as water is to fish. The universal similarity of the gallery interior explains why embiids display no great range of general form and habits, in spite of great evolutionary age evidenced by diversity of male genitalic characters.

The galleries are coverways for all embiid activity except dispersal. They provide predetermined routes of rapid escape from predators, such as ants, which most often are

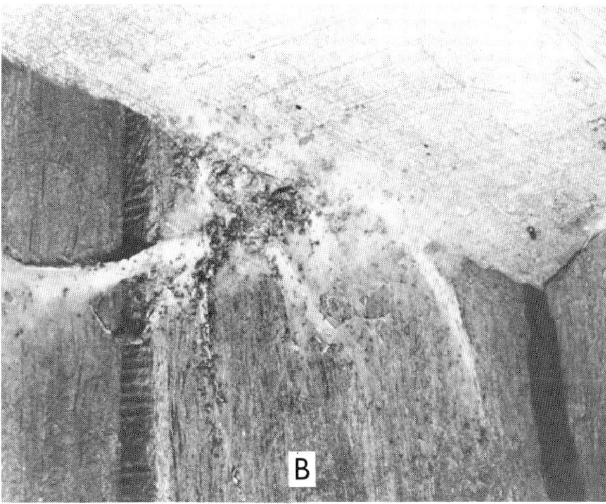

Fig. 23.5. Galleries of: A, *Aposthonia* sp. of *approximans* group, Oligotomidae, extending up from soil retreat to lichen food source on exposed rock surface on desert hillside, W.A.; B, *Notoligotoma hardyi*, Notoligotomidae, female nest on old fence near Perth, W.A. [Photos by E. S. Ross]

encountered at the periphery of the galleries. At such times an embiid darts backward with great rapidity, while the predator makes a usually unsuccessful pursuit outside the gallery. The exceptionally large depressor muscles of the hind tibiae motivate this reverse movement, while highly sensitive cerci function as tactile caudal 'eyes'.

The food of embiids is entirely vegetable, consisting of outer bark, dead leaves, living moss, and lichens. A ready source of food is likely to be present wherever embiids choose to live on the basis of other ecological factors. They are easily cultured in tubes or jars containing habitat material.

Originally embiids were alate in both sexes, but the stiff wings of ancestral forms must have slowed reverse movement by producing friction or snagging against the gallery wall. Through selection these became flexible and able to fold back at any point to reverse the axis of friction. Wings of modern embiids, when laterally extended for flight, are temporarily stiffened by blood pressure in the elongate, sac-like R_1. To produce eggs and guard the young necessitates a prolongation of life in females, and consequent longer exposure to the selective pressures of gallery life. Thus, possession of wings by females was disadvantageous in spite of their flexibility. It may be that all modern adult female embiids are wingless as a result of selection for arrested development at a stage comparable to third-instar nymphs, and neoteny would also have led to elimination of protruding genitalia. Size increase continued, however, and females are generally larger and more robust than males.

Aptery, either by neoteny or by brachypterism, occurs in males of certain species, especially those inhabiting arid or marginal environments. In some genera alate males are unknown, but usually only certain species of a genus are apterous, while in others both apterous and alate males occur in the same population or brood.

Embiids are often regarded as subsocial insects. Actually, they are simply gregarious, without castes or division of labour. In some species cannibalism occurs unless each nymph leaves the parent colony and develops in an independent niche. Males of such species are usually consumed by the female after mating. However, a typical colony consists of one or more adult females living in the midst of their broods which create and share a common labyrinth of galleries. Adult males are short-lived, and usually take no food. Their distinctive mandibles are adapted for grasping the head of the female during copulation.

Reproduction. After mating, the female starts a new colony, or an extension of an old one, by utilizing a protected portion of the colony as a retreat and depositing a layer of eggs on a silk-enclosed surface. Often the eggs are coated by, or embedded in, a hardened paste of masticated bark or leaf fragments. In some species they are naked, and laid singly or in clusters moved about in the gallery in the manner of ant eggs. The parent female guards her eggs and newly hatched young, but these soon spin small galleries to fit their body size.

Natural Enemies. As long as embiids remain in their galleries they are relatively safe from free-roving predators, but the predation hazard greatly increases during even short movements in the open to seek new niches. Within the galleries the embiids have a variety of natural enemies. Scelionid wasps parasitize the eggs; ectophagous larvae of bethyloid wasps (Sclerogibbidae) parasitize developing nymphs; tachinid fly larvae and eugregarine Sporozoa are internal parasites; and, in certain regions, tiny cimicoid Hemiptera (*Embiophila*, Microphysidae) infest the galleries and suck blood of the embiids. There is, as well, the usual hazard of mites and of viral and bacterial disease.

Economic Significance. Because of their preference for dead vegetable food and uncultivated habitats, embiids are of practically no economic importance. Only occasional minor economic records, such as the webbing of grape bunches or colonization of a sugar refinery, have been reported. A few species, particularly Indian species of *Oligotoma*, have become widespread through ancient and modern commerce.

Special Features of the Australian Fauna

Embioptera are best represented on the great continental land masses of the tropics. Perhaps because dispersal is slow and limited by aptery of females, the order is sporadically distributed and very poorly represented on islands. Even such insect-rich islands as New Guinea and the Philippines have only related species-complexes of a single genus (*Aposthonia*). Australia, however, though peripheral to Embioptera centres, has an interesting embiid fauna developed from three basic stocks, which were undoubtedly gained by way of land connections with south-eastern Asia, while perhaps a species or two moved southward from New Guinea by way of a Torres Strait connection. Two species have recently been introduced in commerce. Apparently, the order has not reached New Zealand, and it is absent from Chile and Patagonia. As may be expected in a tropically centred group, there is no evidence of faunal exchange through southern connections.

CLASSIFICATION

Order EMBIOPTERA (65 Australian spp.)

Clothodidae (0)
Embiidae (0)
1. Notoligotomidae (6)

Embonychidae (0)
Anisembiidae (0)
2. Australembiidae (32)

Teratembiidae (0)
3. Oligotomidae (27)

There are also several new family categories to be established in other faunas of the world.

Due to character reduction in females resulting from neoteny, the classification of embiids must be based on adult males. The most important characters are head structure, hind basitarsal sole-papillae, and, above all, the complex abdominal terminalia. Females and nymphs without associated adult males can be identified only when the fauna of the particular area is well known. Many of the species-complexes of the Australian fauna tend to form geographic races. Specimens are preserved best in vials of 70 per cent alcohol. Carefully cleared slide preparations often are required for close study.

Key to Families of Embioptera Known in Australia—Adult Males

1. Left cercus with terminal segment normal, clearly delimited by a membranous joint, basal segment often inwardly lobed but never echinulate; hind basitarsus with only 1 (terminal) ventral papilla .... **Oligotomidae**
   Left cercus with terminal segment partially or entirely combined with basal segment, the composite segment with one or more echinulate inner areas or lobes; hind basitarsus usually with 2 ventral papillae (sole-bladders) 2
2. Usually winged; if apterous, basal segment of right cercus is elongate and similar to the terminal segment; sternum 9 of abdomen (hypandrium) is continuously produced caudally as a subgenital lobe (HP) .... **Notoligotomidae**
   Always apterous; basal segment of right cercus very short, hemispherical; sternum 9 with caudal margin transverse, and with a nearly completely detached, large, triangular sclerite (HP?) on its right caudal half .... **Australembiidae**

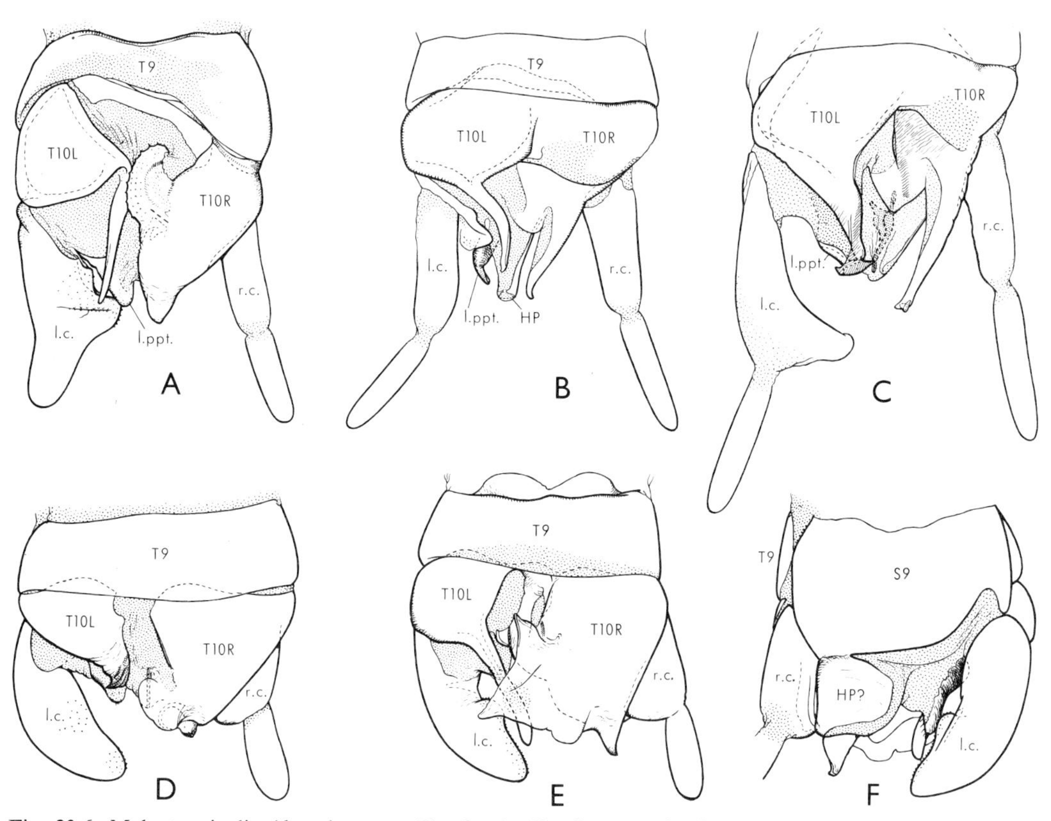

Fig. 23.6. Male terminalia (dorsal, except F) of: A, *Notoligotoma hardyi*, Notoligotomidae; B, *Oligotoma nigra*, Oligotomidae; C, *Aposthonia glauerti*, Oligotomidae; D, *Australembia incompta*, Australembiidae; E, *Metoligotoma illawarrae*, Australembiidae; F, *M. illawarrae*, ventral.
T10L, *T10R*, left and right hemitergites of T10; *l.c.*, *r.c.*, left and right cerci; *l.ppt.*, left paraproct (the right paraproct is atrophied); *HP*, caudal process of S9 (hypandrium).

1. Notoligotomidae. This family, as recently redefined (Ross, 1963), may be represented in both the Old and New World tropics. The one Australian genus, *Notoligotoma* (Fig. 23.1A), is most closely related to *Ptilocerembia* of south-eastern Asia. *N. hardyi* Fried. and a complex of related species or races occur in south-western Australia, usually on the under-surfaces of exfoliated rock slabs on granitic outcrops of arid regions. In Perth, *hardyi* is common in crevices of old board fences in residential areas (Fig. 23.5B). The large, pale, winged males fly to light during the first rains of the cold season in May. Species or races related to *hardyi* are also common in dry bark or on rock surfaces in the drier zones of northern Queensland. Lichens are the preferred food of *Notoligotoma*, and the species never use soil as a habitat. *N. nitens* Davis occurs as a complex of races from Victoria northward into Queensland, inhabiting lichens of ledges, bark flakes, and old fence posts. Unlike pale *hardyi*, *nitens* adults are glossy jet-black; the males are either apterous or alate.

2. Australembiidae. This recently defined family, known only from eastern Australia and Tasmania, is the only peculiarly Australian element of the order. Both males and females are entirely apterous. The most important family character is the semi-detached, triangular, ventral lobe of the right caudal margin of sternum 9 in the male. The male cranial and mouth-part anatomy (Fig. 23.2) is also unusual. The left cercus of the male is always 1-segmented, and the rudiment of the terminal segment is not projected as a lobe as in *Notoligotoma*. Species of *Australembia* have only one hind basitarsal papilla, and the males have massive maxillary palpi and a peculiar, elongate submentum. The 2 species, *rileyi* Davis and *incompta* Ross, are found in leaf litter in savannah zones of north Queensland. *Metoligotoma* (Figs. 23.1B, 7) is a large complex of closely related species and races with robust apterous males similar in appearance to those of *Australembia*, but having 2 hind basitarsal papillae, normal palpi, and usually a soft, transverse submentum. All species live in leaf litter, and range from the central Queensland coast southward into Tasmania. Most occur east of the Great Dividing Range, and one species has been found above 2,000 m in the Australian Alps.

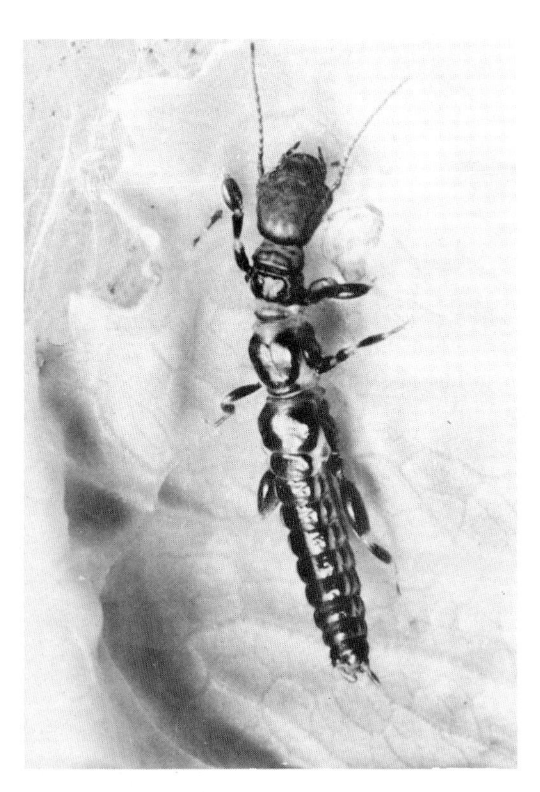

Fig. 23.7. *Metoligotoma* sp. of *reducta* complex, Australembiidae, ♂-length *ca* 10 mm.
[Photo by E. S. Ross]

3. Oligotomidae. This family includes Australia's most widespread embiid genus, *Aposthonia*, which also occurs throughout tropical Asia and the eastern Pacific area. In eastern Australia the principal species complex is *gurneyi* (Frogg.), found mostly in bark, post and rock crevices, and amongst lichens. The genus has its greatest diversity of species in W.A. A small black species in north Queensland is one of the smallest of the order, averaging only 5 mm in length. Many western species live in soil crevices, with no surface galleries evident except after rains. Males frequently fly to light. *Oligotoma* is an Indian genus represented by the introduced species *saundersii* (Westw.) in Queensland and *nigra* (Hagen) from inland N.S.W. Both may be expected to extend their range steadily in Australia, as they have in many other regions of the world.

24

PSOCOPTERA

(Psocids, booklice)

by C. N. SMITHERS

Small, free-living, exopterygote Neoptera, with large, mobile head, filiform antennae, and bulbous postclypeus; mandibles asymmetrical; maxillae with rod-shaped lacinia; labial palpi reduced; wings membranous, usually held roofwise over abdomen, venation reduced, brachyptery and aptery frequent; tarsi 2- or 3-segmented in adults, 2-segmented in nymphs; cerci absent.

This is an order of about 1,700 described species arranged in about 200 genera. Psocoptera are found in all regions. They range from less than 1 to almost 10 mm in length, and have a characteristic appearance due mainly to their having a round, mobile head, long antennae, enlarged pterothorax, and the wings held roofwise over the abdomen (Fig. 24.1A). Most species are winged as adults, but alary polymorphism occurs, and brachyptery or aptery in one or both sexes is common. Their relationships are not clear, but their nearest living relatives appear to be the Phthiraptera–Mallophaga. Both groups have a hypopharynx of peculiar form, but fossil evidence to link them is lacking. The Psocoptera would seem to have been derived from primitive hemipteroid stock. Publications on the order up to 1964 have been annotated by Smithers (1965c).

Anatomy of Adult

Head (Fig. 24.1B). Large and mobile, with distinct epicranial suture; clypeus divided into narrow transverse anteclypeus and characteristically bulbous postclypeus; frons small. Compound eyes usually strongly convex, sometimes reduced to groups of ommatidia (e.g. *Liposcelis*); 3 ocelli present in winged forms (usually absent in apterous forms), grouped in most families on a tubercle, widely separated in some (e.g. Lepidopsocidae). Antennae filiform, usually 13-segmented, segments sometimes very numerous (e.g. Lepidopsocidae, Trogiidae); scape and pedicel short, remaining segments elongate. Labrum simple. Mandibles asymmetrical, with large, ridged molar area and a toothed incisor edge. Maxillae without differentiated cardo; stipes with a broad, fleshy galea strengthened by complex sclerotizations; lacinia (Fig. 24.1D) modified into an elongate, strongly sclerotized rod, proximally sunken well into head capsule, apically variously toothed; palpi 4-segmented. Labium with sclerotized mentum; prementum divided; paraglossae membranous, flanking minute glossa; palpi reduced, 1- or 2-segmented. Hypopharynx with extremity of lingua bearing two superlinguae; lingua partially thickened ventrally into two oval lingual sclerites, each connected to a median

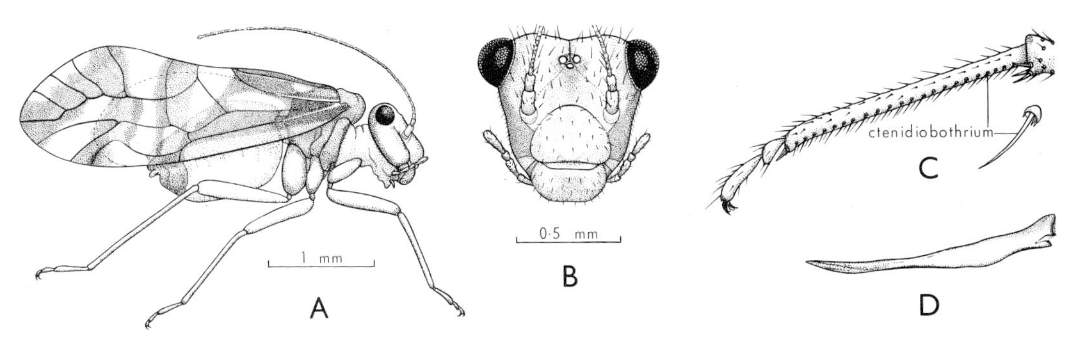

Fig. 24.1. *Pentacladus eucalypti*, Elipsocidae: A, ♂, lateral; B, head of same, anterior; C, hind tarsus and an enlarged ctenidiobothrium. D, lacinia of *Myopsocus griseipennis*, Myopsocidae. [B. Rankin]

sitophore sclerite by a fine filament.

Thorax. Prothorax reduced in winged forms; pterothorax well developed, the terga divided into a scutum and scutellum, behind which lies the postnotum. In apterous forms terga of meso- and metathorax sometimes fused, without subdivision. Pleura developed in accordance with powers of flight, reduced in flattened apterous forms. Sterna reduced in winged forms, broad in flattened apterous forms. Normally two pairs of spiracles.

Legs. Usually slender, similar; in *Liposcelis* the femora are strongly dilated. Hind coxae in many families bear on their inner surfaces a supposed stridulatory organ (Pearman's organ) consisting of a small rugose dome and an adjacent membranous area of integument (tympan or mirror). Trochanters without movable articulation with femur. Tibiae long, cylindrical, apically spurred, carrying ctenidiobothria. Tarsi 2- or 3-segmented; at least 1st segment usually with row of ctenidiobothria (Fig. 24.1c); pretarsus with 2 apical claws, toothed or not, and a variously formed pulvillus; empodia lacking.

Wings (Fig. 24.2). Membranous, hind wings smaller than fore wings, both often reduced or absent; at rest usually held roofwise over the body with the hind margins uppermost; coupled both in flight and at rest. Membrane usually bare, except for pterostigma, in some families scaled (e.g. Lepidopsocidae); veins and margins bare or setose. Venation of fore wing reduced; Sc reduced; pterostigma present, bounded behind by R_1; R and M usually 3-branched; M fused with Rs for a length, meeting it at a point, or

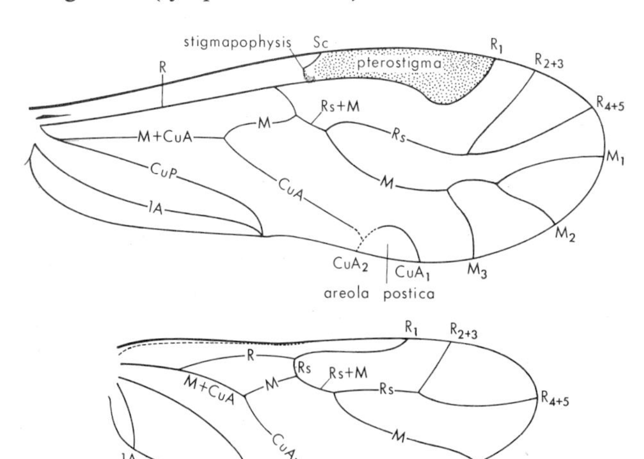

Fig. 24.2. Wing venation of *Caecilius* sp. Caeciliidae.

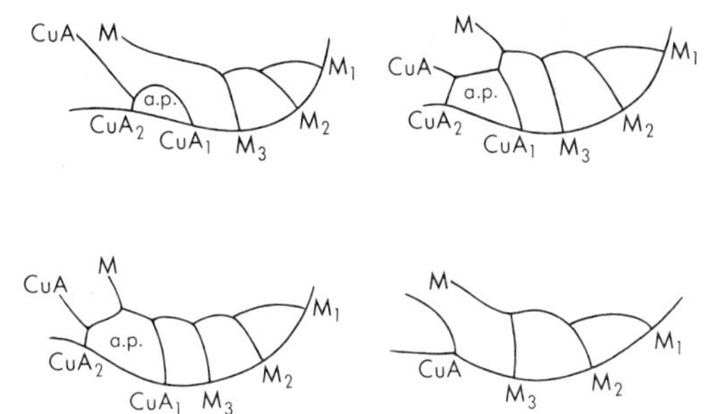

Fig. 24.3. Various relationships between M and CuA in the fore wing, with CuA dividing to form an areola postica *(a.p.)* or not.

joined to it by a cross-vein; M and CuA fused in basal part of wing; CuA usually forked distally, the cell between the branches *(areola postica)* being a characteristic feature of the psocopteran wing (Fig. 24.3); M frequently fused with apex of areola postica, or joined to it by a cross-vein, or meeting it in a point (closed discoidal cell); CuP (analis of Enderlein) usually finer than other veins, less often setose, runs free to margin in primitive forms, meets margin at same point *(nodulus)* as 1A in advanced forms; only one anal vein (axillaris of Enderlein) present, except in Amphientomidae and extinct families. Hind wing with venation further reduced; M and CuA usually not branched. Venational aberrations are frequent, and departures from the basic plan occur in some families, either by loss (especially of CuA_1) or additional branching. The Lower Permian Psocoptera (Fig. 8.13B) had a more generalized venation and other primitive features.

Abdomen. Nine-segmented, terminating in a dorsal epiproct and a pair of lateral paraprocts; paraprocts of winged forms usually each with a field of sensory setae (trichobothria). Cerci never present. Usually 8 pairs of spiracles. Sternum 9 of male (hypandrium) well developed (Fig. 24.4), lying ventral to the phallosome, usually simple, sometimes complexly ornamented with sclerotized structures (e.g. Psocidae). The phallosome consists of two 'parameres' ('external parameres' of some authors), which are sclerotized and free distally, joined basally, and flank the aedeagus ('internal parameres' of some authors). Within the framework so formed lies the expanded and eversible end of the ejaculatory duct (penial bulb) of which the walls may be sclerotized in a complex manner (e.g. Peripsocidae). Sternum 7 of female forming well-developed subgenital plate (Fig. 24.5). Ovipositor of 3 pairs of valves: the gonapophyses of segment 8 (ventral valves) which are usually elongate and pointed; and two pairs of appendages of segment 9, the dorsal valves, usually long and broader than the ventral valves, and the external valves, which are usually short, broad, and setose. Reduction of some or all of the valves occurs in varying degree, and they may be absent (some *Archipsocus* spp.).

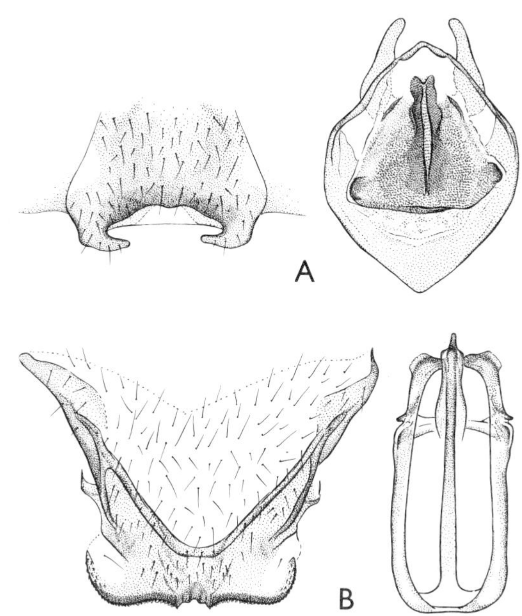

Fig. 24.4. Male hypandrium (left) and phallosome (right) of: A, *Pentacladus eucalypti*; B, *Myopsocus griseipennis*. [B. Rankin]

Internal Anatomy. Oesophagus elongate; mid gut wide, convoluted, leading into short hind gut; 4 Malpighian tubes. A pair of long, tubular, ventral labial glands function as salivary glands, and a pair of variously formed dorsal glands as silk glands. Nervous system concentrated; meso- and metathoracic ganglia fused, and a single small abdominal ganglion adjacent to that in the pterothorax. Two large nerves and their

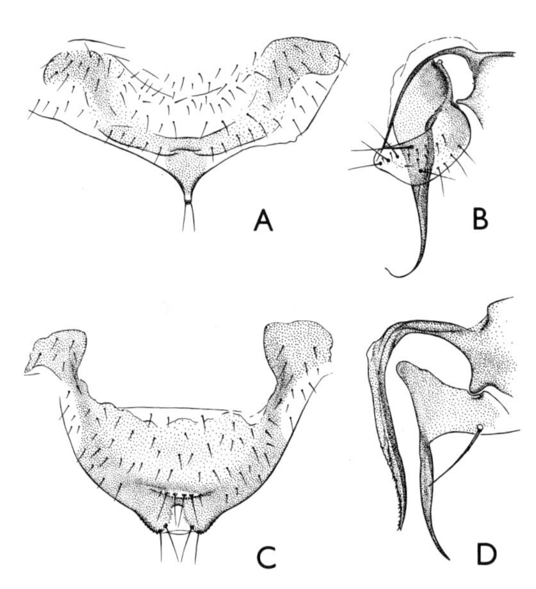

Fig. 24.5. A, subgenital plate of ♀ *Myopsocus griseipennis*; B, gonapophyses of same; C, subgenital plate of ♀ *Pentacladus eucalypti*; D, gonapophyses of ♀ *Caecilius* sp. [B. Rankin]

branches from the abdominal ganglion serve the abdomen, except for segment 1 which is served by a pair of small nerves. Testes usually 3-lobed, sometimes spherical or fusiform; vasa deferentia lead to large, complex seminal vesicles which secrete spermatophore material; ejaculatory duct short, but broadens distally to form penial bulb of phallosome. Ovaries of 3-5 polytrophic ovarioles opening into common median duct via short transverse oviducts; gonopore behind sternum 7; a spermatheca opens on sternum 9 by a duct of variable length, the opening sometimes having characteristic adjacent sclerotizations.

Immature Stages

The eggs are ellipsoidal, ovoid, or oblong, and the chorion may be sculptured or smooth. Development of the embryo has been followed in only a few species. Hatching is achieved with the aid of an egg-burster on the frontal region of the embryonic cuticle, which is immediately shed. On hatching, the nymph is generally like the adult, but always has 2-segmented tarsi, relatively shorter antennae, lacks ocelli, and has equal thoracic segments. There are normally 6 nymphal instars, but the number may vary, especially in polymorphic species. Wing-bud development is apparent from the 2nd instar, and rudiments of the external genitalia may be discernible in the final nymphal instar.

Biology

Psocids are found on the foliage or branches of trees and shrubs, on or under bark, on fences and walls, in leaf litter, under stones, on rocks, in caves, in human habitations, and in stored products. Some species occur in several habitats. They feed on unicellular algae, lichens, fungal hyphae, spores, and fragments of plant or insect tissue; *Liposcelis bostrychophilus* Bad. has been reared on yeast media. In the field they are sometimes taken with a sweep net, but are best collected by beating foliage and stems of trees and shrubs. They should be captured by means of an aspirator, and transferred directly into 70–80 per cent alcohol. Specimens should be handled carefully, as antennae and legs are easily lost.

Various degrees of intraspecific association are found, some species occurring in loose groups apparently brought into proximity of each other because of attraction to food source or other environmental factors. In other cases nymphs remain in close physical contact, the groups reassembling after forced dispersal of the members; the adults of such species are usually solitary. Small groups of nymphs or adults are sometimes found under communal webs, the size of web depending on the species; in *Archipsocus* the webs may be of spectacular proportions covering the trunks and branches of large trees. Nymphs are sometimes rendered inconspicuous by means of particles of debris adhering to glandular body hairs; other nymphs and adults may resemble their backgrounds by virtue of colour pattern.

The coxal (Pearman's) organ is presumably stridulatory in function. The ticking noise frequently described is, however, caused by the underside of the apex of the abdomen being struck against the surface on which the insect is standing.

Polymorphism is fairly common in some families, the usual form involving loss or reduction of wings in the female, but loss of wings in the male alone and equal reduction in both sexes are also known. Control of polymorphism appears in some species to be at least in part environmental, and loss or reduction of wings is frequently associated with loss of ocelli, trichobothria, and coxal organ, and retention of duplex setae in the adult.

Reproduction. Copulation is usually preceded by a nuptial dance, the male facing the female, after which he intrudes himself backwards under her from in front. Spermatozoa are transferred in a spermatophore which may be of complex form (e.g. in *Lepinotus*). Eggs are laid singly or in groups on or under bark, or on leaves, usually on the lower surface and frequently adjacent to a vein. They may be covered with silk or an encrustation of debris. Viviparity occurs in *Archipsocus*, and obligatory parthenogenesis is frequent. Males are rare in some species, and species are known which are parthenogenetic in some parts of their range but not in others; facultative parthenogenesis also occurs.

Natural Enemies. Psocids are preyed upon by spiders, pseudoscorpions, neuropterous larvae, ants, reduviids, wasps, and thrips. They are attacked by parasitic nematodes and entomophagous fungi; the gut usually contains protozoa. Mymarid parasites (*Alaptus*, Hymenoptera) and cimicoid predators (Hemiptera) have been known to destroy eggs.

Economic Significance. Psocids are not of great economic importance, although species associated with stored products sometimes develop enormous populations. Their occurrence seems to be secondary, poor storage and infestation by pests rendering conditions suitable for them (Champ and Smithers, 1965). They occasionally occur in large numbers in houses, where they are a nuisance rather than destructive. Neglected insect collections may be ruined by psocids (usually *Liposcelis* spp).

Special Features of the Australian Fauna

Any remarks on the faunal relationships of the Australian Psocoptera must be tentative, as the fauna of this region is, as yet, poorly known. Moreover, psocids are easily dispersed by air currents, and have been found in samples of the drifting insect population of the upper air (Thornton and Harrell, 1965), whilst some of the species that are common in stored products and human habitations have been widely distributed by man.

Two groups of families appear to be unrepresented (Amphientometae and Epipsocetae), and all families that are present are also found elsewhere. There are certain recognizable elements in the Australian fauna. There are the cosmopolitan species, some of which are associated with man and his products (e.g. *Liposcelis*), whereas others occur naturally (e.g. *Ectopsocus*). A second element consists of species having relationships to species of other southern areas (e.g.

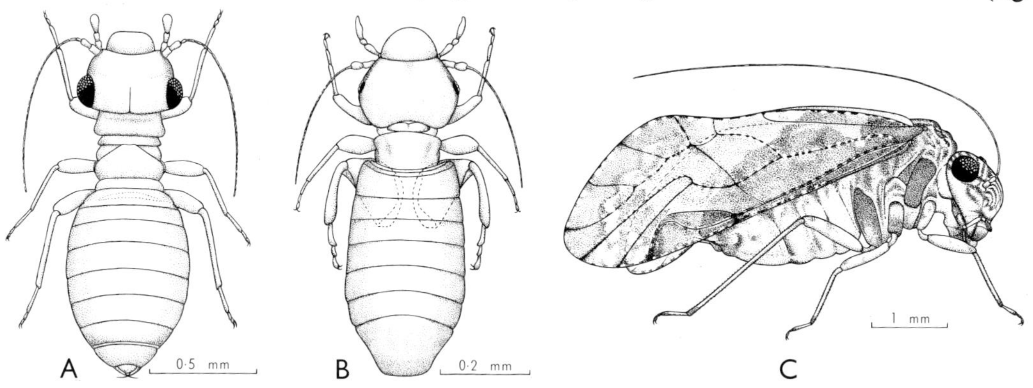

Fig. 24.6. A, *Lepinotus reticulatus*, Trogiidae; B, *Liposcelis bostrychophilus*, Liposcelidae; C, *Myopsocus griseipennis*, Myopsocidae. [B. Rankin]

Propsocus). An interesting genus is *Sphaeropsocus*, with a species in Tasmania and S.A., one in the Argentine, and a third recorded from Baltic amber. Some of the northern species represent a more or less tropical, world-encircling element (e.g. Archipsocidae), whereas others are more definitely related to Papuan forms (e.g. *Calopsocus*). Finally, there is an element consisting of widespread endemic Australian species belonging to groups represented beyond Australia.

CLASSIFICATION

Order PSOCOPTERA (120 Australian spp.)

Suborder TROGIOMORPHA (13)

Atropetae (11)
1. Lepidopsocidae (6)
2. Trogiidae (4)
3. Psoquillidae (1)

Psocatropetae (2)
4. Psyllipsocidae (2)
   Prionoglaridae (0)

Suborder TROCTOMORPHA (8)

Nanopsocetae (8)
5. Liposcelidae (6)
6. Pachytroctidae (1)
7. Sphaeropsocidae (1)

Amphientometae (0)

Suborder PSOCOMORPHA (99)

Homilopsocidea (49)
8. Elipsocidae (7)
9. Philotarsidae (10)
   Mesopsocidae (0)
10. Lachesillidae (1)
11. Peripsocidae (20)
12. Hemipsocidae (3)
13. Pseudocaeciliidae (3)
14. Trichopsocidae (3)
15. Archipsocidae (2)
    Psoculidae (0)

Epipsocetae (0)

Caecilietae (27)
16. Caeciliidae (23)
    Amphipsocidae (0)
17. Stenopsocidae (3)
    Polypsocidae (0)
18. Calopsocidae (1)

Psocetae (23)
19. Psilopsocidae (1)
20. Myopsocidae (6)
21. Psocidae (16)
    Thyrsophoridae (0)

The classification adopted here is that of Badonnel (1951), which is a combination of the classifications of Pearman and Roesler in both of which a range of characters was used. Older classifications were less indicative of true relationships, due partly to convergence which became apparent when more characters were used. The suprafamily groups do not correspond with the superfamilies of other orders, and it is best to regard them merely as convenient categories until the many genera requiring further study can be reassessed. This applies especially to the Homilopsocidea.

Some colour loss of psocids is inevitable in alcohol, but dry specimens become very brittle and difficult to study. Critical study requires at least some dissection, usually of mouth-parts and genitalia; small species are sometimes mounted whole.

KEY TO THE SUBORDERS OF PSOCOPTERA—ADULTS

1. Antennae more than 20-segmented, never secondarily annulated; tarsi 3-segmented; pterostigma not thickened, or absent; paraprocts with strong posterior spine .... TROGIOMORPHA
   Antennae usually 13-segmented, if 15- to 17-segmented, some segments are secondarily annulated; tarsi 2- or 3-segmented; pterostigma thickened or not; paraprocts without strong posterior spine 2
2. Antennae 12- to 17-segmented, some flagellar segments secondarily annulated; tarsi 3-segmented; pterostigma not thickened .... TROCTOMORPHA
   Antennae usually 13-segmented; tarsi 2- or 3-segmented, if latter, then flagellar segments not secondarily annulated; pterostigma thickened .... PSOCOMORPHA

Suborder TROGIOMORPHA

Antenna with up to 50 or more segments; filaments of hypopharynx separated for whole length; the four labial glands with a common duct; ♀ gonapophyses reduced.

Key to the Families of Trogiomorpha Known in Australia

1. Head long and vertical; maxillary palp without sensillum on 2nd segment; CuP and 1A end together at wing margin (nodulus) .... PSOCATROPETAE. **Psyllipsocidae**
   Head short and transverse; inner side of 2nd segment of maxillary palp with sensillum; in winged forms CuP and 1A end separately at wing margin .... ATROPETAE. 2
2. Claws with preapical tooth; body and wings usually scaled; if scales absent, fore wings acuminate .... **Lepidopsocidae**
   Claws not toothed; body and wings never scaled; fore wing rounded or absent 3
3. Fore wing well developed or shortened, but always with veins; M_3 arises basal to the forking of Rs; hind wing usually reduced .... **Psoquillidae**
   Fore wing present only as veinless flap, or absent; hind wing absent .... **Trogiidae**

1. Lepidopsocidae. The only Australian family in which the body and wings are scaly; in the THYLACELLINAE scales are absent, but the wings are acuminate. These psocids have the appearance of small moths, and are found in leaf litter and on bark, sometimes in fairly dry situations. [Smithers, 1965a.]

2. Trogiidae (Fig. 24.6A). Several of the common indoor species of world-wide distribution belong to this family, e.g. *Lepinotus inquilinus* Heyd. They are apterous, or have small, easily detached wing rudiments; the thoracic nota are not subdivided, and ocelli are absent. [Smithers, 1965b.]

3. Psoquillidae. The widespread *Psoquilla marginepunctata* Hagen, in which the wings are dark with marginal semicircular hyaline areas, has been found in stored products in Queensland.

4. Psyllipsocidae. Often pale forms, with long legs, and capable of jumping, they sometimes have reduced wings. Several representatives are found in caves and in buildings. The widespread *Psyllipsocus ramburii* Selys-Long. is polymorphic, with brachypterous, pale, almost blind forms occurring in caves and houses, and winged or brachypterous pigmented forms with larger eyes in houses and elsewhere. Wing development appears to be affected by population density and temperature conditions.

Suborder TROCTOMORPHA

Antennae with less than 20 segments and with secondary annulations; filaments of hypopharynx separated only near their distal ends; ♀ gonapophyses of various forms.

Key to the Families of Troctomorpha Known in Australia

1. Wings, when present, flat, fore wings with complete venation. [In both alate and apterous forms eyes seated near vertex; thoracic sterna narrow, without cilia; hind femora not basally widened] .... **Pachytroctidae**
   Wings, when present, with incomplete venation, lacking terminal branching. [In all apterous forms meso- and metathorax indistinguishably fused] 2
2. In alate forms both fore and hind wings present; eyes near vertex. In apterous forms eyes remote from vertex, each consisting of 2 large elements alone or preceded by 6 or fewer smaller ocelloids. Pronotum with lobar divisions; thoracic sterna broad and bearing cilia; hind femora broad basally .... **Liposcelidae**
   In alate forms only fore wings present, convex, elytriform. In all forms eyes remote from vertex, composed of few ocelloids, none greatly enlarged; pronotum simple; thoracic sterna narrow, without cilia; hind femora not widened basally .... **Sphaeropsocidae**

5. Liposcelidae (Fig. 24.6B). Small, flattened species with enlarged hind femora; when wings are present, venation is reduced. Species of *Liposcelis* sometimes occur in large numbers in stored products, houses, ships, and merchandise stores; many species are of world-wide occurrence, the traditional 'booklouse' usually being one or other species of this genus. They also cause damage to insect collections, and may occur in houses in sufficient numbers to be a nuisance. Members of this family are commonly found on or under bark.

6. Pachytroctidae. Frequently found in or under bark or in vegetable debris, the integument of many species is characteristically sculptured.

7. Sphaeropsocidae. Small species with elytriform fore wings; found in grass tussocks in Tasmania and S.A.

Suborder PSOCOMORPHA

Antennae with 13 segments, without secondary annulations; filaments of hypopharynx only partly separated; ♀ gonapophyses various. Includes the more specialized lines of development within the order and about 75 per cent of the described species.

Key to the Families of Psocomorpha Known in Australia

1. Labial palp broadly triangular, laterally diverging; lacinia narrowed towards apex, which is without small teeth but may be divided .... CAECILIETAE. 2
   Labial palp short and appressed, somewhat semicircular; lacinia not narrowed towards apex, which has many small teeth 4

2(1). Vertex sharp; venation reticulate in distal half of fore wing; CuA_2 a little shorter than CuA_1; 1A in hind wing almost as long as CuP .... **Calopsocidae**
   Vertex rounded; venation not reticulate; CuA_2 very short; 1A in hind wing about half as long as CuP 3

3(2). CuA_1 fused with M, or joined to it by a cross-vein .... **Stenopsocidae**
   Areola postica free .... **Caeciliidae**

4(1). Areola postica free or absent; ♀♀ often brachypterous or apterous, but without glandular hairs on head .... most HOMILOPSOCIDEA and Psilopsocidae. 5
   CuA_1 fused with M in fully winged forms; ♀♀ occasionally brachypterous, and then always with glandular hairs on head . most PSOCETAE and Hemipsocidae, some Elipsocidae. 13

5(4). Tarsi 3-segmented 6
   Tarsi 2-segmented 8

6(5). Fore and hind wings bare; areola postica extending well basal to junction of CuA_2 with wing margin .... **Psilopsocidae**
   Fore and hind wings with at least some marginal setae even in brachypterous forms; areola postica normal 7

7(6). Hind wing setose along entire margin; always macropterous; ♂ hypandrium strongly sclerotized; ♀ subgenital plate with median posterior lobe .... **Philotarsidae**
   Hind wing with, at most, setae on margin between R_{2+3} and R_{4+5}; brachyptery common; ♂ hypandrium lightly sclerotized; ♀ subgenital plate usually bilobed ...... **Elipsocidae** (pt)

8(5). Areola postica absent .... **Peripsocidae**
   Areola postica present (sometimes indistinct in Archipsocidae) 9

9(8). Wing membrane setose. [Fore wings more or less reduced, veins evanescent in distal part] .... **Archipsocidae**
   Wings, apart from veins and margin, at most with hairs on pterostigma 10

10(9). Wings without hairs .... **Lachesillidae**
   Wings with hairs on margins and veins 11

11(10). Distal parts of veins in fore wing with more than a single row of setae .... **Pseudocaeciliidae**
   Distal parts of veins in fore wing with only one row of setae, or apterous 12

12(11). Claws with preapical tooth; subgenital plate with median posterior lobe; hind wing with marginal setae only between R_{2+3} and R_{4+5} .... **Elipsocidae** (pt)
   Claws without preapical tooth; subgenital plate simple; hind wing with marginal setae as far as CuP .... **Trichopsocidae**

13(4). Tarsi 2-segmented 14
   Tarsi 3-segmented 15

14(13). M of fore wing 3-branched, or brachypterous, or apterous .... **Psocidae**
   M of fore wing 2-branched; brachyptery or aptery not known .... **Hemipsocidae**

15(13). Fore wing bare; wing pattern of numerous, small, irregular, confluent dark areas. **Myopsocidae**
Fore wing with at least some small, fine setae on veins and margin; pigmentation in a bold pattern of large hyaline and coloured areas, or without pattern (see couplets 7 and 12) .... **Elipsocidae** (pt)

8. Elipsocidae. (Figs. 24.1A–C). A world-wide family, with representatives living in a variety of situations. The nymphs are without glandular setae, and females usually have a bilobed subgenital plate and a complete set of gonapophyses, the dorsal valve being apically divided. Most species are dull-coloured, but the wings are marked in a bold pattern in *Propsocus* and *Pentacladus*. The venation is usually of the *Caecilius* type (Fig. 24.2).

9. Philotarsidae. Very similar to the Elipsocidae, but with the wings more hairy, the female subgenital plate with a median lobe, and the nymphs with glandular setae. Members of this family are more frequently found on fresh foliage than are the elipsocids.

10. Lachesillidae. Venation of the *Caecilius* type, wings bare, and tarsi 2-segmented; terminal structures of male abdomen complex, phallosome simple; female gonapophyses strongly reduced. Species of this family are found mainly in dried leaves and leaf litter. They have been known to occur in large numbers indoors, and occasionally swarm at dusk.

11. Peripsocidae. The species are immediately recognizable by the absence of areola postica of fore wing. ECTOPSOCINAE, with more or less rectangular pterostigma, usually frequent dried leaves, *Ectopsocus briggsi* McLachl. being almost world-wide in distribution. PERIPSOCINAE, with normal pterostigma, are mainly bark dwellers.

12. Hemipsocidae. Distinguishable by 2-branched M and areola postica connected to M by cross-vein. Many species have conspicuous black spots at bases of wing setae, and become very active when disturbed, fluttering and tumbling in a characteristic manner. They are mainly leaf-litter and dried-leaf inhabitants.

13. Pseudocaeciliidae. Venation as in *Caecilius*, but with elongate pterostigma and areola postica; wings with long setae. Usually found on fresh foliage or twigs.

14. Trichopsocidae. Similar to Pseudocaeciliidae, but with normal pterostigma and areola postica.

15. Archipsocidae. Wings often reduced, and usually with indistinct venation, membrane setose; in viviparous species gonapophyses are absent, in others reduced. These insects live in colonies under sheets of webbing, large trees sometimes being enshrouded.

16. Caeciliidae. A large family, world-wide in distribution, usually found on fresh foliage. Gonapophyses reduced.

17. Stenopsocidae. Genitalia are reduced as in the Caeciliidae, but the fore wing has cross-veins between pterostigma and Rs and between areola postica and M. *Taeniostigma perkinsi* Banks, probably the largest Australian psocid, is common on leaves in eastern Queensland and N.S.W.

18. Calopsocidae. The fore wings in this family have a complex of supplementary veins. The single known Australian species from Queensland is brightly coloured.

19. Psilopsocidae. The only Australian species of this small family so far known was found near Sydney. The nymphs are peculiar in that the apex of the abdomen is strongly sclerotized, with the epiproct and paraprocts in a ventral position forming a cover to the anus.

20. Myopsocidae (Fig. 24.6C). Large, common, mottled-winged species, found on bark and paling fences, where they feed on algae and fungi. [Smithers, 1964.]

21. Psocidae. A large family, in which the areola postica is joined to M for a length; fore wings bare. Most species are bark dwellers and of sombre colour, the many genera exhibiting a wide range of genitalic characters.

25

PHTHIRAPTERA

(Lice)

by J. H. CALABY

Apterous, dorsoventrally flattened, exopterygote Neoptera, with mouth-parts mandibulate or piercing and sucking. Ectoparasitic on birds and mammals; entire life spent on hosts.

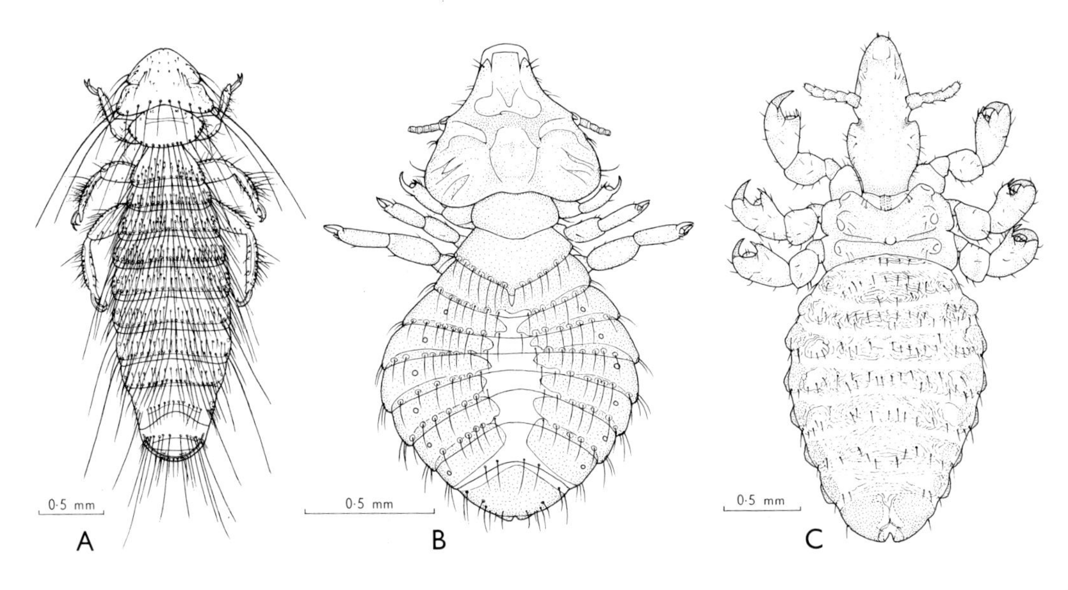

Fig. 25.1. Representative lice: A, *Menacanthus stramineus*, Mallophaga-Amblycera-Menoponidae, ♂ from *Gallus gallus*; B, *Saemundssonia platygaster*, Mallophaga-Ischnocera-Philopteridae, ♀ from *Charadrius bicinctus*; C, *Haematopinus asini*, Anoplura-Haematopinidae, ♀ from *Equus caballus*. [F. Knight]

This highly specialized order was undoubtedly derived from free-living psocopteroid ancestors. It is divided into three very distinct suborders, the Mallophaga (chewing or biting lice), Rhynchophthirina (elephant lice), and Anoplura (sucking lice). Approximately 3,000 species are known. There is a great diversity in size and body form (Fig. 25.1). The adults range in length from less than 0·5 to 10 mm (males are smaller than females).

They are highly modified for their parasitic existence, the most characteristic features being flattening of the body, complete aptery, and adaptation of the tarsi for clinging to the feathers or hair of the host. Most species are well sclerotized, many are deeply pigmented, and some have a characteristic pattern. In some groups the body is covered with setae, but in others it is relatively bare.

The Rhynchophthirina are an anomalous group, which includes only the genus *Haematomyzus* parasitizing African and Asian elephants and the African wart-hog; they are not considered in this chapter. General literature on the other two suborders includes Hopkins and Clay (1952) and Rothschild and Clay (1952) on Mallophaga, Hopkins (1949) on Mallophaga and Anoplura of mammals, and Ferris (1951) on Anoplura.

Anatomy of Adult

Head. Sessile on prothorax, long axis in same plane as rest of body; in Mallophaga flat and hollow, with sutures and thickened internal apodemes, both of which provide useful generic characters. Eyes reduced or absent; ocelli absent. Antennae 3- to 5-segmented; in Amblycera usually capitate and

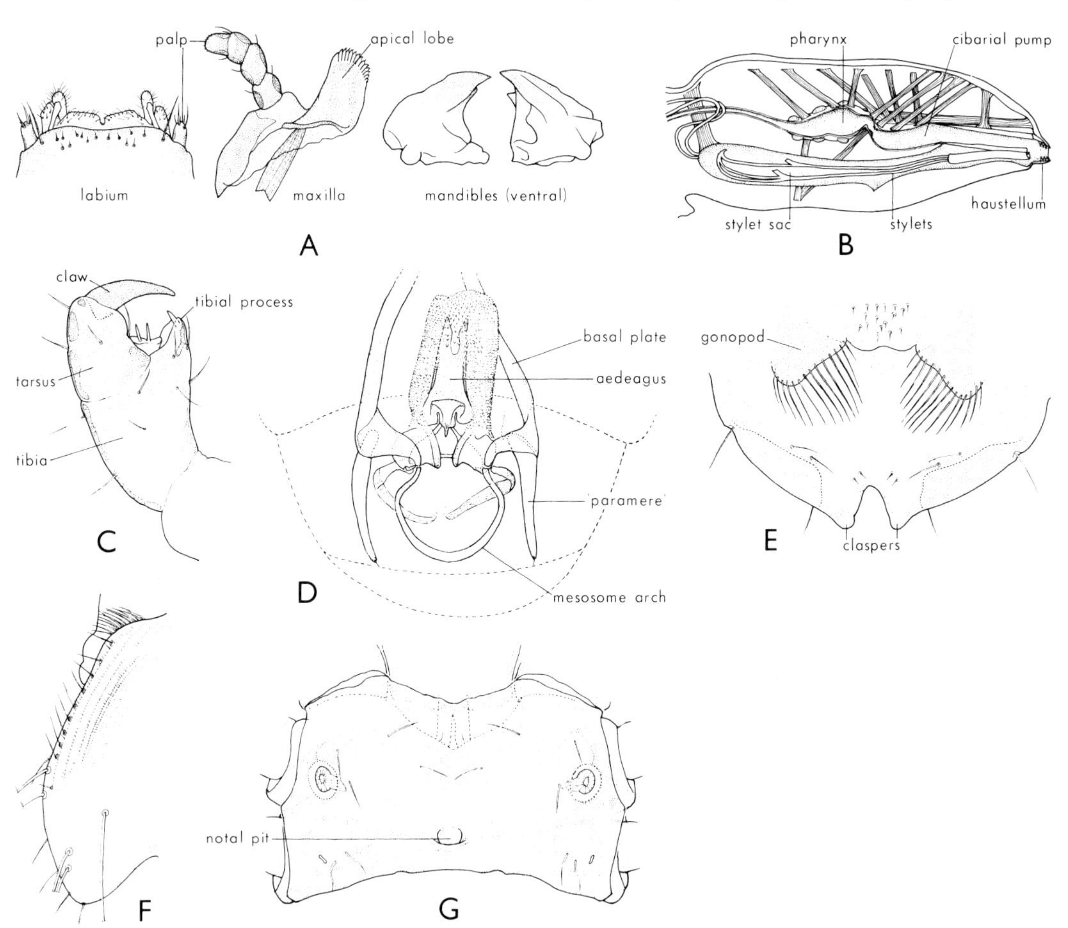

Fig. 25.2. A, mouth-parts of *Laemobothrion* sp., Mallophaga (after Snodgrass, 1905); B, median sagittal section of head of *Pediculus humanus*, Anoplura (after Buxton, 1947); C, fore leg of *Haematopinus asini*, Anoplura; D, ♂ genitalia of *Paraheterodoxus insignis*, Mallophaga; E, ♀ terminalia of *Haematopinus asini*; F, side of head of *Laemobothrion* sp., showing rows of minute projections posterior to eye; G, thoracic nota of *Haematopinus asini*, showing notal pit. [F. Knight]

lying in a groove; in Ischnocera and Anoplura filiform, exserted, and sometimes modified in the male as clasping organs. Mouth-parts mandibulate in Mallophaga (Fig. 25.2A), and differ in their insertion in the two divisions; maxillary palpi 2- to 4-segmented in Amblycera, absent in Ischnocera. In Anoplura anterior part of head is an unjointed, more or less conical or rounded beak (Fig. 25.2B); the mouth-parts consist of a small soft proboscis (haustellum) with small internal teeth, which is eversible to grip the host during feeding, and 3 stylets, the piercing organs, accommodated in the trophic sac opening off the ventral side of the cibarium; maxillary palpi absent.

Thorax. Prothorax almost always free, meso- and metathorax often imperfectly separated in Mallophaga; all segments fused in Anoplura.

Legs. Well developed, especially stout in the Anoplura. In Mallophaga tarsus 1- or 2-segmented, bearing a pair of claws in most families, but single claws in two mammal-infesting families; in Anoplura tarsus 1-segmented or weakly 2-segmented, with single large claw working against a tibial process (Fig. 25.2C).

Abdomen. Number of visible segments varies from 8 to 10, 9 in the Anoplura. Cerci absent. Male has a median eversible aedeagus; anatomy of male genitalia (Fig. 25.2D) important in taxonomy. Ovipositor absent, but there is a pair of short gonopods in Anoplura (Fig. 25.2E) and some Mallophaga which are concerned with grasping the hair of the host and placing eggs on it. In females of many species last segment terminates in 2 posterior lobes or claspers.

Internal Anatomy. In Mallophaga there is a well-developed crop and a large mid gut; hind gut usually short, but may be almost as long as mid gut. In Anoplura crop and proventriculus are undeveloped, and mid gut is large; the cibarium and pharynx and their dilator muscles form a powerful sucking pump. There are 4 Malpighian tubes and 4 or more, usually 6, rectal glands. Nervous system highly specialized. Tracheal system of Mallophaga, to which that of Anoplura is generally similar, has been studied by Harrison (1915); Webb (1946) considers that spiracular structure is important in phylogenetic studies. The female reproductive system includes an accessory gland to provide cement for egg attachment.

Immature Stages

The eggs (Fig. 25.3) are elongate oval, commonly whitish; in some species ornamented with surface sculpturing or plume-like processes associated with the operculum; large

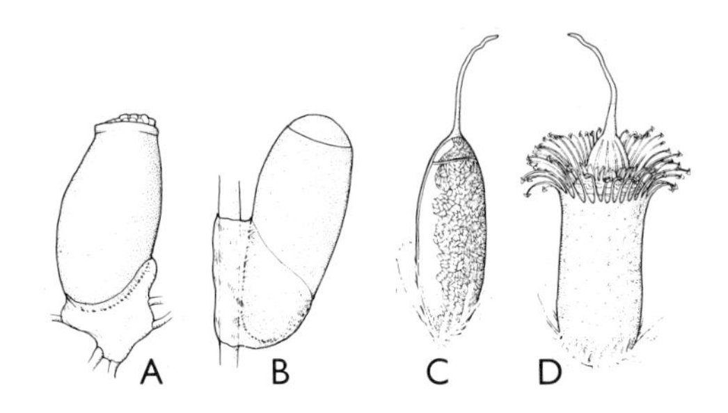

Fig. 25.3. Eggs of Phthiraptera: A, *Pediculus humanus*, Anoplura, from *Homo sapiens* (after Ferris, 1951); B, *Haematopinus suis*, Anoplura, from *Sus scrofa* (after Ferris, 1951); C, *Brueelia* sp., Mallophaga-Philopteridae, from a passerine bird (after Richter, in Balter, 1968); D, *Kelerimenopon* sp., Mallophaga-Menoponidae, from an Australian megapode (after Richter, in Balter, 1968).

relative to size of host, varying from less than 1 to nearly 2 mm in length. The nymphs resemble the adult generally, but are much smaller and unpigmented in first instar, becoming successively larger and darker after each ecdysis.

Biology

Adults. In general, Mallophaga feed on feathers (usually down or the downy parts) and skin-surface detritus, and the Anoplura exclusively on blood. Some species of Mallophaga, however, particularly in the Amblycera, take blood in addition to feathers. The common fowl louse, *Menacanthus stramineus* (Nitsch), habitually punctures developing quills to obtain blood, and also bites through the host's skin.

All orders of birds are parasitized and most orders of mammals; two important exceptions seem to be the monotremes and bats.

The degree of host specificity is generally high (Hopkins, 1949). Because of their host specificity, and the fact that most genera of lice parasitize related hosts, the louse fauna may provide important clues where host relationship is in doubt. The higher classification of birds is unsatisfactory, owing, at least in part, to their apparently rapid evolution and lack of fossils. The lice presumably became established on birds early in their history, and louse evolution has been more conservative than that of their hosts. For example, the flamingoes are usually placed with the storks and herons (Ciconiiformes), but occasionally with the swans, geese, and ducks (Anseriformes), and it may be significant that they share with the Anseriformes 3 genera of Mallophaga not found on the Ciconiiformes. This kind of evidence is not necessarily convincing, however, because the mallophagan fauna of a bird may have been acquired secondarily from another host at some time during its history. The number of species present on a host species varies from as low as one up to 15 in 12 genera and 3 families on a South American tinamid (Clay, 1949). The number of species of lice generally averages 5 or 6 per host species of bird.

In general, lice occupy specific regions on their hosts' bodies, those of mammals probably being less restricted than those of birds. Some are particularly specific in their ecological niche. Perhaps the most bizarre example is provided by the mallophagan genus *Piagetiella*, the species of which are restricted to the interior lining of the bill pouches of pelicans and a few species of cormorants. *P. australis* (Bedf.) on *Pelecanus conspicillatus* is an Australian representative. The lice are firmly attached by their mandibles to the mucous membrane, and apparently their diet consists of blood or mucus. Well-known examples of specific niches are also provided by the human lice and the lice of sheep (p. 122).

Mallophaga are generally very active and Anoplura much slower in their movements. Lice are negatively phototropic, and the numerous sensory setae no doubt help to guide them through the feathers or hair. The antennae, in some groups at least, are important sensory organs. The lice are sensitive to the temperature and smell of their hosts. Their optimum temperature is that of the skin surface, and their ability to withstand extremes is low. The insulating properties of the feathers or hair ensure that the habitat is relatively uniform in temperature and humidity. As the temperature of a dying host decreases, the lice migrate to the outside of the plumage or pelage, and their movements become very slow. They survive for only a few days off the host.

Transfer between individual hosts takes place by contact only: during brooding or nursing of young, during copulation, at roosts or communal sleeping places of gregarious species, or in common dust baths. Accidental transfers take place from prey to predator, and probably some species of genera which are common to predator and possible prey species originally arose on the predator by such means. Phoresis—the transport of Mallophaga by other insects, particularly hippoboscid flies—is probably a means of transfer in some instances (Bequaert, 1953).

The numbers of lice present on individual hosts varies from nil to many thousands. In general, small hosts have few—small birds usually less than 10—and large hosts have larger numbers. Some figures for mammals are given by Hopkins (1949), who found that in a number of species nymphs were roughly twice as numerous as adults. Little is known of the longevity of adult lice. There is some evidence that females live longer than males. The average adult life of the two species infesting man is about a month. The maximum life of the female of *Damalinia bovis* (L.) has been given as 42 days (Matthysse, 1944). The maximum adult lives of males and females of *Haematopinus eurysternus* (Nitsch) are 10 and 16 days (Roberts, 1952). Some individuals of the highly specialized *Lepidophthirus macrorhini* End. survive for several months while its host, the elephant seal, is at sea.

Reproduction. In natural populations of lice females are usually in excess, and in some species males may be very rare (see Hopkins,

1949). Matthysse has shown that *D. bovis* is facultatively parthenogenetic, and that males are most numerous in rapidly increasing populations. Pairing occurs frequently in *Pediculus humanus* L., and one male may fertilize many females (Buxton, 1947). Although *P. humanus* may lay scattered and unattached eggs, lice usually lay their eggs in groups in favoured sites. The eggs are cemented to the feathers or hairs (Fig. 5.23). A common egg-laying site of wing lice of birds is along the grooves between the barbs of flight feathers close to the shaft. The eggs are laid end to end and in rows. The head lice of birds lay their eggs on the feathers of the head or neck close to the skin, where they escape the bill of the preening bird. In blood-sucking species a blood meal is apparently necessary for maturation of the eggs.

Oviposition behaviour, the distribution of eggs, and influence of environmental factors have been studied by Murray (1957a–d, 1960a, b). In experimental situations a *Damalinia ovis* (Schrank) female moved to the warm end of a temperature gradient some time before laying eggs. Here it rested with its head towards the warm end while abdominal contractions took place. About five minutes before egg-laying, the louse suddenly turned around so that the abdomen pointed in the direction of the warmest temperature. The gonopods were raised, and moved around until the wool fibre was caught and held against the abdomen. After a few seconds, cement was secreted, then all of the egg except the cap was expelled. After a short time the louse walked along the fibre, thus assisting the expulsion of the cap end of the egg. The oviposition behaviour of other species was broadly similar. In some species the gonopods are also used as moulds for the cement, and hence attachments may be characteristic for a species.

Egg-laying was inhibited by very high humidities and absence of suitable fibres, but the presence of light had no effect. Gravid lice were attracted to other ovipositing lice. *D. ovis* and *Linognathus pedalis* (Osborn) laid eggs only within a narrow temperature range, and in *L. pedalis* development of the eggs was completed only within a narrow range. Very high humidities also caused hatching failure. The distribution of eggs of *D. ovis* on the body of the host is determined by the optimum temperature zone, which is regulated by skin temperature, skin topography, thickness of air blanket trapped in the fleece, and the ambient temperature. Populations of some species, such as the sheep foot louse, vary with seasonal temperatures.

The number of eggs laid varies with the species. *P. humanus* may lay as many as 300 and *Damalinia bovis* up to 30. Only one egg develops at a time in *D. ovis* (Murray, 1957b). Fully engorged *Linognathus stenopsis* (Burm.) laid 2 eggs a day for 3 days before requiring another blood meal (Murray, 1957a), and *Columbicola columbae* (L.) laid one egg every 2 or 3 days (M. Martin, 1934). On the other hand, females of *Lepidophthirus macrorhini* laid up to 22 eggs per day when the skin temperature of the host seal was maintained at 25–30°C, and the average daily production was 6–9 (Murray and Nicholls, 1965).

Life History. Hatching behaviour has been described for a few species, e.g. *C. columbae* by Martin (1934). When the pharate nymph was ready to hatch, it sucked in air through its mouth, and the air passed through the alimentary tract and accumulated inside the shell behind the nymph. The air pressure soon became so intense that the operculum was forced open, and the nymph freed itself during the next 20 minutes by muscular contractions, abdominal expansion, and further sucking of air. Food was not taken for several hours after hatching. *P. humanus* and some other Anoplura have hatching devices, a pair of upcurved teeth and several pairs of blades, on a rigid area of embryonic cuticle on the front of the head. At the appropriate time, the blades probably pierce the vitelline membrane, and the teeth force up the cap of the egg (Buxton, 1947).

There are 3 nymphal instars. Life history data have been determined for some economically important species, and some representative figures are given in Table 25.1. The incubation period in *C. columbae* under

TABLE 25.1
Average Life Cycles of some Species of Lice under Optimal Conditions

Stage (days)	*Damalinia ovis*		*Linognathus pedalis* (Scott, 1950; minima)	*Haematopinus eurysternus* (Roberts, 1952)	*Pediculus humanus* (Buxton, 1947)
	Optimum laboratory conditions (Scott, 1952)	On sheep (Scott, 1952)			
Incubation	8–10	9–10	17	12	6
1st instar	8–9	7	7	4	} 8–9
2nd instar	5–6	5	7	4	
3rd instar	8	9	7	4	
Pre-oviposition	4	3	5	4	1–2
Egg to egg	36	34	43	28	16

optimal laboratory conditions was 3–5 days (Martin, 1934).

Natural Enemies. The principal enemies of lice are their hosts. Lice are certainly killed by the hosts' scratching, grooming, and preening, and population control by grooming has been demonstrated experimentally. Dust-bathing in birds probably helps to rid them of their lice, and the habit may have evolved as a response to ectoparasitic irritation. The curious phenomenon of 'anting' in birds may also be a response to louse irritation. Fragments of lice have been found in the gut of other lice, and it is possible that predation of this nature may play a part in population regulation.

Economic Significance. Lice are known to transmit the organisms responsible for several diseases of man and other animals. The most important is epidemic typhus in man, transmitted by *P. humanus*. This infection has fortunately been absent from Australia since the early days of settlement. Murine typhus may be transmitted among rats by *Polyplax spinulosa* (Burm.), but the principal vector is a flea (p. 651). Other blood parasites of rats and dogs may be transmitted by lice, and *Trichodectes canis* (De G.) may serve as an intermediate host of the dog tapeworm *Dipylidium caninum*. The virus of rabbit myxomatosis can be transmitted mechanically by the rabbit louse, *Haemodipsus ventricosus* (Denny). Heavy infestations with lice cause considerable production losses in the wool, meat, dairying, and poultry industries, and blood-sucking species may even kill very young animals. The veterinary importance of lice in Australia has been discussed by Roberts (1952).

Special Features of the Australian Fauna

The Phthiraptera are among the least known of Australian insects. From a consideration of the bird and mammal fauna, a conservative estimate of the louse fauna would be 3,000 species, yet less than 180 native species have been recorded. One reason for this may be that the collection of lice involves the tedious examination of dead or captured vertebrate hosts, an activity not often indulged in by the general entomologist.

As in other zoogeographical regions, the great majority of bird lice belong to the large families Menoponidae and Philopteridae. The other bird-infesting families are small and parasitize a limited range of hosts, and one is restricted to a single order of South American birds. As the same genera of lice usually parasitize whole orders or several orders of birds, it is not surprising that the only endemic genera so far discovered occur on the emu (the only endemic order of birds) and on marsupials. The family Boopiidae is almost restricted to Australian and Papuan marsupials. It is to be expected that there would be few native species of Anoplura, as the only available hosts are rodents and seals, and only a few have been reported, all in a cosmopolitan genus found on rodents. The other groups of native warm-blooded vertebrates—birds, monotremes, marsupials, and bats—are known, or believed, not to harbour Anoplura.

The Australian louse fauna may be divided into four categories: (1) endemic species from endemic host species (the great majority); (2) widespread species occurring on hosts that occur also in other parts of the world (e.g. *Tyto alba* and many species of sea-birds, particularly petrels); (3) species which are apparently not very host specific and occur on native hosts, but which are not specifically distinguishable from lice occurring on related hosts in other parts of the world (e.g. *Saemundssonia africana* Timm. on *Vanellus novaehollandiae* is only subspecifically distinct from the louse on a related African plover; a louse found on *Corvus mellori* in Tasmania is indistinguishable from *Colpocephalum fregili* Denny found on a European corvid, *Pyrrhocorax pyrrhocorax*); (4) introduced species parasitizing domestic and feral introduced birds and mammals, including man, among which are most of the Anoplura.

CLASSIFICATION

Order PHTHIRAPTERA (208 Australian spp.)

Suborder MALLOPHAGA (186)

Division AMBLYCERA (80)

1. Menoponidae (45)	3. Ricinidae (1)	Trimenoponidae (0)
2. Laemobothriidae (3)	4. Boopiidae (31)	Gyropidae (0)

Division ISCHNOCERA (106)

5. Trichodectidae (6)	6. Philopteridae (100)	Heptapsogasteridae (0)

Suborder RHYNCHOPHTHIRINA (0)

Haematomyzidae (0)

Suborder ANOPLURA (22)

7. Pediculidae (2)	Neolinognathidae (0)	10. Haematopinidae (4)
8. Linognathidae (6)	9. Hoplopleuridae (10)	Echinophthiriidae (0)

The higher classification adopted here is that of Weber (1954). The families of Mallophaga are those recognized by Hopkins and Clay (1952), and of Anoplura those in Ferris (1951). There is no agreement among authors on the classification of lice, not even on whether they should be included in one or two orders (p. 164). In any case, there are four very distinct kinds of lice, which are classified here into 3 suborders and 2 divisions, although the relationships between them are obscure (see Clay, 1949). The two suborders known in Australia may be distinguished by the following key.

Head relatively large (Figs. 25.1A,B); mouth-parts mandibulate; on birds and mammals .... MALLOPHAGA

Head relatively small (Fig. 25.1C); mouth-parts developed into an unjointed more or less pointed beak, or rounded anteriorly, without mandibles; on mammals .... ANOPLURA

Suborder MALLOPHAGA

Key to the Families of Mallophaga Known in Australia

1. Antennae concealed in grooves, generally capitate, usually 4-segmented, 3rd segment wineglass-shaped; mandibles horizontal; maxillary palpi 2- to 4-segmented; meso- and metathorax usually separate .... AMBLYCERA. 2

   Antennae not concealed, usually filiform, 3- to 5-segmented; mandibles vertical; maxillary palpi absent; meso- and metathorax usually fused .... ISCHNOCERA. 5

2. Mesonotum with 2 protuberances, each bearing a spine-like seta; spiracles on apparent segments 2–7; apparent terga 1–3 with a fine trichobothrium on each side, or, if these are absent, maxillary palpi with less than 4 segments .... **Boopiidae**

   Without this combination of characters 3

3. Abdominal segment 1 fused to thorax; spiracles on apparent segments 2–7. [Head elongate (in *Ricinus*); base of antennae widely removed posteriorly from base of maxillary palpi] ... **Ricinidae**
   Without this combination of characters 4
4. Head elongate, with strong swelling of lateral margins in front of eye, followed by approximately parallel lateral margins to horizontal (flattened or concave) anterior margin; base of antennae widely removed posteriorly from base of maxillary palpi; lateral area posterior to eye with rows of minute projections (Fig. 25.2F); hind femur and some sternites with dense patches of microtrichia .... **Laemobothriidae**
   Without this combination of characters .... **Menoponidae**
5. Antennae 3-segmented; tarsi single-clawed; on introduced mammals .... **Trichodectidae**
   Antennae 5-segmented; tarsi with paired claws; on birds .... **Philopteridae**

Division AMBLYCERA

1. Menoponidae (Fig. 25.1A). This family, much the largest of the division, is worldwide in distribution, and occurs on all orders of birds that have been sufficiently studied. Approximately 35 native species in about 20 genera have been recorded from Australian birds. Perhaps a further 10 species have been recorded from domestic poultry and introduced feral birds. Roberts (1952) lists those recorded from poultry and the domestic pigeon, among the more important being *Menacanthus stramineus* and *Menopon gallinae* (L.) on fowls and *Colpocephalum turbinatum* Denny on pigeons.

2. Laemobothriidae (Fig. 25.4A). The only included genus, *Laemobothrion*, is parasitic on rails, storks, and hawks. A few specimens have been collected from Australian rails and a hawk, but none has been determined to species. The species present on the coot (*Fulica atra*) appears to be *L. atrum* (Nitsch) which occurs on the same host in Europe.

3. Ricinidae (Fig. 25.4B). This family consists of 2 genera, one on Neotropical humming-birds and the other (*Ricinus*) parasitizing small passerines of several families. Its occurrence in Australia rests on an undetermined species of *Ricinus* taken from a Tasmanian thornbill (*Acanthiza*). These lice seem incongruously large relative to the size of their hosts.

4. Boopiidae (Fig. 25.4C). With one intriguing exception, this family of 7 genera and over 30 species is confined to Australian and Papuan marsupials. The last revision was that of Werneck and Thompson (1940). Identification of hosts has often been unreliable, and most of the species are known only from captive hosts. A new revision (Kéler, in preparation) is being based on material collected from wild-caught hosts. All families of Australian marsupials carry these lice, except the Notoryctidae and Phalangeridae (including the koala). Their absence from the possums may only be a reflection of lack of collection, but large numbers of common species such as *Trichosurus vulpecula* have been searched. The 'non-marsupial' boopiid is *Heterodoxus spiniger* (End.), which occurs on domestic dogs in many parts of Australia, Africa, Asia, and the Americas. There are good records from the jackal and coyote. The species is close to, but distinct from, *H. longitarsus* (Piag.) found on *Macropus giganteus*. *Heterodoxus* is otherwise restricted to Macropodidae, and, although there are no unequivocal records of *H. spiniger* from this family, it is probable that it is a macropodid parasite which has become transferred to dogs, perhaps via the dingo, and transported around the world on domestic dogs.

It is worth noting that one genus of a South American family of Amblycera, the Trimenoponidae, is parasitic on marsupials, both didelphids and caenolestids. This family appears to be closely related to the Boopiidae. Two species (*Gyropus ovalis* Burm. and *Gliricola porcelli* (Schrank)) of another South American family, the Gyropidae, have been recorded in Australia on captive guinea-pigs, their natural host.

Division ISCHNOCERA

5. Trichodectidae (Fig. 25.4D). A family of 13 genera and many species parasitizing a great variety of mammals in most parts of the world. Several have been introduced into

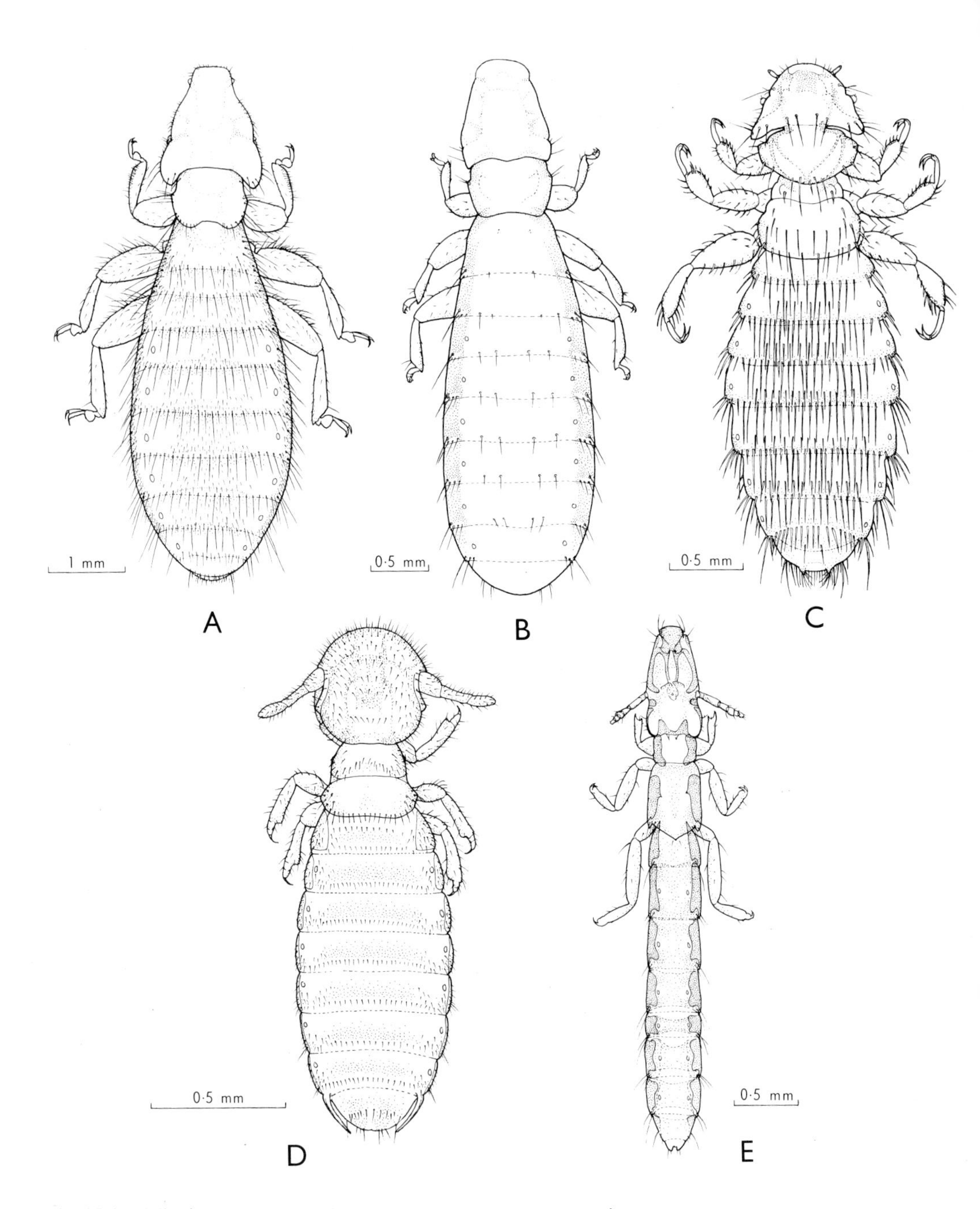

Fig. 25.4. Mallophaga: A, *Laemobothrion tinnunculi*, Laemobothriidae, ♂ from *Falco tinnunculus* (exotic); B, *Ricinus ernstlangi*, Ricinidae, ♀ from *Turdus merula* (exotic); C, *Paraheterodoxus insignis*, Boopiidae, ♀ from *Aepyprymnus rufescens*; D, *Damalinia ovis*, Trichodectidae, ♀ from *Ovis aries*; E, *Halipeurus diversus*, Philopteridae, ♀ from *Puffinus tenuirostris*. [F. Knight]

Australia with their domestic hosts: *Damalinia bovis* (cattle), *D. ovis* (sheep), *D. caprae* (Gurlt) (goat), *D. equi* (Denny) (horse), *Trichodectes canis* (dog, including the dingo), *Felicola subrostratus* (Burm.) (cat).

6. Philopteridae (Figs. 25.1B, 4E). This is the largest family of lice, and contains numerous genera and species found on all orders of birds in all parts of the world. About 90 native species in about 45 genera have been recorded from Australia, including the one endemic genus of the family (*Dahlemhornia* from the emu, *Dromaius novaehollandiae*). Several introduced species have been recorded from domestic poultry and introduced feral birds. Roberts (1952) lists those from domestic poultry and the pigeon, among the more important being *Lipeurus caponis* (L.), *Cuclotogaster heterographus* (Nitsch), *Goniodes dissimilis* Denny, and *Goniocotes gallinae* (De G.) on fowls, and *Anatoecus dentatus* (Scop.) and *Anaticola crassicornis* (Scop.) on ducks. *Columbicola columbae* is common on the domestic pigeon.

Suborder ANOPLURA

Key to the Families of Anoplura Known in Australia

1. Body densely clothed with thick setae which are sometimes modified into scales; abdomen never with sclerotized tergal, paratergal, or sternal plates; only on seals .... (Echinophthiriidae)
   Body less densely clad with thinner setae, arranged in rows and rarely modified into scales; never on seals 2
2. Abdominal paratergal plates absent; abdomen almost invariably membranous except for genital region .... **Linognathidae**
   Abdominal paratergal plates present on at least one segment; abdominal tergal and sternal plates frequently present 3
3. Abdominal paratergal plates with an apical margin which projects freely from the body .... **Hoplopleuridae**
   Paratergal plates without freely projecting apical margin 4
4. Abdominal cuticle finely wrinkled; dorsum of thorax with a conspicuous central notal pit (Fig. 25.2G) .... **Haematopinidae**
   Abdominal cuticle unwrinkled; dorsum of thorax without a conspicuous central notal pit .... **Pediculidae**

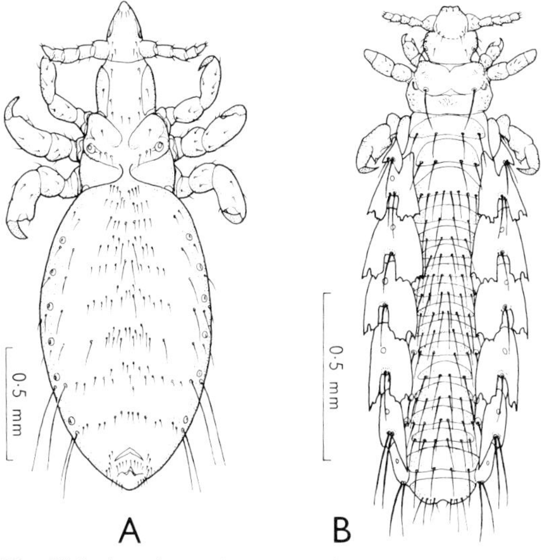

Fig. 25.5. Anoplura: A, *Linognathus vituli*, Linognathidae, ♂ from *Bos taurus*; B, *Hoplopleura calabyi*, Hoplopleuridae, ♀ from *Pseudomys higginsi*. F. Knight]

7. Pediculidae (Fig. 5.22D). This family is parasitic on primates. *Pediculus* occurs on man and other hominoids and the Neotropical monkeys. Both the human louse (*P. humanus*) and the crab louse *Pthirus pubis* (L.) are present in Australia.

8. Linognathidae (Fig. 25.5A). The genus *Linognathus* parasitizes canids and ruminants, and the species present in Australia are all parasites of domestic mammals: *L. setosus* (Olf.) on dogs, including the dingo, *L. pedalis* and *L. ovillus* (Neum.) on sheep, *L. vituli* (L.) on cattle, and *L. stenopsis* on goats. *Solenopotes capillatus* End. occurs on cattle.

9. Hoplopleuridae (Fig. 25.5B). The family contains many genera, and the chief hosts are rodents. In Australia one genus occurs on the rabbit and 2 others on rodents. The common rabbit louse is *Haemodipsus ventricosus*. The hare louse (*H. lyriocephalus* (Burm.), the type species of the genus) has not been recorded, but may occur. It differs from *H. ventricosus* in that abdominal paratergal plates are lacking; in the key it would come out as Linognathidae, where it may well belong (Ferris, 1951). Two introduced species of *Polyplax* infest rats and mice, and *P. spinulosa*, which is common on *Rattus rattus* and *R. norvegicus*, is also found on some native species of *Rattus* in settled areas. The genus *Hoplopleura* contains the only native Anoplura: *H. bidentata* (Neum.) from *Hydromys chrysogaster*, *H. calabyi* Johns. from *Pseudomys higginsi* in Tasmania, and 4 undescribed species from other native rodents.* *H. pacifica* Ewing, a common parasite on *R. rattus* and *R. exulans* in the Pacific, is common on *R. rattus* in Australia.

* These four (and a fifth not named) have been described by H.-J. Kuhn and H. W. Ludwig, *Ann. Mag. nat. Hist.* (13) **9**: 657–74, 1966.

10. Haematopinidae (Fig. 25.1c). The species in this family are best known as parasites of ungulates, and those present in Australia are important pests of domestic stock: *Haematopinus eurysternus* (= *H. quadripertusus* Fahren.) on cattle, *H. suis* (L.) on pigs, and *H. asini* (L.) on horses. *H. tuberculatus* (Burm.) is common on buffalo *(Bubalus bubalis)* in northern Australia, and is occasionally found on cattle.

Echinophthiriidae. The 4 genera in this family are parasites on seals. None appears to have been recorded in Australia, but, as a number of species of seals occur here as breeding species or vagrants, the lice certainly await collection.

ACKNOWLEDGMENT. I am greatly indebted to Dr Theresa Clay, British Museum (Natural History), London, for a critical review of the manuscript and for providing the key to the Amblycera.

26

HEMIPTERA

(Bugs, leafhoppers, etc.)

by T. E. WOODWARD, J. W. EVANS, *and* V. F. EASTOP

Exopterygote Neoptera which feed by suction, the mouth-parts consisting of hinged stylets—mandibles and maxillae—resting in a dorsally grooved rostrate labium; suction canal and salivary canal both lying between the maxillary stylets; maxillary and labial palps absent; usually with two pairs of wings, the fore wings usually completely or in part of harder consistency than the hind wings; cerci lacking. Metamorphosis usually gradual.

The Hemiptera are the dominant group of exopterygote insects. They range in length from about 1 to 90 mm, and comprise insects with a great range of different structural features and occupying a wide range of different environments. Most are phytophagous. Except for male coccoids and some aphidoids, which do not feed, the most characteristic feature is the structure of the mouth-parts.

This is an old order, probably originating in the Carboniferous (p. 170). By the Upper Permian, at least the Cicadelloidea, Cercopoidea, and Psylloidea were represented (Evans, 1964). During the Mesozoic, further extensive diversification occurred in both suborders in association with the emergence of the flowering plants. The hemipterous mouth-parts are highly efficient for extracting the liquid contents of plants, and their basic structure has remained unaltered, despite all other modifications that have occurred; it provides, in fact, the oldest as well as the most generally reliable ordinal character. It is noteworthy that the Homoptera do not include any carnivorous or aquatic groups. It seems possible that adoption of invertebrate blood sucking marked the primary divergence of Heteroptera from homopterous ancestors, and that plant-sucking habits in Heteroptera are secondary. Moreover, adaptation to aquatic life by some Heteroptera probably depended on their first having become carnivorous.

An excellent general account of the Hemiptera has been given by Poisson and Pesson (1951) and of the Australian fauna by Tillyard (1926b).

Anatomy of Adult

Head (Fig. 26.1). Opisthognathous and without a gula in Homoptera; more or less prognathous with a sclerotized gula in most, but not all, Heteroptera. Antenna usually with only a few segments, commonly 4–5 in Heteroptera, 3–10 in Homoptera; antennifers

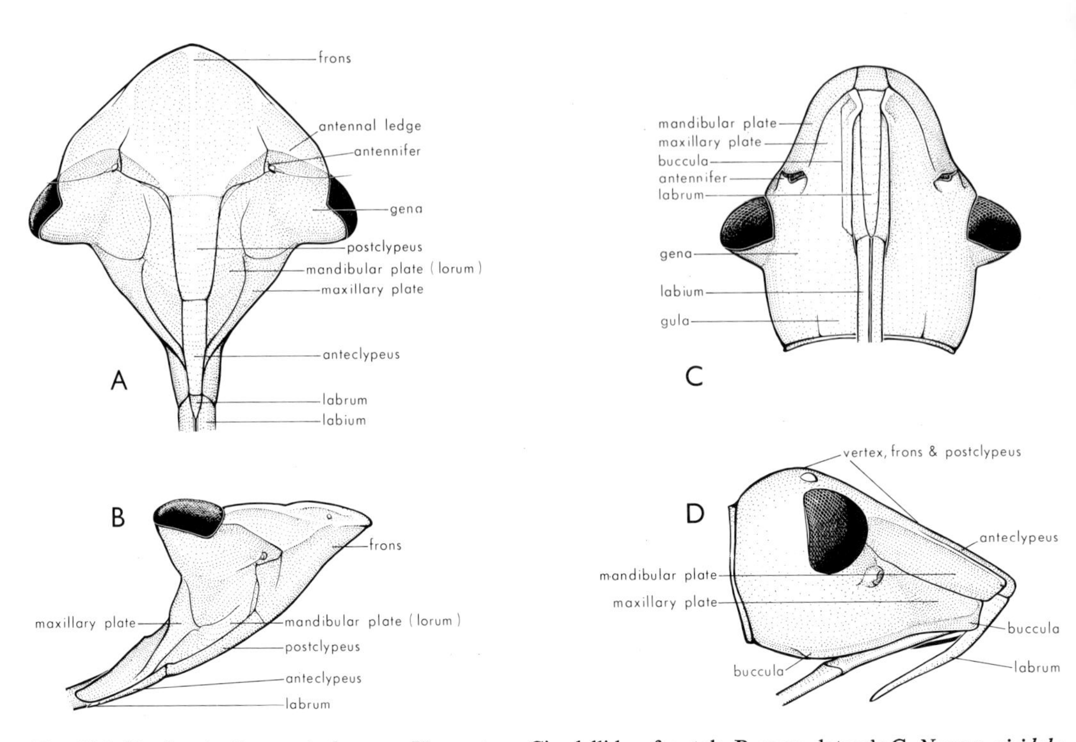

Fig. 26.1. Heads: A, *Stenocotis depressa*, Homoptera–Cicadellidae, frontal; B, same, lateral; C, *Nezara viridula*, Heteroptera–Pentatomidae, ventral; D, same, lateral. [S. Curtis]

usually well developed. The lateral compound eyes are of various sizes and shapes, occasionally absent. There may be 2 ocelli, or 3 in some Homoptera, or ocelli may be lacking.

The mouth-parts (Fig. 26.2) consist of 2 pairs of sclerotized flexible stylets, the mandibular and the maxillary stylets, lying in a morphologically dorsal groove in a 1- to 4-segmented labium. The mandibular stylets, which lie externally to the maxillary ones, are apically serrated. Longitudinal grooves on the inner surface of each maxillary stylet form two channels when the stylets are apposed. During feeding, salivary secretions are pumped down the ventral channel, and the usually liquid food (particulate in Corixidae) is sucked up the dorsal one. The mandibular stylets are internally attached, by means of transverse levers, to the 'mandibular plates' *(paraclypei)*. These plates are continuous medially with the ventral surface of the sucking pump, which lies above the hypopharynx containing the salivary syringe. The maxillary stylets are likewise internally associated with the 'maxillary plates', which lie external to the mandibular plates and are usually continuous posteriorly with the genae. In most Heteroptera there is a pair of ventral flanges, the *bucculae*, one on each side of the first segment of the resting labium. The clypeus, particularly in Homoptera, is frequently divided into two parts by a transverse fold. The labrum is small and usually narrowly triangular, sometimes apically rounded. The postclypeus, to which the dilator muscles of the sucking-pump are internally attached, may be separated from the frontal region of the head by the frontoclypeal sulcus, or the sulcus may be obsolete and the two sclerites continuous.

Thorax (Fig. 26.3). Pronotum large in Heteroptera; variously developed in Homoptera, in Membracidae extended over the abdomen and often provided with dilatations

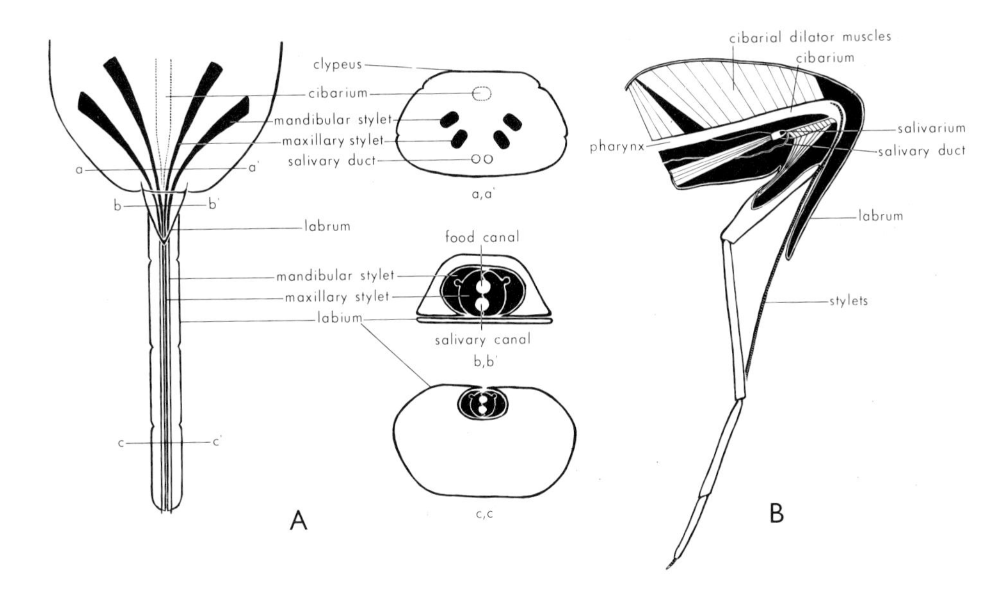

Fig. 26.2. Mouth-parts, diagrammatic: A, general relationships and transverse sections through a–a′ clypeal, b–b′ labral, and c–c′ apical labial regions; B, longitudinal section showing relationships of cibarium, pharynx, and salivarium. [S. Curtis]

and processes. Mesonotum large in both suborders, and usually with all the primary subdivisions. In Heteroptera the mesoscutellum (usually referred to simply as the 'scutellum') is large and usually triangular or subtriangular; in some Pentatomoidea it extends to the posterior end of the abdomen, largely or completely covering the wings at rest. In most Fulgoroidea the mesonotum bears a pair of tegulae.

Although the propleura are evident in peloridiids, psyllids, and most Auchenorrhyncha, and usually divided into epimeron and episternum, in most other Homoptera they are more or less fused with the pronotum or prosternum. The mesopleura are also often fused with the mesonotum, or the pleural sutures may be lost. The metapleura are relatively well developed in Auchenorrhyncha, psyllids, and aleyrodids, in association with the musculature of the hind legs, although in the Cicadelloidea they are obscured by the enlarged hind coxae. In Heteroptera the propleuron is largely effaced by the ventral growth of the pronotum; the meso- and metapleura are well developed,

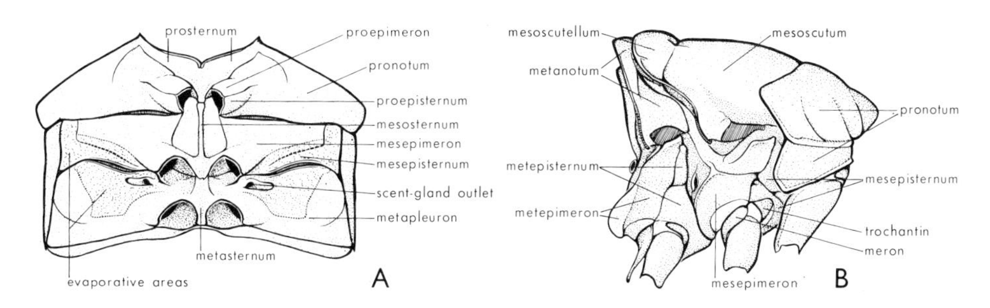

Fig. 26.3. Thorax: A, ventral aspect of *Nezara viridula*, Heteroptera–Pentatomidae; B, lateral aspect of *Cystosoma saundersi*, Homoptera–Cicadidae. [S. Curtis]

the episternum of both being large and overlapping and more or less obscuring the epimeron. The metepisterna in the adults of most families bear the openings of repugnatorial glands and are often provided with evaporative areas (p. 396). Meso- and metathoracic spiracles present in both suborders. The sterna are small.

Legs. Usually all legs cursorial and rather similar. The fore legs of some predacious Heteroptera are raptorial, e.g. emesine reduviids, gelastocorids, naucorids, belostomatids, and nepids. The fore legs of corixids are short, and the tarsus modified as a hair-fringed scoop or *pala* for particle feeding. The Gerridae, Hydrometridae, and Mesoveliidae have long, very slender legs for moving over the water surface. Many of the Hydrocorisae have at least the hind legs modified for swimming, the tibiae in particular often flattened and with fringes of long hairs. In the sedentary females of many Coccoidea the legs are vestigial or absent. Many Homoptera have the hind legs adapted for leaping.

Coxae either mobile or fixed; in Cicadelloidea the hind coxae are immobile, and form transverse plates occupying most of the ventral surface of the metathorax. Trochanters usually distinct, but sometimes (e.g. many Aradidae) fused with the femora. Maximum number of tarsomeres 3, but reduced to 2 or 1 in sternorrhynchous Homoptera and several families of Heteroptera. There are usually paired pretarsal claws, but a single claw in Coccoidea, while many Hydrocorisae lack claws on the fore or hind legs or both. Paired pretarsal processes between the claws may be present or absent. They are usually referred to as 'pulvilli' or 'pseudarolia', in the Miridae as 'arolia'; they may not be homologous structures in all groups. In Miridae the term 'pseudarolia' is applied to a more lateral pair of processes present in some subfamilies at the bases of the claws.

Wings. Differ considerably in the two suborders, but nevertheless share certain basic features in common. In the Heteroptera (Fig. 26.4), the fore wing usually, but not invariably, has 3 distinct parts. These consist, proximally, of the anterior *corium* and smaller posterior *clavus*, which are separated by the *claval suture*, and distally of the membranous portion of the wing, the *membrane*. In some groups, in particular the Miridae and Anthocoridae, there may be one, or two, additional separately defined areas. One of these, the *embolium*, consists of that part of the corium which lies anterior to R + M. The other, the *cuneus*, is the distal part of the embolium, which is separated from the remainder by the *costal fracture*, a short transverse line of weakness in the fore wings of many Heteroptera.

The fore wings of Homoptera usually have a clavus. Unlike those of the Heteroptera, they are usually of homogeneous texture, although the apical part of the wing is sometimes more membranous than the proximal part. Furthermore, the membrane, as in certain cicadas, may be separated from the basal half of the wing by a transverse division, the *nodal line*. The hind wings of some Homoptera, for instance many Cicadoidea, have a *marginal vein*, which is lacking in those of the Heteroptera. The venation of the Homoptera can be observed in its most complete condition in the fore wings of *Tettigarcta* (Fig. 26.4C). In most other Homoptera and in all Heteroptera the venation is reduced to a varying extent, although secondary vein proliferation has taken place in some insects in both suborders.

Alary polymorphism is widespread. Sometimes both sexes are dimorphic or polymorphic, as with many gerrids, delphacids, and aphids. In other groups, as in many Coccoidea, only the female is wingless. In many Aphidoidea most generations are apterous, and there is seasonal production of winged generations in response to environmental factors (p. 421). In some of these groups aptery is associated with rapid growth and reproduction and the development of winged generations with dispersal. Some groups are consistently flightless, and this condition may obtain at the specific, generic, or higher taxonomic level. The flightlessness of all Cimicidae and Polyctenidae is related to their ectoparasitic habits. Brachyptery and aptery

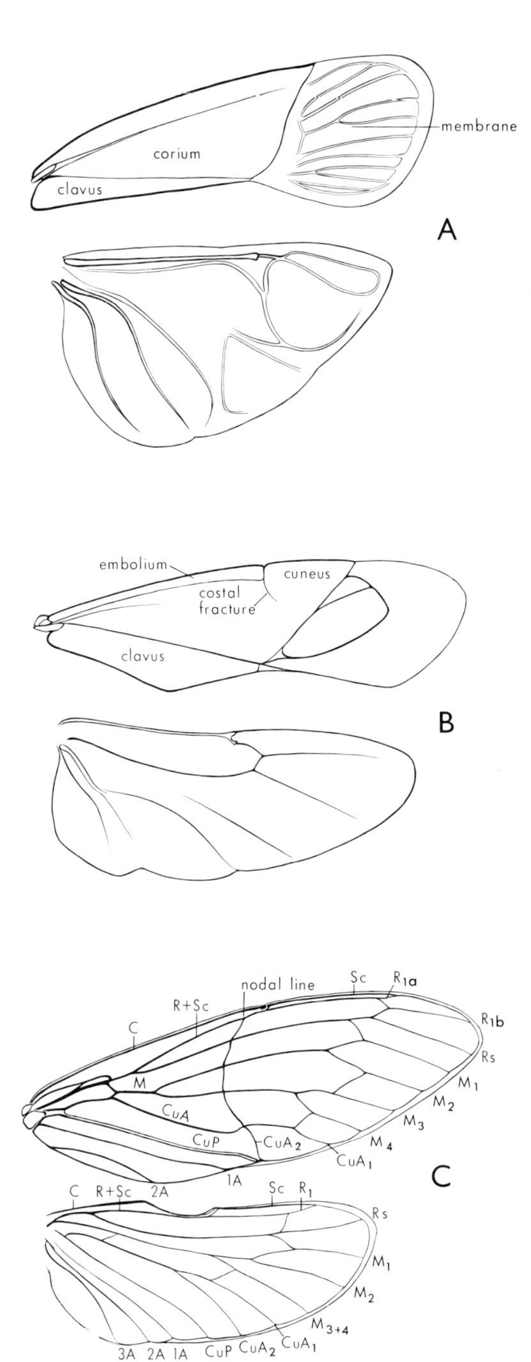

Fig. 26.4. Wings: A, *Nezara viridula*, Heteroptera–Pentatomidae; B, *Megacoelum* sp., Heteroptera–Miridae; C, *Tettigarcta tomentosa*, Homoptera–Tettigarctidae. [S. Curtis]

are common among ground- and litter-dwelling bugs, such as enicocephalids, anthocorids, and many groups of lygaeids. Aptery is also frequent in the fungus-feeding Aradidae living under bark and in other secluded situations, while the related Termitaphididae, inhabitants of termite nests, are all apterous.

Abdomen. There is considerable diversity of structure. Often segment 1 (and sometimes 2) is reduced and closely associated with the metathorax. Usually most of the pregenital segments in Heteroptera have lateral flanges (laterotergites or paratergites) attached by membrane or sutures to the main tergal plates; sometimes a narrower, more medial series of inner laterotergites is also present (e.g. Fig. 26.53A). The entire flange on each side is known as the *connexivum.* The connexiva allow greater distension of the abdomen for food intake or egg production. Segment 10 in both sexes is usually modified as the anal tube *(proctiger)*, and in some families this incorporates also what appears to be the reduced segment 11.

In the Peloridiidae, many psyllids, and most families of auchenorrhynchous Homoptera and of Heteroptera, there are 8 pairs of spiracles. Some psyllids are reported to have them reduced to 7 or even 3 pairs. Most Aphidoidea have 7 pairs of spiracles, those of segment 8 being usually absent, while one or more of the other pairs may be atrophied. In the aleyrodids only an anterior and posterior pair are open, there is no continuity between the thoracic and abdominal tracheae, and there are no transverse tracheal connectives. The nymphs and adult females of most Coccoidea have only the 2 pairs of thoracic spiracles supplying the whole body, but in some of the more primitive families there are also abdominal spiracles, e.g. on segments 2–8 in Ortheziidae and Margarodidae.

The usual positions of the abdominal spiracles in the Heteroptera are in the membrane between the metathorax and reduced segment 1 and on segments 2–8. The spiracles are usually ventrolateral. In the Lygaeidae some or all may be dorsolateral, and their arrangement is a useful taxonomic character at the subfamily and tribal levels. Aquatic

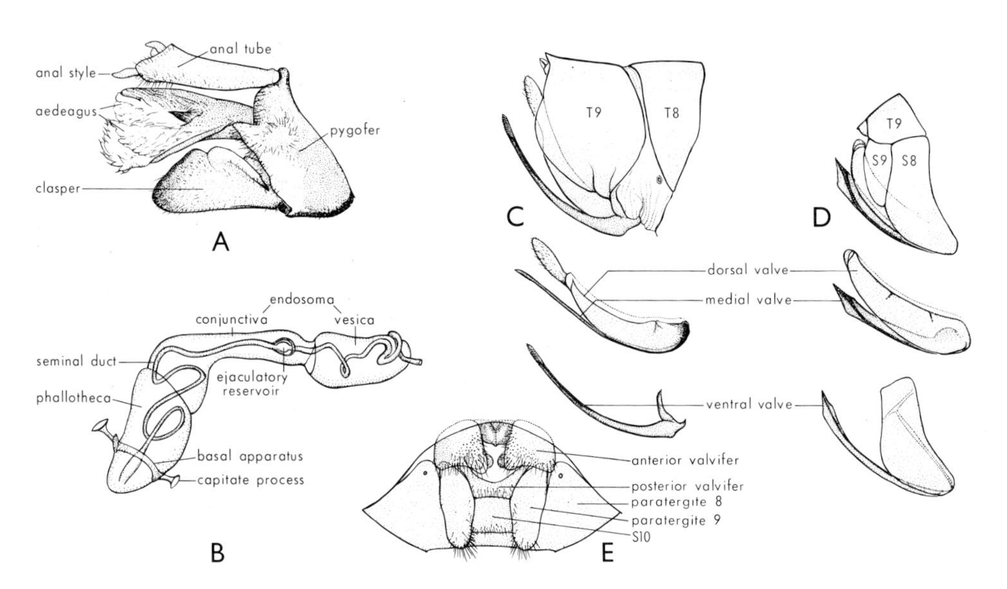

Fig. 26.5. A, ♂ terminalia of *Desudaba psittacus*, Homoptera–Fulgoridae; B, aedeagus of *Nysius clevelandensis*, Heteroptera–Lygaeidae; C, ovipositor of *Cystosoma saundersi*, Homoptera–Cicadidae, and valves separated; D, same of *Megacoelum* sp., Heteroptera–Miridae; E, flat ovipositor of *Nezara viridula*, Heteroptera–Pentatomidae. [S. Curtis]

bugs show various modifications. In nymphs of Nepidae the spiracles of segments 2–8 lie in hair-lined ventrolateral grooves functioning as air-channels. In adult nepids most respiratory exchange is through the last pair, at the base of the siphon; the spiracles of segments 2, 3, and 7 are closed, and those of 4, 5, and 6 closely associated with hydrostatic sense organs. Notonectids also have hair-lined respiratory channels; there are no spiracles on segment 8, inspiration is through those on segment 7, and expiration mainly through the thoracic spiracles. Corixids surface in such a way as to take in air through the first abdominal spiracles.

GENITALIA (Figs. 26.5A–E). Segment 9 of the male is usually modified (but not in Sternorrhyncha) as a more or less capsule-like *pygofer* (pygophor). This generally bears a well-developed aedeagus of varied and often complex structure, and a pair of claspers or 'harpagones', which are called parameres by some authors and gonostyli by others. The claspers may be asymmetrical, as in many Cimicoidea and Corixidae and some Notonectidae and Naucoridae. Claspers are absent in some, e.g. Aphidoidea, Coccoidea, most Enicocephalidae, some Thaumastocoridae.

The presence or absence and structure of the ovipositor in the female are usually related to the mode of oviposition. A well-developed, more or less laciniate ovipositor is present in most groups of Homoptera and in such Heteroptera as Miridae, Nabidae, and Notonectidae, which usually insert their eggs, or the ends of them, into plant tissues, and in others, such as most Lygaeoidea, which bury them in the soil or other substrate or in inflorescences. Three pairs of valvulae are present in the homopterous ovipositor, though the 2nd valvulae are often fused (e.g. Cicadoidea, Psyllidae, Aleyrodidae). In the Heteroptera the 3rd valvulae are usually reduced or absent. Many Heteroptera (e.g. most Pentatomoidea and Coreoidea) attach their eggs to the surfaces of plants, etc., and the ovipositor is short and flattened and the valvulae reduced. In some Cicadelloidea, although an ovipositor is developed, the eggs are laid in groups on plant tissue. In the Coccoidea and most Aphidoidea the eggs are laid superficially, and there is no valved ovipositor.

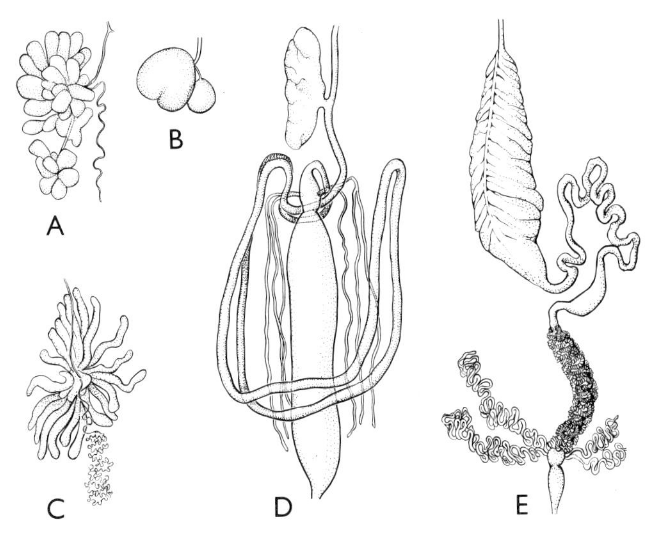

Fig. 26.6. Digestive system: A, salivary glands of *Cystosoma saundersi*, Homoptera–Cicadidae; B, same of *Macrosiphum rosae*, Homoptera–Aphididae; C, same of *Tectocoris diophthalmus*, Heteroptera–Scutelleridae; D, alimentary canal of *Dardus* sp., Homoptera–Eurybrachidae; E, same of *Tectocoris diophthalmus*. [S. Curtis]

Internal Anatomy. The sucking pump is the cibarium (Fig. 26.2B); the pharynx is small, and there is no true crop. Pharyngeal teeth occur in corixids and a few other Heteroptera. The salivarium is specialized as a salivary pump, and the anterior end of its outlet duct projects into the base of the lumen of the maxillary salivary canal. Paired salivary glands (Figs. 26.6A–C) lie in the thorax; their ducts unite anteriorly and enter the pump by a single canal. In Heteroptera each gland consists of a principal gland, usually bi- or multilobed, and an accessory gland, which may be tubular as in Pentatomomorpha, or vesicular as in Cimicomorpha and Hydrocorisae (Southwood, 1955). The condition of the salivary glands in Homoptera is more variable; there are usually 2 or 3 pairs.

The mid gut (Figs. 26.6D,E) is frequently differentiated into several regions, including an anterior crop-like dilatation and a posterior tubular region, the last in many Pentatomomorpha having caeca which commonly house symbiotic bacteria. Hemiptera that suck plant sap from the phloem have developed various mechanisms for water excretion. Among the Homoptera, the Cicadoidea, Cercopoidea, and many Cicadelloidea and Sternorrhyncha have a filter chamber (p. 63) in which the epithelium of a more anterior part of the gut is in close association with a more posterior part. Sometimes the Malpighian tubes are involved in this complex. Surplus water and relatively small molecules, such as sugars and amino acids, are thus removed directly through the hind gut as 'honey-dew'. The Fulgoroidea have no filter chamber, but surrounding the coiled mid gut is a membranous sheath which Goodchild (1966) has suggested prevents dilution of the blood. He has also given evidence for considering that the original function of the mid-gut caeca in sap-sucking Pentatomomorpha was water excretion. There are usually 4 Malpighian tubes, which are sometimes united basally in pairs; occasionally only 2 or 3 are present. Aphidoidea lack Malpighian tubes, their function in nitrogenous excretion being assumed by the fat body and mid-gut epithelium.

The ventral ganglia are always concentrated, the number of centres ranging from 3 (suboesophageal, prothoracic, and a fused mass representing the other 2 thoracic and all the abdominal ganglia) to one.

The reproductive system is shown in Figure 26.7. So far as is known, the ovarioles are acrotrophic; among Heteroptera (Pendergrast, 1957) the number is constant for each species, ranging between 2 and 8 in each ovary, with 7 the commonest, whereas the Homoptera display much greater variation (1 to over 100), and some Aphidoidea have considerable intraspecific variation. Usually the ovarioles are arranged in 'rosette' fashion at the anterior end of the lateral oviduct; occasionally there is a linear and probably more primitive arrangement. There is commonly a single spermatheca, which may be simple or complex with flanges for attachment of pumping muscles; in some groups there are 2 or 3 spermathecae, in others none. A bursa copulatrix sometimes functionally replaces the spermatheca. The number of follicles in each testis varies from 1 to 8. Accessory glands are present in both sexes.

Karyotype. The chromosomes are holocentric. The haploid number of autosomes is usually between 5 and 16, although up to 26 have been recorded in some Heteroptera and as few as 2–4 in some Coccoidea and Aphidoidea, with 4 the commonest number in aphids. The usual sex-chromosome mechanism is XY or XO in Heteroptera and XO in Homoptera, but X is multiple in some members of both suborders. Coccoidea display some peculiar sex-determining mechanisms (p. 80).

Immature Stages

Egg (Fig. 26.8). In Homoptera usually ovoid with a simple chorion. In Aleyrodidae there is a stalk-like pedicle, which attaches to the food plant and from which water is absorbed. The eggs of many Heteroptera have an operculum, particularly well developed in Cimicomorpha. At the cephalic end of heteropterous eggs are usually hollow chorionic processes of varying length and shape which, as in Cimicomorpha, may be true micropyles for sperm entry, or pseudomicropyles for gaseous exchange, or both types

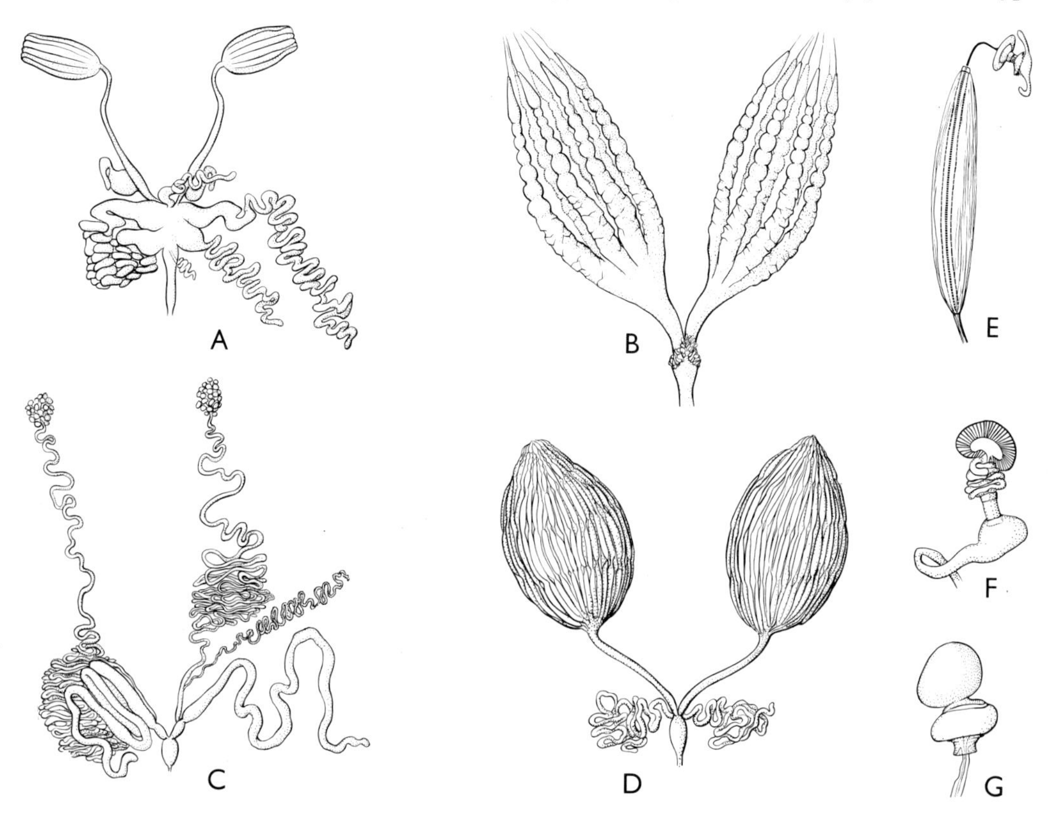

Fig. 26.7. Reproductive system: A, ♂, B, ♀ *Tectocoris diophthalmus*, Heteroptera–Scutelleridae; C, ♂, D, ♀ *Cicadetta* sp., Homoptera–Cicadidae. Spermathecae: E, *Nezara viridula*, Pentatomidae; F, *Mictis profana*, Coreidae; G, *Metacanthus pluto*, Berytidae. [S. Curtis]

together, or, as in Pentatomomorpha, processes of dual function with a central sperm canal surrounded by a spongy respiratory layer (Southwood, 1956).

Nymph (Fig. 26.9). Most Hemiptera are typically hemimetabolous, having nymphs similar to the adult in general structure and type of feeding. They usually resemble their adults closely when the latter are wingless or brachypterous. Thus, with some experience it is possible to place most nymphs at least to family level. However, they usually have fewer tarsal segments and often fewer antennal segments than the adults, and lack ocelli. Because of these and other differences and the lack of fully developed wings, they can rarely be classified by means of keys devised for adults. Keys to the nymphs of the commoner families of Heteroptera have been given by Jordan (1951) and Leston and Scudder (1956). In some groups the nymphs depart more markedly than usual from the adult habits, and show corresponding structural adaptations. The best known of these are the Cicadoidea, with subterranean, root-sucking nymphs having fossorial fore legs.

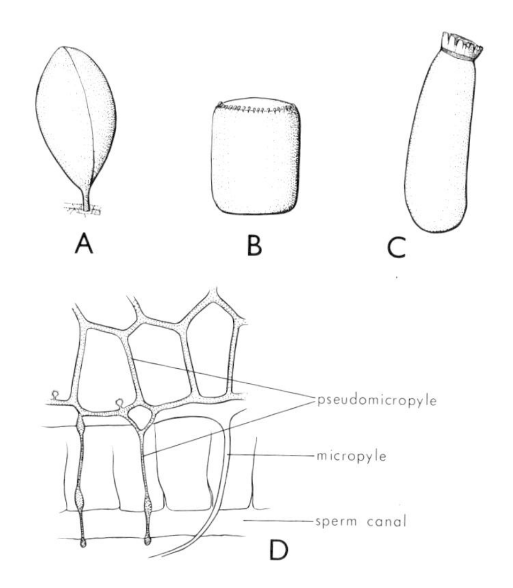

Fig. 26.8. Eggs of: A, *Cardiaspina albicollaris*, Psyllidae, height 0·35 mm; B, *Nezara viridula*, Pentatomidae, height 1 mm; C, *Pristhesancus papuensis*, Reduviidae, height 2 mm; D, surface view of base of micropylar region of *Rhinocoris*, Reduviidae (after Southwood, 1956). [S. Curtis]

Most Heteroptera have 5 nymphal instars, although some have 3, 4, or 6; the number in Homoptera is more variable, from 3 to 7. A more or less inactive instar, commonly referred to as the 'pupa', precedes the adult instar of aleyrodids and winged male coccoids; the latter also have a 'prepupal' instar.

Biology

Adults. All Homoptera and many Heteroptera are phytophagous. All succulent parts of the plant—roots, stems, leaves, flowers, and

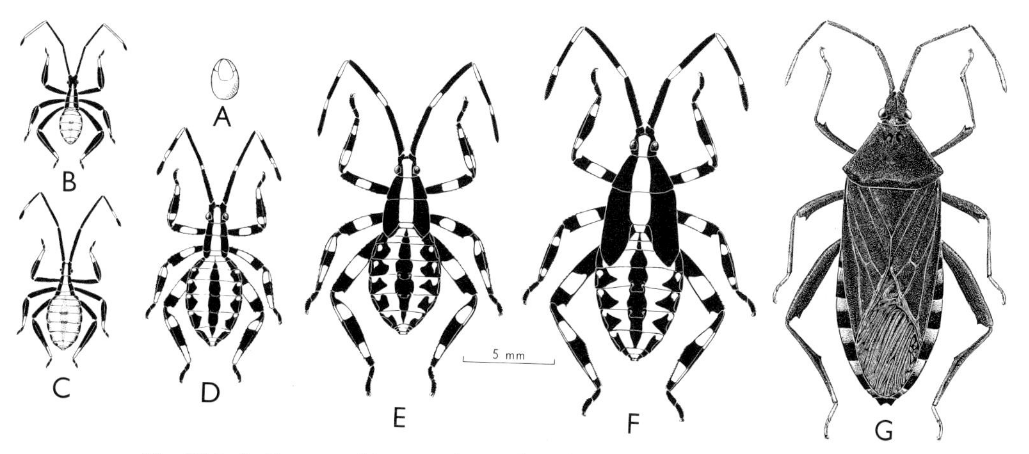

Fig. 26.9. A–G, egg and instars of *Amorbus alternatus*, Heteroptera–Coreidae. [S. Curtis]

fruit—are fed on by diverse members of the order. Most Lygaeidae appear to feed on seeds, either on the plant or, as in many Rhyparochrominae, those dropped on the ground.

There are no aquatic Homoptera. Among the Heteroptera, however, are found varying degrees of adaptation to the aquatic environment. The Ochteroidea are littoral and the Saldoidea mainly so, although some inhabit damp meadows, whereas others are intertidal and are submerged at high tide. Most of the Gerroidea are adapted for walking or skating over the surface of water, and some gerrids (Halobatinae) and veliids (Haloveliinae) are marine; a few mesoveliids are wingless inhabitants of wet moss at high altitudes. In the Notonectoidea and Corixoidea, in which all nymphal instars live under water, the adults have most of their activities there, including feeding and reproduction, coming to the surface to replenish their air supply, and usually being capable of flight for dispersal. With the exception of most Corixidae, which are particle and algal feeders, all these littoral and aquatic groups are predacious.

Predacious habits are very common among Heteroptera, and it is possible that the prognathous condition characteristic of this suborder originated in association with the development of predation. In addition to the groups mentioned above, the Enicocephalidae, Dipsocoroidea, and Reduvioidea are entirely predacious. The Cimicoidea display an interesting range of feeding habits. The Thaumastocoridae are apparently phytophagous. The Miridae include plant-sucking species, facultative predators, and obligate predators. The Nabidae and Anthocoridae are predacious on other arthropods, while one anthocorid, *Lyctocoris campestris* (F.), is a facultative blood-sucker of mammals. The Cimicidae are all obligate blood-suckers of mammals or birds, and the Polyctenidae are ectoparasites on bats. Among the Lygaeidae there are a few predacious groups (notably the Geocorinae), and in the Pentatomidae the Asopinae.

Sound production is common. In Heteroptera (Fig. 26.10) structures on various parts of the body are used for stridulation, e.g., apex of labium rubbing against a transversely striated prosternal groove (many Reduviidae), tubercles on hind femora and strigose areas on abdominal sterna (some Pentatomoidea), dorsal abdominal files and teeth on ventral surface of hind wings (Pentatomoidea and Lygaeoidea), spines of fore femora and margin of clypeus (males of some Corixidae). The males of many auchenorrhynchous Homoptera produce sound by a pair of tymbals at the base of the abdomen. These are best developed in male cicadids, in which the lateral cavity containing the tymbal is in close association with a ventral chamber containing the auditory tympanum (Fig. 2.25).

Repugnatorial glands in many families of Heteroptera produce substances repellent to predators and toxic to arthropods if they penetrate the integument. In the adult they usually open ventrally on the thorax, often by spout-like processes, and the secretion is spread and volatilized on evaporative areas of the metapleura and often also on smaller areas of the mesopleura (Fig. 26.11). These areas have a granular appearance produced

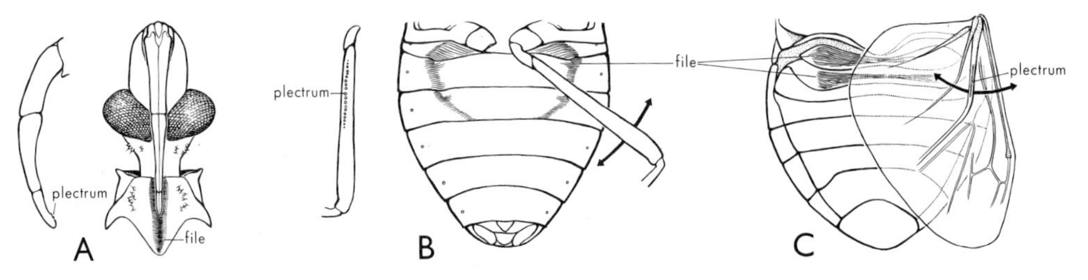

Fig. 26.10. Stridulatory mechanisms of Heteroptera: A, *Oncocephalus confusus*, Reduviidae; B, *Oncocoris punctatus*, Pentatomidae; C, *Aethus indicus*, Cydnidae. [S. Curtis]

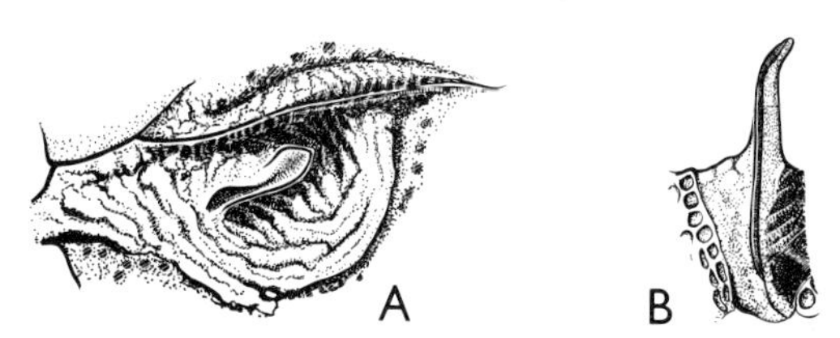

Fig. 26.11. A, evaporative area of *Dictyotus caenosus*, Pentatomidae; B, scent-gland outlet of *Metacanthus pluto*, Berytidae. [S. Curtis]

by numerous, small, more or less mushroom-shaped, cuticular outgrowths which hold the secretions and prevent them from spreading to other parts of the body. Often the secretions can be squirted at the attacker; sometimes they are wiped on to predators, such as ants, by the legs. The composition of the secretions is complex, and is characteristic of particular groups of bugs (Gilby and Waterhouse, 1965). In some reduviids the saliva is used in a similar way, being directed by 'spitting' (Edwards, 1961).

Many species with repugnatorial secretions, especially the larger ones among the Lygaeoidea and Pentatomoidea, and many Reduviidae which have painful 'bites', are aposematically coloured, most commonly in red and black (Plate 3). Groups of these species, often including members of different families, share similar colour patterns which seem to be synaposematic (Mullerian mimicry, p. 131). *Leptocorisa* spp. (Alydidae) appear to be models for pseudaposematic mimics in the beetle genus *Lygesis* (Cerambycidae). Numbers of lygaeids, especially nymphs, and mirids are ant mimics; the smaller mirids are particularly difficult to distinguish from ants in the field, mingling with them and resembling them in movements as well as in shape and colour. Procrypsis is common in both suborders: many plant-feeders are green, stem-feeders are often grey or brown, often with a disruptive, mottled or streaked coloration, and many with graminaceous hosts are linear; many ground- and bark-dwellers and seed-feeders are brown or black.

Many Homoptera are protected by copious wax secretions in the form of a powdery layer, filaments, or thin sheets. Some produce honey-dew, and the relationships between these and the ants that often attend them have been discussed by Way (1963).

Reproduction. Although the majority of Hemiptera are oviparous, there are numerous examples of ovoviviparity, as in most generations of Aphididae. Among the Coccoidea are oviparous, ovoviviparous, and viviparous species. Polyctenids are all viviparous. Hermaphroditism occurs in the Californian race of *Icerya purchasi* Mask. (p. 80). Haploid arrhenotokous parthenogenesis is found in some aleyrodids and coccoids, and cyclical parthenogenesis in aphidids.

In most Heteroptera copulation begins with the male above the female and clasping her thorax with his fore legs. Sometimes this position is maintained throughout coitus, as in most aquatic species. With most terrestrial bugs, however, after the aedeagus is inserted the male leaves the back of the female, and the pair assume an end-to-end position facing in opposite directions. The female drags the male with her when she walks. The males of many mirids do not completely leave the back of the female, but assume a dorsolateral position parallel or at an angle to her, or with head and thorax directed upwards. The prostemmine nabids copulate with their ventral surfaces apposed. Copulation is

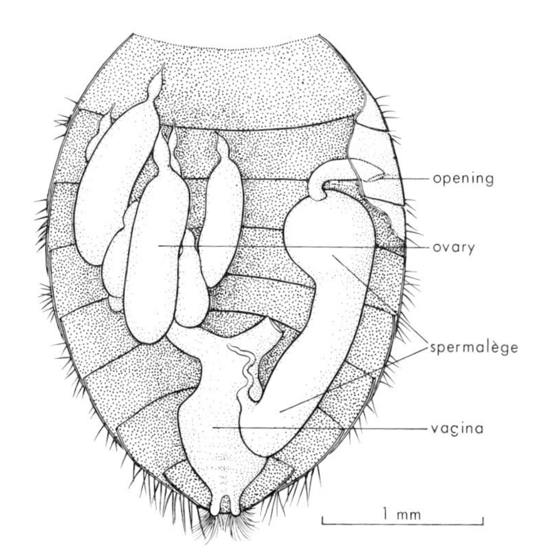

Fig. 26.12. Reproductive system of *Stricticimex brevispinosus*, Cimicidae, with right ovary removed (after Carayon, 1964).

side-by-side in most homopterous superfamilies. In fulgoroids it is end-to-end, with the male upside down.

Anomalous methods of insemination are common in Cimicoidea (Carayon, 1964). In some Nabidae normal intragenital insemination occurs. In others there is haemocoelic insemination following perforation of the vaginal wall by the aedeagus and introduction of the sperm into the haemocoele, either directly or first into a pouch, which either ruptures later, or has a cord that conducts the sperm to the proximity of the oviducal wall. In Cimicidae, Polyctenidae, and many Anthocoridae the males never copulate with the female genital tract, but either inject the sperm directly into the haemocoele after perforating the abdominal integument, usually with the clasper, or introduce it into a pouch opening to the exterior remote from the female genital aperture. Such a pouch (and that in the vagina of the nabids) is termed by Carayon a 'spermalège' (Fig. 26.12). In *Cimex*, in which this structure was first discovered, it has commonly been known as the 'organ of Ribaga' or the 'organ of Berlese'.

Life History. Eggs may be glued to the surface of food plants, usually in groups (e.g. most Pentatomoidea and Coreoidea), embedded in plant tissues or between closely adjacent parts of plants (as in most Cimicomorpha and Notonectoidea and many Lygaeoidea and Homoptera), laid in the soil or surface litter (some Lygaeoidea), or in crevices or on objects near the host (as in Cimicidae). In Corixidae and Hydrometridae the egg has an attachment disc at its posterior pole joined to the chorion by a short or long stalk. In a few Heteroptera the eggs are glued to the dorsum of the male (e.g. some Belostomatidae). A sclerotized cephalic egg-burster occurs on the embryonic cuticle of aphidids and many Heteroptera; it usually remains attached to the chorion after hatching.

Parental care is shown by several species; e.g. in the scutellerid *Tectocoris diophthalmus* (Thunb.) and the tessaratomid *Erga longitudinalis* (Westw.) the female stands over the eggs and tilts her body to cover any exposed in the direction of attack. The nymphs of many Hemiptera are gregarious in the first instar; after this they disperse in most Heteroptera. In many phloem-feeding Homoptera, particularly those attended by ants, all nymphal instars are gregarious and sometimes also the adults. Sedentary Homoptera, such as Aphidoidea and Coccoidea, commonly build up large and dense local populations. Protective secretions similar to those of the adults are usually also produced by the nymphs. The repugnatorial secretions of heteropterous nymphs are released from openings on the abdominal dorsum.

Feeding habits of the nymphs are, in general, similar to those of the adults, although with predators the size of the prey is more limited, and the nymphs of many phytophagous species have different preferences for food plant or parts of the plant from those of the adults (Puchkov, 1956). The nymphs of some groups (Cicadoidea, *Oliarus* spp. in the Cixiidae) are subterranean root suckers, whereas the adults feed on other parts of the plant. Gall-forming is commonest among the Coccoidea; some psyllids and aphidoids and a few tingids also form galls.

Natural Enemies. In addition to vertebrate predators, Hemiptera are attacked by a variety of insect and arachnid predators and insect parasitoids. Predators of hemipterous eggs include other Hemiptera (particularly the nymphs and adults of anthocorids and mirids) and predacious mites. Nymphs and adults are attacked by predacious insects, particularly Hemiptera and Coleoptera, and by arachnids such as spiders and scorpions. Aphidids are especially prone to attack by syrphid and neuropterous larvae (especially hemerobiids), and mealy bugs and soft scales by coccinellids. Caterpillars of some cosmopterigids, stathmopodids, and pyralids are predatory on coccoids. Some Gorytini (Sphecidae) provision their nests with cicadids. Because of their ubiquity in the habitats frequented by most Hemiptera, ants, although they protect many of the Homoptera that excrete honey-dew, must be important predators of the early stages of many other hemipterous groups.

There are many chalcidoid parasites, particularly Encyrtidae, Eulophidae, Elasminae, Trichogrammatidae, and Mymaridae. Dryinids are mostly ectoparasitic on auchenorrhynchous Homoptera. Braconid parasitization of both Homoptera and Heteroptera is common. Tachinids most frequently parasitize the stouter-bodied bugs, such as pentatomoids and coreids. Strepsiptera parasitize some pentatomids and many auchenorrhynchous Homoptera; some of the latter are also infested by first-instar larvae of species of *Cyclotorna* (Zygaenoidea), although these seem to do little or no harm. Mermithid and other nematodes have been recorded, the latter from aquatic Heteroptera. Fungal attack seems most common in bugs exposed to humid conditions.

Economic Significance. Both suborders include many pests, native and introduced, of crop plants, all parts of which are attacked by different species. Damage is sometimes caused by loss of sap, resulting in stunting, distortion, or wilting, especially when large populations of phloem-feeders, such as aphids, occur on young growing parts of the plant. Some Hemiptera, especially the mesophyll-feeders among Miridae and Tingidae, inject toxic saliva which causes necrosis of plant tissues; apart from the immediate damage, entry may be provided for pathogenic fungi. Seed-suckers, e.g. many Lygaeidae, have a similar effect upon seeds, breaking down and liquefying the contents preparatory to their ingestion. Leaves may be covered by sooty mould growing on honey-dew, particularly of coccoids. Many Coreoidea, Lygaeidae, and Pentatomidae suck the softer tissues of fruits. The cotton-stainers of the families Pyrrhocoridae and Lygaeidae stain and damage the fibres. The method of feeding of Hemiptera has led to many of them becoming vectors of plant viruses, particularly among the Aphididae, Cicadellidae, and Miridae. Some cicadids damage trees and shrubs, both by feeding and by slitting the bark for oviposition. Triatomine reduviids transmit trypanosomiasis in tropical America; but the only obligate blood-sucker attacking man in Australia is *Cimex lectularius* L., and it is not a proven vector of any disease. *Lyctocoris campestris* occasionally attacks man, cattle, and horses.

On the other hand, predacious Hemiptera must play a part in regulating the populations of pest arthropods. Thus, many anthocorids and mirids (e.g. *Tytthus mundulus* (Bredd.)) are predacious on eggs of delphacids, aphids, other phytophagous insects, and red spider mites (Tetranychidae). Nabids and reduviids also probably have some importance in this respect. Some phytophagous Hemiptera have been used with varying success against weeds, e.g. the tingid *Teleonemia scrupulosa* Stål against lantana and the coreid *Chelinidea tabulata* Burm. against prickly pear. A few Hemiptera or their products are sometimes used as human food in some parts of the world, e.g. the sugary manna of coccoids, corixid eggs in Mexico, pentatomids (usually roasted) in parts of New Guinea. Cochineal dye and natural shellac, though not produced commercially in Australia, are derived from coccoids.

Special Features of the Australian Fauna

The hemipterous fauna has strong relationships with the Oriental fauna in the wide sense, including that of New Guinea, Melanesia, and adjacent areas. The northern element includes: (a) forms conspecific with, or closely related to, species occurring in these areas, evidently having entered Australia during and since the Pleistocene; (b) older groups with less close relationships to the present Oriental fauna. Both are most strongly represented in the Torresian province, and extend more diffusely into other parts of Australia. Some groups, apparently now centred in Melanesia, extend into Australia, particularly the tropical and subtropical north and north-east. A typical example is the lygaeid tribe Targaremini, which, in isolation, has become the dominant tribe of lygaeids in New Zealand.

The Bassian element is best exemplified by the moss-feeding Peloridiidae, with a discontinuous circum-Antarctic distribution, and in Australia not found north of the N.S.W.–Queensland border ranges. The period of

Tertiary isolation also resulted in the development of several endemic or nearly endemic groups associated with groups of plants, such as *Eucalyptus*, which radiated at this time. The eurymelids are a notable example, and Evans (in Leeper, 1962) also noted several subfamilies and tribes of Cicadelloidea.

An enigmatic distribution (p. 194) is shown by the flightless Orgeriinae (Dictyopharidae), which have a wide global distribution in arid and semi-arid regions, but are not recorded from the wet tropics. In Australia they are represented by a single known species recorded only from the summits of several remanent volcanic plugs north of Brisbane, where the ecological conditions are harsh. It is thus geographically remote from any overseas centre of distribution, its nearest relatives probably being in South Africa.

CLASSIFICATION

Order HEMIPTERA (3,661 Australian spp.)

Suborder HOMOPTERA (2,084)

PELORIDIOIDEA (8)
1. Peloridiidae (8)

FULGOROIDEA (405)
2. Cixiidae (42)
   Tettigometridae (0)
3. Delphacidae (53)
4. Eurybrachyidae (44)
5. Fulgoridae (18)
6. Achilidae (23)
7. Tropiduchidae (7)
8. Issidae (29)
9. Lophopidae (3)
   Acanaloniidae (0)
10. Ricaniidae (29)
11. Flatidae (82)
12. Derbidae (49)
    Achilixiidae (0)
13. Meenoplidae (8)
    Kinnaridae (0)
14. Dictyopharidae (10)
    Gengidae (0)
15. Nogodinidae (8)
    Hypochthonellidae (0)

CERCOPOIDEA (30)
16. Cercopidae (9)
17. Aphrophoridae (12)
18. Machaerotidae (9)

CICADOIDEA (197)
19. Tettigarctidae (2)
20. Cicadidae (195)

CICADELLOIDEA (639)
    Aetalionidae (0)
21. Cicadellidae (500)
    Hylicidae (0)
22. Eurymelidae (84)
23. Membracidae (55)

PSYLLOIDEA (144)
24. Psyllidae (144)

APHIDOIDEA (118)
25. Aphididae (102)
26. Pemphigidae (12)
27. Adelgidae (1)
28. Phylloxeridae (3)

ALEYRODOIDEA (24)
29. Aleyrodidae (24)

COCCOIDEA (519)
30. Margarodidae (25)
31. Ortheziidae (2)
32. Eriococcidae (157)
33. Aclerdidae (1)
    Stictococcidae (0)
34. Asterolecaniidae (28)
    Phenacoleachiidae (0)
35. Pseudococcidae (38)
36. Coccidae (65)
37. Lacciferidae (6)
38. Halimococcidae (1)
    Conchaspididae (0)
39. Diaspididae (196)

Suborder HETEROPTERA (1,577)

ENICOCEPHALOIDEA (6)
40. Enicocephalidae (6)

DIPSOCOROIDEA (5)
41. Dipsocoridae (1)
42. Schizopteridae (4)

CIMICOIDEA (148)
43. Cimicidae (1)
44. Polyctenidae (2)
45. Anthocoridae (24)
    Microphysidae (0)
46. Nabidae (11)
47. Thaumastocoridae (10)
    Joppeicidae (0)
48. Miridae (100)

TINGOIDEA (139)
49. Tingidae (139)

REDUVIOIDEA (244)
50. Reduviidae (244)

SALDOIDEA (11)
51. Saldidae (10)
52. Leptopodidae (1)
    Leotichiidae (0)

ARADOIDEA (129)
53. Aradidae (128)
54. Termitaphididae (1)

COREOIDEA (73)
55. Coreidae (50)
56. Alydidae (14)
57. Rhopalidae (5)
58. Hyocephalidae (4)
    *Stenocephalidae (0)

LYGAEOIDEA (234)
59. Lygaeidae (205)
60. Piesmidae (4)
61. Colobathristidae (1)
62. Berytidae (7)
63. Largidae (5)
64. Pyrrhocoridae (12)

* J. A. Grant (personal communication) has reported the occurrence of 2 species in Australia. The family has been discussed by G. G. E. Scudder (*Proc. R. ent. Soc. Lond.* (A) **32:** 147–58, 1957) and C. W. Schaefer (*Ann. ent. Soc. Am.* **57:** 670–84, 1964).

PENTATOMOIDEA (459)
65. Tessaratomidae (16)
66. Dinidoridae (1)
67. Scutelleridae (38)
68. Plataspidae (16)
69. Aphylidae (2)
70. Lestoniidae (1)
71. Urostylidae (1)
Phloeidae (0)
72. Cydnidae (36)
73. Acanthosomatidae (54)
74. Pentatomidae (294)
GERROIDEA (38)
75. Gerridae (15)
76. Veliidae (14)
77. Hydrometridae (5)
78. Hebridae (2)
79. Mesoveliidae (2)
OCHTEROIDEA (22)
80. Ochteridae (2)
81. Gelastocoridae (20)
NOTONECTOIDEA (45)
82. Notonectidae (30)
83. Pleidae (2)
Helotrephidae (0)
84. Nepidae (5)
85. Naucoridae (4)
86. Belostomatidae (4)
CORIXOIDEA (24)
87. Corixidae (24)

KEY TO THE SUBORDERS OF HEMIPTERA—ADULTS AND NYMPHS

Fore wings, when developed, in form of tegmina of more or less uniform texture, without sharp differentiation into corium and membrane, and usually held roofwise over abdomen; insertion of labium close to prosternum, without an intervening sclerotized gula (Fig. 26.1A); never truly aquatic .... HOMOPTERA (p. 401)

Fore wings, when fully developed, usually in form of hemelytra with basal thickened corium and apical membrane, usually folding flat, or nearly so, on abdomen and with apices widely overlapping; usually (almost always in adults in which fore wings are absent or not differentiated into corium and membrane) insertion of labium remote from prosternum, with a sclerotized gular region intervening (Fig. 26.1C); includes several aquatic groups .... HETEROPTERA (p. 431)

Some Heteroptera (Corixidae, Notonectidae, Pleidae) have a homopteran (opisthognathous) type of head without a well-developed gula; others have fore wings of essentially homogeneous texture (Colobathristidae, Dipsocoroidea, Enicocephalidae, Tingidae, Gerridae, Veliidae, some Reduviidae). But almost all of the former group have hemelytra as adults, the nymphs have a general structure similar to that of the adults, and all are aquatic. The second group has the heteropteran type of head with a gula. It follows that the possession of only one of the three heteropteran characteristics is sufficient to exclude a hemipteron from the Homoptera.

Suborder HOMOPTERA

The Homoptera are customarily separated into 3 divisions: Coleorrhyncha, containing only the Peloridioidea; Sternorrhyncha, comprising the Psylloidea, Aleyrodoidea, Aphidoidea, and Coccoidea; and Auchenorrhyncha, comprising the Fulgoroidea, Cercopoidea, Cicadelloidea, and Cicadoidea. The Membracidae, often regarded as equivalent in status to the superfamilies listed above, are here included in the Cicadelloidea. While this arrangement expresses natural relationships, in so far as it indicates that there have been three separate lines of descent from an original protohomopteran stem, it is not satisfactory in other respects. This is because it tends to over-emphasize certain affinities at the expense of others. For example, the Peloridioidea, Psylloidea, and Fulgoroidea are all early derivations from the Protohomoptera, and are, in some respects, more closely interrelated than the Coccoidea are related to the Psylloidea, or the Cicadelloidea to the Fulgoroidea. The phylogeny of the Homoptera has been discussed by Evans (1963).

Key to the Superfamilies of Homoptera

1. Tarsi with 1 or 2 segments, occasionally absent 2
   Tarsi with 3 segments (Auchenorrhyncha) 6
2. Fore wings with many closed cells formed by raised veins (Fig.26. 13A); with well-developed lateral pronotal expansions (Coleorrhyncha) .... PELORIDIOIDEA (p. 402)
   Fore wings, when present, without or with few closed cells, wing veins not much raised above surface of wing lamina; usually without conspicuous lateral pronotal expansions (Sternorrhyncha) 3

3. Adults always with well-developed legs; tarsi of 2 segments of about equal length; antennae 7- to 10-segmented; ♂♂ and ♀♀ always alate, often present in about equal proportions, and generally similar to one another, although the sexes may differ somewhat in size and pigmentation. [Mostly living on leaves or twigs of trees and shrubs] 4
   Legs sometimes absent or much reduced; tarsi, when present, either of 1 segment, or 2-segmented with basal segment small and usually triangular; antennae 1- to 13-segmented; adults often apterous; ♀♀ often much more common than ♂♂; adult ♂♂ may be devoid of wings and/or mouth-parts. [Many species on trees and shrubs, but also many others on herbaceous plants, often concealed] 5
4. Fore wings (Fig. 26.14) of rather harder consistency than hind wings and with evident veins, both M and Cu usually forked, clavus present; antennae usually 10-segmented . PSYLLOIDEA (p. 418)
   Fore wings membranous (Fig. 26.15) with obscure reduced venation, without clavus; antennae 7-segmented. [Body wax-dusted] .... ALEYRODOIDEA (p. 424)
5. Tarsi, when present, usually 1-segmented and bearing only a single claw; antennae 1- to 13-segmented; ♀♀ always neoteinic, apterous; ♂♂ winged or apterous, when winged dipterous. [Scale insects and mealy bugs] .... COCCOIDEA (p. 425)
   Tarsi usually present and 2-segmented, sometimes only 1-segmented, rarely atrophied; claws paired; antennae 1- to 6- but most commonly 5- to 6-segmented, last segment often with an elongate processus terminalis; ♂♂ and ♀♀ either winged or apterous; when winged, hind wing small, usually with 2 oblique veins, venation characteristic (Fig. 26.16). [Abdomen often with a pair of siphunculi varying from barely discernible pores to black cylinders nearly as long as body] .... APHIDOIDEA (p. 420)
6. Tegula almost always present on mesothorax; mid coxae elongate, widely separated; pedicel of antenna enlarged, often bulbous, with numerous wart-like sensilla (Figs. 26.17A, 18) .... FULGOROIDEA (p. 403)
   Tegula lacking; mid coxae short, not widely separated; pedicel of antenna not or scarcely thicker than scape, without wart-like sensilla 7
7. Three ocelli on crown; fore femora thickened .... CICADOIDEA (p. 410)
   Ocelli 2 and variable in position, or absent; fore femora not thickened 8
8. Hind tibiae short, cylindrical, with 1 or 2 strong spines (Fig. 26.28B); hind coxae short, conical .... CERCOPOIDEA (p. 409)
   Hind tibiae long, angular, with a few, or, more usually, numerous spines, often in conspicuous rows; hind coxae transverse, plate-like (Fig. 26.31F) .... CICADELLOIDEA (p. 413)

Superfamily PELORIDIIOIDEA

1. Peloridiidae (Fig. 26.13). A southern-hemisphere group of small, flattened, cryptically coloured Homoptera which live in saturated moss and liverworts. At present 20 species, in 10 genera, are known from the world, of which 7 species in 4 genera have been recorded from Australia. The remainder occur in New Zealand and cool-temperate South America. The Australian species are *Hemiodoecus leai* China (Tasmania, Victoria, N.S.W.), *H. wilsoni* Evans (Victoria), *Hackeria veitchi* (Hacker) (south Queensland and northern N.S.W.), *Hemiodoecellus fidelis* (Evans) (Tasmania), *H. donnae* (Woodw.) (Victoria), and 3 species of *Howeria* on Lord Howe I.

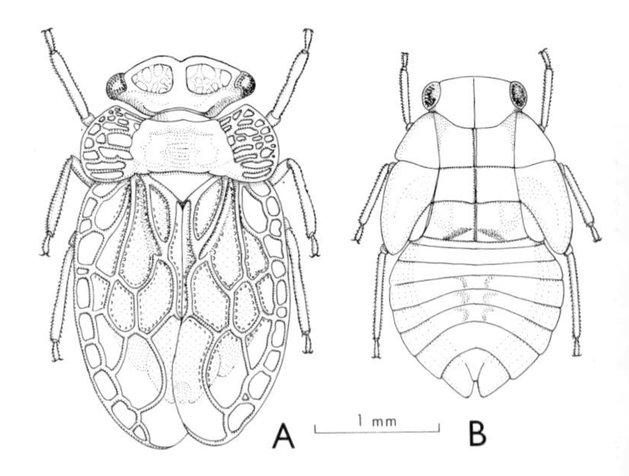

Fig. 26.13. *Hemiodoecellus fidelis*, Peloridiidae: A, adult; B, last-instar nymph. [F. Evans]

Adults are approximately 3 mm long and greenish brown in colour. Head with 2 large oval cells, or areolae, and prominent lateral eyes. Pronotum large, with extensive lateral paranota, which may have a few large cells or numerous smaller ones. Fore wings lying horizontally over abdomen and likewise with a considerable number of cells, the veins raised in relief; hind wings absent in known Australian peloridiids; ovipositor with 3 pairs of valves; gut without a filter chamber. The nymphs are flattened, yellow, oval in outline, and remarkable in later instars for having both dorsal and ventral sutures separating the developing wing pads from the rest of the thorax. These sutures extend on to the prothorax, and their presence and the nature of the paranotal tracheation suggest that the pronotal paranota of the adults are serially homologous with wings; similar pronotal expansions were present in certain Palaeozoic and Mesozoic Homoptera. [China, 1962; Evans, 1939; Pendergrast, 1962.]

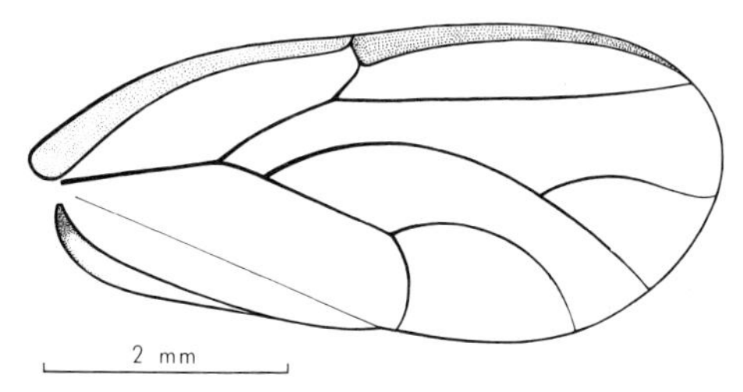

Fig. 26.14. Wing venation of *Psylla acaciaedecurrentes*, Psyllidae. [T. Nolan]

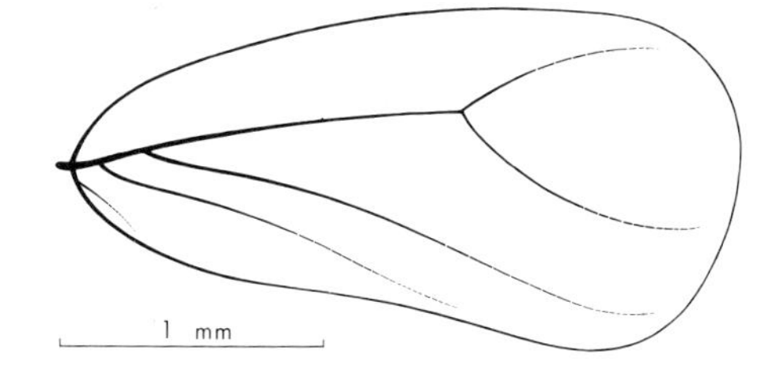

Fig. 26.15. Wing venation of *Synaleurodicus* sp., Aleyrodidae. [T. Nolan]

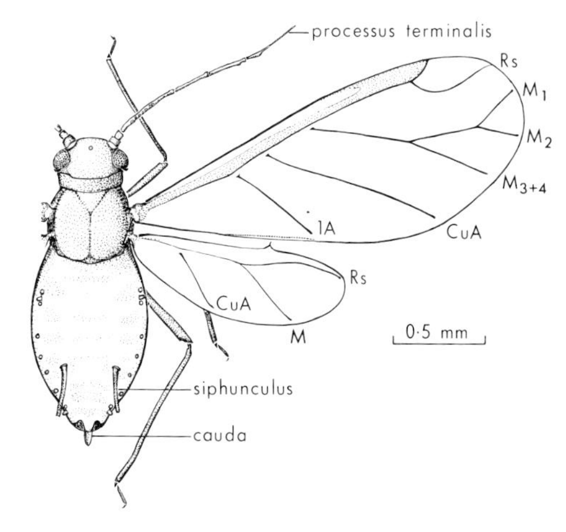

Fig. 26.16. *Aphis acaenovinae*, Aphididae, viviparous ♀ from *Acaena ovina*. [S. Curtis]

Superfamily FULGOROIDEA

Very diverse plant-hoppers, mostly phloem-feeders on flowering plants, but some fungus-feeders. Head sometimes narrowly produced anteriorly; clypeus not extending posteriorly to between the eyes in adults, and separated from the frontal region by a ridge marking the position of the former fronto-clypeal sulcus; antennae varied in shape, usually with an enlarged pedicel bearing conspicuous sensory organs; ocelli 2, occasionally 3, the lateral ocelli on the sides of the face adjacent to eyes and antennae; tegulae present on mesothorax (except in orgeriine dictyopharids); anal veins of tegmen apically confluent; hind wing without a marginal vein. The nymphs have an abundance of sensory pits, and some secrete wax which exudes in the form of long abdominal filaments. Similar filaments occur also in some adults. Wade (1960) has given a species index to the Fulgoroidea; more than 9,200 have now been described.

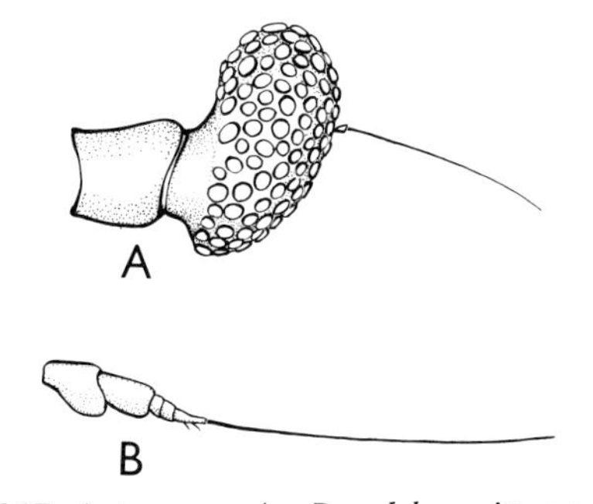

Fig. 26.17. Antennae: A, *Desudaba psittacus*, Fulgoridae; B, *Stenocotis depressa*, Cicadellidae. [S. Curtis]

Key to the Families of Fulgoroidea Known in Australia

1. Hind tibiae with a large movable spur at apex (Figs. 26.18A,B) .... **Delphacidae**
   Hind tibiae without a large movable spur 2
2(1). Apex of 2nd segment of hind tarsi truncate or emarginate, with a ventral row of 3 or more small dark spines 3
   2nd segment of hind tarsi very small, apex rounded or pointed, without spines, or with only 1 at each side 8
3(2). One or both claval (anal) veins granulate; apical segment of rostrum much longer than wide .... **Meenoplidae**
   Claval veins not granulate, or, if so, granules small, similar to those on other veins; or apical segment of rostrum short, about as wide as long 4
4(3). Anal area of hind wings reticulate. [Clypeus with lateral carinae; head often prolonged] .... **Fulgoridae**
   Anal area of hind wings not reticulate 5
5(4). Apical segment of rostrum short, almost as wide as long .... **Derbidae**
   Apical segment of rostrum distinctly longer than wide 6
6(5). Claval vein entering apex of clavus. [Horizontally flattened forms with membranes overlapping] .... **Achilidae**
   Claval veins not reaching apex of clavus 7
7(6). Head prolonged in front, or, if not, frons with 2 or 3 carinae (in addition to lateral carinae), or the tegulae absent and claval suture obscure; no median ocellus .... **Dictyopharidae**
   Head usually not prolonged; if prolonged, frons with only a median carina or none (excluding lateral margins); tegulae and claval suture always distinct; median ocellus often present .... **Cixiidae**
8(2). Apex of 2nd segment of hind tarsi with a spine at each side. [Claval vein nearly always ending in apex of clavus] 9
   Apex of 2nd segment of hind tarsi without spines 12
9(8). Posterior angle of mesonotum restricted off by a groove or fine line and at same level as main anterior division .... **Tropiduchidae**
   Posterior angle not restricted off, or, if so, by a broad trough and at different level from main anterior division 10
10(9). With a cross-veined precostal area, but without granules on clavus; clypeus nearly always with lateral (marginal) carinae, if without them, base of costa not strongly bent. [Wings mostly transparent] .... **Nogodinidae**
   Without cross-veined area, or, if so, either clavus granulate, or clypeus without lateral carinae and base of costa strongly bent 11
11(10). With a cross-veined precostal area, and either the clavus granulate, or base of costa strongly bent. [Green or brown] .... **Flatidae**
   Clavus not granulate and base of costa gradually curved, not strongly bent .... **Issidae**
12(8). Tegmina very steeply tectiform, with apical margin broad, as long as anal margin; clavus always long, reaching almost to anal angle. [With cross-veined precostal area] ... **Ricaniidae**
   Tegmina less steeply tectiform, with apical margin narrower, much shorter than anal margin; clavus sometimes short and ending far from anal angle 13
13(12). Frons wider than long, with sides angular, but with no other carinae; clypeus without lateral carinae .... **Eurybrachyidae**
   Frons longer than wide, with carinae in addition to angulate margins; clypeus with lateral carinae .... **Lophopidae**

2. Cixiidae (Figs. 26.18J, 19). These, regarded as among the most primitive fulgoroids, are the only ones in which all 3 ocelli are sometimes found. There are over 1,000 species, widely distributed throughout the world. Of the 15 genera in Australia, *Oliarus* and *Gonyphlepsia* contain more than half the species. Cixiid nymphs are primarily root-feeders. Hacker (1925) described the life history of *Oliarus felis* Kirk., the nymphs of which live underground on grass and may be submerged by brackish water at high tide.

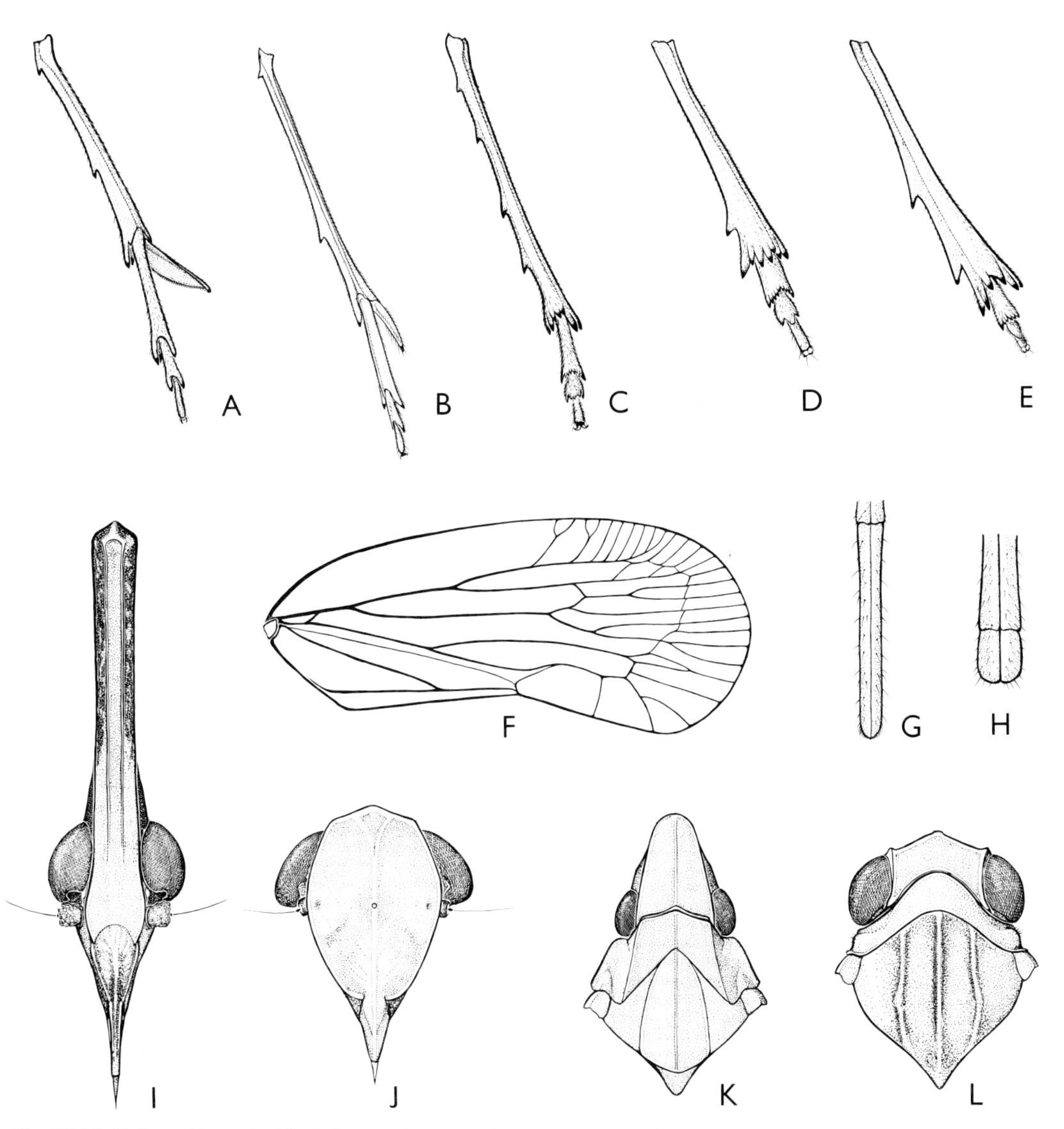

Fig. 26.18. Fulgoroidea: A, hind tibia and tarsus of *Perkinsiella saccharicida*, Delphacidae; B, same of *Ugyops* sp., Delphacidae; C, same, ventral, of *Desudaba psittacus*, Fulgoridae; D, same of *Massila sicca*, Flatidae; E, same of *Scolypopa australis*, Ricaniidae; F, tegmen of *Achilus flammeus*, Achilidae; G, apex of rostrum of *Desudaba psittacus*, Fulgoridae; H, same of *Zoraida* sp., Derbidae; I, facial view of head of *Thanatodictya praeferrata*, Dictyopharidae; J, same of *Oliarus lubra*, Cixiidae; K, mesonotum of *Kallitambinia australis*, Tropiduchidae; L, same of *Salona panorpaepennis*, Nogodinidae. [S. Curtis]

Nymphs of other species have been found in ants' nests. [Fennah, 1956.]

3. Delphacidae (Araeopidae; Figs. 26.18A, B, 20). Mostly small, from less than 2 to 9 mm in length to tip of fore wings, with a large mobile spur at end of hind tibiae. Several species are vectors of virus diseases of cultivated Gramineae. *Perkinsiella saccharicida* Kirk., a native of north Queensland, is a pest of sugar cane which was introduced into Hawaii. It is a vector of Fiji disease. *Peregrinus maidis* Ashm. is known as a

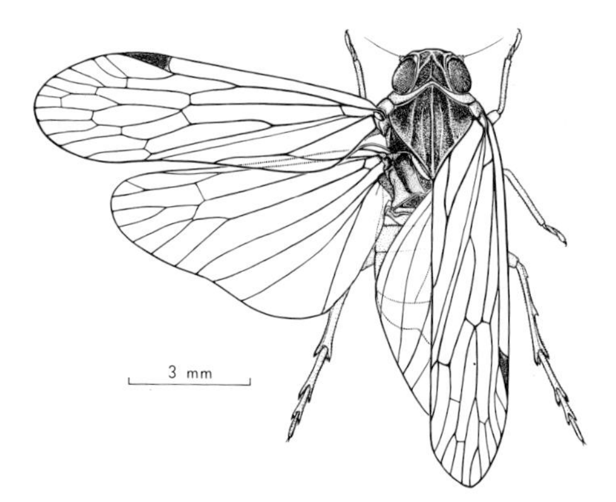

Fig. 26.19. *Oliarus lubra*, Cixiidae. [S. Curtis]

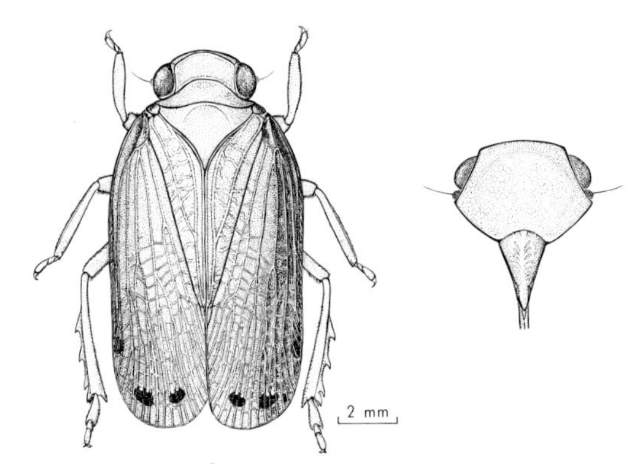

Fig. 26.21. *Platybrachys maculipennis*, Eurybrachyidae, and facial view of head. [S. Curtis]

vector of maize stripe virus in America, and *Delphacodes striatella* Fall. is recorded as a vector of rice dwarf virus in Japan. More than 1,300 species occur in the world. [Fennah, 1956, 1965.]

4. Eurybrachyidae (Plate 3, D, E; Figs. 26.6D, 21). Mostly broad, squat insects of fairly large size (wing-span 16–60 mm, length to tip of wings 7–35 mm) and mottled brown and red, yellow, or orange coloration; some are black. Some (e.g. *Platybrachys* spp.) superficially resemble fulgorids, but are easily distinguished by the very broad frons. They are predominantly Australian, Oriental, and Ethiopian, with over 170 species. The females of some Australian species oviposit on the trunks of *Eucalyptus*, and the nymphs move further up the tree to feed. Hacker (1924) described the life history of *Platybrachys leucostigma* (Walk.).

5. Fulgoridae (lantern flies; Plate 3,A; Figs. 26.5A, 17A, 18C,G, 22). Medium-sized to large species, mostly tropical, most Australian species with a wing-span of 25–40 mm. Clavus open distally, running full length of wing, with numerous cross-veins between anal veins. *Desudaba maculata* Dist. has dark brown tegmina with red spots, a span of about 38 mm, and bases of hind wings brilliant crimson. *D. psittacus* Walk. has a green abdomen and no red spot on the tegmina. *Eurinopsyche* includes several peculiar brownish species with a process projecting in front of the head like an elongate snout.

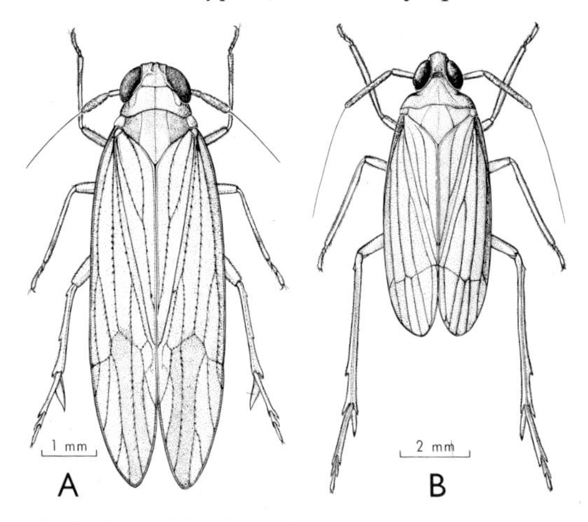

Fig. 26.20. Delphacidae: A, *Perkinsiella saccharicida*; B, *Ugyops* sp. [S. Curtis]

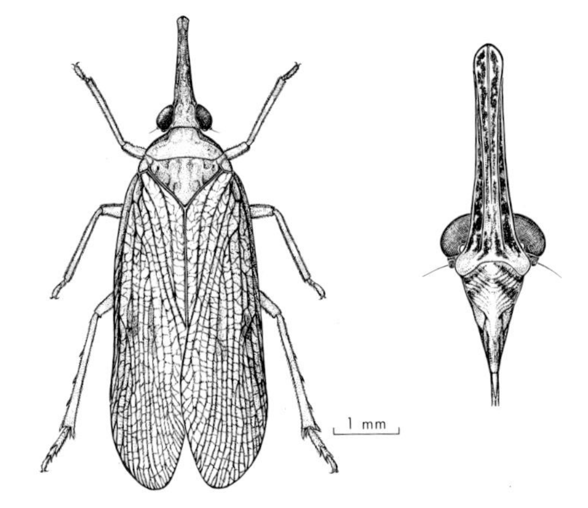

Fig. 26.22. *Eurinopsyche arborea*, Fulgoridae, and facial view of head. [S. Curtis]

Many species inhabit Queensland, but *Eurystheus dilatatus* Westw. occurs around Perth in W.A. The world fauna is about 650 species.

6. Achilidae (Plate 3,B; Fig. 26.18F). Widely distributed in all parts of the world except the Arctic and Antarctic, with about 350 described species. Peculiar in having the tegmina folded flat and partially overlapping at rest. Clavus rather short, ending a little beyond half-way along the tegmen, and making a distinct angle with it; wing-span 7–26 mm, length to tip of wings 4–14 mm. Most Australian species are brown to black; some are green. The nymphs of the bright red *Achilus flammeus* Kirby live under bark or in cavities in dead wood. [Fennah, 1950.]

7. Tropiduchidae (Figs. 26.18K, 23). Mostly tropical and subtropical, with largely clear wings of 10–33 mm span. The world total is about 330 species. The beautiful *Ossa venusta* Kirk. and *O. formosa* Kirk. occur in Queensland.

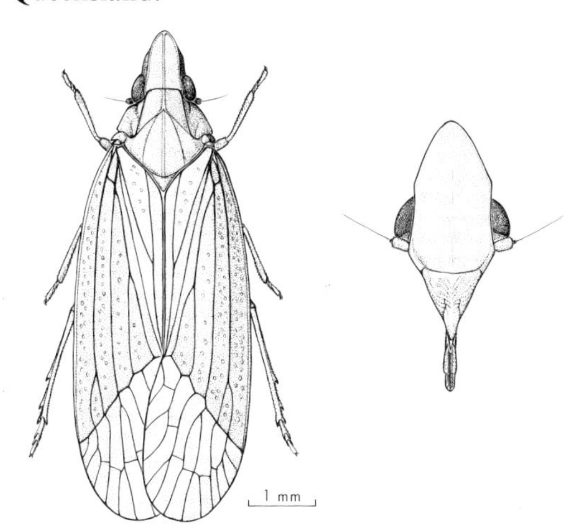

Fig. 26.23. *Kallitambinia australis*, Tropiduchidae, and facial view of head. [S. Curtis]

8. Issidae (Fig. 26.24). The species are mostly dull coloured, with peculiar tegmina, often shortened or convex and of roughened consistency. Over 1,000 species are known. [Fennah, 1954.]

9. Lophopidae (Fig. 26.25). Medium-sized, with a wing-span of 12–28 mm and length to tip of wings 7–14 mm. The world fauna comprises over 120 species, confined to the tropics and warm temperate regions. *Lophops saccharicida* Kirk. has yellowish brown tegmina with darker stripes, and is found on sugar cane and grasses in Queensland.

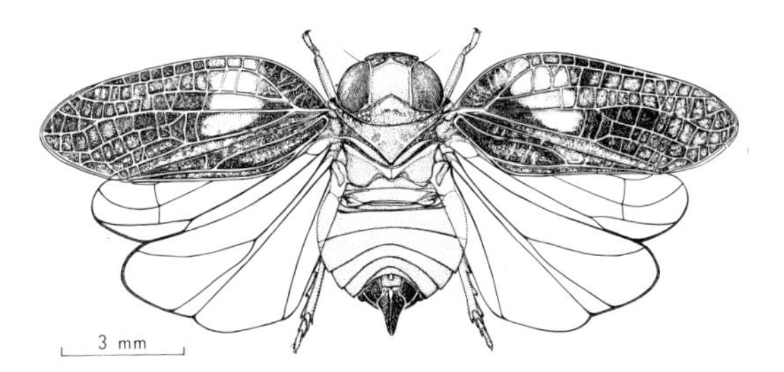

Fig. 26.24. *Chlamydopteryx vulturnis*, Issidae. [S. Curtis]

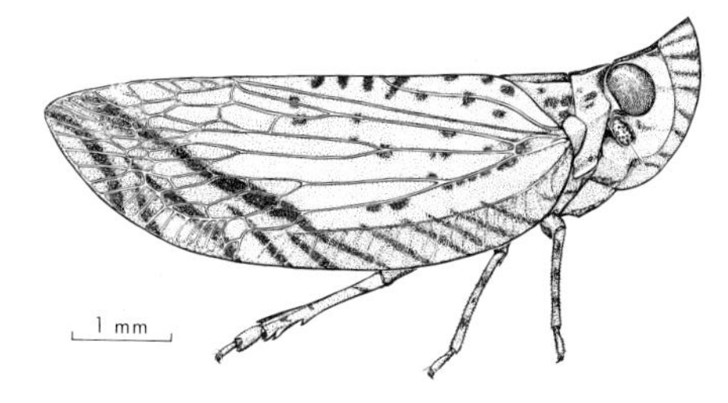

Fig. 26.25. *Lophops saccharicida*, Lophopidae. [S. Curtis]

10. Ricaniidae (Figs. 26.18E, 26A). Medium-sized to rather large; wing-span 12–40 mm; length to tip of wings 7–17 mm. There are over 360 species, mostly from the warmer parts of the Old World. *Scolypopa australis* Walk. sometimes does considerable damage to passion vines, and also attacks other plants. Nymphs and adults sit in rows along the stems. The tegmina are hyaline and heavily barred with black along the costal and distal margins.

11. Flatidae (Plate 3,C; Figs. 26.18D, 26B). Regarded as the most specialized of all fulgoroids; they occur mainly in the tropics and subtropics; more than 1,000 species are known. Their broad triangular tegmina are usually coloured and opaque, and folded to form a steep roof over the body. Myers (1922) gave an account of the bright green *Siphanta acuta* Walk. which feeds on the leaves of *Eucalyptus*. Another common species is the bluish grey *Sephena cinerea* Kirk.

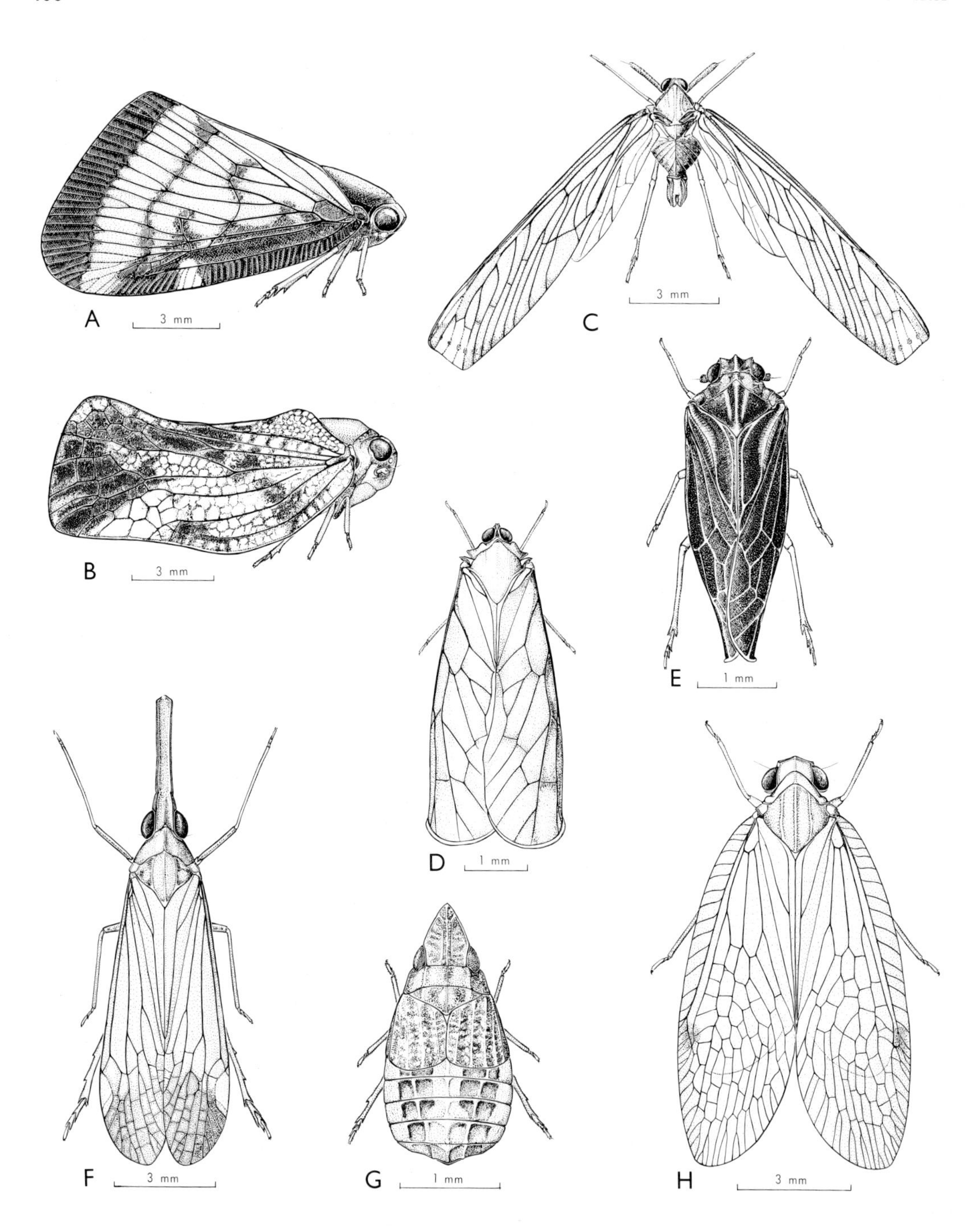

Fig. 26.26. A, *Scolypopa australis*, Ricaniidae; B, *Massila sicca*, Flatidae; C, *Zoraida* sp., Derbidae–Zoraidini; D, unidentified sp. of Derbidae–Otiocerini; E, *Phaconeura froggatti*, Meenoplidae; F, *Thanatodictya praeferrata*, Dictyopharidae; G, *Austrorgerius collinae*, Dictyopharidae; H, *Salona panorpaepennis*, Nogodinidae. [S.Curtis]

12. Derbidae (Figs. 26.18H, 26C,D). Small to medium-sized, often with striking patterns of yellow, brown, or cream; feeding on fungi or phanerogams. *Rhotana chrysonoe* Kirk., with rather broad yellowish wings, occurs in rain forests. The body of *Zoraida* is very short, but the wings exceedingly long and narrow. There are about 800 species widely distributed throughout the tropics. [Fennah, 1952.]

13. Meenoplidae (Fig. 26.26E). Mostly rather small and inconspicuous, with a wing-span of 5–15 mm, and confined to the eastern hemisphere, occurring mainly in the tropics and subtropics. There are about 80 described species. All Australian species belong to NISIINAE; there are 6 spp. of *Phaconeura* and 2 of *Nisia*, all except *P. froggatti* Kirk. (N.S.W.) being recorded only from Queensland. [Muir, 1925; Woodward, 1957.]

14. Dictyopharidae (Figs. 26.18I, 26F,G). Medium-sized, usually with the head strongly prolonged forwards and the wings rather narrow. The 540 described species are widely distributed throughout the world. The Australian fauna includes several species of *Thanatodictya*, with brown markings on the tegmina, and green species of *Hasta*. The nymph-like ORGERIINAE, all brachypterous and flightless and lacking tegulae, are wide-spread in arid and semi-arid parts of the world; only one endemic genus and species, *Austrorgerius collinae* Woodw., has been recorded from Australia (Woodward, 1960), though others may possibly occur in the arid interior.

15. Nogodinidae (Figs. 26.18L, 26H). Mostly clear-winged, with a 15–30 mm wing-span. The world fauna of over 120 species is mostly tropical and subtropical. Species of *Salona* and *Nurunderia* occur in Queensland.

Superfamily CERCOPOIDEA

Many cercopoids, known as 'froghoppers' or 'spittle bugs', superficially resemble cicadelloids or 'leafhoppers', but may be distinguished by the characters given in the key (p. 402). They differ also in constancy of cephalic characteristics, in venational features of the tegmen, and in the biology of their nymphs and their associated adaptive structures. Head variously shaped, sometimes narrowly produced anteriorly; paired ocelli always on the crown; tegmen usually with a short Sc; M usually, though not invariably, basally incorporated in the same vein as CuA, and never proximally associated with R; hind coxae short, conical, laterally dilated.

Nymphs of Cercopidae and Aphrophoridae live enclosed in 'spittle' (Fig. 26.27A); those of the Machaerotidae inhabit calcareous tubes (Fig. 26.27B) situated on their food plants, in which they live immersed in their liquid excretions. Nymphs in all 3 families have a ventral abdominal channel, which terminates anteriorly at the opening of the large mesothoracic spiracles. In spittle-making forms this channel becomes closed when the terga of the two sides of the abdomen are brought together. In tube-making forms, it is permanently closed by a membrane, and in some species abdominal terga 5 and 6 are modified so as to form an operculum with which the opening of the tube can be closed. Spittle is formed by air being taken into the ventral abdominal channel and then expelled

Fig. 26.27. A, 'spittle' of cercopid nymphs; B, tubes and nymph of *Chaetophyes* sp., Machaerotidae. [Photos by P. L. Grant and C. Lourandos]

through a film of anal excretions, thus forming bubbles. Doubtless spittle and tube formation are both associated with prevention of excessive evaporation (see Marshall, 1966).

Although the Cercopoidea are sometimes separated into 4 families, the 3 following are the ones most usually recognized. [Evans, 1966.]

Key to the Families of Cercopoidea

1. Eyes about as long as wide; hind margin of pronotum straight or slightly curved .... **Cercopidae**
   Eyes longer than wide; hind margin of pronotum W-shaped 2
2. Pronotum flat, sometimes anteriorly declivous; tegmen without or with a very small appendix .... **Aphrophoridae**
   Pronotum convex; tegmen with a wide appendix continuing around its apex .... **Machaerotidae**

16. Cercopidae (Plate 3,F; Figs. 26.27A, 28E). World-wide, particularly richly represented in the tropics, and including the largest and most handsome of all cercopoids. The largest species in Australia is *Megastethodon urvillei* Le Pellet. & Serv., up to 19 mm in length, which is found in the north-east and New Guinea. Another striking though considerably smaller species, *Aufidus trifasciatus* Stål, which is orange and black, is likewise restricted to northern Australia and the islands to the north. Other well-known representatives include *Petyllis deprivata* (Walk.), a brown insect with a fine gold pubescence, which has wide distribution, and species of *Tonnoiria* which occur in Tasmania and at high altitudes on the mainland.

17. Aphrophoridae (Figs. 26.28A,B). The largest family of the Cercopoidea and of universal distribution. The 2 most abundant and widely distributed Australian aphrophorids are *Philagra parva* Don. and *Bathylus albicinctus* (Erich.). The former has a long, narrowly produced head, and is associated particularly with *Casuarina* and *Acacia*. The latter is beetle-like, brown and white in colour, and feeds on very numerous plants.

18. Machaerotidae (Figs. 26.27B, 28C,D). Restricted to the Oriental region, tropical Africa, and Australia. The most frequently encountered Australian species is *Chaetophyes compacta* (Walk.). Both sexes have shiny tegmina with a crumpled appearance, those of females, which are larger than the males, being brown, whereas the males are black. MACHAEROTINAE occur in Australia only in the north; they have the scutellum considerably enlarged and produced posteriorly into a spine-like process. Evans (1940) discussed the tube-building nymphs. [Maa 1963a.]

Superfamily CICADOIDEA

Cicadas are abundant insects in all countries having a warm climate, including Australia. Their most notable characteristics are their ability to produce a large volume of sound by means of special organs and the possession of associated organs for sound reception. On the dorsal surface of abdominal segment 1 of the males there is a pair of ridged prominences, the *tymbals*, which may be partly or entirely exposed or concealed beneath an exoskeletal fold. Sound is produced as a succession of short pulses by rapid vibration of the tymbals (p. 53). The auditory *tympana* are situated anteriorly on the ventral surface of the abdomen of both males and females, but are best developed in males. They are usually concealed beneath a pair of large plates, the *opercula*, formed from extension of the metathoracic epimera. Each tympanum, which is known as a 'mirror' and consists of a thin, delicate membrane, has a sclerotized process connecting it with the auditory capsule, a hemispherical swelling on abdominal segment 2.

It was formerly supposed that tymbal sound production was confined to male cicadas among insects. It is now known (Ossiannilsson, 1949) that tymbals are of widespread occurrence within the Homoptera and are found also in representatives of certain

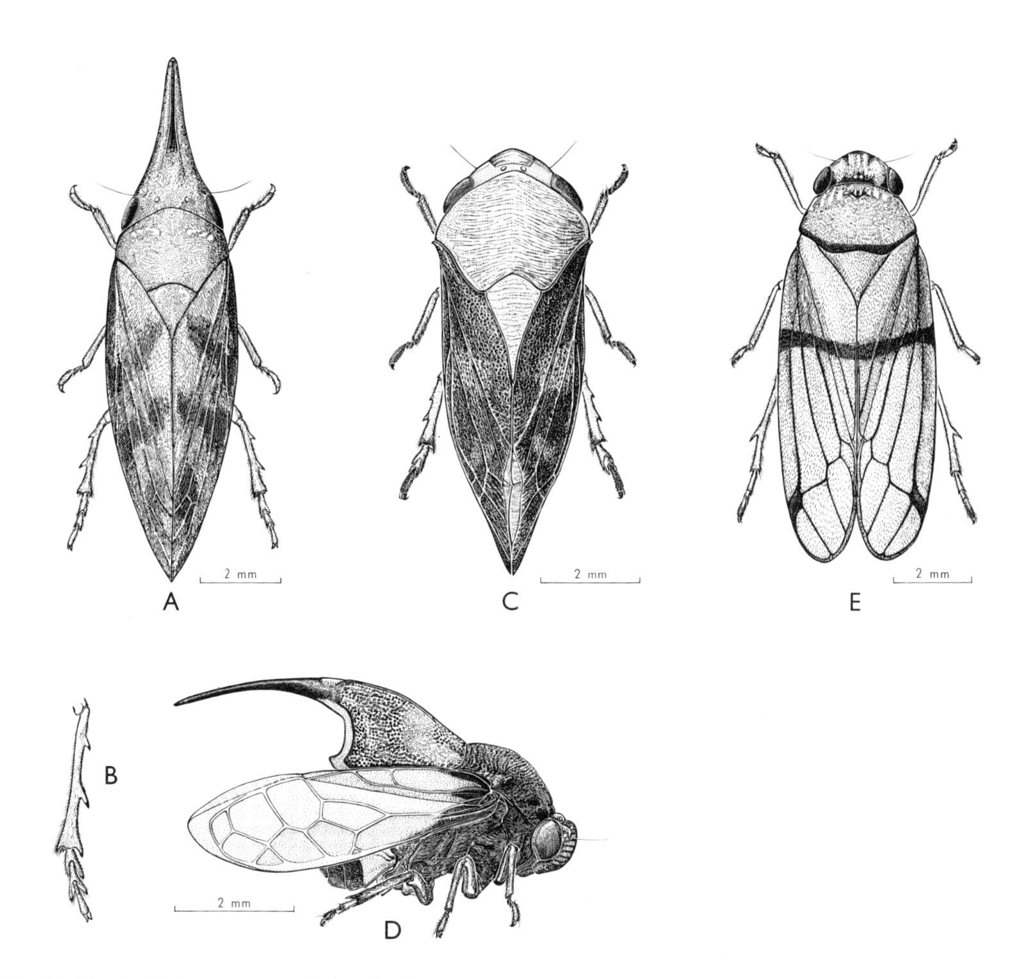

Fig. 26.28. A, *Philagra parva*, Aphrophoridae; B, hind tibia and tarsus of same; C, *Chaetophyes compacta*, Machaerotidae; D, *Machaerota finitima*, Machaerotidae; E, *Aufidus trifasciatus*, Cercopidae. [S. Curtis]

other orders. Furthermore, in Homoptera they are not confined to the male, though usually better developed in males than in females. Complex auditory organs like those of the Cicadidae are, however, lacking in other Homoptera.

As in related groups of Homoptera, eggs are inserted in slits cut in branches and twigs by the saw-like ovipositor of the female. On hatching, young cicadas fall to the ground, and burrow into the soil with the aid of their fore legs which are modified for digging (Fig. 26.30C). The entire nymphal existence is then spent below ground, and may last for several years. On reaching maturity, the nymphs leave the soil, and the adults emerge from the nymphal exuviae, which are common sights on the trunks of trees. A brood of the same age will emerge at the one time, resulting in cyclical appearance of adults.

Cicadas, when regarded as a family group, are usually separated into 3 subfamilies; but the differentiating characteristics are of uncertain phylogenetic significance, so it is preferable to disregard this classification and instead to recognize 2 families. [Burns, 1957; Evans, 1941; Myers, 1929; Pringle, 1957b.]

Key to the Families of Cicadoidea

Tymbals present in both sexes .... **Tettigarctidae**
Tymbals lacking in ♀♀ .... **Cicadidae**

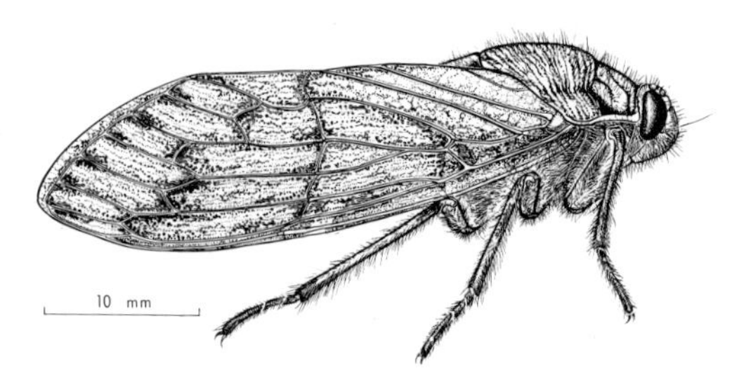

Fig. 26.29. *Tettigarcta tomentosa*, Tettigarctidae. [S. Curtis]

19. Tettigarctidae (Figs. 26.4C, 29). The sole living representatives are confined to Australia, but extinct forms have been recorded from Mesozoic and Tertiary strata in the northern hemisphere. There are 2 species, *Tettigarcta tomentosa* White confined to Tasmania and *T. crinita* Dist. from south-eastern Australia, both associated with an alpine environment. At one time it was supposed that *Tettigarcta* was incapable of sound production, but it is now known that both sexes have tymbals and associated tymbal muscles. Tympana, however, are lacking. *Tettigarcta* differs from other cicadas in many features. The insects are pilose, and have a complete pattern of tegminal venation with a well-developed nodal line, tarsal empodia, and harpagones in the male genitalia. The nodal line is associated with the mechanics of flight and enables the wing to be apically flexed in a downward direction. It is present in the fore wings of nearly all cicadas, though usually inconspicuous.

20. Cicadidae (Plate 3,G; Figs. 26.3B, 5C, 6A, 7C,D, 30). Many striking cicadas occur in Australia, several being so well known that they have acquired common names. Among

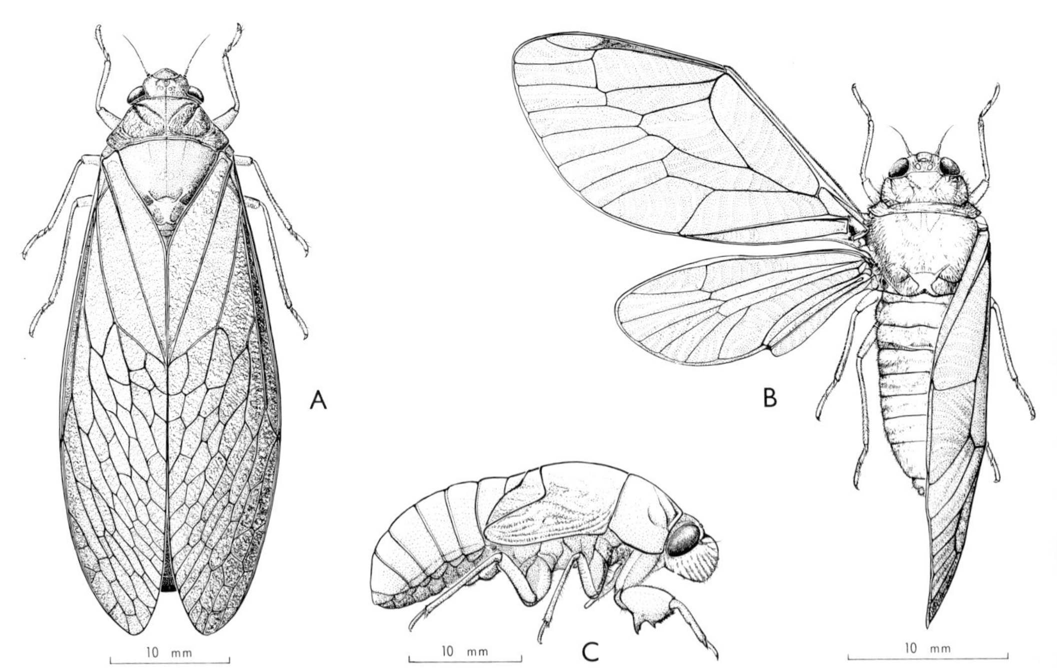

Fig. 26.30. Cicadidae: A, *Cystosoma saundersi*, ♂; B, *Froggattoides typicus*; C, *Macrotristria angularis*, nymph. [S. Curtis]

these are the 'double drummer', *Thopha saccata* (F.), the 'floury miller', *Abricta curvicosta* Germ., and the 'green Monday', *Cyclochila australasiae* (Don.). The largest and best-known of the bladder cicadas is *Cystosoma saundersi* Westw., restricted to southern Queensland and northern N.S.W. Particularly striking species are the 'bent-wing cicadas', *Froggattoides typicus* Dist., and *Lembeja brunneosa* Dist. The latter, which occurs in north Queensland, has some of its abdominal terga modified to form a hood concealing the genital segments. The genus *Cicadetta* has the largest number of species, many of which have a very restricted distribution.

Superfamily CICADELLOIDEA

A large, diverse superfamily, represented in Australia by 3 families, of which the Eurymelidae are endemic, the Cicadellidae cosmopolitan, and the Membracidae widespread though not world-wide. Leafhoppers differ from other superfamilies in having a reduced tentorium, in venational features, in the transverse hind coxae, and in the armature of the hind tibiae (Fig. 26.31). They feed almost exclusively on flowering plants, the Australian species predominantly on trees; some eurymelids are root-feeders. [Evans, 1966.]

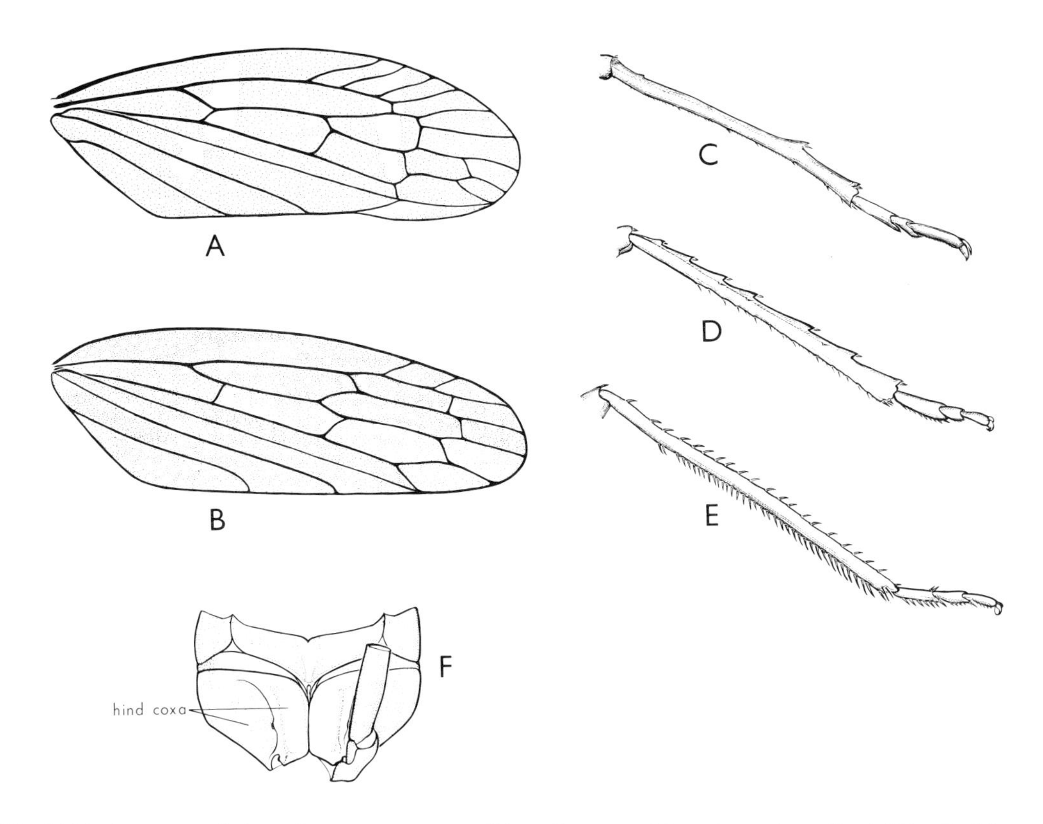

Fig. 26.31. A, tegmen of *Australoscopus* sp., Eurymelidae; B, same of *Putoniessa nigra*, Cicadellidae–Ledrinae; C, hind tibia and tarsus of *Eurymela fenestrata*, Eurymelidae; D, same of *Stenocotis depressa*, Cicadellidae–Ledrinae; E, same of *Cicadella spectra*, Cicadellidae–Cicadellinae; F, ventral aspect of metathorax of *Cicadella spectra*. [S. Curtis]

Key to the Families of Cicadelloidea Known in Australia

1. Pronotum considerably enlarged, sometimes grotesque .... **Membracidae**
   Pronotum usually of normal proportions; if enlarged, then scutellum completely exposed 2
2. Ocelli on ventral surface of head; tegmen with M_{1+2} extending to apex, Rs absent; hind tibiae, except in myrmecophiles, with 1 or several spines on prominent bases .... **Eurymelidae**
   Ocelli ventral, marginal, or dorsal; tegmen usually with M_{1+2} apically fused with Rs; hind tibiae with numerous spines, which, except in Ledrinae, are not on prominent bases .... **Cicadellidae**

21. Cicadellidae (Figs. 26.1A,B, 17B, 31B, D–F, 32, 33). An unusual feature of the rich Australian fauna is the high proportion of arboreal species and the paucity of forms feeding on herbaceous plants. Of the 15 subfamilies represented in Australia, only the Austroagalloidinae are endemic, and the AGALLIINAE are represented by a single species, *Austroagallia torrida* Evans, of uncertain geographical origin. The classification of the family was treated by Evans (1947).

Two of the 4 known tribes of the ULOPINAE are represented by numerous species, all of which, so far as is known, show alary dimorphism. The Ulopini, which comprise relict forms of probable Mesozoic origin, retain a structural feature in their heads lacking in other leafhoppers: separation of the maxillary plates from the genae by a transverse suture. The Cephalelini, which occur also in South Africa and New Zealand, include forms resembling long, narrow, plant seeds.

Two of the tribes of LEDRINAE in Australia are endemic. The third, the Ledrini, which is represented also in the Oriental, Ethiopian, and Palaearctic regions, includes the largest Australian leafhopper, *Ledromorpha planirostris* (Don.), 23–28 mm long. The Stenocotini, which have paper-thin nymphs frequently found on the trunks of eucalypts, comprise several genera, of which the best known is *Stenocotis*. *S. depressa* (Walk.) is sexually dimorphic, and has ocelli situated in lateral pits on the anterior margin of the head. The Thymbrini likewise have several genera; the largest species, brown with triangular heads, belong to *Rhotidus*.

The HECALINAE comprise 2 tribes, of which the Hecalini are represented in Australia by a few species of mostly northern, grass-feeding leafhoppers. The Paradorydiini are mostly small, narrow, and elongate, superficially resembling the Cephalelini but with marginal instead of dorsal ocelli. APHRODINAE are represented by 2 genera of Aphrodini, *Kosmiopelix* belonging to the Indo-Malayan element, and *Euacanthella* apparently to the southern element. Some species of both genera have brachypterous and fully-winged forms.

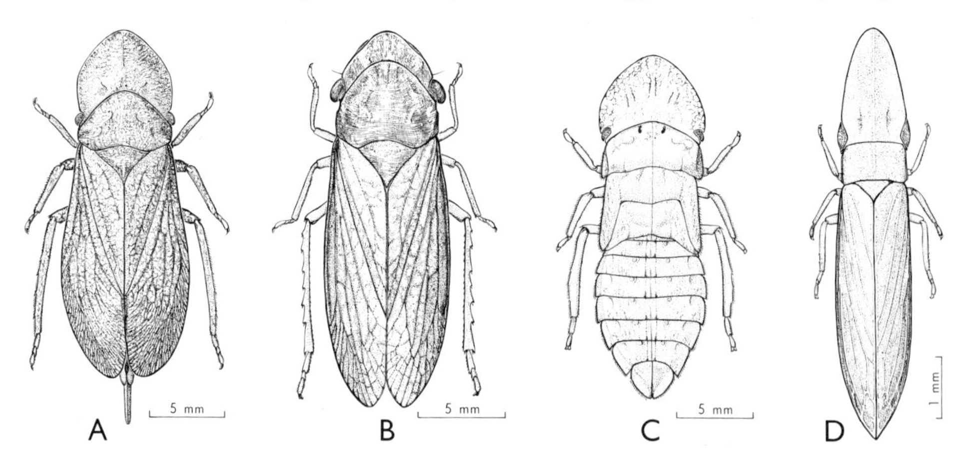

Fig 26.32. Cicadellidae: A, *Ledromorpha planirostris*, Ledrinae; B, *Stenocotis depressa*, Ledrinae, ♀; C, nymph of *L. planirostris*; D, *Cephalelus ianthe*, Ulopinae. [S. Curtis]

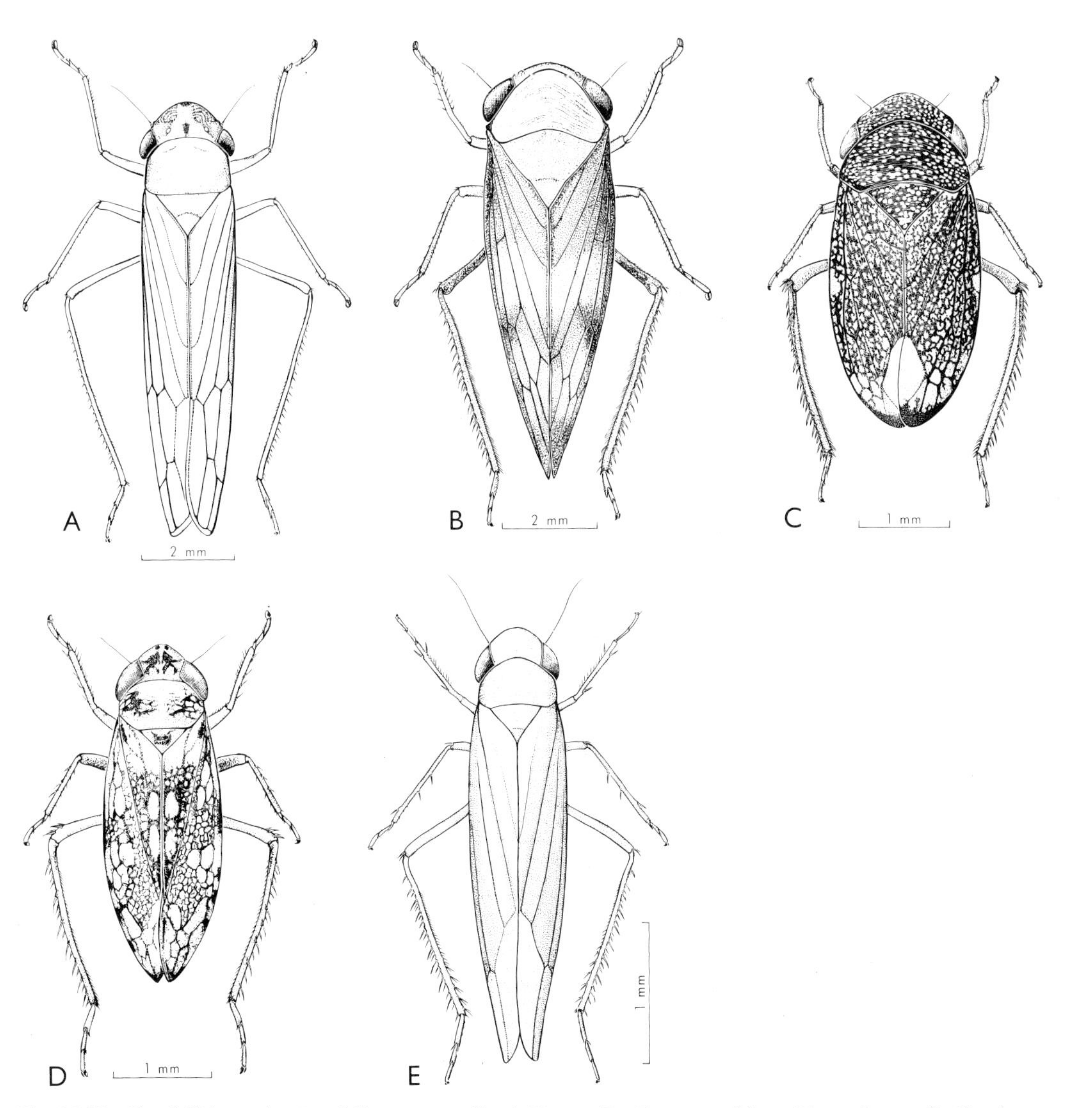

Fig. 26.33. Cicadellidae: A, *Cicadella spectra*, Cicadellinae; B, *Tartessus fulvus*, Tartessinae; C, *Neodartus vaedulcis*, Penthimiinae; D, *Orosius argentatus*, Deltocephalinae; E, *Erythroneura ix*, Typhlocybinae. [S. Curtis]

The CICADELLINAE, of world-wide distribution but most abundantly represented in the Neotropical region, are represented in Australia by a group of small bluish-black and yellow insects of wide occurrence, a few species confined to north-east Queensland, and *Cicadella spectra* (Dist.) which occurs in all countries with a tropical or subtropical climate. The NIRVANINAE are essentially an Oriental group. Most of the Australian species are found in Queensland, but a few, such as *Occinirvana eborea* Evans which feeds on *Casuarina* and is confined to W.A., occur elsewhere. Australia has an exceptionally rich fauna of MACROPSINAE, a world-wide subfamily. Notable species are *Stenoscopus drummondi* Evans, which is by far the largest known macropsine, and *Stenopsoides turneri* Evans, which has the pronotum enlarged and narrowly produced in front of the head. Both

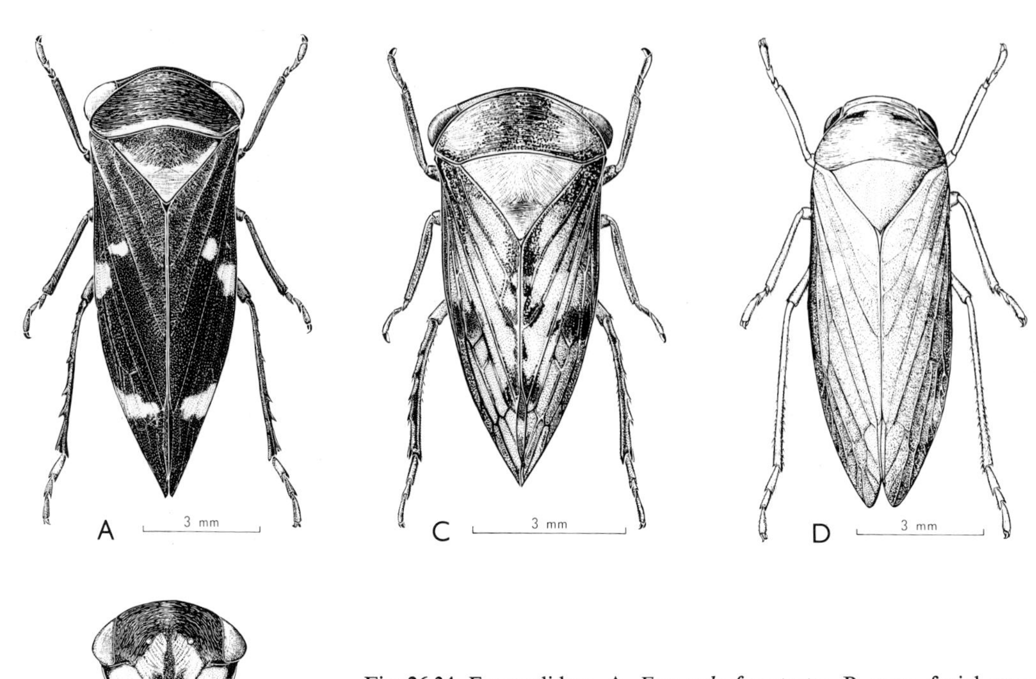

Fig. 26.34. Eurymelidae: A, *Eurymela fenestrata*; B, same, facial aspect of head; C, *Ipo pellucida*; D, *Pogonoscopus lenis*. [S. Curtis]

are confined to south-western Australia.

The AUSTROAGALLOIDINAE include only *Austroagalloides*, the species of which exhibit marked sexual dimorphism. Superficially they resemble IDIOCERINAE, which are world wide and abundantly represented in Australia. The COELIDIINAE, which are commonest in the tropical regions of both the Old and the New World, have been recorded in Australia only from north-eastern Queensland, where they are represented by a few species of *Tharra*. They may be recognized by their unusually long antennae and somewhat fulgoroid appearance. In the TARTESSINAE all species in 3 out of the 4 described genera are confined to Australia. Most species of the fourth genus, *Tartessus*, are also Australian, but a few occur in the Oriental region and on various Pacific islands.

Three tribes of JASSINAE are known in Australia. The cosmopolitan Jassini are represented by numerous species of *Batrachomorphus*. The Trocnadini and Reuplemmelini are endemic tribes, and the species differ from those of Jassini in shape of head and position of ocelli. The PENTHIMIINAE include a seemingly diverse group of leafhoppers. Some, such as small species of *Neodartus* and *Vulturnus*, are largely confined to the north-east; others, such as *Ectopiocephalus australis* (Walk.) and *Chinaella* spp., occur in the low-rainfall inland areas of the continent.

The DELTOCEPHALINAE (Euscelinae) are an extensive group of universal distribution but most abundant in the Nearctic. While some endemic genera undoubtedly occur, it is possible that most of the species in Australia belong to cosmopolitan genera and are of comparatively recent establishment. One such species in the Deltocephalini is *Orosius argentatus* (Evans), a well-known vector of plant virus diseases. The Selenocephalini are represented by a few species confined to

Queensland. Some Australian representatives of the XESTOCEPHALINAE, which are small, oval, and brown with yellowish markings, are myrmecophiles. The TYPHLOCYBINAE are minute, yellow, green, white, or red leafhoppers with reduced wing venation. Apart from species of *Austroasca*, they have been little studied in Australia (Lower, 1951). A well-known introduced species that feeds on the leaves of apple and hawthorn is *Typhlocyba froggatti* Baker.

22. Eurymelidae (Plate 3,H; Figs. 26.31A,C, 34). The only family of Cicadelloidea restricted to a single geographical area, as, apart from Australia, it is sparsely represented only in New Guinea and New Caledonia. Many species are brightly coloured, or predominantly black; others, which have a drab colour pattern, superficially resemble some cicadellids. The male genitalia are distinctive. Unlike cicadellids and membracids, in which the aedeagus is basally continuous with a connective joining the claspers, the aedeagus of eurymelids is dorsally situated and lacks association with the claspers. Eurymelids are readily recognized by their widely flattened face, which has a diamond-shaped frontoclypeus against which the ocelli are posteriorly adjacent. They are always ant-attended, and some species of *Pogonoscopus* and related genera live in the nests of ants. The nymphs, and to some extent the adults, are gregarious, and the nymphs do not jump if disturbed, though the adults do so like other leafhoppers. The largest species belong to *Eurymela* and *Eurymelops*. While most

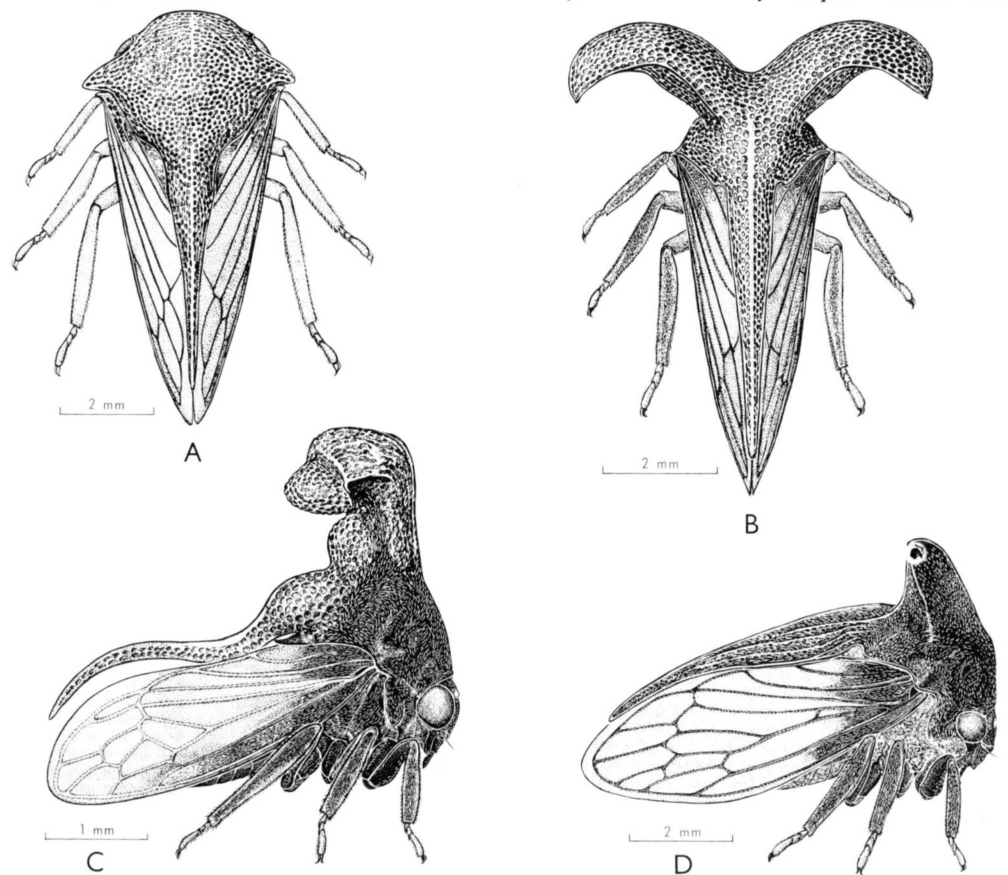

Fig. 26.35. Membracidae: A, *Sertorius australis*; B, *Eufrenchia falcata*; C, *Eutryonia monstrifer*; D, *Eufairmairia fraternus*. [S. Curtis]

eurymelids feed on eucalypts, a few feed on other plants, in particular on *Casuarina*. [Evans, 1931, 1946.]

23. Membracidae (Fig. 26.35). On account of their enlarged pronotum the Membracidae are usually regarded as equally distinctive as the other higher categories of the Homoptera. They are included here in the Cicadelloidea, because they share with the leafhoppers many characteristics in which they differ from the Cercopoidea and Cicadoidea, including cephalic characters, such as possession of a reduced tentorium, venational features, and pretarsal structure. Moreover, some Australian macropsine cicadellids have an enlarged pronotum and resemble membracids in other respects. The nymphs, which are ant-attended and usually gregarious, have lateral abdominal processes, and tergite 9, which is long and tubular, enfolds the 10th or anal segment, which is extrusible.

The most frequently encountered species, *Sextius virescens* (Fairm.), is green, and, like many other Australian membracids, feeds on acacias. Other well-known species belong to *Acanthucus*, *Eufairmairia*, and *Sertorius*, and are black and brown. As well as a pair of lateral pronotal processes, species of *Acanthucus* have a narrow elevated median one. The largest Australian membracid is *Eufairmairia giganticus* (Goding), up to 12 mm in length. *Sertorius australis* (Fairm.) has a convex pronotum with very small acute lateral processes. The most bizarre species are *Lubra spinicornis* (Walk.) and *Eutryonia monstrifer* (Walk.). The former has a pair of apically inflated, strap-shaped pronotal processes, and the latter a single vertical column which is dorsally swollen and laterally wide. [Metcalf and Wade, 1965.]

Superfamily PSYLLOIDEA

24. Psyllidae (Figs. 26.8A, 14, 36). The described world fauna is about 1,250 species in 150 genera. Most of the Australian psyllids are native species associated with native plants. With few exceptions, they are confined to one host species or to a closely related group of hosts. Some, notably *Phellopsylla*, *Cometopsylla*, and *Glycaspis*, are attended by ants. In other parts of the world several species are regarded as pests because of their toxic saliva or gall-making activities. However, most records of psyllids as virus vectors seem to be in error, the damage being caused by the saliva.* Four subfamilies have been recognized in the Australian fauna, with the Spondyliaspinae the best represented. The Psyllinae, Triozinae, and Ciriacreminae have distinct relationships with the fauna of the northern hemisphere and the Indo-Malayan area. It is possible that some undescribed species should be placed in the Aphalarinae. [K. Moore, 1961; Taylor, 1962.]

Key to the Subfamilies of Psyllidae Known in Australia

1. Hind coxae with meracanthus lacking, or small and blunt .... SPONDYLIASPINAE
   Hind coxae with distinct conical meracanthus (Fig. 26.36E) 2
2. M, Cu, and R of fore wing arising at same point, or nearly so, from basal vein .... TRIOZINAE
   M and Cu with a common petiole 3
3. Vertex with median suture; Rs usually long; genae either developed into anterior processes, or swollen .... PSYLLINAE
   Vertex more or less cleft in front, occasionally without median suture; Rs usually short; genae not developed into anterior processes .... CIRIACREMINAE

The SPONDYLIASPINAE contain 113 described species in 14 genera, and there are at least 150 undescribed species. Except for *Syncarpiolyma* and some *Eucalyptolyma* and *Ctenarytaina*, the Australian species occur on *Eucalyptus*. In most species the nymphs build a characteristic scale or test *(lerp)* on leaves or small twigs, under which all the immature stages shelter. Exceptions are *Ctenarytaina*, most species of *Eucalyptolyma*, 2 or 3 undescribed genera, and most species of *Phellopsylla*. The free-living species, which usually secrete copious white flocculent material, are generally found on the young succulent tips, though most *Phellopsylla* feed on the smooth bark of eucalypts. Lerp

* But see A. P. D. McClean and P. C. J. Oberholzer *S. Afr. J. agric. Sci.* **8**: 297–8, 1965.

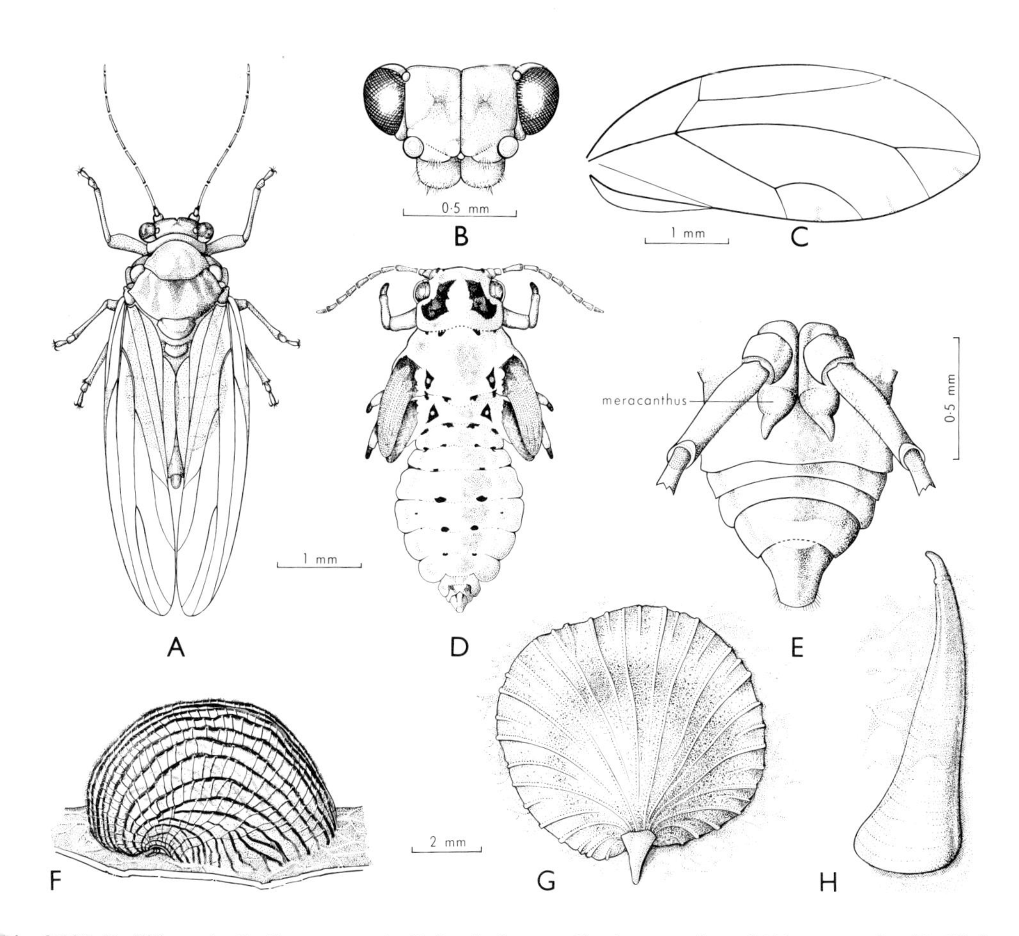

Fig. 26.36. Psyllidae: A, *Creiis costatus*, ♂; B, head of same; C, wing venation of *Trioza eugeniae*; D, 5th-instar nymph of *Creiis costatus*; E, hind legs and abdomen of *Psylla acaciaedecurrentes*, ventral; F, lerp of *Cardiaspina textrix*; G, lerp of *Creiis periculosa*; H, lerp of *Creiis costatus*. [T. Nolan]

formation has probably evolved because of the need to protect the nymphs from desiccation. The lerps take many forms, and they are sometimes associated with curling of sides or tips of the leaves to provide added protection. The most elaborate structures occur in species of *Cardiaspina*, in which they are flat and lace-like or intricately woven in the form of inverted baskets (Taylor, 1962). Most species of *Glycaspis* and *Lasiopsylla* are lerp-builders, but some form large bubble-shaped galls with an orifice at the base plugged with the same waxy or sugary material as is used by other species to build lerps. All species of *Cardiaspina* cause death of the leaf tissues where the nymphs are feeding, and infested trees may be virtually defoliated. Some species of *Glycaspis* also occur in very high numbers and cause debilitation, sometimes death, of the trees (K. Moore, 1961). *Ctenarytaina eucalypti* (Mask.), which has been accidentally introduced into New Zealand, South Africa, and England, is a pest on the growing tips of young blue gum trees.

Only 16 species of Psyllinae have so far been described in 5 genera, the major one being *Psylla* (= *Acizzia*). All are free-living, mostly on succulent growing tips, though some (e.g. *P. acaciaedecurrentes* Frogg. and *Aconopsylla sterculiae* (Frogg.)) occur in colonies on twigs. It is common to find two species together on various species of *Acacia*, and at least one may be found on many Australian trees and shrubs other than *Eucalyptus*. *P. acaciaebaileyanae* Frogg. has been recorded in South Africa, and several species occur in New Zealand on hosts introduced from Australia.

The TRIOZINAE are represented by 11 described species in 3 genera. Most *Trioza* spp. are similar in habits to *Psylla* spp., but many form shallow galls on the leaves of their host plants, which include a wide range of trees and shrubs; a few cause leaf-curling and distortion. All species of *Schedotrioza* form large galls on *Eucalyptus*. The nymphs are completely enclosed, and depend on the splitting of the galls to emerge as adults. *Aacanthocnema* occurs on *Casuarina* (with at least one other undescribed genus) and on some Proteaceae.

The CIRIACREMINAE, represented by only 4 described species in 3 genera, are poorly known, because most species occur in rain forest where little collecting has been done. The 2 known species of *Mycopsylla* live underneath solidified latex exuded from the leaves as a result of their feeding on *Ficus* spp. *Protyora sterculiae* (Frogg.) retains long, white, waxy filaments at the end of its abdomen, hence the name 'star psylla' given to it by Froggatt.

Superfamily APHIDOIDEA

There has been no generally accepted family classification for the last fifty years. Some authors have recognized only one family, others up to 9; 4 are recognized here. The Aphidoidea are rather poorly represented in Australia, with only 118, mostly immigrants, out of a world fauna of about 3,600 species.

About 200 species have been recorded as vectors of plant viruses, and one, *Myzus persicae* (Sulz.), as a vector of 108 virus diseases. Kennedy *et al.* (1962) have summarized the literature on virus-vector specificity. There is little information on the damage done to crops by average aphid populations. Large populations in dry weather cause plants to wilt, and even moderate colonies may considerably reduce yield. The saliva of some aphids is toxic to some plants, and may cause discoloration or distortion of the leaves and shortening of the internodes. Many Aphididae form pseudo-galls, but few species form true closed galls, which are more commonly caused by Pemphigidae and Adelgidae.

Abdominal wax-secreting organs, the *cornicles* or *siphunculi* (Fig. 26.16), are a peculiarity of many aphids. Aphidoidea lack Malpighian tubes. Nutrition is apparently aided by the presence of intracellular symbionts in mycetomes; some are said to be acetobacters capable of synthesizing amino acids from the nitrogen of the air. A standard work on Aphidoidea is by Börner and Heinze (1957). Eastop (1966) gave an account of Australian aphids, Kennedy and Stroyan (1959) a review of the literature, and Auclair (1963) reviewed aphid feeding and nutrition.

Key to the Families of Aphidoidea

1. Parthenogenetic forms viviparous, the body contents consisting largely of embryos; Rs present in fore wings; antennae of alatae usually 5- or 6-segmented, of apterae 4- to 6-segmented; siphuncul often present; cauda often developed (Figs. 26.16, 37, 38A-C) 2
   All forms oviparous; Rs absent from fore wing; antennae short, 2- to 5-segmented; siphuncul absent; cauda not evident (Figs. 26.38D-F) 3
2. Antennae usually slender, at least a quarter as long as body in apterae and usually relatively longe in alatae, 4- to 6-, usually 6-segmented, terminal segment usually with a short thick base and a long slender processus terminalis (Fig. 26.38A), but the latter sometimes only one-tenth to half a long as base, secondary sense organs ('rhinaria') usually round or oval, sometimes transversely elongate; large compound eyes usually present in all forms, but sometimes reduced to 3 separate lenses ('triommatidea') in apterous and immature forms; M of fore wing usually twice branched

sometimes once branched; head usually free, rarely fused with prothorax; siphunculi usually elongate; cauda usually well developed, often elongate .... **Aphididae**

Antennae short, one-20th to one-fifth as long as body in apterae, one- to two-fifths in alatae, 1- to 6-, often 5-segmented, processus terminalis short, usually much less than half as long as base, only about half as long when base unusually short, secondary rhinaria often annular, strongly transverse, sometimes round or oval; alatae with large compound eyes, but apterae and immature forms with only triommatidea; M of fore wing once branched or simple; head and prothorax of apterae sometimes fused; siphunculi rarely more than sclerotized rings, sometimes absent; cauda usually broadly rounded, sometimes weakly knobbed, never elongate .... **Pemphigidae**

3. Antennae of apterae with 2 and of alatae with 3 primary rhinaria; fore wing with CuA and CuP separated at base; wings held roofwise in repose; sexuales with mouth-parts; on Coniferae .... **Adelgidae**

Antennae of apterae with 1 and of alatae with 2 primary rhinaria; CuA and CuP of fore wing with a common base; wings held horizontally in repose; sexuales arostrate; on dicotyledons .... **Phylloxeridae**

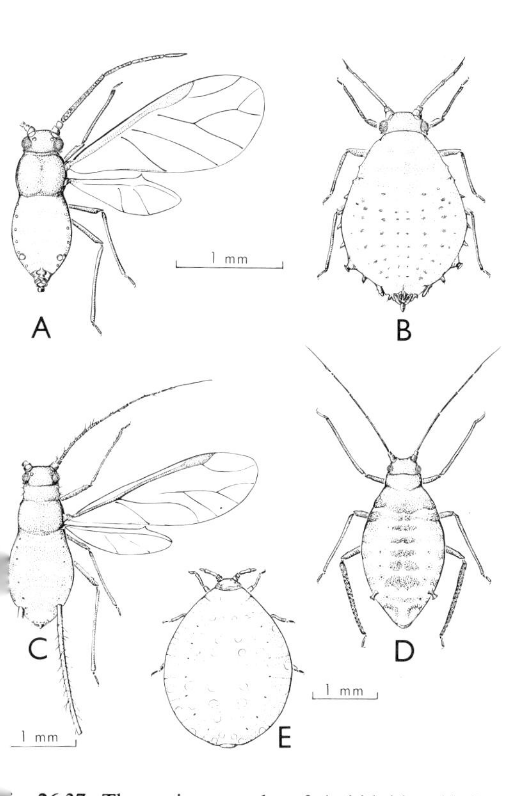

Fig. 26.37. The major morphs of Aphidoidea (A–D Aphididae, E Pemphigidae): A, ♂ of *Neophyllaphis gingerensis* from *Podocarpus alpina*; B, apterous viviparous ♀ of *Sensoriaphis tasmaniae* from *Nothofagus Cunninghamii*; C, alate viviparous ♀ of *Greenidea ficicola* from *Ficus Hillyi*; D, oviparous ♀ of *Kallistaphis basalis* from *Betula alba*; E, fundatrix of *Pemphigus bursarius* from *Populus nigra*. [S. Curtis]

25. Aphididae (Figs. 26.6B, 16, 37A–D, 38A). About 46 species have apparently been introduced to Australia from Europe on the host plants to which they are specific, and another 4 originated in America. There are 30 Oriental species, which occur in south-east Asia, and probably reached Australia without the assistance of Europeans. Another 13 are now circumtropical and are of uncertain origin, as are 5 which have spread over much of the world and appeared in Australia in the last 30 years. Eight species are probably endemic and specific to native plants.

Most aphid species are polymorphic (Figs. 5.45, 26.37). Males are usually rare and produced only in the autumn, or absent; they may be apterous or winged, some species having both forms. Females may be apterous or winged, oviparous or viviparous, and parthenogenetic or require fertilization. Only oviparous females, which are usually apterous, mate. The resultant overwintering eggs always give rise to viviparous females which are usually apterous. This form is anatomically distinguishable from later generations, and is known as the 'fundatrix', The progeny of the fundatrix give rise to a succession of viviparous generations which, depending on the species of aphid and condition of the host plant, may be mostly apterous or mostly alate. Commonly a number of generations mostly of apterae is followed by one generation consisting mostly of alatae, which in turn is followed by further apterous generations.

Except in those groups in which 'apterae viviparae' are unknown, it is rare for the progeny of alates to be alate.

In aphids with alternating hosts, the fundatrix is apterous, but the second, third, or rarely later generation is alate. The alatae, known as 'spring migrantes' or 'fundatrigeniae' fly to the summer host, and deposit young which develop into apterae viviparae. After a number of generations of these apterous, parthenogenetic, viviparous 'exules', a generation of alate exules may occur which fly to other secondary hosts of either the same or a different species from that on which the alate exule matured. The progeny of the alate exule are apterous exules and, after another few generations of apterae, generations containing alatae known as 'gynoparae' occur. The gynoparae transfer to the primary host, and their progeny develop into the 'oviparae'. At the same time the aphids remaining on the secondary host are producing alate males which will fly to the primary host to copulate with the oviparae, after which the overwintering eggs are laid. Each ovipara lays a number of eggs. Thus alate males are a necessity for alternating Aphididae, in contrast to the alternating Pemphigidae and Adelgidae in which the males are apterous. Although some Aphididae have numerous secondary hosts, it is rare for more than two or three species of plants to be suitable as primary hosts, and these usually belong to the same genus.

Host alternation occurs only in the Aphidinae, and was apparently evolved in the continental-type climates of the Holarctic region, in response to the unsuitability as food of some hosts during the summer and of most host plants during the winter. In the milder climatic conditions of much of Australia, either host alternation does not occur, or only some populations alternate, whereas others remain on either the primary or the secondary hosts, or treat them both as secondary hosts, producing parthenogenetic viviparous generations on whichever host happens to be suitable at the time. Decreasing daylight length is the most important factor influencing oviparae production, and decreasing temperature for male production. In the Holarctic region these two factors ensure that males and females are produced together, but under different climatic conditions this very mechanism may ensure that the males are produced at a different time o

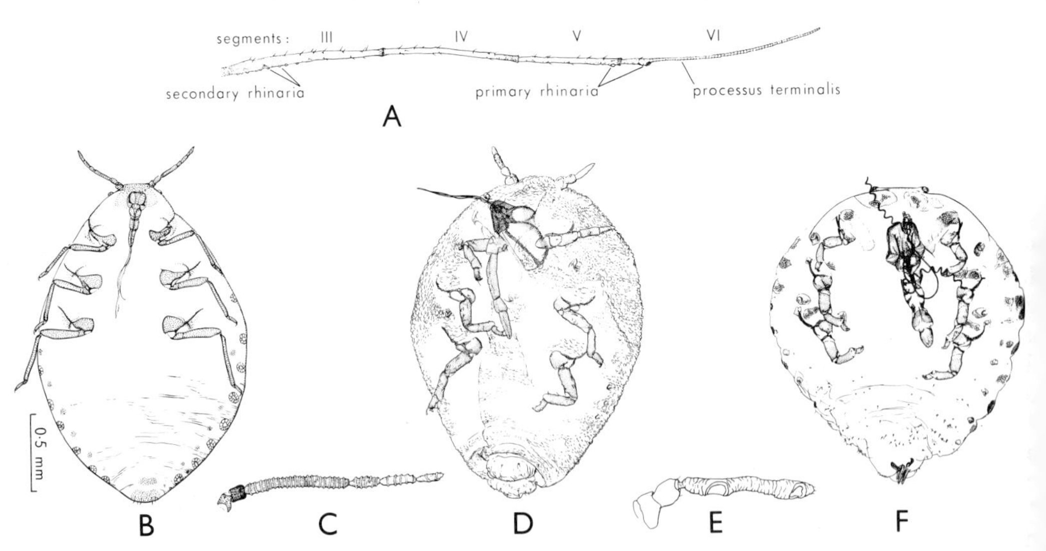

Fig. 26.38. A, antenna of *Macrosiphum rosae*, Aphididae, aptera vivipara; B, *Eriosoma lanigerum*, Pemphigidae, aptera; C, antenna of alata; D, *Viteus vitifoliae*, Phylloxeridae, aptera; E, antenna of alata; F, *Pineus pini*, Adelgidae, aptera. [All except B from Eastop, 1966, by permission of the publisher]

year from the females with which they should mate. Although host alternation occurs in a number of species introduced to Australia, probably very few individuals ever give rise to a sexual generation here, most reproduction being parthenogenetic. The control of polymorphism in Aphidoidea has been reviewed by Lees (1966) and Hille Ris Lambers (1966), who discussed the effects of population density, photoperiod (either directly or through the food plant), temperature, and influence of the fundatrix.

Most aphids have 7 or more generations a year, the duration of each depending mainly on temperature, and varying from about 2 weeks to 2 months. Viviparae are recorded as producing from 3 to 176 young. The first generation may reproduce only 3 days after birth under exceptionally favourable conditions, but 5–22 days is more usual. Some aphids occur in dense colonies of several hundred on a small leaf or several thousand clustered along a stem, whereas others live more evenly scattered over the host plant. The size of aphid populations changes rapidly, and Hughes (1963) has discussed population problems.

Many aphids are specific to one plant or to a few closely related plants; others have a rather wider range including a number of related genera. Host alternation normally entails living on two botanically distant plants, and the generations of aphids produced on the primary and secondary hosts are often anatomically distinguishable. Aphids may differ anatomically when reared on different hosts when alternation does not occur, e.g. the banana aphid, which looks so different when reared on Araceae that it was described as a distinct species. There can be considerable anatomical variation between individuals not only of the same species but of the same clone of an aphid, depending on temperature and food plant. Colonies reared at high temperatures tend to be small, pale, and short-haired. Some features, such as length of 3rd antennal segment, may be correlated with body length. Others, associated with locomotion and feeding, such as length of the tarsi and apical rostral segment, remain almost constant over a large range of body size, but eventually decrease or increase considerably, the critical body sizes often differing for the two characters.

Parthenogenetic reproduction offers a method of perpetuating abnormalities, and thus colonies have been found with extraordinary characters, such as branched siphunculi. When the sexuales are apterous, particularly in root-living aphids, these abnormal populations may be perpetuated indefinitely.

Aphids are known as 'greenfly' or 'blackfly' when they live exposed on leaves, shoots, or buds. Many other species live concealed in unfolded or crumpled leaves, on the undersides of the lower leaves touching the ground, on the stem or trunk, especially at the base, on roots, or under earthen shelters constructed by ants. Some live on Gramineae which are periodically submerged.

Many species of aphids are attended by ants. This association varies from the casual collection of honey-dew from the plant by the ants to a very close association, in which the ants solicit honey-dew from the aphids, carry them about, build shelters over them, collect their eggs, and generally earn the analogy with human beings and cattle. The closeness of the association varies not only with the species involved, but also with the season of the year, some aphids being attended assiduously in the spring and early summer but hardly at all in the autumn.

The APHIDINAE include many species of economic importance. *Macrosiphum euphorbiae* (Thomas), *Myzus ornatus* Laing, *M. persicae* (Sulz.), *M. ascalonicus* Don., *Neomyzus circumflexus* (Buckt.), and *Aphis gossypii* Glov. have a very wide host range. *A. craccivora* Koch is most common on Leguminosae and *A. spiraecola* (Patch) on Rosaceae, Compositae, and Umbelliferae. *Brevicoryne brassicae* (L.) and *Lipaphis erysimi* Kalt. are pests of crucifers. *Cavariella aegopodii* (Scop.) has *Salix* as a primary host and various Umbelliferae as secondary hosts. *Chaetosiphon fragaefolii* (Cock.) is the vector of the principal strawberry viruses. The green or pinkish brown *Macrosiphum rosae* (L.) occurs conspicuously on rose twigs. *Elatobium abietinum* (Walk.)

causes leaf fall of some *Picea* species, while other spruces are apparently immune. *Idiopterus nephrelepidis* Davis and *Shinjia pteridifoliae* (Shinji) occur on ferns. *Macrosiphum miscanthi* Tak. occurs on Gramineae and some dicotyledons. Of the 2 species of black citrus aphid, *Toxoptera aurantii* (Boy. de Fonsc.), occurring on many shrubs, possesses a functional stridulating apparatus (Eastop, 1952), and *T. citricidus* (Kirk.), occurring mostly on Rutaceae, is the principal vector of Tristeza Citrus virus. *Hyalopterus pruni* (Geoff.) is the mealy plum aphid. *Longiunguis sacchari* (Zehntn.) is a pest of sugar cane, and *Rhopalosiphum maidis* (Fitch) and *R. padi* (L.) feed on a wide range of Gramineae and are vectors of barley yellow dwarf virus. *R. rufiabdominalis* (Sas.) occurs on the subterranean parts of both cereals and dicotyledons.

The DREPANOSIPHINAE mostly attack dicotyledonous trees and shrubs or grasses and sedges. Host alternation does not occur. *Sensoriaphis*, on *Nothophagus*, is known only from Tasmania and New Zealand. Only 2 CHAITOPHORINAE, both feeding on *Acer*, occur in Australia. The GREENIDEINAE include *Anomalaphis* and *Meringosiphon*, two genera known only from Australia; close relatives occur in south-east Asia.

26. Pemphigidae (Eriosomatidae; Figs. 26.37E, 38B,C). Two subfamilies occur in Australia. The HORMAPHIDINAE include *Oregma* and *Cerataphis*, the latter with 3 species on palms and orchids. Most PEMPHIGINAE have been introduced from Europe, although *Eriosoma lanigerum* (Hausm.), the woolly aphid of apples, apparently originated in America. *E. pyricola* (B. & D.) is the pear-root aphid. *Pemphigus bursarius* (L.) alternates between *Populus* and the roots of several Compositae. *Aploneura lentisci* (Pass.) and *Geoica lucifuga* (Zehntn.) live on the roots of grasses and sugar cane, and *Smynthurodes betae* Westw. on the roots of dicotyledons.

Most Pemphiginae alternate between primary and secondary hosts to which they are specific. The fundatrix is apterous, and the second or third generation is alate. In late spring or early summer these alatae fly to the secondary hosts, on which a few generations of apterous exules are produced. Alate exules are rare, and are usually 'sexuparae' which return to the primary host where their progeny develop into small, arostrate, apterous males and females. After copulation only one egg develops. In some species the cycle has been secondarily simplified by the loss of the primary or secondary host.

27. Adelgidae (Fig. 26.38F). The primary host of all known alternating Adelgidae is *Picea* (spruce), and the secondary hosts are other conifers. The identity of the species of *Pineus* living on *Pinus radiata* in Australia is uncertain, but it is probably *P. pini* (Macq.).

28. Phylloxeridae (Figs. 26.38D,E). The species known in Australia are introduced. *Viteus vitifoliae* (Fitch) (= *Phylloxera vastatrix*) is a pest of grapes. Host alternation is unknown in this family.

Superfamily ALEYRODOIDEA

29. Aleyrodidae (white flies; Figs. 26.15, 39). About 1,100 species in 90 genera have been described, but only 20 apparently endemic and 4 introduced or cosmopolitan species are known from Australia. Adults are small (about 3 mm wing-span) and covered with a powdery wax secretion. Phylogenetically the Aleyrodidae may be degenerate psyllids. This is suggested both by the general morphology and by such behaviour as the side-by-side copulation in which psyllids and aleyrodids differ from aphids and coccids. The *vasiform orifice*, a characteristic structure of aleyrodids, is a large dorsal opening on the last abdominal segment of all instars. It receives the anus, is partly covered by an operculum, and contains a tongue-like process, the *lingula*. The honey-dew accumulates there in globules, and is flicked off by the lingula, which seems to have the same function in both nymphs and adults as the cauda of some adult aphids. Aleyrodidae are unusual, in that most of the present-day classification is based on a study of the cutic of the last nymphal instar, the so-called 'pupa case'. This has resulted in a knowledge of the host plants of many species, and a paucity of knowledge of the anatomy of the adults (but see Weber, 1935).

Immature Aleyrodidae live mostly on the undersides of the leaves of angiosperms, and a few genera are associated with ferns; there is no record from gymnosperms. Both sexual and parthenogenetic reproduction occurs, and in some species populations consisting mainly of males or mainly of females may be found. The unfertilized eggs of some species develop into males, whereas fertilized eggs give rise to both sexes, but parthenogenetic female-producing races apparently exist in other species. A female may lay up to about 200 eggs attached to leaves by a pedicle, in some species in clusters, in others irregularly scattered. The first-instar nymph wanders for a few hours, and then attaches itself to the leaf. All the following immature instars are spent in the same place. Nymphs are ovoid in outline and dorsoventrally flattened; legs and antennae atrophy after the first ecdysis. The usual number of nymphal stages is 4, and the cast skins often remain attached to the dorsum of the nymph. During the last nymphal instar ('pupa'), feeding ceases and adult appendages develop within sheaths. The wax covering the adults is secreted on to plates by large ventrolateral glands, and is transferred to the wings, antennae, etc., by the hind tibiae which bear a row of 'comb hairs', absent in *Neomaskellia* in which the wax plates are little developed.

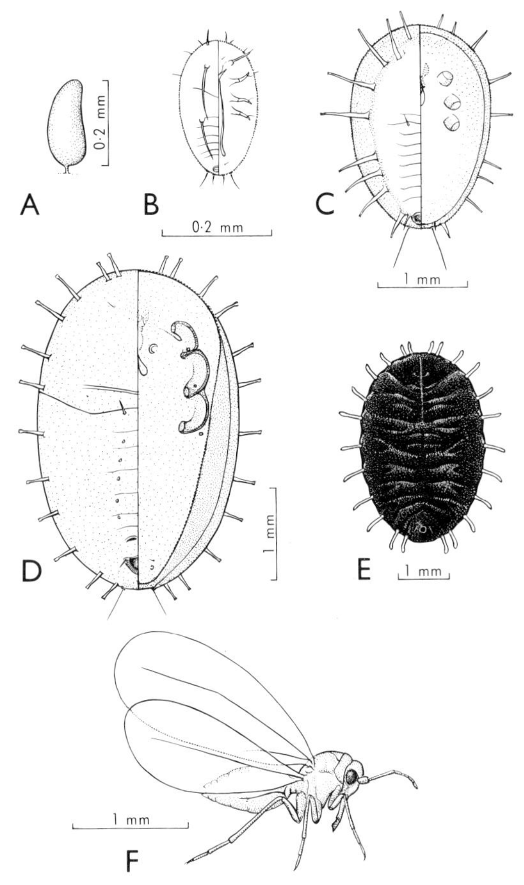

Fig. 26.39. Development of *Neomaskellia eucalypti*, Aleyrodidae: A, egg; B, 1st instar ('crawler'); C, older nymph; D, 'pupa'; E, 'pupa' dorsal; F, *Aleyrodes* sp., ♀. In B–D, dorsal view in left half, ventral in right. [S. Curtis]

The introduced species are *Trialeurodes vaporariorum* (Westw.), which occurs on numerous plants, *Aleurodicus destructor* Mackie (= *alboflocossa* Frogg.), known in Australia only from *Banksia* and undetermined trees although elsewhere a pest of coconut, *Pealius azaleae* (Baker & Moles) on *Rhododendron*, and *Neomaskellia bergii* Mask. on sugar cane. One endemic species, *Synaleurodicus hakeae* Solomon, living on *Hakea*, belongs to the Udamoselinae, and there are endemic species in 10 genera of Aleyrodinae. [Dumbleton, 1956.]

Superfamily COCCOIDEA

About 6,000 species of scale insects and mealy bugs have been described, and about 10 per cent of these have been found in Australia. Froggatt (1921) catalogued the species then known from Australia, and Brookes (1957) gave an account of the introduced Coccoidea of South Australia*. Beardsley (1966) has reviewed the coccoids of Micronesia.

---

* See also H. M. Brookes, *Trans. R. Soc. S. Aust.* **88:** 15–20, 1964.

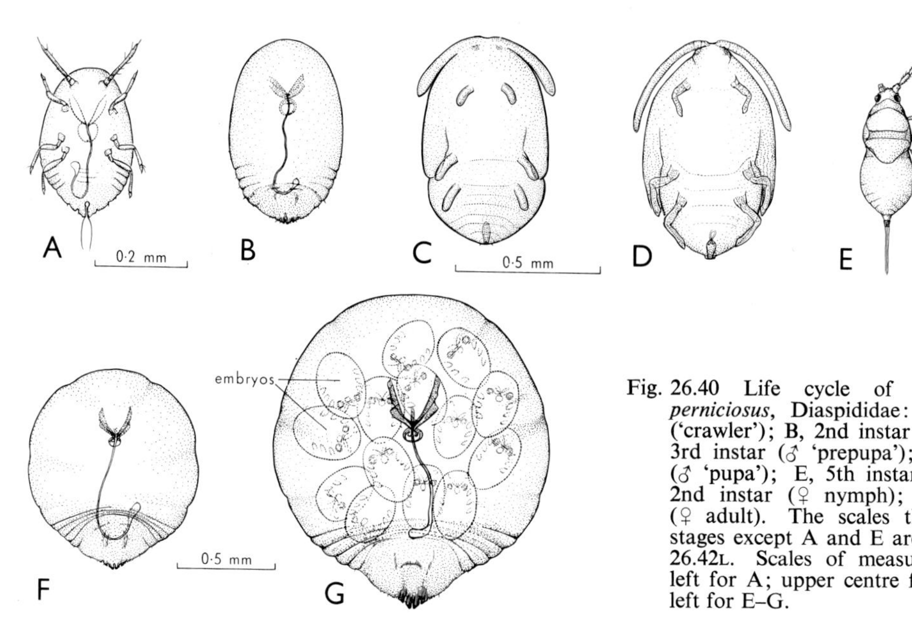
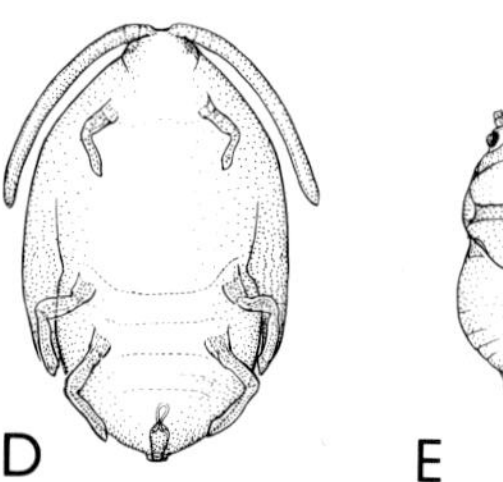
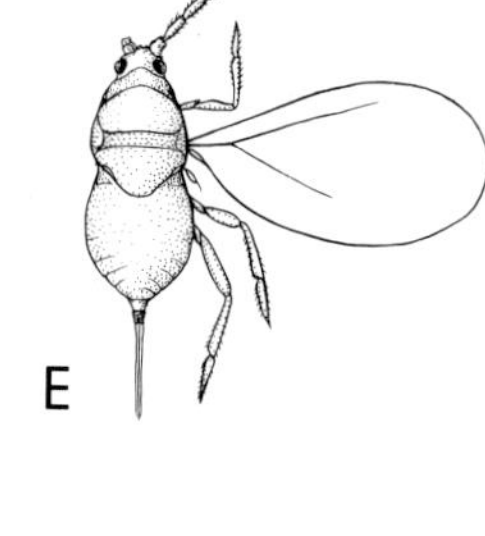

Fig. 26.40 Life cycle of *Quadraspidiotus perniciosus*, Diaspididae: A, 1st instar ('crawler'); B, 2nd instar (♂ nymph); C, 3rd instar (♂ 'prepupa'); D, 4th instar (♂ 'pupa'); E, 5th instar (♂ adult); F, 2nd instar (♀ nymph); G, 3rd instar (♀ adult). The scales that conceal all stages except A and E are shown in Fig. 26.42L. Scales of measurement: upper left for A; upper centre for B–D; lower left for E–G. [S. Curtis]

Reproduction (Fig. 26.40) is oviparous, ovoviviparous, or viviparous, and sometimes parthenogenetic. Eggs are usually laid beneath the body or the scale of the female or in a wax ovisac. First-instar nymphs ('crawlers') are mobile, with functional legs, but later nymphal instars and the adult female may be sedentary, with the legs reduced or lost and their very long stylets remaining inserted into one part of the food plant. The adult female is neoteinic, wingless, and protected by a soft or hard scale or a mealy secretion. The adult male is apterous or 2-winged, with the hind wings reduced to pseudohalteres; the mouth-parts are atrophied. In the last 2 immature instars of the male ('prepupa' and 'pupa'), wing pads are developed in alate forms.

Key to the Families of Coccoidea Known in Australia—Nymphs and Females

[By D. J. Williams, Commonwealth Institute of Entomology, London]

1. Abdominal spiracles present 2
   Abdominal spiracles absent 3
2. Anal ring with a pore band and 6 anal-ring setae; terminal segment of antenna with a stout apical seta .... **Ortheziidae**
   Anal opening from an apical to a distinctly dorsal position, sometimes surrounded by a cluster of pores and setae, but never with cellular anal ring and setae; antennae ranging from flat plates with one or more setae to a normal 11-segmented type with numerous setae but without a single stout apical seta .... **Margarodidae**
3. Adult ♀ with a pair of pore-bearing brachial plates on thorax; anterior spiracles much larger than posterior; with a median sclerotized spine on dorsal surface of abdomen. [Abdomen prolonged and bearing an anal ring with 10 setae surrounded by a ring of points and fimbriations] .... **Lacciferidae**
   Without brachial plates; spiracles not differing in size; dorsal spine absent 4
4. With a distinct pair of triangular or subtriangular plates forming an operculum situated at base of an anal cleft .... **Coccidae**
   Without a pair of plates forming an operculum at base of anal cleft 5

5. With geminate or 8-shaped pores situated somewhere on the body, rarely lacking in adult ♀, and then present in at least 1st instar; antennae usually reduced to tubercles, sometimes 8-segmented; legs absent in adult ♀ or represented by small tubercles .... **Asterolecaniidae**
   Without this combination of characters 6
6. With at least one of the following characters: dorsal ostioles present, a ventral circulus between 4th and 5th segments, antennae present in adult ♀, 5- to 9-segmented or reduced to 2-segmented tubercles, often with terminal segment swollen and elongate in relation to penultimate segment; cerarii usually present at least on anal lobe; usually with trilocular pores and tubular ducts with inner end truncate or slightly convex. [Mealy bugs] .... **Pseudococcidae**
   Without a combination of dorsal ostioles, circuli, trilocular pores, or with the terminal segment of antenna slightly swollen and elongate 7
7. Adult ♀ with terminal abdominal segments fused into a pygidium surrounding the anal ring, which is never setigerous. [Legs lacking in all stages except 1st; antenna in adult ♀ always reduced to a tubercle; body of adult ♀ either secreting a covering scale which incorporates the larval skins, or completely enclosed within exuviae of 2nd instar] 8
   Adult ♀ without a pygidium 9
8. Pygidium of adult ♀ with dorsal ducts and a marginal fringe of lobes and plates or gland spines; usually secreting a scale, sometimes enclosed within exuviae of 2nd instar possessing a similar type of pygidium .... **Diaspididae**
   Adult ♀ with simple membranous pygidium without lobes and plates or gland spines; always enclosed within exuviae of 2nd instar, the posterior end of which forms a flat anal plate or operculum surrounded by a heavily sclerotized rim .... **Halimococcidae**
9. Adult ♀ with posterior end of body forming a short anal cleft, and with a short telescoping anal tube extending into the body and bearing at its inner extremity an anal ring with about 10 setae or seta-like projections; dorsal part of anal-tube structure formed into a triangular or oval plate bearing about 10 setae; anal plate sometimes cleft at apex but this cleft never extending entirely to base; inner extremity of tubular ducts either slightly expanded and forming a shallow cup and laterally prolonged into a small filament, or prolonged into several minute filaments . **Aclerdidae**
   Anus a mere slit, or well developed and surrounded by a cellular and setigerous anal ring; without a triangular or oval plate forming the dorsal part of the anal-tube structure, or, if there is any structure resembling a plate or cauda, then it is without setae; body often with well-developed anal lobes; tubular ducts, when present, with inner end expanded and reflexed to form a distinct cup .... **Eriococcidae**

30. Margarodidae (Monophlebidae; Figs. 26.41C,D, 42A,B). The Australian species are placed in about 14 genera, including *Auloicerya* and *Monophlebulus*; many of the species have been collected from Australian plants. *Icerya purchasi* Mask., the cottony cushion scale, is a wattle-feeding species, apparently of Australian origin, but now a world-wide pest of *Citrus*. The predatory Australian coccinellid *Rodolia cardinalis* (Muls.) has been used for its control (p. 593). The Margarodidae contain some of the largest coccoids, some females having a body length up to 35 mm. The cysts of *Eumargarodes* and *Promargarodes* and of some related non-Australian genera are known as 'ground pearls', which are used as beads in some parts of the world. They are an additional immature stage capable of long periods of quiescence during unfavourable conditions. The females emerging from the ground pearls have greatly enlarged fore legs modified for digging, but the other legs are normal. Legs and antennae are present in the first instar of Margarodini, absent from the middle instars, and appear again in the adults. The males of *Callipappus* can be taken flying about in the sunlight in search of the female on a stump or branch a short distance above the ground. From their bright red tints and beautiful tails of wax filaments the males are popularly called 'bird of paradise flies'. [Morrison, 1928.]

31. Ortheziidae (Fig. 26.41A, B). Both species are apparently introduced. *Newsteadia floccosa* (Westw.) is usually found in damp shady places, and *Orthezia insignis* Browne feeds on the aerial parts of many dicotyledons.

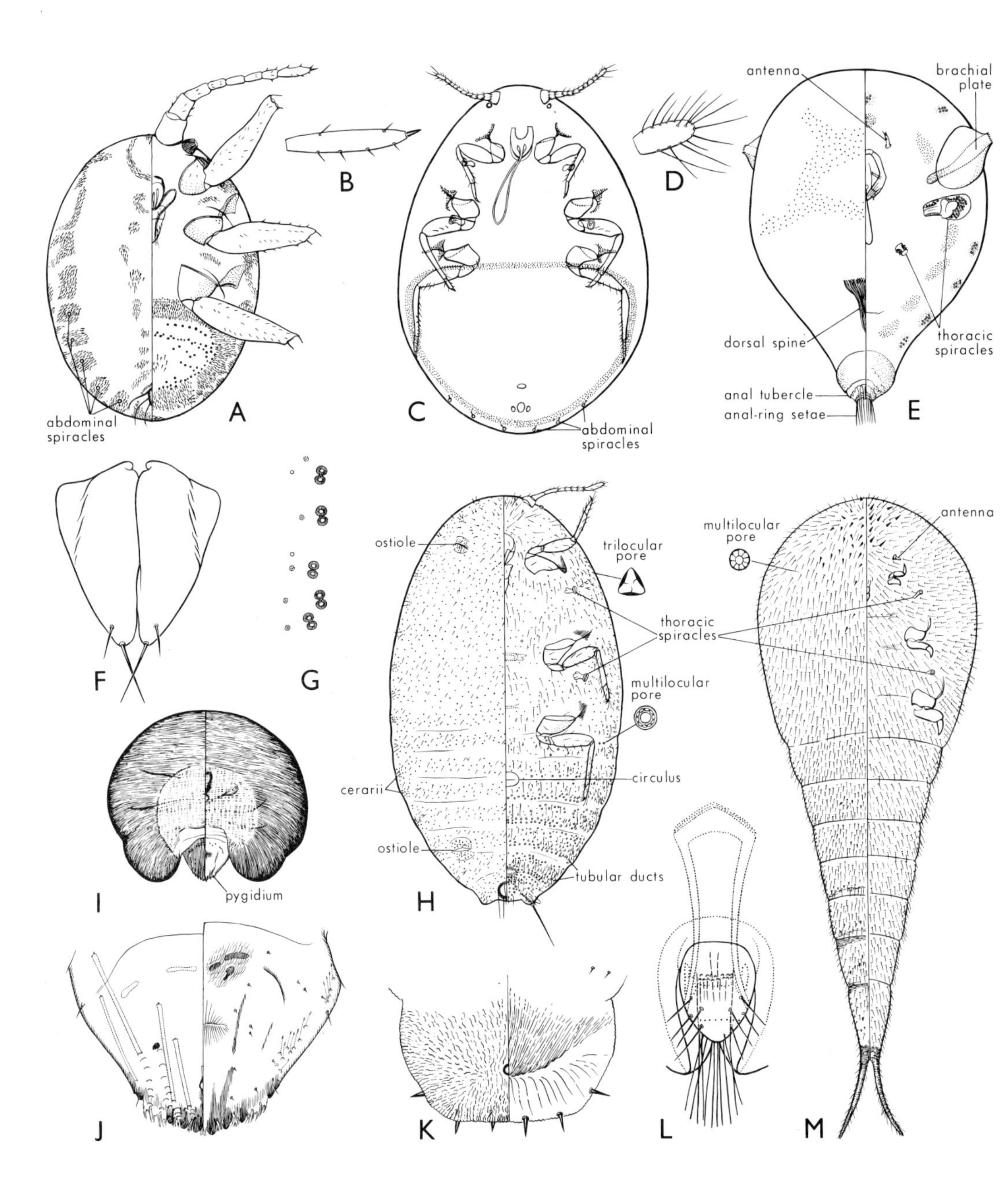

Fig. 26.41. Coccoidea (slide preparations of ♀♀): A, *Orthezia insignis*, Ortheziidae, whole insect (after Morrison, 1925); B, same, terminal antennal segment; C, *Icerya seychellarum*, Margarodidae, whole insect (after Morrison, 1928); D, same, terminal antennal segment; E, *Austrotachardia angulata*, Lacciferidae, whole insect (after J. Chamberlin, 1923); F, *Ceroplastes rubens*, Coccidae, operculum (after Ferris, in Zimmerman, 1948); G, *Asterolecanium hakeae*, Asterolecaniidae, geminate pores (after Russell, 1941); H, *Planococcus citri*, Pseudococcidae, whole insect (after Ezzat and McConnell, 1956); I, *Aonidiella aurantii*, Diaspididae, whole insect (after McKenzie, 1938); J, same, pygidium; K, *Colobopyga coperniciae*, Halimococcidae, pygidium (after Ferris, 1952); L, *Aclerda takahashii*, Aclerdidae, anal cleft (after McConnell, 1953); M, *Apiomorpha frenchi*, Eriococcidae, whole insect (after Ferris, 1957a—and see also Fig. 5.14).

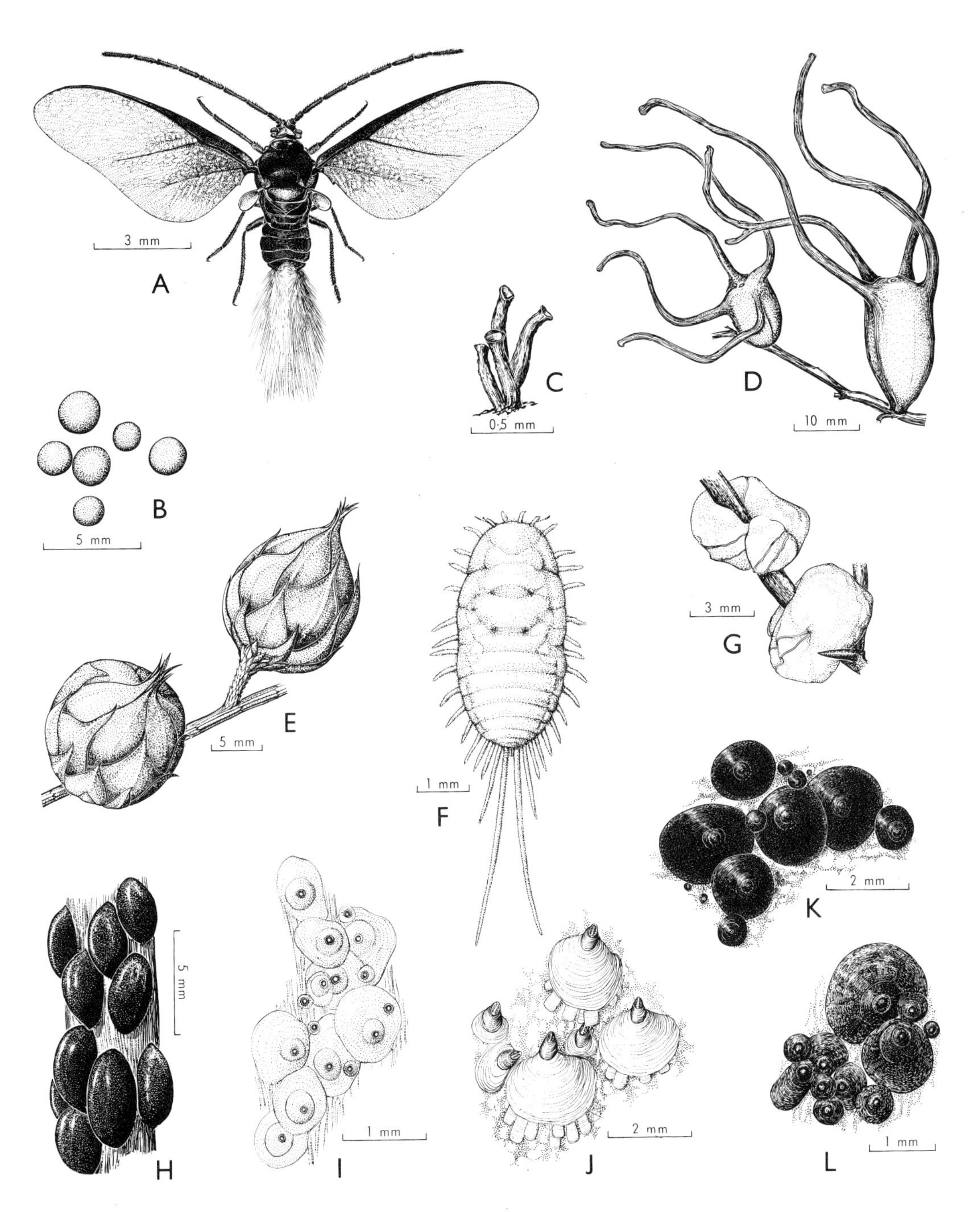

Fig. 26.42. Coccoidea: A, male of *Callipappus* sp., Margarodidae; B, cysts of *Eumargarodes* sp., Margarodidae; C, ♂ gall of *Apiomorpha munita*, Eriococcidae; D, ♀ gall of same; E, gall of *Cylindrococcus spiniferus*, Eriococcidae; F, *Pseudococcus longispinus*, ♀, Pseudococcidae; G, *Ceroplastes destructor*, ♀, Coccidae; H, scales of *Saissetia nigra*, Coccidae. I–L, diaspidid scales: I, *Aonidiella aurantii*; J, *Berlesaspis spinifera*; K, *Chrysomphalus ficus*; L, *Quadraspidiotus perniciosus*. [S. Curtis]

Ortheziids secrete a waxen ovisac, which may be longer than the rest of the body. They are sometimes known as marsupial coccids because the eggs hatch while the insect is still mobile. [Morrison, 1952.]

32. Eriococcidae (Figs. 26.41M, 42C–E). These insects, particularly well represented in Australia, are of diverse form, and have a wide host range. One group contains scales with felted wax coverings: *Rhizococcus* spp. on *Acacia* and *Casuarina*, *Gossyparia* spp. on *Casuarina* and *Eucalyptus*, and *Eriococcus* spp. on a great variety of plants. *E. coriaceus* Mask. is a pest of many species of *Eucalyptus*, and Dumbleton (1940) described attempts to control it in New Zealand with Australian parasites. *Dactylopius coccus* Costa is a Mexican species from which cochineal is prepared. *Dactylopius* spp. were introduced into Australia in an attempt to control prickly pear (F. Wilson, 1960). Most of the Australian species originally described as *Dactylopius* are Pseudococcidae. Another group, erroneously referred to as the Idiococcinae, contains *Sphaerococcopsis* and *Olliffia* which live on the bark of *Eucalyptus*, sometimes forming pits. *Ourococcus* spp. live on the bark of both *Eucalyptus* and *Casuarina*. Some *Sphaerococcus* spp. deform the branches of *Acacia*, and others live on *Eucalyptus*, *Melaleuca*, and *Casuarina*, some species forming galls on *Casuarina*, as do all known *Cylindrococcus*. Some endemic eriococcids have been referred to the Brachyscelinae or Apiomorphinae by some authors. *Apiomorpha*, *Ascelis*, *Cystococcus*, and *Opisthoscelis* form galls on the stems, twigs, or leaves of *Eucalyptus*. The males of *Apiomorpha* and *Opisthoscelis* form differently shaped galls from those of the female, but the male nymphs of *Ascelis* remain in the parent gall and do not produce a gall on their own. [Ferris, 1957a,b; Hoy, 1962.]

33. Aclerdidae (Fig. 26.41L). The only Australian representative is *Aclerda sellahispanica* Lind. from W.A., and is of uncertain identity. *A. takahashii* Kuw. occurs on sugar cane in some parts of the world, and might be found in Queensland.

34. Asterolecaniidae (Fig. 26.41G). This family contains scale insects living on a variety of plants. The antennae of the females may be reduced, and the legs present, vestigial, or absent. They may be covered in a felted sac enclosed with a hard waxy test, or they may form galls.

35. Pseudococcidae (Figs. 26.41H, 42F). Many species of true mealy bugs occur on native Australian plants. Several live concealed on Gramineae; those of some economic importance include *Antonina graminis* (Mask.), *Ripersiella rumicis* (Mask.), and *Tridiscus distichlii* (Ferris). Other pest species are *Dysmicoccus brevipes* (Cock.) on pineapples and *Planococcus citri* (Risso) on *Citrus*.

36. Coccidae (Lecaniidae; Figs. 26.41F, 42G,H). These are scale insects which may be either naked or covered with wax secretions. Many of the Australian species live on native plants, but others, such as *Coccus hesperidum* L., *Paralecanium expansum* (Green), *Saissetia coffeae* (Walk.), *S. nigra* (Nietn.), and *S. oleae* (Bern.), are introduced species. *Coccus* and *Pulvinaria* are widely distributed genera, and *Saissetia* spp. are injurious to many cultivated plants. Smith (1944) gave an account of *S. nigra* and Blumberg (1935) described the biology of *Ceroplastes rubens* Mask. *C. destructor* Newst. is a common pest of many shrubs and trees. *C. ceriferus* (And.) and *Ericerus pela* Sign. yield wax on a small commercial scale in Asia.

37. Lacciferidae (Fig. 26.41E). Most of the known Australian species belong to the genus *Austrotachardia*, and live on native plants. The Oriental *Laccifer lacca* (Kerr) secretes the stick-lac from which shellac is prepared. Males of lacciferids have been recorded with 2-segmented tarsi, a small basal segment being present. [Kapur, 1958.]

38. Halimococcidae (Fig. 26.41K). The only known Australian representative is *Colobopyga kewensis* (Newst.) from Lord Howe I. Other species described from Australia are now regarded as belonging to the Diaspididae (Brown and McKenzie, 1962).

39. Diaspididae (Figs. 26.40, 41I,J, 42I–L). The hard or armoured scales include many introduced species living on crops of economic importance. There is also a rich endemic fauna, including at least 10 common genera. *Aonidiella aurantii* (Mask.) and *A. citrina* (Coq.) occur on *Citrus*. Attempts have been made to control the former by introducing the encyrtid *Comperiella bifasciata* How. Another encyrtid, *Aphytis diaspidis* How., has also been recorded as exercising some control. *Aspidiotus nerii* Bouché (= *A. hederae* auctt.) occurs on many trees and shrubs. *Chrysomphalus ficus* Ashm. and *Lepidosaphes beckii* (Newm.) are pests of *Citrus* in Queensland. *L. ulmi* (L.) is a pest of apples. *Parlatoria proteus* Curt. is of world-wide distribution on date palms, *Citrus*, and other plants. *Quadraspidiotus ostreaeformis* (Curt.), *Q. pyri* (Licht.), and *Q. perniciosus* (Comst.), the San José scale, are pests of apples and other fruit trees. *Unaspis citri* (Comst.) is said to be injurious to *Citrus* only in neglected orchards. [Brimblecombe, 1962; S. Brown, 1965; Ghauri, 1962.]

Suborder HETEROPTERA

The Heteroptera are commonly separated into 3 divisions: the Hydrocorisae, with short antennae wholly or mostly concealed beneath the head, and comprising the littoral Ochteroidea and the aquatic Notonectoidea and Corixoidea, which can move underwater as adults and have entirely underwater nymphs; the Amphibicorisae, including only the Gerroidea, which are mostly adapted to life on the surface of water; and the Geocorisae, comprising all the other superfamilies, mostly terrestrial, but including some littoral and semiaquatic forms. The Geocorisae and Amphibicorisae have often been united as the Gymnocerata, and the Hydrocorisae called the Cryptocerata. There are still differences of opinion on the relationships of the superfamilies, and in some instances on their constitution and limits and the family or subfamily status of the included groups. Leston, Pendergrast and Southwood (1954) have distinguished two major groups of Geocorisae, the Pentatomomorpha (Pentatomoidea, Coreoidea, Lygaeoidea, Aradoidea) and the Cimicomorpha (Cimicoidea, Tingoidea, Reduvioidea). There is some doubt as to the position of the Saldoidea, Enicocephaloidea, and Dipsocoroidea in this scheme. A general account of the Heteroptera is given by Poisson and Pesson (1951); their biology has been treated by N. Miller (1956) and Southwood and Leston (1959); China and Miller (1959) have given keys to the families and subfamilies of the world.*

* R. H. Cobben (*Evolutionary Trends in Heteroptera Part I*, 475 pp. (Wageningen, Netherlands: Centre for Agricultural Publishing and Documentation, 1968)) has published a new arrangement which may dispose of the uncertainties indicated above.

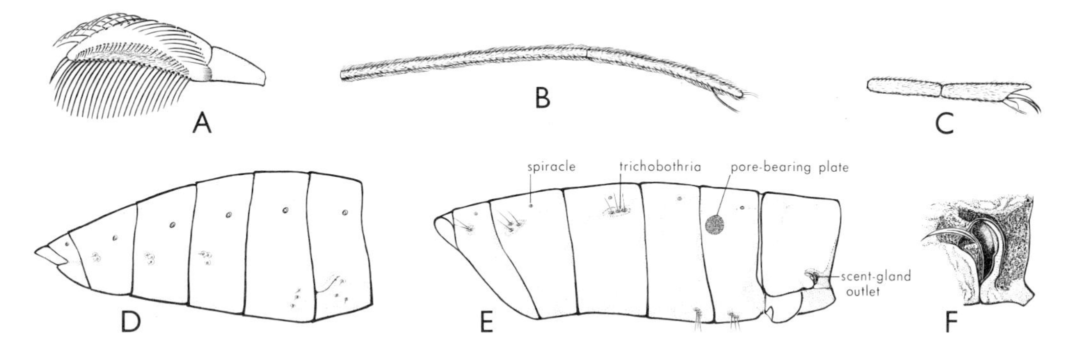

Fig. 26.43. A, pala of *Agraptocorixa eurynome*, Corixidae; B, hind tarsus of *Limnogonus luctuosus*, Gerridae; C, same of *Microvelia mjobergi*, Veliidae; D, abdomen of *Mictis profana*, Coreidae, showing groups of trichobothria; E, abdomen of *Maevius indecora*, Hyocephalidae; F, scent-gland opening of *Maevius indecora*. [S. Curtis]

Key to the Superfamilies of Heteroptera—Adults

1\. Antennae usually entirely concealed beneath, and shorter than, head (less than half visible dorsally in Ochteridae) 2
Antennae exposed and fully visible from above, as long as, usually longer than, head; terrestrial or semiaquatic (Gymnocerata) 5

2(1). Body flat, ovoid, with marginal laminae; without wings or eyes; in termites' nests (Termitaphididae) .... ARADOIDEA (pt., p. 442)
Body not with all characters as above; mostly aquatic (underwater), sometimes semiaquatic (Hydrocorisae = Cryptocerata) 3

3(2). Ocelli present; riparian species .... OCHTEROIDEA (p. 454)
Ocelli absent (except in *Diaprepocoris* (Corixidae) which has palae and other corixid characters as in couplet 4); aquatic species 4

4(3). Fore tarsi usually scoop-like (palae), with a row of long hairs, never with 2 claws (Fig. 26.43A); head not set into prothorax, but slightly overlapping pronotum; dorsal surface of body never strongly arched .... CORIXOIDEA (p. 457)
Fore tarsi not scoop-like, sometimes with 2 claws; head normally set in prothorax; dorsal surface of body sometimes strongly arched .... NOTONECTOIDEA (p. 455)

5(1). Claws of at least fore legs inserted well before apex of last tarsal segment, which is usually more or less cleft longitudinally (cleft sometimes difficult to see); living on surface of water .... GERROIDEA (pt., p. 452)
Claws all apical; last tarsal segment not cleft; habits varied 6

6(5). Head linear, several times longer than wide, somewhat dilated in front, as long as whole thorax, eyes remote from base; body and legs very slender (Hydrometridae) .... GERROIDEA (pt., p. 453)
Head no more than about twice as long as wide, shorter than thorax including scutellum; body and legs of varied form 7

7(6). Pronotum entirely or almost covering scutellum; ocelli absent; hemelytra, and often thorax, with a raised, lace-like reticulation enclosing small cells; tarsi 2-segmented. [Less than 5 mm long] .... TINGOIDEA (p. 438)
Pronotum usually not covering scutellum, but, if it does, then ocelli present and/or tarsi 3-segmented 8

8(7). Flightless, ovoid ectoparasites of vertebrates; fore wings represented by short coriaceous scales (Cimicidae, Polyctenidae) .... CIMICOIDEA (pt., p. 435)
Not flightless ectoparasites with the above form 9

9(8). Body flattened dorsoventrally; when hemelytra developed, sides of abdomen usually extending well beyond them; tarsi 2-segmented but robust (Aradidae) ...... ARADOIDEA (pt., p. 441)
Body rarely much flattened; if it is, then tarsi either 3-segmented or very short and several times as slender as tibiae 10

10(9). Antennae with apical 2 segments very slender, with long erect hairs, and together more than twice as long as the short, thick scape and pedicel, segment 3 swollen at base .... DIPSOCOROIDEA (p. 434)
Antennae not as above 11

11(10). Hemelytra with a cuneus .... CIMICOIDEA (pt., p. 435)
Hemelytra without a cuneus 12

12(11). Tarsi 2-segmented, very small, several times as slender as tibiae (Thaumastocoridae) .... CIMICOIDEA (pt., p. 437)
Tarsi usually 3-segmented, if 2-segmented, not as slender as above 13

13(12). Labium with basal segment, even in repose, well separated from underside of head, either by permanent curvature of basal segment, or by ventral projection of apex of head 14
Labium with basal segment straight and apex of head not ventrally projecting, so that in resting condition labium lies closely parallel to head and prothorax 18

14(13). Corium of hemelytra in winged forms with strong, raised, dark veins enclosing 2 elongate cells; membrane without veins, often torn off in copulation; apterous forms common; tarsi 3-

segmented, basal segment very small, much shorter than 2nd; antennae 4-segmented; antennae and legs long and slender; body elongate-oval, yellowish or greenish; living on surface of water or in wet places (Mesoveliidae) .... GERROIDEA (pt., p. 453)

Corium without such veins and cells; membrane with or without veins; when tarsi 3-segmented, 2nd segment small, shorter than 1st; antennae of 4 or 5 segments; form and habits varied. 15

15(14). Ocelli placed between eyes; head broader than long, with large, prominent eyes; mainly riparian, littoral, or intertidal .... SALDOIDEA (p. 440)

Ocelli placed behind eyes, rarely absent; head longer than broad; eyes rarely unusually prominent; habitat usually terrestrial, sometimes on or near water 16

16(15). Fore tarsi 1-segmented, mid and hind 2-segmented; head constricted, then globularly swollen behind eyes; fore wings entirely membranous .... ENICOCEPHALOIDEA (p. 434)

All tarsi usually 3-segmented, sometimes all 2-segmented; head not constricted and globularly swollen behind eyes; fore wings usually with distinct corium, sometimes entirely membranous or entirely coriaceous 17

17(16). Labium short, stout, apparently 3-segmented; prosternum with median stridulatory groove extending anterior to fore coxae (most Reduviidae) .... REDUVIOIDEA (pt., p. 439)

Labium long and slender, usually 4-segmented; prosternum without a stridulatory groove (Nabidae) .... CIMICOIDEA (pt., p. 436)

18(13). Scutellum reaching apex of clavus or beyond, sometimes covering abdomen .... PENTATOMOIDEA (p. 449)

Scutellum not reaching apex of clavus 19

19(18). Prosternum with median stridulatory groove extending anterior to fore coxae; hemelytron with 2 or 3 large, very long, closed cells occupying most of membrane (some Reduviidae) ... . REDUVIOIDEA (pt., p. 439)

Prosternum usually without a median groove anterior to fore coxae; if a groove present, membrane of hemelytron without such large closed cells 20

20(19). Tarsi 2-segmented, basal segment very small; very small species, length up to 2 mm; membrane of hemelytron without veins; the apical 2 antennal segments much more slender than the others; on surface or near edge of water (Hebridae) .... GERROIDEA (pt., p. 453)

Tarsi usually 3-segmented (2nd short), when 2-segmented, insect either much more than 2 mm long, or apical antennal segment stout; membrane usually with veins, sometimes without; terrestrial 21

21(20). Abdominal trichobothria absent; ♂ claspers asymmetrical; brachypterous forms with membranes reduced or absent (flightless anthocorids and mirids only) . CIMICOIDEA (pt., p. 435)

Abdominal trichobothria present; ♂ claspers symmetrical; usually macropterous, sometimes brachypterous 22

22(21). Ocelli absent; insects more than 5 mm long, and with red or yellow ground colour .... LYGAEOIDEA (pt., p. 444)

Ocelli usually present (absent only in a few brachypterous forms which are brown or dark and less than 5 mm long) 23

23(22). Membrane of hemelytron, when developed, with at least 7 and usually many more main longitudinal veins; brachypters with antennifers inserted dorsolaterally, dorsal to the line between centre of eyes and apex of head .... COREOIDEA (pt., p. 442)

Membrane of hemelytron with no more than 6 and usually fewer main longitudinal veins; brachypters with antennifers inserted on or ventral to the line between centre of eyes and apex of head 24

24(23). Base of abdomen beneath with an ovoid pore-bearing plate on each side (Fig. 26.43E); a bristle-like process from opening of metapleural scent gland (Fig. 26.43F); membrane of macropters with 4 main veins connected by cross-veins to form large basal cells (Hyocephalidae) .... COREOIDEA (pt., p. 442)

Base of abdomen without such ventral pore-bearing plates; without a bristle-like process from opening of metapleural scent gland; membrane of macropters usually with 5 main veins, sometimes fewer, rarely with basal cells .... LYGAEOIDEA (pt., p 444).

Superfamily ENICOCEPHALOIDEA

40. Enicocephalidae (Fig. 26.44). The enicocephalids have often been included in the Reduvioidea, but the present tendency is to regard them as a rather isolated family which has probably arisen from near the base of the heteropteran stem. These bugs, of small to medium size, are, so far as known, predacious on a wide range of small insects. Both flightless and macropterous species occur on the forest floor, under stones, and in rotting logs, while winged forms are sometimes seen at dusk in large swarms resembling those of midges. The adults are unusual among terrestrial bugs in lacking ventral thoracic openings to the scent glands; instead they retain the median dorsal nymphal opening on tergum 4 of the abdomen. Fore wings entirely membranous. The eggs have a thin, simple, permeable chorion, and hatch only in a high atmospheric humidity. Described Australian species belong to *Henschiella* and *Didymocephalus*, both in ENICOCEPHALINAE. [Jeannel, 1941; Usinger, 1945.]

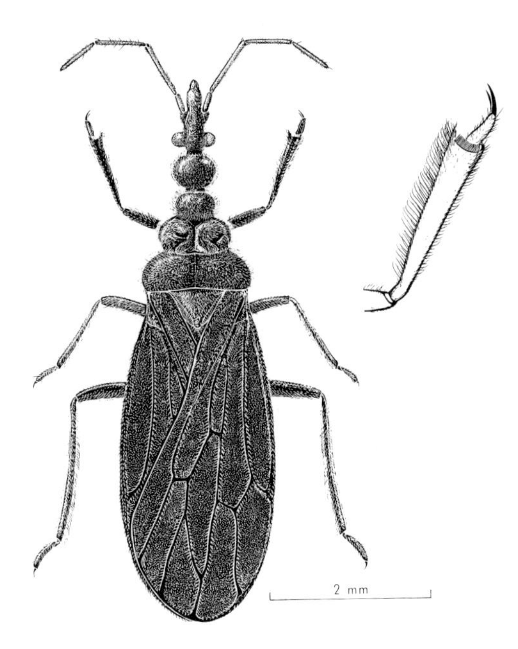

Fig. 26.44. *Oncylocotis* sp., Enicocephalidae, and fore tibia and tarsus enlarged. [S. Curtis]

Superfamily DIPSOCOROIDEA

Another apparently primitive group of rather uncertain affinities. It includes small, predacious bugs living in moist places, such as among leaf litter and moss and in ants' nests. The Schizopteridae are often included as a subfamily of the Dipsocoridae, though the present tendency is to separate them. [Gross, 1951.]

41. Dipsocoridae (Cryptostemmatidae; Fig. 26.45A). Head porrect, or at least not strongly deflexed, more or less conical; hemelytron usually with cuneal fracture.

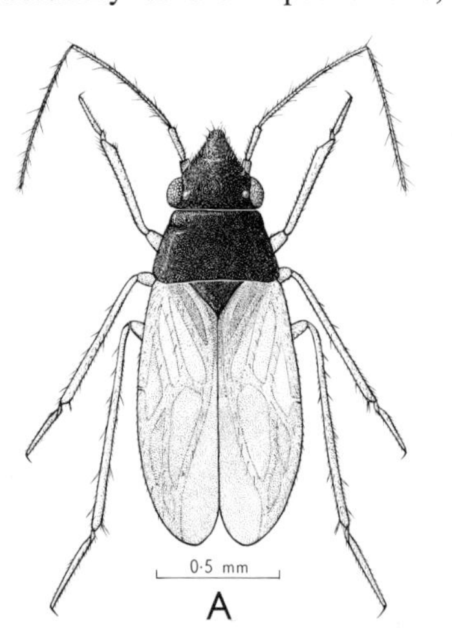

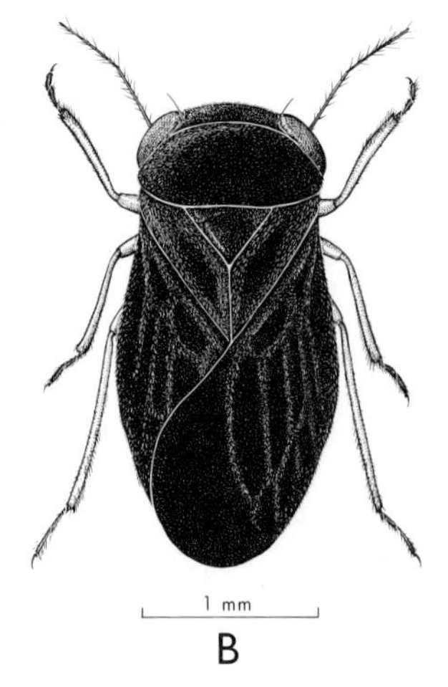

Fig. 26.45. A, *Ceratocombus australiensis*, Dipsocoridae; B, undescribed genus of Schizopteridae. [S. Curtis]

Represented by *Ceratocombus (Xylonannus) australiensis* Gross, with a wide distribution in Australia.

42. Schizopteridae (Fig. 26.45B). Head strongly deflexed, broad at apex; hemelytron without cuneal fracture. The described Australian species belong to the endemic genera *Dictyonannus*, *Pachyplagia*, and *Pachyplagioides*.

Superfamily CIMICOIDEA

There is a strong tendency to blood-sucking in this superfamily, although the Thaumastocoridae and possibly the majority of the Miridae are phytophagous; many Miridae, however, are facultative or obligatory suckers of the blood or eggs of other insects. It is possible that blood-sucking is a primitive habit of the cimicomorphs (p. 396). There is also a tendency for haemocoelic fertilization, strongest in the Cimicidae, but appearing also in some Anthocoridae (Lyctocorinae) and Nabidae.

Key to the Families of Cimicoidea Known in Australia

1. With ctenidia on head or thorax; eyes absent; micropterous ectoparasites of bats ... **Polyctenidae**
   Without ctenidia; eyes present; wing development and habits varied 2
2. Body oval, much flattened; flightless, with fore wings reduced to short, transverse scales; pronotum with flattened lateral expansions behind eyes; clypeus apically broadened; obligatory blood-suckers of vertebrates .... **Cimicidae**
   Body, fore wings, pronotum, and clypeus not as above; not obligatory blood-suckers 3
3. Tarsi 2-segmented, very small, several times as slender as tibiae; tibiae with apical process almost as long as tarsi. [Small lygaeid-like bugs with pedunculate or subpedunculate eyes] .... **Thaumastocoridae**
   Tarsi 3-segmented; tarsi and tibiae not as above 4
4. Ocelli absent; hemelytron (except in some brachypters) with a cuneus; membrane of hemelytron with 1 or 2 closed cells near cuneus (Fig. 26.4B) .... **Miridae**
   Ocelli present; hemelytron with or without a cuneus; membrane either with no basal cells, or with 3 or more large basal cells 5
5. Hemelytron, except in some brachypters, with a cuneus; labium apparently 3-segmented; ♂ claspers asymmetrical; body less than 6 mm long .... **Anthocoridae**
   Hemelytron never with a cuneus; labium obviously 4-segmented; ♂ claspers symmetrical; body more than 6 mm long .... **Nabidae**

43. Cimicidae (Figs. 26.12, 46A). All are flightless ectoparasites, feeding on the blood of mammals or birds.* The only species recorded from Australia is the cosmopolitan bed-bug of man, *Cimex lectularius* L., introduced since white settlement. The female has a thick-walled spermalège opening on abdominal sternum 5, which protects the viscera from the enlarged left clasper of the male and reduces bleeding when this penetrates. The sperms proceed by way of the haemocoele to the oviducts, where the eggs are fertilized.

* See R. L. Usinger. *Monograph of Cimicidae (Hemiptera-Heteroptera)*. Thomas Say Foundation Publ. no. 7, 585 pp. (College Park, Md.: Ent. Soc. Am., 1966).

44. Polyctenidae (Fig. 5.22B). Small ectoparasites of bats, lacking eyes, ocelli, and hind wings; fore wings short and coriaceous. There are comb-like rows of flattened spines (ctenidia), variable in number and position, on the body. They are viviparous, the young being born at an advanced stage, and differing markedly from the adults; 2 ecdyses occur after birth. The Australian species are *Adroctenes magnus* Maa, parasitic on *Taphozous*, and *Eoctenes intermedius* (Speiser), on *Hipposideros diadema*, both from north Queensland (Maa, 1964).

45. Anthocoridae (Fig. 26.46B). Most feed on insect eggs or blood, but the cosmopolitan *Lyctocoris campestris* has been recorded in several countries as a facultative though

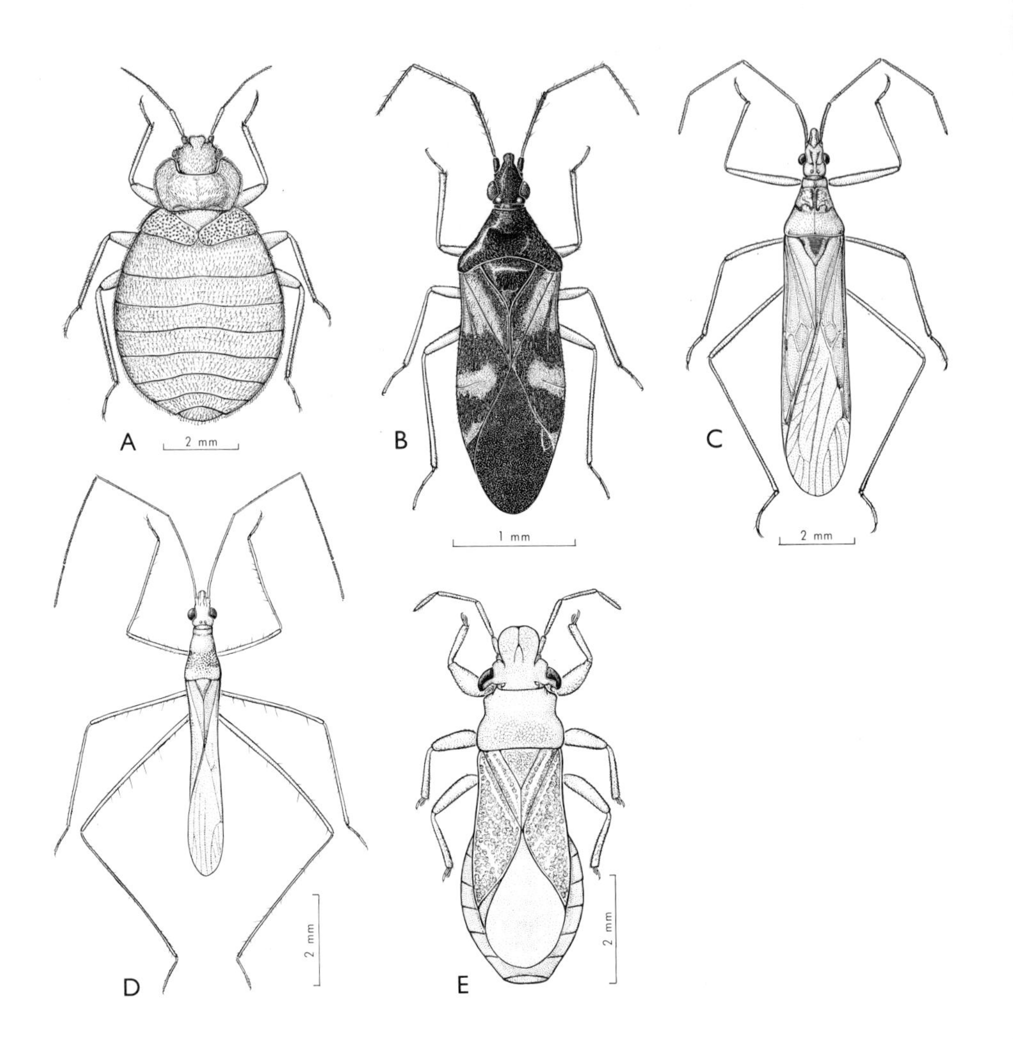

Fig. 26.46 A, *Cimex lectularius*, Cimicidae; B, *Plochiocorella elongata*, Anthocoridae; C, *Nabis capsiformis*, Nabidae; D, *Arbela telomi*, Nabidae; E, *Thaumastocoris australicus*, Thaumastocoridae. [S. Curtis]

persistent blood-sucker of man and other mammals. The anthocorids are now often regarded as a subfamily of Cimicidae. Most Australian species are small, brown or black, about 2–4 mm long, and occur commonly among leaf litter and other vegetable debris. Some, particularly Anthocorinae, are taken by sweeping plants, and others, e.g. species of *Cardiastethus*, occur under bark. The LYCTOCORINAE are most closely related to Cimicidae, and are represented by 7 species in 4 genera; the introduced *Xylocoris flavipes* (Reut.) often occurs in large numbers in peanuts and stored grain, feeding on the early stages of pest species. The ANTHOCORINAE are poorly represented here, with 2 species of *Orius* and 1 of *Anthocoris*; in the DUFOURIELLINAE we have 14 species in 7 genera, including 5 of *Cardiastethus* and 4 of *Physopleurella*. [Gross, 1954–7.]

46. Nabidae (Figs. 26.46C,D). Another predacious group, formerly included in the Reduvioidea. Recent work, including the discovery by Carayon (1954) of haemocoelic fertilization, has led to its transfer to the Cimicoidea. The eggs are inserted into the tissues of plants. NABINAE include several native species of *Nabis* and the cosmopolitan *N. capsiformis* Germ., while the PROSTEMMINAE are represented by *Alloeorhynchus flavolimbatus* Kirk., dark brown with yellowish markings, and the smaller *A. queenslandicus* Gross, as well as species of *Phorticus*. GORPINAE are represented by 2 species of *Gorpis*.

47. Thaumastocoridae (Fig. 26.46E). Formerly placed in or near the Lygaeidae, this small family was the subject of detailed study by Drake and Slater (1957), who demonstrated cimicoid relationships. The species are apparently phytophagous; both nymphs and adults of *Thaumastocoris australicus* Kirk. have been found on *Acacia*, while *T. hackeri* Drake & Slater has been collected from *Elaeocarpus obovatus*, *Baclozygum depressum* Bergr. and *B. brevipilosum* Rose from *Eucalyptus trachyphloia*, and *Onymocoris hackeri* Drake & Slater from *Banksia* (Rose, 1965).

48. Miridae (Figs. 26.4B, 5D, 47). This, one of the largest families in the Heteroptera, is predominantly phytophagous, although many species are known to be at least facultative blood- or egg-suckers of other insects. Commonest colours are green and brown, though some are attractively marked.

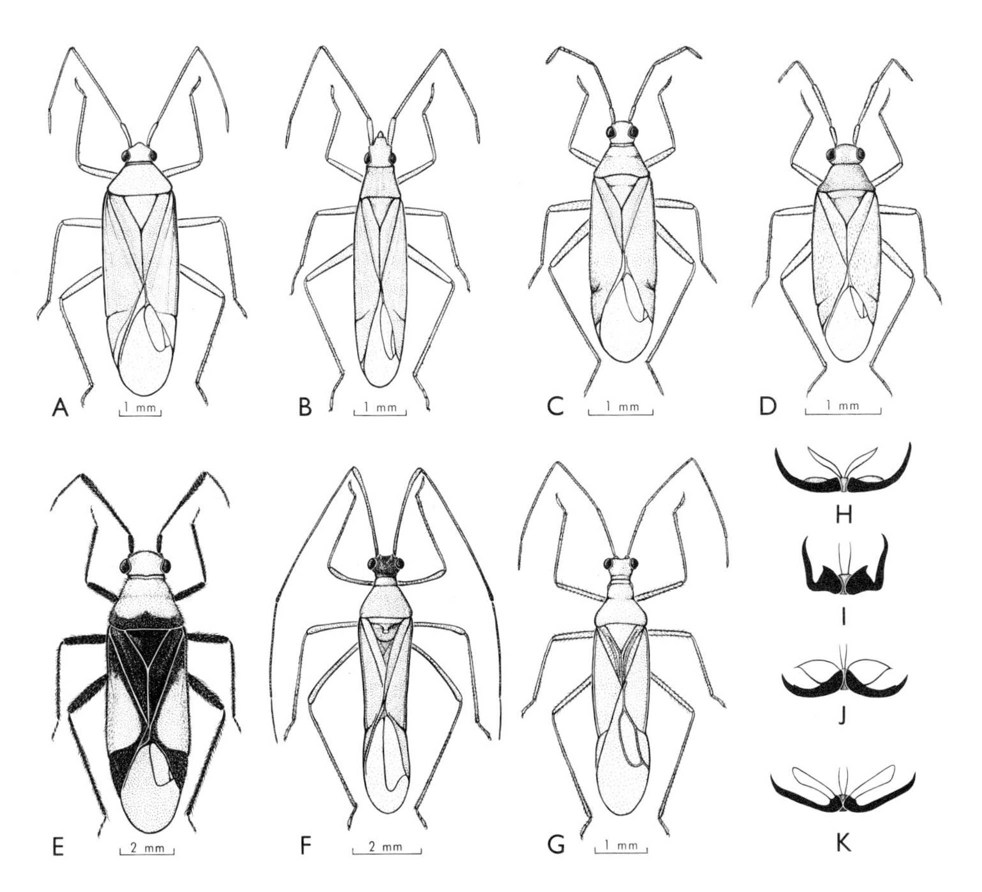

Fig. 26.47. Miridae: A, *Megacoelum modestum*; B, *Trigonotylus* sp.; C, *Cyrtopeltis tenuis*; D, *Tytthus mundulus*; E, *Trilaccus nigroruber*; F, *Helopeltis australiae*; G, *Felisacus glabratus*. H–K, pretarsal structures of: H, *Trigonotylus* sp., Mirinae; I, *Trilaccus nigroruber*, Deraeocorinae; J, *Helopeltis australiae*, Bryocorinae; K, *Cyrtopeltis tenuis*, Phylinae. [S. Curtis]

The elongate eggs are inserted into or between plant tissues by the long ovipositor. Mirids are still poorly known, and the estimate of species will probably be multiplied several-fold. They are fragile, and are best collected alive, a few in a tube, and killed shortly before mounting; if killed together with other insects, or if allowed to stick to condensed moisture, etc., hairs, scales, and antennal segments, which provide important taxonomic characters, are easily lost. Legs should not be gummed down, as the pretarsal structures are critical in classification. [J. Carvalho, 1955–60.]

The MIRINAE include about 50 described species in Australia. The widespread genus *Lygus (sens. lat.)* is well represented, with numbers of undescribed species; many overseas species are of economic importance. *Megacoelum modestum* Dist. damages a variety of crops. The narrow, linear, mainly grass-feeding Stenodemini are represented by 7 genera, including the widespread *Megaloceroea* and *Trigonotylus* and the predominant endemic genus *Zanessa*. The large subfamily ORTHOTYLINAE is poorly known in Australia, with 11 described species, 5 of which belong to the ant-mimicking tribe Pilophorini. Another large subfamily, the PHYLINAE, has 16 Australian species. Two south-western species of Dicyphini, *Cyrtopeltis droserae* China and *C. russelli* China, are adapted for survival on the sticky leaves of sundews *(Drosera)*, and suck the blood of insects trapped by these plants (p. 120). *Tytthus mundulus* (Phylini), occurring in Queensland and other parts of the south-western Pacific, sucks the eggs of several species of Delphacidae, and was successfully introduced into Hawaii to help in controlling the sugar-cane leafhopper, *Perkinsiella saccharicida*. It is dark-bodied, about 3·5 mm long, with pale green hemelytra. The ant-like Hallodapini are poorly represented.

In the DERAEOCORINAE, the entirely predacious Deraeocorini are represented by 7 recorded species, but many remain undescribed; they are brown or black and, especially the females, rather stout-bodied. The Australian Saturniomirini include only 4 species of *Trilaccus* and 1 of *Imogen*. Black and red species of *Trilaccus* are sometimes collected in large numbers from potato and cape gooseberry plants in Queensland. The CYLAPINAE are represented by only 2 recorded species. There are 9 described species of BRYOCORINAE, all but 3 belonging to the Monaloniini, parallel-sided mirids to which at least 7 undescribed species also belong. *Felisacus glabratus* (Motsch.) (Bryocorini) is slender, shining green to straw-brown, and occurs on ferns. *Helopeltis australiae* Kirk. and *Eucerocoris suspectus* Dist. (Monaloniini) belong to genera with species in other countries attacking tea, cacao, coffee, and cinchona, and might become pests here if these crops were established on a large scale. *Helopeltis* is peculiar in having a long, apically knobbed spine projecting from the scutellum. The Isometopinae, which are exceptional in possessing ocelli, and are sometimes regarded as a separate family, are not recorded from Australia.

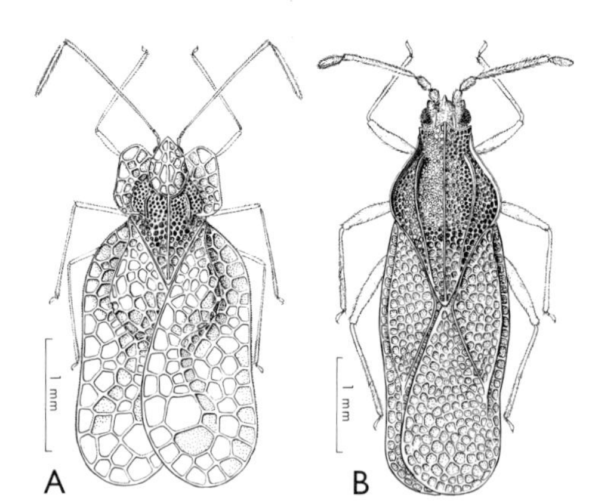

Fig. 26.48. Tingidae: A, *Stephanitis queenslandensis*; B, *Teleonemia scrupulosa*. [S. Curtis]

Superfamily TINGOIDEA

49. Tingidae (Fig. 26.48). Known as 'lace bugs' because of the raised, reticulate venation of the hardened fore wings, the tingids are phytophagous; a few produce galls. Some are pests of cultivated plants, e.g. *Stephanitis queenslandensis* Hacker and *S. pyrioides*

(Scott) have been reported as damaging azaleas. The Central American species, *Teleonemia scrupulosa* Stål, was introduced (via Fiji) to attack lantana, and is now common in Queensland and northern N.S.W. Several authors have suggested that this family has reduviid relationships, while Carayon (1954) indicates both reduviid and mirid affinities. Drake and Davis (1960) have grouped the Tingidae and Miridae in a superfamily Miroidea. [Drake and Ruhoff, 1960.]

Superfamily REDUVIOIDEA

50. Reduviidae (assassin bugs; Plate 3,I–K; Figs. 26.8C,D, 10A, 49, 50). A large, predacious group, the subfamily and tribal classification of which is still in flux (Davis, 1966). It has been variously divided into up to about 30 subfamilies, less than half of which have been recorded from this country. No species sucking vertebrate blood is known from Australia, but several, particularly the larger ones, are reported as inflicting painful 'bites' on humans. The salivary secretions responsible have the basic functions of paralysing and killing the arthropod prey and breaking down its tissues. Most reduviids can produce sound by rubbing the short labium against a transversely striated stridulatory groove on the prosternum.

Members of the large subfamily HARPACTOCORINAE are mostly diurnal and found on plants. One of the best known is the 'bee-killer', *Pristhesancus papuensis* Stål, brown and up to 25 mm long. *Gminatus nigroscutellatus* Bredd. and *Havinthus rufovarius* Bergr. are up to 17 mm in length, the former being reddish brown with the legs, scutellum, and ventral surface black, the latter variably marked in red and black, including the legs. There are several species of *Veledella*, 8–11 mm long, reddish brown and black, with the posterior angles of the pronotum spined and numerous erect spines on the head and anterior lobe of the pronotum, and of *Catasphactes*, about the same size but less reddish and without spines on the head and pronotum. *Euagoras bispinosus* (F.) is about 15 mm long, with the anterior and posterior angles of the pronotum spined and the legs and antennae long and slender. Also in this subfamily are the micropterous species *Austrocoranus mundus* Miller, *Dicranucoris canberrae* Miller, and *D. tasmaniae* Miller, the last collected in tussocks.

The PIRATINAE are mostly nocturnal, ground-living species with the pronotum constricted well behind the middle, fore femora usually swollen, and fore coxae large and flattened on the outer surface. *Pirates* includes several robust, black or dark brown species, usually with a yellowish patch at the base of the hemelytron. *Ectomocoris decoratus* Stål, an orange and black species with both macropterous and apterous forms, has been reported as 'biting' forestry workers in Queensland, both on trees and on the ground.

The TRIBELOCEPHALINAE are dull coloured,

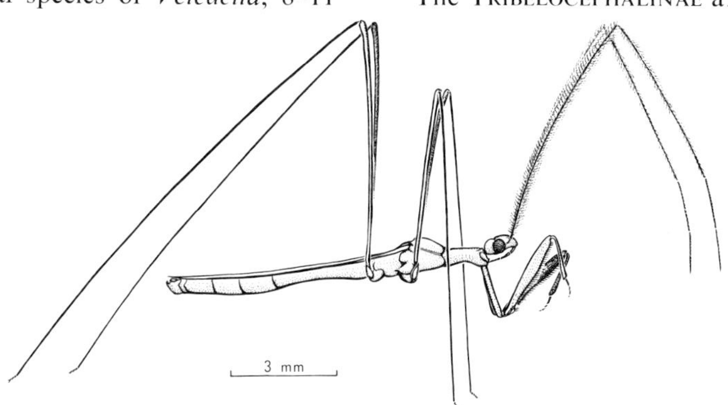

Fig. 26.49. *Ploiaria regina*, Reduviidae.
[S. Curtis]

without ocelli, and with the corium narrow and elongate but the membrane very broad; antennal scape thick and longer than head; dorsal surface of body often densely tomentose. There are several brown, flattened species of *Opisthoplatys*. The STENOPODINAE include elongate, mostly pale brown species of *Sastrapada*, *Oncocephalus*, and *Thodelmus*; the antennae and legs are slender, and the head elongate and cylindrical. The HOLOPTILINAE include several species of *Ptilocnemus* with numerous long hairs on the antennae, legs, and abdomen, particularly conspicuous on the hind tibiae. A ventral trichome, the elevated opening of a gland with an associated tuft of setae, is usually present at the base of the abdomen. Its secretion is reported to attract and anaesthetize ants, on which the holoptilines prey.

EMESINAE have the body and legs slender, antennae long, fore legs raptorial, and hemelytra long and narrow; they usually lack ocelli. They are predatory on small insects, such as Diptera, and are themselves often gnat-like in appearance. Wygodzinsky (1966) listed 44 Australian species. The genera best represented are *Ploiaria* and *Stenolemus*. One of the commonest species is the cosmopolitan *Empicoris rubromaculatus* (Blackb.) with a red costal spot.

Superfamily SALDOIDEA

As at present constituted, this superfamily comprises only the Saldidae, Leptopodidae, and Leotichiidae, a small family recorded from caves in Burma and Malaya. Cobben (1959) has revised the classification of the Saldidae, and forecast the erection of a new family for *Omania*; he supports the ideas of China (1955) on the relationship of the Saldoidea to the Gerroidea, and thus their transfer from the Geocorisae to the Amphibicorisae. Gupta (1963), on the other hand, agrees with Scudder (1959) and others in regarding them as Geocorisae; on this concept they are either cimicomorphs or have arisen from close to the ancestral stock of both cimicomorphs and pentatomomorphs.

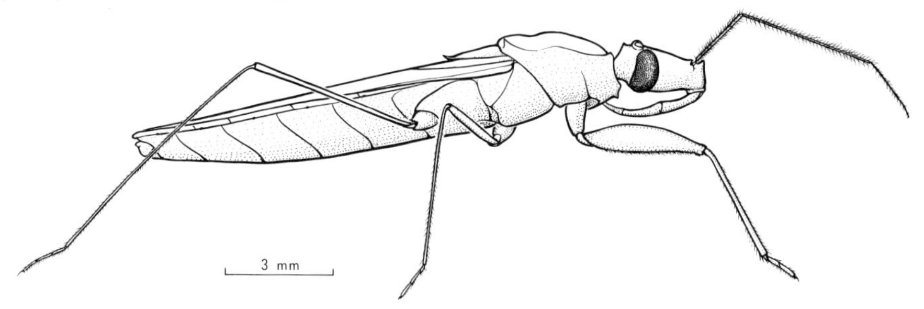

Fig. 26.50. *Oncocephalus confusus*, Reduviidae.
[S. Curtis]

Key to the Families of Saldoidea Known in Australia

Labium long, reaching or passing mid coxae; labium and fore femora without long spines; ocelli not situated on a tubercle; eyes prominent, but not pedunculate .... **Saldidae**

Labium short, not passing prosternum; 1st and often 2nd segment of labium and the fore femora with very long spines; ocelli situated close together on a median tubercle; eyes set on short, broad peduncles .... **Leptopodidae**

51. Saldidae (shore bugs; Figs. 26.51A–C). Saldids are dark coloured, more or less ovoid in outline, and capable of leaping. They are usually very fast-running and difficult to catch, commonly progressing in a series of short, darting flights close to the ground. Most are littoral, beside either fresh or brackish water (streams, lakes, estuaries) and some along the sea shore. Others live in damp fields and boggy areas. There are varying degrees of adaptation to submergence by tides. The subfamily CHILOXANTHINAE of Cobben (1959) is represented here only by *Pentacora salina* (Bergr.) from W.A. and *P.*

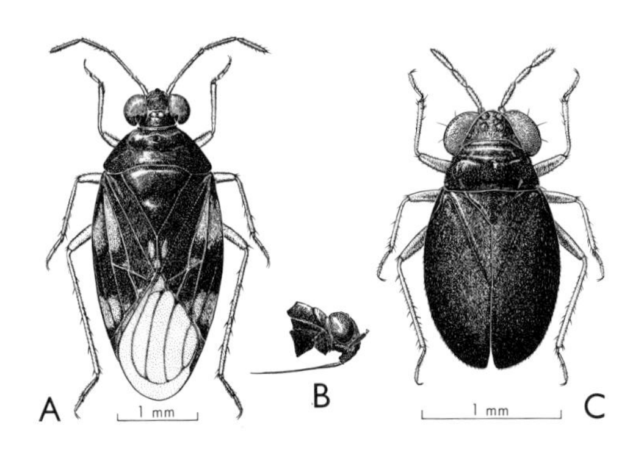

Fig. 26.51. Saldidae: A, *Saldula brevicornis*; B, same, lateral aspect of head; C, *Omania marksae*. [S. Curtis]

leucographa Rimes from S.A., while the SALDINAE (in Cobben's sense) include 4 Australian species of *Saldula*. The aberrant genus *Omania* includes small, beetle-like, flightless species less than 2 mm long living in the intertidal zone, and so far recorded from Arabia, Japan, Australia and Samoa, though doubtless widely distributed in the warmer parts of the world. The Australian species, *O. marksae* Woodw., is known from the Great Barrier Reef; during high tides adults and nymphs retreat into fine crevices and channels in coral rocks (Woodward, 1958). [Drake and Hoberlandt, 1950; Rimes, 1951.]

52. Leptopodidae. Known only from the eastern hemisphere and mainly tropical and subtropical, these bugs frequent the drier parts of rocks in streams. Only one Australian species is recorded, *Valleriola wilsonae* Drake (Fig. 26.52), which is widespread in eastern Queensland.

Superfamily ARADOIDEA*

Key to the Families of Aradoidea

Eyes absent; antennae very short, completely concealed beneath head; wings always absent; body fringed by a row of short, peg-like processes; associated with termites .... **Termitaphididae**

Eyes present; antennae clearly visible from above; wings present or absent; body not fringed with such processes; commonly found beneath bark .... **Aradidae**

53. Aradidae (flat bugs, bark bugs; Fig. 26.53). Usinger and Matsuda (1959) recognize 8 subfamilies, all of which are now known in Australia. Some authors have removed several of these to a separate family Meziridae (= Dysodiidae). Aptery is very common, and about half of the genera and a third of the species in Australia are apterous. Aradids are commonly found under the bark of rotting logs and others on fungi, but some, particularly apterous species, occur in the forest-floor litter. The maxillary and mandibular stylets are extremely long and basally coiled in a spiral within the head, an adaptation for feeding on mycelia.

The primitive ISODERMINAE include only *Isodermus*, with a typically 'Antarctic' distribution—1 species in Australia (including Tasmania), 1 in southern South America, and 2 in New Zealand. This group has the unusual tendency to break off the wings, the hemelytra having a transverse line of weakness between the poorly differentiated corium and membrane near the level of the apex of the scutellum. The species are polished, strongly

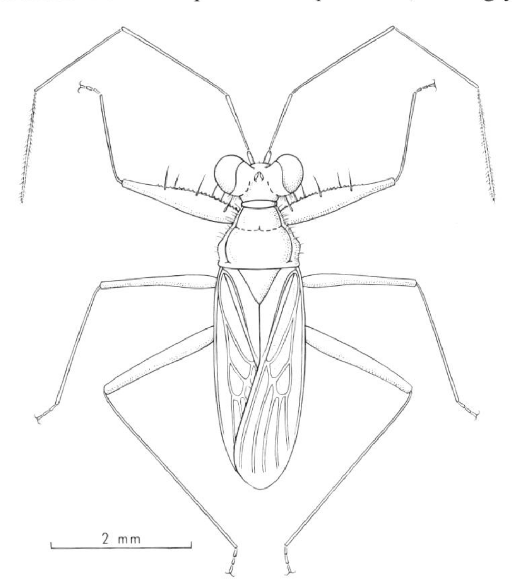

Fig. 26.52. *Valleriola wilsonae*, Leptopodidae. [T. Nolan]

* Morphology and relationships discussed by R. Kumar, *Ann. ent. Soc. Am.* **60**: 17–25, 1967.

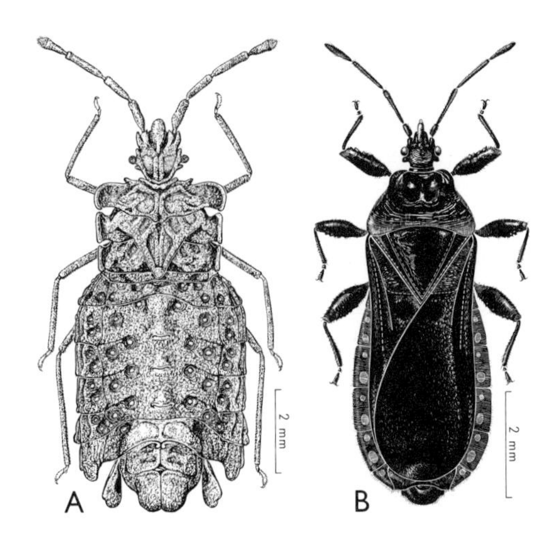

Fig. 26.53. Aradidae: A, *Kumaressa scutellata*; B, *Isodermus planus*. [S. Curtis, T. Nolan]

flattened, and the ventral surface of the head lacks bucculae and a rostral groove or atrium. PROSYMPIESTINAE have a similar distribution; in Australia there are 5 species of *Prosympiestus*. The CHINAMYERSIINAE are known only from New Zealand and Australia, *Kumaressa scutellata* Monteith being described from Queensland (Monteith, 1966). The ARADINAE are world-wide, with nearly 200 species, at present all included in *Aradus*; only 3 are known in Australia. The widespread CALISIINAE and ANEURINAE are represented by 5 and 4 species respectively. CARVENTINAE are particularly abundant in the tropics; 4 species in 3 apterous genera occur in Queensland.* The large subfamily MEZIRINAE, occurring in all zoogeographical regions, has 7 genera and 16 species known from Australia. The Aneurinae, Carventinae, and Mezirinae differ from other subfamilies in having large genae which often extend forward beyond the apex of the clypeus. [Kormilev, 1965.]

54. Termitaphididae (Fig. 26.54). A small family of small, flattened, apterous insects superficially resembling wingless Homoptera; recorded only from termites' nests. Little is known of their habits. Although possessing long, coiled stylets like those of aradids, they differ in lacking eyes and dorsal abdominal scent-gland openings, and in having the margins of the body laminate and bearing a row of small processes. The Australian species is *Termitaradus australiensis* Mjöb.

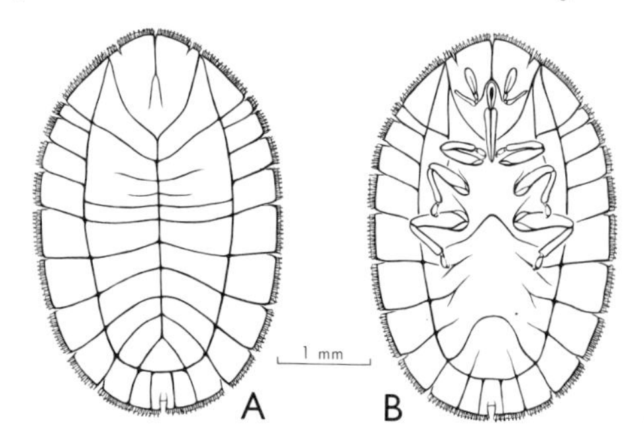

Fig. 26.54. *Termitaradus australiensis*, Termitaphididae: A, dorsal; B, ventral. [S. Curtis]

* For additional species, see G. B. Monteith, *Proc. R. ent. Soc. Lond.* (B) **36:** 50–60, 1967.

Superfamily COREOIDEA

There is not as yet general agreement on the limits between the Coreoidea and the Lygaeoidea, nor on the families that should be included and the higher classification of the Coreoidea. These problems have been discussed by Scudder (1963), Schaefer (1965), and Kumar (1966).

Key to the Families of Coreoidea Known in Australia

1. Membrane of hemelytron in macropters with 4 main veins connected by cross-veins to form large basal cells; base of abdomen beneath with an ovoid pore-bearing plate on each side (Fig. 26.43F); opening of metapleural scent gland with a bristle-like process; antennifers inserted lateroventral to a line between centre of eye and apex of head .... **Hyocephalidae**
   Membrane of hemelytron in macropters with 6 or usually more longitudinal veins, rarely with basal cells; base of abdomen without ovoid pore-bearing plates; opening of metapleural scent gland rarely with a bristle-like process; antennifers inserted dorsal to a line between centre of eye and apex of head 2

2. Metapleural scent gland openings with peritremes (spouts) usually obsolete, but, if visible, each leading into 2 divergent grooves; abdominal tergum 5 (4th visible) constricted in mid line; corium often with a large hyaline area (lacking in most Australian species) .... **Rhopalidae**
   Metapleural scent gland openings with distinct peritremes; posterior margins of abdominal terga 4 and 5 produced posteriorly in mid line; corium without a large hyaline area 3
3. Stout, robust species with relatively thick legs; bucculae extending to, and usually behind, level of antennifers; head usually much less than half as wide as base of pronotum .... **Coreidae**
   Narrow, elongate species with relatively long, slender legs; bucculae very short, not extending behind, and rarely reaching level of, antennifers; head more than half as wide as base of pronotum .... **Alydidae**

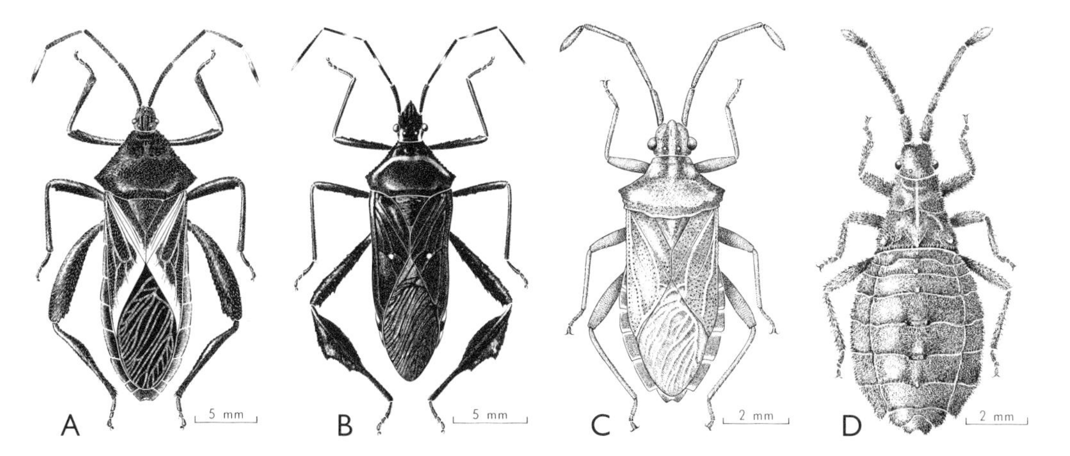

Fig. 26.55. Coreidae: A, *Mictis profana*; B, *Leptoglossus bidentatus*; C, *Cletus* sp.; D, *Agriopocoris* sp. [T. Nolan]

55. Coreidae (Plate 3,Z; Figs. 26.9, 43D, 55). One of the largest genera is *Amorbus*, with large, robust, brown species in which, as commonly in the family, the males have much stouter hind femora than the females and the hind tibiae expanded or spined; they feed on eucalypts. *Mictis* is of similar facies, the 'crusader bug', *M. profana* (F.), of some economic importance, being easily recognized by the yellowish markings forming a cross on the closed hemelytra. Other common species in the northern half of the country are *Cletus similis* Blöte and *C. bipunctatus* Herr.-Sch., brown, with the apical antennal segment fusiform and the posterior angles of the pronotum acutely produced outward; the former is about 7–8 mm long, the latter 10–12 mm. *Pomponatius* includes several species with the apex of the abdomen acutely produced. In *Agriopocoris* are several micropterous, nymph-like species with broad flattened abdomens of ovoid outline.

Leptoglossus bidentatus Montr. damages cucurbits, pawpaw, passion fruit, and citrus. It is black, with a transverse red line across the pronotum, red bands on the antennae, and red patches on the ventral surface of the body; the hind femora have large flattened expansions. Several species of *Amblypelta* (fruit-spotting bugs) are important pests of a wide range of fruits. They are brown, usually shining, about 10–15 mm long, and nearly parallel-sided with acutely produced posterior pronotal angles. The false cotton stainer, *Aulacosternum nigrorubrum* Dall., of similar size and general form, is red or brownish red with black membranes and appendages. The American species, *Chelinidea tabulata* Burm., introduced to attack prickly pear, is light brown, with dark brown membranes and whitish veins on the corium, and the antennal segments flattened and fluted.

56. Alydidae (Figs. 26.56A,B). *Leptocorisa acuta* Thunb., *L. oratorius* (F.), and *Mutusca*

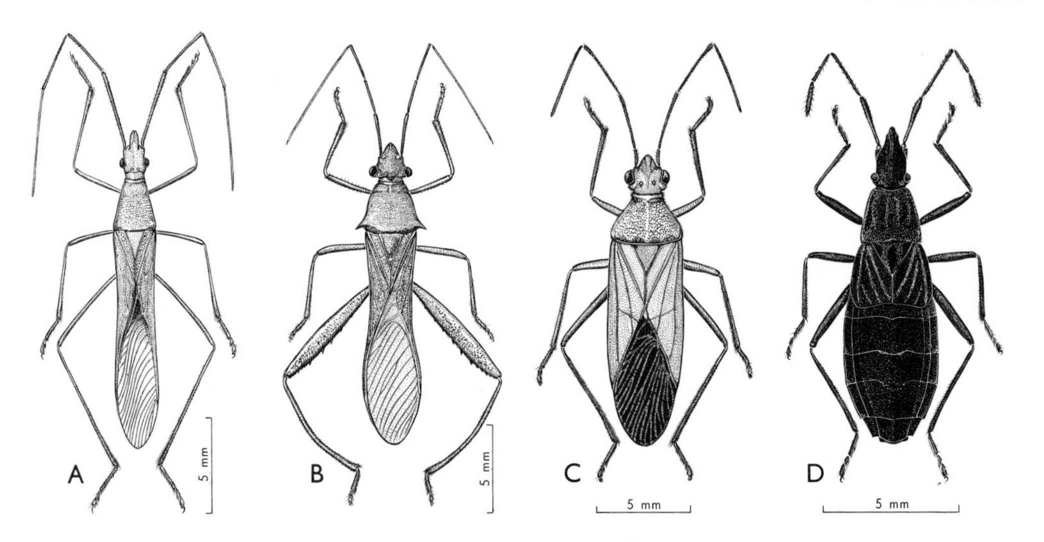

Fig. 26.56. A, *Leptocorisa acuta*, Alydidae; B, *Riptortus serripes*, Alydidae; C, *Leptocoris rufomarginata*, Rhopalidae; D, *Maevius indecora*, Hyocephalidae. [S. Curtis]

brevicornis Dall. are very narrow and linear and about 15 mm long; the first 2 are common on grasses, and are pests of rice (I. Ahmad, 1965). *Noliphus erythrocephalus* Stål, another common species, has the dorsal surface of the abdomen red with black connexival bars and the posterior angles of the pronotum strongly spined. *Riptortus serripes* (F.), the pod-sucking bug, is a pest especially of beans; it is dark brown with yellow lateral streaks, and has the posterior pronotal angles spined and the hind femora with a row of spines.

57. Rhopalidae (Fig. 26.56c). *Leptocoris* includes 4 Australian species, yellowish brown to purplish red, of elongate ovoid form, and ranging from about 9 to 29 mm in length; the corium lacks a large hyaline area. The widely distributed *Liorhyssus hyalinus* (F.) has been introduced. [Gross, 1960.]

58. Hyocephalidae (Figs. 26.43E,F, 56D). A small, entirely Australian family of brown or black species of lygaeid-like facies, some macropterous and some brachypterous. Most belong to *Hyocephalus*; they are seed-feeders.

Superfamily LYGAEOIDEA

Key to the Families of Lygaeoidea

1. Ocelli absent; membrane of hemelytron in macropters with at least 7, usually many more, main longitudinal veins usually arising from 2 basal cells; insects more than 5 mm long and with red or yellow ground colour 2
   Ocelli present (except in a few brachypterous forms which are brown or dark and less than 5 mm long); membrane of hemelytron in macropters with no more than 6, usually fewer, main longitudinal veins which rarely arise from basal cells; size and colour various, sometimes with red ground colour, but more commonly brown 3
2. ♀ with long ovipositor and abdominal sternum 7 (6th visible) cleft in mid-line to receive it (Fig. 26.57A) .... **Largidae**
   ♀ with short ovipositor and abdominal sternum 7 without median cleft .... **Pyrrhocoridae**
3. Legs and antennae long and exceedingly slender, thread-like; apices of femora and usually antennal scape slightly but appreciably swollen; apical antennal segment short, relatively thick, spindle-shaped; body linear (Fig. 26.62B) .... **Berytidae**
   Legs and antennae robust to slender, but legs never exceedingly slender and thread-like, nor with apices of femora swollen; if antennae exceedingly slender and thread-like, then scape not swollen at apex, and apical segment long and slender; form of body varied 4

4. Antennae longer than body, thread-like, all segments very long and slender; wings long and narrow; membrane of hemelytron with veins absent or obsolescent; abdomen narrowed at base .... **Colobathristidae**
   Antennae not longer than body nor thread-like; form of wings and abdomen varied 5
5. Tarsi 2-segmented; body less than 5 mm long; pronotum and corium with deep punctures; a few tingid-like species on *Acacia* .... **Piesmidae**
   Tarsi 3-segmented; size and punctation varied; a large and abundant family .... **Lygaeidae**

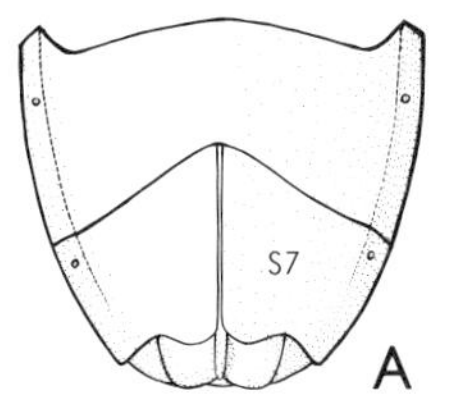

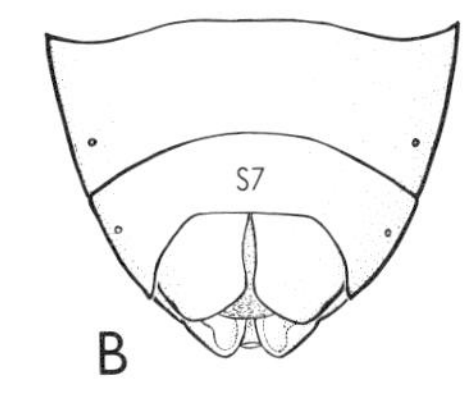

Fig. 26.57. Apex of abdomen of ♀, ventral: A, *Physopelta famelica*, Largidae; B, *Dindymus versicolor*, Pyrrhocoridae. [S. Curtis]

59. Lygaeidae (Plate 3,V,W; Figs, 26.5B, 58–60). This large family appears to be predominantly seed-feeding (Sweet, 1960). Slater (1964) catalogued the Lygaeidae of the world, and China and Miller (1959) dealt with the subfamily classification.

The dominant subfamily RHYPAROCHROMINAE, with over 100 species in Australia, is distinguished by the presence of cephalic trichobothria near the eyes, and, with few exceptions, by the suture between abdominal sterna 4 and 5 being curved forward at the sides and not reaching the lateral margins; fore femora more or less strongly incrassate and usually with ventral spines (Fig. 26.58). Probably most feed on ripe seeds that have dropped to the ground, so they are often overlooked. Brachyptery is common, and there are several examples of ant mimicry, particularly by nymphs. Among the larger genera in Australia are *Myocara* (12 spp.), *Pachybrachius* (*sens. lat.*, 8 spp.), *Daerlac* (6 spp.), and *Dieuches* (13 spp.). *Paromius pallidus* (Montr.), widespread in the Oriental and Pacific areas, is a very common species found on grasses. *Brentiscerus australis* (Bergr.), a widely distributed species 2·5–4·5

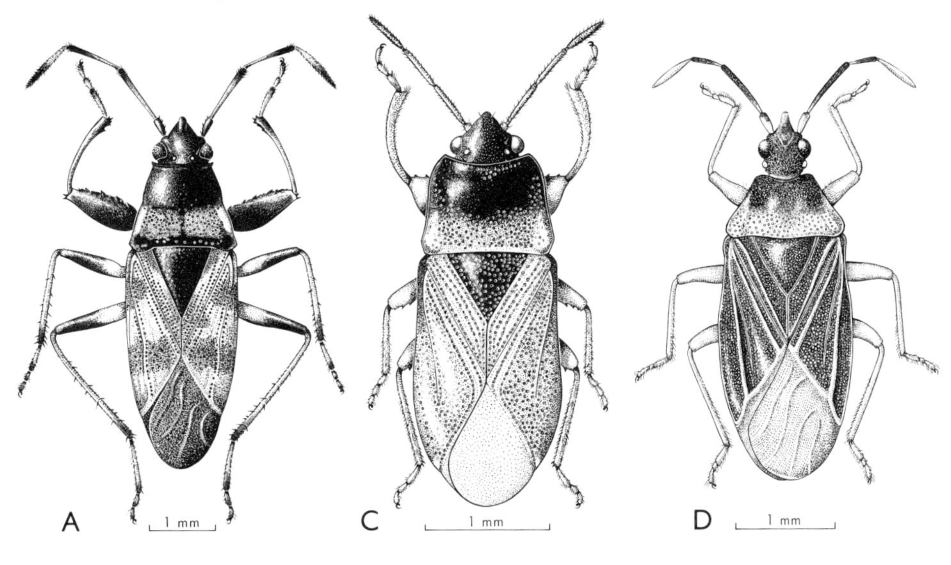

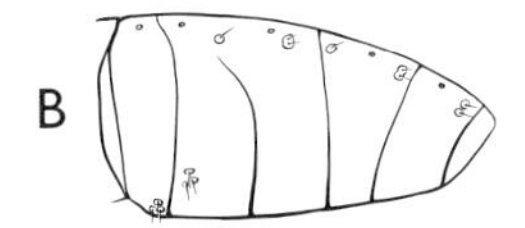

Fig. 26.58. Lygaeidae–Rhyparochrominae: A, *Euander lacertosus*; B, abdomen of same, lateral; C, *Plinthisus* sp.; D, *Clerada* sp. [T. Nolan]

mm long, is one of the commonest lygaeids of leaf litter; it belongs to the Drymini. *Euander lacertosus* (Erich.), a widespread Australian species, has been reported as attacking strawberries in Tasmania and A.C.T. *Clerada*, widely distributed in the tropics and subtropics, with 5 species recorded from Australia, includes brown, flattened species some of which are known to be ectoparasitic on phalangers and other mammals. *Plinthisus* (2 spp.) includes small, dark species unusual among rhyparochromines in having a complete and nearly straight suture between abdominal sterna 4 and 5. [Ashlock, 1964; Gross, 1965; Gross and Scudder, 1963; Slater and Sweet, 1961.*]

The Lygaeinae and Cyminae have all the abdominal spiracles dorsal, on the laterotergites. LYGAEINAE (39 spp.) are mostly moderate-sized to large and often conspicuously coloured, with red and black predominating; membrane of hemelytron with a closed basal cell. *Arocatus* has 3 species; the commonest, *A. rusticus* (Stål), occurs also in New Zealand. *Astacops*, also with 3 species, has protuberant eyes. Other genera are *Graptostethus*, *Oncopeltus*, *Scopiastes*, *Spilostethus*, *Caenocoris*, and *Melanerythrus*. A species of some economic importance is *Oncopeltus sordidus* (Dall.), attacking cotton in Queensland. CYMINAE, which are included in part in the Berytidae by Southwood and Leston (1959), are small brown species; only 8, in 5 genera, are known from Australia, including 4 of *Ontiscus*. The usual food plants seem to be rushes and sedges.

Most GEOCORINAE (18 spp.) belong to the widespread genera *Germalus* and *Geocoris*. Head at least as wide as base of pronotum, with prominent, pedunculate or subpedunculate eyes; first 3 pairs of abdominal spiracles dorsal, the other 3 pairs ventral. So far as is known, geocorines are predacious; they are sometimes caught by sweeping, but are more commonly found on the ground beneath herbage and leaf litter. Only 3 species of OXYCARENINAE are recorded from Australia, all belonging to *Oxycarenus*, and none of them endemic. They are small, black and white bugs feeding by preference on malvaceous plants, including cotton, *Malva*, and hibiscus, and sometimes occurring in large numbers. They may also damage orchard fruits and other crops, especially in dry seasons.

The BLISSINAE are represented by 8 species, the biology of which is poorly known; more will probably be found, particularly in arid and semi-arid areas. They occur on grasses, especially in the bases of tussocks and in the leaf axils. Most blissines are narrow and parallel-sided, and brachyptery is common. Particularly in the broader species, the abdomen often extends far lateral to the folded wings, giving a rather aradid-like facies. Several overseas species of *Blissus* are pests of graminaceous crops; in North America they are known as 'chinch bugs'. The closely related subfamily SLATELLERINAE is monotypic; the one species, *Slaterellus hackeri* Drake & Davis, is Australian (Drake and Davis, 1959).

ORSILLINAE are represented by 3 species of the cosmopolitan genus *Nysius*, all of some economic importance, and with a wide host range, including many crop plants. Large numbers of individuals, including eggs and nymphs, often occur on seed-heads of composites, such as thistles. *N. vinitor* Bergr., the Rutherglen bug, occurs throughout Australia, although *N. clevelandensis* Evans is much commoner in coastal Queensland; *N. turneri* Evans is restricted to Tasmania. PACHYGRONTHINAE, with a mainly tropical and subtropical distribution, include 10 Australian species in 7 genera, 4 of these endemic (Slater, 1955, 1962). Food plants, so far as known, are grasses, sedges, and rushes. Among the commoner Australian species are *Opistholeptus vulturnus* (Kirk.), *Stenophyella macreta* Horvath, and *Pachygrontha austrina* Kirk. They are heavily punctate species with all the abdominal spiracles ventral and the fore femora strongly swollen and spined.

The ISCHNORHYNCHINAE (6 spp.) are small, subovoid lygaeids, with a covering of short sericeous hairs and all abdominal spiracles dorsal (Scudder, 1962). What little is known

* See also M. H. Sweet, *Ann. ent. Soc. Am.* **60**: 208–26, 1967.

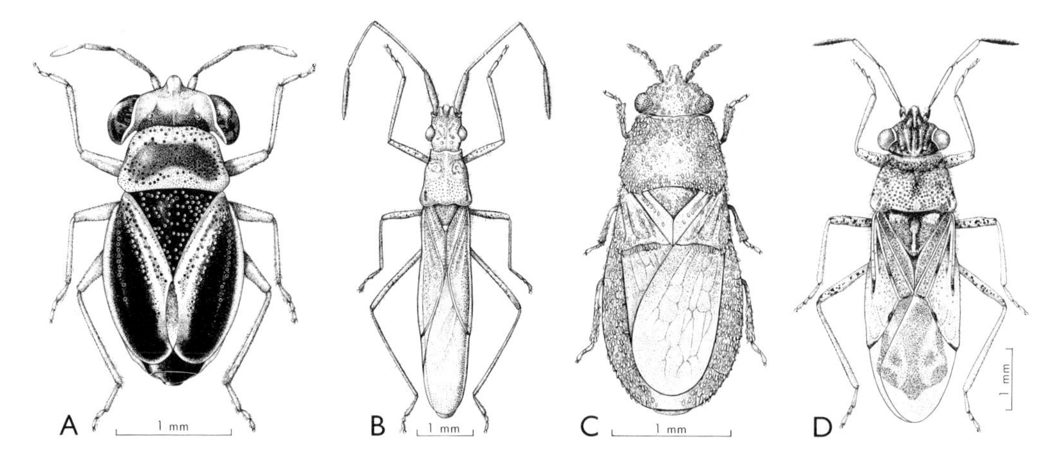

Fig. 26.59. Lygaeidae: A, *Geocoris elegantulus*, Geocorinae; B, *Ontiscus* sp., Cyminae; C, *Slaterellus hackeri*, Slatellerinae; D, *Nysius clevelandensis*, Orsillinae. [T. Nolan]

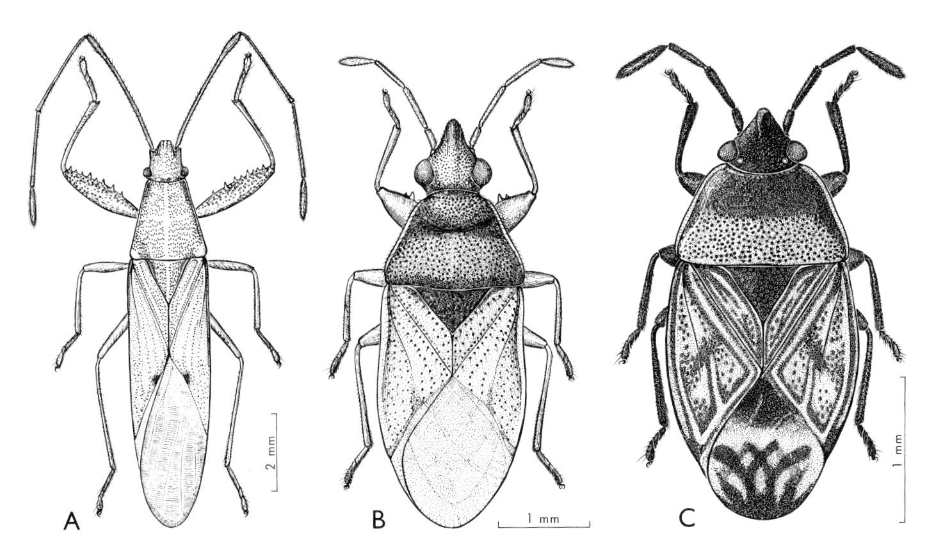

Fig. 26.60. Lygaeidae: A, *Pachygrontha austrina*, Pachygronthinae; B, *Crompus opacus*, Ischnorhynchinae; C, *Dilompus robustus*, Artheneinae. [S. Curtis]

of the host relationships indicates that they feed on a wide range of dicotyledons, but the only host records in Australia are of Myrtaceae (*Callistemon, Leptospermum, Metrosideros*). Two of the most widely distributed species are *Crompus oculatus* Stål and *C. opacus* Scudder. The HETEROGASTRINAE (4 spp.) have all abdominal spiracles ventral, and the membrane of the hemelytron has basal cells; the intersternal sutures of the abdomen of the female are often curved far forward in the mid line. The ARTHENEINAE have only one species in this country, *Dilompus robustus* Scudder, representing an endemic tribe (Slater, Woodward and Sweet, 1962). *Trisecus pictus* Bergr. is our only representative of the IDIOSTOLINAE, which should probably be regarded as a separate family (Schaefer, 1966).*

* A new species of *Trisecus* and a new genus and species remain to be described.

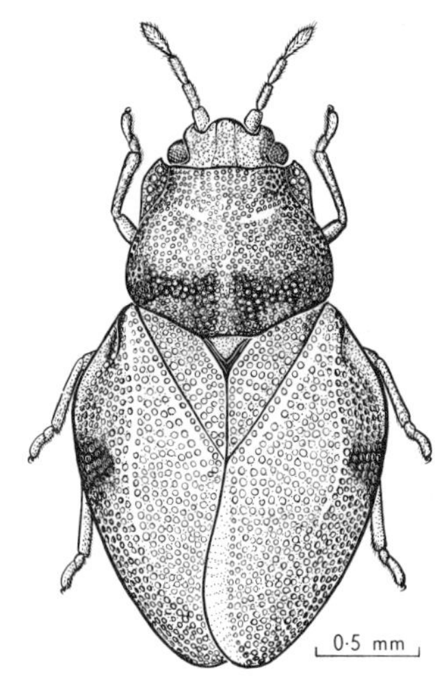

Fig. 26.61. *Mcateella* sp., Piesmidae. [S. Curtis]

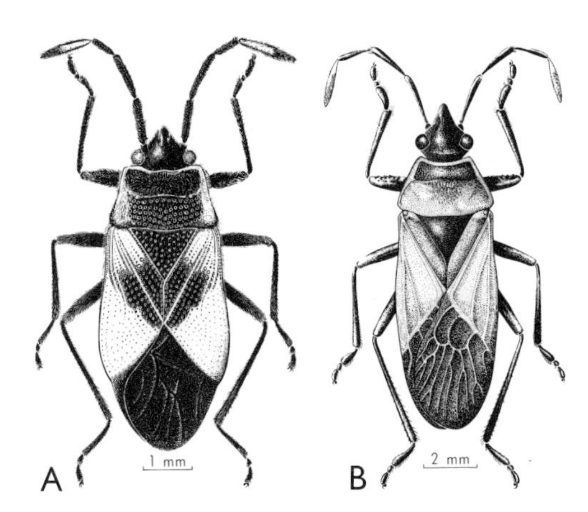

Fig. 26.63. A, *Delacampius militaris*, Largidae; B, *Dindymus versicolor*, Pyrrhocoridae. [T. Nolan]

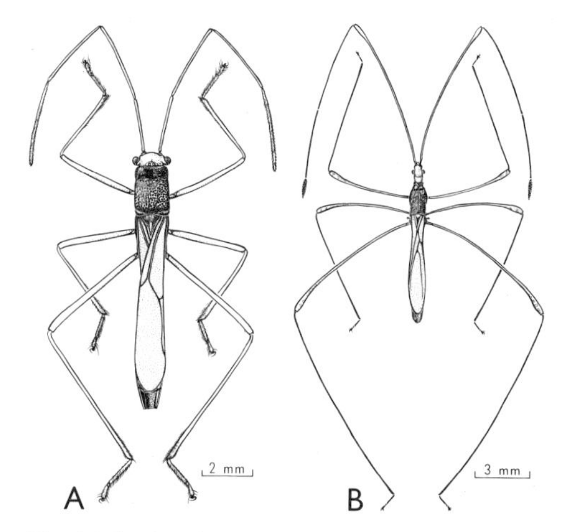

Fig. 26.62. A, *Phaenacantha australiae*, Colobathristidae; B, *Metacanthus pluto*, Berytidae. [S. Curtis]

60. Piesmidae (Fig. 26.61). Though superficially resembling tingids, morphological studies have shown that the Piesmidae are related to the Lygaeidae, and they are now usually placed in the Lygaeoidea. Only one of the 3 included genera occurs in Australia, the endemic *Mcateella* with 4 species. Their only known host plants are species of *Acacia*. [Drake and Davis, 1958.]

61. Colobathristidae. *Phaenacantha australiae* Kirk. (Fig. 26.62A), the only recorded Australian species, is restricted to the tropical north, where it occurs in large numbers on grasses, including sugar cane. [Štys, 1966.]

62. Berytidae (Neididae; Figs. 26.11B, 62B). The stilt bugs are phytophagous, but little is known of the biology of the Australian species. [Gross, 1950.]

63. Largidae (Plate 3,Y; Figs. 26.57A, 63A). The best-known species is *Physopelta famelica* Stål, up to about 16 mm long, orange-red, with a large black spot on the disc of each corium, and the membrane and apex of the corium also black. A much smaller black and red species, *Delacampius militaris* Dist. occurs in New Guinea and north Queensland. So far as known, the Largidae are phytophagous.

64. Pyrrhocoridae (Plate 3,X; Figs. 26.57B, 63B). *Dysdercus sidae* Montr., the cotton stainer, stains the lint and damages the seed of cotton, and also infests other Malvaceae. It is yellowish brown to orange, with a red head and the following black: scutellum, membrane, a small spot on each corium, and usually a transverse band just behind the pale anterior margin of the pronotum. There are several species of *Dindymus*. *D. versicolor* (Herr.-Sch.) damages deciduous fruit; it is red and black but without central corial spots. The Pyrrhocoridae and Largidae are sometimes grouped in a superfamily Pyrrhocoroidea.

Superfamily PENTATOMOIDEA

The shield bugs or stink bugs are mostly phytophagous, although some are predacious. There is not yet general agreement on their classification, some groups which are separated as families by some authors being regarded as subfamilies of Pentatomidae by others. [Kirkaldy, 1909; McDonald, 1966.]

Key to the Families of Pentatomoidea Known in Australia

1. Pronotum with large, rounded, posterolateral lobes; meso- and metanotum visible in lateral view beneath base of hemelytron; scutellum large, covering apex of abdomen. [A few species of small, convex bugs] .... **Aphylidae**
   Pronotum without large posterolateral lobes; meso- and metanotum not visible in lateral view; scutellum varied 2
2. Tibiae with numerous strong spines; apices of mid and hind coxae fringed with closely set, rigid setae. [Ovoid, black or brown bugs] .... **Cydnidae**
   Tibiae without numerous strong spines; apices of mid and hind coxae not fringed with closely set, rigid setae 3
3. Antennifers cylindrical, set on lateral margins of head, and clearly visible from above; ocelli close together, sometimes vestigial. [Green bugs of coreid-like facies] .... **Urostylidae**
   Antennifers not cylindrical, set below lateral margin of head, and not or only apically visible from above; ocelli usually well separated 4
4. Head, pronotum, and part of costal margins of hemelytra strongly laminately expanded and recurved to give a tortoise-like appearance; abdomen with a ventral pair of disc-shaped organs; tarsi 2-segmented. [One described species about 3·5 mm long] .... **Lestoniidae**
   Head, pronotum, and costal margins of hemelytra not strongly laminately expanded and recurved; not tortoise-like; abdomen without ventral disc-shaped organs; tarsi 2- or 3-segmented 5
5. Scutellum very large, convex, more or less U-shaped, completely or almost completely covering abdomen and the wings when at rest 6
   Scutellum triangular or subtriangular, not covering hemelytra and abdominal region 8
6. Hemelytra much longer than abdomen, in resting position folded between corium and membrane to fit beneath scutellum; abdominal sterna with a straight, black, transverse sulcus on each side; short species, usually with dorsal surface very convex and shining and more or less truncate posteriorly .... **Plataspidae**
   Hemelytra not or scarcely longer than abdomen, not folded between corium and membrane; abdominal sterna without a straight, black, transverse sulcus on each side; form varied but not posteriorly truncate 7
7. Lateral margins of pronotum without slender, pointed, anterior projections extending to near the eyes; hind wings with a hamus in basal cell; prosternum with a deep median sulcus .... **Scutelleridae**
   Lateral margins of pronotum with slender, pointed, anterior projections extending to near eyes; hind wings without a hamus; prosternum without a deep median sulcus (Podopinae) .... **Pentatomidae** (pt.)
8. Spiracles of abdominal segment 2 (1st visible) fully exposed, removed from posterior margins of metapleura 9
   Spiracles of abdominal segment 2 concealed by metapleura (usually entirely hidden, occasionally barely visible above their margins) 10
9. Membrane of hemelytron reticulately veined; antennae 4-segmented in the one Australian species .... **Dinidoridae**
   Membrane of hemelytron not reticulately veined; antennae usually 5-segmented (3rd segment very short), occasionally 4-segmented .... **Tessaratomidae**
10. Tarsi 2-segmented; abdominal sternum 3 (2nd visible) with a long, spine-like anterior process; mesosternum with a median carina .... **Acanthosomatidae**
    Tarsi usually 3-segmented, rarely 2-segmented; abdominal segment 3 without a long, spine-like process, at most with a small anterior tubercle; mesosternum without a median carina .... **Pentatomidae** (pt.)

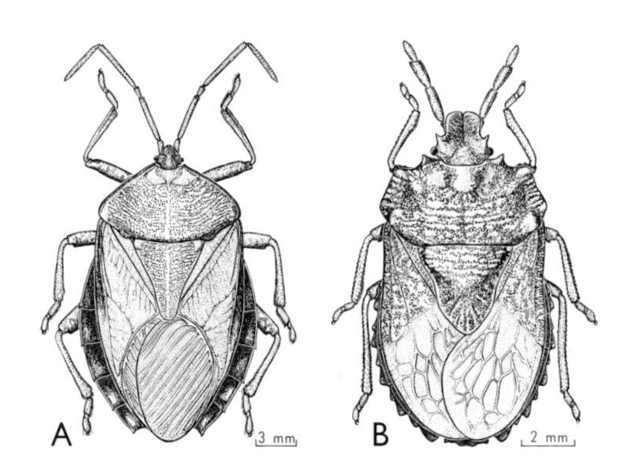

Fig. 26.64. A, *Musgraveia sulciventris*, Tessaratomidae; B, *Megymenum insulare*, Dinidoridae. [S. Curtis]

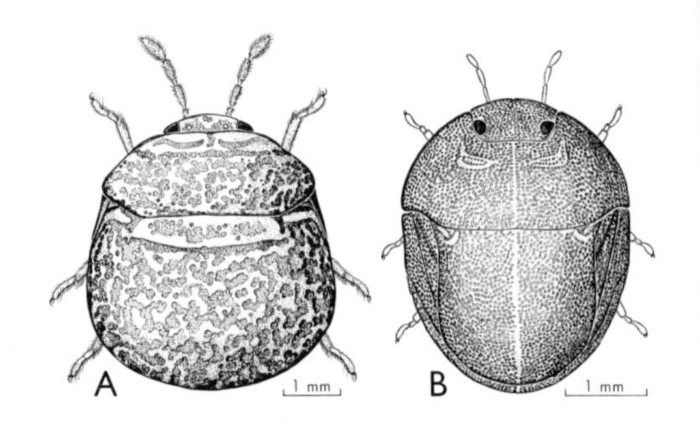

Fig. 26.65. A, *Coptosoma* sp., Plataspidae; B, *Lestonia haustorifera*, Lestoniidae. [S. Curtis]

65. Tessaratomidae (Plate 3,S,T; Fig. 26.64A). The best-known species is the bronze orange bug, *Musgraveia sulciventris* (Stål), over 20 mm long and with flattened, ovoid nymphs. Its native hosts are Rutaceae, but it has become a pest of citrus. *Stilida indecora* Stål is a paler brown species of similar size. *Lyramorpha* has 2 species in eastern Australia; the posterior angles of the abdomen are strongly produced. *Erga longitudinalis* (Westw.), 12–16 mm long, brown with paler connexiva and a pale median band on head, pronotum, and scutellum, feeds on *Lonchocarpus blackii*, a vine common along the banks of creeks.

66. Dinidoridae (Fig. 26.64B). Predominantly Oriental and Ethiopian, the family is represented in Australia only by *Megymenum insulare* Westw., which feeds on the stems, leaf stalks, and young fruit of pumpkins, etc. It is a brown, not very active species 10–14 mm in length.

67. Scutelleridae (Plate 3,P,Q,U; Figs. 26.6C,E, 7A,B). Predominantly tropical and subtropical in Australia, although *Choerocoris paganus* (F.) ranges from northern Queensland to South Australia and Tasmania, and *Coleotichus* spp. extend to arid regions. *Tectocoris diophthalmus* (Thunb.), the 'harlequin bug' of Queensland, attacks the bolls of cotton, and is common on other malvaceous plants. It is extremely variable in colour, from almost wholly yellow to orange-yellow with black patches, while the males are often red with extensive dark metallic blue markings. *Cantao parentum* (White) is of about the same size (up to 20 mm long), but narrower and orange-brown with 2 black spots on the pronotum and 8 on the scutellum. *Calliphara imperialis* (F.) is shining orange-red with a metallic green head. *Lampromicra* includes several smaller species, mostly with a ground colour of metallic green or blue. *Scutiphora pedicellata* (Kirby), widespread in the eastern States, is a metallic green species with red markings on the pronotum and base of the scutellum and black patches on the scutellum.

68. Plataspidae (Coptosomatidae, Brachyplatidae; Fig. 26.65A). Convex, usually shining bugs, mainly Oriental and Ethiopian. Most of our species belong to the large genus *Coptosoma*, and are restricted to the north; they have been reported to attack cow-peas in Queensland.

69. Aphylidae. Restricted to Australia, and known only from 2 species of *Aphylum*. They resemble Plataspidae, but differ in the characters given in the key and in the dull surface and the strongly punctate pronotum and scutellum.

70. Lestoniidae (Fig. 26.65B). The only described species is *Lestonia haustorifera* China from N.S.W. It is small, light brown, and tortoise-like; China (1963) has suggested that the ventral abdominal discs function in adhering to some smooth surface, such as eucalypt leaves or bark. A species also occurs in W.A.

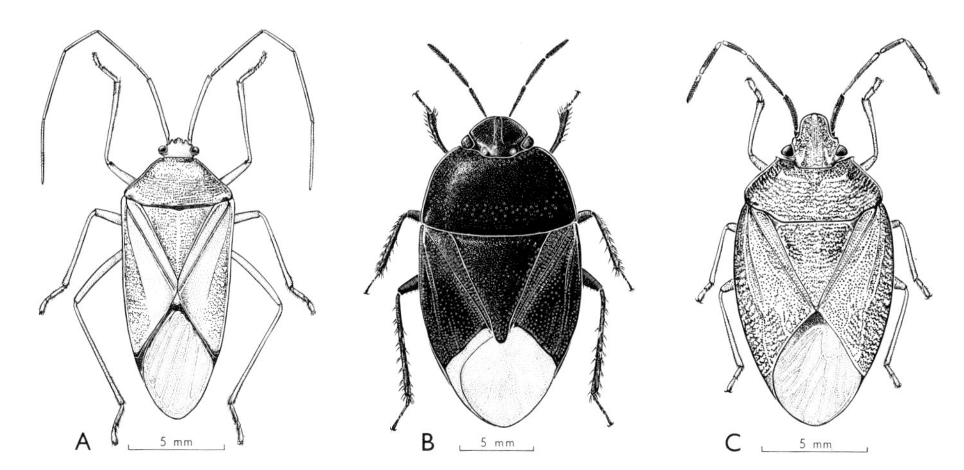

Fig. 26.66. A, *Urolabida histrionica*, Urostylidae; B, *Adrisa* sp., Cydnidae; C, *Amphaces* sp., Acanthosomatidae. [S. Curtis]

71. Urostylidae (Fig. 26.66A). These bugs superficially resemble greenish coreids; little is known of their biology.

72. Cydnidae (Figs. 26.10C, 66B). Burrowing bugs, usually shining black or brown; most live in the soil, but they are often attracted in large numbers to light. Species of *Adrisa*, *Cydnus*, and *Geotomus* occur in Australia. Little is known of their habits, but many are probably root-feeders.

73. Acanthosomatidae (Plate 3,R; Fig. 26.66C). The Australian species are small to medium-sized; one of the most widely distributed is *Anischys luteovaria* (Westw.). The best represented genera are *Andriscus*, *Amphaces*, and *Stictocarenus*, all restricted or almost restricted to Australia.

74. Pentatomidae (Plate 3,L–O; Figs. 26.1C,D, 3A, 4A, 5E, 8B, 10B, 11A, 67). In the PENTATOMINAE, the Pentatomini include several species each of *Cephaloplatus*, *Dictyotus* (brown), and *Ocirrhoe* and *Cuspicona* (mostly greenish); all about 10 mm long. *Cuspicona simplex* Walk., the green potato bug, attacks a range of crops. *Agonoscelis rutila* (F.), the horehound bug, is up to 12 mm long and conspicuously marked in red and black. *Nezara viridula* (L.), the green vegetable bug, is an introduced pest about 15 mm long, attacking crops such as beans, tomatoes, and lucerne. *Glaucias amyoti* (Dall.), a native green bug, is superficially rather similar to *Nezara*, but with a shinier surface, and lacking the 3 small yellowish

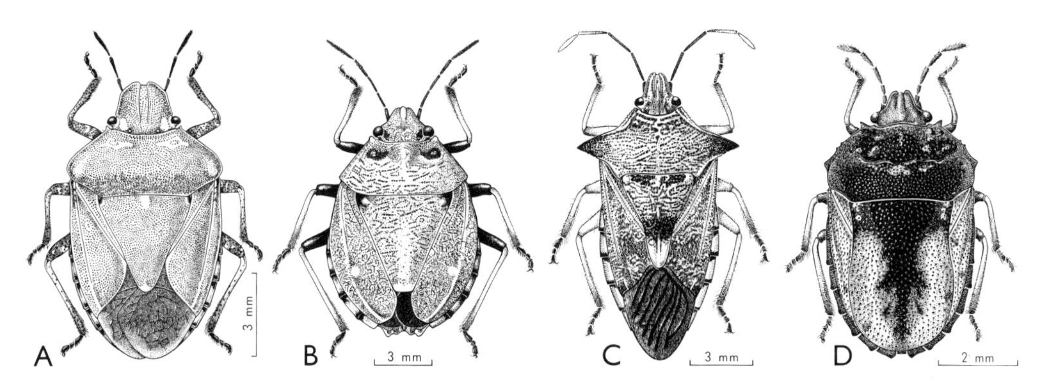

Fig. 26.67. Pentatomidae: A, *Dictyotus caenosus*; B, *Hypogomphus bufo*; C, *Oechalia schellenbergii*; D. *Coracanthella geophila*. [T. Nolan]

spots on the anterior margin of the scutellum and the small black spot on each of its anterolateral angles. *Biprorulus bibax* Bredd., the spined citrus bug, is about 15–20 mm long, with the posterior pronotal angles strongly spined. There are several species of *Vitellus* and *Morna*, of similar facies, but 9–13 mm long, and not of economic importance.

Halyini are well represented, many of the Australian members being associated with eucalypts, the nymphs especially often being found under their bark. The species are of medium to large size. The predominant genera here are *Eumecopus* and *Poecilometis*. Most species have a brown ground colour variously marked. *Notius* includes a few dark, rather flattened species, of which *N. depressus* Dall. is the most widely distributed. *Hypogomphus bufo* (Westw.) is broadly ovoid and strongly convex above and below, with the hemelytral membranes vestigial. The Diemeniini include 14 genera in Australia, the largest and most widely distributed being *Oncocoris*. There are 10 genera of Sciocorini.

The ASOPINAE are predacious on caterpillars and other insects, and have the basal labial segment very stout. They are poorly represented in Australia, the commonest species being *Cermatulus nasalis* (Westw.) and *Oechalia schellenbergii* (Guér.–Mén.), about 9–13 mm long. Both are brown with the apex of the scutellum paler; the latter species has the posterior angles of the pronotum acutely produced and the apex of the scutellum narrowly pointed. The PODOPINAE are small and dull, with a large scutellum which often covers the abdomen and much of the hemelytra. Australian genera include *Testrica*, *Protestrica*, and *Deroploa*. [Musgrave, 1930.]

Superfamily GERROIDEA

The Gerroidea all feed on live or dead animals. Most live on the surface of water, some in moist places usually near the edges of water. [Hale, 1925, 1926.]

Key to the Families of Gerroidea

1. Head several times longer than wide, narrow, but somewhat dilated in front, with eyes about half way along its length .... **Hydrometridae**
   Head no more than about twice as long as wide, not dilated in front, with eyes at or near base ... 2
2. Claws of at least fore legs inserted well before apex of last tarsal segment 3
   Claws of all legs apical 4
3. Legs extremely long and slender; hind femora extending beyond apex of abdomen; mid coxae closer to hind than to fore coxae; medial margins of eyes usually sinuate .... **Gerridae**
   Legs usually shorter, with hind femora not or scarcely exceeding apex of abdomen; mid coxae usually nearly equidistant between fore and hind coxae; medial margins of eyes straight ... **Veliidae**
4. Tarsi 2-segmented (basal segment very small); small, broad bugs 2 mm or less in length; a median longitudinal groove between bucculae extending the length of the head. .... **Hebridae**
   Tarsi 3-segmented (basal segment very small); elongate oval bugs over 2 mm in length; without a median longitudinal groove extending the length of the head .... **Mesoveliidae**

75. Gerridae (Figs. 26.43B, 68). The true water striders skate rapidly on the surface and most live on fresh water. Alary dimorphism is common. The HALOBATINAE and HERMATOBATINAE, with extremely short abdomens, are marine; probably 9 species of *Halobates* occur around the Australian coasts, and 3 of *Hermatobates* off the north and north-west coasts. A few TREPOBATINAE are found in the north. The remaining species all belong to the GERRINAE; they include several species of *Limnogonus* and one each of *Gerris* and *Tenagogerris*. Large brown species of *Tenagogonus* occur in the tropical north. [Hungerford and Matsuda, 1960; Matsuda, 1960.]

76. Veliidae (Figs. 26.43C, 69). *Rhagovelia australica* Kirk. and at least 8 species of *Microvelia* live on the surface of fresh water, while *Halovelia maritima* Bergr. is marine. The species of *Microvelia* are small and dark, about 1·7–3·7 mm long; in some only the

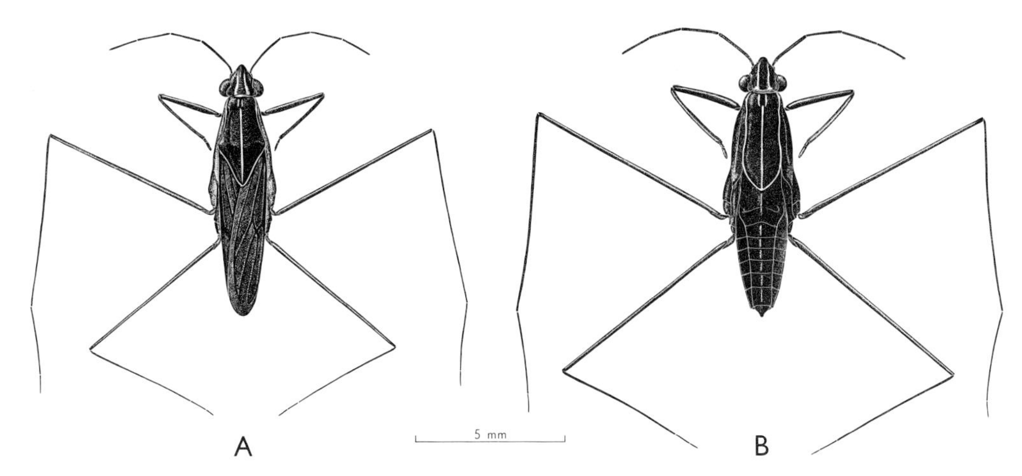

Fig. 26.68 *Limnogonus luctuosus*, Gerridae: A, macropterous; B, apterous.
[S. Curtis]

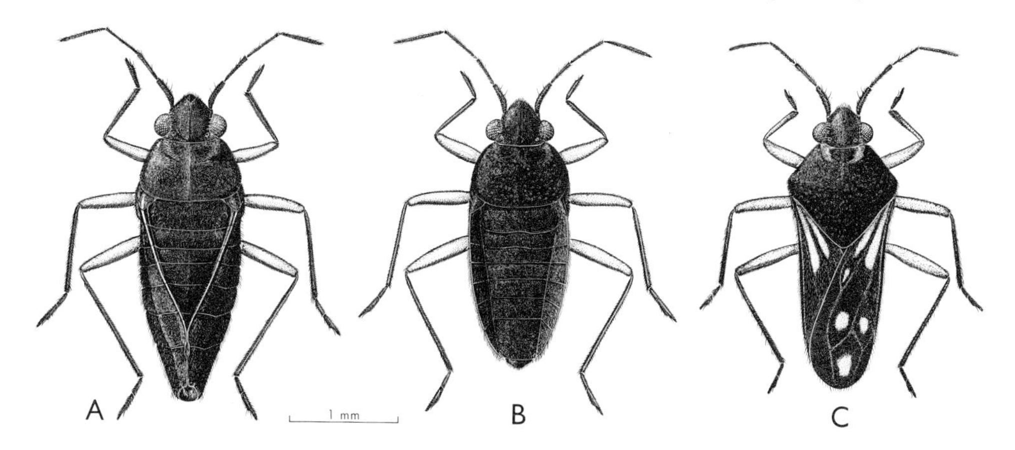

Fig. 26.69. *Microvelia mjobergi*, Veliidae: A, ♀, apterous; B, ♂, apterous; C, ♂, macropterous.
[S. Curtis]

macropterous form is known, in some only the apterous, and others are dimorphic. They are often found swimming in large groups, particularly in sheltered side pools or along the edges of bodies of water.

77. Hydrometridae (Fig. 26.70A). The slender water striders are slow moving, and commonly found by dragging amongst vegetation such as reeds in shallow water. *Hydrometra hoplogastra* Hale and *H. feta* Hale, about 13–15 mm long, are restricted to the north; *H. strigosa* Skuse is smaller, and widely distributed. *H. risbeci* Hungerf. occurs also in New Caledonia and New Zealand.

78. Hebridae (Fig. 26.70B). *Hebrus axillaris* Horvath is a small, stout bug between 1 and 2 mm long, found near the edges of fresh water. *Merragata hackeri* Hungerf. occurs in Queensland.

79. Mesoveliidae (Fig. 26.70C). The common species is *Mesovelia hungerfordi* Hale, living on the surface of quiet water, particularly at edges overhung by vegetation or among floating water plants. It exhibits alary dimorphism, with the apterous form predominating. The membranes of the hemelytra of macropters are often torn off, apparently at copulation.

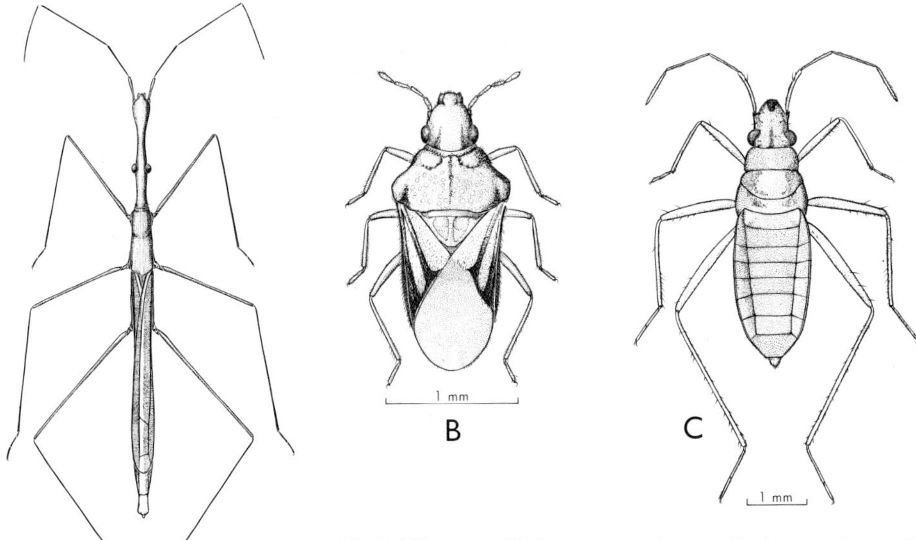

Fig. 26.70. A, *Hydrometra strigosa*, Hydrometridae; B, *Merragata hackeri*, Hebridae; C, *Mesovelia hungerfordi*, Mesoveliidae. [S. Curtis]

Superfamily OCHTEROIDEA

Key to the Families of Ochteroidea

Fore legs not raptorial, similar in form to other legs; labium long, reaching at least to hind coxae; apical part of antennae usually visible from above .... **Ochteridae**

Fore legs raptorial, with broadened femur; labium short, not reaching behind fore coxae; antennae not visible from above, held in grooves beneath eyes .... **Gelastocoridae**

80. Ochteridae (Fig. 26.71A). The species are littoral, particularly common on sandy stretches, and apparently all are predacious. The nymphs often occur just under the surface of shallow water; their dorsal surface is frequently covered with small grains of sand. *Ochterus australica* Jacz. is dark brown with paler pronotal flanges, and 3–5 mm in length. *Megochterus* is of similar facies, but about 9 mm long.

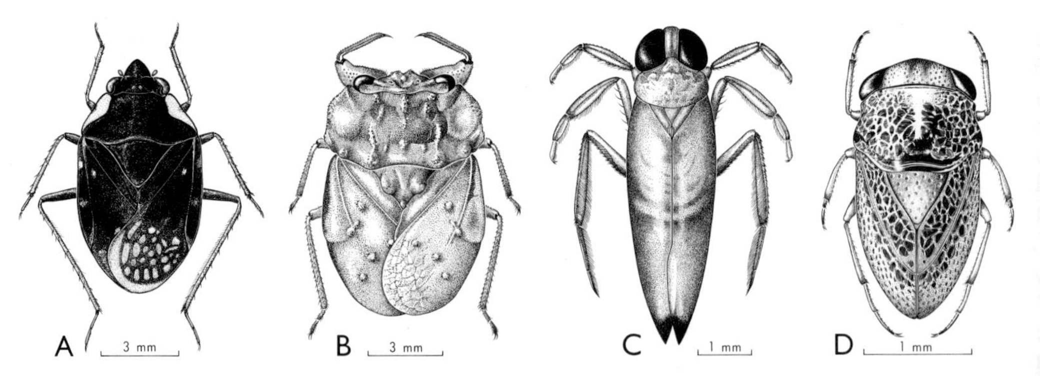

Fig. 26.71. A, *Ochterus marginatus*, Ochteridae; B, *Nerthra nudata*, Gelastocoridae; C, *Anisops inconstans*, Notonectidae; D, *Plea* sp., Pleidae. [S. Curtis, T. Nolan]

81. Gelastocoridae (toad bugs; Fig.26.71B). Todd (1960) placed all the Australian species in *Nerthra* (NERTHRINAE). Their characteristic habitat is on muddy or sandy banks or shores, where their colour merges with the background; they may also crawl beneath the water. However, some species are found long distances from water, particularly in forests, under leaves or debris, or burrowing in the soil. Even macropterous specimens have not been seen to fly, but both nymphs and adults can hop actively and pounce on their prey, which apparently includes a wide range of insects.

Superfamily NOTONECTOIDEA

The Notonectoidea are predacious. Together with the corixids, they are the most completely adapted of the bugs to the aquatic environment, the nymphs spending all their time under water and the adults most of their time. The adults, however, require atmospheric oxygen, which they obtain either as bubbles by surfacing or by protruding respiratory siphons through the surface film. Most species are capable of flight for dispersal. The families included here are sometimes grouped into 2 or 3 separate superfamilies.

Key to the Families of Notonectoidea Known in Australia

1. Apex of abdomen with a respiratory siphon composed of 2 filaments about as long as body . **Nepidae**
   Without a long siphon 2
2. Body ovoid or shield-shaped, not unusually convex above; fore legs raptorial, with femur thickened and grooved to receive curved tibia and tarsus 3
   Body very strongly convex above; fore legs not raptorial 4
3. Fore femora greatly thickened and flattened, nearly as deep as long; membrane of hemelytron without veins .... **Naucoridae**
   Fore femora less thickened, several times as long as deep; membrane of hemelytron with veins ... . **Belostomatidae**
4. Hind legs long, oar-like, with a single claw; hind tibiae flattened, with a fringe of long hairs; abdomen with a broad median ventral keel; more than 4 mm long .... **Notonectidae**
   Hind legs short, with 2 claws; hind tibiae cylindrical, without long hairs; abdomen with a fine laminate ventral keel on segments 2–6; less than 4 mm long .... **Pleidae**

82. Notonectidae (Fig. 26.71C). The backswimmers are easily recognized by their convexly keeled dorsal surface, a feature related to their upside-down swimming position. The abdomen has a mid-ventral keel, with a hair-covered groove on each side for trapping air. The eyes are large, and the hind legs natatorial with fringes of swimming hairs. The large genus *Notonecta* is represented here by only one certain species, *N. handlirschi* Kirk. in W.A. *Anisops*, another large and cosmopolitan genus, has more than 20 species in Australia (Lansbury, 1964). There are 4 species of *Enithares* and one of *Nychia*.

83. Pleidae (Fig. 26.71D). Small backswimmers, with the head and pronotum partially fused, and the membranes of the hemelytra vestigial or absent. Both Australian species belong to the genus *Plea*.

84. Nepidae (Figs. 26.72A,B). The NEPINAE are represented by *Laccotrephes tristis* (Stål), and the more slender RANATRINAE (sometimes separated as a family) by 3 species of *Ranatra* and *Austronepa angusta* (Hale) (Menke and Stange, 1964*). They spend most of their time among the mud and waterweeds of ponds, lakes, and streams, preying on arthropods and small aquatic vertebrates, which they capture with their raptorial fore legs. Respiration is by means of the long respiratory siphon, the insect either remaining suspended by it from the surface film or pushing it through the surface while clinging to underwater vegetation.

85. Naucoridae (Fig. 26.72C). Superficially

* See also I. Lansbury, *Aust. J. Zool.* **15:** 641–9, 1967.

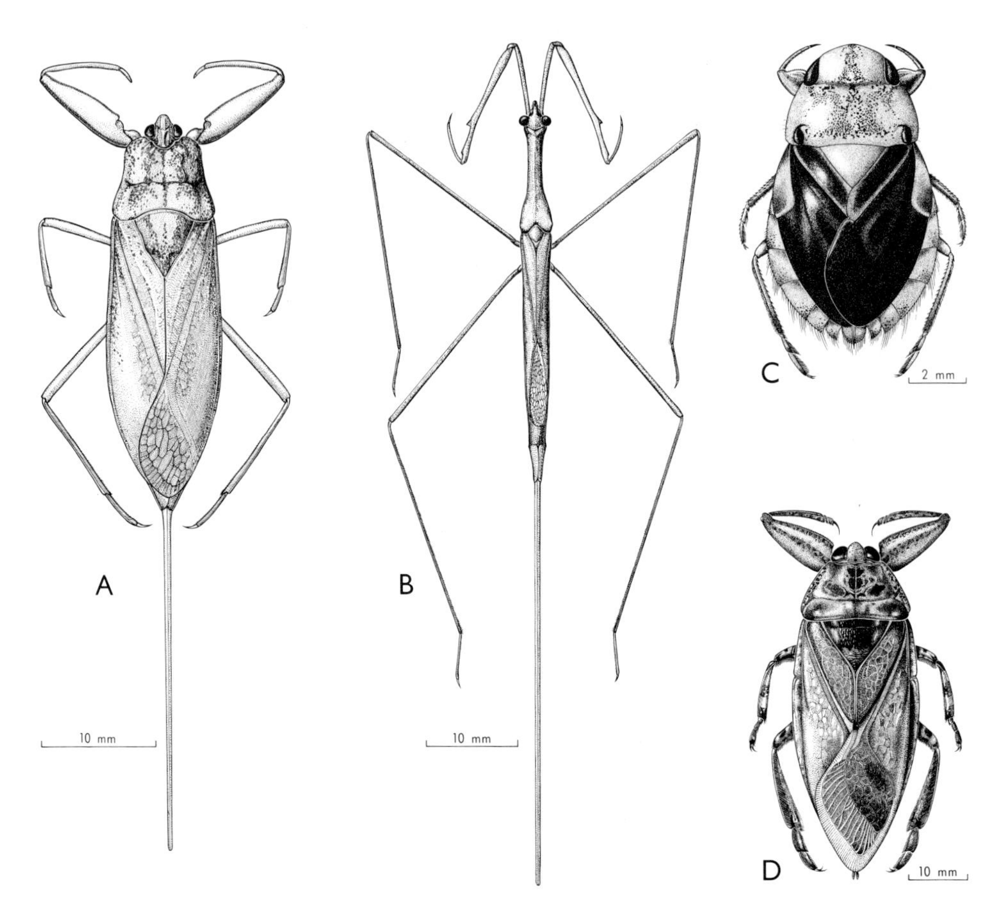

Fig. 26.72. A, *Laccotrephes tristis*, Nepidae; B, *Ranatra dispar*, Nepidae; C, *Naucoris congrex*, Naucoridae; D, *Lethocerus insulanus*, Belostomatidae. [S. Curtis, T. Nolan]

like small belostomatids, and also with raptorial fore legs. The only genera recorded in Australia are *Naucoris* and *Aphelocheirus*.

86. Belostomatidae (Figs. 26.72D,73). The giant water bugs, although entirely aquatic as nymphs and spending much of their time under water as adults, are active fliers in the adult stage, and are sometimes caught in numbers at light. Menke (1960) recorded 2 species of *Lethocerus* from Australia, *L. insulanus* (Mont.) and *L. distinctifemur* Menke, and suspected that earlier records of *L. indicus* (Le Pelet. & Serv.), a south-east Asian species, may refer to *L. insulanus*. These are large species (50–70 mm long), which can kill and feed on tadpoles and small fish as well as aquatic insects. *Diplonychus* (= *Sphaerodema*) *rusticus* (F.) and *D. eques* (Dufour) are smaller species (16–20 mm long); the female lays her eggs in a large batch on the dorsum of the male, where they are carried around and doubtless protected from insect predators. [Lauck and Menke, 1961.]

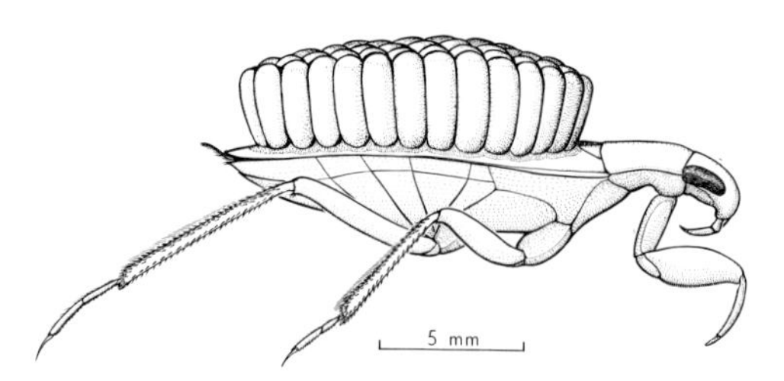

Fig. 26.73. *Diplonychus rusticus*, Belostomatidae, ♂ carrying eggs. [S. Curtis]

Superfamily CORIXOIDEA

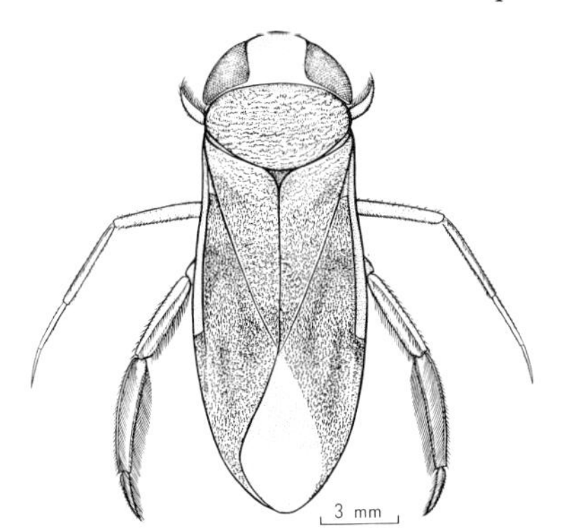

Fig. 26.74. *Agraptocorixa eurynome*, Corixidae. [S. Curtis]

87. Corixidae (Figs. 26.43A, 74). The water-boatmen are mostly phytophagous, usually having the fore tarsi modified as scoop-like palae for gathering bottom detritus that includes small algae. Some, including many of the Australian species, are at least partly carnivorous. The mid legs are used mainly for clinging, whereas the hind legs are modified for swimming. The eggs of corixids are very characteristic, being oval, with a short, more or less conical operculum at the cephalic end, and attached to the substrate at the posterior end by either an elongate stalk or a flattened disc which is sometimes shortly stalked. *Agraptocorixa* has 4 species in Australia, and *Sigara (Tropocorixa)* 5. *Micronecta* includes 14 described species, 2·5–5mm long (Chen, 1965). *Diaprepocoris*, restricted to Australia, and the allied genus *Corixanecta* in New Zealand, are exceptional in possessing ocelli (Hale, 1924).

ACKNOWLEDGMENTS. We are grateful to the following for valuable information that has been incorporated in the text: Dr W. E. China, formerly British Museum (Natural History), London; Mr L. J. Dumbleton, Crop Research Division, D.S.I.R., New Zealand; Mr J. Grant, formerly British Museum (Natural History), London; Mr G. F. Gross, South Australian Museum, Adelaide; Mr R. G. Fennah, Commonwealth Institute of Entomology, London; Mr L. A. Mound, British Museum (Natural History), London; Mr K. L. Taylor, Division of Entomology, CSIRO, Canberra; Dr D. J. Williams, Commonwealth Institute of Entomology, London.

27

THYSANOPTERA

(*Thrips*)

by E. M. REED

Small, usually slender, exopterygote Neoptera, with asymmetrical rasping and piercing mouth-parts; tarsi with eversible apical bladders; wings, when present, short, narrow, with reduced venation and wide marginal fringe; cerci absent. Metamorphosis gradual, but with two or three pre-imaginal resting stages.

Thrips (Fig. 27.4) are small insects, *Idolothrips spectrum* Hal., which attains a length of 12 mm, being amongst the largest known species. Asymmetry of the mouth-parts and possession of eversible tarsal bladders are characteristic features of the order, so there is little difficulty in placing its representatives correctly. Modern evidence supports the view that they are nearly related to the Hemiptera, in which group they were originally placed by Linnaeus. About 4,000 species have been recognized in the world. The Australian fauna is not well known, and such favourable habitats as grassland and woodland litter are largely unexplored.

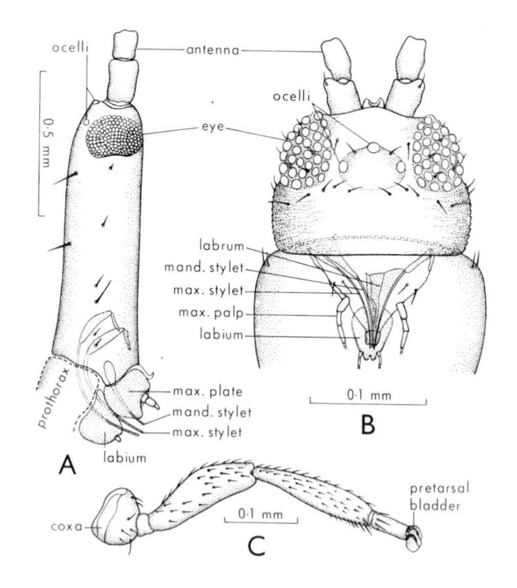

Fig. 27.1. A, head and mouth-parts of *Idolothrips spectrum*, Tubulifera, lateral; B, head of *Isoneurothrips australis*, Terebrantia, dorsal, with mouth-parts seen through prothorax; C, hind leg of *I. australis*. [B. Rankin] *mand.*, mandibular; *max.*, maxillary.

Anatomy of Adult

Head (Fig. 27.1). Broad-based, hypognathous or opisthognathous; capsule devoid of sutures, striated; genae often spiny in the Tubulifera. Compound eyes of various size; facets usually large, but often reduced and separated in cryptic forms. Three ocelli present in winged adults only. Antennae of

4 to 9 segments, moniliform or filiform, with several segments carrying spine-like sense organs, and sensory areas on the more basal segments. Mouth-parts arranged as a ventrally oriented cone formed by the labrum and labium, with the piercing elements, comprising paired maxillary stylets and a single (left) functional mandible, emerging at the apex. Maxillary palpi 2- to 8-segmented. Labial palpi 1- to 4-segmented.

Thorax. Prothorax with conspicuous notum, sometimes divided at the posterior angles by epimeral sutures. Pterothorax variously developed according to the presence or absence of wings. Prosternum marked by the development of paired and unpaired plates; prepectus present only in Tubulifera, prospinasternum only in Terebrantia. Meso- and metathoracic spiracles present.

Legs. Ambulatory. Fore legs often modified, reaching an extreme of development in the crab-like *Carcinothrips*. Tarsi 1- or 2-segmented, with the distal segment possessing an eversible bladder (Fig. 27.1) which is used for adhesion to varied surfaces.

Wings. Membranous, strap-like, varying considerably in length, with a marginal fringe of long filaments. They are carried unfolded along the abdomen while at rest. The venation is considerably reduced, but at least one longitudinal vein reaches the apex in the Terebrantia. A scale in the fore wing, the rudiment of the ano-jugal area, has curved setae that attach to the costa of the hind wing during flight.

Abdomen. Ten-segmented, often dorso-ventrally compressed. Terminal segment conical in Terebrantia, drawn out into a conspicuous tube in the Tubulifera (Figs. 27.2, 4); cerci absent. Functional spiracles present on segments 1 and 8 only. Gonopore of male between segments 9 and 10; genitalia concealed at rest, symmetrical, with one or two pairs of 'parameres' in Terebrantia, greatly reduced in Tubulifera (Priesner, in Tuxen, 1956). Females of Terebrantia with gonopore between segments 8 and 9, and two pairs of valves (arising from the same segments) forming the serrated ovipositor; in

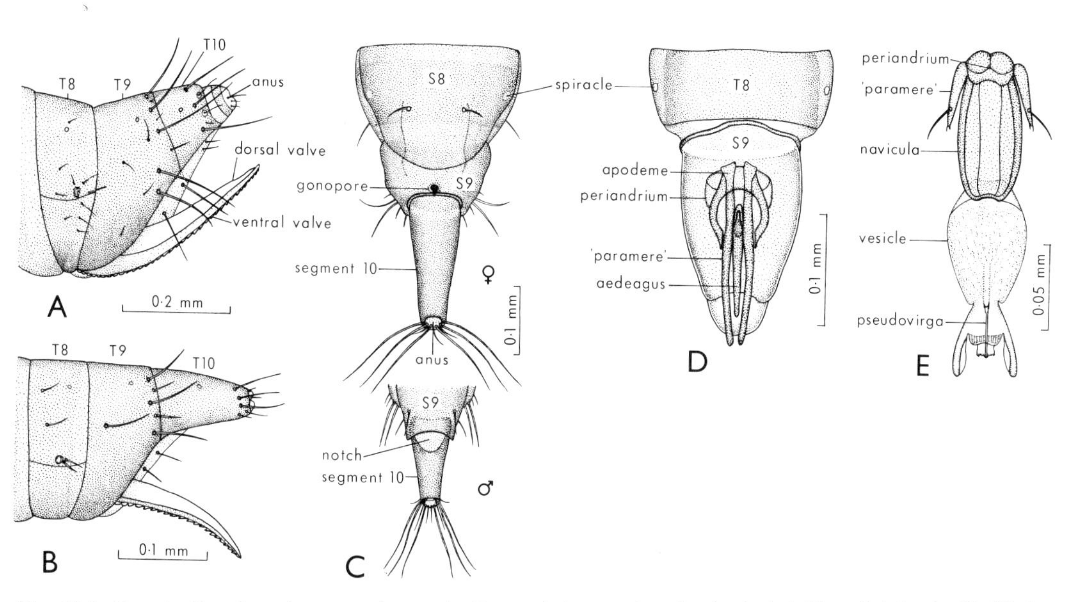

Fig. 27.2. Terminalia, cleared preparations: A, *Desmothrips tenuicornis*, Aeolothripidae, ♀ lateral; B, *Thrips imaginis*, Thripidae, ♀ lateral (Merothripidae, Terebrantia with reduced ovipositor, also occur in Australia—found recently by Mr L. A. Mound); C, *Oncothrips tepperi*, Phlaeothripidae, ♀ and ♂ ventral; D, *Desmothrips steeleae*, Aeolothripidae, ♂ in resting position, with T9 and T10 removed, penile vesicle not shown; E, undetermined sp. of Phlaeothripidae–Megathripinae, ♂ with vesicle extruded. [B. Rankin]

Tubulifera the gonopore is between segments 9 and 10, and there is no ovipositor.

Internal Anatomy. There is a cibarial sucking pump, usually two pairs of salivary glands, an extensive mid gut divided into two parts, 4 Malpighian tubes, and usually 4 rectal papillae. The nervous system is concentrated, the sub-oesophageal and prothoracic ganglia being fused, and the abdominal ganglia concentrated in segment 1. In the female each ovary consists of 4 panoistic ovarioles. In the male there is a pair of fusiform testes and one or two pairs of large accessory glands.

Immature Stages

Egg. The eggs are surprisingly large in relation to the body size of the female, more or less kidney-shaped in the Terebrantia, usually oval in the Tubulifera. The embryology is characterized by precocious rupture of the chorion.

Nymph (Fig. 27.3). The most notable feature in the postembryonic development is the occurrence of two (sometimes three) resting stages in which a significant amount of metamorphosis occurs (Snodgrass, 1954). The first and second instars are like the adults, but with transparent cuticle (often overlying dense pigment in the Tubulifera), reduced number of antennal segments, reduced eyes, no ocelli, and no wings. Wing rudiments appear in the first resting ('propupal') stage in Terebrantia, not until the second ('pupal') stage in Tubulifera. The antennae may be reduced in the 'propupa' and, they are characteristically turned backward in the 'pupa'. There is some histolysis and reconstruction of the cells of the gut, nervous system, body muscles, and especially of the muscles of the head, in these stages.

The development shows a striking trend, also foreshadowed in a few Homoptera, towards a holometabolous type of metamorphosis, and the resemblance is accentuated by the occurrence of a 'pupal' cocoon in some of the Aeolothripidae. It does not indicate relationship with the endopterygote

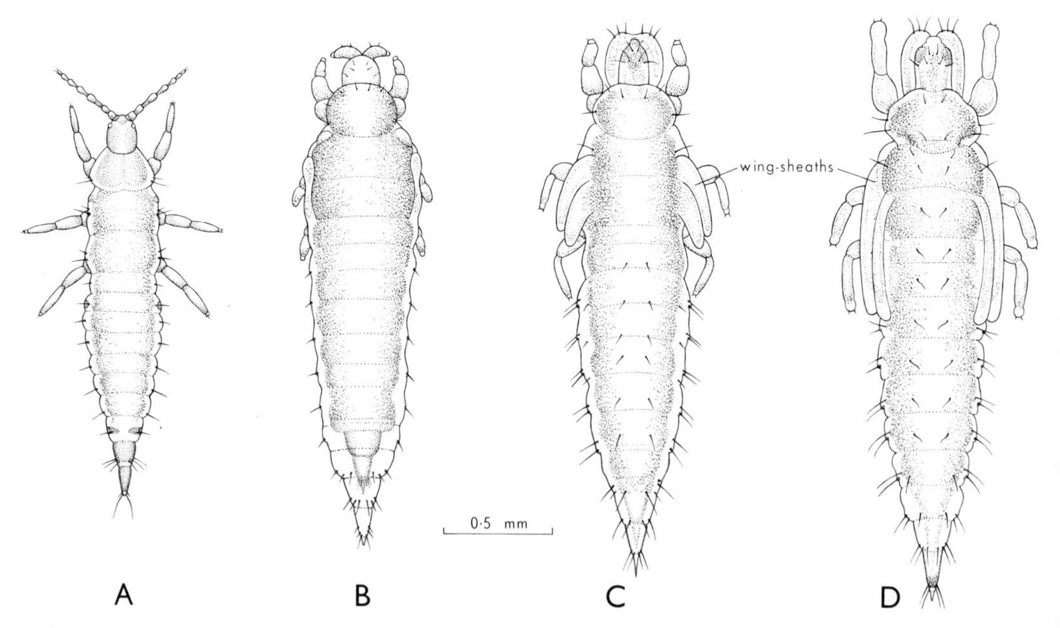

Fig. 27.3. Immature stages of Tubulifera (A, C, D *Teuchothrips* sp., B unidentified): A, stage II (active nymph); B, stage III ('1st pupal' or 'propupa'); C, stage IV ('2nd pupal'); D, stage V ('3rd pupal'). The anterior abdominal spiracle is on segment 2 in the nymph, on segment 1 in the adult. [B. Rankin]

orders, but it does provide an example of the kind of parallel evolution that might be expected to occur with such an advantageous acquisition (p. 104).

Biology

Adults. A wide range of food materials is utilized by thrips, with the representatives of both suborders showing somewhat different preferences. Wildflower blossoms are particularly favoured by the more active terebrant species, whereas the Tubulifera commonly hide in debris and feed on fungi, or combine phytophagy with the cryptic habit as gall formers. Several species are predacious on a variety of small insects and mites. A few are found in the debris collected between eucalypt leaves webbed together by moth larvae, and some cryptic forms occur beneath flaking bark. Woodland litter yields a fauna marked by apterous forms, such as *Amphibolothrips*. The inland species are predominantly gall formers associated with the softer foliage of fodder trees, such as *Geijera* and *Acacia aneura*. Morphs are common, including several anatomical categories analogous to those of the social orders. Adults may lack wings, in which case ocelli are also lacking. Thrips populations may be exceedingly dense, and the flower heads of some native grasses are infested so heavily at times as to suggest physical crowding. Parasites and predators appear to be of little consequence in regulating their abundance, and population studies on *Thrips imaginis* Bagn. have shown weather to be a major factor influencing fluctuations in the numbers of that species (Andrewartha and Birch, 1954).

Reproduction and Immature Stages. Most thrips lay fertilized eggs, but parthenogenesis may occur and result in production of males or females only. *Frankliniella schultzei* Trybom produces males if mated, females if unmated. Viviparity has been reported in several species of the *Idolothrips* group. The oviposition period extends over many days. In the Terebrantia the eggs are laid singly in a slit cut into host tissue by the ovipositor, whereas the Tubulifera scatter theirs on surfaces near where they feed. During warm weather hatching takes place in about 7 days, the pale-coloured nymphs feeding alongside the adults for another 10 days or so before the onset of quiescence. The 'pupal' stages are usually cryptic, and readily identified by the prominent wing buds (Fig. 27.3). Pupation often occurs within an earthen cell located several inches below the soil surface.

Natural Enemies. Several species of Hymenoptera, including *Thripoctenus* spp., confine their attacks to thrips. Other natural enemies, such as predacious bugs, spiders, and fungi, attack insects generally.

Economic Significance. It is difficult to gauge the economic importance of such a poorly studied group, but only a few species are proven pests. The absence of tubular mouth-parts precludes the deep sucking characteristic of the Hemiptera, so that feeding by phytophagous thrips is more or less confined to soft recent growth. Floral parts, when favoured, often sustain crucial attacks on the sexual components, preventing fruit development and set of seed. Many fruit and vegetable crops sustain serious injury from the feeding of all stages on the ovaries, anthers, young buds, and ripe fruit. In addition to feeding lacerations, damage also results from faecal deposits on growing fruit and injury to the sexual components by the ovipositor.

Some species are vectors of various viruses, bacteria, and fungi. Sakimura (1947) considered that all the viruses transmitted by thrips were varieties of tomato spotted wilt virus, and *Frankliniella schultzei* is a vector of this virus in Australia. Wheat rust spores have been observed on the bodies of captured individuals.

Special Features of the Australian Fauna

The east-coastal area has provided most species, with Queensland and Victoria accounting for almost two-thirds of the recognized fauna. The gall-forming thrips are almost restricted to Indonesia and Australia.

CLASSIFICATION

Order THYSANOPTERA (287 Australian spp.)

Suborder TEREBRANTIA (99)

1. Aeolothripidae (13) Merothripidae (0)
2. Thripidae (86) Heterothripidae (0)

Suborder TUBULIFERA (188)

3. Phlaeothripidae (188)

The classification adopted is essentially that of Priesner (1949). Stannard (1957) has discussed the phylogeny of the Tubulifera. The numbers of Australian species are based on Kelly and Mayne (1934), and may be drastically altered when the order is revised.

Key to the Families of Thysanoptera Known in Australia

1. ♀♀ without ovipositor; terminal abdominal segment tubular in both sexes; wings overlapping along the dorsum when at rest .... TUBULIFERA. **Phlaeothripidae**

   ♀♀ with saw-like ovipositor and apex of abdomen conical; ♂♂ with apex of abdomen rounded; wings carried side by side along the dorsum when at rest .... TEREBRANTIA 2
2. ♀♀ with point of ovipositor directed dorsally; wings broad, rounded at apex; antenna 9-segmented .... **Aeolothripidae**

   ♀♀ with point of ovipositor directed ventrally; wings narrow, usually pointed at apex; antennae 6- to 9-segmented .... **Thripidae**

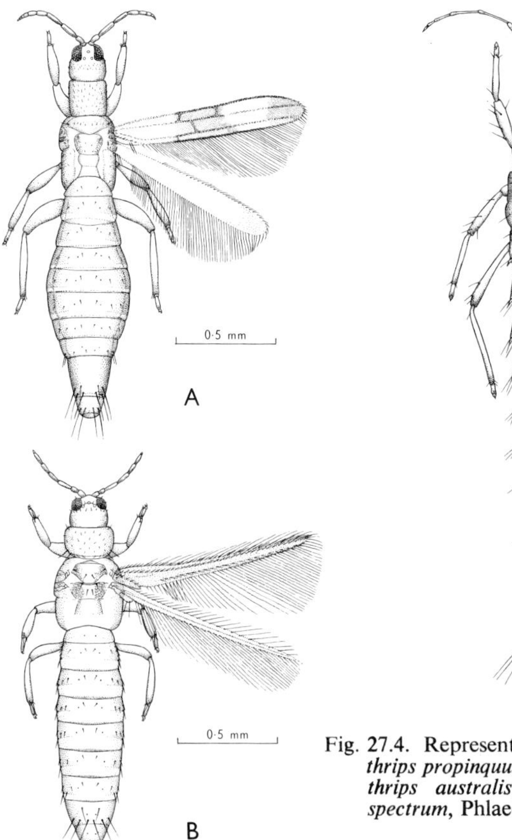

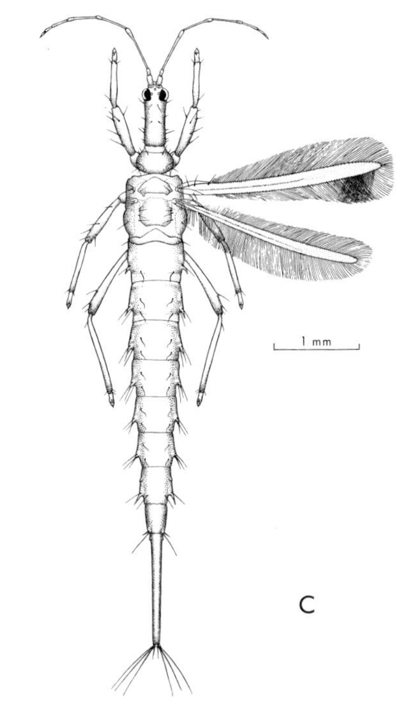

Fig. 27.4. Representative thrips, dorsal: A, *Desmothrips propinquus*, Aeolothripidae, ♀; B, *Isoneurothrips australis*, Thripidae, ♀; C, *Idolothrips spectrum*, Phlaeothripidae, ♂. [B. Rankin]

Suborder TEREBRANTIA

1. Aeolothripidae. A widely distributed family containing the more primitive members of the order. Wings large and usually conspicuously veined, often with reduced fringe area; palpi and antennae elongate. Five genera are represented in Australia. *Desmothrips* has antennal segments 5–9 united, and wings often marked with dark patches. *D. australis* Bagn. is found on a variety of grasses and blossoms, and is probably predacious (Vevers Steele, 1935).*

2. Thripidae. A large family containing most species of the suborder. They have 6–9 antennal segments, with a distal group forming a compact style; wings are usually narrow and tapering. The HELIOTHRIPINAE include species with deep polygonal reticulation on parts of the body. Six genera are represented. *Heliothrips* has a finely pointed style. *H. haemorrhoidalis* Bouché is a cosmopolitan pest commonly encountered on ornamental shrubs. *Australothrips* has antennae as in *Heliothrips*, but has small projections around the prothoracic margin. *A. bicolor* Bagn. is found in wildflower blossoms.

The THRIPINAE embrace all the species not marked by deep polygonal reticulation, and include a heterogeneous assemblage of genera. *Frankliniella* and *Taeniothrips* are both large, world-wide genera containing many species. *T. brevicornis* (Bagn.) shows a decided preference for dandelion flowers. The species of *Anaphothrips* are often apterous, with a general tendency towards reduction in the numbers of wing and body bristles. *A. striatus* Osb. is an immigrant species found mainly on grasses, and has been implicated in the sterility of cereals. *Thrips* is one of the largest genera in the order, with 7 antennal segments and lacking prominent bristles on the anterior angles of the pronotum. *T. imaginis* is an endemic species, which feeds on a wide range of wild and cultivated flowers; apple crops have been seriously damaged at times (Vevers Steele, 1935). *T. tabaci* Lind. is a cosmopolitan pest with catholic tastes; cotton, tobacco, and onion crops are often severely attacked. *Isoneurothrips* is similar to *Thrips* in thoracic chaetotaxy, but with antennal segment 6 much longer in proportion to 7. *I. australis* Bagn. is commonly found in *Eucalyptus* flowers, possibly feeding on nectar. *Chaetanaphothrips* and *Pseudanaphothrips* are closely related to *Anaphothrips*. *Ch. signipennis* (Bagn.) and *Ch. bilongilineatus* (Gir.) cause rust-coloured blemishes on banana. *Limothrips* has lateral projections on segment 3 of the antenna, *Chirothrips* on segment 2. *L. cerealium* Hal. is an immigrant pest of cereals. *Ch. atricorpus* Gir. is a native species found in grasses and in sorghum seed heads.

Suborder TUBULIFERA

3. Phlaeothripidae. This family contains some 300 genera distinguished from those of the Terebrantia by numerous ecological as well as anatomical features. The species are generally larger, more darkly coloured, and fly less readily. Reduced activity is associated with gregariousness, the exploitation of litter, and gross development of parts of the body (e.g. the fore legs). Parthenogenesis is rare, and there are three 'pupal' instars. Maxillary palpi are always 2-segmented, and the stylets reach an extraordinary length in some species.

Phlaeothrips is a large genus, to which many nondescript species have been assigned. Many are fungal-feeders found in a wide variety of habitats, such as under bark, in leaf litter, and on dead wood. *Horistothrips* and *Liothrips* were formerly included in the Liothripinae, which embraced medium-sized species with a sharply pointed mouth-cone and smooth cheek region. *Horistothrips* are mostly cryptic species found under bark. The Australian species referred to *Liothrips* are found on leaf surfaces, in flowers, and in galls. *Teuchothrips* is a large genus composed of several morphological groupings. *T. pittosporiicola* Bagn. is commonly found on

* L. A. Mound (*Bull. Br. Mus. nat. Hist.* Ent. **20**: 43–74, 1967) now recognizes 19 Australian species in 7 genera.

Pittosporum leaves in the Sydney area. *Haplothrips*, like *Phlaeothrips*, cannot be separated from other genera by the presence of any single outstanding feature. The eyes are generally larger, and most species have the fore wings constricted near the middle (as do several species of other genera). Few habitats have not been exploited by members of this genus; sweeping grass usually produces a few individuals. Both *Froggattothrips* and *Rhopalothripoides* contain minute species found on *Acacia*. *Nesothrips* has the head rounded laterally and often with small spines. In *N. dimidiatus* (Hood) the eyes are prolonged ventrally. This species has the habit of curling the tail over the abdomen like a staphylinid beetle. A large number of diverse forms have been referred to *Cryptothrips*, in which the head is usually rectangular and the mouth-cone broadly rounded.

The larger species found in the warmer regions of the world are well represented in Australia, with *Idolothrips* providing the giants of the group. The members of this genus are common on fallen decaying eucalypt branches. *Kleothrips* and *Elaphrothrips* are related genera which, like *Idolothrips*, have the head projecting before the eyes and carrying the anterior ocellus some distance forward.

Galls take the form of a bulbous growth, as produced on *Casuarina* stems by *Thaumatothrips*, or, as in *Eugynothrips* on *Smilax*, a leaf blade becomes distorted through overgrowth and curling. *Kladothrips* is the dominant gall-forming genus, with the members confined to species of *Acacia*. *Moultonia* and *Choleothrips* are found in *Geijera* leaf galls. Eucalypts are not attacked.

ACKNOWLEDGMENT. I am grateful to Mr L. A. Mound, British Museum (Natural History), London, for his comments on the manuscript and help with the illustrations.

28

MEGALOPTERA

(Alderflies)

by E. F. RIEK

Mandibulate endopterygote Neoptera, in which the mouth-parts are not produced into a beak, and with two pairs of slightly dissimilar functional wings. Larvae aquatic; with well-developed mandibulate mouth-parts; 3-segmented labial palpi; well-developed prothorax; functional legs; and lateral abdominal gills. Pupae decticous and exarate.

This is a small order, with less than 300 described species, but it is of special interest because it is considered to be the most primitive of the endopterygote orders. It has a wide distribution throughout the temperate regions, though a few species occur in the tropics. Many species are of large size, with a wing-span up to 16 cm.

Australia has less than twenty, mostly large species (Riek, 1954a). They resemble broad-winged lacewings when in flight, but can be distinguished from them when at rest by the soft, very flexible abdomen. They are also similar to soft-bodied beetles, but the fore wings are functional in flight and have a clearly defined venation. Adults can be distinguished from most Mecoptera by CuA, which is normally forked in both wings (simple in the fore wing of one Australian genus), and from most Neuroptera by the absence of end-twigging.

Although formerly considered as part of an enlarged order Neuroptera, the very distinctive aquatic larvae do not have the mouth-parts modified into a sucking beak and have lateral abdominal gills. On the other hand, separation of the larvae (Fig. 28.6) from some coleopterous larvae (Gyrinidae) is not easy. The labial palpi are 2-segmented or less in Coleoptera, whereas they are 3-segmented in Megaloptera (Fig. 28.4). However, in Gyrinidae they appear to be 3-segmented through deep emargination of the labium. The labrum is distinct in Megaloptera, but not in Gyrinidae.

Anatomy of Adult

Head (Fig. 28.2). Prognathous, with prominent, bulging eyes; postocular region prominent, sometimes (in non-Australian species) expanded and with a forwardly directed marginal spine. Three ocelli well developed (Corydalidae) or absent (Sialidae). Antenna shorter than the fore wing, with the first one or two segments enlarged; variable in form, typically a simple tapering flagellum,

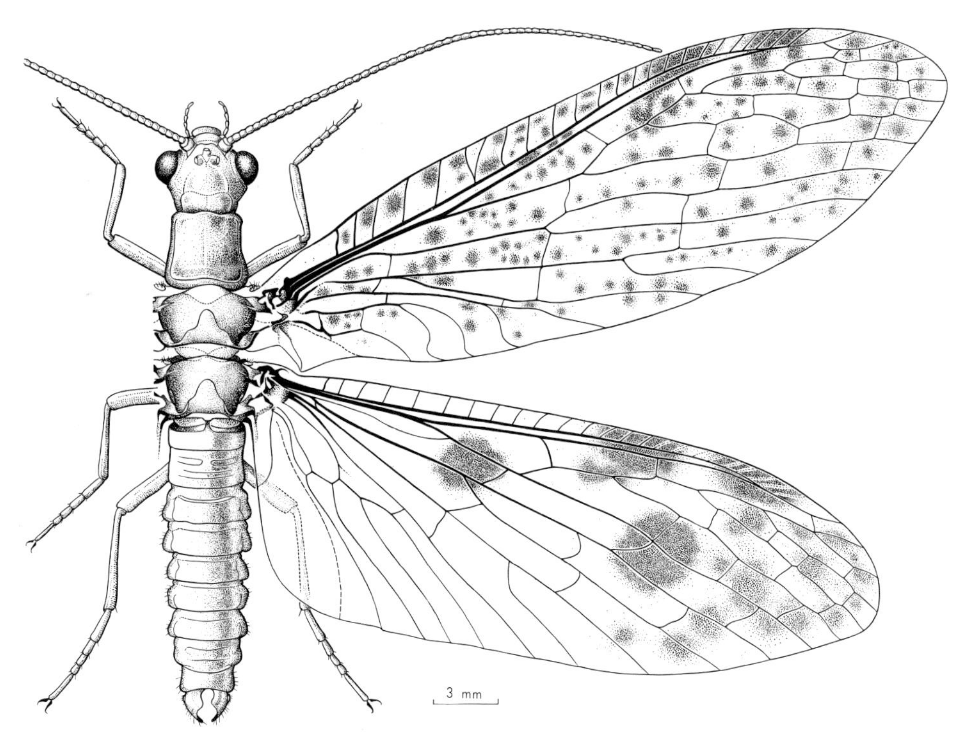

Fig. 28.1. *Archichauliodes* sp., Corydalidae.
[T. Binder]

but sometimes moniliform or serrate or even flabellate in the male. Mouth-parts well developed, with strong mandibles; maxillary palpi 5-segmented; labial palpi 3-segmented; gular plate exposed. The males of some exotic Corydalidae have the mandibles greatly enlarged into formidable tusks.

Thorax. All segments well developed, freely movable. Pronotum large, subquadrate. Mesothorax and metathorax with distinct postnota and normal spiracles.

Legs. Short, with short coxae; mid and hind coxae enlarged at base above; tarsi 5-segmented, ending in 2 simple claws and without a pulvillus. The laterally and ventrally expanded 4th tarsal segment of Sialidae functions as a pulvillus.

Wings (Fig. 28.5). Membranous, dissimilar, the hind wing with an expanded anal field. Venation complete, end-twigging absent or only slightly developed. Sc and R_1 fused towards apex; R_{2+3} sometimes pectinately branched; M of fore wing usually 2-branched, but 3- or 4-branched in Sialidae; CuA nearly always branched in both fore and hind wings. In Corydalidae, the jugal lobe of the fore wing is distinct, and has a well-developed jugal vein. The humeral region of the hind wing is without special frenulate hairs, so that the wings are coupled merely by the jugal lobe of the fore wing overlying the humeral region of the hind wing. The jugal lobe is not developed in Sialidae, though a faint jugal vein is present.

Abdomen. Very soft and flexible, segment 1 with some sclerotization. Spiracles on segments 1–8. The male (Figs. 28.3A,B) has tergite 9 generally hooded, with the 'gonapophysis' articulating with its lower lateral margin, except in Chauliodinae (Corydalidae); sternite 9 forming a variously shaped subgenital plate; tergite 10 forming a pair of anal claspers of varying shape, their outer surface with a group of sensilla ('trichobothria') forming a more or less convex wart, or the group of sensilla borne on a separate cercus;

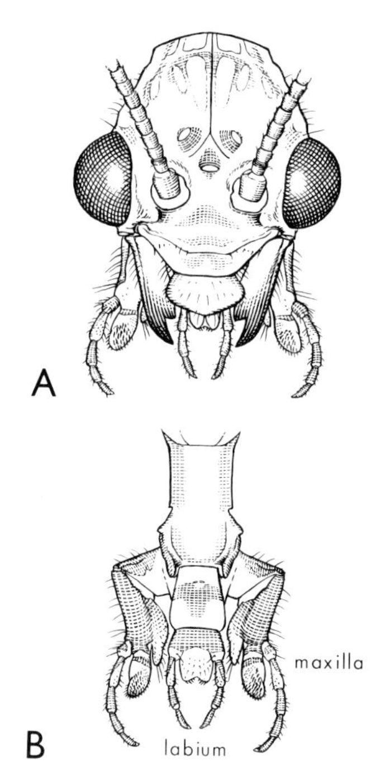

Fig. 28.2. Head of *Archichauliodes* sp.: A, frontal; B, maxillae and labium, ventral. [M. Quick]

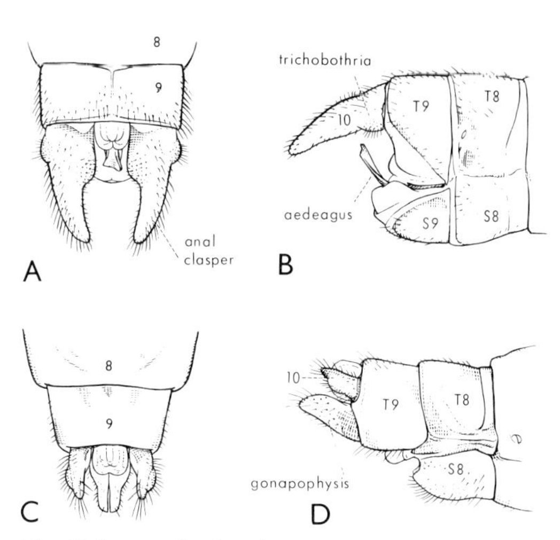

Fig. 28.3. Terminalia of *Archichauliodes* sp.: A, ♂, dorsal; B, ♂, lateral; C, ♀, dorsal; D, ♀, lateral. [M. Quick]

sternite 10 generally membranous, though occasionally it forms a sclerotized plate; aedeagus hinged to the lower basal angle of tergite 9 and either simple or paired. The female (Figs. 28.3C,D) has sternite 8 sometimes produced apically as a subgenital plate; tergite 9 large and produced downwards laterally, sternite 9 membranous; the pair of 'gonapophyses' attached to the lower angles of tergite 9; segment 10 forming a pair of cercoid anal plates of varying form, each with a group of sensilla. The terminalia have the tergites less sclerotized in Sialidae.

Internal Anatomy. Very inadequately investigated. The digestive system is provided with a median dorsal food reservoir; Malpighian tubes normally 8 (sometimes 6). The ventral nervous system consists of 3 thoracic and generally 7 abdominal ganglia. The ovaries consist of a variable number of usually polytrophic ovarioles.

Immature Stages

Egg. The eggs of *Archichauliodes* are cylindrical. They are about 0·7 mm high, attached by one end, and the other bears a white, elongate, terminally clubbed micropyle which is about a quarter as long as the egg. This white projection gives the egg masses a rather frosted appearance. The micropyle is not conspicuous in *Sialis*.

Larva (Fig. 28.6). Elongate and flattened. Head (Figs. 28.4A,B) prognathous, quadrangular, with well-developed mouth-parts; mandibles stout; maxillae elongate, each sometimes appearing to end in two 5-segmented palpi; labial palpi 3-segmented; gular plate distinct, large. Eyes with separate stemmata (6 in *Archichauliodes*). Antennae relatively well developed, with 4 elongate segments. Prothorax large, subquadrate, pronotum and prosternum both heavily sclerotized; mesonotum and metanotum transverse, sclerotized. Legs with long, unsegmented tarsi ending in 2 claws. Abdomen bearing lateral filamentous gills. In Sialidae, segments 1–7 bear segmented gills with long setae. In Corydalidae, segments 1–8 bear long, simple or imperfectly segmented, tapering gills with a few very short setae. There may be a tuft of accessory gill filaments at the base of each, except on segment 8. Segments 1–8 with spiracles, that of 8 sometimes raised on a long stalk in Corydalidae. Segment 9 without processes.

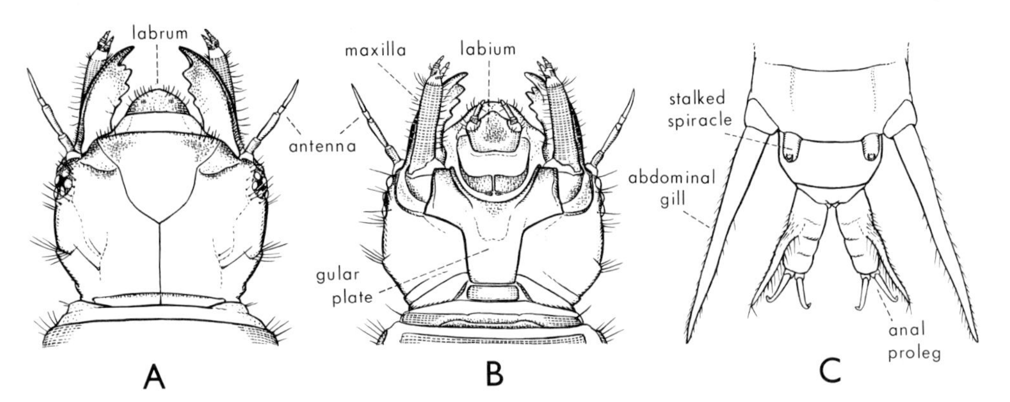

Fig. 28.4. Larva of *Archichauliodes* sp.: A, head, dorsal; B, head, ventral; C, apex of abdomen, dorsal. [M. Quick]

Segment 10 prolonged and tapering in Sialidae, very short in Corydalidae and bearing a pair of large prolegs, each of which has a lateral filament near its middle and ends in two large claws (Fig. 28.4c).

Pupa. Active, decticous, exarate, very like that of Coleoptera, except for the presence of functional mandibles. The head is similar to that of the larva, apart from the development of the antennae and compound eyes.

Biology

Adults. Megaloptera are among the commonest of stream insects, and they are generally associated with clear, cold waters. They have not been observed to feed. Adults of *Archichauliodes* are common along such streams in summer and early autumn. They fly actively when the light intensity is low, especially in the late afternoon and early evening, but do not move far from the streams. In dull weather they may be on the wing throughout the day. Their flight is awkward, rather slow and irregular, but can be swift when they are disturbed. They also have a gliding flight high over the stream. The significance of this type of flight, in which the wings are outspread, slightly swept back, and raised above the horizontal, is unknown. At first it could be mistaken as a preliminary to egg-laying, similar to that seen in some Plecoptera, but the eggs are deposited on fixed objects, and both sexes exhibit this type of flight, though it is more often observed in the male. During flight they occasionally dip on to the surface of the water, and appear to rise again with a little difficulty. Again, this action is not associated with egg-laying. The adults rest on the stems of vegetation margining the stream, especially in clumps of *Leptospermum*. The wings are folded rather flat over the abdomen, and the anal area of the hind wing is folded under the remainder of the wing. On bright days they are easily collected by hand, as they make little effort to fly when disturbed. On handling they exude a white or pinkish fluid from the anus.

The Sialidae frequent slow-flowing streams with muddy bottoms. Nothing is known of the biology of the Australian species. They, too, are attracted to light, or can be beaten from the foliage margining the streams. They are not conspicuous insects, and when they fly at dusk, as does *Archichauliodes*, they can be confused with the similar-sized caddisflies.

Reproduction. Copulation takes place at rest on twigs and branches of shrubs close to or overhanging the stream. The pair face in opposite directions along the twig. Females normally lay in the late afternoon and early evening; but on overcast days they can be observed laying at all times, and even when there is a change to bright sunlight they continue to lay. The eggs are deposited in large, single-layered clusters of 1,000 to 3,000 or more, on rocks, stumps, or dead shrubs overhanging the stream, in positions that are protected from the sun during the hottest part of the day. The same sites are used for many seasons. Many individuals aggregate at this time, and their conjoined single-layered egg masses may cover many square centimetres. Females ovipositing later lay their eggs adjoining those already deposited, rarely on top of them. The aggregates of egg masses resemble lichens encrusting the rocks. The egg colour varies somewhat with individual females. When the eggs are freshly laid, they may be nearly white, but vary to a pale brown. After a short period they darken somewhat, the darkest egg masses being almost black. They hatch in 2 to 4 weeks.

Immature Stages. On hatching, the larvae of *Archichauliodes* (length about 1 mm, width of head capsule 0·4 mm) drop to the stream below, where they sink to the bottom and conceal themselves under rocks and debris. Older larvae occur in all situations in the stream. They are actively predacious. The number of larval instars is apparently large, for the head-capsule width of a sample falls into a large number of distinct sizes (Tillyard recorded only 4 instars). The rate of development is not known; but there are many immature larvae of various sizes in the stream when the first adults are on the wing in summer, so that growth probably extends over 2 or more years. Larval development in some extra-Australian species is known to extend over 3 years.

When the larvae are full-grown, they migrate to the stream margins during periods of high water level in spring and early summer, and remain under stones and logs, often at a considerable distance above the normal water level, when the water recedes. There they construct simple pupal cells in the soil, and may remain in the active larval stage for several months, though pupation usually takes place in early summer. Under unfavourable conditions, fully grown larvae have been kept without food for upwards of 12 months before pupation and successful eclosion of the adults.

Not all the maturing larvae leave the water at the one time, but they continue to do so following each marked rise in water level throughout spring and early summer, so that there is often a clear distinction between those that have pupated and are almost ready to emerge and those that are still larvae in their pupal cells. Larvae select their pupation

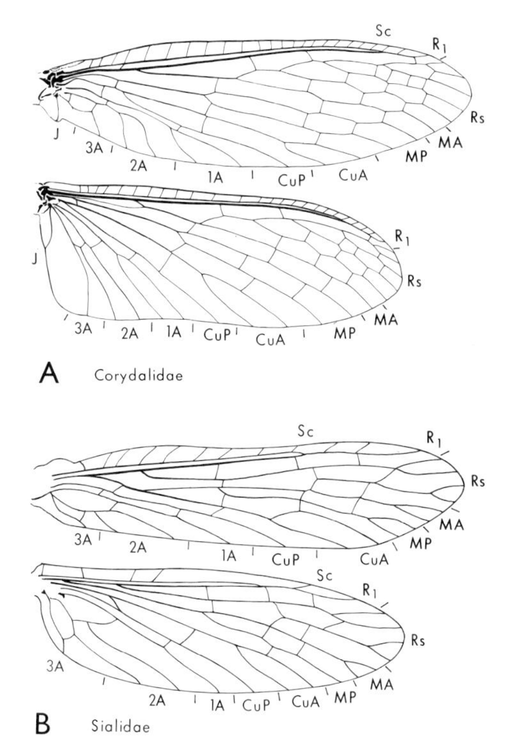

Fig. 28.5. Wings: A, *Archichauliodes* sp.; B, *Austrosialis* sp. **[M. Quick]**

sites carefully. The stones, or occasionally logs, must rest on soil, rather than on sand or gravel, to ensure retention of adequate moisture. Stones at the highest levels covered by floods are not suitable, even if they rest on soil, as such situations dry out too rapidly, and are usually invaded by ants. When one locates larvae or pupae in their cells under stones, they usually occur in numbers, and the same situations are used in following years unless there has been great disturbance in the stream bed.

The pupal stage lasts from 2 to 4 weeks. The legs are all free, and the insect is able to walk actively and use its mandibles in defence. The wing-pads darken before the emergence of the adult.

Natural Enemies. Adults are captured on the wing at dusk by birds, but most predation takes place when the females are egg-laying, when frogs often take a heavy toll. Both adult and larval anthicid beetles are recorded as predators of the exposed egg-masses of corydalids. Eggs are parasitized by Trichogrammatidae (Hymenoptera). Larvae have not been examined for parasites.

Economic Significance. The larvae are predatory on other stream animals, and form an important link in the food chain of large fish. The adults, too, are readily taken by large fish when they dip on to the surface of the stream.

Special Features of the Australian Fauna

The larger Corydalidae are all generalized types, not unlike those of South America and South Africa. *Archichauliodes* is recorded from Australia, Chile, and New Zealand. *Protochauliodes* occurs also on the western side of South and North America. The Sialidae are represented by the endemic *Austrosialis* in which the wings are longer and narrower than in *Sialis*, but it is similar to *Protosialis* from South America and related genera from South Africa and India. The Australian distribution is predominantly in the upper reaches of the east coast streams and the mountains of the south-east, with one sialid in Tasmania and one corydalid in south-western Australia.

CLASSIFICATION

Order MEGALOPTERA (16 Australian spp.)

1. Corydalidae (14)
2. Sialidae (2)

Some authors regard the Raphidioptera, which do not occur in Australia, as a suborder of Megaloptera, but they are best treated as a separate order. The Corydalidae might well be divided into two families, Corydalidae and Chauliodidae.

Keys to the Families of Megaloptera

ADULTS

Three ocelli; 4th tarsal segment simple; insects of medium to large size (fore wing 35–85 mm) .... **Corydalidae**

Ocelli absent; 4th tarsal segment prominently bilobed; insects of rather small size (fore wing 8–12 mm) .... **Sialidae**

LARVAE

Eight pairs of lateral abdominal gills; apical segment of abdomen with a pair of prolegs (Fig. 28.6B) .... **Corydalidae**

Seven pairs of lateral abdominal gills; apex of abdomen drawn out into a long terminal filament (Fig. 28.6A) .... **Sialidae**

1. Sialidae. Nothing is known of the biology of the Australian species. The adults are attracted to lights set up close to the streams, or they may be swept from foliage bordering the stream. They are like lacewings in their general facies and manner of flight. There is a single species of *Austrosialis* in Tasmania and another on the mainland ranging from coastal N.S.W. to south-eastern Queensland.

2. Corydalidae. This widspread family is well represented in Australia. All the Australian species are placed in the subfamily CHAULIODINAE, tribe Chauliodini (Kimmins, 1954), in which the head of the adult is rather triangular and without a forwardly directed posterolateral spine and there are only 3 cross-veins between R and Rs. The females are slightly larger than the males, and the apices of their wings are a little more pointed. The 9 species of *Archichauliodes* are inhabitants of cold-water streams from Victoria to north Queensland. Except for one species in the Kosciusko area, the adults all have 4 or 5 rarely 3, large spots on the hind wing, in addition to an area of smaller apical spots. *Protochauliodes*, with 4 species, occurs in coastal N.S.W. and south-eastern Queensland. It lacks large spots on the hind wing, but otherwise is distinguished from *Archichauliodes* only by the anal vcins of the wings. There is a small species of an undescribed genus in south-western W.A.

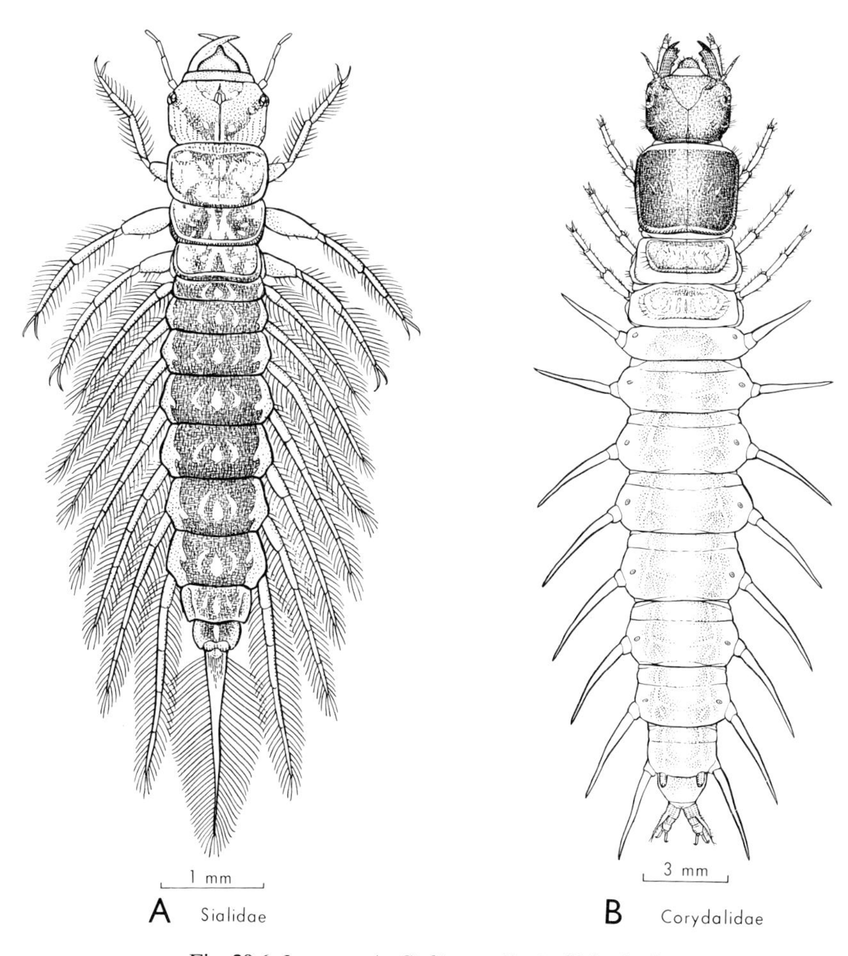

Fig. 28.6. Larvae: A, *Sialis* sp.; B, *Archichauliodes* sp.
[M. Quick]

29

NEUROPTERA

(Lacewings)

by E. F. RIEK

Endopterygote Neoptera with mouth-parts of the simple biting type; antennae multisegmented, well developed, often conspicuous; compound eyes always present; ocelli usually absent; generally two pairs of large, equal or subequal, many-veined wings, often divided into many small cells by numerous cross-veins; main veins usually with end-twiggings. Larvae with distinctive sucking jaws and modified alimentary canal. Pupation in a silken cocoon; pupae decticous.

The Neuroptera form one of the most archaic orders of endopterygote Neoptera. Adult lacewings (Figs. 29.1, 2) vary from minute insects with a wing-span of 5 mm to large species with a wing-span of more than 120 mm. Many are highly coloured and attractively patterned, and many have dense, rather long hairs on the body (Ascalaphidae and some Myrmeleontidae) and sometimes on the wings; a few are moth-like in appearance (Ithonidae). Some are swift fliers that resemble Odonata on the wing (Stilbopterygidae and some Ascalaphidae), but most have a slow, irregular flight, and in these the wing-coupling mechanism, when present, is not efficient. Sexual dimorphism is usually slight, but it is strongly marked in some Chrysopidae, in Stilbopterygidae, and in some Ascalaphidae. Most lacewings may be recognized by the end twigging of the main veins and the roofwise manner in which the wings are held at rest.

This is a small order, with some representatives in all the major zoogeographical regions, though they are more abundant in the tropics than in temperate regions. Adults occur in most terrestrial ecological zones, perhaps most commonly in the warmer, drier parts of the country. The order has been reviewed by Berland and Grassé (1951), the early stages discussed by Withycombe (1925), and the Australian species described by Esben-Petersen in 1914–29 (see Esben-Petersen, 1926–29), Tillyard in 1916–19, and Kimmins (1939). Main references to 1925 are in Tillyard (1926b).

Anatomy of Adult

Head (Fig. 29.3A). Usually transverse, but with clypeus and labrum elongated in Nemopteridae and to a less extent in Mantispidae; hind margin deeply excavated in Ascalaphidae. Compound eyes large. Ocelli

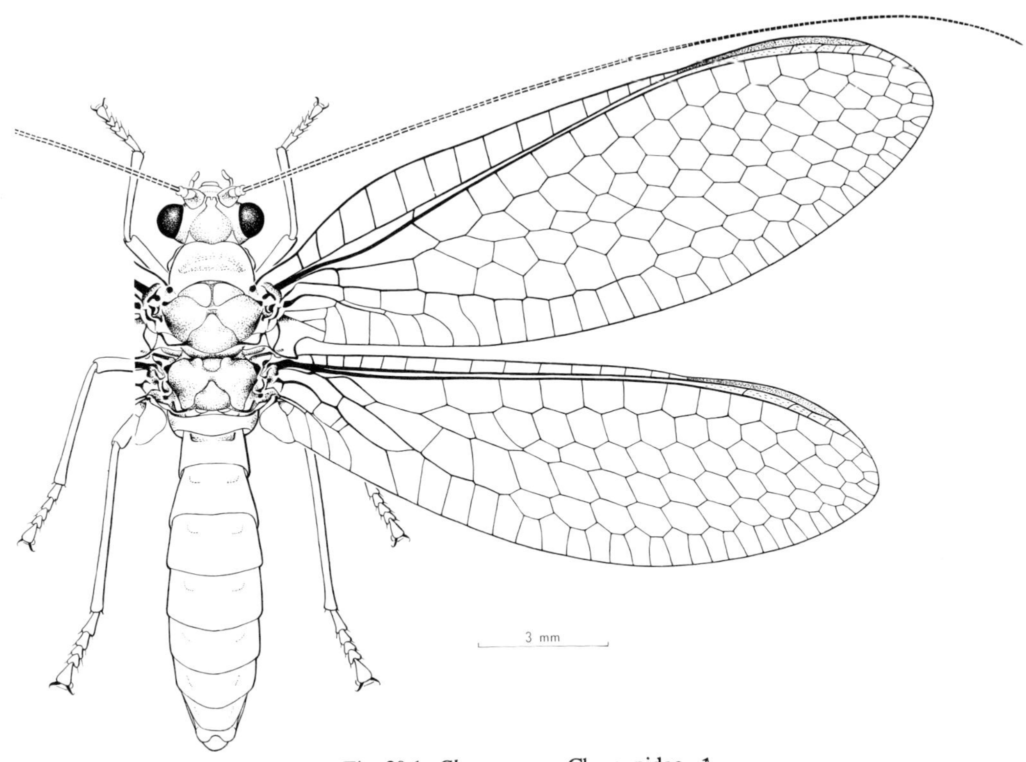

Fig. 29.1. *Chrysopa* sp., Chrysopidae, ♂.
[T. Binder]

usually absent; present, though sometimes reduced, in Osmylidae. Antennae multi-segmented, of varied form, often conspicuous; scape large, and the following one or two segments sometimes enlarged. In the primitive forms, the antennae are whip-like and tapering, and may be very long. In more advanced families, there is some thickening, and usually shortening, of the antenna, which may be uniform throughout or slightly flattened. The apex is thickened or clubbed in Myrmeleontidae, Stilbopterygidae, and Ascalaphidae. Mouth-parts simple; maxillary palp (Fig. 29.3C) 5-segmented in all except one exotic species; labial palp (Fig. 29.3B) 2- or more usually 3-segmented; ligula reduced to a median, sometimes slightly bilobed process, or absent. The apical segment of both the maxillary and labial palpi is sensory in function, and one or both is usually enlarged or swollen.

Thorax. Prothorax freely movable, varying from transverse (Ascalaphidae) to very long and narrow (Mantispidae, Fig. 29.3D). Mesothorax and metathorax both well developed.

Legs. Mostly cursorial, but fore legs raptorial in Mantispidae and some exotic Berothidae. Fore coxae often elongate, partly subdivided in some Mantispidae. Hind trochanter usually produced at apex on the inner side, except in Myrmeleontoidea. Tibiae with or without apical spurs. Tarsi 5-segmented, ending in 2 claws except in the fore leg of some Mantispidae. In some Myrmeleontidae the claws are very long, in others they bear one or 2 ventral teeth of various form, and in Mantispidae they may be flattened and bear up to 5 or 6 terminal spines. Empodium simple or divided into 2 lobes.

Wings (Figs. 29.7, 8). Generally two pairs of equal or subequal, many-veined wings; but the hind wings of Nemopteridae (Fig. 29.8G) are very long and narrow, quite unlike

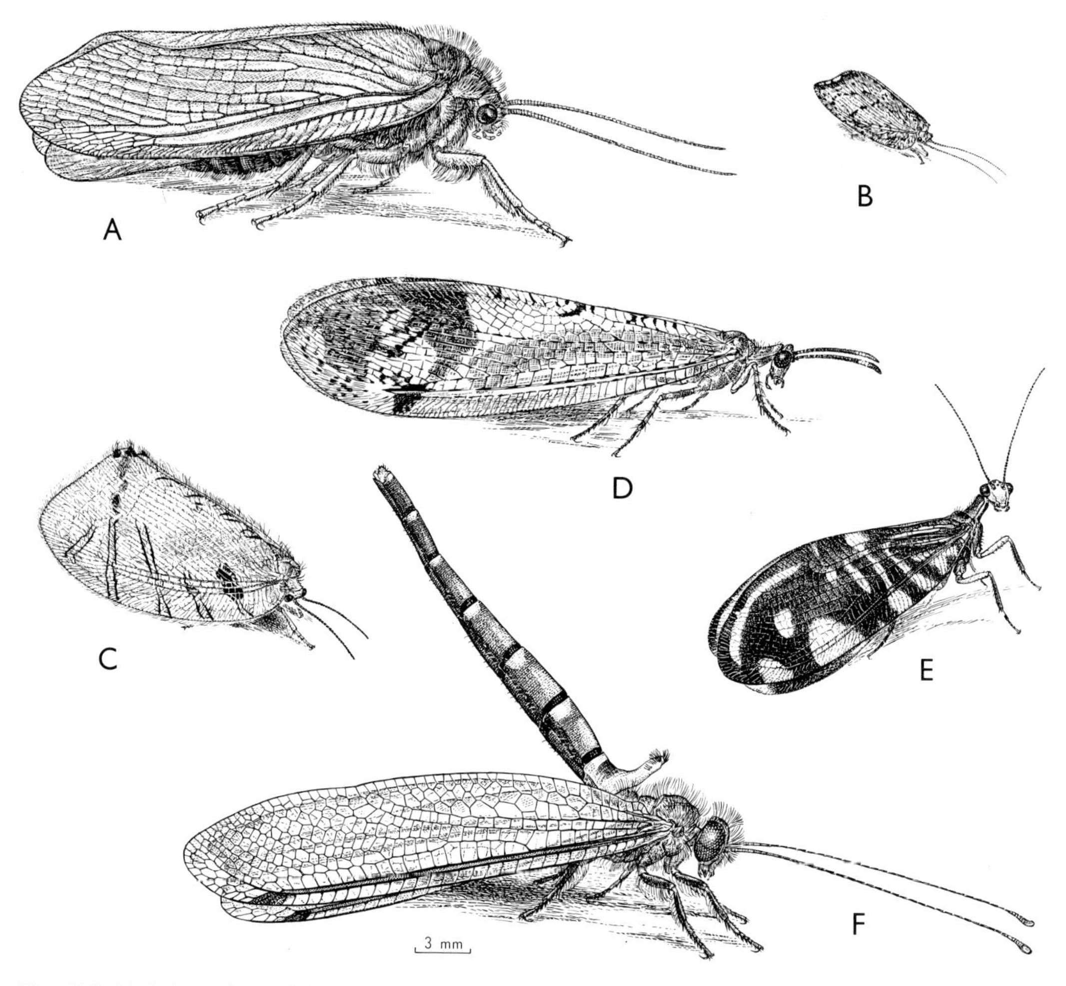

Fig. 29.2. Adults in resting position: A, *Megalithone megacerca*, Ithonidae; B, *Drepanacra humilis*, Hemerobiidae; C, *Psychopsis mimica*, Psychopsidae; D, *Glenoleon pulchellus*, Myrmeleontidae; E, *Porismus strigatus*, Osmylidae; F, *Acmonotus magnus*, Ascalaphidae. [M. Quick]

the short, rounded fore wings. The hind wing of Ithonidae (Fig. 29.7A) is more expanded in the anal region than is the fore wing, but in all other Neuroptera the wings are homonomous, or the hind wing is reduced as in *Notiobiella* (Hemerobiidae). In the exotic *Psectra diptera* (Hemerobiidae) the hind wing is sometimes reduced to a minute basal scale. There is one brachypterous coniopterygid. In a few families there are few branches to the wing veins and few cross-veins. This may possibly be a primitive condition (Coniopterygidae, Fig. 29.7E), or due to reduction (Sisyridae, Fig. 29.7G). Costal cross-veins are present in the fore wing, except in Coniopterygidae, and there is usually at least some indication of a pterostigma. Fusion between parts of some of the branches of the main veins produces the distinctive zigzagged venation of the Chrysopidae (Figs. 29.8B,C).

Nygmata, or sensory spots, occur in the wings of some species. They are situated between the posterior two branches of the apparent Rs, rarely behind the posterior branch, and often also closer to the base between Rs and M. There may be one between the stems of R and M almost at the

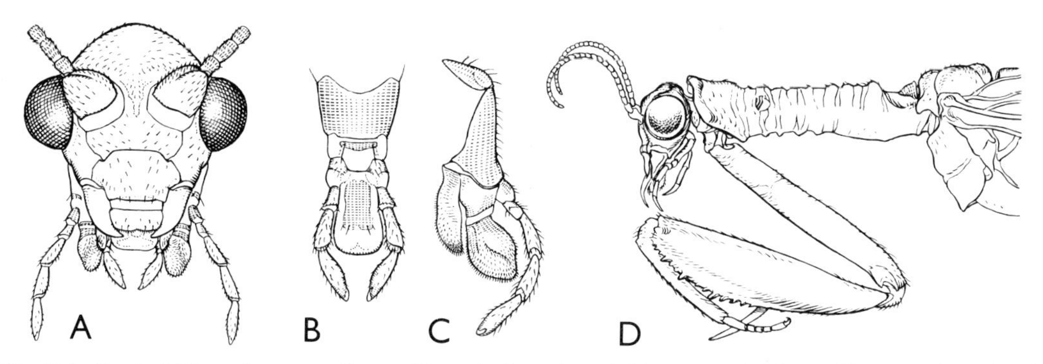

Fig. 29.3. Head of *Dictyochrysa* sp., Chrysopidae: A, frontal; B, labium, ventral; C, maxilla, ventral. D, head and prothorax of *Campion* sp., Mantispidae, lateral. [M. Quick]

base of the wing. The margin of the wing sometimes has a fringe of separate hairs, or there may be *trichosors*, which are thickenings of the margin bearing several hairs. When fully developed, there is a single trichosor between each pair of vein-endings and a similar thickening at the end of each branch of the vein. The wing membrane may be more or less covered with minute microtrichia. A more or less developed *clavate process* with a dense hair tuft occurs at the base of the posterior margin of the hind wing in the males of some genera of Stilbopterygidae and Myrmeleontidae.

WING-COUPLING. Simple, except in Coniopterygidae. In all Hemerobiidae there is a frenulum type of coupling. A well-developed jugal lobe at the base of the fore wing couples with a basal process bearing a few somewhat enlarged bristles from the costal margin of the hind wing. There are similar, though not always well developed, processes in Sisyridae, Berothidae, Osmylidae, most Mantispidae, and some Chrysopidae (*Dictyochrysa*, *Triplochrysa*). In most other families the jugal lobe is reduced, and there is a thickening of the wing margin at the end of the anal vein. In some ascalaphids this thickened zone may be developed into a spur-like process. The thickening is best developed in strong fliers, such as the Stilbopterygidae and Ascalaphidae. In these species the wings are held in position during flight by opposing pressures.

The Coniopterygidae have a distinctive type of wing-coupling. The anterior margin of the hind wing bears a series of several

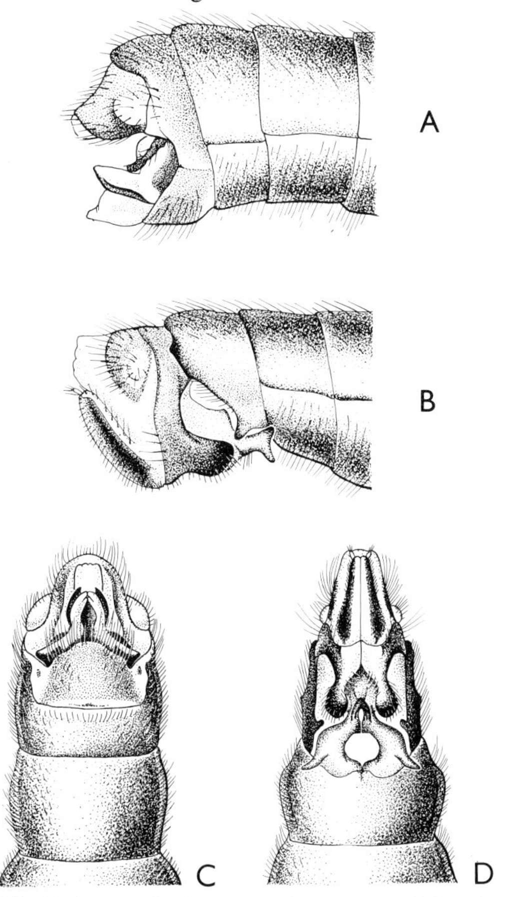

Fig. 29.4. Terminalia of *Nymphes* sp., Nymphidae: A, ♂, lateral; B, ♀, lateral; C, ♂, ventral; D, ♀ ventral. [R. Ewins]

hamuli-like bristles, which interlock with the hind margin of the fore wing in the anal region, and a series of hooks from the jugal region of the fore wing interlock with the basal part of R in the hind wing.

Abdomen. Ten-segmented, except in Chrysopidae in which only 9 are distinguishable. Sternite 1 is reduced. In the males of some Ascalaphidae there is a dorsal process on segment 2. In the males of Australian Stilbopterygidae there is a lengthening of segments 3–5 which may be variously swollen and produced. The terminalia are considerably modified in both sexes (Fig. 29.4), and the papers by Acker (1960) and Tjeder (1954) should be consulted for details. A long ovipositor occurs only in Dilaridae and in some Mantispidae and Berothidae.

Internal Anatomy. No better known than that of Megaloptera. Digestive system provided with a median dorsal food reservoir; Malpighian tubes usually 8, but 6 in Coniopterygidae. Three thoracic and usually 7 abdominal ganglia. In the male an aedeagus is developed, and a spermatophore is usually deposited (absent in Coniopterygidae and many, if not most, Chrysopidae); in the female the ovaries consist of a variable number of usually polytrophic ovarioles.

Immature Stages

Egg. Varies from elongate oval to almost spherical. The micropyle is well marked and usually raised (not distinguishable in *Croce*). Myrmeleontidae and Ascalaphidae have a micropyle-like apparatus at each end of the egg. The chorion is usually ornamented and sculptured.

Larva. The general form (Figs. 29.9–13) is that of an active predacious larva, more or less carabiform, with peculiarly modified

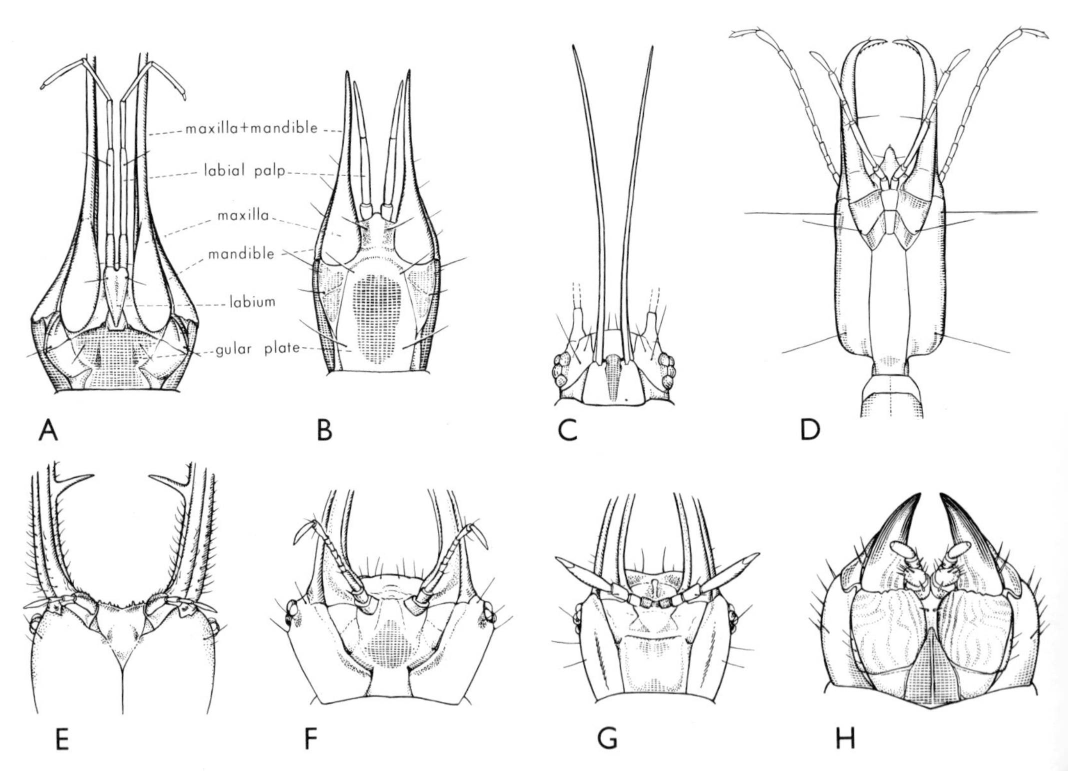

Fig. 29.5. Heads of larvae, ventral: A, *Porismus strigatus*, Osmylidae; B, a berothid; C, *Sisyra* sp., Sisyridae; D, *Austroneurorthus* sp., Neurorthidae; E, a myrmeleontid; F, *Chrysopa* sp., Chrysopidae; G, a hemerobiid; H, *Ithone* sp., Ithonidae. [M. Quick]

jaws (Fig. 29.5) and alimentary canal. The tarsi are 1-segmented, and there are usually 2 tarsal claws. The mid-gut region is separated from the hind gut, so that no solid excrement can be passed during larval life. The neuropterous sucking mouth-parts are formed by the combined mandibles and maxillae, each pair of which fits together to form a hollow sucking tube. The maxilla is always a single lobe (galea or lacinia), and there is no trace of a maxillary palp. Labial palpi are usually well developed, but may be absent (Sisyridae). The labrum is reduced, except in Coniopterygidae. The abdomen often ends in one or 2 suction discs. In some families the discs contain hooked spicules or crochets similar to those that occur in Lepidoptera, and in Osmylidae they are identical in appearance with the last abdominal prolegs of Lepidoptera.

Pupa. Always decticous and exarate. It lies in a silken cocoon, with the head and the terminal abdominal segments bent ventrally. The pupal mandibles are strongly sclerotized, sharply pointed, and usually finely toothed internally. They normally cross in the mid-line, and may be asymmetrical.

Biology

Adults. The Australian species are found from desert regions to cold subalpine stream margins depending mainly on the habits of the larvae. Thus, Sisyridae are always collected close to slow-flowing or standing fresh waters, and the same applies to those Osmylidae with semiaquatic larvae *(Kempynus)*, but other osmylids occur in open forest country *(Porismus, Stenosmylus)*. Psychopsidae occur in the wetter coastal zone, but the Berothidae, which also have bark-dwelling larvae, are more abundant in drier regions. The free-moving Chrysopidae, Hemerobiidae, and Coniopterygidae have a more general distribution wherever aphids, psyllids, and other slow-moving, soft-bodied insects are likely to be abundant. Ithonidae and Myrmeleontidae that lay their eggs in sandy situations occur wherever these conditions are available, so that there are coastal as well as inland species. The ithonid larvae are adapted to damp conditions, but the myrmeleontids tend to seek out drier situations, such as caves and overhangs and dry patches under fallen logs.

Many lacewings give off a distinctive odour when handled. Sometimes this may be due to the food on which they live, rather than to any specific scent glands, but *Nymphes myrmeleonides* produces a musk-like odour from an eversible gland opening between abdominal sternites 6 and 7. *Myiodactylus* gives off an ant-like smell that is also a little like rancid coconut. The males of Mantispidae have a scent gland opening between tergites 5 and 6. A scent gland occurs on the lateral portion of abdominal segment 1 in Dendroleontinae (Myrmeleontidae). The larvae of *Ithone* give off a strong odour of citronella. A stridulatory mechanism between the abdomen and hind femora has been described in both sexes of an exotic chrysopid (Adams, 1962). Modified microtrichia at or close to the anal margin of the fore wing, and sometimes the hind wing, of both sexes of most neuropterons are very probably associated with stridulation.*

Reproduction. Massed mating flights occur in the Ithonidae, but such flights are unusual in other families in which adults emerge over a longer period. Many species are attracted to light, and flights may be spectacular under warm conditions prior to rain. Many osmylids aggregate at rest in rock crevices at stream margins.

The eggs may be laid singly in sand or soil (Ithonidae, Myrmeleontidae, Nemopteridae), or they may be cemented by one side to some substrate, either singly or in clusters. Some eggs are stalked, and may be laid singly, in clusters, or in lines of several to many eggs. In *Nymphes* (Fig. 29.6) the double row of stalked eggs forms a distinctive U-shape. Stalked eggs occur in Chrysopidae, Berothidae, Mantispidae, and Nymphidae. Escape from the egg is usually effected with an egg-breaker, by means of which a longitudinal rent, or a circular slit to form a lid, is made in the chorion. The egg-breaker consists of a small longitudinal, sclerotized ridge, single or

* E. F. Riek, *Aust. J. Zool.* **15:** 337–48, 1967.

Fig. 29.6. Eggs of *Nymphes myrmeleonides*. [Photo by D. J. Lee]

double, lying over the labrum and clypeus of the embryo. In some genera the eggs seem to split along a line of weakness without the help of an egg-breaker. In the myrmeleontoid type of egg, there is apparently a circular line of weakness which yields to pressure from within to produce the lid.

Immature Stages. There seem to be consistently 3 larval instars in most families. Tillyard (1922a), however, reported 5 in *Ithone fusca*, and 4 have been recorded for two species of Coniopterygidae. Up to 12 ecdyses have been described in an American dilarid, but this may have been due to the conditions under which the larvae were reared.

The larvae are of very diverse form and habits, though most are active predacious types. The soil-burrowing ithonid larvae resemble the curl grubs (scarab larvae) on which they feed. The grub-like mantispine larva is a 'parasite' in the egg-capsules of spiders. The aquatic sisyrid larvae are 'external' feeders on freshwater sponges. They have extremely long, thin mouth-parts, but lack labial palpi, and possess paired ventral abdominal gills. The elongate osmylid larvae with long needle-like jaws occur in damp situations bordering streams, or else under the bark of trees, where they may be associated with the flattened but elongate larvae of Berothidae and Psychopsidae. The chrysopid and hemerobiid larvae scavenge on trees and shrubs for the sap-sucking insects and mites on which they prey. Some chrysopid larvae carry detritus attached to hooked hairs on their backs. The ant-lion types of larvae of the Ascalaphidae and Nymphidae occur in litter, or externally on the bark of trees, where they are camouflaged by the litter they carry or by the markings on the body. A flattened, disc-like, nymphid type of larva is found amongst foliage. The myrmeleontid ant-lions burrow in loose, sandy soil in which some of them construct pit-traps.

At the end of larval life, portions of the Malpighian tubes are said to secrete silk, which is spun through the anus in making the cocoon for pupation. Commonly the pharate adult emerges completely from the cocoon before eclosion. As a rule the pupal mandibles are used to tear an irregular rent in the wall of the cocoon, but in Chrysopidae a circular lid is cut. Some of the rupturing of the wall of the cocoon may be caused by flexion and extension of the pharate adult. Myrmeleontidae pupating in the soil work their way up to the surface before emerging, but Ithonidae emerge from the pupal skin while still in the soil. Some overwintering *Chrysopa* pupae fail to emerge until the second year.

Natural Enemies. Representatives of the major superfamilies of parasitic Hymenoptera are recorded from neuropterous hosts. They generally attack the larvae and prepupae, but egg and pupal parasites are also recorded. Larvae of Cleridae (Coleoptera) are recorded as predators of neuropteran cocoons. As the adult lacewings are mainly crepuscular or nocturnal, they are preyed on by bats but rarely by birds, though the day-time

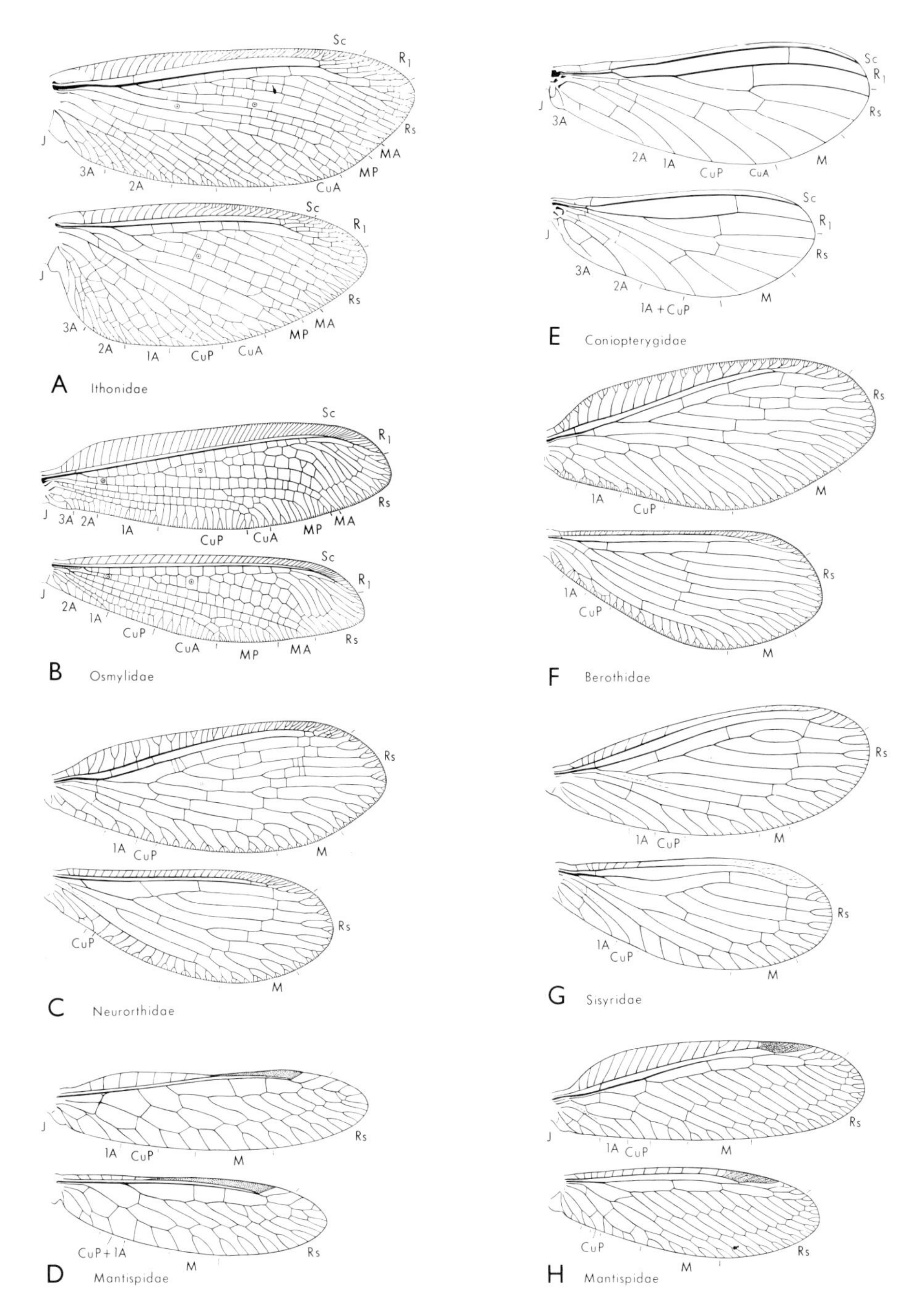

Fig. 29.7. Wings: A, *Ithone* sp.; B, *Eidoporismus* sp.; C, *Austroneurorthus* sp.; D, *Mantispa* sp.; E, *Neosemidalis* sp.; F, *Spermophorella* sp.; G, *Sisyra* sp.; H, *Ditaxis* sp. [M. Quick]

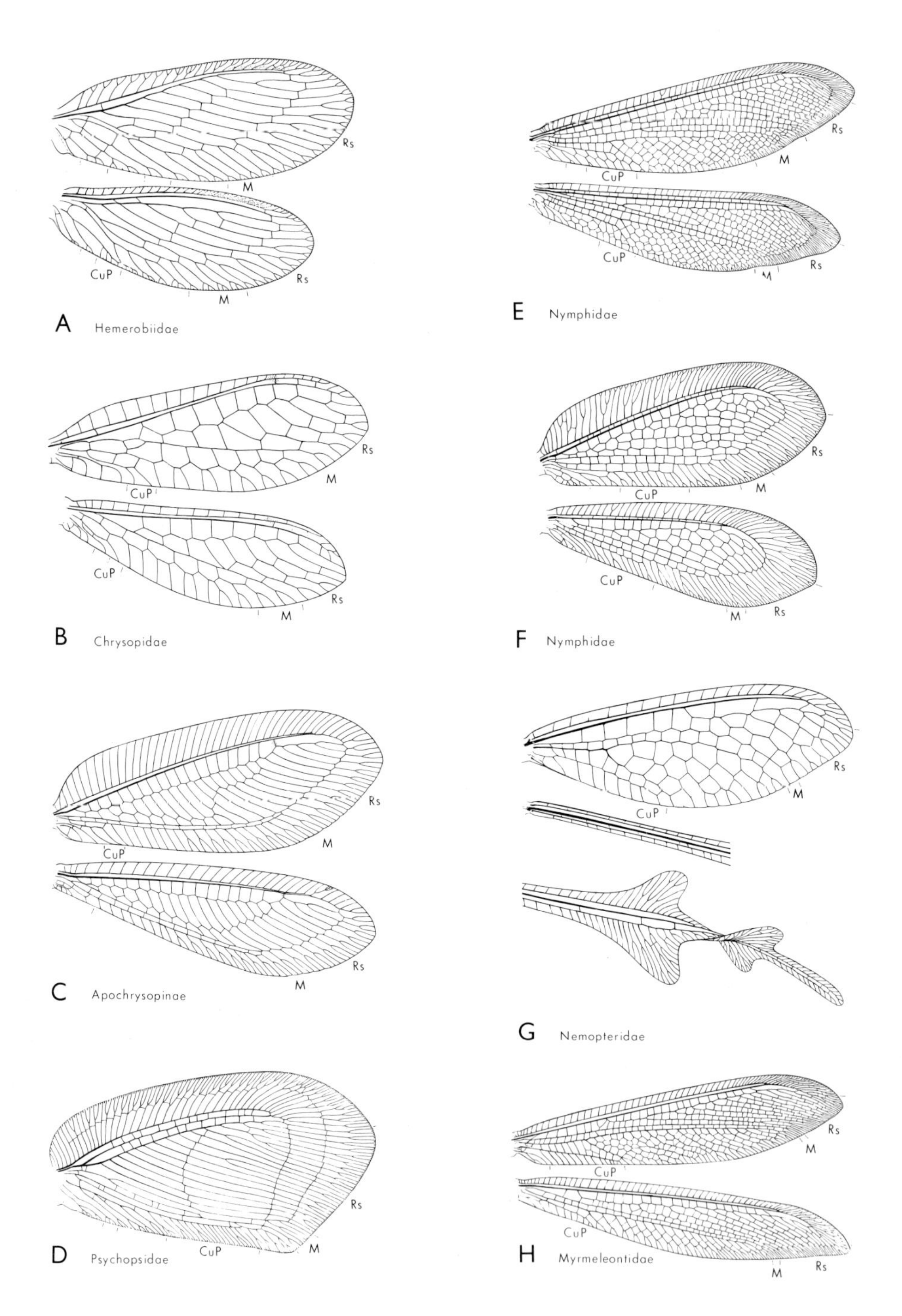

Fig. 29.8. Wings: A, *Psychobiella* sp.; B, *Chrysopa* sp.; C, *Oligochrysa* sp.; D, *Psychopsis* sp.; E, *Nymphes* sp.; F, *Myiodactylus* sp.; G, *Chasmoptera* sp.; H, *Eidoleon* sp. [M. Quick]

aggregations of ithonids may be attacked. When adult ithonids settle on tree trunks during mating flights, they are attacked by spiders, several large ants *(Myrmecia, Iridomyrmex, Ectatomma)*, and frogs.

Economic Significance. There is no order of insects, with the exception of the Hymenoptera, whose members are so generally beneficial to mankind as the Neuroptera. Hemerobiidae, Chrysopidae, and Coniopterygidae feed, both as larvae and adults, on a wide variety of small sap-sucking insects, such as aphids, psyllids, and scale insects, and also on mites and lepidopteran eggs. Many mantispids feed as larvae in the egg-capsules of spiders; while as adults they prey upon a wide variety of insects that live amongst foliage or visit blossoms, and they are often abundant on flowering eucalypts. Ithonids have been used in attempted biological control of some pest species of curl grubs. In certain areas, adult ithonids can be a nuisance for a week or so, when they aggregate in mating swarms that may shelter in houses during the day. Although sisyrid larvae feed on freshwater sponges, they are too small, both in numbers and size, to exert any significant control on sponges that can be pests in water reticulation systems. Osmylid larvae feed on chironomid and other dipterous larvae breeding at stream margins.

Special Features of the Australian Fauna

Australia has a great variety of Neuroptera. Many of the archaic groups are well represented, and a few specialized families are either restricted to, or best developed in, the region. There is considerable similarity between the Neuroptera of Australia and South America. If the Asian *Rapisma* is excluded, the moth-like Ithonidae occur only in Australia and North America. The Nymphidae, which possibly preserve the line of development leading to the higher Myrmeleontoidea, occur only in Australia, Lord Howe I., and New Guinea. The Stilbopterygidae, a small, specialized group close to the Myrmeleontidae, occur only in Australia and South America *(Albardia)*. The Osmylidae, which have a long fossil history, are widely distributed, but the more primitive subfamilies are best represented in Australia, New Zealand, and South America. Another archaic family, the Psychopsidae, which is well represented in Australia, occurs elsewhere only in South Africa and in the Oriental region. Only two small families, Dilaridae and Polystoechotidae, are not represented. The Dilaridae have a Holarctic distribution, and the Polystoechotidae are recorded only from North and South America.

Most families are widely distributed in Australia, but Myrmeleontidae and Ascalaphidae are more abundant in the drier regions, and there are possibly more species recorded from north-western Australia than from any other part of the country. Very few lacewings are recorded from Tasmania: three species of Osmylidae (*Kempynus* and *Odeosmylus*) and one each of Hemerobiidae *(Micromus)*, Myrmeleontidae *(Acanthaclisis)* and Coniopterygidae *(Perisemidalis)* are known to occur. The few species found in the subalpine areas of south-eastern Australia are mainly Osmylidae; the stream-frequenting *Kempynus* is found above the tree-line, and the bark-dwelling *Porismus* and *Oedosmylus* almost to the upper limits of trees. *Nymphes* also reaches these altitudes. The Psychopsidae, Coniopterygidae, and Nymphidae are most abundant in the higher-rainfall areas along the east coast.

CLASSIFICATION

Order NEUROPTERA (396 Australian spp.)

CONIOPTERYGOIDEA (27)
1. Coniopterygidae (27)

OSMYLOIDEA (38)
2. Ithonidae (8)
   Dilaridae (0)
3. Osmylidae (28)
   Polystoechotidae (0)
4. Neurorthidae (2)

MANTISPOIDEA (61)
5. Berothidae (19)
6. Sisyridae (9)
7. Mantispidae (33)

HEMEROBIOIDEA (84)
8. Hemerobiidae (20)
9. Chrysopidae (52)
10. Psychopsidae (12)

MYRMELEONTOIDEA (186)
11. Nymphidae (21)
12. Stilbopterygidae (7)
13. Myrmeleontidae (129)
14. Ascalaphidae (22)
15. Nemopteridae (7)

The wing venation of the adults and the head structure of the third-instar larvae have formed the main bases for the higher classification of the order. The arrangement presented here departs from those more generally accepted, in that the Mantispoidea are split off from the Osmyloidea, the Ithonidae are included in the restricted Osmyloidea, the Nymphidae are included in the Myrmeleontoidea, and the Myiodactylidae in the Nymphidae. Both the Osmyloidea and Mantispoidea have straight-jawed larvae, but the clearly defined mentum of the Osmyloidea bears a pair of stout bristles and nygmata are present in the adult wings. In adult characters the Mantispoidea are more like the Hemerobioidea. The Hemerobioidea and Myrmeleontoidea have larvae with curved jaws. The Coniopterygoidea differ so markedly from the other superfamilies that it might be preferable to raise them to ordinal rank on a combination of the distinctive mouthparts and reduced Malpighian tubes of the larvae and the wing venation of the adults.

Keys to the Families of Neuroptera Known in Australia

ADULTS

1. Costal area with, at most, 1 or 2 basal cross-veins, including humeral; no pterostigma; very small insects, fore wing 2–5 mm; wings and most parts of body covered with a waxy, mealy secretion; wing-coupling by 2 sets of hamuli-like curved bristles .... CONIOPTERYGOIDEA. **Coniopterygidae**
   Costal area with more numerous cross-veins and a more or less defined pterostigma; generally larger insects; wings and body without a mealy secretion; wing-coupling, when well developed, of the frenulate type 2
2(1). Nygmata present between the posterior 2 branches of Rs, occasionally behind the posterior branch, and usually between Rs and M closer to the base .... OSMYLOIDEA. 3
   Nygmata not present between the posterior 2 branches of Rs 5
3(2). Ocelli present .... **Osmylidae**
   Ocelli absent 4
4(3). Cross-veins numerous; venation dense .... **Ithonidae**
   Cross-veins reduced, never more than 2 gradate series .... **Neurorthidae**
5(2). Sc and R_1 connected by a cross-vein towards apex (appearing fused in some Chrysopidae, but then antenna longer than fore wing; cross-vein very short or Sc and R_1 fused for a considerable distance in some Mantispidae, but then the pterostigma is discrete); CuA of fore wing not forming a large convex triangular area; trichosors usually present 6
   Sc and R_1 clearly fused at apex for a considerable distance; CuA of fore wing usually forming a large convex triangular area; trichosors absent, except in Nymphidae .... MYRMELEONTOIDEA. 11
6(5). Numerous cross-veins between R_1 and Rs, or Rs arising on more than one stem .... HEMEROBIOIDEA. 7
   One to 4, rarely 5, cross-veins between R_1 and Rs .... MANTISPOIDEA. 9
7(6). Rs of fore wing arising on 2 or more stems. [Trichosors present] .... **Hemerobiidae**
   Rs of fore wing arising on 1 stem 8
8(7). Vena triplica present; trichosors present. [Basal nygma between the stems of R and M] .... **Psychopsidae**
   No vena triplica; trichosors absent, but with a marginal fringe of hairs .... **Chrysopidae**
9(6). Fore legs raptorial; trichosors absent in Mantispinae (present but irregular in cubital and anal fields in Platymantispinae) .... **Mantispidae**
   Fore legs cursorial; trichosors present and usually regular 10
10(9). First antennal segment distinctly longer than wide; costal cross-veins forked in fore wing .... **Berothidae**
   First antennal segment only about as long as wide; costal cross-veins simple in fore wing .... **Sisyridae**
11(5). Trichosors present, but irregular in the cubital and anal fields .... **Nymphidae**
   Trichosors absent 12

12(11). Hind wing greatly lengthened and narrowed at least over basal half, only 1–3 simple longitudinal veins there; face lengthened into a rostrum; antenna of practically uniform width throughout, tapering slightly to apex .... **Nemopteridae**
Hind wing at most slightly narrowed at base, with several longitudinal veins; face not lengthened into a rostrum; antennae fusiform, or more often clubbed at apex 13
13(12). Antenna more than half as long as fore wing, distinctly clubbed; eye often partly divided by a horizontal groove .... **Ascalaphidae**
Antenna less than half as long as fore wing, club varied in development; eyes not partly divided 14
14(13). Antenna distinctly clubbed at apex; 1A of hind wing with numerous pectinate branches, CuA appearing simple, connected below to the long apparent CuP by simple cross-veins .... **Stilbopterygidae**
Antenna slightly clubbed at apex or subfusiform; 1A of hind wing with at most 3–4 pectinate branches, or, if with more branches, then CuA distinctly forked and only its lower branch connected to CuP .... **Myrmeleontidae**

THIRD-INSTAR LARVAE (The larvae of Stilbopterygidae are not known.)
1. Very small; labial palpi 2-segmented, the terminal segment greatly enlarged; antenna 2-segmented (Fig. 29.9) .... **Coniopterygidae**
Generally larger; labial palpi and antenna with more than 2 segments (labial palpi absent in Sisyridae) 2
2(1). Gular plate of head not covered by ventral extensions of genae (Figs. 29.5A,B); no large internal teeth on mandible 3
Gular plate of head covered by ventral extensions of the genae meeting in mid-ventral line (Fig. 29.5E); generally with at least one large internal tooth on mandible 11
3(2). Body resembling a curl grub (Fig. 29.10A); all tibiae and tarsi fused. [Head capsule much smaller than pronotum; eyes absent; tarsal claws double, very large; empodium absent; antenna 5-segmented; palpi 4-segmented] .... **Ithonidae**
Body not resembling a curl grub, though grub-like in the parasitic Mantispidae; tibiae and tarsi separate. [Eyes present, though sometimes quite reduced] 4
4(3). Parasitic in egg-capsules of spiders (Fig. 29.11A). [Mandibles short; tarsi with one claw] .... **Mantispidae**
Free-living, or external parasites of freshwater sponges 5
5(4). Tarsi with 1 claw, empodium absent; palpi absent; abdomen with 7 pairs of ventral respiratory filaments; living in freshwater sponges (Figs. 29.10C,D). [Mouth-parts greatly elongate] .... **Sisyridae**
Tarsi with 2 claws, empodium present; palpi present; without ventral gills; terrestrial, or at most semiaquatic 6
6(5). Mandible and maxilla straight, usually long and thin, at most inwardly curved at apex 7
Mandible and maxilla inwardly curved 9
7(6). Mandible and maxilla inwardly curved at apex (Figs. 29.11F,G) .... **Neurorthidae**
Mandible and maxilla straight 8
8(7). Base of maxilla markedly expanded; antenna 5-segmented (Figs. 29.10G,H). [Empodium trumpet-shaped] .... **Berothidae**
Base of maxilla only moderately expanded, similar to mandible; antenna (and palpi) with a greater number of segments (Figs. 29.10E,F) .... **Osmylidae**
9(6). Antenna composed of 10 distinct segments which are usually longer than wide; labial palpi 5-segmented. [Dorsal surface of abdomen bare; antenna only about as long as mandible] .... **Psychopsidae**
Antenna divided into a large number of small ill-defined segments, or with one segment greatly elongated; labial palpi with more than 5 segments, or segmentation ill defined 10
10(9). Empodium trumpet-shaped (Fig. 29.11D); dorsal surface of abdomen often with long hairs covered with debris (Fig. 29.11C) .... **Chrysopidae**
Empodium not trumpet-shaped (except possibly in first instar); dorsal surface of abdomen with very short hairs, appearing bare (Fig. 29.11H) .... **Hemerobiidae**

11(2). Hind tibia and tarsus not fused to form a single segment 12
Hind tibia and tarsus fused to form a single segment. [Mandible usually with 3 enlarged internal teeth; palpi 4-segmented] 13

12(11). Mandible without internal teeth, or with several small teeth; palpi 3-segmented (Fig. 29.12A). .... **Nemopteridae**
Mandible with 1 internal tooth; palpi 4-segmented (Fig. 29.12B). [Lateral body processes well developed] .... **Nymphidae**

13(11). Lateral body processes absent or weakly developed; hind margin of head not markedly bilobed; hind leg with enlarged, forwardly directed claws (Fig. 29.12D) .... **Myrmeleontidae**
Lateral body processes well developed, 1 on each abdominal segment and 2 on each of mesothorax and metathorax; hind margin of head strikingly bilobed; hind tarsal claws normal (Fig. 29.12C) .... **Ascalaphidae**

Superfamily CONIOPTERYGOIDEA

The Coniopterygidae have many distinctive features. The venation (Fig. 29.7E) is quite reduced and superficially psocid-like. Costal cross-veins, which are present in all other families, are absent, and there is no indication of a pterostigma. The wing-coupling is distinctive, with hamuli-like hooks at the base of both wings. The adult male has a distinctive reproductive system: an aedeagus has developed (or persisted), and a spermatophore is not deposited as in most other Neuroptera. The larvae have only 6 Malpighian tubes instead of the usual 8.

1. Coniopterygidae. Very small species, with fore wing 2–5 mm. Body, wings, and often legs and basal portion of antennae covered by a whitish or greyish meal secreted by wax glands situated on the head, thorax, and abdomen. The insect rubs some of the wax off its body with its legs, and distributes it over the wings. Antennae short, moniliform, with scape enlarged and pedicel somewhat elongate (the first 3 segments are enlarged in some males). Ocelli absent. Prothorax short. Wings generally longer than the body, with rounded apex; Rs 2-branched.

The oval, slightly flattened eggs (about 200 per female) are usually laid singly on leaves or twigs. They hatch in 1 to 3 weeks, and the larvae are arboreal. There are 3 or 4 instars. The larvae (Fig. 29.9) are short, fusiform, and the head is extremely small, with almost straight, needle-like jaws. In some species the labrum is large and completely covers the mandibles. The larvae pupate in a cocoon constructed in bark crevices or on leaves.

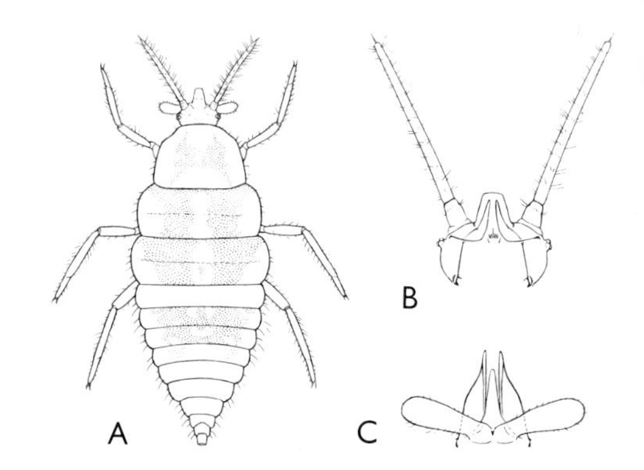

Fig. 29.9. Coniopterygid larva: A, dorsal; B, head, ventral, maxilla and labium removed; C, maxilla and labium, ventral. [M. Quick]

There are more than 100 described species, distributed in all regions. Six genera are known from Australia, *Parasemidalis* (11 spp.) being the commonest and *Spiloconis*, with a fore wing up to 5 mm long, the giant of the family. The others are *Heteroconis* (4 spp.), *Coniopteryx* (5 spp.), *Neosemidalis* (3 spp.), and *Cryptoscenea* (2 spp.). The Coniopterygidae are a valuable family, for they feed both as larvae and adults on psyllids and scale insects, as well as on mites.

Superfamily OSMYLOIDEA

Nygmata are present on the wings. The Ithonidae are stout, moth-like species. The Osmylidae are the only Neuroptera with ocelli. The Neurorthidae resemble Sisyridae and Hemerobiidae.

2. Ithonidae. A small, archaic family almost restricted to the Australian region

Adults (Fig. 29.2A) large, stoutly built, and superficially resembling the smaller and duller hepialid moths. Wings (Fig. 29.7A) broad, expanding 40–65 mm, and generally uniformly infuscated, but with a distinct pattern in *Varnia*. Males with enlarged claspers, and females with a distinctive, ploughshare-like ovipositor used for depositing the eggs in sandy soil. The large (1·7 mm), smooth eggs are laid singly, and hatch in about 30 days. There are at least 5 larval instars, and the life cycle occupies two years. The larva (Figs. 29.10A,B) is a white, blind, melolonthid-like grub, with short, straight jaws and the mandibles smaller than the maxillae.

There are 3 Australian genera. *Ithone* (4 spp.) is widely distributed in low-lying, sandy, coastal areas. *Megalithone* (2 spp.), found in higher country in eastern N.S.W. and south-eastern Queensland, is distinguished by the greatly expanded claspers in the males and the dense covering of long fine hairs in both sexes; the emergence period is short, and the adults swarm in great numbers in mating flights at night, hiding in crevices during the day. The 2 species of *Varnia* have patterned wings and a greater development of branched veinlets and cross-veins in the costal space than in *Ithone*; they are found in desert country, and little is known of their habits.

3. Osmylidae. Attractive lacewings of medium to large size, with a wing-span of 30–55 mm. Antennae filamentous, setose, multisegmented, short, less than half length of fore wing. Three ocelli present, but not always distinct. No evident wing-coupling mechanism for the rather broad wings; venation (Fig. 29.7B) generalized, and nygmata present. The males of some species have scent glands that open dorsolaterally between abdominal segments 8 and 9 as eversible, rather long sacs. The female terminalia are much more conspicuous than those of the male.

The elongate-oval eggs are laid attached by one surface in short, curved rows of up to 12 eggs touching side by side, either on leaves or stones in the vicinity of water, or else on tree trunks. The older larvae (Fig. 29.10E) are long, fusiform, and dark or mottled in colour. The body bears only a few long bristles. The long, tapering maxillae and mandibles (Fig. 29.10F) are finely serrated on the inner side. During cocoon formation, the jaws are broken off close to their bases, and the broken-off portions are retained within the cocoon. The food of semiaquatic larvae consists mainly of nematoceran larvae (Chironomidae), but the bark-dwelling larvae feed on a variety of insects. Adults mostly appear in the autumn. The family has a wide, discontinuous distribution, with about 100 described species. There are 5 subfamilies in Australia.

The only Australian species of PORISMINAE, *Porismus strigatus* (Fig. 29.2E), has many cross-veins between Sc and R_1 and a large triangular fork at the apex of CuA. It occurs over most of the eastern States, including S.A. The larvae live under the bark of eucalypts, and adults emerge in the late summer and autumn.

The STENOSMYLINAE (15 spp.) are the common Australian osmylids, mostly of medium size, with a wing-span about 40 mm, a few up to 55 mm. Most have clear or nearly clear wings, but *Euporismus albatrox* and undescribed species of *Oedosmylus* are attractively patterned. There are 4 Australian genera, and the subfamily occurs also in South America. The known larvae are found under the bark of eucalypts, where they also pupate.

The KALOSMYLINAE (10 spp.) are relatively large species (wing-span 40–55 mm) with attractively patterned wings, found along streams, especially at the edge of rain forest or in open subalpine areas. The fore wing may be entire, or variously emarginate and lobed on the caudal margin. The adults cluster on overhanging rocks and tree trunks projecting from the streams. The larvae are semiaquatic, and are found under stones and litter bordering the stream. There are 2 Australian genera, one of which occurs also in New Zealand.

The single species of SPILOSMYLINAE, *Conchylosmylus triseriatus* from north Australia, can be distinguished from all other

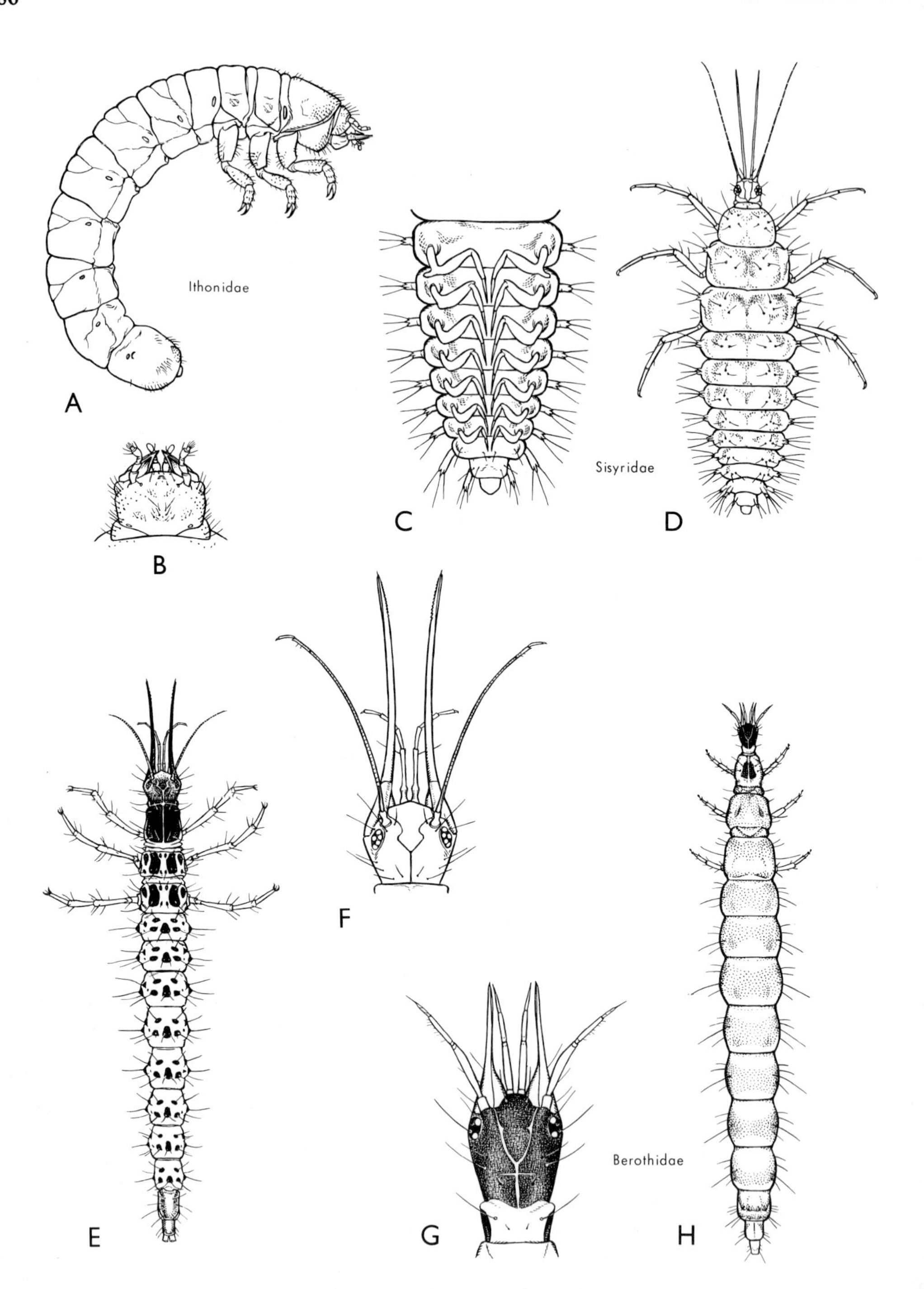

Fig. 29.10. Larvae: A, B, *Ithone* sp., lateral, and head, dorsal; C, D, *Sisyra* sp., abdomen, ventral, and whole larva, dorsal; E, F, *Kempynus* sp., dorsal and head enlarged; G, H, a berothid, dorsal and head enlarged. **[M. Quick]**

Australian osmylids by the presence of a rounded, bead-like thickening or embossed spot on the hind margin of the fore wing. There is also one species of PROTOSMYLINAE, *Eidoporismus pulchellus* from the Sydney area; it is similar in appearance to *Oedosmylus*, but differs in 1A not extending beyond the level of origin of first branch of Rs in either wing.

4. Neurorthidae. The small, delicate species of *Austroneurorthus* are generally included in a distinct subfamily of Sisyridae. The group differs from typical Sisyridae, however, in having nygmata on the wings (Fig. 29.7C) and slightly forked costal cross-veins in the fore wing. The immature stages are not known with certainty, but an unusual larva (Figs. 29.11F,G) found in moist litter is tentatively referred to the family. The adults occur only in damp situations and along the margins of dense forest, and their larvae are therefore unlikely to be parasitic on freshwater sponges, as are the Sisyridae.

Superfamily MANTISPOIDEA

Nygmata are absent from the wings, and there are few cross-veins between R and Rs. Mantispidae have raptorial fore legs; Sisyridae are very small, with reduced venation; Berothidae have rather hairy wings due to long macrotrichia on the veins.

5. Berothidae. Delicate, small to medium-sized lacewings, with fore wing 6–13 mm. Antennae shorter than fore wing, setose, multisegmented, moniliform, with the scape enlarged. Apical segment of labial palp generally without palpimacula (present in *Trichoberotha*). Prothorax usually at least as long as wide. Wings (Fig. 29.7F) subequal and variable in shape, with Sc terminating distally either on R_1, or on the wing margin and joined to R_1 by a cross-vein. Outer margin of fore wing sometimes more or less incised below the apex, but most Australian species have more rounded wings. Specialized scale-like hairs occur on the wings and sometimes fore coxae or pronotum of the females of many species, but they are always absent on the males. The females often have a pair of long hypocaudae on the lateral gonapophyses.

Life histories are not well known. The elongate-oval, stalked eggs of *Spermophorella* are laid in patches, and resemble those of *Chrysopa*. The larvae (Figs. 29.10G,H) occur under bark on eucalypts; they have straight, needle-like mouth-parts, with a very broad basal portion to the maxilla which distinguishes them from all other neuropterous larvae. Berothidae have a wide but discontinuous distribution. The Australian species are divided among 6 genera, of which *Spermophorella*, with broadly rounded speckled wings, and *Stenobiella*, with narrow strap-like wings, are the commonest. Many of the species are found in drier inland areas.

6. Sisyridae. Very small (fore wing 3·5–10 mm), hemerobiid-like lacewings, found only near bodies of fresh water in which sponges grow. The adults are nocturnal or crepuscular, and are readily attracted to light. Antennae multisegmented, moniliform, about half as long as fore wing (Fig. 29.7G), with scape enlarged. Prothorax wider than long.

Females lay about 50 eggs similar to those of *Hemerobius*, but deposited singly or in clusters on leaves and twigs overhanging water. The egg masses are covered with a layer of white silk strands. Incubation lasts about 2 weeks. The newly hatched larvae drop to the water, break through the surface film, swim like a *Cyclops*, and are carried along until they come in contact with a sponge, on the fluids of which they then feed. The larvae (Figs. 29.10C,D) leave the water to pupate on a tree or similar situation above ground level, often at a considerable distance away. They spin cocoons, hibernate as prepupae, and pupate in spring or early summer. There is only one generation a year. The family has a wide distribution, with about 30 described species. *Sisyra* (8 spp.) is world-wide, and *Sisyrina* (1 sp.) occurs in India and Australia.

7. Mantispidae. Small to medium-sized (wing-span 10–50 mm), distinctive, mantis-like species, with long pronotum, raptorial fore legs (Fig. 29.3D), greatly elongate fore coxae, and long, narrow, subequal wings (Figs. 29.7D,H) in which there is a conspicuous pterostigma. Antennae short in all genera, without obvious thickening in the more generalized *Calomantispa* and *Ditaxis*, but somewhat thickened on their whole length in *Mantispa* and *Campion*, and obviously so in *Euclimacea*. Fore tarsi highly modified and shortened in some genera. Wings with 1 or 2 gradate series of cross-veins, and a well-developed, margined, jugal lobe usually present. *Ditaxis* (Fig. 29.7H), apart from the raptorial modifications, is quite similar to the Berothidae, but the distinction between the two families is more marked in other Australian genera in which there is only one gradate series of cross-veins (Fig. 29.7D).

The Platymantispinae (parasitic in the nests of vespoid wasps) are not known in Australia, but the MANTISPINAE, which have a 5-segmented fore tarsus and lack a strongly produced ovipositor, are represented by 8 genera, some of which have brightly patterned wings. The minute, smooth, oval, white eggs, with very short stalks, are laid in large clusters of up to 400 on the bark of trees. They hatch in 2 to 3 weeks, and the larvae (Fig. 29.11A) disappear into crevices in the bark, where they later search for egg-sacs of spiders, in which the older, more grub-like larvae feed and pupate (McKeown and Mincham, 1948).

Superfamily HEMEROBIOIDEA

Rs arises on two or more stems in the fore wing of Hemerobiidae, a condition that occurs elsewhere only in some Ithonidae (and Dilaridae). The wings of Chrysopidae are mostly clear and glassy, whereas those of Psychopsidae are very broad and hairy and usually have a distinctive pattern.

8. Hemerobiidae. Rather small, delicate lacewings, the Australian species with a wing-span of 10–20 mm. Antennae multisegmented, moniliform, tapering, usually nearly as long as the fore wing, scape enlarged. Prothorax much broader than long. Wings (Fig. 29.8A) normally subequal, but hind wing reduced in *Notiobiella*; Sc terminating on the costal margin; and a frenulum type of wing-coupling present in all genera. The presence or absence of the recurrent humeral vein divides the family into two groups of genera. When it is present, the wings are usually broad; when it is absent, they are long and narrow, as in *Micromus*.

Most hemerobiids feign death when disturbed during the day. Adults are crepuscular or nocturnal, and live for several months. They are predacious, feeding on the same prey as the larvae, and the jaws are adapted for chewing solid food. Copulation usually occurs at night with the pair arranged in a linear position. The elongate-oval eggs, 0·5–1 mm long, are laid attached by their sides to leaves and bark. In most genera the chorion is lightly sculptured, but it is smooth in *Micromus*. Eggs usually hatch in 1 to 2 weeks, but in some species they overwinter. The number of eggs is usually small, but up to 500 are recorded. The smooth, fusiform larvae (Fig. 29.11H) do not carry detritus. They are very active, run rapidly, and the terminal segment of the abdomen is often used as an aid in walking, especially when climbing. The larval period is usually short. The prepupa, in its cocoon under bark or on the lower side of a leaf, lasts for several days to as long as several months, and many species overwinter in this stage. Consequently the number of broods in a year varies considerably, depending on species and locality.

The family is world-wide in distribution, mainly in temperate regions. It is represented in Australia by 10 genera, which are found almost exclusively in the eastern coastal and montane areas. Two of the commonest species are the eastern *Drepanacra humilis* (Fig. 29.2B), with broad, attractively but variably patterned wings spanning up to 18 mm, and the Australia-wide *Micromus tasmaniae*, which has relatively long narrow wings.

9. Chrysopidae. The family is divided into 2 very distinct subfamilies. *Oligochrysa lutea*, the only Australian species of APOCHRYSOPINAE, is a large, delicate lacewing, with broadly rounded wings (Fig. 29.8C) spanning about 40 mm, and very long antennae. All the longitudinal veins appear to arise from very close to the base, and M is apparently unforked in the fore wing. This is the only

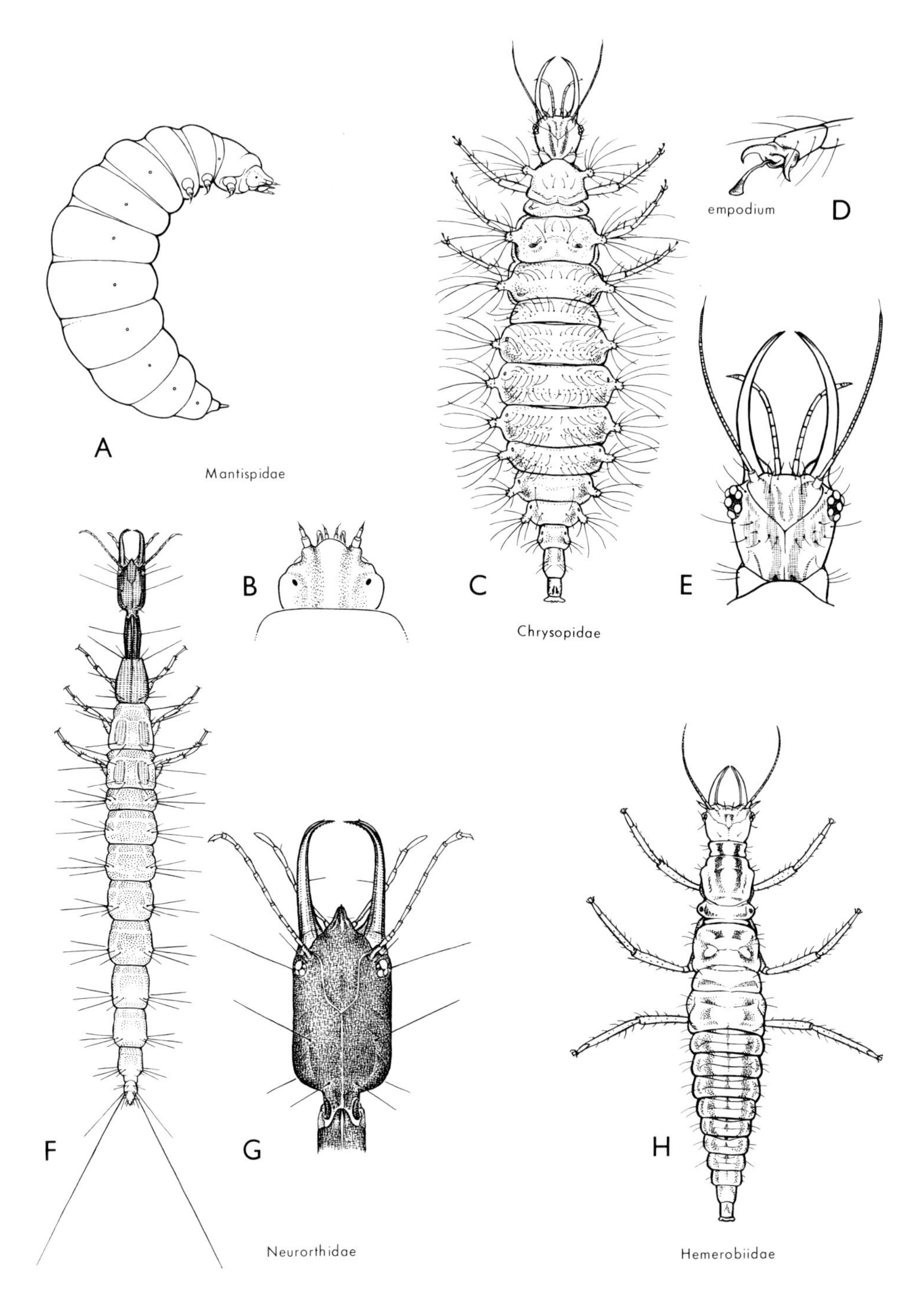

Fig. 29.11. Larvae: A, B, *Mantispa* sp., first instar, lateral, and head enlarged, dorsal; C–E, *Chrysopa* sp., dorsal, apex of leg, head enlarged; F, G, a neurorthid, dorsal and head enlarged; H, a hemerobiid, dorsal. [M. Quick, T. Binder]

genus in which Sc and R_1 fuse towards the apex of the wing. The costal space is markedly expanded over its whole length. *O. lutea* is found in coastal Queensland and N.S.W.

The CHRYSOPINAE are a group of easily recognized, but not so easily diagnosed, green to yellowish species, of medium size, with clear, glassy wings (Fig. 29.8B), lightly patterned in *Glenochrysa*. Antennae as long as, or somewhat longer than, fore wing, slightly shorter in one genus. Venation distinctive, with many of the branches appearing fused to produce a number of cellules; jugal lobe of fore wing usually very reduced; trichosors absent. One tibial spur, except in *Dictyochrysa* (2 spurs). The eggs are stalked, and placed at random or in groups or rows. They are quite large, conspicuous because of their white or green colour, and the stalk is twice to several times as long as the egg. They hatch in 1 to 2 weeks. The larvae (Fig. 29.11C) are active, and some species cover themselves with detritus, usually the sucked-out bodies of their victims. The pupal cocoon is constructed in crevices, and covered with the same detritus as the larva. Adults occur throughout the year in many parts of the country, and there may be several generations in a year. There are 8 Australian genera, with *Chrysopa* the commonest.

10. Psychopsidae. Attractive species, of medium to large size, with very broad, hairy, often strikingly patterned wings (Fig. 29.2C) expanding 25–60 mm. Antennae short, much less than half length of fore wing, with a slight tendency to thicken towards apex. Jugal lobe clearly developed, though sometimes small. Two tibial spurs. The family differs from other hemerobioid families mostly in the development of a vena triplica and in the shortening of the antennae. Broad wings with a markedly expanded costal space also occur in both the Chrysopidae and Hemerobiidae, and abundant longitudinal veins in some Hemerobiidae.

During the day the adults rest concealed and motionless on the under surface of a leaf or hidden away in debris; they are active only at night. Pairing takes place at rest with the bodies of the pair parallel and facing in the same direction. Adults live for 1 to 2 months. Females deposit at least 50 oval eggs, which are laid separately and attached by one side. They hatch in 10 to 14 days, and the larvae are found under the thick rough bark of eucalypts. The empodium is trumpet-shaped in all instars, as in Chrysopidae, and the antenna and palp have a small number of clearly defined segments. The larval instars are all of long duration, and the complete life cycle takes two years. The larvae pupate in crevices in the bark, in cocoons consisting of an outer loose envelope and an inner fine one of silk.

Psychopsidae are best represented in Queensland and N.S.W., where they occur inland as well as on the coast. One species, *Psychopsis mimica* (Fig. 29.2C), is recorded from S.A. and one, *Wernzia maculipennis*, from W.A.; none is known from Tasmania. Five genera are recognized, *Megapsychops* from south Queensland being the most striking. The family occurs also in Burma, Tibet, China, Formosa, and South Africa.

Superfamily MYRMELEONTOIDEA

Australian representatives of the families can be recognized readily by one or two outstanding characters. Ascalaphidae always have very long, clubbed antennae and the eyes divided by a horizontal groove in both sexes. Nemopteridae have very long, narrow hind wings and an elongate face. The large Stilbopterygidae, with short, distinctly clubbed antennae, can be distinguished from Myrmeleontidae by the sigmoidally curved CuP in the hind wing, and in the male by the dorsal processes on one or more of segments 3–5 of the abdomen. Some Nymphidae look like Myrmeleontidae, but the antennae taper at apex, whereas others resemble osmylids.

11. Nymphidae. This small but very interesting family is confined to Australia, Tasmania, Lord Howe I., and New Guinea. The species are most common along the east coast, but there are some in northern Australia and in south-western W.A. The

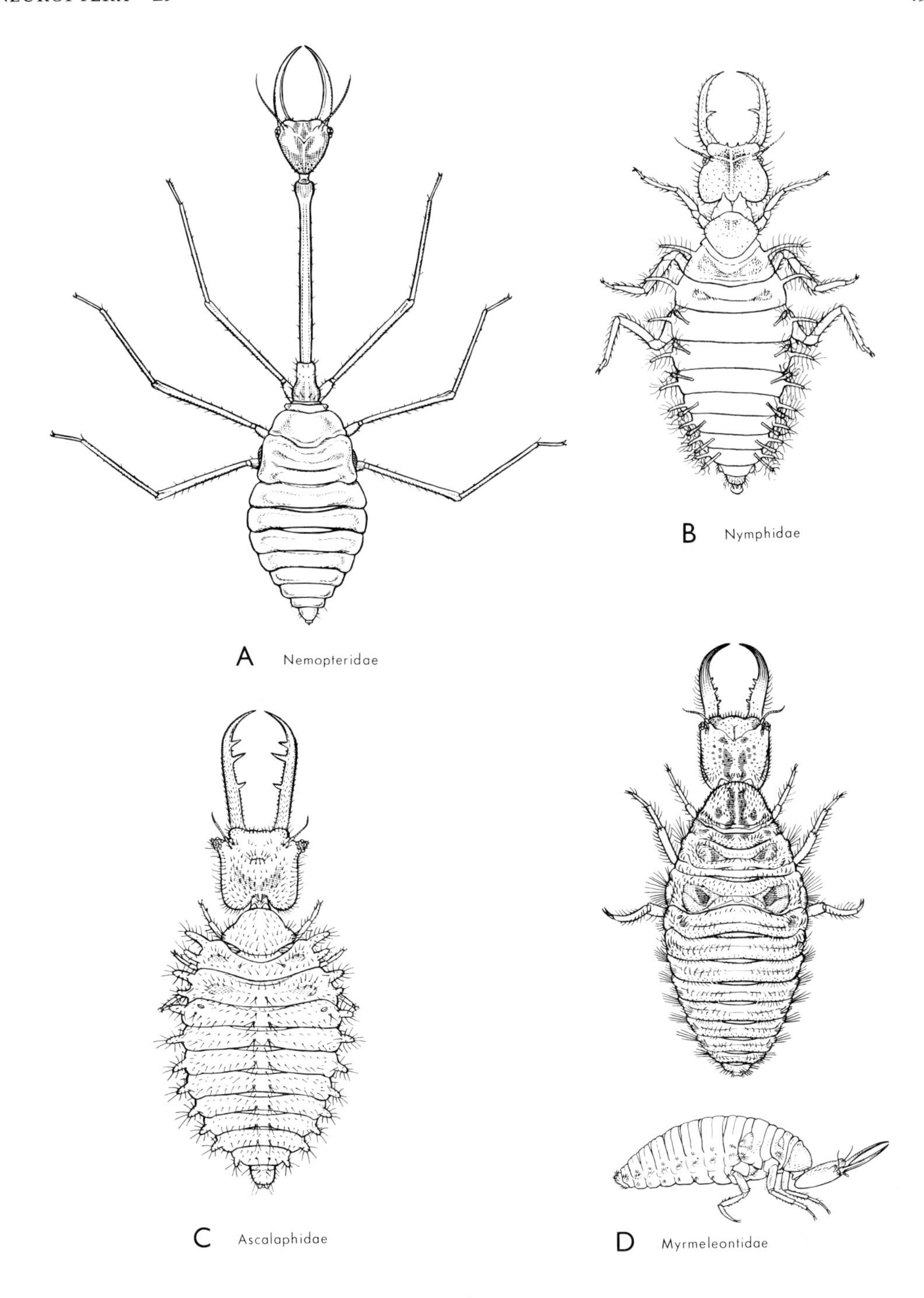

Fig. 29.12. Larvae of Myrmeleontoidea. [M. Quick]

species are mostly large, with broad clear wings. In adult characters the family is transitional between Psychopsidae and Myrmeleontidae, *Myiodactylus* (Fig. 29.8F) being like a narrow-winged psychopsid and *Nymphes* (Fig. 29.8E) like a myrmeleontid in which the antenna has not been thickened into a club-like structure. The antenna is more tapering at apex in *Nymphes myrmeleonides* than in other species of the family.

The eggs of *Nymphes* are laid on long stalks in characteristic U-shaped groups of 30 or more alternating in two directions at right angles to each other. The first-instar larvae are of the myrmeleontid type, but with only a single internal tooth on the enlarged mandibles and well-developed lateral and dorsal body processes. Later instars (Fig. 29.12B) are said to live in rubbish and on rotten logs, and to cover their bodies with particles of rubbish so that only the head and jaws are exposed. The mature larvae are said to pupate in sand in a cocoon similar to that of Myrmeleontidae.

There are 5 genera, *Nymphes* (5 spp.) being the commonest. It occurs over the whole east coast and Tasmania, with one undescribed species in south-western W.A. All the species are light coloured, except for the rather small western species. The only common species, *N. myrmeleonides*, with a wing-span of 60–70 mm, has clear wings apart from a lightly infuscated stigma and a white apical semicrescentic marking; it is widespread in eastern Australia, and is the only nymphid known from Tasmania. Adults are commonly found in damp situations bordering streams, often in association with acacias.

12. Stilbopterygidae. A small family confined to Australia and South America. They are amongst the largest of the lacewings, with a wing-span up to 110 mm, long narrow wings, and a long thin body. They are closely related to the Ascalaphidae. They have sometimes been considered as part of the Myrmeleontidae, or as a short-antennaed subfamily of Ascalaphidae. There are 2 Australian genera: *Stilbopteryx* (4 spp.), with the anterior border and apex of the wings darkened, and a well-developed clavate process with a dense hair tuft at the base of the posterior margin of the hind wing in the males; and an undescribed genus (3 spp.), with almost clear wings and the process much reduced.

13. Myrmeleontidae. The dominant family of lacewings; mostly large species, with long, narrow wings (Fig. 29.8H), generally lightly patterned, but a few species are amongst the most attractive lacewings. Abdomen long and thin, but the folded wings usually extending well beyond its apex. Antennae usually short, thickened, and expanded at apex into a club, occasionally almost half as long as fore wing, and rarely (*Ceratoleon brevicornis*, Dendroleontinae) short and subfusiform, widest in middle. Venation characteristic; in fore wing apparent CuA forking about its middle, and its branches demarcating a large triangular sector; this area sometimes less clearly defined in hind wing. Tibial spurs may be absent, as in some Nymphidae.

The eggs are laid singly in dry soil, and the typical larva (Fig. 29.12D) is the ant-lion that burrows into dry sand. Some construct conical pits into which other insects fall (p. 129), but others merely burrow in sand and loose soil. The bristles on the body are directed forwards, and the hind legs have fused tibia and tarsus, and enlarged, forwardly-directed claws, enabling the larva to move backwards quickly through the soil. Myrmeleontids are possibly more abundant in the lower-rainfall areas than elsewhere. In higher-rainfall areas the larvae generally occur in the shelter of rock overhangs or under slightly raised logs, where they are protected from rain. Some of the inland species have a wide distribution.

Key to the Subfamilies of Myrmeleontidae Known in Australia

1. 2A and 3A of fore wing not fused, just touching or connected by a cross-vein. [In Australian species, only 1, rarely 2, cross-veins before the origin of Rs in the hind wing; CuA of fore wing forking at or after the origin of Rs] .... DENDROLEONTINAE

2A and 3A of fore wing fused over the middle portion, with 2A bent up at base towards 1A 2
2. In the hind wing, only 1, rarely 2, cross-veins before the origin of Rs. [In Australian genera, CuA of fore wing forking before the origin of Rs] .... MACRONEMURINAE
In the hind wing, 3 or more cross-veins before the origin of Rs 3
3. Body and legs covered with long, dense hairs. [Australian genera with a double row of cells in at least part of the costal space of the fore wing] .... ACANTHACLISINAE
Hairs on body short, those on legs very much shorter than the bristles .... MYRMELEONTINAE

The ACANTHACLISINAE, with 17 species in 3 genera, occur mostly in the northern half of the continent, with 2 species of *Acanthaclisis* extending into southern Australia, and one of these into Tasmania. All the species are large, hairy, rather stoutly built as compared with other myrmeleontids. A clavate process is present at the base of the hind wing in the males, and there is an inter-radial 'Banksian' line in Australian genera, the bent branches of Rs forming a straight line through the middle of the apical part of the wings.

The 2 Australian genera of MYRMELEONTINAE, *Myrmeleon* (15 spp.) and *Callistoleon* (2 spp.), have long, thin body and wings, rather broad pronotum, the Banksian line usually ill defined, and costal veinlets of fore wing simple. The tibial spurs are almost straight, thin, and the apical tarsal segment is not as long as the basal 4 segments combined. The species of *Myrmeleon* have mostly clear wings, with the pterostigma often very white; those of *Callistoleon* have dark spots, mainly in the fore wing.

The MACRONEMURINAE include about 40 species in 14 genera. Most of them are undistinguished, medium-sized myrmeleontids, with very little wing pattern apart from the usual speckling. The large *Distoleon bistrigatus* with a longitudinal dark stripe at the apex of the rather pointed hind wing is widespread.

The DENDROLEONTINAE are another large subfamily, with 46, mostly medium-sized species distributed in 8 genera. The pronotum is rather narrower than in the other subfamilies. *Glenoleon* (Fig. 29.2D) contains some common myrmeleontids with a blotch towards the apex of the hind wing. The large species of *Periclystus* with emarginate hind margins to the wings are amongst the most handsome myrmeleontids. *Froggattisca* is rather similar, but the wings are entire. In *Mossega* the hind wing extends further than the fore wing when folded.

14. Ascalaphidae. Medium-sized to large insects (wing-span 35–70 mm), generally similar to Myrmeleontidae except for the antennal and eye characters. The wings are mostly without colour, except for the pterostigma and sometimes the subcostal space. The males of some genera have a pronounced dorsal process on abdominal tergite 2 (Fig. 29.2F). One undescribed genus has a very hairy body in both sexes. *Suhpalacsa* (13 spp.) is the largest genus. The large eggs are laid in clusters of 40 or more around twigs and grass-stems, or in single rows. The larvae (Fig. 29.12C) are of the myrmeleontid type, but distinguished by the characters given in the key. They live openly on tree trunks or on the ground.

15. Nemopteridae. A small family of very

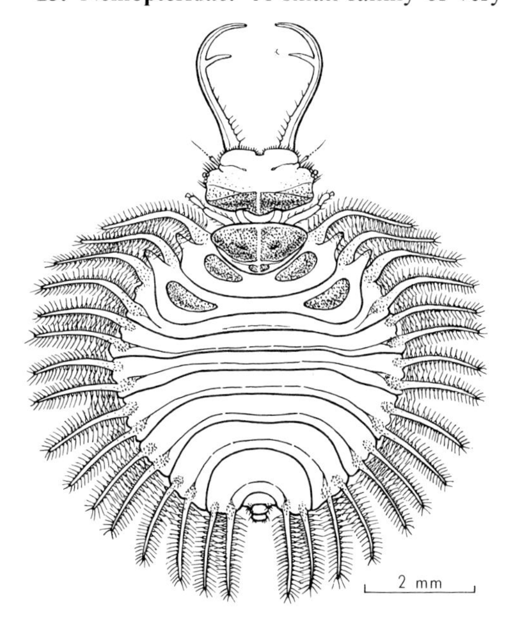

Fig. 29.13. Larva of Nymphidae, myiodactylid type. [R. Ewins]

distinct species, in which the hind wing (Fig. 29.8G) is greatly lengthened and extremely narrow, at least over the basal half, where there are, at most, 3 simple longitudinal veins, apparently Sc, R_1 and CuA. The head is produced into a rostrum. The larvae (Fig. 29.12A) live on the sandy floors of small caves and ledges, preying on other insect larvae. They have a neck of variable length, and in some species the long narrow pronotum gives the larva a grotesque appearance. The legs are often very long. The family is divided into the NEMOPTERINAE, or spoon-winged nemopterids, with 2 species of *Chasmoptera* in south-western W.A., and the CROCINAE, or thread-winged nemopterids, represented by 5 species of *Croce*, delicate insects with white-tipped hind wings, which are widely distributed in the drier parts of the continent.

ACKNOWLEDGMENT. I am grateful to Dr Ellis G. MacLeod, Department of Entomology, University of Illinois, U.S.A., for helpful discussions of the manuscript.

30

COLEOPTERA

(Beetles)

by E. B. BRITTON

Endopterygote Neoptera, with the mesothoracic wings modified into more or less hardened, non-folded, rigid elytra, which meet edge to edge at rest, and partly or wholly cover the hind wings and abdomen; metathoracic wings, when developed, membranous, folded, and alone used for propulsion in flight; mouth-parts mandibulate; prothorax well developed, free, and forming with the head a distinct forebody, contrasted with the hind body formed by the elytra covering the meso- and metathorax and abdomen; mesothorax usually reduced; abdominal sternites more strongly sclerotized than tergites. Larva with or without thoracic legs; with a distinct head capsule, antennae, and mandibulate mouth-parts, but rarely with abdominal prolegs. Pupa adecticous and exarate, rarely obtect.

Estimates of the number of species of Coleoptera so far described in the world vary between 277,000 and 350,000. It is a notable fact that, whereas 70 per cent of the known species of animals are insects, no less than 40 per cent of the species of insects and about one-third of all animal species are beetles. The number of species of Coleoptera known to occur in Australia is greater than 19,000. This total will, however, be considerably exceeded when the fauna is thoroughly studied.

The basic characters of Coleoptera suggest no obvious reason for the remarkable success of this particular insect pattern. It is commonly assumed that the main function of the elytra is to protect the folded hind wings. There are, however, other groups (Orthoptera, Dermaptera, Hemiptera–Heteroptera) in which the more or less sclerotized fore wings serve as covers for the hind wings, and it is perhaps significant that the Coleoptera differ from these in that the abdominal spiracles do not open directly to the exterior but into a space enclosed between the elytra and the abdomen. This can be of importance in the conservation of water, for the opening of the spiracles into a volume of still air beneath the elytra has the effect of reducing the steepness of the humidity gradient from tracheae to external air. It may be noted that beetles that have to resist a very dry environment (desert Tenebrionidae, Curculionidae) are able to seal the elytral cavity very effectively. The protection provided by the elytra also permits the dorsal surface of the abdomen to be largely soft and unsclerotized, which may be important in allowing for the growth of the viscera, including the reproductive organs, without change of external body size and shape. The area of soft integument also assists in the elimination of carbon dioxide (Thorpe, 1928).

The earliest known fossil beetles have been found as elytral imprints in the Upper Permian of Russia and New South Wales (p. 183), and many of the dominant families of the present-day fauna made their appearance during the early Mesozoic. Coleoptera have exploited all possible habitats with the exception of the open sea.

Beetles include some of the largest as well as the smallest insects. The largest are the tropical American *Megasoma elephas* F. and *Dynastes hercules* L. (Scarabaeidae–Dynastinae), the latter attaining a maximum length of 16 cm. The largest Australian beetles are the carabid, *Hyperion schroetteri* Schreib., and the buprestid, *Stigmodera grandis* Don., both of which reach a length of 6 cm. *Haploscapanes barbarossa* F. (Dynastinae) and *Batocera boisduvali* Hope (Cerambycidae) closely approach this length. At the other end of the scale, the smallest beetles are included in the Ptiliidae and Corylophidae. The smallest known beetle is probably the Mexican species *Nanosella fungi* Le C. (0·4 mm long). The smallest species so far discovered in Australia is 0·6 mm in length.

Anatomy of Adult

Head. The head (Fig. 30.1A) is a rigid capsule of very varied shape. It is divided into fairly well-defined areas, which are useful for purposes of description, but have little significance for purposes of homology (p. 8). The frons (front or vertex) is that part of the upper (in prognathous forms) or anterior surface of the head lying between the eyes. This is limited anteriorly by the clypeus (or epistoma), the boundary with the frons being marked by a suture, ridge, or declivity.

The eyes are very varied in size, and may be absent, or so large that they meet above and below (in males of some Lampyridae). In a number of families the eyes are partially divided by a lobe (canthus). In Gyrinidae and in some Lucanidae *(Lissapterus)* the division into an upper and a lower eye is complete. The facets of the compound eye tend to be small and flat in diurnal species, whereas they are coarse and convex in forms that are active in reduced light. A median ocellus occurs on the frons in some Dermestidae (e.g. *Anthrenus*), and there are 2 ocelli on the head between the compound eyes in Staphylinidae–Omaliinae, but 3 ocelli are never present in the adult.

The antennae are normally 11-segmented, and the scape and pedicel are always clearly

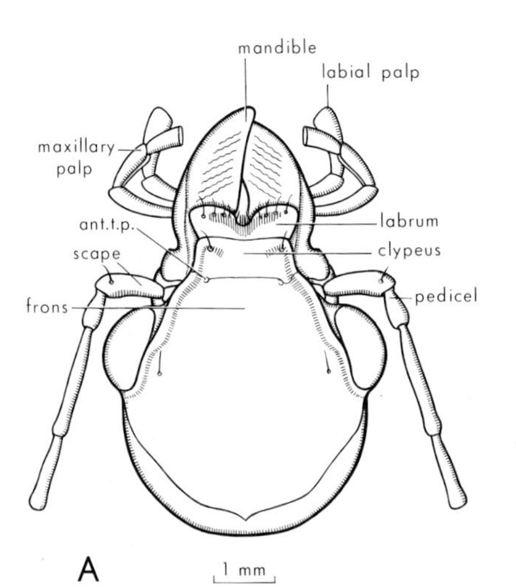

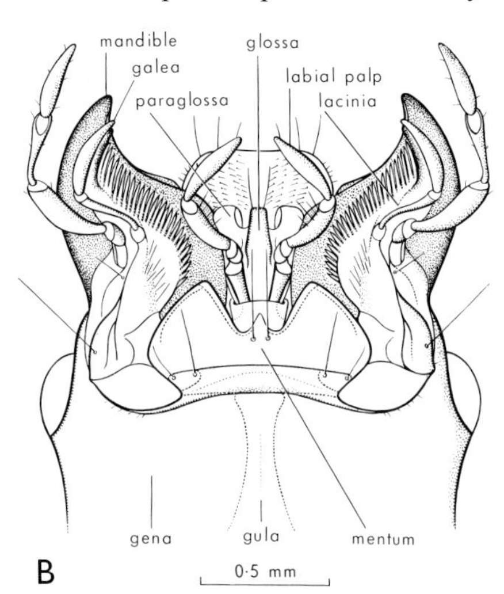

Fig. 30.1. Head: A, *Calosoma schayeri*, Carabidae, dorsal; B, *Hypharpax* sp., Carabidae, ventral. *ant. t. p.*, anterior tentorial pit. [F. Nanninga]

differentiated. The length of the antennae is very varied, from two or three times the length of the body in Cerambycidae to less than the length of the maxillary palpi in Hydrophilidae. Antennal form is equally varied. Antennae may be *filiform* (flagellum linear, cylindrical), *moniliform* (as a string of beads), *geniculate* (elbowed between the elongate scape and the rest of the antenna), *serrate* (saw-like), *pectinate* (saw-like, with longer teeth), *flabellate* (extreme development of pectinate antenna), *clavate* (clubbed), or *lamellate* (with an asymmetrical 3- to 7-segmented club of more or less flattened segments). The antennae are frequently longer (e.g. in Cerambycidae) or more elaborate (e.g. in Elateridae, Meloidae, Eucnemidae) in the male than in the female.

The clypeus is usually trapezoidal, with the labrum attached to its narrower anterior edge. The labrum is commonly visible, but may be membranous or hidden beneath the clypeus as in the Scarabaeidae (Scarabaeinae and Aphodiinae), or fused with the clypeus as in the Curculionidae. The sides of the head, below and behind the eyes as far as the longitudinal gular sutures, are termed the *genae*, and the area between the gular sutures is the *gula*. This area is always without setae. In the Curculionidae the gular sutures are fused.

Articulated to the anterior end of the gula (or of the genae in Curculionidae) is the mentum, a distinct flat plate, the base of the *labium* (Figs. 30.1B,2C). The mentum covers the articulations of the *labial palpi*, which are usually 3-segmented or, more rarely, 2- or even 1-segmented. Connected to its upper surface is the membranous prementum on which the labial palpi are articulated. Between the palpi is a membranous ligula, with the glossae and paraglossae usually represented by lobes. In some forms, including *Hydrophilus* (Hydrophilidae) and *Nicrophorus* (Silphidae), there is a distinct sclerite defined by a suture intervening between the mentum and the gula. This is the submentum, a name which in other beetles has also been applied to the undifferentiated anterior margin of the gula. The gular sutures mark the inflection of the posterior arms of the tentorium.

The cardo of the *maxilla* (Figs. 30.2B, D) on each side is articulated with the anterior end of the posterior arm of the tentorium. On the cardo is hinged the stipes, and to this are articulated the palpifer laterally, bearing the palp, and the galea and lacinia (the 'external' and 'internal lobes' of the maxilla). The galea is often a single piece, but in the

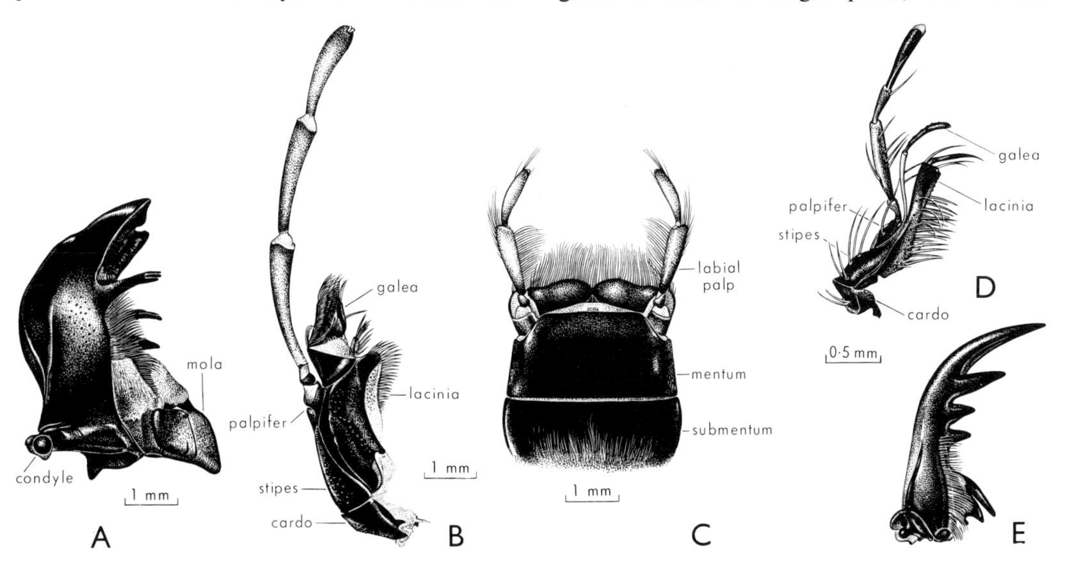

Fig. 30.2. Mouth-parts. *Hydrophilus latipalpus*, Hydrophilidae: A, mandible, ventral; B, maxilla, ventral; C, labium, ventral. *Cicindela semicincta*, Carabidae: D, maxilla, ventral; E, mandible, ventral. [F. Nanninga]

Adephaga it is 2-segmented and palp-like, and the lacinia is often large, blade-like, and spinose. The *maxillary palp* is usually 4-segmented, rarely 5- or 3-segmented, and is greatly developed in the Hydrophilidae ('Palpicornia'). In most Coleoptera the maxillae are adapted for chewing the food, but in some they are modified to assist in imbibing nectar, as in some Meloidae *(Zonitis)*.

The *mandible* (Figs. 30.2A, E) is basically in the form of a tetrahedron, with a dorsal, ventral, and external face, and a triangular base. The mandibles are articulated at the dorsolateral and ventrolateral angles, so that movement is normally transverse and opposed. The dorsal articulation is composed of an acetabulum on the mandible and a condyle on the head, whereas the ventral articulation comprises a condyle on the mandible connecting with an acetabulum on the head. The basal part of the internal ridge of the mandible, when thickened and enlarged, as in Polyphaga, is known as the *mola*. The mola is used for grinding the food, and is particularly well developed in phytophagous forms. In Adephaga there is an accessory tooth, the *retinaculum*, distal to the mola; and in Sphaerioidea and some Hydrophilidae (Fig. 30.2A) there is an articulated tooth below the apical point. The apices of the mandibles are acute and curved in carnivorous forms, and are used for seizing prey. The mandibles attain their maximum development in males of the Lucanidae (e.g. *Phalacrognathus*, *Cacostomus*), in which they are used solely in combat. By contrast, the mandibles in Scarabaeinae are reduced and more or less membranous.

The *tentorium* consists typically of a central transverse connecting piece with three pairs of arms. The anterior arms terminate at the anterior tentorial pits, which are slightly deeper parts of the fronto-clypeal suture. The posterior arms run back to pits on the gular sutures, where they meet the invaginated gular apophyses. The dorsal arms of the tentorium bend forwards and upwards, and terminate at the cranial roof between the upper edge of the eye and the middle of the head (M. Evans, 1961).

Thorax. The *prothorax* in the Coleoptera is always well developed, forming with the head an obvious 'forebody'. It is composed of a single dorsal sclerite, the pronotum, which also forms part of the sides, a ventral prosternum, and lateral proepisterna.

In the Adephaga, Archostemata, and Myxophaga the pleuron on each side is visibly separated from the pronotum and prosternum by notopleural and pleurosternal sutures (Fig. 30.3). The trochantin is a small sclerite, articulating by a single condyle with the coxa, but separated by membrane from

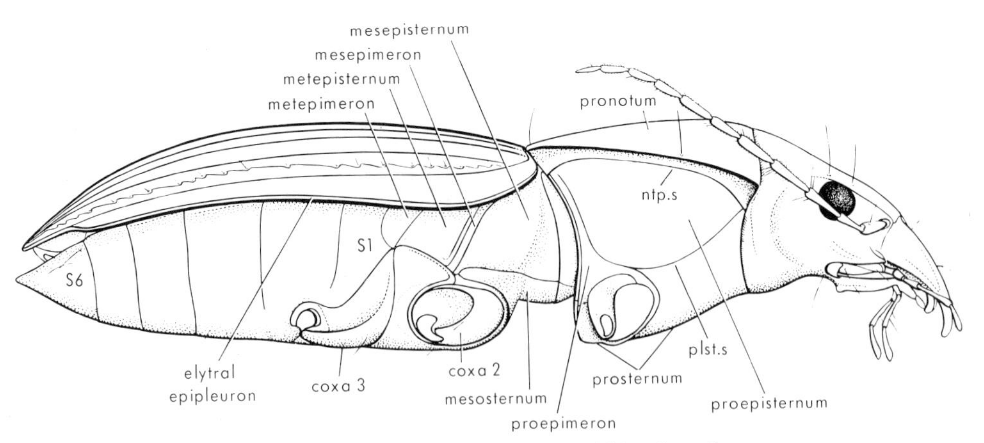

Fig. 30.3. *Notonomus violaceus*, Carabidae, lateral. [F. Nanninga]
ntp. s., notopleural suture; *plst. s.*, pleurosternal suture.

the body of the prothorax. The coxa itself articulates with the pleuron. The notopleural suture is not obvious in the Polyphaga, and it has been commonly assumed that the pleuron and the deflexed part of the notum are fused to form the ventrolateral face of the prothorax, so that the notopleural suture is absent; but, in fact, the ventrolateral face of the prothorax is formed by the deflexed notum alone (Ferris, 1935). The pleuron here is separated from the notum and united with the trochantin, taking with it the coxal articulation. This is indicated by the fact that the trochantin articulates with the coxa at two points (Fig. 30.4B), instead of one as in the Adephaga (Fig. 30.4A). The notopleural suture in Polyphaga is represented by the junction between the deflexed part of the notum and the trochantin (Hinton, 1939a). In the Adephaga the deflexed lateral margin of the pronotum bordering the notopleural sutures on each side is termed the *pronotal epipleuron*, whereas in the Polyphaga the side of the prothorax is the *pronotal hypomeron*.

The prosternum is distinct, being separated from the proepisterna or hypomera by the *pleurosternal* sutures, except in the Curculionoidea in which the prothorax forms a single, ring-shaped sclerite. The prosternum is sometimes (e.g. in Elateridae, Buprestidae, Histeridae) produced forward into a plate which covers the mouth-parts beneath when the head is retracted. Behind the fore coxa on each side there is a narrow transverse sclerite, the *proepimeron*, which is separated by a suture from the proepisternum in the Adephaga. In other Coleoptera it is merely a posterior inner process of the hypomeron. When the inner end of the proepimeron reaches the prosternal process, the coxal cavity is said to be *closed* (Fig. 30.3); otherwise the coxal cavity is *open* (the primitive condition).

The *mesothoracic spiracles* are normally hidden in the cavity between the pro- and mesothorax. When the prothorax is detached, they are seen as large, elongate-oval structures on the part of the intersegmental membrane attached to the back of the prothorax.

The meso- and metathorax are immovably fused together to form the *pterothorax*, which bears the elytra and wings. The mesothorax is reduced by comparison with the metathorax, except in forms in which the flying wings are absent or reduced. The dorsal surface of both meso- and metathorax is divided into an anterior *prescutum*, a median *scutellum*, and lateral *scuta* (the divided scutum). Behind these sclerites in the metathorax is the transverse *postnotum*, which is absent in the mesothorax. The *mesoscutellum* is usually triangular, and is often visible from above between the bases of the elytra. It is known simply as the scutellum.

The ventral surface of the pterothorax has almost always been interpreted as being composed of a median mesosternum in front and metasternum behind, flanked by their respective episterna and epimera. Ferris (1940a, b) has, however, shown that it is highly probable that there is no sternal element in the pterothorax of any neuropteroid insect. On this interpretation, the ventral surface is formed of symmetrical subcoxal pieces which meet and infold together along the median line (the 'discriminal line'), and the sternum is absent, or at most limited to a very narrow area between the sternal apophyseal pits and perhaps along the apex of the internal median ridge.

The mid coxal cavities are bordered in some by the 'mesosternum' anteriorly and medially, the 'metasternum' medially and posteriorly, and the mesepisternum and mesepimeron laterally (e.g. in Hydrophilidae, Buprestidae, Lycidae, etc.). In a second type, each mid coxal cavity is enclosed by the mesosternum, the metasternum, and the mesepimeron, the mesepisternum being excluded. This form of mid coxal cavity is found in the Cupedidae, Carabidae–Paussinae, Noteridae, Dytiscidae–Laccophilinae, Histeridae, Scarabaeidae, Elateridae, Cerambycidae, and Chrysomelidae. In the Cupedidae the metepisternum also adjoins the mid coxal cavity. Rarely (e.g. in *Epilachna*), the mesepisternum forms part of the border of the coxal cavity, and the mesepimeron is excluded. Lastly, both episternum and epimeron may be excluded, the mid coxal cavity being completely enclosed by the meso- and metasternum. This is seen in the Tenebrionidae, Erotylidae, Curculionidae, Brenthidae, and in the Passalidae and Trogidae among the Scarabaeoidea.

The *metathoracic spiracles* are hidden beneath the mesepimera, the posterior edges of which can be raised to expose the spiracles to view (Fig. 30.11). They are sometimes non-functional (e.g. in Carabidae), but when functional are often large.

The hind coxal cavities are usually transverse, separated by the metacoxal apophysis, and completed on the outer side by the metepimeron; they are bounded posteriorly by the abdomen (sternite of segment 2 or 3). The metasternum commonly bears a transverse groove, which marks the invagination of the metathoracic furca to which the muscles of the hind legs are attached.

Legs. The legs are normally adapted for walking or running, but in some families they are modified for burrowing (Scarabaeidae, Bostrychidae, Scolytinae, etc.), for swimming (Dytiscidae, Gyrinidae), or for jumping (Chrysomelidae–Halticinae, and also some Curculionidae, e.g. *Orchestes*). The form and degree of separation of the coxae provide characters of major taxonomic importance, as does the number of segments in the tarsi. The coxae are normally capable of a limited rotary movement, but in Adephaga the hind coxae are immovable. The trochantins are associated with the fore and mid coxae and are visible externally in some Polyphaga, but are present on the metathorax only in Cupedidae.

The trochanter articulates with the coxa, and is largest in the hind legs, in which it sometimes shows sexual characters. The femora of all three pairs of legs are usually rather similar in shape, but the hind femora are usually enlarged in species that jump. The tibiae are usually more or less expanded towards their apices, where they bear combs of spines, 2 of which are specially enlarged and are known as the *spurs.* The fore tibiae are often expanded and toothed on the outer side to assist in digging (Fig. 30.4B).

The tarsi are normally 5-segmented, but the number is reduced to 4 in some large groups and to 3 in others (e.g. Lathridiidae, Pselaphidae); rarely the sexes have different numbers of segments (e.g. Cryptophagidae). In the Chrysomeloidea and Curculionoidea the 4th tarsal segment is very small and is fused to the 5th, so that the tarsus is in effect 4-segmented ('pseudotetramerous' or 'cryptopentamerous'), the 3rd segment being large and usually bilobed. In the families of the Cucujoidea–Heteromera the fore and mid tarsi are 5-segmented and the hind tarsi 4-segmented. In some male Adephaga the segments of the fore tarsi are dilated and clothed beneath with adhesive lamellae. The terminal segment normally bears 2 claws. These may be equal, unequal, or reduced to one, of simple form, dentate, or pectinate, and free or connate (i.e. immovably joined).

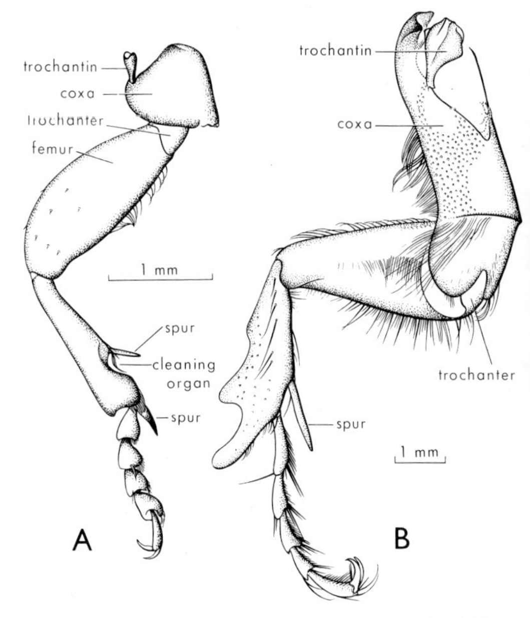

Fig. 30.4. Fore leg: A, *Hypharpax* sp., Carabidae; B, *Colpochila* sp., Scarabaeidae. [F. Nanninga]

Elytra. The elytra are modified mesothoracic wings. They are characteristically rigid, fitting over the abdomen at rest, with their inner edges in contact. Except in the Scarabaeidae–Cetoniinae, the elytra are opened in flight and held at a constant angle to the body, allowing free movement of the wings but playing no significant part in flight.*

The named parts of an elytron are the *disc*, which is the general dorsal surface, the medial *sutural* edge, which meets the corresponding edge of the other elytron, and usually a defined incurved lateral border known as the *epipleuron*. There is a small, 2-headed tubercle at the base, by means of which the elytron articulates with the lateral part of the mesonotum through the axillary sclerites. The articulatory process carries nerves and tracheae, and acts as the point of attachment of the muscles that move the elytron.

The edge-to-edge fitting of the elytra at rest is assisted in many beetles by the engagement of a flange on the edge of one elytron with a groove on the opposed edge (Fig. 30.5A). In wingless forms in which the elytra are

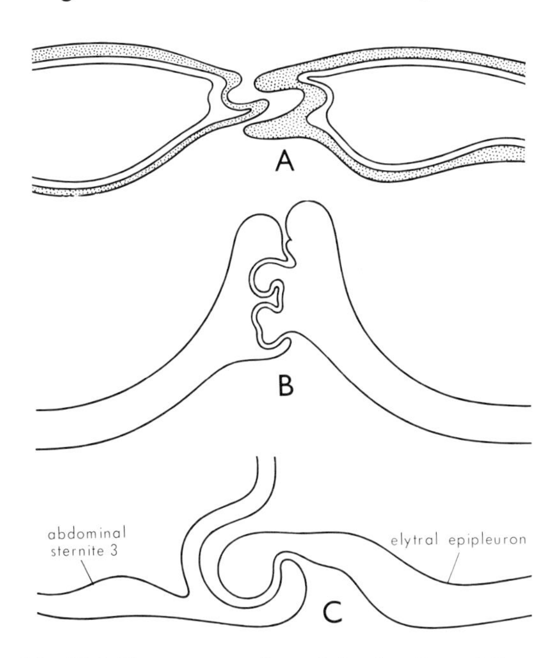

Fig. 30.5. Transverse section of the junction of the elytra: A, *Colpochila* sp., Scarabaeidae; B, *Sympetes* sp., Tenebrionidae; C, transverse section of the junction of elytron and sternite 3. [F. Nanninga]

* It has been shown in *Melolontha* (Demoll, 1918) that removal of 45 per cent of the elytra doubles the speed of flight, suggesting that they actually impede flight.

apparently united along their sutural edges, they are, in fact, not fused but locked together by a mortice and tenon joint (e.g. Fig. 30.5B). Similarly the lateral edge of the elytron may be locked to the basal sternites of the abdomen (Fig. 30.5C).

The incurved epipleura are separated from the disc in most beetles by a longitudinal ridge on each side. The elytral epipleura are, however, absent in some (e.g. Curculionidae, Meloidae).

The disc of each elytron is usually marked by longitudinal *striae*, each of which corresponds to a row of sclerotized pillars connecting the upper and lower faces of the elytra. These, together with the thickness of the cuticle, give the elytra their characteristic rigidity. The basic number of striae is 9, and they are numbered from the stria adjacent to the suture outwards. They are more numerous in some forms (e.g. up to 25 in some Carabidae), or may disappear. The spaces between the striae are known as *elytral intervals*, *interstices*, or *interstriae*. The odd-numbered interstriae appear to correspond with the primitive wing veins. This is indicated by the fact that the 7th and 9th are traversed by nerves and tracheae of the costoradial group, whereas the 1st, 3rd, and 5th receive their nerves and tracheae from the cubito-anal group. In addition, the odd interstriae are frequently distinguished externally in one way or another from the even interstriae (e.g. by convexity, width, sculpture, or the presence of sensory setae). Crowson (1955) has suggested that the elytron was derived from a primitive wing with a series of longitudinal veins and regular cross-veins, by thickening of all the veins and cross-veins, with the result that the enclosed areas of wing membrane were finally obliterated. The cavity between the upper and lower walls of the elytron is lined by epidermis, and contains blood channels, nerves, and tracheae. The elytra exhibit various degrees of shortening in different families, but are rarely completely atrophied.

Wings. The metathoracic wings of Coleoptera, where functional, are nearly always longer than the elytra, and at rest are normally folded longitudinally and transversely so that they can be contained beneath the elytra. The pattern of folding varies considerably in the various families (Forbes, 1926).

The wing is rotated forwards on its base into the flight position by the action of the direct flight muscles. The same action spreads the wing, opening the longitudinal folds, which leads, by virtue of the stiffness and springiness of the wing membrane, to automatic opening of the transverse folds. As a result, the costal edge, which is acutely folded at the hinge at rest, straightens as the wing rotates forwards. At the same time the anal fold is opened, and the whole wing flattens as the membrane tightens. Relaxation of the direct flight muscles allows the wing to fold longitudinally, and this automatically leads to the transverse folding centred on the costal hinge. The final stage of closure is assisted by movements of the abdomen and elytra. The folding lines of the wing are determined by the lines of stiffness produced by adjacent veins. Thus, longitudinal folds normally lie alongside main veins, whereas the angles of the chevron-shaped transverse folds are determined by terminations and weak hinge-points in veins.

In the Archostemata (Fig. 30.15A) and Myxophaga (Fig. 30.15B) the apices of the wings are rolled into a double spiral instead of being folded. In many Buprestidae (e.g. *Stigmodera*), the wings are broad, but not longer than the elytra. They are therefore folded longitudinally, but not transversely, and there is no hinge in the costal edge.

Variations in wing venation are related to the complexity of folding, but, in spite of this, homologies with the venation of other orders can be traced. The basic pattern is most clearly seen in the Adephaga and Archostemata, and is reduced or otherwise partially obscured in the Polyphaga, which are less primitive in this and some other respects. The main veins normally visible are C, Sc, R, and CuA. The first three are crowded together close to the costal edge and often fused throughout the greater part of this length, being recognizable only near the base of the wing. M lies between R and CuA, and is less strongly developed, or abbreviated, or absent. The posterior inner lobe of the wing is occupied by 4 anal veins or by reductions of these.

Two main types of venational pattern are recognized.

1. The *adephagan* type. This is characterized by the presence of a closed cell, the *oblongum*, formed by cross-veins connecting M and CuA (Fig. 30. 6B). An oblongum is also seen in the Archostemata (Fig. 30. 6C). A second common character is the presence of another closed cell formed close behind the costal edge, in the apical half, by branches of R.
2. The *polyphagan* type. In wings of this type, the closed cells are absent. A further subdivision of this group into a staphylinoid type (Fig. 30. 6D) and a cantharoid type (Fig. 30. 6E) has been made, but this appears to lack systematic value. The staphylinoid type shows greatly reduced venation, with no cross-veins, whereas in the cantharoid type, which occurs in a great many families, CuA and M unite distally and continue to the margin as a single vein, and M does not reach the base.

In very small beetles the wings are commonly fringed with long setae (Fig. 30.6A). An extreme example of this is to be seen in the Ptiliidae (Fig. 30.30), the smallest of all beetles, in which the wing surface is greatly reduced and the setae have taken over its function.

Wings in Coleoptera are frequently atrophied to a greater or lesser extent. This phenomenon is common in island and mountain species (Darlington, 1943). The wings may be merely shortened and not folded (brachypterous), or reduced to mere vestiges (micropterous), or completely absent (apterous). Jackson (1928) and Lindroth (1946) have demonstrated that the wing condition is genetically determined, and it seems probable that brachyptery functions as a simple Mendelian dominant with respect to macroptery throughout the Coleoptera. Lindroth has pointed out that, when a wing-dimorphic species invades new areas, the macropterous form will arrive first, and, as it is homozygous, cannot give rise to brachypterous forms, which must therefore arrive separately and more slowly. A distribution

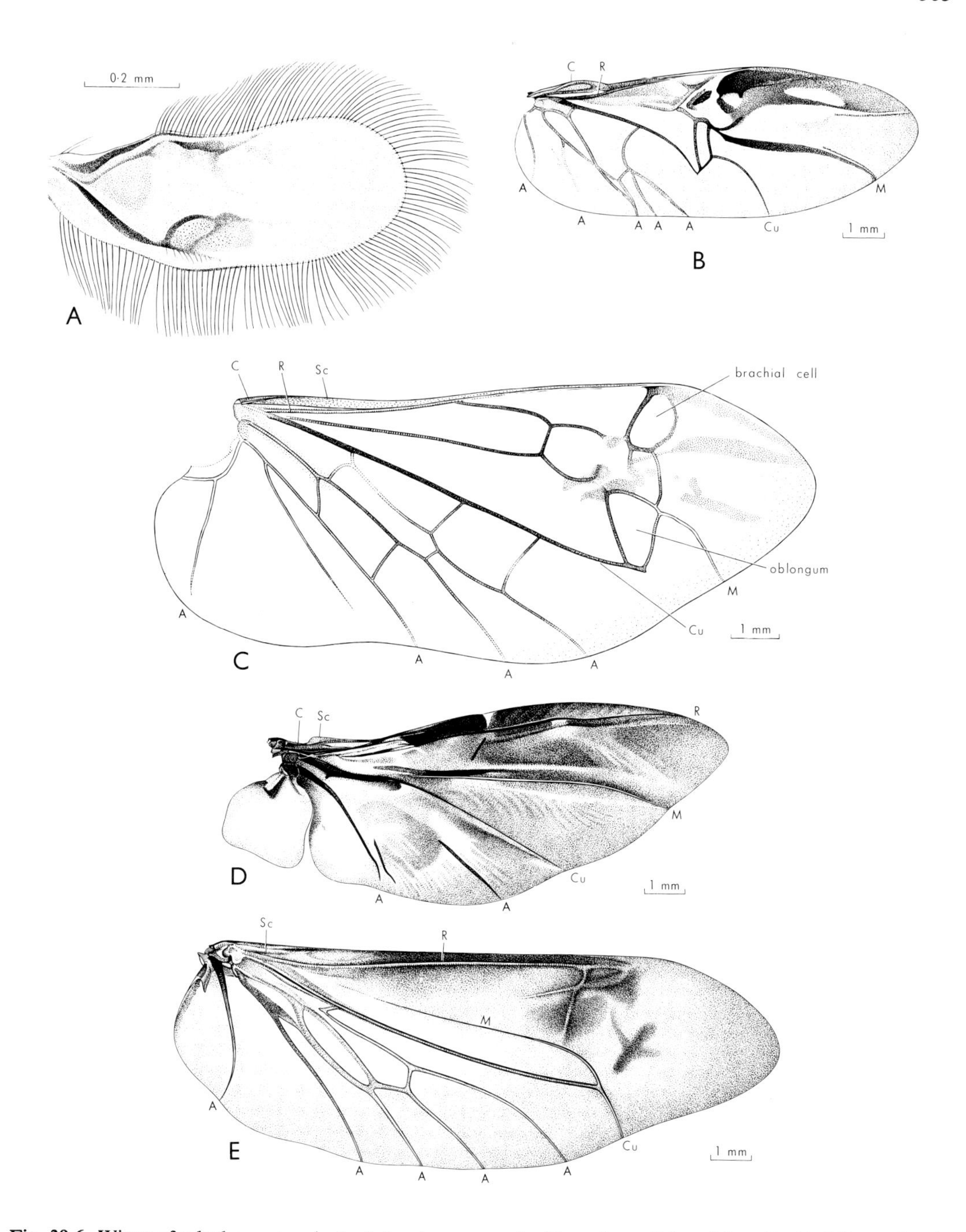

Fig. 30.6. Wings of adephagan type: A, *Sphaerius ovensensis*, Myxophaga–Sphaeriidae; B, *Eudalia macleayi*, Adephaga–Carabidae; C, *Omma varians*, Archostemata–Cupedidae; D, wing of staphylinoid type, *Creophilus erythrocephalus*, Polyphaga–Staphylinidae; E, wing of cantharoid type, *Tanychilus striatus*, Polyphaga–Alleculidae.

[F. Nanninga]

map of the two forms can therefore reflect the history of immigration (Lindroth, 1963).

Some beetles with fully developed wings are unable to fly because of atrophy of the flight muscles, e.g. some Dytiscidae (Jackson, 1956) and *Othnonius* (Melolonthinae). The degeneration of the flight muscles may occur during the life of the adult, so that the insect is able to fly at first, but loses the power of flight as the gonads develop. The flight muscles can even undergo cyclic changes (Chapman and Kinghorn, 1958).

Abdomen. The abdomen is composed of 10 segments in both sexes, but segment 9 is modified as the genital segment and hidden within the body, and 10 is greatly reduced. Eight tergites can be counted on the dorsal surface of the abdomen in most adult Coleoptera; they are more easily traceable in forms with functional wings than in wingless forms in which the tergites are never exposed. In all cases, however, the number of tergites is easily determined by relating them to the spiracles (Fig. 30.11). The last visible tergite is known as the *pygidium*, and is sclerotized, as is the penultimate tergite *(propygidium)* in beetles in which the elytra are shorter than the abdomen.

Segment 9 (the *genital segment*) is invaginated and completely hidden (Fig. 30.7), except in a very few (e.g. Carabidae–Brachininae, Staphylinidae), and segment 8 is sometimes ring-shaped and telescoped inside segment 7 which becomes the apparent terminal segment. Where most fully developed (Adephaga, Staphylinidae, Hydraenidae, Cantharidae), segment 9 surrounds both rectum and aedeagus in the male, whereas in the female it forms the armature at the apex of the genital tube. In many families this segment is more or less reduced. In the male reduction of segment 9 has been thought to leave a Y-shaped ventral piece, or *spiculum gastrale*, but a well-developed segment 9 and a spiculum gastrale can exist together (Fig. 1.31c). Segment 10 *(perianal)* is almost always absent, but possible remains of it exist in male Cupedidae in the form of a subanal plate which lies between the aedeagus and the anus, and in some families (e.g. Elateridae, Lucanidae) in which there is a sclerite following tergite 9.

The number of abdominal sternites (referred to by many coleopterists as *ventrites*) is variable, but always less than the number of terga. Sternite 1, corresponding to tergite 1, is missing in the great majority of Coleoptera. It is visible externally as a separate median intercoxal process and internally on the posterior face of the hind coxal cavity in some Hydraenidae. Otherwise 5 to 7 sternites are visible externally, except in the males of Carabidae–Brachininae, in which 8 are visible. Four types of abdomen are recognized in Coleoptera, marking apparent stages of regression of sternite 2 (Jeannel and Paulian, 1944).

The *hologastrous* type—sternite 2 complete and fully sclerotized, like segment 3. This is seen in some Cantharoidea (e.g. *Chauliognathus*, *Metriorrhynchus*).

The *haplogastrous* type—sternite 2 reduced to a small triangular plate on each side. The first complete sternite is 3 (e.g. Scarabaeidae).

The *adephagous* type—sternite 2 divided into two separate lateral pieces, and is fused with segment 3 (e.g. Carabidae).

The *cryptogastrous* type—sternite 2 membranous and hidden from view in the hind coxal cavity (e.g. in Curculionidae).

In most families there are 8 pairs of functional abdominal spiracles. These usually lie in the membranous pleura or terga, and are covered by the elytra at rest. Those of segment 1 are large, especially in beetles capable of flight. In Cucujoidea, Scarabaeoidea, Chrysomeloidea, and Curculionoidea the spiracles of segment 8 are non-functional, and in Cucujoidea the spiracles of segment 7 (and sometimes 6) may also be non-functional.

The visible sternites may be connected flexibly so that all are capable of movement (e.g. in Cantharidae), but in many families the basal 2 or 3 sternites are *connate* (i.e. immovably joined together), as, for example, in the Adephaga and Curculionidae. Connation is carried to completion in Anthribidae and Attelabidae in which 4 or all sternites are immovably joined.

Male Genitalia. The male copulatory organ, or *aedeagus*, is a development of the

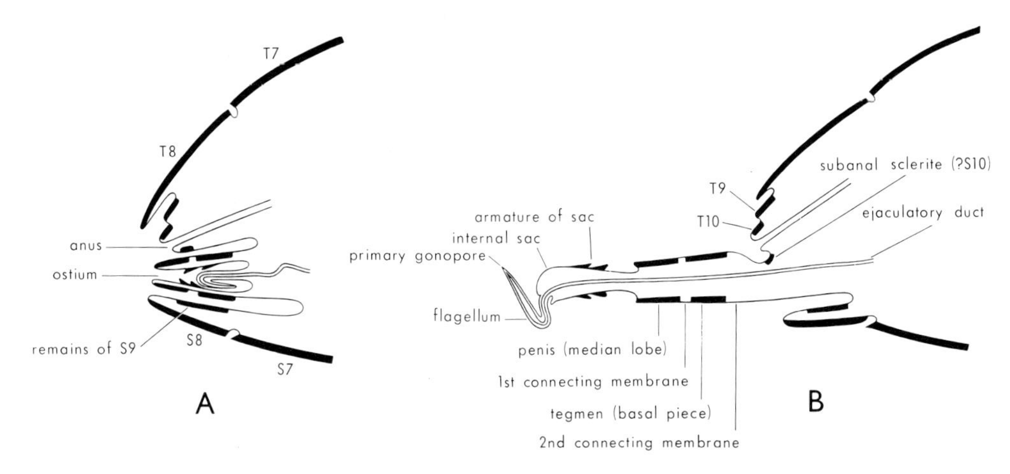

Fig. 30.7. Apex of abdomen of male, diagrammatic sagittal sections: A, with aedeagus retracted; B, aedeagus evaginated. [F. Nanninga]

posterior end of the ejaculatory duct, which opens behind sternite 9 (Fig. 30.7). This part of the duct can be considered to be composed of three successive parts. The true apical part is membranous, and permits extrusion and withdrawal of the second part, which is usually stiffened by strong sclerotization. This second part is composed of a sclerotized tubular *tegmen* connected by a short tubular membrane to the sclerotized tubular *penis* or *median lobe*. Invaginated into the penis at rest is the third specialized section of the duct, the *internal sac*. This is often armed with spines, teeth, etc., which project into the lumen at rest, but are external when the sac is everted, and may assist in holding it in the female genitalia. It is, however, more probable that the teeth serve to rupture the spermatophore (Hinton, in Highnam, 1964). The male genitalia at rest within the body are therefore doubly invaginated, and, when active, are everted in two stages. First the aedeagus is protruded from the body through the genital aperture, and second the internal sac is everted from the penis through the *ostium*. The opening of the ejaculatory duct into the internal sac is the *primary gonopore*, and this is apical when the sac is everted. In many Coleoptera the ejaculatory duct is continued, as the *flagellum* (Fig. 30.9B), into the cavity of the inverted internal sac. When the sac is everted, the flagellum is terminal and enters the aperture of the spermathecal duct of the female. This appears to be a device for economizing sperm, which otherwise would have to reach the spermatheca as the result of contractions of the female duct or by chemotaxis.

In its most primitive form, therefore, the aedeagus is composed of a single basal piece (the *tegmen* or *phallobase*) with a median lobe (penis) attached to its apical end, and it has a pair of *parameres* (which may correspond to the 'gonocoxites' of other endopterygotes—see p.26) attached one on each side (Figs. 30.9A, B). A considerable range of variation of this basic pattern exists. All components—median lobe, basal piece, internal sac, and parameres—may be present, or any part may be reduced or absent. For example, the complete trilobed structure occurs in the Byrrhoidea, whereas in Carabidae, Staphylinidae, Buprestidae, and Elateridae the basal piece is reduced or absent and the parameres articulate with the median lobe. The alternative modification is also common, i.e. parameres present and connected to the apex of the basal piece, median lobe absent or much reduced and contained within the basal piece (as in Scarabaeoidea). Again, the parameres may be absent and the basal piece reduced to form an open half-tube or gutter in which the median lobe can slide. This basal piece may be either above or beneath

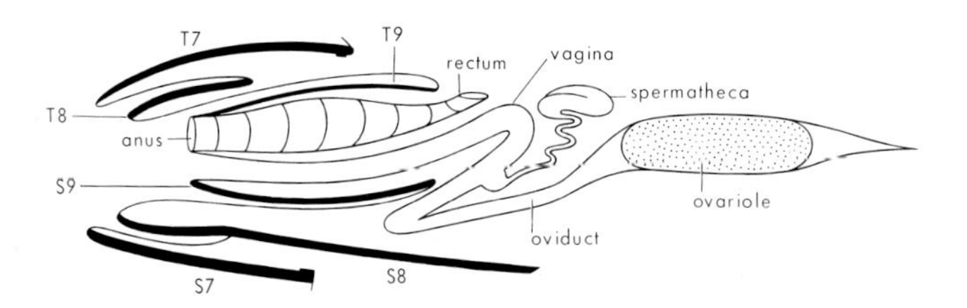

Fig. 30.8. Apex of abdomen of female *Atomaria* sp., Cryptophagidae (diagrammatic sagittal section after M. Evans, 1961).

the median lobe. Finally, the sides of such a dorsal basal piece may be produced into long processes which meet beneath, forming a kind of ring around the median lobe ('*mode en cavalier*' of Jeannel, 1955). This type is found throughout the Chrysomeloidea and in the Cucujoidea.

In many Coleoptera, in particular in the Adephaga and Staphylinoidea, the aedeagus, when at rest within the abdomen, lies on its side, i.e. rotated on its axis through 90°. This is shown by the relative courses of the main aedeagal tracheae. In addition, in some, e.g. Adephaga, during eversion of the aedeagus there is a further rotation through 90° in the same direction, whereas in others the rotation at rest is cancelled by a rotation in the opposite direction. In order to avoid confusion that can result from the use of left and right, dorsal and ventral, these terms should be applied to the aedeagus when everted in the position of activity

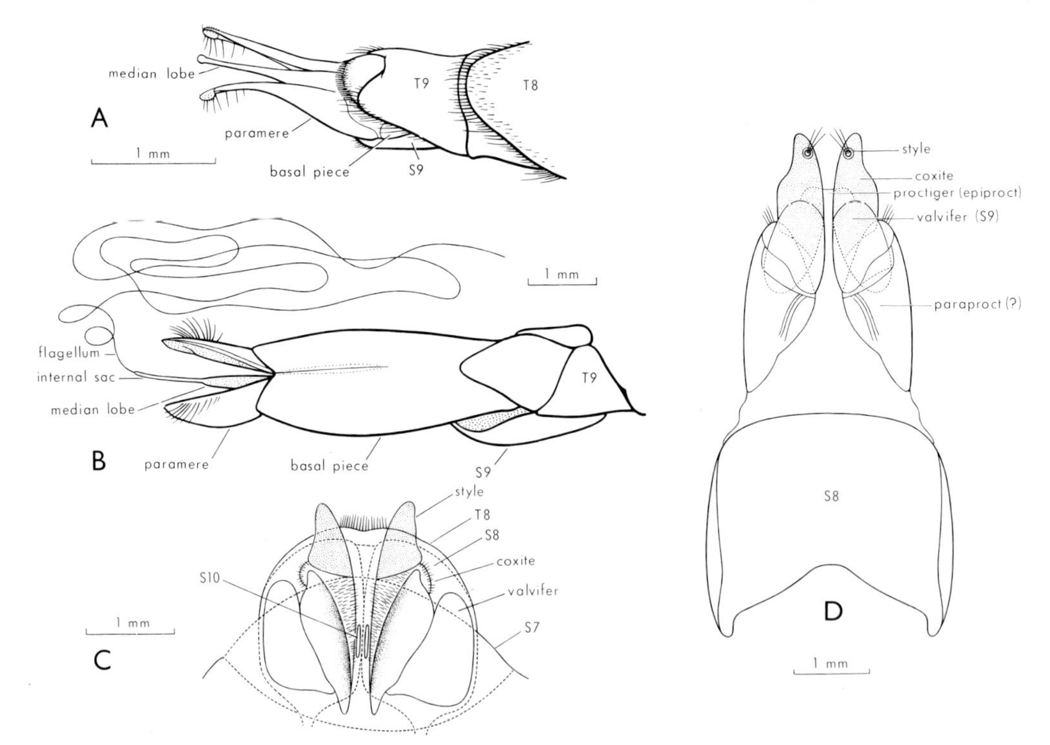

Fig. 30.9. Apex of male abdomen with everted aedeagus: A, *Conoderus* sp., Elateridae; B, *Lamprima aurata*, Lucanidae. Female genitalia, ventral: C, *Calosoma schayeri*, Carabidae, with sternites 7 and 8 removed (shown by broken lines); D, *Ptomaphila lacrymosa*, Silphidae. [F. Nanninga]

(Lindroth and Palmén, in Tuxen, 1956).

The form of the parts of the genitalia and the details of the armature of the internal sac are very varied, and often have considerable taxonomic value.

FEMALE GENITALIA (Fig. 30.8). Segment 9 is largely membranous, and the sternite bears a pair of segmented setose appendages (Fig. 30.9C). Each appendage comprises a sclerotized part of sternite 9, the *valvifer*, to which is articulated a 1- or 2-segmented *coxite* and a terminal setose *style* (Tanner, 1927). The *vulva* is situated between the appendages. An ovipositor may be present, formed either by enlargement of the styles (Dytiscidae), or by the extension of segments 8 and 9 which are telescoped when at rest (e.g. in Cerambycidae).

Internal Anatomy. In the prognathous position, the preoral cavity is formed by the labrum and clypeus above, the hypopharynx and labium below, and the maxillae and mandibles at the sides. The paired salivary glands are often very long, and open by a single duct into the salivarium; they are absent in Adephaga. In most Coleoptera the oesophagus is expanded towards its posterior end to form a crop which is capable of considerable distension. The proventriculus is small, very muscular, bulb-like, and lined with sclerotized ridges or spines. It is well developed in carnivorous and wood-boring forms, but may be absent (e.g. Curculionidae–Rhynchitinae, Attelabidae, Apionidae, etc.). The mid gut varies considerably in length and complexity, from a simple sac to a long, convoluted form as in Scarabaeoidea. Its most characteristic feature is the presence of large numbers of enteric caeca (Fig. 30.10), which may be very long or variable in form. The hind gut is usually convoluted, and a large *rectal pouch* of unknown function opens into it at its junction with the rectum in some Dytiscidae and Silphidae. The ratio of adult gut length to body size is greatest among dung-feeding beetles and least in the Lucanidae, which feed little or not at all.

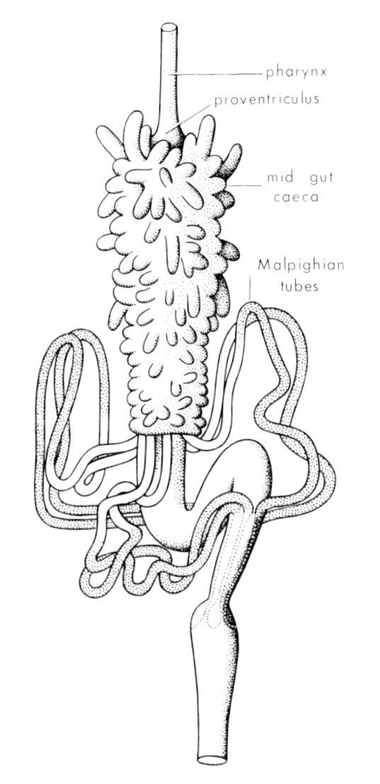

Fig. 30.10. Alimentary canal of *Atomaria* sp., Cryptophagidae (after M. Evans, 1960).

There are 4 or 6 Malpighian tubes. They may all be equal in length, or one pair may be short and the others long, and they occasionally anastomose (e.g. in some Tenebrionidae). The *cryptonephric* condition is common in Coleoptera with 6 tubes, the distal ends of the tubes being applied to, or buried in, the wall of the rectum, instead of remaining free as in other forms (Fig. 30.10).

Paired *anal* or *pygidial* glands occur in most Adephaga and some Staphylinidae. They open near the anus, and, except in a few species, secrete a corrosive, pungent fluid which can be ejected explosively in *Pheropsophus verticalis* Dejean (p. 36). In *Stenus* (Staphylinidae), the species of which live near water, the anal glands have an unusual function. The secretion lowers the surface tension of water, and this serves to propel the insect rapidly to the bank if it should chance to be carried away (Jenkins, 1960).

The number of discrete abdominal ganglia varies from 8 in Elateridae and some Tenebrionidae to 1 in Histeridae, Scarabaeidae, and Curculionidae. This trend reaches its limit in Melolonthinae, in some of which the thoracic and abdominal ganglia are

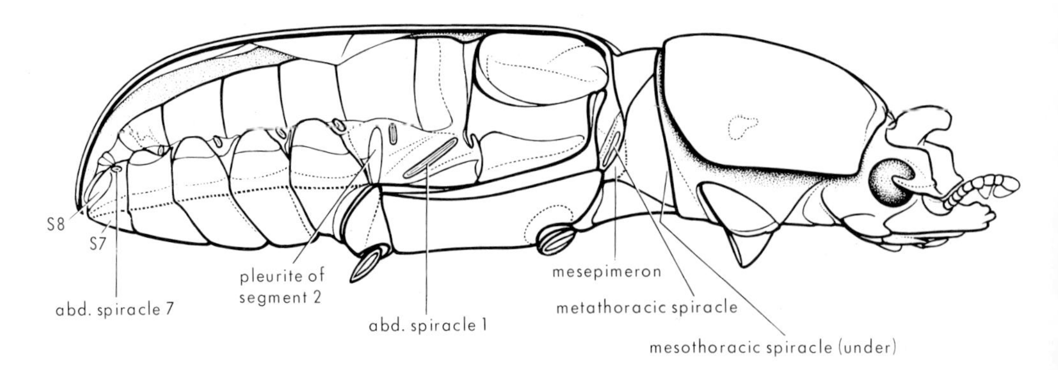

Fig. 30.11. *Aulacocyclus collaris*, Passalidae, right elytron and wing removed to show disposition of spiracles. [F. Nanninga]

fused into a single mass (Peyerimhoff, 1933).

In many flying beetles the tracheal system includes large numbers of air-sacs distributed throughout the body, even including the head. Their main function is to increase the efficiency of ventilation of the tracheal system (p. 39), but their distension also reduces the volume of the haemolymph, resulting in higher concentration of nutrient substances and more efficient supply to the muscles (Wigglesworth, 1963b). It may be noted here that a parallel development of air-sacs has occurred in another group of flying animals, the birds. Air-sacs also provide space which can be occupied by developing organs, such as the ovaries, important in beetles with a rigid exoskeleton. Large air-sacs are found in aquatic beetles and in Carabidae which live in the intertidal zone. It is assumed that these serve mainly for the storage of air.

In the male reproductive system, there are paired accessory glands, and a median ejaculatory duct. The vasa deferentia are sometimes dilated to form vesiculae seminales. The testis in Adephaga consists of a coiled tube, whereas in Polyphaga it usually consists of a number of separate follicles, and when the follicle is single it is not long and coiled (e.g. in *Austrolimnius*; Hinton, 1965). In the female, a *bursa copulatrix* is sometimes present as a modified part of the common oviduct, and there is usually a spermatheca opening into the vagina or bursa copulatrix by a long slender duct. An accessory gland is usually connected with the spermatheca. The ovarioles in Adephaga are polytrophic (nutritive cells between oocytes), whereas in Polyphaga they are acrotrophic (nutritive cells at apex of ovariole).

Immature Stages

Egg. Eggs of Coleoptera appear usually to be simple, ovoid, without surface ornamentation. There are, however, exceptions: the eggs in *Paropsis* (Chrysomelidae) are, for example, ornamented with spines and surface sculpture in a highly specific manner. Hatching is often initiated by special spines (egg-bursters) which are present up to the first larval ecdysis.

Larva. Distinguished from those of other endopterygotes by: ventral abdominal prolegs very rarely present; presence of a well-developed, usually fully sclerotized head capsule; and mouth-parts adapted for chewing (rarely with suctorial mandibles). If without thoracic legs, the head capsule and mouth-parts are either directed ventrally or forwards and downwards, and the body is more or less crescent-shaped, thicker in the middle, in contrast to the larvae of Diptera–Nematocera and of Siphonaptera, in which the head is directed forwards and the body is straight. Curved, legless larvae of Hymenoptera–Apocrita are distinguishable from those of Coleoptera by having the head only slightly sclerotized, or reduced and sunk into the thorax; the larvae of Apocrita also differ by

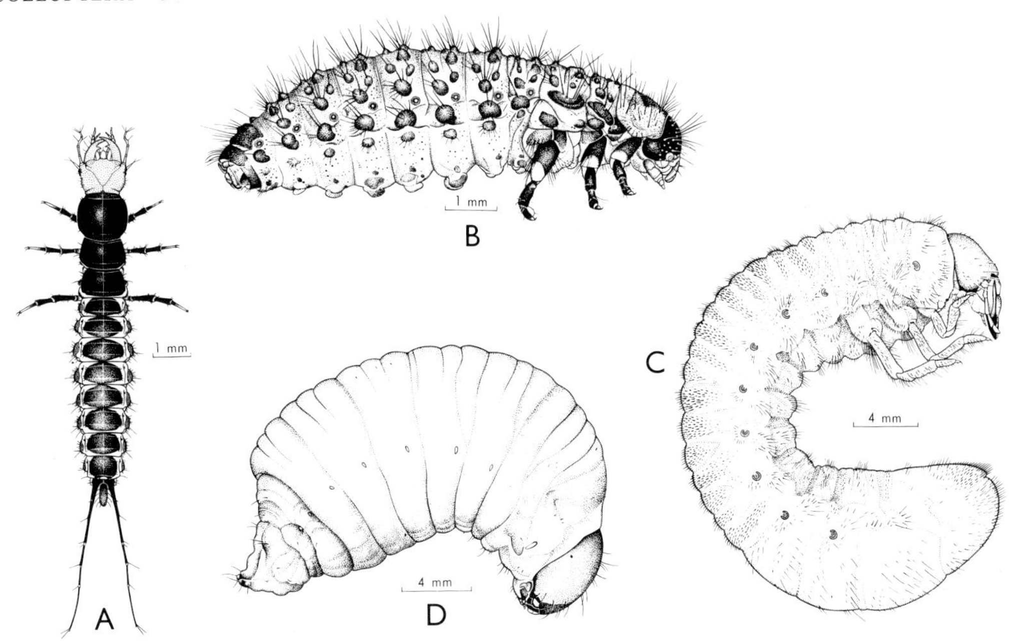

Fig. 30.12. Larval types: A, *Eudalia macleayi*, Carabidae–Odacanthinae; B, *Paropsisterna beata*, Chrysomelidae–Chrysomelinae; C, *Anoplognathus pindarus*, Scarabaeidae–Rutelinae; D, *Trigonotarsus rugosus*, Curculionidae–Rhynchophorinae. [F. Nanninga]

lacking ocelli, and by having antennae in the form of 1-segmented papillae or absent. Larvae of Coleoptera also differ from those of Hymenoptera and Diptera by their lack of labial (salivary) glands. The antennae in the Coleoptera are inserted on the inner side of the anterior ends of the Y-shaped epicranial ecdysial line—in contrast to larvae of Lepidoptera in which they are inserted outside the arms of the line. The mandibles, when suctorial, are distinctly curved (in contrast to larvae of Neuroptera). If the anal segment bears small hooks, then the antennae are 2-segmented (1-segmented in Trichoptera).

The beetle larva has a well-developed head, 3 thoracic segments, and usually 10 abdominal segments (sometimes reduced to 9 or even 8). The thoracic segments are rather similar, although the first is sometimes enlarged and more strongly sclerotized. Abdominal segment 10 is small, surrounds the anus, and sometimes forms a ventral pseudopod. Segment 9 often carries a pair of segmented or unsegmented *urogomphi*, which may be homologous with the cerci of other endopterygotes.

Campodeiform, eruciform, scarabaeiform, and apodous are purely descriptive terms applied to larval types which are associated with different modes of life. *Campodeiform* larvae are active and predatory, with a markedly prognathous head, long thoracic legs, and divergent, setiferous, 1- or many-segmented urogomphi (e.g. larvae of Carabidae (Fig. 30.12A) and Staphylinidae). *Eruciform* larvae are less active, cylindrical larvae, with short legs, and without or with short urogomphi (e.g. Fig. 30.12B). The *scarabaeiform* type is the C-shaped, rather long-legged larva characteristic of Scarabaeidae (Fig. 30.12C), which live buried in soil or rotten wood; whereas the *apodous* type lacks thoracic legs and urogomphi, and has reduced mouth-parts and antennae (e.g. Curculionidae, Fig. 30.12D).

The head may bear up to 6 ocelli (stemmata), or, especially in those larvae that feed in wood or in the soil, ocelli may be absent. The antennae, when well developed (Carabidae), are composed of 3 or 4 segments, and, at the other extreme, are reduced to mere vestiges in the grub-like larvae of Curculionidae. The ecdysial lines of the head consist of a V-shaped epicranial line, open anteriorly, and one or two longitudinal lines beneath (Hinton, 1963a).

The larval tracheal system is usually peripneustic, with 9 pairs of spiracles. The first pair is normally on the anterior margin of the mesothorax, but in some the spiracles migrate on to the prothorax during embryonic development (e.g. in Ptinidae). The remainder are on the first 8 abdominal segments. The number of functional spiracles is invariably reduced in aquatic larvae (Hinton, 1947a). The spiracles show considerable variety of structure. They may be of the simple annular type like those of adults, as in the larvae of Adephaga and many Staphylinoidea, or they may be *cribriform* or *biforous*. In the cribriform type, which occurs in the larvae of the Scarabaeoidea, Dascillidae, and Buprestidae, the original aperture is represented by a crescent-shaped scar, and is closed except during ecdyses. Around this is a broad crescentic plate which is perforated by minute holes, through which air diffuses into the atrial chamber.* The term biforous was originally applied to spiracles which had two external openings. Spiracles of this type are now known to have two or more external apertures. The atrium has two pouches, each of which has its own aperture. The original atrial aperture is closed except during ecdysis.

The legs are 6-segmented in Adephaga (coxa, trochanter, femur, tibia, tarsus, and 1 or 2 claws), whereas in Polyphaga there are 5 or fewer segments, including a single claw, or legs are absent. Jeannel (1949b) believed that the segment following the femur in the larvae of Adephaga had disappeared in Polyphaga. The fact that each claw of the pair in some Carabidae bears a single seta, and the single claws of the larvae of other Carabidae bear 2 setae suggests that the single claw represents 2 fused claws. The terminal segment of the leg of Polyphaga larvae normally bears 2 setae near the middle. This suggests that this segment is homologous with the fused tarsus and claw(s) of Adephaga, and so should be referred to as a *tarsungulus* rather than tibiotarsus.

The illustrated synopsis by Böving and Craighead (1931) is a comprehensive study of beetle larvae.

Pupa. Always adecticous and nearly always exarate (Fig. 30.13B). Obtect pupae are found in Staphylinidae–Staphylininae (Hinton, 1946b). The abdomen usually includes 9 terga and 8 sterna, and the number of spiracles is usually reduced in comparison with the larva. The spiracles of abdominal segment 8 are always non-functional. The head and body bear, as a rule, various prominences and setae, which are peculiar to the pupal stage, and which serve to hold the pupa away from the walls of the pupal cell. Many beetle pupae are equipped with supposed organs of defence against arthropod predators in the shape of the so-called 'gin-traps' (Hinton, 1955a). These are formed by the local sclerotization of opposable edges of adjacent abdominal segments. Median dorsal gin-traps occur in the pupae of Dermestidae (Fig. 30.13A), Scarabaeidae, Cerambycidae, and Coccinellidae, and paired lateral gin-traps are found in many Tenebrionidae and Colydiidae.

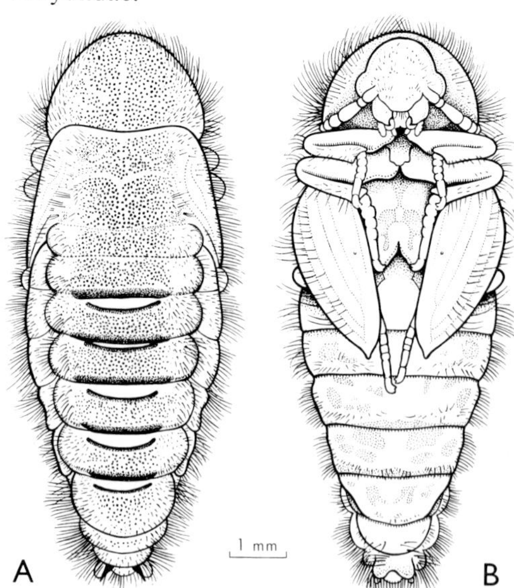

Fig. 30.13. Pupa of *Dermestes maculatus*: A, dorsal; B, ventral. [F. Nanninga]

* H. E. Hinton, *Aust. J. Zool.* **15**: 947–53, 1967.

Some Coleoptera that have a spinose or setose larva pupate within the larval skin, which acts as a protective covering (e.g. *Anthrenus*). In others the larval skin remains attached to the abdomen of the pupa (e.g. Chrysomelidae–Cassidinae). Many Coleoptera pupate in earthen cells below ground or within the food plant. A cocoon is frequently constructed, but little is known of the nature and origin of the material used. Cetoniinae and Passalidae use faecal material, and in some Cerambycidae and Curculionidae (e.g. *Leptopius*) the cocoon is hard and calcified.

Biology

General. Beetles show so many diverse adaptations to such a wide variety of habitats and ways of life that it has not been possible to follow quite the same sequence of subjects as in other chapters. They are found, both as adults and larvae, on plants of all kinds, in flowers, in plant debris, in soil, in moss in swamps, in grass tussocks, under stones and fallen logs, beneath the bark of trees, in galls; in fungi, especially bracket fungi on trees; in hollow stems; in carrion, and in the nests of ants, termites, wasps, bees, birds, and mammals; in stored foodstuffs; in dung; in and on the wood of dead, burnt, or dying trees; in fresh water, from temporary pools to mountain torrents; in brackish water (e.g. salt-lakes); beneath high-water debris on the seashore; in rock clefts in the intertidal zone; in sand, mud, and gravel at the margins of creeks and rivers; and in caves. Cavernicolous species are rare in Australia, and none so far discovered shows any special adaptation to cave life. This suggests that ancient cave faunas here may have become extinct, possibly as the result of desiccation, and that reinvasion of caves has only recently commenced (Moore, 1964).

Although most beetles are winged, their flight is, in general, not sustained, and usually occurs only when the temperature is high.

Most beetles are phytophagous,* but almost all Carabidae, Dytiscidae, and Gyrinidae prey, as both larvae and adults, upon insects and other small animals. The Silphidae, Staphylinidac, Histeridae, Lampyridae, Cantharidae, Cleridae, Meloidae, and Coccinellidae are also very largely carnivorous. Most of these are general feeders, but a few are limited in their choice of prey. The Coccinellidae, for example, feed largely upon Aphididae, Coccidae, and Aleyrodidae; the Lampyridae upon snails and earthworms; and the Meloidae on the eggs of grasshoppers and the larvae of bees. The Rhipiphoridae are the only family in which the larvae of all species are parasitic; otherwise the parasitic habit has been observed in a relatively few representatives of the Carabidae, Staphylinidae, Colydiidae, and Coccinellidae. A few beetles live as ectoparasites on mammals. These include a few species of Leptinidae (*Platypsyllus castoris* on the beaver in Europe and North America) and Staphylinidae–Staphylininae–Amblyopinini (including *Myotyphlus*, recorded as a parasite of rats in Tasmania). The small coprine beetles of the genus *Macropocopris*, although not parasitic, are found on wallabies, clinging to the hair near the anus (p. 123).

Some Coleoptera, both adults and larvae, are adapted for extra-oral digestion. This occurs in the Dytiscidae, Hydrophilidae, Lampyridae, and Drilidae. In the larvae of Dytiscidae each of the long, curved mandibles has a longitudinal, almost tube-like channel, which leads to a slit-like aperture at each side of the otherwise closed preoral cavity. The larva seizes its prey with the mandibles, and injects a secretion of the mid gut which digests the flesh. The liquid products of digestion are taken up through the mandibles by a pumping action of the pharynx. In some Hydrophilidae (e.g. *Hydrophilus*) the digestive secretion is applied externally from the mouth, while the prey is held by the mandibles and masticated above the surface of the water. It is subsequently eaten in the normal way. In the Lampyridae and Drilidae the mandibles of the larva are pierced by a canal, and are used to inject a fluid which serves the double function of paralysing the prey, a snail, and of digesting its tissues. The resulting fluid is

* J. Brooks (1948, 1965a) has recorded the host plants of 308 north Queensland species.

imbibed with the aid of the other mouthparts, and not through the mandibles as in Dytiscidae.

Relationship with Symbiotes. Many Coleoptera harbour aggregations of symbiotic yeasts or bacteria, the type of micro-organism being constant in each species of beetle. The symbiotes may be extracellular in the lumen of the gut or its processes, or they may be intracellular in the cytoplasm of fat-body cells known as *mycetocytes* which may be aggregated into special organs known as *mycetomes* (Koch, 1960). The gut mycetomes are commonly situated around the anterior end of the mid gut, and probably evolved as detached intestinal crypts. Stages in the evolution of crypt to mycetome can, in fact, be seen in the larvae of various species of Anobiidae (Gräbner, 1954). The importance of the symbiotes to the insect is indicated by the fact that they appear in all individuals, larval and adult, of the species in which they occur.

Symbiotes are observed most frequently in forms that feed upon such materials as wood, hair, wool, feathers, humus, and dry cereals. Such forms are included mainly in the Anobiidae, Lyctidae, Bostrychidae, Cucujidae, Lagriidae, Buprestidae, Cerambycidae, and Tenebrionidae. Experimental evidence from insects artificially deprived of their symbiotes appears to confirm the view that, in general, the micro-organisms supply a nutritional requirement (M. Brooks, 1963). The importance of extracellular micro-organisms is less obvious, but flagellates are very abundant in the larvae of some Scarabaeidae. In the larvae of Cetoniinae the hind gut is greatly enlarged and sac-like, and contains an abundant special fauna which is probably concerned with the digestion of cellulose.

The mechanisms by which the symbiotes are transmitted to the next generation of the host are varied. In Anobiidae the symbiotic yeasts contaminate the surface of the egg and are eaten by the emerging larva, whereas in *Lyctus* and *Oryzaephilus* the symbiotes enter the egg before the secretion of the chorion. In *Rhizopertha* the symbiotes invade the testes and pass with the sperm to the female, in which they infect the eggs through the micropyle.

A highly specialized symbiosis is found in 'ambrosia beetles' or 'pinhole borers', which are diverse Curculionidae of the subfamilies Hylesinae (some), Scolytinae (many), and Platypodinae (all). Scolytine bark beetles attack living and newly felled trees, and feed upon the living tissues under the bark—phloem, cambium, and outer sapwood (Baker, 1963). This food is highly nutritious, so that these beetles do not depend on internal symbiotes. The ambrosia beetles among the scolytines have, however, adopted the habit of boring into the wood, and in so doing they introduce the ambrosia fungi, which grow on the tunnel walls and serve as food for larvae and adults. The fungi are always associated with the beetles, and are transmitted from tree to tree by the adult, which often has specialized organs for the purpose. In general, the ambrosia beetles are more numerous than bark beetles in tropical forests, whereas the reverse is true in temperate forests. The Lymexylidae have also, quite independently, evolved the habit of ambrosia culture.

Luminosity. The ability to produce light is rare in insects. Otherwise limited to a few mycetophilid Diptera, it is restricted within the Coleoptera to the Lampyridae, Phengodidae, and Elateridae. Luminous Elateridae are numerous in South America, but are otherwise found only in Fiji (*Photophorus jansoni* Cand.) and the New Hebrides (*P. bakewelli* Cand.). Phengodidae, with about 50 species, are limited to the Americas. The Lampyridae (fireflies), on the other hand, are widely distributed throughout the world, and are represented in Australia by *Luciola* and *Atyphella.* Unlike some other members of the family, most Australian species are winged in both sexes. The light is produced in both sexes by a special organ (p. 60) on the underside of the abdomen. Its function is to bring the sexes together, although eggs and larvae are also somewhat luminous. The light is pale yellowish green in colour, and the amount of heat produced is negligible.

Brightness is apparently controlled by regulation of the oxygen supply via the tracheae.

Sound Production. Stridulation is effected by rubbing a file-like area on one part of the exoskeleton with an adjoining part (the *plectrum*). It is widely distributed throughout the Coleoptera, occurring more commonly in adults than in larvae, and is achieved in a variety of ways in various groups. The stridulatory organ is usually equally developed in both sexes.

Stridulation is well developed in the larvae of Lucanidae, Passalidae, and Geotrupidae. In the larvae of Lucanidae the file, consisting of one or more longitudinal rows of tubercles, is on the coxa of the mid leg, and is rubbed by a plectrum in the form of a ridge on the trochanter of the hind leg. The mechanism is similar in the larvae of Passalidae and Geotrupidae, but the hind leg is reduced to a mere stump the only function of which is to act as a plectrum. Adult Scarabaeidae *(Trox)* stridulate by rubbing a transverse ridge on the abdomen over a file on the underside of each elytron, whereas some Dynastinae have files on the propygidium which can be rubbed against the elytra. Many Cerambycidae produce sound by rubbing the hind edge of the pronotum over a file on the mesonotum; in others the hind femora are rubbed against the edges of the elytra. A file on the dorsal or ventral surface of the head rubbed on an inner ridge of the pronotum is employed by some Endomychidae, Nitidulidae, Erotylidae, Curculionidae–Scolytinae, and Chrysomelidae–Hispinae. In the Heteroceridae the adult beetles stridulate by rubbing a ridge on the hind femora over a file on the first visible sternite.

The functions of stridulation in Coleoptera have not been satisfactorily explained (Arrow, 1942). If it serves as a medium of communication between individuals, appropriate organs of hearing are implied, and these are not obvious. The sound may, however, be secondary to the production of vibration in the substrate, which could have a defensive function. The conversion of the normal hind legs into stridulatory organs in the larvae of Geotrupidae and Passalidae suggests that sound production is of considerable importance to these insects. Geotrupid larvae live solitary lives in burrows provisioned by the adults. It is therefore difficult to imagine any function for the sound other than defence, but it is also difficult to imagine an enemy against which the sound or vibration would be effective. The function of stridulation in Passalidae may be different. These beetles appear to live in family groups, and Ohaus maintained that stridulation by adults and larvae served to keep the family group intact, and that the larvae stridulated when hungry and were fed by the adults. Heymons (1929) has refuted this, but Pearse *et al.* (1936) have shown that, whereas well-grown larvae can be raised independently, newly hatched larvae require food material previously treated by adults. Additional observations on the Australian species are desirable.

Adaptations to Aquatic Life. About 5,000 species (about 1·7 per cent of the world fauna of beetles) are more or less adapted to life in fresh water, and the Australian species total about 415. Almost all species of Amphizoidae, Haliplidae, Noteridae, Dytiscidae, Gyrinidae, Sphaeriidae, Hydraenidae, Dryopidae, Helminthidae, and Georyssidae are adapted for life in water both as adults and larvae. Hydrophilidae also include a large number of aquatic forms. In the Helodidae, Psephenidae, Ptilodactylidae, and Limnichidae, the larvae are truly aquatic, and the adults, although normally terrestrial, live near water which some at least are able to enter. Adults of the 3 Australian species of the curculionid genus *Bagous* live under water, feeding upon aquatic plants, and are able to swim. They are reported to respire with the aid of bubbles of oxygen evolved by plants, or alternatively by tapping the air in plant stems. In addition, a relatively few species of some other families, e.g. Carabidae, Staphylinidae (see p. 546), Heteroceridae, Ptiliidae, live in or on wet mud or gravel close to the water's edge. A few species live in the intertidal zone in other parts of the world, and, although none has yet been recorded, might be expected to occur in Australia, more particularly in Tasmania.

For example, species of Aepini, a tribe of the Carabidae–Trechinae, live in crevices in rocks and under stones and seaweed.

Adults of the adephagan water-beetle families (Amphizoidae, Haliplidae, Noteridae, Dytiscidae, Gyrinidae) take up air at the surface and store it beneath their elytra, whereas many aquatic Polyphaga have areas of the body covered with a very dense hydrofuge pubescence which retains a film of air when the beetle is submerged. This film is continuous with the air beneath the elytra into which the spiracles open. The film of air functions as a gill, taking up oxygen and losing carbon dioxide to the water (Thorpe and Crisp, 1949).

Among aquatic larvae there is a marked tendency for the later instars to be less strictly aquatic than the earlier instars, and the converse is never true. The same tendency is shown by the fact that aquatic adults always have aquatic larvae, but aquatic larvae may have terrestrial adults. The final larval instar has functional spiracles in order to be able to leave the water to pupate, whereas the earlier instars are commonly without functional spiracles, and remain beneath the surface, extracting oxygen from the water by means of abdominal or tracheal gills or the thinner parts of the body cuticle. In the Dytiscidae larvae of all instars come to the surface to obtain air through the terminal abdominal spiracles, and, although the final instar has other functional spiracles, these are not used until the larva leaves the water to pupate. Non-functional spiracles, consisting of a surface scar, are non-functional only in that they are not open to the passage of air; they still have an important function at ecdysis (p. 38).

Aquatic larvae leave the water to pupate, and no Australian Coleoptera are known to have truly aquatic pupae. Noteridae pupate under water, but the pupa is enclosed in a gas-filled cocoon. The only truly aquatic pupae known are those of the Psephenidae–Psephenoidinae. Here the water is in contact with the pupal cuticle and the pupa breathes by means of spiracular gills (Hinton, 1947a).

Reproduction and Life Histories. The antennae of male Coleoptera are often larger or more elaborate than those of the female. This is correlated with the emission of a pheromone by the female, which stimulates the male to fly upwind in search of a mate (e.g. in *Rhopaea*). In copulation the male commonly assumes a position on the back of the female, using his fore tarsi and sometimes the mandibles to hold him in position. The necessity to curve the abdomen down and forwards to reach the apex of the abdomen of the female is often reduced or obviated by a curvature of the aedeagus itself. The median lobe alone is inserted into the vagina, the internal sac being subsequently everted. In some beetles, however, the median lobe is of such a shape that penetration of the vagina is impossible. In these the penetration is accomplished by the everted internal sac alone. In most Coleoptera the males may copulate with more than one female, and the females similarly copulate with more than one male. Monogamy does, however, occur in a few groups, particularly among the coprophagous Scarabaeidae, in which the couple collaborate in building and provisioning the nest. The more lasting relationship, in which the couple remain together and feed their larvae, has been described by Ohaus in Passalidae. Polygamy is very rare, but occurs in certain genera of Scolytinae.

The number of eggs deposited varies greatly throughout the order, from the single large egg of *Sphaerius* to the many thousands of minute eggs produced by Meloidae.

Subsocial behaviour, i.e. parental care of eggs or larvae, is known only in 9 families (Hinton, 1944). An example is provided by the ambrosia beetles, already mentioned, which cultivate a fungus on which the larvae feed. Geotrupidae and Scarabaeidae–Scarabaeinae carry vegetable debris or dung into deep burrows to provide stores of food for the future use of the larvae. In some Scarabaeinae (Scarabaeini–Canthonides, e.g. *Canthonosoma*, *Cephalodesmius*, and the introduced Mexican species *Canthon humectus* Say) the male and female co-operate to form a ball of dung and roll it to a suitable place for burial as a larval store. The subject of parental care in Coleoptera has been reviewed by von Lengerken (1954).

Atypical life histories have evolved in various families, although not all of them have been observed in Australian species.

1. The virtual elimination of the free-living larva as a growing stage has been reported in some European cave beetles, e.g. some *Speonomus* (Silphidae) (Deleurance, 1958). The adult female lays very large eggs, singly, at long intervals. The larva hatches fully grown in the last instar; after building a pupal cell, it goes into diapause without feeding, and subsequently pupates. Although this displacement of the growth stage has so far been reported only in some European cave beetles, it is likely, in view of the apparent rarity of the larvae, that it occurs in cave species elsewhere.
2. Viviparity, with the production of small, fully formed, active larvae, has been described in some European genera of Chrysomelidae, and may occur in some Australian genera.
3. Larval heteromorphosis occurs in the Micromalthidae, Meloidae, and Rhipiphoridae, in which the first instar is active, whereas later instars are grub-like. These larvae are specialized predators or parasites of other insects. Heteromorphosis is also seen in Drilidae and in some Carabidae and Staphylinidae.
4. Paedogenesis has not been observed in any Australian species, but has been described by Mjöberg in *Duliticola*, a genus of Lycidae which is known from the Indo-Malayan subregion. Larviform females occur in the Lampyridae and Drilidae; but they are not neoteinic, being distinguished from those of *Duliticola* by the fact that the females emerge from pupae, and are adult in all respects but for the lack of elytra and wings. In contrast, the female in *Duliticola* has the single-clawed, 5-segmented legs of a polyphagan larva, 4-segmented antennae, and simple eyes, and does not pass through an obvious pupal stage. Paedogenesis also occurs in *Micromalthus* (p. 519).
5. Parthenogenesis appears to be rare in Coleoptera, but does occur in some Curculionidae (e.g. *Otiorrhynchus*, *Listroderes*), Chrysomelidae, and Dermestidae. Pseudogamous parthenogenesis occurs in the triploid form (*P. mobilis* Moore) of *Ptinus clavipes* Panz. (Woodroffe, 1958). This form consists of females only. They reproduce only after mating with males of *P. clavipes*, *P. pusillus*, or *P. fur*. The offspring of these crosses are all female and of the form *mobilis*.

Natural Enemies. Coleoptera are subject to attack by the usual range of predators and parasitic animals. Eggs are parasitized by Chalcidoidea, and larvae are preyed upon by spiders, wasps, and vertebrates, and parasitized by Chalcidoidea, Braconidae, Ichneumonidae, and Tachinidae. Larvae living in the soil are attacked by mites and mermithid worms. Scolioid wasps of the subfamily Thynninae, otherwise known only in South America, and the large rutiliine Tachinidae (Diptera) are common parasites of the soil-dwelling larvae of Scarabaeidae in Australia, and Pyrgotidae (Diptera) parasitize adult Scarabaeidae. Aquatic Coleoptera form an appreciable part of the food of fish, and most amphibia and many reptiles and birds feed to a large extent upon insects of which Coleoptera form a considerable proportion. Beetles are, in addition, subject to disease caused by fungi, bacteria, and viruses, e.g. the Sericesthis iridescent virus (SIV) which was first found attacking larvae of *Sericesthis geminata* Boisd. in northern N.S.W.

Economic Significance. The economic importance of Coleoptera results almost entirely from the fact that some species consume or damage materials of value to man. Coleoptera are of little or no importance in the transmission of diseases of plants, animals, and man, and they do not attack larger animals.

Materials attacked by Coleoptera include food plants of all kinds, trees, and a great variety of natural products, both stored and in use, including foodstuffs, clothing, leather, crude drugs, tobacco, and wood. Some of the beetle pests of stored products are able to breed in a considerable variety of materials, including cereals, dried fruit, spices, and tobacco. Examples are *Stegobium paniceum* L. (Anobiidae) and *Ptinus tectus* Boield. (Ptinidae) which can subsist on a great variety of dry foodstuffs. Some pest species, however, are mainly restricted to particular kinds of products, e.g. *Dermestes* spp. which attack dry animal products of high protein content and *Sitophilus* spp. (Curculionidae) which infest whole grains of wheat, rice, barley, or maize. It is a characteristic of pests of stored materials that they are well adapted to a very dry environment. About 400 species of Coleoptera have been found in stored products. Only a small proportion of these are true pests, in that they feed directly on the stored material. The majority feed upon vegetable debris, or moulds, or are carnivorous, preying upon other insects, and the remainder are of accidental occurrence. Beetles associated with stored products have been reviewed by Busvine (1966), Hinton (1945), and Lepesme (1944).

Species of *Paederus* (Staphylinidae) are known to cause severe human skin lesions ('whip-lash dermatitis', p. 546), and Meloidae and Oedemeridae contain cantharidin, a highly toxic substance which also causes blistering. Deaths have occurred as the result of eating meloids, and drinking liquids in which oedemerids have been accidentally immersed has been known to produce severe symptoms.

A great many species of Coleoptera prey upon other insects and for this reason are indirectly of great value to man. Predatory species have been used for the control of insect pests, as, for example, the introduction of Coccinellidae into California and Hawaii to control scale insects (p. 593). Coleoptera are also of great importance by virtue of the role of many species in breaking down, consuming, or burying plant and animal remains, including dung. Even plant-feeding Coleoptera may be beneficial in controlling noxious weeds, and some species have been employed for this purpose (e.g. *Chrysolina geminata* Suffr. for the control of St John's wort in Victoria).

Special Features of the Australian Fauna

Of the major families, the Curculionidae, the dominant family throughout the world, is well represented in Australia. The world total of Curculionidae is near 60,000 species,* or about 21 per cent of all known species of Coleoptera. The degree of development of the more important families in Australia in comparison with the world fauna is shown in the following table.

Family	*Australian Species* (*No. of species as % of total*)	*World Species* (*No. of species as % of total*)
Curculionidae	21	21
Chrysomelidae	10.5	7
Scarabaeidae	10·5	6
Carabidae	8	9
Tenebrionidae	6	5·5
Cerambycidae	5·5	7
Buprestidae	4	4
Staphylinidae	3·5	9·5

The relative preponderance of the Scarabaeidae in Australia is the result of an abundance of species of Melolonthinae which more than offsets a deficiency of species in the other subfamilies. Tenebrionidae are well represented, as might be expected of a family which flourishes in arid areas elsewhere. Staphylinidae are, on the other hand, poorly represented. This again probably reflects climatic conditions, as this family is particularly abundant in damp climates.

The degree of endemism is high, a great many genera and many tribes being limited to Australia. In the scarabaeid subfamily Melolonthinae, for example, about 80 genera are represented in the fauna, only 2 of which are found elsewhere. One of these, *Lepidiota*, has species widely distributed in India and S.E. Asia, but extends into Australia only as far as north Queensland. This illustrates the limited Indo-Malayan element that has penetrated into Australia in relatively recent times. The Antarctic (Bassian) element is clearly distinguishable. This is best seen in those families in which the classification is soundly established, as, for example, the Carabidae. The Australian representatives of the carabid subfamilies Migadopinae, Broscinae, Trechinae, and Psydrinae have closely related forms in Chile, Patagonia, and New Zealand. Of the 14 genera of Migadopinae, 7 are found in the southern half of

* Estimated from the total number of species (34,500) listed in the *Catalogus Coleopterorum* (1910–1939) and the British Museum card catalogue of species described since the publication of the *Catalogus*.

South America, mostly in Tierra del Fuego and the Falkland Is., 4 are Australian, and 3 occur in New Zealand. Similarly, the curculionoid family Belidae is restricted to Chile (2 genera), Australia (6 genera), and New Zealand (1 genus). Within Australia the species of Antarctic origin are limited very largely to the cooler and wetter parts of the continent, notably the temperate rain forests of the south-east and Tasmania.

CLASSIFICATION

In this Table the first figure within the brackets is the approximate number of species (totals rounded off) known in the world and the second is the number of known Australian species.

Order COLEOPTERA (278,000; 19,219)

Suborder ARCHOSTEMATA (26; 6)

CUPEDOIDEA (26; 6)

1. Cupedidae (25; 6)
Micromalthidae (1; 0)

Suborder MYXOPHAGA (22; 2)

SPHAERIOIDEA (22; 2)

2. Sphaeriidae (11; 2)
Lepiceridae (1; 0)
Hydroscaphidae (9; 0)
Torridincolidae (1; 0)

Suborder ADEPHAGA (30,200; 1,849)

CARABOIDEA (30, 200; 1,849)

3. Rhysodidae (125; 9)
4. Carabidae (25,000; 1,613)
5. Haliplidae (200; 5)
6. Hygrobiidae (4; 2)
Amphizoidae (5; 0)
7. Noteridae (150; 4)
8. Dytiscidae (4,000; 188)
9. Gyrinidae (700; 28)

Suborder POLYPHAGA (247,800; 17,362)

Staphyliniformia

HYDROPHILOIDEA (2,400; 127)
10. Hydraenidae (300; 20)
11. Hydrochidae (69; 19)
12. Spercheidae (17; 1)
13. Georyssidae (23; 3)
14. Hydrophilidae (2,000; 84)
HISTEROIDEA (2,500; 144)
Sphaeritidae (3; 0)
Synteliidae (4; 0)
15. Histeridae (2,500; 144)
STAPHYLINOIDEA (35,100; 1,336)
16. Limulodidae (33; 6)
17. Ptiliidae (300; 51)
Dasyceridae (9; 0)
Leptinidae (7; 0)
18. Anisotomidae (1,100; 27)
19. Scydmaenidae (1,200; 80)
20. Silphidae (200; 3)
21. Scaphidiidae (300; 19)
22. Staphylinidae (27,000; 650)
23. Pselaphidae (5,000; 500)

Scarabaeiformia

SCARABAEOIDEA (18,800; 2,332)
24. Lucanidae (750; 75)
25. Passalidae (490; 34)
26. Trogidae (167; 41)
27. Acanthoceridae (120; 1)
28. Geotrupidae (300; 81)
29. Scarabaeidae (17,000; 2,100)
DASCILLOIDEA (495; 71)
30. Clambidae (61; 11)
*Eucinetidae (24; 0)
31. Helodidae (360; 53)
32. Dascillidae (50; 7)

Elateriformia

BYRRHOIDEA (300; 45)
33. Byrrhidae (270; 44)
34. Nosodendridae (30; 1)
DRYOPOIDEA (1,053; 100)
35. Limnichidae (70; 8)
36. Psephenidae (22; 1)
37. Ptilodactylidae (174; ?)
38. Heteroceridae (150; 9)
Eurypogonidae (45; 0)
*Chelonariidae (42; 0)
Dryopidae (250; 0)
39. Helminthidae (300; 82)
BUPRESTOIDEA (11,500; 800)
40. Buprestidae (11,500; 800)
RHIPICEROIDEA (180; 15)
41. Rhipiceridae (65; 11)
42. Callirhipidae (115; 4)
ELATEROIDEA (8,200; 690)
43. Elateridae (7,000; 608)
44. Eucnemidae (1,000; 78)
Cerophytidae (5; 0)
Perothopidae (3; 0)
45. Trixagidae (200; 4)
CANTHAROIDEA (12,400; 572)
Brachypsectridae (1; 0)
Telegeusidae (3; 0)
Homalisidae (19; 0)
Karumiidae (4; 0)
Drilidae (80; 0)
Phengodidae (50; 0)
46. Lampyridae (1,700; 16)
47. Cantharidae (3,500; 100)
48. Lycidae (3,000; 206)
49. Melyridae (4,000; 250)

Bostrychiformia

DERMESTOIDEA (819; 91)
50. Dermestidae (731; 91)
Derodontidae (10; 0)
Thorictidae (74; 0)
*Sarothriidae (4; 0)
BOSTRYCHOIDEA (2,300; 206)
51. Anobiidae (1,100; 100)
52. Ptinidae (700; 61)
53. Bostrychidae (434; 40)
54. Lyctidae (70; 5)

Cucujiformia

CLEROIDEA (4,000; 327)
55. Trogossitidae (600; 27)
Chaetosomatidae (3; 0)
56. Cleridae (3,400; 300)
Phloeophilidae (1; 0)
LYMEXYLOIDEA (37; 9)
57. Lymexylidae (37; 9)
CUCUJOIDEA (41,100; 3,029)
58. Nitidulidae (2,200; 120)
Smicripidae (6; 0)
59. Rhizophagidae (200; 8)
*Protocucujidae (1; 0)
*Sphindidae (20; 0)
Hypocopridae (8; 0)
60. Passandridae (98; 6)
61. Cucujidae (500; 50)
62. Silvanidae (400; 42)
Helotidae (80; 0)
63. Phycosecidae (4; 3)
*Propalticidae (3; 0)
64. Cryptophagidae (800; 23)
65. Biphyllidae (200; 21)
Byturidae (11; 0)

* These families are now known to be represented in Australia (Crowson, *in litt.*).

66. Languriidae (400; 8)
67. Erotylidae (1,500; 81)
68. Phalacridae (500; 72)
69. Cerylonidae (186; 14)
70. Corylophidae (300; 37)
71. Coccinellidae (5,000; 260)
72. Endomychidae (1,100; 32)
Discolomidae (30; 0)
73. Merophysiidae (77; 1)
74. Aculagnathidae (1; 1)
75. Lathridiidae (520; 34)
76. Ciidae (250; 13)
77. Merycidae (2; 2)
78. Mycetophagidae (200; 6)
79. Colydiidae (1,400; 100)
Pterogeniidae (1; 0)
80. Tenebrionidae (15,000; 1,227)
81. Lagriidae (400; 29)
82. Alleculidae (1,100; 199)
83. Zopheridae (100; 1)
Monommidae (200; 0)
Tetratomidae (19; 0)
Perimylopidae (10; 0)
†Elacatidae (26; 1)
84. Inopeplidae (55; 3)
85. Salpingidae (276; 20)
Cononotidae (3; 0)
86. Mycteridae (63; 10)
87. Hemipeplidae (16; 1)
Trictenotomidae (12; 0)
88. Pythidae (20; 6)
*Pyrochroidae (116; 0)
89. Melandryidae (600; 12)
90. Scraptiidae (200; 16)
91. Mordellidae (650; 123)
92. Rhipiphoridae (250; 57)
93. Oedemeridae (1,500; 84)
Cephaloidae (11; 0)
94. Meloidae (2,000; 60)
95. Anthicidae (1,800; 172)
96. Aderidae (700; 74)
CHRYSOMELOIDEA (41,200; 3,137)
97. Cerambycidae (20,000; 1,040)
98. Bruchidae (1,200; 14)
99. Chrysomelidae (20,000; 2,083)
CURCULIONOIDEA (65,300; 4,331)
100. Nemonychidae (18; 1)
101. Anthribidae (2,400; 57)
102. Belidae (176; 156)
Oxycorynidae (20; 0)
Proterrhinidae (120; 0)
103. Attelabidae (310; 68)
104. Brenthidae (1,230; 33)
105. Apionidae (1,060; 31)
106. Curculionidae (60,000; 3,985)

* These families are now known to be represented in Australia (Crowson, *in litt.*).
† *Othnius delusa* Pasc., recorded from Cairns (J. Brooks, 1965b).

The development of the classification of Coleoptera has been reviewed by Fowler (1912) and by Leng (1920). The primary division into the suborders Adephaga and Polyphaga was proposed by Ganglbauer (1903), and the suborder Archostemata was erected by Kolbe (1908). This was confirmed by the important work of Böving and Craighead (1931) based on larval morphology. Peyerimhoff (1933) followed with a survey of the characters significant in the classification of Coleoptera and a critical evaluation of the system of Böving and Craighead. Jeannel and Paulian (1944) proposed a classification based on the degree of regression of the basal abdominal sternites. This divided the Polyphaga into two 'suborders', Heterogastra (= Symphiogastra of Kolbe) and Haplogastra. The suborder Heterogastra is characterized by having sternite 2 complete or completely lost in the coxal cavity, whereas in Haplogastra (composed of Staphylinoidea, Histeroidea, Hydrophiloidea, and Scarabaeoidea) sternite 2 is visible only as a lateral rudiment and pleural sclerite.

Crowson (1955) synthesized earlier work on the classification, with much original observation, and proposed a system which, with a few minor exceptions, is followed here. He tentatively proposed the establishment of a fourth suborder, the Myxophaga, to include 3 families, each of a very few species, the Lepiceridae, Sphaeriidae, and Hydroscaphidae. The description of a fourth family (Torridincolidae) by Steffan (1964) and the larva of *Sphaerius* (Britton, 1966) has confirmed the validity of the union of these families. The suborder Myxophaga is therefore adopted here, and is characterized on p. 520.

Authorities differ on the position to be assigned to the Strepsiptera. The group has certain striking similarities to the cucujoid family Rhipiphoridae: e.g. both have metathoracic flying wings (in the male); reduced elytra; berry-like eyes; flabellate antennae in the male and simple female antennae; a planidium larva, heteromorphosis, and an endoparasitic habit. On the other hand, Strepsiptera differ from Rhipiphoridae in having isomerous tarsi, in lacking trochanters, and in venation of the hind wing (p. 622). Crowson (1960) has treated them as a superfamily, Stylopoidea, of Coleoptera; Bohart (1941) placed them in a separate order closely related to Coleoptera; Pierce (1936) regarded them as an order far removed from Coleoptera; and Jeannel (1945) thought that they were more closely related to Hymenoptera. In view of this lack of agreement, the group is here treated conservatively as an order separate from Coleoptera.

The great bulk of the Coleoptera remaining after separation of the Archostemata and Myxophaga is clearly divisible into two natural suborders of very different size, the

Adephaga and the Polyphaga. Adephaga are relatively uniform, and include only a single superfamily, the Caraboidea, whereas Polyphaga are divided into 18 superfamilies. These may be grouped into 5 large 'series' which represent the main evolutionary lines of the Coleoptera (Crowson, 1960).

The taxonomic literature on the beetles of the world has been catalogued in the *Catalogus Coleopterorum*, edited by S. Schenkling. This is the work of many authors, and was published in 170 parts between 1910 and 1940 by W. Junk. The catalogue is arranged in families, and lists references to 221,480 species. Supplementary catalogues to some families have been published since 1940.

The sizes of beetles recorded in the following pages refer to the total length of the body including the head.

KEY TO THE SUBORDERS OF COLEOPTERA

1. Prothorax without notopleural sutures (e.g. Fig. 30.11); wings without an oblongum; hind coxae movable, not fused to the metasternum and not dividing the basal abdominal sternite into 2 lateral pieces. [Testes not tubular and coiled; ovarioles acrotrophic; aedeagus usually with a basal piece.] Larva with legs 5-segmented (including a single claw) or absent; mandibles always with a mola .... POLYPHAGA (p. 532)
   Prothorax with visible notopleural sutures (Fig. 30.3); wings usually with an oblongum; hind coxae often fused to the metasternum, usually dividing the basal abdominal sternite into 2 lateral pieces 2
2. Wings at rest not having the distal part spirally rolled; basal abdominal sternite divided into 2 lateral pieces by hind coxae; aedeagus without a basal piece. Larva with legs 5-segmented, plus paired claws; mandibles without a mola; maxillae without a lacinia; labrum absent; urogomphi usually present .... ADEPHAGA (p. 522)
   Wings at rest having the distal part spirally rolled (Fig. 30.15); basal abdominal sternite not divided by the hind coxae; aedeagus with a basal piece. Larva with legs 4- or 5-segmented, plus 1 or 2 claws; mandibles with a mola; maxillae with a lacinia; labrum visible; urogomphi absent 3
3. Antennae filiform (Fig. 30.14); mandibles without an articulated tooth; wings not fringed with long hairs, venation well developed (Fig. 30.6C); length 7–17 mm. Larvae wood-boring .... ARCHOSTEMATA (p. 519)
   Antennae clubbed (Fig. 30.16); left mandible with an articulated tooth; wings fringed with long hairs, venation reduced (Fig. 30.6A); length less than 1 mm. Larvae aquatic .... MYXOPHAGA (p. 520)

Suborder ARCHOSTEMATA

Wing venation with an oblongum (Fig. 30.6C); distal part of the wing spirally rolled in repose (Fig. 30.15A); notopleural sutures visible on the prothorax; hind coxae attached to metasternum, but slightly movable, and not dividing 1st visible sternite; abdomen cryptogastrous (sternites 1 and 2 missing), with 5 visible sternites; tarsal segmentation 5-5-5; testes tubular, coiled, and ovarioles polytrophic as in Adephaga. Larva with legs 5-segmented plus 1 or 2 claws; labrum distinct; without urogomphi.

Includes only 2 families. The systematic position of the Micromalthidae is doubtful. Jeannel and Paulian (1944) and Arnett (1960) consider that the family should be returned to its former association with the Lymexyloidea or Cantharoidea, because *Micromalthus* is said to lack notopleural sutures. Although not recorded from Australia, the Micromalthidae deserve special mention, as the life history of the single species, *Micromalthus debilis* Le C., is the most complicated known in the Insecta (Pringle, 1938). It includes at least five different kinds of larvae, one of which is paedogenetic, producing first-instar caraboid larvae viviparously, or a single egg which gives rise to a curculionoid larva that lives as an ectoparasite on the mother larva. The curculionoid larva pupates and produces an adult male. The second-stage or cerambycoid larva can pupate, and then gives rise to the adult female. Males are haploid and females diploid. The product of the mating of the adult males and females is unknown. *M. debilis* has been recorded from North America, Mexico, Cuba, Brazil, Gibraltar, South Africa, and Hawaii. The distribution shows that the species can be accidentally transported. It feeds in rotten wood, and, in view of its small size (1·5 mm), might easily have escaped noticc in Australia.

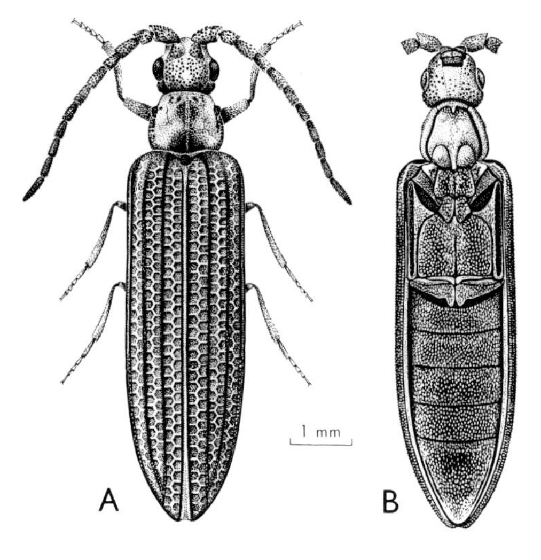

Fig. 30.14. *Cupes mathesoni*, Cupedidae: A, dorsal; B, ventral. [F. Nanninga]

1. Cupedidae (Figs. 30.6C, 14, 15A). Body surface tuberculate; elytral surface sculpture clathrate (lattice-like); metasternum with a transverse suture; size medium (10–18 mm), body elongate; antennae of moderate length, rather thick; tibiae without spurs; aedeagus trilobed. Larvae eruciform, without urogomphi, with short legs and single claws, mandibles with a mola, labrum distinct. Cupedidae occur in all continents, but individuals are, in general, rare. The larvae are borers in rotting wood. Two genera, *Cupes* and *Omma*, occur in Australia. *Cupes* (4 spp.) occurs in south Queensland, N.S.W., and Victoria, and is represented by other species in North America, E. and S. Asia, and S.E. Africa. The present species are the dwindling relics of the most ancient known family of beetles which, on fossil evidence, were apparently common throughout the Mesozoic. [Neboiss, 1960; Crowson, 1962; Atkins, 1963.]

Suborder MYXOPHAGA

Size minute; mandibles with a distinct mola; a supplementary articulated tooth on left mandible only; prothorax with notopleural sutures; wings folded in basal half but rolled apically (Fig. 30.15B), with a distinct oblongum, but venation otherwise reduced; maxilla without a galea; antennae clubbed. Larvae (of the 3 families so far known) aquatic, broadly ovate, with strongly deflexed head, 4 prominent ocelli on each side, a large labrum, mandibles with a mola, maxillae without a galea, tubular spiracular gills, and 5-segmented legs.

The suborder includes only 4 families, each containing one genus. The combination of characters indicates that the group is very ancient, with an origin near that of the Archostemata and Adephaga. The distribution of the families is as follows:

Hydroscaphidae *(Hydroscapha)*: Europe (4 spp.), N. Africa (1 sp.), N. America (1 sp.), India (1 sp.), Burma (1 sp.), Madagascar (1 sp.).

Lepiceridae *(Lepicerus)*: Central America (1 sp.).

Sphaeriidae *(Sphaerius)*: Europe (3 spp.), N. America (4 spp.), India (1 sp.), Madagascar (1 sp.), Australia (2 spp.).

Torridincolidae *(Torridincola)*: Rhodesia (1 sp.).

2. Sphaeriidae (Figs. 30.6A, 15B, 16). Minute (0·8 mm), ovate, strongly convex, shining beetles, with elytra covering the abdomen, and only 3 visible abdominal sternites, the 1st and 3rd of which are long; antennae 11-segmented, inserted under edge of frons, with scape and pedicel swollen, 3rd segment long and slender, 4th and 5th subglobular, 6th–8th short, transverse, and last 3 forming a strong, compact club; mandibles strongly curved, bifurcate at apex, with a basal molar surface and above this is a slender membranous appendage, ciliate on inner edge. In addition, on the left mandible only, above the ciliated appendage there is a slender movable appendage. Maxillae with one lobe only and a 4-segmented palp, the terminal segment of which is short. Labial palpi 3-segmented. Fore coxae transverse and closely placed; notopleural suture present; mid coxae widely separated; hind coxae triangular, almost in contact, each produced behind into a horizontal plate which covers the hind femur when retracted. Fore femur hollowed for reception of the coxa and tibia. Tarsi short, 3-segmented, not lobed, with 2 or 3 long setae between the claws, which are unequal, the inner being longer than the outer. Hind wings fringed with long hairs, venation reduced, but with an obvious closed cell (Fig. 30.6A) between M and Cu homologous with the oblongum in Adephaga. The wing falls into two oblique transverse folds in the basal half, and a longitudinal fold, the longitudinally folded distal half forming a conical roll (Fig. 30.15B).

The Australian species of *Sphaerius* are *S. ovensensis* (Oke) in Victoria and N.S.W. and *S. coenensis* (Oke) in north Queensland. They can be collected, along with Hydraenidae, by stirring wet mud, gravel, and roots from the edges of streams into a pan of water. The adult has no plastron, but stores air beneath the elytra. The larva (Fig. 30.16B), found in the same situation as the adult, is yellowish white, broadly ovate, with stout 5-segmented legs, strongly deflexed triangular head, large labrum covering the mandibles, prominent,

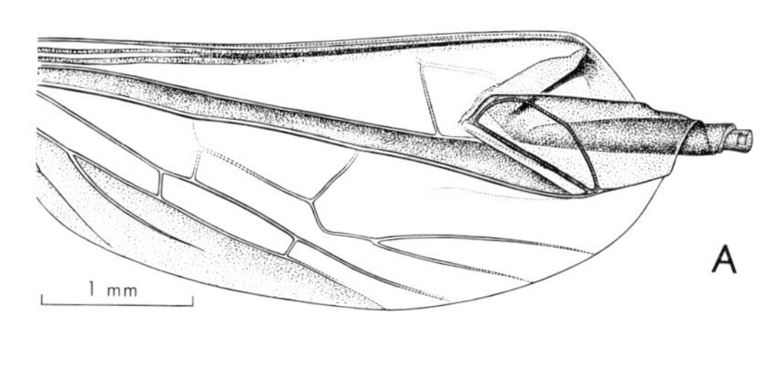

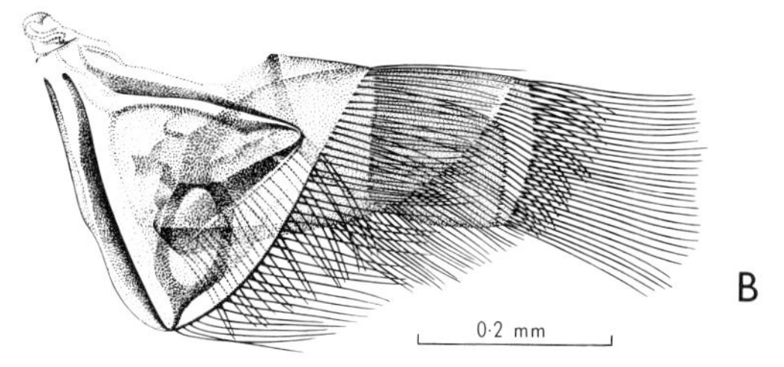

Fig. 30.15. A, *Cupes mathesoni*, Cupedidae, right wing with apical roll; B, *Sphaerius ovensensis*, Sphaeriidae, right wing folded with apical roll. [F. Nanninga]

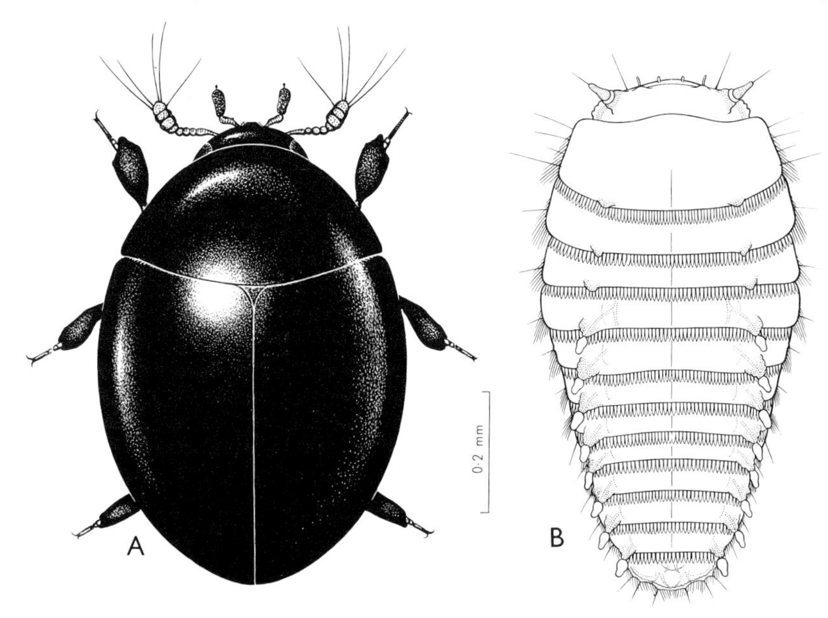

Fig. 30.16. *Sphaerius ovensensis*, Sphaeriidae: A, adult; B, larva. [F. Nanninga]

conical, 2-segmented antennae, and grouped pigmented ocelli. The spiracular tracheae terminate in open-ended thin-walled vesicles (spiracular gills), one pair on each of abdominal segments 1–8 (H. E. Hinton, *Aust. J. Zool.* **15**: 955–9, 1967).

Suborder ADEPHAGA

Wing venation usually with an oblongum; notopleural sutures visible on lateroventral sides of prothorax (Fig. 30.3); hind coxae immovably fixed to metasternum, and completely dividing 1st visible abdominal sternite; 6 abdominal sternites (segments 2–7) normally visible, the first 3 partly fused or immovably connected; tarsal segmentation 5-5-5; genitalia trilobed, comprising median lobe and parameres, basal piece absent. Four Malpighian tubes; testes tubular, coiled inside a membranous sheath; ovarioles polytrophic. Larva with 5-segmented legs, plus 1 or 2 claws; without labrum; mandible without a mola; maxilla without a lacinia; segmented urogomphi usually present. The species are largely predatory in habit.

Superfamily CARABOIDEA

Key to the Families of Caraboidea Known in Australia

(Adapted from Crowson, 1955)

1. Hind coxae not extended laterally to meet the elytra (Fig. 30.3), the junction of the metepimeron and 1st abdominal sternite usually visible; antennae largely pubescent; body with well-developed projecting sensory setae in definite positions; terrestrial 2
   Hind coxae extended laterally to meet the elytra, so that the junction of the metepimeron and 1st abdominal sternite is not visible (Figs. 30.21–23); antennae entirely glabrous or nearly so; body rarely with projecting sensory setae; aquatic 3
2. Metasternum without a transverse suture in front of hind coxae; 1st visible abdominal sternite with a large median portion exposed between hind coxae; antennae relatively thick, moniliform (Fig. 30.17) .... **Rhysodidae** (p. 523)
   Metasternum with a transverse suture in front of hind coxae; 1st visible abdominal sternite with median portion very small or absent; antennae usually slender, and when thick are not moniliform .... **Carabidae** (p. 523)
3. Posteroventral edges of hind coxae produced into large flat plates which conceal basal abdominal sternites; metasternum with a distinct transverse suture; elytra with large punctures in regular striae (Fig. 30.21) .... **Haliplidae** (p. 529)
   Posteroventral edges of hind coxae not produced; if the metasternum has a distinct transverse suture, then the elytra are without regular punctured striae 4
4. Metasternum with a transverse suture; head more exserted; eyes strongly protuberant; hind coxae relatively small (Fig. 30.22B); legs moved alternately in swimming .... **Hygrobiidae** (p. 529)
   Metasternum without a transverse suture; head less exserted; eyes not protruding beyond general outline of head; hind coxae larger; hind legs moved simultaneously in swimming 5
5. Eyes completely divided; antennae very short and thick, pedicel large; abdominal segment 6 visible beyond the elytra; fore legs long, raptorial, mid and hind in the form of short broad paddles (Fig. 30.24D) .... **Gyrinidae** (p. 531)
   Eyes not completely divided; antennae longer, pedicel not unusually large; abdominal segment 8 normally concealed; mid and hind legs much longer and narrower 6
6. Dorsal surface strongly convex; scutellum not visible; ventral surface flattened; hind coxae smaller, with characteristic longitudinal plates which cover the articulations of the trochanters; metasternal–metacoxal suture angulate in middle; metepisterna clearly not reaching the mid coxal cavities; mesepisterna not distinct (Fig. 30.23) .... **Noteridae** (p. 530)
   Dorsal surface hardly more convex than ventral surface; scutellum often visible (Fig. 30.24A); hind coxae large, without such longitudinal plates; metasternal–metacoxal suture arcuate; metepisterna reaching or almost reaching the mid coxal cavities; mesepisterna always distinct .... **Dytiscidae** (p. 531)

Fig. 30.17. *Rhysodes* sp., Rhysodidae. [F. Nanninga]

3. Rhysodidae (Fig. 30.17). Small (4–8 mm), black beetles of elongate, subcylindrical form. Antennae strongly moniliform; head and pronotum with deep longitudinal furrows carrying a yellowish secretion; the furrows are suggestive of a myrmecophilous habit, but no special association with ants has yet been reported; head with a pronounced constriction; legs short, fore tibiae without a cleaning organ; metasternum without a transverse suture; 1st visible abdominal sternite divided by hind coxae into a large triangular median portion and two lateral parts. Larvae eruciform; without urogomphi; anterior edge of head with a median projection; mandibles well developed, but maxillae and labium reduced. Two genera are known. Only *Rhysodes* (9 spp.) is found in Australia. Both larvae and adults live in decaying wood, but no details of their life history or habits are known.

4. Carabidae (ground beetles; Figs. 30.18–20). Antennae usually 11-segmented (10-segmented in Paussinae), filiform or flattened, inserted between eyes and mandibles under a frontal ridge; mandibles and maxillary and labial palpi prominent; prothorax almost always with defined lateral margins; fore and mid coxae globular, the fore coxal cavities open or closed behind; metasternum small, with a transverse and a longitudinal suture making a cross in the middle; legs usually long and slender, cursorial; fore tibiae always (except in Paussinae) with an anterior cleaning organ, sometimes flattened and toothed for digging; elytra almost always striate, usually with characteristic 'fixed' setiferous punctures; wings sometimes absent; abdomen with 6 visible sternites (7 or 8 in Brachininae), the first 3 immovably connected, the basal sternite visible only at the sides.

Carabidae can be placed in three main ecological groups: *geophiles*, which live on the ground, not associated with surface water; *hydrophiles*, living at the edges of streams or ponds or in swamps; and *arboricoles*, living on the trunks of trees or on leaves. According to Darlington (1961), the proportion of geophiles : hydrophiles : arboricoles in the Australian fauna is roughly 2 : 1 : 1.

This is one of the major families, with 232 genera so far known in Australia. Both adults and larvae are usually active and carnivorous. Carabid larvae are almost always campodeiform, with 10-segmented abdomen, and legs each usually with 2 claws; antennae prominent, 3- to 5-segmented; mandibles curved, sharply pointed, with retinaculum; segmented urogomphi present.

The basic classification of the family is now well established. The following key, based on the work of Andrewes (1935), Sloane (1923), and Jeannel (1946), is adapted to fit the Australian fauna.

Key to the Subfamilies of Carabidae Known in Australia

1. Fore tibia with 2 spurs placed level with each other at the apex ventrally, the spurs not separated by the comb-organ, which is sometimes absent (Fig. 30.18C). [Antennae sometimes very broad and flattened (Fig. 30.18A); outer edge of each elytron with a deep notch or prominent flange near the apex; metepimeron visible as a lobe between hind coxa and elytral

epipleuron; mesepimeron forming part of the border of mid coxal cavity between mesosternum and metasternum] .... (ISOCHAETA). PAUSSINAE
Fore tibia with the outer spur placed nearer to the apex than the inner spur, respectively below and above (Fig. 30.4A) or in front of and behind the comb-organ, which is always present 2

2(1). Metepimeron not visible between the posterior edge of the metepisternum and anterior edge of 1st visible abdominal sternite; mesepimeron always forming part of the border of the mid coxae, between the mesosternum and metasternum .... (SIMPLICIA). 3
Metepimeron visible as a lobe on each side attached to posterior edge of metepisternum; mesepimeron usually not reaching the mid coxal cavity 6

3(2). Fore coxal cavities open behind; a single seta above the eye. [Mesepimeron forming part of the border of the mid coxal cavity between the mesosternum and metasternum; parameres elongate and non-setose] 4
Fore coxal cavities closed behind; head with 1 or 2 setae above the eye 5

4(3). Mandibles multidentate, their upper surfaces smooth; clypeus without setiferous punctures. [Head with a deep transverse groove behind eyes; fore tibia with outer apical angle prolonged] ...... CYCHRINAE
Mandibles not toothed, their upper surfaces rugose; clypeus with a setiferous puncture in each anterior angle (Fig. 30.1A) .... CARABINAE

5(3). Clypeus extended laterally in front of antennal insertions; elytra not regularly striate; parameres elongate and joined by a dorsal bridge .... CICINDELINAE
Clypeus not extended laterally in front of antennal insertions; elytra usually regularly striate; parameres not joined by a bridge. [Mandibles usually with a seta on scrobe; a single seta above each eye; apterous; no setae on pronotum; elytra with 10 striae] .... MIGADOPINAE

6(2). Mesepimeron reaching edge of mid coxal cavity between mesosternum and metasternum; body often pedunculate (Fig. 30.19B). [Antennae at rest lie in a posteriorly directed groove below the eye; fore tibiae flattened and toothed for digging; parameres usually setiferous] .... (SCROBIFERA). SCARITINAE
Mesepimeron not reaching edge of mid coxal cavity, which is closed by the meeting of mesosternum and metasternum (Fig. 30.3); body rarely pedunculate 7

7(6). Mandibles with a seta on anterior part of outer face; parameres always elongate and setiferous .... (STYLIFERA). 8
Mandibles with or without a seta on anterior part of outer face; one paramere short and broad, the other usually different in shape, both without setae 14

8(7). Head with 1 setiferous puncture above each eye 9
Head with 2 setiferous punctures above each eye 10

9(8). Body pubescent; prothorax globular, without defined lateral margins; palpi long and slender .... APOTOMINAE
Body not pubescent; prothorax with well-defined lateral margins; palpi normal .... BROSCINAE

10(8). Terminal segment of maxillary palp fusiform (widest in the middle); the marginal series of 12 setiferous punctures not divided into 2 groups; base of elytra with a complete, defined border; tarsi without setae beneath .... POGONINAE
Terminal segment of maxillary palp conical or subulate (very small, cylindrical); the marginal series of 8 or 9 setiferous punctures of elytra divided into an anterior group of 4 and a posterior group of 4 or 5; base of elytra without a complete, defined border; tarsi pubescent beneath 11

11(10). Terminal segment of maxillary palp very small .... BEMBIDIINAE
Terminal segment of maxillary palp of about the same length as penultimate segment 12

12(11). Elytra without an inner longitudinal ridge visible beneath their apical edge; frontal furrows of head extending behind level of posterior edge of eyes (Fig. 30.20A). .... TRECHINAE
Elytra with an inner longitudinal ridge, the end of which is visible beneath the edge of the elytron, level with the apical end of the epipleuron; frontal furrows of head not extending back beyond posterior edge of eyes 13

13(12). Penultimate segment of maxillary palp

setose; fore tarsi of ♂ with 2 basal segments dilated and dentate on inner side .... MERIZODINAE

Penultimate segment of maxillary palp bare; fore tarsi of ♂ not or slightly modified .... PSYDRINAE

14(7). Head without supraorbital setiferous punctures, or with only 1 on each side 15

Head with 2 supraorbital setiferous punctures on each side .... (CONCHIFERA, pt.). 19

15(14). Mandibles with a single seta in the outer hollow (scrobe) towards the apex, or head without a supraorbital seta; aedeagus with left paramere very short and broad, in the form of a half-annulus; ♂ with 7 or 8 visible sternites .... (BALTEIFERA). 16

Mandibles without a seta in the scrobe, and head with a supraorbital seta on each side; aedeagus with left paramere short, flattened, and rounded, right paramere elongate or reduced; ♂ with 7 or fewer visible sternites .... (CONCHIFERA, pt.). 17

16(15). Mandibles with a single seta in the scrobe; head with 1 supraorbital seta on each side; body not flattened; head without a groove on each side for reception of the antennae; femora without cavities for reception of the tibiae ...... BRACHININAE

Mandibles without a seta in the scrobe; head without supraorbital setae; body much depressed and oval in outline (Fig. 30.20B); head with a deep longitudinal groove beneath eye on each side for reception of the antennae at rest; femora each with a longitudinal cavity into which the tibia can be retracted. [In ♂, 7th (6th visible) sternite deeply emarginate, 8th (7th visible) divided into 2 lobes, and 9th (not visible externally) expanded at posterior end into a sclerotized plate] .... PSEUDOMORPHINAE

17(15). Antennae densely setose from middle of 3rd segment onwards ...... HARPALINAE

Antennae densely setose from base of 4th segment onwards 18

18(17). Eighth elytral stria extended at distal end by a deep, broad channel, parallel to apical edge of elytra as far as sutural angle; elytra bare; body black .... OODINAE

Eighth elytral stria normal, not extended around distal margin of elytron; elytra pubescent; body colour green or black, elytra often with yellowish markings .... CHLAENIINAE

19(14). Apices of elytra rounded off together in a single curve or acutely angled at apex, terminal abdominal segment normally covered 20

Apices of elytra transversely or obliquely truncate, terminal abdominal segment visible from above 24

20(19). Palpi long, with terminal segments strongly securiform (axe-shaped) and inserted eccentrically on the penultimate segment. [Head relatively small, with a marked groove around neck, prominent eyes, and deep longitudinal grooves on frons; body pubescent and coarsely punctured]. .... PANAGAEINAE

Palpi not unusually long, terminal segments not securiform 21

21(20). Elytra with sparse, short pubescence on the depressed outer-apical marginal area outside the 8th stria; remainder of elytral surface smooth, bare, shining, with striae very faint or absent .... PERIGONINAE

Elytra without pubescence on outer-apical area outside the 8th stria; striae visible on disc of elytra 22

22(21). Clypeus emarginate and sometimes asymmetrical; a pale membrane present in the emargination, between clypeus and labrum; labrum often deeply emarginate .... LICININAE

Clypeus not emarginate and never asymmetrical; without an obvious pale membrane between clypeus and labrum; labrum not deeply emarginate 23

23(22). Elytra with an inner longitudinal carina the apical end of which is visible below edge of elytron towards apex where the elytral epipleuron ends; mentum with median tooth usually bifid .... PTEROSTICHINAE

Elytra without an inner carina visible externally below edge of elytron towards apex; mentum with a simple median tooth, or without a tooth ...... AGONINAE

24(19). Scape of antenna obviously larger than the remainder; antenna pubescent from base 25

Scape of antenna short; antenna pubescent from segment 4 27

25(24). Antennae stout, scape not longer than

next 2 segments together (*Pogonoglossus*, 2 spp.) .... PHYSOCROTAPHINAE
Antennae slender, scape about as long as next 3 segments together 26

26(25). Prothorax subcylindrical, the lateral edges, if present, not sharp; neck about half as wide as head; 4th tarsal segment bilobed .... DRYPTINAE
Prothorax with normal sharp lateral edges; neck about one-third as wide as head; 4th tarsal segment not bilobed ... . ZUPHIINAE

27(24). Head tapered behind to a narrow neck. [Prothorax subcylindrical with obsolete lateral borders] .... ODACANTHINAE
Head not narrowed to a condyliform neck 28

28(27). Labrum large, smooth, semicircular, practically concealing mandibles. [Body hairy; legs and antennae short and stout] .... HELLUONINAE
Labrum normal, more or less quadrate 29

29(28). Tibiae serrate, fore tibiae dilated at apex; pronotum without lateral setae. [Fourth tarsal segment emarginate and often bilobed; claws usually pectinate] .... ORTHOGONIINAE
Tibiae not serrate; pronotum almost always with 1 lateral seta on each side, usually with 2 30

30(29). Hind tibiae with unusually long spurs .... TETRAGONODERINAE
Hind tibiae with spurs of normal length 31

31(30). Mentum supported on a projecting submentum; head not sharply constricted behind eyes; mandible with an outer hollow face; claws often pectinate beneath .... LEBIINAE
No projecting submentum; head sharply constricted behind eyes; mandible without a hollow face; claws simple .... PENTAGONICINAE

The PAUSSINAE, as defined here, include Ozaenini and Paussini, the former distinguished by the possession of 11-segmented linear antennae. Ozaenini are represented by one genus, *Mystropomus* (3 spp.; Fig. 30.18B); they occur in ground litter in dense rain forest in northern N.S.W. and Queensland. The Australian Paussini include 3

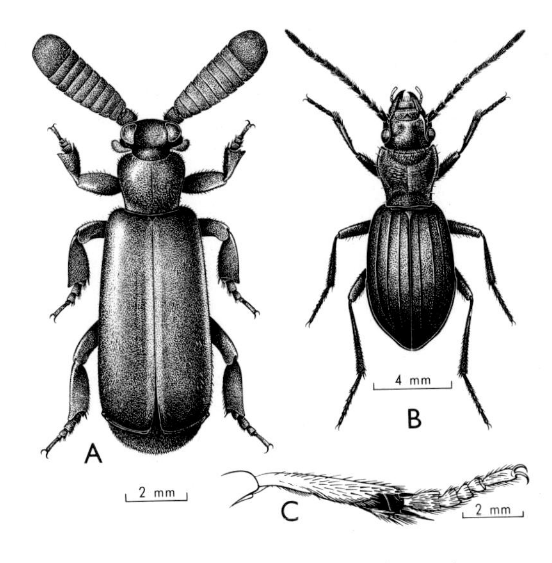

Fig. 30.18. Carabidae–Paussinae: A, *Arthropterus planicollis*; B, *Mystropomus subcostatus*; C, *M. subcostatus*, inner side of fore leg. [F. Nanninga]

genera and 91 species, 89 of which belong to *Arthropterus* (Fig. 30.18A), and are immediately recognizable by their extraordinary antennae. They are dark brown, 5–13 mm long, the antennae with 10 visible segments which are very broad and co-adapted to form a segmented lamina. Paussini are sluggish and generally rare, but occasionally come to lights at night. Little is known of their habits, and very few larvae are known. Most live with ants, but the Australian species are exceptional in that the adults do not have *trichomes* (glands opening beneath special tufts of hairs, sought by ants), and it seems likely that only their larvae are myrmecophilous. The function of the remarkable antennae of the adults remains to be discovered. The adults lack all the fixed tactile setae which appear in other Carabidae.

The CICINDELINAE (tiger beetles) have the head deflexed, with large eyes, broader than the prothorax; clypeus extending laterally beneath the bases of the antennae; labrum largely covering the mandibles, which are long, acute, and curved; and elytra without striae. Six genera are known in Australia: *Cicindela* (29 spp.), *Megacephala* (19 spp.), *Distipsidera* (13 spp.), *Nickerlea* (2 spp.),

Rhysopleura (1 sp.), and *Tricondyla* (1 sp.). The great majority of the species occur in Queensland, N.T., and north-western Australia, only 3 being found in the east and south; none occurs in Tasmania. All are predatory. *Cicindela* spp. inhabit sandy heaths, the margins of creeks, sand-banks, and saltpans; they run and fly very actively in daylight. *C. ypsilon* is found on sea beaches in N.S.W. and Victoria. *Megacephala* spp. (Plate 4, A) are active at night and have not been observed to fly, although some species are winged. All species of *Distipsidera* live as adults on the trunks of trees. *Rhysopleura orbicollis* Sloane is found on mossy tree trunks at Kuranda, north Queensland. The species of *Megacephala* have a striking metallic green or purple colour. *M. crucigera* Macl. has a metallic green head and prothorax and yellowish elytra bearing a dark cross-shaped mark. *Tricondyla aptera* Oliv. is black, slender, wingless, and 20–25 mm in length. It is widely distributed from the Moluccas to the New Hebrides, and extends into north Queensland. The larvae of Cicindelinae live separately in tunnels excavated in the soil, and are highly adapted to this habitat, having the head and pronotum heavily sclerotized, forming a lid which is used to close the aperture of the tunnel. The abdomen is not sclerotized, and tergite 5 bears strong hooks which assist in the movement of the larva in the tunnel. *Tricondyla* is arboreal, and is unusual in that its larva makes its tunnel in wood.

CYCHRINAE (Pamborinae) are represented in Australia by the single genus *Pamborus* (13 spp.) in N.S.W. and Queensland. They are striking insects, 17–35 mm long, with head and pronotum black, and elytra metallic green or blue or black bordered with metallic red. The head is flattened, with hemispherical eyes, and the palpi terminate in securiform segments. Their resemblance to the Cychrinae of the northern hemisphere suggests that they may feed upon snails.

Calosoma, a world-wide genus, is the sole representative of the CARABINAE in Australia. The only Australian species are *C. schayeri* Erich. in the east and south and Tasmania (Plate 4, B), *C. oceanicum* Perroud in the west, and *C. australe* Hope in central Australia. *C. schayeri* is a common species of stout build and bright metallic green colour, which preys upon larvae of Lepidoptera.

The MIGADOPINAE (Jeannel, 1938), a primitive group with 4 genera in Australia, is of considerable interest, because its geographical distribution stronglv suggests a former land connection between Australia, New Zealand, southern Chile, and the Falkland Is. The 5 species are rare, and all occur in mountain forests. Larvae and habits are unknown. The Australian genera are *Stichonotus*, *Calyptogonia*, *Nebriosoma*, and *Decogmus*.

BROSCINAE (Fig. 30.19A) have an unusual distribution covering both temperate zones. Australia has 10 genera and 116 species of a world total of 27 genera and 250 species. The largest genus, *Promecoderus*, includes 54 species, and is widely distributed mainly in dry forest and arid regions. Striking features of the distribution of the subfamily are its absence from Africa and the Oriental region and its paucity in the American continent except in the extreme south. The distribution suggests an Antarctic origin.

SCARITINAE (Plate 4, S; Fig. 30.19B) live in deep burrows in the soil, on sea shores, river

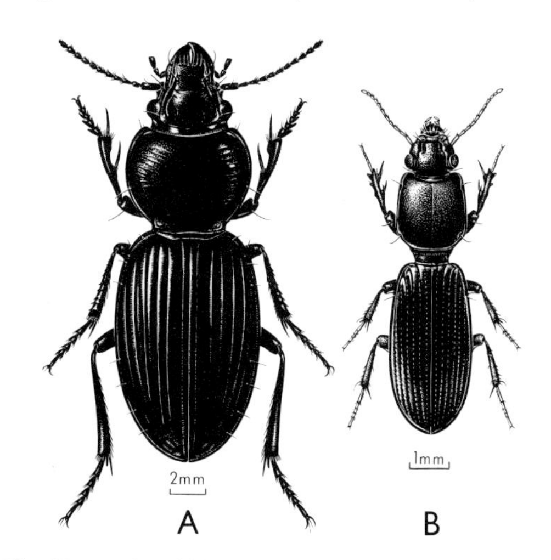

Fig. 30.19. Carabidae: A, *Eurylychnus blagravei*, Broscinae; B, *Clivina basalis*, Scaritinae.
[F. Nanninga]

banks, and in mountains. There are 19 Australian genera and 216 species, including about 120 species of *Carenum*. They are very strongly sclerotized beetles, often metallic in colour, wingless, with elytra joined together, and they are well adapted for burrowing. *Euryscaphus waterhousei* Macl. from central Australia is the giant of the group, a shining black beetle up to 5 cm in length, with mandibles 1 cm long, and broad, heart-shaped elytra. Little is known of the habits or life histories of the Australian species, but elsewhere in the world the Scaritinae are carnivorous and nocturnal.

Among the TRECHINAE (Fig. 30.20A) 6 genera of the tribe Homoloderini (e.g. *Tasmanorites*, *Pogonoschema*, *Sloanella*), which are found in the mountains of south-eastern Australia and Tasmania, are of special interest in that they are clearly of Antarctic origin, having their closest relatives in South America. Eleven genera and 27 species of the subfamily are so far known in Australia. They frequent damp situations, and a few in Europe, South America, and New Zealand live in rock crevices in the intertidal zone, but so far none has been found in this habitat in Australia. The minute *Perileptus constricticeps* Sloane is found in gravel at the edges of streams.

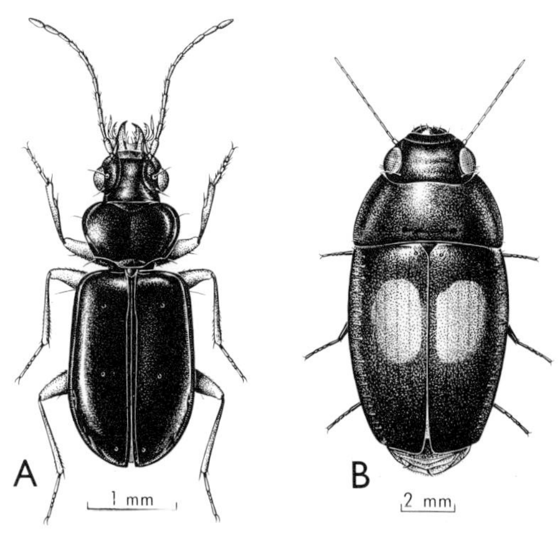

Fig. 30.20. Carabidae: A, *Trechodes bipartitus*, Trechinae; B, *Sphallomorpha bimaculata*, Pseudomorphinae. [F. Nanninga]

The BEMBIDIINAE are numerous in the northern hemisphere, but rather poorly represented in Australia (Sloane, 1921). *Bembidion* and *Tachys*, for example, with over 1,000 and 500 species respectively, are represented in Australia by 6 and 50 species. They are small beetles, which live on sand or gravel beaches along mountain streams or rivers or in boggy places. MERIZODINAE include the Tasmanian cave-dwelling *Idacarabus troglodytes* Lea. [Jeannel, 1940.]

The PSYDRINAE, like the Migadopinae, are probably Antarctic in origin. Eighteen genera and 36 species are so far known in Australia. The subfamily has been reviewed by Moore (1963).

PTEROSTICHINAE are world-wide in distribution, and well developed in the Australian region. Of about 150 known genera and over 5,000 species, 43 genera and 380 species are Australian. The largest genus, *Notonomus* (106 spp.) is almost entirely limited to the wet coastal forests of eastern Australia, the species being notably allopatric. Pterostichinae include some very large forms, such as *Hyperion schroetteri* Schr. and *Catadromus elseyi* White, both 6 cm or more in length, and many attractive metallic species. Pterostichinae are typically ground-dwelling, carnivorous beetles, usually found during the day under stones, logs, etc., especially in damp places. *H. schroetteri* preys upon scarabaeid larvae in rotting wood in the centres of large eucalypts. Moore (1965) has provided a key to the Australian genera of Pterostichinae and Agoninae.

Numerous in the tropics and notably so in New Guinea (Darlington, 1961), AGONINAE are poorly represented in Australia, where 8 genera and 23 species are known (Darlington, 1956). Also relatively poorly represented in Australia (110 spp.) but very extensive elsewhere, the HARPALINAE include the handsome ground beetle *Gnathaphanus pulcher* Dej., with metallic green head and pronotum and reddish elytra, a species common in coastal N.S.W. and Queensland.

The LEBIINAE are small flattened Carabidae, often with yellow and black patterned elytra, and are commonly found beneath loose bark.

They are represented in Australia by 24 genera and 153 species. The largest genus, *Xanthophaea*, includes *X. grandis* Chaud. and *X. vittata* Dej., with yellow elytra longitudinally striped with black, which are common in the south-east. Lebiinae elsewhere are often predators on the larvae and pupae of Chrysomelidae; some are semi-parasitic, but nothing is known of the habits of the Australian forms.

There are about 500 known species of BRACHININAE, but only one, the 'bombardier beetle' *Pheropsophus verticalis* Dej., is known in Australia. This widely distributed insect is found under stones in damp places, and is distinguished by its black elytra marked by 4 yellow spots. When disturbed, it emits a small, visible cloud of vapour from the pygidial glands, accompanied by an audible 'pop' (see p. 36).

PSEUDOMORPHINAE (Fig. 30.20B) are a predominantly Australian subfamily, comprising 5 genera and 96 species. They are very flat beetles of unusually smooth oval contour, dark brown or brown with yellow spots, found under loose bark. They appear to be related to the Brachininae. The larvae of a species of *Sphallomorpha* live in burrows around nests of the meat ant, *Iridomyrmex purpureus* Sm., and prey upon the ants in the manner of cicindeline larvae.

5. Haliplidae (Fig. 30.21). A small but world-wide family of aquatic beetles, recognizable by the fact that the posterior edges of the hind coxae are produced into very large plates which conceal the basal abdominal sternites and hind femora. The adults are boat-shaped, 2–5 mm long, strongly convex, yellowish, with the punctures which mark the elytral striae very obvious and dark; antennae bare, 11-segmented; abdomen with 5 visible sternites; scutellum not visible; mid and hind tarsi with long swimming hairs; metasternum with a transverse suture. The family is represented in Australia by 5 species of *Haliplus*. The adults and larvae are found among aquatic vegetation at the edge of ponds, lakes, or slow streams. The adults usually crawl, but are able to swim, using alternate leg movements, and are said to feed

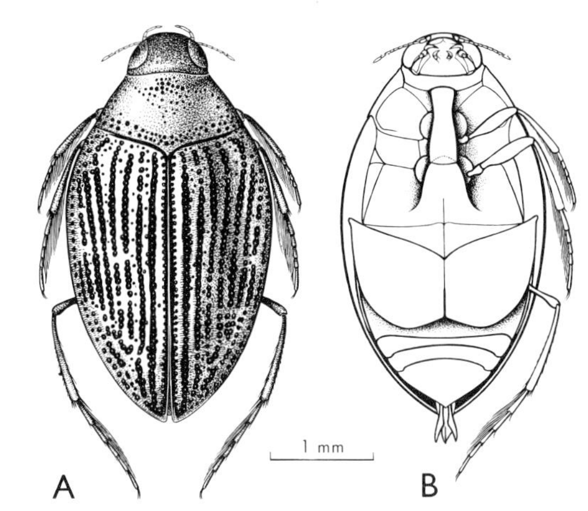

Fig. 30.21. *Haliplus testudo*, Haliplidae: A, dorsal; B, ventral. [F. Nanninga]

both upon algae and animal materials; they breathe air taken at the surface. The larvae breathe dissolved oxygen through the body wall, which is increased in area by segmentally arranged processes. They are predacious; and have short, curved, hollow, suctorial mandibles; 4-segmented antennae; legs with single claws; abdomen with tergal processes, and an elongate segment 10 with cerci. The larval spiracles on the mesothorax and first 8 abdominal segments are open only when the insects leave the water to pupate.

6. Hygrobiidae (Pelobiidae; Fig. 30.22). The body of the adult is stout, oval; antennae bare, 11-segmented; eyes, unlike those of other water beetles, strongly protuberant; metasternum with a transverse suture; legs equipped with swimming-hairs, tarsi longer than the tibiae. Both adults and larvae are carnivorous. The adult swims by the more primitive method of alternate leg movements, like the Haliplidae, and unlike the Dytiscidae, Noteridae, and Gyrinidae. It breathes air stored beneath the elytra at visits to the surface, whereas the larva uses dissolved oxygen, except when it leaves the water to pupate. Functional spiracles are present on the mesothorax and abdominal segments 1 to 8 of the final larval instar only. Larval respiration is assisted by gill filaments on the underside of the body. The larval mandibles are not channelled. The body of the larva is

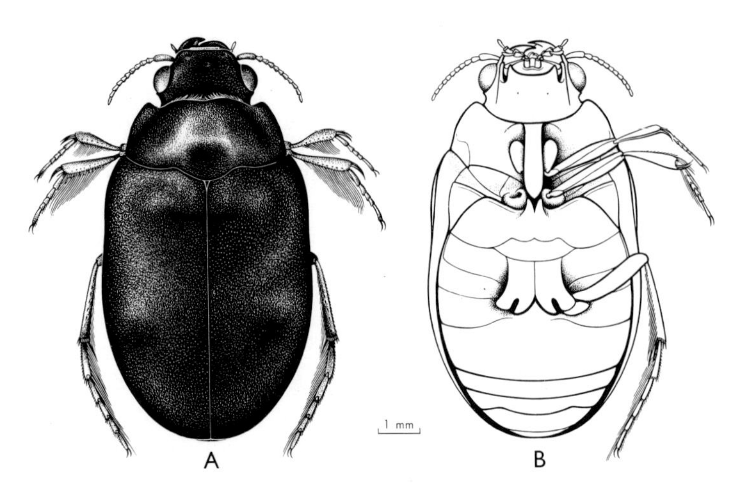

Fig. 30.22. *Hygrobia niger*, Hygrobiidae: A, dorsal; B, ventral. [F. Nanninga]

club-shaped, with enlarged head and prothorax, and narrow abdomen terminated by 3 long filaments. Adult Hygrobiidae stridulate loudly by rubbing the apex of the abdomen on a file on the inside of the elytra. The Australian species live in muddy, stagnant water. Of the 4 known species of the single genus *Hygrobia*, one is found in Victoria (*H. australasiae* Clark), one in Queensland (*H. niger* Clark), one in Tibet, and one in Europe. This relict distribution, combined with the primitive characters of the insects, indicates that the genus is of great age, and it is in some respects related to the ancestral form of the aquatic families of the Adephaga.

7. Noteridae (Fig. 30.23). Distinguished from Dytiscidae by the greater convexity of the dorsal surface, the division of the hind coxae into an inner and an outer part, the angulate metasternal–metacoxal suture, and the fact that the scutellum is not visible (visible in some, not all, Dytiscidae). The inner part of the hind coxae, as seen from below, overlies the articulation of the femur and trochanter, and forms, with the middle of the metasternum, a flat median longitudinal keel; metasternum without a transverse suture. The larvae are distinguished from those of Dytiscidae by the absence of a neck constriction; by the stout mandibles, which have an inner superficial groove instead of a tubular perforation, and which usually have a sharp retinaculum; abdominal segment 8 with 2 short conical urogomphi. They are probably carnivorous, but, unlike Dytiscidae, do not feed by extra-oral digestion. Larvae of some species appear to obtain air by tapping plants with the aid of a pointed apical process, but others obtain air at the surface. The only known pupae (Balfour-Browne and Balfour-Browne, 1940) are contained in air-filled cocoons attached to roots under water. Noteridae inhabit the

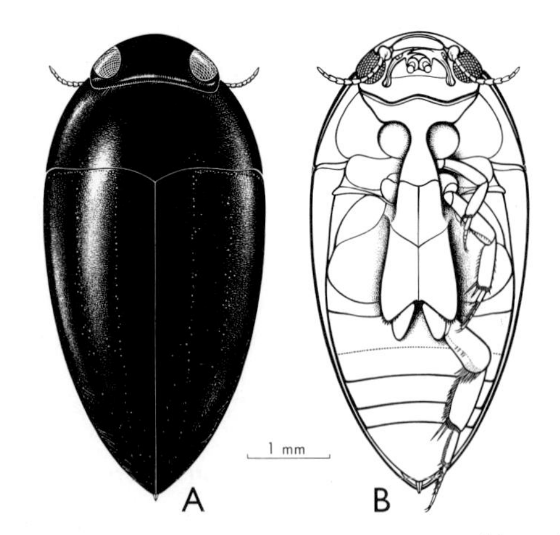

Fig. 30.23. *Hydrocanthus australasiae*, Noteridae: A dorsal; B, ventral. [F. Nanninga

shallow margins of stagnant or slowly running water, and the larvae are found in mud at the bottom. The Australian species are *Hydrocanthus australasiae* Wehnke, *Canthydrus bovillae* Blackb., *Notomicrus tenellus* Clark, and *Hydrocoptus subfasciatus* Sharp.

8. Dytiscidae (Figs. 30.24A–C). A family of predacious water beetles. The adults are smooth and boat-shaped, with flattened, paddle-like hind legs which are moved together in swimming. The dorsal and ventral surfaces of the body are of similar convexity; hind coxae extensive, each bounded anteriorly by a pronounced curve and without a lamella covering the articulation of the trochanter; body without a median longitudinal flat keel; metasternum without a transverse suture; hind tarsi strongly flattened, with a brush of hairs forming an efficient paddle. Adult Dytiscidae vary in length from a few to over 25 mm.

The beetles and their larvae are found in both running and still water. Both larvae and adults take in air periodically at the surface through the terminal pair of spiracles, while the adult also stores air between the abdomen and the elytra. The terminal spiracles alone are functional in the first and second instar larvae. The males in some genera (e.g. *Eretes*) have the first 3 segments of the fore tarsi greatly dilated to form adhesive pads equipped with cup-shaped suckers, which are used to hold the female. Both larvae and adults are carnivorous, preying upon a wide variety of aquatic animals, including molluscs and small fish. The mandibles in the adult are adapted for chewing, and digestion is internal. In the larva, on the other hand, digestion takes place externally (p. 511). The eggs are often deposited in slits made by the ovipositor in the stems of aquatic plants, but this habit is not universal. The habits of the Australian species have not been studied. Pupation takes place in a cell formed by the larva in damp soil near, but out of, the water. Adults are normally capable of prolonged flight, necessary to enable them to migrate to isolated ponds. Flight is crepuscular or nocturnal. The beetles are attracted to lights, and often appear to mistake artificial shining surfaces, such as glass, for water. The commonest Australian species is *Eretes sticticus* L., a yellowish brown beetle about 12 mm in length, which is almost world-wide in distribution. The largest Australian species are *Cybister tripunctatus* Oliv. and *Homeodytes scutellaris* Germ., both dark olive green with yellow margins, and about 25 mm in length.

9. Gyrinidae (whirligig beetles, Figs. 30.24D, E). The adults, smooth and boat-shaped like the Dytiscidae, are distinguished from that family by their habit of swimming mainly on the surface of the water, by the fact that each eye is completely divided into an upper and a lower segment, and by the very short, stout antennae. The largest species, *Macrogyrus striolatus* Guér. from the mountains of N.S.W., is about 17 mm long. All species are very dark blue or olive green in colour. The mid and hind legs are short, broad, flattened, and fringed with long flattened swimming hairs, the whole forming very effective paddles. The fore legs are long and modified for grasping, and, unlike the other legs, are visible from above the body. The fore tarsi are expanded in the males. Metasternum without a transverse suture. The larvae are elongate, with slender, sickle-shaped mandibles perforated by a duct, and digestion is extra-oral. Ten pairs of feathery tracheal gills are borne on the abdomen, and there are no functional spiracles.

Gyrinidae inhabit both still and moving water, and the adults are usually, but not always, active in daylight. They are gregarious, and are well known for the way in which they swim in tight circles on the surface of the water. The adults are mainly, and the larvae entirely, carnivorous. The adults find their food on the surface of the water with the aid of Johnston's organ in the antenna, which is sensitive to surface waves. The larvae live at the bottom, and breathe dissolved oxygen through the gills. As in Dytiscidae, the third-instar larva, which has functional spiracles, leaves the water for pupation. This occurs in a dark-coloured cocoon on land, or on reeds, etc., some centimetres above the surface of the water. The Australian species have been reviewed by Ochs (1949).

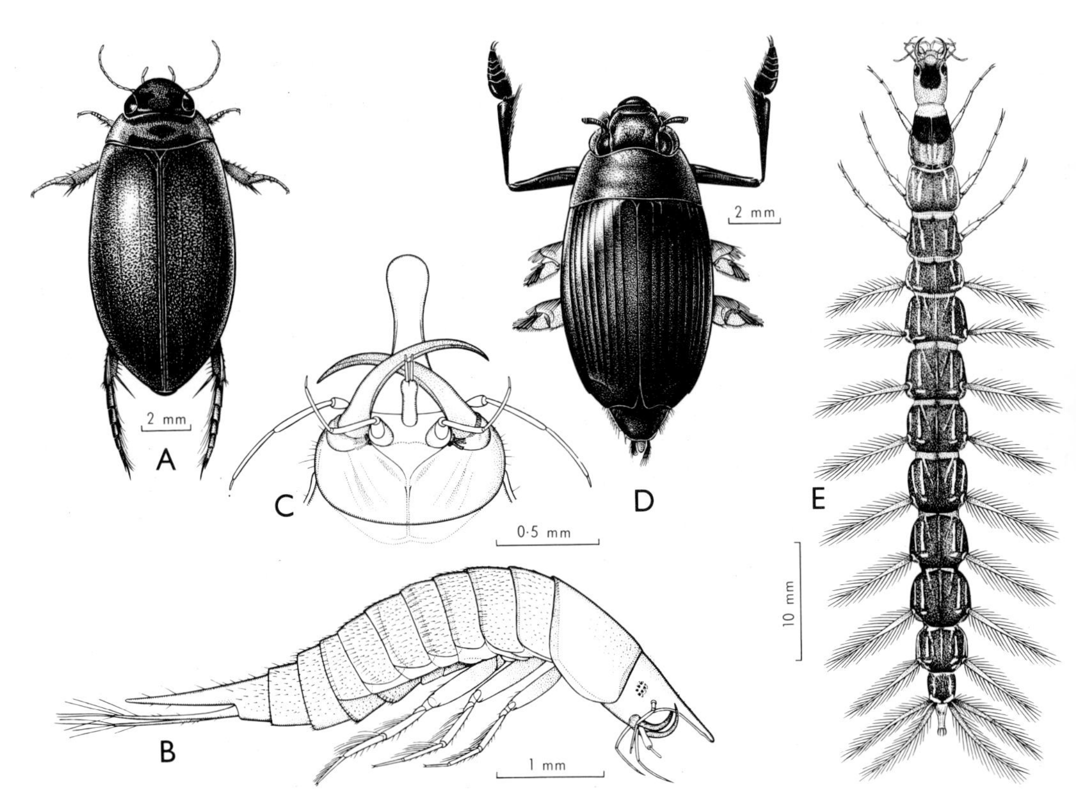

Fig. 30.24. A, *Rhantus pulverosus*, Dytiscidae; B, larva of Dytiscidae–Hydroporinae; C, head of same, ventral; D, *Macrogyrus oblongus*, Gyrinidae; E, larva of *Macrogyrus striolatus*. [F. Nanninga]

Suborder POLYPHAGA

Wing without an oblongum; notopleural sutures not visible on the sides of the prothorax; hind coxae rarely fused to the metasternum, but if so they do not divide the 1st abdominal sternite; if abdomen has 6 sternites, these belong to segments 3 to 8; tarsal segmentation various; basal piece of aedeagus present or absent; Malpighian tubes various; testes not tubular or coiled; ovarioles acrotrophic. Larva: legs with 4 segments plus 1 claw, or less than 4 segments, or vestigial, or absent.

Key to the Superfamilies of Polyphaga

1. Antennae with from 3 to 8 distal segments very asymmetrically expanded, forming a well defined, one-sided, often lamellate club (e.g. Figs. 30.37, 39, 45); head visible from above, not covered by the prothorax; legs usually adapted for digging, the fore tibiae being flattened and with one or more teeth on the outer edge (Fig. 30.4B); tarsal segmentation clearly 5–5–5 .... SCARABAEOIDEA (p. 547)

   Antennae other than with a markedly asymmetrical club formed by the 3 to 8 distal segments; or with a lamellate club, but the club with more than 8 lamellate segments; or with the legs not adapted for digging (some Anobiidae); or head not visible from above; tarsal segmentation sometimes other than 5–5–5 2

2(1). Maxillary palpi prominent (Fig. 30.26), as long as, or longer than, antennae (which are normally held below the eye); antennae short, with an obvious, dull, pubescent club composed of 3 to 5 segments, 6th (more rarely 5th or 4th) segment transverse and laterally angulate or cup-shaped (Fig. 30.25); frons usually with a faint median longitudinal line which forms a Y with the fronto-clypeal suture, or with a deep V-shaped fronto-clypeal suture .... HYDROPHILOIDEA (pt., p. 536)

Maxillary palpi shorter than antennae; antennae otherwise, where clubbed, the segment before the club is not laterally angulate or cup-shaped; head without a median longitudinal line 3

3(2). Elytra short, exposing 2 or more abdominal tergites (Figs. 30.28, 34–36); 3 or more apical tergites sclerotized, or elytra and wings absent 4

Elytra covering all or nearly all of abdomen; if with apex of abdomen exposed, then with at most 2 tergites sclerotized 12

4(3). Wings extending beyond elytra, without transverse folds, or folded only near apex, extended in the resting position to cover half or more of the abdomen; elytra very short, not reaching base of abdomen, or elytra and wings absent (e.g. Fig. 30.66; Plate 6, M, O) 5

Wings, at rest, complexly folded and concealed beneath elytra 7

5(4). Antennae long, reaching at least to base of abdomen; wings with tip obliquely folded; eyes in ♂ not meeting in front .... CHRYSOMELOIDEA (pt., p. 607)

Antennae flabellate, serrate, or filiform, short, apices not reaching to base of prothorax; wings without apical transverse folds, or absent; eyes in ♂ very large, meeting in the mid-line in front 6

6(5). Antennae flabellate, serrate, moniliform, or 2- to 4-segmented; tarsal segmentation 5-5-4 or 4-4-4; wings, where present, longer than abdomen (Figs. 30.93A, B) .... CUCUJOIDEA (Rhipiphoridae, p. 603)

Antennae serrate or filiform; tarsal segmentation 5-5-5; wings shorter than abdomen (Fig. 30.66) .... LYMEXYLOIDEA (pt., p. 579)

7(4). Claws simple; antennae usually clubbed 8

Claws toothed or appendiculate beneath; antennae filiform or serrate 12

8(7). Antennae elbowed, the 3 apical segments forming a compact club; body of characteristic short, broad, compact shape, surface always hard and shining (Fig. 30.28). [Colour dark brown, black, or metallic blue; elytra often with discal striae short or absent] .... HISTEROIDEA (p. 539)

Antennae not elbowed, club when present usually not compact; body often elongate, and rarely very hard and shining 9

9(8). Tarsal segmentation 5-5-4; head very broad and flat (width about one-third greater than length of pronotum) (Fig. 30.88) .... CUCUJOIDEA (Inopeplidae, p. 600)

Tarsal segmentation other than 5-5-4; head not so broad and flat 10

10(9). Abdomen with 6 or 7 visible sternites; antennal club not compact and sharply defined (Fig. 30.35) .... STAPHYLINOIDEA (pt., p. 540)

Abdomen with 5 visible sternites; antennal club compact and sharply defined, or antennae reduced 11

11(10). Abdomen with at least the 3 apical tergites exposed; tarsi with 1 claw or with 2 unequal claws (Fig. 30.36) .... STAPHYLINOIDEA (Pselaphidae, p. 546)

Abdomen with only 2 apical tergites exposed; tarsi with 2 equal claws (Fig. 30.67) .... CUCUJOIDEA (Nitidulidae, p. 584)

12(7, 3). Abdomen with 6 or 7 visible sternites 13

Abdomen with 5, 4, or 3 visible sternites 18

13(12). Tarsal segmentation apparently 4-4-4 or 3-3-3 14

Tarsal segmentation 5-5-5 15

14(13). Tarsal segmentation 4-4-4 (apparently, actually 5-5-5 with the basal segment greatly reduced); terminal segment of maxillary palpi bluntly pointed (Fig. 30.26A) .... HYDROPHILOIDEA (Hydraenidae, p. 537)

Tarsal segmentation 3-3-3; terminal segment of maxillary palpi securiform or sharply pointed (Fig. 30.78) .... CUCUJOIDEA (Coccinellidae, p. 592)

15(13). Antennae with a distinct club 16

Antennae without a club; mostly narrow, flattened beetles 17

16(15). Claws often toothed beneath; penultimate tarsal segment deeply bilobed; antennal club sometimes with more than 3 segments (Fig. 30.65) .... CLEROIDEA (pt., p. 577)
Claws simple; penultimate tarsal segments not bilobed; antennal club of 3 segments .... STAPHYLINOIDEA (pt., p. 540)

17(15). Tarsi slender, as long as, or longer than, tibiae, 4th segment not bilobed; eyes very large in ♂, meeting anteriorly (Fig. 30.66) .... LYMEXYLOIDEA (pt., p. 579)
Tarsi usually stout, shorter than tibiae, 4th segment bilobed or entire at apex; eyes large in ♂ but not meeting anteriorly (Figs. 30.57, 58) .... CANTHAROIDEA (p. 568)

18(12). Hind coxa with a vertical declivity or a concavity, defined by a transverse ridge or covered by a lamina, against or into which the femur can be retracted (Figs. 30.46, 49, 50, 60, 61) 19
Hind coxa without such a steep or hollowed posterior face 29

19(18). Fore coxae obliquely or dorsoventrally elongate (Figs. 30.60, 61), usually projecting; if apparently globular, then the penultimate tarsal segments not bilobed; trochantins usually visible; frons sometimes with a median ocellus. [Fore coxae obscured by the closely applied, flattened fore tibiae in Byrrhidae] 20
Fore coxae globular; penultimate tarsal segments lobed beneath or bilobed; trochantins not obvious; frons never with an ocellus 28

20(19). Empodium between the claws large and bearing numerous setae; antennae very strongly flabellate (♂; Fig. 30.54); fore coxae elongate in the vertical direction .... RHIPICEROIDEA (p. 565)
Empodium, when present, small, bearing 2 or 3 setae; antennae clubbed, serrate, filiform, or flattened; fore coxae obliquely or transversely elongate or globular 21

21(20). Mid coxae widely separated; hind coxae very close at their inner ends (e.g. Fig. 30.49); tarsal segmentation 5-5-5, 4-4-4, or 3-3-3, with no segments bilobed; never with a median ocellus; colour dark brown, black, or metallic green. [Femora and tibiae often flattened and capable of being fitted precisely into cavities to conform with level of undersurface of body] 22
Mid coxae usually close together (e.g. Fig. 30.60), if widely separated then hind coxae also widely separated, or tarsi 4-4-4, with one or more segments lobed beneath; frons often with a median ocellus; colour brown or black, or yellowish and black 24

22(21). Abdomen with 3 visible sternites; tarsal segmentation 3-3-3; body length less than 1 mm (Figs. 30.6A, 16). [Strongly convex, black, shining; wings fringed with setae; 3-segmented antennal club bearing long setae] .... (Myxophaga: Sphaerioidea, p. 520)
Abdomen with 5 visible sternites; tarsal segmentation 5-5-5 or 4-4-4; size larger, length more than 1 mm 23

23(22). Without a defined clypeus .... BYRRHOIDEA (p. 559)
With a well-defined clypeus .... DRYOPOIDEA (Limnichidae, p. 560)

24(21). Penultimate tarsal segments deeply bilobed (Fig. 30.47); tarsal segmentation 5-5-5 or 4-4-4; antennae filiform, serrate, or with a 2-segmented club (Figs. 30.46–48) .... DASCILLOIDEA (p. 557)
Penultimate tarsal segments not bilobed; tarsal segmentation 5-5-5 (basal segment often very small, so segmentation is apparently 4-4-4); antennal club, if present, composed of 3 or more segments 25

25(24). Antennae with a well-marked 3- to 5-segmented club, or with the last 3 segments together longer than rest of antenna; ventral side of prothorax often with lateral hollows for reception of antennae at rest; head sometimes with a median ocellus 26
Antennae otherwise, club, if present, not 3- to 5-segmented nor with the last 3 segments elongate; prothorax without hollows for reception of antennae at rest; head without an ocellus 27

26(25). The 3 distal segments of antennae very long or large, their combined length equal to, or longer than, the other segments. [Prosternum short, so that the anterior aperture of the prothorax faces downwards; pronotum sometimes convex, forming a 'hood' (Figs. 30.61–63)] .... BOSTRYCHOIDEA (pt., p. 573)
The 3 distal segments of antennae not unduly elongate, together not longer than the other segments .... DERMESTOIDEA (p.572)

27(25). Body 10–20 mm long and very slender; abdomen 5 to 12 times as long as the pronotum (Fig. 30.66) .... LYMEXYLOIDEA (pt., p. 579)
Body of normal proportions, length less than 10 mm; abdomen less than 3 times as long as the pronotum .... DRYOPOIDEA (pt., p. 560)

28(19). Abdomen with the 2 basal sternites fused or immovably joined, the suture between them partly obliterated; metasternum with a well-marked transverse suture which curves backwards to meet the posterior edge of the sclerite on each side; prothorax not movable on mesothorax; all tergites of abdomen strongly sclerotized ... BUPRESTOIDEA (p. 564)
Abdomen with suture between basal and 2nd sternites as obvious as those between the other sternites; metasternum without a transverse suture; prothorax usually movable relative to the mesothorax; abdominal tergites much less strongly sclerotized than the exposed parts .... ELATEROIDEA (p. 566)

29(18). Tarsal segmentation 5-5-5 (basal segment often small) 30
Tarsal segmentation 3-3-3, 4-4-4, or 5-5-4 36

30(29). Fore coxae in the form of broad plates which project downwards from the prothorax, curve towards each other at their apices, and almost meet in the middle, covering the short, lightly sclerotized prosternum; head deflexed, covered by the pronotum as viewed from above. [Strongly convex, heavily sclerotized, black, length 1·5–3 mm (Figs. 30.27A, B)] .... HYDROPHILOIDEA (Georyssidae, p. 538)
Fore coxae not in the form of broad plates covering the prosternum; head usually visible from above 31

31(30). Tarsi with proximal segment very long and slender, longer than segments 2-5 together; antennae with elongate scape, short funicle, and a large, broad, flat, 1-segmented club (Fig. 30.112A). [Body notably cylindrical, like that in Bostrychidae] .... CURCULIONOIDEA (Curculionidae–Platypodinae, p. 621)
Tarsi with proximal segment not as long as segments 2-5 together; antennae with a short scape, and club, where present, of more than one segment 32

32(31). Junction of trochanter and femur transverse, at right angles to their axis (Fig. 30.63B); trochanters often unusually long; fore coxae conical, projecting; antennal insertions very close, sometimes confluent .... BOSTRYCHOIDEA (pt., p. 573)
Junction of trochanter and femur oblique (Fig. 30.65B); trochanters small; fore coxae usually not conical and projecting; antennal insertions not unusually close 33

33(32). Tarsi with a conspicuous bisetose empodium between the claws; fore coxae projecting; body usually with many erect bristly setae in addition to the normal clothing of small setae (Figs. 30.64, 65) .... CLEROIDEA (pt., p. 577)
Tarsi with empodium inconspicuous or absent; fore coxae not projecting; body without many erect bristly setae 34

34(33). Body length less than 2 mm; antennal club and large areas of ventral surface covered with very fine, short, dense, velvety, hydrofuge pubescence; scape long and curved to fit around lower side of eye; terminal segment of tarsi always as long as, or longer than, remaining segments together; dorsal surface almost always glabrous, never densely pubescent; coxae of each pair very closely approximated .... HYDROPHILOIDEA (Hydraenidae, p. 537)
Body length often more than 2 mm; antennal club and ventral surface of the body without hydrofuge pubescence; scape of antenna not unusually elongate (shorter than combined length of the remaining segments below the club); terminal segment of tarsi often shorter than remaining segments together; dorsal surface often pubescent; coxae of each pair not always close 35

35(34). Fourth tarsal segment usually clearly visible, but, if very small, then not in a deep concavity of 3rd segment .... CUCUJOIDEA (pt., p. 580, + Lyctidae)
Fourth tarsal segment very small, almost hidden at base of terminal segment and in a deep concavity of 3rd segment (Fig. 30.100B) 40

36(29). Tarsal segmentation 5-5-4 .... CUCUJOIDEA (pt., p. 580)
Tarsal segmentation 4-4-4 or 3-3-3. [Prothorax often without distinct lateral margins and head often with a rostrum] 37

37(36). Tarsal segmentation 4-4-4 or 3-3-3, with no segments bilobed or lobed beneath 38
Tarsal segmentation 4-4-4, with one or more segments bilobed or lobed beneath 39

38(37). Fore coxae strongly projecting, in the form of plates which curve towards each other, almost meeting in the middle and covering the reduced and lightly sclerotized prosternum. [Tarsi actually 5-5-5-segmented, but basal segments very small and hidden] .... HYDROPHILOIDEA (Georyssidae, p. 538)
Fore coxae not in the form of arched plates which cover the prosternum .... CUCUJOIDEA (pt., p. 580)

39(37). Metasternum without a transverse suture; tarsi with segments 1, 2, and 3 deeply bilobed, or with no segments bilobed; terminal segment of labial palpi often securiform; antennae thickened towards apex or clubbed, not elbowed .... CLEROIDEA (pt., p. 577)
Metasternum with a transverse suture near the posterior edge; tarsi with only segment 2 or segment 3 deeply bilobed, or without lobed segments (segment 2 usually triangular); labial palpi not securiform; antennae various, often filiform, or elbowed, or clubbed ... 40

40(39, 35). Head more or less produced forwards to form a rostrum, gular sutures usually confluent, forming a single median suture (Fig. 30.102C); antennae usually elbowed or clubbed, their scapes retractable into grooves on the head ...... CURCULIONOIDEA (pt., p. 613)
Head without a rostrum, 2 gular sutures present; antennae not elbowed, rarely clubbed; never with longitudinal grooves for reception of the scape 41

41(40). Antennae with a pronounced 3-segmented club; fore coxal cavities widely open behind; 3rd tarsal segments with the ventral lobe entire; body elongate .... CUCUJOIDEA (Languriidae, p. 589)
Antennae other than with a strong 3-segmented club; fore coxal cavities almost always closed, if open, then body not elongate; 3rd tarsal segments usually bilobed .... CHRYSOMELOIDEA (pt., p. 607)

Superfamily HYDROPHILOIDEA (Palpicornia)

Mainly aquatic or semiaquatic species, distinguished by elongate maxillary palpi which appear to have taken over the sensory function of the antennae, which in turn are involved in respiration. Antennae short, the segment below the club (rarely the second below) cup-shaped or acutely angulate (Figs. 30.25A–D); the apical 3–5 segments forming a strongly pubescent club, the cup-shaped segment usually the 6th; wings usually with a distinct M–Cu loop (not in Hydraenidae); usually with a Y-shaped line on the frons; 6 Malpighian tubes. Larvae active and usually carnivorous; adults phytophagous.

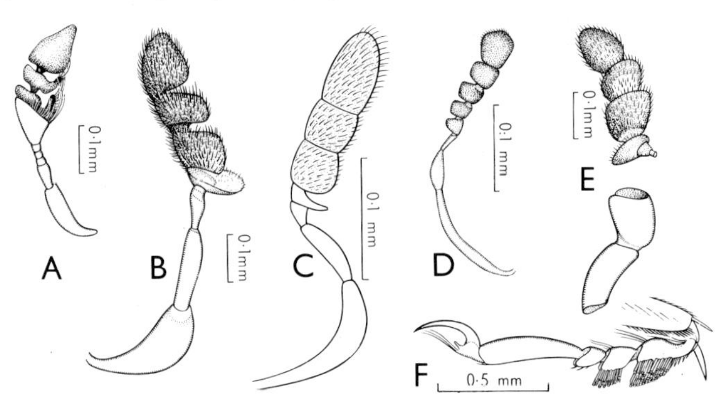

Fig. 30.25. Antennae of adult Hydrophiloidea: A, *Hydrophilus* sp., Hydrophilidae; B, *Berosus australiae* Hydrophilidae; C, *Hydrochus* sp., Hydrochidae; D, *Ochthebius* sp., Hydraenidae; E, *Spercheus platycephalus* Spercheidae, broken to show 3rd segment. F, tarsus of *Berosus australiae*. [F. Nanninga]

Key to the Families of Hydrophiloidea

1. Abdomen with 6 or 7 sternites; antennae with a 5-segmented pubescent club (Fig. 30.25D); tarsi without a setose empodium between the claws; length less than 2 mm ...... **Hydraenidae** (p. 537)
   Abdomen with 5 sternites; antennae with 3 pubescent segments beyond the cupule (Figs. 30.25A–C, E); tarsi with a setose empodium between the claws; length 2 mm or more 2
2. Antennae with 2 or 3 segments before the cupule (Fig .30.25C, E); fore coxal cavities closed behind ... 3
   Antennae usually with 5 well-developed segments before the cupule, sometimes 3 or 4 (Figs. 30.25A,B); fore coxal cavities open behind 4
3. Antennae with pedicel and cupule pubescent, the latter looking like part of the club (Fig. 30.25E); tarsi with a large plurisetose empodium between the claws; body form broad and very convex (Fig. 30.27C) .... **Spercheidae** (p.538)
   Antennae with pubescence limited to the 3-segmented club; tarsi with a normal bisetose empodium between the claws; body form elongate and parallel-sided .... **Hydrochidae** (p. 538)
4. Maxillary palpi shorter than antennae; the 2 basal abdominal sternites connate; fore coxae very large, concealing the sternum; tarsi apparently 4-4-4-segmented. [Length *ca* 2 mm] .... **Georyssidae** (p. 538)
   Maxillary palpi as long as, or longer than, antennae; the 2 basal abdominal sternites not connate; fore coxae not concealing prosternum; tarsi usually distinctly 5-5-5-segmented .... **Hydrophilidae** (p. 539)

10. Hydraenidae (Limnebiidae; Figs. 30.26A, 25D). Very small (*ca* 1·5 mm), elongate beetles, distinguished by the presence of 6 or 7 abdominal sternites and a 5-segmented antennal club. Maxillary palpi longer or shorter than the antennae; tarsal segmentation 5–5–5 or 4–4–4, apical segments long. Larvae active, subcylindrical, with well-developed legs and a 10-segmented abdomen, segment 10 with a pair of curved, downwardly projecting hooks; segment 9 with short 2-segmented urogomphi; 9 pairs of functional annular spiracles. The larvae live close to water, but breathe air, and are easily drowned. Some adult Hydraenidae breathe in water by using a film of air held on hydrofuge pubescence on the ventral surface. Both adults and larvae feed on algae.

The genera *Hydraena*, *Ochthebius*, and *Gymnoclathebius* are so far known in Australia. Hydraenidae are easily overlooked, and it is likely that many more species remain to be found. Nothing is recorded of the habits of the Australian species, but elsewhere species are known to inhabit both running and stagnant water, while some live in water seepages on rock faces and others in waterside gravel, in brackish ditches, marine rock pools (above high tide but in the splash zone), and even in the hypersaline water beneath the crust of salt pans. There is considerable

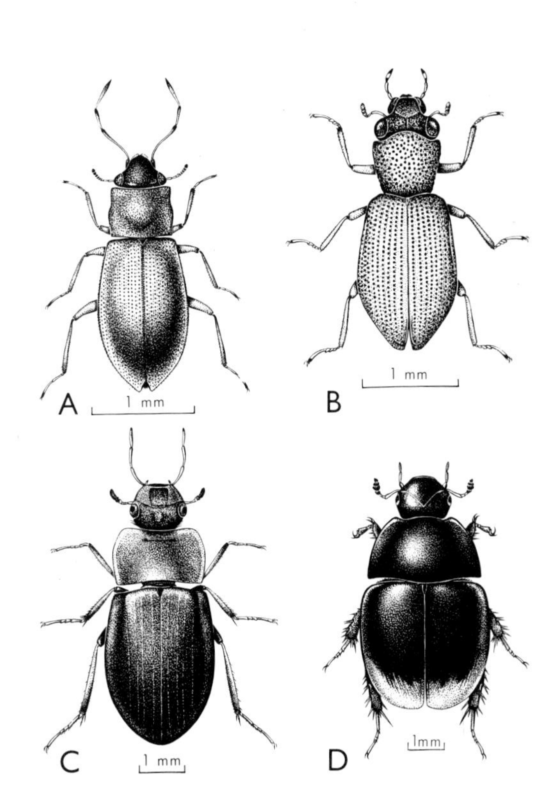

Fig. 30.26. Hydrophiloidea: A, *Hydraena luridipennis*, Hydraenidae; B, *Hydrochus serricollis*, Hydrochidae; C, *Helochares australis*, Hydrophilidae; D, *Sphaeridium dimidiatum*, Hydrophilidae. [F. Nanninga]

doubt concerning the systematic position of the family. It shows affinities with the Staphylinoidea in wing venation, absence of the basal piece of the aedeagus, presence of a distinct segment 9 (tergite and sternite) in the abdomen of the adult, and the annular spiracles of the larvae (biforous in typical Hydrophiloidea).

11. Hydrochidae (Figs. 30.26B, 25C). Small (*ca* 3 mm), dark brown or somewhat metallic beetles of elongate, narrow form, rather like Hydraenidae; antennae with 3 segments below the cupule, which is followed by a densely pubescent 3-segmented club; tarsi 5-5-5-segmented, but with basal segment very small and difficult to see, apical segment large, longer than the rest together; empodium bisetose; eyes hemispherical; scutellum minute; ventral surface and bases of femora clothed with a dense velvety pile. The single genus *Hydrochus* is more or less world-wide in distribution. The beetles are not active, and are found attached to plants in ponds or slow-moving creeks, but little is known of their habits or larvae.

12. Spercheidae (Figs. 30.27C, 25E). Distinguished by the fact that what appears to be the hydrophiloid cupuliform antennal segment is pubescent like the segments of the club. It is the 4th segment, and is preceded by a small cylindrical non-pubescent segment which may, in fact, be homologous with the hydrophiloid cupule. The pedicel is large and pubescent. The family includes the single genus *Spercheus*. *S. platycephalus* Macl. is found in Australia, New Caledonia, and Indonesia. It is a brown beetle, about 4 mm long, with the elytra coarsely punctured between very strong longitudinal ribs. Both adults and larvae of Spercheidae occur in stagnant ponds, and are reported to walk in an inverted position on the underside of the surface film.

13. Georyssidae (Figs. 30.27A, B). Small (*ca* 2 mm), black, stout-bodied beetles with tuberculate sculpture; antennae 7- or 9-segmented, the 2 basal segments stout, and 1 or 3 apical segments forming a short oval club, the insertions of the antennae covered from above by the projecting frons; head deflexed, covered from above by the pronotum; fore coxae very large, filling the gap between head and mesosternum, concealing the lightly sclerotized, very short, transverse prosternum; tarsi 5-5-5-segmented, the basal segment very short and hidden; abdomen with 5 visible sternites, the first 2 connate. Aedeagus simple, trilobed, very like that in Hydrochidae. The beetles are found at the edge of fresh water, in sand or mud. The single genus *Georyssus* is represented throughout a large part of the world. The Australian species are *G. australis* King (eastern Australia), *G. kingi* Macl. (Queensland), and *G. occidentalis* Carter (W.A.).

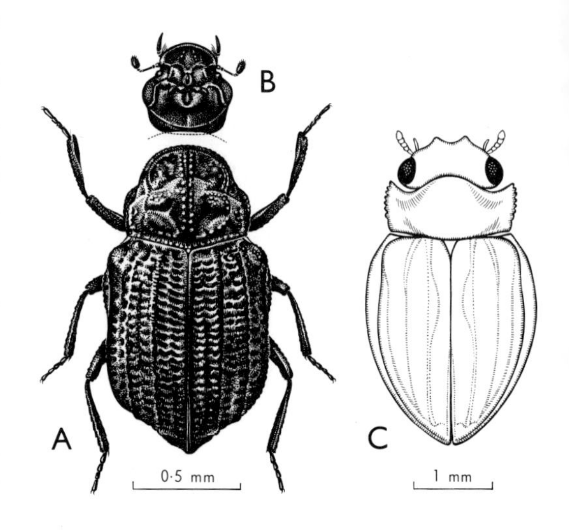

Fig. 30.27. A, *Georyssus australis*, Georyssidae, dorsal; B, same, head; C, *Spercheus mulsanti*, Spercheidae, dorsal. [F. Nanninga]

The systematic position of the family is doubtful. It has commonly been associated with Byrrhidae and Helminthidae, but Crowson (1955) has drawn attention to the somewhat hydrophiloid antenna, concealed antennal insertions, closed fore coxal cavities, absence of a marked prosternal process, reduction of the basal tarsal segment, and hind coxae which lack a cavity or declivity, all characters which tend to indicate a relationship with the hydrophiloid families. In addition, van Emden, in 1956, described a hydrophiloid larva which was claimed, by association, to be that of *Georyssus*. On the other hand, *Georyssus* has short maxillary palpi, a strongly deflexed head, and connate basal abdominal sternites, characters which tend to exclude it from the Hydrophiloidea.

14. Hydrophilidae (Figs. 30.26C, D, 25A, B, F). A world-wide family, of which 19 genera are so far known to occur in Australia. The adults range in size from 1 to 40 mm, and are usually aquatic and of smooth oval form and drab colour, many superficially resembling Dytiscidae, but distinguishable by the elongate, antenna-like maxillary palpi. Abdomen with 5 visible sternites, none connate; tarsal segmentation 5-5-5; a setose empodium between the claws. The adults of the smaller aquatic species are able to renew the reserve of air held on the hydrofuge pile on the ventral surface of the body by making use of bubbles of oxygen produced by the algae on which they feed, but the larger species have to return to the surface to renew their supplies. Air is stored between the elytra and abdomen, and this reservoir communicates with air held on areas of hydrofuge pile on each side of the ventral surface of the thorax and abdomen. The spiracles open into the air reservoir under the elytra. When the insect rises to the surface to renew its air it inclines the body to one side, so that the angle between the head and prothorax on one side is brought into proximity with the surface and the hydrofuge, pilose club of the antenna is used to break the surface film, allowing the submerged air-reservoir to communicate with the atmosphere. *Berosus*, *Hydrobius*, and *Sternolophus* stridulate when alarmed.

The larvae are diverse in form and are usually carnivorous, whereas the adults feed upon vegetable matter. Most are aquatic, but some live in dung, damp soil, or decaying vegetable matter. Aquatic forms are almost all metapneustic, breathing at the surface of the water by means of the 8th spiracles which open into an atrium. Such larvae are easily drowned if unable to reach the surface. The larvae of *Berosus* are apneustic, and have 7 pairs of lateral gills.

The largest Australian species are the shining black *Hydrophilus latipalpus* Cast. and *H. albipes* Cast., both of smooth, oval form, with a strong mid-ventral keel running between all 3 pairs of coxae. *Sternolophus marginicollis* Hope is a common aquatic species occurring in northern, eastern, and southern Australia. *Cercyon haemorrhoidalis* F., a small (3 mm) convex beetle with reddish elytra, is an introduction from Europe, and is common in dung.

Superfamily HISTEROIDEA

Adults are compact, with retractile legs and antennae, and hard, dark-coloured, polished integument. Antenna almost always elbowed, having a long, curved scape followed by 7 small segments and ending in a compact, pilose, 3-segmented club. Elytra never fully 9- or 10-striate, truncate, leaving the propygidium and pygidium exposed. Wings usually with M–Cu loop distinct. Abdomen with 5 visible sternites and 7 tergites, all heavily sclerotized. Fore coxae strongly transverse; mid and hind coxae oval and widely separated. Tarsal segmentation 5-5-5; 6 Malpighian tubes; aedeagus with parameres fused. Larval maxillae without lacinia.

15. Histeridae (Fig. 30.28). The Australian species (in 27 genera) are mostly of flattened, quadrate, compact form, with the

Fig. 30.28. *Hister walkeri*, Histeridae [F. Nanninga]

head deeply sunk into the prothorax. Antennae elbowed, with a pronounced club; integument shining, very strongly sclerotized and black or metallic green (e.g. in *Saprinus*); length 1–16 mm, the largest species being included in the genus *Hololepta*. Wings are well developed; adult Histeridae are able to fly. As far as is known, both adults and larvae are carnivorous, preying upon larvae of other insects, and they are commonly found in carrion, dung, and decaying vegetable matter, or under the bark of dead or dying trees. Some species (e.g. of *Chlamydopsis*) are equipped with trichomes on the elytra and live in ants' nests; others may be termitophilous. Histeridae are also likely to be found in the nests of birds and mammals, in bat guano, and in fungi. The Histeridae found in stored food products have been reviewed by Hinton (1945).

Superfamily STAPHYLINOIDEA

Venation of wings, when present, with M–Cu loop absent; abdomen with at least one, usually more, of the basal tergites membranous, in sharp contrast with the sclerotized posterior segments; antennae not elbowed and with, at most, a loose club; body often elongate, and integument rarely heavily sclerotized; elytra usually abbreviated, usually with more than the 2 apical tergites exposed; abdomen usually very flexible. Larvae with galea and lacinia of maxilla more or less fused. Carnivorous, mould-feeding, sometimes myrmecophilous, very rarely phytophagous.

Key to the Families of Staphylinoidea Known in Australia

1. Length 1·2 mm or less; hind coxae with a very deep cavity into which the femora can be retracted; wings, when present, narrow, with a marginal fringe of very long setae (Fig. 30.30) 2
   Length more than 1·2 mm; hind coxae with, at most, shallow cavities for reception of hind femora; wings not narrow, without a fringe of long, fine setae 3
2. Eyes absent; wings absent; prosternal intercoxal process broad and long, extending back over mesosternum (Fig. 30.29); antennae short, retractable into grooves on lower side of head. [Myrmecophilous] .... **Limulodidae** (p. 541)
   Eyes present; wings present and fringed with long hairs; prosternal intercoxal process narrow, not extending back over mesosternum; antennae not retractable into grooves ...... **Ptiliidae** (p. 541)
3. Elytra covering the whole of the abdomen, or leaving some tergites exposed; antennae usually clubbed; abdomen with at least 3, usually 4 of the basal tergites membranous or of different texture from the apical tergites, not flexible 4
   Elytra abbreviated, leaving at least 3 tergites exposed; antennae usually filiform; abdomen with not more than 2 basal tergites membranous or of different texture from 3rd and succeeding tergites, usually very flexible dorsoventrally 7
4. Last 3 to 5 segments of antenna forming a loose club, and with segment 8 nearly always smaller than segments 7 and 9 (sometimes, e.g. in *Dietta*, so small as to be scarcely recognizable); elytra entire and completely covering the abdomen; fore coxal cavities closed behind. [♂ nearly always with the fore tarsi expanded; 5 mm or less] .... **Anisotomidae** (p. 542)
   If with antennal segment 8 smaller than segments 7 and 9, then elytra truncate; fore coxal cavities open behind 5
5. Fore coxae with trochanters exposed; antennae widely separated; hind coxae very close or contiguous. [Large, flat, carrion beetles, 16–30 mm]. .... **Silphidae** (p. 542)
   Fore coxae with trochanters hidden; antennae inserted on the frons, not widely separated; hind coxae fairly widely separated 6
6. First abdominal sternite as long as sternites 2–4 together; elytra truncate, the last abdominal tergite usually exposed, conical; the 3 basal abdominal tergites membranous, remainder sclerotized; maxillary palpi with penultimate and terminal segments about equal in length; body glabrous and shining. [1–5 mm] .... **Scaphidiidae** (p. 543)
   First abdominal sternite not or hardly longer than 2nd; elytra usually entire, covering whole

abdomen; at least the 4 basal abdominal tergites membranous; maxillary palpi with penultimate segment very large compared with the terminal segment; body pubescent, not shining. [Eyes coarsely granular] .... **Scydmaenidae** (p. 542)

7. Abdomen not flexible dorsoventrally; tarsi 3-3-3-segmented; tarsi with one claw or 2 very unequal claws; abdomen with 5 or 6 visible sternites; antennae usually more or less clubbed; terminal segment of maxillary palpi much larger than the rest; labial palpi 1- or 2-segmented. [Body usually with deep pits at various places, especially on vertex of head and sides of mesosternum] .... **Pselaphidae** (p. 546)

   Abdomen more or less flexible dorsoventrally; tarsal segmentation various; tarsi with 2 equal claws; abdomen with 6 or 7 visible sternites; antennae usually filiform; terminal segment of maxillary palp usually not larger than the penultimate segment; labial palpi 3-segmented .... **Staphylinidae** (p. 544)

16. Limulodidae (Fig. 30.29). Minute (less than 1 mm), reddish-brown beetles with semicircular, convex pronotum which is rather more voluminous than the slightly tapered elytra; abdomen tapered, with 3 segments exposed in life, but highly contractile, often covered by the elytra in dry specimens; body clothed with fine, recumbent, sericeous pubescence; head deflexed; eyes absent; antennae short, clubbed; posterior angles of pronotum produced backwards into acute angles as in Corylophidae (p. 592); fore coxae globular, widely separated, the broad intercoxal process continued back to cover the mesosternum; hind coxae in the form of large plates beneath which the hind femora can be retracted; femora very short and wide, with a cavity for reception of the tibia; tarsi 3-3-3-segmented; abdomen with 6 visible sternites; wings absent. The Australain species belong to the genus *Rodwayia* (6 spp.), and are found under stones, in the nests of ants (*Camponotus*, *Polyrhachis*, *Iridomyrmex*, *Rhytidoponera* spp.) in all States. They are largely ignored by their hosts, and appear to feed on the exudations of the ant larvae. [Seevers and Dybas, 1943.]

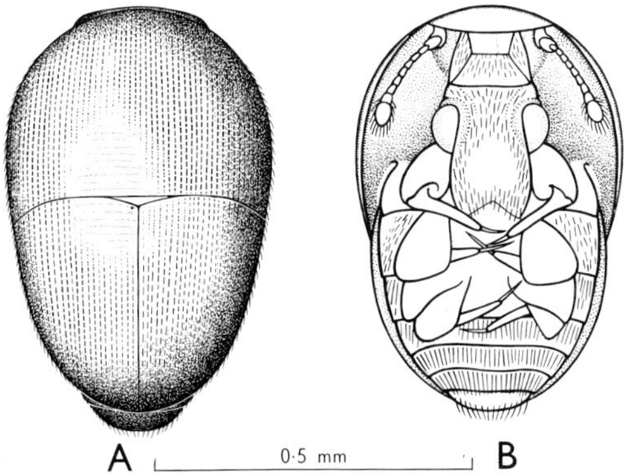

Fig. 30.29. *Rodwayia orientalis*, Limulodidae: A, dorsal; B, ventral. [F. Nanninga]

17. Ptiliidae (Trichopterygidae; Fig. 30.30). Minute beetles, less than 2 mm long, including the smallest known Coleoptera. Characterized by the narrow wings, fringed with very long setae, without venation, the well-developed hind coxal plates which cover the hind femora when these are retracted, and the antennal segments each with a whorl of long setae. Body pubescent; antennae 11-segmented with a loose 3-segmented club; fore coxae globular, close together; tarsal segmentation 3-3-3, the terminal segments very long; elytra truncate or entire; abdomen with 6 or 7 visible sternites. Larvae elongate, with 3-segmented antennae, unsegmented urogomphi, and well-developed thoracic legs. The pupa is obtect. *Actinopteryx fucicola* Allibone is found in decaying seaweed, and is widely distributed on sea coasts throughout the world. Other species occur in mouldy, decaying plant debris, dung, and ants' nests. [Deane, 1932a.]

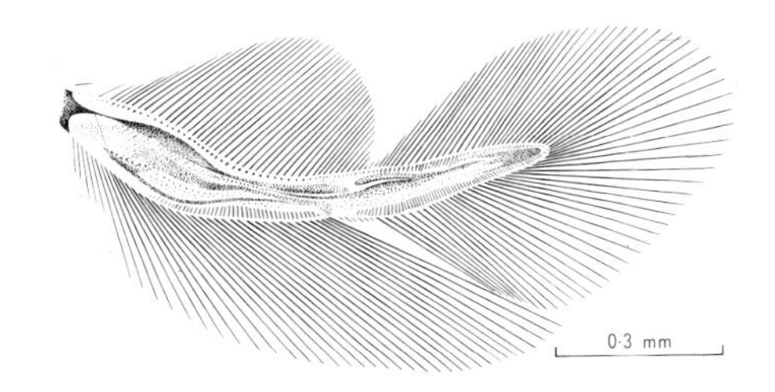

Fig. 30.30. Ptiliidae, wing. [F. Nanninga]

18. Anisotomidae (Leiodidae, Catopidae; Fig. 30.31). The Australian species are small (5 mm or less), elongate-oval, and convex; colour dark brown or reddish brown, almost always pubescent, and distinguished from other staphylinoids by the elytra, which cover the whole of the abdomen, and the antennae, which have a more or less well marked 5-segmented club in which segment 8 is almost always smaller (shorter, or shorter and narrower) than segments 7 and 9. In *Dietta sperata* Sharp, from the north-west, the club is very strongly developed, with segments 7, 9, 10, and 11 disc-shaped and 8 small and hidden, its presence indicated only by the gap between 7 and 9 which is wider than between 9 and 10. Fore coxal cavities closed behind; abdomen with 6 visible sternites; antennae with one or more internal vesicles (presumably chemoreceptors) opening on the distal surface of the 9th and 10th segments; tarsal segmentation normally 5-5-5. The shiny, non-pubescent species (e.g. *Anisotoma* spp.) are found in fungi, while the pubescent forms are often found in carrion, but may also occur in subterranean fungi or in the nests of mammals or insects (e.g. *Myrmecholeva*). In the northern hemisphere there has been a considerable proliferation of species adapted to cave life.

19. Scydmaenidae (Fig. 30.32). Small or minute beetles, distinguished from related families by their 'waisted' body form, complete elytra, general pubescence, and coarse granular eyes. Fore and mid coxae close together, hind coxae widely separated; basal abdominal sternite not exceptionally large, as it is in Scaphidiidae; antennae moniliform or weakly clubbed; penultimate segment of maxillary palp much larger than terminal segment; abdomen with 6 visible sternites; tarsi 5-5-5-segmented; hind wings absent. Larvae and adults occur in damp places, in leaf-mould, moss, or under stones, and are carnivorous. Some live in ants' nests (e.g. *Phagonophana* spp.). Almost nothing is known of the biology of the family. The best-represented genera are *Scydmaenus* and *Phagonophana*.

20. Silphidae (Fig. 30.33). Mostly confined to temperate regions. The 3 Australian species are large (16–40 mm), flattened beetles found in carrion. Distinguished from beetles of related families in Australia by their size; exposed trochanters of the fore coxae; the 4 basal tergites of the abdomen, which are soft, at least on each side of the middle, in contrast to the wholly sclerotized posterior

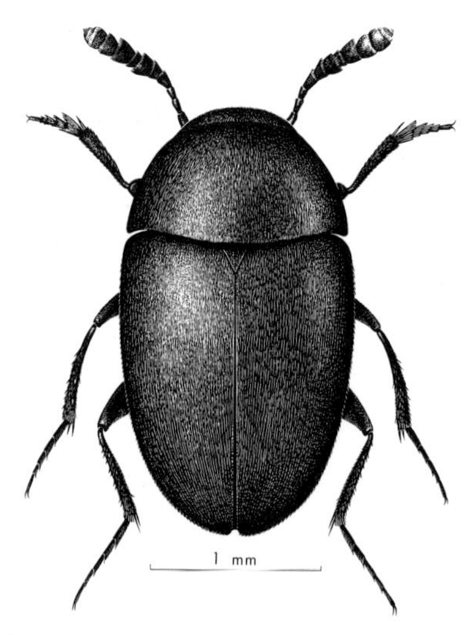

Fig. 30.31. *Choleva australis*, Anisotomidae. [F. Nanninga]

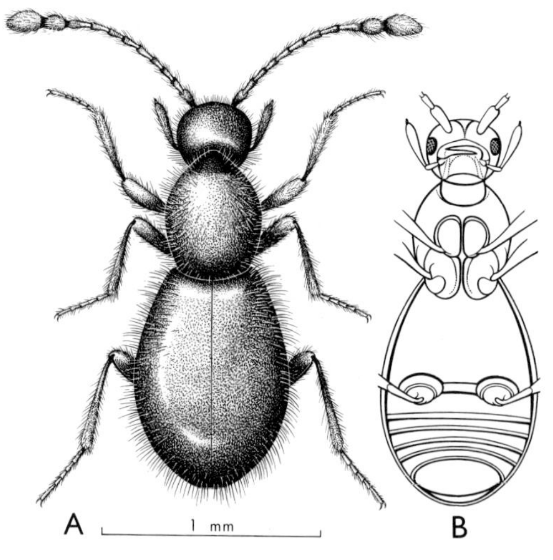

Fig. 30.32. *Heterognathus myrmecophilus*, Scydmaenidae: A, dorsal; B, ventral. [F. Nanninga]

segments; antennae widely separated, with a densely and finely pubescent, 3-segmented club, and segment 8 not smaller than segments 7 and 9, black with the apical 1 or 3 segments bright yellow; fore coxal cavities open behind; fore and hind coxae contiguous; elytra covering the abdomen or not; tarsal segmentation 5-5-5.

Ptomaphila lacrymosa Schreib. and *P. perlata* Kraatz are large, very much flattened beetles with the middle of the body dark brown and the expanded margins of the pronotum and elytra dull yellowish brown. The elytra cover the abdomen completely, and are marked with elongate ribs and tubercles. Both species occur in the south-east and Tasmania. *Diamesus osculans* Vigors is widely distributed from India through Indonesia and New Guinea to Queensland. In this species the elytra are truncate, leaving 4 or 5 abdominal tergites exposed. The general colour is black, the elytra having two dull orange marks. The larvae are broad and depressed, pigmented, and of 'trilobite' form, the segments very transverse with broad lateral plate-like expansions which have acute posterolateral angles. The biology of the Australian species has not been described, but the carrion beetles of the northern hemisphere are known as 'burying-beetles' from their habit of burying the corpses of small animals to serve as food for the larvae. The adults are reported to feed the young larvae. A few Silphidae feed upon snails, and others are pests of garden crops.

21. Scaphidiidae (Fig. 30.34). Small (1·5–5 mm), highly convex, shining, stoutly fusiform beetles, with small head and truncate elytra which, in life, expose the conical, pointed abdominal apex; abdomen with 6 visible sternites, 1st as long as 2–4 together; legs long, tarsi 5-5-5-segmented; antennae 11-segmented, the last 5 segments pubescent

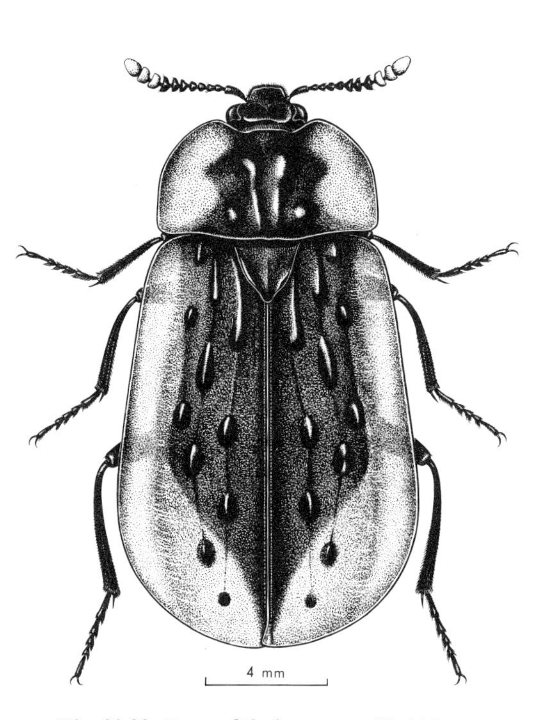

Fig. 30.33. *Ptomaphila lacrymosa*, Silphidae. [F. Nanninga]

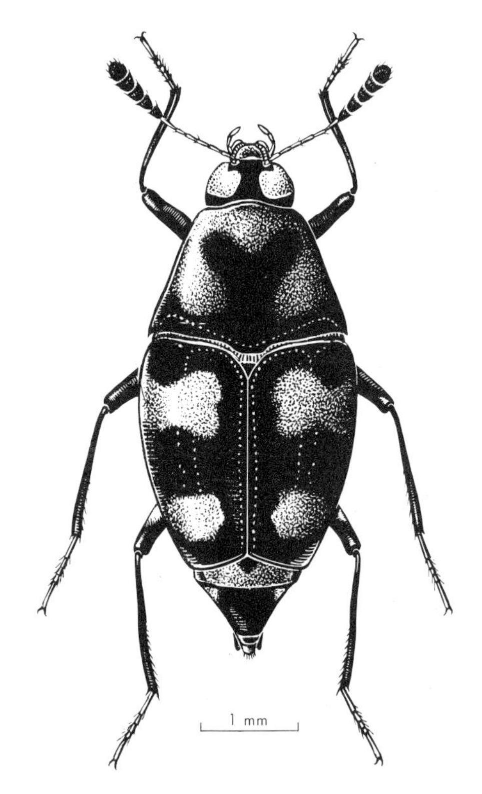

Fig. 30.34. *Scaphidium quadripustulatum*, Scaphidiidae. [F. Nanninga]

and more or less expanded to form a loose club, 8th sometimes smaller than 7th or 9th; mid and hind coxae widely separate, fore coxae contiguous; hind wings well developed. The beetles are found in rotten wood, fungi, among dead leaves, or beneath bark, and the larvae of some are known to feed on fungi. The larvae are fusiform, pigmented, with 3-segmented antennae and 2-segmented urogomphi. The principal Australian genera are *Scaphidium* and *Scaphisoma*.

22. Staphylinidae (rove beetles; Fig. 30.35). Elytra very short, nearly always* leaving more than half of the abdomen exposed. In general, Staphylinidae are elongate, parallel-sided, or fusiform beetles, 1–20 mm in length, with a very flexible abdomen, and usually with functional flying wings, which are packed away in complicated folds beneath the short elytra but can be unfolded for flight with great rapidity. Antennae usually 11-segmented, filiform, moniliform, or with a (usually) ill-defined club; scape long and other segments more or less equal in length; eyes varied in size, sometimes absent; exposed tergites of abdomen usually sclerotized and often laterally margined; 6 or 7 visible sternites; tarsal formula various; aedeagus without a basal piece. Staphylinid larvae are active, with 2-, 3-, or 4-segmented antennae, 10-segmented abdomen, and long, 1- or 2-segmented urogomphi; spiracles simple, annular, present on mesothorax and abdominal segments 1–8.

The Staphylinidae constitute one of the major families, and many species undoubtedly remain to be discovered in Australia. This may be judged from the fact that 950 species are known in the British Isles, with a very much smaller area and more uniform environment. They are found in a wide variety of habitats, e.g. in decaying vegetable matter of all sorts, under stones, at the edges of creeks, in moss, fungi, dung, carrion, seaweed, under bark, in flowers, and in the nests of birds, mammals, Hymenoptera, and termites. Many species are predacious, but the precise feeding habits of the majority are not known.

* An exception is *Sartallus signatus* Sharp (Oxytelinae), a small, yellow, oval species found on beaches under carrion.

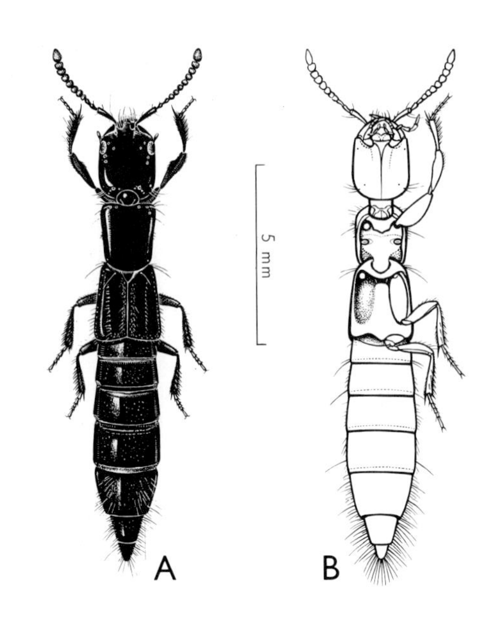

Fig. 30.35. *Eulissus chalcopterus*, Staphylinidae: A, dorsal; B, ventral. [F. Nanninga]

Key to the Subfamilies of Staphylinidae

1. Mesothoracic spiracles (on the intersegmental membrane between pro- and mesothorax) concealed by the triangular proepimera which are fused with the pronotal epipleura at the sides 2
   Mesothoracic spiracles exposed, or, if concealed, then with proepimera free from the pronotal epipleura 7
2. Abdomen with 7 visible sternites; antennae inserted under edge of frons between the anterior end of the eye and the clypeus on each side .... OXYTELINAE
   Abdomen with 6 visible sternites; antennae under prominent anterolateral angles of the frons, or otherwise 3
3. Head with a pair of ocelli level with the posterior edges of the eyes .... OMALIINAE
   Head without a pair of ocelli 4
4. Head with two acute-angled setose processes of the labrum extending forwards from beneath the clypeus, above the mandibles; eyes very large, bulging ...... MEGALOPSIDIINAE
   Head without such processes; eyes rarely very prominent 5

5. Antennae with the 2 terminal segments enlarged, forming a sharply defined club; tarsi 4-4-4-segmented; length 2 mm or less. [Very like Pselaphidae] .... EUAESTHETINAE
   Antennae without a club, or with club of 3 or more segments; tarsi 5-5-5-segmented; length greater than 2 mm........................ 6
6. Eyes not strongly convex, not occupying most of the sides of the head; basal segment of maxillary palp very short .... PAEDERINAE
   Eyes large, bulging, occupying most of the sides of the head; basal segment of maxillary palp elongate .... STENINAE
7. Antennae inserted between eyes, posterior to the level of the anterior ends of the eyes; inner edges of eyes converging anteriorly; elytral epipleura not defined. [Head narrowed anteriorly, width between the outer edges of the bases of the mandibles less than width between the middles of the inner edges of the eyes] .... ALEOCHARINAE
   Antennae inserted in front of eyes, anterior to the level of the anterior ends of the eyes; inner edges of eyes not converging anteriorly; elytral epipleura usually defined by a longitudinal ridge 8
8. Body not fusiform; epipleura of elytra not defined by a longitudinal ridge . STAPHYLININAE
   Body fusiform, the head small and deeply inserted in the prothorax, and the abdomen very strongly tapered; epipleura of elytra defined by a longitudinal ridge .... TACHYPORINAE

The important subfamilies Staphylininae, Aleocharinae, and Tachyporinae are distinguished by the fact that the prothoracic epimeron is reduced so that the cavities of the fore coxae are widely open below and behind, and the large mesothoracic spiracle is visible from below. In the Aleocharinae the antennae, which are usually thickened, are inserted close to the anterior inner edge of the eye, whereas in Staphylininae they are inserted on the front edge of the frons well away from the eye.

The ALEOCHARINAE are small or minute, dull-coloured beetles, the principal genera being *Atheta (= Homolota)*, *Calodera*, *Tachyusa*, and *Gyrophaena*. Species of *Dabra* and *Dabrosoma* are found in ants' nests, and *Coptotermoecia*, *Hetairotermes*, *Termitoecia*, and *Termitaptocinus* live with termites. The larger species of the family belong to the STAPHYLININAE. The fore coxae are conical and elongate, and the tarsi 5-5-5-segmented. *Creophilus erythrocephalus* F., the largest common species, is of striking appearance, 15–20 mm in length, the body black, and the head bright orange with a central black spot and black antennae and mandibles. It is widely distributed, and found commonly in carcasses where it preys upon other insects. When disturbed, this insect assumes a threatening attitude, facing the intruder and raising the head and abdomen. *Actinus imperialis* Fauv. and *A. macleayi* Oll. are large and handsome species from north Queensland. They have the head and pronotum of a metallic coppery green colour and the elytra of deep metallic purple. *Cafius sabulosus* Fauv. is a common littoral species, up to 10 mm in length, with shining black head and pronotum and dark brown elytra and abdomen. It is found beneath piles of seaweed cast up on the beach. The tribe Amblyopinini includes species that are parasitic on mammals in South America and Australia. *Myotyphlus jansoni* (Matthews), the only Australian species, is found in the fur of native species of *Rattus* in Tasmania and Victoria, and has been collected in bat guano in caves (Hamilton-Smith and Adams, 1966). The beetle is about 7 mm long, wingless, with single-faceted eyes and elytra reduced to small scales.

The TACHYPORINAE have a characteristic fusiform shape, with a small rounded head sunk into the prothorax, the latter widened towards the base, elytra usually longer than the metasternum, and the abdomen markedly tapered posteriorly. In Australia the subfamily is represented by *Conosoma* (31 spp.), *Tachyporus* (4 spp.), *Coproporus* (7 spp.), *Tachinomorphus* (3 spp.), and *Tachinus* (1 sp.). The insects are found in decaying wood, dead leaves, etc., where they feed on other insects.

The PAEDERINAE include some species which are common in wet places, such as the gravel and sand at the margins of streams. Species of *Paederus* are narrow in form, 6–10 mm in length, usually with head black, the pronotum shining reddish yellow, very short metallic blue elytra, and a black or red and black abdomen. The mandibles are falcate, coxae strongly projecting, tarsi 5-5-5-segmented, and the terminal segment of the maxillary palp very short (less than one-fifth of the length of the penultimate segment). The larger genera of Paederinae are *Cryptobium*, *Lathrobium*, *Hyperomma*, *Scopaeus*, *Paederus*, and *Pinophilus*.

Species of *Paederus*, when irritated or crushed, produce a fluid that causes severe blistering of the human skin. The phenomenon is known in Africa, the Oriental region, South America, and Australia, and the name 'whip-lash rove beetle' has been applied to the genus in reference to the linear appearance of the lesion, which is the common result of brushing the insect off the skin with the hand. The skin reaction appears one or two days after contact with the beetle, and as a result is rarely associated with its cause (McKeown, 1951). In Australia the species usually involved is the widespread *Paederus cruenticollis* Germ. (Plate 5,J), which lives on the gravel margins of rivers.

The STENINAE include only 2 genera, *Stenus* and *Dianous*, of which only *Stenus* is represented in Australia (25 spp.). *Stenus* is characterized by large, hemispherical eyes (seen otherwise only in *Megalopsidia*), coarse and dense puncturation, and the insertion of the antennae between the eyes. Beetles of this subfamily live in marshes and at the edges of ponds and streams. Their interest lies in the fact that they are the only animals that are able to move over the surface of water by reducing the surface tension behind the body. The resulting 'skimming' movement is very rapid, and the beetle directs itself towards the nearest bank, orientating itself by the contrast of illumination between the edge of the bank and the sky. The ability to move speedily to the bank presumably has survival value in an insect that lives close to water.

The MEGALOPSIDIINAE, represented by 3 species of *Megalopsidia*, resemble the Steninae in having large bulging eyes, but are clearly distinguished by two sharply angled processes which extend forwards from beneath the clypeus. The pronotum is transversely sulcate. OMALIINAE include only 4 genera: *Omalium* (7 spp.), *Amphicroum* (4 spp.), *Phloeonomus* (3 spp.), *Phyllodrepa* (1 sp.). The subfamily is distinguished by the presence of a pair of ocelli on the posterior part of the head. The general form is flat, with elytra covering more of the abdomen than is usual in the family. Another small subfamily, the OXYTELINAE, includes 8 genera of which only *Oxytelus* (40 spp.) and *Bledius* (17 spp.) are noteworthy. Species of *Bledius* are gregarious, live in sand (sea shore, river sandbanks, and sandpits), and have a cylindrical form and digging fore tibiae. *Oxytelus* species are found in decaying vegetable matter, including seaweed. They have a strongly transverse pronotum with longitudinal depressions. The Oxytelinae are unusual in that some species exhibit a degree of parental care of eggs and young larvae (Hinton, 1944).

23. Pselaphidae (Fig. 30.36). Small, reddish or yellowish beetles (length 0·75–3·5 mm), distinguished from all except Staphylinidae by the greatly abbreviated elytra, which leave most of the abdomen exposed. From Staphylinidae, the Pselaphidae are distinguishable by the non-flexible abdomen, which is oval in shape and much stouter than the head and pronotum. Antennae clubbed, and often with less than 11 segments; tarsal formula 3-3-3 or less; abdomen with 5 or 6 visible sternites. Pselaphidae are found in decaying vegetable matter, in forest-floor litter, under bark, in moss, beneath logs and stones, or in the nests of ants and termites. With the possible exception of some myrmecophilous species, the adults and larvae are carnivorous. The CLAVIGERINAE are specialized for life in ants' nests, having highly modified antennae, reduced maxillary palpi, and trichomes in the form of a hollow or hollows at the base of the abdomen surrounded by tufts of yellow hair. The best represented genus of this subfamily is *Articerus*, in which the tarsi are 1-segmented, and the antennae are reduced to the elongate scape which looks as if it had been broken off near the apex. The PSELAPHINAE, in contrast,

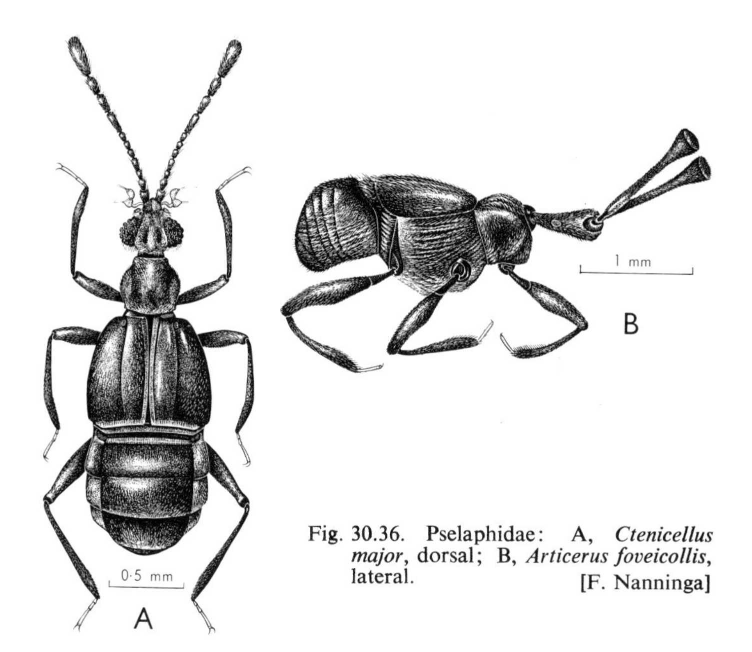

Fig. 30.36. Pselaphidae: A, *Ctenicellus major*, dorsal; B, *Articerus foveicollis*, lateral. [F. Nanninga]

have 11-segmented antennae and enlarged maxillary palpi. Hind wings are usually present, and winged forms often come to lights at night. Little is known of the larvae of Pselaphidae.

Superfamily SCARABAEOIDEA (Lamellicornia)

Antennae non-tactile, short, 7- to 11-segmented, with distal 3–7 segments expanded on inner aspect to form a lamellate or, less commonly, pectinate club. Head small, but often with greatly enlarged mandibles, deeply sunk in large pronotum; both head and pronotum sometimes with conspicuous armature. Fore coxal cavities closed behind. Mesepimeron reaching the mid coxal cavities. Legs fossorial; fore tibiae with single apical spurs and with lateral margins toothed; tarsi 5-segmented. Elytra almost always without a scutellar striole. Wing venation reduced, with M incomplete and forming a loop with CuA; with 1 or 2 detached apical veins between CuA and first complete anal vein; wing folding at conspicuous pterostigma. Abdomen with 5 or 6 visible sternites, the ultimate tergite frequently not concealed by elytra and forming a true pygidium. The 8th abdominal spiracles non-functional, represented by a sclerotized scar on the membrane of the angle between the terminal tergite and sternite. Four Malpighian tubes.

Larvae soft and greyish white, usually of the curl-grub type (Fig. 30.12c); all 3 pairs of legs usually developed; rarely with distinct ocelli; abdomen without urogomphi, and with cribriform spiracles. Subterranean forms feeding on roots or soil organic matter, or associated with rotting wood, fungi, carcasses, or dung. Three larval instars; generation period rarely less than one year, usually 2, sometimes 3 or 4 years.

Within Australia, speciation has been most intense in the north-eastern portion of the continent, with a few notable exceptions. In Tasmania, the fauna is impoverished with the exception of one genus of Lucanidae which has speciated prolifically there. Groups such as the Lucanidae, Passalidae, and ruteline Scarabaeidae, which appear to require moist environments, are conspicuously absent from, or poorly represented in, the drier portions of the continent, including

Western Australia. In these regions, scarabaeoids associated with dung and carcasses (Geotrupidae, Trogidae, and laparostict Scarabaeidae) have proliferated. The phytophagous pleurostict Scarabaeidae (Melolonthinae, Dynastinae, and Rutelinae) have all produced eremophilous forms which show remarkable convergence expressed in sedentariness of the females and great development of the sensory organs of the males, which have poorly developed mouth-parts and probably do not feed. In many of these genera the females have not yet been discovered, and the taxonomy rests precariously upon secondary sexual characters of the male.

The biology of the Scarabaeoidea is most varied (Ritcher, 1958). The larval stages live in environments that provide them with adequate protection from desiccation. Synchronization of emergence of adults is achieved by a diapause occurring at the end of the third instar, the so-called 'prepupal' stage. The adults may feed on the same materials as the larvae, or may be leaf or nectar feeders; in some groups the adults do not feed at all, but rely on fat body carried over from the larval stage to mature their eggs and supply their own energy requirements; such adults are naturally short-lived. A few native scarabaeids have become serious pests of pastures, crops, and shade trees.

Keys to the Families of Scarabaeoidea

ADULTS

1. Segments of antennal club thick, rarely capable of close apposition (Fig. 30.37A); abdomen with 5 visible sternites 2
   Segments of antennal club usually thin, always capable of close co-adaptation (e.g. Fig. 30.39); abdomen usually with 6 visible sternites 3
2. Antennae with long scape, usually elbowed; labrum not conspicuous; mentum complete; mandible without a movable tooth; scutellum present, conspicuous, forming a triangular plate separating the elytra at their bases (Fig. 30.37A); sexual dimorphism marked, mandibles of ♂ usually greatly enlarged; elytra rarely with conspicuous longitudinal striae .... **Lucanidae** (p. 549)
   Antennae not elbowed, curled in repose (Fig. 30.37B); labrum conspicuous; mentum with ligula placed in deep excision; mandibles with a movable tooth on posterodorsal surface; scutellum not separating elytra at their bases; sexual dimorphism slight; elongate, dark brown to black beetles, rather flattened, always with conspicuous elytral striae .... **Passalidae** (p. 549)
3. Antennae 11-segmented, club circular (Fig. 30.39). [More or less hemispherical, yellowish or reddish brown species, often with extensive cephalic or pronotal armature] .... **Geotrupidae** (p. 550)
   Antennae with 10 or fewer segments, club usually elongate, rarely with segments cupuliform 4
4. Body capable of being contracted into the shape of a ball, mouth-parts and legs concealed .... **Acanthoceridae** (p. 550)
   Body not contractile, legs always exposed 5
5. Pygidium concealed by the elytra; head deflexed and mouth-parts partly concealed (Fig. 30.38); abdomen with 5 visible sternites; tarsi with empodium absent or small, not setose .... **Trogidae** (p. 550)
   Pygidium almost always exposed; head not deflexed, mouth-parts visible; abdomen usually with 6 visible sternites; tarsi with setose empodium between the claws (Figs. 30.40–45) .... **Scarabaeidae** (p. 551)

LARVAE

1. Antennae 2-segmented; abdomen cylindrical, neither tapered posteriorly nor curled in repose; concavity of 'prothoracic' spiracles directed anteriorly, those of abdominal spiracles directed posteriorly; metathoracic legs reduced to minute tooth-scrapers working against area of fine ridges on mid coxae. [Living in association with adults in rotting timber] .... **Passalidae**
   Antennae 3- or 4-segmented; larvae usually curled in repose; concavity of 'prothoracic' spiracles directed posteriorly (except in Lucanidae), those of abdominal spiracles directed anteriorly, usually with all 3 pairs of legs equally developed (e.g. Fig. 30.12C) 2

2. Antennae 4-segmented. [Often root-feeders] .... **Scarabaeidae**
Antennae 3-segmented, terminal segment reduced to a small cone. [Living in humus, decaying wood, carcasses, or dung, never root-feeding] 3
3. Hind legs unmodified; abdominal tergites divided into 3 annulets clothed with conspicuous sharp setae and longer hairs; ocelli present near bases of antennae. [Associated with carcasses] .... **Trogidae**
Hind legs modified, femora having a series of teeth capable of movement across an area of fine ridges on mid coxae; abdominal tergites not annulate, without conspicuous hairs or setae; tarsal claws reduced or absent; lacking ocelli near bases of antennae 4
4. Abdomen strongly curved, dilated in central region, mouth approximating to anus; episternal suture usually obliterated; legs slender and devoid of claws; hind legs sometimes reduced to a passalid-like condition. [Usually associated with dung] .... **Geotrupidae**
Abdomen slightly curved, tapering posteriorly; episternal suture always distinct; legs robust and with at least a trace of claws on fore legs. [Associated with decaying timber] .... **Lucanidae**

24. Lucanidae (Plate 4,N,U; Fig. 30.37A). Of moderate to large size (10–50 mm), characterized by the thick lamellae of the antennal club, the scape long so that the antenna appears to be elbowed, scutellum visible, and the abdomen with 5 visible sternites. Colour dull brown or black or metallic; general surface of body without setae; head prognathous, the mandibles sometimes sexually dimorphic. The mandibles of the male exhibit allometry, ranging from a condition scarcely different from that in the female to extreme hypertrophy with increase in body size. Sexual dimorphism in Coleoptera has been reviewed by Arrow (1951).

The 17 genera represented in Australia are distributed among 4 of the subfamilies recognized by Holloway (1960). *Lamprima*, *Hololamprima*, and *Phalacrognathus* are placed in the LAMPRIMINAE, *Ceratognathus* in AESALIINAE, *Syndesus* in the SINODENDRINAE, and the remainder in the LUCANINAE. The classification of Lucanidae has been dealt with by Benesh (1960), and Holloway (1960). Most of our species occur in tropical and subtropical areas of the mainland, although one genus *(Lissotes)* has its headquarters in Tasmania. About half of the genera are endemic. Only 2 species are known from W.A. (*Lamprima micardi* Reiche and *L. varians* Germ.). The magnificent metallic purple and green *Phalacrognathus muelleri* Macl., the most attractive of all Australian Coleoptera, is very rare and restricted to the rain forest of north Queensland, where it is reported to inhabit logs and stumps of *Cedrela australis*. Larvae of the widely distributed golden stag beetle (*Lamprima aurata* F.) can be found in logs of *Casuarina Cunninghamiana* on river banks in N.S.W. Most lucanids are nocturnal, although some Lampriminae are found feeding on eucalypt or wattle leaves in daylight. The genus *Eucarteria* includes several nectar-feeding species. Adults are attracted to lights on warm nights.

25. Passalidae (Fig. 30.37B). Large (20–60 mm), shining black beetles of somewhat flattened form; head prognathous, sometimes with a short median horn; antennae curved, not elbowed; labrum conspicuous;

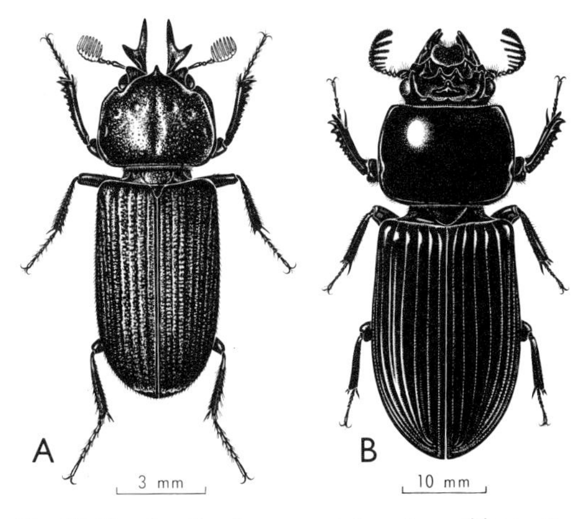

Fig. 30.37. A, *Syndesus cornutus*, Lucanidae; B, *Mastochilus quaestionis*, Passalidae.
[F. Nanninga]

mandibles with a movable tooth; mentum deeply excised; galea hooked; scutellum not visible; elytral striae conspicuous; wings with M–Cu loop absent. Adults stridulate by friction between the wings and the upper surface of the abdomen. Larvae subcylindrical, dull white or greyish. The hind legs are reduced to paw-like stumps, consisting of a very short coxa and a more elongate trochanter, which are used to scrape over a file on the coxa of the mid leg, thereby producing a high-pitched sound.

Passalidae are nearly all tropical. The Australian species are most numerous in the north; half occur in Queensland and 10 in New Guinea. Only one species, *Pharochilus politus* Burm., is known from Tasmania, and there are no Passalidae in W.A. The commonest species in south-eastern Australia are *Pharochilus dilatatus* Dalm. and *Aulacocyclus edentulus* Macl. Adult Passalidae feed on decaying wood, and are said to prepare food for the larvae which are always found in association with adults. [Dibb, 1938.]

26. Trogidae (Fig. 30.38). Length 6–20 mm; body robust, with heavily sculptured dorsal surface and dull, uniform dark colour; head almost wholly concealed from above, deflexed; antennae 10-segmented; wings with M–Cu loop present; abdomen with 5 free sternites; pygidium concealed by the elytra; tarsi with empodium, if present, not setose. Adults stridulate by rubbing the edge of the abdominal segments against the internal margin of the elytra. The greater part of the world's trogid fauna is included in the world-wide genus *Trox*, and a quarter of the species are Australian. The genus appears to flourish in the arid regions of the interior, where many as yet undescribed forms are known to exist. Haaf (1954) has reviewed the Australian species. The larvae are curved, with 3 pairs of well-developed legs with prominent claws; they do not stridulate. The adults are strongly attracted to lights. Adults and larvae feed upon dry animal remains, being one of the last of the succession of insects that invade carcasses. The larvae form vertical burrows in the soil beneath the carcass.

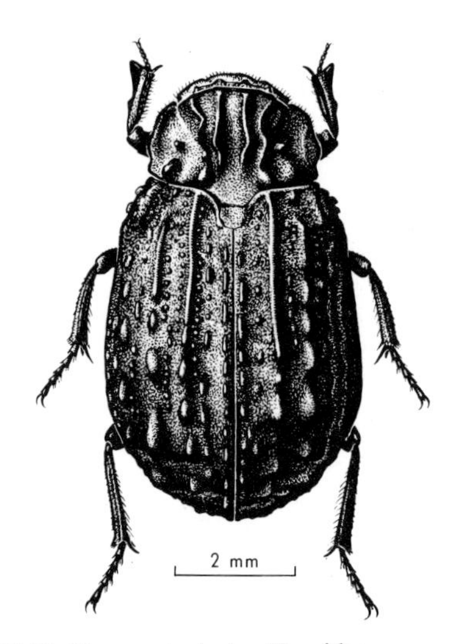

Fig. 30.38. *Trox australasiae*, Trogidae. [F. Nanninga]

27. Acanthoceridae. A small family found mainly in South America, Madagascar, and Malaya, with a few species in New Guinea. A single species (*Pterorthochaetes simplex* Gestro) has been described from Queensland. Acanthoceridae are small, highly convex insects with a broad head and pronotum which can be deflexed (as in Clambidae), so that the mouth-parts, abdomen, and even the legs are completely concealed, the body assuming a pill-like form. All that can be seen from above is the posterior edge of the pronotum, a large triangular scutellum, and the elytra. The puncturation of the surface is annular, each puncture bearing a very short, clavate seta. The first 4 segments of each tarsus are clothed with long hairs. Nothing is known of their biology, although some species have been recorded as living under the bark of fallen trees. The protective adaptation may indicate association with ants.

28. Geotrupidae (Fig. 30.39). Stout, very strongly convex beetles, length 7–27 mm, colour dark reddish brown or, rarely, black. Head prognathous; head and pronotum in male frequently armed with long horns; antennae 11-segmented, with a circular, biconvex club; legs very powerful and heavily armed; scutellum large, obvious.

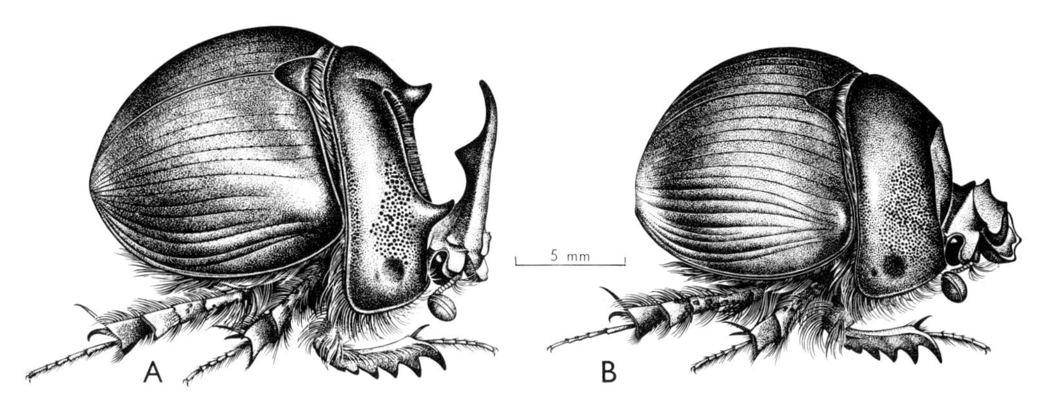

Fig. 30.39. *Blackburnium cavicolle*, Geotrupidae: A, ♂; B, ♀.
[F. Nanninga]

Both adults and larvae stridulate, the adults by rubbing a file on the hind coxae over a sharp ridge on the abdomen, and the larvae by rubbing the mid and hind legs. The hind legs of the larvae are atrophied as in Passalidae.

The family is well represented in all States except Tasmania. All Australian species belong to the BOLBOCERATINAE, which are also represented in Africa and South America. Most Australian species (formerly *Bolboceros* spp.) are now placed in the genus *Blackburnium* (Howden, 1954). The only Geotrupidae known from Tasmania belong to the genus *Elephastomus*. Armature of the head and prothorax is highly developed in the males, and takes the form of tubercles, horns, and complex excavations. Because of the differences between the sexes it is often difficult to associate them with confidence. Although some exotic species feed on subterranean fungi or vegetable matter, all Australian species, as far as is known, are dung-feeders. The adults dig deep burrows beside dung pads and bury large quantities of dung in which the eggs are laid. Some species are very abundant, and heavily grazed pastures are often seen peppered with piles of soil excavated by the adults. No larval stages of Australian species have been found, and their biology has not yet been studied. Adults of all species are attracted to lights at night.

29. Scarabaeidae (Figs. 30.40–45). Usually stout-bodied beetles of moderate to large size; rarely brightly coloured, or small and clothed with scales (Valginae); head not deflexed, mouth-parts visible from below; antennae with 10 or fewer segments, antennal club composed of 3 to 7 movable lamellae; abdomen with 5 or 6 visible sternites; pygidium usually exposed; tarsi with a setose empodium between the claws; fore tarsi sometimes reduced or absent (Scarabaeinae). Larvae white, with well developed, pale reddish brown head and 3 pairs of legs, the hind legs being fully developed and usually longer than the mid legs; antennae 4-segmented. The larvae are rarely able to stridulate, and live concealed, feeding on roots, dung, or decaying vegetable matter. Some are termitophilous.

Key to the Subfamilies of Scarabaeidae Known in Australia

1. Spiracles situated on pleural membrane, covered by the elytra, not visible in intact insects (laparostict Scarabaeidae, Fig. 30.40A) 2
   Spiracles diverging, so that several lie on the abdominal sternites, and at least the last is not covered by the elytra (pleurostict Scarabaeidae, Figs. 30.40B, C) 6
2. Labrum and mandibles clearly visible from above the head 3
   Labrum and mandibles concealed beneath clypeus 5
3. Labrum very conspicuous, as long as mandibles; tarsi slender and very long. [Tropical species] .... ACLOPINAE

Labrum not as long as mandibles, tarsi normal 4

4. Antennae 10-segmented, with proximal segment of club forming a cup into which the distal segments are fitted; mandibles and labrum flattened and strongly produced, the former not toothed .... HYBOSORINAE
Antennae 9-segmented, club not cupuliform; labrum just visible, mandibles not flattened but strongly toothed .... AEGIALIINAE
5. Elytra completely covering the abdomen; hind tibiae with 2 terminal spurs; mid coxae contiguous; scutellum present; more or less elongate beetles, very rarely with cephalic or pronotal armature; fore tarsi of normal development .... APHODIINAE
Elytra leaving the pygidium exposed; hind tibiae with single terminal spurs; mid coxae widely separated; scutellum absent; more or less rounded beetles, usually with complex cephalic and pronotal armature; fore tarsi often reduced or absent .... SCARABAEINAE
6. Line of the last 3 abdominal spiracles diverging feebly from that of the anterior spiracles (Fig. 30.40B); claws of hind legs of equal size, not capable of movement ... MELOLONTHINAE
Line of the last 3 abdominal spiracles diverging strongly (Fig. 30.40C); claws of hind legs sometimes asymmetrical and movable 7
7. Hind tarsal claws unequal, movable, often large; elytra often with a narrow membranous margin beneath the posterolateral edges. [Pronotum without armature; sexual dimorphism slight, most conspicuously expressed in form of clypeus; mostly diurnal species] .... RUTELINAE
Hind tarsal claws equal, immovable; elytra never with a membranous margin 8
8. Fore coxae transverse, not produced ventrally; mandibles usually visible, and metathoracic epimera invisible from above; sexual dimorphism strong, ♂♂ with cephalic tubercles, horns, or complex excavations; mostly nocturnal phytophagous species of dull or uniform coloration .... DYNASTINAE
Fore coxae conical, produced ventrally; mandibles concealed, metathoracic epimera usually visible from above; sexual dimorphism slight; mostly diurnal, nectar-feeding species, usually brilliantly coloured or patterned 9
9. Small species (less than 5 mm in length) clothed with scales; hind coxae widely separated; maxillae with pencil of fine hairs protruding beyond mouth; penultimate abdominal tergite exposed .... VALGINAE
Larger species, without clothing of scales; hind coxae contiguous; maxillae without such hairs; penultimate abdominal tergite concealed .... CETONIINAE

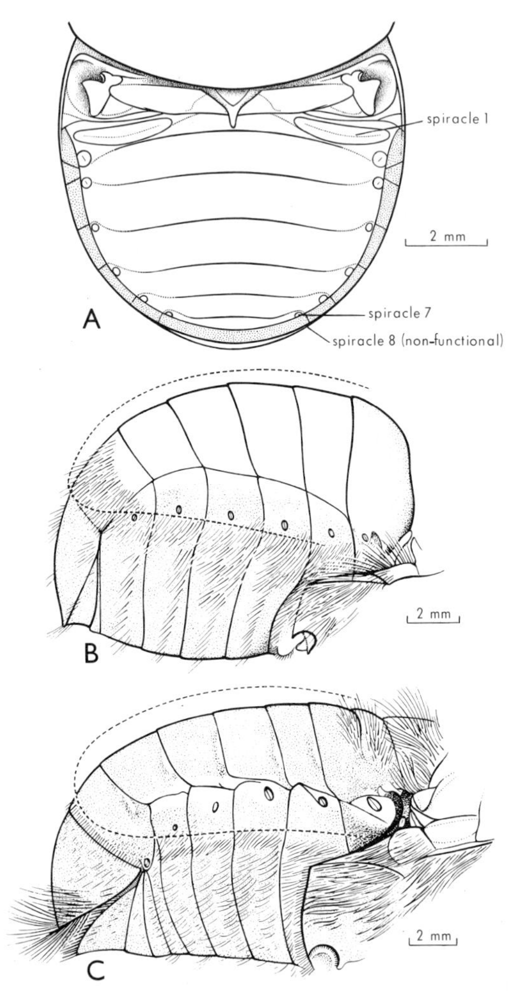

Fig. 30.40. A, laparostict abdomen (*Onthophagus declivis*), dorsal. Pleurostict abdomen, lateral: B, Melolonthinae (*Colpochila* sp.); C, Rutelinae (*Anoplognathus* sp.). [F. Nanninga]

ACLOPINAE (world 14 spp.; Australia 6 spp.; Fig. 30.41) include only 2 genera confined to South America and northern Australia. The Australian species of *Phaenognatha* are very distinctive brown or black insects, with somewhat tapered, separately rounded elytra, and very long legs, the

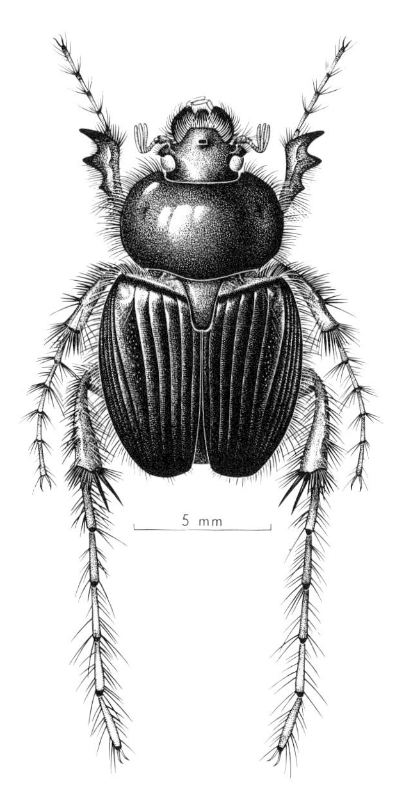

Fig. 30.41. *Phaenognatha* sp., Scarabaeidae–Aclopinae. [F. Nanninga]

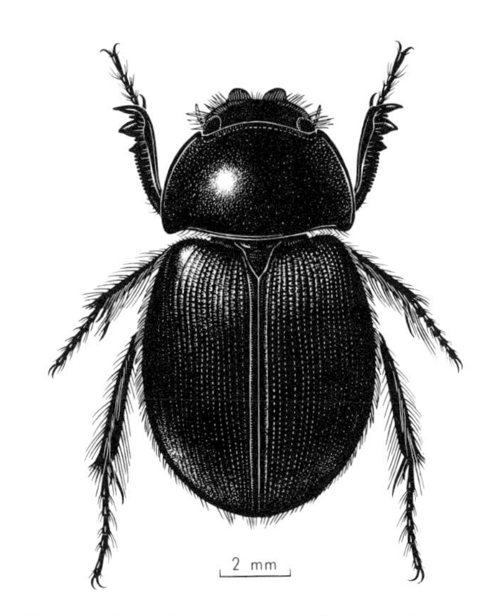

Fig. 30.42. *Phaeochrous hirtipes*, Scarabaeidae–Hybosorinae. [F. Nanninga]

hind tibia and tarsus together being as long as, or longer than, the body. Mandibles broad, concave, plate-like, visible from above on either side of the prominent labrum; abdomen with 6 sternites, small, largely concealed by the large hind coxae; fore coxae very large, occupying the greater part of the underside of the prothorax. Nothing is known of the biology of these insects.

HYBOSORINAE (world 100 spp.; Australia 17 spp.; Fig. 30.42) are 5–12 mm long, shining, dark brown or black beetles, in which the mandibles are sharply curved, with pointed apices directed inwards towards each other, and visible from above beyond the strongly transverse labrum. Antennae 10-segmented, the club resembling that in Geotrupidae, with the proximal segment circular and concave, into which the 2 apical segments are closely fitted; abdomen with 6 visible sternites in male and 5 in female; fore tibia with 3 large and numerous small teeth; tarsi shorter than tibiae; aedeagus strongly asymmetrical. The Australian genera are *Liparochrus*, *Antiochrus*, and *Phaeochrous*. The species are tropical apart from a few in the south-east. The scanty biological data available suggest that hybosorines, like trogids, are associated with carrion.

AEGIALIINAE (world 23 spp.; Australia 1 sp.) all occur in the northern hemisphere, other than *Saprus griffithi* Blackb. from Tasmania. This is a black insect, about 4 mm long, with the general appearance of an aphodiine, but differing by having protruding, pointed mandibles. Nothing is known of its biology.

APHODIINAE (world 1,200 spp.; Australia 60 spp.; Fig. 30.43A) are distinguished by having the pygidium covered, mouth-parts and labrum concealed beneath the fronto-clypeus, hind tibia with 2 apical spurs, fore tarsi normal, and the scutellum visible. The larvae are separable from those of Scarabaeinae by lacking a hump and by having fully developed legs. These beetles are mainly dung-feeders, but the larvae of some species live on organic matter in the soil.

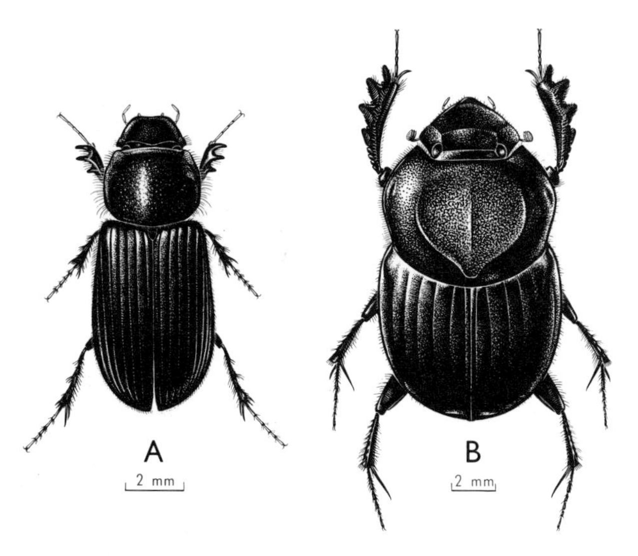

Fig. 30.43. Scarabaeidae: A, *Aphodius howitti*, Aphodiinae; B, *Onthophagus declivis*, Scarabaeinae. [F. Nanninga]

Two species, *Aphodius howitti* Hope and *A. pseudotasmaniae* Given are phytophagous and regarded as major pasture pests in south-eastern Australia and Tasmania. Their larvae form vertical burrows, from which they emerge at night to cut off leaves which they drag into their tunnels for consumption. The dung-feeding aphodiines are beneficial, in that they assist in the disposal of animal droppings. The species are distributed among 11 genera, of which *Aphodius*, *Ataenius*, *Saprosites*, and *Psammobius* are the largest. *Aphodius lividus* Oliv. and *granarius* L. are cosmopolitan.

In the SCARABAEINAE (=Coprinae; 'dung beetles'; world 2,000 spp.; Australia 212 spp.; Fig. 30.43B) the body is broadly oval, colour almost always black, and length 2–23 mm; they are characterized by the exposed pygidium, hind tibiae with only 1 spur, mid coxae widely separated, and scutellum absent. The frontoclypeus is expanded to cover the mouth-parts; mandibles lamelliform, usually membranous with only the outer edge sclerotized; antennae 8- or 9-segmented, club 3-segmented; abdomen with 6 visible sternites; fore tarsi sometimes absent. The larvae have a pronounced hump in the middle of the dorsal surface, and the tarsal claws are minute or absent. Scarabaeinae feed almost exclusively on the faeces of various animals, but no details of the habits of Australian species have been recorded.* The subfamily is divided into 5 tribes, of which 3 are represented in the fauna. The Scarabaeini (e.g. *Temnoplectron*, *Tesserodon*, *Cephalodesmius*, *Canthosoma*) are the dung-ball rollers, commonly recognizable by their long, slender, curved hind tibiae. The beetle uses its shovel-like frontoclypeus with a rotary movement of the body to cut out a piece of dung which it then moulds into a ball. The ball is pushed away from the dung-mass by the beetle and finally buried, either to be eaten by the adult, or stored as food for a future larva. In some species the sexes co-operate in this work. The Sacred Scarab of ancient Egypt belongs to this tribe. The Coprini (e.g. *Coptodactyla*, *Pedaria*, *Macropocopris*) excavate a chamber immediately below the mass of dung. The species of *Macropocopris* are remarkable in that the adults cling to the hair around the anus of wallabies (Arrow, 1920). The Onthophagini

* For a review of the ecology and ethology of Scarabaeinae, see G. Halffter and E. G. Matthews, *Folia ent. méx*, nos 12–14: 1–312, 1966.

(Onthophagus) behave like Geotrupidae, digging deep tunnels beneath the dung. The blind end of each tunnel is packed with dung to serve as food for a larva. [Lea, 1923.]

MELOLONTHINAE ('chafers'; world 9,000 spp.; Australia 1,400 spp.; Fig. 30.44) are 3·5–40 mm long, and usually of reddish brown colour, rarely black or metallic (*Diphucephala*; Plate 5,I). Head with labrum sclerotized and visible externally, rarely fused with the clypeus; abdominal spiracles in the sclerotized upper margins of the sternites, not in the membranes; lines of 6th and 7th spiracles on the two sides not strongly divergent; claws equal, fixed, with or without a tooth beneath; head and prothorax without horn or tubercle. The eggs are large, relatively few, and deposited below the surface of the soil in spring or early summer. The larvae, which are C-shaped, with sclerotized head and well-developed legs, feed on roots and vegetable matter in the soil. The great abundance of the adults of some species suggests that root damage by the larvae must be severe, but the effect on pasture has not yet been assessed. Serious damage is often caused to vegetables, wheat, and sugar cane by melolonthine larvae, which are commonly known as 'white grubs'. The 'grey-back' (*Dermolepida albohirtum* Waterh.) is the most serious pest of sugar cane in north Queensland. Pupation occurs in a cell at some depth in the soil, and the adult, after emergence, remains in the cell until rain softens the soil. This is advantageous in synchronizing the flight of the adults. Adults of crepuscular or nocturnal species shelter in daylight under debris or beneath the surface of the soil. Adults of most species eat leaves, but a few appear to be adapted to feed on nectar. Among the leaf-eaters are some (e.g. *Sericesthis geminata* Boisd., *Liparetrus* spp.) which are serious pests because they defoliate trees. Some species of *Maechidius* (e.g. *M. tibialis* Blackb.) are found in termite mounds. *Diphucephala colaspidoides* Gyll. and *D. aurulenta* Kirby often appear in enormous numbers on flowering trees in the south-east. Also found swarming on flowers in daylight are the slender, brown, or brown and black species of *Phyllotocus*. [Britton, 1957.]

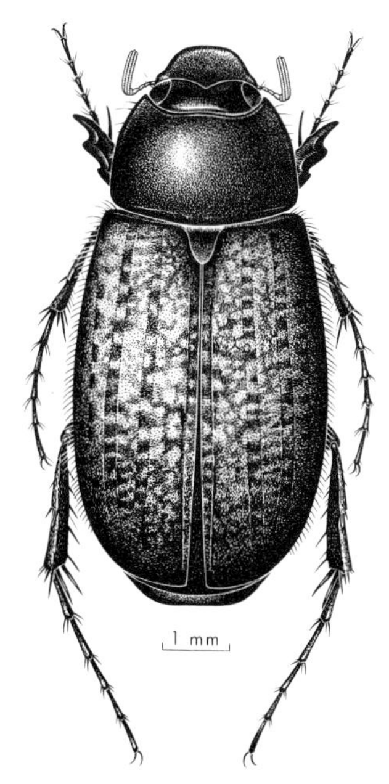

Fig. 30.44. *Sericesthis geminata*, Scarabaeidae–Melolonthinae. [F. Nanninga]

RUTELINAE (world 2,500 spp.; Australia 96 spp.; Plate 4,D–F,L,M) are of moderate to large size (8–35 mm) and stout build; colour usually bright and more or less metallic; abdominal spiracles set in the sclerotized upper part of the sternites, the lines of the posterior spiracles strongly divergent; claws large, unequal, and movable; head and prothorax without horns or tubercles. The larvae are typically scarabaeoid, feeding on roots and soil organic matter. Some species of *Anoplognathus* cause damage to lawns and pastures, and adults of the genus, commonly called 'Christmas beetles', often defoliate eucalypts. Apart from *A. macleayi macleayi* Blackb. from central Australia and a few small, rather dynastine-like species from the N.T., the subfamily is limited to the higher-rainfall coastal areas of eastern Australia. The metallic colouring so common in this subfamily is of interest because, like that in Cetoniinae but unlike that of metallic Melolonthinae *(Diphucephala)* and Lucanidae, the reflected light responsible is circularly polarized. Of the 21 genera known in Australia 19 are endemic. [Carne, 1957a, 1958.]

DYNASTINAE (world 1,400 spp.; Australia 169 spp.; Fig. 30.45) are stout-bodied insects, 10–70 mm long, and mainly black in colour. Spiracles set in the sclerotized upper part of the abdominal sternites; labrum membranous, not visible externally; mandibles partly visible from above; fore coxae transverse; claws equal and fixed; head and pronotum in the male often bearing horns or tubercles which are disproportionately large in large specimens (allometric; Figs. 30.45C, D). The larvae feed upon roots or decayed vegetable matter in the soil. The adults, unlike many Melolonthinae and Rutelinae, do not feed on leaves. They are active at night, and are strongly attracted to lights. The subfamily includes several well-known pest species, including the introduced South African *Heteronychus arator* F. (=*sanctaehelenae* Blanch.), a serious pest of maize in coastal N.S.W. *Adoryphorus couloni* Burm. is a pasture pest in the south-eastern States. The only dynastine with a colour pattern (yellowish brown with black marks) is an Argentine species, *Cyclocephala signaticollis* Burm., which appeared in Sydney in 1947 and has since become firmly established in that area. Species of the largest genus, *Cryptodus*, live in association with ants or termites. *Cryptodus* adults have no projecting armature or setae, and the mouth-parts and antennae are protected by shield-like expansions of the mentum. Other large genera are *Teinogenys*, *Corynophyllus*, *Trissodon*, and *Novapus*. Speciation has been most intense in the north and west of the continent, and only 5 species are known from Tasmania. [Carne, 1957b.]

VALGINAE (world 200 spp.; Australia 17 spp.) are small beetles (5 mm or less) in which the lines of abdominal spiracles are in the sclerotized upper part of the sternites and strongly divergent towards the apex; tarsal claws equal; fore coxae conical; mandibles not visible from above; tarsi

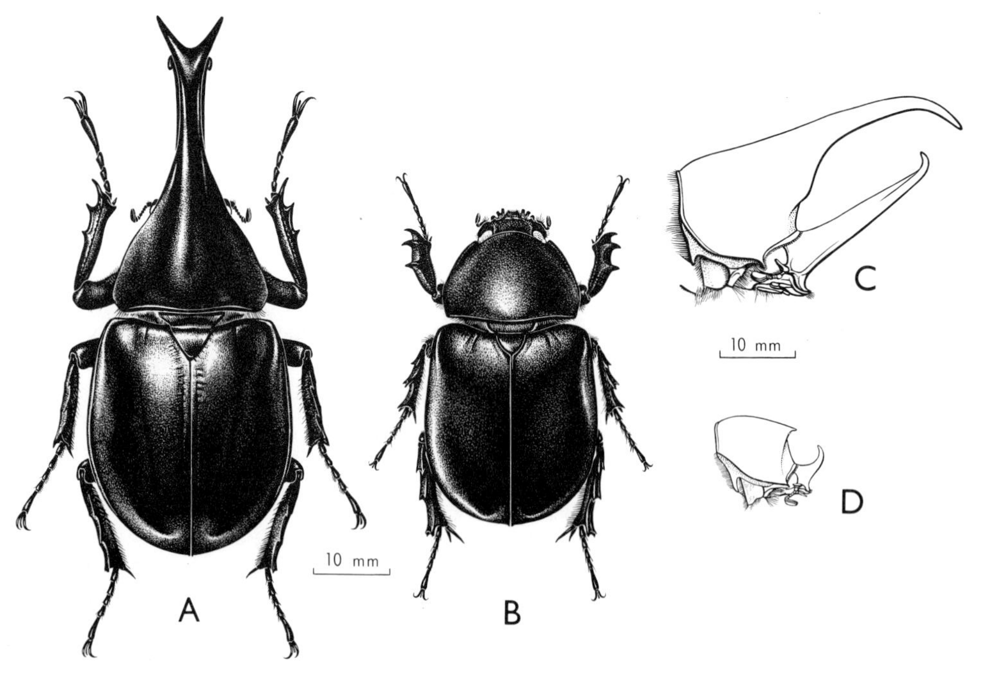

Fig. 30.45. *Xylotrupes gideon*, Scarabaeidae–Dynastinae: A, ♂; B, ♀; C, D, profile of head and prothorax showing variation of form with size in males. [F. Nanninga]

longer than tibiae; body clothed with scales; pygidium and penultimate abdominal tergite not covered by the elytra. All Australian species belong to the endemic genus *Microvalgus*. They are diurnal, and are often seen in great numbers on flowering trees and shrubs. The pencil of fine hairs projecting from the mouth appears to be an adaptation to their nectar-feeding habits. Little is known of their biology, but the larvae may be associated with termites, as large numbers of valgines have been reared from *Cryptotermes* mound material from southern N.S.W. Most of the species are found in eastern Australia, only one being known from the west.

CETONIINAE (world 2,600 spp.; Australia 120 spp.; Plate 4,O,P) are of moderate to large size (8–45 mm) and rather depressed, flattened form. Fore coxae conical; elytra covering the penultimate abdominal tergite but leaving the pygidium exposed; lateral edges of elytra with a concavity behind the shoulder which exposes the metepimeron and the edge of the hind coxa from above; mesepimeron also visible from above in the angle between pronotum and elytra. Most cetoniines have strikingly patterned, metallic or 'enamelled' colouring. All are diurnal and feed on nectar. The larvae appear to feed on humus, being found near rotting logs, in damp litter, compost, etc. They are thick, with relatively small head and legs, and are not curled like most scarabaeid larvae. They are able to move horizontally, and some species make no use of their legs but lie on their backs, progressing with a worm-like peristaltic motion, purchase being obtained by the dense, strong bristles with which the abdomen is clothed. Adult cetoniines are unusual in that the elytra are not raised in flight. There are 32 Australian genera, of which *Lomaptera*, *Dilochrosis*, *Diaphonia*, and *Pseudoclithria* are the largest. [Lea, 1914.]

Superfamily DASCILLOIDEA

Adults with prosternal process very narrow, not received into a cavity on the mesosternum; fore coxae conically projecting, the fore coxal cavities open behind; lateral margins of pronotum always complete, posterior angles not acute or projecting; antennae filiform, serrate, or with a 2-segmented club; tarsi 5-5-5- or 4-4-4-segmented, empodium absent; abdomen with 5 free sternites; hind coxae with transverse excavation behind to receive hind femur; aedeagus trilobed; 6 Malpighian tubes.

Key to the Families of Dascilloidea Known in Australia

1. Tarsi 4-4-4-segmented; head very broad, reflexed against underside of prothorax; antennae 8- to 10-segmented, with a 2-segmented club .... **Clambidae** (p. 557)
   Tarsi 5-5-5-segmented; head narrower; antennae 11-segmented, filiform to serrate 2
2. Tarsi with only 4th segment lobed; head with a fine longitudinal carina below each eye; wings with only 2 distinct anal veins .... **Helodidae** (p. 558)
   Tarsi with segments, 2, 3, and 4 strongly lobed; head without a longitudinal carina below each eye; wings with 5 distinct anal veins .... **Dascillidae** (p. 558)

30. Clambidae (Fig. 30.46). Very small beetles (*ca* 1 mm), distinguishable by their ability to roll into a ball, by their hind coxae which are extended backwards into a thin plate which covers the hind femora when these are retracted, and by the hair-fringed wings. The body form is broadly oval and convex; surface sparsely and conspicuously setose; head strongly deflexed against underside of prothorax; antennae with last 2 segments forming a distinct club; eyes small; mouth-parts reduced; pronotum broad but very short; fore coxae conical, prominent, contiguous, the cavities open behind; mid coxae separate; hind coxae contiguous, with large femoral plates; tibiae without apical spurs; tarsi 4-4-4-segmented; elytra entire; wing venation reduced; abdomen with 5

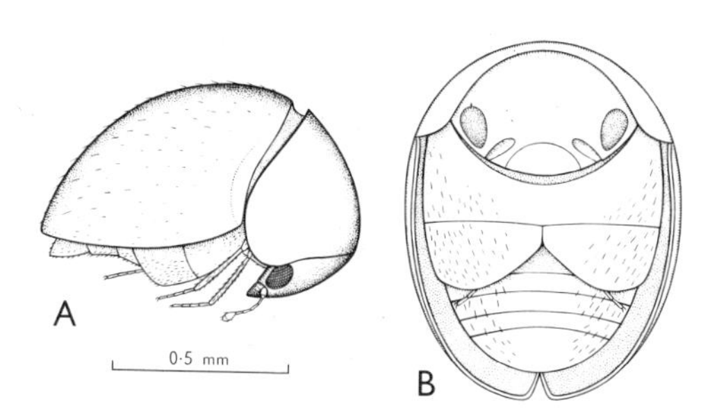

Fig. 30.46. *Clambus simsoni*, Clambidae: A, lateral; B, ventral. [F. Nanninga]

visible sternites. The Australian species are all placed in the genus *Clambus*. Very little is known of their biology, but some species are said to be myrmecophiles. The systematic position of the family is doubtful. It has been tentatively placed in the Dascilloidea by Crowson, because of apparent similarities of the larvae with those of Eucinetidae. In the past the family has been associated with the Anisotomidae.

31. Helodidae (Fig. 30.47). Small (*ca* 3–10 mm), ovoid, convex beetles; head strongly deflexed, not visible from above, with a longitudinal carina below the eye; antennae 11-segmented, long, slender, flattened, scape much stouter than the rest; upper edge of the antennal cavity in front of eye very finely margined; labrum small, quadrate, free; pronotum broad and very short; prosternum very short; fore coxae elongate, conical, slightly separate; mid coxae conical, slightly separate; tarsi 5-5-5-segmented, and with only the 4th segment bilobed; claws simple; femora sometimes swollen; elytral epipleura broad, continued to the apices; wings with 2 anal veins. The larvae, which are somewhat campodeiform, are unique among endopterygote larvae in having long multisegmented antennae. They have anal gills, and are found in both stagnant and running water. The intermediate and last larval instars have functional spiracles on abdominal segment 8 only. *Macrohelodes crassus* Blackb. is a stout-bodied, pale yellow or black-spotted species, which occurs in Tasmania and as far north as the Blue Mountains. Other Australian species are included in the genera *Pseudomicrocara*, *Heterocyphon*, *Macrocyphon*, and *Macrodascillus*. [Armstrong, 1953.]

32. Dascillidae (Fig. 30.48). Small to medium-sized beetles, distinguished from

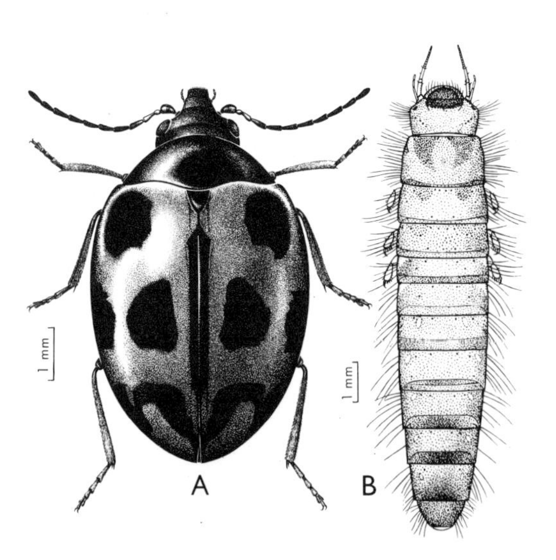

Fig. 30.47. A, *Macrohelodes crassus*, Helodidae, head hidden from above in life; B, helodid larva. [F. Nanninga]

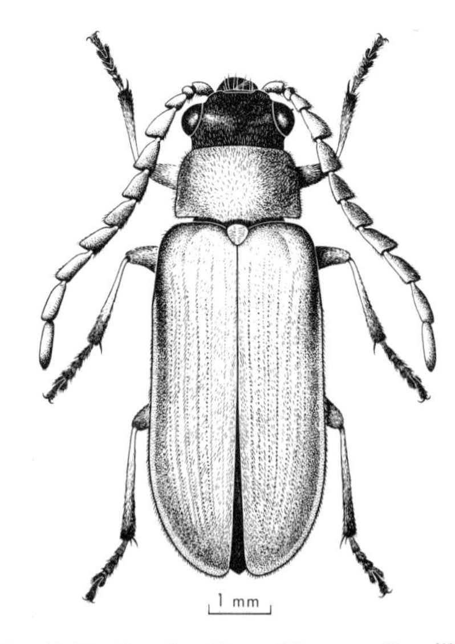

Fig. 30.48. *Notodascillus sublineatus*, Dascillidae. [F. Nanninga]

Helodidae mainly by their more elongate, more parallel-sided form, and by having segments 2, 3, and 4 of the tarsi bilobed; the head is less strongly deflexed and remains visible from above; pronotum transverse; body surface clothed with recumbent pubescence; colour pale brown. The larvae are C-shaped, likc those of Scarabaeidae, and also live in soil. Their antennae are 3-segmented but rather long; ocelli absent; abdominal segment 10 reduced, lying beneath the large segment 9. The Australian species belong to the genera *Notiocyphon*, *Notodascillus*, and *Epilichas* (see footnote on p. 562).

Superfamily BYRRHOIDEA

Small, stout-bodied, very convex beetles, in which the head is strongly deflexed, the frons vertical, and the clypeus not distinguishable. Intercoxal process of prosternum fitted into an anterior emargination of the mesosternum; mid coxae widely separated; hind coxae almost in contact at their inner angles and each with a sharp-edged declivity or hollow for reception of the hind femur; fore coxae transverse, not projecting, with large exposed trochantins; fore coxal cavities confluent and widely open behind; femora and tibiae often capable of being folded into cavities on the underside of the body, and tarsi retractable in grooves on or above the tibia; tarsi 5-5-5-segmented, 3rd segment lobed beneath, 4th small, 5th long, no segments bilobed, and the terminal segments not as long as the other 4 together; aedeagus trilobed; 6 Malpighian tubes.

Most authors have included the Limnichidae in the Byrrhidae as a tribe or subfamily, because external differences between adults of the Byrrhidae and Limnichidae are slight. Hinton (1939b) has, however, shown, on the basis of larval characters and the internal anatomy of the adults, that Limnichidae are more closely related to Dryopidae than are the Helminthidae. They are therefore treated here in the Dryopoidea.

Key to the Families of Byrrhoidea

Antennae gradually widened to the apex; mentum small, hidden between head and prosternum .... **Byrrhidae** (p. 559)

Antennae with a sharply defined 3-segmented club; mentum in form of a horizontal plate, larger than the visible part of the prosternum .... **Nosodendridae** (p. 560)

33. Byrrhidae. Short, stout beetles of dull black or bright metallic green colour. On being disturbed they retract the appendages into close contact with the body or into special cavities, and remain motionless. In typical genera (e.g. *Morychus*) the legs, when folded into the ventral cavities, conform completely with the ventral surface of the body. The tarsi are folded back into line with, and above, the tibiae. The beetles and their larvae are found at the roots of grasses or in damp moss. The larvae of the Australian species are not described, but those known elsewhere are scarabaeiform or onisciform, without urogomphi. The adults of *Pedilophorus* have an attractive metallic colouring, that of *P. gemmatus* Lea (Plate 5,H), metallic green with red bosses on the

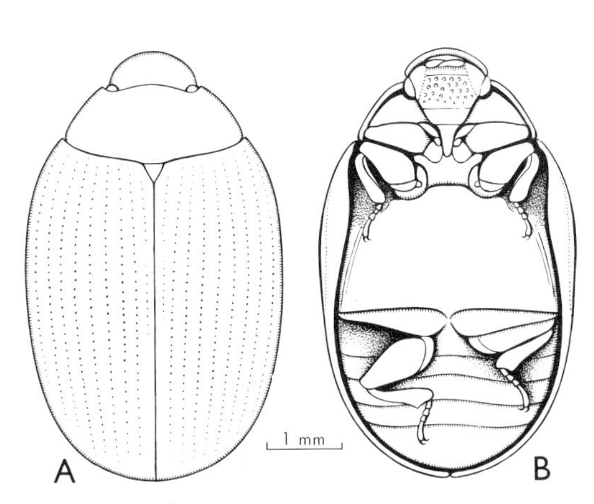

Fig. 30.49. *Nosodendron australicum*, Nosodendridae: A, dorsal; B, ventral. [F. Nanninga]

elytra, being the most striking. [Lea, 1920.]

34. Nosodendridae (Fig. 30.49). Ovoid, convex, shining beetles; legs capable of being folded into cavities on the ventral surface of the body as in Byrrhidae. In contrast with the Byrrhidae, the ventral surface of the head is enlarged, the prosternum correspondingly reduced, and the head is not retractable. Antennae slender, with a well-marked, 3-segmented, tomentose club; the 3rd segment unusually elongate; the club is received at rest in a cavity formed between the fore tibia and the pronotal hypomeron. The family contains only the widely distributed genus *Nosodendron*. Adults and larvae occur under bark. *N. australicum* Lea is found in north Queensland. The family was referred to the Dermestoidea by Crowson, who suggested that the retractable legs and the prosternal process are possibly independently evolved.

Superfamily DRYOPOIDEA

Small or very small beetles, largely of subaquatic habit; body usually pubescent; antennae 11-segmented, filiform or clubbed, or very short and thick; mid coxae usually well separated; fore coxae open behind, and separated by a prosternal process which fits into a depression on the mesosternum; hind coxae with a steep transverse declivity against which the femur can be retracted; head with a distinct fronto-clypeal suture; abdomen with 5 visible sternites; tarsi 4-4-4- or 5-5-5-segmented, segments not bilobed or emarginate, terminal segments often elongate with strong claws; hind margin of pronotum often crenulate. Hinton (1939a) has discussed the relationships of the families. The absence of Dryopidae from the relatively well-collected streams of N.S.W. and Victoria is probably real, but it seems possible that representatives may yet be found in north Queensland.

Key to the Families of Dryopoidea Known in Australia

1. Tarsi 4-4-4-segmented; fore and mid tibiae broad, flattened, spinose along their outer edges; antennae very short, with a broad, 7-segmented, serrate club; mouth directed forwards, mandibles in horizontal plane, visible from above .... **Heteroceridae** (p. 563)
   Tarsi 5-5-5-segmented; fore and mid tibiae not broad and flattened, and not spinose along their outer edges; antennae usually filiform; mouth directed ventrally, mandibles not visible from above 2
2. Posterior edge of pronotum and the opposed basal edge of the elytra not crenulate (Fig. 30.50) ... 3
   Posterior edge of pronotum and sometimes the basal edge of the elytra crenulate (Fig. 30.51A) 5
3. Ventral surface with shallow cavities into which the legs can be folded to conform with the general body surface; hind coxae very closely approximated at their inner ends, mid coxae widely separated. [Body stout, ovoid, with both dorsal and ventral surfaces convex] .... **Limnichidae** (p. 560)
   Ventral surface without cavities for reception of the legs; hind coxae widely separated, as are the mid coxae 4
4. Antennae very short, with 6 or more apical segments forming a pectinate club .... (Dryopidae)
   Antennae slender, never with a pectinate club .... **Helminthidae** (p. 564)
5. Head somewhat rostrate (i.e. clypeus prolonged forwards between unusually close antennal insertions); labrum short; mandibles hidden from above by the clypeus; head without a transverse occipital keel; tarsi with segments not lobed .... **Psephenidae** (p. 561)
   Head not rostrate; clypeus transverse; labrum and mandibles visible; head with a transverse occipital keel; tarsi with 3rd segment lobed beneath .... **Ptilodactylidae** (p. 562)

35. Limnichidae (Fig. 30.50). Very small (*ca* 3 mm), of broadly oval, strongly convex form, clothed with short, dense pubescence. They resemble Byrrhidae in having cavities on the ventral surface into which the legs can be folded. Limnichidae differ externally from Byrrhidae by having a large clypeus separated by a distinct fronto-clypeal suture, and by the fact that the coxites of the female genitalia have no styli. The adults and, no doubt,

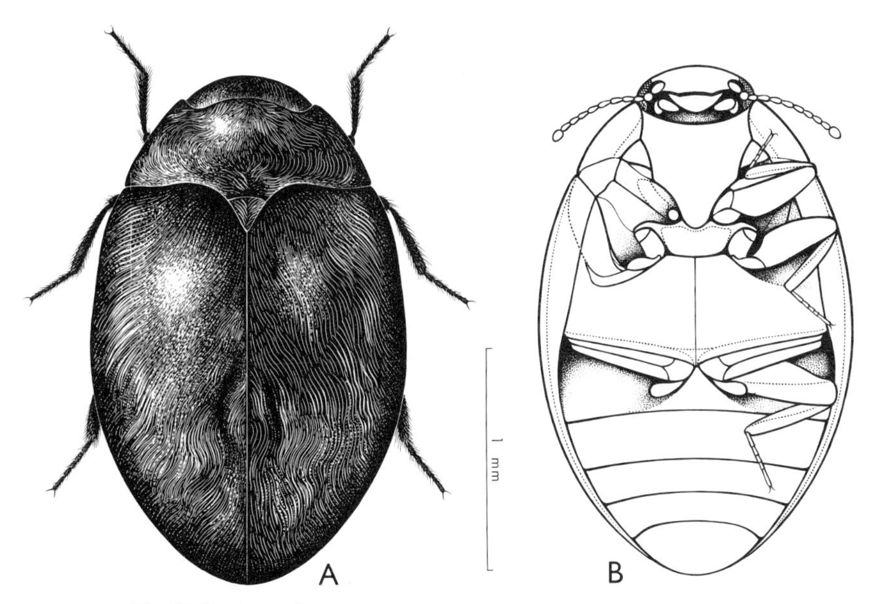

Fig. 30.50. *Limnichus australis*, Limnichidae: A, dorsal; B, ventral. [F. Nanninga]

the larvae also, live on or in mud on the margins of ponds or streams. The Australian species are included in the genera *Limnichus* and *Byrrhinus*. Nothing is known of the larvae of the Australian species.

36. Psephenidae (Fig. 30.51). Small (*ca* 5 mm), oval, flattened beetles, black or dark brown in colour, clothed with very short, fine, dense pubescence. Head deflexed, top of head and eyes visible from above, narrowed between the antennal insertions giving a rostrate appearance; labrum and mandibles concealed beneath clypeus; maxillary palpi with 2nd segment long and apical segment

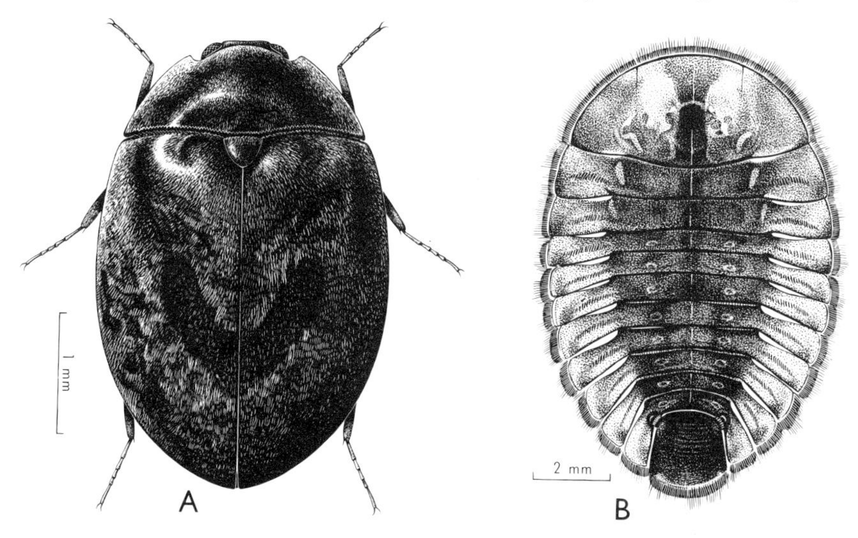

Fig. 30.51. ***Sclerocyphon*** **sp., Psephenidae: A, adult; B, larva. [F. Nanninga]**

flattened and expanded; antennae slender, slightly serrate; pronotum strongly transverse, strongly narrowed anteriorly, margins explanate; prosternum with a median process which is received into a cavity on the mesosternum; fore coxae transverse, separate, but not projecting, with large trochantin visible; hind coxae with a sharp-edged very steep posterior face; abdomen with 5 visible sternites; tibiae without spurs; tarsi with apical segments as long as segments 2, 3, and 4 together; basal edge of the elytra and the posterior edge of the pronotum finely and uniformly crenulate.

Larvae (Fig. 30.51B) of characteristic, flattened, disc shape ('water pennies'), clinging to stones in swiftly running water; head, legs, and gills hidden from above by broad marginal extensions of all segments; colour yellowish brown or dark brown; all outer edges of the marginal extensions of the segments with a uniform fringe of long, straight friction-setae. The only larvae of Australian species so far known belong to the EUBRIINAE, and are characterized (with the Psephenoidinae) by having anal retractile tracheal gills and lacking exposed branched tracheal gills on 4 or more abdominal segments (cf. Psepheninae, Eubrianacinae). Eubriinae also have the early instars apneustic and the final instar with only the spiracles of abdominal segment 8 functional (and cribriform). The anal gills comprise two tufts of doubly bifurcating fine tubes, which are retractable into a cloacal chamber covered by an operculum formed by abdominal sternum 9. There is also a conspicuous brush of setae on each side of tergite 9, close to, and facing, the large spiracle on each side of the posterior edge of segment 8. This 'spiracular brush' serves to keep the spiracle free of silt when it becomes functional as the larva leaves the water to pupate. Psephenid larvae are common in mountain streams in south-eastern Australia, but have not yet been associated with adults. No Australian species of Psephenidae have been described as such, but *Sclerocyphon maculatus* Blackb. which was described as an helodid, is, in fact, a psephenid. [Hinton, 1955b.]

37. Ptilodactylidae (Fig. 30.52). No Australian Ptilodactylidae have been described, but typical ptilodactylid larvae are common in streams in the mountains of N.S.W. (Hinton, *in litt.**). For this reason, a brief account of the family is included here to facilitate its recognition. Adults are about 5 mm long, yellowish brown, with dense clothing of fine pubescence; distinguished by crenulate opposed basal edges of pronotum and elytra, non-rostrate head, and long pectinate antennae; tarsi 5-5-5-segmented, the 4th segment often minute and the 3rd with a large ventral lobe; abdomen with 5 visible sternites, the 5th distinctly emarginate.

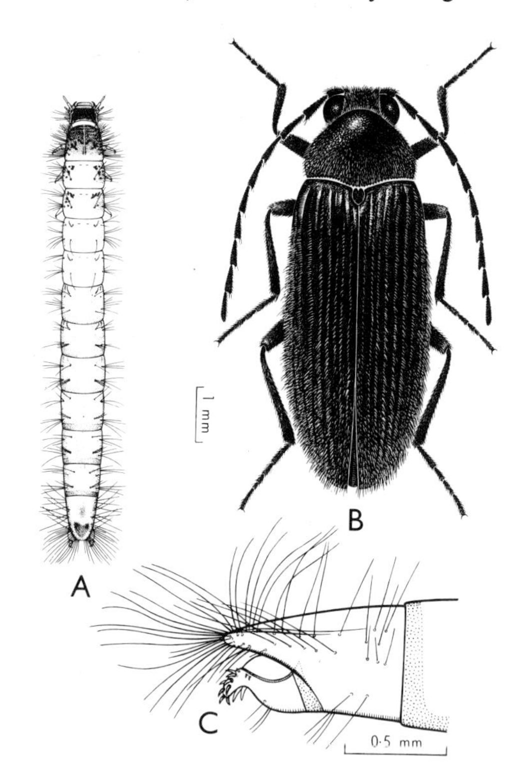

Fig. 30.52. Ptilodactylidae: A, unidentified larva; B, *Epilichas oblongus*, adult; C, larva, apex of abdomen, lateral. [F. Nanninga]

* The species placed by Carter (*Proc. Linn. Soc. N.S.W.* **60**: 186–7, 1935) in *Epilichas* (Dascillidae) are now known to be Ptilodactylidae. They differ from the description given here in having serrate antennae, no lobe on the 3rd tarsal segments, and no emargination of the 5th sternite.

The larvae are elongate and subcylindrical, up to 15 mm long, with conspicuous setae; antennae 3-segmented, conspicuous; a pigment spot on each side of head close to the base of the antenna; prothorax almost as long as the meso- and metathorax together; abdomen 10-segmented, with segment 10 small, situated beneath the terminal segment 9, and with 2 small setiferous projections on each side of the anal cleft; with biforous spiracles on the mesothorax and abdominal segments 1 to 8, or with a pair of ventral tufted tracheal gills on abdominal segments 1 to 8, or with gills confined to segment 9. The larvae are found in water or in damp, decaying wood. The adults occur on vegetation near water. Undescribed species are known from Queensland, N.S.W., Victoria, and Tasmania. Seven species are known from New Guinea.

38. Heteroceridae (Fig. 30.53A). Small (3–4 mm), elongate, densely pubescent beetles; head and pronotum dark brown, elytra with a pattern of dull yellow elongate spots or bands on dark brown. Head directed more or less forwards, so that the flattened mandibles and large labrum are visible from above; antennae very short, with a thick, 7-segmented, serrate club; prothorax strongly transverse; fore coxae transverse, with large visible trochantins; prosternal process narrow; mid coxae separated; hind coxae contiguous; tibiae flattened, with long spines on the outer and apical edges; tarsi 4-4-4-segmented; segments simple, the terminal segment only about half as long as the basal 3 together, claws small and slender; abdomen with 5 visible sternites. Larvae elongate, without urogomphi, and with rather distended thoracic segments; legs well developed; 10 abdominal segments visible.

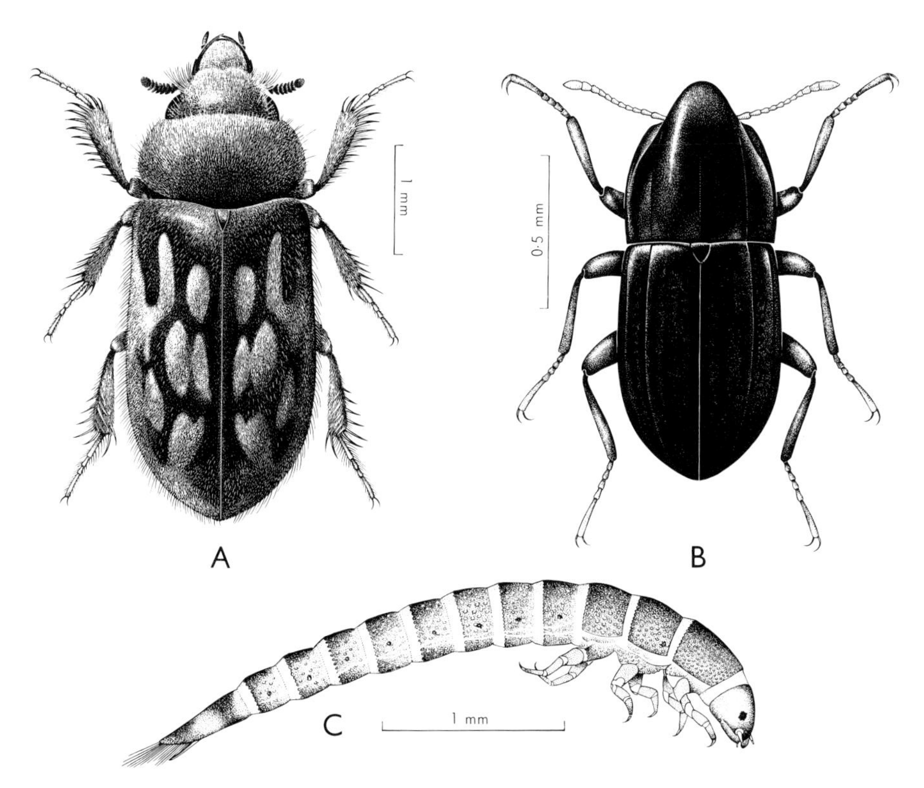

Fig. 30.53. A, *Heterocerus flindersi*, Heteroceridae; B, *Austrolimnius waterhousei*, Helminthidae, adult; C, *A. waterhousei*, larva. [F. Nanninga]

Heterocerids are widely distributed throughout the world, almost all species being in the genus *Heterocerus*. Both adults and larvae tunnel in stiff mud at the sides of ponds and streams, and are phytophagous. The family is represented in Australia by 8 species of *Heterocerus* and the Queensland species *Elythomerus elongatulus* Waterh.

39. Helminthidae (Elmidae; Figs. 30.53B, C). Small (2–5 mm), dark beetles found on stones or weeds in running water. Antennae slender, never with apical segments forming a close, pectinate club (as they do in Dryopidae); mid and hind coxae widely separated; ventral surface without cavities for reception of the folded legs; head deflexed, not rostrate; legs long, with strong tarsi and claws; tarsi 5-5-5-segmented, terminal segment as long as the 4 basal segments together; coxites of female genitalia symmetrical and always with styli. Adults breathe under water by means of a plastron. Larva onisciform (p. 845) or cylindrical, with an abdomen of 9 segments, the posterior part of sternum 9 forming a movable operculum, and 3 tufts of anal retractile gills; last instar with functional spiracles on mesothorax and first 8 abdominal segments. Larvae of Australian species have yet to be described.

Some of the Australian species were at first placed in the Dryopidae. *Austrolimnius* (51 spp.) also occurs in South America (Hinton, 1965). In addition to *Austrolimnius*, the genera *Simsonia*, *Coxelmis*, *Stenelmis*, *Kingolus*, and *Hydrethus* are represented in the fauna. Helminthids live in streams of all kinds, but are more common in those with rocky bottoms, clear water, and high oxygen content. Both larvae and adults are phytophagous, and occur together throughout the year. The larvae leave the water to pupate under stones on the bank of the stream.

Superfamily BUPRESTOIDEA

Elongate, heavily sclerotized, rigid beetles; size range considerable (1·5 to 60 mm). Head vertical and rather deeply sunk into prothorax; antennae 11-segmented, short and serrate; tarsi 5-5-5-segmented, penultimate segments bilobed as seen from above and segments 1 to 4 with a membranous lamella beneath, terminal segments without an empodium; fore coxae globular, the coxal cavities open behind; hind coxae with a backwardly facing concavity for reception of the femur; abdomen with 5 visible sternites, the 2 basal sternites being immovably joined and the suture between them less obvious than those between the other sternites; metasternum with a well-marked transverse suture which is curved back towards the posterior edge on each side; pronotum closely applied to the elytra; elytra striate, scutellar striole present; prothorax not movable on the mesothorax; median process of prosternum received into a cavity formed by the mesosternum or (more commonly) the mesosternum and metasternum; all tergites of abdomen strongly sclerotized; wings short and broad, folded longitudinally, but without the usual transverse apical fold; anal cell, where present, pointed and produced into a single anal vein; aedeagus without a distinct basal piece; Malpighian tubes 6, cryptonephric. Larvae soft-bodied, with prothorax markedly expanded and flattened, hind body relatively slender; without legs; labrum free; spiracles cribriform. This is one of the most clearly defined and uniform superfamilies of Coleoptera.

40. Buprestidae (Plates 4, T, V–ZA, 6, E). Represented throughout the world and particularly abundant in the humid tropics, the Buprestidae (commonly known as 'jewel beetles') include some of the most brilliantly coloured of all insects. The adults are very active in hot weather, and fly readily in sunlight. They are usually to be found on nectar-bearing flowers. The larvae of many species feed in the wood or roots of trees, where they produce characteristic flattened excavations, or in the stems of herbaceous plants. Some of the smaller species are gall-makers (e.g. *Paracephala*, *Ethon* spp.). The larvae are characterized by the relatively wide expansion of the prothorax, small retracted head, and slender hind body. The antennae are very short, and there are no ocelli. Anal processes are known only in *Agrilus* spp. Of 12 recognized subfamilies, 6 are represented in Australia.

Key to the Subfamilies of Buprestidae Known in Australia

1. Antennae with the pores on the toothed segments spread over both surfaces .... CHALCOPHORINAE
   Antennae with the pores on each toothed segment limited to a small fovea 2
2. Surface of scutellum uniformly inclined upwards from beneath edge of pronotum; inner edges of eyes strongly convergent towards top of head; lower anterior edge of fore femur with an angle near the middle .... CHRYSOBOTHRINAE
   Surface of scutellum not inclined, terminating parallel with edge of pronotum; inner edges of eyes not convergent towards top of head; fore tibia without angles 3
3. Basal edge of pronotum straight .... MASTOGENINAE
   Basal edge of pronotum sinuate 4
4. Cavity for reception of prosternal process formed by mesosternum only ... POLYCESTINAE
   Cavity for reception of prosternal process formed by mesosternum at sides and metasternum at bottom 5
5. Labrum largely hidden beneath the clypeus, but, when visible, pale in colour; mouth-parts not produced downwards to form a short rostrum .... BUPRESTINAE
   Labrum prominent and coloured like the clypeus; mouth-parts produced downwards to form a short rostrum .... STIGMODERINAE

Merimna atrata (Castel. & Gory), an all-black, rather flat beetle, of length 15–30 mm, is commonly known as the 'fire beetle' because of its habit of flying into camp fires. Adult buprestids are sometimes found alive in houses, having emerged from structural timber or furniture years after it has been put into use. Cases are known in Europe in which the adult of *Buprestis aurulenta* L. has emerged from wood imported from North America as much as 25 years earlier. *B. aurulenta* has also been recorded in Australia, having emerged from imported timber. The largest genus, *Stigmodera*, which includes about half of the native species, is restricted to Australia. The genus is recognizable by having the mouth-parts produced downwards to form a short rostrum, the pores on the antennae concentrated into foveae on the toothed segments, the labrum long, coloured like the clypeus, the posterior edge of the pronotum sinuate, and the frons not narrowed between the antennal cavities. The range in size in *Stigmodera* is considerable (10–60 mm), and most species are strikingly patterned in various combinations of red, yellow, black, and metallic colours. [Carter, 1929.]

Superfamily RHIPICEROIDEA

Characterized by the presence of a large, multisetose empodium between the claws, and very strongly flabellate antennae in the male. Fore coxae conical, the coxal cavities widely open behind; prosternal process not articulating with the mesosternum; hind coxae with a steep posterior face against which the femur can be retracted; tarsi 5-5-5-segmented; abdomen with 5 visible sternites. The species are distributed mainly in the warmer parts of the world.

Key to the Families of Rhipiceroidea

Tarsal segments 1–4 deeply bilobed, with paired membranous appendages beneath; mandibles prominent, strongly curved; lateral margin on each side of pronotum indicated at the anterior or posterior end by a longitudinal carina; basal abdominal sternite not raised into a keel between hind coxae. Larva with labrum fused to head capsule; mandibles with apex simple and without a mola at base; abdominal segment 9 not operculate; body soft .... **Rhipiceridae** (p. 566)

Tarsal segments without membranous lobes beneath; mandibles short, not projecting; lateral margin on each side of pronotum obsolete, not marked by a carina; basal abdominal sternite raised into a keel between hind coxae. Larva with labrum free; mandibles with apical teeth and a mola at the base; abdominal segment 9 operculate; body cylindrical .... **Callirhipidae** (p. 566)

41. Rhipiceridae (Sandalidae; Fig. 30.54). Easily recognized by the large hairy empodium, the strongly flabellate antennae of the male, and the presence of membranous lobes on the tarsal segments. The beetles are of moderate size (10–25 mm). The beautiful fan-like antennae of the male *Rhipicera* are composed of a large but variable number of segments (21–41 in *Rhipicera mystacina* F.). *R. femorata* Kirby from northern N.S.W. is black, the pronotum and elytra marked with conspicuous spots formed by patches of white setae, and the femora pale yellow with apices black. Nothing is known of the biology of the Australian species, but the larva of an American species (*Sandalus niger* Koch) is known to parasitize the immature stages of cicadas. The Australian species are all included in the genus *Rhipicera*.

42. Callirhipidae. The Australian species are rare, and all are from the rain forest of Queensland. They belong to the genera *Callirhipis* and *Ennometes*, and are dark brown, with distinct longitudinal ribs on the elytra. The antennae are 11-segmented. The only larvae known resemble those of Elateridae, and are found in rotting wood.

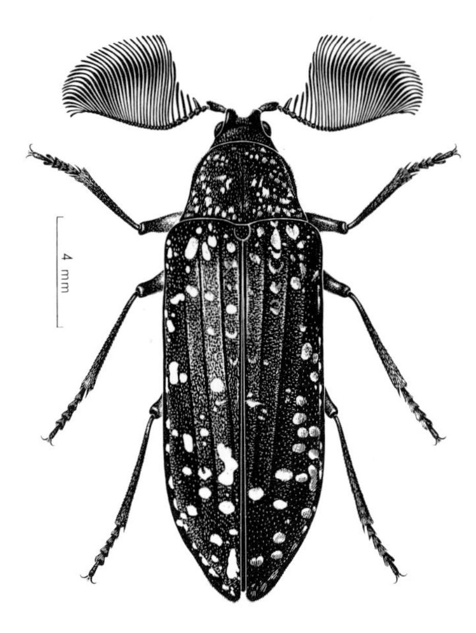

Fig. 30.54. *Rhipicera femorata*, Rhipiceridae, ♂. [F. Nanninga]

Superfamily ELATEROIDEA

Beetles of a characteristic elongate, rather flattened form, with the prosternal process capable of being received into a corresponding cavity on the mesosternum, the two parts usually forming a 'clicking mechanism' which is used by the beetle in righting itself and as a means of escape, the prothorax being freely movable relative to the mesothorax. Hind angles of pronotum acute, projecting posteriorly; antennae usually serrate; metasternum without a transverse suture; fore coxae rounded, trochantins concealed; fore coxal cavities open behind; prosternum long in front of the coxae; hind coxae contiguous, with transverse cavities for reception of hind femora; abdomen with 5 visible sternites; tarsi 5-5-5-segmented; elytra without a scutellar striole; wings with anal cell, where present, truncate apically, giving rise to 2 anal veins; 4 Malpighian tubes; aedeagus trilobed, with a reduced basal piece and free parameres. Larvae with labrum absent or fused to clypeus; without a median epicranial suture.

The Elateroidea and the Buprestoidea together formed the Sternoxia of early authors.

Key to the Families of Elateroidea Known in Australia

1. Labrum not externally visible; antennal insertions situated more than one eye-width in front of eye; abdomen with all sternites connate, sternite 5 not movable on 4 in life ... **Eucnemidae** (p. 567)
   Labrum visible, free; antennae inserted close in front of eyes; abdomen with sternites 1 to 3 or 4 immovably connected, 5 free and movable in life (the membrane connecting 4 and 5 often visible) 2

2. Head not transversely ridged between the eyes; last 3 segments of antenna forming a serrate club, or segments 1–4 of tarsi with long membranous lobes; insects not able to 'click' .... **Trixagidae** (p. 568)
   Head transversely ridged between the eyes; last 3 segments of antenna not forming a club; tarsi simple or with segments 3 and 4 lobed; species able to 'click' .... **Elateridae** (p. 567)

43. Elateridae (Plate 4, H; Fig. 30.55A). Beetles of characteristic elongate form, length 3–54 mm; with a 'click' mechanism formed by the long prosternal process which can be forced suddenly into the corresponding cavity in the mesosternum, causing a sudden movement of the prothorax relative to the hind body. Labrum visible, and head with a sharp transverse ridge between the eyes; antennae serrate, not clubbed, pedicel inserted on the axis of the scape; tarsi simple or with segments 3 and 4 bilobed. Larvae elongate, cylindrical, with legs, colour usually shining reddish brown; labrum fused to clypeus; ocelli vestigial or absent; abdomen 9-segmented, 10 reduced and visible only from below; spiracles on mesothorax and abdominal segments 1–8; short, heavily sclerotized urogomphi often present.

The larvae are phytophagous, feeding upon roots, or carnivorous. Among the phytophagous species are the well-known 'wireworms' which are pests of cereals, grasses, and root crops. *Lacon variabilis* Cand. is a pest of sugar cane in Queensland. The larvae of *Paracalais* are very active, and prey upon the larvae of the larger wood-boring beetles; the adults are large and attractively mottled with longitudinal white, grey, and black marks. [Neboiss, 1956.]

44. Eucnemidae (Melasidae; Fig. 30.55B). Beetles 5–20 mm long, of elateroid form; distinguished from Elateridae by the apparent absence of the labrum; abdomen with sternite 5 not movable on 4 in life (intervening membrane not visible); frons usually without a transverse ridge; antennae 11-segmented, flabellate (in male), moniliform, or filiform, sometimes received into a groove beneath lateral edge of prothorax; pedicel inserted on outer apical angle of the thick scape; prothorax movable relative to mesothorax, but the click mechanism apparently not functional. Larvae white, subcylindrical, flattened, with an unusually small head, and without legs; mouth-parts greatly reduced; mandibles toothed externally; antennae very small; abdomen 9-segmented, with or without

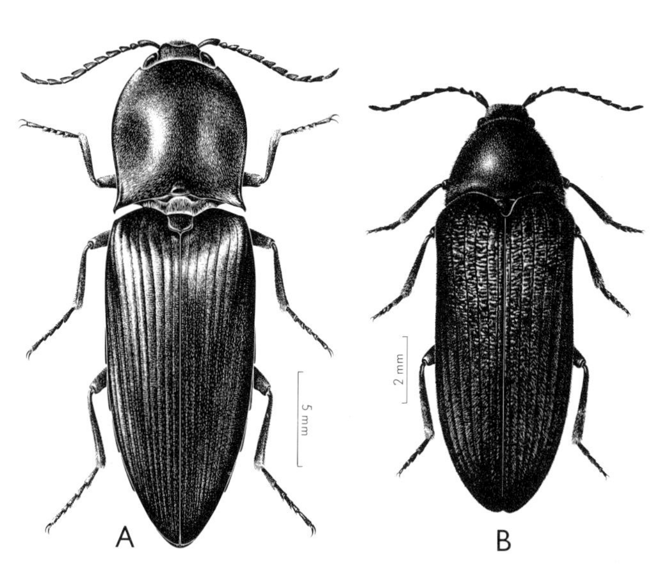

Fig. 30.55. A, *Agrypnus mastersi*, Elateridae; B, *Euryptychus porosus*, Eucnemidae. [F. Nanninga]

small, fixed urogomphi. It seems probable that the larvae are carnivorous, but their mouth-parts are reduced, and the precise method of feeding has not been observed. They are found in wood beneath the bark of trees which have been attacked by other insects, and are rather rare. *Hemiopsida* (15 spp.) is the largest of the 27 genera represented in Australia.

45. Trixagidae (Throscidae). Small beetles (*ca* 4 mm) of elateroid form; distinguished from Eucnemidae by the presence of the labrum, and from Elateridae by the absence of a transverse ridge between the eyes. Prothorax with a deep groove on each side between prosternum and hypomeron, for reception of the antenna; pedicel inserted on the axis of the scape, the last 3 segments usually forming a serrate club; head deflexed, frons convex; tibiae flattened, without spurs; mid and hind tarsi sometimes received into sharply defined grooves on the metasternum and first 2 abdominal sternites respectively (e.g. in *Trixagus*); prosternal process broad and flat, without a 'click' mechanism. Larvae soft-bodied, unlike those of Elateridae, usually with very short legs, head reduced and mandibles biting outwardly as in Eucnemidae, and the terminal segment with 2 short, fixed urogomphi. The larvae are found in warm, fermenting plant debris, and the adults are taken in flowers.

Superfamily CANTHAROIDEA

Integument unusually soft and flexible, body with unusually loose articulation; elytra usually not closely applied to each other at the sutural edges; abdomen with 6 or 7 visible sternites; tarsal segmentation 5-5-5, the tarsi stout, shorter than the tibiae; antennae long, filiform, serrate, or flabellate; fore and mid coxal cavities open behind; the large tubular mesothoracic spiracles often visible between fore and mid coxae (particularly in fresh specimens); hind coxae transverse, with a steep, flat or convex posterior face; 4 free Malpighian tubes. Sexual dimorphism commonly very marked; aedeagus trilobed. Larvae always with channelled mandibles, adapted presumably to extra-oral digestion of insect prey; cuticle usually soft and pubescent; last abdominal segment forming a pygopod.

This superfamily, together with Melyridae, Lymexylidae, Dascillidae, and Cleridae, comprised the 'Malacodermata' of older authors.

Key to the Families of Cantharoidea Known in Australia

1. Luminous organs present on one (in ♀) or two (in ♂) abdominal sternites (visible as pale yellowish white sternites); elytral epipleura broad at the base .... **Lampyridae** (p. 568)
   Luminous organs absent; elytral epipleura narrow or absent 2
2. Tarsi with 4th segment entire, its apex not emarginate; mid coxae obviously separated; labrum visible; lateral edge of metasternum straight; pronotum and elytra strongly ridged; not found on flowers .... **Lycidae** (p. 570)
   Tarsi with 4th segment often bilobed; mid coxae very close together, almost in contact; labrum covered by clypeus; lateral edge of metasternum curved; pronotum and elytra smooth; commonly found on flowers 3
3. Claws simple; labrum not visible, covered by clypeus; body usually clothed with fine pubescence .... **Cantharidae** (p. 569)
   Claws toothed or with an appendage beneath; body usually bearing erect bristles .... **Melyridae** (p. 571)

46. Lampyridae (fireflies; Fig. 30.56). Adults with luminescent organs on the 5th (in female) or 5th and 6th (in male) abdominal sternites; integument rather soft, and head largely covered by the explanate anterior margin of the pronotum as in other cantharoids; body rather flattened, 4–11·5 mm long, with yellow pronotum and brownish elytra. Antennae filiform, 11-segmented, inserted close together between the large,

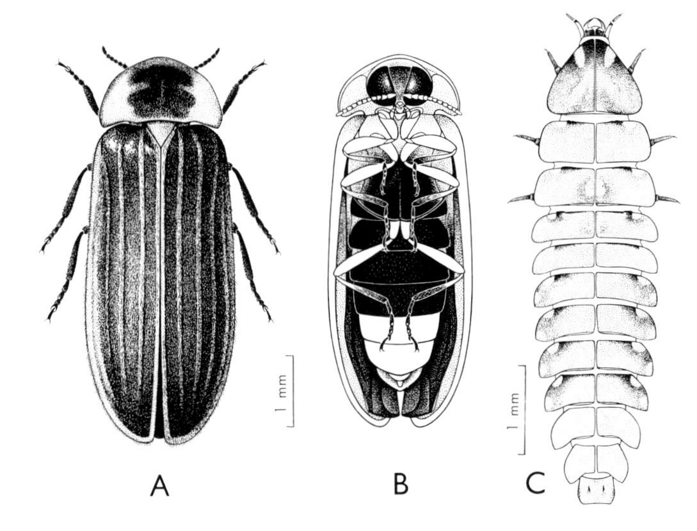

Fig. 30.56. *Atyphella lychnus*, Lampyridae: A, dorsal; B, ventral; C, *Atyphella* sp., larva. [F. Nanninga]

hemispherical eyes, which are much larger in male than in female; mandibles curved, perforated by a canal, and acutely pointed; palpi pubescent, with enlarged, pointed, terminal segments; tibiae flattened, apical spurs absent; elytra with short dense pubescence, epipleura broad at the base; elytra covering the abdomen and wings fully developed in most species of the 2 Australian genera. (In the Palaearctic *Lampyris* the females are wingless, resembling larvae, and are consequently known as 'glow-worms'.) Larvae usually brown, elongate, strongly tapered in front and behind; head small, narrow, and concealed beneath the prothorax; all body segments except the last laterally explanate in *Atyphella* and with a strongly impressed median longitudinal groove; head with a single ocellus on each side; antennae 3-segmented; legs short; abdomen composed of 9 segments; with mesothoracic and 8 abdominal spiracles.

Where observations have been made, eggs, larvae, and pupae of Lampyridae are found to be faintly luminous; in the adults the light is much brighter and is restricted to special organs. The light is emitted in a series of controlled flashes which are assumed to have sexual significance. The males of the Australian species fly in damp forests on dark, still evenings, and the light is emitted in flashes with a frequency of about one per second. The females, although often winged, are not known to fly. The very impressive synchronization of the flashes of a large number of individuals, which has been described for some tropical species, has not been observed in Australia. The adult beetles are said not to feed, but the larvae are all carnivorous, preying mainly upon gastropod molluscs. The prey is paralysed by injecting the secretion of a pair of acinose glands at the anterior end of the fore gut (there are no salivary glands) through the tubular, sharply pointed mandibles. The secretion is also proteolytic, and the larva imbibes the liquefied tissues of the prey with the aid of its maxillae and labium. The Australian species belong to the genera *Luciola* and *Atyphella*. Most are found in the north, but *A. lychnus* Oll. occurs in the Blue Mts in N.S.W. [Lea, 1909.]

47. Cantharidae (soldier beetles; Fig. 30.57A). Soft-bodied, elongate, flattened, parallel-sided beetles, 3–18 mm long, commonly coloured in various combinations of yellow and bluish black, and clothed with fine, fairly dense pubescence. Head deflexed, with oval, entire, convex eyes; antennae inserted in large sockets before and between the eyes, 11-segmented, filiform; labrum largely hidden; mandibles strongly curved

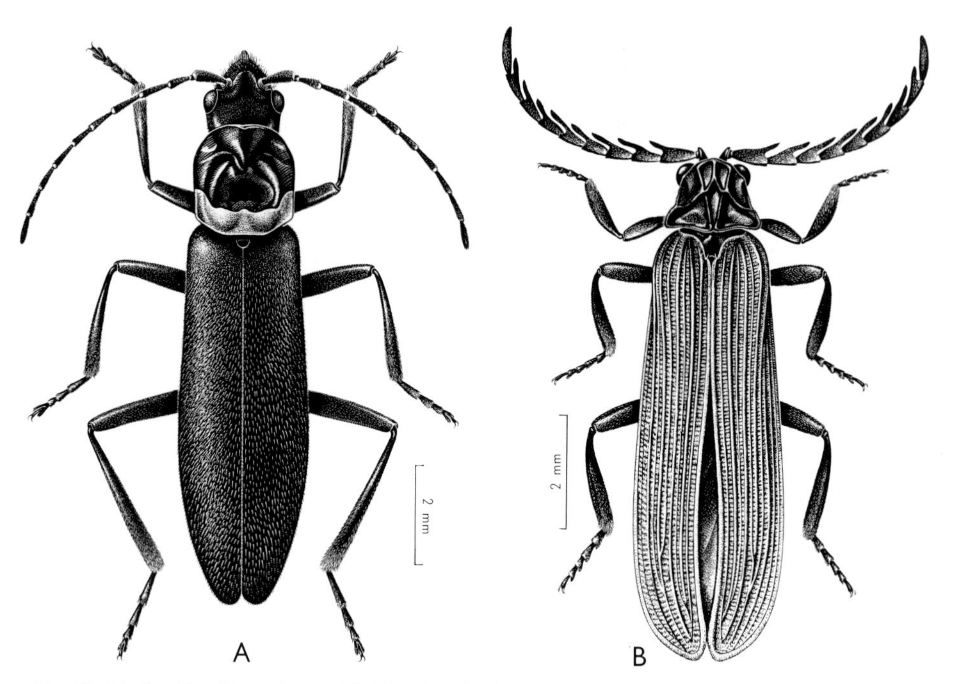

Fig. 30.57. A, *Chauliognathus pulchellus*, Cantharidae; B, *Metriorrhynchus rhipidius*, Lycidae. [F. Nanninga]

and sharply pointed; palpi often with terminal segments securiform; pronotum oval or subquadrate, rather flat with somewhat explanate margins, partly covering the head from above; tarsi with 4th segment bilobed; mid coxae contiguous or nearly so; elytra smooth. Larvae stout, campodeiform, with a characteristic velvety clothing of fine setae, often dark coloured, brown or bluish-black; antennae 3-segmented; one pair of ocelli; mandibles acute, channelled; abdomen 10-segmented, segment 10 small and hidden beneath 9.

Both adults and larvae are free-living and carnivorous, preying upon soft-bodied insects; adults are active and common on flowers. *Chauliognathus pulchellus* Macl. is often very abundant in the mountains of N.S.W. in summer. It is about 12 mm long, blue-black in colour, with the base of the pronotum and apex of the abdomen bright orange. A closely related species, with abbreviated elytra and reduced wings, is found on Lord Howe I. This family was referred to as Lampyridae–Telephorinae by Tillyard.

48. Lycidae (Fig. 30.57B). Beetles 5–20 mm long, with characteristic soft integument, flattened form, and long, narrow, dehiscent elytra; colour usually combinations of reddish yellow and black. Head triangular or rostrate, partly covered by the explanate front margin of the pronotum; antennae 11-segmented, long, thick, flattened, serrate or pectinate, inserted in large, almost contiguous cavities between the eyes; labrum distinct; maxillary palpi 4-segmented, apical segment flattened and expanded; eyes oval, very strongly convex; pronotum with margins explanate and surface divided into areas by ridges; fore and mid coxae conical, projecting, fore coxae with a large triangular trochantin; mid coxae obviously separate; lateral edge of metasternum straight; femora and tibiae flattened, tibiae without apical spurs; tarsal segments expanded apically, but not bilobed or emarginate; elytra elongate, 5 to 8 times as long as pronotum, usually widened apically, surface with 4 marked longitudinal costae. Larvae elongate, tapered to both ends, without velvety pubescence; setae limited to a few on the ventral surface and sides; head small, depressed, with short, stout, 1- or 2-segmented antennae; labrum, clypeus, and frons fused; mandibles divided longitudinally into two portions; abdomen 10-segmented, 10 very small, covered by 9; spiracles on mesothorax and abdominal segments 1–8; without urogomphi.

Both adults and larvae are predatory, the larvae living beneath bark or in the soil. Of the 16 Australian genera, *Metriorrhynchus* (87 spp.; Plate 6, A) comprises species that are distasteful to birds and appear to serve as models for a variety of mimics, including Oedemeridae *(Pseudolycus)*, Cerambycidae *(Eroschema)*, Buprestidae, and Belidae *(Rhinotia)* (Plate 6, B–E), as well as moths, flies, and wasps. [Lea, 1909.]

49. Melyridae (Malachiidae; Fig. 30.58). Small, rather soft-bodied, elongate, flattened beetles; head, pronotum, and elytra usually bearing long erect setae. Antennae 10- or 11-segmented, inserted in large sockets near the anterior edge of the frons, rather distant from the eyes, filiform, serrate, or flabellate, often with marked sexual modification of the scape and pedicel in the male; labrum visible, sclerotized like the clypeus, often separated from it by a broad pale membrane; mandibles curved, pointed, simple or bifid; maxillary palpi with apical segment pointed, fusiform, or truncate; pronotum as in Cantharidae; tarsi with penultimate segment simple or bilobed; claws, short and stout, each with a fleshy appendage beneath; mid coxae nearly contiguous; elytra smooth, not striate, sometimes abbreviated, so that up to 6 abdominal tergites are exposed. Larva subcylindrical, with the last (9th) abdominal segment sclerotized and bearing a pair of short curved urogomphi; head quadrate, usually with 4 pairs of ocelli; mandibles with a prostheca, but without a mola.

Like the Cantharidae, these beetles are almost always carnivorous, and are commonly found on flowers. Many species are

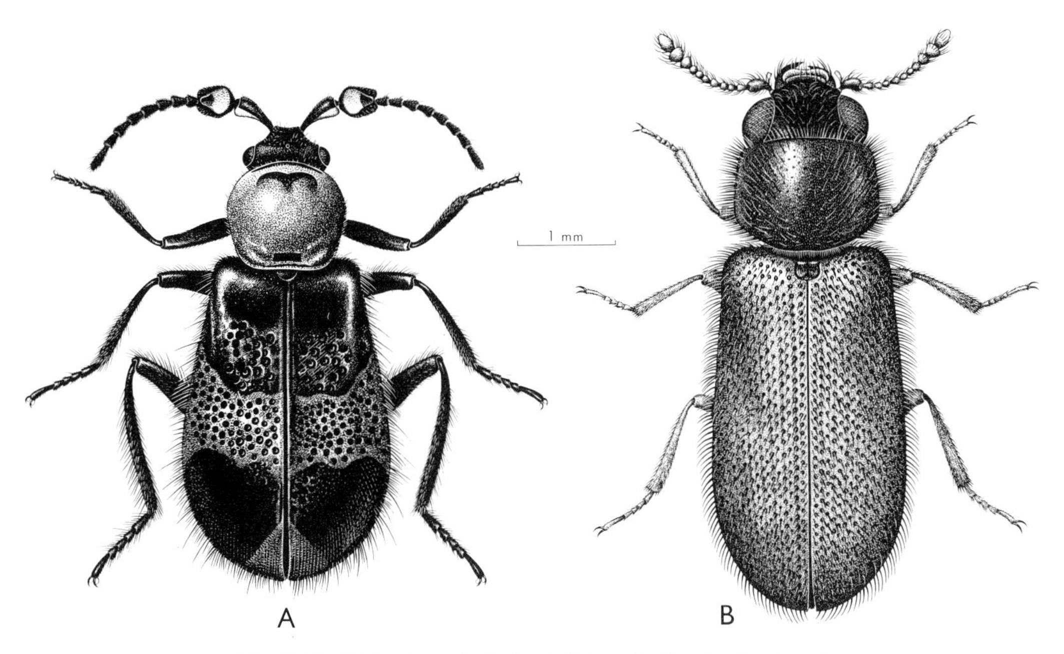

Fig. 30.58. Melyridae: A, *Laius bellulus*; B, *Dasytes fuscipennis*. [F. Nanninga]

brightly coloured, mainly in yellow and metallic blue. Sexual dimorphism is marked. Males of *Laius* have the scape and pedicel of the antennae greatly enlarged, the pedicel usually with a deep channel, possibly used for gripping the antenna or the edge of the pronotum of the female. The fore tibiae of the males are also markedly swollen. Males in *Balanophorus* have strongly flabellate antennae. In *Carphurus* and *Helcogaster* the males have a black comb-like structure on the basal segment of each fore tarsus, while in *Balanophorus* this structure is present in both sexes. In contrast to the normal predatory habit of the family, one species, *Laius cinctus* Redtenb., is a pest of growing rice in N.S.W.

The family is divided into 4 subfamilies. The 2 larger, DASYTINAE and MALACHIINAE, are well represented in our fauna. The Malachiinae are distinguished by having protrusible vesicles at the sides of the prothorax and basal segments of the abdomen. Adult Melyridae (especially Malachiinae) resemble Cantharidae, whereas their larvae appear to be closer to those of the Cleridae. The Melyridae were grouped with Cleridae by Böving and Craighead on larval characters, and Crowson (1955) also holds the view that the adult similarities of the Melyridae and Cantharidae are the result of convergence. Nevertheless the similarities of the adults are such that the Melyridae are more easily recognized as Cantharoidea, and for this reason they are retained in this superfamily. [Lea, 1909.]

Superfamily DERMESTOIDEA

Small, ovoid, compact beetles; hind coxae with a vertical or concave posterior face; antennae with a well-marked 3- to 5-segmented club; head very often with a single large median ocellus; prothorax never hooded over head; posterior edge of pronotum obtusely angled in the middle; tarsi 5-5-5-segmented, segments not lobed; abdomen with 5 visible free sternites; spiracles of abdominal segment 8 functional; fore coxae projecting or transverse; mesepimera forming part of the mid coxal cavities; body frequently densely clothed with hairs or scales; elytra entire, usually not striate; Malpighian tubes cryptonephric and asymmetrically arranged.

50. Dermestidae (Figs. 30.59, 60). Pronotum strongly narrowed anteriorly, the hind angles acute; head deflexed; labume visible; mandibles stout and blunt-poinred; fore coxal cavities open behind; fore cotxa projecting; tarsal segments not lobed; body

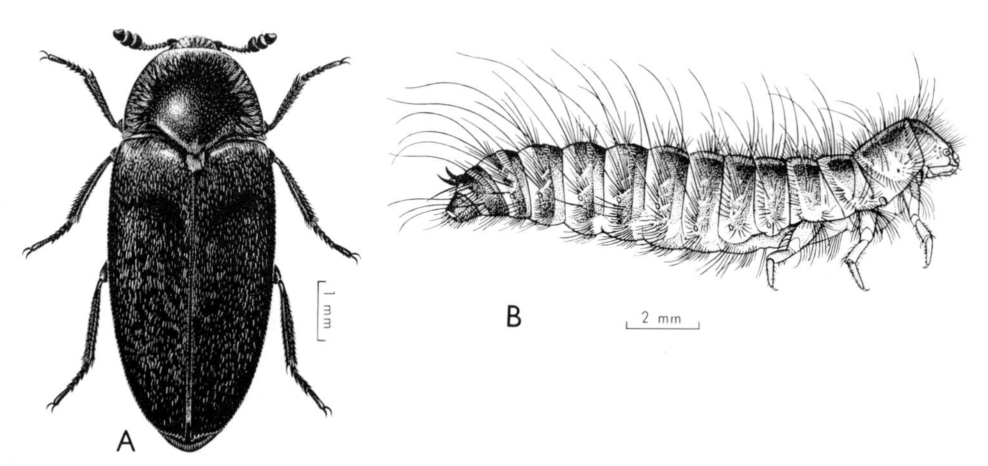

Fig. 30.59. *Dermestes maculatus*, Dermestidae: A, adult; B, larva. [F. Nanninga]

colour usually dark, often with patterns of coloured scales or hairs; elytra entire; aedeagus trilobed, median lobe and parameres long and slender; basal piece short. Larvae subcylindrical, elongate or short, oval, with distinct tergal sclerites; body densely clothed with long setae, some of which may be complex, spear-headed, or clubbed; antenna 3-segmented; 3 to 6 lateral ocelli present; urogomphi present or absent.

Dermestidae feed as larvae on dry animal material of high protein content, e.g. dry carcasses, insect remains, bees', birds', or mammals' nests. Adults are commonly found on flowers. The family includes pests of stored products, such as *Dermestes maculatus* De G., and *D. lardarius* L., which attack hides, furs, and skins, as well as stored foodstuffs such as bacon and cheese. The larvae of *Dermestes*, when about to pupate, commonly leave their food supply and burrow into neighbouring materials, with the result that they damage, and adults are often found in, materials on which the species cannot feed (e.g. wood, cork, cotton, salt). *Trogoderma granarium* Everts is unique among dermestids, in that its larva is a serious pest of grain and cereal products. Like the species of *Dermestes*, it has now spread widely with commerce, but is not yet established in Australia. Species of *Anthrenus* (*A. verbasci* L., *A. scrophulariae* L., *A. vorax* Waterh.) and *Attagenus* (*A. pellio* L., and *A. piceus* Oliv.) are troublesome household pests, because of the damage that their larvae cause to carpets, blankets, and other woollen fabrics. Adults of *Anthrenus* have to leave the house temporarily to feed on flowers and mature their eggs. *Anthrenus* spp., as well as *Trogoderma versicolor* Creutz., are also troublesome as pests of insect collections in museums. Dermestidae live naturally in the debris of birds' nests, and these are an important source of household infestations. Cultures of larvae of *D. maculatus* are often employed in museums in the cleaning of delicate skeletons of small vertebrates. [Hinton, 1945; Armstrong, 1945.]

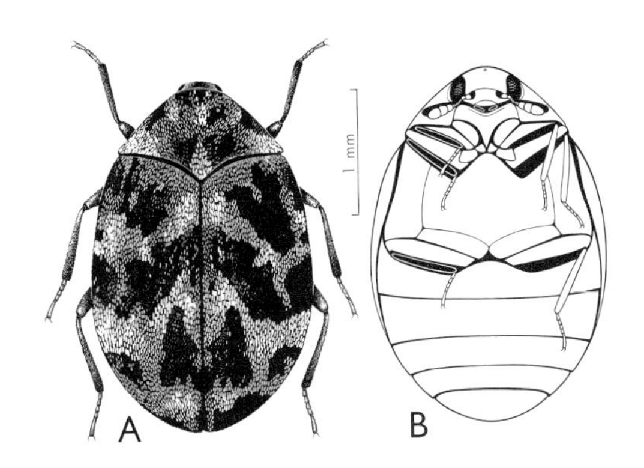

Fig. 30.60. *Anthrenus verbasci*, Dermestidae: A, dorsal; B, ventral. [F. Nanninga]

Superfamily BOSTRYCHOIDEA

Small to medium-sized beetles of elongate, cylindrical shape; with the head small and deflexed; prothorax often with pronotum 'hooded', i.e. strongly convex and with prosternum short, so that the anterior aperture faces downwards (e.g. Fig. 30.63). Head usually concealed by prothorax when viewed from above, usually elongate behind the eyes, and with a median longitudinal suture in the occipital region; antennae varied, filiform, moniliform, clubbed, or with the 3 terminal segments elongate and subequal, together as long as, or longer than, basal part of antenna; hind coxa usually without a vertical or concave posterior face; tarsi 5-5-5-segmented; mesepimera not forming part of the mid coxal cavities, which instead are closed externally by the meso- and metasternum; elytra often very convex, more than semicircular in section, reducing the exposed area of sternites; visible sternites usually 5, sometimes 3 (some Ptinidae); spiracle of abdominal segment 8 functional. Larvae C-shaped, soft, white, without sclerotized dorsal plates or urogomphi; legs present; feeding in wood or dry vegetable or animal materials.

Keys to the Families of Bostrychoidea

ADULTS

1. Hind coxae with transverse concavity for reception of the femur, *or* antennal insertions very close together on the frons; antennae usually 11-segmented, with a loose club; trochanters squarely truncate at apex 2
   Hind coxae without a concavity or obvious steep posterior face against which the femur can be retracted; antennal insertions placed just before the eyes, widely separated; antennae usually with less than 11 segments, and with a 2- or 3-segmented compact club; trochanters obliquely truncate at apex 3
2. Hind coxal cavities with a transverse concavity for reception of the femur; antennal insertions separated by more than the length of the scape .... **Anobiidae** (p. 574)
   Hind coxal cavities without an obvious transverse concavity for reception of the femur; antennal insertions close together, separated by less than the length of the scape .... **Ptinidae** (p. 575)
3. Pronotum strongly convex, the planes of the anterior and posterior apertures of the prothorax nearly at right angles (Fig. 30.63); fore coxae projecting; head not visible from above; 1st visible sternite hardly longer than 2nd or 3rd .... **Bostrychidae** (p. 576)
   Pronotum not strongly convex, the planes of the anterior and posterior apertures of the prothorax not nearly at right angles (Fig. 30.62B); head visible from above; fore coxae globular, not projecting; 1st visible sternite almost as long as 2nd and 3rd together .... **Lyctidae** (p. 577)

LARVAE

1. Head exserted, rounded, with mouth-parts directed downwards; antennae 1- or 2-segmented; setae on body abundant 2
   Head partly buried in prothorax, mouth-parts directed forward; antennae longer, usually 3-segmented; setae on upper side of body scanty 3
2. Anterior spiracles on the posterolateral part of the prothorax or between prothorax and mesothorax; abdominal tergites each with a transverse band of spinules; almost always wood borers .... **Anobiidae**
   Anterior spiracles on the anterolateral part of the prothorax; abdominal tergites without transverse bands of spinules; never wood borers .... **Ptinidae**
3. Eighth abdominal spiracles of about the same size as the rest .... **Bostrychidae**
   Eighth abdominal spiracle much larger than the other spiracles .... **Lyctidae**

51. Anobiidae (Fig. 30.61). Small beetles (2–6 mm) of uniform reddish or dark brown colour; form subcylindrical or ovoid; head covered above by the pronotum; body surface clothed with fine recumbent or semi-erect setae. Antennae 9- to 11-segmented, last 3 segments markedly elongate, inserted in front of eyes, the insertions widely separated; mandibles short, broadly triangular, dentate; labrum distinct but very small; pronotum with anterior edge rounded, posterior edge often continuous with lateral edges, without an obvious angle, the posterolateral edge meeting the anterior edge in a deflexed acute angle or right angle; the planes of the anterior and posterior apertures of the prothorax nearly at right angles, so that the prosternum is very short; fore coxal cavities open behind; fore and mid coxae conical, contiguous, trochantins exposed; hind coxae transverse, with transverse concavities for reception of the retracted femora; trochanters squarely truncate at their apices; tibiae slender, without spurs. The legs can be retracted into the angular cavities between the head and prothorax, prothorax and mesothorax, and behind the declivity of the hind coxae. Species of *Dorcatoma* in addition have oblique grooves on the metasternum and abdomen for reception of the tarsi. Larvae C-shaped, soft, white, with rounded, hypognathous head and 10-segmented abdomen; segment 10 small; antennae 2-segmented, very small.

The larvae are almost all borers in the wood or bark of dead trees. The notorious furniture beetle (*Anobium punctatum* De G.), now cosmopolitan in distribution, is a member of

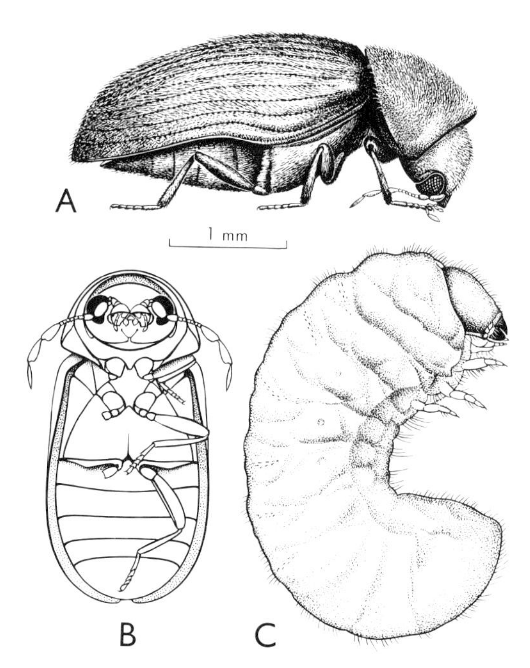

Fig. 30.61. *Stegobium paniceum*, Anobiidae: A, B, adult; C, larva. [F. Nanninga]

the family. The name 'death watch', sometimes applied to this species, belongs in fact to *Xestobium rufovillosum* De G., a large anobiid which does not occur in Australia. *Stegobium paniceum* F. is a highly polyphagous, cosmopolitan pest of drugs, stored food, and other products. This is the biscuit weevil that the sailors of earlier days knocked out of their 'hard tack'. Like *Ptinus tectus* Boield., it is a serious pest of dried plants in herbaria. *Lasioderma serricorne* F. is a pest of stored tobacco. Many Anobiidae have mycetomes in the form of caeca in connection with the anterior part of the gut. The microorganisms from the mycetomes are transmitted in the female by way of the anus to special sacs from which the surface of the eggs is infected (p. 512). [Lea, 1924.]

52. Ptinidae (Fig. 30.62A). Small (2–5 mm) beetles, with head and pronotum narrow in comparison with the rather globular elytra. Head deflexed; antennae 2- to 11-segmented, inserted in close proximity on the frons; labrum distinct; palpi short and slender; pronotum without defined lateral edges or margins, anterior aperture facing downwards; prosternum very short, fore coxal cavities open behind; mesosternum very short, so that the small globular fore and mid coxae are unusually close; legs long; hind coxae without a transverse concavity for reception of the femur; trochanters squarely truncate at their apices. Larvae C-shaped, white, setose; head exserted, hypognathous; antennae small, 2-segmented; abdomen 10-segmented, sometimes with segments 9 and 10 greatly reduced.

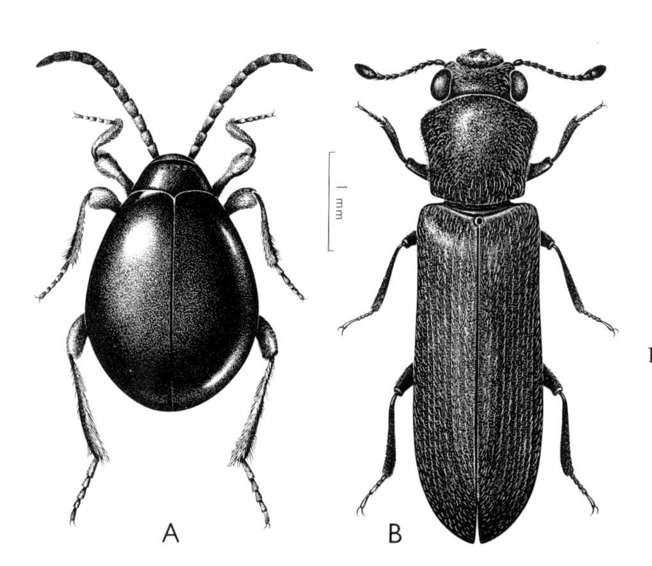

Fig. 30.62. A, *Gibbium psylloides*, Ptinidae; B, *Lyctus brunneus*, Lyctidae. [F. Nanninga]

Ptinidae are not borers, but feed mainly on dry materials of animal and vegetable origin, and for this reason they are common inhabitants of the nests of birds, mammals, and social insects. The family includes a number of species that infest stored foodstuffs, skins, furs, etc., as well as museum specimens. The Australian *Ptinus tectus* is now cosmopolitan and a common household pest. *Niptus hololeucus* Falder., *Gibbium psylloides* Czemp., and *Mezium affine* Boield. are also household pests, introduced into Australia, and originally from the Mediterranean region. Unlike *Ptinus* and *Niptus*, which are densely clothed with recumbent hairs, *Gibbium* and *Mezium* have the elytra extremely convex, highly polished, and bare, giving the insects the appearance of large mites. A number of remarkable genera and species occur in ants' nests (especially of *Iridomyrmex* spp.). Among these, *Paussoptinus laticornis* Lea has broad, flattened, 10-segmented antennae which recall those in Carabidae–Paussinae. *Diphobia*, *Enasiba*, and *Diplocotes* have stout, moniliform, 11-segmented antennae. *Hexaplocotes sulcifrons* Lea has 6-segmented antennae terminated in a large 2-segmented club, whereas in species of *Ectrephes* the antennae are reduced to 2 segments. All but 6 of the Australian species are indigenous. [Hinton, 1941—Ptinidae of economic importance.]

53. Bostrychidae (Fig. 30.63). Cylindrical, strongly sclerotized, black or black and brown beetles, 3–20 mm long. Head directed downwards, not visible from above; anterior aperture of prothorax facing downwards; surface of pronotum usually rugose, sometimes with curved hooks, pronotum without lateral margins; antennae 8- to 10-segmented, with a loose 3- or 4-segmented club, inserted near the eyes, not close together; labrum distinct, small; palpi small and slender; eyes small but strongly convex; prosternum short; fore coxae rounded and projecting, their cavities confluent and open behind; trochantins not visible; mid coxae globular and close together; hind coxae triangular, transverse, without a concavity for reception of the femora; elytra usually with a marked

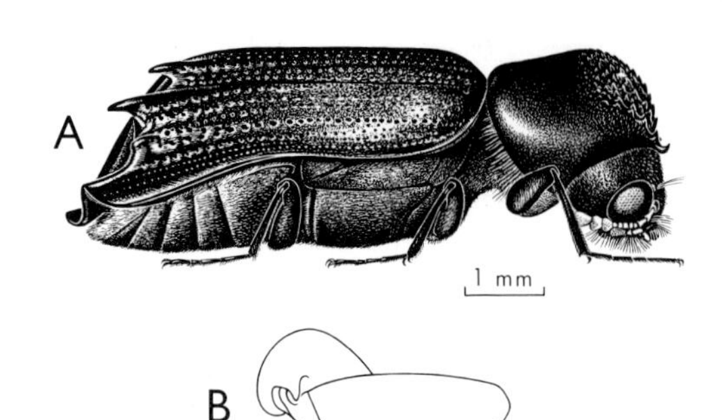

Fig. 30.63. *Bostrychoplites cylindricus*, Bostrychidae: A, adult; B, junction of femur and trochanter. [F. Nanninga]

flat apical declivity and apical spines. Because of their cylindrical form and apical declivity, adult Bostrychidae resemble Scolytinae, but can be readily distinguished by their non-elbowed antennae, tuberculate pronotum, and 5-segmented tarsi. The larvae are wood-borers, and, like the Anobiidae, have mycetomes at the anterior end of the gut. The micro-organisms of the mycetomes are transmitted to the eggs among the sperm of the male. Their activities are limited to moribund or freshly felled trees. Bostrychidae do not have a cellulase, and they make use of starch and sugars in the wood; they are consequently unable to infest seasoned timber.

The largest Australian species is *Bostrychopsis jesuita* F. This species is cylindrical, 10–20 mm long, black, with spinose prothorax. It normally attacks wattle and eucalypt trees, but is known to tunnel in the trunk of unhealthy citrus trees. In contrast to the usual sombre colour of Bostrychidae, *Xylion collaris* Erich. has the pronotum yellowish brown and the elytra black. *Xylobosca bispinosa* Macl. is a very small species (2–3 mm) in which the male is distinguished by having 2 long divergent spines on the apices of the elytra. In this species the male guards the egg-laying female by sealing the entrance to the tunnel with his body. One introduced species, *Rhizopertha dominica* Steph. is now the most serious pest of grain in Australia. *Dinoderus minutus* F. is another widely distributed pest which is recorded in this country. It commonly infests the wood of bamboos, especially those belonging to the genera *Dendrocalamus* and *Phyllostachys*, and is responsible for much damage to basketwork, furniture, etc. made from bamboo. Like *R. dominica*, the species is polyphagous and has been recorded from a wide variety of plant material, including other woods, dried fruit, ginger, cinnamon, avocados, etc. The adult is cylindrical, 2–3 mm long, dark brown, with a rounded tuberculate prothorax covering the head. Bostrychidae are numerous only in the tropics.

54. Lyctidae (Fig. 30.62B). Small, parallel-sided, flattened beetles; head visible from above, not covered by the prothorax; with a transverse constriction just behind the eyes which are prominent, convex; antennae 11-segmented with a 2-segmented club, inserted in front of eyes; clypeus and labrum distinct; mandibles stout, curved, with bifid apices; palpi with terminal segments slender; pronotum with a median depression or fovea; plane of anterior aperture of prothorax at 45° to plane of posterior aperture; prosternum relatively long in front of the coxae; fore coxal cavities closed behind; fore and mid coxae globular, trochantins not visible; metasternum long; elytra parallel-sided, rounded at apex; hind coxae transverse, widely separated. Larvae C-shaped, white; head partly buried in prothorax, with mouthparts directed forwards; antennae 3-segmented; abdomen 10-segmented, terminal segment small; spiracles as in Bostrychidae, but those of segment 8 much larger than the others.

Both larvae and adults feed in dry, dead wood and dry roots of herbaceous plants. A few species, especially the introduced *Lyctus brunneus* Steph., are serious timber pests. *L. brunneus* infests only unseasoned sapwood of hardwoods, which contains a relatively high proportion of starch. The larvae reduce the wood to a soft powder, leaving the surface in the form of a shell which collapses under the slightest pressure (hence the common name 'powder-post beetles'). Lyctidae do not attack the wood of coniferous trees. The family is poorly represented in Australia.

Superfamily CLEROIDEA

Small to medium-sized, elongate beetles; hind coxae without a steep or concave posterior face for reception of the femur; fore coxae transverse, with exposed trochantins; tarsi 5-5-5-segmented (sometimes with basal segment minute or hidden), with segments bilobed or filiform, terminal segment often with a conspicuous bisetose empodium; body often clothed with erect bristles, at least on margins of elytra; abdomen with 5 (rarely 6) visible sternites; spiracles of segment 8 non-functional; elytra almost always covering the abdomen, and, if striate, without a scutellar striole; labial palpi often with terminal segment securiform; antennae 8- to 11-segmented, filiform or clubbed; prosternal process not received into a cavity on the mesosternum; Malpighian tubes 6, cryptonephric, regularly arranged around the gut. Larval prothorax strongly sclerotized; abdominal segment 9 with sclerotized urogomphi arising from a basal sclerite; labrum free; maxillae without an articulating area, and with galea and lacinia fused; mandibles without mola; almost always carnivorous. [Crowson, 1964.]

Key to the Families of Cleroidea Known in Australia

Pronotum with defined lateral margins; body without erect bristles, but sometimes clothed with scales; tarsi with segments not lobed, terminal segment with a bisetose empodium between the claws; fore coxae transverse, not projecting .... **Trogossitidae** (p. 578)

Pronotum with or without defined lateral margins; body with erect bristles; if antennae clubbed, then tarsi with lobed segments; empodium sometimes absent; fore coxae projecting ... **Cleridae** (p. 578)

55. Trogossitidae (Ostomidae, Temnochilidae; Fig. 30.64). Small to moderate-sized beetles of ovoid, rather depressed form. Head visible from above, prognathous; antennae 10- or 11-segmented, with a marked 1- or 3-segmented club; tarsi 5-segmented, but with the basal segment very small and difficult to see; terminal segment with a bisetose empodium; claws simple;

Fig. 30.64. *Leperina turbata*, Trogossitidae. [F. Nanninga]

pronotum with well-marked lateral margins; fore coxae transverse and not projecting; terminal segments of palpi filiform; eyes not prominent, oval or reniform; antennae inserted between eye and mandible; upper surface of body with or without setae, sometimes clothed with flattened setae or with scales. Larvae elongate, subcylindrical or tapered towards the head, whitish; head exserted, with several pairs of ocelli; body with a few setae on most segments; pronotum and abdominal segment 9 sclerotized, the latter with urogomphi which are usually simple. The larvae are predatory, and live beneath bark, in tunnels of wood-boring insects, or in bracket fungi.

Tenebroides mauretanicus L. is a cosmopolitan pest of cereals, cereal products, dried fruit, etc. Its larva is also partly predacious upon the larvae of other stored-product insects. The adults are black, shiny, strongly flattened and with a distinctive 'waist' between prothorax and elytra. They are long-lived and highly resistant to starvation. *Leperina* (10 spp.) includes flattened beetles which are found under bark; they are clothed with scales and setae, and usually have a mottled colouring. The Trogossitidae were formerly classified among the Clavicornia, near the Cucujidae, but were associated with the Cleridae by Böving and Craighead (1931) on larval characters (absence of maxillary articulating area and the mandible without a mola). The family is divided by Crowson (1964) into Trogossitidae and Peltidae (the latter represented by *Egolia variegata* Erich. from Tasmania and Lord Howe I.)

56. Cleridae (Fig. 30.65). Small to moderate-sized, rarely large beetles, of elongate parallel-sided form, clothed with erect setae; colour usually dark, sometimes metallic green, or patterned with red or yellow. Head deflexed; antennae short, 11-segmented, including a poorly defined 3- to 6-segmented club; antennae inserted in front of eyes; labrum short, transverse, the basal membrane exposed; maxillary palpi 4-segmented, terminal segment cylindrical or securiform; labial palpi longer than maxillary palpi, 3-segmented, terminal segment securiform; pronotum convex, usually without lateral margins, sometimes transversely constricted; fore coxae prominent, conical, contiguous, fore coxal cavities closed or open behind; mid coxae not strongly projecting, close together; hind coxae transverse; metasternum convex posteriorly, with a declivity to the hind edge, in front of hind coxae; tarsi 5-5-5-segmented, with 1 to 4 segments bilobed beneath, the lobes continued by membranous extensions beneath, the basal segment often small (short above, longer beneath, obliquely connected to 2nd segment) and partly hidden in the emargination of the apex of the tibia; claws simple; elytra rounded at apices, covering abdomen, except in *Cylidrus* (Tillinae) in which the 3 apical tergites are exposed and sclerotized; abdomen with 5 visible, movable sternites,

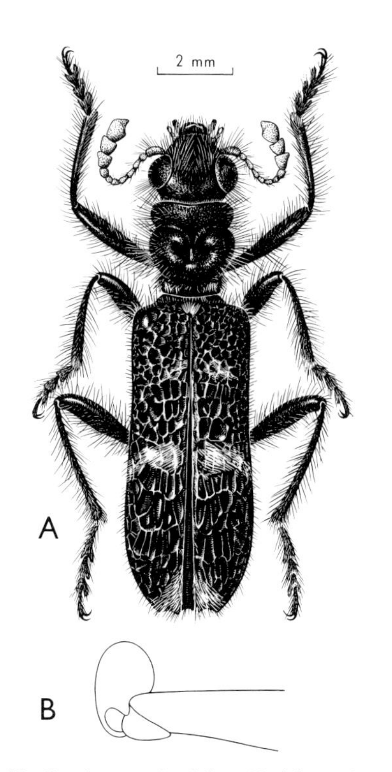

Fig. 30.65. *Scrobiger splendidus*, Cleridae: A, adult; B, junction of femur and trochanter. [F. Nanninga]

except in *Cylidrus* in which there are 6. Larvae elongate, often unusually coloured (e.g. yellow, orange, pink, blue, or brown); head usually flattened dorsally and convex ventrally, parallel-sided, prognathous; antennae 3-segmented; labrum and clypeus distinct; pronotum sclerotized; abdomen 10-segmented, segment 10 reduced, sometimes forming a foot; spiracles on segments 1 to 8; urogomphi branched or unbranched on the sclerotized tergite 9.

Both larvae and adults of almost all Cleridae prey upon other insects, especially insects associated with bark or wood of trees, e.g. Scolytinae. The adults are often found on flowers, and some come to light. The largest Australian species is the dark brown *Eunatalis titana* (Thoms.), which attains a length of 40 mm. Among the attractively patterned species is *Trogodendron fasciculatum* Schreib., an eastern species up to 20 mm long, black, with yellow antennae, 2 lateral yellow spots on each elytron, and, towards the apex, a velvety-black area encircled with grey pubescence. *T. fasciculatum* is found running on tree trunks in hot sunshine. *Phlogistus* includes metallic green and blue species. *Phlogistus eximius* White, 7 mm long, has elytra of metallic purple with a metallic blue spot in the middle. *Necrobia rufipes* De G. has a world-wide distribution, and infests products of high fat or oil content, e.g. ham, cheese, copra. It also preys upon larvae of other stored-product insects, and in Australia is commonly found in carcasses. [Blackburn, 1900.]

Superfamily LYMEXYLOIDEA

Elongate, soft-bodied beetles with 5, 6, or 7 visible sternites; antennae short, filiform, fusiform, or serrate; tarsi 5-5-5-segmented, filiform, as long as, or longer than, the tibiae; coxae narrow and elongate; fore coxal cavities confluent and open behind; mid coxae contiguous; males with eyes very large, meeting in the midline in front, and with drooping, flabellate maxillary palpi. Elytra either long and narrow, covering the wings at rest and all except the terminal abdominal tergite, or very short, not reaching the abdomen.

57. Lymexylidae (Fig. 30.66). The larvae are cylindrical, with short but well-developed legs. The pronotum is enlarged and forms a partial hood over the head. Segment 9 is large, terminal, overlying segment 10, heavily sclerotized or with a pair of hooks. They bore in hard wood of previously weakened trees, and use the sclerotized terminal segment to push the core of debris out of the tunnel. The larvae appear to feed on a fungus, which grows on the walls of the tunnel and is transmitted by the insect. Australian species are placed in the genera *Hylecoetus*,* *Melitomma*, *Lymexylon*, and *Atractocerus*. Those

* The Australian species are not congeneric with *Hylecoetus* of the northern hemisphere and need to be transferred to another genus.

of the first 2 genera have long narrow elytra which conceal the folded wings at rest, whereas in *Atractocerus* the elytra are very short, not extending beyond the thorax. The exposed wings, which fold fanwise, extend to about half the length of the long, flattened, dark brown abdomen. The metathoracic spiracle in *Atractocerus* is unusual in being completely exposed, and tibial spurs are absent. Adults of a species of *Atractocerus* in W.A. are reported to fly in swarms at dusk.

Superfamily CUCUJOIDEA

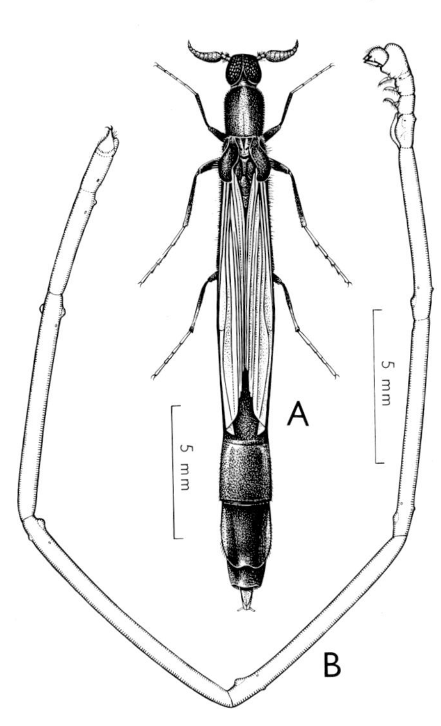

Fig. 30.66. *Atractocerus crassicornis*, Lymexylidae: A, adult; B, larva. [F. Nanninga]

Antennae filiform, thickened apically, or with a 2- or 3-segmented club, never elbowed; hind coxae without a transverse concavity or steep face against which the hind femora can be retracted; abdomen with 5 (rarely 6 or 7) visible sternites; spiracles of segment 8 (and sometimes of segment 7, or 6 and 7) non-functional; tarsi usually 5-5-4- or 3-3-3-segmented, more rarely 4-4-4- or 5-5-5-segmented; if abdomen has 6 or 7 visible sternites, or fore coxae projecting, then tarsi are 5-5-4-segmented; if tarsi are 4-4-4-segmented, the parameres of the aedeagus are not articulated; wings rarely with a spur on the *r–m* cross-vein, in contrast to Chrysomeloidea; if prothorax without side margins or the head rostrate, then the tarsi are 5-5-4-segmented, in contrast to Curculionoidea. Malpighian tubes 6, cryptonephric; aedeagus with a semi-tubular basal piece surmounting a well-developed median lobe.

The superfamily, established by Böving and Craighead and accepted by Peyerimhoff (1933), includes both the Heteromera and Clavicornia of older authors, and its diversity is indicated by the fact that it covers about half the recognized families of Coleoptera. On the other hand, the families, with the exception of Tenebrionidae and Coccinellidae, are small in numbers of species, so that the superfamily contains only about one-tenth of the known species of Coleoptera. Tarsal segmentation is an important character in the classification of Cucujoidea. The segments are often difficult to count because of the reduction of the basal or penultimate segment. A slide preparation of the legs may be necessary in order to permit the segments to be counted with certainty.

Key to the Families of Cucujoidea Known in Australia

In this key provision is made for possible errors in the counting of tarsal segments, and difficult members of other superfamilies, which might be traced in error to Cucujoidea, are included.

1. Tarsal segmentation various, but never 5-5-4; antennae almost always with a well-marked club; fore coxae never projecting; lateral margins of prothorax distinct. (Clavicornia) ...2
   Tarsal segmentation always 5-5-4; antennae usually without, or with a slightly developed club; fore coxae usually projecting; lateral margins of prothorax often absent. (Heteromera) 35

2(1). Tarsi 5-5-5-segmented (4th segment often very small and partly concealed) 3
Tarsal segmentation 4-4-4, 3-3-3, or 2-3-3 15
3(2). Antennae with a club 4
Antennae filiform or moniliform 14
4(3). Epipleura continued to apices of elytra (e.g. Fig. 30.75B) 5
Epipleura terminated before apices of elytra 12
5(4). Body very convex, glossy, non-pubescent, less than 3 mm long; claws appendiculate or with a tooth beneath .... **Phalacridae** (pt., p. 590)
Body elongate and usually pubescent, length often more than 3 mm; claws simple 6
6(5). Upper surface of body without pubescence; elytra often with obvious yellow or red markings, or prothorax of different colour from the elytra 7
Upper surface of body bearing setae; body uniformly brown 9
7(6). Fore coxal cavities open behind .... **Languriidae** (pt., p. 589)
Fore coxal cavities closed behind 8
8(7). Trochantins clearly visible; tarsi with basal segment smaller than 2nd, obscure; 4th segment of same size as 3rd, 5th with a bisetose empodium between the claws .... (Cleroidea–Trogossitidae, pt., p. 578)
Trochantins not visible; tarsi with basal segment as long as, or longer than, 2nd; 4th segment obviously smaller than 3rd, 5th not with an obvious bisetose empodium between the claws .... **Erotylidae** (p. 590)
9(6). Basal sternite of abdomen with 'femoral lines' extending back from inner ends of the coxae; pronotum with raised lines at the sides (Fig. 30.73) .... **Biphyllidae** (p. 589)
Basal sternite of abdomen without femoral lines; pronotum without raised lines at the sides 10
10(9). Mesepimera not reaching the mid coxal cavities; prosternum produced backwards, overlapping anterior edge of mesosternum; fore coxal cavities open or closed behind; pronotum often with obvious anterolateral marginal tubercles (Fig. 30.72) .... **Cryptophagidae** (pt., p. 588)
Mesepimera reaching the mid coxal cavities; prosternum not overlapping mesosternum; fore coxal cavities closed behind; pronotum without anterolateral tubercles 11
11(10). Fore coxae globular; trochanters not visible; length 3·5 mm or less ...... **Silvanidae** (p. 587)
Fore coxae strongly transverse; trochanters visible; length greater than 3·5 mm .... (Cleroidea–Trogossitidae, pt., p. 578)
12(4). Abdomen with at least apical part of last tergite exposed; mid coxal cavities closed by mesepimera between the sterna 13
Abdomen completely covered by the elytra; mid coxal cavities closed by meso- and metasterna only .... (Bostrychoidea–Lyctidae, p. 577)
13(12). Antennae with segments 10 and 11 closely coadapted, forming a spheroidal club; tarsi with no segments lobed .... **Rhizophagidae** (pt., p. 585)
Antennae with a club composed of 3 flattened, strongly transverse, clearly separate segments; tarsi with at least one segment lobed .... **Nitidulidae** (p. 584)
14(3). Basal segment of tarsi much smaller than 2nd segment (Fig. 30.69); maxillae partly concealed from below by forwardly projecting processes of the head capsule; mid coxal cavities closed outwardly by the sterna; antennae moniliform; gular sutures confluent .... **Passandridae** (p. 586)
Basal segment of tarsi about equal in size to 2nd segment (Fig. 30.70B); head capsule without forwardly directed processes below; mid coxal cavities closed by the mesepimera between the sterna; antennae filiform or moniliform; with 2 gular sutures .... **Cucujidae** (pt., p. 586)
15(2). Tarsal segmentation 4-4-4 or 3-4-4 16
Tarsal segmentation 3-3-3 or 2-3-3 30
16(15). Antennae filiform; body flattened .... **Cucujidae** (pt., p. 586)
Antennae with a gradual or well-defined club 17
17(16). Claws appendiculate or toothed beneath; body form rounded or convex 18
Claws simple; body form often elongate 19

18(17, 30). Apical segment of maxillary palp securiform; fore coxae much closer together than hind coxae; mid coxal cavities closed by the mesepimera in addition to the sterna; viewed from side of body the lateral edges of pronotum and elytra form a square or acute angle (Figs. 30.77,78); length often more than 3 mm .... **Coccinellidae** (p. 592)

Apical segment of maxillary palp not securiform; hind coxae closer than fore coxae; mid coxal cavities closed by the sterna only; viewed from side of body the lateral edges of pronotum and elytra form a very obtuse angle (Fig. 30.75); length 3 mm or less .... **Phalacridae** (pt., p. 590)

19(17). Head usually hidden by the semicircular pronotum when viewed from above; abdomen with 6 visible sternites; length less than 3 mm; form broadly ovate, pubescent .... **Corylophidae** (p. 592)

Head visible at least in part, when viewed from above; abdomen with 5 visible sternites; length often more than 3 mm 20

20(19). Mid coxal cavities closed by the meso- and metasternum only 21

Mid coxal cavities closed by the mesepisterna or mesepimera between the meso- and metasternum 26

21(20). Pronotum and elytra clothed with scales; anterior edge of pronotum produced forwards into a semicircular lobe which projects partly over the head (Fig. 30.71B), lateral edges not serrate. [Segments 10 and 11 of antenna fused to form a large rounded club; length 1·5–3 mm] .... **Phycosecidae** (p. 587)

Pronotum and elytra without scales; anterior edge of pronotum without a semicircular lobe, lateral edges sometimes serrate 22

22(21). Ventral surface of head capsule on each side of mentum produced forwards into a long acute process which is visible from above outside the mandible; mandibles multidentate; antennae moniliform with the 3 terminal segments enlarged .... **Cucujidae**–Prostominae (p. 586)

Ventral surface of head capsule without forwardly directed processes; mandibles not multidentate; antennae clubbed 23

23(22). Fore coxal cavities open behind (Fig. 30.82B); elytra smooth, striate, without traces of longitudinal ribs 24

Fore coxal cavities closed behind; elytra usually with longitudinal ribs and the surface dull and rough 25

24(23). Tarsi with 3rd segment broadly lobed beneath; prothorax usually of paler colour than elytra; surface glabrous, elytra with clearly defined punctate striae; length 2–10 mm .... **Languriidae** (pt., p. 589)

Tarsi with 3rd segment not lobed below; prothorax and elytra of the same colour; surface setose, elytra without clearly defined punctate striae; length 1–3 mm .... **Mycetophagidae** (p. 595)

25(23). Basal abdominal sternite with femoral lines (grooves) extending backwards, or backwards and outwards, from inner end of the hind coxal cavity; scape of antenna completely visible from above; body surface smooth and shining .... **Cerylonidae** (p. 590)

Basal abdominal sternite without femoral lines; scape of antenna often partly covered; body surface usually dull and rough, or tuberculate, or with longitudinal ribs on the elytra .... **Colydiidae** (p. 595)

26(20). Antennae with the 3 apical segments slightly thickened; fore coxal cavities closed behind .... **Merycidae** (p. 595)

Antennae with a pronounced club; fore coxal cavities open or closed behind 27

27(26). Elytra truncate, exposing at least one abdominal tergite; fore coxal cavities closed behind; antennal club spherical, composed of 2 closely fitted segments .... **Rhizophagidae** (pt., p. 585)

Elytra completely covering abdomen; fore coxal cavities open or closed behind; antennal club composed of 3–5 segments 28

28(27). Fore coxal cavities closed behind; trochantins obvious .... (Cleroidea–Trogossitidae, pt., p. 578)

Fore coxal cavities open behind; trochantins not visible 29

29(28). Tarsal segments not lobed beneath .... **Ciidae** (p. 595)
Second tarsal segment lobed beneath .... **Endomychidae** (pt., p. 593)

30(15). Claws appendiculate or toothed beneath. [Body rounded or convex] 18
Claws simple 31

31(30). Fore coxal cavities closed behind; mid coxal cavities closed by the mesepimera between the sterna; tarsal segments simple. [Length 3 mm or less] .... **Lathridiidae** (p. 594)
Fore coxal cavities open behind; mid coxal cavities closed by the sterna only; 2nd segment of tarsi bilobed or simple 32

32(31). Second segments of tarsi deeply bilobed; pronotum with an obvious longitudinal groove on each side at base; length 2–7 mm; fungus-feeding species...**Endomychidae** (pt., p. 593)
Tarsal segments simple; pronotum without obvious longitudinal grooves; length less than 2 mm; myrmecophilous species 33

33(32). Labrum and mandibles lanceolate (Fig. 30.80A); maxillary palpi with penultimate segment broadly expanded and flattened, terminal segment very acute and slender; dorsal surface coarsely punctured and clothed with rather sparse, short, erect setae; length 1·5 mm .... **Aculagnathidae** (p. 594)
Labrum fused with clypeus, not lanceolate; mandibles short, curved; terminal segments of maxillary palpi not acute and slender; dorsal surface pubescent or glabrous; length 1·2 mm or less 34

34(33). Eyes absent; dorsal surface with fine sericeous pubescence; pronotum semicircular, larger than elytra; elytra without striae; length less than 1 mm .... (Staphylinoidea–Limulodidae, p. 541)
Eyes present; dorsal surface glabrous; pronotum smaller than elytra; elytra with a longitudinal stria next to the suture; length *ca* 1·2 mm .... **Merophysiidae** (p. 594)

35(1). Elytra truncate, leaving at least 2 abdominal tergites exposed; head very broad and flat (width of head 1·3 to 1·5 times the length of the pronotum; Fig. 30.88); prothorax strongly constricted to the base; antennae filiform; tarsi without lobed segments .... **Inopeplidae** (p. 600)
Elytra covering all, all but one, or none of the abdominal tergites; head less broad and flat; antennae sometimes clubbed or thickened towards apex; tarsi sometimes with one or more segments lobed beneath 36

36(35). Underside of prothorax with cavities for reception of the antennae. [Length 25–35 mm; strongly tuberculate] .... **Zopheridae** (p. 600)
Underside of prothorax without cavities for reception of the antennae 37

37(36). Fore coxal cavities closed behind; abdomen with 3 basal sternites connate, 4th and 5th movable (often recognizable in dried specimens by the membrane between segments 3, 4, and 5) 38
Fore coxal cavities open behind, or, if closed, with all abdominal sternites free 40

38(37). Prothorax without defined lateral margins (Fig. 30.85A); penultimate tarsal segments strongly lobed beneath; fore coxae very close together and somewhat projecting; claws simple .... **Lagriidae** (p. 598)
Prothorax usually with distinct lateral margins; penultimate tarsal segments rarely lobed; fore coxae fairly widely separated and not projecting above prosternum; claws sometimes toothed 39

39(38). Tarsal claws comb-like beneath; sides of frons not produced over antennal insertions; species almost always winged; antennae filiform .... **Alleculidae** (p. 599)
Tarsal claws simple; sides of frons usually produced to cover antennal insertions; species often wingless or with clubbed antennae .... **Tenebrionidae** (p. 596)

40(37). Fore coxae small, spherical, widely separated; mid coxal cavities usually closed outwardly by the sterna only; antennae usually with last 3 segments forming a club 41
Fore coxae transverse or projecting, or with prosternal process between the coxae narrow or absent; antennae filiform or gradually thickened to apex 42

41(40). Antennae filiform or gradually thickened towards apex .... **Cucujidae** (pt., p. 586)
Antennae with a 3-segmented club .... **Cryptophagidae** (pt., p. 588)

42(40). Mid coxal cavities closed by the sterna only 43
Mid coxal cavities closed by the mesepimera between the sterna 45

43(42). Tarsal segments not lobed; apical segment of maxillary palp not securiform .... **Salpingidae** (p. 601)
Tarsi with 1 or 2 segments lobed below; apical segment of maxillary palp often securiform 44

44(43). Tarsi with only penultimate segment lobed below; claws usually appendiculate or toothed below; apical segment of maxillary palp securiform .... **Mycteridae** (p. 601)
Tarsi with the 2 segments preceding the terminal segment lobed below; apical segment of maxillary palp not securiform .... **Hemipeplidae** (p. 601)

45(42). Claws simple 46
Claws toothed, serrate, pectinate, or appendiculate below (e.g. Figs. 30.92, 93D, 95) 51

46(45). Pronotum with defined lateral margins or sharp edges; base of pronotum equal in width to base of elytra 47
Pronotum with or without defined lateral margins; base of pronotum considerably narrower than base of elytra 48

47(46). Tibial spurs serrate or pectinate (Fig. 30.91A); penultimate segment of fore tarsus not lobed below; head not constricted behind eyes; eyes not pubescent...... **Melandryidae** (p. 602)
Tibial spurs pubescent, not serrate or pectinate; penultimate segment of fore tarsus lobed below the terminal segment; head sharply constricted behind eyes; eyes pubescent .... **Scraptiidae** (p. 602)

48(46). Head not sharply constricted behind eyes and not, or only slightly, deflexed; pronotum widest close to anterior end or in middle (Figs. 30.90, 94) 49
Head sharply constricted behind eyes and strongly deflexed; pronotum often widest near base 50

49(48). Tarsi with penultimate segment deeply bilobed; head, pronotum, and elytra uniformly pubescent; antennae usually filiform (Fig. 30.94) .... **Oedemeridae** (p. 605)
Tarsi with penultimate segment not deeply bilobed; body with or without pubescence; antennae slightly thickened towards apex .... **Pythidae** (p. 601)

50(48). Mesepisterna not nearly meeting in front of mesosternum; tarsi with penultimate segment small, antepenultimate segment lobed below; abdomen with the 2 basal sternites connate; eyes hairy (Fig. 30.97) .... **Aderidae** (p. 607)
Mesepisterna meeting in front of mesosternum (Fig. 30.96B); tarsi with penultimate segment lobed and antepenultimate segment simple; all 5 abdominal sternites free; eyes bare .... **Anthicidae** (p. 606)

51(45). Seventh abdominal tergite produced backwards beyond elytra as a long, stout spine; base of prothorax nearly or quite as wide as base of elytra; prothorax with sharp lateral longitudinal margins (Fig. 30.92); antennae thickened apically, slightly serrate; tarsal claws serrate, with bristle-like lobes beneath .... **Mordellidae** (p. 603)
Seventh abdominal tergite not produced apically beyond elytra; base of prothorax usually much narrower than base of elytra, or, if not, then with antennae flabellate in ♂, serrate in ♀; prothorax without sharp lateral longitudinal edges; tarsal claws serrate or pectinate and with or without lobiform appendages 52

52(51). Antennae flabellate (♂) or strongly serrate (♀) (Fig. 30.93); claws serrate or pectinate, but without lobiform appendages; base of prothorax as wide as base of elytra .... **Rhipiphoridae** (p. 603)
Antennae filiform or slightly serrate; claws, if serrate, with long appendages beneath (Fig. 30.95); base of prothorax much narrower than base of elytra. [Head strongly deflexed, neck constricted] .... **Meloidae** (p. 605)

58. Nitidulidae (Fig. 30.67). Small (1–7 mm), brown or black beetles, usually broad and flattened. Head prognathous, often constricted at base of mandibles resulting in a somewhat rostrate appearance; mandibles broad and very strongly curved, with a tooth near apex and a brush of hairs on inner side; clypeus not distinguished from frons; labrum separate, often bilobed; terminal segments of palpi subcylindrical, antennae 11-segmented, slender between the swollen scape and the pronounced club of 3 transverse segments; pronotum transverse, quadrate, sometimes with posterior margin explanate, overlapping base of elytra; fore coxal cavities strongly transverse and broadly closed behind; fore coxae transverse, with trochantins very obvious; mid coxae transverse, separated by less than their width, trochantins obvious, the mesepimeron reaching the coxal cavity; hind coxae transverse, widely separated; legs short, stout, tibiae expanded apically; fore tibiae often serrate externally; tarsi 5-5-5-segmented, the 3 basal segments usually broadly dilated and hairy beneath, 4th very small, 5th elongate, bearing simple claws; scutellum large, broadly triangular; elytra usually truncate, exposing the pygidium, or 2 or 3 tergites; abdomen with 5 visible free sternites. Larvae subcylindrical, white, with small prognathous head; short, pointed or truncate urogomphi often present on a sclerotized area on abdominal segment 9; prothorax often sclerotized on the upper surface; spiracles sometimes carried on short projecting tubes.

Little is known of the biology of Australian species, but habits in other regions are very varied. Some are found in flowers, where they feed mainly on pollen and nectar. Such species are restricted to particular plant species, and their larval life is very short. Other species feed in fungi, and a few are leaf-miners. Pupation of the phytophagous forms occurs in the soil. Some species are predatory, and a few, probably also carnivorous, are myrmecophiles. Twenty-two species have been recorded as occurring in stored food products, but only species of *Carpophilus* appear to be of economic importance. Several species of this genus, notably *C. hemipterus* L. and *C. aterrimus* Macl., are pests of dried fruits, but they also attack ripe fruit in the orchards. *C. hemipterus* is a dark brown beetle, about 3 mm long, with abbreviated elytra, each of which has a large yellow patch. The larva has 2 pairs of small, pointed, brown urogomphi, and the prothorax and abdominal spiracles are brownish and sclerotized in contrast with the rest of the body. *C. aterrimus* is similar in size, but of uniform dark brown colour. Some species of *Brachypeplus* are found in the nests of native bees, where they probably feed on the stored pollen. *Circopes pilistriatus* Macl. is hairy, reddish brown, 3 mm long, with the setae on the elytra arranged in many parallel longitudinal lines; all stages live inside the seed pods of the kurrajong *(Brachychiton)*.

59. Rhizophagidae (Fig. 30.68). Very small beetles (3 mm or less). Head large, triangular, sharply constricted behind the eyes in the

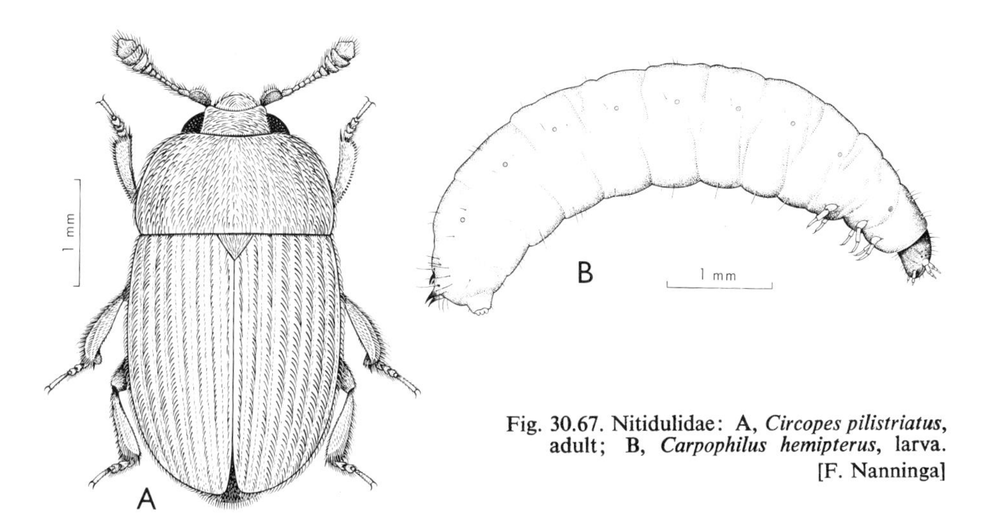

Fig. 30.67. Nitidulidae: A, *Circopes pilistriatus*, adult; B, *Carpophilus hemipterus*, larva. [F. Nanninga]

male; eyes strongly convex; antennae 11-segmented, with segments 10 and 11 closely coadapted, forming a pronounced, spherical club; labrum not distinct from the clypeus; pronotum transverse, but narrower than the head; scutellum narrowly triangular; elytra subparallel, punctate-striate, truncate, the last 1 or 2 tergites of the abdomen exposed; fore coxae small, the cavities broadly closed behind, separated by about half their width; mid coxal cavities ovoid, transverse, separated by about half their width, closed externally by the mesepimera between the sterna; hind coxae transverse, separated; tarsi 5-5-5-segmented in female or 4-4-4-segmented in male, penultimate segment small, segments 1 and 2 broad, with long setae; claws simple; basal abdominal segment larger than the next 2 segments together. The larvae of the Australian species have not been described, and nothing is known of their biology. The beetles are found under bark or in rotten wood. Larvae and adults of the northern-hemisphere *Rhizophagus* live in the tunnels of Scolytinae upon which they prey. The genera *Mimemodes*, *Tristaria*, *Thione* are recorded in Australia.

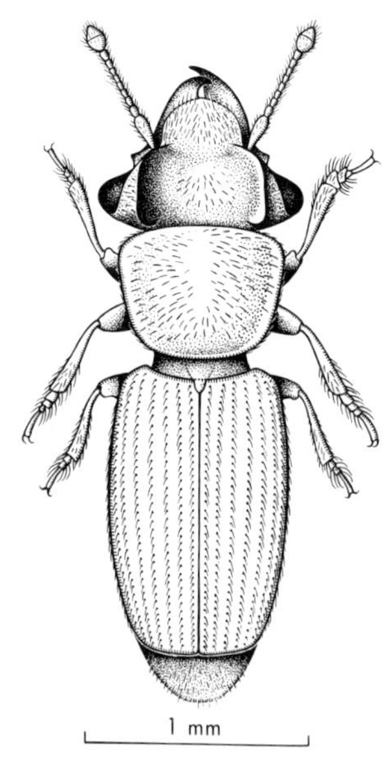

Fig. 30.68. *Mimemodes laticeps*, Rhizophagidae. [F. Nanninga]

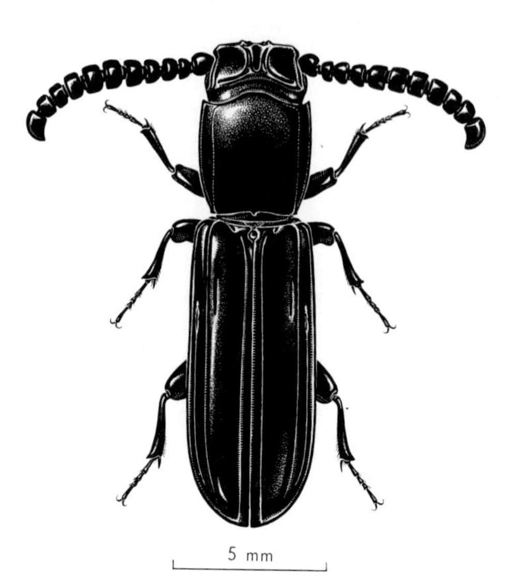

Fig. 30.69. *Hectarthrum heros*, Passandridae. [F. Nanninga]

60. Passandridae (Fig. 30.69). Beetles of moderate size (6–24 mm) and cylindrical form; closely allied to Cucujidae, but distinguishable by having the maxillae partly concealed from below by forwardly projecting gular processes, confluent gular sutures, and mesepimera not reaching the mid coxal cavities; tarsi 5-5-5-segmented, with the basal segment smaller than the 2nd. The adults live in the tunnels of wood-boring insects, and the larvae are ectoparasitic and partly degenerate. *Ancistria* (3 spp.), *Passandra* (2 spp.), and *Hectarthrum* (1 sp.) are represented in the fauna. *H. heros* F. appears to be relatively common in north Queensland. It is a cylindrical, shining black species with thick moniliform antennae.

61. Cucujidae (Fig. 30.70). Length 1·5–25 mm, extremely flattened, head prognathous. Antennae usually long, usually filiform or submoniliform, sometimes more or less clubbed, inserted under frontal ridges; fore coxal cavities open or closed behind, and with or without lateral extensions exposing the trochantins; mesepimera reaching mid coxal cavities or not; hind coxae widely separated; abdomen with 5 visible sternites; elytra flat on the disc, with a sharp lateral declivity outside a ridge which extends back from the shoulder, epipleura usually continuous to the apex; tarsi 5-5-5-, 5-5-4-, or 4-4-4-segmented, the terminal segment long and claws simple. This family is poorly characterized, and is probably in need of further subdivision. Crowson (1955) has erected a family Protocucujidae based on a Chilean species which appears to be the most primitive cucujoid type, showing affinities with the Chrysomeloidea. He mentions an Australian representative of the new family, but this is, as yet, undescribed. It differs from Cucujidae in having 2 segments lobed on each tarsus and a conspicuous bisetose empodium between the claws.

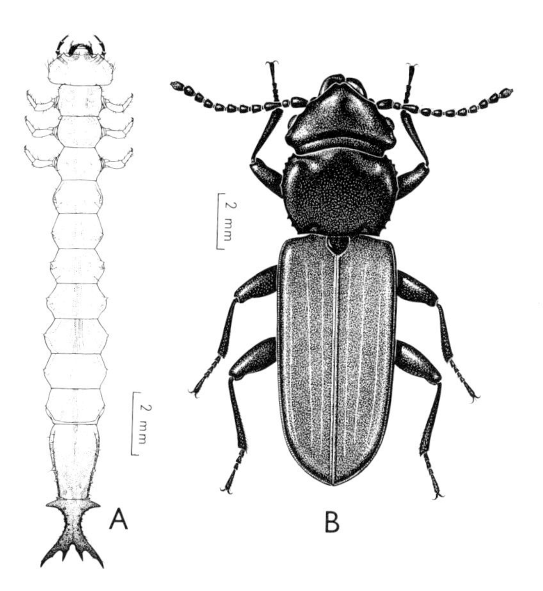

Fig. 30.70. Cucujidae: A, *Platisus moerosus*, larva; B, *Cucujus colonarius*, adult. [F. Nanninga]

The larvae of Cucujidae are of diverse form. Almost nothing is known of the biology of the Australian species. The extraordinary larva of *Platisus moerosus* Pasc. (Fig. 30.70A) is not uncommon under bark in the south-east. *Laemophloeus ferrugineus* Steph. is a cosmopolitan species, commonly occurring in stored foodstuffs, and sometimes sufficiently numerous to be ranked as a pest. The large (25 mm), black *Parandra frenchi* Blackb. is included here, but forms a link with the Cerambycidae, to which it was transferred by Crowson (1955).

62. Silvanidae (Fig. 30.71A). Minute (1·5–3·5 mm), elongate, flattened, yellowish or reddish brown beetles. Head large, elongate, prognathous; antennae inserted beneath frontal ridges, 11–segmented, and thickened towards the apex or with a 3-segmented club; pronotum usually with prominent anterior angles and lateral edges serrate; dorsal surface of body clothed with short recumbent setae; fore coxal cavities closed behind; mid coxae, like fore coxae, globular, spaced by their own width; mesepimera reaching mid coxal cavities; hind coxae widely separated; abdomen with 5 visible sternites; elytra with epipleura continuous to the apex; tarsal segmentation 5-5-5, 3rd segment lobed below; claws simple. Silvanidae are found on plants or, more rarely, under bark and in the tunnels of wood-boring insects. Several cosmopolitan species are associated with dry stored foodstuffs, including *Cryptamorpha desjardinsi* Guér., *Nausibius clavicornis* Klug, *Ahasverus advena* Waltl, and *Cathartus quadricollis* Guér. *Oryzaephilus surinamensis* L. is a pest of a wide variety of stored plant products.

63. Phycosecidae* (Fig. 30.71B). Length

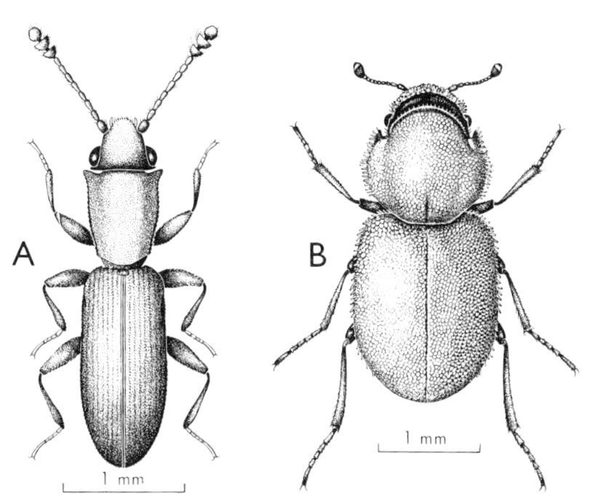

Fig. 30.71. A, *Silvanus unidentatus*, Silvanidae; B, *Phycosecis litoralis*, Phycosecidae. [F. Nanninga]

* The family was transferred to the Cleroidea by Crowson (1964).

1·5–3 mm; body broad, convex, and densely clothed with whitish scales. Head flattened, with only a narrow margin exposed in front of the almost semicircular forward projection of anterior edge of pronotum; antennae 10-segmented, including a marked 1-segmented club; mandibles bidentate at apex and an obtuse tooth near middle, with a setose prostheca but no mola; prothorax with defined lateral edges; fore coxae ovoid, not projecting, their cavities closed behind, anterior angles forwardly projecting, acute, tooth-like; mid coxal cavities closed by the meeting of the sterna; scutellum broadly triangular; elytra with epipleura complete; hind coxae transverse-ovoid, flat, extending laterally about half way to edge of abdomen; abdomen with 5 visible sternites; tarsi 4-4-4-segmented, the segments not lobed. The family includes only 4 species, 3 from Australia and 1 from New Zealand. Adults and larvae are found on sandy beaches above high-water mark, in the remains of dead birds, etc. They appear to be predatory. *Phycosecis litoralis* Pasc., *ammophilus* Lea, and *hilli* Lea occur on the shores of N.S.W., south-western W.A., and north Queensland, respectively.

64. Cryptophagidae (Fig. 30.72). Very small (1–3 mm) pubescent beetles. Head prognathous; antennae moniliform, 11-segmented, including a 3-segmented club; clypeus not distinct from frons; labrum distinct; palpi short, terminal segments subcylindrical; eyes small, strongly convex, coarsely facetted; pronotum quadrate, sometimes with a tooth or tubercle on anterior half of lateral margin; fore coxae globular or ovoid, moderately separated, not projecting, the cavities open behind, and the elongate intercoxal process overlapping the mesosternum; mid coxae globular, moderately separate; hind coxae transverse, narrow, moderately separate; mid coxal cavities closed by the sterna; abdomen with 5 visible sternites, basal segment about twice as long as 2nd; tarsi 5-5-5-segmented (sometimes 5-5-4 in male); elytra with no scutellar striole and epipleura limited to basal half. Larvae subcylindrical, whitish, with exserted head; antennae 3-segmented; labrum visible; ocelli present; abdomen 10-segmented, the terminal segment forming a proleg, 9 with a pair of short, pointed urogomphi.

Like Mycetophagidae, the Cryptophagidae feed upon fungi and mouldy materials, and are sometimes found in flowers and in nests of bees and wasps, birds, and mammals. Cryptophagidae are commonly found in

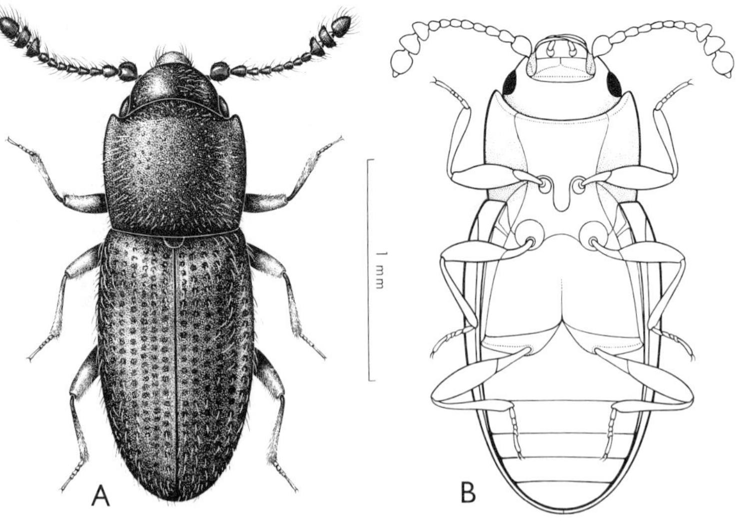

Fig. 30.72. *Cryptophagus gibbipennis*, Cryptophagidae: A, dorsal; B, ventral. [F. Nanninga]

stored foodstuffs, but are always associated with mouldy materials, and are therefore not of economic importance. The beetles also commonly swarm in new houses, where the larvae have fed upon moulds growing on damp plaster. The main genus is *Cryptophagus*. [Hinton, 1945, species in stored products; M. Evans, 1961, morphology of *Atomaria ruficornis* (Marsham).]

65. Biphyllidae (Fig. 30.73). Very small (2–3 mm), rather parallel-sided, with a flattened 3-segmented antennal club. Eyes hemispherical, with very coarse, convex facets; maxillary palpi 4-segmented, terminal segment subcylindrical; labial palpi 3-segmented, terminal segment expanded, flattened, and pointed; pronotum usually with one or more narrow, longitudinal, punctured grooves parallel to the lateral edge on each side; fore coxal cavities broadly closed behind; fore coxae rounded but rather transverse, separated by a narrow, parallel-sided prosternal process; mid coxae globular, separated by about one-third their width; mid coxal cavities closed externally by the sterna; hind coxae transverse and almost contiguous; abdomen with 5 visible sternites, all free, the basal sternite bearing 'femoral lines', fine ridges which run backwards and outwards from the inner ends of the hind coxae (Fig. 30.73B); elytra entire, coarsely punctate-striate, dorsal surface with long erect pubescence and shorter semi-erect pubescence; scutellum transverse; tarsi 5-5-5-segmented, 3rd segment with a membranous lobe beneath, 4th small; claws simple. Little is known of the biology of species of this family; adults and larvae are found in fungi and under bark. *Biphyllus*, *Diplocoelus*, and *Althaesia* occur in Australia.

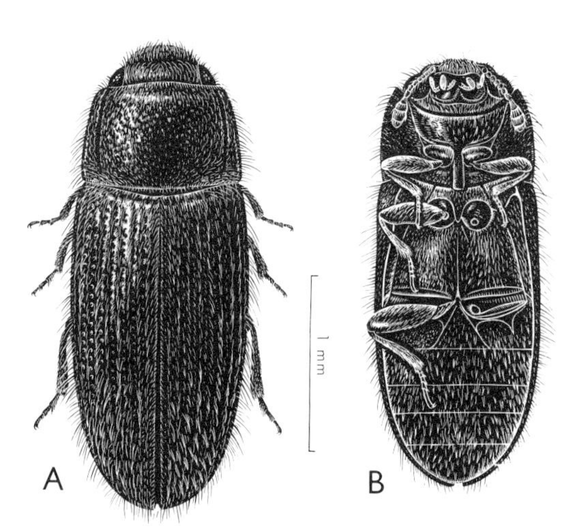

Fig. 30.73. *Diplocoelus punctatus*, Biphyllidae: A, dorsal; B, ventral. [F. Nanninga]

66. Languriidae (Fig. 30.74B). Polished, glabrous beetles of narrow, elongate, cylindrical shape; length 2–10 mm; pitchy black with orange prothorax or elytral marks. Head large; eyes prominent, with coarse, convex facets; frons with a longitudinal groove above each eye; antennae 11-segmented, with a 3- or 4-segmented club; clypeus and labrum distinct; mandibles small with apices bifid; maxillary palpi pointed at apices; terminal segment of labial palpi expanded and truncate; pronotum markedly convex anteriorly, but depressed across the base where there is an impression on each side; base sinuate, produced behind the angles which are square or even acute; prosternum long in front of fore coxae, which are globular and without visible trochantins; coxal cavities open behind; mid coxal cavities closed by the sterna only; hind coxae separated by about the same distance as the mid and fore coxae; abdomen with 5 visible sternites; legs slender, tibiae without spurs; tarsi 5-5-5-segmented, segments 1–3 broadly triangular and densely setose beneath, 4 very small and hidden in the cavity of the 3rd, 5 elongate with simple claws; elytra elongate, tapering, rounded at their apices, with punctate striae, epipleura continuous to the apices; wings present. Larvae white, elongate, cylindrical, with exserted head and well-developed legs; abdomen 10-segmented, with short, stout, curved, pointed urogomphi. Nothing is known of the biology of the Australian species, but elsewhere the larvae of many species are stem-borers. Adults are found on flowers, and may feed on pollen. Five genera are known in Australia, of which *Telmatophilus* (6 spp.) was formerly included in the Cryptophagidae.

67. Erotylidae (Plate 5, B; Fig. 30.74A). Small to large (3–23 mm), ovoid or elongate beetles, always associated with fungi, and frequently with a bright yellow and black pattern or metallic colour. Head prognathous, deeply inserted into the transverse prothorax; antennae 11-segmented, including a short, flattened 3-segmented club; labrum short, transverse; clypeus fused to frons; mandibles strongly curved; palpi with terminal segments fusiform to securiform; prosternum with fore coxal cavities widely separated and broadly closed behind; fore coxae globular, not projecting, trochantins not visible; mid coxal cavities closed laterally by the meso- and metasterna, widely separated; hind coxae widely separated and not reaching the sides of the abdomen laterally; abdomen with 5 visible sternites; tarsi 5-5-5-segmented, 4th usually small; claws simple; elytra entire, with epipleura obvious from shoulder to sutural angle; upper surface pubescent or glabrous. Larvae subcylindrical, whitish to pale brown, with segmentally arranged sclerites carrying spines which may be serrate; urogomphi present, sometimes very long; legs well developed; head hypognathous, with 5 or 6 pairs of ocelli; antennae 3-segmented, long.

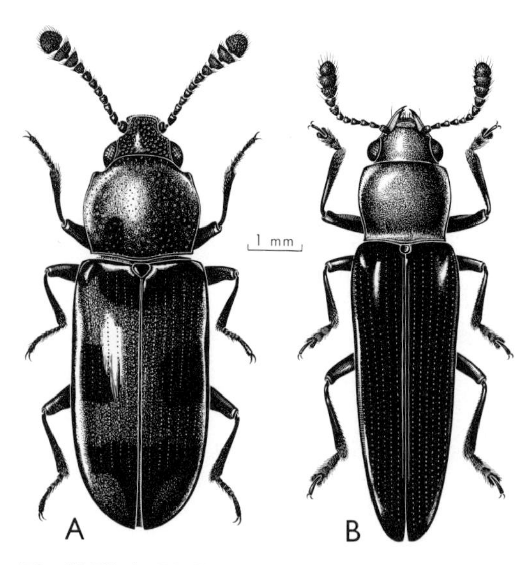

Fig. 30.74. A, *Thallis compta*, Erotylidae; B, *Caenolanguria vulgaris*, Languriidae. [F. Nanninga]

The larvae feed in fleshy fungi or on mycelia in decayed wood. Adults commonly overwinter beneath the bark of trees. The family is widely distributed, but is especially well represented in South America. The important genera in our fauna are *Thallis* and *Episcaphula*. The largest species, *E. hercules* Lea, which occurs in north Queensland, reaches 23 mm in length. Erotylidae are closely related to Languriidae, which have been included in the same family. The larvae of Languriidae are, however, not fungus feeders; their adults have the fore coxal cavities open, and lack the continuous elytral epipleura.

68. Phalacridae (Fig. 30.75). Very small (1–3 mm), broadly oval beetles of strongly convex form, with highly polished, glabrous surface; margins of pronotum and elytra broadly explanate and deflexed; head deflexed but visible from above. Antennae 11-segmented with a 3-segmented club, scape large, triangular, received into a cavity below the head on the inner side of the eye; clypeus not distinct from frons; labrum visible; maxillae with lacinia and galea; fore coxae globular, separate, the cavities open behind; mid coxae separated; hind coxae transverse, nearly contiguous; tarsi 5-5-5-segmented, with 4th segment greatly reduced and 1–3 broad and lobed below; claws appendiculate or with a basal tooth; elytra entire; wings present; abdomen with 5 visible sternites. Larvae rather onisciform, whitish, with depressed head, 3-segmented antennae, and 2 to 5 pairs of ocelli on each side; legs well developed; abdomen 9-segmented; short, curved urogomphi present. Adults and larvae are commonly found in flowers, particularly Compositae, where they eat pollen. Some appear to be associated with fungal smuts of grasses. The more important Australian genera are *Litochrus* (22 spp.), *Parasemus* (11 spp.), and *Phalacrinus* (8 spp.). [Lea, 1932.]

69. Cerylonidae (Fig. 30.76A). Minute beetles (*ca* 2 mm) of broadly ovoid or parallel-sided form, dark reddish brown or with elytra yellow-brown. Antennae with a stout scape and a club with 2 small transverse

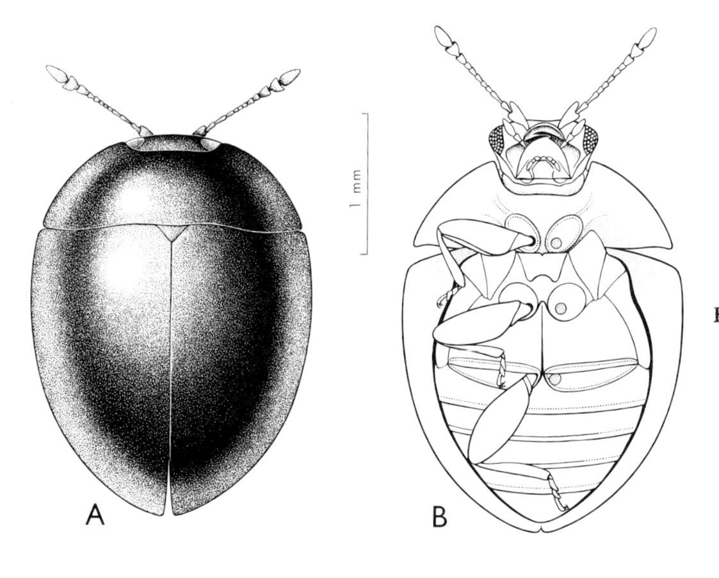

Fig. 30.75. *Phalacrinus rotundus*, Phalacridae: A, dorsal; B, ventral. [F. Nanninga]

basal segments and a large terminal segment; pronotum transverse, with a broad median basal lobe; elytra smooth, punctate-striate; coxae all widely separated; fore coxal cavities closed behind, mid coxal cavities closed by the meeting of the meso- and metasterna; the large basal abdominal sternite with a femoral line on each side extending backwards *(Cerylon)* or backwards and outwards *(Euxestus)* from the inner end of hind coxal cavity; tarsi short, 4-4-4-segmented; claws simple. A small family, the Australian fauna comprising *Ocholissa leai* Grouv. from Tasmania, *Cerylon* (4 spp.), *Euxestus* (7 spp.), and one cosmopolitan species, *Murmidius ovalis* (Beck) which has been recorded in Darwin. As far as is known, Cerylonidae live under bark and in decaying leaves.

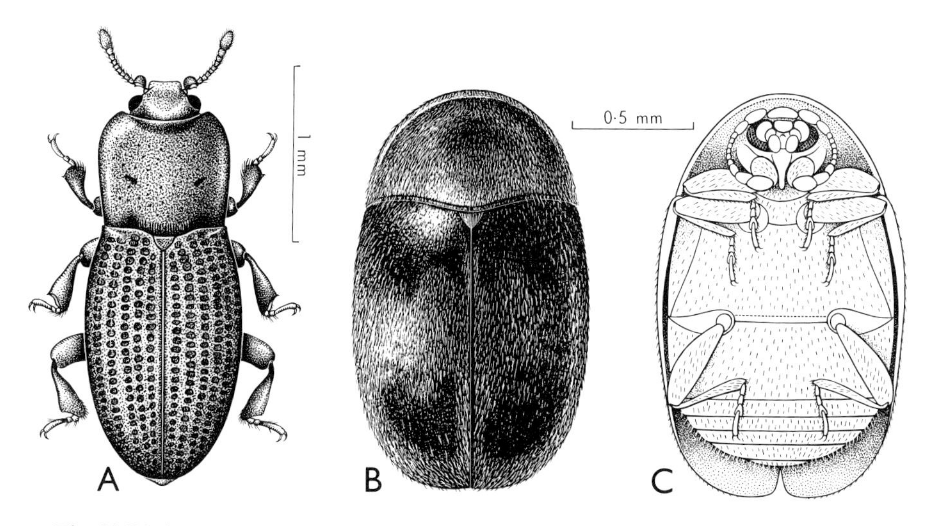

Fig. 30.76. A, *Cerylon alienigerum*, Cerylonidae; B, C, *Clypeaster pulchella*, Corylophidae. [F. Nanninga]

70. Corylophidae (Orthoperidae; Figs. 30.76B, C). Minute beetles (*ca* 1 mm) of broadly ovate, only slightly convex form; pronotum approximately semicircular in outline with angles square or acute, anterior edge often straight, not arched, when viewed from in front; head sometimes completely covered; colour brown, surface uniformly pubescent. Antennae with scape and pedicel enlarged and a loose 3-segmented club; fore coxae close together, mid and hind coxae widely separated; mesosternum very short, metasternum very large, without a median line; abdomen with 6 visible sternites; tarsi 4-4-4-segmented, short, apical segment as long as, or longer than, 1 to 3, 3rd lobed beneath; claws simple; wings narrow, fringed with hairs, with obsolete venation. Corylophidae are found under bark and in rotting wood and decaying vegetable matter, and feed on the spores and hyphae of small fungi. The pupa, as in Coccinellidae, is obtect, and not enclosed in a cocoon. The common genera are *Sericoderus*, *Corylophodes*, and *Clypeaster*. [Deane, 1932b.]

71. Coccinellidae ('ladybirds'; Figs. 30.77, 78). Distinguished by their broadly ovate, highly convex form, 3-segmented tarsi, and toothed claws; length 1–7 mm; colour yellow to black, often bicoloured or spotted, surface usually bare, sometimes finely pubescent. Head deflexed, deeply sunk into prothorax; antennae 11-segmented (rarely 8-segmented), with a 3-segmented club; terminal segment of maxillary palp large, securiform; pronotum with lateral edge usually strongly curved; fore coxae transverse, very close, the cavities usually closed behind; mid coxal cavities more widely separated, closed by the mesepimera (or mesepisterna) between the sterna; hind coxae more widely separated, transverse; abdomen with 5 or 6 visible sternites; basal

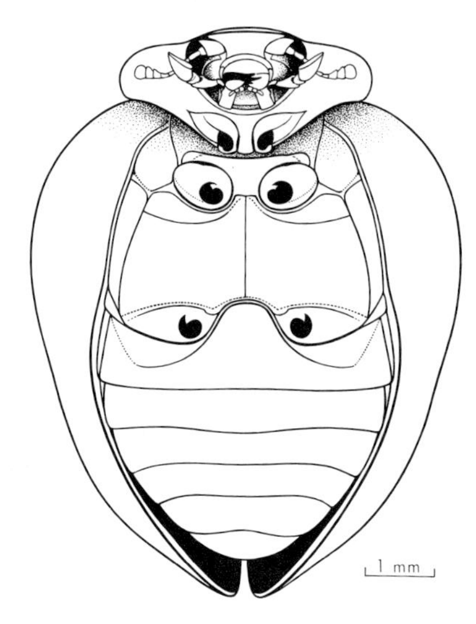

Fig. 30.77. *Epilachna guttatopustulata*, Coccinellidae, ventral. [F. Nanninga]

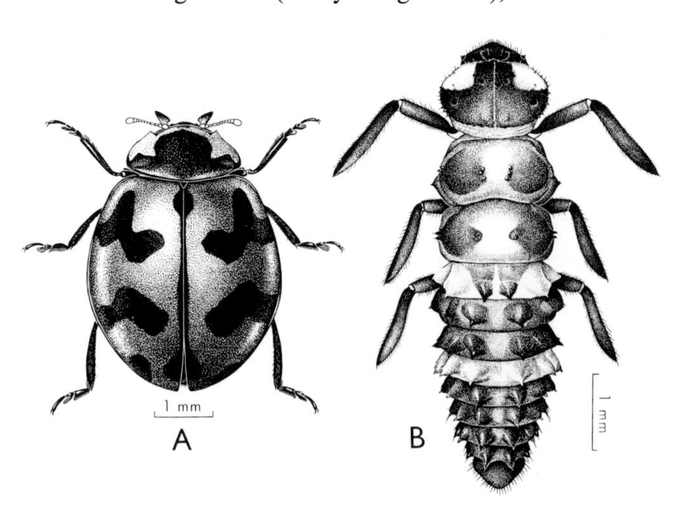

Fig. 30.78. *Coccinella repanda*, Coccinellidae: A, adult; B, larva. [F. Nanninga]

sternite with a curved femoral line extending backwards from the inner end of the coxa; tarsi apparently 3-segmented, actually 4-segmented, the 2 basal segments expanded and densely pilose beneath, 2nd produced into a long lobe beneath, 3rd minute and hidden, 4th long; claws with a tooth beneath or appendiculate. Larvae campodeiform, with body tuberculate or spinose, dark grey commonly spotted with paler colours; urogomphi absent. Pupae obtect, without cocoons, and usually conspicuously coloured. They are attached by the anal end, which is surrounded by the larval skin.

Adults and larvae of most species are carnivorous, preying upon aphids, coccids, or other small insects and mites. The EPILACHNINAE are, however, leaf-eating, and they include some species that rank as pests. *Epilachna 28-punctata* F. and *E. guttatopustulata* F., for example, damage the foliage of potato, tomato, and pumpkin. The carnivorous species are important in the control of scale insects, and some have been used with notable success as agents of biological control. The introduction of *Rodolia cardinalis* Muls., from Australia, for example, saved the Californian citrus industry from destruction by the cottony cushion scale (*Icerya purchasi*). Another Australian species, *Cryptolaemus montrouzieri* Muls., has been similarly employed in Hawaii and California. Its larvae mimic the mealy bugs on which they prey. The commonest of our ladybirds, *Leis conformis* (Boisd.), is bright orange with black spots. Adult Coccinellidae, when alarmed, feign death and discharge drops of yellow blood from the tibio-femoral articulations. This fluid is toxic to vertebrates, and is assumed to be defensive.

Coccinellidae are remarkable for the formation of vast aggregations of adults. Aggregation differs from a cluster of individuals competing for shelter, in that the individuals touch and each has its head hidden beneath a neighbour. The phenomenon is exhibited almost exclusively by species that prey upon aphids. The first step towards aggregation is the accumulation of fat reserves. After cessation of feeding, the beetles migrate towards the most prominent silhouette on their horizon, which may be a mountain peak or, in level country, any prominent object (hypsotactic aggregation). Aggregation may be further concentrated by the odour of dead beetles from earlier years. Its value appears to lie in bringing the sexes together in species in which the ephemeral nature of the food makes a long dormant period necessary. The beetles mate before dispersal, and are then prepared to exploit the rapid growth of the aphid populations. The biology of predacious Coccinellidae has been reviewed by Hagen (1962).

72. Endomychidae (Plate 5, G; Fig. 30.79). Beetles of moderate size (2–8 mm) and convex form, usually found in fungi on trees or feeding on moulds. Antennae with a prominent 3-segmented club; pronotum with anterior angles prominent, so that head is sunk into a concavity, and often with a transverse depression near the base connected with a longitudinal groove on each side of the base; fore coxal cavities open behind; mid coxal cavities bordered by the mesepimera (or mesepisterna) in addition to the sterna; hind coxae transverse, oval, separated by about their own width; abdomen with 5 visible sternites; tibiae without spurs; tarsi 4-4-4-segmented, but 3rd segment minute;

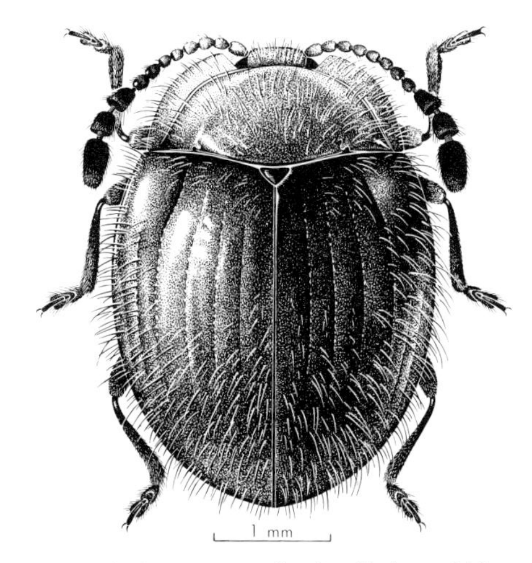

Fig. 30.79. *Stenotarsus pisoniae*, Endomychidae. [F. Nanninga]

basal segment broadly triangular, 2nd broad and with a long lobe beneath, 4th slender, as long as the remainder, claws simple; elytral epipleura continuous almost to the apex. Twelve Australian genera are known, the largest of which is *Stenotarsus*. No Australian larvae have been described.

73. Merophysiidae. The family was erected by Crowson in 1955 to include Merophysiinae and Holoparamecinae formerly classified as Lathridiidae, from which it is distinguished by having open fore coxal cavities, a maxillary lacinia, only 5 pairs of functional abdominal spiracles, and a different facies. *Holoparamecus depressus* Curtis, a cosmopolitan species which feeds on moulds, has been recorded from Australia. It is 1–1·4 mm long, shining brown, elytra with a sutural stria only, and pronotum with 3 short basal longitudinal striae. The larvae, unlike those of the Lathridiidae, have urogomphi.

74. Aculagnathidae (Fig. 30.80A). A family erected by Oke (1932) on a single species, *Aculagnathus mirabilis* Oke, found near Melbourne in nests of the ant *Amblyopone obscurus*. The family is unusual in having the mouth-parts partly modified for piercing and presumably for sucking. Length *ca* 1·5 mm; head without a distinguishable clypeus; labrum sharply pointed; mandibles with a thin styliform process on the outer edge; maxillary and labial palpi with penultimate segments securiform and apical segments sharply pointed and slender; antennae 9-segmented, with large scape and a 3-segmented club composed of a small 7th and large 8th and 9th segments; prothorax broad; fore coxae small, globular, their cavities open behind; mid coxal cavities closed by the sterna only; hind coxal cavities widely separated; dorsal surface coarsely punctured and sparsely setose; tarsi 3-segmented, without lobes; claws simple.

75. Lathridiidae (Fig. 30.80B). Minute (1–3 mm) beetles; pronotum usually much narrower than base of elytra; elytra often coarsely punctured or ribbed; antennae 8- to 11-segmented including a 2- or 3-segmented club; palpi with terminal segments pointed; fore coxal cavities closed behind; mid coxal cavities bordered by the mesepimera in addition to the sterna; tarsi 3-3-3-segmented (males sometimes 2-3-3), segments simple, claws simple; abdomen with 5 or 6 visible sternites. Adults and larvae feed on moulds and in larger fungi and mycetozoa. They are found in plant debris, under stones or bark, in stored foodstuffs affected by damp, and some species live in the nests of ants and termites. A few species are commonly found in new houses, where they feed on moulds that grow on damp plaster. *Lathridius*

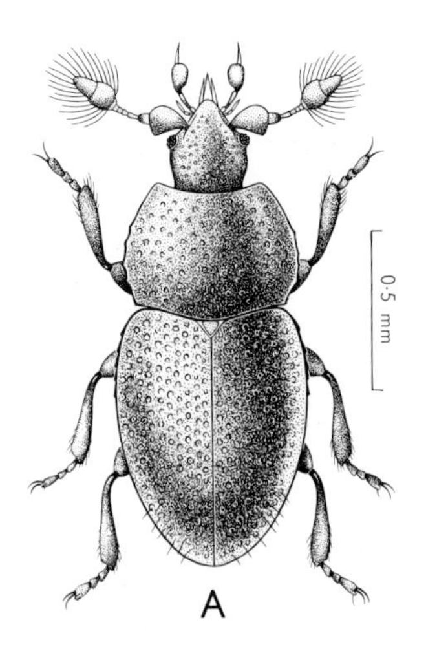

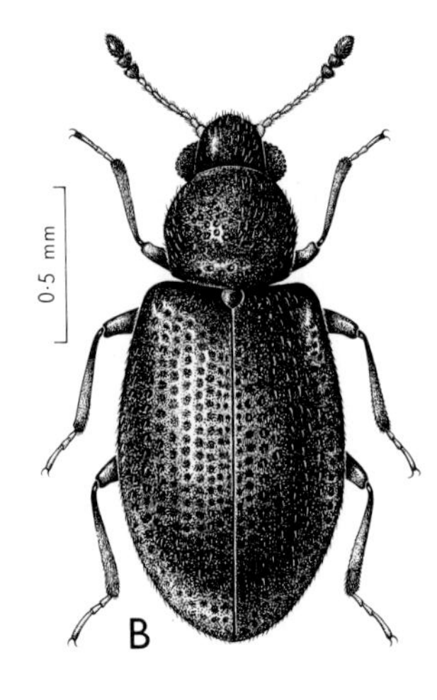

Fig. 30.80. A, *Aculagnathus mirabilis*, Aculagnathidae; B, *Melanophthalma australis*, Lathridiidae. [F. Nanninga]

nodifer Westw., *Coninomus constrictus* Gyll., and *Enicmus minutus* L. have a world-wide distribution.

76. Ciidae (Cisidae, Cioidae; Fig. 30.81). Minute (1–3 mm), cylindrical beetles, found in bracket fungi on trees or in rotting wood. Head deflexed, often largely covered by pronotum, with a frontal ridge between eyes; antennae 8- to 11-segmented with a prominent, 3-segmented, elongate club; maxillae with lacinia, but no galea; terminal segment of palp long and conical; pronotum long, prosternum short, so that the anterior aperture faces forwards and downwards and the head is largely covered from above; pronotum sometimes with anterior edge toothed, with defined lateral margins; fore coxal cavities closed behind; fore coxae projecting; mesepimera reaching the mid coxal cavities; abdomen with 5 visible sternites; tarsi 4-4-4-segmented, segments slender, 1–3 short, 4th long; claws simple; elytra without striae, usually clothed with rather sparse, short, stout setae, often arranged in longitudinal rows. Larvae with upturned, pointed urogomphi. Ciidae may be confused with Curculionidae–Scolytinae, but can be distinguished by the fact that the antennal club has the 3 segments obviously separated, whereas in Scolytinae the club is compact.

77. Merycidae (Fig. 30.82A). A family erected by Crowson (1955) to accommodate the genus *Meryx*, which is composed of 2 Australian species formerly treated as Colydiidae. Length 6–10 mm; dark brown, with pale yellowish, short, recumbent pubescence; head prognathous; antennae 11-segmented, long, slender, with last 3 segments slightly thickened; fore coxae globular, slightly projecting, the cavities closed behind; mesepimera reaching mid coxal cavities; abdomen with 5 visible sternites; tarsi 4-4-4-segmented, the segments not lobed, 1st moderately long, 2 and 3 short, 4 as long as 1–3 together; claws simple. The larvae are unknown. The known species are *M. aequalis* Blackb. and *M. rugosa* Latr., the latter having elytra with a remarkable polygonal pattern of raised ridges.

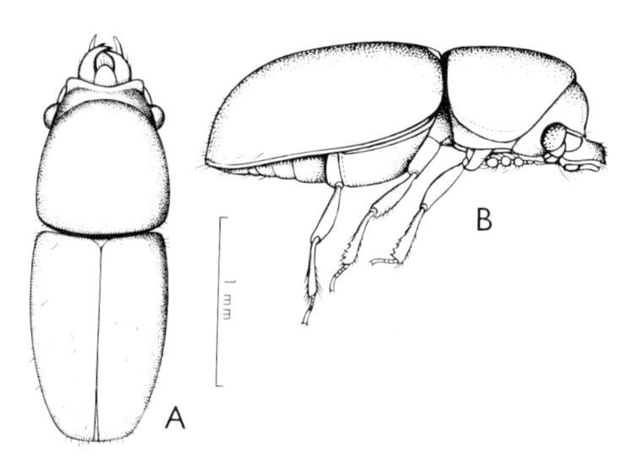

Fig. 30.81. *Octotemnus dilutipes*, Ciidae, ♂. [F. Nanninga]

78. Mycetophagidae (Fig. 30.82B). Very small (1·5–3 mm), broadly ovate beetles of brown or dark brown colour, clothed with semi-erect pubescence. Head triangular, visible from above; eyes with coarse convex facets; antennae 11-segmented with a 2- or 3-segmented club; clypeus and labrum distinct, labrum quadrate; maxillary palpi long, with terminal segment as long as the whole labial palp; pronotum trapezoidal, as wide as elytra at base; fore coxal cavities close together, open behind; fore coxae ovate, with trochantins not visible; mid coxae almost in contact, the coxal cavities closed outwardly by the meso- and metasternum; hind coxae narrow, transverse, contiguous; abdomen with 5 visible sternites; tibiae with terminal spurs; tarsi 4-4-4-segmented or 3-4-4- in the males; 1st and 4th segments longer than 2nd and 3rd, none bilobed; claws simple; elytra entire, punctate-striate, epipleura not reaching the apices. Larvae subcylindrical, pale brown, with few setae; head exserted, with 4 to 6 pairs of ocelli; legs well developed; abdomen 9-segmented; urogomphi present, short, dorsally curved. The adults and larvae feed on fungi, and are found in bracket fungi or mouldy materials. *Typhaea stercorea* L. and *Litargus balteatus* Le C. are cosmopolitan species commonly found in stored foodstuffs affected by mould (Hinton, 1945).

79. Colydiidae (Fig. 30.83). Cylindrical or flattened beetles (2–12 mm), usually dark brown or black, with surface dull and

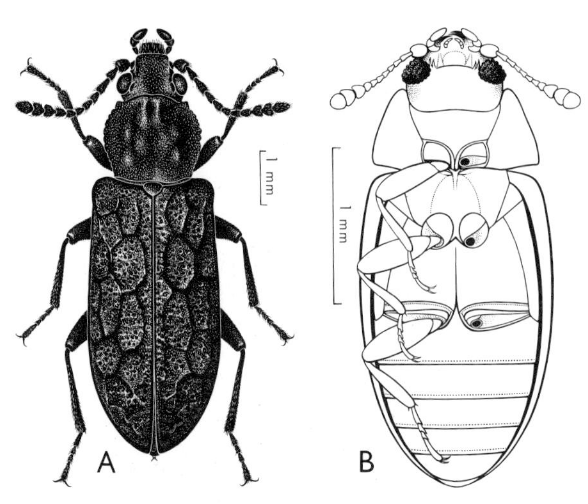

Fig. 30.82. A, *Meryx rugosa*, Merycidae; B, *Typhaea stercorea*, Mycetophagidae, ventral. [F. Nanninga]

coarsely sculptured or with strong longitudinal ridges. Antennae inserted below a frontal margin, with a 2- or 3-segmented, oval club; head often with a channel below eye for reception of the antennae; palpi with apical segments minute; fore coxae globular, cavities closed behind; mid coxal cavities closed by the sterna only; abdomen with 5 visible sternites; tarsal segmentation 4-4-4, none lobed, terminal segment longer than the rest; claws simple. Larvae subcylindrical, with 9-segmented abdomen and backwardly projecting urogomphi. Most Colydiidae live under bark or in decaying wood, and feed on fungi. Some appear to be carnivorous. The larvae of some are said to be ectoparasitic on the larvae or pupae of Cerambycidae and Buprestidae. *Euclarkia costata* Lea and *Kershawia rugiceps* Lea live in ants' nests. The biology of the Australian species is, however, almost entirely unknown. *Deretaphrus* and *Bothrideres* are important Australian genera. [Carter and Zeck, 1937.]

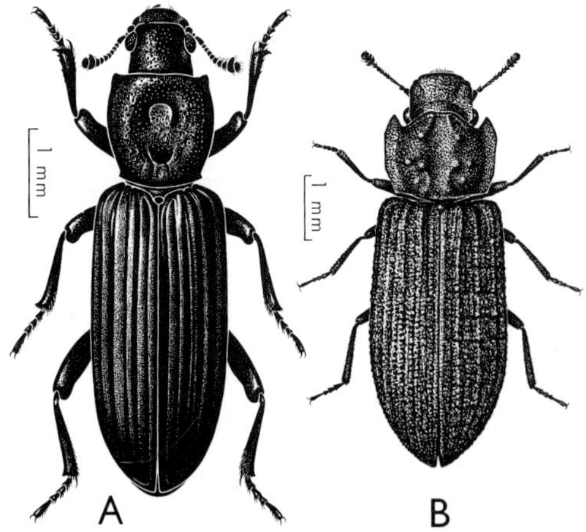

Fig. 30.83. Colydiidae: A, *Bothrideres taeniatus*; B, *Sparactus elongatus*. [F. Nanninga]

80. Tenebrionidae (Plates 4, C, 5, E; Fig. 30.84). One of the major families of Coleoptera, of very varied size and shape and usually dull brown or black colour. Tarsal segmentation always 5-5-4, the segments not lobed beneath, claws simple; abdomen with 5 visible sternites, the 3 basal segments immovably connected; head with sides of frons usually strongly produced over the antennal insertions and eyes deeply indented; antennae usually stout, often moniliform, sometimes clubbed; fore coxae separated and not projecting, the cavities closed behind; prosternum with lateral margins distinct, and without cavities for reception of the antennae; mid coxae usually widely separated, mesepimera reaching the coxae or not; elytra complete, often strongly embracing the sides of the body, often firmly joined together and wings absent. Larvae usually subcylindrical and strongly sclerotized; resembling larvae of Elateridae, but with clypeus and labrum

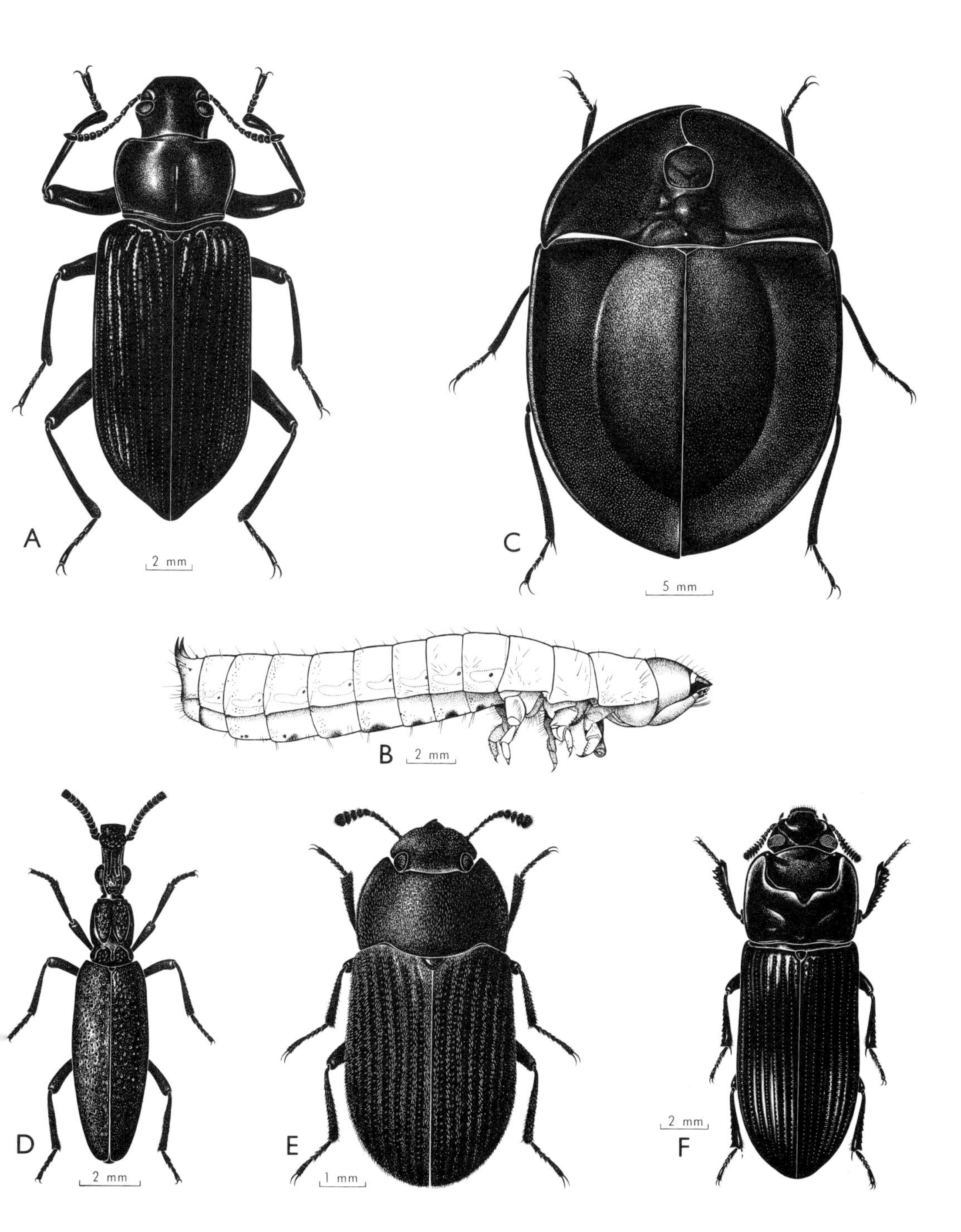

Fig. 30.84. Tenebrionidae: A, *Promethis nigra*, Tenebrioninae; B, *P. nigra*, larva; C, *Helaeus waitei*, Helaeinae; D, *Tretothorax cleistostoma*, Dacoderinae; E, *Gonocephalum meyricki*, Opatrinae; F, *Uloma westwoodi*, Ulominae. [F. Nanninga]

distinct, abdominal segment 9 not subconical, and pleural sutures always present; urogomphi usually present; legs usually short, fore legs sometimes longer than the others.

Adult Tenebrionidae are usually strongly sclerotized and well adapted to resist desiccation, e.g. subelytral cavity largely sealed, communicating with the exterior only by the limited movement of the 2 apical segments of the abdomen. The family is very large, well represented in all the arid parts of the world, and the species are found in a wide variety of habitats. Most are found on the ground, beneath stones or logs, but they also occur under bark, in rotten wood, fungi, birds' nests, and the nests of ants and termites. A few are pests of stored foods. *Sphargeris physodes* Pasc. (PHALERINAE), a rounded, pale yellowish-brown beetle about 5 mm long, lives in moist sand on sea beaches at high water mark. *Caediomorpha heteromera* King, *Caedius sphaeroides* Hope, and *Trachyscelis* spp. are also widely distributed on ocean beaches.

There is a great need for research on the classification of the family. Carter (1926) has provided a catalogue of Australian species with keys to subfamilies and genera. Twenty-eight subfamilies are recognized, and 75 per cent of the genera are endemic. The largest subfamilies are the Helaeinae and Adeliinae, each with more than 200 species. HELAEINAE, which include the remarkable 'pie-dish beetles' (*Helaeus*, *Pterohelaeus*, *Sympetes*), are known only from Australia. In *Helaeus* the flanges of the prothorax meet in front of the head (Fig. 30.84C). *Helaeus* species are wingless, and live on the ground in the dry interior. *Adelium* and *Cardiothorax* are the largest genera of ADELIINAE. *Cardiothorax* (60+ spp.) are attractive beetles with a somewhat heart-shaped pronotum and strongly striate elytra. They occur only in the east, and are found under logs. Species of *Adelium*, when disturbed, evert a pair of glands and emit a defensive odour. The AMARYGMINAE are another large subfamily including the genera *Amarygminus* and *Chalcopterus*, species of which are recognizable by their strongly convex shape and beautiful metallic blue and coppery colour. They are found under bark. Some large attractive species are also found in the CYPHALEINAE. *Prophanes mastersi* Hope is 18 mm long and bright metallic blue.

The ULOMINAE include a number of species which are pests of cereal products and have a world-wide distribution. Among these are *Gnathocerus cornutus* F., *Latheticus oryzae* Waterh., *Palorus ratzeburgii* Wissm., *Alphitobius diaperinus* Panz., *Tribolium castaneum* Herbst, and *T. confusum* Duval. *Tenebrio molitor* L. and *T. obscurus* F. (TENEBRIONINAE) are minor pests of cereal products. Their shining, yellowish brown, cylindrical larvae ('mealworms') are bred and sold by pet shops as food for reptiles and insectivorous birds. Three species of LEIOCHRININAE, *Leiochrodes suturalis* Westw. and 2 species of *Derispia*, are known from Australia. *L. suturalis* is 5 mm long, black, highly convex, shining, and remarkable for its resemblance to a coccinellid. It has been recorded from Sydney north to Cooktown, and also occurs in New Guinea and Amboina; it is found among decaying leaves, and flies to light. The subfamily DACODERINAE,* which has species in California, Mexico, and Colombia, is represented by *Tretothorax cleistostoma* Lea, a remarkable beetle found in ants' nests in Queensland. The pronotum in this species has a pair of tubercles on each side which meet over a transverse cavity, and the head is greatly elongate between eyes and antennae. The 3 genera of AMPHIDORINAE, *Ectyche* and *Micrectyche* (W.A., N.T.) and *Phaennis* (N.S.W., Tasmania), are unusual in the family in being clothed with long erect bristles like those of Cleridae. Species of *Platydema* (DIAPERINAE), small ovoid beetles with red and black elytra, live in birds' nests.

81. Lagriidae (Fig. 30.85). Beetles of moderate size (5–14 mm), characterized by the narrow, approximately cylindrical head and thorax followed by ample elytra which are widest in the apical half; colour yellowish brown, or dark blue, or green, commonly with a metallic reflection. Head prognathous;

* Raised to family rank by J. C. Watt, *Proc. R. ent. Soc. Lond.* (B) **36:** 109–18, 1967.

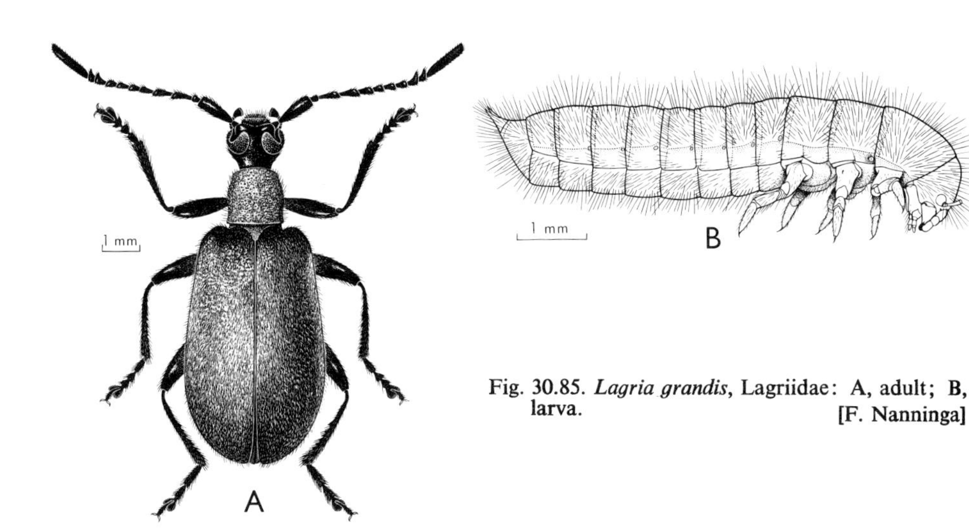

Fig. 30.85. *Lagria grandis*, Lagriidae: A, adult; B, larva. [F. Nanninga]

antennae filiform, 11-segmented, apical segment as long as the 3 or more preceding segments; pronotum subcylindrical with ill-defined margins; fore coxae somewhat projecting, almost contiguous, the cavities closed behind; mid coxae close together, the mesepimera reaching the cavities; legs with tibiae slender, clothed with long setae, without spurs; tarsal segmentation 5-5-4, basal segment long, penultimate segment broadly lobed beneath apical segment and spongy beneath; claws simple; elytra covering abdomen, uniformly punctured and setose, without striae; abdomen with 5 visible sternites. The larvae, which are subcylindrical, sclerotized and pigmented, with short urogomphi, live in decaying wood and vegetable matter. The adults are found beneath bark or on low shrubs, and are said to be phytophagous. Most of the Australian species belong to the Old-World genus *Lagria*. [Armstrong, 1948.]

82. Alleculidae (Cistelidae, Gonaderidae; Fig. 30.86). Beetles of moderate size (4–19 mm), elongate-oval shape, and brown, black, or metallic colour; distinguished by the pectinate tarsal claws, distinctly margined prothorax, terminal segment of antennae not unduly elongate, and fore coxae globular, sunken. Antennae filiform; fore coxal cavities closed; mesepimera reaching mid coxal cavities; abdomen with 5 visible sternites; eyes large; elytra striate; legs slender, long, all tibiae with 2 apical spurs; tarsal segmentation 5-5-4, penultimate segment lobed beneath. Larvae of Australian species are not known, but elsewhere larvae live in rotting wood. The adults are found

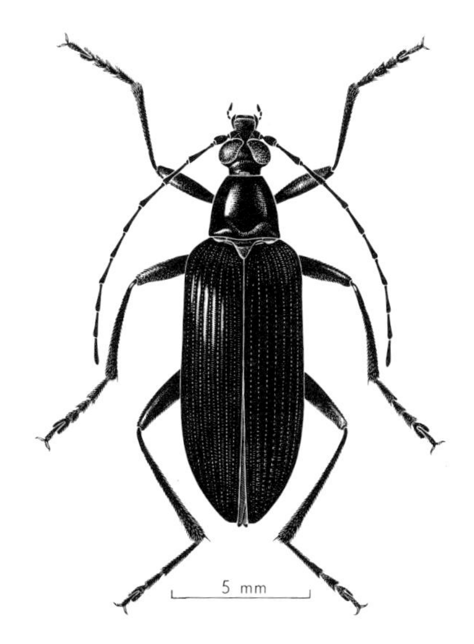

Fig. 30.86. *Tanychilus striatus*, Alleculidae. [F. Nanninga]

under bark or on flowers, and are said to feed on pollen, but very little is known of their biology. The large, shining black *Tanychilus* spp. are common in eastern Australia. *Aethyssius* includes some attractive metallic green and blue species (*A. viridis* Boisd., *A. virescens* Boisd.).

83. Zopheridae (Fig. 30.87). Represented in Australia by *Zopherosis georgei* White (the 'ironbark beetle'). Length 25–30 mm, colour black or dark brown, surface dull, very heavily sclerotized, form parallel-sided and flat, the surface ornamented with deep pits and prominent tubercles. Prothoracic hypomera with deep channels anteriorly for reception of the antennae; head deeply sunk into prothorax, eyes almost hidden; antennae short, stout, with a short, flat, 3-segmented club; fore coxae globular, the cavities closed behind; mid coxae globular, very broadly closed by the meso- and metasterna; abdomen with 5 convex sternites, 5th with a deep impression on each side, 3rd, 4th, and 5th without visible connecting membranes; tarsal segmentation 5-5-4, segments 1–4 of the fore and mid legs and 1–3 of the hind legs short, stout, subequal, terminal segments as long as about 3 other segments; claws simple. The family is small, occurring mainly in North and South America, but with a few species in Africa, E. Asia, and one in Australia. *Z. georgei* is found in northern N.S.W. on or beneath the bark of eucalypts. The larva is unknown.

84. Inopeplidae (Fig. 30.88). Small (*ca* 4 mm), flat beetles. Head larger than prothorax; labrum and clypeus transverse, separated by a broad pale membrane; antennae filiform or slightly moniliform, inserted beneath a preocular shelf; pronotum very strongly tapered towards base; elytra broad and rounded apically, strongly narrowed to base, leaving 2 or 3 abdominal tergites exposed; fore coxae rounded, not projecting, the cavities open

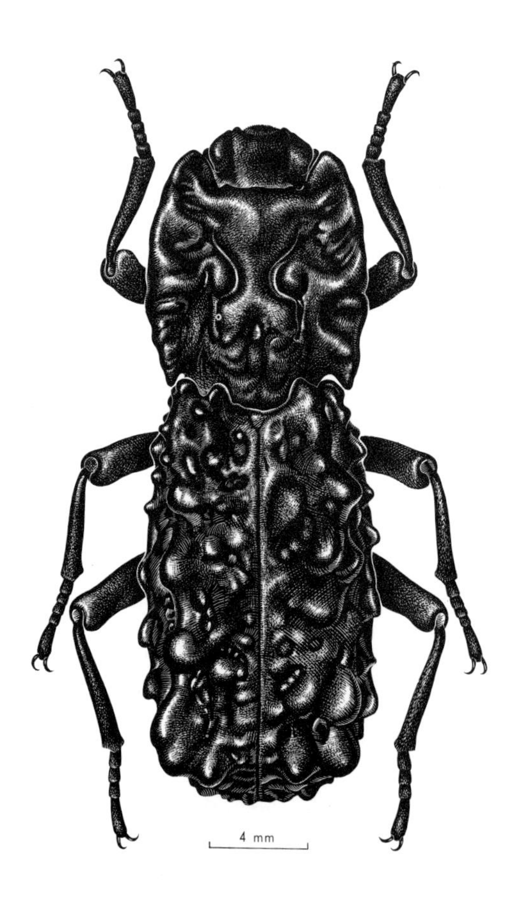

Fig. 30.87. *Zopherosis georgei*, Zopheridae. [F. Nanninga]

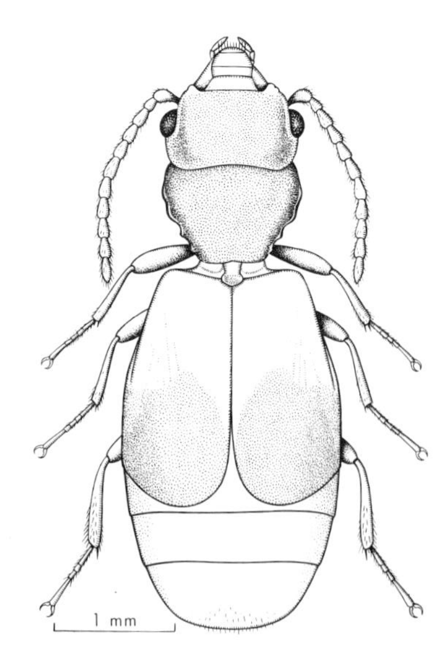

Fig. 30.88. *Inopeplus dimidiatus*, Inopeplidae. [F. Nanninga]

behind; prosternal process broad; mesepimera not reaching mid coxal cavities; mid and hind coxae widely separated; abdomen with all 5 visible sternites free; tarsi 5-5-4-segmented, segments not lobed; claws simple. *Inopeplus* is the only genus. Adults are found beneath bark, but nothing is known of their biology.

85. Salpingidae. Small or minute (1–5 mm) beetles. Head prominent, prognathous, sometimes rostrate; antennae gradually thickened to apex; maxillary palpi with terminal segment not securiform; prothorax contracted towards base; elytra subparallel, covering abdomen; fore coxae projecting, contiguous, the cavities open behind; mesepimera not reaching the mid coxal cavities; abdomen with 5 free sternites; tarsi 5-5-4-segmented, the segments and claws simple. Nothing is known of their biology, other than that adults are found under bark and in moss. Six Australian genera are known, of which *Notosalpingus* (6 spp.), *Neosalpingus* (4 spp.), and *Lissodema* (3 spp.) are the largest. The family is represented in all States.

86. Mycteridae (Fig. 30.89). Small (3–6 mm), elongate, parallel-sided beetles; reddish brown or dark brown, rarely with paler spots on the elytra; body with uniform pale pubescence. Head prognathous; eyes large, hemispherical; antennae filiform, inserted beneath a preocular shelf; terminal segment of maxillary palp obliquely truncate, subsecuriform; prothorax quadrate, without defined lateral margins; fore coxal cavities contiguous, fused, widely open behind; fore coxae conical, projecting; mesepimera not reaching mid coxal cavities; mesepisterna meeting anteriorly (as in Anthicidae); mid coxae almost in contact; hind coxae separated; scutellum subquadrate; elytra covering abdomen: abdominal sternites 1–3 connate; tarsi 5-5-4-segmented, penultimate segment lobed beneath. This small family has a worldwide distribution. *Trichosalpingus* (9 spp.) occurs in south-eastern Australia; *Hybogaster* is known from 1 species in Queensland and 2 in Chile. Mycteridae are found under bark, and the larvae are unknown.

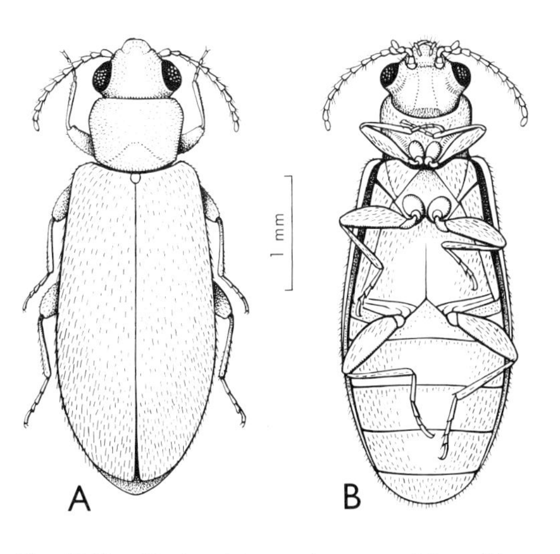

Fig. 30.89. *Trichosalpingus brunneus*, Mycteridae: A, dorsal; B, ventral. [F. Nanninga]

87. Hemipeplidae. Small (*ca* 4 mm), elongate, parallel-sided beetles. Head broad, with prominent eyes; antennae short, gradually widened to apex; maxillary palpi not securiform; prothorax transverse, tapered posteriorly; elytra pubescent, covering abdomen, tapered near apex; fore coxae close together, the cavities closed behind; mesepimera not reaching mid coxal cavities; abdominal sternites freely movable; tarsi 5-5-4-segmented, the 2 segments before the terminal one lobed beneath. Hemipeplidae feed on the leaves of palms. *Hemipeplus* is represented in all continents except Europe. *H. australicus* Arrow, from Darwin, is pale yellow with a reddish-brown apical spot.

88. Pythidae (Boridae; Fig. 30.90). Elongate and parallel-sided or rather flattened beetles; length 3·5–10 mm; body clothed with sparse, erect bristles or bare. Head large, prognathous, with prominent eyes; antennae 11-segmented, the segments gradually increasing in thickness to apex, inserted beneath a shelf-like extension of the frons in front of the eyes; prothorax widest towards apex or about middle, lateral margin absent or very faint and continuous with anterior margin; fore coxae contiguous, projecting, the cavities open behind; mid coxae almost contiguous, the mesepimera reaching the coxal cavities; abdomen covered by the elytra; abdominal

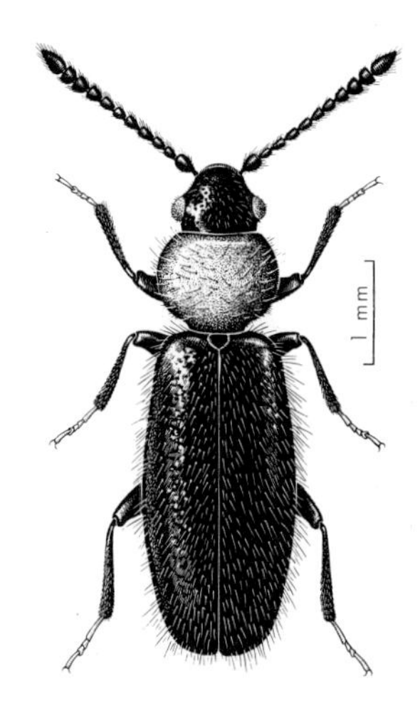

Fig. 30.90. *Temnopalpus bicolor*, Pythidae. [F. Nanninga]

sternites all free; tarsal segments simple, or with penultimate segment lobed beneath, terminal segments relatively long, claws simple. Nothing is known of the larvae or biology of the Australian species, which are included in the genera *Synercticus*, *Paromarteon*, and *Temnopalpus*. *T. bicolor* Blackb. is about 5 mm long, clothed with sparse erect setae, with head and elytra black and prothorax reddish yellow.

89. Melandryidae (Fig. 30.91A). Beetles of small to moderate size (3–15 mm, very variable within species) and elongate-ovate form, with unusually elongate, tapered elytra; hind tarsi and hind tibial spurs very long; body finely and uniformly pubescent; colour brown, or dark brown with yellow spots. Head deflexed, not constricted behind eyes; antennae filiform or thickened towards apex; maxillary palpi very large, the segments broad, terminal segment securiform; eyes vertically elongate, not prominent and not pubescent; pronotum with strongly arched front edge largely covering head, base as wide as base of elytra, lateral margins obliterated anteriorly; fore coxae strongly projecting, contiguous, the cavities closed behind; mesepimera reaching mid coxal cavities; mid coxae close together; abdomen strongly tapered, all sternites freely movable; tibial spurs pectinate; tarsi 5-5-4-segmented, basal segment as long as next 3 together, penultimate segments not lobed below; claws simple. Little is known of the habits of Melandryidae, other than that larvae and adults are associated with decaying wood and woody fungi. Adults may be found under bark, and are said to be active at night. The few Australian species are found near the east coast and in Tasmania, but *Dircaea lignivora* Lea occurs in south-western Australia. *Talayra elongata* (Macl.) is the common species in N.S.W.

90. Scraptiidae (Fig. 30.91B). Size small (2–5 mm); body elongate-oval, yellowish brown, uniformly pubescent. Head deflexed, sharply constricted behind eyes, which are pubescent and cover the sides and part of the ventral surface; antennae long, filiform; terminal segment of maxillary palp strongly securiform; pronotum narrowed anteriorly, with obvious angulate lateral margins, base as wide as the elytra; fore coxae projecting, very long, resembling the femora, widely separated at their bases, the cavities closed behind; mesepimera reaching mid coxal cavities; mid coxae contiguous, greatly elongate, and directed backwards; all abdominal sternites

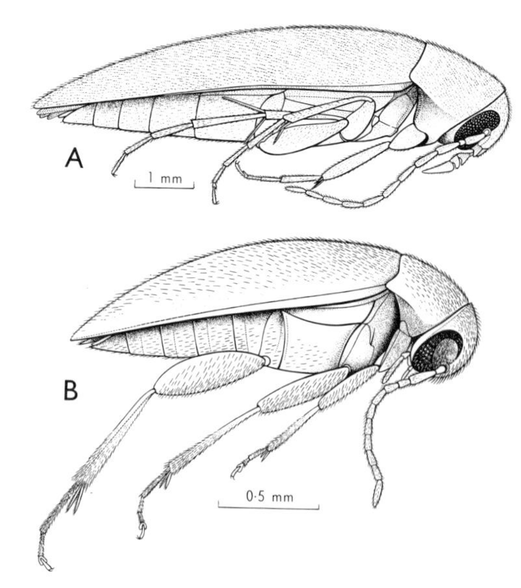

Fig. 30.91. A, *Talayra elongata*, Melandryidae; B, *Scraptia australis*, Scraptiidae. [F. Nanninga]

freely movable; legs very long, tibial spurs pubescent; tarsi 5-5-4-segmented, basal segments very long, penultimate segments very short and lobed beneath; claws very small and simple. *Scraptia* is represented throughout the world. *S. picturata* Champ. from north Queensland is unusual in having the pronotum and elytra patterned in yellowish brown and brown; other species are known from Tasmania. The adults are found on flowers, and the larvae are reported to feed in decaying wood.

91. Mordellidae (Fig. 30.92). Small to medium-sized (2–13 mm), characteristically humped and smoothly tapered beetles; apical tergite of abdomen produced beyond elytra in the form of a long, stout spine; head and prothorax deflexed, so that the head covers the fore coxae and the pro- and mesosterna; body dark, frequently patterned with patches of pale recumbent setae. Head with a sharply angulate constriction behind eyes, the declivity closely applied to the anterior edge of pronotum; antennae with 7 apical segments enlarged, forming an elongate, serrate club; terminal segments of maxillary palpi securiform; prothorax with sharp lateral longitudinal edges, base as wide as base of elytra; fore coxae elongate, projecting (although covered by the head), the cavities closed behind, contiguous but not confluent; prosternum greatly reduced; mesepimera reaching mid coxal cavities, which are contiguous; hind coxae in the form of large plates; all abdominal sternites freely movable; tarsi 5-5-4-segmented, simple, sometimes flattened; claws serrate or pectinate beneath, each with a sclerotized process of equal length beneath. Adults are found on flowers, and are very active. Most of the species belong to the world-wide genera *Mordella* and *Mordellistena*. The larvae are predatory or parasitic, or are leaf- or stem-miners (*Mordellistena* spp.). Mordellids are found in all States.

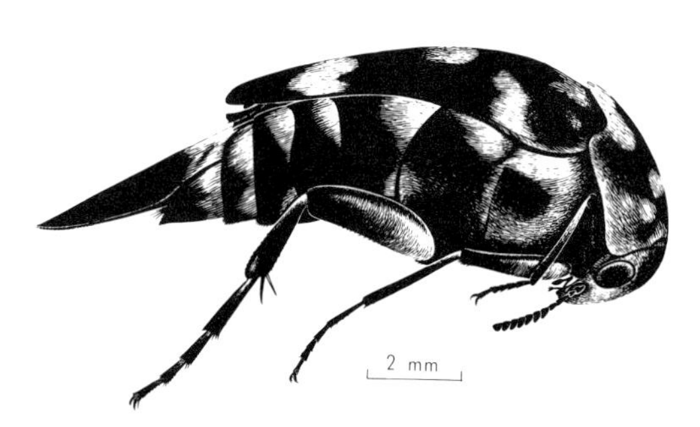

Fig. 30.92. *Mordella leucosticta*, Mordellidae. [F. Nanninga]

92. Rhipiphoridae (Fig. 30.93). Small to large (3–30 mm), humped and tapered beetles; brown or black, elytra sometimes with a yellow or brown and black pattern; usually winged and with entire elytra, but sometimes (in Rhipidiinae) males have greatly reduced elytra but well-developed wings, and females lack both elytra and wings. Head and pronotum deflexed, so that the head rests on the anterior faces of the long, projecting fore coxae; head with a sharp constriction behind eyes, the declivity closely applied to the front edge of the prothorax; antennae usually 11-segmented, with apical 8 or 9 segments flabellate in male, serrate in female (or 11-segmented and moniliform, or 2- to 4-segmented, in the larviform females of Rhipidiinae); maxillary palpi filiform or atrophied; prothorax strongly tapered anteriorly, without defined lateral edges, base as wide as base of elytra; fore coxal cavities open behind; prosternum reduced; mesepisterna reaching mid coxal cavities; mesosternum with a mid-longitudinal carina; abdomen with all 5 visible sternites free, without an apical spine; tarsi normally 5-5-4-segmented (4-4-4-segmented in the larviform females of *Rhipidioides*), segments simple; claws serrate or pectinate, without appendages beneath.

Rhipiphoridae are unique among Coleoptera, in that all the larvae so far known are, at least temporarily, endoparasitic. The adults are free-living and winged (except female Rhipidiinae), and are commonly found on flowers. Some species fly to light at night. Adult life appears to be brief. As is suggested by the flabellate male antennae, males are attracted to females by scent. The larvae of the RHIPIPHORINAE–Macrosiagonini parasitize the larvae of wasps (Scoliidae and Tiphiidae), whereas those of

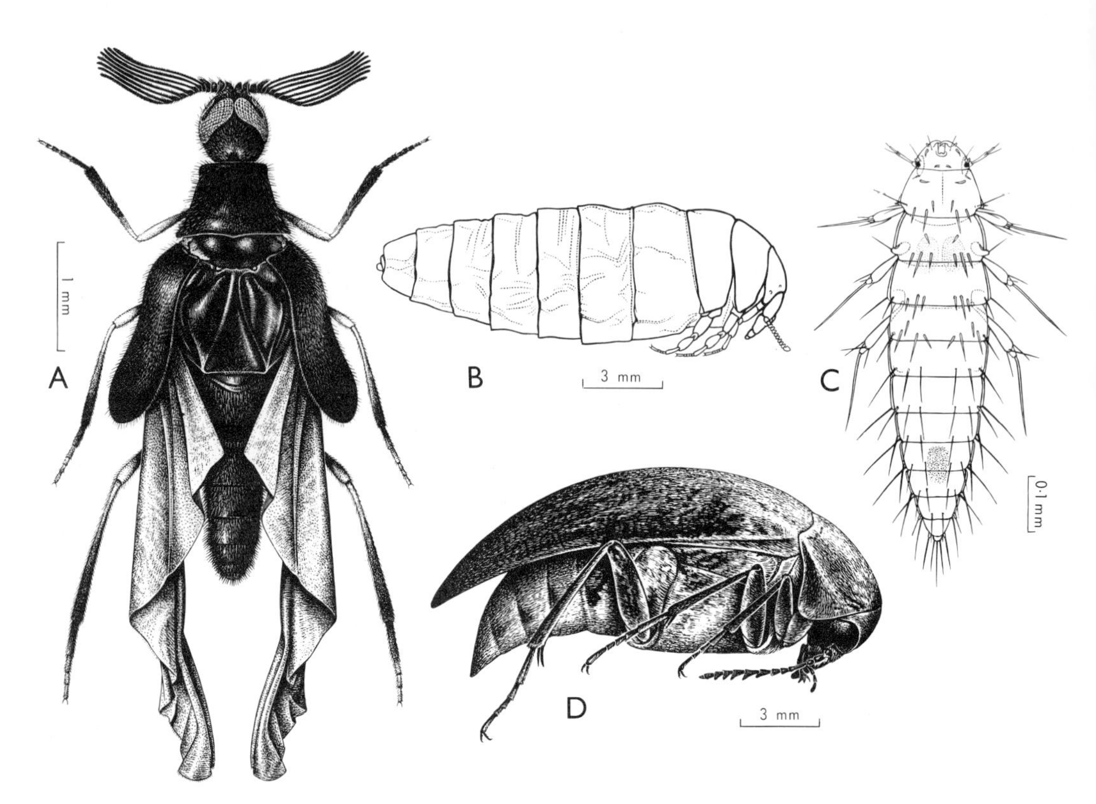

Fig. 30.93. Rhipiphoridae: A, *Rhipidioides rubricatus*, adult ♂; B, *Riekella australis*, adult ♀; C, *Rhipidioides helenae*, planidium; D, *Pelecotomoides marmorata*, ♀. [F. Nanninga]

Rhipiphorini parasitize the larvae of bees. Little is known of the biology of the Pelecotomini, but there is some evidence that they may parasitize beetle larvae. RHIPIDIINAE parasitize cockroaches.

Female Rhipiphoridae lay large numbers (500 to 3,000) of small (0·3–0·7 mm long) eggs, on flowers, or, in Rhipidiinae, on bark, logs, or other places frequented by cockroaches. The eggs hatch into minute, active, heavily sclerotized, elongate-oval larvae with multiple ocelli, long 2- or 3-segmented antennae, and long, fine caudal bristles. This type of larva, named 'planidium' by Wheeler, is found in several orders (p. 106), often as an adaptation to the use of winged insects to provide transport to the vicinity of a host larva. The planidium awaits the arrival of a bee or wasp, whereupon it attaches itself, and is transported to the nest or host larva (e.g. of a scoliid wasp). The low probability of success of this step is offset by the large number of planidia produced. Having located a suitable host larva, the planidium penetrates the cuticle, and proceeds to feed as an endoparasite. In most species observed the endoparasitic habit is limited to the first instar. The greatly distended planidium re-emerges and moults into a legless, tuberculate grub, which curls around the thorax of the host larva and thereafter feeds externally. In Rhipidiinae the larva is endoparasitic throughout its life, and the adult female lacks elytra and wings. Species of *Macrosiagon* in Australia are hyperparasitic on the larvae of the scoliid wasps, *Campsomeris* spp., which, in turn, are parasitic on the larvae of scarabaeid beetles in the soil. [Selander, 1957, classification of the family; Riek, 1955a, Australian Rhipidiinae.]

93. Oedemeridae (Fig. 30.94). Size moderate (5–18 mm); body slender, parallel-sided, with rather soft integument, finely and uniformly pubescent; usually pale yellowish brown, sometimes with a dark pattern or spots. Head small, moderately deflexed, elongate before eyes, not constricted behind, at most as wide as pronotum; antennae long, filiform; palpi long, with apical segments triangular; prothorax subcylindrical, widest near anterior end or in middle; fore coxae elongate, projecting, contiguous, the cavities confluent and widely open behind; mesepimera reaching mid coxal cavities; mid coxae elongate, contiguous; abdomen with 5 free sternites; elytra entire, rounded apically; wings present; legs long, slender; tarsal segmentation 5-5-4, basal segment long, penultimate segments bilobed and spongy beneath; claws simple. As far as is known, the larvae feed in moist, rotting wood, especially driftwood, so that Oedemeridae are most common on the coast, particularly on islands. Adults are found on flowers, and are attracted to light at night. *Nacerdes melanura* L. is cosmopolitan, and breeds in wharves and old ships' timbers. Some species (e.g. *Sessinia* sp.) are known to be toxic and, from the symptoms produced, appear to contain cantharidin. *Lagrioida australis* Champ. is an aberrant form, with slightly clubbed antennae, found on sand beaches in eastern Australia.

94. Meloidae (blister beetles; Plate 6, C; Fig. 30.95). Of moderate size (7–18 mm), with rather soft integument; head and pronotum reddish yellow or black; elytra reddish yellow, or bluish or greenish black. Head large, constricted to a narrow 'neck', strongly deflexed over the large, elongate fore coxae; antennae filiform or moniliform; maxillae with galeae broad and ciliate, adapted for sucking; palpi slender; pronotum widest at base, but narrower than base of elytra, without lateral longitudinal marginal carinae; fore coxal cavities confluent and widely open behind; mesepimera reaching mid coxal cavities; mid coxae elongate and contiguous; legs long, tarsi 5-5-4-segmented, claws pectinate beneath and with a sclerotized blade-like process beneath each claw; elytra entire, rather soft, and often not well fitted together; wings present; abdomen with 5 or 6 free sternites.

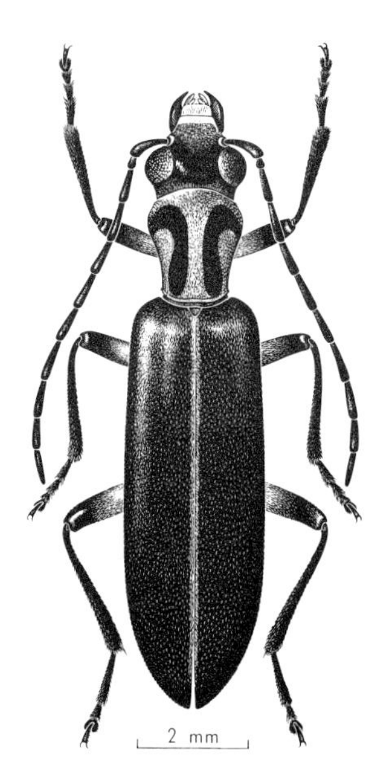

Fig. 30.94. *Copidita mira*, Oedemeridae. [F. Nanninga]

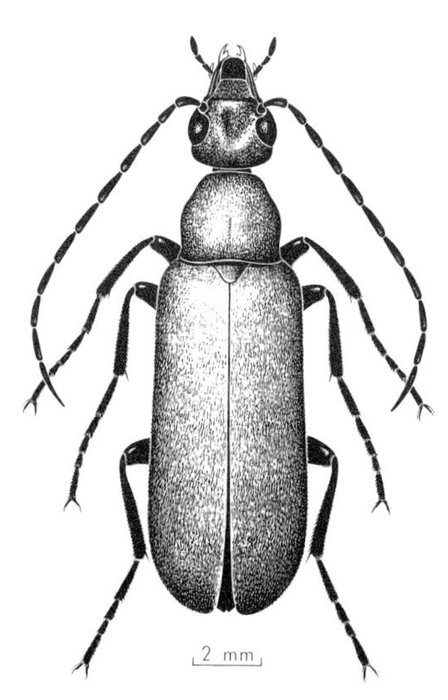

Fig. 30.95. *Zonitis lutea*, Meloidae. [F. Nanninga]

The life histories of Australian meloids are not yet known, and the following notes are based on observations in Europe and North America. The adults are phytophagous, whereas the larvae are parasitic or predacious upon the eggs, honey, and pollen stores of bees or the egg-pods of Orthoptera. Oviposition may be near or remote from the future host. When remote, the number of eggs laid is very large, and the first-stage larvae ascend plants to wait in flowers for the visits of bees, which provide transport to the nest. Meloidae resemble Rhipiphoridae in exhibiting larval heteromorphosis. Six larval stages have been distinguished (Clausen, 1940), namely: triungulin (1st instar), caraboid (2nd instar), scarabaeoid (3rd and 4th instars), coarctate (5th instar), scolytoid (6th instar). The triungulin is a planidium in which the single tarsal claws are flanked by 2 strong setae. The 'coarctate' larva has non-functional mouth-parts, and is the stage in which the meloid passes the winter. The 'scolytoid' stage is a reversion to a feeding, modified scarabaeoid form. Meloids contain cantharidin, a vesicating substance which is highly toxic. Meloidae are represented in all States, the majority of species being found in Queensland. The largest Australian genus is *Zonitis*. [Blackburn, 1899.]

95. Anthicidae (Plate 5, D; Fig. 30.96). Small to moderate (1·5–12 mm); slender; body pubescent, brown, black, or red, elytra often with a pattern of bands or spots. Head strongly deflexed and sharply constricted behind eyes; antennae filiform, moniliform, or thickened apically; maxillary palpi large, terminal segment securiform or pointed; eyes small, not or slightly emarginate; prothorax of about same size as head, subcylindrical or narrowed posteriorly, or waisted, sometimes sharply constricted near anterior end, without carinate or otherwise defined margins, sometimes with a broad, median, horn-like process which curves over the head *(Mecynotarsus)*; fore coxae elongate, projecting, the cavities confluent and usually open behind; mesepimera reaching mid coxal cavities; mesepisterna meeting in front of mesosternum; mid coxae ovoid, very close; hind coxae narrowly or widely separated; elytra entire, without epipleura; wings usually present; abdomen with 5 visible sternites, all free; tarsi 5-5-4-segmented, basal segments of mid and hind tarsi elongate, penultimate segment emarginate and lobed beneath, antepenultimate segment simple; claws simple. The subfamily PEDILINAE (e.g. *Egestria*) has been treated as a family by some authorities. It is distinguished

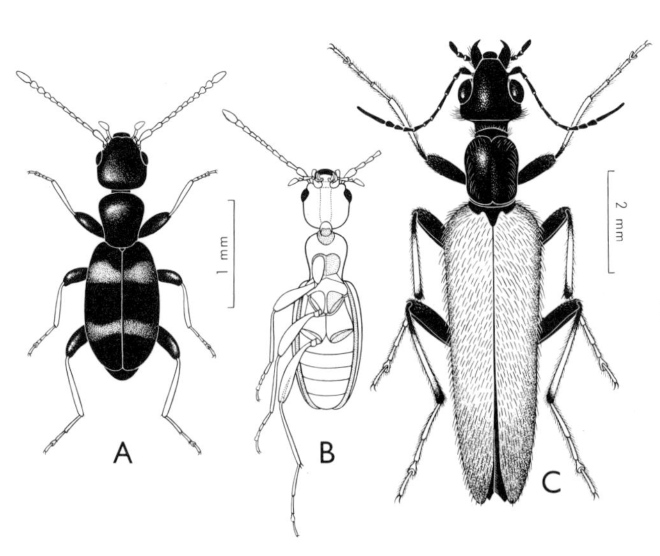

Fig. 30.96. Anthicidae: A, B, *Anthicus australis*; C, *Egestria taeniata*. [F. Nanninga]

by having eyes finely facetted and slightly emarginate in front, larger body size, and hind coxae almost contiguous.

The habitats of Anthicidae are varied. Some species are found in damp plant debris and on flowers, but others are common in arid areas, on sand dunes, at the edges of streams, in the burrows of waterside Staphylinidae (*Bledius* spp.), and in salt marshes. The biology of Australian species has not been studied. Males of some species of *Anthicus* and *Notoxus* are strongly attracted to meloid beetles, even to dead specimens. The attractive agent is the cantharidin contained in Meloidae, which is presumably the same as, or resembles, the sex-attractant substance of these Anthicidae. Many of the genera represented in Australia (e.g. *Anthicus*, *Tomoderus*, *Formicomus*) have world-wide distribution, but a few (e.g. *Lemodes*, *Trichananca*) are endemic. The species of *Lemodes* (Plate 5, D) are a beautiful bright velvety red, sometimes with metallic blue apex and white-tipped antennae. *Anthicus floralis* L., dark brown with the base of the elytra reddish, is cosmopolitan in stored foodstuffs, mainly wheat and dried fruit. [Lea, 1922.]

96. Aderidae (Hylophilidae, Xylophilidae, Euglenidae; Fig. 30.97). Very small (1·5–3 mm) beetles, resembling Anthicidae; body pubescent, yellowish to dark brown, elytra sometimes with paler or darker spots or bands. Head deflexed, constricted at base; eyes coarsely facetted, hairy; antennae filiform or slightly thickened towards apex; apical segment of maxillary palpi securiform; pronotum about as wide as head, narrower than base of elytra, without defined lateral margins; fore coxae projecting, the cavities open behind; mid coxae conical, mesepimera reaching mid coxal cavities; tibiae without spurs; tarsi 5-5-4-segmented, basal segment very long, antepenultimate segment lobed beneath, penultimate segment very small; claws simple; elytra entire; abdomen with 5 visible sternites, the 2 basal sternites immovably joined, the suture between them faint in comparison with sutures between other sternites. The larvae appear to live in rotten wood, but nothing more is known of their biology. The Australian genera are *Aderus*, *Syzeton*, *Syzetoninus*, and *Syzetonellus*. [Pic, 1902.]

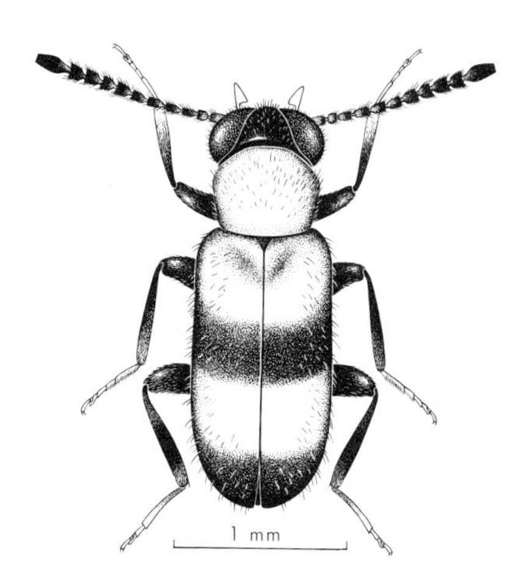

Fig. 30.97. *Syzeton abnormis*, Aderidae. [F. Nanninga]

Superfamily CHRYSOMELOIDEA (Phytophaga)

Tarsi pseudotetramerous, i.e. 5-5-5-segmented, but with penultimate segment very small and hidden in the emargination of the 3rd segment, which is usually deeply bilobed; head not produced into a rostrum, with 2 gular sutures; antennae not elbowed, sometimes very long, very rarely clubbed, never received in channels on the head; metasternum with a transverse suture; hind coxae without a steep declivity or concavity against, or into, which the femora can be retracted; abdomen with 5 visible sternites and 7 pairs of functional spiracles; wings usually with a spur on the *r–m* cross-vein; Malpighian tubes 6, cryptonephric; aedeagus without articulated parameres. Larva usually with legs and a 3-segmented antenna, and without a mandibular mola; habit phytophagous (wood, roots, stems, leaves, seeds). The Chrysomeloidea are most nearly related to the Curculonoidea and, less closely, to the Cucujoidea.

Key to the Families of Chrysomeloidea

1. Head not deeply engaged in prothorax, long behind eyes and downwardly curved, so that it covers the prosternum and fore coxae as viewed from below (Fig. 30.99); basal segment of hind tarsus longer than the other segments together; fore coxae contiguous; antennae thickened, not reaching middle of elytra; body stout, ovoid, compact, length 5 mm or less; elytra truncate (length: width *ca* 1·1 : 1), pygidium large and fully exposed; elytral striae regular, obvious; mentum supported on a pedunculate submentum .... **Bruchidae** (p. 610)
   Head usually deeply set in prothorax, not covering prosternum and fore coxae as viewed from below; basal segment of hind tarsus shorter than the other segments together; fore coxae usually separated; antennae often filiform, and often extending beyond middle of elytra; body usually more elongate; elytra not truncate (length : width more than 1·15 : 1); pygidium rarely wholly exposed; elytral striae often absent; submentum not pedunculate 2
2. Antennae usually more than two-thirds as long as the body and usually inserted on frontal prominences; all tibiae with 2 spurs (sometimes almost obscured by pilosity); often more than 20 mm long (Figs. 30.98A, B) .... **Cerambycidae** (p. 608)
   Antennae rarely more than half as long as the body; if more, then not inserted on frontal prominences; tibiae with less than 2 spurs on one or more legs; not more than 20 mm long .... **Chrysomelidae** (p. 610)

97. Cerambycidae (Longicornia, longhorn beetles; Plates 4, G, J, K, R, 6, B, J, M, O; Fig. 30.98). Small to large (5–60 mm), usually elongate, subcylindrical or flattened, usually pubescent. Prognathous or hypognathous; antennae at least two-thirds the length of the body, often much longer, inserted on prominences in the emargination of the eyes, and capable of being directed backwards above and parallel to the body; palpi with terminal segments subcylindrical or fusiform; pronotum with sharp-edged lateral margins only in Prioninae; fore coxae transverse or globular, separated, the cavities open behind or narrowly closed; mesepimera not reaching mid coxal cavities; a pair of tibial spurs present on all legs; claws usually simple; elytra usually covering abdomen, rarely very short; wings usually present.

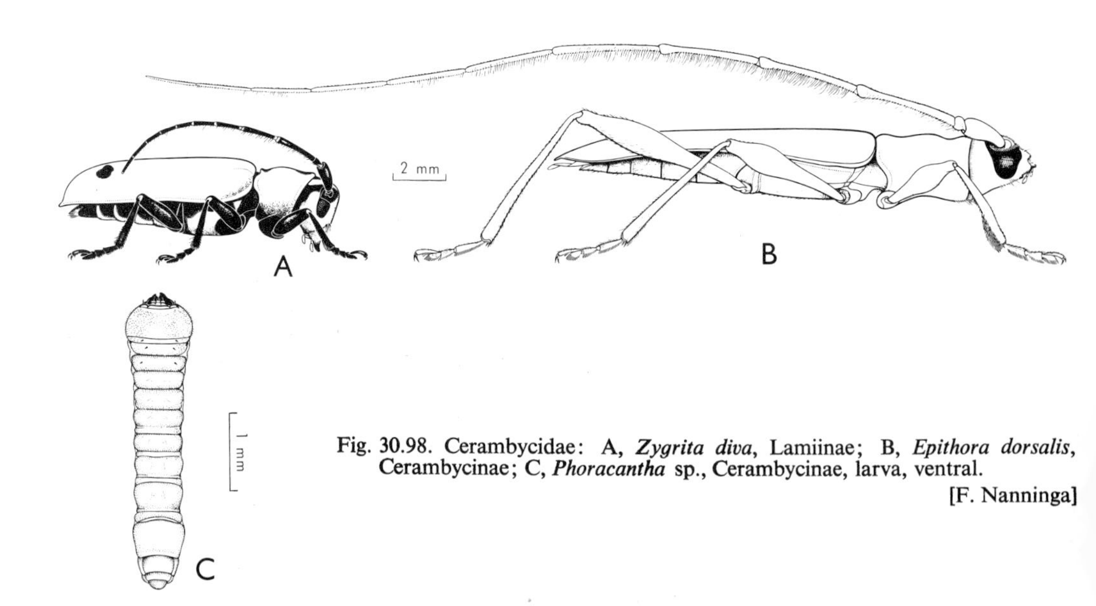

Fig. 30.98. Cerambycidae: A, *Zygrita diva*, Lamiinae; B, *Epithora dorsalis*, Cerambycinae; C, *Phoracantha* sp., Cerambycinae, larva, ventral. [F. Nanninga]

The eggs are laid in cracks in wood or bark or in holes made by the female, usually in dying or dead trees. The larvae bore into the wood, usually close to the bark. They are whitish, cylindrical or flattened grubs, with a small transverse head, and legs very small or absent (Fig. 30.98C). The larvae of Buprestidae, which are found in similar situations, are distinguished by having the thorax more flattened and more sharply and strongly expanded, and the spiracles cribriform, whereas those of Cerambycidae are bilabiate or annular. The larvae of some Cerambycidae live in the stems of herbaceous plants, and others feed in roots. Duffy (1963) has described the known larvae and pupae of Australian Cerambycidae. The colour of adults is very varied, and cryptic coloration is common. The family is divided into 3 subfamilies. They have been regarded as families, but it is now usual to treat them as subfamilies because they are linked by a number of intermediate forms. [McKeown, 1947.]

Key to the Subfamilies of Cerambycidae

1. Prothorax with sharp, well-defined lateral margins, at least towards base; fore coxae strongly transverse, not projecting, the cavities open or narrowly closed behind; mesonotum without stridulatory file; lacinia of maxilla absent or reduced. Larval mandibles with a simple cutting edge .... PRIONINAE
   Prothorax without defined lateral edges; fore coxae usually projecting and less transverse, the cavities closed behind; mesonotum with stridulatory file; lacinia of maxilla usually well developed. Larval mandibles with a gouge-like excavation 2
2. Head only partly deflexed, prognathous (Fig. 30.98B); hind legs obviously longer than mid legs .... CERAMBYCINAE
   Head strongly deflexed, subvertical, hypognathous (Fig. 30.98A); hind legs scarcely longer than mid legs .... LAMIINAE

The PRIONINAE include a number of large, dark brown species characterized by serrate lateral pronotal edges. The common *Eurynassa australis* Boisd. occurs in all States. *Paroplites australis* Erich., of similar size and appearance, feeds as a larva in the trunks of *Banksia* spp. *Rhipidocerus australasiae* Westw. (Plate 4, G), a relatively small species (20 mm) which occurs from N.S.W. to north Queensland, is dull green, and *Phaolus metallicum* Newm. is bright metallic blue. Adult Prioninae are nocturnal, and are commonly found under bark during the day.

The CERAMBYCINAE include many more species than the Prioninae. *Phoracantha semipunctata* F. has a yellow and black pattern on the elytra and a spine on each side of the prothorax, at the apex of each elytron, and on the antennal segments. It attacks *Eucalyptus* spp. in all States, and has been accidentally introduced into Israel. *Uracanthus triangularis* Hope (Plate 4, R) is another common species, long, slender, grey or pale brownish with a shining triangular brown patch on each elytron. Its larvae feed in *Acacia* stems. *U. cryptophagus* Oll., in northern N.S.W. and Queensland, has moved from its native host plant *Citrus australasica* to cultivated *Citrus*, and has caused serious damage in orange orchards. The most serious insect pest of structural timbers, *Hylotrupes bajulus* (L.), has been accidentally introduced on a number of occasions, but has so far been prevented from becoming established. This species, first known from Algeria, is established in Europe, South Africa, New Zealand, and Argentina. It favours the wood of *Pinus* spp., and is particularly destructive in the roof timbers of houses. The species of the *Hesthesis* group (Plate 6) have very short elytra which do not quite cover the metathorax, leaving the abdomen and longitudinally-folded wings exposed. They frequent flowers, and are striking mimics of wasps both in colour and motion. Further examples of probable mimicry are the species of *Sternaderus*, *Eroschema*, and *Pterostenus*, which bear a striking resemblance to Lycidae, and species of *Macrones*, which, with their slender waisted bodies and narrow, almost invisible elytra, look very like sphecid wasps.

Among the LAMIINAE, the dark-coloured, ground-living species of *Athemistus*, *Microtragus*, and *Ceraegidion* closely resemble amycterine weevils which are found in similar habitats. *Batocera boisduvali* Hope, from north Queensland, is one of the largest and most striking of all Australian Coleoptera. It is 50–60 mm in length, with antennae up to 130 mm, and dark grey to black, usually with a few yellow or white spots on the elytra. These beetles, and the related pearl-grey *Rosenbergia megacephala* v. d. Poll, are said to assemble to the sap of damaged native fig trees.

98. Bruchidae (Lariidae, Mylabridae; seed weevils; Fig. 30.99). Small (1·5–5·0 mm), ovoid, compact beetles, clothed with recumbent hairs, and usually with a pattern of spots. Head deflexed, covering prosternum and fore coxae, on a long neck; eyes very deeply emarginate in front; antennae 11-segmented, compressed, gradually thickened from 5th segment towards apex, moderately serrate, short, not reaching beyond shoulders of elytra; clypeus elongate; head sharply constricted below and at sides behind the eyes and submentum; mentum supported on a pedunculate submentum; palpi with terminal segments fusiform; pronotum narrowed in front, with lateral posterior angles obvious; prosternum reduced; fore coxae elongate, projecting, contiguous, cavities closed behind; mid coxae slightly separated, ovoid, cavities closed by the sterna; mesepimera indistinct; elytra short and broad, with very distinct regular longitudinal striae; pygidium large and fully exposed; tarsi with basal segment elongate; hind femora sometimes with a tooth on the posterior edge; claws together at base.

The larvae live in seeds, mainly those of Leguminosae and Palmaceae. The eggs are normally laid on the seeds or seed-pods, and each larva usually completes its development and pupates in a single seed. Bruchidae include some pests of crops and stored foodstuffs, and the mode of life has resulted in ready movement of the pests about the world. *Bruchus pisorum* L. is a common pest of growing peas. The adults feed on the pea plants, and the species does not infest dried peas. *Acanthoscelides obtectus* (Say), on the other hand, attacks peas and beans both dried and in the field.

99. Chrysomelidae (leaf beetles; Plates 4, Q, 5, C, F; Figs. 30.100, 101). Size small to moderate (1·5–22 mm); body form varied, usually rather robust; usually not pubescent; often brightly coloured, spotted, or metallic. Head prognathous or hypognathous, sometimes reflexed beneath; antennae filiform, moniliform, slightly serrate, or thickened apically but without a pronounced club, not extending past middle of body, not inserted in emarginations of eyes, not inserted on prominences, not capable of being directed

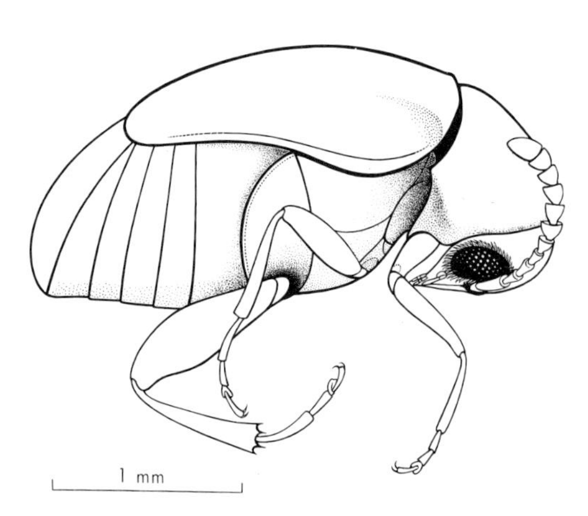

Fig. 30.99. *Bruchus lyndhurstensis*, Bruchidae. [F. Nanninga]

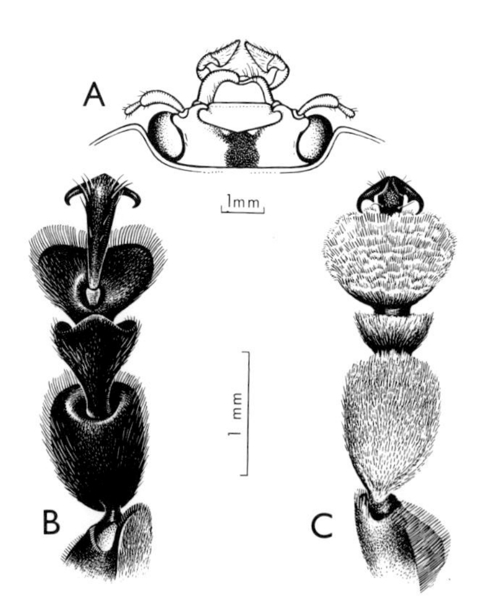

Fig. 30.100. *Sterromela trimaculata*, Chrysomelidae–Chrysomelinae: A, head; B, tarsus, dorsal; C, tarsus, ventral. [F. Nanninga]

backwards over body; eyes round or emarginate; pronotum with or without defined margins; fore coxae globular or transverse, projecting or not, cavities open or closed behind; mesepimera usually reaching mid coxal cavities; a pair of tibial spurs not present on all legs; hind femora sometimes dentate behind; elytra usually covering abdomen, rarely exposing part of the pygidium or abbreviated; wings usually present. About 16 subfamilies are recognized, of which 10 are represented in Australia.

Key to the Subfamilies of Chrysomelidae Known in Australia

1. Antennal insertions widely separated, distance between inner edges of sockets more than twice width of one socket (Fig. 30.100A) 2
   Antennal insertions close together, distance between inner edges of sockets less than twice width of one socket 6
2. Penultimate tarsal segments not bilobed (viewed from below, Fig. 30.100C); fore coxae transverse .... CHRYSOMELINAE
   Penultimate tarsal segments deeply bilobed (viewed from below); fore coxae transverse, projecting, or globular 3
3. Pygidium fully exposed; head deeply sunk into prothorax, the eyes partly covered, hypognathous, frons at right angles to axis of prothorax (Fig. 30.101A); abdominal sternites 3 and 4 much narrower in middle than at sides 4
   Pygidium not fully exposed; head sometimes exserted, the eyes free, prognathous, frons obliquely inclined to axis of prothorax; abdominal sternites 3 and 4 not, or only little, narrower in middle than at sides 5
4. Antennae thickened or serrate in distal half; fore coxae very close .... CLYTRINAE
   Antennae filiform; fore coxae widely separated .... CRYPTOCEPHALINAE
5. Lateral margins of prothorax absent or indistinct; head usually with cruciform grooves; prothorax much narrower than elytra; fore coxae transverse, projecting; hind femora sometimes inflated and with a strong tooth on the hind face .... SAGRINAE
   Lateral margins of prothorax obvious; head without cruciform grooves; prothorax nearly as wide as elytra (Fig. 30.101B); fore coxae usually globular, sometimes transversely ovoid; hind femur not much larger than mid femur, without a tooth behind .... EUMOLPINAE
6. Head covered by the explanate front edge of prothorax, facing downwards; elytra very broad with explanate margins, so that outline of pronotum and elytra is broadly elliptical (Fig. 30.101C) .... CASSIDINAE
   Head visible from above, frons usually facing forwards and upwards; outline of body not broadly elliptical 7
7. Head hypognathous, frontoclypeus and labrum not visible from above; head without cruciform grooves; body very slender or with strong spines or coarse sculpture (Figs. 30.101D, E) .... HISPINAE
   Head prognathous, frontoclypeus and labrum visible from above; head with or without cruciform grooves; body usually not slender, not with spines 8
8. Prothorax without defined lateral longitudinal edges; head with cruciform grooves (Fig. 30.101F) .... CRIOCERINAE
   Prothorax with sharp, defined, lateral longitudinal edges; head without cruciform grooves 9
9. Hind femora much more distended than mid femora; adapted for jumping (Fig. 30.101G) .... HALTICINAE
   Hind and mid femora both slender, adapted for walking (Fig. 30.101H) ...... GALERUCINAE

The CHRYSOMELINAE include the common *Paropsis*, with some hundreds of species, which are very difficult to identify as adults, although their larvae, which are gregarious on eucalypts and wattles, promise to be more easily recognizable. On abdominal segment 8 in *Paropsis* larvae there are paired eversible glands containing hydrocyanic acid, which are used defensively. Adults and larvae in general feed exposed on leaves. *P. atomaria* Oliv. is responsible for extensive damage to planted eucalypts in N.S.W. and Victoria, and *P. bimaculata* Erich. is a serious defoliator of *Eucalyptus regnans* in Tasmania. [Blackburn, 1897–1901.]

The EUMOLPINAE include the widely distributed genus *Rhyparida*. *R. morosa* Jac. feeds on grasses, and attacks sugar cane in Queensland. *Spilopyra sumptuosa* Baly (Plate 5, C) is a beautiful metallic green and purple eumolpine from northern N.S.W. The GALERUCINAE include a number of pests, e.g. *Monolepta australis* (Jac.), a small,

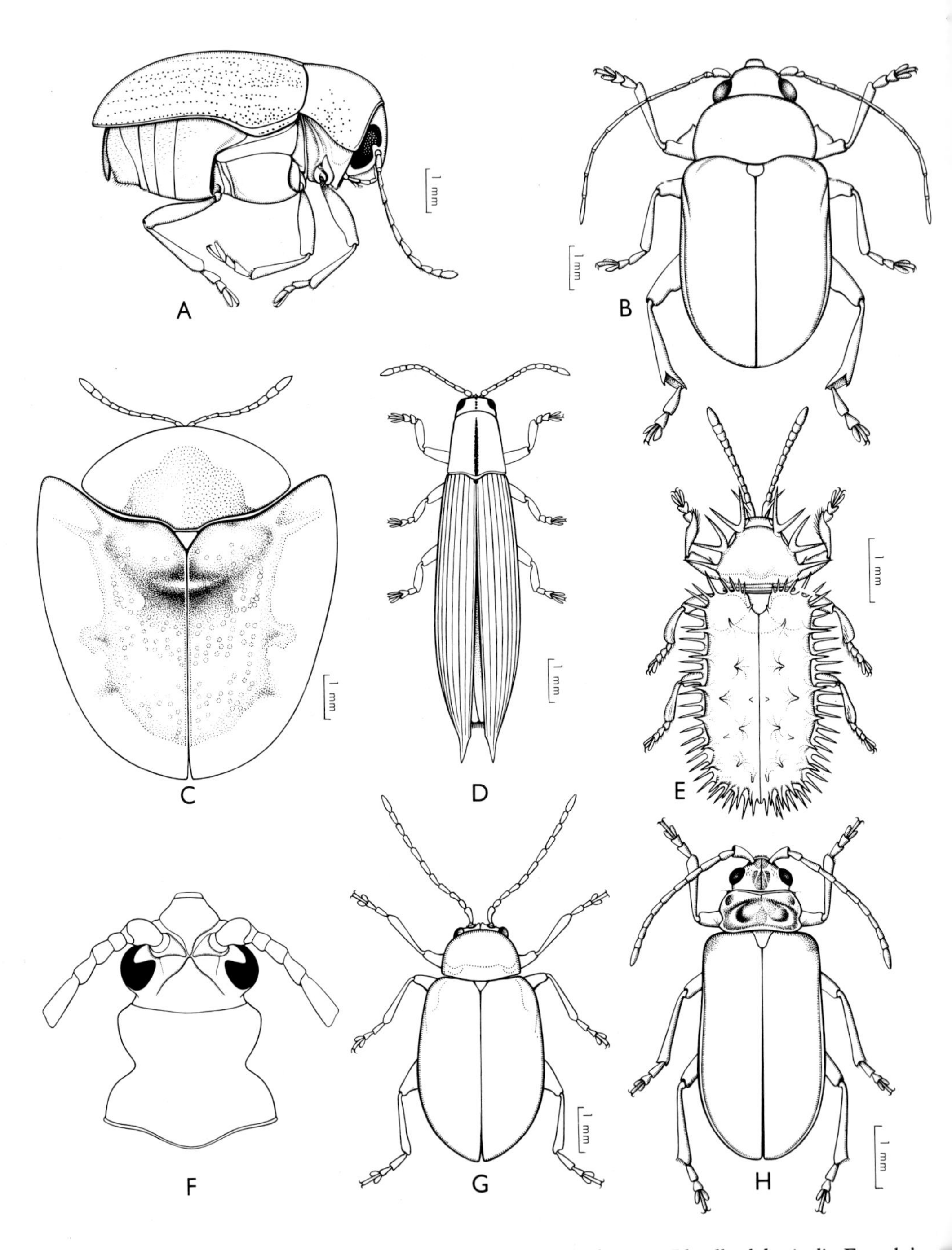

Fig. 30.101. Chrysomelidae: A, *Cryptocephalus carnifex*, Cryptocephalinae; B, *Edusella abdominalis*, Eumolpinae; C, *Meroscalcis selecta*, Cassidinae; D, *Eurispa howitti*, Hispinae; E, *Monochirus multispinosus*, Hispinae; F, *Stethopachys formosa*, Criocerinae; G, *Haltica pagana*, Halticinae; H, *Poneridia australis*, Galerucinae. [F. Nanning]

red-shouldered, pale yellow species which attacks citrus, maize, and stone fruits, and *Aulacophora hilaris* (Boisd.) and *A. abdominalis* (F.), the adults of which destroy the foliage and flowers of cucumbers, melons, and pumpkins, the larvae tunnelling in the stems. *Rupilia ruficollis* Clark (Plate 5, F) from the N.T., is unusual in that it has short elytra.

The HALTICINAE (flea beetles) attack a variety of plants and trees, causing characteristic 'shot-hole' perforations in the leaves. They are small, shining, metallic beetles, noteworthy for their jumping ability. Among the relatively few SAGRINAE are some unusually large species, e.g. the metallic green *Sagra papuana* Jac. (Plate 4, Q) and the dark brown *Megamerus kingi* Macl., both about 20 mm long, from north Queensland. Some sagrines have inflated hind femora, but they are, nevertheless, unable to jump. As far as is known, the larvae live in stem-galls. HISPINAE are characteristically slender beetles whose larvae are leaf-miners or feed inside hollow stems. The relatively few species are mainly tropical. Some are conspicuously spiny, e.g. *Monochirus multispinosus* Germ., a black beetle found on grass in N.S.W.

CASSIDINAE, the 'tortoise beetles', have strongly explanate margins and an almost circular outline. Most of the species are included in *Aspidomorpha*. Cassidines are mainly tropical and occur as far south as northern N.S.W. The larvae are leaf-feeders, and carry their cast skins or faeces over their backs with the aid of a long forked caudal process. The larvae of CRYPTOCEPHALINAE and CLYTRINAE live in close-fitting portable cases built of faecal material. The larva can withdraw into the case, sealing the aperture with its head. *Cryptocephalus*, *Cadmus*, *Ditropidus*, and *Loxopleurus* have numerous Australian species. The CRIOCERINAE are a small subfamily with few Australian species. The adults are notable for their powerful stridulation, produced by rubbing the apices of the elytra over 2 parallel files of minute striae on the last tergite. Some of the larvae conceal themselves under an accumulation of faecal matter. Two Australian species (*Microdonacia*) have been described as Donaciinae, but were subsequently shown to be Eumolpinae (Monrós, 1958). Donaciinae are the only aquatic Chrysomelidae (p. 112), and are almost entirely confined to temperate parts of the northern hemisphere.

Superfamily CURCULIONOIDEA (Rhynchophora)

Tarsi pseudotetramerous, i.e. 5-5-5-segmented, but with penultimate segment very small and hidden in emargination of 3rd segment, which is deeply bilobed; head produced forwards into a rostrum; antennae usually elbowed and usually with a short 1- to 3-segmented club, the (usually) long scape usually received into a channel on the side of the head, which usually has a single median gular suture; prothorax almost always without defined lateral margins; metasternum with a transverse suture near posterior edge; hind coxae without a declivity or concavity against which the femur can be retracted; abdomen with 5 visible sternites and 7 pairs of spiracles; wing venation without a spur on the *r–m* cross-vein; Malpighian tubes 6, cryptonephric; aedeagus with parameres not articulated; body usually very strongly sclerotized and often clothed with scales. Larva almost always without legs; antennae 1- or 2-segmented; urogomphi absent; phytophagous.

Key to the Families of Curculionoidea Known in Australia

1. Maxillary palpi normal, flexible; labrum distinct and separated (Fig. 30.102A) 2
   Maxillary palpi, where visible, short, rigid (Figs. 30.102B, C); labrum not distinct and free 3
2. Rostrum broad, flattened, short or moderately elongate (Fig. 30.103); gular sutures not visible; lateral longitudinal edges of prothorax distinct at least in basal half; mid coxal cavities closed outwardly by the broad junction of meso- and metasterna; first 4 visible abdominal sternites more or less rigidly connected .... **Anthribidae** (p. 614)

Rostrum long; gular sutures paired; prothorax without visible lateral longitudinal edges; mid coxal cavities not, or incompletely, closed by the sterna; all 5 visible abdominal sternites free .... **Nemonychidae** (p. 614)

3. Antennae elbowed or with scape longer than next 3 segments together, with a well-defined 1- to 4-segmented club .... **Curculionidae** (p. 616)

   Antennae not elbowed and with scape shorter than next 3 segments together, with or without a defined 3-segmented club 4

4. Antennae with a well-defined 3-segmented club or a very long apical segment; body not unusually slender; pronotum usually distinctly narrower than elytra 5

   Antennae without a defined 3-segmented club; body unusually long and slender; pronotum as wide as, or only slightly narrower than, elytra. [Rostrum almost straight] 6

5. Rostrum long and uniformly curved downwards; 2nd tarsal segments without acute distal angles; elytra usually dome-shaped .... **Apionidae** (p. 616)

   Rostrum broad and flattened, expanded at apex, or long and slender; 2nd tarsal segment with acute distal angles which embrace 3rd segment; elytra not unusually convex .... **Attelabidae** (p. 615)

6. Antennae gradually thickened towards apex; abdomen with 5 free sternites; gular sutures, if visible, short and paired; elytra with an inner flange parallel with the outer edge (raise the elytron to view) .... **Belidae** (p. 615)

   Antennae not thickened towards apex; abdomen with at least the 2 basal segments connate; head with a single median gular suture; elytra without an inner flange .... **Brenthidae** (p. 615)

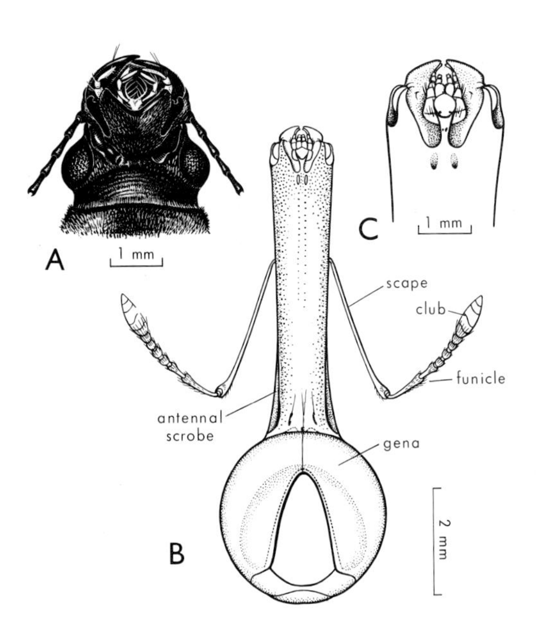

Fig. 30.102. Heads, ventral: A, *Dendrotrogus colligens*, Anthribidae; B, C, *Orthorrhinus cylindrirostris*, Curculionidae. [F. Nanninga]

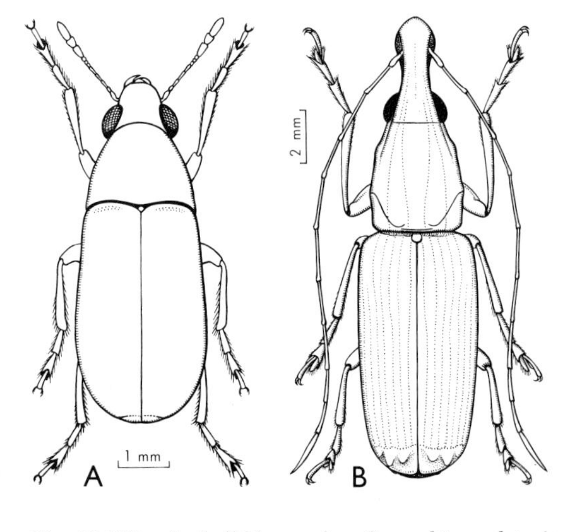

Fig. 30.103. Anthribidae: A, *Caccorhinus lateripunctus*; B, *Ancylotropis waterhousei*. [F. Nanninga]

100. Nemonychidae. Small beetles (*ca* 3 mm), with elongate rostrum, distinct labrum, and paired gular sutures; maxillae with lacinia present; maxillary palpi flexible; prothorax without trace of lateral margins; mid coxal cavities not completely closed by the sterna; abdomen with all 5 sternites freely movable. *Cimberis* (= *Rhinomacer*) *australiae* Lea, from north Queensland, is the only Australian species, but, as the genus is otherwise known only from the northern hemisphere, its generic placing should be verified.

101. Anthribidae (Figs. 30.102A, 103). Beetles 2–15 mm long, with a broad, short or elongate, flattened rostrum; antennae not elbowed, scape shorter than pedicel, short with a 3-segmented club, or elongate and without a club, or with only a slightly marked club; labrum visible; maxillae with lacinia present; maxillary palpi slender, 4-segmented, flexible; mentum large, pedunculate, ligula large, sclerotized; gular sutures not visible; prothorax with lateral margins present on basal half; fore coxal cavities contiguous, the coxae globular; mid coxal cavities broadly closed by the sterna; abdomen with 4 sternites connate and only the apical sternite movable; elytra with an inner flange below the shoulder; pygidium largely exposed (viewed from behind); tarsi with 3rd segment deeply bilobed and embraced by the prominent angles of the 2nd; claws sometimes toothed beneath. Larvae curved, fleshy, widest in middle; head usually exserted, hypognathous; antennae 1-segmented or absent; legs absent or, if present, with reduced segments. The larvae feed in dead wood, fungi, and galls, but almost nothing is known of the habits of Australian species. *Doticus pestilans* Oll. is unusual in that the larva develops in dried apples. *Araecerus fasciculatus* De G. is a pest of coffee and cocoa beans and of spices, especially nutmeg. It has been carried to all parts of the world, but is established only in warmer climates. [Blackburn, 1900.]

102. Belidae (Fig. 30.104A). Beetles 9–20 mm long; body long and slender; rostrum of moderate length, straight; maxillae with palps rigid and lacinia absent; labrum not distinct; 2 short gular sutures present; antennae thickened apically, but without a defined club, inserted between one-third and half the length of the rostrum from its base; prothorax without defined lateral margins; fore coxae large, projecting, contiguous; all abdominal sternites free; elytra unusually elongate, usually little wider than prothorax, with an inner flange parallel to the outer edge. As far as is known the larvae tunnel in the branches of wattles. The adults fly actively in sunlight. Species of *Rhinotia* (Plate 6, D) are orange and black, and appear to mimic the lycid genus *Metriorrhynchus*. The family is largely Australian, a few species only being native to Chile, New Zealand, and New Guinea. *Belus* (*ca* 100 spp.) is the largest genus, and is represented in all parts of Australia. [Lea and Bovie, 1909.]

103. Attelabidae (Fig. 30.104B). Length 1·5–7 mm; rostrum broad, flattened, expanded apically, or long and slender; maxillary palpi rigid; labrum not distinct; antennae not elbowed, with a well-defined 3-segmented club; prothorax conical, without trace of lateral margins, narrower than elytra; fore coxal cavities confluent; fore coxae large, strongly projecting, fore legs longer than the other legs; 2nd tarsal segments with acute apical angles which embrace the base of the 3rd; claws free or fused at base; abdomen with only the apical sternite or with all sternites movable; elytra sometimes with a scutellar striole. The large genera *Euops* (ATTELABINAE) and *Auletobius* (RHYNCHITINAE) occur in Australia. Attelabinae have the tarsal claws fused at the base, and Rhynchitinae have the mandibles dentate on the outer side. The habits of the Australian species are unknown, but elsewhere adult Attelabidae provide for their larvae by making cuts in the leaf on which the egg is laid, so that the leaf forms a roll. In Rhynchitinae the egg is laid in a terminal shoot, which is then ringed, so that it withers and falls to provide food for the larva.

104. Brenthidae (Fig. 30.104C). Elongate, parallel-sided beetles 4–40 mm long; head, including rostrum, elongate and slender, directed forwards, about as long as prothorax; antennae not elbowed, usually short, not thickened towards apex, inserted usually towards middle of rostrum; maxillae without lacinia, with rigid palpi; labrum not distinct; a single median gular suture present; prothorax elongate, conical, without lateral margins; fore coxae circular, slightly projecting, contiguous; metathorax elongate, about as long as abdomen; 2 basal abdominal sternites rigidly connected. The body size is variable within species, and pronounced sexual dimorphism is common. The larvae

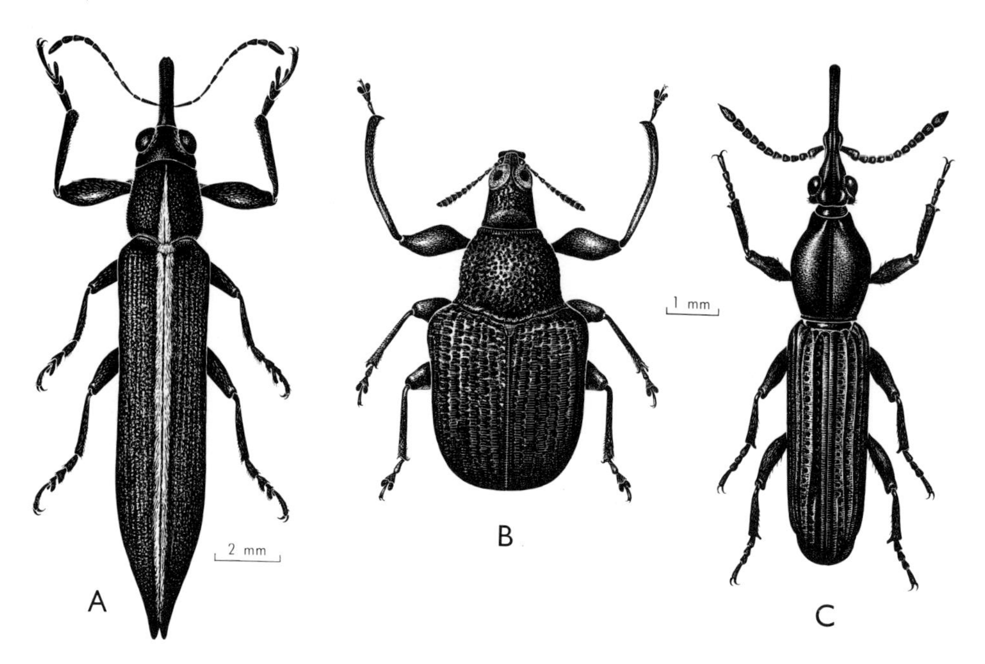

Fig. 30.104. A, *Belus suturalis*, Belidae; B, *Euops falcata*, Attelabidae; C, *Euschizus internatus*, Brenthidae. [F. Nanninga]

are unusual among Curculionoidea in having small thoracic legs. As far as is known they feed in decaying wood. The family as a whole has a tropical distribution, but in Australia extends south to Victoria.

105. Apionidae (Fig. 30.105). Beetles of length 1·5–20 mm, with a long, curved, slender to rather stout rostrum; maxillary palpi rigid, labrum not distinct; antennae not elbowed, scape shorter than next 3 segments together, club 1- to 3-segmented; prothorax without lateral margins, subcylindrical; fore coxal cavities confluent; fore coxae projecting; hind coxal cavities closed by the sterna; abdomen with 2 basal sternites enlarged, fused; elytra inflated, highly convex; tarsi with 3rd segment not embraced by 2nd; claws simple or appendiculate. Australian representatives include some fairly large black weevils of the genus *Eurhynchus* from the south-east, about 50 species of *Apion* (1·5–4 mm long), and the red and black *Cylas formicarius* F., an introduced Indian species which is a pest of sweet potato in eastern Australia.

106. Curculionidae (weevils; Plates 4, I, 5, A; Figs. 30.106–112). Length 1–50 mm; head usually produced into a rostrum at the end of which are the mouth-parts; antennae with a marked, compact, 3- to 4-segmented club, elbowed, scape enlarged and at least as long as next 3 segments; antennae

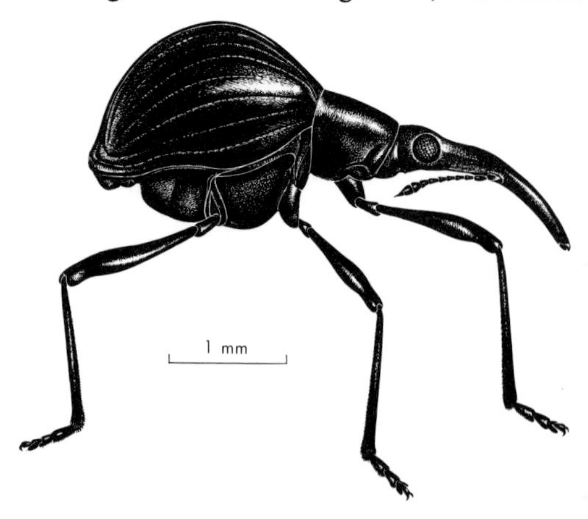

Fig. 30.105. *Apion convexipenne*, Apionidae. [F. Nanninga]

inserted on rostrum in front of eyes, the rostrum usually having an elongate channel for reception (in part) of the scape; labrum absent; mandibles short, powerful, and strongly articulated above and below; maxillary palpi short, 3-segmented, rigid, often concealed beneath mentum; maxillae without a distinct lacinia; head with a single median gular suture; tarsi usually apparently 4-4-4-segmented, the true 4th segment being usually minute in the emargination of the bilobed 3rd, but sometimes (in Platypodinae) with the 3rd not bilobed and the 4th small but clearly visible, rarely with the claw segment absent and the 3rd segment not bilobed; fore coxae contiguous, mid coxae slightly and hind coxae widely separated; abdomen with 5 visible sternites, the first 2 connate; elytra usually entire, but sometimes exposing the pygidium; body very strongly sclerotized, often clothed with scales. Larvae C-shaped, without legs, rarely feeding in an exposed situation; antennae small, 1-segmented. Pupae almost always in the ground or within the food plant.

The classification of the Curculionidae, by far the largest family in the animal kingdom, is in a most unsatisfactory state, and has shown no real improvement since Lacordaire, a hundred years ago. About 100 subfamilies and more than 200 tribes have been proposed, but the evolutionary success of the family has been such that many linking forms survive, making definition difficult. The great size of the family has deterred any comprehensive study, and there is now no more pressing need in the study of Coleoptera than a comprehensive classification of the subfamilies and tribes of Curculionidae.

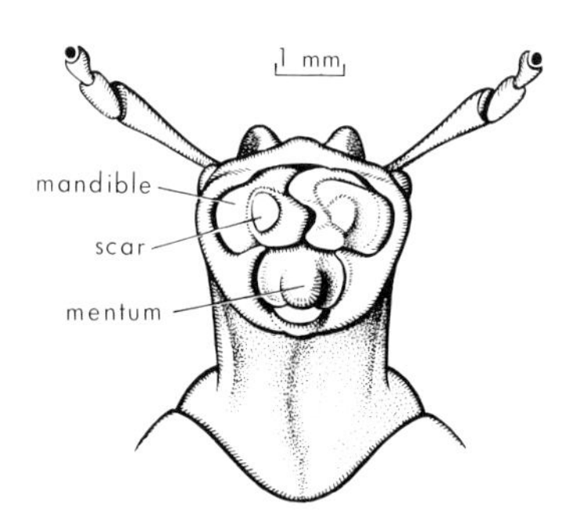

Fig. 30.106. Head of *Leptopius* sp., Curculionidae, ventral. [F. Nanninga]

Key to the Major Subfamilies of Curculionidae Known in Australia

1. Maxillae hidden by mentum (Fig. 30.106); mandibles with a conspicuous scar (Fig. 30.106) on outer surface (left by detachment of a tooth used by adult to escape from pupal cell); rostrum usually short and thick; fore tibiae never with a stout, apical, hook-like process. Larval antenna not projecting; larvae living free in the soil, feeding on roots. (Adelognatha) 2

   Maxillae not completely hidden by mentum; mandibles without a scar; rostrum usually long and slender; fore tibiae sometimes with a stout, apical, hook-like process. Larval antenna with a conical, projecting apex; larvae usually not living free in the soil, not feeding on roots. (Phanerognatha) 4

2(1). Anterior edge of prothorax in front of fore coxae markedly concave, so that the lateral edges appear broadly lobed behind and below the eyes; eyes elongate (Fig. 30.107A) .... LEPTOPIINAE

   Anterior edge of prothorax in front of fore coxae transverse, straight, not concave, so that the lateral edges do not appear lobed below and behind the eyes; eyes approximately circular 3

3(2). Scrobe narrow, elongate, sharply defined, on side of head, not visible from above .... BRACHYDERINAE

   Scrobe wider and less well defined behind, visible from above the head .... OTIORRHYNCHINAE

4(1). Tarsi with 3rd segment bilobed as seen from below (Fig. 30.107B), 4th minute, hidden from above between lobes of 3rd, or with only 3 segments, the last not bilobed, and the claw segment absent ... 5

   Tarsi with 3rd segment not bilobed as seen from below (Fig. 30.107C), at most with an apical cavity, visible from above, in which the small but visible 4th is inserted 20

5(4). Fore coxae contiguous, their cavities usually confluent 6

   Fore coxae separated 15

6(5). Pygidium exposed .... MAGDALINAE, TYCHIINAE, BALANINAE, CIONINAE
Abdomen covered by elytra 7
7(6). Metasternum shorter than 1st abdominal sternite (in mid-line) 8
Metasternum longer than 1st abdominal sternite (in mid-line) 10
8(7). Rostrum very short, broader than long (Fig. 30.109); submentum not pedunculate; mandibles very large, the opposed edges meeting in a straight, vertical line; tarsal segments 1–3 of equal width, lobes of 3rd narrow, vertical (Fig. 30.107B); large (8–30 mm), tuberculate or rough-surfaced, ground-living weevils .... AMYCTERINAE
Rostrum more elongate, longer than broad; submentum pedunculate; mandibles not conspicuous, the opposed edges often toothed; tarsal segments 1–3 not of equal width, lobes of 3rd usually broadly expanded (Fig. 30.107D); length often less than 8 mm 9
9(8). Fore tarsi with 1st and 3rd segments wider than 2nd. [Small, grey-scaled weevils] .... TANYRRHYNCHINAE
Fore tarsi with 1st, 2nd, and 3rd segments of increasing width .... CYLINDRORRHININAE, MOLYTINAE
10(7). Rostrum short and stout, or flattened (Fig. 30.110A), rather quadrate in section 11
Rostrum more elongate, usually cylindrical, round in section 12
11(10). Metasternum about as long as 1st abdominal sternite .... GONIPTERINAE, ATERPINAE
Metasternum about as long as 1st and 2nd abdominal sternites together .... COSSONINAE (pt.)
12(10). Claws fused at their bases (Fig. 30.107E) .... CLEONINAE
Claws separate at their bases, or absent 13
13(12). Body long and slender; head straight, cylindrical, long behind the eyes; elytra fusiform, each terminating in a pointed process (Fig. 30.110B) .... RHADINOSOMINAE
Body not unduly long and slender; head not straight and cylindrical; elytra not terminating in pointed processes 14
14(13). Eyes almost meeting beneath head .... AMALACTINAE
Eyes not nearly meeting beneath head .... ERIRRHININAE (pt.), CURCULIONINAE, DIABATHRARIINAE
15(5). Prosternum and mesosternum with a deep, median, longitudinal cavity passing between the fore coxae for reception of the rostrum (Fig. 30.108) .... CRYPTORRHYNCHINAE
Prothorax without a median, ventral, longitudinal channel 16
16(15). Pygidium exposed 17
Abdomen completely covered 18
17(16). Mesepimeron visible from above in angle between shoulder of elytron and pronotum (Fig. 30.111B) .... CEUTHORRHYNCHINAE, BARIDINAE
Mesepimeron not visible from above .... LAEMOSACCINAE
18(16). Hind femora very strongly swollen .... RAMPHINAE, TRACHODINAE
Hind femora not unduly swollen 19
19(18). Tarsi with basal segment much longer than the other segments together .... ZYGOPINAE
Tarsi with basal segment not longer than the other segments together .... HAPLONYCHINAE, ALCIDINAE
20(4). Rostrum longer than broad 21
Rostrum very short, broader than long 24
21(20). The 2 basal abdominal sternites separated by a well-defined suture, length of metasternum equal to, or less than, their combined length ...... ERIRRHININAE (pt.)
The 2 basal abdominal sternites fused, the suture between them faint or absent in the middle, metasternum longer than their combined length 22
22(21). Pygidium exposed (Fig. 30.111A) .... RHYNCHOPHORINAE
Pygidium covered by the elytra 23
23(22). Eyes almost meeting beneath head; size large (more than 15 mm) ...... SIPALINAE
Eyes small, widely separated beneath head; size small (10 mm or less) .... COSSONINAE (pt.)
24(20). Tarsi with basal segment longer than the other segments together; eyes approximately circular (Fig. 30.112A) .... PLATYPODINAE
Tarsi with basal segment shorter than the other segments together; eyes dorso-ventrally elongate (Fig. 30.112B) .... SCOLYTINAE

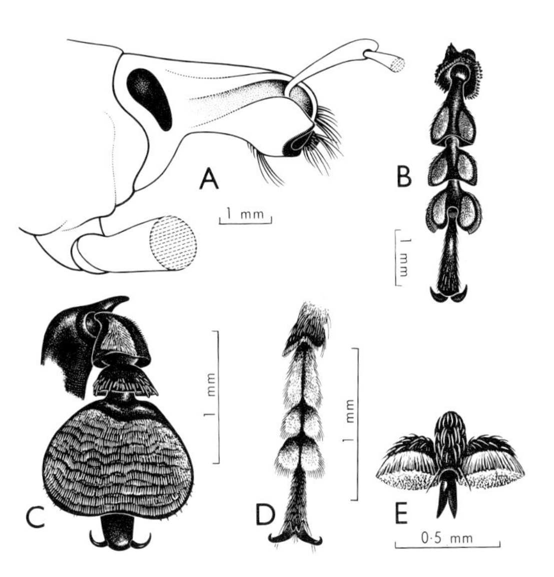

Fig. 30.107. Curculionidae: A, *Leptopius rhizophagus*, Leptopiinae, head and prothorax, lateral; B, *Talaurinus rugifer*, Amycterinae, tarsus, ventral; C, *Diathetes morio*, Rhynchophorinae, tarsus, ventral; D, *Methypora tibialis*, Molytinae, tarsus, ventral; E, *Lixus mastersi*, Cleoninae, end view of fore tarsus. [F. Nanninga]

The largest subfamilies, in numbers of species, are the Cryptorrhynchinae, Erirrhininae, Gonipterinae, Amycterinae, and Leptopiinae. CRYPTORRHYNCHINAE (1,000+ spp.) are typically rather long-legged weevils, with a long, downward and backwardly directed rostrum which can be received into an intercoxal channel. The larvae feed mainly in wood or seeds (Lea, 1898–1913). Among ERIRRHININAE, there are some genera (e.g. *Misophrice*) in which the terminal tarsal segments are absent. Another genus, *Bagous*, is of interest in that both adults and larvae live on plants under water. Among the GONIPTERINAE, *Gonipterus* and *Oxyops* have numerous species that feed on eucalypts (Lea, 1927). *G. scutellatus* Gyll., accidentally introduced from Australia, is a pest of planted eucalypts in South and East Africa, New Zealand, and South America. The AMYCTERINAE (Phalidurinae) contain nearly 500 species of large, tuberculate or spiny, ground-dwelling weevils, confined to Australia and New Zealand. The adults are found beneath stones and logs, but nothing

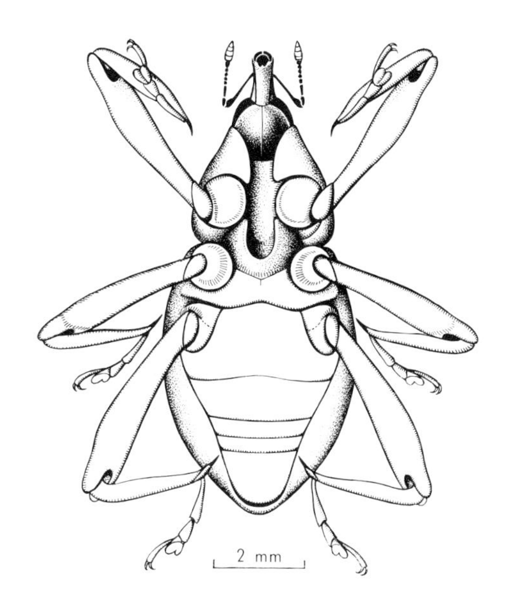

Fig. 30.108. *Paleticus subereus*, Cryptorrhynchinae, ventral. [F. Nanninga]

is known of their life histories. Important genera are *Phalidura*, *Talaurinus*, *Sclerorinus*, *Cubicorrhynchus*, and *Acantholophus* (Ferguson, 1909–23). In *Phalidura* the males have very large, projecting, terminal forceps. ATERPINAE include the common *Chrysolopus spectabilis* (F.) (Plate 5, A), the 'diamond beetle', first taken by Banks at Botany Bay, an insect of striking appearance, coloured by metallic blue or green scales. Its larvae feed in the wood of wattles. The MOLYTINAE include *Syagrius* and *Neosyagrius*, the larvae of which live in the stems of fern fronds. AMALACTINAE, some of which (e.g. *Tranes internatus* Pasc.) feed in the female cones of cycads *(Macrozamia)*, have the general form of Rhynchophorinae.

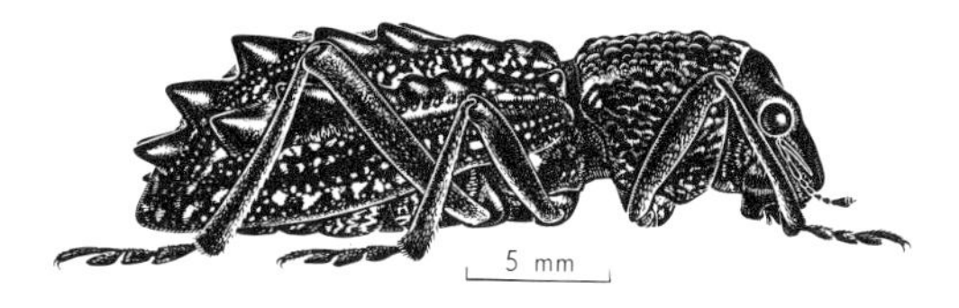

Fig. 30.109. *Macramycterus boisduvali*, Amycterinae, lateral. [F. Nanninga]

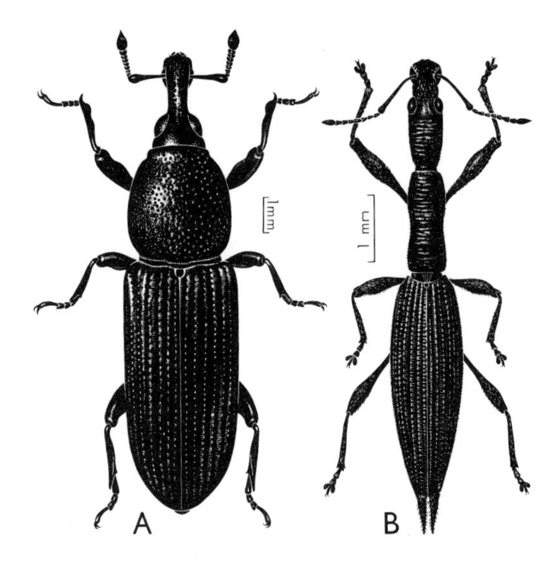

Fig. 30.110. A, *Cossonus simsoni*, Cossoninae; B, *Rhadinosomus lacordairei*, Rhadinosominae. [F. Nanninga]

The CURCULIONINAE (Hylobiinae) include *Eurhamphus fasciculatus* Shuck. (Plate 4, I), the largest Australian weevil, a rare species, the larva of which tunnels in the wood of hoop pine *(Araucaria Cunninghamii)*. The adult is 30–50 mm long, with a very long, straight rostrum and remarkable, erect, dark brown brushes on pronotum and elytra. *Orthorrhinus cylindrirostris* F. is probably the best known and most widely distributed Australian weevil. It is 10–20 mm long, with a long, straight, downwardly-directed rostrum, and is readily recognized by its disproportionately long fore legs and very broad expanded tarsi.

RHADINOSOMINAE (11 spp.) are an endemic subfamily of very characteristic, slender body form (Fig. 30.110B). The head and prothorax are elongate, subcylindrical, and the slender fusiform elytra end in sharp processes. *Rhadinosomus lacordairei* Pasc. is a pest of strawberries in Tasmania. LEPTOPIINAE contain more than 100 species of large, tuberculate, wingless weevils of the genus *Leptopius* (= *Leptops*). They are found in all parts of the continent, and usually feed on the roots and foliage of *Acacia*. The larva of *L. squalidus* Boh. damages the roots of apple trees. *Catasarcus* includes about 60 species of stout, ovoid, or spiny weevils almost entirely restricted to W.A.

The BRACHYDERINAE (Thylacitinae; 4,000 world spp.) are poorly represented. *Prosayleus* (7 spp.), common in the drier inland of N.S.W., is an example. A striking species is *Pantorhytes chrysomelas* Montr., a very hard black weevil with spots of pale green scales, found in north Queensland. *Maleuterpes phytolymus* Oll., 3 mm long, is a root-feeding pest of citrus, and *Graphognathus leucoloma* Boh., the 'white-fringed weevil', is a pest species, originating in Argentina and now recorded from hundreds of host plants. OTIORRHYNCHINAE are represented in Queensland by many small species of *Myllocerus*, distinguished by their clothing of metallic blue or greenish scales. *Otiorrhynchus sulcatus* F., *O. cribricollis* Gyll., and *O. scabrosus* Marsham are introductions from Europe which attack fruit trees and strawberries. RHYNCHOPHORINAE (Calandrinae) are poorly represented, but include 2 large

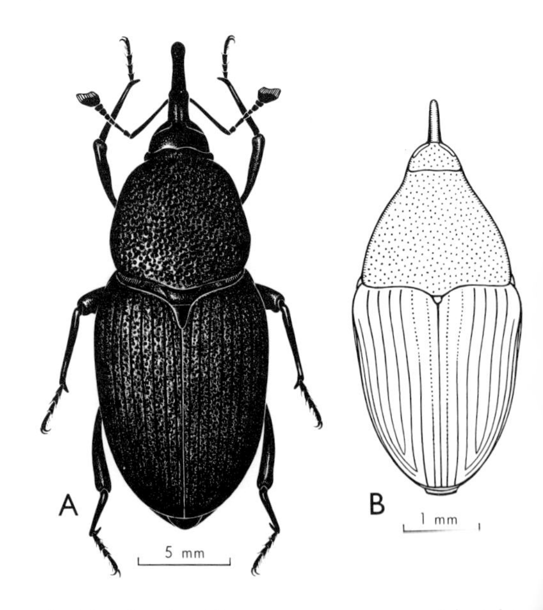

Fig. 30.111. A, *Trigonotarsus rugosus*, Rhynchophorinae; B, *Myctides barbatus*, Baridinae. [F. Nanninga]

species, *Trigonotarsus rugosus* Boisd. (Fig. 30.111A) from N.S.W. and *Iphthimorhinus australasiae* Roelofs from north Queensland, both about 35 mm long. *Sitophilus (= Calandra) granarius* L. and *S. oryzae* L., are widespread pests of grain, notably wheat and rice, and *Cosmopolites sordidus* Germ. is a pest of banana plants in Queensland. The COSSONINAE are small, black, cylindrical weevils, that live mainly in dead wood. Australian cossonines are not numerous, but the subfamily is well represented on islands (e.g. Lord Howe and Norfolk Is.).

SCOLYTINAE and PLATYPODINAE have commonly been separated as families distinct from the Curculionidae. This appears to be based more on habit than on morphology, and a number of other curculionid groups could be separated with as much justification. Species of these subfamilies are commonly known as 'bark beetles'. They are mainly small, cylindrical insects which burrow in trees between the bark and wood. Some species bore into solid wood or roots or twigs and a few attack seed (e.g. of palms). The adults bear a superficial resemblance to Bostrychidae, but are distinguished by their large, compact antennal club. The eggs are deposited in tunnels made by the adults, and the pattern formed by the combined maternal and larval tunnels is often characteristic of the species. The bark beetles include the 'ambrosia beetles' (p. 512) which feed on a fungus grown in the tunnels.

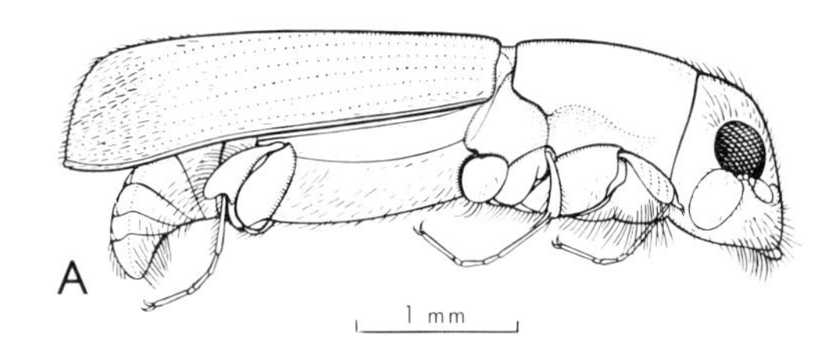

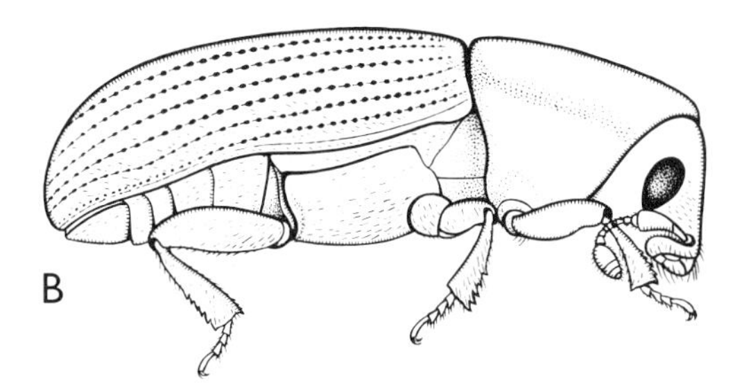

Fig. 30.112. A, *Platypus subgranosus*, Platypodinae; B, *Hylastes ater*, Scolytinae. [F. Nanninga]

ACKNOWLEDGMENT. I am indebted to Dr P. B. Carne, CSIRO, Canberra, for help with the section on the Scarabaeoidea.

31

STREPSIPTERA

by E. F. RIEK

Endopterygote Neoptera with reduced mandibulate mouth-parts, extreme development of the metathorax, reduced prothorax, and without differentiated trochanters. Adult males free-living, with functional hind wings and small elytriform fore wings. Females larviform, viviparous; usually parasitic, in a puparium, and then with secondary progoneate genital apertures. Heteromorphosis during larval growth, the first-instar larvae free-living and active, later instars parasitic.

This is a small order, comprising about 300 described species of remarkably specialized insects. The males (Figs. 31.1, 9–11) are small (1·5–4 mm long), usually black or brown in colour, and have distinctively shaped heads and flabellate antennae. The females of the primitive family Mengeidae (Fig. 31.2) leave their hosts and move about; all others (Figs. 31.3, 8) protrude from the integument of their hosts enclosed in their yellow to brownish puparia. Strepsiptera are known from Thysanura, Blattodea, Mantodea, Orthoptera, Hemiptera, Diptera, and aculeate Hymenoptera. Only 9 Australian species have been described (R. Perkins, 1905b; Lea, 1910; Ogloblin, 1923), but more than 90 are known today.

The Strepsiptera are unique among insects in the absence of a trochanter in the adult leg, and in the development of a brood canal in the puparium of most females. The reduced, elytriform fore wing and markedly expanded hind wing with radial venation in the males give them a considerable superficial resemblance to some rhipiphorid beetles, and the group has been included in the Coleoptera by some authors (see p. 518). However it differs basically from rhipidiine Rhipiphoridae in having the anal region of the wing very expanded, with 2 anal veins and fully developed jugal vein, and the scutellum and postscutellum of the metathorax both very large. The two groups have reached almost the same stage of progressive modification to the parasitic habit, and therefore a close relationship between them is not necessarily implied by their similarities.

Anatomy of Adult

Male (Fig. 31.1)

Head. Distinguished mainly by the large, berry-like eyes with large, well-separated facets. Antenna 4- to 7-segmented, with at least 3rd segment laterally flabellate; thickly

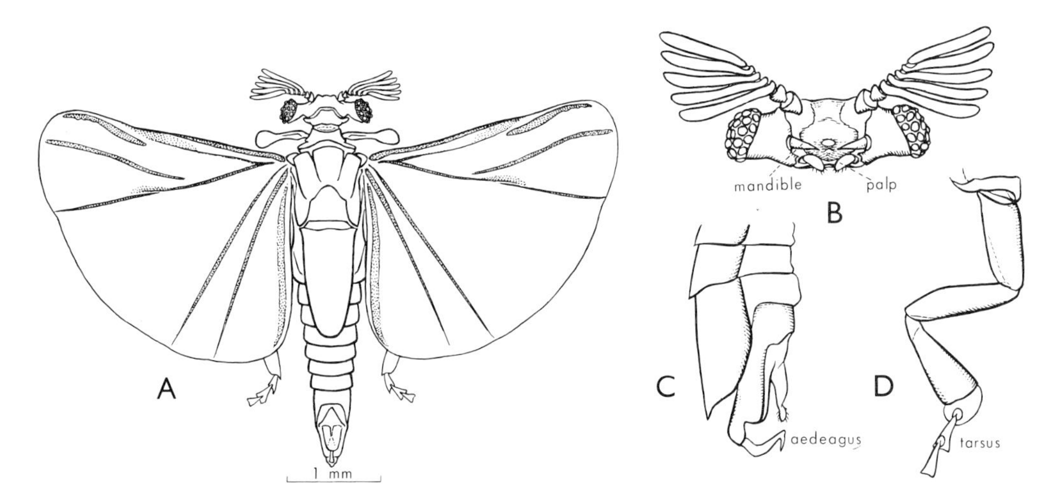

Fig. 31.1. Adult ♂ *Halictophagus* sp., Halictophagidae: A, dorsal; B, head, frontal; C, terminalia, lateral; D, fore leg. [R. Ewins]

covered with relatively large sensilla termed *sensoria*. Males, with few exceptions, have a large round sensorium near the base of the 4th antennal segment, except in Elenchinae, in which fusion has taken place between the 3rd and 4th segments and the sensorium is on the apparent 3rd. In the Mengeidae, *Mengea* and some species of *Mengenilla* possess both labium and labrum in a reduced form, but they are absent in all other species. Mandibles broad and scimitar-shaped, except in Corioxenidae in which they are absent, and in some Myrmecolacidae in which they are thread-like. Maxillae represented by 2-segmented palps; labial palps absent.

Thorax. Prothorax and mesothorax very reduced, but the metathorax is very large.

Legs. Remarkable for their long fore and mid coxae, absence of trochanters, and modified isomerous tarsi. The tarsi of Mengeidae and *Triozocera* are 5-segmented and terminate in 2 simple claws (minute in *Triozocera*). In all other species the tarsi are without claws, and the terminal segments are flattened and furnished with expanded lateral lobes that function as pulvilli. In *Triozocera* the 4th tarsal segment is bilobed and forms the functional termination of the tarsus, with the reduced 5th segment and claws held away from the substrate.

Wings. Much of the unique appearance of male Strepsiptera is due to the wings. The fore wings are reduced, elytriform, and without venation. They are somewhat similar to the halteres of Diptera, and may be analogous structures. The hind wings are remarkable for their reduced radial venation and their great size. There are two open negative folds between the cubital and anal fields with a positive fold in between, followed by a positive and then a negative fold in the anal field. Progressive reduction of the wing venation parallels that of the antennae and tarsi.

Abdomen. Ten-segmented, the basal segments partly concealed beneath the postscutellum and the first apparently fused with the metathorax. In Mengeidae, in which copulation takes place with a free-living female, the aedeagus is nearly straight. In other families it is almost invariably hooked at apex to enable the male to fertilize the partly endoparasitic female. It is straight, however, in *Elenchus* and *Corioxenos*.

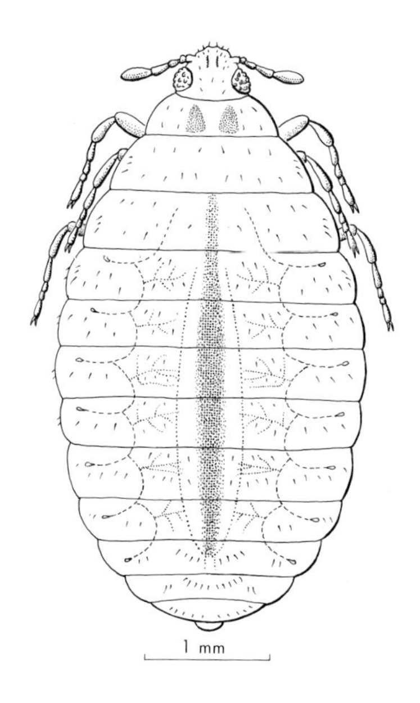

Fig. 31.2. Adult ♀ *Eoxenos laboulbenei*, Mengeidae (after Parker and Smith, 1933).

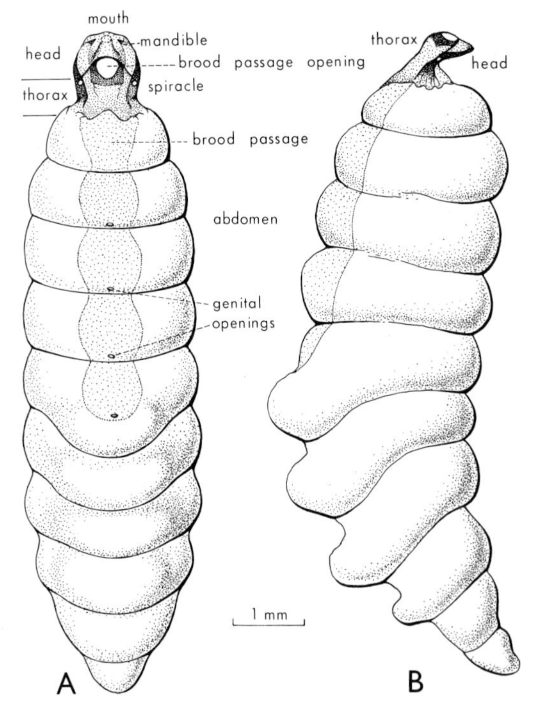

Fig. 31.3. Adult ♀ *Halictophagus* sp., Halictophagidae: A, ventral; B, lateral, dorsal surface to right. [R. Ewins]

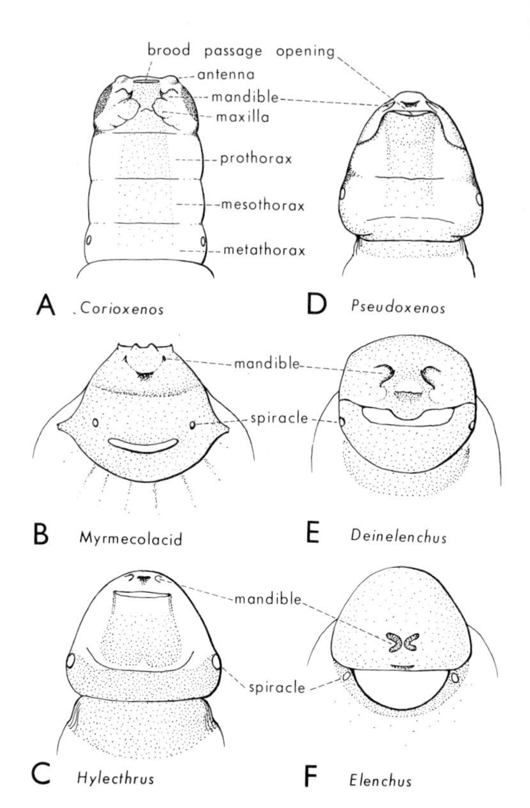

Fig. 31.4. Cephalothorax of ♀♀: A, Corioxenidae; B, myrmecolacid from gryllid; C, D, Stylopidae; E, F, Halictophagidae–Elenchinae. [R. Ewins]

Female (Figs. 31.2, 3)

Female mengeids are free-living, and, apart from the absence of a trochanter, resemble the larviform adults of certain Coleoptera, especially Rhipidiinae. All other female Strepsiptera are partly endoparasitic throughout adult life. Only the head and thorax, which are more or less fused to form a cephalothorax, protrude from the host. Although immovable, the head can be distinguished in most genera, and occasionally the three segments of the thorax are discernible (*Corioxenos* and some Stylopidae). In some genera, including *Hylecthrus* and *Halictoxenos*, there is complete fusion between head and thorax. This description of the external features actually refers to the puparium formed by the last larval cuticle within which the parasitic female is permanently pharate. The puparium, through the role it plays in respiration, reproduction, and food absorption, is an integral part of the parasite.

The head (Fig. 31.4) is relatively small in the cephalothorax of most female Stylopidae; in the Halictophagidae it is often the dominant part of the cephalothorax; and in *Elenchus* (Elenchinae) it greatly overshadows the reduced thorax. The main distinguishing features of the head are the mouth opening, a reduced pair of mandibles, and a pair of rudimentary palps.

The cephalothorax possesses one characteristic structure that is without parallel in the Insecta. A brood-passage is developed in the ventral space between the female and its puparium (Fig. 31.3). It has, typically, a slit-like opening between the head and thorax of the puparium, at which point the true female is reinforced with ridges. It is through this opening that insemination takes place, and the first-instar larvae emerge. At the junction of the cephalothorax and abdomen the integument of the puparium is often thickened and dark-coloured. The abdomen ends blindly, without genitalia, but an anus is usually visible. The secondary genital ducts occur on segments 2 to 5. They are normally single, median, tube-like invaginations with an open inner end. In Stylopidae the number of genital tubes varies from 2 to 5. There are 4 or fewer in Halictophaginae. The genital pores of *Stichotrema* are arranged in three transverse rows of 12 to 14 each. In *Deinelenchus* (Elenchinae) there are two rows of from 3 to 8 pores.

Internal Anatomy

The digestive tract is simple, and ends blindly in both sexes; there are no Malpighian tubes. The ventral nervous system is concentrated into an anterior ganglionic mass in the thorax, comprising all ganglia to the 2nd or 3rd abdominal, and connected by a median nerve cord to a posterior ganglionic centre in abdominal segment 3. The respiratory system opens by 1 or 2 thoracic and up to 7 abdominal spiracles in males; 7 abdominal and possibly 1 or 2 thoracic spiracles in female Mengeidae; and only a single

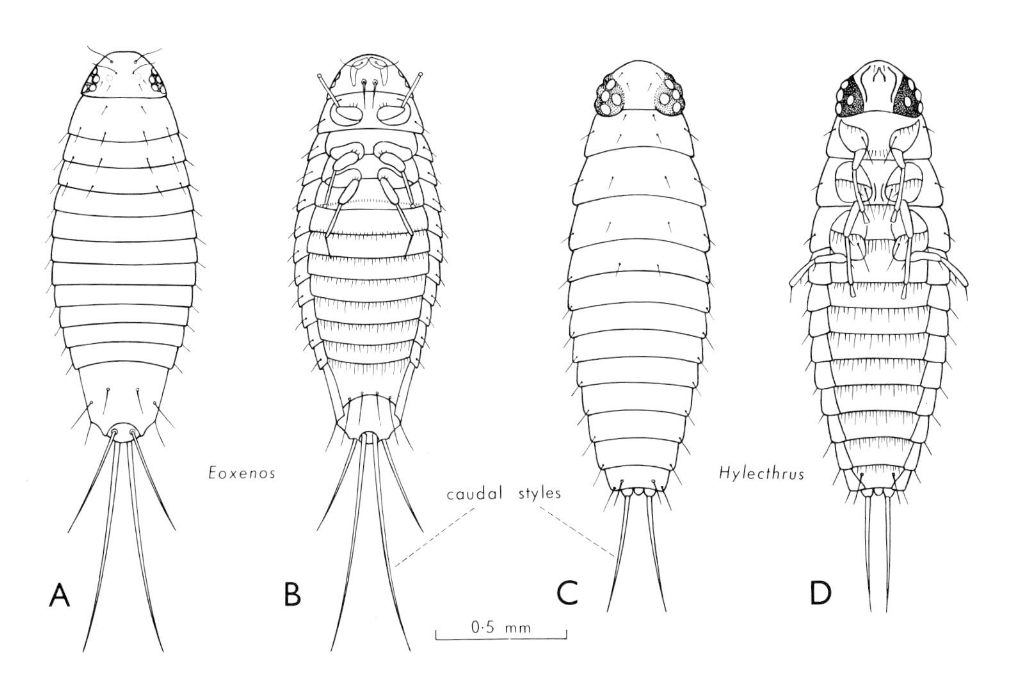

Fig. 31.5. Triungulinids, dorsal and ventral: A, B, Mengeidae; C, D, Stylopidae (after Bohart, 1941).

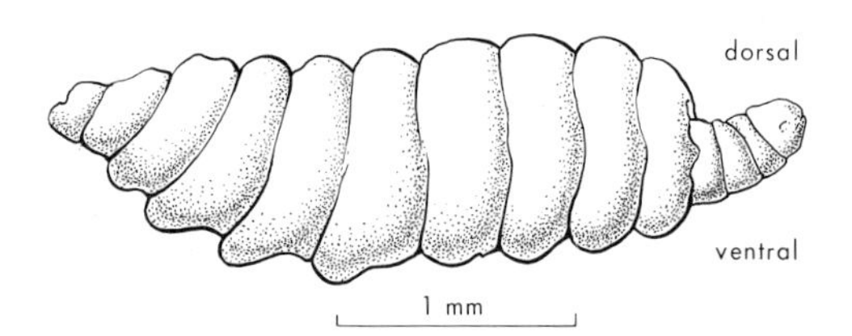

Fig. 31.6. Parasitic larval stage of *Halictophagus* sp. from pentatomid bug. [R. Ewins]

thoracic spiracle in other females (the second pair said to be present in the extra-Australian *Callipharixenos* are probably artifacts). The so-called metathoracic spiracles appear in the head close to the junction between head and prothorax in the second-instar (first parasitic instar) larva, and can be traced throughout larval development to their final metathoracic position. The female reproductive system consists of a pair of simple tubes, which break down before the adult stage, distributing the ova in the haemocoele. The secondary genital ducts are invaginations of the cuticle, and are independent of the puparial cuticle.

Immature Stages

Egg. Microlethical, reproduction being by haemocoelic viviparity (p. 83).

Larvae. The young are born as active forms, known in this order as *triungulinids* (Fig. 31.5). They are without well-formed antennae or mandibles, the maxillae are reduced, and all legs are without a trochanter. Otherwise there is little to distinguish them from various first-instar larvae of Coleoptera. The eyes are composed of 3–5 large, well-separated ocelli located in a pigmented area. The maxillary palps are apparently 3- or possibly 4-segmented. The abdomen ends in a pair of long medial styles and a shorter pair of lateral 'cerci'. The older, parasitic larvae (Fig. 31.6) are legless and grub-like.

Pupa. Adecticous and exarate in the male, the insect emerging by pushing off a cap at the anterior end of the puparium. The male cephalothecae (Fig. 31.7C) are not normally used for specific differentiation, but they preserve impressions of the mandibles, eyes,

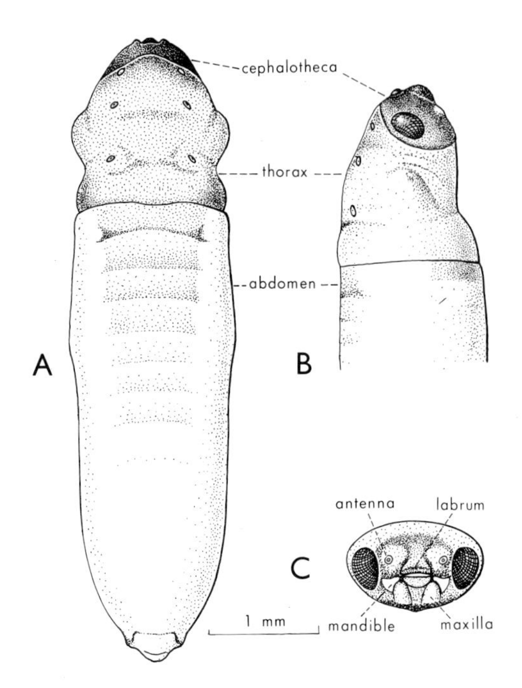

Fig. 31.7. Male puparium: A, dorsal; B, lateral; C, cephalotheca, frontal. [R. Ewins]

palps, and antennae which are specific in their shape. The pupal stage is suppressed in the female, except in Mengeidae.

Biology

The biology of few species of Strepsiptera has been studied in detail.

Males. The males of many species apparently emerge in the morning on bright days (some species of *Halictophagus*), but those of *Elenchus* and *Deinelenchus* have been observed to emerge in the late afternoon. *Mengenilla*, *Triozocera*, and species of the Myrmecolacidae are attracted to light, so they probably emerge in the late afternoon or evening, or possibly their adult life may extend over more than one day. Males fly actively, and are very conspicuous for their size, because the rather milky wings appear white during flight. They fly with the body almost vertical, though the tip of the abdomen is turned up and therefore appears horizontal.

Extrusion of the Parasite (Fig. 31.8). The

position where the parasite is extruded varies in different Hemiptera–Homoptera. The female *Halictophagus* is protruded on the ventral pleural regions between the segments, with the head pointing to the host's posterior end; whereas the male puparia are protruded dorsally below the wings, usually towards the median region of flat-bodied hosts and towards one side in hosts that are convex, as are most jassines. In *Elenchus* and *Deinelenchus*, which also parasitize small Homoptera, there is no definite method of extrusion, though there, too, females tend to the ventral margins and males to the dorsal. The

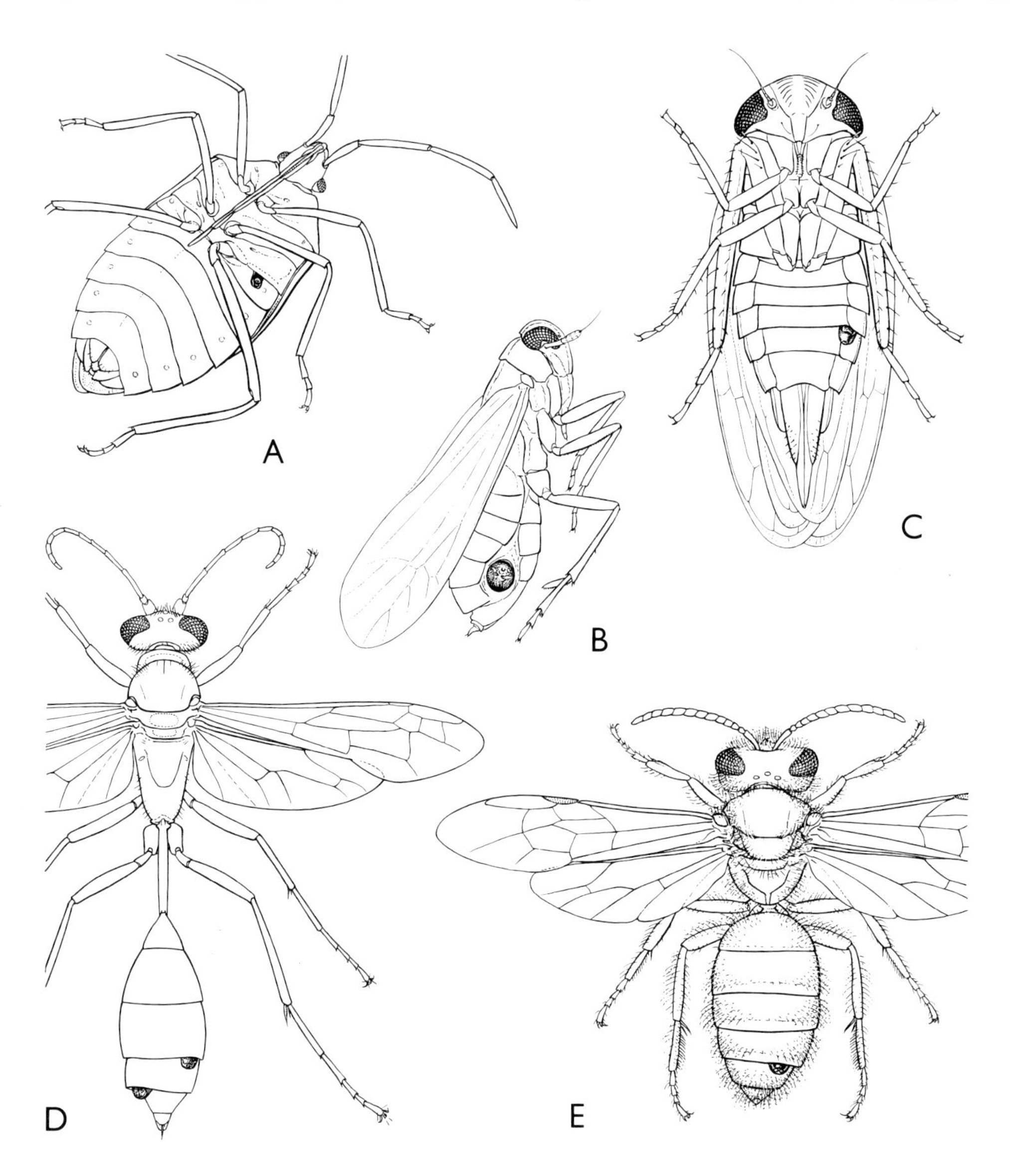

Fig. 31.8. A, *Halictophagus* sp., ♀ in pentatomid bug; B, *Elenchus* sp., ♂ puparium in delphacid bug; C, *Halictophagus* sp., ♀ in jassine bug; D, *Pseudoxenos* sp., ♂ puparium and ♀ in *Sceliphron laetum* (Sphecidae); E, *Hylecthrus* sp., ♀ in *Euryglossa* sp. (Colletidae). [M. Quick]

arrangement in delphacids and flatids is irregular, whereas in eurybrachyids the parasites follow the halictophagid type more closely. The number of parasites present in a host varies greatly. One specimen of the flatid, *Colgar peracuta*, had 7 male puparia and 2 female *Deinelenchus*, but normally only 1 or 2 parasites are present in the one host. Up to 5 female *Halictophagus* have occurred in one specimen of *Platybrachys*. In jassine hosts one normally finds only one parasite, or both a male and a female, but occasionally 2 females occur together.

In Hemiptera–Heteroptera the parasites are protruded from one of two positions. The females of *Halictophagus* are found ventrally and towards the lateral margin at the junction between thorax and abdomen. In the exotic *Corioxenos* the parasites are protruded dorsally between tergites 3 and 4 (apparent 1st and 2nd tergites), with the males often in a median position and the females laterally. There may be from 1 to 4 female *Halictophagus* in the one bug, though usually there are not more than 2. The host of the male is not recorded.

In *Sexava* and gryllids the female parasites are protruded in the soft pleural regions. There may be several in the one host.

In platystomatid Diptera the relatively large parasites are protruded in the soft pleural regions of the abdomen ventral to the large tergites. The male puparium causes great distortion of the host abdomen. There are 1 or 2 parasites. of mixed sex, per host.

In the Hymenoptera both the male and the female parasites are normally extruded dorsally, either in the median area, or more commonly to one side. Occasionally a female will be extruded ventrally, particularly when there is multiple parasitism. In the normal condition, the triungulinids emerge at the lip of the brood-passage opening with their ventral side uppermost, and must turn over to right themselves before they can spring away from the opening or walk on to the dorsal surface of the wasp. The male puparia also lie on their back. There is generally only one parasite per host, but it is not uncommon to find 2 or 3 in *Bembix*. Three female and 2 male parasites were present in one *Sceliphron*, and up to 12 male puparia have been observed in one *Polistes*. Extrusion of the parasite in eumenids, *Bembix*, and *Bembicinus* usually takes place between tergites 3 and 4 or 4 and 5 of the gaster, with females tending to extrude from a more distal segment than the males. In *Sphex* and *Sceliphron* extrusion of the parasites is more usually between tergites 5 and 6, less often between 4 and 5 or 3 and 4, and only rarely between 6 and 7 (in a male host).

Reproduction. Males are relatively strong fliers, and actively seek out the females that remain partly protruded from the parasitized host. Copulation takes place on the host. A virgin female is able to attract males for a long period. In *Halictophagus* sp. parasitizing *Poecilometis strigatus* (a pentatomid bug), this attraction is known to last for more than four weeks, and Kirkpatrick (1937) has recorded attraction over a period of 119 days in *Corioxenos*. In some male Strepsiptera the fore legs are greatly expanded as an aid in grasping the female during copulation. The apparent rarity of males in some species has given rise to the belief that parthenogenesis may occur. Polyembryony is recorded for at least one species.

Life History. The triungulinid finds its way out of the female through one of the genital tubes and along the brood-passage to its opening, which lies just outside the abdomen of the host. One female normally produces more than a thousand triungulinids. If the species is a parasite of the Homoptera the triungulinid must find an homopteran nymph to parasitize. If it is a parasite of the Hymenoptera, it must depend on an unparasitized adult of its host species to carry it back to her brood, where it can attack an early larval instar.

Triungulinids emerge in considerable numbers over a short period—at the rate of about 150 a day in one species of *Pseudoxenos* parasitizing *Sphex*. These triungulinids were about 0·4 mm long, and readily seen with the naked eye. They were quite active, swarming over the body of the host, and able to jump 2–3 cm at a time. They raised the abdomen and tucked the two inner caudal appendages under to form a spring for their take-off, which was generally preceded by raising the anterior portion of the body also, so that the fore and mid legs were free. Triungulinids have been observed to emerge from a female parasite at least a week after the death of the host wasp. Those of some species can live for at least 14 days.

The triungulinid apparently enters the body of the host by enzymatic action. It exudes a fluid from its mouth in considerable quantity, so that it soon becomes completely immersed in its own exudation. The exudate hardens on the outside in 3 to 4 minutes, while, at the same time, it apparently dissolves the cuticle of the wasp larva. The triungulinid is quite active while immersed in the exudate, contracting and expanding its body as though in an effort to break through the softened cuticle of its future host. Once it has penetrated into the host tissues, it wanders aimlessly until it sinks too deeply to be observed. The triungulinid inside the host appears to be considerably enlarged, with the cuticle between its abdominal segments stretched and appearing pale between the bands of normal darkened cuticle.

The triungulinid soon moults, the cast cuticle persisting in the body of the host. The second-instar larva grows rapidly, but there is no clear indication of moulting, although Kirkpatrick (1937) recorded 6 parasitic instars in *Corioxenos*. No recognizable cast skins can be recovered from the body of the host, even though the very much smaller (though more heavily sclerotized) cuticle of the triungulinid is always recoverable. It is possible that there is only one instar between the triungulinid and the pupa, and that the cuticle itself grows, as is the case in the early larvae of rhipiphorids. There is certainly a great increase in size of the female parasite due to stretching of the cuticle during the embryonic development of the triungulinids.

In pentatomid bugs of the genera *Poecilometis* and *Omyta* parasitized by a species of *Halictophagus*, it has been possible to delimit the points of entry of the triungulinids into the adults and nymphs by finding their cast cuticles in fatty tissues just below the host cuticle. Most entered through the intersegmental membranes in the region of the genitalia and apical segments of the abdomen, but some entered ventrally at the junction of thorax and abdomen close to the insertion of the hind legs and less usually at other intersegmental membranes.

Most Strepsiptera overwinter as parasitic larvae within the host. The triungulinids mature and 'hatch' in the late summer and autumn, and infest immature, or less often adult, stages of the host. Most Hemiptera parasitized by Strepsiptera have one generation a year; but in delphacids there may be continuous development throughout the warmer months, and in these one finds all stages of parasitism at any time over this period. The life cycle of the males *(Elenchus)* in the delphacids must be very short, for very immature hosts in which the wing buds are only just discernible may show empty male puparia. Most parasitized species of Homoptera overwinter as early nymphs, but some of the Heteroptera overwinter as adults. In Homoptera the triungulinids parasitize the early nymphs, whereas in Heteroptera it is the adult or late nymphal instars that are attacked. The parasites of bees and wasps normally overwinter in the host larva or 'prepupa'.

Host-parasite Relationships. Infestation with Strepsiptera is known as *stylopization.* In general, one does not notice any striking modification due to parasitism, except that in the Homoptera the wings may be slightly displaced when a male puparium is extruded dorsally. However, in some eurybrachyids there are often modifications in the external genitalia. Some, but not all, parasitized specimens which would be considered as females on wing venation have external genitalia approaching those of the male. Parasitized male hosts show little or no obvious modification. The abdomen of a parasitized specimen is often a deeper yellow than normal at the intersegmental membranes. Internally, there is some atrophy of the reproductive system as well as of most other organs, particularly if the parasite is a female. Richards (1962) found that parasitized female *Paragia decipiens* acquired male characters (in clypeus, scape, orbital spot, etc.), but that males were less altered, though the tubercle of sternite 2 was reduced. He considered that a male parasite produced a greater effect than a female.

Parasitism does not shorten the normal life span of the host. Very often the percentage parasitism rises towards the end of the season through selective survival of parasitized specimens. This survival of host individuals is essential for the female parasite, to permit embryonic development and hatching of the triungulinids. Male parasites emerge early in the season, and the empty puparium remains protruding from the host. It is not known whether normal sexual development of the host is possible after emergence of the male parasite.

The percentage parasitism is often high in localized areas, especially with parasites of Homoptera and certain bees. More than 50 per cent parasitism has been recorded, though it is more usually of the order of 10 to 20 per cent. In sphecid and eumenid Hymenoptera parasitism is usually less than 5 per cent, but on occasion it may rise to as much as 20 per cent in a common host species. This relatively low rate of parasitism is due no doubt to the solitary habits of the hosts. In social Vespidae the parasitism may be high.

Very rarely are stylopized specimens affected by other parasites. Epipyropid larvae (Lepidoptera), on rare occasions, have occurred on *Platybrachys decemmacula* parasitized by either *Halictophagus tryoni* or *Deinelenchus australiensis*. A mermithid nematode has been found in the same host specimen of *Dardus abbreviatus* (Eurybrachyidae) as *Deinelenchus* on one occasion. The two male puparia of the *Deinelenchus* were fully formed, but had not been exserted. In the pentatomid bug, *Poecilometis strigatus*, a species of *Halictophagus* is sometimes associated with tachinid larvae. A single host species is normally attacked by only one species of strepsipteron, although two species (*Halictophagus tryoni* and *Deinelenchus australiensis*) are recorded as common parasites of *Platybrachys*. On only one occasion, however, has an individual host been known to harbour both species.

The degree of host specificity varies, and, in general, that for parasites of the Hymenoptera is rather high, whereas the parasites of the Homoptera are not so specific. The genus *Halictophagus* is recorded from Fulgoroidea, Cicadellidae, Membracidae, and Cercopidae amongst the Homoptera and from Pentatomidae and Coreidae amongst the Heteroptera, while closely related genera occur in grylloid Orthoptera, Blattodea, and Diptera. Even in parasites of the Hymenoptera, there is a tendency for the one species of parasite to occur in all of the closely related species of the host group that are present in the locality. One species of *Pseudoxenos* occurs in 8 species of eumenid wasps of the genera *Odynerus* and *Paralastor*, but it is common in only 2 or 3 of them. *Deinelenchus* parasitizes a number of homopteran hosts of the families Eurybrachyidae, Flatidae, and occasionally Dictyopharidae, Ricaniidae, and Fulgoridae. The Flatidae constitute the normal hosts, and male parasites are abundant in species of this family, whereas in Eurybrachyidae, although parasites are quite common, only females have been found.

Natural Enemies. No parasites of Strepsiptera are known, and there are no important predators.

Economic Significance. It seems probable that Strepsiptera exert some control of the population of the host species, but it is doubtful whether they will ever be a prime factor controlling the abundance of their hosts, except possibly for a few parasites of the Homoptera. Hyperparasites are unknown, and this is one factor in favour of Strepsiptera as control agents. *Elenchus* may effect some control over the density of delphacid populations, and *Stichotrema* has been used in efforts to control the *Sexava* which is a pest of coconuts.

Special Features of the Australian Fauna

The Stylopidae and Halictophagidae are

cosmopolitan, but the other families have a discontinuous distribution. The Mengeidae are recorded only from the Mediterranean region, China, and Australia. The Myrmecolacidae are widely distributed in the Oriental region, New Guinea, and Australia, and also in Africa and South America. The Corioxenidae occur in North Africa, tropical America, Japan, Philippines, New Guinea, and Australia.

In Australia Strepsiptera are more common in the northern half and in the dry interior. The Halictophagidae parasitizing Homoptera are almost limited to the higher-rainfall areas of Queensland and N.S.W. The Stylopidae that parasitize the Sphecinae and the eumenid wasps *(Pseudoxenos)* are mostly tropical and subtropical; those from *Bembix* are widespread in the interior where this sphecid genus is most common; parasites of the Masaridae *(Paragioxenos)* are known only from S.A. and inland N.S.W. Few parasites have been recorded from Australian bees, but they range over the whole continent, including Tasmania. The Mengeidae are also widely distributed; Corioxenidae occur mainly in the interior; and Myrmecolacidae mainly in the north.

CLASSIFICATION

Order STREPSIPTERA (93 Australian spp.)

1. Mengeidae (8)
2. Corioxenidae (6)
3. Myrmecolacidae (3)
4. Stylopidae (53)
5. Halictophagidae (23)

The classification adopted has been modified from Bohart (1941) and E. Carvalho (1956, 1961). The higher classification is based to a large extent on the structure of the males, but the females are important at the generic level.

Keys to the Families of Strepsiptera

MALES

1. Tarsi 5-segmented, the apical segment broad and with large terminal claws; antenna 6- or 7-segmented (Fig. 31.9A) .... **Mengeidae**
   Tarsi with fewer segments and no terminal claws, or, if 5-segmented, the 4th segment bilobed and forming the functional terminal segment, 5th very narrow and the claws minute 2
2. Tarsi 2- or 3-segmented; antenna 4- or 7-segmented (Fig. 31.1) .... **Halictophagidae**
   Tarsi with 4 or 5 segments 3
3. Antenna 7-segmented, the first 4 segments very short, 3rd with a lateral flabellum, 5th to 7th very long (Figs. 31.9C, D) .... **Myrmecolacidae**
   Antenna 5- to 7-segmented, first 4 segments very short, 3rd and 4th laterally flabellate, 5th elongate, 6th and 7th, when present, elongate (Fig. 31.9B) .... **Corioxenidae**
   Antenna 4- to 6-segmented, 3rd segment laterally flabellate, 4th elongate (Fig. 31.10) .... **Stylopidae**

FEMALES

1. Free-living; tarsi 5-segmented, with terminal claws (Fig. 31.2) .... **Mengeidae**
   Parasitic, with reduced cephalothorax protruded from the host; apodous 2
2. Brood-canal opening terminal and dorsal to buccal crest (Fig. 31.4A); parasites of Heteroptera .... **Corioxenidae**
   Brood-canal opening ventral, at junction of head and thorax (Fig. 31.4D) 3
3. Parasites of Hymenoptera .... **Stylopidae**
   Not parasites of Hymenoptera 4
4. Cephalothorax with hook-like projections behind spiracles; with multiple genital apertures on each segment; parasites of Orthoptera .... **Myrmecolacidae**
   Cephalothorax without such projections; with a single genital aperture on each segment, except in *Deinelenchus*; parasites of Homoptera, Heteroptera, Orthoptera, Blattodea, Diptera .... **Halictophagidae**

TRIUNGULINID LARVAE

1. All tarsi long, setiform (though there is a very slight terminal swelling) .... **Halictophagidae**
   Tarsi of at least some legs with pulvilli 2
2. Fore tarsus pulvilliform, mid tarsus short-setiform *(Stichotrema)* .... **Myrmecolacidae**
   Fore and mid tarsi with pulvilli 3

3. Fore and mid tarsi with small round pulvilli (Figs. 31.5A, B) .... **Mengeidae**
   Fore and mid tarsi with large pulvilli 4
4. Pulvilli round or elongate (Figs. 31.5C, D); associated with wasps or bees .... **Stylopidae**
   Pulvilli large and oblong; associated with bugs .... **Corioxenidae**

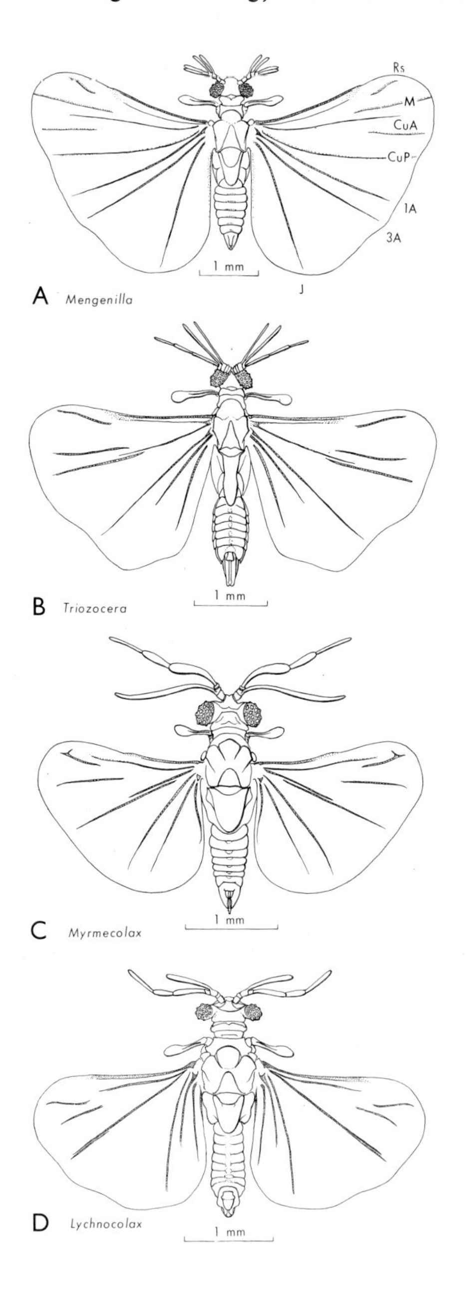

Fig. 31.9. Males of: A, Mengeidae; B, Corioxenidae; C, D, Myrmecolacidae. [R. Ewins]

1. Mengeidae. The most primitive Strepsiptera are placed in this family. The mature females, though larviform, are free-living, and both sexes have tarsi with terminal claws. The aedeagus of the male is not hooked, and this is probably correlated with the more usual type of copulation with a normal type of female. The mandibles are well developed and blade-like; the palps are 2-segmented. *Mengea* has been found only in Baltic amber. *Eoxenos* is recorded from the Mediterranean area. It is an internal parasite of Thysanura, but both sexes pupate under stones and on tree-stumps away from the host. *Mengenilla* (Fig. 31.9A) also occurs in the Mediterranean and in China, and is widespread in Australia (8 spp.) from north Queensland to Canberra, with one species in W.A. Only males have been collected here, at light or enmeshed in spiders' webs, so the host relationship is unknown.

2. Corioxenidae. Male with mandibles apparently always absent. Tarsi 4- or 5-segmented, the 4th segment bilobed and forming the functional end of the tarsus; basitarsus about as long as the following segments combined. The 5th tarsal segment in *Triozocera* is long but very narrow, and the two terminal claws are minute. Female with the brood-canal opening terminal, instead of ventral at the junction of head and thorax. The segmentation of the thorax is distinct. The few hosts recorded are pentatomid, cydnid, and mirid bugs. *Triozocera* (Fig. 31.9B) is widespread in Australia, with 4 species from north Queensland to Canberra, and 2 in W.A.; it also occurs in Japan and New Guinea. The male is more primitive than that of the Ethiopian *Corioxenos* in retaining the 5th tarsal segment, tarsal claws, and an extra anal vein. *Triozocera* is often referred to the Mengeidae on account of its tarsal characters.

3. Myrmecolacidae. The species are known mostly from the males, generally collected at light, and all presumably parasitic on ants (Ogloblin, 1939). Stylopized genera of ants include *Camponotus*, *Pheidole*, *Pseudomyrma*, *Solenopsis*, and *Eciton*. Parasitism by Strepsiptera, as with mermithid worms, produces a change in the behaviour of the ants, which are often found at the tips of grass stems even in bright sunlight. The female parasites apparently develop in Orthoptera (Gryllidae, Gryllotalpidae, Tettigoniidae) and Mantodea. A family Stichotrematidae was based on these females. So far there has been no direct association of males and females. Both male and female parasites are recorded from Australia. Male specimens of *Myrmecolax* (Fig. 31.9C) are known from the northern half of the continent, and males of *Lychnocolax* from Bourke, N.S.W. There is a single recorded undetermined gryllid host from the Armidale area, N.S.W. *Sexava* (Tettigonioidea) is parasitized in New Guinea.

4. Stylopidae. A very large family of species that parasitize bees and wasps. The males are characterized by expanded 4-segmented tarsi, a small prescutum, and an elongate postscutellum. The females are not so easily defined, except that they are parasites of Hymenoptera. The male antenna is 4- to 6-segmented, but only the 3rd segment is laterally flabellate. The females have 2 to 5 genital tubes, and the head and thorax may or may not be completely fused ventrally.

Stylops, a common parasite of *Andrena* and the Panurgini (Andrenidae), with 6-segmented male antenna, is not known in Australia, although there are possibly more species of *Stylops* than of any other genus of Strepsiptera. *Hylecthrus* (7 spp.) occurs in colletid bees of the genera *Euprosopis* (Hylaeinae) and *Euryglossa* (Euryglossinae) from Queensland to Victoria and in W.A. There are 5 species of a related genus in *Paracolletes* (Colletinae). *Halictoxenos* (5 spp.) parasitizes halictid bees (*Halictus* and *Parasphecodes* in Australia); the head of the female is not separated from the thorax ventrally by lateral extensions of the brood-canal opening, and there are 5 genital apertures; the male has 4-segmented antennae, whereas most of the parasites of bees have 5- or 6-segmented male antennae.

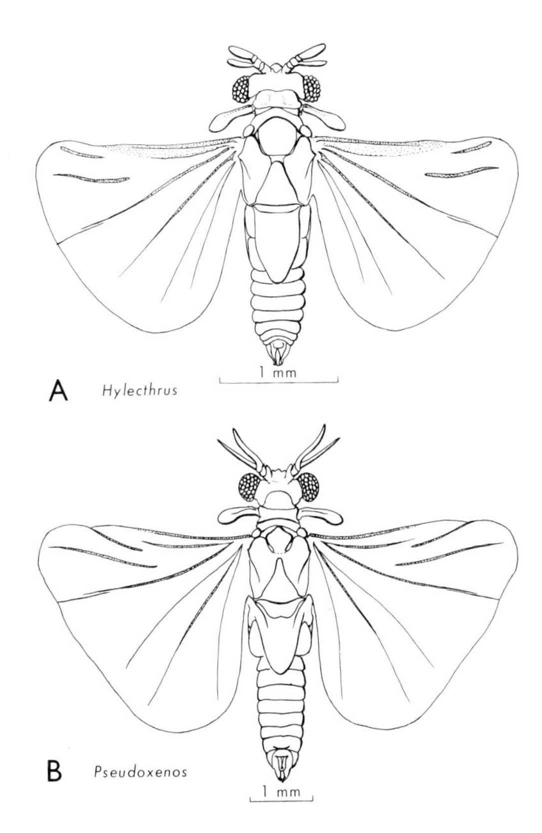

Fig. 31.10. Males of Stylopidae.
[R. Ewins]

The parasites of vespoid and sphecoid wasps have the antenna of the male 4-segmented, with the 3rd and 4th segments flabellate and subequal. Parasites of these wasps are common in the northern half of the continent, but there are few species in the south, and only 2 are known from Tasmania. *Xenos* (3 spp.), a parasite of the social wasps, is not common in Australia. *Ropalidia cabeti* and other species of *Ropalidia* in north Queensland are parasitized. *Polistes variabilis* is parasitized in south-east Queensland, and there is one record of parasitism of *P. tepidus* in southern N.S.W. *Pseudoxenos* (20 spp.; Fig. 31.10B) is the dominant genus in the Australian fauna, parasitizing Sphecinae (*Sphex* and *Sceliphron*) and Eumenidae (mainly '*Odynerus*' and *Paralastor*, but also recorded from *Abispa*,

Rhynchium, *Eumenes*, *Discoelius*, and *Montezumia*). *Paragioxenos* (1 sp.) is a parasite of *Paragia decipiens* in S.A. and its subspecies *aliciae* in N.S.W. There are two undescribed genera, one on *Bembix* (10 spp.) and the other on *Bembicinus* (2 spp.)

5. Halictophagidae. The family is divided into two clearly defined subfamilies which differ in tarsal and antennal segmentation of the males.

The HALICTOPHAGINAE (21 spp.) are common parasites of Homoptera, but are recorded also from Heteroptera, Orthoptera, and Blattodea. Males can be distinguished by their 3-segmented tarsi and 7-segmented antennae. In the females the number of median genital openings on the anterior segments of the abdomen is variable (1–4), though 3 is apparently the usual number. The subfamily is well represented in Queensland, but is almost entirely absent from southern Australia, apart from one species in pentatomid bugs. *Tridactylophagus* is a parasite of *Tridactylus* (Orthoptera–Caelifera). The single recorded species is from India, but there is a second species in Australia. The male antenna has only a single flabellum, and in this respect resembles the Elenchinae. *Stenocranophilus* is a parasite of the Fulgoroidea, whereas the Australian species of *Halictophagus* (Fig. 31.1) are nearly all parasites of the Cicadellidae, although there is one very common species, *H. tryoni*, in eurybrachyids, and another is a parasite of membracids. The male of *H. tryoni* is not known, but the female is a common parasite of *Platybrachys* and *Dardus*. There are also two species in pentatomids in eastern Australia (and one in a coreid in East Africa). There is an undescribed genus parasitizing cockroaches (*Melanozosteria* sp.) in W.A.

The ELENCHINAE (2 spp.) are distinguished from all other Strepsiptera by the combination of 2-segmented tarsi and 4-segmented antennae. The 3rd segment of the antenna is Y-shaped, and most probably developed by fusion of the true 3rd segment with one or more distal segments. This is partly substantiated by the presence of the large sensorium on this segment, for in all other

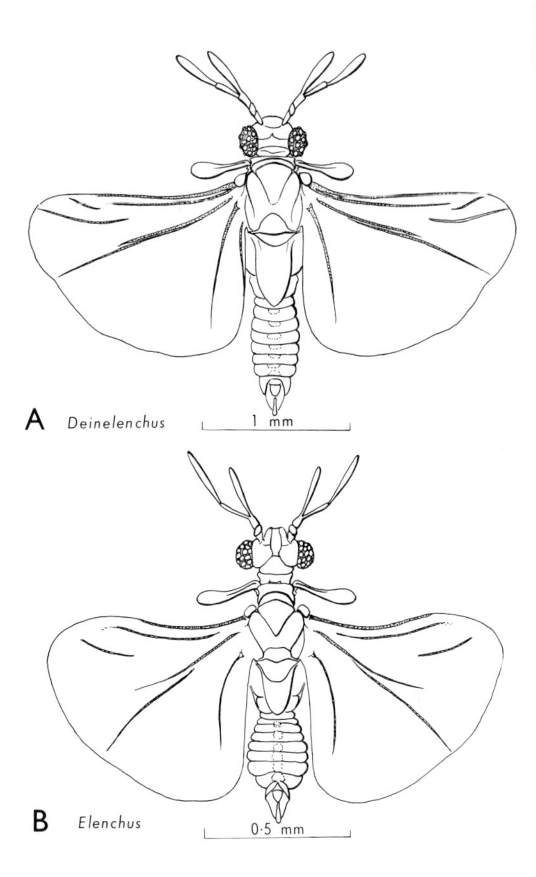

Fig. 31.11. Males of Halictophagidae–Elenchinae. [R. Ewins]

Strepsiptera, except a few in which it is absent, it occurs on the 4th segment. The enlarged brood-canal opening and reduced thorax of the female are diagnostic only of *Elenchus* (Fig. 31.4F).

The subfamily is probably represented in the Australian region by three genera. *Elenchus* is the only generally recognized (and widely distributed) genus. It is a parasite of the Delphacidae *(sens. lat.)*, and the single Australian species is known to occur from Cairns to Canberra; it is common in delphacids inhabiting seedheads of couch grass, *Cynodon dactylon*, at Brisbane. Male exuviae can be found in young nymphs in summer, and the parasite overwinters as a larva in the hibernating delphacid nymphs. *Deinelenchus*, with its strongly hooked male aedeagus, more complex wing venation, and a tendency for the antenna to be 5-segmented, shows characters which link the Elenchinae to the Halictophaginae. *D. australiensis* is a parasite of Eurybrachyidae, Flatidae, Dictyopharidae, Fulgoridae, and Ricaniidae in Queensland. The male is quite large (wing-span 5·5 mm), and the genital apertures of the female are in two transverse rows on segments 2 and 3 of the abdomen. There are two overlapping generations in the year. An undescribed genus parasitizing platystomatid flies in New Guinea should probably be referred to this subfamily though the adult male is unknown.

ACKNOWLEDGMENT. Appreciation is expressed to Professor Richard M. Bohart, University of California, Davis, for a critical review of the manuscript.

32

MECOPTERA

(Scorpion-flies)

by E. F. RIEK

Mandibulate endopterygote Neoptera, in which the head is produced into a rostrum; abdominal tergum 1 fused to the thorax; normally two pairs of subequal wings, with simple venation. Larvae with compound eyes and mouth-parts of a simple mandibulate form. Pupae decticous.

This is a relatively small order, with less than 400 described species. Nearly half the species are placed in the genus *Panorpa*, and there are more than 60 species of *Bittacus* (both extra-Australian genera). Australia has only 16 described species distributed in 8 genera (Riek, 1954b), and few undescribed species are known in collections.

The Choristidae (Fig. 32.1) and Nannochoristidae resemble small lacewings, both in general appearance and in their mode of flight, but the most common species (*Harpobittacus* spp., Fig. 7.8) are more like large tipulid flies, from which they are easily distinguished by having two pairs of wings. *Austromerope* could possibly be mistaken for a small, soft-winged cockroach when at rest, for the rather broad, much-veined wings are folded rather flat, whereas the adults of other Australian species rest with the wings held roofwise over the abdomen. The known terrestrial larvae of Australian genera are caterpillar-like and carnivorous. The aquatic larvae of *Nannochorista* are elongate, wireworm-like forms. The compound eyes of mecopteran larvae are most diagnostic, as this is an unusual development in endopterygote insects, though compound eyes also occur in the larvae of Micropterigidae.

Anatomy of Adult

Head (Fig. 32.2). Elongate; clypeus and labrum produced to form a rostrum, which is often long (especially in Panorpidae). Compound eyes well developed. Normally 3 ocelli (absent in Meropeidae and *Apteropanorpa*). Antenna filiform, many-segmented, often about as long as fore wing (Panorpidae, Choristidae), but sometimes short (Bittacidae); scape and pedicel enlarged. Mandibles elongate, usually with an inner tooth towards apex; reduced in Nannochoristidae. Maxilla with galea, lacinia, and 5-segmented palp, which is sometimes variously enlarged in

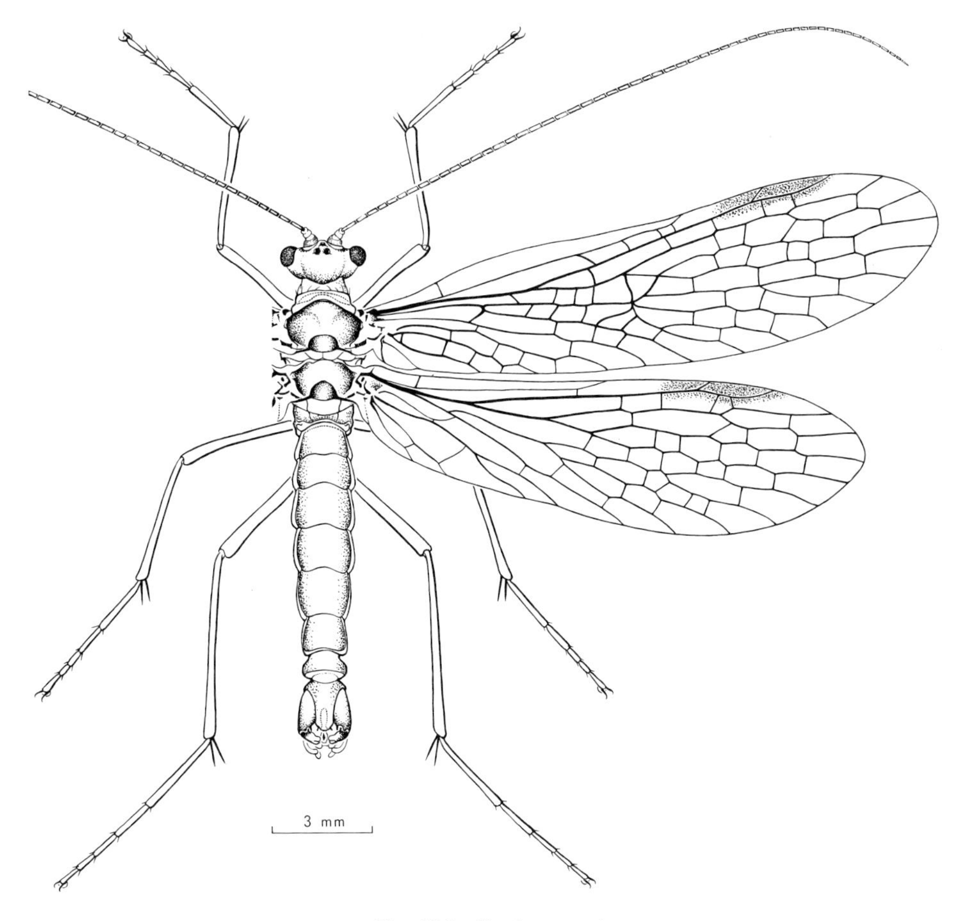

Fig. 32.1. *Chorista* sp., ♂.
[T. Binder]

male. Labium with 2-segmented palps.

Thorax. Lateral cervical sclerite prominent. Prothorax transverse, about as wide as head, with separate sclerotized plates, normally articulating with the mesothorax (fused to it in Bittacidae). Mesothoracic spiracle large, elongate. Mesothorax, metathorax, and abdominal tergum 1 immovably fused. The mesothorax is large, with the scutum, parapsides (p. 871), scutellum, and postnotum clearly defined. Tegula distinct, but not enlarged. Metathorax similar, with a normal spiracle. The fusion of abdominal tergum 1 with the thorax is complete, but often only the anterior portion of the segment is sclerotized, and there is a wide zone of articulation between it and the remainder of the abdomen. It is more fully sclerotized in Choristidae and Nannochoristidae. It is usually quite short, but in the wingless *Apteropanorpa* it is as large as the preceding thoracic segments.

Legs. Long and thin, especially the hind pair. Coxae elongate. Tibiae with 2 apical spurs. Tarsi 5-segmented, ending in 2 claws, except in Bittacidae which have a single claw (Fig. 32.9). Tarsal claws simple *(Apteropanorpa)*, with a ventral comb of a few teeth (Panorpidae, Choristidae), or with an enlarged basal tooth (Nannochoristidae).

Wings (Fig. 32.5). Membranous, subequal, usually elongate. The venation is normally

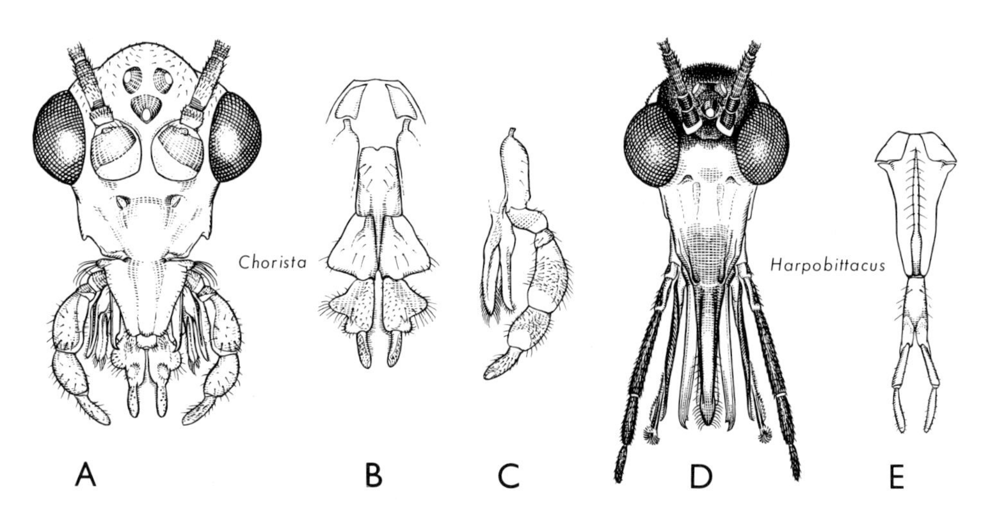

Fig. 32.2. Head and mouth-parts: A–C, Choristidae, head frontal, labium ventral, maxilla ventral; D, E, Bittacidae, head frontal, labium ventral. [M. Quick]

simple, with few branches to the main veins and a complete system of openly spaced cross-veins, but occasionally much branched and reticulate. *Apteropanorpa* (Fig. 32.6) and a few non-Australian species are wingless. Wing-coupling is usually of the frenulate type, with a series of thickened and lengthened hairs from the thickened jugal margin of the fore wing interlocking with 2 stout bristles from the humeral area at the base of the hind wing. In the Bittacidae the wings are narrowed at the base, and there is no coupling between fore and hind wings, although the reduced 3A of the fore wing could be mistaken for a jugal vein, and 1 or 2 stout frenulate bristles persist at the base of the hind wing.

Abdomen. Cylindrical, fusiform. Tergum 1 fused to the metathorax, as indicated above, but the sternite is free, with only a pair of lateral sclerotized plates. The female has 11 clearly defined apparent segments and a pair of terminal, 2-segmented cerci that sometimes appear to be 3-segmented through narrow prolongations of the base (Panorpidae, Choristidae). The male similarly has an apparent 11th segment formed by sclerotized plates surrounding the anus, but the cerci are normally 1-segmented (absent in *Apteropanorpa*).

In the male segment 9 is ring-like (Nannochoristidae, Figs. 32.3E, F), produced dorsally to a pair of claspers (Bittacidae, Figs. 32.3I–K) or produced both dorsally and ventrally (Panorpidae, Choristidae, Figs. 32.3A, B). Gonocoxites (basistyles) bulbous, fused ventrally, separated dorsally except in Nannochoristidae; gonostyle (dististyle) of varied form. Aedeagus tapering, sometimes enlarged and curved upwards and forwards, sometimes with complicated accessory processes (Choristidae). Females (Figs. 32.3C, D, G, H) normally without genital appendages, but an ovipositor is developed in the divergent, non-Australian Boreidae.

Internal Anatomy. The digestive system is unusual; the oesophagus has two dilatations that form what appears to be a muscular pumping apparatus; proventriculus lined with numerous long setae; mid gut large and long; a pair of tubular salivary glands present; 6 Malpighian tubes. The ventral nervous system has 3 thoracic and a reduced number of abdominal ganglia, with 1st abdominal fused with metathoracic. In the male the paired testes join paired vasa deferentia, which open separately into a large median vesicula seminalis; a pair of accessory glands present. In the female each ovary consists of 7–19 polytrophic ovarioles; the two

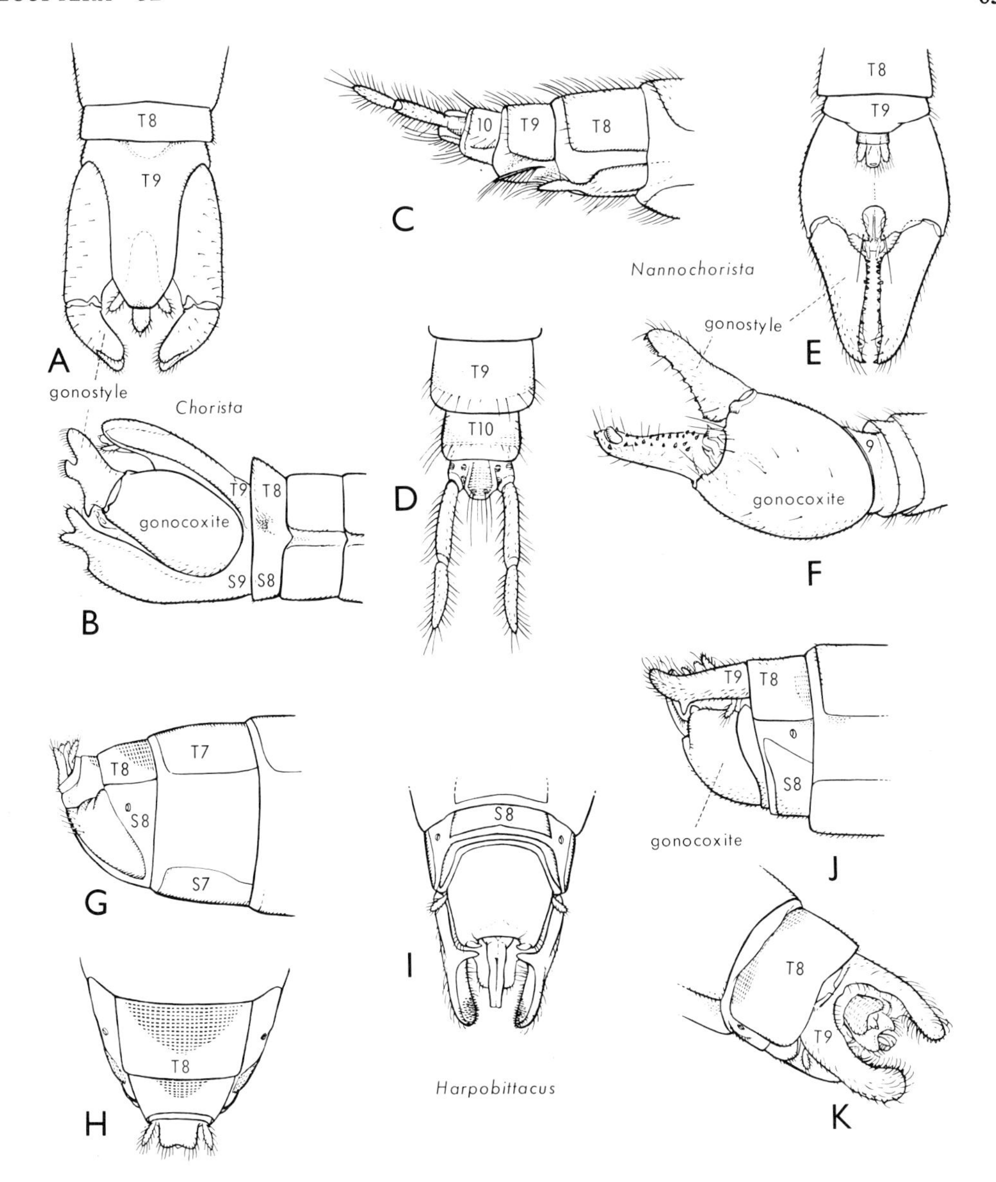

Fig. 32.3. Terminalia: A, B, Choristidae, ♂ dorsal, ♂ lateral; C–F, Nannochoristidae, ♀ lateral, ♀ dorsal, ♂ dorsal, ♂ lateral; G–K, Bittacidae, ♀ lateral, ♀ dorsal, ♂ ventral, ♂ lateral, ♂ dorsal. [M. Quick]

oviducts unite before opening into a genital pouch; spermatheca and accessory glands present.

Immature Stages

Egg. Broadly oval, soft and smooth (Choristidae, Nannochoristidae, Panorpidae), or hard, cuboidal, and with somewhat concave faces which become convex before hatching (Bittacidae). The eggs of *Harpobittacus* have sides of about 0·5 mm. Before hatching they inflate to a sphere measuring more than 1 mm in diameter.

Larva (Figs. 32.7, 8). Caterpillar-like in

most families. The head capsule is well defined, and the division into sclerites clear. There are few facets in the compound eyes of Bittacidae, but a large number in Panorpidae, Choristidae, and Nannochoristidae; a median 'ocellus' is present. Thorax normally with only the pronotum sclerotized, but with all terga sclerotized in Nannochoristidae (Fig. 32.8). The single tarsal claw is small, except in Nannochoristidae. Abdominal prolegs are distinct on segments 1–8. The apex of the abdomen is modified into a suction disc (Fig. 32.7C), except in Nannochoristidae (Fig. 32.8C) in which it ends in a pair of anal hooks; and there are paired dorsal ('excretory?') processes on segments 8 and 9 and a median process on segment 10 in Panorpidae, Choristidae, and Bittacidae. The first-instar larva of *Chorista* has similar but smaller paired processes on the first 7 abdominal segments as well, but they are progressively reduced in the later instars.

The Boreidae have very different, curl-grub-like, herbivorous larvae, with no thoracic sclerotization, no abdominal prolegs (though they are apparently present in the early instars), and no apical suction disc or dorsal processes on segments 8–10 of the abdomen.

Pupa (Fig. 32.4). Decticous and exarate in all Mecoptera.

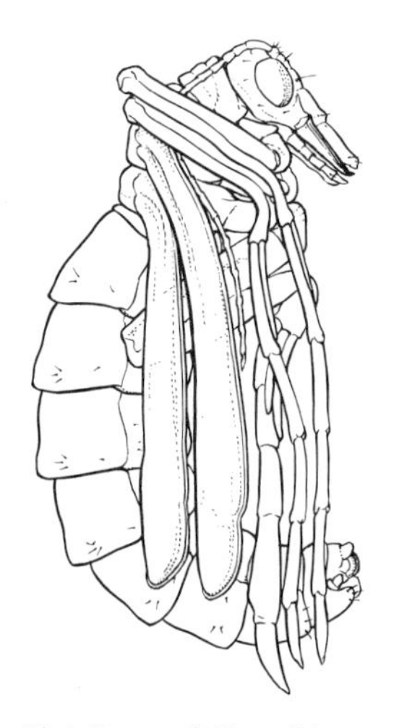

Fig. 32.4. Pupa of *Harpobittacus* sp. [M. Quick]

Biology

Adults. Scorpion-flies generally frequent cool, moist situations, though adults of *Harpobittacus* are able to survive under more severe conditions. Adults of *Nannochorista* are never found far from swampy areas on small mountain streams or along the margins of lakes.

Scorpion-flies are active fliers, and many species have extensive ranges. *Harpobittacus* frequents blossom (often *Leptospermum*) in the spring to suck nectar. The males also prey on soft-bodied insects, such as flies, lampyrid beetles, and honey bees, also caterpillars and soft beetle larvae (e.g. *Paropsis*), and in the autumn they even prey on other scorpion-flies (e.g. *Chorista*). The prey is taken both on the wing and as the adult walks over the blossom. The females apparently do not capture living prey. *Harpobittacus* trails its long legs during leisurely flight, but it is capable of rapid flight when disturbed. *Apteropanorpa* may be able to make hops of a few centimetres, like the flightless exotic *Boreus*.

Harpobittacus adults emerge in spring. The first adults occur in close association with damp areas that often support shrubs of *Leptospermum*. Later they spread out from these areas, but are still more abundant on the blossom where they forage for their prey. They are also common in the autumn in long grass bordering small streams. Adults may live through the summer, though it is possible that there are two generations each year. The autumn-collected adults are, in general, larger and darker than those of spring, and more often have worn or tattered wings. The adults of *Chorista* occur only in autumn, when they may be common in long grass adjoining small streams or in moist gullies. They make short darting flights, and depend to some extent on concealment in grass tussocks and low vegetation for protection. They are sometimes attracted to light. *Nannochorista* adults first appear in late spring, and there is a second peak of abundance in autumn, so that there are probably two generations at least in some localities. The adults of *Chorista* and

Nannochorista have not been observed to feed.

Reproduction. There is a complicated courtship in *Harpobittacus*, matings being most frequent towards midday when temperatures are high. The male secures a prey in the hind legs, subduing it by puncturing it with the mouth-parts. He flies with it held in the trailing hind legs to a resting place in some low shrub, where, hanging by his fore legs, he transfers it to his mouth-parts. Females are attracted by a musty secretion apparently produced by glandular areas behind terga 7 and 8 of the male (Bornemissza, 1966). When one approaches, she is grasped ventrally by his hind and mid legs, so that the ventral surfaces of the two are opposed. There is considerable agitation at this stage. His abdomen then seeks out the tip of hers, its apex is partly turned up towards her dorsal surface, and copulation takes place. The genitalia of the two come to be oriented in opposite directions, with a twist of 90° in the apical section of the male abdomen.

They remain passive for a minute or two, with her hind legs still firmly grasped by his. He then frees his grasp of her, and uses his hind legs to transfer the food from his mouth to the proximity of hers. She grasps it with her fore legs, and commences to feed. Feeding may continue for several minutes. When the female has finished feeding, they separate, she flies off to settle nearby, and he consumes the remaining food for a minute or so before allowing the sucked-out remains to fall to the ground. Finally, he cleans his legs and mouth-parts, using one fore leg as he hangs suspended by the other, and then flies off to forage again.

Only males with food are attractive to the females, and several females may be attracted to the one male at the same time. A male will copulate with more than one female, using the same portion of food, and the second, or even a third, will follow the regular routine, such as remaining passive for a time before commencing to feed again. It would seem that the females depend on this food supply for maturation of the eggs, for they have not been observed to take prey, although they do feed on nectar.

In *Chorista*, copulation takes place with both dorsal surfaces uppermost, with the pair arranged side by side and the abdomen of the male deflected sideways. The mouths of the pair touch, and there is a very evident and persistent exchange of fluids. They are not readily disturbed when copulating. *Nannochorista* apparently behaves similarly, for pairs can be swept from foliage, and remain in copulation in the collecting net.

The eggs of Choristidae are laid in clusters in moist hollows in the soil. Those of Bittacidae apparently are dropped freely. The eggs of Nannochoristidae are laid in a single row of up to 25 eggs, end to end, apparently in fully saturated litter at the bases of grass tussocks in swampy ground at the margins of streams and lakes. The hard eggs of *Harpobittacus* are laid in summer and autumn, and can remain dormant for some time, but hatch in $1\frac{1}{2}$ to 3 months following the rains of early autumn.

Immature Stages. The known larvae of Australian genera are easy to recognize in the field, but most information is available for a species of *Harpobittacus* which was bred from egg to adult in the laboratory by Currie (1932). There were 4 larval instars. Under artificial conditions, the larvae would feed only on dead insects, and started away from any moving object. They could flush the skin with fluid secreted from glands opening on the pronotum, but nevertheless did not live long in a dry atmosphere. They covered the body with excrement, and it seems likely that, in nature, they may use the terminal suction cup to move about on low vegetation and erect the body to simulate a twig when disturbed, as does *Panorpa*. Growth was completed in about a month. The larva then constructed an earthen pupal cell just beneath the surface of the soil, covered the entrance with a lid, and remained in the cell, lying with the head uppermost, for several months during the winter. The subsequent pupal period was relatively short (14 to 50 days).

Natural Enemies. Nothing is known of parasitism or predation on the eggs and larvae of Australian species. Adults of

Chorista are occasionally eaten by *Harpobittacus*. Large asilid flies are often seen feeding on *Harpobittacus*.

Economic Significance. *Harpobittacus* feeds on a wide range of other insects. It is often abundant in heath country, and may be a factor in limiting populations of some Diptera, Coleoptera, and perhaps Lepidoptera. Little is known of the biology of the predatory types of larvae that are usual throughout the order.

Special Features of the Australian Fauna

The Australian Mecoptera exhibit a strong endemism. The Choristidae are known only from Australia, and are absent from Tasmania. The Nannochoristidae occur in New Zealand and South America, as well as in Tasmania and Australia. *Apteropanorpa* has been placed in the Panorpidae, but it could be a derivative from the Choristidae. The Australian Bittacidae belong to distinct endemic genera, except for one that is apparently allied to species in Central America. The single species of Meropeidae is a relict of a group that was apparently widespread in the late Palaeozoic.

The local distribution is essentially cool temperate. No Mecoptera are known from the northern part of the continent or from the dry interior, except in swampy areas. *Harpobittacus* is the common widespread genus that occurs from central Queensland to Tasmania and S.A. and is represented in south-western W.A. Another two genera extend the range of Bittacidae further north in central Queensland. Choristidae range from Victoria to southern Queensland, and Nannochoristidae from Tasmania to northern N.S.W. *Apteropanorpa* has been found only in Tasmania, and *Austromerope* only in south-western W.A.

CLASSIFICATION

Order MECOPTERA (20 Australian spp.)

Suborder PROTOMECOPTERA (1)

1. Meropeidae (1)

Notiothaumidae (0)

Suborder EUMECOPTERA (19)

Boreidae (0)
2. Panorpidae (1)
3. Choristidae (4)
4. Nannochoristidae (3)
5. Bittacidae (11)

The classification adopted is that most commonly used. The Boreidae are considered by Hinton (1958a) to constitute a separate order Neomecoptera. The Choristidae and Nannochoristidae have sometimes been treated as subfamilies of the Panorpidae. Byers (1965) divided Panorpidae into three families: Panorpidae, Panorpodidae, and Apteropanorpidae.

Key to the Families of Mecoptera Known in Australia

ADULTS

1. Tarsi raptorial, with the 5th segment folding down against the 4th, and with a single claw (Fig. 32.9); wings usually subpetiolate (Figs. 32.5E, F); antenna much shorter than fore wing ...... **Bittacidae**
   Tarsi not raptorial, with 2 claws 2
2. Wings with reticulate cross-veins in the costal space (Fig. 32.5D); CuA forked at apex; eyes elongate or reniform. [Ocelli absent] ...... **Meropeidae**
   Wings, when present, with at most a few simple cross-veins in the costal space (Figs. 32.5A–C); CuA simple; eyes more or less rounded 3
3. The only Australian species apterous (Fig. 32.6); ocelli absent. [In winged forms, the costal space of the fore wing narrow and CuA not fused with the main stem of M] ...... **Panorpidae**
   With fully developed wings; ocelli present. [Costal margin of fore wing convex over basal half, especially in Choristidae] 4
4. In the fore wing (Figs. 32.5A, B) CuA only touches the main stem of M; Rs 4-branched, M normally 5-branched; vertex of head raised; sternum 9 of ♂ produced; gonocoxites not meeting in mid line dorsally; gonostyle trifid or blunt at apex (Figs. 32.3A, B) ...... **Choristidae**

In the fore wing (Fig. 32.5C) CuA and the main stem of M coalesce for a longer distance; Rs 3-branched, M 4-branched; vertex of head not raised; sternum 9 of ♂ not produced; gonocoxites meeting in mid-line dorsally; gonostyle tapering at apex (Figs. 32.3E, F) .... **Nannochoristidae**

LARVAE

(The larvae of Meropeidae are not known).

1. Aquatic; terminal segment of abdomen not modified into a suction cup; body very long and thin .... **Nannochoristidae**

   Not aquatic, living in litter, moss, or other encrusting vegetation; abdomen terminating in a suction cup; caterpillar-like 2

2. Terminal segment of antenna not thread-like though thin; eye elongate; [Median dorsal process of abdominal segment 10 and paired processes of 8 and 9 well developed] .... **Panorpidae**

   Terminal segment of antenna thread-like; eye more or less circular 3

3. Median dorsal process of abdominal segment 10 greatly reduced, shorter than hairs on the segment, paired processes of 8 and 9 almost indistinguishable (Fig. 32.7A) .... **Choristidae**

   Median dorsal process of abdominal segment 10 and paired processes on 8 and 9 well developed (Fig. 32.7D) .... **Bittacidae**

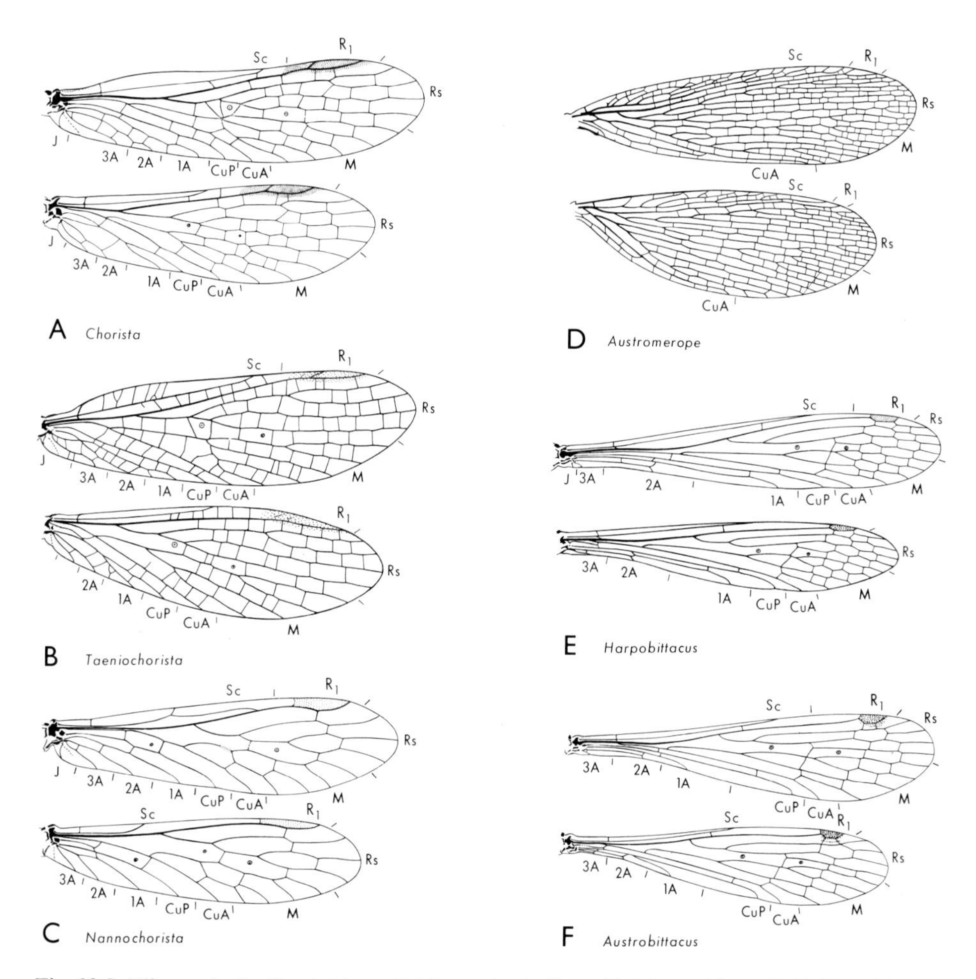

Fig. 32.5. Wings: A, B, Choristidae; C, Nannochoristidae; D, Meropeidae; E, F, Bittacidae. [M. Quick]

Suborder PROTOMECOPTERA

1. Meropeidae. This small family is known only from North America and Australia. *Merope* is restricted to the eastern part of the United States, while *Austromerope poultoni* is known from two specimens taken under log or stone in the bush in W.A. The antennae are much shorter than the fore wing; the male claspers are greatly lengthened; and the wings are folded rather flat over the abdomen.

Suborder EUMECOPTERA

2. Panorpidae. This large family is essentially Holarctic, but it occurs in India and Indonesia. *Apteropanorpa tasmanica* (Fig. 32.6) is isolated from the rest of the family structurally and in distribution; Byers (1965) made it the basis of a separate family. It is apterous in both sexes, and was collected on snow in Tasmania in May and September. Its biology is unknown, though Evans (1942) has described a young larva collected in moss that could possibly belong to it.

3. Choristidae. The wing venation (Figs. 32.5A, B) of this endemic family, including the expanded costal space of the fore wing, provides ready separation from winged Panorpidae. The 2 included genera are superficially similar, but *Chorista* has only one cross-vein in the costal space of the fore wing, whereas *Taeniochorista* has several. *Chorista australis* is widespread over most of the inland areas of the south-eastern portion

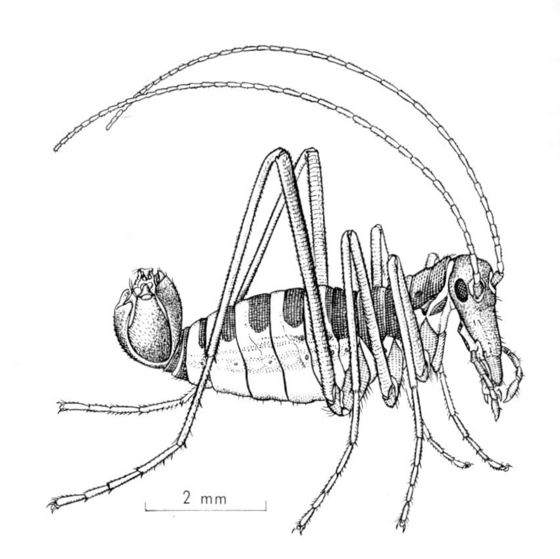

Fig. 32.6. *Apteropanorpa tasmanica*, ♂. [M. Quick]

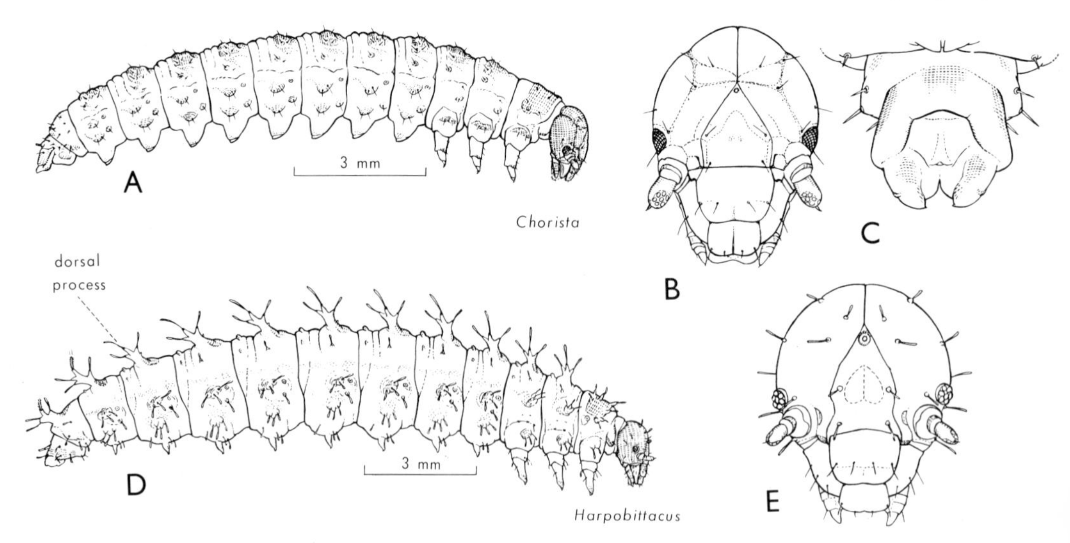

Fig. 32.7. Larvae: A, B, Choristidae; C, apex of abdomen of same, ventral; D, E, Bittacidae. [M. Quick]

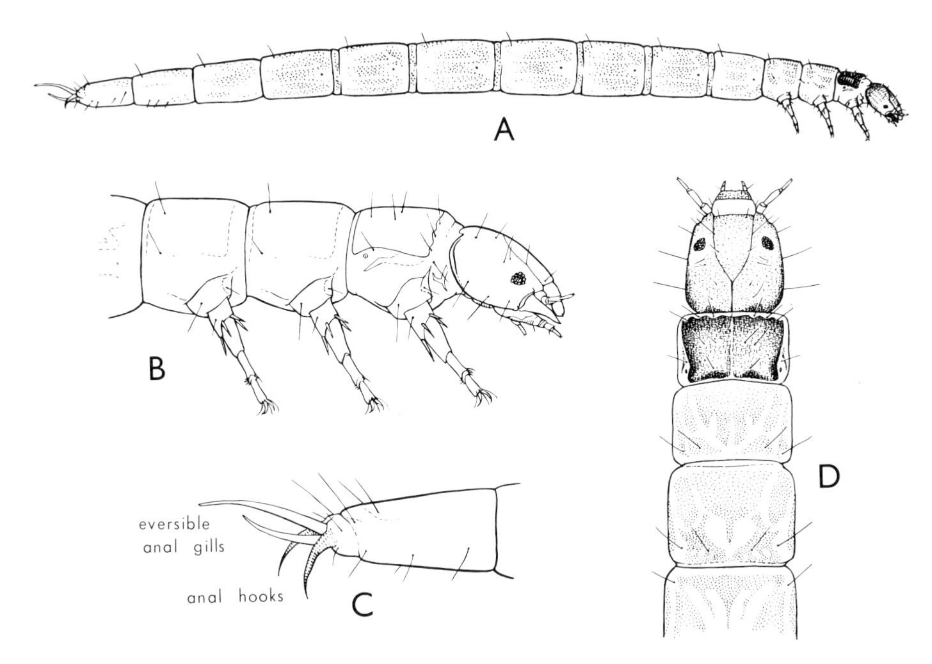

Fig. 32.8. Larva of *Nannochorista* sp.
[M. Quick]

of Australia, with a distinct subspecies in coastal N.S.W. *C. ruficeps* occurs mainly in Victoria. *Taeniochorista pallida* ranges from coastal N.S.W. into south-eastern Queensland, where there is also a second species in the higher areas. Adults occur only in the late summer and autumn. *Chorista* apparently lays its eggs in pockets in damp situations. The almost smooth larvae live in litter and low vegetation. They pupate in simple earthen cells in the soil, and there is apparently a long prepupal period within the cells.

4. Nannochoristidae. A distinctive, southern-hemisphere family of small species (fore wing 5·5–8·5 mm). The head shape and mouth-parts are very different from those of Choristidae and Panorpidae. The postocular region is not enlarged and raised, and the rostrum is very narrow and needle-like. The antennae are considerably shorter than the fore wing, except in one species. The wing venation (Fig. 32.5C) is reduced almost to the simplest possible condition, with Rs only 3-branched. The male terminalia are distinctive. The female subgenital plate is cleft at apex, but not as markedly as in *Chorista*. The larvae are aquatic, and adults only occur close to small streams and lake margins. On the mainland they are most abundant in the high country. There are 3 species of *Nannochorista* in Tasmania, one of which has a mainland subspecies extending from Victoria to Ebor in N.S.W. They differ most in wing venation and in the faint colour pattern.

5. Bittacidae. This is a large, world-wide family, which is distinguished at once by the raptorial tarsi (Fig. 32.9) and slender, usually subpetiolate wings (Figs. 32.5E, F). The 6 species of *Harpobittacus*, with a wing-span about 50 mm, are the most commonly encountered Australian scorpion-flies. They can

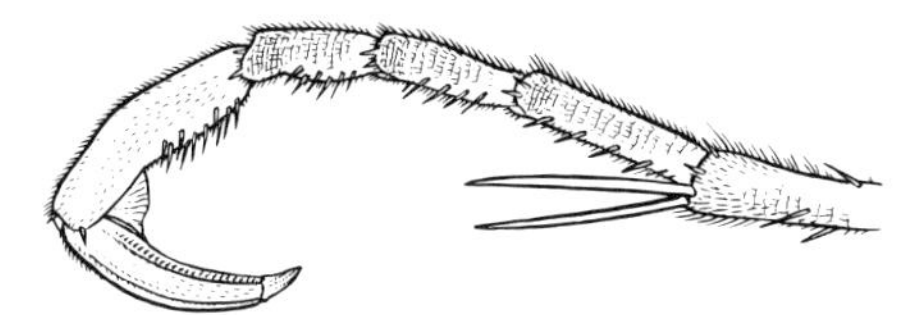

Fig. 32.9. Tarsus of *Harpobittacus* sp.
[M. Quick]

be divided into two species-groups on a combination of the presence or absence of an apical cross-vein between CuP and 1A and the shape of the apex of the hind basitarsus. A species of each group is often found in the one area, but under different ecological conditions, the *nigriceps* group preferring moist conditions often bordering on streams, whereas the *australis* group can survive under dry conditions. The larvae have well-developed branched processes on the body, and are at least 21 mm long when full-grown.

The other 2 Australian genera are smaller, more delicate insects. Two species of *Austrobittacus* occur in coastal Queensland and N.S.W., usually at the margins of rain forest. The single, similar-looking species of *Kalobittacus* occurs in drier situations in central Queensland. This genus is recorded also from central America.

33

SIPHONAPTERA

(Fleas)

by G. M. DUNNET

Apterous, laterally compressed, endopterygote Neoptera, with piercing and sucking mouthparts; ectoparasites of mammals and birds. Larvae apodous and vermiform, usually living in nests of hosts. Pupae adecticous, exarate.

This small order of highly specialized insects includes about 1,370 described species, 53 described and at least 15 undescribed species being known to occur in Australia. The adults (Fig. 33.1) are 1–6 mm long (males usually smaller than females), strongly sclerotized, and have long legs which enable them to leap characteristically. The body is covered with backwardly directed setae and spines, sometimes arranged in combs *(ctenidia)* to facilitate progress through the hair or feathers of the host. Fleas are so modified structurally for their particular kind of parasitic life (Snodgrass, 1946) that their relationships with other orders are difficult to determine (p. 166); and they show equally striking biological adaptations, in that all stages, except perhaps the egg, can withstand unfavourable environmental conditions for remarkably long periods. Their phylogeny, classification, and host relations have been reviewed by Holland (1964).

Anatomy of Adult

Head. Sessile on prothorax; clearly divided laterally by deep antennal groove into frons and occiput. Frons sometimes with small frontal tubercle, and sometimes developed into 'helmet' (*Stephanocircus*, Fig. 7.9). Genal ctenidium may be present. Compound eyes absent, but two atypical lateral ocelli may be large, vestigial, or absent. Antennae 3-segmented, with terminal segment usually consisting of 9 units plus a petiole; lying in a deep groove laterally; usually relatively long in male and used as secondary claspers in copulation. Mouthparts (Fig. 33.1c) adapted for piercing and sucking; mandibles absent; laciniae of maxillae long cutting blades more or less serrated on distal portion; stipites short, broad blades; maxillary and labial palpi well developed, the latter forming a sheath for the laciniae; epipharynx a long stylet.

Thorax. The three segments free, prothorax smallest, metathorax largest in association with the strong development of the

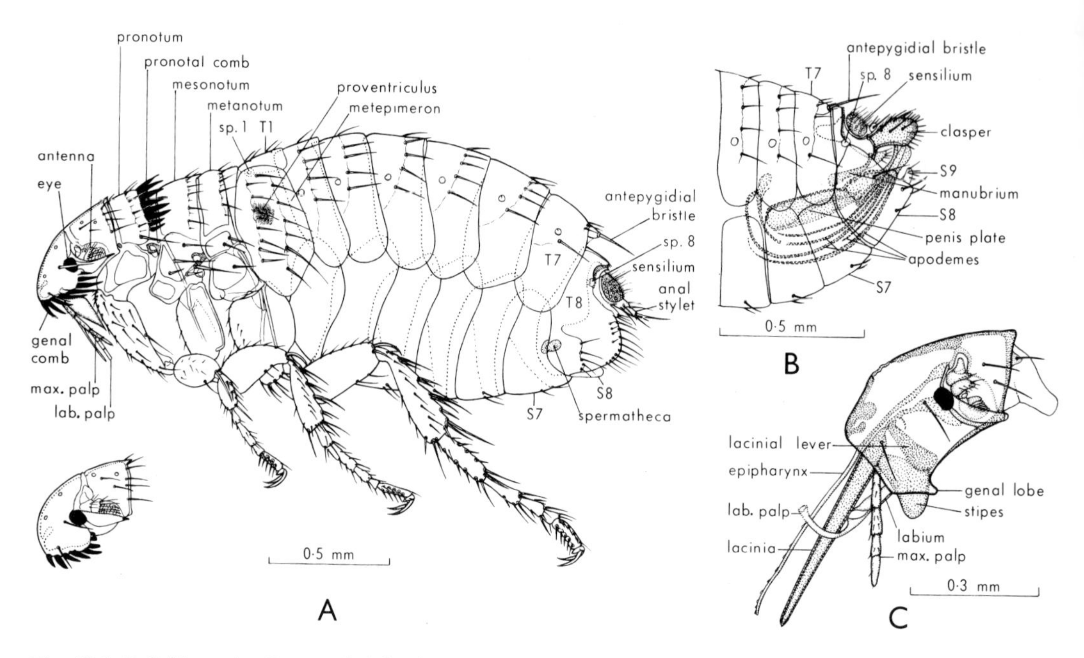

Fig. 33.1. Pulicidae: A, *Ctenocephalides felis*, ♀ (antenna partly hidden by genal process)—inset, head of *Ct. canis*, ♀, to show shape of frons; B, distal abdominal segments of *Ct. felis*, ♂; C, mouth-parts of *Echidnophaga myrmecobii*, ♀. [B. Rankin]
lab., labial; *max.*, maxillary; *sp.*, abdominal spiracle; *comb* = ctenidium.

legs. Prothorax frequently, mesothorax never, and metathorax rarely (in some bat fleas) with a ctenidium. Meso- and metathoracic spiracles present. Arrangement of sclerotized plates and setae is of taxonomic importance.

Legs. Long and well developed. Fore legs, especially coxae, modified for pushing through pelage or plumage of hosts; hind legs greatly enlarged for jumping. Articulating surfaces protected by setae. Fifth tarsal segments terminate in a pair of strong claws for clinging to host.

Abdomen. Ten segments can be distinguished, the posterior segments, especially in male, considerably modified. Terga with rows of setae and sometimes ctenidia. The dorsal *sensilium* occupies position of tergum 10, and consists of a number of sensory pits in an elaborate organ protected by a variable number of antesensilial setae of tergum 7. Spiracles on segments 1–8. In the male, sternum 8 may be large, medium-sized, or small, and tergum 8 is contrastingly small to large; sternum 9 is modified to form an L-shaped clasping organ; and an elaborate clasping organ (tergum 9), consisting of paired manubria and articulating claspers, occupies the posterior tip of the abdomen and encloses the complex aedeagus. In the female, the terminal segments are less modified; sternum 7 is frequently shaped and strengthened in a specific way; the spermatheca (paired in Macropsyllidae) is well sclerotized, and its shape and ducts are characteristic features.

Internal Anatomy. The principal modifications are in the digestive tract, in association with feeding on blood. There is a salivary pump for injecting saliva into the wound, and cibarial and pharyngeal pumps for sucking up the blood; the proventriculus is small, and provided with characteristic, radially arranged, sclerotized spines which may act as a valve or as a triturating mechanism; the stomach is very large. There are 4 Malpighian tubes, and 6 rectal glands. The nervous

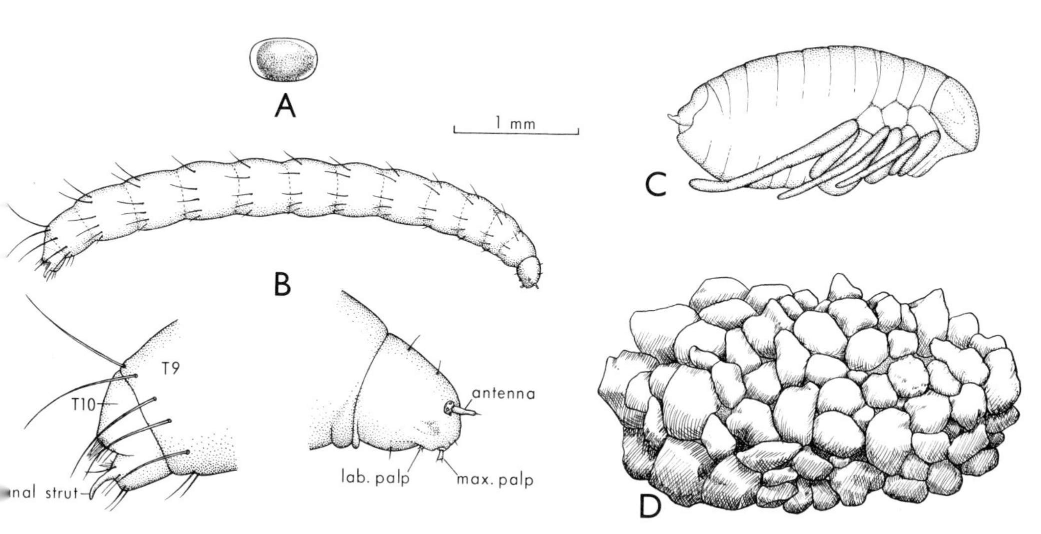

Fig. 33.2. *Ctenocephalides felis*: A, egg; B, larva, with caudal and cranial ends enlarged; C, pupa; D, sand-encrusted cocoon. [B. Rankin]

system is generalized, but with short connectives. The ovaries are panoistic.

Immature Stages

Egg (Fig. 33.2A). Oval, whitish, often glistening; about 0·5 mm long, and therefore large relative to the size of the adult.

Larva (Fig. 33.2B). Vermiform, whitish, rarely parasitic, 4–10 mm long when full-grown. Head usually well developed, with serrated mandibles; maxillae small and brush-like; no eyes; antennae 1-segmented. The 13 body segments, each with many rigid setae, are not clearly differentiated into thoracic and abdominal; last segment with pair of anal struts.

Pupa (Fig. 33.2C). Adecticous and exarate, in a thin, loosely constructed cocoon. The presence of wing-buds in the pupae of several species confirms the origin of fleas from winged ancestors.

Biology

Adults. Both sexes normally live exclusively on blood, but their relationship with their host is not nearly so close as that in the lice. Their larvae are nearly all free-living and the adults parasitic only intermittently, even the stickfast fleas *(Echidnophaga)* spending much of their time off the body of the host. Fleas also quickly leave a dead host and seek a new one, not necessarily of the same species, which accounts for their importance as transmitters of plague. The Pulicoidea are generally more dependent on their hosts than the Ceratophylloidea.

The degree of host specificity varies greatly, from the monoxenous *Bradiopsylla echidnae* (Denny) occurring only on the echidna, to *Echidnophaga myrmecobii* Rothsch. which has 15 known hosts, and *Pygiopsylla hoplia* J. & R. which is recorded from 24 hosts and occurs regularly on a wide variety of marsupials and rodents and even on domestic livestock. Conversely, whereas some hosts (e.g. *Gymnobelideus leadbeateri*) carry only a single species of flea, others support a variety of species (e.g. 11 have been recorded from *Rattus assimilis* and 7 from *Isoodon macrourus*). Some species are primarily nest fleas, occurring only rarely on the body of the active host, and these usually have a reduced chaetotaxy (e.g. *Acedestia chera* Jord. and *Idilla caelebs* Smit). There are relatively few bird fleas (about 100 spp.), all apparently derived as offshoots from diverse groups that evolved on mammals (Holland). In general, these are nest-inhabitants with a life cycle closely linked with the nesting activities of their hosts. Fleas occur on such aquatic mammals as the platypus and *Hydromys*, but not on seals whose ties with the land are brief and mainly intertidal.

Adults may live for a long time. Thus, Bacot (1914) found that *Pulex irritans* L. would survive for 125 days unfed and for more than 500 days if fed on man, *Xenopsylla cheopis* (Rothsch.) might live for more than a year, and other species for somewhat similar periods, observations that explain the not infrequent occurrence of fleas in premises that have been unoccupied for considerable periods. There are few observations on fleas of wild hosts, but Allan (1956) has shown that the European rabbit flea, *Spilopsyllus cuniculi* (Dale), may live for over 10 months in association with its host, and there is evidence that species associated with migratory birds overwinter in the nesting places as larvae, pupae, and adults.

Reproduction. Mating may take place on the host or in the nest, and may be stimulated simply by warmth or by a blood meal, although a blood meal is usually (perhaps always) necessary for maturation of the ovaries. In *S. cuniculi*, which is highly host-specific, the physiological state of the fleas, their feeding sites, maturation of ova, migration to infant rabbits in the nest, mating, and oviposition appear to be governed by the hormone levels, particularly of corticosteroids, in the blood of the host (Mead-Briggs, 1964; Rothschild, 1965). The breeding cycle of the parasite is thus linked closely with that of the rabbit. This kind of linkage has evident selective advantages for specialized nest breeders; it does not occur in *X. cheopis*, which lives on rats with quite different colonizing and breeding habits from those of the rabbit.

Life History. The eggs may be deposited on the body or in the nest of the host. Those deposited on the body are nearly always smooth and soon fall to the ground, whereas those laid in the nest are often sticky and adhere to the substrate. They are usually laid a few at a time, and *P. irritans* may lay a total of more than 400 in the course of its life (Bacot). The incubation period of the egg usually varies from about 2 to 12 days, and the first-instar larva frees itself from the shell by means of a sharp, spinous egg-burster on its head. The immature stages of all species studied require moderately high temperatures (20–30°C) and humidities (70 per cent or more) for development, but the larvae of the domesticated species, which often live in dust on floors and similar situations, may be able to resist adverse conditions better than those that are restricted to nests. Duration of the 3 larval instars varies from about 10–15 days to over 200, depending on the conditions to which they are exposed, and the pupa may require similar periods before emergence. In laboratory cultures, Sharif (1937) reported total developmental periods of 25–65 days for *Nosopsyllus fasciatus* (Bosc) and Kerr (1946) 20–24 days for *Ctenocephalides felis* (Bouché).

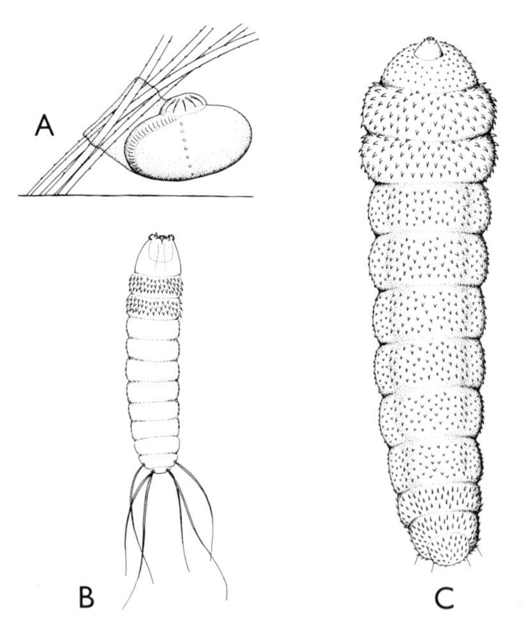

Fig. 33.3. Egg and parasitic larvae of *Uropsylla tasmanica*, Pygiopsyllidae, from *Dasyurus* and *Sarcophilus*: A, egg (length 0·65 mm) attached to hair; B, first-instar larva (body 0·5 mm long); C, full-grown larva (6 mm long) from burrow in skin. [From R. M. Warneke, unpublished]

Flea larvae have occasionally been found on the bodies of their hosts, but the only species with truly parasitic larvae would appear to be *Uropsylla tasmanica* Rothsch., which lives on dasyurids in Tasmania and Victoria (Warneke, unpublished). The eggs (Fig. 33.3A) are cemented to the fur of the host, and the first-instar larvae (Fig. 33.3B) have large mandibles which are used for tearing the chorion and penetrating the skin. Older larvae (Fig. 33.3C) are considerably modified, with very small heads, considerably expanded 2nd and 3rd trunk segments, specialized spines on the body, and probably only the 3 posterior pairs of spiracles functional. They live in burrows extending into the dermis, with only the posterior segments protruding, quickly withdrawing them into the burrow if disturbed. When mature, they drop to the ground, and spin a cocoon in the normal way.

Natural Enemies. No enemies of fleas, other than their hosts, have been recorded in Australia.

Economic Significance. Fleas can be serious domestic pests, because many people become sensitized to their bites, but they are of much greater public health importance as transmitters of disease. *Xenopsylla cheopis* is the principal vector of bubonic plague from rats to man. Australia has been fortunately free from this disease, except for periodic outbreaks in seaports, the last more than forty years ago; sylvatic plague is unknown here. A less dangerous infection, murine typhus, is endemic in country towns, and is also transmitted from rodents by their fleas. They serve, too, as intermediate hosts of the dog tapeworm, *Dipylidium caninum*, and the rodent tapeworm, *Hymenolepis diminuta*, both of which occasionally infect children, of one of the dog filarioids, *Dipetalonema reconditum*, and of the non-pathogenic *Trypanosoma lewisi* of rats. All these parasites have been recorded in Australia. It was thought also that species of *Echidnophaga* might be useful in spreading myxomatosis of rabbits, but they were not found to contribute to the development of epizootics in the field (Fenner and Ratcliffe, 1965). The veterinary importance of fleas in Australia has been reviewed by Roberts (1952).

Special Features of the Australian Fauna

The fleas so far known from Australia fall into three groups: 9 cosmopolitan species of 3 families; 3 species of *Parapsyllus*, a rhopalopsyllid genus derived from South American rodent fleas, now circumpolar parasites of penguins and petrels; and the endemic fauna of 56 species in 6 families dominated by the Pygiopsyllidae. Several of the endemic fleas, closely associated with their marsupial and rodent hosts, have related forms on South American marsupials and rodents, and also in Africa. Endemic genera of bat fleas are also found. Only 3 fleas are known from

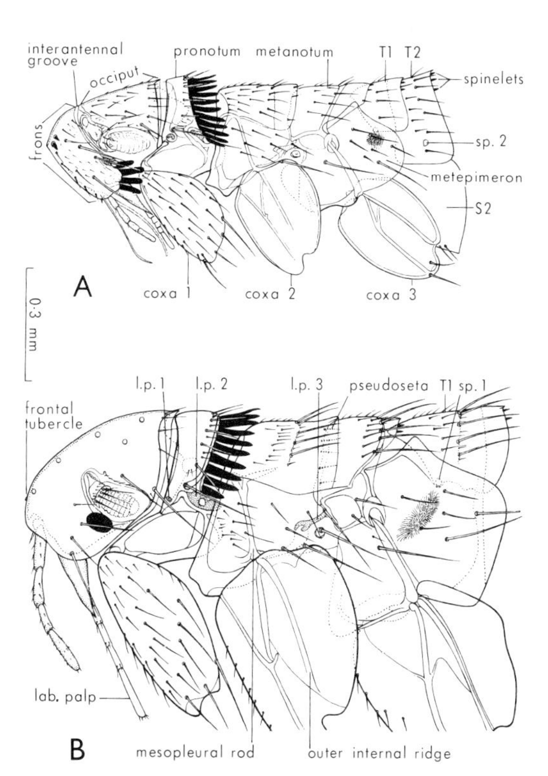

Fig. 33.4. A, *Leptopsylla segnis*, Leptopsyllidae, ♂ from *Rattus lutreolus*—fracticipit; B, *Nosopsyllus fasciatus*, Ceratophyllidae, ♀ from *Rattus rattus*—integricipit. [B. Rankin]

l.p. 1–3, first to third link plates (2 and 3 with associated thoracic spiracles); *sp.*, abdominal spiracle; *pseudosetae* lack sockets.

land birds, the cosmopolitan *Echidnophaga gallinacea* (Westw.) and *Ceratophyllus gallinae* (Schrank) on poultry, and the endemic *Hoogstraalia vandiemeni* Smit found on small passerines in Tasmania. Subspeciation in different parts of their range is a common feature of flea distributions (Holland).

CLASSIFICATION

Order SIPHONAPTERA (68 Australian spp.)

PULICOIDEA (15)
Tungidae (0)
1. Pulicidae (15)

CERATOPHYLLOIDEA (53)
2. Rhopalopsyllidac (3)
Malacopsyllidae (0)
Vermipsyllidae (0)
3. Pygiopsyllidae (29)
Xiphiopsyllidae (0)
Coptopsyllidae (0)
4. Hystrichopsyllidae (2)
5. Stephanocircidae (9)
6. Macropsyllidae (2)
7. Ischnopsyllidae (4)
8. Leptopsyllidae (1)
Amphipsyllidae (0)
Ancistropsyllidae (0)
9. Ceratophyllidae (3)
Hypsophthalmidae (0)

The classification adopted here is that of Hopkins and Rothschild (1962). Differentiation of the superfamilies is not simple, requiring several characters. Well-cleared material is essential for identification, but considerable care in treatment in KOH is necessary with small, pale specimens. No key to the larvae is available.

Key to the Families of Siphonaptera known in Australia

1. Hind coxa with spiniform setae on lower part of inner side; outer internal ridge of mid coxa absent; mesonotum without pseudosetae under the collar; abdominal terga 2–7 with not more than one row of setae; no setae above spiracle of tergum 8; sensilium with 14 pits on each side ... **Pulicidae**
   Hind coxa with no spiniform setae on inner surface; outer internal ridge of mid coxa present; mesonotum with pseudosetae under the collar; terga 2–7 with more than one row of setae (except Hystrichopsyllidae); sensilium with 16 or more pits on each side 2
2. Occiput with conspicuous, single, subdorsal, internal tubercle on each side 3
   Occiput without conspicuous internal tubercle 4
3. Head with conspicuous ctenidium extending along ventral margin and up antennal fossa .... **Macropsyllidae**
   Head with vertical anterior helmet with conspicuous ctenidium, and separate vertical genal ctenidium .... **Stephanocircidae**
4. Fourth link-plate present between basal abdominal sternum and metepimeron. [Antennal fossa closed; no frontal tubercle—except *Lycopsylla*; sensilium exceptionally convex—except *Lycopsylla*, in which there is a small hook-like projection at its anterior end] .... **Pygiopsyllidae**
   No fourth link-plate 5
5. Metanotum without marginal spines or spinelets; genal ctenidium of 4 sharply pointed spines; no setae on lateral surfaces of femora; ♂ with no apodeme extending anteriorly from angle of sternum 9; ♀ with single apical seta on long slender stylet .... **Hystrichopsyllidae**
   Metanotum with spinelets; other characters not as above 6
6. Fracticipit (Fig. 33.4A). [Antennal fossa closed; club of ♂ antenna not extending to propleuron] ... 7
   Integricipit (Fig. 33.4B), with an inter-antennal suture at most indicated 8
7. Head with 2 broad, blunt, preoral spines at anterior ventral angle. [On bats] ...... **Ischnopsyllidae**
   Head with 2 spiniform setae at frontal angle, and 4 blunt spines in vertical genal ctenidium. [Row of spiniform setae on dorsoposterior margin of tibiae] .... **Leptopsyllidae**
8. No ctenidia; lower half of frons with arrow-shaped tubercle sunk in a groove and pointing forward and upward. [Single long antesensilial seta in both sexes; on sea birds] .... **Rhopalopsyllidae**
   Pronotal but no genal ctenidium; frontal tubercle present. [Eye well developed, but no tentorial arch in front of it] .... **Ceratophyllidae**

Superfamily PULICOIDEA

1. Pulicidae (Figs. 33.1, 5A–C). The 4 Australian genera are of medical and veterinary interest, and may be distinguished by the following key.

1. Pleural rod of mesothorax absent 2
   Pleural rod of mesothorax present 3
2. Frons angulate; metanotum much shorter

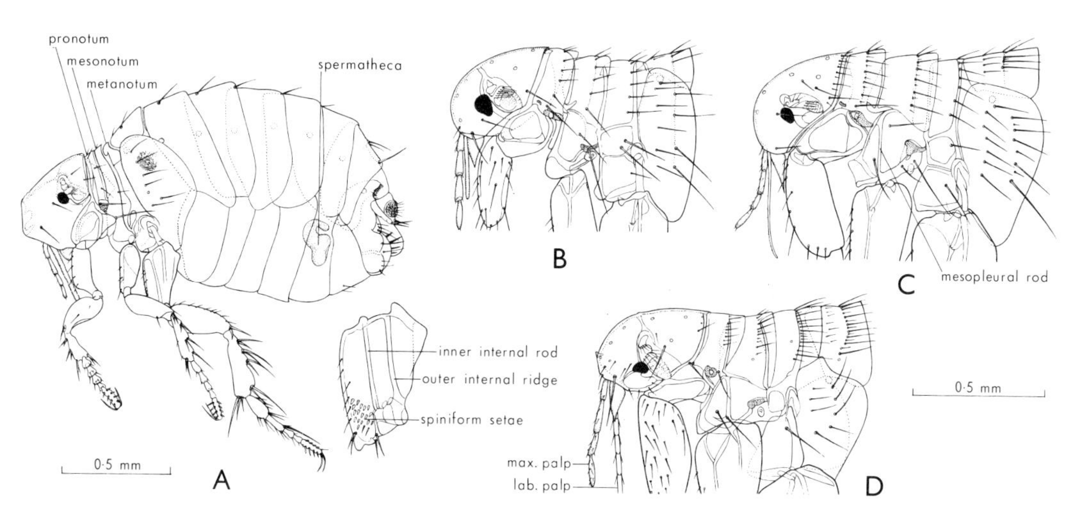

Fig. 33.5. A, *Echidnophaga myrmecobii*, Pulicidae, ♀ from *Trichosurus vulpecula*—inset, inner aspect of hind coxa enlarged; B, *Pulex irritans*, Pulicidae, ♂ from man; C, *Xenopsylla vexabilis meseris*, Pulicidae, ♀ from *Rattus villosissimus*; D, *Parapsyllus taylori*, Rhopalopsyllidae, ♀ from *Puffinus tenuirostris*. Right-hand scale for B–D. [B. Rankin]

than abdominal tergum 1; labial palp membranous .... *Echidnophaga*
Frons smoothly rounded; metanotum nearly as long as abdominal tergum 1; labial palp sclerotized and stiff .... *Pulex*

3. Genal ctenidium of 8 long, curved, sharp spines .... *Ctenocephalides*
No genal or pronotal ctenidium. [Hind coxa narrowing suddenly below the middle of posterior margin] .... *Xenopsylla*

The introduced species, with their normal hosts, are: *Pulex irritans*, man; *Xenopsylla cheopis*, rats; *Ctenocephalides canis* (Curtis), dogs; *C. felis*, cats; and *Echidnophaga gallinacea*, fowls; all overflow on to quite a wide range of casual hosts. *Xenopsylla vexabilis* Jord. is a native species on rodents, mainly in warmer parts of the country, and there are 8 native species of the predominantly Australian 'stickfast fleas', *Echidnophaga myrmecobii* and *E. perilis* Jord. being the best known; Hopkins and Rothschild (1953) have provided a key for their identification.

Superfamily CERATOPHYLLOIDEA

2. Rhopalopsyllidae (Fig. 33.5D). A South American family, represented in Australia by the circumpolar genus *Parapsyllus*, which occurs in the nests of penguins and petrels on the southern coasts from Perth to Sydney.

3. Pygiopsyllidae (Fig. 33.6). This family is primarily Australian, but has representatives in south-east Asia, Africa, and South America. Eight genera have been recorded from Australia. The aberrant *Hoogstraalia*, distinguished by having a genal ctenidium of 10–11 long slender spines, is known only from one rare species in nests of *Acanthornis magnus* and *Sericornis humilus* in Tasmania and another from a thrush in the Philippines. *Notiopsylla*, which parasitizes flying sea birds, has not been recorded, but probably occurs in the Tasmanian area. The other genera are all found on mammals: *Pygiopsylla* on the platypus and a wide range of marsupials and native rodents; *Acanthopsylla* on a variety of terrestrial and arboreal marsupials and rodents, including *Conilurus*; *Stivalius* on native rats in Queensland; and *Choristopsylla* on several different phalangers; but the other 3 genera appear to be more restricted, *Bradiopsylla* to the echidna, *Lycopsylla* to wombats, and *Uropsylla* to *Dasyurus* and *Sarcophilus*. The introduced *Rattus rattus* has become widely established in the canefields and bush, and is included among the hosts of several of the species that feed on rodents.

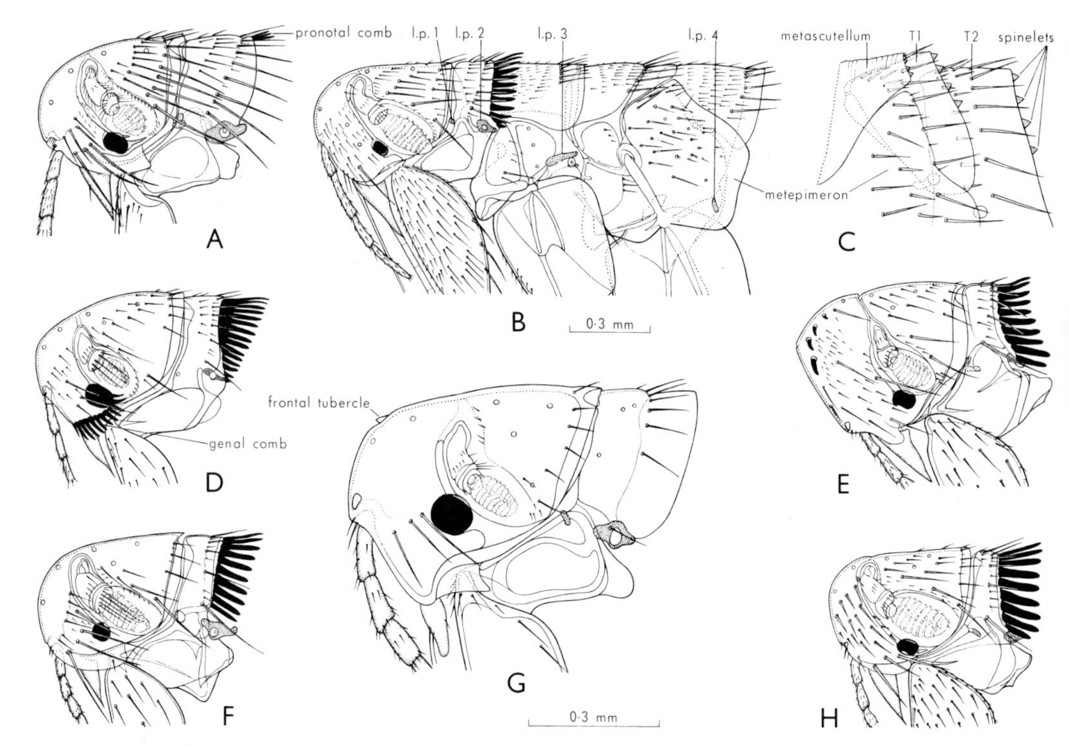

Fig. 33.6. Pygiopsyllidae (upper scale for A–C, lower scale for D–H): A, *Bradiopsylla echidnae*, ♂ from *Tachyglossus aculeatus*; B, *Pygiopsylla rainbowi*, ♂ from *Rattus lutreolus*; C, *Uropsylla tasmanica*, ♀ from *Dasyurus viverrinus*; D, *Hoogstraalia vandiemeni*, ♀ from nest of *Sericornis humilus*; E, *Acanthopsylla scintilla*, ♀ from *Eudromicia lepida*; F, *Choristopsylla ochi*, ♂ from *Trichosurus vulpecula*; G, *Lycopsylla nova*, ♀ from *Vombatus hirsutus*; H, *Stivalius rectus*, ♂ from *Rattus fuscipes assimilis*. [B. Rankin] *l.p. 1–4*, first to fourth link plates.

4. Hystrichopsyllidae (Fig. 33.7A). The subfamily ACEDESTIINAE (see Hopkins and Rothschild, 1962, p. 35) includes only 2 aberrant monotypic genera, which appear to be associated with the nests of marsupials. *Acedestia* is known from 3 females from bandicoots and a potoroo, and *Idilla* from numerous specimens taken from *Antechinus flavipes*.*

5. Stephanocircidae (Fig. 7.9). There are 2 subfamilies: Craneopsyllinae with 7 genera in South America, and STEPHANOCIRCINAE with one genus, *Stephanocircus*, in Australia. Five species of *Stephanocircus* have been described, but at least 4 others are known. They parasitize a wide variety of rodents and marsupials, and at least one species occurs on arboreal phalangers.

6. Macropsyllidae (Figs. 33.7B, C). Exclusively Australian, with 2 monotypic genera. *Macropsylla*, with high convex frons, occurs widely on rodents (also on *Antechinus*); *Stephanopsylla*, with very shallow frons, is known from a rat in the Monte Bello Is. and a phascogale in Victoria.

7. Ischnopsyllidae (Figs. 5.22C, 33.7D). There are 3 genera, *Lagaropsylla*, *Porribius*, and an undescribed genus with strongly sclerotized bands on terga 1–7 of the abdomen. They are widely distributed on a variety of Microchiroptera, but too few specimens are available to permit discussion.

8. Leptopsyllidae (Fig. 33.4A). This family is represented by one cosmopolitan species,

* Hopkins and Rothschild (1966, p. 113) now refer these genera to the Doratopsyllinae.

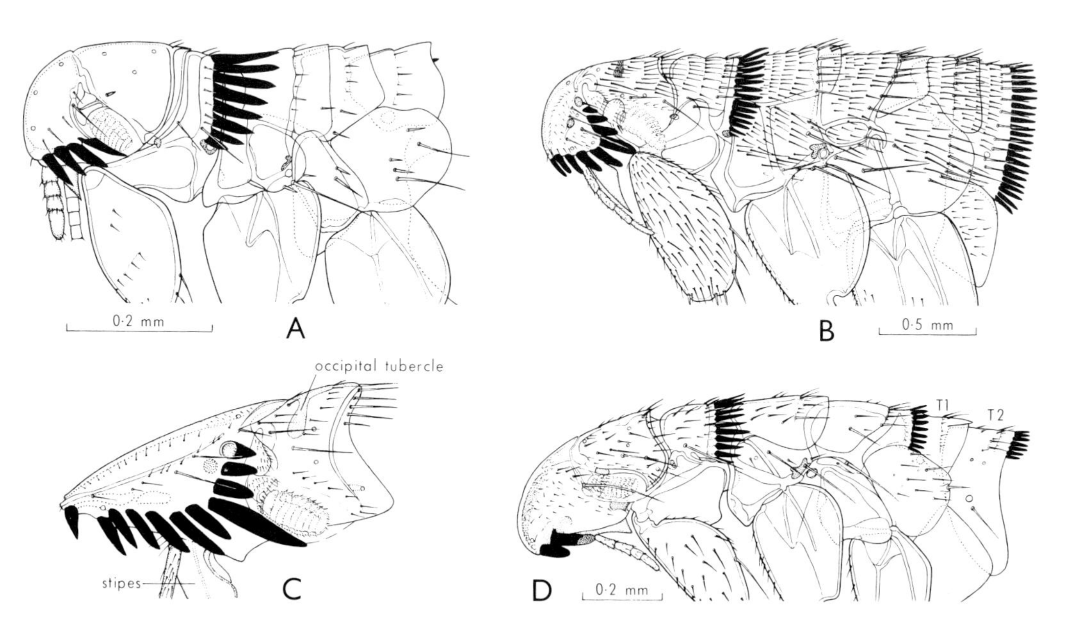

Fig. 33.7. A, *Idilla caelebs*, Hystrichopsyllidae, ♀ from *Antechinus stuartii*; B, *Macropsylla hercules*, Macropsyllidae, ♂ from *Rattus fuscipes assimilis*; C, *Stephanopsylla thomasi*, Macropsyllidae, ♀ from *Pseudomys ferculinus* (after Hopkins and Rothschild, 1956); D, *Porribius bathyllus*, Ischnopsyllidae, ♀ from *Tadarida australis*.
[B. Rankin]

Leptopsylla segnis (Schönh.), which occurs on the introduced house mouse and the black rat, and very occasionally on native rodents and small dasyurids.

9. Ceratophyllidae (Fig. 33.4B). The cosmopolitan *Nosopsyllus fasciatus* and *N. londiniensis* (Rothsch.) occur on introduced mice and rats, and both have been found occasionally on native mammals. The European poultry flea, *Ceratophyllus gallinae*, has been recorded in N.S.W. (Roberts, 1952).

ACKNOWLEDGMENTS. I am greatly indebted to Mr R. M. Warneke, Victorian Fisheries and Wildlife Department, for permission to quote from his unpublished studies of *Uropsylla tasmanica*. I also wish to acknowledge the information and unstinted assistance which I have received from Dr I. M. Mackerras in the preparation of this chapter.

34

DIPTERA

(Flies)

by D. H. COLLESS *and* D. K. McALPINE

Endopterygote Neoptera, with a pair of membranous wings on mesothorax only, the metathoracic pair represented by club-like halteres; prothorax and metathorax greatly reduced; mouth-parts of adults suctorial, often adapted for piercing. Larvae without true legs. Pupae adecticous and obtect or exarate, the latter in a puparium.

The Diptera form one of the larger insect orders, the world total of species, described and undescribed, probably being at least 150,000. Despite a considerable diversity of structure, almost all adults are immediately recognizable by the presence of only one pair of functional wings. The exceptions are the relatively few apterous species, which bear other characters of mouth-parts, thorax, etc., that indicate their true relationships. These characters also distinguish the Diptera from the few other insects (some Ephemeroptera, male Coccoidea) that have only two wings.

The order includes many common and familiar insects: mosquitoes, midges, sand flies, house flies, blow flies, etc. Some are important pests or vectors of disease, but others are beneficial, and, by virtue of their parasitic or predatory habits, play an important role in regulating the populations of many plants and animals that adversely affect human welfare. We may also note our special debt to the inconspicuous *Drosophila*, to which we owe so much of our basic knowledge of cytogenetics and genetic mechanisms.

Important general works include those on anatomy by Séguy (1951), on the larvae and pupae by Hennig (1948–52) and Brauns (1954), on cytology by White (1949) and Boyes (1958), on biology by Séguy (1950, 1951) and Oldroyd (1964), and on the northern-hemisphere faunas by Lindner (1924–), Stone *et al.* (1965), and Curran (1965).

Anatomy of Adult

For reasons that are largely historical, the terminology of the external anatomy of the Diptera is in a confused state, with various special terminologies applying within limited groups of families. Some terms, particularly where the homologies are obscure, are not precisely defined, and their limits are variously drawn by different authors; in fact, a given term may have several different meanings.

The terminology recommended below is based largely on the classical treatise by Crampton (1942), but it has been necessary to make certain arbitrary selections, and to accept a few unfortunate, but long established usages. Generally, many terms seem destined to remain conventional or topographical rather than strictly morphological.

Head (Fig. 34.1). A highly mobile, relatively large capsule, its parts defined principally by reference to the occipital foramen, median ocellus, insertions of antennae, margins of eyes and oral region, and the anterior tentorial pits. The tentorial pits may be poorly, or not at all, developed, particularly in the Cyclorrhapha. They are best seen in the Orthorrhapha, and completely penetrate the head in some Stratiomyidae (e.g. *Boreoides*), the arms of the tentorium forming a pair of hollow tubes.

The *occiput* may be flattened or concave in higher Diptera, with a distinct *median sclerite* (epicephalon, cerebrale). The *vertex* is of rather indefinite extent, but there may be a distinct *vertical triangle*, often raised and defined by grooves, then better termed *ocellar tubercle* (or ocellarium). The median ocellus, when present, marks the dorsal limit of the true frons, which, strictly, includes all the anterior surface down to the clypeus. However, students of the higher Diptera traditionally restrict the term to include only the area dorsal to the antennae. It would be better to use 'front' or 'postfrons' for that area, or include it under the general designation of 'vertex', as is usual in the Nematocera; but past attempts to change the usage have not been successful. The *frons*, in this restricted sense, may be differentiated into a median *interfrons* (frontal stripe or vitta) and *parafrontal* or *fronto-orbital* areas adjacent to the eyes. In the Schizophora, a transverse *ptilinal fissure* crosses just above the antennae and extends down laterally, in the form of an inverted U, towards the clypeus.* It represents the closed lips of the *ptilinum*, an eversible sac used in eclosion, and cuts off a small oval or crescentic

* A transverse groove may occur in other Diptera, but it rarely continues down laterally past the antennae.

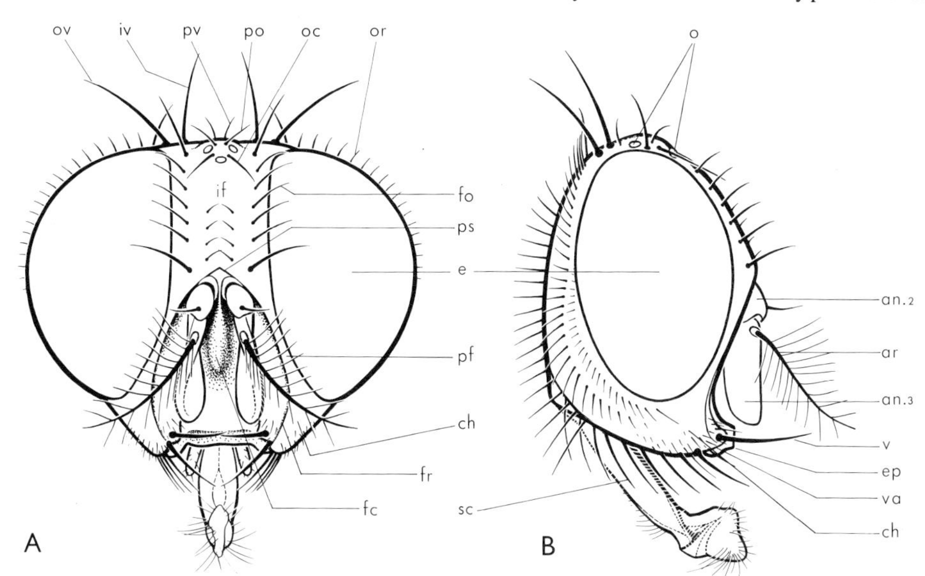

Fig. 34.1. Head of muscoid fly: A, anterior; B, lateral.
[T. Binder]

an. 2, *an. 3*, antennal segments 2 and 3; *ar.*, arista; *ch.*, cheek; *e.*, eye; *ep.*, epistoma; *fc.*, facial carina; *fr.*, facial ridge; *o.*, ocellus; *pf.*, parafacial; *ps.*, ptilinal suture; *va.*, vibrissal angle. Bristles: *fo.*, fronto-orbital; *if.*, interfrontal; *iv.*, inner vertical; *oc.*, ocellar; *or.*, orbital; *ov.*, outer vertical; *po.*, postocellar; *pv.*, postvertical; *sc.*, subcranial; *v.*, vibrissa.

frontal lunule above the bases of the antennae.

The area below the antennae is the *face*, a general term for the anterior frons, which is bounded ventrally by the fronto-clypeal or epistomal suture; or, if this is absent, by the level of the tentorial pits. Some authors include the clypeus under the term 'face', but this is undesirable if the sclerites are at all differentiated. The ventral arms of the ptilinal fissure divide the face into a median *facial plate* and lateral strips, the *parafacials*. The latter term is also applied to the similar, but not homologous, strips between the eyes and the enlarged clypeus in, e.g., Tabanidae. These would be better called *genae*, but that term is often restricted to the more ventral areas, better styled *cheeks* (jowls, buccae) below the level of the eyes. In Cyclorrhapha, the antennae may lie on the face in longitudinal grooves, the *antennal fossae* or *foveae*, separated by a median ridge, the *facial carina*; or the whole face may be sunken. In such cases, the raised lateral margins form the *facial ridges* or *facialia*, which may terminate ventrally in prominences, the *vibrissal angles*, each bearing one or more stout bristles or *vibrissae*. Between the vibrissal angles, there may be differentiated a

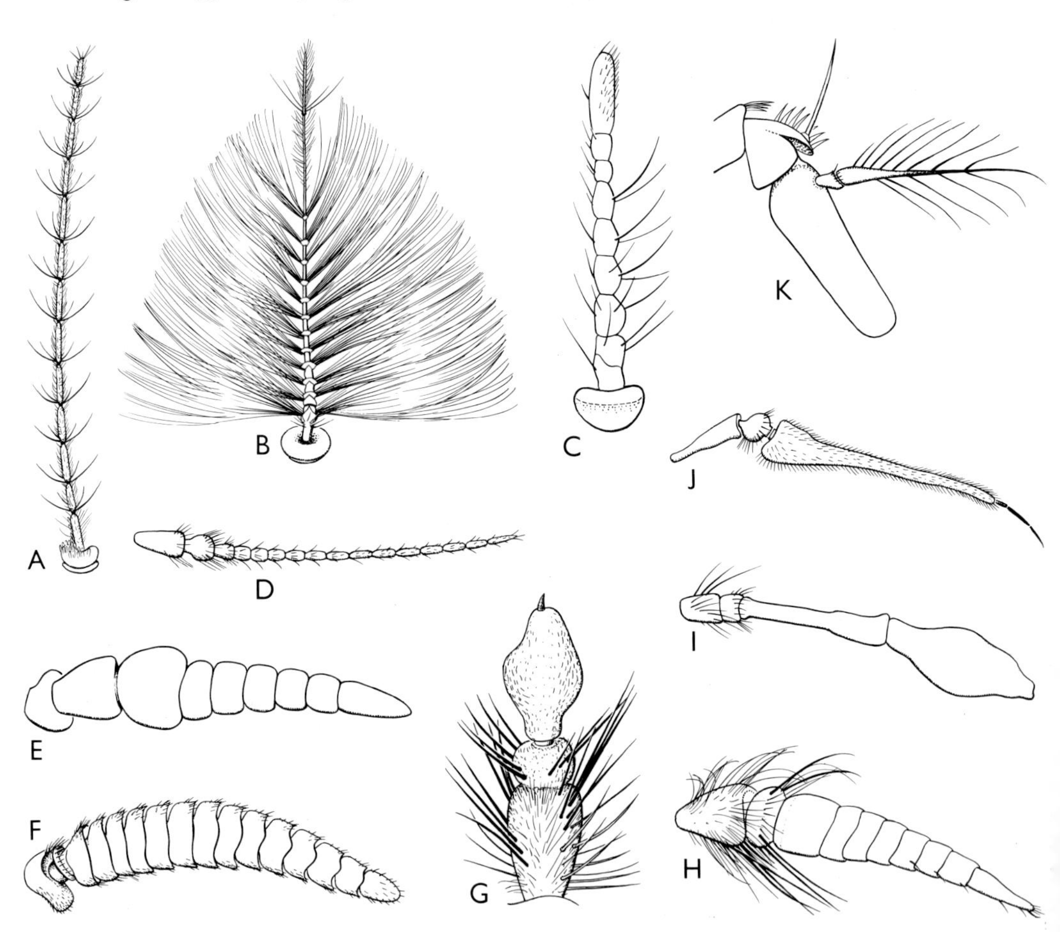

Fig. 34.2. Antennae of various Diptera: A, *Aedes alternans*, Culicidae, ♀; B, *A. alternans*, ♂; C, *Heptagyia tasmaniae*, Chironomidae, ♀; D, *Sylvicola dubius*, Anisopodidae, ♀; E, *Austrosimulium bancrofti*, Simuliidae, ♀; F, *Keroplatus mastersi*, Mycetophilidae, ♂; G, *Apiocera asilica*, Apioceridae, ♀; H, *Scaptia maculiventris*, Tabanidae, ♀; I, *Miltinus viduatus*, Mydaidae, ♂; J, *Xiphandrium pudicum*, Dolichopodidae, ♂; K, *Musca vetustissima*, Muscidae, ♂. [T. Binder]

projecting *epistoma*, which is probably the postclypeus and, strictly, not part of the face. Below the face, the clypeus or anteclypeus ('prelabrum') is usually distinct, sometimes borne on the proboscis. The cavity into which the proboscis is retracted, often miscalled the 'oral cavity', is the *subcranial cavity*, and its margin the *subcranial margin*. The posteroventral region of the head-capsule is formed by the postgenae, often joined by a sclerotized region, the *pseudogula*.

The compound eyes may be dichoptic or holoptic. The holoptic condition is typically seen in males; or in both sexes in families, such as the Acroceridae and Pipunculidae, in which the eyes may occupy most of the surface of the head. In some Nematocera (Sciaridae, Cecidomyiidae) the eyes are connected above the antennae by a narrow line of facets, or *eye-bridge* (Fig. 34.20C). Sometimes each eye is differentiated into dorsal and ventral components by a difference in facet size or by a transverse groove. In the cecidomyiid *Trisopsis*, the components are completely separated, and the dorsal parts contiguous, giving a three-eyed appearance, while in *Perissomma* (Fig. 34.18D) all four parts are completely separated. Many Diptera have small hairs between the facets, som ti ıes producing a marked pilosity over the whole eye. There are typically 3 ocelli, but the median one, or all 3, may be absent.

The antennae (Fig. 34.2) vary considerably in structure, and may exhibit strong sexual dimorphism. The simplest type is filiform, with 16 segments, but the number can be greater or, particularly in the Brachycera, much less. The scape may be rudimentary (e.g. in the Culicidae), while in many Nematocera the pedicel is distinctly enlarged and the remaining, flagellar segments more or less uniform in size and shape. In most higher Diptera, the 3rd segment is enlarged to some degree and the more distal segments reduced and appearing as a mere appendage. When fine and bristle-like, the appendage is termed an *arista*; when stouter and more rigid, a *style*; its segmentation may or may not be apparent. In many Diptera, the pedicel (torus) encloses Johnston's organ, a group of receptors which serve to detect air vibrations (p. 44).

The mouth-parts (Figs. 34.3, 4) are adapted for sucking, and form a more or less elongate *proboscis* or *rostrum*, which usually incorporates the labrum, and sometimes elements of clypeus and even of frons, as in *Elephantomyia* (Tipulidae) and *Neoantlemon* (Mycetophilidae). The labrum (labrum-epipharynx) may be small and flap-like, but it is normally more or less elongated to form the roof of the food canal. The floor of the canal is formed either by the overlapping mandibles or by the elongate, deeply grooved hypopharynx into which the salivary duct opens basally. The mandibles, usually absent in males and often so in females, are found mainly in predatory species as long, piercing stylets. The maxillae have their basal sclerites reduced and fused into the structure of the proboscis. The free portions comprise the slender, elongate lacinia, which is toothed or pointed in most predators, or, in many Nematocera, brush-like at its apex; and the 1- to 5-segmented palp, which may bear *sensory pits* or *plaques*. The labium is the largest of the mouth-parts, and both mentum and prementum may be distinct; in the proboscis of many Cyclorrhapha, the prementum forms a conspicuous ventral sclerite, sometimes called the *theca*. The labial palps are represented by the *labella*, a pair of apical lobes on the prementum. In some lower or predatory forms, they are slender and 2-segmented; but in higher forms, they become a pair of inflated membranous lobes (sometimes fused), with their surfaces traversed by a series of sclerotized canals, the *pseudotracheae*, which may bear *pseudotracheal teeth*, or be replaced by *prestomal teeth*. The piercing type of proboscis found in predators and blood-suckers may have been independently evolved in a number of lineages, although Downes (1958) regards the 'biting' type of mouth-parts as the primitive condition, and 'non-biting' groups as independent, secondary developments.* Usually the labrum, galeae, mandibles, and

* See also J. A. Downes and D. H. Colless, *Nature, Lond.* **214:** 1355–6, 1967.

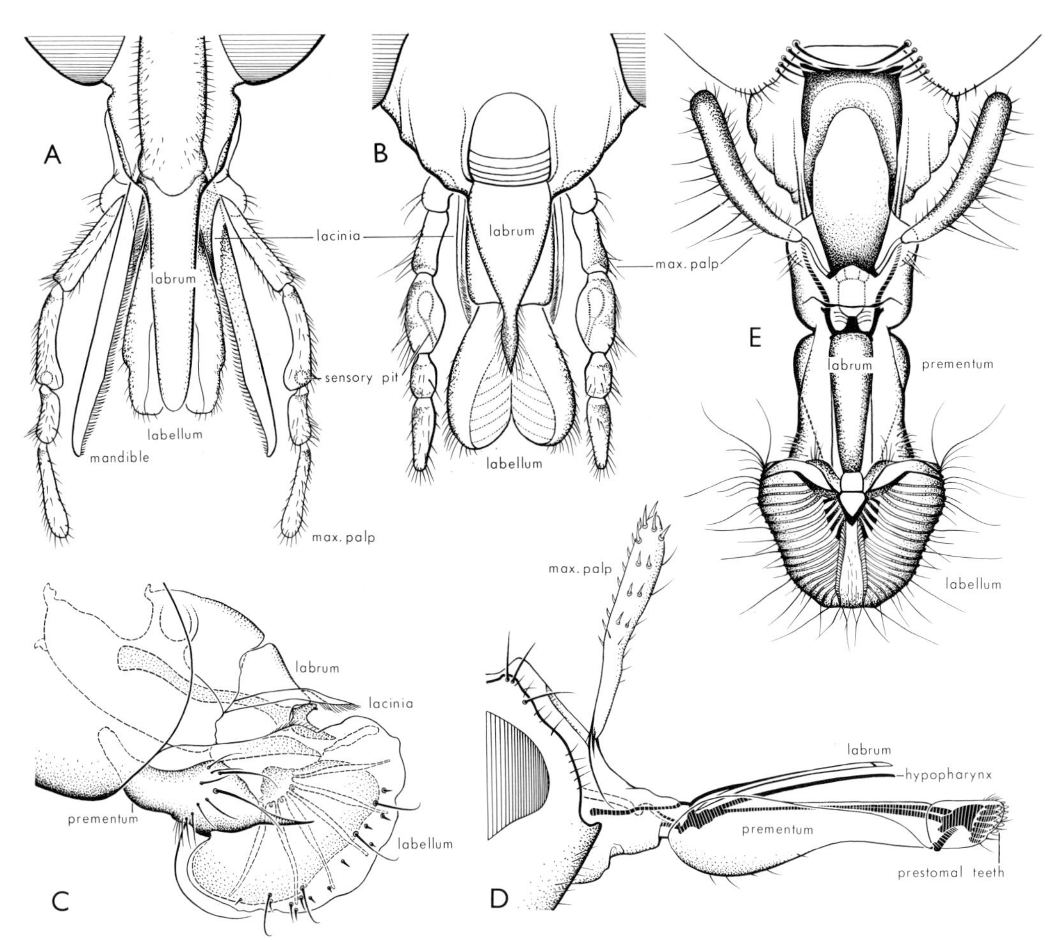

Fig. 34.3. Mouth-parts of various Diptera (all ♀♀): A, *Edwardsina* sp., Blephariceridae, dorsal; B, *Sylvicola dubius*, Anisopodidae, dorsal; C, *Heteropsilopus cingulipes*, Dolichopodidae, lateral, palp not shown; D, *Haematobia exigua*, Muscidae, lateral; E, *Calliphora stygia*, Calliphoridae, dorsal. [T. Binder]

hypopharynx form a set of elongate *stylets*, wrapped within a groove in the fleshy labium (e.g. Culicidae, Fig. 34.4), or the labium itself may also be sclerotized to form a piercing organ (e.g. *Haematobia*, Fig. 34.3D).

Chaetotaxy of the head is shown in Figure 34.1.

Thorax (Fig. 34.5). Both pro- and metathorax are greatly reduced, and the mesothorax correspondingly enlarged to accommodate the muscles of the single pair of wings. The principal thoracic landmarks are the meso- and metathoracic spiracles, and the attachments of wings, halteres, and legs.

The cervix (neck) is a largely membranous area of modified prothorax, and bears at least one lateral pair of *cervical sclerites*. The pronotum is usually clearly divided into anterior and posterior parts; the *anterior pronotum* is best developed in lower Diptera (e.g. Tipulidae, Bibionidae) and may form a pair of prominent *pronotal lobes*, whereas the *posterior pronotum*, better developed in higher forms, is usually intimately associated with the mesonotum to form the *humeri* or *humeral calli*. Laterally, the episternum *(propleuron)* is normally distinct, but the epimeron may merge completely with the mesopleuron. The principal sternal sclerite is the basisternum *(prosternum)*, lying a little

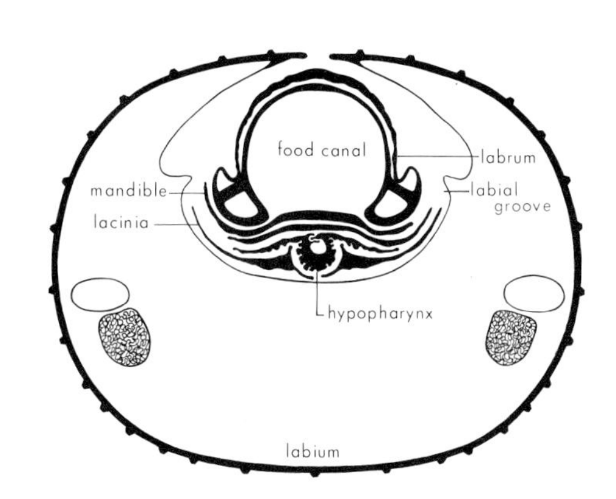

Fig. 34.4. Transverse section towards apex of proboscis of *Aedes aegypti*, Culicidae (from electron micrograph prepared by B. K. Filshie). [T. Binder]

anterior to the bases of the coxae. It may connect with the episternum via a narrow *precoxal bridge* (e.g. Culicidae, Bibionidae), or the two sclerites may be broadly fused (e.g. certain Empididae and Therevidae).

Enlargement of the mesothorax has resulted in distortion, and the homologies of its sclerites are not readily apparent. The mesonotum is, strictly, its entire dorsal surface, but the term is traditionally restricted to the larger, anterior, wing-bearing plate (*eunotum*, or alinotum) only. The eunotum (often called 'scutum') is frequently divided by a transverse sulcus or 'suture' (V-shaped in Tipulidae) into 'prescutum' and 'scutum' (again not strictly correct morphologically), between the lateral parts of which there may be a triangular, sunken *notopleural area.* The *humeral pits* lie anteriorly on each side. In Nematocera and many Orthorrhapha, a narrow lateral sclerite, the *paratergite* (a vestige of the true prescutum), is differentiated in front of the wings, while, in Brachycera, the posterolateral angles of the scutum form the *postalar calli.* The *scutellum*, a clearly defined, rounded or triangular lobe, bounds the eunotum posteriorly, and is sometimes preceded by a small ridge, originally termed proscutellum, but for obvious reasons better termed *prescutellum.*

Even more confusing terminologies have been applied to the parts below and behind the scutellum. They probably represent the intersegmental acrotergite, and should preferably be termed the *postnotum*; 'postscutellum' (see below) has also been used, and even (quite incorrectly) 'metanotum'. The postnotum is divided longitudinally into a median *mediotergite* and lateral *pleurotergites* or laterotergites; the latter (sometimes miscalled 'metapleura') extend down to the metathoracic spiracle and may be divided into a dorsal *anatergite* and ventral *katatergite.* The anterior region of the mediotergite is usually concave but, particularly in Tachinidae, may be differentiated as a convex, transverse ridge or lobe, the *subscutellum* (often called 'postscutellum', but conflicting usage of the term makes it undesirable).

The mesothoracic spiracle lies towards the dorsal margin of the pleuron, near the junction of pronotum and mesonotum. The mesopleuron is divided by the vertical *pleural suture*, which runs more or less directly from wing base to coxa in most Nematocera, but becomes markedly zigzag in higher forms. It may include a distinct *midpleural pit.* The episternum comprises a dorsal *anepisternum* (mesopleuron) and ventral katepisternum or *sternopleuron*, the former often divided by a vertical *anepisternal cleft*, which may lie posteriorly, very close to the true pleural suture. The two sternopleura are often contiguous ventrally, fusing with or displacing much of the small sternum. The epimeron commonly has its dorsal anepimeron or *pteropleuron* clearly demarcated; the katepimeron is usually small, or fused with the detached meron of the coxa to form a composite *meropleuron*; but the meron is distinct in some Nematocera. Both meron and meropleuron have been miscalled 'metasternum', or included amongst the various meanings of 'hypopleuron', a term that has lost even its topographic usefulness and should be discarded.

The metathorax may be so reduced as to be almost vestigial. The metanotum forms a narrow, transverse strip (best developed in some Psychodidae), with the halteres arising

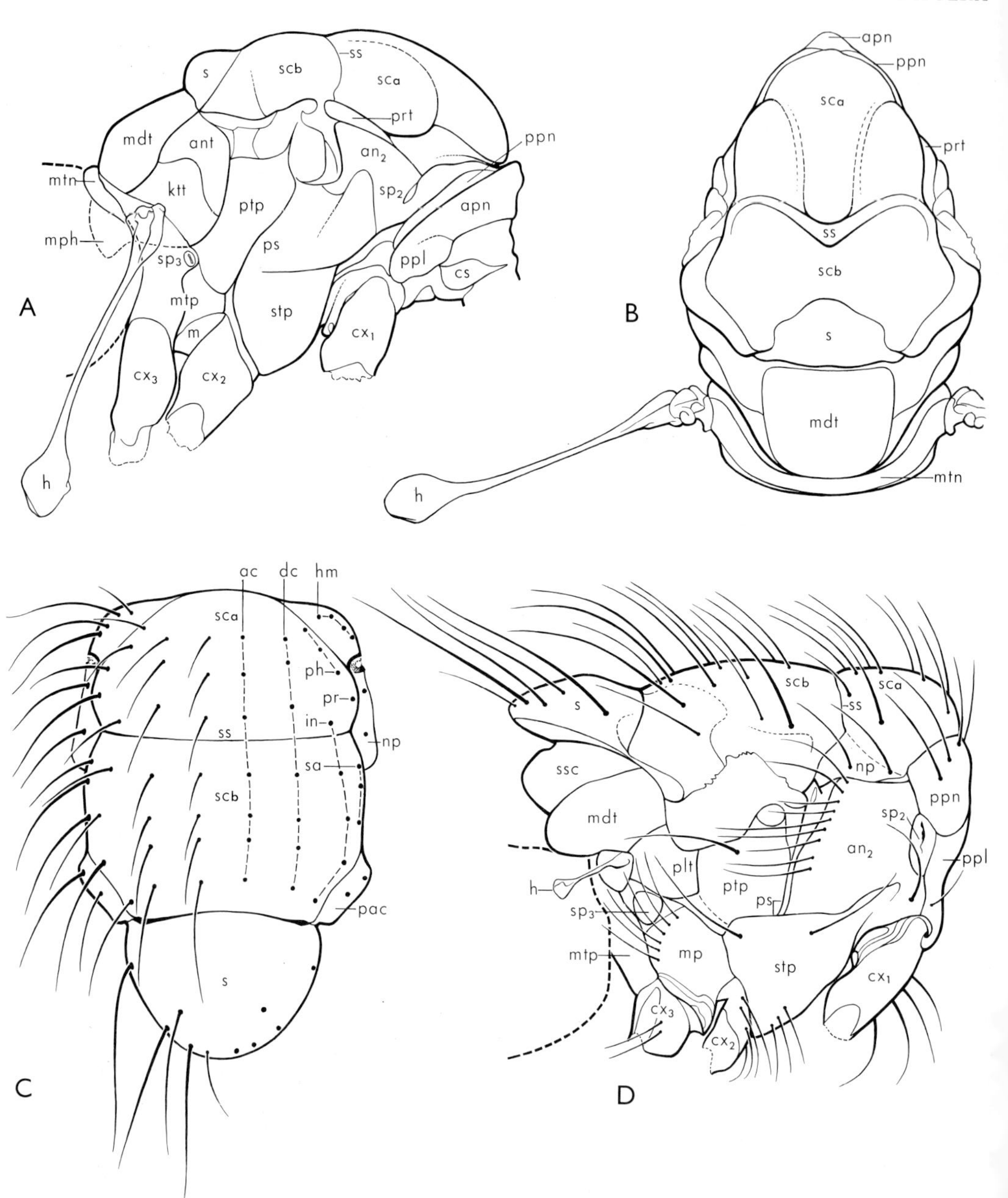

Fig. 34.5. Thoracic structure and chaetotaxy: A, B, Tipulidae, lateral and dorsal; C, D, Tachinidae, dorsal and lateral. [T. Binder]
an., anepisternite (mesopleuron); *ant.*, anatergite; *apn.*, anterior pronotum; *cs.*, cervical sclerite; *cx.*, coxa; *h.*, haltere; *ktt.*, katatergite; *m.*, meron; *mdt.*, mediotergite; *mp.*, meropleuron; *mph.*, mesopostphragma; *mtp.*, metapleuron; *mtn.*, metanotum; *np.*, notopleuron; *pac.*, postalar callus; *plt.*, pleurotergite; *ppl.*, propleuron; *ppn.*, posterior pronotum; *prt.*, paratergite; *ps.*, pleural suture; *ptp.*, pteropleuron; *s.* scutellum; sc_a, sc_b, presutural and postsutural parts of scutum or mesonotum; *sp.*, spiracle; *ss.*, scutal or mesonotal suture; *ssc.*, subscutellum; *stp.*, sternopleuron. Subscript numbers 1–3 indicate pro-, meso-, and metathorax, respectively. Bristles (in C): *ac.*, acrostichal; *dc.*, dorsocentral; *hm.*, humeral; *in.*, intraalar; *ph.*, posthumeral; *pr.*, presutural; *sa.*, supraalar.

from its lateral margins. The true metapleuron, lying below and behind the metathoracic spiracle, is not always clearly separated from the mesopleural sclerites, but when moderately developed, as in some Nematocera, it may be divided into anepisternum, etc. It, or its katepisternum, has also been included under the outworn label of 'hypopleuron'. The true metasternum is usually greatly reduced, but may be distinct.

The thoracic endoskeleton is well developed to support the powerful flight muscles. Although largely ignored by taxonomists, it offers some useful characters. Its principal structures are: the paired sternal apophyses, arising from the sternal plates and, on meso- and metathorax, fused basally to form complex *furcae*; the *pleural arms*, arising from the mesopleural suture; and the mesothoracic *prephragma* and *postphragma*. The prephragma, at the anterior margin of the mesonotum, is often small or vestigial, but the postphragma is well developed as an internal, posterior extension of the postnotum. Its primitive bilobed condition is preserved in most Tipulomorpha and Culicomorpha, but in most other families it forms a convex plate, almost completely closing the thoracic cavity; in some, e.g. Scatopsidae, it projects deeply into the abdomen.

The chaetotaxy of the thorax is shown in Figure 34.5. The traditional term 'mesopleural bristle' is retained with some reluctance.

Legs (Fig. 34.6). There is great diversity of leg structure in the Diptera, and there are examples of striking modifications on all segments but the trochanter. The adaptive significance, if any, is often obscure (e.g. the male coxal processes in some *Mycomya* spp., Mycetophilidae).

The coxae are usually rather small, though the fore coxae are often lengthened and sometimes swollen, while all three are characteristically elongate in the Mycetophilidae. In almost all Diptera, the articulation of the fore legs is quite different from that of the mid and hind legs, due to the greater mobility of the fore coxae. The mid coxa may possess a distinct meron, but it is normally fused into the pleuron, particularly in higher Diptera; mobility may be restored by transverse division of the remaining eucoxa into *basicoxa* and *disticoxa*. The trochanter is usually small.

The femur and tibia are the longest segments of the leg, and normally subequal, with the femur rather stout and flattened and the tibia more slim and cylindrical. In some Empididae and Ephydridae the fore femur and tibia are modified in an apparently raptorial, mantid-like fashion, while a few species in various families have the mid or hind legs thus modified. In many families, particularly of Nematocera, the tibiae may bear *apical combs* of close-set setulae and/or one or two articulated *spurs*.

The tarsi typically consist of 5 segments, which may be variously modified, principally by thickening or flattening; the basal segment (*basitarsus* or metatarsus) and, very rarely, other segments may also bear apical combs. The distal segment bears a pair of claws, or *ungues*, usually simple but sometimes toothed. Beneath the claws there may be a pair of pad-like *pulvilli* and/or a median *empodium* (Fig. 34.6), the latter often bristle-like but sometimes, e.g. in Bibionidae and Tabanoidea, pulvilliform. Tipulidae and Trichoceridae may have a rather similar (but not homologous) median lobe, the *arolium*. Many flies secrete on the pulvilli and pulvilliform empodium an adhesive substance that enables them to cling to smooth surfaces.

Wings (Fig. 34.7). The functional wings are borne on the mesothorax, the metathoracic pair having become reduced to small, club-like structures, the *halteres* (Fig. 34.5). The original four-winged condition is partially restored in certain *Drosophila* mutants, and is still recognizable in some pupae and primitive Nematocera. The halteres vibrate rapidly during flight, and are believed to act as gyroscopic sense organs of balance, precessional forces being detected by sensilla near their bases.

The base of the fore wing includes a series of axillary sclerites, of which the most conspicuous are the tegula *(epaulet)* at the extreme

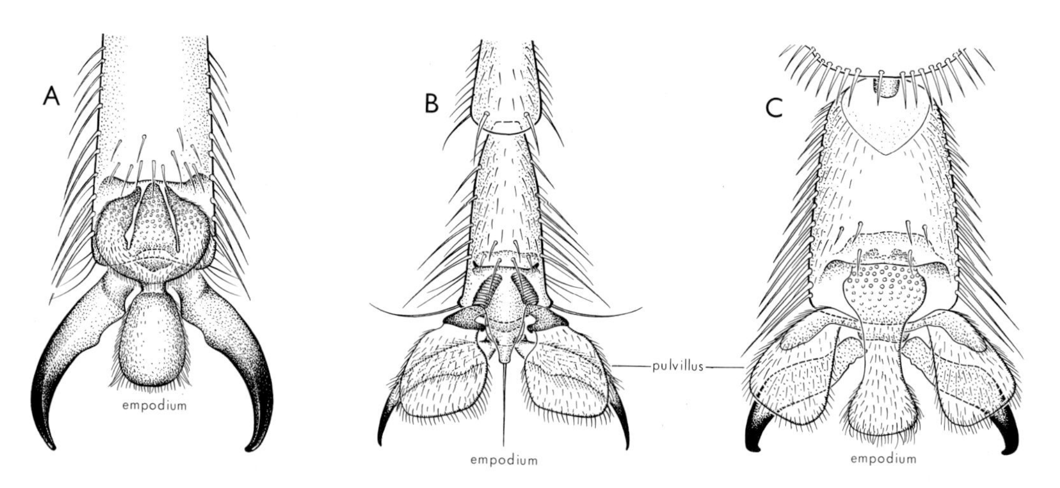

Fig. 34.6. Tarsal claws and associated structures: A, *Clytocosmus helmsi*, Tipulidae; B, *Musca domestica*, Muscidae; C, *Altermetoponia rubriceps*, Stratiomyidae. [T. Binder]

base of the anterior margin, and the adjacent, more distal humeral plate *(basicosta, subepaulet)*. Posteriorly, the membrane may have up to three basal lobes; the *upper* and *lower calypters* or squamae, which are folded one above the other; and the more distal *alula* or axillary lobe which, when differentiated, is marked off by a deep *axillary incision*. The lower calypter is immobile, with its margin continuing on to the notum; it may be vestigial, particularly in Nematocera, but in many Tabanoidea and Cyclorrhapha it is greatly enlarged, roofing over the hollow in which the halteres lie. Distal to the axillary incision there may be a distinct anal lobe. The membrane of the wing may be hyaline or may bear distinctive colour patterns. The disposition of micro- and macrotrichia on the membrane, and of the latter on the wing veins, are important taxonomic characters.

The hypothetical venation of the primitive dipteron is shown in Figure 34.7A and its modifications in subsequent figures. As the older Loew–Williston and Schiner notations have been extensively used by dipterists, their equivalents in the modified Comstock–Needham system are shown in Table 34.1. Several interpretations have been proposed for the radial and cubital fields (Alexander, 1929; Hennig, 1954), but we prefer here to follow the more conservative interpretations shown. The anomalous convexity of the posterior branches of Rs and M possibly arose through incorporation in them of MA and a primitive anterior branch of CuA respectively, while M, as used here, is strictly MP.

Amongst the more characteristic features of the primitive venation are: (a) the 'kink', constriction, or sulcus in R close to the level of the humeral veinlet, marking off a basal section, the *stem-vein*; (b) the transverse fold, often vein-like, forming a *brace* ('arculus') between M or Cu and the apex of the stem-vein; (c) the well defined 'discal' ('median') cell; (d) the reduction of CuP to little more than a concave fold; (e) the presence of only two anal veins. The primitive pattern is most closely approached by the Tanyderidae, and the most characteristic lines of specialization are: (a) shortening of the costa to end near or proximal to the wing apex; also, development of constrictions or breaks towards its base; (b) loss of branching and the apical portion of Sc; (c) reduction in branching of Rs and/or basal shift of its origin; R_{2+3} is normally branched only in the more primitive Nematocera; (d) loss of *i-m* and hence of the discal cell; (e) reduction of branching of M, or loss of its basal portion, or detachment of M_{3+4} which may appear to fork from

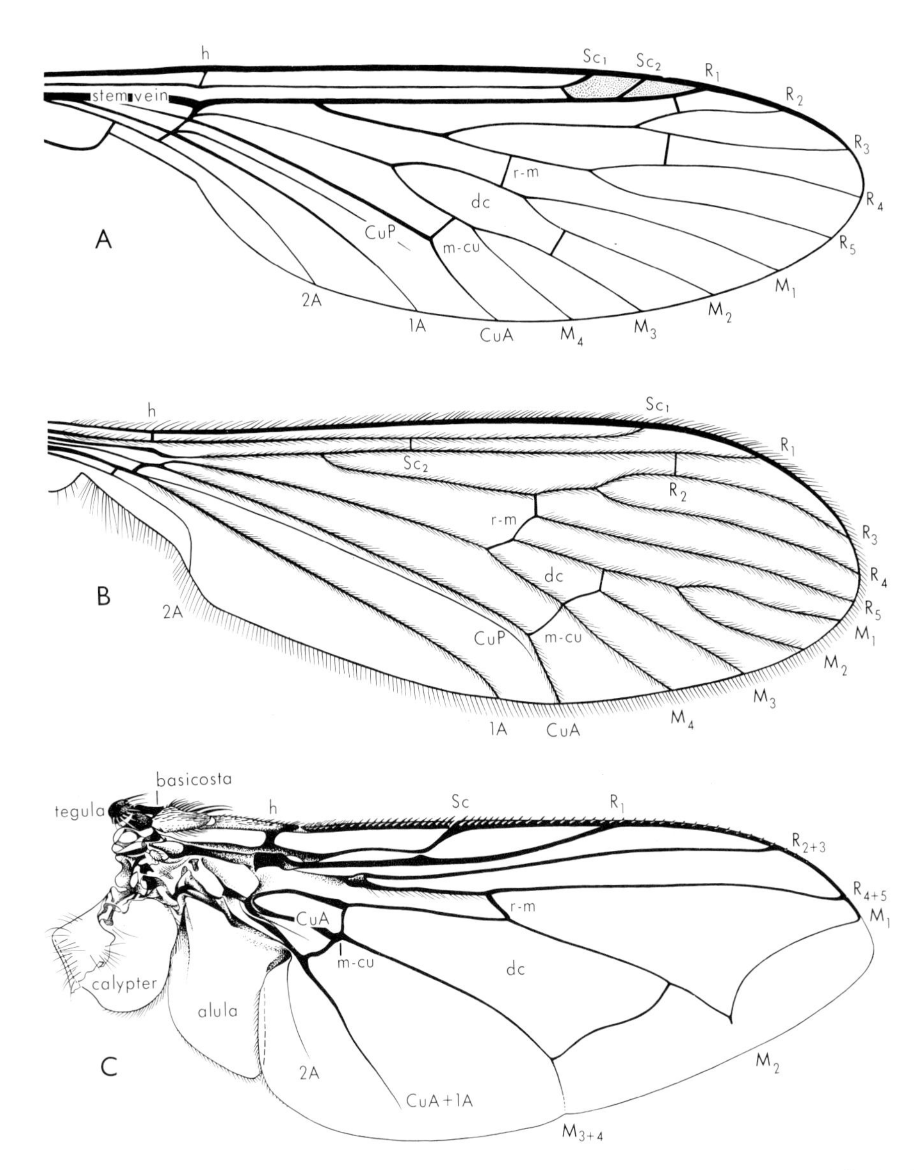

Fig. 34.7. Wing venation: A, reconstruction of hypothetical ancestor of Diptera (cf. Fig. 1.18A); B, tipuloid pattern, *Nothotrichocera cingulata*, Trichoceridae; C, muscoid pattern, *Chrysomya mallochi*, Calliphoridae. Standard notation for veins; *dc.*, discal cell. [T. Binder]

CuA; (f) shortening of CuA, and flexion of its apex to end in 1A; (g) shortening or loss of 2A, and sometimes 1A also—2A is very short, rudimentary, or absent in all but Tipulidae. Almost all these trends are towards reduction of the venation, particularly in the posterior field, and are presumably correlated with improvements in the dynamics of two-winged flight.

Abdomen. The principal landmarks are: the spiracles, borne within, or just below, the lateral margins of the tergites; the male

TABLE 34.1

Comparison of Systems of Wing Venation Applied to the Diptera

Veins

Modified Comstock–Needham (Fig. 34.7)	Loew–Williston	Schiner
C	Costa	Costa
Sc	Auxiliary	Mediastinal
R_1	1st longitudinal	Subcostal
R_2, R_3	2nd longitudinal	Radial
R_4, R_5	3rd longitudinal	Cubital
M_1, M_2	4th longitudinal	Discal
M_3, M_4 (Cu_1, Cu_{1a})	5th longitudinal (branches)	
CuA (Cu_1)	5th longitudinal (base)	Postical (incl. *m-cu*)
CuP (Cu_2)	—	—
1A	6th longitudinal	Anal
2A	Axillary	Axillary

Cells

Modified Comstock–Needham (Fig. 34.7)	Loew–Williston	Schiner
C	Costal	Costal
Sc	Subcostal	Mediastinal
R	1st basal	1st basal
R_1	Marginal	Subcostal
R_3	1st submarginal	1st cubital
R_4	2nd submarginal	2nd cubital
R_5	1st posterior	1st posterior
M	2nd basal	2nd basal
Discal (median, 1st M_2)	Discal	Discoidal
M_1	2nd posterior	2nd posterior
M_2	3rd posterior	3rd posterior
M_3	4th posterior	4th posterior
M_4	5th posterior	5th posterior or postical
CuP (CuA+CuP)	Anal	3rd basal or anal
1A	Axillary	Axillary

and female genital orifices; the cerci; and the anus. Primitively, there are 11 segments, but 10 and 11 are generally fused to form a composite *proctiger* (or, simply, segment 10) which bears cerci and anus. In higher Diptera, there is also a progressive tendency towards reduction of segment 1 and fusion of tergites 1 and 2. Most Diptera show some degree of telescoping of the terminal segments, but in higher forms, particularly Cyclorrhapha, the more posterior segments are clearly differentiated as a slender *postabdomen*, retracted within, or curved beneath, the broader *preabdomen*. The postabdomen may be strongly modified, particularly in males. The convenient, general designation *terminalia* is applied to the terminal complex of modified genital and anal segments *plus* any adjacent segments that show modifications for copulation or oviposition; it may include the entire postabdomen. 'Hypopygium' and 'genitalia' should not be used as synonyms of 'terminalia'; their meanings are much more restricted.

FEMALE TERMINALIA (Fig. 34.8). These structures are usually relatively simple. In many families, segments 6 and 7 are reduced and, together with the more posterior segments, form a telescopic, eversible, tubular ovipositor (Fig. 34.8B). However, in a few (e.g. Agromyzidae, Tephritidae), segment 7 forms an ovipositor sheath, which is stout, rigid, and permanently exserted, while the terminal segments are adapted for piercing (Fig. 34.8C). In many Nematocera and Orthorrhapha, sternite 8 is produced posteriorly and may be bilobed; in the complex ovipositor of many Tipulidae, those lobes are produced to form the *sternal* (hypogynial) *valves* ('gonapophyses'), the *tergal valves* being formed by the cerci (Fig. 34.8A). The oviduct opens behind sternite 8, and its distal portion may bear a rod-like or more complex sclerotization, the *vaginal apodeme* or *furca*. Behind the genital orifice, there may be a series of small sclerites, derived mainly from sternite 9, which are of taxonomic importance in some families. In many Orthorrhapha, tergite 9 is longitudinally divided into a pair of hemitergites, the *acanthophorites*, which bear strong spines (Fig. 34.8D); in Cyclorrhapha, it tends to unite with the proctiger. The latter structure shows signs of segmental differentiation in

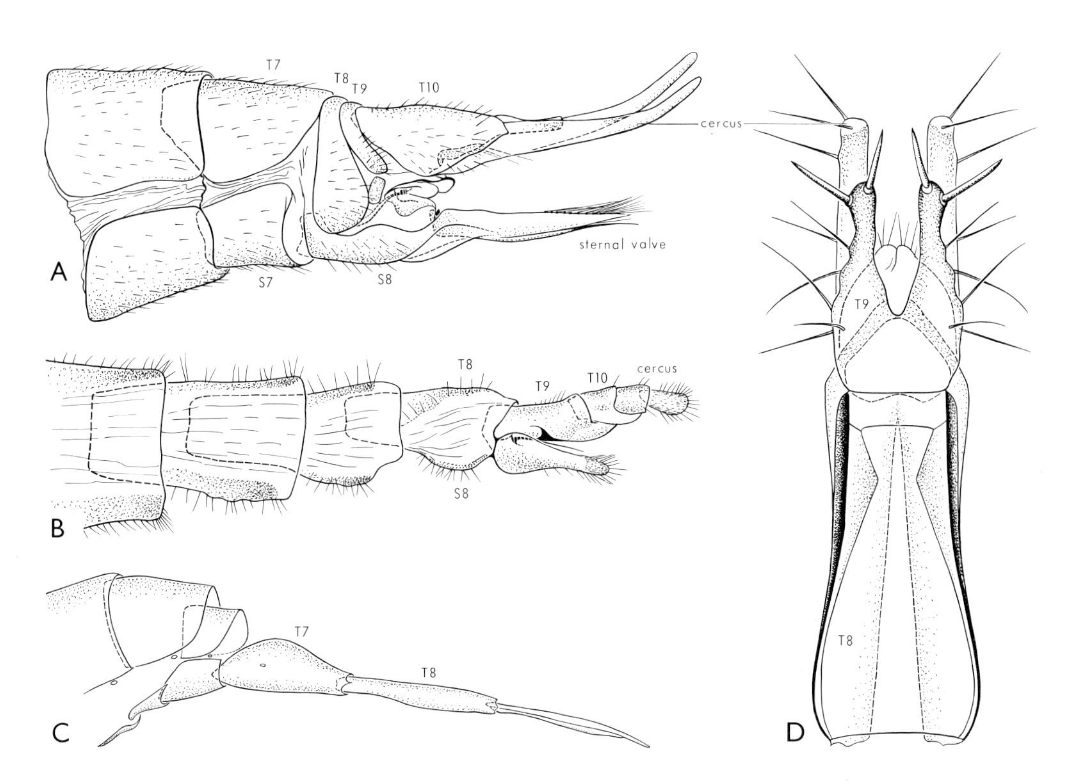

Fig. 34.8. Female teminalia: A, *Gynoplistia* sp., Tipulidae, lateral; B, *Sciara* sp., Sciaridae, lateral; C, *Dacus tryoni*, Tephritidae, lateral; D, *Heteropsilopus cingulipes*, Dolichopodidae, dorsal. [T. Binder]

some primitive forms, but normally appears as a single unit, composed of a dorsal *supra-anal plate* (epiproct) and a ventral *subanal plate* (hypoproct, postgenital plate); it bears the cerci, primitively 2-segmented, but sometimes reduced to tiny lobes.

MALE TERMINALIA (Fig. 34.9). These structures are often complicated, and the terminology of the parts has become exceedingly confused (see Crampton, 1942; Zumpt and Heinz, 1950; van Emden and Hennig, in Tuxen, 1956). The basic pattern is best seen in the more primitive Nematocera (Fig. 34.9A), in which homologies are fairly clear; but, ascending through the Orthorrhapha (Fig. 34.9C), considerable modifications occur until, in the Cyclorrhapha (Fig. 34.9B), homologies become obscure and are still disputed. In the simplest types, modifications are largely restricted to segment 9 and the proctiger, apart from reductions associated with telescoping of segments 7 and 8, and occasional development of lobes, spines, etc. ventrally on segment 8 (pregenital segment, *protandrium*). Segment 9 (genital segment, *andrium*) is considerably modified; its tergite *(epandrium)* may be reduced or enlarged, sometimes bearing articulated processes, the *surstyli*, while the sternite *(hypandrium)* bears the usually forcipate *coxites* (basistyle, sidepiece), each with an apical *style* (dististyle, clasper), and the median copulatory organ, the *aedeagus*. The coxites may bear various accessory lobes or appendages (e.g. the *claspettes* of Culicidae) while the style may be double (as in Tipulidae) or branched. Sternite 9 is often greatly reduced, and may be fused with the coxites.

The name 'aedeagus' seems the best available general term for the copulatory

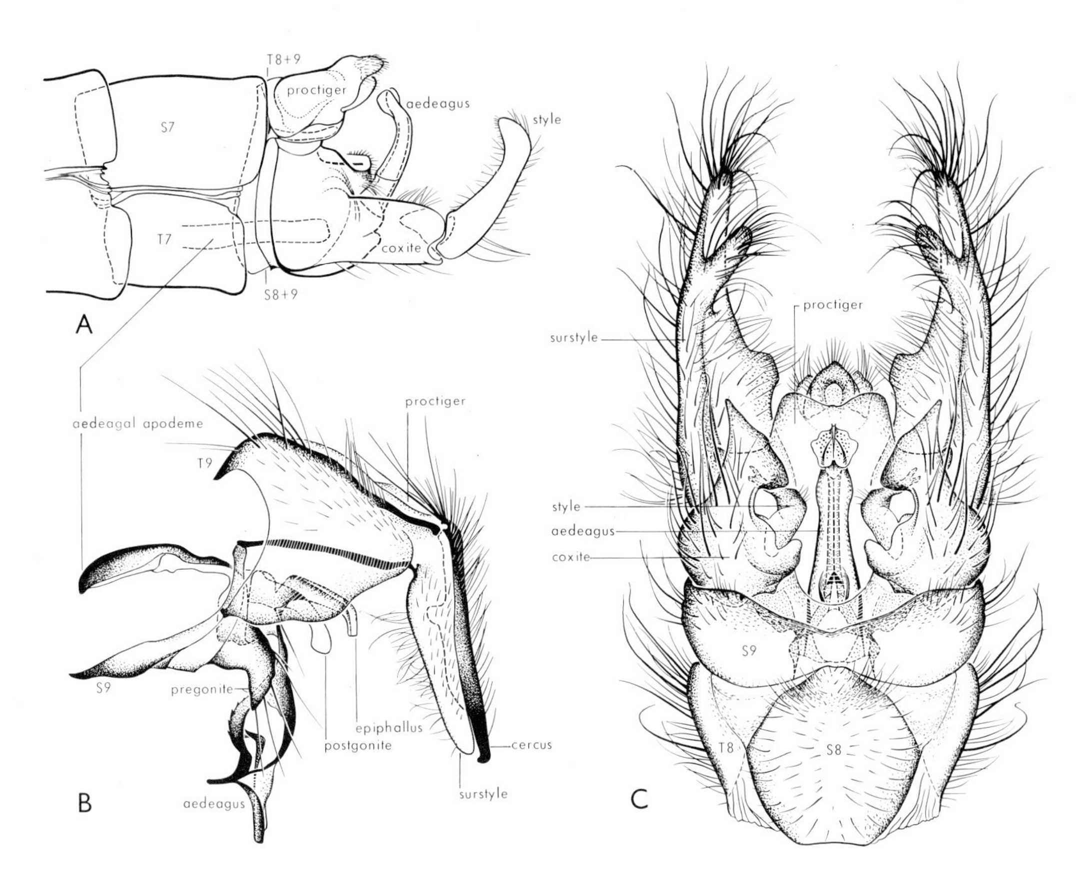

Fig. 34.9. Male terminalia: A, *Eutanyderus wilsoni*, Tanyderidae, lateral, ventral surface uppermost; B, *Calliphora stygia*, Calliphoridae, lateral; C, *Neoaratus hercules*, Asilidae, ventral. [T. Binder]

organ (also called phallosome, mesosome, penis, etc.). This structure exhibits an enormous variety of forms, from a membranous or sclerotized lobe or tube, up to the complex 'phallosome' of some Culicidae. This is due, in part, to incorporation of a variety of elements, including intersegmental sclerotizations and internal lobes of the 'parameres' (Abul-Nasr, 1950). The latter may not be the structures so designated in Coleoptera and accepted by Snodgrass (1957), and are best referred to as parandrites (p. 26); they may remain distinct (e.g. in *Chironomus* and *Trichocera*) as appendages flanking the aedeagus. The proctiger usually forms a distinct unit bearing a pair of 1-segmented cerci.

In some Nematocera and Orthorrhapha, segment 9 and the proctiger are temporarily or permanently rotated through 180°, so that true ventral structures lie in a dorsal position. Such a condition is readily recognized by reference to the proctiger and aedeagus. A somewhat similar rotation occurs in certain Empididae, but only through about 90°, and associated with considerable asymmetry of the parts. In Dolichopodidae and Cyclorrhapha, the rotation has proceeded through 360°, restoring the original relations of anus and genital structures, a process called *circumversion*. The 'torsion', which takes place in the pupa, may affect the entire postabdomen, causing a sinistral displacement of the sternites, with sternite 8 usually adopting a dorsal position (laterally, to the left of segment 9, in Dolichopodidae). In some forms, however, external symmetry is restored (e.g. Drosophiloidea). The circumverted terminalia are nearly always folded forward ventrally against the preceding sternites.

In these higher Diptera (Fig. 34.9B), sternite 5 often bears a prominent pair of *pregenital* ('copulatory') *lobes*, while the coxites are greatly reduced or absent. Dorsally, the displaced sternites of segments 7 and 8 may fuse to form a composite protandrium. The most prominent appendages are the surstyli (lateral valves, edita) of tergite 9 and the cerci (medial valves), the latter being sometimes partly fused. Sternite 9 is sometimes reduced, but may be inflected to form a *hypandrial apodeme*. The aedeagus exhibits a variety of bizarre forms (e.g. the coiled, ribbon-like form in Tephritoidea), and is often differentiated into *basiphallus* (phallobase) and *distiphallus* (phallus, aedeagus), the former bearing a posterior, spine-like *epiphallus* (gonacanthus, spinus titillatorius). At the base of the aedeagus, there are often two pairs of small appendages, the *pregonites* and *postgonites* (anterior and posterior 'gonapophyses'), which are presumably derived from the coxites or parandrites; also, internally, there is usually a strong *aedeagal apodeme*.

Internal Anatomy. The alimentary canal has the buccal cavity dilated and lined with several small sclerites; attached dilator muscles operate the cavity as a pump to draw up fluid through the mouth-parts. In the blood-sucking Tabanidae, Culicidae, and Psychodidae *(Phlebotomus)*, a similar *pharyngeal pump* is also present, and the sclerites of the two chambers in Culicidae and *Phlebotomus* form a *bucco-pharyngeal armature* which is of taxonomic importance. The tubular oesophagus bears usually one, but sometimes three, characteristic diverticula, of which the ventral one ('crop') forms a large, distensible sac lying mainly in the abdomen. It functions as a primary or secondary food reservoir, its contents passing slowly back to the gut; in mosquitoes, it stores only liquids other than blood, which passes directly through the gut. The mid gut is a simple sac in the lower Diptera, but highly convoluted in Cyclorrhapha. The hind gut terminates in a dilated rectum bearing 2, 4, or 6 papillae. There are generally 4 Malpighian tubes (2 in *Culicoides*, 5 in *Psychoda* and Culicidae). The salivary glands usually lie in the thorax, but may extend into the abdomen (e.g. in *Musca*); they are normally elongate and tubular, and may be branched (e.g. the trilobed glands of the Culicidae).

The nervous system shows a broad evolutionary sequence, from the generalized Nematocera with 3 thoracic and 7 abdominal ganglia (8 in the exotic Nymphomyiidae) to the higher Cyclorrhapha with all ventral ganglia fused; other groups show a variety of intermediate conditions. The respiratory system is notable for the development of greatly dilated air-sacs, mainly in the abdomen, and particularly prominent in the Cyclorrhapha.

In the female, the ovaries are paired, and each comprises from one up to more than a hundred polytrophic ovarioles, while up to 4 spermathecae and a series of accessory glands open into the oviduct. In ovoviviparous and viviparous groups, the number of ovarioles is small, and the oviduct is dilated to form a *uterus*; in viviparous groups, the accessory glands secrete a milky fluid that nourishes the larva *in utero*. In the male, the testes are normally small, compact, ovoid bodies, often deeply pigmented; in 'pupiparous' groups, they take the form of compactly coiled, thread-like tubules. Their vasa deferentia join to form the ejaculatory duct, with which is often associated a muscular ejaculatory sac; paired accessory glands may also be present. A conspicuous sclerite, the *ejaculatory apodeme*, is often attached to the ejaculatory duct.

Cytogenetics. Compared with other orders, the Diptera have, in general, very few chromosomes. Most have from 3 to 6 pairs (usually 6 in higher forms), and none is known with more than 10 pairs; the homologous chromosomes generally form pairs during mitosis ('somatic pairing'). The male sex is almost always heterogametic, and X and Y chromosomes are commonly distinct, but in at least three lineages (limoniine Tipulidae, Psychodidae, and Culicidae) they are barely, or not at all, apparent, presumably through fusion with autosomes (e.g. Breland, 1961). In Nematocera, the X and Y chromosomes do not form bivalents at meiosis ('distance-pairing'), this probably being the primitive condition, but the Brachycera mostly show normal pairing. Another very distinctive feature is found in Bibionomorpha and Brachycera, which form no chiasmata during meiosis in the male, while the Cecidomyiidae and Sciaridae exhibit several striking bizarre features (p. 79). A remarkable type of sex determination in the Phoridae has been described by Mainx (1964).

A characteristic and apparently unique feature of dipterous larvae, also described in chapter 3, is the presence of giant, polytene chromosomes (p. 75) in the cells of the salivary glands, and, to a lesser extent, in other tissues. They rarely occur in adult tissues, but have been found in developing pulvilli (Whitten, 1964) and Malpighian tubes (Bush, personal communication). They seem to occur in most families, but useful preparations have so far been demonstrated in only a few, their development being apparently affected by larval nutrition. Such chromosomes have provided a powerful tool for genetic and cytotaxonomic research in the Diptera (e.g. J. Martin, 1963; Basrur, 1962; Frizzi and Holstein, 1956).

Immature Stages

Egg. Generally small and elongate-oval. Being usually deposited in moist situations, the outer wall is normally pale and relatively thin. However, some have a strongly sclerotized chorion, which may be sculptured or developed into a plastron (Fig. 4.4B). Eggs are usually deposited singly or in irregular masses, but may be laid in well-defined *rafts* or *rosettes* (e.g. some Culicidae), or in gelatinous ribbons (Chironomidae).

Larva. (Figs. 34.10, 12–14). The usual number of instars is 4, though more are recorded in some groups, whereas the Cyclorrhapha have the 4th instar suppressed. The notes below refer mainly to the mature larva.

Body form is variable, but usually more or less elongate, and cylindrical or dorsoventrally flattened. Macroscopically, the integument may appear smooth, but it usually bears rows of microscopic spines, etc.; these and any setae that may be present provide important taxonomic characters. Segmentation is normally distinct, into head, 3 thoracic, and 8 or 9 abdominal segments, but the apparent number may be reduced by fusion (e.g. Blephariceridae, Lonchopteridae) or increased by secondary division (e.g. Anisopodidae, Therevidae). True segmented legs are never present, but prothoracic and/or abdominal prolegs occur in some families (Figs. 34.13G, H). In certain parasitic species of Nemestrinidae, Acroceridae, Bombyliidae, Sarcophagidae, and Tachinidae, the active first-instar larva is a planidium (p. 106).

The Nematocera are generally *eucephalic* (Fig. 34.10A), with a distinct head capsule formed from a median dorsal plate, the *cephalic apotome* ('frontoclypeus' of authors), separated by the *epicranial* 'suture' from the *lateral*, or *epicranial, plates* (or genae); the latter curve under to form the sclerotized ventral wall of the capsule, and may meet or even fuse along the mid-line. The capsule is small and very weak in Cecidomyiidae, while the Tipulidae (Fig. 34.12C) have the more posterior parts weakly sclerotized and the head partially, or even completely, retracted into the prothorax. A rather similar *hemicephalic* condition is typical of most Orthorrhapha, in which the posterior portion of the head is retracted within the thorax, and usually represented by a dorsal pair of *metacephalic rods* or *plates*, a pair of *tentorial rods*, and a *pharyngeal skeleton* (Fig. 34.10B). The Cyclorrhapha are all *acephalic*, without any indication of an external head skeleton (Fig. 34.10C).

Many Nematocera have distinct, segmented antennae, and most have normal, chewing mouth-parts, with the mandibles hinged to move in the horizontal plane. In some

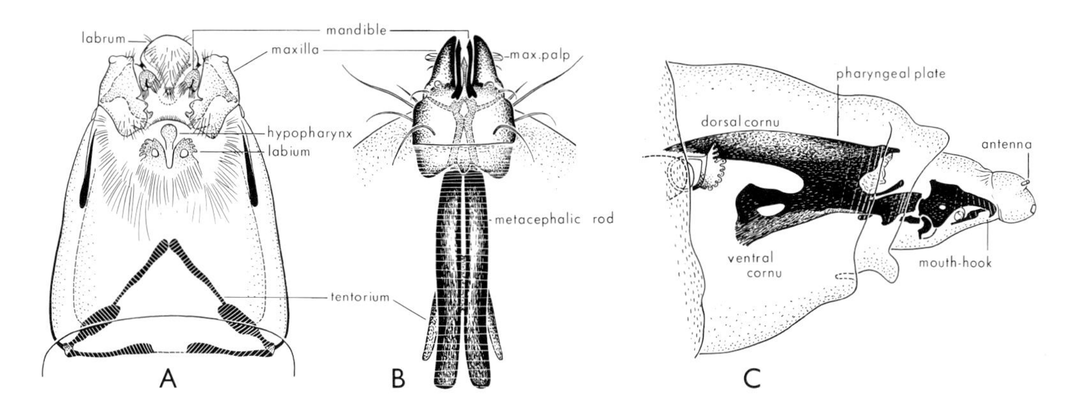

Fig. 34.10. Larval heads and mouth-parts: A, *Sylvicola dubius*, Anisopodidae, ventral; B, unidentified sp. of Asilidae, dorsal; C, *Calliphora vicina*, Calliphoridae, lateral. [T. Binder]

families, the labrum bears a pair of ventral appendage-like structures, the *premandibles*, which may take the form of toothed plates or processes (e.g. Trichoceridae, Chironomidae), or mobile, brush-like organs (the *mouth-brushes*, or *cephalic fans* of Culicidae, Simuliidae, etc.; Fig. 34.13F). Ventral to the mouth opening, there may be a median toothed plate, commonly called 'mentum' or 'submentum' (e.g. in Culicidae), but now believed to belong to the head capsule proper and better termed the *hypostomium* (Crosskey, 1960). In the Orthorrhapha, the mandibles move in the vertical plane and are modified to sickle- or hook-like structures, whereas the maxillae, and sometimes antennae are fairly well developed. In the Cyclorrhapha, the antennae are rudimentary and the mouth-parts and associated pharyngeal structures form a characteristic *cephalopharyngeal* skeleton (Fig. 34.10C), to the parts of which a variety of terminologies have been applied (Sanjean, 1957). The principal parts are the paired, anterior *mouth-hooks** and the sclerotized areas of the pharynx: the latter usually as a pair of *pharyngeal plates*, each produced posteriorly into a *dorsal* and a *ventral cornu*.

Spiracles, functional or rudimentary, are present on the pro- and metathorax and 8 abdominal segments, the arrangement of *functional* spiracles (pp. 38-9) varying considerably, usually between families, but also between instars of a single species (for details of spiracular structure, etc., see Keilin, 1944). Mature larvae of the more primitive families are often holopneustic (e.g. Bibionidae, Stratiomyidae), or peripneustic, lacking only the metathoracic spiracle (e.g. Cecidomyiidae) or the posterior abdominal pair as well (e.g. Sciaridae). However, some are metapneustic (e.g. Culicidae, Tipulidae), amphipneustic (e.g. Anisopodidae), propneustic (e.g. some Mycetophilidae), or apneustic (e.g. Thaumaleidae, Chironomidae). With the exception of some Stratiomyidae, the Brachycera are typically amphipneustic. In all but the asiloid families, the spiracles of the posterior pair tend to be approximated on the penultimate or ultimate segment; in some, they lie in a deep cleft or depression (Nemestrinidae, Sarcophagidae), at the apex of a breathing tube or *siphon* (Psychodidae, Culicidae—Fig. 34.12F,H), or even on paired individual siphons (Scatopsidae—Fig. 34.14G). The siphon (e.g. in *Mansonia*, Culicidae) may be modified to take air from tissues of aquatic plants.

Internally, the alimentary canal is a relatively simple tube, though greatly convoluted in most Cyclorrhapha. The oesophagus projects into the mid gut, to form a valve or *cardia*, immediately behind which lies a series of enteric caeca; Cyclorrhapha may also have an oesophageal diverticulum,

* Although long believed to represent the mandibles, there is strong evidence that the mouth-hooks are, in fact, modified maxillae (Menees, 1962).

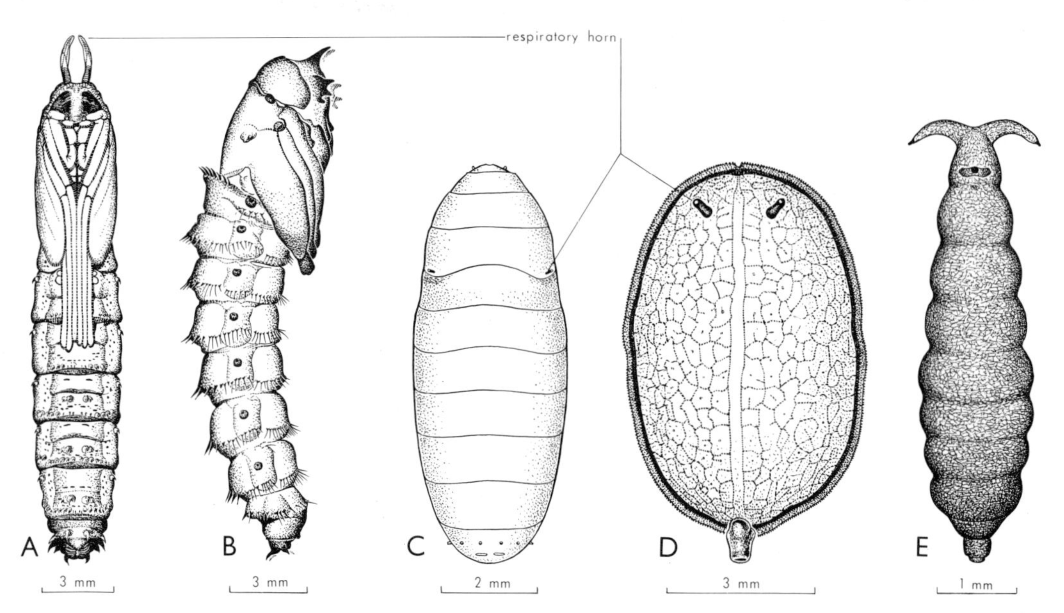

Fig. 34.11. Pupae and puparia: A, *Ischnotoma rubriventris*, Tipulidae, ventral; B, unidentified sp. of Asilidae, lateral; C, *Lucilia cuprina*, Calliphoridae, puparium; D, *Microdon* sp., Syrphidae, puparium; E, unidentified muscid, puparium. [T. Binder]

as in the adult. There are normally 4 or 5 Malpighian tubes. Salivary glands are always prominent, usually tubular and elongate, and sometimes very large (e.g. in *Mycetophila*). The nervous system tends to resemble that of the adult, the Nematocera with ganglia in most segments, the Cyclorrhapha with a single ganglionic mass. The brain may be in the head capsule, if present, or in the thorax. The tracheal system consists principally of two dorsal and two lateral longitudinal trunks, with a largely segmental series of dorsal anastomoses and lateral connectives; details of the pattern provide characters of considerable taxonomic and phylogenetic importance (Whitten, 1963).

Pupa (Fig. 34.11). Dipterous pupae are usually immobile, though capable of rapid swimming movements in the Culicidae; the sluggish movement seen in many families is said to be that of the pharate adult (Hinton, 1946b). All are adecticous, and in the Nematocera and Orthorrhapha almost all are obtect, the adhering sheaths of the legs and wings being termed *pedothecae* and *pterothecae*, respectively. The Cyclorrhapha are secondarily exarate and coarctate, with their delicate integument protected within a usually globular or barrel-shaped *puparium* (Figs. 34.11C–E). This is formed from the retained, hardened skin of the 3rd larval instar, and is lined with a delicate moulting membrane (Hinton, 1958c). Functionally transitional types of puparia occur; in Stratiomyidae and Perissommatidae, the delicate pupa, more or less exarate in the former but obtect in the latter, is enclosed within the unmodified last larval skin; in Scatopsidae, a normal, sclerotized, obtect pupa is almost completely enclosed in the last larval skin; and in a number of nematocerous pupae (e.g. Tipulidae, Chironomidae), the larval skin remains attached to the posterior segments of the abdomen.

In many families, the obtect pupa is enclosed in a silken cocoon, the walls often incorporating substrate materials. Such pupae may bear sharp anterior processes ('cocoon cutters'), while in all groups, the integument often bears spines and tubercles that have various functions during eclosion of the adult. These furnish important taxonomic characters, as do the relative lengths and positions of the appendage sheaths; e.g.

in Bibionidae, Psychodidae, and several other families the pedothecae lie one above the other, instead of side by side.

Although functional spiracles may be present on the abdomen, most pupae respire mainly or entirely by the 'prothoracic' pair (as in the larva, actually mesothoracic spiracles that have migrated forward). These may be sessile, but frequently bear long, projecting extensions, the *respiratory horns* or *trumpets* (Fig. 34.11A), which, although usually more or less rod-like, may be filamentous and branched. In aquatic forms, they may act as gills, operating via a plastron (e.g. Hinton, 1962a), or, as in the larvae, may take air from plant tissues. In the coarctate pupae of the Cyclorrhapha, most acalyptrates have the prothoracic spiracles internal, enclosed within the puparium; other Cyclorrhapha have, in addition to the internal spiracles, a pair of respiratory horns formed by extensions of the spiracular chambers. These, in a most remarkable fashion, come to penetrate, and protrude through, special thin 'windows' in the first abdominal segment of the puparium (Keilin, 1944; Roddy, 1955).

Biology

Adults. Adult Diptera are free-living and ubiquitous, found in aerial plankton, in practically all terrestrial habitats, and even, in the case of the extraordinary chironomid *Pontomyia*, beneath the sea. At least some families form significant elements of the fauna in desert regions, intertidal zones, rain forests, and snowfields. Their often considerable powers of flight make them generally wide-ranging within broadly defined habitats, although some are more restricted, e.g. the ectoparasites that live mainly on the host's skin. Generally, the majority of species are associated with forest or water.

Except for a few groups with non-functional mouth-parts (e.g. Oestridae), most adults are polyphagous, but take only the liquid food for which their mouth-parts are specially adapted. Such foods include free water; a variety of animal and plant secretions; products of decomposition of organic matter; soluble solids, which are first liquefied by salivary secretions; and the tissue fluids of other animals. The last refers to the large and important group which prey upon, or parasitize, other arthropods and vertebrates. Most of these species extract the host's body fluids by inserting a sharp proboscis, although the Dolichopodidae and Ephydridae, for instance, 'masticate' their prey between their labella, while some Cyclorrhapha merely lacerate the skin with their prestomal teeth and suck up the exuding fluids.

Reproduction. Almost all Diptera are normally bisexual; parthenogenesis is rare (e.g. in certain Psychodidae, Chironomidae), while paedogenesis occurs in some Cecidomyiidae. Mating commonly takes place, or at least may continue, on the wing, and pairs are sometimes captured *in copula*. In some groups, particularly in Nematocera, it is usually preceded by formation of dancing swarms of males, while various forms of elaborate courtship procedure occur in the Empididae (Melander, 1927) and other families. The female may be periodically inseminated by a number of males, but in some species she mates only once in her lifetime. Peacock and Erickson (1965) have recently demonstrated that in *Drosophila melanogaster* Meig. half of the spermatozoa are regularly non-functional, and the phenomenon may well be present in other families. Sex ratios are usually normal, but *Chrysomya rufifacies* (Macq.) (Calliphoridae) is exceptional, in that individual females produce either all male, or all female, offspring throughout their lives (Norris, in Keast *et al.*, 1959). A similar phenomenon probably occurs in some Australian Sciaridae (Metz, 1938).

Development of the ova is usually rapid, being complete in a matter of days after eclosion (or minutes in the chironomid *Polypedilum nubifer* (Skuse)), and may be independent of insemination. It may, however, require a prior meal of protein (some Calliphoridae) or blood (*anautogenous*, as opposed to *autogenous*, Culicidae). The eggs (or larvae) are deposited in, on, or near a substrate suitable for larval development, the gravid female being quite precisely guided by physical and chemical stimuli from the environment. Some Cyclorrhapha are 'larviparous' (ovoviviparous), the egg hatching in the oviduct shortly before deposition; or 'ovo-larviparous', hatching immediately after deposition. The 'pupiparous' families (Hippoboscidae, Streblidae, and Nycteribiidae) and the African *Glossina* show the even more advanced condition of adenotrophic viviparity (p. 83). In many, if not most, Diptera, the maturation and release of the eggs leave visible traces, such as stretching of the ovarian tracheae or oviduct, or *follicular relics* within the ovarioles. These have made possible a number of very useful techniques for age determination (e.g. Detinova, 1962).

Immature Stages. Dipterous eggs usually hatch fairly rapidly—diapause is known in some groups, but seems to be uncommon. Larval habitats are usually in moist situations: in mud, soil, etc.; in decomposing organic matter; in plant or animal tissues; or in free water. Apart from parasitic forms, most larvae crawl or swim actively in the breeding medium, although those of the Simuliidae and Blephariceridae are more sedentary, and live anchored to stones, etc., in flowing water; also many aquatic Chironomidae inhabit small silken tubes in bottom silt or attached to debris. Some larvae are predators or parasites, devouring the tissues of their hosts, whereas the remainder feed on fragments of organic matter. Noteworthy amongst feeding devices are the mouth-brushes of Culicidae and Simuliidae, which strain food particles from the surrounding water; the aquatic Stratiomyidae appear to have an analogous, but not homologous, organ.

Larval respiration is commonly direct from the atmosphere, even in liquid media. Many aquatic forms surface for air, though a few (e.g. Culicidae, Tipulidae, and Syrphidae) take air from plant tissues; others (particularly Chironomidae, Simuliidae, and Ceratopogonidae) absorb dissolved oxygen directly from the water, a process assisted in certain bottom-dwelling Chironomidae by the presence of haemoglobin in their blood. The endoparasitic larvae form an analogous series; many take air through perforations in the host's integument or tracheae (e.g. Tachinidae); others use air in plant tissues (e.g. leaf-miners); while yet others draw directly on oxygen dissolved in the host's blood (e.g. early stages of various families).

Pupation may occur in the larval breeding medium (e.g. in most aquatic forms and soil-dwellers), but in many families, both free-living and parasitic, the mature larvae leave their very moist habitat to pupate in relatively drier sites near by, usually in the soil. This applies even to some aquatic forms (e.g. Tipulidae). The cyclorrhaphous pupae (and a few others) are protected within puparia, while many Nematocera (e.g. Simuliidae, Mycetophilidae) spin silken cocoons. Before eclosion, obtect 'pupae' (pharate adults) may first wriggle free from any surrounding cocoon (or puparium, in Perissommatidae); aquatic forms commonly come to the surface (e.g. Chironomidae), but adult Simuliidae emerge from the pupal shell *in situ*, beneath the surface.

The pharate adult bursts from the pupal skin or puparium usually by the pressure of swallowed air. In obtect pupae, it emerges from a longitudinal slit on the pupal thorax and sometimes abdomen. In coarctate forms, the anterior end of the puparium is pushed off or open, and the pushing process is facilitated by the *ptilinum*, a balloon-like sac of elastic cuticle, which can be protruded from above the bases of the antennae. Although the ptilinum is a characteristic feature of the Schizophora, at least some Aschiza (e.g. Syrphidae) have an analogous elasticity of the frontal cuticle that seems to function in the same way. The process of eclosion is reviewed by Hinton (1946b).

Natural Enemies. Diptera, like most insects, are attacked by a wide range of predatory animals. Bats, birds, reptiles, amphibia, dragonflies, and spiders consume large numbers of adults, particularly of 'swarming' species, while sphecoid wasps use them to provision their nurseries; fishes, birds (to some extent), Odonata, Coleoptera, and Hemiptera feed on larvae and pupae of aquatic Diptera, while both adults and immature stages are attacked by other Diptera—adult Empididae, Asilidae, Dolichopodidae, etc., and the predatory larvae of many families, e.g. Muscidae. Fish (*Gambusia* and other genera) have even been utilized as controlling agents for mosquito larvae.

Apart from such generalized predation, Diptera support a wide range of parasites. Their ectoparasites are mainly Acarina, particularly Hydrachnoidea, which are common on aquatic species, but probably cause little mortality; also, epibiont algae, fungi, and protozoa (such as *Vorticella* spp.) may cause mortality in aquatic larvae. Their endoparasites have been more thoroughly studied, because of their potential use in biological control; a brief outline of the principal groups is given below.

VIRUSES. Little investigated; 'Tipula Iridescent Virus' is an example.

BACTERIA. Also little investigated, though known to cause mortality in aquatic larvae.

FUNGI. A number of specialized parasites are known; e.g. *Empusa* spp. of Cyclorrhapha, and *Coelomomyces* almost entirely in larval Culicidae.

PROTOZOA. Apart from commensal species, pathogenic flagellates, gregarines, and microsporidia are known from larvae and, to a lesser extent, adults of aquatic species.

HELMINTHS. Mainly nematodes; mermithids are rather common in some Culicomorpha (e.g. see p. 691).

INSECTA. Hymenoptera: Mainly in larvae and pupae (occasionally in eggs) of Cyclorrhapha, though a few are known from Orthorrhapha and Nematocera (Stammer, 1956). Several species have been tested as biological control agents, but only *Opius* spp. against the Queensland fruit fly has shown any promise. A variety of small wasps are known from higher Diptera; Braconidae (*Opius* spp. and various Alysiinae) parasitize acalyptrate Schizophora, the Dacnusini particularly in the leaf-mining Agromyzidae; while predatory syrphids are parasitized by an ichneumonoid (*Diplazon* sp.) and several Encyrtidae. Hyperparasitic species of Trigonalidae and Perilampinae are known from Tachinidae, and of Encyrtidae from Acroceridae.

Diptera: Relatively few known, mainly Tachinidae and Bombyliidae.

Economic Significance. The Diptera outrank all other orders of insects in medical and veterinary importance. Blood-sucking species are directly responsible for the transmission of malaria, filariasis, leishmaniasis, trypanosomiasis (sleeping sickness), and a wide range of arboviruses, including yellow fever, dengue, and various encephalitides. Most of the vector species are mosquitoes, but others are found in the Simuliidae, Ceratopogonidae, Muscidae–Glossininae, and the Tabanidae. Similarly transmitted are various protozoa, helminths, and viruses that cause disease in domestic animals and birds; e.g. trypanosomiasis (surra) in horses, the blue-tongue virus of sheep, and the avian malarias. These diseases of man and animals are mostly absent, or of limited occurrence, in Australia; but in some areas the bites of mosquitoes, sand flies, black flies, and horse flies can create significant problems through physiological side effects and economic effects on land use.

Apart from the biting flies, the house fly (*Musca domestica* L.) is a cosmopolitan vector of enteric diseases, and larvae of other Cyclorrhapha (principally Calliphoridae), infest the tissues of livestock and occasionally man, producing a condition known as *myiasis*. 'Blow-fly strike' of sheep costs Australian industry millions of dollars annually, and we live under the threat that *Chrysomya bezziana* Vill., which causes a similar and very serious myiasis in cattle, may be introduced from New Guinea. Of lesser importance are the 'bot flies' (Oestridae, Gasterophilidae), whose larvae are internal parasites of sheep, horses, and kangaroos, and the ectoparasitic Hippoboscidae (e.g. the sheep ked).

Cultivated plants probably suffer less from dipterous pests in Australia than in many other countries, but the Queensland and Mediterranean fruit flies (Tephritidae) are serious pests of fruit. Also, leaf-miners and stem-borers (Agromyzidae, Chloropidae, etc.) although of little significance now, may yet increase in importance with the development of tropical agriculture in the north.

Against this record can be set the numerous

examples of beneficial species. Various Diptera play at least a secondary role as pollinators, some with highly specialized behaviour, e.g. *Forcipomyia* spp. (p. 691) and a mycetophilid (*Mycomya* sp.) that pollinates a ground orchid. The order has been little used in Australia for planned biological control, although an introduced cecidomyiid, *Zeuxidiplosis giardi* (Kieff), is of some importance in combating the weed, St John's wort; but we have numerous native predators and parasites which play at least some part in checking the activities of potential or actual pests. Even the biting flies, particularly the mosquitoes, have been of use as vectors of the myxoma virus which so dramatically reduced our rabbit plague. We may also instance the Tachinidae, Bombyliidae, and Nemestrinidae, whose larvae are endoparasites of other insects, including many pest species; and the larvae of Syrphidae and adults of Dolichopodidae, Empididae, Asilidae, and others, which feed largely on arthropod prey; also, the phytophagous larvae of Cecidomyiidae which assist in keeping in check plants, such as *Hakea* spp., that have shown potentialities as pests when introduced abroad. Generally, this aspect of the ecology of our native Diptera has been little investigated.

Special Features of the Australian Fauna

The Australian dipterous fauna may be regarded as more or less 'normal' in size—its probable total of some 7,000–8,000 species is about 5 per cent of the estimated world total, as is (approximately) the whole Australian insect fauna, and that figure also represents the proportion of habitable land (i.e. excluding Antarctica) provided by the the Australian continent. Most families occur here, those missing being mainly small and specialized, principally acalyptrate Schizophora, of which some may yet be discovered (as were the Cypselosomatidae in 1964). However, the absence of the Ptychopteridae is noteworthy; they are an ancient group, occurring in all other regions, and their absence poses an interesting problem in biogeography.

The zoogeography of the Australian fauna has been discussed most recently by Paramonov (in Keast *et al.*, 1959), Gressitt (1961), and, to some extent, Hennig (1960). The accepted biogeographic provinces are reflected in the distributions of dipterous species, but the Eyrean is not very distinct, and is characterized mainly by the absence of groups found elsewehere. The most distinct elements of the fauna can be classified on the basis of their world distribution and probable origins, as follows.

(A) Groups of obscure origin: (1) Endemic forms without obviously close relatives elsewhere, possibly relicts of ancient lineages; e.g. *Exeretoneura* (Nemestrinidae). (2) Primitive forms with wide, but disjunct, world distributions, presumably relicts of an ancient fauna; e.g. *Nemopalpus* (Psychodidae), *Olbiogaster* (Anisopodidae). (3) Other forms with enigmatic distributions; e.g. *Culiseta* (Culicidae), *Microphorella* (Empididae). (4) Cosmopolitan forms of such widespread distribution, perhaps by wind currents, that their origins are no longer discernible; e.g. various Chironomidae and Cecidomyiidae. The foregoing groups form a relatively small part of the fauna, and with increasing knowledge may yet be satisfactorily placed in the division below.

(B) Groups showing evidence of probable origins: (1) 'Indo-Malayan' (northern); mostly recent groups, with close relatives in the Indo-Malayan subregion; they comprise numerous groups, particularly of Cyclorrhapha, principally Torresian in distribution, but with members penetrating throughout the continent. (2) 'Antarctic' (southern); almost all belonging to the Nematocera and Orthorrhapha, with near relatives in New Zealand, southern South America, and, in a few cases, in South Africa; most are characteristically Bassian in distribution, but some penetrate well into the tropics, particularly at high altitudes. (3) Recent immigrants; these include two curious cases of immigrants from South America, *Tricharaea brevicornis* (Wied.) (Sarcophagidae) and *Prosopantrum flavifrons* (Tonn. and Mall.) (Heleomyzidae).

It is doubtful if older and younger northern

elements can yet be distinguished in the Diptera, apart from noting that some forms seem to have been derived from more ancient stocks than others. The southern element, found in so many floral and faunal groups, is very marked, and almost every family of Nematocera and Orthorrhapha offers examples of the 'Antarctic' distribution. In the Cyclorrhapha, the reverse is equally marked, and there are very few such groups.

The existence of the clear-cut 'Antarctic' elements seems most satisfactorily explained by the theory of continental drift (chapter 9). Cases of recent immigration from South America notwithstanding, it is difficult to give much credence to alternative hypotheses; particularly those that regard the Australian and South American 'Antarctic' faunas as nothing more than terminal isolates from a once Holarctic distribution. The resemblances are just too close and too numerous, and it is relevant to note that the Tertiary fauna of the northern hemisphere seems to lack all the typically 'Antarctic' groups of Nematocera and Orthorrhapha (Colless, 1964).

CLASSIFICATION

Order DIPTERA (6,256 Australian spp.)

Suborder NEMATOCERA (1,836)

Division TIPULOMORPHA (696)

1. Tipulidae (691)
Ptychopteridae (0)
2. Trichoceridae (5)

Division PSYCHODOMORPHA (73)

3. Tanyderidae (6)
4. Psychodidae (67)

Division CULICOMORPHA (608)

5. Dixidae (11)
6. Culicidae (228)
7. Chironomidae (129)
8. Ceratopogonidae (197)
9. Simuliidae (34)
10. Thaumaleidae (9)

Division BIBIONOMORPHA (459)

11. Blephariceridae (11)
Deuterophlebiidae (0)
12. Anisopodidae (4)
Pachyneuridae (0)
Axymyiidae (0)
13. Perissommatidae (4)
14. Scatopsidae (21)
Nymphomyiidae (0)
Hyperoscelididae (0)
15. Bibionidae (25)
16. Cecidomyiidae (111)
17. Sciaridae (61)
18. Mycetophilidae (222)

Suborder BRACHYCERA (4,420)

Division ORTHORRHAPHA (1,811)

TABANOIDEA (519)
19. Pelecorhynchidae (28)
20. Rhagionidae (65)
21. Tabanidae (240)
Coenomyiidae (0)
Pantophthalmidae (0)
22. Stratiomyidae (100)
23. Xylomyidae (1)
Xylophagidae (0)
24. Nemestrinidae (56)
25. Acroceridae (29)

ASILOIDEA (1,077)
26. Therevidae (162)
27. Scenopinidae (13)
28. Asilidae (380)
29. Apioceridae (75)
30. Mydaidae (35)
31. Bombyliidae (412)

EMPIDOIDEA (215)
32. Empididae (88)
33. Dolichopodidae (127)

Division CYCLORRHAPHA (2,609)

Series *Aschiza* (301)

LONCHOPTEROIDEA (1)
34. Lonchopteridae (1)

PHOROIDEA (102)
35. Platypezidae (11)
36. Sciadoceridae (1)
37. Phoridae (90)

SYRPHOIDEA (198)
38. Pipunculidae (30)
39. Syrphidae (168)

Series *Schizophora* (2,308)

CONOPOIDEA (75)
40. Conopidae (75)
TEPHRITOIDEA (298)
41. Otitidae (3)
Richardiidae (0)
42. Platystomatidae (96)
43. Pyrgotidae (69)
44. Tephritidae (130)
Tachiniscidae (0)
MICROPEZOIDEA (16)
45. Pseudopomyzidae (1)
46. Cypselosomatidae (1)
47. Neriidae (3)
48. Micropezidae (11)
Megamerinidae (0)
TANYPEZOIDEA (4)
Nothybidae (0)
Diopsidae (0)
49. Tanypezidae (2)
50. Psilidae (2)
SCIOMYZOIDEA (268)
Helcomyzidae (0)
Ropalomeridae (0)
51. Sepsidae (8)
Dryomyzidae (0)
52. Sciomyzidae (20)
53. Chamaemyiidae (10)
54. Lauxaniidae (230)
Celyphidae (0)
HELEOMYZOIDEA (94)
55. Coelopidae (6)
56. Heleomyzidae (55)
57. Sphaeroceridae (30)
58. Chyromyidae (3)
Somatiidae (0)
OPOMYZOIDEA (143)
Pallopteridae (0)
59. Lonchaeidae (21)
Neottiophilidae (0)
60. Piophilidae (11)
Opomyzidae (0)
61. Clusiidae (28)
62. Odiniidae (2)
63. Agromyzidae (57)
64. Fergusoninidae (23)
65. Carnidae (1)
Acartophthalmidae (0)
ASTEIOIDEA (26)
66. Teratomyzidae (9)
Periscelididae (0)
Aulacigastridae (0)
67. Anthomyzidae (12)
68. Asteiidae (5)
DROSOPHILOIDEA (399)
Camillidae (0)
69. Ephydridae (55)
Diastatidae (0)
*Curtonotidae (0)
70. Drosophilidae (70)
71. Milichiidae (14)
72. Cryptochetidae (3)
73. Tethinidae (8)
74. Canaceidae (8)
75. Chloropidae (240)
76. Braulidae (1)
MUSCOIDEA (985)
Scatophagidae (0)
Mormotomyiidae (0)
77. Gasterophilidae (3)
78. Anthomyiidae (6)
79. Muscidae (200)
80. Calliphoridae (135)
81. Sarcophagidae (66)
82. Tachinidae (520)
83. Oestridae (2)
84. Hippoboscidae (30)
85. Streblidae (7)
86. Nycteribiidae (16)

* See p. 729, footnote.

A stable phylogenetic classification has not yet been attained in the Diptera, and the evidence of morphology, palaeontology, and cytology has been variously adduced to support systems that differ in details (e.g. Hennig, 1954; Rohdendorf, 1964; Boyes, 1958). The broad picture is, however, fairly clear. One difficulty lies in the long period during which the order has been actively evolving, with extinction of many lineages, variable divergence in the remainder, and active radiation in the more recently successful. Some nematocerous families date back for 200 million years, whereas some muscoid families seem to be still radiating and *in statu nascendi*. Thus, what we conveniently recognize as families occupy a great diversity of evolutionary levels, and are separated by a corresponding diversity of phenetic 'gaps'.

The classification used here is conservative at the family level, and unnecessary splitting has been avoided; but at the same time we have accepted families such as the Sciaridae and Carnidae on grounds of distinctive morphology and probable phylogeny. The recency of certain nomenclatural decisions has made it necessary to give synonyms of many family names, particularly the 'Meigen 1800' names (Internat. Comm. Zool. Nomencl., 1963). The category of superfamily has been used to indicate probably monophyletic family groups in the Brachycera, but precise definition has often proved difficult. In the Nematocera, the category has not been applied at all; the great diversity within the groups recognized warrants a higher category, and the available evidence for further subdivision is equivocal or leads to a cumbersome number of monotypic groups.

The keys have been designed primarily for convenience in identification and make no pretence to be 'natural'. For this reason, the Cyclorrhapha Aschiza have been included in a single key with the Orthorrhapha.

KEYS TO THE MAJOR GROUPS OF DIPTERA

ADULTS

1. Antenna relatively simple, often filiform and longer than thorax, with scape, pedicel, and a flagellum of 6–14 segments, rarely more or less (Figs. 34.2A–F); maxillary palps usually with 3–5 segments (rarely reduced); vein CuA rarely converging towards 1A (except in Bibionidae), never meeting it;

discal cell often absent; mesopleural suture following a roughly straight or wavy line from wing root to mid coxa (except in some Psychodidae and Scatopsidae). Mostly rather slender flies .... NEMATOCERA (p. 679)

Antenna usually short, with less than 6 segments (except in some Orthorrhapha), often highly modified, terminating in narrow style or bristle-like arista (Figs. 34.2G–K); maxillary palps with 1 or 2 segments; vein CuA converging towards 1A, often meeting it; discal cell usually present; mesopleural suture sharply angled where it meets sternopleuron, and often similarly angled posteriorly around that sclerite. Often more stoutly built flies .... BRACHYCERA. 2

2. Ptilinal fissure absent, the sutures, if any, at sides of face not confluent above antennae (except in a few Syrphidae); R_{4+5} sometimes branched; CuA often long, reaching wing margin or joining 1A near its apex .... Orthorrhapha (p. 699) and Cyclorrhapha Aschiza (p. 712)

Ptilinal fissure present above bases of antennae, continuing down sides of face as an inverted U; R_{4+5} unbranched; CuA usually short and joining 1A well back from its apex, usually towards its base .... Cyclorrhapha Schizophora (p. 714)

MATURE LARVAE

1. Head capsule usually well formed, complete, or with deep incisions posteriorly (may be considerably reduced or modified in Cecidomyiidae, Tipulidae, and Blephariceridae); mandibles usually of chewing type, toothed, opposable, and moving in horizontal plane of head (Fig. 34.10A) .... NEMATOCERA (p. 679)

Head capsule absent or incomplete posteriorly*; if incomplete, with a strong internal skeleton of paired rods retracted within the thorax; 'mandibles' (p. 671) usually hook- or sickle shaped, moving in the vertical plane of head .... BRACHYCERA. 2

2. Head capsule partially developed, more or less retracted within prothorax, the anterior portion with a distinct dorsal sclerite, which bears the usually well-formed antennae; posterior portion with an internal skeleton of longitudinal rods; mandibles usually sickle-shaped, not attached to a cephalopharyngeal skeleton (Fig. 34.10B) .... Orthorrhapha (p. 699)

Head capsule not developed, without dorsal sclerotization; antennae absent, or poorly developed, borne on membranous areas; mandibles replaced by 'mouth-hooks', attached to a characteristic cephalopharyngeal skeleton (Fig. 34.10C) .... Cyclorrhapha (p. 711)

* Stratiomyidae, with an apparently complete head capsule, have paired anterior incisions on the dorsal surface, flanking a sharp, beak-like labrum.

Suborder NEMATOCERA

Mostly small, rather delicate flies, with the general characteristics given in the key; also with X and Y chromosomes not forming bivalents at spermatogenesis (except in Ptychopteridae), or sex-chromosomes apparently absent. Generally, the Nematocera most closely resemble the ancestral Diptera and include the oldest families, some known from Jurassic fossils; some, e.g. the Culicidae, show considerable specialization, and have probably undergone relatively recent radiations.

Keys to the Families of Nematocera Known in Australia

ADULTS

1. Wings with a secondary network of vein-like markings; slender, long-legged flies (Figs. 34.18B, C) .... **Blephariceridae** (p. 694)

Wings without such markings 2

2(1). Discal cell and ocelli both present 3

Discal cell and/or ocelli absent 5

3(2). R_{2+3} forked, R_2 ending in costa; eyes divided laterally into completely separated dorsal and ventral components; small flies (Figs. 34.18D, E) .... **Perissommatidae** (p. 695)

R_2 absent, or ending in R_1; eyes normal, at most differentiated into contiguous areas of different facet size 4

4(3). Delicate, long-legged flies; mesonotum with V-shaped transverse suture; 2A short but strong, curved down to meet wing margin (Fig. 34.7B) .... **Trichoceridae*** (p. 687)
Stoutly-built flies; mesonotum without such suture; 2A weak, not reaching wing margin (Fig. 34.18A) .... **Anisopodidae** (p. 695)
5(2). Mesonotum with complete V-shaped transverse suture; wing with 2 complete anal veins reaching margin; long-legged flies (Fig. 34.15) .... **Tipulidae*** (p. 685)
Mesonotum without V-shaped suture (indicated, but incomplete centrally in Tanyderidae); wing with at most a single complete anal vein reaching margin 6
6(5). Discal cell present; rather large, long-legged flies, with patterned wings (Fig. 34.16A) .... **Tanyderidae** (p. 687)
Discal cell absent 7
7(6). Small to minute, delicate flies, with elongate, usually moniliform antennae; wings with Rs unbranched, usually hairy, with few main veins and few or no cross-veins; eyes bridged above the antennal sockets; tibiae without spurs; coxae usually not elongate (Fig. 34.20E) .... **Cecidomyiidae** (p. 696)
Not with the above combination of characters 8
8(7). Ocelli present 9
Ocelli absent 13
9(8). Mid and hind tibiae with 1 or 2 distinct apical spurs 10
Mid and hind tibiae without apical spurs 12
10(9). Antennae arising from near or below level of lower margins of eyes; empodium and pulvilli strongly and equally developed, forming a triple pad beneath the tarsal claws; coxae normal (Fig. 34.19A) .... **Bibionidae** (p. 695)
Antennae arising from near or above level of centre of eyes; pulvilli at most weakly developed; coxae more or less elongate 11
11(10). Eyes usually connected dorsally above the antennae; a distinct mid-pleural pit usually present below the wing-root; wing with characteristic venation, *r-m* in line with apical portion of Rs, M_{3+4} arising near base of wing; tibiae without strong spines (Figs. 34.20C, D) .... **Sciaridae** (p. 697)
Eyes not connected dorsally; mid-pleural pit indistinct or absent; wing venation not as above, and/or tibiae with strong spines (Figs. 34.20A, B) .... **Mycetophilidae** (p. 698)
12(9). Rs unbranched; radial veins very strong, other veins much weaker (Fig. 34.19B) .... **Scatopsidae** (p. 695)
Rs forked .... (Hyperoscelididae)
13(8). Wings broadly ovate, often pointed; M always 4-branched, Rs sometimes 4-branched, cross-veins usually restricted to basal third of wing; mostly small, very hairy, moth-like flies (Fig. 34.16B) .... **Psychodidae** (p. 687)
Both M and Rs with 3 or fewer branches 14
14(13). Rs and M each with 3 branches 15
Rs, and/or M, with fewer than 3 branches 16
15(14). R_{2+3} strongly arched, not in line with main stem of Rs; wing veins with sparse setulae; mouth-parts not forming an elongate proboscis (Fig. 34.17H) .... **Dixidae** (p. 689)
R_{2+3} not arched, more or less in line with main stem of Rs; wing veins clothed with scales; most species with mouth-parts forming an elongate proboscis (Figs. 34.17C, D) .... **Culicidae** (p. 689)
16(14). R_2 short, vertical; R_3 distinctly arched on its basal half; antennae short in both sexes, apical part of flagellum abruptly narrowed, fine (Fig. 34.17F) .... **Thaumaleidae** (p. 693)
Radial veins and antennae otherwise; R_2 usually vestigial or absent 17
17(16). Wings very broad, with large anal lobe; M_{3+4} arising at extreme base of wing; small, stout, biting flies (Fig. 34.17E) .... **Simuliidae** (p. 693)
Wings mostly longer and narrower, with small anal lobe or none; M_{3+4} forking from CuA at about centre of wing 18

* Shrivelled specimens of Trichoceridae, in which the ocelli may be overlooked, will key to Tipulidae, but may be recognized by the very short, curved 2A.

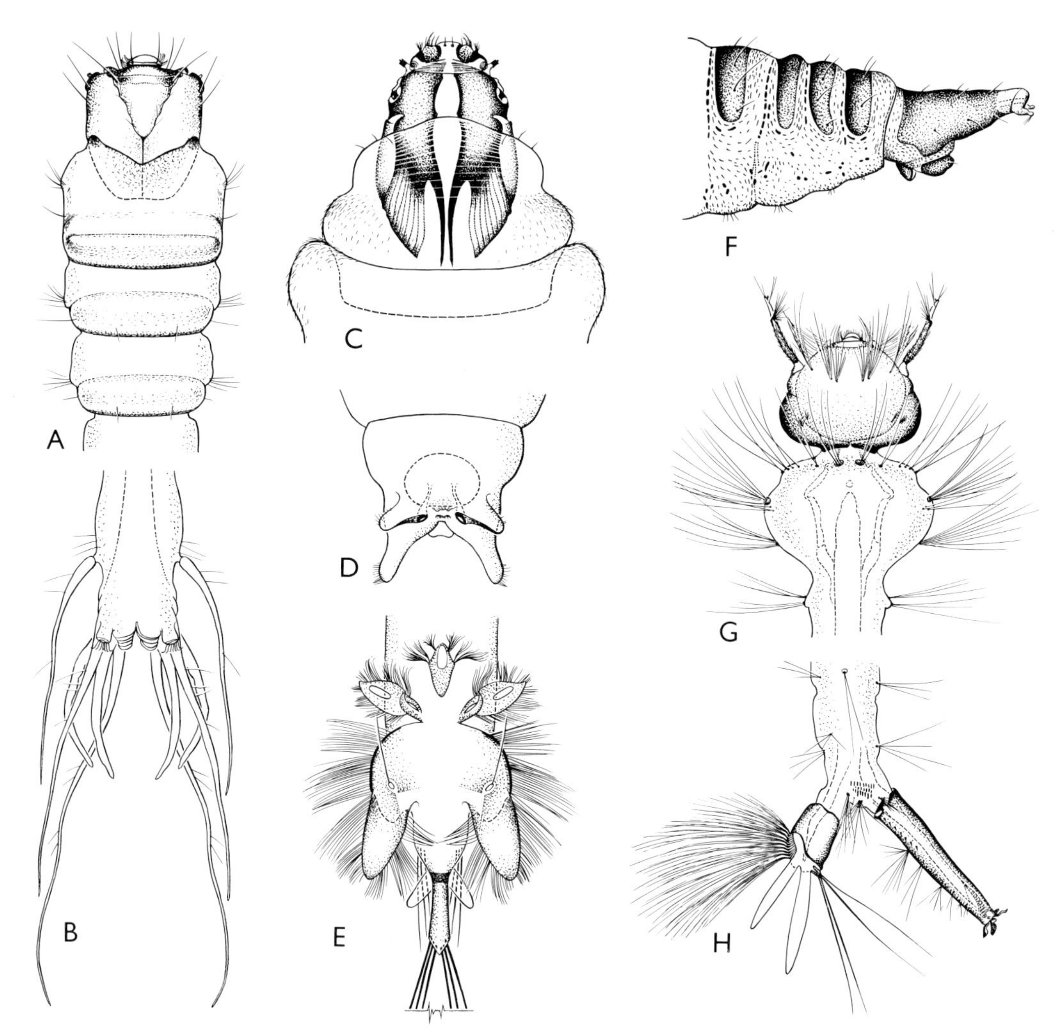

Fig. 34.12. Larvae of Nematocera (F and H lateral, remainder dorsal): A, *Eutanyderus wilsoni*, Tanyderidae, head and thorax; B, same, end of abdomen; C, *Dolichopeza* sp., Tipulidae, head and prothorax; D, *Trichocera annulata*, Trichoceridae, end of abdomen; E, *Dixa nicholsoni*, Dixidae, end of abdomen; F, unidentified sp. of Psychodidae, end of abdomen; G, *Culex pipiens australicus*, Culicidae, head and thorax (minor setae omitted); H, same, end of abdomen. [T. Binder]

18(17). Radial veins short, usually meeting costa well before apex of wing, and enclosing 1 or 2 radial cells; M_1 and M_2 both present; mouth-parts of piercing type (Fig. 34.17G) .... **Ceratopogonidae** (p. 691)

Radial veins longer, R_{4+5} almost always ending near apex of wing; M_{1+2} unbranched; mouth-parts almost always non-piercing (Figs. 34.17A, B) .... **Chironomidae** (p. 690)

MATURE LARVAE

1. Very small larvae, with 13 postcephalic segments and a tiny, much reduced head capsule; mouth-parts rudimentary; antennae distinct; mature larvae usually with a longitudinal sclerotized strip ('sternal spatula') ventrally on thorax; peripneustic; terrestrial or in plant galls (Fig. 34.14H) .... **Cecidomyiidae** (p. 696)

Not with the above combination of characters 2

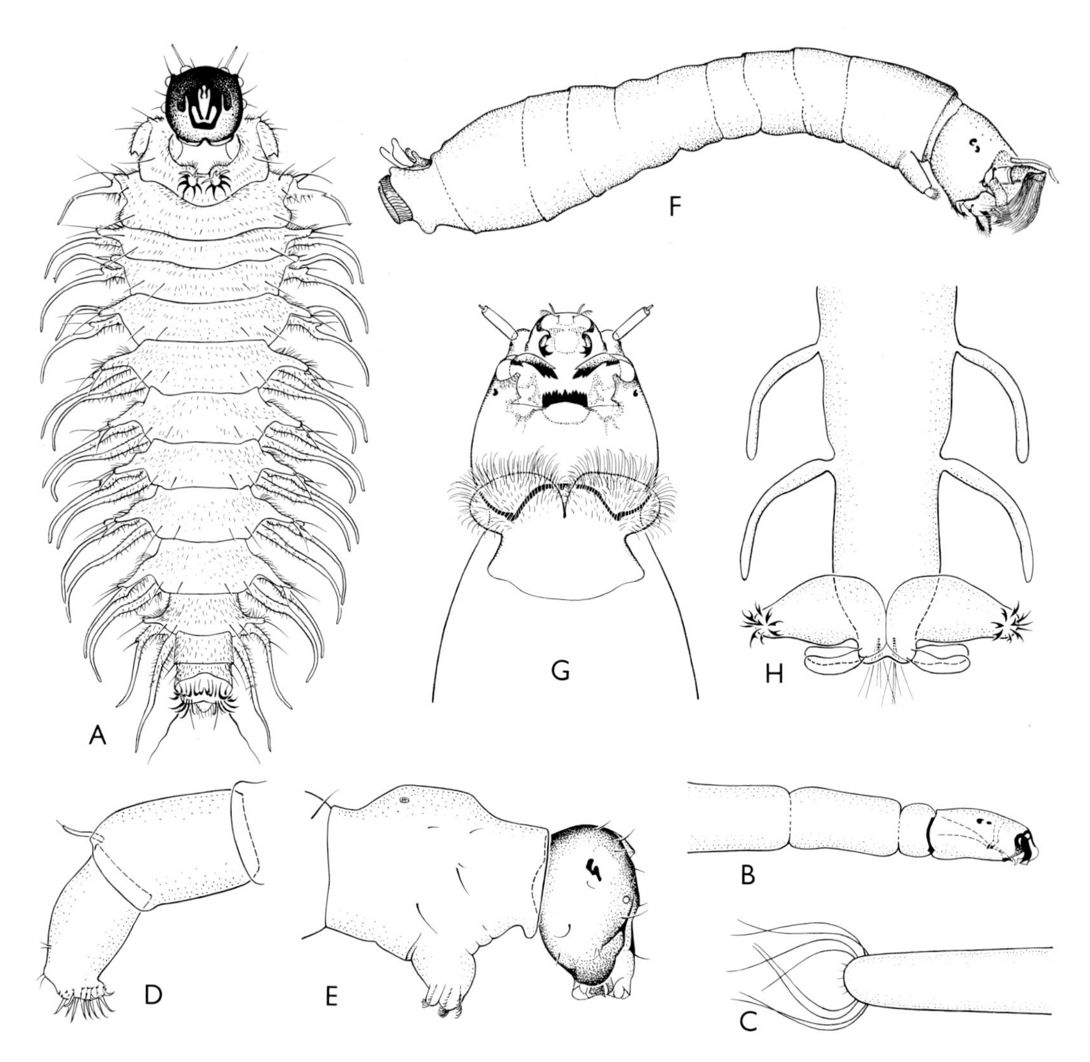

Fig. 34.13. Larvae of Nematocera: A, *Atrichopogon* sp., Ceratopogonidae, ventral; B, *Culicoides angularis*, Ceratopogonidae, head and thorax; C, same, apex of abdomen, lateral; D, *Austrothaumalea* sp., Thaumaleidae, apex of abdomen, lateral; E, same, head and thorax; F, *Simulium ornatipes*, Simuliidae; G, *Chironomus tepperi*, Chironomidae, head and thorax, ventral; H, same, apex of abdomen, ventral.

[T. Binder]

2(1). Thorax, first 2 abdominal segments, and sometimes head capsule fused into a single large mass; the 5 or 6 apparent segments of abdomen separated by deep constrictions; ventrally with a conspicuous median row of sucking discs; attached to rocks, etc., in running water (Fig. 34.14A) .... **Blephariceridae** (p. 694)

Not with the above combination of characters 3

3(2). Head more or less retracted within thorax, sometimes completely so, capsule with the strongly sclerotized areas deeply incised posteriorly, or even reduced to little more than a series of longitudinal rods and plates; apex of abdomen often with radiating protuberances; terrestrial or aquatic (Fig. 34.12C) .... **Tipulidae** (p. 685)

Head capsule well formed, completely sclerotized dorsally, more or less exserted 4

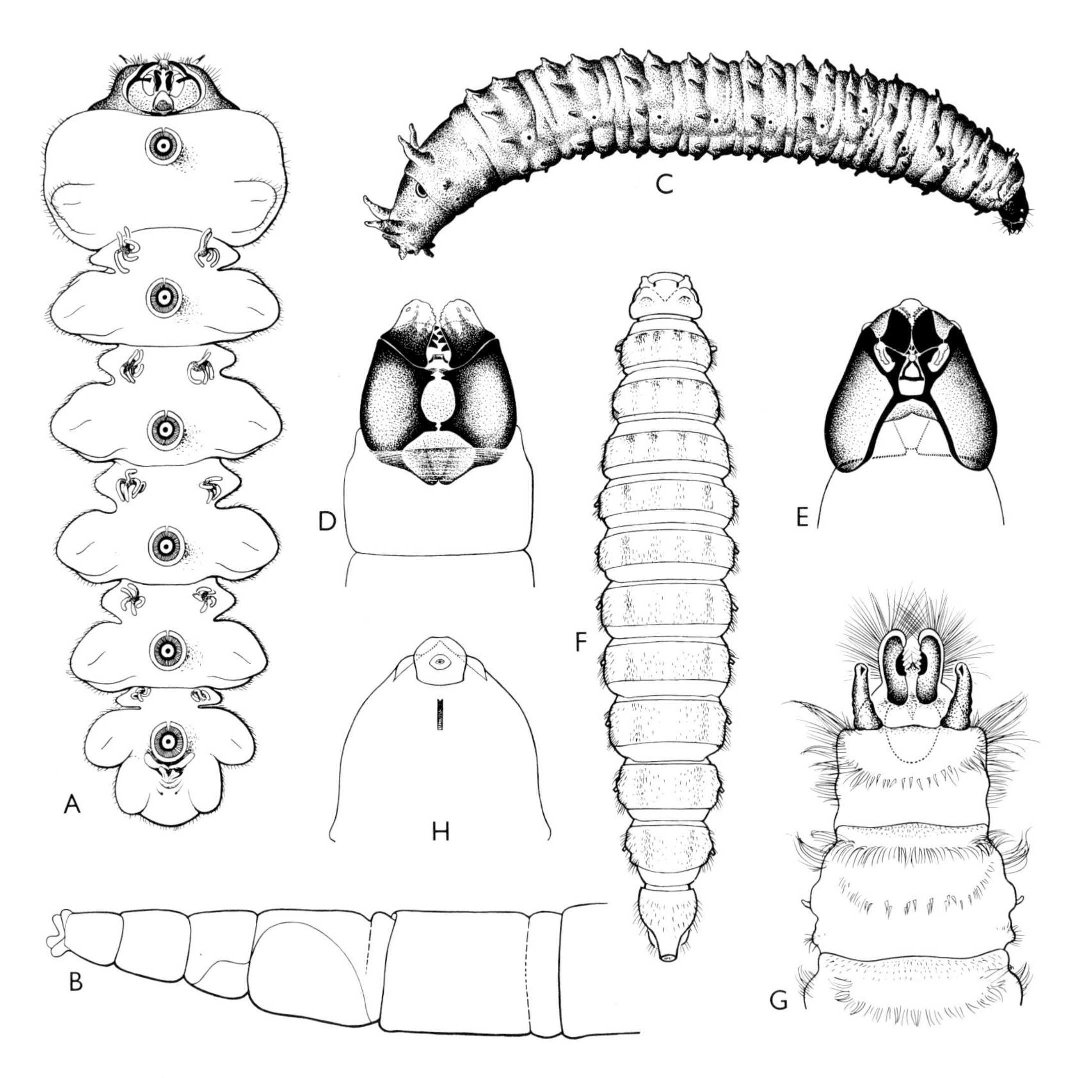

Fig. 34.14. Larvae of Nematocera: A, *Edwardsina ferrugiana*, Blephariceridae, ventral; B, *Sylvicola dubius*, Anisopodidae, apex of abdomen, lateral; C, unidentified sp. of Bibionidae; D, *Sciara* sp., Sciaridae, head, ventral; E, *Exechia* sp., Mycetophilidae, head, ventral; F, *Perissomma fusca*, Perissommatidae, dorsal; G, *Scatopse fuscipes*, Scatopsidae, apex of abdomen, dorsal; H, *Zeuxidiplosis giardi*, Cecidomyiidae, head and thorax, ventral. [T. Binder]

4(3). Abdomen with a club-like shape produced by dilated posterior segments, the apical segment with a sucker-like disc bearing concentric circles of tiny hooks; thorax with a median ventral proleg; mouth-brushes conspicuous, fan-like; attached to stones etc., in running water (Fig. 34.13F) .... **Simuliidae** (p. 693)

Not with the above combination of characters 5

5(4). Thorax with segments completely fused into a single dilated mass; metapneustic, posterior spiracles typically at apex of a stout, sclerotized siphon or approximated on a complex dorsal plate; head large, highly mobile; aquatic, very active (Figs. 34.12G, H) .... **Culicidae** (p. 689)

Thorax with distinct segmentation 6

6(5). Abdominal segments 1 and 2 each with a pair of prolegs, which bear apical rows of short, curved, bristles; apex of abdomen with a pair of projecting plates fringed with long hairs and a short rod tipped with long bristles; aquatic (Fig. 34.12E) .... **Dixidae** (p. 689)
Abdominal segments 1 and 2 without paired prolegs 7

7(6). Small larvae; peripneustic, the posterior spiracles borne on a pair of short, tubular, sclerotized siphons; terrestrial (Fig. 34.14G) .... **Scatopsidae** (p. 695)
Posterior spiracles, if present, sessile or borne on a single siphon 8

8(7). Small larvae, completely sclerotized when mature; segmentation very distinct, simple; peripneustic, the posterior spiracles borne on a short, stout, tapering siphon; antennae rudimentary; head capsule indented laterally, not smoothly rounded; terrestrial (Fig. 34.14F) .... **Perissommatidae** (p. 695)
Not with the above combination of characters 9

9(8). Small larvae; amphipneustic or metapneustic; body with marked secondary segmentation, indicated on at least some segments by narrow, transverse, sclerotized strips; and/or posterior spiracles borne on a short, stout, tapering siphon; or apex of abdomen with several very long hairs *(Phlebotomus)*; or 'louse-like' larvae, flattened, with long, 4-segmented antennae and dorsal plates fringed with fine processes *(Sycorax)*; terrestrial or semiaquatic (Fig. 34.12F) .... **Psychodidae*** (p. 687)
Not with the above combination of characters 10

10(9). Head projecting ventrally, at about a right angle to thorax, with short hairs and/or knob-like protuberances; antennae rudimentary; prothorax with median pseudopod ventrally; apex of abdomen with semicircle or cluster of hooklets; amphipneustic, thoracic spiracles distinct; aquatic (Figs. 34.13D, E) .... **Thaumaleidae** (p. 693)
Not with the above combination of characters 11

11(10). Apneustic; often with prolegs, bearing apical spines or hooklets, on prothorax and sometimes on apex of abdomen also; aquatic 12
Spiracles distinct on at least prothorax and apex of abdomen; prolegs absent; terrestrial ... 14

12(11). Anterior margin of head truncate, almost rectangular; thorax and abdomen with a dorsal and a ventral protuberance ('creeping welts') on most segments; apex of abdomen with 8 long gill-filaments, 4 blunt, 4 longer and tapering; a pair of the latter type on the penultimate segment also (Figs. 34.12A, B) .... **Tanyderidae** (p. 687)
Without the above combination of characters 13

13(12). Usually with a pair of partially or completely fused prolegs on thorax and a pair at apex of abdomen; body hairs, if present, normal; head without conspicuous internal pharyngeal skeleton (Figs. 34.13G, H) .... **Chironomidae** (p. 690)
Without prolegs or, if present, posterior pair completely fused; or head with conspicuous internal pharyngeal skeleton; or body hairs rising from sclerotized tubercles or plaques; sometimes extremely elongate (Figs. 34.13A–C) .... **Ceratopogonidae** (p. 691)

14(11). Integument finely pilose; thorax and abdomen with marked secondary segmentation; amphipneustic, posterior spiracles surrounded by 4 protuberances fringed with hairs, or set in a circular depression fringed with long hairs (Fig. 34.12D) .... **Trichoceridae** (p. 687)
Not with the above combination of characters 15

15(14). Abdomen with the main portion of each segment separated from its neighbour by a narrow, ring-like secondary segment; amphipneustic; body cylindrical, smooth, without conspicuous hairs or tubercles (Fig. 34.14B) .... **Anisopodidae** (p. 695)
Not with the above combination of characters 16

16(15). Integument with conspicuous hairs or fleshy processes and usually dark, rather 'leathery'; metathoracic spiracle usually distinct (Fig. 34.14C) .... **Bibionidae** (p. 695)
Integument smooth, usually pale; no metathoracic spiracle 17

17(16). Head black, shiny, centre of ventral surface with epicranial plates almost, but not quite, meeting at apices of a pair of short, blunt projections; abdomen without transverse rows of tiny spines ventrally (Fig. 34.14D) .... **Sciaridae** (p. 697)

* Larvae of *Nemopalpus* and *Trichomyia* are not sufficiently known for inclusion, and Australian *Sycorax* may not conform with the above description.

Head otherwise, centre of ventral surface with epicranial plates broadly, or not at all, approximated, or joined by a complete narrow bridge; abdomen often with paired, transverse rows of tiny spines ventrally (Fig. 34.14E) .... **Mycetophilidae** (p. 698)

Division TIPULOMORPHA

Long-legged and long-winged flies, mostly with a generalized venation, including 2 distinct anal veins (except in Ptychopteridae, which are doubtfully placed here) and Rs often 4-branched. They retain the primitive V-shaped suture and bilobed postphragma of the mesonotum, have distinguishable sex-chromosomes (except in limoniine Tipulidae), and form chiasmata at spermatogenesis. Larvae are mostly aquatic or semiaquatic, and metapneustic or amphipneustic.

1. Tipulidae (crane flies, daddy-long-legs; Figs. 34. 5A, B, 6A, 8A, 11A, 12C, 15). An immense family, cosmopolitan in distribution, with a world total of over 11,000 known species. In Australia, as elsewhere, it is by far the richest in species of all families of Diptera. Most Tipulidae are readily recognized by their slim build and long, unusually brittle legs, though the Trichoceridae and Tanyderidae have a very similar appearance. They are amongst the most generalized of Diptera, with a strong, V-shaped mesonotal suture and 2 complete anal veins, but lack ocelli and sometimes the discal cell. The interpretation of the venation (Fig. 34.15) is that proposed by Alexander (1929). In some of the larger species, the sexes differ strikingly in colour pattern, while the females of some species have vestigial wings. There is remarkable variation in size, from *Semnotes imperatoria* Westw. (Fig. 34.15F), one of our largest Diptera, with a wing-span of 75 mm, down to tiny, midge-like species of *Tasiocera*, expanding only 6–8 mm.

The family is well represented in Australia; most species have been described from the mountainous areas of the south-east, and no doubt many more await recognition. They are notable lovers of moisture, usually found resting on foliage, overhanging banks, etc., in damp shady places. Some occur in very narrowly restricted habitats; e.g. at least one species appears to rest by preference on spiders' webs. Many of the smaller species are crepuscular, while a number of species, large and small, are attracted to light. The larvae (Fig. 34.12C) are hemicephalic, and found either in water or, more commonly, in wet soil or decomposing vegetable matter.

The following key will place most specimens in the subfamilies currently recognized (some authors regard them as separate families). It must be noted that there are exceptions to many of the individual characters given.

Key to the Subfamilies of Tipulidae

1. Last segment of palp almost always elongate, whiplash-like (Fig. 34.15A); antenna with not more than 15 segments (usually not more than 13); Sc usually ending in R; *m-cu* meeting M_{3+4} close to its fork or more distally on M_4; mostly larger species .... TIPULINAE
   Last segment of palp short, subequal to the others (Fig. 34.15B); antenna with less than 13 segments *(Eriocera)* or with 14–39 segments; wings with Sc ending in costa; *m-cu* usually joining M at, before, or only a short distance beyond its fork; mostly smaller species or of moderate size 2
2. Wing with *r-m* joining Rs at or before the fork; R_{1+2+3} apparently with a long fusion back from the margin .... CYLINDROTOMINAE
   Wing with *r-m* joining R_{4+5} or R_5 beyond the fork of Rs (except in *Helius*); radial veins without such apparent fusion ...... LIMONIINAE

Australia is rich in endemic genera of TIPULINAE (8 out of 19), including the large, handsome, and rather rare species of *Clytocosmus* (Plate 5, K). Several genera occur also in New Zealand, South America, and some in South Africa; whereas *Tipula* and *Nephrotoma* are world-wide genera which have invaded the continent only in the extreme north. *Plusiomyia*, *Phymatopsis*, and *Macromastix* include a number of species with subapterous females.

The CYLINDROTOMINAE are represented only by the endemic *Stilbadocerodes*. The vast subfamily LIMONIINAE, however, is

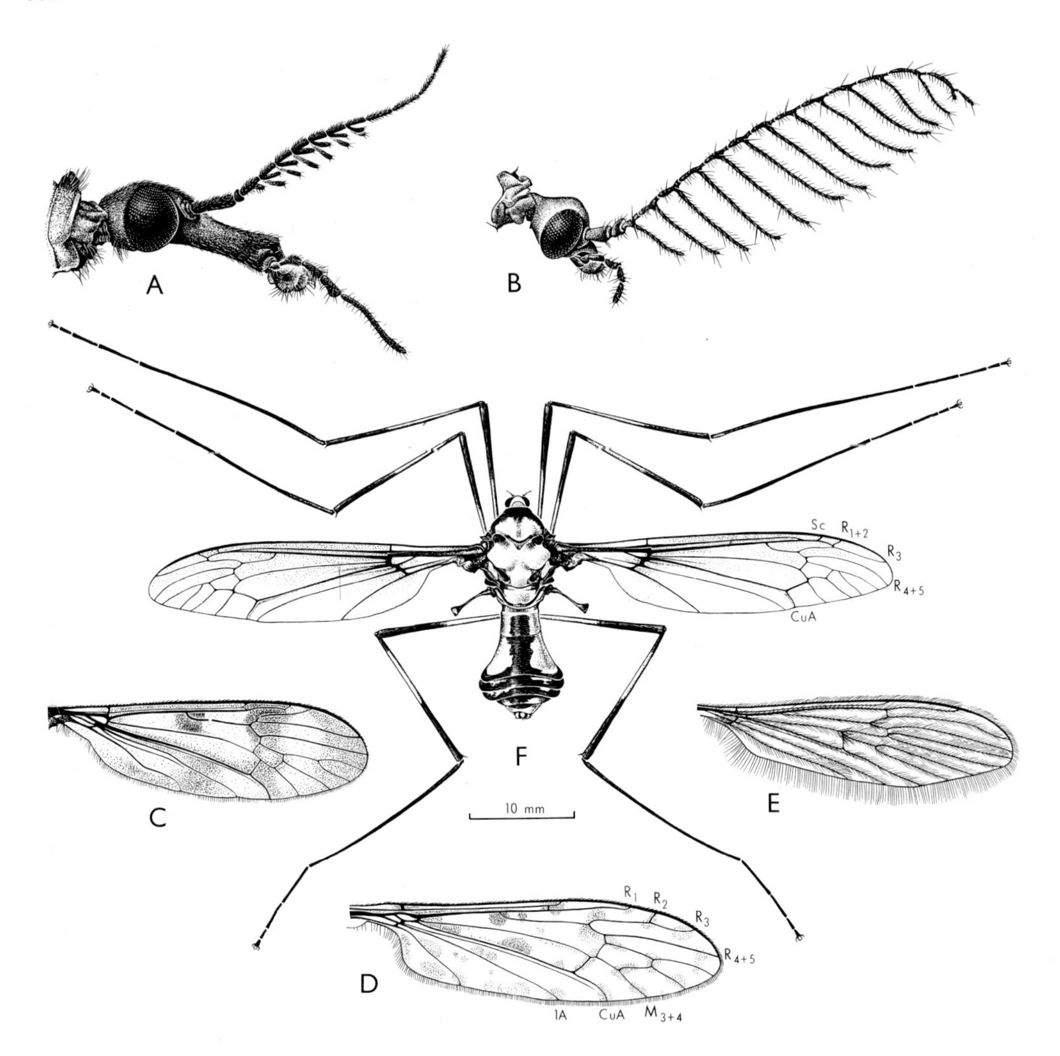

Fig. 34.15. Tipulidae: A, *Ptilogyna ramicornis*, ♂; B, *Paralimnophila setulicornis*, ♂; C, *Gynoplistia apicalis*; D, *Ischnothrix australasiae*; E, *Erioptera* sp.; F, *Semnotes imperatoria*. [T. Binder]

represented here by 5 tribes. Of these, the Lechriini have the single genus *Lechria*, which is entirely Austro-Malayan, while the more or less cosmopolitan Pediciini have only 2 species of *Pedicia (Tricyphona)* recorded from this country. In the Limoniini, the great genus *Limonia* has several characteristic subgenera, notably *Dicranomyia* and *Geranomyia*. *Idioglochina* has 2 northern species which, like those found in the Pacific islands, are probably marine, while *Tonnoiromyia* is of particular zoogeographical interest, with 2 species in south-eastern Australia and one in Chile.

In the Hexatomini, which resemble some Trichoceridae in having small tibial spurs, there are 5 endemic genera and a number showing an 'Antarctic' distribution; of the latter, *Gynoplistia* (Fig. 34.15C) and *Ischnothrix* (Fig. 34.15D) are most characteristic. Of the cosmopolitan genera, there is *Limnophila*, with many Australian species, and the aptly-named *Elephantomyia*, which has a long, superficially mosquito-like proboscis.

Both *Limnophila* and *Gynoplistia* include species with subapterous females, and, in the latter, subapterous males also.

The Eriopterini are mostly small flies, many of which lack a discal cell in the wing. They are most abundantly represented here by the genus *Erioptera* (=*Molophilus*; Fig. 34.15E), with about 150 species, some with subapterous females. The genus is also well developed in New Zealand and Chile. *Tasiocera* is another characteristic Australian genus with a somewhat similar 'Antarctic' distribution. [Alexander, 1932, and numerous short papers, 1923 to present; Paramonov, 1953e.]

2. Trichoceridae (Petauristidae; winter crane flies; Figs. 34.7B, 12D). A small, primitive family, including only 4 genera. Superficially, they resemble the Tipulidae, but they have ocelli and a eucephalic larva. The morphology of the immature stages has led some authors to ally them with the Anisopodidae. All are small, delicate, long-legged, brownish flies, with a wing-span of about 10–12 mm, and, like the Tipulidae, are found in damp, shady places. They are so far known only from the cooler areas of south-eastern Australia, and are commonly found only in the colder months of the year. Most appear to be crepuscular or nocturnal in habit, and they are often taken in light traps. The larvae occur principally in decomposing vegetable matter, and several species have been reared from rotting mushrooms.

Trichocera, a Holarctic genus with small tibial spurs, is represented by the introduced European *T. annulata* Meig. *Paracladura*, found in various countries around the Pacific, has a single undescribed Australian species, recognizable by the extremely short basitarsus. There are also 4 species of *Nothotrichocera* (Fig. 34.7B), with normal tarsi and no tibial spurs; the genus is known elsewhere only from the Subantarctic Campbell and Auckland Is. [Alexander, 1926.]

Division PSYCHODOMORPHA

Small to large flies with generalized wing venation; Rs and M both 4-branched (except in *Trichomyia* and *Sycorax*, with Rs 3-branched). Mesonotum without complete, V-shaped suture; postphragma partly fused, but with distinguishable lobes; metathorax usually better developed than in other Diptera. As far as known, cytogenetic details resemble those of the Culicomorpha. Although poorly represented as fossils, they appear to represent a distinct and probably very ancient lineage.

3. Tanyderidae (Figs. 34.9A, 12A, B, 16A). An essentially Australasian family, which is amongst the most primitive in the Diptera. In general facies, Tanyderidae resemble the Tipulidae, being long-legged, about 20–40 mm in wing-span, with patterned wings, and usually elongate mouth-parts. However, structural details ally them with the Psychodidae. Their primitive nature is indicated principally by the strongly developed pronotum, presence of tibial spurs, and the wing venation, which has the free tip of Sc_2 preserved (in *Nothoderus*), all branches of Rs and M present and parallel, and an elongate discal cell near the centre of the wing. However, vein 2A is reduced to a relic at the wing base.

Adults are usually found together with Tipulidae in moist forest habitats. The only known Australian larva (*Eutanyderus* sp.) is eucephalic and apneustic, with a series of long gill-filaments at the apex of the abdomen; two of them arise from small prolegs (Fig. 34.12B). It bores in the surface layers of submerged rotting logs in alpine streams. All the Australian species are rare, some known from single specimens only. There are 2 endemic genera, *Nothoderus* (1 sp.) and *Eutanyderus* (2 spp.), the latter closely related to a South American genus. The other 3 species fall in *Radinoderus*, an essentially tropical genus; *R. occidentalis* (Alex.) is particularly interesting in being recorded only from south-western W.A. [Alexander, 1938; Colless, 1962c; Hinton, 1966.]

4. Psychodidae (moth flies; Figs. 34.12F, 16B). A cosmopolitan family of small flies,

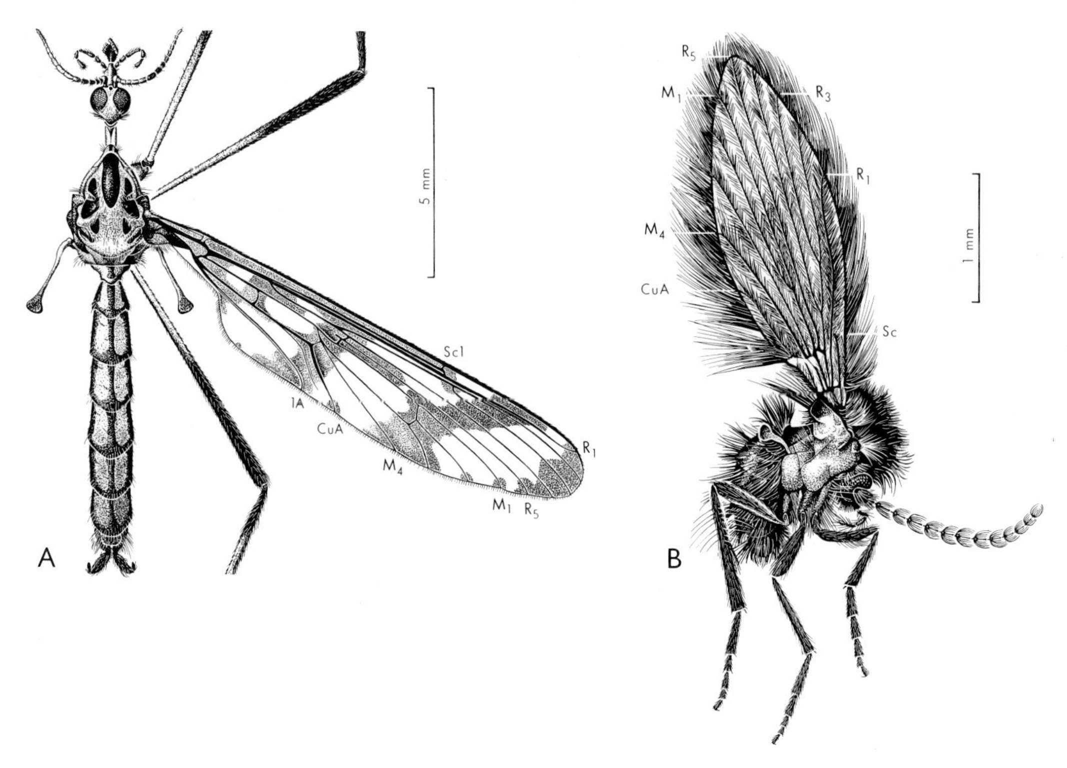

Fig. 34.16. Psychodomorpha: A, *Eutanyderus oreonympha*, Tanyderidae, ♂; B, *Atrichobrunnettia alternata*, Psychodidae, ♀. [T. Binder]

with wing-span rarely exceeding 8 mm, usually much less. In most, the wings are broadly ovate, pointed, hairy, and folded roofwise over the abdomen; this and the hairy body together give them a moth-like appearance. In all but a few genera, both Rs and M are 4-branched and the anal veins greatly reduced.

Adults frequent moist, shady places, but are rarely seen on the wing; at least some are nocturnal and come to light. They are most abundant in early summer, but *Psychoda* is also common during the winter. Most are short-lived (1–2 days) and do not feed, except for members of the blood-sucking genus *Phlebotomus*. In certain countries, where they are often called 'sand flies', species of *Phlebotomus* are important vectors of kala-azar, other forms of leishmaniasis, and the viruses of 'sand-fly fever'. They are relatively uncommon in Australia, and not known to transmit disease here. Psychodid larvae feed on decomposing organic matter, usually at the edges of freshwater habitats or in rotting vegetation, dung, etc., although *Trichomyia* probably bores in wood. *Pericoma* and *Telmatoscopus* have a long larval life and require permanent moisture, but the shorter-lived *Psychoda* can exploit more temporary habitats. *Psychoda alternata* Say is sometimes a pest of sewage-treatment plants and is often seen in bathrooms. The Australian fauna is not yet well known, but most genera occur here, including such rarities as *Nemopalpus*, *Sycorax*, and *Trichomyia*. Those commonly encountered are *Psychoda*, *Pericoma*, *Brunettia*, and *Telmatoscopus*. [Lee, Reye and Dyce, 1962; Duckhouse, 1965.]

Division CULICOMORPHA

Mostly small, delicate flies (except for the more stoutly built Simuliidae and Thaumaleidae), lacking the V-shaped mesonotal suture, and with rather specialized venation. Almost all have 3 or 4 pairs of chromosomes, and the X and Y chromosomes are poorly, if at all, distinguishable. The larvae are almost all aquatic.

5. Dixidae (Figs. 34.12E, 17H). A small, cosmopolitan family, sometimes treated as a subfamily of the Culicidae; there are 3 genera, one of which occurs in Australia. All are small, with wing length about 3–4 mm, shiny integument, and characteristic venation. The adults rest on rocks or vegetation close to streams, generally under forest cover. The larvae are aquatic, usually on vegetation, at the edge of flowing water, but may occasionally be seen swimming in the surface film with a characteristic jerky motion. Only the posterior spiracles are functional, with a complex apparatus of valves, which allows them to be closed off. The Australian species, so far known only from the south-east, all belong to the large genus *Dixa*. There are 6 or 7 species of the subgenus *Nothodixa*, which also occurs in New Zealand and Patagonia; these have *r-m* placed well before the fork of Rs. The others belong to the cosmopolitan *Paradixa*, with longer antennae and *r-m* at or after the fork. [Tonnoir, 1923.]

6. Culicidae (mosquitoes; Figs. 34. 2A, B, 12G, H, 17C, D). A large cosmopolitan family, with characteristic venation and with scales (sometimes very narrow) along the veins and posterior margin. Most species have elongate mouth-parts, forming the typical 'proboscis' of the true mosquitoes; its visible portion is formed by the fleshy labium wrapped around the fine stylets which, in blood-sucking species, are used to pierce the host's skin. Males are more slender than females and, with rare exceptions, have strongly plumose, bushy antennae. Males are not blood-suckers, but females usually require a blood meal before their eggs can mature. There is considerable host specificity, and not all will normally attack man. Many take blood from other animals, and some feed mainly on birds, reptiles, or frogs; even fish ('mud-skippers', *Periophthalmus* sp.) are attacked. Mating usually takes place on the wing, and eggs are laid on free water or wet surfaces, both activities often involving highly specialized behaviour. Adults normally rest by day in shady, humid sites, flight activity commencing at sundown, but some species are active by day in shady places.

The larvae are all aquatic, taking in atmospheric oxygen at the surface by means of spiracles which are borne dorsally on abdominal segment 8; these may be sessile (Anophelini) or at the apex of a sclerotized tube, the 'siphon' (Fig. 34.12H). One genus *(Mansonia)* takes air from the tissues of aquatic plants through a specially adapted, piercing siphon. Both larvae and pupae are very active swimmers, and dive rapidly when disturbed. All species show at least some restriction to a 'preferred' larval habitat, which is selected by the ovipositing female: in permanent ground pools and streams, usually amongst emergent vegetation; in transient pools; or in a variety of container-habitats, some highly specialized (e.g. tree-holes, crab-holes, plant axils). All groups may include saline habitats; e.g. *Aedes australis* (Erich.) breeds in concentrated sea-water in seaside rock-pools; some species (e.g. *Aedes aegypti* (L.)) are domestic, breeding mainly in containers around human dwellings.

Certain species are important vectors of disease (p. 122). In Australia, malaria is transmitted mainly by *Anopheles farauti* Lav.; filariasis by *Culex fatigans* Wied.; dengue fever by *Aedes aegypti*; and other arboviruses by several species, of which *Culex annulirostris* Sk. appears to be the most important. Several also transmit myxomatosis of rabbits, and so perform a valuable economic function. Apart from disease, mosquitoes can be a serious pest, disturbing the comfort of humans and affecting the productivity of domestic animals. Some major pest species are *Culex fatigans* (domestic), *C. annulirostris* (widespread), *Aedes vigilax* (Sk.) (coastal, breeding in brackish swamps), and *A. notoscriptus* (Sk.) (a widespread container-breeder); other species may be very abundant in particular localities.

Key to the Subfamilies and Tribes of Culicidae

1. Mouth-parts at most slightly elongate; larval antennae prehensile, with long, strong, apical spines .... CHAOBORINAE
   Mouth-parts elongate, forming a long proboscis (Fig. 34.17C); larval antennae not prehensile. (Mosquitoes) .... CULICINAE. 2
2. Abdomen with few or no scales, at least on the sternites; ♂ palps clubbed apically, ♀ palps usually about as long as proboscis; larval spiracles sessile .... Anophelini
   Abdomen completely clothed with scales; ♂ palps not clubbed, ♀ palps much shorter than proboscis; larva with siphon 3
3. Proboscis with apical half ventrally recurved; scutellum evenly rounded; large, metallic species; larval mouth-brushes prehensile, each composed of 10 stout rods ... Toxorhynchitini
   Proboscis more or less straight; scutellum with 3 lobes; larval mouth-brushes rarely prehensile, with about 30 or more hairs .... Culicini

The CHAOBORINAE (sometimes treated as a family) are a small distinctive group, with some 8 Australian species in 4 genera; the widespread *Chaoborus* is common. Their larvae are all predacious, occurring in a variety of habitats; one species of *Corethrella* breeds in pitcher-plants.

The Anophelini have only 2 Australian genera. The tropical *Bironella* is rare, but *Anopheles* has about 15 species, mostly tropical; *A. annulipes* Walk., with spotted wings (Fig. 34.17D), occurs throughout the continent. Anopheline larvae are readily recognized in the field by their habit of lying suspended horizontally from the surface film; the adults' resting and feeding attitude is also characteristic, with proboscis and body in a straight line at an angle to the surface. The Toxorhynchitini (formerly Megarhinini) include only 3 Australian species. The adults are large, handsome, with metallic coloration, and do not suck blood; the larvae are predacious, mainly on other mosquito larvae. The Culicini comprise the bulk of our mosquito fauna, with *Culex* and *Aedes* as the dominant genera; *Aedes* is represented by 11 subgenera, dominated by *Ochlerotatus* (41 spp.) and *Finlaya* (34 spp.). There are 9 other genera, including the curious plant-axil-breeding *Malaya*, whose adults are fed by ants.

The bulk of the Australian fauna is derived from the tropics to the north. Many New Guinea species extend into the Cape York area, while *Mansonia uniformis* Theob. and *Aedes lineatopennis* (Ludl.) range as far west as Africa. The domestic species, *C. fatigans* and *A. aegypti*, also are widely distributed in other regions. No genus is endemic, but there are several endemic subgenera, e.g. *Aedes (Chaetocruiomyia)*, and a number of characteristically Australian species-groups, particularly in the south. *Culiseta*, with 12 species and 2 endemic subgenera, is of particular interest, since, apart from one species in Africa and one in New Zealand, the genus is otherwise almost completely Holarctic. [Stone, Knight and Starcke, 1959 (references to Dobrotworsky, Lee, Mackerras, Marks, Woodhill); Clements, 1963; Dobrotworsky, 1965; Fenner and Ratcliffe, 1965.]

7. Chironomidae (Tendipedidae; midges; Figs, 34. 2C, 13G, H, 17A, B). A large, cosmopolitan family, diverse in form but mostly small, delicate flies, some superficially resembling mosquitoes. They range in size from *Chironomus alternans* Walk., wing length 7·5 mm, down to the minute *Orthosmittia reyei* Freem., with wing length of only 0·8 mm. Many are brownish or black, but green, reddish, and yellow species occur, and the wings are sometimes hairy, and sometimes with dark markings. In almost all species, the male antennae are strongly plumose (Fig. 34.17A), and some show sexual dimorphism in colour pattern.

Adults are common, particularly in the vicinity of bodies of water. They are mainly crepuscular or nocturnal, often forming mating swarms at sundown, and can occur in such enormous numbers as to cause considerable annoyance around lights on warm evenings. Several species *(Clunio, Telmatogeton)* are marine, the adults occurring on intertidal rocks; in *Clunio*, the female is wingless. Also, the extraordinary submarine genus *Pontomyia* (p. 114) has an apterous, vermiform female and the male wings reduced to paddle-like structures. In a number of genera, females are frequently parasitized by mermithid worms, usually resulting in partial reversal of their sex.

The larvae are, with few exceptions, aquatic, living either buried in the bottom debris or free on vegetation, etc.; many of the former live enclosed in a gelatinous tube coated with particles of debris. One species damages rice seedlings in N.S.W., principally through physical disturbance of the roots. The tube-dwellers include the 'bloodworms' (*Chironomus* spp.), whose colour is due to haemoglobin in the haemolymph. Cryptobiotic species are known (p. 112). Typically, the larvae bear a pair of prolegs on the prothoracic and anal segments, and a pair of papillae, each with a tuft of hairs, on the anal segment.

Key to the Subfamilies of Chironomidae

1. True base of M_{3+4} (*m-cu* of authors) present 2
   True base of M_{3+4} absent (Fig. 34.17B) 3
2. R_{2+3} forked (or crowded out in some small species with hairy wings) .... TANYPODINAE
   R_{2+3} simple and distinct *(Heptagyia)* .... ORTHOCLADIINAE (pt.)
   R_{2+3} absent, R_1 and R_{4+5} well separated .... PODONOMINAE
3. Anterior basitarsus shorter than the tibia; ♂ styles folded inwards ... ORTHOCLADIINAE (pt.)
   Anterior basitarsus longer than, or very occasionally equal to, the tibia; ♂ styles directed rigidly backwards ...... CHIRONOMINAE

Fifty-one genera are so far known from Australia, with the Orthocladiinae and Tanypodinae more numerous in the colder areas of the south-east, and the Chironominae more abundant in the hotter regions. No doubt many species await recognition, since some are now known that are distinguished mainly by cytological features (e.g. Martin, 1963). Three elements can be recognized in the fauna. (a) Typical members of well-known cosmopolitan genera (e.g. *Chironomus*, *Polypedilum*, *Tanytarsus*); the majority of species fall in this group, including 8 which are apparently identical with forms known from other regions (e.g. the Palaearctic and Oriental *Polypedilum nubifer*). (b) A small group of probably primitive genera, which have an 'Antarctic' distribution (e.g. *Austrocladius*, common during the winter, and also found in New Zealand, Patagonia, and south Chile). (c) Eight endemic genera and one *(Harrisius)* known also from New Zealand; these include only about 16 species. There is also an extraordinary undescribed genus from W.A., with piercing mouth-parts resembling those of the Ceratopogonidae. [Freeman, 1961; Martin, 1963, 1964; Edward, 1963. Also L. Brundin, *K. svenska VetenskAkad. Handl.* (4) **11**: 1–472, 1966.]

8. Ceratopogonidae (Heleidae; sand flies, biting midges; Figs. 34.13A–C, 17G). A widespread family of small to minute blood-sucking flies, not to be confused with the phlebotomine 'sandflies' (Psychodidae). A few genera, e.g. *Johannsenomyia*, have species expanding up to 5 mm, but most are much smaller. All have somewhat elongate, piercing mouth-parts, usually associated with a predatory or blood-sucking habit, and many are notable pests of vertebrates. Other animals also are attacked, including insects; e.g. *Pterobosca* spp. are found on the wings of dragonflies, while *Culicoides anophelis* Edw. and its relatives are secondary blood-suckers, feeding on serum from the abdomens of engorged mosquitoes. Those most commonly encountered in Australia are the pest species of *Culicoides*, *Leptoconops*, and *Styloconops*, whose attacks cause severe annoyance and, particularly in the tidal zone, may render areas barely fit for human habitation. Some species are proved vectors of parasitic worms in other countries; e.g. *Onchocerca* spp. of cattle and horses, and several minor filariases of man. However, in Australia they are so far incriminated only as causing an allergic dermatitis of horses. Not all are predatory; some frequent flowers, and the cocoa plant is apparently dependent upon species of *Forcipomyia* for pollination. Apart from those taken biting, adults sometimes enter light traps in large numbers, and

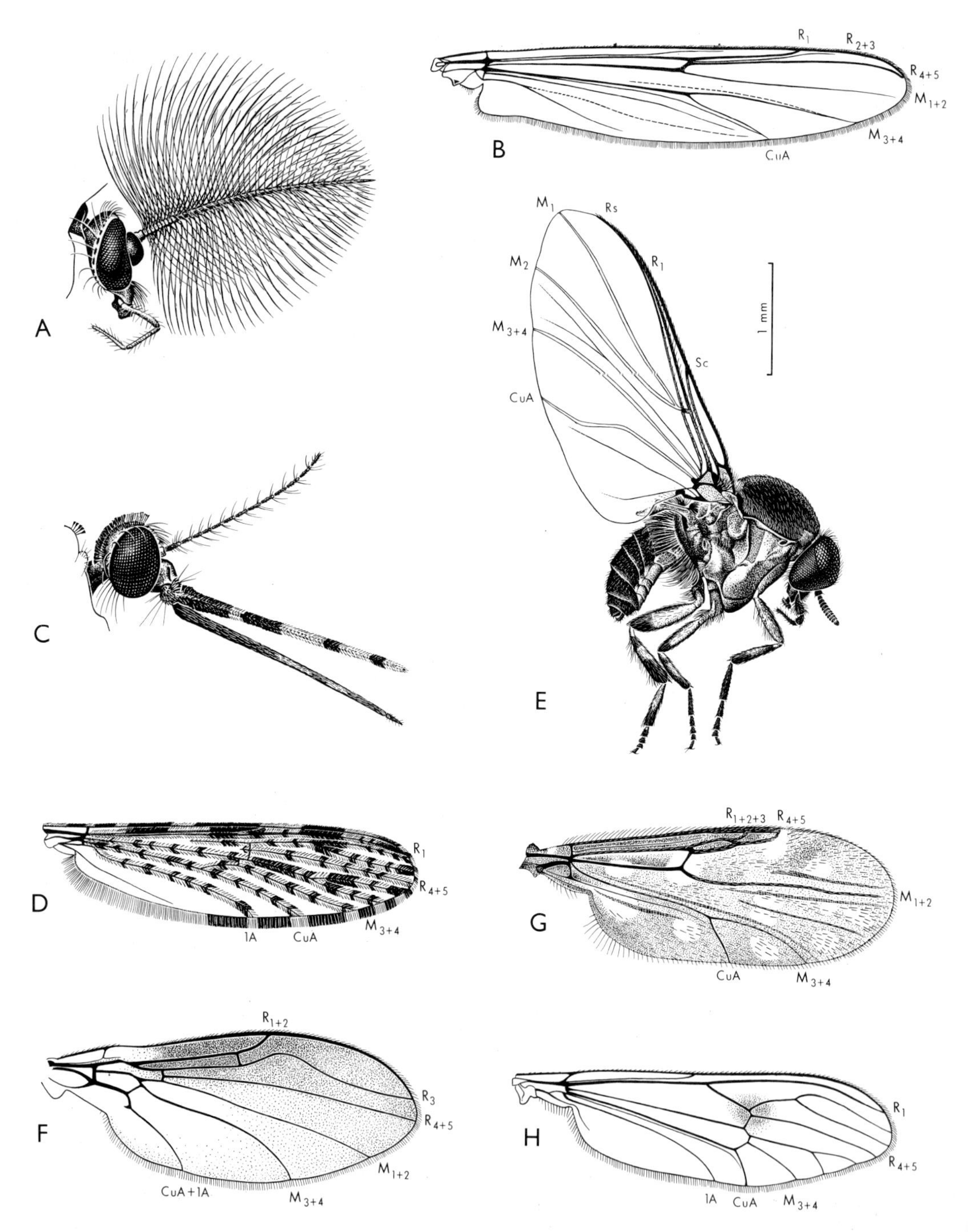

Fig. 34.17. Culicomorpha: A, *Chironomus alternans*, Chironomidae, ♂, head, lateral; B, same, wing; C, ***Anopheles annulipes***, Culicidae, ♀, head, lateral; D, same, wing; E, ***Simulium ornatipes***, Simuliidae, ♂; F, ***Austrothaumalea*** sp., Thaumaleidae; G, *Culicoides antennalis*, Ceratopogonidae; H, *Dixa nicholsoni*, Dixidae. [T. Binder]

many can be swept from vegetation. The tiny larvae are more elusive, living mainly in mud or debris in or around ground-water or container habitats, or in rotting vegetation; some occur in tree-holes and rock-pools. A number of species, including some notable pests, breed in saline or brackish habitats.

The general composition of the Australian fauna is now fairly clear. The dominant genus is *Culicoides* (Fig. 34.17G), while 29 others are known to occur here, the richest in species being *Dasyhelea*, *Forcipomyia*, and *Atrichopogon*. The ease of dispersal of these small flies is reflected in the fact that at least 15 of our genera, including the 4 above, have world-wide distributions. However, *Austroconops* is apparently endemic; *Macrurohelea*, *Acanthohelea*, and *Paradasyhelea* also occur in the Neotropical region; while *Xenohelea* and *Dibezzia* (Oriental and Ethiopian regions) are examples of a northern element. [References listed by Lee, Reye and Dyce, 1962; Reye, 1964; Smee, 1966.]

9. Simuliidae (Melusinidae; black flies; Figs. 34.2E, 13F, 17E). A cosmopolitan family of biting flies, readily recognized by their stout build and wings with large anal lobe and characteristic venation. All are relatively small, of wing length 2·5–3·5 mm, and most are dark in colour. At least some species require a blood meal for maturation of the eggs, and *Austrosimulium pestilens* M. & M. is a vicious pest of man and animals in Queensland; *A. bancrofti* (Tayl.) is a widespread feeder on stock, and other species are troublesome in Tasmania and southern Australia. In general, however, adults are infrequently encountered, usually swept from vegetation near streams, or taken in light traps. Larvae and pupae are more readily found, being aquatic, attached to stationary objects in running water only; there are specific differences in rate of flow tolerated and type of substratum preferred. The larva and pharate pupa (Hinton, 1958b) can move about with a caterpillar-like motion, and breathe by means of anal gills, whereas the pupa and pharate adult have thoracic gill-tufts and are enclosed in small silken cocoons.

The 3 genera found here present in miniature much of the zoogeographical picture of the Australian fauna as a whole. *Cnephia* and, possibly, *Austrosimulium* show the 'Antarctic' pattern; both have related species in South America, and the latter in New Zealand also (but see Dumbleton, 1963). However, *Simulium* appears to be derived from the north; one group of species occurs only in the northern half of the continent and in New Guinea, with related species in Indonesia, while *S. ornatipes* Sk. (Fig. 34.17E) occurs throughout Australia and extends into New Guinea and New Caledonia. [References listed by Lee, Reye and Dyce, 1962.]

10. Thaumaleidae (Orphnephilidae; Figs. 34.13D, E, 17F). A small family of stoutly built flies, the Australian species with a wingspan of only 5–10 mm. Their systematic position is puzzling. Most authorities place them near the Chironomidae, but, cytogenetically, they seem closer to the Bibionomorpha. Little is known of their biology. Adults are occasionally swept from wet rocks or vegetation near streams in wet forest, and are immediately recognizable by the wing-venation. The wings also have a peculiar tendency to fold downwards across a transverse line of weakness near the apex of Sc. The larvae are aquatic, superficially resembling Chironomidae. The family is of considerable zoogeographic interest. The Australian species, which are still undescribed, belong mostly to *Austrothaumalea*, to which also belong the species of New Zealand and southern South America. One other species appears to belong to the South African genus, *Afrothaumalea*.

Division BIBIONOMORPHA

Mostly rather stoutly built flies, but some very small; characterized by the more or less complete fusion of the lobes of the postphragma, which is often strongly developed and may project well into the base of the abdomen; also by the absence of chiasma

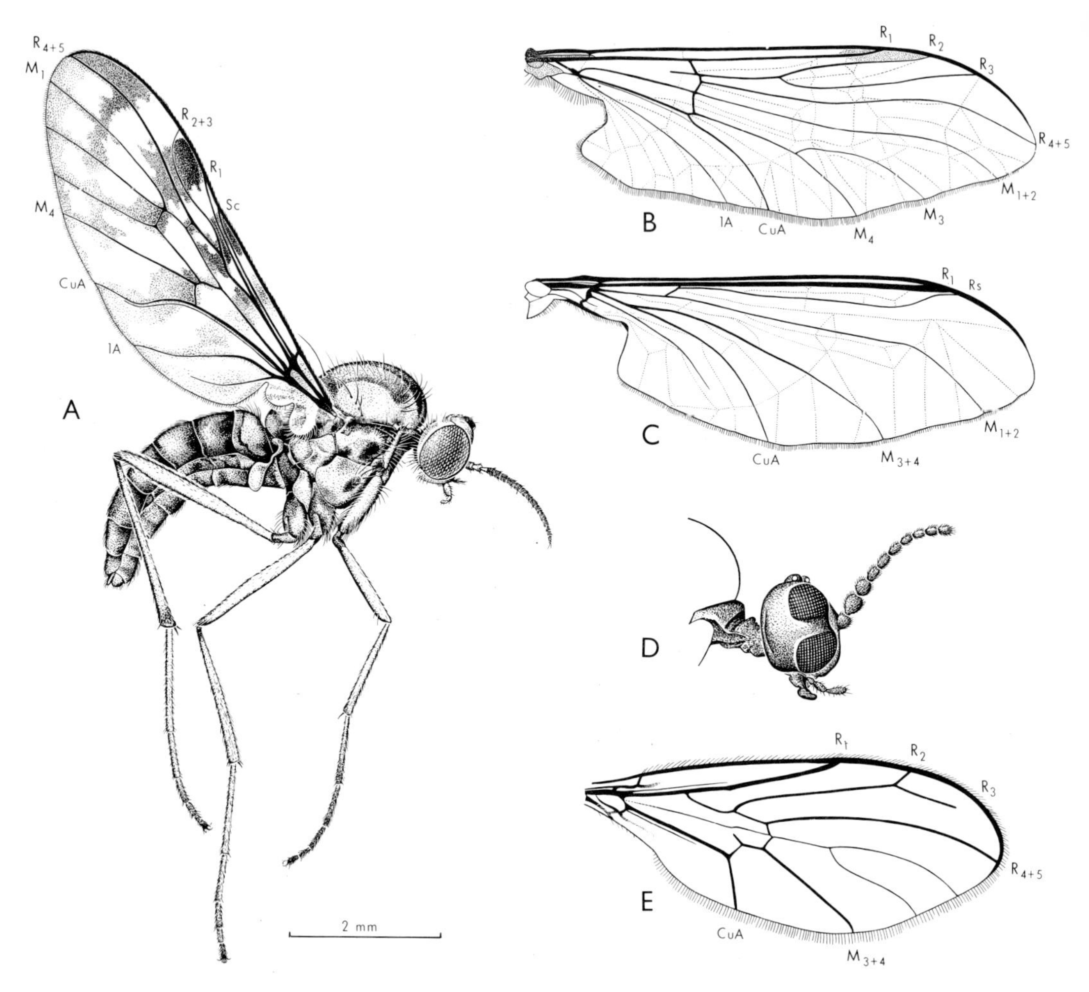

Fig. 34.18. Bibionomorpha: A, *Sylvicola dubius*, Anisopodidae; B, *Edwardsina* sp., Blephariceridae; C, *Apistomyia tonnoiri*, Blephariceridae; D, *Perissomma fusca*, Perissommatidae, head, lateral; E, same, wing. [T. Binder]

formation during spermatogenesis. The venation tends to be reduced, with Rs at most 3-branched and the basal portion of M often missing. Larvae are mostly terrestrial, with the exception of highly specialized aquatic groups, such as the Blephariceridae. Certain family groupings can be discerned, e.g. Mycetophilidae—Sciaridae—Cecidomyiidae, but no satisfactory scheme of superfamilies is yet apparent. The Blephariceridae stand well apart from the other families, but seem best placed here on grounds of morphology and cytogenetics. The inclusion of the exotic Deuterophlebiidae is quite tentative.

11. Blephariceridae (Figs. 34.3A, 14A, 18B, C). An archaic, cosmopolitan family of unusual interest, both zoogeographically and in the unique immature stages. The relationship to other families is obscure, and they are here included in the Bibionomorpha on grounds of the fused postphragma and cytogenetic features; some authorities ally them with the Culicomorpha, but they also show points of resemblance to the Tanyderidae.

The adults are long-legged, not unlike Tipulidae, and *Edwardsina* has a complete, V-shaped mesonotal suture; but their wings have a characteristic network of pseudo-venation (Figs. 34.18B, C), produced by folding within the pupal wing-cases. In life, the wings are held out at right angles to the body. The mouth-parts are elongate, adapted for predation, the female (but not the male) having long, spear-like mandibles. Adults are found in the vicinity of fast-flowing streams and waterfalls; some species tend to congregate on rocks close to the water, but others are most elusive and rarely seen. Both larvae and pupae are aquatic, attaching themselves to stones in fast-flowing water, often in the splash-zone of cascades. The larvae have a particularly characteristic appearance (Fig. 34.14A).

Australian species fall in 2 of the 4 subfamilies. The primitive EDWARDSININAE have 8 species of *Edwardsina* in the highlands of south-eastern Australia. In the more specialized APISTOMYIINAE, there are 2 species of *Apistomyia*, one northern and one southern and a single species of *Neocurupira*. Both *Edwardsina* and *Neocurupira* are 'Antarctic' elements, the former occurring also in Chile and the latter in New Zealand, but *Apistomyia* is derived from the north, with related species in Europe and Asia. [Alexander, 1958.]

12. Anisopodidae (Rhyphidae, Phryneidae, Sylvicolidae; Figs. 34.2D, 3B, 10A, 14B, 18A). A rather small family, often postulated as most closely resembling the ancestors of the Brachycera. All are stoutly built flies of moderate size, and the common genus, *Sylvicola*, has attractively mottled wings. *S. dubius* Macq. is very common in temperate Australia during the cooler months, usually in moist forest, but it also occurs in semi-domestic situations. All species breed in decomposing organic matter, *Olbiogaster* being apparently restricted to rotting wood. *Sylvicola* larvae, which frequent more liquid media, are extremely active, and their posterior spiracles can be retracted and sealed off during submergence. The family is poorly represented here, with only 3 species of *Sylvicola* and one (rare and undescribed) of *Olbiogaster*. Those of the cosmopolitan *Sylvicola* are possibly of 'Antarctic' origin, one of them occurring also in New Zealand, but *Olbiogaster*, known from Queensland, has a distinct, pantropical distribution. [Fuller, 1935.]

13. Perissommatidae (Figs. 34.14F, 18D, E). A very small family including only the 5 known species of *Perissomma* (3 undescribed). All are small, sluggish flies with a characteristic venation and three species have mottled wings. The eyes are remarkable in being divided into separated dorsal and ventral components. Adults have been found during the winter months only, and are rarely encountered, except for *P. fusca* Colless, which breeds in large numbers in decomposing fungi in certain pine plantations. The larva pupates within the unmodified larval skin. The genus is 'Antarctic', occurring in wet sclerophyll and rain forests in south-eastern Australia, while an undescribed species is known from Chile. [Colless, 1962a.]

14. Scatopsidae (Figs. 34.14G, 19B). Small to minute flies, predominantly black in colour, with 1-segmented palps and the radial veins thickened and dark. They are not commonly encountered or collected, except for the introduced *Scatopse notata* (L.) and *S. fuscipes* Meig. which thrive in peri-domestic habitats. At least some of our species are confined to wet forest. Larvae occur in rotting vegetable matter, dung, etc., and the pupae are enclosed in the last larval skin, only their branched respiratory horns protruding. The Australian fauna remains practically unknown. Only 4 species are described, including the 2 immigrants, but species of *Scatopse*, *Colbostema*, *Rhegmoclema*, and other genera are known to occur here. [Skuse, 1889, 1890b; Womersley, 1950.]

15. Bibionidae (Figs. 34.14C, 19A). Flies of small to moderate size, some species of *Plecia* expanding 20–25 mm. The structure of the head is particularly characteristic, and males are all holoptic, almost always with each eye differentiated into a large anterodorsal section and small posteroventral section. Adults are rather sluggish and poor fliers, and most species inhabit forest. Some are active

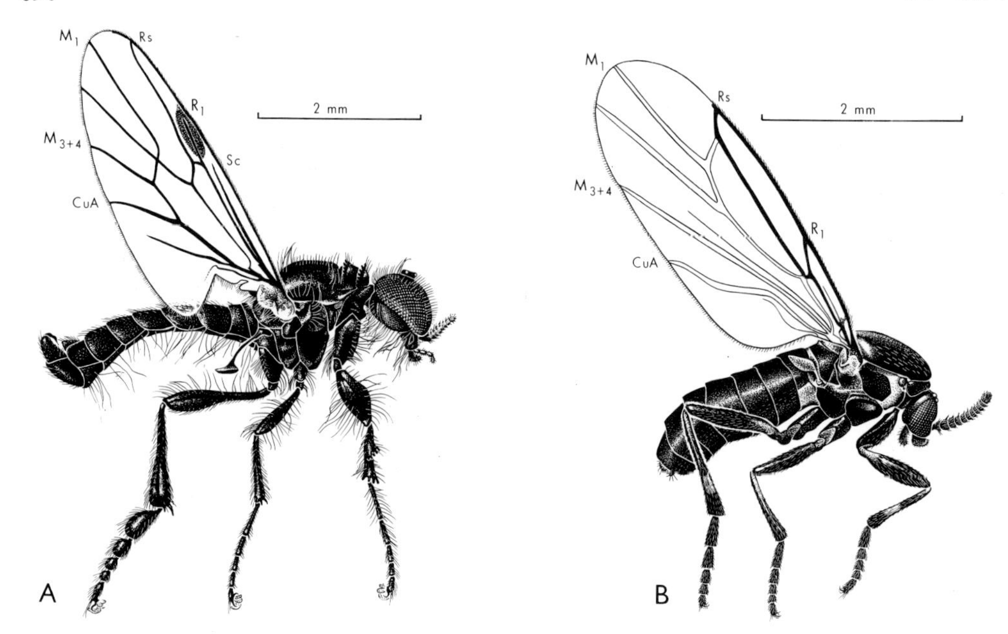

Fig. 34.19. Bibionomorpha: A, *Dilophus* sp., Bibionidae; B, *Scatopse notata*, Scatopsidae. [T. Binder]

by day, but flight activity is probably mainly crepuscular and nocturnal; they are common in light traps. The larvae occur in decomposing vegetation and in the soil; some feed on plant roots, but probably on dead tissue. In Australia, *Bibio imitator* Walk. sometimes occurs in very large numbers in garden soil, but does not attack living plants.

Fourteen species are described from Australia, and at least 10 others occur here. The tropical genus *Plecia* (Plate 6, ZA), with smoky wings, usually orange thorax, and Rs branched, is dominant in the north, though the small *P. dimidiata* Macq. is mainly southern. *Dilophus* (=*Philia*, auct.; Fig. 34.19A), which has 2 transverse rows of blunt teeth on the thorax, is dominant in southern areas (as also in New Zealand and southern South America). *Bibio*, with fore tibia produced into a large apical spine, is represented here by *B. imitator*, which, like many *Dilophus*, has pale brown females and dark males. An interesting recent discovery is *Enicoscolus*, from north Queensland, which has shortened costa and M_{3+4} not connected with M_{1+2} it is otherwise known only from Mexico. [D. Hardy, 1951, 1958, 1962.]

16. Cecidomyiidae (Itonididae; gall midges; Figs. 34.14H, 20E). A large, cosmopolitan family of small to minute flies, most with delicate hairy wings and reduced venation (*Lestremia* resembles the Sciaridae in venation, but has no tibial spurs). The adults are ubiquitous, and their small size and active flight habits make them very susceptible to dissemination by air currents. Many show a remarkable tendency to congregate unharmed on spiders' webs. Many larvae, when mature, have a characteristic, longitudinal, sclerotized 'sternal spatula' on the prothoracic sternum. Most live in galls or other deformities in living plants, or are scavengers in decomposing organic matter; some are paedogenetic (Fig. 4.2). Various exotic species are predacious on aphids, etc., while a few are endoparasites, or live as inquilines in insect galls; no doubt species with similar habits occur here also. In other countries, the plant parasites include some very destructive crop pests, e.g. *Mayetiola destructor* (Say), the hessian fly of wheat; but, as far as is known, the sorghum midge, *Contarinia sorghicola* (Coq.), is the only serious pest so far established in Australia.

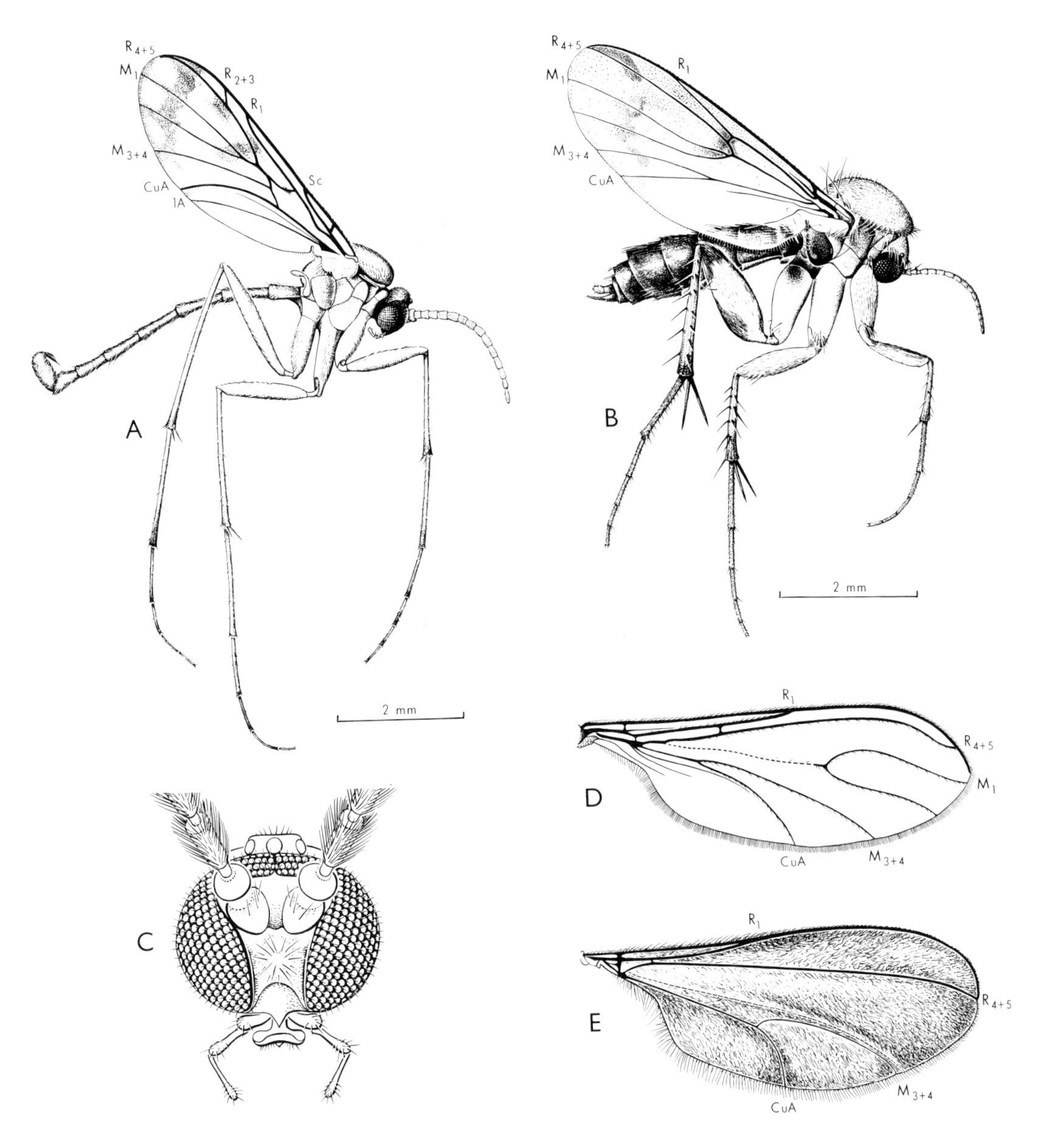

Fig. 34.20. A, *Orfelia fenestralis*, Mycetophilidae, ♂; B, *Mycetophila propria*, Mycetophilidae, ♀; C, *Sciara* sp., Sciaridae, frontal view of head; D, same, wing; E, unidentified sp. of Cecidomyiidae. [T. Binder]

As against this, *Zeuxidiplosis giardi* is a beneficial species introduced for the control of St John's wort. Almost all our known species were described by Skuse (1888a, 1890a), who gives an interesting account of their biology; they have been little studied since then. Most of the well-known genera occur here, and we probably have a number of immigrant species, but the aberrant *Ipomyia* is apparently endemic (Colless, 1965).

17. Sciaridae (Lycoriidae; Figs. 34.8B, 14D, 20C, D). An extremely widespread family, with members adapted to a wide variety of climates. Their uniformity of structure makes taxonomic treatment very difficult, and our fauna badly needs revising, using modern generic concepts (Tuomikoski, 1960). Sometimes treated as a subfamily of Mycetophilidae, they resemble that family in adult habits, but at least some are much more active by day in more open situations. They are, however, common in light traps, some of the rain-forest species being taken in very large numbers. One apterous species of *Austrosciara* is an inquiline in termite nests. The larvae, usually with pale body and shiny black head-capsule, tend to be gregarious, and are commonly found in rotting vegetable matter or highly organic soils; a species of *Bradysia* is reputed to attack roots of plants in greenhouses. The Australian fauna probably includes most of the described genera and possibly some immigrant species. The aberrant genera *Colonomyia* and *Ohakunea* are currently lodged in this family, on somewhat dubious grounds. The former is otherwise known only from the mountains of New Guinea, but the latter has one closely related species in southern Chile and another, less closely related, in New Zealand. [Skuse, 1888b, 1890a; Colless, 1962b.]

18. Mycetophilidae (Fungivoridae; fungus gnats; Figs. 34.2F, 14E, 20A, B). A large family, widely distributed throughout the world in both tropics and temperate regions. Despite variability in wing venation, it forms a well-knit group, with considerable uniformity in structure and biology; e.g. the constricted base of the abdomen and the larval habitat. Numbers, both of individuals and of species, are highest in wet forest, particularly in temperate climates, but some are quite abundant in the moister parts of open savannah and heath. Adults, sometimes in enormous numbers, are commonly taken by sweeping low vegetation. They also enter light traps, and flight activity seems to be mainly crepuscular or nocturnal. Males of *Ceroplatus (Heteropterna)* have the remarkable habit of hanging from spider-webs, particularly in caves. The larvae are mostly peripneustic, and usually found associated with fungi, either inside the fruiting bodies, or externally in webs or mucilaginous tubes. The more primitive genera include one known endoparasite of land planarians, the remarkable *Planarivora* (Hickman, 1965), and a few predators, e.g. the self-luminescent 'glow-worms' *(Arachnocampa)*, which trap their prey in hanging mucilaginous threads. Glow-worms are not uncommon in certain of our caves, but we have nothing to match the spectacular display seen in the Waitomo Cave, New Zealand.

Key to the Subfamilies of Mycetophilidae

1. M_{3+4} connected with M_{1+2} by an apparent cross-vein (*m-cu* of some authors, actually the true base of M_{3+4}) 2
   M_{3+4} appearing to fork from CuA, not connected to M_{1+2} 5
2. R_4 long, almost as long as R_5, and strongly curved .... DITOMYIINAE
   R_4 short and only slightly curved, or absent 3
3. Rs and M_{1+2} fused over a short distance .... CEROPLATINAE
   These veins not fused, cross-vein *r-m* distinct 4
4. M with distinct basal section, *m-cu* well basal to *r-m* .... BOLITOPHILINAE
   Base of M absent, *m-cu* more or less in line with *r-m* .... DIADOCIDIINAE
5. Mouth-parts greatly elongate, forming a mosquito-like proboscis; Rs appearing to arise at base of wing; basal part of M absent .... LYGISTORRHININAE
   Mouth-parts not so; Rs branching from R_1 well beyond base of wing (except in *Allactoneura*), or its base absent 6
6. Prothorax large, without strong bristles; base of M absent; tibial setulae arranged in lines, fore tibia very short; palpal segment 2 attached subapically on segment 1 .... MANOTINAE*
   Without this combination of characters 7
7. Microtrichia of wing arranged in distinct lines, at least near posterior margin, macrotrichia absent; tibial setulae also arranged in lines; ocelli touching eye margins ... MYCETOPHILINAE
   Microtrichia of wing irregularly arranged, macrotrichia sometimes present; tibial setulae irregularly arranged, and/or ocelli distant from eye margins .... SCIOPHILINAE

Unpublished studies show that all subfamilies are represented in Australia, and the

* Excluding *Allactoneura*, which we prefer to place in the Sciophilinae.

number of species is about the same as in New Zealand; but the fauna more closely resembles that of southern South America in having more genera and fewer species per genus. The only noticeably large genera are *Orfelia (=Platyura)*, *Mycomya*, and *Mycetophila*. Four distinct elements can be discerned. (a) Some 10 to 15 probably endemic genera and subgenera, e.g. *Antriadophila* and the remarkable wasp-mimics, *Nicholsonomyia* and *Pseudalyssiinia*. (b) A distinct, but not well known, Indo-Malayan element mainly in the north, e.g. *Allactoneura*, *Epicypta (=Delopsis)*. (c) A very distinct 'Antarctic' element, e.g. *Australosymmerus* (New Zealand and southern South America), *Paramacrocera* (New Zealand), and *Paraleia* (southern South America). (d) Cosmopolitan genera, some with species-groups belonging in the two preceding categories. Present information gives the impression that (c) and (d) form the bulk of our fauna, but further collecting in northern areas may modify this view. [Tonnoir, 1929.]

Suborder BRACHYCERA

Mostly stouter and larger flies than the Nematocera, with the characteristics given in the key. They are clearly derived from bibionomorph-type ancestors, having the fused postphragma and generally similar venational traits. As far as known, they also lack chiasmata in the male, but the X and Y chromosomes form bivalents at spermatogenesis (except in the aberrant Hippoboscidae and a few XO forms). Larvae are generally terrestrial, though many (probably secondarily) aquatic groups are known. It is not yet possible to present keys to the larvae based on Australian material.

Division ORTHORRHAPHA

Separated from the Cyclorrhapha principally by the normal, obtect pupa (enclosed in the unmodified larval skin in Stratiomyidae); also by the form of the male terminalia, which are, at most, only partly rotated and not flexed under the apex of the abdomen (except in Dolichopodidae). It includes the more primitive families of Brachycera, many with a rather generalized venation, with Rs 3-branched and M up to 4-branched. In many families, the adults and/or larvae are predators, or the larvae endoparasites.

Key to the Families of Orthorrhapha and Aschiza Known in Australia

1. Empodium pulvilliform, i.e. 3 subequal pads below the tarsal claws (Fig. 34.6C); CuA and 1A separate, or meeting at an acute angle rather close to wing margin 2
   Empodium bristle-like or absent, i.e. at most 2 well-developed pads below the tarsal claws, or, if a median pad present, CuA and 1A meeting far from wing margin, not at an acute angle ... 8

2(1). Head very small, eyes holoptic in both sexes; thorax and abdomen greatly inflated; calypters very large; wing venation specialized (Fig. 34.22B) .... **Acroceridae** (p. 704)
   Not such flies 3

3(2). Wing with a strong, composite, oblique vein; branches of M parallel with posterior margin; cerci of ♀♀ usually produced into a long 'ovipositor' (Fig. 34.22A) ... **Nemestrinidae** (p. 704)
   Wing venation and cerci of ♀♀ otherwise 4

4(3). R_4 and R_5 divergent, the former ending before, the latter far behind, wing apex; cell R_4 relatively short and broad 5
   R_4 and R_5 of different conformation; cell R_4 relatively long and narrow 6

5(4). Vein 1A strongly sinuate, well separated from CuA; calypters undeveloped (Fig. 34.21G) .... **Pelecorhynchidae** (p. 701)
   Vein 1A almost straight, almost always fused distally with CuA; calypters almost always well developed (Fig. 34.21F) .... **Tabanidae** (p. 701)

6(4). Cell M_3 closed. [Flagellum of antenna 8-annulate; Rs ending in wing apex] .... **Xylomyidae** (p. 703)
   Cell M_3 open or absent 7

7(6). Prosternum with broad precoxal bridges; Rs almost always originating opposite or distal to (rarely, slightly basal to) the junction of *m-cu* and CuA; frequently with radial veins crowded anteriorly, discal cell compact (rarely absent), and some branches of M weak or abbreviated distally (Figs. 34.21A–D) .... **Stratiomyidae** (p. 702)

Prosternum isolated, without precoxal bridges; Rs originating far basal to junction of *m-cu* and CuA; radial veins and discal cell otherwise (Fig. 34.21E) .... **Rhagionidae** (p. 701)

8(1). CuA long, reaching wing margin or joining 1A at less than a quarter of its length back from the wing margin; cell CuP markedly longer than cell M 9

CuA absent, vestigial, or short, joining 1A at more than a quarter of its length back from the wing margin*; cell CuP often little, if at all, longer than cell M 15

9(8). Vein M_1 turned forwards to meet R_{4+5}, which is unbranched; either vein, or both, often sinuous; often with a vein-like fold or thickening ('vena spuria') between Rs and M, and/or terminal branches of M joining to form an 'ambient vein' parallel to margin; antenna with dorsal arista, except in a few wasp-mimics (Figs. 34.27A, B) .... **Syrphidae** (p. 713)

Venation otherwise, and/or antenna with terminal style 10

10(9). Head at least as large as thorax and subspherical, bounded mainly by the enormous eyes; antenna with dorsal arista; R_{4+5} unbranched (Figs. 34.27C, D) ...... **Pipunculidae** (p. 713)

Head not as above, and/or antenna with terminal style, and/or R_{4+5} branched 11

11(10). R_5 and M_1 terminating anteriorly before apex of wing; at most 3 veins terminating posterior to apex. [Antenna swollen apically] 12

R_5 terminating posterior to wing apex (except in a few species with R_{4+5} unbranched); almost always with 4 or more veins terminating posterior to apex 13

12(11). Antennal club petiolate, the preceding segment narrow and elongate; M with only one branch reaching margin posterior to wing apex; palps vestigial (Figs. 34.2I, 23E)... **Mydaidae** (p. 707)

Antennal club more or less sessile, the preceding segment short; M often with 2 branches reaching or approaching margin posterior to wing apex; palpi distinct, often expanded apically (Figs. 34.2G, 23F) .... **Apioceridae** (p. 707)

13(11). Ocellar tubercle set in a distinct notch or depressed area between the well-separated, bulging eyes; face relatively long, more or less vertical, with long bristles forming a 'moustache'-like tuft or line on at least the lower margin; proboscis a stout, horny beak. [Head never subspherical; antennae usually set above centre of anterior surface of head—Figs. 34.23A, B] .... **Asilidae** (p. 705)

Frons at most slightly depressed; eyes often holoptic; face usually short and receding or sharply protruding; proboscis with fleshy labella and/or thin and elongate 14

14(13). M with 3 (rarely 2) branches; R_{2+3} and R_4 often strongly curved distally, meeting costa at about a right angle; often woolly flies (Fig. 34.24) .... **Bombyliidae** (p. 707)

M with 4 branches, M_3 and M_4 usually converging or meeting; R_{2+3} and/or R_4 less strongly curved; usually sparsely haired flies (Fig. 34.23C) .... **Therevidae** (p. 705)

15(8). Wings rather lanceolate, pointed, with R_{4+5} terminating at apex and cross-veins confined to base; main veins, except Sc and R_3, with a dorsal row of black setulae (Fig. 34.26A) .... **Lonchopteridae** (p. 712)

Wing otherwise 16

16(15). Radial veins strongly thickened, terminating along with costa at about centre of anterior margin; other veins much weaker, more or less parallel, running obliquely across wing; antenna apparently 1-segmented, globular or discoid, with dorsal arista; head and forwardly-projecting palps with strong, spiny bristles (Fig. 34.26E) .... **Phoridae** (p. 712)

Not with the above combination of characters 17

17(16). M_1 and M_2 incomplete basally; alula and calypters with long plumose hairs; basal half of wing with large area devoid of microtrichia (Fig. 34.26B) .... **Sciadoceridae** (p. 712)

Not with the above combination of characters 18

18(17). Hind tarsi modified, either the basal segment or most segments expanded and flattened, or the 4th segment with a prominent, acute projection on inner or outer side; wings broad, cell

* Scenopinidae (Fig. 34.23D) and a few Empididae (Fig. 34.25D) have CuA rather long. The latter are immediately recognizable by the humped mesothorax and/or swollen, spinose hind femora; the former have the characters given in couplet 19.

CuA acute distally (Figs. 34.26C, D) .... **Platypezidae** (p. 712)
Hind tarsi rarely modified; if modified, cell CuA not acute apically 19

19(18). Antenna without apical style or arista, sometimes bifid at tip; CuA not recurved apically, meeting 1A at one-quarter to one-third its length back from wing margin; R_{4+5} branched; M_1 often curved forward to meet or approach R_5; mouth-parts non-predatory (Fig. 34.23D) .... **Scenopinidae** (p. 705)
Not with the above combination of characters 20

20(19). Rs arising from R very close to level of humeral vein, distal to it by no more than the length of that vein; at most 2 separate veins from apex of discal cell to wing margin; cell M confluent with discal cell (base of M_{3+4} missing); mouth-parts with strong, opposable labella (Figs. 34.3C, 25A, B) .... **Dolichopodidae** (p. 711)
Origin of Rs well distal to level of humeral vein, and/or 3 separate veins from apex of discal cell to wing margin; cell M almost always separated from discal cell by base of M_{3+4}; mouth-parts usually a piercing beak (Figs. 34.25C–E) .... **Empididae** (p. 709)

Superfamily TABANOIDEA

Characterized principally by the enlarged, pad-like empodium, which resembles the pulvilli and, with them, forms a triple pad beneath the tarsal claws. The venation is rather generalized (except in Acroceridae), with CuA reaching the wing margin or meeting 1A near its apex. Larvae are aquatic or terrestrial, and some are endoparasites; they have the posterior pair of spiracles approximated on the terminal segment.

19. Pelecorhynchidae (Fig. 34.21G). This family contains the single primitive genus *Pelecorhynchus*,* which was placed in the Tabanidae by early authors, and in the Coenomyiidae by Steyskal (1953). The species are of robust build, 10–20 mm long, usually ornate, with 8-segmented antennal flagellum and hatchet-shaped labella. The females have 2-segmented cerci, and reduced mandibles; they are not blood-suckers. The larvae and pupae are more like those of Rhagionidae than of Tabanidae.

The species are confined to cool temperate parts of southern Chile and eastern Australia. Adults have been taken on *Leptospermum* flowers, or hovering in the air, in the mountains during the summer; a few species appear in coastal districts in the spring. Larvae have been collected in the damp margins of swampy areas, where they feed on earthworms and possibly other soft-bodied animals; they move to drier levels to pupate. [Mackerras and Mackerras, 1953.]

20. Rhagionidae (Leptidae; Fig. 34.21E). A world-wide family of rather small flies (largest Australian species 10 mm long), with sparse hairs, sometimes bristle-like, on body and legs. The primitively 8-segmented antennal flagellum shows progressive reduction to a swollen basal segment and attenuated, often unsegmented style; at least some of the tibiae have apical spurs (a distinction from primitive Stratiomyidae); the cerci of the female are 2-segmented (*Atherimorpha*, *Chrysopilus*, *Spaniopsis*) or 1-segmented (*Atherix*, *Austroleptis*, *Dasyomma*); and the male terminalia are basically simple, though showing remarkable variation even within genera.

Most Australian Rhagionidae may be taken by sweeping among vegetation in sheltered, rather damp places, where they are believed to prey on other insects; they are particularly common in spring. The species of *Spaniopsis* suck blood, and some may be minor pests in eastern N.S.W. Larvae of *Chrysopilus* were found in damp soil, and *Austroleptis rhyphoides* Hardy has been reared from rotting wood. Of the 6 Australian genera, *Austroleptis*, *Dasyomma*, and *Atherimorpha* occur also in South America (and *Atherimorpha* in South Africa), whereas *Chrysopilus*, *Atherix* (recently found in Queensland), and possibly *Spaniopsis* belong to a later northern element. [Paramonov, 1962.]

21. Tabanidae (March flies*, horse flies;

*The North American *Bequaertomyia*, formerly placed here, belongs to the Coenomyiidae *sens. str.* (Mackerras, personal communication).

* In Europe and North America, the term 'March fly' is applied to the Bibionidae.

Figs. 34.2H, 21F). A large and important family, well studied because of its medical and veterinary importance. The Australian species vary from small (6 mm long) to large (20 mm), and all are distinguished by having large eyes, segmented antennal flagellum, strong proboscis and labella, smooth bodies which are more or less hairy and often show dense pleural tufts, at least mid tibiae with apical spurs, and 1-segmented cerci in the female. Division of the family into subfamilies has been based primarily on characters of the terminalia, but most species can be placed correctly on external characters.

Key to the Subfamilies of Tabanidae

1. Ocelli rudimentary or absent; hind tibiae without spurs; antennal flagellum a consolidated basal plate and 4-annulate style .... TABANINAE
   Ocelli fully developed; hind tibiae with paired apical spurs (occasionally small, rarely absent); antennal flagellum varied 2
2. Antennal flagellum usually 8-annulate; if with fewer annuli, Sc is bare below and R_4 has a strong appendix; frons of ♀ almost always without a shining callus .... PANGONIINAE
   Antennal flagellum usually with a basal plate and 4-annulate style, occasionally with fewer divisions; Sc usually setulose below; R_4 usually without appendix; frons of ♀ almost always with a shining callus, which is occasionally linear .... CHRYSOPINAE

Most of the Australian species suck blood, but some species of *Scaptia* (Plate 5, N) appear to be exclusively flower-feeders, and the most primitive Pangoniini seem to live obscure lives among low vegetation; the females of a few of these have atrophied mandibles. The commonest genera in southern Australia are *Scaptia*, thickset Pangoniinae with hairy eyes, and *Dasybasis*, hairy-eyed Tabaninae with bare basicosta and R_4 appendiculate; in the north, *Tabanus*, with setulose basicosta, *Cydistomyia*, bare-eyed Tabaninae with bare basicosta and R_4 without appendix, and *Mesomyia*, *Cydistomyia*-like Chrysopinae, are common. Some of the blood-sucking species are pests of man and stock, and one northern species is known to produce sensitization, with severe reaction to subsequent bites. However, none of our species occurs in such vast numbers as those of the Subarctic zone of the northern hemisphere. None is known to transmit disease in this country, although Tabanidae transmit loaiasis of man in Africa and trypanosome infections of stock in the Old World tropics.

The breeding places of the Australian species are extremely varied: in classical situations among floating vegetation in open swamps; in the mud of rivers; in more or less damp soil, among rotting vegetation; in dry or beach sand; or even in rot-holes in the trunks of *Casuarina*. The more primitive genera have a predominantly southern distribution in Australia, and are related to groups in New Zealand, South America, and South Africa. Some old genera and subgenera, however, have a more northern distribution, and they are connected with corresponding African groups by a discontinuous band of species that extends around the Indian Ocean and also into the Pacific. The most highly evolved genera *(Chrysops, Tabanus)* appear to have become distributed from northern centres of evolution along with the eutherian mammals. [Mackerras, 1956–61.]

22. Stratiomyidae (Figs. 34.6C, 21A–D). A cosmopolitan family, containing a wide range of forms, from among the most primitive in the suborder to others that are highly specialized; but most members can be recognized by the characteristic venation. The antennal flagellum is often elongate, and usually annulate; the proboscis short and fleshy; the tibiae without apical spurs (except in some Beridinae); the cerci of the female 2-segmented; and the male terminalia of rather simple form (Fig. 1.31A).

The family has not yet been studied critically in this country, but it can be seen that, as in the Tabanidae, the older elements show relationships with South America, and the more specialized ones with the Oriental region. The elongate, soft-bodied CHIROMYZINAE, including *Boreoides* which has apterous females (Fig. 34.21A), are not uncommon in the south-east, especially in higher country. *Neoexaireta spinigera* (Wied.) is common on window-panes. The slightly metallic species

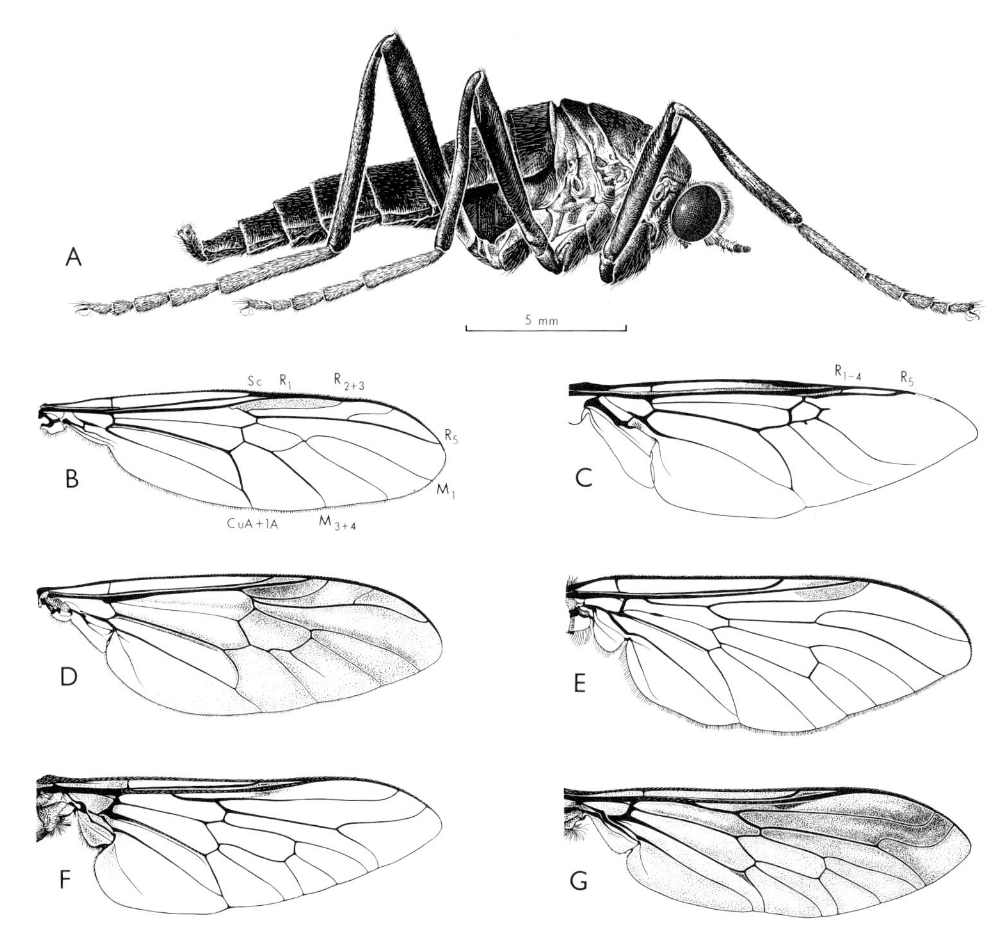

Fig. 34.21. Tabanoidea: A, *Boreoides subulatus*, Stratiomyidae, ♀; B, same, ♂ wing; C, *Odontomyia scutellata*, Stratiomyidae; D, *Neoexaireta spinigera*, Stratiomyidae; E, *Chrysopilus gratiosus*, Rhagionidae; F, *Scaptia auriflua*, Tabanidae; G, *Pelecorhynchus fulvus*, Pelecorhynchidae. [T. Binder]

of *Odontomyia* are the most common representatives of the family; they may be collected in numbers in any swampy area, both coastal and montane, settled on low vegetation or hovering in the air over open patches. A few genera contain some remarkably perfect wasp mimics (Plate 6, I; see also Nicholson, 1927). Larvae of Stratiomyidae are distinctive, being elongate, somewhat flattened, with permanently exserted heads, and densely shagreened cuticles. Some are aquatic, but many are found in damp soil or rotting vegetation; *Altermetoponia rubriceps* (Macq.) breeds abundantly in soil, and is a serious pest of pastures and sugar cane in Queensland. The pupa remains enclosed within the last larval skin, a characteristic shared only with Xylomyidae in the superfamily. [G. Hardy, 1959.]

23. Xylomyidae. A small family, of scattered world distribution, related to Stratiomyidae, and distinguished in Australia from the Rhagionidae by the swollen hind femora, closed cell M_3, and complex, strongly down-turned male terminalia. *Solva* has been recorded from New Guinea and

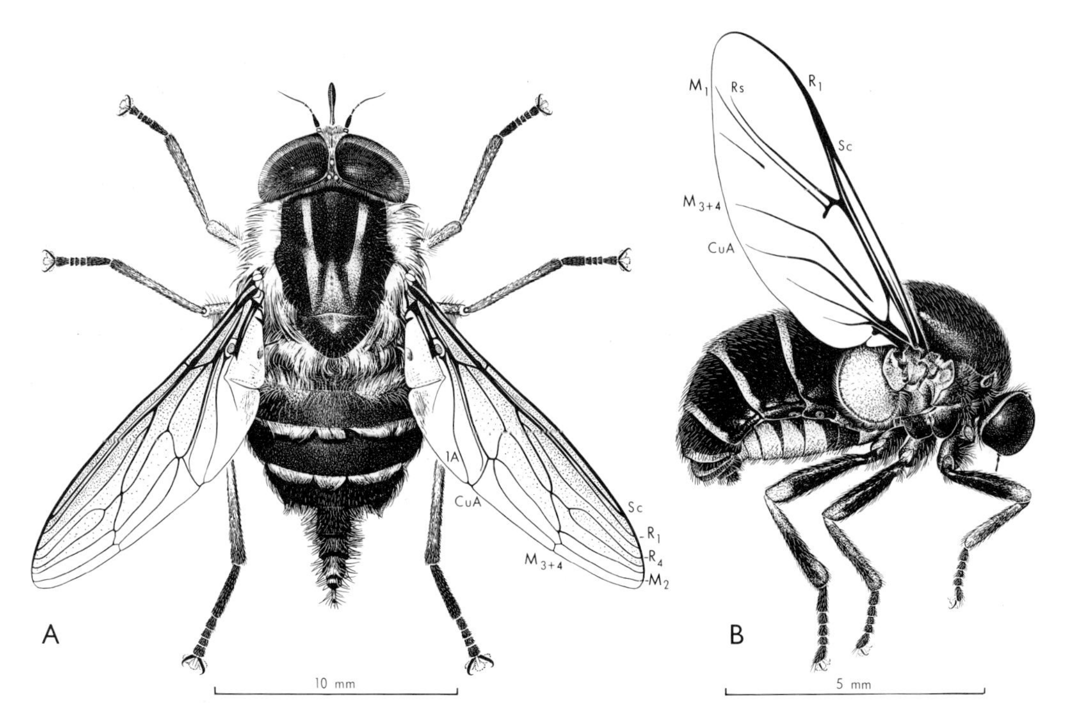

Fig. 34.22. Tabanoidea: A, *Trichophthalma laetilinea*, Nemestrinidae; B, *Ogcodes basilis*, Acroceridae. [T. Binder]

recently found in north Queensland.

24. Nemestrinidae (Fig. 34.22A). Except for *Exeretoneura*, which may be wrongly placed here, this is a compact family, easily recognized by the wing venation. Most are large to medium-sized, compactly built flies, usually hairy but never bristly, apical spurs on some tibiae, and the abdomen conical, the 6th and subsequent segments in the female forming a tubular, retractile 'ovipositor', with a pair of strikingly elongate, 1-segmented cerci.

The medium-sized to large, often rather ornate species of *Trichophthalma* are much the most abundant. They are most often taken when feeding on blossom or hovering motionless over open sunny patches, darting away when disturbed, but usually returning quickly to the same spot. The larvae are parasitic, and those of *Trichopsidea oestracea* Westw. are found in grasshoppers. Of the 7 genera definitely recognized in Australia, the very primitive *Exeretoneura* is endemic, *Cyclopsidea* is a more specialized endemic element, *Trichophthalma* is shared with southern South America, and the others are recent derivatives from the north. [Paramonov, 1953c.]

25. Acroceridae (Cyrtidae; Fig. 34.22B). A small family of grotesquely inflated flies, remarkable for having the eyes holoptic in both sexes (apparently fused in *Pterodontia*), contact being below the long antennae in *Panops* and the wasp-like *Leucopsina* (Plate 6, Y), and above the very short antennae in *Pterodontia* and *Ogcodes*. Consequently, the sexes cannot usually be distinguished without examining the terminalia, which are also considerably modified, with those of the male rotated from 90 to 180°. Most of the species are rare, but occasionally *Ogcodes* has been taken in considerable numbers. The early stages are parasitic in spiders, but no life histories have been worked out in detail in Australia. *Leucopsina* is endemic, *Panops* occurs also in South America, while

Pterodontia and *Ogcodes* are cosmopolitan. [Paramonov, 1957d.]

Superfamily ASILOIDEA

Generally similar to the Tabanoidea, but empodium bristle-like, reduced, or absent. The larvae are mostly terrestrial, and many are predatory, or, in the Bombyliidae, endoparasites. Their posterior abdominal spiracles are placed laterally on the penultimate segment.

26. Therevidae (Fig. 34.23C). A cosmopolitan family of flies of moderate to small size, many with prettily patterned wings and/or silvery pubescent markings on the body; many species show marked sexual dimorphism. Some resemble small Asilidae, others Rhagionidae or Apioceridae, while many are wasp mimics (Plate 6, W). The antennae are sometimes very distinctive, with a greatly elongate or thickened scape. Adults frequent a wide variety of habitats, often in rather dry situations, e.g. sand dunes or beaches; they are common on tree trunks. Little is known of their habits; they are reputed to be predacious, but this seems unlikely. The larvae are smooth and vermiform, with rather well developed head; the abdomen is secondarily divided into some 16 apparent segments, and terminates in a pair of tiny pseudopods. They are found mainly in sand or soil, and are predacious (English, 1950). The prepupal larva lies in the soil in a characteristically curved attitude.

Our genera show an extraordinary degree of endemism. Of 33 so far recognized (11 still undescribed), all but 4 are so far known only from Australia. *Anabarrhynchus*, one of our commonest genera, is also dominant in New Zealand, and it, or close relatives, also occur in South America and Madagascar. *Ectinorrhynchus* is also recorded from New Zealand and (somewhat doubtfully) South Africa and South America. These two genera appear to be rather ancient elements, of the 'Antarctic' type, but the cosmopolitan *Psilocephala* (and *Phycus*, if it really occurs here) presumably entered by a northern route. The large number of endemic genera reflects the great structural diversity found within the world fauna. This is perhaps related in some fashion to the wasp-mimicking habit, and tends to obscure relationships. However, there is a strong indication of a perhaps ancient radiation in Australia from ancestors whose origins are no longer clear. [Mann, 1928–33; Paramonov, 1950b.]

27. Scenopinidae (window flies; Fig. 34.23D). A small but widespread family of rather small, dark flies; 7 genera are known from Australia, 3 of them apparently endemic. In addition to the key characters, all but *Scenopinula* have M_{1+2} converging distally with Rs, and the veins meet to form a closed cell in several genera. Adults are uncommon; those of the rarer genera have been found on vegetation, but *Scenopinus* spp. are most commonly collected on window-panes. The larvae have a secondarily segmented abdomen, like small Therevidae. They occur in rotting vegetation, etc., and have been found in birds' nests; in *Scenopinus* at least, they are probably predacious. *S. fenestralis* L. is said to breed in houses, preying on larvae of moths, carpet beetles, etc., which may explain why this European species has now spread to many countries of the world. [Paramonov, 1955e.]

28. Asilidae (robber flies; Figs. 34.9C, 10B, 11B, 23A, B). A very large family of predatory flies. The shape of the head, with its prominent, well separated eyes, is very characteristic, while most species are bristly, with strong legs, short, stout thorax, and long, but stout, tapering abdomen; though some have the abdomen narrow and elongate. The venation is rather generalized, with both R and M 4-branched, specialization being towards apical fusion of R_{2+3} with R_1, M_3 with M_4, or CuP with 1A. Most asilids are of moderate size, but there are tiny species of *Stichopogon* with wing length of only 1·7 mm, while *Phellus* and *Blepharotes* include the giants of Australian Diptera, expanding up to 75 mm. Some species, particularly in the Dasypogoninae *(sens. lat.)*, have orange or yellow markings on wings or body (Plate 6, H, L), and are effective mimics of sphecoid and vespoid wasps.

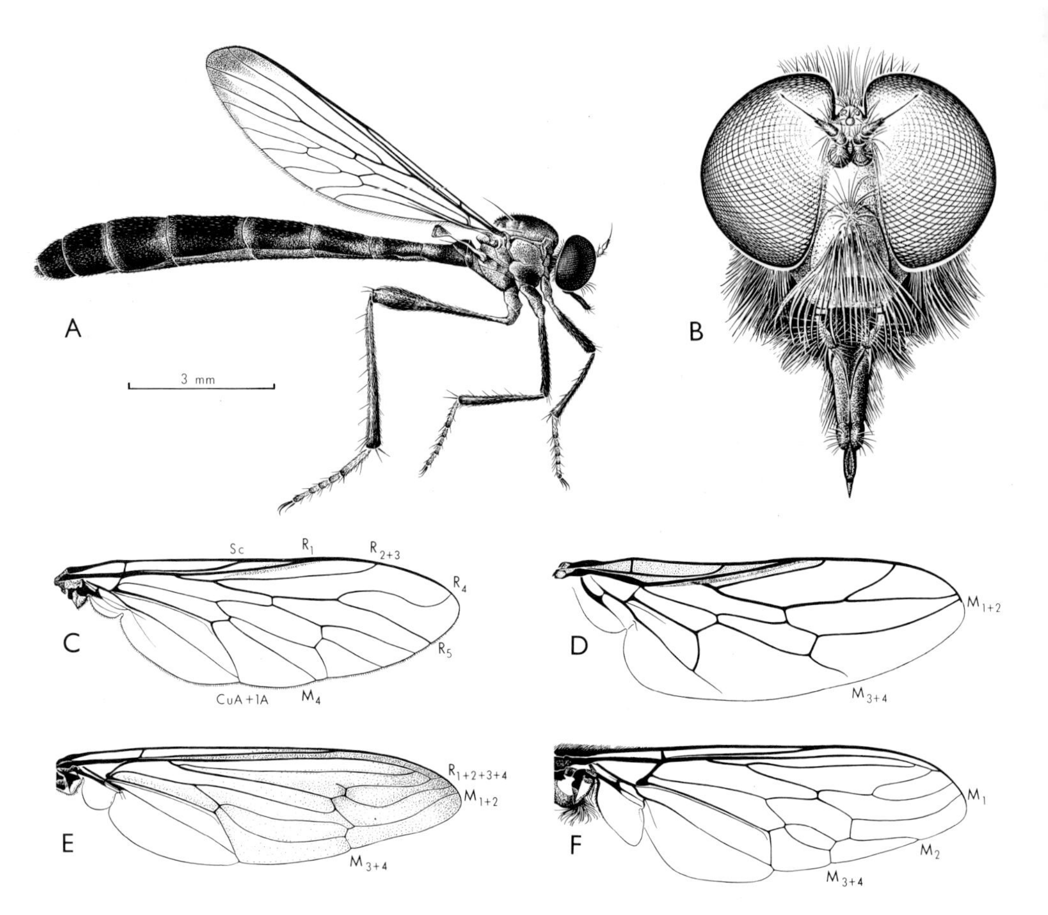

Fig. 34.23. Asiloidea: A, *Leptogaster* sp., Asilidae, ♀; B, *Neoaratus hercules*, Asilidae, head, anterior view; C, *Anabarrhynchus calceatus*, Therevidae; D, *Scenopinus glabrifrons*, Scenopinidae; E, *Miltinus viduatus*, Mydaidae; F, *Apiocera asilica*, Apioceridae. [T. Binder]

The adults, which live mainly in open forest country, are aggressive predators, feeding particularly on Diptera and Hymenoptera, but attacking almost all insects, including dragonflies; even the hand of the collector is not immune! The prey is usually caught on the wing, held by the powerful legs, and the body juices sucked out via the sharp proboscis. It seems that powerful enzymes are injected, since the victim may be reduced to a mere shell. Eggs are laid in the soil, or attached to foliage or bark. The larvae are cylindrical, elongate, tapered at each end, and with a small, distinct head. They live in soil or rotting wood, etc., and, although often regarded as predacious, at least some species are not so.

Four subfamilies may be recognized (from 2 to 6 or more, according to various authorities). The ASILINAE, with 1-segmented palp, slender antennal style, and cell R_1 closed and apically petiolate, are well represented here by species of *Neoaratus*, *Cerdistus*, *Ommatius*, and others, including the giant species of *Blepharotes*. The LEPTOGASTRINAE, which differ in having cell R_1 open to the costal margin, include, in Australia, only the delicate, elongate species of *Leptogaster*. The LAPHRIINAE and DASYPOGONINAE have 2-segmented palps and a thickened terminal style, or none, on the antenna. The former, with cell R_1 closed, comprise mainly the relatively stout, usually shiny metallic, species of *Laphria* (Plate 5, M), some of which are wasp-mimics. The Dasypogoninae, with cell R_1 usually open, or with a strong terminal spine on the fore tibia, are the dominant group, including over half of the known Australian fauna. Generally, the Australian asilid fauna is very characteristic, with some 80 per cent of its genera (e.g. *Chrysopogon*) endemic; although some are clearly related to those of other faunas. Most of the remaining 20 per cent are more or less cosmopolitan, e.g. *Laphria* and *Leptogaster*, but there are some notable links with the Neotropical region, e.g. *Bathypogon* and the tribe Phellini. [Hull, 1962 (refs. to papers by Hardy and Paramonov); Paramonov, 1964a–c.]

29. Apioceridae (Figs. 34.2G, 23F). A small family of mostly rather large flies, somewhat resembling Asilidae and Therevidae, but rather short winged, with R_4, R_5, and M_1 all turned up distally to end before the wing apex. The abdomen is usually broad at the base, tapering distally, and often has grey markings. Adults are strong, noisy fliers, and occupy a wide range of habitats, from sea beaches and desert to forests at high altitudes; at least half of our known species are associated with relatively arid conditions. Although essentially flower-feeders, they are usually found resting on the ground. The immature stages are known only for *Apiocera maritima* Hardy, which breeds in beach sand (English, 1947). Its larva is possibly carnivorous, somewhat resembling an asilid, but with long penultimate abdominal segment.

The world fauna has a curiously disjunct distribution, possibly due to considerable extinction during past epochs. Some two-thirds of known species are Australian, and none is known from the Palaearctic Region or New Zealand. There are 2 Australian genera; *Apiocera* (68 spp.) with M_1 ending on the wing margin, and the endemic *Neorhaphiomydas* (7 spp.) with both M_1 and M_2 fused apically with the radial veins. The latter has a close relative in Chile. [Paramonov, 1953d, 1961c.]

30. Mydaidae (Figs. 34.2I, 23E). A small family of handsome flies, of moderate to very large size (the Neotropical *Mydas heros* Phil. is one of the world's largest flies). They bear a superficial resemblance to Asilidae and Apioceridae, but are recognizable by the long, clubbed antennae and the venation. Many are wasp mimics, e.g. *Diochlistus auripennis* Westw. which resemble the large, orange-winged Pompilidae. Mydaids are not very common, but occupy a wide variety of habitats, usually in open country. *Miltinus viduatus* (Westw.) is widespread, found both in the humid eastern States and the dry interior, but most species seem to be much more restricted in range and habitat. Adults may be taken visiting flowers, but it is thought that at least some are predators. The larvae are said to be predacious, and are recorded in other countries as feeding on beetle larvae in rotting wood; they have been little studied in Australia.

The Australian fauna has a striking distribution, resembling that of the Apioceridae. There are 2 genera, both endemic: *Diochlistus* with M_2 present, and *Miltinus* with M_2 absent. *Diochlistus* is found in most States, but best represented in south Queensland by 6 of the 12 known species. *Miltinus*, however, is found mainly in the more arid areas, occurring around, and presumably in, the desert interior. The world fauna generally seems to be an old one, adapted to hot and/or arid climates, and has probably suffered much extinction due to climatic changes. [Paramonov, 1955c, 1961b.]

31. Bombyliidae (Fig. 34.24). A very large, cosmopolitan family of usually stoutly built flies, of small to large size, and with very characteristic venation. In Australia, wing-spans vary from over 70 mm in *Comptosia lateralis* (Newm.) down to some 3·5 mm in the tiny *Pachyneres australis* Mall. A few, e.g. *Systropus flavoornatus* Rob. (Plate 6, S), are remarkable wasp mimics; in others, mostly members of the subfamily Bombyliinae, the stout, hairy body and long, thin proboscis, together with their flight habits, have earned them the vernacular name 'bee flies'.

The adults favour warm, sunny localities. Although occurring throughout the continent, they form a particularly characteristic element of the fauna in the more arid climates. Most have a strong, hovering flight, and are commonly taken hovering above, or resting on, blossom or patches of bare earth. The larvae are believed to be all parasites in eggs or larvae of other insects. Little is known about Australian species, but some are known to parasitize Hymenoptera and Lepidoptera, and several have been reared from Diptera (Asilidae, Therevidae, and Mydaidae) and Neuroptera. Nothing appears to be known of the larval biology of our dominant genus, *Comptosia*, but it is thought that some species of this and other genera attack the egg-masses of grasshoppers and locusts.

Ten subfamilies are known to occur in Australia, but less than half of our species have been described. The dominant subfamily, with some 170 species, is the LOMATIINAE, in which the occiput has a deep central cavity, the posterior eye-margins are indented, and Rs forks well before *r-m*. Most species belong to the great genus *Comptosia*, an 'Antarctic' element, closely related to *Lyophlaeba* of South America (Paramonov, unpublished). There is also a small northern element represented by several species of *Petrorossia*. The EXOPROSOPINAE, which resemble the Lomatiinae, but with Rs forking close to *r-m*, are well represented, mainly by the cosmopolitan genera *Ligyra (Hyperalonia auct.)*, *Exoprosopa*, and *Villa*. The first two belong to the northern element, but *Villa* is abundant in Tasmania, where the other two are rare or absent. The ANTHRACINAE, another northern element, are represented by some 20 species of *Anthrax*, which differ from the Exoprosopinae in having a pencil of hairs at the tip of the antenna.

The subfamily BOMBYLIINAE includes some 80 Australian species, principally of *Bombylius*,

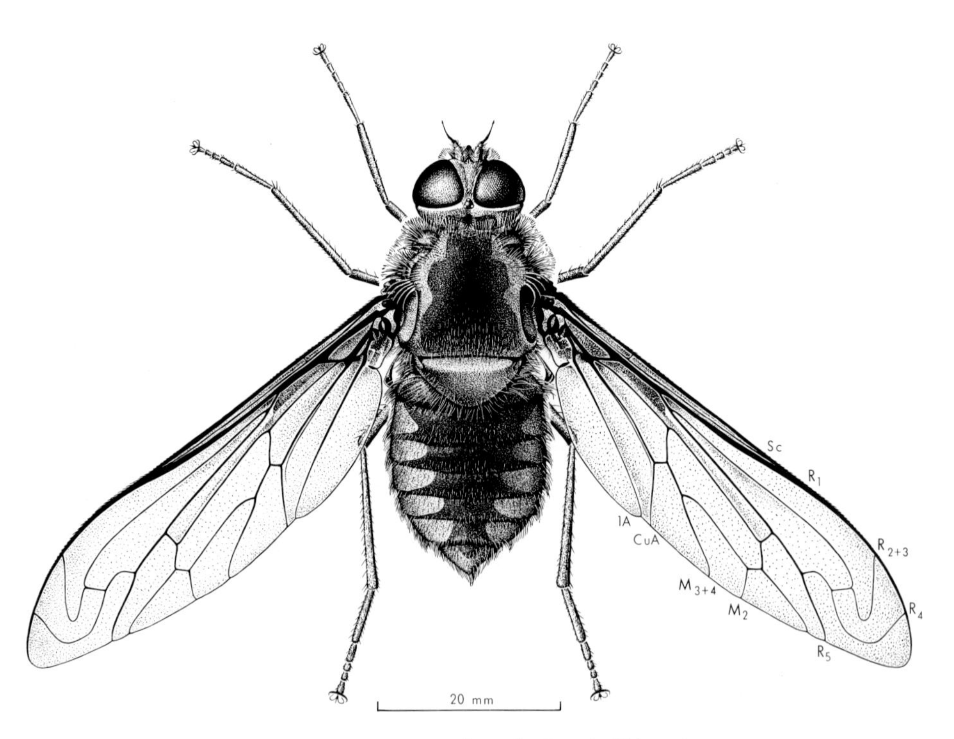

Fig. 34.24. *Comptosia lateralis*, Bombyliidae, ♀.
[T. Binder]

with smaller numbers of *Sisyromyia*, *Systoechus*, and other genera. Most are stout and hairy, with long, slender proboscis and 1-segmented palp; the abdomen is broad and oval, the tibiae spinose, and, in many, vein M_1 meets R_5 before the wing margin. *Sisyromyia* is endemic, but the subfamily as a whole is of northern origin. The PHTHIRIINAE form a smaller, rather heterogeneous group, somewhat resembling the Bombyliinae, but with smooth tibiae, or 2-segmented palp, or elongate body; *Geron* has M only 2-branched. *Acreotrichus* and *Marmosoma* are endemic, *Eclimus* may be 'Antarctic' in origin, whereas *Geron* is apparently of northern origin. *Phthiria* is almost cosmopolitan, but apparently absent from the Oriental region. The rest of the Australian fauna consists of a few species in the primitive, wasp-mimicking SYSTROPINAE and the extremely rare TOXOPHORINAE; several in the endemic *Neosardus* (CYLLENIINAE) and *Eusurbus* (USIINAE); and a few small to minute species of CYRTOSIINAE, which have R_{4+5} unbranched. [Roberts, 1928–9; G. Hardy, 1933b, 1941; Paramonov, 1950c, 1951a, 1953a. Also S. J. Paramonov, *Aust. J. Zool.* **15**: 123–44, 1967.]

Superfamily EMPIDOIDEA

Characterized principally by the shortening of CuA, which is usually recurved to end in the basal half of 1A, or may even be obsolescent. The male terminalia are generally very complex structures, partly rotated and asymmetrical in hybotine Empididae, and circumverted in the fashion typical of the Cyclorrhapha in the Dolichopodidae. As far as is known, most adults and larvae are predatory, and the larvae are terrestrial or aquatic.

32. Empididae (Figs. 34.25C–E). A very large family of flies, of moderate to minute size. Despite considerable diversity, they form a clear-cut group, recognized mainly by the venation and the general predatory appearance. In most genera, the proboscis is elongate to some extent and adapted for piercing, though a few have 'chewing labella', as in the Dolichopodidae. The male terminalia are often highly complex, and, in the Tachydromiinae and Hybotinae, asymmetrical, and the legs may be greatly modified; e.g. the extraordinary mid legs of males of *Tomia* (Paramonov, 1961a). Probably most adults are predacious on smaller arthropods. They frequent moist places, particularly along streams, and are commonly collected by sweeping vegetation; many are taken in light traps, and a few are found visiting flowers. Species of *Hilara* are often to be seen dancing swiftly close to the surface of small ponds, whereas some other genera form dancing swarms in the air. However, the Tachydromiinae and Hemerodromiinae are mainly terrestrial and rarely fly. Swarms consist largely of males, and swarming forms part of their often complex mating activity; some have very elaborate courtship behaviour, offering captured prey or a bubble of frothy secretion as a lure to attract females (Melander, 1927). Little is known of the larvae; some are recorded from decaying vegetation, whereas others are aquatic.

Five subfamilies are here recognized, the Clinocerinae being included in the Hemerodromiinae and the Ocydromiinae in the Hybotinae.

Key to the Subfamilies of Empididae

1. Discal cell absent; R_{2+3}, R_{4+5}, M_{1+2}, and M_{3+4} all unbranched; cell CuP absent, or, if present *(Tachydromia)*, mid legs raptorial; fore legs not raptorial .... TACHYDROMIINAE
   Discal cell usually present, and/or one or more of the above veins branched, and/or fore legs raptorial 2
2. Third antennal segment very large, with 2nd segment inserted thumb-like on its inner side towards the base; cell CuP indistinct .... CERATOMERINAE
   Antenna not as above; cell CuP often complete and distinct 3
3. Wings relatively long and narrow, anal lobe not developed or very weak; thorax usually rather elongate, mid coxae vertically beneath scutellum or postnotum; fore legs often raptorial, with elongate coxae and strong spines on underside of femur .... HEMERODROMIINAE
   Wings relatively broader, with distinct, rounded anal lobe, and/or thorax more

globular in shape; mid coxae vertically beneath scutum; fore legs rarely raptorial 4

4. Distal part of CuA not in line with *m-cu*, and/or not strongly recurved into 1A, often weak at apex and not quite reaching 1A; R_{4+5} unbranched, and, in known Australian species, M with at most 2 branches reaching margin of wing; cell CuP sometimes longer than cell M .... HYBOTINAE

Distal part of CuA usually more or less in line with *m-cu* and strongly recurved into 1A; M with 3 branches reaching margin and/or R_{4+5} branched; cell CuP shorter than cell M .... EMPIDINAE

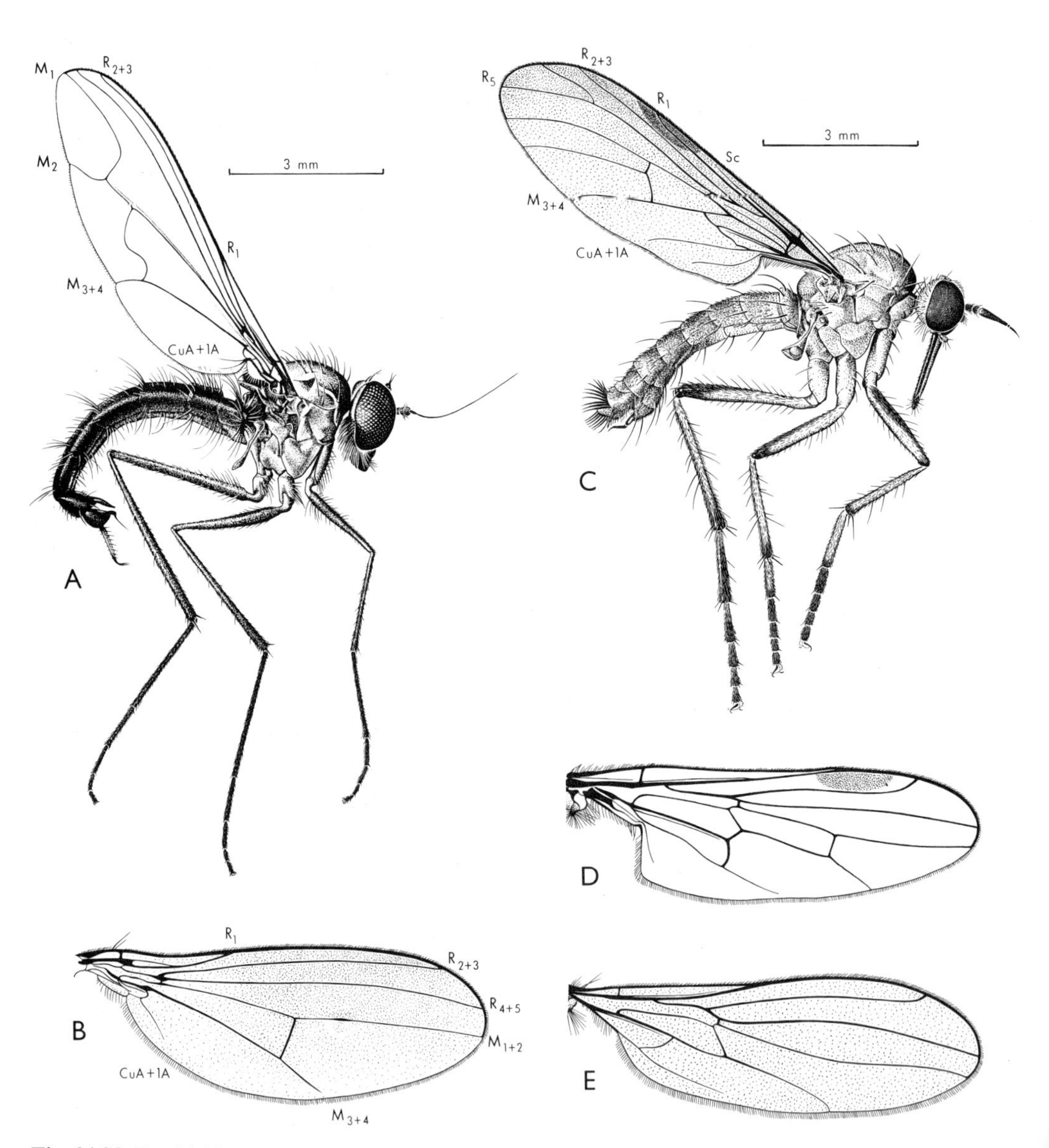

Fig. 34.25. Empidoidea: A, *Heteropsilopus cingulipes*, Dolichopodidae, ♂; B, *Sympycnus allectorius*, Dolichopodidae; C, *Empis* sp., Empididae, ♂; D, *Syneches* sp., Empididae; E, *Tachydromia* sp., Empididae. [T. Binder]

At least half our species await description, but the family seems most abundant in the cooler southern and eastern States, though Tachydromiinae are very common in the north. All subfamilies are represented, dominated by the Empidinae, with many species in the great world-wide genera *Empis* and *Hilara*. There is a distinct 'Antarctic' element, e.g., *Ceratomerus*, *Proagomyia*, *Asymphyloptera*, and *Atrichopleura* (one species closely resembling the S. American *A. caesia* Collin); while a little-studied northern element is represented by various species of Hybotinae and Tachydromiinae. The curious genus *Microphorella*, which probably requires a separate subfamily, is apparently annectant with the Dolichopodidae (Colless, 1963). [G. Hardy, 1930, 1934; Malloch, 1930.]

33. Dolichopodidae (Figs. 34.2J, 3C, 8D, 25A, B). A large family of flies, apparently a specialized offshoot of the Empididae. Most are of rather slender build and moderate to small size; *Heteropsilopus cingulipes* (Walk.) (Fig. 34.25A) expands about 17 mm, but most are much smaller. The thoracic integument is usually metallic, and often bluish, greenish, or bronzy in hue. In many genera, the venation resembles that of the muscoid families. The mouth-parts are predacious, but developed in an unusual fashion: only rarely elongate, rather fleshy, with the labella forming a pair of longitudinally opposed lips, their edges sharp and sometimes set with small spines (Fig. 34.3C). The antennae usually have a long, fine arista, but are sometimes strikingly modified; e.g. *Syntormon*, which shows a remarkable parallel with the empidid *Ceratomerus*. Some Sciapodinae show strong sexual dimorphism, with bizarre modifications of the male wing. Adults are very common, and often to be seen on foliage, tree trunks, etc., or on mud or sand beside bodies of water, fresh or salt; species of *Hydrophorus* are water-striders. The Sciapodinae are common in exposed, sunny situations, but others, e.g. *Sympycnus*, are largely restricted to wet forest. All seem to be predacious; some prey on smaller arthropods, and certain Sciapodinae are useful predators on pest species of aphid; many others probably feed on small oligochaete worms in mud, sand, etc. The larvae resemble those of the Empididae; exotic species have been reported from mud, rotting vegetation, under bark, etc., and at least some are predacious.

The Australian fauna is fairly well known in outline, though not in detail. Twenty-eight genera are recorded, with about 60 per cent of described species falling in two dominant subfamilies. The SCIAPODINAE are mostly rather large, brilliant green or blue species, with excavated frons and M_{1+2} usually branched; *Heteropsilopus cingulipes*, *H. ingenuus* (Erichs.), and *Sciapus connexus* (Walk.) are common widespread species, the two last with banded wings. The CAMPSICNEMINAE are mostly small, semi-metallic species, including *Sympycnus* (usually dark and lacking acrostichal bristles) and *Chrysotimus* (often green, with biserial acrostichals and a depressed area in front of the scutellum). Six other subfamilies are represented. There is a large element of northern origin, e.g. *Neurogona*, *Paraclius*, and *Medetera*, which are found mainly in northern areas. The Sciapodinae are also essentially tropical, though they are widely distributed in Australia. On the other hand, *Sympycnus* may be an 'Antarctic' element, being a dominant genus in the faunas of New Zealand and southern South America, although also distributed well into the tropics to the north of Australia. [G. Hardy, 1958.]

Division CYCLORRHAPHA

Characterized principally by the coarctate pupa, which remains permanently pharate within a puparium formed from the larval skin; also by the circumverted male terminalia, which are flexed forwards ventrally beneath the preceding segments and usually lack recognizable coxites. The larvae are generally of the acephalic 'maggot' type and mainly terrestrial, typically in decaying organic matter, though true aquatic forms occur; many are phytophagous, parasitic, or predacious.

Series Aschiza

A rather heterogeneous assemblage of families, and possibly polyphyletic; distinguished mainly by the absence of the ptilinal fissure. Some families show features annectant with the Orthorrhapha, and all are included in the key on pp. 699–701.

Superfamily LONCHOPTEROIDEA (Anatriata)

The most significant characters are found in the immature stages. The larvae retain maxillae and a vestigial head capsule which is not withdrawn into an atrium, yet the pupa is coarctate as in the higher groups of Cyclorrhapha. The male postabdomen has not been adequately investigated, but is presumably circumverted.

34. Lonchopteridae (Musidoridae; Fig. 34.26A). Slender flies with well developed bristling on head and thorax and characteristic wing venation; CuA + 1A joined distally to M_4 in female only. The only Australian species, *Lonchoptera furcata* (Fall.) (= *dubia* Curran), of this monogeneric family has perhaps been introduced from Europe or America. It reproduces parthenogenetically (Stalker, 1956), and males are extremely rare. The larvae live in decaying vegetation.

Superfamily PHOROIDEA

Adults and larvae resembling those of the next superfamily in many characters, but adults usually with terminal arista and cell CuP more or less shortened. Preabdomen of male consisting of 6 segments, the postabdomen circumverted, at least in Sciadoceridae and presumably in the other families.

35. Platypezidae (Clythiidae; Figs. 34.26C, D). Rather small flies, somewhat resembling Phoridae in life, because of the humped thorax and jerky method of running. The hind tarsi are always modified, and often conspicuously compressed and dilated. The larvae are ovoid and flattened, with lateral projections on the segments; they live in fungi so far as known. *Platypeza*, with M_{1+2} forked beyond the discal cell, is the commonest genus. *Agathomyia* is similar, but with M_{1+2} simple. The smoke fly, *Microsania australis* Collart, may be seen flying in the smoke of camp fires in large numbers in cool localities; it has very weak venation and the discal cell open. *Ironomyia* is an aberrant endemic genus with M_2 arising from the apex of the discal cell.* [Tonnoir, 1925; Collart, 1938.]

36. Sciadoceridae (Fig. 34.26B). The venation is intermediate between that of the Platypezidae and the more specialized venation of the Phoridae. The male has only one vestigial sclerite between abdominal segment 6 and the terminalia, which are not deflexed. The Australian species, *Sciadocera rufomaculata* White, occurs in wet forest from Tasmania to central Queensland, and also in New Zealand. The only other recorded species of the family are the recent *Archiphora patagonica* (Schmitz) from Patagonia and *A. robusta* (Meunier), a fossil from Baltic amber (Hennig, 1964).* *S. rufomaculata* has been reared in the laboratory (Fuller, 1934a) and may be a carrion-breeder, but the early stages have not been found in the field.

37. Phoridae (Fig. 34.26E). A very aberrant family of small to minute flies, with a characteristic hunchbacked appearance due to the relatively large thorax. The wings are folded flat over the abdomen, and the legs, particularly the hind femora, are strongly developed. Adults are common, usually seen running with a quick, jerky motion upon foliage or litter, etc. Some have apterous or brachypterous females, which, in at least one species, are transported by the male during copulation. Two species with apterous females and one normal species live as inquilines in termite nests, while another is recorded from ants' nests. Larval habits vary greatly. Many are scavengers in carrion and other decomposing organic matter, and some species will oviposit and develop satisfactorily in bacterial cultures on ordinary agar plates (Riek, unpublished). Others are probably endoparasites, while at least one

* In a recent paper, J. F. McAlpine and J. E. H. Martin (*Can. Ent.* **98**: 527–44, 1966) have described further fossil species of Sciadoceridae and proposed family status for *Ironomyia*.

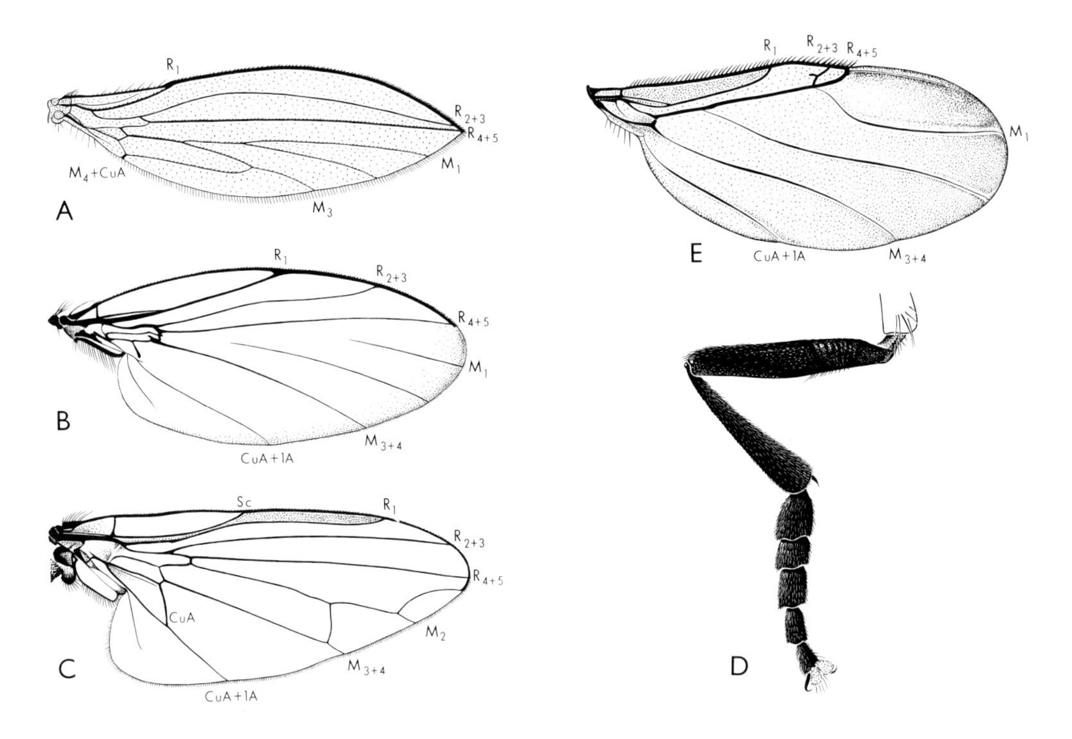

Fig. 34.26. Lonchopteroidea and Phoroidea: A, *Lonchoptera furcata*, Lonchopteridae; B, *Sciadocera rufomaculata*, Sciadoceridae; C, *Platypeza* sp., Platypezidae; D, same, ♂ hind leg; E, unidentified sp. of Phoridae. [T. Binder]

species lives as an ectoparasite attached to the neck of ants of the genus *Camponotus* (Riek, unpublished).

Superfamily SYRPHOIDEA

This group approaches the Schizophora more closely than do any other of the Aschiza. They differ from Schizophora in the absence of a ptilinal fissure (but see p. 674), and from all but a few of that group in the very long cell CuP. Though very variable, the larvae are basically of the 'muscoid' type.

38. Pipunculidae (Dorilaidae; Figs. 34.27C, D). Rather small flies with remarkable powers of hovering in confined spaces. The larvae are endoparasites of Homoptera. The commonest genus is *Pipunculus* (=*Dorilas*), with pigmented pterostigma and venation complete except for the absence of M_2. *Tomosvaryella* is similar, but without pterostigma. *Nephrocerus* with M_{1+2} forked, *Beckerias* with vein 1A obsolescent, and *Chalarus* with discal cell and M_1 also incomplete, occur in eastern Australia. [D. Hardy, 1964.]

39. Syrphidae (hover flies; Figs. 34.11D, 27A, B). A common, widespread family of flies, many with characteristic yellow markings on the body. Some are stoutly built, closely resembling bees or muscoid flies; others (e.g. *Cerioides*, *Baccha*) are quite remarkable wasp mimics, with waisted abdomen and appropriate markings (Plate 6, U; Fig. 5.37). *Psilota rubra* Klöcker even flies in association with the bee that it mimics.

The adults are swift fliers, and many species habitually hover, apparently motionless, in the air; hence their vernacular name of 'hover fly'. In warm, sunny weather, they are a common sight, either on the wing or visiting blossom, and they are probably of importance as pollinators of plants. Certain

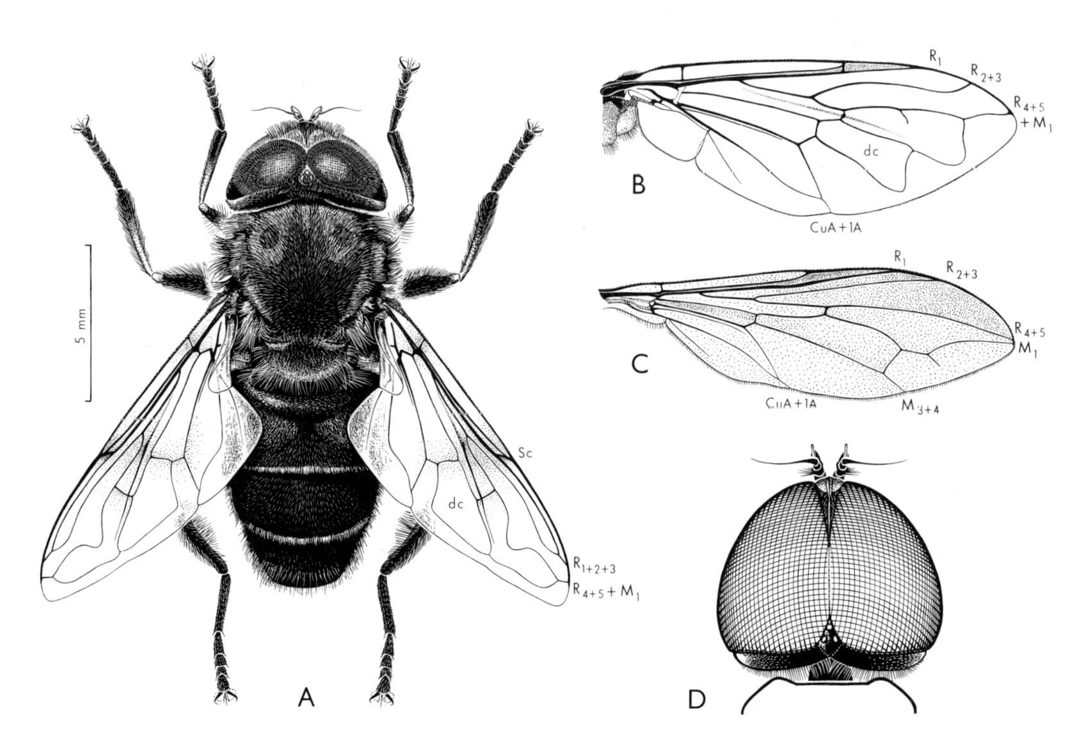

Fig. 34.27. Syrphoidea: A, *Eristalis tenax*, Syrphidae; B, *Microdon modestus*, Syrphidae; C, *Nephrocerus* sp., Pipunculidae, wing; D, same, ♂ head, dorsal view. [T. Binder]

larvae (*Syrphus* spp.) are also beneficial, being important predators of aphids, but a few species of *Eumerus* are injurious to plant bulbs. Most larvae are maggot-like in appearance, living in rotting vegetation or fruit, or in liquid media; amongst the last are the 'rat-tailed maggots' of *Eristalis* spp., which are common in drains. The extraordinary larvae and puparia of *Microdon* spp. (Fig. 34.11D) are oval and convex dorsally, soft and flat ventrally, and were originally described as molluscs! They are usually found in ants' nests, where the larvae live as scavengers.

In Australia, the family seems to be rather poor in species, and may be a relatively recent element of the fauna. Quite a few of our species (some 25, in 9 genera) also occur in other countries. Some are obviously recent immigrants (e.g. the cosmopolitan 'drone fly', *Eristalis tenax* L.), but others belong to wide-ranging species, or perhaps species-complexes, principally Oriental in distribution. The dominant genera in Australia are *Cerioides*, *Microdon*, *Syrphus (sens. lat.)*, *Eristalis*, and *Psilota*, the last being poorly represented in other countries but relatively abundant here. [G. Hardy, 1933a; Paramonov, 1955b, d, 1956a, 1957b.]

Series Schizophora

A clearly demarcated group, including the most recently evolved families of Diptera, some of them probably still actively radiating; characterized by the presence of the ptilinal fissure and the venation (Fig. 34.7C) which is rather uniform throughout the group.

Key to the Families of Schizophora known in Australia

1. Coxae little or not separated (except Braulidae with mesonotum short, resembling abdominal segments); not parasitic on vertebrates in adult stage 2
Coxae of mid and hind legs widely separated; mesonotum large, thorax depressed; parasitic on mammals and birds 57

2(1). Mouth-parts vestigial, subcranial cavity reduced to a small, roughly circular orifice; latero-facial sutures approximated, the broad facialia bare, or with fine golden hairs 3
Mouth-parts and subcranial cavity normal (except in a few rare Tachinidae with dark, spiny bristles on the facialia) 4

3(2). M_1 curved forwards to meet R_5; mesonotum with short, rather sparse hairs (Fig. 34.36D) .... **Oestridae** (p. 737)
M_1 approximately straight, ending in wing margin; mesonotum with long, dense hairs (Fig. 34.35D,E) .... **Gasterophilidae** (p. 732)

4(2). Transverse suture of mesoscutum broadly interrupted medially, or, if complete, incurved lower fronto-orbital bristles and vibrissae absent; lower calypter absent or vestigial (except in many Platystomatidae) .... Acalyptrate families. 5
Transverse suture of mesoscutum complete; incurved lower fronto-orbital bristles present; vibrissae present, or several strong bristles on vibrissal angle; lower calypter well developed (except in some Anthomyiidae) .... Calyptrate families. 53

5(4). Wings absent; mesonotum short and resembling the abdominal tergites; scutellum absent .... **Braulidae** (p. 732)
Wings usually present; mesonotum large; scutellum present 6

6(5). Sc complete, separate from R_1 throughout 7
Sc either indistinct distally, or not ending in costa independently of R_1, or joined to R_1 by sclerotization of the intervening region 26

7(6). Occiput broadly flattened, head very closely fitted to thorax; antennae decumbent, 3rd segment discoid; tarsi with terminal segment triangular and wider than other segments; principally sea-shore species (Figs. 34.30C, D) .... **Coelopidae** (p. 724)
Not with the above combination of characters 8

8(7). Metathoracic spiracle with one or more fine bristles on lower margin; face with a row of bristles on each side, from which the vibrissae are usually not well differentiated; palps vestigial; ant-like flies with abdomen constricted basally (Fig. 34.29F) ... **Sepsidae** (p. 723)
No setulae on lower margin of metathoracic spiracle; other characters not all as above ... 9

9(8). One or rarely 2 pairs of outstanding vibrissae; costa broken or weakened at end of Sc ... 10
Vibrissae absent, or vibrissal angle with a row of undifferentiated setulae (except in a few Lauxaniidae with complete costa) 12

10(9). Second antennal segment with angular projection near middle of outer side of distal margin; postvertical bristles, when present, divergent; face largely desclerotized and continuous with lining of subcranial cavity below (Fig. 34.31D) .... **Clusiidae** (p. 726)
Second antennal segment without angular projection on outer side; postvertical bristles, when present, convergent; face evenly sclerotized, with well-defined lower margin ... 11

11(10). Two strong mesopleural bristles; arista plumose *(Axinota)* .... (Curtonotidae)
Mesopleural bristle rarely present; if present, arista not plumose (Figs. 34.30A, B) .... **Heleomyzidae** (pt., p. 724)

12(9). Tibiae with preapical dorsal bristles; costa unbroken 13
Tibiae without preapical dorsal bristles; or, if these are present, costa broken at end of Sc 15

13(12). CuA + 1A discontinued well before margin; mesopleural bristle present; postvertical bristles convergent or, rarely, absent (Figs. 34.30F, G) .... **Lauxaniidae** (p. 724)
CuA + 1A discernible almost to margin; mesopleural bristle absent (except in some Helcomyzidae); postvertical bristles usually divergent or parallel 14

14(13). Prosternum with well-developed precoxal bridges; sea-shore species ...... (Helcomyzidae)
Prosternum without precoxal bridges, rarely inhabiting sea-shore (Figs. 34. 29D, E) .... **Sciomyzidae** (p. 724)

15(12). The two scutellar bristles arising from apices of long spines; cell M and discal cell broadly confluent. [Eyes and antennae generally borne on lateral processes of head in both sexes] .... (Diopsidae)
Scutellar bristles not at apices of long spines (except in a few Platystomatidae); cell M nearly always closed 16

16(15). Body very elongate; legs, or at least the hind pair, abnormally long; R_5 and M_1 converging or fused distally; R_1 not setulose, ending before middle of wing; proboscis short; a projecting prong on outer surface of mid coxa 17
Not as above; if elongate, with very long legs and R_5 and M_1 convergent, then mid coxal prong is absent and R_1 ends beyond middle of wing 19

17(16). Hind femur enlarged, with a double row of ventral spines; prothorax with broad precoxal bridges *(Gobrya)* .... ('Megamerinidae')
Mid and hind femora both elongate, the latter usually without ventral spines; prothorax without precoxal bridges 18

18(17). Arista dorsal; fore legs much shorter than other legs and widely separated from them (Fig. 34.29A) .... **Micropezidae** (p. 722)
Arista terminal or almost so; fore legs as long as others or almost so, not widely separated from the mid legs (Fig. 34.29B) .... **Neriidae** (p. 722)

19(16). Cell CuP long, acute, the free section of vein CuA never indented nor angulate; or, if cell CuP short *(Stylogaster)*, then proboscis very slender and much longer than head; no mid coxal prong; R_1 ending beyond middle of wing (Figs. 34.28A, B) ... **Conopidae** (p. 719)
If cell CuP long and acute, then vein CuA angulate or indented; proboscis stout, usually shorter than head; mid coxal prong generally present 20

20(19). Free section of CuA recurved; R_1 bare; wings without markings (except Pallopteridae). 21
Free section of CuA angulate, so that cell CuP is acutely produced; or R_1 setulose above; wings usually with dark markings 23

21(20). Costa not broken at end of Sc; if fronto-orbital bristles well developed, then convergent postverticals also present (Fig. 34.30E) .... **Chamaemyiidae** (p. 724)
Costa broken at end of Sc; postvertical bristles never convergent; one distinct fronto-orbital bristle 22

22(21). Anepisternite with posterior marginal and upper anterior bristles; colour shining black, blue, or green (Figs. 34.31E, F) .... **Lonchaeidae** (p. 726)
Anepisternite without upper anterior bristles; colour brown or fulvous ... (Pallopteridae)

23(20). Incurved lower fronto-orbital bristles present; costa broken at end of Sc (Fig. 34.28G) .... **Tephritidae** (pt., p. 721)
No incurved lower fronto-orbital bristles; costa not broken at end of Sc 24

24(23). Vein R_1 bare, or setulose on distal half only .... **Otitidae** (p. 719)
Vein R_1 setulose above for almost its entire length 25

25(24, 39). Cell CuP acutely produced, vein CuA indented, or, if straight, ocelli are absent; ♀ with abdominal segment 7 enlarged, forming a conical or cylindrical ovipositor sheath (Fig. 34.28F) .... **Pyrgotidae** (p. 721)
Cell CuP not acutely produced, vein CuA usually recurved; ocelli distinct; ♀ with abdominal segment 7 forming an inconspicuous, usually flattened ovipositor sheath (Fig. 34.28D) .... **Platystomatidae** (p. 720)

26(6). Hind basitarsus much swollen, or shortened and compressed (Figs. 34.31B, C) .... **Sphaeroceridae** (p. 726)
Hind basitarsus similar to fore basitarsus (except in a few Ephydridae that have mantid-like fore legs) 27

27(26). Lower fronto-orbital bristles incurved, upper fronto-orbitals otherwise directed 28
No incurved lower fronto-orbital bristles 34

28(27). Fold representing distal part of Sc abruptly bent forwards distally to meet costa almost at right angles; cell CuP usually acutely produced; no vibrissae (Fig. 34.28G) .... **Tephritidae** (pt., p. 721)
Distal vestige of Sc close to R_1; cell CuP not angularly produced; vibrissae present 29

29(28). Cell CuP absent or open distally; arista plumose (some *Stenomicra*, etc.) .... **Anthomyzidae** (pt., p. 729)
Cell CuP closed, arista not plumose 30

30(29). CuA + 1A absent; postverticals convergent or parallel (Fig. 34.34A) .... **Milichiidae** (p. 730)
CuA + 1A extending well beyond cell CuP, though sometimes weak 31

31(30). Mesopleural bristle present; postvertical bristles divergent; ♀ with elongate ovipositor (Fig. 34.32C) .... **Agromyzidae** (p. 726)
Mesopleural bristle absent, or postverticals convergent and ovipositor absent 32

32(31). Fore and hind tibiae with short preapical dorsal bristles; postvertical bristles divergent .... **Odiniidae** (p. 726)
Tibiae without preapical dorsal bristles; postvertical bristles convergent or more or less parallel 33

33(32). Postvertical bristles strongly convergent; costa complete or broken only at end of Sc .... **Chyromyidae** (pt., p. 726)
Postvertical bristles more or less parallel; costa broken near humeral vein and also at end of Sc (Figs. 34.32E, F) .... **Carnidae** (p. 727)

34(27). Face convex or protuberant; cell CuP incomplete; cell M confluent with discal cell; arista often with long hairs above but at most pubescent below; postvertical bristles divergent or absent (Figs. 34.33A, B) .... **Ephydridae** (p. 729)
Not as above; if face convex or protuberant, then cell CuP closed, or postvertical bristles convergent, or arista bipectinate (plumose) 35

35(34). Cell CuP and vein CuA + 1A absent. [Cell M confluent with discal cell] 36
Cell CuP at least partly enclosed, or vein CuA + 1A distinct in part 37

36(35). Propleuron with lateral part flat and separated from the transverse anterior part by a vertical carina; M_1 usually at least slightly bent at junction with cross-vein *im*, and not notably converging with R_5 towards apex; costa broken just before end of R_1 (except in *Pemphigonotus*) (Fig. 34.34G) .... **Chloropidae** (p. 731)
Propleuron sloping inwards anteriorly, without vertical carina; M_1 not at all bent at junction with *im* (which may be absent), thereafter curving forward to converge with R_5; costa unbroken (Fig. 34.32B) .... **Asteiidae** (pt., p. 729)

37(35). Antennae very short, lying in deep pits level with lower margin of eye; fronto-orbital bristles, when present, directed outwards; ptilinal fissure highly arched, reaching well above antennae; postvertical bristles divergent or parallel; costa unbroken (Fig. 34.32D) .... **Fergusoninidae** (p. 727)
Antennae and ptilinal fissure not as above 38

38(37). M_4 completely absent beyond discal cell; 1 outwardly directed and 2 proclinate fronto-orbital bristles; postvertical bristles strong, divergent; mesopleural and sternopleural bristles absent (Fig. 34.28E) .... **Cypselosomatidae** (p. 722)
M_4 at least faintly discernible beyond discal cell; without the above combination of bristle characters 39

39(38). R_1 setulose above; abdominal segment 7 of ♀ forming an ovipositor sheath 25
R_1 not setulose; no such ovipositor sheath 40

40(39). Four or 5 dorsocentral bristles, at least 1 in front of suture, postvertical bristles convergent or parallel, widely spaced; anepisternite bristled; CuA + 1A absent beyond cell CuP; habitat sea shore or estuarine (except *Pelomyia* with 1 fronto-orbital) (Figs. 34.34E, F) .... **Tethinidae** (p. 730)
Either with fewer dorsocentral bristles, or postverticals divergent, or anepisternite bare, or CuA + 1A developed 41

41(40). Mid tibiae with distinct preapical dorsal bristles; all fronto-orbital bristles reclinate, or, if 1 proclinate, then there are 2 preapical dorsal bristles and 1 anterior bristle on mid tibiae; postvertical bristles convergent; vibrissae present; prosternum without precoxal bridges .... **Heleomyzidae** (pt., p. 724)
Tibiae without preapical dorsal bristles, except many Drosophilidae which have a proclinate fronto-orbital bristle, only 1 preapical bristle on mid tibiae, and prosternum with precoxal bridges 42

42(41). Sc not obsolete apically, either terminating in R_1 or joined to it apically by sclerotization of the intermediate region. [Postvertical bristles more or less divergent] 43
Sc obsolescent apically, free from R_1 44

43(42). CuA + 1A not extending beyond cell CuP; 2A often distinct; subcranial cavity enlarged; sea-shore species (Figs. 34.34C, D) .... **Canaceidae** (p. 730)
CuA + 1A well developed; 2A vestigial; subcranial cavity not enlarged (Fig. 34.31A) .... **Piophilidae** (p. 726)

44(42). Arista of antenna absent, or minute and terminal on the very large 3rd segment; vertical, ocellar, and fronto-orbital bristles not differentiated (Fig. 34.33C) .... **Cryptochetidae** (p. 730)
Arista well developed, or, if vestigial, replaced by a long, haired process 45

45(44). One proclinate and one or 2 reclinate fronto-orbital bristles (except in a few rare Drosophilidae which have precoxal bridges on prothorax, no presutural bristle, and postverticals convergent) 46
No strong proclinate fronto-orbital bristles; if presutural bristle absent, then precoxal bridges also absent, or postverticals divergent 47

46(45). Postvertical bristles absent, or closely placed and divergent; presutural bristle absent; costa not broken at end of fold representing Sc, but sometimes distinctly narrowed at this point (*Cyamops*, etc.) (Figs. 34.32G, H) .... **Anthomyzidae** (pt., p. 729)
Postvertical bristles convergent, or, if parallel, then widely separated and presutural bristle present; costa clearly broken at end of subcostal fold (Figs. 34.33D, E) .... **Drosophilidae*** (p. 729)

47(45). Sternopleural and presutural bristles absent; fronto-orbital bristles short and weak. [No vibrissae] 48
Sternopleural and usually presutural bristles present; at least one pair of strong, reclinate fronto-orbital bristles 50

48(47). Cells M, CuP, and discal open distally; head exceptionally flattened *(Nothoasteia)* .... **Asteiidae** (pt., p. 729)
All the above cells closed; head not flattened 49

49(48). Free section of CuA strongly recurved; distal vestige of Sc and costal fracture close to R_1 (Fig. 34.29C) .... **Tanypezidae** (p. 722)
Free section of CuA straight; distal vestige of Sc bent forward to end in costal fracture well before end of R_1 .... **Psilidae** (p. 722)

50(47). Three fronto-orbital bristles; mesopleural bristle present. [Postvertical bristles crossed] ... . **Chyromyidae** (pt., p. 726)
One or 2 fronto-orbital bristles; mesopleural bristle generally absent 51

51(50). Four pairs of dorsocentral bristles .... **Pseudopomyzidae** (p. 722)
At most 3 pairs of dorsocentral bristles 52

52(51). Cross-veins *r-m* and *im* very close together; CuA + 1A very long but weak; antenna porrect, with broad 3rd segment and pubescent arista (Fig. 34.32A) .... **Teratomyzidae** (p. 729)
These cross-veins well separated; CuA + 1A shortened; antenna with narrow, deflexed 3rd segment and plumose arista .... **Anthomyzidae** (pt., p. 729)

53(4). Meropleuron usually bare or with weak hairs; if bristled, then pteropleuron bare and M_1 not distinctly bent forward 54
Meropleuron with row or group of bristles and pteropleuron with one or more bristles, or occasionally either meropleuron or pteropleuron with long dense hairs only; M_1 almost always strongly bent forward distally 55

54(53). CuA + 1A reaching wing margin, though often faint distally; lower calypter not longer than upper calypter (Fig. 34.35C) .... **Anthomyiidae** (p. 732)
CuA + 1A not reaching margin; lower calypter nearly always longer than upper calypter (Figs. 34.35A, B) .... **Muscidae** (p. 732)

* A few Milichiidae key out here, but have interfrontal bristles, a row of incurved fronto-orbital setulae on each side, and costa very deeply incised at end of Sc.

55(53). Subscutellum prominent, as a transverse, rounded, convex lobe beneath scutellum and above normal postnotum (rather weak in Palpostomatini, which have the arista bare or pubescent and labella often with a pair of palp-like processes posteriorly); anterior lappet of metathoracic spiracles without tuft of long hairs (Fig. 34.36C) ...... **Tachinidae** (p. 737)

Postnotum normal, without distinct, rounded subscutellum dorsally; or, if subscutellum distinct (blue or green, metallic species), anterior lappet of metathoracic spiracle with tuft of long, fine hairs towards dorsal margin 56

56(55). Colour non-metallic, thorax usually with longitudinal dark stripes; external posthumeral bristle, if present, never distinctly lateral to presutural bristle; M_1 angled at a point usually nearer to apex of discal cell than to wing margin; posterior spiracle not enlarged, set posterior to line of meropleural bristles; if arista bare, base of R without setae posteriorly (Fig. 34.36A) .... **Sarcophagidae** (p. 735)

Colour metallic; or external posthumeral bristle distinctly lateral to presutural bristle; or M_1 angled at a point nearer to wing margin than to apex of discal cell; or posterior spiracle very large, partly projecting beyond main axis of line of meropleural bristles; if arista bare, base of R with setae posteriorly (Fig. 34.36B) ...... **Calliphoridae** (p. 734)

57(1). Head small and inserted dorsally on thorax; wingless; parasites of bats (Fig. 34.37B) ... . **Nycteribiidae** (p. 738)

Head larger, not inserted dorsally on thorax 58

58(57). Palps appressed, sheathing the proboscis, not broadened; head usually closely fitted to thorax; when wings present, the stronger veins crowded towards costa; parasites of birds and mammals other than bats (Fig. 34.37A) .... **Hippoboscidae** (p. 738)

Palps not appressed and sheathing, extremely broad; head not closely fitted to thorax; veins not crowded towards costa in winged forms; parasites of bats (Fig. 34.37C) .... **Streblidae** (p. 738)

Superfamily CONOPOIDEA

This is apparently the most primitive group of the Schizophora. The venation resembles that of the Syrphoidea in having M_1 approximated to or fused with R_5 distally, cell CuP long and acute (except in *Stylogaster*), and often a vestige of the vena spuria present. The male postabdomen is quite symmetrical.

40. Conopidae (Figs. 34.28A, B). A rather small family, but unusual in that most of its members are, to some degree, wasp mimics, some very effectively so (Plate 6, Q). Many superficially resemble syrphids of the genus *Cerioides*, even in venation, but have a well-developed ptilinum. *Stylogaster* has an aberrant venation, but is recognizable by the long, thin proboscis (present in many conopids) and the elongate ovipositor. Most conopids have both male and female terminalia enlarged and conspicuous, those of the female often opposed or covered by a large lobe developed from sternite 5. Adults are slow fliers, and are usually taken on blossom, often accompanying the wasps which they mimic. Larvae of Australian species presumably resemble those of other countries, which are recorded as endoparasites of adult Hymenoptera, Diptera, and orthopteroids. The 16 Australian genera include the widely distributed *Conops*, *Myopa*, *Physocephala*, *Stylogaster*, and *Thecophora*; the remainder are mostly endemic, though species of *Microconops* are also found in Chile. [Camras, 1961.]

Superfamily TEPHRITOIDEA (Otitoidea)

Vein R_1 setulose (except in Richardiidae and some Otitidae), cell CuP usually acutely produced (except in Richardiidae and Platystomatidae). Male with abdominal segment 6 vestigial or absent, the aedeagus very long and coiled, often looped around the terminal segments. Female with abdominal segment 7 more or less enlarged to form an ovipositor sheath, the subsequent segments forming an elongate, usually pointed ovipositor (except in Pyrgotidae). The Richardiidae (Neotropical) and Tachiniscidae (Neotropical, Ethiopian) have not been found in Australia. The classification of the superfamily has been discussed by Steyskal (1961).

41. Otitidae (Ortalidae, Ulidiidae). Male with aedeagal apodeme simple; female with abdominal segment 6 well developed. The Australina species belong to the subfamily ULIDIINAE, having vein R_1 bare and aedeagus not setulose. *Physiphora aenea* (F.) is a common introduced species, with metallic, green-black body colour. Its larvae live in decaying vegetable refuse.

42. Platystomatidae (Figs. 34.28C, D). Abdominal segment 6 of female vestigial; costa usually broken beyond humeral vein, but not at end of Sc. In this and the next two families, the aedeagal apodeme ('fultella') has a pair of lateral arms which pivot on sternite 9. The larvae occur in living and dead vegetable matter, and have also been found in graves in Europe. A larva of *Euprosopia*

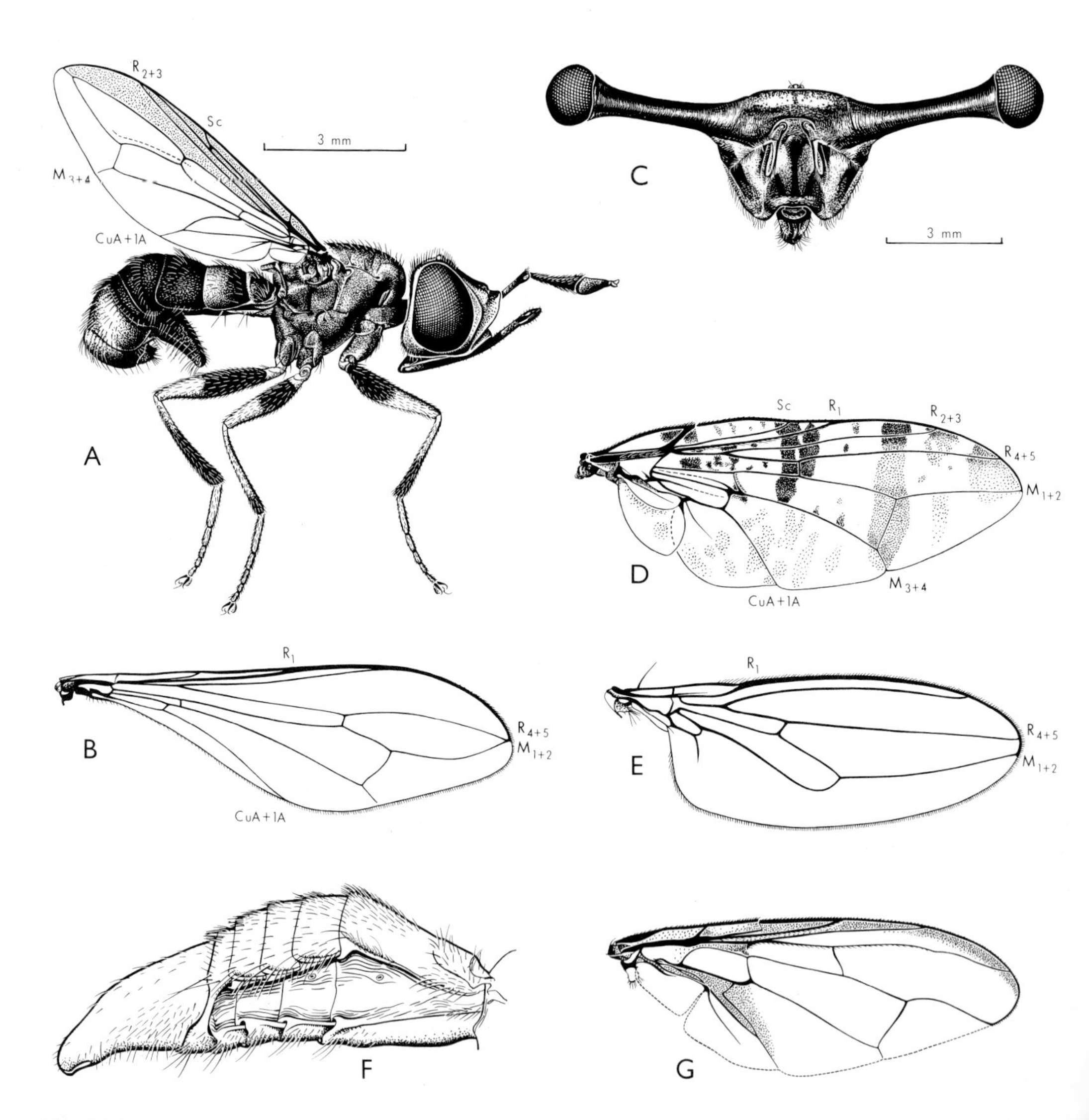

Fig. 34.28. A, *Conops splendidus*, Conopidae, ♀; B, *Stylogaster* sp., Conopidae; C, *Achias australis*, Platystomatidae, ♂, front view of head; D, *Euprosopia tenuicornis*, Platystomatidae; E, *Cypselosoma australis*, Cypselosomatidae; F, *Adapsilia illingworthana*, Pyrgotidae, abdomen; G, *Dacus tryoni*, Tephritidae. [T. Binder]

has been found in soil near Sydney, eating a curl-grub pupa (Scarabaeidae). Once included in the Otitidae, this group is now considered just as distinct as Pyrgotidae and Tephritidae. In Australia, the dominant genera are *Rivellia*, *Duomyia*, *Euprosopia*, and *Lamprogaster* (Plate 5, L). The almost cosmopolitan *Rivellia* includes small species, usually black and with transverse bands on the wings. They occur in most forested parts of Australia, but few have yet been described. *Duomyia* is almost exclusively Australian, though one species occurs in Chile and another in Fiji. The species are widely distributed through the continent in open country. *Achias*, *Lamprogaster*, and *Euprosopia* are found mainly in wet tropical and subtropical areas, though the last two are represented in the highlands of south-eastern Australia. *Achias*, with 3 species in north Queensland and others in New Guinea, usually has the eyes on stalks or lateral extensions of the head in the males (Fig. 34.28C); the females are much less highly modified. [Malloch, 1928, 1929, 1930, 1939; Paramonov, 1957c.]

43. Pyrgotidae (Fig. 34.28F). The most consistent distinctive character is the reduction of the ovipositor; segment 8 is mostly membranous, and the piercing apparatus simple and thorn-like or reduced to a small sclerite; 2nd antennal segment usually tapered basally. Other characters, such as absence of ocelli and of mid coxal prong, are inconsistent in Australian species. As far as known, the larvae are parasites of larval or adult Scarabaeidae. Several Australian species have been observed apparently ovipositing on adults of *Phyllotocus* and *Anoplognathus*. Though often taken in light traps, the adults of some species are also active in the daytime. [Paramonov, 1958b, c.]

44. Tephritidae (Trypetidae, Trypaneidae, Trupaneidae; fruit flies; Figs. 34.8C, 28G). Sc usually bent forward distally to meet the costa almost at right-angles; abdominal segment 6 of female well developed. A large, cosmopolitan family, best represented in the tropics.

The numerous species of *Dacus* are fruit-eating in the larval stage. There are several subgenera (sometimes called genera). The Queensland fruit fly, *Dacus (Strumeta) tryoni* (Frogg.), is by far the most serious pest species in eastern Australia, attacking a wide range of cultivated and native fruits. Other destructive species in Queensland are *Dacus (Austrodacus) cucumis* French, principally in cucumbers and other cucurbits, and *Dacus (Afrodacus) jarvisi* (Tryon) in a variety of fruits. *Dirioxa pornia* (Walk.) (=*Rioxa musae* (Frogg.)), of N.S.W. and Queensland, is of little importance because it usually attacks damaged fruit. The larvae of *Termitorioxa termitoxena* (Bezzi), in the N.T., live in termite galleries in tree trunks. The Mediterranean fruit fly, *Ceratitis capitata* (Wied.), an introduced species, attacks oranges and a wide variety of other cultivated fruits; once a serious pest in N.S.W., it has not been seen there since 1941, but is still of importance in W.A. *Procecidochares utilis* Stone is a Mexican species, introduced into Queensland and N.S.W. to control the weed *Eupatorium adenophorum*; eggs are laid in the young shoots which subsequently become galled by the larvae. The Bathurst burr seed fly, *Euaresta bullans* (Wied.), which is common in N.S.W., also occurs in North and South America, and Europe. Its larvae eat the seeds of the burr *(Xanthium spinosum)* and probably of other plants. The larvae of *Tephritis*, *Trupanea*, and *Stylia* live in the flower-heads of Compositae. [Malloch, 1939; May, 1965.]

Superfamily MICROPEZOIDEA

Legs often very elongate. Prosternum with a strongly sclerotized posterior part which is broadly continuous with mesosternal region; often also with a median anterior plate; precoxal bridges absent. Preapical tibial bristles absent. Male postabdomen with ventral and often almost symmetrical sternite 6, sternite 7 ventral or sublateral. Epandrium (p. 667) often elongate, channelled ventrally to receive the basiphallus which is elongate, rigid, usually with longitudinal struts, the distiphallus usually folded along its anterior side. Female abdomen with segment 7 elongate, forming an ovipositor sheath, the more distal segments telescopic, not forming a piercing organ but with small, separate cerci. The mainly Oriental Megamerinidae do not occur in Australia, though *Gobrya*, erroneously referred to that family, occurs in New Guinea.

45. Pseudopomyzidae. Medium-sized to minute flies resembling Heleomyzidae in cephalic chaetotaxy, but clearly belonging to this superfamily from the structure of the prosternum and postabdomen of both sexes. The family includes the genera *Pseudopomyza* (= *Heluscolia*) from Europe, New Zealand and Australia, and *Heloclusia* from Chile and Argentina.

46. Cypselosomatidae (Fig. 34.28E). Legs not attenuated; vibrissae and ocellar bristles well developed; 4–6 dorsocentral bristles; mid tibia with strong dorsal bristles; male with almost symmetrical postabdomen, and no pregenital lobes on sternite 5. This small family is represented in Australia by a species of *Cypselosoma*, the larvae of which live in bat guano in caves, sometimes in very large numbers. It is known elsewhere only from Formosa, the Philippines, Indonesia, and New Guinea. [D. McAlpine, 1966.]

47. Neriidae (Fig. 34.29B). Head elongate, postvertical bristles convergent; 2nd antennal segment with terminal finger-like process on inner side; costa usually without distinct break; CuA+1A not reaching margin; pregenital lobes of male absent; protandrium (p. 669) asymmetrical; segment 9 elongate. The family is widely distributed, but principally tropical. The banana-stalk fly, *Telostylinus bivittatus* Cress., occurs in southern Queensland and N.S.W. The larvae have been found in the decaying ends of cut banana bunches and in other decaying vegetable matter.

48. Micropezidae (Tylidae; Fig. 34.29A). Ocellar bristles absent or minute; postverticals usually divergent or parallel; 2nd antennal segment without process; costa unbroken; vein CuA straight; male usually with pregenital lobes on abdominal sternite 5, sternite 6 ventral, 7 lateral. The larvae live in decaying wood and other vegetable matter. Two subfamilies occur in Australia: CALOBATINAE (Trepidariinae) with 1 or 2 distinct sternopleural bristles and ocelli situated near vertex; and TAENIAPTERINAE with a fascicle of numerous, long sternopleural bristles and ocelli well in front of vertex. The two are sometimes accorded family rank. The Calobatinae include the dominant Australian genus, *Metopochetus*, distinguished by the occipital tubercle just above the neck. It occurs from New Guinea to Tasmania. The other genera of the subfamily, *Trepidarioides* and *Gongylocephala*, are tropical and subtropical only. *Mimegralla* is the only Australian genus of Taeniapterinae.

Superfamily TANYPEZOIDEA (Nothyboidea)

Generally elongate flies. Vibrissae absent; postverticals, when present, divergent. Prosternum with broad precoxal bridges (except Psilidae, which may not belong here). Male postabdomen with well-developed segment 6, the sternite ventrally placed. Female usually with terminal segments elongate, but no distinct ovipositor sheath or piercing organ. Non-Australian families are Nothybidae (Oriental) and Diopsidae, which reach New Guinea.

49. Tanypezidae (Fig. 34.29C). Costa broken near end of R_1; postvertical bristles divergent; mesopleural bristle present; face desclerotized medially; male with protandrial sternites asymmetrical and without articulated surstyli. The genus *Strongylophthalmyia* is principally Oriental, with one European and 2 undetermined Australian species occurring mainly in rain forest. The male of the commoner Australian species has the arista reduced and replaced by an outgrowth of the 3rd segment (Fig. 34.29C). The genus, which differs from typical tanypezids in its incomplete subcosta, was once included in the Psilidae.

50. Psilidae. Costa broken well before end of R_1, Sc discontinued opposite the break; face sclerotized; pleural bristles absent; prosternum without precoxal bridges; male postabdomen symmetrical, with or without surstyli. The genus *Chyliza* occurs in north Queensland and Victoria. The Australian

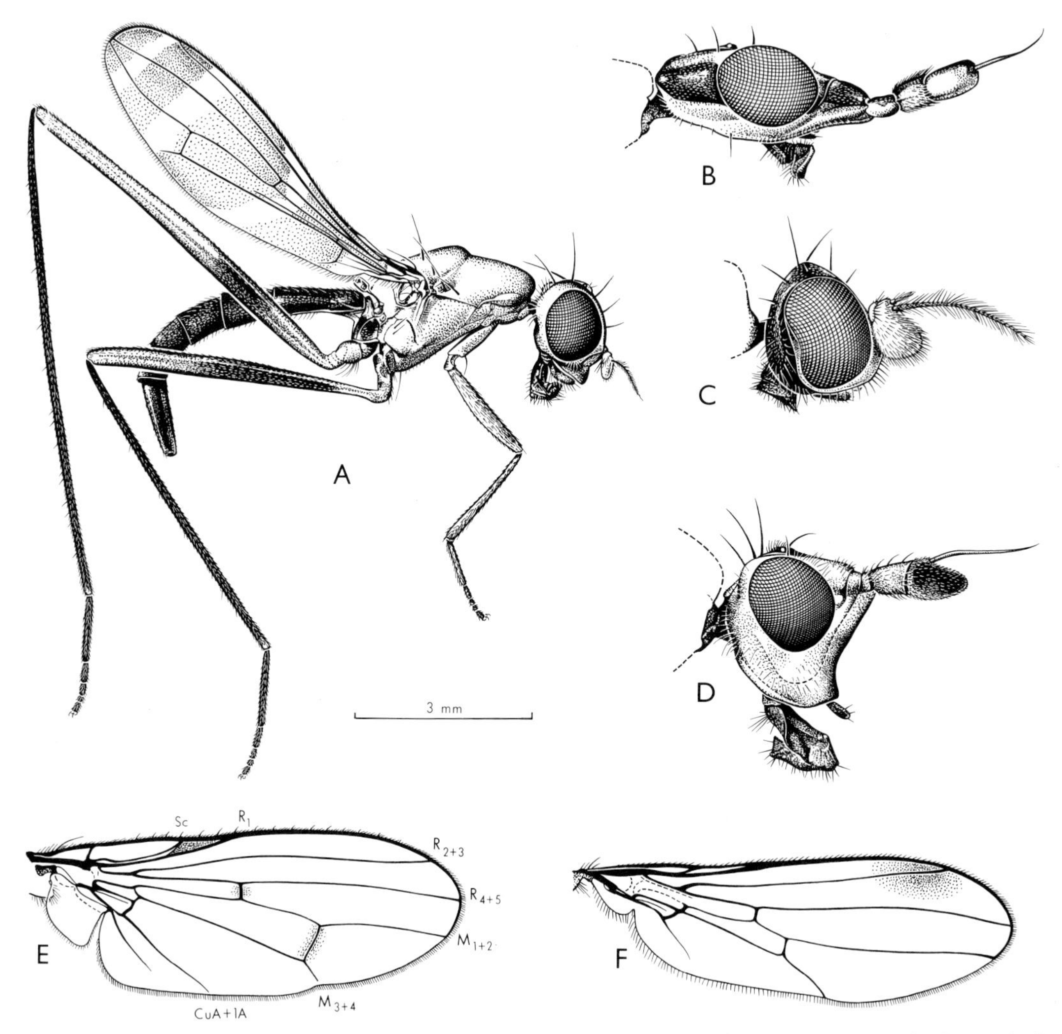

Fig. 34.29. A, *Metopochetus tenuipes*, Micropezidae, ♀; B, *Telostylinus bivittatus*, Neriidae; C, *Strongylophthalmyia* sp., Tanypezidae, ♂; D, *Dichaetophora* sp., Sciomyzidae; E, *Helosciomyza* sp., Sciomyzidae; F, *Australosepsis niveipennis*, Sepsidae. [T. Binder]

record of *Loxocera*, with very long 3rd antennal segment, requires confirmation.

Superfamily SCIOMYZOIDEA

Vibrissae absent (except in many Sepsidae); tibiae usually with preapical dorsal bristles. Costa unbroken; Sc complete and separate from R_1; cell CuP short, not angularly produced. Male with sternites 6 and 7, when present, laterally displaced; tergite 6 abbreviated or absent. Female postabdomen not modified to form an ovipositor, the cerci usually separate. This superfamily may be polyphyletic, as most of its characters are merely the primitive basic ones of the Schizophora. The Dryomyzidae (Holarctic), Rhopalomeridae (Neotropical), Helcomyzidae (Europe, North and South America, New Zealand), and Celyphidae (Ethiopian, Oriental) have not been found in Australia.

51. Sepsidae (Fig. 34.29F). Small, ant-like flies with the habit of continuously waving the wings when at rest. The larvae live in mammal faeces or vertebrate carcasses, and the adults often swarm in the vicinity of these. [Steyskal, 1949.]

52. Sciomyzidae (Tetanoceridae; Figs. 34.29D, E). The larvae of typical forms are predators or parasites of freshwater and terrestrial snails. The free-living larvae and floating puparia of *Dichaetophora* (SCIOMYZINAE) are common in ponds or dams containing snails. *Helosciomyza* and *Xenosciomyza* (HELOSCIOMYZINAE) differ in having spaced costal spines and in other characters; both also occur in New Zealand. Larvae of *Xenosciomyza* have been found in agaric fungi in South Australia. [Malloch, 1928.]

53. Chamaemyiidae (Ochthiphilidae; Fig. 34.30E). Small flies; postvertical bristles convergent or absent; at most 2 fronto-orbital bristles, both reclinate. The larvae are predators of coccids and psyllids. The genera *Pseudoleucopis*, *Chaetoleucopis*, *Chamaemyia*, and *Leucopis* occur in Australia, the first two being endemic. They have vein CuA+1A much abbreviated, and the male postabdomen almost symmetrical. In *Gayomyia*, from Australia and Chile, CuA+1A is long and curved and the protandrium very asymmetrical; it probably belongs, with some other Neotropical forms, to the subfamily EURYCHOROMYIINAE. [Malloch, 1930; J. McAlpine, 1960.]

54. Lauxaniidae (Sapromyzidae; Figs. 34.30F, G). This is one of the largest and commonest of the 'acalyptrate' families, the adults occurring in a wide range of habitats: mangrove swamps, sand dunes, grasslands, all forest types, and on the summits of such mountains as Kosciusko and Wellington. The larvae, on the other hand, are very little known, and it has been suggested that they are mainly saprophagous. A species of *Sapromyza* has been reared from the nest of *Acanthornis magnus* in Tasmania. Malloch has given a key to the Australian genera, but there remain many unrecorded genera and species. [Malloch, 1927, 1928.]

Superfamily HELEOMYZOIDEA

Basic characters principally as in Sciomyzoidea, with the following modifications: postvertical bristles convergent (except in a few Coelopidae); vibrissa or a group of bristles or hairs present on vibrissal angle; costa broken or weakened at end of Sc (except in Coelopidae and some Chyromyidae); prosternum without precoxal bridges; female postabdomen not forming an elongate ovipositor, cerci usually free. Many of the basic characters are modified or lost in specialized forms, e.g. the postvertical bristles may be lost, Sc may be incomplete, preapical tibial bristles may be absent. In addition to the families mentioned below, the little known Neotropical Somatiidae may belong to this superfamily.

55. Coelopidae (Phycodromidae; Figs. 34.30C, D). Vibrissae usually replaced by a group of hairs or bristles; costa entire; Sc well separated from R_1; free section of vein CuA straight; CuA+1A reaching wing margin; protandrium of primitive type, strongly twisted. The larvae live on decaying, stranded kelp and sea grass *(Zostera)*, and the adults generally occur on beaches, though they have been found inland. Most of the species are confined to southern Australia, where *Chaetocoelopa sydneyensis* (Schin.) is the commonest species; *Dasycoelopa australis* Mall. is known only from northern N.S.W. and Queensland.

56. Heleomyzidae (including Trixoscelidae, Rhinotoridae; Figs. 34.30A, B). Postvertical bristles convergent (often absent in *Cairnsimyia*); fronto-orbital bristles usually reclinate; sternopleural and presutural bristles usually present; tibiae usually with preapical dorsal bristles; costa weakened or broken at end of Sc; cells CuP and M closed; male postabdomen with sternites 6 and 7 laterally displaced or absent (sternite 6 median, symmetrical in *Waterhouseia*, which may require a separate family). The larvae of *Tapeigaster* live in fungi, whereas those of *Cairnsimyia robusta* (Walk.) have been found in the burrows of longicorn beetles in a fig tree. The family occurs principally in temperate forests.

The typical northern-hemisphere forms, with CuA+1A reaching margin and strong, spaced costal spines, are represented in

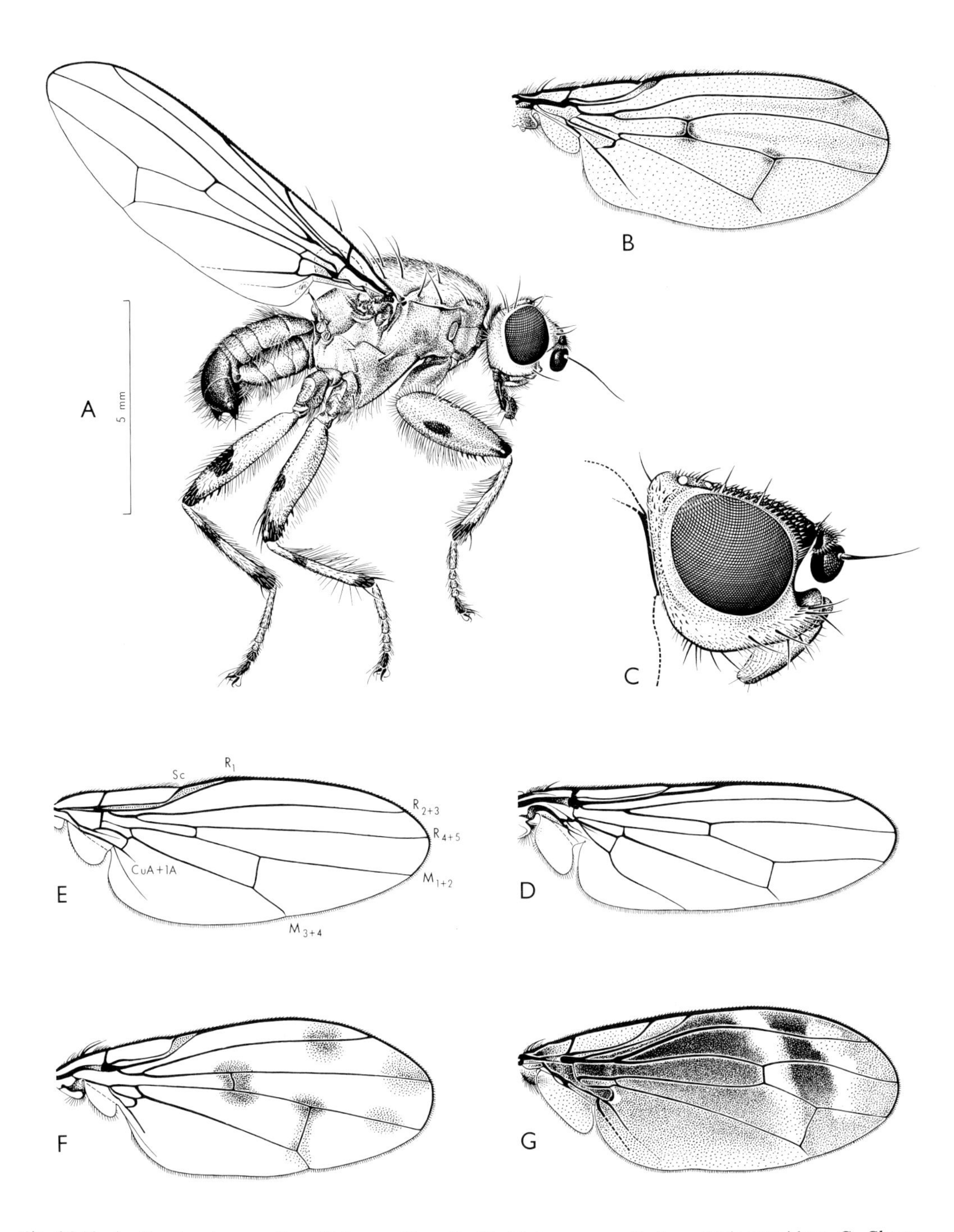

Fig. 34.30. A, *Tapeigaster annulipes*, Heleomyzidae, ♂; B, *Diplogeomyza diaphora*, Heleomyzidae; C, *Chaetocoelopa sydneyensis*, Coelopidae, head; D, same, wing; E, *Pseudoleucopis fasciventris*, Chamaemyiidae; F, *Homoneura proximella*, Lauxaniidae; G, *Depressa striatipennis*, Lauxaniidae. [T. Binder]

southern Australia by 2 introduced species, *Pseudoleria pectinata* (Loew) and *Oecothea fenestralis* (Fall.). The other Australian genera deviate from the typical forms to a varying extent, making the family limits obscure. The dominant genera, *Tapeigaster* and *Diplogeomyza*, are endemic. *Tapeigaster*, with no spaced costal spines, CuA + 1A long, and only 1 or 2 pairs of dorsocentral bristles, contains large, stout species; it is a phylogenetically isolated genus of doubtful systematic position. *Diplogeomyza*, with 4 dorsocentrals and 2 long preapical bristles on the mid tibia, is related to the New Zealand genus *Allophylopsis*.

57. Sphaeroceridae (Borboridae, Cypselidae; Figs. 34.31B, C). Small to very minute flies with distinct vibrissae, often found on animal dung or other organic matter in which the larvae live. *Leptocera*, with cells M and CuP open, is the dominant genus, and it is possible that many Australian species will be later recognized. *Sphaerocera* and *Copromyza* have these cells closed, the former without, the latter with, scutellar bristles. *S. curvipes* Latr. is an immigrant from Europe.

58. Chyromyidae (Chiromyiidae). Very small but stout flies; fronto-orbital bristles reclinate, or the foremost somewhat sloping inwards; vibrissae often not differentiated from cheek bristles; preapical tibial bristles absent; cells M and CuP complete; vein CuA + 1A not reaching margin; male postabdomen symmetrical. Adults of *Aphaniosoma* spp. have been found in the flowers of *Hibiscus* and closely related plants, while larvae of one species have been collected in bat guano in caves.

Superfamily OPOMYZOIDEA (Pallopteroidea)

Postvertical bristles divergent or almost parallel. Costa broken at end of Sc (except in Fergusoninidae and Acartophthalmidae); cells M and CuP complete. Male postabdomen with sternite 6 asymmetrical, 7 laterally displaced, or with these sclerites lost and symmetry restored (Agromyzidae, Fergusoninidae, some Lonchaeidae). Female postabdominal segments elongate, often forming an ovipositor, the cerci usually fused. Besides the families occurring in Australia, the Neottiophilidae (Holarctic), Opomyzidae and perhaps Acartophthalmidae (Holarctic), and Pallopteridae (New Zealand and most temperate regions) belong to this superfamily. Only the first 5 families listed on p. 678 are closely related to one another, the others being doubtfully placed here.

59. Lonchaeidae (Figs. 34.31E, F). Stout flies of moderate to small size; females with an ovipositor resembling that of Otitoidea. *Lamprolonchaea brouniana* (Bezzi), the metallic-green tomato fly, is very common; its larvae often live in damaged tomatoes. [Malloch, 1928; J. McAlpine, 1964.]

60. Piophilidae (Thyreophoridae; Fig. 34.31A). Postvertical bristles well developed; at most 2 weak fronto-orbital bristles; vibrissae strong; mesopleural bristle absent. The larvae live in dead animal matter. Those of the introduced *Piophila casei* (L.) ('cheese skippers') live in cheese, preserved meats, and old carrion. *Protopiophila australis* Harrison, differing in having 4 instead of 2 pairs of dorsocentral bristles, is apparently endemic to Australasia. *Piophilosoma (=Chaetopiophila)*, an endemic genus, has been placed in the Thyreophoridae, a group scarcely deserving family rank; its larvae live in dried carrion. [Paramonov, 1954b.]

61. Clusiidae (Heteroneuridae, Clusiodidae; Fig. 34.31D). Second antennal segment with angular projection on outer side; male postabdomen asymmetrical, with tergite 6 and sternite 8 present. The species are principally found in wet forest country. The larvae live in moist rotting wood. [D. McAlpine, 1960.]

62. Odiniidae. Related to Agromyzidae, but distinguished by the presence of short preapical tibial bristles and the less developed ovipositor. The Australian species are undescribed.

63. Agromyzidae (Fig. 34.32C). Mostly small or minute flies. The larvae are leaf- or stem-miners and gall-makers. Larvae of several species occur on cultivated plants, and are important economically, whilst those of *Ophiomyia lantanae* (Frogg.) live in the

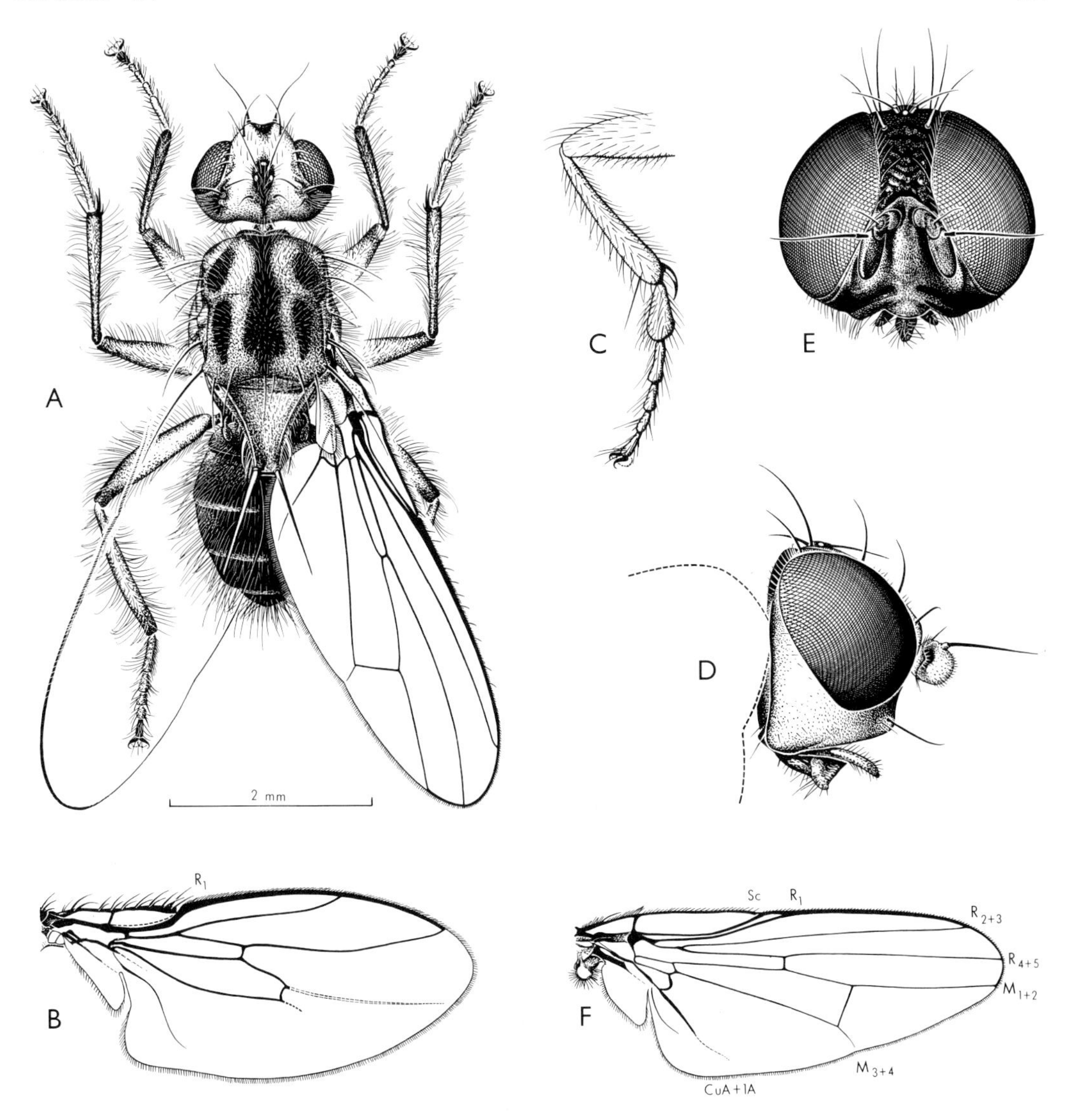

Fig. 34.31. A, *Piophilosoma scutellata*, Piophilidae, ♂; B, *Leptocera* sp., Sphaeroceridae; C, *Copromyza* sp., Sphaeroceridae, hind leg; D, *Heteromeringia norrisi*, Clusiidae; E, *Lamprolonchaea brouniana*, Lonchaeidae; F, same, wing. [T. Binder]

developing fruit of lantana and reduce the quantity of seed set by that weed. [Spencer, 1963.]

64. Fergusoninidae (Fig. 34.32D). The family, established by Hennig (1958), includes only the interesting endemic Australian genus *Fergusonina*, which is possibly only a specialized agromyzid. Larvae live in leaf, bud, and stem galls on various species of *Eucalyptus*, in a remarkable association with nematodes of the genus *Anguillulina (Fergusobia)* (p. 116). As the galling may prevent flower production and setting of seed, some species are of importance to the honey and timber industries. [Currie, 1937.]

65. Carnidae (Figs. 34.32E, F). Though often included in Milichiidae, this group is evidently not referable to Drosophiloidea.

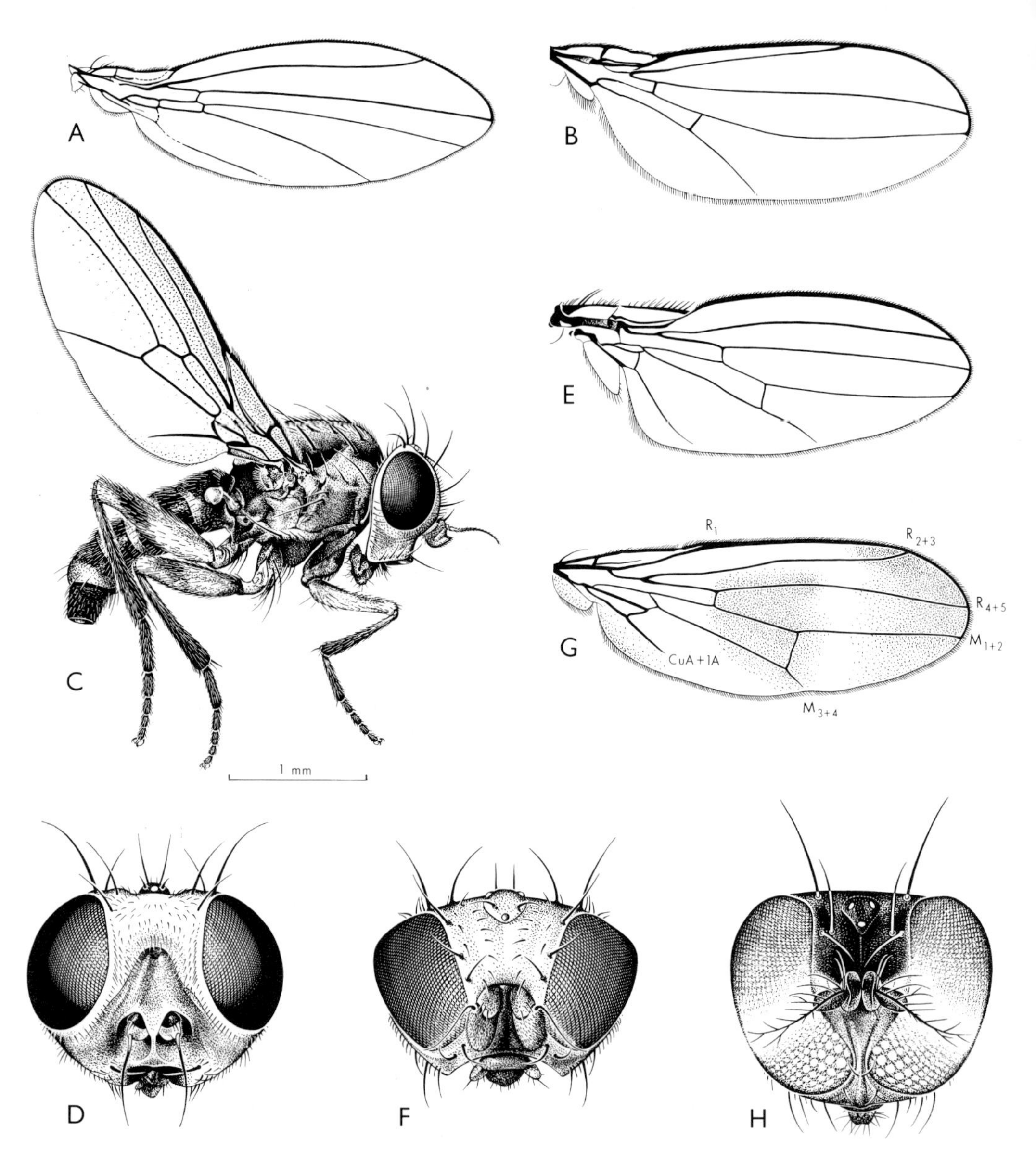

Fig. 34.32. A, *Teratoptera* sp., Teratomyzidae; B, *Leiomyza nitidula*, Asteiidae; C, *Cerodontha robusta*, Agromyzidae, ♀; D, *Fergusonina* sp., Fergusoninidae; E, *Australimyza* sp., Carnidae, wing; F, same, head; G, *Cyamops* sp., Anthomyzidae, wing; H, same, head. [T. Binder]

Specialized genera such as *Meoneura* resemble Milichiidae through reduction of the protandrium and wing venation, but may generally be distinguished by having CuA + 1A more distinct than 2A. In more primitive forms *(Hemeromyia, Australimyza)* CuA is very long, and male sternites 6 and 7 are asymmetrically placed. The predominantly Holarctic family is represented in Australia by a minute species of *Australimyza*, occurring on stranded debris on the shores of sheltered inlets.

Superfamily ASTEIOIDEA

Small or minute elongate flies. Fronto-orbitals usually strongly developed; postverticals, when well developed, divergent (except some Anthomyzidae); antennae often with 3rd segment deflexed at an angle to 2nd; arista often long-plumose. Prosternum usually broad; sometimes with weak precoxal bridges; legs not notably elongate; preapical tibial bristles absent. Wings usually narrow, Sc incomplete. Protandrium usually retaining signs of at least two sclerotized segments. Female postabdomen short, without piercing organ. The Aulacigastridae and Periscelididae are confined to the Holarctic and Neotropical regions.

66. Teratomyzidae (Fig. 34.32A). Elongate flies. Face sclerotized; 1 strong fronto-orbital; postverticals small, slightly divergent to slightly convergent; vibrissae present; antennae porrect, with segment 3 broad, subcircular; arista not plumose. Prosternum reduced, without precoxal bridges. Wings narrow; costa broken at end of Sc, weakened near humeral vein, not reaching M_1; cross-veins *r-m* and *im* close together before middle of wing; cell CuA complete; vein CuA + 1A very long. Protandrium symmetrical, segment 6 with complete tergite and sternite. The family includes *Teratomyza* (New Zealand), *Teratoptera* (Chile and Australia), and other genera.

67. Anthomyzidae (Figs. 34.32G, H). Small, slender flies; vibrissae present. The members of this family are so diverse as to render it difficult to define, and it is possible that it includes more than one group of family value. Some genera are rather close to the Aulacigastridae. The typical genera have the face largely desclerotized and the postverticals, when present, convergent. They are principally Holarctic, and only *Amygdalops* reaches Australia. The remaining genera have the face entirely sclerotized and the postverticals, when present, divergent. Adults are frequently associated with Araceae, and the larvae of an undescribed species live in the fruiting spathes of *Alocasia*. The widely distributed genera *Cyamops* and *Stenomicra* are represented in Australia, and there are 2 apparently endemic undescribed genera. [Sabrosky, 1965.]

68. Asteiidae (Fig. 34.32B). Small or minute flies with reduced venation. The larvae do not seem to have been recorded. The genera *Leiomyza*, *Sigaloessa* (= *Crepidohamma*), *Asteia*, and *Nothoasteia* occur in Australia, the last being endemic.

Superfamily DROSOPHILOIDEA

Postvertical bristles convergent, or replaced by the divergent postocellar bristles (Canaceidae, Ephydridae), or absent; vibrissae usually present. Prosternum with precoxal bridges (lost in some Milichiidae). Costa broken at end of Sc, which is usually obsolete distally or fused with R_1; CuA + 1A shortened or absent. Male postabdomen symmetrical, with reduced segmentation. Female postabdominal segments short, cerci usually free. The group includes the Milichioidea and Drosophiloidea of Hennig (1958), with the addition of the Chloropidae and Cryptochetidae. The Camillidae, Diastatidae, and Curtonotidae are not known in Australia, though the curtonotid genus *Axinota* is common in New Guinea.*

69. Ephydridae (Figs. 34.33A, B). These flies are often found near water, both salt and fresh, but certain species of *Hydrellia* and *Scatella* occur on grasslands and lawns far from water. The larvae are mainly aquatic, or live within the stems and shoots of plants, often of aquatic plants. *Hydrellia tritici* Coq. is generally the most abundant fly in extra-tropical pastures, and is associated with both land and freshwater plants. [Malloch, 1925, 1928; Cresson, 1948.]

70. Drosophilidae (Figs. 34.33D, E). Distinguished mainly by the position of the proclinate fronto-orbital bristle near the eye, and absence of a mesopleural bristle. The larvae of most species are fungivorous, some eating yeasts growing in decaying fruit. Although many species of *Drosophila* are known (see Mather, 1960), the other genera in Australia have been little studied. A key to the Pacific genera (now incomplete) was given by M. Wheeler (1952).

* Discovered in north Queensland since the above was written.

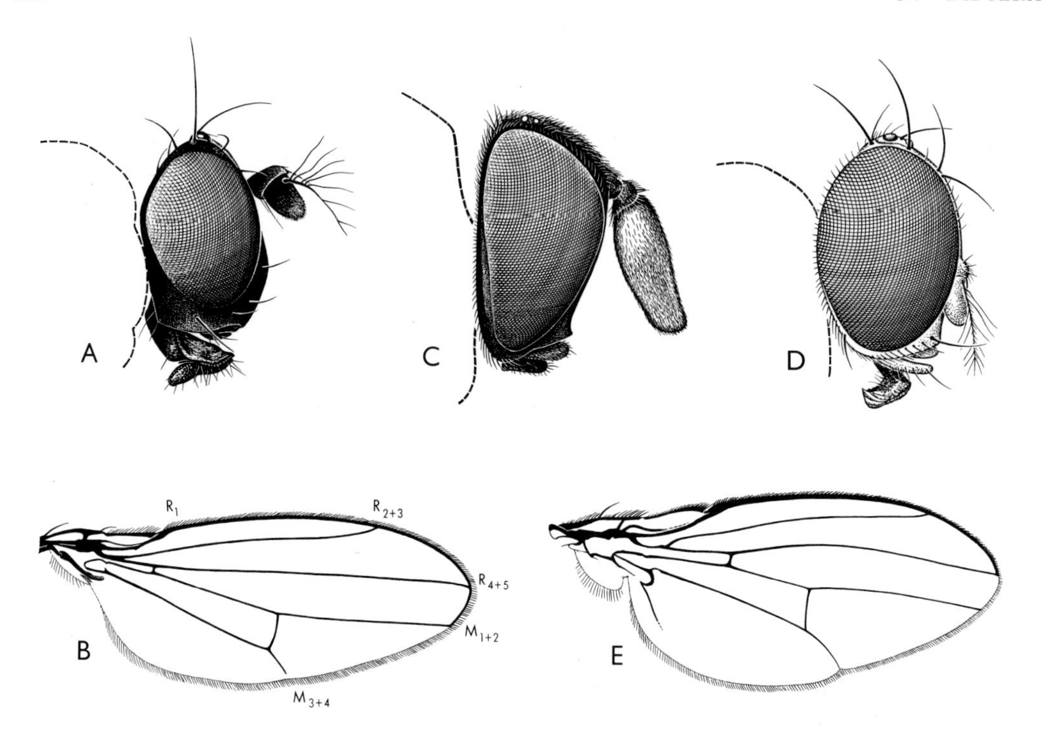

Fig. 34.33. A, *Hydrellia tritici*, Ephydridae, head; B, same, wing; C, *Cryptochetum monophlebi*, Cryptochetidae; D, *Leucophenga niveifasciata*, Drosophilidae, head; E, same, wing. [T. Binder]

71. Milichiidae (Phyllomyzidae; Figs. 34.34A, B). Postvertical bristles convergent or parallel; lower fronto-orbital bristles incurved (sometimes reduced to hairs), upper ones otherwise directed; vibrissae present; prosternal bridges present, or the prosternum reduced; costa twice broken; cells M and CuP closed; protandrium consisting of a single sclerite situated dorsally. Larval habitats are varied; in Australia the larva of *Milichia piscivora* Mall. has been found in dead fish and that of *Milichiella* sp. in soil on rotting wood.

72. Cryptochetidae (Fig. 34.33C). Small, stout flies; costa twice broken; cell M confluent with discal cell; vein 2A strongly developed, CuA+1A absent. *Cryptochetum* is the only genus. The larvae are endoparasites of mealy bugs of the family Margarodidae, and some are considered to be of economic importance. *C. monophlebi* (Sk.) and *C. iceryae* (Will.) are Australian species which were introduced to California many years ago to control *Icerya purchasi* Mask. [Malloch, 1927.]

73. Tethinidae (Figs. 34.34E, F). Very small flies; postvertical bristles usually convergent, often a pair of divergent postocellar bristles in front of them; 1 to 4 fronto-orbital bristles reclinate or sloping outwards, often also an inner row of incurved setulae on orbits; vibrissae present; anepisternite with one or more posterior bristles and an upwardly directed bristle near upper margin; costa broken only at end of subcosta. Larvae unknown, but those of *Tethina* may have similar habits to Coelopidae. *Dasyrhicnoessa* is abundant in mangroves, whereas *Tethina* is found on stranded kelp and sea grass *(Zostera)* on beaches or on adjacent sand dunes.

74. Canaceidae (Figs. 34.34C, D). Closely related to Tethinidae, differing in the absence of convergent postvertical bristles. Fronto-orbital bristles usually flexed outwards over eyes; vibrissae usually weak or inserted

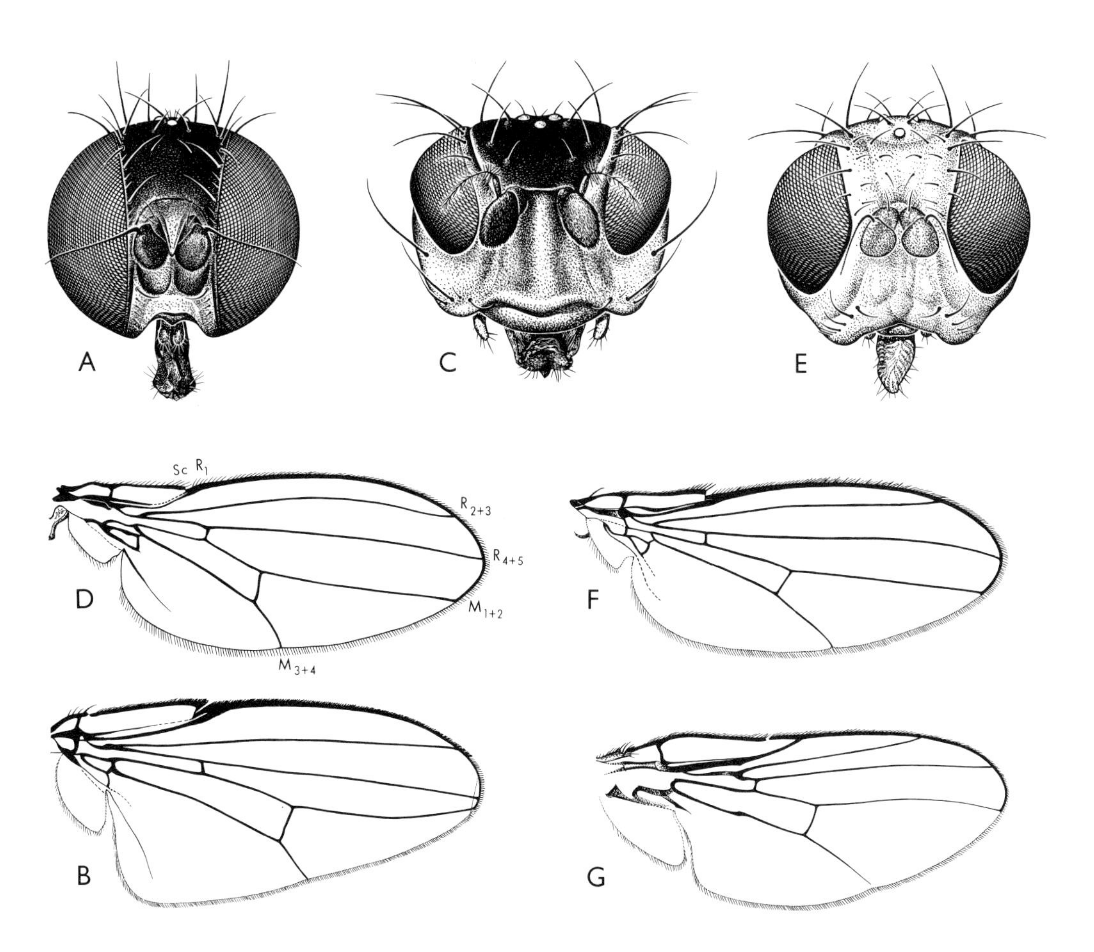

Fig. 34.34. A, *Milichiella* sp., Milichiidae, head; B, same, wing; C, *Canace albiceps*, Canaceidae, head; D, same, wing; E, *Tethina nigriseta*, Tethinidae, head; F, same, wing; G, *Batrachomyia* sp., Chloropidae. [T. Binder]

behind vibrissal angle; face with a bulbous carina; male with compound penultimate abdominal segment (segments 6–8) at least as long as segment 5 (shorter or absent in Tethinidae); size small or minute. Except for a few foreign species inhabiting mountain streams, the family is confined to the sea shore and estuaries, the adults and larvae frequenting the intertidal zone. The genera *Canace*, *Nocticanace*, *Trichocanace*, *Xanthocanace*, and *Chaetocanace* are represented in Australia. [Wirth, 1964.]

75. Chloropidae (Oscinidae; Fig. 34.34G). Postvertical bristles usually convergent (subparallel in some Chloropinae); fronto-orbital bristles usually weak, never with distinct proclinate and reclinate elements; prosternum with well-developed precoxal bridges; mesopleural bristle absent. The adults are of almost ubiquitous occurrence, and the larvae inhabit a wide range of habitats, though still little known. Larvae of numerous species feed within the young shoots and stems of grasses and other plants.

In the N.T., larvae of *Anatrichus* are associated with rice, but it is uncertain whether they damage rice or prey on rice-eating insects. *Lioscinella flavoapicalis* (Mall.) has been reared from both new and rotting swedes. Larvae of *Batrachomyia* live beneath the skin of frogs, feeding on blood. Forms with somewhat similar adults live as larvae in spiders' egg-cocoons, while *Botanobia tonnoiri* Mall. has been reared from cocoon masses of sawflies and oothecae of mantids. Larvae of *Botanobia luteohirta* (Mall.) are recorded by Evans (1931) as eating the eggs of *Eurymela distincta* Sign. (Homoptera), and some species of *Lasiopleura* are parasites or hyperparasites within the burrows of fossorial Hymenoptera. *Cadrema nigricornis flavus* (Thoms.) has been reared from stranded marine molluscs. [Sabrosky, 1955.]

76. Braulidae. Wingless, flattened, highly specialized flies, with tarsi broadened distally. The relationships are still obscure, but it seems certain that there is no connection with the Phoridae or with the pupiparous muscoid families. The bee louse, *Braula coeca* Nitz., a widely distributed species, has been introduced into Tasmania. Adults are usually found upon honey bees, whilst the larvae live in cells of that species. They pupate without any modification of the larval cuticle, though it is not shed.

Superfamily MUSCOIDEA

Includes all the 'calyptrate' families of Schizophora and several greatly modified, but probably related, parasitic families (Hippoboscidae, etc.).

77. Gasterophilidae (Figs. 34.35D, E). A small family, represented in Australia only by 3 introduced species of horse bot fly, *Gasterophilus intestinalis* De G., *G. nasalis* (L.), and *G. haemorrhoidalis* (L.). The adults have vestigial mouth-parts, as in the Oestridae (also called bot flies, and with similar habits), but M_1 is not curved forwards to meet R_{4+5}. All 3 species are brown and rather bee-like, and *G. intestinalis* has patterned wings. The eggs are attached to hairs on the host's head or body, and the young larvae make their way to the horse's mouth, or are picked up by licking. They spend a period burrowing in the epithelial tissues of the mouth and tongue, and then pass to the stomach, where they attach to the mucous membrane, but the effect on the host appears usually to be slight. Eventually, they pass out in the faeces and pupate in the soil. Very rarely, the young larvae cause a 'creeping eruption' in human skin.

78. Anthomyiidae (Fig. 34.35C). M_1 not curved forward distally. The complete vein CuA + 1A occurs otherwise among the calyptrate families only in the Scatophagidae, an exotic group often included in the Anthomyiidae. The larvae are mainly phytophagous. The larvae of a species of *Fucellia* live in stranded seaweed on beaches. *Hylemyia platura* (Meig.) is a cosmopolitan pest species, its larvae damaging seedlings of onions, beans, and other plants.

79. Muscidae (Figs. 34.2K, 3D, 6B, 35A, B). A large and variable family, with many species of economic and medical importance.

Key to the Subfamilies of Muscidae Known in Australia

1. Palps usually broadly spatulate or, if otherwise, parafacials haired over their whole length; pteropleuron haired .... LISPINAE
   Palps not spatulate, but sometimes apically thickened; pteropleuron bare or, if haired, parafacials bare above 2
2. Proboscis very elongate, rigid and sclerotized. [M_1 curved forward distally] ...... STOMOXINAE
   Proboscis shorter, folding into subcranial cavity 3
3. Lower calypter very broad, with almost straight, transverse, posterior margin; M_1 strongly bent forward from near middle of distal section .... MUSCINAE
   Lower calypter narrow, with rounded apex; M_1 curved forward near apex, or straight ... 4
4. Fold representing vein 2A curved forward to end beyond apex of CuA + 1A but in a straight line with it .... FANNIINAE
   Fold representing 2A not much curved forward 5
5. Meropleuron with one or more bristles in a vertical row; frons of ♂ not narrowed .... EGINIINAE
   Meropleuron at most with a few hairs, in which case the frons is narrowed in ♂ 6
6. Sternopleuron with one bristle which is nearly equidistant from the 2 upper bristles; lower bristle of proepimeral (stigmatal) group curved downwards, or very weak and horizontal .... COENOSIINAE
   Sternopleuron otherwise bristled, or, if as above, then lower proepimeral bristle curved

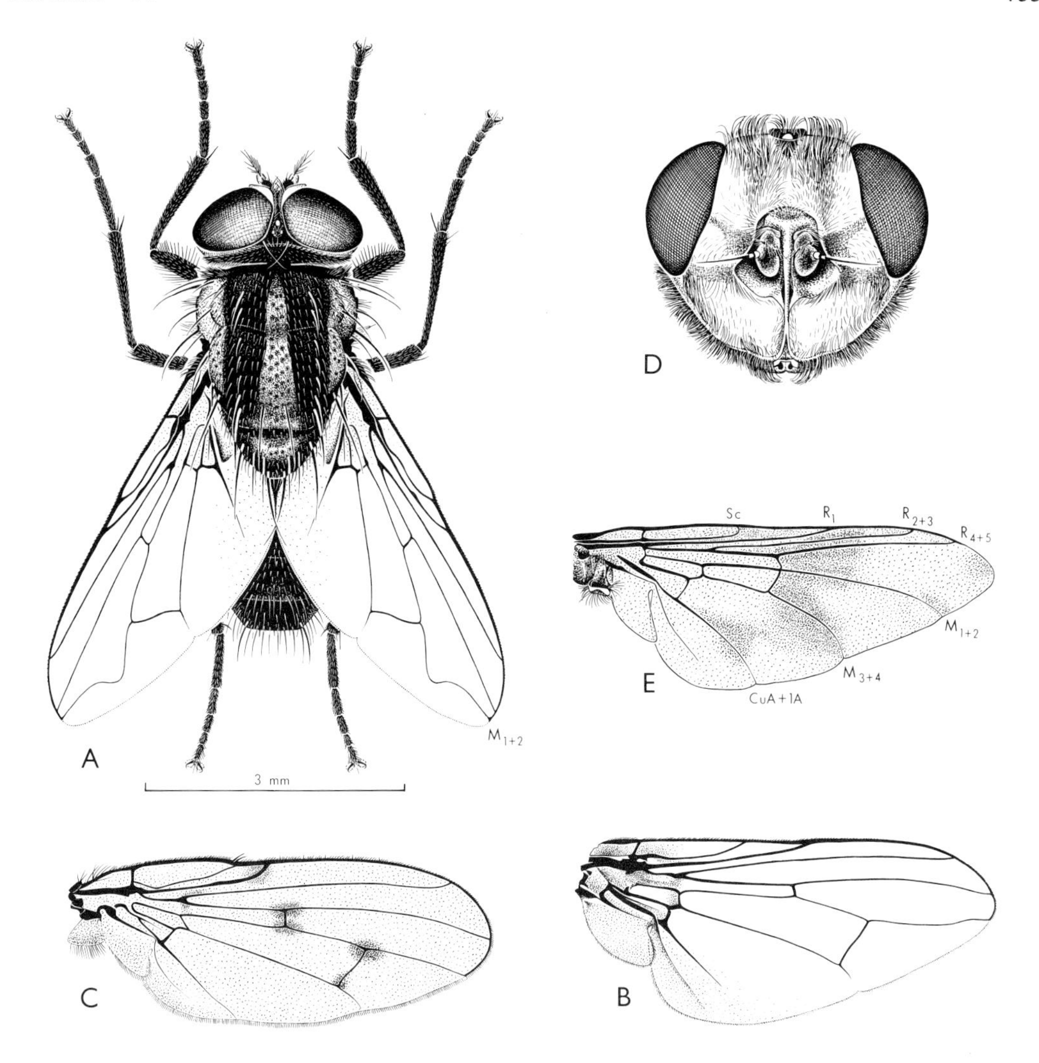

Fig. 34.35. A, *Musca vetustissima*, Muscidae, ♂; B, *Ophyra rostrata*, Muscidae; C, *Hylemyia deceptiva*, Anthomyiidae; D, *Gasterophilus intestinalis*, Gasterophilidae, head; E, same, wing. [T. Binder]

upwards or absent, never curved downwards .... PHAONIINAE

The FANNIINAE include the lesser house fly, *Fannia canicularis* (L.); the adults occur in numbers in houses, and the larvae live in organic refuse. It is not comparable with the house fly either as a threat to health or as a nuisance. A few small undescribed species inhabiting mountain forests appear to belong to the small subfamily EGINIINAE. Adults of the LISPINAE generally occur near fresh or salt water; the larvae are probably mainly aquatic and the adults predacious. The COENOSIINAE contain the genera *Coenosia*, *Pygophora*, and *Atherigona*, which are common in grasslands and gardens. The adults, of some forms at least, are predacious on smaller insects. Available records of larvae suggest mainly phytophagous habits. Those of *Atherigona* have been found in decaying fruit and boring in the stems of such grain crops as millet.

In the large subfamily PHAONIINAE, the larvae are probably mainly predacious on other insect larvae, particularly those of Diptera. They occur in very diverse habitats. The larvae of *Limnophora nigriorbitalis* Mall. live in moss at the margins of streams, where they are frequently submerged. Those of other species live in decaying fruit and other vegetable matter, manure, and carrion. The predacious larvae of *Ophyra* also occur in carrion.

The most familiar species of MUSCINAE is the house fly, *Musca domestica*, which is recorded as carrying many kinds of pathogenic bacteria affecting man and domestic animals. The adults, which tend to congregate in human dwellings, are attracted to both wet and dry organic foods, including much that is to be eaten by man. The larvae live in excrement of various animals and other organic refuse. The bush fly, *Musca (Byomya) vetustissima* Walk. (Fig. 34.35A), is a constant worry to man and domestic animals in the open air during summer. The larvae of *Passeromyia* occur in nests of many species of birds, either as parasites sucking the blood of the nestlings, or merely as scavengers; the taxonomic status of these ecologically different forms has not been established.

The STOMOXINAE include blood-sucking flies with the peristomial teeth of the labella adapted for piercing (Fig. 34.3D). The cosmopolitan stable fly, *Stomoxys calcitrans* (L.), bites man and domestic animals, particularly horses. Its larvae normally live in horse manure mixed with straw, etc. The buffalo fly, *Haematobia exigua* (de Meij.), which was introduced from the Oriental region, is now a major pest of cattle and horses in the northern half of Australia. [Malloch, 1923, 1924, 1925, 1926; Paramonov, 1960a.]

80. Calliphoridae (blow flies, bluebottles; Figs. 34.3E, 7C, 9B, 10C, 11C, 36B). A large, cosmopolitan family of flies, mostly stoutly built and of moderate size; almost all have the antennal arista plumose. They vary rather considerably in details of structure, and several well-marked subfamilies can be recognized.

Key to the Subfamilies of Calliphoridae Known in Australia

1. Anterior lappet of metathoracic spiracle with tuft of long, fine hairs towards dorsal margin; subscutellum distinct; face almost always with strong carina between antennae .... AMENIINAE
   Anterior lappet of metathoracic spiracle bare, or without long hairs; subscutellum and facial carina rudimentary or absent 2
2. Stem-vein (extreme base of R) with row of hairs posteriorly 3
   Stem-vein bare 4
3. Lower calypter haired on upper surface; no external posthumeral bristle; arista strongly plumose .... CHRYSOMYINAE
   Lower calypter bare on upper surface; usually with a posthumeral bristle placed lateral to level of presutural bristle; arista rather weakly plumose or pubescent only .... RHINIINAE
4. With an external posthumeral bristle placed lateral to level of presutural bristle; sternopleural bristles usually 2 : 1 (if 1 : 1, 3 notopleural bristles present) .... CALLIPHORINAE
   Without any such external posthumeral bristle; sternopleural bristles 1 : 1 .... POLLENIINAE
   Without any such external posthumeral bristle, and sternopleural bristles 2 : 1 *(Euphumosia)*; or with an external posthumeral bristle and sternopleural bristles 1 : 1 *(Caiusa)*; 2 notopleural bristles present. [Northern area only] .... PHUMOSIINAE

Adult calliphorids are ubiquitous, flying mainly by day, although a few species appear in light traps. They are strongly attracted to moisture, and feed mainly on nectar, honeydew, and other sweet liquids, and on the liquid products of organic decomposition; the last provide the proteins essential for egg maturation. Reproduction is oviparous or ovoviviparous. Many species, particularly in the CALLIPHORINAE and CHRYSOMYINAE, breed in carrion (p. 124), although at least some species of *Calliphora* parasitize earthworms (Fuller, 1933). By contrast, the POLLENIINAE are probably all parasites of earthworms, the larvae of at least some RHINIINAE live in the nests of ants or termites, and the AMENIINAE are apparently parasites of land snails.

Although their numbers sometimes reach

plague proportions, blow flies are far less important than other groups of flies as domestic nuisances or vectors of disease. However, the larvae of some species are responsible for the cutaneous myiasis of sheep, known as 'blow-fly strike', which annually causes direct and indirect losses of millions of dollars to Australian industry. This myiasis is an extension of the carrion-breeding habit, and is initiated by only a few species, the 'primary flies'. These are attracted to oviposit on areas of the sheep's body, usually where persistent wetting of the wool by urine, sweat, or rain has caused 'scalding' of the skin and superficial bacterial infection. The younger larvae feed on serous exudates, etc., but the older larvae attack the living tissues, producing extensive lesions and sometimes death of the sheep. Once the strike is established, other 'secondary' species may be attracted to extend the infestation, while a group of 'tertiary flies' is also recognized, whose larvae occur in healing scabs. In Australia, some 80 per cent of primary strikes are caused by the introduced, metallic green sheep blow fly *Lucilia cuprina* (Wied.). The related *L. sericata* (Meig.), of importance in some other countries, also occurs here, but is practically innocuous. The other important primary flies are native species of *Calliphora*, notably *C. augur* (F.) in south-eastern Australia and *C. nociva* Hardy, which replaces it in the drier inland and western areas. The most important secondary fly is *Chrysomya rufifacies*, whose 'hairy maggots' not only compete with, but actively prey upon, the primary maggots. A similar type of succession occurs in carrion, but with the primary flies playing a much smaller role.

The family is well developed in Australia. Some 4 per cent of our species are immigrants, while the Phumosiinae, Ameniinae, Rhiniinae, and Chrysomyinae tend to be distributed mainly in the north, and show other evidence of relatively recent entry from that direction. By contrast, *Calliphora* (66 spp.) and *Pollenia* (29 spp.) seem to have undergone considerable radiation in the more humid, temperate areas; their origins, and relationships with other faunas, remain obscure. The striking metallic green and blue Ameniinae (Frontispiece), hitherto regarded as prosenine Tachinidae, have recently been transferred to the Calliphoridae (Crosskey, 1965). [G. Hardy, 1940; Norris, 1965; Paramonov, 1957a, 1960b.]

81. Sarcophagidae (flesh flies; Fig. 34.36A). A cosmopolitan family, treated by some authors as a subfamily of Calliphoridae, but forming a fairly compact group defined by the characters given in the key. The larvae of most species have the posterior spiracles set in a deep pit or chamber. Two subfamilies occur in Australia:

Arista plumose; thorax with 3 broad, longitudinal, dark stripes (very faint in *Tricharaea*); abdomen with changing tessellated pattern of silvery-grey and black patches; usually 4 notopleural bristles; sternopleural bristles 1 : 1 : 1 or 2 : 1 .... SARCOPHAGINAE

Arista bare, or nearly so; thoracic pattern variable; abdomen usually with silvery-grey, pruinose bands; only 2 notopleural bristles; sternopleural bristles 1 : 2–4 .... MILTOGRAMMATINAE

The MILTOGRAMMATINAE are not common, and, although widely distributed, most specimens seem to come from the drier inland areas. As far as is known, the females are viviparous, and the larvae live as food-parasites in wasps' nests, generally those of fossorial species. Seven genera, 4 of them endemic, are recorded from Australia, but *Protomiltogramma* seems to be by far the commonest.

The SARCOPHAGINAE are ubiquitous, commonly seen on foliage, etc., or around carrion. Almost all species are viviparous, usually depositing their larvae on some type of decomposing organic matter, and in northern areas certain species are common primary or secondary breeders in carrion. However, the family is very rarely involved in myiasis, animal or human, in Australia, although species of considerable medical importance occur in other countries. The genus *Blaesoxipha* is interesting, in that the known larvae are all parasites of Acrididae; other genera also have been bred from insects and snails, but at least some of these are necrophagous,

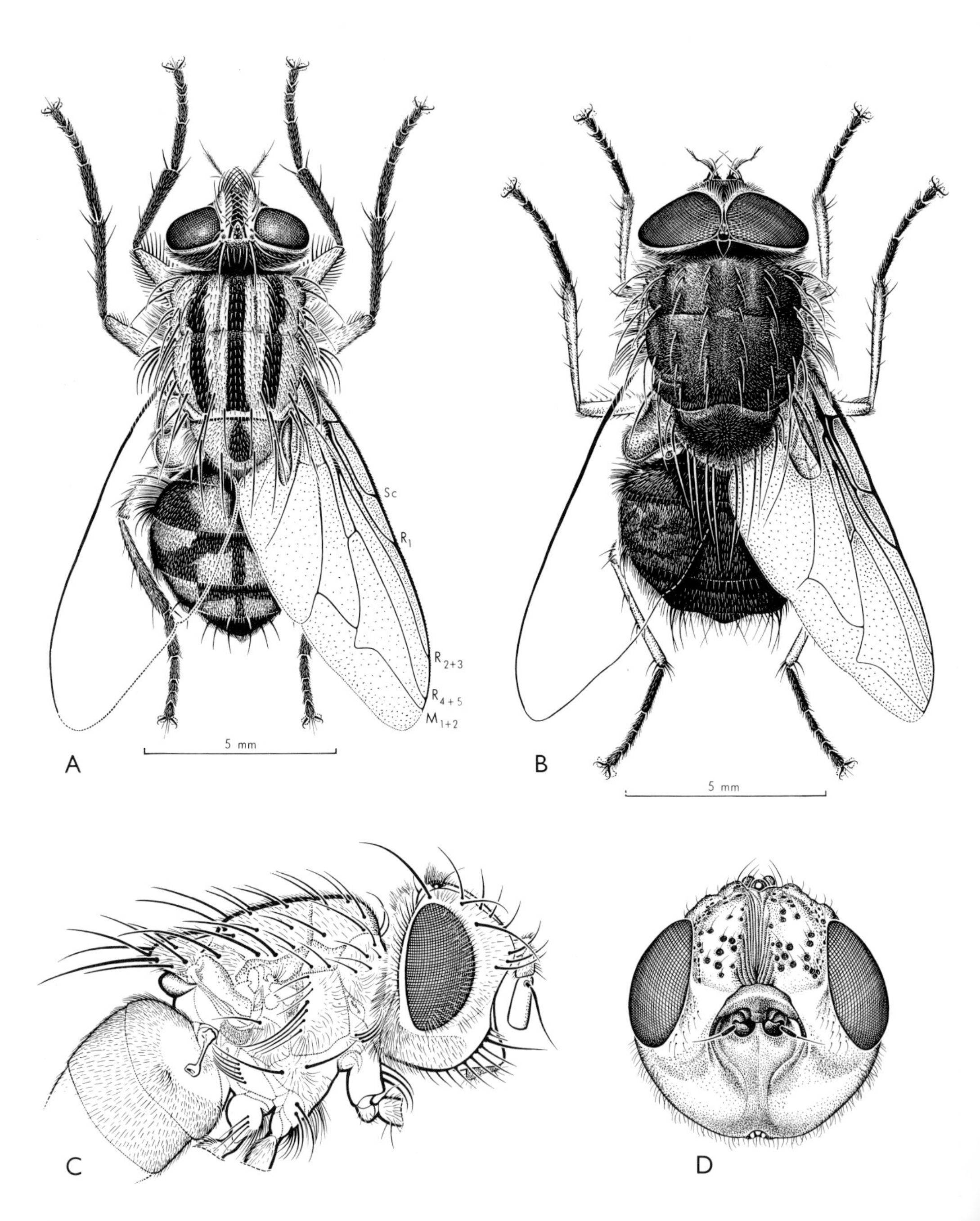

Fig. 34.36. A, *Tricholioproctia hardyi*, Sarcophagidae, ♀; B, *Calliphora stygia*, Calliphoridae, ♂; C, *Chaetophthalmus* sp., Tachinidae, ♀; D, *Oestrus ovis*, Oestridae. [T. Binder]

rather than truly parasitic. The dominant genera in Australia are *Tricholioproctia* and *Parasarcophaga*, which together make up more than half our fauna in this subfamily. A number of species are extralimital, e.g. the cosmopolitan *P. crassipalpis* (Macq.), and *Tricharaea brevicornis*, a presumably recent immigrant from South America, now widespread in south-eastern Australia. [Souza Lopes, 1959; Malloch, 1930.]

82. Tachinidae (Figs. 34.5C, D, 36C). An immense and taxonomically difficult family, subdivided by various authorities into from three to over sixty smaller families—current practice is to recognize only one, defined mainly by the venation and the usually strongly developed subscutellum. Most species are stout-bodied and strongly bristled, ranging in size from small species about half as big as a house fly, up to the large, showy species of *Rutilia* with wing-spans of over 35 mm. Some, e.g. *Chaetophthalmus*, superficially resemble calliphorids or sarcophagids, whereas a few genera (e.g. *Cylindromyia*) are more elongate and include wasp mimics.

The adults are ubiquitous, while their larvae are all endoparasitic in other arthropods, principally insects; a possible exception is *Chrysopasta elegans* (Macq.), reported to live as an inquiline in termites' nests. Reproduction is oviparous or ovoviviparous, but life histories vary considerably in details. Some species lay their eggs or larvae directly on, or rarely in, the host, or deposit them in places frequented by the host, which is then sought out and penetrated by the young larva. Others lay numerous, tiny, 'microtype' eggs on the host's food plant; if ingested, they hatch and the larvae penetrate via the host's gut wall. During some part of their growth, the larvae are usually connected with the external atmosphere through a perforation in the body wall or a trachea, and the apex of the abdomen becomes enclosed there in a sclerotized 'funnel' formed by reaction of the host tissue. As a rule, the host eventually succumbs, and the parasite leaves to pupate in the soil; though species are known which pupate within the host, or, by feeding on non-vital tissues and retaining metabolic wastes within their bodies, leave the host while it is still alive.

The family includes parasites of many orders, but principally Lepidoptera, Coleoptera, Hemiptera, and Orthoptera, including many pest species. No doubt, tachinids play an important role as natural regulators of insect numbers, though little seems to be known of their real importance in this respect. They have been employed as biological control agents in some countries, but only to a very limited extent in Australia.

Although over 400 species have been recorded from Australia, our fauna is still very imperfectly known. Of some 150 genera, more than half are known from single species only—perhaps partly because of over-splitting of genera, but also through unrecognized synonymy and general lack of study. The family is probably still actively radiating, and its members present an extraordinary range of taxonomic characters. This creates formidable problems in the delimitation of groups at all levels. Some six, not very well defined, subfamilies are currently accepted by students of the family. The PHASIINAE, parasitic in Coleoptera and Hemiptera, often lack strong bristles on the abdomen, and include *Hyalomyia* with peculiar, short, broad wings, the long-bodied species of *Cylindromyia*, and the characteristically Australian *Palpostoma*, a parasite of melolonthine scarabs. The PROSENINAE (='Dexiinae') usually have a strong facial carina and/or plumose arista, and are mainly parasites of Coleoptera. They include the Oriental–Australasian tribe Rutiliini, with many very large, often brilliantly coloured species (Plate 5, O, P). The genus *Rutilia* is very well developed in Australia, with about 115 species, probably most of them parasites of scarabaeid beetle larvae. Other Proseninae include the long-legged species of *Rhynchodexia*, etc., common on tree trunks in sunny weather, The remaining subfamilies include many diverse forms, parasitic mainly in Lepidoptera and Coleoptera. [Malloch, 1928, 1929, 1930, 1931, 1932, 1934, 1936; Paramonov, 1950a, 1951c, 1953b, 1954a, 1955a, 1956b, 1958a, 1959a-b.]

83. Oestridae (Fig. 34.36D). A small family of bot flies, related to the Tachinidae, but with vestigial mouth-parts and, in Australian genera, M_1 meeting R_{4+5} well short of the wing margin. The larvae are parasitic in the nasal and respiratory passages of mammals. The introduced sheep bot fly, *Oestrus ovis* L., is widespread in Australia, but apparently causes no serious disease. The female deposits young larvae around the sheep's nostrils; they develop in the nasal sinuses, and are sneezed out when mature, to pupate in the soil. Man is sometimes infested, particularly in the eyes, in sheep-raising countries, but such cases are extremely rare in Australia. There is only one known native oestrid, *Tracheomyia macropi* (Frogg.), whose larvae infest the trachea of the red kangaroo *(Megaleia rufa)*; the same, or a related, species occurs also in the euro *(Osphranter robustus)*. Little is known of its life history, but infestations can be common, with rates of up to 30 per cent in some populations of *M. rufa* (Mykytowycz, 1963).

84. Hippoboscidae (louse flies, wallaby flies, keds; Fig. 34.37A). A family of greatly modified, blood-sucking flies, perhaps related to the blood-sucking muscids. The adults live as ectoparasites amongst the hair or feathers of mammals and birds, and, with their flattened bodies, porrect mouth-parts, and robust legs, have a characteristic louse- or tick-like appearance. Although only occasionally used in flight, the wings are well developed in most species, with the strong veins concentrated anteriorly. In several exotic genera, they are shed when the fly becomes established on its host, whereas other species have them reduced or, in *Melophagus*, vestigial. Reproduction is by adenotrophic viviparity, and the mature larvae are usually deposited away from the host; though in the sheep ked (*Melophagus ovinus* (L.); Fig. 5.22A) they are laid and pupate in the hosts' wool.

The large subfamily ORNITHOMYIINAE and the smaller ORNITHOICINAE, both represented in Australia, are restricted to avian hosts; the other four subfamilies, which include only some 30 per cent of hippoboscid species, are almost exclusively parasites of mammals other than bats. The ORTHOLFERSIINAE, with some 4 species of *Ortholfersia* and one of *Austrolfersia*, occur only in Australia, as parasites of wallabies and, doubtfully, of kangaroos. Our species of HIPPOBOSCINAE and MELOPHAGINAE are all introduced. *Hippobosca equina* L., occasionally imported on horses and cattle, does not appear to be established here, but *Melophagus ovinus*, often called the 'sheep tick' is common in cooler parts of the continent, where heavy infestations sometimes cause losses through anaemia in the sheep and staining of the wool. [Maa, 1963b.]

85. Streblidae (Fig. 34.37C). A small family of blood-sucking ectoparasites of bats, widely distributed in the tropics and subtropics. Their origins are obscure, and the New World subfamilies may have been independently evolved. Streblids show many features associated with the ectoparasitic habit; piercing mouth-parts, enlarged legs and claws, reduced sensory apparatus, specialized reproduction, etc.: but they have tended to retain functional wings, presumably used in finding a host. There are usually only some 2–8 individuals per bat, and most species show a marked association with one family, and occasionally one species, of bat. Of the Old World subfamilies, the NYCTERIBOSCINAE are free-living in the host's fur, but in the ASCODIPTERINAE the females embed themselves in the subcutaneous tissue, and there degenerate to a sac-like organism enclosed in a cyst. Both reproduce by adenotrophic viviparity, the larvae developing one at a time and pupating on the floor of the roosting site. Both subfamilies occur in the warmer parts of Australia. The two most common species are extralimital, the large *Brachytarsina amboinensis* (Rond.) being widely distributed in the Oriental region, and the small *B. minuta* Jobling extending to New Guinea and the Solomon Is. [Paramonov, 1951b; Jobling, 1951; Maa, 1965.]

86. Nycteribiidae (Fig. 34.37B). A small but widespread family of blood-sucking ectoparasites, somewhat spider-like in appearance, and of obscure origins. As in the Streblidae, the adults inhabit the fur of bats, but their structure is more specialized. Legs

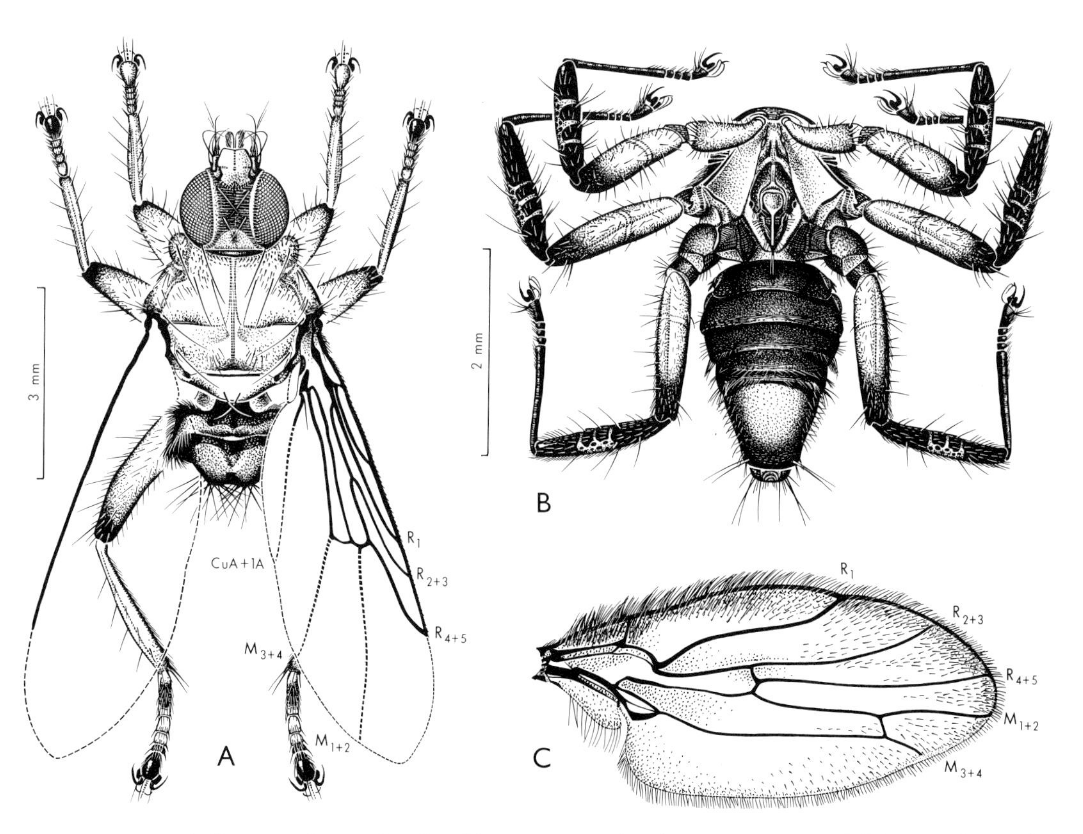

Fig. 34.37. A, *Ortholfersia macleayi*, Hippoboscidae, ♂; B, *Cyclopodia australis*, Nycteribiidae, ♀; C, *Brachytarsina uniformis*, Streblidae. [T. Binder]

and claws are very strongly developed, eyes reduced or absent, etc., but all vestiges of wings have been lost, and the body is markedly flattened—particularly the thorax, which has an antero-dorsal groove in which the extraordinary, backwardly-flexed head lies when at rest. They resemble the Streblidae also in mode of reproduction, except that the larvae are actively glued to the wall of the bat's roost by the larvipositing female. The puparium is oval and glossy black, and emergence of the adult is apparently postponed until triggered by warmth or physical contact of a roosting bat in its immediate vicinity. Many species appear to be host specific, but, as also in the Streblidae, this is probably at least partly due to the isolated roosting habits of some bats. However, the restriction of the very large species of *Cyclopodia* to the fruit bats may be an ancient adaptation. The Australian species belong in some 6 genera, and are found wherever suitable hosts occur. Amongst the better known species are *Basilia falcozi* (Musgrave), widely distributed within Australia on cave-dwelling vespertilionid bats, and *Cyclopodia albertisii* Rond., found on flying-foxes (*Pteropus* spp.). The latter is also widely distributed on islands to the north of Australia. [Theodor, 1956, 1959; Maa, 1965.]

ACKNOWLEDGMENTS. We are indebted to many colleagues for assistance with the sections on individual families; to Dr I. M. Mackerras (Tabanoidea); to Dr S. J. Paramonov,* whose unpublished manuscripts supplied much of the

* *Obit.* 22 November 1967.

information concerning Therevidae, Bombyliidae, and Calliphoridae, and who also assisted with details of other families; and to Drs E. N. Marks and N. V. Dobrotworsky (Culicidae), Mr D. J. Lee (Ceratopogonidae), Mr P. Aitken (Streblidae and Nycteribiidae), and Dr D. A. Duckhouse (Psychodidae), who, in each case, supplied much of the information given under the respective families. We have also been greatly assisted by other colleagues, too numerous to name, who read the manuscript and provided many suggestions that were incorporated in the text.

35

TRICHOPTERA

(Caddis-flies, caddises)

by E. F. RIEK

Endopterygote Neoptera with reduced mouth-parts; fore wing with anal veins looped; CuA forked in both wings; body and wings covered with hairs and at times with some scales. Larvae aquatic, with a pair of terminal abdominal prolegs; often living in portable cases. Pupae decticous.

Trichoptera resemble small, rather hairy moths, from which they differ in details of wing venation and in the structure of the mouth-parts. The larvae differ in having only one pair of abdominal prolegs and in their general aquatic habits. This is a relatively small order of stream-frequenting insects, with an estimated 5,000 species. Adult caddises are generally of small to moderate size (1·5–40 mm in length), drab-coloured, and cryptic. The length of the antennae and the shape of the wings are apparently correlated with the strength of the wing-coupling and the powers of flight. The strongest fliers have strongly coupled wings, a narrow fore wing and expanded hind wing, and the antennae are usually much longer than the fore wing. The principal recent papers on the Australasian fauna are those by Mosely and Kimmins (1953), Neboiss (1958, 1959), and Kimmins (1962).

Anatomy of Adult

Head (Fig. 35.1). Compound eyes well developed, sometimes almost meeting at the vertex (male Odontoceridae); 3 ocelli present or absent, only 2 in some Hydroptilidae. Antennae filiform, setaceous, many-segmented, about as long as, to much longer than, fore wing; scape sometimes greatly enlarged, pedicel somewhat enlarged. Plumed antennae occur in both sexes of one Australian species of Philorheithridae, and the middle segments are thickened in the males of some Hydroptilidae. Mouth-parts weak and specialized for ingestion of liquid foods; mandibles vestigial; galea small or absent. Maxillary palp typically 5-segmented, the terminal segment sometimes divided, sometimes long, flexible, and multiarticulated; segmentation reduced in the males of some families. Development of the apical segments is partly correlated with the presence or absence of a larval case. Labial palp 3-segmented, the apical segment sometimes long, flexible, or multiarticulated. Eversible scent glands, 'pilifers', and other processes

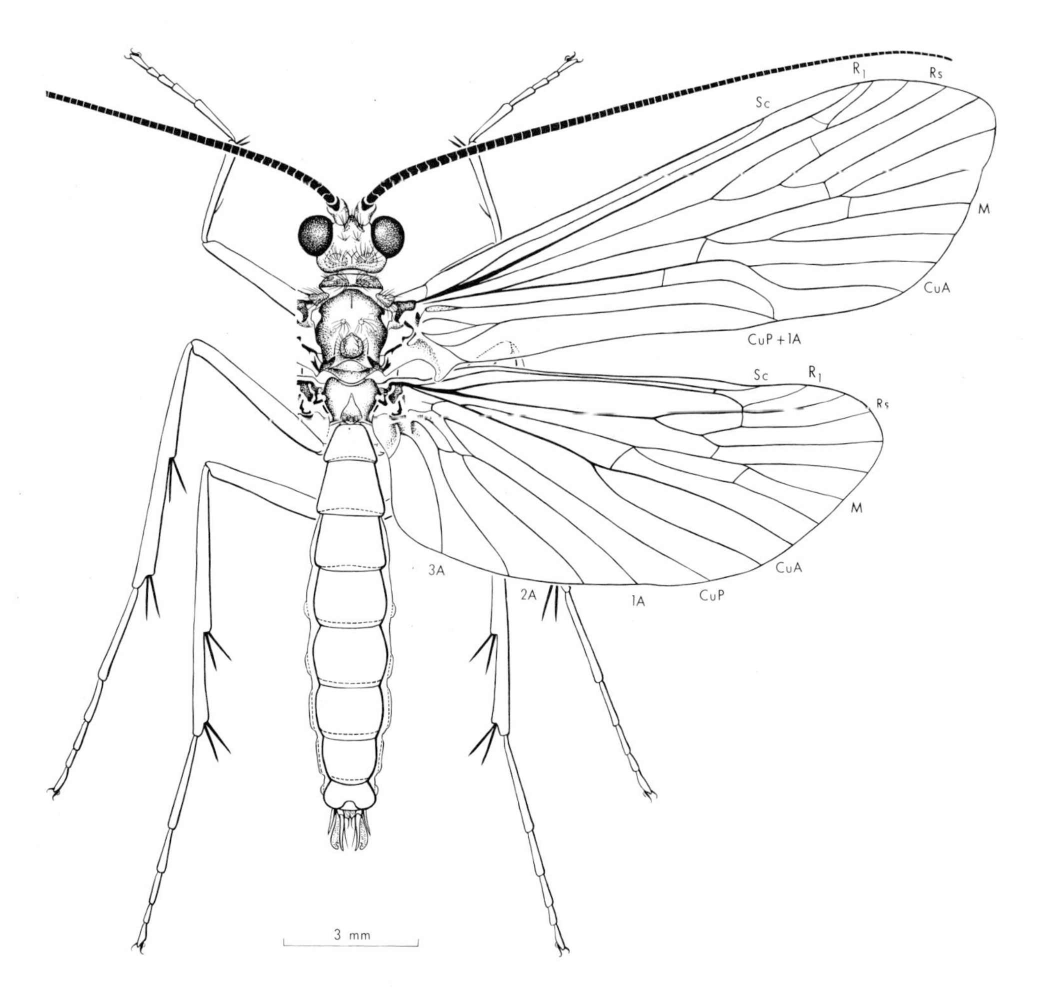

Fig. 35.1. *Stenopsychodes* sp., Psychomyiidae, ♂.
[T. Binder]

are sometimes present on vertex of head or between bases of antennae. Raised setiferous warts often occur above antennae and at hind margin of head, and there are occasionally large scales on the antennal scape or maxillary palp. In Plectrotarsidae and some Psychomyiidae *(Stenopsychodes)* the mouthparts are produced to form a rostrum.

Thorax. All three segments distinct. Prothorax short, meso- and metathorax well developed. Mesonotum with scutum and scutellum defined; scutum with or without a pair of anterior warts, or with a pair of longitudinal lines of setiferous punctures; scutellum with a large median wart or a pair of warts (Fig. 35.3).

Legs. Strongly built, cursorial, usually long, especially the hind leg. Coxae long. Fore femur sometimes with dense spines on anterior margin; mid femur occasionally with a row of stout bristles on outer surface. Tibiae with or without long spines, with varying numbers of apical and preapical spurs, usually in pairs, occasionally single; sometimes with long hair-fringes (male *Anisocentropus*). Tarsi 5-segmented, with ventral spines on each segment, those at apex sometimes enlarged and aligned; claws double, without teeth; hind tarsus elongate in Leptoceridae.

Wings (Figs. 35.1, 4). Two pairs of sub-equal wings, reduced in a few species, especially in females; held roofwise over the body when at rest, though nearly flat in Hydroptilidae and the Tasmanian Goeridae. Fore wing slightly coriaceous, hind wing flimsy, shorter but usually broader than fore wing. Venation simple. Basically, CuA is always forked toward apex, but forking sometimes obscured by fusion between portions of veins; end-twigging absent; pterostigma not clearly developed, except in Rhyacophilidae; cross-veins few or almost entirely absent; basal stems of Rs and M sometimes reduced. In the fore wing each anal vein is looped on to the one in front, so that only one anal vein reaches the wing margin, but the anal veins are occasionally absent (Goeridae). The wings are clothed with hairs, sometimes thickened at the tips, and occasionally with areas of scales along some of the veins. Nygma ('corneous spot') usually present at the base of cell R_4 ('apical fork 2') and in cell M ('thyridial') in both wings (cf. many Neuroptera). Wings of males sometimes with longitudinal folds in various positions, and the venation then considerably modified; those of females always without folds.

WING-COUPLING (Fig. 35.4). There are two basic methods of coupling the wings in the order, and each has been modified in various ways. The simplest is of rather general occurrence in the non-case-making families, but occurs also in Limnephilidae and Sericostomatidae. There is overlapping of the wing margins, with or without some strengthening of a marginal fringe of macrotrichia on the anterior border of the hind wing. The jugal lobe of the fore wing is developed, and there may be a corresponding enlargement of a few basal macrotrichia on the hind wing. In the Helicopsychidae, Philanisidae, Helicophidae, and Hydroptilidae there is an increase in the strength and number of the curved macrotrichia on the anterior margin of the hind wing, with a corresponding reduction in the size of the jugal lobe, and in some Hydroptilidae there is a brush of stout macrotrichia ventrally on CuA of the fore wing which strengthens the coupling. In Plectrotarsidae hooked macrotrichia occur over the basal half of the anterior margin of the hind wing; but the basal two macrotrichia are straight, and greatly enlarged to form a frenulum interlocking with the well-developed jugal lobe.

The Oestropsinae (Hydropsychidae) have a totally different method of coupling the wings, and this has been modified in various ways in the Odontoceridae, Philorheithridae, Leptoceridae, and Calamoceratidae. There is a ventral longitudinal ridge behind the anal veins of the fore wing. Enlarged macrotrichia on the anterior margin of the hind wing curve over this ridge from below upwards, and interlock the wings. There is also a dorsal series of curved macrotrichia on portion of R_1 of the hind wing interlocking with a marginal fringe of the fore wing. The jugal lobe is not developed, and the marginal zone of the anal field of the fore wing is wide and covered with setae. In Odontoceridae and Philorheithridae enlarged curved macrotrichia are developed on the longitudinal ridge of the fore wing, and interlock with the short curved macrotrichia of the marginal fringe of the hind wing. The marginal zone of the anal field of the fore wing is narrower, and there is a seta-free area which is variously thickened and often produced to form a knob. These knobs, which are covered with file-like microtrichia and overlap when the wings are folded, are most probably modifications for stridulation.

In Leptoceridae, the modified macrotrichia on the hind wing are limited to a middle section of the anterior margin, and are developed into hamuli. They interlock with the remnant of the ventral ridge of the fore wing, where it meets the wing margin at the apex of the anal field. This portion is reflexed, so that it forms a very secure coupling with the strongly hooked hamuli. The Calamoceratidae have a coupling of a modified leptocerid type, with the enlarged curved macrotrichia of the hind wing situated on the upper surface of R_1 towards its distal extremity where it approaches close to the wing margin. They are not hooked enough to be termed hamuli, and they interlock with portion of the anal margin of the fore wing.

Abdomen. Segmentation distinct; male with 9 clearly defined segments, 10 incorporated in the terminalia; female usually with 10 defined segments. Sternum 1 not differentiated. Spiracles on segments 1–7. Backwardly directed ventral processes may occur on one or more of segments 6–8, and there may be a pair of lateral filaments arising from the pleural region of segment 5. Female with genital opening between sterna 8 and 9, and often a pair of small cerci on segment 10. The terminal segments are sometimes modified to form a long ovipositor (glossosomatine Rhyacophilidae and Philanisidae).

Internal Anatomy. Generalized, apparently with few distinctive features; 6 Malpighian tubes; numerous polytrophic ovarioles.

Immature Stages

Egg. Small, more or less spherical (about 0·3 mm in diameter in *Plectrotarsus*); usually pale in colour, but sometimes blue or green.

Larva. 'Campodeioid' or 'eruciform', with many intermediate forms (Figs. 35.6–10). The head and thorax may be heavily sclerotized, but the abdomen is soft and often conspicuously coloured a soft green, yellow, cream, or white.

Head with very small antennae, two patches of black stemmata, and chewing mouth-parts. Clypeus fused to frontal region; labrum well developed, usually more or less retractile beneath the clypeus by means of an infolding of the membranous part of the anteclypeus. The ventral cleavage system may be a straight line extending from the occipital foramen to the submentum, two parallel lines, or a Y-shaped line. The postmentum consists of a distinct mentum and submentum, the mental part always bearing 2 innervated setae, although secondary setae also occur in Hydropsychidae. Maxillary lobe undivided; stipes subdivided; labial palp 1- or 2-segmented, absent in some Polycentropinae (Psychomyiidae).

The three thoracic segments distinct. Fore legs generally the shortest and stoutest, and used more for holding prey and case-making than for walking. Legs with the normal 5 segments, but some are subdivided; trochanter possibly in two sections in all species; mid and hind femora may be divided, and occasionally the tibiae and tarsi. The mid tibia and tarsus are occasionally fused, less commonly also in the fore leg. Abdomen with 9 discrete segments. Ventral prolegs absent, but the abdomen ends in a pair of terminal prolegs which bear hooks for anchoring the larva in its case or to the substrate. Functional spiracles absent, but many case-making larvae and a few non-case-makers have non-retractile tracheal gills on most abdominal segments and a few species also on the thorax. They are usually simple, but may be arranged in tufts, or branched. Small, finger-like, eversible anal gills also occur. In larvae that have no tracheal gills respiratory exchange takes place through the general body surface. The lateral abdominal hair fringe present in some species is used to produce a current of water through the case.

CONSTRUCTION OF THE CASE. The larvae of many families construct characteristic cases (Fig. 35.2), which they carry around with them; those of a few families live in silken retreats, or are free-living. Non-case-making larvae, in 4 families, appear to be the most primitive. In most respects the family Hydropsychidae contains the most generalized type of larva (Figs. 35.6D–I). All thoracic terga are sclerotized; the prosternum has a sclerotized bar; the legs are unmodified; abdominal tergum 9 is unmodified; and the prolegs on the last abdominal segment have simple claws. These larvae have developed ventral abdominal gills. In the other 3 families (Figs. 35.6A–C, 7) only the pronotum is sclerotized, and ventral abdominal gills are absent. The prosternum is sclerotized in Rhyacophilidae, but not in Philopotamidae and Psychomyiidae.

Hydropsychidae construct a net at the

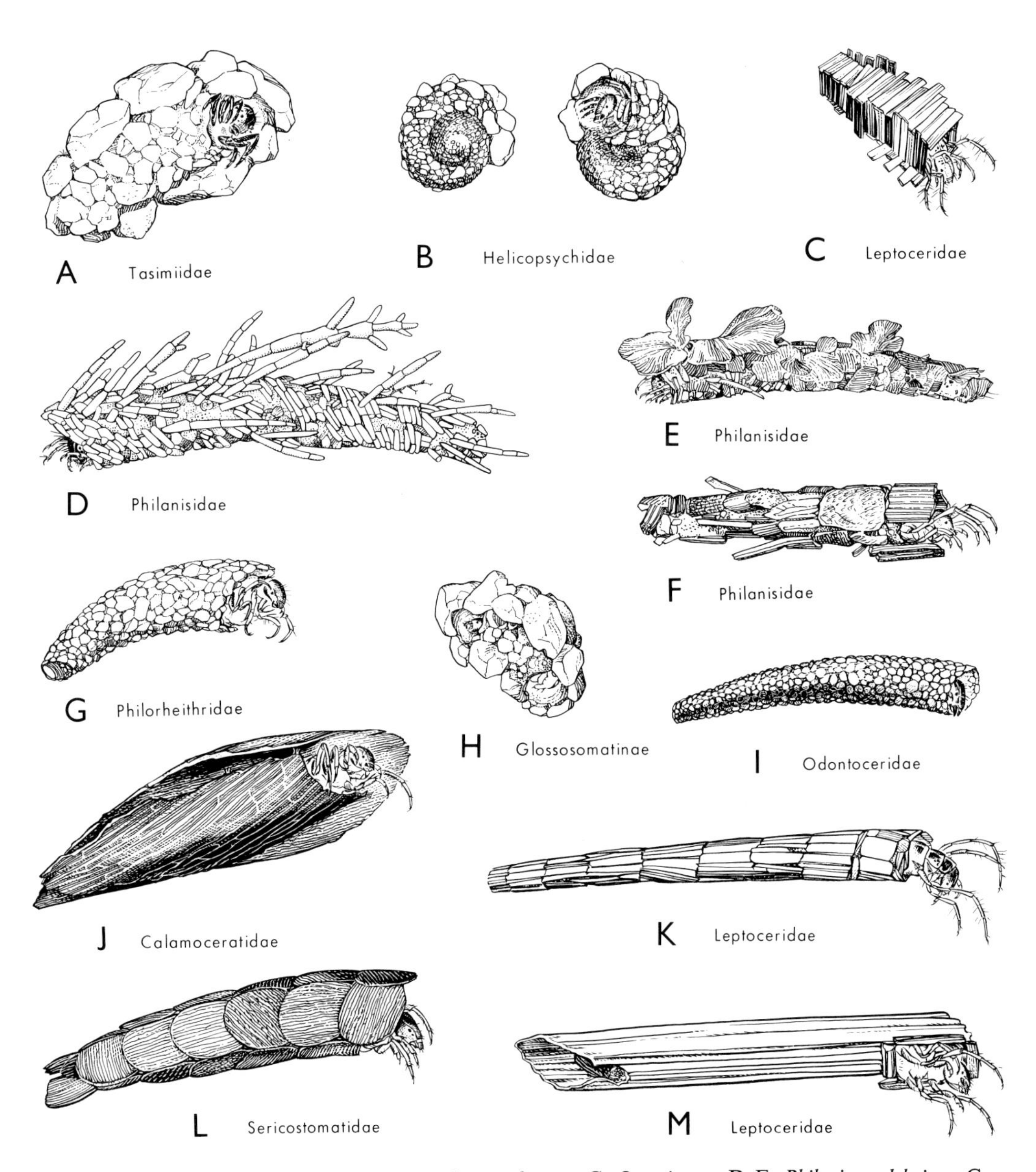

Fig. 35.2. Larval cases: A, *Tasimia* sp.; B, *Helicopsyche* sp.; C, *Oecetis* sp.; D–F, *Philanisus plebeius*; G, a philorheithrid; H, *Synagapetus* sp.; I, *Marilia* sp.; J, *Anisocentropus* sp.; K, a leptocerid; L, *Caenota* sp.; M, a leptocerid. [M. Quick]

entrance to their retreat. Most Rhyacophilidae are free-living, but a transition to case-making is seen in the Glossosomatinae which construct simple saddle-cases. Philopotamidae and Psychomyiidae live in silken, finger-like tubes. The Hydroptilidae are free-living in the early instars, but older larvae construct cases that are usually flattened, purse-like structures open at both ends.

A number of modifications of the free-living type of larva are correlated with the case-making habit. There is partial fusion of the prolegs to give the appearance of a tenth abdominal segment; a lateral comb of fine spicules is usually developed on segment 8; there are swollen areas or tubercles, usually 3, lateral and mid-dorsal, on segment 1; the hind leg is lengthened; and dorsal, as well as ventral, abdominal gills are commonly developed.

Case-making families have apparently arisen more than once from different types of non-case-making ancestors. There is a close similarity between the larvae of the Glossosomatinae, Limnephilidae, and Tasimiidae. The saddle-type cases of Glossosomatinae cover only the middle portion of the body, the opening for the apex of the abdomen being transverse and towards the ventral side. In Tasimiidae the case is similar, but the opening is terminal and partly closed. Limnephilidae have a tubular type of straight or slightly curved case. In Glossosomatinae the abdominal prolegs are free, but they are fused in Limnephilidae and Tasimiidae. In the last two families the larva maintains a grip on the case by tubercles on abdominal segment 1, but there is no lateral comb on 8.

The other types of larvae that live in cases appear to have arisen from different ancestors. In all of them there are both tubercles on abdominal segment 1 and a lateral comb of fine spicules on 8. The larvae of the Odontoceridae, which make tapering tubular cases of sand grains, are very rhyacophilid-like. In contrast, the larvae of the remaining families are more like those of the Hydropsychidae. Thus, the larvae of Sericostomatidae differ most from Hydropsychidae in the shape of the clypeus. Their cases are straight or nearly so, usually tubular and composed of sand grains, but occasionally flattened and constructed of plant tissue. The Helicopsychidae are distinguished chiefly in characters that are apparently associated with the helical shape of their cases of sand grains. The larvae of Leptoceridae and Calamoceratidae have a distinct lateral line and fringe on the abdomen and the hind legs greatly lengthened. They are closely related, but their Australian representatives differ considerably from each other in association with the type of case they construct. Those of Calamoceratidae *(Anisocentropus)* build a flat case of two pieces of leaf, whereas leptocerid larvae live in very long, thin, straight cases, circular or square in section, and usually made of spirally incorporated fragments of plant tissues, although a few use sand grains or merely a hollow twig. The larvae of these two families are possibly the most highly evolved within the order.

Pupa. Decticous and exarate, either free in the larval case, or within a gelatinous cocoon in a specially made case of sand grains. The characteristically shaped, powerful mandibles are used to cut a way out of the cocoon and case.

Biology

Adults. Caddis-flies are found mainly near fresh water, either running streams or the standing water of lakes and ponds, more especially in the colder regions. Several species, especially Leptoceridae, occur in sedge-grown ponds and swamps. There is one marine species, *Philanisus plebeius* (p. 763), which breeds in rock pools close to low-tide mark on the open coasts of southern Australia and New Zealand. *Symphitoneuria wheeleri* (Leptoceridae) in south-western Australia breeds in lakes in which the salinity is as much as three times that of sea-water.

Adults are mainly crepuscular or nocturnal in habits, although some species, especially *Asmicridea* spp. and many Leptoceridae, fly in swarms over the water or at its margins during the day. Adults of some species move away from their breeding areas during the heat of summer, when the water dries up. They shelter in dense vegetation, where the humidity is relatively high, and have been collected at distances of 2 km from permanent water. Although the mouth-parts are reduced, it would appear that the adult caddis-fly normally has the ability to take up water or nectar, and so may live for a number of weeks under suitable conditions. Adults of many species may be swept or beaten from the foliage margining the stream or pond during the day. Their movements, whether in flight or running, are rapid and jerky. During the day many seek cool, dark, damp situations, such as the undersurfaces of fallen timber or overhanging rocks, or under bridges crossing the stream. Most adult caddises occur during the warmer months, but some specimens can be collected at all times of the year. The small numbers of adults occurring in winter are mostly Rhyacophilidae, but there are also a few Hydropsychidae *(Cheumatopsyche)*.

Reproduction. Most caddis-flies mate at dusk and in the early evening, but others have mating flights during the day. Mating pairs of *Asmicridea* spp. and Leptoceridae can be seen during the day on vegetation overhanging or margining the streams. Copulation apparently takes place on the wing, but the pairs then settle and rest facing in opposite directions.

The eggs are deposited in or near the water, either as strings or in a mass. The females of many families (non-case-making families especially) enter the water to attach their eggs to submerged objects. Others (Limnephilidae and probably Plectrotarsidae) may deposit their egg masses on various objects above the water. The female of *Plectrotarsus* in captivity covers the egg mass with hairs detached from the undersurface of the wings and the marginal fringe of the fore wings. The hairs remain adhering to the sticky egg mass when the female moves away. *Philanisus* has a long ovipositor that is probably used for inserting the eggs into crevices at the bases of tufts of calcareous algae. The number of eggs produced varies considerably. There are about 150 in the egg mass of *Plectrotarsus*, which measures about 8 × 6 mm by 5 mm high. Caddis-fly eggs usually hatch quickly, though up to 2 months are required for those of *Plectrotarsus*. There may be several generations in a year, but more usually only one.

Immature Stages. There are 6 or 7 instars. Most case-making larvae are unable to swim, but some Leptoceridae and Phryganeidae can swim well, especially those species with cases made from fine plant material. Their hind legs are modified for swimming by the development of brushes of long setae. Larvae are generally collected by turning over stones, logs, and other objects in the water, or from the foliage of water plants. A few species live exposed on sand or gravel, or lie partly buried. In streams they occur most abundantly in the swift water flowing over rocky or tree-strewn bottoms. In general, individual species are restricted to particular habitats, and show a preference for certain types of water and kinds of bottom.

Pupation takes place in the larval case, which is attached at the head end to some support, and partly closed at the posterior or free end. In the non-case-making larvae special pupal cells of sand grains are constructed. The Glossosomatinae remove the lower surface of the saddle-case, and usually attach the upper portion to a rock surface. Pupation in the non-case-makers and the Glossosomatinae is within a semitransparent, smooth, silken cocoon, but in the case-making species the pupa lies free within the old larval case. This is the only order in which it is normal for pupation to take place in water, though aquatic pupae do occur in other orders. The 'pupae' (pharate adults) cut their way out of the pupal cell or case, and swim to the surface by means of the usually well-developed swimming brushes on the mid tarsus of the pupal leg. Eclosion takes place on the surface film, and the adult is able to fly immediately.

Pupae and their pupal cases or cocoons should be preserved with their contents intact, because 'pupae' can be identified if the male genitalia of the pharate adult are well developed. It is then possible to associate captured adults with larval cases, and sometimes with larvae based on an examination of the cast last larval skin.

Natural Enemies. The larvae are preyed on by fish and by other insect larvae, such as dytiscids. There are non-Australian records of small Hymenoptera parasitizing the larvae and pupae. Water mites attack larvae in their cases. As most adults fly only at night, they are attacked only by nocturnal predators. Bats are sometimes active round lights to which caddises are attracted. Birds attack some day-flying species *(Stenopsychodes)*, but others *(Asmicridea)* are generally ignored. Some adults are devoured by frogs as they return to the water for egg-laying.

Economic Significance. Caddis-flies are a most important group in many food chains of streams. Their larvae are mostly omnivorous, though Hydropsychidae, in particular, feed mainly on simple plant tissue. Many of the non-case-makers are carnivorous at times. Some Rhyacophilidae have well-developed raptorial fore legs which enable them to grasp larger prey, but in the other families the prey consists of the minute organisms caught on their nets or silken tubes. Caddis larvae and pupae, in turn, form a staple item in the diet of many freshwater fish, especially trout, which devour the 'pupae' as they rise through the water or transform on the surface. Some leptocerid larvae are pests in paddy fields, where they eat the young shoots of rice plants.

Special Features of the Australian Fauna

The Australian fauna contains representatives of all but one of the major, more universally recognized families. The mainly Holarctic Phryganeidae are absent, and the Limnephilidae, which have a similar general distribution, are represented by few species. Some families, including the 4 non-case-making families and the Hydroptilidae, Leptoceridae, Calamoceratidae, and Sericostomatidae *(sensu lato)*, are almost worldwide, whereas the Odontoceridae and Beraeidae have a more discontinuous distribution. The dominant families in Australia are the Leptoceridae, Hydropsychidae, and Rhyacophilidae, but in Tasmania there are many species of Sericostomatidae.

There is a distinctive element in the fauna. The marine Philanisidae and the cold-adapted Plectrotarsidae, Tasimiidae, Philorheithridae, and Helicophidae are either restricted to Australia or occur elsewhere only in New Zealand. There seems no reason to doubt that these families evolved within Australia, and they possibly represent developments from ancestral forms that are known to have occurred in the Australian Upper Permian and Triassic.

Some strong fliers of the less cold-adapted species (e.g. in the Leptoceridae and Calamoceratidae) are possibly recent introductions. The greater part of the caddis-fly fauna of the northern and the drier parts of the continent shows close relationships with the Asiatic fauna. This is especially so with *Anisocentropus*, the only genus of Calamoceratidae in Australia, and with *Atriplectides* of the Leptoceridae, one species of which is considered to range from Asia to New Zealand.

CLASSIFICATION

Order TRICHOPTERA (260 Australian spp.)

1. Hydropsychidae (27)
2. Rhyacophilidae (35)
3. Philopotamidae (14)
4. Psychomyiidae (33)
   Stenopsychidae (0)
5. Hydroptilidae (7)
6. Limnephilidae (2)
   Phryganeidae (0)
7. Plectrotarsidae (5)
8. Goeridae (1)
   Lepidostomatidae (0)
   Brachycentridae (0)
   Kitagamiidae (0)
9. Tasimiidae (4)
10. Calamoceratidae (5)
11. Leptoceridae (65)
    Molannidae (0)
12. Odontoceridae (7)
13. Philorheithridae (18)
14. Helicopsychidae (5)
15. Sericostomatidae (26)
16. Beraeidae (3)
17. Philanisidae (1)
18. Helicophidae (2)

The order is usually divided into two suborders (Aequipalpia and Inaequipalpia) on the structure of the male maxillary palp, but this is considered to be an unnatural division, and has not been adopted here. The family classification has been modified from H. Ross (1944, 1956), and includes also the endemic families Plectrotarsidae, Tasimiidae, Philorheithridae, and Helicophidae. Mosely and Kimmins (1953) separated Polycentropodidae from Psychomyiidae, and combined Helicopsychidae with Sericostomatidae.

The adults of some Australian groups are relatively easy to recognize. Very long antennae occur only in Leptoceridae, Calamoceratidae, Odontoceridae, and Oestropsinae (Hydropsychidae). The Plectrotarsidae are moth-like species with thick antennae and usually some metallic colouring. Species of *Asmicridea* and *Smicridea* (Hydropsychidae) are like small white moths. *Stenopsychodes* spp. (Psychomyiidae) have black and orange patterns on the wings, while most species of *Anisocentropus* (Calamoceratidae) have an orange basal half to the distinctly triangular fore wing. Ocelli are present in Rhyacophilidae, Plectrotarsidae, Limnephilidae, and Philopotamidae. In most Glossosomatinae (Rhyacophilidae) the mid tibia and tarsus of the female are dilated, and the abdomen drawn out into a long ovipositor. *Philanisus* is the only marine caddis-fly.

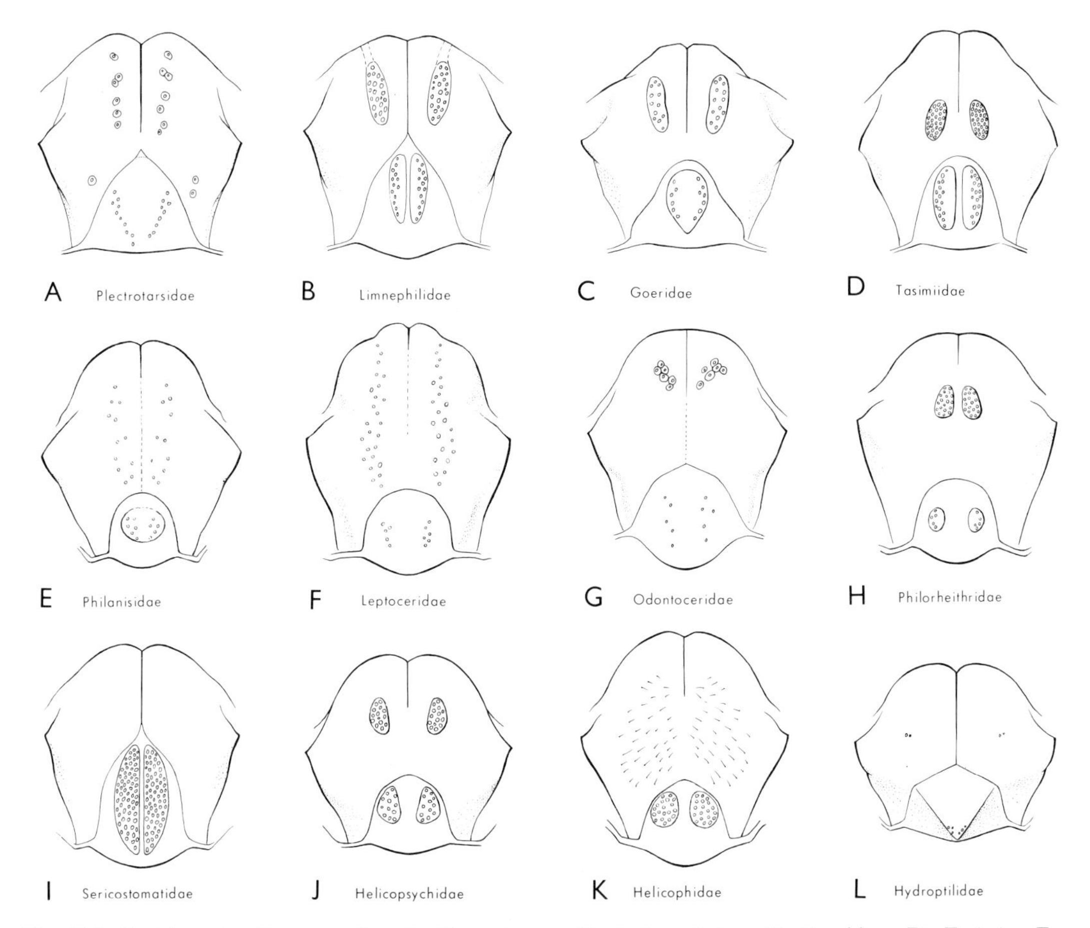

Fig. 35.3. Scutal warts, diagrammatic: A, *Plectrotarsus*; B, *Archaeophylax*; C, Goeridae; D, *Tasimia*; E, *Philanisus*; F, *Triplectides*; G, *Marilia*; H, *Aphilorheithrus*; I, *Caenota*; J, *Helicopsyche*; K, *Helicopha*; L, *Hydroptila*. [R. Ewins]

Similar distinction can be made in the larvae. Raptorial fore legs occur only in the Hydrobiosinae (Rhyacophilidae). Among other non-case-making larvae, Hydropsychidae have abundant abdominal gills, and the Philopotamidae have a distinctive labrum, much broadened at apex. Saddle-cases are found only in the Glossosomatinae (Rhyacophilidae), and purse-cases only in the Hydroptilidae. The cases characteristic of other families are shown in Figure 35.2.

Keys to the Families and Subfamilies of Trichoptera Known in Australia

ADULTS (Fig. 35.3)

1. Mesoscutellum with posterior portion forming a triangular flat area with steep sides; mesoscutum without warts; size small, not more than 6 mm long; wings long, narrow, fringes very long, those of hind wing usually longer than breadth of wing; antenna shorter than fore wing. [Maxillary palp of both sexes 5-segmented, terminal segment simple] .... **Hydroptilidae**
   Either mesoscutellum evenly convex without a triangular posterior portion set off by sharp sides, or mesoscutum with warts; generally larger, 5–40 mm; fringes not longer than breadth of wings; antenna generally as long as, or longer than, the fore wing 2

2(1). Terminal segment of maxillary palp multiarticulated, flexible, generally longer than all the other segments combined 3
   Terminal segment of maxillary palp not multiarticulated, mostly not flexible, only slightly or not at all longer than any other segment except when segmentation of palp is reduced 8

3(2). Mesoscutum with a pair of warts, sometimes with setiferous punctures as well. [Fore tibia with 2 or 3 spurs; R_1 simple or branched in fore wing; ocelli absent; jugal lobe well developed] .... **Psychomyiidae**. 4
   Mesoscutum with 2 lines of setiferous punctures, which are sometimes reduced, but never with a pair of warts. [Fore tibia with 2 spurs or less, never with a subapical spur; R_1 simple in fore wing] 6

4(3). R_1 forked at apex in fore wing .... ECNOMINAE
   R_1 simple in fore wing 5

5(4). Spurs 3, 4, 4. [Fork 1 present in fore wing, except in *Nyctiophylax*] ...... POLYCENTROPODINAE
   Spurs 2, 4, 4. [Fork 1 absent in fore wing] .... PSYCHOMYIINAE

6(3). Ocelli present .... **Philopotamidae**
   Ocelli absent .... **Hydropsychidae**. 7

7(6). Fore wing without a ventral longitudinal ridge behind the anal veins, usually with a well-developed jugal lobe (reduced in *Cheumatopsyche*); antenna shorter than fore wing .... HYDROPSYCHINAE
   Fore wing with a ventral longitudinal ridge just behind the anal veins, jugal lobe not developed; antenna much longer than fore wing .... MACRONEMATINAE

8(2). Ocelli present. [Fore wing with very distinct jugal lobe] 9
   Ocelli absent 12

9(8). First segment of maxillary palp short, subquadrate; hind tibia greatly lengthened, almost twice as long as hind femur. [Mid and hind tibiae with 4 spurs] .... **Rhyacophilidae.** 10
   First segment of maxillary palp distinctly longer than wide; hind tibia only slightly longer than hind femur. [Second segment of maxillary palp elongate, much longer than 1st; fore tibia with 1 spur; tibiae and tarsi with long stout spines] 11

10(9). Second segment of maxillary palp short; fore tibia with 1–3 spurs .... GLOSSOSOMATINAE
   Second segment of maxillary palp elongate; fore tibia with 2–3 spurs .... HYDROBIOSINAE

11(9). Scutum with a posterolateral wart on a level with the scutellum. [Hind wing with apex of R_1 strongly deflected towards R_2, often fusing with it] .... **Plectrotarsidae**
   Scutum without such wart .... **Limnephilidae**

12(8). Mesoscutum with a pair of irregular lines of large setate spots extending full length of scutum; wing-coupling by hamuli or curved macrotrichia developed over the middle portion of anterior margin of the hind wing, sometimes reaching almost to wing base but becoming less hooked. [Scutellum rounded anteriorly] 13

Without this combination of characters; mesoscutum with or without a pair of warts, sometimes with dense fine setae; or with simple wing-coupling if a pair of irregular lines of setate spots extend the full length of scutum 14

13(12). Median cell closed in fore wing; hind tarsus not much longer than mid tarsus; antenna only slightly longer than fore wing; curved macrotrichia on portion of upper surface of the radius of hind wing. [Maxillary palp 6-segmented in Australian species] .... **Calamoceratidae**

Median cell open in fore wing; hind tarsus much longer than mid tarsus; antenna much longer than fore wing, up to 3 times as long; hamuli or a row of hooked macrotrichia on portion of costal margin of hind wing .... **Leptoceridae**

14(12). Fore wing with a ventral longitudinal ridge at or close to the hind margin, bearing a row or rows of enlarged curved macrotrichia which interlock with the marginal fringe of the hind wing to form the wing-coupling. [Second segment of maxillary palp long; jugal lobe not developed; scutellum rounded anteriorly] 15

Wing-coupling by long hairs developed on the overlapping margins of the wings, or by hamuli on portion of the anterior margin of the hind wing interlocking with a marginal fringe on the fore wing 16

15(14). First segment of maxillary palp without an enlargement at apex; 'corneous area' on anal margin of fore wing small .... **Odontoceridae**

First segment of maxillary palp produced at apex medially to a rounded knob; but, if knob is indistinct, 'corneous area' on anal margin of fore wing strongly produced .... **Philorheithridae**

16(14). Scutum with 2 lines of setate punctures. [Mid tibia longer than femur] .... **Philanisidae**

Scutum without 2 lines of setate punctures, though sometimes with dense fine punctures 17

17(16). Scutellum with a large median wart, extending almost its full length and sometimes partly divided in mid-line. [Scutum with 2 elongate warts] .... **Goeridae**

Scutellum with 2 distinct warts 18

18(17). Scutum with 2 warts separated from the mid-line. [Jugal lobe reduced] 19

Scutum without warts, or with 2 warts almost touching 20

19(18). Scutellum rounded anteriorly; hamuli developed on anterior margin of hind wing; fore tibia with second spur reduced .... **Helicopsychidae**

Scutellum tapering anteriorly; wing-coupling by enlarged hairs, at most only slightly curved; fore tibia with 2 well-developed spurs .... **Tasimiidae**

20(18). Wing-coupling by hamuli over basal half of fore margin of hind wing, with the distal ones more closely grouped; jugal lobe reduced; scutum with dense, fine, setate punctures .... **Helicophidae**

Wing-coupling by enlarged hairs, at most only slightly curved; jugal lobe well developed; scutum without dense setate punctures 21

21(20). Mid femur and tibia with a lateral row of stout black bristles .... **Beraeidae**

At least mid femur without such bristles .... **Sericostomatidae**

LARVAE (families only; larvae of Plectrotarsidae and Helicophidae are not known.)

1. Terminal abdominal prolegs not fused to form an apparent segment 10 (Figs. 35.6, 7); tubercles absent on abdominal segment 1 2

Basal segments of terminal abdominal prolegs fused to form an apparent segment 10 (Figs. 35.9, 10); 2 or 3 tubercles present on abdominal segment 1. [Case-making larvae] 6

2(1). Abdomen swollen, much higher or wider than thorax (without gills); very small larvae, usually living in silken purse-like cases; (Figs. 35.7D,E). [All thoracic nota and sterna sclerotized] .... **Hydroptilidae**

Abdomen only slightly wider than thorax; larvae without portable cases, except saddle-cases of Glossosomatinae 3

3(2). Tracheal gills abundant, branched, on venter of abdomen; all terga of thorax fully sclerotized; (Figs. 35.6D–I). [Prosternum with distinct spur-like trochantin] .... **Hydropsychidae**

Tracheal gills absent on venter of abdomen, rarely with lateral gill tufts; only pronotum sclerotized 4

4(3). A sclerotized shield present on dorsum of abdominal segment 9; prosternum entirely sclerotized in Australian species; trochantin reduced; (Figs. 35.6A–C, 8D–F) .... **Rhyacophilidae**

Without a sclerotized shield on dorsum of abdominal segment 9; prosternum only partly sclerotized; trochantin spur-like, sometimes small 5

5(4). Labrum expanded distally, entirely membranous; prosternum mostly not sclerotized; with only a minute spur-like trochantin; (Figs. 35.7F–H) .... **Philopotamidae**

Labrum shorter, entirely sclerotized; prosternum with a sclerotized band; with a distinct spur-like trochantin; (Figs. 35.7A–C) .... **Psychomyiidae**

6(1). Abdomen without a lateral comb of fine spicules on segment 8, or with a few spicules above the lateral fringe line. [Trochantin very reduced] 7

Abdomen with a lateral comb of fine spicules on segment 8 (sometimes very pale in Sericostomatidae, several irregular rows in Philanisidae) 9

7(6). Prosternum without a forwardly-directed median spur or horn; abdominal sternum 1 with a median pair of bristles. [Mesonotum not divided into sclerotized plates; (Figs. 35.8A–C)] .... **Tasimiidae**

Prosternum with a forwardly-directed median spur or horn (not always pigmented); abdominal sternum 1 with several bristles 8

8(7). Mesonotum with 2 pairs of sclerotized plates .... **Goeridae**

Mesonotum not divided into plates; (Figs. 35.8G–I) .... **Limnephilidae**

9(6). Several irregular rows of spicules on segment 8; marine; (Figs. 35.10A–C). [Metasternum with a transverse row of bristles; abdominal sternum 1 with several bristles; abdomen without a lateral fringe] .... **Philanisidae**

A single row of spicules on segment 8; freshwater 10

10(9). Anal hooks formed of a single large tooth; prosternum fully sclerotized in Australian species. [Hind leg only slightly longer than mid leg] .... **Odontoceridae**

Anal hooks with at least a small second dorsal tooth; prosternum not heavily sclerotized .. 11

11(10). Mid tibia and tarsus fused; (Figs. 35.9G–I) .... **Philorheithridae**

Mid tibia and tarsus not fused 12

12(11). Metasternum with a transverse row of at least 6 or a dense zone of stout bristles; (Figs.35.9D–F) .... **Leptoceridae**

Metasternum with not more than a median pair of bristles 13

13(12). Labrum with a distinct transverse row of numerous stout setae across the middle of the dorsal surface; abdomen with a lateral fringe of long hairs; (Figs. 35.9A–C) ...... **Calamoceratidae**

Labrum with only a few bristles across the middle; abdomen without a distinct lateral fringe. 14

14(13). Second tooth of anal hook comb-like; (Figs. 35.10G–J); larvae living in helical cases of sand grains .... **Helicopsychidae**

Teeth of anal hook not comb-like; larval cases not helical in shape 15

15(14). Hind tarsal claw as long as tarsus .... **Beraeidae**

Hind tarsal claw much shorter, similar to or smaller than that of other legs; (Figs. 35.10D–F) .... **Sericostomatidae**

1. Hydropsychidae. A large family found mostly in clear flowing streams. The adults emerge mainly in summer and autumn. The family contains possibly the most primitive of all caddis-flies in both larval and adult characters. It is divided into 4 subfamilies, 2 of which occur in Australia. These differ in the adults as stated in the key and also in the segmentation of the maxillary palp.

MACRONEMATINAE (4 spp.) have very long antennae, and in general appearance resemble the Leptoceridae, from which they are readily distinguished by the multiarticulated terminal segment of the maxillary palp. The one genus, *Macronema*, is widely distributed in the cooler streams of the east coast.

HYDROPSYCHINAE (23 spp.), including Diplectroninae which differ in hind-wing venation, are found only in streams where there are sections of rapids. Most breed in the fast-flowing coastal and subalpine streams, but a few species of *Cheumatopsyche* occur on sections of the slow-flowing inland streams. Some of the species are well known, because of their attractive appearance and day-flying habits. *Asmicridea edwardsi*, the 'white moth', is widely distributed in eastern Australia and Tasmania; *A. grisea*, the 'Shannon moth',

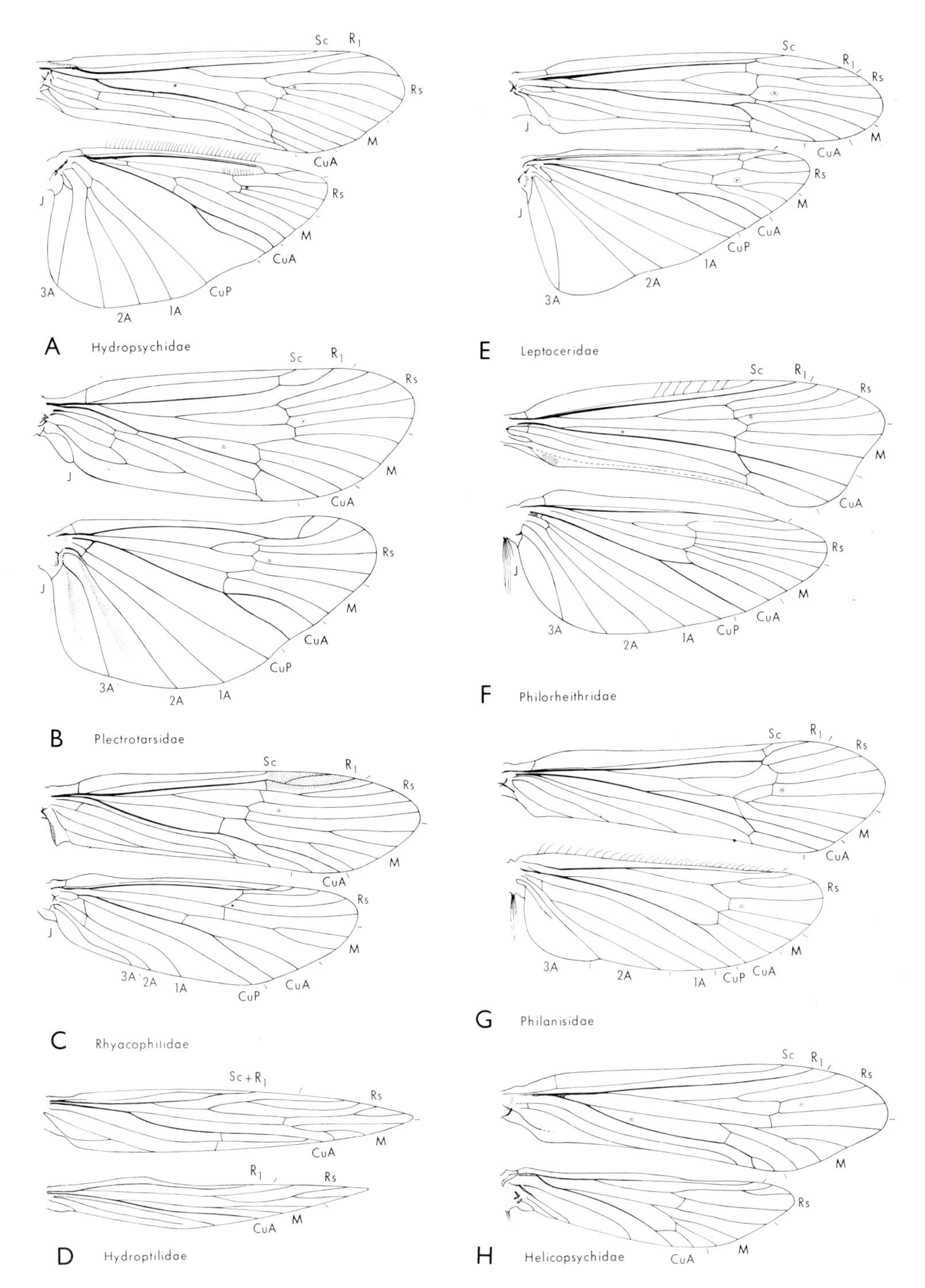

Fig. 35.4. Wings: A, *Macronema* sp.; B, *Plectrotarsus* sp.; C, *Taschorema* sp.; D, *Hydroptila* sp.; E, *Triplectides* sp.; F, *Aphilorheithrus* sp.; G, *Philanisus plebeius*; H, *Helicopsyche* sp. [M. Quick]

is abundant in parts of Tasmania. The males of both fly in bright sunlight on hot steamy days, in swarms in the lee of bushes bordering the stream. The females shelter in the bushes, and are rarely seen on the wing.

The Hydropsychidae have primitive larvae (Fig. 35.6D,G), which do not build cases, but erect a wide net directly in front of a tube-like retreat attached to irregularities on rocks or pieces of wood, usually in a swift-flowing current. They have numerous ventral and lateral branched abdominal gills, and the whole body is covered with a dense mat of short, dark hairs.

2. Rhyacophilidae. A large family of mostly cold-adapted species that occur only in clear cold-water streams; they are well represented in Tasmania. Adults are found on the wing throughout the year, though they are less abundant in midwinter and mid-summer. The species are generally widely distributed. The wings are usually of a generalized elongate-oval shape. The family is divided into 3 subfamilies, 2 of which occur in Australia.

The HYDROBIOSINAE (29 spp.) contain the larger, more commonly collected Australian species, with upright, rather thickened hairs on portions of some veins, the long setae on wings and body giving the adults a rather hairy appearance. There are several well-defined genera, based mainly on differences in wing venation. *Taschorema* contains most of the larger species (wing-span 15–35 mm), some of which have mottled wings, whereas others are of a general dark colour. The genus is represented in Tasmania and W.A., as well as over most of the eastern States. The males have a small modified area of setae on the undersurface of the hind wing surrounding an elongate 'scent gland' situated between CuP and 1A. In addition, there is a short line of scales on 1A and 2A in most species, but in others there are long dense hairs in these positions. Spooned, scale-like hairs occur over most of the fore wing in the male of *T. rieki*.

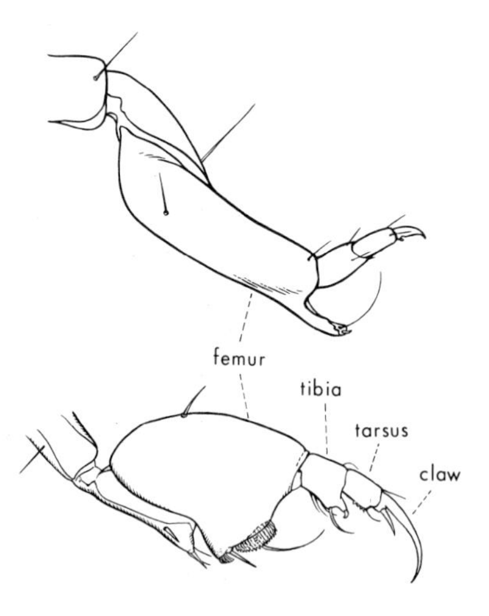

Fig. 35.5. Types of fore leg in larvae of Hydrobiosinae (Rhyacophilidae). [M. Quick]

The GLOSSOSOMATINAE (6 spp.) are very small, dull-coloured caddis-flies. The species of *Agapetus* have a wing-span of 7·5–9·5 mm, and are dark coloured. They could be confused with Hydroptilidae in size, but have a more expanded apex to the fore wing and the marginal fringes are not exceedingly long. The mid tibia and tarsus of the females are generally considerably dilated. The first 2 segments of the maxillary palp are subequal. The species are mostly found in the swift-flowing waters of mountain streams and rivulets of Tasmania and south-eastern Australia.

The free-living larvae of the HYDROBIOSINAE (Fig. 35.6A) are predacious, mainly on other aquatic insects. They are long and thin, pale in colour, with only the head and pronotum sclerotized. Body smooth, with a few very long, bristle-like hairs. Prosternum heavily sclerotized, without a spur-like process. Head elongate. Abdominal tergum 9 with a sclerotized shield bearing a few (usually 6) large hairs. Abdominal proleg with the apical segment fully sclerotized, with 2–3 apical hairs, the claw simple. Fore leg unusual for the order, enlarged, chelate or subchelate (Fig. 35.5), the chela formed by a spur or process from the femur articulating with the combined tibia-tarsus-claw. Tarsal claws of mid and hind legs not enlarged. Maxillary palp with 3rd and 4th segments long. Labrum rather bare, with only 2 enlarged upright

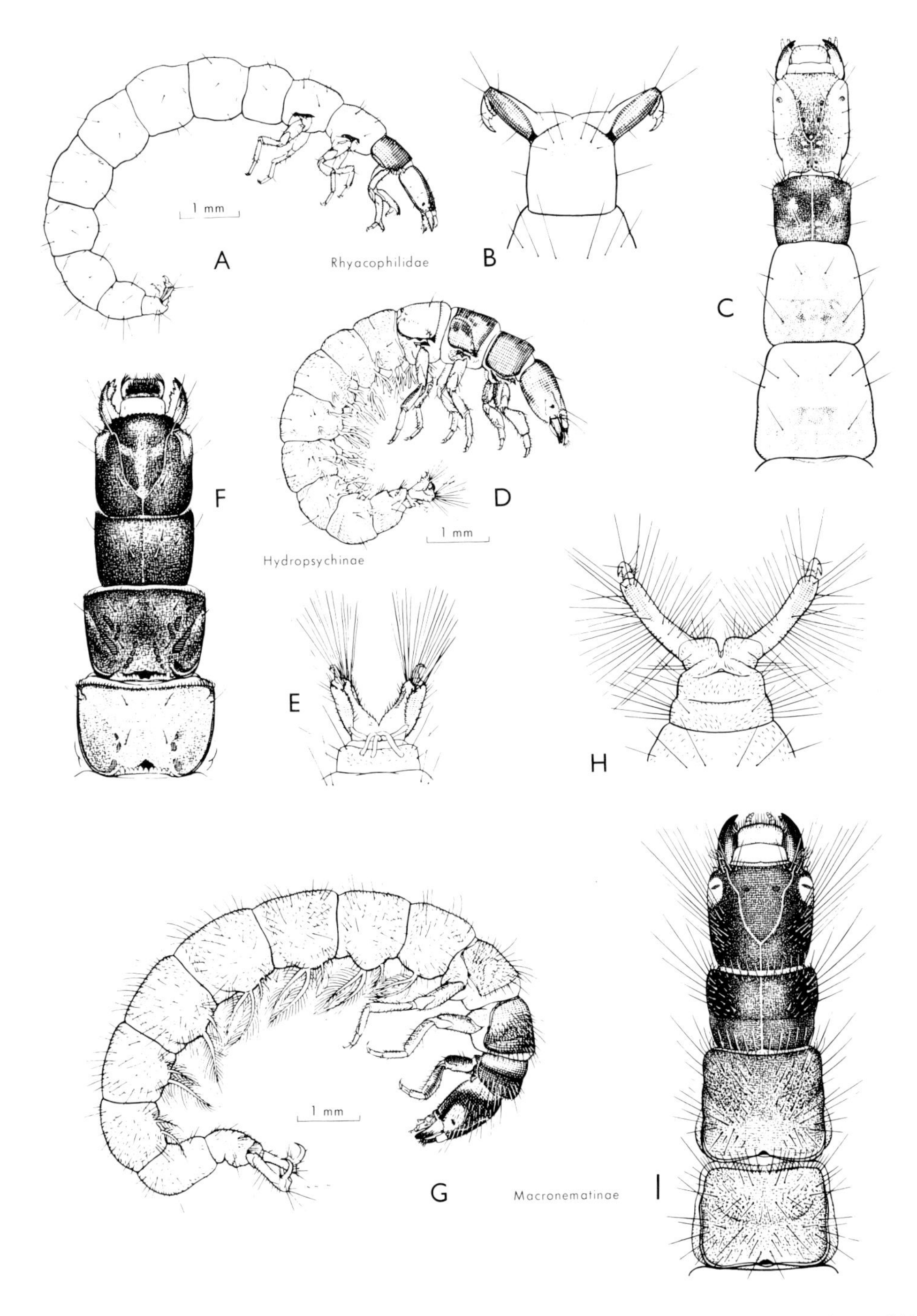

Fig. 35.6. Larvae: A–C, Hydrobiosinae (Rhyacophilidae); D–F, Hydropsychinae (Hydropsychidae); G–I, Macronematinae (Hydropsychidae). [M. Quick]

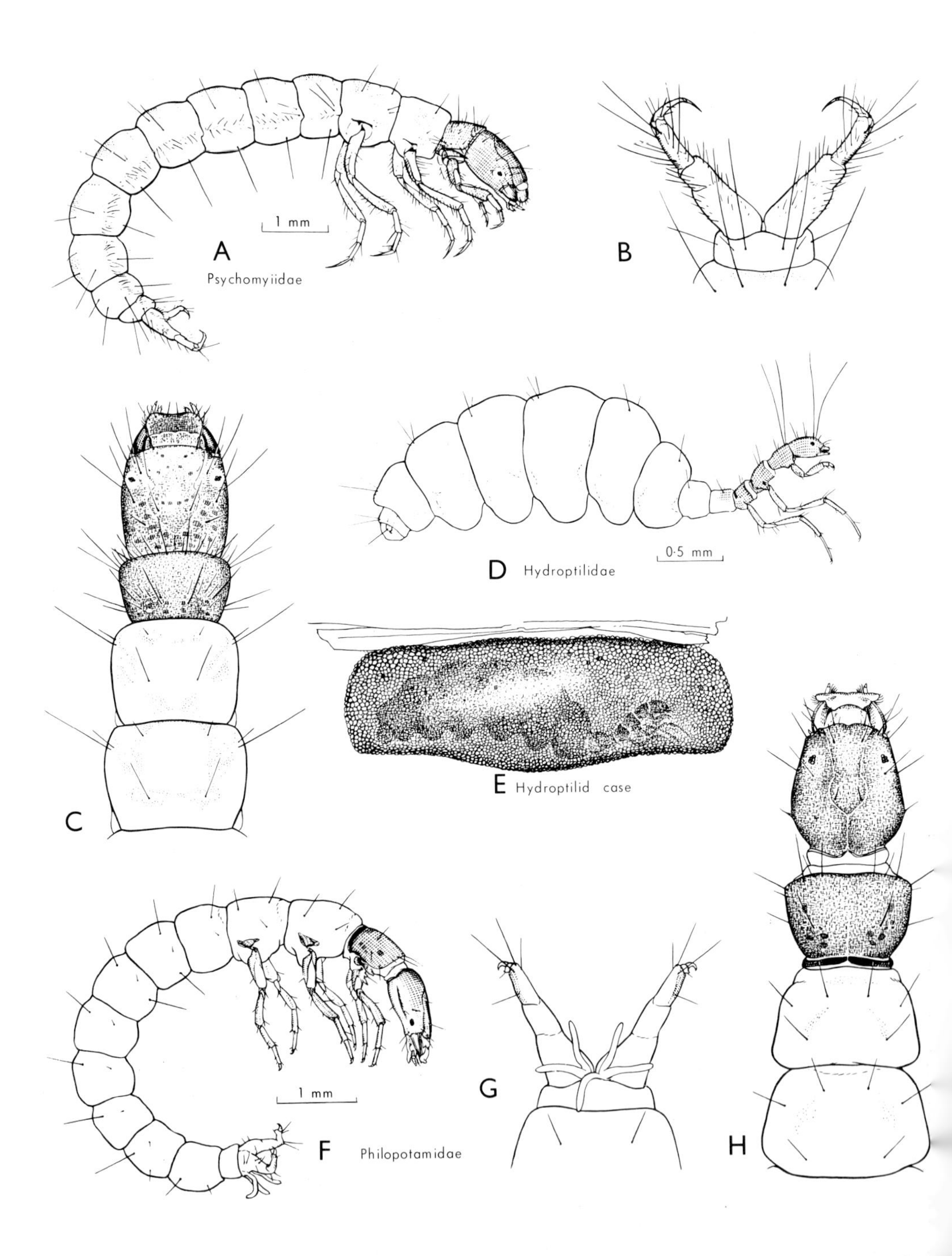

Fig. 35.7. Larvae: A–C, Psychomyiidae; D, Hydroptilidae; E, hydroptilid larva in case prior to pupation F–H, Philopotamidae. [M. Quick

hairs. Antenna not distinguishable. The larvae form a pupal cell of cemented sand grains, within which they spin a smooth silk cocoon that has a tanned appearance.

The larvae of the GLOSSOSOMATINAE (Fig. 35.8D) live in convex saddle-cases made of sand grains (Fig. 35.2H), the flat lower surface being constructed from smaller grains than the upper. They occur usually on rocks in the quieter sections of the stream, along with Hydroptilidae. They are all very small and of generalized appearance. The pupal cell is formed by cutting away the sand grains from the lower side of the case, and attaching the upper part directly to the substrate.

3. Philopotamidae. The Australian species are all small, with dark grey to almost black wings (span 9·5–18 mm). One species, *Chimarra uranka*, has bright orange hairs on head and anterior thorax. The adults mostly emerge in summer, but some appear in early spring or in autumn. *Chimarra* has an almost world-wide distribution, and contains more species than the rest of the family combined. *Hydrobiosella* is confined to Australia and New Zealand.

The larvae (Fig. 35.7F), smooth-looking, slim, and usually of a distinctive yellow colour, are found mostly in clear rapid streams. They are similar in general appearance to rhyacophilid larvae (Hydrobiosinae), but have a distinctive labrum, and are without a sclerotized shield on abdominal tergum 9. The flimsy silken tubes or sac-like nets in which they live are covered with aggregations of fine silt. These collapse when withdrawn from the water, and look like pieces of brown slime.

4. Psychomyiidae. This is a large family, closely allied to, but more specialized than, the Philopotamidae, from which the adults differ mainly in the absence of ocelli. Most of the Australian species occur in the southern portion of the continent, but there are representatives of many of the genera in Queensland. The only representatives in Tasmania and W.A. belong to *Ecnomus* and *Ecnomina*. The species vary widely in habits. *Stenopsychodes* and *Plectrocnemia* breed only in clear mountain streams. Some species of *Ecnomus* occur in slower-flowing, warmer streams, whereas others are found only in cold streams. *E. continentalis* is widely distributed in the inland, but some of the coastal species have very restricted distributions. The large species of *Stenopsychodes* (wing-span up to 45 mm) are amongst the most attractive Australian caddis-flies. Most have golden fore wings heavily splashed or speckled in black. The remaining genera contain smaller, mostly drab species. In *Plectrocnemia* and *Polyplectropus* there is a golden iridescence on the fore wing. The species of *Nyctiophylax*, *Ecnomus*, and *Ecnomina* are all very small (wing-span as little as 9 mm).

The pale larvae (Fig. 35.7A) are similar to those of the Philopotamidae, apart from the labrum and the shorter head. They occur in all types of habitats from standing waters to rapid cold streams. Some larvae pupate in woven cells incorporating coarse sand grains and covered usually with slime-like silken strands, attached under rocks in quiet portions of pools. Pupae have simple, fine, curved, very long mandibles, crossed at base and with apex extending to the opposite eye. There is a pair of long, simple gills on each pleural margin of the abdominal segments.

5. Hydroptilidae. The adults usually occur in dense numbers during summer. These very small species (wing-span 6–11 mm) are readily recognized from other caddises, for the wings are folded rather flat over the abdomen in a moth-like manner. The narrow wings have very long marginal fringes. Wing-coupling is by curved macrotrichia along the anterior margin of the hind wing. Jugal lobe not developed in fore wing.

The larvae live in ponds, in quiet backwaters and areas of deep silt. Most larvae, at least of Australian species, construct elongate-oval, laterally compressed, purse-like cases (Fig. 35.7E) of 'chitinous' material, the outer surface being finely shagreened through incorporation of fine sand grains. During larval life the case, open at both ends, is apparently opened periodically on the lower margin to allow increase in size. When mature, the larva attaches the case to the

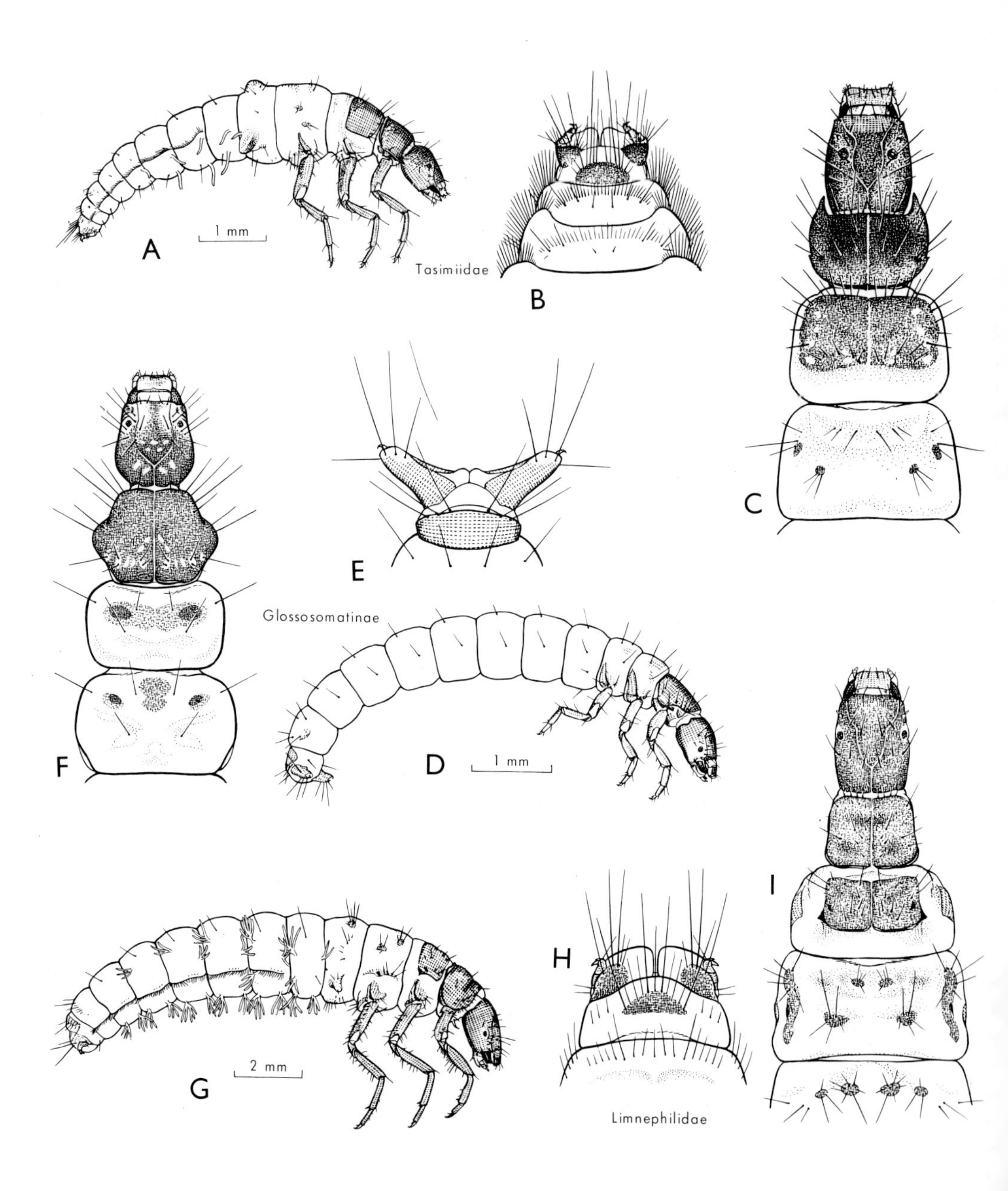

Fig. 35.8. Larvae: A–C, *Tasimia* sp.; D–F, *Synagapetus* sp. (Rhyacophilidae); G–I, *Archaeophylax* sp
[M. Quick

substrate either at the lower side of each end by a few silken strands or by a silken stalk, turns over with its ventral surface towards the dorsal side of the case, and pupates. Cases are only about 3·5 mm long.

6. Limnephilidae. A family widespread in the northern hemisphere and of very varied habits, but the 2 rather large species of *Archaeophylax* (wing-span up to 30 mm), occurring in Victoria and Tasmania, are limited to clear mountain streams. The adults have long stout spines on the tibiae and tarsi, and the fore femur sometimes bears an area of short stout bristles on its anterior border. Ocelli well developed. Maxillary palp with reduced segmentation in male. Face slightly produced.

The larvae (Fig. 35.8G) live under stones in calm pools of otherwise rapidly flowing, small, stony streams. They construct straight or slightly bent tubular cases of sand grains. The larvae are not unlike those of the Glossosomatinae (Rhyacophilidae), apart from a more definite adaptation to the case-making habit. The abdominal prolegs are fused, as in other true case-makers, but abdominal segment 8 has not developed the lateral row of spicules for gripping the wall of the case. There is no dorsal tubercle on abdominal segment 1. Gills have developed on the abdomen, as often happens with case-makers.

7. Plectrotarsidae. A small family restricted to the subalpine areas of south-eastern Australia and Tasmania, with one species in the higher-rainfall areas of south-western W.A. Structurally, the adults differ from Limnephilidae only in the development of warts on the dorsal thorax, details of the hind-wing venation, the slightly thickened and flattened antennae, and the development of a frenulum-type wing-coupling, but they are very distinctive in appearance. They are the most moth-like of all Australian caddises. Two species of *Plectrotarsus* have attractively coloured wings with many iridescent sheens, spanning 11–15 mm, whereas the fore wing of the western species, *P. minor*, is black with large white marks. Adults occur along cold, clear streams and lakes. Only first-instar larvae are known with certainty.

8. Goeridae. There is a large, undescribed species in Tasmania that folds the wings flat over the abdomen, and is further distinguished by the absence of anal veins in the fore wing. The hind wing has an expanded anal field. Scutum with 2 elongated warts, separated from the mid line; scutellum tapering, and with a single median wart sometimes indistinctly divided. Preapical spurs of mid tibiae hairy; mid and hind tibiae with 4 spurs. Jugal lobe very reduced. Maxillary palp of male with reduced segmentation. Its larva is not known.

9. Tasimiidae. The small species of *Tasimia*, wing-span 15–20 mm, are widespread in cold mountain streams. Scutum with 2 rounded warts; scutellum with 2 elongated warts. Mid and hind tibiae with 4 spurs, preapical spurs hairy, those of hind tibia situated well beyond the middle. Maxillary palp of male 4-segmented. Hind wing without an expanded anal area. Jugal lobe of fore wing reduced. The larva (Fig. 35.8A), similar to that of Limnephilidae but prosternum without a median horn, constructs a saddle-like case closed at apex (Fig. 35.2A).

10. Calamoceratidae. Species of the only Australian genus, *Anisocentropus*, occur throughout the eastern coastal area and Tasmania. They are not known from W.A., but have been recorded from Darwin. The genus is probably a relatively recent immigrant from the north. Three species have the basal half of the distinctly triangular fore wing an orange colour; one *(latifascia)* is dark grey with some suggestion of the orange pattern in some specimens; and *banghaasi*, from north Australia and extra-Australian localities, has areas of iridescent metallic colouring at the base and white markings on the distal half of the fore wing. Wing-span usually between 16 and 20 mm. Antennae clearly somewhat longer than fore wing. Median cell closed in fore wing. Mesoscutum with 2 longitudinal lines of setiferous punctures. The only Australian genus with a 6-segmented maxillary palp, and the segments are elongate.

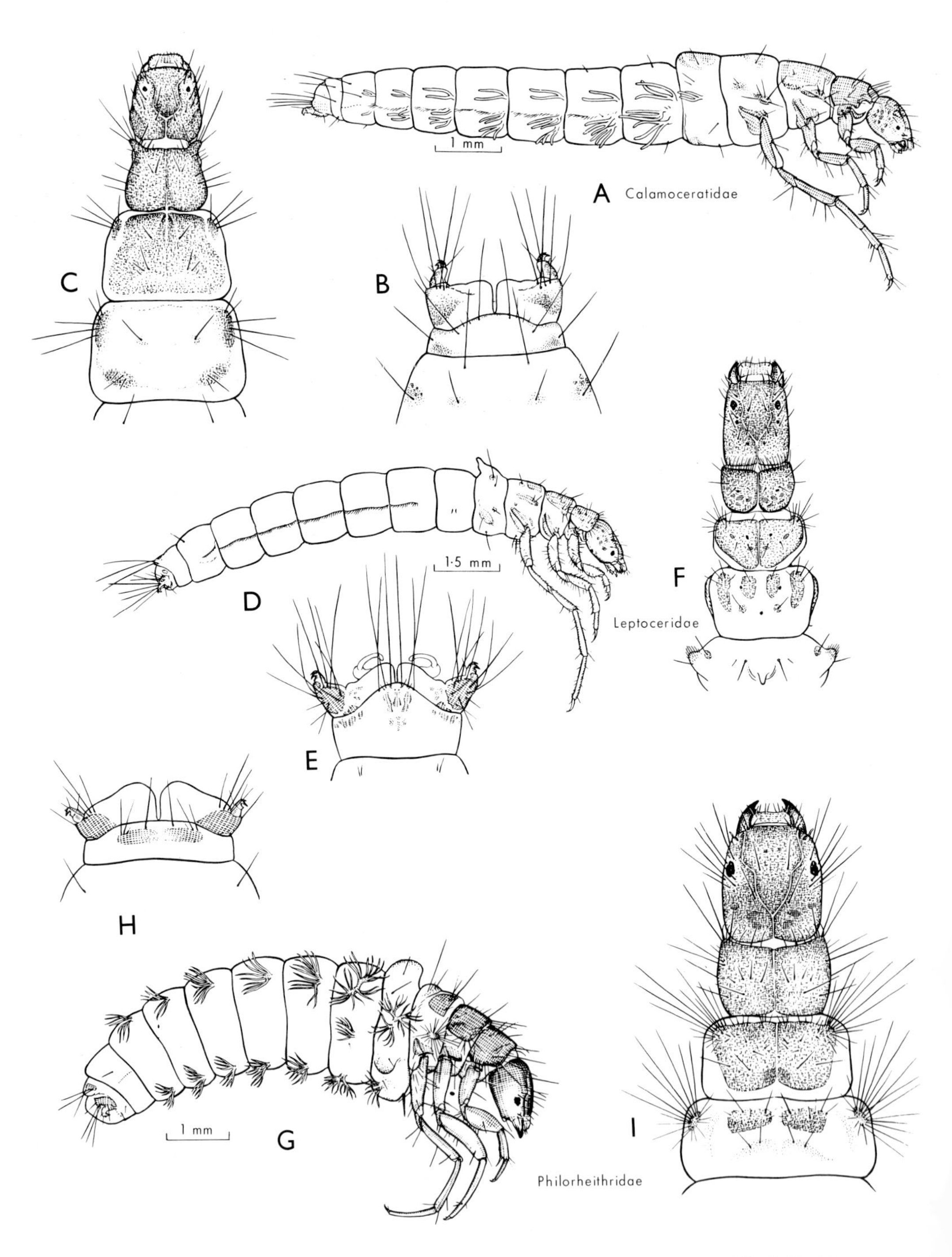

Fig. 35.9. Larvae: A–C, *Anisocentropus* sp.; D–F, Leptoceridae; G–I, Philorheithridae. [M. Quick]

The larvae (Fig. 35.9A) occur in swamps and peaty backwaters, as well as in cool, clear mountain streams. *Anisocentropus* constructs a flat, elongate-oval case (Fig. 35.2J) of two portions of leaves bound together, the upper piece somewhat larger and overlapping the lower, especially at the head end. In general appearance the larvae are not unlike those of Hydropsychidae apart from modifications due to case-making.

11. Leptoceridae. All the species are distinguished by the very long antennae, which may be more than three times as long as the fore wing, and relatively longer in males than in females. In Australia such long antennae occur elsewhere only in the Macronematinae (Hydropsychidae) and to a less extent in some Odontoceridae. Many of the species are of considerable size (wing-span up to 40 mm), but others are quite minute. The fore wings are very long and narrow, but the hind wing may be expanded and three or more times as wide as the fore wing. Males have more expanded hind wings than females. Several species of *Oecetis* have distinctive areas of scales on the wings. *Triaenodes* has the antennal scape greatly enlarged, as long as the width of the head. Many of the species are active day-fliers, but, as only a few aggregate during flight, they are rarely conspicuous. The large species of *Triplectides* and related genera are common along stream margins, or flying in an irregular jerky pattern over sedge-grown swamps in the afternoon. The small, almost black *Leptorussa darlingtoni* is often seen flying in numbers over small ponds. The dark *Triplexina nigra* swarms along the margins of some high Tasmanian lakes.

Representatives of this family breed in a wide variety of habitats. Some occur in swift-flowing mountain streams. Others breed in the standing water of sedge-grown swamps, both in cool southern waters and the warm waters of the coast or inland. The pale yellow-brown *Notalina fulva* and the dark brown slightly speckled *Oecetis pechana* are amongst the few species that breed in the standing waters of the dry inland. The larvae (Fig. 35.9D) generally construct long, thin, tubular cases (Figs. 35.2K,M), which often commence with small sand grains, but later change over to pieces of plant tissue incorporated in a spiral fashion, rarely irregularly. Sometimes they are made entirely of sand grains. A few use hollow twigs or grass stems for their cases, and others appear actually to hollow out the twig. Still others form box-like tubular cases from short sections of twigs (*Oecetis*; Fig. 35.2C).

12. Odontoceridae. The species occur only in clear cold-water streams. The adults are active mostly in summer. *Marilia* has the antennae considerably longer than the fore wing, narrow, obliquely truncate wings, and large eyes, especially in the male. The wings normally exhibit faint patterns in greys and browns, but in *Barynema* they are blackish with a white or orange bar about the middle. The larvae construct tubular cases of sand grains (Fig. 35.2I).

13. Philorheithridae. The species—of Australia, Tasmania, and New Zealand—have a wrinkling of the tips of the fore wing when the wings are folded. In *Austrheithrus* there is a well-developed longitudinal fold. *Aphilorheithrus* forms a connecting link between Philorheithridae and Odontoceridae. The scutellum is somewhat produced anteriorly, and the thickening on the anal margin of the fore wing is not produced. Maxillary palp normally 5-segmented in both sexes, but segments reduced to 2 in the males of *Ausmanthrus*, 3 in *Austrheithrus*, and 4 in *Tasmanthrus*; 1st segment short and enlarged on the medial side at apex where it bears a tuft of enlarged bristles, 2nd and 3rd long and stout. *Pilifers*, finger-like processes bearing sensory hairs arising from above the base of the antennae, occur in the males of several genera (*Aphilorheithrus*, *Kosrheithrus*, *Tasmanthrus*, and *Ausmanthrus*). The antennae of the males and the anal margin of the fore wing in both sexes are modified for a possible stridulatory function, and they are plumed in both sexes of one genus.

The larvae (Fig. 35.9G) occur in all types of cool flowing water, sometimes in the quieter pools. They construct slightly curved, leptocerid-like cases of sand grains (Fig. 35.2G).

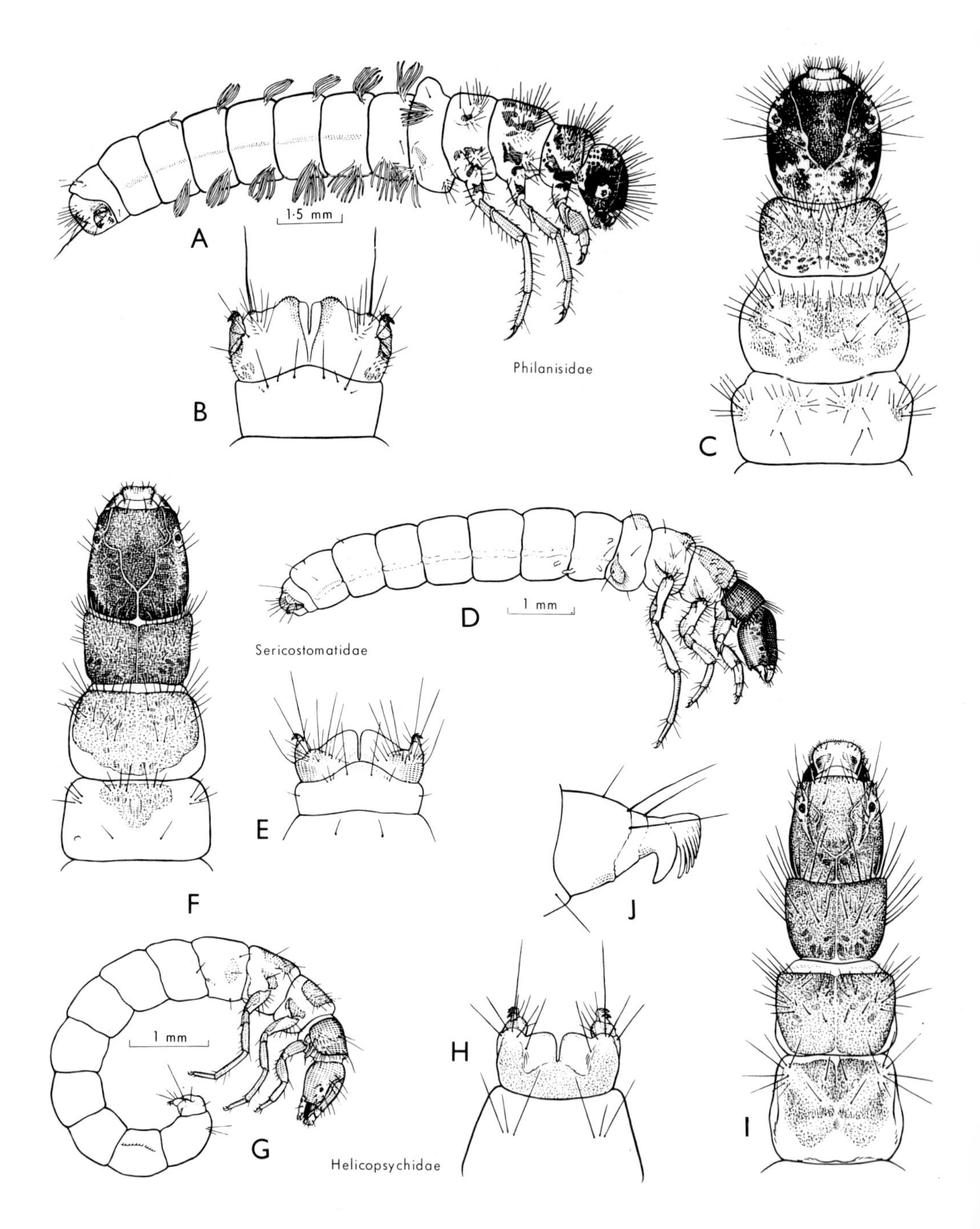

Fig. 35.10. Larvae: A–C, *Philanisus plebeius*; D–F, Sericostomatidae; G–I, *Helicopsyche* sp.; J, apical proleg of *Helicopsyche*. [M. Quick]

Some larvae burrow in the sandy bottom of the stream. There is a general similarity to the larvae of Rhyacophilidae (Hydrobiosinae). In both families the prosternum is fully sclerotized, and the spur-like trochantin is very reduced.

14. Helicopsychidae. Small, grey to black species, with a wing-span about 10 mm. Wings long and thin, fore wing rather triangular. *Helicopsyche* is the only recognizable Australian genus. Adults occur only along clear mountain streams. They differ from Sericostomatidae in thoracic structure and wing-coupling. Scutellum rounded anteriorly, with two large warts. Scutum with two rounded warts, sometimes small. Jugal lobe of fore wing reduced, not forming a re-entrant angle with the anal lobe, and a well-developed row of hamuli over the basal half of the costal border of the hind wing. They are also similar in many ways to the Hydropsychinae (Hydropsychidae), differing most in the structure of the palps, reduced venation, and more highly developed wing-coupling.

The larvae (Fig. 35.10G) are found mainly on rocks in the quieter waters of clear mountain streams. They construct snail-like helical tubes of sand grains (Fig. 35.2B). They are similar to those of Sericostomatidae, apart from the anal hook and the spiral twist to the body. Upper tooth of anal hook flattened and comb-like, with parallel grooves on its upper surface. The labrum is prominent.

15. Sericostomatidae. This large family is well represented in the southern parts of the mainland and in Tasmania, where the species are restricted mostly to the colder waters. They are generally small, but there are a few larger ones (wing-span 7–30 mm). The adults have a general rhyacophilid-like appearance, due to the relatively long hairs that clothe the body and wings. The males often have grotesque enlargements of the maxillary palps and large frontal processes covered with dense, long hairs; the fore legs may be margined with similar long hairs. Scutum without warts or with 2 warts almost touching. Scutellum tapering anteriorly, with 2 elongated warts. Jugal lobe of fore wing well developed and forming a re-entrant angle with the anal lobe. Jugal lobe and more heavily sclerotized portion of anal lobe without setae. Wings of males often with folds and scale-like areas, especially in the anal area. Antennae only about as long as the fore wing, their bases sometimes with scale-like or enlarged hairs. Maxillary palp of male with reduced segmentation.

The larvae (Fig. 35.10D) occur in all types of situations. Larval cases (Fig. 35.2L) consist of almost straight tubes of sand grains, or may be constructed of an upper and lower biserial row of rounded fragments of leaves. The larvae are similar to the Hydropsychinae in having a short, somewhat flattened head and the hind leg only slightly lengthened. The prosternum has a well-developed trochantin.

16. Beraeidae. The genus *Alloecella* from Victoria has been placed in this family. The 2 species are small (fore wing 5–5·5 mm). The family is defined in the adult on the structure of the mid and hind tarsus, in which each segment has a group of 4 black spines at apex. *Antipodoecia turneri*, originally placed in the Sericostomatidae on the reduced segmentation of the palp, is transferred to the Beraeidae on its leg structure. This very small species (wing-span 7 mm) occurs from Ebor to Kangaroo Valley in N.S.W.

17. Philanisidae. The single species, *Philanisus plebeius*, breeds in rock pools, mainly close to low-tide mark, on rocky headlands on the open sea-coast of south-eastern Australia and New Zealand. It is the only known marine caddis-fly. The slightly curved, tubular case of the larva (Fig. 35.10A) is usually covered with coralline fragments (Figs. 35.2D–F). The medium-sized adults (wing-span about 17 mm) are not unlike Hydropsychidae in general appearance. Scutellum rounded anteriorly, with 2 reduced warts; scutum with a few setiferous punctures. Mid tibia longer than femur. Hairs on head of male spooned at tips. Second segment of maxillary palp of male produced on one side of apex to a rounded knob. The adults rarely move from the headlands, and are generally found sheltering in the lee of projecting rocks close to water level. They apparently require the high humidity of the sea spray for survival.

18. Helicophidae. The small species of *Helicopha*, wing-span 11–15 mm, are known to occur from Ebor, N.S.W., to Tasmania. Scutum without warts, but with numerous fine setae; scutellum tapering anteriorly, with 2 warts. Wing-coupling by hamuli over the basal half of the fore wing, with the distal ones more closely grouped. Jugal lobe of fore wing reduced, and anal region without a bare area at the wing margin. Hind wing with a large vein-free area discally. Maxillary palp of male 5-segmented. This small family has a distinctive scutum as well as wing venation. The wing-coupling is similar to that of the Helicopsychidae. The larva is unknown.

ACKNOWLEDGMENT. I am grateful to Mr A. Neboiss, National Museum of Victoria, Melbourne, for helpful discussions of the manuscript.

36

LEPIDOPTERA

(Moths and butterflies)

by I. F. B. COMMON

Haustellate, or rarely mandibulate, endopterygote Neoptera, with two pairs of membranous wings, clothed on both surfaces with usually overlapping scales. Larvae eruciform, peripneustic, or rarely holopneustic. Pupae rarely decticous, usually adecticous and obtect.

The Lepidoptera are one of the largest insect orders, including some 10,000 described Australian species, with possibly half as many again still to be named. They range in size from tiny leaf-miners, with wings expanding barely 3 mm, to the huge saturniids and cossids expanding some 25 cm. Their colouring and elegance have given them considerable popular appeal, while the destructive qualities of the larvae of many species establish the economic importance of the order.

Members of this order are readily distinguished from other panorpoid orders having two pairs of wings by the clothing of usually broad, overlapping scales on the head, body, and appendages of the adult. The wing venation of the primitive Zeugloptera and Dachnonypha approaches that of certain Trichoptera, but M_4 rarely occurs as a separate vein terminating on the wing margin. An epiphysis is present on the fore tibia of most of these archaic Lepidoptera, as in all but a few of the more specialized families of the order, whereas this structure does not occur in other orders. The modification of the galeae into a haustellum or proboscis, found widely in the Lepidoptera, occurs in no other order.

Anatomy of Adult

Head (Fig. 36.1). Compound eyes large, rounded, often with erect hairs between the facets; ocelli when present paired, one above each eye; both sexes sometimes with paired sensory organs, *chaetosemata* (Jordan, 1923; Fig. 36.1A). Frons and clypeus sometimes separated by an epistomal suture, or not differentiated (frontoclypeus); genae narrow; labrum a narrow, pointed, or transverse plate, with laterally projecting *pilifers* and median projection (epipharynx). Antennae anterior to ocelli, many-segmented, usually partly clothed with scales, but sometimes naked; scape sometimes with tuft of scales, or with anterior pecten of stiff, hair-like scales, or laterally expanded and concave beneath forming an *eye-cap* (Fig. 36.16A); flagellum in male usually more specialized than in female, varying greatly in structure,

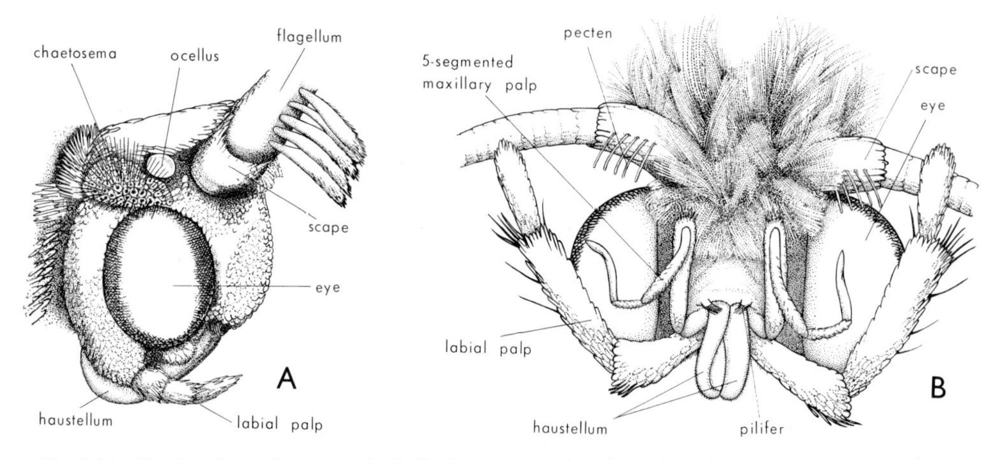

Fig. 36.1. Head and mouth-parts: A, *Pollanisus*, Zygaenidae, lateral; B, *Monopis*, Tineidae, ventral. [B. Rankin]

moniliform, filiform, ciliate, clubbed, dentate, serrate, lamellate, or pectinate.

Mouth-parts (Fig. 36.1B) variable. Mandibles usually absent, rarely functional, dentate (Zeugloptera, Fig. 36.12B) or non-dentate (Dachnonypha, Fig. 36.12E), in both furnished with well-developed abductor and adductor muscles, or vestigial (a few families). Maxillae with laciniae present (Zeugloptera, some Dachnonypha, Figs. 36.12C, F), rudimentary or absent; galeae often greatly elongated, usually grooved internally, and fastened together by interlocking hooks and spines to form a tubular *haustellum* (proboscis) through which liquid food may be drawn. Haustellum often with dense overlapping scales towards base; it is composed of a large number of sclerotized rings joined together by membranous bands, each half containing a nerve, a trachea, and two sets of muscles which enable it to be coiled beneath the head when not in use. Maxillary

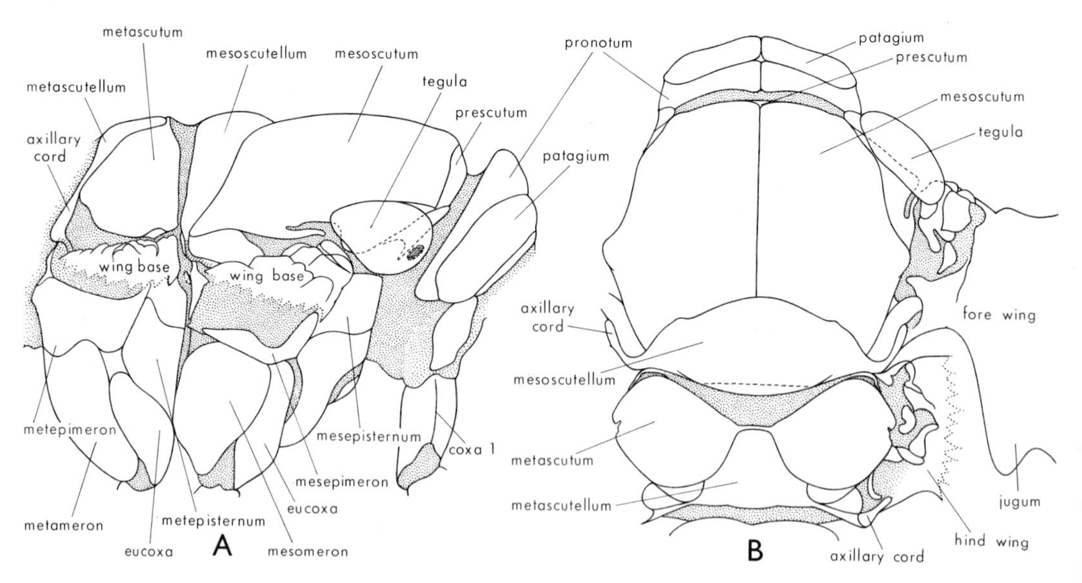

Fig. 36.2. Thorax of *Oncopera*, Hepialidae: A, lateral; B, dorsal. [B. Rankin]

palpi usually present and clothed with scales, often 5-segmented and folded in more primitive families, or reduced to 4, 3, 2, or 1 segment, or absent. Labium small, usually with well-developed, 3-segmented or rarely 2- or 4-segmented palpi, clothed with scales, the apical segment bearing a depressed or invaginated sensory area.

Thorax (Fig. 36.2). Prothorax usually small and membranous, often with a pair of articulated dorsal plates, the *patagia*. Mesothorax large, with small prescutum, prominent scutum, and smaller scutellum; *tegulae* uniformly well developed, each arising from a special tegular plate of the notum and supported by a tegular arm carried by the base of the pleural wing process. Metathorax only slightly smaller than mesothorax in primitive suborders, much smaller in the Ditrysia.

In Notodontoidea and Noctuoidea the metathorax carries a pair of auditory tympanal organs situated in cavities between the epimeron and the postnotum (Fig. 36.5A; Kiriakoff, 1963; Dethier, 1963). The transparent and iridescent tympanic membrane, to the inner surface of which 2 chordotonal sensilla are attached, covers a large tracheal sac. The tympanum is usually separated from the membranous dorsoposterior portion of the epimeron by a small sclerite, the *epaulette* or nodular sclerite, which varies greatly in shape between species. At the base of the abdomen, posterior to each tympanal cavity, there is usually a counter-tympanal cavity, which is expanded laterally to form a *hood* covering the tympanal cavity posteriorly (Figs. 36.5A, B).

Legs (Fig. 36.3). Usually well developed for walking and clothed with scales; fore legs reduced in some Papilionoidea, hind legs in some Geometridae, all degenerate in females of some Psychidae. Coxa relatively immobile; trochanter small. Femur strong, often bearing long hairs. Fore tibia usually with leaf-like, or spur-like, basally articulating *epiphysis* on inner surface, perhaps used for cleaning antennae or haustellum, sometimes with spines, bristles, long hairs, or prominent scales; mid tibia usually with pair of apical articulating spurs; hind tibia usually with 2 pairs of spurs, one pair apical, the other medial; the number of spurs on the fore, mid, and hind legs is shown conventionally by a formula (e.g. 0-2-4); males sometimes with expansible tuft or pencil of scent-scales on hind tibia, rarely on mid tibia. Tarsi 5-segmented, usually spined; apical segment

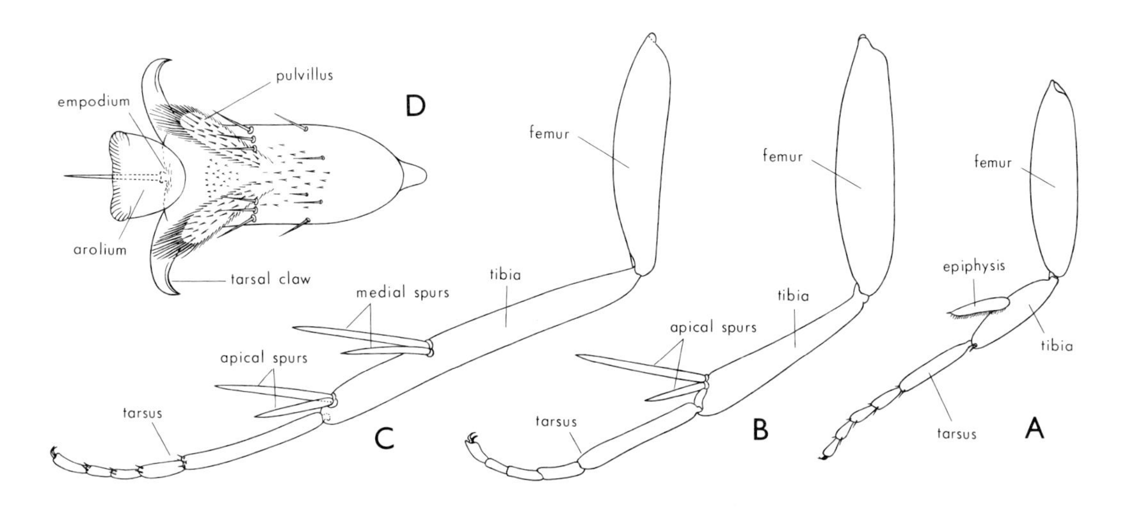

Fig. 36.3. Legs of *Epiphyas*, Tortricidae: A, fore; B, mid; C, hind; D, apical fore tarsal segment, ventral. [B. Rankin]

with a pair of lateral articulated curved claws, simple, with a tooth (e.g. Drepanidae), bifid (Pieridae), or reduced, usually with a median arolium and empodium, and sometimes with well-developed lateral pulvilli (Fig. 36.3D).

Wings. Membranous; both surfaces usually clothed with overlapping, usually broad, flattened macrotrichia *(scales)*; each with a short pedicel, inserted in a minute socket on wing membrane, its surface bearing many fine longitudinal ridges or striae, which often produce iridescent colours, pigments usually present within the scale. In males of some groups specialized scales *(androconia)*, which help to diffuse scents secreted by associated glands, scattered over wings, or in well-defined sex brands or patches. Microtrichia or *aculeae* also present in Zeugloptera, Dachnonypha, and most Monotrysia. The sides of the triangular lepidopterous wing are referred to as the *costa*, the *termen* (outer margin), and the *inner margin* (posterior margin or dorsum).

The wings in the females of a few Cossidae, Psychidae, Oecophoridae, Geometridae, Anthelidae, Lymantriidae, and Arctiidae are reduced and non-functional, or absent. On certain Subantarctic islands brachyptery occurs in both sexes in several families.

The articular sclerites of the lepidopterous wing base differ in some respects from the generalized type. As well as the usual anterior and posterior notal processes, an intermediate notal process is developed. Its size in the fore wing is correlated with that of the first axillary sclerite, which increases as one passes from the Zeugloptera, Dachnonypha, and Monotrysia to the Ditrysia. In the fore wing of Trichoptera and all Lepidoptera, except the Hepialoidea, the second axillary sclerite consists of two separate parts, dorsal and ventral, sometimes joined by thin cuticle. In the hind wing, however, it is a single sclerite, as it is in both wings of Mecoptera and Neuroptera. Unlike other panorpoid orders, all Lepidoptera, including the Zeugloptera, have a well-developed tergopleural apodeme. Other features of the wing bases of Lepidoptera are discussed by Sharplin (1963–4).

VENATION. The most outstanding features are its simplicity and apparent uncompromising reduction from the homoneurous venation of the Zeugloptera, the Dachnonypha, and the Hepialoidea to the heteroneurous condition of the more advanced Lepidoptera. The occasional presence of an extra vein (a 2-branched Rs) in the hind wing of Gracillariidae led Busck to suggest that this family may be transitional, but other characters lend little support; a similar aberration sometimes occurs in Elachistidae and other families.

The most generalized lepidopterous venation of the Zeugloptera and the Dachnonypha (Figs. 36.11A, B) closely resembles that of the more primitive Trichoptera. In both wings a humeral vein is present, Sc and R_1 may each have 2 branches, Rs has 4 branches, M has 3 branches, CuP is present, and the three anal veins are distally fused. Few cross-veins are retained. A homoneurous venation also occurs in the Hepialoidea, but here CuP is usually reduced.

The heteroneurous venation, found in the Nepticuloidea and Incurvarioidea of the Monotrysia and in the whole of the Ditrysia, shows a trend towards simplicity produced by the fusion or loss of veins. The most generalized type of heteroneurous venation may well be exhibited by the Cossidae (Figs. 36.17A, B; Turner, 1947), in which the stem of R_{4+5} *(chorda)* forms an accessory cell (cell R_3) in the fore wing, M is well developed and 2-branched, forming a second cell (cell M_2), and CuP is present in both wings. In the fore wing two anal veins are present but fused distally; in the hind wing Sc is fused, at least distally, with R_1, Rs is unbranched, and three anal veins are present, although 1A and 2A practically coalesce. This venation is closely approached by the more primitive Psychidae (e.g. *Trigonocyttara*). The first step in reduction is the loss of M and the chorda, resulting in the formation of one large 'discal' cell. Further reduction varies between the superfamilies, but generally takes the form of fusion or coalescence of the peripheral veins. In the fore wing a narrow accessory cell *(areole)* may arise by the partial fusion of two or more radial veins. In the most advanced superfamilies CuP has atrophied in both fore and hind wing, but in the Tortricoidea, Tineoidea, Yponomeutoidea, Gelechioidea, Copromorphoidea, Zygaenoidea, and Castnioidea CuP has usually persisted, at least towards the wing margin.

In Lepidoptera the amount of blood circulating in the wing veins and the force of the current are much less than in most other orders. This is due to the large tracheae which almost fill the lumen of the main veins and many of their branches. The number of haemocytes in the wing is similarly limited. Pulsatile organs in the thorax maintain the circulation, the blood flowing outwards in the costal, radial, and median veins, and returning through the cubital and anal veins (Arnold, 1964).

WING-COUPLING. Coupling is effected in several different ways. In the Zeugloptera, Dachnonypha, and Hepialoidea, a *jugum (fibula)* projects from the inner margin of the fore wing (Figs. 36.2B, 11, 13A–C), and lies above the base of the hind wing when the wings are extended, but folds beneath the fore wing when the wings are at rest. Sometimes a few small frenular bristles are also present, arising near the base of C of the hind wing, or a series of more distal costal bristles *(pseudofrenulum)* which press upon the anal area beneath the fore wing. Costal bristles are absent in Hepialidae, in some genera of which the jugum is long and thin.

A reduced jugum or jugal area persists in the remainder of the Monotrysia and in the Cossoidea, Tortricoidea, Tineoidea, Yponomeutoidea, Gelechioidea, Copromorphoidea, Castnioidea, Zygaenoidea, and Pyraloidea (Sharplin, 1963). In these forms, as in the rest of the Ditrysia, wing-coupling is effected by a true *frenulum* arising from a thickened frenulum-base at the humeral angle of the hind wing, and held beneath the fore wing by a *retinaculum* (Fig. 36.4). The frenulum is composed of a single composite bristle in males and one to many bristles in females. In males, the retinaculum is a subcubital membranous hook, usually short and broad in the more primitive groups and arising between C and Sc, or on a spur of Sc,

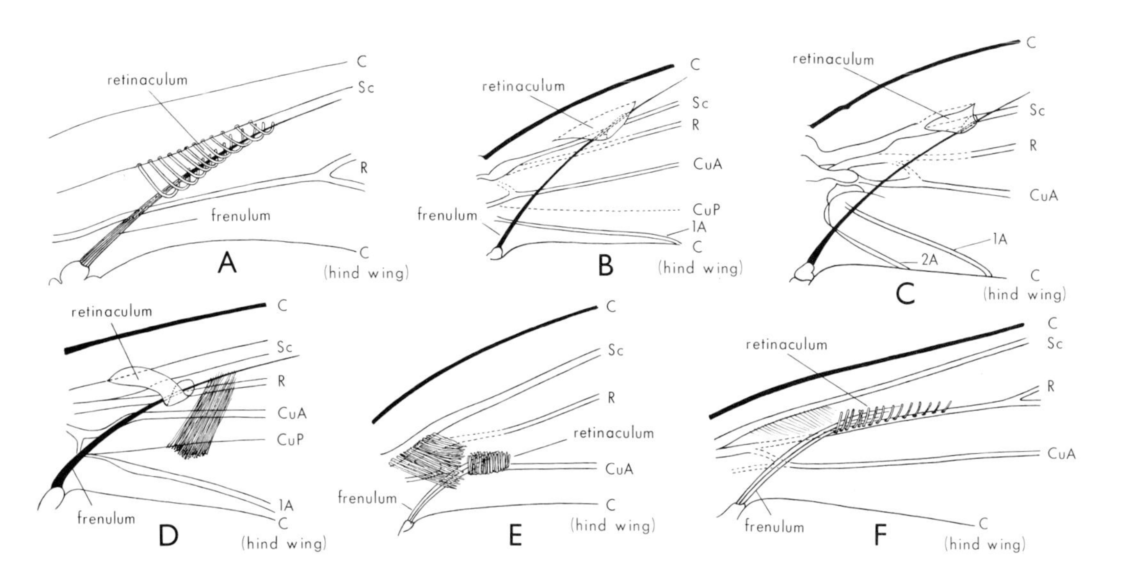

Fig. 36.4. Wing bases, ventral, showing retinaculum and frenulum: A, *Nepticula* ♂, Nepticulidae; B, *Narycia* ♂, Psychidae; C, *Barea* ♂, Oecophoridae; D, *Urisephita* ♂, Pyralidae; E, *Barea* ♀, Oecophoridae; F, *Phthorimaea* ♀, Gelechiidae. [B. Rankin]

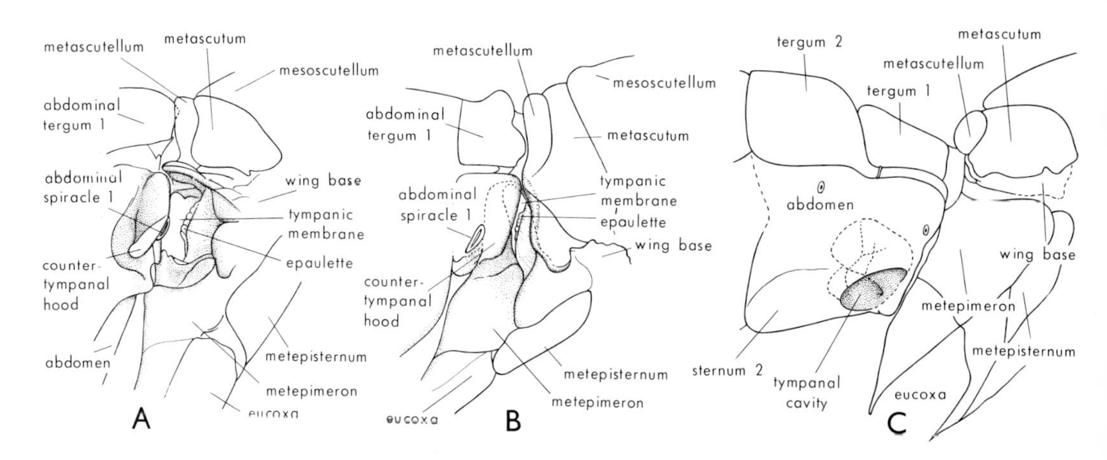

Fig. 36.5. Tympanal organs: A, *Agrotis*, Noctuidae, thoracic, showing postspiracular hood of counter-tympanal cavity; B, *Amsacta*, Arctiidae, thoracic, showing prespiracular hood of counter-tympanal cavity; C, *Chlorocoma*, Geometridae, abdominal. [B. Rankin]

becoming narrow, more elongate, and arising on Sc in the more advanced forms. In females, it usually consists of one or two series of stiff scales, arising near Sc *(subcostal)* or near CuA *(subcubital)*, but occasionally R *(radial)*; some males also have a subdorsal retinaculum of hairs or scales near CuA. In the Nepticuloidea the retinaculum is a series of hooks arranged along the down-folded base of C. In some Nepticuloidea, Incurvarioidea, and Tineoidea pseudofrenular bristles may also be present. A few families of the Ditrysia, in which the frenulum is reduced or lost, feature the *amplexiform* method of wing-coupling. Here a great enlargement of the humeral area of the hind wing is substantially overlapped by the fore wing, and the former may then be strengthened by the development of one or more short humeral veins.

Abdomen. Ten-segmented, but segments 7–10 may be greatly modified by the structures of the genitalia, and the sternum of segment 1 is lacking. Functional spiracles on segments 1–7. In the Pyraloidea and the Geometroidea a pair of tympanal organs is found in the abdomen (Fig. 36.5C), usually laterally near the base. Their detailed structure and taxonomic significance require further study. The dorsal surface of the abdomen in certain families of Yponomeutoidea, Gelechioidea, Copromorphoidea, Pyraloidea, and Sphingoidea bears scales which are modified into short, often flattened, spines. The abdomen, especially in males, may also contain eversible or other sensory organs. The anus in the male is situated at the end of an anal tube projecting from segment 9, and in the female on segments 9–10 at the posterior extremity of the abdomen. Cerci are entirely lost.

Reproductive Organs. In both sexes of most Lepidoptera the genital organs provide characters of great value in the separation of species, and often for the characterization of genera and higher taxa. The external genitalic structures are derived from the integument of abdominal segments 7 to 10. The internal genitalic structures, especially in the female, are also important, and their study normally necessitates the removal and dissection of the whole abdomen. The homologies of the complex lepidopterous genitalia are not well understood, even between some of the families within the order (Klots, in Tuxen, 1956).

MALE (Fig. 36.6). Tergum 9 and parts of 10 usually form a hood-like *tegumen*, from which arise various usually paired processes of the genitalia. Posteriorly, part of tergum

10 forms a mid-dorsal *uncus*, which is usually simple and hook-like, but may be divided, modified, or even absent. In certain groups tergum 8 may bear an uncus-like process. Ventrally, sternum 9 forms a U-shaped *vinculum*, the upper ends of which articulate with the ventral extremities of the tegumen. Mid-ventrally, the vinculum may be produced anteriorly within the body cavity to form a *saccus*. In some families paired, lightly sclerotized, often hairy *socii* arise from the posterior margin of the tegumen, beneath the uncus. Beneath the socii, a pair of processes derived from sternum 10 articulates with the posterior margin of the tegumen; these together form the *gnathos* and are often fused medially. The anal tube, between the gnathos and the ventral surface of the tegumen and uncus, may also be variously sclerotized.

The posterior end of the abdomen is closed by a membranous *diaphragma*, through which the *aedeagus* projects. Sclerotized portions of the diaphragma above, beneath, and around the aedeagus are called respectively the *fultura superior*, the *fultura inferior*, and the *anellus*. The fultura superior at times takes the form of a transverse band known as the *transtilla*, while the fultura inferior may be represented by a shield-shaped plate *(juxta)* to which the aedeagus is hinged. The invaginated distal end of the aedeagus forms an inner tube (endophallus) known as the *vesica*, which is everted during copulation and penetrates the female bursa copulatrix. The exterior of the everted vesica often bears sclerotized areas or spines, which may be shed in the bursa copulatrix. The paired clasping organs or *valvae* usually articulate with the tegumen and vinculum, and may be derived from the coxites, styli, or parandrites of segment 9. Each valva is a flattened sac, open proximally, and may be simple and rounded, or may assume a complicated shape and bear a variety of often complex structures and spines.

The testes lie adjacent to the intestine just beneath terga 5 and 6. Each usually consists of 4 follicles, but in *Nemophora* the number is increased to 20. The follicles are normally compressed together, but occasionally (some Hepialidae) they remain separate. The two testes are either separate from one another, each contained in a separate 'scrotum' (Zeugloptera, some Bombycoidea, some Papilionoidea), or more usually fused and contained in a common 'scrotum'. Each slender *vas deferens* opens into a *ductus*

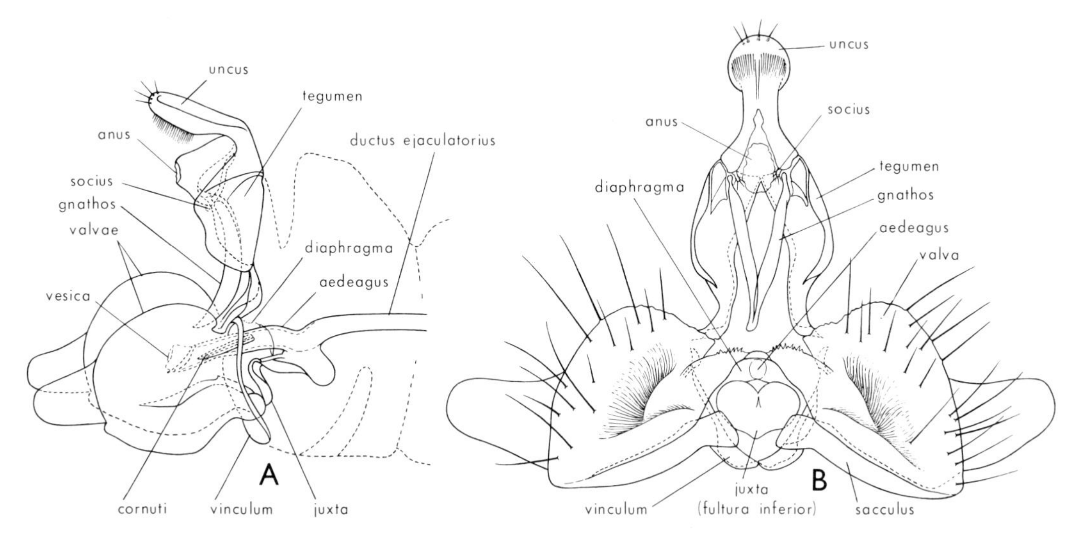

Fig. 36.6. Male genitalia of *Epiphyas*, Tortricidae: A, lateral; B, posteroventral. [B. Rankin]

ejaculatorius, the pair uniting to form a single duct leading to the aedeagus. Spermatozoa are stored both in dilatations of the vasa deferentia, the *vesiculae seminales*, and in the paired ductus ejaculatorii. The latter also store the secretion of the paired filamentous *accessory glands* which join them. The accessory secretion provides a medium for the spermatozoa in the spermatophore, and also initiates peristaltic contractions in the muscles of the female ductus seminalis. The simple distal ductus ejaculatorius includes two secretory sections in which the spermatophore is produced. In primitive Lepidoptera the spermatophore may disintegrate soon after deposition in the female genital tract, but in advanced forms, such as Noctuidae, it is very durable and contains chitin (Callahan and Cascio, 1963).

FEMALE. Two main types of genitalia (Fig. 36.7), the *monotrysian* and the *ditrysian*, are found in female Lepidoptera, and the higher classification reflects these fundamental differences. The monotrysian type occurs throughout the Zeugloptera and the Dachnonypha, and the ditrysian throughout the Ditrysia. The monotrysian type is also found in the Monotrysia, but here a somewhat intermediate type, the *exoporian*, also occurs.

The external genital aperture (vulva) is situated on the fused segments 9–10 at the posterior extremity of the abdomen. It is usually flanked by a pair of soft hairy lobes, the *papillae anales*, but these may be heavily sclerotized and modified for the insertion of the eggs into plant tissues or into crevices, and the terminal segments of the abdomen may be extensile and serve as an ovipositor. The muscles operating the terminal segments of the female abdomen are inserted in paired sclerotized apodemes, which extend forward from segment 8 *(apophyses anteriores)* and from the bases of the papillae anales on segment 9–10 *(apophyses posteriores)*. In monotrysian genitalia (Fig. 36.7A) a single terminal genital opening serves for both copulation and oviposition. In ditrysian genitalia (Figs. 36.7B, C) an additional copulatory opening, the *ostium bursae*, is situated mid-ventrally on sternum 8. Exoporian genitalia, found in the Hepialoidea, also have a separate copulatory opening, but this is situated close to the terminal genital opening on segment 9–10.

In most female Lepidoptera each ovary has 4 polytrophic ovarioles, but up to 20 are known in a few Incurvariidae and Aegeriidae. The egg passes from the ovary through an *oviductus* to the *oviductus communis*, and thence to the invaginated genital chamber *(vagina)*, the anterior part of which is often enlarged and termed the *vestibulum*. The egg is fertilized in the vagina before passing to the genital aperture. At copulation the spermatozoa, contained in a spermatophore, are received into an elongate sac, the *bursa copulatrix*. From there they may pass direct to the vagina (monotrysian genitalia), or through a lateral duct, the *ductus seminalis*, to the vestibulum and thence to the *receptaculum seminis* or *spermatheca*, where they are stored before finally returning to the vagina to fertilize the ova (ditrysian genitalia). In exoporian genitalia the spermatozoa pass out from the external opening of the bursa copulatrix, the ostium bursae, usually along a mid-ventral groove to enter the vagina through the genital aperture.

Three sections of the bursa copulatrix are recognized, the ostium, *ductus*, and *corpus bursae*. The ostium bursae is often surrounded by a sclerotized area *(sterigma)*, which is sometimes formed into a projecting tube. The ductus bursae and the corpus bursae may be membranous or variously sclerotized, and the inner wall of the corpus bursae frequently bears one or more sclerotized *signa*, usually in the form of spines or of dentate or roughened patches. In the Ditrysia the orifice of the spermatophore approximates the point in the corpus bursae or ductus bursae from which the ductus seminalis leads to the vestibulum. Before reaching the vestibulum the spermatozoa are sometimes stored temporarily in a diverticulum of the ductus seminalis, the *bulla seminalis*. Paired accessory glands open by a common duct into the vagina. Their secretion is thought by Callahan and Cascio (1963) to

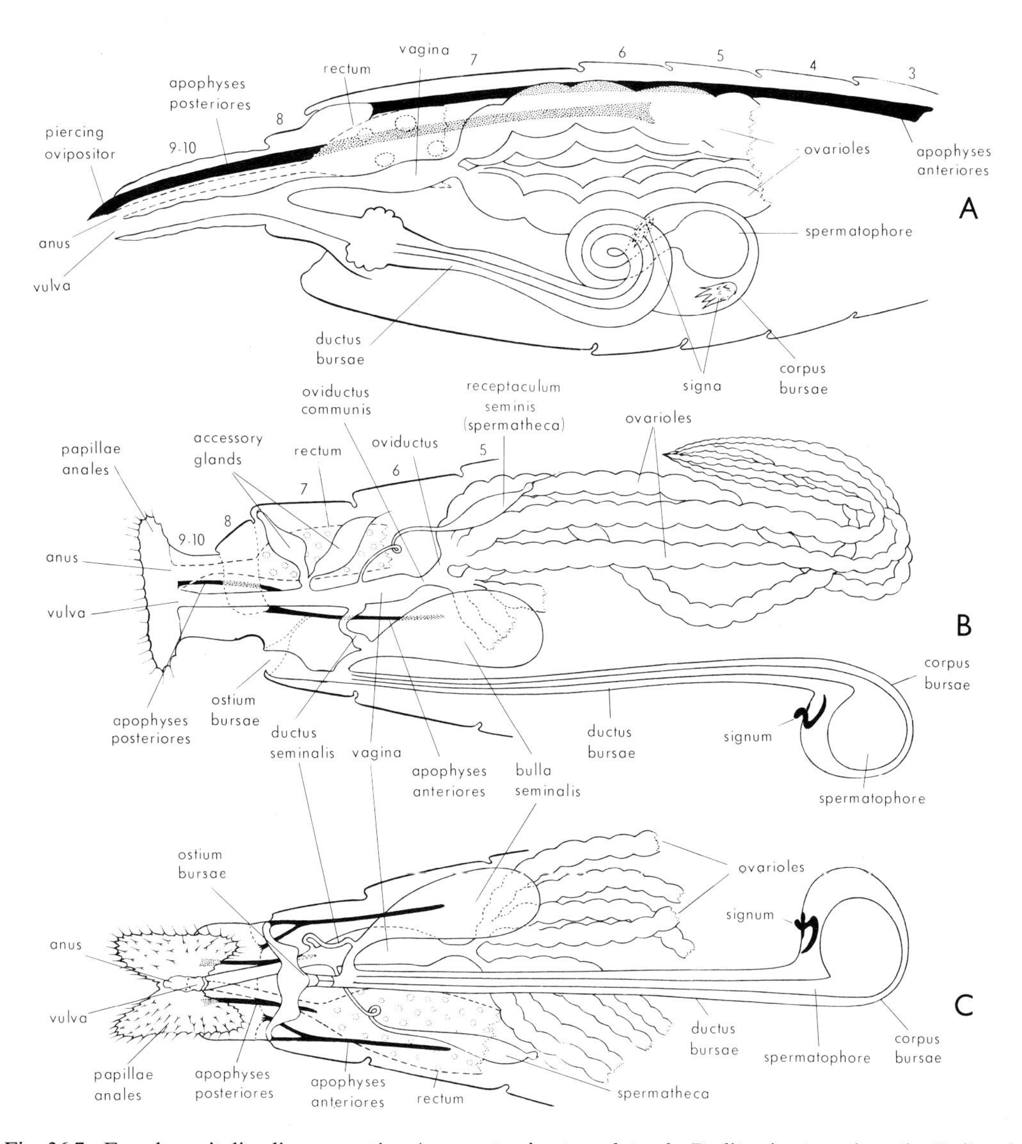

Fig. 36.7. Female genitalia, diagrammatic: A, monotrysian type, lateral; B, ditrysian type, lateral; C, ditrysian type, ventral. [B. Rankin]

function not as an adhesive for the eggs but as a medium for the motile passage of the spermatozoa across the vagina to the spermatheca.

Internal Anatomy. In adult Lepidoptera with a functional haustellum, liquids pass through it to the pharynx, where a partial vacuum is created by muscular dilation of the pharyngeal walls. A pharyngeal valve is then closed, and, by contraction of the pharynx, the liquid continues its passage backwards into the long slender oesophagus. Posteriorly the oesophagus is usually dilated to form a crop. The crop may be a symmetrical dilatation (Hepialidae, Cossidae, some Psychidae, some Tineidae), a lateral dilatation (some Tineidae, Zygaenidae), or a diverticulum connected to the oesophagus by a short narrow duct (most Lepidoptera). A tubular chitinous peritrophic membrane occurs in the mid gut of several families (D. Waterhouse, 1953). Usually 6 Malpighian tubes, in two groups of 3, enter at either side of the anterior end of the hind gut through a pair of excretory chambers. They are reduced to 2 in *Tineola*, *Monopis*, and *Galleria*. The central nervous system of the adult contains fewer ganglia than that of the larva. In the Hepialidae there are 3 thoracic and 5 abdominal ganglia; in the Zeugloptera, some Tineidae, Cossidae, Zygaenidae, and a few other moths, the 4th and 5th abdominal ganglia are fused; in the remainder the ganglia of the meso- and metathorax are fused, and there are 4 abdominal ganglia. The reproductive system has been described above.

Immature Stages

Egg. Two types of egg may be distinguished in the Lepidoptera: a *flat* type and an *upright* type. The flat egg is asymmetrical in horizontal section, the long axis is usually horizontal, and the micropyle is at one end. The upright egg is symmetrical in horizontal section, the long axis is vertical, and the micropyle is at the top. The chorion may be relatively smooth, it may bear a regular or irregular sculptured pattern, or, as in most eggs, especially in the upright type, prominent ribs, usually with cross-ribbing as well. The micropyle is often surrounded by a rosette-like pattern of radiating ribs (Döring, 1955).

Larva. The larva, or caterpillar, exhibits a clear division into head, thorax, and 10-segmented abdomen. Both hypognathous and prognathous forms occur, depending to a large extent upon their feeding behaviour.

The head (Figs. 36.8A–C) is a heavily sclerotized capsule with a very large occipital foramen, and is usually strengthened dorsally by a λ-shaped internal ridge, represented on the dorsal surface by median and lateral *adfrontal* sutures (Hinton, 1947b). Laterally, beyond the adfrontal sutures, there is usually a pair of ecdysial lines, which represent the

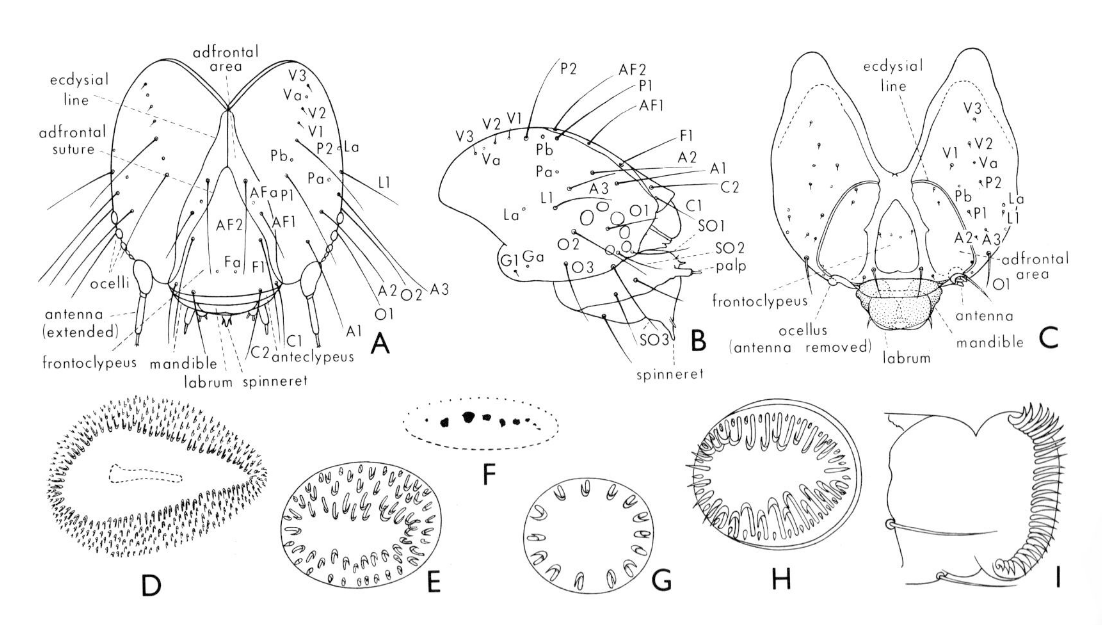

Fig. 36.8. Larvae: A, B, head of *Epiphyas*, Tortricidae, dorsal and lateral; C, head of *Eriocrania*, Eriocraniidae. D–I, arrangement of crochets: D, multiserial (Hepialidae); E, multiserial (Yponomeutidae); F, transverse band (Incurvariidae); G, uniordinal circle (Carposinidae); H, triordinal circle (Pyralidae); I, uniordinal mesoseries (Noctuidae). [B. Rankin]

dorsal lines of cleavage at ecdysis and normally appear only in late or last instars. The triangular area between the lateral adfrontal sutures is a composite frontoclypeus, bordered laterally by the two adfrontal areas. Anterior to the frontoclypeus is the *anteclypeus*, articulating with which is the labrum. The mandibles are normally well developed and dentate, but are modified in the sap-feeding instars of leaf-miners. The maxillae have well differentiated cardo and stipes and 3-segmented palpi. The submentum, mentum, and prementum can be distinguished in the labium, the last usually carrying a median *spinneret* and a pair of minute, lateral, usually 2-segmented palpi. The antennae are usually short, 3-segmented, and bear several sensilla (Dethier, 1941). They are usually situated lateral to the anterior extremity of the adfrontal sutures. Posterior to the antenna is a group of usually 6 'ocelli' (i.e. stemmata, p. 45) which may be reduced in number or absent. Each thoracic segment bears a pair of 5-segmented legs, terminating in a single claw. Occasionally they are modified, reduced, or lost. The prothorax carries a large dorsal sclerotized area, the *prothoracic shield*, and a pair of lateral spiracles.

Spiracles usually occur also on segments 1 to 8 of the abdomen. Leg-like processes *(prolegs)* are usually borne on segments 3 to 6 *(ventral prolegs)* and on 10 *(anal prolegs)*. The truncate end of each proleg, the *planta*, usually bears small sclerotized hooks or *crochets*, the form and arrangement of which are of taxonomic importance (Figs. 36.8D–I). The crochets may be all of one length *(uniordinal)*, of two or three alternating lengths (*bi-* or *triordinal*) or of many lengths *(multiordinal)*. In archaic families they are more usually arranged in a complete circle or ellipse, sometimes in two or more series; or the circle may be broken ('penellipse'). Occasionally they are arranged in transverse bands, and in the more specialized arboreal larvae in a longitudinal medial row *(mesoseries)*, the outer crochets of each circle having been reduced or lost, or in a combined mesoseries and reduced *lateroseries*. The prolegs may be reduced in size or number, and, especially in some of the leaf-miners, may be lost altogether.

The head and body of the larva bear *setae* and *punctures*. *Primary* setae or punctures are almost invariably present in the first instar, and are fairly constant in number and distribution. A few *subprimary* setae, also with definite positions, appear in the second and later instars. The primary setae are mostly long and tactile in function, but there are a few *microscopic* primary setae, thought to be proprioceptors. These occur near the anterior margin of each segment, and near the posterior margin of the prothorax. In many families, especially those with concealed larvae, further setae are not acquired. In others much more generally distributed *secondary* setae appear, and may form tufts or pencils, which may be grouped on sclerotized

TABLE 36.1

Nomenclature and Usual Distribution of the Setae of the Thorax and Abdomen in Larvae of Lepidoptera (Hinton, 1946a).

Tactile Setae	Thorax I	Thorax II	Thorax III	Abdomen 1 – 9
XD1	+	–	–	–
XD2	+	–		–
D1	+	+	+	+
D2	+	+	+	+
SD1	+	+	+	+
SD2	+	+	+	+*
L1	+	+	+	+
L2	+	+	+	+†
L3	±	+	+	+†
SV1	+	+	+	+
SV2	+	±	±	±
SV3	–	–	–	±‡
V1	+	+	+	+
Microscopic or Proprioceptors				
MXD1	±	–	–	–
MD1	–	+	+	+
MD2	–	–	±	±
MSD1	–	+	+	–
MSD2	–	+	+	–
MV1	–	+	+	–
MV2	+	+	+	–
MV3	+	+	+	+

* Always absent on 9.
† Often absent on 9.
‡ Always absent on 8 and 9.

plates or various elevations and processes of the body wall; they are sometimes found even on the mandibles. In some families, such as the Arctiidae, the primary setae are replaced by *verrucae*, tufts of secondary setae borne on raised plates. In others, such as the Lymantriidae, secondary setae are present in the first instar obscuring the primary setae, and later pencils of setae borne on flat plates *(verricules)* may be present.

Pioneering work on the homology of larval setae has been done by a series of workers, notably by Fracker (1915). Perhaps the most generally adopted nomenclature (Table 36.1) is that proposed by Hinton (1946a). The setal patterns (chaetotaxy) are used widely for the identification of larvae and contribute to the higher classification of the Lepidoptera. The distribution of the setae on each segment is usually plotted on a rectangular setal map, which represents the flattened integument extending from the mid-dorsal to the mid-ventral lines (Fig. 36.9).

INTERNAL ANATOMY. The alimentary canal of the lepidopterous larva is a relatively simple tube with few convolutions. The oesophagus is short, the mid gut seldom has caeca or diverticula, and the hind gut is very short. A peritrophic membrane is present in the mid gut. Normally 6 Malpighian tubes, in two groups of 3, open proximally on either side into a small excretory chamber leading to the hind gut. Paired labial silk glands, homologous with the salivary glands in other insects, lead through a common duct to the spinneret, within which lies the spinning apparatus. Paired accessory glands usually open into the silk ducts near their anterior end. The tube-like silk glands are often extremely long, especially in the Bombycoidea. Paired mandibular glands, which function as salivary glands, usually open on the inner side of each mandible near the base. The central nervous system consists of the brain, suboesophageal ganglion, 3 thoracic ganglia joined by paired connectives, and 7 or 8 abdominal ganglia joined by single (fused) connectives. The gonads are situated in abdominal segment 5, on each side of the dorsal vessel, and are visible through the integument in male larvae of many less advanced families.

Pupa. The term pupa is used in this chapter in the conventional sense. It refers to the entire insect from the time the last larval skin was shed until the shedding of the pupal cuticle by the emerging adult.

The most extensive study of lepidopterous pupae is that of Mosher (1916). The various components of a pupa are demarcated by sutures (Fig. 36.10), some of which may be lacking in the more specialized families. The head, thorax, and abdomen can be readily distinguished. In the head the fronto-clypeal and clypeo-labral sutures are usually not distinct. Posterolateral projections of

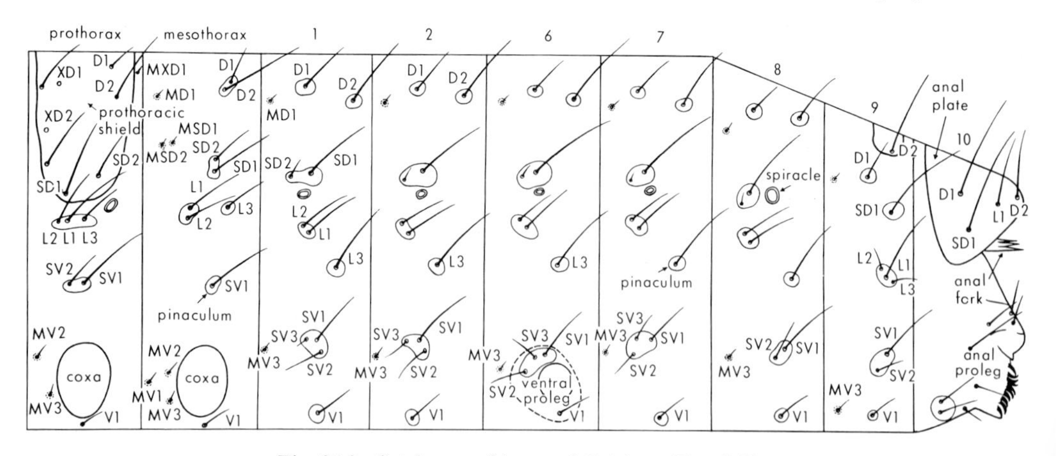

Fig. 36.9. Setal map of larva of *Epiphyas*, Tortricidae.
[B. Rankin]

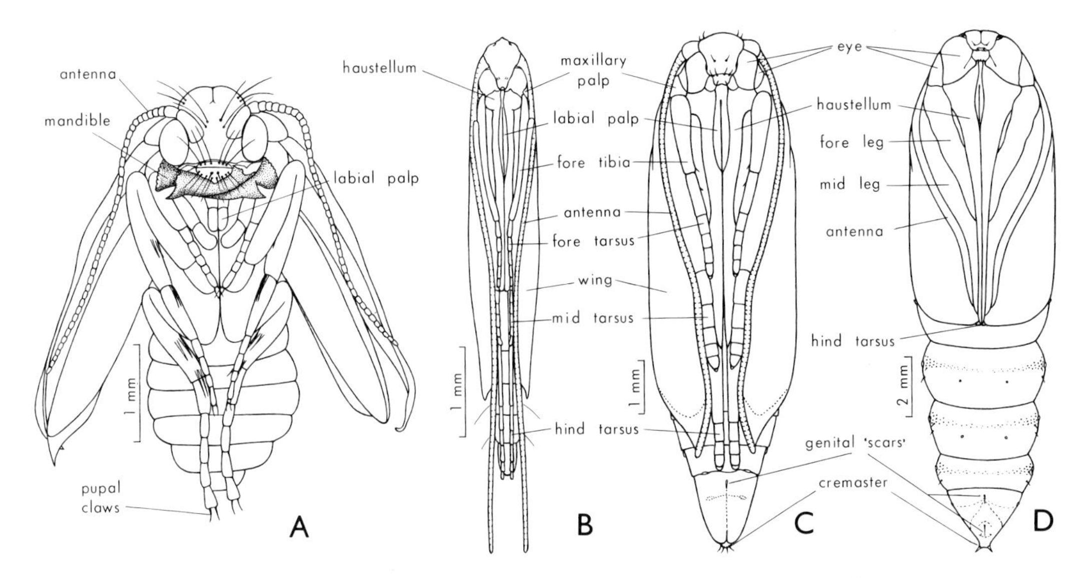

Fig. 36.10. Pupae, ventral. A, decticous exarate (*Agathiphaga*, Eriocraniidae, with wings and antennae displaced). B–D, adecticous obtect: B, *Acrocercops*, Gracillariidae; C, *Yponomeuta*, Yponomeutidae; D, *Persectania*, Noctuidae. [B. Rankin]

the labrum, the *pilifers*, are recognizable in a few Yponomeutoidea, most Pyraloidea, and in Pterophoroidea, Hesperioidea, and Papilionoidea. Functional mandibles are found only in the Zeugloptera and Dachnonypha (pupae decticae), and are operated by the muscles of the future adult to cut open the cocoon or to assist escape (Fig. 36.10A). Small non-functional mandibles, adjacent to the posterolateral angles of the labrum, are present in the pupae of many families (pupae adecticae, p. 105). Maxillary palpi are frequently present near the posterior margin of the eyes. The labial palpi are usually visible, but may be concealed by the haustellum. The head of the pupa often bears a pointed projection which aids its escape from the cocoon.

The three thoracic segments are visible dorsally, and are of nearly equal length in the generalized species; in the remainder, the mesothorax is the longest. One pair of spiracles is visible, between pro- and mesothorax. Primitive forms have exarate pupae, whereas the appendages are glued to each other and to the body to various degrees in the obtect pupae of more specialized forms. In some species with brachypterous females the wings of the female pupa show little reduction, but in *Orgyia* and the more specialized Psychidae they are greatly reduced.

Of the 10 abdominal segments, independent movement is never possible in the last 3. The number of free or movable segments is greatest in the most primitive forms. Segments 1 to 7 are free in the Zeugloptera and the Dachnonypha, and 2 to 7 or 2 to 6 in most of the Monotrysia and in some of the lower Ditrysia. The male usually has one more movable segment than the female. In the higher Ditrysia only 3, or even fewer, abdominal segments are movable. Pupae with more than 3 movable segments are sometimes known as 'pupae incompletae'. Segments 1–8 of the abdomen have spiracles, but those on 1 are seldom visible, and those on 8 are non-functional. Segment 10 is sometimes modified to form the *cremaster*, when it may bear a series of usually hooked setae used to attach the pupa. The cremaster is seldom present in the more generalized forms. Segment 10 also bears a circular or slit-like 'scar' which represents the anus. The genital openings of the ditrysian adult are represented by 'scars' on sternum 9 in the male and sterna 8 and 9 in the female. The mobile abdomen of the Monotrysia and the more primitive Ditrysia is often armed with backwardly directed dorsal spines, which force the pupa forward when the abdomen is wriggled. These pupae partially emerge from the cocoon or larval tunnel just before ecdysis. In the more advanced forms ecdysis occurs within the cocoon or shelter, and adult emergence is facilitated in various ways. The cocoon often has a weak area, or a slit-like or funnel-like opening at one end, which permits the adult to escape easily (e.g. *Anthela*). The newly emerged adult sometimes regurgitates a fluid which softens the silk *(Cerura)*, sometimes cutting a hole through the softened silk by means of a sharp hook on the first axillary sclerite of the fore wing *(Antheraea)*.

Biology

The remarkable success of the Lepidoptera seems to have been associated directly with the development of the higher plants, especially the angiosperms, in the Upper Cretaceous and Tertiary. A predominantly phytophagous group, it has demonstrated its great plasticity and adaptability in exploiting most parts of the plant.

Adults. The evolution of the haustellum probably contributed substantially to the success of the Lepidoptera. Not only did it enable the adults to ingest water freely, thus increasing their resistance to desiccation, but it also permitted utilization of the carbohydrates contained in nectar, which the lepidopteron can convert to, and store as, fat. In some species carbohydrate food is essential before egg maturation can ensue; in others it greatly increases egg output; in most it increases longevity. Sometimes the ingestion of carbohydrates is necessary before mating can occur. On the other hand, feeding is often not possible in forms with greatly reduced mouth-parts, and the eggs are ready for deposition when the adult sheds the pupal cuticle.

At rest the wings are folded in a variety of ways. The ancestral method of folding the wings roofwise above the abdomen, with the hind wing covered by the fore wing (Fig. 36.26A), is characteristic of most moths. The slope of the fore wing in such species varies from almost horizontal to almost vertical. Many exceptions to this resting position may be noted. In many Geometridae and some other groups, the wings are expanded flat and appressed to the substrate; in most Papilionoidea they are held back to back above the body; in other groups they may be variously rolled, twisted, or raised in repose.

Flight is not possible unless the temperature of the flight muscles is high enough, and therefore moths often vibrate their wings preparatory to flight. Diurnal species, such as butterflies, usually depend on the heat of the sun to raise their temperature sufficiently. A body temperature at least 10°C higher than that of the surrounding air is reached during flight in the hepialid *Trictena* (Tindale, 1935), and up to 18°C higher in sphingids (Dorsett, 1962).

Apart from normal dispersal movements, migratory flights of varying magnitude have long been observed in Lepidoptera (Williams, 1958). In Australia, the most conspicuous migrants are butterflies, especially the pierids *Anaphaeis java* (Sparrman) and *Catopsilia* spp., the nymphalid *Vanessa kershawi* (McCoy), and the hesperiid *Badamia exclamationis* (F.). The noctuid moth, *Agrotis infusa* (Boisd.) is known to undertake a remarkable annual two-way migration (p. 108).

Variation due to either environmental or genetic factors is common in Lepidoptera. Seasonal variation, for example, is found in *A. infusa*: a sombre spring form with dark hind wings and a much paler autumn and winter form with whitish hind wings (Common, 1958a). Many caterpillars are kentromorphic (p. 139), a green or pale-coloured phase being characteristic of a low population level and a conspicuously striped (probably aposematic) or dark-coloured phase of a high population level. Geographical variation, which may be of either environmental or genetic origin, is well demonstrated in the butterflies, especially the Satyrinae. Two genetically distinct racial groups of *Tisiphone abeona* (Don.) have been recognized in coastal south-eastern Australia, with an extremely variable population occupying a small zone of contact between them (G. Waterhouse, 1928). Sexual dimorphism is marked in many Lepidoptera, with the sexes of such extreme examples as *Epitymbia cosmota* (Meyr.) having been described not only as separate species, but as distinct genera. The problem of balanced polymorphism in Lepidoptera has received little attention in Australia, but the two forms of *Catopsilia pyranthe* (L.), with either black or pink antennae, may be an example.

Reproduction. Male moths may be attracted to females of the same species by visual stimuli or by pheromones. The latter are thought to be effective over considerable distances, but there is some evidence that far infrared radiation may also be important in bringing the sexes together (Callahan, 1965). When the male has perceived the female, mating may be preceded by the emission of male pheromones which evoke the mating response in the female. These may be secreted by special glands of the abdomen, legs, or wings, androconia and extensile brushes of scales aiding in their dissemination.

Olfactory stimuli provided by the essential oils of the food plant are believed to induce oviposition. The average number of eggs laid varies from a few dozen to many thousands; more than 18,000 are laid by *Xyleutes* and *Abantiades*. The eggs may be deposited singly or in batches, near or on the food plant; the oviposition habits are characteristic of the species and often of the genus. Some of the Hepialidae scatter their eggs at random during flight *(Hepialus)*, others in loose masses on the ground (some *Oncopera*); *Ptilomacra senex* Walk. (Cossidae) arranges its eggs in a cylindrical pattern around small twigs; *Epiphyas postvittana* (Walk.) and many other Tortricidae lay their scale-like eggs slightly overlapping one another in batches; the Limacodidae also lay scale-like eggs; *Persectania ewingii* (Westw.) and other hadenine Noctuidae insert their eggs with their wedge-shaped papillae anales beneath the leaf sheaths of grasses; and the Incurvariidae and many Eriocraniidae insert theirs into plant tissues with their sclerotized cutting ovipositors. In some species of *Monopis* (Tineidae) the fertilized eggs are retained in a grossly enlarged vagina until the embryo is mature; hatching follows immediately the eggs are laid.

Parthenogenesis, as a regular feature of the life cycle, has been reported only in special races of European Psychidae. Occasionally, however, it occurs in individuals of species which normally reproduce bisexually (e.g. Pieridae, *Galleria*).

Immature Stages. In some species the newly hatched larva makes its first meal of the chorion. Many leaf-miners tunnel direct from the egg into the leaf tissue beneath (e.g. Nepticulidae). Although most lepidopterous larvae are phytophagous, a few are carnivorous, feeding on the egg masses of other Lepidoptera *(Titanoceros)* and of spiders *(Stathmopoda)*, on living ant larvae (some Lycaenidae, Cyclotornidae), or on scale insects *(Batrachedra, Stathmopoda, Catoblemma)*. Cannibalism is not uncommon amongst phytophagous larvae reared in confined spaces (e.g. *Heliothis*, *Agrotis*, Lycaenidae), but probably does not occur commonly in nature unless populations are exceptionally dense. Some larvae feed on materials of animal origin, such as wool, keratin, etc. (e.g. *Tineola*), digestion being permitted by the secretion of an enzyme which breaks the disulphide linkage of the wool molecule. A few species (Epipyropidae, early stages of Cyclotornidae) are ectoparasitic on leafhoppers.

There are few plant habitats that have not been exploited by Lepidoptera, and every part of the plant provides food: the roots, trunk, bark, branches, twigs, leaves, buds, flowers, fruits, seeds, galls, and fallen seeds, fruit, or leaves. Larvae that feed in concealed situations, such as the borers, leaf- and bark-miners, leaf-tiers, and case-bearers, usually belong to the more primitive families, whereas the exposed feeders, especially those that feed during the day, usually belong to the more advanced families.

Many who live in inland Australia have observed that Lepidoptera at times are among the most abundant insects in dry country. Under such conditions, diurnal species, such as butterflies, are usually scarce. The most abundant are those whose larvae feed in concealment and so avoid desiccation. The adults are usually on the wing for only a few days, especially just after rain, during which time they mate and lay their eggs. Of special note are the Hepialidae and Cossidae, the larvae of which either tunnel in roots, or feed externally on roots which tap subterranean supplies of water. The Oecophoridae, Xyloryctidae, and Gelechiidae are other families with larvae tunnelling in stems, galls, seed capsules, or living in concealed shelters, which have effectively exploited the inland. Certain other species utilize the rapidly responding vegetation which follows sporadic rains, emerging as adults, laying their eggs, and rapidly completing their larval development, before entering the soil and remaining quiescent as larvae or pupae until further rain falls. Included here are many species in the Noctuidae, Geometridae, and Pyralidae which feed on herbs.

Although concealment in itself provides some measure of protection from desiccation, rapid changes in temperature, and other physical hazards, it by no means protects the larva from potential parasites and predators. Even vertebrates, such as the black cockatoo, are able to detect and extract wood-boring larvae and pupae from deep within tree trunks.

Larvae that feed in exposed situations have developed a remarkable range of adaptations which lessen potential hazards. Procryptic concealment, by means of green, grey, and brown coloration, resembling that of the leaves, bark, and other plant parts with which the insect associates, aids it to escape the attention of predators. The twig-like larvae of many Geometridae, and the larvae of Lasiocampidae merging with the branch to which they closely appress their hairs, are good examples. Once detected, the larva may then rely on an alternative mechanism to startle, or escape from, its enemy. Papilionid larvae display brightly coloured and pungent-smelling osmeteria; *Entometa* (lasiocampid) larvae arch their thorax and expose black, elliptical, mouth-shaped areas; many larvae rear up and strike at an intruder with the head, at the same time often regurgitating a brightly coloured fluid; and others feign death and drop to the ground or lower themselves rapidly on a thread of silk. Similar mechanisms are exhibited by adult moths. When disturbed, many species suddenly display brightly coloured hind wings which were concealed beneath the procryptic fore wings while the insect was at rest. Adults of some Arctiidae and several other families feign death when handled, and *Rhodogastria* (Arctiidae) emits a pungent, yellow, frothy liquid from glands at the bases of the wings. During flight Pyralidae and Noctuidae respond to the ultrasonic cries of bats by making vigorous evasive movements (Belton, 1962; Dethier, 1963).

Aposematic or warning coloration, which might be associated with distasteful or injurious properties, is common in the diurnal larvae of several lepidopterous families. The brightly coloured bands found in *Danaus* and *Euploea* (Nymphalidae) and many Agaristidae probably discourage predators. Some of these species may be genuinely distasteful, whereas others benefit from the protection afforded by similar coloration. Day-flying moths frequently have aposematic colour patterns, some, such as the Amatidae and Aegeriidae, having orange and black bands on the abdomen and partially transparent wings resembling wasps. The orange and black body and transparent wings of the sphingid *Cephonodes kingi* (Macl.) suggest a large bumble bee, while the brown, yellow, and black markings of the small stathmopodid *Snellenia lineata* (Walk.), reinforced by its behaviour at flowers, bear a striking resemblance to the distasteful *Metriorrhynchus* beetles which also visit flowers (p. 132). The larva of *Homodes* (Noctuidae) has a series of lateral, clubbed, thread-like processes which are constantly waved as it moves, giving it a startling resemblance to two of the *Oecophylla* ants which frequent the same host plant.

Hairy caterpillars generally seem to be distasteful to predators, although cuckoos and certain other birds are able to eat them with impunity. Modified setae of various kinds produce a stinging sensation if they contact the human skin, while others cause dermal rashes or urticaria. In *Doratifera* (Limacodidae) paired dorsal processes contain eversible tufts of brightly coloured stinging hairs, which resemble minute sea-anemones. Contact produces a sensation similar to a mild nettle sting. Hairy caterpillars of several families are able to cause skin rashes in humans. No Australian Lasiocampidae are known to do this, but in the northern hemisphere *Malacosoma* is notorious. Amongst the Bombycoidea, *Anthela nicothoe* (Boisd.) (Anthelidae) and the gregarious *Panacela lewinae* (Lew.) (Eupterotidae) can both cause skin rashes. Both in Australia and Europe the processionary caterpillars of the subfamily Thaumetopoeinae (Notodontidae) are known to have urticating properties. The gregarious *Ochrogaster contraria* (Walk.) is a good example. In Australia urticaria is most usually caused by larvae of the lymantriids *Euproctis edwardsi* (Newm.) and *Leptocneria reducta* (Walk.). Australian Arctiidae apparently do not have serious urticating properties, although rashes caused by *Spilosoma glatignyi* (le Guillou) have been reported. The method by which larval setae cause urticaria is not well understood, and the problem is complicated by the occurrence of human allergies.

Natural Enemies. At any stage in its life the lepidopteron may fall prey to a great variety of predators and parasites. Perhaps the greatest loss of eggs is due to parasitism by Chalcidoidea. The larvae provide food for mites, spiders, wasps (especially Eumenidae and Vespidae), and a great number of vertebrates, especially birds. The fat-rich larvae of certain Cossidae were important items in the diet of the Aborigines. Both larvae and pupae are prone to parasitism by Mermithidae (Nematoda), Chalcidoidea, Braconidae, Ichneumonidae, and Tachinidae. Bacterial and viral diseases at times take their toll of both, drastically reducing populations. Adult moths and butterflies are preyed upon mainly by vertebrates, birds probably destroying the greatest numbers. Many moths, especially those which shelter in the leaf litter, have parasitic mites attached to their antennae and other parts of their bodies, while some Noctuidae (e.g. *Pseudaletia convecta* (Walk.)) harbour colonies of mites within their tympanal organs (Treat, 1955). Bogong moths are parasitized in their aestivation camps by the mermithids *Amphimermis bogongae* Welch and *Hexamermis cavicola* Welch, and were also used as food by the Aborigines.

Economic Significance. Being predominantly phytophagous, the Lepidoptera contribute substantially to the economic losses suffered by man's agricultural crop plants. In Australia the majority of lepidopterous pests are native species, some of which occur naturally also in New Guinea and south-east Asia. Of special note are the underground grass grubs (Hepialidae), borers (Hepialidae, Cossidae, Aegeriidae, Xyloryctidae), leaf-miners (Gracillariidae), leaf-rollers (Tortricidae), leaf-tiers and web-worms (Pyralidae), forest defoliators (Limacodidae, Geometridae, Eupterotidae, Nolidae), and cutworms and armyworms (Noctuidae). The introduced species are mainly cosmopolitan pests of stored foodstuffs and fibres (Tineidae, Oecophoridae, Gelechiidae, Pyralidae), and of orchard (Tortricidae) and vegetable crops (Yponomeutidae, Gelechiidae, Pieridae).

Only a few Lepidoptera are obviously beneficial to man. Species which prey on coccids (*Batrachedra*, *Stathmopoda*, *Catoblemma*) and parasitic species (Epipyropidae) which attack leafhoppers presumably assist in the control of these pests. *Titanoceros* destroys the egg masses of the bag-shelter moth (*Ochrogaster*), a minor pest of shade and fodder trees. Other species, such as those that assist in reducing leaf litter to humus, are probably beneficial in a less tangible way. A few exotic Lepidoptera have been introduced to assist in the control of major introduced weeds. Of these the most outstanding has been the South American pyralid *Cactoblastis* which effectively controlled prickly pear (*Opuntia* spp.) in Queensland and N.S.W. (p. 116).

Special Features of the Australian Fauna

The major families of Lepidoptera are all represented in Australia, but the relative abundance of the species in some groups differs markedly from that of other continents. Some sections represent a relict fauna, probably largely of Oriental origin, upon which a much later fauna of similar origin has been superimposed in several waves.

Archaic families, such as the Hepialidae, Cossidae, and Castniidae, have reached a great degree of development, but whether this radiation took place in the absence of more advanced groups in the continent, or in competition with those groups in response to increasing aridity, is not clear. The larvae of each of these families are adapted to feed within stems and roots or deep in the soil externally on roots, which would greatly favour their persistence in an arid environment. Concealed feeding and an outstanding adaptation to a dominant flora of *Eucalyptus* and *Acacia* have probably also contributed to the abundance of species in the Oecophoridae and the Xyloryctidae. The last two families comprise more than a quarter of the fauna, and are especially abundant in the *Eucalyptus* forests and heathlands in both coastal and inland areas.

Groups which have reached Australia more recently, and which are still most abundant in the north, include the Aegeriidae, Timyridae, Papilionidae, Pieridae, Uraniidae, Sphingidae, Lymantriidae, Hypsidae, and Amatidae, together with large sections of the Gelechiidae, Pyralidae, Geometridae, Notodontidae, and Noctuidae. Nevertheless, some of these families contain large blocks of endemic genera, notably the Oenochrominae (Geometridae) and the Thaumetopoeinae (Notodontidae). The Anthelidae, Cyclotornidae, and Carthaeidae are confined to the Australian region.

CLASSIFICATION

Order LEPIDOPTERA (11,221 Australian spp.)

Suborder ZEUGLOPTERA (6)

MICROPTERIGOIDEA (6)

1. Micropterigidae (6)

Suborder DACHNONYPHA (2)

ERIOCRANIOIDEA (2)

2. Eriocraniidae (2)

Neopseustidae (0)

Mnesarchaeidae (0)

Suborder MONOTRYSIA (254)

HEPIALOIDEA (112)

3. Protothеoridae (1)
4. Palaeosetidae (2)
5. Hepialidae (109)

NEPTICULOIDEA (75)

6. Nepticulidae (55)
7. Opostegidae (20)

INCURVARIOIDEA (67)

8. Incurvariidae (35)

Prodoxidae (0)

9. Heliozelidae (32)

Tischeriidae (0)

Suborder DITRYSIA (10,959)

COSSOIDEA (99)

10. Cossidae (99)

Metarbelidae (0)

TORTRICOIDEA (822)

11. Tortricidae (809)
12. Phaloniidae (13)

TINEOIDEA (678)

Pseudarbelidae (0)

Arrhenophanidae (0)

13. Psychidae (152)
14. Tineidae (146)
15. Lyonetiidae (128)
16. Phyllocnistidae (30)
17. Gracillariidae (222)

YPONOMEUTOIDEA (317)

18. Aegeriidae (19)
19. Glyphipterigidae (123)
20. Douglasiidae (1)
21. Heliodinidae (1)
22. Yponomeutidae (158)
23. Epermeniidae (15)

GELECHIOIDEA (4,062)

24. Coleophoridae (4)
25. Agonoxenidae (1)
26. Elachistidae (22)
27. Scythridae (14)
28. Stathmopodidae (137)
29. Oecophoridae (2,445)
30. Ethmiidae (12)
31. Timyridae (87)
32. Cosmopterigidae (394)

Metachandidae (0)

Anomologidae (0)

Pterolonchidae (0)

33. Blastobasidae (11)
34. Xyloryctidae (418)
35. Stenomidae (34)
36. Gelechiidae (483)

*Physoptilidae (0)

* Previously known from 2 Indian species; an undescribed species has recently been taken at Cape York.

Strepsimanidae (0)
COPROMORPHOIDEA (78)
37. Copromorphidae (7)
38. Alucitidae (13)
39. Carposinidae (58)
CASTNIOIDEA (29)
40. Castniidae (29)
ZYGAENOIDEA (114)
Heterogynidae (0)
41. Zygaenidae (21)
Chrysopolomidae (0)
Megalopygidae (0)
42. Cyclotornidae (5)
43. Epipyropidae (8)
44. Limacodidae (80)
PYRALOIDEA (1,155)
45. Hyblaeidae (4)
46. Thyrididae (49)
47. Tineodidae (10)
48. Oxychirotidae (2)
49. Pyralidae (1,090)
PTEROPHOROIDEA (33)
50. Pterophoridae (33)
HESPERIOIDEA (106)
51. Hesperiidae (106)
Megathymidae (0)
PAPILIONOIDEA (253)
52. Papilionidae (18)
53. Pieridae (31)
54. Nymphalidae (79)
55. Libytheidae (1)
56. Lycaenidae (124)
GEOMETROIDEA (1,244)
57. Drepanidae (5)
Thyatiridae (0)
58. Geometridae (1,203)
59. Uraniidae (9)
60. Epiplemidae (27)
Axiidae (0)
Sematuridae (0)
CALLIDULOIDEA (1?)
61. Callidulidae (1?)
Pterothysanidae (0)
BOMBYCOIDEA (172)
Endromidae (0)
62. Lasiocampidae (76)
63. Anthelidae (73)
64. Eupterotidae (9)
Lacosomidae (0)
Bombycidae (0)
Lemoniidae (0)
Brahmaeidae (0)
65. Carthaeidae (1)
Oxytenidae (0)
Cercophanidae (0)
66. Saturniidae (13)
Ratardidae (0)
SPHINGOIDEA (58)
67. Sphingidae (58)
NOTODONTOIDEA (89)
Dioptidae (0)
68. Notodontidae (89)
Thyretidae (0)
NOCTUOIDEA (1,650)
69. Lymantriidae (74)
70. Arctiidae (248)
71. Amatidae (54)
72. Hypsidae (8)
73. Nolidae (91)
74. Noctuidae (1,138)
75. Agaristidae (37)

The Lepidoptera are here divided into the four suborders, Zeugloptera, Dachnonypha, Monotrysia, and Ditrysia. The first three include only about 2 per cent of the known species and comprise the original Monotrysia of Börner, who divided the order in 1939 into the Monotrysia and the Ditrysia, according to the occurrence of one or two genital openings in the female. Hinton (1946a) separated two primitive groups, the Zeugloptera and the Dachnonypha, from Börner's Monotrysia, for the first of which (following Chapman in 1917) he proposed ordinal status. The Zeugloptera, he considered, differed basically from the Dachnonypha by the structure of the adult and of the larva. The discovery of *Agathiphaga* by Dumbleton (1952), which Hinton (1958a) referred to the Dachnonypha, reduced the number of adult structures by which the Zeugloptera could be distinguished from the Dachnonypha. The larvae of the Zeugloptera now provide the main characters separating them from other Lepidoptera.

KEYS TO THE SUBORDERS OF LEPIDOPTERA

ADULTS

1. Venation of fore and hind wings similar; hind wing with 10 or usually more veins, excluding anals, reaching margin 2
   Venation of fore and hind wings dissimilar; hind wing with 8 or usually fewer veins, excluding anals, reaching margin 4
2. Maxillary palpi minute or vestigial .... MONOTRYSIA (pt., p. 787)
   Maxillary palpi 5-segmented, folded (3-segmented in *Mnesarchaea*) 3
3. Mid tibia without spurs, but with apical bristles .... ZEUGLOPTERA (p. 784)
   Mid tibia with at least one spur .... DACHNONYPHA (p. 785)
4. Maxillary palpi conspicuous, 5-segmented, folded 5
   Maxillary palpi inconspicuous (if folded), porrect, or vestigial 6
5. Fore wing with venation reduced to 3 or 4 unbranched veins, or R branched and strongly curved downwards; if venation of fore wing not so, hind wing with $Sc+R_1$ remote from Rs and R_1 strong and forming a basal fork with Sc .... MONOTRYSIA (pt., p. 787)
   Fore wing with venation not so; hind wing with $Sc+R_1$ not remote from Rs, and R_1 either completely fused with Sc, or much weaker than Sc and not joining it close to base .... DITRYSIA (pt., p. 793)

6. Small (less than 18 mm expanse) diurnal species; if head with smooth shining scales, then ocelli absent, labial palpi very short, drooping, and wings lanceolate; if head with erect hair-scales, then antennae much longer than fore wing, which has brilliant metallic markings .... MONOTRYSIA (pt., p. 787)
   Small to very large (3 to 250 mm expanse) species; if very small and with smooth shining scales on head, then not with lanceolate wings and very short drooping labial palpi, or, if so, ocelli present; if erect hair scales on head, then not with both very long antennae and metallic markings on fore wing .... DITRYSIA (pt., p. 793)

LARVAE

1. Head with 5 conjoined ocelli on each side; antennae long; thoracic legs 3-segmented .... ZEUGLOPTERA (p. 784)
   Head never with 5 ocelli on each side, when more than 1 ocellus, then well spaced; antennae short; thoracic legs 5-segmented or absent 2
2. All abdominal prolegs absent or vestigial, without crochets; thoracic legs usually absent 3
   At least one pair of abdominal prolegs present, if reduced, always indicated by crochets; thoracic legs usually present 6
3. Thoracic legs present .... DITRYSIA (pt., p. 793)
   Thoracic legs absent 4
4. Head with ecdysial line joining anterior margin well lateral to adfrontal suture and behind antennae, or adfrontal suture absent .... DACHNONYPHA (p. 785)
   Head with ecdysial line joining adfrontal suture near anterior margin 5
5. Frontoclypeus narrower posteriorly; paired ventral protuberances sometimes present on meso- and metathorax and on abdominal segments 2–7; body sometimes very slender and elongate, about ten times as long as wide; ocelli 1 or 2 on each side, or vestigial .... MONOTRYSIA (pt., p. 787)
   Frontoclypeus wider posteriorly, or extending upwards to the vertex; paired ventral protuberances absent; body not very slender; ocelli 1 or more on each side .... DITRYSIA (pt., p. 793)
6. Anal prolegs absent 7
   Anal prolegs present 8
7. Prolegs on abdominal segments 3–6 reduced, but indicated by one or more uniserial or multiserial transverse bands of crochets; larvae leaf-miners or case-bearers ...... MONOTRYSIA (pt., p. 787)
   Prolegs on abdominal segments 3–6 prominent, with crochets uniordinal or biordinal in a mesoseries or both a mesoseries and reduced lateroseries; larvae external feeders . DITRYSIA (pt., p. 793)
8. Crochets in a multiserial complete ellipse; maxillary palpi with 3 free segments; 6 ocelli present, the 4 anterior ones arranged in an oblong group .... MONOTRYSIA (pt., p. 787)
   Crochets otherwise; maxillary palpi with 2 free segments; if 6 ocelli present, then 4 anterior ones arranged in a semicircle .... DITRYSIA (pt., p. 793)

Suborder ZEUGLOPTERA

Very small, diurnal, metallic moths; mandibles present, dentate, functional (Fig. 36.12B); lacinia erect, functional (Fig. 36.12C); galea unspecialized; hypopharynx forming special triturating basket for grinding pollen grains; wings aculeate; fore and hind wing with similar venation; jugum present, frenular bristles often present at base of hind wing, frenulum absent; tegulae well developed; female with bursa copulatrix opening into the vagina which, with the rectum, forms a cloaca, with a single opening on sternum 9–10, without apophyses. Larva with prolegs represented by a pair of conical processes, without muscles, on abdominal segments 1–8; thoracic legs with fused coxa, trochanter, and femur; antennae long; adfrontal sutures absent; ocelli 5, grouped. Pupa decticous; mandibles not hypertrophied.

The Micropterigidae are usually accepted as archaic Lepidoptera, with which they share more characters than with other panorpoid orders. These include in the adult the presence of the lacinia, the unspecialized galea, and functional mandibles (as in

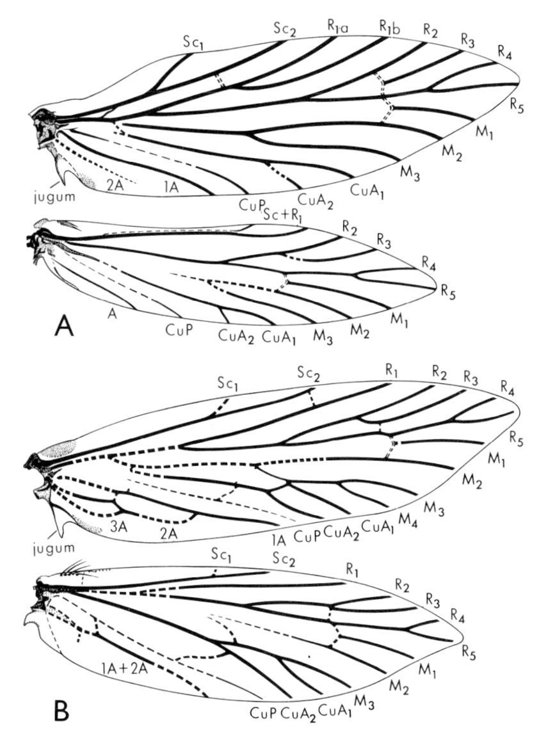

Fig. 36.11. Wing venation of Zeugloptera and Dachnonypha: A, *Sabatinca*, Micropterigidae; B, *Agathiphaga*, Eriocraniidae. [J. Wedgbrow]

Agathiphaga), 2 ocelli (never 3), an apical invaginated sensory area of the labial palpi, a tergopleural apodeme, equally membranous wings, the clothing of regularly overlapping broad striated scales on the body, legs, and wings, well-developed tegulae, and an epiphysis of the fore tibiae. In the pupa the wing tracheation resembles that of the adult. As in other Lepidoptera, the wings of the adult lack a separate vein M_4 running to the wing margin, a thyridium, and an anal plate, and, as in *Agathiphaga*, the larval head is without adfrontal sutures.

The suborder retains many archaic characters, but has clearly diverged from the lepidopterous stem. Specialized characters of the larva include the 5 grouped ocelli, the long antennae, the 3-segmented thoracic legs, and the specialized, sometimes scale-like setae. In both sexes the adult too has evolved curious organs, perhaps sensory in function, on abdominal segment 5. Somewhat similar organs are found on this segment in *Agathiphaga* and other Dachnonypha.

1. Micropterigidae. Head (Fig. 36.12A) rough-haired; ocelli present; antennae moniliform, submoniliform, or filiform; maxillary palpi 5-segmented; labial palpi 2- to 4-segmented; epiphysis often present, tibial spurs 0-0-4, mid tibiae with terminal spines; fore wing (Fig. 36.11A) with humeral vein present, Sc forked, R_1 sometimes forked, chorda present, M forked, CuP present, 3A apically confluent with 1A, or 2A and 3A greatly reduced; cross-veins present towards base between CuP, 1A, and 2A; hind wing with R_1 confluent with Sc_2 in Australian species, base of R_1 lost, 2A and 3A absent; male with sternum and tergum 9 sometimes fused. Egg ovoid, with blunt spines. Larva slug-like, ventral abdominal processes in young larva each with an apical claw. Pupa with pointed mandibles used for opening the cocoon.

The adults are difficult to see when they fly in patchy sunlight and shade, but are sometimes taken at light or when sweeping foliage. They are attracted to flowers to feed on the pollen. The minute larvae were thought to feed on mosses and liverworts, but recent work suggests that some may be detritus feeders. They pupate in a strong, parchment-like, oval cocoon. Australian species of *Sabatinca* (6 spp.) differ from those in New Zealand by the forking of R_1 and the reduction of 2A and 3A in the fore wing, and by the 4-segmented labial palpi. *S. calliplaca* Meyr. (Fig. 36.14A) and another species are from rain forest in southern Queensland and N.S.W. Three others, including *S. sterops* Turn. and *S. porphyrodes* Turn., occur in rain forest in northern Queensland, and another in temperate rain forest in western Tasmania. Careful collecting, especially in rain forest, may well disclose further species.

Suborder DACHNONYPHA

Very small, usually diurnal moths; mandibles present or absent; lacinia sometimes present (Figs. 36.12F, G); galea unspecialized (Fig. 36.12F) or elongated into short functional haustellum (Fig. 36.12G); wings aculeate, fore and hind wing with similar venation; jugum usually projecting, frenular bristles usually present, frenulum absent; tegulae well developed; male usually with pair of short anterolateral apodemes on sternum 8 or 9; female with single genital opening on sternum 9–10. Larva apodous; cleavage line terminating on anterior margin of head behind antenna, with seta A_1 included in adfrontal area, or adfrontal sutures absent. Pupa decticous; mandibles hypertrophied.

Of the 3 families in the suborder, the Mnesarchaeidae are confined to New Zealand, the Neopseustidae contain 3 species in India and Formosa, and the Eriocraniidae are mainly Holarctic, but include 2 species from Australia and one from Fiji. The early stages are known only for the Eriocraniidae. As in the Zeugloptera, the suborder exhibits many archaic characters, especially in the pupa and adult. These two suborders, alone in the Lepidoptera, feature decticous pupae. Primitive characters of adult Dachnonypha include the presence of lacinia and functional mandibles *(Agathiphaga)*, the reduction of which in other genera is accompanied by the elongation of the galea into a haustellum. The larvae are specialized leaf-miners, or, in *Agathiphaga*, specialized for life within a seed.

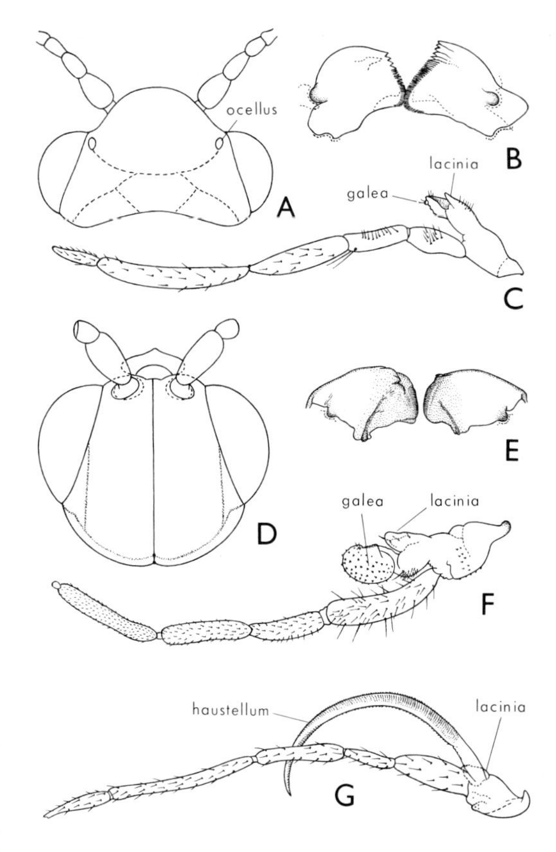

Fig. 36.12. A–C, head, mandibles, maxilla of *Sabatinca*, Micropterigidae; D–F, head, mandibles, maxilla of *Agathiphaga*, Eriocraniidae; G, maxilla of *Eriocrania*, Eriocraniidae. [B. Rankin]

Key to the Families of Eriocranioidea

1. Jugum not projecting from wing margin .... (Neopseustidae)
   Jugum projecting distinctly from wing margin 2
2. Maxillary palpi 3-segmented; mandibles absent; R_2 and R_3 coincident .... (Mnesarchaeidae)
   Maxillary palpi 5-segmented; mandibles present; R_2 and R_3 separate or stalked ...... **Eriocraniidae**

2. Eriocraniidae. Head (Fig. 36.12D) rough-haired; ocelli present or absent; antennae filiform; mandibles (Fig. 36.12E) present, with abductor and adductor muscles; maxillary palpi long, 5-segmented, folded; labial palpi 3- or 4-segmented: fore tibia sometimes with epiphysis or short apical spur, tibial spurs 0-1-3, 0-1-4, 0-2-3, or 1-4-4; fore wing (Fig. 36.11B) with humeral vein sometimes indistinct, Sc simple or forked, CuP present, 3A apically confluent with 2A, which is apically confluent with 1A, cross-veins present towards base, between CuP, 1A, and 2A; hind wing with Sc simple or forked, R_1 usually forked, CuP present; male with paired ventral processes on abdominal segment 5; female ovipositor sometimes heavily sclerotized and adapted for piercing. Larva with one pair of ocelli, or ocelli reduced to two pigmented spots; in leaf mines in angiosperms or in seeds of gymnosperms. Pupa (Fig. 36.10A) with mandibles asymmetrical, tarsi sometimes with paired claws; in a stout cocoon of silk and earthen particles in soil, or in an oval cell in seed of *Agathis*.

The adults of Holarctic genera are diurnal and fly during only a brief period of the year. The species are local in occurrence, and their mines may be restricted to the foliage of a few small trees. The adult behaviour of the primitive *Agathiphaga queenslandensis* Dumbleton* (Fig. 36.14C) is not known. Its mature larva is found in a hard resinous cell within the seed of kauri pines, *Agathis robusta* in southern Queensland and *A. palmerstoni* in northern Queensland (Dumbleton, 1952). It has minute setae and dense spinules; the head lacks adfrontal sutures and is flexed beneath the thorax. Before diapause begins the larva excavates near one end of the cell an oval emergence hole which, however, does not penetrate the exterior of the seed. An undescribed genus and species from Canberra, known only from the adult, is more closely related to Holarctic genera.

*Separated as family Agathiphagidae by N. P. Kristensen (*Ent. Meddr.* **35**: 341-5, 1967).

Suborder MONOTRYSIA

Very small to very large moths; without mandibles or laciniae; galeae usually produced into short haustellum; labial palpi 2- or 3-segmented; wings usually aculeate; venation of fore and hind wing similar, or venation reduced in hind wing or in both wings; female with one or two genital openings on sternum 9–10. Pupa adecticous, obtect.

The suborder apparently lacks homogeneity, containing both homoneurous and heteroneurous forms. Its three superfamilies are associated primarily because of the preponderance of archaic features exhibited, when compared with the Ditrysia, and it is probable that none of them originated directly from the Dachnonypha. Archaic characters shared with the latter include the presence of aculeae over much of the wing surface, and the absence of a separate copulatory aperture on sternum 8 in the female. Other primitive characters which have persisted in the Monotrysia, though not confined to this suborder, are the projecting jugum found in the Hepialoidea, and in a reduced form in the remainder, the homoneurous venation of the Hepialoidea, and the 5-segmented maxillary palpi of the Nepticuloidea and some Incurvarioidea. If the Monotrysia as a whole have not originated directly in the Dachnonypha, it is equally probable that neither the Nepticuloidea nor the Incurvarioidea originated in the Hepialoidea (Hinton, 1946a). The absence of ocelli, the vestigial maxillary palpi, and frequently reduced labial palpi in the adult, together with the highly modified pupa, show that the Hepialoidea have diverged greatly from the main stream of development in the Monotrysia. The Nepticuloidea and the Incurvarioidea, on the other hand, probably had a common origin, their individual divergence occurring after that of the Hepialoidea.

Key to the Superfamilies and Australian Families of Monotrysia

1. Venation of fore and hind wings similar; ♀ with 2 genital openings on sternum 9–10 .... HEPIALOIDEA. 2
   Venation of fore and hind wings dissimilar; ♀ with a single genital opening on sternum 9–10 ... 4
2. Hind tibia with 2 pairs of spurs .... **Prototheoridae** (p. 789)
   Hind tibia with 1 apical spur, or spurs absent 3
3. Both wings with M simple in discal cell .... **Palaeosetidae** (p. 789)
   Both wings with M forked in discal cell .... **Hepialidae** (p. 789)
4. Scape of antenna expanded into broad eye-cap; fore wing usually without closed discal cell .... NEPTICULOIDEA. 5
   Scape of antenna not expanded; fore wing with closed discal cell .... INCURVARIOIDEA. 6

5. Venation of fore wing reduced to 3 or 4 longitudinal, unbranched veins .... **Opostegidae** (p. 792)
   Venation of fore wing reduced, but branched (R strongly curved downwards) .... **Nepticulidae** (p. 792)
6. Head with smooth appressed scales; wings not aculeate; venation of hind wing greatly reduced .... **Heliozelidae** (p. 793)
   Head rough-haired; wings aculeate; venation of hind wing not greatly reduced .... **Incurvariidae** (p. 792)

Superfamily HEPIALOIDEA

Small to very large; antennae short; ocelli minute or absent; haustellum vestigial or absent; maxillary palpi minute or absent; labial palpi 2- or 3-segmented; wings usually aculeate, fore wing with humeral vein, Sc simple or forked, CuP present, at least in basal half; hind wing with venation similar to fore wing, Sc usually simple, M simple or forked, CuP present or vestigial; female genitalia of exoporian type. Pupa protruded from larval tunnel before ecdysis.

The superfamily includes those Monotrysia with primitive homoneurous venation and exoporian female genitalia. The Prototheoridae and Palaeosetidae represent small, restricted and specialized, Old-World offshoots from the hepialoid stem. The Hepialidae themselves show considerable specialization,

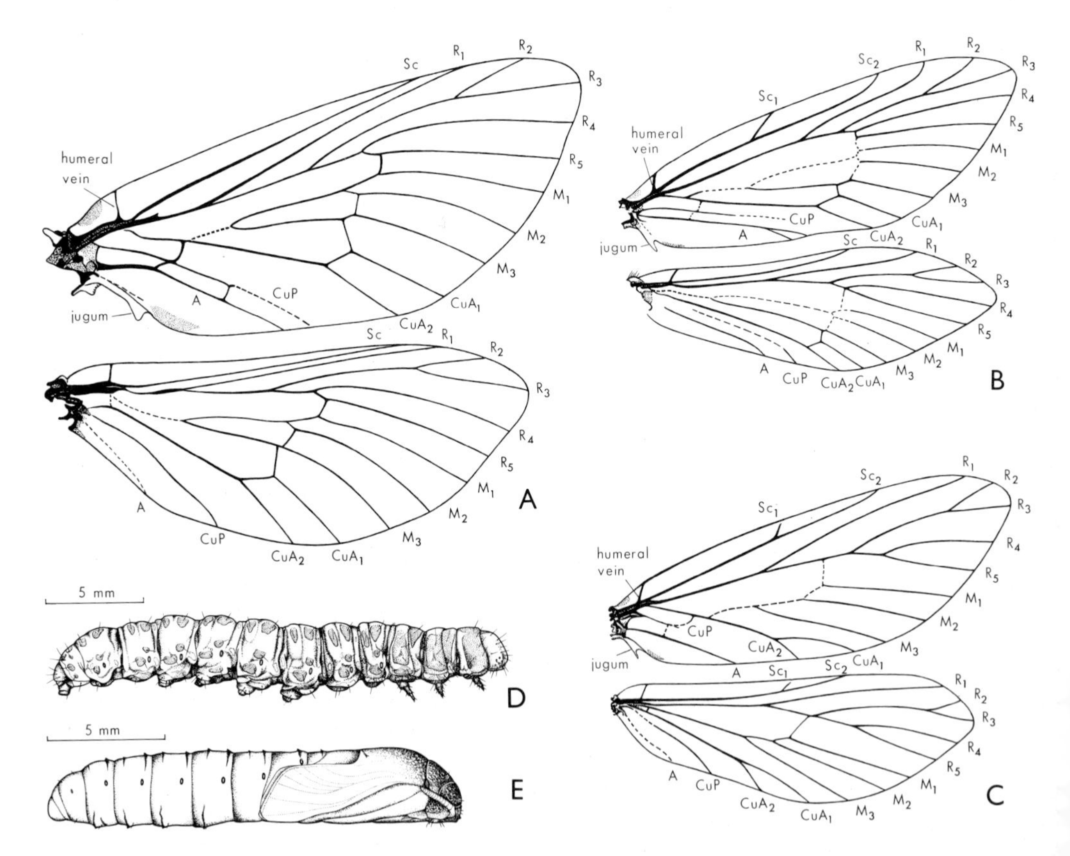

Fig. 36.13. Hepialoidea: A, wing venation of *Oncopera*, Hepialidae; B, wing venation of *Anomoses*, Prototheoridae; C, wing venation of *Palaeoses*, Palaeosetidae; D, E, larva and pupa of *Aenetus*, Hepialidae. [T. Binder, J. Wedgbrow]

and have one of the most advanced incomplete pupae known in the Lepidoptera (Hinton, 1946a).

3. Prototheoridae. Small; ocelli minute; antennal segments simple, hairy; haustellum very short; maxillary palpi minute, 2- or 3-segmented; labial palpi 3-segmented; epiphysis usually present; tibial spurs 0-2-4; wings with basal aculeae only; jugum and frenular bristles present; fore wing (Fig. 36.13 B) with Sc forked, CuP present; hind wing with Sc usually simple, R_1 simple, CuP present, 1A and 2A confluent, or 1A forming oblique cross-vein between CuP and 2A.

Containing only 8 species from South Africa and Australia, this small family may be less specialized than others in the super-family. None of the known species exceeds 27 mm in wing expanse. *Anomoses hylecoetes* Turn. (Fig. 36.14B), referred to a separate family Anomosetidae by Turner (1922), but to the Prototheoridae by Philpott (1928), is a dull brown species from rain forest in southern Queensland and northern N.S.W., which at rest resembles a small hepialid. The life history of none of the species is known.

4. Palaeosetidae. Small; ocelli absent; antennal segments simple, hairy; haustellum absent; maxillary palpi absent; labial palpi 2- or 3-segmented; epiphysis present; tibial spurs 0-1-1 or absent; wings aculeate; pterostigma often present in both wings; jugum small; fore wing (Fig. 36.13C) with Sc simple or forked, M simple, arising from CuA, M_4 absent in *Palaeoses*, basal half of CuP present, one anal vein; hind wing with Sc simple or forked, M simple, from base of discal cell, M_4 absent in *Palaeoses*, CuP present, one anal vein.

The family (Turner, 1922) has reduced mouth-parts, slightly reduced fore-wing venation, and tibial spurs either reduced or absent. *Palaeoses scholastica* Turn. (Fig. 36.14D) is a small, dark brown species less than 18 mm in wing-span, taken by sweeping foliage in rain forest in southern Queensland. *Ogygioses* is from Formosa and *Genustes* from Assam. The early stages of none of the species are known.

5. Hepialidae. Small to very large; ocelli absent; antennal segments angular, dentate, bipectinate or tripectinate; haustellum vestigial or absent; maxillary palpi minute, 1- to 5-segmented; labial palpi short, 2- or 3-segmented; epiphysis sometimes reduced; tibial spurs absent; wings aculeate; jugum present, frenular bristles absent; fore wing (Fig. 36.13A) with humeral vein strong, Sc simple or forked, R_2 and R_3 stalked, M forked, often arising from CuA; base of CuP present, with cross-veins to CuA and 1A; hind wing with humeral vein strong, Sc simple or rarely forked, R_2 and R_3 stalked, M forked from base of discal cell. Egg nearly spherical. Larva (Fig. 36.13D) long, cylindrical, with 6 ocelli, 3-segmented maxillary palpi, thoracic legs and abdominal prolegs present, crochets multiserial, in an ellipse. Pupa (Fig. 36.13E) long, cylindrical, with rudimentary mandibles; maxillary palpi absent; abdominal segments each with 2 series of spines, segments 2–6 movable; cremaster absent; appendages glued to body wall.

The family (Tindale, 1932–64) occurs in all regions, but reaches its greatest development in Australia where it includes many large, fast-flying species. *Zelotypia* (1 sp.) and *Aenetus* (16 spp.) contain beautiful insects, whose larvae tunnel down vertically in the stems of living shrubs and trees. They feed on the bark regrowth around the entrance to the tunnel, where they form a vestibule covered with a webbing of silk, wood particles, and excrement. Pupation occurs at the end of the tunnel, the entrance being previously closed with a silken wad or membrane. The larvae of the large, orange-brown *Z. stacyi* Scott (Fig. 36.14I) tunnel in *Eucalyptus* in eastern N.S.W. *A. lignivorus* (Lew.) and *A. eximius* (Scott) (Plate 8, A) are sexually dimorphic species from the coast and tablelands of south-eastern Australia. The former, with pale green male and green and red female, attacks various small trees, including *Eucalyptus*, *Acacia*, and also *Lantana*. The latter has a green and pink female; the larvae form a tunnel often 50 cm long in the stems and larger roots of *Acmena*, *Glochidion*, and other small trees.

Fig. 36.14. A, *Sabatinca calliplaca*, Micropterigidae; B, *Anomoses hylecoetes*, Prototheoridae; C, *Agathiphaga queenslandensis*, Eriocraniidae; D, *Palaeoses scholastica*, Palaeosetidae; E, *Nepticula anazona*, Nepticulidae; F, *Opostega gephyraea*, Opostegidae; G, *Oxycanus diremptus*, Hepialidae; H, *Heliozela prodela*, Heliozelidae; I, *Zelotypia stacyi*, J, *Abantiades magnificus*, Hepialidae. Scales: A–F, H = 1 mm; G, I, J = 10 mm.

[Photos by C. Lourandos]

The larvae of *Trictena* (3 spp.) and of *Abantiades* (14 spp.) live in vertical tunnels in the soil, and feed externally on the roots of *Eucalyptus*. Their life history is well adapted to the aridity of inland Australia, where great flights of the moths occur after a fall of rain. Natural mortality of the early stages is very high; a single female of *A. magnificus* (Lucas) (Fig. 36.14J) lays more than 18,000 eggs. Vertical tunnels in the soil are also formed by the larvae of *Oncopera* (12 spp.) and *Oxycanus* (44 spp.), some species of which emerge at night to feed on the foliage of herbaceous plants growing near the entrance of the tunnel. A few are pasture pests. The larvae of *Oxycanus diremptus* (Walk.) (Fig. 36.14G) feed mainly on leaf litter on the soil surface.

Superfamily NEPTICULOIDEA

Very small; ocelli absent; antennae often thickened, scape expanded and concave beneath, forming eye-cap; haustellum short or rudimentary; maxillary palpi 5-segmented, folded; labial palpi short, 3-segmented, usually drooping; epiphysis absent; tibial spurs 0-2-4; hind tibiae with long bristles; wings aculeate, at least at base; both wings with venation reduced; female with short cloaca and fleshy ovipositor. Pupa protruded from cocoon at ecdysis.

The superfamily retains many primitive features, notably the monotrysian genitalia, aculeate wings, and folded 5-segmented maxillary palpi. Specializations include the reduced venation, the leaf-mining habit, and the modified larval structure. The adults are among the smallest Lepidoptera.

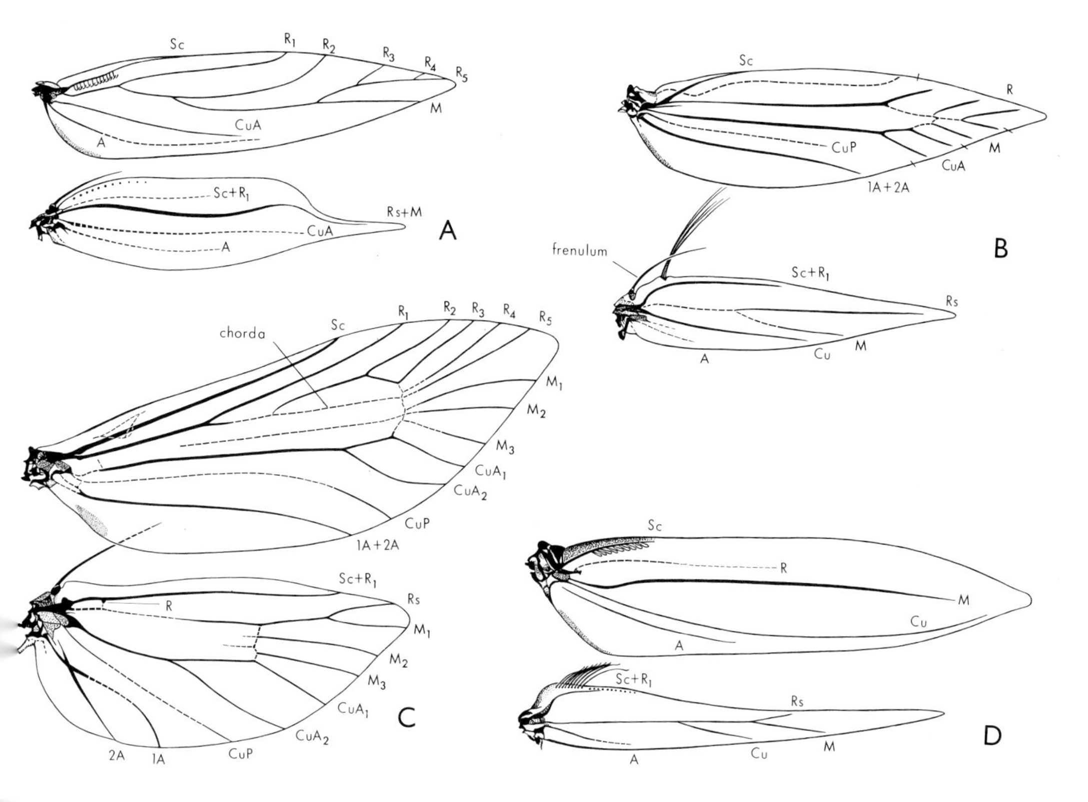

Fig. 36.15. Wing venation of Nepticuloidea and Incurvarioidea; A, *Nepticula*, Nepticulidae; B, *Heliozela*, Heliozelidae; C, *Nemophora*, Incurvariidae; D, *Opostega*, Opostegidae. [T. Binder]

6. Nepticulidae. Head (Fig. 36.16A) with erect hair-scales; antennae often thickened in male; haustellum very short or rudimentary; labial palpi usually drooping; wings aculeate; frenulum simple in male, a series of pseudofrenular bristles in female; fore wing (Fig. 36.15A) with R strongly curved downwards in middle. Egg oval, flattened, convex. Larva a leaf-miner without segmented thoracic legs or crochets, ventral paired leg-like processes on meso- and metathorax and abdominal segments 2–7; head with single ocellus on each side. Pupa with abdominal segments 1–7 movable.

In Australia, only 15 species of these tiny dark moths have been described, all in the genus *Nepticula*. They are, however, numerous, and the fauna may well exceed 200 species. They fly erratically, and run rapidly over the leaves of their food plants, upon which the eggs are laid singly. The larva tunnels direct from the egg into the leaf, forming a slender tortuous mine, which later expands gradually or abruptly into a blotch mine. The form of the mine is usually characteristic of each species. At maturity the larva leaves the mine and spins an oval cocoon of silk amongst detritus. *N. anazona* Meyr. (Fig. 36.14E) is black with a silver transverse band on the fore wing; the larva mines in the leaves of *Tristania suaveolens*.

7. Opostegidae. Head smooth except for tuft of long hair-scales on vertex; antennae often stout; haustellum very short; labial palpi drooping; hind tibiae with long stiff bristles; wings aculeate at base; fore wing (Fig. 36.15D) with venation reduced to 4 or 5 unbranched veins; hind wing lanceolate, with a series of pseudofrenular bristles. Larva apodous, without ventral processes; head with single rudimentary ocellus on each side.

The adults, which are usually white with a darker pattern, are relatively numerous in the tropics. The larvae mine in leaves and bark, but little is known about the life histories of the Australian species. *Opostega gephyraea* Meyr. (Fig. 36.14F) comes from Queensland.

Superfamily INCURVARIOIDEA

Very small to small; ocelli absent; haustellum short, scaled near base; labial palpi 3-segmented; hind wing with strong frenulum; female with long cloaca and sclerotized ovipositor for piercing plant tissues to insert eggs. Larva apodous or with short prolegs, crochets in transverse rows, anal prolegs and crochets always absent; leaf-miners, later cutting out flattened cases used as portable shelters or for pupation; pupa protruded from case at ecdysis.

The Tischeriidae and Prodoxidae are not known from Australia. The two other families are superficially dissimilar, but clearly had a common origin. The female genitalia and the structure and habits of the larvae indicate a real relationship. The Incurvariidae, with complete heteroneurous venation and 5-segmented folded maxillary palpi, at least in the less specialized genera, are evidently the more primitive family.

8. Incurvariidae. Head (Fig. 36.16B) rough-haired; antennae short or much longer than wings, especially in male; maxillary palpi 5-segmented and folded, or 2-segmented, or rudimentary; labial palpi bristly; wings aculeate, venation not or only slightly reduced; fore wing (Fig. 36.15C) with chorda present. Larva with ventral prolegs present, usually short, broad, each with one or more transverse rows of crochets, rarely reduced; head with 6 ocelli on each side.

Two divergent subfamilies are recognized, the Incurvariinae and the Adelinae. The Australian INCURVARIINAE have previously been included in the Tineidae, and may well exceed 50 species. The moths are very small, with sombre colours, have 5-segmented folded maxillary palpi, and short antennae. The larvae mine in leaves, and later cut out oval or irregular, flattened cases, in which they pupate, or which they use as shelters while they feed within new mines which they excavate in the leaf. *'Tinea' phaulopter*a Meyr. mines in the leaves of *Banksia serrata* and *'T'. nectaria* Meyr. cuts ou

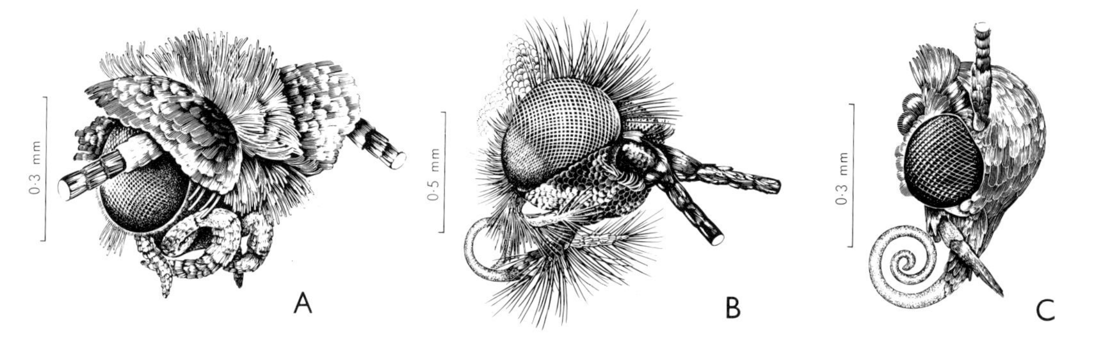

Fig. 36.16. Heads: A, *Nepticula*, Nepticulidae; B, *Nemophora*, Incurvariidae; C, *Heliozela*, Heliozelidae. [F. Nanninga]

very irregular flat cases from successively larger mines it makes in the leaves of *Eucalyptus* seedlings.

The ADELINAE have reduced maxillary palpi, and the males have very large eyes and the longest known antennae in the Lepidoptera. *Nemophora* (9 spp.) have brilliant metallic coloration, and fly in sunshine, resting on flowers in which the first-instar larvae feed. The larva of *N. topazias* (Meyr.) later forms a case from the flower parts of *Acacia* and feeds on fallen flowers on the ground. Pupation takes place in this case. Adults of *N. sparsella* (Walk.) (Plate 7, I) are found around *Bursaria* flowers. The larval case occurs within the dry seed capsules.

9. Heliozelidae. Head (Fig. 36.16C) smooth-scaled, shining; antennae short; maxillary palpi minute, 5-segmented; labial palpi slender, drooping; epiphysis present; tibial spurs 0-2-4; wings (Fig. 36.15B) lanceolate, not aculeate, venation reduced in both wings; fore wing with R_5 and M_1 stalked, R_5 to costa. Larva apodous in all instars, sometimes with colourless discs; prolegs on segments 4–6 sometimes represented by crochets; head with 2 ocelli on each side.

The moths, which often have metallic scales, fly in sunshine and rest on flowers. The larvae are leaf-miners, finally cutting out flat oval cases from the mine, which either drop to the ground, or are attached by silk amongst detritus before pupation. *Heliozela prodela* Meyr. (Fig. 36.14H) mines in young terminal leaves of *Eucalyptus*.

Suborder DITRYSIA

Very small to very large; ocelli and chaetosemata present or absent; mandibles absent; galeae usually produced into haustellum; maxillary palpi 1- to 5-segmented or vestigial; labial palpi 3-segmented, rarely 2-segmented; wings rarely with a few aculeae, venation reduced in hind wing or in both wings; female with two genital openings, the ostium bursae on sternum 8 and the genital aperture on sternum 9–10. Pupa adecticous, obtect.

The suborder contains the bulk of the Lepidoptera, with a great diversity of forms having in common the complex genital system of the female. Wing-coupling is never of the jugate type, being either frenulate or amplexiform. In some of the more specialized superfamilies, such as the Bombycoidea, Papilionoidea, and Hesperioidea, the frenulum has often been lost. Aculeae have largely disappeared, persisting only in a few of the more primitive Tineoidea and then in restricted areas of the wing membrane. A heteroneurous venation distinguishes the Ditrysia from the Zeugloptera, Dachnonypha, and the Hepialoidea. The reduction of the venation in the hind wing, and sometimes also in the fore wing, reaches extravagant limits in some of the narrow-winged species.

Although mandibles are generally absent (vestigial in a few families) the primitive 5-segmented folded maxillary palpi are retained in some of the Tineidae and Lyonetiidae. Reduction, or even loss, of the maxillary palpi is characteristic of most superfamilies. The haustellum has become highly developed in some of the more specialized families which ingest sugary liquids. In some the haustellum has become secondarily degenerate or even entirely lost.

Organs for the dissemination of pheromones and production of sounds, together with complex organs of sensory perception, have also evolved in the more advanced superfamilies. Thus, in such groups as the Pyraloidea, Hesperioidea, Papilionoidea, and Noctuoidea, specialized scent brushes and scales are often found, especially in the males. Females in many families, such as the Saturniidae and the Lymantriidae, produce special sex odours which serve to attract the males. Chaetosemata have developed, probably independently, in several superfamilies, but their function is not understood. Tympanal organs have evolved, likewise apparently independently, in the Pyraloidea, Geometroidea, and Noctuoidea. Moths in these groups are very sensitive to sound. Primitive tympanal organs occur in some Cossidae. Special sound-producing mechanisms have developed in the males of a few species of Noctuidae and Agaristidae. The whistling sounds produced are probably significant in bringing the sexes together.

Key to the Superfamilies of Ditrysia

1. Functional wings present 2
   Apterous, or with one or both wings greatly reduced, non-functional (♀♀ only) 33

2(1). Small to very small (usually less than 20 mm expanse); wings very narrow, sometimes with apex of hind wing produced and termen emarginate, or each wing with single deep cleft, venation sometimes reduced; length of marginal scales of hind wing approaching or exceeding its width 3
   Small to large (usually more than 10 mm expanse); wings broader, termen of hind wing not emarginate, venation usually not greatly reduced; or wings deeply cleft with hind wing divided into 3 or more plumes; length of marginal scales of hind wing much less than its width 6

3(2). Hind wing with $Sc+R_1$ diverging from Rs at base, then fusing with Rs from towards end to well beyond discal cell; or each wing with deep terminal cleft ... PYRALOIDEA (pt., p. 833)
   Hind wing with $Sc+R_1$ either completely separate from Rs, or Sc joined to discal cell by a short, slightly oblique R_1 4

4(3). Maxillary palpi small, folded over base of haustellum, which has dense imbricated scales at least near base .... GELECHIOIDEA (pt., p. 815)
   Maxillary palpi either large, 5-segmented, and folded; or small and not folded over base of haustellum, which is naked 5

5(4). Scape of antenna expanded and concave beneath to form eye-cap; or maxillary palpi long, folded, 5-segmented; or hind tibiae with long stiff dorsal bristles; or, if hind tibiae smooth-scaled, then hind wing linear or linear-lanceolate, much narrower than fore wing .... TINEOIDEA (pt., p. 802)
   Scape of antenna not expanded; maxillary palpi never 5-segmented, folded; hind tibiae and tarsi surrounded by long stiff bristles, or with whorls of bristles at apices of segments, or with long loose hair scales above, or, if smooth-scaled, hind wing lanceolate, ovate-lanceolate, or ovate, nearly as broad as fore wing .... YPONOMEUTOIDEA (pt., p. 809)

6(2). Wings deeply cleft, hind wing divided into 3 or more plumes 7
   Wings not cleft 8

7(6). Hind wing divided into 3 plumes .... PTEROPHOROIDEA (p. 838)
   Hind wing divided into 6 or 7 plumes .... COPROMORPHOIDEA (pt., p. 826)

8(6). M present as a tubular vein in discal cell of either wing 9
   M vestigial or absent from discal cell of both wings 14

9(8). M strong and forked in discal cell of both wings; fore wing with 5 branches of R and usually with chorda .... COSSOIDEA (pt., p. 796)
M either not forked in discal cell of one wing, or, if forked in both wings, fore wing either without chorda or R with only 4 branches 10
10(9). Antennae terminating in a distinct rounded club .... CASTNIOIDEA (p. 828)
Antennae not terminating in a club 11
11(10). Hind wing with marginal scales near anal angle scarcely longer than near apex 12
Hind wing with marginal scales near anal angle much longer than near apex 13
12(11). Fore wing with CuP, 1A, and 2A fusing before margin; or mid and hind tibiae each with a single apical spur .... TINEOIDEA (pt., p. 802)
Fore wing with CuP separate from 1A and 2A; mid and hind tibiae with apical spurs paired or absent .... ZYGAENOIDEA (p. 828)
13(11). Labial palpi porrect and beak-like, apical segment usually turned down .... TORTRICOIDEA (pt., p. 798)
Labial palpi not porrect or beak-like, apical segment not turned down ... TINEOIDEA (pt., p. 802)
14(8). Hind wing with $Sc+R_1$ approaching, approximating to, or fusing with, Rs for a short distance beyond discal cell; CuP rarely present in fore wing, but, if so, $Sc+R_1$ fused with Rs beyond cell 15
Hind wing with $Sc+R_1$ not approaching Rs beyond discal cell; or if approaching Rs before or near end of discal cell, CuP present in fore wing 18
15(14). Wings narrow or of medium width, not broadly rounded, and fore wing not falcate; fore wing of ♂ with membranous hooked retinaculum, hind wing with frenulum usually strong...16
Wings very broad, often rounded or fore wing falcate; fore wing of ♂ without membranous retinaculum, hind wing with frenulum weak or absent 17
16(15). Large (usually more than 50 mm expanse), robust, with stout, fusiform body; tympanal organs absent; maxillary palpi reduced to one very small segment; hind wing with CuP absent .... SPHINGOIDEA (pt., p. 855)
Small to medium-sized (usually less than 50 mm expanse), body moderate to slender, not fusiform; tympanal organs usually present at base of abdomen; maxillary palpi 4-segmented, prominent; hind wing with CuP present .... PYRALOIDEA (pt., p. 833)
17(15). Chaetosemata large, prominent; antennae simple, slightly thickened distally; tympanal organs absent; fore wing not falcate .... CALLIDULOIDEA (p. 851)
Chaetosemata very small or absent; antenna bipectinate in both sexes; tympanal organs present; fore wing usually falcate .... GEOMETROIDEA (pt., p. 846)
18(14). Haustellum clothed with dense imbricated scales, at least near base; maxillary palpi small, almost always folded over base of haustellum; ocelli small or absent .... GELECHIOIDEA (pt., p. 815)
Haustellum naked, or, if sparse scales present near base, then maxillary palpi not folded over base and ocelli prominent 19
19(18). Marginal scales on hind wing long and flexible, those near anal angle much longer than near apex 20
Marginal scales on hind wing short and firm, those near anal angle scarcely longer than near apex 23
20(19). Labial palpi ascending or drooping, not porrect or beak-like; if ascending, then not appressed to face 21
Labial palpi porrect and beak-like, with apical segment usually short and turned downwards; if ascending, then palpi short, appressed to face, with apical segment short, blunt 22
21(20). Hind wing with strong pecten on CuA; if not, fore wing with tufts of raised scales; ocelli and chaetosemata absent .... COPROMORPHOIDEA (pt., p. 826)
Hind wing without pecten on CuA; fore wing smoothly scaled; ocelli and chaetosemata never both present .... YPONOMEUTOIDEA (pt., p. 809)
22(20). Chaetosemata absent; hind wing with 1 and often 2 branches of M absent; ocelli absent .... COPROMORPHOIDEA (pt., p. 826)
Chaetosemata present; hind wing with 3 branches of M present, or if 1 absent, ocelli almost always present .... TORTRICOIDEA (pt., p. 798)

23(19). Length of fore wing at least 4 times width; wings partly hyaline and coupled together by series of interlocking recurved spines as well as by frenulum ... YPONOMEUTOIDEA (pt., p. 809)
Fore wing broader; if partly hyaline, then wings coupled together only by frenulum 24
24(23). Tympanal organs present in thorax or abdomen 25
Tympanal organs absent 27
25(24). Tympanal organs in abdomen; chaetosemata present ...... GEOMETROIDEA (pt., p. 846)
Tympanal organs in metathorax, counter-tympanal cavities usually at base of abdomen; chaetosemata absent 26
26(25). Fore wing with M_2 arising nearer to CuA than to R, or M_2 absent .... NOCTUOIDEA (pt., p. 859)
Fore wing with M_2 not arising nearer to CuA than to R .... NOTODONTOIDEA (p. 856)
27(24). Hind wing with $Sc+R_1$ coincident with Rs (i.e. $Sc+R_1$ apparently absent); abdomen boldly ringed in orange and black .... NOCTUOIDEA (pt., p. 859)
Hind wing with $Sc+R_1$ distinct from Rs; abdomen not boldly ringed in orange and black ... 28
28(27). Body stout, long, fusiform; wings narrow, termen very oblique. ... SPHINGOIDEA (pt., p. 855)
Body slender or, if stout, not long and fusiform; wings broad 29
29(28). Antennae gradually or abruptly clubbed, tip sometimes hooked 30
Antennae not clubbed or hooked 31
30(29). Bases of antennae widely separated; veins arising separately from discal cell in both wings .... HESPERIOIDEA (p. 839)
Bases of antennae approximated; at least some peripheral veins stalked .... PAPILIONOIDEA (p. 840)
31(29). Ocelli present .... PYRALOIDEA (pt., p. 833)
Ocelli absent 32
32(31). Hind wing with $Sc+R_1$ not approximated to Rs beyond discal cell; humeral area often expanded; frenulum often absent .... BOMBYCOIDEA (pt., p. 851)
Hind wing with $Sc+R_1$ approximated to Rs beyond discal cell; humeral area not expanded; frenulum present .... PYRALOIDEA (pt., p. 833)
33(1). Legs greatly reduced, non-functional; moth not leaving larval case, at least until after oviposition .... TINEOIDEA (pt., p. 802)
Legs normally developed, functional; if larva a case-bearer, then moth emerging normally from pupal skin and larval case 34
34(33). Both wings much shorter than in ♂, but of same general shape, with similar venation; M strong and forked in discal cell of both wings .... COSSOIDEA (pt., p. 796)
Hind wing vestigial and fore wing greatly shortened or vestigial; or apterous 35
35(34). Labial palpi ascending, apical segment recurved, slender, acute; haustellum present, scaled near base; hind wing much more reduced than fore wing. ...... GELECHIOIDEA (pt., p. 815)
Labial palpi not so; haustellum rudimentary or absent; both fore and hind wings equally reduced, or absent 36
36(35). Chaetosemata present; reduced wings with overlapping scales and erect bristles, or abdomen clothed with long bristles .... GEOMETROIDEA (pt., p. 846)
Chaetosemata absent; vestigial wings and abdomen with overlapping scales or woolly hair-scales, without erect bristles 37
37(36). Wings reduced to sclerotized lobes; thorax and base of abdomen clothed with appressed thin hair-scales, sometimes with notched apices; tarsi with dense, stout spines beneath .... BOMBYCOIDEA (pt., p. 851)
Wings reduced to crumpled vestiges; thorax and base of abdomen not so clothed; tarsi without noticeable spines beneath 38
38(37). Body clothed with normal but rather sparse scales .... TINEOIDEA (pt., p. 802)
Body clothed with long woolly hair-scales .... NOCTUOIDEA (pt., p. 859)

Superfamily COSSOIDEA

10. Cossidae. Small to very large; ocelli rarely present; chaetosemata absent; antennae usually bipectinate, rarely lamellate or simple in male; haustellum very short, naked, or absent; maxillary palpi minute, 1- or 2-segmented; labial palpi short or moderate; epiphysis usually present; tibial spurs 0-2-4 or absent; wings (Figs. 36.17A, B) strong, narrow, frenulum in female 2 to several bristles, rectinaculum in male often between costa and Sc; fore wing with M strong and forked in discal cell, chorda usually present, CuP present; hind wing with M usually forked, CuP sometimes reduced. Egg of flat type in Zeuzerinae, upright in Cossinae. Larva (Fig. 36.19A) stout, prothorax with large sclerotized shield and 3 prespiracular (L) setae; crochets usually in a circle, biordinal or triordinal; wood-boring. Pupa (Fig. 36.19B) long, cylindrical, abdomen spined, segments 3–7 movable in male, 3–6 in female, cremaster absent; protruded from tunnel at ecdysis.

The family is usually regarded as one of the most primitive of the Ditrysia (Turner, 1947). The venation suggests a relationship to some of the most primitive Tineoidea, whereas the chaetotaxy of the larvae and the male genitalia suggest a relationship to the Tortricoidea. The Psychidae on the one hand and the Tortricidae on the other may have been derived from cossoid stock. Nevertheless, the reduced mouth-parts, the loss of the ocelli, the strongly bipectinate antennae, the loss of tibial spurs, and the upright egg of the Cossinae, show that the Cossidae have been subject to much specialization.

In Australia the family (Turner, 1945) contains numerous fast-flying grey species. The larvae bore in heartwood or in the larger roots of living trees, shrubs, or woody herbs, form galleries beneath the bark, or rarely tunnel in soil, feeding externally on roots. Pupation occurs in a chamber within the tunnel.

The COSSINAE have tibial spurs. The male antennae of *Ptilomacra* (1 sp.), *Macrocyttara* (2 spp.), *Culama* (8 spp.), *Charmoses* (1 sp.), and *Idioses* (1 sp.) are bipectinate to the apex. The larvae of *Culama* and *Macrocyttara* excavate galleries beneath the bark of trees, those of *M. expressa* (Lucas) (Fig. 36.18B) tunnelling gregariously in the trunks of mangroves *(Excaecaria agallocha)* in Queensland. The fore wing of *Idioses* lacks the chorda and the male has a pterostigma. In the hind wing M is simple, as it also is in *Charmoses*. In *Archaeoses* (2 spp.) and *Dudgeonea* (3 spp.), the segments of the male antennae are thickened by the development of closely appressed lamellae. *Dudgeonea* is sparsely distributed from Africa and Madagascar to India, New Guinea, and Australia. With *Chilecomadia* from Chile and *Pseudocossus* from Madagascar, *Dudgeonea* is remarkable for the occurrence of primitive tympanal organs at the base of the abdomen. *D. actinias* Turn. (Plate 8, D), as in other species, is reddish brown with a pattern of silver spots. The larvae are said to tunnel in trees. *Idioses littleri* Turn., from eastern Australia, and *Pseudocossus* are the only cossids known to have retained ocelli.

In the ZEUZERINAE, the large genus *Xyleutes* (69 spp.), though not confined to Australia, is best developed here. The male antennae are bipectinate only in the basal half, and tibial spurs are absent, but the mid and hind tibiae each bears a pair of stout apical spines. The species range in wingspan from about 30 mm in males of *X. amphiplecta* Turn. (Fig. 36.18E) to about 230 mm in the females of *X. boisduvali* Roths. and *X. affinis* Roths. (Fig. 36.18A), with abdomens measuring up to 70 mm in length and 22 mm in diameter. Vast numbers of eggs are laid; those deposited by a single female of one of the larger species, *X. durvillei* Herr-Sch., exceeded 18,000. The mass of small yellowish eggs is covered with a glutinous secretion, beneath which the newly hatched larvae live for a day or two before dispersing. The first-instar larvae spin great quantities of silk, and Kalshoven (1965) stated that they are dispersed by wind when they lower themselves on silken strands. The larvae of *X. durvillei* tunnel in the larger roots of *Acacia*, but many, such as *X. boisduvali* and *X. affinis*, bore singly in the trunks of *Eucalyptus*, taking 2 and possibly 3 years to reach maturity. Excrement ejected through a small opening betrays the presence of a larva. Before pupation, the tunnel to the ejection hole is enlarged, and

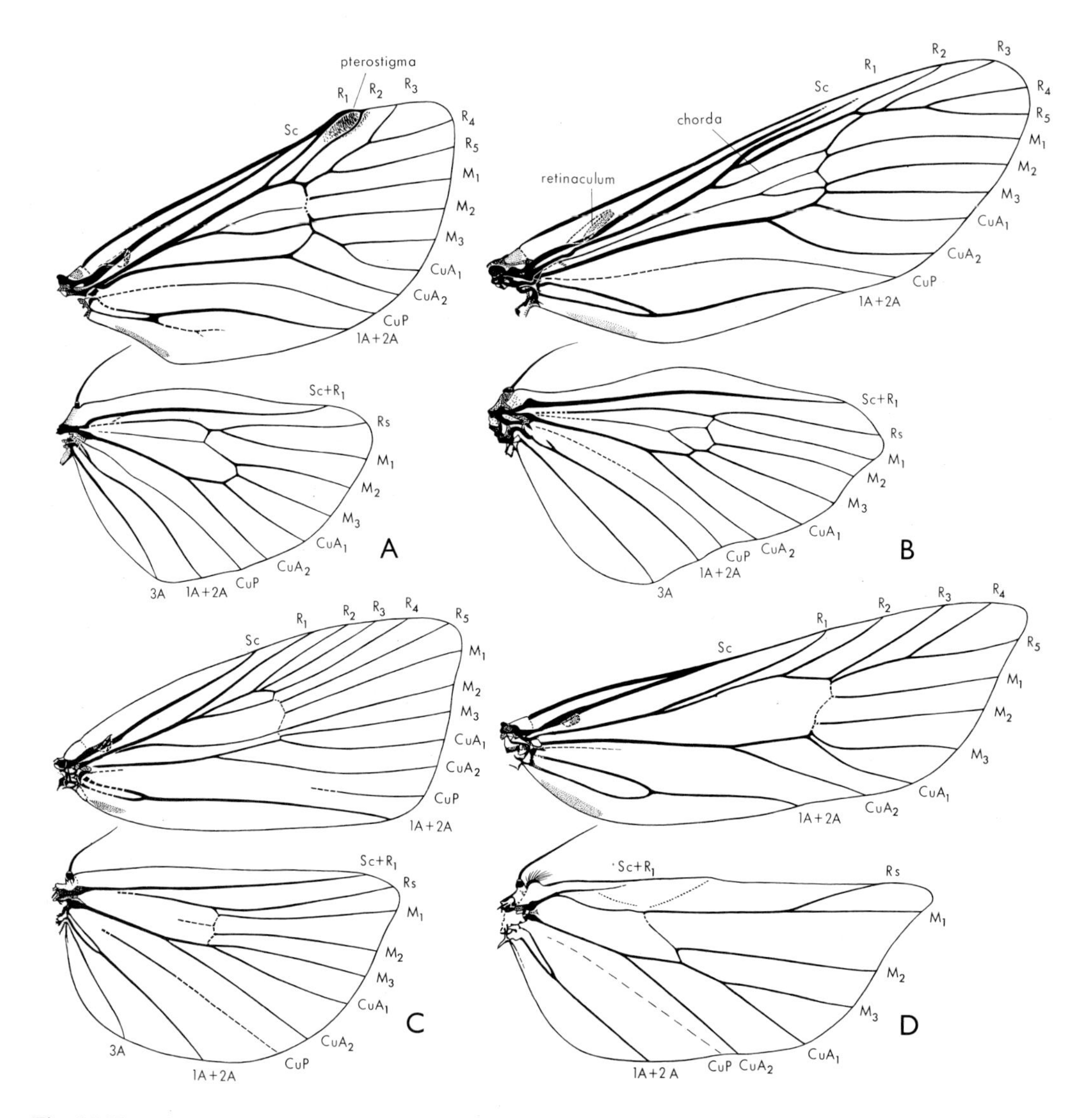

Fig. 36.17. Wing venation of Cossoidea and Tortricoidea: A, *Idioses*, Cossidae; B, *Xyleutes*, Cossidae; C, *Tanychaeta*, Tortricidae; D, *Clysiana*, Phaloniidae. [T. Binder]

the bark covering it almost severed. Pupation takes place in the tunnel, sometimes in a chamber of chewed wood, and a felted pad which is usually fixed in the tunnel is pushed out by the emerging pupa before ecdysis. The larva of the small *X. leucomochla* Turn., which forms a silk-lined tunnel in the soil and feeds externally on the roots of *Acacia ligulata* in inland South Australia, is the true witchety grub of the Aborigines.

Superfamily TORTRICOIDEA

Small; head usually rough-scaled; ocelli and chaetosemata usually present; haustellum naked; maxillary palpi very small, 2- to 4-segmented, or vestigial; labial palpi

Fig. 36.18. A, *Xyleutes affinis*, B, *Macrocyttara expressa*, Cossidae; C, *Arotrophora siniocosma*, D, *Palaeotoma styphelana*, Tortricidae; E, *Xyleutes amphiplecta*, Cossidae; F, *Anisogona simana*, G, *Olethreutes melanocycla*, H, *Crocidosema plebeiana*, Tortricidae; I, *Trigonocyttara clandestina*, J, *Cebysa leucoteles*, K, *Clania tenuis*, Psychidae. Scales: A = 10 mm; B–K = 2 mm. [Photos by C. Lourandos]

usually porrect; epiphysis present; tibial spurs 0-2-4; fore wing often with costal fold in male, M sometimes present, simple, in discal cell, CuP usually present; hind wing usually as broad as, or broader than, fore wing, CuA sometimes with basal pecten of hairs, CuP usually present; abdomen not spined. Larva with 3 prespiracular (L) setae on prothorax, crochets in a circle; concealed feeders, mostly in shelters formed by tying adjacent mature or terminal leaves, or boring in fruits, occasionally leaf-mining, or tunnelling in bark, galls, flower spikes, or seeds. Pupa with dorsal abdominal spines and often a cremaster; in larval shelter, protruded from shelter at ecdysis.

This large and homogeneous group has many archaic characters in common with the Tineoidea on the one hand and the Cossoidea on the other. Some authors have included the Cossidae in the Tortricoidea (Forbes, 1924), whereas others have regarded them as only a section of an enlarged Tineoidea (Tillyard, 1926b). The more primitive family Tortricidae is well represented in the Australian fauna; the Phaloniidae include few Australian species.

Key to the Families of Tortricoidea

Fore wing with CuA_2 from before three-fourths of discal cell, CuP rarely absent ... **Tortricidae** (p. 800)
Fore wing with CuA_2 from beyond three-fourths of discal cell, CuP absent ...... **Phaloniidae** (p. 801)

11. Tortricidae. Head (Fig. 36.19E) usually rough-scaled; ocelli usually present; chaetosemata present; maxillary palpi 2- to 4-segmented; labial palpi short to very long, usually porrect, rarely ascending, apical segment usually short, obtuse; fore wing (Fig. 36.17C) with costa often strongly arched, sometimes with costal fold in male, chorda and M sometimes present, CuA_2 arising before three-fourths of discal cell, CuP rarely absent as a tubular vein, at least near margin; hind wing sometimes with pecten of hairs on CuA, CuP present. Egg flat, scale-like, laid singly or in batches. Larva often with anal fork, crochets uniordinal, biordinal, or triordinal; joining leaves or shoots, rarely leaf-miners in early instars, or tunnelling in flower spikes, fruits, or galls. Pupa (Figs. 36.19C, D) with abdomen spined, cremaster or anal segment with hooked spines, antennae reaching nearly to tip of wings; in larval shelter, protruded at ecdysis.

The family is well represented in Australia by the two subfamilies Tortricinae and Olethreutinae. The former are more abundant in southern Australia, where endemic genera predominate; the latter are numerous in northern Australia.

The most primitive TORTRICINAE (Common, 1965) have retained M in the discal cell and the larvae have uniordinal crochets. The larvae of *Arotrophora* (8 spp.) tunnel in the flower spikes of *Banksia*. *A. siniocosma* Turn. (Fig. 36.18C) is found in *B. paludosa* in the south-east. Galls formed on *Eucalyptus* by *Apiomorpha* (Coccoidea) and other insects are tunnelled by the larvae of *Palaeotoma styphelana* Meyr. (Fig. 36.18D), whereas the larvae of *Proselena annosana* Meyr. are leaf-miners in *Bursaria spinosa*. *Trymalitis optima* Meyr. and related species have long hair-scales on the tibiae. At rest the fore legs are extended in front, and the tip of the fore wing is bent abruptly upwards at a point where the venation is modified.

The larvae of many species feed on dead *Eucalyptus* leaves, thus helping to reduce the forest leaf litter to humus. *Epitymbia cosmota* (Meyr.) and *Anisogona simana* Meyr. (Fig. 36.18F) are sexually dimorphic examples of this group. The larvae of the most specialized Tortricinae feed between joined living leaves, and are usually green in colour. Those of *Epiphyas postvittana* (Walk.) and *Acropolitis rudisana* (Walk.) have a wide range of food plants, including exotic ornamental and fruit trees. The former is a serious orchard pest, and was introduced to New Zealand, Hawaii, and Britain. *Merophyas divulsana* (Walk.), which attacks mainly herbaceous plants, is a pest of lucerne. Most species have solitary larvae, but the early stages of the large *Cryptoptila australana* (Lew.) feed gregariously in a webbing on *Tieghemopanax sambucifolius*.

The OLETHREUTINAE also include some notable pests, such as *Cydia pomonella* (L.) and *C. molesta* (Busck.), which attack pome and stone fruits. *Olethreutes* (48 spp.) contains mostly rain-forest species such as *O. melanocycla* (Turn.) (Fig. 36.18G) from the north-east. The larvae of *O. euryphaea* (Turn.), however, from the east coast join leaves of *Banksia integrifolia*. *Cryptophlebia ombrodelta* (Low.) is a minor pest of *Macadamia*, but the larvae also feed within the seed pods of *Acacia* and other trees. The cosmopolitan genus *Bactra* (10 spp.) includes drab species, such as *B. ablabes* Turn., which feed on sedges. *Epinotia lantana* Busck, introduced to assist in the control of *Lantana*, now occurs widely along the eastern coast wherever its host plant grows. Malvaceous weed and crop plants are attacked by the larvae of *Crocidosema plebeiana* Zell. (Fig. 36.18H), a species which also occurs widely abroad. Galls on *Acacia decurrens* are tunnelled by the larvae of *Eucosma triangulana* Meyr. and the terminal leaves of *Hibbertia obtusifolia* are tied by the larvae of *Acroclita erythrana* Meyr. The antennae in males of *Spilonota* (42 spp.) are deeply notched near the base. *S. macropetana* Meyr. and *S. infensa* Meyr. are common species with larvae tying leaves of *Eucalyptus*.

12. Phaloniidae (Cochylidae). Head rough-scaled; ocelli and chaetosemata usually present; antennae simple, ciliated; haustellum short or absent; maxillary palpi reduced to minute papillae; labial palpi long, porrect; hind tibiae with long hair-scales; fore wing (Fig. 36.17D) with M sometimes present in discal cell, chorda often present, R_4 and R_5 sometimes stalked, other veins separate, CuA_2 arising beyond three-fourths of discal cell, CuP usually absent as a tubular vein; hind wing as broad as fore wing, Sc joined to Rs by R_1, Rs and M_1 approximated or stalked, CuA_2 arising well before lower angle of discal cell, CuP weak or absent. Larva with ventral prolegs small, crochets uniordinal, in a circle; case-bearing or tunnelling in flower or seed heads, or in stems. Pupa with dorsal spines on abdomen, sometimes in 2 rows, or very short

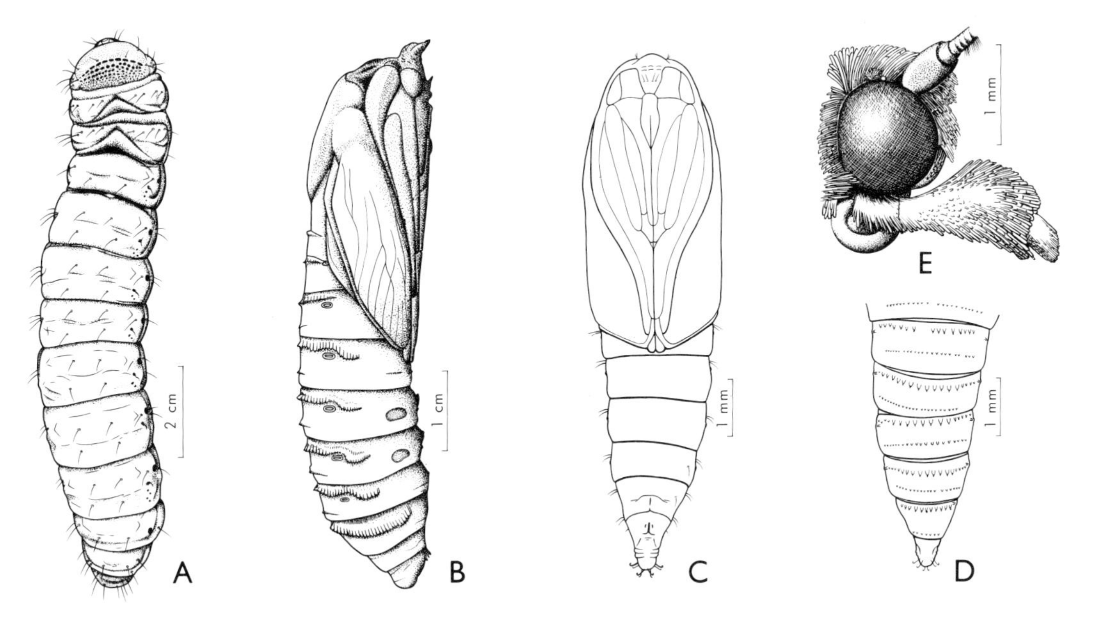

Fig. 36.19. A, B, larva and pupa of *Xyleutes*, Cossidae; C, D, pupa, ventral and dorsal, of *Epiphyas*, Tortricidae; E, head of *Epiphyas*, lateral. [J. Wedgbrow]

in transverse area on each segment; in larval case, or in cocoon, usually protruded at ecdysis.

Few species occur in Australia. *Clysiana* (1 sp.) is a small Oriental and Papuan genus, with a single species, *C. acrographa* (Turn.), from coastal Queensland and N.S.W. *Heliocosma* (7 spp.) is probably misplaced here. The larva, which feeds on flowers of daisies, *Ranunculus*, and clover, constructs a flimsy portable case of the flower parts. Before pupating within, it suspends the case by a silken thread up to 2 cm in length; the pupa is not protruded at ecdysis. *H. incongruana* (Walk.) occurs widely in south-eastern Australia.

Superfamily TINEOIDEA

Very small to medium-sized; ocelli present or absent; chaetosemata absent; scape sometimes forming eye-cap; haustellum present, naked, or reduced; maxillary palpi 1- to 5-segmented, folded, porrect or ascending, not folded over base of haustellum; labial palpi usually short, drooping or ascending, sometimes bristly, apical segment usually not acute; epiphysis rarely absent; tibial spurs 0-2-4, rarely reduced; M retained in discal cell and CuP strong in primitive forms, venation often reduced, especially in hind wings of narrow-winged species; retinaculum in female subcubital; abdomen without dorsal spining. Larva with 3 prespiracular (L) setae on prothorax (only 2 in *Scardia*), crochets when present in circle, ellipse, pseudoellipse, or transverse band, setae L1 and L2 of abdomen remote; concealed feeders, in portable cases, tunnels, silken tubes, and leaf mines. Pupa usually with dorsal abdominal spines, in male usually with segments 4–7, 4–8, or 5–8 movable; in larval shelter, in silken cocoon, or suspended amid strands of silk, except for some female Psychidae protruded from cocoon or shelter at ecdysis.

This is clearly the most primitive of the 'tineoid' superfamilies. Archaic forms retain M in the discal cell, have only a slightly reduced heteroneurous venation, and sometimes have 5-segmented folded maxillary palpi. The reduced mouth-parts in the Psychidae, the reduced venation in many of the narrow-winged Lyonetiidae, Phyllocnistidae, and Gracillariidae, and the form of the pupa in some of the Lyonetiidae, are specialized conditions.

Key to the Families of Tineoidea Known in Australia

1. Apterous or brachypterous (♀♀ only) .... **Psychidae** (pt., p. 803)
   Fully winged 2
2. Head with erect hair-scales, at least on vertex, or forming a posterior and lateral fringe 3
   Head entirely covered with appressed scales 6
3. Head with roughened hair-scales or with a tuft of hair above, frons smooth-scaled; if head entirely rough-haired, tip of wing bent abruptly upwards .... **Lyonetiidae** (pt., p. 805)
   Head smooth-scaled except for posterior and lateral fringe of long hairs above; if head entirely rough-haired, tip of wing not bent abruptly upwards 4
4. Head smooth-scaled except for fringe of erect hair above .... **Gracillariidae** (pt., p. 808)
   Head entirely rough-haired 5
5. Fore wing with M present and often forked in discal cell, chorda often present, R_5 when present ending at apex or on termen; scape of antenna without pecten; ♀ abdomen with terminal scale-tuft .... **Psychidae** (pt., p. 803)
   Fore wing with M usually vestigial or absent in discal cell; if present, R_5 ending on costa; scape of antenna often with pecten; ♀ abdomen without terminal scale-tuft .... **Tineidae** (p. 804)
6. Scape of antenna expanded and concave beneath. [Species with ground colour shining white] ... 7
   Scape of antenna not expanded 8
7. Hind tibiae with long hair-scales above and below .... **Lyonetiidae** (pt., p. 805)
   Hind tibiae with dorsal row of stout bristles .... **Phyllocnistidae** (p. 808)
8. Maxillary palpi long, folded, 5-segmented .... **Lyonetiidae** (pt., p. 805)
   Maxillary palpi short, porrect, 4-segmented or (rarely) minute .... **Gracillariidae** (pt., p. 808)

13. Psychidae. Small to medium-sized; head with roughened hairs; ocelli present or absent; antennae often bipectinate to apex; haustellum and maxillary palpi rudimentary or absent; labial palpi short or rudimentary; epiphysis usually present, sometimes very long; tibial spurs 0-2-4, 0-1-1, or absent; fore wing (Figs. 36.20A, B) with M usually present in discal cell, often forked, chorda sometimes present; CuP usually separate or partly fused with 1A+2A or connected by a cross-vein; hind wing with M usually present in discal cell, sometimes forked, R_1 sometimes joining Rs and Sc, sometimes with one or more short branches from $Sc+R_1$ to costa, CuP often present; females fully winged, brachypterous, or apterous; abdomen in male often grossly extensible, in female with anal tuft of long hair. Larva case-bearing; case with anterior opening through which larva feeds, and posterior opening through which excrement is ejected (Fig. 36.21); crochets in lateral penellipse, uniordinal. Pupa (Figs. 36.23A, B) with dorsal abdominal spines, in female eyes and appendages sometimes greatly reduced or absent; in larval case, partially protruded at ecdysis or, in females, sometimes remaining in case.

Australia is rich in primitive species, often with fully-winged females, sometimes referred to the TALEPORIINAE. They have relatively simple venation, with M usually present and often forked in the discal cell of both wings. The fore wings often lack one vein, and the tibial spurs are 0-2-4. The larvae feed on many angiosperms, but a few feed on gymnosperms, lichens, and mosses. Certain exotic species feed on dead insects.

Ctenocompa (1 sp.), *Narycia* (71 spp.), and *Trigonocyttara* (1 sp.) have winged females. The larva of the stout-bodied *T. clandestina* Turn. (Fig. 36.18I) forms a cylindrical case

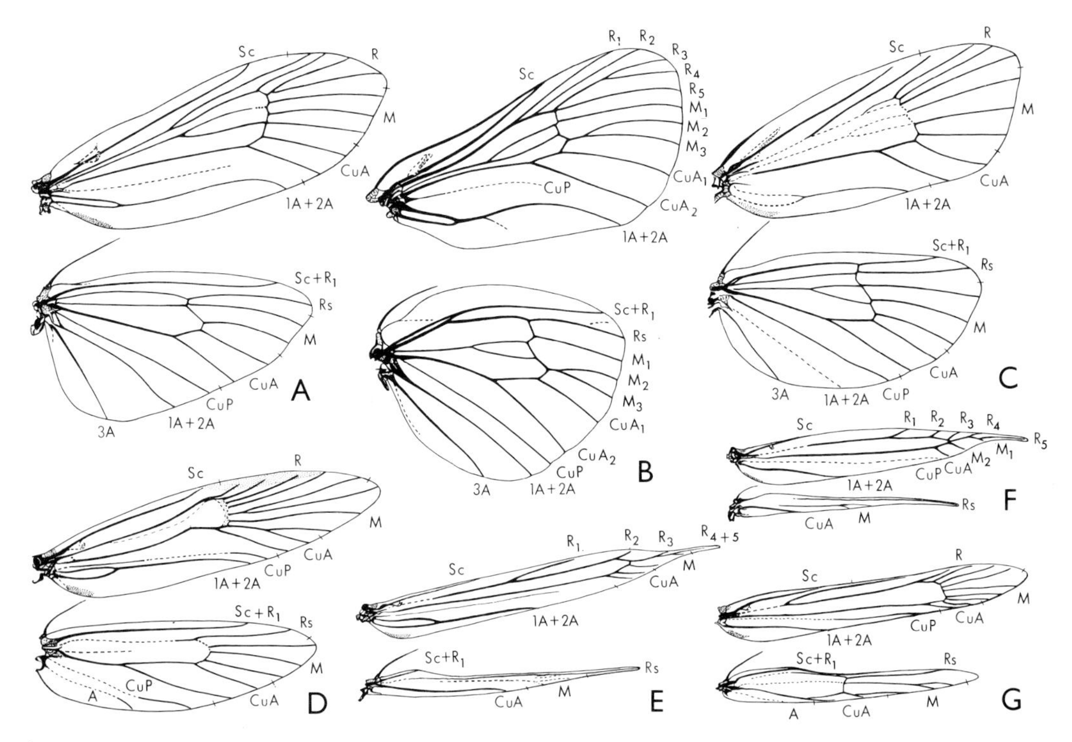

Fig. 36.20. Wing venation of Tineoidea: A, *Lepidoscia*, Psychidae; B, *Lomera*, Psychidae; C, *Ardiosteres*, Tineidae; D, *Monopis*, Tineidae; E, *Lyonetia*, Lyonetiidae; F, *Phyllocnistis*, Phyllocnistidae; G, *Macarostola*, Gracillariidae. [B. Rankin]

(Fig. 36.21C) with one or two twigs attached lengthwise. It feeds on *Eucalyptus*, *Acacia*, the introduced *Pinus radiata*, and several exotic ornamentals. In *Lepidoscia arctiella* (Walk.) the female has vestigial wings, but emerges normally from the pupal cuticle which, as in the males of all species, partially protrudes from the larval case at ecdysis. Fertilization takes place near or on the slender, tapering case (Fig. 36.21B), which is ornamented with spirally-arranged short lengths of twig. The brilliant metallic green and orange female of *Cebysa leucoteles* Walk. (Fig. 36.18J) has short, non-functional wings. An unfertilized female soon attracts several eagerly fluttering males, even in bright sunshine. The larvae form soft sacks (Fig. 36.21A) tapering at each end and ornamented with tiny pieces of the lichens upon which the larvae feed.

In the more specialized PSYCHINAE (Turner, 1947) the males have short, thinly-scaled wings, and fly rapidly. The enlarged costal area of the hind wing is usually strengthened by one or more short branches from $Sc+R_1$. The females are apterous, and the other appendages are greatly reduced. They are fertilized within the larval case, and the eggs, mixed with hair shed from the tip of the female abdomen, are laid within the pupal cuticle from which the female usually partially, but sometimes completely, emerges in the process. Spent females sometimes drop from the lower opening of the case, through which the young larvae later descend on silken threads. In some species wind drift of larvae suspended from strands of silk greatly aids dispersal. Larvae at first run rapidly on their thoracic legs, with the abdomen held high in the air. Later they construct a small conical case, which is gradually enlarged as they grow.

In *Lomera* (16 spp.) the venation is little reduced, CuP is not fused with 1A+2A in the fore wing, and the mid and hind tibiae each has a single apical spur. The epiphysis and basal segment of the fore tarsus are very long. The most specialized genera include *Clania* (9 spp.), *Hyalarcta* (2 spp.), and *Elinostola* (4 spp.). The last two have lost the tibial spurs and one or more veins, and *Hyalarcta* has also lost the epiphysis. *H. nigrescens* (Doubld.) has almost transparent wings; the larva constructs a whitish silken case, tapering at each end, with prominent ribs, but without ornamentation (Fig. 36.21E). The food-plant is *Eucalyptus*. The cylindrical larval case of *C. tenuis* Rosenstock (Fig. 36.18K) is ornamented with irregularly arranged lengths of twig from *Leptospermum* or other food-plant (Fig. 36.21D).

14. Tineidae. Small; head (Fig. 36.24A) with rough hair-scales; ocelli absent; antennae usually simple, rarely dentate or

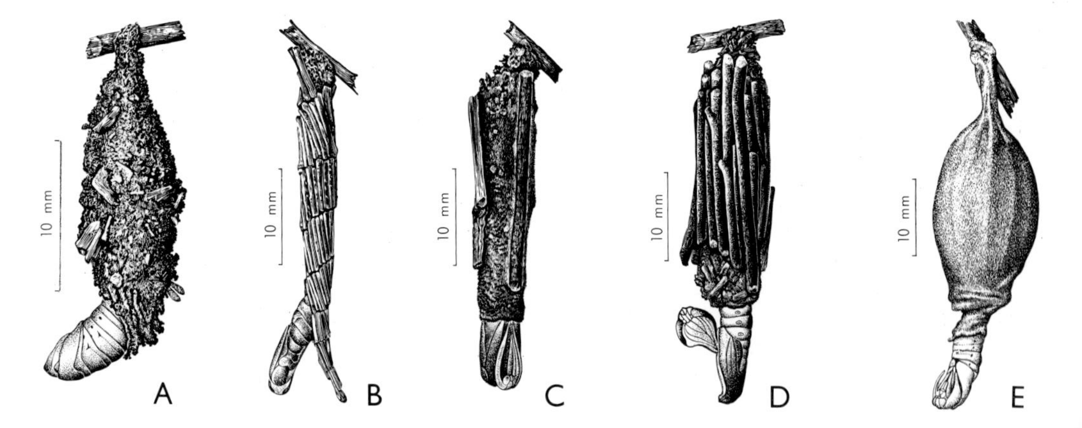

Fig. 36.21. Larval cases of Psychidae: A, *Cebysa*; B, *Lepidoscia*; C, *Trigonocyttara*; D, *Clania*; E, *Hyalarcta*. [F. Nanninga]

bipectinate, scape usually with pecten; haustellum short or absent; maxillary palpi 5-segmented and folded, or reduced; labial palpi drooping, porrect, or ascending, 2nd segment usually with lateral bristles; epiphysis present, tibial spurs 0-2-4, hind tibiae never smooth-scaled; fore wing (Figs. 36.20C, D) with M and chorda often present in discal cell, M sometimes forked, R_5 usually to costa or apex. Eggs oval. Larva sometimes case-bearing, usually feeding on dried animal or vegetable matter. Pupa (Fig. 36.23C) with dorsal abdominal spines, segments 4–8 movable in male, 4–7 in female; in larval shelter or cocoon.

The rough hairy head, bristly labial palpi, long folded maxillary palpi, when present, primitive venation, and rough-scaled hind tibiae, are useful distinguishing characters. Cosmopolitan genera include pests of animal fibres and grain. Several endemic genera occur in Australia.

Ardiosteres (9 spp.), with dentate antennae, and *Iphierga* (12 spp.), with bipectinate antennae, often have transverse wing patterns in black and white or black and orange. The 2nd segment of the labial palpi has long projecting scales, but lacks bristles. The case-bearing larvae of *I. macarista* Turn. (Fig. 36.22A) occur in meat ants' nests, and are thought to be scavengers. *Mesopherna* (8 spp.) has the bristly labial palpi of more specialized genera.

The widespread *Scardia* (2 spp.), *Tinea* (67 spp.), and *Monopis* (12 spp.), together with *Moerarchis* (6 spp.) and *Mimoscopa* (2 spp.), are more specialized. These usually have narrower wings, and M is often reduced or absent from the discal cell of one or both wings. Forking or loss of veins is common. Larvae of *Scardia*, which tunnel in fungi, are exceptional in having only 2 prespiracular (L) setae on the prothorax. In *Monopis* (Fig. 36.26A) there is a small transparent area in the discal cell of the fore wing. In some species, such as *M. chrysogramma* (Low.), the fertilized eggs are retained in the greatly enlarged vagina until just before hatching. The larvae are case-bearers, and feed on animal fibres. Endemic species of *Tinea* web leaves of trees or shrubs together or feed beneath loose bark. The larvae of *T. diaphora* Meyr. (Fig. 36.22B) are found beneath bark of *Eucalyptus*, and those of *T. tryphera* Meyr. live in a shelter formed by joining adjacent leaves of *Persoonia*. Introduced species include the well-known case-bearing clothes moth, *T. pellionella* (L.). The cosmopolitan *Nemapogon granella* (L.) and *Niditinea fuscipunctella* (Haw.) are pests of grain and other foodstuffs. In the clothes moth *Tineola bisselliella* (Humm.) digestion of wool is mediated by mid-gut enzymes.

Harmaclona is a small but widespread genus which Bradley (1953) referred to the Arrhenophanidae. Superficially the moths resemble Tineidae, with long narrow wings, primitive venation, rough-scaled head, dense pecten on the scape, and labial palpi with bristle-like projecting scales. The costa of the hind wing bears a series of stiff specialized scales, which hook beneath the down-curved inner margin of the fore wing, and assist in wing-coupling. Similar species occur in eastern Australia, but their generic and family status have not yet been studied.

15. Lyonetiidae. Very small; head usually with rough hairs, at least on crown; ocelli present or absent; scape usually forming an eye-cap; haustellum present; maxillary palpi 5-segmented, folded, or short, or vestigial; labial palpi short, porrect or drooping; epiphysis present, tibial spurs 0-2-4, hind tibiae with long hair-scales; fore wing (Fig. 36.20E) usually narrow, apex tapering, sometimes tip bent upwards, venation often reduced, R_5 to costa, CuP present; hind wing lanceolate or linear, fringe longer than width of wing, venation reduced, $Sc+R_1$ short, Rs to apex, discal cell open, CuA separate, simple, anal veins reduced. Larva with normal or reduced thoracic legs and prolegs, crochets uniordinal, prolegs sometimes absent only in early instars; feeding on refuse, or mining in leaves, bark, stems, or galls. Pupa (Fig. 36.23D) often with appendages glued down and abdominal segments immovable, maxillary and labial palpi not visible, abdomen usually lacking dorsal spines; in elliptical cocoon, sometimes

Fig. 36.22. A, *Iphierga macarista*, B, *Tinea diaphora*, Tineidae; C, *Comodica acontistes*, D, *Lyonetia sulfuratella*, Lyonetiidae; E, *Phyllocnistis diaugella*, Phyllocnistidae; F, *Acrocercops antimima*, G, *A. calicella*, Gracillariidae; H, *Simaethis chionodesma*, I, *Choreutis lampadias*, Glyphipterigidae; J, *Yponomeuta paurodes*, K, *Lactura suffusa*, L, *Amphithera heteroleuca*, Yponomeutidae. Scales = 2 mm. [Photos by C. Lourandos]

within mine or tunnel, or suspended amid strands of silk.

The family as here defined may not be homogeneous. *Comodica* (11 spp.) and *Erechthias* (21 spp.), separated by some authors as the Erechthiidae, are akin to the Tineidae. The adults have ocelli and 5-segmented folded maxillary palpi, and the wing tips are bent upwards at a sharp angle. The larvae tunnel in galls, bark, or stems, and the abdomen of the pupa has dorsal spines. The larva of *C. acontistes* (Meyr.) (Fig. 36.22C) tunnels in the flower stems of grass-trees *(Xanthorrhoea)*, and that of *C. mystacinella* (Walk.) bores in galls growing on *Exocarpos* and *Acacia*, and in the gall-like swellings of apples injured by woolly aphid. The larva of the black and white *Erechthias symmacha* (Meyr.) tunnels in the stringy bark of *Eucalyptus delegatensis*.

Opogona (28 spp.), with smooth-scaled head, is sometimes separated as the Oinophilidae. The frons is very oblique, ocelli are absent, the maxillary palpi are folded and 5-segmented, and the scape is long and slender. *O. comptella* (Walk.) has been reared from larvae tunnelling in the bark of a dead ornamental tree; *O. glycyphaga* Meyr. injures sugar cane and bananas.

The remaining genera constitute the Lyonetiidae *sensu stricto*. *Bucculatrix* (9 spp.; Braun, 1963) lacks ocelli, and has reduced maxillary palpi. The abdomen of the pupa is spined, and some segments are movable. Pupation occurs in a fusiform ribbed cocoon. The native *B. gossypii* Turn. attacks cotton in northern Australia. During the first 2 instars and part of the 3rd it is an apodous leaf-miner; it then spins an oval shelter on the leaf surface, in which it moults to the 4th instar. This older larva feeds exposed on the surface, skeletonizing the leaf. It has normal thoracic legs, and ventral prolegs, each with 2 transverse bands of uniordinal crochets.

The most specialized genera include *Bedellia* (1 sp.), *Lyonetia* (9 spp.), and *Leucoptera* (18 spp.). They are without ocelli, and the maxillary palpi are reduced or absent. The larvae are leaf-miners. In *Bedellia* the prolegs on abdominal segments 3 and 6 are somewhat reduced; in *Lyonetia* and *Leucoptera* prolegs are absent in the earlier instars, and the thoracic legs may be more or less reduced in later instars. The pupae are suspended by the posterior end amid several strands of silk. As in the unrelated Elachistidae, Pterophoridae, and Papilionoidea, in which exposed pupae also occur, the pupae are often angular and patterned. The appendages are glued down, and the abdomen is unspined, without movable segments. *Bedellia somnulentella* (Zell.) is a cosmopolitan pest of sweet potato *(Convolvulus)*, and *Lyonetia sulfuratella* (Meyr.) (Fig. 36.22D) mines in the young terminal leaves of *Banksia*.

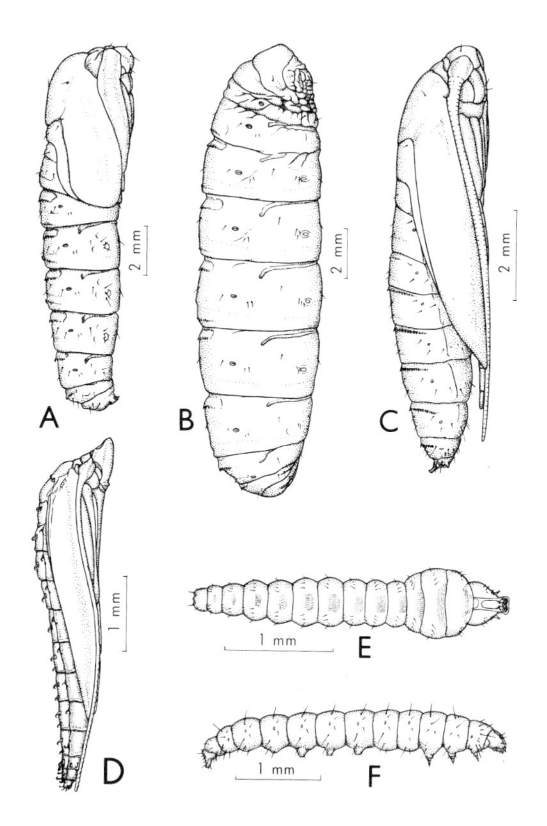

Fig. 36.23. A, B, pupa, ♂ and ♀, of *Lomera*, Psychidae; C, pupa of *Monopis*, Tineidae; D, pupa of *Lyonetia*, Lyonetiidae; E, F, early and final instar larvae of *Acrocercops*, Gracillariidae. [B. Rankin]

16. Phyllocnistidae. Very small; head smooth-scaled; ocelli absent; scape sometimes forming small eye-cap, without pecten; haustellum present; maxillary palpi very short; labial palpi slender, porrect or ascending; epiphysis present, tibial spurs 0-2-4, hind tibia with dorsal row of bristles; fore wing (Fig. 36.20F) lanceolate, apex tapering, R_1 from beyond middle of wing, R_5 to apex, M_3 and CuA_1 absent. Larva an apodous leaf-miner: first 3 instars flattened, with blade-like mandibles, sap-feeding; 4th instar cylindrical, with reduced head, non-feeding; all instars have paired retractile protuberances on meso- and metathorax and on each of abdominal segments 1–7, 2 pairs on 8. Pupa with antennae and hind legs extending beyond wing tips; abdominal segments 4–7 movable in male, 4–6 in female, 3–7 with dorsal pits and sclerotized setae.

The family contains tiny, usually shining white species, with distinctive wing pattern, and with a prominent row of dorsal bristles on the hind tibiae, as in some Gracillariidae. However, the maxillary palpi are shorter, the venation more reduced, and the non-feeding apodous last-instar larva is significant. The larvae produce silvery serpentine mines in the epidermis of leaves, superficially suggesting the trails of snails. Pupation occurs in a silk-lined chamber formed at the end of the mine by partially folding in the sides.

All the known species belong to the genus *Phyllocnistis*. *P. diaugella* Meyr. (Fig. 36.22E) is common in coastal eastern Australia, often covering the upper surface of the leaves of *Breynia oblongifolia* and *Phyllanthus* with its mines.

17. Gracillariidae. Very small, slender; head (Fig. 36.24B) usually smooth-scaled; ocelli absent; antennae nearly as long as, or longer than, fore wing, simple, scape slender, usually without pecten; haustellum present; maxillary palpi slender, porrect or ascending, 4-segmented; labial palpi slender, porrect or ascending; epiphysis present, tibial spurs 0-2-4, hind tibiae smooth or with dorsal row of bristles, never hairy, fore and mid tibiae sometimes thickened with scales; fore wing (Fig. 36.20G) lanceolate or linear-lanceolate, R_5 to costa, some veins often lost, 1A+2A without basal fork; hind wing lanceolate or linear, venation often reduced. Larva (Figs. 36.23E, F) leaf- or gall-mining, at least in early stages, with heteromorphosis; early instars flattened, with large blade-like mandibles, sap-feeding, thoracic legs sometimes lost, ventral prolegs more or less reduced, always absent from segment 6; after second or third ecdysis larva feeds on parenchyma, body cylindrical, thoracic legs present, ventral prolegs on abdominal segments

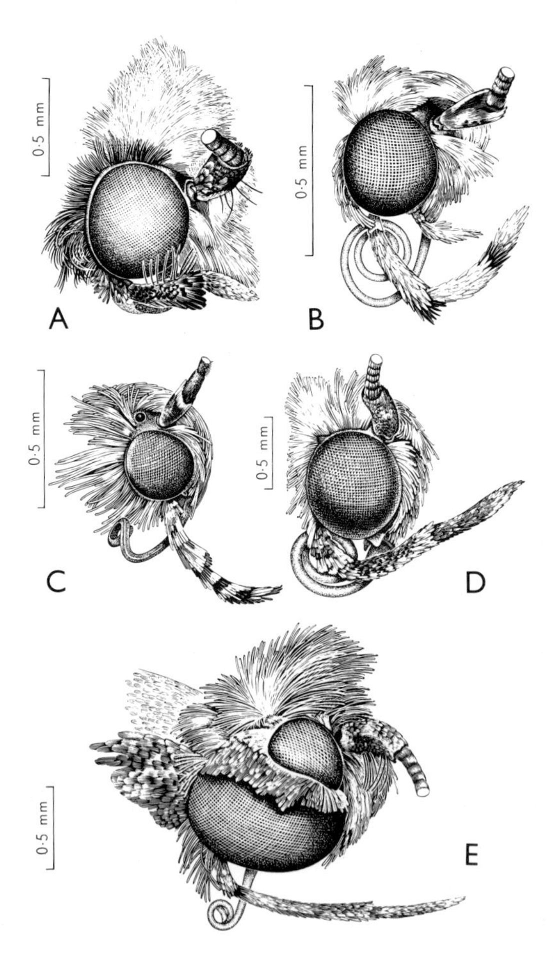

Fig. 36.24. Heads, lateral: A, *Monopis*, Tineidae; B, *Acrocercops*, Gracillariidae; C, *Glyphipterix*, Glyphipterigidae; D, E, *Yponomeuta*, *Amphithera*, Yponomeutidae. [F. Nanninga]

3–5. Pupa with antennae and haustellum extending beyond wing tips; abdomen with fine scattered dorsal spining, segments 5–8 movable in male, 5–7 in female; usually in flattened cocoon.

The family (Vári, 1961) contains elegant, tiny moths, which usually rest with the anterior part of the body raised at a steep angle and the fore and mid legs prominently displayed (Fig. 36.26B). In *Caloptilia* (35 spp.) the larva is a leaf-miner at first, later cutting and rolling the edge of the leaf to form a shelter within which the larva feeds and usually pupates. The introduced *C. azaleella* Brants attacks azaleas in eastern Australia. The mature larvae of the endemic *Macarostola formosa* (Staint.) (Plate 7, A) leaves its rolled leaf shelter on *Acmena smithii* to pupate in an oval flattened cocoon. In *Caloptilia* the fore and mid legs are ornamented with tufts of scales, but they are smooth in *Macarostola*.

Most species in the composite genus *Acrocercops* (98 spp.) produce blister-like mines in leaves, pupating in an oval flattened cocoon, either inside or outside the mine. The cocoon is usually ornamented with tiny froth-like globules, the presence and number of which are of specific significance. The hind tibia of the adult has a series of stiff dorsal bristles. *A. antimima* Turn. (Fig. 36.22F), on *Lomatia myricoides* is one of several similar species which attack Proteaceae. *A. plebeia* Turn. is a minor pest of *Acacia podalyrifolia*, and several species, such as *A. calicella* (Staint.) (Fig. 36.22G), mine the leaves of *Eucalyptus*. *A. tricuneatella* (Meyr.) produces elongate mines in the leaves of the reed *Typha*.

In Australia *Lithocolletis* (11 spp.) contains tiny orange and white species on Malvaceae, and black and white species on Leguminosae. The head is smooth-scaled with a fringe of erect hair-scales above, and the maxillary palpi are short. The larva forms a small mine in which the silk lining causes the epidermis to contract to form a characteristic cell. Several mines may occur in a single leaf. *L. aglaozona* Meyr. mines in *Kennedya*, *Glycine*, and sometimes in French beans.

Superfamily YPONOMEUTOIDEA

Ocelli often prominent, or absent; chaetosemata usually absent; scape sometimes with pecten; haustellum usually naked; maxillary palpi small, 3- or 4-segmented, or reduced, or rudimentary, porrect or drooping, rarely folded over base of haustellum; labial palpi short, drooping, or moderate, ascending, apical segment usually not acute; epiphysis present; tibial spurs 0-2-4; fore wing with M rarely retained in discal cell, chorda and pterostigma sometimes present, venation sometimes reduced; retinaculum in female sometimes a subcostal scale-tuft, hook, or series of recurved bristles; abdomen sometimes with inconspicuous dorsal spining. Larva with 3 prespiracular (L) setae on prothorax (only 2 in Epermeniidae), crochets in a circle, pseudocircle, or irregular mesoseries, sometimes in 2 bands; boring in stems, mining in leaves, feeding beneath a slight webbing, or in a more extensive webbing, occasionally feeding exposed. Pupa often with dorsal abdominal spines, in male with segments 3 or 4 to 7 movable; in larval gallery or in fusiform or oval silken, sometimes network or parchment-like cocoon, usually protruded from cocoon or shelter at ecdysis.

The superfamily retains many relatively primitive characters, such as the mobile and well-spined pupa of the Aegeriidae and the prominent ocelli in the Aegeriidae, Glyphipterigidae, Douglasiidae, and Heliodinidae *(sensu stricto)*. Narrow wings and reduced venation in the Douglasiidae and the highly specialized adult form, pattern, and behaviour of the Aegeriidae are probably specializations. In some Yponomeutidae and Glyphipterigidae chaetosemata have appeared, and the pupa of the Epermeniidae seems to be the most advanced of the superfamily. The larval habits of the latter, however, and the network cocoon in some *Epermenia*, suggest a relationship to the Yponomeutidae.

Most of the families are well represented in Australia, especially the genus *Glyphipterix* and the brightly coloured *Lactura* group of the Yponomeutidae. The Heliodinidae and the Douglasiidae, on the other hand, include only one Australian species each.

Key to the Families of Yponomeutoidea

1. Wings partly hyaline, coupled by series of recurved bristles as well as by frenulum .... **Aegeriidae** (p. 810)
   Wings usually fully scaled, coupled only by frenulum 2
2. Hind tibia and tarsus with generally distributed dense stiff bristles .... **Epermeniidae** (p. 813)
   Hind tibia and tarsus not so 3
3. Hind tarsus with whorls of stiff bristles at apices of segments, tibia with median and apical whorls of scales or bristles .... **Heliodinidae** (p. 812)
   Hind tibia and tarsus not bristled, tibia either smooth-scaled, or with long fine hair-scales 4
4. Ocelli small or absent; if ocelli present, maxillary palpi porrect, easily seen without dissection .... **Yponomeutidae** (p. 812)
   Ocelli prominent; maxillary palpi very small and obscure, not clearly porrect 5
5. Hind wing with venation reduced, discal cell open, Rs at or near long axis of wing .... **Douglasiidae** (p. 812)
   Hind wing with venation not greatly reduced, discal cell closed, Rs nearer to C than to inner margin .... **Glyphipterigidae** (p. 810)

18. Aegeriidae (Sesiidae). Small; head smooth-scaled; ocelli prominent; chaetosemata absent; antennae gradually thickened and then tapered to apex, simple or pectinate, apex usually with minute tuft of bristles, scape without pecten; haustellum naked; maxillary palpi minute; labial palpi ascending; hind tibiae with long hair-scales or long tufts or bristles at spurs; fore wing (Fig. 36.25A) long, narrow, partially hyaline, CuP reduced or absent, 1A+2A in longitudinal downward fold which fits into upward fold of hind wing, each bearing a series of recurved spines which interlock and help couple wings, retinaculum in female a broad subcostal hook; hind wing partially hyaline, Sc and Rs parallel, anastomosing distally, concealed in costal fold, CuP present or vestigial, frenulum in female simple. Larva with crochets in 2 transverse uniordinal rows; tunnelling in trunk, bark, or roots of trees or herbs, or in galls. Pupa with 2 rows of dorsal abdominal spines, ventral spines on last segment, 3–7 movable in male, 3–6 in female; protruded from gallery at ecdysis.

The brightly coloured moths bear a striking resemblance to wasps. Although diurnal, their rapid flight makes them difficult to see. Most of the species are from Queensland, with only a couple from N.S.W. and one from the N.T. The larva of the introduced *Synanthedon tipuliformis* (Clerck) tunnels in the stems of currants and gooseberries in the south-east and in Tasmania. *S. chrysophanes* (Meyr.) (Plate 7, D) has been reared from the bark of *Alphitonia excelsa*. *Diapyra igniflua* (Lucas) tunnels in the bark of quandong.

19. Glyphipterigidae. Small; head (Fig. 36.24C) usually smooth-scaled; ocelli prominent; chaetosemata rarely present, scape without pecten; haustellum usually naked, sometimes sparsely scaled near base; maxillary palpi very small, 3- or 4-segmented, or reduced to 1 segment; labial palpi usually ascending, smooth-scaled, or 2nd segment tufted; fore wing (Fig. 36.25B) often with chorda, pterostigma often present, M sometimes present in discal cell, R_5 to apex or termen, CuP present near margin, female usually with only a subcostal retinaculum; hind wing with R_1 usually joining Sc and Rs, Rs and M_1 separate, parallel or divergent, CuP often reduced, frenulum in female usually of 1 or 2 bristles. Larva with crochets uniordinal in a circle, abdominal segments with L1 and L2 approximated; usually feeding beneath slight webbing, rarely gregarious in dense web, on leaves, amongst

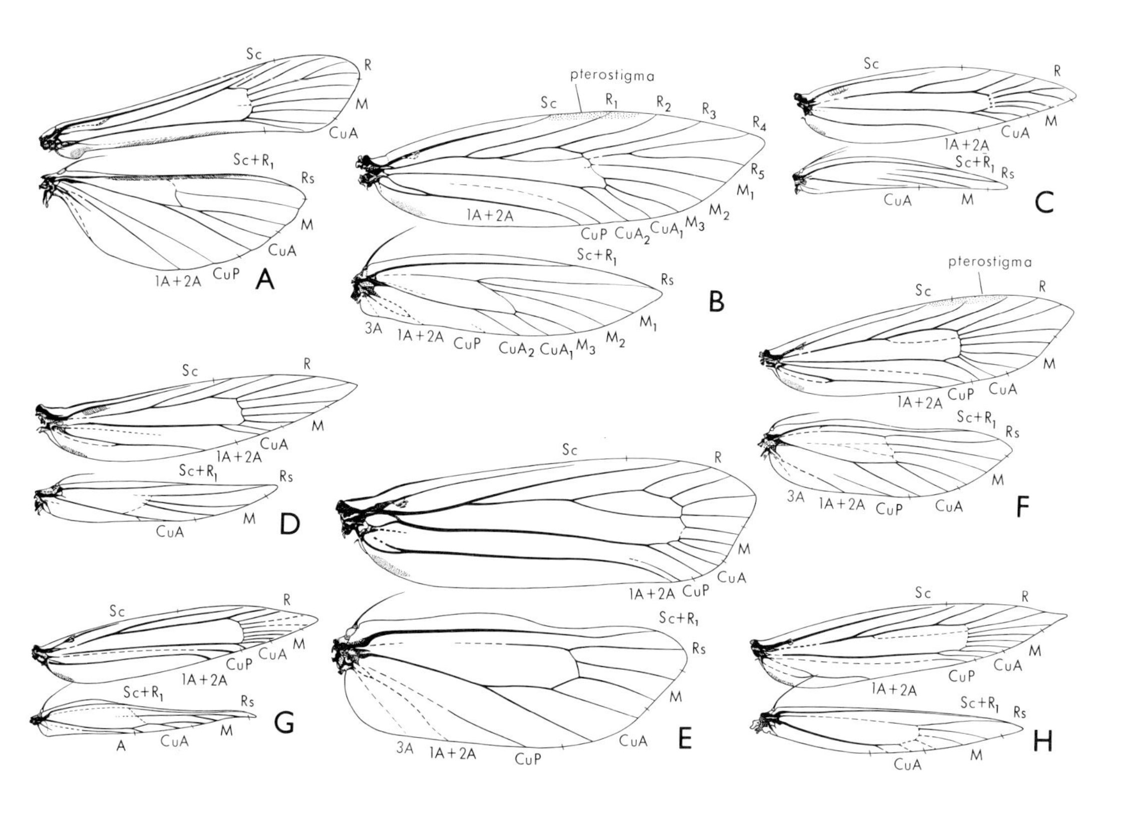

Fig. 36.25. Wing venation of Yponomeutoidea: A, *Synanthedon*, Aegeriidae; B, *Glyphipterix*, Glyphipterigidae; C, *Tinagma* ♀, Douglasiidae; D, *Heliodines* ♀, Heliodinidae; E–G, *Yponomeuta*, *Plutella*, *Argyresthia*, Yponomeutidae; H, *Epermenia*, Epermeniidae. [J. Wedgbrow]

seeds, or tunnelling into shoots. Pupa with dorsal abdominal spines, segments 3–7 movable in male, 3–6 in female, in silken cocoon, partially protruded at ecdysis.

The family contains mostly diurnal species with bright metallic or orange markings often displayed in various ways by the resting adult. It features a relatively unspecialized venation, often with pterostigma in fore wing, prominent ocelli, and female usually with subcostal retinaculum and frenulum of 1 or 2 bristles.

By far the largest Australian genus is *Glyphipterix* (57 spp.) in which the fore wings have brilliant metallic markings, and the maxillary palpi are 3- or 4-segmented. The moths fly in damp places, and the larvae feed amongst the seeds or in the shoots of Juncaceae. *G. gemmipunctella* (Walk.) has the basal half of the hind wing bright orange. *Simaethis* (10 spp.) has larvae feeding on *Ficus*. The moths have broad wings, some scaling at the base of the haustellum, 1-segmented maxillary palpi, and the labial palpi truncate or the apical segment short and blunt. At rest the wings are raised and curled. *S. chionodesma* Low. (Fig. 36.22H) has a broad orange band on the fore wing. The larva of *S. ophiosema* Low. makes a silken shelter in a fold of a leaf, and pupates in a fusiform cocoon of dense white silk. The base of the haustellum is also scaled and the maxillary palpi are 1-segmented in the cosmopolitan genus *Choreutis* (3 spp.). *C. lampadias* Meyr. (Fig. 36.22I) is common at high altitudes in the south-east. *Homadaula* (4 spp.), also with very short maxillary palpi, contains grey species found in the drier areas of the south and west. The larvae of *H. myriospila* Meyr. from W.A. feed gregariously in a dense web on the phyllodes of *Acacia*.

20. Douglasiidae. Very small; head smooth-scaled; ocelli prominent; chaetosemata absent; haustellum naked; scape without pecten; maxillary palpi rudimentary; labial palpi short, drooping; hind tibiae with long hair-scales; fore wing (Fig. 36.25C) lanceolate, R with 4 or 5 branches to costa, R_5 and M_1 stalked, CuP weak or absent, female with subcostal retinaculum; hind wing lanceolate, frenulum in female of 1 or 2 bristles, Rs at or near long axis of wing, venation reduced, CuA free. Larva fusiform, prolegs small, crochets absent, abdomen with L1 and L2 approximated; mining in leaves *(Tinagma)*, tunnelling in stems amongst flowers *(Douglasia)*. Pupa in stem *(Douglasia)*.

Tinagma leucanthes Meyr. is a tiny black species with two transverse white bands on the fore wing, known from Sydney and Stradbroke I. The prominent ocelli, unscaled haustellum, and the venation make recognition easy. The larvae are unknown, but the moths are said to rest with their wings partly raised.

21. Heliodinidae. Very small; head smooth-scaled; ocelli prominent; chaetosemata absent; haustellum naked; scape without pecten; maxillary palpi very small; labial palpi short, drooping or recurved; hind legs with whorls of stiff bristles at bases of spurs and apices of tarsi; fore wing (Fig. 36.25D) lanceolate, R_5 to termen, CuP absent, female with subcostal retinaculum of recurved bristles; hind wing lanceolate, female frenulum of 2 bristles, venation often reduced, CuP vestigial. Egg upright. Larva with prolegs long, slender, crochets few in circle, abdomen with L1 and L2 approximated; in a slight web on foliage, or within fruits. Pupa with dorsal spines on abdominal segments 5–7, 3–7 movable in male, 3–6 in female, spiracles on conical protuberances; in open-mesh cocoon, protruded at ecdysis.

Included here are those genera, such as *Heliodines* (1 sp.) and *Schreckensteinia*, with naked haustellum, prominent ocelli, female with subcostal retinaculum and frenulum of one bristle, and pupa that is protruded from the cocoon at ecdysis. *Stathmopoda* and related genera, usually associated with *Heliodines* mainly because of the similar habit of raising and displaying the hind legs, are here referred to the Gelechioidea. The family apparently includes few Australian species. *H. princeps* Meyr. (Plate 7, C), from north Queensland, has orange and dark brown fore wings ornamented with brilliant metallic spots.

22. Yponomeutidae. Very small to small; head (Figs. 36.24D, E) smooth-scaled or with long hair-scales; ocelli rarely present; chaetosemata sometimes present; scape often with pecten; haustellum naked; maxillary palpi porrect, 1- to 4-segmented, or rudimentary; labial palpi ascending, or short and drooping; hind tibiae usually smooth-scaled; fore wing (Figs. 36.25E–G) often with pterostigma, R_5 to termen, chorda and M

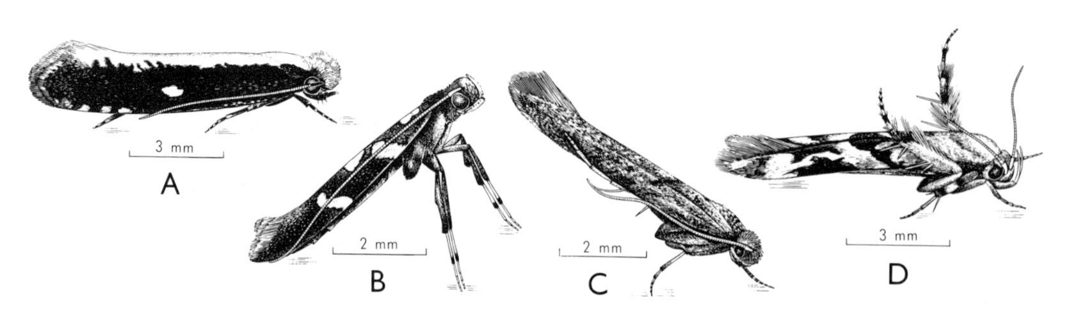

Fig. 36.26. Adults, lateral: A, *Monopis*, Tineidae; B, *Caloptilia*, Gracillariidae; C, *Zelleria*, Yponomeutidae; D, *Stathmopoda*, Stathmopodidae. [F. Nanninga]

sometimes present, CuP present; hind wing with Rs and M_1 usually well separated, M_1 and M_2 sometimes stalked, CuP present; abdomen sometimes with fine dorsal spining. Larva with crochets in a circle, pseudocircle, or irregular mesoseries, uni- or multiserial, uniordinal or biordinal; abdominal setae L1 remote from L2; beneath slight webbing on leaves or flowers, gregarious in webs, or tunnelling in shoots, mining in leaves, or feeding exposed. Pupa with maxillary palpi present, pilifers sometimes indicated, cremaster sometimes present; in fusiform or oval, sometimes network cocoon, protruded at ecdysis.

The family contains 4 subfamilies sometimes treated as families. The PLUTELLINAE have porrect 4-segmented maxillary palpi, and many rest with porrect antennae. Ocelli are present in *Plutella* (3 spp.); the larvae of the cosmopolitan *P. xylostella* (L.) skeletonize the foliage of Cruciferae, and pupate in an open-mesh cocoon. The native *Prays nephelomima* Meyr. and *P. parilis* Turn. damage *Citrus* flowers in N.S.W. and Queensland.

The YPONOMEUTINAE, with 1- or 2-segmented maxillary palpi, include many brightly coloured tropical and subtropical species. *Yponomeuta* (5 spp.) has fore wings with numerous black spots; the larvae, which live gregariously in webs, have multiserial crochets. *Y. paurodes* Meyr. (Fig. 36.22J) and *Y. pustulellus* Walk. are from eastern Queensland and N.S.W. *Atteva* (5 spp.) is unusual because chaetosemata are present, the larvae have the MD group bisetose on segments 1 to 8 of the abdomen, and the pupa has pilifers. In eastern Australia the larvae of *A. niphocosma* Turn. (Plate 7, L) occur gregariously in webs. Indian species have similar habits, feeding on *Ailanthus*. A related group, including *Lactura* (22 spp.), *Anticrates* (5 spp.), and *Thyridectis* (1 sp.), also have chaetosemata in the adult and pilifers in the pupa. The striking slug-like larva of *L. suffusa* (Walk.) (Fig. 36.22K), patterned in blue-grey and white, with yellow tubercles, feeds exposed on the leaves of *Sideroxylon australe*, and pupates in a stiff cocoon. It has white eversible lateral lobes, and is covered with sticky secretion which enables it to adhere to anything it contacts. This could minimize the accidental removal of larvae by wind or other mechanical means. *T. psephonoma* Meyr. also has slug-like larvae suggestive of Limacodidae. *Imma* (12 spp.) and *Burlacena* (1 sp.) are usually referred to the Glyphipterigidae, but lack ocelli, and have chaetosemata and, in the female, a subcubital retinaculum. The larva of neither genus is known, but Meyrick stated that *Imma* spins an open network cocoon, similar to that of *Plutella*.

The AMPHITHERINAE contain elegant species with long antennae and 1-segmented maxillary palpi. The male of the rain-forest species *Amphithera heteroleuca* (Turn.) (Fig. 36.22L) is remarkable for its large eyes which are divided horizontally by a scaled band (Fig. 36.24E). In the female the eyes have a deep posterior indentation. The larvae of *Thereutis* (9 spp.) and *Macarangela* (4 spp.) are leaf-miners at first, but later feed exposed. *M. leucochrysa* Meyr. (Plate 7, F) is white with an orange stripe; the larvae feed on *Banksia paludosa*.

The ARGYRESTHIINAE include *Argyresthia notoleuca* (Turn.) from north Queensland, and *Zelleria* (20 spp.). They rest with head depressed and body raised at a steep angle (Fig. 36.26C) and, although the wings are narrow, the venation is not greatly reduced. The larva of *Z. euthysema* Turn. lives in a slight web on *Melichrus urceolatus*. The small grey *Ogmograptus scribula* Meyr. is perhaps also referable to the Yponomeutidae. Its larva produces conspicuous scribble-like mines (Fig. 5.16) in the living bark of smooth-barked *Eucalyptus*, such as *E. pauciflora* and *E. micrantha*. Its stiff cocoon, from which the pupa is protruded at ecdysis, resembles that of *Bucculatrix* (Lyonetiidae).

23. Epermeniidae. Very small; head smooth-scaled; ocelli and chaetosemata absent; scape with pecten; haustellum naked; maxillary palpi 3-segmented, folded over base of haustellum; labial palpi recurved; hind tibiae with long stiff bristles above and beneath; fore wing (Fig. 36.25H)

Fig. 36.27. A, *Epermenia eurybias*, Epermeniidae; B, *Coleophora ochroneura*, Coleophoridae; C, *Agonoxena phoenicia*, Agonoxenidae; D, *Cosmiotes aphanta*, Elachistidae; E, *Scythris rhabducha*, Scythridae; F, *Stathmopoda callichrysa*, Stathmopodidae; G, *Hypertropha tortriciformis*, H, *Thema chlorochyta*, I, *Hippomacha callista*, J, *Philobota productella*, Oecophoridae; K, *Ethmia sphaerosticha*, Ethmiidae; L, *Crocanthes prasinopis*, Timyridae. Scales = 2 mm. [Photos by C. Lourandos]

lanceolate, small scale tufts on inner margin, chorda often present, R_5 to termen, CuP present near margin; hind wing lanceolate, Rs and M_1 sometimes stalked, other veins separate. Larva with 2 prespiracular setae (L) on prothorax, crochets in uniordinal circle, L1 and L2 of abdomen approximated; usually mines in leaves at first, then feeds externally in slight web. Pupa with abdominal segments 4–7 movable in male, 4–6 in female, 2 deep lateral pits on 9; in silken cocoon, not protruded at ecdysis.

Easily recognized by the stiff bristles of the hind tibiae, and the scale tufts along the inner margin of the forc wing, the Australian species are placed in *Epermenia*. The presence of 2 instead of 3 prespiracular setae on the larval prothorax is exceptional in the superfamily. The pupa is also unusual in not being protruded from the cocoon at ecdysis. The small green larvae of *E. eurybias* Meyr. (Fig. 36.27A) may be beaten from *Exocarpos* upon which they feed.

Superfamily GELECHIOIDEA

Ocelli present or absent; chaetosemata absent; scape often with pecten; haustellum with dense imbricated scales towards base, rarely reduced; maxillary palpi small, 4-segmented, folded over base of haustellum, rarely reduced; labial palpi usually recurved, apical segment often exceeding vertex, usually tapering, acute; epiphysis present; tibial spurs 0-2-4; fore wing with M rarely present in discal cell, chorda vestigial or absent, pterostigma sometimes present, retinaculum in female normally subcubital; hind wing with venation sometimes reduced, CuA rarely with basal pecten of hairs, CuP sometimes absent, female frenulum usually with 3 bristles; abdomen often with prominent dorsal spining. Larva with 3 prespiracular (L) setae on prothorax, crochets in a circle, ellipse, or rarely in 2 transverse rows, abdominal setae L1 and L2 approximated; concealed feeders, case-bearing, tunnelling in stems or fruits, leaf-mining, joining foliage, or feeding beneath silken shelter. Pupa without dorsal abdominal spines, normally segments 5–7 movable in male, 5–6 in female; usually in larval shelter, not protruded from shelter or cocoon at ecdysis.

Most species can be easily placed by the densely scaled haustellum, the base of which is clasped by the small 4-segmented maxillary palpi. In the Coleophoridae, Agonoxenidae, Elachistidae, and a few Xyloryctidae, the maxillary palpi are reduced. *Stathmopoda* and related genera, usually referred to the Heliodinidae, are here referred to the Stathmopodidae.

This is by far the largest superfamily in Australia, and more than half of it consists of Oecophoridae. The Xyloryctidae, Gelechiidae, and Cosmopterigidae are also large families containing many endemic genera. At the other extreme are the Coleophoridae and Agonoxenidae.

Key to the Families of Gelechioidea Known in Australia

1. Hind wing narrow, lanceolate, linear-lanceolate, or linear, narrower than its fringe 2
   Hind wing ovate-lanceolate, ovate, or trapezoidal with sinuate or emarginate termen 9

2(1). Hind tibia with medial and terminal whorls of bristles, or, if with dense dorsal bristles, the hind tarsus with whorls of bristles at apices of segments. [Hind legs raised in repose] .... **Stathmopodidae** (p. 818)
   Hind tibia and tarsus not so 3

3(2). Fore wing with R_2 approximated at base to upper angle of discal cell, remote from R_1 4
   Fore wing with R_2 from well before upper angle of discal cell 5

4(3). Fore wing with R_5 to costa, pterostigma present; scape usually expanded; ♀ with both subcubital and radial retinacula .... **Blastobasidae** (p. 822)
   Fore wing with R_5 to termen, pterostigma absent; scape not expanded; ♀ with only subcubital retinaculum .... **Scythridae** (p. 818)

5(3). Hind wing with Rs remote from $Sc+R_1$, at or near longitudinal axis of wing 6
Hind wing with Rs nearer to $Sc+R_1$, not axial 7
6(5). Frons strongly oblique; labial palpi flattened; antennae porrect in repose .... **Agonoxenidae** (p. 816)
Frons vertical; labial palpi not flattened; antennae turned back in repose .... **Elachistidae** (p. 817)
7(5). Maxillary palpi rudimentary, not folded over base of haustellum; abdominal terga each bearing 2 small longitudinal patches of short spines. [Antennae porrect in repose] .... **Coleophoridae** (p. 816)
Maxillary palpi distinctly folded over base of haustellum; abdominal terga with spines more generally distributed or absent 8
8(7). Hind wing with Rs and M_1 separate, parallel .... **Oecophoridae** (pt., p. 819)
Hind wing with Rs and M_1 approximated, connate (arising from one point on the discal cell), or stalked .... **Cosmopterigidae** (p. 821)
9(1). CuP absent as a tubular vein from both wings; hind wing often with emarginate termen; ♀ with radial retinaculum of specialized scales .... **Gelechiidae** (p. 824)
CuP present as a tubular vein at least in hind wing; termen of hind wing not emarginate; ♀ with subcubital retinaculum of hair-scales 10
10(9). Hind wing with Rs and M_1 well separated at base, parallel or slightly divergent 11
Hind wing with Rs and M_1 closely approximated at base, connate, or stalked, divergent 12
11(10). Hind wing with M_2 usually not arising nearer to M_1 than to M_3; if M_2 nearer to M_1 or equidistant from M_1 and M_3, then abdomen with dorsal spining ...... **Oecophoridae** (pt., p. 819)
Hind wing with M_2 arising nearer to M_1 than to M_3, or rarely equidistant from M_1 and M_3; abdomen without dorsal spining .... **Ethmiidae** (p. 820)
12(10). Hind wing with $Sc+R_1$ approaching Rs near or before end of discal cell ... **Stenomidae** (p. 824)
Hind wing with $Sc+R_1$ diverging from Rs well before end of discal cell 13
13(12). Fore wing with CuA_2 arising well before lower angle of discal cell ...... **Xyloryctidae** (p. 822)
Fore wing with CuA_2 arising close to lower angle of discal cell 14
14(13). Antennae often longer than fore wing and thickened, at least in ♂; dorsum of abdomen often spined; gnathos in ♂ not spinose .... **Timyridae** (p. 820)
Antennae much shorter than fore wing, not thickened in ♂, but often ciliate; dorsum of abdomen not spined; gnathos in ♂ spinose .... **Oecophoridae** (pt., p. 819)

24. Coleophoridae. Very small; head smooth-scaled; ocelli absent; scape usually with scale tuft; maxillary palpi minute, 2-segmented; labial palpi long, recurved; hind tibiae with roughened hair-scales; fore wing (Fig. 36.28A) narrow, elongate, without transverse markings or pterostigma, at least 1 vein absent; hind wing linear-lanceolate, usually at least 1 vein absent; abdomen with terga 1–7 each with 2 elongate patches of spines; valva in male partly divided, gnathos elongate, with spinose apical knob. Larva in first instar with ventral prolegs sometimes absent, but anal prolegs present; in later instars prolegs may be absent from segment 6, or from 3–6; a leaf-miner in first instar, then a case-bearer, feeding externally on leaves or flowers, or mining in leaves. Pupa in larval case.

The adults rest with antennae porrect. The few endemic species are referred to *Coleophora*, but their early stages are not known. *C. ochroneura* (Low.) (Fig. 36.27B), with grey wings and veins outlined with ochreous, is not uncommon in the south-east. The bronzy green *C. alcyonipenella* Koll. from Europe is established in south-eastern Australia, Tasmania, and New Zealand, where the larva feeds on the flowers of white clover, greatly reducing seed production.

25. Agonoxenidae. Very small; head (Fig. 36.29A) smooth-scaled; ocelli absent; frons very oblique; scape without pecten; maxillary palpi 1-segmented; labial palpi recurved, apical segment flattened; hind tibiae with long hair-scales; fore wing (Fig. 36.28B) lanceolate, without pterostigma, R_5 and M_1 stalked, R_5 to costa, 1 vein absent;

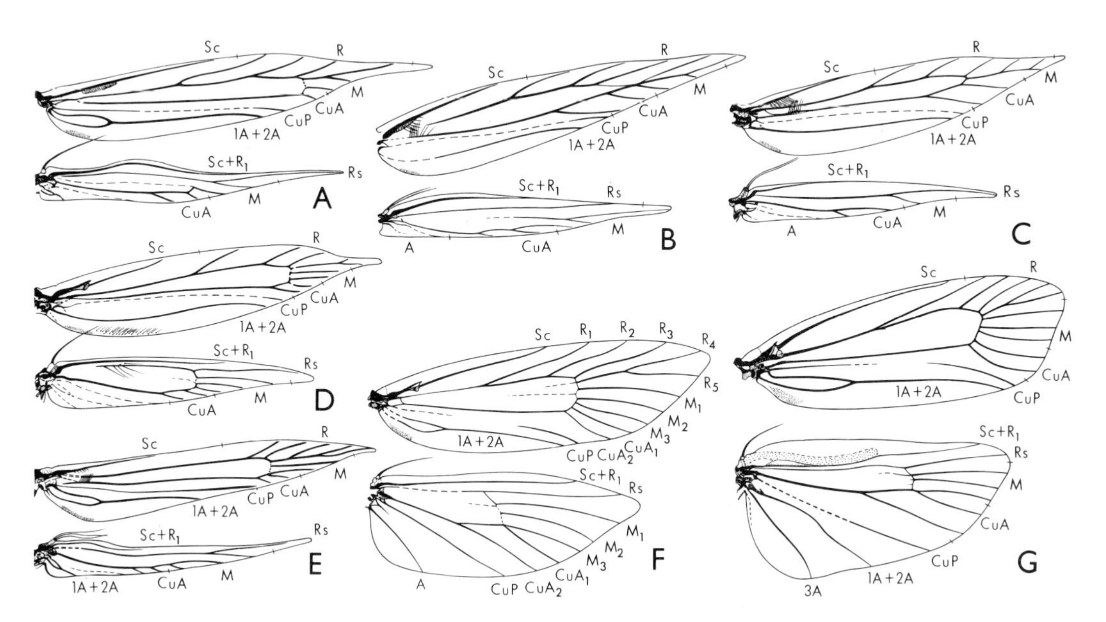

Fig. 36.28. Wing venation of Gelechioidea: A, *Coleophora*, Coleophoridae; B, *Agonoxena*, Agonoxenidae; C, *Elachista*, Elachistidae; D, *Scythris*, Scythridae; E, *Stathmopoda*, Stathmopodidae; F, *Philobota*, Oecophoridae; G, *Ethmia*, Ethmiidae. [J. Wedgbrow]

hind wing linear-lanceolate, $Sc+R_1$ remote from Rs, Rs and M_1 remote, M_3 to CuA_2 very short, or 2 veins absent, discal cell open; female with frenulum of 2 bristles; abdomen without dorsal spines; male gnathos a spinose knob, female with signum a pair of dentate patches. Larva with crochets irregularly uniordinal in a circle; beneath a flimsy silken web. Pupa flattened, antennae exceeding end of wings, abdominal segment 9 with pair of ventral processes, each with a series of short apical bristles; within double-walled silken cocoon.

The reduced maxillary palpi, the position of Rs in the hind wing, the female frenulum, and the male genitalia indicate a relationship to less specialized species of Elachistidae. The larva of *Agonoxena phoenicia* Bradley (Fig. 36.27C), from north Queensland, feeds beneath the leaves of the palm *Archontophoenix alexandrae*, and pupates in a fusiform, white, silken cocoon spun in the larval shelter. The related *A. argaula* Meyr. is a pest of coconut palms in Fiji and other Pacific islands. The adult runs about rapidly, and rests with the wings tightly folded along the body, the antennae porrected, and the fore legs extended in front.

26. Elachistidae. Very small; head smooth-scaled; ocelli present or absent; scape usually with pecten; haustellum rarely absent; maxillary palpi 1- or 2-segmented; labial palpi recurved, porrect or drooping; hind tibiae with long hair-scales; fore wing (Fig. 36.28C) lanceolate, R_5 to costa; hind wing lanceolate, $Sc+R_1$ remote from Rs which extends in nearly straight line through or near axis of wing, discal cell usually closed; abdomen rarely with dorsal spining; male gnathos usually 1 or 2 spinose knobs. Larva with crochets uniordinal, in broken circle, or transverse band, or prolegs absent; leaf- or stem-miners, usually in grasses or sedges. Pupa roughened, usually exposed, attached by cremaster, sometimes in flimsy cocoon.

The adults can usually be recognized by the axial condition of Rs of the hind wing, and the spinose gnathos in the male. The larva of *Cosmiotes aphanta* (Turn.) (Fig. 36.27D) pupates in a flimsy cocoon. It mines in the leaves of grasses, and has been reared from wheat. The known pupae of other Australian species, referred to *Elachista*, are without a cocoon.

27. Scythridae. Very small; head (Fig. 36.29B) smooth-scaled; ocelli sometimes present; scape sometimes with pecten; maxillary palpi 4-segmented; labial palpi recurved, smooth-scaled; hind tibiae with long hair-scales; fore wing (Fig. 36.28D) lanceolate, R_1 from beyond middle of discal cell but remote from R_2, R_4 and R_5 stalked, R_5 to termen, 1 or 2 veins absent, CuP present near margin; hind wing lanceolate, Rs and M_1 parallel; abdomen without spines. Larva with tufted hair, crochets bi- or triordinal in a circle; spins silken webbing or gallery. Pupa with antennae not reaching wing tips, wings extending to abdominal segment 7, spiracles tubular, setae mostly hooked.

The Australian species are referred to *Scythris*. They are blackish above and pale beneath, sometimes with a white longitudinal stripe on the fore wing, as in *S. rhabducha* Meyr. (Fig. 36.27E). Their larvae are not known, but the adults are common amongst grass or amongst *Eucalyptus* leaf litter.

28. Stathmopodidae.* Very small; head (Fig. 36.33A) smooth-scaled, shining; ocelli absent; scape usually without pecten; maxillary palpi 1- to 4-segmented, usually folded over base of haustellum; labial palpi recurved, slender, smooth; hind tibiae and tarsi with whorls of stiff bristles or rarely with dense dorsal bristles; fore wing (Fig. 36.28E) lanceolate or linear, venation usually complete; hind wing narrower than fore wing, discal cell sometimes open, venation reduced, female frenulum of 2 or 3 bristles; abdomen often with dorsal spining. Larva with crochets uniordinal in a circle; predatory on scale insects, spiders' eggs, or tunnelling in leaves, bark, flower spikes, fruit or galls,

* =Tinaegeriidae.

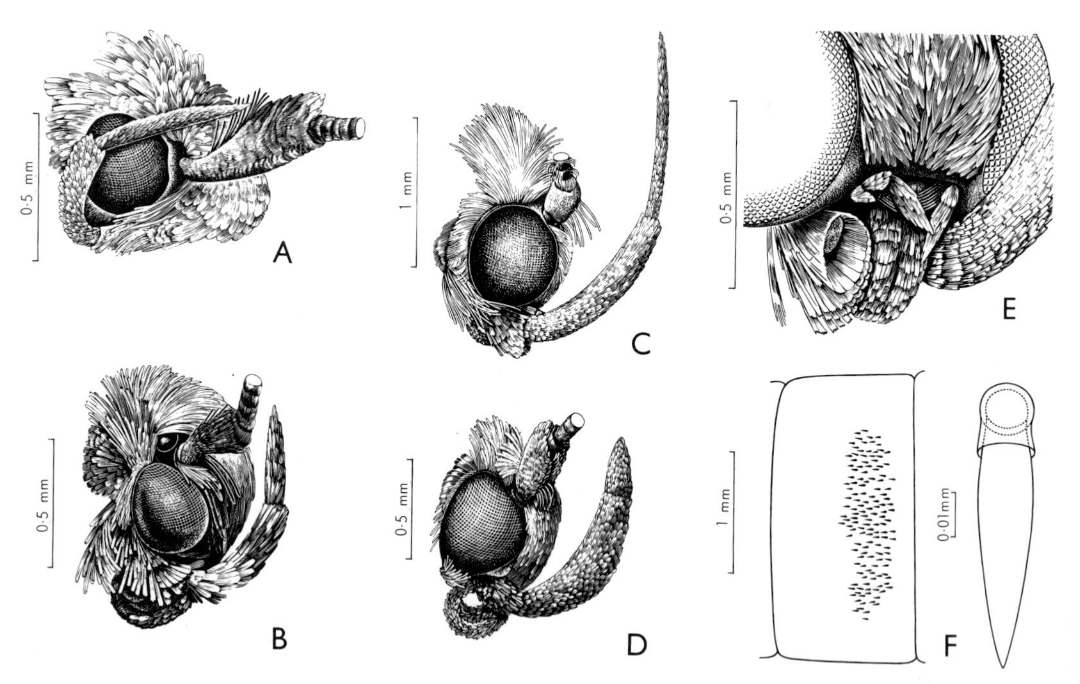

Fig. 36.29. Heads, lateral: A, *Agonoxena*, Agonoxenidae; B, *Scythris*, Scythridae; C, *Philobota*, Oecophoridae; D, *Blastobasis*, Blastobasidae; E, anterolateral view showing maxillary palpi of Oecophoridae. F, dorsal spining of 4th abdominal segment and one enlarged spine (modified scale) in *Philobota*, Oecophoridae. [F. Nanninga]

or amongst fallen leaves. Pupa in larval gallery or cocoon.

The moths may be recognized by the bristles on the hind legs, arranged in whorls, at least at the apex of each tarsal segment. Most rest with the hind legs raised (Fig. 36.26D). *Snellenia* (4 spp.) and *Pseudaegeria* (4 spp.) contain handsome diurnal species which visit flowers. *S. lineata* (Walk.) (Plates 6, F, 7, E) has thickened antennae and enormous labial palpi; it bears a close resemblance to *Metriorrhynchus* (Coleoptera) which feed at the same flowers. The larvae of the red and black *P. phlogina* Turn. tunnel in the bark of the woody climber *Ventilago viminalis* (Rhamnaceae) in inland Queensland. *Vanacela* (3 spp.), *Hieromantis* (2 spp.), and *Calicotis* (4 spp.) have the scape forming an eye-cap, which in *Vanacela* has a pecten. Most Australian species belong to *Stathmopoda* (60 spp.), in which the abdomen is spined dorsally. The larvae of *S. chalcotypa* Meyr., *S. cephalaea* Meyr., and *S. callichrysa* Low. (Fig. 36.27F) tunnel in rust galls on *Acacia decurrens*, and *S. crocophanes* Meyr. has been reared from a loquat *(Photinia)* fruit. *S. melanochra* Meyr. larvae are predators on *Eriococcus* scales on *Eucalyptus*, and those of *S. arachnophthora* Turn. feed in spiders' egg-sacs.

29. Oecophoridae. Small; head (Fig. 36.29C) usually smooth-scaled; ocelli rarely present; antennae simple, ciliate, rarely pectinate, scape often with pecten; maxillary palpi 4-segmented, folded over base of haustellum (Fig. 36.29E); labial palpi recurved, often exceeding vertex, apical segment slender, tapering, acute; hind tibiae with long hair-scales; fore wing (Fig. 36.28F) without pterostigma, R_4 and R_5 stalked, never more than 1 vein lost, CuP present at least towards margin; hind wing with Sc and Rs separate and parallel, R_1 often present, Rs and M_1 well separated at base, parallel or divergent, rarely connate and divergent, M_2 usually arising nearer to M_3 than to M_1, CuP present in outer half; female sometimes brachypterous; abdomen often with dorsal spining (Fig. 36.29F); male gnathos sometimes spinose. Larva with crochets in biordinal ellipse, rarely triordinal; in a portable case, tunnelling in wood or in flowers or galls, joining leaves, amongst detritus, or in tunnels in soil. Pupa with large maxillary palpi, abdominal segments 4–6 movable; in larval shelter or exposed and attached by truncate posterior end.

The family seems homogeneous, and has reached a remarkable degree of development in Australia (Turner, 1932–47). Some genera, such as *Cryptolechia* (30 spp.), *Thudaca* (18 spp.), *Tonica* (2 spp.). and *Peritorneuta* (7 spp.), have an erect naked pupa with truncate abdomen attached to a pad of silk. The gnathos in the male is usually spinose. The larvae are usually leaf-tiers, but in *Tonica effractella* (Snellen), from northern Australia, they tunnel in the stems of cotton and of *Brachychiton paradoxum.*

Eupselia (46 spp.) and *Hypertropha* (6 spp.) belong to a related group with similar pupae (Fig. 36.30A) but distinct genitalia. In *Eupselia* the larvae are at first leaf-miners in *Eucalyptus*, later using the mine as a shelter while they feed on the adjacent leaf tissue. *E. satrapella* Meyr. is one of a complex of similar species with orange and fuscous wings. The larvae of *Hypertropha* form a tubular network shelter of silk and excrement pellets on the under side of *Eucalyptus* leaves, emerging to feed on the surrounding leaf. *H. tortriciformis* (Guen.) (Fig. 36.27G) has metallic markings on the fore wing, with orange and black hind wings.

The abdomen in the preceding genera is smooth, but in most Australian species the dorsum is densely spinose. The male usually has a beak-like gnathos. The larval habits are diverse, but many feed on leaf debris on the forest floor. Some species of *Garrha* (141 spp.) cut oval flattened cases from fallen leaves; other genera, such as *Thema* (42 spp.), join dead leaves on the ground or in tree crevices. *G. carnea* (Zell.) and *T. chlorochyta* (Meyr.) (Fig, 36.27H) are good examples. Living *Eucalyptus* leaves provide food for many species, some of which construct portable cases, and others spin shelters between adjacent leaves or live in tubular shelters amongst the twigs. *Cormotypa fascialis* (F.) forms a flat case of two fragments of leaf, usually on juvenile foliage near the ground. A short length of hollowed-out twig is used as a portable case (Fig. 36.30B) by *Hippomacha callista* (Meyr.) (Fig. 36.27I), the open end being blocked by the modified head capsule of the larva. In *Sphaerelictis hepialella* (Walk.) the case resembles a snail shell, made from a series of successively larger, spirally arranged, overlapping fragments of eucalypt leaf (Fig. 36.30C). Some species, such as *Heliocausta hemiteles* Meyr., live gregariously in silken tubes spun in a bunch of tied leaves. *Wingia rectiorella* (Walk.) (Plate 7, B) spins a solitary silken shelter on *Leptospermum*.

Few native Oecophoridae are known to be pests, but *Philobota productella* (Walk.) (Fig. 36.27J) and related species damage grass pastures in the south-east. The larvae form short vertical tunnels in the soil, emerging to feed on grass. *Barea consignatella* Walk. and related species sometimes damage sapwood in damp structural timbers. Introduced stored-product pests include *Hoffmanophila pseudospretella* (Staint.) and *Endrosis sarcitrella* (Steph.).

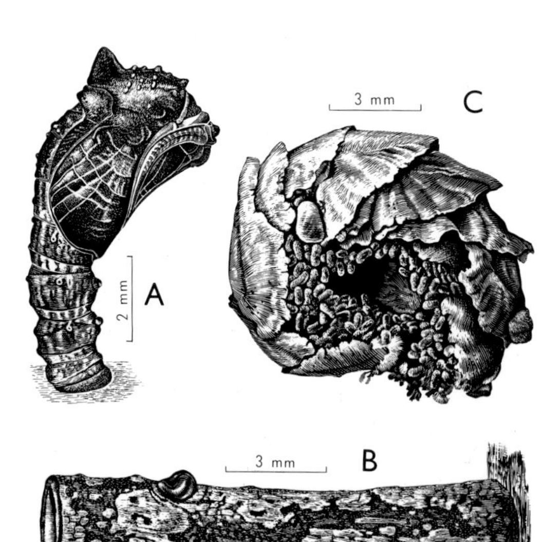

Fig. 36.30. Oecophoridae: A, pupa of *Hypertropha*; B, C, larval cases of *Hippomacha* and *Sphaerelictis*. [F. Nanninga]

30. Ethmiidae. Small; head smooth-scaled; ocelli absent; maxillary palpi 4-segmented, folded over base of haustellum; labial palpi recurved; hind tibiae with long hair-scales; fore wing without pterostigma, R_4 and R_5 stalked, R_5 usually to costa, CuP present at least towards margin, female with subcubital retinaculum of scales; hind wing (Fig. 36.28G) with Rs and M_1 well separated at base, divergent, M_2 not arising nearer to M_3 than to M_1, CuP present; abdomen not spined; male valva with distal lobe and with costal and dorsal areas defined by sutures. Larva with crochets in single biordinal band; usually beneath light webbing on Boraginaceae. Pupa with a pair of ventral processes on abdominal segment 9, each with a series of hooked apical bristles.

In Australia the family is mainly tropical. The species resemble Oecophoridae, but are distinguished by the characteristic male genitalia and usually by the position of M_2 in the hind wing. The fore wings often have bold black spots on a white or grey background. In the rain-forest species, *Ethmia sphaerosticha* (Meyr.) (Fig. 36.27K), the male antenna bears a large expansible tuft of long silky hairs. In other *Ethmia* (11 spp.) the antennae are simple in both sexes. *E. thoraea* Meyr., with grey fore wings and orange hind wings, has a small pecten on the scape. *E. clytodoxa* Turn. and *E. hemadelpha* (Low.) are black-spotted white species. Larvae of the latter feed in northern Australia on *Ehretia saligna* (Boraginaceae). *E. heliomela* Low., with bronzy black fore wings and orange hind wings, lacks conspicuous spots; M_2 does not arise as close to M_1 in the hind wing as in other Australian species.

31. Timyridae. Small; head (Fig. 36.33B) smooth-scaled; ocelli absent; antennae often thickened in male, nearly as long as or longer than fore wing; maxillary palpi 4-segmented, folded over base of haustellum; labial palpi recurved, in male sometimes modified; fore wing (Fig. 36.31A) without pterostigma, R_3 and R_4 or R_4 and R_5 stalked, R_5 usually to costa, CuA_1 and CuA_2 usually stalked, CuP sometimes present; hind wing with Rs and M_1 connate or stalked, CuP present; venation

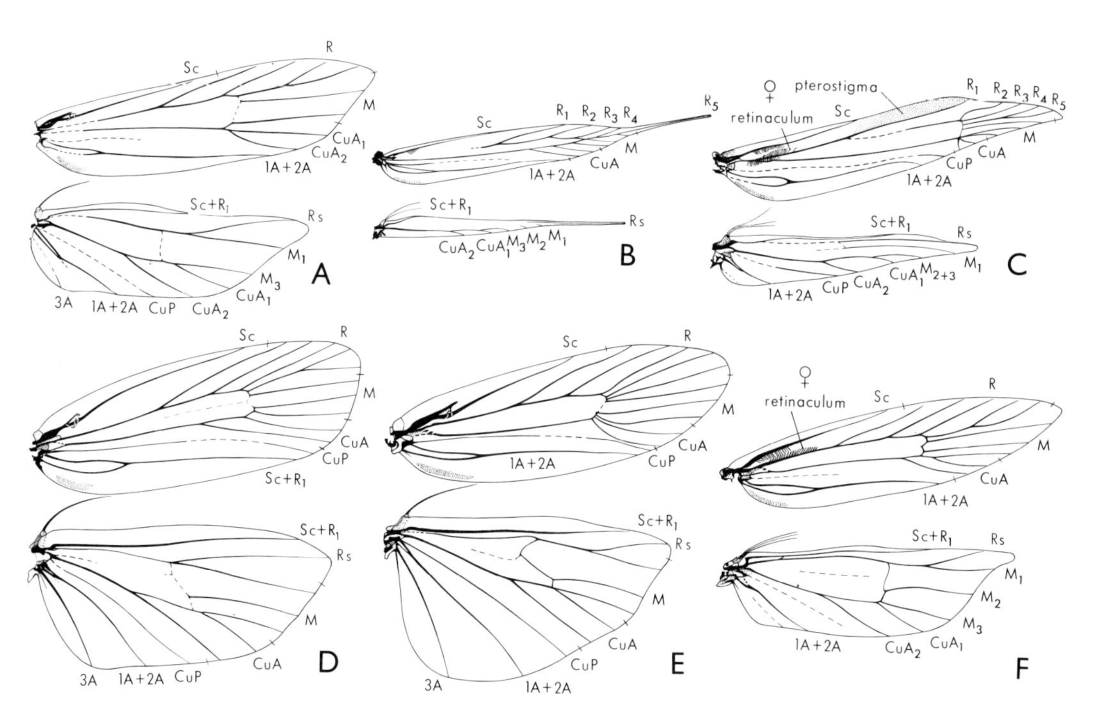

Fig. 36.31. Wing venation of Gelechioidea: A, *Crocanthes*, Timyridae; B, *Cosmopterix*, Cosmopterigidae; C, *Blastobasis*, Blastobasidae; D, *Xylorycta*, Xyloryctidae; E, *Agriophara*, Stenomidae; F, *Stegasta*, Gelechiidae. [J. Wedgbrow]

may be reduced; abdomen often with dorsal spining. Larva not studied; probably feeds on dead leaf litter. Pupa not studied.

The family was separated from the Gelechiidae by Clarke (1955). The presence of CuP often in both wings, the dorsal spining of the abdomen, and the genitalia indicate a relation ship to the Oecophoridae. The moth rests with head directed downwards and antennae held porrect. Most species occur in the Oriental and Australian regions, being mainly tropical and subtropical. *Crocanthes* (25 spp.) contains bright yellow, orange, or pink species with dorsally spined abdomen. The males of some species, such as *C. prasinopis* Meyr. (Fig. 36.27L), have segments 2 and 3 of the labial palpi fused to form a long, weak, partly coiled segment. *Lecithocera* (65 spp.) includes dark species which shelter amongst the leaf litter. *L. micromela* (Low.) from the south-east is common in dry grassy places.

32. Cosmopterigidae. Very small to small; head smooth-scaled; ocelli present or absent; antennae simple or ciliated, scape long and slender, pecten sometimes present; maxillary palpi 3- or 4-segmented, folded over base of haustellum; labial palpi long, recurved; hind tibiae with long hair-scales; fore wing (Fig. 36.31B) linear to lanceolate, without pterostigma, R_2 arising well before end of discal cell, R_4 and R_5 stalked, R_5 to costa, CuP often present towards margin; hind wing linear or lanceolate, Sc separate from Rs, R_1 sometimes present, discal cell sometimes open, venation often reduced; abdomen without dorsal spines; male genitalia often asymmetrical. Larva with crochets in circle, usually biordinal; mining in leaves, boring in stems and seeds, tying leaves, forming galls, or predatory on scale insects. Pupa with cremaster of stout spines.

The family has been divided into 3 families (Hodges, 1962), mainly on the basis of genitalia. They are treated here as subfamilies Cosmopteriginae, Momphinae, and Walshiinae, the last of which has not been recorded from Australia.

The COSMOPTERIGINAE have asymmetrical male genitalia, which lack the uncus. *Cosmopterix* (10 spp.) contains minute blackish species with orange and bright metallic markings. The larvae mine in leaves of grasses and other plants. *C. mimetis* Meyr. occurs north from Sydney. *Limnaecia* (53 spp.) often have blackish fore wings and transverse white or orange markings, as in *L. cirrhozona* Turn. (Fig. 36.32A). The genus was based on the widely distributed *L. phragmitella* Staint., a slender ochreous species with larva burrowing amongst the seeds of *Typha angustifolia*. In *Labdia* (56 spp.) the hind wings are narrower. *L. leucombra* (Meyr.) is one of several species with delicate stripes on the fore wing. The larvae of the blackish *L. semnostola* (Meyr.) mine in the tips of the young phyllodes of *Acacia implexa* and other wattles. The pink larvae of *Sathrobrota rileyi* (Wals.) are scavengers in damaged cotton bolls and sorghum heads in northern Australia and North America.

The male genitalia of the MOMPHINAE are symmetrical, and the fore wings are usually narrower than in Cosmopteriginae. The best-known genus is *Batrachedra* (35 spp.). The larvae of *B. arenosella* (Walk.) are predators on scale insects, feeding beneath a silk webbing, and pupating in a flat, elliptical cocoon.

33. Blastobasidae. Very small; head (Fig. 36.29D) smooth-scaled; ocelli absent; scape often dilated and concave beneath, with dense pecten of scales; maxillary palpi 4-segmented, folded over base of haustellum; labial palpi recurved, apical segment in male often thick and blunt; hind tibiae with long hair-scales; fore wing (Fig. 36.31C) lanceolate, pterostigma present, R_2 arising near upper angle of discal cell, R_4 and R_5 stalked, R_5 to costa, CuP present near margin, female with subradial retinaculum of specialized scales extending across cell to base of CuA; hind wing lanceolate, $Sc+R_1$ fused with Rs near base, Rs and M_1 parallel, CuP present; abdomen with dorsal spining. Larva with crochets uniordinal, ocelli reduced; feeding on fallen woody fruits, seeds, and dry refuse.

The family has a world-wide distribution; in Australia it is best represented in the north-east. The dull grey moths may be recognized by the pterostigma and the position of R_2 in the fore wing, and the fusion of $Sc+R_1$ and Rs in the hind wing. The fused subradial and subcubital retinaculum of the female is not found elsewhere. *Blastobasis tarda* Meyr. (Fig. 36.32B) has a broadly dilated scape, dense pecten, and blunt apical segment of the labial palpi in the male. *B. sarcophaga* Meyr., the larvae of which tunnel in fallen palm fruits, lacks the last two characters.

34. Xyloryctidae. Small to medium-sized; head (Fig. 36.33C) smooth-scaled; ocelli usually absent; antennae in male simple, ciliated, or pectinated, scape without pecten; maxillary palpi 3- or 4-segmented, usually folded over base of haustellum; labial palpi recurved; hind tibiae with long hair-scales; fore wing (Fig. 36.31D) with R_4 rarely to termen, CuA_2 arising well before lower angle of discal cell, CuP present; hind wing often broader than fore wing, $Sc+R_1$ usually separate from and diverging from Rs well before upper angle of discal cell, R_1 sometimes present, Rs and M_1 usually approximated, connate, or stalked, CuP present; abdomen usually with dorsal spining. Larva with crochets biordinal, in circle or ellipse; feeding on lichens, tying leaves, feeding in shelter beneath bark, tunnelling in bark or stems of trees, often dragging leaves to the entrance for food.

The family is well developed in Australia, where species of up to 75 mm wing expanse occur. The adults are seldom seen during the day, but come to light at night. Many are distinctively marked, and some are sexually dimorphic.

The maxillary palpi have 4 segments in most genera, including *Lichenaula* (56 spp.), *Procometis* (20 spp.), *Telecrates* (4 spp.), *Scieropepla* (17 spp.), *Catoryctis* (20 spp.), and *Uzucha* (2 spp.). The relationships of

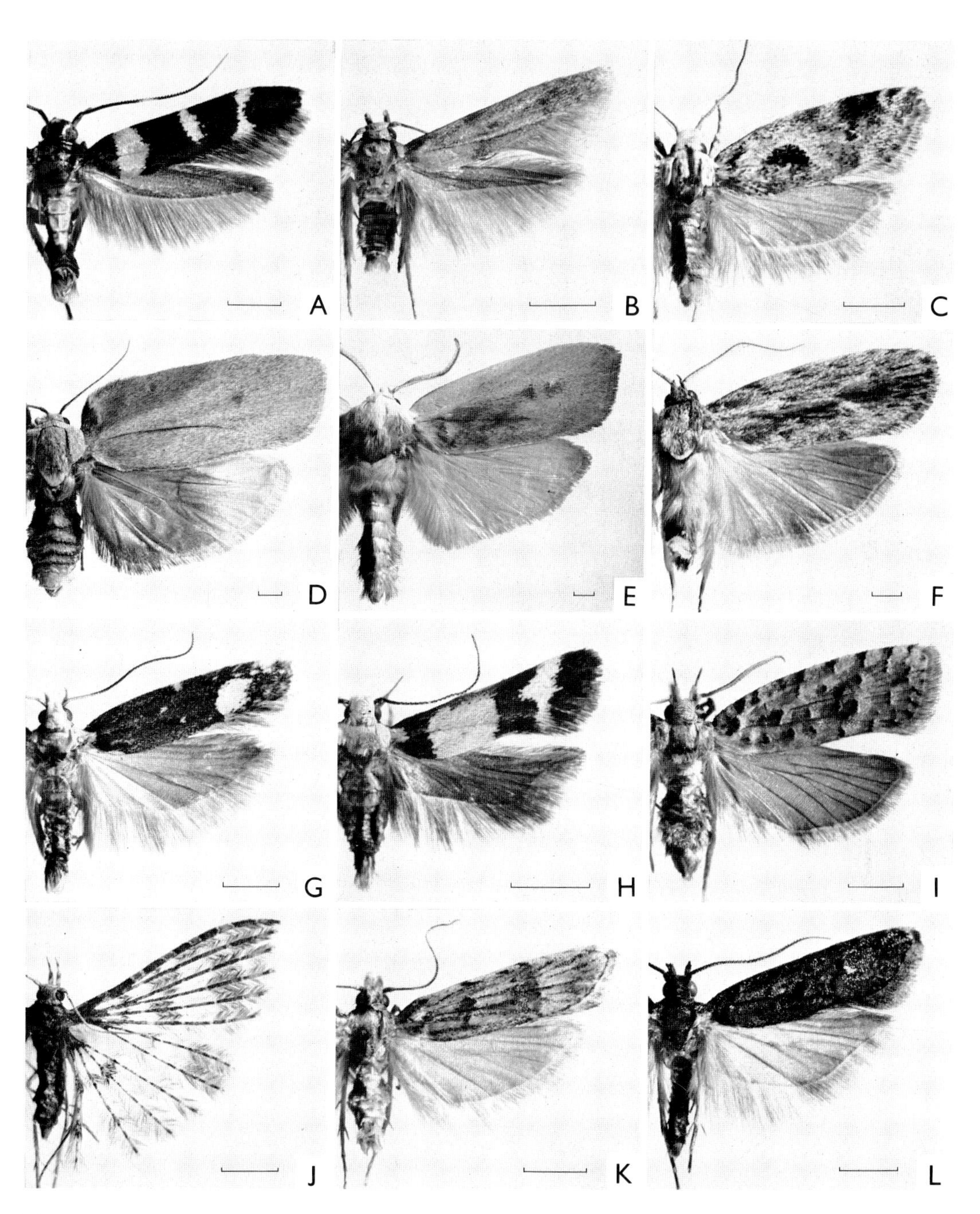

Fig. 36.32. A, *Limnaecia cirrhozona*, Cosmopterigidae; B, *Blastobasis tarda*, Blastobasidae; C, *Lichenaula choriodes*, D, *Uzucha humeralis*, E, *Cryptophasa rubescens*, Xyloryctidae; F, *Agriophara platyscia*, Stenomidae; G, *Protolechia deltodes*, H, *Decatopseustis xanthastis*, Gelechiidae; I, *Phycomorpha prasinochroa*, Copromorphidae; J, *Alucita phricodes*, Alucitidae; K, *Carposina neurophorella*, L, *Bondia nigella*, Carposinidae. Scales = 2 mm. [Photos by C. Lourandos]

these genera are not yet understood. The larvae of some, such as *L. choriodes* Meyr. (Fig. 36.32C) and *L. lichenea* Meyr., feed upon lichens growing on fences and rocks, sheltering in a gallery of silk and refuse particles. *Procometis* bears a pencil of long hair-scales on the costa of the hind wing in the male. The larvae of *P. bisulcata* Meyr. form a vertical tunnel in the soil, emerging at night through a flexible, silken, soil-encrusted tube to feed on terrestrial lichens, pieces of which are stored in a rounded chamber just beneath the soil surface. *Scieropepla* larvae tunnel in flower spikes, usually of *Banksia*, but those of *S. typhicola* Meyr. burrow amongst the seeds of *Typha*. *Banksia* flower spikes are also tunnelled by larvae of *Chalarotona intabescens* Meyr. *Telecrates laetiorella* (Walk.) (Plate 7, H) larvae form a webbing gallery and feed on the inner bark of *Eucalyptus*. A conspicuous gallery of silk and bark particles is constructed by the larva of *Uzucha humeralis* Walk. (Fig. 36.32D), which feeds on the surface of the bark of *Angophora* and smooth-barked *Eucalyptus*. The adult has the base of the costa in the fore wing strongly arched, very short labial palpi, and smoothly scaled hind tibiae.

In *Neodrepta* (7 spp.) and *Xylorycta* (93 spp.) the maxillary palpi are 3-segmented, and the male antennae are ciliated. Many of the species are shining white. The larva of *N. luteotactella* (Walk.) lives either in a webbing shelter amongst twigs and leaves or in a short tunnel in a twig or the woody fruits of Proteaceae, including *Banksia* and *Hakea*, and is a pest of *Macadamia*. *X. strigata* (Lew.) is white with a broad fuscous longitudinal stripe on the fore wing. The larva tunnels in the branches of *Banksia serrata* and *B. integrifolia*, and of *Lambertia formosa*, feeding on leaves which it drags to the entrance of the tunnel. This habit is common in *Cryptophasa* (19 spp.), which contains the giants of the family, with maxillary palpi of 3–4 segments and male antennae usually bipectinate. *C. rubescens* Lew. (Fig. 36.32E) tunnels in the stems of *Acacia*, covering the entrance with a web of silk, and feeding on leaves it drags to the tunnel. By contrast, *C. melanostigma* (Wall.) feeds on the bark of many native and exotic trees, often ring-barking them. Its main native host is *Acacia*, but it attacks citrus, stone and pome fruits, figs, and ornamentals.

35. Stenomidae. Small; head smooth-scaled; ocelli absent; antennae ciliated in male, scape without pecten; maxillary palpi 4-segmented, folded over base of haustellum; labial palpi recurved; hind tibiae with long hair-scales; fore wing (Fig. 36.31E) with R_4 and R_5 usually separate, R_5 to apex, CuA arising near lower angle of discal cell, CuP present; hind wing broader than fore wing, $Sc+R_1$ approaching Rs, or Rs curved upwards, near upper angle of cell, Rs and M_1 usually stalked, CuP present; abdomen without dorsal spining; uncus in male usually long and slender, valvae often complex with large scale-like setae. Larva with crochets in biordinal pseudocircle or circle; in leaf mine or between joined leaves, or (exotic species) boring in shrubs and trees. Pupa short, stout, abdomen with scattered hairs and looped cremaster bristles; between joined leaves on ground, or in larval shelter.

The family is numerous in the Neotropical region, but is represented in other regions by *Agriophara* (31 Australian spp.) and a few smaller genera. The larvae of *Agriophara* at first excavate linear mines in *Eucalyptus* leaves, later feeding between joined leaves. The pupae are found in a shelter between fallen leaves, and can produce a clearly audible clicking sound by tapping the side of the shelter. *A. platyscia* Low. (Fig. 36.32F) is one of many procryptic grey species.

36. Gelechiidae. Small; head (Fig. 36.33D) smooth-scaled; ocelli often present; antennae simple or shortly ciliated, scape seldom with pecten; maxillary palpi 4-segmented, folded over base of haustellum; labial palpi recurved, 2nd segment often tufted beneath or rough-scaled; hind tibiae with long hair-scales; fore wing (Fig. 36.31F) without pterostigma, R_4 and R_5 usually stalked, R_5 to costa, CuP absent, female with subradial retinaculum a row of strongly curved scales on R; hind wing usually more or less trapezoidal, with termen sinuate or emarginate,

CuA sometimes with basal pecten, R_1 present, Rs and M_1 usually approximated at base or stalked, M_3 and CuA_1 connate or stalked, rarely separate, CuP usually absent; abdomen rarely with dorsal spining. Larva with crochets biordinal, in ellipse or in 2 transverse rows, thoracic legs occasionally reduced, prolegs rarely absent; joining leaves, feeding on seeds, or mining in leaves or stems. Pupa with maxillary palpi present, cremaster present; in silken cocoon in larval shelter or amongst detritus on the ground.

Though numerous in Australia, the Gelechiidae are greatly outnumbered by the Oecophoridae. Most genera may be recognized by the trapezoidal hind wing, usually with emarginate termen, and by the condition of Rs and M_1 in the hind wing. In *Protolechia* the hind wing is not trapezoidal, Rs and M_1 are nearly parallel, and the abdominal terga are minutely spinose anteriorly. All our genera lack CuP in both wings, and have the specialized retinaculum in the female. *Symmoca* and allied Holarctic genera are sometimes separated as the Symmocidae.

Anarsia (7 spp.), *Gaesa* (3 spp.), and *Brachiacma* (1 sp.) have a large ventral scale-tuft on the 2nd segment of the labial palpi. In *Anarsia* the apical segment is very short in the male, whereas it is of normal length in the female. The larvae feed between leaflets or in galls on *Acacia*. *A. molybdota* Meyr., on *A. decurrens*, occurs widely in the south. The adults of *Gaesa* are sombre insects found amongst the deep leaf litter of the denser eastern forests. The larvae of *G. capnites* (Meyr.) feed on the foliage of *Acronychia*, and the adults sometimes occur in vast numbers. The widely distributed *Brachiacma palpigera* Wals. is common in eastern Australia, the larvae feeding in the seed pods of Leguminosae.

The largest endemic genus is *Protolechia* (139 spp.), with larvae feeding between joined leaves, usually of *Eucalyptus*. Pupation occurs in a silk-lined cell between fallen leaves.

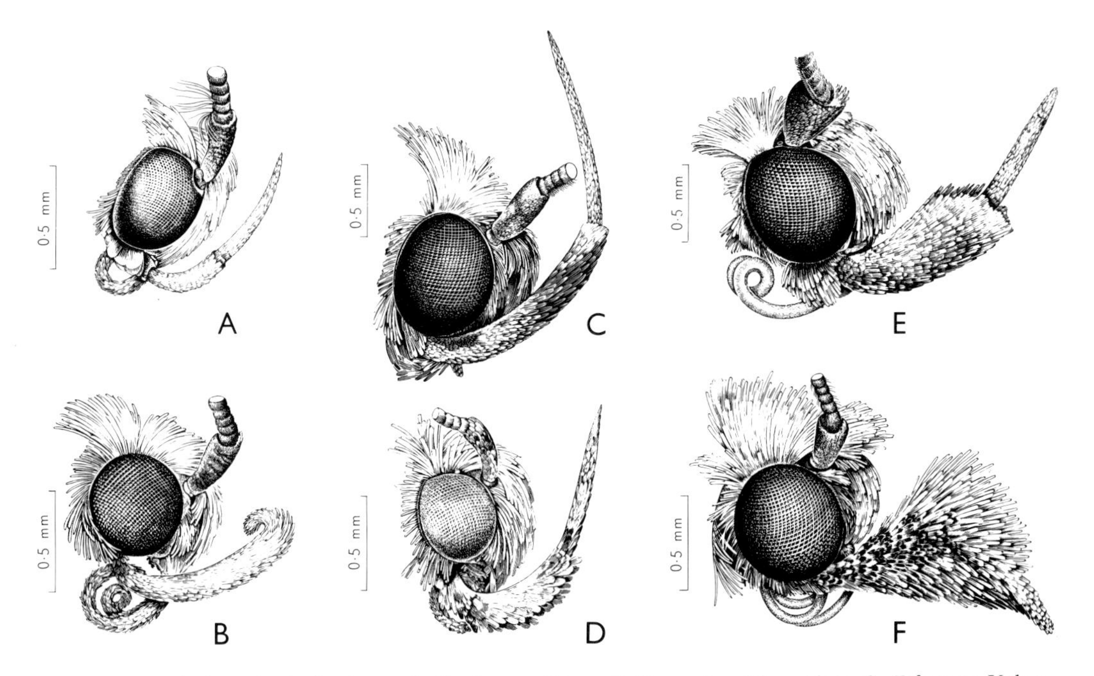

Fig. 36.33. Heads, lateral: A, *Stathmopoda*, Stathmopodidae; B, *Crocanthes*, Timyridae; C, *Xylorycta*, Xyloryctidae; D, *Phthorimaea*, Gelechiidae; E, *Phycomorpha*, Copromorphidae; F, *Carposina* ♂, Carposinidae. [F. Nanninga]

The procryptic adults rest on *Eucalyptus* trunks. *P. aversella* (Walk.) occurs widely in the south, while *P. deltodes* (Low.) (Fig. 36.32G) is found in Victoria and N.S.W.

Phthorimaea (1 sp.) and *Scrobipalpa* (14 spp.) include three pests of Solanaceae. *P. operculella* (Zell.) is a leaf-miner in tobacco and potatoes, and tunnels potato tubers. *S. heliopa* (Low.) also attacks tobacco, and *S. plaesiosema* (Turn.) tunnels in stems of tomatoes.

The larvae of *Pectinophora* (3 spp.) and *Pexicopia* (20 spp.) feed in the seed capsules of Malvaceae (Common, 1958b). A pecten of at least 2 or 3 hair-scales is present on the scape. *Pectinophora gossypiella* (Saund.) damages cotton bolls in north-western Australia and the N.T., and *P. scutigera* (Holdaway) is a minor cotton pest in Queensland. *Hibiscus* is the native host plant. *Pexicopia nephelombra* (Meyr.) flies commonly amongst *Sida* and other malvaceous weeds in sub-coastal Queensland, often with the related *Decatopseustis xanthastis* (Low.) (Fig. 36.32H) which lacks the antennal pecten.

Thiotricha (18 spp.) contains slender, elegantly-patterned, satin-white species, in which the male antennae have long cilia, and the fore wing lacks 2 veins. *T. parthenica* Meyr. has ochreous yellow wings; its larva constructs a portable case of withered leaf particles, and bores into the lower surface of the leaves of *Grevillea punicea*. *Atasthalistis* (3 spp.) and *Onebala* (6 spp.) are Indo-Malayan and Papuan genera, with *A. ochreoviridella* (Pag.) (Plate 7, K) and *O. hibisci* (Staint.) in northern Australia. In *Apatetris* (27 spp.) the scape has a strong pecten, the apical segment of the labial palpi is very short, and the termen of the hind wing is deeply emarginate. The larva of *A. poliopasta* (Turn.), from the east coast, forms a linear mine in leaves of *Lomandra longifolia*, and pupates in a white cocoon in the leaf litter. A pecten is also present on the scape of *Sitotroga* (1 sp.); *S. cerealella* (Oliv.) is a cosmopolitan pest of stored grain.

Superfamily COPROMORPHOIDEA

Small; head smooth-scaled; ocelli present or absent; chaetosemata absent; scape without pecten; haustellum naked; maxillary palpi very small; labial palpi recurved or porrect; epiphysis present, tibial spurs 0-2-4; fore wing often with raised scale-tufts; CuA of hind wing often with basal pecten of hairs; wings sometimes divided into 6 or more plumes. Larva with 2 prespiracular (L) setae on prothorax, crochets uniordinal in a circle, spiracle on abdominal segment 8 larger and more dorsally placed than in other larvae; tunnelling in flowers, fruits, stems, galls. Pupa without abdominal spines; in larval tunnel or cocoon, not protruded at ecdysis.

The systematic position of the 3 families included here has for long remained in doubt. The Carposinidae have been referred to either the Tineoidea *(sensu lato)* or the Tortricoidea, but were raised to superfamily rank by Diakonoff (1961). Meyrick (1928) associated the Carposinidae, Copromorphidae, and Alucitidae in the Copromorphoidea. Turner (1947) treated the first two as subfamilies of the one family Copromorphidae. The genitalia suggest a gelechioid relationship, but the naked haustellum and the paired prespiracular setae of the larval prothorax at once distinguish them.

Key to the Families of Copromorphoidea

1. Fore wing deeply divided into 6 plumes, hind wing divided into 6 or 7 plumes ... **Alucitidae** (p. 827)
   Fore wing and hind wing entire 2
2. Hind wing with all 3 branches of M present .... **Copromorphidae** (p. 826)
   Hind wing with 1, or more usually 2, branches of M absent .... **Carposinidae** (p. 827)

37. Copromorphidae. Ocelli present or absent; antennae simple or unipectinate, rarely bipectinate; maxillary palpi 1-segmented or 4-segmented and clasping base of naked haustellum; labial palpi recurved (Fig. 36.33E) or porrect; hind tibiae smooth or with long hair-scales; fore wing (Fig. 36.34A) usually with raised scale-tufts, CuP vestigial; hind wing with Rs and M_1 parallel, CuA usually with basal pecten, CuP present. Larva tunnels in leaf veins, twigs, and fruits. Pupa in larval tunnel.

The family is represented in eastern Australia by *Copromorpha* (4 spp.), *Phycomorpha* (1 sp.), and *Osidryas* (2 spp.). In the first two the labial palpi are recurved, the thorax bears a posterior crest, the fore wing is tufted, and the hind wing has a cubital pecten. *Osidryas* has long porrect labial palpi, and is without thoracic crest, scale-tufts on the fore wing, and cubital pecten. The larvae of the mossy green *P. prasinochroa* (Meyr.) (Fig. 36.32I) attack *Ficus stenocarpa*, *F. carica*, and cultivated figs in central N.S.W. They bore in the fruits, fruit stems, shoots, twigs, and leaf veins.

38. Alucitidae (Orneodidae). Ocelli present or absent; antennae simple; maxillary palpi 3- to 5-segmented, or absent; labial palpi moderate to long, recurved or porrect; hind tibiae with rough scales or bristles above; fore wing (Fig. 36.34C) divided into 6 plumes, sometimes with narrow costal fold in male; hind wing with 6 or 7 plumes; wings rarely entire (South American species); abdomen with dorsal transverse patches of short spines. Larva tunnels in flowers, buds, and shoots, sometimes producing galls. Pupa in silken cocoon on ground, or in gall.

The family is easily recognized by the broad, plumed wings. The Australian species are all referred to *Alucita*, with each wing divided into 6 plumes. They occur in the north and east. In *A. phricodes* Meyr. (Fig. 36.32J) the larva burrows into *Bignonia* and *Tecoma* flowers. *A. pygmaea* Meyr. is a tiny grey species, and *A. xanthodes* Meyr. from Queensland is orange with a slender costal fold in the male.

39. Carposinidae. Ocelli absent; antennae densely ciliated in male; maxillary palpi

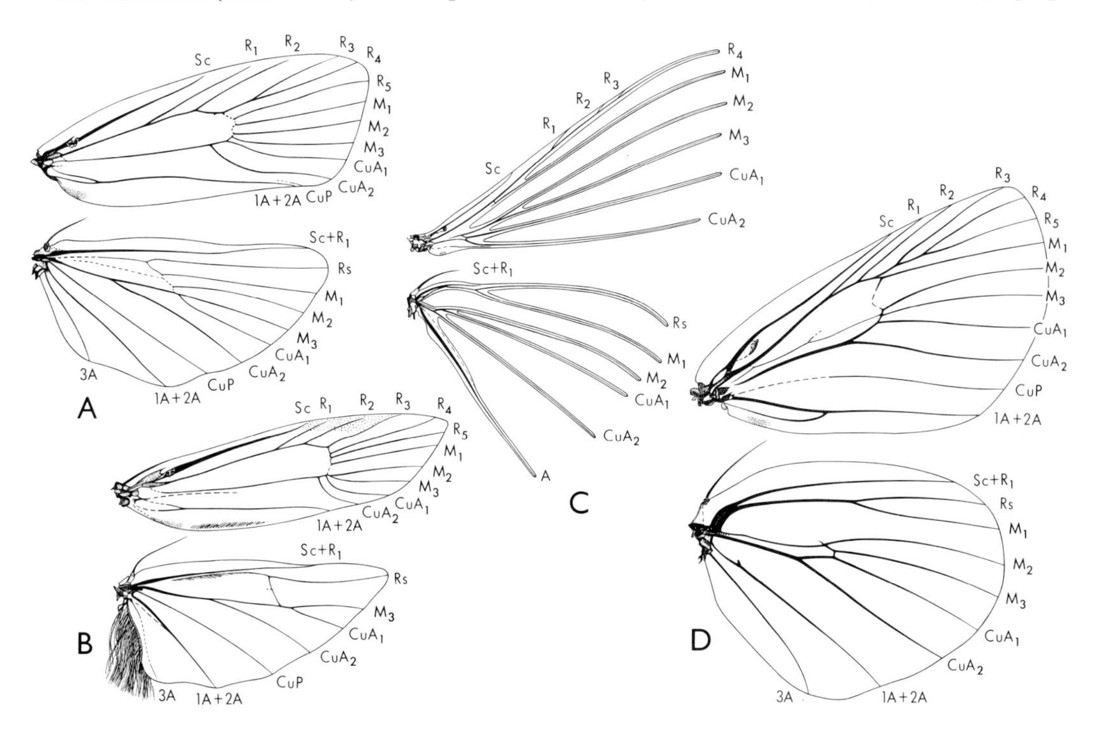

Fig. 36.34. Wing venation of Copromorphoidea and Castnioidea: A, *Copromorpha*, Copromorphidae; B, *Carposina*, Carposinidae; C, *Alucita*, Alucitidae; D, *Synemon*, Castniidae. [J. Wedgbrow]

1-segmented; labial palpi in male recurved, or apical segment porrect (Fig. 36.33F), in female long, porrect; hind tibiae with long hair-scales; fore wing (Fig. 36.34B) with raised scale-tufts, R_5 to termen, widely separate from R_4 at base, M_2, M_3, CuA_1, and CuA_2 approximated from lower angle of cell, CuP absent; hind wing with 1 and usually 2 branches of M absent, CuA often with basal pecten, CuP developed. Larva tunnels in living bark, fruit, or galls. Pupa in cocoon in soil, in bark, or in crevice.

Australian species of the family are most abundant in the sclerophyll forests of the south. The moths usually rest on tree trunks during the day, and the black species of *Bondia* are well adapted to concealment on fire-blackened trunks. *Sosineura* (1 sp.) lacks only one branch of M from the hind wing. At rest *S. mimica* (Low.) resembles a bird-dropping. The hind wing of the male is without a pecten, but bears a large patch of modified scales. Most species are referred to *Carposina* (30 spp.), in which a cubital pecten is present, often modified in the male to form a large expansible tuft of hair-scales. *C. neurophorella* (Meyr.) (Fig. 36.32K) has ochreous fore wings with veins outlined in dark grey. The larvae are thought to tunnel in the bark of *Casuarina*. In *Paramorpha* (11 spp.) and *Bondia* (9 spp.) the hind wing is without a cubital pecten. The larvae of *B. nigella* Newm. (Fig. 36.32L) tunnel in the necrotic bark of *Eucalyptus*, just above the entrance to the tunnels of *Aenetus* (Hepialidae) larvae.

Superfamily CASTNIOIDEA

40. Castniidae. Medium-sized; ocelli large; chaetosemata absent; antennae with flagellum smooth, broadly clubbed apically; haustellum present or reduced; maxillary palpi very small, 2- to 4-segmented; labial palpi short, ascending, apical segment short; epiphysis slender, spine-like, tibial spurs 0-2-4, hind tibiae smooth-scaled; wings broad; fore wing (Fig. 36.34D) with M present in discal cell, chorda sometimes present, CuP present or absent, 1A + 2A with long basal fork; hind wing with M present in discal cell, $Sc + R_1$ separate from, or shortly fused with, Rs near base, discal cell open, CuP present or absent, 2 anal veins. Egg elongate, with longitudinal ridges. Larva boring in stems, or in soil feeding on roots of sedges. Pupa with 1 or 2 transverse rows of dorsal spines on abdomen; at end of larval tunnel, protruded from tunnel at ecdysis.

The family includes only the Neotropical *Castnia*, the Indo-Malayan *Tascina*, and the Australian *Synemon* (29 spp.). *Synemon* contains diurnal species with strongly clubbed antennae, procryptic fore wings, and brightly patterned hind wings. They occur throughout the continent, but are most common in the south-west. Some are sexually dimorphic in pattern and size, and the females have a long ovipositor. In the fore wing CuP is well developed, but is absent from the hind wing. The moths fly rapidly, close to the ground, in grass- or heathland, or in forest clearings, where they may be locally common. They settle on the ground with the fore wings covering the hind wings. Much remains to be learnt about the life history, but the larvae of *S. sophia* (White) tunnel in the ground and feed on the roots of a sedge (*Lepidosperma*), pupating at the end of the silk-lined tunnel (Tindale, 1928b). *S. magnifica* Strand (Plate 8, B) from the Blue Mts is one of the finest species. *S. laeta* Walk. (Fig. 36.36A) from the east and *S. icaria* Feld. from the south-west have narrower fore wings, with R_2 and R_3 stalked, and a short areole formed by R_3 fusing for a short distance with the stalk of R_{4+5}.

Superfamily ZYGAENOIDEA

Maxillary palpi very small or absent; both wings with M present or vestigial in discal cell, CuP present, hind wing with $Sc + R_1$ not approximated to Rs beyond discal cell; tympanal organs absent. Larva stout, sluggish, feeding exposed on foliage, or

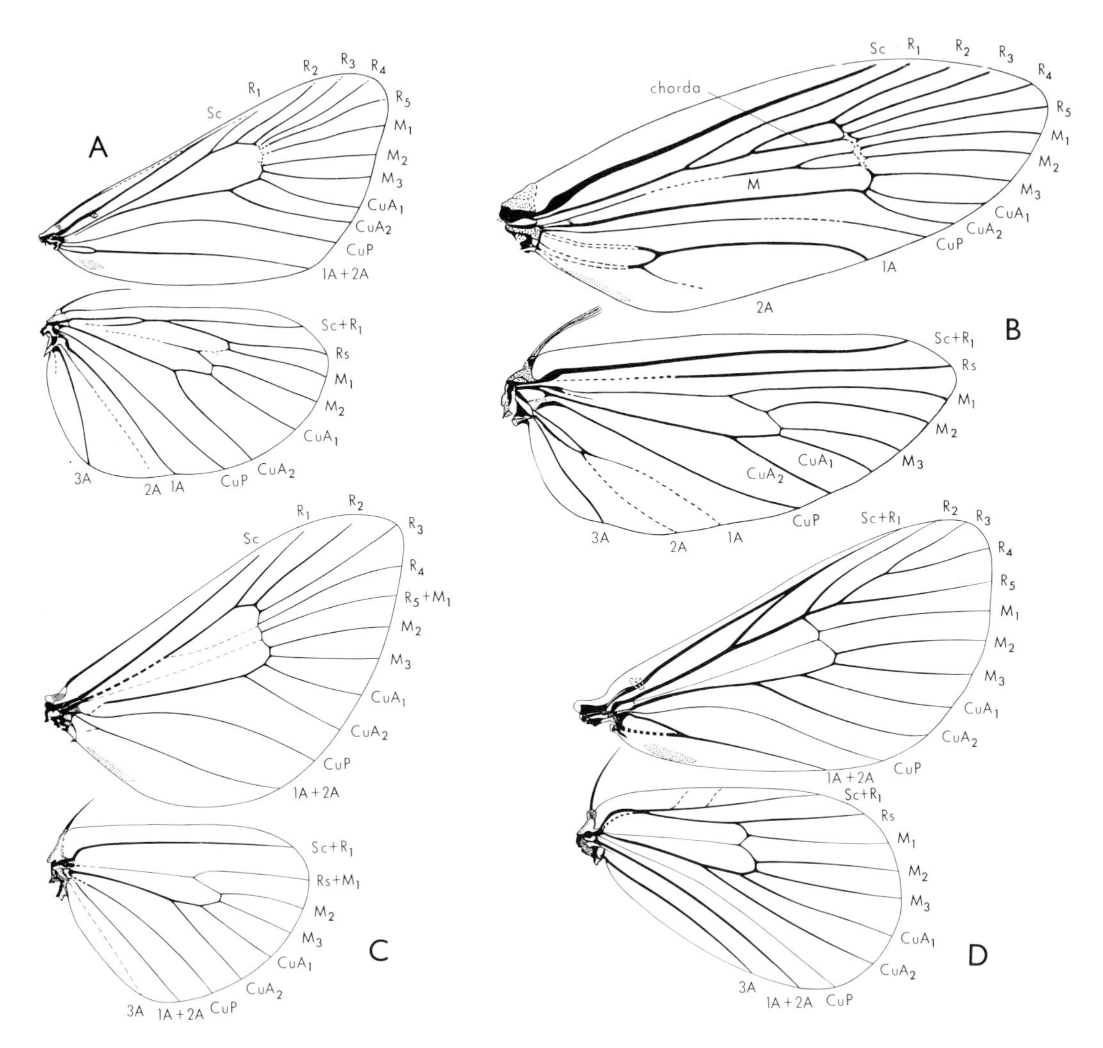

Fig. 36.35. Wing venation of Zygaenoidea: A, *Pollanisus*, Zygaenidae; B, *Cyclotorna*, Cyclotornidae; C, *Heteropsyche*, Epipyropidae; D, *Pseudanapaea*, Limacodidae. [T. Binder, J. Wedgbrow]

ectoparasitic on Homoptera, or in ants' nests. Pupa lightly sclerotized, appendages largely free, abdomen with fine dorsal spines, with spiracles visible on segment 1 (except Epipyropidae); in firm cocoon, protruded at ecdysis.

Primitive features of the superfamily are the presence of M in the discal cell and of CuP, and the generalized pupa which is protruded from the cocoon at ecdysis. The Zygaenidae have a well-developed haustellum and chaetosemata, and usually ocelli, and have sometimes acquired a wasp-like appearance. The other families have lost the haustellum and ocelli. The Epipyropidae have also lost the tibial spurs and, like the Cyclotornidae, have developed a highly specialized parasitic biology. The Limacodidae show remarkable specialization in the larva, which has lost the prolegs and has developed stinging hairs and aposematic coloration in some genera.

Key to the Families of Zygaenoidea Known in Australia

1. Head with chaetosemata large .... **Zygaenidae** (p. 830)
   Head with chaetosemata absent 2
2. Fore wing with R_3, R_4, and R_5 stalked .... **Limacodidae** (p. 832)
   Fore wing with all veins arising separately from discal cell 3
3. Fore wing with 1A and 2A fused in middle, distal ends free; tibial spurs 0-2-4 .... **Cyclotornidae** (p. 830)
   Fore wing with 1A and 2A forming a basal fork; tibial spurs absent .... **Epipyropidae** (p. 832)

41. Zygaenidae, Small; ocelli usually present; chaetosemata large (Fig. 36.1A); antennae dilated, bipectinate in male; haustellum naked; maxillary palpi 1- or 2-segmented; labial palpi short, ascending; epiphysis present or absent; tibial spurs 0–2–2 or 0–2–4, hind tibiae smooth; fore wing (Fig. 36.35A) with R_3 and R_4 sometimes stalked, other veins separate, chorda absent, CuP present; hind wing with Sc + R_1 shortly fused with Rs beyond middle of discal cell, R_1 present about middle of discal cell, CuP present, 3 anal veins. Larva with short dense secondary setae, prothorax sometimes with paired gland-like protuberances, crochets in uniordinal mesoseries. Pupa in strong elongate cocoon.

The family (Turner, 1926) may be recognized by the presence of CuP in both wings, the thickened antennae, and the prominent chaetosemata and ocelli. The adults fly and feed at flowers in the daytime. In *Thyrassia* (1 sp.) the epiphysis is present, but tibial spurs are absent. *T. inconcinna* Swinh. (Fig. 36.36B), from north Queensland, resembles a small amatid. The epiphysis is also present in *Homophylotis* (1 sp.), in which only apical spurs are present on mid and hind tibiae, as in *Onceropyga* (1 sp.), *Hestiochora* (4 spp.), and *Pollanisus* (12 spp.; Plate 7, N). The last three genera lack an epiphysis. The females of *Hestiochora* and *Pollanisus* have a large dense anal tuft. The larva of the common wasp-like species, *H. tricolor* (Walk.) (Plate 6, ZC), erodes the surface of the leaves of *Syncarpia.*

42. Cyclotornidae. Small; head smooth-scaled; ocelli and chaetosemata absent; antennae simple, thickened, scape with pecten; haustellum and maxillary palpi absent; labial palpi very short; epiphysis absent, tibial spurs 0–2–4, hind tibiae with long hair-scales; fore wing (Fig. 36.35B) with chorda, M forked in discal cell, R_5 to apex, all veins separate, CuP present, 2A running separately to margin from basal fork of 1A+2A; hind wing with M simple in discal cell, R_1 absent, CuP present, 2A running separately towards margin from fork of 1A + 2A. Larva heteromorphic, mature larva (Fig. 36.37D) with long antennae, apical adhesive pads on thoracic legs, crochets in uniordinal mesoseries, and fine secondary setae; at first parasitic on Homoptera, later predatory on ant larvae. Pupa in oval, white, silken cocoon.

Confined to Australia, the stout-bodied, grey moths may be recognized by their reduced mouth-parts, smooth head, primitive venation, and especially by the condition of 2A in both wings. The remarkable life history of *Cyclotorna monocentra* Meyr. (Fig. 36.36C) was described by F. P. Dodd in 1912. The eggs are laid on the twigs and bark of a tree already infested with Cicadellidae (Homoptera), which are usually attended by ants. The rounded, flat, first-instar larva (Fig. 36.37E) is an external parasite of a leafhopper. It later spins an oval, flat shelter beneath bark, in which it moults to a brightly coloured, flattened larva with retractile head and body segments laterally expanded and pointed (Fig. 36.37D). This is carried by an ant to its nest, where the moth larva feeds upon the ant larvae, yielding an anal secretion eagerly devoured by the ants. At maturity, the larva spins its cocoon beneath loose bark on a nearby tree. In the smaller *C. egena* Meyr., lacking CuA_1 in the hind wing, the first-stage larvae parasitize Psyllidae. In some species early ecdyses occur in shelters spun on the body of the leafhopper host.

Fig. 36.36. A, *Synemon laeta*, Castniidae; B, *Thyrassia inconcinna*, Zygaenidae; C, *Cyclotorna monocentra*, Cyclotornidae; D, *Heteropsyche melanochroma*, Epipyropidae; E, *Doratifera vulnerans*, Limacodidae; F, *Hyblaea puera*, Hyblaeidae; G, *Oxycophina theorina*, H, *Striglina pyrrhata*, Thyrididae; I, *Tanycnema anomala*, J, *Tineodes adactylalis*, Tineodidae; K, *Oxychirota paradoxa*, L, *Cenoloba obliteralis*, Oxychirotidae. Scales = 2 mm.
[Photos by C. Lourandos]

43. Epipyropidae. Very small; antennae short, bipectinate to apex in both sexes; ocelli, chaetosemata, haustellum, and maxillary palpi absent; labial palpi minute, drooping; epiphysis and tibial spurs absent; wings (Fig. 36.35C) broad; fore wing without retinaculum, all veins arising separately from discal cell, M present or vestigial in cell, chorda present, CuP present near margin; hind wing with frenulum simple in both sexes, Sc + R_1 well separate from Rs, R_1 sometimes present, CuP and 1 or 2 anal veins present. Eggs disc-like. Larva 'subcampodeiform' at first, with large head and thorax, later strongly convex, covered with white waxy secretion, crochets uniordinal in a circle, antennae long, ocelli closely grouped, mandibles long, tapering, toothed; ectoparasitic upon Fulgoridae, Cicadellidae, and Cicadidae. Pupa in white, usually rosette-shaped cocoon (Fig. 36.37A).

The eggs are laid singly or in clusters on the food plants of the homopteran hosts. The first-instar larvae seek actively for a host, and often adopt an erect questing posture as they await its approach. After the first ecdysis they are incapable of parasitizing a second host, should they be dislodged. The feeding habits of the larva have not been determined precisely; its effect upon the host varies, but in some species the host succumbs. In the north and east of Australia the hosts of *Heteropsyche* (7 spp.) include *Platybrachys* spp. on *Eucalyptus* and *Scolypopa australis* on passion vines. *H. melanochroma* Perk. (Fig. 36.36D) is one of several similar species.

44. Limacodidae. Small to medium-sized; ocelli and chaetosemata absent; antennae bipectinate in male, at least in basal half; haustellum very small or absent; maxillary palpi 1- to 3-segmented, or absent; labial palpi short, 2- or 3-segmented; epiphysis absent, tibial spurs 0–2–2 or 0–2–4, rarely absent; wings broad (Fig. 36.35D); fore wing with M present in discal cell, sometimes forked, chorda absent, R_3, R_4, and R_5 stalked, CuP present; hind wing with M present in discal cell, rarely forked, Sc + R_1 fused with Rs near base, or connected to Rs by R_1, CuP present. Eggs scale-like. Larva (Fig. 36.37C) with head retracted, antennae long, thoracic legs reduced, prolegs absent, sometimes ventral suckers on abdominal segments 1–7, setae often modified, with stinging hairs, or setae reduced. Pupa in oval or pyriform cocoon (Fig. 36.37B), with circular lid-like opening at one end.

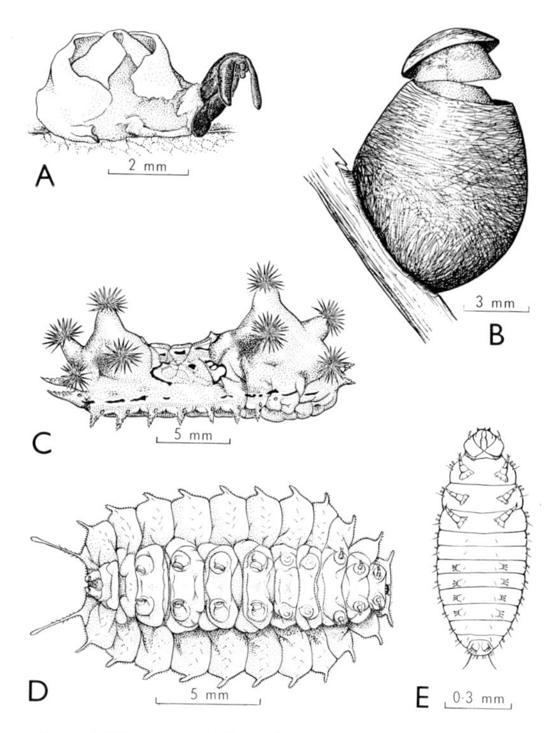

Fig. 36.37. Zygaenoidea: A, cocoon of *Heteropsyche*, Epipyropidae; B, C, cocoon and larva of *Doratifera*, Limacodidae; D, E, final and first instar larvae of *Cyclotorna*, Cyclotornidae. [S. Curtis]

The family (Turner, 1926) is generally distributed, but with most species in the tropics. The legs and thick body are clothed with long, dense hair-scales. In *Comana* (20 spp.) M is forked in the discal cell of the fore wing, and the larvae lack stinging hairs. *C. miltocosma* (Turn.) (Plate 7, J) is a handsome species from the north. The variable black and white *C. fasciata* (Walk.) occurs widely in the north and east, where the green slug-like larvae feed on Proteaceae. The best known species belong to *Doratifera* (8 spp.), in which M is simple in the discal cell of both wings, and the hind tibiae are without median spurs. The brightly coloured larvae feed exposed on foliage, and have paired dorsal protuberances which are crowned with eversible tufts of stinging hairs. The pyriform cocoons are attached to twigs or bark. The larvae of *D. vulnerans* (Lew.) (Fig. 36.36E) feed on fruit and ornamental trees as well as *Eucalyptus*.

Superfamily PYRALOIDEA

Maxillary palpi present, usually 2- to 4-segmented; labial palpi porrect, beak-like, or ascending; legs usually long, slender, with epiphysis and prominent tibial spurs; wings entire or, rarely, each with a deep terminal cleft, M absent from discal cell, fore wing with CuP usually absent, hind wing with $Sc + R_1$ usually approximated to, or shortly fused with, Rs beyond discal cell, CuA sometimes with basal pecten of hairs, CuP present or absent. Egg of flat type. Larva with 2 prespiracular (L) setae on the prothorax; concealed feeders. Pupa usually with maxillary palpi and distinct pilifers; in silken cocoon or in larval gallery, not protruded at ecdysis.

The superfamily may not be completely homogeneous. The Hyblaeidae were referred here by Forbes (1933) on the basis of pupal characters, but Koning and Roepke (1949) and Singh (1956) disagreed. With the Thyrididae, the family is here regarded as a primitive member of the Pyraloidea. In the hind wing of the Hyblaeidae, as in some Thyrididae and Pyralidae, Sc does not fuse with Rs beyond the discal cell, but in some genera of Thyrididae these veins show the more usual pyraloid condition. Many features are common to the remaining families. However, abdominal tympanal organs, found universally in the Pyralidae, do not occur elsewhere in the superfamily. With few exceptions the Pyralidae are also distinguished by the densely scaled base of the haustellum. Whalley (1961) suggested that *Oxychirota* is a tineoid genus, but most characters of the adult, including its resting posture and dorsal spining of the abdomen, indicate a relationship to *Cenoloba* and to the Tineodidae. Its immature stages are not known, but in *Cenoloba* the larva has 2 prespiracular setae on the prothorax, and the pupa has both maxillary palpi and pilifers as in most Pyraloidea.

Key to the Families of Pyraloidea

1. Hind wing with CuP present as a tubular vein, at least towards margin 2
   Hind wing with CuP vestigial or absent, not present as a tubular vein 3
2. Hind wing with $Sc + R_1$ approximated to, or fused with, Rs for a short distance beyond discal cell; haustellum rarely naked; tympanal organs present at base of abdomen ...... **Pyralidae** (p. 835)
   Hind wing with $Sc + R_1$ not approximated to Rs beyond discal cell; haustellum naked; tympanal organs absent .... **Hyblaeidae** (p. 833)
3. Wings medium to broad; hind wing with M_2 arising nearer to M_3 than to M_1, M_1 nearer to Rs than to M_2 at base .... **Thyrididae** (p. 834)
   Wings narrow; hind wing with M_2, when present, arising equidistant from M_1 and M_3, or, if nearer M_3, then M_1 nearer to M_2 than to Rs at base 4
4. Wings almost linear, or each deeply divided into 2 plumes; neither wing with tubular anal veins .... **Oxychirotidae** (p. 835)
   Wings not linear but entire; fore wing with 1 anal, hind wing with 2 anal veins .... **Tineodidae** (p. 835)

45. Hyblaeidae. Medium-sized; ocelli present; chaetosemata absent; antennae simple; haustellum strong, naked; maxillary palpi short, 3- or 4-segmented, with long scales; labial palpi porrect, beak-like; tibial spurs 0–2–0 in male, 0–2–4 in female, hind tibia in male with long projecting apical lobe and hair pencil which folds into a specialized bladder-like appendage from the hind coxa; fore wing (Fig. 36.38A) with retinaculum in male a long slender hook, all veins separate, M_2 arising nearer to M_3 than to M_1, CuP absent, 1A strong, 2A sinuous, not reaching margin; hind wing with Sc fused with Rs near base of discal cell, thence divergent, M_2 arising nearer to M_3 than to M_1, CuP a weak tubular vein in outer one-third, 2 anal veins; abdomen without tympanal organs; uncus in male simple or trifurcate. Larva without secondary setae, crochets triordinal in a circle; in silken gallery between joined leaves. Pupa stout, with distinct maxillary palpi, pilifers not defined, antennae short, epicranial suture distinct, cremaster present; in silken cocoon amongst foliage, fallen leaves, or in soil.

Hyblaea, the only genus, was once referred to the Noctuidae. The absence of tympanal organs, the prominent maxillary palpi, the beak-like labial palpi, the behaviour of the larvae, and the structural characters of both larvae and pupae suggest a position for the family in the Pyraloidea. The larvae of *H. puera* Cram. (Fig. 36.36F) feed on Verbenaceae, defoliating teak in Java and New Guinea, and in northern Australia have been reared from *Vitex trifolia*. The uncus of the male is trifurcate, whereas in *H. ibidias* Turn. it is simple.

46. Thyrididae. Small to large; ocelli usually and chaetosemata always absent; antennae simple, dentate, or pectinate; haustellum present, naked, or absent; maxillary palpi minute, 1- or 2-segmented; labial palpi porrect or recurved, sometimes very short or only 2-segmented; tibial spurs 0–2–2, 0–2–3, or 0–2–4; fore wing (Fig. 36.38B) with retinaculum in male a long slender hook, all veins often separate, CuP absent, 2A vestigial; hind wing with all veins separate, Sc sometimes connected to Rs by R_1 and approximated to, or fused with, Rs beyond discal cell, M_2 arising nearer to M_3 than to M_1, CuP vestigial, 2 anal veins;

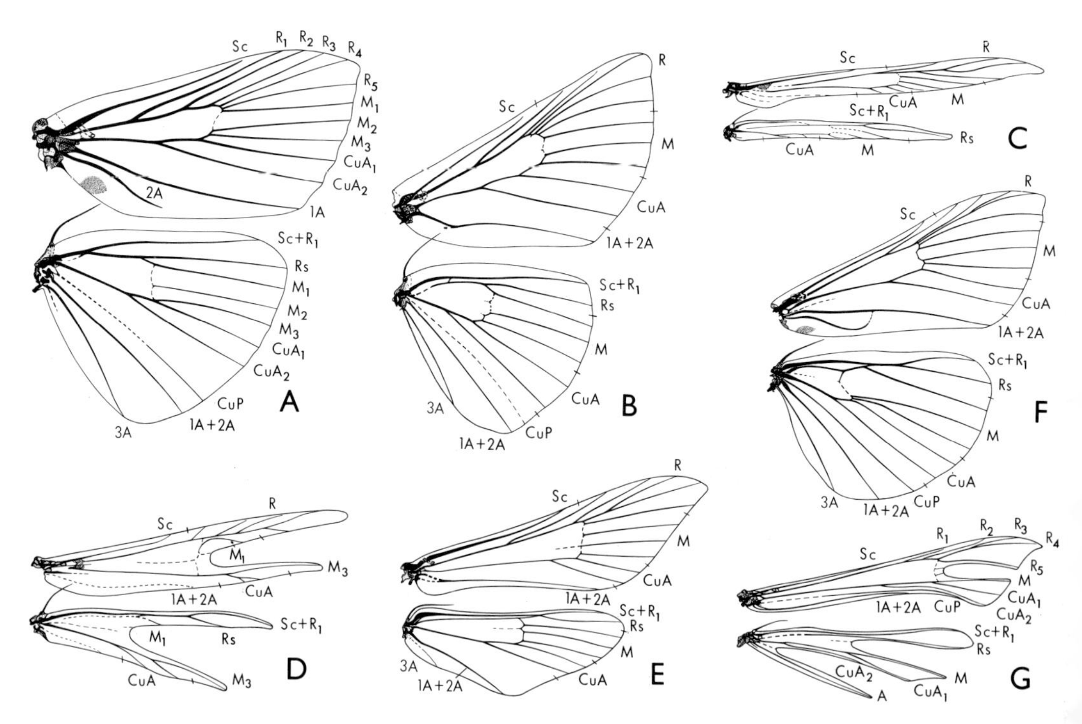

Fig. 36.38. Wing venation of Pyraloidea and Pterophoroidea: A, *Hyblaea*, Hyblaeidae; B, *Striglina*, Thyrididae; C, D, *Oxychirota*, *Cenoloba*, Oxychirotidae; E, *Tineodes*, Tineodidae; F, *Urisephita*, Pyralidae; G, *Platyptilia*, Pterophoridae. [J. Wedgbrow]

abdomen without tympanal organs. Larva with crochets biordinal, in circle or ellipse, sometimes broadly interrupted; tunnelling in twigs and stems, sometimes producing swellings. Pupa with maxillary palpi and pilifers defined.

The family (Whalley, 1964) forms a compact group, most members of which have Sc + R_1 of the hind wing approximated to, but not fusing with, Rs beyond the discal cell. These include large broad-winged species, such as *Oxycophina theorina* (Meyr.) (Fig. 36.36G) from north Queensland, with female wing-span of 6·5 cm. *Striglina* (8 spp.) contains smaller species, such as the variable *S. pyrrhata* (Walk.) (Fig. 36.36H), sometimes rich pink in colour, and the small *S. centiginosa* Lucas with tornus of the fore wing excavated. *Rhodoneura tetragonata* (Walk.) is an Indo-Malayan species which also occurs in New Guinea and eastern Australia. *Addaea* (7 spp.) contains small pyralid-like species, such as *A. pusilla* (Butl.), in which Sc + R_1 fuses with Rs beyond the discal cell of the hind wing.

47. Tineodidae. Small; ocelli present or absent; chaetosemata absent; antennae simple, ciliated; haustellum naked; maxillary palpi porrect, 4-segmented; labial palpi long, porrect; legs long, slender, tibial spurs 0–2–4, sometimes 0–0–2; wings (Fig. 36.38E) entire; fore wing sometimes without 1 branch of R, M_1, M_2, and M_3 well separated, usually parallel, CuP absent, 1A + 2A forming small basal fork; hind wing with Sc + R_1 approximated to, or fusing with, Rs beyond discal cell, M_1, M_2, and M_3 well separated, parallel, or rarely M_2 nearer to M_3 at base, CuP absent, 2 anal veins; tympanal organs absent, abdominal terga with anterior transverse patches of short spines. Larva and pupa between joined leaves.

The family occurs in India, New Guinea, and Australia, and includes species with relatively narrow and curiously shaped wings and long legs. *Tanycnema anomala* Turn. (Fig. 36.36I), from Queensland and New Guinea, superficially resembles an agdistine pterophorid, but has 4-segmented maxillary palpi and is without specialized scales along CuA beneath the hind wing. The larvae of *Euthrausta holophaea* (Turn.) join leaves of *Glochidion ferdinandi* in coastal Queensland and N.S.W. *Tineodes adactylalis* Guen. (Fig. 36.36J) occurs widely in the south-east and south-west.

48. Oxychirotidae. Very small; ocelli and chaetosemata absent; antennae simple, ciliated; haustellum naked; maxillary palpi 4-segmented; labial palpi porrect; legs long, slender, tibial spurs 0–2–4; wings (Figs. 36.38C, D) linear, or each with a deep terminal cleft; fore wing sometimes without 1 branch of R, M_1, M_2, and M_3 well separated and parallel, or M_2 absent, CuP absent, 1A + 2A vestigial or absent; hind wing with Sc + R_1 fused with Rs from before to well beyond end of discal cell, M_1, M_2, and M_3 well separated, parallel, or M_2 absent, CuP and anal veins absent; tympanal organs absent, abdominal terga with anterior transverse patches of short spines. Larva with crochets uniordinal in penellipse. Pupa with maxillary palpi and pilifers; in silken cocoon.

The species are characterized by narrow wings, reduced venation, and long fringe. As in the Tineodidae, CuP is absent from both wings, and the dorsum of the abdomen is similarly spined. However, the anal veins of the hind wings are lost. The wings of *Oxychirota paradoxa* Meyr. (Fig. 36.36K) are almost linear, whereas in *Cenoloba obliteralis* (Walk.) (Fig. 36.36L) each wing has a deep terminal cleft. Both species occur along the east coast. *Oxychirota* is also known from Christmas I. and Ceylon. The larvae of *C. obliteralis* tunnel in the developing cotyledons of fallen seeds of the mangrove *Avicennia marina*.

49. Pyralidae. Small to large; ocelli and chaetosemata present or absent; antennae simple and ciliated, rarely uni- or bipectinate; haustellum densely scaled near base, rarely naked, sometimes reduced; maxillary palpi usually 4-segmented, sometimes 2- or 3-segmented; labial palpi porrect, beak-like, or ascending, especially in male, rarely reduced; tibial spurs 0–2–4; fore wing (Fig. 36.38F) with R_3 and R_4 stalked or coincident, M_2

approximated to M_3 at base, CuP usually absent, 1A and 2A usually forming basal fork; hind wing with Sc + R_1 approximated to, or shortly fused with, Rs beyond discal cell, M_2 approximated to M_3 at base, CuA sometimes with basal pecten, CuP usually present, anal area large, with 2 anal veins, frenulum in female sometimes simple; tympanal organs present at base of abdomen. Larva with crochets usually bi- or triordinal, in a circle or penellipse, rarely uniordinal in 2 transverse bands; in shelters of webbed leaves or shoots, or tunnels in shoots, stems, seed heads, fruits, or galls, or in silken galleries amongst mosses, herbaceous plants, or fallen leaves, or in shelters or cases amongst aquatic plants in fresh water, or in stored products, or in nests of Hymenoptera, rarely predacious on Coccidae. Pupa with pilifers defined, antennae long; in silken cocoon usually in larval shelter.

This is a large and ubiquitous family, with species adapted to diverse terrestrial and aquatic habitats. Many are pests of cultivated plants and stored products. Several subfamilies are recognized, and some authors treat them as families.

The SCHOENOBIINAE have chaetosemata, and CuP is present as a tubular vein near the margin of the fore wing. Many are white and rest during the day amongst grasses, rushes, and sedges, in which the larvae tunnel (Common, 1960). The females often have a large anal tuft of hair-scales used to cover the egg masses. *Scirpophaga* (8 spp.) includes several sexually dimorphic species, such as *S. imparella* (Meyr.) (Fig. 36.39A) which is found amongst *Eleocharis sphacelata*. *Tryporyza innotata* (Walk.) attacks rice in northern Australia and south-east Asia.

The SCOPARIINAE also have chaetosemata. *Eudonia* (43 spp.) is the largest genus, with larvae often feeding on mosses. *E. aphrodes* (Meyr.) is one of several procryptic species from temperate rain forest in the south-east. The closely related CRAMBINAE (Bleszynski and Collins, 1962) are also numerous in the south, the predominant genus being *Hednota* (51 spp.). The hind wing bears a pecten of hairs on CuA, and the prominent maxillary palpi are dilated apically by scales. Chaetosemata may be present or absent. The larvae feed on grasses, living in tubes of silk in the crown of the tussocks. The wings are often striped longitudinally, making the moths inconspicuous as they rest amongst grass. Some, such as *H. pedionoma* (Meyr.) (Fig. 36.39B), are pests of grass pasture, especially in the south-west. *Chilo suppressalis* (Walk.) is a rice-stem borer in the N.T. and in south-east Asia.

The NYMPHULINAE, with chaetosemata present, are widely distributed, especially in the north, with larvae adapted for aquatic life. Respiration may be cutaneous in the early larval stages, but later is permitted by filamentous tracheal gills, with or without functional spiracles.

The PYRAUSTINAE, which lack chaetosemata, are most numerous in the tropical north. Most rest with their wings partly expanded, usually on the underside of leaves, and are readily flushed during the day. The larvae usually live in webbed shelters amongst foliage. Iridescent colours are common, as in the pale blue '*Margaronia*' *glauculalis* (Guen.) (Plate 7, M). '*M.*' *canthusalis* (Walk.) (Fig. 36.39C) has a marbled pattern of brown, with white spots. Many species are pests of agricultural crops. *Hymenia recurvalis* (F.) attacks the leaves of beet, '*Margaronia*' *tolumnialis* (Walk.) damages the foliage of figs, *Phakellura indica* (Saund.) is a pest of Cucurbitaceae, *Maruca testulalis* (Geyer) bores in beans, *Sceliodes cordalis* (Doubl.) is a pest of egg-fruit, *Dichocrocis punctiferalis* (Guen.) attacks peaches, pawpaws, sorghum, and castor, and *Psara licarsisalis* (Walk.) is a pest of *Paspalum* pastures and lawns.

The GALLERIINAE also lack chaetosemata, but have a cubital pecten. In males the labial palpi are usually short. The larvae of *Hylaletis latro* (Zell.) tunnel in the flower spikes of grass-trees (*Xanthorrhoea*). Those of *Callionyma sarcodes* Meyr. (Fig. 36.39D) are pink, and occur beneath loose bark on *Eucalyptus* trunks. The stored-product pest *Corcyra cephalonica* (Staint.) and the wax moths *Galleria mellonella* (L.) and *Achroia*

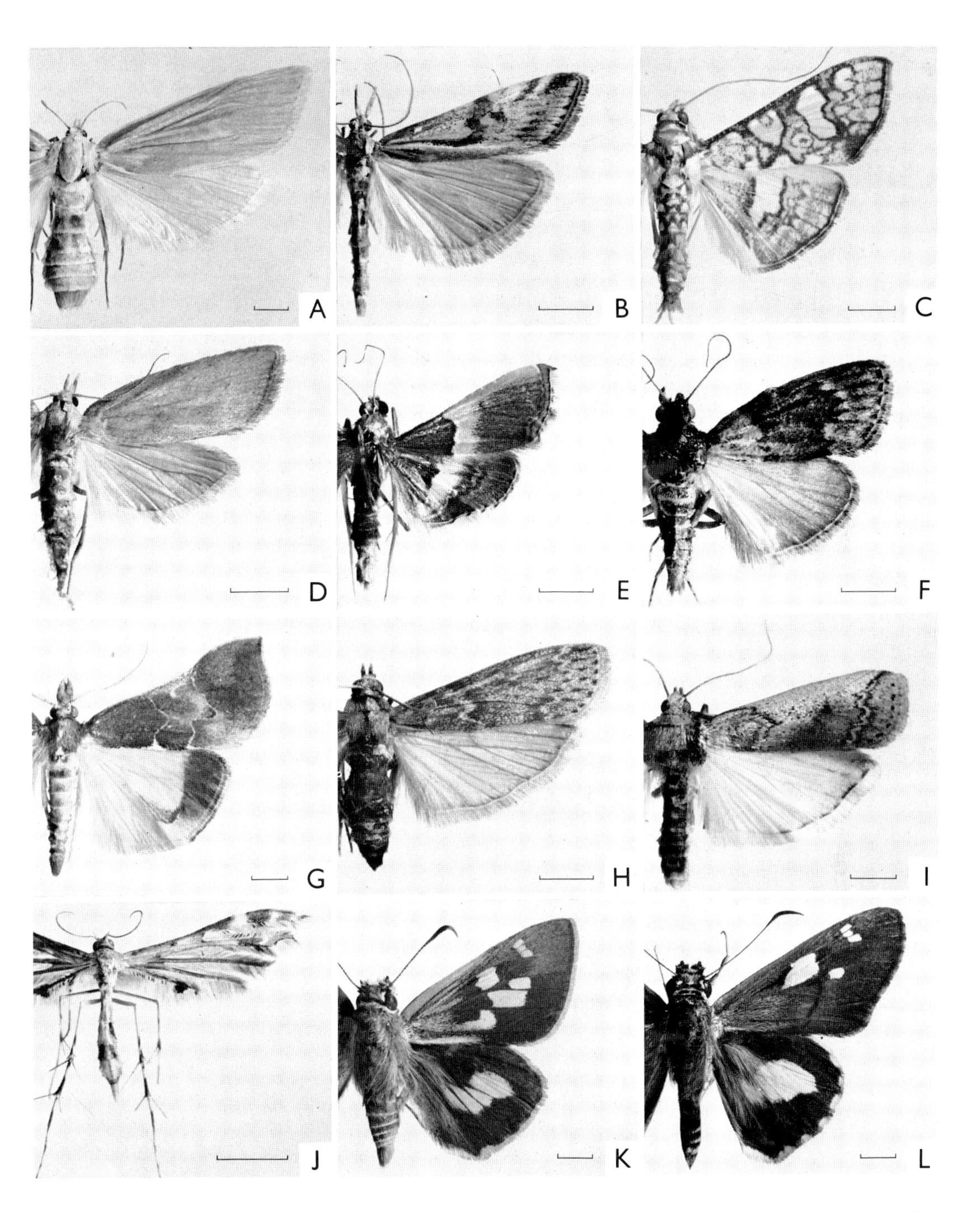

Fig. 36.39. A, *Scirpophaga imparella*, B, *Hednota pedionoma*, C, '*Margaronia*' *canthusalis*, D, *Callionyma sarcodes*, E, *Endotricha mesenterialis*, F, *Epipaschia nauplialis*, G, *Drymiarcha exanthes*, H, *Hypsipyla robusta*, I, *Cactoblastis cactorum*, Pyralidae; J, *Sphenarches anisodactylis*, Pterophoridae; K, *Trapezites eliena*, L, *Hesperilla mastersi*, Hesperiidae. Scales = 3 mm. [Photos by C. Lourandos]

grisella (F.) are cosmopolitan species.

The PYRALINAE, with chaetosemata present, include some brilliant tropical species such as *Hypsidia erythropsalis* Rothsch. Most species are procryptic, and rest with their wings partly outspread and the tip of the abdomen curled upwards between them. In *Endotricha* (14 spp.) the tegulae of the males bear very long scales. *E. mesenterialis* (Walk.) (Fig. 36.39E) is common in Queensland. Several introduced pests of stored products, such as *Pyralis farinalis* (L.) and *Aglossa caprealis* (Hübn.), are also included here. The EPIPASCHIINAE are well represented in Australia, mainly by species in *Epipaschia* (16 spp.) and *Macalla* (23 spp.). The larvae of *E. nauplialis* (Walk.) (Fig. 36.39F) spin silken galleries amongst fallen leaves, upon which they feed. *Drymiarcha exanthes* Meyr. (Fig. 36.39G) is one of few CHRYSAUGINAE.

The PHYCITINAE are most numerous in the tropics, and include many pest species. The fore wing lacks R_5, the hind wing has a cubital pecten, chaetosemata are present, and the labial palpi usually curve upwards. The larvae of *Etiella behrii* Zell. attack lucerne grown for seed in South Australia. *Hypsipyla robusta* (Moore) (Fig. 36.39H) is a pest of regenerating red cedar, *Toona australis*, in N.S.W. and Queensland, tunnelling in the terminal shoots. Introduced pests of stored products include *Ephestia kuehniella* (Zell.), *E. elutella* (Hübn.), *E. cautella* (Walk.), *E. figulilella* (Greg.), and *Plodia interpunctella* (Hübn.). Perhaps the most famous pyralid is *Cactoblastis cactorum* (Berg) (Fig. 36.39I) which was successfully introduced to control prickly pear.

Superfamily PTEROPHOROIDEA

50. Pterophoridae. Small; ocelli and chaetosemata absent; antennae simple, ciliated in male; haustellum naked; maxillary palpi minute, 1-segmented; labial palpi porrect, sometimes small, slender; legs long, slender, epiphysis present, tibial spurs 0–2–4, prominent; fore wing (Fig. 36.38G) cleft into 2, 3, or 4 lobes or plumes, or rarely entire, R_1 often absent, R_3 and R_4 usually stalked, M_1 and M_2 usually very short and weak, CuP present, 1A + 2A usually simple; hind wing deeply cleft into 3 plumes, or rarely entire, underside with double series of specialized scales along CuA, CuP present or reduced, 1 anal vein; tympanal organs absent. Egg of flat type, oval, smooth. Larva with 2 or 3 prespiracular (L) setae on prothorax, usually with numerous secondary setae, ventral prolegs very long, crochets uniordinal in a mesoseries; sometimes leaf-mining at first, later exposed, or in stem. Pupa smooth or spined, maxillary palpi absent, pilifers present; attached by cremaster, usually without cocoon, or in larval tunnel.

The superfamily differs from the Pyraloidea by the reduced maxillary palpi, by the larva which has long ventral prolegs, crochets in a mesoseries, and usually numerous secondary setae, and by the unusual pupa which lacks maxillary palpi. However, as in most Pyraloidea, pilifers are defined in the pupa. The adults fly weakly, and rest with wings outspread and legs prominently displayed. In the Australian species the fore wing is divided into 2 or 3 lobes and the hind wing into 3 plumes; the Agdistinae, with entire wings, have not been recorded. Several Australian species are also widely distributed abroad.

In the PLATYPTILIINAE the second lobe of the hind wing contains 3 veins and there is 1 anal vein. The larva of *Sphenarches anisodactylis* (Walk.) (Fig. 36.39J) bears modified scale-like as well as normal setae, and feeds on the flower-buds and pods of *Dolichos*; the pupa is spiny with many setae. *Trichoptilus wahlbergi* (Zell.) also occurs in India and Japan, where the larva feeds on *Oxalis* (Yano, 1963). The pantropical species *Lantanophaga pusillidactyla* (Walk.) has larvae feeding on flowers of *Lantana* and may assist in its control. The PTEROPHORINAE have only 2 veins in the second lobe of the hind wing, but there are 2 anal veins.

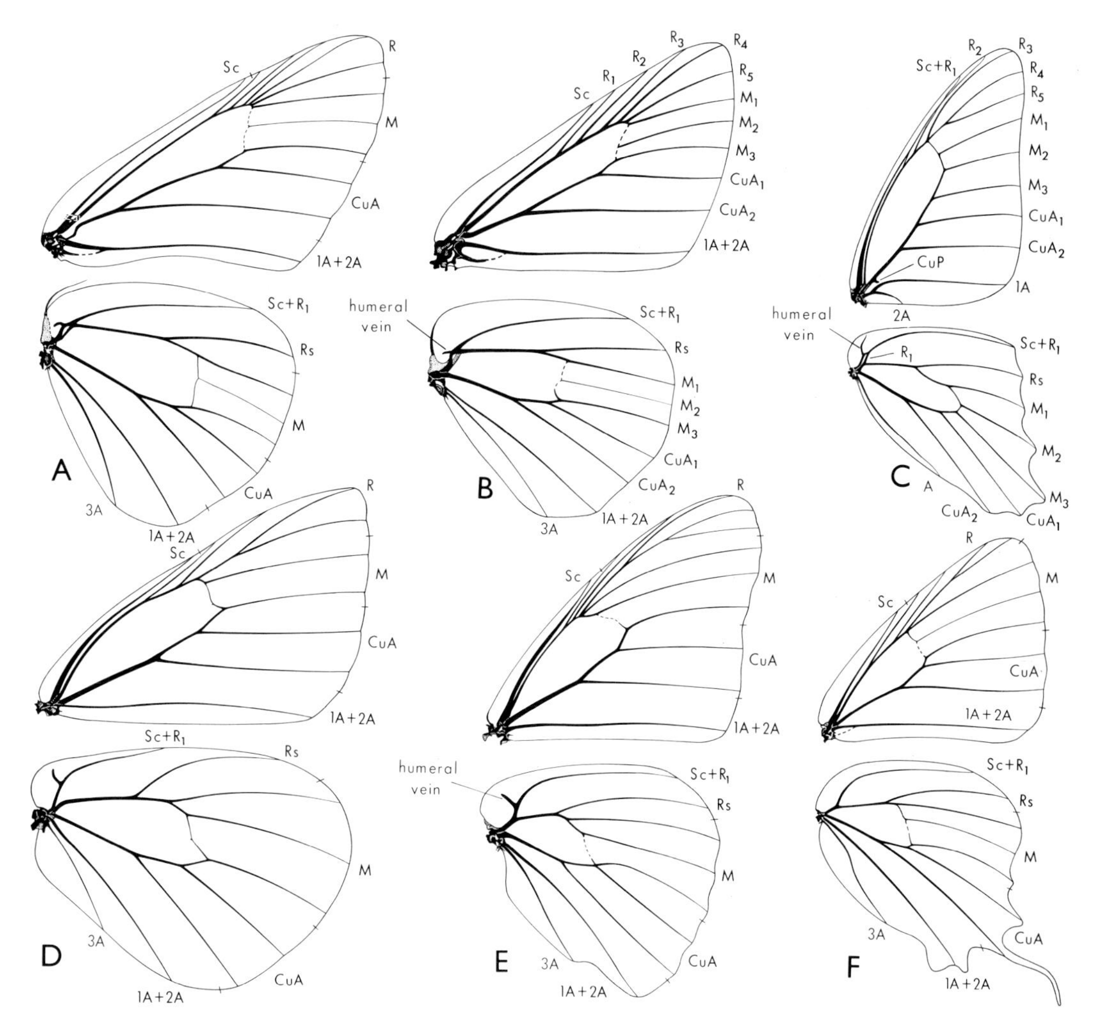

Fig. 36.40. Wing venation of Hesperioidea and Papilionoidea: A, *Euschemon* ♂, Hesperiidae; B, *Trapezites*, Hesperiidae; C, *Graphium*, Papilionidae; D, *Pieris*, Pieridae; E, *Vanessa*, Nymphalidae; F, *Jalmenus*, Lycaenidae. [J. Wedgbrow]

Superfamily HESPERIOIDEA

51. Hesperiidae. Small to medium-sized; ocelli absent; chaetosemata present; antennae widely separated at base, scape with scale-tuft, flagellum gradually dilated apically to form club, usually with hooked tip; haustellum naked; maxillary palpi absent; labial palpi ascending; epiphysis present, tibial spurs usually 0–0–4 or 0–0–2; fore wing (Figs. 36.40A, B) without retinaculum (except male *Euschemon*), veins arising separately from discal cell, CuP absent, 1A + 2A simple or forming basal fork; hind wing with frenulum absent (except male *Euschemon*), humeral vein usually present, Sc connected to Rs near base by R_1, veins arising separately from discal cell, CuP absent, 2 anal veins. Egg of upright type, smooth or with vertical ribs. Larva (Fig. 36.41A) without obvious primary setae, but with fine short hair, constricted behind large head, crochets multiordinal in a circle; in silk-lined shelter between joined leaves or in longitudinally rolled leaf. Pupa (Fig. 36.41B) in larval shelter, attached by cremaster and usually by a central silken girdle.

This homogeneous family (Waterhouse and Lyell, 1914) differs from the Papilionoidea in having the antennae set widely apart at the base and the peripheral veins in the wings not stalked. Most species are diurnal, but a few are crepuscular. Their flight is rapid and rather jerky, and in repose the wings may be held back to back, sometimes with the hind wings depressed, or extended flat. The larvae of the most archaic species feed on dicotyledons, whereas those of the more specialized feed on monocotyledons.

The COELIADINAE (6 spp.) and the PYRGINAE (10 spp.) include large, robust species which rest with wings outspread, often on the underside of a leaf. *Euschemon rafflesia* (Macl.) (Plate 8, N) is remarkable because of the presence of a frenulum and retinaculum in the male. The brightly coloured larva feeds and pupates between joined leaves of *Wilkiea macrophylla* in rain forest. *Chaetocneme* (4 spp.) contains crepuscular species, often with bright red eyes. *C. beata* (Hewitson) occurs as far south as Wollongong.

The TRAPEZITINAE (54 spp.) are mainly Australian, and are most numerous in the south. The larvae feed on sedges, grasses, and other monocotyledons, usually sheltering between two or three leaves joined with silk. In *Trapezites* (11 spp.) the larvae feed at night on *Lomandra*, hiding during the day at the base of the plant. *T. eliena* (Hewitson) (Fig. 36.39K), with a large white spot on the underside of the hind wing, occurs in coastal eastern Australia. *Toxidia* (9 spp.) includes dull brown species, which also occur in the tropics. The larvae of *T. doubledayi* (Feld.) and other species feed on grasses. Sword-grass *(Gahnia)* is the food plant of *Hesperilla* (11 spp.), the finest of which is *H. mastersi* Waterh. (Fig. 36.39L) mainly from coastal N.S.W.

The HESPERIINAE (36 spp.) are mostly tropical, with larvae feeding on coarse grasses and palms. *Taractrocera* (5 spp.) and *Ocybadistes* (4 spp.) contain small orange and black species frequenting long grass. The antennal club of the former is spoon-shaped. *O. walkeri* Heron extends from Indonesia to Australia. *Telicota* (8 spp.) has similar but larger species. *Pelopidas lyelli* (Rothsch.) is one of several rather similar greenish grey-brown species from northern Australia.

Superfamily PAPILIONOIDEA

Ocelli absent; chaetosemata prominent; antennae approximated at base, clubbed distally, without apical hook; haustellum naked; maxillary palpi minute, 1-segmented, or absent; CuP absent from both wings, frenulum absent; tympanal organs absent. Egg of upright type. Larva with crochets in a circle in first instar, usually in a mesoseries in later instars. Pupa usually exposed and attached at posterior end to a pad of silk, and often with central silken girdle.

The superfamily (Ehrlich, 1958), with the Hesperioidea, constitutes the butterflies, or Rhopalocera. Their relationship to the remaining Lepidoptera is controversial, but it is now usually agreed that they deserve no more than superfamily rank, and that the name Rhopalocera has no special significance. Some authors regard the Papilionoidea as the most advanced Lepidoptera, and without doubt they are highly specialized. Nevertheless, such archaic characters as the presence of a short basal section of CuP and a separate 2A are retained in the fore wing of the Papilionidae. Ontogenetic evidence for the archaic origin of the superfamily is provided by the larvae which, in the first instar, have crochets in a complete circle, whereas in later instars they are progressively modified to a mesoseries (Hinton, 1952). In the Bombycoidea, Sphingoidea, Notodontoidea, and Noctuoidea, on the other hand, the crochets of all instars are arranged in the more specialized mesoseries.

Four of the 5 families are well represented in Australia, especially in the tropics. New Guinea and Oriental genera predominate, but endemic genera occur in the Satyrinae and Lycaenidae.

Key to the Families of Papilionoidea

1. Fore tibia with epiphysis; fore wing usually with transverse section of CuP near base; hind wing with 1 anal vein .... **Papilionidae** (p. 841)
   Fore tibia without epiphysis; fore wing without transverse section of CuP; hind wing usually with 2, rarely with 1 anal vein 2
2. Tarsal claws bifid; fore leg in ♂ fully developed and functional; abdomen without sclerotized prespiracular bar in basal segment .... **Pieridae** (p. 842)
   Tarsal claws rarely bifid; fore leg in ♂ reduced; abdomen with a sclerotized prespiracular bar in basal segment 3
3. Eyes notched or emarginate at base of antennae, or at least eye and edge of antennal socket contiguous; fore leg usually reduced but functional in ♂, seldom brush-like, one or both tarsal claws sometimes absent, in ♀ not reduced, fully functional, with tarsal claws .... **Lycaenidae** (p. 845)
   Eyes not notched or emarginate at base of antennae, eye and edge of antennal socket not contiguous; fore leg abbreviated, non-functional, without tarsal claws, often brush-like in both sexes (except ♀ Libytheidae) 4
4. Labial palpi very long, porrect, beak-like; fore leg in ♀ fully developed, with pair of tarsal claws; patagia not prominent or rounded, sclerotized only laterally .... **Libytheidae** (p.845)
   Labial palpi not very long and porrect, usually ascending; fore leg in ♀ abbreviated, non-functional, without tarsal claws, sometimes with tarsi clubbed; patagia prominent, rounded, fully sclerotized .... **Nymphalidae** (p. 843)

52. Papilionidae. Large; antennae short; maxillary palpi reduced to tiny projection; labial palpi appressed to frons; epiphysis present, fore leg fully developed, tibial spurs 0–0–2, tarsal claws usually simple; fore wing (Fig. 36.40C) with R_4 and R_5 usually stalked, CuP usually present near base (often said to be a cross-vein between CuA and 1A), 2 anal veins, 2A free at base or stalked with 1A, terminating on inner margin; hind wing with humeral vein present, Sc connected to Rs near base by R_1, 1 anal vein. Egg nearly spherical, smooth or with raised protuberances, laid singly. Larva (Fig. 36.41C) stout, often with paired fleshy dorsal processes, thorax sometimes humped; a forked eversible osmeterium (Frontispiece) which emits a pungent odour can be extruded from dorsum of prothorax. Pupa (Fig. 36.41D) exposed, attached by cremaster to a silken pad and by a central girdle of silk, head sometimes bifid, or thorax with a median horn.

The Australian species belong to the 3 tribes Leptocircini, Papilionini, and Troidini of the PAPILIONINAE (Munroe, 1960), and are chiefly tropical and subtropical. The Leptocircini include *Protographium* (1 sp.) and *Graphium* (6 spp.). *P. leosthenes* (Doubl.) has long straight tails to the hind wings; its pupa lacks the prominent median thoracic horn found in *Graphium*. The pale blue and black *G. sarpedon* (L.) (Fig. 36.42A) is one of several tailless species. The larva feeds on camphor laurel *(Camphora officinalis)*, *Geijera*, and *Daphnandra*. The procryptic green larva with slender yellow lateral stripe may be detected, like other papilionid larvae, by the pungent odour emitted by the osmeterium. The green and black *G. macleayanus* (Leach), with clubbed tails, occurs along the eastern tablelands, and is the only species found in Tasmania.

The Papilionini have larvae with a series of paired dorsal fleshy spines, especially while young. They feed mainly on Rutaceae. *Papilio* (7 spp.), sometimes with a tail to the hind wing, is the only Australian genus. The large, sexually dimorphic *P. aegeus* Don. and the smaller *P. anactus* Macl. attack native Rutaceae and cultivated *Citrus*. The brilliant blue and black *P. ulysses* (L.) (Plate 8, K) occurs in New Guinea and north Queensland. The larvae are green with white markings, and feed on *Euodia*. The widely distributed Old-World species, *P. demoleus* L., is abundant in inland Australia. The larva is exceptional in feeding mainly on a legume, *Psoralea*, although it also attacks *Citrus*.

The larvae of the Troidini, which feed on *Aristolochia*, have a series of soft dorsal

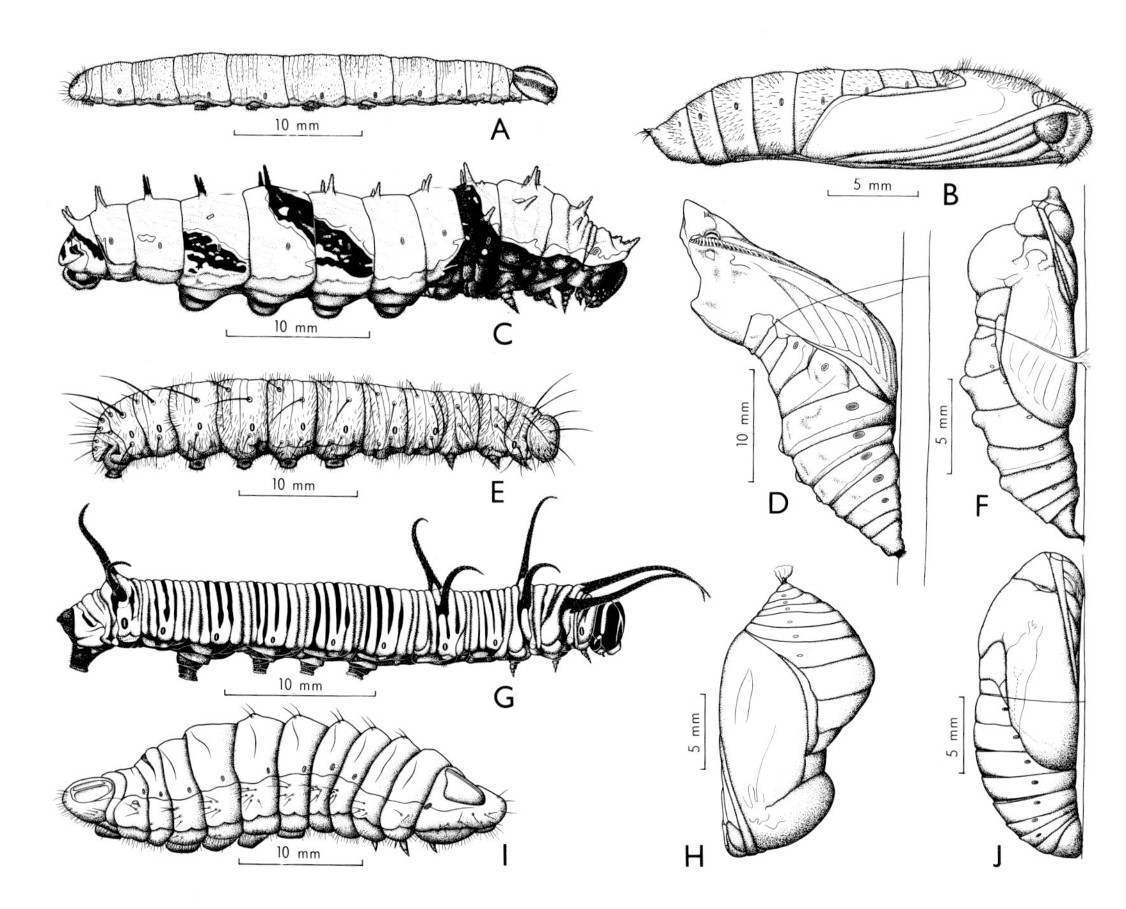

Fig. 36.41. Larvae and pupae of: A, B, *Cephrenes*, Hesperiidae; C, D, *Papilio*, Papilionidae; E, F, *Delias*, Pieridae; G, H, *Euploea*, Nymphalidae; I, J, *Ogyris*, Lycaenidae. [J. Wedgbrow]

protuberances. In *Cressida cressida* (F.) (Fig. 36.42B) the fore wing is largely hyaline, especially in the female, and the flight is slow. The largest and finest species is the sexually dimorphic *Ornithoptera priamus* (L.), with green, black, and gold male and much larger black, white, and yellow female. There are three Australian races, the smallest and dullest of which occurs as far south as the Clarence River.

53. Pieridae. Medium-sized; maxillary palpi absent; labial palpi ascending, appressed to frons; epiphysis absent, fore leg fully developed, tibiae spined, tibial spurs 0–2–2, tarsal claws bifid; fore wing (Fig. 36.40D) usually lacking at least 1 branch of R, 1A + 2A simple; hind wing with humeral vein present, vestigial, or absent, $Sc + R_1$ diverging from Rs at base, 2 anal veins. Egg vertically fusiform, with vertical and horizontal ribs; laid singly or in groups. Larva (Fig. 36.41E) cylindrical, with short fine secondary setae, crochets multiordinal; sometimes gregarious. Pupa (Fig. 36.41F) angular, often with spines or ridges; attached by cremaster and central silken girdle.

This sun-loving family is best represented in the tropics. In Australia there are 2 subfamilies, the Coliadinae (11 spp.), in which the humeral vein in the hind wing is reduced or absent, and the Pierinae (20 spp.), in which the humeral vein is present. The larvae of the former usually feed on Leguminosae, whereas those of the Pierinae feed on Capparidaceae, Loranthaceae, and Cruciferae.

In the COLIADINAE *Catopsilia* (3 spp.) includes tropical, migratory species which also occur in the Oriental region. The larvae feed on *Cassia*. The white *C. pyranthe* (L.) and the lemon-yellow *C. pomona* (F.) occur in two forms, one with black antennae and the other with pink. *Eurema* (8 spp.) contains much smaller, usually sulphur-yellow species, which fly close to the ground. The larvae of *E. smilax* (Don.), found throughout the continent, feed on *Cassia* and *Neptunia*, whereas those of *E. hecabe* (L.) (Fig. 36.42C) are found on *Indigofera* and *Sesbania*, as well as *Breynia* (Euphorbiaceae).

Among the PIERINAE, *Delias* (8 spp.) has bright red and yellow markings beneath the wings. The larvae are dark with fine whitish hairs, and feed on mistletoes. Those of *D. harpalyce* (Don.) are gregarious, and many of the shining black spiny pupae are attached to a communal web spun amongst foliage. *D. aganippe* (Don.) (Fig. 36.42D) occurs in the south-west and the south-east, and *D. nigrina* (F.) is found from Cape York to southern N.S.W. The larvae of *Anaphaeis java* (Sparrman) defoliate *Capparis*, and the adults often take part in great migratory flights. *Elodina* (4 spp.) contains smaller species seldom recorded as far south as Sydney. Their larvae also feed on *Capparis*. The introduced *Pieris rapae* (L.) occurs in the south-east and south-west and in Tasmania; the larva damages crucifers.

54. Nymphalidae. Small to large; maxillary palpi 1-segmented; labial palpi ascending; epiphysis absent, fore legs modified and non-functional in both sexes, tibial spurs 0–2–2; fore wing (Fig. 36.40E) with all branches of R present, 1A + 2A simple or forming a basal fork; hind wing with humeral vein usually present, Sc + R_1 diverging from Rs near base, 2 anal veins. Egg taller than diameter, with vertical and horizontal ribs, or nearly spherical and sometimes nearly smooth. Larva with long paired filaments (Fig. 36.41G), branching spines, or with fine hairs and bifid anal segment or horned head; crochets multiordinal. Pupa (Fig. 36.41H) suspended by cremaster or, rarely, loose on ground beneath debris.

The Nymphalidae are characterized in both sexes by the modified fore legs, often clothed with long hair. Seven subfamilies occur in Australia. The DANAINAE include large, conspicuous forms thought to be distasteful to birds and other predators. They are tenacious of life, and can survive physical injury of an order fatal to other Papilionoidea. Scent glands and expansible hair pencils are present in the males. The larvae have 2 or more long fleshy dorsal filaments, and are usually aposematically marked with bright transverse bands. They usually feed on Apocynaceae, Asclepiadaceae, and Moraceae. The smooth pupae often have brilliantly reflective colours. The large migratory *Danaus plexippus* (L.) was first seen in numbers in Australia about 1870, and now occurs wherever its introduced milkweed host plants, *Asclepias* and *Calotropis*, are established. Its extensive two-way migrations and gregarious hibernation in North America are not a feature of its ecology here, although winter aggregations do occur in N.S.W. and South Australia. *D. chrysippus* (L.) (Fig. 36.42E) is a smaller native species found throughout the continent. *Euploea* (10 spp.) is mainly tropical, but *E. core* (Cram.) extends south to Victoria. Its larva feeds on *Ficus*, *Mandevillea*, and oleander, and the pupa is brilliantly patterned in silver or gold.

The SATYRINAE include endemic genera in the south and a few Oriental genera in the north. The larvae feed on grasses and sedges, and have a bifid anal segment. The adults are usually orange and black, with eye-spots, and the bases of the veins are often swollen. *Melanitis leda* (Drury) is mainly crepuscular in habits; the larvae feed on *Imperata* and even sugar cane. *Hypocysta* (6 spp.) contains small grass-frequenting species. Endemic genera include *Heteronympha* (7 spp.), *Argynnina* (2 spp.), *Oreixenica* (6 spp.), and *Geitoneura* (3 spp.). *H. merope* (F.) is a ubiquitous species in the south-east and south-west. The most remarkable satyrine is

Fig. 36.42. A, *Graphium sarpedon*, B, *Cressida cressida*, Papilionidae; C, *Eurema hecabe*, D, *Delias aganippe*, Pieridae; E, *Danaus chrysippus*, F, *Tisiphone abeona*, G, *Polyura pyrrhus*, H, *Vanessa kershawi*, Nymphalidae; I, *Libythea geoffroyi*, Libytheidae; J, *Liphyra brassolis*, K, *Jalmenus evagoras*, L, *Paralucia aurifer*, Lycaenidae; M, *Oreta jaspidia*, Drepanidae; N, *Ciampa arietaria*, O, *Thalaina selenaea*, P, *Hypobapta eugramma*, Geometridae. Scales = 5 mm. [Photos by C. Lourandos]

Tisiphone abeona (Don.) (Frontispiece; Fig. 36.42F) in which several geographical forms occur between southern Queensland and S.A. (p. 779). Its larvae feed on *Gahnia*.

Polyura pyrrhus (L.) (Fig. 36.42G) is the only Australian species of CHARAXINAE. The adults fly rapidly, and feed on fermenting juices. The larvae have 4 horns on the head, and feed on many trees, including *Acacia*. The larvae of the NYMPHALINAE have branched spines. Those of the common migratory *Vanessa kershawi* (McCoy) (Fig. 36.42H) feed on Compositae, and those of *V. itea* (F.) on nettles. The subfamily contains many fine tropical species, such as the red and biack *Cethosia chrysippe* (F.) (Plate 8, G). *Hypolimnas* (3 spp.) is also northern. The female of *H. misippus* (L.) resembles *Danaus chrysippus* (L.), a species thought to be distasteful to birds; it may thereby derive some protection. The ACRAEINAE, richly developed in Africa, have only one Australian species, *Acraea andromacha* (F.), with larvae feeding on *Passiflora*.

55. Libytheidae. Medium-sized; maxillary palpi vestigial; labial palpi long, porrect, beak-like; epiphysis absent, fore leg reduced in male, non-functional, normal in female; tibial spurs 0–2–2; fore wing produced and truncate apically, 1A + 2A forming basal fork; hind wing with humeral vein, $Sc + R_1$ diverging from Rs near base, 2 anal veins. Egg fusiform, with vertical and horizontal ribs. Larva resembling a pierid, cylindrical, covered with minute secondary hairs, head small. Pupa smooth, suspended by cremaster.

Closely related to the Nymphalidae, the family occurs sparsely in most regions. *Libythea geoffroyi* Godt. (Fig. 36.42I), found in the far north, is the only Australian species. The larvae are not known from Australia.

56. Lycaenidae. Small to large; eyes often emarginate at base of antenna, or contiguous with antennal socket, often hairy; maxillary palpi absent; labial palpi ascending; epiphysis absent, fore leg in male somewhat reduced, normal in female, tibial spurs 0–2–2; fore wing (Fig. 36.40F) often lacking 1 or 2 branches of R, 1A + 2A sometimes forming basal fork; hind wing usually without humeral vein, $Sc + R_1$ diverging from Rs near base, 2 anal veins, margin often produced into 1 or more tails; gnathos of male a pair of slender curved hooks. Egg with diameter usually greater than height, or nearly spherical, often densely pitted or with projections. Larva (Fig. 36.41I) usually onisciform (i.e. shaped like a wood-louse) with retractile head, sometimes with dense short hairs, abdomen often with medial dorsal gland on segment 7 and a pair of dorsolateral eversible organs on 8 (Hinton, 1951), prolegs with median fleshy lobe, more or less interrupting the row of crochets. Pupa (Fig. 36.41J) usually attached at anal end and by central silken girdle, sometimes lying loose, cremaster absent.

The family is predominantly tropical. The LIPHYRINAE are represented by the large, crepuscular, orange and black *Liphyra brassolis* Westw. (Fig. 36.42J) from the north. The larva is found in the arboreal nests of *Oecophylla*, where it feeds on the ant larvae and pupae. It has a hard integument which can resist any attack from the ants and within which it pupates.

The LYCAENINAE are the dominant Australian subfamily. The fore wing lacks 1 or 2 branches of R, and the hind wing may bear 1 or 2 slender tails. The wings are usually some shade of blue, but are sometimes orange. The larvae are usually attended by ants, their dorsal glands secreting a substance sought by a single or at most a few species of ants. Several tropical species, such as *Hypolycaena phorbas* (F.) are attended by *Oecophylla*. The larvae of *Jalmenus* (7 spp.) feed openly during the day on foliage of *Acacia*. Those of the pale blue *J. evagoras* (Don.) (Fig. 36.42K) are each attended by many small black *Iridomyrmex*; they pupate gregariously on a communal web spun amongst the twigs. *Ogyris* (13 spp.) contains brilliant blue or purple species with larvae feeding on mistletoes. The nocturnal larvae hide during the day under bark, in holes or crevices, or in the attendant ants' nests. The larvae and pupae of *O. genoveva* Hewitson (Plate 8, E) occur in *Camponotus* nests at the foot of a eucalypt infested with mistletoe.

Hypochrysops (15 spp.) have shining purple, blue, or orange wings, with metallic green and red markings beneath. The larvae of the purple *H. ignita* (Leach) feed on *Acacia* and several other trees, while those of the orange *H. apelles* (F.) (Plate 8, C) feed on mangroves. *Paralucia aurifer* (Blanchard) (Fig. 36.42L) has larvae which feed at night on *Bursaria spinosa*, hiding during the day in ants' nests at the base of the food plant. The small *Neolucia* (5 spp.) fly close to the ground near their food plants. The larva of *N. agricola* (Westw.), brownish with chequered wing margins, feeds in spring on flowers of Leguminosae. The ubiquitous *Zizeeria otis labradus* (Godt.) also feeds on legumes, including lucerne and French beans.

Superfamily GEOMETROIDEA

Ocelli usually absent; chaetosemata usually present; maxillary palpi minute or vestigial; haustellum naked; epiphysis usually present, tibial spurs sometimes reduced; wings usually broad; fore wing with M_2 rarely arising nearer to M_3 than to M_1, CuP absent, 1A + 2A often forming a basal fork; hind wing with frenulum usually present, CuP absent; abdominal tympanal organs present (except Sematuridae). Egg of flat type. Larvae rarely with secondary setae, crochets biordinal, in a mesoseries, rarely in uniordinal circle on anterior prolegs, ventral or anal prolegs often reduced or absent.

Except for the Drepanidae, the Australian species possess chaetosemata. M_2 of the fore wing nearly always arises either nearer to M_1 or equidistant from M_1 and M_3. Many species which rest with wings outspread have the fore-wing pattern continued on the hind wing. In some the wings are held back to back above the body, as in the Papilionoidea. The less specialized species often have a reduced hind-wing pattern, and the wings are folded roofwise above the body. The superfamily is noted for procryptic wing patterns, and for remarkable larval adaptations which aid concealment.

Key to the Families of Geometroidea Known in Australia

1. Apterous or brachypterous (♀♀ only) .... **Geometridae** (pt., p. 847)
   Fully winged 2
2. Hind wing with $Sc+R_1$ approximated to, or fusing with, Rs for a short distance beyond discal cell .... **Drepanidae** (p. 846)
   Hind wing with $Sc+R_1$ remote from Rs beyond discal cell 3
3. Hind wing with $Sc+R_1$ approximated to, or partly fusing with, Rs, or Sc connected to Rs by R_1, before end of discal cell; fore wing with R_5 not remote from R_4 and usually stalked with it, or R_5 arising from areole .... **Geometridae** (pt., p. 847)
   Hind wing with $Sc+R_1$ diverging from Rs at base; fore wing with R_5 remote from R_4, R_5 stalked with M_1 or approximated to M_1 at base, areole absent 4
4. Hind wing with 2 anal veins, frenulum well developed, humeral angle not greatly expanded, humeral vein vestigial or absent .... **Epiplemidae** (p. 851)
   Hind wing with 1 anal vein, frenulum reduced or absent, humeral angle greatly expanded, humeral vein present .... **Uraniidae** (p. 849)

57. Drepanidae. Small to medium-sized; ocelli absent; chaetosemata reduced or absent; antennae pectinate, lamellate, or simple; haustellum vestigial or absent; maxillary palpi vestigial; labial palpi small, slender, ascending; epiphysis usually present, tibial spurs 0–2–4, 0–0–4, 0–2–2; wings usually broad; fore wing (Fig. 36.43A) usually produced apically, areole sometimes present, M_2 arising nearer to M_3 than to M_1, 1A + 2A usually forked at base; hind wing with frenulum present or absent, $Sc + R_1$ approximated to, or shortly fusing with, Rs beyond discal cell, M_2 approximated to M_3 at base, 1 anal vein; basal segment of abdomen with dorsolateral tympanal organs opening posteriorly. Larva slender, without anal prolegs, posterior extremity produced

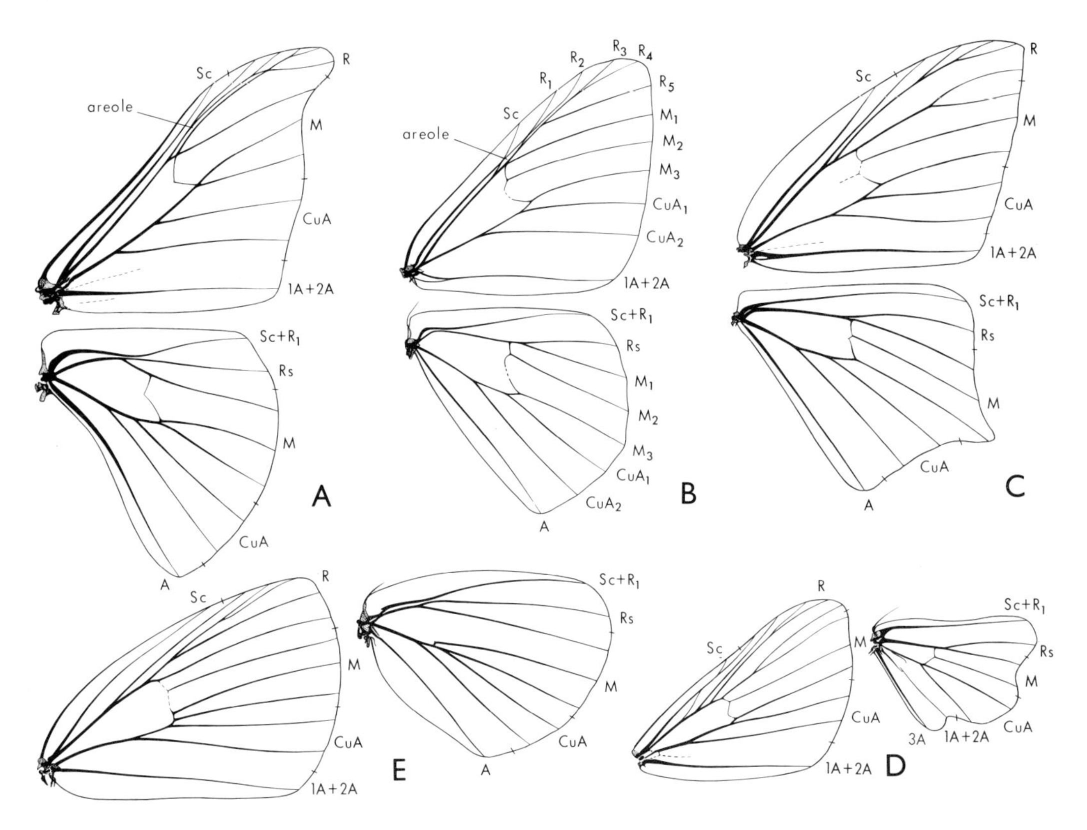

Fig. 36.43. Wing venation of Geometroidea and Calliduloidea: A, *Oreta*, Drepanidae; B, *Chlorocoma*, Geometridae; C, *Aploschema*, Uraniidae; D, *Lobogethes*, Epiplemidae; E, *Cleis*, Callidulidae.

[J. Wedgbrow]

and raised in repose. Pupa in cocoon amongst fallen leaves.

This mainly Old-World tropical family (A. Watson, 1965; Turner, 1926) is represented in eastern Australia by the DREPANINAE and ORETINAE. Their life history here is not known. *Oreta jaspidea* (Warren) (Fig. 36.42M) occurs from Cape York to Mackay.

58. Geometridae. Small to large; ocelli usually absent; chaetosemata present; antennae simple or pectinate; haustellum present; maxillary palpi 1- or rarely 2-segmented; labial palpi rarely long, usually ascending; epiphysis present, tibial spurs usually 0–2–4; wings (Fig. 36.43B) broad, sometimes reduced in females, often with 1 or 2 areoles; fore wing with R_4 and R_5 stalked, M_2 usually not arising nearer to M_3 than to M_1, 1A + 2A forming basal fork; hind wing usually with frenulum, costa and Sc usually strongly bent near base, sometimes with humeral vein from angle of Sc to base of frenulum, Sc approximated to, or fusing with, Rs near base, or joined to Rs by R_1, and then divergent, M_2 not arising nearer to M_3 than to M_1 or absent, anal area narrow, 1 or 2 anal veins; basal segment of abdomen with tympanal organs in pair of deep ventrolateral cavities opening anteriorly. Larva (Figs. 36.44A–C) with anterior 2 or 3 pairs of ventral prolegs usually reduced or absent, when not reduced sometimes bearing crochets in a uniordinal circle; often twig-like, or occasionally in loose shelter of leaves, feeding on foliage. Pupa (Fig. 36.44D) with cremaster well developed, setae often hooked; usually in flimsy cocoon in debris or in soil.

The abdominal tympanal organs, the chaetosemata, the wing shape, and the venation will distinguish most genera. In some of the more primitive, however, neither the base of Sc nor the costa of the hind wing is strongly bent basally. The reduced prolegs result in the larvae progressing with a looping motion. Many provide striking examples of protective resemblance to plant parts.

Six subfamilies are recognized (Forbes, 1948). Australia is rich in primitive forms, but the Archiearinae (Brephinae *auct.*) are not recorded. The Oenochrominae are very numerous, especially in the south. In sub-alpine areas the Larentiinae are abundant, whereas in the tropics the Sterrhinae and the green Hemitheinae are abundant.

The OENOCHROMINAE include 2 large groups (Turner, 1929–30), the primitive *Oenochroma* group with stout bodies, and the *Taxeotis–Nearcha* group with slender bodies. A few genera fit into neither group. The larvae of *Oenochroma* (26 spp.) have 2 pairs of ventral prolegs and, where known, feed on Proteaceae. *O. vinaria* Guen. (Fig. 36.44B) feed on *Grevillea*, *Banksia*, and *Hakea*, and *O. subustaria* (Walk.) and *O. phyllomorpha* (Low.) on *Persoonia*. Adults of *Circopetes* (1 sp.) and *Onychopsis* (1 sp.) rest with wings and abdomens twisted, and resemble withered leaves. The larvae of *O. lutosaria* (Feld.), which feed on *Eucalyptus*, are slender with a pair of long forked dorsal processes on the prothorax. *Lissomma* (29 spp.) includes species with strongly sclerotized frontal projections in the adults.

The ENNOMINAE (Boarmiinae) have lost M_2 as a tubular vein in the hind wing, and in the fore wing an areole is seldom present and R_3, R_4, and R_5 are stalked. *Mnesampela* (5 spp.) has much in common with the Archiearinae. The larvae (Fig. 36.44A) have 4 pairs of ventral prolegs, those on segments 3–5 in the first instar, and on 3 in the last instar, bearing uniordinal crochets in a circle. In *M. privata* (Guen.), which is a pest of seedling blue gums *(Eucalyptus)*, and *M. lenaea* Meyr. two or three larvae are often found together in a shelter formed by joining two or more leaves of the food plant with silk. In *M. fucata* (Feld.) the larva feeds exposed on *Eucalyptus* leaves. Other Ennominae are also of economic importance. *Ciampa arietaria* Guen. (Fig. 36.42N) is a minor pest of pastures in the south-west, where the larvae feed on *Erodium* and other dicotyledons, and *Chlenias* spp. attack *Pinus radiata* in Victoria and S.A. *Thalaina* (7 spp.) includes satin-white species, such as *T. selenaea* (Doubl.) (Fig. 36.42O), the larvae of which feed on *Acacia*. Many genera have procryptic wing patterns of wavy transverse lines, and rest on bark. In the males an oval, blister-like *fovea* is often present near the base of the fore wing. The larvae have only one pair of ventral prolegs; those of *Boarmia luxaria* (Guen.) feed on *Leptospermum*.

The HEMITHEINAE contain 2 major groups, a more primitive one with thick bodies and wings bearing a grey or lichen-like pattern, and a more specialized group of slender-bodied species, usually with green wings. In the fore wing R_1 is free from R_2, usually arising separately from the discal cell and fusing for a short distance with Sc. *Hypobapta eugramma* (Low.) (Fig. 36.42P) lacks medial spurs of the hind tibiae, and the stick-like larvae with pointed heads feed on *Eucalyptus*. *Chlorocoma assimilis* (Lucas) and *Eucyclodes pieroides* (Walk.) (Plate 8, L) are fine green species. The larvae of both feed on *Acacia*, those of *C. assimilis* (Fig. 36.44C) being slender, green, and twig-like, whereas in *E. pieroides* the body segments are broadly flanged, resembling the leaflets of *Acacia*. The latter also feed on avocado, cherry, guava, and rose.

The STERRHINAE are small and slender-bodied, usually pale in colour, with delicate wavy pattern. One or two areoles are present in the fore wing, situated before the end of the discal cell, and Sc is free from R; in the hind wing Sc + R_1 fuses with Rs only near the base. Little is known of their larvae in Australia, although some species, such as *Scopula rubraria* (Doubl.), are common.

The LARENTIINAE have a distinct wing pattern of wavy lines. As in the Sterrhinae, one or two areoles are present in the fore wing, and Sc is free from R. However, in the hind wing Sc + R_1 usually fuses with Rs for

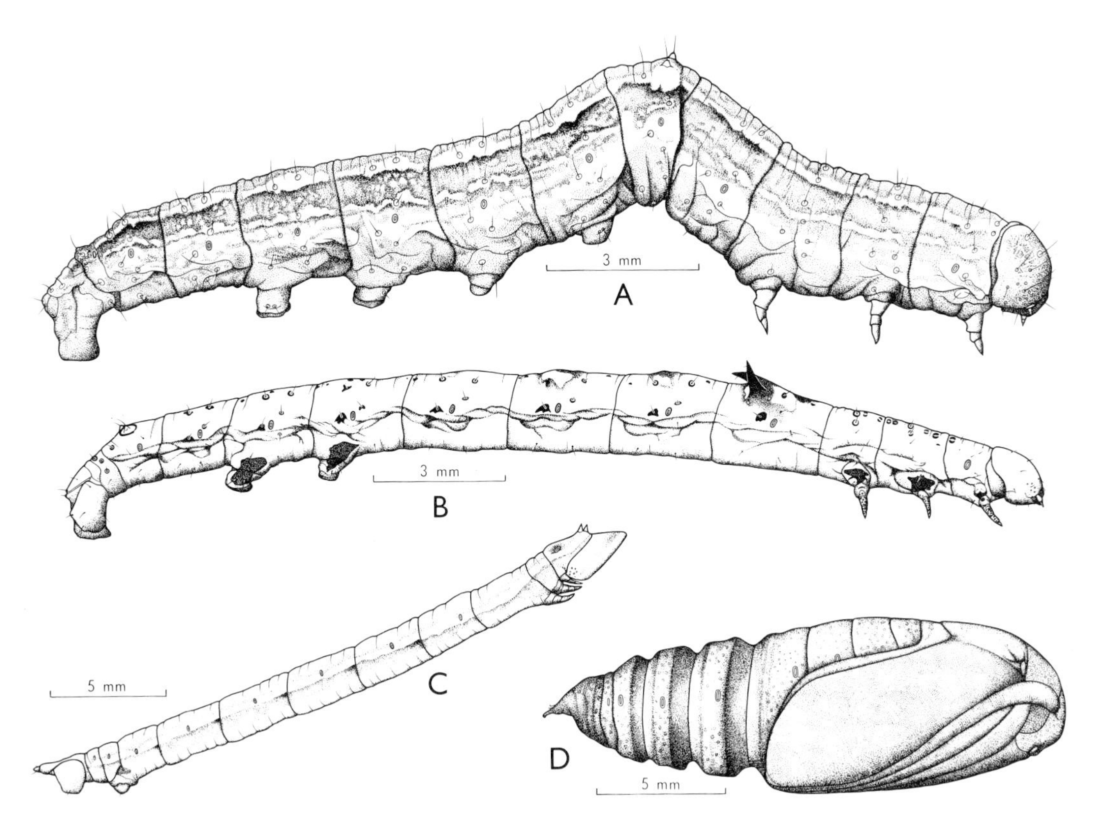

Fig. 36.44. Larvae of Geometridae: A, *Mnesampela*; B, *Oenochroma*; C, *Chlorocoma*. D, pupa of *Melanodes*, Geometridae. [J. Wedgbrow]

at least half the length of the discal cell. Occasionally, as in *Sauris* (13 spp.), the hind wing of the male is modified. The larvae of *Chloroclystis approximata* (Walk.) (Fig. 36.45A) and *C. laticostata* (Walk)., which normally feed on the buds and flowers of *Acacia*, are minor pests of apple. *Poecilasthena pulchraria* (Doubl.) is one of several pale green species with translucent wings; the larva feeds on *Monotoca*.

59. Uraniidae. Small to large; ocelli absent; chaetosemata large; antennae thickened in male, sometimes shortly dentate; haustellum present; maxillary palpi 1-segmented; labial palpi ascending; tibial spurs 0–2–3 or 0–2–4; fore wing (Fig. 36.43C) without areole, R_5 widely separate from R_4, R_5 and M_1 stalked or approximated at base, M_2 never arising nearer to M_3 than to M_1, 1A + 2A forming basal fork; hind wing with frenulum vestigial or absent, humeral angle greatly expanded, usually with weak humeral vein, Sc + R_1 diverging from Rs at base, 1 anal vein, margin of wing produced, at least at apex of M_3; abdomen in male with lateral tympanal organs on tergum 2 opening posteriorly, in female ventral at base of abdomen. Larva with prolegs normal. Pupa in loose silken cocoon.

Mainly tropical, the family includes the URANIINAE (2 spp.) and MICRONIINAE (7 spp.). The former contains large, tailed species which resemble papilionid butterflies. The frenulum is absent. The diurnal *Alcides zodiaca* (Butl.) (Plate 8, I) occurs in rain forest in north Queensland and New Guinea. The moths feed at flowers, rest with wings outspread, and sometimes, towards evening,

Fig. 36.45. A, *Chloroclystis approximata*, Geometridae; B, *Acropteris nanula*, Uraniidae; C, *Lobogethes interrupta*, Epiplemidae; D, *Entometa sobria*, Lasiocampidae; E, *Chelepteryx collesi*, F, *Chenaula heliaspis*, Anthelidae; G, *Antheraea helena*, Saturniidae; H, *Anthela excellens*, Anthelidae; I, *Eupterote expansa*, Eupterotidae; J, *Carthaea saturnioides*, Carthaeidae; K, *Pinara cana*, Lasiocampidae; L, *Munychryia senicula*, M, *Pterolocera amplicornis*, Anthelidae; N, *Panacela lewinae*, Eupterotidae. Scales = 5 mm.

[Photos by C. Lourandos]

fly to and fro high above the forest. The nocturnal *Nyctalemon patroclus* (L.), with a much longer tail, is found in New Guinea and at Cape York. The Microniinae are delicate, nocturnal, white species with pencilled markings and hind wing produced to an acute angle at the end of M_3; a vestigial frenulum is sometimes present. They rest with wings outspread, and have a slow, laboured flight. In the fore wing of *Micronia* (1 sp.) R_5 and M_1 are separate, and CuA_1 and CuA_2 are stalked. In other Australian genera R_5 and M_1 are stalked in the fore wing. CuA_1 and CuA_2 are also stalked in *Acropteris* (3 spp.), but separate in *Aploschema* (1 sp.) and *Urapteroides* (1 sp.). *Aploschema discata* (Warren) and *Acropteris nanula* (Warren) (Fig. 36.45B) are from southern Queensland and N.S.W.

60. Epiplemidae. Small; ocelli absent; chaetosemata large; antennae dentate in male; haustellum present; maxillary palpi 1-segmented; labial palpi short, subascending; tibial spurs 0–2–4; fore wing (Fig. 36.43D) without areole, R_5 widely separate from R_4, R_5 and M_1 stalked or approximated at base, M_2 not arising nearer to M_3 than to M_1, 1A + 2A simple; hind wing with frenulum, humeral angle not greatly expanded, humeral vein vestigial or absent, $Sc + R_1$ diverging from Rs at base, 2 anal veins, wing margin irregular; abdominal tympanal organs as in Uraniidae. Larva with normal prolegs, first and last spiracles twice as large as others; early stages in a communal web, later feeding exposed. Pupa in cocoon.

Although closely related to the Uraniidae, the species are much smaller and darker, and rest with the fore and hind wings separated. The hind wing is usually more or less folded along the abdomen, whereas the fore wing is either rolled, curled, or extended flat. Little is known of the early stages, but *Lobogethes interrupta* Warren (Fig. 36.45C) has been reared from *Canthium oleifolium*. The family is mainly tropical in distribution.

Superfamily CALLIDULOIDEA

61. Callidulidae. Medium-sized; ocelli absent; chaetosemata large; antennae simple, increasing slightly in diameter towards apex; haustellum present, naked; maxillary palpi vestigial; labial palpi strong, ascending; epiphysis present, tibial spurs 0–2–4; wings (Fig. 36.43E) broad; fore wing without retinaculum, R_2, R_3, and R_4 stalked, without areole, discal cell short, M_2 arising nearer to M_3 than to M_1, CuP absent, 1A + 2A simple; hind wing with small frenulum, broad humeral area, $Sc + R_1$ bent near base, separate from Rs but approaching it beyond discal cell, discal cell short, open, M_2 arising nearer to M_3 than to M_1, CuP absent, 1 anal vein; tympanal organs absent. Larva and pupa unknown.

The adults are diurnal, and usually dark brownish black in colour with orange markings. In flight, and at rest with wings held back to back, they resemble small butterflies. The family is mainly Oriental and Papuan, but *Cleis scotti* (Macl.) was based on a single specimen said to be from Cape York. Its identity requires confirmation.

Superfamily BOMBYCOIDEA

Medium-sized to large; ocelli and chaetosemata absent; antennae bipectinate, at least in male; maxillary palpi small or absent; wings broad; body stout, clothed with long hair-scales; tympanal organs absent. Egg of flat type, smooth. Larva with dense secondary setae, usually with verrucae, or with sparse setae, or with *scoli* (fleshy processes bearing secondary setae), often with dorsal projection on abdominal segment 8, crochets usually biordinal in a mesoseries; feeding exposed in daytime. Pupa stout, usually in cocoon of silk sometimes mixed with larval hairs.

Broad wings and usually strongly pectinate antennae are features of the superfamily, which also displays a progressive reduction or loss of many characters found elsewhere. Ocelli, chaetosemata, and tympanal organs are never present. The frenulum and retinaculum have been lost in Lasiocampidae, in female Anthelidae, and in Saturniidae, but the humeral area of the hind wing is often expanded, permitting amplexiform wing-coupling. The haustellum is rudimentary or lost in all but a few of the least specialized Anthclidac, Eupterotidae, and Saturniidae, and in the Carthaeidae. The larvae of Anthelidae, Eupterotidae, and many Lasiocampidae have long, dense, secondary setae, unlike Bombycidae, Carthaeidae, and Saturniidae. Most species spin strong silken cocoons, and the silk spun by the commercial silk worm, *Bombyx mori* L., and some Saturniidae has long been exploited by man.

Key to the Families of Bombycoidea Known in Australia

1. Wings vestigial (♀♀ only) .... **Anthelidae** (pt., p. 853)
   Fully winged 2
2. M_2 arising nearer to M_3 than to M_1 in both wings 3
   M_2 not arising nearer to M_3 than to M_1 in both wings 4
3. Frenulum and retinaculum present in ♂, absent in ♀; fore wing with 1 or 2 areoles; Sc in hind wing without short costal branches .... **Anthelidae** (pt., p. 853)
   Frenulum and retinaculum absent in both sexes; fore wing without areole; Sc in hind wing with 1 or more short costal branches .... **Lasiocampidae** (p. 852)
4. Hind wing with 1 anal vein, humeral angle not thickened, frenulum absent; R in fore wing with 3 branches .... **Saturniidae** (p. 855)
   Hind wing with 2 anal veins, humeral angle with thickened frenulum base, frenulum often present; R in fore wing with 4 or 5 branches 5
5. Hind wing with Sc not connected to discal cell by R_1. Mature larva with conspicuous dense secondary setae .... **Eupterotidae** (p. 854)
   Hind wing with Sc connected to discal cell by R_1. Mature larva smooth, with inconspicuous secondary setae 6
6. Haustellum strong; labial palpi prominent; frenulum well developed in both sexes, retinaculum present in ♂ .... **Carthaeidae** (p. 854)
   Haustellum absent; labial palpi very small; frenulum rudimentary, retinaculum absent .... (Bombycidae)

62. Lasiocampidae. Antennae bipectinate in both sexes; haustellum and maxillary palpi absent; labial palpi porrect, often beak-like, with chaetosema-like sense organ; epiphysis present in male, reduced or absent in female, tibial spurs very short, 0–2–2; fore wing (Fig. 36.46A) without retinaculum, R_2 and R_3 stalked, without areole; hind wing without frenulum, humeral angle expanded, Sc fused with Rs near base, or connected to Rs by R_1, Sc with 1 or more short branches to costa, CuP absent, 2 anal veins. Larva sometimes with dorsal protuberances, usually with dense secondary setae, sometimes in verrucae, pencils, or tufts, setae never branched, sometimes scale-like, crochets biordinal, simple. Pupa often hairy, epicranial suture present, sometimes with hooked anal setae; in stiff silken cocoon.

The family is world-wide in distribution, but absent from New Zealand. The Australian genera (Turner, 1924) are mainly endemic. The moths are stout-bodied, hairy, and sexually dimorphic, with fast-flying males and larger sluggish females. The larvae lie along twigs, to which they appress their dense lateral hairs, effectively disguising their presence. The white, parchment-like cocoons are often spun amongst the foliage of the food plant.

Perna (3 spp.) has short labial palpi, hairy eyes, and a broad 'sub-costal' cell in the hind wing, formed by R_1 joining Sc and Rs beyond the end of the discal cell. The larva of *P. exposita* (Lew.) feeds on *Casuarina*, spinning its white cocoon between branchlets. *Entometa* (8 spp.) contains the largest species, with smooth eyes, long and beak-like labial palpi, and Sc in the hind wing fused with Rs near its base. The dark larvae usually have a

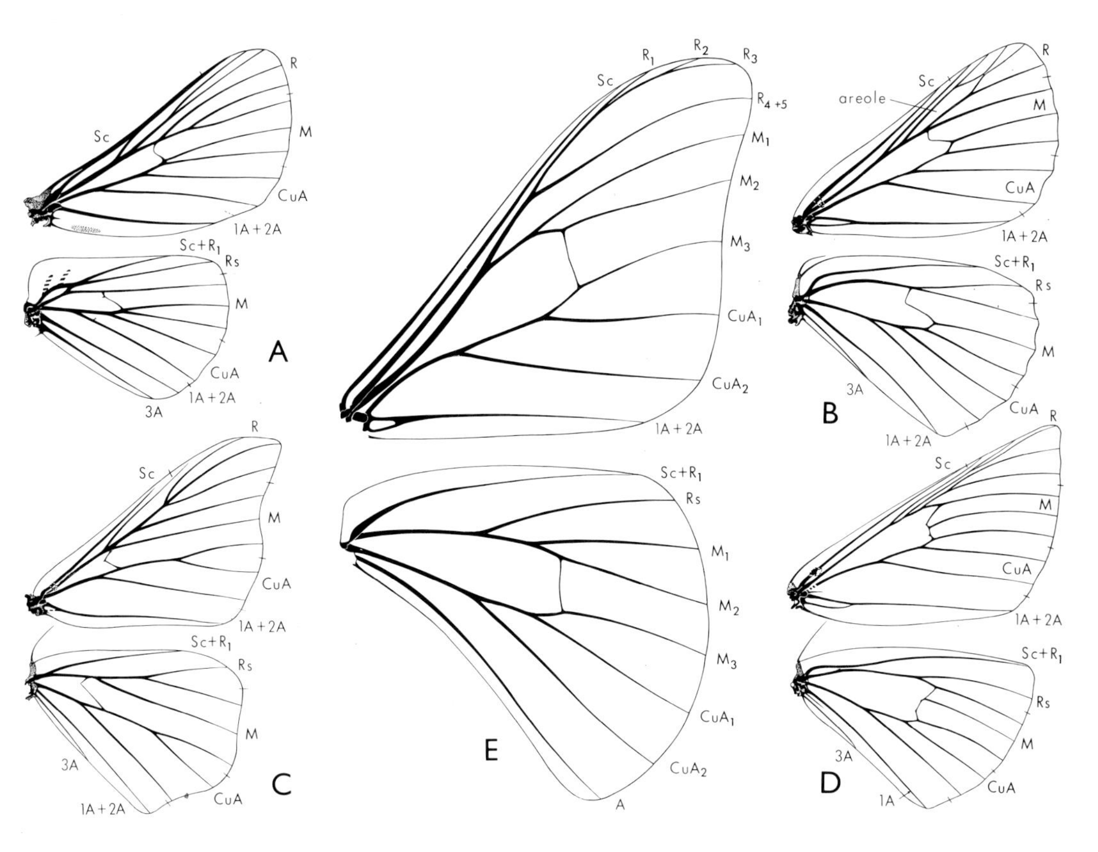

Fig. 36.46. Wing venation of Bombycoidea: A, *Porela*, Lasiocampidae; B, *Anthela*, Anthelidae; C, *Panacela*, Eupterotidae; D, *Carthaea*, Carthaeidae; E, *Antheraea*, Saturniidae. [J. Wedgbrow]

prominent projection on abdominal segment 8 and a pair of erectile dorsal protuberances on the metathorax; they feed on *Eucalyptus*. *E. sobria* (Walk.) (Fig. 36.45D) is from the east and south-east. In *Crexa* (8 spp.) the males have largely hyaline wings, whereas in *Pinara* (4 spp.) they are dark brown with an orange patch on the hind wing. The larvae of *C. acedesta* Turn. feed on mistletoe and *Exocarpos*, and those of *P. cana* Walk. (Fig. 36.45K) on *Eucalyptus*.

63. Anthelidae. Antennae bipectinate to apex in male, usually bipectinate or pectinate in female; haustellum rarely present; maxillary palpi vestigial; labial palpi porrect; epiphysis present in male, reduced or absent in female, tibial spurs short, 0–2–4 or 0–2–2; females rarely with vestigial wings; fore wing (Fig. 36.46B) with retinaculum in male, 1 or 2 areoles usually present, M_2 arising nearer to M_3 than to M_1, CuP absent, 1A + 2A forming basal fork; hind wing with functional frenulum in male, in female with only thickened frenulum base, Sc + R_1 separate from Rs, or Sc connected to Rs by R_1, M_2 as in fore wing, CuP absent, 2 anal veins. Larva with verrucae and dense, branched or roughened setae, verrucae of abdominal segment 1 arranged differently from those of 2–8, crochets biordinal or multiordinal, simple. Pupa in double-walled cocoon of silk often mixed with larval hairs.

The family (Turner, 1921) is a conspicuous element of the Australian fauna, but elsewhere is known only from New Guinea. Turner originally separated it from the Lymantriidae by the structure of the areole, but the absence of tympanal organs at once distinguishes it from the Noctuoidea. Most genera have lost the haustellum and pilifer bristles, but they persist in *Munychryia* (1 sp.) in which median spurs are also present on the hind tibiae. *M. senicula* Walk. (Fig. 36.45L) occurs widely. The hind tibiae also have median spurs in *Chelepteryx* (2 spp.), which includes the large *C. collesi* Gray (Fig. 36.45E), females of which reach 18 cm in wing-span. The *Eucalyptus*-feeding larvae are covered with tufts of stout setae, which are thrust through the silken fusiform cocoon during its construction. If the larvae or cocoons are handled, these bristles readily enter the skin and cause mechanical irritation.

Anthela (51 spp.) lacks median spurs on the hind tibiae. Some of the species are extremely variable in colour and pattern, especially in the male. Thus in *A. nicothoe* (Boisd.) the males may be brown, yellow, or red, and in *A. excellens* (Walk.) (Fig. 36.45H) they vary from yellowish green through yellow to bright orange-red. A heavily blotched form in this and other species is fairly common. *Acacia dealbata* is the normal food plant of *A. nicothoe*, but at times both it and *A. excellens* damage *Pinus radiata*.

In *Chenaula* (1 sp.) the frenulum is reduced in the male, and the tornus of the hind wing is produced. *C. heliaspis* (Meyr.) (Fig. 36.45F) is sexually dimorphic, with reddish brown male and pale grey female. The larva feeds on *Eucalyptus*, but also attacks *Pinus radiata*. In *Pterolocera* (5 spp.) the male antennae have very long pectinations, whereas the female antennae are simple. It is the only genus in which females have vestigial wings; the strong legs have rudimentary tibial spurs. The larva of *P. amplicornis* Walk. (Fig. 36.45M) feeds on grasses, and spins a flask-shaped cocoon in a vertical shaft in the soil, with the emergence exit just below the surface.

64. Eupterotidae. Antennae bipectinate to apex in both sexes; haustellum very weak or absent; maxillary palpi absent; labial palpi short; epiphysis sometimes present, tibial spurs 0–2–2; fore wing (Fig. 36.46C) usually with retinaculum, areole absent, M_2 arising nearer to M_1 than to M_3, CuP absent; hind wing with frenulum usually present, functional, Sc and Rs well separated, usually connected by R_1, M_2 as in fore wing, CuP usually absent, 2 anal veins. Larva with dense secondary setae, often branched, dorsal verrucae of abdominal segment 1 similar to those of 2–8, crochets biordinal, shorter series with subapical spur or dentate. Pupa in flimsy cocoon of silk mixed with larval hairs.

This chiefly Old-World family (Forbes, 1955) includes 4 genera in Australia. In the Indo-Malayan *Eupterote* (1 sp.), and in *Cotana* (3 spp.) which also occurs in New Guinea, a large epiphysis is present in the male, but absent in the female. *Eupterote* lacks the frenulum and retinaculum found in *Cotana*. *E. expansa* (Lucas) (Fig. 36.45I) is a large anthelid-like species from north Queensland. *Panacela* (4 spp.) and *Gastridiota* (1 sp.) are endemic genera with frenulum and retinaculum in both sexes, but without epiphysis; the tornus of the hind wing is slightly produced. The larvae of *P. lewinae* (Lew.) (Fig. 36.45N) live gregariously in communal shelters of silk, spun in the branches of *Exocarpos*, *Eucalyptus*, and other trees, and sometimes cause extensive defoliation. They occasionally damage *Pinus radiata*. The larval hairs are capable of inflicting skin rashes in humans.

65. Carthaeidae. Antennae bipectinate to apex in male, dentate in female, flagellum scaled; haustellum strong; maxillary palpi 3-segmented; labial palpi strong, ascending; epiphysis present; tibial spurs 0–2–4, apices bare; fore wing (Fig. 36.46D) with retinaculum in male, R_2 and R_3 stalked, M_2 arising about equidistant between M_1 and M_3, CuP absent; hind wing with frenulum strong in male, about 10 short bristles in female, Sc approaching Rs before one-half of discal cell where R_1 joins Rs and Sc, M_2 as in fore wing, CuP absent, 1A and 3A present, 2A vestigial. Larva in early instars with scoli and dorsal horn on abdominal segment 8, mature larva (Fig. 36.49A) with minute setae and with slight hump on segment 8, crochets biordinal, simple. Pupa with cremaster bearing a group of hooked setae; in flimsy cocoon on ground.

The single fine species, *Carthaea saturnioides* Walk. (Fig. 36.45J) from south-western Australia (Common, 1966), is grey with a rose-flushed hind wing. When disturbed suddenly, the insect displays its eye-spots by depressing the head, protracting the fore wings, and moving the hind wings rhythmically. The handsome orange and brown larva, with lateral eye-spots, feeds exposed in the early summer on the young foliage of *Dryandra cirsioides* and *D. lewardiana* (Proteaceae).

66. Saturniidae. Antennae short, naked, bipectinate to apex in both sexes, each segment usually with 4 pectinations; haustellum usually absent; maxillary palpi vestigial; labial palpi small; epiphysis present, tibial spurs 0–2–2 or absent, short, apices bare; wings very broad; fore wing (Fig. 36.46E) without retinaculum, areole absent, R usually with 3, never more than 4 branches, M_2 arising nearer to M_1 than to M_3, CuP absent; hind wing without frenulum or thickened frenulum base, humeral angle expanded, Sc + R_1 diverging from Rs, M_2 as in fore wing, CuP absent, 1 anal vein. Larva (Fig. 36.49B) with scoli, at least in early instars, abdominal segment 8 with dorsal projection, crochets uniordinal, simple. Pupa with cremaster simple, when present; in tough silken cocoon.

The family (Michener, 1952) includes some of the largest and finest moths, with stout hairy bodies, small in proportion to the broad wings, and prominent eye-spots. The 3 Australian genera (Turner, 1922) also occur in the Oriental or Papuan areas. *Attacus* (1 sp.) and *Coscinocera* (1 sp.) are without tibial spurs, and the discal cell is open in both wings. *A. dohertyi* (Rothsch.) occurs at Darwin and in Indonesia. The tornal area of the hind wing in *C. hercules* (Miskin) (Plate 8, F), a striking species from north Queensland and New Guinea, is produced into a long tail, much broader in the female.

In *Antheraea* (11 spp.) an epiphysis is present in both sexes, and the mid and hind tibiae each has an apical pair of short stout spurs. The first axillary sclerite of the fore wing is armed with a strong distal hook, used by the freshly emerged moth to open the cocoon (p. 778). *A. eucalypti* Scott is from eastern Australia, and *A. helena* (White) (Fig. 36.45G) from both the south-east and south-west. The mature larva of *A. eucalypti* (Fig. 36.49B) is bluish green, with a pale lateral stripe and red and blue scoli; that of *A. helena* is green with a broad pink lateral stripe, and lacks scoli. Both feed on the foliage of *Eucalyptus*, and attach their oval grey cocoons to the bark. The larvae of the larger *A. loranthi* Luc. feed gregariously on mistletoe growing on *Eucalyptus*; the cocoons are spun in a mass on the butt of the mistletoe.

Superfamily SPHINGOIDEA

67. Sphingidae. Large; ocelli and chaetosemata absent; antennae usually thickened, sometimes clavate or hooked apically, ciliate, serrate, or shortly pectinate in male, simple in female; haustellum usually strong, often long; maxillary palpi 1-segmented; labial palpi thick, ascending, appressed to frons, with unscaled areas on inner surface, basal segment often with patch of sensory hairs on inner surface; epiphysis present, tibial spurs 0–2–4 or rarely 0–2–2; fore wing (Fig. 36.47A) long, narrow, CuP absent, one other vein usually absent, 1A + 2A forked at base, retinaculum rarely absent; hind wing much shorter than fore wing, with anal lobe, frenulum usually strong, Sc connected to Rs by R_1, Sc + R_1 approaching Rs beyond discal cell, CuP absent, 2 anal veins; tympanal organs absent; abdomen large, fusiform, posterior margins of segments usually with flattened spines or stiff modified scales. Egg of flat type. Larva (Fig. 36.49C) without conspicuous setae, abdominal segment 8 with spine-like dorsal horn, sometimes rudimentary in final instar, crochets biordinal, in a meso-series; feeding exposed in daytime. Pupa (Fig. 36.49F) fusiform, cremaster prominent, rarely armed; in cell in soil, or in flimsy cocoon amongst detritus.

This mainly tropical family (Rothschild and Jordan, 1903) includes fast-flying species, often with long haustellum used for ingesting

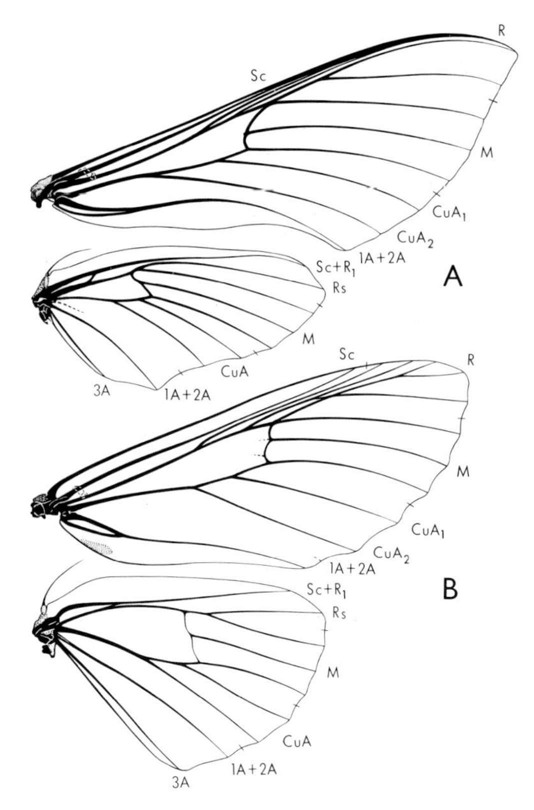

Fig. 36.47. Wing venation of Sphingoidea and Notodontoidea: A, *Hippotion*, Sphingidae; B, *Sorama*, Notodontidae. [J. Wedgbrow]

nectar as the insect hovers before flowers. Of the genera occurring in Australia, 7 are endemic, and the others occur also in New Guinea and the Orient; a few of the species are almost cosmopolitan.

The ACHERONTIINAE and AMBULICINAE lack the area of sensory hairs on the basal segment of the labial palpi. The former includes the large grey *Psilogramma menephron* (Cram.) (Fig. 36.48B) ranging from the eastern Palaearctic to Australia. The larvae often feed on privet *(Ligustrum)* and jasmine, and the haustellum in the pupa is free (Fig. 36.49F). In the cosmopolitan *Herse* (2 spp.), the pupal haustellum is also free, but distally recurved, whereas in the endemic *Coenotes* (1 sp.) it is glued down. *C. eremophilae* (Lucas) occurs widely in the northern inland, where its larvae sometimes defoliate *Eremophila mitchelli.* The Ambulicinae include the endemic *Metamimas* (1 sp.) and *Coequosa* (1 sp.) in which the haustellum is short and stout. The retinaculum is absent, the frenulum is vestigial, and in *Coequosa* the hind tibiae lack medial spurs. The mature larva of *C. triangularis* (Don.), which feeds mainly on *Persoonia*, lacks a posterior horn, but has lateral eye-spots above the enormous anal claspers.

The day-flying *Cephonodes* (4 spp.) is included in the SESIINAE; the wings are largely hyaline, the antennae are clubbed and hooked, and the male genitalia asymmetrical. As they fly in sunshine with an expanded anal tuft of long hair-scales, they resemble large bumble bees. The larvae of *C. kingi* (Macl.) feed on gardenia. The PHILAMPELINAE occur mainly in the north. *Chromis erotus* (Cram.) is found from India to Australia; the larvae, which feed on virginia creeper, have 8 lateral eye-spots. *Cizara ardeniae* (Lew.) (Fig. 36.48E), with rich green fore wings, occurs along the east coast, where the larvae have been recorded on *Embothrium*, *Cissus*, and *Grevillea*. *Macroglossum* (14 spp.) contains many Indo-Malayan and Papuan species. In flight they resemble bumble bees but, unlike *Cephonodes*, the wings are scaled and the abdomen bears lateral scale-tufts. *M. hirundo* Boisd. (Fig. 36.48A), with orange hind wings, is found in Queensland.

The CHOEROCAMPINAE include the widely distributed *Celerio* (1 sp.), *Hippotion* (5 spp.), and *Theretra* (11 spp.). *C. lineata livornicoides* (Lucas) is common inland. The small *H. scrofa* (L.) (Fig. 36.48C) occurs throughout the continent and Tasmania, where the larvae feed on *Ipomaea*. *H. celerio* (L.) ranges with little variation over most of the Old World. The larva, with large eye-spots on abdominal segments 1 and 2, feeds on grape vines.

Superfamily NOTODONTOIDEA

68. Notodontidae. Medium-sized to large; ocelli usually reduced or absent; chaetosemata absent; antennae usually bipectinate in male, pectinate or simple in female;

Fig. 36.48. A, *Macroglossum hirundo*, B, *Psilogramma menephron*, C, *Hippotion scrofa*, Sphingidae; D, *Cerura australis*, Notodontidae; E, *Cizara ardeniae*, Sphingidae; F, *Epicoma melanosticta*, G, *Hylaeora dilucida*, Notodontidae; H, *Iropoca rotundata*, I, *Orgyia athlophora*, J, *Euproctis edwardsi*, Lymantriidae; K, *Nyctemera amica*, L, *Spilosoma canescens*, Arctiidae. Scales = 5 mm. [Photos by C. Lourandos]

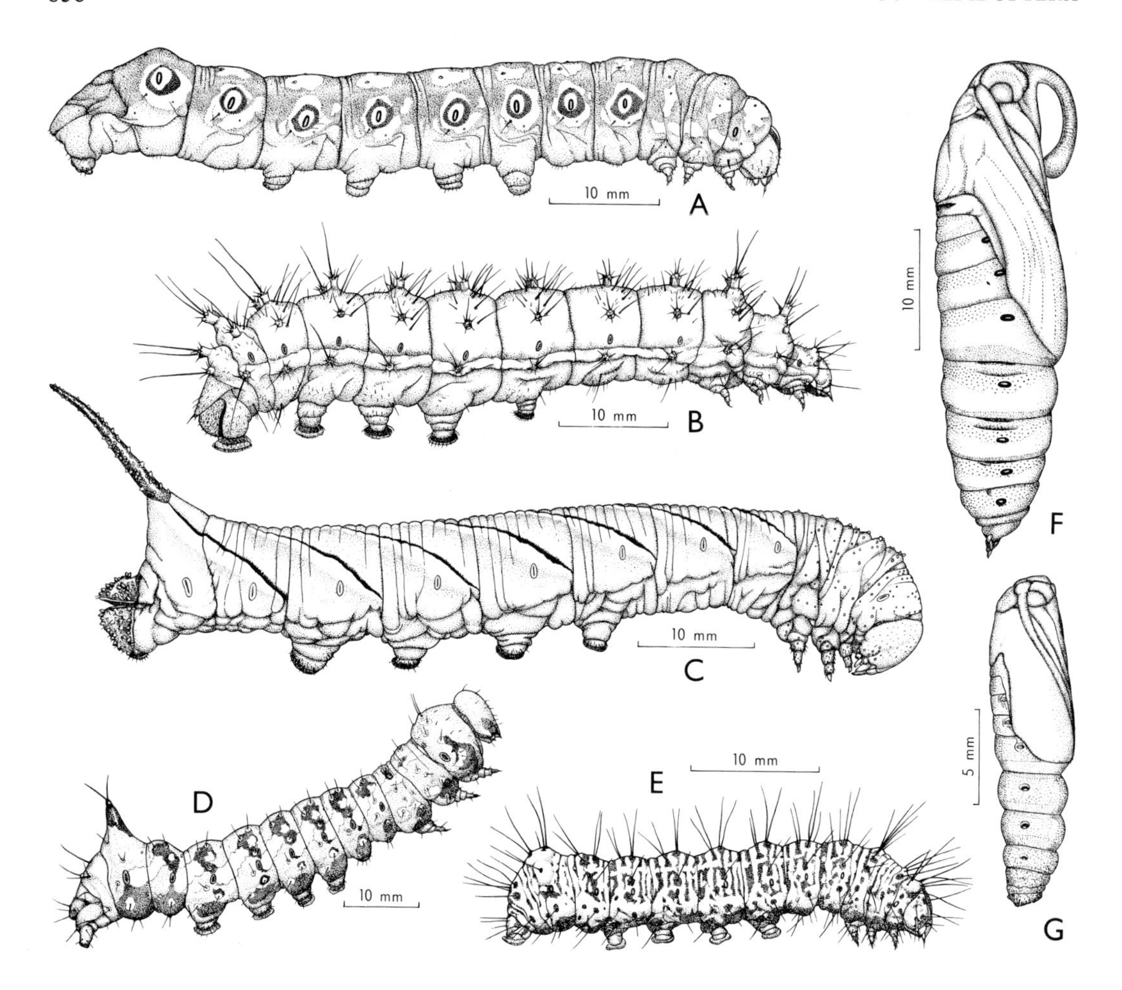

Fig. 36.49. Larvae: A, *Carthaea*, Carthaeidae; B, *Antheraea*, Saturniidae; C, *Psilogramma*, Sphingidae; D, *Danima*, Notodontidae; E, *Phalaenoides*, Agaristidae. Pupae: F, *Psilogramma*, Sphingidae; G, *Phalaenoides*, Agaristidae. [B. Rankin]

haustellum strong, reduced, or absent; maxillary palpi very small, 2- to 4-segmented, or absent; labial palpi porrect, ascending, or rudimentary; epiphysis present in male, absent in female, tibial spurs 0–2–4 or 0–2–2; fore wing (Fig. 36.47B) often with areole, M_2 not arising nearer to M_3 than to M_1, CuP absent, 1A + 2A sometimes forming basal fork; hind wing with frenulum, Sc approximated to, or shortly fused with, Rs, or connected to Rs by R_1, M_2 as in fore wing, often weak, CuP absent, 2 anal veins; metathorax with tympanal organs, tympanum directed ventrally; abdomen in female sometimes with dense anal tuft. Eggs hemispherical or nearly spherical, of upright type. Larva (Fig. 36.49D) with spines, fleshy tubercles, humps, or verrucae, sometimes with dense secondary setae, metathorax and abdominal segment 1 each with 2 MD setae, prothorax sometimes with a forked eversible organ beneath head, anal prolegs often reduced, crochets uniordinal in a mesoseries, rarely biordinal. Pupa with M_2 of fore wing, when distinguishable, as in adult.

The superfamily differs from the Noctuoidea primarily by the ventrally directed tympanum, the position of M_2 in the fore wing of the adult and pupa, and the presence of 2 MD setae on abdominal segment 1 of the larva. Only the one family (Turner, 1922) occurs in Australia. The NOTODONTINAE usually have well-developed labial palpi and haustellum, and larva with reduced or modified anal prolegs raised in repose. *Hylaeora* (3 spp.) and *Neola* (2 spp.) are endemic genera with crested head and thorax. The larvae of *H. dilucida* Feld. (Fig. 36.48G) and *H. eucalypti* Doubl. feed on *Eucalyptus*, and those of *N. semiaurata* Walk. on *Acacia*. *Danima banksiae* (Lew.), found in the south and east, feeds on *Banksia*, *Hakea*, and *Grevillea*. *Cerura* (2 spp.) is a world-wide genus with remarkable, brightly coloured larvae with a pair of long posterior filaments replacing the anal prolegs. *C. australis* Scott (Fig. 36.48D) is from subtropical rain forest, with larva on *Scalopia brownii*; the hard cocoon is spun in an excavation in the bark.

The THAUMETOPOEINAE have lost the haustellum, and the labial palpi are greatly reduced. Except for the female of the aberrant *Oenosanda* (1 sp.), the antennae are bipectinate. The females of all species have a large dense anal tuft of hair-scales. The larvae exhibit processionary behaviour, and have verrucae and normally developed anal prolegs. In *O. boisduvalii* Newm. they lack the dense secondary hair found in other genera, and the crochets are biordinal, an exception in the superfamily. They shelter beneath loose bark on *Eucalyptus*, feeding at night on the foliage. *Ochrogaster contraria* (Walk.) has gregarious larvae living in large silken bags, partly filled with excrement and cast skins, spun amongst the branches of *Acacia* and other trees. At night they move in processions to feed, and often defoliate the tree. The larval hairs can cause irritating skin rashes. Less noticeable processionary behaviour is displayed by *Epicoma* (14 spp.) and *Trichetra* (3 spp.). *E. melanosticta* (Don.) (Fig. 36.48F) occurs widely in the east and south, the food plant being *Leptospermum*. In the white male of *T. sparshalli* (Curt.) the tegulae and tip of the abdomen both bear very long hair-scales.

Superfamily NOCTUOIDEA

Small to large; ocelli present or absent; chaetosemata absent; haustellum often strong, sometimes reduced or absent; maxillary palpi minute, 1-segmented, or absent; fore wing with M_2, when present, arising nearer to M_3 than to M_1, CuP absent; hind wing with frenulum, CuP absent, 2 anal veins; metathorax with tympanal organs, rarely reduced or lost *(Amata)*, tympanum directed obliquely posteriorly; abdomen with basal counter-tympanal cavities, usually with prominent hood, with abdominal spiracle 1 either on anterior face (hood postspiracular) or at posteroventral angle (hood prespiracular; Figs. 36.5A, B). Larva sometimes with dense secondary setae, crochets uniordinal, rarely biordinal (a few Noctuidae), in a mesoseries.

The superfamily is remarkably homogeneous, and the families, though long recognized, are difficult to define. Metathoracic tympanal organs directed obliquely backwards occur in all but some Amatidae, in which they are believed to be secondarily lost. There are 2 family groups (Kiriakoff, 1963): the Arctiidae and Lymantriidae have a prespiracular counter-tympanal hood, and the remainder have the counter-tympanal hood, when developed, postspiracular.

Key to the Families of Noctuoidea

1. Fully winged 2
   Abbreviated crumpled wings (♀♀ only) 10
2. Hind wing with $Sc+R_1$ coincident with Rs (i.e. $Sc+R_1$ apparently absent); wasp-like species with boldly ringed abdomen ..... **Amatidae** (p. 862)
   Hind wing with $Sc+R_1$ separate from Rs, at least beyond discal cell; species not wasp-like ... 3

3. Hind wing with M_2 arising nearer to M_3 than to M_1 4
   Hind wing with M_2 not arising nearer to M_3 than to M_1 9
4. Hind wing with Sc sharply diverging from Rs at base and then fusing with Rs for a short distance near middle of discal cell, or connected to Rs by R_1 towards middle of discal cell (Figs. 36.50A, B) 5
   Hind wing with Sc not sharply diverging from Rs at base, but either approximated to Rs near base and then fused with Rs for a short distance, or fused with Rs from base to near or beyond middle of discal cell (Figs. 36.50C–G) 6
5. Haustellum very weak or absent; ocelli absent; fore wing with 1A + 2A not forming basal fork; ♀ abdomen with large compact terminal tuft .... **Lymantriidae** (pt., p. 860)
   Haustellum strong; ocelli present; fore wing with 1A + 2A forming basal fork; ♀ abdomen without terminal tuft .... **Hypsidae** (p. 862)
6. Hind wing with $Sc + R_1$ approximated to Rs near base and then fused with Rs for a short distance, but not beyond one-third of discal cell; $Sc + R_1$ not swollen at base (Fig. 36.50F) .... **Noctuidae** (pt., p. 864)
   Hind wing with $Sc + R_1$ fused with Rs from base to near or even beyond middle of discal cell; if $Sc + R_1$ separate from, but approximated to, Rs near base and then fused with Rs to one-third of discal cell or less, then $Sc + R_1$ swollen at base (Figs. 36.50C–E, G) 7
7. Hood of counter-tympanal cavity at base of abdomen prespiracular (Fig. 36.5B) .... **Arctiidae** (pt., p. 861)
   Hood of counter-tympanal cavity at base of abdomen postspiracular (Fig. 36.5A) 8
8. Fore wing with raised scales, in tufts or lines .... **Nolidae** (p. 862)
   Fore wing without raised scales .... **Noctuidae** (pt., p. 864)
9. Antennae usually thickened distally and often slightly hooked, rarely bipectinate in ♂; large counter-tympanal cavities at base of abdomen visible laterodorsally; mainly diurnal species .... **Agaristidae** (p. 866)
   Antennae not thickened nor hooked distally; counter-tympanal cavities at base of abdomen not readily visible laterodorsally; mainly nocturnal species .... **Noctuidae** (pt., p. 864)
10. Antennae simple; integument of abdomen and thorax whitish, with distinct brownish black dorsal pattern .... **Arctiidae** (pt., p. 861)
    Antennae with rudimentary pectinations; integument not distinctly patterned .... **Lymantriidae** (pt., p. 860)

69. Lymantriidae. Small to large; ocelli absent; antennae bipectinate to apex in male and usually in female; haustellum usually absent; maxillary palpi 1-segmented or absent; epiphysis present in male, reduced or absent in female, tibial spurs short, 0–2–4, rarely 0–2–2; females sometimes brachypterous, flightless; fore wing (Fig. 36.52A) often with areole, R_3 and R_4 stalked from discal cell or areole, 1A + 2A simple; hind wing with Sc diverging from Rs at base but approximated to, and usually connected with, Rs by R_1 towards middle of discal cell (Fig. 36.50B), rarely fusing with Rs for short distance, Rs and M_1 usually stalked, M_2 arising nearer to M_3 than to M_1; thorax and abdomen densely hairy; abdomen with counter-tympanal hood prespiracular, in female with dense anal tuft. Egg often hemispherical, rounded, or subcylindrical, laid in cluster, usually covered with hair-scales from anal tuft. Larva with dense tufted secondary setae, often with 4 long dense dorsal tufts or with hair pencils, 2 coloured dorsal glands on abdominal segments 6 and 7. Pupa stout, hairy, in silken cocoon incorporating larval hairs.

The family (Turner, 1921) is best developed in the tropics, and has few endemic genera. Most species can be distinguished by the absence of the haustellum, by the position of M_2 and of Sc in the hind wing, and by the prespiracular counter-tympanal hood.

In *Iropoca* (1 sp.) and *Orgyia* (3 spp.) there is an areole in the fore wing and females have abbreviated crumpled wings. *Orgyia* lacks medial spurs on the hind tibiae. *I. rotundata* (Walk.) (Fig. 36.48H) feeds on *Eucalyptus*, and the cocoons are spun beneath loose bark. *O. anartoides* (Walk.) from the

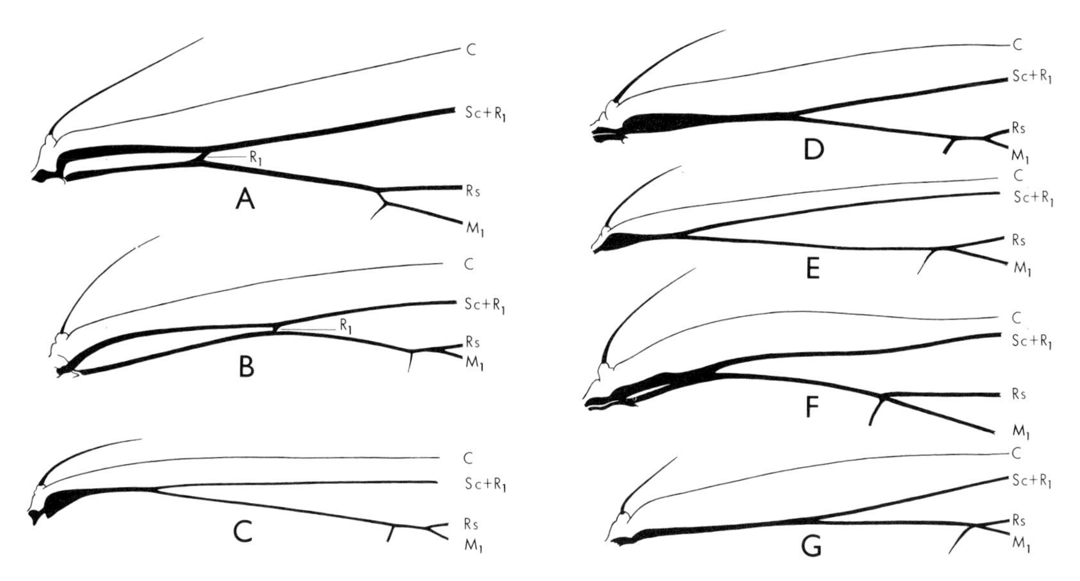

Fig. 36.50. Base of hind wing in Noctuoidea: A, *Asota*, Hypsidae; B, *Euproctis*, Lymantriidae; C–E, *Nyctemera*, *Amsacta*, *Termessa*, Arctiidae; F, G, *Heliothis*, *Earias*, Noctuidae. [B. Rankin]

south-east and *O. athlophora* Turn. (Fig. 36.48I) from the south-west are polyphagous pests of garden plants. The principal native food plant of the former is *Acacia*. Most other genera lack an areole. The larvae of *Leptocneria* (2 spp.) are urticating; *L. reducta* (Walk.) defoliates white cedar *(Melia)* in Queensland and N.S.W. *Euproctis edwardsi* (Newm.) (Fig. 36.48J) also causes urticaria, and feeds on mistletoes, sheltering in crevices and beneath loose bark. To *Euproctis* (21 spp.) also belong smaller species, such as the orange *E. lucifuga* (Lucas) from Queensland. This genus and *Porthesia* (13 spp.) are widely distributed abroad. The latter includes small white or yellow species, such as *P. paradoxa* (Butl.) from eastern Australia. The endemic *Icta* (2 spp.) has narrow elongate fore wings and short hind wings in the male, and short functionless wings in the female. The larvae of *I. fulviceps* Walk. from N.S.W. feed on *Eucalyptus* foliage and shelter beneath loose bark.

70. Arctiidae. Small to medium-sized; ocelli present or absent; antennae usually bipectinate or ciliate in male, simple in female; haustellum usually reduced; maxillary palpi 1-segmented; labial palpi short; epiphysis present, tibial spurs 0–2–4 or 0–2–2; fore wing (Fig. 36.52B) often without areole, 1A + 2A simple or with small basal fork; hind wing with Sc + R_1 sometimes swollen at base, usually fused with Rs to near middle of discal cell, then divergent (Figs. 36.50C–E), M_2 arising nearer to M_3 than to M_1; abdomen with counter-tympanal hood prespiracular. Egg usually hemispherical, with raised network on surface, usually laid in cluster. Larva usually with dense secondary setae, in tufts from verrucae, meso- and metathorax with 2 verrucae above spiracle, crochets abruptly shorter at end of each mesoseries. Pupa glabrous, cremaster weak or absent; in a cocoon of felted larval hairs with little silk.

The family is widely distributed. In the NYCTEMERINAE and ARCTIINAE ocelli are present and Sc + R_1 in the hind wing is swollen basally. To the former belong *Nyctemera* (4 spp.) and *Exitelica* (1 sp.), mostly northern, but *N. amica* (White) (Fig. 36.48K) is widespread in the east and south. Its larvae feed on *Senecio* and cineraria. The Arctiinae (Turner, 1940) contain few Australian species, usually black and white with red or orange markings. *Argina cribraria* (Clerck) from northern Australia and the Orient is rich orange with black spots. The larvae feed within the seed pods of *Crotolaria*. The cosmopolitan *Utetheisa* (3 spp.) includes *U. lotrix* (Cram.), feeding on *Crotolaria*, and *U. pulchelloides* Hamps., feeding on Boraginaceae. *Spilosoma* (6 spp.) and *Amsacta* (5 spp.) have densely hairy larvae which usually eat herbaceous plants. *S. canescens* (Butl.) (Fig. 36.48L) also attacks woody ornamentals.

The LITHOSIINAE, which lack ocelli, are smaller, slender-bodied species, usually patterned in orange or red and black and white. There are many endemic genera. The larvae often feed on lichens. *Scoliacma bicolora* (Boisd.) is a variable red and black species from the south-east, whereas the much larger *Oenistis entella* (Cram.) (Plate 7, G) occurs in rain forest in north Queensland and New Guinea. *Halone coryphaea* Hamps. is usually found resting on lichen-covered rocks, the procryptic fore wings covering the orange hind wings. *Thallarcha* (30 spp.) includes delicately patterned forms, such as *T. jocularis* (Rosen.) (Fig. 36.51A) from the south-east. In *Xanthodule* (2 spp.) the female has rudimentary wings. *X. semiochrea* Butl. feeds on several trees and shrubs, including *Eucalyptus*. *Termessa* (12 spp.) contains orange and black species, but also the delicate white *T. nivosa* (Walk.) with larvae found beneath loose bark on *Eucalyptus*.

71. Amatidae (Syntomidae, Ctenuchidae). Small to medium-sized; ocelli present or absent; antennae simple, thickened, dentate, or bipectinate; haustellum present; maxillary palpi 1-segmented; labial palpi small; epiphysis present, tibial spurs 0–2–4; tympanal organs often reduced or lost; fore wing (Fig. 36.52C) narrow, retinaculum sometimes absent, areole absent, 1A + 2A with small basal fork; hind wing small, Sc coincident with R_1 and Rs; abdomen with counter-tympanal hood, when present, postspiracular. Larva with dense secondary setae on verrucae, meso- and metathorax with only 1 verruca above spiracle. Pupa in cocoon of felted larval hairs and silk.

A tropical family, the Amatidae are represented in Australia by *Euchromia* (4 spp.), *Eressa* (7 spp.), *Amata* (38 spp.), and *Ceryx* (5 spp.). They are diurnal and, with narrow fore wings, small hind wings, and ringed abdomen, rather resemble wasps. *Euchromia creusa* (L.) (Plate 8, J) is a brilliant species from Cape York. *Amata* contains confusing black and orange species, in which the male retinaculum, ocelli, and tympanal organs have been lost. The larvae are dark reddish black. Some, such as *A. aperta* (Walk.), feed on living foliage, whereas others, such as *A. trigonophora* (Turn.) (Fig. 36.51B), feed on fallen flowers and leaves.

72. Hypsidae. Medium-sized; ocelli present; antennae ciliate in male; haustellum present; maxillary palpi 1-segmented; labial palpi ascending, apical segment long, slender, erect, smooth; epiphysis present, tibial spurs 0–2–4; fore wing (Fig. 36.52D) sometimes without retinaculum, R_3 and R_4 stalked from areole, 1A + 2A forming basal fork; hind wing with Sc connected to Rs by R_1 before middle of discal cell (Fig. 36.50A), M_2 arising nearer to M_3 than to M_1; abdomen with counter-tympanal hood postspiracular. Eggs laid in cluster. Larva noctuid-like. Pupa in flimsy cocoon.

Four genera are represented in Australia: *Digama* (1 sp.), *Agape* (1 sp.), *Asota* (5 spp.), and *Neochera* (1 sp.). The small *D. marmorea* Butl. is common in the north, with larvae on *Carissa ovata* (Apocynaceae). The larger *Agape chloropyga* (Walk.) (Plate 8, H) lacks a retinaculum in the male. The larvae of *Asota* feed on *Ficus*; when young several rest together beneath a leaf, but later they are found singly. The dark brown larvae of *A. iodamia* (Herr.–Sch.) (Fig. 36.51C) have short white setae, and are found on *F. macrophylla*.

73. Nolidae. Small; ocelli absent; antennae bipectinate or ciliate in male, scape with anterior tuft; haustellum present; maxillary palpi 1-segmented; labial palpi medium to long, porrect or ascending; epiphysis present, tibial spurs 0–2–4; fore wing (Fig. 36.52E) with scales more or less raised and 3 or 4 raised scale-tufts, areole small or absent, 1 or 2 radial veins often lost, 1A + 2A simple; hind wing with Sc + R_1

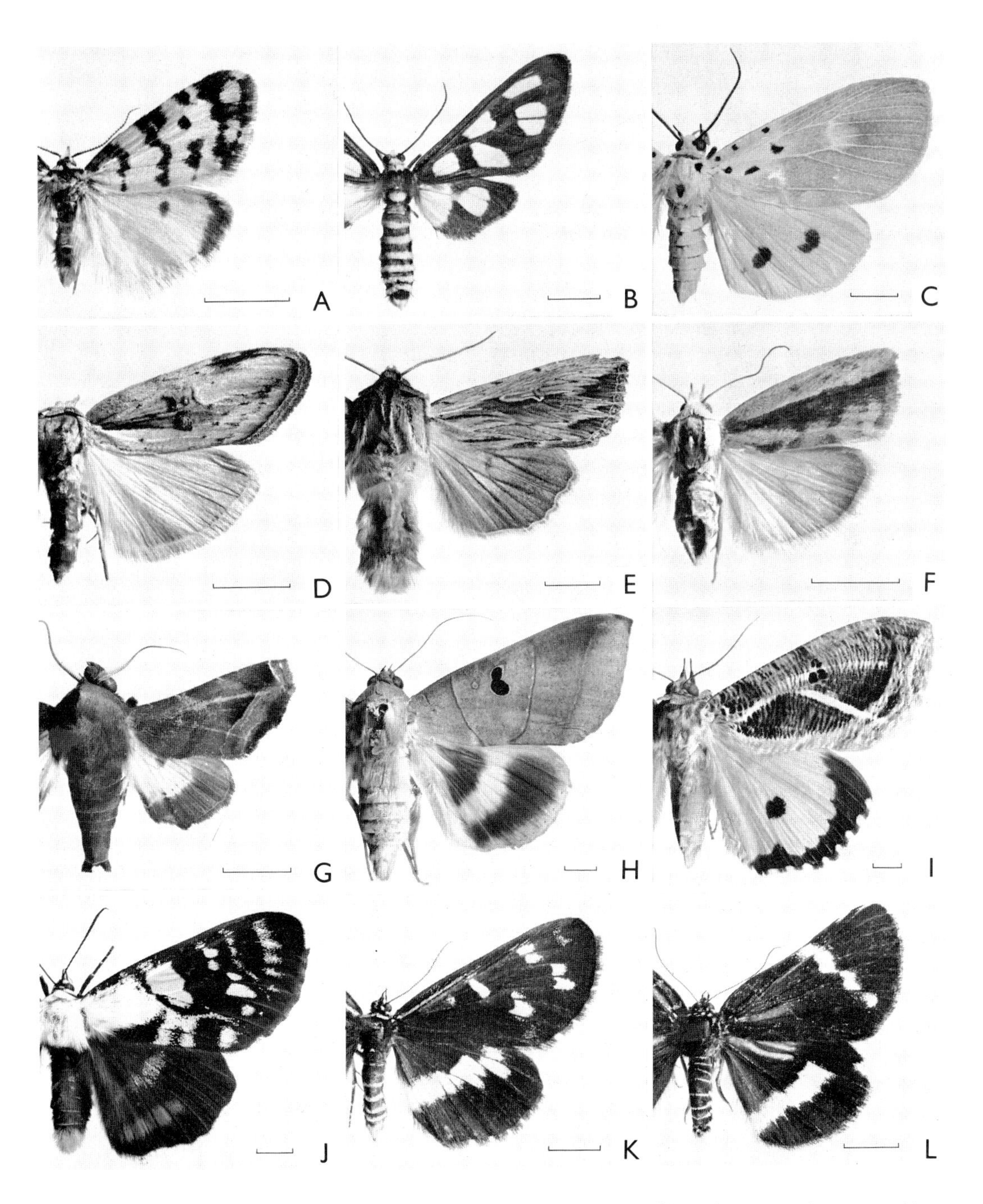

Fig. 36.51. A, *Thallarcha jocularis*, Arctiidae; B, *Amata trigonophora*, Amatidae; C, *Asota iodamia*, Hypsidae; D, *Aquita tactalis*, Nolidae; E, *Persectania ewingii*, F, *Earias huegeli*, G, *Bombotelia jocosatrix*, H, *Anua coronata*, I, *Othreis materna*, Noctuidae; J, *Agarista agricola*, K, *Cruria donovani*, L, *Idalima tetrapleura*, Agaristidae. Scales = 5 mm. [Photos by C. Lourandos]

fusing with Rs from base to near middle of discal cell, M_2 arising nearer to M_3 than to M_1, or M_3 absent; abdomen with dorsal tuft on segment 1, counter-tympanal hood poorly developed. Larva with dense secondary setae on verrucae, prolegs absent from abdominal segment 3. Pupa in stiff cocoon, often boat-shaped.

The family (Turner, 1944) is distinguished from the Noctuidae and Arctiidae by the scale-tufts on the fore wing, and usually from the former also by the condition of Sc + R_1 in the hind wing. The larvae resemble those of Arctiidae, but lack the first pair of ventral prolegs in all instars. The cocoon is similar to that of the Sarrothripinae in the Noctuidae. *Aquita* (3 spp.) and *Uraba* (4 spp.) have 5 branches of R in the fore wing. *A. tactalis* (Walk.) (Fig. 36.51D), with ciliate male antennae, is found amongst *Leptospermum*. *U. lugens* Walk., with bipectinate male antennae, often defoliates *Eucalyptus*. The head capsules of the earlier instars are stacked upon the prothorax of the older larva. Most of the Australian species are referred to *Nola* (81 spp.), in which the fore wing lacks 1 or 2 branches of R. The small, stout, green larvae of *N. lechriopa* Hamps. feed on *Melichrus* (Epacridaceae).

74. Noctuidae. Small to large; ocelli usually present; antennae pectinate, dentate, or simple; haustellum usually strong; maxillary palpi 1-segmented; labial palpi porrect or ascending; epiphysis present, tibial spurs 0–2–4, tibiae and tarsi sometimes spined; fore wing (Fig. 36.52F) usually with areole, 1A + 2A forming basal fork; hind wing with Sc + R_1 shortly fused with Rs near base, rarely fused to about one-half of discal cell (Figs. 36.50 F, G), M_2 weak and arising nearer to M_1 than to M_3 (trifid), or strong and arising nearer to M_3 than to M_1 (quadrifid); abdomen with counter-tympanal hood postspiracular. Egg usually domed, with vertical ribs, laid singly or in clusters. Larva usually without secondary setae, rarely with dense secondary setae on verrucae, crochets uniordinal, rarely biordinal, not abruptly shorter at ends of mesoseries; mostly phytophagous, sometimes stem-boring, or predacious on Coccidae. Pupa in cell in soil, or in silken cocoon not incorporating larval hairs.

The family (Forbes, 1954) is one of the most specialized. In the hind wing Sc + R_1 is not swollen basally, and in most species has a short subbasal fusion with Rs. In a few genera, such as *Earias*, this fusion may extend to the middle of the discal cell, as in the Arctiidae and Nolidae. The postspiracular counter-tympanal hood, however, will separate these from the Arctiidae, and the smooth fore wings will distinguish them from the Nolidae. Throughout the family the genitalia provide excellent characters for separating genera and species.

In the HELIOTHIDINAE, NOCTUINAE, HADENINAE, CUCULLIINAE, AMPHIPYRINAE, and ACRONICTINAE the hind wing is trifid, with M_2 arising nearer to M_1 than to M_3. The first two, with mid and usually hind tibiae spined, include many pests of crops. Most pupate in the soil. *Heliothis punctigera* Wall. and *H. armigera* (Hübn.) attack cotton, lucerne, linseed, tomatoes, and maize. The chief cutworm pests are *Agrotis munda* Walk., *A. ipsilon* (Hufn.), and *A. infusa* (Boisd.), the larvae hiding in soil or beneath debris, and damaging plants at night. *A. infusa* is mainly a winter pest, the adults aestivating in the mountains of the south-east (p. 108). The Hadeninae, with hairy eyes, include *Persectania ewingii* (Westw.) (Fig. 36.51E) and *Pseudaletia convecta* (Walk.), the main armyworm pests of grasses and cereals. The Cuculliinae and Amphipyrinae have naked eyes, but in the former they have long marginal hair-scales which curl partly over the eye. The Amphipyrinae include the 3 widespread Old-World pests *Spodoptera litura* (F.), *S. mauritia* (Boisd.), and *S. exempta* (Walk.). The first attacks cotton in the north-west, and the others are pasture pests in the north and east. *Bathytricha* (3 spp.) has larvae boring in grasses. *B. truncata* (Walk.) damages rice in the Riverina and sugar cane in the north.

The WESTERMANNIINAE are neither trifid nor quadrifid, having M_2, when present in the hind wing, equidistant from M_1 and M_3. *Earias* (8 spp.) contains small green and

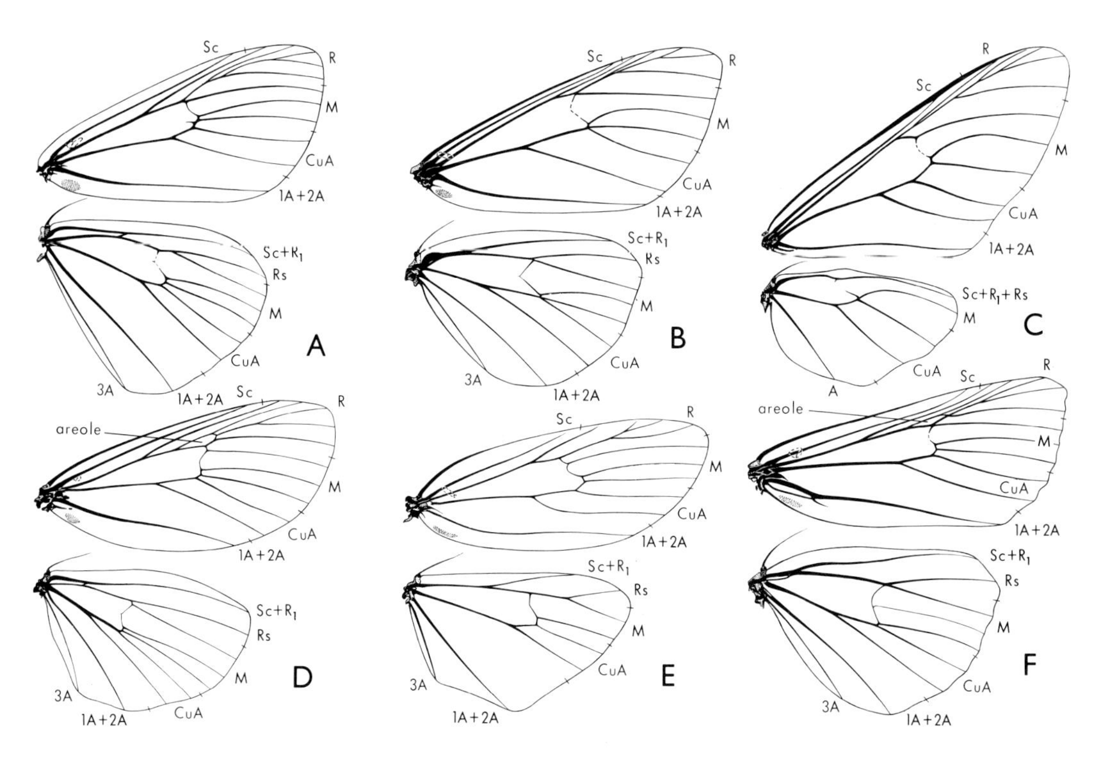

Fig. 36.52. Wing venation of Noctuoidea: A, *Leptocneria*, Lymantriidae; B, *Spilosoma*, Arctiidae; C, *Amata*, Amatidae; D, *Asota*, Hypsidae; E, *Nola*, Nolidae; F, *Persectania*, Noctuidae. [J. Wedgbrow]

cream moths, with larvae in the seed capsules of Malvaceae, pupating in a boat-shaped silken cocoon. The Australia-wide *E. huegeli* Rog. (Fig. 36.51F) and the northern *E. vitella* (F.) are pests of cotton. The ERASTRIINAE contain small species, often with larvae lacking ventral prolegs on abdominal segments 3 and 4. The larvae of *Catoblemma* (12 spp.), however, which are predacious on scale insects, retain all ventral prolegs. *C. dubia* (Butl.) is a common species.

The EUTELIINAE, SARROTHRIPINAE, PLUSIINAE, CATOCALINAE, OPHIDERINAE, and HYPENINAE are the main quadrifid subfamilies. The larvae usually pupate in a silken cocoon. The tropical Sarrothripinae have a long thin apical segment of the labial palpi, and the cocoons have a vertical slit-like exit at one end. Few species of Euteliinae occur in Australia. They rest with wings extended and partly rolled, and the abdomen is curled upwards. The larvae of *Bombotelia jocosatrix* (Guen.) (Fig. 36.51G) bore in the shoots of mango. Most Plusiinae have a large thoracic crest, abdomen with dorsal scale-tufts, and silver markings on the fore wings. The larvae are semi-loopers, with first pair of ventral prolegs reduced or lost and the remainder with biordinal crochets, an exception in the superfamily. *Plusia argentifera* Guen. and *P. chalcites* (Esper) are pests of dicotyledonous crops. The larvae of the Catocalinae, with 1 or 2 pairs of ventral prolegs reduced or lost, taper towards each end, and walk with a looping action. Those of the large *Anua coronata* (F.) (Fig. 36.51H) sometimes feed on *Quisqualis* in Queensland. The Ophiderinae include many tropical and subtropical species with large upwardly-curved labial palpi. The first pair of larval prolegs is reduced. The adults are attracted to fermenting fruits. Some, such as the targe

Othreis fullonia (Clerck) and *O. materna* (L.) (Fig. 36.51I), damage *Citrus* and other tropical fruits by penetrating the rind with their stout haustellum. The Hypeninae contain many obscure tropical species, sometimes with very large labial palpi. The ventral prolegs are also reduced.

75. Agaristidae. Medium-sized; ocelli present; antennae thickened towards apex, often slightly hooked, or simple; haustellum strong; maxillary palpi 1-segmented; labial palpi ascending, apical segment slender, porrect; epiphysis present, tibial spurs 0–2–4; fore wing usually with areole, 1A + 2A forming basal fork; hind wing with Sc + R_1 shortly fused with Rs near base of discal cell, M_2 arising nearer to M_1 than to M_3; abdomen with counter-tympanal cavities large, visible laterodorsally. Egg domed, ribbed, laid singly. Larva (Fig. 36.49E) with posterior hump, without secondary setae. Pupa (Fig. 36.49G) heavily sclerotized, rugose; in a cell in bark or dead wood, or in soil.

The family differs from the Noctuidae by the larger counter-tympanal cavities and usually clavate antennae. Most are diurnal, and brightly patterned in black, yellow, and other colours. The larvae, also aposematically banded in black and yellow or white, with orange or red markings, are exposed feeders.

Agarista agricola (Don.) (Fig. 36.51J) is brightly coloured, with larvae boldly banded in black and white and feeding on *Cissus*. The larvae of several species feed on *Hibbertia*, including *Phalaenoides glycine* Lew., *Eutrichopidia latina* (Don.), *Cruria donovani* (Boisd.) (Fig. 36.51K), and *Periscepta polysticta* (Butl.). *P. glycine* is also a pest of grape vines. The males of *Hecatesia* (3 spp.), the whistling moths, have a sound-producing device. A ribbed hyaline area lies adjacent to a large knob on the costa of each fore wing, and in flight the two knobs can be struck against one another above the body, flexing the ribbed membrane of the wings. The rapid succession of clicking sounds thus produced resembles whistling. *H. fenestrata* Boisd. (Plate 8, M) is common in the east. In *Idalima tetrapleura* (Meyr.) (Fig 36.51L) a similar sound is made when the grooved file-like basal segment of the hind tarsus rubs against the enlarged veins of the underside of the wings.

ACKNOWLEDGMENTS. The author wishes to thank many colleagues for helpful discussions and comments, especially Messrs J. D. Bradley, D. S. Fletcher, W. H. T. Tams, A. Watson, P. E. S. Whalley, and Dr I. W. B. Nye, of the British Museum (Natural History), London, Dr J. F. Gates Clarke, of the U.S. National Museum, Washington, Mr J. S. Dugdale of D.S.I.R., New Zealand, and Mr M. S. Upton, Division of Entomology, CSIRO, Canberra.

37

HYMENOPTERA

(Wasps, bees, ants)

by E. F. RIEK

(Bees by Charles D. Michener; ants by W. L. Brown Jr and R. W. Taylor)

Endopterygote Neoptera with mandibulate mouth-parts and a characteristic reduced venation; wings coupled by hamuli; segment 1 of abdomen usually without sternite and in close association with metathorax; a marked constriction between segments 1 and 2 in all but the most primitive species and a few secondarily specialized ones. Larva a caterpillar in most Symphyta; apodous, usually maggot-like, in Apocrita. Pupa adecticous and generally exarate.

The Hymenoptera are an abundant, ubiquitous, highly specialized, and highly successful order of insects that rivals the Diptera and Lepidoptera in number of species, and probably exceeds both in variety of adaptive radiations. It is distinguished among endopterygotes, not only by the general characters of the thorax, wings, and abdomen, but by the remarkable developments of the ovipositor in the female and the universality of male haploidy. It includes two divergent evolutionary lines: the primitive, relatively unobtrusive, phytophagous Symphyta; and the much more specialized and abundant Apocrita with their diverse adaptations to parasitism, predation, exploitation of the nectar and pollen resources of the flowering plants, and highly organized forms of social life. On the other hand, only a few Apocrita subsist on the tissues of plants (as gall-formers), and only a very few parasitic species have become aquatic. Some Hymenoptera are pests, but no other order contains so many that are of benefit to mankind.

Anatomy of Adults

Head (Fig. 37.1C–E). Characteristically hypognathous, often extremely mobile, and connected to the thorax by a thin neck. Clypeus and labrum usually distinct, though the labrum may be hidden under the clypeus, which is not defined in some parasitic groups. Compound eyes almost always large, and sometimes strongly convergent to holoptic in males; the eyes have atrophied to single facets in certain Formicoidea; sometimes

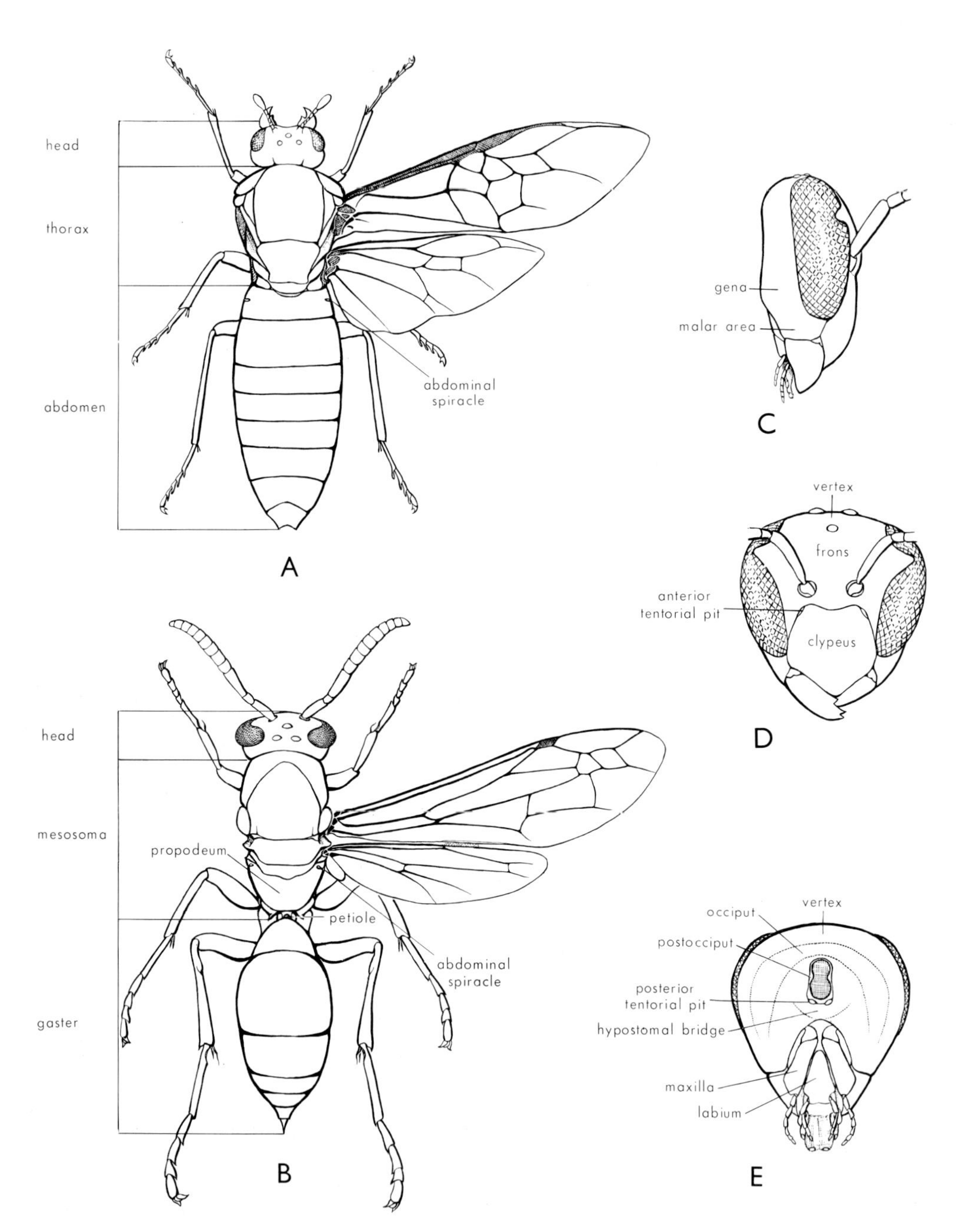

Fig. 37.1. Adults: **A**, Symphyta (*Perga* sp., Pergidae); **B**, Apocrita (*Polistes* sp., Vespidae). C–E, head of *Polistes*: **C**, lateral; **D**, frontal; **E**, posterior. [S. Curtis]

numerous hairs arise between the facets. Three ocelli commonly present, but sometimes reduced or aborted (some Sphecidae and workers of many ants).

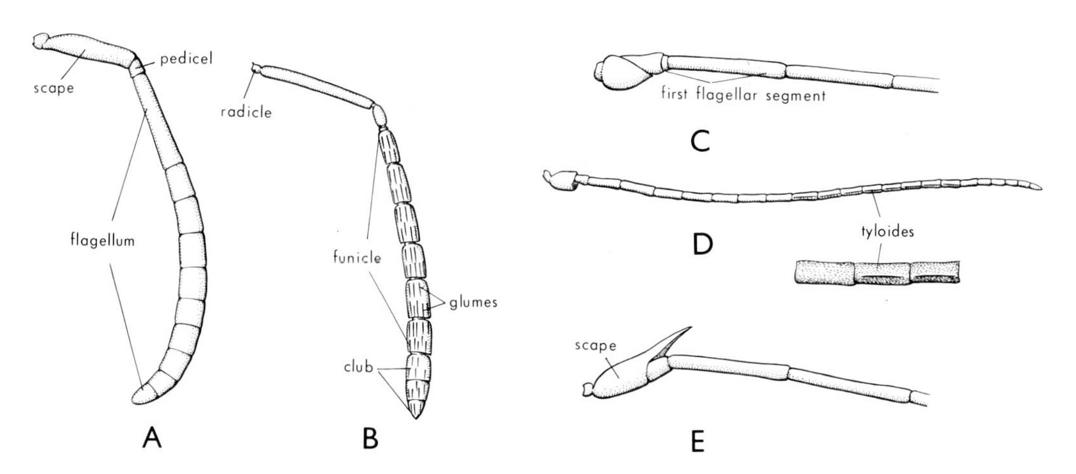

Fig. 37.2. Antennae: A, long scape in Vespidae; B, radicle, club, and glumes in Chalcidoidea; C, divided 1st flagellar segment in Ichneumonidae; D, tyloides in Trigonalidae; E, produced scape in Austroserphidae. [S. Curtis]

Antennae (Fig. 37.2) very variable in Symphyta and the more primitive Apocrita; pronounced sexual dimorphism common. The number of segments is variable, with a tendency to reduction in the more highly evolved forms. In the higher Symphyta there are typically 9 segments; in Sphecoidea, Apoidea, Vespoidea, and Scolioidea there are almost constantly 13 in the male and 12 in the female. With reduction in overall size associated with parasitic habits, there is a tendency for the antennae to become clubbed. When segmentation and length of the antenna are reduced, it is often elbowed *(geniculate)*, the scape elongate, its narrowed base *(radicle)* conspicuous, and the pedicel small; the basal slender portion of the flagellum is the *funicle*, the distal stouter part is the *club*. (In ants the pedicel plus flagellum is called the funiculus.) The combination of long scape and elbowed antenna occurs in most species of Scolioidea, Chrysidoidea, Chalcidoidea, Proctotrupoidea, and in the highly evolved predatory and melliferous species. The surface of many of the antennal segments is covered with sense organs of various types. The males of Trigonalidae and many Ichneumonidae have a large longitudinal keel or sensory patch *(tyloides)* on the ventral side of several of the flagellar segments. Longitudinal ridges *(glumes)* occur on the flagellar segments of many Chalcidoidea, Cynipoidea, and Proctotrupoidea.

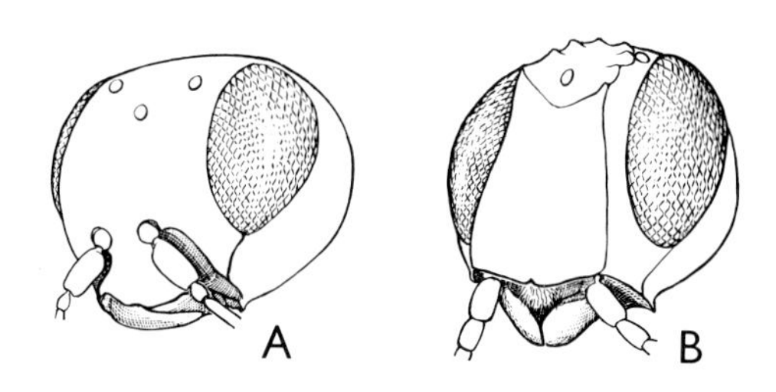

Fig. 37.3. Subantennal groove in: A, *Megalyra* sp., Megalyridae; B, *Guiglia* sp., Orussidae. [S. Curtis]

Most adults rest with the antennae directed forwards with the scape either directed forwards or dorsally. Others fold the flagellum back against the scape. In Platygasteridae the antenna is folded in a closed Z. In some parasitic groups in which the antennae are long, they are folded back dorsally over the thorax when the species must pass through some obstruction, or during emergence from the cocoon or the host body. Megalyridae rest with the long antennae folded back ventrally between the legs, whereas Orussidae rest with them

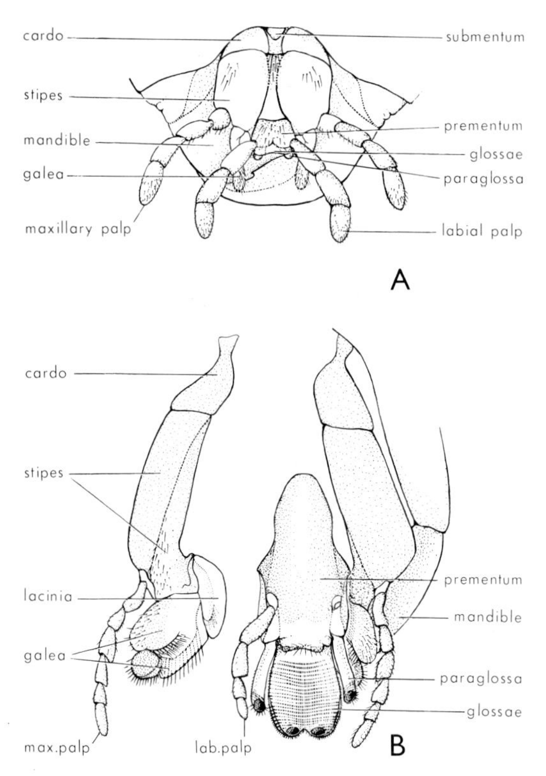

Fig. 37.4. Mouth-parts: A, *Perga* sp., Pergidae; B, *Polistes* sp., Vespidae. [S. Curtis]

directed laterally. In both these families the scape is directed laterally and somewhat ventrally, and lies in a *subantennal groove* (Fig. 37.3).

The mouth-parts (Fig. 37.4) exhibit a wide range of specialization, from the generalized biting type of Symphyta and primitive Apocrita to the combined sucking and chewing type of Apoidea. Mandibles always present; used mainly to enable the adults to cut their way out of the pupal cell (or body of the host in some parasitic species); but in predatory and melliferous types they are used to a large extent in constructing the nest or in the collection of material for its construction. The mouth-parts consist of the normal components in the more generalized forms, but there has been some reduction in certain parasitic groups. In Apoidea the glossa becomes progressively lengthened in correlation with the habit of collecting nectar, and the other mouth-parts become lengthened to produce a proboscis (rostrum). The evolution of the rostrum can be traced in the Apoidea from the simple condition in Colletidae to the highly specialized Apidae, in which it is folded down beneath the head when not in use. The *proboscidial fossa* is the deep groove on the under side of the head in which the proboscis is folded in repose. The *hypostomal carinae* margin this fossa, and turn laterally toward the bases of the mandibles at their anterior ends.

Thorax. Characterized by the close association of abdominal segment 1 with the metathorax, and its complete incorporation into the thorax (Figs. 37.5, 6) as the *propodeum* in all but the most primitive species. In most Symphyta it is still clearly part of the abdomen, and has undergone little change. In Apocrita it is incorporated in the thorax, but its origin is indicated by the presence of the first pair of abdominal spiracles on the propodeum. The peculiar thoracic tagma in Apocrita is variously called the 'thorax', or *mesosoma*, or *alitrunk* (in ants).

The pronotum is dissociated from the remainder of the prothorax in all Hymenoptera. It is hinged laterally to the anterior region of the mesothorax, and appears as an integral part of the mesothorax in the more advanced families. Its lateral lobes are extended ventrally, and the lower ends may meet, so that the pronotum then forms a complete collar about the front of the mesothorax (Proctotrupidae, Gasteruptiidae). The remainder of the prothorax is freely movable in relation to the rest of the thorax; it serves for attachment both of the fore legs and of the head.

There is a close relationship between the development of the pronotum and the ability to fly, though this may be modified in burrowing species. Pompilidae have a well-developed pronotum, and are mostly relatively weak fliers that are unable to transport their prey in flight. Sphecoidea are strong fliers that transport their prey in flight, and increased ability is correlated with an increase in the relative size of the scutum

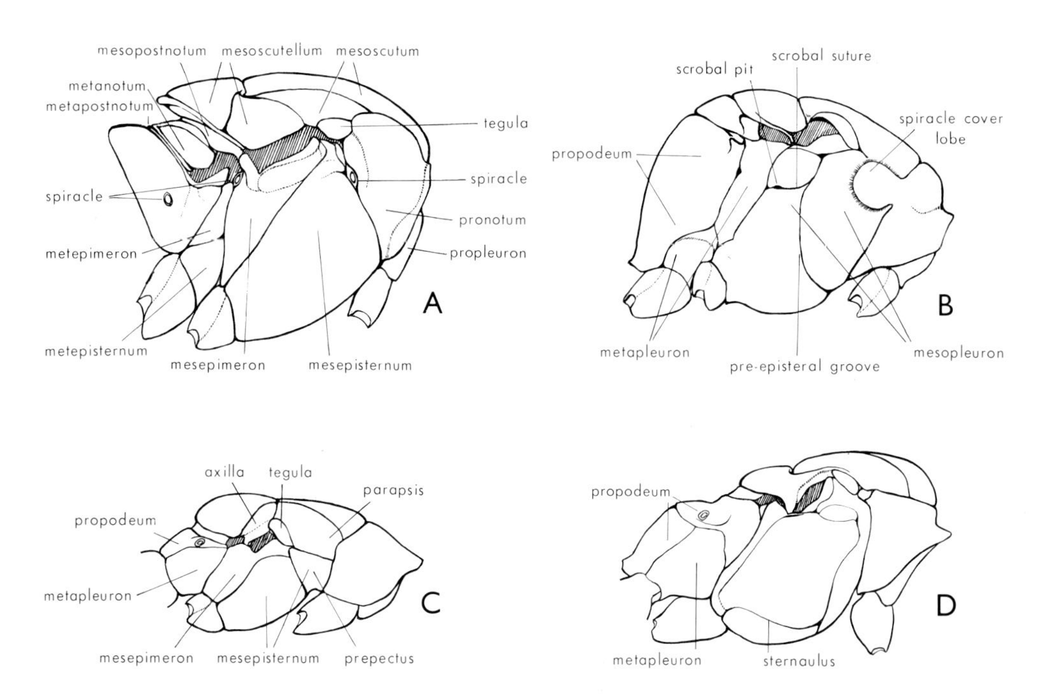

Fig. 37.5. Lateral thorax: A, *Perga* sp., Symphyta-Pergidae; B, *Hyleoides* sp., Apoidea-Colletidae; C, *Thaumasura* sp., Chalcidoidea-Pteromalidae; D, *Ceratomansa* sp., Ichneumonidae. [S. Curtis]

at the expense of the pronotum. Vespoidea are similar to Sphecoidea, but have developed along a different line. Scolioidea are strong fliers, though they have retained a large pronotum in association with the fossorial habits of the females in search of hosts. The males often carry the females in flight. Increase in the mesothorax and its associated muscles has been accomplished by enlargement of the pleural region. Most Bethyloidea have a thorax similar to that of primitive Pompilidae, but there has been reduction in the size of the pronotum in those species in which the wingless female is transported by the male.

In Symphyta the 'prothoracic' spiracle is not covered by a lateral lobe of the pronotum. In Megalyridae there has been fusion of the spiracular plate with the lateral pronotum, so that the spiracle appears to arise from within the pronotum; this occurs also in Proctotrupidae and Pelecinidae. The *spiracle cover lobe* (pronotal tubercle, Fig. 37.5B) is margined with close fine hairs in Trigonalidae and all the more advanced Apocrita. The marginal fringe is well developed in the fossorial species.

The mesonotum (Fig. 37.6) is divided into anterior scutum and posterior scutellum. The lateral portions of the scutum are separated by grooves *(notaulices)* to form the *parapsides*, which sometimes show a longitudinal *parapsidal line* or furrow. Each notaulix is said to be *percurrent* when it extends from the anterior margin of the scutum to the scuto-scutellar suture. A pair of *admedian lines* is usually evident anteriorly and a *median mesoscutal line* between them (Daly, 1964). The *axillae* are formed from the lateral scutellum. The metanotum is reduced to a single transverse sclerite carrying the hind wings; while the metapostnotum in all the higher forms is invaginated under the anterior region of the propodeum, and usually invisible except for lateral vestiges, though it is often well developed in Pompilidae. The *cenchri* of most Symphyta are a pair of protrusions from the anterolateral

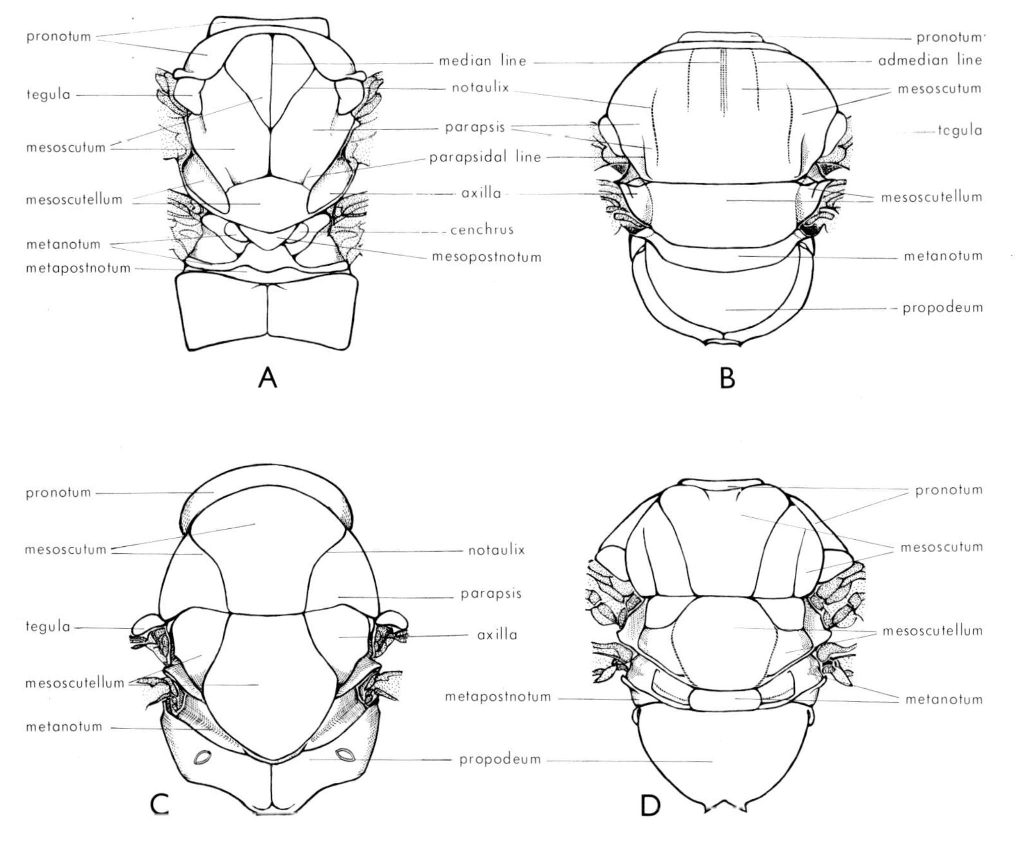

Fig. 37.6. Dorsal thorax and propodeum: A, Symphyta-Tenthredinidae; B, *Sphecius* sp., Sphecidae; C, *Agamerion* sp., Pteromalidae; D, *Taeniogonalos* sp., Trigonalidae. [S. Curtis]

metanotum (Fig. 37.6A). *Piliferous lobes* on the anterior angles of the metanotum, recalling cenchri, occur in a few Ichneumonidae (Kerrich, 1952).

There is no clear differentiation between sternal and pleural region in the mesothorax of most Hymenoptera. In some Symphyta, however, there is a distinct suture ventral to the articulation of the mid coxa, which is apparently the sternopleural suture. In some Ichneumonidae there is a groove *(sternaulus)* in the corresponding position (Fig. 37.5D). The mesosternum is apparently distinct in some Chalcidoidea. The pleuron is usually divided by an oblique pleural 'suture' into an anterior episternum and a posterior epimeron. The *prepectus* is the differentiated anterior portion of the mesepisternum. It often forms a conspicuous plate on the lateral thorax lying between the pronotum and the mesepisternum. Normally only the lateral part of the plate, which is often continuous across the venter, is visible. The prepectus is best developed in Chalcidoidea, but there is a well-developed prepectus in most Chrysidoidea, some Bethyloidea, and Scolioidea. The metapleuron is usually a simple plate. It may be covered by the enlarged mesopleuron in some Encyrtidae.

Legs. A *trochantellus* (Fig. 37.7A) is distinct in most Hymenoptera, at least in some legs, but it is large and more freely articulated mainly in Symphyta and in parasitic Apocrita. It is the proximal end of the femur, though it

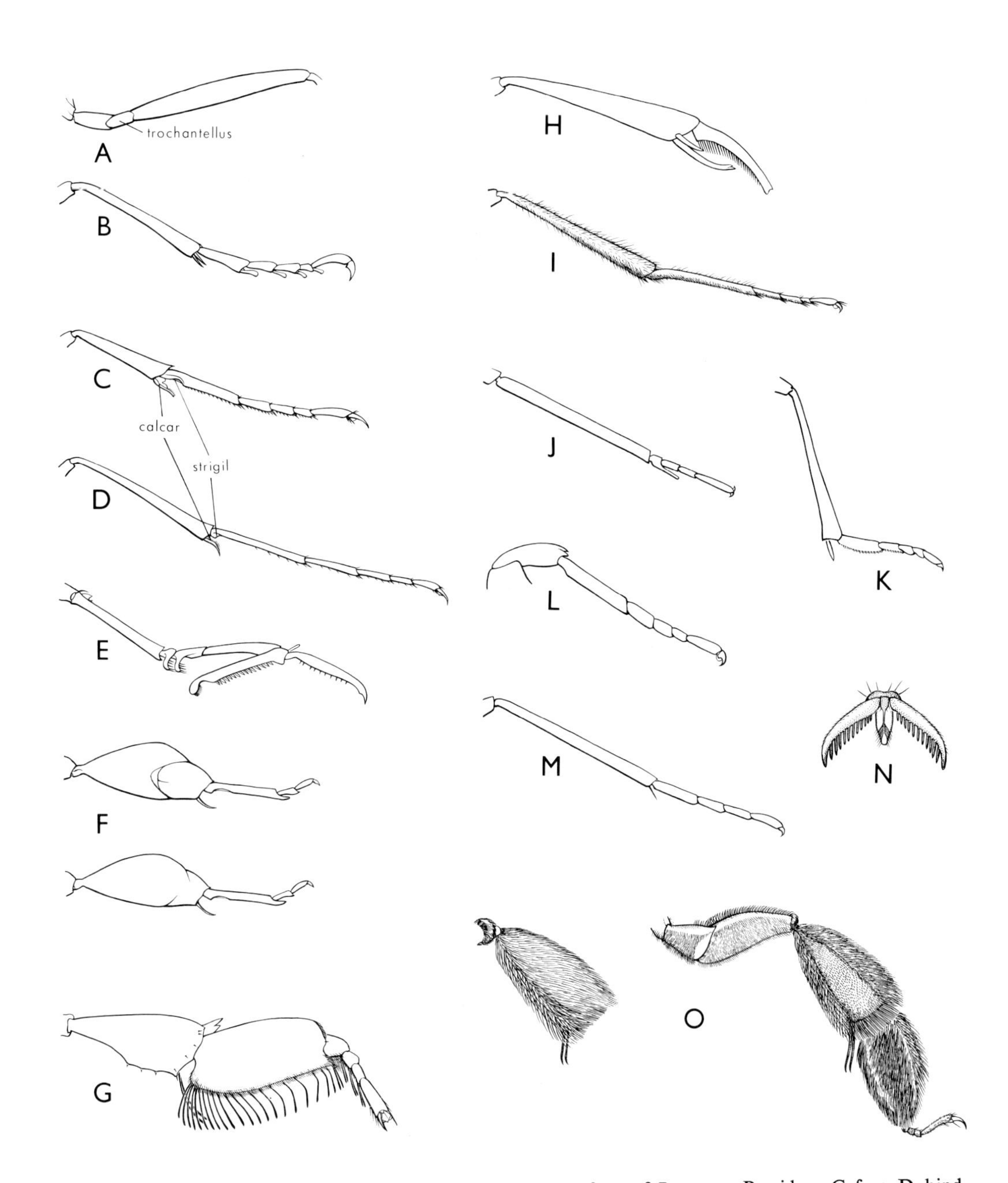

Fig. 37.7. Legs: A, hind, of *Ceratomansa* sp., Ichneumonidae; B, fore, of *Perga* sp., Pergidae; C, fore, D, hind, of *Sceliphron* sp., Sphecidae; E, fore, of *Paradryinus* sp., Dryinidae, ♀; F, fore, of *Guiglia* sp., Orussidae, ♀; G, fore, of *Bembix* sp., Sphecidae, ♂; H, fore, of Ceraphronidae; I, hind, of *Megalyra* sp., Megalyridae; J, hind, of *Ibalia* sp., Ibaliidae; K, mid, of Encyrtidae; L, fore, of *Pleistodontes* sp., Agaonidae; M, fore, of *Elasmus* sp., Eulophidae; N, claws of *Netelia* sp., Ichneumonidae; O, hind leg of *Amegilla* sp., Anthophoridae. [S. Curtis]

appears as an apparent second segment of the trochanter.

The fore tibia (Figs. 37.7B) has two articulated, simple and similar, apical spurs in the most primitive Symphyta, including Argidae and most Pergidae, but a few Pergidae have only a single, simple, straight spur. In the more advanced Symphyta (e.g. Tenthredinidae) the anterior of the two spurs is curved and modified into a *calcar*, and then the basitarsus may have a poorly developed emargination bearing a special comb of hairs (a *strigil*). In the most advanced Symphyta (Orussidae, Siricidae, Cephidae) there is only a single spur, modified into a calcar, which may be expanded and margined with a row of close hairs and fits against a poor to well-developed strigil on the basitarsus. A well-developed calcar and strigil occur in most Apocrita, forming a cleaning mechanism for the antenna, which is passed through the opening between them. Ceraphronidae are unique amongst the Apocrita in possessing 2 fore tibial spurs. The fore tibia is without an apical spur in Agaonidae; and in some other chalcidoid families the spur is straight and simple. The fore legs are also used to clean the eyes and face, while the mouth-parts clean the fore legs. Some Eulophidae, which have a short, straight fore tibial spur, use both legs together to clean each antenna, and there is a definite cleaning of the tibial spur and surrounding areas by the mouth-parts.

In most of the highly evolved superfamilies (Sphecoidea, Apoidea, Vespoidea, Formicoidea, Pompiloidea, Bethyloidea) a calcar is also developed from the inner hind tibial spur, or both spurs are variously modified (Fig. 37.7D). Rarely, spurs may be absent on the hind leg (some Apoidea and Formicoidea). There is also a dense brush of hairs on the inner surface of the tibia towards its apex and sometimes on the basitarsus. Even when special developments are lacking, the hind legs are used to clean the gaster, wings, and mid legs, and the hind legs are cleaned by rubbing and scraping one against the other. Many Formicoidea and Apoidea have a calcar-like development on all legs.

The tarsi are normally 5-segmented, but 3-segmented in Trichogrammatidae, and 4-segmented in some other Chalcidoidea; an arolium is normally present between the claws. The tarsal claws are simple, or variously toothed and spined. In most female Dryinidae the fore tarsus is raptorial (Fig. 37.7E).

The mid legs of many Chalcidoidea (Encyrtidae) are modified for jumping; the tibial spur is enlarged, and pads of special bristles are developed on the ventral surface of each tarsal segment, especially the basitarsus (Fig. 37.7K). A *pecten*, or marginal row of special, long, flattened bristles on the lateral margin of the fore tarsus, especially the basitarsus, occurs in many burrowing species (Sphecoidea, Pompiloidea, Scolioidea). The tarsal segments too may be broadened and flattened. A pecten and broadening of the tarsal segments is very pronounced in the males of some species of *Bembix* (Fig. 37.7G). A flattened basitibial plate is present on the outer side of the base of the hind tibia of many bees, presumably for support as the bee moves about in its burrow in soil. A similar plate is present in *Cerceris* (Sphecidae).

In most Apoidea the hind legs are adapted for pollen carrying (Fig. 37.7O). The hind tibia and basitarsus are more or less dilated and bear long hairs on their outer surface. When the hairs are distributed over the whole surface this pollen brush is referred to as the *scopa*, but when they are reduced to a fringe surrounding a broad concavity the whole is termed a *corbicula* or *pollen basket*. Pollen-carrying hairs (or a scopa) may be developed on various other areas of the body. Thus, a well-developed scopa occurs on the venter of the gaster in Megachilidae. A brush-like patch of dense hairs, also called a scopa, is developed on the hind coxae of some female Ichneumonidae (Heinrich, 1961).

Wings (Figs. 37.8, 9). Relatively stiff and glassy, but usually covered with fine macrotrichia, and with areas of dense microtrichia in the more primitive Symphyta. The hind wing is shorter and usually narrower than the fore wing. A well-developed anal venation in the hind wing occurs mainly in some Symphyta. The anal field may be defined by an apical

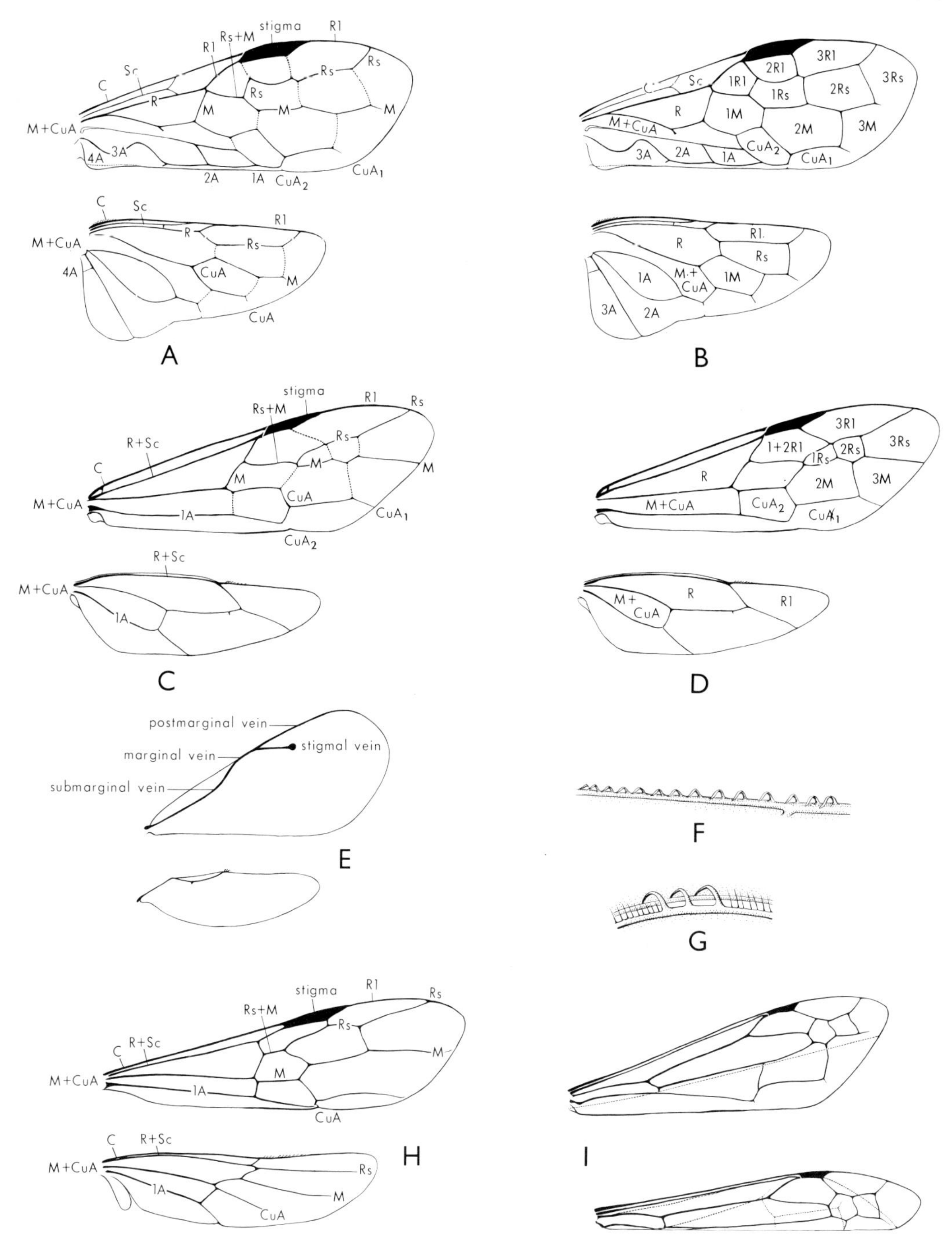

Fig. 37.8. Wings: A, B, *Pamphilius*, Symphyta-Tenthredinidae, A, veins, B, cells; C, D, *Taeniogonalos*, Apocrita-Trigonalidae, C, veins, D, cells; E, terminology in Chalcidoidea *(Nasonia)*; F, hamuli for wing-coupling in *Polistes*, Vespidae; G, wing-coupling in Braconidae; H, wings of *Myrmecia*, Formicidae; I, wing folding in *Polistes*. [S. Curtis]

emargination. It may also be subdivided, and then the basal lobe is referred to as a 'jugal' lobe. Apterous and brachypterous species are common. The workers of all ants are apterous. Wingless males occur in some Chalcidoidea (especially fig-inhabiting species), but it is usually females that are wingless (e.g. in all Mutillidae and Thynninae). When wings are absent, it is sometimes difficult to place a species in a family, except in such characteristic groups as Formicoidea, Mutillidae, and to a less extent Thynninae.

The venation is very reduced, and most of its characteristic appearance is due to fusion of parts of the veins. Homologies can be recognized readily in primitive forms (e.g. Fig. 37.8A). The terminology adopted here has been modified slightly from that of H. Ross (1936). With reduction in overall size there has been a corresponding reduction in wing venation, and some of the smallest species are devoid of venation. For any given size, parasitic species usually have a more reduced venation than predatory or melliferous species.

The fore and hind wings are strongly coupled in flight. The wings are held together by a row of hooks or *hamuli* over a limited portion of the costal margin of the hind wing, which catch on the folded posterior margin of the fore wing at the apex of the anal field. There is a correlation between number of hamuli and size of the species. In very small parasitic species there are normally only 2–3 hamuli, often raised on a slight projection (e.g. in Chalcidoidea and Proctotrupoidea), whereas there are 3–4 hamuli, not raised on a projection, in some Bethyloidea, Evanioidea, and Ichneumonoidea and up to many in the larger wasps and bees.

Hooked macrotrichia may occur over most of the fore margin of the hind wing between the zone of definite hamuli and the wing base. These are usually openly spaced, but they may be aggregated into a zone towards the wing base, or limited to one strong hamulus. When this grouping of secondary hamuli is developed, the fore margin of the hind wing is decidedly convex at this point. A secondary wing-coupling is best developed in primitive forms, in which there is a tendency for the coupling to extend from base to the hamuli with a break in the middle. It is also better developed when the fore margin of the hind wing is curved, and when there is a large number of true hamuli. Braconidae exhibit a very primitive condition, with the marginal macrotrichia well developed but not hooked. There is a well-developed secondary coupling in Trigonalidae and many Ichneumonidae, but in other ichneumonids the marginal macrotrichia are only slightly hooked. A secondary coupling occurs in many Scolioidea, Bethyloidea, and Pompiloidea, but it is usually absent in Sphecoidea, Apoidea, Vespoidea, Formicoidea, and Evanioidea. The wing-coupling may be so strong that the wings do not separate even when they are reflexed to the resting position. This necessitates longitudinal folding of the fore wing, cutting across the venation and dividing the wing into two more or less equal parts. Such wing folding occurs in three unrelated groups: many Vespoidea, all Gasteruptiidae

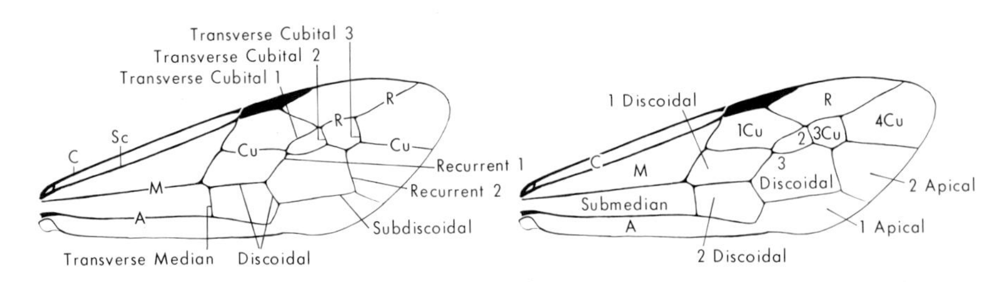

Fig. 37.9. Wing, other terminologies (*Taeniogonalos*, Trigonalidae). [S. Curtis]

(Evanioidea), and all Leucospidinae (Chalcidoidea).

Abdomen (Fig. 37.10). Segment 1 is normally without a sternite, though a reduced sternite is recognized in some Symphyta, and Short (1959b) recognizes a very narrow sternite in most Apocrita. The tergite is generally similar to the following tergites in Symphyta, but in Apocrita it forms the propodeum (p. 870). There is no constriction (or at most a slight one) between segments 1 and 2 in Symphyta, but in Apocrita a marked constriction almost always separates the propodeum from the remainder of the abdomen. The resulting shortened abdominal tagma is variously termed the abdomen, *gaster*, or *metasoma*. In ants the term 'gaster' is applied to the portion behind the waist (i.e. behind the petiole or petiole plus postpetiole).

The narrow constricted zone *(petiole)* at the base of the gaster may be so short that it is visible only when the gaster is deflexed (e.g. honey bee or *Bembix*), or it may be extremely long and thin (e.g. *Sceliphron* and *Ammophila*). Various components may be present in the more pronounced petioles. In the primitive condition it consists of the narrowed base of the 1st segment of the gaster with components of both tergite and sternite. Sometimes one or other of these components may be absent. In Chalcidoidea the petiole is often very reduced and may be overlooked, or the gaster may be secondarily sessile (some Aphelininae and Mymaridae). In some Ichneumonoidea the sternite may be divided. In some Sphecoidea there may appear to be 2 petiolar segments (consisting of displaced sternite followed by the tergite of the 1st gastral segment). There are 2 petiolar segments in many ants. The *glymmae* are lateral foveae between the base and spiracles of the petiolar segment. The *gastrocoeli* are deepenings at both sides of the base of the 2nd gastral tergite.

All 10 abdominal segments can be distinguished in Tenthredinoidea. The number of visible segments in the gaster of Apocrita varies greatly. As a rule, in the higher groups there are 6 tergites in the female and 7–8 in the male, but in some parasitic groups there may be as few as 4, or the basal segments may be fused into an apparent single segment. There is considerable variation in the number of sternites, though 6 is usual in females of Apocrita, sometimes reduced through fusion of the basal 2 (Megalyridae), 3 (Stephanidae, Proctotrupidae), or 4 (some Heloridae). All sternites are fused with tergites in Proctotrupidae and 2–5 fused in Pelecinidae. Abdominal spiracles are very variable. They are usually present on segments 1–8 in Symphyta (1–3 in Orussidae). In Apocrita a pair is always present on the propodeum;

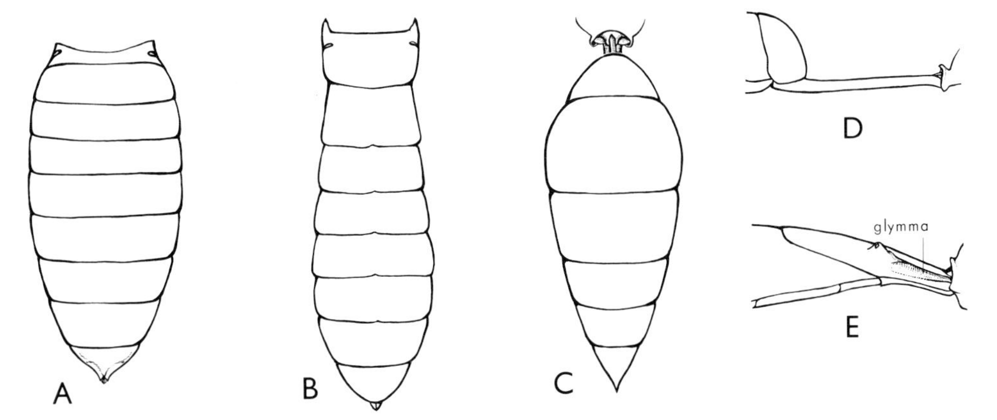

Fig. 37.10. Abdomen: A, *Perga* sp., Symphyta-Pergidae; B, *Guiglia* sp., Symphyta-Orussidae; C, *Polistes* sp., Apocrita-Vespidae; D, gaster of *Sceliphron* sp., Sphecidae, lateral; E, gaster of *Netelia* sp., Ichneumonidae, lateral. [S. Curtis]

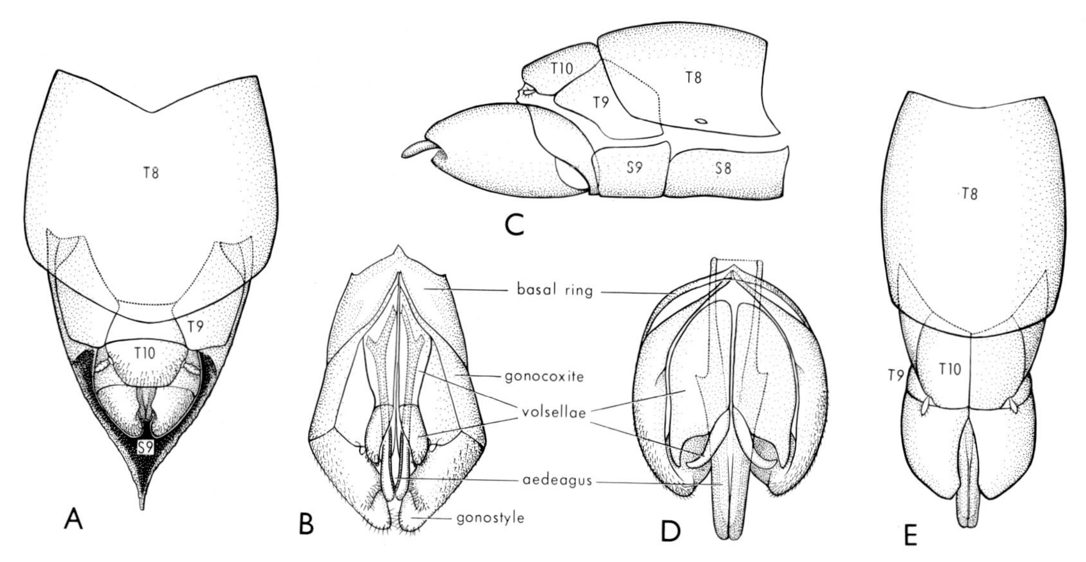

Fig. 37.11. Male terminalia: A, B, Symphyta (*Sirex noctilio*, Siricidae); C–E, Apocrita (*Rhyssa* sp., Ichneumonidae). [S. Curtis]

the gaster has spiracles on 1st–7th segments in some (Ichneumonoidea, Formicoidea), variously reduced in others, on 7th only in Chalcidoidea and Cynipoidea, and none in most Proctotrupoidea.

The gaster normally bears an apical pair of 1-segmented cerci *(pygostyles)*, or they are reduced to a patch of sensilla (trichobothria) on the composite morphological tergite 9 + 10, which is hidden in the predatory species through retraction of the apical segments. The pygostyles are better developed in males than in females. A *pygidial plate*, usually flat and surrounded by a carina or line and sometimes produced as an apical projection, is present on 6th gastral tergite in females, and 7th in males, of some Apoidea, Sphecoidea, and Scolioidea.

The male terminalia (Fig. 37.11) are usually large. The general structure of the phallus is constant throughout, though special types of modification are characteristic of super-family groups. It arises from the sternum of morphological segment 9 as a pair of phallic lobes which subdivide and later coalesce at the base. The lateral lobes develop into parameres (= gonocoxites, p. 26); the composite median lobe becomes the aedeagus; and ventral lobes become the *volsellae*, each of which may be produced into 2 lobes at apex; a distinct basal ring is also set off (Snodgrass, 1941).

The ovipositor (Fig. 37.12) is very highly developed. Basically (in most Symphyta) it is used for sawing, but is often modified for piercing or stinging. Morphologically it is composed of 3 pairs of valves which arise from a similar number of abdominal processes in the larva, with 1 pair on segment 8 (7th gastral) and 2 pairs on 9. Those of segment 8 develop into the *stylets*. The middle pair of processes fuse to form the dorsally placed *stylet-sheath*, or remain separate and form a pair of sheaths. The terminal pair give rise to the palp-like process ('ovipositor valves'). The stylets and stylet-sheath form the actual sting or *terebra* of Apocrita. The stylets are grooved along their entire length. The apices of the stylets and their sheath are usually provided with forwardly directed barbs.

The ovipositor of Hymenoptera is remarkable for the diversity of its secondary functions. In some parasitic species it functions as a sense organ by which the female is able to

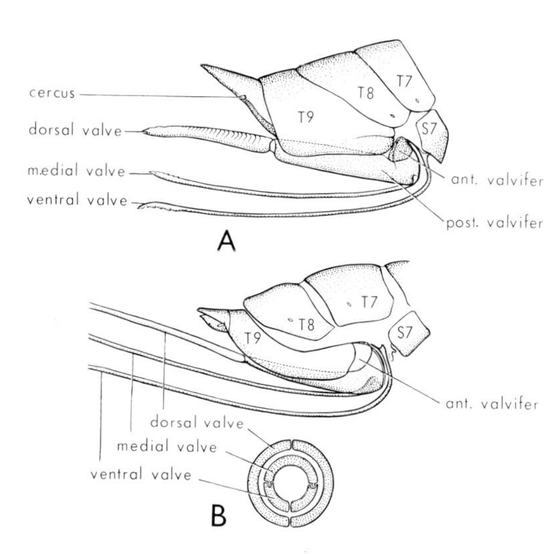

Fig. 37.12. Female terminalia: A, Symphyta (*Sirex noctilio*, Siricidae); B, Apocrita (*Rhyssa* sp., Ichneumonidae). [S. Curtis]

recognize an object as a host. It is used not only to deposit an egg in the host, but as a hypodermic needle to inject paralysing fluids prior to oviposition, or as a weapon of defence in social species. Perhaps its most extraordinary use is that of a food-procuring instrument. Females that need additional protein for egg production use it to tap the host's body for nutrient fluids. If the host is concealed, either within its host plant or its cocoon, the female uses the ovipositor as a mould for construction of a pipe-line through which she imbibes the host's fluids.

Internal Anatomy. Digestive system reasonably uniform. The oesophagus dilates into a thin-walled crop or honey-stomach that serves as a reservoir for imbibed liquids in Sphecoidea, Vespoidea, and Apoidea. In Formicoidea there is an infrabuccal chamber opening into the mouth cavity that serves a variety of functions (Eisner and Happ, 1962). Malpighian tubes variable, 6–250, usually numerous, usually less numerous in parasitic than in predatory and melliferous species, few in ants, moderate number in Symphyta. Three thoracic ganglia, the meso- and metathoracic occasionally fused; 9 abdominal ganglia are said to be present in some Symphyta; there are 6 in many Apocrita, but fewer in many parasitic species, and only one in some Chalcidoidea. In the male, testes paired in Symphyta and a few Apocrita; accessory glands usually large and sac-like. In the female, the ovaries consist of a variable number (sometimes numerous, occasionally single) of polytrophic ovarioles; a median spermatheca is generally present.

Immature Stages

Egg. Usually ovoid or sausage-shaped, and in the parasitic groups commonly provided with a pedicle (Fig. 37.13), which is often long in Cynipoidea. Stalked eggs occur also in some Chalcidoidea and Proctotrupoidea, but the pedicle is poorly developed in Ichneumonoidea. A pedicle may occur in eggs that are laid externally as well as in those which are laid within the host. Its function is often obscure, though in some cases it serves as a respiratory funnel. Microlecithal eggs that develop polyembryonically (p. 84) occur in Chalcidoidea (Encyrtidae), Proctotrupoidea (Platygasteridae), Ichneumonoidea (Braconidae), and Bethyloidea (Dryinidae). The number of individuals produced may be small (*Platygaster*), but in *Litomastix* (Encyrtidae) it may be as high as several thousand.

Larva. The typical larva has a well-developed head, 3 thoracic, and usually 9 or 10 abdominal segments. In Symphyta (Figs. 37.14A–D) the head is strongly sclerotized, and there are powerful biting mouth-parts. In sawfly larvae that feed externally there are 3 pairs of thoracic legs, and usually 6 to 8 pairs of abdominal prolegs (typically on

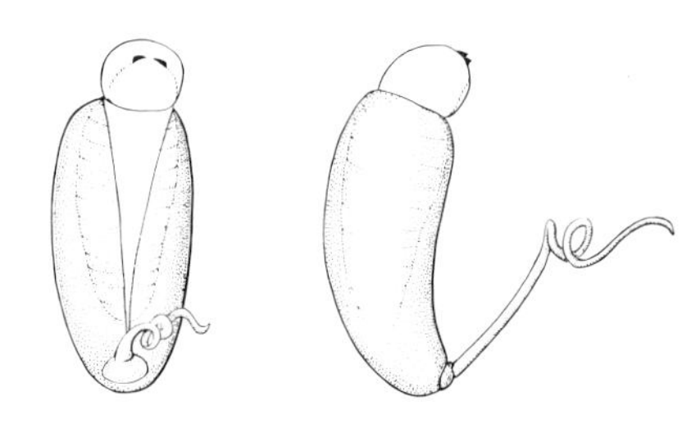

Fig. 37.13. Egg of *Netelia* sp., Ichneumonidae. [S. Curtis]

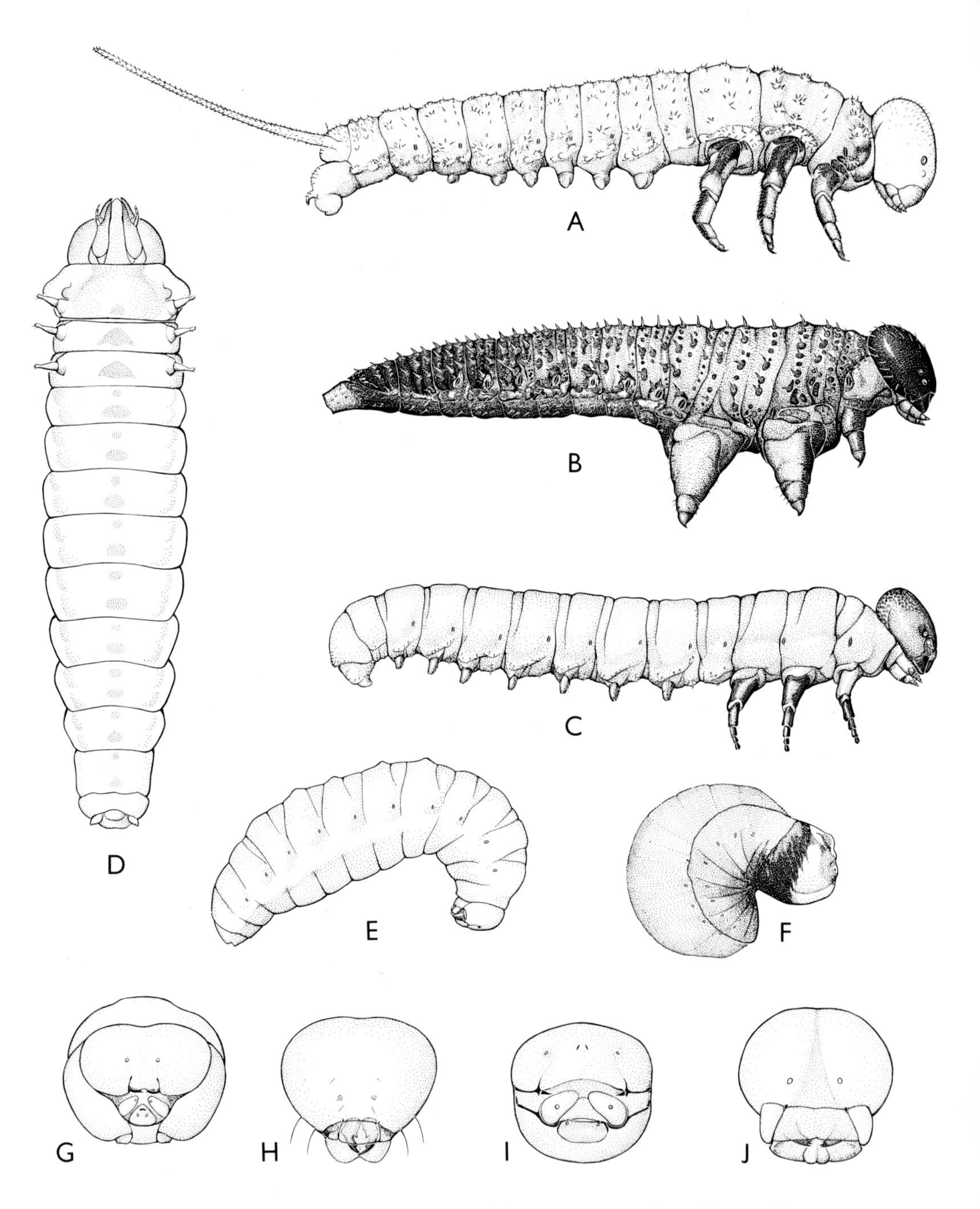

Fig. 37.14. Larvae, A–D Symphyta, E–J Apocrita: A, *Philomastix* sp., Pergidae-Philomastiginae; B, *Perga* sp., Pergidae-Perginae; C, *Zenarge turneri*, Argidae; D, *Phylacteophaga* sp., Pergidae-Phylacteophaginae; E, *Paralastor* sp., Eumenidae; F, unidentified Dryinidae. Heads, frontal: G, *Myrmecomimesis* sp., Cleptidae; H, unidentified Platygasteridae; I, *Westwoodiella* sp., Ichneumonidae; J. *Taeniogonalos* sp., Trigonalidae. [S. Curtis]

segments 2–7 and 10) none of which bears crochets. Larvae that bore into wood or tunnel in stems and leaf-mining species have reduced legs and no abdominal prolegs.

The larvae of Apocrita (Figs. 37.14E–J) are apodous and generally maggot-like. The head is less strongly sclerotized than in Symphyta, but the mandibles and certain bands associated with the mouth-parts are usually defined. In certain parasitic forms the head is greatly reduced and retracted into the prothorax. Definite stemmata are absent, and the antennae are never more than short sensory papillae. The mandibles are dentate, sickle-shaped, or simple pointed spines with broad flattened bases. Spiracles are usually present on the prothorax and first 8 abdominal segments. In almost all Apocrita the stomach ends in a blind sac, and does not communicate with the hind gut until the final instar. The faeces are evacuated just prior to pupation. Michener (1953) has given a key to some groups of hymenopterous larvae.

Heteromorphosis, with various primary larval forms some of which have received special names, occurs in many parasitic species. In the Hymenoptera the planidium (Fig. 37.15A) is an active larva, somewhat like a triungulinid, but provided with spine-like locomotory processes instead of legs. It occurs in certain Chalcidoidea (*Perilampus*, *Orasema*, *Leucospis*, and in modified form in *Spalangia*). In *Agriotypus* (Ichneumonoidea) the form is somewhat like a planidium, but the apex of the abdomen is drawn out into 2 long caudal styles. The *caudate* type has the last abdominal segment produced into a fleshy tail. It occurs in certain Ichneumonidae and Braconidae and in a few Chalcidoidea and Proctotrupoidea. *Cyclopoid* or *naupliiform* types occur in certain Proctotrupoidea. A type with a large caudal vesicle occurs in *Apanteles* and *Microgaster* (Braconidae). In Trigonalidae the early larval instars are vermiform and with distinct segmentation; and the third-instar larva has a greatly enlarged head bearing very stout mandibles. The final-instar larva in all cases is the maggot-like type characteristic of Apocrita.

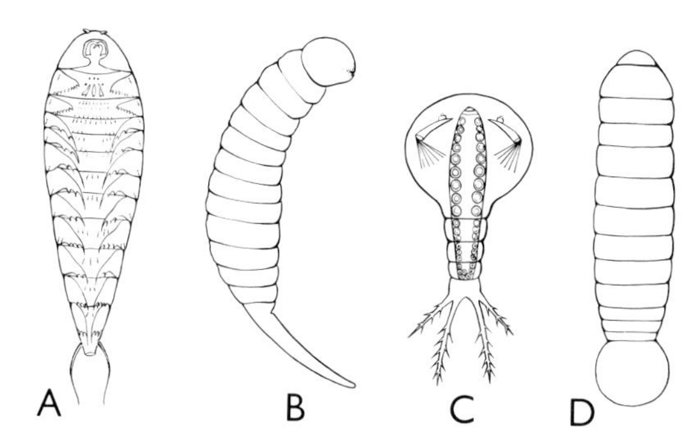

Fig. 37.15. Heteromorphic larvae of Apocrita: A, planidium; B, caudate type; C, cyclopoid type; D, vesicle-bearing type (A–C after Clausen, 1940, D after Imms in Richards and Davies, 1960).

Pupa. Always adecticous and usually exarate. A cocoon spun with silk from the labial glands, though often only slightly developed, is of frequent occurrence, but is absent in Chalcidoidea, Cynipoidea, most Apoidea, and many Formicoidea (a cocoon is formed in *Euplectrus* of the Chalcidoidea, in which it is said to be secreted by the Malpighian tubes). In many Vespoidea the cocoon is little more than a silken lining to the larval cell. The larva of *Polistes* spins the sealing cap to the cell.

Biology

Adults. Hymenoptera can be found in all kinds of situations: in and on the ground, on vegetation, especially at flowers, hawking for prey, drinking or gathering mud at the edges of pools or streams, and a few exotic Mymaridae even swimming under water in search of dragonfly eggs to parasitize. Most adults seem to feed to some extent on nectar or honey-dew. A few sawflies are regularly predacious on other insects. Many parasitic species suck the juices of their hosts as well as oviposit in them (p. 879).

Males of several kinds of bees are known to establish territories which they defend primarily against other males of the same species. Carpenter bees, however, may defend their territory against such intruders as other bees, flies, dragonflies, and sparrows. In some andrenid bees in the northern hemisphere territories involving flowering plants are guarded against all other insect intruders except females of their own species. Guarded plants include those visited by females for nectar or pollen, as well as plants that they were not observed to visit at any time.

Many adult Hymenoptera play an important biological role in pollinating flowering plants, and there are some remarkable mutual adaptations to this end. Thus, males of *Lissopimpla excelsa* (Ichneumonidae, p. 898) are attracted to 'copulate' with flowers of a number of species of *Cryptostylis* and so pollinate these ground-dwelling orchids. Specimens of *Pseudagenia* (Pompilidae), *Larra* (Sphecidae), and *Stilbum* (Chrysididae) have also been observed with pollinia on their legs.

Solitary wasps that construct linear nests in hollow twigs and similar situations where the tunnel is too narrow for the adult wasp to turn in it, employ a digital communication system that subsequently conveys information to the pupating wasp larva regarding the location of the exit from the burrow. The female wasp makes the cell walls asymmetrical in relation to the nest's sole exit, and the larvae respond to this asymmetry by orienting towards the exit when about to pupate (Cooper, 1957). Several important functions depend on odours. Some are sex attractants (Kullenberg, 1956); others are related to territorial behaviour, as in bumble bees; and others are used in orientation, as with trailing by ants. Many venoms are proteinaceous, as in the honey bee, vespids, and some ants, and there is a wide range of additional materials in ants, some species of which produce formic or related aliphatic acids and aldehydes or terpenoid ketones and lactones. The highest levels of both communication and defence occur in the social groups.

Stridulation is probably of widespread occurrence. Many sawflies stridulate by rubbing an area of modified microtrichia at or close to the anal margin of the fore wing over the cenchri. In Perginae the caudal margin of the cenchrus fuses with the metanotum to form a more efficient resonator. Stridulation in species of *Pseudoperga* is distinctly audible. Both sexes of Mutillidae stridulate by rubbing a transversely striated area at the basal margin of the 3rd tergite of the gaster against a sharp median apical ridge on the underside of the 2nd tergite. The wasp accomplishes this by moving the 3rd tergite in and out of the 2nd segment. Many ichneumonids, especially *Metopius*, stridulate loudly. The mud-dauber wasp, *Sceliphron laetum*, stridulates by movement between the overlapping surfaces of the pronotum and mesonotum (or possibly pronotum on propleuron). When stridulating, the whole head, including the opened mandibles, vibrates. The antennae are vibrated through a very small angle at a rate so fast that the amplitude of the movement becomes visible, the distal segments sometimes moving through a much greater arc than the main portion of the flagellum. At the same time, the folded wings are crossed slightly, and they too are vibrated rapidly. If one antenna is held during stridulation the other will continue to vibrate, and if the tips of both antennae are held the basal portions vibrate.

Reproduction. Males are always haploid, but production of females by facultative or cyclical diploid parthenogenesis (p. 80) also occurs. The cherry slug, *Caliroa*, for example, is able to breed from females only, and males are not known in Australia. In Cynipinae with alternating generations, on the other hand, one type of female produces both males and females parthenogenetically, whereas the other, after fertilization, produces females only. In the type of biparental reproduction that is much more general in the order, however, males are produced from ova that have not been fertilized as they pass down the genital tract and females from those that have. Fertilization, and consequently the sex of the offspring, is controlled by regulating the release of sperm from the spermatheca, which itself is subject to the influence of external stimuli. The sex-ratio can thus be regulated to meet the requirements of the species, and remarkably varied ratios are encountered in nature. In some species there is a difference in oviposition habits depending on whether a fertilized or unfertilized egg is laid. In *Coccophagus ochraceus*, for example, the female deposits fertilized eggs within the body of the host scale and unfertilized eggs externally. Preferential stimulation of the spermatheca occurs also in the honey bee, in which the work of the hive is done by diploid (but physiologically sterile) females and the haploid males are 'drones'.

Life Histories. The Hymenoptera may be divided into four broad groups on the feeding habits of the larvae: the primarily phytophagous Symphyta; the great group of parasitic Apocrita, with a few secondarily phytophagous offshoots; the non-social predatory Apocrita feeding on prey provided by their parents; and the usually social Apocrita that develop on specially prepared food. Only in the Symphyta do the larvae have to fend for themselves like those of most other insects.

The eggs of most Symphyta are deposited in leaf tissue, but in Siricoidea they are deposited into woody tissue, or into the frass-filled tunnels of beetle larvae, by means of the long ovipositor. The larvae of Tenthredinoidea feed externally on leaf and other plant tissue, though a few mine in leaves or stems or form galls. The Siricoidea have larvae that feed deep in wood, though they are apparently unable to digest woody tissue unless it is partly broken down by fungal attack. Most Siricoidea attack coniferous trees and tunnel the wood themselves, but at least some Orussidae ingest only the frass left in the tunnels made in angiosperms by cerambycid beetle larvae. These orussid larvae occur only some distance from the active end of the tunnels, where the frass has had time to become completely riddled with fungal mycelia. They may come into contact with beetle larvae that have been killed by parasitic ichneumonids and aulacids, but are not themselves parasitic.

The most primitive Apocrita are parasitoids (p. 124). The adult stage is not closely associated with the host, which is nearly always killed by the larva. If it is not killed, its reproductive cycle is interfered with, often to the extent of sterility should it reach maturity or should the adult be parasitized. The eggs are normally deposited either on or in the host, and the larvae usually develop in the same situations, though early development may often be internal and later development external. A few parasites are heteromorphic, and lay eggs which produce an active larva that finds its own host. In the Trigonalidae the eggs are deposited in leaf tissue just as with most Symphyta, but they do not hatch unless they are first ingested by the host larva (within which the trigonalid larvae have mostly evolved into hyperparasites). The parasitic habit most probably arose in this way. A few groups have developed the gall-forming habit, notably the Cynipidae and some Chalcidoidea. These species oviposit into plant tissue, usually very young tissue, which proliferates around the egg and developing larva, and the larva feeds on the new tissue. There are genera in the Chalcidoidea (e.g. *Eurytoma*) that contain both parasitic and gall-forming species.

Egg deposition in many parasitic Hymenoptera involves the passage of the egg through a fine ovipositor, and eggs that contain yolk in amounts sufficient for the development of the embryo ('yolk-replete eggs') are compressed during their passage. Since development may begin before they leave the genital tract, these eggs must be deposited promptly, before embryonic development prevents their extrusion through the ovipositor.

A few parasitic species have developed the ability to retain the egg until it is nearly or quite ready to hatch. In most Ichneumonidae the egg passes down the very narrow ovipositor and undergoes considerable compression and elongation in the process, due to elasticity in the chorion. In the ectoparasitic *Tryphon* and allied genera only the knob or anchor at the apex of the egg-stalk passes down the ovipositor channel, and the body of the egg is external to it. Up to 17 eggs may be extruded and carried in a cluster at the apex of the ovipositor, where they undergo development. Some further time elapses after deposition on the host before they hatch. In *Polyblastus* and *Netelia* the egg is retained on the ovipositor and partly hatches, but the larva does not leave the split chorion. The egg is attached to the host by its stalked end with the anchor beneath the cuticle. The larva then protrudes its head and attaches externally to the host.

The capacity to regulate egg-disposal is a very important factor in the economy of parasitic Hymenoptera. It is based on two different phenomena: ovisorption, or the ovarian resorption of yolk-replete eggs; and the storage of yolk-deficient microlecithal eggs in the genital tract (Flanders, 1962). In microlecithal eggs development is initiated by immersion in the body fluids of the host, a trophamnion is formed, and polyembryony ensues (p. 879). Since a microlecithal egg undergoes little if any constriction in its passage down the ovipositor, the time required for its deposition is less than that for a yolk-replete egg.

With many parasitic Hymenoptera host preference is determined to a large extent by the habitat and general characteristics of the host, and this sometimes gives an erroneous impression of host specificity. Apparently the host specificity of species with many potential hosts is largely an effect of the host-habitat finding process, whereas with species that have few potential hosts it is largely an effect of host suitability.

In the predatory species the egg is deposited on or close to the paralysed prey, and the larva consumes the food provided. Pompilidae show a transition from parasitic to predatory habits. Some of the most primitive species sting a single host spider, and deposit an egg externally on it. The host soon recovers and leads an active life for a time. Other species inactivate a single host in its own retreat, while others provision a cell with a single host. Still others store several victims in the cell, deposit a single egg on or close to them, and the larva feeds on the stored food, consuming more than one host individual as it grows. Most Sphecoidea also store several hosts, but others (e.g. *Bembix*) practise progressive provisioning, in which they provide additional food at intervals for the growing larva. Similarly, some Vespoidea store several intact hosts in their cells, whereas others masticate the food for the larvae.

In the social species the larvae are fed throughout development on food provided by adults, usually by the parent initially, but by a sterile worker caste in developed colonies. Social vespoids use food of animal origin, though a very few species are melliferous (i.e. pollen and nectar gatherers). Bees use pollen and nectar in place of animal food, and some ants utilize seeds, fungi, and other plant foods. These groups could thus be regarded as secondarily phytophagous. Some of them have become socially parasitic (p. 127), in that they utilize the food supply of other, usually closely related species, including Apoidea, Sphecoidea, Pompiloidea, and Formicoidea.

Natural Enemies. Regulation of populations in Hymenoptera is often through availability of nesting sites and food, though parasitic Hymenoptera and Bombyliidae (Diptera) are important natural enemies. Fungal, bacterial, viral, and protozoal diseases are often important in social species. A number of species are stylopized by Strepsiptera, but usually the percentage parasitism is low. Stylopization is limited to the Sphecoidea, Apoidea, Vespoidea, and Formicoidea. Mermithid worms occur in some Apoidea and Formicoidea. Both these types of parasites produce partial or complete sterility of the host. Some rhipiphorid and meloid beetles parasitize various Vespoidea and Apoidea. Commensal mites occur in some Vespoidea and Apoidea, but their role in the biology of the species is unknown. Several families of mites contain species that are parasites of solitary wasps and bees (Krombein, 1961), especially on the more or less helpless immature stages. Ants are an important item in the diet of some mammals, reptiles, and frogs.

Economic Significance. There is no group of insects so beneficial to man as the Hymenoptera. The parasitic and predatory wasps exercise a marked control on the abundance of other insects and certain other arthropods, including a variety of pests. The bees are most important pollinators of plants, especially trees, so that without their nectar and pollen-collecting activities much of our forested land would be bare and our orchard trees would not bear fruit. They also pollinate pasture plants and crops such as lucerne. Their pollinating activities are, indeed, more important than their commercial value as producers of honey.

A relatively few species are harmful. Sawfly larvae eat the foliage of plants, and sometimes cause complete defoliation, but this rarely kills the plant except in combination with attack by other insects. The wood-boring *Sirex noctilio* causes damage to pine plantations. Some parasitic Hymenoptera are secondary parasites, and so mitigate beneficial effects of the primary parasite. A few parasites attack hosts that are beneficial to man. Some ants are pests, and some tend sap-sucking insects that are harmful to crops.

Special Features of the Australian Fauna

As most Hymenoptera are active fliers, or are so small that they are readily dispersed by air currents, it is surprising that there are any distinctive elements in Australian Hymenoptera. The Symphyta are more distinct than the Apocrita. The native symphytan fauna consists mainly of Pergidae, which are almost limited to Australia and South America. In Australia there has been a marked radiation within the family to produce a number of leaf-feeding types as well as the leaf-mining Phylacteophaginae. The more highly evolved Tenthredinidae and Argidae are represented by a few endemic species, mainly limited to the north, and Siricoidea are represented by a few Xiphydriidae and Orussidae.

An outstanding feature of the Apocrita is the very poor representation of the gall-forming Cynipinae. Their ecological niche has been filled by a marked development of gall-forming Chalcidoidea. Megalyridae, one of the most distinct and primitive families of Apocrita, are well represented. The most primitive Trigonalidae have persisted in Australia. Thynninae, a very successful parasitic group of the Scolioidea, are dominant in Australia and South America. Few primitive or distinct types of the predatory wasps, which are mostly very active fliers, are present. Australia has the most distinctive continental bee fauna in the world, characterized by the marked radiation of the Colletidae, the most primitive family of bees, and the absence or rarity of several groups that are otherwise almost world-wide. Andrenidae are absent, and there is only one melittid (in north Queensland). Most Australian bees have their closest relationships with Asiatic forms. The Australian ants include many old endemic genera, better represented in the south than in the north. Much of the fauna is best explained by a genus-by-genus accretion from Asia via the East Indies, mainly through New Guinea and Cape York, with a lesser route through the Indies to north-western Australia. Due no doubt to its relative isolation, Australia has also accumulated a fauna rich in endemic lineages derived from the old endemic genera.

CLASSIFICATION

Order HYMENOPTERA (8,834 Australian spp.)

Suborder SYMPHYTA (166)

SIRICOIDEA (16)
1. Orussidae (9)
2. Xiphydriidae (6)
3. Siricidae (1)
   Syntexidae (0)

MEGALODONTOIDEA (0)
   Megalodontidae (0)

TENTHREDINOIDEA (150)
4. Tenthredinidae (3)
   Cimbicidae (0)
5. Argidae (11)
   Blasticotomidae (0)
   Diprionidae (0)
6. Pergidae (136)

XYELOIDEA (0)
   Xyelidae (0)
   Pamphilidae (0)

CEPHOIDEA (0)
   Cephidae (0)

Suborder APOCRITA (8,668)

MEGALYROIDEA (36)
7. Megalyridae (29)
8. Stephanidae (7)
TRIGONALOIDEA (12)
9. Trigonalidae (12)
ICHNEUMONOIDEA (390)
10. Ichneumonidae (190)
11. Braconidae (200)
Agriotypidae (0)
EVANIOIDEA (216)
12. Evaniidae (26)
13. Aulacidae (31)
14. Gasteruptiidae (159)
PROCTOTRUPOIDEA (615)
15. Heloridae (11)
16. Proctotrupidae (22)
17. Austroserphidae (3)
18. Pelecinidae (1)
19. Ceraphronidae (84)
20. Platygasteridae (50)
21. Scelionidae (333)
22. Diapriidae (107)
23. Loboscelidiidae (4)
CYNIPOIDEA (67)
24. Liopteridae (3)
25. Ibaliidae (3)
26. Figitidae (2)
27. Cynipidae (59)
CHALCIDOIDEA (2,791)
28. Agaonidae (21)
29. Trichogrammatidae (100)
30. Eulophidae (508)
31. Mymaridae (150)
32. Chalcididae (231)
33. Eurytomidae (194)
34. Torymidae (181)
35. Pteromalidae (435)
36. Encyrtidae (971)
CHRYSIDOIDEA (42)
37. Chrysididae (42)
BETHYLOIDEA (165)
38. Bethylidae (78)
39. Sclerogibbidae (3)
Sierolomorphidae (0)
40. Dryinidae (64)
41. Embolemidae (3)
42. Cleptidae (17)
POMPILOIDEA (122)
43. Rhopalosomatidae (1)
44. Pompilidae (121)
SCOLIOIDEA (733)
45. Scoliidae (25)
46. Mutillidae (197)
47. Tiphiidae (511)
Sapygidae (0)
Plumariidae (0)
VESPOIDEA (281)
48. Masaridae (25)
49. Vespidae (14)
50. Eumenidae (242)
SPHECOIDEA (442)
51. Ampulicidae (21)
52. Sphecidae (421)
APOIDEA (1,656)
53. Colletidae (851)
54. Halictidae (422)
Andrenidae (0)
55. Melittidae (1)
Fideliidae (0)
56. Megachilidae (175)
57. Anthophoridae (193)
58. Apidae (14)
FORMICOIDEA (1,100)
59. Formicidae (1,100)

KEY TO THE SUBORDERS OF HYMENOPTERA

Abdomen broadly sessile at its base and without a marked constriction, though hinged, between segments 1 and 2; thorax with 2 pairs of spiracles, neither visible dorsally; cenchri present except in Cephidae. Larva with segmented legs except in a few tunnelling and leaf-mining species; antenna and maxillary and labial palps distinctly several-segmented, or, if 1-segmented, then apex of abdomen with a median sclerotized process or legs indicated by sclerotized discs .... SYMPHYTA (p. 887)

Abdomen with tergum 1 incorporated in thorax, and with a marked constriction and hinge between segments 1 and 2 or with a large thoracic phragma extending into the gaster; mesosoma with 3 pairs of spiracles, the pair on the propodeum conspicuous and usually distinctly visible dorsally; cenchri absent. Larva always without legs; antenna and maxillary and labial palps 1-segmented or absent; apex of abdomen not sclerotized .... APOCRITA (p. 892)

The Symphyta are clearly the most primitive Hymenoptera on the basis of both larval and adult structures. In the adult there is usually a well-developed anal field in the hind wing and, at most, only a slight constriction between segments 1 and 2 of the abdomen. The venation is the most complete within the order, but forms with a markedly reduced venation do occur. The great majority of species deposit their eggs in plant tissue, and the resultant caterpillar-like larvae feed externally on the plant tissue.

From the phytophagous Symphyta have arisen the basically parasitoid Apocrita, which are distinguished as adults from Symphyta only by the narrowing of the junction between gaster and propodeum, and, in general, by reduction of the anal field of the hind wing. The parasitic habit could have arisen from orussid or cephoid forms or from both. The orussid stem apparently gave rise only to the Megalyroidea. *Megalyra* differs from the orussids mainly in the petiolate gaster. The thorax and head are closely similar, with the basal portion of the antenna directed latero-ventrally below the eye instead of in the more usual direction dorsally or anteriorly. In *Megalyra* the ovipositor is exserted, whereas in the orussids it is coiled within the body.

The cephoid stem is considered to have produced the trigonaloid and ichneumonoid types, from which the remainder of the Apocrita can be traced. The trigonalid type of parasitic development may be the most primitive. Like sawflies, the trigonalids insert their eggs into plant tissue. But the eggs are very small, and, instead of hatching and the larvae feeding on the plant, they do not hatch until they are swallowed by phytophagous insect larvae in which the trigonalid larvae have become parasitic, though some are now hyperparasites. In the next stage in evolution of the parasitic habit, the parasite actually deposits its egg on or in the host. This enables it to produce a smaller number of larger eggs, and, by eliminating the hazards of chance ingestion of the eggs, increases the survival rate of the offspring. The larvae feed on the host, but cause its death only when they have completed their own development.

The first predators differed from their parasitoid ancestors only in that they stored the food in some special site, but one prey still served as food for one host. Subsequently, smaller food was stored, and this required the accumulation of several prey to feed the one developing predator, which led in turn to the more complex methods of provisioning described on p. 884. Predatory groups include some Pompiloidea, Vespoidea, and Sphecoidea. Social behaviour, probably the highest evolutionary development in the order, and often an associated change to food of plant origin, arose independently at least three times: in the primarily predatory Vespoidea; in the Apoidea from sphecoid ancestors; and in the Formicoidea from scolioid ancestors. Some Vespoidea, Sphecoidea, Apoidea, and Formicoidea are secondarily parasitoid, and some have become parasitic as inquilines feeding on the food reserves of the hosts or fed progressively in the case of the social ants.

Suborder SYMPHYTA

Adult with abdomen broadly sessile at its base; tergum 1 distinct, though closely associated with the metanotum, its sternum not usually distinguishable; no marked constriction between segments 1 and 2 (a slight constriction in Orussidae and Cephidae, but not nearly as pronounced as in most Apocrita). A prepectus is defined, but sometimes very small. Cenchri (Fig. 37.6A) protrude from the metanotum, and there is a corresponding 'scaly patch' of modified microtrichia on the anal field of the fore wing, except in Cephidae. These structures have a stridulatory function, at least in Pergidae, in which the piliferous lobes are developed into resonators. Fore wing with a second anal cell, sometimes incomplete; hind wing with basal field strongly developed, and usually with a large, emarginate anal field (small in Cephidae); nygmata (p. 474) present in various areas of the wings. Larva with well-developed head capsule, and usually with well-developed legs and abdominal prolegs which lack crochets (both reduced in tunnelling and leaf-mining species, and the prolegs reduced in some Pergidae).

The higher classification has been reviewed by H. Ross (1937) and Benson (1938b). Apart from Pergidae, the suborder is poorly represented in Australia. Cephoidea, which have no cenchri, the single fore tibial spur developed into a calcar, venation approaching that of Apocrita, and larvae that are stem borers in grasses and rosaceous plants, are absent. Xyeloidea and Megalodontoidea, which contain possibly the most primitive sawflies, are also not recorded. Two families of Siricoidea, Xiphydriidae and Orussidae, have endemic, though not common, species, and there is only an introduced species of Siricidae. Tenthredinoidea are represented mainly by Pergidae, which occur throughout; the few Tenthredinidae are confined to the north, apart from one introduced pest species; and the few Argidae occur mainly in south-western W.A., apart from one widespread species that feeds on *Callitris* and an introduced miner in *Portulaca*.

Key to Families of Symphyta Known in Australia

1. Mesoscutellum completely separated from scutum by a suture and the axillae defined (Fig. 37.16A); subantennal grooves present for reception of the basal segments of the antennae (Fig. 37.3B) ... . SIRICOIDEA. 2

   Mesoscutellum not separated from scutum laterally and axillae not defined anteriorly (Fig. 37.16B); no subantennal grooves .... TENTHREDINOIDEA. 4

2. Antennae inserted on the ventral side of the head, below the lower margins of the eyes and below the apparent clypeus; hind wing without closed discoidal or cubital cell (Fig. 37.17E); eyes normally finely hairy .... **Orussidae** (p. 888)
   Antennae inserted well above the clypeus and on the anterior aspect of the head; hind wing usually with at least one closed cubital or discoidal cell; eyes bare 3
3. Neck long (cervical sclerites viewed from the side appear longer than broad); last abdominal segment without a horn-like projection; maxillary palp 4-segmented (Fig. 37.17F) .... **Xiphydriidae** (p. 890)
   Neck short (cervical sclerites viewed from the side appear broader than long); last abdominal segment with a horn-like projection; maxillary palp 1-segmented (Fig. 37.17G) ...... **Siricidae** (p. 890)
4. Radial cell of fore wing crossed by a vein; fore tibia with 2 spurs, one developed into a calcar (Fig. 37.17C) .... **Tenthredinidae** (p. 890)
   Radial cell of fore wing not crossed by a vein; fore tibia without spur developed into a calcar 5
5. Third antennal segment very long, antenna only 3-segmented (3rd segment may be bifid); fore tibia with 2 spurs in Australian species (Fig. 37.17H) .... **Argidae** (p. 890)
   Third antennal segment not abnormally long, antenna with more than 3 segments; fore tibia usually with 2 well-developed spurs, 1 only in *Phylacteophaga* and *Pergula* (Fig. 37.17I) ... **Pergidae** (p. 890)

Superfamily SIRICOIDEA

The female ovipositor is adapted for boring, rather than sawing as in Tenthredinoidea. The larvae, which have greatly reduced legs and no abdominal prolegs, are all wood-borers or live in wood.

1. Orussidae (Fig. 37.17E). The only family of Symphyta with hairy eyes. Antenna 11-segmented in male; 10-segmented in female, with penultimate segment swollen and longer than any other segment, and apical segment peg-like with truncate apex. Vertex of head with a zone of rasp-like tubercles (similar to those in *Eusandalum* and Stephanidae); mandibles very stout and gouge-like. Fore leg of female (Fig. 37.7F) strikingly modified, with a 'jointing' in the swollen tibia, and 3-segmented tarsus with the apex of the long basitarsus produced on one side; this possibly acts as an auditory mechanism for the female in her search for oviposition sites. Venation reduced mainly to pigmented bands; microtrichia dense; wing-coupling by 3–4 hamuli, with others more openly spaced towards the wing base. Larva legless; without marked constriction between head and thorax; mouth-parts reduced; each thoracic and abdominal segment dorsally with a lateral zone of backwardly-directed spines.

Orussidae differ from other Siricoidea mainly in the development of a long ovipositor, which is spirally twisted within the body in the adult, although it is external in the pupa and reaches forward over the abdomen as far as the front of the head. They are sometimes separated in a distinct suborder, mainly on this development and on their presumed parasitic habits. Adults are chalcid-like both in superficial appearance and in behaviour. They frequent areas of firm, dry, dead wood, and move up and down the surface with a jerky ant-like gait. At times they sway from side to side with the antennae vibrating rapidly, and appear to pivot about the apex of the abdomen. They can leap agilely. They rest with the antenna directed laterally and with apex on the substrate. The Australian species are relatively small (5–13 mm), and the body is usually subcylindrical, but very flattened in one genus.

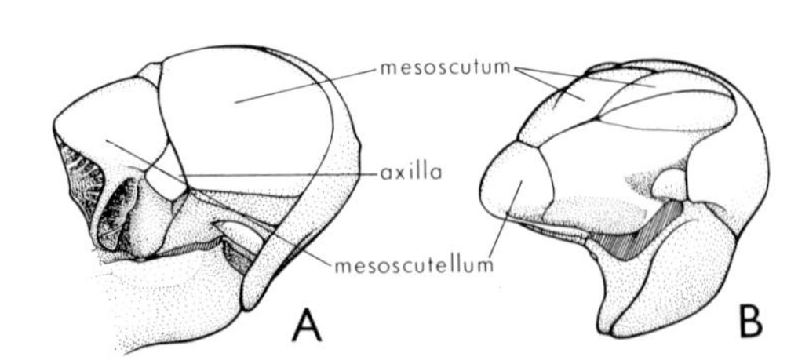

Fig. 37.16. Mesoscutellum and axillae: A, Siricoidea (*Guiglia* sp., Orussidae); B, Tenthredinoidea (*Pterygophorus* sp., Pergidae). [S. Curtis]

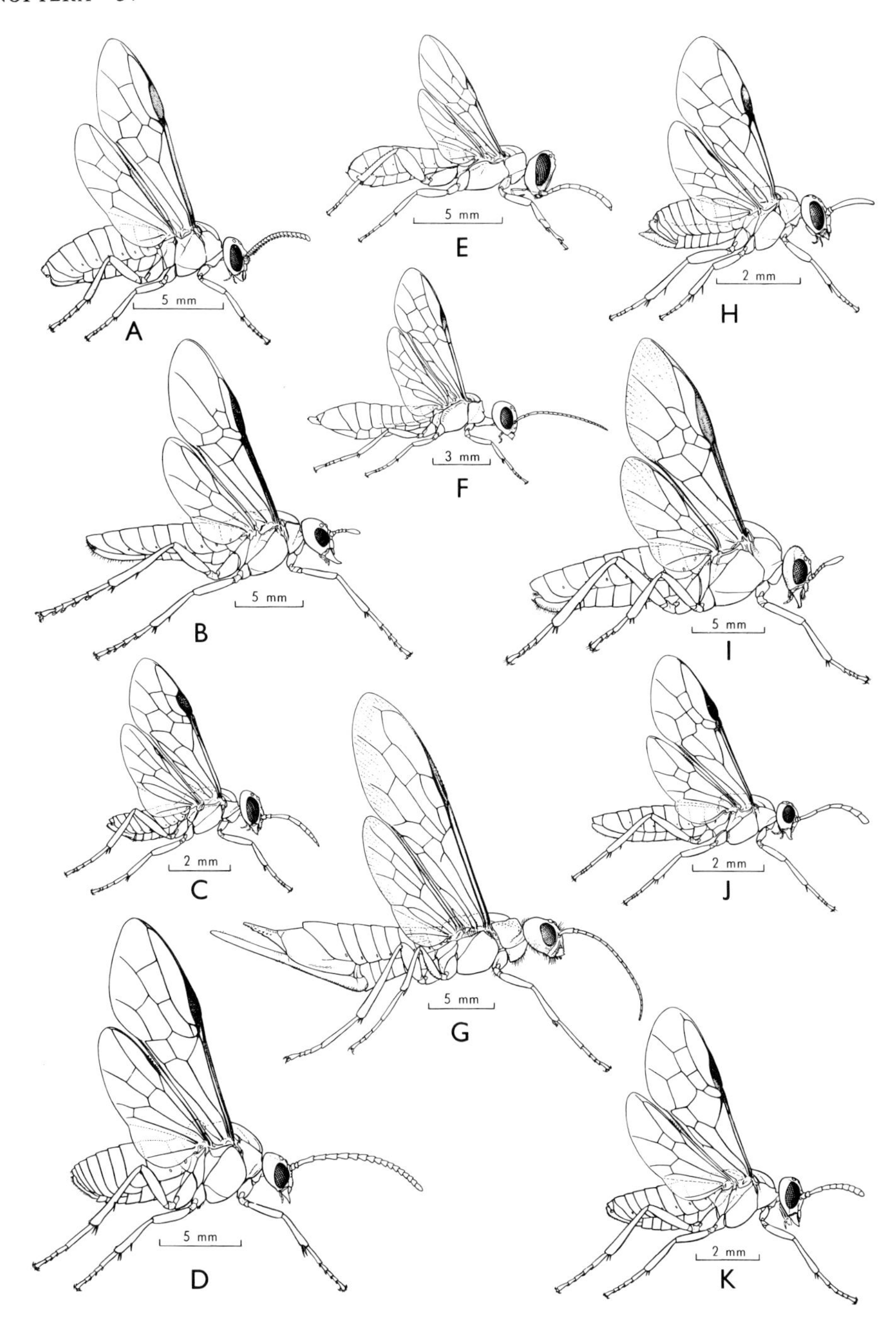

Fig. 37.17. Symphyta: A, *Pterygophorus* sp., Pergidae-Pterygophorinae; B, *Pseudoperga* sp., Pergidae-Perginae; C, *Caliroa cerasi*, Tenthredinidae; D, *Philomastix* sp., Pergidae-Philomastiginae; E, *Guiglia sericatus*, Orussidae; F, *Austrocyrta australiensis*, Xiphydriidae; G, *Sirex noctilio*, Siricidae; H, *Schizocerella pilicornis*, Argidae; I, *Perga affinis*, Pergidae-Perginae; J, *Phylacteophaga froggatti*, Pergidae-Phylacteophaginae; K, *Clarissa* sp., Pergidae-Euryinae. [M. Quick]

Little is known of the biology of orussids, but they are generally considered to be parasites of wood-boring beetle larvae, such as Cerambycidae and Buprestidae. It is possible that they feed only on the frass in the tunnels of such larvae, where fungi and other organisms associated with decomposition of the frass may be a requisite to larval development. Oviposition in one Australian species takes place only into fungus-affected, frass-filled tunnels some distance from an actively tunnelling beetle larva. *Orussobaius wilsoni* has been considered as a possible parasite of *Melobasis purpurescens*, a buprestid attacking acacias. *Orussobaius* adults are associated with acacias and sometimes eucalypts, while *Guiglia* adults frequent eucalypts and the allied *Angophora*. [Riek, 1955c].

2. Xiphydriidae (Fig. 37.17F). There are less than 100 species in this small, world-wide family. Host-plant associations are unknown, except for a few Holarctic species whose larvae, with vestigial legs, bore in deciduous angiosperms. Australian species have a body length of 7–11 mm. The rather long thin adults, with a long neck and a slightly exserted ovipositor, resemble a small *Sirex*. The wing venation is fully developed and the antennae 13–27 segmented. *Austrocyrta* is allied to South American genera. *Rhysacephala* (5 spp.) occurs also on Aru Is. [Riek, 1955i.]

3. Siricidae (Fig. 37.17G). Large wasps with conspicuous colouring, sometimes metallic. The abdomen usually ends in an upturned spine or horn, often conspicuous in females. The ovipositor is slightly but obviously exserted. Females normally oviposit in dead or weakened trees, boring holes through the bark and depositing a single egg in each hole. The larva tunnels into the wood, and causes considerable damage, particularly because of the symbiotic association between the larvae and rot-producing fungi. The introduced *Sirex noctilio* damages *Pinus radiata* and related coniferous trees in south-eastern Australia and Tasmania.

Superfamily TENTHREDINOIDEA

Females are provided with a saw-like ovipositor. The eggs are generally inserted into leaf tissue or woody twigs, but in a few they are only partly enclosed. The larvae are phytophagous, mainly on trees and larger shrubs, but a few on ferns and horsetails. Larvae often bear a close general resemblance to the caterpillars of Lepidoptera. They usually have well-developed legs and abdominal prolegs, but these are reduced or absent in leaf-mining and stem-boring species, and abdominal prolegs are absent in some Pergidae. The larvae of *Caliroa* are slug-like, for the body is obscured by a dark-coloured slime or exudation. Pupation generally takes place in a silken cocoon in soil.

4. Tenthredinidae (Fig. 37.17C). The small black introduced pear and cherry 'slug', *Caliroa cerasi*, is a pest of some fruit trees and ornamental shrubs. The larvae skeletonize the mature leaves, and pupate in the soil under the affected trees. *Cheilophleps xantha*, a small (8 mm) species with orange abdomen, occurs in N.S.W. *Senoclidia furva* is a New Guinea species also recorded from Darwin. [Benson, 1938a.]

5. Argidae (Fig. 37.17H). This widespread family is poorly represented. Most species belong to distinctive endemic groups. Larvae of *Zenarge turneri* (Fig. 37.14C), a large (10 mm) species with orange abdomen, feed on *Callitris* spp. and cause part defoliation; they also feed on introduced *Cupressus*. There are two generations a year in coastal areas, but emergences are irregular inland. The larvae of *Trichorhachus* (3 spp.) feed on smoke-bushes (*Conospermum* spp., Proteaceae) in W.A. The small (7 mm) adults are a bright metallic blue. Nothing is known of the biology of the widespread *Antargidium* (5 spp.) or the north Queensland *Styphelarge*. The introduced *Schizocerella pilicornis*, a small (5 mm) dark species, mines the leaves of the introduced *Portulaca oleracea* in eastern Australia. [Benson, 1962.]

6. Pergidae. This large family includes species of very diverse habits. The best known Australian sawflies are the large species of *Perga*, *Pergagrapta*, and *Pseudoperga* whose larvae feed on eucalypts and sometimes cause complete defoliation. The family occurs mainly in Australia and South America, but the faunas are distinct in general at the subfamily level. All species of *Ancyloneura* (Euryinae), Perreyiinae, a subfamily not recorded from Australia, and species of *Perga*, *Lophyrotoma*, and *Acanthoperga* are known from New Guinea. The host plants of many species remain unknown. Perginae and Phylacteophaginae are apparently restricted to *Eucalyptus* and the closely related *Angophora* and *Tristania*. Pterygophorinae occur on *Melaleuca* and *Callistemon*, but there are also records from *Eucalyptus*, *Acacia*, *Angophora*, the introduced *Rumex*, and *Emex*. *Philomastix* feeds on wild raspberry, and *Pteryperga galla* is recorded from *Elaeocarpus cyaneus*, a common small tree in south-eastern Queensland. [Benson, 1939, 1965.]

Key to the Subfamilies of Pergidae Known in Australia

1. Fore tibia with 1 apical spur 2
   Fore tibia with 2 apical spurs 3
2. Hind tibia with a preapical spine, sometimes reduced but its insertion obvious .... PHYLACTEOPHAGINAE
   Hind tibia without a preapical spine .... PERGULINAE
3. Hind tibia with a preapical spine 4
   Hind tibia without a preapical spine 7
4. Scutellum without lobes or paired projections behind 5
   Scutellum either bilobed, or with a backward projection from each hind angle 6
5. 'Propodeum' strongly sclerotized all over .... PTERYPERGINAE
   'Propodeum' strongly emarginate behind, only the front margin and sides sclerotized .... STYRACOTECHYINAE
6. Costa strongly swollen so that it touches R at least apically; anal vein of fore wing straight and close to the anal fold throughout; antenna generally less than 13-segmented .... PERGINAE
   Costa not swollen and much narrower than the intercostal area; anal vein of fore wing bent away from the anal fold apically; antenna with 13 or more segments (and serrate) .... PHILOMASTIGINAE
7. Cenchri small, the distance between them nearly twice as great as the transverse width of one; anal cell absent in fore wing. [Metanotum large, about the same width as scutellum] .... PTERYGOPHORINAE
   Cenchri large, the distance between them at most about as great as the breadth of one; anal cell of fore wing petiolate though it may be open apically .... EURYINAE

The larvae of PHYLACTEOPHAGINAE (5 spp.; Figs. 37.14D, 17J) form blotch-mines on the almost mature leaves of various *Eucalyptus* and *Tristania* species. They are amongst the very few leaf-mining sawflies in Australia. They occur over most of the east coast and Tasmania and in inland N.S.W. and Victoria. In colder regions there are two distinct generations a year, but in coastal N.S.W. and Queensland there is more overlapping of generations. The spring brood is more intensely coloured and larger than the autumn brood. [Riek, 1955b.]

PERGINAE (59 spp.; Figs. 37.17B, I) are mostly large, up to 25 mm in length. The larvae (Fig. 37.14B) are the common large species that occur on eucalypts. The eggs are inserted in batches in leaf tissue, and this causes a characteristic swelling of the leaf surface. The larvae aggregate during non-feeding periods, and are then most conspicuous, especially when the trees are partly defoliated. They pupate in the soil at the bases of the affected trees. Adults emerge over 2 or more seasons. *Pseudoperga* stands guard over the egg-pod and for some period over the immature larvae, and stridulates loudly, with vibration of the partly extended wings, when disturbed or handled during this period.

PHILOMASTIGINAE (3 spp.; Fig. 37.17D) are similar in size to Perginae, but have long conspicuous antennae and attractively patterned wings. Larvae (Fig. 37.14A) have long cercus-like processes on abdominal tergum 9. PTERYGOPHORINAE (18 spp.; Fig. 37.17A) are attractive, soft-bodied sawflies of medium size, often strongly marked with orange and with partly or completely infuscated wings. Adults are often found resting individually on grass stems in the early morning or during dull weather. Males have attractively plumed antennae. Sexual dimorphism is marked. EURYINAE (47 spp.; Fig. 37.17K) are very small (5–7 mm), soft-bodied sawflies common in early summer on blossom, especially the small, tubular, blue flowers of *Wahlenbergia*, but also attracted to eucalypt and other flowers. Food plants and biology unknown. They are mostly black or dull-coloured, but the species of *Eurys* have metallic hues. Plumed antennae occur in the males of *Polyclonus*. *Ancyloneura* occurs also in New Guinea, but the other genera are restricted to Australia. Each of the remaining 3 subfamilies is represented by a single, uncommon species. Males of *Pteryperga galla* are distinguished by biflabellate antennae.

Suborder APOCRITA

Segment 1 of the abdomen incorporated in the thorax to form the propodeum, which always possesses a pair of large spiracles. There is almost invariably a marked constriction between the propodeum and the gaster or metasoma; but in some Aphelininae, Trichogrammatidae, and Mymaridae (Chalcidoidea) the gaster is broadly sessile, the widening of its basal attachments being correlated with the development of a phragma extending back into it. Larvae apodous and nearly always with reduced head capsule.

Key to the Superfamilies of Apocrita

1. Hind tibia with 1 or 2 spurs (without spur in a few Ichneumonidae), but none modified for preening (slight enlargement of one spur in some Scolioidea); sometimes a dense brush of enlarged hairs developed on the basitarsus and apex of tibia; parasitic 2
   When hind tibial spur or spurs present, one (the medial when there are 2) is modified into a calcar through development of a comb of hairs or teeth on its inner (tarsal) margin; a corresponding strigil or brush is developed on the slightly emarginate basitarsus (Fig. 37.7D); if both spurs are of non-simple form or lacking (in some Apoidea and Formicoidea), then at least some body hairs are plumed and the hind basitarsus widened, or first 1 or 2 segments of metasoma form a node; mostly not parasitic 10

2(1). Subantennal groove present for reception of basal segments of antenna (Fig. 37.3A). [A dense preening brush on inner surface at apex of hind tibia and on ventromedial surface of basitarsus (Fig. 37.7I), similar to that in Sphecoidea; spiracle cover lobe of pronotum with a marginal fringe of hairs, or, when spiracle is enclosed, then margined with fine hairs] .... MEGALYROIDEA (p. 893)
   No subantennal groove 3

3(2). Pronotum with at least a lateral spiracle cover lobe reaching back to tegula, or, if wingless, lobe margined with close fine hairs 4
   Pronotum not reaching back to tegula (almost so in some Leucospidinae, in which tegula is lengthened, and in some Mymaridae) (Fig. 37.5C) 9

4(3). Spiracle cover lobe of pronotum not margined with close fine hairs 5
   Spiracle cover lobe of pronotum margined with close fine hairs (poorly developed and more openly spaced in Aulacidae, but then fore-wing venation well developed and hind wing without closed cells) 7

5(4). Lateral pronotum not vertically grooved for reception of fore femur, though with a groove close to and parallel with anterior margin. [A grooved recess for the fore femur usually occurs on lower anterior portion of mesopleuron—not always defined, especially in Ibaliidae, and absent in Oberthuerellinae; antenna not elbowed] .... CYNIPOIDEA (p. 910)
   Lateral pronotum vertically grooved for reception of fore femur 6

6(5). Costal cell absent, but distinct marginal vein from base and a pterostigma; hind wing with at least one closed basal cell (except in Aphidiinae) .... ICHNEUMONOIDEA (p. 896)
   Costal cell present, or venation greatly reduced; hind wing without closed cells (see also Bethyloidea) .... PROCTOTRUPOIDEA (p. 904)

7(4). Pronotum with well-developed dorsal surface in median area, reaching back above tegula as well as below (pronotum reduced in nocturnal ♂ Mutillidae). [Hind-wing venation well developed] .... SCOLIOIDEA (p. 931)

Pronotum without a dorsal surface in median area, and not reaching back above tegulae 8
8(7). Hind wing without closed cells (see also Megalyroidea) .... EVANIOIDEA (p. 902)
Hind wing with closed basal cells .... TRIGONALOIDEA (p. 894)
9(3). Fore wing with closed basal cells; pronotum separated from tegula by lateral lobe of scutum (see also Bethyloidea and Formicoidea) .... CHRYSIDOIDEA (p. 925)
Fore wing with venation very reduced, without fully formed cells; pronotum separated from tegula by prepectus (though sometimes in part by lateral lobe of scutum) .... CHALCIDOIDEA (p. 913)
10(1). Lateral pronotum and mesopleuron (or prepectus) overlapping and with considerable free movement, lower portion of pronotal lobe rounded. [Secondary fusion between pronotum and prepectus in a few Dryinidae, but then antennae 10-segmented].......................... 11
Lateral pronotum and mesopleuron meeting with carinate margins and with very little free movement between them, lower portion of pronotal lobe tapering, usually to a point 12
11(10). Spiracle cover lobe of pronotum margined with close fine hairs; hind wing with at least one closed basal cell (see also Scolioidea) .... POMPILOIDEA (p. 928)
Spiracle cover lobe of pronotum not margined with close fine hairs; hind wing without closed basal cells (except in Sierolomorphidae). [Prepectus defined] ...... BETHYLOIDEA (p. 925)
12(10). A well-developed wingless worker caste present; posteroventral corners of thorax each with a metapleural gland (Fig. 37.37); first 1 or 2 segments of metasoma nodiform, sharply marked off from remainder .... FORMICOIDEA (p. 951)
Workers, when present, with wings; metapleural glands lacking; 1st segment of metasoma not often constricted to a node, though sometimes petiolate 13
13(12). Posterior lateral lobes of pronotum reaching back to and ending above tegula, and always angulate .... VESPOIDEA (p. 934)
Posterior lateral lobes of pronotum not reaching back to and ending below tegula, lobe rounded and limited to the spiracle cover 14
14(13). At least a few branched hairs on body; frequently social species. [Hind basitarsus wider than following segments] .... APOIDEA (p. 943)
Hairs on body not branched; non-social species .... SPHECOIDEA (p. 937)

Superfamily MEGALYROIDEA

Megalyridae and Stephanidae have a distinctive head and hind-leg structure which distinguish them from Ichneumonoidea. They differ from most Ichneumonoidea also in their reduced wing venation, whereas similarity in structure of the gaster is correlated with similar parasitic habits. The preening mechanism on the hind leg, better developed in Megalyridae than in Stephanidae, consists of a brush of dense short hairs on the medial surface of the tibia, especially towards its apex, and a similar development ventrally on the long basitarsus. In Stephanidae the brush on the tibia is limited to a very short apical area. The multisegmented antennae and wing venation of Stephanidae suggest a close relationship to Braconidae, but the head has a very different structure, with the crown of teeth on the vertex, the shape of mandible, and subantennal grooves (and venation) reminiscent of Orussidae.

Key to the Families of Megalyroidea

Gaster subsessile; antenna 14-segmented .... **Megalyridae** (p. 893)
Gaster with a long petiole; antenna with more than 14 segments .... **Stephanidae** (p. 894)

7. Megalyridae. Short, stocky species with subsessile gaster and very long ovipositor (Fig. 37.18A). Length of body 4–22 mm; ovipositor sometimes more than 5 times as long as body. Body and legs covered with relatively long hairs. The wings generally have a dark transverse infuscation in the middle, but entirely deeply infuscated in a few of the largest species; males may have clear wings. Antennal segments long and narrow, and tucked under along the venter when adults are feeding or at rest. The scutum has a median longitudinal groove mainly in the large species of *Megalyra*. There are several genera (undescribed) of MEGALYRINAE in Australia. *Megalyra* occurs also in New Guinea, Ceram and Sumatra. Dinaspinae, with a closed radial cell in the fore wing and a post-ocular carina on the head are recorded from South Africa, Java, and the Philippines. The biology of one species of *Megalyra* has been studied by Rodd (1951). As a female walks along in search of an oviposition site, gently tapping with the tips of the outspread antennae, there is an agitated slight raising and lowering of the folded wings. Oviposition takes place through the frass plug of beetle galleries, through cracks in the dry timber, or through sound timber. The larval life is apparently of short duration, and the larvae are said to be parasitic on beetle larvae in wattles and eucalypts. [Froggatt, 1906; R. Turner, 1916.]

8. Stephanidae. Adults (Fig. 37.18B) have a long thin body, thin multisegmented antennae, spherical head crested much as in Orussidae, long neck, petiolate gaster, long hind coxa, swollen hind femur, and a very long ovipositor. Venation reduced, but with distinct costal cell in the fore wing. Length 4·5–13 mm in Australian species. In general appearance they resemble Spathiinae (Braconidae). The hind tibia and tarsus are modified, except in the most primitive genera that retain a 5-segmented tarsus. The distal portion of the tibia is swollen, there is a tendency for it to be articulated to the basal part, and the tarsus, with reduced segmentation, has the basitarsus lengthened and produced on one side at apex. These modified hind legs probably form vibratory receptors. Species are most common in the tropics. Adults are usually collected on dead timber. The larvae are parasites of wood-boring coleopterous larvae; Australian species have been bred from bostrychid larvae in *Angophora* (Rodd, 1951) and from beetle-infested acacia twigs.

Superfamily TRIGONALOIDEA

Trigonalids have long been considered an archaic group close to the evolutionary point of separation of the 'terebrant' (parasitoid) and 'aculeate' (predatory) stems. They have developed a specialized method of depositing the eggs, and the prothoracic spiracle is covered by a lobe of the pronotum; but in all other respects they are ancestral to the other parasitic families and, through them, to the predatory and melliferous families. In the parasitic groups the tendency has been towards reduction in size and venation, but the predatory groups and the Apoidea and certain Vespoidea remain large and retain a well-developed venation. In general body form trigonalids resemble Sphecoidea, and, apart from the narrowed base to the gaster, resemble Pamphilidae amongst the Symphyta, though in venation they are more like Cephidae.

9. Trigonalidae. A small family of medium-sized wasps, mostly hyperparasites of sawflies through an ichneumonoid or tachinid primary parasite, but at least one Australian species of *Taeniogonalos* is a primary parasite of sawflies (Raff, 1934). Australian species have also been bred as hyperparasites from lepidopterous cocoons (Anthelidae, Geometridae, and ?Notodontidae). Species also parasitize Vespoidea, reaching the final host through leaf-eating caterpillars masticated as food by the wasps in feeding their larvae (Clausen, 1940).

The adults (Fig. 37.18C) are of stout build, with the gaster convex both above and below so that the ovipositor appears to issue from its apex. Antennae with 15 to more than 20 segments; costal cell present in fore wing; hind wing with distinct venation, 2 closed cells, and anal lobe vestigial; trochantellus usually free; mandibles broad, with stout apical teeth. The apical sternite 6, channelled and produced to a point, is used to slit the leaf tissue for insertion of the ovipositor. The very small eggs (0·15 × 0·07 mm) in *Taeniogonalos venatoria* are deposited in eucalypt leaves. During egg-laying, the female sits on the upper surface of the leaf

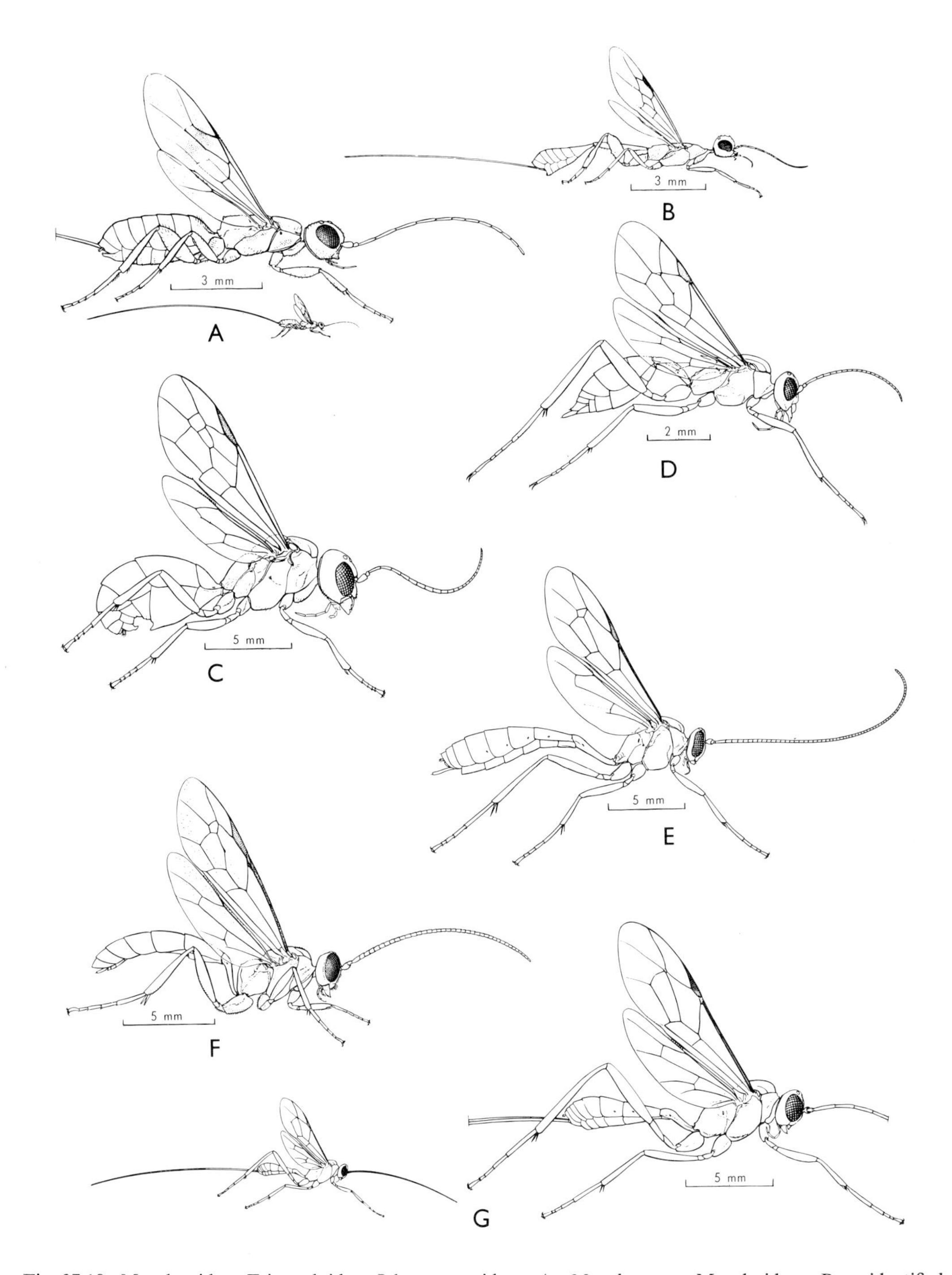

Fig. 37.18. Megalyroidea, Trigonaloidea, Ichneumonoidea: A, *Megalyra* sp., Megalyridae; B, unidentified Stephanidae; C, *Taeniogonalos* sp., Trigonalidae; D, *Agriotypus* sp., Agriotypidae (exotic); E–G, Ichneumonidae: Ophioninae, Ichneumoninae, Cryptinae. [M. Quick]

at right angles to the margin, passes the apex of the gaster over the margin, and makes a slit on the lower side of the leaf. The egg is deposited just below the slit and just beneath the leaf epidermis. Eggs are laid only in leaves that are just reaching maturity, and several may be deposited round the margins of one leaf. The eggs are ingested with plant tissue by sawfly larvae or lepidopterous caterpillars.

There are 2 Australian genera. In *Taeniogonalos*, which ranges to the islands to the north, the 2nd sternite of the gaster of the female has a process that is used during oviposition to anchor the basal part of the gaster to the upper surface of the smooth eucalypt leaf. The species are rarely less than 10 mm long, and best distinguished from one another by the secondary sexual characters of the females. In the endemic *Mimelogonalos*, the 2nd gastral sternite of the female lacks processes; the species are smaller (8 mm or less), and most readily distinguished by differences in colour. Adults of Australian species are normally bred from sawfly cocoons or collected on eucalypt foliage in summer and autumn during oviposition. Each species has a short emergence period, and adults are short lived. [Riek, 1954c, 1962,a,b.]

Superfamily ICHNEUMONOIDEA

The constituents of this superfamily have varied, but it is usually limited to the families Ichneumonidae, Braconidae, and Agriotypidae, in which the costal cell of the fore wing is absent or very reduced, and this arrangement is accepted here. It may be enlarged to include Megalyridae and Stephanidae and also Evaniidae, Gasteruptiidae, and Aulacidae. Agriotypidae, which have a stout aculeate-type of gaster and parasitize caddis larvae (Goeridae) in their cases (Fisher, 1932), are sometimes combined with Ichneumonidae, and the Aphidiinae are sometimes separated from Braconidae.

Antennae generally long and multisegmented, scape short, pedicel relatively large, and the antenna not elbowed; segments reduced to as few as 13 in females of some small Braconidae. In most Ichneumonidae and Braconidae the ventral surface of the gaster remains rather soft and flexible, so that it can be curved markedly as an aid in oviposition; but it is more heavily sclerotized in some Blacinae (Braconidae) in which the ovipositor is very short and stout. The trochantellus is large and jointed to the femur, except in at least the fore leg of Metopiinae (Ichneumonidae). This extra joint facilitates raising of the body and insertion of the ovipositor into the host. The distal sternites are shortened, and the ovipositor appears to issue before the apex of the gaster in all but the Agriotypidae, although it issues almost at the apex in some Blacinae and other braconids with reduced segmentation in the antennae. The fused Rs + M section of the venation is absent in the fore wing of Braconidae. The distal *m–cu* cross-vein is present in Ichneumonidae, but absent in Braconidae. There are 2 closed basal cells in the hind wing of Braconidae and Ichneumonidae, except in a few species of Braconidae with reduced venation, such as the Aphidiinae.

Key to the Families of Ichneumonoidea Known in Australia

Fore wing with 1 or without recurrent vein (Fig. 37.19); hind wing with median cell not extending to the base of the marginal vein. [Gastral tergites 2 and 3 immovably united except in Aphidiinae] .... **Braconidae** (p. 900)

Fore wing with 2 recurrent veins (Fig. 37.18G), very rarely only 1; hind wing with median cell elongated beyond the base of the marginal vein .... **Ichneumonidae** (p. 896)

10. Ichneumonidae (Figs. 37.18E–G). A large, diverse, and highly successful family of wasps that parasitize the larvae and pupae of endopterygote insects, immature and adult spiders, and the egg-sacs of spiders and pseudoscorpions. Host selection is sometimes very specific, but more often parasitization includes a variety of species that may be exposed to attack in the habitat explored by the ovipositing female. Most species prefer habitats with high humidity, and in drier areas individuals are more abundant along streams or during seasons of greater moisture, though some may depend on dew for survival. Adults are numerous in subalpine habitats during the summer, especially *Labium* and Ophioninae. In many habitats species are common in spring, but fade out in the hot dry summer period.

Most Ichneumonidae are of medium size; none is minute, but a few are very large. The family is characterized mainly by its wing venation. There are structural features in the head capsule of the larva (Fig. 37.14I) that appear to be connected with whether it is an ectoparasite or an endoparasite. The antenna in most ectoparasites is papilliform, whereas in most endoparasites it is reduced and disc-like. The labral sclerite is present and the mandible is often toothed in ectoparasites. Characters of the larvae indicate relationships similar to those based on adults. [Baltazar, 1964; Short, 1952, 1959a.]

Key to the Major Subfamilies of Ichneumonidae Known in Australia

1. First tergite of gaster petiolate (narrowest before base) and bent downwards towards apex; spiracle of 1st segment usually beyond mid-point (very rarely in the middle)......... 2
   First segment of gaster sessile, or, if subpetiolate, either straight or regularly curved and flattened in section; spiracle of 1st segment at or before mid-point 5
2. First segment of gaster narrowing at apex, round in section over basal half (and smooth); areolet of fore wing petiolate, rhomb-shaped, or absent (Fig. 37.18E) .... OPHIONINAE
   First segment of gaster depressed, widening or parallel-sided at apex; areolet of fore wing usually 5-sided or absent, never petiolate, rarely quadrilateral or rhomb-shaped 3
3. Areolet rhomb-shaped; ♂ cerci elongate, thin; spiracle of petiolar segment about mid-point. [First tergite with large glymma; tarsal claws pectinate] .... MESOCHORINAE
   Areolet not rhomb-shaped; ♂ cerci short and stout; spiracle of petiolar segment beyond mid-point 4
4. Mesosternum not separated from mesopleuron by a line or furrow (sternaulus absent); spiracles of petiolar segment often further from each other than from caudal margin of the segment; ovipositor usually not obviously exserted (Fig. 37.18F) .... ICHNEUMONINAE
   Mesosternum separated from mesopleuron by a sigmoidally curved line or furrow, usually extending more than half length of mesopleuron; spiracles of petiolar segment often closer to each other than to caudal margin of the segment; ovipositor usually distinctly exserted (Fig. 37.18G). [Areolet usually pentagonal] .... CRYPTINAE
5. Gaster subpetiolate, compressed. [Tarsal claws pectinate at least on fore leg] ....TRYPHONINAE
   Gaster sessile or subsessile, depressed 6
6. Fore trochantellus not defined; clypeus not separated from face by a groove. [Medial margin of eye emarginate; areolet usually rhomb-shaped, sometimes absent; gaster with subparallel sides] .... METOPIINAE
   Fore trochantellus defined; clypeus separated from face by a more or less distinct groove ... 7
7. Tergites of gaster each with a pair of oblique furrows, sometimes indistinct; ovipositor usually strongly exserted; areolet usually triangular or absent .... PIMPLINAE
   Tergites without such furrows; ovipositor not obviously exserted; areolet usually absent .... DIPLAZONINAE

There are only a few Australian species of ICHNEUMONINAE, and all are internal parasites of Lepidoptera. They oviposit in the host larva, less frequently in the pupa, and always emerge from the pupa (Heinrich, 1961). Host specificity is high. Sexual dimorphism is marked. One very common, easily recognized species, *Ichneumon promissorius*, with the basal half of the gaster orange and the apical half black banded with white, occurs throughout Australia. *Labium*, sometimes placed in Ichneumoninae, is best referred to a separate tribe of Xoridinae or Pimplinae. The head is elongate, the antennae short and with a tendency to be clubbed. The petiolar segment of the gaster, which is inserted rather high on the propodeum, is almost regularly slightly curved, and the spiracles are slightly beyond its middle. The species are cold-adapted, and occur in great numbers in the subalpine regions throughout summer. They are more generally distributed in Tasmania than on the mainland.

CRYPTINAE attack a wide variety of hosts. Many are occasional or obligatory secondary parasites. Most are ectoparasites, but some emerge from the host pupa. They are typically parasitic in cocoons, usually lepidopterous, but also of sawflies, braconids, ichneumonids, Neuroptera, egg-sacs of spiders, puparia of Diptera, and nests of wasps. A few parasitize pupae in stems, and a few attack borers in wood. *Gambrus stokesi* is a parasite of the codling moth. *Ceratomansa* spp. have been reared from the hard ovoid cocoons of Limacodidae. In *Gotra* several adults are often bred from one large lepidopterous cocoon; some species have an ovipositor considerably longer than the body. Poecilocryptini, small, mainly yellow species limited to the Australian region, are reared from soft galls on the new growth of acacias and eucalypts. Hemitelini may be secondary parasites. The large Mesostenini are the most conspicuous of all ichneumonids in the tropics.

Ichneumoninae and Cryptinae are generally rather large, stoutly built species, and the antennae often show a transverse white band about the middle. In general, the Australian species of these subfamilies differ in body colouring and length of the ovipositor. Many Cryptinae are black with white markings and the ovipositor is clearly exserted, whereas many Ichneumoninae are marked with orange or red and the ovipositor is not prominent.

PIMPLINAE are common. Many are parasites of wood-boring Coleoptera, but others parasitize large Lepidoptera. A few (Polysphinctini) are external parasites on the bodies of spiders or parasitize spider egg-sacs. The species are usually recognized by the oblique furrows, sometimes indistinct, that occur on the tergites of the gaster. The ovipositor is usually strongly exserted; but it is quite short and stout in the Echthromorphini, in which the tarsal claws are distinctly enlarged and which parasitize the pupae and 'prepupae' of Lepidoptera with little host specificity, so that adults are of very variable size. *Echthromorpha intricatoria*, one of the most common ichneumonids, emits a pungent odour when handled. *Theronia viridicans* is a widespread Pimplini with a bright green gaster. Pseudocopulation by males of *Lissopimpla* with terrestrial orchids of the genus *Cryptostylis* (Frontispiece) is a well-known phenomenon. It is apparently a response to scent, and is attended by protrusion of the male claspers as in normal copulation. The evidence from clusters of pollinia attached to the wasp's gaster indicates that it may be performed repeatedly (Coleman, 1938). Rhyssini have a long ovipositor for parasitizing wood-boring Coleoptera and Siricidae; they have a body length up to 25 mm and an overall length, including antennae and ovipositor, of 60 mm.

OPHIONINAE, which are very common, are characterized by a long petiolar segment to the long compressed gaster, and long hind legs, usually very conspicuous because of the pale tarsi, held straight out behind when in flight; antennae very long; ovipositor usually short and inconspicuous. They are often quite large, with a body length up to 30 mm, but some species of *Hymenobosmina* are only 2·5 mm long. The number of antennal segments is sometimes reduced to as few as 14 in the female (a few more in the male). One species has a toothed hind femur. The small delicate Cremastini parasitize mostly microlepidopterous larvae in leaf rolls. Many frequent drier habitats than are usual for the family. Tersilochini include parasites of vegetable weevils. One species of *Temelucha* parasitizes Lepidoptera attacking lucerne seed-pods. *Idechthis* is a common parasite of moth-infested grains and dried fruits in warehouses. *Hymenobosmina* parasitizes cabbage moth. The larger *Therion* has been bred from Anthelidae and *Enicospilus* from large Lepidoptera (?Notodontidae). [Kerrich, 1959, 1961.]

TRYPHONINAE are a small group, apart from the abundant, widespread species of *Netelia* (= *Paniscus*). They are ectoparasites mostly of lepidopterous larvae, but some parasitize sawfly larvae. The eggs are attached to the skin of the host by a stalk or other structure.

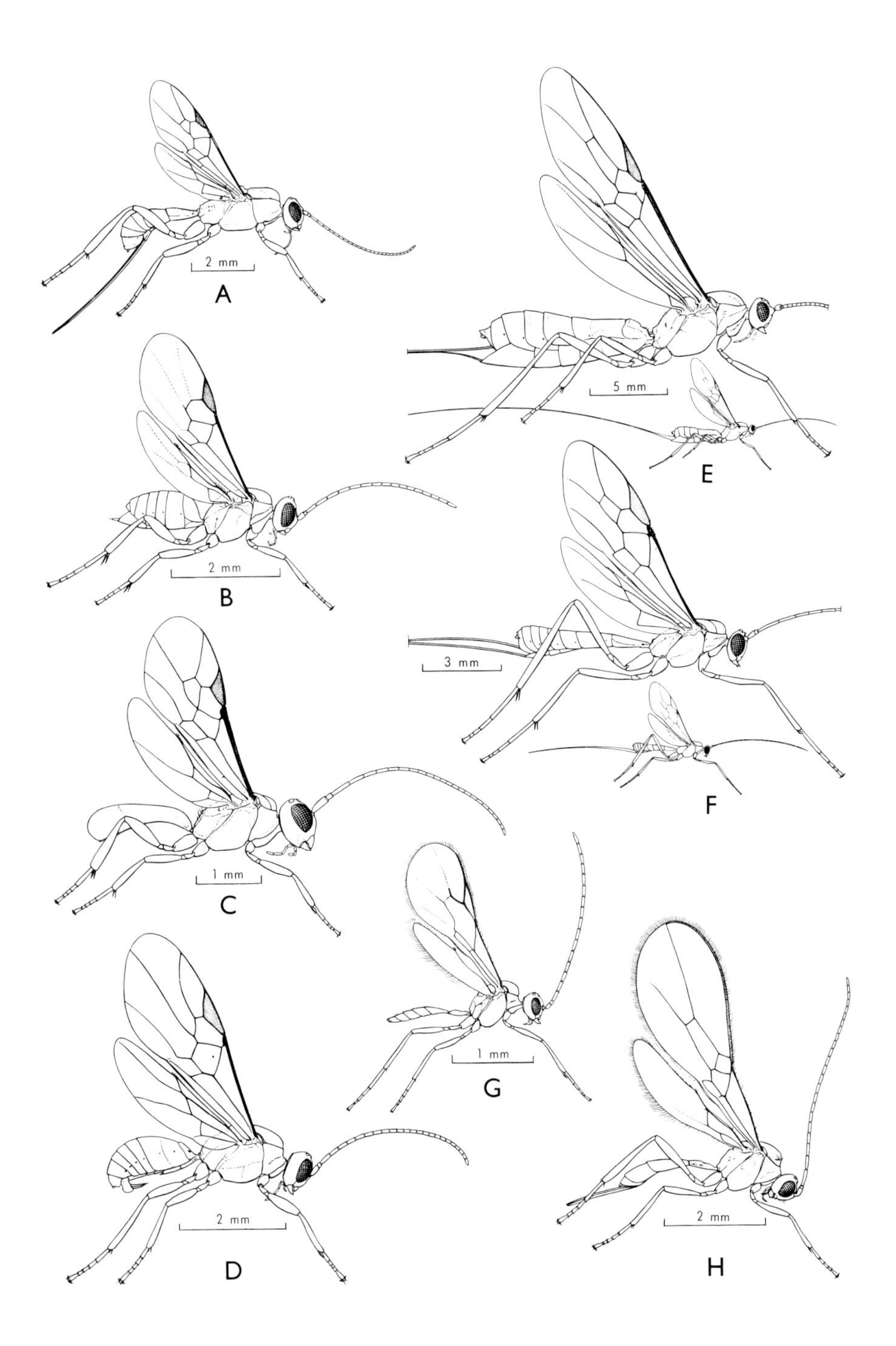

Fig. 37.19. Ichneumonoidea-Braconidae: A, Helconinae; B, Microgastrinae; C, Cheloninae; D, Blacinae; E, Braconinae; F, Macrocentrinae; G, Aphidiinae; H, Alysiinae. [M. Quick]

They are retained externally on the ovipositor for some time before deposition, and in a few cases hatch during this period and are deposited as first-instar larvae. In *Netelia* the egg is attached by a spirally twisted stalk (Fig. 37.13). Most of the larval development occurs after the host spins its cocoon or makes its pupal cell in the ground. The species of *Netelia* are of moderate size, thin, with compressed gaster and long legs and usually red-brown colour. They are the best known Australian Ichneumonidae, for the adults, with very large ocelli, are crepuscular and attracted to light. The females are amongst the few Ichneumonidae that can inflict a perceptible sting. [Kerrich, 1936, 1952.]

MESOCHORINAE occupy an isolated position on both adult and larval characters. The cerci of the male are elongate and could be mistaken for ovipositor sheaths. There are only 6 or 7 small (5–8 mm) species of *Mesochorus* and *Cidaphus* in Australia. All are believed to be both endoparasitic and secondary parasites; one species of *Mesochorus* has been reared as a hyperparasite from cocoons of *Perga*.

METOPIINAE are endoparasites of Lepidoptera; they apparently parasitize the early larvae, but emerge from the host pupae, a single parasite per host. The strongly convex face with the antennae appearing to be inserted on a frontal shelf is diagnostic of all but *Metopius*, which has a flat, margined face produced to a point between the antennal insertions. The world-wide *Metopius* is represented by 3 species, all with black and yellow banded gasters and a body length of 12–15 mm, but most species of other genera are much smaller. In *Metopius* and some other genera the trochantellus of the fore and mid legs is not separated from the femur. Specimens of *Metopius* stridulate with a high-pitched, sawfly-like buzz when handled. Species of *Carria* are common in the high-rainfall areas of the east coast and south-western corner of W.A.

DIPLAZONINAE are endoparasites of Syrphidae (Diptera). They oviposit into the egg or young host larva, and emerge from the puparium. They are a rather isolated group. The adults are short and stocky, with a flattened sessile gaster which is slightly widened in the middle. The short, almost hidden ovipositor is directed dorsally. *Diplazon laetatorius*, 5–6 mm long, with gaster and legs mostly orange, is the only common species in Australia; it ranges from Queensland to S.A. and W.A.

11. Braconidae. Distinguished from Ichneumonidae by the absence of the second recurrent vein (or distal *m–cu* cross-vein; Fig. 37.19), though a few species in which this vein is absent have been referred to the Ichneumonidae on their general facies. In a few species there is a very narrow costal cell visible over at least the distal half (e.g. Helconinae), and in many a minute cell is distinguishable just before the pterostigma. Many small species have reduced antennal segmentation, and in a few (Pambolinae and some species of *Monoctonus*–Aphidiinae) there are only 13 segments in the female antenna. In some Cheloninae, Blacinae, and Neoneurinae the pterostigma appears double through development of a large prestigma (such as occurs in some Bethylidae). In *Ctenistes* (Blacinae) the gaster is heavily sclerotized ventrally, and therefore remains convex when dry. The hind coxa is enlarged in Microgastrinae and Cardiochilinae. [Baltazar, 1962; Parrott, 1953.]

Key to the Major Subfamilies of Braconidae

1. Mandibles widely separated, teeth curving outwards (exodont braconids; Fig. 37.20A) .... ALYSIINAE, DACNUSINAE
   Mandibles normal, their apices opposed and meeting when closed 2
2. Venation greatly reduced (Fig. 37.19G), hind wing without a closed basal cell; gaster subpetiolate and with all segments freely movable .... APHIDIINAE
   Venation not as reduced, or gaster subsessile; hind wing with at least one closed basal cell; at least basal 2 segments of gaster not freely movable 3
3. Clypeus semicircularly emarginate below, and forming with the mandibles a more or less circular opening or cavity (cyclostome braconids; Fig. 37.20B) 4

Clypeus not emarginate below, or at most with a broad shallow emargination (Fig. 37.20C) 6
4. Gaster distinctly petiolate .... SPATHIINAE
Gaster subsessile 5
5. Occiput not carinate (Fig. 37.19E) .... BRACONINAE, EXOTHECINAE
Occiput carinate .... DORYCTINAE, RHOGADINAE
6. Gaster with tergites fused to form a rigid dorsal shield, sutures absent or indicated only by fine grooves 7
Gaster with tergites separated by distinct sutures, all beyond the 2nd freely movable 8
7. Fore wing with 3 cubital cells (Fig. 37.19C) .... CHELONINAE
Fore wing with 2 cubital cells .... TRIASPIDINAE
8. Gaster distinctly petiolate .... METEORINAE, EUPHORINAE
Gaster subsessile 9
9. Venation greatly reduced (Fig. 37.19B)........ MICROGASTRINAE
Venation not greatly reduced 10
10. Radial cell very narrow, proctotrupid-like .... AGATHIDIINAE
Radial cell longer (Fig. 37.19F) .... MACROCENTRINAE, OPIINAE

The cyclostome group of subfamilies includes the typical subfamily BRACONINAE containing many, often relatively large species with a body length up to 15 mm and an ovipositor that may be twice as long; the wings are generally dark, and the body orange and black (Plate 6, Z, ZB). They parasitize a wide variety of hosts, including Lepidoptera, sawflies, and chalcid and psyllid galls. EXOTHECINAE are mostly large species with infuscated wings, and are like Braconinae except in venation. DORYCTINAE are long thin species, up to 13 mm, with ovipositors of the same length. They are external parasites of coleopterous larvae. Some parasitize buprestid larvae in acacias and *Callitris*, and one has been bred from the fern weevil. RHOGADINAE are somewhat smaller than Doryctinae, and the ovipositor is quite short; they parasitize Lepidoptera. SPATHIINAE could be mistaken for Stephanidae. The ovipositor is long, and there are a number of wingless species. They are apparently external parasites of coleopterous larvae.

The tergites of the gaster are fused to form a rigid shield in Cheloninae and Triaspidinae. The large subfamily CHELONINAE, parasites of lepidopterous larvae, contains stout species with a length up to 7–8 mm and usually coarsely sculptured body. In *Chelonus* there are no grooves across the gaster, but they are distinct in *Ascogaster*. TRIASPIDINAE resemble Cheloninae, but the ovipositor is quite obvious and often as long as the body; they are parasites of coleopterous larvae.

MICROGASTRINAE are internal parasites of lepidopterous larvae. They leave the host to pupate and spin a cocoon close to, or on the body of, the host. There are often several to several hundred parasites per host. The species of *Apanteles* are amongst the most commonly observed Braconidae. They are usually dull, dark-coloured, with a short, inconspicuous ovipositor, and easily recognized by the characteristic reduced venation. The fore wing of the AGATHIDIINAE has a very narrow radial cell lying along the stigma, and the remainder of the venation is very irregular. The pterostigma and radial cell are very proctotrupid-like. They are usually attractive species that can be mistaken for Braconinae. They parasitize lepidopterous larvae.

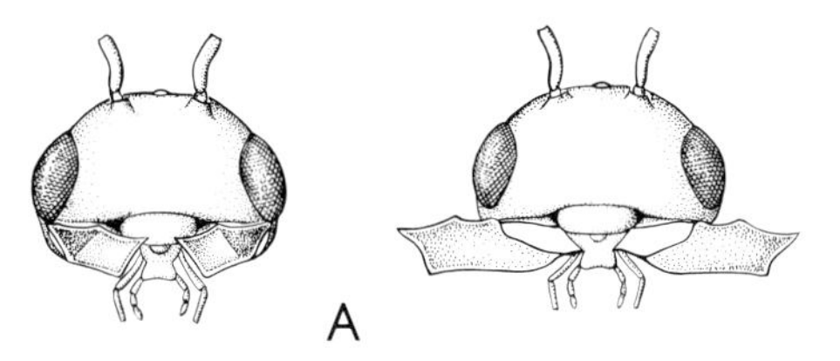

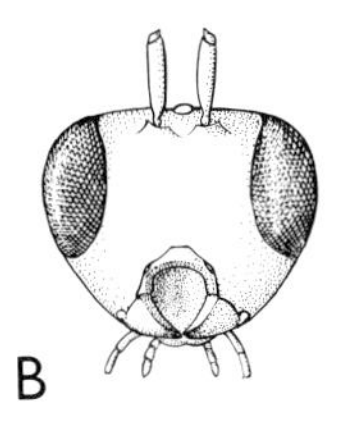

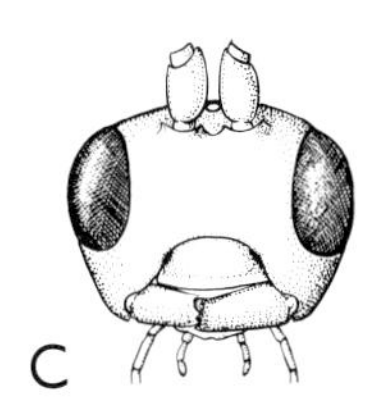

Fig. 37.20. Braconidae, heads, frontal: A, exodont type; B, cyclostome type; C, normal type. [S. Curtis]

The gaster is petiolate in Aphidiinae, Meteorinae, and Euphorinae, though it is not always distinctly so. The APHIDIINAE (Starý, 1958) are all parasites of aphids. The venation is quite reduced except in one or two small genera, and the hind wing is without distinctly closed basal cells. This is the only subfamily of the Braconidae in which the 2nd and 3rd tergites of the gaster are not fused. Some species of METEORINAE have the general appearance of Braconinae. A large (8 mm), dark-winged, black species of *Aridelus* with red head parasitizes *Paropsis* larvae (Chrysomelidae). A rather solid globular gaster follows the distinct petiole in the Australian species of EUPHORINAE.

MACROCENTRINAE are long thin species, up to 12 mm, with long legs and very long ovipositor. In general appearance they resemble ichneumonids. They parasitize mainly lepidopterous larvae. OPIINAE are mostly very small species that parasitize agromyzid and other small Diptera. HELCONINAE (Fig. 37.19A) are mostly parasites of coleopterous larvae. *Austrohelcon* is ichneumonid-like in appearance, but it has the typical braconid venation apart from indications of a costal cell in the fore wing. In *Helconidea* the hind femur has a large blunt tooth below at the middle of its length, an unusual character in Ichneumonoidea. Many diverse species are placed in the BLACINAE (Fig. 37.19D). One has an unusual apple-green body and pale blue eyes. *Centistes* has a biconvex gaster and a short stout ovipositor.

In the exodont subfamilies Alysiinae and Dacnusinae the mandibles face outwards. There are very few Australian species of ALYSIINAE. *Alysia manducator* was introduced for the control of sheep blow flies, and will attack *Lucilia*, *Calliphora*, and *Sarcophaga*. It oviposits in the almost full-grown maggots, and a single parasite emerges from the puparium. DACNUSINAE are distinguished from Alysiinae by the short radial cell. Most Australian species have a very long narrow stigma. They are small to very small species that are common parasites of agromyzid leaf miners, but they have also been bred from larvae of *Platypeza* (Diptera).

Superfamily EVANIOIDEA

The families placed here have been grouped with the Ichneumonoidea or Proctotrupoidea, but they are separated on the distinct spiracle cover lobe of the pronotum, development of a jugal lobe in the fore wing (small in Gasteruptiidae), and the forwardly bowed marginal vein of the hind wing that extends to the hamuli. Hind-wing venation very reduced, and no closed cells apart from a narrow costal cell. Gaster inserted high on the propodeum, well above the insertions of the coxae, particularly in Evaniidae and Gasteruptiidae; ovipositor strongly exserted in Aulacidae and some Gasteruptiidae. Antennae 13-segmented in all males and in females of Evaniidae, but 14-segmented in female Aulacidae and Gasteruptiidae. Aulacidae have the most primitive fore-wing venation, which differs from that of Trigonalidae in the presence of only one *r–m* cross-vein. In Gasteruptiidae there is some modification in the size of the basal cells, and loss of both *r–m* cross-veins and of the *m–cu* cross-vein. In Evaniidae the venation has been modified by increase in the size of one of the basal cells, but it is otherwise similar to that of Gasteruptiidae, though there may be indications of one *r–m* cross-vein.

Key to the Families of Evanioidea

Fore wing with complete venation; hind wing with a very small anal notch; trochantellus clearly defined ventrally; first 2 tergites of gaster fused (Fig. 37.21B) .... **Aulacidae** (p. 903)

Fore wing with apical venation reduced; hind wing without obvious anal notch, but with a deep 'jugal' notch; trochantellus defined ventrally on mid and hind legs; first 2 tergites of gaster not fused (Fig. 37.21A) .... **Evaniidae** (p. 903)

Fore-wing venation modified; hind wing without anal or 'jugal' notch; trochantellus not distinctly defined on any leg; first 2 tergites of gaster not fused, partly hinged (Figs. 37.21C–E) .... **Gasteruptiidae** (p. 904)

12. Evaniidae. Short, stout-bodied species (Fig. 37.21A) parasitic in the oothecae of Blattodea. They differ from all but a few Hymenoptera in having the gaster attached near the top of the propodeum instead of down near the coxae. The small gaster, with a long petiole, is laterally compressed and cynipoid in appearance. Braconid-like venation in the fore wing; hind wing with a very deep notch separating an anal lobe, which could be confused as a jugal lobe as it lies in the basal half of the anal field. All coxae grooved above for reception of the trochanter, and all trochanters very long. Length of body 3–15 mm in Australian species. The species are usually black, but a few have orange antennae, and a few have the body partly reddish brown. Species occur over the whole of Australia and Tasmania, but are more common in coastal areas. Adults are usually collected from tree trunks, generally on smooth white bark in bright sunlight. [Hedicke, 1939.]

13. Aulacidae. Medium-sized, body length 7–20 mm in Australian species. Adults (Fig. 37.21B) resemble Ichneumonidae, from which they can be distinguished by the presence of a distinct costal cell in the fore wing and very reduced hind-wing venation. They are parasitic on wood-boring Coleoptera and Xiphydriidae (Symphyta). Adults have been bred from beetle-infested acacias and eucalypts. *Aulacus fusicornis* has been bred from longicorn larvae, and *Aulacostethus apicalis* from larvae of *Piesarthrus marginellus*. The wings of the female are vibrated in the 'alert' position as she searches for an oviposition

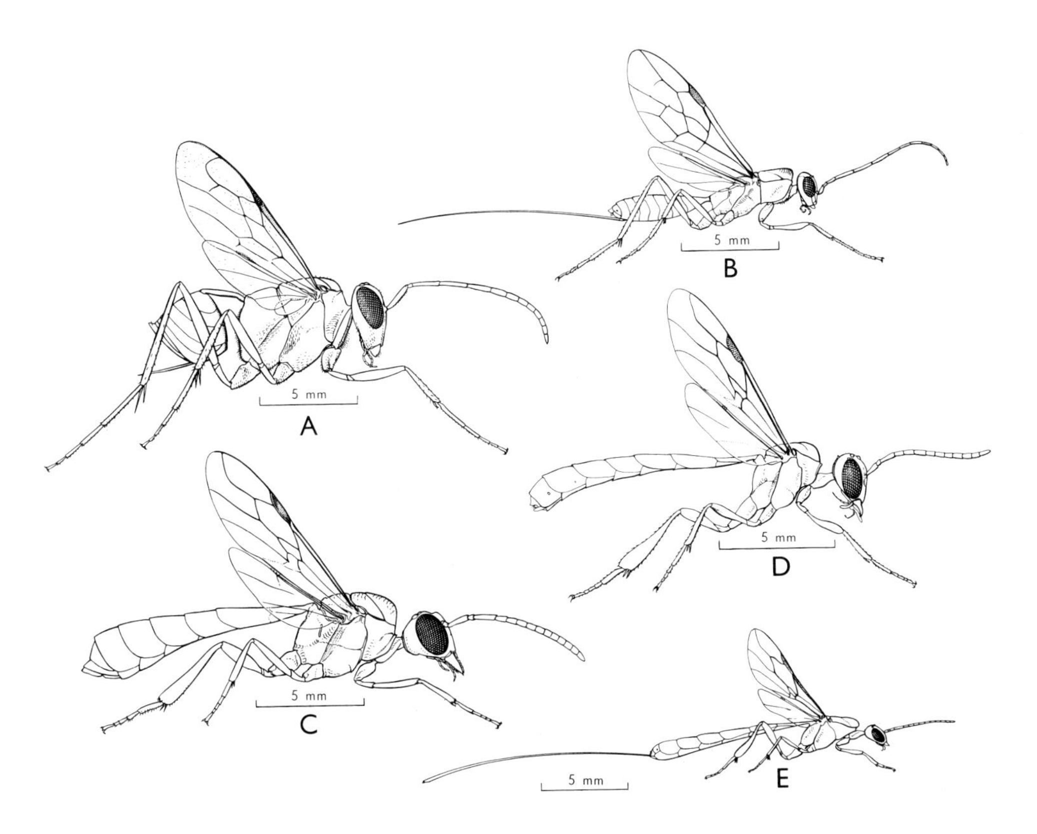

Fig. 37.21. Evanioidea: A, *Evania* sp., Evaniidae; B, *Aulacostethus* sp., Aulacidae; C, *Hyptiogaster* sp., Gasteruptiidae; D, *Gasteruption* sp., Gasteruptiidae, ♂; E, *Gasteruption* sp., ♀. [M. Quick]

site, and the apex of the gaster is raised and lowered in unison with the wing action. The hind coxae of the females are usually produced and grooved on the medial side to control the ovipositor during oviposition, the valves playing no part in the process. The metasternum is strongly produced so that the gaster appears to be inserted high on the propodeum; this is correlated with the role played by the hind coxae in oviposition. [Crosskey, 1953b.]

14. Gasteruptiidae. Long thin wasps parasitic in the nests of bees and wasps nesting in wood, rotten logs, and in the ground; also in some mud nests. Adults (Figs. 37.21C–E) are commonly collected flying around stumps and dead tree trunks, clay banks, or other nesting sites of their hosts. They also visit flowers and young foliage. Few Australian species have been bred: *Gasteruption* spp. from *Callomelitta* spp. (Colletinae) nesting in rotten logs; from the resinous cells of a megachilid bee; and from the mud cells of *Sceliphron. Hyptiogaster* spp. have been collected in close association with ground-nesting bees. The wings remain coupled, so that the fore wing is folded longitudinally at rest. The gaster is inserted very high on the propodeum, and the 1st tergite is partly free, hinged at its apex, and lateral regions creased to allow greater flexibility of the gaster during oviposition.

The species can be divided into 2 groups on wing venation, which is less modified in the *Hyptiogaster* group than in the *Gasteruption* group. *Gasteruption* (120 spp.) occurs in all regions. Australian species are 6–26 mm long, with the ovipositor up to 40 mm in some of the larger species, and are distinguished by the long thin body and distinctly exserted ovipositor of the female. In the *Hyptiogaster* group (30 spp.), Australian species are 4–22 mm long, and the ovipositor is usually short, hidden, and directed dorsally, but it is distinctly exserted in *Odontofoenus* which is more like *Gasteruption.* In the more generalized *Hyptiogaster* the hind trochanter is without a transverse groove, the face without a median carina, the first funicle longer than the second and longer than the scape, and the ovipositor not obviously exserted. [Crosskey, 1953a; Pasteels, 1957.]

Superfamily PROCTOTRUPOIDEA

A large superfamily of small parasitic species, most of which are easily distinguished from Chalcidoidea by the structure of the pronotum and position of the 'prothoracic' spiracle. In all but a few of the more primitive species the venation is greatly reduced or absent, and the wing-coupling is by 2–3 hamuli on a projection from the hind wing. The antennae are usually geniculate, with a large, prominent scape, and, as in all small parasites, there is a shortening in length and reduction in segmentation as compared with most larger parasites. Mid tibia with 2 spurs, except in Scelionidae and Platygasteridae; fore tibia with 2 spurs in Ceraphronidae (alone amongst Apocrita). Proctotrupoid species are mostly black, heavily sclerotized, and rather smooth and shining (except in Scelionidae). Many are egg-parasites, but host relationships are most varied. The primitive species resemble Braconidae, but the whole gaster is lengthened, so that the ovipositor issues from its apex and the sheaths are very small. Heloridae, especially *Austronia,* possess many primitive features. *Austronia* is very ichneumonoid in facies, but there is a costal cell in the fore wing, and the hind-wing venation is very reduced; the ovipositor appears to arise some distance from the apex, though the sheaths are not exserted; and the trochantellus is free in all legs.

Key to the Families of Proctotrupoidea

1. Antenna not geniculate; scape small and usually much shorter than 1st funicle segment 2
   Antenna geniculate and scape long and prominent, or scape produced to a point at apex 4
2. Fore wing with only 1 basal cell (the costal). [With large pterostigma and short transverse radial cell; trochantellus not free on any leg] .... **Proctotrupidae** (p. 905)
   Fore wing with 2 or 3 closed basal cells 3

3. Radial cell of fore wing triangular, not extending to apex; trochantellus free, usually on all legs, but at least on hind leg .... **Heloridae** (p. 905)
   Radial cell parallel-sided, extending to apex; trochantellus not free on any leg .... **Pelecinidae** (p. 907)
4. Scape not long, produced on one side to a point (Fig. 37.2E); fore wing with 3 basal cells, large pterostigma, and large radial cell .... **Austroserphidae** (p. 905)
   Scape long, not strongly produced at apex, antenna distinctly geniculate; fore-wing venation otherwise, though sometimes with basal cells or large pterostigma 5
5. Mid (and hind) tibia with 2 spurs 6
   Mid (and hind) tibia with 1 spur 8
6. Fore tibia with 2 apical spurs, one simple, the other developed into a calcar (Fig. 37.7H). [Fore wing with distinct marginal vein, sometimes with a large pterostigma, and with a well-developed stigmal vein] .... **Ceraphronidae** (p. 907)
   Fore tibia with only one apical spur (calcar) 7
7. Fore wing without a marginal vein, but with distinct venation and at least one closed cell (Fig. 37.22J) ....**Loboscelidiidae** (p. 910)
   Fore wing with both marginal and submarginal veins, or venation quite reduced (Figs. 37.23I–N) ....**Diapriidae** (p. 909)
8. Fore wing with distinct, though sometimes short, marginal and stigmal veins (Figs. 37.23A–G) .... **Scelionidae** (p. 907)
   Venation absent, or reduced to submarginal vein which is generally slightly knobbed at apex (Figs. 37.22H,I) .... **Platygasteridae** (p. 907)

15. Heloridae. This small family contains a number of distinctive genera that are united mainly on wing venation, which is somewhat braconid-like, with a narrow radial cell similar to that in some Braconidae but more diagnostic of Proctotrupidae. The hind wing may have a distinct venation, though closed basal cells are absent. The family possibly forms the base stem of the Proctotrupoidea, with many characters of a braconid-like ancestor. *Monomachus* (Figs. 37.22A, B), which occurs also in South America, is a parasite of *Boreoides* spp. (Diptera–Stratiomyidae). Adults emerge from the fully grown larva (or puparium) of the host; one parasite only from each small male host, but several from each large female host. Sexual dimorphism strongly marked; female (15–20 mm) with a very long gaster, thin and tapering and without clearly defined petiole, and antenna 15-segmented; male with distinct long petiole and short, slightly clubbed gaster, and very long, 14-segmented antennae. *Austronia* (5 mm) is short and braconid-like, with a distinct short petiole (Fig. 37.22C). Nothing is known of its biology. Extra-Australian genera have been bred from a variety of hosts: Neuroptera (Chrysopidae), eucnemid larvae (Coleoptera), and Tenthredinidae. [Riek, 1955g.]

16. Proctotrupidae. A small family of mostly small (3–6, but up to 12 mm), shining black species; host relationships mostly unknown; mostly collected by sweeping in late summer and autumn. *Proctotrupes janthinae* has been bred from larvae of *Thallis janthina* (Coleoptera–Erotylidae); and *P. syagrii* from the fern weevil, *Syagrius fulvitarsis*, which attacks the stems of many ferns, including tree-ferns and bracken. The fore-wing venation is distinctive (Fig. 37.22E); the stout ovipositor of the female appears as a prolongation of the apex of the gaster. Six genera are represented, but most species belong in the widely distributed *Cryptoserphus* (Fig. 37.22F). [Masner, 1961b; Riek, 1955d.]

17. Austroserphidae (Fig. 37.22D). A small, distinct, endemic family of unknown host relationships. *Austroserphus* (5–7 mm) has the fore trochantellus defined and a longer than wide petiole to the gaster, whereas the trochantellus is not distinct and the petiole is short in *Acanthoserphus*. The scape is excavated below at apex so that the remainder of the antenna can be folded back ventrally, and then the two sharp points of the scapes project anteriorly (Fig. 37.2E). The fore-wing

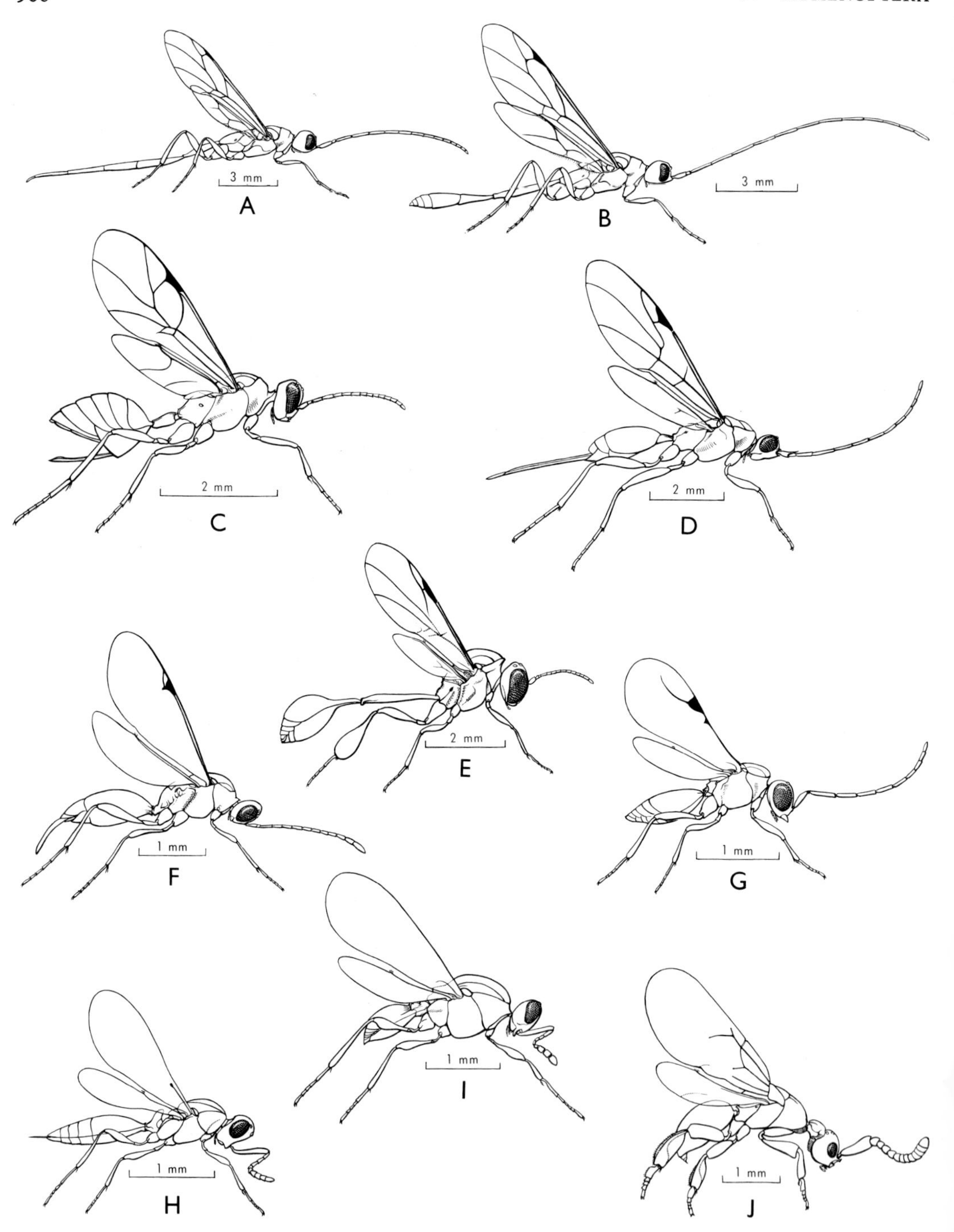

Fig. 37.22. Proctotrupoidea: A, *Monomachus* sp., Heloridae, ♀; B, *Monomachus* sp., ♂; C, *Austronia rubrithorax*, Heloridae; D, *Austroserphus* sp., Austroserphidae; E, undetermined genus of Proctotrupidae; F, *Cryptoserphus* sp., Proctotrupidae; G, *Conostigmus* sp., Ceraphronidae; H, *Brachinostemma* sp., Platygasteridae; I, unidentified Platygasteridae-Platygasterinae; J, *Loboscelidia* sp., Loboscelidiidae. [S. Curtis]

venation is similar to that of Heloridae except for the shape of the radial cell. [Riek, 1955d.]

18. Pelecinidae. There is a record of an introduced species of *Pelecinus* from the N.T. The large shining black body, reduced wing venation, and extremely long thin gaster distinguish species of the family from all other Hymenoptera.

19. Ceraphronidae. Adults (Fig. 37.22G) are mostly swept or obtained from soil and litter, but because of their very small size (0·5–2·8, mostly 1–1·5 mm) they are not commonly collected. Host relationships are not known, but they are probably parasites or hyperparasites of Diptera. They are generally stocky black species with hairy eyes; usually winged, but a few are wingless. Fore wing generally with a distinctive large pterostigma and usually with prestigma, but the marginal vein extending to the wing base is a diagnostic feature; antenna 9- to 11-segmented, geniculate, with a long scape; the only family of Apocrita with 2 fore tibial spurs (Fig. 37.7H). [Masner and Dessart, 1967.]

20. Platygasteridae. All very small, from less than 1 to 2·5 mm, mostly shining black, with legs and antennae variously coloured. The PLATYGASTERINAE are parasites of cecidomyiid galls and of woody galls on eucalypts. The INOSTEMMINAE are parasites of the eggs of Fulgoroidea (Homoptera) and Chrysomelidae (*Paropsis*), but species have also been bred from galls (?cecidomyiid) on *Melaleuca*. *Tetrabaeus* is a gregarious internal parasite of the larvae of crabronine wasps. Adults (Figs. 37.22H, I) usually winged, but some females brachypterous; venation reduced to a submarginal vein that usually ends in a slight thickening (Inostemminae) or absent (Platygasterinae). Scape large, flattened, and grooved below for reception of the basal segments of the funicle. The scape is directed caudally under the body when at rest, and the whole of the funicle and club remains hidden, with the club bent back along the funicle so that the antenna forms a closed Z (Fig. 37.22I). The structure of the antenna distinguishes these species from those Scelionidae in which the venation is reduced. Hind tarsus 5-segmented in Inostemminae and Platygasterinae, but only 4-segmented in Iphitrachelinae. [Kieffer, 1926.]

21. Scelionidae. These small species parasitize the eggs of a wide range of insects and other arthropods. Most adults are collected from leaf litter. *Scelio* can be collected on bare patches of ground in summer, especially on light soils. A few genera, such as *Oxytelia* and *Prosapegus*, are found on leaves, whereas others occur in very decayed wood on the ground.

Key to the Subfamilies of Scelionidae

1. Gaster with lateral margins rounded (Fig. 37.23C) .... TELENOMINAE
   Gaster with lateral margins carinate 2
2. Marginal vein very long, much longer than the stigmal (Fig. 37.23D). [Stigmal vein short; postmarginal absent] .... TELEASINAE
   Marginal vein shorter 3
3. Antenna of ♀ usually 7-segmented, rarely 6- or 8-segmented, ending in an enlarged, solid club; of ♂ 12-segmented, rarely 11-segmented (Figs. 37.23A, B) .... BAEINAE
   Antenna 12-segmented in both sexes, rarely 10- or 11-segmented (Figs. 37.23E–H) .... SCELIONINAE

TELENOMINAE (40 spp.; Fig. 37.23C) are reared mainly from the eggs of bugs and Lepidoptera, but are also recorded from tabanid eggs. The common species, placed in *Trissolcus* and *Aholcus*, are short, stocky, shining black insects, 0·7–2 mm long. Australian species of *Trissolcus* parasitize bug eggs, and *Aholcus* lepidopterous eggs; *Telenomus* and *Phanurus* are also bred from lepidopterous eggs. In *Telenomus* from the large eggs of the saturniid, *Antheraea helena*, there are several parasites per egg, but normally there is one parasite per host egg. [Masner, 1958.]

Adult TELEASINAE (57 spp.) are very similar to one another in general facies (Fig. 37.23D), and are usually found in damp situations in summer. They are possibly all parasites of coleopterous eggs, but little is known of their host associations. The sexes may differ markedly in colour and sculpture, but only a few species have brachypterous females. *Trimorus* (43 spp.) is the dominant genus in Australia. Usually only males are collected.

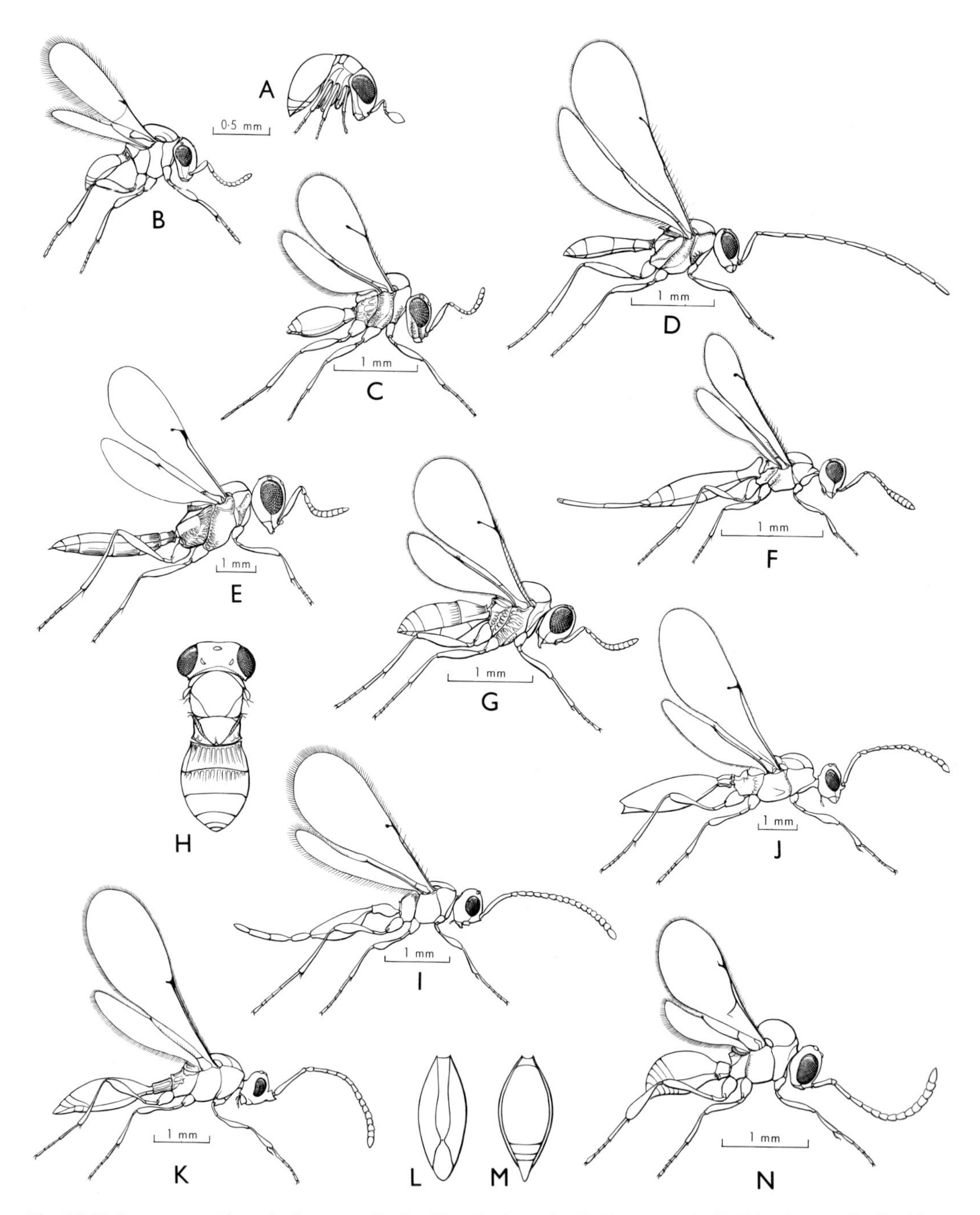

Fig. 37.23. Proctotrupoidea: A, *Baeus* sp., Scelionidae-Baeinae, ♀; B, *Baeus* sp., ♂; C, *Trissolcus* sp., Scelionidae-Telenominae; D, *Trimorus* sp., Scelionidae-Teleasinae, ♂; E, *Scelio* sp., Scelionidae-Scelioninae; F, *Baryconus* sp., Scelionidae-Scelioninae; G, *Hadronotus* sp., Scelionidae-Scelioninae; H, same, dorsal; I, *Stylaclista* sp., Diapriidae-Belytinae; J, *Hemilexomyia* sp., Diapriidae-Diapriinae; K, *Betyla* sp., Diapriidae-Ambositrinae; L, gaster of Diapriinae, ventral; M, gaster of Ambositrinae, ventral; N, *Ismarus* sp., Diapriidae-Ismarinae. [S. Curtis]

The long thin antennae and the median spine of the metanotum of the common species enable recognition of this subfamily as readily as does the distinctive venation. [Dodd, 1930a.]

BAEINAE (45 spp.; Figs. 37.23A, B) are most commonly collected under bark or on the webs of spiders; a few are parasites of Homoptera. They are usually 0·7–1 mm long, all less than 2 mm. The antennae of the female are the most distinctive feature separating the subfamily from Scelioninae; but there is no enlarged club in males, so they are not so easily distinguished. The wingless females of *Baeus* (Fig. 37.23A) are the most easily recognized members of the subfamily. Their nearly spherical shape, with the gaster composed mainly of one greatly enlarged, convex segment, combined with very small size and reduced segmentation of the thorax, make them most distinct. The males, of about the same size, are normal winged forms. The manner in which the female can parasitize the spider eggs enclosed in their tough capsule is not known. Parasitized eggs overwinter, and adults emerge in summer. *Ceratobaeus* and *Odontacolus* have a dorsal process on the 1st tergite of the gaster, and this usually reaches forwards to the scutellum. *Dissacolus* contains many undistinguished species.

The large subfamily SCELIONINAE (191 spp.) contains mostly stout, strongly ornamented, heavily sclerotized species, some of which are the largest in the family. Length 1–6·5 mm. There is marked variation in venation, but the subfamily is distinguished by having a short marginal vein, fully segmented female antenna, and margined gaster. *Scelio* (31 spp.; Fig. 37.23E), *Oxyscelio* (32 spp.), *Hadronotus* (30 spp.), *Macroteleia* (5 spp.), and *Prosapegus* (14 spp.) contain common large species. The last two include mostly long, slender species. There is a considerable size range in the short, stocky, coarsely sculptured species of *Hadronotus* (Figs. 37.23G, H). The species are all egg parasites: *Scelio* from acridoid eggs; *Hoploteleia* from tettigoniids; *Baryconus* (36 spp.) from gryllids; *Embidobia* from embiids; and *Hadronotus* from pentatomids. The host relationships of some of the smaller genera remain unknown. [Dodd, 1927, 1930b, 1939.]

22. Diapriidae. These small species (1–6 mm) are widespread and common in the higher-rainfall zones, where they are most commonly collected by sweeping in moist fern gullies. All have 2 spurs on the mid tibia, but antennal segmentation and the number of free gastral segments are quite variable. The subfamilies (Kieffer, 1916) are sometimes difficult to define; Ambositrinae are the most easily recognized.

Key to the Subfamilies of Diapriidae

1. Second sternite of gaster short, with concave apical margin (Fig. 37.23M). [Lateral margins of gaster carinate, with a distinct doublure or overfold] .... AMBOSITRINAE
   Second sternite of gaster very large (Fig. 37.23L) 2
2. Lateral margins of gaster with a distinct doublure; the gastral tergites fused into a carapace but retaining the sutures between tergites; antennae not inserted on a distinct frontal shelf (Fig. 37.23N) .... ISMARINAE
   Otherwise 3
3. Antenna of ♀ 15-segmented, rarely 14- and exceptionally 12- or 13-segmented, with no abrupt club; of ♂ 14-segmented, usually with 3rd segment modified (Fig. 37.23I) .... BELYTINAE
   Antenna of ♀ 11- to 13-segmented, rarely 14-segmented, with a more or less abrupt club; of ♂ 13- or 14-segmented, the 4th segment modified, rarely the 3rd or no segments modified (Fig. 37.23J) .... DIAPRIINAE

DIAPRIINAE (75 spp.) are usually collected by sweeping in grass or low vegetation or as they oviposit on carrion. They parasitize the larvae and pupae of various insects, mainly Diptera, but also Coleoptera and Lepidoptera; some are hyperparasites of tachinids. Adults found in ants' nests are considered to be commensals. Length of body 1–6 mm; antennae inserted on a frontal shelf; mouthparts directed caudally; gaster distinctly petiolate. Most species are winged in both sexes, but wingless and brachypterous females are known, and there are reduced thread-like wings in the males of a few species. Venation reduced, usually at most with submarginal,

marginal, and stigmal veins; often limited to basal third of wing; hind wing without a closed basal cell; wings usually clear or lightly infuscated (fore wing patterned in one species of *Propentapria*). The very small, rather glabrous species of *Trichopria* and *Phaenopria* parasitize dung- and carrion-breeding Diptera. *Hemilexomyia*, a common parasite of carrion-breeding Diptera, is a giant for the subfamily; the stout 6 mm body has the gaster composed mainly of one tubular segment. *Spilomicrus* is a common parasite of *Calliphora*. A related genus is a hyper-parasite of a tachinid parasitizing sawflies (*Perga*). The head is elongate and grotesquely ornamented in *Galesus*, with the scape flattened and irregular, and the mandible enlarged and 3-toothed. *Neurogalesus* also has an elongate head and very pronounced frontal projection.

BELYTINAE (15 spp.) are usually collected in damp situations, such as overhanging banks and in the vegetation bordering streams. Where life histories are known, the species parasitize dipterous larvae. Many Australian species are probably parasites of Mycetophilidae. Wing venation most complete in the *Xenotoma* group, in which there is a closed radial cell in the fore wing; hind wing often with a partly closed basal cell; wings clear, except in a species of *Xenotoma* in which apex of fore wing is darkened. Eyes usually hairy; antennae always inserted on a frontal shelf; gaster distinctly petiolate. In *Stylaclista* the apical 4 segments of the gaster are narrow, tubular, and together much longer than the normal-shaped basal segment. There is usually a large median pit at the base of the scutellum, but in *Aclista* there are 2 pits. ISMARINAE (2 spp.) differ from Belytinae in their stocky build, with the antennae inserted on an indistinct frontal shelf, bare eyes, and distinctive gaster.

AMBOSITRINAE (15 spp.) are usually collected in damp situations in association with Belytinae. The distinct doublure of the lateral margin of the gaster is almost diagnostic, though it does occur in Ismarinae (Fig. 37.23N) and a few Belytinae. The gaster of the males is very disc-like (Fig. 37.23M). Most species are fully winged in both sexes, but brachypterous females occur in *Betyla*, and a few species of *Neobetyla* are almost wingless. The wings are normally clear, but the whole fore wing is patterned in *Prosoxylabis*. There is a most unusual undescribed genus, related to *Neobetyla*, with grotesquely thickened fore wings which form a cover fitting closely over the propodeum, petiole, and base of the gaster; the apex of the fore wing is emarginate, exposing the apical segments of the gaster. Ocelli are absent, and the vertex has a pair of short sharp horns. *Betyla* contains a number of very common species. [Masner, 1961a.]

23. Loboscelidiidae. The family is recorded from the Oriental region (Maa and Yoshimoto, 1961). Undescribed Australian species have been reared from the eggs of phasmatids, possibly *Acrophylla* sp., and one specimen is recorded from the nest of *Ectatomma metallicum*, the green-head ant. Both sexes are fully winged. The body is mainly smooth and shining; it has a most grotesque appearance, with occipital flanges and projections and expanded scape and legs; the tegulae are very large. The fore wing is without a costal vein and pterostigma, and the glossy wing membrane is irregularly wrinkled (Fig. 37.22J).

Superfamily CYNIPOIDEA

The superfamily is not well represented. Most known species are Eucoilinae (parasites of the puparia of Diptera) and Charipinae (secondary parasites of aphids). The gall-forming Cynipinae are represented by few endemic species, but one introduced species is very common (gall-forming Hymenoptera in Australia are mostly Chalcidoidea). Most Australian cynipoids have been collected by sweeping. No parasites of Cynipoidea are recorded from Australia, but a chalcidoid parasite of the introduced *Aylax hypochoeridis* is present in Europe.

Insects of small to medium size (1–6, with a

few to 25 mm). Antenna of female usually 13- never more than 19- or less than 11-segmented, never geniculate, usually filiform, but in parasitic forms several of the terminal segments often form a club; antenna of male usually 14- or 15-segmented, 3rd (rarely 4th) often elongate and bent. The pronotum reaches back to the tegulae. Fore wing without an anterior marginal vein from the base and without a true pterostigma, typically with a characteristic radial cell. Ovipositor usually concealed within the gaster. Gaster with sternites well developed but almost covered by lateral prolongations of the tergites; apical sternite (hypopygium) much enlarged. Mid tibia with 2 apical spurs. Subfamilies, though more clearly diagnosed than the usually accepted families, are defined more on their biology than on clearly defined structural differences. This is particularly true of Cynipinae and Figitinae. [Weld, 1952.]

Key to the Families and Subfamilies of Cynipoidea Known in Australia

1. Radial cell at least 9 times as long as broad; hind basitarsus twice as long as segments 2 to 5 combined (Fig. 37.24A) .... **Ibaliidae** (p. 911). IBALIINAE
   Radial cell not 9 times as long as broad; hind basitarsus less than twice as long as segments 2 to 5 combined 2
2. Largest tergite of gaster (lateral view) 4, 5, or 6, at least 2 short tergites preceding the large tergite (Fig. 37.24C) .... **Liopteridae** (p. 911). MESOCYNIPINAE
   Largest tergite of gaster (lateral view) 2 or 3, never more than 1 short tergite (often none) preceding the large tergite 3
3. Scutellum with a characteristic raised 'cup' on the disc (Fig. 37.24G) .... **Cynipidae** (p. 913). EUCOILINAE
   Scutellum without such a cup 4
4. Body without sculpture; scutellum smooth, rarely with basal pits; length under 2 mm (Fig. 37.24I) .... **Cynipidae**. CHARIPINAE
   Body sculptured unless mesoscutum and scutellum are fused without a suture; length usually greater than 2 mm 5
5. Gaster with distinct petiole; body when viewed from above distinctly wedge-shaped; tergite 2 longer than 3 along dorsal margin (Fig. 37.24B) .... **Figitidae** (p. 911). ANACHARITINAE
   Gaster sessile or nearly so 6
6. Tergite 2 usually much smaller than 3 and saddle-shaped; hypopygium of ♀ ending without caudal spine (Fig. 37.24D) .... **Figitidae**. FIGITINAE
   Tergite 2 or fused 2+3 usually large; hypopygium of ♀ ending in a caudal spine (Fig. 37.24H) .... **Cynipidae**. CYNIPINAE

24. Liopteridae. MESOCYNIPINAE (Fig. 37.24C) are recorded from the Oriental region and Australia, Oberthuerellinae from Africa, and Liopterinae from the Neotropical region. Biology unknown. Wing venation well developed for the superfamily, but not as complete as in Ibaliidae. Australian species (5–8 mm) are of stout build and with coarsely ornamented thorax. [Hedicke and Kerrich, 1940.]

25. Ibaliidae. *Ibalia* (Fig. 37.24A) contains relatively large species (up to 15 mm) with the most complete venation in Cynipoidea; gaster almost sessile, that of the female strongly compressed; a peg-like spur present on the 2nd segment of hind tarsus of both sexes. Represented in Australia only by introduced species parasitic on *Sirex noctilio.*

26. Figitidae. There are 3 subfamilies. Aspicerinae, bred from the puparia of Syrphidae (Diptera) are not recorded. FIGITINAE, parasitic on other, usually smaller Diptera, are represented by an undescribed species of *Xyalophora* (Fig. 37.24D). *Thrasorus*, which is usually referred to Figitinae, is transferred to Cynipidae, and the Australofigitinae, based on a single Australian species, are an aberrant group related to *Thrasorus*. ANACHARITINAE parasitize only hemerobioid lacewings. The one common

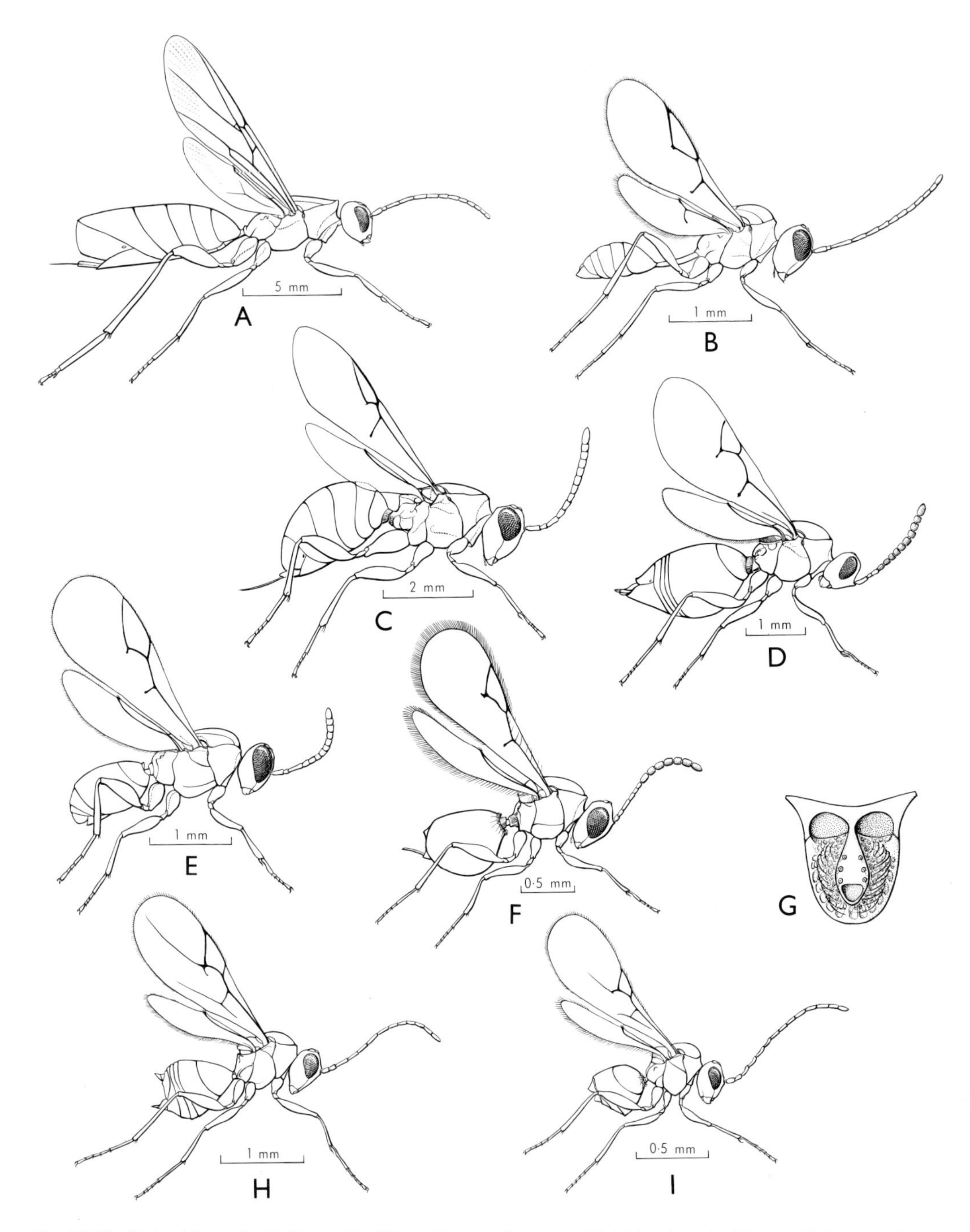

Fig. 37.24. Cynipoidea: A, *Ibalia* sp., Ibaliidae; B, *Anacharis* sp., Figitidae-Anacharitinae; C, *Mesocynips* sp., Liopteridae-Mesocynipinae; D, *Xyalophora* sp., Figitidae-Figitinae; E, *Thrasorus* sp., Cynipidae-Cynipinae; F, *Heptamerocera* sp., Cynipidae-Eucoilinae; G, cup of Eucoilinae; H, *Aylax hypochoeridis*, Cynipidae-Cynipinae; I, *Charips australiae*, Cynipidae-Charipinae. [S. Curtis]

species of *Anacharis* (2 spp.; Fig. 37.24B) is distinguished from all other Australian Cynipoidea by the very long petiole to the gaster. [Weld, 1952.]

27. Cynipidae. The family is divided into 4 subfamilies, of which all but the African Pycnostigmatinae are recorded. CYNIPINAE (9 spp.), gall-formers or inquilines in galls, are represented by at least 2 introduced species and the endemic *Thrasorus* (7 spp.; Fig. 37.24E), which has been reared from brachyscelidiphagine pteromalid galls on eucalypts and acacias where it is probably an inquiline. *Andricus* sp. forms galls on introduced oaks. *Aylax hypochoeridis* (Fig. 37.24H), gall-former in the flower stems of the introduced 'dandelion', *Hypochoeris radicata*, is possibly the most common cynipoid in Australia.

CHARIPINAE (9 spp.) are hyperparasitic on aphids through aphidiine braconids. Three genera, all widely distributed, are recorded. *Charips australiae* (Fig. 37.24I) and *Alloxysta bifoveata* are common hyperparasites of cabbage aphids in eastern Australia. EUCOILINAE (40 spp.) have been reared from the puparia of Calliphoridae, Sphaeroceridae, Drosophilidae, and other cyclorrhaphous Diptera. They can be recognized by the raised cup on the scutellum. Many belong to endemic genera, but most common species are placed in widely distributed genera, such as *Eucoila* (Kerrich and Quinlan, 1960) in which the wing membrane is without the usual pubescence, *Eucoilidea* in which the hair-ring is absent from the base of the gaster, and *Pseudeucoila (Miomera)* in which the female antenna is 12-segmented. *Kleidotoma*, *Trybliographa*, and *Rhoptromeris* are other common genera, divided into subgenera on the structure of the club in the female antenna. *Thoreauella*, an endemic genus with reduced cup on the scutellum, links this subfamily structurally with Charipinae. Its biology is unknown. [Weld, 1952.]

Superfamily CHALCIDOIDEA

A very large group of mostly small to minute parasitic or phytophagous species (mostly gall-formers), united mainly on the structure of the pronotum and the very reduced venation. The pronotum does not reach back to the tegula, and is separated from it usually by the prepectus, but when the prepectus is very reduced the two may be separated mainly by the lateral lobe of the mesonotum. Chalcidoidea and Chrysidoidea are the only groups of Hymenoptera, apart from a few Bethyloidea, in which the pronotum does not reach the tegula. It almost reaches the tegula in Leucospidinae (Chalcididae), in which the tegula has a long, narrow, forward projection, and in some Mymaridae. The prepectus is fused with the pronotum in many Eucharitinae (Pteromalidae), giving a false impression that the pronotum reaches the tegula. Chalcidoidea are the only Hymenoptera in which the 'prothoracic' spiracle is situated at or above the level of the tegula.

The species are of very diverse structure, but there is invariably only 1 mid tibial spur. Tarsi 5-segmented in the majority; but 3-segmented in Trichogrammatidae; 4-segmented in most Eulophidae and in some Mymaridae, Aphelininae, and Encyrtinae. Agaonidae do not have a fore tibial spur, and there are two types in other species, correlated with the manner in which the antennae are cleaned. Most have a curved calcar on the tibia and a strigil ventrally on the basitarsus. Eulophidae and Trichogrammatidae have a short, straight spur, and the strigil forms an oblique brush of setae on the medial surface of the basitarsus. These brushes are used together to clean each antenna, at least in some Eulophidae.

Many chalcidoids jump very agilely, and those with the best jumping ability have a modified mid leg: the tibial spur is often enlarged, and the plantar surface of the tarsus, especially basitarsus, develops stout bristles that grip the substrate. Correlated with this ability to jump there is an increase in the size of the mesothoracic muscles, and this results in inflation of the lateral thorax, more especially of the mesepimeron. Inflation is pronounced in many Encyrtidae and all stages occur in Aphelininae. Both mid and hind legs are enlarged in a few (Elasminae and Myiocneminae), and then the legs and lateral thorax are smooth and flattened. The hind femur is enlarged and toothed below in Chalcididae, some Torymidae, and some Pteromalidae, and there is a single tooth-like projection in some Torymidae and Pteromalidae. Enlarged scale-like hairs are present on the body of some species, and similar but smaller hairs are of widespread occurrence. Enlarged bristles are present towards the apex of the femora posteriorly, especially in Eulophidae. Many species have bright metallic colouring, but because of small size they are not conspicuous. They range from 0·2 to more than 30 mm, but the great majority are between 1 and 3 mm. [Girault, 1912–16; Peck, Bouček and Hoffer, 1964.]

The family classification is unsatisfactory. That presented differs from previous ones in reducing many groups to subfamily rank. The primary divisions are based on the manner in which the antennae are cleaned and on the ability to jump. Most species can be placed by the following key.

Key to the Families and Major Subfamilies of Chalcidoidea

1. Fore wing folded longitudinally when at rest. [Hind femur enlarged and toothed below] .... **Chalcididae.** LEUCOSPIDINAE (p. 918)
   Fore wing not folded 2
2(1). Hind femur greatly enlarged and toothed below, teeth sometimes fine and limited to a short distal zone; hind tibia curved around the femur 3
   Hind femur sometimes enlarged, but not regularly toothed below, and tibia not distinctly curved 5
3(2). Axillae distinctly advanced; prepectus large, not impressed. [Hind tibia with 2 spurs] .... **Pteromalidae.** CHALCEDECTINAE (p. 921)
   Axillae at most only slightly advanced; prepectus sometimes large but impressed 4
4(3). Basal tergites of gaster emarginate in mid-line; notaulices only slightly impressed, very widely separated. [Ovipositor usually strongly exserted] ...... **Torymidae.** PODAGRIONINAE (p. 919)
   Basal tergites of gaster not emarginate; notaulices clearly percurrent .... most **Chalcididae** (p. 918)
5(2). Mesopleuron completely inflated, without impressed lines, grooves, or pits ... most **Encyrtidae.** 6
   Mesopleuron with at least a mesopleural 'suture', usually with impressed femoral furrow ... 7
6(5). Mesosternum elongate, mid coxae widely separated from fore coxae. [Notaulices usually distinctly impressed, never entirely absent] ...... EUPELMINAE (p. 924)
   Mesosternum short, mid coxae usually widely separated from hind coxae .... ENCYRTINAE (p. 923)
7(5). Tarsi 3-segmented (see also wingless ♂♂ of Agaonidae and Sycophaginae) .... **Trichogrammatidae** (p. 917)
   Tarsi 4- or 5-segmented 8
8(7). Fore and hind tibiae very short, much shorter than femora; fore tibia without an apical spur. [Mouth-parts of ♀ with a backwardly-directed radula-like process] ...... **Agaonidae** (p. 915)
   Fore and hind tibiae about as long as femora; fore tibia with an apical spur 9
9(8). Venation greatly reduced and limited to basal third of wing, not extending beyond level of wing-coupling, stigmal vein always absent. [Hind wing clearly stalked at base] .... **Mymaridae** (p. 918)
   Venation otherwise, stigmal vein normally distinct 10
10(9). Spur of fore tibia short and straight; tarsi 4-segmented ...... **Eulophidae.** 11
   Spur of fore tibia curved and often large; tarsi almost invariably 5-segmented 12
11(10). Hind coxa greatly enlarged and flattened; lateral thorax without impressed femoral furrow, but with mesopleural 'suture' ...... ELASMINAE (p. 917)
   Hind coxa not greatly enlarged; lateral thorax with femoral furrow indicated .... other **Eulophidae** (p. 917)

12(10). Mandibles large, falcate. [Pronotum very reduced at mid-line; body usually metallic] .... **Pteromalidae.** EUCHARITINAE (p. 920)
Mandibles not falcate though sometimes large 13
13(12). Mesepimeron with caudal margin deeply excised. [Basal tergites of gaster emarginate at mid-line; mesepimeron with an impressed line] .... **Torymidae.** TORYMINAE (p. 919)
Mesepimeron with caudal margin entire or only slightly emarginate 14
14(13). Gaster with transverse rows of very large, deep punctures. [Hind tibial spurs thickened and one enlarged; occiput carinate; lateral thorax shining and prepectus small] .... **Pteromalidae.** ORMYRINAE (p. 921)
Gaster without such punctures 15
15(14). First 2 tergites of gaster fused above but with the line of junction retained, free laterally, gaster consisting mainly of these 2 segments. [Body often metallic blue, green, or purple; pronotum carinate at declivity] .... **Pteromalidae.** PERILAMPINAE (p. 921)
If basal tergites are fused, then of different form 16
16(15). Body mostly glabrous, flattened; mesopleuron shining, without distinct femoral furrow but with an irregular pit or pits. [Antenna without a ring-segment; 1 hind tibial spur; stigmal vein well developed] .... **Pteromalidae.** SPALANGIINAE (p. 921)
Otherwise 17
17(16). Cerci large, longer than wide. [Basal tergites of gaster usually emarginate at mid-line; occipital carina usually distinct; ovipositor usually strongly exserted] .... most **Torymidae** (p. 919)
Cerci small, often indistinct 18
18(17). Gaster broadly sessile, without pronounced constriction at junction with propodeum. [Fore wing with long marginal, short stigmal, and no postmarginal vein; notaulices distinct] .. .. **Encyrtidae.** APHELININAE (p. 924)
Gaster distinctly constricted at base, but appearing sessile when the petiole is short 19
19(18). Antenna 7- to 10-segmented, inserted near mouth .... **Pteromalidae.** EUNOTINAE (p. 923)
Antenna 11- to 13-segmented, inserted distinctly above the mouth 20
20(19). Axillae distinctly advanced; prepectus not impressed, caudal margin straight or sloping forward; eyes hairy .... **Pteromalidae.** CLEONYMINAE (p. 921)
Axillae not obviously advanced; prepectus normally impressed, caudal margin normally concave; eyes rarely distinctly hairy 21
21(20). Inner margins of axillae much closer than inner margins of notaulices, often meeting at mid-line; hind coxa large. [Mesepimeron not impressed] .... **Pteromalidae.** BRACHYSCELIDIPHAGINAE (p. 923)
Inner margins of axillae not much closer than inner margins of notaulices, or hind coxa small . 22
22(21). Pronotum large and with simple caudal margin; genal carina well developed, at least at mandible; hind coxa small; propodeum usually concave in mid-line; antennae 11-segmented (club counted as 3 segments), with one ring-segment ... **Eurytomidae.** EURYTOMINAE (p. 919)
Otherwise 23
23(22). Notaulices deep and percurrent; hind coxa large ... **Pteromalidae.** MISCOGASTERINAE (p. 923)
Notaulices shallow, not percurrent, indicated only anteriorly; hind coxa not usually enlarged .... **Pteromalidae.** PTEROMALINAE (p. 923)

28. Agaonidae. All species of this small family live in the 'fruits' of certain figs, where they act as caprification and pollinating agents. The sexes are markedly dimorphic. Males are wingless and rarely emerge from the 'fruits'. Females are all small, 1–3 mm, though the ovipositor may be exserted for 1 mm. Fore and hind legs enlarged, with peculiar shortened tibiae; mid legs weak; fore wings broad. There is a spine-like process on the 1st funicle segment, and the antenna is 11-segmented, or 12-segmented when a ring-segment is distinguishable, though the 3-segmented club may show indistinct segmentation. Agaonidae are the only Hymenoptera in which there is no fore tibial spur. The species are often parasitized by Sycophaginae (Torymidae), some of which also have wingless males. Many belong to the endemic genus *Pleistodontes* (Fig. 37.25A). Different species occur in different species of figs. *P. froggatti*, from the Moreton Bay fig, is one of the largest and most common species. Unless this insect is present the fig

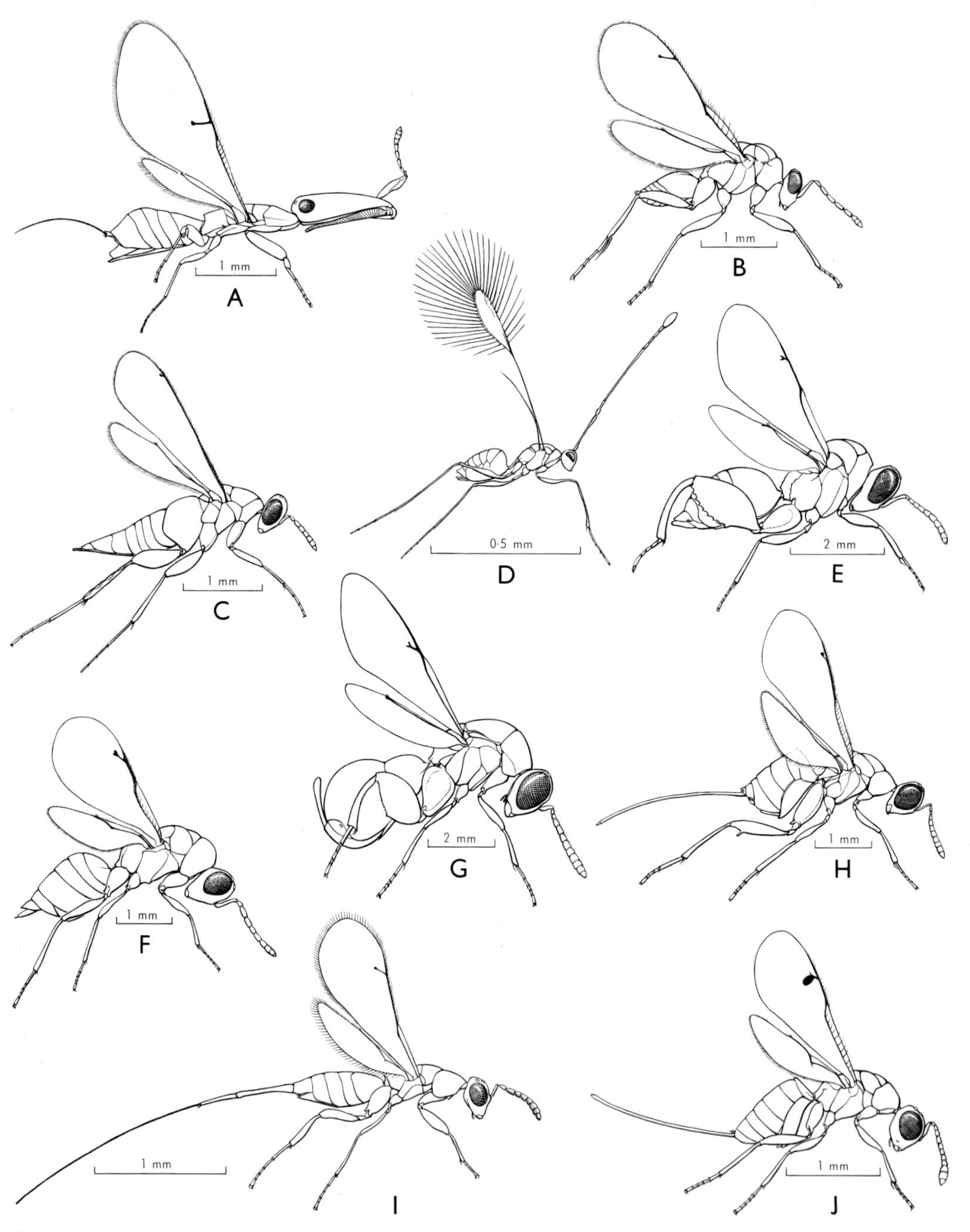

Fig. 37.25. Chalcidoidea: A, *Pleistodontes* sp., Agaonidae, ♀; B, *Euplectrus* sp., Eulophidae-Euplectrinae; C, *Elasmus* sp., Eulophidae-Elasminae; D, *Mymar* sp., Mymaridae; E, *Brachymeria* sp., Chalcididae-Chalcidinae; F, *Eurytoma* sp., Eurytomidae-Eurytominae; G, *Epexoclaenoides bicinctus*, Chalcididae-Leucospidinae; H, *Diomorus* sp., Torymidae-Toryminae; I, *Philotrypesis* sp., Torymidae-Sycophaginae; J, *Megastigmus* sp., Torymidae-Megastigminae. [S. Curtis]

does not produce mature 'fruits'. *Blastophaga psenes* was introduced for caprification of Capri figs, but other varieties of commercial figs are self-pollinating. [Grandi, 1961.]

29. Trichogrammatidae. Minute species, mostly 0·5–1 mm, with very reduced venation and long marginal fringes; marginal vein short, stigmal vein short or sometimes absent, postmarginal always absent; venation never reaching beyond middle of wing; pubescence of fore wing often arranged in lines. Fore leg with modified strigil; fore tibial spur straight. They are the only Chalcidoidea with 3-segmented tarsi, apart from a few wingless males found in figs. All are egg parasites, and attack a wide range of small insects.

30. Eulophidae. A large family of very diverse form and habits united on the structure of the fore leg. Length 0·6–4, mostly 1–3 mm. The pronotum retains a considerable degree of movement with the mesothorax. The large prepectus is usually slightly convex caudally and has impressed margins, especially above, except in Elasminae. The antennae may be thickened throughout the length of the flagellum, or there may be a well-developed club. Fore tibial spur short and straight, and basitarsus with the strigil modified into an oblique comb at the base medially, though this is not always easily seen; mid tibial spur often, but not invariably, enlarged; hind tibial spurs greatly enlarged in Euplectrinae; tarsi 4-segmented in most (5-segmented only in female Tetracneminae). There is an enlarged curved bristle at the apex laterally of the fore and mid femora, though they are not always well developed in Tetrastichinae. The cuticle of many of the smallest species is soft, so that they collapse dorsoventrally when dried. The family has been divided into a number of lesser groups on a combination of wing venation, development of notaulices, and attachment of gaster.

TETRASTICHINAE are usually reared from lepidopterous and coleopterous pupae, less frequently from acrocerid (dipterous) pupae and lepidopterous eggs; also as hyperparasites of lepidopterous larvae through ichneumonoid or tachinid primary parasites; from seed capsules of eucalypts, from galls on eucalypts and acacias, and from tephritid and cecidomyiid galls. EUPLECTRINAE (Fig. 37.25B) are parasites of lepidopterous larvae. They are either external feeders, or bore their way out and complete development externally on approaching maturity. They spin small silken cocoons (the only chalcidoids to do so) either on or close to the host larva. *Euplectrus agaristae* often parasitizes larvae of the vine moth. In all species of this subfamily one at least of the hind tibial spurs is greatly enlarged. EULOPHINAE, in which the notaulices are incomplete, are smooth, often elongate species (up to 3 mm). They are possibly all hyperparasites. Eulophini, with 2 hind tibial spurs, parasitize Lepidoptera. Hemiptarsenini, with 1 hind tibial spur and usually long basitarsus, are often reared from leaf-mining agromyzids.

ENTEDONINAE, in which the notaulices are broadened posteriorly and there is 1 hind tibial spur, are mostly shining species with an irregular surface to the thorax and a few large bristles. Entedonini have a petiolate gaster. The species of *Epientedon* are parasites (or hyperparasites) of *Phylacteophaga*, the leaf-mining sawfly on eucalypts. ELACHERTINAE are often similar in appearance to Tetrastichinae, but differ in venation and have less advanced axillae. *Rhicnopeltella*, and related genera reared from galls, are relatively large, robust species that are most readily separated from Brachyscelidiphaginae (Pteromalidae) by their 4-segmented tarsi. ELASMINAE (Riek, 1967) are small (1–2·7 mm), phorid-like species of slim form, tapering gaster, and rather polished appearance. Most are patterned in yellow or orange and have pale legs, but some are shining black. Both mid and hind legs are enlarged, and the hind coxa is extremely large. The long wing with very long marginal and minute stigmal veins, mesopleuron, and prepectus are also characteristic. The hind tibia of most has rows of hairs arranged to form rhomb-shaped areas. The subfamily is based mainly on *Elasmus* (Fig. 37.25C) the species of which are parasitic on the larvae of Lepidoptera; some are reared as hyperparasites from ichneumonoid cocoons. One species is recorded from wasps' nests *(Polistes)*.

31. Mymaridae. Mostly small (0·5–2 mm), delicate species with characteristic very reduced venation; distinguished from Trichogrammatidae by the 4- or 5-segmented tarsi and well-developed strigil and calcar on the fore leg. Fore-wing venation usually limited to the basal third, stigmal vein always absent. Wings usually with a long marginal fringe which is often longer than the width of the hind wing. The wings may be thread-like apart from an apical enlargement, as in *Mymar* (Fig. 37.25D). The pronotum almost reaches the tegula in some species, but is widely separated from it in *Polynema.* Legs often very long, thin, not enlarged. The femoral furrow is not impressed, except caudally in some, and the lateral thorax, though smooth, is not inflated. The ovipositor is sometimes extremely long, and is then brought forward as a loop under the thorax, or even extends further forward than the head. The species are all egg parasites of other insects. Many are collected by sweeping. *Polynema* has a long-petiolate gaster similar to that of the cynipoid *Anacharis.* [Annecke and Doutt, 1961.]

32. Chalcididae. The species, often large, are distinguished by the enlarged hind leg with the femur toothed below, though there is a similar development of the hind leg of Podagrioninae (Torymidae) and Chalcodectinae (Pteromalidae). Pronotum usually closely associated with mesothorax, and emarginate anteriorly to receive posterior portion of head; 'mesosternum' grooved anteriorly to receive fore coxa and trochanter. The body is heavily sclerotized, and the species often partly roll up and feign death when disturbed. The 2 dominant genera in Australia are *Brachymeria* (Fig. 37.25E), in which the teeth of the hind femur are large and the gaster usually not obviously produced, and *Antrocephalus*, in which the teeth are fine and comb-like and the apex of the gaster is usually produced in the female.

LEUCOSPIDINAE (12 spp.) are relatively large (up to 15 mm), and distinguished from all other Chalcidoidea by the fore wing being folded longitudinally at rest through retention of the wing-coupling. Hind tibia not grooved above for the tarsus, and normally produced distally into a process that bears one of the tibial spurs at its apex, the other spur arising normally. Gaster usually more or less laterally compressed, with rounded apex; ovipositor reaching partly or entirely round the apex, sometimes extending dorsally as far as the scutellum. Little is known of the biology of Australian species. They are usually attracted to the extra-floral nectaries of young foliage or to blossom. *Epexoclaenoides bicinctus* (Fig. 37.25G) is a parasite of *Pison* (Sphecidae).

CHALCIDINAE (76 spp.) are divided into 2 tribes on the development of the gaster. Chalcidini, in which the petiole is long, are well represented in South and Central America, but there is only one species of *Chalcis* in Australia. The mid tibia is without an apical spur in this genus. Brachymerini are of diverse form. In *Megalocolus* the gaster is produced into a stout caudal style. *Chalcidellia ucalegon*, a parasite of *Euthyrrhinus meditabundus*, a weevil pest of mango trees in north Australia, has the scutum produced anteriorly. *Mirochalcis* has elaborately produced hind legs. *Brachymeria* (65 spp.) comprises stoutly built, mostly large (2·5–7 mm) species, which parasitize a wide range of hosts, mostly Lepidoptera (emerging from the cocoon or naked pupa), but also dipterous puparia and ichneumonid cocoons. The apex of the gaster is rarely produced, but it is long in females of *B. pomonae*, a parasite of codling moth. Adults of *Brachymeria* are often attracted to young foliage, and males to smooth-barked tree trunks in bright sunlight.

DIRHININAE (11 spp.) are flattened species with the head produced into 2 horns and the gaster petiolate. They are pupal parasites of Diptera. They differ most obviously from the similarly flattened Spalangiinae (Pteromalidae), also pupal parasites of Diptera, in the enlarged, toothed, hind femur. *Dirhinus ruficornis* parasitizes several common Calliphoridae, and other species have been recorded from Tephritidae. EPITRANINAE (7 spp.) have a long petiole, a characteristic flange or spine on the hind tibia, and antennae inserted on a frontal shelf. *Neoanacryptus piceiventris* has been reared from *Cydia molesta* (Lepidoptera).

The HALTICHELLINAE (125 spp.) include many unusual species. The hind tibia is grooved above at apex for the base of the tarsus and not produced into an apical spine, or the spine is very short and the tibia has the normal 2 apical spurs. Species of *Trichoxenia* are all large and have prominent tufts of silvery or golden hairs which give them a rather bee-like appearance. *Uga colliscutella*, with a grotesquely produced scutellum is a parasite of coccinellid larvae. *Antrocephalus* (53 spp.) is a widespread genus parasitizing mainly lepidopterous pupae, but with species recorded from cockroach oothecae and dipterous puparia. Each species usually attacks a wide range of similarly-sized hosts. More than a dozen species have been recorded from the introduced *Cactoblastis*, which is the only known host of *Irichohalticella* (which has numerous longitudinal striae on the rather flattened 1st tergite of the gaster) and related genera.

33. Eurytomidae. The species differ from Chalcididae mainly in the non-toothed and usually small hind femur, and from more generalized Pteromalidae in the structure of pronotum and propodeum. They are non-metallic, except for one small genus. The family is divided into 3 subfamilies, of which the Heimbrinae, distinguished by an unusual gaster, do not occur in Australia. RILEYINAE (4 spp.) have 13-segmented antennae, with 2 or 3 ring segments. The gaster is cynipid-like, with the very short basal tergites followed by 1 or 2 large tergites. The prepectus is sometimes almost completely hidden. One Australian species has been bred from cecidomyiid galls on *Santalum lanceolatum*. EURYTOMINAE (190 spp.) have 1 ring segment and the antenna never more than 11-segmented. *Eurytoma* (130 spp.; Fig. 37.25F) generally has a cynipoid facies with a highly polished, laterally compressed gaster. Most of the Australian species form galls on eucalypts, acacias, and other native trees; some are parasitic on the gall-forming Brachyscelidiphaginae; a few form galls in seeds; and a few are parasites of Lepidoptera (Limacodidae, Psychidae) and Diptera. In *Eudecatoma* (7 spp.) the marginal vein is thickened and appears rather like a pterostigma. Species are bred from cecidomyiid and cynipid galls, but the only common Australian species are parasites of Brachyscelidiphaginae.

34. Torymidae. First tergite of gaster usually emarginate in mid-line, and ovipositor exserted, often markedly so (not very obvious in some Monodontomerinae); 2 hind tibial spurs in all species except Podagrioninae, but 2nd spur very reduced in Megastigminae; hind tibia not produced or grooved above at apex for the tarsus, except in Podagrioninae. There is considerable freedom of movement between pronotum and mesothorax.

PODAGRIONINAE (20 spp.) are distinguished by an enlarged hind femur, toothed below, and an arcuate hind tibia that is produced into a long curved spine at apex and without obvious tibial spurs, or there may be one weak spur on the tarsal surface of the spine. Hind coxa elongate, except in *Iridophaga*. First segment of gaster deeply emarginate caudally, and 2nd segment normally hidden under 1st; otherwise the species, about 3 mm long, are rather like Chalcidini. Ovipositor very long, sometimes more than twice as long as body. The species are parasites of mantid oothecae, with one to several per egg and a single generation per year. In *Pachytomoides* and *Pachytomoidella* the club is greatly enlarged (solid and longer than the combined funicle segments in *Pachytomoidella*), and the sensory flattening on its lower side is prominent. *Podagrionella* has an obviously petiolate gaster, and the adults emerge from the sides of the eggs of the naked oothecae. *Iridophaga* has a thorax rather like that of Pteromalinae.

TORYMINAE (6 spp.) are distinguished by the excised dorsoposterior margin of the mesepimeron (unusual in Chalcidoidea, but seen also in some Spalangiinae). The species are metallic or black, and have a cross furrow on the scutellum. In *Ecdamua* and *Australtorymus* the gaster is petiolate, and in *Ecdamua*, which is recorded also from India and Africa, the ovipositor is extremely long. *Diomorus* (Fig. 37.25H), with an enlarged tooth-like projection below on the hind femur, is represented by one species in north Australia. The introduced *Syntomaspis varians*, with simple hind femur, is the most commonly collected species. The only Australian rearing record of the subfamily is from galls on *Apophyllum*.

SYCOPHAGINAE (23 spp.) are parasites of the fig-inhabiting Agaonidae. Length, excluding the usually very long ovipositor, 1·5–2·5 mm. The body is obviously flattened. The fore and hind legs are large and the mid legs weak (as in Agaonidae). Most Australian species belong to wide-ranging genera, such as *Philotrypesis*, *Sycoryctes*, and *Apocrypta*. Many genera can be bred together from one species of fig, and none is apparently parasitic on only one genus of fig insect. The males are wingless and of grotesque appearance in *Sycoryctes* and *Philotrypesis*, but winged in *Pseudidarnes* and *Metidarnes*. The apex of the female gaster is produced into a long tubular style in *Sycoryctes*, while there is a 2-segmented style in *Philotrypesis* (Fig. 37.25I). In the metallic green-black *Pseudidarnes*, which is much larger and stouter than most species, the gaster is obviously petiolate. It resembles some Monodontomerinae, but has a long stigmal vein, and the axillae are not advanced. The pronotum is reduced and not clearly visible from above in *Idarnes*. In *Paracolystichus* and *Micranisa* the ovipositor is short and the axillae slightly advanced.

In MEGASTIGMINAE (100 spp.) the stigmal knob is much enlarged and the axillae are not advanced. The 2nd hind tibial spur is often quite reduced, and may be overlooked in *Megastigmus*. Length 1–13 mm. Most species are yellow or straw-coloured, but some are metallic. Australian species of *Megastigmus* (27 spp.; Fig. 37.25J) and *Epimegastigmus* (28 spp.), which differs from *Megastigmus* mainly in having a distinct median longitudinal sulcus from the anterior margin of the scutellum, are gall-formers on a wide range of plants, most commonly on eucalypts and acacias. Some are apparently inquilines in galls, and a few are parasites of tephritid gall-makers. The large *Bootania gigantea*, body up to 13 mm and ovipositor 25 mm, has clearly defined venation and whorls of long hairs on the male antenna. *Epibootania* forms galls on endemic *Eremocitrus*.

MONODONTOMERINAE (23 spp.) are characterized mainly by the absence of diagnostic features. They differ from Sycophaginae in having advanced axillae, the scutellum strongly raised at apex, and the whole thorax more convex from side to side. The body form resembles that of a pteromalid, but the ovipositor is usually exserted. The males, in particular, can be confused with Miscogasterinae, but the emarginate apex of the 1st tergite (and often the 2nd) of the gaster distinguishes all Australian species. They also have well-developed cerci. The species are mostly gall-formers or parasites of gall-formers, mostly on acacias and eucalypts, on which they form small round red galls; they also form galls on *Callitris*. A few are parasites of tephritid gall-makers, and one has been bred from the fruits of *Eremophila*. In *Macrodontomerus* (20 spp.) the hind femur normally has a tooth below towards the apex; the species are metallic green, with the ovipositor about as long as the body. The gaster is laterally compressed in *Ditropinotella* and the ovipositor directed dorsally.

35. Pteromalidae. There is little agreement on classification of this large and difficult family, apart from a few rather distinct groups that are sometimes treated as separate families.

EUCHARITINAE (60 spp.) all have large falcate mandibles. Pronotum normally reduced dorsally in correlation with the very mobile head; coxae all small; first 1 or 2 segments of gaster large; axillae usually completely fused and forming a transverse sclerite between mesoscutum and scutellum. They are of very diverse and often bizarre form, and nearly all are metallic. As far as is known, all are parasites of ants, *Myrmecia*, *Ectatomma*, *Camponotus*, and *Pheidole* being common hosts. The wasps emerge from the ant pupal cocoons, and may be seen flying over ants' nests on sunny days. The genera fall into 2 groups. In *Orasema* and *Eucharomorpha* there is a ring segment in the antenna, and the prepectus is large, triangular, and vertically elongate. In the other group, which includes most genera, the antenna is without a ring segment, and the prepectus is either fused with pronotum or quite reduced; when present it is horizontally elongate. In *Tricoryna*, small black species reared from *Ectatomma*, the scape is little longer than the pedicel, and the gaster of the male is composed mainly of one large tergite. The common, large (up to 10 mm), usually metallic green, blue, or coppery species of *Epimetagea* (Fig. 37.26A) are parasites of *Myrmecia*; there is marked dimorphism in size of gaster and length of antenna. In *Schizaspidia* the scutellum is markedly produced and forked over the gaster, and the males have distinctly flabellate antennae.

PERILAMPINAE (20 spp.) are specialized chalcidoids in which the gaster is composed largely of the partly fused first 2 tergites (hinged dorsally but free laterally). They are of short stout build (length 1·5–5·5 mm) and metallic appearance. Mandibles stout, each prolonged into a stout tooth, but not sickle-shaped as in Eucharitinae; coxae all small; a ring segment present in the antenna; pronotum with a well-developed dorsal surface, the anterior declivous portion usually concave. Many of the species are parasites of lepidopterous larvae, or of tachinids and ichneumonoids parasitizing lepidopterous and coleopterous larvae. Most are placed in *Perilampus* (Fig. 37.26B) in which the gaster is sessile. *Euperilampus scutellatus* has the scutellum strongly produced and tapering and the post-marginal vein long. [Riek, 1966.]

ORMYRINAE (9 spp.) are small (1·7–2·2 mm), metallic, green to almost black species, with a characteristic ornamentation of the gaster, but otherwise pteromaline-like. Gaster tapering in female, but ovipositor not exserted; femoral furrow rather glabrous; prepectus small and almost hidden; hind coxae enlarged, and axillae very slightly advanced; one hind tibial spur long and very stout; basal segments of gaster with straight caudal margins; antennae 13-segmented, with 2 ring segments, but in *Ormyrus* the second ring segment considerably larger than first. The species are parasites of gall-forming chalcids. One rather glabrous species has been bred from a seed-gall-former in figs, other species from galls on *Apophyllum*, *Eremocitrus*, *Melaleuca*, and *Casuarina*.

LEPTOFOENINAE (1 sp.) are very large thin species with the facies of the Gasteruptiidae. Their biology is unknown. *Amotura* (2 spp.) (CHALCEDECTINAE) has a hind leg similar to that of the Haltichellinae (Chalcididae), but is distinguished by the advanced axillae. An enlarged hind leg occurs in some CLEONYMINAE, such as *Agamerion* (Fig. 37.26C), but the femur is never regularly toothed below. The bright metallic, stocky species of *Agamerion*, very common on tree trunks, are parasites of cockroach oothecae. *Cameronella* has the apex of the gaster dilated into 3 leaf-like expansions in the female; the prepectus has a concave posterior margin, and the notaulices end in the advanced point of the axilla and not closer to the mid-line as is usual. The species are bred from *Apiomorpha* galls on eucalypts. In *Paraheydenia*, bred from beetle-infested acacia twigs, the fore femur is greatly swollen and bears a series of teeth below. The pronotum is very large and saddle-like, as in *Oodera* which has in addition an unusual development of the thoracic grooves and sutures. *Marxiana* has a series of tubercles on the head like those of *Eusandalum* (Eupelminae) and Orussidae. *Thaumasura* and related genera are parasites of longicorn and curculionid larvae. The numerous species have long thin legs and a strongly exserted ovipositor. Many are very large (more than 30 mm including ovipositor). In *Cleonymus* the apical funicle segment is produced on one side almost to the apex of the club; the small (up to 5 mm) species are bred from beetle-infested twigs. [Heqvist, 1961.]

SPALANGIINAE (20 spp.) are distinguished by their rather smooth, flattened, elongate

Fig. 37.26. Chalcidoidea: A, *Epimetagea* sp., Pteromalidae-Eucharitinae; B, *Perilampus* sp., Pteromalidae-Perilampinae; C, *Agamerion* sp., Pteromalidae-Cleonyminae; D, *Spalangia* sp., Pteromalidae-Spalangiinae; E, *Nasonia vitripennis*, Pteromalidae-Pteromalinae; F, *Trichilogaster* sp., Pteromalidae-Brachyscelidiphaginae; G, *Psyllaephagus* sp., Encyrtidae-Encyrtinae; H, *Eupelmus* sp., Encyrtidae-Eupelminae; I, *Tanaostigmodes* sp., Encyrtidae-Tanaostigmodinae; J, *Centrodora* sp., Encyrtidae-Aphelininae. [S. Curtis]

bodies, long, forwardly-directed heads, and ill-defined femoral furrow. *Spalangia* (Fig. 37.26D) is a common parasite of dipterous puparia. *Cerocephala* is a parasite of xylophagous beetles, whereas *Theocolax* is a parasite of *Anobium* attacking stored products. The DIPARINAE (5 spp.) have the propodeum produced to a distinct neck; the notaulices are deeply impressed. The species of *Australolelaps* have a glabrous body, one greatly enlarged hind tibial spur, and the apex of the gaster produced into a style in the female. *Dipara* has a few very large bristles on the head and dorsal thorax.

PTEROMALINAE (130 spp.) have shallow incomplete notaulices and, in the great majority of species, only 1 hind tibial spur. The gaster is subsessile in Pteromalini, but obviously petiolate in Sphegigasterini; Merisini have no spiracular grooves on the propodeum, but many are most readily distinguished by the obliquely truncate antennal club. Pteromalini are mostly parasites of lepidopterous pupae. *Dibrachys* is a common parasite of *Galleria mellonella* in honey comb. *Nasonia vitripennis* (Fig. 37.26E) is a common parasite of *Lucilia* and *Chrysomya*. Some species are hyperparasites. Many Sphegigasterini are egg parasites of Lepidoptera, Chrysomelidae, and spiders, but some are reared from agromyzid leaf-miners. *Neopolycystus* is a common parasite of *Paropsis* eggs on eucalypts. The marginal vein is thickened and the head rather large in *Pachyneuron*, all species of which are hyperparasites.

MISCOGASTERINAE (100 spp.) differ from Pteromalinae in having deeply impressed notaulices, 2 hind tibial spurs, and slightly advanced axillae. The 2nd hind tibial spur is sometimes reduced, as in *Coelocyba*, common elongate yellow species from galls on eucalypts. The gaster is obviously petiolate in Miscogasterini and subsessile in Trigonoderini. EUNOTINAE differ from Tridyminae in possessing a menisciform head and low insertion of the antennae, in which segmentation is reduced.

BRACHYSCELIDIPHAGINAE (60 spp.), which are mainly Australian, differ from Tridyminae in the strong development of the axillae, which touch in the mid-line, or at least are much closer together than the inner margins of the notaulices. First tergite of gaster emarginate in mid-line, but ovipositor not strongly exserted; development of the pronotum variable, but it is almost as wide as the mesothorax. In a few genera, including *Decatomothorax* from galls on *Brachychiton*, the pronotal collar is large and *Eurytoma*-like. Other genera, especially *Lisseurytoma* which forms bud-like galls on *Casuarina*, have an obviously petiolate gaster as in Miscogasterini. *Trichilogaster* (Fig. 37.26F) forms apple-galls on acacias. *Brachyscelidiphaga* has been reared from *Apiomorpha* galls on eucalypts where it is possibly an inquiline.

36. Encyrtidae. This large family embraces all species in which the mid leg is modified for jumping: the tibial spur is enlarged, and the tarsal segments have rows or patches of modified setae on the ventral (plantar) surface (Fig. 37.7K) except in the most primitive species. The mesopleuron is completely swollen in all except Thysaninae, most Aphelininae, and males of *Eupelmus (sens. lat.)*, but even in these there is some enlargement, and the lower portion of the mesepisternum is produced and swollen. The prepectus is swollen and convex in frontal view in Eupelminae and Tanaostigmodinae, and, particularly in the latter, produced forward to cover the lateral lobe of the pronotum, a character that does not occur in other Chalcidoidea. It is flat in the other subfamilies, and both its posterior and dorsal margins overlap the adjoining structures to some extent, especially the anterior margin of the mesepimeron. The hind tibia has 2 spurs, but the 2nd is always decidedly reduced to virtually absent.

The ENCYRTINAE (600 spp.) are a large group of small to minute species (5–0·5 mm), of diverse habits, but mostly parasites or hyperparasites of scale insects. The subfamily has been subdivided on the structure of the mandible, which is edentate or bears one to several teeth, and some of the upper teeth may be united to form a straight cutting edge. This character is difficult to see without dissection, and a more satisfactory subdivision is based on the development of metapleuron and malar groove of the head. Notaulices almost invariably absent, and when present never distinctly impressed; antenna without a ring segment, usually with 6 funicle segments (less often 5 or 7), and the club often solid; male antenna often of grotesque form; marginal vein short to punctiform. *Psyllaephagus* (Fig. 37.26G), a large and diverse genus of relatively large species, parasitizes lerp-forming psyllids. *Metaphycus* is a common parasite of scale insects. *Ooencyrtus* and related genera are egg parasites. *Hunterellus* parasitizes nymphal ticks. *Tachinaephagus* is a common parasite of large dipterous puparia. *Quaylea* is hyperparasitic on other encyrtids. *Copidosoma* and related genera are polyembryonic. *Paralitomastix kohleri*, which oviposits in the eggs of potato moth and emerges from the larva after it has spun a cocoon, produces up to 100 individuals from one egg, but more than 1,000 specimens of *Litomastix* may be reared from one noctuid larva. *Neocladia*, with leaf-like expansions on the hind leg and 1-toothed mandibles, is common on smooth-barked tree trunks.

EUPELMINAE (213 spp.) are similar to Encyrtinae, but the mid leg occupies a more normal position close to the hind leg, and the fore and mid legs are widely separated. The species are of medium to rather large size, with *Metapelma* up to 9 mm, including an ovipositor of 2 mm. The mesonotum is large and often characteristically sunken. The mesepimeron is greatly enlarged and convexly swollen in all except the males of *Eupelmus (sens. lat.)*, which are similar to male Cleonyminae, but distinguished by the more inflated prepectus and mesepisternum and the single hind tibial spur. *Eupelmus* (131 spp.; Fig. 37.26H) is a primary parasite of insect eggs. *Metapelma*, which differs mainly in having a flange-like expansion on the hind leg, is a parasite of lepidopterous and possibly coleopterous larvae. *Calosota*, with very small axillae limited to the lateral margins, and *Eusandalum*, with rows of rasp-like tubercles on the head and the prepectus very large, are associated with borer-infested timbers. *Neoanastatus*, which has a very long tibial spur and a longitudinal groove on the small scutellum, has been bred from galls on *Atalantia glauca* and native limes. *Eueupelmus*, with a cornuted head, is associated with *Cactoblastis*. *Eupelminus*, which contains species with subapterous females, is apparently a hyperparasite. The scape is widened in *Cerambycobius* and a few species of both *Eupelmus* and *Eusandalum*. A few species of *Eupelmus* have enlarged scale-like hairs.

TANAOSTIGMODINAE (8 spp.; Fig. 37.26I) are stout, stocky species, 2–3 mm, recognized at once by the greatly swollen prepectus. The notaulices meet in the mid-line before the caudal margin of the scutum, and then usually continue as a single median groove. The male antenna often has whorls of long hairs, and is sometimes flabellate. Some species of *Tanaostigmodes* have enlarged scale-like hairs. They are apparently all gall-formers. One Australian species has been bred from seed-galls on *Bossaea* (Leguminosae).

APHELININAE (130 spp.) are very small to minute; mostly parasites of scale insects, mealy bugs, and aphids, but *Centrodora* (Fig. 37.26J) and related genera are egg parasites, and *Marietta* and related genera are hyperparasites. The mid leg is not as well developed as in other Encyrtidae, so that the mesopleuron may have an impressed vertical line or show more definite indications of the femoral furrow, but it is entirely inflated in *Physcus*. Prepectus not overlapping anterior margin of mesopleuron; notaulices clearly impressed, and widely separated caudally; axillae usually obviously advanced, but not, or only slightly, in genera in which the lateral thorax is decidedly inflated; marginal vein long and stigmal sessile. *Aphelinus mali* is a common parasite of woolly aphids on apples.

THYSANINAE (10 spp.) are very small (0·5–1 mm), shining black, flattened species with reduced segmentation in the antenna and axillae fused with the scutellum. The mid tibial spur has a fringe of long stout spines on the inner margin. The prepectus is large but flat, and the mesopleuron is inflated. The mid tibia has a number of stout bristles. These most distinct species are hyperparasites; one has been bred from *Psyllaephagus* parasitizing lerp-forming psyllids.

Superfamily CHRYSIDOIDEA

Prothorax freely movable, overlapping the prepectus, but not reaching back to the tegulae; hind tibial spurs and basitarsus not modified for preening, but a dense brush over the whole medial surface of tibia and basitarsus; spiracle cover lobe of pronotum margined with close fine hairs, but they are not always obvious; fore-wing venation reduced apically; hind wing without closed cells, but sometimes with distinct longitudinal veins. There is thus considerable similarity to the Chalcidoidea, in which the venation is even more reduced. Antennae inserted near mouth, elbowed, the scape long; gaster concave below, with 2–4, rarely 5, visible tergites; body usually metallic and with coarse sculpture. The Cleptidae, which are generally placed in Chrysidoidea, are transferred to Bethyloidea on the structure of the hind leg and pronotal lobe, though the pronotum does not reach back to the tegula as is usual in Bethyloidea.

37. Chrysididae (Fig. 37.27I). Australian species are mainly green and blue (Plate 5, V), but some have purple or dull copper colouring. They are external 'parasites', mainly of mud-dauber wasps (Sphecoidea and Vespoidea). Development occurs only after the host larva has consumed the food in its cell; when full-grown, the parasite larva spins a cocoon within the cell of the host. The most commonly bred species are those that parasitize the large mud cells of *Sceliphron* and *Abispa*. The adults are able to roll themselves up into a ball for protection against the jaws of the host wasp, and the body is heavily sclerotized apart from the soft concave venter. The apex of the gaster is tubular and telescoped. Both sexes winged, the sexes not easily distinguished; males are not commonly collected. Wing venation well developed, except apically; costal cell linear, pterostigma very small. Apparent apex of gaster sometimes produced into a varying number of spines. Pronotum separated from tegula by a lateral lobe of mesoscutum, and not by the large prepectus. Adults (length 3–22 mm) are common on dead tree trunks and around old timber buildings. Some visit flowers, and pollinia have been found attached to the legs of *Stilbum*, large species common only in the north. The somewhat smaller species of *Pyria*, which also have a produced metanotum, are widespread. The small species of *Holochrysis*, in which the apex of the gaster is entire, are most common. They are conspicuous because of their green colour.

Superfamily BETHYLOIDEA

Closely related to, but generally much smaller than, Pompiloidea. Their venation is reduced in correlation with their small size and parasitic habits, the non-Australian Sierolomorphidae having the most complete venation in the superfamily. The species are all parasites, mainly ectoparasites. There is considerable variation both in structure and general appearance. Bethylidae, the only common family, resemble Tiphiidae and some of the more primitive Pompilidae. Both sexes are winged in some genera, but others have wingless females that are distinguished from the wingless females of *Eirone* (Tiphiidae–Thynninae) only on the development of the hind tibial spurs and basitarsus. Winged female Dryinidae resemble Ampulicidae (Sphecoidea), apart from differences in structure of the prothorax and in venation; wingless females, which are not transported by the males, often resemble ants, especially those that attend their cicadelloid and fulgoroid hosts. Australian Cleptidae also have wingless females that are somewhat ant-like; the winged males are more like Chrysididae, and the family has generally been referred to the Chrysidoidea.

Key to the Families of Bethyloidea Known in Australia

1. Antenna with 17–40 segments (Fig. 37.27C). [Head elongate; antennae inserted near mouth under a pronounced frontal shelf; ♀ wingless; pterostigma linear in ♂] .... **Sclerogibbidae** (p. 926)
   Antenna 10- to 13-segmented 2

2. Antenna inserted high on the forwardly produced face (Fig. 37.27B). [♂ antenna 10-segmented, ♀ 13-segmented; 7 visible tergites in gaster of ♂] .... **Embolemidae** (p. 928)
   Antenna inserted near mouth, or, if high on the face, then venation not attaining the costal margin. 3
3. Antenna 10-segmented in both sexes; fore tarsus of ♀ usually chelate (Figs. 37.7E,27D,E). [Six or 7 visible tergites] .... **Dryinidae** (p. 926)
   Antenna 12- or 13-segmented, rarely 11-segmented; fore tarsus simple 4
4. Pronotum reaching back to tegula; 7–8 visible tergites (Fig. 37.27A) .... **Bethylidae** (p. 926)
   Pronotum not reaching back to tegula, but separated by a lobe of the scutum; 5–6, rarely 4, visible tergites (Figs. 37.27F–H) .... **Cleptidae** (p. 928)

38. Bethylidae (Fig. 37.27A). External parasites of small larval Lepidoptera and Coleoptera. Habits diverse, intermediate between true parasites and the fossorial families. Usually no true nest is made, but the prey may be dragged to a sheltered position (as occurs in the more primitive Pompilidae). Unlike fossorial wasps, more than one egg may be laid on each host. The egg is placed externally, and the mature larva spins a cocoon. Females may be fully-winged, brachypterous, or wingless, and dimorphic and polymorphic species are known; ocelli absent in some wingless females. Some species with wingless females *(Propristocera)* copulate in a similar manner to Thynninae. Adults of the common species are collected on tree trunks or from the leaf-litter at their bases. They occur throughout the year, and there may be more than one generation a year in some species; development in *Goniozus* may be as short as 3 weeks. The family is well represented in most parts of Australia. The species vary in size from 1·5 mm *(Cephalonomia)* to 10 mm *(Rhabdepyris)*, but are generally 3–4 mm. *Eupsenella, Perisierola, Sierola*, and *Goniozus* are parasites of small Lepidoptera; *Cephalonomia* and *Parasclerodermus* are probably parasites of Coleoptera; *Rhabdepyris* and *Epyris* are associated with ants (*Camponotus* and *Ectatomma*). [Richards, 1939a.]

39. Sclerogibbidae (Fig. 37.27C). A widespread family with few described species. The strikingly distinct sexes, with wingless females, have been associated in only one species. The only reared species are ectoparasites of nymphal Embioptera. The larva feeds externally, attached dorsally between the prothorax and mesothorax; it spins a cocoon after the manner of Bethylidae. [Richards, 1939b.]

40. Dryinidae (Figs. 37.27D, E). Small species (1·5–10 mm), parasitic on Homoptera, mainly Fulgoroidea and Cicadelloidea, mostly ectoparasites on the abdomen of the nymph or adult, but at least one species of *Aphelopus* is endoparasitic and polyembryonic. The position of the parasite varies, and one to several may be present. During development the larva is enclosed in an external gall-like sac *(thylacium)*, usually black, smooth, shining and composed of cast larval skins. Pupation takes place either in soil or litter or on the food-plant of the host (in an oval white cocoon). Species of very diverse form are grouped on the 10-segmented antennae of both sexes and the chelate fore tarsus of the female (except in *Aphelopus*); the fore tarsi of males are not chelate. The fore wing has a reduced braconid-like venation, but with a distinct costal cell; the hind-wing venation is limited to the marginal vein. The chela (Fig. 37.7E) is formed by an elongate lobe from the 5th tarsal segment and an enlarged claw. The female uses it to capture an adult or nymphal host, which she stings to insensibility before laying an egg externally. The host soon recovers and resumes normal activity.

APHELOPINAE and ANTEONINAE, in which both sexes are usually fully winged, are stocky insects not unlike Pemphredoninae (Sphecidae). In Aphelopinae (sometimes separated from the Dryinidae) the fore leg of the female is not chelate, the prepectus fused with the lateral pronotum, and the scape small. In Anteoninae the fore leg has a large coxa, elongate trochanter, and expanded femur, the prepectus is large and clearly defined *(Chelogynus)* or secondarily fused

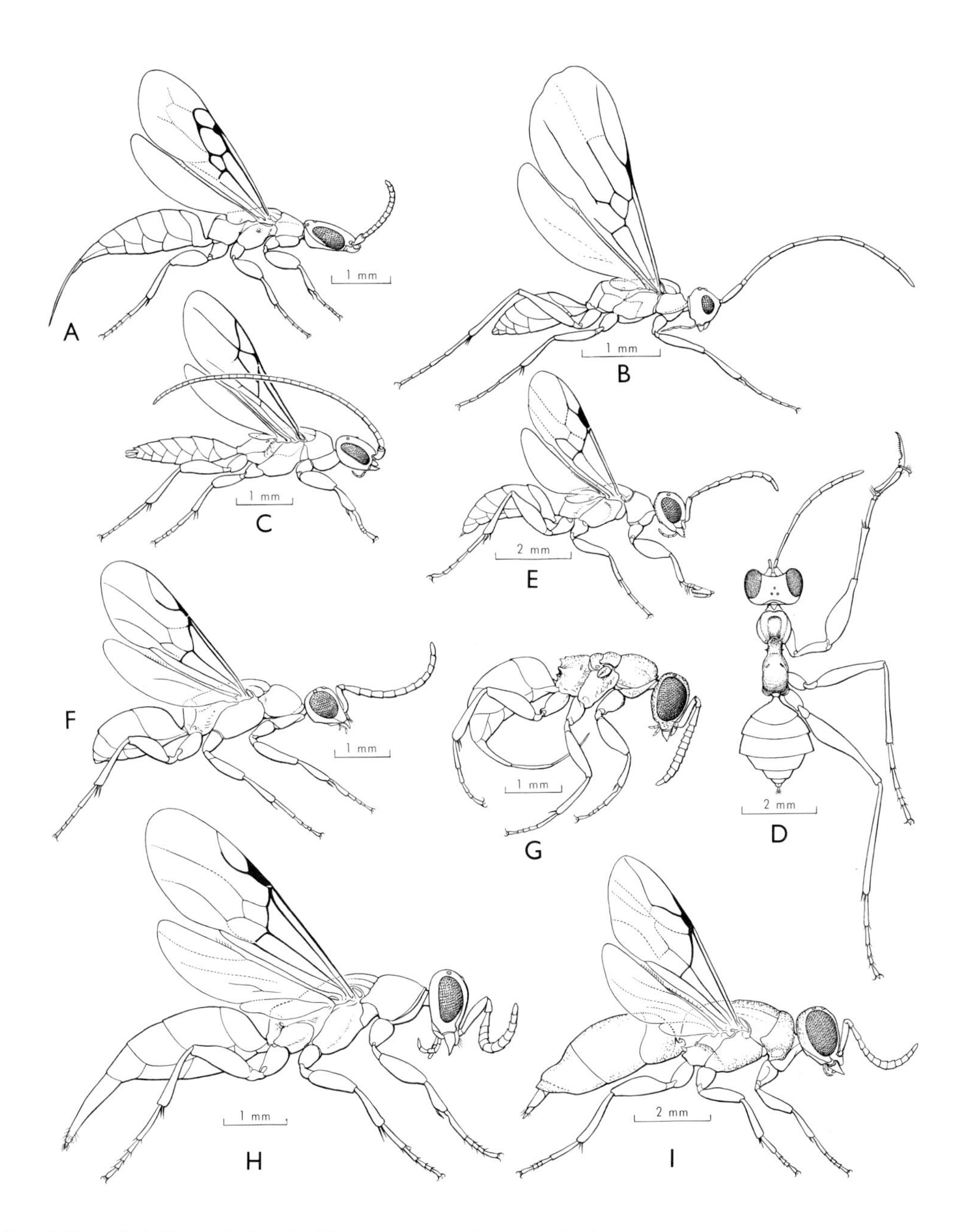

Fig. 37.27. Bethyloidea and Chrysidoidea: A, *Eupsenella* sp., Bethylidae; B, *Embolemus* sp., Embolemidae; C, genus near *Sclerogibba*, Sclerogibbidae; D, *Chalcogonatopus* sp., Dryinidae, ♀; E, *Chelogynus* sp., Dryinidae, ♀; F, *Myrmecomimesis* sp., Cleptidae, ♂; G, *Myrmecomimesis* sp., ♀; H, *Cleptes* sp., Cleptidae, ♀; I, *Holochrysis* sp., Chrysididae. [T. Binder]

with the pronotum *(Anteonella)*, and the scape is long (as in Sphecoidea). In GONATOPODINAE and DRYININAE the females have a fore leg similar to that of Anteoninae but with a more elongate trochanter, the prepectus is well defined, the pronotum does not extend back to the tegula, and the scape is short. Wingless females are often long, slender, rather ant-like, and quite unlike the winged males. Adult dryinids, which occur in summer, are an inconspicuous element in the fauna, but both sexes of some species *(Chelogynus)* and the males of others are attracted to lights. *Chelogynus* and *Anteonella* are the more normally encountered genera on tree trunks, while many of the wingless females of Gonatopodinae *(Pseudogonatopus, Chalcogonatopus, Haplogonatopus)* occur in litter where they may be mistaken for ants. [R. Perkins, 1905a; Richards, 1939a, 1953.]

41. Embolemidae (Fig. 37.27B). The host relationships are not known, but males have been taken in sweepings from moist gullies. The sexes are so unlike that they were referred to different genera. The venation of the male fore wing is well developed, with a narrow pterostigma; females are wingless. The head shape of the male is distinctive, with the antennae inserted high on the prominently produced face.

42. Cleptidae (Figs. 37.27F–H). The family is divided into 2 (or more) subfamilies. Cleptinae, parasitic on sawfly 'prepupae' within their cocoons, do not occur in Australia. The Australian species, egg parasites of Phasmatodea (Riek, 1955e), are placed in the widely distributed AMISEGINAE. Some species of *Myrmecomimesis* (16 spp.), recorded from the eastern States and W.A., parasitize the eggs of species of *Podacanthus* and *Didymuria.* Sexual dimorphism is marked: males are fully winged, but not active fliers; and the rarely seen brachypterous females, with fore wings reduced to minute pads, forage in litter in search of phasmatid eggs. Females of extra-Australian genera are sometimes fully winged. [Krombein, 1957.]

Superfamily POMPILOIDEA

In many respects the Pompiloidea are intermediate in structure between Scolioidea and Sphecoidea. All three are fossorial, and each has the prothorax modified, though in a different manner, to meet the burrowing requirement. The pronotum is closely associated with the mesothorax in all Sphecoidea and some Scolioidea (Scoliidae and Mutillidae), but there is considerable freedom of movement between these two structures in Pompiloidea and primitive Scolioidea (Tiphiidae). The Pompiloidea have a special mechanism on the hind legs for brushing the wings and body, similar to that of Sphecoidea, especially in the females. The Rhopalosomatidae are usually referred to Scolioidea, but are here included in Pompiloidea on the development of a preening calcar and strigil on the hind leg.

Key to the Families of Pompiloidea

Mesopleuron with a 'horizontal' (scrobal) groove .... **Pompilidae** (p. 928)
Mesopleuron without a 'horizontal' groove .... **Rhopalosomatidae** (p. 928)

43. Rhopalosomatidae. *Harpagocryptus australiae*, described as a dryinid, and subsequently referred to Bethylidae and also with some doubt to Pompilidae, may belong in this family (Reid, 1939; Gurney, 1953). The larva formed 'a sac on the sides of the abdomen of small crickets' (Gryllidae).

44. Pompilidae (Fig. 37.28). Distinguished from all other fossorial wasps by the structure of the prothorax combined with a well-developed hind-wing venation. The more primitive parasitic species are not readily distinguished from primitive Bethyloidea. There is considerable freedom of movement of the posterolateral margin of the pronotum over the anterior portion of the mesopleuron, and the lower end of the lobe is somewhat rounded, allowing considerable flexion of head and pronotum. The hind wing has a well-developed 'jugal' lobe, except in brachypterous species. Secondary hamuli, generally 4 and easily distinguished, occur towards the base of the hind wing.

All species are parasites or predators of spiders. Some use only one spider to provision the nest, which is usually in the ground. Most prepare only one cell to each burrow, but in some species of *Priocnemis* several cells may be prepared. Some first construct the cell and then provision it, but many do not construct the burrow until after the prey is captured. Many Macromerini construct mud cells; those of *Pseudagenia* are placed under bark or in other sheltered situations. Some species do not construct a nest, but insensitize the spider in its own retreat and deposit an egg on it (some Australian Aporini and Pompilini). In some Planicepinae and in *Platyderes collaris* paralysis of the spider is very light and soon wears off, and the larva develops on the active spider, later pupating in a silken cocoon within the spider's retreat. *Ceropales* is a 'parasite' that deposits its egg on the prey of other pompilids before it is placed in the nest. *Deuteragenia* has a peculiar tuft of bristles on the maxilla, which are used for collecting spider's web used in the closure of the nest. Similar bristles occur also in other genera, in which, however, the nesting habits have not been observed.

Many females walk quickly with their wings flicking rapidly when hunting for prey. When prey is secured, the wasp usually walks backwards, grasping the spider with her mandibles by the base of the hind legs, and, unless it is very large, holds it rather high in the air. Sometimes the wasp may turn round and proceed forwards if the prey is small, or may even make small hopping flights with it. Some species deposit the prey and make exploratory trips in the direction of the nest, but others proceed in an almost straight line directly to the nest. In those species in which the nest is not constructed until the prey is secured, it may be hidden or temporarily buried until the nest is prepared. [Richards and Hamm, 1939; Townes, 1957.]

Key to the Subfamilies of Pompilidae

1. Fore femur greatly enlarged; eyes hairy in ♀ (Fig. 37.28A). [Medial margin of eyes almost parallel, closest above antennae or at vertex; femora without apical spines; hind tibia without spines above or at apex; 2nd sternite of gaster without a transverse sulcus; empodium very small] .... PLANICEPINAE
   Fore femur not enlarged (somewhat enlarged in some species of *Aporus*); eyes smooth, or at most with a few minute hairs 2
2. Eyes deeply emarginate medially above the mid-point (Fig. 37.28F). [Antenna short and stout, scape expanded below; vertex distinctly raised; hind tarsal claws strongly hooked; hind coxa greatly enlarged, twice as long as mid coxa; femora without apical spines; 2nd sternite of gaster without a transverse sulcus] .... CEROPALINAE
   Eyes with medial margin at most slightly sigmoid, closest either at vertex or below at clypeus 3
3. Metapostnotum small or not visible in mid-line, its caudal margin irregular (not straight); femora each with a single enlarged spine at apex anteriorly, rarely with additional small spines (rarely reduced on hind femur and, more rarely, on mid femur too, but then 2nd and 3rd radial cells united) (Fig. 37.28H). [Apical tarsal segment usually with a median row of spines below] .... POMPILINAE
   Metapostnotum clearly developed in mid-line (sometimes small in *Chirodamus*), and with more or less straight caudal margin; femora without spines at apex anteriorly, or, if spines are present, there is never regularly one large spine on each femur (fore femur may have only one spine) (Fig. 37.28D) ...... PEPSINAE

The PEPSINAE (62 spp.) are divided into tribes mainly on the structure of the gaster. Macromerini (15 spp.; Plate 6, V; Fig. 37.28C) are readily distinguished from all other pompilids by the subpetiolate gaster. The Australian species, of medium to large size (7·5–24 mm), almost invariably have red antennae and infuscated wings, sometimes with 2 transverse fasciae. The males, in particular, frequent smooth tree trunks in bright sunlight. Pepsini (39 spp.) include the common, large, orange, ground-dwelling pompilids. *Cryptocheilus* (7 spp.; Fig. 37.28D) contains large (up to 35 mm), orange and black species that are amongst the most

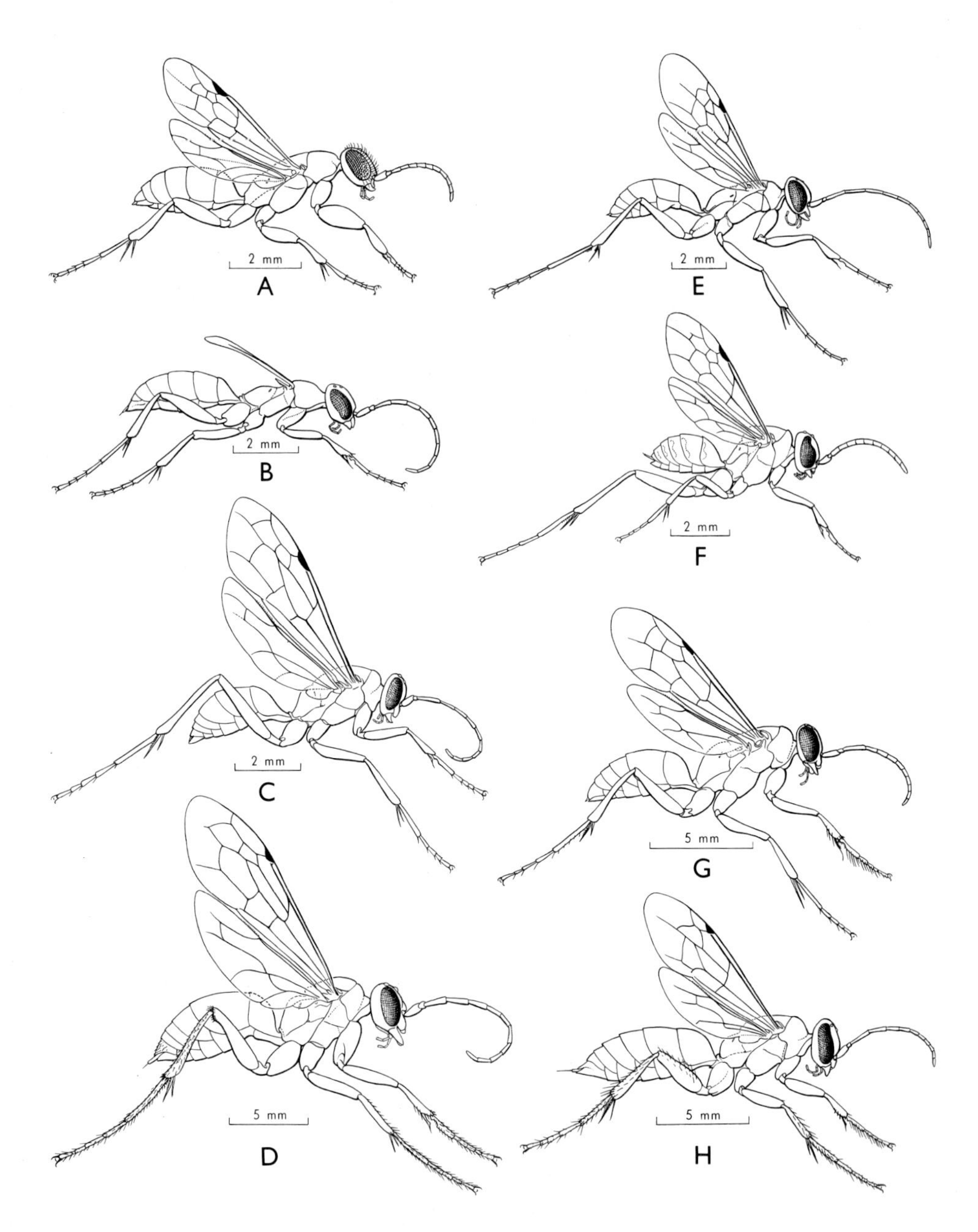

Fig. 37.28. Pompiloidea-Pompilidae: A, unidentified Planicepinae; B, brachypterous Pepsini; C, *Phanagenia* sp., Pepsinae-Macromerini; D, *Cryptocheilus* sp., Pepsinae-Pepsini; E, unidentified Pepsini; F, *Ceropales* sp., Ceropalinae; G, *Pompilus* sp., Pompilinae-Pompilini; H, *Aporus* sp., Pompilinae-Aporini. [T. Binder]

common pompilids; they carry their prey to burrows in soil. *Chirodamus* (30 spp.) are more stoutly built. Most are large (up to 28 mm), and many are marked with orange and black, but there are also dark species. The female antenna is 13-segmented in one species (usually 12-segmented in the females of higher Hymenoptera). *Priocnemis* (9 spp.) contains mostly small species with 2 dark fasciae on the fore wing. One Australian pepsine has very reduced wings and the general facies of an ant (Fig. 37.28B). The Notocyphini and Minageniini, in which the 2nd sternite of the gaster of the female is without a transverse groove, are not recorded, though both occur in the Oriental region.

POMPILINAE (46 spp.; Plate 5, T) are divided into 2 tribes on venation, but in other respects there is great similarity. The 2nd and 3rd radial cells are fused in Aporini, but separate in Pompilini. This large subfamily contains small to medium-sized species (3·5–19 mm), mostly black, with infuscated wings, silvery grey bands on the gaster, and silvery markings on the legs and body. There is often considerable size variation (from 3·5 mm in small males to 13 mm in large females of a single species). Pompilini are mainly predators of web-spinning spiders. Most nest in soil and store the gallery with a single spider. The nests of one species of *Pompilus* are often very common in coastal sand dunes. Many Australian Aporini attack spiders under bark, and pupate in the spider's retreat. Others are said to attack trap-door spiders, and to utilize the spider burrow as a nest. *Platyderes collaris*, an iridescent purple-black species with a large orange pronotum and black wings, cannot be confused with any other Australian pompilid. *Batozonellus* (8 spp.) has the metanotum larger than in most genera of Pompilini, and the tibiae have additional small spines at the apex. The males could be mistaken for ichneumonids.

CEROPALINAE (1 sp.) are 'parasitic' on other pompilids. The female *Ceropales* inserts an egg in the stigma at the base of the metasoma of the prey of another pompilid before it is placed in the burrow. The biology of the unnamed Australian species, which occurs from inland N.S.W. to W.A., is unknown. PLANICEPINAE (6 spp.) could be mistaken for large Bethylidae or Sierolomorphidae, but they have a more complete, generalized venation and lack notaulices on the meso-scutum. They are small (up to 10 mm), dark species, with a very well developed pronotum.

Superfamily SCOLIOIDEA

Scolioidea are mostly stoutly built, heavily sclerotized, rather hairy, fossorial species of medium to large size. They are all parasitic; many parasitize scarab larvae in the soil, but mutillids are mostly parasitic on ground-dwelling sphecoid and vespoid wasps and bees. No calcar development occurs on the hind tibial spur of Australian species, and the body remains free of debris mainly through the smoothness of its surface and the presence of numerous coarse hairs. Hairs are often finer in the non-burrowing males than in females. The nocturnal males of some exotic Tiphiidae and Mutillidae have extremely large ocelli. The Scolioidea are generally considered to be the ancestral group from which the Sphecoidea–Apoidea, Formicoidea, and Vespoidea arose. They are also the ancestral group of Bethyloidea, in which the parasitic habit is more highly developed and size reduced.

Key to the Families and Subfamilies of Scolioidea Known in Australia

1. Winged 2
   Wingless (♀♀ only) 6
2. Hind wing with notched jugal (and usually anal) lobe; pronotum not angulate above tegula (Fig. 37.29E) .... **Tiphiidae** (p. 932). 3
   Hind wing without 'jugal' lobe; pronotum angulate above tegula, less so in Mutillidae (Fig. 37.29J). 5
3. Mid tibia with 1 apical spur; fore wing with 2 submarginal cells; tegula elongate ...... TIPHIINAE
   Mid tibia with 2 apical spurs; fore wing with 3 submarginal cells; tegula short 4

4. Antennae arising from simple sockets; ♀ winged .... ANTHOBOSCINAE
   Antennae arising from beneath a frontal ridge; ♀ wingless (Figs. 37.29A, B) .... THYNNINAE
5. Anal lobe of hind wing notched (with pre-axillary incision); mid tibia with 1 spur; apex of wing with close pseudovenation; meso- and metasterna together forming a flat plate, which is divided by a transverse, more or less sinuous suture, and overlies the bases of the mid and hind coxae; ♀ winged (Fig. 37.29J) .... **Scoliidae** (p. 932)
   Anal lobe of hind wing not notched; mid tibia with 2 spurs; apex of wing with microtrichia; meso- and metasterna not forming such a plate; ♀ wingless (Figs. 37.29F, G) .... **Mutillidae** (p. 932)
6. Thorax with distinct segmentation .... **Tiphiidae.** THYNNINAE
   Thorax without distinct segmentation dorsally .... **Mutillidae**

45. Scoliidae (Fig. 37.29J). Mostly large (9–36 mm), stout-bodied, densely hairy species that parasitize scarab larvae in soil. The host is insensitized before the egg is placed externally on its abdomen (transversely on the 3rd segment). The apex of the wings develops a secondary pseudovenation of ridges and grooves not unlike the dense venation of some Blattodea. Eyes emarginate; tarsal claws simple; sexual dimorphism often marked, but both sexes winged and heavily sclerotized. Males have longer, stouter antennae and much thinner bodies than females. *Trisciloa ferruginea*, covered with dense rich golden hair, is one of the largest Australian wasps. Some of the common species of *Campsomeris* (11 spp.) are orange and yellow with golden hairs, but others are dark. [Betrem, 1928, 1933.]

46. Mutillidae (Figs. 37.29F, G). Due to the difficulty in associating sexes of the 'velvet ants', there may be over-estimation of the number of Australian species, which are numerous in the tropics. Males fly with the wingless female in copula. All mutillids are characterized by a heavily sclerotized integument. Females are ant-like, but most are readily separated from ants by a 'felt line' on the 2nd gastral segment. They occur in litter, or are seen running at the ends of grass stems or twigs, or running like ants on tree trunks. Males are often attracted to smooth tree trunks in bright sunlight. Mutillids parasitize sphecoid and vespoid wasps and social and solitary bees, and emerge from the host cocoon. They generally parasitize ground-nesting species, but have also been reared from mud cells of *Sceliphron* and *Pison*. They have been recorded from puparia of *Glossina* (Diptera) in Africa. Most Australian species have been referred to the widespread, very diverse *Ephutomorpha*, in which there are many species-groups. Females range in size from 3 to 23 mm. [André, 1895.]

47. Tiphiidae. Only 3 of the 8 subfamilies are recorded in Australia. Both sexes are winged in the primitive Anthoboscinae and Tiphiinae, but the females of Thynninae are wingless. The loss of wing is correlated with the strongly developed fossorial habits of the females, as evidenced by the development of the pecten on the fore tarsus. Both sexes feed on nectar or honey-dew. ANTHOBOSCINAE (13 spp.) occur mainly in Australia, the Ethiopian region, and South America. The Australian species of *Anthobosca* (Fig. 37.29D), length 6·5–26 mm, mostly black and shining, are very similar in general appearance to small Scoliidae, but lack dense hairs on the body. TIPHIINAE are represented by a single species of the cosmopolitan *Tiphia* (Fig. 37.29C) in north Queensland. They are mostly small to medium-sized, shining black species, rather like *Anthobosca*, but with only 1 mid tibial spur and large tegula. They parasitize scarab larvae, and some species lay more than one egg on each host.

THYNNINAE (497 spp.; Figs. 37.29A, B) are well represented in Australia and South America, and occur also in Lord Howe I., New Guinea, Solomon Is., New Caledonia, Celebes, Philippines (Luzon), and North America. They parasitize scarab larvae, though *Diamma* (Plate 5, R) is said to parasitize mole crickets. In females the legs are modified for burrowing to enable them to reach the soil-dwelling host larva on which they lay an egg; the thorax is divided into 3 regions by sutures; and ocelli are absent in

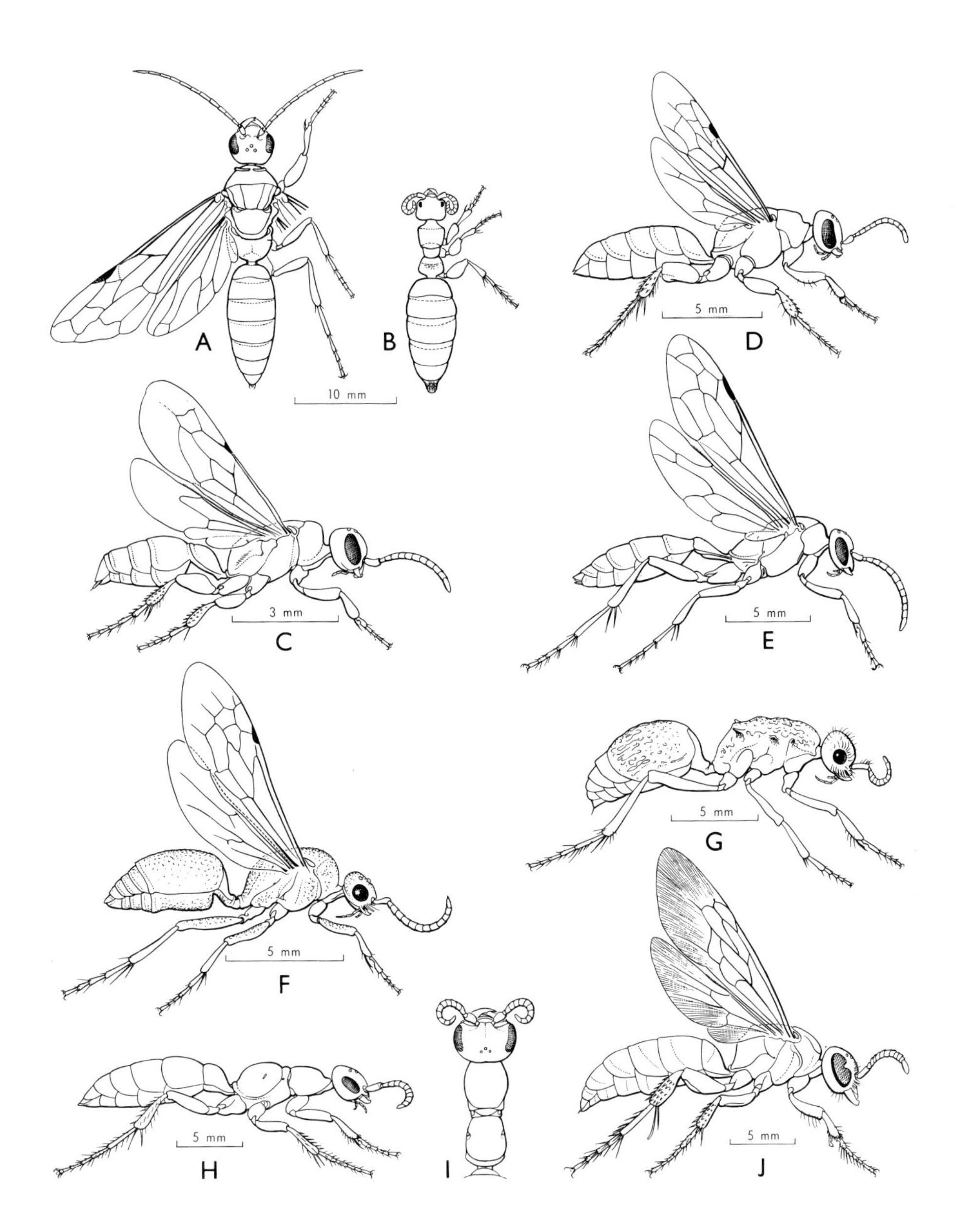

Fig. 37.29. Scolioidea: A, *Hemithynnus* sp., Tiphiidae-Thynninae, ♂; B, *Hemithynnus* sp., ♀; C, *Tiphia* sp., Tiphiidae-Tiphiinae; D, *Anthobosca* sp., Tiphiidae-Anthoboscinae; E, *Diamma bicolor*, Tiphiidae-Thynninae, ♂; F, *Ephutomorpha* sp., Mutillidae, ♂; G, *E. rugicollis*, ♀; H, *Diamma bicolor*, ♀; I, same, dorsal; J, *Campsomeris* sp., Scoliidae. [T. Binder]

Australian species except those of *Diamma* (Figs. 37.29E, H, I). There are normally 2 mid tibial spurs in females as well as in males, but *Eirone* has only 1 spur in the female and the 2nd tooth of the tarsal claw is very reduced. Adults feed on secretions from aphids and scale insects, as well as on nectar. Females (2·2–30 mm long) are smaller than the males, except in *Diamma*, and many males have a highly developed copulatory mechanism for transporting the females in flight to the feeding sites. The females of some species are not transported, but are fed by the male either with regurgitated food or with food stored in excavations beneath the head. *Diamma* females are very active runners and metallic green or purple in colour, hence the name 'blue ant'; they inflict a very painful sting. Females of other species are without metallic colouring. They are mostly stout-bodied, with the large gaster ending in an obliquely truncate pygidium, and the 2nd tergite distinctively banded and ornamented. [Given, 1954a, b.]

Superfamily VESPOIDEA

Distinguished from all other Hymenoptera on the combined structure of the pronotum and well-developed hind-wing venation, although Scolioidea have a similar but less advanced type of prothorax. The larvae are fed chiefly on food of animal origin (insect larvae), except that some Masaridae provision their cells with pollen and nectar (Richards, 1962). Masaridae are also exceptional in a number of structural features, including possession of a relatively weak wing-coupling and absence of fusion between the 1st gastral tergite and sternite. In Vespidae and Eumenidae, the eyes are deeply emarginate; fore wings folded at rest; trochantellus defined on mid femur; tergite and sternite of 1st gastral segment fused at least basally (except in some Eumeninae), and, when a petiole is present, it is formed from this combined tergite and sternite; posterior segments of gaster retractable within 2nd segment. More generally, the hind wing has a straight fore margin, and there are no secondary hamuli towards the base; median (1st submarginal) cell of fore wing greatly lengthened in Australian species; calcar spur of hind tibia curved at apex, and strigil well defined. The following simplified key is valid for Australian species only.

Key to the Families of Vespoidea

1. Fore wing not folded longitudinally when at rest, and with 2 submarginal cells ... **Masaridae** (p. 934)
   Fore wing folded longitudinally when at rest, and with 3 submarginal cells 2
2. Tarsal claws simple; tegula not margined; social species .... **Vespidae** (p. 936)
   Tarsal claws bifid; tegula with raised margin (indistinct in some *Odynerus*); solitary species .... **Eumenidae** (p. 936)

48. Masaridae (Figs. 37.30A, B). Mostly medium-sized (9–24 mm), stout-bodied, solitary wasps, readily mistaken for eumenids. The wing-coupling is by a large number of small hamuli, and the fore wing is not folded at rest except in a few exotic species. Eyes at most slightly emarginate medially in Australian species; mandibles not crossing; tarsal claws usually toothed (almost bifid in some *Paragia*); no suture marking off a trochantellus, except in the hind femur of *Euparagia*; 'jugal' lobe of hind wing usually small and notched; hind tibial spur calcar-like, or forked at apex; usually 2 mid tibial spurs. Of the 3 subfamilies, only the MASARINAE are represented by several genera of Paragiini. Masarinae sometimes construct hard mud cells attached to rocks and trees, in groups or singly, and provision them with pollen and nectar. Some Paragiini make burrows in the ground, often erecting a chimney over the mouth. The food of Australian species is unknown, but many are regular visitors to blossom (*Hakea*, *Leptospermum*, *Eucalyptus*, and *Goodenia*). They are usually strongly marked with orange, and

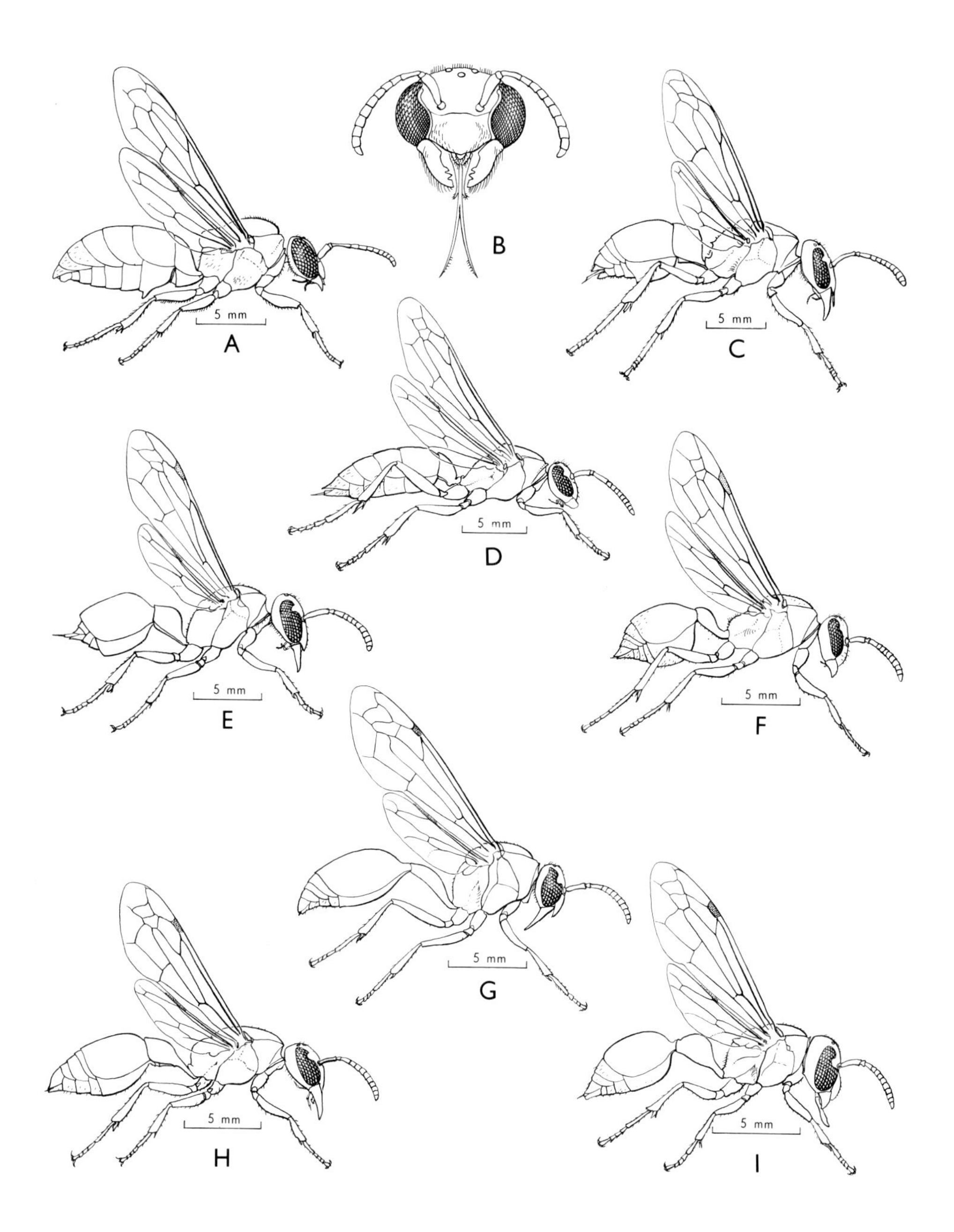

Fig. 37.30. Vespoidea: A, *Paragia* sp., Masaridae; B, *Rolandia* sp., Masaridae, head, frontal; C, *Abispa* sp., Eumenidae-Eumeninae; D, *Polistes* sp., Vespidae-Polistinae; E, *Paralastor* sp., Eumenidae-Eumeninae; F, *Ropalidia* sp., Vespidae-Ropalidiinae; G, *Eumenes* sp., Eumenidae-Eumeninae; H, *Acarozumia amaliae*, Eumenidae-Eumeninae; I, *Pseudozethus* sp., Eumenidae-Zethinae. [R. Ewins]

are found in localized habitats mainly in low-rainfall parts of the country. *Paragia decipiens* is the only species known to be stylopized. [Richards, 1962.]

49. Vespidae (Figs. 37.30D, F). Social, predatory, and more or less melliferous species, or less often inquilines of allied social species. The fore wing is folded longitudinally at rest, except in Stenogastrinae. Mandibles usually short and broad, with their apices overlapping only slightly; trochantellus marked off by a distinct suture on mid and often on fore femur; 2 tibial spurs on mid leg; tarsal claws usually simple; hind wing generally with a 'jugal' lobe. Only Polistinae and Ropalidiinae are known in Australia, but the mainly northern-hemisphere Vespinae, the mainly South American Polybiinae, and the Oriental Stenogastrinae in which the petiole of the gaster is extremely long and thin could possibly occur in the north. The European wasp *Vespula germanica* has become established.

In the Australian Polistinae the nest is formed of a single horizontal comb of hexagonal 'paper' cells attached to a support by a pedicel, or there may be more than one attachment when it is built in an enclosed space. In Ropalidiinae the cells usually form a vertical biserial line, or less commonly there may be many horizontal tiers of cells enclosed in an outer papery envelope (*Ropalidia cabeti*). The larvae are fed chiefly on masticated caterpillars, and there is progressive provisioning. The queens hibernate in cold climates, but the nests may be perennial. There are continuous generations in the tropics. Adults generally rest at night on the upper surface of the comb or on the surrounding support.

POLISTINAE (6 spp.; Fig. 37.30D) are not recorded from Tasmania, but the cosmopolitan genus *Polistes* is widespread, especially in the tropics. Body length 11–26 mm. The many geographical races of the large *P. tepidus* change strikingly in colour in passing from islands to the north of Australia to localities in southern N.S.W. This species often nests in hollow logs. The smaller species, red-brown with yellow markings, have lightly infuscated wings that are sometimes dark at the apices. *P. tasmaniensis* is widespread, and has been introduced into W.A. It is common under the eaves of houses, where it can be a nuisance because of its painful sting, but it is a beneficial species preying on a wide variety of caterpillars. In southern N.S.W. only females overwinter, on top of the old nest, though males are common throughout the summer and autumn. There is no obvious differentiation into queens and workers (p. 136) in any Australian species.

ROPALIDIINAE (8 spp.; Fig. 37.30F) occur in the Ethiopian, Oriental, and Australian regions. They are distinguished from Polistinae by the petiolate gaster. They are relatively small (7–10 mm), and are common in the tropics, though one species extends south to Victoria. *R. cabeti*, of north Queensland, is a small, mostly yellow species, whose large paper-covered nests on the lower, rain-protected side of the branch of a tree blend well with their surroundings. Some of the north Queensland species are stylopized. [van der Vecht, 1962.]

50. Eumenidae (Figs. 37.30C, E, G–I). These solitary, predatory wasps store caterpillars, or less often larvae of sawflies or beetles. The wing-coupling is very strong, and the fore wing is consequently folded longitudinally when at rest. Trochantellus marked off by a distinct suture on the mid and often on the fore femur; usually only 1 mid tibial spur (rarely none or 2); tarsal claws normally bifid, rarely toothed or simple; 'jugal' lobe of hind wing always present as an oval or subcircular area. The 2 major subfamilies, both well represented, are separated mainly on the structure of the mandibles; but all Australian species of Zethinae have a distinctly petiolate gaster, whereas in Eumeninae the gaster is subsessile except in *Eumenes* and *Ectopioglossa*.

In ZETHINAE (27 spp.; Fig. 37.30I) the mandibles are short and not or only slightly crossed at the apices. The species construct nests of various types, and some show an approach to social habits. Wood cavities may be utilized, or several females may construct a colony of cells of vegetable fibres or leaves pasted together with a resinous material. The 14 species of *Pseudozethus* (Plate 6, N), generally distributed in the eastern States with one in W.A., are mostly large (12–25 mm) and rather stout apart from the long petiole. They are mostly black and rather shining, with the gaster marked with yellow or orange. The 10, long (11–15 mm), thin species of *Ischnocoelia* are widespread, mostly in the interior; most are largely orange. *Macrocalymma* (1 sp.), which occurs from Victoria to north Queensland, has the tegula greatly developed, with its apex reaching slightly beyond the posterior margin of the scutellum.

In EUMENINAE (215 spp.; Plate 6, K, T; Figs. 37.30C, E, G) the mandibles are crossed at their apices; they are usually long, but are short in *Ectopioglossa*. *Abispa* often builds its large mud cells in houses in Queensland, with one species extending into southern N.S.W. Other genera make mud cells in soil (*Stenodynerus*), or line burrows in wood with mud. All species apparently construct a chimney to the cell while it is being provisioned, and all apparently use caterpillars as prey. *Eumenes* (9 spp.), the only common genus in which the gaster is distinctly petiolate, are all rather large (up to 30 mm). Their jug-like mud nests are common on buildings in inland areas. *Ectopioglossa australiensis* of north Queensland resembles a *Eumenes*, except for the relatively short mandibles. The 3 species of *Abispa* are all very large (23–32 mm) and stout-bodied. *Eudiscoelius gilberti* is one of the few metallic (green) species.

Acarozumia amaliae (Fig. 37.30H) and related species have a median caudal pouch on the propodeum, in which the resting stage (hypopi) of symbiotic saproglyphid mites shelter. In *Stenodyneroides* there are 2 pouches in this position, a pair under the posterolateral projection of the propodeum, a pouch at the junction of the 1st and 2nd gastral tergites, and one at the base of the 2nd gastral sternite. Infestation of the wasp takes place in the nest when the pupal skin is shed. The mites do not seem to affect the host adversely. The species of Saproglyphidae are nearly always host specific (Krombein, 1961). Even more complex relationships are known with other mites that enter the genital chamber of the female wasp, and pass to the cell as the egg is laid.

The common medium-sized species are placed in *Odynerus* (71 spp.) and *Paralastor* (111 spp.). These 2 genera can almost invariably be separated on wing venation, and *Odynerus* is sometimes subdivided into a large number of genera. In one species-group of *Paralastor* the gaster is subpetiolate, and these species can be confused with *Ropalidia* (Vespidae). [Soika, 1962; van der Vecht, 1959, 1963.]

Superfamily SPHECOIDEA

Fossorial wasps that prey on other insects and a few other arthropods, such as spiders and Collembola; a few are inquilines in the nests of other sphecoids. Distinguished from all other Hymenoptera on the combined structure of the pronotum and simple body hairs. Each species tends to specialize, hunting down a particular prey, and disregarding others of similar size even when more readily available. Some restrict their predation to a single species or genus, and only a few take victims from more than one order of arthropods. The nature of the prey gathered is as characteristic of the wasp as the anatomical features that distinguish it from other wasps.

The adaptation of a particular wasp to its prey presents one of the most intriguing problems in the study of behaviour. Solitary wasps are predators of a rather special kind. Only a few take prey as food for themselves. For the most part the adults of all species feed on nectar, or ripe fruit, or honey-dew secreted by aphids and other plant-sucking insects. The males, in fact, are not predators at all, and feed exclusively on plant exudates. Only the females take prey, and they do so primarily to feed their larvae. In their behaviour the solitary wasps foreshadow the more elaborate 'larva-nurturing' of the social ants, bees, and vespids, all of which apparently arose independently from solitary wasps.

The size of the prey relative to that of the

wasp is of significance in the evolutionary sequence. Since primitive wasps generally place only a single victim in each brood cell, they must take prey as large as, or larger than, themselves if their larvae are to have food to reach full size. In this feature they betray their more immediate derivation from the parasitoid Hymenoptera, which are invariably smaller than their hosts. The more primitive predatory wasps accordingly drag their prey over the ground, grasping the victim in their mandibles. They cannot cover much ground by this method, and consequently nest close to the habitat of their prey. Thus, wasps of the genus *Prionoyx* nest in bare spots in open country where their grasshopper prey abounds. On the other hand, *Sphex*, which also hunts grasshoppers, is able to transport them considerable distances in flight, because it takes smaller specimens and provisions each cell with several of them.

Those species that carry the prey in the mandibles must put it down while they scrape out the entrance to their nest, or else leave the nest open all the time. Either of these actions exposes the prey or the contents of the cell to the attacks of parasites. The more advanced species grasp the prey in either or both the mid and hind legs, thereby leaving the mandibles and fore legs free. They close the nest entrance when they leave, and are able to open it readily when they return while still holding their prey. The ability is well developed in Nyssoninae. Some Nyssoninae (Bembicini) have progressed further, and practise progressive provisioning, opening the cell at intervals and supplying new prey to the developing larva. The known prey of the sphecoid wasps is shown in the following table.

Predator	*Prey*
Ampulicidae	Blattodea.
Sphecidae	
ASTATINAE	Hemiptera–Heteroptera.
SPHECINAE	Spiders; Orthoptera; lepidopterous larvae; sawfly larvae.
LARRINAE	Spiders; Orthoptera; Heteroptera; adult Hymenoptera; Blattodea; adult Diptera.
TRYPOXYLONINAE	Spiders.
PEMPHREDONINAE	Collembola; Hemiptera–Homoptera; Thysanoptera.
CRABRONINAE	Homoptera; Heteroptera; Psocoptera; adult Diptera; adult Hymenoptera; adult Coleoptera; and other adult insects.
NYSSONINAE	
Nyssonini	(Inquilines.)
Gorytini	Homoptera.
Stizini	Orthoptera; Homoptera.
Bembicini	Heteroptera; adult Diptera; adult Lepidoptera.
CERCERINAE	Adult Coleoptera; adult Hymenoptera.
PHILANTHINAE	Adult Hymenoptera. [Not Australian.]

Key to the Families and Subfamilies of Sphecoidea Known in Australia

1. Notaulices well developed; hind wing usually without 'jugal' lobe; fore wing with 2 interradial cross-veins unless venation is reduced (Fig. 37.31A) .... **Ampulicidae** (p. 940)
   Notaulices absent or ill-defined; hind wing with 'jugal' lobe; fore wing with only 1 interradial cross-vein (except in Astatinae) .... **Sphecidae** (p. 940). 2
2. Fore wing with 2 interradial cross-veins (radial cell distinctly appendiculate); 2 mid tibial spurs .. .. ASTATINAE
   Fore wing with 1 interradial cross-vein; 1 or 2 mid tibial spurs 3
3. Hind wing with distinct 2nd anal vein, well separated from 1A (Figs. 37.31B, C). [Mid tibia with 2 spurs; 'jugal' lobe of hind wing without notch] .... SPHECINAE
   Hind wing without distinct 2nd anal vein 4
4. Labrum visible beyond the clypeus, or mid tibia with 2 spurs (Figs. 37.31D–H) ..... NYSSONINAE
   Labrum not protruding (or only very slightly so) and mid tibia with 1 spur 5
5. Hind wing with 'median' cell shorter than 'costal' cell (Fig. 37.32A) .... CRABRONINAE
   Hind wing with 'median' cell longer than 'costal' cell (Fig. 37.32B) 6
6. Eyes deeply emarginate (Fig. 37.32C) .... TRYPOXYLONINAE
   Eyes not or very slightly emarginate 7
7. Pterostigma enlarged (Fig. 37.32B) .... PEMPHREDONINAE
   Pterostigma little wider than the combined C+R width 8
8. Gaster with first segment short, node-like (Fig. 37.32D) .... CERCERINAE
   Gaster subsessile (Fig. 37.31I) .... LARRINAE

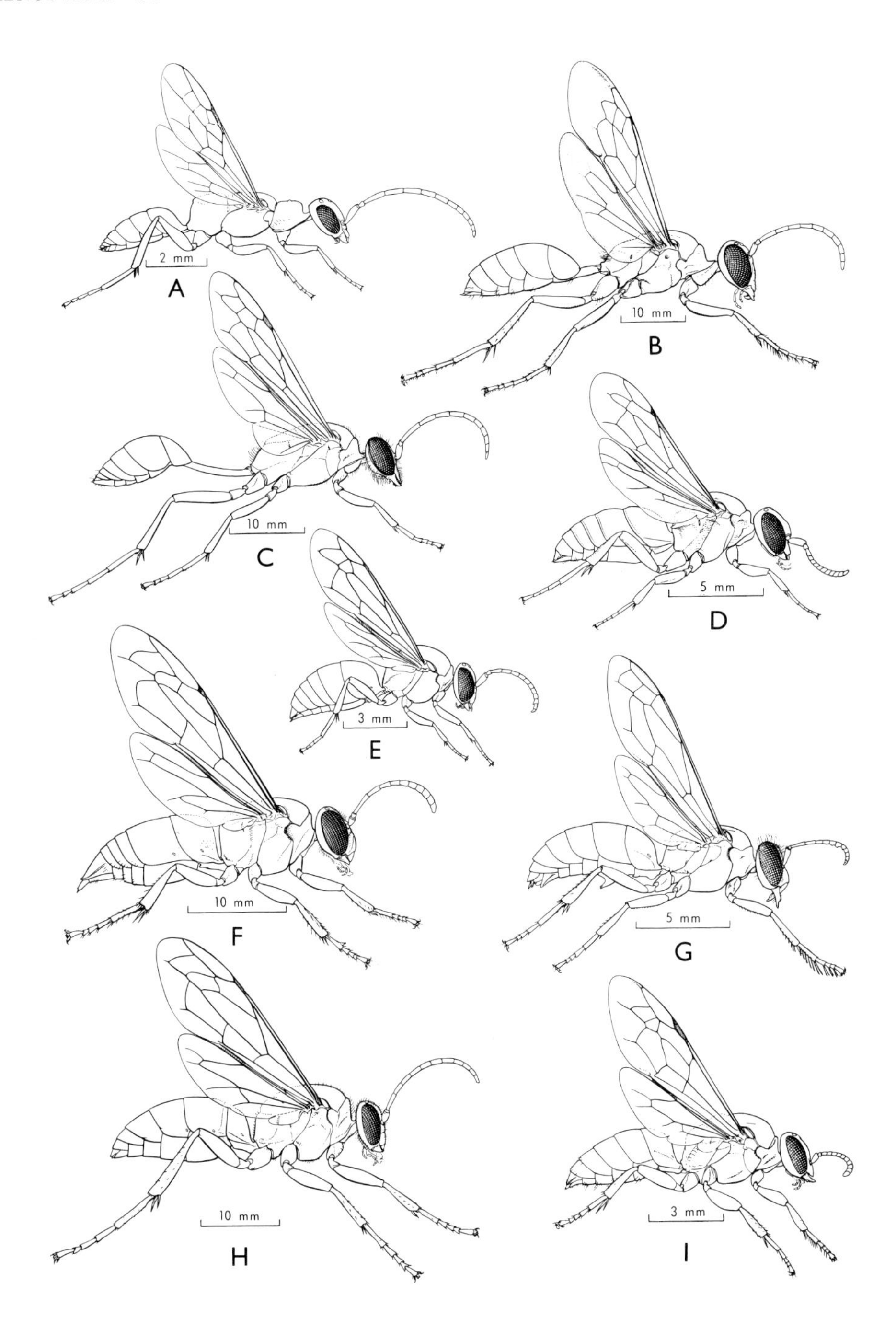

Fig. 37.31. Sphecoidea: A, *Aphelotoma* sp., Ampulicidae; B, *Sphex* sp., Sphecidae-Sphecinae; C, *Sceliphron laetum*, Sphecidae-Sphecinae; D, *Nysson* sp., Sphecidae-Nyssoninae-Nyssonini; E, *Bembicinus* sp., Sphecidae-Nyssoninae-Stizini; F, *Sphecius pectoralis*, Sphecidae-Nyssoninae-Gorytini; G, *Bembix* sp., Sphecidae-Nyssoninae-Bembicini; H, *Exeirus lateritius*, Sphecidae-Nyssoninae-Gorytini; I, *Sericophorus* sp., Sphecidae-Larrinae-Miscophini. [M. Quick]

51. Ampulicidae. Easily distinguished from other Sphecoidea by the absence of a separated 'jugal' lobe in the hind wing (Fig. 37.31A). The metasternum is strongly produced and forked in *Ampulex*, but not markedly so in Australian ampulicids. The pronotum is long, and hind femur basally swollen or clavate. Australian species are all small (4·5–15, mostly 7–8 mm), with females slightly larger than males; and all are non-metallic and mostly rather hairy. Adults are common on tree trunks. Males are attracted to patches of sunlight on living, smooth-barked eucalypts, but females prefer dead trees where they hunt for their cockroach prey. *Aphelotoma* (16 spp.), the only common genus, is Australia-wide. There is marked sexual dimorphism in the colour of the legs and antennae. [Riek, 1955f.]

52. Sphecidae. The family, divided into a large number of subfamilies, contains the great majority of large predatory wasps that are important in regulating the populations of a wide variety of insects and other arthropods. All the well-recognized subfamilies are present except Philanthinae, but there is only one species of the primitive Astatinae. The subfamilies are often treated as families.

ASTATINAE. The single species of *Astata* described from north Australia may be incorrectly assigned to this genus and subfamily, in which the radial cell of the fore wing is very distinctly appendiculate and the mid tibia has 2 apical spurs. [Parker, 1962.]

SPHECINAE (47 spp.) are all rather large, and the gaster is clearly petiolate; in *Ammophila* the petiole is extremely long. The subfamily, however, is distinguished by the well-developed 2nd anal vein in the hind wing, the legs appear relatively long, and the body is not stout. The 3 tribes present differ considerably in nesting habits and in the type of prey used in provisioning. Sphecini provision nests in the ground or in old tunnels in timber with Orthoptera; Ammophilini, which also nest in the ground, utilize lepidopterous larvae; Sceliphronini construct clay cells in sheltered situations and provision with spiders.

Sphecini (33 spp.) are mostly large, black, rather stout-bodied species with a shining gaster. The thorax often has long, dense, silvery, gold, or black pubescence, and the wings are sometimes infuscated with black or orange. Body length usually 15–34 mm, most commonly about 25 mm, though small males of *Prionony globosus* may be only 8 mm long (females up to 17 mm). They are very strong fliers, widespread in drier areas. Adults of both sexes are territorial and pugnacious when nesting. *Sphex* (25 spp.; Fig. 37.31B) provisions with Tettigoniidae; *Isodontia* (5 spp.) with Gryllidae or Tettigoniidae; and *Prionony* (3 spp.) with Acridoidea. The prey may be larger than the predator, and it is then straddled and dragged along the ground. The pecten, or digging comb, on the fore tarsus is usually well developed. Species of *Sphex* are often stylopized.

Ammophilini (10 spp.) generally have a longer and thinner body than Sphecini, but *Podalonia suspiciosa* has a body form, especially in the female, resembling that of *Sphex*. It is one of the most widespread sphecoid wasps, occurring in sandy areas, especially river banks, over all Australia and Tasmania. The basal half of the gaster is orange in both sexes. There is marked sexual dimorphism in the large, very long, thin species of *Ammophila* (9 spp.). Females are usually strongly marked with orange on legs and gaster, whereas males are normally mostly black. Length of body 14–33 mm; both petiole and 1st segment of gaster very long and thin. The species are common in the inland and in dry areas. *Ammophila* normally nests in soil rather than sand.

Sceliphronini (4 spp.) are the common long-bodied mud-daubers of coastal regions (the long-petiolate mud-daubers of the inland area are usually the stouter-bodied species of *Eumenes*). *Sceliphron laetum* (Plate 6, R; Fig. 37.31C), with prominent yellow body markings, is common in houses. It ranges to New Guinea, and there is a darker subspecies in the south-west of W.A. Length 18–30 mm. The mud cells have a rather smooth finish, whereas those of the smaller *S. formosum* have a coarse finish of raised ridges. The dark coloured American species *S. caementarium* is established in south-eastern Queensland. One species of metallic blue *Chalybion*, length 15 mm, reaches north Queensland. Both endemic species of *Sceliphron* are often stylopized. [Bohart and Menke, 1963.]

NYSSONINAE (100 spp.) are distinguished from the following subfamilies by the presence, with odd exceptions, of 2 mid tibial spurs, and from Sphecinae on wing venation and absence of a distinct petiole (Figs. 37.31D–H). The labrum protrudes beyond the clypeus, with few exceptions (*Miscothyris*, *Ammatomus*). This large subfamily of moderate to very large species is divided into a number of tribes. Many species nest in the soil, but Nyssonini are inquilines in the nests of other sphecoid wasps and of bees.

Nyssonini (12 spp.; Fig. 37.31D) are rather dark and dull due to a coarse ornamentation; length 5–9 mm. Australian species are characterized by the development of a strong posterolateral spine on the propodeum. *Nysson* was recorded as an inquiline in the cells of *Sericophorus* (Larrinae) by Rayment (1953).

Gorytini (24 spp.) are of medium size (7–17 mm), with *Sphecius pectoralis* (Fig. 37.31F) 20–28 mm and *Exeirus lateritius* (Plate 5, Q; Fig. 37.31H) 24–36 mm long. These 2 species, the largest Australian sphecoids, provision their cells in the ground with cicadas. *S. pectoralis* is a smooth-looking, mostly dull orange wasp, whereas *E. lateritius* has a black, rather hairy thorax and base of gaster and relatively long hind legs. *Miscothyris* (5 spp.; Plate 6, X) provisions with Cicadelloidea, but the prey of *Gorytes* (15 spp.) is not recorded, though adults are most commonly collected at blossom.

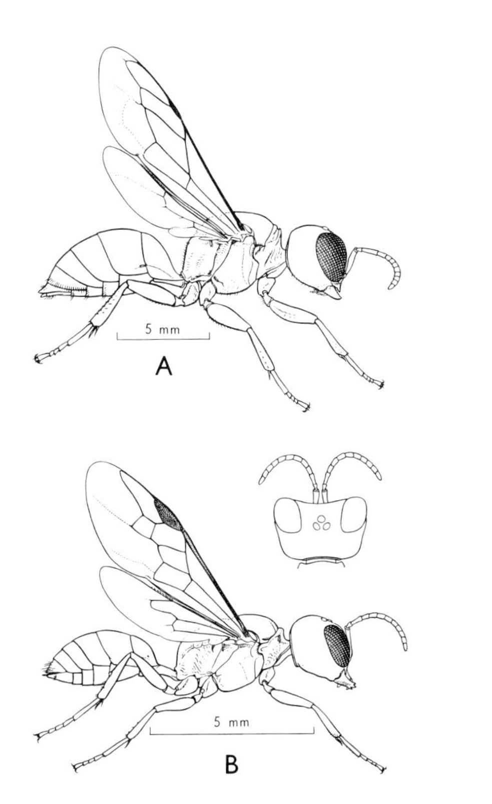

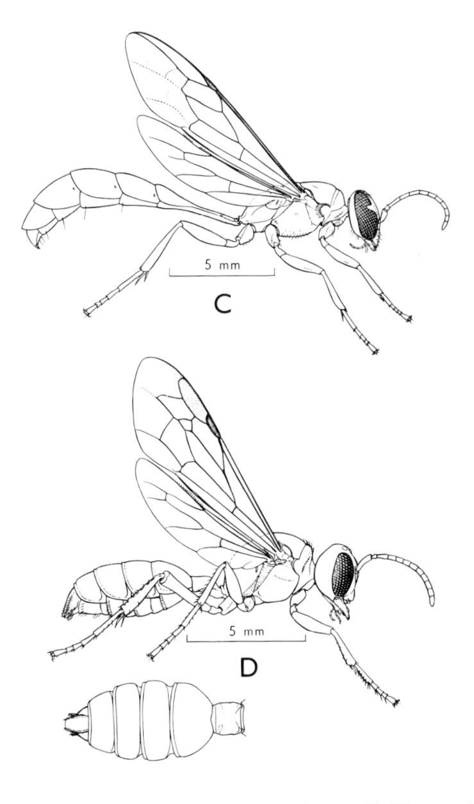

Fig. 37.32. Sphecidae: A, *Williamsita* sp., Crabroninae; B, *Paracrabro* sp., Pemphredoninae; C, *Trypoxylon* sp., Trypoxyloninae; D, *Cerceris* sp., Cercerinae. [M. Quick]

Australian Stizini (14 spp.) are less than 11 mm long, and all included in *Bembicinus* (Fig. 37.31E), which has a hook-like process on the 11th antennal segment of the male. The species provision their nests in the ground with Diptera, and often live in communities in sandy banks; they are most abundant in drier areas. They are sometimes stylopized. *Xanthochroism*, in which males only are entirely sandy yellow instead of the normal black with yellow markings, is common in an undescribed Australian species of *Bembicinus*.

Bembicini (50 spp.) construct nests in the soil, usually in colonies. Many provision with Diptera, though a few use other insects. Most species practise progressive provisioning. Most are relatively large (12–25 mm), stocky species with the gaster fitting close against the propodeum. They have a very rapid, pugnacious, noisy flight, and return rapidly to the spot from which they are disturbed. *Bembix* (Fig. 37.31G), the only genus in Australia, has the median ocellus in the form of an inverted V-shaped slit. In males the 2nd sternite is almost invariably produced (though there is often marked individual variation in the development of this process), and the fore tarsus is markedly expanded and spooned in a few large species (Fig. 37.7G). The Australian species are generally black with pale transverse bands on the gaster, but a few are orange. They are often stylopized, and also parasitized by bombyliids, conopids, and mutillids. [Beaumont, 1954.]

CRABRONINAE (56 spp.) are readily distinguished on the reduced fore-wing venation and enlarged rather square head, but the characteristic hind-wing venation is more diagnostic; the 'jugal' lobe of the hind wing is very small. Oxybelini are not recorded. Some Crabronini nest in soil, others in holes in wood, and many kinds of insects are used as prey (Homoptera, Heteroptera, Psocoptera, Diptera, Hymenoptera, Coleoptera). Australian species are of medium to small size (3·5–18, mostly 7–12 mm), and mostly black or have orange markings on the gaster and legs. Adults are collected at the nesting sites (of a few to several individual burrows) or on blossom and young foliage. *Podagritus tricolor*, one of the larger, long-bodied species of a common genus, provisions its nest in the ground with a wide range of small adult Diptera. Some species of *Rhopalum*, the other common genus, nest in deserted beetle holes in trees. *Williamsita* (Plate 6, P; Fig. 37.32A) contains common species with a broad gaster. [Leclercq, 1954, 1957.]

The deeply emarginate eyes are diagnostic of TRYPOXYLONINAE (54 spp.; Fig. 37.32C). There are two very different body forms in the subfamily. *Trypoxylon* (2 spp.) has a long thin gaster and resembles a small *Sphex*; whereas *Pison* (52 spp.) has the broad body form of typical Larrinae. Australian species are of medium to small size and mostly dark colour, but some have gold or silver pubescence, and some a ferruginous gaster. They construct mud cells in many situations, often in the old mud nests of other wasps, and provision with spiders.

PEMPHREDONINAE (29 spp.) are mostly small to very small species, with an enlarged pterostigma (Fig. 37.32B), and are mistaken for small, smooth-bodied bees, for they are regular visitors to blossom. Psenini, with more complete fore-wing venation than Pemphredonini, occur only in the extreme north, with one of the 2 species of *Diodontus* also in New Guinea. Pemphredonini (27 spp.) are widespread in Australia and Tasmania. They often nest in holes in wood, and provision with Collembola, Thysanoptera or Homoptera. *Harpactophilus* (5 spp.), of medium size and stout build, the body coarsely ornamented and the gaster ferruginous, occurs in Queensland, with one species in Mysol. *Austrostigmus* (7 spp.) utilizes small deserted beetle holes in dead trees and chalcidoid emergence holes in *Apiomorpha* galls as nesting sites. *Austrostigmus* and *Spilomena* (10 spp.) are often minute (1·6–8 mm), and the adults are most commonly collected on blossom along with small bees. *Paracrabro* (length 6–7 mm), the single species ranging from Kosciusko to Tasmania, has a distinctly petiolate gaster and large, square, crabronine-like head.

CERCERINAE (35 spp. of *Cerceris*) are easily recognized by the beaded gaster with the 1st segment forming a petiole (Fig. 37.32D). Head large and wide; eyes entire, and not slightly emarginate as in the exotic Philanthinae; gaster ending in a flat oblique pygidium; apex of hind femur expanded and apically truncate, and the ball-and-socket joint of the hind coxa enlarged. The species, of small to medium size (5–25, mostly 8–12 mm), are often coloured orange and black. Some provision their cells in the ground with small beetles.

LARRINAE (99 spp.) are mostly of very generalized body form, with few distinctive features apart from the development of the ocelli (Fig. 37.33) on which the subfamily is divided into tribes. The ocelli are normal in Miscophini, but in Larrini and Tachytini the lateral ocelli are modified in different ways. In Larrini the face is produced below the median ocellus, but it is flat in Tachytini.

Larrini (22 spp.) are mostly tropical, but *Notogonia australis* occurs in Tasmania. *Larra* (3 spp.), black with highly polished gaster, preys on Gryllotalpidae. *Liris* (1 sp.), with bright golden pubescence and golden wings, preys on crickets. *Notogonia* (15 spp.) provisions with tettigoniids. Tachytini (36 spp.) are mostly tropical, but at least 2 species of *Tachysphex* occur in Tasmania. They are mostly of medium size, but *T. pugnator* may have a body length of only 5–6 mm. The species are mostly dark coloured, but some have bands of silvery pubescence, and a few have a pale gaster or golden pubescence. *Tachytes* (13 spp.) preys on Acridoidea, whereas *Tachysphex* (23 spp.) also uses Blattodea and Mantodea. Miscophini (43 spp.) are mostly tropical, but *Sericophorus relucens* is a common widespread species of the interior. *Sericophorus* (14 spp.; Fig. 37.31I; Rayment, 1955) and the closely related *Zoyphium* (10 spp.) nest in the ground and provision with Diptera. Most of the species of *Sericophorus*, apart from *relucens*, have a metallic colouring. The very small (2·5–6 mm) black species of *Nitela* (7 spp.), with rather long pronotum and reduced venation, are generally collected on tree trunks.

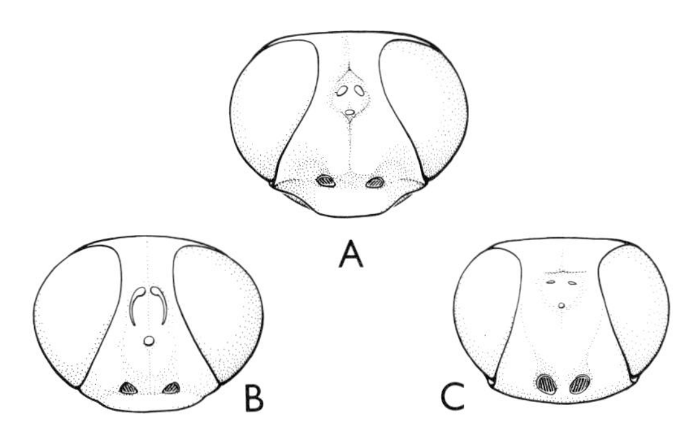

Fig. 37.33. Sphecidae-Larrinae, ocellar region: A, *Tachysphex* sp.; B, *Tachytes* sp.; C, *Larra* sp. [S. Curtis]

Superfamily APOIDEA

by CHARLES D. MICHENER

Apoidea are basically sphecoid wasps that use pollen rather than insect or spider prey as a protein source for their larvae; most of them are not social. Some of the primitive bees are very similar to Sphecoidea, but can be distinguished by the presence of at least a few branched hairs on the body (often difficult to see) and by having the hind basitarsi wider than the following segments. Bees also differ from sphecoids in having the last 2 gastral terga of the female completely divided to form hemitergites associated with the sting apparatus. These structures are hidden by the preceding terga so that they can be seen only on dissection. The larvae differ from those of sphecoids, not only in their feeding habits, but almost always also in the presence of only 1 papilla (the palp) on the maxilla instead of 2. The bees constitute a group of some 20,000 species in the world, ranging from forms only 2 mm long to gigantic ones attaining 39 mm. In Australia there is an enormous number of undescribed species, and an estimate of the total number of species would be about 3,000.

The great majority of bees are solitary, i.e., each female makes her own nest without the co-operation of others, and there are no separate queen and worker castes, for each female must engage in the activities of both worker and queen if progeny are to be produced. The nest typically consists of a burrow in the soil, in wood, or in a pithy stem, from which lateral burrows extend to the cells, or which is itself subdivided into cells. Some solitary bees, however, construct cells of resin, mud, or other materials in protected or exposed situations. Normally each cell is completed, provisioned with pollen and honey, an egg laid in it, and sealed, before the next cell is begun; but in many Halictidae this sequence is broken, and new cells are started before their predecessors are completed. The mother ordinarily dies before her progeny reach maturity and emerge from their cells.

Many bees provide for their young in the same way, and are usually called solitary, although they nest in aggregations; sometimes there are thousands of nest burrows, each made by a single female, in a few square metres of soil. Still others live in small, more or less social groups consisting of two or several females in a single nest. Finally, a few bees live in very large colonies with clearly differentiated female castes of worker and queen (p. 136). The nesting behaviour of Australian bees has been studied by Hacker, Rayment, Michener, and a few others (references in Michener, 1965b), but it has not yet received the attention that it deserves.

There are some socially parasitic (inquiline, cleptoparasitic, or 'cuckoo') bees that do not make their own nests. The larvae feed on provisions provided by the host, which is always some other species of bee. Cuckoo bees are scarcer in Australia than in any other region, perhaps because so many of the Australian bees are Colletidae, a family containing few if any parasitic species. The best known of the few Australian genera of social parasites is *Thyreus* (Anthophoridae–Melectini), which parasitizes *Amegilla* (Anthophoridae–Anthophorini). In *Thyreus*, *Nomada*, and *Coelioxys* the first-stage larva has a large, heavily sclerotized head with huge sharp mandibles for destroying the egg or young larva of the host. These structures have doubtless arisen independently in each of the three groups.

Bees are dependent on nectar from flowers as their chief source of carbohydrates and on pollen as their source of proteins. Many species obtain nectar from a wide variety of flowers, and some also gather pollen from many kinds of flowers. However, some are restricted in their pollen collecting to particular kinds of flowers. Since most of the pollen that is gathered is used in provisioning cells for larvae, an activity in which only the females engage, it is primarily the females that show the restriction in the kinds of flowers they visit. Even in these species, the males and females gathering nectar may visit a variety of flowers. Bees that gather pollen from many kinds of flowers are called *polylectic*, whereas those that gather from only a few species of related flowers are called *oligolectic* (Linsley, 1958). The numerous hylaeine and euryglossine colletids carry pollen in the crop with the nectar instead of among scopal hairs on the legs and body. For such bees it is not easy to determine the degree of oligolecty, since one never knows whether a female is gathering pollen or merely sucking nectar.

Australia is the only continent where most bees are largely dependent on a single family of plants, the Myrtaceae. Genera attractive to bees include *Angophora*, *Baeckea*, *Callistemon*, *Eucalyptus*, *Eugenia*, *Leptospermum*, *Melaleuca*, and *Tristania*. A list of the chief honey sources of southern Queensland included 47 Myrtaceae, 10 introduced weeds and crops, and only 17 species of native plants in all other families (Blake and Roff, 1958). Most bees oligolectic on Myrtaceae will collect pollen from whatever members of that family are in bloom in the area; the occurrence of narrower oligolecty has not been verified. Many polylectic forms also include Myrtaceae among their pollen sources.

Some examples of other oligolectic relationships are: the subgenus *Cladocerapis* of

Leioproctus (Colletidae) gathers pollen only from *Persoonia*; certain species of *Tricho-colletes* and other bees seem restricted to small yellow legumes, such as *Daviesia*; certain species of Colletidae and Halictidae appear to be restricted to yellow Compositae, and some may possibly be limited within that group; an interesting group of unrelated small black Colletidae and Halictidae is restricted to *Wahlenbergia*; *Leioproctus moretonianus* appears to be restricted to Goodeniaceae; the species of *Lithurge* are restricted to large Malvaceae, such as native species of *Hibiscus*.

Although very diversified in size, structure, and appearance, the bees are not readily divisible into a few sharply defined major groups of equivalent status. For this reason their classification has never reached the stage of general agreement. Some workers would place them all in a single family, the Apidae, whereas others divide them into from two to many families. The classification adopted here is that of Michener (1965b), which contains definitions of taxa down to subgenera, an annotated list of Australian species, and a bibliography. The characters for the families of bees are often difficult to appreciate, or are in the mouth-parts which are usually folded and hidden in dried specimens. Therefore an artificial key to genera, using characters easier to see, has been given in the same paper.

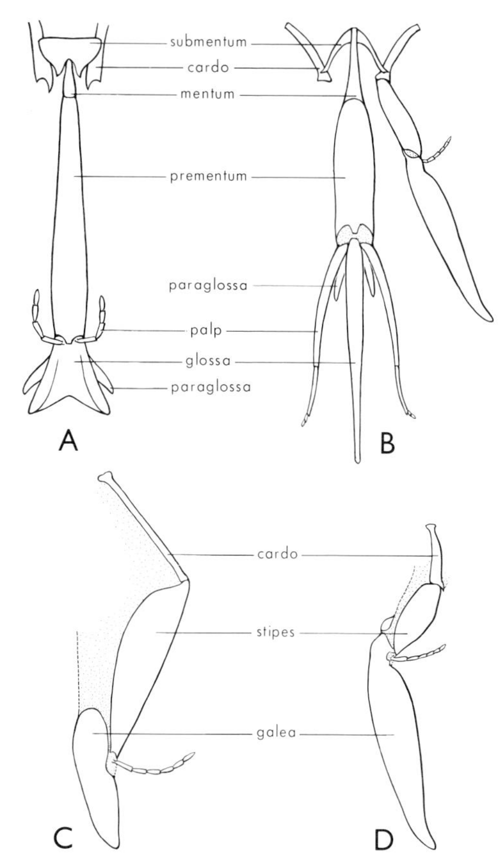

Fig. 37.34. Apoidea, mouth-parts (after Michener, 1965b): A, labium and maxillary cardines of a colletid; B, labium and maxilla of an anthophorid; C, maxilla of a colletid; D, maxilla of an anthophorid.

Key to the Families of Apoidea Known in Australia

1. Labial palp with first 2 segments elongate, sheath-like, and flattened, in strong contrast to distal 2 segments; galea with postpalpal part greatly elongated, usually longer than stipes (Fig. 37.34B); glossa elongate and pointed, usually longer than prementum; mid coxa over two-thirds as long as distance from its base to base of hind wing; pre-episternal groove or suture absent below scrobal suture. ('Long-tongued bees') 2

   Labial palp with segments similar and subcylindrical; galea with postpalpal part much shorter than stipes (Figs. 37.34A, C); glossa shorter than prementum; mid coxa (or at least exposed part) much shorter than distance from its base to base of hind wing; pre-episternal groove usually present below scrobal suture (Fig. 37.5B). ('Short-tongued bees') 4

2. Labrum longer than broad; subantennal suture directed toward outer margin of antennal socket; 2 submarginal cells (Fig. 37.36D); scopa, when present, on gastral sterna ... **Megachilidae** (p. 948)

   Labrum broader than long; subantennal suture directed toward inner margin of antennal socket; 3 submarginal cells (Fig. 37.36A; distal veins weak and no distinct submarginal cells in *Trigona* (Apidae), 2 cells in some Ceratinini); scopa, when present, on hind tibia 3

3. Hind tibial spurs absent (except in *Bombus*, p. 951); scopa of ♀ forming a corbicula on hind tibia (except in queens); inner apical margin of hind tibia (except in queens) provided with a comb of

stiff setae; pygidial plate absent; distance between costal ends of the 2 recurrent veins (if present) nearly twice as great as length of 2nd recurrent vein and longer than 1st recurrent. Social species with queen and worker castes and large colonies .... **Apidae** (p. 951)

Hind tibial spurs present; scopa of ♀ not forming a corbicula; inner apical margin of hind tibia without a comb of stiff setae; pygidial plate present or absent; distance between costal ends of the 2 recurrent veins (*2nd may be absent*) less than twice as great as length of 2nd recurrent vein and shorter than, or equal to, 1st recurrent. Non-social species without ♀ castes .... **Anthophoridae** (p. 949)

4. Inner hind tibial spur of ♀ greatly broadened basally so that the whole spur is crescentic, the concave side finely ciliate; pre-episternal suture absent below scrobal suture; submentum V-shaped. [Northern Queensland and northward] .... **Melittidae** (p. 948)

Inner hind tibial spur not broadened basally and not crescentic; pre-episternal suture present below scrobal suture (except in Stenotritinae and *Hesperocolletes* of Colletidae); submentum absent or in the form of a plate, but not V-shaped 5

5. Glossa broad, emarginate, truncate, or rounded apically in all ♀♀ and most ♂♂, pointed in a few ♂♂; submentum present, sometimes only weakly sclerotized; mentum usually recognizable although often weakly sclerotized; galea not evenly tapering to pointed base, prepalpal portion shorter than postpalpal portion .... **Colletidae** (p. 946)

Glossa pointed apically; submentum and mentum absent or scarcely recognizable and not sclerotized; galea elongate prepalpally and evenly tapering to pointed base, prepalpal portion usually as long as postpalpal portion .... **Halictidae** (p. 948)

53. Colletidae. This family is better represented in Australia than in any other region. Colletidae are unique among bees in that they line their cells, whether they are irregular and in pre-existing burrows or beautifully regular in shape, with a translucent or transparent cellophane-like material applied by the broad glossa of the female (see Rayment, 1935). The family contains no social species and, at least in Australia, no known parasitic ones. Colletidae vary from large, robust, hairy bees to some of the smallest and most slender in the world. Some of the diversity is reflected in the following key to subfamilies. It is quite possible that the Stenotritinae and perhaps other subfamilies do not belong in the family.

Key to the Subfamilies of Colletidae Known in Australia

1. With 3 submarginal cells, or, if with 2, 2nd about as long as 1st; jugal lobe of hind wing not greatly (if at all) exceeding cross-vein *cu-a* (except in *Callomelitta*); ♀♀ with scopa, and with pygidial plate broadened basally; relatively hairy forms 2

With 1 or usually 2 submarginal cells, 2nd much shorter than 1st (except in *Hyleoides* and a few uncommon species of *Palaeorhiza*, etc.); jugal lobe of hind wing much exceeding *cu-a*; ♀♀ without scopa, and pygidial plate, when present, usually narrow and more or less parallel-sided; relatively bare or short-haired forms 3

2. Pre-episternal groove absent below scrobal groove; glossa short, blunt, apex rounded; 1st flagellar segment longer than scape .... STENOTRITINAE

Pre-episternal groove extending well below scrobal groove except in *Hesperocolletes*; glossa truncate or bilobed; 1st flagellar segment shorter than scape .... COLLETINAE

3. Pygidial plate of ♀ present, a slender, more or less parallel-sided band; basitibial plate usually recognizable, although often demarcated only by a few tubercles, or only the apex indicated by a tubercle; anterior face of 1st gastral tergum with longitudinal median groove .... EURYGLOSSINAE

Pygidial plate absent, or, in the few species that possess it, broad basally rather than parallel-sided; basitibial plate absent; anterior face of 1st gastral tergum usually without longitudinal median groove .... HYLAEINAE

The world-wide COLLETINAE are moderate-sized, hairy bees, which transport pollen externally on the scopa. The only Australian tribe, which is very large, is the Paracolletini. These bees nest in soil or occasionally in rotting wood. Each cell is at the end of a lateral burrow radiating from the main nest burrow. Included genera are *Callomelitta*, *Hesperocolletes*, *Leioproctus*, *Neopasiphae*, *Paracolletes*, and *Trichocolletes*.

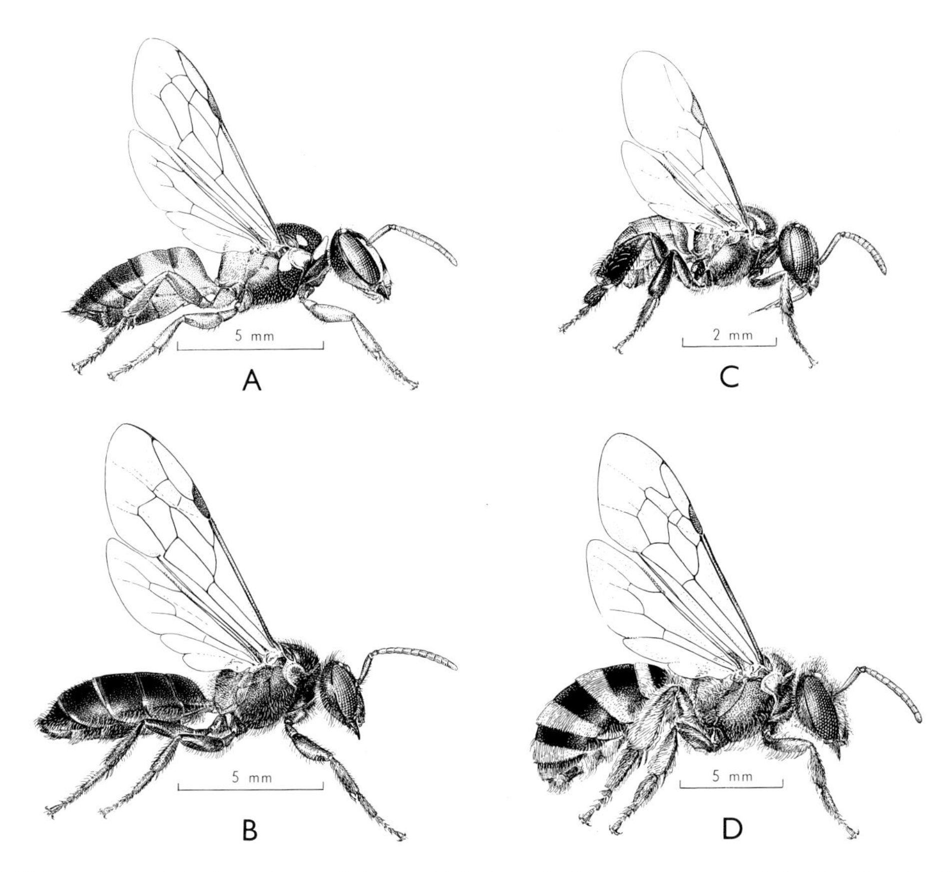

Fig. 37.35. Apoidea: A, *Hylaeus elegans*, Colletidae-Hylaeinae; B, *Euryglossa depressa*, Colletidae-Euryglossinae; C, *Trigona essingtoni*, Apidae; D, *Nomia australica*, Halictidae. [M. Quick]

STENOTRITINAE are moderately large to very large, strongly hairy bees found only in Australia. A distinctive feature is the location of ocelli low on the frons, nearer to the antennal bases than to the posterior margin of the vertex. Nests of this subfamily are unknown, and should be searched for because their architecture and the larvae may shed light on the relationships of the group. The only genera are *Ctenocolletes* and *Stenotritus*.

The endemic subfamily EURYGLOSSINAE (Fig. 37.35B) consists of moderate-sized to minute bees, usually sparsely hairy, and lacking a scopa of pollen-carrying hairs. The pollen is carried to the nest in the crop. The only similar group is the Hylaeinae, from which the Euryglossinae differ not only in the characters indicated in the key but also in having a usually broad face with the clypeus not extending much above the level of the tentorial pits. The Euryglossinae mostly nest in the soil, but some inhabit holes in old wood. Neither social nor parasitic species are known. Included genera are *Euryglossa*, *Euryglossina*, *Pachyprosopis*, and several less well known groups.

HYLAEINAE are also moderate-sized to minute, sparsely hairy bees, usually dark with yellow or white markings on the face and sometimes elsewhere. Most are more slender than the Euryglossinae, and the face and particularly the clypeus are commonly rather elongate, the latter extending well above the level of the anterior tentorial pits. Pollen is carried to the nest in the crop. Nests of this subfamily are usually made in burrows in pithy stems, or in holes made by beetles or other burrowing insects in woody stems or in logs and stumps. Some exotic species, and perhaps some Australian ones, nest in the ground. The cells are placed end to end in the burrow, merely by dividing it by transverse partitions into a series of cells. Parasitic species are not known in Australia, and social species are not known anywhere. However, several young adults of some Australian species may be present in a single nest before they disperse to establish new ones, and one may even be provisioning cells while her sisters are still present in the nest. Males of some genera have the glossa pointed, unlike any other Colletidae. Australian genera are *Heterapoides*, *Hylaeus* (Fig. 37.35A), *Hyleoides* (Plate 6, G; Fig. 37.36F), *Meroglossa*, *Palaeorhiza*, and several less common genera.

54. Halictidae. A large, cosmopolitan family of moderate-sized to small bees, black or greenish, sometimes with the metasoma red. The pubescence is moderately abundant, and (except in the parasitic *Sphecodes*) forms a pollen-collecting scopa on the hind femora, coxae, and sometimes on the sides of the propodeum and on the under surface of the gaster. The nest burrows, usually in the ground but occasionally in rotting wood, are often branched, and lead to cells which are either at the ends of lateral burrows or radiate from branch or main burrows. The cells are lined with a thin layer of wax-like material, and the provisions are in the form of a slightly flattened ball lying on, or attached to, the lower, flatter surface of the cell. The egg is placed on top of this ball of pollen. (For details, see Sakagami and Michener, 1962.)

Many Australian species of Halictidae are not solitary, but several individuals live together in the nest. So far as known, this is always a matter of several fertilized egg-laying individuals jointly occupying a single nest burrow (Michener, 1960). In other continents, however, some species have small colonies with workers and one or a few queens. There are several parasitic genera elsewhere in the world, and two species of *Sphecodes* presumably parasitic in the nests of *Nomia* or *Lasioglossum* reach northern Australia. Australian genera include *Nomia* (Fig. 37.35D; commonly put in a separate subfamily NOMIINAE), *Nomioides*, *Homalictus*, and *Lasioglossum*. The generic classification is not stabilized. Australian species falling in the last two groups have commonly been placed in *Halictus*, except that some of the species of *Lasioglossum* (Fig. 37.36B) have often been placed in a separate genus *Parasphecodes*.

55. Melittidae. This family has the type of mentum and submentum found in the long-tongued families, but short glossa and labial palps as in Halictidae and Andrenidae. It is a small family, divided into 4 subfamilies, of which the CTENOPLECTRINAE are represented by a single species of *Ctenoplectra* in north Queensland.

56. Megachilidae. This is one of the most readily recognized families because of the characteristic form, with a large head which is usually well developed in the genal and occipital regions. There are always 2 submarginal cells, with the 2nd as long, or nearly as long, as the 1st. The scopa is on the gastral sterna, not on the legs (absent in *Coelioxys*). Basitibial and pygidial plates are absent or nearly so (except that males of *Lithurge* have a pygidial plate). No members of this family are social. Some are parasitic in the nests of other megachilids; in Australia *Coelioxys* inhabits the nests of *Megachile*.

Although some megachilids make their own burrows, most use pre-existing burrows or hollows or construct exposed nests. The family is noteworthy for the use of foreign materials carried to the nests from other locations. *Megachile* (Fig. 37.36D) makes its cells of pieces of leaves (commonly rose leaves) neatly cut and carried to the nest burrow. *Chalicodoma* uses resin and (in other continents at least) mud, pebbles, and other materials. Such nests are usually placed in small natural cavities. Nests of Australian *Anthidiellum* are not known, but elsewhere this genus makes exposed nests entirely of resin on the surfaces of stones, etc. Larvae of this family spin strong cocoons before pupation, unlike those of the short-tongued families which usually spin no cocoons.

Almost all Australian megachilids are Megachilini, included genera being *Chalicodoma*, *Coelioxys*, and *Megachile*. *Anthidiellum*, restricted in Australia to the north and distinguished by having yellow integumental markings, belongs to the Anthidiini. *Lithurge* is currently placed in a separate subfamily LITHURGINAE because it exhibits certain primitive characteristics, such as a long jugal lobe on the hind wing and remnants of a pygidial plate.

57. Anthophoridae. This family includes small and sparsely haired to large and densely hairy bees. The pollen-carrying scopa (absent in parasitic forms) is restricted to the hind tibiae and basitarsi. The family is large and diversified in other continents, but poorly represented in Australia by 3 subfamilies which differ greatly from one another in behaviour.

Key to the Subfamilies of Anthophoridae Known in Australia

1. Pygidial plate absent or represented by apical spine usually hidden in dense pubescence; clypeus not strongly protuberant, lateral parts seen from below but little bent back and not parallel to long axis of body ...... XYLOCOPINAE
   Pygidial plate present in ♀; clypeus strongly protuberant, so that, seen from below, lateral parts are bent back parallel to long axis of body 2
2. Marginal cell longer than distance from its apex to wing tip, pointed on wing margin; stigma large, extending well into marginal cell beyond base of R; small, slender bees .... NOMADINAE
   Marginal cell shorter than distance from its apex to wing tip, apex of cell rounded and separated from wing margin; stigma small, not extending into marginal cell beyond base of R; large, robust species ... ANTHOPHORINAE

Numerous tribes of NOMADINAE occur on other continents. Only Nomadini reach Australia, represented by the single genus *Nomada* in Queensland. All species of the subfamily are parasitic and have no scopa. The most closely related non-parasitic bees are probably in the tropical American Exomalopsini of the Anthophorinae. Australian species are rather small, slender, wasp-like insects with short, sparse pubescence.

ANTHOPHORINAE are most diversified in the Americas, less so in Eurasia and Africa, and only 2 tribes reach Australia. Anthophorini (Fig. 37.36A) are represented only by the large, robust, hairy, often banded bees of the genus *Amegilla* (formerly included in *Anthophora*). They often nest in aggregations, either in flat ground or in banks of clay or soil, but each female makes her own burrow and cells. Each cell is lined with a layer of wax-like material, and the provisions are semiliquid, not forming a firm mass as in the Halictidae or the Xylocopinae. The pupae are not enclosed in cocoons. The Melectini (Fig. 37.36E) contain *Thyreus (= Crocisa)*, robust bees without much long hair and without a scopa, but adorned with striking blue or white patches of appressed hairs. They are social parasites in the nests of *Amegilla*.

The XYLOCOPINAE include 2 very different looking tribes. Xylocopini, represented by *Xylocopa* and *Lestis* (Plate 5, U), are large to very large, robust, hairy bees, black or metallic blue or green in colour, often with part of the pubescence pale. They excavate large burrows in sound wood, dry rotten wood, or even pithy stems such as the flower spikes of *Xanthorrhoea*. The cells are placed end to end in these burrows, and are unlined. The pollen mass is firm, loaf-like, and the egg is laid on its surface.

The Ceratinini are small, slender, sparsely haired bees, black or rarely metallic greenish, sometimes with a red metasoma. *Ceratina* is world-wide, and represented in Australia by only one species found in Queensland and N.S.W. It nests in pithy stems, and constructs series of cells basically similar to those of the Xylocopini. The other Australian genera, *Allodapula*, *Exoneura* (Fig. 37.36C), and their relatives, also nest in pithy stems or in burrows made by other insects in stems or wood. These two genera and their close relatives found in Asia and Africa are unique among the bees in that the nests are not divided into cells, the immature stages being reared together and fed progressively in the

Fig. 37.36. Apoidea: A, *Amegilla cingulata*, Anthophoridae-Anthophorinae-Anthophorini; B, *Lasioglossum altichum*, Halictidae; C, *Exoneura bicolor*, Anthophoridae-Xylocopinae- Ceratinini; D, *Megachile chrysopyga*, Megachilidae; E, *Thyreus nitidulus*, Anthophoridae-Anthophorinae-Melectini; F, *Hyleoides concinna*, Colletidae-Hylaeinae. [M. Quick]

nest burrow. The larvae thus occupy an environment very different from that of most bee larvae, and are provided with head and body projections totally different from those of other species of bees. In nearly all the species there is a primitive social organization. Young adult females remain in the nests and care for their own younger brothers and sisters, and there is a weak caste system, queens and workers being anatomically alike externally, although the queens average larger than the workers. Queens mate, lay eggs, and are relatively long-lived, whereas workers usually do not mate, usually do not lay eggs, and probably are relatively short-lived. The nature of this primitive social organization is little known (Michener, 1965a).

Inquilina is closely related to *Exoneura*. It differs primarily by the reduction in the scopa, and lives as a social parasite in the nests of certain species of *Exoneura*. There are also species of *Allodapula* which appear to be social parasites in the nests of other species of the same genus (Michener, 1961a).

58. Apidae. This family contains all of the highly social bees and the tropical American Euglossini, in which pollen is carried in a corbicula. It consists of 2 subfamilies. The BOMBINAE are large, robust, usually hairy bees with the hind tibial spurs present. There are no Australian species, and efforts to introduce European species of *Bombus* (bumble bees) have been unsuccessful, although several are established in New Zealand. The APINAE are represented by the introduced *Apis mellifera* and native species of *Trigona*.

A. mellifera is of commercial importance as a honey producer and as a pollinator in many parts of the continent. It visits a wide range of native flowers, and has escaped from hives into hollow trees, holes in banks and cliffs, and other nesting sites. The combs of cells are made of wax, and are vertical, hanging down from some support in the nesting cavity. Honey and pollen are stored in cells similar to brood cells. The female castes are very different from one another, the queen lacking a corbicula and being unable to survive without the accompanying workers. *A. mellifera* is easily distinguished from almost all other Australian bees by the densely hairy surfaces of the eyes.

Trigona (Fig. 37.35C) includes the small, dark, 'native bees' of the northern half of the continent. Like honey bees, they exist in large colonies consisting of a structurally strikingly differentiated queen without corbiculae, and thousands of workers. Unlike honey bees, they are unable to sting. They place their brood cells either in clusters (Australian species of the subgenus *Plebeia*) or in horizontal combs with the cells opening upward (subgenus *Tetragona*), unlike the vertical combs of *Apis*. They store honey and pollen not in cells like brood cells but in large wax pots totally different from the brood cells. The nests are almost always in hollow trees. Establishment of new nests is not by the departure of the old queen and a swarm of workers as in *Apis*, but is a gradual process in which workers from a nest locate a new site, carry nesting materials (wax, resin) there, construct a new nest, and even carry provisions to it. Ultimately a young queen goes there and establishes herself with a group of workers, but interchange between the new and the old nest may go on for some weeks after the arrival of the young queen at the new nest. Old queens are so heavy and swollen that they cannot fly, and hence could not depart to establish new nests as in *Apis*. In *Trigona* queens are produced in special large queen cells, as in *Apis*. The nests of Australian species have been described by Michener (1961b).

Superfamily FORMICOIDEA

by W. L. BROWN JR *and* R. W. TAYLOR*

59. Formicidae. Superfamily and family coextensive, consisting entirely of social species, each usually having a (winged) male, (deciduously winged or wingless) female, and (wingless, neuter-female) worker castes. All

* This study was supported in part by grants nos G–23680 and GB–1634 from the U.S. National Science Foundation.

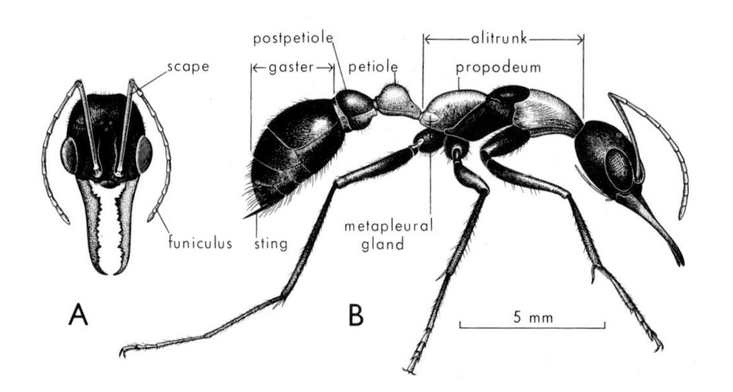

Fig. 37.37. *Myrmecia nigrocincta*, Formicidae-Myrmeciinae, worker: A, head, frontal; B, whole insect, lateral. [F. Nanninga]

forms have a nodiform, binodal, or scale-like 'waist' consisting of the modified true abdominal segment 2 or 2 + 3. With the exception of a few rare parasitic forms and the males of army ants (Dorylinae), all ants have a large metapleural gland with external bulla and a small orifice, opening on each side of the metathorax at its lower posterior corners (Fig. 37.37).

Ants have a characteristic facies, and are normally confused (among non-mimics) only with female mutillid or thynnine wasps and certain other wingless Hymenoptera, which, however, lack the peculiar waist and metapleural glands. There are perhaps 10,000 known, valid species of the world already in collections; approximately a tenth of these occur in Australia. The world tally may expand to 15,000 as collecting is improved and extended; ultimately the Australian count may exceed 1,500 species.

The female castes are the female proper, or *queen*, and the *worker*. The worker caste may be subdivided into *soldier*, *media*, *minor*, etc. phases. Female caste differences are phenotypic, and arise during ontogeny due to the interaction of an intrinsic system of physiological thresholds with variation in amount or composition of the larval food (and pheromones?), and in some cases, of the nutriment stored in the egg.

The virgin queen bears wings, has a well-developed pterothorax and flight muscles, large compound eyes and 3 ocelli, and is usually larger than workers of the same species. Her gaster is relatively bulky and contains well-developed ovaries and fat body.

The worker lacks wings, and has greatly reduced pterothoracic structure. The only flexible or semi-flexible joint lies between the pro- and mesothorax; even this is soldered solid in Myrmicinae, many Ponerinae, and some genera of other subfamilies. The metanotum may be obsolete, or represented by a narrow impressed piece with paired spiracles (*Myrmecia*, some Formicinae) or by a transverse metanotal groove. Head as in queen, but usually with smaller compound eyes (sometimes absent) and ocelli reduced or absent; mouth-parts and antennae are usually the same, and function similarly, in both worker and queen.

The soldier phase (Figs. 37.38C, D), when present, is larger than the other workers, and has a disproportionately large head, sometimes absolutely larger even than that of the queen. Worker minor and soldier are either connected by a series of intermediate forms (*Anisopheidole*, *Pheidologeton*, most *Camponotus* and *Melophorus*), or else normally lack the intermediate forms ('dimorphism' of *Pheidole* and *Oligomyrmex*). The soldiers apparently function mainly as defenders of the colony against attack, especially by other ants.

Various wingless intermediates between queens and workers occur in some species. These are the *ergatogynes* (ergatoids), which may accompany true queens in the nest, or replace them as the reproductive caste. Ergatogynes serving as functional queens are unusually frequent among Australian species,

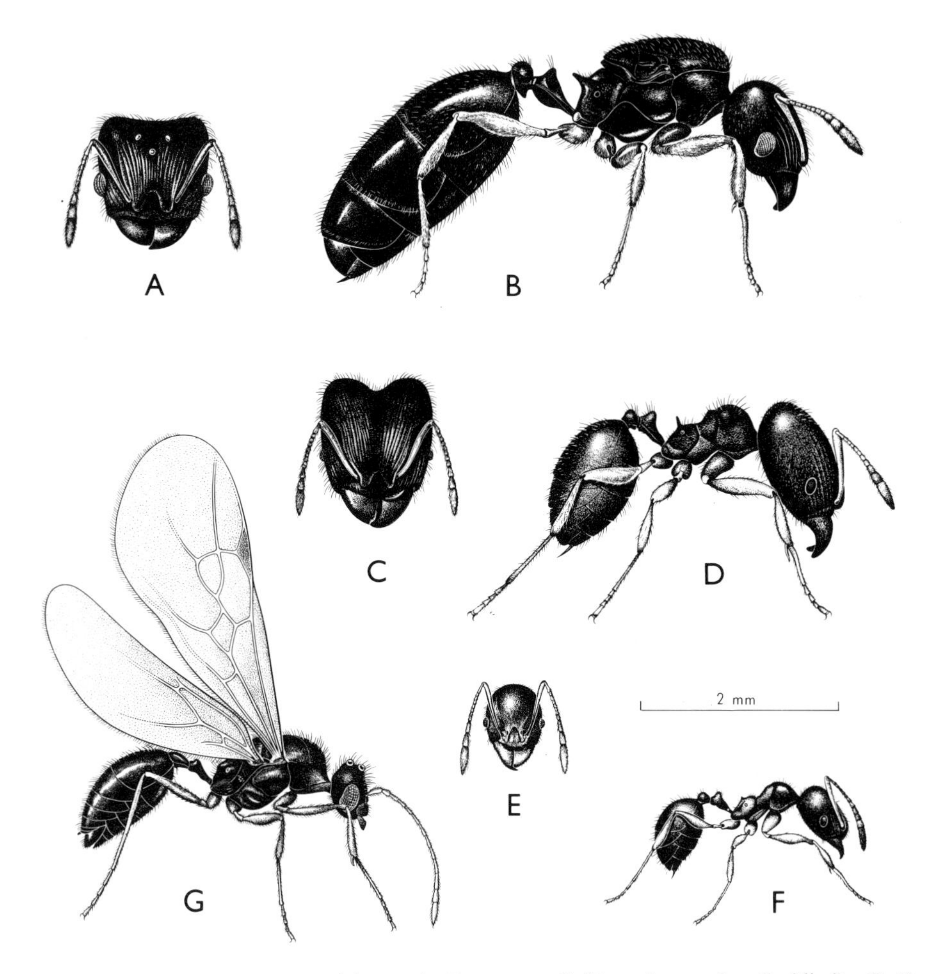

Fig. 37.38. Castes of *Pheidole* sp., Myrmicinae: A, B, queen; C, D, major worker ('soldier'); E, F, minor worker; G, male. [F. Nanninga]

and are common in *Myrmecia*, *Heteroponera*, *Leptogenys*, *Sphinctomyrmex*, *Mayriella*, and *Leptomyrmex*. In many *Rhytidoponera*, *Diacamma*, and *Leptogenys* neither queen nor differentiated ergatogyne is known, and the queen function is assumed by workers capable of egg production and normal fertilization with sperm storage. In some genera, such as *Aenictus*, *Leptanilla*, and the ponerine *Onychomyrmex*, which probably all follow a group-predatory, nomadic existence (i.e. are army ants), the reproductive female is a bizarre type of ergatogyne with enlarged, ovary-filled gaster, and is referred to as a *dichthadiigyne*.

Males are usually permanently winged, with relatively small heads and large compound eyes and ocelli; their mandibles are much modified, and often reduced and apparently functionless. In most species the males are larger than the workers and smaller than the queens. Male genitalia are unremarkable in most ants, and are easily homologized with the basic pattern in aculeate Hymenoptera (Snodgrass, 1941). The terminalia seem promising as discriminatory characters within *Myrmecia*, *Amblyopone*, *Bothroponera*, and the dolichoderine genera. In a few scattered species of *Hypoponera*, *Cardiocondyla*, and *Technomyrmex* there are ergatoid males, wingless and with more or less worker-like characteristics.

Both female and male sexes of ants share a primitive number of palpal segments (6 maxillary, 4 labial) with other old hymenopteran families, but this formula is reduced to as low as 1, 1 in some specialized ant genera. The primitive ant antennal count is 13 in the male and 12 in the female, but this number is often reduced in either or both sexes; queens and workers of some Dacetini (Myrmicinae) have as few as 4 segments, but male numbers do not fall so low. The worker-queen antennae, and sometimes those of males, are elbowed, with the scape forming the elongate basal segment. Ant wing venation has undergone convergent reduction in several advanced lines in different subfamilies (Figs. 37.8H, 38G; Brown and Nutting, 1950). In general, veins or parts of veins are lost in lineages that have undergone reduction in body size; reversal of the size trend does not lead to replacement of regular venational elements.

Internally, adult ants are remarkable chiefly for (1) the development of the crop, and especially of its sclerotic posterior valve, the proventriculus, which is elaborate in the Dolichoderinae and Formicinae, and varies widely to furnish valuable tribal or even generic characters (Eisner, 1957); and (2) the great variety and prominence of glands that open externally in various parts of the body, most of them apparently secreting pheromones—substances that are primary media of social and sexual communication (Pavan and Ronchetti, 1955; E. Wilson, 1963).

Larvae of ants (Fig. 37.39) share the white, grub-like, apodous condition of other aculeate larvae, and are dependent on the adult female castes as food providers. So radically reduced are most larval characters in even the primitive ants that the main evolutionary opportunities for morphological change clearly have lain in adding new structures or elaborating old ones. Significant trends in different lineages involve the elaboration of tubercles or specialized hairs that fasten the larvae to nest walls or ceiling, or clump them together for quick transport by the workers. The number of larval instars is not known for any ant species, but there are probably 3. Most primitive ants, and most Formicinae, spin a pupal cocoon; but the pupae are naked in Myrmicinae, Pseudomyrmecinae, and Dolichoderinae, as well as in several groups of Dorylinae and scattered species in other subfamilies.

Winged males and queens are usually produced in the colony at a time of the year that is fixed for the species. Where males and queens are both winged, a mass nuptial flight or flights may occur, during which mating takes place. After copulation the male wanders off and dies, while the queen divests herself of her wings and seeks a place to start a new colony. In the higher genera the wing muscles are converted into fat body that serves as a food supply during the lean time of nest foundation, when the queen remains enclosed in a small chamber. In some primitive genera the queen leaves the chamber to hunt from time to time. In forms having ergatogynes, and in some winged species, the nuptial flight is much modified, and may be reduced to a precopulatory promenade on the nest surface or nearby vegetation. In genera such as *Rhytidoponera* and the army ants males leave their own nests and fly to other nests, where mating takes place.

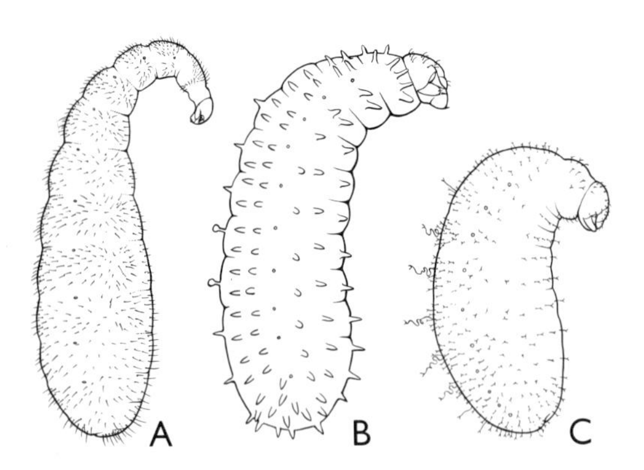

Fig. 37.39. Larvae of ants: A, *Amblyopone* sp., Ponerinae; B, *Hypoponera* sp., Ponerinae; C, *Pheidole* sp., Myrmicinae. [S. Curtis]

Established nests may contain a single dealate, egg-laying queen, or there may be several to many queens present. In the latter case colony reproduction may occur by simple fission or 'swarming', or by a kind of budding process.

A few ants in different groups found their nests by entering colonies of another species and living there as social parasites or inquilines. In the so-called temporary social parasites the host queen is killed, either by the inquiline queen or by host workers, and a mixed nest is formed of host workers with the inquilinous queen and her worker offspring; such a nest gradually loses its host workers and becomes a pure independent society of the inquilinous species. Probably a number of Australian ants are temporary parasites; *Rhoptromyrmex* and some *Bothriomyrmex* are likely examples. Permanent social parasites have specialized to the extent that the parasitic species requires the continued presence of the host species in order to survive and reproduce. Such parasites commonly lack a worker caste altogether; well-known Australian examples are *Myrmecia inquilina* (hosts *M. nigriceps* and *M. vindex*) and *Strumigenys xenos* (host *S. perplexa*). The special type of permanent parasitism known as *dulosis*, or slave-making, in which workers of the 'slave' component in the mixed colony are constantly replenished by the addition of pupae acquired through mass raids on neighbouring pure nests of the host species, is so far not known to occur in Australia.

Conclusive generalizations on orientation and communication in ants are not possible at this time. The known methods of communication are: (1) olfactory, through the medium of volatile pheromones, secreted by specific glands upon a particular stimulus and evoking a particular response, such as alarm behaviour or trail-following; (2) gustatory, particularly during food exchange by regurgitation from the crop or stomach; (3) auditory, including tapping, stridulation, and other sounds transmitted mainly through the substrate; (4) mechanical, including antennal stroking, body recoil, etc.; and (5) visual, in those forms with well-developed eyes.

Ants usually live in more or less permanent nests, excavated in the soil or in wood, or utilizing pre-existing cavities in plants or in rocks. The great majority of Australian ants inhabit the ground layers, in chambers deep in the soil or under rocks or other objects. Many species, particularly those of arid areas, construct mounds or disc nests, often with a large entrance, and a surface cover of pebbles, charcoal, etc. (e.g. Fig. 5.39). In forests, rotting logs and small fragments of rotting wood are favoured nesting sites. Although a number of Australian ants nest in twigs (in *Camponotus*, *Crematogaster*, *Myrmecorhynchus*, *Tetraponera*, and *Metapone*), this habit is best developed in the northern areas, and is not as common as in some parts of the world. Other tree-dwelling ants belong to *Polyrhachis*, which often build silken or plant-fibre nests on leaves, twigs, or bark; to *Podomyrma*, which frequently choose beetle burrows in solid wood; to *Oecophylla*, the weaver ants, which build nests of leaves joined by sheets of larval silk (Fig. 5.38); to *Myrmecia mjobergi*, a nester in epiphytic ferns, and to *Iridomyrmex cordatus* and the *Pheidole* species that live in the hollows of the tubers of the epiphytic 'ant-house plants' (*Myrmecodia* and *Hydnophytum*). Aside from these ant plants, and perhaps a few other less well understood plants also belonging to the Indo-Malayan intrusive floral element of the north, symbiotic relations between ants and plants apparently are not well developed in Australia.

Australian ants vary by taxa from highly specific predators (Cerapachyini on other ants, *Discothyrea* on spider eggs, some *Leptogenys* on slaters, *Metapone* on termites, most *Strumigenys* on entomobryoid Collembola, etc.) to scavengers and near-omnivores. Some genera of Myrmicinae, notably *Pheidole*, *Chelaner*, and *Meranoplus*, have species that depend on seed-harvesting for much of their food. Many genera attend homopteran insects for their honey-dew, and some ants, for instance the subterranean *Acropyga*, are obligatory tenders of symbiotic root-feeding plant lice. Forms such as the bulldog and jumper ants (*Myrmecia*; Frontispiece and Plate 5, S) feed largely on nectar and honey-dew as adults; the queen and youngest larvae subsist on worker-laid 'trophic' eggs, and the older larvae are fed on insect prey. Regurgitative feeding of the queen and brood is common in the higher genera.

Some formicine ants in arid areas (*Melophorus bagoti*, *Camponotus* spp.) store regurgitated honey-dew and nectar in the enormously distended crops of special large 'honeypot' workers (Figs. 5.5, 35) that remain in deep chambers of the nest. While this is obviously a mechanism of food storage, its relationship to special needs during poor seasons has never been properly studied.

Foraging in ants seems to involve two main kinds of orientation that are followed by different taxa: (1) by individuals learning restricted feeding areas, to which they return repeatedly; and (2) by trail-following, in which foraging is induced and directed by a chemical trail laid by food-laden workers returning to the nest. Foraging in many taxa is rather strictly either by day or by night, but in others it seems to be controlled more by temperature or humidity, without apparent regard for light conditions.

Many predators commonly thought to eat ants, such as *Myrmecobius fasciatus* and *Tachyglossus aculeatus*, actually feed primarily on termites and eat true ants less often. A number of Australian birds are frequent ant feeders, including the magpies (*Gymnorhina*), the coachwhip bird (*Psophodes crepitans*), and especially the thickhead (*Pachycephala pectoralis*). Various agamid lizards feed extensively on ants in the arid lands; of these *Moloch horridus* is a specialist predator of certain desert *Iridomyrmex*. Frogs (e.g. *Pseudophryne corroboree*) and spiders take many ants in Australia, as possibly also do the burrowing typhlopid snakes. Ant-lions (Neuroptera) exact their formicid toll. Many arthropods live as inquilines in ants' nests and feed on ant larvae or adults, or on the secretions, the food stores, or the cast-off wastes of ants, in such diverse orders as: Collembola, Thysanura, Hemiptera, Neuroptera, Coleoptera (Carabidae–Paussinae, Pselaphidae, Ptiliidae, Scydmaenidae, Staphylinidae, Histeridae, Scarabaeidae, Ptinidae, Limulodidae, Clambidae, Dermestidae, Silvanidae, Tenebrionidae, Aculagnathidae, Colydiidae, Brenthidae), Lepidoptera (Lycaenidae), Diptera (Phoridae, Syrphidae, Milichiidae), Hymenoptera (Diapriidae, Bethylidae), and various mites. Ants specializing in predation on other ants probably include all species of cerapachyines and at least 2 of *Melophorus*; army ants of the genus *Aenictus* also belong here.

Ants are host to various parasites, including ascomycete fungi (*Cordyceps* and *Laboulbenia*), juvenile mermithid nematodes, and the metacercarial stage of a sheep fluke, *Dicrocoelium dendriticum* (Brangham, 1959). Internal and external parasitoid insects attacking ants are certain phorid flies, Strepsiptera, and eucharitine Pteromalidae (p. 920), among others.

Ant colonies may contain as few as ten adults, or up to millions, but colony size in most Australian species is under 2,000 workers. Colonies may, however, be extremely numerous in tropical forest, savannah woodland, or semi-desert like that covering so much of Australia, and it is not uncommon to find 20 or 30 nests in a single rotten log, or even two or three colonies of different species under the same stone. Upwards of 150 species have been found in a few square km of lowland rain forest in New Guinea, and similar counts could no doubt be made in north Queensland.

Their ubiquity, their sheer abundance, and their generally high rates of activity make ants in Australia one of the most important animal groups in the environmental system of energy flow. Ants tend to be active through the seasons, and many of them are rather general feeders. As predators and scavengers of other arthropods, their operations are probably broadly beneficial, and the distributions of many other kinds of insects, including especially sawflies, predatory beetles, and termites, are strongly influenced, or determined, by the presence or absence of ants in their

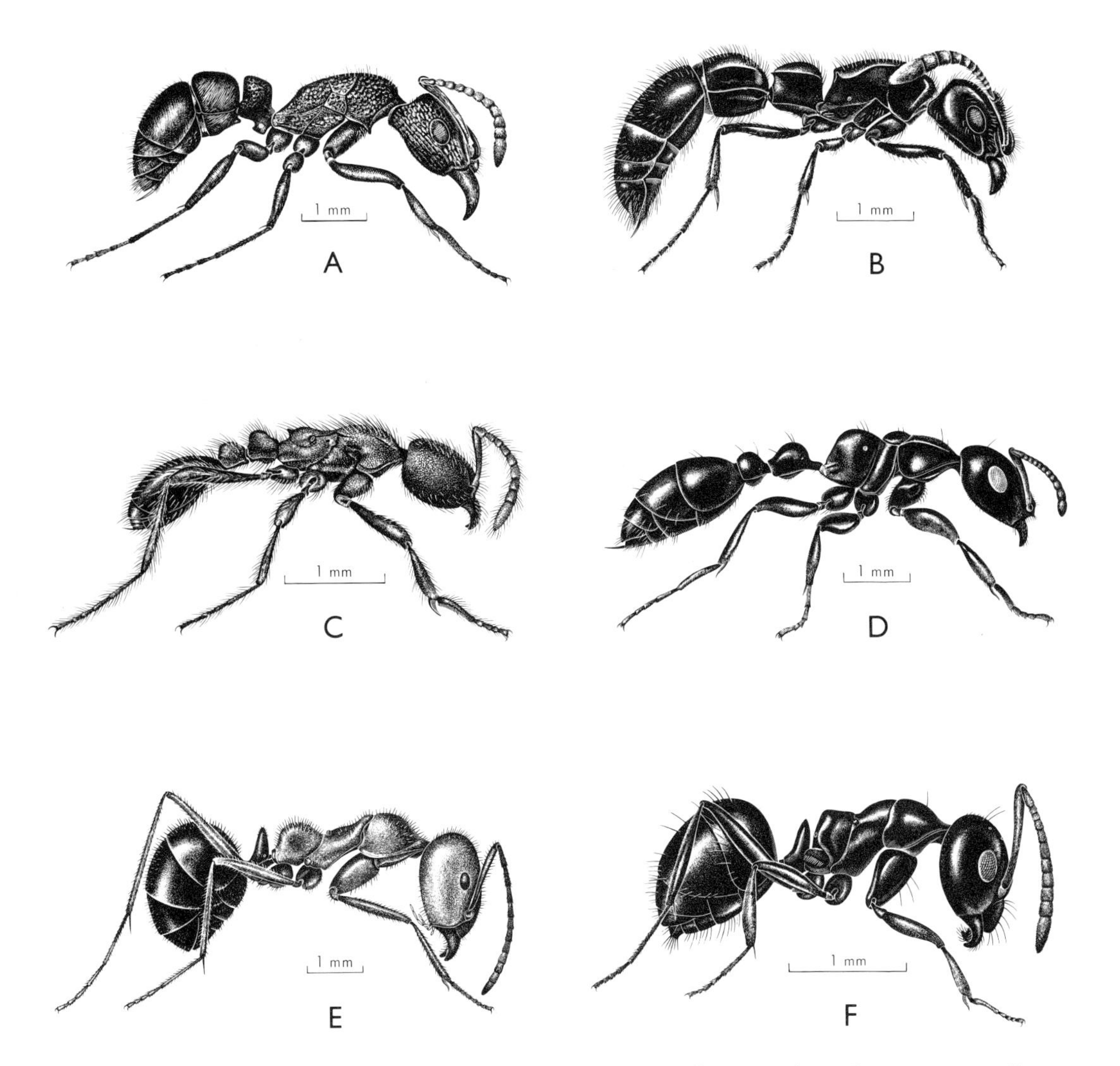

Fig. 37.40. Some representative ants, workers: A, *Rhytidoponera metallica*, Ponerinae; B, *Lioponera* sp., Ponerinae-Cerapachyini; C, *Aenictus* sp., Dorylinae; D, *Tetraponera* sp., Pseudomyrmecinae; E, *Iridomyrmex purpureus*, Dolichoderinae; F, *Prolasius* sp., Formicinae. [F. Nanninga]

niches. A little appreciated influence of ants is their role in the vertical mixing and aeration of the soil. Even in temperate North America, with a much poorer complement of ants, transport by ants of particles upward from lower horizons has been found to be of importance in soil formation (see also p. 284).

Occasional harvester species may cause some damage by attacking crop seed-beds or collecting the seeds of forest trees, but the main agricultural damage done by ants is in connection with their guarding and transporting of aphids and coccids harmful to vegetable crops and orchards. Indeed, several kinds of scale insects are a serious threat only when tended by ants. Australians as a rule feel the baneful influence of ants most directly in the home, where introduced species such as the Argentine ant (*Iridomyrmex humilis*), black house ant (*Technomyrmex albipes*), and hospital ant (*Monomorium pharaonis*) are the main pests in the south, while *Pheidole megacephala*, the fire ant (*Solenopsis geminata*), and the ghost ant (*Tapinoma melanocephalum*) are familiar nuisances in tropical Australia. Bush pests include the ferociously stinging bulldog ants and jumpers of the genus *Myrmecia*, the mainly nocturnal sugar ants (*Camponotus* spp.), and the meat ants (*Iridomyrmex* spp.). The last-mentioned species can be a serious pest around homes and food-processing plants.

All of the 8 ant subfamilies occur in Australia. The MYRMECIINAE (*Myrmecia* and *Nothomyrmecia*) include the most primitive living ants, and are restricted to Australia and New Caledonia. This subfamily is represented elsewhere by the extinct Oligocene genera *Prionomyrmex* (Baltic amber) and *Ameghinoa* (Argentina). *Nothomyrmecia* is the most archaic known living ant. Among the other major subfamilies, PONERINAE (including Cerapachyini) are exceptionally well developed, but DORYLINAE and PSEUDOMYRMECINAE are represented only by a few Indo-Malayan elements in the tropical north.

[W. Brown, 1953, 1955, 1958, 1960; Clark, 1930; Emery, 1910–25; Ettershank, 1966; W. Wheeler, 1918.]

Key to the Subfamilies of Formicidae—Workers

(Where there are two figures in the square brackets, the first is the number of genera and the second the number of species known in Australia.)

1. Gaster attached to alitrunk by a waist consisting of 2 more or less reduced and nodiform segments (Figs. 37.37,38) 2
   Gaster attached to alitrunk by a waist consisting of a single reduced segment, which may be nodiform, erect or inclined scale-like, or even prostrate and more or less hidden by overhanging gaster (Figs. 37.40A,E,F) 7
2. Larger species, total length usually exceeding 8 mm, with large eyes and long, slender, serially dentate mandibles. [*Myrmecia*; 65 spp.; Fig. 37.37] .... MYRMECIINAE (pt.)
   Smaller species, or, if largest workers or soldiers are over 8 mm, eyes very small or absent and/or mandibles not elongate 3
3. Pygidium flattened at apex, the flattened part bordered with denticles arranged in rows on each side; underside of head with a strong carina on each side running forward from posterior corner. [Cerapachyini–*Lioponera* (Fig. 37.40B), *Syscia*, *Sphinctomyrmex*; 55 spp.] .... PONERINAE (pt.)
   Pygidium simple; no distinct carinae on underside of head extending forward from posterior corners 4
4. Eyes lacking; small to minute army ants. [Queens dichthadiiform, i.e. blind, wingless, with simple alitrunk, gaster long and bulky] 5
   Eyes large to minute, but normally present in Australian forms; not army ants 6
5. Extremely minute (under 2·5 mm) slender yellow species, with 12-segmented antennae. [Rare and subterranean; *Leptanilla*; 1 sp.] .... LEPTANILLINAE
   Small species, but usually over 2·5 mm long; antennae 10-segmented. [*Aenictus*; 3 spp.; Fig. 37.40C] .... DORYLINAE
6. Tarsal claws toothed; tibial spurs of mid and hind legs distinctly pectinate; Australian species very slender, black, with large eyes; inhabiting hollow twigs and similar plant cavities. [*Tetraponera*; 2 spp.; Fig. 37.40D] .... PSEUDOMYRMECINAE
   Tarsal claws simple; tibial spurs of mid and hind legs, when present, simple or at most very indistinctly pectinate; body form and nesting habits very diverse (Fig. 37.38). [37, 284; e.g. *Podomyrma*, *Crematogaster*, *Meranoplus*, *Strumigenys*] .... MYRMICINAE
7. Sting well developed and functional, usually extended and visible in dead specimens 8
   Sting absent, or vestigial and not extensible 9
8. A large (10 mm or more), tawny yellow species with large convex eyes set at the middle of the sides of the head; long, robust mandibles with finely dentate inner margins; palpi segmented 6,4; alitrunk and petiole much as in *Myrmecia*; body with abundant stiffly erect hairs; tarsal claws toothed. [Presumably in south-western W.A.; *Nothomyrmecia*; 1 sp.] .... MYRMECIINAE (pt.)
   Disagreeing with some or all of the above characters. [28, 200; e.g. *Amblyopone*, *Rhytidoponera* (Fig. 37.40A), *Leptogenys*, *Odontomachus*] .... PONERINAE (pt.)
9. Seventh sternite rolled into a short ventro-apical cone with a round apical orifice (with or without a coronula of minute hairs) that serves as a nozzle for a defensive acid spray; not to be confused with the cloacal orifice, which is more dorsal and normally hidden (Fig. 37.40F).

[21, 370; e.g. *Melophorus, Oecophylla, Camponotus, Polyrhachis*] .... FORMICINAE
Gastric apex without such an acid-spray-ejecting cone (dried specimens may sometimes have the 7th sternite distorted into a more or less conical shape, but then still without a circular orifice); defensive secretion in the form of a viscous fluid, ejected through a slit-like orifice. [10, 120; e.g. *Leptomyrmex, Iridomyrmex* (Fig. 37.40E), *Technomyrmex*] .... DOLICHODERINAE

ACKNOWLEDGMENTS. E. F. Riek is grateful to Dr Gerald I. Stage and Dr Howell V. Daly, University of California, Berkeley, U.S.A., for helpful discussions of the manuscript and anatomical details, respectively.

REFERENCES

This is not a bibliography: it merely provides a selected entry into main literature. Musgrave (1932) published a bibliography of Australian entomology to 1930, and copies of his second, unpublished (uncorrected) volume, covering the period 1931–58, are held by the Australian Museum, Sydney, the Australian Academy of Science, Canberra, and the Division of Entomology, CSIRO, Canberra. World literature from 1864 is covered by the *Zoological Record*, issued annually by the Zoological Society of London. In the following list, abbreviations are, where possible, those given in the 4th edition of *The World List of Scientific Periodicals* (London: Butterworths, 1963–5), and the numbers in square brackets are the chapters in which the references are cited.

Abul-Nasr, S. E. (1950). Structure and development of the reproductive system of some species of Nematocera (order Diptera: suborder Nematocera). *Phil. Trans. R. Soc.* (B) **234:** 339–96. [34]

Acker, T. S. (1960). The comparative morphology of the male terminalia of Neuroptera (Insecta). *Microentomology* **24:** 25–83. [29]

Adams, P. A. (1962). A stridulatory structure in Chrysopidae (Neuroptera). *Pan-Pacif. Ent.* **38:** 178–80. [29]

Ahmad, I. (1965). The Leptocorisinae (Heteroptera: Alydidae) of the world. *Bull. Br. Mus. nat. Hist.* Ent. Suppl. **5:** 1–156. [26]

Ahmad, M. (1950). The phylogeny of termite genera based on imago-worker mandibles. *Bull. Am. Mus. nat. Hist.* **95** (2): 37–86. [15]

Alexander, C. P. (1926). The Trichoceridae of Australia (Diptera). *Proc. Linn. Soc. N.S.W.* **51:** 299–304. [34]

—— (1929). A comparison of the systems of nomenclature that have been applied to the radial field of the wing in the Diptera. *Trans. 4th int. Congr. Ent.*: 700–7. [34]

—— (1932). A review of the Tipulidae of Australia (Diptera). I. Introduction; Historical; Distribution. Subfamily Tipulinae: *Clytocosmus* Skuse. *Proc. Linn. Soc. N.S.W.* **57:** 1–23. [34]

—— (1938). New or little known Tipulidae from eastern Asia (Diptera). XXXVII. *Philipp. J. Sci.* **66:** 221–59. [34]

—— (1958). Geographical distribution of the net-winged midges (Blepharoceridae, Diptera). *Proc. 10th int. Congr. Ent.* **1:** 813–28. [34]

Alexander, R. D. (1962). The role of behavioral study in cricket classification. *Syst. Zool.* **11:** 53–72. [21]

Allan, R. M. (1956). A study of the populations of the rabbit flea *Spilopsyllus cuniculi* (Dale) on the wild rabbit *Oryctolagus cuniculus* in north-east Scotland. *Proc. R. ent. Soc. Lond.* (A) **31:** 145–52. [33]

Ander, K. (1934). Über die Gattung *Cylindracheta* und ihre systematische Stellung (Orthoptera Saltatoria). *Ark. Zool.* **26A** (21): 1–16. [21]

Andersen, S. O., and Weis-Fogh, T. (1964). Resilin. A rubberlike protein in arthropod cuticle. *Adv. Insect Physiol.* **2:** 1–65. [2]

André, E. (1895). Mutillides d'Australie nouvelles ou imparfaitement connues. *Mém. Soc. zool. Fr.* **8:** 475–517. [37]

Andrewartha, H. G., and Birch, L. C. (1954). *The Distribution and Abundance of Animals*, 782 pp. (University of Chicago Press). [5, 27]

Andrewes, H. E. (1935). *The Fauna of British India, Including Ceylon and Burma.* Coleoptera. Carabidae II, Harpalinae I, 323 pp. (London: Taylor & Francis). [30]

Annecke, D. P., and Doutt, R. L. (1961). The genera of the Mymaridae. Hymenoptera: Chalcidoidea. *Entomology Mem. Dep. agric. tech. Serv. Repub. S. Afr.* **5**: 1–71. [37]

Armstrong, J. W. T. (1945). On Australian Dermestidae. Part IV. *Proc. Linn. Soc. N.S.W.* **70**: 47–52. [30]

—— (1948). On Australian Coleoptera. Part I. *Proc. Linn. Soc. N.S.W.* **72**: 292–8. [30]

—— (1953). On Australian Helodidae (Coleoptera). I. *Proc. Linn. Soc. N.S.W.* **78**: 19–32. [30]

Arnett, R. H. (1960–2). *The Beetles of the United States (A Manual for Identification)*, 1,112 pp. (Washington, D.C.: Catholic University of America Press—issued in 6 parts). [30]

Arnold, J. W. (1964). Blood circulation in insect wings. *Mem. ent. Soc. Can.* **38**: 1–48. [36]

Arrow, G. J. (1920). A peculiar new genus of Australian beetles. *Ann. Mag. nat. Hist.* (9) **6**: 434–7. [30]

—— (1942). The origin of stridulation in beetles. *Proc. R. ent. Soc. Lond.* (A) **17**: 83–6. [30]

—— (1951). *Horned Beetles, A Study of the Fantastic in Nature*, 154 pp. (Den Haag: Junk). [30]

Ashlock, P. D. (1964). Two new tribes of Rhyparochrominae: a re-evaluation of the Lethaeini (Hemiptera-Heteroptera: Lygaeidae). *Ann. ent. Soc. Am.* **57**: 414–22. [26]

Atkins, M. D. (1963). The Cupedidae of the world. *Can. Ent.* **95**: 140–62. [30]

Auclair, J. L. (1963). Aphid feeding and nutrition. *A. Rev. Ent.* **8**: 439–90. [26]

Bacot, H. W. (1914). A study of the bionomics of the common rat fleas and other species associated with human habitations, with special reference to the influence of temperature and humidity at various periods of the life history of the insect. *J. Hyg., Camb.* **13** (Plague Suppl.): 447–654. [33]

Badonnel, A. (1951). Ordre des Psocoptères. Psocoptera Psoquillae Latreille, 1810. In P.-P. Grassé (ed.), *Traité de Zoologie: Anatomie, Systématique, Biologie*, Vol. X, pp. 1301–40. (Paris: Masson). [24]

Baker, J. M. (1963). Ambrosia beetles and their fungi, with particular reference to *Platypus cylindrus* Fab. *Symp. Soc. gen. Microbiol.* **13**: 232–65. [30]

Balfour-Browne, F., and Balfour-Browne, J. (1940). An outline of the habits of the water-beetle, *Noterus capricornis* Herbst (Coleopt.). *Proc. R. ent. Soc. Lond.* (A) **15**: 105–12. [30]

Baltazar, C. R. (1962, 1964). The genera of parasitic Hymenoptera in the Philippines. Parts 1, 2. *Pacif. Insects* **4**: 737–71; **6**: 15–67. [37]

Balter, R. S. (1968). Lice egg morphology as a guide to taxonomy. *Med. biol. Illust.* **18**: 94–5. [25, fig.]

Barth, R. (1954). Untersuchungen an den Tarsaldrüsen von *Embolyntha batesi* MacLachlan, 1877 (Embioidea). *Zool. Jb.* **74**: 172–88. [23]

Basrur, P. K. (1962). The salivary gland chromosomes of seven species of *Prosimulium* (Diptera: Simuliidae) from Alaska and British Columbia. *Can. J. Zool.* **40**: 1019–33. [34]

Basrur, V. R., and Rothfels, K. H. (1959). Triploidy in natural populations of the black fly *Cnephia mutata* (Malloch). *Can. J. Zool.* **37**: 571–89. [3]

Beament, J. W. L. (1964). The active transport and passive movement of water in insects. *Adv. Insect Physiol.* **2**: 67–129. [2]

——, Noble-Nesbitt, J., and Watson, J. A. L. (1964). The waterproofing mechanism of arthropods. III. Cuticular permeability in the firebrat, *Thermobia domestica* (Packard). *J. exp. Biol.* **41**: 323–30. [11]

Beardsley, J. W. (1966). Homoptera: Coccoidea. *Insects Micronesia* **6** (7): 377–562. [26]

Beaumont, J. de (1954). Remarques sur la systématique des Nyssoninae paléarctiques (Hym. Sphecid.). *Revue suisse Zool.* **61**: 283–322. [37]

Beer, G. R. de (1954). *Archaeopteryx* and evolution. *Advmt Sci., Lond.* **11**: 160–70. [7]

Beermann, W. (1962). Riesenchromosomen. *Protoplasmatologia* **6** (D): 1–161. [3]

——, and Clever, U. (1964). Chromosome puffs. *Scient. Am.* **210** (4): 50–8. [2]

Beier, M. (1935). Mantodea. Fam. Mantidae. *Genera Insect.* **201**: 1–10; **203**: 1–146 (+3 pp. supplement, 1937). [16]

—— (1955). Ordnung: Saltatoptera m. (Saltatoria Latreille 1817). *Bronn's Kl. Ordn. Tierreichs* (5) (3) **6**: 34–304. [21]

—— (1957). Ordnung Cheleutoptera Crampton 1915 (Phasmida Leach 1815). *Bronn's Kl. Ordn. Tierreichs* (5) (3) **6**: 305–454. [22]

—— (1964). Blattopteroidea. Ordnung Mantodea Burmeister 1838 (*Raptoriae* Latreille 1802; *Mantoidea* Handlirsch 1903; *Mantidea* auct.). *Bronn's Kl. Ordn. Tierreichs* (5) (3) **6**: 849–970. [16]

Belton, P. (1962). Responses to sound in pyralid moths. *Nature, Lond.* **196**: 1188–9. [36]

Benesh, B. (1960). Lucanidea. *Coleoptm Cat.*, Suppl. Pt 8, 178 pp. (Den Haag: Junk). [30]

Benson, R. B. (1938a). Sawflies of the family Tenthredinidae in Australia, with the description of a new genus and species (Hymenoptera, Symphyta). *Ann. Mag. nat. Hist.* (11) **2**: 236–9. [37]

—— (1938b). On the classification of sawflies (Hymenoptera Symphyta). *Trans. R. ent. Soc. Lond.* **87**: 353–84. [37]

—— (1939). A revision of the Australian sawflies of the genus *Perga* Leach, *sens. lat.* (Hymenoptera Symphyta). *Aust. Zool.* **9**: 324–57. [37]

—— (1962). The affinities of the Australian Argidae (Hymenoptera). *Ann. Mag. nat. Hist.* (13) **5**: 631–5. [37]

—— (1965). Some new pergid sawflies from New Guinea (Hymenoptera Symphyta). *Ann. Mag. nat. Hist.* (13) **8**: 45–9. [37]

Bequaert, J. C. (1953). The Hippoboscidae or louse-flies (Diptera) of mammals and birds. Part I. Structure, physiology and natural history. *Entomologica am.* **32** (n.s.): 1–209. [25]

Berland, L., and Grassé, P.-P. (1951). Super-ordre des Névroptéroides. Neuropteroidea Handlirsch, 1903. In P.-P. Grassé (ed.), *Traité de Zoologie: Anatomie, Systématique, Biologie*, Vol. X, pp. 3–69. (Paris: Masson). [29]

Betrem, J. G. (1928). Monographie der indo-australischen Scoliiden (Hym. Acul.) mit zoogeographischen Betrachtungen. *Treubia* **9** Suppl.: 1–388. [37]

—— (1933). Die Scoliiden der indoaustralischen und palaearktischen Region aus dem Staatlichen Museum für Tierkunde zu Dresden (Hym.). *Stettin. ent. Ztg* **94**: 236–63. [37]

Bigelow, R. S., and Cochaux, P. S. A. (1962). Intersterility and diapause differences between geographical populations of *Teleogryllus commodus* (Walker) (Orthoptera: Gryllidae). *Aust. J. Zool.* **10**: 360–6. [21]

BLACKBURN, T. (1897–1901). Revision of the genus *Paropsis*. Parts I–VI. *Proc. Linn. Soc. N.S.W.* **21:** 637–93; **22:** 166–89; **23:** 218–63, 656–700; **24:** 482–521; **26:** 159–96. [30]

—— (1899, 1900). Further notes on Australian Coleoptera, with descriptions of new genera and species. *Trans. R. Soc. S. Aust.* **23:** 22–101; **24:** 113–69. [30]

BLACKITH, R. E. (1957). Polymorphism in some Australian locusts and grasshoppers. *Biometrics* **13:** 183–96. [21]

——, and BLACKITH, R. M. (1966). The anatomy and physiology of the morabine grasshoppers: I, II. *Aust. J. Zool.* **14:** 31–48, 1035–71. [21]

BLAKE, S. T., and ROFF, C. (1958). *The Honey Flora of South-Eastern Queensland*, 199 pp. (Brisbane: Govt Printer). [37]

BLESZYŃSKI, S., and COLLINS, R. J. (1962). A short catalogue of the world species of the family Crambidae (Lepidoptera). *Acta zool. cracov.* **7:** 197–389. [36]

BLUMBERG, B. (1935). The life cycle and seasonal history of *Ceroplastes rubens*. *Proc. R. Soc. Qd* **46:** 18–32. [26]

BOHART, R. M. (1941). A revision of the Strepsiptera with special reference to the species of North America. *Univ. Calif. Publs Ent.* **7** (6): 91–159. [30, 31]

——, and MENKE, A. S. (1963). A reclassification of the Sphecinae with a revision of the Nearctic species of the tribes Sceliphronini and Sphecini (Hymenoptera, Sphecidae). *Univ. Calif. Publs Ent.* **30:** 91–181. [37]

BOLIVAR Y PIELTAIN, C., and CORONADO-G, L. (1963). Estudio de un nuevo *Zorotypus* proviento de la Región Amazónica Peruana (Ins., Zoraptera). *Ciencia, Méx.* **22** (4): 93–100. [17]

BONGARDT, J. (1903). Beiträge zur Kenntnis der Leuchtorgane einheimischer Lampyriden. *Z. wiss. Zool.* **75:** 1–45. [2, fig.]

BORNEMISSZA, G. F. (1957). The first Projapygidae from Western Australia, with some additional notes on the family and its allies. *West. Aust. Nat.* **6:** 76–9. [10]

—— (1966). Observations on the hunting and mating behaviour of two species of scorpion flies (Bittacidae: Mecoptera). *Aust. J. Zool.* **14:** 371–82. [32]

BÖRNER, C., and HEINZE, K. (1957). Aphidina—Aphidoidea. Blattlause, plantlice (aphids), pucerons (aphides). *Handb. PflKrankh.* (5th edn) **5** (4): 1–402. [26]

BÖVING, A. G., and CRAIGHEAD, F. C. (1931). *An Illustrated Synopsis of the Principal Larval Forms of the Order Coleoptera*, 351 pp. (Brooklyn, N.Y.: Brooklyn ent. Soc.). [30]

BOYES, J. W. (1958). Chromosomes in classification of Diptera. *Proc. 10th int. Congr. Ent.* **2:** 899–906. [34]

BRADLEY, J. D. (1953). On *Harmaclona* Busck, 1914 (Lep., Arrhenophanidae), with descriptions of a new species, of a supplementary wing-coupling device present, and some new synonymy. *Entomologist* **86:** 61–6. [36]

BRANGHAM, A. N. (1959). Ant vectors in the life-cycle of the lesser liver fluke, *Dicrocoelium dendriticum* (Rudolphi 1819) (Trematoda: Dicrocoeliidae). *Entomologist's Gaz.* **10:** 111–31. [37]

BRAUN, A. F. (1963). The genus *Bucculatrix* in America north of Mexico (Microlepidoptera). *Mem. Am. ent. Soc.* **18:** 1–208. [36]

BRAUNS, A. (1954). *Terricole Dipterenlarven*, 179 pp., and *Puppen terricoler Dipterenlarven*, 156 pp. (Berlin: Musterschmidt). [34]

BRELAND, O. P. (1961). Studies on the chromosomes of mosquitoes. *Ann. ent. Soc. Am.* **54:** 360–75. [34]

BRIMBLECOMBE, A. R. (1962). Studies of the Coccoidea. 13. The genera *Aonidiella*, *Chrysomphalus* and *Quadraspidiotus* in Queensland. *Qd J. agric. Sci.* **19:** 403–23. [26]

BRINCK, P. (1956). Reproductive system and mating in Plecoptera. I, II. *Opusc. ent.* **21:** 57–127. [1]

—— (1957). Reproductive system and mating in Ephemeroptera. *Opusc. ent.* **22:** 1–37. [1]

BRITTON, E. B. (1957). *A Revision of the Australian Chafers (Coleoptera: Scarabaeidae: Melolonthinae)*, Vol. 1, 185 pp. (London: Br. Mus. nat. Hist.). [30]

—— (1966). On the larva of *Sphaerius* and the systematic position of the Sphaeriidae (Coleoptera). *Aust. J. Zool.* **14:** 1193–8. [30]

BROOKES, H. M. (1957). The Coccoidea (Homoptera) naturalized in South Australia: an annotated list. *Trans. R. Soc. S. Aust.* **80:** 81–90. [26]

BROOKS, J. G. (1948). North Queensland Coleoptera and their food plants. Parts 1, 2. *N. Qd Nat.* **15:** 26–9; **16:** 6–7. [30]

—— (1965a). North Queensland Coleoptera, their food and host plants (continued). Part 3. *N. Qd Nat.* **32:** 29–30. [30]

—— (1965b). New records of Coleoptera in Australia. *J. ent. Soc. Qd* **4:** 85. [30]

BROOKS, M. A. (1963). Symbiosis and aposymbiosis in arthropods. *Symp. Soc. gen. Microbiol.* **13:** 200–31. [30]

BROWN, S. W. (1965). Chromosomal survey of the armored and palm scale insects (Coccoïdea: Diaspididae and Phoenicococcidae). *Hilgardia* **36:** 189–294. [3, 26]

——, and BENNETT, F. D. (1957). On sex determination in the diaspine scale *Pseudaulacaspis pentagona* (Targ.) (Coccoïdea). *Genetics, Princeton* **42:** 510–23. [3]

——, and MCKENZIE, H. L. (1962). Evolutionary patterns in the armored scale insects and their allies (Homoptera: Coccoidea: Diaspididae, Phoenicococcidae, and Asterolecaniidae). *Hilgardia* **33:** 141–70-A. [26]

BROWN, W. L. (1953). Revisionary notes on the ant genus *Myrmecia* of Australia. *Bull. Mus. comp. Zool. Harv.* **111:** 1–35. [37]

—— (1955). A revision of the Australian ant genus *Notoncus* Emery, with notes on the other genera of Melophorini. *Bull. Mus. comp. Zool. Harv.* **113:** 471–94. [37]

—— (1958). Contributions towards a reclassification of the Formicidae. II. Tribe Ectatommini (Hymenoptera). *Bull. Mus. comp. Zool. Harv.* **118:** 175–362. [37]

—— (1960). Contributions towards a reclassification of the Formicidae. III. Tribe Amblyoponini (Hymenoptera). *Bull. Mus. comp. Zool. Harv.* **122:** 143–230. [37]

——, and NUTTING, W. L. (1950). Wing venation and the phylogeny of the Formicidae (Hymenoptera). *Trans. Am. ent. Soc.* **75:** 113–32. [37]

BRUES, C. T., MELANDER, A. L., and CARPENTER, F. M. (1954). Classification of insects. Keys to the living and extinct families of insects, and to the living families of other terrestrial arthropods. *Bull. Mus. comp. Zool. Harv.* **108:** 1–917. [8]

BRUNDIN, L. (1965). On the real nature of trans-antarctic relationships. *Evolution, Lancaster, Pa.* **19:** 496–505. [9]

BUCK, R. C., and HULL, D. L. (1966). The logical structure of the Linnaean hierarchy. *Syst. Zool.* **15:** 97–111. [6]

BULLOCK, T. H., and HORRIDGE, G. A. (1965). *Structure and Function in the Nervous System of Invertebrates*, 2 vols, 1,719 pp. (San Francisco: Freeman). [2]

BURKS, B. D. (1953). The mayflies, or Ephemeroptera, of Illinois. *Bull. Ill. St. nat. Hist. Surv.* **26** (1): 1–216. [12]

BURNS, A. N. (1957). Check list of Australian Cicadidae. *Ent. Arb. Mus. Georg Frey* **8:** 609–78. [26]

——, and NEBOISS, A. (1957). Two new species of Plecoptera from Victoria. *Mem. natn. Mus. Vict.* **21:** 92–100. [20]

BURR, M. (1915–16). On the male genital armature of the Dermaptera. Pt. 1: Protodermaptera (except Psalidae). Pt. 2: Psalidae. Pt. 3: Eudermaptera. *Jl R. microsc. Soc.* **1915:** 413–47, 521–46; **1916:** 1–18. [19]

BURTT, E. T., and CATTON, W. T. (1962). A diffraction theory of insect vision. I. An experimental investigation of visual acuity and image formation in the compound eyes of three species of insects. *Proc. R. Soc.* (B) **157:** 53–82. [2]

BUSNEL, R.-G. (ed.) (1955). *Colloque sur l'Acoustique des Orthoptères*, 448 pp. (Paris: Inst. natn. Rech. agron.). [21]

BUSVINE, J. R. (1966). *The Biology and Control of Insect Pests of Medical and Domestic Importance*, 2nd edn, 467 pp. (London: Methuen). [30]

BUXTON, P. A. (1947). *The Louse. An Account of the Lice which Infest Man, their Medical Importance and Control*, 2nd edn, 164 pp. (London: Edward Arnold). [25]

BYERS, G. W. (1965). Families and genera of Mecoptera. *Proc. 12th int. Congr. Ent.*: 123. [32]

CALLAHAN, P. S. (1965). Intermediate and far infra-red sensing of nocturnal insects. Part I. *Ann. ent. Soc. Am.* **58:** 727–45. [36]

——, and CASCIO, T. (1963). Histology of the reproductive tracts and transmission of sperm in the corn earworm, *Heliothis zea*. *Ann. ent. Soc. Am.* **56:** 535–56. [36]

CAMPBELL, K. G. (1961). The effects of forest fires on three species of stick insects (Phasmatidae Phasmatodea) occurring in plagues in forest areas of south-eastern Australia. *Proc. Linn. Soc. N.S.W.* **86:** 112–21. [22]

CAMRAS, S. (1961). Notes on Australian flies of the family Conopidae. *Proc. U.S. natn. Mus.* **113:** 61–76. [34]

CARAYON, J. (1954). Organes assumant les fonctions de la spermathèque chez divers Hétéroptères. *Bull. Soc. zool. Fr.* **79:** 189–97. [26]

—— (1964). Les aberrations sexuelles 'normalisées' de certains Hémiptères Cimicoidea. In A. Brion et H. Ey (eds), *Psychiatrie Animale*, pp. 283–94. (Paris: Desclée de Brouwer). [26]

CAREY, S. W. (ed.) (1958). *Continental Drift: A Symposium*, 363 pp. (Hobart: University of Tasmania). [9]

CARNE, P. B. (1957a). A revision of the ruteline genus *Anoplognathus* Leach (Coleoptera: Scarabaeidae). *Aust. J. Zool.* **5:** 88–143. [30]

—— (1957b). *A Systematic Revision of the Australian Dynastinae (Coleoptera: Scarabaeidae)*, 284 pp. (Melbourne: Commonw. scient. ind. Res. Org.). [30]

—— (1958). A review of the Australian Rutelinae (Coleoptera: Scarabaeidae). *Aust. J. Zool.* **6:** 162–240. [30]

CARPENTER, F. M. (1930). The Lower Permian insects of Kansas. Part 1. Introduction and the order Mecoptera. *Bull. Mus. comp. Zool. Harv.* **70:** 69–101. [8, fig.]

—— (1931). The Lower Permian insects of Kansas. Part 2. The orders Palaeodictyoptera, Protodonata, and Odonata. *Am. J. Sci.* **21:** 97–139. [8, fig.]

—— (1932). The Lower Permian insects of Kansas. Part 5. Psocoptera and additions to the Homoptera. *Am. J. Sci.* **24:** 1–22. [8, fig.]

—— (1933). The Lower Permian insects of Kansas. Part 6. Delopteridae, Protelytroptera, Plectoptera and a new collection of Protodonata, Odonata, Megasecoptera, Homoptera, and Psocoptera. *Proc. Am. Acad. Arts Sci.* **68:** 411–503. [8, fig.]

—— (1935). The Lower Permian insects of Kansas. Part 7. The order Protoperlaria. *Proc. Am. Acad. Arts Sci.* **70:** 103–46. [8, fig.]

—— (1943). The Lower Permian insects of Kansas. Part 9. The orders Neuroptera, Raphidiodea, Caloneurodea and Protorthoptera (Probnisidae), with additional Protodonata and Megasecoptera. *Proc. Am. Acad. Arts Sci.* **75:** 55–84. [8, fig.]

—— (1947). Lower Permian insects from Oklahoma. Part 1. Introduction and the orders Megasecoptera, Protodonata, and Odonata. *Proc. Am. Acad. Arts Sci.* **76:** 25–54. [8, fig.]

CARRICK, R. (1959). The food and feeding habits of the straw-necked ibis, *Threskiornis spinicollis* (Jameson), and the white ibis, *T. molucca* (Cuvier), in Australia. *CSIRO Wildl. Res.* **4:** 69–92. [21]

CARTER, H. J. (1926). A check list of the Australian Tenebrionidae. *Aust. Zool.* **4:** 117–63, 280, 294. [30]

—— (1929). A check list of the Australian Buprestidae. *Aust. Zool.* **5:** 265–304. [30]

——, and ZECK, E. H. (1937). A monograph of the Australian Colydiidae. *Proc. Linn. Soc. N.S.W.* **62:** 181–208. [30]

CARVALHO, E. L. de (1956). Primeira contribuição para o estudo dos estrepsípteros angolenses (Insecta Strepsiptera). *Publções cult. Co. Diam. Angola* **29:** 11–54. [31]

—— (1961). Tabela para a determinação dos géneros de estrepsípteros (Insecta). *Garcia de Orto* **9:** 691–8. [31]

CARVALHO, J. C. M. (1955). Chaves para os gêneros de Mirídeos do mundo (Hemiptera). *Bolm Mus. para. Emilio Goeldi Hist. nat. Ethnogr.* **11 (2):** 5–151. [26]

—— (1957–60). A catalogue of the Miridae of the world. Pts 1–5. *Archos Mus. nac., Rio de J.* **44:** 1–158; **45:** 1–216; **47:** 1–161; **48:** 1–384; **51:** 1–194. [26]

CASIMIR, M. (1962). History of outbreaks of the Australian plague locust, *Chortoicetes terminifera* (Walk.), between 1933 and 1959, and analysis of the influence of rainfall on these outbreaks. *Aust. J. agric. Res.* **13:** 674–700. [21]

Chamberlin, J. C. (1923). A systematic monograph of the Tachardiinae or lac insects (Coccidae). *Bull. ent. Res.* **14:** 147–212. [26, fig.]

Chamberlin, R. V. (1920). The Myriapoda of the Australian region. *Bull. Mus. comp. Zool. Harv.* **64:** 1–269. [7]

Champ, B., and Smithers, C. N. (1965). Insects and mites associated with stored products in Queensland. 1. Psocoptera. *Qd J. agric. Sci.* **22:** 259–62. [24]

Chapman, J. A., and Kinghorn, J. M. (1958). Studies of flight and attack activity of the ambrosia beetle, *Trypodendron lineatum* (Oliv.), and other scolytids. *Can. Ent.* **90:** 362–72. [30]

Cheesman, L. E. (1951). Old mountains of New Guinea. *Nature, Lond.* **168:** 597. [9]

Chen, L.-c. (1965). A revision of *Micronecta* of Australia and Melanesia (Heteroptera: Corixidae). *Kans. Univ. Sci. Bull.* **46:** 147–65. [26]

Child, C. M. (1894). Ein bisher wenig beachtetes antennales Sinnesorgan der Insekten mit besonderer Berücksichtigung der Culiciden und Chironomiden. *Z. wiss. Zool.* **58:** 475–528. [2, fig.]

China, W. E. (1955). The evolution of the water bugs. *Bull. natn. Inst. Sci. India* (New Delhi) **7:** 91–103. [26]

—— (1962). South American Peloridiidae (Hemiptera–Homoptera: Coleorrhyncha). *Trans. R. ent. Soc. Lond.* **114:** 131–61. [26]

—— (1963). *Lestonia haustorifera* China (Hemiptera: Lestoniidae)—a correction. *J. ent. Soc. Qd* **2:** 67–8. [26]

——, and Miller, N. C. E. (1959). Check-list and keys to the families and subfamilies of the Hemiptera–Heteroptera. *Bull. Br. Mus. nat. Hist.* Ent. **8:** 3–45. [26]

Chopard, L. (1938). La biologie des Orthoptères. *Encycl. ent.* (A) **20:** 1–541. [16, 21, 22]

—— (1949a). Ordre des Dictyoptères Leach, 1818 (= Blattaeformia Werner, 1906; = Oothecaria Karny, 1915). In P.-P. Grassé (ed.), *Traité de Zoologie: Anatomie, Systématique, Biologie*, Vol. IX, pp. 355–407. (Paris: Masson). [16]

—— (1949b). Ordre des Orthoptères Latreille 1793—Olivier 1789. In P.-P. Grassé (ed.), *Traité de Zoologie*, Vol. IX, pp. 617–722. [21]

—— (1949c). Ordre des Dermaptères Leach 1817. In P.-P. Grassé (ed.), *Traité de Zoologie*, Vol. IX, pp. 745–70. [19]

—— (1951). A revision of the Australian Grylloidea. *Rec. S. Aust. Mus.* **9:** 397–533. [21]

—— (1961). Les divisions du genre *Gryllus* basées sur l'étude de l'appareil copulateur (Orth. Gryllidae). *Eos, Madr.* **37:** 267–87. [21]

Christiansen, K. (1964). Bionomics of Collembola. *A. Rev. Ent.* **9:** 147–78. [10]

Claassen, P. W. (1931). *Plecoptera Nymphs of America North of Mexico*. Thomas Say Foundation Publ. No. 3, 199 pp. (Springfield, Ill.: Thomas). [20]

Clark, J. (1930). The Australian ants of the genus *Dolichoderus* (Formicidae). Subgenus *Hypoclinea* Mayr. *Aust. Zool.* **6:** 252–68. [37]

Clarke, J. F. G. (1955–65). *Catalogue of the Type Specimens of Microlepidoptera in the British Museum (Natural History) Described by Edward Meyrick*. Vol. I, 332 pp.; II, 531pp.; III, 600 pp.; IV, 521 pp.; V, 581 pp. (London: Br. Mus. nat. Hist.). [36]

Clausen, C. P. (1940). *Entomophagous Insects*, 688 pp. (New York: McGraw-Hill). [5, 30, 37]

Clay, T. (1949). Some problems in the evolution of a group of ectoparasites. *Evolution, Lancaster, Pa.* **3:** 279–99. [25]

Clements, A. N. (1963). *The Physiology of Mosquitoes*, 393 pp. (Oxford: Pergamon Press). [34]

Cobben, R. H. (1959). Notes on the classification of Saldidae with the description of a new species from Spain. *Zoöl. Meded., Leiden* **36:** 303–16. [26]

Coleman, E. (1928). Pollination of an Australian orchid by the male ichneumonid, *Lissopimpla semipunctata*, Kirby. *Trans. ent. Soc. Lond.* **76:** 533–9. [5]

—— (1938). Further observations on the pseudocopulation of the male *Lissopimpla semipunctata* Kirby (Hymenoptera parasitica) with the Australian orchid, *Cryptostylis leptochila* F. v. M. *Proc. R. ent. Soc. Lond.* (A) **13:** 82–3. [37]

Collart, A. (1938). Description d'un *Microsania* nouveau d'Australie (Diptera: Platypezidae). *Bull. Mus. r. Hist. nat. Belg.* **14:** 1–4. [34]

Colless, D. H. (1962a). A new Australian genus and family of Diptera (Nematocera: Perissommatidae). *Aust. J. Zool.* **10:** 519–35. [34]

—— (1962b). New species of *Ohakunea* Edwards and a related new genus with notes on the relationships of *Heterotricha* Loew. (Diptera). *Proc. Linn. Soc. N.S.W.* **87:** 303–8. [34]

—— (1962c). Notes on Australasian Tanyderidae, with description of a new species of *Radinoderus* Handl. (Diptera). *Proc. Linn. Soc. N.S.W.* **87:** 309–11. [34]

—— (1963). An Australian species of *Microphorella* (Diptera: Empididae), with notes on the phylogenetic significance of the genus. *Proc. Linn. Soc. N.S.W.* **88:** 320–3. [34]

—— (1964). Fossil Diptera and continental drift. *Nature, Lond.* **204:** 402–3. [9, 34]

—— (1965). An extraordinary Australian cecidomyiid, possibly related to the Nymphomyiidae (Diptera, Nematocera). *Proc. R. ent. Soc. Lond.* (B) **34:** 145–9. [34]

Common, I. F. B. (1948). The yellow-winged locust, *Gastrimargus musicus* Fabr., in central Queensland. *Qd J. agric. Sci.* **5:** 153–219. [21]

—— (1954). A study of the ecology of the adult bogong moth, *Agrotis infusa* (Boisd.) (Lepidoptera: Noctuidae), with special reference to its behaviour during migration and aestivation. *Aust. J. Zool.* **2:** 223–63. [5]

—— (1958a). The Australian cutworms of the genus *Agrotis* (Lepidoptera: Noctuidae). *Aust. J. Zool.* **6:** 69–88. [36]

—— (1958b). A revision of the pink bollworms of cotton (*Pectinophora* Busck (Lepidoptera: Gelechiidae)) and related genera in Australia. *Aust. J. Zool.* **6:** 268–306. [36]

—— (1960). A revision of the Australian stem borers hitherto referred to *Schoenobius* and *Scirpophaga* (Lepidoptera: Pyralidae, Schoenobiinae). *Aust. J. Zool.* **8:** 307–47. [36]

—— (1965). A revision of the Australian Tortricini, Schoenotenini, and Chlidanotini (Lepidoptera: Tortricidae: Tortricinae). *Aust. J. Zool.* **13:** 613–726. [36]

—— (1966). A new family of Bombycoidea (Lepidoptera) based on *Carthaea saturnioides* Walker from Western Australia. *J. ent. Soc. Qd* **5:** 29–36. [36]

Cooper, K. W. (1957). Biology of eumenine wasps. V. Digital communication in wasps. *J. exp. Zool.* **134:** 469–513. [37]

Corbet, P. S. (1962). *A Biology of Dragonflies*, 247 pp. (London: Witherby). [13]

Cott, H. B. (1940). *Adaptive Coloration in Animals*, 508 pp. (London: Methuen). [21]

Crampton, G. C. (1938). The structures called parameres in male insects. *Bull. Brooklyn ent. Soc.* **33:** 16–24. [1]

—— (1942). The external morphology of the Diptera. In Guide to the insects of Connecticut. Part VI. The Diptera or true flies of Connecticut, fasc. 1. *Bull. Conn. St. geol. nat. Hist. Surv.* **64:** 10–165. [34]

Crane, J. (1952). A comparative study of innate defensive behavior in Trinidad mantids (Orthoptera, Mantoidea). *Zoologica, N.Y.* **37:** 259–93. [16]

Cresson, E. T. (1948). A systematic annotated arrangement of the genera and species of the Indo-Australian Ephydridae (Diptera). II. The subfamily Notiphilinae and supplement to Part I on the subfamily Psilopinae. *Trans. Am. ent. Soc.* **74:** 1–28. [34]

Crosskey, R. W. (1953a). A revision of the genus *Hyptiogaster* Kieffer (Hymenoptera: Gasteruptionidae), with descriptions of two new genera and three new species. *Trans. R. ent. Soc. Lond.* **104:** 347–84. [37]

—— (1953b). Two new species of *Aulacostethus* Philippi, and a new species of *Aulacus* Jurine, from Australia, together with a key to the Australian species of *Aulacostethus* (Hymenoptera: Aulacidae). *Ann. Mag. nat. Hist.* (12) **6:** 758–66. [37]

—— (1960). A taxonomic study of the larvae of West African Simuliidae (Diptera: Nematocera) with comments on the morphology of the larval black-fly head. *Bull. Br. Mus. nat. Hist.* Ent. **10:** 1–74. [34]

—— (1965). A systematic revision of the Ameniinae (Diptera: Calliphoridae). *Bull. Br. Mus. nat. Hist.* Ent. **16:** 35–140. [34]

Crowson, R. A. (1955). *The Natural Classification of the Families of Coleoptera*, 187 pp. (London: Nathaniel Lloyd). [30]

—— (1960). The phylogeny of Coleoptera. *A. Rev. Ent.* **5:** 111–34. [7, 30]

—— (1962). Observations on the beetle family Cupedidae, with descriptions of two new fossil forms and a key to the recent genera. *Ann. Mag. nat. Hist.* (13) **5:** 147–57. [30]

—— (1964). A review of the classification of Cleroidea (Coleoptera), with descriptions of two new genera of Peltidae and of several new larval types. *Trans. R. ent. Soc. Lond.* **116:** 275–327. [30]

Curran, C. H. (1965). *The Families and Genera of North American Diptera*, 2nd edn, 515 pp. (Woodhaven, N.Y.: Tripp). [34]

Currie, G. A. (1932). Some notes on the biology and morphology of the immature stages of *Harpobittacus tillyardi* (order Mecoptera). *Proc. Linn. Soc. N.S.W.* **57:** 116–22. [32]

—— (1937). Galls on Eucalyptus trees. A new type of association between flies and nematodes. *Proc. Linn. Soc. N.S.W.* **62:** 147–74. [5, 34]

Da Cunha, A. B. (1960). Chromosomal variation and adaptation in insects. *A. Rev. Ent.* **5:** 85–110. [3]

Daly, H. V. (1964). Skeleto-muscular morphogenesis of the thorax and wings of the honey bee *Apis mellifera* (Hymenoptera: Apidae). *Univ. Calif. Publs Ent.* **39:** 1–77. [37]

Darlington, P. J. (1943). Carabidae of mountains and islands: data on the evolution of isolated faunas, and on atrophy of wings. *Ecol. Monogr.* **13:** 37–61. [30]

—— (1956). Australian carabid beetles. III. Notes on the Agonini. *Psyche, Camb.* **63:** 1–10. [30]

—— (1957). *Zoogeography: The Geographical Distribution of Animals*, 675 pp. (New York: Wiley). [9]

—— (1961). Australian carabid beetles. V. Transition of wet forest faunas from New Guinea to Tasmania. *Psyche, Camb.* **68:** 1–24. [30]

—— (1965). *Biogeography of the Southern End of the World*, 236 pp. (Cambridge, Mass.: Harvard University Press). [9]

Davey, K. G. (1965). *Reproduction in the Insects*, 96 pp. (Edinburgh: Oliver & Boyd). [2]

Davis, C. (1936–44). Studies in Australian Embioptera. Parts I–VII. *Proc. Linn. Soc. N.S.W.* **61:** 229–53, 254–8; **63:** 226–72; **65:** 155–60; **67:** 331–4; **68:** 65–6; **69:** 16–20. [23]

—— (1944a). Revision of the Embioptera of Western Australia. *J. Proc. R. Soc. West. Aust.* **28:** 139–47. [23]

Davis, N. T. (1966). Contributions to the morphology and phylogeny of the Reduvioidea (Hemiptera: Heteroptera). Part III. The male and female genitalia. *Ann. ent. Soc. Am.* **59:** 911–24. [26]

Day, M. F. (1950). The histology of a very large insect, *Macropanesthia rhinocerus* Sauss. (Blattidae). *Aust. J. scient. Res.* (B) **3:** 61–75. [14]

Deane, C. (1932a). Trichopterygidae of Australia and adjacent islands. *Proc. Linn. Soc. N.S.W.* **57:** 181–96. [30]

—— (1932b). New species of Corylophidae (Coleoptera). *Proc. Linn. Soc. N.S.W.* **57:** 332–7. [30]

Delany, M. J. (1957). Life histories in the Thysanura. *Acta zool. cracov.* **2:** 61–90. [11]

Deleurance, S. (1958). La contraction du cycle évolutif des Coléoptères Bathysciinae et Trechinae en milieu souterrain. *C. r. hebd. Séanc. Acad. Sci., Paris* **247:** 752–3. [30]

Demoll, R. (1918). *Der Flug der Insekten und der Vögel*, 67 pp. (Jena: Fischer). [30]

Demoulin, G. (1955). Ephéméroptères nouveaux ou rares du Chili. *Bull. Inst. r. Sci. nat. Belg.* **31** (22): 1–15. [12]

—— (1958). Nouveau schema de classification des Archodonates et des Ephéméroptères. *Bull. Inst. r. Sci. nat. Belg.* **34** (27): 1–19. [12]

Dempster, J. P. (1963). The population dynamics of grasshoppers and locusts. *Biol. Rev.* **38:** 490–529. [21]

Denis, R. (1949). Super-ordre des Ectotrophes. In P.-P. Grassé (ed.), *Traité de Zoologie: Anatomie, Systématique, Biologie*, Vol. IX, pp. 209–75 (Paris: Masson). [11]

Dethier, V. G. (1941). The antennae of lepidopterous larvae. *Bull. Mus. comp. Zool. Harv.* **87:** 455–507. [36]

—— (1963). *The Physiology of Insect Senses*, 266 pp. (London: Methuen). [2, 36]

Detinova, T. S. (1962). Age-grouping methods in Diptera of medical importance, with special reference to some vectors of malaria. *Monograph Ser. W.H.O.* **47:** 1–216. [34]

DIAKONOFF, A. (1961). Taxonomy of the higher groups of the Tortricoidea. *Trans. 11th int. Congr. Ent.* **1:** 124–6. [36]

DIBB, J. R. (1938). Synopsis of Australian Passalidae (Coleoptera). *Trans. R. ent. Soc. Lond.* **87:** 103–24. [30]

DIRSH, V. M. (1956). The phallic complex in Acridoidea (Orthoptera) in relation to taxonomy. *Trans. R. ent. Soc. Lond.* **108:** 223–356. [21]

—— (1961). A preliminary revision of the families and subfamilies of Acridoidea (Orthoptera, Insecta). *Bull. Br. Mus. nat. Hist.* Ent. **10:** 351–419. [21]

DOBROTWORSKY, N. V. (1965). *The Mosquitoes of Victoria (Diptera, Culicidae)*, 237 pp. (Melbourne University Press). [34]

DOBZHANSKY, T. (1951). *Genetics and the Origin of Species*, 3rd edn, 364 pp. (New York: Columbia University Press). [6]

DODD, A. P. (1927). The genus *Scelio* Latreille in Australia (Hymenoptera: Proctotrypoidea). *Proc. R. Soc. Qd* **38:** 127–75. [37]

—— (1930a). A revision of the Australian Teleasinae (Hymenoptera: Proctotrypoidea). *Proc. Linn. Soc. N.S.W.* **55:** 41–91. [37]

—— (1930b). The genus *Oxyscelio* Kieffer, its synonymy and species, with a description of one new genus (Hymenoptera: Proctotrypoidea). *Proc. R. Soc. Qd* **42:** 71–81. [37]

—— (1939). Hymenopterous parasites of Embioptera. *Proc. Linn. Soc. N.S.W.* **64:** 338–44. [37]

DOHERTY, R. L. (1964). A review of recent studies of arthropod-borne viruses in Queensland. *J. med. Ent.* **1:** 158–65. [5]

DÖRING, E. (1955). *Zur Morphologie der Schmetterlingseier*, 154 pp, (Berlin: Akademie-Verlag). [36]

DORSETT, D. A. (1962). Preparation for flight by hawk moths. *J. exp. Biol.* **39:** 579–88. [36]

DOWNES, J. A. (1958). The feeding habits of biting flies and their significance in classification. *A. Rev. Ent.* **3:** 249–66. [34]

DRAKE, C. J., and DAVIS, N. T. (1958). The morphology and systematics of the Piesmatidae (Hemiptera), with keys to world genera and American species. *Ann. ent. Soc. Am.* **51:** 567–81. [26]

——, and —— (1959). A new subfamily, genus, and species of Lygaeidae (Hemiptera–Heteroptera) from Australia. *J. Wash. Acad. Sci.* **49:** 19–26. [26]

——, and —— (1960). The morphology, phylogeny, and higher classification of the family Tingidae, including the description of a new genus and species of the subfamily Vianaidinae (Hemiptera: Heteroptera). *Entomologica am.* **39** (n.s.): 1–100. [26]

——, and HOBERLANDT, L. (1950). Catalogue of genera and species of Saldidae (Hemiptera). *Sb. ent. Odd. nár. Mus. Praze* **26** (376): 1–12. [26]

——, and RUHOFF, F. A. (1960). Lace-bug genera of the world (Hemiptera: Tingidae). *Proc. U.S. natn. Mus.* **112:** 1–107. [26]

——, and SLATER, J. A. (1957). The phylogeny and systematics of the family Thaumastocoridae (Hemiptera: Heteroptera). *Ann. ent. Soc. Am.* **50:** 353–70. [26]

DUCKHOUSE, D. A. (1965). Psychodidae (Diptera, Nematocera) of southern Australia, subfamilies Bruchomyiinae and Trichomyiinae. *Trans. R. ent. Soc. Lond.* **117:** 329–43. [34]

DUFFY, E. A. J. (1963). *A Monograph of the Immature Stages of Australasian Timber Beetles (Cerambycidae)*, 235 pp. (London: Br. Mus. nat. Hist.). [30]

DUMBLETON, L. J. (1940). Australian parasites of *Eriococcus coriaceus* Maskell. *N.Z. Jl Sci. Technol.* **22:** 102A–8A. [26]

—— (1952). A new genus of seed-infesting micropterygid moths. *Pacif. Sci.* **6:** 17–29. [36]

—— (1956). The Australian Aleyrodidae (Hemiptera–Homoptera). *Proc. Linn. Soc. N.S.W.* **81:** 159–83. [26]

—— (1963). The classification and distribution of the Simuliidae (Diptera) with particular reference to the genus *Austrosimulium*. *N.Z. Jl Sci.* **6:** 320–57. [34]

DUPORTE, E. M. (1957). The comparative morphology of the insect head. *A. Rev. Ent.* **2:** 55–70. [1]

EASTOP, V. F. (1952). A sound production mechanism in the Aphididae and the generic position of the species possessing it. *Entomologist* **85:** 57–61. [26]

—— (1966). A taxonomic study of Australian Aphidoidea (Homoptera). *Aust. J. Zool.* **14:** 399–592. [26]

EDMUNDS, G. F. (1962). The principles applied in determining the hierarchic level of the higher categories of Ephemeroptera. *Syst. Zool.* **11:** 22–31. [12]

——, and TRAVER, J. R. (1954). An outline of a reclassification of the Ephemeroptera. *Proc. ent. Soc. Wash.* **56:** 236–40. [12]

EDWARD, D. H. D. (1963). The biology of a parthenogenetic species of *Lundstroemia* (Diptera: Chironomidae), with descriptions of the immature stages. *Proc. R. ent. Soc. Lond.* (A) **38:** 165–70. [3, 34]

EDWARDS, J. S. (1961). The action and composition of the saliva of an assassin bug *Platymeris rhadamanthus* Gaerst. (Hemiptera, Reduviidae). *J. exp. Biol.* **38:** 61–77. [26]

EHRLICH, P. R. (1958). The comparative morphology, phylogeny and higher classification of the butterflies (Lepidoptera: Papilionoidea). *Kans. Univ. Sci. Bull.* **39:** 305–70. [36]

——, and DAVIDSON, S. E. (1961). The internal anatomy of the monarch butterfly, *Danaus plexippus* L. (Lepidoptera: Nymphalidae). *Microentomology* **24:** 87–133. [2, fig.]

EISNER, T. (1957). A comparative morphological study of the proventriculus of ants (Hymenoptera, Formicidae). *Bull. Mus. comp. Zool. Harv.* **116:** 439–90. [37]

——, and HAPP, G. M. (1962). The infrabuccal pocket of a formicine ant: a social filtration device. *Psyche, Camb.* **69:** 107–16. [37]

EMDEN, F. I. VAN (1946). Egg-bursters in some more families of polyphagous beetles and some general remarks on egg-bursters. *Proc. R. ent. Soc. Lond.* (A) **21:** 89–97. [4]

EMERSON, A. E. (1955). Geographical origins and dispersions of termite genera. *Fieldiana, Zool.* **37:** 465–521. [15]

—— (1960). New genera of termites related to *Subulitermes* from the Oriental, Malagasy, and Australian regions (Isoptera, Termitidae, Nasutitermitinae). *Am. Mus. Novit.* **1986:** 1–28. [15]

EMERY, C. (1910–25). Hymenoptera. Fam. Formicidae. *Genera Insect.*: 1910, subfam. Dorylinae, **102:** 1–34; 1911, subfam. Ponerinae, **118:** 1–125; 1912, subfam. Dolichoderinae, **137:** 1–50; 1921–22, subfam. Myrmicinae, **174:** 1–397; 1925, subfam. Formicinae, **183:** 1–302. [37]

English, K. M. I. (1947). Notes on the morphology and biology of *Apiocera maritima* Hardy (Diptera, Apioceridae). *Proc. Linn. Soc. N.S.W.* **71:** 296–302. [34]
—— (1950). Notes on the morphology and biology of *Anabarrhynchus fasciatus* Macq. and other Australian Therevidae (Diptera, Therevidae). *Proc. Linn. Soc. N.S.W.* **75:** 345–59. [34]
Ergene, S. (1953). Homochrome Farbanpassungen bei *Mantis religiosa*. *Z. vergl. Physiol.* **35:** 36–41. [16]
Esben-Petersen, P. (1926–9). Australian Neuroptera. Parts V(*sic*)–VI. *Qd Nat.* **6:** 11–12; **7:** 31–5. [29]
Ettershank, G. (1966). A generic revision of the world Myrmicinae related to *Solenopsis* and *Pheidologeton* (Hymenoptera: Formicidae). *Aust. J. Zool.* **14:** 73–171. [37]
Evans, J. W. (1931). Notes on the biology and morphology of the Eurymelinae (Cicadelloidea, Homoptera). *Proc. Linn. Soc. N.S.W.* **56:** 210–26. [26, 34]
—— (1939). The morphology of the thorax of the Peloridiidae (Homopt.). *Proc. R. ent. Soc. Lond.* (B) **8:** 143–50. [26]
—— (1940). Tube-building cercopids (Homoptera, Machaerotidae). *Trans. R. Soc. S. Aust.* **64:** 70–5. [26]
—— (1941). The morphology of *Tettigarcta tomentosa* White, (Homoptera, Cicadidae). *Pap. Proc. R. Soc. Tasm.* **1940:** 35–49. [26]
—— (1942). The morphology of *Nannochorista maculipennis* Tillyard (Mecoptera). *Trans. R. Soc. S. Aust.* **66:** 218–25. [32]
—— (1946–7). A natural classification of leaf-hoppers (Homoptera, Jassoidea). Parts 2, 3. *Trans. R. ent. Soc. Lond.* **97:** 39–54; **98:** 105–271. [26]
—— (1956). Palaeozoic and Mesozoic Hemiptera (Insecta). *Aust. J. Zool.* **4:** 165–258. [8, fig.]
—— (1963). The phylogeny of the Homoptera. *A. Rev. Ent.* **8:** 77–94. [8, 26]
—— (1964). The periods of origin and diversification of the superfamilies of the Homoptera–Auchenorhyncha (Insecta) as determined by a study of the wings of Palaeozoic and Mesozoic fossils. *Proc. Linn. Soc. Lond.* **175:** 171–81. [8, 26]
—— (1966). The leafhoppers and froghoppers of Australia and New Zealand (Homoptera: Cicadelloidea and Cercopoidea). *Mem. Aust. Mus.* **12:** 1–347. [26]
Evans, M. E. G. (1961). On the muscular and reproductive systems of *Atomaria ruficornis* (Marsham) (Coleoptera, Cryptophagidae). *Trans. R. Soc. Edinb.* **64:** 297–399. [30]
Ezzat, Y. M., and McConnell, H. S. (1956). A classification of the mealybug tribe Planococcini (Coccoidea–Homoptera). *Bull. Md agric. Exp. Stn* **A-84:** 3–108. (26, fig.]
Fennah, R. G. (1950). A generic revision of Achilidae (Homoptera: Fulgoroidea) with descriptions of new species. *Bull. Br. Mus. nat. Hist.* Ent. **1:** 3–170. [26]
—— (1952). On the generic classification of Derbidae (Fulgoroidea), with descriptions of new Neotropical species. *Trans. R. ent. Soc. Lond.* **103:** 109–70. [26]
—— (1954). The higher classification of the family Issidae (Homoptera: Fulgoroidea) with descriptions of new species. *Trans. R. ent. Soc. Lond.* **105:** 455–74. [26]
—— (1956). Homoptera: Fulgoroidea. *Insects Micronesia* **6** (3): 1–211. [26]
—— (1965). Delphacidae from Australia and New Zealand. Homoptera: Fulgoroidea. *Bull. Br. Mus. nat. Hist.* Ent. **17:** 3–59. [26]
Fenner, F., and Ratcliffe, F. N. (1965). *Myxomatosis*, 379 pp. (Cambridge University Press). [5, 33, 34]
Ferguson, E. W. (1909–23). Revision of the Amycterides (Coleoptera). Parts I–VIII. *Proc. Linn. Soc. N.S.W.* **34:** 524–85; **37:** 83–135; **38:** 340–94; **39:** 217–52; **40:** 685–718, 759–805; **41:** 422–52; **46:** 19–75, 393–406; **48:** 381–435. [30]
Ferris, G. F. (1935). The prothoracic pleurites of Coleoptera. *Ent. News* **46:** 63–8. [30]
—— (1940a). The morphology of *Plega signata* (Hagen) (Neuroptera: Mantispidae). *Microentomology* **5:** 33–56. [30]
—— (1940b). The myth of the thoracic sternites of insects. *Microentomology* **5:** 87–90. [30]
—— (1951). The sucking lice. *Mem. Pacif. Cst ent. Soc.* **1:** 1–320. [25]
—— (1952). Some miscellaneous Coccoidea (Insecta: Homoptera). *Microentomology* **17:** 1–5. [26, fig.]
—— (1957a). Notes on some little known genera of the Coccoidea (Homoptera). *Microentomology* **22:** 59–79. [26]
—— (1957b). A review of the family Eriococcidae (Insecta: Coccoidea). *Microentomology* **22:** 81–9. [26]
Fisher, K. (1932). *Agriotypus armatus* (Walk.) (Hymenoptera) and its relations with its hosts. *Proc. zool. Soc. Lond.* **1932:** 451–61. [37]
Flanders, S. E. (1962). The parasitic Hymenoptera: specialists in population regulation. *Can. Ent.* **94:** 1133–47. [37]
Flower, J. W. (1964). On the origin of flight in insects. *J. Insect Physiol.* **10:** 81–8. [7]
Forbes, W. T. M. (1924, 1948, 1954, 1960). Lepidoptera of New York and neighbouring States. Parts I–IV. *Mem. Cornell Univ. agric. Exp. Stn* **68** (1923): 1–729; **274:** 1–263; **329:** 1–433; **371:** 1–188. [36]
—— (1926). The wing folding patterns of the Coleoptera. *Jl N.Y. ent. Soc.* **34:** 42–68, 91–138. [30]
—— (1933). The pupa of *Hyblaea* (Lepidoptera, Hyblaeidae). *Ann. ent. Soc. Am.* **26:** 490. [36]
—— (1955). The subdivision of the Eupterotidae (Lepidoptera). *Tijdschr. Ent.* **98:** 85–132. [36]
Fowler, W. W. (1912). *The Fauna of British India, Including Ceylon and Burma.* Coleoptera. General Introduction and Cicindelidae and Paussidae, 529 pp. (London: Taylor & Francis). [30]
Fracker, S. B. (1915). The classification of lepidopterous larvae (2nd edn, 1930). *Illinois biol. Monogr.* **2** (1): 1–161. [36]
Fraser, F. C. (1957). *A Reclassification of the Order Odonata*, 133 pp. (Sydney: R. zool. Soc. N.S.W.). [13]
—— (1960). *A Handbook of the Dragonflies of Australasia with Keys for the Identification of all Species*, 67 pp. (Sydney: R. zool. Soc. N.S.W.). [13]
Freeman, P. (1961). The Chironomidae (Diptera) of Australia. *Aust. J. Zool.* **9:** 611–737. [34]
Frisch, K. von (1954). *The Dancing Bees: An Account of the Life and Senses of the Honey Bee*, translated by D. Ilse, 182 pp. (New York: Harcourt, Brace). [2, 5]

FRISON, T. H. (1942). Studies of North American Plecoptera with special reference to the fauna of Illinois. *Bull. Ill. St. nat. Hist. Surv.* **22:** 235–355. [20, fig.]

FRIZZI, G., and HOLSTEIN, M. (1956). Étude cytogénétique d'*Anopheles gambiae*. *Bull. Wld Hlth Org.* **15:** 425–35. [34]

FROGGATT, W. W. (1906). Notes on the hymenopterous genus *Megalyra* Westw., with descriptions of new species. *Proc. Linn. Soc. N.S.W.* **31:** 399–407 [37]

—— (1921). A descriptive catalogue of the scale insects ('Coccidae') of Australia. Part ii. *Sci. Bull. Dep. Agric. N.S.W.* **18:** 1–159. [26]

FULLER, M. E. (1933). The life history of *Onesia accepta* Malloch (Diptera, Calliphoridae). *Parasitology* **25:** 342–52. [34]

—— (1934a). The early stages of *Sciadocera rufomaculata* White (Dipt. Phoridae). *Proc. Linn. Soc. N.S.W.* **59:** 9–15. [34]

—— (1934b). The insect inhabitants of carrion: a study in animal ecology. *Bull. Coun. scient. ind. Res., Melb.* **82:** 1–62. [5]

—— (1935). Notes on Australasian Anisopodidae (Diptera). *Proc. Linn. Soc. N.S.W.* **60:** 291–302. [34]

GANGLBAUER, L. (1903). Die neueren Classification en der Koleopteren nach Sharp, Lameere und Kolbe. *Münch. koleopt. Z.* **1:** 271–319. [30]

GAY, F. J. (ed.) (1966). Scientific and common names of insects and allied forms occurring in Australia. *Bull. Commonw. scient. ind. Res. Org.* **285:** 1–52. [Intro.]

GEITLER, L. (1953). Endomitose und endomitotische Polyploidisierung. *Protoplasmatologia* **6** (C): 1–89. [3]

GHAURI, M. S. K. (1962). *The Morphology and Taxonomy of Male Scale Insects (Homoptera: Coccoidea)*, 221 pp. (London: Br. Mus. nat. Hist.). [26]

GIGLIO-TOS, E. (1927). Orthoptera: Mantidae. *Tierreich* **50:** 1–707. [16]

GILBY, A. R. (1965). Lipids and their metabolism in insects. *A. Rev. Ent.* **10:** 141–60. [2]

——, and WATERHOUSE, D. F. (1965). The composition of the scent of the green vegetable bug, *Nezara viridula*. *Proc. R. Soc.* (B) **162:** 105–20. [26]

GILES, E. T. (1963). The comparative external morphology and affinities of the Dermaptera. *Trans. R. ent. Soc. Lond.* **115:** 95–164. [7, 19]

—— (1965). The alimentary canal of *Anisolabis littorea* (White) (Dermaptera: Labiduridae), with special reference to the peritrophic membrane. *Trans. R. Soc. N.Z.* (Zool.) **6:** 87–101. [19]

GILMOUR, D. (1965). *The Metabolism of Insects*, 195 pp. (Edinburgh: Oliver & Boyd). [2]

GIRAULT, A. A. (1912–16). Australian Hymenoptera Chalcidoidea. Parts I–XIV, General Supplement. *Mem. Qd Mus.* **1:** 66–189; **2:** 101–334; **3:** 142–346; **4:** 1–365; **5:** 205–30. [37]

GISIN, H. (1960). *Collembolenfauna Europas*, 312 pp. (Geneva: Mus. Hist. nat.). [10]

GIVEN, B. B. (1954a). A catalogue of the Thynninae (Tiphiidae, Hymenoptera) of Australia and adjacent areas. *Bull. N.Z. Dep. scient. ind. Res.* **109:** 1–89. [37]

—— (1954b). Evolutionary trends in the Thynninae (Hymenoptera: Tiphiidae) with special reference to feeding habits of Australian species. *Trans. R. ent. Soc. Lond.* **105:** 1–10. [37]

GOODCHILD, A. J. P. (1966). Evolution of the alimentary canal in the Hemiptera. *Biol. Rev.* **41:** 97–140. [26]

GRÄBNER, K.-E. (1954). Vergleichend morphologische und physiologische Studien an Anobiiden- und Cerambyciden-Symbionten. *Z. Morph. Ökol. Tiere* **2:** 471–528. [30]

GRANDI, G. (1961). The hymenopterous insects of the superfamily Chalcidoidea developing within the receptacles of figs. Their life-history, symbioses and morphological adaptations. *Boll. Ist. Ent. Univ. Bologna* **26:** I–XIII. [37]

GRAY, E. G. (1960). The fine structure of the insect ear. *Phil. Trans. R. Soc.* (B) **243:** 75–94. [2, fig.]

GREAVES, T. (1947). The control of silverfish and the German cockroach. *J. Coun. scient. ind. Res. Aust.* **20:** 425–33. [11]

GRESSITT, J. L. (1958). Zoogeography of insects. *A. Rev. Ent.* **3:** 207–30. [9]

—— (1961). Problems in the zoogeography of Pacific and Antarctic insects. *Pacif. Insects Monogr.* **2:** 1–94. [5, 9, 34]

—— (ed.) (1963). *Pacific Basin Biogeography. A Symposium*, 563 pp. (Honolulu: Bishop Museum Press). [9]

GROSS, G. F. (1950). The stilt-bugs (Heteroptera–Neididae) of the Australian and New Zealand regions. *Rec. S. Aust. Mus.* **9:** 313–26. [26]

—— (1951). On three new genera and species of Schizopterinae (Heteroptera–Cryptostemmatidae) from Australia. *Rec. S. Aust. Mus.* **9:** 535–44. [26]

—— (1954–7). A revision of the flower bugs (Heteroptera Anthocoridae) of the Australian and adjacent Pacific regions. Parts 1–3. *Rec. S. Aust. Mus.* **11:** 129–64, 409–22; **13:** 131–42. [26]

—— (1960). A revision of the genus *Leptocoris* Hahn (Heteroptera: Coreidae: Rhopalinae) from the Indo-Pacific and Australian regions. *Rec. S. Aust. Mus.* **13:** 403–51. [26]

—— (1965). A revision of the Australian and New Guinea Drymini (Heteroptera–Lygaeidae). *Rec. S. Aust. Mus.* **15:** 39–78. [26]

——, and SCUDDER, G. G. E. (1963). The Australian Rhyparochromini (Hemiptera: Lygaeidae). *Rec. S. Aust. Mus.* **14:** 427–69. [26]

GUNTHER, K (1953). Über die taxonomische Gliederung und die geographische Verbreitung der Insektenordnung der Phasmatodea. *Beitr. Ent.* **3:** 541–63. [22]

GUPTA, A. P. (1963). A consideration of the systematic position of the Saldidae and Mesoveliidae (Hemiptera: Heteroptera). *Proc. ent. Soc. Wash.* **65:** 31–8. [26]

GURNEY, A. B. (1938). A synopsis of the order Zoraptera, with notes on the biology of *Zorotypus hubbardi* Caudell. *Proc. ent. Soc. Wash.* **40:** 57–87. [17]

—— (1947). Notes on some remarkable Australasian walkingsticks, including a synopsis of the genus *Extatosoma* (Orthoptera: Phasmatidae). *Ann. ent. Soc. Am.* **40:** 373–96. [22]

—— (1948). The taxonomy and distribution of the Grylloblattidae (Orthoptera). *Proc. ent. Soc. Wash.* **50:** 86–102. [18]

—— (1953). Notes on the biology and immature stages of a cricket parasite of the genus *Rhopalosoma*. *Proc. U.S. natn. Mus.* **103:** 19–34. [37]

—— (1961). Further advances in the taxonomy and distribution of the Grylloblattidae (Orthoptera) *Proc. biol. Soc. Wash.* **74:** 67–76. [18]

HAAF, E. (1954). Die australischen Arten der Gattung *Trox* (Col. Scarab.). 3. Beitrag zur Kenntnis der Subfam. Troginae. *Ent. Arb. Mus. Georg Frey* **5:** 691–740. [30]

HACKER, H. (1924). Field notes on *Platybrachys*, etc. (Homoptera). *Mem. Qd Mus.* **8:** 37–42. [26]

—— (1925). The life history of *Oliarus felis* Kirk. (Homoptera). *Mem. Qd Mus.* **8:** 113–14. [26]

HACKMAN, R. H., and GOLDBERG, M. (1960). Composition of the oothecae of three Orthoptera. *J. Insect Physiol.* **5:** 73–8. [16]

HADLINGTON, P., and SHIPP, E. (1961). Diapause and parthenogenesis in the eggs of three species of Phasmatodea. *Proc. Linn. Soc. N.S.W.* **86:** 268–79. [22]

HAGAN, H. R. (1951). *Embryology of the Viviparous Insects*, 472 pp. (New York: Ronald Press). [4]

HAGEN, K. S. (1962). Biology and ecology of predacious Coccinellidae. *A. Rev. Ent.* **7:** 289–326. [30]

HALE, H. M. (1924). Studies in Australian aquatic Hemiptera. No. IV. The corixid genus *Diaprepocoris. Trans. R. Soc. S. Aust.* **48:** 7–9. [26]

—— (1925). Results of Dr. E. Mjöberg's Swedish scientific expeditions to Australia 1910–1913. 44. The aquatic and semi-aquatic Hemiptera. *Ark. Zool.* **17A** (20): 1–19. [26]

—— (1926). Studies in Australian aquatic Hemiptera No. VII. *Rec. S. Aust. Mus.* **3:** 195–217. [26]

HAMILTON-SMITH, E., and ADAMS, D. J. H. (1966). The alleged obligate ectoparasitism of *Myotyphlus jansoni* (Matthews) (Coleoptera: Staphylinidae). *J. ent. Soc. Qd* **5:** 44–5. [30]

HARDY, D. E. (1951). Studies in Pacific Bibionidae (Diptera). Part II: Genus *Philia* Meigen. *Proc. Hawaii. ent. Soc.* **14:** 257–75. [34]

—— (1958). The *Plecia* of the Pacific and Southeast Asia (Bibionidae–Diptera). *Pacif. Sci.* **12:** 185–220. [34]

—— (1962). A remarkable new bibionid fly from Australia (Diptera: Bibionidae). *Pacif. Insects* **4:** 783–5. [34]

—— (1964). A re-study of the Perkins types of Australian Pipunculidae (Diptera) and the type of *Pipunculus vitiensis* Muir from Fiji. *Aust. J. Zool.* **12:** 84–125. [34]

HARDY, G. H. (1930). Australian Empididae. *Aust. Zool.* **6:** 237–50. [34]

—— (1933a). Notes on Australian Syrphinae (Diptera). *Proc. R. Soc. Qd* **45:** 12–19. [34]

—— (1933b). Miscellaneous notes on Australian Diptera. I. *Proc. Linn. Soc. N.S.W.* **58:** 408–20. [34]

—— (1934). Miscellaneous notes on Australian Diptera. II. *Proc. Linn. Soc. N.S.W.* **59:** 173–8. [34]

—— (1940). Notes on Australian Muscoidea, V. Calliphoridae. *Proc. R. Soc. Qd* **51:** 133–46. [34]

—— (1941). Miscellaneous notes on Australian Diptera. VIII. *Proc. Linn. Soc. N.S.W.* **66:** 223–33. [34]

—— (1958). The Diptera of Katoomba. Part 2. Leptidae and Dolichopodidae. *Proc. Linn. Soc. N.S.W.* **83:** 291–302. [34]

—— (1959). Diptera of Katoomba. Part 3. Stratiomyiidae and Tachinidae. *Proc. Linn. Soc. N.S.W.* **84:** 209–17. [34]

HARKER, J. E. (1950). Australian Ephemeroptera. Part I. Taxonomy of New South Wales species and evaluation of taxonomic characters. *Proc. Linn. Soc. N.S.W.* **75:** 1–34. [12]

—— (1954). The Ephemeroptera of eastern Australia. *Trans. R. ent. Soc. Lond.* **105:** 241–68. [12]

HARRIS, W. V. (1961). *Termites: Their Recognition and Control*, 187 pp. (London: Longmans). [15]

HARRISON, L. (1915). The respiratory system of Mallophaga. *Parasitology* **8:** 101–27. [25]

—— (1928). The composition and origins of the Australian fauna, with special reference to the Wegener hypothesis. *Rep. Australas. Ass. Advmt Sci.* **18:** 332–96. [9]

HASKELL, P. T. (1962). Sensory factors influencing phase change in locusts. *Colloques int. Cent. natn. Rech. scient.* **114:** 145–63. [5]

HEBARD, M. (1943). Australian Blattidae of the subfamilies Chorisoneurinae and Ectobiinae (Orthoptera). *Monogr. Acad. nat. Sci. Philad.* **4:** 1–129. [14]

HEDICKE, H. (1939). *Hymenopterorum Catalogus.* Pars 9: Evaniidae, 50 pp. ('s-Gravenhage: Junk). [37]

——, and KERRICH, G. J. (1940). A revision of the family Liopteridae (Hymenopt., Cynipoidea). *Trans. R. ent. Soc. Lond.* **90:** 177–225. [37]

HEINRICH, G. H. (1961). Synopsis of Nearctic Ichneumoninae Stenopneusticae with particular reference to the northeastern region (Hymenoptera). Part I. *Can. Ent.* **92,** Suppl. 15: 1–87. [37]

HENNIG, W. (1948–52). *Die Larvenformen der Dipteren.* Vol. 1 (1948), 186 pp.; 2 (1950), 458 pp.; 3 (1952), 628 pp. (Berlin: Akademie-Verlag). [34]

—— (1954). Flügelgeäder und System der Dipteren unter Berücksichtigung der aus dem Mesozoikum beschriebenen Fossilien. *Beitr. Ent.* **4:** 245–388. [8, 34]

—— (1957). Systematik und Phylogenese. In *Bericht über die Hundertjahrfeier der Deutschen Entomologischen Gesellschaft Berlin 30. September bis 5. Oktober 1956*, pp. 50–71. (Berlin: Akademie-Verlag). [6]

—— (1958). Die Familien der Diptera Schizophora und ihre phylogenetischen Verwandtschaftbeziehungen. *Beitr. Ent.* **8:** 505–688. [34]

—— (1960). Die Dipteren-Fauna von Neuseeland als systematisches und tiergeographisches Problem. *Beitr. Ent.* **10:** 221–329. [9, 34]

—— (1962). Veränderungen am phylogenetischen System der Insekten seit 1953. In *Bericht über die 9. Wanderversammlung Deutscher Entomologen 6. bis 8. Juni 1961 in Berlin.* Tagungsberichte No. 45, pp. 29–42. (Deutsche Akademie der Landwirtschaftwissenschaften zu Berlin). [7]

—— (1964). Die Dipteren-Familie Sciadoceridae im Baltischen Bernstein (Diptera: Cyclorrhapha Aschiza). *Stuttg. Beitr. Naturk.* **127:** 1–10. [34]

—— (1965). Phylogenetic systematics. *A. Rev. Ent.* **10:** 97–116. [6]

HEQVIST, K.-J. (1961). Notes on Cleonymidae (Hym. Chalcidoidea). I. *Ent. Tidskr.* **82:** 91–110. [37]

HESS, W. N. (1922). Origin and development of the light-organs of *Photurus pennsylvanica* de Geer. *J. Morph.* **36:** 245–77. [2, fig.]

HEYMONS, R. (1929). Über die Biologie der Passaluskäfer. *Z. Morph. Ökol. Tiere* **16:** 74–100. [30]

HICKMAN, V. V. (1965). On *Planarivora insignis* gen. et sp. n. (Diptera: Mycetophilidae), whose larval stages are parasitic in land planarians. *Pap. Proc. R. Soc. Tasm.* **99**: 1–8. [34]

HIGHNAM, K. C. (ed.) (1964). *Insect Reproduction*, 120 pp. (London: Symposium No. 2, R. ent. Soc. Lond.). [1, 30]

HILL, G. F. (1942). *Termites (Isoptera) from the Australian Region*, 479 pp. (Melbourne: Coun. scient. ind. Res. Aust.). [15]

HILLE RIS LAMBERS, D. (1966). Polymorphism in Aphididae. *A. Rev. Ent.* **11**: 47–78. [26]

HINCKS, W. D. (1954). Report from Professor T. Gislén's expedition to Australia in 1951–1952. 8. Dermaptera. *Acta Univ. lund.* **50** (4): 1–10. [19]

HINTON, H. E. (1939a). An inquiry into the natural classification of the Dryopoidea, based partly on a study of their internal anatomy (Col.). *Trans. R. ent. Soc. Lond.* **89**: 133–84. [30]

—— (1939b). A contribution to a classification of the Limnichidae (Coleoptera). *Entomologist* **72**: 181–6. [30]

—— (1941). The Ptinidae of economic importance. *Bull. ent. Res.* **31**: 331–81. [30]

—— (1944). Some general remarks on sub-social beetles, with notes on the biology of the staphylinid, *Platystethus arenarius* (Fourcroy). *Proc. R. ent. Soc. Lond.* (A) **19**: 115–28. [30]

—— (1945). *A Monograph of the Beetles Associated with Stored Products*, Vol. 1, 443 pp. (London: Br. Mus. nat. Hist.). [30]

—— (1946a). On the homology and nomenclature of the setae of the lepidopterous larvae, with some notes on the phylogeny of the Lepidoptera. *Trans. R. ent. Soc. Lond.* **97**: 1–37. [36]

—— (1946b). A new classification of insect pupae. *Proc. zool. Soc. Lond.* **116**: 282–328. [4, 30, 34]

—— (1946c). Concealed phases in the metamorphosis of insects. *Nature, Lond.* **157**: 552–3. [4]

—— (1947a). On the reduction of functional spiracles in the aquatic larvae of the Holometabola, with notes on the moulting process of spiracles. *Trans. R. ent. Soc. Lond.* **98**: 449–73. [30]

—— (1947b). The dorsal cranial areas of caterpillars. *Ann. Mag. nat. Hist.* (11) **14**: 843–52. [36]

—— (1951). Myrmecophilous Lycaenidae and other Lepidoptera—a summary. *Proc. Trans. S. Lond. ent. nat. Hist. Soc.* **1949–50**: 111–75. [36]

—— (1952). The structure of the larval prolegs of the Lepidoptera and their value in the classification of the major groups. *Lepid. News* **6**: 1–6. [36]

—— (1953). Some adaptations of insects to environments that are alternately dry and flooded, with some notes on the habits of the Stratiomyidae. *Trans. Soc. Br. Ent.* **11**: 209–27. [5]

—— (1955a). Protective devices of endopterygote pupae. *Trans. Soc. Br. Ent.* **12**: 49–92. [30]

—— (1955b). On the respiratory adaptations, biology, and taxonomy of the Psephenidae, with notes on some related families (Coleoptera). *Proc. zool. Soc. Lond.* **125**: 543–68. [30]

—— (1957). Some little known respiratory adaptations. *Sci. Prog., Lond.* **45**: 692–700. [2]

—— (1958a). The phylogeny of the panorpoid orders. *A. Rev. Ent.* **3**: 181–206. [1, 7, 32, 36]

—— (1958b). The pupa of the fly *Simulium* feeds and spins its own cocoon. *Entomologist's mon. Mag.* **94**: 14–16. [4, 34]

—— (1958c). Concealed phases in the metamorphosis of insects. *Sci. Prog., Lond.* **46**: 260–75. [4, 5, 34]

—— (1959). How the indirect flight muscles of insects grow. *Sci. Prog., Lond.* **47**: 321–33. [4]

—— (1960a). The ways in which insects change colour. *Sci. Prog., Lond.* **48**: 341–50. [5]

—— (1960b). Cryptobiosis in the larva of *Polypedilum vanderplanki* Hint. (Chironomidae). *J. Insect Physiol.* **5**: 286–300. [5]

—— (1962a). The structure and function of the spiracular gills of *Deuterophlebia* (Deuterophlebiidae) in relation to those of other Diptera. *Proc. zool. Soc. Lond.* **138**: 111–22. [34]

—— (1962b). Respiratory systems of insect egg-shells. *Sci. Prog., Lond.* **50**: 96–113. [2, 4]

—— (1963a). The ventral ecdysial lines of the head of endopterygote larvae. *Trans. R. ent. Soc. Lond.* **115**: 39–61. [1, 30]

—— (1963b). The origin and function of the pupal stage. *Proc. R. ent. Soc. Lond.* (A) **38**: 77–85. [4]

—— (1965). A revision of the Australian species of *Austrolimnius* (Coleoptera: Elmidae). *Aust. J. Zool.* **13**: 97–172. [9, 30]

—— (1966). The spiracular gill of the fly *Eutanyderus* (Tanyderidae). *Aust. J. Zool.* **14**: 365–9. [34]

—— (1967). The respiratory system of the egg-shell of the common housefly. *J. Insect Physiol.* **13**: 647–51. [4, fig.]

HODGES, R. W. (1962). A revision of the Cosmopterigidae of America north of Mexico, with a definition of the Momphidae and Walshiidae (Lepidoptera: Gelechioidea). *Entomologica am.* (n.s.) **42**: 1–171. [36]

HOLLAND, G. P. (1964). Evolution, classification, and host relationships of Siphonaptera. *A. Rev. Ent.* **9**: 123–46. [33]

HOLLOWAY, B. A. (1960). Taxonomy and phylogeny in the Lucanidae (Insecta: Coleoptera). *Rec. Dom. Mus., Wellington* **3**: 321–65. [30]

HOPKINS, G. H. E. (1949). The host-associations of the lice of mammals. *Proc. zool. Soc. Lond.* **119**: 387–604. [25]

——, and CLAY, T. (1952). *A Check List of the Genera & Species of Mallophaga*, 362 pp. (London: Br. Mus. nat. Hist.). [25]

——, and ROTHSCHILD, M. (1953–66). *An Illustrated Catalogue of the Rothschild Collection of Fleas (Siphonaptera) in the British Museum (Natural History)*. Vol. I, 361 pp.; II, 445 pp.; III, 560 pp.; IV, 549 pp. (London: Br. Mus. nat. Hist.). [33]

HOWDEN, H. F. (1954). Notes on Australian beetles in the tribe Bolboceratini formerly in the genus *Bolboceras*. *Proc. Linn. Soc. N.S.W.* **79**: 142–4. [30]

HOY, J. M. (1962). Eriococcidae (Homoptera: Coccoidea) of New Zealand. *Bull. N.Z. Dep. scient. ind. Res.* **146**: 1–219. [26]

HUGHES, R. D. (1963). Population dynamics of the cabbage aphid, *Brevicoryne brassicae* (L.). *J. Anim. Ecol.* **32**: 393–424. [26]

HUGHES-SCHRADER, S. (1948). Cytology of coccids (Coccoïdea–Homoptera). *Adv. Genet.* **2**: 127–203. [3]

—— (1959). On the cytotaxonomy of phasmids (Phasmatodea). *Chromosoma* **10**: 268–77. [22]

HULL, F. M. (1962). Robber flies of the world. The genera of the family Asilidae. *Bull. U.S. natn. Mus.* **224**: 1–907. [34]

HUNGERFORD, H. B., and MATSUDA, R. (1960). Keys to subfamilies, tribes, genera and subgenera of the Gerridae of the world. *Kans. Univ. Sci. Bull.* **41**: 3–23. [26]

HUXLEY, J. S. (1958). Evolutionary processes and taxonomy with special reference to grades. *Uppsala Univ. Årsskr.* **1958** (6): 21–39. [7]

ILLIES, J. (1960). Archiperlaria, eine neue Unterordnung der Plecopteren. (Revision der familien Eustheniidae und Diamphipnoidae) (Plecoptera). *Beitr. Ent.* **10**: 661–97. [20]

—— (1965). Phylogeny and zoogeography of the Plecoptera. *A. Rev. Ent.* **10**: 117–40. [9, 20]

International Commission for Zoological Nomenclature (1963). Opinion 678. The suppression under the plenary powers of the pamphlet published by Meigen, 1800. *Bull. zool. Nom.* **20**: 339–42. [34]

International Congresses of Zoology (1961–4). *International Code of Zoological Nomenclature adopted by the XV International Congress of Zoology*, 176 pp. (London: International Trust for Zoological Nomenclature). [6]

JACKSON, D. (1928). The inheritance of long and short wings in the weevil *Sitona hispidula*, with a discussion of wing reduction among beetles. *Trans. R. Soc. Edinb.* **55**: 665–735. [30]

—— (1956). Observations on flying and flightless water beetles. *J. Linn. Soc.* Zool. **43**: 18–42. [30]

JEANNEL, R. (1938). Les Migadopides (Coleoptera Adephaga), une lignée subantarctique. *Revue fr. Ent.* **5**: 1–55. [30]

—— (1940). Croisière du Bougainville aux îles australes françaises. III. Coléoptères. *Mém. Mus. natn. Hist. nat., Paris* **14**: 63–201. [30]

—— (1941). Les hénicocéphalides. Monographie d'un groupe d'hémiptères hématophages. *Annls Soc. ent. Fr.* **110**: 273–368. [26]

—— (1945). Sur la position systématique des Strepsiptères. *Revue fr. Ent.* **11**: 111–18. [30]

—— (1946). Coléoptères carabiques de la région malgache (première partie). *Faune Emp. fr.* **6**: 1–372. [30]

—— (1949a). Classification et phylogénie des insectes, *and* Les insectes fossiles. In P.-P. Grassé (ed.), *Traité de Zoologie: Anatomie, Systématique, Biologie*, Vol. IX, pp. 3–17, 18–85. (Paris: Masson). [7, 8]

—— [and PAULIAN, R.] (1949b). Ordre des Coléoptères (Coleoptera Linné, 1758). In P.-P. Grassé (ed.), *Traité de Zoologie*, Vol. IX, pp. 771–1077. [30]

—— (1955). L'Édéage. Initiation au recherches sur la systématique des Coléoptères. *Publs Mus. natn. Hist. nat.* **16**: 1–155. [30]

——, and PAULIAN, R. (1944). Morphologie abdominale des Coléoptères et systématique de l'ordre. *Revue fr. Ent.* **11**: 65–110. [30]

JENKIN, P. M., and HINTON, H. E. (1966). Apolysis in arthropod moulting cycles. *Nature, Lond.* **211**: 871. [4]

JENKINS, M. F. (1960). On the method by which *Stenus* and *Dianous* (Coleoptera: Staphylinidae) return to the banks of a pool. *Trans. R. ent. Soc. Lond.* **112**: 1–14. [30]

JOBLING, B. (1951). A record of the Streblidae from the Philippines and other Pacific islands, including morphology of the abdomen, host-parasite relationship and geographical distribution, and with descriptions of five new species (Diptera). *Trans. R. ent. Soc. Lond.* **102**: 211–46. [34]

JOHANNSEN, O. A., and BUTT, F. H. (1941). *Embryology of Insects and Myriapods*, 462 pp. (New York: McGraw-Hill). [4]

JONES, J. C. (1962). Current concepts concerning insect hemocytes. *Am. Zool.* **2**: 209–46. [2]

JORDAN, K. (1923). On a sensory organ found on the head of many Lepidoptera. *Novit. zool.* **30**: 155–8. [36]

JORDAN, K. H. C. (1951). Bestimmungstabellen der Familien von Wanzenlarven. *Zool. Anz.* **147**: 24–31. [26]

JUDD, W. W. (1948). A comparative study of the proventriculus of orthopteroid insects with reference to its use in taxonomy. *Can. J. Res.* **26** (D): 93–161. [16]

KALSHOVEN, L. G. E. (1965). Notes on some injurious Lepidoptera from Java. *Tijdschr. Ent.* **108**: 73–93. [36]

KAPUR, A. P. (1958). *A Catalogue of Lac Insects (Lacciferidae, Hemiptera)*, 47 pp. (Ranchi: Indian Lac Cess Committee). [26]

KARLSON, P. (1963). Chemistry and biochemistry of insect hormones. *Angew. Chem.* (Internat. edn) **2**: 175–82. [2]

——, and BUTENANDT, A. (1959). Pheromones (ectohormones) in insects. *A. Rev. Ent.* **4**: 39–58. [2]

KEAST, A., CROCKER, R. L., and CHRISTIAN, C. S. (eds) (1959). Biogeography and ecology in Australia. *Monographiae biol.* **8**: 1–640. (Den Haag: Junk). [5, 9, 15, 21, 34]

KEILIN, D. (1944). Respiratory systems and respiratory adaptations in larvae and pupae of Diptera. *Parasitology* **36**: 1–66. [34]

KELLY, R., and MAYNE, R. J. B. (1934). *The Australian Thrips: A Monograph of the Order Thysanoptera in Australia: Including a Classification, with Concise Descriptions, of the Known Australian Species*, 81 pp. (Sydney: Australasian Medical Publishing Co.). [27]

KENNEDY, J. S. (ed.) (1961). *Insect Polymorphism*, 115 pp. (London: Symposium No. 1, R. ent. Soc. Lond.). [5]

——, DAY, M. F., and EASTOP, V. F. (1962). *A Conspectus of Aphids as Vectors of Plant Viruses*, 114 pp. (London: Commonw. Inst. Ent.). [26]

——, and STROYAN, H. L. G. (1959). Biology of aphids. *A. Rev. Ent.* **4**: 139–60. [26]

KERR, R. W. (1946). Control of fleas. Laboratory experiments with DDT and certain other insecticides. *J. Coun. scient. ind. Res. Aust.* **19**: 233–40. [33]

KERR, W. E. (1962). Genetics of sex determination. *A. Rev. Ent.* **7**: 157–76. [3]

KERRICH, G. J. (1936). Notes on larviposition in *Polyblastus* (Hym. Ichn. Tryphoninae). *Proc. R. ent. Soc. Lond.* (A) **11**: 108–10. [37]

—— (1952). A review, and a revision in greater part, of the Cteniscini of the Old World. (Hym. Ichneumonidae). *Bull. Br. Mus. nat. Hist.* Ent. **2**: 307–459. [37]

—— (1959). Description of new cremastine Ichneumonidae (Hym.) from Australia, New Zealand and Thailand, with a consideration of the generic categories. *Ann. Mag. nat. Hist.* (13) **2**: 48–64. [37]

—— (1961). A study of the tersilochine parasites of vegetable weevils of the genus *Listroderes*. (Hym., Ichneumonidae). *Eos, Madr.* **37**: 497–503. [37]

——, and QUINLAN, J. (1960). Studies on eucoiline Cynipoidea (Hym.). *Opusc. ent.* **25**: 179–96. [37]

KEY, K. H. L. (1954). *The Taxonomy, Phases, and Distribution of the Genera* Chortoicetes *Brunn. and* Austroicetes *Uv. (Orthoptera: Acrididae)*, 237 pp.

(Canberra: Commonw. scient. ind. Res. Org.). [5, 21]
—— (1957). Kentromorphic phases in three species of Phasmatodea. *Aust. J. Zool.* **5:** 247–84. [22]
—— (1958). Research on the Australian locust and grasshopper problem. *Proc. 10th int. Congr. Ent.* **3:** 63–7. [21]
—— (1960). Proposed addition of certain generic and specific names in the family Phasmatidae (Class Insecta, Order Phasmatodea) to the official lists and indexes. *Bull. zool. Nom.* **17:** 235–40. [22]
——, and DAY, M. F. (1954a). A temperature-controlled physiological colour response in the grasshopper *Kosciuscola tristis* Sjöst. (Orthoptera: Acrididae). *Aust. J. Zool.* **2:** 309–39. [21]
——, and —— (1954b). The physiological mechanism of colour change in the grasshopper *Kosciuscola tristis* Sjöst. (Orthoptera: Acrididae). *Aust. J. Zool.* **2:** 340–63. [21]
KIEFFER, J. J. (1916). Diapriidae. *Tierreich* **44:** 1–627. [37]
—— (1926). Scelionidae. *Tierreich* **48:** 1–885. [37]
KILBY, B. A. (1963). The biochemistry of the insect fat body. *Adv. Insect Physiol.* **1:** 111–74. [2]
KIMMINS, D. E. (1939). A review of the genera of the Psychopsidae (Neuroptera), with a description of a new species. *Ann. Mag. nat. Hist.* (11) **4:** 144–53. [29]
—— (1951). A revision of the Australian and Tasmanian Gripopterygidae and Nemouridae (Plecoptera). *Bull. Br. Mus. nat. Hist.* Ent. **2:** 45–93. [20]
—— (1954). A new genus and some new species of the Chauliodini (Megaloptera), with notes on certain previously described species. *Bull. Br. Mus. nat. Hist.* Ent. **3:** 417–44. [28]
—— (1962). Miss L. E. Cheesman's expeditions to New Guinea. Trichoptera. *Bull. Br. Mus. nat. Hist.* Ent. **11:** 99–187. [35]
KING, L. C. (1962). *The Morphology of the Earth: A Study and Synthesis of World Scenery*, 699 pp. (Edinburgh: Oliver & Boyd). [9]
KIRBY, W. F. (1906, 1910). *A Synonymic Catalogue of Orthoptera.* Vol. II. Orthoptera Saltatoria. Part I. (Achetidae et Phasgonuridae), 562 pp.; Vol. III. Orthoptera Saltatoria. Part II. (Locustidae vel Acridiidae.), 674 pp. (London: Br. Mus.). [21]
KIRIAKOFF, S. G. (1963). The tympanic structures of the Lepidoptera and the taxonomy of the order. *J. Lepid. Soc.* **17:** 1–6. [36]
KIRKALDY, G. W. (1909). *Catalogus Hemipterorum (Heteropterorum).* Vol. I, Cimicidae, 392 pp. (Berlin: Felix Dames). [26]
KIRKPATRICK, T. W. (1937). Studies on the ecology of coffee plantations in East Africa. II. The autecology of *Antestia* spp. (Pentatomidae) with a particular account of a strepsipterous parasite. *Trans. R. ent. Soc. Lond.* **86:** 247–343. [31]
KOCH, A. (1960). Intracellular symbiosis in insects. *A. Rev. Microbiol.* **14:** 121–39. [30]
KOLBE, H. (1908). Mein System der Coleopteren. *Z. wiss. InsektBiol.* **4:** 116–23, 153–62, 219–26, 246–51, 286–94, 389–400. [30]
KONING, H. S. DE, and ROEPKE, W. (1949). Remarks on the morphology of the teak moth, *Hyblaea puera* Cr. (Lep. Hyblaeidae). *Treubia* **20:** 25–30. [36]
KORMILEV, N. A. (1965). Notes on Australian Aradidae (Hemiptera: Heteroptera) with descriptions of new genera and species. *Proc. R. Soc. Qd* **77:** 11–35. [26]
KRISHNA, K. (1961). A generic revision and phylogenetic study of the family Kalotermitidae (Isoptera). *Bull. Am. Mus. nat. Hist.* **122:** 303–408. [15]
KROMBEIN, K. V. (1957). A generic review of the Amiseginae, a group of phasmatid egg parasites, and notes on the Adelphinae (Hymenoptera, Bethyloidea, Chrysididae). *Trans. Am. ent. Soc.* **82:** 147–215. [37]
—— (1961). Some symbiotic relations between saproglyphid mites and solitary vespid wasps. *J. Wash. Acad. Sci.* **51:** 89–93. [37]
KULLENBERG, B. (1956). Field experiments with chemical sexual attractants on aculeate Hymenoptera males. I. *Zool. Bidr. Upps.* **31:** 253–354. [37]
KUMAR, R. (1966). Studies on the biology, immature stages, and relative growth of some Australian bugs of the superfamily Coreoidea (Hemiptera: Heteroptera). *Aust. J. Zool.* **14:** 895–991. [26]
LANSBURY, I. (1964). The genus *Anisops* in Australia (Hemiptera: Notonectidae). Part I. *J. ent. Soc. Qd* **3:** 52–65. [26]
LAUCK, D. R., and MENKE, A. S. (1961). The higher classification of the Belostomatidae (Hemiptera). *Ann. ent. Soc. Am.* **54:** 644–57. [26]
LAURENTIAUX, D. (1953). Classe des Insectes. In J. Piveteau (ed.), *Traité de Paléontologie*, Vol. III, pp. 397–527. (Paris: Masson). [8]
LEA, A. M. (1898–1913). Revision of the Australian Curculionidae belonging to the subfamily Cryptorhynchides. Parts I–XII. *Proc. Linn. Soc. N.S.W.* **22:** 449–513; **23:** 178–217; **24:** 200–70, 522–46; **27:** 408–42; **28:** 643–79; **30:** 235–58; **32:** 400–30; **33:** 701–32; **34:** 593–635; **37:** 602–16; **38:** 451–89. [30]
—— (1909). Revision of the Australian and Tasmanian Malacodermidae. *Trans. ent. Soc. Lond.* **1909:** 45–251. [30]
—— (1910). On a new genus of Stylopidae from Australia. *Trans. ent. Soc. Lond.* **1910:** 514–16. [31]
—— (1914). Notes on Australian Cetonides; with a list of species and descriptions of some new ones. *Trans. R. Soc. S. Aust.* **38:** 132–218. [30]
—— (1920). On Australian Coleoptera. Family Byrrhidae. *Rec. S. Aust. Mus.* **1:** 273–90. [30]
—— (1922). On Australian Anthicidae (Coleoptera). *Proc. Linn. Soc. N.S.W.* **47:** 471–512. [30]
—— (1923). Australian dung beetles of the sub-family Coprides. *Rec. S. Aust. Mus.* **2:** 353–96. [30]
—— (1924). On Australian Anobiides (Coleoptera). *Trans. R. Soc. S. Aust.* **48:** 15–64. [30]
—— (1927). Australian Curculionidae of the subfamily Gonipterides. *Proc. R. Soc. Vict.* **39:** 76–112. [30]
—— (1932). The Phalacridae (Coleoptera) of Australia and New Guinea. *Rec. S. Aust. Mus.* **4:** 433–81. [30]
——, and BOVIE, A. (1909). Coleoptera. Fam. Curculionidae. Subfam. Belinae. *Genera Insect.* **91:** 1–13. [30]
LECLERCQ, J. (1954). *Monographie Systématique, Phylogénétique et Zoogéographique des Hyménoptères Crabroniens*, 371 pp. (Liège: 'Lejeunia'). [37]
—— (1957). Le genre *Rhopalum* (Kirby, 1829) en Australie (Hym. Sphecidae, Crabroninae). *Bull. Annls Soc. r. ent. Belg.* **93:** 177–232. [37]

LECOMTE, J. (1964). Animal sexuality. In *The Animal World*, pp. 49–52 (London: Thomas Nelson). [1, fig.]
LEE, D. J., CRUST, M., and SABROSKY, C. W. (1956). The Australasian Diptera of J. R. Malloch. *Proc. Linn. Soc. N.S.W.* **80**: 289–342. [See Malloch, 1923–41]
——, REYE, E. J., and DYCE, A. L. (1962). 'Sandflies' as possible vectors of disease in domesticated animals in Australia. *Proc. Linn. Soc. N.S.W.* **87**: 364–76. [34]
LEEPER, G. W. (ed.) (1962). *The Evolution of Living Organisms: A Symposium . . . held in Melbourne, December 1959*, 459 pp. (Melbourne University Press). [3, 9, 26]
LEES, A. D. (1966). The control of polymorphism in aphids. *Adv. Insect Physiol.* **3**: 207–77. [26]
LENG, C. W. (1920). *Catalogue of the Coleoptera of America, North of Mexico*, 470 pp. (Mount Vernon, N.Y.: Sherman). [30]
LENGERKEN, H. VON (1954). *Die Brutfürsorge- und Brutpflegeinstinkte der Käfer*, 383 pp. (Leipzig: Akad. Verlagsgesellschaft). [30]
LEPESME, P. (1944). Les Coléoptères des denrées alimentaires et des produits industriels entreposés. *Encycl. ent.* (A) **22**: 1–335. [30]
LESTON, D., PENDERGRAST, J. G., and SOUTHWOOD, T. R. E. (1954). Classification of the terrestrial Heteroptera (Geocorisae). *Nature, Lond.* **174**: 91–2. [26]
——, and SCUDDER, G. G. E. (1956). A key to larvae of the families of British Hemiptera–Heteroptera. *Entomologist* **89**: 223–31. [26]
LIEFTINCK, M. A. (1949). The dragonflies (Odonata) of New Guinea and neighbouring islands. Part VII. *Nova Guinea* (n.s.) **5**: 1–271 (see pp. 236–8). [13]
—— (1951). Odonata of the 1948 Archbold Cape York Expedition, with a list of the dragonflies from the Peninsula. *Am. Mus. Novit.* **1488**: 1–46. [13]
LINDNER, E. (1924–). Series *Die Fliegen der Palaearktischen Region.* (Stuttgart: Schweizerbart'sche Verlag). [34]
LINDROTH, C. H. (1946). Inheritance of wing dimorphism in *Pterostichus anthracinus* Ill. *Hereditas* **32**: 37–40. [30]
—— (1963). The fauna history of Newfoundland illustrated by carabid beetles. *Opusc. ent.* Suppl. **23**: 1–112. [30]
LINK, E. (1909). Über die Stirnaugen der hemimetabolen Insekten. *Zool. Jb.* (Anat.) **27**: 281–376. [2, fig.]
LINSLEY, E. G. (1958). The ecology of solitary bees. *Hilgardia* **27**: 543–99. [37]
LLOYD, F. E. (1942). *The Carnivorous Plants*, 352 pp. (Waltham, Mass.: Chronica Botanica). [5]
LOCKE, M. (1961). Pore canals and related structures in insect cuticle. *J. biophys. biochem. Cytol.* **10**: 589–618. [2, fig.]
—— (1965). Permeability of insect cuticle to water and lipids. *Science, N.Y.* **147**: 295–8. [2]
LOWER, H. F. (1951). A revision of Australian species previously referred to the genus *Empoasca* (Cicadellidae, Homoptera). *Proc. Linn. Soc. N.S.W.* **76**: 190–221. [26]
MAA, T. C. (1963a). A review of the Machaerotidae (Hemiptera: Cercopoidea). *Pacif. Insects Monogr.* **5**: 1–166. [26]
—— (1963b). Genera and species of Hippoboscidae (Diptera): types, synonymy, habitats and natural groupings. *Pacif. Insects Monogr.* **6**: 1–186. [34]
—— (1964). A review of the Old World Polyctenidae (Hemiptera: Cimicoidea). *Pacif. Insects* **6**: 494–516. [26]
—— (1965). An interim world list of batflies (Diptera: Nycteribiidae and Streblidae). *J. med. Ent.* **1**: 377–86. [34]
——, and YOSHIMOTO, C. M. (1961). Loboscelidiidae, a new family of Hymenoptera. *Pacif. Insects* **3**: 523–48. [37]
MCALPINE, D. K. (1960). A review of Australian species of Clusiidae (Diptera, Acalyptrata). *Rec. Aust. Mus.* **25**: 63–94. [34]
—— (1966). Description and biology of an Australian species of Cypselosomatidae (Diptera), with a discussion of family relationships. *Aust. J. Zool.* **14**: 673–85. [34]
MCALPINE, J. F. (1960). A new species of *Leucopis (Leucopella)* from Chile and a key to the world genera and subgenera of Chamaemyiidae (Diptera). *Can. Ent.* **92**: 51–8. [34]
—— (1964). Descriptions of new Lonchaeidae (Diptera). I and II. *Can. Ent.* **96**: 661–757. [34]
MACBRIDE, E. W. (1914). *Text-Book of Embryology* (W. Heape, ed.), Vol. I, Invertebrata, 692 pp. (London: Macmillan). [4, fig.]
MCCONNELL, H. S. (1954). A classification of the coccid family Aclerdidae (Coccoidea, Homoptera). *Bull. Md agric. Exp. Stn* **A-75**: 1–121. [26, fig.]
MCDONALD, F. J. D. (1966). The genitalia of North American Pentatomoidea (Hemiptera: Heteroptera). *Quaest. ent.* **2**: 7–150. [26]
MCELROY, W. D., and SELIGER, H. H. (1962). Mechanism of action of firefly luciferase. *Fedn Proc. Fedn Am. Socs exp. Biol.* **21**: 1006–12. [2]
MCKENZIE, H. L. (1938). The genus *Aonidiella* (Homoptera; Coccoidea; Diaspididae). *Microentomology* **3**: 1–36. [26, fig.]
MCKEOWN, K. C. (1937). New fossil insect wings (Protohemiptera, family Mesotitanidae). *Rec. Aust. Mus.* **20**: 31–7. [8, fig.]
—— (1942). *Australian Insects: An Introductory Handbook*, 304 pp. (Sydney: R. zool. Soc. N.S.W.). [5]
—— (1947). Catalogue of the Cerambycidae (Coleoptera) of Australia. *Mem. Aust. Mus.* **10**: 1–190. [30]
—— (1951). Dermatitis apparently caused by a staphylinid beetle in Australia. *Med. J. Aust.* 1951, **2**: 772–3. [30]
——, and MINCHAM, V. H. (1948). The biology of an Australian mantispid (*Mantispa vittata* Guérin). *Aust. Zool.* **11**: 207–24. [29]
MACKERRAS, I. M. (1950). Marine insects. *Proc. R. Soc. Qd* **61**: 19–29. [5]
—— (1956–61). The Tabanidae (Diptera) of Australia. I–IV. *Aust. J. Zool.* **4**: 376–443; **8**: 1–152; **9**: 827–906. [34]
—— (1964). The Tabanidae (Diptera) of New Guinea. *Pacif. Insects* **6**: 69–210. [9]
——, and MACKERRAS, M. J. (1953). A new species of *Pelecorhynchus* (Diptera, Tabanoidea) from the Dorrigo Plateau, New South Wales. *Proc. Linn. Soc. N.S.W.* **78**: 38–40. [34]
MACKERRAS, M. J. (1965–6). Australian Blattidae (Blattodea). I–V. *Aust. J. Zool.* **13**: 841–927; **14**: 305–63. [Also **15**: 593–618, 1207–98, 1967; **16**: 237–331, 1968]. [14]

——, and MACKERRAS, I. M. (1948a). *Salmonella* infections in Australian cockroaches. *Aust. J. Sci.* **10:** 115. [14]

——, and —— (1948b). Simuliidae (Diptera) from Queensland. *Aust. J. scient. Res.* (B) **1:** 231–70. [5]

MCKITTRICK, F. A. (1964). Evolutionary studies of cockroaches. *Mem. Cornell Univ. agric. Exp. Stn* **389:** 1–197. [14]

MAINX, F. (1964). The genetics of *Megaselia scalaris* Loew (Phoridae): a new type of sex determination in Diptera. *Am. Nat.* **98:** 415–30. [34]

MAKINO, S. (1951). *An Atlas of the Chromosome Numbers in Animals*, 2nd edn, 290 pp. (Ames: Iowa State College Press). [3]

MALLOCH, J. R. (1923–41). See Lee, Crust and Sabrosky (1956) for classified and general bibliographies (pp. 295 and 340, respectively). [34]

MANN, J. S. (1928–33). Revisional notes on Australian Therevidae. Parts 1–3. *Aust. Zool.* **5:** 151–94; **6:** 17–49; **7:** 325–44. [34]

MANNA, G. K. (1958). Cytology and inter-relationships between various groups of Heteroptera. *Proc. 10th int. Congr. Ent.* **2:** 919–34. [3]

MANTON, S. M. (1958). Habits of life and evolution of body design in Arthropoda. *J. Linn. Soc.* Zool. **44,** Bot. **56:** 58–72. [7]

—— (1964). Mandibular mechanisms and the evolution of arthropods. *Phil. Trans. R. Soc.* (B) **247:** 1–183. [7, 10, 11]

MARKS, E. P., and LAWSON, F. A. (1962). A comparative study of the dictyopteran ovipositor. *J. Morph.* **111:** 139–71. [16]

MARSHALL, A. T. (1966). Histochemical studies on a mucocomplex in the Malpighian tubules of cercopid larvae. *J. Insect Physiol.* **12:** 925–32. [26]

MARTIN, J. (1962). Interrelation of inversion systems in the midge *Chironomus intertinctus* (Diptera: Nematocera). I. A sex-linked inversion. *Aust. J. biol. Sci.* **15:** 666–73. [3]

—— (1963). The cytology and larval morphology of the Victorian representatives of the subgenus *Kiefferulus* of the genus *Chironomus* (Diptera: Nematocera). *Aust. J. Zool.* **11:** 301–22. [3, 34]

—— (1964). Morphological differences between *Chironomus intertinctus* Skuse and *C. paratinctus*, sp. nov., with descriptions and a key to the subgenus *Kiefferulus* (Diptera: Nematocera). *Aust. J. Zool.* **12:** 279–87. [34]

MARTIN, M. (1934). Life history and habits of the pigeon louse (*Columbicola columbae* (Linnaeus)). *Can. Ent.* **66:** 6–16. [25]

MARTYNOV, A. V. (1925). To the knowledge of fossil insects from Jurassic beds in Turkestan. *Izv. ross. Akad. Nauk* (6) **19:** 233–46, 569–98, 753–62. [8, fig.]

—— (1928a). A new fossil form of Phasmatodea from Galkino (Turkestan) and on Mesozoic phasmids in general. *Ann. Mag. nat. Hist.* (10) **1:** 319–28. [8, fig.]

—— (1928b). Permian fossil insects of north-east Europe. *Trudy geol. Muz.* **4:** 1–118. [8, fig.]

—— (1930). The interpretation of the wing venation and tracheation of the Odonata and Agnatha. (Translated, with an introductory note, by F. M. Carpenter.) *Psyche, Camb.* **37:** 245–80. (Original in *Russk. ent. Obozr.* **18:** 145–74, 1924). [1]

MARTYNOVA, O. M. (1952). Permian Neuroptera of the USSR. (In Russian.) *Trudy paleont. Inst.* **40:** 197–237. [8, fig.]

—— (1958). New insects from Permian and Mesozoic deposits of the USSR. (In Russian.) *Mater. Osnov. Paleont.* **2:** 69–94. [8, fig.]

—— (1961). Palaeoentomology. *A. Rev. Ent.* **6:** 285–94. [7, 8]

MASNER, L. (1958). Some problems of the taxonomy of the subfamily Telenominae (Hym. Scelionidae). *Trans. 1st int. Conf. Insect Path. biol. Control*, Praha, 1958: 375–82. [37]

—— (1961a). Ambositrinae, a new subfamily of Diapriidae from Madagascar and central Africa (Hymenoptera Proctotrupoidea). *Mém. Inst. scient. Madagascar* (E) **12:** 289–95. [37]

—— (1961b). Proctotrupidae. Key to the genera of the world (Hymenoptera Proctotrupoidea). *Parc National de L'Upemba*, Fasc. **60** (4): 37–47. (Bruxelles: Hayez). [37]

——, and DESSART, P. (1967). La réclassification des catégories taxonomiques supérieures des Ceraphronoidea (Hymenoptera). *Bull. Inst. r. Sci. nat. Belg.* **43** (22): 1–33. [37]

MATHER, W. B. (1955–6). The genus *Drosophila* (Diptera) in eastern Queensland. I–IV. *Aust. J. Zool.* **3:** 545–82; **4:** 65–97. [3]

—— (1960). Additions to the *Drosophila* fauna of Australia. *Pap. Dep. Zool. Univ. Qd* **1:** 229–39. [34]

—— (1963). Patterns of chromosomal polymorphism in *Drosophila rubida*. *Am. Nat.* **97:** 59–63. [3]

MATSUDA, R. (1960). Morphology, evolution and a classification of the Gerridae (Hemiptera–Heteroptera). *Kans. Univ. Sci. Bull.* **41:** 25–632. [26]

MATTHYSSE, J. G. (1944). Biology of the cattle biting louse and notes on cattle sucking lice. *J. econ. Ent.* **37:** 436–42. [25]

MAY, A. W. S. (1965). New species and records of Dacinae (Diptera: Trypetidae) from northern Australia. *J. ent. Soc. Qd* **4:** 58–66. [34]

MAYR, E. (1942). *Systematics and the Origin of Species from the Viewpoint of a Zoologist*, 334 pp. (New York: Columbia University Press). [6]

—— (1963). *Animal Species and Evolution*, 797 pp. (Cambridge, Mass.: Harvard University Press). [6, 9]

MEAD-BRIGGS, A. R. (1964). The reproductive biology of the rabbit flea *Spilopsyllus cuniculi* (Dale) and the dependence of this species upon the breeding of its host. *J. exp. Biol.* **41:** 371–402. [33]

MELANDER, A. L. (1927). Diptera. Fam. Empididae. *Genera Insect.* **185:** 1–434. [34]

MENEES, J. H. (1962). The skeletal elements of the gnathocephalon and its appendages in the larvae of higher Diptera. *Ann. ent. Soc. Am.* **55:** 607–16. [34]

MENKE, A. S. (1960). A review of the genus *Lethocerus* (Hemiptera: Belostomatidae) in the eastern hemisphere with the description of a new species from Australia. *Aust. J. Zool.* **8:** 285–8. [26]

——, and STANGE, L. A. (1964). A new genus of Nepidae from Australia with notes on the higher classification of the family. *Proc. R. Soc. Qd* **75:** 67–72. [26]

METCALF, Z. P., and WADE, V. (1965). *General Catalogue of the Homoptera.* Fasc. I, Suppl. Membracoidea, 2 vols, 1,552 pp. (Raleigh: North Carolina State University). [26]

METZ, C. W. (1938). Chromosome behavior, inheritance and sex determination in *Sciara*. *Am. Nat.* **72:** 485–520. [34]

MEYRICK, E. (1928). *A Revised Handbook of British Lepidoptera*, 914 pp. (London: Watkins & Doncaster). [36]
MICHENER, C. D. (1952). The Saturniidae (Lepidoptera) of the western hemisphere. Morphology, phylogeny, and classification. *Bull. Am. Mus. nat. Hist.* **98**: 337–501. [36]
—— (1953). Comparative morphological and systematic studies of bee larvae with a key to the families of hymenopterous larvae. *Kans. Univ. Sci. Bull.* **35**: 987–1102. [37]
—— (1960). Notes on the biology and supposed parthenogenesis of halictine bees from the Australian region. *J. Kans. ent. Soc.* **33**: 85–96. [37]
—— (1961a). Probable parasitism among Australian bees of the genus *Allodapula*. *Ann. ent. Soc. Am.* **54**: 532–4. [37]
—— (1961b). Observations on the nests and behavior of *Trigona* in Australia and New Guinea (Hymenoptera, Apidae). *Am. Mus. Novit.* **2026**: 1–46. [37]
—— (1965a). The life cycle and social organization of bees of the genus *Exoneura* and their parasite, *Inquilina* (Hymenoptera: Xylocopinae). *Kans. Univ. Sci. Bull.* **46**: 317–58. [37]
—— (1965b). Classification of the bees of the Australian and South Pacific regions. *Bull. Am. Mus. nat. Hist.* **130**: 1–362. [37]
——, and MICHENER, M. H. (1951). *American Social Insects*, 267 pp. (New York: van Nostrand). [5]
MILLER, D. (1956). Bibliography of New Zealand entomology 1775–1952 (with annotations). *Bull. N.Z. Dep. scient. ind. Res.* **120**: 1–492. [Intro.]
MILLER, N. C. E. (1956). *The Biology of the Heteroptera*, 162 pp. (London: Leonard Hill). [5, 26]
MONRÓS, F. (1958). Die Gattung *Microdonacia* Blackburn (Col. Chrysomelidae). *Ent. Arb. Mus. Georg Frey* **9**: 742–9. [30]
MONTEITH, G. B. (1966). A new genus of Chinamyersiinae (Heteroptera: Aradidae) from Australia, with notes on its relationships and male genitalia. *J. ent. Soc. Qd* **5**: 46–50. [26]
MOORE, B. P. (1963). Studies on Australian Carabidae (Coleoptera). 3. The Psydrinae. *Trans. R. ent. Soc. Lond.* **115**: 277–90. [30]
—— (1964). Present-day cave beetle fauna in Australia: a pointer to past climatic change. *Helictite* **3** (1): 3–9. [5, 30]
—— (1965). Studies on Australian Carabidae (Coleoptera). 4—The Pterostichinae. *Trans. R. ent. Soc. Lond.* **117**: 1–32. [30]
—— (1966). Isolation of the scent-trail pheromone of an Australian termite. *Nature, Lond.* **211**: 746–7. [15]
MOORE, K. M. (1961). Observations on some Australian forest insects. 7–9. *Proc. Linn. Soc. N.S.W.* **86**: 128–67, 185–202. [26]
MOREAU, R. E. (1963). Vicissitudes of the African biomes in the late Pleistocene. *Proc. zool. Soc. Lond.* **141**: 395–421. [9]
MORRISON, H. (1925). Classification of scale insects of the subfamily Ortheziinae. *J. agric. Res.* **30**: 97–154. [26, fig.]
—— (1928). A classification of the higher groups and genera of the coccid family Margarodidae. *Tech. Bull. U.S. Dep. Agric.* **52**: 1–240. [26]
—— (1952). Classification of the Ortheziidae. Supplement to classification of scale insects of the subfamily Ortheziinae. *Tech. Bull. U.S. Dep. Agric.* **1052**: 1–80. [26]
MOSELY, M. E., and KIMMINS, D. E. (1953). *The Trichoptera (Caddis-flies) of Australia and New Zealand*, 550 pp. (London: Br. Mus. nat. Hist.). [35]
MOSHER, E. (1916). A classification of the Lepidoptera based on characters of the pupa. *Bull. Ill. St. Lab. nat. Hist.* **12** (2): 14–159. [36]
MUIR, F. (1925). On the genera of Cixiidae, Meenoplidae and Kinnaridae (Fulgoroidae, Homoptera). *Pan-Pacif. Ent.* **1**: 97–110, 156–63. [26]
MULLER, H. J. (1938). The remaking of chromosomes. *Collecting Net* **13**: 181–95 & 198. Reprinted in H. J. Muller, *Studies in Genetics*, pp. 384–408. (Bloomington: Indiana University Press, 1962). [3]
MUNROE, E. (1960). The classification of the Papilionidae (Lepidoptera). *Can. Ent.* Suppl. **17**: 1–51. [36]
—— (1965). Zoogeography of insects and allied groups. *A. Rev. Ent.* **10**: 325–44. [9]
MURRAY, M. D. (1957a–d). The distribution of the eggs of mammalian lice on their hosts. I–IV. *Aust. J. Zool.* **5**: 13–29, 172–87. [25]
—— (1960a, b). The ecology of lice on sheep. I–II. *Aust. J. Zool.* **8**: 349–62. [25]
——, and NICHOLLS, D. G. (1965). Studies on the ectoparasites of seals and penguins. I. The ecology of the louse *Lepidophthirus macrorhini* Enderlein on the southern elephant seal, *Mirounga leonina* (L.). *Aust. J. Zool.* **13**: 437–54. [25]
——, SMITH, M. S. R., and SOUCEK, Z. (1965). Studies on the ectoparasites of seals and penguins. II. The ecology of the louse *Antarctophthirus ogmorhini* Enderlein on the Weddell seal, *Leptonychotes weddelli* Lesson. *Aust. J. Zool.* **13**: 761–71. [5]
MUSGRAVE, A. (1930). Contributions to the knowledge of Australian Hemiptera. No. II. A revision of the subfamily Graphosomatinae (family Pentatomidae). *Rec. Aust. Mus.* **17**: 317–41. [26]
—— (1932). *Bibliography of Australian Entomology 1775–1930 with Biographical Notes on Authors and Collectors*, 380 pp. (Sydney: R. zool. Soc. N.S.W.). [Intro.]
MYERS, J. G. (1922). Life-history of *Siphanta acuta* (Walk.), the large green plant-hopper. *N.Z. Jl Sci. Technol.* **5**: 256–63. [26]
—— (1929). *Insect Singers: A Natural History of the Cicadas*, 304 pp. (London: Routledge). [26]
MYKYTOWYCZ, R. (1963). Occurrence of bot-fly larvae *Tracheomyia macropi* Froggatt (Diptera: Oestridae) in wild red kangaroos, *Megaleia rufa* (Desmarest). *Proc. Linn. Soc. N.S.W.* **88**: 307–12. [34]
NEAVE, S. A. (1939–50). *Nomenclator Zoologicus: A List of the Names of Genera and Subgenera in Zoology from the Tenth Edition of Linnaeus 1758 to the End of 1935*. Vol. I, 957 pp.; II, 1,025 pp.; III, 1,065 pp.; IV, 758 pp.; V (without subtitle), *1936–1945*, 308 pp. [VI (without subtitle), *1946–1955*, 329 pp., by M. A. Edwards and A.T. Hopwood, publ. 1966.] (London: Zool. Soc. Lond.). [6]
NEBOISS, A. (1956). A check list of Australian Elateridae (Coleoptera). *Mem. natn. Mus. Vict.* **22** (2): 1–75. [30]
—— (1958). Larva and pupa of an Australian limnephilid (Trichoptera). *Proc. R. Soc. Vict.* **70**: 163–8. [35]
—— (1959). New caddis fly genus from Tasmania. (Trichoptera: Plectrotarsidae). *Mem. natn. Mus. Vict.* **24**: 91–6. [35]

—— (1960). On the family Cupedidae, Coleoptera. *Proc. R. Soc. Vict.* **72:** 12–20. [30]

NEEDHAM, J. G., TRAVER, J. R., and HSU, Y.-C. (1935). *The Biology of Mayflies with a Systematic Account of North American Species*, 759 pp. (Ithaca, N.Y.: Comstock). [12]

NICHOLSON, A. J. (1927). A new theory of mimicry in insects. *Aust. Zool.* **5:** 10–104. [5, 34]

NORRIS, K. R. (1965). The bionomics of blow flies. *A. Rev. Ent.* **10:** 47–68. [34]

—— (1966). *The Collection and Preservation of Insects*, 34 pp. (Brisbane: Handbook No. 1, Aust. ent. Soc.). [Intro., 3]

NUTTING, W. L. (1951). A comparative anatomical study of the heart and accessory structures of the orthopteroid insects. *J. Morph.* **89:** 501–97. [2, fig.]

OCHS, G. (1949). Revision of Australian Gyrinidae. *Rec. Aust. Mus.* **22:** 171–99. [30]

O'FARRELL, A. F. (1964). On physiological colour change in some Australian Odonata. *J. ent. Soc. Aust. (N.S.W.)* **1:** 5–12. [5, 13]

OGLOBLIN, A. A. (1923). Two new Strepsiptera from materials of National Museum of Natural History in Prague. *Sb. ent. Odd. nár. Mus. Praze* **1:** 45–7. [31]

—— (1939). The Strepsiptera parasites of ants. *Proc. 7th int. Congr. Ent.* **2:** 1277–84. [31]

OKE, C. (1932). Aculagnathidae. A new family of Coleoptera. *Proc. R. Soc. Vict.* **44:** 22–4. [30]

OLDROYD, H. (1964). *The Natural History of Flies*, 324 pp. (London: Weidenfeld & Nicholson). [34]

OSSIANNILSSON, F. (1949). Insect drummers. A study on the morphology and function of the sound-producing organ of Swedish Homoptera Auchenorrhyncha with notes on their sound production. *Opusc. ent.* Suppl. **10:** 1–145. [26]

PACLT, J. (1956). *Biologie der primär flugellosen Insekten*, 258 pp. (Jena: Gustav Fischer). [10, 11]

—— (1957). Diplura. *Genera Insect.* **212:** 1–123. [10]

—— (1963). Thysanura. Fam. Nicoletiidae. *Genera Insect.* **216:** 1–58. [11]

PAGÉS, J. (1952). Diploures Japygidés de Nouvelle-Zélande. *Rec. Canterbury Mus.* **6:** 149–62. [10]

PARAMONOV, S. J. (1950–64). Series 'Notes on Australian Diptera', Parts II–XL, in *Ann. Mag. nat. Hist.*: 1950a, Pt. II, (12) **3:** 519–25; 1950b, Pt. III, (12) **3:** 525–9; 1950c, Pt. IV, (12) **3:** 529–33; 1951a, Pt. VI, (12) **4:** 745–52; 1951b, Pt. VII, (12) **4:** 752–60; 1951c, Pt. VIII, (12) **4:** 761–79; 1953a, Pt. XI, (12) **6:** 204–5; 1953b, Pt. XII, (12) **6:** 206–8; 1954a, Pt. XIII, (12) **7:** 275–83; 1954b, Pt. XV, (12) **7:** 292–7; 1955a, Pt. XVI, (12) **8:** 125–30; 1955b, Pt. XVII, (12) **8:** 130–3; 1955c, Pt. XVIII, (12) **8:** 134–5; 1955d, Pt. XIX, (12) **8:** 135–44; 1956a, Pt. XXII, (12) **9:** 812–16; 1957a, Pt. XXIII, (12) **10:** 52–62; 1957b, Pt. XXIV, (12) **10:** 125–8; 1957c, Pt. XXV, (12) **10:** 779–81; 1958a, Pt. XXVII, (13) **1:** 594–8; 1958b, Pt. XXVIII, (13) **1:** 598–600; 1959a, Pt. XXIX, (13) **2:** 691–6; 1959b, Pt. XXX, (13) **2:** 696–704; 1960a, Pt. XXXI, (13) **3:** 505–12; 1961a, Pt. XXXIII, (13) **4:** 100–2; 1961b, Pt. XXXIV, (13) **4:** 103–7; 1961c, Pt. XXXV, (13) **4:** 107–10; 1964a, Pt. XXXVIII, (13) **7:** 151–3; 1964b, Pt. XXXIX, (13) **7:** 153–7; 1964c, Pt. XL, (13) **7:** 157–9. [34]

—— (1953c). A review of Australian Nemestrinidae (Diptera). *Aust. J. Zool.* **1:** 242–90. [34]

—— (1953d). A review of Australian Apioceridae (Diptera). *Aust. J. Zool.* **1:** 449–536. [34]

—— (1953e). A new species of *Clytocosmus* (Tipulidae). *Encycl. ent.* (Dipt.) **11:** 51–5. [34]

—— (1955e). A review of Australian Scenopinidae (Diptera). *Aust. J. Zool.* **3:** 634–53. [34]

—— (1956b). A review of the Australian species of *Cylindromyia* Meigen and *Saralba* Walker (Tachinidae: Diptera). *Aust. J. Zool.* **4:** 358–75. [34]

—— (1957d). A review of Australian Acroceridae (Diptera). *Aust. J. Zool.* **5:** 521–46. [34]

—— (1958c). A review of Australian Pyrgotidae (Diptera). *Aust. J. Zool.* **6:** 89–137. [34]

—— (1960b). A review of the genus *Euphumosia* Malloch (Diptera, Calliphoridae). *Nova Guinea*, n.s., **10** (Zool. 1): 1–13. [34]

—— (1962). A review of Australian Leptidae (Diptera). *Aust. J. Zool.* **10:** 113–69. [34]

PARKER, F. D. (1962). On the subfamily Astatinae, with a systematic study of the genus *Astata* of America north of Mexico (Hymenoptera: Sphecidae). *Ann. ent. Soc. Am.* **55:** 643–59. [37]

PARKER, H. L., and SMITH, H. D. (1933). Additional notes on the strepsipteron *Eoxenos laboulbenei* Peyerimhoff. *Ann. ent. Soc. Am.* **26:** 217–31. [31, fig.]

PARROTT, A. W. (1953). A systematic catalogue of Australian Braconidae. *Pacif. Sci.* **7:** 193–218. [37]

PASTEELS, J. (1957). Revision du genre *Gasteruption* (Hymenoptera, Evanoidea, Gasteruptionidae). Espèces australiennes. *Mém. Inst. r. Sci. nat. Belg.* (2) **56:** 1–125. [37]

PAVAN, M., and RONCHETTI, G. (1955). Studia sulla morfologia esterna e anatomia interna dell' operaia di *Iridomyrmex humilis* Mayr, e ricerche chimiche e biologiche sulla iridomyrmecina. *Atti Soc. ital. Sci. nat.* **94:** 379–477. [37]

PEACOCK, W. J., and ERICKSON, J. (1965). Segregation-distortion and regularly nonfunctional products of spermatogenesis in *Drosophila melanogaster*. *Genetics, Princeton* **51:** 313–28. [34]

PEARSE, A. S., PATTERSON, M. T., RANKIN, J. S., and WHARTON, G. W. (1936). The ecology of *Passalus cornutus* Fabricius, a beetle which lives in rotting logs. *Ecol. Monogr.* **6:** 455–90. [30]

PECK, O., BOUČEK, Z., and HOFFER, A. (1964). Keys to the Chalcidoidea of Czechoslovakia (Insecta: Hymenoptera). *Mem. ent. Soc. Can.* **34:** 1–120. [37]

PENDERGRAST, J. G. (1957). Studies on the reproductive organs of the Heteroptera with a consideration of their bearing on classification. *Trans. R. ent. Soc. Lond.* **109:** 1–63. [26]

—— (1962). The internal anatomy of the Peloridiidae (Homoptera: Coleorrhyncha). *Trans. R. ent. Soc. Lond.* **114:** 49–65. [26]

PERKINS, F. A. (1958). Australian Plecoptera. Part 1. Genus *Trinotoperla* Tillyard. *Pap. Dep. Ent. Univ. Qd* **1:** 85–100. [20]

PERKINS, R. C. L. (1905a). Leaf-hoppers and their natural enemies (Pt. I. Dryinidae). *Bull. Hawaiian Sug. Plrs' Ass. Exp. Stn* **1:** 1–69. [37]

—— (1905b). Leaf-hoppers and their natural enemies (Pt. III. Stylopidae). *Bull. Hawaiian Sug. Plrs' Ass. Exp. Stn* **1:** 86–111. [31]

PEYERIMHOFF, P. DE (1933). Les larves des Coléoptères d'après A. Böving et F. C. Craighead et les grands critériums de l'ordre. *Annls Soc. ent. Fr.* **102:** 77–106. [30]

PHILPOTT, A. (1928). On the systematic position of *Anomoses* (Lepidoptera Homoneura). *Trans. ent. Soc. Lond.* **76:** 93–6. [36]

PIC, M. (1902). Coleoptera Heteromera. Fam. Hylophilidae. *Genera Insect.* **8:** 1–14. [30]

PIERCE, E. D. (1936). The position of the Strepsiptera in the classification of insects. *Ent. News* **47:** 257–63. [30]

PLUMSTEAD, E. P. (1962). Fossil floras of Antarctica. In *Trans-Antarctic Expedition 1955–1958.* Sci. Rep. No. 9, Geology 2, 154 pp. (London: Trans-Antarctic Expedition Committee). [8, 9]

POISSON, R., and PESSON, P. (1951). Super-ordre des Hémiptéroïdes (Hemiptera Linné, 1758, Rhynchota Burmeister, 1835). In P.-P. Grassé (ed.), *Traité de Zoologie: Anatomie, Systématique, Biologie*, Vol. X, pp. 1385–1803. (Paris: Masson). [26]

POPE, P. (1953). Studies of the life histories of some Queensland Blattidae (Orthoptera). Parts 1 and 2. *Proc. R. Soc. Qd* **63:** 23–59. [14]

POPHAM, E. J. (1965). The functional morphology of the reproductive organs of the common earwig *(Forficula auricularia)* and other Dermaptera with reference to the natural classification of the order. *J. Zool.* **146:** 1–43. [19]

PRIESNER, H. (1949). Genera Thysanopterorum. Keys for the identification of the genera of the order Thysanoptera. *Bull. Soc. Fouad I. Ent.* **33:** 31–157. [27]

PRINGLE, J. A. (1938). A contribution to the knowledge of *Micromalthus debilis* LeC. (Coleoptera). *Trans. R. ent. Soc. Lond.* **87:** 271–86. [30]

PRINGLE, J. W. S. (1957a). *Insect Flight.* Cambridge Monographs in Experimental Biology, No. 9, 133 pp. (Cambridge University Press). [2]

—— (1957b). The structure and evolution of the organs of sound-production in cicadas. *Proc. Linn. Soc. Lond.* **167:** 144–59. [2, 26]

PUCHKOV, V. G. (1956). Principal trophical groups of Heteroptera and the change of their feeding during their life history. (In Russian, with English summary in Suppl. 1, p. 5.) *Zool. Zh.* **35:** 32–44. [26]

RAFF, J. W. (1934). Observations on saw-flies of the genus *Perga*, with notes on some reared primary parasites of the families Trigonalidae, Ichneumonidae, and Tachinidae. *Proc. R. Soc. Vict.* **47:** 54–77. [37]

RAGGE, D. R. (1955a). *The Wing-venation of the Orthoptera Saltatoria. With Notes on Dictyopteran Wing-venation*, 159 pp. (London: Br. Mus. nat. Hist.). [16, 21]

—— (1955b). The wing-venation of the order Phasmida. *Trans. R. ent. Soc. Lond.* **106:** 375–92. [22]

RASNITSYN, A. P. (1964). New Triassic Hymenoptera of Central Asia. (In Russian.) *Paleont. Zh.* **1964** (1): 88–96. [8]

RATCLIFFE, F. N., GAY, F. J., and GREAVES, T. (1952). *Australian Termites. The Biology, Recognition, and Economic Importance of the Common Species*, 124 pp. (Melbourne: Commonw. scient. ind. Res. Org.). [15]

RAYMENT, T. (1935). *A Cluster of Bees: Sixty Essays on the Life-histories of Australian Bees, with Specific Descriptions of over 100 New Species, and an Introduction by Professor E. F. Phillips, D.Ph., Cornell University, U.S.A.*, 752 pp. (Sydney: Endeavour Press). [37]

—— (1953). New bees and wasps. Part XXI. *Victorian Nat.* **70:** 123–7. [37]

—— (1955). Taxonomy, morphology and biology of sericophorine wasps. *Mem. natn. Mus. Vict.* **19:** 11–105. [37]

READSHAW, J. L. (1965). A theory of phasmatid outbreak release. *Aust. J. Zool.* **13:** 475–90. [22]

REHN, J. A. G. (1952–7). *The Grasshoppers and Locusts (Acridoidea) of Australia.* I. Families Tetrigidae and Eumastacidae, 326 pp.; II. Family Acrididae (Subfamily Pyrgomorphinae), 270 pp.; III. Family Acrididae: Subfamily Cyrtacanthacridinae. Tribes Oxyini, Spathosternini, and Praxibulini, 273 pp. (Melbourne: Commonw. scient. ind. Res. Org.). [21]

REID, J. A. (1939). On the relationship of the hymenopterous genus *Olixon* and its allies, to the Pompilidae (Hym.). *Proc. R. ent. Soc. Lond.* (B) **8:** 95–102. [37]

REMINGTON, C. L. (1954). The suprageneric classification of the order Thysanura (Insecta). *Ann. ent. Soc. Am.* **47:** 277–86. [11]

—— (1955). The 'Apterygota'. In *A Century of Progress in the Natural Sciences, 1853–1953*, pp. 495–505. (San Francisco: Calif. Acad. Sci.). [7]

REYE, E. J. (1964). The problems of biting midges (Diptera: Ceratopogonidae) in Queensland. *J. ent. Soc. Qd* **3:** 1–6. [34]

RICHARDS, O. W. (1939a). The British Bethylidae (s.l.) (Hymenoptera). *Trans. R. ent. Soc. Lond.* **89:** 185–344. [37]

—— (1939b). The Bethylidae subfamily Sclerogibbinae (Hymenoptera). *Proc. R. ent. Soc. Lond.* (B) **8:** 211–23. [37]

—— (1953). The classification of the Dryinidae (Hym.) with descriptions of new species. *Trans. R. ent. Soc. Lond.* **104:** 51–70. [37]

—— (1962). *A Revisional Study of the Masarid Wasps (Hymenoptera, Vespoidea)*, 294 pp. (London: Br. Mus. nat. Hist.). [31, 37]

——, and DAVIES, R. G. (1960). *A. D. Imms: A General Textbook of Entomology*, 9th edn, 886 pp. (London: Methuen). [Intro., 1, 4, 7, 37 fig.]

——, and HAMM, A. H. (1939). The biology of the British Pompilidae (Hymenoptera). *Trans. Soc. Br. Ent.* **6:** 51–114. [37]

RICKER, W. E. (1950). Some evolutionary trends in Plecoptera. *Proc. Indiana Acad. Sci.* **59:** 197–209. [20]

—— (1952). Systematic studies in Plecoptera. *Indiana Univ. Publs Sci. Ser.* **18:** 1–200. [20]

RIEK, E. F. (1953). Fossil mecopteroid insects from the Upper Permian of New South Wales. *Rec. Aust. Mus.* **23:** 55–87. [8, fig.]

—— (1954a). The Australian Megaloptera or alderflies. *Aust. J. Zool.* **2:** 131–42. [28]

—— (1954b). The Australian Mecoptera or scorpionflies. *Aust. J. Zool.* **2:** 143–68. [32]

—— (1954c). Australian Trigonalidae (Hymenoptera: Ichneumonoidea). *Aust. J. Zool.* **2:** 296–307. [37]

—— (1954d). Further Triassic insects from Brookvale, N.S.W. (orders Orthoptera Saltatoria, Protorthoptera, Perlaria). *Rec. Aust. Mus.* **23:** 161–8. [8, fig.]

—— (1955a). The Australian rhipidiine parasites of cockroaches (Coleoptera: Rhipiphoridae). *Aust. J. Zool.* **3:** 71–94. [14, 30]

—— (1955b). Australian leaf-mining sawflies of the genus *Phylacteophaga* (Hymenoptera: Tenthredinidae). *Aust. J. Zool.* **3:** 95–8. [37]

—— (1955c). The Australian sawflies of the family Orussidae (Hymenoptera, Symphyta). *Aust. J. Zool.* **3**: 99–105. [37]
—— (1955d). Australian wasps of the family Proctotrupidae (Hymenoptera: Proctotrupoidea). *Aust. J. Zool.* **3**: 106–17. [37]
—— (1955e). Australian cleptid (Hymenoptera: Chrysidoidea) egg parasites of Cresmododea (Phasmodea). *Aust. J. Zool.* **3**: 118–30. [37]
—— (1955f). Australian Ampulicidae (Hymenoptera: Sphecoidea). *Aust. J. Zool.* **3**: 131–44. [37]
—— (1955g). Australian Heloridae, including Monomachidae (Hymenoptera). *Aust. J. Zool.* **3**: 258–65. [37]
—— (1955h). Revision of the Australian mayflies (Ephemeroptera). 1. Subfamily Siphlonurinae. *Aust. J. Zool.* **3**: 266–80. [12]
—— (1955i). The Australian Xiphydriidae (Hymenoptera, Symphyta). *Aust. J. Zool.* **3**: 281–5. [37]
—— (1955j). Fossil insects from the Triassic beds at Mt. Crosby, Queensland. *Aust. J. Zool.* **3**: 654–91. [8, fig.]
—— (1956). A re-examination of the mecopteroid and orthopteroid fossils (Insecta) from the Triassic beds at Denmark Hill, Queensland, with descriptions of further specimens. *Aust. J. Zool.* **4**: 98–110. [8, fig.]
—— (1962a). A new species of trigonalid wasp parasitic on the sawfly *Perga affinis* Kirby (Hymenoptera). *Proc. Linn. Soc. N.S.W.* **87**: 92–5. [37]
—— (1962b). A trigonalid wasp (Hymenoptera, Trigonalidae) from an anthelid cocoon (Lepidoptera, Anthelidae). *Proc. Linn. Soc. N.S.W.* **87**: 148–50. [37]
—— (1962c). Fossil insects from the Triassic at Hobart, Tasmania. *Pap. Proc. R. Soc. Tasm.* **96**: 39–40. [8]
—— (1963). An Australian mayfly of the family Ephemerellidae (Ephemeroptera). *J. ent. Soc. Qd* **2**: 48–50. [12]
—— (1966). Australian Hymenoptera Chalcidoidea. Family Pteromalidae, subfamily Perilampinae. *Aust. J. Zool.* **14**: 1207–36. [37]
—— (1967). Australian Hymenoptera Chalcidoidea. Family Eulophidae, subfamily Elasminae. *Aust. J. Zool.* **15**: 145–99. [37]
RIMES, G. D. (1951). Some new and little-known shore-bugs (Heteroptera-Saldidae) from the Australian region. *Trans. R. Soc. S. Aust.* **74**: 135–45. [26]
RITCHER, P. O. (1958). Biology of Scarabaeidae. *A. Rev. Ent.* **3**: 311–34. [30]
ROBERTS, F. H. S. (1928–9). A revision of the Australian Bombyliidae (Diptera). Parts I–III. *Proc. Linn. Soc. N.S.W.* **53**: 90–144, 413–55; **54**: 553–83. [34]
—— (1952). *Insects Affecting Livestock with Special Reference to Important Species Occurring in Australia*, 267 pp. (Sydney: Angus & Robertson). [5, 25, 33]
ROCKSTEIN, M. (ed.) (1964). *The Physiology of Insecta.* Vol. I, 640 pp.; II (1965), 905 pp.; III, 692 pp. (New York: Academic Press). [2]
RODD, N. W. (1951). Some observations on the biology of Stephanidae and Megalyridae (Hymenoptera). *Aust. Zool.* **11**: 341–6. [37]
RODDY, L. R. (1955). A morphological study of the respiratory horns associated with the puparia of some Diptera, especially *Ophyra anescens* (Wied.). *Ann. ent. Soc. Am.* **48**: 407–15. [34]
ROEDER, K. D. (ed.) (1953). *Insect Physiology*, 1,100 pp. (New York: Wiley). [2]
ROHDENDORF, B. B. (1961). Description of the first winged insect from the Devonian beds of the Timan. *Ent. Rev. Wash.* (translation of *Ent. Obozr.*) **40**: 260–2. [8]
—— (1964). Historical development of the dipterous insects. (In Russian.) *Trudy palaeont. Inst.* **100**: 1–311. [34]
——, BECKER-MIGDISOVA, H. E., MARTYNOVA, O. M., and SHAROV, A. G. (1961). Palaeozoic insects of the Kusznetsk Basin. (In Russian.) *Trudy paleont. Inst.* **85**: 1–705. [8, fig.]
RÖLLER, H., DAHM, K. H., SWEELY, C. C., and TROST, B. M. (1967). The structure of the juvenile hormone. *Angew. Chem.* (Internat. edn) **6**: 179–80. [2]
ROSE, H. A. (1965). Two new species of Thaumastocoridae (Hemiptera: Heteroptera) from Australia. *Proc. R. ent. Soc. Lond.* (B) **34**: 141–4. [26]
ROSS, E. S. (1963). The families of Australian Embioptera, with descriptions of a new family, genus, and species. *Wasmann J. Biol.* **21**: 121–36. [23]
ROSS, H. H. (1936). The ancestry and wing venation of the Hymenoptera. *Ann. ent. Soc. Am.* **29**: 99–111. [37]
—— (1937). A generic classification of the Nearctic sawflies (Hymenoptera, Symphyta). *Illinois biol. Monogr.* **15** (2): 1–173. [37]
—— (1944). The caddis flies, or Trichoptera, of Illinois. *Bull. Ill. St. nat. Hist. Surv.* **23**: 1–326. [35]
—— (1955). The evolution of the insect orders. *Ent. News* **66**: 197–208. [7]
—— (1956). *Evolution and Classification of the Mountain Caddisflies*, 213 pp. (Urbana: University of Illinois Press). [35]
ROTH, L. M., and EISNER, T. (1962). Chemical defences of arthropods. *A. Rev. Ent.* **7**: 107–36. [2]
——, and WILLIS, E. R. (1954). The reproduction of cockroaches. *Smithson. misc. Collns* **122** (12): 1–49. [1, fig.]
——, and —— (1960). The biotic associations of cockroaches. *Smithson. misc. Collns* **141**: 1–470. [14]
——, and —— (1961). A study of bisexual and parthenogenetic strains of *Pycnoscelus surinamensis* (Blattaria: Epilamprinae). *Ann. ent. Soc. Am.* **54**: 12–25. [14]
ROTHSCHILD, M. (1965). The rabbit flea and hormones. *Endeavour* **24**: 162–8. [5, 33]
——, and CLAY, T. (1952). *Fleas, Flukes & Cuckoos: A Study of Bird Parasites*, 304 pp. (London: Collins). [25]
ROTHSCHILD, W. and JORDAN, K. (1903). A revision of the lepidopterous family Sphingidae. *Novit. zool.* **9**, Suppl.: 1–972 (in 2 vols). [36]
RUNCORN, S. K. (ed.) (1962). *Continental Drift*, 338 pp. (New York: Academic Press). [9]
—— (1963). Satellite gravity measurements and convection in the mantle. *Nature, Lond.* **200**: 628–30. [9]
RUSSELL, L. M. (1941). A classification of the scale insect genus *Asterolecanium*. *Misc. Publs U.S. Dep. Agric.* **424**: 1–322. [26, fig.]
SABROSKY, C. W. (1955). Notes and descriptions of Australian Chloropidae (Diptera). *Proc. Linn. Soc. N.S.W.* **79**: 182–92. [34]

—— (1965). Diptera from Nepal. Asiatic species of the genus *Stenomicra* (Diptera: Anthomyzidae). *Bull. Br. Mus. nat. Hist.* Ent. **17:** 209–18. [34]

SAKAGAMI, S. F., and MICHENER, C. D. (1962). *The Nest Architecture of the Sweat Bees (Halictinae): A Comparative Study of Behavior*, 135 pp. (Lawrence: University of Kansas Press). [37]

SAKIMURA, K. (1947). Thrips in relation to gall-forming and plant disease transmission: a review. *Proc. Hawaii. ent. Soc.* **13:** 59–95. [27]

SALMON, J. T. (1964). An index to the Collembola. *Bull. R. Soc. N.Z.* **7:** 1–644 (2 vols). [10]

SANJEAN, J. (1957). Taxonomic studies of *Sarcophaga* larvae of New York, with notes on the adults. *Mem. Cornell Univ. agric. Exp. Stn* **349:** 1–115. [34]

SAVAGE, A. A. (1956). The development of the Malpighian tubules of *Schistocerca gregaria* (Orthoptera). *Q. Jl microsc. Sci.* **97:** 599–615. [4, fig.]

SCHAEFER, C. W. (1965). The morphology and higher classification of the Coreoidea (Hemiptera–Heteroptera). Part III. The families Rhopalidae, Alydidae, and Coreidae. *Misc. Publs ent. Soc. Am.* **5:** 1–76. [26]

—— (1966). The morphology and higher systematics of the Idiostolinae (Hemiptera: Lygaeidae). *Ann. ent. Soc. Am.* **59:** 602–13. [26]

SCHELLER, U. (1961). A review of the Australian Symphyla (Myriapoda). *Aust. J. Zool.* **9:** 140–71. [7]

SCHMIDT, K. P. (1954). Faunal realms, regions, and provinces. *Q. Rev. Biol.* **29:** 322–31. [9]

SCHNEIDERMAN, H. A., and GILBERT, L. I. (1964). Control of growth and development in insects. *Science, N.Y.* **143:** 325–33. [2]

SCHWARZBACH, M. (1963). *Climates of the Past: An Introduction to Palaeoclimatology* (translated and edited by R. O. Muir), 328 pp. (London: van Nostrand). [9]

SCOTT, M. T. (1950). Observations on the bionomics of *Linognathus pedalis*. *Aust. J. agric. Res.* **1:** 465–70. [25]

—— (1952). Observations on the bionomics of the sheep body louse *(Damalinia ovis)*. *Aust. J. agric. Res.* **3:** 60–7. [25]

SCUDDER, G. G. E. (1959). The female genitalia of the Heteroptera: morphology and bearing on classification. *Trans. R. ent. Soc. Lond.* **111:** 405–67. [26]

—— (1961). The comparative morphology of the insect ovipositor. *Trans. R. ent. Soc. Lond.* **113:** 25–40. [1]

—— (1962). The Ischnorhynchinae of the world (Hemiptera: Lygaeidae). *Trans. R. ent. Soc. Lond.* **114:** 163–94. [26]

—— (1963). Adult abdominal characters in the lygaeoid-coreoid complex of the Heteroptera, and the classification of the group. *Can. J. Zool.* **41:** 1–14. [26]

SEEVERS, C. H., and DYBAS, H. S. (1943). A synopsis of the Limulodidae (Coleoptera): a new family proposed for myrmecophiles of the subfamilies Limulodinae (Ptiliidae) and Cephaloplectinae (Staphylinidae). *Ann. ent. Soc. Am.* **36:** 546–86. [30]

SÉGUY, E. (1950). La biologie des Diptères. *Encycl. ent.* (A) **26:** 1–609. [34]

—— (1951). Ordre des Diptères (Diptera Linné, 1758). In P.-P. Grassé (ed.), *Traité de Zoologie: Anatomie, Systématique, Biologie*, Vol. X, pp. 449–744. (Paris: Masson). [34]

SEILER, J. (1963). Untersuchungen über die Entstehung der Parthenogenese bei *Solenobia triquetrella* F.R. (Lepidoptera, Psychidae). IV. *Z. VererbLehre* **94:** 29–66. [3]

SELANDER, R. B. (1957). The systematic position of the genus *Nephrites* and the phylogenetic relationships of the higher groups of Rhipiphoridae (Coleoptera). *Ann. ent. Soc. Am.* **50:** 88–103. [30]

SHARIF, M. (1937). On the life history and the biology of the rat-flea, *Nosopsyllus fasciatus* (Bosc). *Parasitology* **29:** 225–38. [33]

SHAROV, A. G. (1957). Peculiar Palaeozoic wingless insects of the new order Monura (Insecta, Apterygota). (In Russian.) *Dokl. Akad. Nauk. SSSR* **115** (4): 795–8. [8]

—— (1966). *Basic Arthropodan Stock: With Special Reference to Insects*, 271 pp. (Oxford: Pergamon Press). [7, 11]

SHARPLIN, J. (1963–4). Wing base structure in Lepidoptera. I–III. *Can. Ent.* **95:** 1024–50, 1121–45; **96:** 943–9. [36]

SHAW, J., and STOBBART, R. H. (1963). Osmotic and ionic regulation in insects. *Adv. Insect Physiol.* **1:** 315–99. [2]

SHORT, J. R. T. (1952). The morphology of the head of larval Hymenoptera with special reference to the head of the Ichneumonoidea, including a classification of the final instar larvae of the Braconidae. *Trans. R. ent. Soc. Lond.* **103:** 27–84. [37]

—— (1959a). A description and classification of the final instar larvae of the Ichneumonidae (Insecta, Hymenoptera). *Proc. U.S. natn. Mus.* **110:** 391–511. [37]

—— (1959b). On the skeleto-muscular mechanisms of the anterior abdominal segments of certain adult Hymenoptera. *Trans. R. ent. Soc. Lond.* **111:** 175–203. [37]

SILVESTRI, F. (1947). On some Japygidae in the Museum of Comparative Zoölogy (Dicellura). *Psyche, Camb.* **54:** 209–29. [10]

SIMPSON, G. G. (1961). *Principles of Animal Taxonomy*, 247 pp. (New York: Columbia University Press). [6, 7]

SINGH, B. (1956). Description and systematic position of larva and pupa of the teak defoliator, *Hyblaea puera* Cramer (Insecta, Lepidoptera, Hyblaeidae). *Indian Forest Rec.* n.s., Ent. **9:** 1–16. [36]

SKUSE, F. A. A. (1888–90). Series 'Diptera of Australia', Parts I–IV and Supplements I and II, in *Proc. Linn. Soc. N.S.W.* (2nd ser.): 1888a, Part I, **3:** 17–145; 1888b, II, **3:** 657–726; 1889, IV, **3:** 1363–86; 1890a, Suppl. I, **5:** 373–412; 1890b, Suppl. II, **5:** 595–640. [34]

SLATER, J. A. (1955). A revision of the subfamily Pachygronthinae of the world (Hemiptera: Lygaeidae). *Philipp. J. Sci.* **84:** 1–160. [26]

—— (1962). *Darwinocoris*, a new genus of Pachygronthinae from Australia, with a description of the type species *D. australicus* sp. n. (Hemiptera: Lygaeidae). *J. ent. Soc. Qd* **1:** 44–5. [26]

—— (1964). *A Catalogue of the Lygaeidae of the World*, 2 vols, 1,668 pp. (Storrs: University of Connecticut). [26]

——, and SWEET, M. H. (1961). A contribution to the higher classification of the Megalonotinae (Hemiptera: Lygaeidae). *Ann. ent. Soc. Am.* **54:** 203–9. [26]

——, Woodward, T. E., and Sweet, M. H. (1962). A contribution to the classification of the Lygaeidae, with the description of a new genus from New Zealand (Hemiptera: Heteroptera). *Ann. ent. Soc. Am.* **55**: 597–605. [26]

Slifer, E. H., Prestage, J. J., and Beams, H. W. (1959). The chemoreceptors and other sense organs on the antennal flagellum of the grasshopper (Orthoptera; Acrididae). *J. Morph.* **105**: 145–91. [2, fig.]

Sloane, T. G. (1921). Revisional notes on Australian Carabidae. Part vi. Tribe Bembidiini. *Proc. Linn. Soc. N.S.W.* **46**: 192–208. [30]

—— (1923). The classification of the family Carabidae. *Trans. ent. Soc. Lond.* **1923**: 234–50, A–C. [30]

Smart, J. (1963). Explosive evolution and the phylogeny of insects. *Proc. Linn. Soc. Lond.* **174**: 125–6. [7]

Smee, L. (1966). A revision of the subfamily Leptoconopinae Noé (Diptera: Ceratopogonidae) in Australasia. *Aust. J. Zool.* **14**: 993–1025. [34]

Smith, R. H. (1944). Bionomics and control of the nigra scale, *Saissetia nigra*. *Hilgardia* **16**: 225–88. [26]

Smith, S. G. (1953). Chromosome numbers of Coleoptera. *Heredity, Lond.* **7**: 31–48. [3]

—— (1960). Chromosome numbers of Coleoptera. II. *Can. J. Genet. Cytol.* **2**: 66–88. [3]

Smithers, C. N. (1964). The Myopsocidae (Psocoptera) of Australia. *Proc. R. ent. Soc. Lond.* (B) **33**: 133–8. [24]

—— (1965a). The Lepidopsocidae (Psocoptera) of Australia. *J. ent. Soc. Qd* **4**: 72–8. [24]

—— (1965b). The Trogiidae (Psocoptera) of Australia. *J. ent. Soc. Qd* **4**: 79. [24]

—— (1965c). A bibliography of the Psocoptera (Insecta). *Aust. Zool.* **13**: 137–209. [24]

Snodgrass, R. E. (1905). A revision of the mouthparts of the Corrodentia and of the Mallophaga. *Trans. Am. ent. Soc.* **31**: 297–307. [25, fig.]

—— (1935). *Principles of Insect Morphology*, 667 pp. (New York: McGraw-Hill). [1, 2, 4]

—— (1937). The male genitalia of orthopteroid insects. *Smithson. misc. Collns* **96** (5): 1–107. [1, 16, 21, 22]

—— (1941). The male genitalia of Hymenoptera. *Smithson. misc. Collns* **99** (14): 1–86. [37]

—— (1946). The skeletal anatomy of fleas (Siphonaptera). *Smithson. misc. Collns* **104** (18): 1–89. [33]

—— (1952). *A Textbook of Arthropod Anatomy*, 363 pp. (Ithaca, N.Y.: Comstock). [1, 7, 10, 14]

—— (1954). Insect metamorphosis. *Smithson. misc. Collns* **122** (9): 1–124. [4, 27]

—— (1957). A revised interpretation of the external reproductive organs of male insects. *Smithson. misc. Collns* **135** (6): 1–60. [1, 34]

Snyder, T. E. (1949). Catalog of the termites (Isoptera) of the world. *Smithson. misc. Collns* **112**: 1–490. [15]

—— (1956). Annotated, subject-heading bibliography of termites 1350 B.C. to A.D. 1954. *Smithson. misc. Collns* **130**: 1–305. [15]

—— (1961). Supplement to the annotated, subject-heading bibliography of termites 1955 to 1960. *Smithson. misc. Collns* **143** (3): 1–137. [15]

Soika, A. G. (1962). Gli *Odynerus* sensu antiquo del continente australiano e della Tasmania. *Boll. Mus. civ. Stor. nat. Venezia* **14**: 57–202. [37]

Sokal, R. R., and Sneath, P. H. A. (1963). *Principles of Numerical Taxonomy*, 359 pp. (San Francisco: Freeman). [6]

Southwood, T. R. E. (1955). The morphology of the salivary glands of terrestrial Heteroptera (Geocorisae) and its bearing on classification. *Tijdschr. Ent.* **98**: 77–84. [26]

—— (1956). The structure of the eggs of the terrestrial Heteroptera and its relationship to the classification of the group. *Trans. R. ent. Soc. Lond.* **108**: 163–221. [26]

——, and Leston, D. (1959). *Land and Water Bugs of the British Isles*, 436 pp. (London: Frederick Warne). [26]

Souza Lopes, H. de (1959). A revision of Australian Sarcophagidae (Diptera). *Studia ent.* **2** (1–4): 33–67. [34]

Spencer, K. A. (1963). The Australian Agromyzidae (Diptera, Insecta). *Rec. Aust. Mus.* **25**: 305–54. [34]

Stalker, H. D. (1956). On the evolution of parthenogenesis in *Lonchoptera* (Diptera). *Evolution, Lancaster, Pa.* **10**: 345–59. [34]

Stammer, H. J. (1956). Die Parasiten der Bibioniden. *Proc. 14th int. Congr. Zool.*: 349–58. [34]

Stannard, L. J. (1957). The phylogeny and classification of the North American genera of the suborder Tubulifera (Thysanoptera). *Illinois biol. Monogr.* **25**: 1–200. [27]

Starý, P. (1958). A taxonomic revision of some aphidiin genera with remarks on the subfamily Aphidiinae. *Sb. faun. Prací ent. Odd. nár. Mus. Praze* **3**: 53–96. [37]

Steffan, A. W. (1964). Torridincolidae, coleopterorum nova familia e regione aethiopica. *Ent. Z. Frankf. a. M.* **74**: 193–200. [30]

Steinhaus, E. A. (ed.) (1963). *Insect Pathology: An Advanced Treatise*. Vol. 1, 661 pp.; 2, 689 pp. (New York: Academic Press). [5]

Steyskal, G. C. (1949). Sepsidae from the Australasian Region (Diptera). *Pan-Pacif. Ent.* **25**: 161–71. [34]

—— (1953). A suggested classification of the lower brachycerous Diptera. *Ann. ent. Soc. Am.* **46**: 237–42. [34]

—— (1961). The genera of Platystomatidae and Otitidae known to occur in America north of Mexico (Diptera, Acalyptratae). *Ann. ent. Soc. Am.* **54**: 401–10. [34]

Stone, A., Knight, K. L., and Starcke, H. (1959). *A Synoptic Catalog of the Mosquitoes of the World (Diptera, Culicidae)*. Thomas Say Foundation, Vol. VI, 358 pp. (Washington: Ent. Soc. Am.). [34]

——, Sabrosky, C. W., Wirth, W. W., Foote, R. H., and Coulson, J. R. (eds) (1965). *A Catalogue of the Diptera of America North of Mexico*, 1,696 pp. (Washington: U.S. Govt Printing Office). [34]

Stone, W. S. (1962). The dominance of natural selection and the reality of superspecies (species groups) in the evolution of *Drosophila*. *Univ. Tex. Publs* **6205**: 507–37. [3]

Stys, P. (1966). Morphology of the wings, abdomen and genitalia of *Phaenacantha australiae* Kirk. (Heteroptera, Colobathristidae) and notes on the phylogeny of the family. *Acta ent. bohemoslov.* **63**: 266–80. [26]

Suomalainen, E. (1962). Significance of parthenogenesis in the evolution of insects. *A. Rev. Ent.* **7**: 349–66. [3]

Sweet, M. H. (1960). The seed bugs: a contribution to the feeding habits of the Lygaeidae (Hemiptera: Heteroptera). *Ann. ent. Soc. Am.* **53**: 317–21. [26]

Sweetman, H. L. (1958). *The Principles of Biological Control*, 560 pp. (Iowa: Brown). [5]

Tanner, V. M. (1927). A preliminary study of the genitalia of female Coleoptera. *Trans. Am. ent. Soc.* **53**: 5–50. [30]

Taylor, K. L. (1962). The Australian genera *Cardiaspina* Crawford and *Hyalinaspis* Taylor (Homoptera: Psyllidae). *Aust. J. Zool.* **10**: 307–48. [26]

Theodor, O. (1956). On the genus *Tripselia* and the group of *Basilia bathybothyra* (Nycteribiidae, Diptera). *Parasitology* **46**: 353–94. [34]

—— (1959). A revision of the genus *Cyclopodia* (Nycteribiidae, Diptera). *Parasitology* **49**: 242–308. [34]

Thornton, I. W. B., and Harrell, J. C. (1965). Air-borne Psocoptera trapped on ships and aircraft, 2—Pacific ship trappings, 1963–64. *Pacif. Insects* **7**: 700–2. [24]

Thorpe, W. H. (1928). Elimination of carbon dioxide in the Insecta. *Science, N.Y.* **68**: 433–4. [30]

—— (1950). Plastron respiration in aquatic insects. *Biol. Rev.* **25**: 344–90. [2]

——, and Crisp, D. J. (1949). Studies on plastron respiration. IV. Plastron respiration in the Coleoptera. *J. exp. Biol.* **26**: 219–60. [30]

Tiegs, O. W. (1940). The embryology and affinities of the Symphyla, based on a study of *Hanseniella agilis*. *Q. Jl microsc. Sci.* **82**: 1–225. [7]

—— (1945). The post-embryonic development of *Hanseniella agilis* (Symphyla). *Q. Jl microsc. Sci.* **85**: 191–328. [7]

——, and Manton, S. M. (1958). The evolution of the Arthropoda. *Biol. Rev.* **33**: 255–337. [1, 7]

Tillyard, R. J. (1918–19). The panorpoid complex. Parts 1–3. *Proc. Linn. Soc. N.S.W.* **43**: 265–319, 626–57; **44**: 533–718. [1, 7]

—— (1919a). A fossil insect wing belonging to the new order Paramecoptera, ancestral to the Trichoptera and Lepidoptera, from the Upper Coal-Measures of Newcastle, N.S.W. *Proc. Linn. Soc. N.S.W.* **44**: 231–56. [8, fig.]

—— (1921a). Revision of the family Eustheniidae (order Perlaria) with descriptions of new genera and species. *Proc. Linn. Soc. N.S.W.* **46**: 221–36. [20]

—— (1921b). Mesozoic insects of Queensland. No. 8. Hemiptera Homoptera (contd.). The genus *Mesogereon*; with a discussion of its relationship with the Jurassic Palaeontinidae. *Proc. Linn. Soc. N.S.W.* **46**: 270–84. [8, fig.]

—— (1922a). The life-history of the Australian moth-lacewing, *Ithone fusca*, Newman (order Neuroptera Planipennia). *Bull. ent. Res.* **13**: 205–23. [29]

—— (1922b). Mesozoic insects of Queensland. No. 9. Orthoptera, and additions to the Protorthoptera, Odonata, Hemiptera and Planipennia. *Proc. Linn. Soc. N.S.W.* **47**: 447–70. [8, fig.]

—— (1923a). The Embioptera or web-spinners of Western Australia. *J. Proc. R. Soc. West. Aust.* **9** (1): 61–8. [23]

—— (1923b). Mesozoic insects of Queensland. No. 10. Summary of the Upper Triassic insect fauna of Ipswich, Q. (With an appendix describing new Hemiptera and Planipennia.) *Proc. Linn. Soc. N.S.W.* **48**: 481–98. [8, fig.]

—— (1926a). Upper Permian insects of New South Wales. Part i. Introduction and the order Hemiptera. *Proc. Linn. Soc. N.S.W.* **51**: 1–30. [8, fig.]

—— (1926b). *The Insects of Australia and New Zealand*, 560 pp. (Sydney: Angus & Robertson). [Intro., 17, 26, 29, 36]

—— (1928). Some remarks on the Devonian fossil insects from the Rhynie chert beds, Old Red Sandstone. *Trans. ent. Soc. Lond.* **76**: 65–71. [8]

—— (1937). Kansas Permian insects. Part 20. The cockroaches, or order Blattaria. Part I. *Am. J. Sci.* **34**: 169–202. [8, fig.]

Tindale, N. B. (1923–4). Review of Australian Mantidae, I, II. *Rec. S. Aust. Mus.* **2**: 425–57, 547–52. [16]

—— (1928a). Australasian mole-crickets of the family Gryllotalpidae (Orthoptera). *Rec. S. Aust. Mus.* **4**: 1–42. [21]

—— (1928b). Preliminary note on the life history of *Synemon* (Lepidoptera, fam. Castniidae). *Rec. S. Aust. Mus.* **4**: 143–4. [36]

—— (1932–64). Revision of the Australian ghost moths (Lepidoptera Homoneura, family Hepialidae). Parts I–VIII. *Rec. S. Aust. Mus.* **4**: 497–536; **5**: 13–43, 275–332; **7**: 15–46, 151–68; **11**: 307–44; **13**: 157–97; **14**: 663–8. [36]

Tjeder, B. (1954). Genital structures and terminology in the order Neuroptera. *Ent. Meddr* **27**: 23–40. [29]

Todd, E. L. (1960). The Gelastocoridae of Australia (Hemiptera). *Pacif. Insects* **2**: 171–94. [26]

Tokunaga, M. (1932). Morphological and biological studies on a new marine chironomid fly, *Pontomyia pacifica*, from Japan. Part I. Morphology and taxonomy. *Mem. Coll. Agric. Kyoto Univ.* **19** (Ent. Ser. **3**): 1–56. [5]

Tonnoir, A. L. (1923). Australian Dixidae. (Dipt.). *Pap. Proc. R. Soc. Tasm.* **1923**: 58–71. [34]

—— (1925). Australian Platypezidae (Diptera). *Rec. Aust. Mus.* **14**: 306–12. [34]

—— (1929). Australian Mycetophilidae. Synopsis of the genera. *Proc. Linn. Soc. N.S.W.* **54**: 584–614. [34]

Townes, H. (1957). Nearctic wasps of the subfamilies Pepsinae and Ceropalinae. *Bull. U.S. natn. Mus.* **209**: 1–286. [37]

Treat, A. E. (1955). Distribution of the moth ear mite *(Myrmonyssus phalaenodectes)*. *Lepid. News* **9**: 55–8. [36]

Tuomikoski, R. (1960). Zur Kenntnis der Sciariden (Dipt.) Finnlands. *Suomal. eläin- ja kasvit. Seur. van. eläin. Julk.* **21** (4): 1–164. [34]

Turner, A. J. (1921–47). See bibliography in I. M. Mackerras, *Proc. R. Soc. Qd* **60**: 69–87, 1949. [36]

Turner, R. E. (1916). Two new species of the hymenopterous genus *Megalyra* Westw. *Ann. Mag. nat. Hist.* (8) **17**: 246–7. [37]

Tuxen, S. L. (ed.) (1956). *Taxonomist's Glossary of Genitalia in Insects*, 284 pp. (Copenhagen: Ejnar Munksgaard). [1, 10, 17, 27, 30, 34, 36]

—— (1958). Relationships of Protura. *Proc. 10th int. Congr. Ent.* **1**: 493–7. [7]

—— (1959). The phylogenetic significance of entognathy in entognathous apterygotes. *Smithson. misc. Collns* **137** (Snodgrass Anniversary Volume): 379–416. [7]

—— (1967). Australian Protura, their phylogeny and zoogeography. *Z. zool. Syst. Evolut.-forsch.* **5**: 1–53. [10]

USINGER, R. L. (1945). Classification of the Enicocephalidae (Hemiptera, Reduvioidea). *Ann. ent. Soc. Am.* **38:** 321–42. [26]

——, and MATSUDA, R. (1959). *Classification of the Aradidae (Hemiptera–Heteroptera)*, 410 pp. (London: Br. Mus. nat. Hist.). [26]

UVAROV, B. (1966). *Grasshoppers and Locusts: A Handbook of General Acridology*, Vol. 1, 481 pp. (Cambridge University Press). [21]

VÁRI, L. (1961). South African Lepidoptera, Vol. I, Lithocolletidae. *Transv. Mus. Mem.* **12:** 1–238. [36]

VECHT, J. VAN DER (1959). On *Eumenes arcuatus* (Fabricius) and some allied Indo-Australian wasps (Hymenoptera, Vespidae). *Zool. Verh., Leiden* **41:** 1–71. [37]

—— (1962). The Indo-Australian species of the genus *Ropalidia (Icaria)* (Hymenoptera, Vespidae) (Second Part). *Zool. Verh., Leiden* **57:** 1–72. [37]

—— (1963). Studies on Indo-Australian and East-Asiatic Eumenidae (Hymenoptera, Vespoidea). *Zool. Verh., Leiden* **60:** 1–116. [37]

VEVERS STEELE, H. (1935). Thrips investigation: some common Thysanoptera in Australia. *Pamph. Coun. scient. ind. Res. Aust.* **54:** 1–59. [27]

WADE, V. (1960). *General Catalogue of the Homoptera.* Species Index. Fasc. IV, Fulgoroidea, 78 pp. (Raleigh: North Carolina State College). [26]

WALKER, E. M. (1937). *Grylloblatta*, a living fossil. *Trans. R. Soc. Can.* Sect. V **26:** 1–10. [18]

WALLACE, A. R. (1876). *The Geographical Distribution of Animals: With a Study of the Relations of Living and Extinct Faunas as Elucidating the Past Changes of the Earth's Surface.* Vol. I, 503 pp.; II, 607 pp. (London: Macmillan—reprinted by Hafner, New York, 1962). [9]

WATERHOUSE, D. F. (1953). The occurrence and significance of the peritrophic membrane, with special reference to adult Lepidoptera and Diptera. *Aust. J. Zool.* **1:** 299–318. [36]

—— (1957). Digestion in insects. *A. Rev. Ent.* **2:** 1–18. [2]

WATERHOUSE, G. A. (1928). A second monograph of the genus *Tisiphone* Hübner. *Aust. Zool.* **5:** 217–40. [36]

——, and LYELL, G. (1914). *The Butterflies of Australia: A Monograph of the Australian Rhopalocera*, 239 pp. (Sydney: Angus & Robertson). [36]

WATSON, A. (1965). A revision of the Ethiopian Drepanidae (Lepidoptera). *Bull. Br. Mus. nat. Hist.* Ent. Suppl. **3:** 1–177. [36]

WATSON, J. A. L. (1962). *The Dragonflies (Odonata) of South-Western Australia*, 72 pp. (Perth: Handbook No. 7, W.A. Naturalists' Club). [13]

—— (1965). The endocrine system of the lepismatid Thysanura and its phylogenetic implications. *Proc. 12th int. Congr. Ent.*: 144. [11]

WAY, M. J. (1963). Mutualism between ants and honeydew-producing Homoptera. *A. Rev. Ent.* **8:** 307–44. [26]

WEBB, J. E. (1946). Spiracle structure as a guide to the phylogenetic relationships of the Anoplura (biting and sucking lice), with notes on the affinities of the mammalian hosts. *Proc. zool. Soc. Lond.* **116:** 49–119. [25]

WEBER, H. (1933). *Lehrbuch der Entomologie*, 726 pp. (Jena: Gustav Fischer). [2, fig.]

—— (1935). Der Bau der Imago der Aleurodinen. Ein Beitrag zur vergleichenden Morphologie des Insektenkörpers. *Zoologica, Stuttg.* **33** (89): 1–71 [26]

—— (1954). *Grundriss der Insektenkunde*, 3rd edn, 428 pp. (Stuttgart: Gustav Fischer). [25]

WEIS-FOGH, T. (1961). Power in flapping flight. In J. A. Ramsay and V. B. Wigglesworth (eds), *The Cell and the Organism*, pp. 283–300. (Cambridge University Press). [2]

WELD, L. H. (1952). *Cynipoidea (Hym.) 1905–1950 . . .* , 351 pp. (Ann Arbor, Mich.: Mimeo., privately printed). [37]

WERNECK, F. L., and THOMPSON, G. B. (1940). Sur les mallophages des marsupiaux d'Australie (Mallophaga: Boopidae). *Mems Inst. Oswaldo Cruz* **35:** 411–55. [25]

WHALLEY, P. E. S. (1961). A change of status and a redefinition of the subfamily Endotrichinae (Lep. Pyralidae) with the description of a new genus. *Ann. Mag. nat. Hist.* (13) **3:** 733–6. [36]

—— (1964). Catalogue of the world genera of the Thyrididae (Lepidoptera) with type selection and synonymy. *Ann. Mag. nat. Hist.* (13) **7:** 115–27. [36]

WHEELER, M. R. (1952). A key to the genera of Drosophilidae of the Pacific Islands (Diptera). *Proc. Hawaii. ent. Soc.* **14:** 421–3. [34]

WHEELER, W. M. (1918). The Australian ants of the ponerine tribe Cerapachyini. *Proc. Am. Acad. Arts Sci.* **53:** 215–65. [37]

WHITE, M. J. D. (1949). Cytological evidence on the phylogeny and classification of the Diptera. *Evolution, Lancaster, Pa.* **3:** 252–61. [34]

—— (1951). Cytogenetics of orthopteroid insects. *Adv. Genet.* **4:** 267–330. [16, 21, 22]

—— (1957). Cytogenetics and systematic entomology. *A. Rev. Ent.* **2:** 71–90. [3, 7]

—— (1962a). A unique type of sex chromosome mechanism in an Australian mantid. *Evolution, Lancaster, Pa.* **16:** 75–85. [3, fig.]

—— (1962b). Genetic adaptation. *Aust. J. Sci.* **25:** 179–86. [21]

—— (1965). Sex chromosomes and meiotic mechanisms in some African and Australian mantids. *Chromosoma* **16:** 521–47. [16]

——, BLACKITH, R. E., BLACKITH, R. M., and CHENEY, J. (1967). Cytogenetics of the *viatica* group of morabine grasshoppers. I. The 'coastal' species. *Aust. J. Zool.* **15:** 263–302. [3]

——, CHENEY, J., and KEY, K. H. L. (1963). A parthenogenetic species of grasshopper with complex structural heterozygosity (Orthoptera: Acridoidea). *Aust. J. Zool.* **11:** 1–19. [3]

——, and CHINNICK, L. J. (1957). Cytogenetics of the grasshopper *Moraba scurra.* III. Distribution of the 15- and 17-chromosome races. *Aust. J. Zool.* **5:** 338–47. [3]

——, and KEY, K. H. L. (1957). A cytotaxonomic study of the *pusilla* group of species in the genus *Austroicetes* Uv. (Orthoptera: Acrididae). *Aust. J. Zool.* **5:** 56–87. [3]

WHITING, P. W. (1945). The evolution of male haploidy. *Q. Rev. Biol.* **20:** 231–60. [3]

WHITTEN, J. M. (1963). The tracheal pattern and body segmentation in the blepharocerid larva. *Proc. R. ent. Soc. Lond.* (A) **38:** 39–44. [34]

—— (1964). Giant polytene chromosomes in hypodermal cells of developing footpads of dipteran pupae. *Science, N.Y.* **143:** 1437–8. [34]

WIDDOWS, R. E., and WICK, J. R. (1959). Morphology of the reproductive system of *Tetrix arenosa angusta* (Hancock) (Orthoptera, Tetrigidae). *Proc. Iowa Acad. Sci.* **66**: 484–503. [21]

WIGGLESWORTH, V. B. (1954). *The Physiology of Insect Metamorphosis*. Cambridge Monographs in Experimental Biology, No. 1, 152 pp. (Cambridge University Press). [2, 4]

—— (1963a). Origin of wings in insects. *Nature, Lond.* **197**: 97–8. [1, 7]

—— (1963b). A further function of the air sacs in some insects. *Nature, Lond.* **198**: 106. [30]

—— (1964). The hormonal regulation of growth and reproduction in insects. *Adv. Insect Physiol.* **2**: 247–336. [2]

—— (1965). *The Principles of Insect Physiology*, 6th edn, 741 pp. (London: Methuen). [2, 11]

——, and others (1963). Discussion: The origin of flight in insects. *Proc. R. ent. Soc. Lond.* (C) **28**: 23–32. [7]

WILLIAMS, C. B. (1958). *Insect Migration*, 235 pp. (London: Collins). [36]

WILSON, E. O. (1963). The social biology of ants. *A. Rev. Ent.* **8**: 345–68. [37]

—— (1965). Chemical communication in the social insects. *Science, N.Y.* **149**: 1064–71. [5]

WILSON, F. (1960). A review of the biological control of insects and weeds in Australia and Australian New Guinea. *Tech. Commun. Commonw. Inst. biol. Control* **1**: 1–102. [5, 26]

WILSON, J. Tuzo (1963). Continental drift. *Scient. Am.* **208** (4): 86–100. [9]

WIRTH, W. W. (1964). New species and records of the genus *Trichocanace* Wirth (Diptera, Canaceidae). *Pacif. Insects* **6**: 225–7. [34]

WISE, K. A. J. (1964). New records of Collembola and Acarina in Antarctica. *Pacif. Insects* **6**: 522–3. [5]

WITHYCOMBE, C. L. (1925). Some aspects of the biology and morphology of the Neuroptera. With special reference to the immature stages and their possible phylogenetic significance. *Trans. ent. Soc. Lond.* **1924**: 303–411. [29]

WOMERSLEY, H. (1939). *Primitive Insects of South Australia*, 322 pp. (Adelaide: Govt Printer). [10, 11]

—— (1945). New species of Diplura (Insecta Apterygota) from Australia and New Guinea. *Trans. R. Soc. S. Aust.* **69**: 223–8. [10]

—— (1950). On the female of the dipteron *Scatopse aptera* Womersley 1942. *Rec. S. Aust. Mus.* **9**: 331–2. [34]

WOODROFFE, G. E. (1958). The mode of reproduction of *Ptinus clavipes* Panzer form *mobilis* Moore (=*P. latro* auct.) (Coleoptera: Ptinidae). *Proc. R. ent. Soc. Lond.* (A) **33**: 25–30. [30]

WOODWARD, T. E. (1957). Studies on Queensland Hemiptera. Part II. Meenoplidae (Fulgoroidea). *Pap. Dep. Ent. Univ. Qd* **1**: 57–70. [26]

—— (1958). Studies on Queensland Hemiptera. Part III. A remarkable new intertidal saldid. *Pap. Dep. Ent. Univ. Qd* **1**: 101–10. [26]

—— (1960). Studies on Queensland Hemiptera. Part IV. The first record of Orgeriinae (Fulgoroidea; Dictyopharidae) from Australia. *Pap. Dep. Ent. Univ. Qd* **1**: 149–56. [26]

WOOTTON, R. J. (1963). Actinoscytinidae (Hemiptera–Heteroptera) from the Upper Triassic of Queensland. *Ann. Mag. nat. Hist.* (13) **6**: 249–55. [8]

WYGODZINSKY, P. (1948). Redescription of *Nesomachilis maoricus* Tillyard, 1924, with notes on the family Machilidae (Thysanura). *Dom. Mus. Rec. Ent.* **1**: 69–78. [11]

—— (1955). Thysanura. In B. Hanström *et al.* (eds), *South African Animal Life*, Vol. 2, pp. 83–190. (Stockholm: Almqvist & Wiksell). [11]

—— (1961). On a surviving representative of the Lepidotrichidae (Thysanura). *Ann. ent. Soc. Am.* **54**: 621–7. [11]

—— (1966). A monograph of the Emesinae (Reduviidae, Hemiptera). *Bull. Am. Mus. nat. Hist.* **133**: 1–614. [26]

WYNNE-EDWARDS, V. C. (1962). *Animal Dispersion in Relation to Social Behaviour*, 653 pp. (Edinburgh: Oliver & Boyd). [21]

YANO, K. (1963). Taxonomic and biological studies of Pterophoridae of Japan (Lepidoptera). *Pacif. Insects* **5**: 65–209. [36]

ZEUNER, F. E. (1939). *Fossil Orthoptera Ensifera*, 2 vols, 321 pp. (London: Br. Mus. nat. Hist.). [18]

—— (1959). Jurassic beetles from Grahamland, Antarctica. *Palaeontology* **1**: 407–9. [8]

ZIMMERMAN, E. C. (1948). Homoptera: Sternorhyncha. *Insects Hawaii* **5**: 1–464. [26, fig.]

ZUMPT, F., and HEINZ, H. J. (1950). Studies on the sexual armature of Diptera. II.—A contribution to the study of the morphology and homology of the male terminalia of *Calliphora* and *Sarcophaga* (Dipt., Calliphoridae). *Entomologist's mon. Mag.* **86**: 207–16. [34]

INDEX

Principal entries of taxa are indicated by bold face page numbers; coloured plates by 'Pl.'; synonyms by cross-reference. The names of plants are distinguished by '(Fungi)' or '(Bot.)', as appropriate; of animals other than hexapods by phylum, class, or order. Economically significant insects can be traced through the common name of the host or product.